Centrifugal Pumps

Johann Friedrich Gülich

Centrifugal Pumps

Fourth Edition

Springer

Johann Friedrich Gülich
Villeneuve, Switzerland

ISBN 978-3-030-14790-7 ISBN 978-3-030-14788-4 (eBook)
https://doi.org/10.1007/978-3-030-14788-4

This Springer imprint is published by the registered company Springer Nature Switzerland AG.
The registered company address is: Gewerbestrasse 11, 6330 Cham, Switzerland

Preface

Life is linked to liquid transport, and so are vital segments of economy. Pumping devices - be it the human heart, a boiler feeder or the cooling water pump of a motorcar - are always part of a more or less complex system where pump failure can lead to severe consequences. To select, operate or design a pump, some understanding of the system is helpful, if not essential. Depending on the application, a centrifugal pump can be a simple device which could be built in a garage with a minimum of know-how - or a high-tech machine requiring advanced skills, sophisticated engineering and extensive testing. When attempting to describe the state of the art in hydraulic engineering of centrifugal pumps, the focus is necessarily on the high-tech side rather than on less-demanding services even though these make up the majority of pump applications.

Centrifugal pump technology involves a broad spectrum of flow phenomena which have a profound impact on design and operation through the achieved efficiency, the stability of the head-capacity characteristic, vibration, noise, component failure due to fatigue, as well as material damage caused by cavitation, hydro-abrasive wear or erosion corrosion. Operation and life cycle costs of pumping equipment depend to a large extent on how well these phenomena and the interaction of the pump with the system are understood.

This book endeavors to describe pump hydraulic phenomena in their broadest sense in a format highly relevant for the pump engineer involved in pump design and selection, operation and troubleshooting. Emphasis is on physical mechanisms, practical application and engineering correlations for real flow phenomena, rather than on mathematical treatment and theories of inviscid flow.

Vibrations and noise remain among the most persistent pump problems encountered by manufacturers and operators alike. Major additions in the present 4th English edition focus on this theme. Specifically, additional topics discussed in the 4th edition are as follows:

1. Chapter 10.14 "Vertical-pump vibrations" has been added; it focuses on large cooling water pumps with power of up to 4000 kW. Time and again, the approach flow through the large intake structures induces vibrations in spite of

sump model testing done to eliminate vortices. The case histories reported suggest that Froude model testing is not sufficient to guarantee trouble-free operation since the inertia effects associated with large Reynolds numbers are not sufficiently modeled. Non-uniform approach flows create forces on the pump body, and correlations are offered to estimate this type of excitation forces.

2. With the object of helping with vibration diagnostics, Chap. 10 has been again significantly enhanced by a more detailed discussion of various unsteady flow phenomena and their impact on hydraulic excitation forces. Case histories are provided which afford additional insight into excitation mechanisms. Experimental data on lateral excitation forces and pressure pulsations at blade passing frequency are presented in a generic format which facilitates application to other projects.
3. New correlations to estimate the rotordynamic coefficients of annular seals and the hydraulic impeller interaction are provided in Chap. 10.6.
4. With ever-increasing computing power, steady or unsteady numerical flow calculations of entire pumps have meantime become state of the art. Chapter 8 has been revised accordingly. A statistical evaluation of the discrepancies between test and CFD has been done with the object of assessing the accuracy which can be expected from numerical flow calculations.
5. In Chap. 7, a new correlation for estimating the required NPSH of inducers from geometric parameters has been developed and methods for designing twin volutes for multistage pumps and vaned inlet chambers are provided.
6. A new correlation for estimating the radial pump performance in two-phase flow has been developed, Chap. 13. The viscosity anomality linked to emulsions is illustrated by some test data.
7. Impeller stresses and failure modes have been discussed in more detail and the correlations developed for estimating the risk of failure.
8. Some additions were made in most of the chapters, and some printing errors were corrected. The below table gives a detailed list of new items.

Villeneuve, Switzerland Johann Friedrich Gülich
January 2019

Supplements to the 4th Edition

For readers familiar with the 3rd (2014-) edition, the following table might be helpful to get a quick overview of the topics which have been added in the present edition.

Chapter	Topic	Figs., Tables, Eqs.
1.6	Pressure recovery in curved diffusers	Fig. 1.24, Eq. (1.54)
3.2.6	Calculation of friction factors, and optimization, of annular seals with deep serrations	Figs. 3.14-1 to 3.14-3, Table 3.7(4)
4.5.1	Design of inlet rings and low-flow impellers: remedies for over-sized impeller inlets causing noise and vibrations Exponents for trimming impellers	Eqs. (4.15a), (4.17a)
5.2.6	Additional test data on impeller inlet recirculation	Eq. (5.7c), Table D5.1, Fig. 5.16
5.6.6	Casing grooves for suppression of instabilities in the Q-H-curve of semi-open impellers (semi-axial)	Fig. 5.43-1
5.9	Some comments on the limits of understanding 3D-flow phenomena	
6.2.4	Some test data on the limit of the suction specific speed	Eq. (6.8a)
6.3.1	Effect of seal clearances on $NPSH_3$	
6.8.2.3	Unsteady volute cavitation	Fig. 6.47
6.8.3	Inlet rings for reducing cavitation problems during partload operation	Figs. 6.48 and 6.49
7.2.4	Suction impeller design with flat pressure distributions	Table D7.1-1, Fig. 7.10-1
7.3.4	"Double-acting" impellers with specific speeds of $n_q < 10$: two additional concepts reported in the literature	
7.5	Selection of speed and head for large semi-axial cooling water pumps	Eqs. (7.11a) and (7.11b)

(continued)

(continued)

7.7.3	Inducer performance prediction from geometry	Eq. (7.23a), Fig. 7.37-1
7.7.4	Inducer design systematic	
7.8.2	Multistage pumps with double volutes	Figs. 7.42-1 to 7.42-3
7.13	Design of vaned inlet casings for multistage pumps	Figs. 7.51-1 to 7.51-4
7.14	Update of impeller design according an analytical procedure	Table 7.10, Fig. 7.53
8.6.3	Unsteady CFD calculations update	
8.8.4	Accuracy of numerical performance calculations	Table 8.3, Figs. 8.22 to 8.24
8.8.5	CFD set-up parameters	Table 8.4
9.2.1	Static pressure at the impeller outlet for axial force calculations	Figs. 9.14-1 to 9.14-3
9.3.4	Radial forces in double volute	Fig. 9.30-1
9.3.7	Influence of the approach flow on radial forces: new test data	Fig. 9.32
10.6.2	Additional method for calculating the rotordynamic reaction force coefficients for annular seals	Table 10.4(1)
10.6.3	A set of correlations for calculating the rotordynamic coefficients for the hydraulic impeller interaction has been developed based on test data	Table 10.4(2)
10.6.6	Two case histories of super-synchronous rotor instabilities are discussed	
10.7.1	This chapter has been thoroughly revised: (1) Guidelines for staggering double-entry impellers; (2) Rotor/stator interaction: effect of incidence at diffuser or volute; (3) Test data for pressure pulsations at blade passing frequency; (4) Alternate stall in vaned diffuser; (5) Test data for lateral excitation forces at blade passing frequency	Figs. 10.25-1, 10.25-2, 10.26-1 and 10.28-1
10.7.3	Vibrations caused by cavitation in a pump with double-entry impeller Vibrations caused by inlet recirculation	Figs. 10.32-1 to 10.32-4
10.12.4	Trailing edge design for reduced vortex shedding	
10.14	Extensive discussion of vertical-pump vibrations and problems related to unsteady and non-uniform approach flow through the intake structure	Figs. 10.65 to 10.78 Tables 10.14, 10.15
13.1.5	Performance impairment when pumping emulsions of immiscible liquids	Fig. 13.12-1
13.2.4	A new model is presented for estimating the head reduction when pumping gas-liquid mixtures	Eqs. (13.19b to d) Figs. 13.23 and 13.24
14.1.2	Impeller fractures - analysis and prediction	Figs. 14.7-1 to 14.7-4 Eqs. (14.2a) to (14.2e)

Acknowledgements

The 1st English edition owed its existence to the initiative and sponsoring of the management of Sulzer Pumps. For this, I am most grateful to Dr. A. Schachenmann who initiated the project, R. Paley and Dr. R. Gerdes.

The book benefited from the help which I got from many colleagues at Sulzer Pumps in the USA and the UK to whom my sincere thanks are extended. Most important were the reviews of the English text. M. Cropper was instrumental in this activity. S. Bradshaw, R. Davey, Dr. J. Daly, D. Eddy, M. Hall, Dr. A. Kumar, P. Sandford, D. Townson and C. Whilde reviewed individual chapters.

J. H. Timcke meticulously checked most of the 1st English edition for consistency with the 2nd German edition and made many suggestions for making the text and figures easier to understand. Mrs. H. Kirchmeier helped with the figures and computer problems. Last but not least, my wife Rosemarie Gülich was a tremendous help in checking and improving the final text.

I am grateful to various individuals who provided me with the literature and granted permission for using figures: Prof. Dr.-Ing. F. Avellan, Dr. M. Farhat, Dr. O. Braun and Dr. S. Berten from the Ecole Polytechnique Lausanne; Prof. Dr.-Ing. D. H. Hellmann, Prof. Dr.-Ing. M. Böhle and Dr.-Ing. H. Roclawski from the Technical University of Kaiserslautern; Prof. Dr.-Ing. G. Kosyna, Prof. Dr.-Ing. habil. U. Stark, Mrs. Dr.-Ing. I. Goltz, Mrs. P. Perez, Dr.-Ing. H. Saathoff from the Technical University of Braunschweig; C. H. van den Berg, MTI Holland; Prof. Dr.-Ing. H. Wurm, A. Töws, Wilo, Dortmund; Dr. P. Dupont, Sulzer; U. Diekmann, Wilo-Emu; A. Nicklas, Sterling Fluid Systems, T. Folsche, CP Pumpen AG; and H. Bugdayci, IHC Merwede. I am especially grateful R. Palgrave for sharing with me a number of instructive photographs of pump components which have been subject to abrasive wear.

The 1st and 2nd editions benefited from the reviews of individual chapters provided by: Dr.-Ing. G. Scheuerer, ANSYS München; Dr. P. Heimgartner, W. Bolliger, W. Schöffler, Dr.-Ing. W. Wesche, Dr. P. Dupont, Dr. S. Berten, G. Caviola, E. Leibundgut, T. Felix, A. Frei, E. Kläui, W. Handloser (all from Sulzer); and J. H. Timcke.

The following organizations or individuals kindly granted permission for using figures:

- Sulzer Pumps Ltd, Winterthur
- T. McCloskey, Electric Power Research Institute, Palo Alto, CA, USA
- VDMA, Frankfurt
- VDI Verlag, Düsseldorf
- Mr. J. Falcimaigne, Institut Français du Pétrole, Paris
- ASME New York

The appropriate references are given in the figure captions.

Symbols, Abbreviations, Definitions

Hints for the reader: The text is written according to US-English spelling rules. As is customary in English publications, the decimal *point* is used.

Nomenclature: Unfortunately, there is no commonly accepted nomenclature and use of technical terms. As far as possible, I have consulted various standards as to the most accepted terms. The reader is referred to the extensive list of symbols given below. For easy reference, this list defines the chapters, tables or equations where the respective symbols are introduced. A number of subscripts from the German original were left unchanged since replacing them by meaningful English abbreviations involved too much of a risk of overlooking some items which are used throughout the text and the equations.

Conventions: Equations, tables and figures are numbered by chapter. The geometrical dimensions of impellers, diffusers and volutes are defined in Table 0.2.

To improve the readability, simplified expressions have sometimes been used (e.g., "volute" instead of "volute casing"). In order to avoid monotonous repetition of technical terms, synonyms are (sparingly) employed.

Formulae frequently used in practice were gathered in tables which present the sequence of calculation steps required to solve a specific problem. These tables help to find information quickly without looking through a lot of text; they also facilitate programming. The equations presented in the tables are labeled by "T". For example, Eq. (T3.5.8) refers to equation 8 in Table 3.5. Most of the tables are labeled as "Table 6.1," for example. Some "data tables" are referred to as "Table D6.1" for instance; this subterfuge was made necessary by the layout of the German editions which contained "tables" and "plates."

Mathematical expressions: Empirical data in the literature are frequently presented in graphical form. In most cases, such data are given in this book in the form of approximate equations in order to ease programming and to save space.

For reasons of simplicity, the upper limit of a sum is not specified when there can be no doubt about the variable. For example, $\sum_{st} P_{RR}$ stands for

$\sum_{i=1}^{i=z_{st}} P_{RR,i}$ and represents the sum of the disk friction losses in all stages of a multistage pump.

An equation of the form y = a × exp(b) stands for y = a × e^b, where "e" is the base of the natural logarithm.

The symbol ~ is used for "proportional to"; for example, $P_{RR} \sim d_2^5$ stands for "the disk friction loss is proportional to the fifth power of the impeller diameter."

Frequent reference is made to the specific speed n_q which is always calculated with n in rpm, Q in m^3/s and H in m. For conversion to other units, refer to Table D2.1 or Table 3.4. For simplicity, the specific speed n_q is treated as a dimensionless variable even though this is not true.

Many diagrams were calculated with MS Excel which has limited capabilities for graphic layout. For example, 1E + 03 stands for 10^3; curve legends cannot show symbols or subscripts. The sketches should not be understood as technical drawings. Equations in the text are written for clarity with multiplier sign, i.e., a × b (instead of a b). This is not done in the numbered equations.

Literature: There is a general bibliography quoted as [B.1], [B.2], etc., while standards are quoted as [N.1], [N.2], etc. The bulk of the literature is linked to the individual chapters. This eases the search for the literature on a specific topic. The roughly 800 quotations provided represent only 1 % (order of magnitude) if not less of the relevant literature. This statement applies to all topics treated in this book. The literature quoted was selected with the objectives: (1) to provide the sources of specific data or information; (2) to back up a statement; (3) to refer the reader to more details on the particular investigation or topic; (4) to provide reference to the literature in neighboring fields. In spite of these criteria, the selection of the literature quoted is to some extent coincidental.

In order to improve the readability, facts which represent the state of the art are not backed up systematically by quoting the literature where they may have been reported. In many cases, it would be difficult to ascertain where such facts were published for the first time.

Patents: Possible patents on any devices or design features are not necessarily mentioned. Omission of such mention should not be construed so as to imply that such devices or features are free for use to everybody.

Disclaimer of warranty and exclusion of liabilities: In spite of careful checking text, equations and figures, neither the Springer book company nor the author:

- makes any warranty or representation whatsoever, express or implied, (A) with respect of the use of any information, apparatus, method, process or similar item disclosed in this book including merchantability and fitness for practical purpose, or (B) that such use does not infringe or interfere with privately owned rights, including intellectual property, or (C) that this book is suitable to any particular user's circumstances; or

- assumes responsibility for any damage or other liability whatsoever (including consequential damage) resulting from the use of any information, apparatus, method, process or similar item disclosed in this book.

In this context, it should well be noted that much of the published information on pump hydraulic design is empirical in nature. The information has been gathered from tests on specific pumps. Applying such information to new designs harbors uncertainties which are difficult to asses and to quantify.

Finally, it should be noted that the technological focus in the various sectors of the pump industry is quite different. Low-head pumps produced in vast quantities are designed and manufactured to other criteria than engineered high-energy pumps. This implies that the recommendations and design rules given in this book cannot be applied indistinctly to all types of pumps. Notably, issues of standardization and manufacturing are not addressed in this text to any extent.

The recommendations and application limits are considered to be mostly on the safe side. There are certainly pumps on the market and in service with proven operation records, which exceed some of the limits given in this book.

Symbols, Abbreviations and Definitions

Unless otherwise noted all equations are written in consistent units (SI-System). Most symbols are defined in the following. As appropriate, the equation or chapter is quoted where the symbol has been defined or introduced. Vectors in the text and in equations are printed as bold characters. Symbols with local significance only are defined in the text.

The following tables may help in understanding the physical meaning of various parameters of prime importance:

- Table 0.1 and 0.2: Geometric dimensions of the flow channels, flow angles and velocities
- Table 2.2: Head and net positive suction head (NPSH)
- Tables 3.1 and 3.2: Velocity triangles
- Table 3.4: Model laws and dimensionless parameters

A	area, cross section	Chapter or equation
A [%]	elongation at rupture (ductility of material)	Chap. 14
A	amplitude	Chap. 10
A_{1q}	impeller inlet throat area (trapezoidal: $A_{1q} = a_1 \times b_1$)	
A_{2q}	area between vanes at impeller outlet ($A_{2q} = a_2 \times b_2$)	
A_{3q}	diffuser/volute inlet throat area ($A_{3q} = a_3 \times b_3$)	
a	distance between vanes (subscript 1 to 6)	Table 0.2
a	sound velocity in a pipe	Eq. (10.17)

(continued)

(continued)

a_o	sound velocity in the fluid	Eq. (10.17)
a_L	sound velocity in the casing material	Eq. (T6.1.7)
BEP	best efficiency point	
b	acceleration	
b	width of channel in the meridional section	
b_2	impeller outlet width; if double-entry, per impeller *side*	
b_{2tot} (b_{2ges})	impeller outlet width including shrouds	Eq. (9.6)
b_{ks}	solid-borne noise acceleration	Eq. (10.6)
CNL	cavitation sound pressure	Table 6.1
CV	solid-borne noise as RMS of acceleration	
CV*	dimensionless solid-borne noise acceleration	$CV^* = CV \times d_1/u_1^2$
c	absolute velocity	Chap. 1.1
c	rotor damping coefficient	Eq. (10.7)
c_A	axial thrust reduction coefficient	Eq. (9.4), Table 9.1
c_A	approach-flow velocity in pump sump	Chap. 10.14
c_d	flow velocity in discharge nozzle	
c_{Fe}	concentration of iron ions	Eq. (14.7), Table 14.7
c_{3q}	average velocity in diffuser throat	$c_{3q} = Q_{Le}/(z_{Le} \times A_{3q})$
c_c	cross-coupled damping	Eq. (10.7)
c_{eq}	roughness equivalence factor	Eq. (1.36b)
c_f	friction coefficient of a flat plate	Eq. (1.33)
c_p	pressure recovery coefficient	Eqs. (1.11), (1.40), (T9.1.5)
c_p	specific heat at constant pressure	Chap. 13.2
c_{ph}	phase velocity	Chap. 10.7.1
c_s	flow velocity in suction nozzle	
c_s	concentration of solids	Table 14.16
$c_{s,eq}$	equivalent concentration of solids	Table 14.16
c_T	velocity at inlet to suction bell	Eq. (11.15)
c_v	solid volume concentration	Table 13.5
CWP	cooling water pump	
D	damping coefficient	Chap. 10.6.5
D, d	diameter	
D_{fz}	diffusion factor	Table 7.5
DR	liquid/gas density ratio $DR \equiv \rho^* = \rho'/\rho''$	Chap. 13.2
D_T	inlet diameter of suction bell	Fig. (11.20)
d_{3q}	equivalent diameter of volute throat	Eq. (T7.7.7)
d_b	arithmetic average of diameters at impeller or diffuser e.g., $d_{1b} = 0.5\ (d_1 + d_{1i})$; defined such that: $A_1 = \pi \times d_{1b} \times b_1$	Table 0.2
d_d	inner diameter of discharge nozzle	
d_k	ball passage (sewage or dredge pumps)	Chap. 7.4
d_m	geometric average of diameters at impeller or diffuser, e.g., $d_{1m} = \sqrt{0.5(d_{1a}^2 + d_{1i}^2)}$	Table 0.2
d_n	hub diameter	
d_D	diameter at shaft seal	Table 9.1
d_s	inner diameter of suction nozzle	
d_s	diameter of solid particles	Table 14.16

(continued)

(continued)

E	Young's modulus of elasticity	
E_R	maximum erosion rate (at location of highest metal loss)	Table 6.1
$E_{R,a}$	metal loss rate in mm/a	Tables 14.7 and 14.16
e	vane thickness	Table 0.2
F	force	
F_{ax}	axial force ("axial thrust")	
F_{Dsp}	radial thrust correction factor for double volutes, Fig. 9.18 Table 9.4	
F_R	radial force ("radial thrust")	Eq. (9.6), Table 9.4
Fr	Froude number	Chaps. 10.14, 11.7.3
F_r, F_t	radial and tangential forces on rotor	Eq. (10.8)
F_{cor}	corrosion factor	Table 6.1
F_{Mat}	material factor for cavitation	Table 6.1, for abrasion: Table 14.16
f	frequency	
f_{EB}	natural frequency at operational speed	Chap. 10.6.5
f_{e1}	natural frequency	Chap. 10.6.2
f_{kr}	critical speed (as frequency)	Chap. 10.6.5
f_L	influence of leakage flow on disk friction	Eq. (T3.6.7), Table 3.6
f_n	frequency of rotation $f_n = n/60$	
f_q	impeller eyes per impeller: single-entry $f_q = 1$; double-entry $f_q = 2$	
f_H	correction factor for head (roughness, viscosity)	Eq. (3.32)
f_Q	correction factor for flow rate (viscosity)	Eq. (3.32)
f_R	roughness influence on disk friction	Eq. (T3.6.6)
f_{RS}	frequency of rotating stall	
f_η	correction factor for efficiency (roughness, viscosity)	Eq. (3.31)
g	acceleration due to gravity ($g = 9.81\ m/s^2$, rounded)	Appendix 17.4
H	head *per stage*	Tables 2.2 and 3.3
H_{Mat}	hardness of material	Table 14.16
H_s	hardness of solid particle	Table 14.16
H_{tot}	total head of a multistage pump	Table 2.2
H_p	static pressure rise in impeller	Eq. (T3.3.8)
h_{tot}	total enthalpy	Eq. (1.4)
h	casing wall thickness (at location of accelerometer)	Table 6.1
h_D	casing cover wall thickness	Table 6.1
I_{ac}	acoustic intensity	Table 6.1
I_{Ref}	reference value for intensity	Table 6.1
i	incidence (i = blade angle minus flow angle)	Table 3.1
J_{sp}	integral over diffuser or volute throat area	Eqs. (3.15), (4.13)
k	rotation of fluid in impeller sidewall gap $k = \beta/\omega$	Eq. (9.1), Table 9.1
k_E, k_z	rotation of fluid at inlet to impeller sidewall gap	Fig. 9.1
k	stiffness	Eq. (10.7)
k_c	cross-coupled stiffness	Eq. (10.7)
k_n	blockage caused by hub: $k_n = 1 - d_n^2/d_1^2$	
k_R	radial thrust coefficient (steady component)	Eq. (9.6)
$k_{R,D}$	radial thrust coefficient referred to d_2 (steady)	Table 9.7

(continued)

(continued)

$k_{R,dyn}$	dynamic (unsteady) radial thrust coefficient	Table 9.7
$k_{R,tot}$	total radial thrust coefficient (steady and unsteady)	Table 9.7
$k_{R,o}$	radial thrust coefficient (steady) for operation at Q = 0	Table 9.7
k_{Ru}	radial thrust coefficient (steady)	Eq. (9.7)
k_{RR}	disk friction coefficient	Table 3.6
L	length	
L_{PA}	A-type sound pressure level	Table 10.4
L_{Dam}	damage length	
L_{cav}	cavity length	
M	torque	
m	difference of impeller and diffuser periodicity	Chap. 10.7.1
m	mass coefficient	Eq. (10.7)
$\dot{m}$	mass flow rate	
m_c	cross-coupled mass	Eq. (10.7)
NPSH	net positive suction head	
$NPSH_A$	net positive suction head available	Tables 2.2 and 6.2
$NPSH_i$	net positive suction head required for cavitation inception	
$NPSH_R$	net positive suction head required according to a specific cavitation criterion	Chaps. 6.2.2, 6.2.5, 6.3
$NPSH_x$	net positive suction head required for operation with x-percent head drop	Chap. 6.2.2
NL	fluid-borne sound pressure as RMS value; $NL^* = 2NL/(\rho \times u_1)$	
NL_o	background sound pressure	Chap. 6.5
n	rotational speed (revolutions per minute)	
$n^{(s)}$	rotational speed (revolutions per second)	
n_N	nominal speed	
n_q	specific speed [rpm, m^3/s, m]	Tables D2.1 and 3.4, Chap. 3.4
n_{ss}	suction specific speed [rpm, m^3/s, m]	Chap. 6.2.4, Table 3.4
P	power; without subscript: power at coupling	
P_i	inner power	Table 3.5
P_m	mechanical power losses	Table 3.5
P_u	useful power transferred to fluid $P_u = \rho \times g \times H_{tot} \times Q$	Table 3.5
P_{RR}	disk friction power	Tables 3.5 and 3.6
P_{ER}	specific erosion power $P_{ER} = U_R \times E_R$	Table 6.1
P_{er}	disk friction power loss caused by balance device	Table 3.6
P_{s3}	power loss dissipated in interstage seal	Tables 3.5 and 3.7 (1)
PI	pitting index	Eq. (14.8)
p	static pressure	
p	periodicity	Chap. 10.7.1
p_{amb}	ambient pressure at location of pump installation (usually atmospheric pressure)	
p_e	pressure above liquid level in suction reservoir	Table 2.2
p_g	gas pressure (partial pressure)	Appendix 17.3
p_i	implosion pressure	Table 6.1
p_v	vapor pressure	

(continued)

(continued)

Q	flow rate, volumetric flow	
Q_{La}	flow rate through impeller: $Q_{La} = Q + Q_{sp} + Q_E + Q_h = Q/\eta_v$	
Q_{Le}	flow rate through diffuser: $Q_{Le} = Q + Q_{s3} + Q_E$	
Q_E	flow rate through axial thrust balancing device	
Q_{CB}	flow rate through center bush, Figs. 2.9 and 2.10	back-to-back pumps
Q_h	flow rate through auxiliaries (mostly zero)	
Q_R	rated flow or nominal flow rate	Chap. 15
Q_{sp}	leakage flow rate through seal at impeller inlet	Tables 3.5 and 3.7 (1)
Q_{s3}	leakage flow rate through interstage seal	Tables 3.5 and 3.7 (1)
q^*	flow rate referred to flow rate at best efficiency point	$q^* \equiv Q/Q_{opt}$
R, r	radius or gas constant	Table 13.3
R_G	degree of reaction	Chap. 3.2
Re	Reynolds number, channel: $Re = c \times D_h/n$; plate or blade: $Re = w \times L/n$	
Ro	Rossby number	Chap. 5.2
R_m	tensile strength	
RMS	root mean square	
r_{3q}	equivalent radius of volute throat area	Table 7.7
S	submergence	Chap. 11.7.3
S	sound absorbing surface of inlet casing	Table 6.1
SG	specific gravity; $SG \equiv \rho/\rho_{Ref}$ with $\rho_{Ref} = 1000$ kg/m^3	
S_{Str}	Strouhal number	Table 10.13
s	*radial* clearance	Eq. (3.12), Figs. 3.12, 3.15 Table 3.7 (1) and (2)
s_{ax}	axial distance between impeller shrouds and casing	Fig. 9.1
T	temperature	
t	time	
t	pitch	$t = \pi \times d/z_{La}$ (or z_{Le})
t_{ax}	axial casing part in impeller sidewall gap	Fig. 9.1
U	wetted perimeter (of a pipe or channel)	
U_R	ultimate resilience: $U_R = R_m^2/(2 \times E)$	Table 6.1
u	circumferential velocity	$u = \pi \times d \times n/60$
V	volume	
w	relative velocity	
w_{1q}	average velocity in impeller throat area	$w_{1q} = Q_{La}/(z_{La} \times A_{1q})$
x	dimensionless radius	$x = r/r_2$ Table 9.1
x	gas (or vapor) mass content; mass concentration of solids	Chap. 13
x_D	mass concentration of dissolved gas	Appendix 17.3
x_{ov}	overlap at impeller/diffuser side disks	Fig. 9.1
Y	specific work	$Y = g \times H$
$Y_{sch} \equiv Y_{th}$	specific work done by the impeller blades: $Y_{th} = g \times H_{th}$	Table 3.3
$Y_{th\infty}$	specific work done by the impeller blades with vane congruent flow	
y^+	dimensionless distance from the wall	Table 8.1
Z	real gas factor	Table 13.3

(continued)

(continued)

Z_h	hydraulic losses (impeller: Z_{La} diffuser: Z_{Le})	
z	height coordinate	
z_{La}	number of impeller blades	
z_{Le}	number of diffuser vanes (volute: number of cutwaters)	
z_R	number of return vanes	
z_{pp}	number of pumps operating in parallel	
z_{st}	number of stages (in Fig. 2.9, 2.10: $z_{st} = z_{st1} + z_{st2}$)	
z_{st1}	number of stages in first group of stages	back-to-back pumps
z_{st2}	number of stages in second group of stages	back-to-back pumps
z_{VLe}	number of vanes of prerotation control device	
$\alpha \equiv$ GVF	gas content, gas volume fraction, void fraction	Table 13.2
α	angle between direction of circumferential and absolute velocity	
α_k	notch factor	Eq. (T14.1.7)
α_T	total absorption coefficient	Table 6.1
β	angle between relative velocity vector and the negative direction of circumferential velocity	
β	angular velocity of fluid between impeller and casing	Chap. 9.1
β	mass transfer coefficient	Chap. 14.3, Table 14.8
γ	impeller discharge coefficient ("slip factor")	Table 3.2
δ^*	displacement thickness	Eq. (1.18)
Δp^*_d	pressure pulsations(dimensionless)	Eq. (10.1)
Δp_a	amplitude of pressure pulsations	Chap. 10.2.6
Δp_{p-p}	pressure pulsations measured peak-to-peak	Chap. 10.2.6
ε	angle in polar coordinate system	
ε	equivalent sand roughness	Chap. 1.5.2
ε_{sp}	wrap angle of the inner volute (for double volutes)	Table 0.2
ζ	loss coefficient (with subscript La, Le, Sp etc.)	Table 3.8
ζ_a	lift coefficient	Tables 7.1 and 7.4
ζ_w	drag coefficient	Table 7.4
η_{vol}, η_v	volumetric efficiency	Eq. (T3.5.9)
η	overall efficiency (at coupling)	Eq. (T3.5.3)
η_i	inner efficiency	Eq. (T3.5.5)
η_h	hydraulic efficiency	Eq. (T3.5.8) and Table 3.8
η_D	diffuser efficiency	Eq. (1.43)
η_{st}	stage efficiency	Eq. (T3.5.7)
θ_u	similarity parameter for cavitation erosion	Table 6.1
ϑ	diffuser opening angle	Eq. (1.42)
κ	exponent of isentropic expansion/compression	Table 13.3 (2)
λ	angle between vanes and side disks (impeller or diffuser)	Table 0.1
λ	power coefficient	Table 3.4
λ	wave length	Table 10.12
λ_c, λ_w	coefficient for NPSH calculation	Eq. (6.10)
λ_R	friction coefficient for pipes and channels	Eq. (1.36)
μ	dynamic viscosity: $\mu = \rho \times \nu$	
ν	kinematic viscosity: $\nu = \mu/\rho$	

(continued)

(continued)

ν	hub ratio	$n = d_n/d_{1a}$
ν_1, ν_2	vibration orders, natural numbers (1, 2, 3, ….)	
ξ	hydraulic vane loading according to [7.2]	Table 7.1
ρ	density	
ρ''	density of gas or saturated vapor	
ρ_{mat}	density of material	
ρ_p	density of casing material	
ρ_s	density of solids suspended in the fluids	Chaps. 13.4, 14.5
σ	cavitation coefficient (same subscripts as NPSH)	Table 3.4
σ	mechanical stress	Chap. 14.1
τ	blade blockage factor	Table 0.1
τ	shear stress	
φ	flow coefficient	Table 3.4
φ_{sp}	flow coefficient of impeller sidewall gap	Table 9.1
ψ	head coefficient	Table 3.4
ψ_p	pressure coefficient of static pressure rise in impeller	Table 3.3
Ω	orbit- (vibration-)circular frequency	Chap. 10.6.2
Ω_{limit}	orbit frequency of stability limit	Eq. (10.9)
ω	angular rotor velocity	
ω_E	circular natural frequency	Chap. 10
ω_s	universal specific speed	Tables D2.1 and 3.4

Subscripts, Superscripts and Abbreviations

Sequence of calculations: in the pumping mode the fluid flows from station 1 to 6, in the turbine mode from 6 to 1:

1	impeller blade leading edge (low pressure)	
2	impeller blade trailing edge (high pressure)	
3	diffuser vane leading edge or volute cutwater	
4	diffuser vane trailing edge	
5	return vane leading edge	
6	return vane trailing edge	
A	plant	
a	plant, executed pump, prototype	
al	allowable	
ax	axial	
a, m, i	outer, mean, inner streamline	
B	blade angle (impeller, diffuser, volute cutwater)	
BEP	best efficiency point	Chap. 2.2.4
BPF	impeller blade passing frequency $z_{La} \times f_n$	Chap. 10
cor	corrosion	
CP	column pipe of a vertical pump	
DE	drive end (coupling end of a pump shaft)	
Ds (FS)	front shroud	

(continued)

(continued)

d	discharge nozzle	
ER	erosion	
eff	effective	
FS	front shroud	
Gap A	radial gap between impeller shrouds and diffuser side plate	Table 0.2 (2)
Gap B	radial gap between impeller blades and diffuser vanes or volute cutwater	Table 0.2 (2)
Gap E, Gap F	axial gap between impeller (rear and front) shrouds and casing	Table 0.2 (2)
GVF	gas volume fraction, void fraction	Table 13.2
ISR	impeller side room (impeller sidewall gaps E, F) Table 0.2	Chap. 9.1
h	hydraulic	
L	runaway in turbine mode (M = 0)	Chap. 12
La	impeller	
Le	diffuser	
LE	leading edge	
M	model	
m	meridional component	
max	maximum	
min	minimum	
mix	two-phase mixture	
NDE	non-drive end (of a pump shaft)	
o	shut-off operation (Q = 0)	
opt	operation at maximum (best) efficiency (BEP)	
P	pumping mode	Chap. 12
PS (DS)	pressure surface (pressure side)	
pol	polytropic	Chap. 13.2
q	average velocity calculated from continuity (to be distinguished from velocity vector)	
RB	onset of recirculation	
Rec (Rez)	recirculation	
Ref	reference value	
RR	disk friction	Chap. 3.6.1, Table 3.6
RS	rear shroud or hub	
RSI	rotor–stator interaction	Chap. 10.7.1
r	Radial	
s	inlet or suction nozzle	
s	solid particle	Chap. 13.4
sch	blade or vane	
SF	shockless flow (zero incidence)	Eq. (T3.1.10)
SM	separation margin	Chap. 10.14
Sp	volute	
sp	annular seal, leakage flow	
SPL	single-phase liquid	
SS	suction surface (suction side)	
st	stage	
stat	static	

(continued)

(continued)

T	turbine mode	Chap. 12
TE	trailing edge	
TP	two-phase	
Ts	(RS) rear shroud or hub	
th	theoretical flow conditions (flow without losses)	
tot	total (total pressure = static pressure + stagnation pressure)	
u	circumferential component	
v	loss	
v	viscous fluid	Chap. 13
VPF	diffuser vane passing frequency $z_{Le} \times f_n$	Chap. 10
w	water	Chap. 13.1
w	resistance curve in turbine mode locked rotor (n = 0)	Chap. 12
zul (al)	allowable	

The following are superscripts

′	with blade blockage	Tables 0.1 and 3.1
*	dimensionless quantity: all dimensions are referred to d_2 e.g., $b_2^* = b_2/d_2$, velocities are referred to u_2, e.g., $w_1^* = w_1/u_2$	
′	liquid phase	Chaps. 6 and 13
″	gaseous phase	Chaps. 6 and 13

Table 0.1 Dimensions and flow parameters

	Location, main dimensions	Blade blockage	Flow parameters	Vane angles
Impeller: z_{La}	Inlet: $d_{1a}, d_{1m}, d_{1i}, d_n, a_1, e_1$	without	$u_{1a}, u_{1m}, u_{1i}, c_{1m}, c_{1u}, c_1, w_1, \alpha_1, \beta_1$	$\beta_{1B,a}$, β_{1B}, $\beta_{1B,i}$
		$\tau_1 = \dfrac{1}{1-\dfrac{z_{La} e_1}{\pi\, d_1 \sin\beta_{1B} \sin\lambda_{La}}}$	$c_{1m}', c_{1u}, w_1', c_1', \alpha_1', \beta_1'$, $w_{1q} = Q_{La}/(z_{La} A_{1q})$	
	Outlet: $d_{2a}, d_{2m}, d_{2i}, b_2, a_2, e_2, e$	$\tau_2 = \dfrac{1}{1-\dfrac{z_{La} e_2}{\pi\, d_2 \sin\beta_{2B} \sin\lambda_{La}}}$	$c_{2m}', c_{2u}, c_2', w_{2u}, w_2', \alpha_2', \beta_2'$	$\beta_{2B,a}$, β_{2B}, $\beta_{2B,i}$
		without	$u_{2a}, u_{2m}, u_{2i}, c_{2m}, c_{2u}, c_2, w_{2u}, w_2, \alpha_2, \beta_2$	
Diffuser or volute: z_{Le}	Inlet: d_3, b_3, a_3, e_3, $A_{3q} = a_3\, b_3$	without	$c_{3m}, c_{3u}, c_3, \alpha_3$	$\alpha_{3B,a}$, α_{3B}, $\alpha_{3B,i}$
		$\tau_3 = \dfrac{1}{1-\dfrac{z_{Le}\, e_3}{\pi\, d_3 \sin\alpha_{3B} \sin\lambda_{Le}}}$	$c_{3m}', c_{3u}, c_3', \alpha_3'$, $c_{3q} = Q_{Le}/(z_{Le} A_{3q})$	
	Outlet: d_4, b_4, a_4, e_4, $A_4 = a_4\, b_4$	$c_4 = \dfrac{Q_{Le}}{z_{Le}\, b_4\, a_4}$		$\alpha_{4B,a}$, α_{4B}, $\alpha_{4B,i}$
Return vanes: z_R	Inlet: d_5, b_5, a_5, e_5	without	$c_{5m}, c_{5u}, c_5, \alpha_5$	$\alpha_{5B,a}$, α_{5B}, $\alpha_{5B,i}$
		$\tau_5 = \dfrac{1}{1-\dfrac{z_R \cdot e_5}{\pi\, d_5 \sin\alpha_{5B}}}$	$c_{5m}', c_{5u}, c_5', \alpha_5'$	
	Outlet: d_6, b_6, a_6, e_6	$\tau_6 = \dfrac{1}{1-\dfrac{z_R e_6}{\pi\, d_6 \sin\alpha_{6B}}}$	$c_{6m}', c_{6u}, c_6', \alpha_6'$	$\alpha_{6B,a}$, α_{6B}, $\alpha_{6B,i}$
		without	$c_{6m}, c_{6u}, c_6, \alpha_6$	

Note:
All flow parameters can be supplemented by subscripts a, m or i in order to define the streamline e.g. $c_{1m,a}, \beta_{1a}, \beta_{1i}$
Without special subscript: $u_1 \equiv u_{1a}$ and $d_1 \equiv d_{1a}$ as well as $d_2 \equiv d_{2a}$, if $d_{2a} = d_{2i}$
The meridional velocity components are equal in relative and absolute system: $w_m = c_m$
Circumferential velocity components c_u and w_u are not influenced by blade blockage.

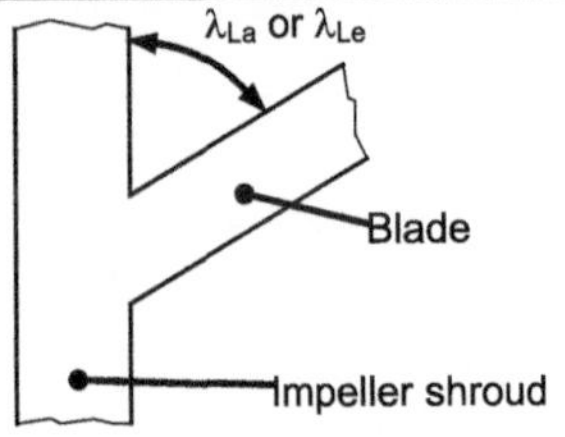

Influence of blade inclination on blade blockage (twisted blades)

Table 0.2 (1) Geometric dimensions

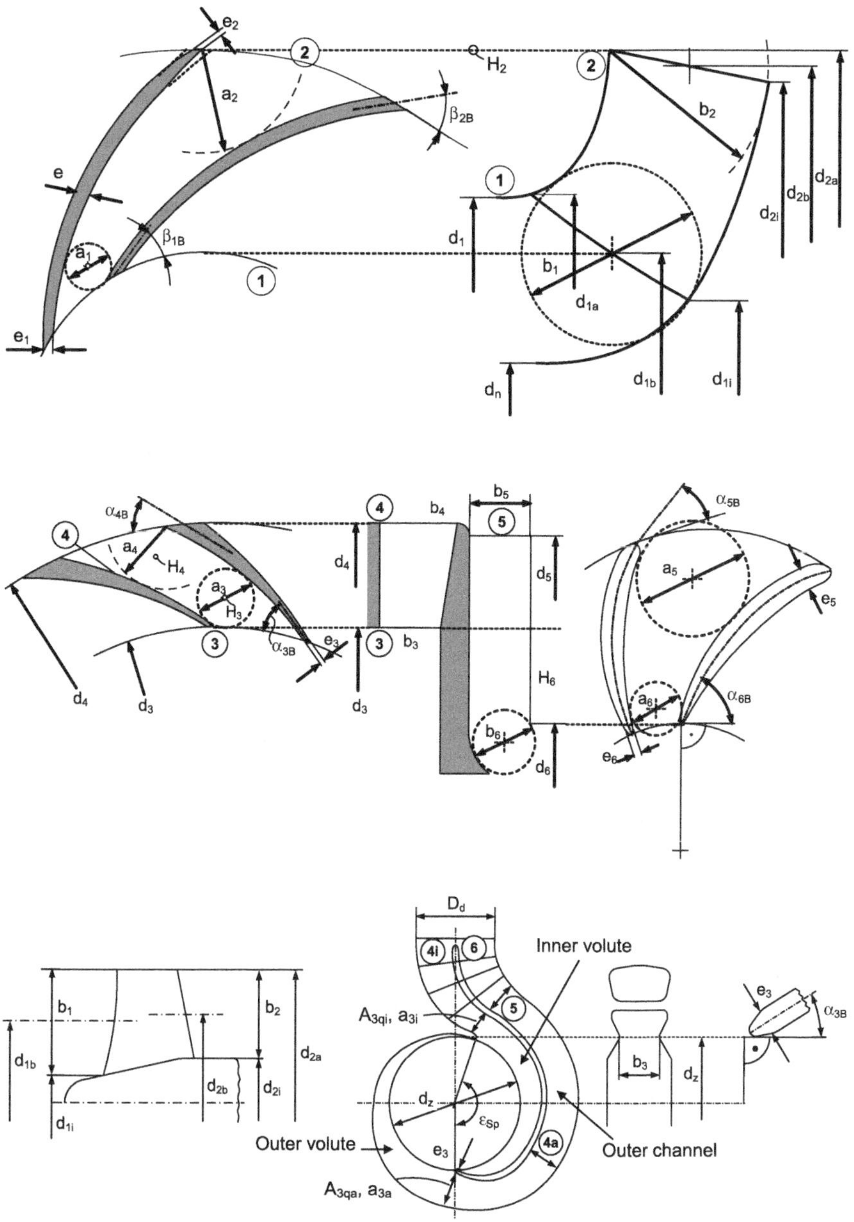

Table 0.2 (2) Geometric dimensions

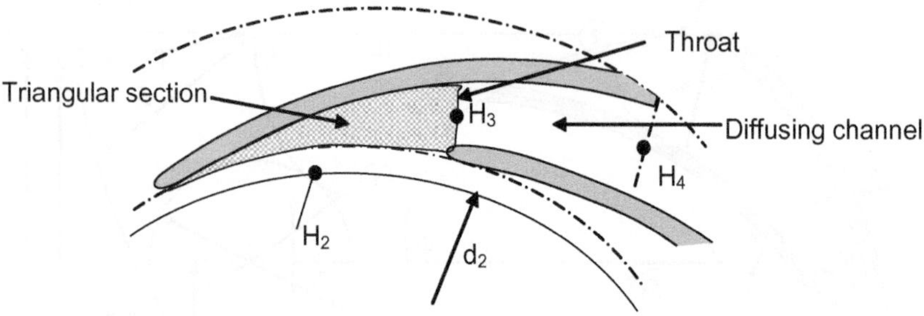

Impeller sidewall gaps and diffuser inlet geometry for multistage pumps

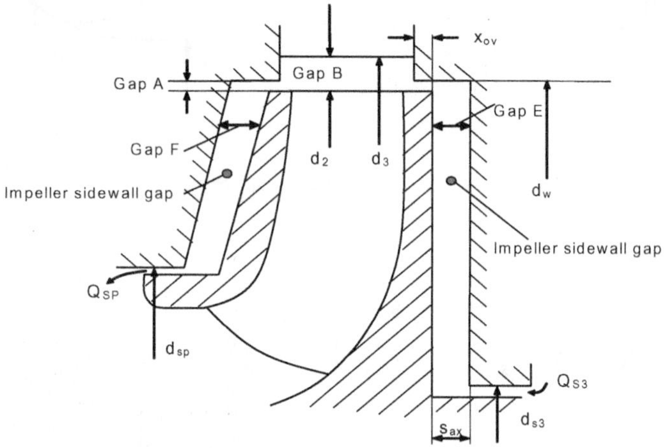

Diffuser inlet geometry with chamfer, for multistage pumps, Chap. 10 [61]

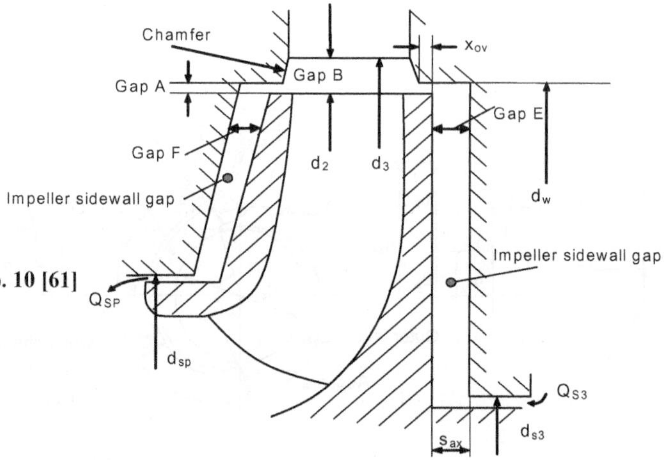

Pump Styles

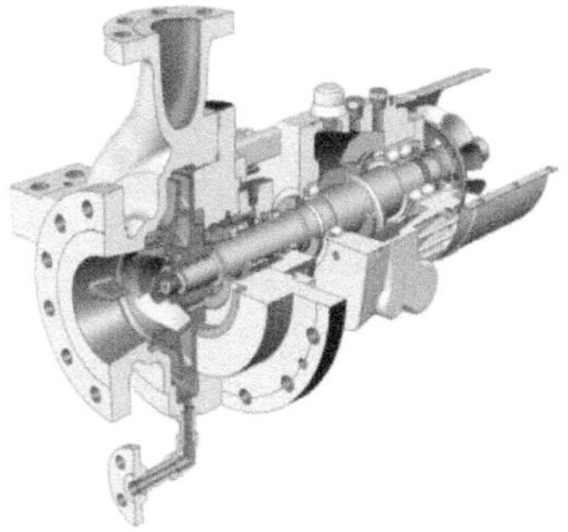

OHH—Overhung Single Stage
ISO 13709 (API 610) Type OH2

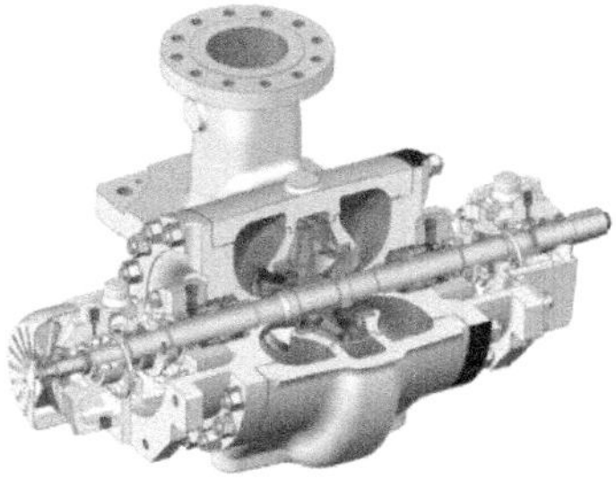

BBS—Between Bearings Single Stage
ISO 13709 (API 610) Type BB2

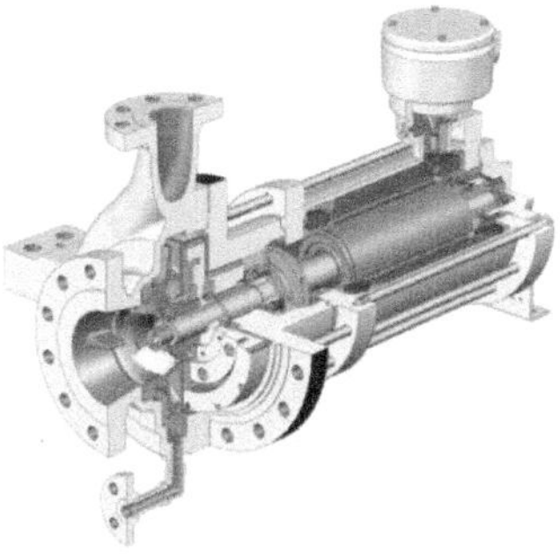

OHC—Overhung Single Stage
with Canned Motor
API 685

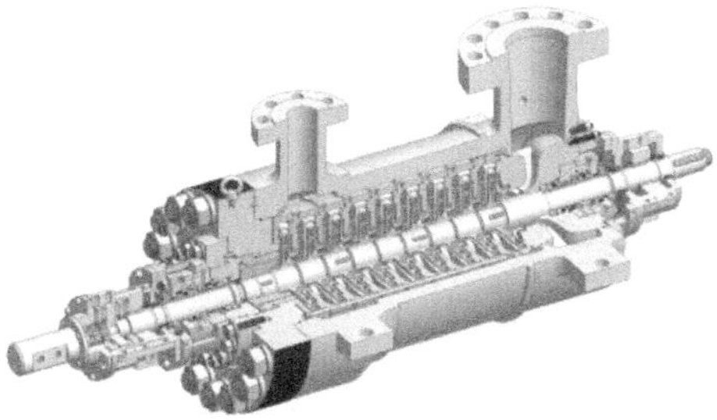

GSG—Inline or Opposed Impeller
Diffuser Barrel Type
ISO 13709 (API 610) Type BB5

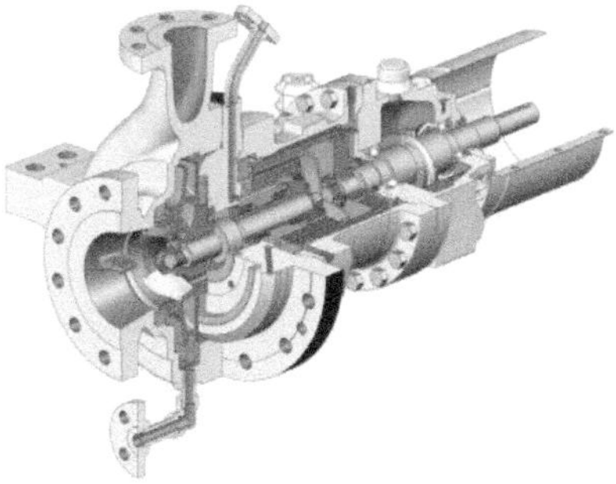

OHM—Overhung Single Stage
with Magnetic Coupling, API 685

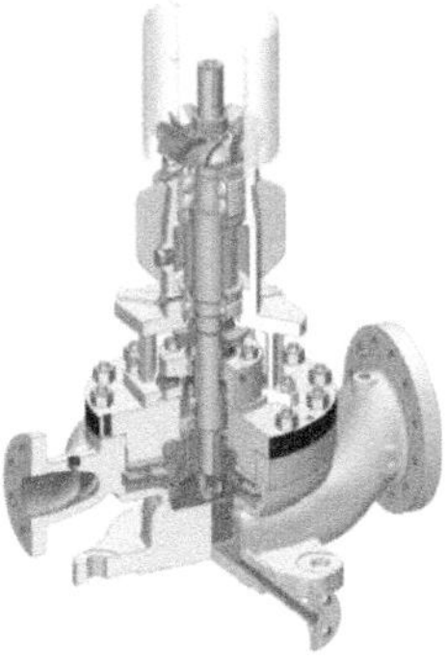

OHV—Vertical Inline
ISO 13709 (API 610) Type OH3

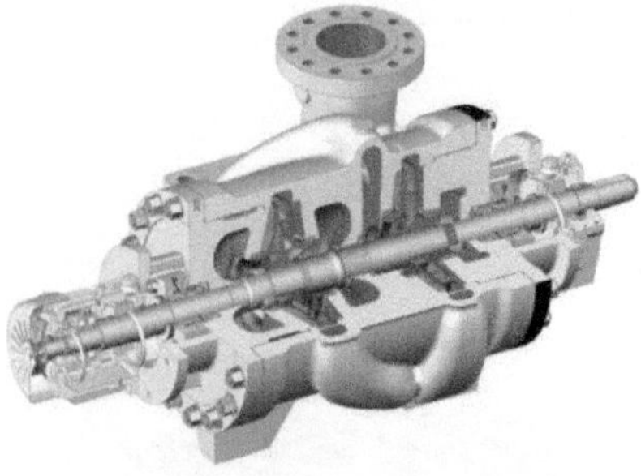

BBT-D—Between Bearings
Two-Stage Double Suction
ISO 13709 (API 610) Type BB2

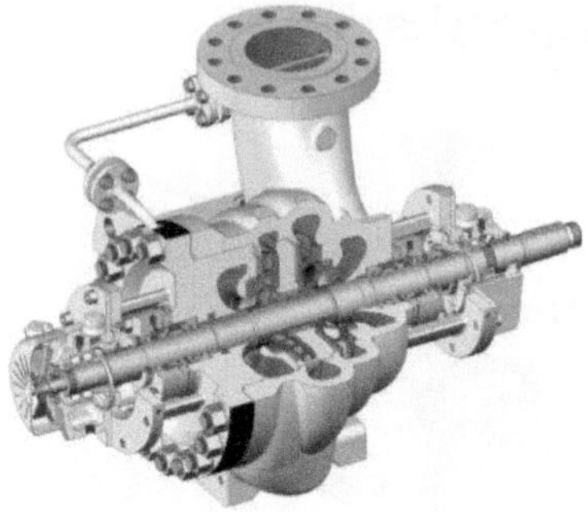

BBT—Between Bearings Two-Stage
ISO 13709 (API 610) Type BB2

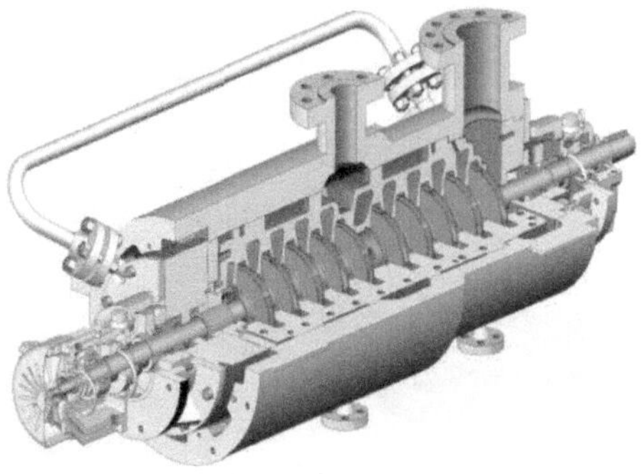

CP—Dual Volute Opposed
Impeller Barrel Type
ISO 13709 (API 610) Type
BB5

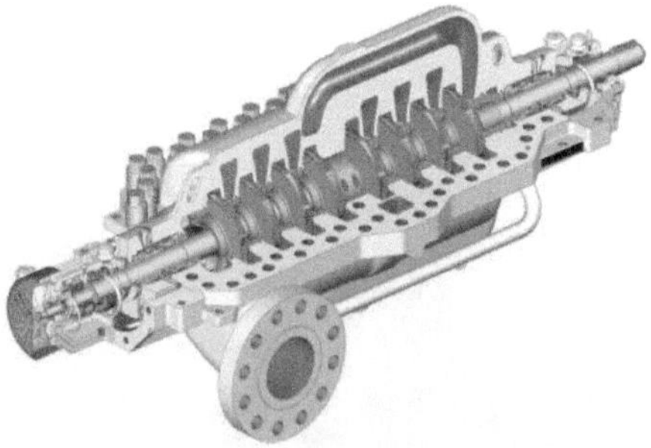

MSD—Dual Volute Opposed
Impeller Horizontally Split
ISO 13709 (API 610) Type BB3

Contents

List of Tables

List of Data Tables

Chapter 1
Fluid Dynamic Principles

The nearly inexhaustible variety of flow phenomena - from the flow through blood vessels, the flow in centrifugal pumps to global weather events - is based on a few basic physical laws only. In this chapter these will be briefly reviewed and their general nature illuminated. Emphasis will be on the phenomena which are of special interest and significance to the pump engineer. Basic knowledge of the terms of fluid dynamics is taken for granted. There is a wealth of textbooks and handbooks on fluid dynamics, only a few can be quoted: [1–3, 6, 7, 11, 15, 18, 19].

1.1 Flow in the Absolute and Relative Reference Frame

In turbo machine design a flow described in fixed coordinates is called an "absolute" movement, while a flow in the rotating reference frame is termed "relative". The flow in the relative reference frame corresponds to the movement that would be seen by a co-rotating observer. A point on a rotating disk is stationary in the relative reference frame, while it describes a circle in the absolute system. If a mass moves radially outward in a guide on a rotating disk, it follows a *straight* path in the relative reference frame, while it describes a *spiral-shaped* movement in the absolute reference frame. When transforming a movement from the absolute to the relative system, the centrifugal and Coriolis forces must be introduced. The absolute acceleration b_{abs} is then obtained as a vectorial sum from relative, centrifugal and Coriolis acceleration as:[1]

$$\mathbf{b}_{abs} = d\mathbf{w}/dt - \omega^2\mathbf{r} + 2(\omega \times \mathbf{w})$$

The velocity conditions are described by three vectors: the peripheral velocity $\mathbf{u} = \omega \times \mathbf{r}$, the relative velocity $\mathbf{w}$ and the absolute velocity $\mathbf{c}$ which is obtained

[1] Vectors are printed in bold characters.

J. F. Gülich, *Centrifugal Pumps*,
https://doi.org/10.1007/978-3-030-14788-4_1

Fig. 1.1 Vector diagram

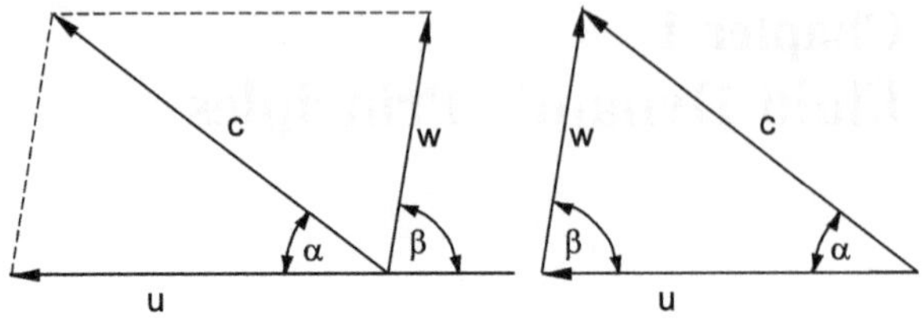

through the *vectorial* addition of **u** and **w**: **c** = **u** + **w**. In turbo machine design this addition is usually presented graphically by "velocity triangles" according to Fig. 1.1 (see Chap. 3.1).

1.2 Conservation Equations

The conservation laws for mass, energy and momentum form the actual basis of fluid dynamics. These describe the (not strictly derivable) observation that neither mass nor energy nor momentum can be destroyed or created in a closed system or control volume. Accordingly, the following balance equation applies to each of the above quantities X:

$$X_1 - X_2 + \frac{\Delta X}{\Delta t} + Z = 0 \tag{1.1}$$

In Eq. (1.1) X_1 stands for the input, X_2 is the output, $\Delta X/\Delta t$ is the change in the control volume over time step Δt, and Z is an additional quantity (supplied or removed mass, heat or work). In this general form, the conservation laws apply to steady and unsteady processes of any complexity with or without losses. Only steady processes with $\Delta X/\Delta t = 0$ will be discussed in the following. If no mass, work or heat is supplied or removed (i.e. Z = 0), *Input* = *Output* applies.

The consistent application of these seemingly trivial balance equations is often the only means of quantitatively treating complex problems without resorting to experiments. By applying the conservation laws to an infinitesimal volume element of a flowing fluid, partial differential equations are obtained. These completely describe the three-dimensional flow field (continuity and Navier-Stokes equations) which, in general, cannot be solved analytically but only numerically.

1.2.1 Conservation of Mass

To formulate the conservation laws, consider any control volume according to Fig. 1.2 which can be a streamline, a pipe or a machine. At the inlet cross section with the control surface A_1, fluid enters with the velocity c_1 and the density ρ_1. The

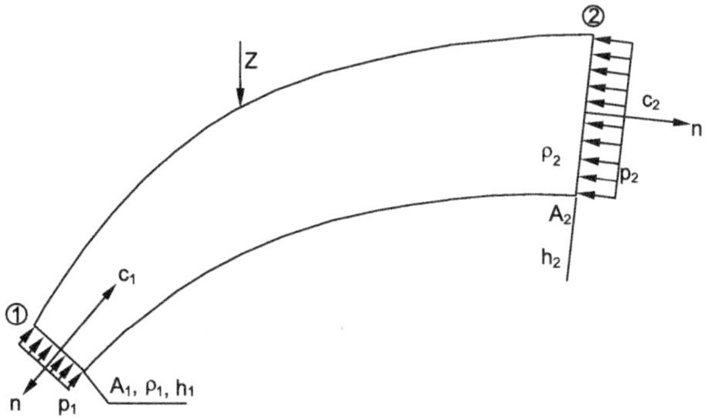

Fig. 1.2 Conservation of mass, energy and momentum

corresponding quantities at the outlet control surface are described by the subscript 2. The conservation of mass according to Eq. (1.1) yields with Z = 0:

$$\dot{m} = \rho_1 A_1 c_1 = \rho_2 A_2 c_2 = \text{constant} \tag{1.2}$$

and for incompressible flow (constant density): $A_1 \times c_1 = A_2 \times c_2$. This is the continuity equation. It states that the incoming and outgoing mass flows are identical in magnitude for a given control volume under steady conditions.

1.2.2 Conservation of Energy

The conservation of energy is described by the first law of thermodynamics if the sum of input or output thermal power P_w and mechanical power P is substituted for Z:

$$\dot{m}_1 h_{Tot,1} + \dot{m}_2 h_{Tot,2} + P_w + P = 0 \tag{1.3}$$

In Eq. (1.3) h_{Tot} is the total enthalpy [B.3]; it is equal to the sum of the internal energy per unit mass U, the static pressure energy p/ρ, the kinetic energy ½c^2 and the potential energy $g \times z$:

$$h_{Tot} = U + \frac{p}{\rho} + \frac{c^2}{2} + gz \tag{1.4}$$

If the mass flows at the inlet and outlet of the control volume are equal (i.e. $(\dot{m}_2 = -\dot{m}_1 = \dot{m})$, the power of a turbo machine without heat exchange ($P_w = 0$) is obtained as the product of the mass flow and the difference of the total enthalpies at the machine inlet and outlet:

$$P_i = \dot{m}\left(h_{Tot,2} - h_{Tot,1}\right) \tag{1.5}$$

P_i is the internal power; it is the sum of the mechanical energy transmitted to the fluid and all the losses which heat up the fluid, see Chap. 3.5. For exact calculations on high-pressure pumps, e.g. the thermometric determination of efficiency, as well as for all turbo machines with compressible flows, the generally applicable form of Eq. (1.5) must be utilized. The enthalpy difference is obtained from Eqs. (1.4) and (1.5) as follows:

$$\Delta h_{Tot} = \frac{P_i}{\dot{m}} = U_2 - U_1 + \frac{p_2 - p_1}{\rho} + \frac{c_2^2 - c_1^2}{2} + g(z_2 - z_1) \qquad (1.6)$$

Since the exchange of heat with the environment has been neglected, the change of the internal energy U is obtained solely from the heating due to losses within the control volume or the machine. This is true for incompressible flow only; for liquids it is thus possible to set $(U_2 - U_1) = \Delta p_v/\rho$.

Equation (1.6) follows from the first law of thermodynamics; it describes the energy on a streamline without external work transmission ($\Delta h_{Tot} = 0$):

$$p_1 + \frac{\rho}{2} c_1^2 + \rho\, g\, z_1 = p_2 + \frac{\rho}{2} c_2^2 + \rho\, g\, z_2 + \Delta p_v + \rho \int_{s_1}^{s_2} \frac{\partial c}{\partial t} ds \qquad (1.7)$$

This is Bernoulli's equation for incompressible flows. Since all flow processes are inevitably affected by losses, it includes the loss element Δp_v. Bernoulli's equation must only be employed along streamlines or closed channels, because the exchange of mass and energy with the environment or adjacent streamlines is assumed to be zero. To be able to use Eq. (1.7) for unsteady processes, it has been expanded by the term on the very right, which is obtained from Eq. (1.22). With $\partial c/\partial t \neq 0$, integration follows the path from position s_1 to s_2.

In the above form Eqs. (1.5) and (1.6) comprise *all* losses which lead to the heating of the fluid. If in Eq. (1.6) only the *hydraulic* losses (generated by the flow through the machine) are substituted for $(U_2 - U_1) = \Delta p_v/\rho$, the total enthalpy difference Δh_{Tot} corresponds to the theoretical work Y_{th} of the pump:

$$Y_{th} = \frac{\Delta p_v}{\rho} + \frac{p_2 - p_1}{\rho} + \frac{c_2^2 - c_1^2}{2} + g(z_2 - z_1) \qquad (1.8)$$

The quantity Y_{th} thus constitutes the work transmitted to the pumped medium per unit mass. Usually Y_{th} is largely converted into useful work, while the losses ($\Delta p_v/\rho$) lead to a - generally negligible - heating of the fluid.

If we set $U_2 = U_1$, Eq. (1.6) yields the specific useful (isentropic) work $\Delta h_{is} = Y = g \times H$ of a pump which constitutes the increase of the total pressure created by the pump, see Chap. 2.2.

1.2.3 Conservation of Momentum

According to Newton's 2nd law of mechanics the change of the momentum ($\rho \times Q \times c$) of a mass over time is equal to the vectorial sum of all volume and surface forces acting on the mass. Such are: (1) pressure forces $p_1 \times A_1$ and $p_2 \times A_2$ at the boundaries of the control volume, (2) external forces on fixed walls F_w, (3) gravity or other accelerations acting on a mass ("body forces") and (4) friction forces due to wall shear stresses F_τ. Considering a control volume as per Fig. 1.2, the momentum conservation for steady incompressible flows can be described by:

$$(p_1 + \rho c_1^2)A_1\mathbf{n}_1 + (p_2 + \rho c_2^2)A_2\mathbf{n}_2 = \mathbf{F}_{vol} + \mathbf{F}_w + \mathbf{F}_\tau \tag{1.9}$$

$\mathbf{n}_1$ and $\mathbf{n}_2$ are unit vectors directed *outward* normal to the areas A_1 and A_2. If the volume flow rate is explicitly introduced into Eq. (1.9), we get:

$$p_1A_1\mathbf{n}_1 + \rho Qc_1\mathbf{n}_1 + p_2A_2\mathbf{n}_2 + \rho Qc_2\mathbf{n}_2 = \mathbf{F}_{vol} + \mathbf{F}_w + \mathbf{F}_\tau \tag{1.10}$$

When applying the conservation of momentum in the form of Eqs. (1.9) or (1.10), the following must be observed: (a) The equations are valid for steady incompressible flow with uniform velocity and pressure distributions on the areas A_1 and A_2. (b) A_1 and A_2 must be perpendicular to the velocity vectors. The momentum forces are *always directed towards the control volume.* (c) The signs of the outward-directed unit vectors must be treated carefully: $\mathbf{c}_1 = -c_1 \times \mathbf{n}_1$ and $\mathbf{c}_2 = c_2 \times \mathbf{n}_2$. (d) All terms are vectors and must be added according to the rules of vector calculation. (e) The successful application of momentum conservation equations often hinges on the appropriate choice of the control volume which must allow the quantification of pressures and velocities at the control surfaces. (f) Avoid generating indefinable forces, when a control surface is placed through structures.

Consider a **sudden channel expansion** according to Fig. 1.3 ("Carnot shock") as an example for the application of the conservation of momentum and Bernoulli's equation. Control surface 1 is positioned immediately downstream of the step; here pressure p_1 is acting over the *entire* cross section A_2, since the same pressure prevails in the jet and in the wake. Control surface 2 is selected downstream far enough for the flow to be uniform again. No external forces are acting normal to the

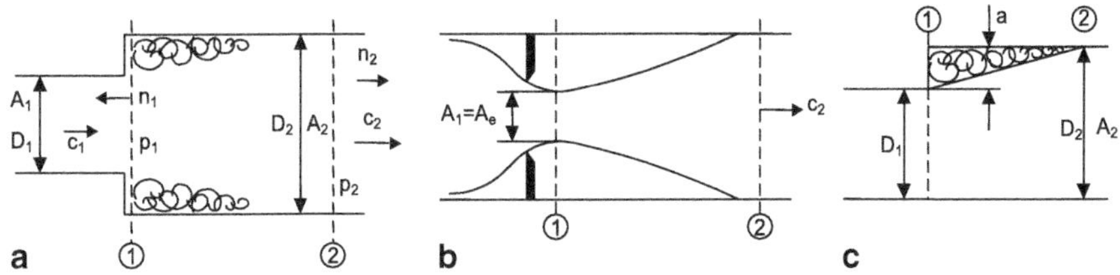

Fig. 1.3 Sudden expansion. **a** To both sides, **b** orifice, **c** to one side

channel walls: $F_w = 0$. Neglecting the gravity ($F_{vol} = 0$) and the force F_τ due to wall shear stresses (which are not transmitted in the wake), the pressure recovery according to Eq. (1.10) is obtained from:

$$p_2 A_2 + \rho Q c_2 - p_1 A_1 - \rho Q c_1 = 0$$

Solving for the pressure recovery we get:

$$p_2 - p_1 = c_p \frac{\rho}{2} c_1{}^2 \quad \text{where} \quad c_p = 2\frac{A_1}{A_2}\left(1 - \frac{A_1}{A_2}\right) \tag{1.11}$$

From Eq. (1.7) it is then possible to calculate the pressure loss using Eq. (1.11):

$$\Delta p_v = \frac{\rho}{2}(c_1 - c_2)^2 = \zeta_1 \frac{\rho}{2} c_1{}^2 \quad \text{where} \quad \zeta_1 = \left(1 - \frac{A_1}{A_2}\right)^2 \tag{1.12}$$

According to [1] the pressure recovery measured reaches approximately 95% of the theoretical values calculated from Eq. (1.11).

To get an essentially uniform flow velocity through exchange of momentum, a certain length of channel downstream of the expansion is required. It is given by:

$$\frac{L}{D_2} \approx b\left(1 - \frac{D_1}{D_2}\right) \tag{1.12a}$$

For an expansion to *both sides* according to Fig. 1.3a, b the factor b = 10 applies, while b = 20 must be used for a *one-sided* expansion according to Fig. 1.3c. Both values of b correspond to an angle of expansion of just under 3° which "limits" the wake (these relationships were derived from data in [1]). Equations (1.11) and (1.12) can also be employed for orifices or other cases with a jet contraction, if A_1 is substituted for the *contracted* cross section A_e (also termed "vena contracta") as indicated in Fig. 1.3b. The contracted cross section is calculated from $A_e = \mu \times A_1$; the contraction coefficient for sharp-edged orifices is approximately $\mu = 0.61$, if $A_1 \ll A_2$.

Conservation of angular momentum: Another consequence of Newton's 2nd law of mechanics is the conservation of the angular momentum (or the "moment of momentum") which is of fundamental significance to all turbo machines. Accordingly, the change of the angular momentum is equal to the sum of the external moments. Angular momentums $\rho \times Q \times r \times c_u$ at inlet and outlet, an external torque M and friction moments due to shear stresses M_τ are acting on an impeller or a diffuser. Since no pressure forces are created on cylindrical surfaces in the circumferential direction, it is possible to write Eq. (1.10) as:

$$\rho Q(c_{2u} r_2 - c_{1u} r_1) = M + M_\tau \tag{1.13}$$

This is Euler's turbine equation where c_{2u} is the circumferential component of the flow velocity at the outlet of the control volume, r_2 is the outer radius of the impeller and c_{1u} and r_1 are the corresponding quantities at the inlet. As will be explained in Chap. 3.2, the specific work done by an impeller is obtained from Eq. (1.13) with $M \times \omega = P$ and neglecting M_τ:

$$Y_{th} = c_{2u}\, u_2 - c_{1u}\, u_1 \tag{1.14}$$

Substituting Eq. (1.14) in Eq. (1.8), Bernoulli's equation in the relative reference frame is obtained (with $z_1 = z_2$). In the derivation, $c^2 = w^2 - u^2 + 2 \times u \times c_u$ was used which follows from the velocity triangles, Chap. 3.1:

$$p_1 + \frac{\rho}{2} {w_1}^2 - \frac{\rho}{2} {u_1}^2 = p_2 + \frac{\rho}{2} {w_2}^2 - \frac{\rho}{2} {u_2}^2 + \Delta p_v \tag{1.15}$$

This consideration shows that, with an incompressible flow, the conservation of energy is always satisfied simultaneously with the conservation of momentum.

If no external moments are active - i.e. M and M_τ in Eq. (1.13) are zero - conserving the angular momentum requires $\mathbf{c_u \times r = constant}$ for a flow free of forces. This relationship is of fundamental significance to all flows moving with tangential velocity components in radial direction. Examples are the flow exiting from an impeller into a collector, the flow in radial inlet chambers, in pump sumps, and, in general terms, the creation and movement of vortices (e.g. a tornado).

1.3 Boundary Layers, Boundary Layer Control

In technically relevant fluid processes, the flow occurs relative to stationary walls (e.g. pipelines or channels) or moving structures (such as air foils or impellers). Although the fluid near the wall comprises only a fraction of the flow field, it largely determines losses and velocity distributions in or around a component. In classical fluid dynamics the flow field is described by an "inviscid" main flow (which is considered as being free of friction) and a boundary layer flow capturing the processes near solid walls. In this concept it is assumed that the relative velocity is zero immediately at the wall ("no-slip condition") and that no gradients of the static pressure occur perpendicular to the wall within the boundary layer ($\partial p/\partial y = 0$). Consequently, the core flow imparts the pressure to the boundary layer and determines the pressure distribution in the flow field under consideration. All streamlines have the same pressure gradient $\partial p/\partial x$ in the flow direction but possess different amounts of kinetic energy.

Owing to the no-slip condition, the fluid adhering to a stationary wall (e.g. a pipe or a diffuser) has an *absolute* velocity of zero. In contrast, on the walls of a rotating impeller it has a *relative* velocity of zero. Accordingly, a fluid particle adhering to the walls or blades of an impeller moves with the absolute velocity $c_u = u = \omega \times r$.

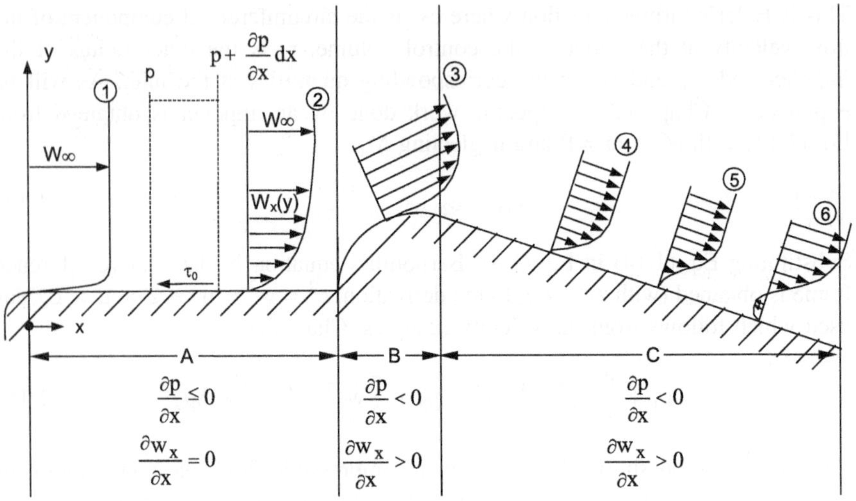

Fig. 1.4 Boundary layers with positive and negative pressure gradients

Apart from molecular diffusion no exchange between the individual streamlines takes place in a *laminar* flow. In contrast, mixing movements perpendicular to the flow occur in *turbulent* flows in which eddies of finite magnitude ensure the lateral transport. This exchange of momentum caused by turbulence largely determines the thickness of the turbulent boundary layer and the velocity distribution over the channel cross section. In this way, the fluid layers near the wall are supplied with energy through increased turbulence, resulting in a reduced boundary layer thickness and a fuller velocity profile.

The velocity distribution in a pipe of radius R with developed turbulent flow is approximately given by the following formula:

$$\frac{w}{w_{max}} = \left(1 - \frac{r}{R}\right)^{x} \quad \text{and} \quad \frac{w_{max}}{w_m} = \frac{(y+1)(2y+1)}{2y^2} \tag{1.16}$$

In Eq. (1.16) it is $x = \lambda_R{}^{0.5}$ and $y = \lambda_R{}^{-0.5}$. This formula takes into account the effects of the Reynolds number and roughness since the friction coefficient λ_R depends on both these quantities. Wall roughness reduces the velocity in the boundary layer, thereby creating higher hydraulic losses (Chap. 1.5.1). The boundary layer thickness increases with the roughness; the hydraulic losses grow and the wall shear stresses drop with the boundary layer thickness.

Figure 1.4 depicts schematically the development of a two-dimensional boundary layer in a channel. Suppose the channel has a constant cross section in domain A, thus w = constant and $\partial w_x/\partial x = 0$. At the inlet the boundary layer thickness tends to zero and grows with increasing flow path length. The boundary layer is laminar at the inlet. Provided that the main flow is turbulent, the boundary layer, too, becomes turbulent after a certain length of the flow path while a laminar sub-layer remains.

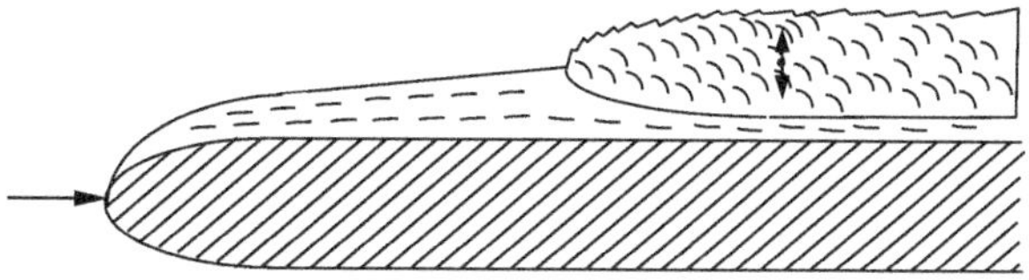

Fig. 1.5 Boundary layer transition from laminar to turbulent

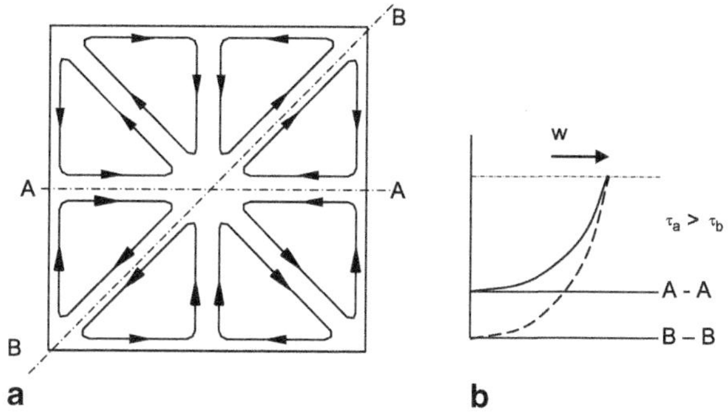

Fig. 1.6 Secondary flow in a square channel **a** cross section, **b** velocity distributions in sections A-A and B-B (diagonal)

The transition from a laminar to a turbulent boundary layer (depicted in Fig. 1.5) depends on: Reynolds number, surface roughness, turbulence of the outer flow, pressure gradient and wall curvature.

If the flow is accelerated (domain B in Fig. 1.4), $\partial w_x/\partial x > 0$ and $\partial p/\partial x < 0$ apply according to Eq. (1.7). The boundary layer thickness then decreases; in accelerated flow correspondingly lower hydraulic losses are experienced. The maximum shear stress on the wall occurs at $\partial p/\partial x < 0$; fuller velocity profiles are obtained with increasing acceleration. Accelerating inlet chambers therefore produce more uniform inflows to the impeller of a pump.

In decelerated flow (Fig. 1.4, domain C) $\partial w_x/\partial x < 0$ and accordingly $\partial p/\partial x > 0$ applies. The boundary layer thickness grows and the wall shear stress drops; it becomes zero at a certain deceleration (profile 5 in Fig. 1.4). At this point the flow separates from the wall (it "stalls"), resulting in recirculation in the boundary layer further downstream (profile 6). Due to the separation the main flow contracts and accelerates in accordance with the reduced cross section. Downstream it intermixes with the wake through the exchange of momentum. These processes result in correspondingly high hydraulic losses which generate a drop of the static pressure (refer to Chap. 1.6).

If the channel cross section is other than circular, the velocity distribution is not rotational-symmetric. Similar to Eq. (1.16), different velocity profiles are expected in sections A-A and B-B (diagonal) of a square channel (Fig. 1.6). Consequently, the wall shear stresses change over the channel perimeter. The shear stresses are lowest in the diagonal B-B and greatest in section A-A. Since the static pressure (with flow in straight channels) is constant over the cross section, the equation:

$$\Delta p = \tau U\, \Delta L = \rho\, c_f \frac{w^2}{2} U \Delta L \quad (U = \text{wetted perimeter}) \tag{1.17}$$

can only be satisfied if fluid is transported through compensatory flows normal to the channel axis from locations with low shear stress to zones with higher shear stress. Such secondary flows occur, for example, in triangular or square channels (Fig. 1.6), hence also in the impeller or diffuser channels. Vortices ("corner vortices") develop in the corners of a channel, between blades and shroud or between pillar and baseplate. Their effect can be observed when abrasive particles are carried with the flow giving rise to erosion, Chap. 14.5.

Corner vortices can also play a role in cavitation (Chap. 6), since the pressure in the vortex core drops relative to the pressure in the main flow, Chap. 1.4.2. The velocity components of the secondary flow attain approximately 1 to 2% of the velocity in mid channel [11]. By rounding the corners the losses can be somewhat reduced.

There are a number of methods of **boundary layer control** that aim at reducing the flow resistance and retarding stall: (1) The point of separation in diffusers or on air foils can be shifted downstream through boundary layer extraction. (2) By injecting fluid, energy can be supplied to the boundary layer or separation enforced. (3) Boundary layer fences may be designed to change the flow direction in the boundary layer in order to prevent the development of large vortices, or to subject the boundary layer flow to the same deflection as the main flow. (4) By means of very fine longitudinal grooves it is possible to influence the vortex structure in the boundary layer in a way that the flow resistance is reduced by several percent. (5) The admixture of friction-reducing agents has a similar effect by reducing the thickness of the boundary layer. (6) Turbulence or swirl generators can be employed to supply energy to the fluid near the wall through the exchange of momentum. In this manner the boundary layer thickness, hence losses, as well as the separation tendency can be reduced.

The boundary layers in decelerated flow through pumps are mostly three-dimensional. In the impeller they are also subjected to centrifugal and Coriolis forces. Such complex conditions are not amenable to analytical treatment. However, if an estimation of the boundary layer thickness is desired, the conditions on a flat plate without pressure gradients may be used for want of better information. To this end the displacement thickness δ* may be used; it is defined by Eq. (1.18):

$$\delta^* \equiv \int_0^{\delta} \left(1 - \frac{w_x}{w_\infty}\right) dy \qquad (1.18)$$

The integration is performed up to the boundary layer thickness δ which is defined by the point where the local velocity reaches 99% of the mean velocity of the core flow. The displacement thickness with laminar or turbulent flow over a hydraulically smooth plate is obtained from:

$$\delta^*_{lam} = \frac{1.72x}{\sqrt{Re_x}} \qquad \delta^*_{turb} = \frac{0.0174x}{Re_x^{0.139}} \qquad \text{with} \quad Re_x = \frac{w_x x}{\nu} \qquad (1.19a, b, c)$$

Here x is the length of the flow path calculated from the leading edge of the plate or component. Depending on the turbulence of the upstream flow and the roughness of the plate, the change from a laminar to a turbulent boundary layer occurs in the range of $2 \times 10^4 < Re_x < 2 \times 10^6$, refer also to Eq. (1.33b). These equations show that the boundary layer thickness grows from the leading edge at x = 0 with the length of the flow path and the viscosity of the fluid. The above formulae can also be used for the inlet into a channel or other components, provided the boundary layer thickness is significantly less than half the hydraulic diameter of the channel so that the boundary layers of opposite walls do not exert a noticeable effect on each other. Consider for example an impeller with an outlet width of $b_2 = 20$ mm and a blade length of L = x = 250 mm delivering cold water at a relative velocity of $w_x = 24$ m/s. From Eqs. (1.19b and c) we find at the impeller outlet: $Re_x = 6 \times 10^6$ and $\delta^*_{turb} = 0.5$ mm. Since δ^* is much smaller than b_2 no appreciable interaction of rear and front shroud boundary layers would be expected. Boundary layer considerations based on Eqs. (1.19a to c) would therefore be relevant as a first approximation.

The inlet length L_e necessary to establish fully developed flow in pipes can be determined from Eq. (1.19d), [3] (in laminar flow L_e is much longer than in turbulent flow):

$$\frac{L_e}{D_h} = 14.2 \log Re - 46 \quad \text{valid for } Re > 10^4 \quad \text{with} \quad Re = \frac{cD_h}{\nu} \qquad (1.19d)$$

1.4 Flow on Curved Streamlines

1.4.1 Equilibrium of Forces

According to Newton's law of motion a body remains in *uniform, straight-line* motion (or at rest) if it is not compelled by external forces to change this condition. To move a body or a fluid particle on a *curved* path, a force must consequently be exerted on that body. To visualize this fact, consider a point mass m on a rotating disk; the mass is held by a spring and is able to move in a radial guide Fig. 1.7. If the disk rotates with the angular velocity ω, the spring is tensioned so that a radius r_g is obtained where the spring force $K \times \Delta r$ is equal to the centrifugal force $m \times \omega^2 \times r_g$. The spring force that compels the mass to move along a circular path is directed radially inward; it is therefore called *centripetal* force. The spring force is opposed by the mass force $m \times \omega^2 \times r_g$ (action equals reaction) which is called *centrifugal* force. (These relationships can be easily visualized by manually spinning a body on the end of a string.)

Let us examine now the more general motion of a fluid particle on an arbitrary 3-dimensionally curved streamline: at any point and at any time it is possible to select a path element ds in the direction of the local velocity vector such that a plane according to Fig. 1.8 is spread through the momentary curvature radius r and the path element ds. A fluid element of mass dm is subjected to forces in the flow

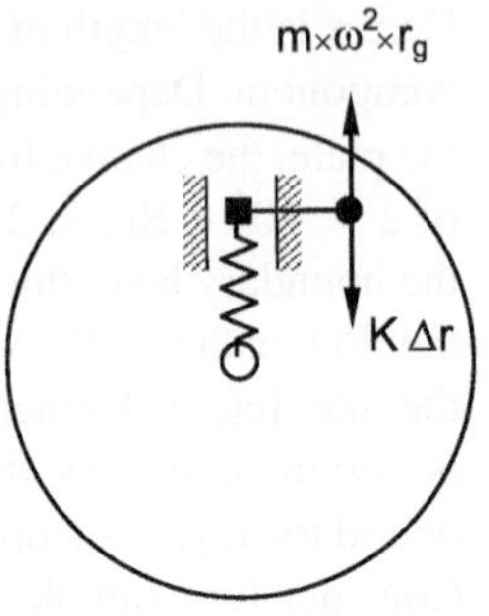

Fig. 1.7 Movement of a point mass on a curved path

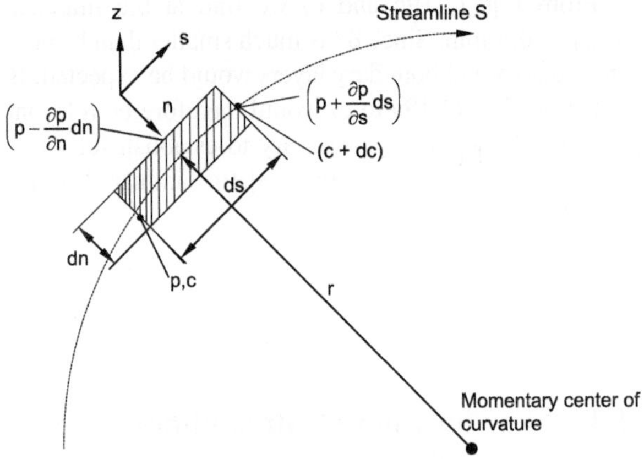

Fig. 1.8 Equilibrium of forces acting on a fluid element

direction and perpendicular to the streamline. Newton's law F = m × dc/dt forms the basis for calculating these forces.

First consider the forces acting in the direction of the motion. Since the velocity c(t, s) is a function of time and space, the change in velocity is:

$$dc = \frac{\partial c}{\partial t}dt + \frac{\partial c}{\partial s}ds \tag{1.20}$$

Taking into account that ds/dt = c, the acceleration becomes:

$$\frac{dc}{dt} = \frac{\partial c}{\partial t} + \frac{\partial c}{\partial s}\frac{ds}{dt} = \frac{\partial c}{\partial t} + c\frac{\partial c}{\partial s} \tag{1.21}$$

The fluid element is exposed to the pressure forces according to Fig. 1.8 and the gravity dm g × ∂z/∂s; from Eq. (1.21) it follows:

$$\frac{\partial c}{\partial t} + c\frac{\partial c}{\partial s} + \frac{1}{\rho}\frac{\partial p}{\partial s} + g\frac{\partial z}{\partial s} = 0 \tag{1.22}$$

The integration of Eq. (1.22) yields Bernoulli's equation (1.7) without the loss term.
Similar to the above procedure we can write for $c_n(s, t)$ normal to the streamline:

$$\frac{dc_n}{dt} = \frac{\partial c_n}{\partial t} + \frac{\partial c_n}{\partial s}\frac{ds}{dt} \tag{1.23}$$

With ds/dt = c and $\partial c_n/\partial s$ = c/r the acceleration normal to the streamline becomes:

$$\frac{dc_n}{dt} = \frac{\partial c_n}{\partial t} + \frac{c^2}{r} \tag{1.24}$$

Finally the equilibrium of forces perpendicular to the streamline is:

$$\frac{\partial c_n}{\partial t} + \frac{c^2}{r} + g\frac{\partial z}{\partial n} + \frac{1}{\rho}\frac{\partial p}{\partial n} = 0 \tag{1.25}$$

If the effect of gravity is negligible, we get for steady flow (with dn = −dr):

$$\frac{dp}{dr} = \rho\frac{c^2}{r} \tag{1.26}$$

These equations, which apply unconditionally to loss-affected flows, demonstrate that a flow on a curved path is *always* coupled to pressure gradients *perpendicular to the flow direction* such that the pressure decreases *from the outside to the inside* towards the *momentary* center of the streamline curvature. This pressure differential provides the centripetal force which makes flow on a curved path possible in the first place, keeping the body force (centrifugal force) acting on the fluid element at equilibrium. Conversely, there follows the equally fundamental fact that transversal pressure gradients *cannot exist in a straight flow*. Consequently, a flat plate in a *parallel* flow creates a resistance (or "drag"), but is not subjected to any force perpendicular to the flow direction.

In the general case, a flow around a body creates a force perpendicular to the flow direction, which is called hydrodynamic lift, as well as a drag opposing the flow. Lift can be generated only if pressure gradients are present perpendicular to the flow direction. As deduced above, this is only the case when the flow around the body moves on curved paths. Aerodynamic lift due to curved streamlines is generated when the flow around a body is asymmetric - either because that body is inclined against the incident flow or because it has an asymmetric shape.

Consider an air foil according to Fig. 1.9 in a control volume, which is large enough so that the same atmospheric pressure p_0 prevails on all of its boundaries. For the sake of illustration assume the lower side of the air foil to be essentially flat and parallel to the approaching flow w_0, while the upper side of the wing is strongly curved. Underneath the air foil an almost undisturbed parallel flow must be assumed (except for the leading edge). Above the wing the flow is accelerated due to the displacement effect, causing the static pressure to drop locally from p_0 to p

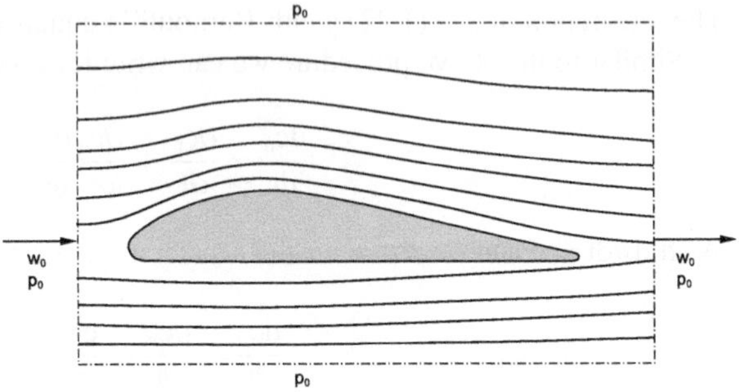

Fig. 1.9 Flow around an air foil

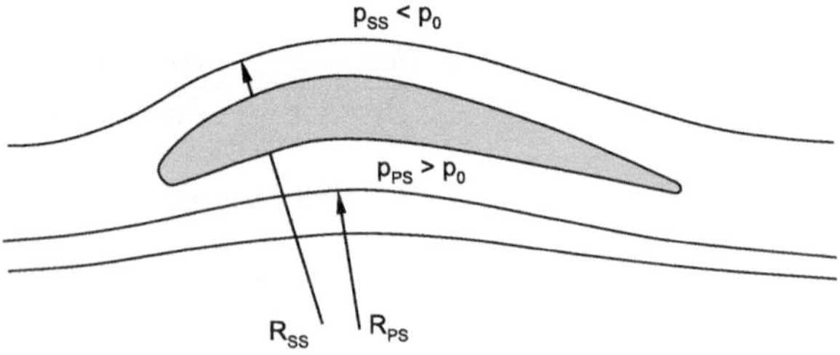

Fig. 1.10 Cambered air foil

according to Eq. (1.7). The differential between p and p_0 can only be maintained when the streamlines are sufficiently curved; refer also to [B.27]. This means that the possible pressure differentials adjust in a way that Eqs. (1.25) and (1.22) are satisfied (a friction term is added in the latter). The integral of the pressure distribution over upper and lower surface of the air foil yields the hydrodynamic lift.

On the cambered air foil of Fig. 1.10 there is $p_{PS} > p_0$ on the lower wing surface, while $p_{SS} < p_0$ is found on the upper surface, Eq. (1.26). Of course, these considerations are not restricted to the specific shapes of the air foils shown in Fig. 1.9 or Fig. 1.10, but apply quite generally to inclined plates and all possible shapes of wings, blades or structures: the curvature of the streamlines creates pressure gradients perpendicular to the main flow, and this generates the hydrodynamic lift. Equation (1.25) therefore is of fundamental significance for understanding the origin of lift forces on air foils or on turbo machine blades. Hydrodynamic lift is frequently described as a circulation superimposed on a parallel flow according to the law of Kutta-Joukowsky where inviscid flow is assumed. This approach may be of service for the mathematical treatment. However, the physical meaning is less obvious, since the circulation cannot be measured in reality.

In a turbo machine impeller the *absolute* flow always occurs along curved paths whose shape is determined by the kinematics of the flow around the rotating blades. This is even true for an impeller with many closely-spaced radial blades. Here the pressure distribution in the impeller is basically established in accordance with Eq. (1.25). These relationships also explain why the pressure in the impeller increases from the inside to the outside, why the fluid in a pump *flows from the lower to the higher pressure* and why the pressure increases even though the absolute flow in the impeller is accelerated.

1.4.2 Forced and Free Vortices

Forced vortex: Consider the movement along concentric circular paths as it occurs when a liquid-filled cylindrical vessel rotates around its axis. Here the liquid rotates like a solid body with $c_u = u = \omega \times r$ (assuming stationary conditions). Hence, the integration of Eq. (1.26) produces the pressure distribution in the vessel as follows:

$$p - p_0 = \frac{\rho}{2}\omega^2(r^2 - r_0^2) = \frac{\rho}{2}c_{u,0}{}^2\left(\frac{r^2}{r_0{}^2} - 1\right) \tag{1.27}$$

In the relative reference frame (viewed by the co-rotating observer), no flow is present in this case, the fluid is at rest. In the absolute reference frame, it constitutes a rotational flow described as a "forced vortex". In the sidewall gaps between the impeller and the casing, for example, a forced vortex according to Eq. (1.27) is created, when there is no leakage flow through the sidewall gap, Chap. 9.1.

Free vortex: If the flow follows the conservation of angular momentum, the pressure distribution is obtained after integration when $c_u \times r = c_{u,0} \times r_0$ = constant is substituted in Eq. (1.26):

$$p - p_0 = \frac{\rho}{2}c_{u,0}{}^2\left(1 - \frac{r_0{}^2}{r^2}\right) \tag{1.28}$$

In this case the pressure from the inside to the outside increases less strongly than according to Eq. (1.27). The flow is non-rotational and is called a "free vortex".

Air-drawing vortices: Consider a vortex in a tank whose dimensions are large compared to the dimensions of that vortex. Measure the velocity $c_{u,o}$ at any radius r_o of the vortex where a pressure p_0 is assumed to prevail. In a large liquid-filled tank this is usually a free vortex. However, in the vortex center, at radius $r = 0$, Eq. (1.28) would yield an infinitely large drop in pressure. Since this is physically impossible, a forced vortex with ω = constant is established in the core of the vortex. Hence a vortex in a real fluid consists of a core in which forced rotation occurs and an outer flow which follows the laws of the free vortex. This combined vortex is called "Rankine vortex". The model of the Rankine vortex approximately describes the flow conditions in a tornado [3] or in vortices which can lead to problems in pump inlets (vortices are discussed in detail in Chap. 11.7.3).

The formulae for the Rankine vortex are summarized in Table 1.1 where r_k is the core radius that defines the boundary between forced and free vortex. These formulae were evaluated in Fig. 1.11: the lower part of the figure shows the liquid

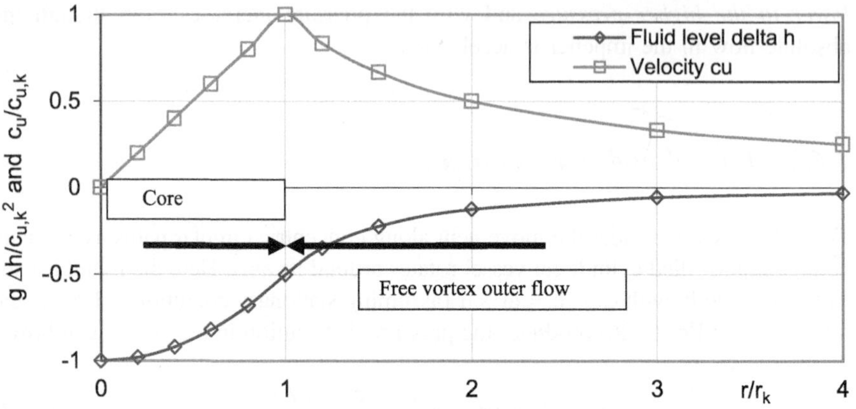

Fig. 1.11 Velocity profile and liquid level in a Rankine vortex

Table 1.1 Rankine vortex

Domain	Vortex core with radius r_k	Outer domain	Eq.
	Forced vortex	Free vortex	
	$0 < r < r_k$	$r > r_k$	
Velocity distribution	$c_u = \omega r = c_{u,k} \frac{r}{r_k}$	$c_u = \frac{\Gamma}{2 \pi r} = c_{u,k} \frac{r_k}{r}$	1.1.1
Circulation	$\Gamma = 2 \pi r c_u = 2 \pi r^2 \omega$	$\Gamma = 2 \pi r_k c_{u,k}$	1.1.2
Vorticity Ω = rot $\mathbf{c_u}$	$\Omega = 2 \omega$ with $\omega = c_{u,k}/r_k$	$\Omega = 0$	1.1.3
Pressure distribution	$p - p_0 = \frac{\rho}{2} \omega^2 (r^2 - r_k^2) = \frac{\rho}{2} c_{u,k}^2 \left(\frac{r^2}{r_k^2} - 1\right)$	$p - p_0 = \frac{\rho}{2} c_{u,k}^2 \left(1 - \frac{r_k^2}{r^2}\right)$	1.1.4
Liquid level depression	$\frac{2 g \Delta h}{c_{u,k}^2} = -1.0$	$\frac{2 g \Delta h}{c_{u,k}^2} = 1.0$ for $r \to \infty$	1.1.5
Total liquid level depression	Theoretical: $\frac{2 g \Delta h}{c_{u,k}^2} = 2.0$	real: $\frac{2 g \Delta h}{c_{u,k}^2} = 0.6$ to 0.72	1.1.6

level depression in a vortex that can be observed in flows with a free surface in a tank or a river. The velocity distribution is plotted in the upper graph.

According to Table 1.1, the pressure (or liquid level) is lowered by the same amount in the vortex core and in the outer flow. The velocity and pressure or level curves in the graph are theoretical, since viscosity effects in real flows yield flatter velocity gradients at the transition from the core to the outer flow. The real liquid level depression is only about 30 to 36% of the theoretical value,

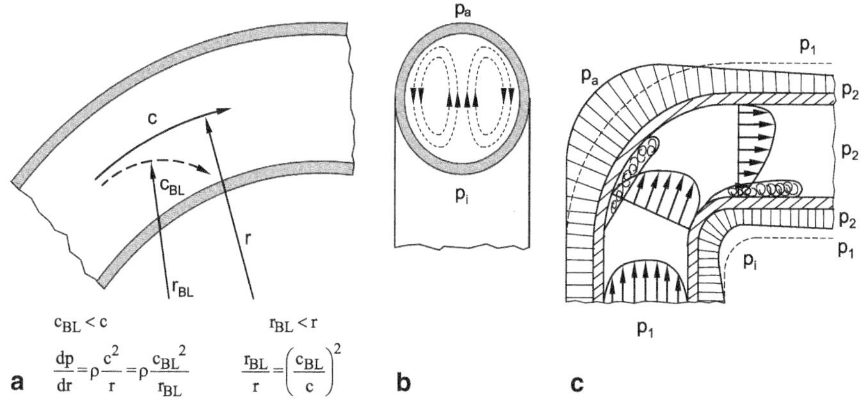

Fig. 1.12 Flow in a bend: **a** origin of secondary flow (subscript BL refers to the boundary layer), **b** secondary flow, **c** pressure and velocity distributions

i.e. $\Delta h = (0.3\text{–}0.36)c_{u,k}^2/g$, [9]. Because of its dimensionless presentation Fig. 1.11 is universally applicable.

1.4.3 Flow in Curved Channels

The pressure in a curved channel decreases from the outside to the inside in the direction of the center of curvature according to Eq. (1.26). This pressure gradient imparts to the fluid the necessary centripetal acceleration that allows motion on a curved path. The flow velocity near the wall (in the boundary layer) is lower than in the center of the channel, but the pressure gradient perpendicular to the streamlines is imposed by the main flow. Therefore, the boundary layer flow must follow a *narrower* radius than the main flow in a way that the equilibrium according to Eq. (1.26) is established. The boundary layer flow is therefore deflected towards the inside as indicated in Fig. 1.12a. Satisfying the continuity condition, fluid in the center of the channel is transported to the outside so that the secondary flow sketched in Fig. 1.12b is obtained. In a section through the channel this flow appears as a dual vortex superimposing itself on the main flow and resulting in a spiral-shaped flow path. Note that the main flow is deflected to the outside *against* a positive pressure gradient.

The development of the secondary flow can also be described as follows: Because of their greater velocity the fluid particles in the center of the channel are subjected to greater centrifugal forces than the slower flowing particles near the wall. Center flow particles are consequently deflected to the outside; for continuity reasons fluid will flow back to the inside through the boundary layer.

Since the formation of the secondary flow requires a certain flow path length, a velocity distribution similar to the angular momentum conservation of

$c_u \times r$ = constant initially forms in the inlet region of the bend. There, the velocity maximum is thus found near the inner streamline. Further in the bend, the secondary flow transports fluid towards the outer streamline, where the velocity maximum is found at the outlet of the bend, Fig. 1.12c. In sharp bends flow separation occurs in the inlet domain on the outer streamline since the fluid has to travel against a positive pressure gradient [$p_a > p_1$ according to Eq. (1.26)]. Separation likewise occurs on the inner streamline in the outlet region since the fluid flows against a positive pressure gradient here too ($p_i < p_2$), see Fig. 1.12c. In terms of pressure and velocity distribution the effect of the bend is already noticeable at approximately $L = 1 \times D$ upstream. Since a swirling flow is usually quite stable, uniform pressure and velocity distributions are only found after a distance of 50 to 70 pipe diameters downstream of a sharp bend. This fact has to be taken into account when doing experiments. The secondary flow has an important effect on the velocity distribution in stationary or rotating curved channels (impellers). These flow mechanisms basically apply to all curved channels, while the shape of the channel cross section exerts an additional influence.

An example of the frequently dominating influence of the secondary flow can be observed along brooks and rivers: sand is deposited on the *inside* of the bend and eroded from the outside creating meanders (a superficial consideration might expect sand deposits on the outer streamline due to centrifugal forces).

The dual vortices created by the secondary flow are unsteady: they change their shape and size as a function of time, see Chap. 10.12.4.

For $Re = c \times D/\nu > 5 \times 10^5$ the pressure differential between outer and inner streamline of a bend can be calculated directly from Eq. (1.26) if $\Delta r = R_a - R_i = D$ is substituted and R_m is the radius of curvature of the center streamline:

$$p_a - p_i = \frac{D}{R_m}\rho c^2 \quad D = \text{pipe diameter} \tag{1.29}$$

In a bend with $R_m/D = 2$ the pressure differential between outer and inner streamline is therefore equal to the stagnation pressure. In the range $4000 < Re < 5 \times 10^5$ the pressure differential obtained from Eq. (1.29) must be multiplied by the factor $(5 \times 10^5/Re)^{0.17}$.

1.5 Pressure Losses

Energy losses in a flow system are caused by friction and flow separation. Stall creates particularly high losses caused by the mixing of stalled fluid with the non-separated flow. In practice, information for the pressure loss calculation is often required; data of loss coefficients for frequently used components are compiled in Table 1.4, comprehensive collections of empirical loss coefficients can be found in the manuals [3, 6, 11].

1.5.1 Friction Losses (Skin Friction)

Energy losses in a flow caused by non-separated boundary layers are termed friction resistances. The velocity gradients in the boundary layer create shear stresses according to:

$$\tau = \rho\,(\nu + \nu_t)\frac{dw}{dy} \tag{1.30}$$

While the kinematic viscosity ν is a fluid property, the "eddy viscosity" ν_t depends on the intensity and structure of the turbulence. In laminar flow it is $\nu_t = 0$, while $\nu_t \gg \nu$ applies to fully developed turbulent flow. Since Eq. (1.30) cannot be easily evaluated in practice, the wall shear stresses are represented by friction coefficients c_f.

$$\tau_0 = c_f \frac{\rho}{2} w^2 \tag{1.31}$$

The wall shear stress is a friction force dF_τ per unit area between flow and wetted wall element dA, which amounts to $dF_\tau = \tau_0 \times dA$. The power dissipated is:

$$dP_d = w dF_\tau = {}^1\!/\!_2 \rho\, c_f w^3 dA \tag{1.32}$$

The friction coefficients c_f must be obtained from tests (or boundary layer calculations). For estimating purposes the friction coefficients of flat plates are frequently used. They are presented in Fig. 1.13 as a function of the Reynolds number and the relative roughness, [3]. For turbulent flows they can also be approximately determined from Eq. (1.33) which is valid in the range of $10^5 < Re_L < 10^8$ and $0 < \varepsilon/L < 10^{-3}$ (L = length of the plate).

$$c_f = \frac{0.136}{\left\{-\log\left(0.2\frac{\varepsilon}{L} + \frac{12.5}{Re_L}\right)\right\}^{2.15}} \quad \text{with} \quad Re_L = \frac{wL}{\nu} \tag{1.33}$$

For laminar flows in the range of $0.01 < Re < Re_{crit}$ the friction coefficients can be calculated from Eq. (1.33a), [4]:

$$c_f = c_{f,lam} = \frac{2.65}{Re_L^{0.875}} - \frac{2}{8 Re_L + 0.016/Re_L} + \frac{1.328}{\sqrt{Re_L}} \tag{1.33a}$$

The transition from laminar to turbulent flow occurs at the critical Reynolds number Re_{crit} according to Eq. (1.33b); Re_{crit} depends on the turbulence intensity Tu and the roughness.

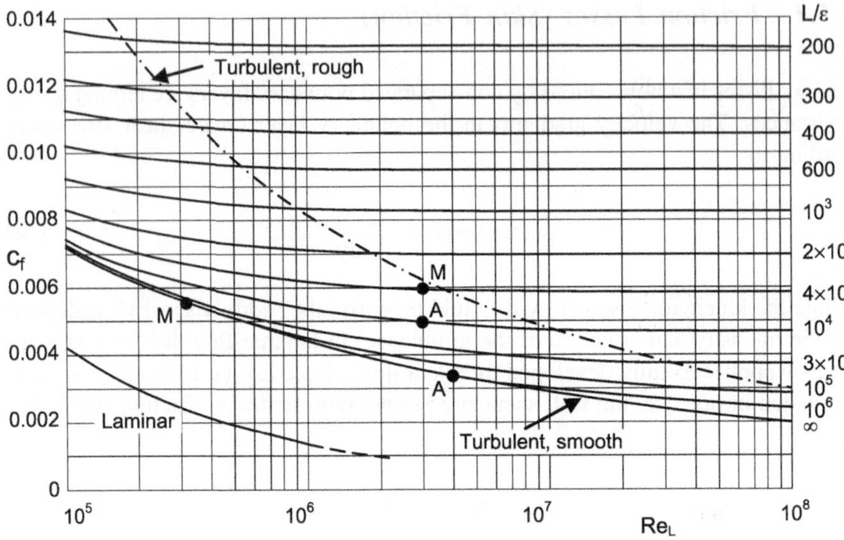

Fig. 1.13 Friction coefficients c_f of flat plates in parallel flow

$$Re_{crit} = \frac{3 \times 10^6}{1 + 10^4 Tu^{1.7}} \quad \text{valid for Tu} < 0.1 \tag{1.33b}$$

Equations (1.33), (1.33a) and Fig. 1.13 supply the *overall* resistance coefficient of a plate of length L. The *local* resistance coefficients, which depend on the local boundary layer thickness, are approximately 10 to 30% smaller (dependent on Re_L and L/ε). The drag force on *one side* of a plate with width b amounts to:

$$F_w = {}^1\!/\!_2\, \rho\, w_\infty^2 c_f b L \tag{1.33c}$$

In a pipe of length L with a constant cross section, the integral of Eq. (1.32) over the wetted surface leads to $A = \pi \times D \times L$ and a power dissipation of:

$$P_d = c_f \frac{\rho}{2} w^3 \pi D L \tag{1.34}$$

Dissipated power and pressure loss are related by $P_d = \Delta p_v \times Q$. With $Q = (\pi/4) \times D^2 \times w$ the well-known equation for the pressure losses in pipes is obtained:

$$\Delta p_v = 4 c_f \frac{L}{D} \frac{\rho}{2} w^2 = \lambda_R \frac{L}{D} \frac{\rho}{2} w^2 \tag{1.35}$$

Consequently, the relationship $\lambda_R = 4 \times c_f$ exists between the pipe friction coefficient λ_R and the shear stress coefficient c_f. Equation (1.35) is also obtained from the conservation of momentum given by Eq. (1.9) which leads to $(\pi/4) \times D^2 \times \Delta p_v = \pi \times \tau \times D \times L$ and with Eq. (1.31) again to Eq. (1.35).

Friction coefficients for turbulent flow through pipes and channels can be calculated from Eq. (1.36) or read from the graph in Table 1.6 (at end of this chapter).

$$\lambda_R = \frac{0.31}{\left\{\log\left(0.135\frac{\varepsilon}{D_h} + \frac{6.5}{Re}\right)\right\}^2} \quad \text{with} \quad Re = \frac{wD_h}{\nu} \tag{1.36}$$

The range of validity is: $4000 < Re < 10^8$ and $0 < \varepsilon/D_h < 0.05$. Equation (1.36) can be used for pipes with circular cross section ($D_h = D$) and channels of any cross section with $D_h = 4 \times A/U$ (U = wetted perimeter of a channel with cross section A). It applies to fully developed flow, i.e. long pipes with about $L/D_h > 50$. For short channels with $L/D_h > 5$ it is better to use the friction coefficients of the flat plate according to Eq. (1.33) together with Eqs. (1.34) or (1.35).

Equation (1.35) can be employed only to calculate the pressure losses due to friction on channel walls. To do so, the velocity averaged over the cross section must be substituted. In the general case Eq. (1.32) should be used. An example is the calculation of the energy loss caused by a plate placed in a channel in parallel flow; in this case it is not possible to assign a specific flow rate to the disturbance caused by the plate.

The friction coefficients for pipes or plates apply to flows without pressure gradients in the direction of flow. If such gradients occur, the friction coefficient c_f has to be modified or substituted by a dissipation coefficient c_d. In accelerated flow the thinner boundary layers lead to $c_d < c_f$, whereas $c_d > c_f$ applies to decelerated flow with thicker boundary layers (refer also to Table 3.8).

1.5.2 Influence of Roughness on Friction Losses

Experience shows that roughness increases the flow resistance in turbulent flow. However, this is only the case if the roughness elements protrude beyond the laminar sub-layer. In laminar flow the roughness has no influence on the resistance since there is no exchange of momentum across the flow. If all roughness peaks remain within the laminar sub-layer, the wall in the prevailing flow conditions is considered "hydraulically smooth", meaning a minimum value of the friction coefficient. With growing Reynolds number, the boundary layer thickness decreases and the permissible roughness drops as well. If the roughness peaks are significantly greater than the thickness of the laminar sub-layer, the wall is "hydraulically rough". Vortex shedding from the roughness peaks creates form drag due to the exchange of momentum with the main flow. In the fully rough domain the losses become independent of the Reynolds number and increase with the square of the flow velocity. In the transition region between hydraulically smooth and rough only

Table D1.1 Roughness limits ε

Hydraulically smooth	Transition from smooth to rough	Hydraulically rough
$\varepsilon < \frac{100\nu}{w}$	$100\frac{\nu}{w} < \varepsilon < 1000\frac{\nu}{w}$	$\varepsilon > 1000\frac{\nu}{w}$

the high roughness peaks protrude from the laminar sub-layer into the turbulent domain. Only these active peaks increase the flow resistance which now depends on the Reynolds number *and* the roughness. Table D1.1 indicates the roughness limits ε, separating the three domains discussed.

According to Table D1.1, the limit for hydraulically rough flow conditions is obtained from $Re = 1000 \times L/\varepsilon$; the corresponding curve is shown in Fig. 1.13.

Table D1.1 applies to flat plates. In the absence of more specific data this table can be used to evaluate other components with external or internal flows. Often the goal of such an analysis is to determine what level of manufacturing effort is justified to minimize the losses. The above relationships demonstrate that the surface finish must be improved with increasing flow velocity and/or decreasing kinematic viscosity. Because of the thinner boundary layer, the surface finish should be higher at the inlet of a component than near the outlet. Since a higher degree of turbulence must be generally assumed in pumps (than on flat plates in parallel flow), the roughness limits in pumps tend to be lower than could be expected from Table D1.1. This is particularly true for diffusers or volutes.

The increase in resistance brought about by the roughness depends on the roughness depth and the number of roughness peaks per unit area or their distance (pitch). Basic measurements of friction coefficients on plates and pipes were conducted on test surfaces to which sand grains were glued. Using different grain sizes, it was possible to determine the impact of various roughness heights on the flow resistance, see for instance [12]. The sand grains were glued by means of varnish and covered with another coat of varnish to prevent washing-off. This produced a *regular* surface structure. The resulting roughness depth ε_{max} equaled the grain diameter, since gaps were present between the individual grains. The roughness coefficients thus determined apply to this *specific surface structure* only, which is called "sand roughness" ε (frequently called k_s). The sand roughness is described by the grain diameter d_s ($\varepsilon \equiv k_s = d_s$).

In the transition domain between hydraulically smooth and hydraulically rough, this type of regular roughness leads to a minimum in the curve $\lambda_R = (Re)$ brought about by the fact that the varnish-coated grains do not generate an appreciable form drag when their tops protrude only slightly from the laminar sub-layer.

While the sand roughness possesses a uniform surface structure, *technical* (i.e. ground, cast or machined) surfaces have an *irregular roughness* which can be characterized, for example, by the maximum roughness depth ε_{max} ("technical roughness"). As the Reynolds number increases, the boundary layer thickness is reduced and more roughness peaks gradually rise through the laminar sub-layer. With such surfaces, the friction coefficients continuously fall from hydraulically smooth to the fully rough domain. This behavior is found in Fig. 1.13 and in the graph in

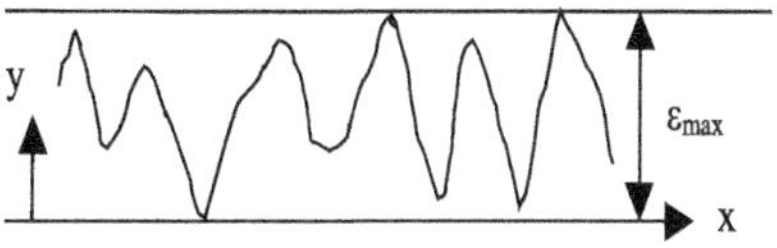

Fig. 1.14 Definition of the maximum roughness depth ε_{max}

Table 1.6. Technically rough surfaces do not exhibit a minimum in the function λ_R = (Re) in the transition region, as it is observed with regular surface structures.

The roughness depth ε_{max} according to Fig. 1.14 is obtained from the envelope of all roughness peaks and troughs. Since the measurement of the effective roughness of technical surfaces is quite involved, surface standards (e.g. "Rugotest") are frequently used in practice. The roughness is determined by touching the surface and comparing with roughness specimens (plates with different roughnesses). The roughness is then defined by the classes N1, N2, N3 etc. The roughness increases by a factor of two from one class to the next. The measured roughness is characterized as arithmetic mean roughness ε_a (corresponding to CLA = center line average or AA = arithmetic average); ε_a is defined by (Eq. 1.36a):

$$\varepsilon_a = \frac{1}{L}\int_0^L |y|\, dx \tag{1.36a}$$

The relationship between the maximum roughness depth ε_{max} and the arithmetic mean roughness ε_a is typically $\varepsilon_{max} = (5\text{–}7) \times \varepsilon_a$. On average we can expect approximately: $\varepsilon_{max} = 6 \times \varepsilon_a$, [5].

If the friction coefficient λ_R is determined from a pressure loss measurement in a pipe with a given roughness depth ε_{max}, the equivalent sand roughness ε can be calculated by means of Eq. (1.36). From this analysis an "equivalence factor" c_{eq} is obtained as the ratio $\varepsilon_{max}/\varepsilon$:

$$c_{eq} \equiv \frac{\varepsilon_{max}}{\varepsilon} \tag{1.36b}$$

Dividing the maximum roughness depth ε_{max} of any given surface by the equivalence factor, we get that equivalent sand roughness which leads to the same friction coefficient as the measurements which were used as a basis for Eq. (1.36).

The equivalence factor depends on the structure of the roughness - i.e. the machining process - and the orientation of the finishing marks relative to the flow direction; c_{eq} can therefore vary within wide boundaries. Of particular influence in this regard is the number of roughness peaks per unit area as shown in Fig. 1.15.

To determine the friction coefficient c_f or λ_R of a component with the roughness ε_a or ε_{max}, the equivalent sand roughness ε must be calculated from Eq. (1.36c):

$$\varepsilon = \frac{\varepsilon_{max}}{c_{eq}} \quad \text{or} \quad \text{with } \varepsilon_{max} = 6\,\varepsilon_a \quad \text{from } \varepsilon = \frac{6\varepsilon_a}{c_{eq}} \tag{1.36c}$$

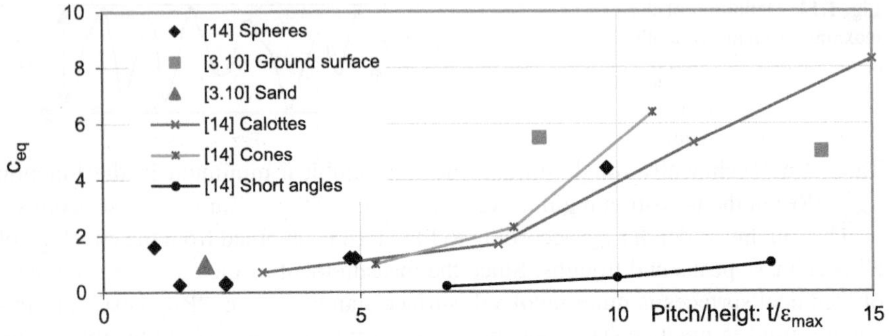

Fig. 1.15 Equivalence factor c_{eq} as function of the roughness density ([14]: Schlichting; [3.10]: Fukuda [20])

Table D1.2 Roughness equivalence factors c_{eq}

	c_{eq}
Manufacturing marks perpendicular to the flow direction	2.6
Manufacturing marks parallel to the flow direction	5
Drawn metal tubes	2–2.6
Smooth coating (e.g. paint)	0.63

Table D1.3 Roughness classes

Roughness class	Arithmetic average roughness	Maximum roughness (Fig. 1.14)	Equivalent sand roughness
	ε_a (μm)	ε_{max} (μm)	ε (μm)
N5	0.4	2.4	1
N6	0.8	4.8	2
N7	1.6	9.6	4
N8	3.2	19	8
N9	6.3	38	16
N10	12.5	75	32
N11	25	150	64
N12	50	300	128
N13	100	600	256

With that calculated equivalent sand roughness, use Eq. (1.33) and Fig. 1.13 or Eq. (1.36) or Table 1.6 to determine the friction coefficient.

Scanty information for c_{eq} is available in the literature. Table D1.2 provides some data, see also [5].

Table D1.3 shows the roughness values ε_a of the surface standards, the maximum roughness values ε_{max} obtained with Eq. (1.36c) and the equivalent sand roughness ε (ε_a is the upper limit of the respective roughness class; the values in Table D1.3 are rounded).

The diagrams for resistance coefficients with sand roughness and technical roughness differ only in the domain of transition from smooth to rough; therefore, they yield identical resistance coefficients in the fully rough region for any given

Table D1.4 Equivalent sand roughness ε

	ε (mm)
Glass, coatings, plastic, drawn metal tubes, polished surfaces	0.001–0.002
Drawn steel pipes, new	0.02–0.1
Drawn steel pipes, lightly rusted	0.15–1
Steel pipes, severely rusted or with deposits	1–3
Cast iron pipes and pump components	0.3–1
Concrete pipes	1–3

value of ε/d. In order to determine the friction coefficient from Eqs. (1.33) or (1.36), we must use that "technical roughness" ε which leads to the same resistance coefficient as the equivalent sand roughness. Some examples are listed in Table D1.4. These values originate from test results which can be found in the literature, for instance in [3, 6, 14].

One of the main uncertainties when calculating the friction losses of turbulent flows lies in determining the physically relevant roughness: if the roughness is assessed incorrectly by the factor of 2, the uncertainty of the loss calculation is approximately 15 to 35%. The impact of the roughness on pump efficiencies is covered in Chap. 3.10.

The flow resistance is greatly increased through vortex shedding caused by *regular* roughness patterns which are perpendicular to the direction of flow. This type of roughness can be generated, for example, by magnetite deposits in boilers where it creates a marked increase of pressure loss ("ripple roughness"). The phenomenon has also been observed in pipelines with ripple-type deposits.

1.5.3 Losses Due to Vortex Dissipation (Form Drag)

While static pressure can be converted into kinetic energy without a major loss (accelerated flow), the reverse process of converting kinetic energy into static pressure involves far greater losses. The reason for this is that in real flows the velocity distributions are mostly non-uniform and subject to further distortion upon deceleration. Non-uniform flow generates losses by turbulent dissipation through exchange of momentum between the streamlines. Such pressure losses may be called "form" or "mixing" losses. These types of losses develop, for example, in flows through curved channels, valves, pipe branches, diffusers, impellers and collectors of turbo machines or in flows around air foils, vehicles, or any type of structure. Because of the complicated three-dimensional flow patterns such losses cannot be predicted theoretically. Either empirical pressure loss coefficients are used for estimation (see for instance [3, 6, 11]) or numerical methods employed. Table 1.4 shows the pressure loss coefficients of some commonly used components.

Flow separations and secondary flows increase the non-uniformity of a velocity distribution, and consequently the energy losses, due to mixing through exchange of momentum. Flow separation and recirculation result in particularly high losses

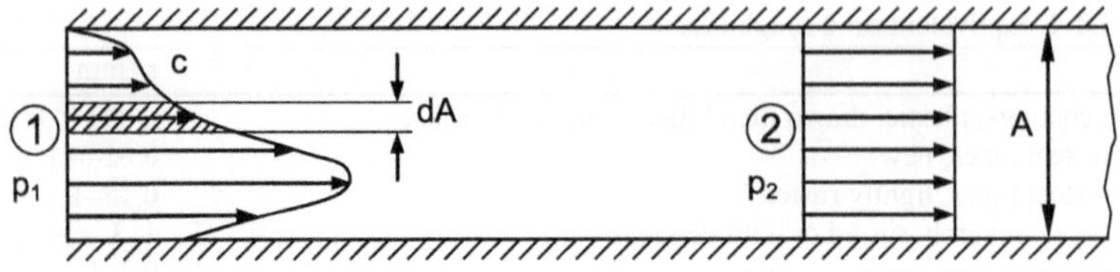

Fig. 1.16 Equalization of a non-uniform velocity distribution

since the kinetic energy in the wake tends to zero while the main flow - amplified by the cross-sectional blockage through the wake - has a large amount of kinetic energy. To minimize these types of energy losses, the extension of a stalled zone must be kept as small as possible. An example of this is the rounding at the inlet of a pipe by means of which flow contraction is reduced or prevented.

Friction losses - except for fully rough flow - depend on the Reynolds number; *form losses* (mixing losses) in fully turbulent flow above $Re > 10^4$ frequently display no noticeable Reynolds dependency. Exceptions are structures where the location of separation - and consequently the extension of the wake - depends on the Reynolds number. Examples of this are the flow around a sphere or a cylinder. In the case of blunt bodies or flow separation on sharp edges, stall largely occurs independently of the Reynolds number.

Mixing losses due to non-uniform velocity distributions form the main source of energy losses in pumps - especially at high specific speeds. It is hardly possible to predict such losses theoretically.

In channel flows a *non*-uniform velocity distribution always contains a greater amount of kinetic energy than a uniform profile (assuming constant flow rate and cross section). Only a fraction of the difference in kinetic energy between non-uniform and uniform distributions can be recovered for useful work, since losses in kinetic energy are inherent to all mixing processes (mixing increases entropy).

The transition of a non-uniform to a uniform velocity distribution can be approximately calculated by the conservation of energy and momentum - refer also to the calculation of a sudden expansion according to Chap. 1.2.3, Eqs. (1.11) and (1.12). To this end, consider a channel of constant cross section as shown in Fig. 1.16 with a non-uniform velocity distribution at the inlet. Suppose the channel is straight which means that the static pressure p_1 is constant over the channel cross section (refer to Chap. 1.4.1). After a distance necessary for velocity equalization, a pressure p_2 and a constant velocity c_{av} are established; c_{av} is obtained from continuity, Eq. (1.2), through $dQ = c \times dA$ and $c_{av} = Q/A = \int c \times dA/A$. Friction effects are not taken into account in the following [these would have to be estimated from Eqs. (1.32) to (1.34)]. The conservation of momentum according to Eq. (1.10) yields for the control surfaces 1 and 2:

$$p_2 A + \rho Q c_{av} - \int p_1 dA - \rho \int c dQ = 0$$

The coefficient of pressure recovery due to the exchange of momentum becomes:

$$c_p \equiv \frac{p_2 - p_1}{\frac{\rho}{2} c_{av}^2} = 2\left\{\int \left(\frac{c}{c_{av}}\right)^2 \frac{dA}{A} - 1\right\} \tag{1.37}$$

Designating the mixing loss as Δp_{mix}, we obtain from the conservation of energy Eq. (1.7): $\int (p_1 + ½\rho \times c^2)\, dQ = (p_2 + ½\rho \times c_{av}^2 + \Delta p_{mix}) \times Q$. The loss coefficient with Eq. (1.37) thus becomes:

$$\zeta_{mix} \equiv \frac{\Delta p_{mix}}{\frac{\rho}{2} c_{av}^2} = \int \left(\frac{c}{c_{av}}\right)^3 \frac{dA}{A} - c_p - 1 \tag{1.38}$$

Equations (1.37) and (1.38) can be evaluated as integrals over the channel cross section. Similarly it is possible to substitute the integrals by summations, if two (or more) flows with constant velocities c_1 and c_2 of the components $\Delta A_1/A$ and $\Delta A_2/A$ are mixed. In CFD calculations the summation is done over the elements in the control surfaces.

1.6 Diffusers

Diffusers serve to decelerate a flow and consequently convert kinetic energy into potential energy (or static pressure). Diffusing elements are therefore very important components in pump design. Consider a diffuser according to Fig. 1.17, the inlet and outlet cross sections of which are called 1 and 2 respectively. Suppose that a component is arranged downstream of the diffuser. Let us include the hydraulic losses into the consideration since the task frequently consists of optimizing the overall system between 1 and 3. Bernoulli's equation (1.7) is first applied to the actual diffuser between cross sections 1 and 2. If a pressure loss coefficient $\zeta_{1\text{-}2}$ is introduced, Eq. (1.7) can be solved for the rise of the static pressure in the diffuser, yielding Eq. (1.39):

$$\frac{p_2 - p_1}{\frac{\rho}{2} {c_1}^2} = 1 - \frac{{c_2}^2}{{c_1}^2} - \frac{\Delta p_{v,1\text{-}2}}{\frac{\rho}{2} {c_1}^2} = 1 - \frac{1}{A_R^2} - \zeta_{1\text{-}2} \tag{1.39}$$

where $A_R = A_2/A_1$ is the area or deceleration ratio of the diffuser.

The pressure recovery of the diffuser is usually identified by a dimensionless coefficient c_p:

$$c_p \equiv \frac{p_2 - p_1}{\frac{\rho}{2} {c_1}^2} = 1 - \frac{1}{A_R^2} - \zeta_{1\text{-}2} = c_{p_{id}} - \zeta_{1\text{-}2} \tag{1.40}$$

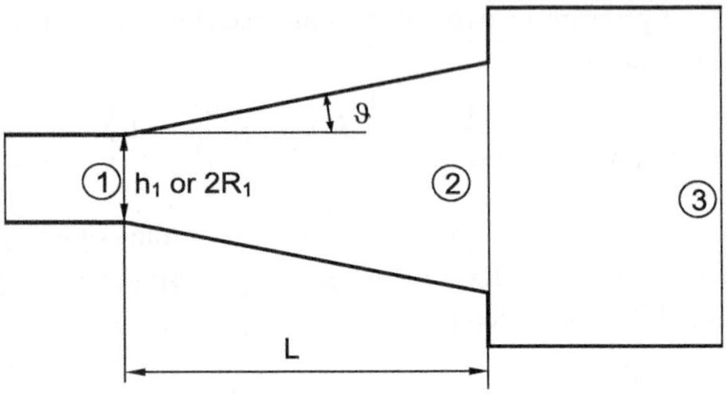

Fig. 1.17 Diffuser

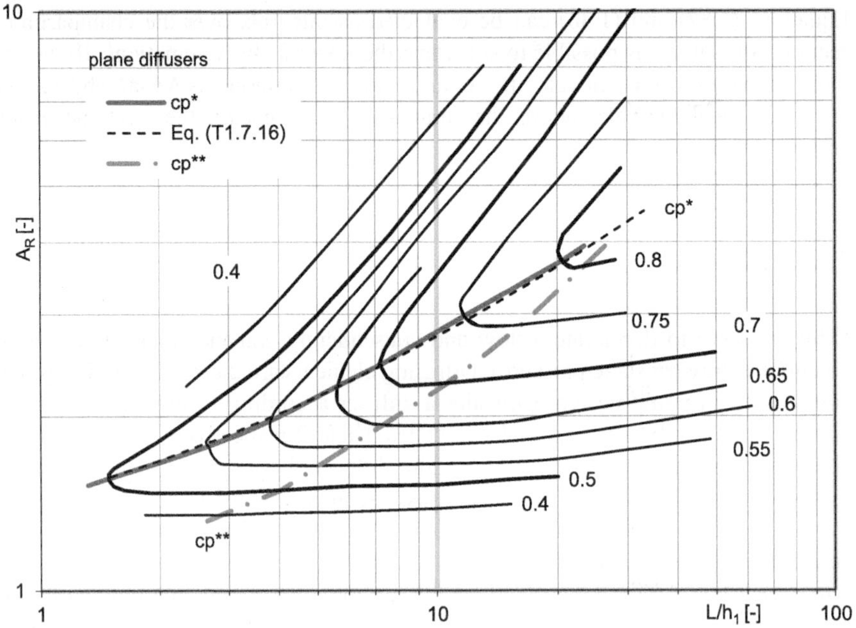

Fig. 1.18 Pressure recovery in *plane* diffusers (Note: In previous editions (A_R-1) was plotted on the ordinate rather than the present A_R) (inlet boundary layer blockage 1.5%) [13]

$c_{p,id} = 1 - 1/A_R^2$ is the pressure recovery in an ideal (loss-free) diffuser. The pressure recovery in a real diffuser cannot be theoretically calculated because of the unavoidable energy losses described by $\Delta p_v = ½\rho \times \zeta_{1\text{-}2} \times c_1^2$. It is necessary to resort to test results as presented in Fig. 1.18 for plane diffusers and in Fig. 1.19 for conical diffusers. These figures show families of curves with c_p as parameter while the area ratio is plotted on the ordinate and the diffuser length on the abscissa. The curves c_p* furnish the optimum area ratio for a specified the diffuser length, while c_p** provides the optimum length for a given area ratio. These optimums correspond to the maximum pressure recovery in the diffuser under the given constraints.

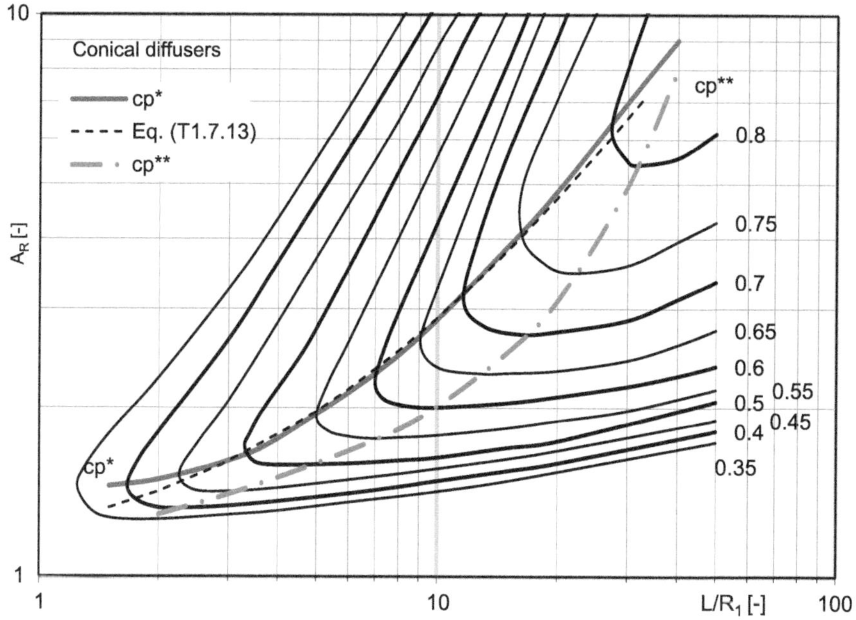

Fig. 1.19 Pressure recovery in *conical* diffusers (Note: In previous editions (A_R-1) was plotted on the ordinate rather than the present A_R), [10]

Figures 1.18 and 1.19 apply to diffusers with a straight axis. The pressure recovery in curved diffusers is inferior as discussed below.

If a certain deceleration ratio is specified, it is possible to determine the diffuser length required for maximum pressure recovery from Fig. 1.18 or Fig. 1.19. In practice diffusers are often designed for a specified length with $A_R < 3$ to 4 since area ratios higher than 4 will increase the energy recovery by less than 2 to 3% only.

Frequently the task requires optimizing a diffuser together with a downstream component the pressure loss of which is described by Eq. (1.41):

$$\Delta p_{v,2\text{-}3} = \zeta_{2\text{-}3} \frac{\rho}{2} c_2^{\,2} \tag{1.41}$$

Consequently, the following loss coefficient is obtained for the system between the cross sections 1 and 3:

$$\zeta_{1\text{-}3} = 1 - c_p - \frac{1}{A_R^2}(1 - \zeta_{2\text{-}3}) \tag{1.42}$$

If the total kinetic energy at the diffuser outlet is counted as useful, $\zeta_{2\text{-}3} = 0$ must be substituted; if it is entirely lost, because the flow exits into a large plenum, we have $\zeta_{2\text{-}3} = 1$ and $\zeta_{1\text{-}3} = 1 - c_p$.

The quality of a diffuser is sometimes described by the "diffuser efficiency" which is defined by Eq. (1.43).

$$\eta_D \equiv \frac{c_p}{c_{p_{id}}} = \frac{c_p}{1 - \frac{1}{A_R^2}} \tag{1.43}$$

As discussed in Chap. 1.3 (Fig. 1.4), the boundary layer thickness grows with increasing flow path length when the fluid flows against a positive pressure gradient. Excessive flow deceleration causes the flow to separate from the walls. Depending on the magnitude of the deceleration, four flow regimes may be distinguished in a diffuser, as shown in Fig. 1.20 according to [8] and [B.3]:

1. Below a deceleration defined by Eq. (1.44) or curve a-a the flow is attached, domain A in Fig. 1.20.
2. If the boundary layer exceeds a certain thickness, local intermittent stall occurs, domain B in Fig. 1.20.
3. With even greater deceleration the flow fully separates on one side, domain C.
4. Finally, with excessive divergence of the channel walls, flow separation occurs on both sides so that the fluid shoots through the diffuser like a jet, domain D.

$$\vartheta_{al} = 16.5^\circ \sqrt{\frac{R_1}{L}} \tag{1.44}$$

The allowable diffusion angle ϑ_{al} according to Eq. (1.44) applies only to favorable inflow conditions. It is important to recognize that the allowable diffusion angle of a diffuser is by no means a universal constant (an opinion still frequently voiced) but that it is strongly dependent on the length of the diffuser. The length of the flow path has a strong impact on the development of the boundary layer thickness and consequently on possible stall. In addition to this, the diffuser flow critically depends on the conditions at the inlet because velocity distribution, boundary layer thickness and turbulence determine the boundary layer development in the diffuser. The thinner the boundary layer and the higher the turbulence levels at the inlet, the lower are the losses and the tendency to stall in a diffuser. *Thick boundary layers are unfavorable to the diffuser flow and lead to premature stall.* Thick boundary layers are generated in long pipes or in non-uniform velocity distributions which may be created by an upstream bend for example. Conversely, a swirl at the diffuser inlet can have a favorable impact on the diffuser performance since the exchange of momentum between main flow and boundary layer is promoted through centrifugal forces.[2] An unsteady approach flow, such as prevailing in a pump at the diffuser inlet, has also a favorable effect on the boundary layer development, hence on the pressure recovery. This is presumably due to the fact that the unsteady flow increases the turbulence level. A higher turbulence in the diffuser enhances the exchange of momentum

[2]An example is the flow in a diffusing discharge nozzle following a volute casing, Chap. 7.8.2.

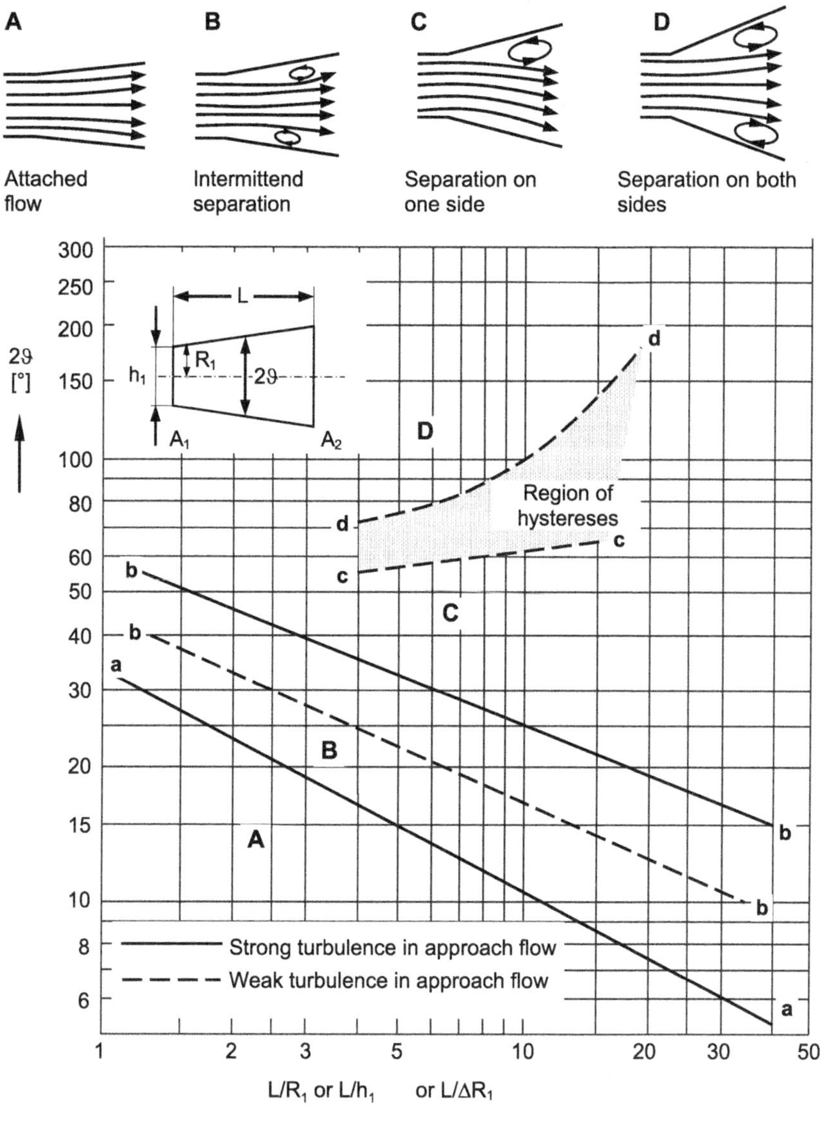

Fig. 1.20 Flow regimes in diffusers

between near-wall and core flows. The thinner the boundary layers, the smaller are the losses and the wider is the allowable diffusion angle. Test data on the influence of the boundary layer on diffuser performance are shown in Fig. 1.25.

If two diffusers are arranged in series, they must not be evaluated separately according to Fig. 1.18 or Fig. 1.19 and Eq. (1.44), but the total deceleration of both diffusers combined must be considered, because this is relevant for the boundary layer growth.

If an excessively expanding diffuser is sub-divided by intermediate walls or by vanes, flow separation can be suppressed and pressure recovery improved. A baffle or resistance arranged downstream of the diffuser reduces the separation tendency too. The diffuser flow can be improved by boundary layer extraction or injection, but because of the associated design complexity and cost such measures are justified only in special cases.

A ratio $b/h_1 = 1$ of diffuser width b to inlet height h_1 yields the maximum pressure recovery in a plane diffuser. In the range $0.8 < b/h_1 < 2$ only minor impairment is incurred compared with this optimum.

For a diffuser with any shape of cross section, an equivalent inlet radius $R_{1,eq}$ and an equivalent opening angle ϑ_{eq} are calculated from Eq. (1.45). Subsequently the diffuser is treated like a conical diffuser according to Fig. 1.19.

$$R_{1,eq} = \sqrt{\frac{A_1}{\pi}} \quad \text{and} \quad \tan\vartheta_{eq} = \frac{R_{1,eq}}{L}\left\{\sqrt{\frac{A_2}{A_1}} - 1\right\} \tag{1.45}$$

Table 1.7 (at the end of this chapter) provides formulae for diffuser design. Note also the formulae and graphs for the optimum diffuser opening angle as function of the non-dimensional diffuser length for the optimum area ratio, Eqs. (T1.7.15 and T1.7.18), see Figs. 1.21 to 1.23.

Curved diffusers: The pressure recovery in curved diffusers has been measured in [17]. Test conditions: Conical diffusers with circular cross section; inlet diameter $d_1 = 100$ mm, $A_R = 4$, $L/R_1 = 715/50 = 14.3$, opening angle $2\vartheta = 8°$; curved diffusers with deflection angles of $\lambda = 15°$ and 30°; inlet Re = 3×10^4 to 8×10^5. The deflection angle is measured at the axis between outlet and inlet (i.e. straight diffuser: $\lambda = 0$; 90°-bend: $\lambda = 90°$).

From the tests in [17], correction factors f_{cp} have been derived by which the c_p-values from Figs. 1.18 or 1.19 can be multiplied for estimating the performance of curved diffusers. The results are shown in Fig. 1.24; they can be approximated by Eq. (1.46):

$$f_{c_p} \equiv \frac{c_{p,curved}}{c_{p,straight}} = \left(1 - 0.13\frac{\lambda}{30°}\right) f(Re) \tag{1.46}$$

The term f(Re) accounts for the influence of the Reynolds number at the diffuser inlet:

For: $Re < 7 \times 10^4$ calculate: $f(Re) = \left(\frac{Re}{7\times10^4}\right)^{0.15}$; for $Re > 7 \times 10^4$ set: $f(Re) = 1.0$

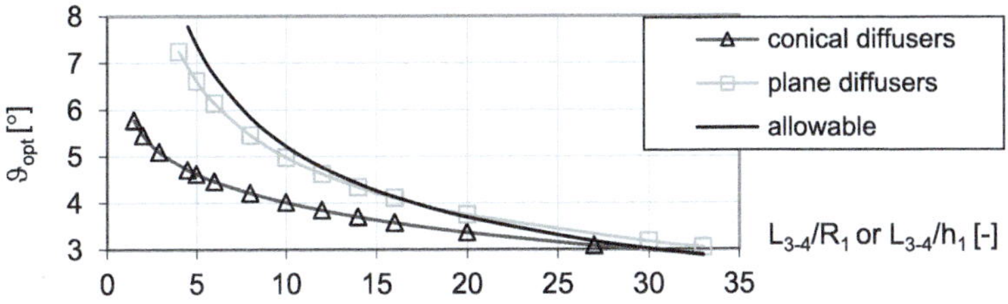

Fig. 1.21 Optimum diffuser opening angle

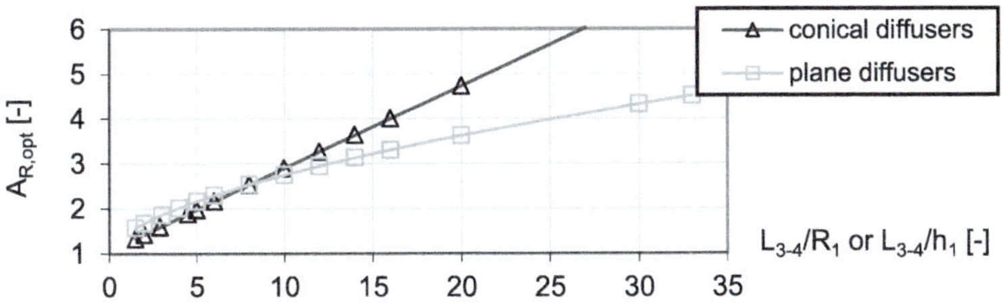

Fig. 1.22 Optimum diffuser area ratio

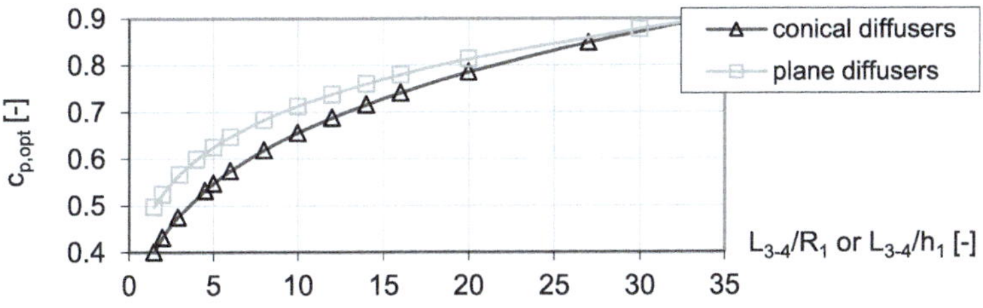

Fig. 1.23 Optimum pressure recovery in diffuser

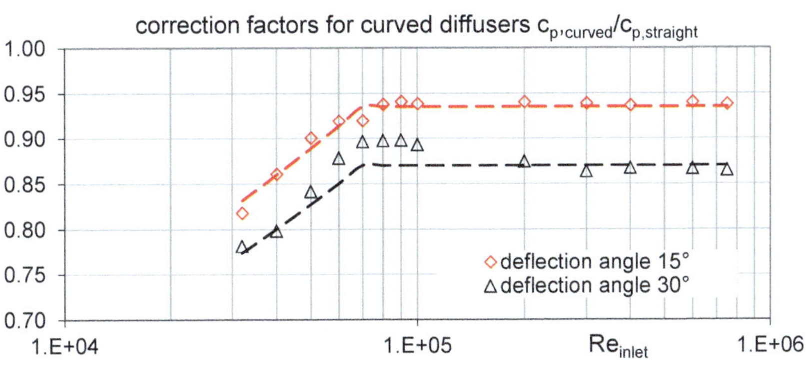

Fig. 1.24 Correction factors for curved diffusers derived from tests in [17]

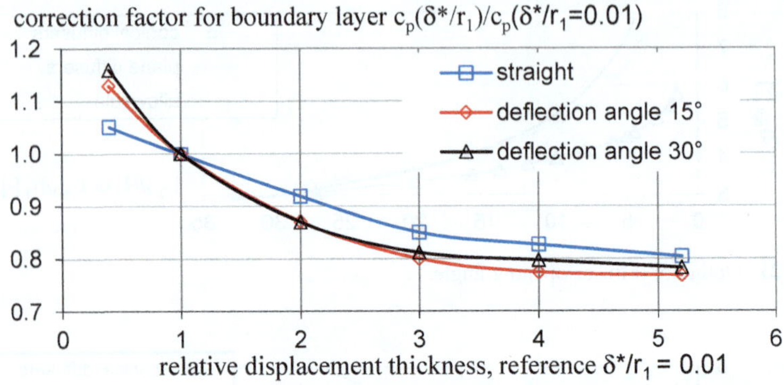

Fig. 1.25 Correction factors for influence of boundary displacement thickness on pressure recovery in straight and curved diffusers; derived from tests in [14, 17]

Above $Re > 7 \times 10^4$ the correction factor is independent of the Reynolds number. Extrapolation of the deflection angles up to 45° should be possible with reasonable confidence.

The influence of the boundary layer on diffuser performance has been measured also in [17]. Correction factors were calculated from these tests which allow to estimate the effect of the boundary layer at the diffuser inlet on the pressure recovery when the relative displacement thickness δ^*/r_1 is different from the reference value $\delta^*/r_1 = 0.01$. The result is shown in Fig. 1.25. It is noteworthy that the correction factors are quite similar for straight and curved diffusers when $\delta^*/r_1 > 0.01$.

1.7 Submerged Jets

A liquid or a gas generates a "jet" when it enters with a finite velocity into a fluid-filled plenum. The interaction between the jet and the static medium is relevant in many applications. We distinguish between:

1. Fluid jets with a phase boundary liquid/gas (e.g. a fountain)
2. Jets without phase boundary i.e. gas/gas or liquid/liquid, called "submerged jets". Examples are the injection of air into a room or the inflow of water through a nozzle or a pipe into a tank, provided that the jet is sufficiently submerged below the liquid surface level.

The mixing of submerged jets with the fluid in a plenum is significant for the evaluation of inlet bays, test stand pits, suction vessels (Chap. 11.7.3) or the annular seal leakages entering upstream of an impeller inlet.

Table 1.2 Submerged turbulent jets

Valid for: $Re = \frac{d_o c_o}{\nu} > 3000$ and $10 < \frac{x}{r_o} < 200$

		Plane Jet	Plane wall Jet	Round Jet	Eq.
Core length with $c_m = c_o$		$x_1 = 12\ b_0$	$x_1 = 6.4\ b_0$	$x_1 = 10\ r_0$	1.2.1
Center-line velocity	Only for $x > x_1$	$c_m = 3.4 c_o \sqrt{\frac{b_o}{x}}$	$c_m = 3.5 c_o \sqrt{\frac{b_o}{x}}$	$c_m = 12 \frac{r_o}{x} c_o$	1.2.2
Velocity distribution δ = boundary layer thickness		$c = c_m e^{-57\left(\frac{y}{x}\right)^2}$	$c = c_m e^{-0.693\left(\frac{y-\delta}{b_0}\right)^2}$ $\delta = 0.014x$	$c = c_m e^{-94\left(\frac{r}{x}\right)^2}$	1.2.3
Width or radius at which the velocity drops to 50% of the value at the jet centerline		$b_{50} = 0.11\ x$	$b_{50} = 0.068\ x$	$r_{50} = 0.086\ x$	1.2.4
Volume flow rate entrained in jet; flow rate at origin is Q_o		$Q = 0.44 Q_o \sqrt{\frac{x}{b_o}}$	$Q = 0.248 Q_o \frac{x}{b_o}$	$Q = 0.16 Q_o \frac{x}{r_o} Q_o$	1.2.5

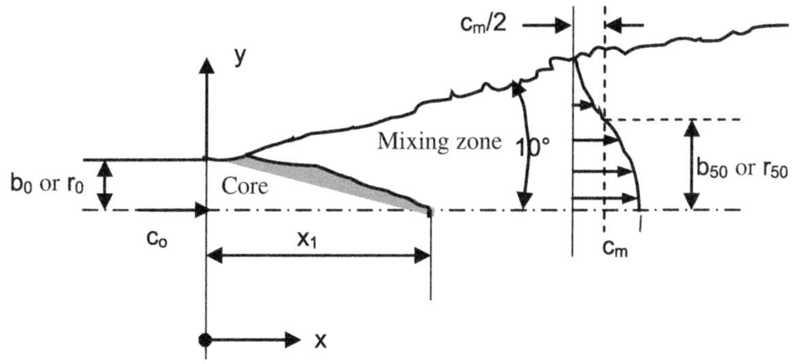

Consider a fluid entering a tank with the velocity c_o from a pipe and let the dimensions of the tank be large relative to the pipe diameter d_o. Suppose the pipe leads into the tank below the liquid level. At the pipe exit a fluid jet is created which mixes with the content of the tank through turbulence. With increasing distance x from the nozzle, the jet expands with an opening angle of approximately $2 \times 10°$ and its velocity decreases, see figure in Table 1.2. In the center of the jet, its core, the nozzle velocity c_o is initially maintained over a certain distance x_1. Subsequently, the velocity c_m in the center drops with growing distance x for distances $x > x_1$. The jet drags fluid along so that its flow rate increases with increasing x. The formulae in Table 1.2 can be used to calculate the expansion and the velocity decrease of plane and round jets, as well as jets entering parallel to a wall.

1.8 Equalization of Non-uniform Velocity Profiles

If a flow of non-uniform velocity distribution meets with a flow resistance evenly distributed over the channel cross section (e.g. a perforated baffle or a strainer), the velocity differences downstream of the resistance are reduced. The equalization mechanism can be qualitatively explained as follows: The local pressure loss created by the resistance is $\Delta p(y) = \frac{1}{2}\rho \times \zeta_A \times c^2$ (local velocity: $c = f(y)$). This would lead to a pressure which is non-uniform over the channel width y. In a straight flow, however, the static pressure is inevitably constant over the cross section (Chap. 1.4.1). For this reason, cross-flows are generated which tend to achieve that condition. Such cross-flows tend to smooth out the velocity distribution.

A non-uniform velocity profile can also be equalized by accelerating the flow. This effect is put to advantage, for instance, when designing the radial inlet casings of multistage or double-entry pumps, Chap. 7.13. For the same reason, the approach flow to a vertical pump is directed to the impeller through a suction bell (bellmouth) or accelerating bends in order to improve the flow at the impeller inlet. Velocity *fluctuations* are also reduced in accelerated flow. Consequently, flow acceleration reduces the turbulence level. Nozzles with high rates of acceleration are therefore used to generate a flow that has very low turbulence.

The reduction of the excess velocities can be estimated according to the formulae given in Table 1.3. To derive these formulae, Bernoulli's equation is written between

Table 1.3 Equalization of non-uniform velocity distributions by flow resistances or acceleration

			Eq.
Average velocity and pressure drop	$c_q = \frac{Q}{A}$	$\Delta p = \frac{\rho}{2}\zeta_A\, c_q^2$	1.3.1
Local excess velocities or deficits	$\Delta c_1 \equiv c_{1max} - c_q$ $\Delta c_2 \equiv c_{2max} - c_q$	$\frac{c_{2,max}}{c_q} = 1 + \frac{\Delta c_2}{\Delta c_1}\left\{\frac{c_{1,max}}{c_q} - 1\right\}$	1.3.2
Reduction of excess velocities by *flow resistance*	$\frac{\Delta c_2}{\Delta c_1} = \frac{1}{(1+\zeta_A)^n}$	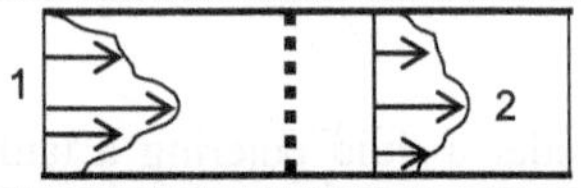	1.3.3
Reduction of excess velocities by *flow acceleration*	$\frac{\Delta c_2}{\Delta c_1} = \frac{A_2}{A_1}$	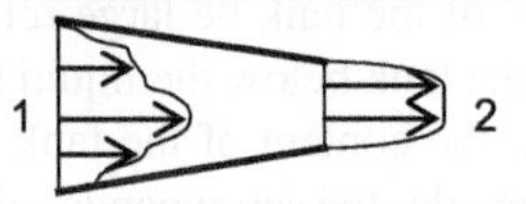	1.3.4
Pressure loss coefficient ζ_A referred to average *approach* velocity c_q	$\zeta_A = \zeta\frac{A_2^2}{A_1^2} = \frac{\zeta}{f^2}$	$f \equiv \frac{A_1}{A_2}$ as defined in Table 1.4	1.3.5

Subscripts: 1 = inlet; 2 = outlet
n = number of flow resistances in series
The pressure loss coefficients given in Table 1.4 for sharp-edged or rounded thick orifice plates can be applied to screens, perrorated plates and similar elements. Note that the resistance coefficients in Table 1.4 are referred to the throat area; such coefficients must be converted by Eq. (1.3.5) to get values which are referred to the average approach velocity.

Table 1.4 Pressure loss coefficients		$\Delta p = \zeta \frac{\rho}{2} w_1^2$	$f \equiv \frac{A_1}{A_2}$
Inlet to pipe or nozzle *without* end wall	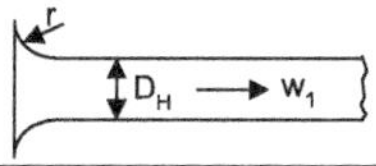	$\zeta = e^{-17\frac{r}{D_H}}$	$0 \le \frac{r}{D_H} \le 0.2$
Rounded inlet *with* end wall	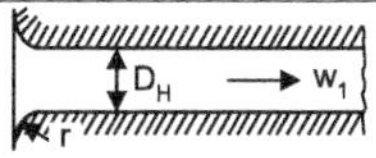	$\zeta = 0.5\, e^{-14\frac{r}{D_H}}$	$0 \le \frac{r}{D_H} \le 0.2$
Inlet *with* end wall, chamfered with 45°	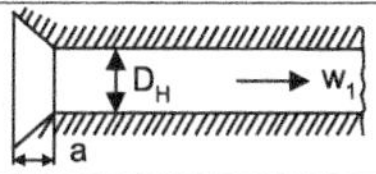	$\zeta = 0.5\, e^{-4.6\frac{a}{D_H}}$	$0 \le \frac{a}{D_H} \le 0.15$
Contraction	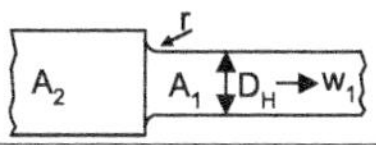	$\zeta = 0.5\,(1-f)\, e^{-14\frac{r}{D_H}}$	$0 \le \frac{r}{D_H} \le 0.2$
Contraction chamfered with 45°	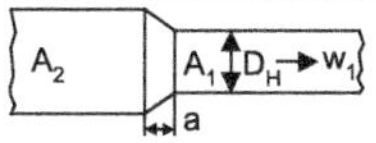	$\zeta = 0.5\,(1-f)\, e^{-4.6\frac{a}{D_H}}$	$0 \le \frac{a}{D_H} \le 0.15$
Sudden expansion	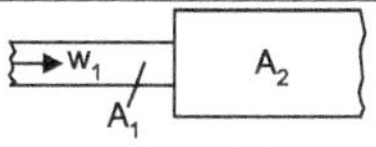	$\zeta = (1-f)^2$	
Sharp-edged *thin* orifice	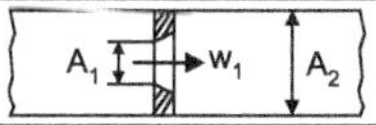	$\zeta = \left\{0.7\sqrt{1-f} + 1 - f\right\}^2$	
Sharp-edged *thick* orifice $L/D_H = 1$	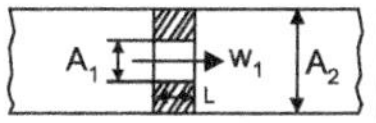	$\zeta = (1-f)\,\left\{1.5 - f + 0.24\,\sqrt{1-f}\right\}$	
Rounded orifice	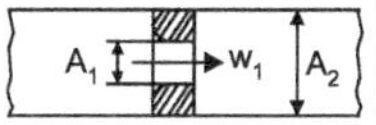	$\zeta = (1-f)\left\{1 - f + \zeta' + 2\sqrt{\zeta'\,(1-f)}\right\}$	$\zeta' = 0.5\, e^{-14\frac{r}{D_H}}$
Sharp knees	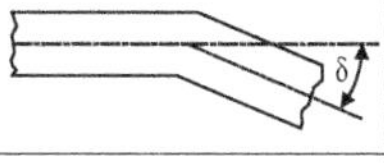	$\zeta = 1.2\left(\frac{\delta°}{90°}\right)^2$	
Bends	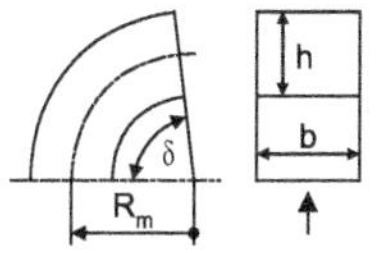	$\zeta = F_R\ F_F\ \frac{0.23}{\left(\frac{R_m}{D_H}\right)^x}\sqrt{\frac{\delta°}{90°}}$	see below
Forked pipe	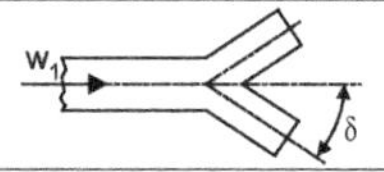	see below	

Bends:

$\frac{R_m}{D_H}$	x
0.5 to 1	2.5
> 1	0.5

Influence of roughness		F_R	**Cross section**	F_F
Hydraulically smooth:		1	Circle, square	1.0
$\frac{\varepsilon}{D_H} > 10^{-3}$	$Re > 4 \times 10^4$	2	Rectangle b/h = 2 to 4	0.4

Forked pipe:

δ	15°	22.5°	30°	45°	60°	90°
ζ	0.15	0.23	0.3	0.7	1.0	1.4

Suction strainer with foot valve	$\zeta = 2.2$ to 2.5	ζ is referred to the velocity calculated with the nominal diameter of the valve.
Gate valves	$\zeta = 0.1$ to 0.5 $(\div 1)$	
DIN-valve	$\zeta = 3$ to 6	
Check valve	$\zeta = 6$	
Butterfly valve, fully open	$\zeta = 0.2$	

inlet cross section A_1 and outlet cross section A_2 for one fictitious streamline with $c_{1max} = c_q + \Delta c_1$ and another streamline with $c_{1min} = c_q - \Delta c_1$, [16]. Since the cross flow is neglected in the derivation, the formulae in Table 1.3 are only an approximation.

The streamline with the greatest excess velocity is decelerated upstream of the resistance; this effect increases with the loss coefficient ζ of the resistance. Therefore, each resistance should not exceed $\zeta \approx 1$. Instead of one large resistance, it is much more efficient to install several resistances in series to achieve an optimum equalization.

Perforated plates, screens, bars or other structures installed for velocity equalization generate relatively uniform turbulence structures whose fineness depends on the size of the mesh used.

1.9 Flow Distribution in Parallel Channels, Piping Networks

Generally a pipe network may consist of various apparatus, fittings and valves connected in series and/or in parallel. To calculate (see Table 1.5) networks of any complexity, a schematic flow diagram is drawn showing all components by groups of resistances connected in parallel or in series. Each component is described by a pressure loss coefficient ζ_i and a cross section A_i, so that the pressure loss in any component is $\Delta p_i = ½\rho \times \zeta_i \times (Q_i/A_i)^2$.

If the components are connected **in series**, the same flow rate Q flows through each component; the pressure losses of all components add up to yield the system pressure loss according to $\Delta p = \sum \Delta p_i$. An arbitrary reference cross section A_{RS} is selected for all parts of the network containing series-connected components. The components located in the considered branch of the network can then be combined into one resistance ζ_{RS} according to Eq. (T1.5.1).

All components which are connected **in parallel** are subject to the same pressure differential, and the flows Q_i are added to give the total flow rate $Q = \sum Q_i$. The sum of the individual cross sections is selected as the reference cross section according to Eq. (T1.5.5). Owing to these relationships, it is possible to calculate a resultant resistance coefficient ζ_{PS} of the system or subsystem, Eq. (1.5.8). Subsequently the individual flow rates through each of the parallel branches can be determined (Table 1.5).

With all of these calculations, careful attention must be paid to ensure that the resistance coefficient is defined for the reference velocity to be used in each case (i.e. ζ_1 relative to Q_1/A_1 etc.). Complex networks are combined step-by-step according to the formulae given, until a resultant total resistance is obtained which allows the determination of the total pressure loss and the total flow rate through the system. Text continued below Table 1.7.

Table 1.5 Calculation of pipe networks		Eq.
Pipe sections connected in series: All pipe sections or components receive the same flow rate Q	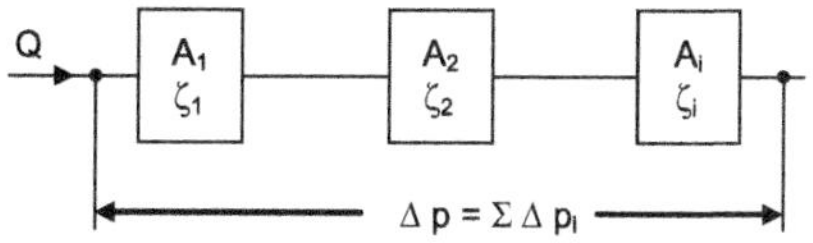	
Resulting overall resistance	$\zeta_{RS} = \sum_i \zeta_i \left(\frac{A_{RS}}{A_i} \right)^2$ — A_{RS} = selected reference cross section	1.5.1
Pressure drop of individual pipe sections or components	$\Delta p_i = \frac{\rho}{2} \zeta_i \left(\frac{Q}{A_i} \right)^2$	1.5.2
Total pressure drop	$\Delta p = \frac{\rho}{2} \zeta_{RS} \left(\frac{Q}{A_{RS}} \right)^2$	1.5.3
Flow rate for a given (imposed) pressure drop Δp	$Q = A_{RS} \sqrt{\frac{2 \Delta p}{\rho \zeta_{RS}}}$	1.5.4
Pipe sections connected in parallel: All pipe sections or components are subject to the same pressure drop; individual flow rates establish themselves to fulfill this condition.	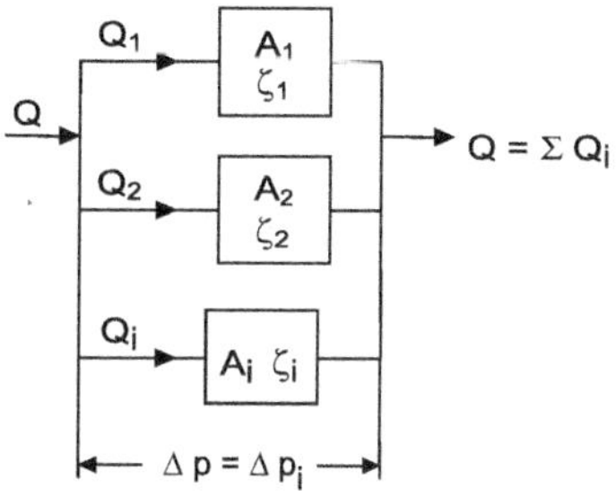	
Reference cross section	$A_{PS} = \sum A_i$	1.5.5
Total flow rate	$Q = \sum Q_i$	1.5.6
Pressure drop of individual pipe sections or components	$\Delta p = \frac{\rho}{2} \zeta_1 \left(\frac{Q_1}{A_1} \right)^2 = \ldots\ldots\ldots = \frac{\rho}{2} \zeta_i \left(\frac{Q_i}{A_i} \right)^2$	1.5.7
Resulting overall resistance	$\zeta_{PS} = \frac{1}{\left\{ \Sigma \frac{A_i}{A_{PS} \sqrt{\zeta_i}} \right\}^2}$	1.5.8
Total pressure drop	$\Delta p = \frac{\rho}{2} \zeta_{PS} \left(\frac{Q}{A_{PS}} \right)^2$	1.5.9
Flow rate for a given (imposed) pressure drop Δp	$Q = A_{PS} \sqrt{\frac{2 \Delta p}{\rho \zeta_{PS}}}$	1.5.10
Flow rate through individual (parallel) pipe sections	$\frac{Q_i}{Q} = \frac{A_i}{A_{PS}} \sqrt{\frac{\zeta_{PS}}{\zeta_i}}$	1.5.11

Table 1.6 Friction losses of pipes and flat plates

Pipe or channel A = cross section U = wetted perimeter L = length	Hydraulic diameter	$D_h = \frac{4\,A}{U}$	Reynolds number	$Re = \frac{w\,D_h}{\nu}$
	Laminar: Re < 2300	$\lambda_R = \frac{64}{Re}$		
	Turbulent, Eq. (1.36) $4000 < Re < 10^8$ $0 < \varepsilon/D_h < 0.05$	$\lambda_R = \frac{0.31}{\left\{\log\left(0.135\frac{\varepsilon}{D_h}+\frac{6.5}{Re}\right)\right\}^2}$		
	Pressure drop	$\Delta p_v = \lambda_R \frac{L}{D_h}\frac{\rho}{2}w^2$		

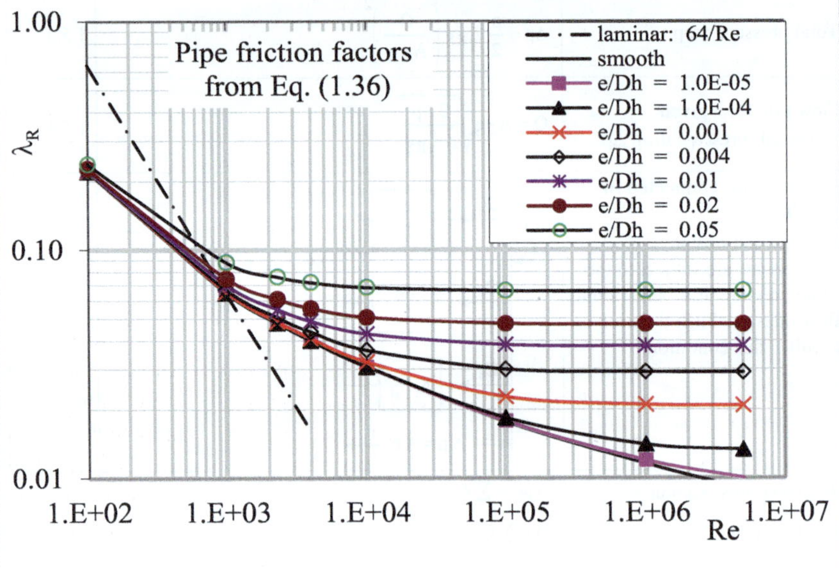

Flat plate of length L	Reynolds number	$Re_L = \frac{wL}{\nu}$
	Laminar: $Re_L < 10^5$ or use Eq. (1.33a)	$c_{f,lam} = \frac{1.328}{\sqrt{Re_L}}$
	Turbulent $10^5 < Re_L < 10^8$ $0 < \varepsilon/L < 10^{-3}$	$c_f = \frac{0.136}{\left\{-\log\left(0.2\frac{\varepsilon}{L}+\frac{12.5}{Re}\right)\right\}^{2.15}}$
	Drag on *one side* of a flat plate of width b and length L	$F_w = ½\,\rho\,w_\infty^2\,c_f\,b\,L$
	c_f is the friction coefficient for the total drag (the local drag coefficient depends on the boundary layer thickness)	

Table 1.7 (1) Diffuser design		Eq.
area ratio	$A_R = \frac{A_2}{A_1}$	1.7.1
allowable opening angle per side for uniform approach flow	$\vartheta_{al} = 16.5° \sqrt{\frac{R_1}{L}}$	1.7.2
pressure recovery coefficient	$c_p \equiv \frac{p_2 - p_1}{\frac{\rho}{2} c_1^2} = 1 - \frac{1}{A_R^2} - \zeta_{1-2} = c_{p_{id}} - \zeta_{1-2}$	1.7.3
pressure recovery in a diffuser	$\frac{p_2 - p_1}{\frac{\rho}{2} c_1^2} = 1 - \frac{c_2^2}{c_1^2} - \frac{\Delta p_{v,1-2}}{\frac{\rho}{2} c_1^2} = 1 - \frac{1}{A_R^2} - \zeta_{1-2}$	1.7.4
pressure loss in diffuser if exit energy is considered useful	$\zeta_{1-2} \equiv \frac{\Delta p_v}{\rho \frac{c_1^2}{2}} = 1 - c_p - \frac{1}{A_R^2}$	1.7.5
pressure loss in diffuser plus exit pressure loss	$\zeta_{1-3} \equiv \frac{\Delta p_v}{\rho \frac{c_1^2}{2}} = 1 - c_p - \frac{1}{A_R^2}(1 - \zeta_{2-3})$	1.7.6
pressure loss if diffuser discharges into a plenum (exit velocity energy dissipated)	$\zeta_{1-2} \equiv \frac{\Delta p_v}{\rho \frac{c_1^2}{2}} = 1 - c_p$	1.7.7
plane diffusers	$c_p = f(A_R, L/h_1)$ from Fig. 1.18	1.7.8
conical diffusers	$c_p = f(A_R, L/R_1)$ from Fig. 1.19	1.7.9
diffusers of arbitrary geometry	$c_p = f(A_R, L/R_{1,eq})$ from Fig. 1.19	1.7.10
equivalent inlet radius	$R_{1,eq} = \sqrt{\frac{A_1}{\pi}}$	1.7.11
equivalent opening angle	$\tan \vartheta_{eq} = \frac{R_{1,eq}}{L}\left\{\sqrt{\frac{A_2}{A_1}} - 1\right\}$	1.7.12
Conical diffusers		
Optimum area ratio for a given diffuser length L	$A_{R,opt} = 1.05 + 0.184 \frac{L}{R_{eq}}$	1.7.13
Pressure recovery in optimum diffuser	$c_{p,opt} = 0.36 \left(\frac{L}{R_{eq}}\right)^{0.26}$	1.7.14
Opening angle per side in optimum diffuser	$\vartheta_{opt} = a\tan\left\{\frac{R_{eq}}{L}\left(\sqrt{A_{R,opt}} - 1\right)\right\}$	1.7.15

Table 1.7 (2) Diffuser design		
Plane diffusers		Eq.
Optimum area ratio for a given diffuser length L	$A_{R,opt} = 1.0 + 0.45\left(\frac{L}{h_1}\right)^{0.588}$	1.7.16
Pressure recovery in optimum diffuser	$c_{p,opt} = 0.46\left(\frac{L}{h_1}\right)^{0.19}$	1.7.17
Opening angle per side in optimum diffuser	$\vartheta_{opt} = \mathrm{a}\tan\left\{\frac{h_1}{2L}\left(A_{R,opt} - 1\right)\right\}$	1.7.18

The methods given in Table 1.5 not only serve for the calculation of piping systems but have a more general significance. Combining series-connected flow resistances as per Table 1.5 is always indicated, if the pressure differential is given and the flow rate is wanted. An example is the calculation of the leakage flow rate through a stepped annular seal according to Table 3.7(1) using Eq. (T3.7.6), which has been derived that in way.

Table 1.5 can also be used to estimate how the flow rate of an impeller is distributed to the individual channels of a double volute according to their different flow resistances, Chap. 7.8.2.

References

1. Ackeret, J.: Aspects of internal flow. In: Fluid Mechanics of Internal Flow. Elsevier Science, Amsterdam (1967)
2. Albring, W.: Angewandte Strömungslehre, 6th edn. Akademie, Berlin (1990)
3. Blevins, R.D.: Applied Fluid Dynamics Handbook. Van Nostrand Reinhold, New York (1984)
4. Churchill, S.W.: Viscous Flows. The Practical Use of Theory. Butterworth Series in Chemical Engineering (1988). ISBN:0-409-95185-4
5. Grein, H.: Einige Bemerkungen über die Oberflächenrauheit der benetzten Komponenten hydraulischer Großmaschinen, Nr. 1, pp. 34–40. Escher Wyss Mitteilungen (1975)
6. Idelchick, I.E.: Handbook of Hydraulic Resistance, 3rd edn. CRC Press, Boca Raton (1994)
7. Johnson, R.W.: The Handbook of Fluid Dynamics. CRC Press LLC, Boca Raton (1998)
8. Kline, S.J.: On the nature of stall. ASME J. Basic Eng. **81**, 305–320 (1959)
9. Knauss, J. (Hrsg): Swirling Flow Problems at Intakes. IAHR Hydraulic structures design manual. AA Balkema, Rotterdam (1987). ISBN:90 6191 643 7
10. McDonald, A.T., Fox, R.W.: An experimental investigation of incompressible flow in conical diffusers. Int. J. Mech. Sci. **8**, 125–139 (1966)
11. Miller, D.S.: Internal Flow Systems, 2nd edn. BHRA, Cranfield (1990)
12. Nikuradse, J.: Strömungsgesetze in rauhen Rohren. VDI Forschungsheft **361** (1933)
13. Reneau, L.R., et al.: Performance and design of straight, two-dimensional diffusers. ASME J. Basic Eng. **89**, 141–150 (1967)

14. Schlichting, H.: Grenzschicht-Theorie, 8th edn. Braun, Karlsruhe (1982)
15. Siekmann, H.E.: Strömungslehre Grundlagen. Springer, Berlin (2000)
16. Siekmann, H.E.: Strömungslehre für den Maschinenbau. Springer, Berlin (2000)
17. Sprenger, H.: Experimentelle Untersuchungen an geraden und gekrümmten Diffusoren. Diss, ETH Zürich (1959)
18. Truckenbrodt, E.: Fluidmechanik, 4th edn. Springer, Berlin (1996)
19. White, F.M.: Fluid Mechanics, 2nd edn. McGraw Hill, New York (1986)
20. Fukuda, H.: The effect of runner surface roughness on the performance of a Francis turbine. Bull. JSME **7**(4), 346–356 (1964)

Chapter 2
Pump Types and Performance Data

Centrifugal pumps are used for transporting liquids by raising a specified volume flow to a specified pressure level. Pump performance at a given rotor speed is described by the rate of flow delivered, the pressure rise achieved, the power absorbed at the coupling, the efficiency and the NPSH. Depending on the application a broad variety of pump types are offered on the market. All of these have at least one impeller and a collector where most of the kinetic energy at the impeller outlet is converted into static pressure. Different forms of impellers, diffusers, volutes and inlet casings are available to build radial, semi-axial or axial, single- or multistage, pumps mounted in horizontal or vertical position - as most suitable for the specific application. For pump selection refer also to Chap. 15.

2.1 Basic Principles and Components

Centrifugal pumps are turbomachines used for transporting liquids by raising a specified volume flow to a specified pressure level. The energy transfer in turbomachines is invariably based on hydrodynamic processes for which characteristically all pressure and energy differences are proportional to the square of the circumferential rotor speed. By contrast, positive displacement pumps (e.g. piston pumps) essentially deliver the same volume V_{stroke} at each stroke independently of flow velocity or rotor speed n. The flow rate then becomes $Q = n \times V_{stroke}$; the pressure rise results solely from the imposed back pressure.

A centrifugal pump according to Fig. 2.1 is essentially composed of a casing, a bearing housing, the pump shaft and an impeller. The liquid to be pumped flows through the suction nozzle to the impeller. The overhung impeller mounted on the shaft is driven via a coupling by a motor. The impeller transfers the energy necessary to transport the fluid and accelerates it in the circumferential direction. This causes the static pressure to increase in accordance with kinetics, because the fluid flow follows a curved path (Chap. 1.4.1). The fluid exiting the impeller is

J. F. Gülich, *Centrifugal Pumps*,
https://doi.org/10.1007/978-3-030-14788-4_2

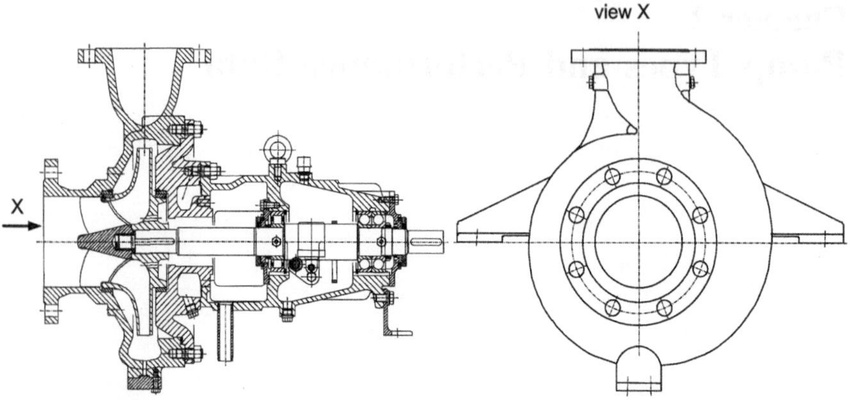

Fig. 2.1 Single-stage volute pump with bearing frame, Sulzer Pumps

decelerated in the volute and the following diffuser in order to utilize the greatest possible part of the kinetic energy at the impeller outlet for increasing the static pressure. The diffuser forms the discharge nozzle.

A shaft seal, e.g. a stuffing box or a mechanical seal, prevents the liquid from escaping into the environment or the bearing housing (the shaft seal is not represented in Fig. 2.1). As shown in Fig. 2.1, an inducer may be added at the impeller inlet for improving the suction performance (Chap. 7.7). However, most applications do not use an inducer.

Impeller and casing are separated by a narrow annular seal through which some leakage flows back from the impeller outlet to the inlet. A second annular seal on the rear shroud serves the purpose of counterbalancing the axial forces acting on the impeller front and rear shrouds. The leakage through this seal flows back into the suction chamber through "axial thrust balance holes" which are drilled into the rear shroud.

The impeller can be described by the hub, the rear shroud, the blades transferring energy to the fluid and the front shroud. In some applications the front shroud is omitted. In this case the impeller is termed "semi-open".

Figure 2.2 shows the meridional section and the plan view of an impeller. The leading face of the blade of the rotating impeller experiences the highest pressure for a given radius. It is called pressure surface or pressure side. The opposite blade surface with the lower pressure accordingly is the suction surface or suction side. When looking into the impeller eye we see the suction surface. Therefore, it is sometimes called the "visible blade face" or the "lower blade face", whilst the pressure surface, not visible from the impeller eye, is called the "upper blade face". These terms are ambiguous and should be avoided. Also defined in Fig. 2.2 are the leading edge LE and the trailing edge TE of the blade. Figure 2.3 shows an impeller designed by means of a 3D-CAD program (the front shroud is removed).

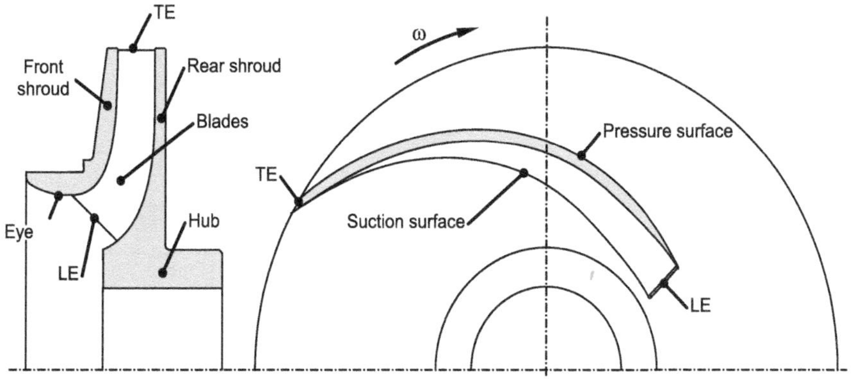

Fig. 2.2 Meridional section and plan view of a radial impeller, *LE*: Leading edge, *TE*: Trailing edge

Fig. 2.3 Radial impeller $n_q = 85$, 3D-model, Sulzer Pumps

According to the performance and application requirements a wealth of different pump types is available which can be classified by different aspects. Table 2.1 Gives an overview of the different types of impellers, diffusing elements, inlet casings and combinations of these elements. The following points A to G relate to the corresponding frames in Table 2.1:

A. Depending on the direction of the flow at the impeller exit we distinguish radial, semi-axial and axial impellers. Accordingly the terms radial, semi-axial and axial pumps are used; the latter are also called "propeller pumps".
B. Impellers with a front shroud are called "closed impellers", those without a front shroud are termed "semi-open impellers" and those with large cut-outs in the rear shroud are designated as "open impellers".
C. According to the flow direction at the diffuser inlet there are radial, semi-axial and axial diffusers. Vaneless diffusers are rarely built.

Table 2.1 Hydraulic pump components and arrangements

Component or feature	**Radial**	**Semi-axial**	**Axial**
(A) Impeller form: Characteristic: flow direction at impeller exit			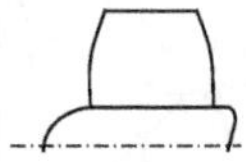
(B) Impeller type	Closed (shrouded) 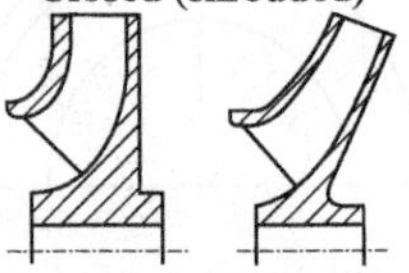	Semi-open 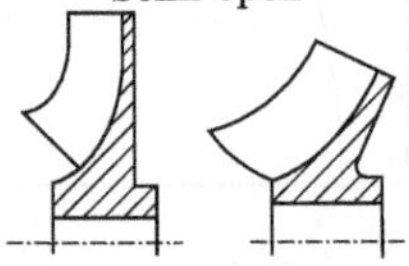	Open 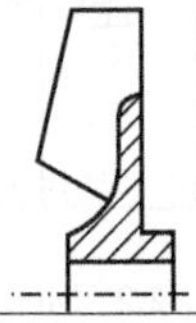
(C) Diffuser Characteristic: flow direction at diffuser inlet	Radial 	Semi-axial 	
(D) Outlet casing	Single-volute Double-volute 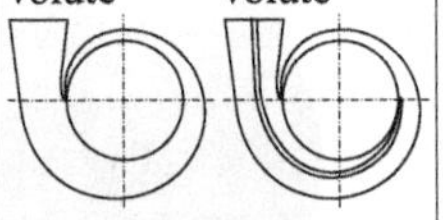	Concentric annulus 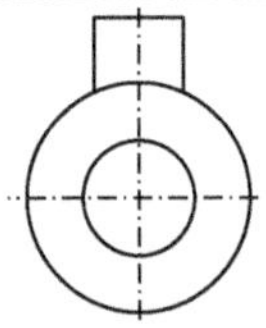	Concentric annulus plus volute 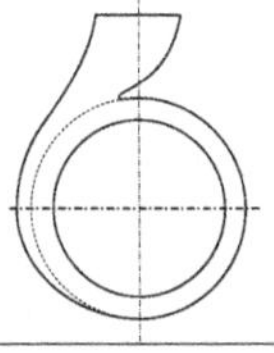
(E) Inlet casing	Annular inlet chamber 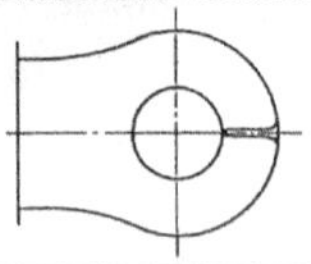	Symmetric inlet 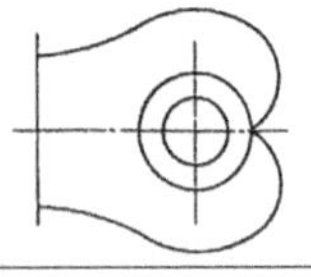	Asymmetric inlet 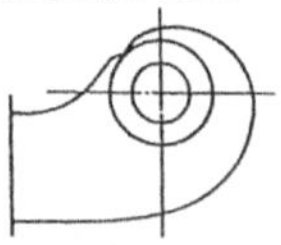
(F) Impellers in series: multistage pumps	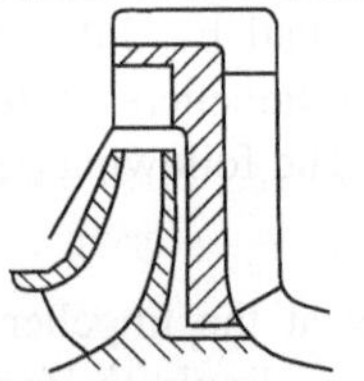	Multistage semi-axial pumps see Fig. 2.14.	
(G) Impellers in parallel: double-entry impellers	Single-stage 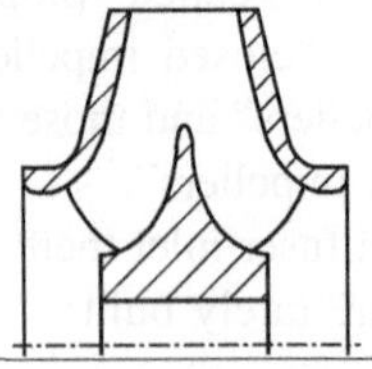	Multistage double-entry pumps see Fig. 2.13.	

D. The most frequent type of diffusing element for a single-stage pump is a volute casing. Sometimes there is a concentric annulus or a combination of annulus and volute.
E. Most single-stage, single-entry pumps have an axial inlet nozzle as in Fig. 2.1. Inline pumps and between-bearing impellers require radial inlet chambers (Figs. 2.5, 2.6, 2.7, 2.8, 2.9, 2.10, 2.11, 2.12 and 2.13). Vertical pumps are often fed via a suction bell (also called "bellmouth") from a sump in a wet pit installation (see Fig. 2.15 and Chap. 11.7.3).

F. If the pressure generated by one impeller is insufficient, several impellers are arranged in series resulting in a "multistage" radial or semi-axial pump. In that type of pump the diffusers include return vanes, which direct the fluid to the impeller of the subsequent stage. Multistage pumps may be equipped with double volute casings instead of diffusers; in that case the fluid is directed to the subsequent stage through appropriately shaped channels.
G. Double-entry radial impellers are used when greater flow rates must be transported. Double-entry pumps may be built as single-stage or multistage.

The components and characteristics shown in Table 2.1 may be combined in many ways for the optimization of pumps for different requirements. Performance data, design specifications, fabrication methods, installation and operation conditions all have an impact on that optimization.

Because of their complicated, three-dimensionally curved surfaces, impellers, volute casings and diffusers are usually fabricated as castings. In some applications impellers and diffusers are manufactured by NC-milling. Small pumps are often made from plastics. There are even impellers and diffusers made entirely from sheet metal. Although simplified hydraulic shapes may be required to ease their manufacturing, the efficiencies achieved with sheet metal pumps are quite good because of the smooth hydraulic surfaces and the small vane blockage.

2.2 Performance Data

The performance data of a centrifugal pump are described by:

- the flow rate Q which is normally defined as the useful volume flow through the discharge nozzle
- the specific work Y or the head $H = Y/g$
- the power consumption P at the pump coupling ("brake horsepower")
- the efficiency η at the pump coupling
- the net positive suction head NPSH at the pump inlet, or the net positive suction energy $NPSE = g \times NPSH$.

In addition to these data, the speed n of the pump rotor is indispensable.

2.2.1 Specific Work, Head

The specific work Y is the total *useful* energy transmitted by the pump to the fluid per unit of mass. Y is measured between the suction and the discharge nozzle. Y is identical to the total useful (isentropic) enthalpy rise Δh_{tot}. In incompressible flow we have $Y = \Delta h_{tot} = \Delta p_{tot}/\rho$ (see Chap. 1.2.2). In practice the head $H = Y/g$ is commonly used (also termed "total dynamic head"). It has to be comprehended as specific *energy unit* (or specific work):

$$Y = \Delta h_{tot} = \frac{p_{2,tot} - p_{1,tot}}{\rho} = g\,H \tag{2.1}$$

The total pressure according to Eq. (1.7) consists of the static (or system) pressure p, the pressure corresponding to the geodetic head $\rho \times g \times z$ and the stagnation pressure $½\rho \times c^2$. With reference to Table 2.2, the total dynamic head of a pump measured between the suction and the discharge nozzles results from the difference of the total pressures expressed as heads $H = H_d - H_s$ (subscript d = discharge nozzle; s = suction nozzle).

$$H = \frac{p_d - p_s}{\rho g} + z_d - z_s + \frac{{c_d}^2 - {c_s}^2}{2g} \tag{2.2}$$

In this equation all energies are expressed as "energy heads": the static pressure heads $p/(g \times \rho)$ measurable at the suction or the discharge nozzle, the potential energy z and velocity heads $c^2/(2g)$. Head and specific work are *independent* of the density or the type of the medium. Thus a pump (in theory) produces the same head when transporting water, mercury or air. But by no means does it create the same pressure rise $\Delta p = \rho \times g \times H$ that could be measured by a manometer. *All pressure differences, powers, forces and stresses are proportional to the density.*

Table 2.2 shows how the different components that make up the head must be taken into account in a measurement or a calculation. The plane of reference should be chosen at the shaft centerline in the case of horizontal pumps. For pumps with a vertical or inclined shaft, the intersecting point of the shaft axis with a horizontal line drawn through the center of the inlet of the upper suction impeller is used as plane of reference, [N.1]. Manometer readings at other levels should be corrected to the plane of reference, [N.2]. As only pressure *differences* are taken into account for calculating the head, values of absolute pressure or gage pressure (i.e. pressure above atmospheric) may be used in Eqs. (T2.2.1 to 2.2.6).

In order to ensure the specified volumetric flow rate through a given pumping plant, the pump must deliver a certain head which is called the required head H_A of the plant. It is calculated from Bernoulli's equation taking into account all head losses in the system (except losses in the pump), see Table 2.2, Eq. (T2.2.6). During steady operation the head of the pump equals the required head of the plant: $H = H_A$.

2.2.2 *Net Positive Suction Head, NPSH*

When the pressure in a liquid drops below the vapor pressure, a portion of the fluid will evaporate. Excess velocities due to the flow around the blade leading edge cause a local pressure drop, which may lead to such partial evaporation. This phenomenon is called "cavitation"; it is discussed in detail in Chap. 6. Extensive cavitation can impair the performance or even interrupt the flow. Therefore, the approach flow conditions at the suction nozzle are an important criterion for the layout and selection of a pump. The relevant parameter is the "net positive suction energy" NPSE or the "net positive suction head" NPSH. It is defined as the absolute suction head $H_{s,abs}$ minus the vapor pressure expressed as head $p_v/(\rho \times g)$. Table 2.2 gives the equations and definitions. We distinguish between the (usually measured) *NPSH of the pump* which is necessary in order to fully or partially suppress cavitation ("NPSH required" or $NPSH_R$) and the *NPSH available in the plant* ($NPSH_A$). Since the vapor pressure p_v is given in the water/steam tables as an absolute pressure, absolute pressures must be inserted into Eqs. (T2.2.7 and 8) for calculating the NPSH.

From Bernoulli's equation we can calculate the absolute pressure at the highest point of the impeller situated at a distance "a" above the rotor axis. This pressure must never be allowed to fall below a level at which an unacceptably large volume of vapor would form at the impeller inlet due to cavitation. Any given pump has its required $NPSH_R = f(Q)$ which corresponds to a *specific amount of cavitation*. The $NPSH_A$ or the liquid level z_e which is necessary for the plant to operate properly must be calculated from the condition $NPSH_A > NPSH_R$, Eq. (T2.2.10), (Chap. 6).

- If z_e calculated from Eq. (T2.2.10) results as *negative*, this value is the *maximum admissible geodetic suction lift*: $|z_{s,geo,max}| = -z_e$.
- If Eq. (T2.2.10) yields a *positive* value, the pump needs a geodetic suction head, which means that the liquid level must be *above* the pump. As demonstrated by Eq. (T2.2.10), this always applies to the pumping of saturated (boiling) liquids, because in that case $p_{e,abs} = p_v$.

With the exception of some special pump types, centrifugal pumps must be filled with liquid for start-up; they are not "self-priming" and thus cannot evacuate the air from the suction pipe.

Types of self-priming pumps include side channel and liquid-ring pumps. Radial impellers are sometimes combined with one side-channel stage in order to allow self-priming of radial pumps, Chap. 2.3.4.

2.2.3 *Power and Efficiency*

Since the specific work represents the energy transferred per unit mass, the *useful power* P_u of a pump is obtained by multiplying the transported mass flow $\dot{m} = \rho \times Q$ by the specific work Y:

Table 2.2 Total dynamic head and net positive suction head (NPSH)

	Pump	Eq.	Plant	Eq.
Head at inlet	$H_s = \frac{p_s}{\rho g} + z_s + \frac{c_s^2}{2g} = H_e - H_{v,s}$	2.2.1	$H_e = \frac{p_e}{\rho g} + z_e + \frac{c_e^2}{2g}$	2.2.2
Head at outlet	$H_d = \frac{p_d}{\rho g} + z_d + \frac{c_d^2}{2g} = H_a + H_{v,d}$	2.2.3	$H_a = \frac{p_a}{\rho g} + z_a + \frac{c_a^2}{2g}$	2.2.4
Total dynamic head	$H_{tot} = H_d - H_s$		$H = H_d - H_s = H_a - H_e + H_{v,d} + H_{v,s}$	
	$H_{tot} = \frac{p_d - p_s}{\rho g} + z_d - z_s + \frac{c_d^2 - c_s^2}{2g}$	2.2.5	$H_A = \frac{p_a - p_e}{\rho g} + z_a - z_e + \frac{c_a^2 - c_e^2}{2g} + H_{v,d} + H_{v,s}$	2.2.6
Total dynamic head above vapor pressure p_v with $p_{abs} = p_{amb} + p_s$	$NPSH = H_s + (p_{amb} - p_v)/(\rho \times g)$		$NPSH_A = H_s + (p_{amb} - p_v)/(\rho \times g)$	
	$NPSH = \frac{p_{s,abs} - p_v}{\rho g} + z_s + \frac{c_s^2}{2g}$	2.2.7	$NPSH_A = \frac{p_{e,abs} - p_v}{\rho g} + z_e + \frac{c_e^2}{2g} - H_{v,s}$	2.2.8
Static pressure in suction nozzle			$p_{s,abs} = p_{e,abs} + \rho g (z_e - z_S - H_{v,s}) + \frac{\rho}{2}(c_e^2 - c_s^2)$	2.2.9
Maximum allowable geodetic suction head (z_e negative) Minimum required geodetic suction head (z_e positive)			$z_e = NPSH_R + H_{v,s} - \frac{p_{e,abs} - p_v}{\rho g} - \frac{c_e^2}{2g} + a$	2.2.10

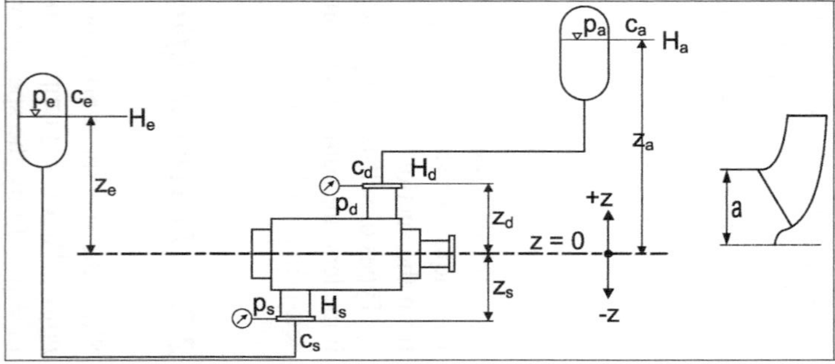

- Note sign (positive or negative) of all levels z
- $H_{v,s}$, $H_{v,d}$ head losses in suction and discharge pipes respectively
- p_s, p_d static pressures measured at suction and discharge branch resp.
- p_e, p_a static pressures above suction and discharge liquid levels, resp.

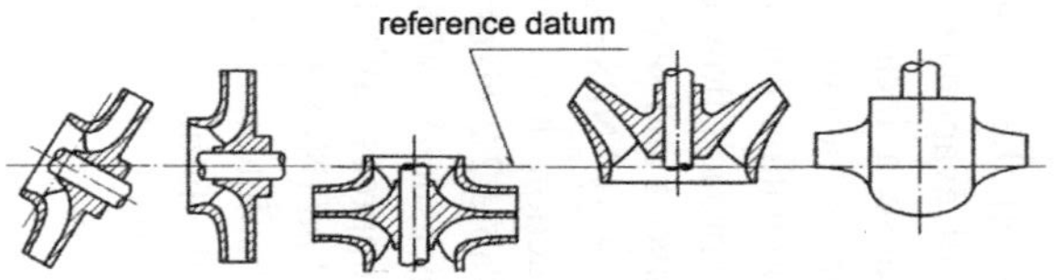

$$P_u = \rho\, Y\, Q = \rho\, g\, H\, Q = Q\, \Delta p \tag{2.3}$$

The power P needed at the coupling is greater than the useful power because it includes all losses of the pump. The ratio of both values is the pump's efficiency η:

$$\eta = \frac{P_u}{P} = \frac{\rho\, g\, H\, Q}{P} \tag{2.4}$$

2.2.4 Pump Characteristics

When the flow rate of a pump varies, the head, the power consumption and the efficiency change too. Plotting these quantities against the flow rate we obtain the "pump characteristics", Fig. 2.4. At a certain flow rate the pump efficiency has a maximum value called the "best efficiency point" (BEP). The pump is designed for this BEP which is characterized by Q_{opt}, H_{opt}, P_{opt} and η_{opt} at a given speed.

The operation point of a pump is invariably where the head generated by the pump equals the head required by the plant: $H = H_A$. In other words, it is at the intersection of the pump characteristic with the system characteristic, refer to Fig. 2.4 and for a detailed discussion to Chap. 11.1.

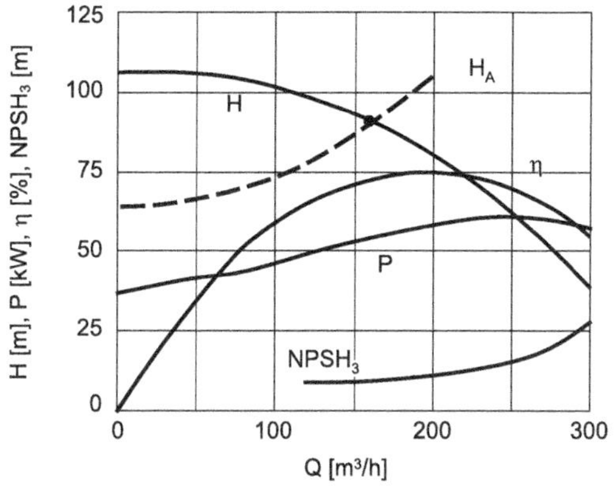

Fig. 2.4 Pump characteristics and system characteristic H_A (dashed line)

2.3 Pump Types and Their Applications

2.3.1 *Overview*

Centrifugal pumps are of eminent technical and economic importance in many areas of life and industry (the world market volume for centrifugal pumps is in the order of 20 billion US$ per year). Their application range comprises small pumps of a few watts, such as central heating circulation pumps or cooling water pumps for automobiles, as well as 60-MW storage pumps and pump turbines with more than 250 MW when operating in the pumping mode.

The term "centrifugal pumps" comprises radial, semi-axial and axial pumps, but also side channel, peripheral and liquid-ring pumps whose working principles are fundamentally different from that of the first group.

Centrifugal pumps in the narrow sense are designed for flow rates from 0.001 to 60 m^3/s, heads of 1 to 5000 m and speeds from a few hundred to about 30,000 revolutions per minute. These values are intended to illustrate the broad range of applications; they do not define the absolute limits of actual or future pumps.

Any pump application is characterized by the flow rate Q_{opt}, the head H_{opt} and the rotor speed n. To a large extent these parameters determine which impeller type and pump design to choose. As will be shown in Chap. 3.4, these three performance parameters are interrelated by the "specific speed" n_q, N_s or ω_s which is of great importance for the selection and design of a pump. Equation (2.5) in Table. D2.1 defines n_q, N_s and ω_s. In Europe n_q is most popular, while N_s is defined in US customary units. The truly non-dimensional quantity ω_s should be preferred for theoretical considerations or for deriving general equations, but is not yet used in practice by the majority of pump engineers. This is because the available documentation is often prepared with either n_q or N_s. Unfortunately, a variety of other definitions of the specific speed can be found in the literature.

$H_{st} = H_{tot}/z_{st}$ is the head per stage where z_{st} is the number of stages. f_q is the number of impeller entries, that is $f_q = 1$ for single-entry and $f_q = 2$ for double-entry impellers. The choice of radial, semi-axial or axial impellers depends on the specific speed as well as on the pump type. A pump with medium specific speed (e.g. $n_q = 60$) may be built with a radial or a semi-axial impeller depending on the pump type which is most economical for the expected application. From Eq. (2.5) it can be seen that:

- For supplying small flow rates at high pressures, pumps with low specific speeds are required. Below $n_q < 20$ the efficiency drops rapidly with diminishing specific speed (Chap. 3.9, Figs. 3.26–3.32). Therefore, the lower economic limit for a centrifugal pump application is generally $n_q = 5$ to 8 for small pumps, but $n_q = 10$ to 15 for larger performances, depending on the application and the importance of keeping energy costs low. Below this limit the speed must be increased, or a multistage pump must be used raising thus the specific speed as per Eq. (2.5). Otherwise another pump type must be selected.

Table D2.1 Definitions of specific speeds

1. Common in Europe, most used throughout this book	2. US customary units	3. Truly dimensionless representation	Eq.
$n_q = n\frac{\sqrt{Q_{opt}/f_q}}{H_{opt}^{0.75}}$	$N_s = n\frac{\sqrt{Q_{opt}/f_q}}{H_{opt}^{0.75}} = 51.6\,n_q$	$\omega_s = \frac{\omega\sqrt{Q_{opt}/f_q}}{(g\,H_{opt})^{0.75}} = \frac{n_q}{52.9}$	2.5
n in rpm Q in m^3/s H in m	n in rpm Q in gpm H in ft	ω in 1/s Q in m^3/s H in m	

Notes: (1) H_{opt} is the *head per stage*: $H_{opt} = H_{tot,opt} / z_{st}$
(2) In the US, the specific speed of double-entry pumps is usually calculated with the *total* flow rate; in this case the factor f_q must be omitted from Eq. (2.5).

- Special pumps with radial blades or Pitot-tube pumps may be built with specific speeds as low as $n_q = 1$, albeit with low efficiencies, Chap. 7.3.
- Large flow rates and low heads require pumps with high specific speeds. The upper limit for economical operation is typically in the region of $n_q = 350$ to 450. Above that limit, costs and hydraulic losses may become unacceptably high. But the economical application is not determined by the specific speed alone. Approach flow conditions and the pump size must also be taken into consideration when selecting the type of pump. If a given application would require an excessively high specific speed, the flow rate must be divided among two or more parallel pumps in order to reduce unit size and n_q. Distributing the flow to several units may prove advantageous in view of the driver cost, plant layout and standby requirements to maintain operation in case of a pump failure.

Table 2.3 lists the different impeller types and their properties in order to illustrate the influence of the specific speed on the design and the range of performance. Explanations to Table 2.3:

- The head per stage at BEP given in the table represents approximately the feasible maximum. Attainable peripheral speed, head per stage and head coefficient fall with increasing specific speed.
- The head coefficient is defined as $\psi_{opt} = 2 \times g \times H_{opt,st}/u_2^2$ (Chap. 3.4).
- The efficiency of a pump depends not only on the specific speed but also on the pump type, design and absolute size. For extremely small pumps it might even be less than the lower limit listed in Table 2.3, refer to Chap. 3.
- Radial, semi-axial and axial pumps are built in a range of $7 < n_q < 400$ for a wealth of universal applications.
- Multistage pumps are commonly built as horizontal units with radial impellers, while in vertical machines often semi-axial impellers are used. Their specific speeds do not generally exceed $n_q = 50$, because the multistage design yields better efficiencies below that figure.

Table 2.3 Pump types

n_q	Type	Impeller shape	Maximum $H_{st,opt}$	ψ_{opt}	η_{opt} [%]	
< 0.5	Piston pumps	Positive displacement pumps	Limited by mechanical constraints		85 - 95	
< 2	Gear pumps				75 - 90	
2-10	Screw pumps				65 - 85	Also for gas/liquid mixtures
0.5-4	Peripheral pumps	Figure 2.22	400 m	5 - 15	30 - 35	Single and multistage
2-11	Side channel pumps	Figure 2.21	250 m	3 - 10	34 - 47	
1-10	Friction pumps	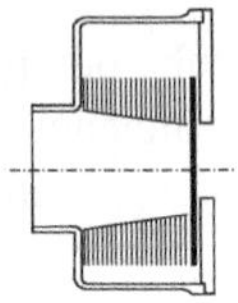		0.5	25 - 35	
7-30	Radial pumps	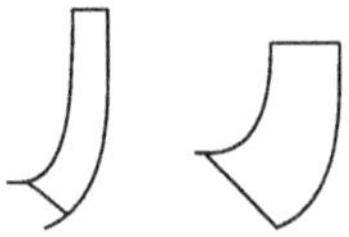	800 m (1200)	1 to 1.2	40 - 88	Below $n_q < 10$ usually small pumps only In most cases: $H_{st,opt} < 250$ m
50			400 m	0.9	70 - 92	
100		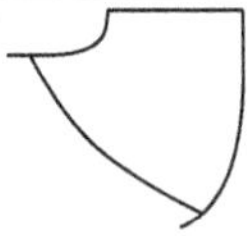	60 m	0.65	60 - 88	$n_q = 100$ is essentially the upper limit for radial impellers
35	Semi-axial-pumps		100 m	1	70 - 90	For $n_q < 50$ often multistage For $n_q > 75$ seldom multistage
160		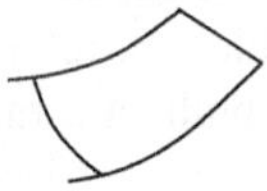	20 m	0.4	75 - 90	For $n_q > 100$ only single-stage
160 to 400	Axial pumps	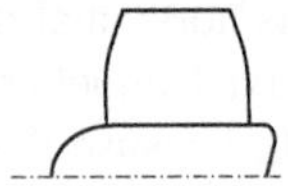	2 m to 15 m	0.4 to 0.1	70 - 88	Flow rates up to 60 m^3/s, only single-stage

The numerical values given in Table 2.3 are intended to illustrate typical application ranges. Extremes beyond these limits could probably be found for all of the examples. Very high heads per stage are only feasible in a limited range of specific speeds (typically $25 < n_q < 40$), Chap. 14.1.

2.3.2 *Classification of Pumps and Applications*

Because of their great number of types and applications, pumps can be classified according to a variety of aspects. The enumeration below provides just a few examples of the many often overlapping concepts for the classification of pumps.

- Operation principle: centrifugal, side channel, positive displacement pumps
- Typical properties such as "self-priming"
- Pump type as described by the flow direction at the impeller outlet: radial, semi-axial or axial
- Pump type as described by the collector: volute or diffuser pumps
- Pump type as described by the design: single-stage, multistage, single-entry, double-entry, horizontal, vertical pumps, wet motor pumps, submerged pumps, "seal-less" pumps (i.e. without shaft seal) with magnetic coupling, or canned motor pumps

  Medium: pumps for the transport of drinking-water, boiler feed water, condensate, sewage, pulp, slurries, acids, oil, or gas/liquid mixtures. Application: e.g. cooling water pump, feed pump, booster pump.

Centrifugal pumps are used in so many different domains and applications that it would by impractical to enumerate them all. Table 2.4 gives examples of some applications of eminent economic significance; some particular requirements are listed too.

In many systems, the pump is only one component in a complex process and its failure means serious economic losses - think for example of the boiler feedpumps in a large power plant. Therefore, the demand for maximum operating reliability is a characteristic feature of many pump specifications. This is true not only for special pumps but also for mass-produced items like water cooling pumps for automobiles or central heating circulation pumps.

2.3.3 *Pump Types*

The following section presents some widely-used pump types. They serve as examples, since a complete discussion would be altogether impractical.

Single-stage, single-entry pumps with volute casings: This group is by far the most common pump type. The basic concept is shown in Fig. 2.1; it is used in various ways for countless applications. The principal dimensions of frequently

Table 2.4 Pump applications

Type or application	Typical pump types	Power class P [kW]	Characteristics, Applications, Special requirements
„Universal" or standard pumps	Single-stage radial	2 - 200	Low investment costs
Process pumps	1-stage, radial	10 - 300	High reliability; often designed according to special standard, e.g. [N. 6] zero leakage, explosion protection, safety
	Multistage, radial or semi-axial	50 - 1000	
Cooling water pumps for central power stations	Single-stage radial or semi-axial	500 - 3000	High specific speed vertical pumps
Boiler feedpumps	Multistage radial	100 - 2000	Industrial power generation
Boiler feedpumps for central power stations		5000 to 45'000	High-speed machines with booster pump, special design features to avoid vibration and cavitation problems
Injection pumps		1000 to 20'000	Injection of water to increase oil production
Pipeline pumps			Far-distance transport of drinking-water or oil
Mine pumps		500 - 3000	Mine de-watering Abrasive wear due to sand-laden water
Irrigation or de-watering pumps	Axial, semi-axial	200 - 2000	Large flow rate variations due to rainfall, abrasive wear due to sand-laden water
Sewage pumps	Radial, semi-axial, axial	10 - 1000	Large flow passages; insensitive to blockage by foreign matter
Marine pumps		1 - 1000	Low space requirements; most vertical inline pumps; low NPSH-values
Submerged pumps		5 - 500	Sewage, de-watering; insensitive to ingress of water
Gas-liquid mixtures		10 - 5000	Process industry, oil production; special hydraulic shapes
Seal-less pumps	Single-stage, radial	5 - 250	For dangerous or toxic fluids (leakage-free pumps) with canned motor or magnetic coupling
Dredge pumps	Single-stage, radial	200 - 5000	Large flow passages to pass foreign matter, protection against abrasive wear
Slurry pumps	1-stage, radial	50 - 1000	Protection against abrasive wear
Rocket pumps	Radial or semi-axial	1000	Extreme tip speeds, short pump life; with inducer
Food industry	All	1 - 50	Extreme cleanness, no ingress of lubricants, protection of process fluid
Medical services (e.g. blood pumps)	All	< 0.1	Extreme cleanness and reliability, protection of process fluid

used single-stage volute pumps have been defined in various standards, e.g. EN 733 (water standard pumps) and ISO 2858 (chemical standard pumps) and several ANSI-standards.

Radial pumps of this type are designed for specific speeds in the range of $n_q = 7$ to 100 typically. The flow rates extend from a few liters per minute to several cubic meters per second in large cooling water pumps for thermal power stations or in water storage pumps for hydroelectric plants. There are radial pumps in horizontal or in vertical arrangement with an axial approach flow, but there are also "inline pumps" with the suction and discharge nozzles in co-axial arrangement, Fig. 2.5.

Single-stage pumps with volute casings are common in many branches of industry. Water supply, sewage, chemical processing (including hydrocarbon processing) and power plants are important areas of application. They are also produced in great numbers as small pumps, e.g. for pumping cooling water in motor vehicles or as central heating circulation pumps in homes and buildings. Severe demands with respect to reliability and noise emission are made on these rather inconspicuous pumps which have impeller diameters typically below 100 mm. Even though their power consumption is 30 to 200 W only, their efficiency or energy consumption is not without impact on the economy. Their market volume in Europe is 5 million pumps per year. Supposing an average power of 60 W per pump, this corresponds to the output of a 300-MW power plant, [1].

Small inline pumps are often built as close-coupled pumps, where the impeller is mounted on the end of the motor shaft. The pump shown in Fig. 2.5 is built as a wet

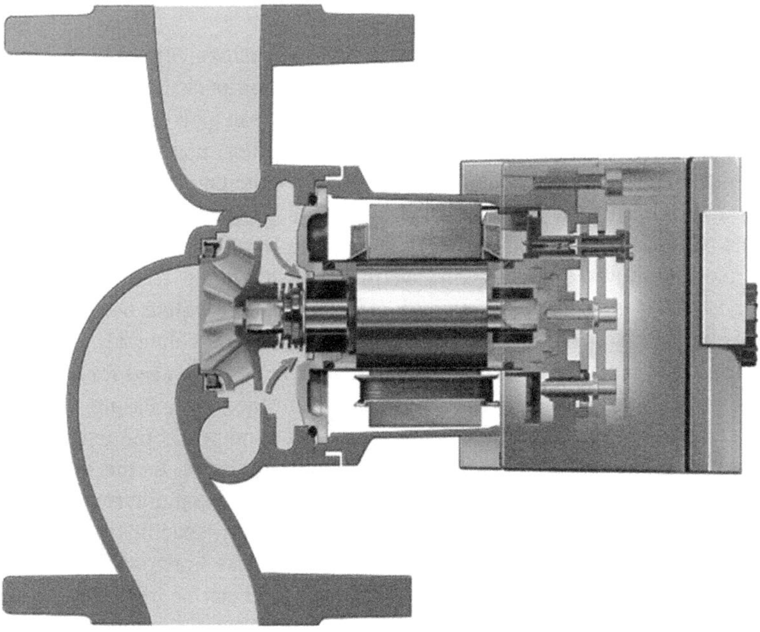

Fig. 2.5 Circulating pump for central heating installations, WILO AG

(or "canned") motor design; also known as "hermetic pumps". The motor and bearings are exposed to, and cooled by, the liquid pumped; no shaft seal is required. The stator windings of the motor are protected from the fluid by a can; likewise the rotor windings are protected by a sleeve. Some liquid is passed through the gap between rotor and stator in order to remove the heat created by the motor losses and fluid friction. The bearings are lubricated by the liquid pumped. Leakage losses at the impeller inlet are reduced by a floating seal ring. The pump is mounted between flanges in the pipeline; no additional support is required. Electronic controls are integrated into the motor block.

Not only radial but also semi-axial impellers up to about $n_q = 140$ are used in pumps with volute casings. Small pumps with low specific speeds can be built also with an annular casing instead of a volute. For pumping liquids with a high content of solids or gas there are a range of special pump designs with various types of closed or open impellers to choose from. Figure 7.17 shows some special impeller designs and Fig. 7.21 shows the section of a sewage pump.

When pumping high-temperature fluids, cooling must be provided for the shaft seal (see the cooling box and cooling water connections in Fig. 2.1). Moreover, the bearing housing and the pump casing should be supported at the level of the shaft centerline in order to minimize any negative effects resulting from thermal deformation. The objective is to avoid a driver misalignment and a distortion of shaft and casing resulting in a possible seizure of the impeller ring on the stationary wear ring.

The range of materials used, depending on the medium to be pumped and the application, comprises cast iron, carbon steel, bronzes and all kinds of stainless steel as well as plastics, glass and ceramics.

Single-stage, double-entry pumps with volute casings, Fig. 2.6: Pumps with double-entry impellers require distinctly lower suction pressures (or $NPSH_A$) than single-entry pumps for delivering a specified flow rate at a given speed, Chap. 6.2.4 . Also, the efficiencies of double-entry pumps are often superior to those obtained with single entry pumps (at $n_q < 40$). With a double-entry design the hydraulic forces acting axially on the impeller are to a great extent balanced for symmetry reasons (Chap. 9.2). Thus a smaller axial bearing is sufficient.

Figure 2.6 shows a double-entry vertical booster pump feeding a pump in a pipeline for the transport of drinking-water. The double volute used considerably reduces the hydraulic radial forces as compared to a single volute design (Chap. 9.3). As the lower radial bearing is water lubricated, there is no need for a lower shaft seal. This bearing is cooled with water taken from the volute casing.

The upper radial bearing is a plain journal bearing while the axial forces are carried by a tilting-pad thrust bearing. Both are oil lubricated. As this is a low speed pump with a large shaft diameter, a stuffing box is a good solution for the shaft seal.

Single-stage, double-entry pumps are designed for specific speeds between approximately $n_q = 10$ to 100. Mostly they are built as horizontal units with the suction and discharge nozzles in line. Thus, the pump can be installed in a straight pipeline without elbows or additional fittings. For large-bore piping this is an important cost advantage if the installation concept is compatible with this

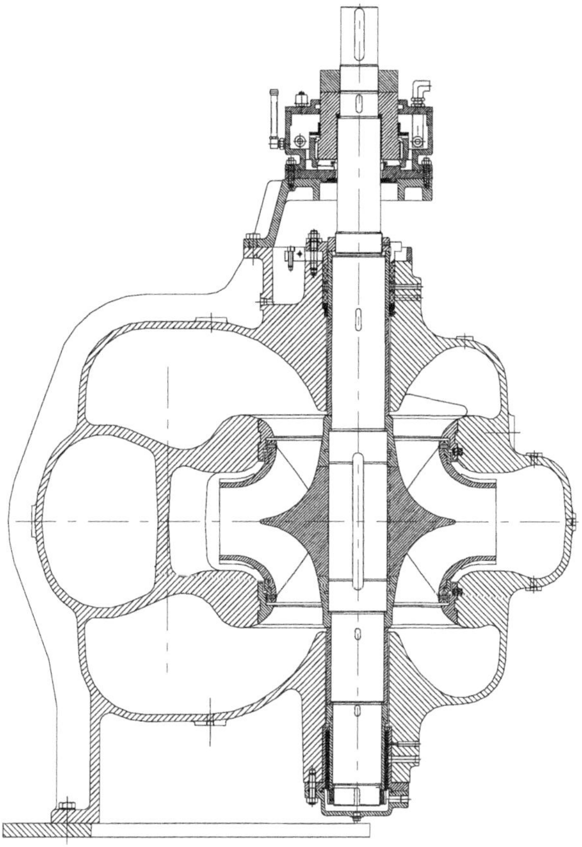

Fig. 2.6 Double-entry vertical pump, Sulzer Pumps, n = 596 rpm, Q_{opt} = 2.4 m^3/s, H_{opt} = 48 m, P_{opt} = 1250 kW, n_q = 36

arrangement. For cold liquids (nonetheless up to 130 °C) double-entry pumps are mostly built as "axial-split" designs. In this type, a flange splits the pump casing through the rotor axis and allows the removal of the complete rotor assembly after lifting off the upper half of the casing. Discharge and suction nozzles are below the split flange and do not hinder the opening of the pump. For high temperatures, high pressures and for chemical processes, single-stage, double-entry pumps are also designed with barrel casings and one or two casing covers which support the bearings and house the shaft seals. The design concept is similar to Fig. 2.12.

Multistage, single-entry pumps: If the specific speed for a required duty falls to a level where the efficiency would be too low and, consequently, the energy costs unacceptably high, multistage pumps are chosen (if the speed cannot be raised). Likewise, it is often necessary to limit the head per stage by using a multistage concept for reasons of mechanical design. Whereas all stages normally deliver the same volumetric flow, the heads of all the stages arranged in series add up to the total head. The maximum possible number of stages is essentially limited by vibration problems. One reason for this is that increasing the rotor length will cause the "critical speed" of the pump shaft to drop. Another reason is that the forces which excite vibrations grow with the square of the circumferential speed u_2 of the impellers: the greater u_2, the higher is the danger of forced as well as self-excited vibrations. As a result, the number of allowable stages decreases with growing circumferential speed.

Multistage pumps are built in Europe as a rule with diffusers, whereas in North America double volutes are often used for many multistage applications. Figure 2.7 shows a 6-stage segmental pump (also called "ring section pump") with diffusers. Usually the fluid enters through an inlet casing with a radially oriented suction nozzle. However, there is also a pump concept with an axial suction nozzle and, consequently, a product-lubricated bearing on the suction side. The pump shown in Fig. 2.7 has product-lubricated bearings. This feature favors a compact design resulting in reduced rotor length, space, weight and costs, [2].

The impeller of the first stage can be designed with a greater inlet diameter than the impellers of the following ("series") stages in order to reduce the NPSH required, Fig. 2.7. The exit casing, which may be in the form of an annulus or a volute, leads the fluid from the last stage into the discharge nozzle.

The stage casings are arranged between the inlet and the exit casings, which are braced by strong tie bolts. Each stage casing comprises an impeller and a diffuser which is centered in the pressure-bearing casing element. Return vanes guide the fluid from the diffuser outlet to the impeller inlet of the subsequent stage. The bearing housings are bolted via flanges to the inlet and exit casings.

Barrel-type multistage diffuser pumps are built for extreme pressures, Fig. 2.8. Again, the diffusers, with integral return vanes, are centered in stage casings which are bolted to each other. The result is an assembly unit (a "cartridge") comprising the casing cover, the rotor, the bearing housings and the inlet casing insert.

The cartridge can be removed from the barrel casing by disassembling the coupling and a small retention ring on the drive-end. The design shown in Fig. 2.8

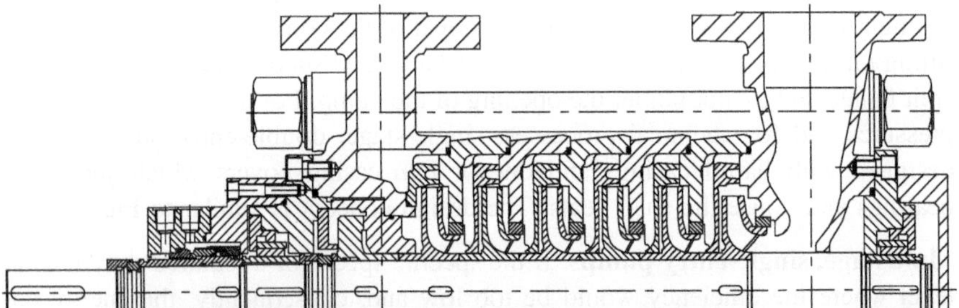

Fig. 2.7 Multistage segmental pump with suction impeller and balance disk, Sulzer Pumps

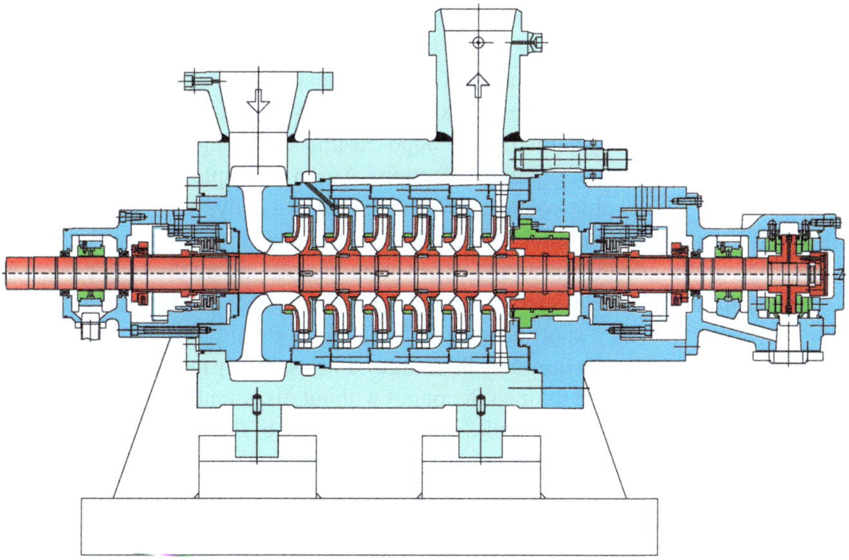

Fig. 2.8 Barrel-type, high-pressure boiler feedpump, Sulzer Pumps n = 5800 rpm, H = 3900 m, P = 34,000 kW. The cartridge assembly is shaded

is typical for boiler feedpumps in fossil-fueled power plants of several hundred MW. The pump shown is of the same type as described in [3]. Similar pumps are used for water injection duties in oil fields for boosting the oil production.

The inlet nozzle leads the water to an annular chamber from which it flows through a vaned insert to the suction impeller. The radial vanes of the insert take out virtually all circumferential components of the inlet velocity and thus ensure a uniform flow to the impeller. The vaned insert also provides a mechanically strong and symmetric connection between the stage casings and the drive end cover.

Multistage pumps may be equipped with an "intermediate take-off" withdrawing a fraction of the flow rate[1]. An example is an intermediate take-off on a boiler feedpump for controlling the steam temperature upstream of the re-heater by water injection. In this case, the flow rate through all stages is *not* the same; this fact must be taken into account in the pump layout. Some boiler feedpumps are equipped with a "kicker stage", which consists of a small additional impeller mounted behind the last stage. It is designed for a fraction of the discharge flow rate with the purpose of controlling the steam temperature upstream or downstream of the super-heater of a fossil-fired boiler.

In multistage pumps the hydraulic axial forces of all stages add up to a considerable "axial thrust" on the rotor that cannot be carried by a reasonably sized

[1]In Fig. 2.8 the intermediate take-off is from the second stage, but it can be arranged at any other stage as well.

thrust bearing. For this reason, multistage pumps usually have some device for axial thrust balancing, Chap. 9.2. The pump in Fig. 2.7 is designed with a balance disk, whereas the barrel-type pump in Fig. 2.8 has a balance piston.

Another possibility to balance the axial thrust is the "back-to-back" arrangement of the impellers in two groups (also: "opposed-impeller design"); Chap. 15 gives some recommendations for pump selection. As an example, Fig. 2.9 shows a 12-stage pump with double volutes. The casing is split axially, making it possible to remove the complete rotor after lifting off the upper half of the casing. The volutes are cast integrally into the lower and upper casing halves, as are the long crossovers leading from the first to the second stage group. The volutes discharge via diffusers into overflow channels ("short crossovers") leading the fluid to the impeller of the subsequent stage. The suction and discharge nozzles are in the lower half of the casing so that it is possible to open the pump without disassembling the piping. The discharge nozzle is in the middle of the pump. A center bushing in the middle of the rotor controls the leakage Q_{CB} from the second to the first group of stages. Another bushing is necessary at the inlet of the second stage group for balancing the axial thrust and relieving the pressure at the shaft seal. The castings for the casing in Fig. 2.9 are quite complicated and a hindrance to standardization.

The back-to-back arrangement requires a smaller thrust bearing (right-side in Figs. 2.9 and 2.10) than inline impellers as per Fig. 2.8. The leakage losses through the bushings in the center and at the inlet to the second stage group tend to be smaller than in inline pumps. By choosing back-to-back pumps, it is thus sometimes possible to improve the efficiency as compared to pumps with inline impellers. This is particularly true with low specific speeds. The center bushing is very favorable with respect to rotor dynamics since it provides additional support (stiffness and damping) in the middle of the bearing span (Chap. 10).

Figure 2.10 shows a back-to-back arrangement of a 12-stage diffuser pump for seawater injection. The performance data are given in the figure caption, note the head of 5300 m. Similar to Fig. 2.8 it is a cartridge design. Between the two stage groups there is a centerpiece with crossover channels. One set of channels, shown below the center bush, leads the water from the first stage group to the second. Another set of channels (shown above the center bush) directs the fluid from the second stage group to an annular chamber and the discharge nozzle. The inlet

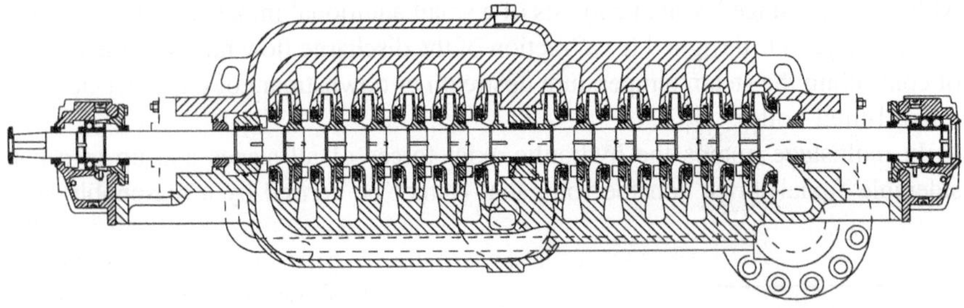

Fig. 2.9 Multistage, axial-split pipeline pump with double volutes. Sulzer Pumps

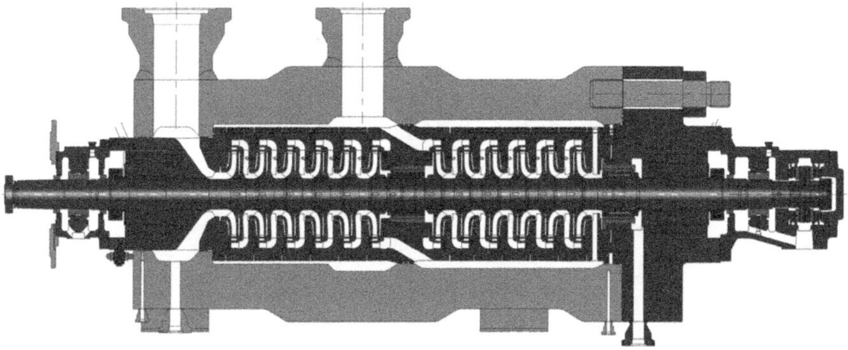

Fig. 2.10 High-pressure sea water injection pump. Sulzer Pumps; rated conditions: n = 6000 rpm, Q = 400 m^3/h, H = 5300 m, P = 7700 kW, n_q = 21

nozzle (left-side in Fig. 2.10), directs the water into an annulus from where it flows through a vaned insert to the suction impeller (as in Fig. 2.8).

Multistage pumps of the types shown in Figs. 2.7, 2.8, 2.9 and 2.10 can also be built with a double-entry impeller as shown in Fig. 2.11.

This arrangement can be advantageous if a direct drive (e.g. 3600 rpm) is to be used, since it allows a higher speed (smaller pump) for a specified available $NPSH_A$. The double-entry suction impeller discharges into a diffuser and subsequently into an annular casing. Several channels lead the fluid from there to the second stage. Similar to Fig. 2.10, the stages are arranged back-to-back in a cartridge which is mounted from the non-drive end "NDE" into the barrel (from the left in Fig. 2.11). The heavy cover on the NDE is a "twist-lock" design, which avoids the large bolting required for a conventional design as in Fig. 2.10. The back-to-back design is used for two-stage pumps too, Fig. 9.19.

Figure 2.12 shows a "radial split" process pump with a double-entry suction stage and a single-entry second stage. Both impellers can be dismantled through casing covers arranged on either side of the pump. The double volutes and crossovers are cast into the casing.

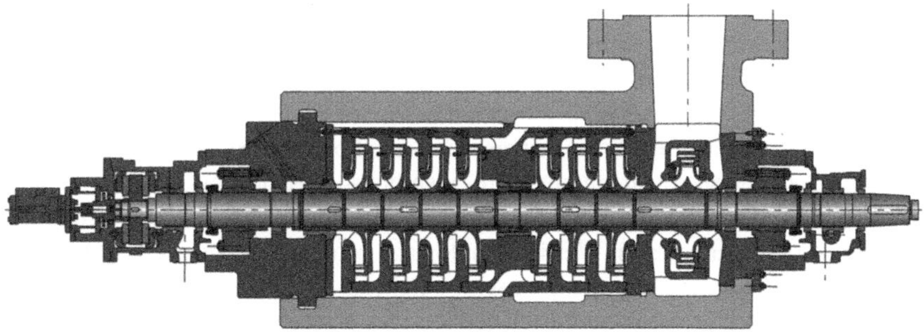

Fig. 2.11 High-pressure back-to-back barrel pump with a double-entry suction impeller. Sulzer Pumps (discharge nozzle not in drawing plane)

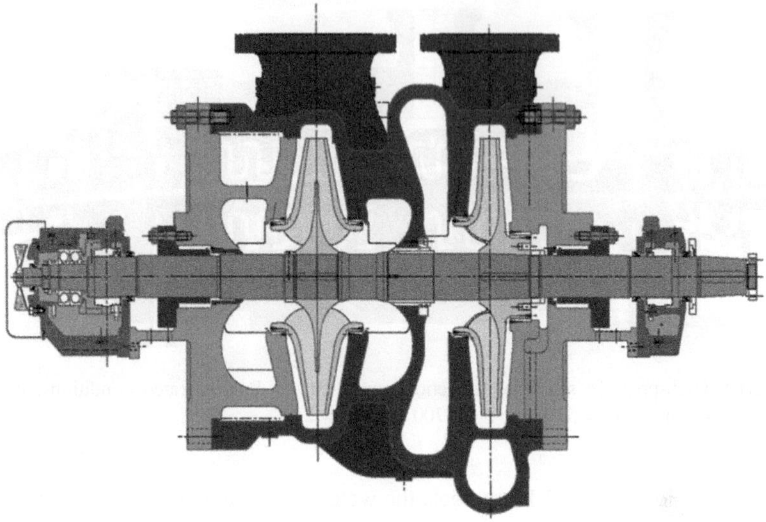

Fig. 2.12 Two-stage process pump with double-entry suction impeller, Sulzer Pumps

Multistage, double-entry pumps: 2- or 3-stage pumps may also be built as double-entry designs. Figure 2.13 shows a two-stage water transport pump with this concept. The fluid is distributed by a forked pipe to both suction nozzles and from there via the inlet casings to both the single-entry suction impellers (which operate in parallel). The water then flows through the diffusers and return channels to the second, double-entry stage. This delivers via a double volute into the discharge nozzle which is arranged inline with the entry to the forked pipe. The double-entry design reduces the $NPSH_A$ requirements (as compared to the single-entry pump) making it an economical solution with regard to investment costs. Moreover, it achieves a better efficiency (and therefore lower energy costs) because no axial thrust balance device is needed. The volumetric and mechanical efficiencies are accordingly higher, since there is no leakage through a balance device and the axial bearing is smaller. The slightly more expensive design of a multistage, double-entry pump thus pays off quickly in large pumping plants due to reduced civil engineering and energy cost and also due to very good reliability (i.e. lower maintenance cost).

Semi-axial pumps: The semi-axial concept helps to reduce the outside diameter of the diffuser in comparison to the volute casing of a radial design. That is why vertical bore-hole pumps for water supply which require, for economic reasons, the smallest possible diameter are mostly multistage, semi-axial pumps. For low flow rates in this kind of application, even pumps with specific speeds as low as approximately $n_q = 20$ are designed with semi-axial diffusers. The impellers hardly vary from those employed in a radial design. Vertical, multistage process or condensate pumps with low specific speeds are often built as semi-axial can-type pumps. Figure 2.14 shows such a design with suction and discharge nozzles arranged inline. The diffusers are cast into the stage casings which are bolted together to form the pump unit. The fluid

Fig. 2.13 Two-stage, double-entry water transport pump, Sulzer Pumps $n = 1490$ rpm, $Q_{opt} = 1.7\ m^3/s$, $H_{opt} = 465$ m, $P_{opt} = 8600$ kW, $n_q = 23$

exits from the last stage into a column pipe. The pump assembly is fitted into a tank (a "can") from where the fluid enters the suction impeller via a suction bell. Axial thrust reduction is accomplished by balance holes and an annular seal on the impeller rear shroud, similar to Fig. 2.1. A design similar to Fig. 2.14 uses volutes with 3 or 4 channels instead of vaned diffusers.

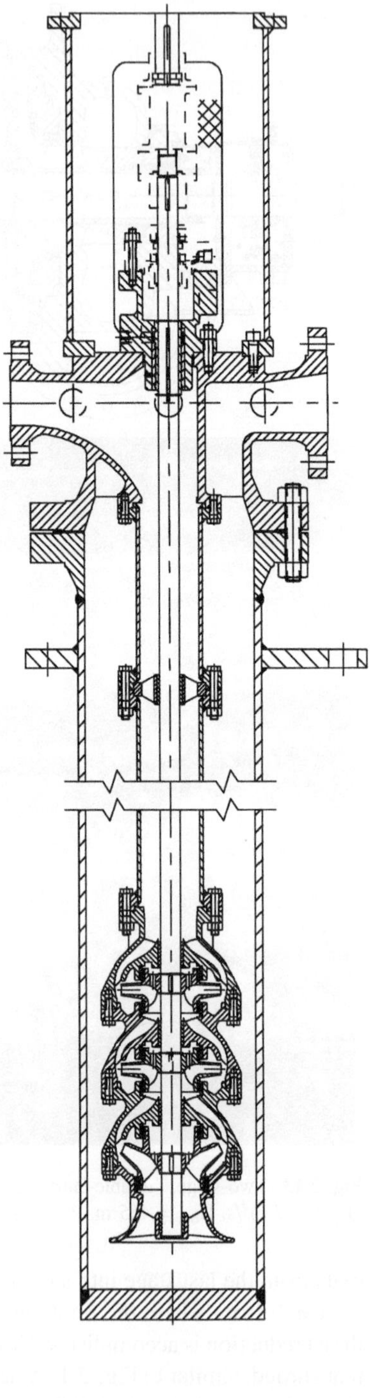

Fig. 2.14 Multistage, semiaxial process can-type pump, Sulzer Pumps

Semi-axial pumps have optimum hydraulic behavior at specific speeds ranging from $n_q = 40$ to 170. Vertical pumps of these specific speeds are used for transporting drinking or cooling water and for irrigation or drainage. They are often installed in intake basins as "wet pit installations". Figure 2.15a shows such a pump with a specific speed of $n_q = 150$ and a suction bell for wet pit installation.

The diffuser discharges into a column pipe in which the shaft is supported by product-lubricated bearings. For low heads (and consequently high specific speeds) special attention has to be paid to good hydraulic design of the discharge bend, since it causes pressure losses of several per cent of the head and thus has an impact on the energy costs.

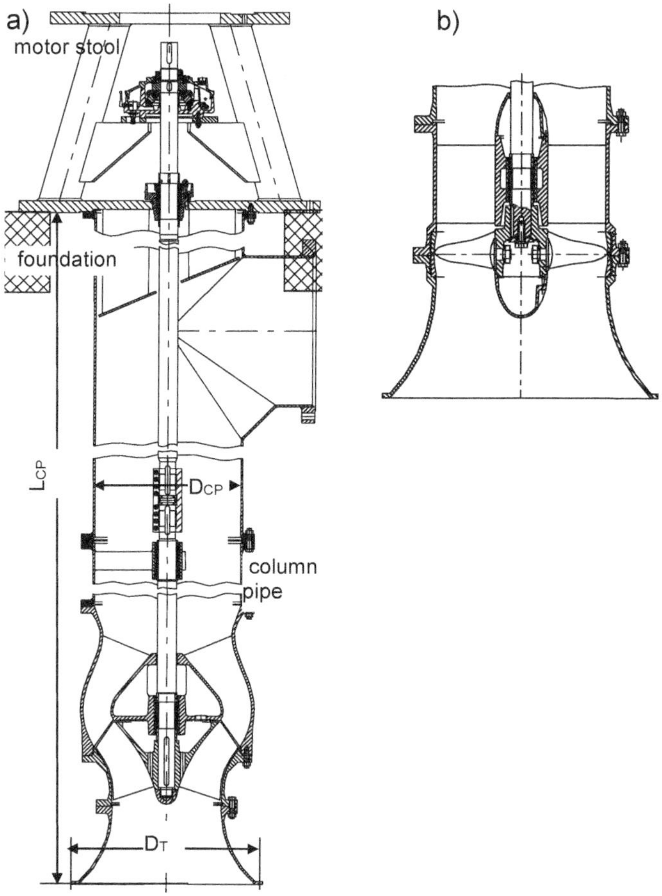

Fig. 2.15 Single-stage vertical pump for wet pit installation: **a** semi-axial pump, **b** axial pump, Sulzer Pumps

Axial pumps: For specific speeds above $n_q > 170$ axial or ("propeller") pumps are used, Fig. 2.15b. The flow through impeller and diffuser is purely in axial direction. The mechanical design is very similar to that of semi-axial pumps.

Seal-less pumps: Pumps without shaft seals ("seal-less pumps") are used when hazardous or noxious substances must be prevented absolutely from leaking to the environment. The technology of seal-less pumps is extensively discussed in [B.14].

One way to eliminate the need for shaft seals is the "canned" or "wet motor" design as shown in Fig. 2.5. With small pumps this very compact design significantly reduces installation costs, thus off-setting the higher costs of the canned motor.

Using a magnetic coupling eliminates the risk of the motor windings coming into contact with the liquid in case of a defect in the can which protects the motor (Fig. 2.16).

In a pump with a magnetic coupling the torque is transmitted from the shaft to the impeller - without any mechanical contact - via a permanent synchronous coupling. The pump casing is sealed on the drive end against the environment by a hermetically tight can which separates the two halves of the magnetic coupling. By this arrangement the need for the sensitive shaft seal and any risk of leakage are avoided.

Figure 2.17 shows a mono-block pump with a magnetic coupling with the motor flanged to the pump. The magnetic coupling consists of two magnetic rotors facing one another. The rotor which is connected to the motor is not in contact with the medium. The second, wet, magnetic rotor forms a unit with the impeller. In the pump shown in Fig. 2.17 the impeller is rotating around a stationary axis supported

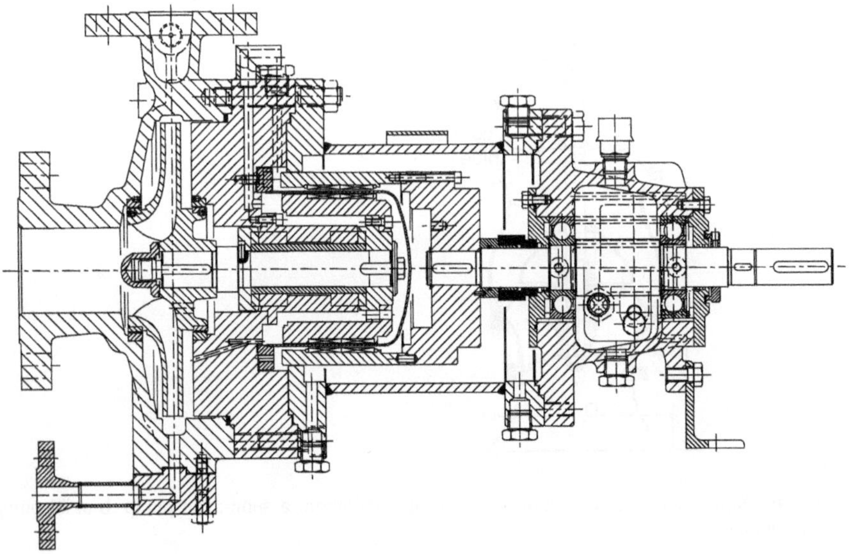

Fig. 2.16 Seal-less process pump driven by a magnetic coupling, Sulzer Pumps

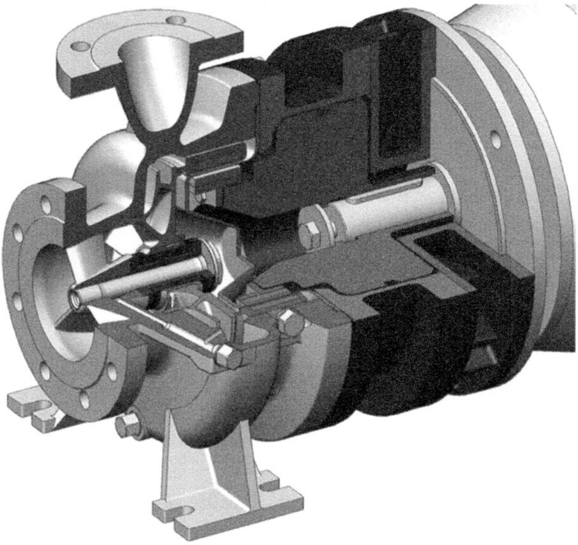

Fig. 2.17 Standard metal process pump with magnetic coupling, close-coupled design, CP Pumpen AG

by ribs in the suction nozzle. The stationary bearing components are fixed to this axis by means of a locking device. There is no additional bearing housing. This arrangement permits a very compact design needing only a few components.

Both of the rotors are equipped with permanent magnets which are arranged with alternating polarities on both of the rotors, Fig. 2.18. The torque is transmitted synchronously (without any slip) by the lines of magnetic flux which act through the air-filled gap between the driving rotor and the can, through the wall thickness of the can and through the liquid-filled clearance between the can and the driven rotor. The tangential force required to transmit the torque is due to the driving rotor preceding the driven rotor by a small angle $\Delta\varepsilon$ (compared to the stand still position). This angle is roughly proportional to the torque. If the maximum possible torque is exceeded, the coupling will slip.

Figure 2.18 shows one possible design of a magnetic coupling. Here, the torque is transmitted from inside to outside. The rotor on the drive end is fixed to the shaft by means of a flywheel. The flywheel will absorb the torque peaks generated during start-up of a standard electric motor. It is thus possible for the magnetic coupling to be designed for just the nominal torque of the motor. Unnecessarily high eddy currents losses are avoided that way.

Using rare-earth permanent magnets (mostly cobalt-samarium), which conserve their magnetic properties up to 350 °C allows magnetic couplings to be manufactured with torques up to 5000 N m (and more). The air-filled gap between the driving rotor and the can virtually does not cause any energy losses; nevertheless, it is to be kept as small as possible (concentration of magnetic flux lines). The radial clearance width is about 1 to 2% of the can radius. If the can consists of a

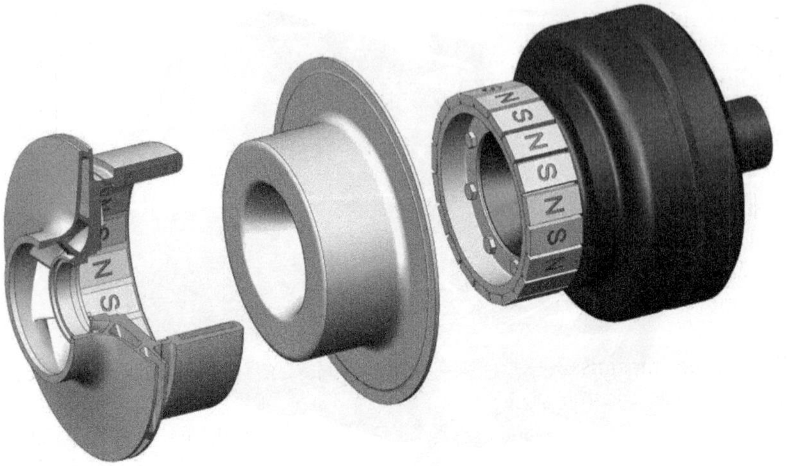

Fig. 2.18 Magnetic coupling, driven rotor with impeller (*left*), can, driving rotor (*right*), CP Pumpen AG

conducting material (e.g. metal), eddy currents are induced in the can by the rotation of the magnetic field. By the losses due to these eddy currents and by the power dissipated through fluid friction the fluid is heated up in the clearance. For removing this heat, a small part of the fluid pumped is diverted through the clearance between rotor and can towards the impeller eye.

The power loss induced by eddy currents is - depending on material and wall thickness of the can - in the range of 0 to 10% of the power transmitted. By designing the can as thin as possible or using a material of high electric resistance (e.g. plastic or ceramic) the induction of eddy currents can be reduced or totally avoided. If the can consists of a non-conducting material, no losses other than those caused by fluid friction and by the cooling flow occur in the magnetic coupling.

Further to the losses by eddy currents, there are losses caused by liquid friction between the can and the rotating magnets wetted by the fluid. The radial width of the fluid-filled gap is about 2 to 3% of the can radius. The power loss (P_{RZ}) can be estimated by Eq. (T3.6.4). If the magnetic force created per surface unit of the magnets is τ_{mag}, the imparted tangential force F_t is proportional to the cylindrical surface of the rotor: $F_t = 2 \times \pi \times R \times L \times \tau_{mag}$. The ratio of the power P_{RZ} dissipated by fluid friction on the rotor to the power transmitted P can be obtained from:

$$\frac{P_{RZ}}{P} = \frac{a\, k_{RZ}\, \rho\, \omega^2 R^2}{2\, \pi\, \tau_{mag}} \tag{2.5}$$

In Eq. (2.5) set a = 2 if there are two wetted gaps, one between can and rotating magnet and the other between rotor and casing. This is the case, if the driving rotor is placed (as in Fig. 2.17) within the impeller magnet. However, if the torque is transmitted from outside to inside, a = 1 is to be used, since the fluid inside the can is supposed to essentially follow the rotation.

It follows from Eq. (2.5) that, with regard to liquid friction, the efficiency of the magnetic coupling is improved by choosing (within the constraints of the mechanical design) magnets as long as possible and with the smallest feasible radius. In addition to the friction of the cylinder surface P_{RZ}, there are other wetted rotor surfaces to be taken into account. Rotor friction losses P_{RZ}/P are expected to be between 1 and 4% of the power - depending on the pump design. For this estimation of losses the value for the BEP is to be used: $\tau_{mag} = P/(2\pi\omega R^2\ L)$. The maximum transferrable torque can be expected in the range of $\tau_{mag,max} = 60{,}000$ to $70{,}000\ N/m^2$.

The torque transmission from outside to inside causes less friction losses than transmission from inside to outside since, in the latter arrangement, the surfaces rotating in the fluid are arranged on a larger diameter, see Figs. 2.16 and 2.17. However, the torque transmission from outside to inside can be advantageous with respect to safety and operational aspects, e.g. venting of vertical pumps, more efficient flushing and cooling of the can if the fluid pumped is close to the boiling point or reducing abrasion on the can when the fluid contains solids. In case of bearing damage debris from the bearing materials are prevented from entering the can. The easy access from the outside to the magnets allows a heating via a jacket integrated in the pump casing if the fluid is close to the melting point. For special requirements or small pumps, these advantages can well override the disadvantage of increased friction losses.

Since the losses inherent to a mechanical shaft seal are avoided, part of the losses caused by the magnetic coupling are compensated. Overall, depending on pump size and design, efficiency losses of 1 to 2% (maxim up to 5%) are to be expected as compared with a traditional process pump.

The impeller bearing is usually a self-lubricated hydrodynamic journal bearing, which can absorb radial and axial forces. By means of pump-out vanes or balancing holes applied to the impeller the axial forces acting on the bearing can be minimized, Chap. 9. The journal bearing of the pump in Fig. 2.17 is supplied with working fluid via a lubricating channel diverting some fluid from the discharge nozzle towards the bearing.

The lubricating properties can vary widely from one medium to the next. Consequently, the design of the lubricating channels must take into account different working fluids and various operating conditions of the pump.

Given an impeller with an optimum axial force balance and suitable lubricating channels, continuous operation without wear can be achieved. During start-up or shut-down of the pump, however, momentary disruption of the lubricating film may occur. Damage can be prevented by using suitable materials. Bearing materials frequently used are sintered silicium carbide (SSiC) or hard metal. If there is a risk of dry running, friction reducing coatings can in addition be applied to the sliding surfaces of the bearing components.

Being hermetically sealed, pumps with magnetic couplings are often used in the chemical industry, for instance for the transport of acids or caustic solutions.

Sufficiently corrosion-resistant materials such as high-alloy stainless steels, nickel alloys, or titanium are necessary in many applications.

As an alternative, pumps with magnetic coupling can also be made from plastic materials or from metals with plastic coatings. Especially when plastics such as PTFE, PVDF or PFA are employed, an almost universal chemical resistance can be achieved.

Submersible pumps: Submerging the pump and motor into the liquid to be transported, reduces the complexity and cost of the pumping plant - in particular if long column pipes would be involved. One of the designs available uses oil-filled motors with a mechanical seal preventing oil leakage into the pumped fluid. Small sewage pumps are predominantly built as submersible pumps. These can be installed or withdrawn from the pit, from above the sewage level, by means of a chain without other manipulation for connecting the pump to the discharge pipe. Figure 2.19 shows a submersible sewage pump with an oil-filled motor. The oil is cooled by a heat exchanger which can be distinguished in the figure just above the impeller rear shroud. The oil is circulated by an impeller arranged below the motor rotor. An external pipe leads the oil from the heat exchanger to the top of the motor. A special oil is used which does not harm the environment in case of a leakage. To avoid the oil circuit, other designs use water cooling of the motor casing; an air gap separates the rotor and stator windings of the motor in this design.

Fig. 2.19 Submersible sewage pump, Wilo-Emu SE

2.3.4 Special Pump Types

Vortex pumps: In vortex pumps according to Fig. 2.20, there is a large axial distance between the casing front wall and the open impeller so that the flow path from the suction to the discharge nozzle is nearly unrestricted. The impeller, equipped with radial or backward-curved blades, generates a recirculating (or "vortex") flow due to centrifugal forces. The incoming fluid is thus forced into a strong rotation. The centrifugal pressure field developed in this way transports the liquid, together with possible inclusions of solids or gas, into the discharge nozzle. As the fluid rotating outside the impeller is moved only by exchange of momentum, there are additional losses due to turbulent dissipation in comparison to centrifugal pumps which transport the fluid in a more regular flow through the impeller channels. Consequently, the efficiencies of vortex pumps are about 30% lower than those of normal centrifugal pumps of similar sizes and specific speeds, see also Chap. 7.4.

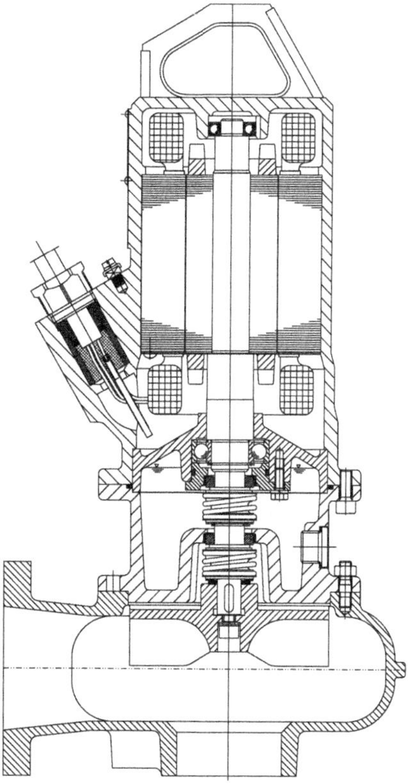

Fig. 2.20 Submersible vortex pump for sewage applications, Wilo-Emu SE

Typical parameters of vortex pumps are: specific speeds n_q = 10 to 80; efficiencies η_{opt} = 0.34 to 0.55; pressure coefficients at BEP ψ_{opt} = 0.2 to 0.6; at shutoff ψ_o = 0.8 to 1.3; impeller blade outlet angles β_{2B} = 30 to 65° or radial blades with 90°; ratio of casing width to impeller outlet width b_k/b_2 = 0.58 to 1.5; ratio of casing diameter to impeller outer diameter d_a/d_2 = 1.25 to 2.1, data from [4].

As their impeller is only slightly exposed to contact with the medium, vortex pumps are fit for pumping liquids with all sorts of impurities including textiles, fibers, abrasive solids and gases (without danger of choking or clogging). These pumps are also convenient for transporting delicate products like fish, vegetables, or crystals. Small vortex pumps are applied quite commonly for pumping sewage.

Friction pumps: The impeller of a friction pump is composed of smooth disks arranged close together by means of bolts and held apart by spacers (see the sketch in Table 2.3). The disks are set into rotation by the driver. The pumping action is due to the centrifugal forces and shear stresses created within the boundary layers on these disks.

The fluid adhering directly to the disks moves with the circumferential speed of the rotor (in the absolute reference frame). Because of laminar and turbulent shear stresses, the liquid between the disks is dragged along, set in rotation and transported outwards by centrifugal forces. The flow regime between the disks may either be laminar or turbulent. The transition takes place in the range of Re_b = 1200 to 2300 ($Re_b = r \times b \times \omega/\nu$ with b signifying the distance between the disks). The head coefficient reaches nearly ψ = 2 at very low flow close to shut-off. With increasing flow rate it drops rapidly, because the momentum transferable by shear stresses is limited. The relation $\psi = 2 - 0.06\ n_q$ gives a rough approximation for the value of the head coefficient, though this depends on the distance between the disks. The efficiencies are in the region of 25 to 35%.

Friction pumps are built with one or several disks operating in parallel. Because of the smooth flow without blades these pumps are very quiet and largely free from vibration and cavitation problems. For that reason they may be considered when quiet operation is of prime importance. Friction pumps are rather insensitive to abrasive wear by solid particles contained in the pumpage. More details concerning friction pumps and literature may be found in [5–7, 11].

Side channel pumps: In side channel and peripheral pumps the energy transfer from the impeller to the fluid results from exchange of momentum between the flow in the impeller and in the stator channel. Due to centrifugal forces in the impeller, a circulating flow is generated between the impeller and the casing channel, see Fig. 2.21.

This circulation strongly intensifies the exchange of momentum because the circumferential speeds of fluid and impeller are nearly the same, but significantly higher than in the channel. The fluid flows back from the side channel into the impeller in a spiral path. This happens several times over the perimeter; the more often this occurs, the higher are energy transfer and pressure rise. The number of spiraling loops grows when the flow rate is reduced. The energy transfer therefore

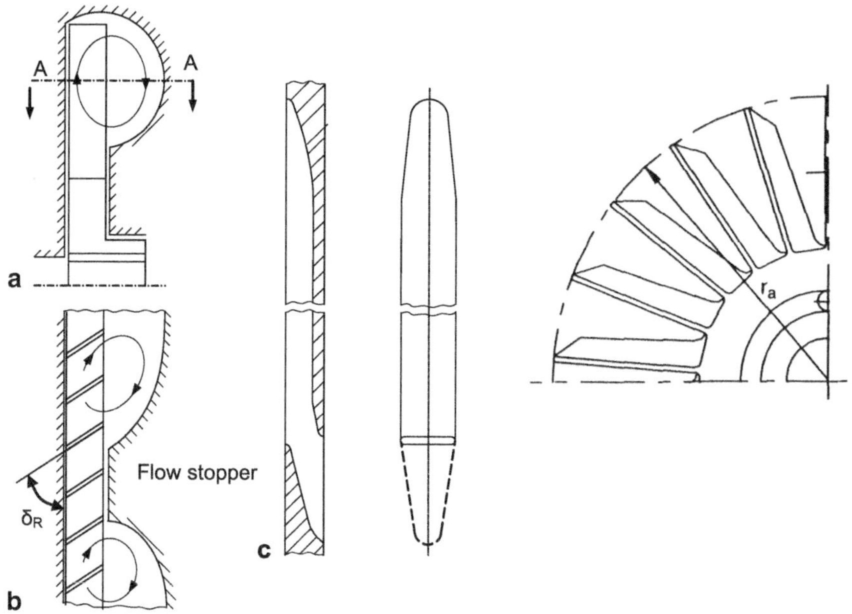

Fig. 2.21 Side channel pump. **a** Meridional section; **b** section A with flow stopper; **c** development of side channel

rises strongly at partload. As the fluid travels through the side channel, the pressure increases from the inlet opening to the outlet port. Both are separated by a flow stopper, which makes the pressure build-up possible. The spiraling flow induces an intensive exchange of momentum resulting in far higher head coefficients than could be obtained with radial impellers (Table 2.3). At partload, head coefficients above $\psi = 10$ are common. The above energy transfer mechanisms can be described as a "partial admission" pump. Detailed flow studies by laser velocimetry confirm this flow model, [8].

The optimum number of impeller blades is between 22 and 26. The straight blades are radial, or they may be inclined backward or forward ($\beta_{2B} = 70°$ to $140°$); they are profiled and may be raked as in Fig. 2.21 (where the rake angle is $\delta_R < 90°$). However, commonly radial blades without a rake are used. The side channel wraps around 270° to 320° of the circumference.

Disadvantages include the relatively low efficiency and the high noise level, the blade passing frequency (rotational frequency times number of blades) being particularly noticeable. Both disadvantages are caused by the recirculating flow and the associated losses. The shape of the flow stopper has an important influence on the noise level. These pumps are also sensitive to cavitation damage. For this reason, the tip speed is limited to about 35 m/s for pumping water.

The axial clearances between impeller and casing should not exceed 0.05 to 0.15 mm. Therefore, the impeller can move axially on the shaft in most designs. Because of the small clearances these pumps are sensitive to abrasive wear.

In addition to the layout with one side channel, shown in Fig. 2.21, there are also pumps with side channels placed on both sides of the impeller in order to double the flow rate.

The vortex flow created within the impeller and the casing channel is able to entrain gases. Side channel pumps are therefore self-priming, i.e. they can evacuate the suction pipe, provided some liquid is left in the pump when starting. Side channel pumps are thus capable of pumping liquids with relatively high gas content generating pressure ratios of $1.6 < p_d/p_s < 35$.

Peripheral pumps according to Fig. 2.22 are similar to side channel pumps as far as the design and working principle are concerned. The difference between both types is essentially limited to the annular chamber which in peripheral pumps is placed not only laterally next to the impeller but also around its outer diameter.

Peripheral pumps are built with up to three sets of blades machined on different diameters of the same disk, forming thus a "3-stage" design with heads up to 2000 m and flow rates up to 80 m^3/h. The impeller being symmetric, axial thrust is balanced. More details concerning side channel and peripheral pumps can be found in [B.8 and B.13].

Liquid-ring pumps are self-priming if sufficient liquid is available in the pump casing. A rotor with radial blades is placed in the pump casing in eccentric position, Fig. 2.23. The rotation of the impeller causes a fluid ring to form at the outer perimeter of the casing due to centrifugal forces. As the fluid ring is concentric with the casing, the cell volumes formed by the blades vary along the circumference. The fluid ring seals the clearance between casing and impeller. Where the cells are largest, gas is drawn in via channels connected to the suction nozzle. The gas exits via slots into the discharge nozzle, where the cells are smallest. The volumetric flow rate depends on the cell size and the speed. Considering its working principle, this machine is a positive displacement rather than a centrifugal pump.

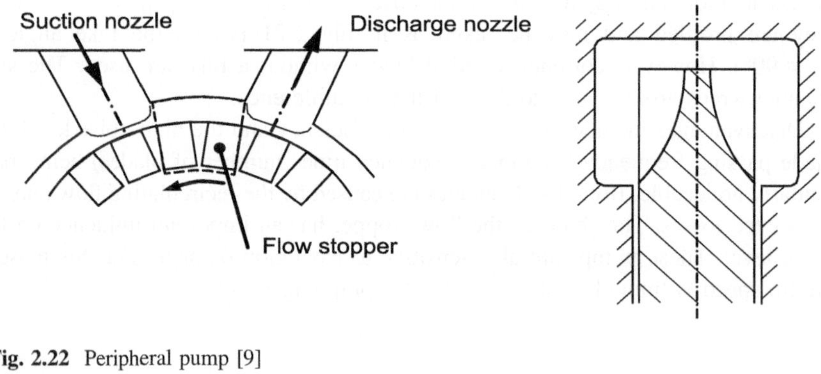

Fig. 2.22 Peripheral pump [9]

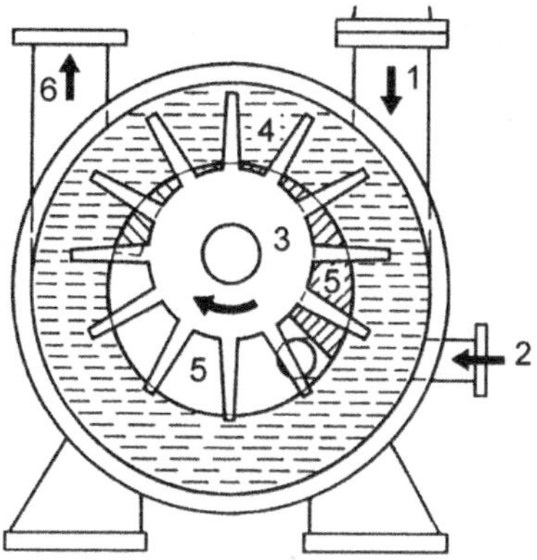

Fig. 2.23 Working principle of a liquid-ring pump. *1* gas inlet, *2* ring fluid entry, *3* impeller, *4* ring fluid, *5* sickle form cavity, *6* gas and ring fluid outlet, [10]

Liquid-ring pumps are useful for processes requiring the pumping (at least sporadically) of liquids containing gas. According to [10] they are employed most often as vacuum pumps in the chemical process industry. Liquid-ring pumps also can transport pure liquids. However, they perform poorly in this application because the cell volumes get narrower towards the exit.

Pitot Tube Pumps, Fig. 2.24: The liquid to be pumped enters via a suction nozzle into a *rotating casing*. Consequently, the fluid rotates with the casing as a forced vortex. Due to centrifugal forces, the static pressure in the casing increases with the square of the circumferential speed according to Eq. (1.27). Some "slip" occurs between the circumferential speed of the casing and the fluid. The slip increases

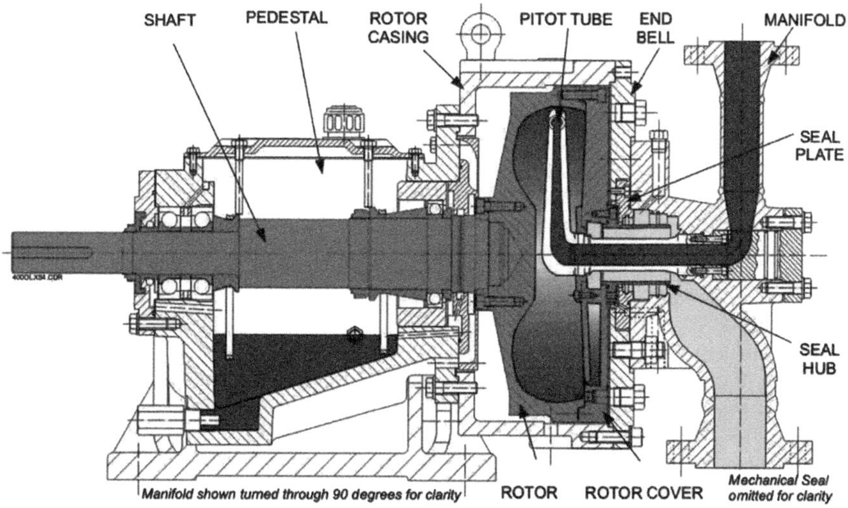

Fig. 2.24 Pitot-tube pump, Sterling Fluid Systems

with growing flow rate. A *stationary* Pitot tube within the rotating casing draws the pumped liquid from the casing and conducts it to the discharge branch. At the inlet of the Pitot tube, the fluid has the absolute velocity $c_{2u} = \gamma \times u_2$, the static head $c_{2u}^2/(2\,g)$ and the dynamic head $c_{2u}^2/(2\,g)$; γ is the slip factor. The fluid is decelerated in a diffuser built into the Pitot tube and led to the discharge nozzle. The theoretical head coefficient would thus be $\psi = 2.0$; due to the slip and losses $\psi_{opt} = 1.9$ may be achieved.

The casing side walls may be equipped with rudimentary vanes, slots or channels to decrease the slip and to improve shut-off head, head rise and stability. The rotating casing eliminates the need of a mechanical seal on the discharge side; there is only one mechanical seal on the suction side.

Since the casing rotates in ambient air, disk friction losses are low. Therefore efficiencies are relatively high in spite of the low specific speeds. Typical performance ranges can be described by the following data: shut-off head up to 2000 m; efficiencies: 50% for $n_q = 2$; 62% for $n_q = 5$; flow rates at BEP up to 100 m^3/h; head rise $H_o/H_{opt} = 1.2$ to 1.3.

Most likely the stationary Pitot tube, which is exposed to the circumferential speed in the rotating casing, is profiled in order to reduce losses and vortex shedding. Since there is no bladed impeller, pressure pulsations are low. Partload recirculation does not occur and minimum flow control valves are not needed.

References

1. Genster, A.: Erhöhung des Wirkungsgrades bei kleinen Heizungsumwälzpumpen. Pumpentagung Karlsruhe, A2–3 (1996)
2. Leibundgut, E.: Mediumgeschmierte Pumpenlager. Techn. Rev. Sulzer. **3**, 8–11 (1998)
3. Laux, C.H., et al.: New size of boiler feedpump in the fossil-fired power stations of the VEAG. ASME Paper 94-JPGC-PWR-47 (1994)
4. Surek, D.: Einfluß der Schaufelgeometrie auf den Kennliniengradienten von Seitenkanalmaschinen. Forsch. Ing. **64**, 173–182 (1998)
5. Roddy, P.J.: Characteristics of a multiple disk pump with turbulent rotor flow. M.Sc. Thesis, Mechanical Engineering, Department of Texas A&M University (1985)
6. Osterwalder, J.: Experimentelle Untersuchungen an Reibungspumpen im turbulenten Strömungsbereich. Pumpentagung Karlsruhe, K10 (1978)
7. Köhler, M.: Die Strömung durch das Spaltelement einer Reibungspumpe. Dissertation, TU Karlsruhe (1969)
8. Heilmann, C.: Strömungsentwicklung längs der Peripherie eines Seitenkanalverdichters. Dissertation, Tu Berlin, Mensch & Buch Verlag, Berlin (2005)
9. Tonn, E.: Zur Berechnung von Peripheralpumpen. Konstruktion **44**, 64–70 (1992)
10. Sobieszek, T.: Umweltfreundliches Evakuieren von Gasen und Dämpfen. Techn. Rundschau. Sulzer. **4**, 27–29 (1986)
11. Straub, H.: Experimentelle Untersuchungen an Reibungspumpen im turbulenten Strömungsbereich. Dissertaion, TH Darmstadt (1983)

Chapter 3
Pump Hydraulics and Physical Concepts

3.1 One-Dimensional Calculation with Velocity Triangles

The objective of the flow calculation for a centrifugal pump is to determine the main dimensions and blade angles of the impeller and the collector for a specified pumping task. To do so, delivery flow, head and speed for the calculation point (usually the best efficiency point) must be available. For this calculation secondary flows and uneven velocity distributions in the impeller and diffusing elements are ignored and the real flow is replaced by an idealized one-dimensional flow (streamline theory). The one-dimensional flow theory is essential for the understanding of the basic flow processes and for the initial design of the impeller, the volute casing or diffuser and the return vanes - even if these components are subsequently optimized by means of numerical methods.

The hydraulic calculation always follows the course of the fluid through the machine. In this process, we consider the inlet and outlet of each component in the flow path. This results in the calculation stations 1 to 6 below. The impeller, diffuser and volute casing with the respective main dimensions, angles and calculation stations are shown in Tables 0.1 and 0.2.

Multistage diffuser pumps:	Volute casing pumps:
(1) Impeller inlet (LE)	(1) Impeller inlet (LE)
(2) Impeller outlet (TE)	(2) Impeller outlet (TE)
(3) Diffuser inlet (LE)	(3) Volute inlet
(4) Diffuser outlet (TE)	(4) Volute outlet
(5) Return vane inlet (LE)	(5) Outer volute
(6) Return vane outlet (TE)	(6) Discharge nozzle

LE means the leading edge and TE the trailing edge of the blades or vanes.

J. F. Gülich, *Centrifugal Pumps*,
https://doi.org/10.1007/978-3-030-14788-4_3

In the case of volutes, the calculation stations 5 and 6 are only required for the outer channel of double volutes, (Table 0.2).

Pump calculation follows the direction of pressure increase (ignoring the minor pressure drop in the inlet nozzle or inlet casing and in the return vanes if applicable). Likewise, turbine operation of a pump (Chap. 12) is calculated in the direction of the flow, but follows the course of diminishing pressure. The numbering of the calculation stations is retained: the turbine calculation sequence therefore is from station 6 to 1. The following shall be determined:

- Circumferential speeds: $u = \omega \times r = \pi \times d \times n/60 = \pi \times d \times n^{(s)}$ (n in rpm; $n^{(s)}$ in s)
- Absolute velocities: c
- Relative velocities: w
- Angle in the absolute reference frame: α
- Angle in the relative reference frame (rotating reference frame): β
- Blade or vane angles on a component: subscript B. Unless otherwise specified this means the *camber* angle of the impeller blades, the diffuser vanes, the return vanes or the volute cutwater.
- Passage width between the blades or vanes: a (shortest distance, Table 0.2)
- Width in the meridional section: b
- Meridional velocity components: subscript m, where: $c_m = Q_{La}/(f_q \times \pi \times d \times b)$; note that $w_m = c_m$ applies.
- Circumferential velocity components: subscript u, where $u = c_u + W_u$
- c_u is counted as positive when acting in the direction of u, negative when opposite to the u-direction
- w_u is counted as positive if acting opposite to the direction of u
- Outer, mean and inner streamlines are described by the subscripts a, m, i; for example $c_{1m,m}$, $\beta_{1B,a}$. The subscript m always refers to the streamline which starts or ends on the inlet or outlet diameter calculated as the *geometric* mean value (d_{1m} and d_{2m} on the impeller and d_{3m} and d_{4m} on the diffuser). The one-dimensional calculation assumes representative conditions on the mean streamlines defined in this way. At low specific speeds frequently only the outer and inner streamlines are considered while three, five or more streamlines are advisable at high specific speeds.
- When the fluid flow enters or leaves a cascade there is a blockage effect due to the finite blade thickness. The flow velocity in the meridional section undergoes (arithmetically) a sudden change at these points. It is, therefore, possible to define velocities with and without blockage at any calculation station; quantities with blockage are described by a raised stroke, e.g. c_{1m}'.
- Two velocities must be distinguished at the inlet of a blading: (1) the velocity vector, which is calculated from the velocity triangle (vector diagram); (2) the mean velocity resulting from the continuity equation. This is given the subscript q to avoid confusion; hence $c_q = Q/A$ and $w_q = Q/A$, where A is the local cross section under consideration.

- The flow rate of the impeller Q_{La} consists of the useful flow rate Q plus leakage through the annular seal at the impeller inlet Q_{sp} and the balance flow Q_E, resulting in $Q_{La} = Q + Q_{sp} + Q_E$. In addition, there may be an intermediate take-off in the case of multistage pumps.

The flow around a blade is considered from the view of an "observer sitting in the blade". Consequently, the *relative* velocities are relevant for the impeller, while the *absolute* velocities are used for the volute or diffuser calculation. The relationship between circumferential impeller speed u, relative velocity w and absolute velocity c is obtained from the rules of vector addition which can be illustrated as velocity triangles (refer also to Chap. 1.1). The equations describing the geometry of triangles therefore apply to the calculation of all velocities, their components in circumferential or meridional direction and the angles α and β.

Inlet triangle: Consider the velocity relationships on the impeller inlet shown in Table 3.1. The meridional velocity immediately in front of the impeller blade leading edges is $c_{1m} = Q_{La}/A_1$, where A_1 is calculated in accordance with the position of the leading edge, Eq. (T3.1.2). Immediately after the leading edges the meridional velocity is increased to $c_{1m}' = \tau \times c_{1m}$ due to the blade blockage. The latter is obtained from $\tau_1 = t_1/\{t_1 - e_1/(\sin \beta_{1B} \times \sin \lambda_{La})\}$, Eq. (T3.1.7), using the sketch in Table 3.1 (with the pitch of the blades $t_1 = \pi d_1/z_{La}$).[1] If the blades are not perpendicular to the front shroud ($\lambda_{La} \neq 90°$), additional fluid is displaced, since the blade blockage obviously becomes larger than with $\lambda_{La} = 90°$. This influence is taken into account by λ_{La} as shown by the sketch in Table 0.1. The *circumferential* components of the absolute or relative velocity are not affected by the blockage, as follows from the conservation of angular momentum.

The fluid flow to the impeller is mostly axial ($\alpha_1 = 90°$). The circumferential component of the absolute inflow velocity is therefore $c_{1u} = 0$. However, if a device for pre-rotation control is installed or when the inlet casing or the return vanes generate an inflow with $\alpha_1 \neq 90°$, the circumferential component is obtained from Eq. (T3.1.3). Figure 3.1 shows the inlet triangles for non-swirling inflow $\alpha_1 = 90°$, pre-rotation $\alpha_1 < 90°$ ("pre-swirl") and counter-rotation $\alpha_1 > 90°$. It can be seen that the approach flow angle β_1 of the blades is increased by pre-swirl and reduced by counter-swirl. As a result of the blade blockage c_{1m} grows to c_{1m}' so that the approach flow angle increases from β_1 to β_1'.

The difference between blade angle β_{1B} and flow angle β_1' is known as incidence: $i'_1 = \beta_{1B} - \beta_1'$. If the incidence is zero, the blade has only a displacement effect on the flow; local excess velocities are correspondingly low. They grow with increasing incidence until the flow separates, since incident flow generates a circulation around the leading edge at $i_1 \neq 0$, Fig. 3.1d. The flow angles without and with blockage are calculated from Eqs. (T3. 1.6 and 3.1.8).

---

[1]With profiled blades, the effect of the blockage may not be easily defined in certain conditions. Generally, e_1 must be selected where the cross section is narrowest relative to the next blade. The effect of the angle λ (if any) must be considered iteratively.

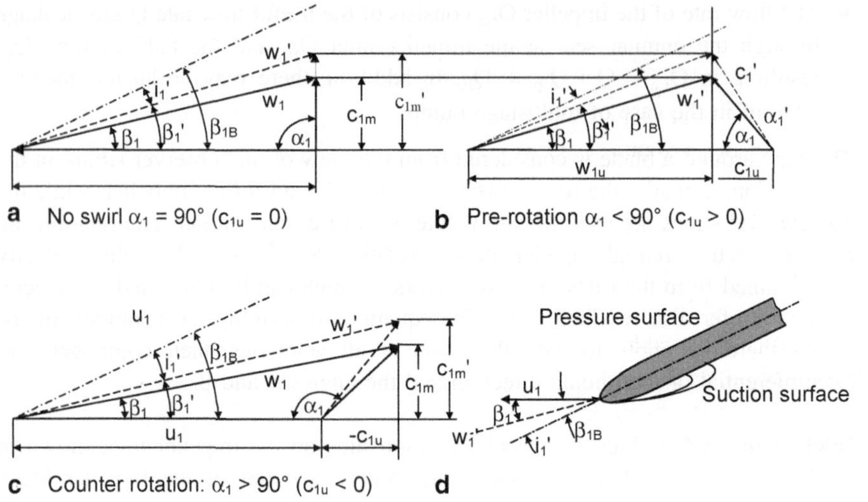

Fig. 3.1 Velocity triangles at the impeller inlet

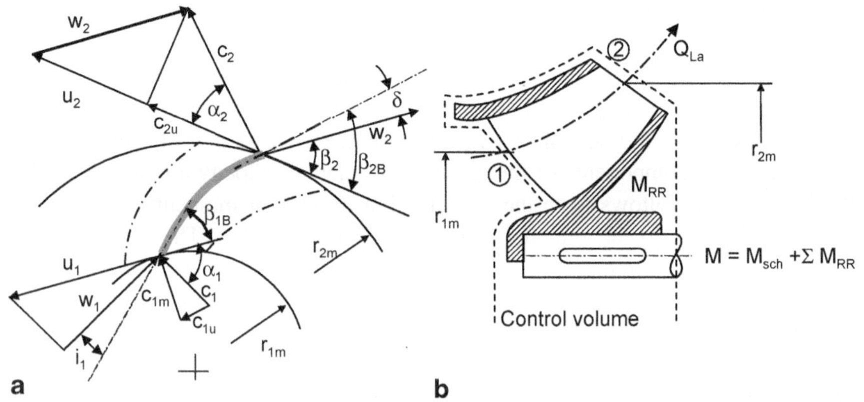

Fig. 3.2 Balance of angular momentums acting on the impeller. **a** Velocity vectors; **b** control volume

With a certain flow rate (i.e. a specific c_{1m}') blade and flow angles are identical ($\beta_{1B} = \beta_1'$) and the incidence becomes zero. This flow situation, called "shockless entry" or "shock-free", is calculated from the condition $\tan\beta_1' = \tan\beta_{1B}$ from Eq. (T3.1.10), where $\varphi_{1,SF} = c_{1m}'_{SF}/u_1 = \tan\beta_{1B}/\tau_1$ is the flow coefficient introduced in Chap. 3.4. If the approach flow angle drops below the blade angle ($i_1 > 0$), the stagnation point is situated on the pressure surface of the blade. If the pump flow rate exceeds the value of the shockless entry, the incidence is negative and the stagnation point is located on the blade suction surface.

Outlet triangle: The velocity relationships at the impeller outlet are shown in Fig. 3.2a and Table 3.2. The meridional velocity downstream of the impeller is obtained from Eq. (T3.2.2). Blade blockage is still present immediately upstream of the impeller outlet and the velocity is correspondingly greater than downstream of the trailing edge: $c_{2m}' = c_{2m} \times \tau_2$, Eq. (T3.2.3). Again, the blockage does not affect the *circumferential* component. The absolute velocity c_2 and outflow angle α_2 are relevant for the design of the diffusing elements, Eqs. (T3.2.10 and 3.2.13).

3.2 Energy Transfer in the Impeller, Specific Work and Head

The head H measurable between discharge and suction nozzles has been defined in Table 2.2 as the useful total head difference generated by the pump. However, Eq. (T2.2.5) does not specify how the impeller converts the power applied to the rotor into pumping work.

To calculate the energy transfer from the impeller to the fluid the conservation of momentum according to Eq. (1.10) can be used, since it establishes a relationship between flow quantities and external forces. If suitable control surfaces are selected, the momentum balance allows a global statement without knowledge of the flow details within the control volume.

The control volume is selected according to Fig. 3.2b around the impeller as presented in the meridional section. The following quantities are applied to its boundary:

- The flow Q_{La} enters the impeller through area 1 with the angular momentum $\rho \times Q_{La} \times r_{1m} \times c_{1u}$.
- The flow leaves the impeller through area 2 with the angular momentum $\rho \times Q_{La} \times r_{2m} \times c_{2u}$.
- The control surface cuts through the shaft. Therefore, external forces in the form of $M = M_{sch} + \sum M_{RR}$ must be applied to this interface.
- Viscous shear stresses cause friction moments M_{RR} on front and rear shrouds that can be estimated according to Chap. 3.6.1. But these need not be considered when calculating the moment M_{sch} acting on the blades.
- Turbulent shear stresses occur in the areas 1 and 2 in circumferential direction, since the control surfaces are not positioned perpendicularly to the velocity vectors c_1 and c_2. These shear stresses generate a moment $M\tau$, which is neglected when calculating according to the streamline theory. From numerical calculations it can be inferred that M_τ amounts to typically 1% of the moment transmitted by the impeller when no recirculation is present[2].

The static pressures in the areas 1 and 2 do not generate any forces in the circumferential direction and therefore do not enter the momentum balance. Likewise, the radial velocity components do not contribute to the angular momentum acting

[2]Recirculation normally occurs at partload below $q^* = 0.5$ to 0.7 (Chap. 5). It may, however, occur even near the best efficiency point when extremely wide impellers are used (for example in dredge pumps).

on the blades; consequently, only the circumferential components c_{1u} and c_{2u} need to be considered. When applying one-dimensional streamline theory, the non-uniform velocity distributions at the control surfaces are replaced by representative average velocities. They generate exactly the same angular momentum on the mean radii r_{1m}, r_{2m} as the integral over the real flow distributions. Subject to all these prerequisites, the conservation of momentum provides the angular momentum acting on the blades in the form of "Euler's turbine equation":

$$M_{sch} = \rho\, Q_{La}\, (r_{2m}\, c_{2u} - r_{1m}\, c_{1u}) \tag{3.1}$$

The mean radii are defined in such a way that they divide inlet and outlet cross sections into two areas which both receive identical flow rates (where c_m is assumed constant over the cross section):

$$r_{1m} = \sqrt{\frac{1}{2}\left(r_{1a}^2 + r_{1i}^2\right)} \quad \text{and} \quad r_{2m} = \sqrt{\frac{1}{2}\left(r_{2a}^2 + r_{2i}^2\right)} \tag{3.2}$$

M_{sch} is the torque which must be applied to the shaft in order to create the flow conditions shown in Fig. 3.2a. According to Newton's law ("action equals reaction"), M_{sch} is also equal to the moment transmitted to the fluid. At an angular velocity ω of the shaft the corresponding driver power is (with $u = \omega \times r$):

$$P_{sch} = M_{sch}\omega = \rho\, Q_{La}\, (u_{2m}\, c_{2u} - u_{1m}\, c_{1u}) \tag{3.3}$$

The specific work produced by the blades is obtained from dividing P_{sch} by the mass flow rate $\dot{m}$ passing through the impeller ($\dot{m} = Q_{La}\, \rho$), Chap. 1.2.2:

$$Y_{sch} = Y_{th} = \frac{P_{sch}}{\rho\, Q_{La}} = u_{2m}\, c_{2u} - u_{1m}\, c_{1u} \tag{3.4}$$

As demonstrated by Eqs. (3.1), (3.3) and (3.4) a pre-swirl ($\alpha_1 < 90°$) reduces blade moment, power consumption and head while a counter-swirl increases these values. Since the specific work of the blades according to Eq. (3.4) does not include any losses (although it applies to flows with losses!), it is also called "theoretical specific work". Accordingly, $H_{th} = Y_{sch}/g$ is the "theoretical head".

The energy Y_{sch} transmitted to the fluid comprises both the useful energy in the discharge nozzle Y according to Eq. (2.1) and the hydraulic losses (see Chap. 1.2.2 and Eq. (1.8)). The following therefore applies: $Y_{sch} = Y + g \times \Sigma Z_h$. These losses are described by the "hydraulic efficiency" η_h defined by Eq. (3.5):

$$\eta_h = \frac{Y}{Y_{sch}} = \frac{H}{H_{th}} = \frac{H}{H + \Sigma\, Z_h} \tag{3.5}$$

The hydraulic efficiency includes all hydraulic losses between the suction and discharge nozzles, i.e. in the inlet, impeller, diffuser and discharge casing. Details concerning η_h can be found in Chap. 3.9, Eq. (3.28a) and Figs. 3.30, 3.31 and 3.32.

From geometrical relationships of the velocity triangles $u \times c_u = \frac{1}{2}(u^2 + c^2 + w^2)$ is obtained. Inserting this expression into Eq. (3.4) produces Eq. (T3.3.2). According to this equation, the specific work consists of three components: centrifugal component $u_2^2 - u_1^2$, deceleration of the relative velocity $w_2^2 - w_1^2$ and acceleration of the absolute velocity $c_2^2 - c_1^2$. The impeller outlet velocity c_2 is decelerated in the collector. There, its kinetic energy is for the most part converted into static pressure. With reference to Eq. (1.15) the first two terms stand for the increase of the static pressure in the impeller plus the impeller losses, Eq. (T3.3.3).

The energy Y_{sch} transmitted to the fluid by the impeller causes a total pressure increase $Y_{tot,La}$ at the impeller outlet. A fraction of Y_{sch} is dissipated in the impeller in the form of losses. Consequently $Y_{sch} = Y_{tot,La} + g \times Z_{La}$ (Chap. 1.2.2) applies. Y_{tot} is the *useful* hydraulic energy present at the impeller outlet (less the inlet energy at the suction nozzle). The specific work $Y = g \times H$ measured between discharge and suction nozzles is obtained as $Y = Y_{tot,La} - g \times Z_{Le}$ when the losses in the diffuser are subtracted. The relationships between specific work and the increase of the total pressure and the static pressure are summarized in Table 3.3. The increase of the static pressure in the impeller H_p (expressed as static head) is obtained from Bernoulli's equation in the relative reference frame, Eq. (1.15). The ratio of the static head to the head of the pump, $R_G = H_p/H$, is called the "degree of reaction". Table 3.3 also includes equations which describe the energy transfer in the relative reference frame.

According to Eq. (3.4) the blade forces create a flow condition which is described only by the circumferential components of the absolute velocity at the impeller inlet and outlet. The static pressures need not be considered. According to Chap. 1.4.1 the static pressures result from the streamline curvature imposed by the blades on the fluid in such a way that they maintain equilibrium with the centrifugal forces.

As will be discussed in Chap. 5, recirculations occur at partload. Their angular momentums acting on the impeller inlet and outlet are *not* included in the above equations. *It is most important to note that one-dimensional pump theory as discussed above applies to pump operation without recirculation only. Ignoring this fundamental fact often leads to erroneous conclusions.* This problem will be discussed in greater detail in Chap. 8.4.

3.3 Flow Deflection Caused by the Blades. Slip Factor

As mentioned above, only the average velocities prevailing at the chosen control surfaces are considered when applying the theorem of conservation of momentum. The complicated flow conditions in the interior of the control volume are ignored. The conservation of momentum consequently fails to show the manner in which the flow observed at the control surfaces was generated; and how the impeller, its outlet

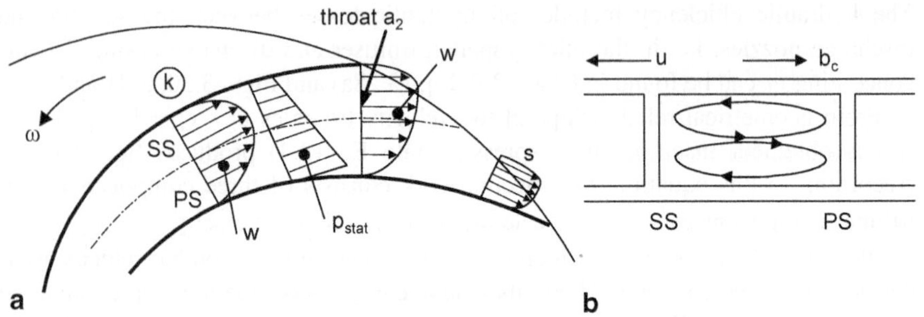

Fig. 3.3 Slip phenomenon. **a** Flow between the blades; **b** Secondary flow

width and blade angles must be designed so that these average velocities actually occur at given values of Q_{La} and ω.

The blade moment results from blade forces which can be imagined as an integral of the pressure and shear stress distributions over the blade surface - much like the lift of a wing is integrated from its pressure distribution. If a blade (or a wing) is to generate a net force, this integral obviously cannot be zero, i.e. greater pressures must be present on the blade pressure surface than on the suction surface. Since the pressure distributions in a turbo machine solely result from the velocity distributions around the blades, it must be concluded that different flow conditions are present on blade pressure and suction surfaces. Hence the flow is not able to follow the blades exactly (the flow is not "blade-congruent"). The work transfer is only made possible through a deviation between the blade and flow angles. In Fig. 3.2a, the flow angle β_2 is therefore represented smaller than the blade angle β_{2B}. The described phenomenon is quantified by the "slip factor" or by the "deviation angle" $\delta = \beta_{2B} - \beta_2$. Both terms implicitly assume the idea of a blade-congruent flow and consider the deviation of the real flow from the blade outlet angle.

As will be discussed in detail in Chap. 5, the flow distribution at the impeller outlet - and consequently the mean flow angle and the slip factor - results from a complex equilibrium of forces. The deviation of the real flow from the blade-congruent flow is essentially influenced by the following mechanisms:

- The velocity differences between the pressure and suction surfaces of the blades are induced by the transfer of work, Fig. 3.3a, Profile k.
- The Coriolis acceleration b_c is opposite to the direction of rotation and causes a secondary flow which transports fluid towards the pressure surface, thus reducing the flow angle β (refer to Fig. 3.2a and Chap. 5).
- Immediately downstream of the trailing edge, the differences in static pressure acting on the blade pressure and suction surfaces vanish, because pressure differences in the free flow can only be maintained through different streamline curvature. The velocity distribution already adapts in the triangular section downstream of the throat at the impeller outlet in a way that this outflow condition is satisfied and that the circulation around the trailing edge is not excessive. Ahead of the throat, in the actual impeller channel, the flow is guided more effectively

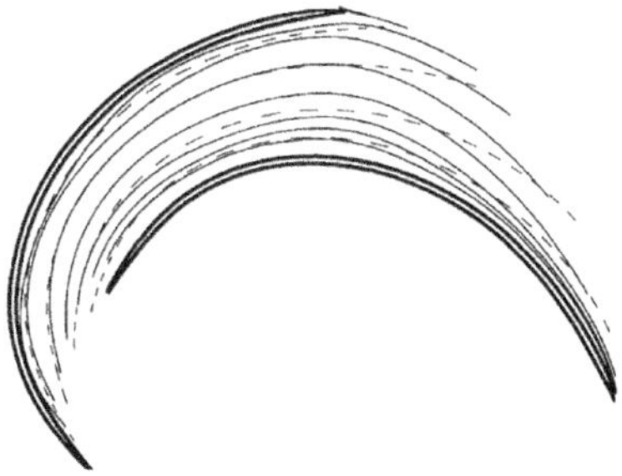

Fig. 3.4 Streamline of relative flow in a radial impeller (Navier-Stokes calculation). The broken lines correspond to blade-congruent flow

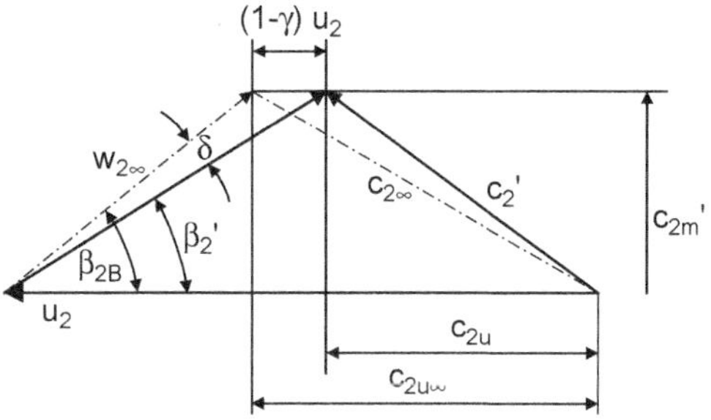

Fig. 3.5 Slip and deviation angle

and deviates less from the blade angle (refer to profiles k and s in Fig. 3.3a). Figure 3.4 illustrates these effects by the paths of the relative velocity calculated with a 3D-Navier-Stokes program. Up to the throat a_2 the flow paths are almost congruent with the blade contour marked by broken lines; in the inclined section after a_2 the flow paths curve in the direction of the pressure surface and the flow angle β is correspondingly reduced toward the impeller outlet. W*ith backward curved blades, the slip factor is to a large extent created in the triangular section at the impeller outlet.* Conversely, with radial blades (having inlet and outlet angles of 90°) the slip factor is primarily caused by the Coriolis force.

Since these flow processes cannot be calculated by simple means, empirical data have to be used when calculating the flow outlet angles according to the streamline theory. Figure 3.5 shows the relevant outlet triangle immediately ahead of the trailing edge (i.e. with blockage), where subscript ∞ is used for blade-congruent flow. Known are u_2, c_{2m}' from Q_{La} and the blade outlet angle β_{2B}. The difference between $c_{2u\infty}$ and c_{2u} is defined as:

$$c_{2u\infty} - c_{2u} = (1 - \gamma)\, u_2 \tag{3.6}$$

Here γ is defined as slip factor, while the quantity $(1 - \gamma)$ is the "slip". Therefore $\gamma = 1.0$ means blade-congruent flow. The smaller γ, the greater is the deviation between flow and blade angles.

With Eq. (3.6) and $\tan \beta_{2B} = c_{2m}'/(u_2 - c_{2u\infty})$ the following definition equation for the slip factor is obtained:

$$\gamma \equiv 1 - \frac{c_{2u\infty} - c_{2u}}{u_2} = \frac{c_{2u}}{u_2} + \frac{c_{2m}\,\tau_2}{u_2 \tan \beta_{2B}} \tag{3.7}$$

To obtain data for the impeller design, the coefficient γ has to be calculated from test data and correlated with geometrical quantities. This calculation is performed at the best efficiency point or in its vicinity. In this context, the measured values of Q, H, P, η are known, and c_{1u} can be calculated from Eq. (T3.1.3) in case of $\alpha_1 \neq 90°$. The hydraulic efficiency η_h is determined by loss analysis according to Chaps. 3.6 and 3.7 so that $H_{th} = H/\eta_h$ is also available. The circumferential component of the absolute velocity c_{2u} at the impeller outlet is obtained from Eq. (T3.2.8). This value can be used to calculate the coefficient γ from Eq. (3.7).

A modified form of the Wiesner equation [44] is used for correlating the data obtained in this way. Based on the calculations by Busemann [3], Wiesner developed a formula for the prediction of slip factors and compared it with measurements on compressors and pumps. Busemann's calculations assumed frictionless flow, zero blade thickness, two-dimensional blades between parallel shrouds and constant blade angle from inlet to outlet in the shape of a logarithmic spiral. To adapt the slip factor correlation to a wider database of pumps, the full blade thickness was used as blockage for the calculation and the correction factor f_1 was introduced. This evaluation produced Eq. (T3.2.6) for slip factor prediction. The correlation reflects the test results of γ for radial impellers with a standard deviation of about $\pm 4\%$ which implies a 95%-confidence limit of approximately $\pm 8\%$. Since the tolerances of the slip factor are emphasized in the calculation of the head, Eq. (T3.3.7), considerable uncertainties must be expected in the performance prediction according to the streamline theory, unless more accurate test data of similar impellers are available. Even an attempt to improve the accuracy of the correlation through additional parameters such as b_2^* or d_1^* has failed so far. Although these parameters have some influence, it is not sufficiently systematic and, consequently, obscured through additional 3D-effects of the flow.

Guided by the idea that a long blade would improve flow guidance and thereby reduce the slip, Pfleiderer developed a slip factor formula incorporating the static moment of the streamline (projected into the meridional section) [B.1]. However, with the analyzed pumps no reduction of the standard deviation below the mentioned level of $\pm 4\%$ was obtained. These findings (at first glance surprising) are explained thus: as mentioned above, the flow guidance in the actual blade channel is quite good so that the length of the blade in this area has only a minor influence. The slip mainly occurs in the "triangle" between a_2 and d_2. Slip is determined to a far greater degree by Coriolis forces and the development of the blade angles and cross sections than by the blade length projected into the meridional section.

The uncertainties inherent to the streamline theory are due to the fact that the 3-dimensional impeller flow cannot be described by a one-dimensional approach - regardless of the number of geometrical parameters taken into account. The main sources of the uncertainties are discussed below:

- The calculation of the slip factor according to Eqs. (T3.2.4–6) only takes into account the parameters outlet angle β_{2B}, blade number z_{La}, blade blockage and indirectly the inlet diameter d_{1m}*. However, the entire blade and channel development determines the velocity distribution at the impeller outlet and therefore also the integral which constitutes the mean value c_{2u} in Eq. (3.4).
- The secondary flow in the impeller increases with growing outlet width, which tends to increase the deviation between the blade and flow angles. Impellers with large relative outlet widths b_2* generate very uneven outflow profiles. As mentioned above, the streamline theory breaks down when there is flow recirculation, and so does the slip factor concept. This may be the case for extremely wide impellers.
- To evaluate the work transfer in the impeller, not only the nominal outlet angle must be used, but the whole blade development near the impeller exit needs to be considered. To be relevant for slip and head prediction, β_{2B} should be approximately constant over an extended range - for example, up to the cross section at a_2. The blade distance a_2 at the throat according to Table 0.2 is an important dimension in the design and for checking a casting. An angle β_{a2} can be defined by Eq. (3.8):

$$\beta_{a2} = \text{arc sin}\frac{a_2}{t_2} = \text{arc sin}\frac{a_2\, z_{La}}{\pi d_2} \tag{3.8}$$

- The smaller the ratio $\tan\beta_{a2}/\tan\beta_{2B}$, the lower is the head and the greater the deviation from the slip factor calculated with β_{2B} according to Eq. (T3.2.6). This is demonstrated by the measurements in [42] where the outlet width a_2 was varied in a wide range, while inlet and outlet angles were kept constant. The resulting Q-H-curves varied from very steep to very flat. Applying any of the published slip factor formulae would completely fail to predict these results.
- Profiling the trailing edge has an effect on the impeller flow and, consequently, the head achieved. Depending on the trailing edge profile, various definitions are possible for the outlet angle, since camber angle β_{2B}, pressure side angle $\beta_{2B,DS}$ and suction side angle $\beta_{2B,SS}$ can be quite different. In the case of un-profiled blades, all three angles are approximately identical. With symmetrical profiling the camber angle constitutes a representative mean value. Conversely, with increased under-filing the camber angle is locally significantly greater than the angle on the pressure side. Finally, with profiling on the pressure side $\beta_{2B} < \beta_{2B,SS}$ applies, (Fig. 3.6).
- The head coefficient tends to decrease with increasing impeller inlet diameter (at otherwise identical design parameters). This fact is not sufficiently represented in the slip factor formula which predicts an influence of the inlet diameter only above some limiting value.
- The pumping action of the shrouds (discussed in Chap. 3.10.3) increases the work done by the impeller - and thus the slip factor. The incidence at the impeller inlet also has an effect on the flow turning achieved by the blades.

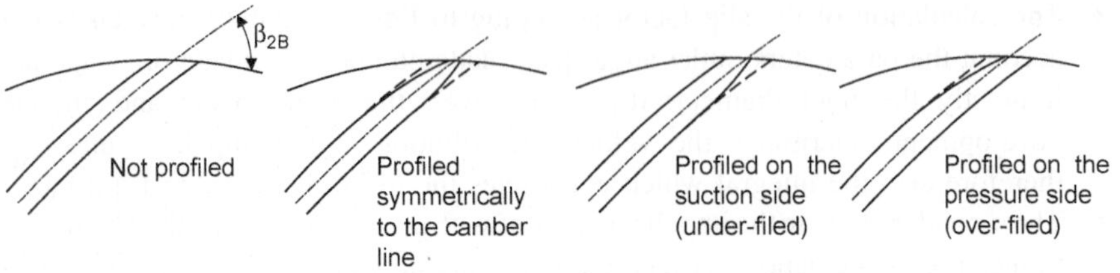

Fig. 3.6 Shapes of impeller blade trailing edges

- In summary, the meridional section, specifically the variation of the width b(r), and the blade angle development $\beta_B(r)$ or $\beta_B(L)$ have an impact on the flow deceleration in the impeller channels and thus on the head. Such effects cannot be taken into account by 1D-theory and the slip factor concept, see also Chaps. 7.2.2 and 7.14.

3.4 Dimensionless Coefficients, Similarity Laws and Specific Speed

Turbulent flows in complex geometries cannot be described accurately by simple analytical means. In practice, such flows are treated with similarity characteristics ("model laws") and dimensionless coefficients by means of which test results can be generalized and applied for prediction purposes in new applications. This approach even applies to geometries as simple as a straight pipe, where Reynolds number, relative roughness and friction coefficients are used for pressure drop prediction, Chap. 1.5.1.

Model tests and their conversion to other speeds and pump sizes consequently form a most important basis in pump design and application. A principle prerequisite for applying similarity laws are geometrical and dynamic similarity. *Geometric* similarity is fulfilled when *all* dimensions of the parts *conducting the flow* of two machines (subscript "a" for the prototype, subscript "M" for the model) are scaled to the same ratio $\lambda = D_a/D_M$. *Dynamic* similarity requires identical values of Euler number, Reynolds number, and Froude number for prototype and model. Which specific similarity characteristics have to be fulfilled depends on the physical process being examined. In pump design, it is only necessary to maintain identical Euler numbers in model and prototype. The effect of the Reynolds number is generally minor and is therefore only taken into account for detailed loss investigations, Chap. 3.10.[3]

If gravity exercises a substantial effect, the similarity according to Froude must be maintained. This applies to flows with a free surface, for example when modeling pump sumps, Chap. 11.

[3]When pumping oils or other media of high viscosity, head and efficiency are severely reduced compared with the pumping of water, Chap. 13.1.

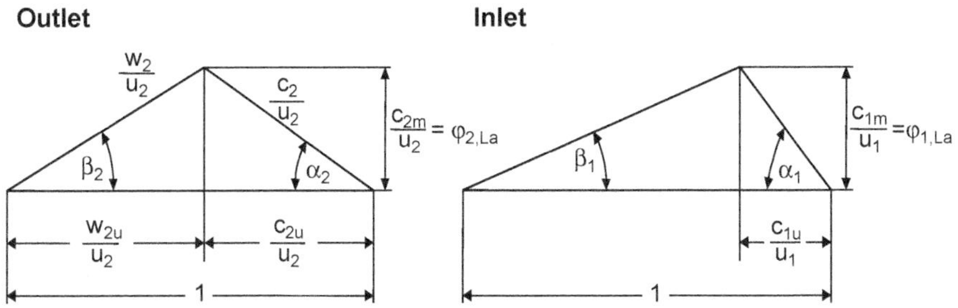

Fig. 3.7 Velocity triangles in dimensionless representation

Similarity or model laws can be developed according to various methods, [1, 12, 45]. Without going into the details of similarity theory, the model laws for pumps (and turbines) can be derived directly from the velocity triangles. To this end, consider the dimensionless inlet and outlet triangles where all velocities are referred to the circumferential velocities u_1 and u_2 respectively, Fig. 3.7. When doing so, all flow angles remain the same. Thus the representation according to Fig. 3.7 is independent of the circumferential velocity - and consequently independent of impeller diameter and speed. Since geometrical similarity is an indispensable requirement, also u_1 and u_2 are in a fixed relationship. All velocities are proportional to $n \times d$. *For geometrically similar impellers the velocity triangles at the inlet and outlet are determined completely by the quantities* c_m/u *and* c_u/u *alone* (Fig. 3.7).

Since the meridional velocity is proportional to the volumetric flow rate, the ratio c_m/u describes the influence of the flow rate. Consequently it is possible to define two flow coefficients: $\varphi_1 = c_{1m}/u_1$ for the impeller inlet and $\varphi_{2,La} = c_{2m}/u_2$ for the impeller outlet, Eqs. (T3.4.6) and (3.4.7). With geometrically similar impellers the same kinematic flow conditions prevail at a given flow coefficient regardless of impeller diameter and speed. The flow coefficients strictly refer to the volumetric flow rate $Q_{La} = Q/\eta_v$ passing through the impeller. Often, however, the flow coefficients are calculated with the useful flow rate Q. This is a good approximation, provided that the volumetric efficiency is not very different between model and prototype.

Since the quantities c_{2u}/u_2 and c_{2m}/u_2 according to Fig. 3.7 are identical in model and prototype because of identical kinematic conditions, it follows from Eq. (3.7) that the slip factor is the same in both machines. From Eq. (T3.3.1) it can be deduced further that the specific work in model and prototype is proportional to the square of the circumferential speed. The following head coefficient is therefore defined: $\psi = 2 \times Y/u_2^2$, Eq. (T3.4.8). $\psi = f(\varphi)$ is also a dimensionless characteristic which is independent of speed and impeller diameter.

Solving Eq. (T3.4.6) for the flow Q and Eq. (T3.4.8) for the head H and inserting the equations obtained in $P_{st} = \rho \times g \times H_{st} \times Q/\eta$ according to Eq. (2.4), a power coefficient λ according to Eq. (T3.4.9) can be defined.

The following proportionalities for geometrically similar pumps result directly from Eqs. (T3.4.6 to T3.4.9): $Q \sim n \times d^3; H \sim n^2 \times d^2$ and $P \sim n^3 \times d^5$. The similarity or model laws between model and prototype according to Eqs. (T3.4.1 to T3.4.4) are obtained from these relationships.

Since all forces are proportional to the product of pressure × area, the following also applies: $F \sim n^2 \times d^4$, Eq. (T3.4.5).

The above relationships presuppose the same efficiencies in model and prototype. If this prerequisite is not satisfied with sufficient accuracy, corrections have to be made by means of an appropriate loss analysis according to Chaps. 3.6–3.10. The similarity conditions for cavitating flows are covered in Chap. 6.2.3.

Specific speed: According to Eqs. (T3.4.8 and T3.4.2) we can write for the head:

$$H = \psi \frac{u_2^2}{2g} = k_1\, n^2\, d_2^2 \psi \tag{3.9}$$

Likewise, Eqs. (T3.4.6 and T3.4.1) yield for the flow rate:

$$Q = \varphi_2\, \pi\, d_2^2\, b_2^*\, u_2 = k_2\, \varphi_2\, b_2^*\, n\, d_2^3 \tag{3.10}$$

where k_1 and k_2 are constants. Eliminating the impeller diameter from these relationships, Eq. (3.11) is obtained after several transformations (written for the best efficiency point):

$$\frac{\sqrt{k_2}\sqrt{\varphi_2 b_2^*}}{k_1^{0.75}\,\psi_{opt}^{0.75}} = n\,\frac{\sqrt{Q_{opt}}}{H_{opt}^{0.75}} = \text{ISC} \tag{3.11}$$

The left side of this equation is a constant "ISC" for geometrically similar pumps; it depends on neither the impeller diameter nor the speed. The middle term describes the relationship between speed, volumetric flow and head of the pump.

Considering the wide range of applications of centrifugal pumps with their various impeller shapes (radial, semi-axial, axial, Table 2.1) described in Chap. 2, such a relationship between the flow characteristics Q, H and n is of fundamental significance for the selection of a suitable type of pump for a given application.

The quantity ISC (impeller shape coefficient) is therefore termed "specific speed" or "type number". Its relevance is explained by way of Fig. 3.8. Consider an impeller at a given speed n_x, providing the flow rate Q_x. If the impeller is designed with various outer diameters or trailing edge shapes (1 to 6 in Fig. 3.8), the head decreases roughly with the square of the diameter according to Eq. (3.9). Since n_x and Q_x are constant, the relationship between Q, H and n changes; this means that the specific speed increases. The shape of the impeller in Fig. 3.8 changes from radial impellers (1 to 4a) to semi-axial impellers. Finally an almost axial impeller is

Fig. 3.8 Development of the impeller shape with the specific speed

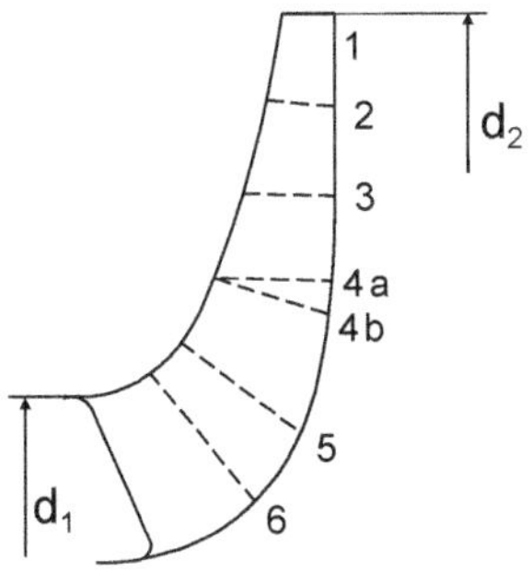

achieved with trailing edge 6. As can also be seen from Fig. 3.8, the relationships b_2/d_2 and d_1/d_2 increase with growing specific speed.

The specific speed must invariably be calculated with the performance data at the best efficiency point and the head per stage. With double-entry impellers the specific speed is defined in Europe with the flow rate *per impeller side*, Eq. (T3.4.15). However, in North America the total flow rate is used (omitting the factor f_q, refer also to Chap. 2.2). Unfortunately n_q is not dimensionless. To obtain a dimensionless characteristic it is possible to divide Q_{opt} by the reference quantity $Q_1 = 1\ m^3/s$ and likewise introduce $H_1 = 1$ m and $n_1 = 1$ rpm. Formulated thus, n_q is a dimensionless figure. In practice, n_q is considered as dimensionless without explicitly writing the reference quantities. A really dimensionless characteristic is ω_s according to Eq. (T3.4.16).

The specific speed permits a comparison of geometrically *dissimilar* impellers. Hence it can be used for comparing pumps from different vendors for a given application. But the specific speed provides only a rough indication concerning the hydraulic features, configuration and impeller shape of a pump. Depending on the application, a pump with given data n, Q_{opt} and H_{opt} - consequently with a given specific speed - can be equipped with impellers of quite different shapes. Examples are: (1) Sewage or dredge pumps typically have impeller outlet widths twice as large as pumps for non-clogging fluids with identical specific speed. (2) Depending on the required suction capability, the ratio of impeller inlet to outlet diameter d_1/d_2 can vary appreciably for a given n_q, (Chap. 6). (3) The head coefficient can be varied in relatively wide limits in order to obtain flat or steep characteristics without great effect on the specific speed. In this way the head rise H_o/H_{opt} can be adapted to specific requirements.

3.5 Power Balance and Efficiencies

Losses arise whenever a fluid flows through (or components move in) a machine. The useful power P_u introduced in Chap. 2.2 is therefore always smaller than the power P supplied at the pump shaft. The sum of all losses $\Sigma P_v = P - P_u$ is dissipated into heat. Apart from the hydraulic losses according to Eq. (3.5) additional losses occur in a pump ("secondary losses") so that altogether the following sources of loss must be considered:

1. Mechanical losses P_m in bearings and shaft seals. These do not generally result in a heating of the fluid and are therefore termed *external* losses.
2. Volumetric losses are due to all leakages which are pumped by the *impeller*. These include: (a) The leakage Q_{sp} through the annular seal at the impeller inlet. (b) The leakage Q_E through the device for axial thrust balancing. These leakages (Q_{S2} in Fig. 3.16) either occur through balance holes in single-stage or multistage machines with individual impeller balancing or through central balance devices such as pistons or disks (Q_E in Fig. 3.16), refer also to Chap. 9.2. In special cases additional fluid (Q_h) may be circulated within the pump which is branched off for auxiliary purposes such as feeding a hydrostatic bearing, flushing, sealing or cooling. Such flows must be taken into account when calculating volumetric efficiency and pump performance. To pump all of these leakages the power $P_L = \rho \times g \times H_{th} \times \left(Q_{sp} + Q_E z_{st1}/z_{st} + Q_{CB} z_{st2}/z_{st} + Q_h\right) = \rho \times g \times H_{th} \times Q/\eta_v$ must be supplied by the impeller, where η_v is the volumetric efficiency, Eq. (T3.5.9). Back-to-back pumps (as shown in Figs. 2.9 and 2.10) comprise two groups of stages: the first group has z_{st1} stages which are opposed by the second group with z_{st2} stages. From the 1st group there is a balance flow rate Q_E over the piston to the external balance line. Between both groups a center bush breaks down the pressure difference between the last stage and the first group of stages; the flow rate is Q_{CB}. The equations in Table 3.5 apply to both in-line and back-to-back pumps. For inline pumps set $z_{st1} = z_{st}$ and $z_{st2} = 0$.
3. Disk friction losses P_{RR} are generated on rear and front shrouds of the impellers which rotate in the fluid as hydraulically smooth or rough disks.
4. Similar friction losses P_{er} are created by the components of axial thrust balance devices (balance piston or disk, Chap. 9.2).
5. In multistage pumps leakages also occur in the "interstage seal", i.e. the annular seal separating two stages, Fig. 3.16. These leakages do not flow through the impeller; consequently, they are not imparted the full blade work. They constitute a throttling loss P_{s3} which should be evaluated with only about 40% of the head, Chap. 3.6.3.
6. Hydraulic losses due to friction and turbulent dissipation in all components between suction and discharge nozzle are covered by the hydraulic efficiency η_h (Chap. 3.7). The dissipated power is $P_{vh} = \rho \times g \times H \times Q \times (1/\eta_h - 1)$.
7. Fluid recirculation at partload creates high losses P_{Rec} due to an exchange of momentum between stalled and non-separated fluid zones. If correctly designed, these are zero near the best efficiency point (and above). When operating against a closed valve or at a low flow rate, they make up the greatest part of the power consumption (Chaps. 4 and 5).

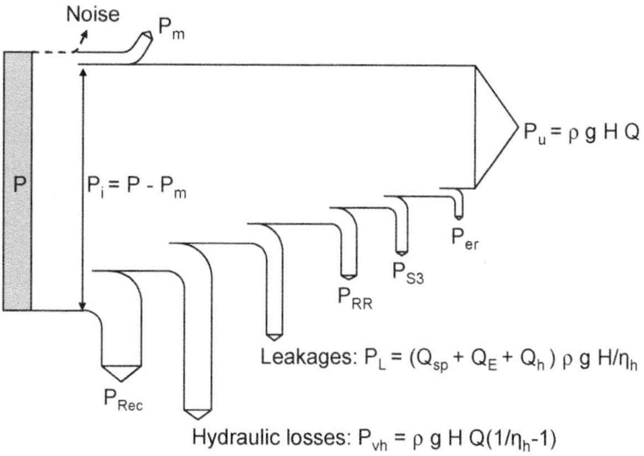

Fig. 3.9 Power balance of a pump

The loss components 2 to 7 are generated within the pump. In the case of the balancing flow this applies only if Q_E is returned into the suction nozzle. All these losses are called "internal losses" because they heat up the fluid in the machine.

The overall power balance of a pump according to Eq. (T3.5.1) is depicted in Fig. 3.9.[4] Table 3.5 contains the relevant equations and rules of thumb to estimate the secondary losses.

In addition, the following efficiencies are used: the mechanical efficiency according to Eq. (T3.5.6); the internal efficiency according to Eq. (T3.5.5) and the volumetric efficiency according to Eq. (T3.5.9).

The secondary losses are discussed in Chap. 3.6 and the hydraulic losses in Chap. 3.8. The calculation of all types of losses is subject to uncertainties of ± 20 to 30%. If the individual loss under consideration amounts to only a few percent of the coupling power - as with pumps having more than 80% overall efficiency - such an uncertainty is not too serious. However, in the case of small pumps and low specific speeds it is significant. Loss analysis is necessary not only for new developments; it can also be very helpful in determining the cause of a performance deficit observed in a plant or on the test stand. When performing this task, often the focus is not so much on the absolute value of the calculated loss; the goal is rather to determine trends caused by a deviation from design or to evaluate the effect of an intended or observed modification. In such cases, consistent consideration of the losses is more important than the choice between different calculation methods.

[4]The sound power radiated as air-, fluid-, and solid-borne noise is negligibly small compared to all other losses.

3.6 Calculation of Secondary Losses

In the following, disk friction losses and leakage losses are considered as separate phenomena that can be measured in special test stands outside, or even inside, a pump. However, a close interaction is in effect between the main flow at the impeller outlet, the flow in the impeller sidewall gaps, leakage through annular seals and the disk friction losses. The interaction is difficult to capture either experimentally or numerically (this topic is discussed extensively in Chap. 9.1).

3.6.1 Disk Friction Losses

When a circular disk or a cylinder rotates in a fluid, shear stresses corresponding to the local friction coefficient c_f occur on its surface, Chap. 1.5.1. On a disk rotating in an extended *stationary* fluid (without the influence of a casing) the shear stress is $\tau = ½\rho \times c_f \times u^2$ with $u = \omega \times R$, Eq. (1.31). The friction force on a surface element $dA = 2\pi \times r \times dr$ is then $dF = 2\pi \times \tau \times r \times dr$ and the torque exercised by friction becomes: $dM = r \times dF = \pi \times \rho \times c_f \times \omega^2 \times r^4 \times dr$. The friction power *per side* of the disk $P_{RR} = \omega \times M$ is obtained from the integral $P_{RR} = \omega \times \int dM$ (between inner radius r_1 and outer radius r_2) as $P_{RR} = (\pi/5) \times \rho \times c_f \times \omega^3 \times {r_2}^5 \times \left(1 - {r_1}^5/{r_2}^5\right)$. The friction coefficient c_f depends on the Reynolds number and the surface roughness. It has roughly the same magnitude as found on a plate in parallel flow, Eq. (1.33).

In the case of a rotating cylinder (with the length L and radius r_2) no integration is necessary since the radius is constant. In the same way as above the power $P_{RZ} = \pi \times \rho \times c_f \times \omega^3 \times {r_2}^4 \times L$ is obtained.

In agreement with the similarity laws in Table 3.4, the disk friction power increases with the third power of the speed and the fifth power of the diameter (assuming geometrical similarity L/r_2 = constant and neglecting the influence of the Reynolds number).

If the body rotates in a casing (as is the case in a pump) the velocity distribution between casing and rotating body depends on the distance between the impeller shroud and the casing wall as well as on the boundary layers which form on the stationary and rotating surfaces. A core flow with approximately $c_u = ½\omega \times r$ is obtained (in other words, $u = \omega \times r$ can no longer be assumed). In the case of turbulent flow, the power absorbed by a disk in a casing therefore amounts to roughly half of the power of a free disk rotating in a stationary fluid.[5]

The disk friction losses of impeller shrouds or of cylindrical bodies are calculated according to Eqs. (T3.6.2 and 3.6.4). The friction coefficients k_{RR} and k_{RZ}

[5]This experimental finding follows from data in [35]; it cannot be deduced by inserting half the angular velocity in the equation given earlier.

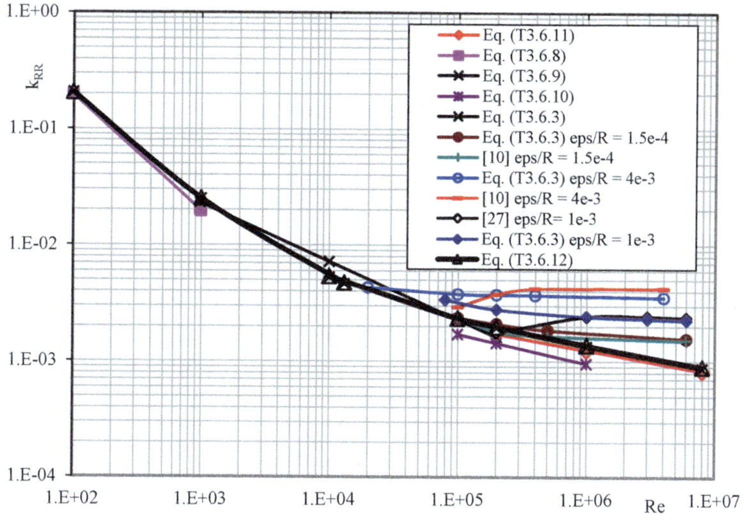

Fig. 3.10 Disk friction coefficients for smooth and rough surfaces calculated for $s_{ax}/R = 0.08$ (without leakage through the impeller sidewall gap); ε = roughness

required for this purpose are determined from correlations derived from experiments. All relevant equations are compiled in Table 3.6. They were derived from measurements on hydraulically smooth disks.[6]

The equations for turbulent flows were expanded by correction factors f_R and f_L by means of which the effect of the roughness and the leakage flow through the impeller sidewall gaps can be estimated. The increase of the friction on rough disks compared with hydraulically smooth disks is determined as follows: the friction coefficient c_f is calculated for the rough surface and for the hydraulically smooth surface $c_{f,o}(\varepsilon = 0)$ from Eq. (1.33) with $Re = u \times r_2/\nu$ and ε/r_2. This produces the correction factor $f_R = c_f/c_{f,o}$ according to Eq. (T3.6.6).

Equation (T3.6.3) produces similar friction coefficients as Eqs. (T3.6.8 to T3.6.11), Fig. 3.10. However, Eq. (T3.6.3) has the advantage that the entire range from laminar to turbulent can be covered with a single equation without any unsteadiness at the transitions.

Numerous tests confirm that the axial casing clearance s_{ax} has only a very small effect on the disk friction; Eq. (T3.6.11) overestimates this effect. With s_{ax} tending to zero it yields too small and with s_{ax} tending to infinity too large friction coefficients. In contrast, Eq. (T3.6.3) produces sensible values even with extreme axial clearances s_{ax}. With s_{ax} tending towards infinity Eq. (T3.6.3) supplies factors

[6]Equations (T3.6.3) and (T3.6.5) are based on [33] and [25], Eqs. (T3.6.8) to (T3.6.11) on [6], Eq. (T3.6.12) on [14].

roughly corresponding to a free disk whose friction factors are given by $k_{RR} = 0.0365/Re^{0.2}$, [35]. Refer to Table 3.6 for further definition.

In [14] a method was developed to calculate disk friction losses by taking into account the fluid rotation in the impeller sidewall gap, Eqs. (T3.6.12 and T3.6.13). This equation covers the effects of the impeller sidewall gap geometry as well as the roughness of the casing and the impeller shrouds. It is useful particularly when the casing and the impeller disk have different roughness values. For a detailed discussion refer to Chap. 9.1.

The various correlations from Table 3.6 are compared in Fig. 3.10. It can be seen that in particular Eqs. (T3.6.3) and (T3.6.12) are equivalent and suitably reflect Eqs. (T3.6.8 to T3.6.11). The effect of the roughness is appropriately covered by Eq. (T3.6.6).

Literature overviews concerning disk friction losses can be found in [14, 17, 34]; the basic theory is treated in [35]. When comparing with other sources, the different definitions of the torque coefficients must be considered (frequently the coefficients are defined in a way that they comprise the torque or the friction power for both sides of the disk).

The disk friction losses in a pump depend on the following parameters:

1. **Reynolds number:** As with the flow over a plate or through a pipe, the friction coefficient drops with increasing Reynolds number. When pumping water, the flow in the impeller sidewall gap is generally turbulent. In the case of oils or other fluids with high viscosity the flow becomes laminar and the disk friction increases greatly (Chap. 13.1).
2. **Roughness of the rotating disk:** The roughness of the stationary or the rotating surface increases the friction power, provided the roughness peaks protrude from the boundary layer thickness. In this context, grooves in circumferential direction such as generated by machining the shrouds on a lathe are less harmful than an unstructured roughness as typically found on a cast surface, [11]. According to the measurements in [11] the disk friction of a machined disk with $\varepsilon_{max} = 120$ µm was practically identical in magnitude to that of a polished disk with $\varepsilon \approx 0$. Machined disks can therefore be calculated in general with $\varepsilon \approx 0$. For cast or sandblasted disks $\varepsilon = \varepsilon_{max}/c_{eq}$ can be assumed. The equivalence factor is $c_{eq} = 2.6$ and ε_{max} is the maximum roughness height, Chap. 3.10. Disk friction losses are minimized if hydraulically smooth surface conditions are achieved. To this end, the permissible roughness that must not be exceeded can be evaluated according to Table D1.1 where $w = c_u \approx ½ \times \omega \times r$ must be employed. The roughness problem is also discussed in Chap. 3.10.
3. **Roughness of the casing wall:** When rotor and casing have the same roughness the rotation of the fluid in the impeller sidewall gap is independent of the roughness, Eq. (T3.6.13). However, if $\varepsilon_{impeller} \neq \varepsilon_{casing}$, the roughness has an effect on the rotation and, consequently, the disk friction. With $\varepsilon_{impeller} > \varepsilon_{casing}$ the rotation increases, while it is reduced if $\varepsilon_{impeller} < \varepsilon_{casing}$. At the same time the disk friction changes as per Eqs. (T3.6.12) and (T3.6.13). For a more detailed calculation refer to Chap. 9.1.

4. **Axial sidewall gap s_{ax}:** With a very small distance between impeller shroud and casing $s_{ax} < \delta$ (δ = boundary layer thickness) the velocity distribution in the axial sidewall gap is linear (Couette flow) and the mean velocity is approximately $\frac{1}{2}\omega \times r$. When increasing the axial sidewall gap, the friction loss drops to a minimum (which has no practical significance since it cannot be executed in reality) before it rises again. As already mentioned, the effect of the axial sidewall clearance is minimal in turbulent flow, while it is significant in laminar flow. The larger the wetted casing surfaces of the sidewall gap, the slower is the fluid rotation.
5. **Shape of the casing and size of the impeller sidewall gap:** have an effect on the rotation of the fluid and consequently the disk friction. *Complicated shapes of the casing, large casing surfaces, ribs and other structures reducing fluid rotation increase the disk friction losses in a way similar to rough surfaces*, Chap. 9.1.
6. **Influencing the boundary layer:** The turbulence structure of the boundary layer can be influenced by fine grooves in the flow direction. The friction coefficient could thus be lowered by up to 10% in the tests reported in [2]. However, producing grooves of the necessary fineness and with sharp tips is expensive and scarcely feasible in practice.
7. **Leakage flow:** through the impeller sidewall gap (Chap. 9.1): A radially inward-directed leakage introduces an angular momentum of $\rho \times Q_{sp} \times c_{2u} \times r_2$ in the impeller sidewall gap which reduces the disk friction, if $c_{2u}/u_2 > 0.5$. Conversely, a radially outward-directed leakage flow decelerates the fluid rotation in the impeller sidewall gap (if it has no or little pre-rotation) so that the disk friction increases. The effect of leakage can be estimated by the correction factor according to Eq. (T3.6.7)[7] which was derived from the calculations in Chap. 9.1. It roughly agrees with the measurements in [38]. In practice, the effects of radially inward leakage (reducing the disk friction) and roughness (increasing disk friction) compensate to some extend so that $f_L = f_R = 1.0$ can often be used for simplification.
8. **Partload recirculation:** slows down the fluid rotation if it enters the impeller sidewall gap with low circumferential velocity, Fig. 5.30. The disk friction increases as demonstrated by Fig. 9.10.
9. **Exchange of momentum:** An exchange of momentum and, consequently, an interaction occur between main flow and impeller sidewall gap flow, Chap. 9.1.

The last three parameters can severely influence the disk friction loss and render its calculation quite uncertain, especially with large leakage flows and at partload. Even at the best efficiency point, the tolerance of the calculated disk friction loss is estimated to be about ±25%. The physical mechanisms of sidewall gap flow are

[7]The impeller sidewall gap integration according to Table 9.1 is recommended for more thorough investigations.

treated comprehensively in Chap. 9.1, since they are very important for calculating the axial forces acting on the impeller

When determining the disk friction losses in a pump, the contributions of all surfaces rotating in the fluid must be calculated using the equations from Table 3.6. These include impeller front and rear shrouds, their cylindrical end faces, annular seals on the impeller and the balance device if any. In the case of radial impellers, friction on the shrouds by far exceeds the contribution of all other components. The size of the impeller inlet diameter has little impact since $(r_1/r_2)^5 \ll 1$, Eq. (T3.6.2).

For typical radial impellers the disk friction loss can be calculated from the useful power P_u and the quantities n_q, Re and ψ_{opt}. This leads to Eq. (T3.5.12) for turbulent flows, or (more generally) to Eq. (T3.5.13) for laminar and turbulent flows. In these equations the contributions of front and rear shrouds, including cylindrical surfaces and annular seals, have been taken into account for typical impeller geometries. The equations show:

- The share of disk friction losses in the power consumption of a pump drops exponentially with increasing specific speed and increasing head coefficient. With low specific speeds disk friction is the main source of loss: at $n_q = 10$, $\psi_{opt} = 1$ and $Re = 10^7$ the disk friction is approximately 30% of the useful power of the impeller.
- With low specific speeds the head coefficient must be selected as high as possible in order to improve the efficiency (a high head coefficient leads to a reduced impeller diameter and consequently less disk friction, since $P_{RR} \sim d_2^5$).

3.6.2 *Leakage Losses Through Annular Seals*

Close running clearances ("annular seals") between impeller and casing limit the leakage from the impeller outlet to the inlet, see Figs. 3.15, 3.16 and Table 3.7. Any leakage reduces the pump efficiency. Since the *entire* mechanical energy transferred by the impeller to the leakage flow (i.e. the increase of the static head and the kinetic energy) is throttled in the seal and converted into heat, one percent of leakage flow also means an efficiency loss of one percent. Leakage flows likewise occur on axial thrust balance devices in which the entire differential head of the pump is throttled to suction pressure (here, too, the overall efficiency is reduced by one percent for every percent of leakage).

The annular seal consists of a case ring and a rotating inner cylinder. The radial clearance s between the rings is small compared to the radius of the rotating parts ($s \ll r_{sp}$). Due to the pressure difference across the seal, an axial flow velocity c_{ax} is generated. With the rotor at rest, this axial flow can be treated according to the laws of channel flow if the hydraulic diameter $d_h = 2 \times s$ is used.

Through the rotation of the inner cylinder a circumferential flow is superimposed on the axial flow. To describe these flow conditions two Reynolds numbers are required: Re for the axial and Re_u for the circumferential flow. Both definitions are given in Table 3.7, Eq. (T3.7.9). The transition from laminar to turbulent occurs at a Reynolds number Re* = 2000, which is formed with the mean vectorial velocity from c_{ax} and $u_{sp}/2$, as given by Eq. (T3.7.10).

The circumferential velocity c_u in the annular seal depends on the pre-rotation at the annular seal inlet, $k_{in} = c_u/(\omega \times r)$; c_u develops in the annular seal with increasing flow path length z according to Eq. (T3.11a), [5]. It reaches an asymptotic value given by $k_{out} = c_u/(\omega \times r) = 0.5$ for long annular seals.

$$k \equiv \frac{c_u}{u_{sp}} = 0.5 + (k_{in} - 0.5)\exp\left\{-\frac{\lambda\, z}{4\, s}\left(1 + \frac{0.75}{1 + 4\left[\frac{Re}{Re_u}\right]^2}\right)\right\} \qquad (3.11a)$$

In laminar flow the velocity distribution in circumferential direction is linear with $c_u = 0$ at the stator and $c_l = \omega \times r_{sp} = u_{sp}$ at the rotor. Since c_u decreases from the rotor to the stator with growing radius, the centrifugal forces, too, decrease from rotor to stator. Consequently, the velocity distribution is inherently unstable. Above a specific circumferential velocity "Taylor vortices" are generated in the circumferential direction. These manifest as pairs of vortices with opposite swirl structures. The stability criterion of the vortex development is determined by the Taylor number Ta: $Ta = u_{sp} \times s/\nu \times (s/r_{sp})^{0.5} = Re_u/2 \times (s/r_{sp})^{0.5}$, [35]. Taylor vortices occur at Ta > 41.3. However, the flow remains laminar up to approximately Ta = 400 with an adequately small axial Reynolds number Re. In laminar flow the resistance coefficient increases considerably due to the Taylor vortices (by a factor of 2 to 3) refer also to [35, 34].

In stable laminar flow (Ta < 41.3) the rotation has no noticeable effect on the friction coefficient λ which is calculated as λ = 96/Re for a concentric gap which is confirmed by the tests in Fig. 3.11 with $Re_u = 0$. Conversely, in turbulent flow λ depends on the ratio of the circumferential to the axial Reynolds numbers Re_u/Re. As follows from Eq. (T3.7.10) the flow is always turbulent for $R_{eu} > 4000$, even if there is no axial flow (Re = 0). Equation (T3.7.14) can be used to estimate the friction coefficient for flows with Re < 2000.

In practical applications, the transition between laminar and turbulent cannot be related clearly to either the criterion Re* = 2000 or the Taylor number. The transition between laminar and turbulent flow depends on the roughness and surface structure (e.g. serrations), the axial Reynolds number Re and the circumferential Reynolds number Re_u. As yet, there are no reliable correlations to describe this transition. Test data in pipes and on flat plates demonstrate that the friction factors in laminar flow usually cannot be lower than in turbulent flow at any given axial Reynolds number. In order to avoid a discontinuity in the curve λ = f(Re) it is therefore suggested to calculate all friction factors by taking the maximum values of the laminar and turbulent flows using the logic: $\lambda = \text{Max}(\lambda_{lam}; \lambda_{tur})$.

An exception to the above statement are surfaces with a *uniform* roughness where a local minimum can be observed in the curve λ = f(Re). Surfaces with technical roughness do not show this behavior. This minimum should therefore be

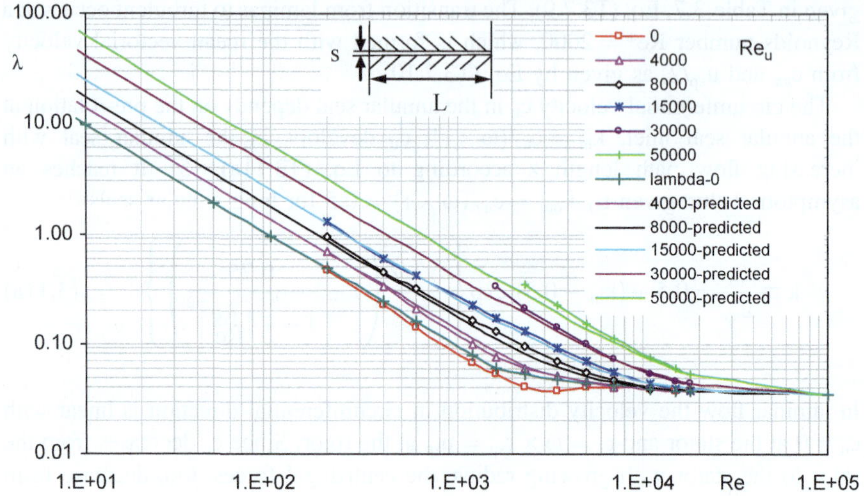

Fig. 3.11 Friction coefficients of smooth annular seals, test data from [36, 41] ε/s = 0.01; L/s = 413; s = 0.315 mm. The curves labeled "predicted" were calculated from Eqs. (T3.7.11) to (T3.7.14)

ignored. Structured surfaces (grooves, serrations and honeycomb) have dimensions greater than the turbulence length scales. Therefore, they do not show the same behavior as a uniform roughness.

To calculate the annular seal leakages, experimentally determined friction coefficients are used for turbulent flow. Figure 3.11 shows measurements on a straight, plain annular seal, Fig. 3.12 an example for a straight serrated seal with large grooves [41] (see Fig. 3.15 for various seal geometries). Figure 3.11 was supplemented by tests from [36] for Re_u = 30,000 and 50,000. As with all turbulent flows, the resistance of an annular seal depends on the roughness of the wall. In the case of small clearances this dependency is particularly high since the *relative* roughness ε/d_h is high due to the small hydraulic diameter.

The required high-pressure differentials and problems of measuring accuracy aggravate the experimental determination of the friction coefficients of annular seals in a wide range of Reynolds numbers. It is therefore frequently necessary to extrapolate the available test data.

The resistance law for the flow in channels, Eq. (1.36), is suitable for extrapolation since the entire range of turbulent flows from hydraulically smooth to hydraulically rough is described by it. The influence of the roughness in this context is expressed by ε/s. This better describes the fine structure of the flow in the annular seal than ε/(2 s) and reflects more accurately the test results at least in this case.[8]

[8]It is open to debate whether ε/s or ε/2 s is more relevant for small clearances. Undoubtedly the hydraulic diameter is d_h = 2 s. Considering the structure of the flow one may rather tend intuitively to select ε/s, which could be viewed as a local obstruction to the flow.

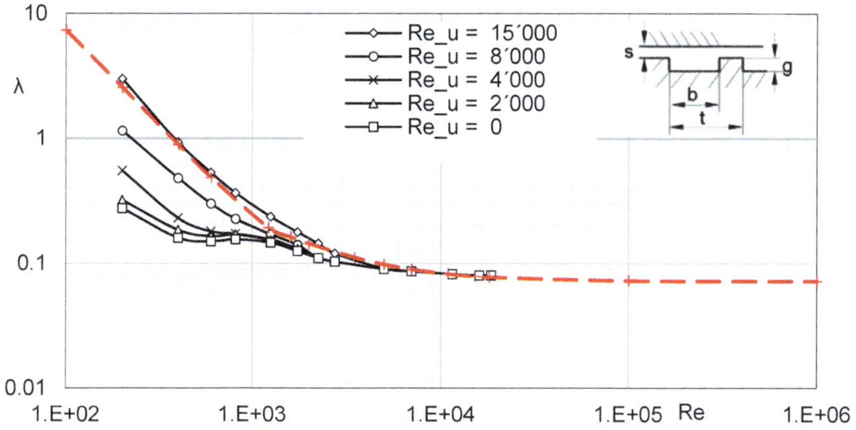

Fig. 3.12 Friction coefficients of serrated annular seal, test data from [41]; s = 0.31 mm; b/t = 0.7; t/s = 16.1; g/s = 3.2

The resistance coefficients calculated in this way apply to a stationary rotor (u = 0) and are therefore termed λ_o, Eq. (T3.7.12). The effect of the rotation in turbulent flow is covered by an experimentally determined factor λ/λ_o, Eq. (T3.7.13), which harbors considerable uncertainties. As can be derived from Eq. (3.11b), $Re_u/Re < 2$ mostly applies to the annular seals on the impeller. Consequently, the correction for the influence of the rotation according to Eq. (T3.7.13) is below 25% in most applications so that this uncertainty is often not really serious.

$$\frac{Re_u}{Re} = d^*_{sp}\sqrt{\frac{\zeta_{EA} + \lambda\frac{L}{2S}}{R_G\,\psi - k^2(1 - d^{*\,2}_{sp})}} \qquad (3.11b)$$

The test data from Fig. 3.11 were corrected for the rotation influence according to Eq. (T3.7.13). In this way λ_o was obtained which can be compared with Eq. (T3.7.12). The results of this calculation demonstrate that the Eqs. (T3.7.12 to T3.7.13) are suitable for extrapolating the test data. This even applies in the laminar range, Fig. 3.13.

The friction coefficients of smooth annular seals and of seals with coarse or fine serrations and isotropic patterns can be estimated from Fig. 3.14. In addition to the friction loss, inlet and outlet losses must be considered in an annular seal. These are covered by ζ_E and ζ_A or by the sum of these coefficients $\zeta_{EA} = \zeta_E + \zeta_A$. Without rotation, $\zeta_E = 0.5$ and $\zeta_A = 1$ would apply to genuinely sharp-edged inlets. As a result of the rotation both values drop so that $\zeta_{EA} = 1$ to 1.2 can be used. The existing tests exhibit a major scatter since the inlet and outlet losses depend on Re and Re_u and even a tiny radius or chamfer at the seal inlet severely reduces the coefficient ζ_E. For example, if the clearance is s = 0.3 mm, a radius of 0.2 mm produces $r/d_h = 0.33$ and reduces the inlet loss coefficient from $\zeta_E = 0.5$ to $\zeta_E = 0.03$ (see Table 1.4).

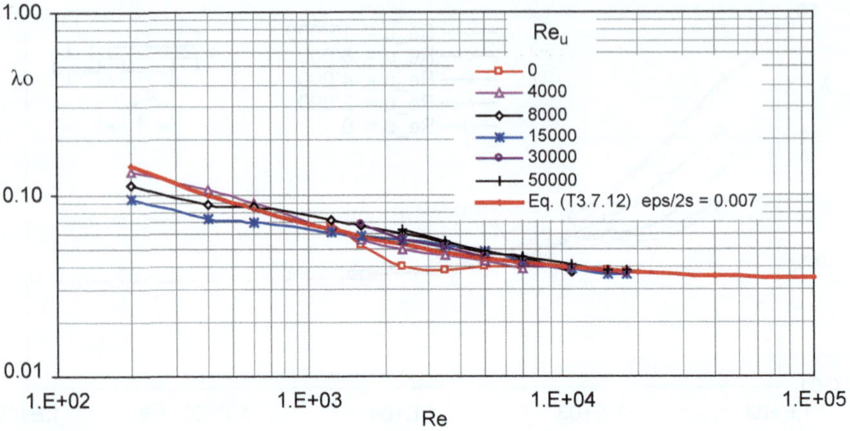

Fig. 3.13 Friction coefficients of smooth annular seals with $u_{sp} = 0$ according to Fig. 3.11 (eps stands for the sand roughness ε)

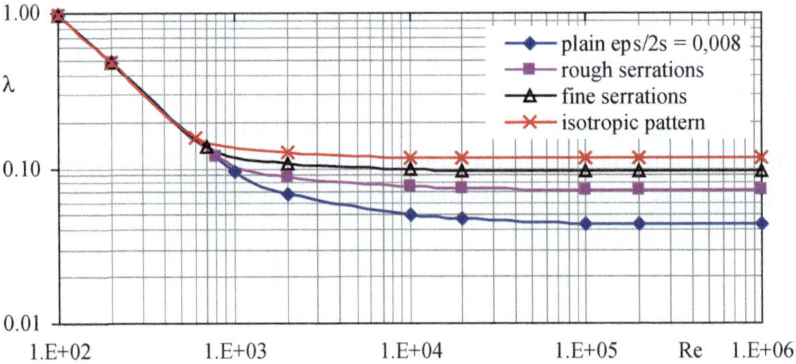

Fig. 3.14 Friction coefficients of annular seals, calculated from Table 3.7 for $Re_u/Re = 2$

In the case of plain seals (without serrations) the greatest possible roughness should be aimed at. However, there are practical limits to this, since the close manufacturing tolerances required for the seal diameters cannot be achieved with very rough surfaces. To reduce the leakage flow, grooves and isotropic patterns have been developed which will increase the resistance coefficient. Large grooves were examined in [41]; fine and coarse grooves in [9]; honeycomb patterns in [4]. The effect of such types of structures is based on increased energy dissipation through turbulent exchange of momentum between the leakage flow and the fluid in the grooves or recesses. A sort of "form resistance" is obtained which causes the transition from hydraulically smooth to fully rough flow being shifted towards lower Reynolds numbers than with a seal having no grooves. For Re > 10,000 it is therefore possible with such geometries, as a first approximation, to assume

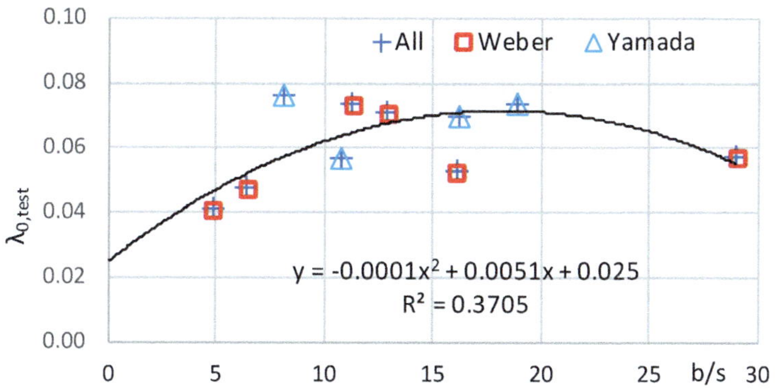

Fig. 3.14.1 Friction factors of serrated seals from test data for $Re_u = 0$

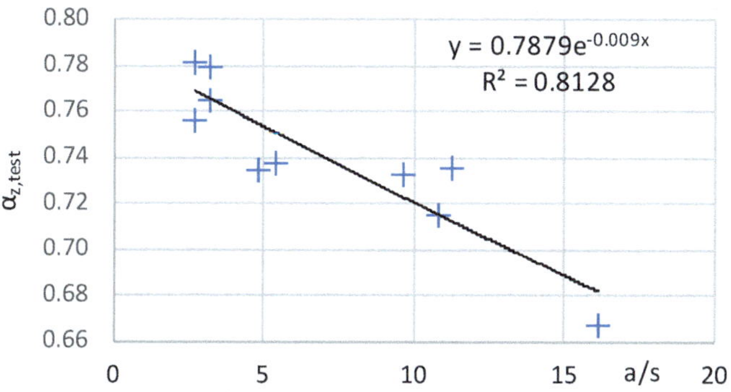

Fig. 3.14.2 Jet contraction factor

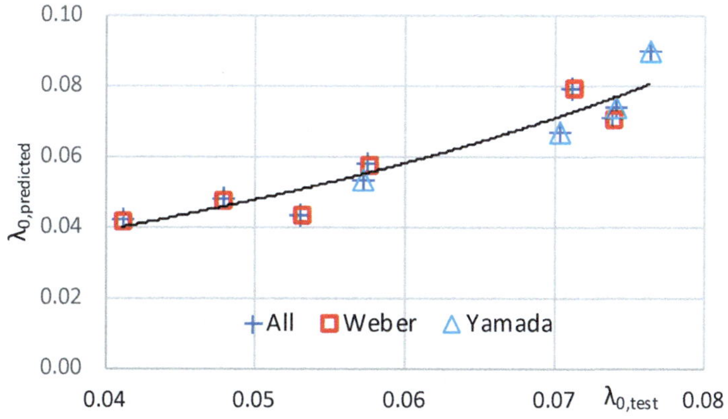

Fig. 3.14.3 Serrated-seal friction factors calculated from Eq. (T3.7.46) compared to test data derived from [41, 47]

Re-independent resistance coefficients for the calculation. As explained above, the influence of rotation on λ/λ_o is relatively small. In many cases it is possible to describe the attainable level of resistance coefficients for Re > 10,000 as follows:

Honeycomb patterns: $\lambda = 0.1-0.18$; Fine grooves: $\lambda = 0.07-0.09$

Hole patterns: $\lambda = 0.07-0.1$; Coarse grooves: $\lambda = 0.04-0.1$

These are only rough indications since the effective resistance coefficients sensitively depend on the surface structure.

Friction factors of seals with deep serrations (g/s > 2):

Grooves with width "b" and smooth portions ("lands") with a = t − b make up the seal; the nominal clearance is "s" (or s_0), see sketch in Fig. 3.12 or Table 3.7(4). A method has been developed for predicting the friction coefficients as function of the groove dimensions; it is based on test data in [41, 47]. The physical model involves the following assumptions and steps: (1) When the fluid enters the gap width s, there is a streamline (or "jet") contraction (similar to the flow through a sharp orifice). The velocity in the resulting vena contracta is $c_{VC} = c_{ax}/\alpha_z$ where c_{ax} is the nominal velocity in the seal $c_{ax} = Q_{sp}/(\pi d_{sp} s)$. (2) When entering a groove, the fluid expands like a plane *wall jet* (formulae given in Table 1.3). (3) The loss in a groove is calculated from Eq. (1.12); i.e. as a shock loss created by decelerating the flow from c_{VC} to c_b where c_b is assumed to half the center-line velocity of the expanding jet at the length x = b. (4) An empirical factor β has been introduced to account for effects not captured by the jet model. (5) The friction loss on the lands is calculated as a plain seal with λ from Eq. (T3.7.12). (6) The friction loss is increased due to the additional surface given by the depth of the grooves. Admittedly, the effect is small and the assumption open to debate, but it is an attempt to enter the groove depth into the model. Overall, the results are very little influenced. (7) The resulting formula for calculating the equivalent friction factor $\lambda_{0,eq}$ of the serrated seal is given by Eq. (T3.7.46). The formula ensures a smooth transition between serrated and plain seals (where a = t and b = g = 0). (8) Eq. (T3.7.46) represents the friction factor for the seal *without rotation* at high Reynolds numbers where the curve levels off as seen in Fig. 3.12. The effect of rotation is subsequently introduced by Eq. (T3.7.13); the dashed curve in Fig. 3.12 shows the curve calculated in this way from Table 3.7(4). (9) The published friction factors have been analyzed by these steps: (A) Test result λ was divided by the factor λ/λ_0 from Eq. (T3.7.13); these test data are plotted in Fig. 3.14.1 as function of the groove width b/s. (B) The friction factor λ_{smooth} for the un-serrated seal was taken from the applicable tests (its influence is moderate). (C) The contraction factor was calculated by solving Eq. (T3.7.46) for α_z. The result is plotted in Fig. 3.14.2 as function of the land a/s. The effect of the contraction increases with increasing a/s (surprisingly?). The resulting correlation, Eq. (T3.7.45) was introduced in Eq. (T3.46) to calculate $\lambda_{0,eq,test}$. (D) A slight adaptation of the empirical factor β was useful to improve the data fit.

The resulting friction factors are compared in Fig. 3.14.3 to the test data. In view of the many uncertainties (which include the reading of the data from the published graphs), the standard deviation of 9.5% seems acceptable.

Results, findings and recommendations:

- The groove width ratio b/s is relevant for the expansion of the jet. The optimum seems in the range of b/s = 8 to 20, Fig. 3.7.14-2. Considering that the "core length" of the wall jet is $x_1/s = 6.4$ (Table 1.2), the groove width b should be greater than x_1 in order to enable the jet to sufficiently expand. Above b/s > 20, there are too few grooves in a given seal length and the dissipation of kinetic energy per unit length of the seal diminishes.
- The width of the grooves should be as large as possible b/t = 0.75 to 0.9 in order to maximize dissipation by vortices. The minimum land "a" is influenced by manufacturing and mechanical reasons as well as wear of the seal. The minimum land may be $a_{min} = 0.4$ mm.
- Jet expansion in the groove demands a sufficiently large expansion angle ε. With tan ε = g/b and assuming a minimum ε = 10° (see Chap. 1.7) one gets $(g/b)_{min} = 0.18$. The optimum from test data in the literature seems near g/b = 0.25 and g/s = 3 to 6. If the pump is designed for operation at double the new clearances, selecting g/s = 5 to 6 will still provide high friction factors at worn conditions. Too deep grooves do not increase the friction factors because the friction on the outer wall of the groove is then reduced.
- The grooves can be machined with an angle other than 90° to the rotor axis in order to use a slightly trapezoidal tool for cutting the grooves.

The correlations cover a fair range of tested geometries as shown in Table 3.7(4). However, there may be geometries with finer grooves which generate higher flow resistances.

Annular seals are designed in a variety of shapes according to Fig. 3.15:

- Straight plain annular seals require the least manufacturing effort.
- Stepped annular seals with 1 to 2 stages or intermediate chambers offer increased resistance since a head corresponding roughly to one stagnation pressure is dissipated in one chamber: $\zeta_k = 1$ to 1.3. Annular seal length and diameter can possibly be better optimized with stepped seals.
- Z-shaped annular seals provide a slightly higher resistance coefficient ζ_k than stepped seals.
- The multiple seal (labyrinth seal) is the most complex design solution. With this type of seal it must be noted that clearance "a" is designed large compared with clearance s to avoid rotor instabilities (Chap. 10).
- Radial and inclined or diagonal seals are designed according to Chap. 3.6.5.

The required seal clearance depends on considerations of mechanical design. The main objective is to prevent contact between rotor and stator. Design criteria are:

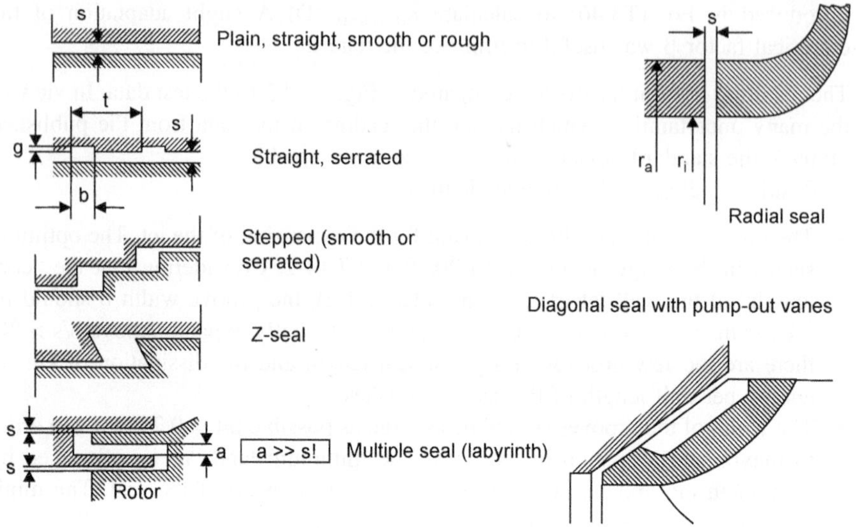

Fig. 3.15 Types of seals

the shaft bending under the rotor weight and radial thrust, possible thermal deformations of rotor and casing and the tendency of the materials used to seize.

The diametrical clearance $\Delta d = 2 \times s$ is in the order of:

$$\frac{\Delta d}{d_{sp}} = 0.004\left\{\frac{d_{Rref}}{d_{sp}}\right\}^{0.53} \quad \text{with} \quad d_{Ref} = 100\,\text{mm} \tag{3.12}$$

Equation (3.12) roughly corresponds to the minimum clearance values from [N.6] *plus* tolerances according to half the tolerance fields H7/h7. Executed clearances as a rule are in the range of ±30% of the clearances calculated from Eq. (3.12). Some pump manufacturers use internal standards, for example defined as 75% [46] to 80% of clearances from [N.6]. In [N.6] an allowance of 0.125 mm is specified for materials with a high tendency to seize and in pumps for media above 260 °C. This allowance is added to the clearances according to Eq. (3.12).

When starting pumps for hot media without preheating, the differential heating of rotor and stator must be considered. The impeller wear rings heat up faster than the casing because of the thin material thicknesses involved. The clearances are thus reduced due to thermal expansion.

All types of seals can be designed either with smooth surfaces or with many different types of grooves or isotropic patterns (honeycomb or hole patterns) in order to increase the flow resistance. A "smooth" surface in terms of manufacturing is not the same as "hydraulically smooth". The roughness must always be considered when calculating the leakage loss, provided the flow is turbulent. For example, machining a seal with a clearance of 0.4 mm to quality N7 (i.e. a sand

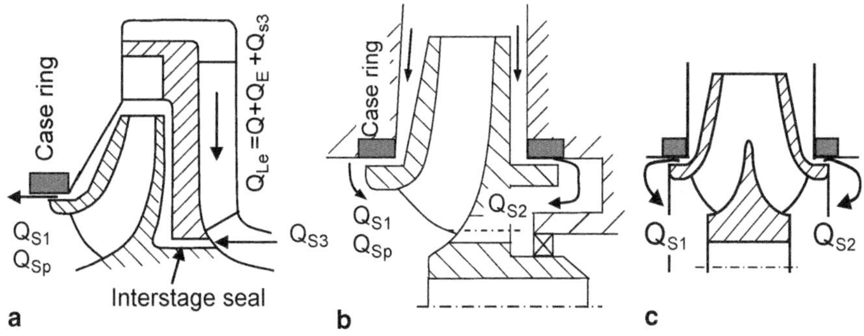

Fig. 3.16 Leakage flows: **a** multistage pump; **b** impeller with balance holes; **c** double-entry impeller

roughness of 4 μm) gives a relative roughness of $\varepsilon/s = 0.01$, which is hydraulically very rough, see the graph in Table 1.6).

The seal diameter is chosen as small as possible, the seal length is typically $L_{sp} = (0.1 \text{ to } 0.14) \times d_{sp}$.

Calculation procedure for leakage losses (Table. 3.7)**:** To start with, the pressure difference acting over the seal must be established. The static pressure H_p prevailing at the impeller outlet can be calculated from Eq. (T.3.7.1). In many cases H_p can be estimated accurately enough with the help of the degree of reaction from $H_p = R_G \times H$. For moderate specific speeds $R_G = 0.75$ is a good assumption. With radially inward directed leakages (Q_{s1} and Q_{s2} in Fig. 3.16) the pressure between the impeller outlet and the seal drops due to the rotation of the fluid in the impeller sidewall gap (Chap. 9.1). This is described by the rotation factor $k = \beta/\omega$ (β is the angular velocity of the fluid).

As per Eq. (1.27) the pressure reduction in the impeller sidewall gap can be calculated from $\Delta p = \frac{1}{2}\rho \times \beta^2 \times \left(r_2^2 - r_{sp}^2\right) = \frac{1}{2}\rho \times k^2 \times u_2^2 \times \left(1 - d_{sp}^{*\,2}\right)$; Eq. (T3.7.3) is obtained for determining the pressure difference over the seal at the impeller inlet.

The greater the fluid rotation (i.e. the greater k) the greater is the pressure drop in the sidewall gap and the smaller is the pressure difference over the seal and the resulting leakage flow. Since k increases with growing leakage (Chap. 9.1), a *radially inward* leakage depends on itself and partly limits itself.

However, if the leakage Q_{s3} flows *radially outward* as in the case of the interstage seals of multistage pumps, the pressure differential due to the rotation in the impeller sidewall gap is added according to Eq. (T3.7.4) to the pressure recovery in the diffuser $\Delta H_{Le} = H - H_p = (1 - R_G) \times H$. Again the leakage tends to limit itself, since a large flow through the sidewall gap has the effect of reducing the fluid rotation or the pumping action of the rear shroud and thus the pressure difference across the seal.

The rotation factor k can be estimated in different ways: (1) according to Eq. (T3.7.2) as a function of the Reynolds number and the seal geometry; (2) by reading from Figs. 9.8 or 9.9; (3) according to Eq. (T9.1.4); (4) by the calculation according to Table 9.1.

The leakage through the axial thrust balance device of a multistage pump (Chap. 9.2.3) is generally returned to the suction nozzle. In that case the head generated by (z_{st} − 1) stages, plus the static head calculated from Eq. (T3.7.3), is throttled by the balance device. The balancing flow must therefore be calculated with the head difference defined by Eq. (T3.7.5).

The axial velocity c_{ax} in the annular seal has to be calculated iteratively according to Eq. (T3.7.6) since the friction coefficient depends on the Reynolds number (i.e. on c_{ax}). Stepped seals or multiple seals are calculated as resistances connected in series according to Chap. 1.9. To this end, the resistances of the individual steps or seals are added as per Eq. (T3.7.6).

Pumping grooves: Instead of *annular* grooves it is also possible to machine *threads* on the rotating internal cylinder which work against the pressure to be sealed ("pumping grooves"). The leakage flow decreases with falling c_{ax}/u_{sp}. Pumping grooves always give less leakage than plain seals, however they make sense only when $c_{ax}/u_{sp} < 0.7$. Annular grooves similar to Fig. 3.12 are more effective above this limit. The ratio of seal length to clearance should be above $L_{sp}/s = 50$. Pumping grooves are geometrically similar if the following parameters are identical: pitch angle α, groove width "a" to pitch (a + b): a/(a + b), groove depth h to clearance s : h/s and pitch to clearance: (a + b)/s. For given values of these parameters the friction coefficient is independent of the clearance.

Since the selection of favorable parameters depends on the ratio of the axial velocity in the seal to the circumferential speed of the rotor, an iterative calculation is necessary. Table 3.7(3) supplies all the required information and equations derived from tests in [18]. Equations (T3.7.39) and (T3.7.40) apply to the geometric parameters specified in Eqs. (T3.7.35) to (T3.7.38) from which a deviation of ±10% is possible without Eqs. (T3.7.39) and (T3.7.40) losing their validity.

Assessment and optimization of the annular seal leakage:

- The seal clearance is the most important parameter since the leakage flow increases approximately with the 1.5th power of the clearance. Seemingly minor deviations from the assumed dimensions have a noticeable effect on the efficiency at low specific speeds. In the case of a pump with $n_q = 15$, for example, a clearance change by 0.03 mm from 0.25 to 0.28 mm means a reduction of efficiency of one percentage point. Usually the clearance calculated from the *nominal* diameters corresponds to the *minimum* clearance. The (statistically) expected *average* clearance is larger than this minimum. Therefore, leakage calculations should be based on an average clearance which is calculated by adding half the tolerance field to the minimum clearance.
- The clearance has practically no effect on the friction coefficient, provided the relative roughness ε/s is used in the calculation. For Re > 10,000, the roughness

must always be considered for leakage flows, since the actual ratio of ε/s is usually quite large.

- When optimizing the seal geometry in terms of shape (Fig. 3.15), seal length and surface structure, the rotor-dynamic characteristics must be considered in addition to the efficiency (Chap. 10).
- With turbulent flow an eccentricity of the inner cylinder can be neglected. In contrast, with laminar flow the resistance of a seal drops with increasing eccentricity, Eq. (T3.7.11).
- The efficiency impairment caused by the seal leakage decreases with growing specific speed since the leakage flow is referred to an increasingly larger useful flow rate (Q_{sp}/Q_{opt}) and because the pressure difference over the seal drops. Equation (T3.5.10) provides a relationship for estimating the leakage. Above $n_q > 60$ the seal loss is roughly 1% or below (and 2% if balance holes are provided).
- Close interaction exists between leakage flow, pressure distribution in the impeller sidewall gap, disk friction and main flow (Chap. 9.1). With unusually high leakage, i.e. with clearances appreciably above those calculated from Eq. (3.12), the calculations should be done according to Chap. 9.1 (Table 9.1).

Because of the many influencing factors and the limited amount of relevant test data, leakage calculations are subject to uncertainties of approximately ±30% due to the following factors:

- Turbulence and roughness structure (and their interaction)
- Pressure difference over the seal (impeller sidewall gap flow)
- Effective clearance which can vary over the circumference and the length because of manufacturing tolerances.
- The actual clearance can be influenced by: (1) the stator being at a slightly different temperature from the rotor; (2) differences in material implying dissimilar coefficients of thermal expansion between rotor and stator; (3) centrifugal force: the widening of the rotating part reduces the clearance
- Stator deformation under load
- Inlet and outlet loss coefficients ζ_E and ζ_A depend on the pre-rotation at the seal entry. The edges are nominally sharp but even tiny rounding or chamfering will reduce the loss coefficients.

3.6.3 Power Loss Caused by the Inter-stage Seal

The leakage flow through seals placed between the stages of a multistage pump is calculated according to Table 3.7(1). However, the relative loss of efficiency is smaller than with leakages around the impeller, because the head increase in the diffuser only amounts to about a quarter of the work of the impeller. Consequently, the loss obtained by throttling the leakage flow Q_{s3} over the head differential ΔH_{s3}

is calculated as power loss in this seal: $P_{s3} = \rho \times g \times Q_{s3} \times \Delta H_{s3}$. This loss is *not* incorporated into the volumetric efficiency. It scarcely depends on the operating point of the pump. The ratio P_{s3}/P_u decreases with growing specific speed. Equation (T3.5.11a) serves to estimate this loss. Often this approximation is sufficient. For more accuracy ΔH_{s3} is calculated from Eq. (T3.7.4), the leakage according to Eqs. (T3.7.7–13) and the power loss P_{s3} from Eq. (T3.5.11a).

3.6.4 Leakage Loss of Radial or Diagonal Seals

Seals with radial or diagonal flow (Fig. 3.15) are used for centrifugal pumps operating with fluids which are laden with abrasive solids. Such designs avoid flow deflections upstream or downstream of the seal which would result in local metal loss due to abrasion.

The flow in a radial or diagonal seal is affected by centrifugal forces generated in the seal through the circumferential component of the flow velocity. This effect is covered by the rotation factor k as in Chap. 3.6.2. The pressure difference Δp present across the seal is suitably corrected with an additional element $\frac{1}{2}\rho \times (k \times \omega \times r_a)^2$. If the flow is from the inside to the outside, this term is positive and the leakage increases accordingly. Conversely, if the fluid flows radially inwards, centrifugal forces oppose the flow. The pumping action in the seal thus reduces the leakage.

The leakage loss of radial-flow seals can be calculated according to [40]:

1. The pressure difference across the seal is determined from Table 3.7(1) on the analogy of Eqs. (T3.7.1 to T3.7.4).
2. The friction coefficients $\lambda = f(Re, \varepsilon)$ must be determined iteratively from Eqs. (T3.7.12) and (T3.7.13) in the same way as for axial seals. The Reynolds numbers are calculated with radial and circumferential velocities, which are determined as the geometric averages between inlet and outlet, Eq. (T3.7.21). For this reason the radius ratio $a_r = r_a/r_i$ appears in this equation.
3. The radial velocity in the seal and the leakage are obtained from Table 3.7(2), Eqs. (3.7.22) and (T3.7.23).

When pumping abrasive suspensions, the clearances can increase considerably during service because of wear. The consequent decrease in head and efficiency is not only caused by the volumetric loss. Large leakage flows also generate a pre-swirl at the impeller inlet and thereby reduce the head according to the Euler equation, Table 3.3. Strong pre-swirl has an impact on cavitation so that $NPSH_R$ is apt to increase appreciably. The leakage flow can be reduced through pump-out vanes (as shown in Fig. 3.15). The effect of the leakage on head, efficiency and $NPSH_R$ can be roughly estimated according to Eqs. (T3.7.18 to T3.7.20).[9]

[9] In order to derive these relationships, the measurements in [21] were recalculated into a more general form.

3.6.5 Leakage Losses in Open Impellers

In the case of open or semi-open impellers (see Table 2.1) fluid flows from the pressure to the suction surface of the blades through the clearance between casing and blades. In the process, the kinetic energy of the leakage flow is largely dissipated and the main flow deflection is reduced (the slip factor dropping accordingly). Consequently, efficiency and head suffer with growing clearance. However, the kinetic energy of the leakage flow is fed into the low-energy zone on the suction surface (the "wake", as per Chap. 5). This can have a favorable effect on the stability of the head-capacity characteristic of a semi-axial pump. The pumping of free gas is also facilitated due to the mixing of the leakage flow with the main flow, because the risk of gas accumulations on the suction surface of the blades is reduced (Chap. 13.2). The leakage flow reinforces the secondary flow in the channel, imparting greater three-dimensionality to the flow in the impeller.

Very large clearances allow some fluid to flow back from the volute into the gap between the blades and the casing and consequently back into the impeller. The conditions are then similar to those in a vortex pump according to Fig. 2.20. The sensitivity of the pump to a clearance change depends on the circumferential component of the fluid flowing back from the volute. This interaction with the collector further contributes to the uncertainty of predicting the drop in performance due to the leakage.

Because of the interaction between the leakage and the main flow, a strict separation of the leakage losses from the hydraulic losses and the friction on the casing wall is not possible. In the case of small clearances ($s/b_2 < 0.01$), such as are used in semi-axial pumps with high specific speeds, it therefore appears sensible not to determine the leakage losses separately but to include them in the hydraulic efficiency.

The influence of the clearance on head, power consumption and efficiency was measured in several experiments; [7, 8, 22, 29] are referenced as recent works. According to these investigations the hydraulic behavior of pumps with open impellers can be characterized as follows:

- The head of a pump drops in a linear fashion with increasing clearance according to $(H_{(s=0)} - H_{(s)})/H_{(s=0)} = g \times s$. The subscript s = 0 describes the conditions extrapolated to zero clearance and subscript s the behavior with finite clearance. The gradient g describing the decrease in head greatly depends on the impeller design. There is no perfect correlation between gradient g and the impeller parameters since the three-dimensional flow in the impeller cannot be described with simple geometrical parameters. Different impellers react to an increase in clearance with quite different sensitivities. Performance impairment is mitigated when the number of blades and/or the outlet angle is increased. Impellers with a large number of radial blades and $\beta_{2B} = 90°$ therefore react less sensitively to the clearance in spite of the low specific speed. The insensitivity of blades with large outlet angles can be explained by the velocity in the gap having a high circumferential component whose energy contributes to the flow deflection. In the case of small blade angles, in contrast, the leakage flow has a radially inward directed component opposing the main flow. From the previously mentioned test results it is possible to derive a correlation, Eq. (T3.7.15), for the head decrease. Equation (T3.7.15) describes the above effects qualitatively quite well, but has a considerable scatter.

- Because of the lower flow deflection, the power consumption also drops with increasing clearance. The same applies to the efficiency which drops more than the power consumption. These effects are demonstrated by Eqs. (T3.7.16 and 3.7.17) according to which the efficiency reduction amounts to approximately 2/3 of the head decrease. Accordingly, the relative power-consumption diminishes by roughly 1/3 of the head reduction.
- The head decreases throughout the load range. The relative impairment increases with growing flow rate. The shut-off head drops approximately half as much as the head at the best efficiency point.
- The best efficiency point is shifted towards smaller flow rates with increasing clearance.

According to [B.1] it is possible to very roughly estimate the effect of the clearance on head and efficiency for radial, semi-axial and axial impellers from:

$$\frac{\Delta\eta}{\eta} = \frac{\Delta H}{H} = \frac{\Delta Q}{Q} = \frac{s}{b_2} \tag{3.13}$$

Semi-axial pumps with $n_q > 70$ to 100 generally have semi-open impellers (without a front shroud) because a large-diameter shroud would generate high disk friction losses and considerably impair the efficiency (typically by 3%).

3.6.6 Mechanical Losses

The mechanical losses P_m are generated by the radial bearings, the axial bearing and the shaft seals. Occasionally P_m includes auxiliary equipment driven by the pump shaft. These losses depend on the design of the pump, i.e. on the selection of anti-friction versus journal bearings or stuffing boxes versus mechanical seals.

The mechanical efficiency of large pumps is around 99.5% or even above. In contrast, the mechanical losses of small pumps (say below 5 kW) can use up a considerable portion of the coupling power. Examples are process pumps which are often equipped with dual mechanical seals.

The mechanical losses can be estimated according to Eq. (T3.5.6a) if more accurate information is not available. But usually it is best to consult the documentation of the manufacturer of the mechanical equipment. Exact analysis is indicated in the case of special designs such as wet motors or magnetic couplings.

When pump performance tests are done at a speed lower than that in the plant, it must be taken into account that the mechanical losses have a lower exponent of the speed than the power of the pump: $P_m \sim n^x$ with x = 1.3 to 1.8 is expected rather than $P \sim n^3$. In other words, the mechanical efficiency *of a given pump* decreases with the speed.[10]

3.7 Basic Hydraulic Calculations of Collectors

At the impeller outlet the fluid has the circumferential velocity $c_{2u} = c_2 \times \cos \alpha_2$ and the specific kinetic energy $E_{kin} = \frac{1}{2}c_2^2$. Up to moderate specific speeds, $\cos \alpha_2$ is close to 1.0. Equation (T3.3.9) yields $c_{2u}/u_2 = \frac{1}{2}\psi_{th}$ for swirl-free inflow. The theoretical specific work is $Y_{th} = \frac{1}{2}\psi_{th} \times u_2^2$. Using these expressions, we can derive the kinetic energy at the collector inlet $E_{kin}/Y_{th} = \psi_{th}/4$ as a fraction of the energy transferred by the impeller. An acceptable efficiency can consequently be achieved only if the kinetic energy available at the impeller outlet is effectively decelerated in the diffusing elements. The collector design attempts to convert the greatest possible part of E_{kin} into static pressure.[11]

In Table 2.1 the following diffusing elements were introduced: volute casings, vaned or un-vaned diffusers or concentric annular chambers, all of which can be designed by relying on similar principles.

According to Newton's law of inertia, a mass retains its momentum if not prevented by external forces to do so. As a consequence, the fluid has the tendency to maintain its angular momentum (its "swirl") $\rho \times Q \times r_2 \times c_{2u}$ downstream of the impeller unless it is influenced by structures or wall friction. In an essentially vaneless casing the flow thus follows the law $c_u \times r = c_{2u} \times r_2$ = constant (apart from friction effects). The circumferential component of the flow velocity in any element of the casing cross section is therefore calculated from $c_u = c_{2u} \times r_2/r$. The diffusing elements have to be designed so that the flow field in the collector is compatible with the conservation of angular momentum. If the angular momentum is conserved, rotational symmetry around the impeller is achieved and interaction between the flows in the collector and impeller is minimized.

The diffusing elements are dimensioned for the flow rate $Q_{Le} = Q_{opt} + Q_{s3} + Q_E$ where Q_E is the volumetric flow through a device for axial thrust balancing and Q_{s3} the leakage of the inter-stage seal.

[10] The mentioned exponent does not imply a contradiction to Eq. (T3.5.6a) which refers to different pumps with optimally adjusted mechanical equipment, while the scaling of the mechanical losses P_m of a given pump assumes given mechanical components which are over-dimensioned at reduced speed.

[11] Exceptions are axial pumps with very high specific speeds, which under certain conditions, have to be designed without diffusing elements, Chap. 7.6.6.

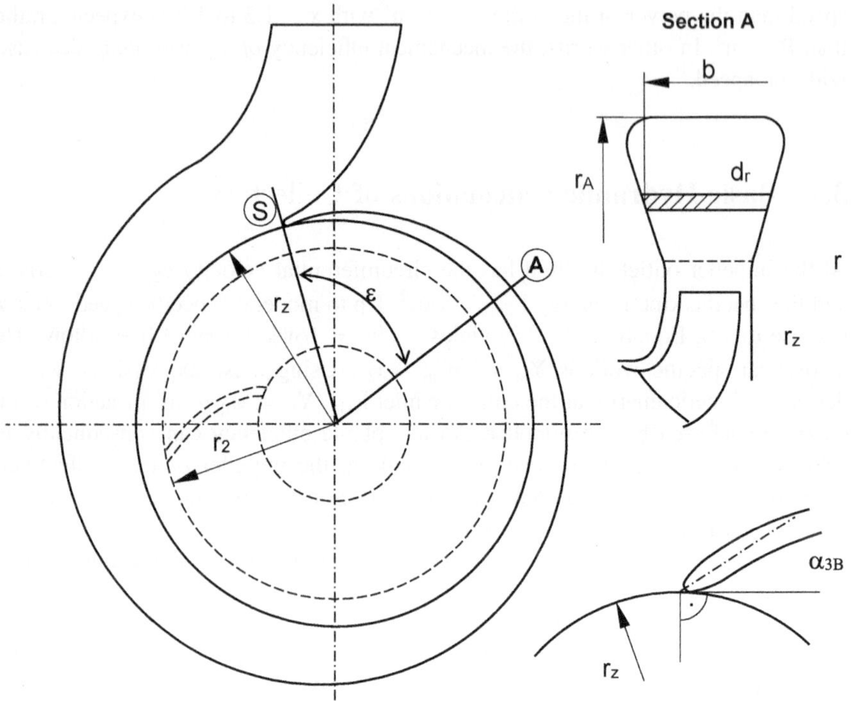

Fig. 3.17 Volute casing

Volute casing: Consider any cross section of a volute located at the circumferential position ε°, Fig. 3.17. The fluid $dQ = c_u \times b \times dr = c_{2u} \times r_2/r \times b \times dr$ flows through an area element $dA = b \times dr$ on the radius r in the section A. The flow rate through the entire cross section is obtained by integration $Q(\varepsilon) = c_{2u} \times r_2 \times \int b/r \times dr$. From the start of the volute (point S in Fig. 3.17) to the cross section A, the impeller discharges the flow $Q(\varepsilon) = Q_{opt}\, \varepsilon°/360°$ in the design point. In order to allow the fluid to flow according to $c_u \times r = c_{2u} \times r_2 =$ constant, cross section A must therefore be designed in such a way that Eq. (3.14) is satisfied:

$$\int_{r_z}^{r_A} \frac{b}{r}\,dr = \frac{Q_{opt}\,\varepsilon°}{360°\, c_{2u}\, r_2} \tag{3.14}$$

Experience confirms that the losses become approximately minimal if a volute casing is dimensioned so that the fluid at the design point is able to flow in accordance with the angular momentum conservation, see also Chap. 4.2. This means that the cross sections of the volute casing - and most importantly the end cross section or "throat area" - have to be selected according to Eq. (3.14). The throat area then follows from Eq. (3.15):

$$Q = z_{Le} \int c_{2u} \frac{r_2}{r} b\, dr \quad \text{or} \quad Q = c_{2u}\, r_2\, J_{SP} \quad \text{with} \quad J_{SP} = z_{Le} \int_{r_z}^{r_A} \frac{b}{r} dr \tag{3.15}$$

z_{Le} is the number of partial volutes forming the casing. Most frequently used are: single volutes with $\varepsilon = 360°$ wrap angle (i.e. $z_{Le} = 1$) and double volutes with 180° each and $z_{Le} = 2$. Occasionally three or four volutes are used to reduce the size of the casing. Double volutes are also designed with wrap angles $\varepsilon_{sp} < 180°$ where $z_{Le} = 360°/\varepsilon_{sp}$ must be used accordingly for each partial volute.

The swirl $r_2 \times c_{2u}$ at the volute inlet is obtained from Eqs. (T3.2.7) or (T3.7.8). If a vaned or vaneless diffuser is installed upstream of the volute, its outlet swirl $c_{4u} \times r_4$ must be inserted into Eqs. (3.14) or (3.15).

The volute cutwater forms a camber angle α_{3B} with the circumferential direction which must be selected in accordance with the approach flow angle. It is obtained from Eqs. (3.16) to (3.19) that apply to both volutes and diffusers. The incidence is selected in the range of $i_3 = \pm 3°$; for volutes set $\tau_3 = 1.0$.

The meridional velocity c_{2m} at the impeller outlet cannot expand suddenly to the inlet width b_3– in particular if the ratio b_3/b_2 is large. Nevertheless, b_3 is used in Eq. (3.16) with the goal to design the diffusor vanes or cutwaters with low inlet angles. This is advantageous for partload operation in that the tendency to flow separation at the cutwater is reduced.

Meridional component with blockage (Table 0.1); for volutes set $\tau_3 = 1.0$.

$$c_{3m}' = \frac{Q_{Le}\, \tau_3}{\pi\, d_3\, b_3} \tag{3.16}$$

circumferential component according $(d_z = d_3)$: to conservation of angular momentum

$$c_{3u} = c_{2u} \frac{d_2}{d_3} \tag{3.17}$$

approach angle with blockage:

$$\tan \alpha_3' = \frac{c_{3m}}{c_{3u}} \tag{3.18}$$

vane inlet angle:

$$\alpha_{3B} = \alpha_3' + i_3 \tag{3.19}$$

It is advisable to shape the cutwater as an elliptical profile, which is less sensitive to changes in incidence (or pump flow rate). The profile can be applied either symmetrically as in Fig. 3.17 or be made asymmetric to obtain small incidences at partload. Any profiling must be made very short to keep stresses in the cutwater within acceptable levels. Not much profiling is possible with small pumps.

The diffuser which usually follows the actual volute must be designed according to Chap. 1.6. For hydraulic and design reasons, a distance is required between the impeller outlet and the cutwater. In the space thus created the fluid flows according to the laws of the vaneless diffuser, but the strong secondary flow created in the volute has an impact on the wall-near flow.

Diffuser: As shown by Fig. 3.18, a diffuser consists of z_{Le} vanes forming triangular inlet sections and closed channels ("diffusing channels" as per Table 0.2), which can be calculated according to Chap. 1.6. The flow through the triangular inlet section can be imagined to behave in a similar way as a partial-volute. The smaller the number of diffuser vanes, the more this is the case. When the number of diffuser vanes is large the flow does not truly follow the relationship $c_u \times r = c_{2u} \times r_2 =$ constant. The diffuser throat area at a_3 cannot be designed strictly according to Eq. (3.15). Nevertheless Eq. (3.15) forms a physically sound basis to calculate the throat area, if an appropriate correction factor f_{a3} is applied, Chap. 7. For a given inlet width b_3, the diffuser inlet area can thus be determined from Eq. (3.20).

$$a_3 = f_{a3}\frac{d_3}{2}\left\{\exp\frac{2\,Q_{Le}}{z_{Le}\,b_3\,d_2\,c_{2u}} - 1\right\} \tag{3.20}$$

This relationship corresponds to the integral of Eq. (3.15) for a volute with rectangular cross section. As for the rest, the diffuser can be calculated according to similar principles as a volute: (1) Blade angle α_{3B} according to Eqs. (3.16) to (3.19) where $i_3 = \pm 3°$; (2) Diffusing channel according to Chap. 1.6; (3) Profiling as described above (for details see Chap. 7).

Vaneless diffuser: In a vaneless diffuser, depicted in Fig. 3.19, the flow is able to develop according to the angular momentum conservation and the pressure increases in radial direction following Eq. (1.28). The meridional component is calculated from continuity as per Eq. (3.21).

$$c_m = \frac{Q_{Le}}{2\,\pi\,r\,b} = \frac{c_{3m}\,r_2\,b_3}{r\,b} \tag{3.21}$$

Setting $c_u = c_{2u} \times r_2/r$ and neglecting friction, the flow angle is obtained from the following equation:

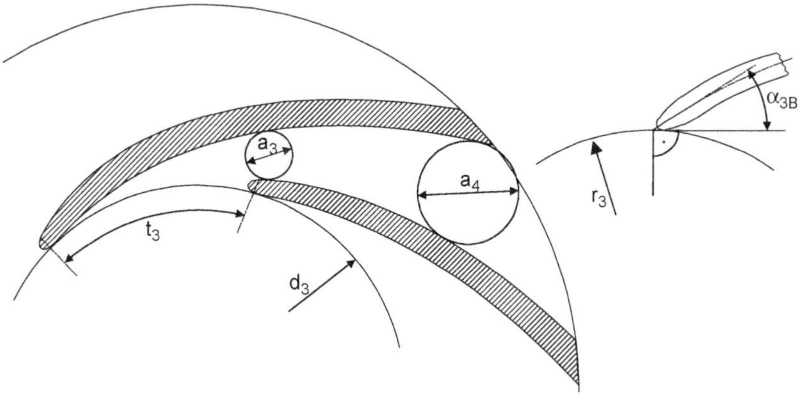

Fig. 3.18 Vaned diffuser

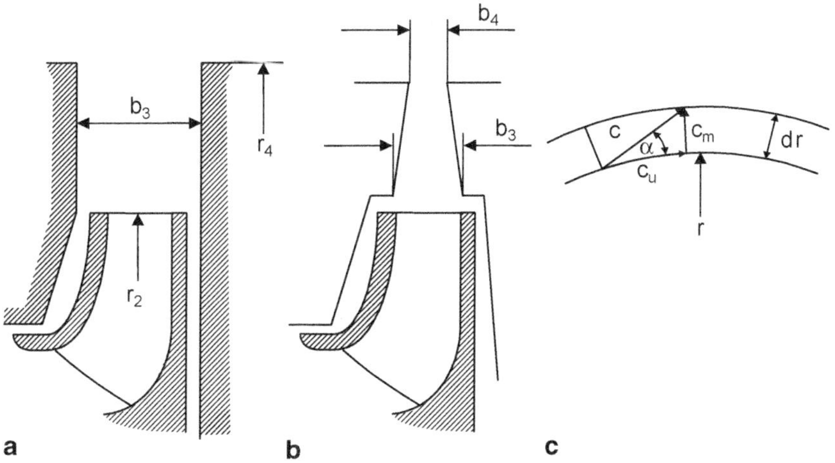

Fig. 3.19 Vaneless diffuser. **a** parallel walls; **b** conical walls; **c** velocity vectors

$$\tan\alpha = \frac{c_m}{c_u} = \frac{c_{3m} b_3}{c_{2u}\, b} = \tan\alpha_3\, \frac{b_3}{b} \tag{3.22}$$

Equation (3.22) shows that the flow angle remains constant in a vaneless diffuser of constant width as long as friction is neglected.

In reality, the circumferential component c_u drops due to wall friction in comparison with the angular momentum conservation employed without shear stresses. As a result, the flow angle grows with increasing radius. According to [B.1] this follows from the relationship:

$$\frac{c_u r}{c_{2u} r_2} = \left\{1 + \frac{c_f\, r_2}{b_3 \tan \alpha_3}\left(\frac{r}{r_2} - 1\right)\right\}^{-1} \tag{3.23}$$

The flow angle in viscous flow can be deduced from Eq. (3.23) leading to:

$$\tan \alpha_4 = \tan \alpha_3 + c_f\, \frac{r_2}{b_3}\left(\frac{r_4}{r_2} - 1\right) \tag{3.24}$$

The meridional velocity results from continuity and is therefore not affected by the friction.

The friction losses in a vaneless diffuser of constant width can be determined approximately as follows: In a ring element (as shown in Fig. 3.19c) the velocity is $c = c_u/\cos \alpha = c_{2u} \times r_2/(r \times \cos \alpha)$ and the wall shear stress is $\tau = \frac{1}{2}\rho \times c_f \times c^2$. Thus, the power $dP_d = c \times dF = \frac{1}{2}\rho \times c_f \times c^3 \times dA$ is dissipated, Chap. 1.5. Integrating dP_d from r_2 to r_4 with $\zeta_{LR} = 2 \times P_d/(\rho \times Q \times u_2{}^2)$ the coefficient of the friction losses ζ_{LR} according to Eq. (T3.8.23) is obtained which comprises *both* sides of the vaneless diffuser.

Equation (T3.8.23) shows that the friction losses *diminish with growing flow rate* since the circumferential component c_u drops. At the same time the flow angle α increases and the friction path length is shortened. For this reason, the best efficiency point shifts towards a higher flow rate if a vaneless diffuser is employed in place of a volute or a diffuser.

There is an optimum flow angle at the impeller outlet that has to be realized in order to place the efficiency optimum at the specified design flow rate. This means that the width b_2 or b_3 cannot be selected arbitrarily.

Since c_u drops with increasing flow, the pressure recovery drops as well. Hence a vaneless diffuser is a stabilizing element - for as long as the flow does not separate. The smaller the outflow angle α_2 from the impeller, the greater the losses become in the vaneless diffuser. Consequently, a vaneless diffuser yields poor efficiencies with low specific speeds. If the vaneless diffuser is designed so that its width decreases from the inside to the outside ($b_4 < b_3$), the flow angle according to Eq. (3.22) grows with increasing radius. Flow path length and losses diminish accordingly.

If the vaneless diffuser width b_3 is designed larger than the impeller outlet width b_2, shock losses occur due to deceleration of the meridional velocity from c_{2m} to c_{3m} which can be calculated according to Eq. (1.12): $2g \times \Delta H = (c_{2m}{}' - c_{3m})^2$. The corresponding loss coefficient is shown as Eq. (T3.8.24) in Table 3.8; it includes the effect of the blade blockage at the impeller outlet. This sudden deceleration creates a minor pressure recovery that can be calculated from Eq. (1.11). *According to the principle of angular momentum conservation the circumferential component c_{2u} is not affected by the expansion from b_2 to b_3.*

Volute casings and diffusers behave like vaneless diffusers with regard to the expansion from b_2 to b_3.

Annular casing: Annular casings are sometimes used for pumps with low specific speeds or for small pumps. The fluid is discharged from the annular casing via a diffuser. The annular casing differs from the vaneless diffuser in that it is noticeably wider than the impeller outlet and its radial extension accordingly smaller. The deceleration of the meridional velocity from c_{2m} to c_{3m} is high, promoting a vigorous secondary flow and exchange of momentum between the annular casing and the impeller. Since the cross sections in the concentric casing do not correspond to a design according to angular momentum conservation, no circumferential symmetry is obtained at the design point. Consequently, large radial forces are generated as discussed in Chap. 9.3.

3.8 Hydraulic Losses

The hydraulic losses in a pump reduce the useful head according to $H = \eta_h \times H_{th}$ or $H = H_{th} - Z_E - Z_{La} - Z_{Le} - Z_{sp} - Z_A$. They include losses in all components located between the suction and the discharge nozzles, i.e. losses in the inlet casing (Z_E), the impellers (Z_{La}), diffusers (Z_{Le}), volutes (Z_{sp}) and the outlet casing Z_A. Hydraulic losses are generated through friction and vortex dissipation (Chap. 1.5.2):

1. Skin friction losses are the result of shear stresses in the boundary layers on solid structures. Skin friction, which depends on the Reynolds number and the roughness, is important in thin boundary layers and attached flow. These conditions are found in accelerating flows. Friction, Reynolds number and the roughness have less impact on losses in decelerated or separated flow.
2. Flow deceleration leads to thick boundary layers which make the velocity distribution less uniform. In non-uniform flow the exchange of momentum between different streamlines increases due to eddies of various scales. Large eddies break up into small-scale turbulence which finally dissipates into increased molecular movement leading to a slight heating of the fluid.

In the way described under item 2 and in Chap. 1.5.3, non-uniform flows have an altogether profound impact on the pump's hydraulic efficiency. Non-uniform flows are generated by various mechanisms:

- The work transfer by the blades is inherently linked to non-uniform flow over the pitch of the blades. The non-uniformity of the flow increases with the blade loading (i.e. with increasing pressure coefficient and decreasing blade length).
- Incident flow creates local zones with decelerated flow or even separation on the blades ("shock-losses"). This can occur with or without cavitation.
- A sudden deceleration of the flow, when the impeller or diffuser throat area does not match with the approach flow velocity vector (w_{1q}/w_1 and c_{3q}/c_2), leads necessarily to a non-uniform flow distribution in the channel. The flow mechanism loosely resembles the Carnot-shock of a sudden expansion, Fig. 1.3.

- Any kind of curved flow path, notably the curvature of the meridional section and of the blades, creates a non-uniform flow as illustrated by Fig. 1.12.
- Flow separation implies zones of stalled fluid with local recirculation. The stalled fluid blocks part of the cross section available to the flow. The un-stalled fluid is therefore accelerated and may form a sort of jet. High losses are generated through the strong exchange of momentum between the jet-like through-flow and the stalled zone.
- Secondary flows are generated by Coriolis forces, streamline curvature (Fig. 1.12) and centrifugal forces. These effects are discussed in detail in Chap. 5. Another type of secondary flow is created by different boundary layer flows on the blades and the side walls as illustrated by Fig. 1.6. The strength of the secondary flow (or the cross-flow) may be thought of as establishing itself according to the *principle of least resistance*.
- The mixing of the leakage through the seal at the impeller inlet creates a non-uniform flow near the front shroud in terms of the c_m-distribution, local pre-swirl and incidence angles.
- Wake flows downstream of blades or ribs. The wake of the blades is increased by low energy fluid near the suction side of the impeller blades.
- When the boundary layer flow on the hub or shroud enters the impeller blading (or a vaned diffuser) it is decelerated upon approaching the leading edge. Due to the blockage caused by the blade and by stagnating fluid, liquid is displaced away from the wall and side-ways into the channel. Thereby a "horseshoe" vortex is formed, Fig. 3.20. It has some blockage effect, increases non-uniformity of the flow and can induce cavitation and abrasive erosion. Some photographs of erosion caused by horseshoe vortices in sand-laden water) on volute cutwaters and impeller blades are shown in Chap. 14.5.6.
- In semi-open impellers the leakage between the blades and the casing interacts with the main flow in the impeller channels.

These phenomena can hardly be assessed by applying simplified models. For this reason, the hydraulic efficiency is mostly determined from the power balance of a measured pump according to Eq. (T3.5.8). This requires calculating the secondary losses by means of the correlations given in Chap. 3.6. Thus, the hydraulic efficiency is only known once the pump has been built and its performance measured. For designing a pump, it is necessary to rely on earlier tests or correlations which allow a reasonable estimate of η_h.

The hydraulic efficiency calculated from the power balance does not allow any statements as to the contribution of individual pump components to the losses. To answer this question, it can be useful to estimate the loss in individual components. Such calculations have an empirical character since the three-dimensional velocity distributions in the impeller and diffusing elements, which determine both friction and turbulence losses, cannot be described by simple models. Estimations of this type are only meaningful near the best efficiency point. *With pronounced partload recirculation they fail entirely* if the exchange of momentum of the recirculating fluids is not taken into account.

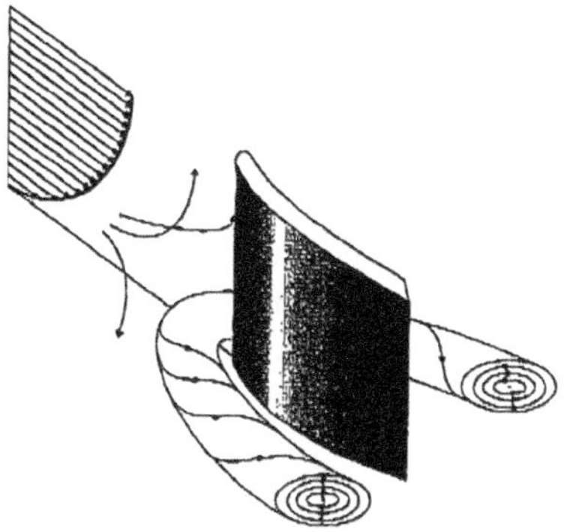

Fig. 3.20 Formation of a "horseshoe" vortex

The **loss model for impellers** according to Table 3.8(1) comprises the following steps and assumptions:

- Equation (T3.8.2) defines an average velocity in the impeller channel which is formed by the throat areas that are defined by a_1 and a_2 at inlet and outlet.
- Equation (T3.8.3) supplies, as a function of the Reynolds number and the roughness, a friction coefficient corresponding to a flat plate in parallel flow. The plate model is preferred to the channel model since undeveloped flows must be assumed because of the short impeller blade lengths. Equation (T3.8.3) includes the roughness. Assessing its effect is frequently required in practice in order to evaluate possible efficiency improvements through additional finishing of the impeller channels.
- Equation (T3.8.5) utilises a dissipation coefficient. According to [B.3] the amount of 0.0015 is added to the friction coefficient since thicker boundary layers, hence greater losses, occur in decelerated flow than on a flat plate in longitudinal flow. The value obtained in this way is further multiplied by an empirical factor containing the relative impeller outlet width. According to this equation, the losses increase considerably with increasing values of b_2^* or specific speed. The empirical factor can thus be interpreted as the effect of uneven velocity distributions and secondary flow.
- Equation (T3.8.6) produces the loss coefficient for the impeller which comprises the friction, deceleration and turbulence effects.
- By means of Eq. (T3.8.9) it is possible to estimate the shock losses at the impeller inlet. This equation describes the deceleration of the vector of the mean inflow velocity w_{1m} (from the velocity triangle) to the velocity w_{1q} in the throat area A_{1q}, calculated from Eq. (T3.8.8). The shock loss in this case is assumed at 30% of a Carnot shock, which supplies plausible results for moderate decelerations. This relationship should not be used for $w_{1q}/w_{1m} < 0.6$.

An alternative model for the friction losses is formulated in Table 3.8(1), Eqs. (T3.8.11–3.8.14). All of the wetted surfaces of the impeller channels, consisting of front and rear shrouds as well as suction and pressure surfaces of the blades, are taken into account by Eq. (T3.8.11). The mean value of the vectors of the relative velocity at inlet and outlet is used as the mean velocity, Eq. (T3.8.12). The dissipated friction power is obtained according to Eq. (1.34) from Eq. (T3.8.13) and the loss coefficient from Eq. (T3.8.14). This calculation corresponds to the procedure for the loss estimation of volute casings (see below). According to Eq. (T3.8.14) the friction losses in the impeller rise with falling specific speed, which is *not* confirmed by tests and numerical calculations.

The **loss model for diffusers** according to Table 3.8 (2) comprises the following steps and assumptions:

- Mixing losses due to an inflow with a non-uniform velocity profile are *not* considered.
- Equation (T3.8.15): calculation of the velocity vector at the impeller outlet
- Equation (T3.8.16): velocity in the diffuser throat area
- Equation (T3.8.17): determination of the pressure recovery coefficient c_p in the diffusing channel. For plane diffusers use Fig. 1.18. For other diffuser types calculate an equivalent conical diffuser from Eq. (1.45) and determine c_p from Fig. 1.19.
- The friction losses between the impeller outlet and the diffuser throat area can be estimated according to Chap. 1.5.1 from Eqs. (1.32) to (1.35) by calculating the energy dissipated through wall shear stresses on the blades and the sidewalls. After several transformations this results in Eq. (T3.8.18).
- Equation (T3.8.19) supplies the pressure loss coefficient for the diffuser including the return channel. If a volute is provided downstream of the diffuser (i.e. in the absence of a return channel), its loss coefficient ζ_{ov} must be set to zero. The first term in Eq. (T3.8.19) takes into account the loss due to the deceleration from c_2 to c_{3q}. The explanations for Eq. (T3.8.9) apply accordingly. The actual flow deceleration is described by the term $1 - c_p - 1/A_R^2$ (Chap. 1.6).
- The losses of the return channels greatly depend on their configuration. With optimal flow design $\zeta_{ov} = 0.2$ could be attainable for a continuous channel, while a whole stagnation pressure $c_4^2/2g$ could be lost with an unfavorable design (i.e. $\zeta_{ov} = 1.0$).

The **loss model for volute casings** including diffuser and pressure nozzle according to Table 3.8 (2) comprises the following steps and assumptions:

- Mixing losses resulting from an inflow with a non-uniform velocity profile have *not* been considered.
- Since the flow rate in the volute casing varies around the circumference, the dissipation due to friction should be calculated according to Eq. (1.34). This leads to Eq. (T3.8.21), which is written as the sum over all elements ΔA of the wetted surface of the volute casing, since the calculation is as a rule performed section by section. The velocity in the volute changes over the cross section according to $c_u \times r$ = constant and over the circumference. The division of the

surface elements must be performed accordingly, depending on the desired accuracy. As with the impeller, the friction coefficient is calculated according to the flat plate model and increased by the value of 0.0015 for decelerated flow. In the case of double volutes, the integration (or summation) is performed for both partial volutes. This sum covers the friction effective on all wetted surfaces; it applies to the *entire* flow rate of the pump (i.e. both partial volutes).

- The actual volute is followed by a diffuser/pressure nozzle. The diffuser is treated similarly to Eq. (T3.8.17): an equivalent conical diffuser is calculated as shown in Chap. 1.6 and the pressure recovery coefficient is determined from Fig. 1.19. The pressure loss in the diffuser is then obtained from Eq. (T3.8.22).
- In the case of double volutes, pressure losses in the outer channel are added to the outer volute. The outer channel can be calculated as a bend according to Table 1.4. If cross sectional expansions are involved, it is calculated with additional diffuser losses according to Fig. 1.19. Since outer and inner volutes have different resistances, the flow distribution over both partial volutes does not correspond to the theoretical value obtained from their wrap angles. The individual flow rates through both volute channels can be determined by applying the concept of parallel flow resistances described in Table 1.5. By means of such loss considerations it is also possible to design the outer channel so that approximately the same flow rates are established in both the outer and inner volutes.
- In multistage pumps the fluid moves from the volute into an overflow channel which directs the liquid to the following stage. Because of the severe curvature of that channel, separations and additional losses occur [13]. The deflection losses can be estimated according to Table 1.4.
- The friction losses in a vaneless diffuser are calculated from Eq. (T3.8.23). This is done with the impeller outlet flow angle α_2 – rather than the angle α_3 from Eq. (3.18) - because the flow cannot expand suddenly from b_2 to b_3.
- The shock losses at the impeller outlet are obtained from Eq. (T3.8.24). These also occur in volute casings. In the case of diffusers, the expansion from b_2 to b_3 is implicitly included in Eq. (T3.8.19). The effect of blade blockage can be considered in a similar way as in Eq. (T3.8.24) by inserting c_2' for c_2.
- Since volute casing, outer channels and diffuser/pressure nozzles can be designed in different ways, the loss calculation sketched above should be adapted to suit the respective requirements.
- The deceleration and shock-losses for off-design conditions can be estimated from the test data discussed in Chap. 4.1.3, which yield the correlation given by Eq. (T3.8.25). In this equation the diffuser (or volute) loss is correlated successfully against the deceleration ratio c_{3q}/c_2. The details of the correlations are discussed in Chap. 4.1.3. When applying Eq. (T3.8.25), the diffuser (or volute) losses are first calculated or the BEP according to Table 3.8 (2); this yields $\zeta_{Le,opt}$ which can be used subsequently as $\zeta_{Le,min}$.

The losses in the pump inlet are usually insignificant. They can generally be neglected in end-suction pumps. If an extended pipe section, a suction bell or a similar component is located between the measuring point in the suction pipe and the impeller inlet, the pressure losses can be calculated according to the customary rules, Tables 1.4 and 1.6. Typical inlet casings of between-bearing pumps have pressure loss coefficients of $\zeta_E = 0.15$ to 0.4 based on c_{1m} (hence $Z_E = \zeta_E c_{1m}^2/2g$), depending on the actual design. According to the measurements in [B.21], the inlet losses decrease with the square of the deceleration of the fluid in the inlet casing following the relationship:

$$\zeta_E = 0.75\left(\frac{d_1^2 - d_n^2}{d_s^2}\right) \tag{3.25}$$

Outlet losses need not be considered in many pumps since the losses in the diffuser/ discharge nozzle are included in the calculation of the volute casing. The losses in the outlet chambers of multistage pumps greatly depend on their design. Typically they are around 1 to 2% of the head of *one* stage.

The procedure described in Table 3.8 has been applied to single-stage end-suction volute pumps in a range of $n_q = 7$ to 85. All pumps were designed for $Q_{opt} = 0.1\ m^3/s$ in a way that they had the same impeller inlet diameter. The factor for decelerating flow was set to zero in Eqs. (T3.8.5) and (T3.8.21). The resultant loss coefficients are plotted in Fig. 3.21. The losses in the volute and the discharge nozzle increase steeply as the specific speed is reduced, while the impeller losses increase almost linearly with n_q.

The hydraulic efficiencies calculated are shown in Fig. 3.22 compared to the values obtained from the statistical data according to Eq. (T3.9.1). Both curves show the same tendency with a flat optimum in the range of $n_q = 30$ to 60.

In view of the scatter of the statistical data, the agreement between the predicted and statistical hydraulic efficiencies appears to be quite good. The same applies to higher design flow rates, while the discrepancies increase at lower flow rates.

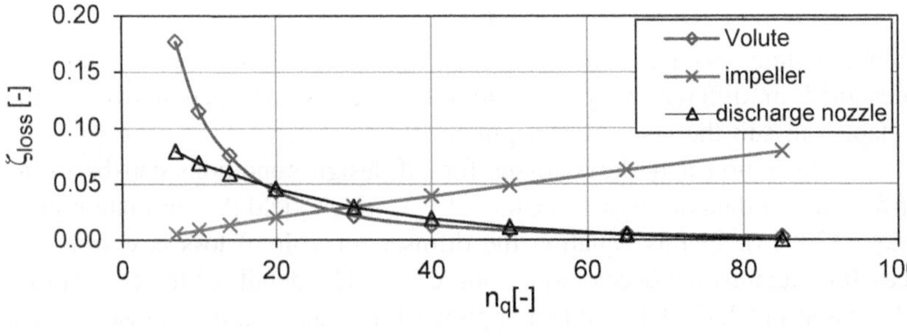

Fig. 3.21 Loss coefficients calculated from Table 3.8

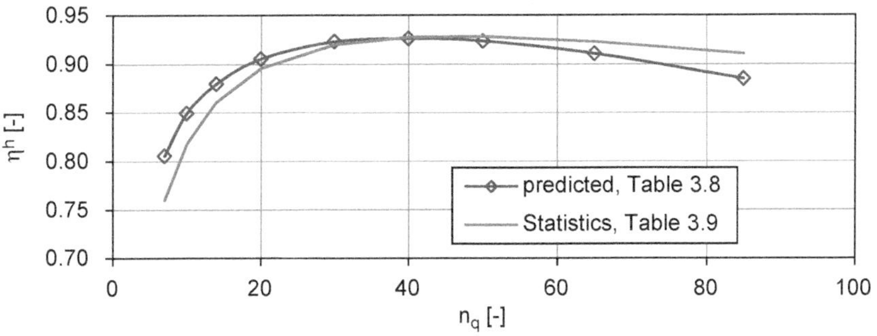

Fig. 3.22 Hydraulic efficiencies calculated from Table 3.8

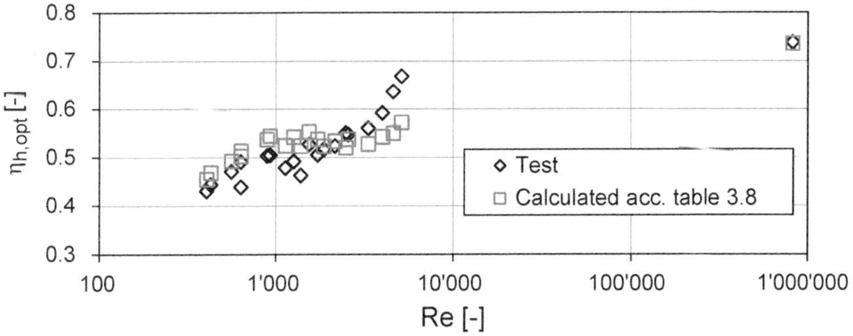

Fig. 3.23 Hydraulic efficiencies tested with various viscosities and speeds compared to calculation according to Table 3.8

Tests with a single-stage end-suction volute pump at various speeds and with viscosities up to 620 cSt are shown in Fig. 3.23 together with the calculation according to Table 3.8. These calculations were done at the actual BEPs measured with viscous flow. The points at about Re = 800,000 represent the tests with water.

There are some problems with this comparison: (1) The determination of the hydraulic efficiency of a pump with $n_q = 7$ in viscous flow is quite uncertain; (2) The calculation of the friction factor in the transition region between laminar and turbulent flow is uncertain; (3) no data on diffuser performance were available at low Reynolds numbers. Nevertheless, the results suggest that Table 3.8 may be used to estimate the hydraulic efficiency in viscous flow. The reason is that flow at low Reynolds numbers is dominated by friction effects which - to a large extent - form the basis for the calculation in Table 3.8.

The good agreement between the curves in Fig. 3.22 is considered to be to some extent coincidental because the loss models in Table 3.8 are not specifically adapted to the flow through pumps.

3.9 Statistical Data of Pressure Coefficients, Efficiencies and Losses

Based on numerous measurements in all types of centrifugal pumps, head coefficients and efficiencies were published which allow quite a reliable evaluation of the achievable values. As with all statistics, the figures supply mean values and mean errors but do not permit any statement on possible deviation of the *individual* measurement which, in principle, can be of any magnitude.

Figure 3.24 furnishes the pressure coefficient ψ_{opt} as a function of the specific speed. Here, ψ_{opt} is always based on the outlet diameter d_{2a} at the *outer* streamline. In contrast, ψ_i is based on the outlet diameter at the *inner* streamline d_{2i} according to [19]. The hydrodynamic blade loading at the hub limits the achievable pressure coefficient with semi-axial and axial impellers.

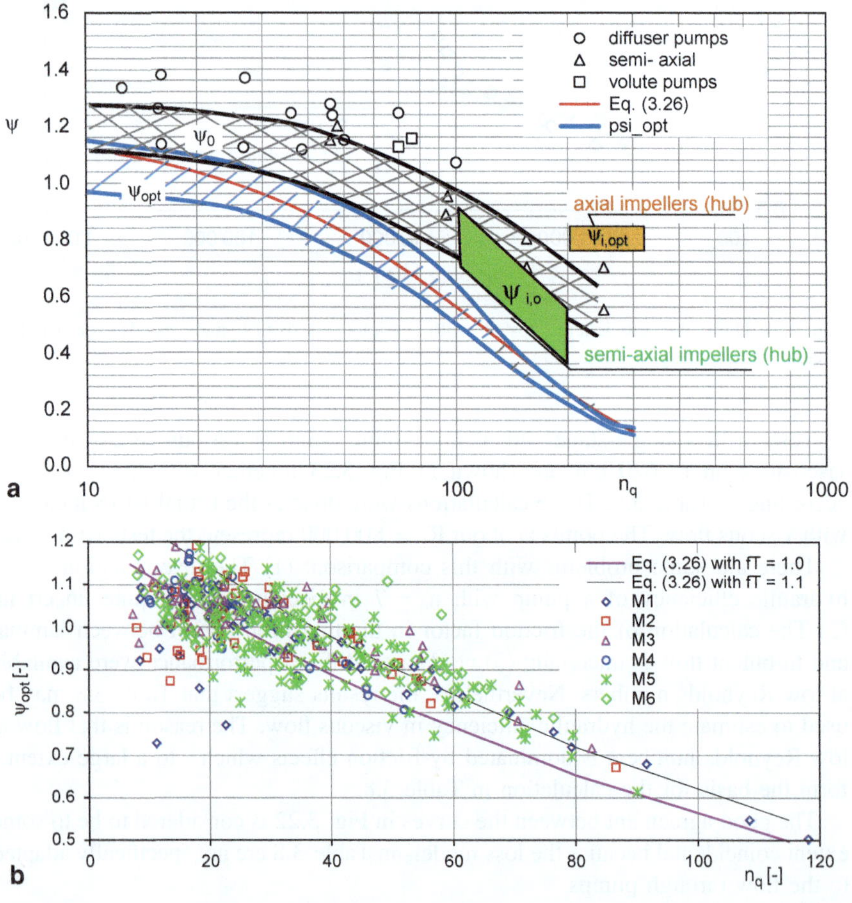

Fig. 3.24 Pressure coefficients: **a** general tendencies (values for hub according to [19]); **b** ψ_{opt} for single-stage volute pumps from various pump manufacturers

The drop of the achievable pressure coefficient with increasing specific speed is largely due to the fact that the centrifugal component of the head according to Eq. (T3.3.2) drops with growing n_q and becomes zero in axial pumps.

Analytical functions for the pressure coefficients according to Fig. 3.24 for the best efficiency point and for Q = 0 are provided by Eqs. (3.26–3.28).

The pressure coefficients to be selected to achieve a competitive design depend on the pump type and application. While Fig. 3.24a gives general trends, Fig. 3.24b shows an evaluation of the BEP pressure coefficients of various pump vendors (M1 to M6) for single-stage volute pumps with axial inlet or as inline pumps. It is seen that these low-head pumps (which include standard pumps according to EN733) are designed with head coefficients higher than calculated from Eq. (3.26) with $f_T = 1.0$. Rather $f_T = 1.1$ would be suitable for this application. A similar evaluation of data from various pumps vendors revealed that $f_T = 1.0$ is a suitable selection for multistage pumps.

Selecting high ψ_{opt} yields flatter characteristics and the risk of an instability of the Q-H-curve increases. The more so, the higher the value of ψ_{opt} is chosen. If a steep characteristic is required, ψ_{opt} is selected near the lower limit curve (or even below). In addition, Fig. 3.24 contains pressure coefficients ψ_o for operation against a closed valve, Chap. 4.

Pressure coefficients for designing sewage pumps (single-, two- or three channel impellers) and vortex pump are given Chap. 7.4.

$$\mathbf{BEP}: \ \psi_{opt} = 1.21\ f_T\ e^{-0.77 n_q/n_{q,Ref}} = 1.21\ f_T\ e^{-0.408\ \omega_s} \quad n_{q,Ref} = 100 \tag{3.26}$$

$$\mathbf{Q = 0}: \quad \textbf{diffuser pumps}: \psi_o = 1.31\ e^{-0.3 n_q/n_{q,Ref}} \quad n_{q,Ref} = 100 \tag{3.27}$$

$$\mathbf{Q = 0}: \ \textbf{volute pumps}: \ \psi_o = 1.25\ e^{-0.3 n_q/n_{q,Ref}} \quad n_{q,Ref} = 100 \tag{3.28}$$

The velocity components $c_{2u}{}^*$ and $w_{2u}{}^*$ can be calculated on the basis of the curve $\psi = f(n_q)$ given in Fig. 3.24, Eq. (T3.2.8) and Eq. (T3.2.11). If an inflow angle is specified, it is possible to determine the relative velocity $w_1{}^*$ at the impeller inlet. Such data can be useful for estimations without details of the pump having to be known or calculated.

Figure 3.25 provides this type of data; the assumptions made for calculating the graph are given in the figure caption. In addition, the mean relative velocity $w_{sp}{}^* = \left(c_{ax}{}^2 + {}^1\!/\!{}_4 u_{sp}{}^2\right)^{0.5}$ in annular seals has been plotted to ease the estimation of abrasive wear as per Chap. 14.5. With $\alpha_1 = 90°$ we have $d_1{}^* \approx w_1{}^*$ since the inlet flow angle is usually small. Hence the impeller inlet diameter can be estimated from $d_1 = d_2 \times d_1{}^*$ when the outer impeller diameter has been determined.

In Fig. 3.26, overall efficiencies of single-stage radial pumps are shown as functions of the specific speed with the design flow rate as parameter. The overall efficiency reaches a maximum in the range from $n_q = 40$ to 50 for a given pump size. The maximum is caused by two tendencies: (1) To the left of the maximum the

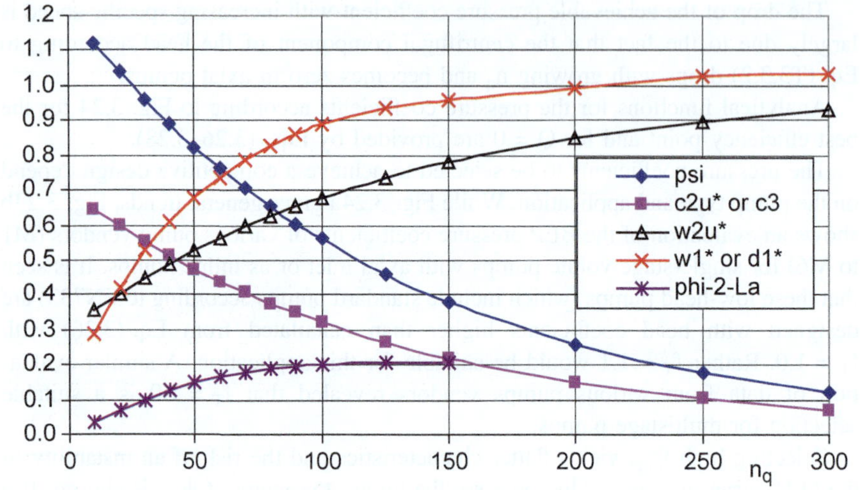

Fig. 3.25 Velocities at impeller inlet and outlet and in annular seals, calculated with $\eta_h = 0.87$, $k_n = 1$, $\beta_{1a} = 15°$, clearance s from Eq. (3.12), leakage flow rate from Eq. (T3.5.10). All velocities are referred to u_2. For $\alpha_1 = 90°$ it is $d_1^* \approx w_1^*$

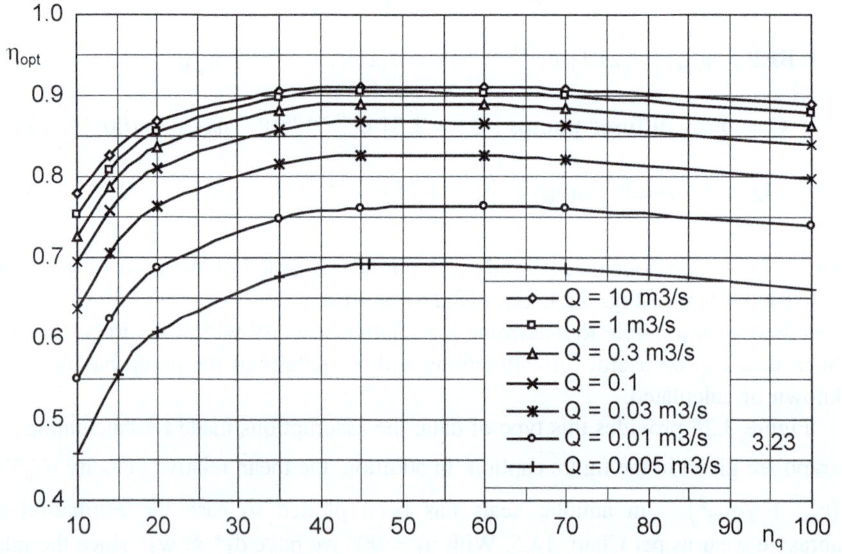

Fig. 3.26 Efficiencies of single-stage, single-entry, radial pumps

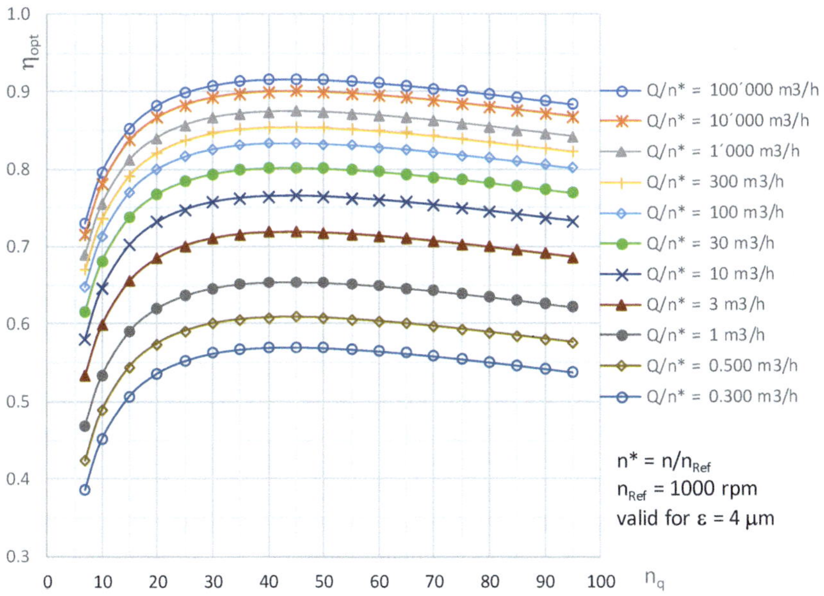

Fig. 3.26.1 Efficiencies of single-stage, single-entry, radial pumps, calculated from Eq. T3.9.1A; parameter Q_{opt}/n^*

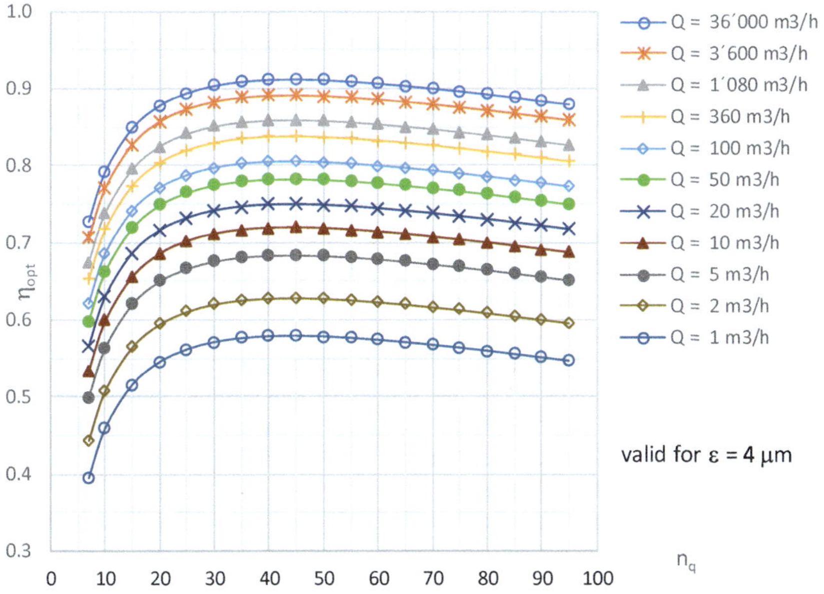

Fig. 3.26.2 Efficiencies of single-stage, single-entry, radial pumps, calculated from Eq. T3.9.1A; parameter Q_{opt}

secondary losses increase exponentially with falling n_q according to Eqs. (T3.5.10 to 3.5.12). (2) At high specific speeds ($n_q > 70$) the hydraulic losses increase with growing n_q due to increasing mixing losses mainly caused by non-uniform flow distributions over the blade height and by secondary flows.

As shown by Fig. 3.26, the overall efficiency increases with the flow rate. This dependency is caused by the influence of pump size and speed since the flow rate follows the proportionality $Q \sim n \times d_2^3$ (Chap. 3.4). In the case of very small pumps or very low flow rates, mechanical losses also have a strong impact, Chap. 3.6.4.

The efficiencies according to Fig. 3.26 apply to impellers without axial thrust balancing. If balance holes are provided, the efficiency must be corrected according to Eq. (T3.9.9).

An alternative to Eq. T3.9.1 and Fig. 3.26 is provided as Eq. T3.9.1A and Fig. 3.26.1. This formula has been adapted from [B.15]. It has the advantage that the influence of the surface roughness is included. The parameter in the graph is Q_{opt}/n^* rather than Q_{opt}. But plotting a graph with Q_{opt} as parameter gives very similar results if reasonable speeds are assumed. This is demonstrated by Fig. 3.26.2 which may be more convenient to use.

Equation T3.9.1A is expected to offer a better estimate than Eq. T3.9.1 for small pumps with $Q_{opt} < 50\ m^3/h$. However, any efficiency predictions from statistical data are uncertain for small pumps, because of the strong influence of the surface roughness; and, because the mechanical losses can take a large portion of the power at the coupling.

The efficiencies of semi-axial and axial pumps are shown in Fig. 3.27. Above $n_q > 70$ the efficiencies of semi-axial impellers are slightly superior to those of radial pumps. This is caused by differences in the shape of the impellers: Radial impellers have stronger front shroud curvatures than semi-axial impellers. Hence flow distributions over the blade height become more non-uniform with increasing specific speed, which leads to higher turbulent dissipation.

The efficiencies in Figs. 3.26 and 3.27 were calculated according to Eqs. (T3.9.1 and T3.9.3). These empirical equations show that the losses in a wide range of pump sizes and specific speeds cannot be covered by simple power laws - a fact that must be taken into account when converting the efficiencies from a model to a prototype with larger diameter and/or speed, see Chap. 3.10.

The efficiencies of multistage industrial pumps according to Fig. 3.28 and Eq. (T3.9.2) are generally below the values of single-stage pumps because of additional losses in the return channels. However, in the case of storage pumps which are built less compact than pumps for industrial applications similar efficiencies as in Fig. 3.26 are achieved.

Double-entry pumps according to Fig. 3.29 and Eq. (T3.9.4) with $n_q < 40$ achieve higher efficiencies than single-entry impellers *of the same specific speed*, since the disk friction losses, referred to the power, are only half as high. Furthermore, the losses in the volute casing, which has been designed for twice the flow rate, turn out to be smaller.

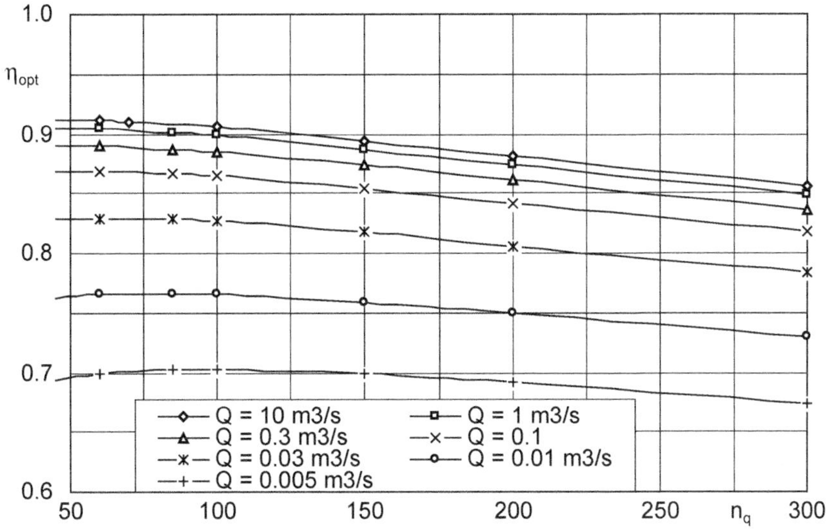

Fig. 3.27 Efficiencies of semi-axial and axial pumps

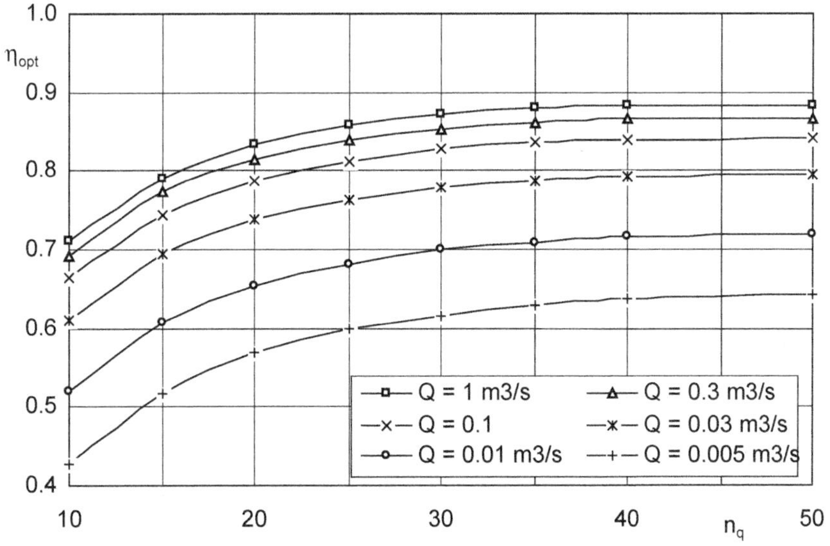

Fig. 3.28 Efficiencies of multistage, single-entry, radial pumps

The hydraulic efficiencies given in Figs. 3.30, 3.31 and 3.32 and Eqs. (T3.9.6) to (T3.9.8) also depend on the specific speeds and the flow rates (or pump sizes and speeds). The maxima are less pronounced than with the overall efficiency. They are caused by two tendencies: (1) The friction and deceleration losses in the diffusing

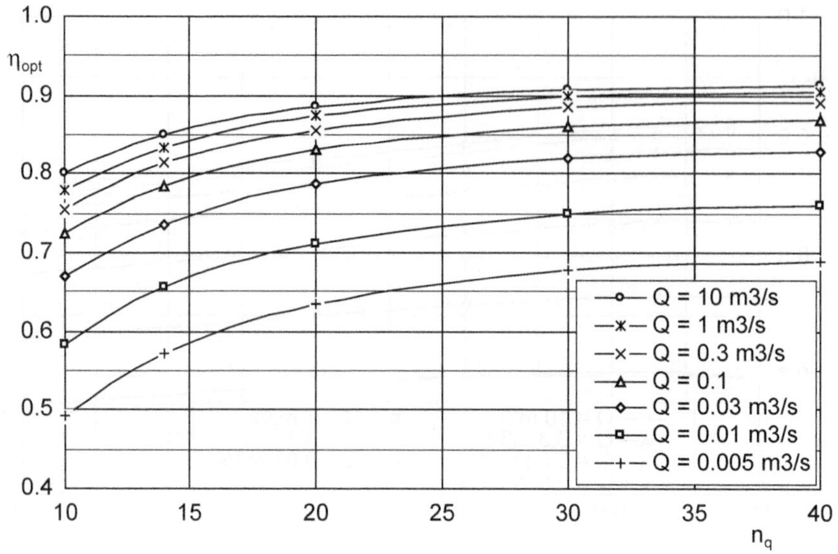

Fig. 3.29 Efficiencies of single-stage pumps with double-entry impeller. Q corresponds to Q_{opt} *of the pump*, while n_q is calculated with the flow *per impeller side*, Eq. (T3.4.15)

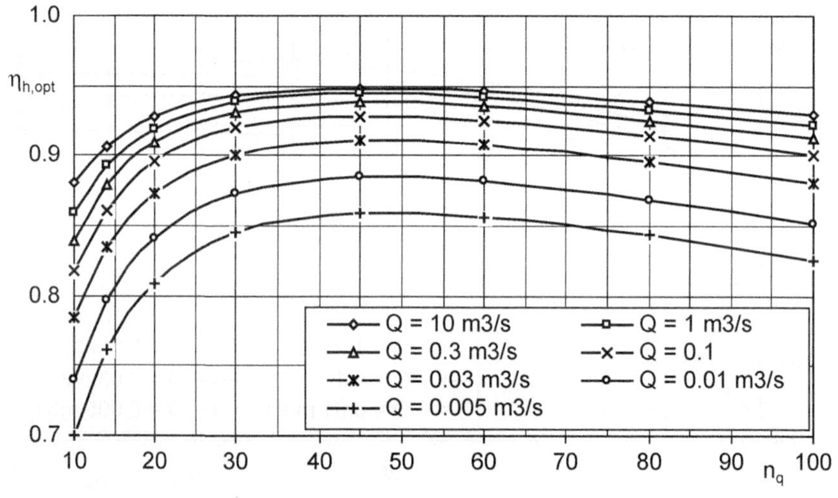

Fig. 3.30 Hydraulic efficiencies of single-stage, single-entry, radial pumps

elements increase strongly at low specific speeds; (2) increasing mixing losses are responsible for the drop of the hydraulic efficiency above n_q = 70 (as in the case of the overall efficiency).

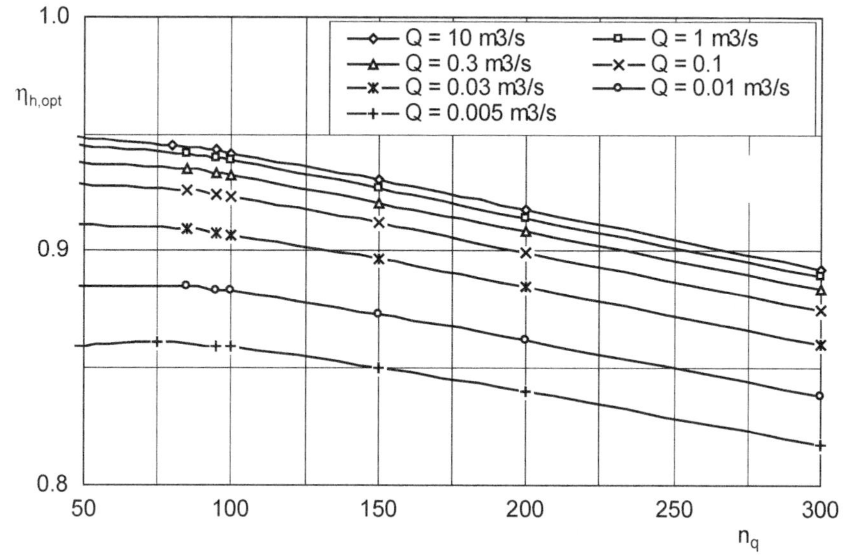

Fig. 3.31 Hydraulic efficiencies of semi-axial and axial pumps

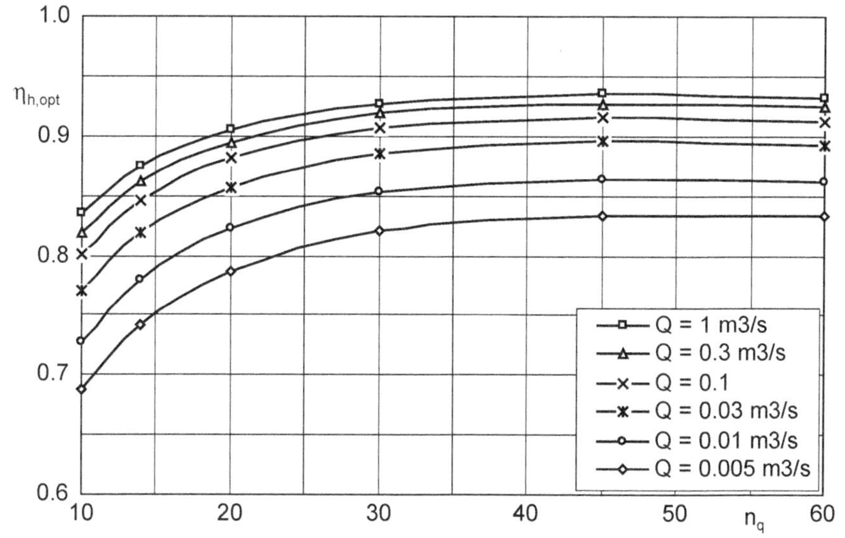

Fig. 3.32 Hydraulic efficiencies of multistage, single-entry, radial pumps

The hydraulic efficiency of very small pumps decreases due to increased *relative* roughness, since the absolute roughness remains more or less constant (it is given by the casting process). In addition, the narrow channels cannot be dressed well. Finally, the economic incentive to improve the efficiency of small pumps is less significant.

When it is difficult to make a reasonable estimate of the hydraulic efficiency, Eq. (3.28a) may be used. It can be particularly helpful with small pumps and at partload.

$$\eta_h = \frac{1}{2}(1+\eta) \tag{3.28a}$$

Equation (3.28a) supplies realistic estimates even at Q = 0 and at partload.[12]

The efficiencies determined according to Table 3.9 or Figs. 3.26, 3.27, 3.28, 3.29, 3.30, 3.31 and 3.32 are subject to a scatter in the magnitude of $\Delta\eta = \pm 0.2 \times (1 - \eta)$. Deviations can be due to the quality of the hydraulic design, seal clearances, axial thrust balancing and mechanical equipment. If sufficient test data of one or several pump ranges are available, the coefficients of the equations in Table 3.9 can be easily adjusted so that the data can be correlated with a smaller standard deviation.

Figure 3.33 shows how the secondary losses and the hydraulic losses in the impeller and the diffusing elements depend on the specific speed. These figures illustrate tendencies and relationships but can also be useful for a rough evaluation. At $n_q < 25$, for example, an optimum design and careful dressing of the diffusing elements is most important. At $n_q < 20$, seal clearance and geometry have to be optimized and the disk friction has to be minimized (as far as these can be influenced). At $n_q > 50$, however, the impeller losses deserve greater attention because of the non-uniform flow distribution. Test data discussed in Chap. 4.1.3 confirm the trends and magnitudes presented by Fig. 3.33.

Since the losses depend on size, type and surface finish of the pump, Fig. 3.33 is obviously unsuitable for an exact loss analysis, for which Chap. 3.6 and Tables 3.5–3.9 should be used.

In order to asses the efficiency potential of a pump, the concept of an "achievable efficiency" may be used. It constitutes the efficiency which would be expected if the pump under consideration were designed and finished in an optimum way. An "achievable efficiency" can be defined with various assumptions and methods depending on the goals of the investigation and the issues at stake. The information on achievable efficiencies can be quite misleading, if mechanical constraints (e.g. safe seal clearances), reliability issues and the impact on costs are disregarded. Therefore, it may often be more useful to perform an analysis as to how sensitively the efficiency reacts to all the parameters which determine the hydraulic and the secondary losses.

Based on the information in Tables 3.5–3.8 the achievable efficiency can be determined (for example) with the following assumptions:

(1) Calculate the disk friction loss as hydraulically smooth assuming that the lower limit is at 75% of the calculated loss.

[12]The relationship $\eta_h = \eta^{0.5}$ which is sometimes employed does not furnish any useful values either at very low overall efficiencies (small pumps) or at low partload.

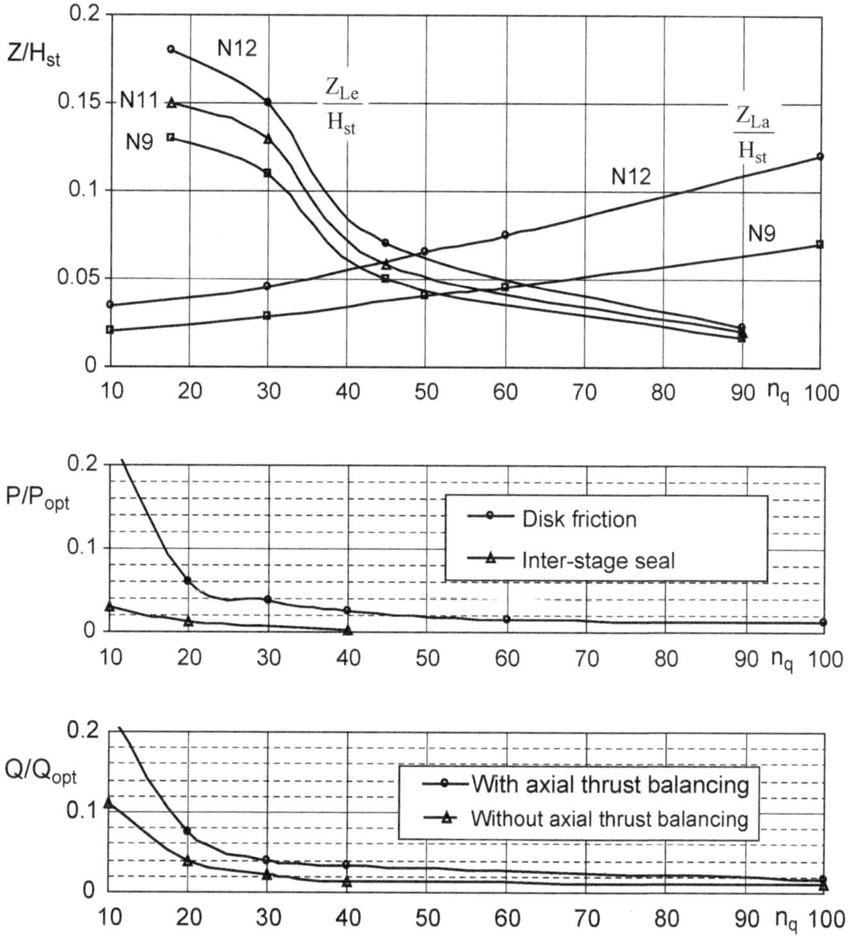

Fig. 3.33 Various losses as functions of the specific speed

(2) Determine the leakages for the minimum seal clearances and assume only 70% of the calculated value as the lower limit of the correlations used. In this context, it is also possible to examine the effect of a different annular seal design: seal length, stepped seal, grooves or isotropic pattern.
(3) The hydraulic losses are calculated for hydraulically smooth surfaces and $c_d = c_f$ is inserted in Eq. (T3.8.6). This means the impeller is considered to be without the detrimental effects of deceleration and secondary flow.
(4) Shock losses in the impeller and diffuser are also assumed as zero.

Theoretical investigations concerning the maximum achievable efficiency of single-stage volute casing pumps are discussed in [23]. This topic is also discussed in detail in [B.21]. The results from [23] can be reflected through Eq. (T3.9.5) if the values for η_{opt} from Fig. 3.26 or from Eq. (T3.9.1) are inserted. The "achievable efficiency" depends on the design of the pump. In practice it is greatly influenced by considerations of mechanical design, cost and reliability. Careful analysis of all losses and the potential for improvement of an actual pump should therefore follow a comparison with theoretical achievable values.

3.10 Influence of Roughness and Reynolds Number

3.10.1 Overview

The losses in a centrifugal pump, and consequently its efficiency, depend on the roughness of the components and the Reynolds number Re. Re covers the influences of pump size, speed and viscosity. The knowledge of these relationships is of great practical importance for:

1. Determining the efficiency of the prototype (subscript "a") $\eta_a = f(Re_a, \varepsilon_a/d_a)$ based on the efficiency $\eta_M = f(Re_M, \varepsilon_M/d_M)$ measured in the model test (subscript "M").
2. Determination of the efficiency expected in the plant based on the efficiency measured in the works test at a lower speed and/or a different viscosity. A typical application would be a boiler feed pump which was measured on the test stand at 1500 rpm with water of 20 °C but is operated at 6000 rpm with water of 180 °C in the power station.
3. Estimation of the characteristics for operation with fluids of high viscosity (e.g. such as oil with 3000×10^{-6} m²/s) based on tests with cold water.
4. Evaluation of the expected efficiency gain upon reduction of the roughness.
5. Influence of the roughness in various components (impeller, diffuser, impeller sidewalls and casing) on the efficiency. Or the answer to the question where, in a specific case, additional effort promises the greatest chance of success to improve the efficiency. These effects depend on the specific speed n_q and the pump type.

In the following it is generally assumed that the characteristics and efficiency $\eta_a = f(Re_a, \varepsilon_a/d_a)$ of an *application* (subscript "a") or prototype are to be determined based on a *model* or *basic* test (subscript "M") with known (measured) efficiency $\eta_M = f(Re_M, \varepsilon_M/d_M)$ and known characteristics (the "basic test" could, of course, also be a sales curve for example).

Points 1 and 2 are extensively covered in the literature under keywords such as "efficiency scaling", "efficiency majoration" or "efficiency step-up". Published formulae for efficiency scaling suffer from a lack of generality because they do not cover the effects of the specific speed and of the roughness. The special aspects of pumping high-viscosity media are covered in Chap. 13.1. On the basis of a simplified loss analysis it is possible to tackle the five tasks mentioned above through a common approach which is discussed in Chap. 3.10.3, while the comparison with test data is documented in [16].

3.10.2 *Efficiency Scaling*

Efficiency scaling means the prediction of the efficiency of a given pump ("prototype") from a test done with a *geometrically similar* "model" with different size and/or Reynolds number.

As has been shown in Chaps. 3.6 and 3.8 the disk friction, the leakage losses and a portion of the hydraulic losses (and therefore also the hydraulic and the overall efficiencies) depend on the Reynolds number and the relative roughness. There are different possibilities for evaluating this Reynolds influence or for scaling the efficiencies:

1. Determine the effective mechanical losses and the secondary losses separately for model and prototype according to Tables 3.6 and 3.7. Likewise, calculate the *difference* in hydraulic efficiency between model and prototype according to Table 3.9. Add this difference to the hydraulic efficiency of the model. In this way it is possible to determine the power consumption of the machine according to Eq. (T3.5.1) and its efficiency according to Eq. (T3.5.3). Note that the mechanical losses often differ between model and prototype since different bearings and shaft seals are employed (as mentioned $P_m \sim n^x$ with x = 1.3 to 1.8 is applicable and not $P_m \sim n^3$ as for scaling the internal power of the pump).
2. The separate evaluation of all losses according to Chap. 3.10.3 largely corresponds to the procedure suggested under item 1. Scaling the efficiency in this way, is presumably the most accurate method at present, since all relevant design- and plant-specific differences between model and prototype can be taken into account.
3. Proceed as under (1), but calculate the *differences* in the hydraulic losses according to Table 3.8 as caused by the differences in the Reynolds numbers and/or in the relative roughness of model and prototype. Note that the absolute values of the hydraulic losses calculated would be too inaccurate since the mixing losses cannot be evaluated adequately.
4. Use of "majoration formula" (see below).
5. Figures 3.26, 3.27, 3.28 and 3.29 and the equations in Table 3.9 provide a relation between overall efficiency and flow rate Q_{opt}. Since $Q \sim n \times d^3$, such a representation indirectly covers the effect of the pump size and the speed on the efficiency. If the viscosity is constant, this means that the effects of the Reynolds number ($Re \sim n \times d^2$) and the relative roughness are taken into account, provided that the absolute roughnesses of model and prototype are identical. Consequently, the efficiency can be scaled on the basis of Table 3.9 by employing $\eta_a = \eta_M + \Delta\eta$. $\Delta\eta$ is the difference between the efficiencies calculated for model and prototype according to Eqs. (T3.9.1–3.9.4). When proceeding thus, the effect of surface roughness can be taken into account by Eq. (T3.9.1A).
6. The hydraulic efficiencies of model and prototype can be established using numerical methods. The rest of the procedure is the same as in step (1).

Majoration equations: Various equations for efficiency scaling have been published. Overviews can be found in [20, 30–32, 43]. The more relevant equations for *internal* efficiency η_i allow a division into "scalable" (i.e. Reynolds-dependent) and non-scalable losses. Disk friction losses and that portion of hydraulic losses which is caused by friction are "scalable". Both losses fall with rising Reynolds number and both drop if the relative roughness ε/d_h decreases. These effects make up the efficiency scaling in its proper sense. However, the leakage losses have an opposite tendency: they increase with Re and drop with growing relative roughness ε/d_h.

As is shown in Figs. 3.26, 3.27, 3.28, 3.29, 3.30, 3.31, 3.32 and Table 3.9, changes in efficiency and loss depend on the specific speed and the pump size. Doubling the design flow rate gives rise to quite different efficiency corrections depending on specific speed and design flow rate. The published efficiency scaling formulae don't do justice to this fact (or only incompletely so). In the light of these considerations it is quite sensible to scale efficiencies according to Table 3.9 as described under point (5).

As mentioned before, the hydraulic losses are composed of friction and mixing losses: Only the friction losses are primarily scalable. Different scaling laws result depending on how the Reynolds numbers and the relative roughnesses of the model and prototype are related with respect to hydraulically smooth or rough behavior. If the surfaces of model and prototype are hydraulically smooth, the scaling follows the relation Re^x, and the exponent drops with increasing Re number. Impeller diameter, speed and viscosity enter into the Reynolds number in the same way. In contrast, if both machines are situated in the fully rough range, the relative roughness has an impact on the efficiency while the Reynolds number has no effect. These relationships become clear from Fig. 1.13, see the points marked M and A. *Majoration formulae with constant exponents of the Reynolds number can offer an approximation only if roughness effects are weak.*

This difficulty can be avoided by the approach implied by Eqs. (T3.9.11 and 3.9.12): the friction coefficients $c_f = f(Re, \varepsilon/L)$ are determined from Eq. (1.33) or Fig. 1.13 for the model and prototype and the hydraulic efficiency is scaled by means of the ratio $c_{f,a}/c_{f,M}$ as given by Eq. (T3.9.11). The coefficient $\zeta_{R,M}$ comprises all friction losses in the pump (especially in the impeller and the diffusing elements). The friction losses can be estimated from the given correlation $\zeta_{R,M}/\psi$, which was derived from calculating various pumps according to Table 3.8. Alternatively, $\zeta_{R,M}/\psi$ can be determined for the case to be examined by means of the equations in Table 3.8.

The disk friction can be treated in a similar manner: The friction coefficients k_{RR} are determined as a function of the Reynolds number and roughness for the model and prototype from Eq. (T3.6.3). Next, the internal efficiency is scaled according to Eq. (T3.9.12) in which the effect of the disk friction and the hydraulic losses are averaged. When scaling the internal and the hydraulic efficiencies, the distribution of the scalable and the Reynolds-independent losses is estimated as a function of the specific speed from Eq. (T3.9.12). Finally, Eq. (T3.9.12) can also be applied to the overall efficiency if the difference in the mechanical losses is neglected.

Equation (T3.9.13) provides the scaling formula according to [N.5]; here, an efficiency η_{hR} which comprises the hydraulic and the disk friction losses is defined and scaled. Annular seal losses and mechanical losses are calculated separately for the model and the prototype and are considered in $\eta_{hR,M}$ and $\eta_{hR,a}$.

The scaling of efficiencies from the model to the prototype harbors a number of uncertainties, which are discussed in Chap. 3.10.3. As a rule, a careful loss analysis according to Chap. 3.10.3 will supply the most accurate prediction, but scaling formulae - such as Eq. (T3.9.11A) - have their justification for rapid estimations. They are important, for example, when contractual formulae for evaluating the guaranteed efficiencies have to be agreed upon between the pump buyer and the manufacturer since the scaling procedure must apply equally to all project bidders.

3.10.3 Calculation of the Efficiency from Loss Analysis

A complete set of equations (Table 3.10) is developed in this chapter by means of which the effect of Reynolds number and roughness on efficiency can be estimated. The process applies to turbulent and laminar flow over hydraulically smooth or fully rough surfaces and covers. It smoothly covers all transitions between the different flow regimes.

The efficiency conversion takes into account the hydraulic and volumetric losses as well as disk friction. These losses depend on the Reynolds number, the roughness and the specific speed in different ways. All the necessary equations are compiled in Table 3.10.

Factors for efficiency conversion: The examination is performed for the best efficiency point. Power consumption and efficiency can be expressed in simplified form (without recirculation) according to Eq. (T3.5.1) as follows:

$$P = \frac{\rho\, g\, H\, z_{st}\, Q}{\eta_{vol}\, \eta_h} + P_{RR} + P_m \qquad \eta = \frac{P_u}{P} = \frac{\rho\, g\, H\, z_{st}\, Q}{P} \tag{3.29}$$

In Eq. (3.29) the effect of the interstage seals in multistage pumps is taken into account by a weighted volumetric efficiency defined by Eq. (T3.10.4). Furthermore, the disk friction power P_{RR} in Eq. (3.29) comprises the sum of *all* components of the pump rotor, i.e. it includes the friction at the axial thrust balance device (P_{er}). The efficiency according to Eq. (3.29) can then be expressed by:

$$\eta = \frac{\eta_{vol}\, \eta_h}{1 + \eta_{vol}\eta_h \left(\frac{P_{RR}}{P_u} + \frac{P_m}{P_u}\right)} \tag{3.30}$$

The quantities η_{vol}, η_h and P_{RR} depend on the Reynolds number and the roughness, while P_m/P_u is assumed to be a constant. To recalculate the pump characteristics from the base or the model ("M") to the application ("a"), the factors according to Eq. (3.31) are introduced. These factors *only* take into account the *deviations* from the model laws. Changes in speed and/or impeller diameter, if any, have to be calculated from Table 3.4 before this analysis.

$$f_\eta = \frac{\eta_a}{\eta_M} \quad f_{\eta h} = \frac{\eta_{h,a}}{\eta_{h,M}} \quad f_{\eta vol} = \frac{\eta_{vol,a}}{\eta_{vol,M}} \quad f_{\eta m} = \frac{\eta_{m,a}}{\eta_{m,M}} \tag{3.31}$$

$$f_Q = \frac{\varphi_a}{\varphi_M} \quad f_H = \frac{\psi_a}{\psi_M} \quad \text{or} \quad f_Q = \frac{Q_a}{Q_M} \quad f_H = \frac{H_a}{H_M} \tag{3.32}$$

The disk friction losses in the prototype and in the model amount to:

$$P_{RR,a} = k_{RR,a}\, \rho_a\, \omega^3\, r_2^5\, f_{geo} \quad \text{respectively} \quad P_{RR,M} = k_{RR,M}\, \rho_M\, \omega^3\, r_2^5 f_{geo} \tag{3.32a}$$

The corresponding useful power is $P_{u,a} = \rho_a \times g \times H_a \times Q_a$ and $P_{u,M} = \rho_M \times g \times H_M \times Q_M$.

With f_Q and f_H from Eq. (3.32), it follows that $P_{u,A} = P_{u,M} \times f_H \times f_Q \times \rho_a/\rho_M$. With these relationships the following is obtained:

$$\left(\frac{P_{RR}}{P_u}\right)_a = \left(\frac{P_{RR}}{P_u}\right)_M \frac{k_{RR,a}}{f_Q\, f_H\, k_{RR,M}} \quad \text{with} \quad \frac{P_{u,M}}{P_{u,a}} = \frac{1}{f_Q\, f_H} \tag{3.33}$$

If Eq. (3.30) is written for the application ("a") and the model ("M") and the resulting equations are divided by one another, the multiplier for the efficiency is obtained by virtue of Eqs. (3.31–3.33) as follows:

$$f_\eta \equiv \frac{\eta_a}{\eta_M} = \frac{f_{\eta h}\, f_{\eta vol}\left[1 + \left\{\left(\frac{P_{RR}}{P_u}\right)_M + \left(\frac{P_m}{P_u}\right)_M\right\} \eta_{vol,M}\, \eta_{h,M}\right]}{1 + \left\{\left(\frac{P_{RR}}{P_u}\right)_M \frac{k_{RR,a}}{k_{RR,M}} + \left(\frac{P_m}{P_u}\right)_M \frac{\rho_M\, f_{\eta m}}{\rho_a}\right\} \frac{f_{\eta h}\, f_{\eta vol}}{f_H\, f_Q}\, \eta_{vol,M}\, \eta_{h,M}} \tag{3.34}$$

This equation covers all the losses. If the geometry of the annular seals is known, the Re-dependency of the volumetric efficiency can be calculated according to Table 3.7 and taken into account in the form of $f_{\eta vol}$. The following assumptions and simplifications were made: (1) $f_{\eta vol} = f_{\eta m} = 1.0$; $\rho_a = \rho_M$; $f_H = f_{\eta h}$. (2) When pumping viscous fluids with viscosities above (50 to 100) $\times\ 10^{-6}$ m²/s the best efficiency point moves more or less along the volute characteristic (Chap. 4.2). Hence, we can set $f_Q = f_H = f_{\eta h} < 1$ (this is discussed in detail in Chap. 13). Under these conditions a somewhat simplified equation for the efficiency conversion is obtained, Eq. (T3.10.20).

Factors for head conversion: The assumption $f_H = f_{\eta h}$ is necessary since the factor f_H (in contrast to f_η) cannot be determined directly from a test, as shown below.

Exceptions are tests with a viscosity sufficiently high so that the following interference effects are masked:

1. With varying Reynolds number, the seal leakage changes, resulting in a slight shift of the Q-H curve since the flow rate through the impeller $Q_{La} = Q/\eta_{vol}$ changes, Fig. 4.29.
2. The impeller shrouds have a pumping effect similar to that of a friction pump. Boundary layer fluid with $c_{2u} \approx u_2$, flung off the impeller shrouds, contributes to the transfer of energy. The relative contribution of this effect grows with decreasing specific speed, falling Reynolds number and increasing roughness.
3. An *increase* in head was also repeatedly measured when the roughness of the impeller channels was increased. Greater roughness implies a deceleration of the relative velocity near solid walls and a thickening of the boundary layer. A lower *relative* velocity however means a higher *absolute* velocity and consequently an increase of the slip factor and the theoretical head. This increase of H_{th} can (but not always does) exceed the additional losses brought about by the roughness. The same mechanism is active if the viscosity is increased.

Some test results relating to this increase in head are quoted below:

- Double-entry impeller $n_q = 10$, a change of roughness from $\varepsilon = 0.025$ mm to 0.87 mm in the channels and the outside of the impeller shrouds [39]: head increase $f_H = 1.1$ with efficiency loss $f_\eta = 0.84$
- Single-entry impeller, $n_q = 7$, a change of roughness from $\varepsilon = 3.7$ μm to 46 μm [15]: increase in head $f_H = 1.01$
- Even with a semi-axial impeller $n_q = 135$ a slight increase in head was measured with increased roughness.
- Tests in [24] showed an increase in head with viscosities up to 45×10^{-6} m²/s; a drop in head was only measured at 100×10^{-6} m²/s. However, this effect cannot be separated from the shift of the Q-H-curve caused by a reduced seal leakage.
- A slight increase in head was observed also in the measurements in [26] and [37] on single-stage pumps with $n_q = 12$ and 20 respectively.

The higher the quality of the hydraulic design (i.e. the impeller operating with thin boundary layers and without separation), the greater is the effect of the roughness on ψ_{th}. In contrast to this, a low-quality hydraulic design with separated flow is less affected by the surface roughness.

Neither the shift of the Q-H-curve due to a change in the annular seal flow, nor the pumping action of the impeller shrouds, nor the change of the slip factor due to changes in the boundary layer thickness or the secondary flows in the impeller can be separated from the friction losses. Attempting to determine the factor f_H directly from a test is therefore too inaccurate.

Hydraulic losses: As explained in Chap. 3.8, the hydraulic losses consist of friction $\zeta_R = f(Re, \varepsilon)$ and mixing losses ζ_M. All of the friction and mixing losses from inlet, impeller and diffusing elements are included in these two quantities. The theoretical head coefficient in the prototype and in the model can therefore be written as follows (refer also to Table 3.8):

$$\psi_{th} = \psi_M + \zeta_{R,M} + \zeta_{M,M} = \psi_a + \zeta_{R,a} + \zeta_{M,a} \tag{3.35}$$

In Eq. (3.35) it is assumed that the slip factor, and consequently the theoretical head, remain roughly constant, refer also to Chap. 13.1. If the blade work remains constant, the ratio f_H of the heads must be equal to the ratio of the hydraulic efficiencies, resulting in $f_H = f_{\eta h}$.

If the mixing losses are assumed to be independent of the Reynolds number, $\zeta_{M,M} = \zeta_{M,a}$ in Eq. (3.35) cancel so that a relation between ψ_a and ψ_M in the form of the multiplier $f_{\eta h}$ can be derived:

$$f_{\eta h} = \frac{\eta_{h,a}}{\eta_{h,M}} = 1 - \frac{\zeta_{R,M}}{\psi_M}\left(\frac{\zeta_{R,a}}{\zeta_{R,M}} - 1\right) = 1 - \frac{\zeta_{R,M}}{\psi_M}\left(\frac{c_{f,a}}{c_{f,M}} - 1\right) \tag{3.36}$$

Equation (3.36) applies at *constant* flow rate.

If the fraction $\zeta_{R,M}/\psi_M$ of the friction losses in the head is known, Eq. (3.36) can be used to estimate how the head will change with a different roughness and/or Reynolds number. To do so, it is assumed that the friction losses $\zeta_{R,a}/\zeta_{R,M}$ are proportional to the corresponding friction coefficients $c_{f,a}/c_{f,M}$. These friction coefficients depend on the Reynolds number and the roughness. They can be calculated for turbulent and laminar flows according to Eqs. (1.33 and 1.33a).

The portion of the friction losses in the hydraulic losses can be estimated for a given pump according to Table 3.8 and Chap. 3.8. It depends on the specific speed, the pump type and the geometry of the hydraulic components. An estimation of $\zeta_{R,M}/\psi_M$ according to Table 3.8 is complicated and afflicted with uncertainties. For this reason an empirical approach according to Eq. (T3.10.18) was developed on the basis of a large number of tests, [15].

Secondary losses: The disk friction losses are calculated according to Table 3.6 and Chap. 9.1, from Eq. (T3.10.9). The influence of different roughnesses on the casing walls and impeller shrouds can be captured by means of Eq. (T3.10.8). Expeller vanes on the rear shroud, if any, are taken into account by Eq. (T3.10.10), the effect of the leakage through the impeller sidewall gap by Eq. (T3.10.12), the roughness of the impeller disks by Eq. (T3.10.7) and the heating of the fluid in the impeller sidewall gap in case of very high viscosity ($\nu > 400 \times 10^{-6}$ m²/s) by the empirical factor f_{therm}, which is discussed in Chap. 13.1. The volumetric losses are estimated from Eq. (T3.10.4) and the mechanical losses from Eq. (T3.10.14).

Roughness: According to Eq. (T3.10.5) and Chap. 1.5.2 the equivalence factor $c_{eq} = 2.6$ is used to assess the roughness of the various components while the maximum roughness is assumed to be 6-times the average roughness. For simplification the Reynolds number for impeller shrouds and hydraulic channels is calculated with r_2. The effect of these assumptions is mitigated because only the ratios of friction coefficients occur in Eqs. (T3.10.8, 3.10.19 and 3.10.20). The roughnesses for impeller and collector are averaged according to the empirical expression given by Eq. (T3.10.6). In doing so, a weighted average of the roughness is calculated which depends on the specific speed, [15].

In [15] a collection of 32 tests with different roughnesses and Reynolds numbers were analyzed according to the procedure given in Table 3.10. The empirical coefficients for calculating the relevant mean roughness from Eq. (T3.10.6) and the component of the scalable hydraulic losses from Eq. (T3.10.18) were optimized in order to minimize the standard deviation. The investigation in [15] covers the range: $n_q = 7$ to 135; $d_2 = 180$ to 405 mm; $u_2 = 22$ to 113 m/s; n = 1200 to 7000 rpm; T = 20 to 160 °C; Re = 2.5×10^6 to 9.1×10^7; equivalent sand roughness $\varepsilon = 1$ to 130 μm and average roughness $\varepsilon_{CLA} = 0.4$ to 75 μm.

The standard deviation between measurement and calculation for f_η is ±1.0% and for $f_{\eta h}$ it is ±1.5%. The scatter increases with decreasing specific speed. A variation of the equivalence factor c_{eq} yielded no improvement; $c_{eq} = 2.6$ is therefore a useful assumption.

The relationship $a_\varepsilon = 0.98 - 0.0012\, n_q\, f_q^{0.5}$ found for weighting the roughness in the impeller and collector shows that the effect of the roughness in the impeller is quite weak in comparison with the diffusing elements, even at high specific speeds. The portion of scalable losses also drops with increasing specific speed, Eq. (T3.10.18). Both functions are shown in Fig. 3.34. Equation (T3.10.18) supplies quite similar values as the expression (1-V) in Eq. (T3.9.12).

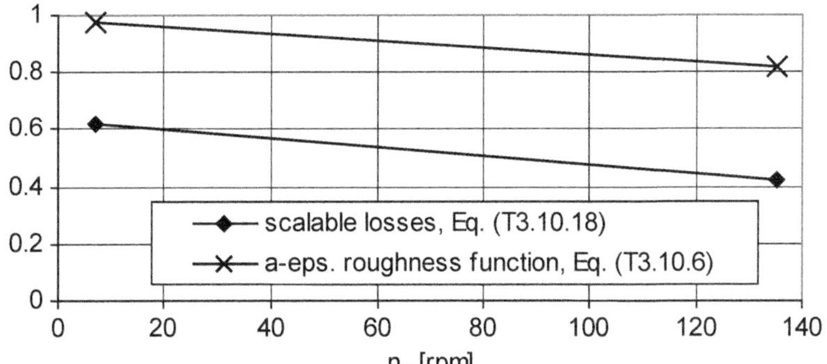

Fig. 3.34 Weighting the roughness a_ε acc. to Eq. (T3.10.6) and fraction of scalable losses acc. to Eq. (T3.10.18): $a_1\text{-}b_1 \times n_q \times f_q^{0.5}$

Uncertainties of efficiency calculation: Depending on the task set (efficiency scaling, assessment of roughness effects or calculation for highly viscous media), determining the efficiency is subject to a series of uncertainties:

1. The scaling concerns minor differences between two (relative to the scaling amount) large numbers. Even if the efficiency is measured with great accuracy in the test, the uncertainty in the efficiency *differences* $\Delta\eta = \eta_a - \eta_M$ determined on model and prototype is considerable; more so since η_a and η_M generally are measured with different instruments in different test circuits. The experimental verification of scaling formulae is therefore very difficult.
2. The casting and manufacturing tolerances of model and prototype are different.
3. The annular seal clearances are usually selected as functions of the size. They are not always strictly geometrically similar to the model. Even the seal design may not be geometrically similar.
4. For economic reasons, minor design differences are often unavoidable. They can be caused, for example, by the annular seal geometry, the impeller fixation on the shaft, the impeller sidewall gaps, or the inlet or outlet chamber.
5. The calculation of the disk friction losses and annular seal leakages is quite uncertain.

Like turbulence, the roughness calls for a statistical rather than a deterministic description; *the fluid-dynamical relevant quantification of the roughness is one of the main difficulties in determining the efficiency*:

- The effect of roughness on the flow cannot be determined, not even by accurate measurement of the surfaces, which is rarely attempted in practice.
- Frequently the roughness varies locally in the various channels depending on the accessibility for dressing or machining.
- The losses due to roughness are caused by the interactions of the velocity profile near the wall and the turbulence with the roughness elevations. The fine structures of roughness and turbulence are, consequently, responsible for the losses. The interaction between roughness and turbulence is determined on the one hand by the size, shape and number of roughness elevations and, on the other hand, by the size and frequency of the turbulence eddies near the wall. The turbulence structure depends on the local velocity distribution as given by accelerated or decelerated flow, Coriolis and centrifugal forces and flow separation (refer also to Fig. 1.15, Chap. 1.5.2 and [16]). An example of the interaction between roughness and turbulence is the increase in pressure drop caused by the "ripple roughness" quoted in Chap. 1.5.2. Another example is the reduction of the disk friction through fine grooves in circumferential direction (Chap. 3.6.1).
- If the pressure decreases in the flow direction ($dp/dx < 0$), the losses increase with growing wall shear stress (or roughness). In decelerated flow, however, with $dp/dx > 0$ (e.g. in a diffuser) the wall shear stress is reduced with increasing roughness and becomes $\tau_w \approx 0$ during separation. The "friction losses"

therefore decrease with increasing roughness, while the "mixing losses" and the overall losses increase. The concept of distinguishing between "friction" and "form or mixing losses" therefore does not make much sense in decelerated flow.

3.11 Minimization of Losses

In practice, the efficiency achievable is influenced by the boundary conditions imposed by the specific application and the mechanical design. Pump size, the required suction specific speed, demands as to the shape of the Q-H-curve and the manufacturing costs also have a considerable impact. Any attempt at increasing the efficiency of a pump which already reflects the state of the art has to rely on introducing at several locations rather small improvements which sum up to yield the desired effect. To minimize losses, it is necessary to first analyze the secondary losses and the hydraulic losses of the existing pump or the design at hand using Tables 3.5–3.8. At the same time these tables serve to quantitatively evaluate the potential of improvement by possible modifications.

Hydraulic losses (refer also to Chaps. 7 and 8):

- A uniform approach flow reduces the losses in the impeller.
- Optimize blade loading and pressure distribution. Excessively long blades cause friction losses, too short blades mixing losses.
- An excessively large impeller outlet width causes mixing losses.
- An excessively large inlet diameter causes increased mixing, friction and leakage losses through the annular seal and impairs the partload performance.
- In the case of diffusers and volute casings, the diffuser must be optimized according to Chap. 1.6.
- Carefully optimize the flow deceleration in diffusing channels: (1) Design for steady, efficient deceleration downstream of the diffuser or volute throat area where velocities (and friction losses) are high; (2) Avoid a sudden increase of the diffuser opening angles near the diffuser outlet where boundary layers are thick and at the brink of stall.
- Avoid flow separation, sharp curvature and unsteadiness in the development of curvature, angles and flow areas.
- Reduce roughness so that hydraulically smooth characteristics are obtained.
- When replacing an oversized impeller in an existing pump by a retrofit impeller adapted to lower flow, the volute or diffuser remains too large (unless it would be modified too). The increase in efficiency expected from the original performance curves will not materialize.

Secondary losses (refer also to Chap. 3.6**):**

- Disk friction: Avoid ribs or cavities in the impeller sidewall gaps since these would slow down the rotation of the fluid. Avoid rough walls and large impeller sidewall gap volumes.
- At low specific speeds the disk friction losses can be noticeably reduced by selecting a high head coefficient, but the effects on the Q-H-curve according to Chaps. 4 and 5 have to be considered.
- Annular seal losses: A reduction of the annular seal clearance would be most beneficial, but the reliability of the machine must not be jeopardized under any circumstances. If very tight clearances are selected, it must be ensured that no damage can result from contact between rotor and stator. The risk of seizure can be mitigated by the correct choice of materials or the surface structure of the annular seal, Chap. 14.4.4. Tight seal clearances that rapidly wear in operation are of little benefit to the operator. Fine grooves, hole or honeycomb patterns reduce the annular seal loss. Increasing the length of the annular seal usually brings only a moderate reduction in leakage.
- Flow through the axial thrust balance device: In the case of multistage pumps a balance disk is superior to a balance piston in terms of efficiency. However, a balance piston is less sensitive and therefore in many applications it is more reliable than a disk.
- Mechanical losses: Anti-friction bearings cause fewer losses than plain bearings. Mechanical seals use up less power than stuffing boxes. However design aspects usually determine the choice of these elements.

3.12 Compendium of Equations for Hydraulic Calculations

All calculations discussed in Chap. 3 can be performed using the following tables. Together with the list of symbols and the dimensions and quantities defined in Tables 0.1 and 0.2, these tables are largely self-explanatory.

Table 3.1 Velocity triangle at impeller inlet		Eq.
Quantities given or selected	n, Q_{La}, d_2, d_1, d_n, b_2, z_{La}, α_1, e_1, e, η_h	---
Circumferential speed	$u_1 = \pi\, d_1\, n/60$	3.1.1
Meridional component of the absolute velocity b_1 see sketch at bottom of table	$c_{1m} = \dfrac{Q_{La}}{f_q\, A_1}$ $A_1 = \dfrac{\pi}{4}(d_1^2 - d_n^2)$ $A_1 = \pi\, d_{1b}\, b_1$	3.1.2
Circumferential component of absolute velocity	$c_{1u} = \dfrac{c_{1m}}{\tan\alpha_1}$	3.1.3
Relative velocity	$w_1 = \sqrt{c_{1m}^2 + (u_1 - c_{1u})^2}$	3.1.4
Flow coefficient	$\varphi_1 = c_{1m} / u_1$	3.1.5
Flow angle without blockage	$\beta_1 = \arctan \dfrac{c_{1m}}{u_1 - c_{1u}}$	3.1.6
Blade blockage	$\tau_1 = \left\{1 - \dfrac{z_{La}\, e_1}{\pi\, d_1 \sin\beta_{1B} \sin\lambda_{La}}\right\}^{-1}$	3.1.7
Flow angle with blockage	$\beta_1' = \arctan \dfrac{c_{1m}\,\tau_1}{u_1 - c_{1u}}$	3.1.8
Blade angle for selected incidence, i_1 defined by: $i_1' \equiv \beta_{1B} - \beta_1'$	$\beta_{1B} = \beta_1' + i_1'$	3.1.9
Shockless inlet defined by $\beta_1' = \beta_{1B}$ giving c_{1m}	$\varphi_{1,SF} = \left(1 - \dfrac{c_{1u}}{u_1}\right)\dfrac{\tan\beta_{1B}}{\tau_1}$	3.1.10

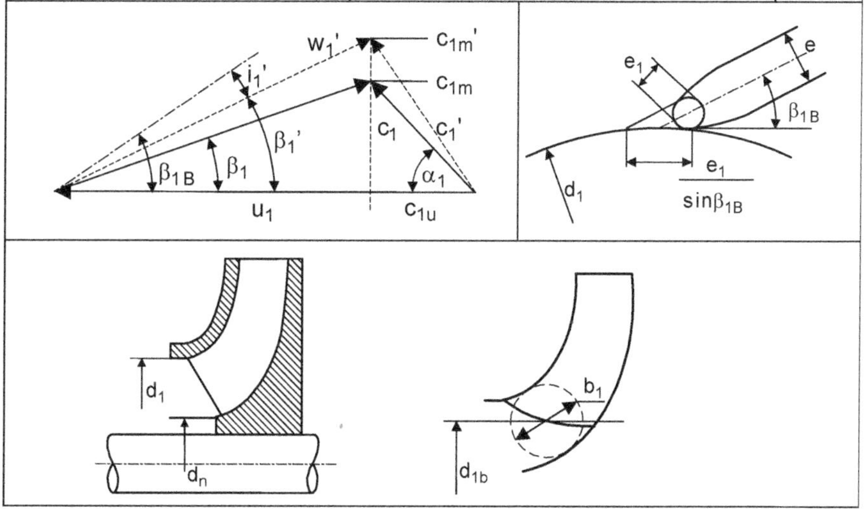

Table 3.2 Velocity triangle at impeller outlet			Eq.
Circumferential speed		$u_2 = \pi\, d_2 n/60$	3.2.1
Meridional component of absolute velocity $A_2 = \pi\, d_{2b}\, b_2$		$c_{2m} = \frac{Q_{La}}{f_q\, A_2}$ $\quad \varphi_{2,La} = \frac{c_{2m}}{u_2}$	3.2.2
Blade blockage		$\tau_2 = \left\{1 - \frac{e\, z_{La}}{\pi\, d_2\, \sin\beta_{2B}\, \sin\lambda_{La}}\right\}^{-1}$	3.2.3
Influence of impeller inlet diameter on slip factor $k_w = 1$ for $d_{1m}^* \le \varepsilon_{Lim}$		$\varepsilon_{Lim} = \exp\left\{-\frac{8.16\,\sin\beta_{2B}}{z_{La}}\right\}$	3.2.4
		$k_w = 1 - \left(\frac{d_{1m}^* - \varepsilon_{Lim}}{1 - \varepsilon_{Lim}}\right)^3$	3.2.5
Slip factor for $z_{La} \ge 3$ Radial: $f_1 = 0{,}98$ Semi-axial: $f_1 = 1{,}02 + 1{,}2 \cdot 10^{-3}\,(n_q - 50)$		$\gamma = f_1 \left(1 - \frac{\sqrt{\sin\beta_{2B}}}{z_{La}^{\ 0.7}}\right) k_w$	3.2.6
Circumferential component of absolute velocity	*Prediction*	$c_{2u} = u_2 \left(\gamma - \frac{c_{2m}\ \tau_2}{u_2 \tan\beta_{2B}}\right)$	3.2.7
	Calculation from measured head H	$c_{2u} = \frac{g\, H}{\eta_h\, u_2} + \frac{u_{1m}\, c_{1u}}{u_2}$	3.2.8
Slip factor from test		$\gamma = \frac{c_{2u}}{u_2} + \frac{\varphi_{2,La}\ \tau_2}{\tan\beta_{2B}}$	3.2.9
Absolute velocity		$c_2 = \sqrt{c_{2m}^{\ 2} + c_{2u}^{\ 2}}$	3.2.10
Relative velocity		$w_{2u} = u_2 - c_{2u}$	3.2.11
		$w_2 = \sqrt{c_{2m}^{\ 2} + w_{2u}^{\ 2}}$	3.2.12
Absolute outlet angle without blockage		$\alpha_2 = \arctan c_{2m}/c_{2u}$	3.2.13
Relative outlet angle with blockage		$\beta_2' = \arctan c_{2m}\, \tau_2 / w_{2u}$	3.2.14
Relative outlet angle without blockage		$\beta_2 = \arctan c_{2m} / w_{2u}$	3.2.15
Deviation angle		$\delta' = \beta_{2B} - \beta_2'$ or: $\delta = \beta_{2B} - \beta_2$	3.2.16

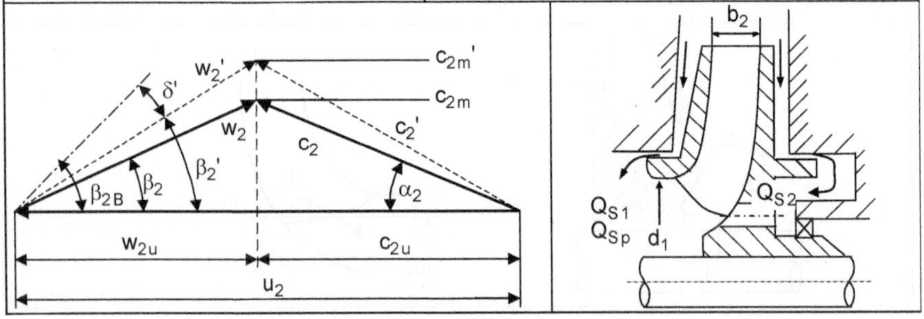

Table 3.3 Work transfer in impeller		Eq.
Specific work of blades Y_{sch} (no losses) Theoretical head: $H_{th} = Y_{th}/g$	$Y_{sch} = g\,H_{th} = u_2\,c_{2u} - u_1\,c_{1u} = u_2^2\left(\frac{c_{2u}}{u_2} - d^*_{1m}\frac{c_{1u}}{u_2}\right)$	3.3.1
	$Y_{sch} = \frac{1}{2}\left(u_2^2 - u_1^2 + w_1^2 - w_2^2 + c_2^2 - c_1^2\right)$	3.3.2
Specific work related to increase of static pressure in impeller (without losses)	$Y_{P,th} = g\,H_{P,th} = \frac{1}{2}\left(u_2^2 - u_1^2 + w_1^2 - w_2^2\right) = \frac{p_2 - p_1}{\rho} + g\,Z_{La}$	3.3.3
	$Y_{P,th} \equiv g\,H_{p,th} = u_2\,c_{2u} - u_1\,c_{1u} - \frac{1}{2}\left(c_2^2 - c_1^2\right)$	3.3.4
Rise of total pressure (enthalpy) in impeller	$Y_{tot,La} \equiv \Delta h_{tot} = \frac{p_2 - p_1}{\rho} + \frac{1}{2}\left(c_2^2 - c_1^2\right) = \frac{p_{2tot} - p_{1tot}}{\rho}$	3.3.5
Impeller losses	$g\,Z_{La} = Y_{sch} - Y_{tot,La} = Y_{P,th} - \frac{p_2 - p_1}{\rho}$	3.3.6
Head	$H = \frac{\eta_h\,u_2^2}{g}\left\{\gamma - \frac{Q_{La}}{f_q\,A_2\,u_2\,\tan\beta_{2B}}\left[\tau_2 + \frac{A_2\,d^*_{1m}\tan\beta_{2B}}{A_1\tan\alpha_1}\right]\right\}$	3.3.7
Static head rise in impeller	$H_P = \frac{1}{2g}\left(u_2^2 - u_1^2 + w_1^2 - w_2^2\right) - Z_{La}$	3.3.8
Theoretical head coefficient (without losses)	$\psi_{th} = 2\left(\frac{c_{2u}}{u_2} - d^*_{1m}\frac{c_{1u}}{u_2}\right)$	3.3.9
	$\psi_{th} = 2\left\{\gamma - \frac{\varphi_{2,La}}{\tan\beta_{2B}}\left[\tau_2 + \frac{A_2\,d^*_{1m}\tan\beta_{2B}}{A_1\tan\alpha_1}\right]\right\}$	3.3.10
Head coefficient	$\psi = 2\eta_h\left\{\gamma - \frac{\varphi_{2,La}}{\tan\beta_{2B}}\left[\tau_2 + \frac{A_2\,d^*_{1m}\tan\beta_{2B}}{A_1\tan\alpha_1}\right]\right\}$	3.3.11
Static head coefficient	$\psi_P = 1 - \frac{u_1^2}{u_2^2} + \frac{w_1^2}{u_2^2} - \frac{w_2^2}{u_2^2} - \zeta_{La}$	3.3.12
Specific work of impeller in relative system (without losses)	$Y_{sch} = u_2^2 - u_1^2 - \left(u_2\,w_{2u} - u_1\,w_{1u}\right)$	3.3.13
Total enthalpy rise in relative system	$Y_{tot,La} = \frac{p_2 - p_1}{\rho} + \frac{1}{2}\left\{w_{2m}^2 - w_{1m}^2 + (u_2 - w_{2u})^2 - (u_1 - w_{1u})^2\right\}$ $w_{2m} = c_{2m}$ $\quad$ $w_{1m} = c_{1m}$	3.3.14
u, c, w are referred either to the average stream line (geometric or mass average) or to any stream line selected (a, b, c, m, i) for calculation on a stream line.		
Relation between Q_{opt} and specific speed	$Q_{opt} = f_q\,\frac{\omega_s^2\,\psi_{opt}^{1.5}}{2\sqrt{2}}\,\omega\,r_2^3 = f_q\,n_q^2\left(\frac{\psi_{opt}}{2\,g}\right)^{1.5}\left(\frac{\pi}{60}\right)^3 n\,d_2^3$	3.3.15

Table 3.4 Scaling laws and dimensionless coefficients

Scaling from model (subscript: M) to prototype (subscript: a)				Eq.
Scaling laws	Flow rate	$Q_a = Q_M \frac{n_a}{n_M}\left(\frac{D_a}{D_M}\right)^3 \frac{\eta_{v.a}}{\eta_{v.M}}$	1. Approximation: $\eta_{v.a} = \eta_{v.M}$	3.4.1
	Head	$H_a = H_M \left(\frac{n_a}{n_M}\right)^2 \left(\frac{D_a}{D_M}\right)^2 \frac{z_{st.a}}{z_{st.M}} \frac{\eta_{h.a}}{\eta_{h.M}}$	1. Approximation: $\eta_{h.a} = \eta_{h.M}$	3.4.2
	NPSH	$NPSH_a = NPSH_M \left(\frac{n_a}{n_M}\right)^2 \left(\frac{D_a}{D_M}\right)^2$		3.4.3
	Power	$P_a = P_M \left(\frac{n_a}{n_M}\right)^3 \left(\frac{D_a}{D_M}\right)^5 \frac{\rho_a}{\rho_M} \frac{z_{st.a}}{z_{st.M}} \frac{\eta_M}{\eta_a}$	1. Approximation: $\eta_a = \eta_M$	3.4.4
	Forces	$F_a = F_M \left(\frac{n_a}{n_M}\right)^2 \left(\frac{D_a}{D_M}\right)^4 \frac{\rho_a}{\rho_M}$		3.4.5
Dimensionless coefficients	Flow coefficient (outlet)	$\varphi_2 = \frac{Q/f_q}{\pi\, d_{2b}\, b_2\, u_2} = \frac{\psi^{1.5}}{b_2^*}\left(\frac{n_q}{316}\right)^2$	$\varphi_{2.La} = \frac{Q_{La}/f_q}{\pi\, d_{2b}\, b_2\, u_2} = \frac{c_{2m}}{u_2}$	3.4.6
	Flow coefficient (inlet)	$\varphi_1 = \frac{4 Q_{La}}{f_q\, \pi\, (d_1^2 - d_n^2)\, u_1} = \frac{4\, b_2^*\, \varphi_{2.La}}{d_1^{*3}\, k_n} = \frac{c_{1m}}{u_1}$		3.4.7
	Head coefficient	$\psi = \frac{2\,g\,H}{u_2^2} = \frac{2\,Y}{u_2^2}$		3.4.8
	Power coefficient with b_2	$\lambda = \frac{2\,P}{f_q\, \rho\, \pi\, z_{st}\, d_{2b}\, b_2\, u_{2a}^3} = \frac{\varphi_2\, \psi}{\eta}$		3.4.9
	Power coefficient with d_2	$\lambda_D = \frac{2\,P}{z_{st}\, \rho\, d_2^2\, u_2^3} = \lambda\, \pi\, b_2^*\, f_q$		3.4.10
	Cavitation coefficient	$\sigma = \frac{2g\, NPSH}{u_1^2}$		3.4.11
	Reynolds number	$Re = \frac{c\,D}{\nu}$; $Re = \frac{w\,L}{\nu}$	$Re_u = \frac{u\,R}{\nu}$; $Re_u = \frac{2\,s\,u_{sp}}{\nu}$	3.4.12
	Froude no. Fr Euler no. Eu	$Fr = \frac{c}{\sqrt{g\,D}}$	$Eu = \frac{2\Delta p}{\rho\, c^2}$	3.4.13 3.4.14
Specific speeds	Specific speed	$n_q = n \frac{\sqrt{Q_{opt}/f_q}}{H_{opt}^{0.75}} = 315.6 \frac{\sqrt{\varphi_{2opt}\, b_2^*}}{\psi_{opt}^{0.75}} = 52.9\, \omega_s$	n [rpm] Q [m³/s] H [m]	3.4.15
	"Type number" DIN 24260 n [rpm]	$n_q = 5.55\, n \frac{\sqrt{Q_{opt}/f_q}}{(g\, H_{opt})^{0.75}}$	$\omega_s = \frac{\omega\sqrt{Q_{opt}/f_q}}{(g\, H_{opt})^{0.75}} = \frac{n_q}{52.9}$	3.4.16
	Suction specific speed	$n_{ss} = n \frac{\sqrt{Q_{opt}/f_q}}{NPSH_{opt}^{0.75}} = 157.8 \frac{\sqrt{\varphi_{1opt}\, k_n}}{\sigma_{opt}^{0.75}}$	n [rpm] Q [m³/s] NPSH [m]	3.4.17
	USA-units	$n_q = 0.0194\, N_s$; $n_{ss} = 0.0194\, N_{ss}$	N_s. N_{ss} in: Q [gpm], H [ft]	3.4.18

Table 3.5 Power balance, efficiencies and losses		Eq.
Reference values	$Q_{Ref} = 1 m^3/s \quad n_{Ref} = 1500$ min	
Power balance	$P = \sum_{st} \frac{\rho\, g\, H\, Q}{\eta_v\, \eta_h} + \sum_{st} P_{RR} + \sum P_{S3} + P_m + P_{er} + P_{Rez}$	3.5.1
Useful power P_u	$P_u = \rho\, g\, H_{tot}\, Q$	3.5.2
Overall efficiency (at coupling) η	$\eta = \frac{P_u}{P} = \frac{\rho\, g\, H_{tot}\, Q}{P}$ $\eta = \frac{\eta_v\, \eta_h\, \rho g\, H_{tot}\, Q}{\rho\, g\, H_{tot}\, Q + \eta_v\, \eta_h\, (\sum_{st} P_{RR} + \sum P_{S3} + P_m + P_{er} + P_{Rez})}$	3.5.3
Inner power P_i	$P_i = P - P_m$	3.5.4
Inner efficiency η_i	$\eta_i = \frac{\rho\, g\, H_{tot} Q}{P_i} = \frac{\eta}{\eta_m}$	3.5.5
Mechanical efficiency η_m	$\eta_m = 1 - \frac{P_m}{P}$	3.5.6
Mechanical losses P_m	$\frac{P_m}{P_{opt}} = 0.0045 \left(\frac{Q_{Ref}}{Q}\right)^{0.4} \left(\frac{n_{Ref}}{n}\right)^{0.3}$	3.5.6a
Stage efficiency	$\eta_{st} = \frac{\rho\, g\, (H_{tot} + Z_{EA})(Q + \frac{z_{st1}}{z_{st}} Q_E + \frac{z_{st2}}{z_{st}} Q_{CB})}{P - P_m - P_{er}}$	3.5.7
Calculation of hydraulic efficiency from test ($P_{Rec} = 0$)	$\eta_h = \frac{\rho\, g\, H_{tot} (Q + Q_{sp} + \frac{z_{st1}}{z_{st}} Q_E + \frac{z_{st2}}{z_{st}} Q_{CB} + Q_h)}{P - \sum P_{RR} - \sum P_{s3} - P_m - P_{er}} = \frac{1 + \frac{Q_{sp}}{Q} + \frac{z_{st1} Q_E}{z_{st} Q} + \frac{z_{st2} Q_{CB}}{z_{st} Q} + \frac{Q_h}{Q}}{\frac{1}{\eta} - \frac{\sum P_{RR} + \sum P_{s3} + P_m + P_{er}}{P_u}}$	3.5.8
Volumetric efficiency η_v	$\eta_v = \frac{Q}{Q + Q_{sp} + \frac{z_{st1}}{z_{st}} Q_E + \frac{z_{st2}}{z_{st}} Q_{CB} + Q_h} = \frac{1}{1 + \frac{Q_{sp}}{Q} + \frac{z_{st1} Q_E}{z_{st} Q} + \frac{z_{st2} Q_{CB}}{z_{st} Q} + \frac{Q_h}{Q}}$	3.5.9
Leakage loss at impeller (API-clearance)	$\frac{Q_{sp}}{Q_{opt}} = \frac{a\, z_H}{n_q^{\,m}}$ n_q / a / m: < 27 / 4.1 / 1.6; ≥ 27 / 0.15 / 0.6 With balance: Holes: $z_H = 2$; Else: $z_H = 1$	3.5.10
Leakage loss of inter-stage seal	$\frac{Q_{S3}}{Q_{opt}} = \frac{5.5}{n_q^{\,1.8}}$ $\frac{\Delta H_{S3}}{H_{opt}} \approx 0.4$	3.5.11
Power loss of inter-stage seal	$P_{S3} = \rho\, g\, \Delta H_{S3} Q_{S3}$ $\frac{P_{S3}}{P_{u,st,opt}} = \frac{2.2}{n_q^{\,1.8}}$	3.5.11a
Disk friction power loss of radial impellers. $f_{R,La}$ from Eq. (T3.6.6); f_L from Eq.(T3.6,7)	$\frac{P_{RR}}{P_{u,opt}} = \frac{770\, f_{R,La}\, f_L}{n_q^2\, \psi_{opt}^{\,2.5}\, Re^{0.2}\, f_q}$ Turbulent: $Re > 10^5$	3.5.12
	$\frac{P_{RR}}{P_{u,opt}} = \frac{38500\, k_{RR}}{n_q^2\, \psi_{opt}^{2.5}\, f_q}$ General	3.5.13

Table 3.6 Friction losses of rotating disks or cylinders

				Eq.
Reynolds number	$Re = \frac{u\,R}{\nu} = \frac{\omega\,R^2}{\nu}$	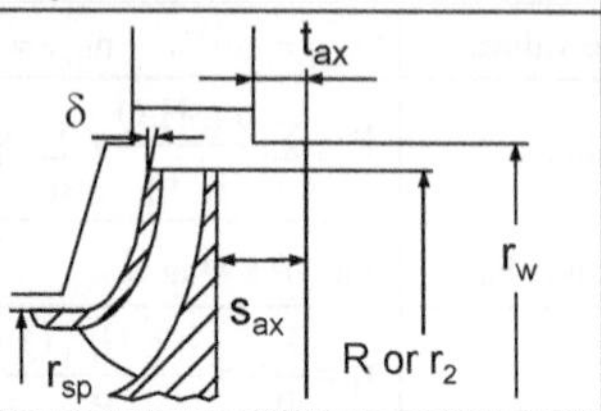		3.6.1
Friction power *per side* of a rotating **disk** for $\delta < 45°$	$P_{RR} = \frac{k_{RR}}{\cos\delta}\,\rho\,\omega^3 R^5 \left\{1 - \left(\frac{R_n}{R}\right)^5\right\}$			3.6.2
Valid for $Re > 10$	$k_{RR} = \frac{\pi\,R}{2 Re\, s_{ax}} + \frac{0.02}{Re^{0.2}} \cdot \frac{1 + \frac{s_{ax}}{R}}{1 + \frac{s_{ax}}{2\,R}}\, f_L\, f_{R,La}$			3.6.3
Friction power of a rotating **cylinder**; valid for $Re > 10$	$P_{RZ} = k_{RZ}\,\rho\,\omega^3 R^4 L$		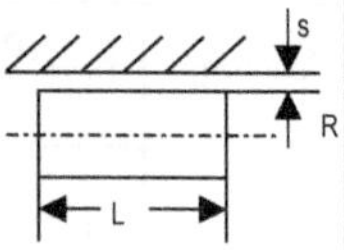	3.6.4
	$k_{RZ} = \frac{2\,\pi\,R}{Re\,s} + \frac{0{,}075}{Re^{0{,}2}} \cdot \frac{1 + \frac{S}{R}}{1 + \frac{S}{2R}} \cdot f_R$			3.6.5
Influence of roughness ε to be calculated with roughness of rotating disk	$f_{R,La} \equiv \frac{k_{RR}(\varepsilon)}{k_{RR}(\varepsilon = 0)} = \left\{ \frac{\log \frac{12.5}{Re}}{\log\left(0.2\frac{\varepsilon}{r_2} + \frac{12.5}{Re}\right)} \right\}^{2.15}$		The roughness of the casing is not taken into account here, but in Eq. (3.6.13). $\varepsilon = \varepsilon_{max}/c_{eq}$	3.6.6
Influence of leakage valid for: $r_{sp}/r_2 > 0{,}3$ und $k_E \approx 0{,}5$	$f_L \equiv \frac{k_{RR}(Q_{sp})}{k_{RR}(Q_{sp} = 0)} = \exp\left\{-350\,\varphi_{sp}\left(\left[\frac{r_2}{r_{sp}}\right]^a - 1\right)\right\}$		Leakage direction: *radially inwards: φ_{sp} positive; a = 1.0 *radially outwards: φ_{sp} negative; a = 0.75	3.6.7
Alternatives to Eq. (T3.6.3) hydraulically smooth; without leakage; analogous to [6]	$k_{RR} = \frac{\pi\,R}{2 Re\, s_{ax}}$	$Re_{lam} \le 8.7\left(\frac{Sax}{R}\right)^{-1.87}$	Laminar, merged boundary layers	3.6.8
	$k_{RR} = \frac{0.925}{Re^{0.5}}\left(\frac{Sax}{R}\right)^{0,1}$	$Re_{lam} < Re < 2 \times 10^5$	Laminar, separated boundary layers	3.6.9
	$k_{RR} = \frac{0.02}{Re^{0.25}}\left(\frac{R}{s_{ax}}\right)^{1/6}$	$10^5 < Re < 10^6$	Turbulent, merged boundary layers	3.6.10
	$k_{RR} = \frac{0.0255}{Re^{0.2}}\left(\frac{s_{ax}}{R}\right)^{0.1}$	$Re > 2{\times}10^5$	Turbulent, separated boundary layers	3.6.11
Taking into account all factors which influence rotation in the sidewall gap [14]	$k_{RR} = \frac{\pi\,R}{2 Re\, s_{ax}} + \frac{0.0625}{Re^{0.2}}\left(1 - k_o\right)^{1.75} f_{R,La}\, f_L$		Valid for $Re > 10$	3.6.12
	$k_o = \frac{1}{1 + \left(\frac{r_w}{r_2}\right)^2 \sqrt{\left(\frac{r_w}{r_2} + 5\frac{t_{ax}}{r_2}\right)\frac{c_{f,casing}}{c_{f,impeller}}}}$		c_f from Eq. (1.33)	3.6.13

Table 3.7 (1) Leakage losses I: Annular seals			Eq.
Static pressure rise created by impeller	$H_p = \frac{u_2^2 - u_1^2 + w_1^2 - w_2^2}{2g} - Z_{La}$	$H_p = R_G\,H$ $R_G \approx 0{,}75$ for $n_q < 40$	3.7.1
Rotation factor k	Q_{sp} radially inward: $k = 0.9\,y_{sp}^{0.087}$ Q_{sp} radially outward: $k = 0.24\,y_{sp}^{-0.096}$	$y_{sp} = Re_{u_2}^{0.3}\,\frac{s\,d_{sp}}{d_2^2}\sqrt{\frac{s}{L_{sp}}}$	3.7.2
Pressure difference across seal if leakage flows radially *inwards*	$\Delta H_{sp} = H_P - k^2\,\frac{u_2^2}{2g}\left\{1 - \frac{d_{sp}^2}{d_2^2}\right\}$	$Re_{u_2} = \frac{u_2\,r_2}{\nu}$	3.7.3
Pressure difference across seal if leakage flows radially *outwards*	$\Delta H_{S3} = H - H_P + k^2\,\frac{u_2^2}{2g}\left\{1 - \frac{d_{S3}^2}{d_2^2}\right\}$		3.7.4
Pressure difference across balance device	$\Delta H_{EK} = (z_{ST} - 1)H + H_p - k^2\,\frac{u_2^2}{2g}\left\{1 - \frac{d_{EK}^2}{d_2^2}\right\}$		3.7.5
Axial velocity in gap ΔH_{sp}, ΔH_{S3}, or ΔH_{EK} i = number of chambers	$c_{ax} = \sqrt{\dfrac{2g\,\Delta H}{\zeta_{EA} + \lambda\,\frac{L_{sp}}{2s} + \sum_i \left(\frac{d_{sp}}{d_{si}}\right)^2 \left(\frac{s}{s_i}\right)^2 \left\{\zeta_K + \lambda_i\,\frac{L_i}{2s_i}\right\}}}$		3.7.6
Leakage flow rate	$Q_{sp} = \pi\,d_{sp}\,s\,c_{ax}$		3.7.7
Loss coefficients	ζ_{EA} = 1 to 1.2 inlet + outlet loss; ζ_K = 1 to 1.3 loss per chamber		3.7.8
Reynolds numbers	$Re_u = \frac{2\,s\,u_{sp}}{\nu}$ $\quad u_{sp} = \frac{\pi\,d_{SP}\,n}{60}$ $\quad Re = \frac{2\,s\,c_{ax}}{\nu}$		3.7.9
Transition laminar/turbulent at $Re^* = 2000$	$Re^* = \sqrt{Re^2 + \frac{1}{4}Re_u^2} = Re\sqrt{1 + \frac{1}{4}\left(\frac{u_{sp}}{c_{ax}}\right)^2}$		3.7.10
Friction coefficient for laminar flow, use for $Re < 2000$ ($u_{sp} = 0$)	$\lambda_0 = \frac{96}{Re}\left(1 - 0.6\,\frac{e_x}{s}\right)$	e_x = eccentricity	3.7.11
Friction coefficient for turbulent flow, $Re > 2000$ ($u_{sp} = 0$)	$\lambda_0 = \frac{0.31}{\left\{\log\left(A + \frac{6.5}{Re}\right)\right\}^2}$	Seal surface: A Rough: 0.135 ε/s Serrated: 0.005 to 0.01 Isotropic pattern: 0.01 to 0.03	3.7.12
Transition laminar-turb.	use $\lambda = Max(\lambda_{lam}\,;\,\lambda_{tur})$ in order to avoid discontinuity		
Influence of rotation for turbulent flow $Re > 2000$	$\frac{\lambda}{\lambda_0} = \left\{1 + 0.19\left(\frac{Re_u}{Re}\right)^2\right\}^{0.375}$		3.7.13
Influence of rotation for flow with $Re < 2000$	$\frac{\lambda}{\lambda_0} = \left\{1 + 0.2\left(\frac{Re_u}{2000}\right)^2\right\}^{1.03}$		3.7.14

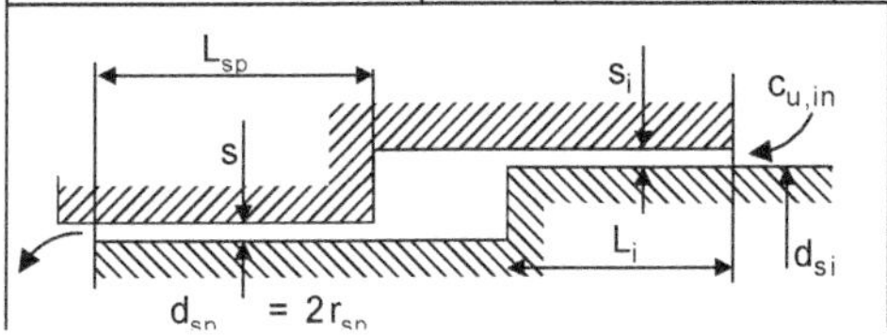

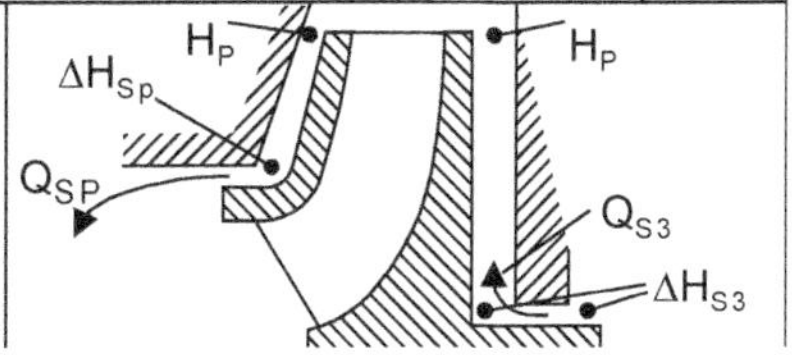

Table 3.7 (2) Leakage losses II: Radial or diagonal gaps, open impellers					Eq.
Semi-open impellers	Head loss	$\frac{\psi_{(s=o)}-\psi_{(s)}}{\psi_{(s=o)}} = \frac{2.5\frac{s}{d_2}}{\sqrt{b_2^*(1-d_1^*)z_{La}}\left(\frac{e}{t_2}\right)^{0.2} n_q^{0.1}(\sin\beta_2)^{1.2}(\sin\beta_1)^{0.4}}$			3.7.15
	Efficiency loss	$\frac{\eta_{(s=o)}-\eta_{(s)}}{\eta_{(s=o)}} = \frac{2}{3}\frac{\psi_{(s=o)}-\psi_{(s)}}{\psi_{(s=o)}}$		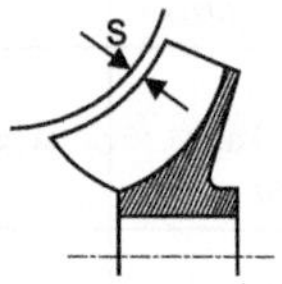	3.7.16
	$\Delta\eta$ according to [B.16]	$\Delta\eta = 0.3\frac{s}{b_2-s}$	for $\frac{s}{b_2-s} < 0.13$		3.7.16a
	Reduction of power consumption	$\frac{\lambda_{(s=o)}-\lambda_{(s)}}{\lambda_{(s=o)}} = \frac{1}{3}\frac{\psi_{(s=o)}-\psi_{(s)}}{\psi_{(s=o)}}$			3.7.17
Diagonal gap	Head loss	$\frac{\psi_{(s=o)}-\psi_{(s)}}{\psi_{(s=o)}} = (2.3 \text{ to } 3.3)\frac{s}{b_2}$		Upper value for smooth gap; lower value for gap with pump-out vanes	3.7.18
	Efficiency loss	$\frac{\eta_{(s=o)}-\eta_{(s)}}{\eta_{(s=o)}} = (1.6 \text{ to } 1.9)\frac{s}{b_2}$			3.7.19
	Increase of $NPSH_3$	$\frac{\sigma_{(s)}-\sigma_{(s=o)}}{\sigma_{(s=o)}} = (15 \text{ to } 40)\frac{s}{b_2}$		see. Fig. 3.15	3.7.20
Radi-al gap	Reynolds numbers	$Re = \frac{2\,s\,c_i}{\nu}\sqrt{a_r}$	$Re_u = \frac{2\,s\,\omega\,r_a}{\nu}\sqrt{a_r}$	$a_r \equiv \frac{r_i}{r_a}$	3.7.21
	Friction coefficient	$\lambda = f(Re, Re_u, \varepsilon)$ according to Table 3.7 (1)			
	Pressure difference across gap	according to Table 3.7 (1)			
	Constants	Leakage flow rate	Rotation faktor k: $Re < 10^4$	$Re > 10^4$	
		Radially inwards	0.4	0.5	
		Radially outwards	0.5	0.6	
	Radial velocity in gap at inner radius r_i 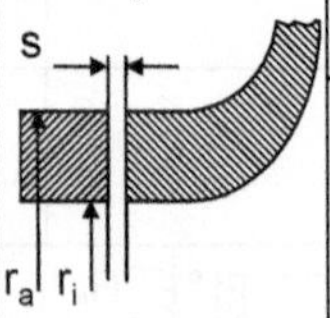	Leakage radially inwards	$c_i = \sqrt{\frac{\frac{2\Delta p}{\rho} - k^2\omega^2 r_a^2(1-a_r^2)}{\zeta_E\,a_r^2 + \frac{\lambda\,r_i}{2\,s}(1-a_r)+1}}$		3.7.22
		Leakage radially outwards	$c_i = \sqrt{\frac{\frac{2\Delta p}{\rho} + k^2\omega^2 r_a^2(1-a_r^2)}{\zeta_E + \frac{\lambda\,r_i}{2\,s}(1-a_r)+a_r^2}}$		3.7.23
	Leakage flow	$Q_{sp} = 2\pi\,r_i\,s\,c_i$			3.7.24

Table 3.7 (3) Screw-type pumping grooves in turbulent flow, [18]

Pumping grooves should be selected only if: $\frac{c_{ax}}{u_{sp}} < 0.7$ and $\frac{L_{sp}}{s} > 50$	The selection of the geometric parameters depends on $\frac{c_{ax}}{u_{sp}}$			
Given quantities	$d_{sp}, L_{sp}, n, \Delta H, s, \rho, \nu$			Eq.
Axial velocity in gap Estimation for first iteration: $\lambda = 0.06$	$c_{ax} = \sqrt{\dfrac{2g\ \Delta H}{\zeta_{EA} + \lambda \dfrac{L_{sp}}{2s}}}$			3.7.30
Loss coefficient	$\zeta_{EA} = 1.5$ for inlet and outlet loss			3.7.31
Reynolds numbers	$Re_u = \dfrac{2\, s\, u_{sp}}{\nu}$ $\quad u_{sp} = \dfrac{\pi\, d_{SP}\, n}{60}$ $\quad Re = \dfrac{2\, s\, c_{ax}}{\nu}$			3.7.32
Flow is turbulent if	$\dfrac{u_{sp}\, s}{\nu} > 700$			3.7.33
Number of threads (integer)	$z = \dfrac{\pi\, d_{SP} \sin\alpha}{a+b} = \dfrac{\pi\, d_{SP} \sin\alpha}{\left(\dfrac{a+b}{s}\right) s}$			3.7.34
Inclination angle $\alpha = 7$ to 13° (up to 15°)	$\alpha[°] = 13 - 8\dfrac{c_{ax}}{u_{SP}}$	Eqs. (3.7.39. and (3.7.40) are valid for the parameters given in the column on the right	$\alpha = 10°$	3.7.35
Depth of grooves h/s = 2,5 to 4	$\dfrac{h}{s} = 1.2\left(\dfrac{c_{ax}}{u_{SP}}\right)^{-0,68}$		$\dfrac{h}{s} = 2.5$	3.7.36
Width of grooves	$\dfrac{a}{a+b} = 0.5$ to 0.7		$\dfrac{a}{a+b} = 0.5$	3.7.36
Check that	$\dfrac{a+b}{s} = 15$ to 20		$\dfrac{a+b}{s} = 17$	3.7.38
Friction coefficient for $u_{sp} = 0$ in turbulent flow	$\lambda_0 = \dfrac{0.32}{\left\{\log\left(0.0022 + \dfrac{20}{Re}\right)\right\}^2}$	Range tested: Re = 400 to 21'000 $Re_u = 0$ to 37'000		3.7.39
Friction coefficient with rotation	$\lambda = \lambda_0 \left\{1 + k\dfrac{Re_u}{Re}\right\}^2$	$k = 0.0243 Re^{0.2255}$		3.7.40
With the friction coefficient from Eq. (3.7.40) the axial velocity c_{ax} is calculated iteratively from Eq. (3.7.30)				
Leakage flow rate	$Q_{sp} = \pi\, d_{sp}\, s\, c_{ax}$			3.7.41

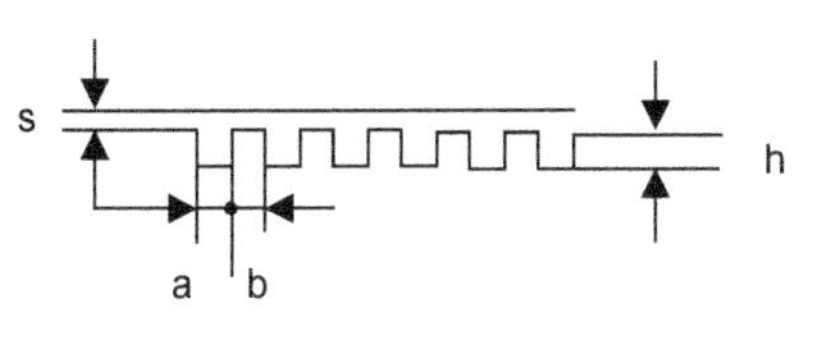

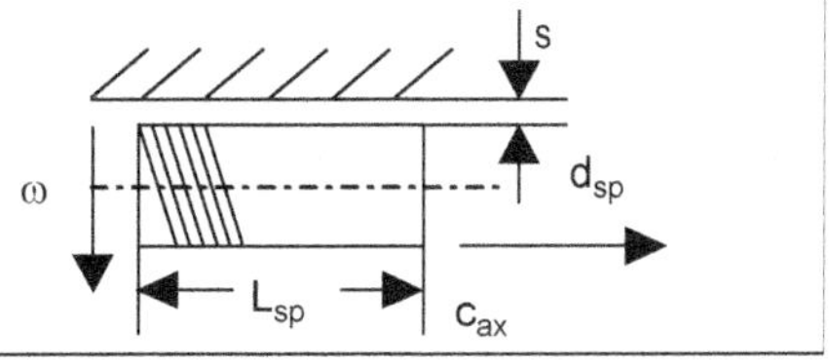

Table 3.7(4) Serrated annular seals		
Land: a = t - b	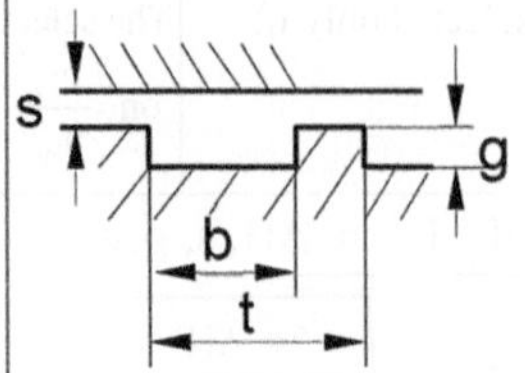	Eq.
Pressure loss of seal without ζ_{EA} Eq. (1.12)	$\Delta p_v = \frac{\rho}{2}(c_{VC} - c_b)^2 = \frac{\rho}{2} c_{ax}^2 \frac{\lambda_{0,eq} L_{sp}}{2\,s}$	3.7.42
Maximum velocity with jet contraction α_z	$c_{VC} = \frac{c_{ax}}{\alpha_z}$	3.7.43
Velocity at end of groove at length x = b	$c_b = 1.75\,\beta\left(\frac{s}{b}\right)^{0.5} c_{ax}$	3.7.44
Jet contraction	$\alpha_z = 0.79 \times e^{-0.001(a/s)}$	3.7.45
Equivalent friction factor for seal: $\lambda_{0,eq} = \frac{a}{t}\left(1+2\frac{g}{t}\right)\lambda_{smooth} + 2\left(\frac{s}{t}\right)\left(\frac{b}{t}\right)\left\{\frac{1}{\alpha_z} - 1.75\beta\left(\frac{s}{b}\right)^{0.5}\right\}^2$ Empirical factor $\beta = 0.87$		3.7.46
Recommendations	g/s = **3** to 6 g/b = 0.2 ….**0.25** ….0.3 b/t = 0.7 ….**0.8** ……0.9	

Test parameters covered by the correlations

		min.	max.
s	mm	0.3	0.9
g	mm	1.0	5.0
b	mm	1.5	17.5
t	mm	5.0	20.0
a = t - b	mm	1.0	15.0
b/s	-	4.8	29.0
b/t	-	0.3	0.9
t/s	-	10.8	32.2
g/s	-	3.2	5.4
g/b		0.1	1.0
a/s	-	2.7	16.2

Table 3.8 (1) Hydraulic losses in impeller			Eq.
Hydraulic efficiency	$\eta_h = \frac{Y}{Y_{sch}} = \frac{H}{H_{th}} = \frac{H}{H + Z_{La} + Z_{Le} + Z_{EA}} = \frac{\psi}{\psi + \zeta_{La} + \zeta_{Le} + \zeta_{EA}}$		3.8.1
Average relative velocity in impeller channel	$w_{av} = \frac{2Q_{La}}{f_q\, z_{La}\,(a_2\, b_2 + A_{1q})}$		3.8.2
Friction coefficient (ε = roughness)	$Re = \frac{w_{av}\, L_{sch}}{\nu}$	$c_f = \frac{0.136}{\left\{-\log\left(0.2\frac{\varepsilon}{L_{sch}} + \frac{12.5}{Re}\right)\right\}^{2.15}}$	3.8.3
Hydraulic diameter	$D_h = \frac{2\,(a_2\, b_2 + A_{1q})}{a_1 + b_1 + a_2 + b_2}$		3.8.4
Dissipation coefficient	$c_d = (c_f + 0.0015)\,(1{,}1 + 4b_2{}^*)$		3.8.5
Friction and mixing losses	$\zeta_{La,R} \equiv 2g\,\frac{Z_{La,R}}{u_2{}^2} = 4c_d\,\frac{L_{sch}}{D_h}\left(\frac{w_{av}}{u_2}\right)^2$		3.8.6
Relative velocity vector	$w_{1m} = \sqrt{c_{1m}{}^2 + (u_1 - c_{1u})^2}$		3.8.7
Velocity in impeller throat	$w_{1q} = \frac{Q_{La}}{f_q\; z_{La}\; A_{1q}}$		3.8.8
Shock loss at impeller inlet	$\zeta_{La,C} \equiv 2g\,\frac{Z_{La,C}}{u_2{}^2} = 0.3\left(\frac{w_{1m} - w_{1q}}{u_2}\right)^2$ only for: $\frac{w_{1q}}{w_{1m}} > 0.65$		3.8.9
Impeller loss	$\zeta_{La} = \zeta_{La,R} + \zeta_{La,C}$		3.8.10
Alternative calculation with wetted surface and c_f instead of c_d			
Wetted surface in *radial impellers* per impeller side	$A_{ben} = \frac{\pi}{4}\left(d_{2a}^2 + d_{2i}^2 - d_1^2 - d_n^2\right) + z_{La}\, L_{sch}\,(b_1 + b_2)$		3.8.11
Average relative velocity	$\overline{w} = 0.5\,(w_1 + w_2)$		3.8.12
Dissipated power per impeller side	$P_d = \frac{\rho}{2} c_f\; \overline{w}^3 A_{ben} = \rho\; g\; Q\; Z_{La,R}$		3.8.13
Friction losses	$\zeta_{La,R} \equiv 2g\,\frac{Z_{La,R}}{u_2{}^2} = \frac{c_f\, A_{ben}}{\varphi_2\, A_2}\left(\frac{\overline{w}}{u_2}\right)^3$		3.8.14

Table 3.8 (2) Hydraulic losses in diffuser or volute					Eq.
Hydraulic efficiency		$\eta_h = \frac{Y}{Y_{sch}} = \frac{H}{H_{th}} = \frac{H}{H + Z_{La} + Z_{Le} + Z_{EA}} = \frac{\psi}{\psi + \zeta_{La} + \zeta_{Le} + \zeta_{EA}}$			3.8.1
Diffuser	Velocity vector at impeller outlet	$c_{2u} = \frac{gH}{\eta_h u_2} + d^*_{1m} c_{1u}$	$c_{2m} = \frac{Q_{La}}{\pi d_2 b_2}$	$c_2 = \sqrt{c_{2u}^2 + c_{2m}^2}$	3.8.15
Diffuser	Flow velocity in throat	$c_{3q} = \frac{Q + Q_E}{z_{Le} a_3 b_3}$			3.8.16
Diffuser	Diffuser c_p from Fig. 1.18 or 1.19	$A_R = \frac{a_4 b_4}{a_3 b_3}$	$\frac{L_{3-4}}{R_1} = L_{3-4}\sqrt{\frac{\pi}{a_3 b_3}}$	$c_p = f\left(A_R, \frac{L_{3-4}}{R_1}\right)$	3.8.17
Diffuser	Friction losses in inlet region	$\zeta_{2\text{-}3} \equiv \frac{2g Z_{2\text{-}3}}{u_2^2} = (c_f + 0.0015)\left(a_3^* + b_3^*\right)\frac{\pi^3\left(\varphi_2 b_2^*\right)^2}{8\left(z_{Le} a_3^* b_3^*\right)^3}\left(1 + \frac{c_2}{c_{3q}}\right)^3$			3.8.18
Diffuser	Diffuser loss including overflow channel	$\zeta_{Le} \equiv \frac{2g Z_{Le}}{u_2^2} = \zeta_{2-3} + \left(\frac{c_{3q}}{u_2}\right)^2\left\{0.3\left(\frac{c_2}{c_{3q}} - 1\right)^2 + 1 - c_p - \frac{1 - \zeta_{ov}}{A_R^2}\right\}$			3.8.19
Diffuser	Overflow channel (alone)	$\Delta\zeta_R = \zeta_{ov}\frac{c_4^2}{u_2^2}$	$\zeta_{ov} = 0.2$ to 1.5		3.8.20
Volute	Friction loss ΔA= wetted surface	$\zeta_{Sp,R} = \frac{2gZ_{Sp}}{u_2^2} = \frac{2 P_d}{\rho Q u_2^2} = \frac{1}{Q u_2^2}\Sigma\left(c_f + 0.0015\right)c^3\Delta A$			3.8.21
Volute	Diffuser/discharge nozzle (c_x = diffuser inlet velocity)	$\zeta_{Sp,D} = \frac{c_x^2}{u_2^2}\left(1 - c_p - \frac{1}{A_R^2}\right)$			3.8.22
Volute	Losses of volute including discharge nozzle	$\zeta_{Sp} = \zeta_{Sp,R} + \zeta_{Sp,D} + \zeta_{LS} + (\zeta_{Le} - \zeta_{Le,min})$ ζ_{LS} from Eq. (T3.8.24)			3.8.22a
Vaneless diffuser	Friction in vaneless diffuser with constant width	$\zeta_{LR} = \frac{2 c_f r_2}{b_3 \sin\alpha_3 \cos^2\alpha_3}\left(\frac{c_{2u}}{u_2}\right)^2\left(1 - \frac{r_2}{r_4}\right)$			3.8.23
Vaneless diffuser	Shock losses	$\zeta_{LS} = \varphi_{2,La}^2\left(\tau_2 - \frac{b_2}{b_3}\right)^2$			3.8.24
Deceleration and shock losses in vaned diffusers and volutes		$\zeta_{Le} - \zeta_{Le,min} = a + b\frac{c_{3q}}{c_2} + c\left(\frac{c_{3q}}{c_2}\right)^2$	volutes $c = 0.63$ $x_o = 0.87$ $a = c\,x_o^2$ $b = -2\,x_o\,c$	diffusers $c = 0.65$ $x_o = 0.93$ $a = c\,x_o^2$ $b = -2\,x_o\,c$	3.8.25
Add losses of vaneless space between impeller and diffuser or volute.					

Table 3.9 Statistical efficiency data and efficiency scaling

	Pump type	n_q	$Q_{Ref} = 1\ m^3/s$ Valid for $Q \geq 0.005\ m^3/s$		Eq.
Overall efficiency	Exponent		$m = 0.1\ a \left(\frac{Q_{Ref}}{Q}\right)^{0.15} \left(\frac{45}{n_q}\right)^{0.06}$	$Q \leq 1\ m^3/s$: $a = 1$; $Q > 1\ m^3/s$: $a = 0.5$	
	Radial pumps single-stage	≤ 100	$\eta_{opt} = 1 - 0.095\left(\frac{Q_{Ref}}{Q}\right)^m - 0.3\left\{0.35 - \log\frac{n_q}{23}\right\}^2 \left(\frac{Q_{Ref}}{Q}\right)^{0.05}$		3.9.1
			$\eta_{opt} = 0.94 - 0.05\left(\frac{Q_{Ref}}{Q}\frac{n}{n_{Ref}}X\right)^{0.213} - 0.29\left\{\log\frac{n_{q,Ref}}{n_q}\right\}^2$		3.9.1A
			$Q_{Ref} = 1\ m^3/s = 3600\ m^3/h$; $n_{Ref} = 1000$ rpm; $n_{q,Ref} = 44$		
			Roughness influence: $X = \left(\frac{\varepsilon}{\varepsilon_{Ref}}\right)^2$ Use equivalent sand roughness with $\varepsilon_{Ref} = 4\ \mu m$ corresponding to N7 in Table D1.3		
	Radial pumps multistage	≤ 60	$\eta_{opt} = 1 - 0.116\left(\frac{Q_{Ref}}{Q}\right)^m - 0.4\left\{0.26 - \log\frac{n_q}{25}\right\}^2$		3.9.2
	Semi-axial and axial pumps	≥ 45	$\eta_{opt} = 1 - 0.095\left(\frac{Q_{Ref}}{Q}\right)^m - 0.09\left\{\log\frac{n_q}{45}\right\}^{2.5}$		3.9.3
	Pumps with double-entry impellers	≤ 50	$\eta_{opt} = 1 - 0.095\left(\frac{Q_{Ref}}{Q}\right)^m - 0.35\left\{0.35 - \log\frac{n_q}{17.7}\right\}^2 \left(\frac{Q_{Ref}}{Q}\right)^{0.05}$		3.9.4
	Theoretically achievable efficiency		$\eta_{th,er} = \eta_{opt} + 0.35\left(\frac{Q_{Ref}}{Q}\right)^{0.08}(1 - \eta_{opt})$		3.9.5
Hydraulic efficiency	Exponent		$m = 0.08\ a \left(\frac{Q_{Ref}}{Q}\right)^{0.15} \left(\frac{45}{n_q}\right)^{0.06}$	$Q \leq 1\ m^3/s$: $a = 1$; $Q > 1\ m^3/s$: $a = 0.5$	
	Radial pumps single-stage	≤ 100	$\eta_{h,opt} = 1 - 0.055\left(\frac{Q_{Ref}}{Q}\right)^m - 0.2\left\{0.26 - \log\frac{n_q}{25}\right\}^2 \left(\frac{Q_{Ref}}{Q}\right)^{0.1}$		3.9.6
	Semi-axial and axial pumps	≥ 45	$\eta_{h,opt} = 1 - 0.055\left(\frac{Q_{Ref}}{Q}\right)^m - 0.09\left\{\log\frac{n_q}{45}\right\}^{2.5}$		3.9.7
	Radial pumps multistage	≤ 60	$\eta_{h,opt} = 1 - 0.065\left(\frac{Q_{Ref}}{Q}\right)^m - 0.23\left\{0.3 - \log\frac{n_q}{23}\right\}^2 \left(\frac{Q_{Ref}}{Q}\right)^{0.05}$		3.9.8
Loss due to balance holes			$\Delta\eta_{EL} = 0.018\left(\frac{25}{n_q}\right)^{1.6}$ $n_q < 40$	$n_q > 40$: $\Delta\eta_{EL} = 0.01$	3.9.9
Uncertainty of η			$\Delta\eta_{Tol} = \pm 0.2\,(1 - \eta_{opt})$		3.9.10
Efficiency scaling	Scaling of hydraulic efficiency		$\frac{\eta_{h,a}}{\eta_{h,M}} = \left\{1 - \eta_{h,M}\frac{\zeta_{R,M}}{\psi}\left(1 - \frac{c_{f,a}}{c_{f,M}}\right)\right\}^{-1}$	$\frac{\zeta_{R,M}}{\psi} = b\left(\frac{12}{n_q}\right)^{0,83}$ $b = 0.06$ to 0.1	3.9.11
	Quick estimate of overall efficiency		$\eta_a = 1 - (1 - \eta_M)\left(\frac{n_M}{n_a}\frac{\nu_a}{\nu_M}\right)^{0.07}\left(\frac{d_{2,M}}{d_{2,a}}\right)^{0.2}$		3.9.11A
	Scaling of stage efficiency		$\frac{1 - \eta_{St,a}}{1 - \eta_{St,M}} = V + \frac{1-V}{2}\left(\frac{c_{f,a}}{c_{f,M}} + \frac{k_{RR,a}}{k_{RR,M}}\right)$	$V = 0.3 + 0.4\frac{n_q}{200}$	3.9.12
	Efficiency scaling according to [N.5]:		$\eta_{hR} = \frac{\rho\, g\, H_{tot}(Q + Q_{sp} + Q_E)}{P - P_m} = \frac{\eta}{\eta_v \eta_m}$	To be calculated for model (M) and prototype (a). $\eta_a = \eta_{hR,a} + \Delta\eta$. η_{hR} includes hydraulic and disk friction losses	3.9.13
			$\Delta\eta = (0.4\text{ to }0.6)(1 - \eta_{hR,M})\left[1 - \left(\frac{Re_M}{Re_a}\right)^{0.2}\right]$		

Table 3.10 (1) Influence of roughness and Reynolds number on efficiency

			Eq.
Quantities given for "model", baseline or water	n_M, $Q_{opt,M}$, $H_{opt,M}$, $d_{2,M}$, η_M, ν_M, roughness ε_M of the different components		
Quantities given for "prototype" or viscous fluid	n_a, $Q_{opt,a}$, $d_{2,a}$, ν_a, roughness ε_a of the different components		
Specific speed	$\omega_s = \omega \frac{\sqrt{Q_{opt}/f_q}}{\left(g H_{opt}\right)^{0,75}} = \frac{n_q}{52.9}$	ω_s is dimensionless; n_q is with rpm, m^3/s, m	3.10.1
Reynolds number	$Re = \frac{u_2 r_2}{\nu} = \frac{\omega r_2^2}{\nu}$	Re is used for hydraulic channels and disk friction	3.10.2
Estimated leakage flow, if no details of labyrinths are available	$\frac{Q_{sp}}{Q_{opt}} = \frac{4.1}{n_q^{1.6}} = \frac{7.16\times10^{-3}}{\omega_s^{1.6}}$	Can be used for Q_{sp} and Q_E as first approximation	3.10.3
Volumetric efficiency use Eq. (T3.5.9) for back-to-back	$\eta_{vol} = \frac{Q}{Q + Q_{sp} + Q_E + Q_{s3}\frac{\Delta H_{s3}(z_{st}-1)}{H_{st} z_{st}}}$	$\Delta H_{s3}/H_{st} = 0.4$ is a good estimation	3.10.4
Equivalent (sand) roughness $c_{eq} = 2.6$	$\varepsilon = \frac{\varepsilon_{max}}{c_{eq}} = \frac{6\,\varepsilon_{CLA}}{c_{eq}}$ with $\varepsilon_{max} = 6\,\varepsilon_{CLA}$	Used in Eqs. (3.10.7 and 3.10.16)	3.10.5
Average roughness for hydraulic channels	$\varepsilon_{av,h} = (1 - a_\varepsilon)\,\varepsilon_{La} + a_\varepsilon\,\varepsilon_{Le}$	$a_\varepsilon = 0.98 - 0.0012\, n_q \sqrt{f_q}$	3.10.6
Roughness effect on disk friction, for turbulent flow only	$f_R = \frac{C_{f,rough}}{C_{f,smooth}} = \left\{\frac{\log\frac{12.5}{Re}}{\log\left(0.2\frac{\varepsilon}{r_2} + \frac{12.5}{Re}\right)}\right\}^{2.15}$	Calculated with equivalent sand roughness $\varepsilon_{impeller}$ and ε_{casing}	3.10.7
Rotation of fluid with zero leakage ($Q_{sp} = 0$)	$k_o = \frac{c_u}{\omega r} = \frac{1}{1 + \left(\frac{r_w}{r_2}\right)^2 \sqrt{\left(\frac{r_w}{r_2} + 5\frac{t_{ax}}{r_2}\right)\frac{f_{R,casing}}{f_{R,impeller}}}}$	Open sidewall gaps: $r_w = r_2$; $t_{ax} = 0$	3.10.8
Disk friction coefficient	$k_{RR} = \left[\frac{\pi r_2}{2\, Re\, s_{ax}} + \frac{0.0625}{Re^{0.2}}(1 - k_o)^{1.75} f_{R,imp} f_L\, f_{RS}\right] f_{therm}$		3.10.9
Effect of pump-out vanes	$f_{RS} = 0.63 + 0.6\, d_{RS}/d_2$	d_{RS} = o.d. of pump-out vanes; if no pump-out vanes: $f_{RS} = 1.0$	3.10.10
Thermal effect on disk friction (empirical)	$f_{therm} = \exp\left\{-2\times10^{-5}\left(\frac{\nu}{\nu_{Ref}}\right)^{1.34}\right\}$	$\nu_{Ref} = 10^{-6}\, m^2/s$ see Chap. 13.1	3.10.11
Influence of leakage, valid for: $r_{sp}/r_2 > 0,3$	$f_L = \exp\left\{-350\,\varphi_{sp}\left(\left[\frac{r_2}{r_{sp}}\right]^a - 1\right)\right\}$	* radially inwards: φ_{sp} positive; a = 1,0 * radially outwards: φ_{sp} negative; a = 0,75	3.10.12

Table 3.10 (2) Influence of roughness and Reynolds number on efficiency			Eq.
Disk friction $f_{geo} \approx 1.22$ for typical radial impellers	$\left(\frac{P_{RR}}{P_u}\right)_M = \frac{8\sqrt{2}\, k_{RR,M}\, f_{geo}}{\omega_s^2\, \psi_{opt}^{2.5}\, f_q} = \frac{31680\, k_{RR,M}\, f_{geo}}{n_q^2\, \psi_{opt}^{2.5}\, f_q}$		3.10.13
Mechanical losses	$\frac{P_m}{P_u} = \frac{0.0045}{\eta}\left(\frac{Q_{Ref}}{Q}\right)^{0.4}\left(\frac{n_{Ref}}{n}\right)^{0.3}$	$Q_{Ref} = 1$ m3/s $n_{Ref} = 1500$ rpm	3.10.14
Hydraulic efficiency of model or baseline	$\eta_{h,M} = \frac{\eta}{\eta_{vol}\left\{1-\eta\left[\left(\frac{P_{RR}}{P_u}\right)_M + \frac{P_m}{P_u}\right]\right\}}$	Calculated from measured overall efficiency and loss analysis for model and prototype	3.10.15
Friction coefficient of hydraulic channels for turbulent flow with $Re > Re_{crit}$	$c_f = \frac{0.136}{\left\{-\log\left(0.2\frac{\varepsilon}{r_2} + \frac{12.5}{Re}\right)\right\}^{2.15}}$ $Re_{crit} = \frac{3\times 10^6}{1+10^4\, Tu^{1.7}}$	c_f is calculated with ε_{av} from Eq. (3.10.6) Tu = turbulence level; valid for Tu < 0.1	3.10.16
Friction coefficient of hydraulic channels for laminar flow	$c_{f,lam} = \frac{2.65}{Re^{0.875}} - \frac{2}{8\, Re + 0.016/Re} + \frac{1.328}{\sqrt{Re}}$	$0.01 < Re < Re_{crit}$	3.10.17
Fraction of losses dependent on roughness and Reynolds number	$\frac{\zeta_{R,M}}{\psi_{opt}} = \left\{\frac{1}{\eta_{h,M}} - 1\right\}\left(a_1 - b_1\, n_q \sqrt{f_q}\right)$	$a_1 = 0.635$ $b_1 = 0.0016$	3.10.18
Correction factor for hydraulic efficiency	$f_{\eta_h,opt} = 1 - \frac{\zeta_{R,M}}{\psi_{opt}}\left(\frac{c_{f,a}}{c_{f,M}} - 1\right)$	$c_{f,a}$ und $c_{f,M}$ from Eq. (3.10.16)	3.10.19
Correction factor for efficiency at: $f_{\eta vol} = 1.0$ $f_{\eta m} = 1.0$; $\rho_a = \rho_M$ $f_Q = f_H = f_{\eta h}$	$f_\eta \equiv \frac{\eta_a}{\eta_M} = \frac{f_{\eta h}\left[1+\left\{\left(\frac{P_{RR}}{P_u}\right)_M + \left(\frac{P_m}{P_u}\right)_M\right\}\eta_{vol,M}\,\eta_{h,M}\right]}{1+\left\{\left(\frac{P_{RR}}{P_u}\right)_M \frac{k_{RR,a}}{k_{RR,M}} + \left(\frac{P_m}{P_u}\right)_M\right\}\frac{\eta_{vol,M}\,\eta_{h,M}}{f_{\eta h}}}$		3.10.20
Efficiency Q_{opt}	$\eta_a = f_\eta\, \eta_M$	f_η from Eq. (3.10.20)	3.10.21
Hydraulic efficiency	$\eta_{ha} = f_{\eta h}\, \eta_{hM}$ $f_{\eta h}$ from Eq. (3.10.19) $f_Q = f_H = f_{\eta h}$ for highly viscous fluids (Chap. 13.1)		3.10.22
The above efficiency scaling procedure is to be followed for the best efficiency point. The factor f_η is then applied to the entire curve: $\eta_a(Q_a) = f_\eta\, \eta_M(Q_M)$.			
f_{geo} represents the ratio of the total disk friction losses of an impeller (on *all* rotating surfaces) to those of the flat sides of front and rear shroud			

References

1. Baker, W.E., et al.: Similarity Methods in Engineering Dynamics (Revised edn). Elsevier, Amsterdam (1991)
2. Brodersen, S.: Reduzierung der Scheibenreibung bei Strömungsmaschinen. Forsch Ing Wes **59**, 184–186 (1993)
3. Busemann, A.: Das Förderhöhenverhältnis radialer Kreiselpumpen mit logarithmisch-spiraligen Schaufeln. ZAMM **8**, 5 (1928)
4. Childs, D.W., et al.: Annular honeycomb seal test results for leakage and rotordynamic coefficients. ASME Paper 88-Trib-35
5. Childs, D.W.: Dynamic analysis of turbulent annular seals based on Hirs' lubrication equation. ASME 82-Lub-41 (1982)
6. Dailey, J.W., Nece, R.E.: Chamber dimension effects on frictional resistance of enclosed rotating disks. ASME J Basic Eng. **82**, 217–232 (1960)
7. Engeda, A., et al.: Correlation of tip clearance effects to impeller geometry and fluid dynamics. ASME Paper 88-GT-92 (1988)
8. Engeda, A.: Untersuchungen an Kreiselpumpen mit offenen und geschlossenen Laufrädern im Pumpen- und Turbinenbetrieb. Diss. TU Hannover, Deutschland (1987
9. Florjancic, S: Annular seals of high energy centrifugal pumps: A new theory and full scale measurement of rotordynamic coefficients and hydraulic friction factors. Diss. ETH Zürich (1990)
10. Fukuda, H.: The effect of runner surface roughness on the performance of a Francis turbine. Bulletin JSME. **7**(4), 346–356 (1964)
11. Geis, H.: Experimentelle Untersuchungen der Radseitenverluste von Hochdruck-Wasserturbinen radialer Bauart. Diss. TH Darmstadt, Deutschland (1985)
12. Görtler, H.: Dimensionsanalyse. Springer, Berlin (1975)
13. Graf, E., et al.: Three-dimensional analysis in a multi-stage pump crossover diffuser. In: ASME Winter Annual Meeting, pp. 22–29 (1990)
14. Gülich, J.F.: Disk friction losses of closed turbomachine impellers. Forsch Ing. Wes. **68**, 87–97 (2003)
15. Gülich, J.F.: Effect of Reynolds-number and surface roughness on the efficiency of centrifugal pumps. ASME J. Fluids Eng. **125**(4), 670–679 (2003)
16. Gülich, J.F.: Pumping highly viscous fluids with centrifugal pumps. World Pumps, 395/396, Aug/Sept. 1999
17. Hamkins, C.P.: The surface flow angle in rotating flow: Application to the centrifugal impeller side gap. Diss. TU Kaiserslautern, Shaker, Aachen, Deutschland (2000)
18. Henning, H.: Experimentelle Untersuchungen an als Drosseln arbeitenden Gewinderillendichtungen für hydraulische Strömungsmaschinen. Diss. TU Braunschweig, Deutschland (1979)
19. Hergt, P.: Hydraulic design of rotodynamic pumps. In: Rada Krishna (Hrsg) Hydraulic Design of Hydraulic Machinery. Avebury, Aldershot (1997)
20. Hippe, L.: Wirkungsgradaufwertung bei Radialpumpen unter Berücksichtigung des Rauheitseinflusses. Diss. TH Darmstadt, Deutschland (1984)
21. Kosyna, G., Lünzmann, H.: Experimental investigations on the influence of leakage flow in centrifugal pumps with diagonal clearance gap. ImechE Paper C439/010 (1992)
22. Lauer, J., et al.: Tip clearance sensitivity of centrifugal pumps with semi-open impellers. ASME Paper FEDSM97-3366 (1997)
23. Lauer, J., Stoffel, B.: Theoretische Untersuchungen zum maximal erreichbaren Wirkungsgrad von Kreiselpumpen. Industriepumpen+Kompressoren. **3**(4), 222–228 (1997)
24. Li, W.G.: The "sudden-rising head" effect in centrifugal oil pumps. World Pumps. **409**, 34–36 (Oct, 2000)
25. Linneken, H.: Der Radreibungsverlust, insbesondere bei Turbomaschinen. AEG Mitt **47**(1/2), 49–55 (1957)

26. Münch, A.: Untersuchungen zum Wirkungsgradpotential von Kreiselpumpen. Diss. TU Darmstadt, Deutschland (1999)
27. Nece, R.E., Dailey, J.W.: Roughness effects on induced flow and frictional resistance of enclosed rotating disks. ASME J. Basic Eng. **82**, 553–562 (1960)
28. Nemdili, A.: Einzelverluste von Kreiselpumpen mit spezifischen Drehzahlen von $n_q = 15$ bis 35 rpm. Diss. Uni Kaiserslautern, SAM Forschungsberichte, Bd. 1 (2000)
29. Ni, L.: Modellierung der Spaltverluste bei halboffenen Pumpenlaufrädern. Fortschrittber VDI Reihe. **7**, 269 (1995)
30. Osterwalder, J., Hippe, L.: Betrachtungen zur Aufwertung von Serienpumpen. VDI Ber. **424**, 1–17 (1981)
31. Osterwalder, J., Hippe, L.: Guidelines for efficiency scaling process of hydraulic turbo-machines with different technical roughnesses of flow passages. J. Hydraul. Res. **22** (2), 77–102 (1984)
32. Osterwalder, J.: Efficiency scale-up for hydraulic turbo-machines with due consideration of surface roughness. J. Hydraul. Res. **16**(1), 55–76 (1978)
33. Pantell, K.: Versuche über Scheibenreibung. Forsch Ing Wes. **16**, 97–108 (1949/1950)
34. Schilling, R., et al.: Strömung und Verluste in drei wichtigen Elementen radialer Kreiselpumpen. Strömungsmech Strömungsmasch. **16** (1974)
35. Schlichting, H: Grenzschicht-Theorie. 8. Aufl, Braun, Karlsruhe (1982)
36. Stampa, B.: Experimentelle Untersuchungen an axial durchströmten Ringspalten. Diss. TU Braunschweig, Deutschland (1971)
37. Tamm, A., Eikmeier, L., Stoffel, B.: The influences of surface roughness on head, power input and efficiency of centrifugal pumps. In: Proceedings of the XXIst IAHR Symposium on Hydraulic Machinery and Systems, Lausanne (2002)
38. Thomae, H., Stucki, R.: Axialschub bei mehrstufigen Pumpen. Techn Rundschau Sulzer **3**, 185–190 (1970)
39. Varley, F.A.: Effects of impeller design and surface roughness on the performance of centrifugal pumps. Proc. Inst. Mech. Eng. **175**(21), 955–969 (1961)
40. Wagner, W.: Experimentelle Untersuchungen an radial durchströmten Spaltdichtungen. Diss. TU Braunschweig, Deutschland (1972)
41. Weber, D.: Experimentelle Untersuchungen an axial durchströmten kreisringförmigen Spaltdichtungen für Kreiselpumpen. Diss. TU Braunschweig, Deutschland (1971)
42. Welz, E.: Der Einfluß der Laufschaufelform auf Förderhöhe und Wirkungsgrad der Kreiselpumpen. Schweizer Archiv. 114–126 (April 1966)
43. Wiesner, F.J.: A new appraisal of Reynolds-number effects on centrifugal compressor performance. ASME J. Eng. Power. **101**, 384–396 (1979)
44. Wiesner, F.J.: A review of slip factors in centrifugal impellers. ASME J. Eng. Power **89**, 558–566 (1967)
45. Zierep, J.: Ähnlichkeitsgesetze u. Modellregeln der Strömungslehre. Braun, Karlsruhe (1971)
46. Liebner, T., Cowan, D., Bradshaw, S.: Influence of wear ring geometry on suction performance. In: 32nd Pump User‘s Symposium (2016)
47. Yamada, Y.: On the pressure loss of flow between rotating co-axial cylinders with rectangular grooves. Bulletin JSME **5**, 642–651 (1962)

Chapter 4
Performance Characteristics

Following the changing operational requirements, practically all pumps temporarily operate away from the design point which is defined by $q^* \equiv Q/Q_{opt} = 1$. Overload corresponds to $q^* > 1$, while operation at $q^* < 1$ is called "partload". The pump characteristics describe the behavior of head, power consumption and efficiency as functions of the flow rate (the behavior of NPSH = f(Q) is discussed in Chap. 6). The shape of these performance curves over the range from shut-off (or zero flow Q = 0) to the maximum possible flow rate is important for the operational behavior of the pump in the plant - for instance when operating in parallel or during start-up (see Chap. 1.1). The majority of applications require a Q-H-curve steadily falling with increasing flow rate, i.e. $\partial H/\partial Q < 0$. This is termed a "stable characteristic". In contrast, if the Q-H-curve has a range with $\partial H/\partial Q > 0$, the characteristic is said to be "unstable". Unstable or flat Q-H-curves can cause problems in parallel operation or with a flat system characteristic, see Chap. 1.1.

The characteristics are measured on the test stand by throttling the discharge valve in order to obtain different flow rates corresponding to the valve openings. At a given speed, unique values of head and power are established at every specific flow rate. The resulting curve H = f (Q) is called "head-capacity curve" or "Q-H-curve".

As a rule, the impeller and collector are designed so that neither flow separation nor recirculation occurs at the best efficiency point (except special designs such as some sewage or dredge pumps). However, at low partload - typically at $q^* < 0.4$ to 0.75 and especially at Q = 0 - recirculations occur at the impeller inlet and outlet. They largely determine the shape of the partload characteristics. These relationships and the measures for influencing the shape of the performance curves are extensively discussed in Chap. 5.

J. F. Gülich, *Centrifugal Pumps*,
https://doi.org/10.1007/978-3-030-14788-4_4

4.1 Head-Capacity Characteristic and Power Consumption

4.1.1 Theoretical Head Curve (Without Losses)

As discussed in Chap. 3, the specific blade work (or H_{th}) is obtained from the difference of the angular momentums at impeller inlet and outlet, Eq. (T3.3.1). The circumferential component c_{2u} of the absolute velocity required for this calculation is obtained from the outlet triangle shown for various flow rates in dimensionless representation in Fig. 4.1 (see also Table 3.2). The circumferential component of the absolute velocity c_{2u} drops with increasing c_{2m} or increasing flow rate at a given blade outlet angle and given slip factor. Suppose vector two in Fig. 4.1 corresponds to the design point, vector 1 to partload and vector 3 to overload. At vector 4 no angular momentum is transmitted to the absolute flow since $c_{2u} = 0$. Finally, at vector 5 c_{2u} is negative and the pump works as a brake or a reverse running turbine (Chap. 12). If there is no pre-rotation (that is $\alpha_1 = 90°$), the relation $\psi_{th} = 2 \times c_{2u}/u_2$ applies. In this most frequent case, the dimensionless velocity triangle according to Fig. 4.1 represents the relationship $\psi_{th} = f(\varphi)$ or, in other words, the dimensionless *theoretical* head characteristic. According to Eq. (T3.3.10) $\psi_{th} = f(\varphi)$ can be expressed by the following equation:

$$\psi_{th} = 2\left(\gamma - K_1 \frac{\varphi_{2,La}}{\tan \beta_{2B}}\right) \text{ with } K_1 \equiv \tau_2 + \frac{A_2\, d^*_{1m} \tan \beta_{2B}}{A_1 \tan \alpha_1} \tag{4.1}$$

The theoretical characteristic can also be formulated by using the *flow* angle β_2 instead of the *blade* angle β_{2B}. To this end, the expressions $c_{2u} = u_2 - w_{2u}$ and $w_{2u}/u_2 = \varphi_{2,La} \times \tau_2 \times \cot \beta_2$ are substituted in Eq. (T3.3.9) (from the outlet triangle). We obtain thus for $\alpha_1 = 90°$:

$$\psi_{th} = 2\left(1 - \tau_2 \frac{\varphi_{2.La}}{\tan \beta_2}\right) \quad \text{and} \quad \cot \beta_2 = \cot \beta_{2B} + \frac{1 - \gamma}{\varphi_{2,La} \tau_2} \tag{4.2}$$

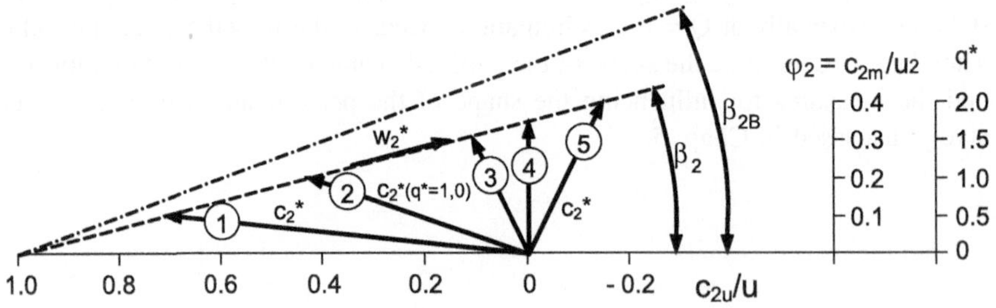

Fig. 4.1 Dimensionless impeller outlet velocity triangle for varying flow, φ_2 or $q^* \sim \varphi_2$

Because of the complex flow conditions in the impeller (Chaps. 5 and 3) both the flow angles and the slip factor generally depend on the flow rate. However as a first approximation, the slip factor γ can be assumed as independent of the flow rate in the range without partload recirculation. In this case, the theoretical characteristic according to Eq. (4.1) is obtained as a linear function of the flow rate (see Fig. 8.11 for a CFD calculation and test data). Experience shows that the outflow angle β_2 in the vicinity of the design point is almost constant so that Eq. (4.2) represents a virtually linear theoretical characteristic (as in Fig. 4.1).

The constant in Eq. (4.1) reduces to $K_1 = \tau_2$ (and is then close to 1.0) if there is no pre-swirl, i.e. at $\alpha_1 = 90°$. If the fluid approaches the impeller with $\alpha_1 \neq 90°$, the inlet triangle must be considered too, Eq. (T3.1.3).

The theoretical head reaches the value $\psi_{th,o} = 2 \times \gamma$ at $Q_{La} = 0$, but the slip factor cannot be predicted when recirculations are present. At Q = 0 the slip factor γ_o is expected between the value at BEP and 1.0, that is $\gamma_{opt} < \gamma_o < 1.0$.

At $Q_{La} = 0$ we therefore find $2\gamma < \psi_{th,o} < 2.0$. The actual value $\psi_{th,o}$ can scarcely be predicted since the effects of recirculation are difficult to quantify. Equation (4.1) further shows that ψ_{th} becomes zero at a flow coefficient $\varphi_{2,max}$:

$$\varphi_{2,max} = \frac{\gamma}{K_1} \tan \beta_{2B} \tag{4.3}$$

However, a pump cannot operate up to this flow rate since in most plants it would run into full cavitation at much smaller flows (Chap. 6). This statement primarily applies to pumps with low or medium specific speeds. Axial pumps can often operate up to the flow where H = 0 (any impeller pumping a gas could do so as well).

Figure 4.2a shows the theoretical characteristic in the form $\psi_{th} = f(\varphi_{La})$ for the three cases $\beta_{2B} < 90°$, $\beta_{2B} = 90°$ and $\beta_{2B} > 90°$. The broken straight lines stand for $\gamma = 1$, i.e. blade-congruent flow. The solid lines take into account the slip factor γ which frequently is in a range from 0.7 to 0.8. The curves with constant slip run parallel with the straight line for $\gamma = 1$ if no recirculation occurs. Most pumps have backward-swept blades with outlet angles in the range $\beta_{2B} = 15$ to 45° (predominantly 20 to 35°). An intersection with the abscissa is obtained according to $\varphi_{2,max}$ as per Eq. (4.3). The smaller the impeller blade outlet angle, the steeper is the characteristic and the lower $\varphi_{2,max}$.

Impellers with straight radial blades ($\beta_{1B} = \beta_{2B} = 90°$) are sometimes selected for small pumps with low specific speeds (Chap. 7.3.3). In this case the theoretical characteristic $\psi_{th} = f(\varphi)$ runs parallel with the abscissa. Blades with $\beta_{2B} > 90°$ are not used in pumps since the characteristic would be inherently unstable. Exceptions to this are side channel and peripheral pumps, Chap. 2.3.4.

The theoretical power consumption is obtained from $P_{th} = \rho \times g \times H_{th} \times Q_{La}$ or with Eq. (4.1) as a dimensionless expression according to Eq. (T3.4.9) as follows:

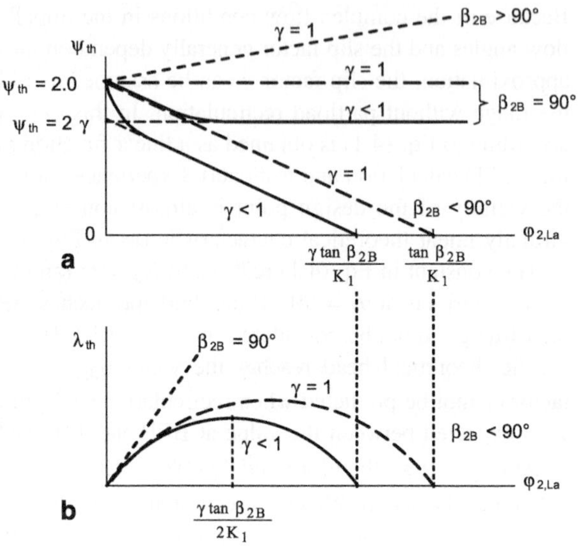

Fig. 4.2 Theoretical head and power characteristics (without losses or partload recirculation). **a** Pressure coefficient. **b** Power coefficient

$$\lambda_{th} = \varphi_{2,La}\psi_{th} = 2\left(\gamma\varphi_{2,La} - K_1 \frac{\varphi_{2,La}^2}{\tan \beta_{2B}}\right) \tag{4.4}$$

With backward-swept blades ($\beta_{2B} < 90°$) the theoretical power curve $\lambda_{th} = f(\varphi_{2,La})$ therefore has the shape of a parabola with zero points at $\varphi_{2,La} = 0$ and $\varphi_{2,La} = \varphi_{2,max}$ corresponding to Eq. (4.3), Fig. 4.2b. Differentiating Eq. (4.4) the maximum of the power consumption is obtained at $\varphi_{2,La} = 0.5 \times \varphi_{2,max}$ as follows:

$$\lambda_{th,max} = \frac{\gamma^2 \tan \beta_{2B}}{2K_1} \tag{4.5}$$

Equation (4.4) shows that the theoretical power increases as a linear function of the flow at $\beta_{2B} = 90°$ (and with the square of the flow at $\beta_{2B} > 90°$).

4.1.2 Real Characteristics with Losses

According to Chap. 3, Eq. (T3.3.7), the real characteristic is obtained by deducting the hydraulic losses caused by friction and vortex dissipation from the theoretical head. It is frequently postulated that the friction losses grow quadratically with the flow rate and that the losses at the best efficiency point caused by an incorrect approach flow ("shock losses") are zero. The friction losses at q* = 1 would then amount to $Z_{r,opt} = H_{th,opt} - H_{opt} = (1 - \eta_{h,opt})H_{th,opt}$. They increase with the square of the flow rate according to $Z_r = (1 - \eta_{h,opt})H_{th,opt}q^{*2}$.

It is thus assumed that the velocities in the pump increase proportionally with the flow rate. However, this hypothesis is fulfilled only conditionally: (1) With the usual approach flow angles of below 20° the relative velocity at the impeller inlet is independent of the flow rate within a few percent. (2) The vector of the relative velocity at the impeller outlet increases less than proportionally with the flow rate. This is shown by Eqs. (T3.2.7 and 3.2.11) according to which the expression $w_2{}^* \approx w_{2u}{}^* = 1 - \gamma + \varphi_{2,La} \times \tau_2 \times \cot \beta_{2B}$ applies. The measurements evaluated in Fig. 5.15 confirm this statement taking into account that $w_{2u} = u_2 - c_{2u}$. (3) The mean relative velocity in the impeller $w_{av} = \frac{1}{2}(w_1 + w_2)$ changes little with the flow rate. (4) The absolute velocity at the collector inlet *drops* with increasing flow while the velocity in the diffuser downstream of the throat area is proportional to the flow rate. The losses in the inlet casing, in the return channels and in the outlet casing of multistage pumps also increase with the square of the flow rate.

The losses due to incorrect approach flow are normally low at the design point. They increase at off-design conditions with $(q^* - 1)^x$, where x = 2 is usually assumed. These losses were usually described in the literature as "shock losses" because the incorrect approach flow to the impeller blades and diffuser vanes was considered to be the cause. As is shown in Chap. 5, the losses due to vortex dissipation (which include the actual shock losses) are predominantly caused by non-uniform flow distributions in the impeller and collector channels. This is particularly true for pumps with high specific speeds. At $n_q < 40$, mainly the pressure recovery in the diffuser is impaired by non-uniform approach flows resulting in corresponding losses. Even at the best efficiency point, the flow distribution is far from uniform so that $Z_{opt} = H_{th,opt} - H_{opt}$ in fact already includes friction *and* vortex dissipation losses. The latter are even dominant at high specific speeds (above $n_q = 50$ to 60) because the flow channels in the impeller and collector are short compared to their hydraulic diameter (L/D_h is small) so that the proportion of friction losses is low.

At the onset of partload recirculation at the impeller inlet and/or outlet, the vortex dissipation losses increase substantially due to exchange of momentum between the recirculating fluid and the net through-flow. As yet these recirculation phenomena have not been treated either theoretically or numerically in a general form; empirical coefficients can be used at best. Therefore, a theoretical calculation of these losses - and consequently of the real characteristic - does not yield satisfactory results. Empirical methods have been proposed but not found general acceptance. Figure 4.3 is a qualitative illustration of the real characteristic H(Q) as obtained from the theoretical characteristic $H_{th}(Q)$ by deducting all friction and vortex dissipation losses.

With $q^* \gg 1$ the hydraulic losses due to flow separation and high velocities increase "exponentially" in the throat areas of the impeller and collector. This is why the flow rate at zero head H = 0 shifts to values far below those calculated from Eq. (4.3). This applies particularly to pumps with low specific speeds where

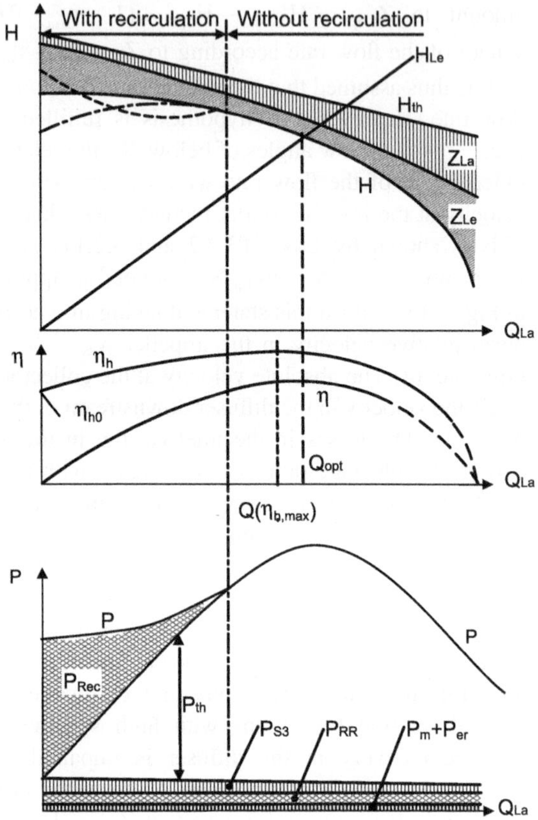

Fig. 4.3 Influence of losses and recirculation on pump characteristics

the flow frequently cavitates just a little above the best efficiency point in the throat area of the collector. At high specific speeds the overload range is also limited since low head coefficients and high flow coefficients have to be selected. Therefore, the range between $\varphi_{2,opt}$ and $\varphi_{2,max}$ inevitably becomes narrower than calculated from Eq. (4.1).

The real power consumption of the pump is obtained from the theoretical power $P_{th} = \rho \times g \times H_{th} \times Q_{La}$ plus the secondary losses discussed in Chap. 3.6 (Fig. 4.3). While the leakage losses are included in Q_{La}, the other secondary losses - P_{RR}, P_m, P_{s3} and P_{er} according to Eq. (T3.5.1) - can be assumed to be approximately independent of the flow rate. The difference between the curves P and P_{th} therefore consists of the secondary losses and the recirculation power P_{Rec} after the onset of recirculation.

In some pump applications it is advantageous for the point of maximum power consumption to be located at the best efficiency point, i.e. at q* = 1. The pump is specified to have a "non-overloading" characteristic. Since the power maximum according to Eq. (4.3) is at $\varphi(_{P_{max}}) = {}^1\!/_2 \times \varphi_{2,max}$, the angle β_{2B} required to achieve a non-overloading characteristic can be derived from the equation:

$$\beta_{2B} = \arctan \frac{2K_1}{\gamma} \varphi_{opt} \tag{4.6}$$

As Eq. (4.6) shows, the blade angle to be executed depends on the selected flow coefficient, hence also on the impeller outlet width, the blade number and the slip factor. Equation (4.6) must therefore be solved iteratively. To obtain a non-overloading power curve, high flow coefficients and small values for blade number, outlet width and angle must be selected. As a consequence a steep Q-H-curve is obtained. This is borne out by Eqs. (4.1 and 4.6) which yield $\psi_{th} = \gamma$. Quite unfavorable conditions are therefore obtained at low and medium specific speeds in that the pump size gets large and the efficiency is low. Therefore non-overloading characteristics at low specific speeds are demanded only in special cases, when the driver power at high flow must be limited.

4.1.3 Component Characteristics

By measuring the static pressure H_2 at the impeller outlet, the losses in the impeller and collector can be determined approximately. In this way the behavior of the losses as a function of the flow rate can be assessed. The losses in the volute and in the downstream diffuser can be determined through additional measurement of the static pressure H_3 in the volute throat area.

In the case of multistage pumps, the static pressures are measured in the diffuser throat area at H_3, at the diffuser outlet H_4 and in the return channel H_6 if applicable (see Fig. 4.4 and sketches in Table 2) to determine the pressure recovery in the collector and the individual losses. The static pressure rise H_p in the impeller, the pressure recovery $(H_3 - H_2)$ in the triangular section at the diffuser inlet and in the entire diffuser $(H_6 - H_2)$ can be obtained from these measurements. Such investigations show the contribution of the individual pump components to the pressure generation (hence to the characteristic's stability, see Chap. 5) and to the hydraulic losses.[1]

The analysis of this type of pressure measurements within a pump is performed in the following steps:

1. The hydraulic efficiency η_h is obtained from Eq. (T3.5.8) for which the secondary losses according to Tables 3.5 to 3.7 have to be determined. The calculation according to Eq. (T3.5.8) is only true in regions where recirculations

[1]However, the accuracy of the measurements is often limited since the pressure distribution over the impeller circumference varies at partload. While this variation is low in diffusers, it is high in volute casings where several pressure tappings are required to obtain useful results. In addition, the impeller and diffuser losses are obtained as a small difference between two large numbers and are therefore afflicted with high relative errors.

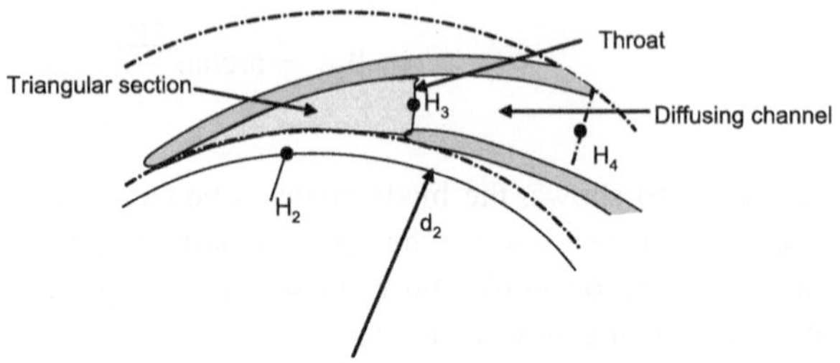

Fig. 4.4 Pressure tappings H_2 to H_4, triangular section at diffuser inlet, throat area at H_3

are absent since these cannot be taken into account in the balance of angular momentums without knowing the recirculating flow rates and their velocities (otherwise unrealistically high values would be obtained for c_{2u} in the working range with recirculation). At low partload it is therefore better to assume a (constant) slip factor $\gamma < 1$.

2. Equation (T3.2.8) allows to calculate c_{2u} for a given η_h. All components of the inlet and outlet triangles can be determined from Tables 3.1 and 3.2.
3. The impeller losses are obtained as the difference of $H_{p,th}$ and the measured increase of the static pressure $H_p = H_2 - H_1$ according to Eq. (T3.3.8) from:

$$Z_{La} = \frac{1}{2g}\left(u_2^2 - u_1^2 + w_1^2 - w_2^2\right) - H_p \tag{4.7}$$

4. The diffuser losses are obtained by applying Bernoulli's Eq. (1.7) between impeller and diffuser outlet:

$$Z_{Le} = \frac{1}{2g}\left(c_2^2 - c_6^2\right) - (H_6 - H_2) \tag{4.8}$$

where $H_6 - H_2$ is the static pressure increase measured in the diffuser.

Three examples shall serve to explain the results of internal pressure measurements in the following. Figure 4.5 shows tests on diffuser pumps with specific speeds of $n_q = 22$ and 33. Here, all pressure heads (rendered dimensionless by $u_2^2/2\,g$) are plotted against q*. The degree of reaction at the best efficiency point is near $R_G = 0.75$. While the head coefficient flattens towards Q = 0, the static head H_p created by the impeller increases significantly below q* = 0.3 for both pumps. As explained in Chap. 5, this is due to recirculation at the impeller inlet. In the triangular section at the diffuser inlet a major pressure increase ψ_{2-3} is recorded at q* < 1. With $n_q = 33$ it amounts to more than 10% of the shut-off head and with $n_q = 22$ it even reaches 17%. Fully developed recirculation at the impeller outlet thus produces an increase in pressure in the diffuser which develops *in the triangular section at the inlet.* The pressure recovery in the actual diffuser ($H_4 - H_3$) only becomes effective at q* > 1; it increases subsequently with the square of the flow rate. The diffuser losses ξ_{Le} increase markedly towards Q = 0. In contrast, the impeller losses ξ_{La} at q* < 1 depend very little on the flow rate; no increase towards

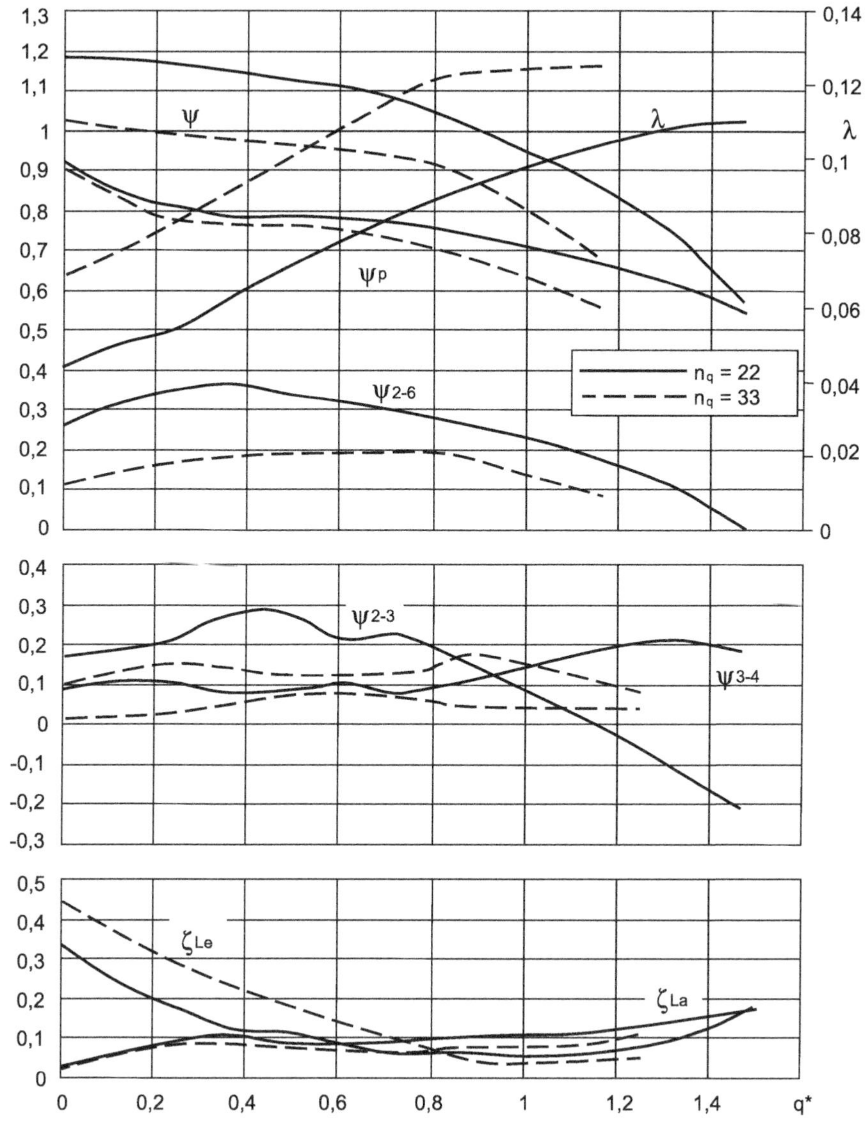

Fig. 4.5 Component characteristics of diffuser pumps with n_q = 22 and 33. [1]

Q = 0 is observed as proposed by the classical shock-loss concept. This is because the recirculating fluid induces a strong pre-rotation.

Figure 4.6a shows measurements on an impeller n_q = 33 which was initially combined with a diffuser and subsequently with a double volute. Note that the volute has a significantly weaker interaction with the impeller flow than the diffuser. This is demonstrated by the lower power consumption and lower static impeller

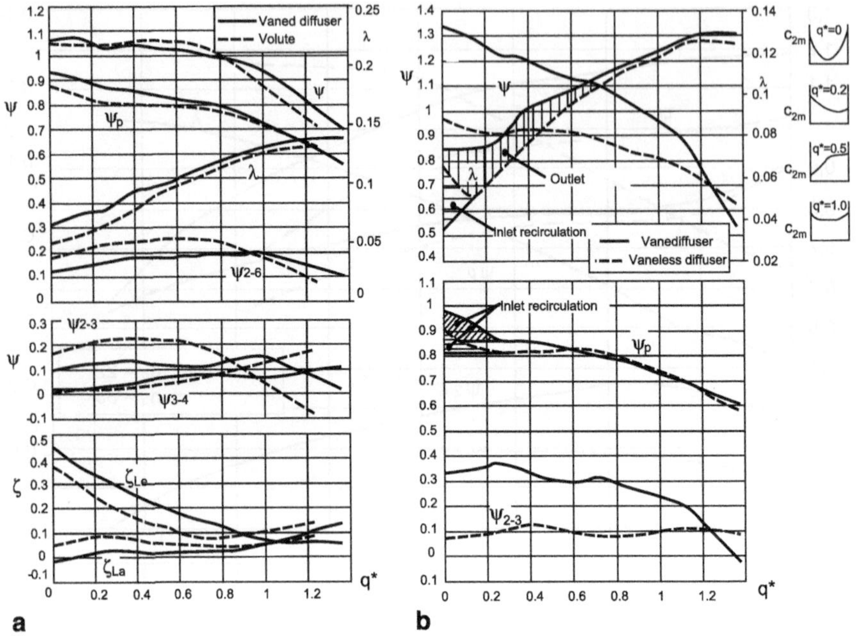

Fig. 4.6 Influence of collector on component characteristics with identical impeller, $n_q = 33$. **a** Comparison between volute and diffuser, [1]. **b** Comparison between vaneless and vaned diffuser. [B.20]

head ψ_p at partload. It is also important to realize that flow separation in the volute occurs similarly to that in the diffuser, as demonstrated by the curves ψ_{2-3} and ψ_{2-6} (the throat area of the volute was approximately 10% smaller than in the diffuser) and that a pressure increase also takes place in the volute even at shut-off. Again, the losses in the volute increase towards Q = 0 while no major shock losses can be detected at the impeller inlet. The pressure recovery in the actual diffuser ψ_{3-4} following the throat area is very small at low flow. It increases approximately with the square of the flow rate as in a diffuser in a pipeline. These findings are also confirmed by the measurements reported in [5].

In a volute casing with only one or two cutwaters at a considerable distance from the impeller outlet, losses caused by an incorrect approach flow are expected to be considerably lower than in a diffuser, where 8 to 12 vanes are arranged at a short distance from the impeller blades. Despite of this, the losses in both collectors are quite similar (Fig. 4.6a). This analysis demonstrates that flow separations due to major deceleration (from c_2 to c_{3q}) in the triangular section of the diffuser or in the volute are far more responsible for the losses in the collector than shock losses caused by an incorrect approach flow (i.e. high incidence).

Figure 4.6b shows measurements on an impeller $n_q = 33$ combined with a smooth vaneless diffuser (broken curves) and a vaned diffuser (solid lines). The interaction of the vaneless diffuser with the impeller and the pressure increase in the vaneless diffuser at Q = 0 are considerably less than with a vaned diffuser or a volute. Consequently, the overall characteristic and the power consumption differ significantly in both tests.

The tests in Figs. 4.5 and 4.6a, b in essence apply qualitatively to all centrifugal pumps. Additional examples can be found in Figs. 5.30, 5.31 and 5.34 as well as in [B.20], [4] and [5]. Quite similar component characteristics were found in measurements on radial compressors, [2]. The evaluation of a large number of component characteristics shows:

- The recirculation from the collector has an effect on the static pressure increase in the impeller.
- The overall Q-H-characteristic becomes flat in the flow range where the curve of the pressure increase ψ_{2-3} in the triangular section upstream of the diffuser throat or in the volute flattens.
- "Shock losses" resulting from incorrect approach flow (incidence) at partload are difficult to verify on impellers since the flow largely adapts to the blades due to the pre-rotation induced by recirculation.
- Components with $\partial H/\partial Q < 0$ are generally stabilizing. This applies to the static pressure increase in the impeller and the triangular section of the diffuser as long as the flow does not stall. Conversely, components with $\partial H/\partial Q > 0$ have a destabilizing effect on the overall characteristic. This always includes the diffusing channel ψ_{3-4} following the throat area of the collector.

Figures 4.7 and 4.8 show the losses determined from internal pressure measurements and plotted against the flow rate q* for 9 diffuser and volute pumps with specific speeds from $n_q = 7$ to 160. The character of the curves is similar to those in Figs. 4.5 and 4.6. These evaluations reveal the following features which are typical for a large number of similar measurements:

- The losses in the collector at BEP significantly drop with rising specific speed because c_2/u_2 decreases and less energy is converted with increasing n_q, Fig. 4.9.
- Tendency-wise, the impeller losses at the best efficiency point *increase* with the specific speed albeit with a large scatter, Fig. 4.9. The primary reason for this is that vortex dissipation losses and secondary flows grow with increasing blade height due to uneven velocity and pressure distributions.
- At off-design conditions, the diffuser or volute losses $\zeta_{Le} = f(q^*)$ increase approximately in the form of a parabola from a minimum near the BEP. This observation would bear out the idea of "shock losses". However, the losses involved may be less caused by an incorrect incidence to the diffuser vanes or the volute cutwater; they rather result from the cross sections in the collector not matching the velocity vector at the impeller outlet: $c_{3q} = Q/(z_{Le} \times a_3 \times b_3) \ll c_2$. This means a sudden flow deceleration. It simultaneously causes both a significant pressure increase H_3-H_2 in the triangular section at the diffuser inlet and major vortex dissipation losses as soon as the

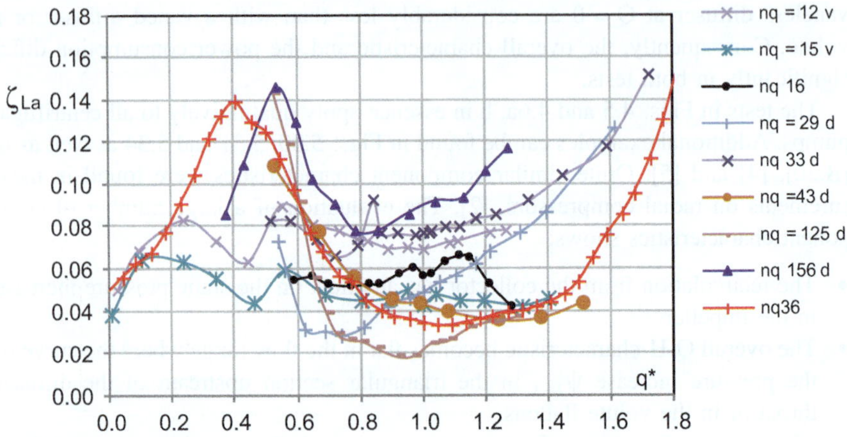

Fig. 4.7 Impeller losses with various specific speeds

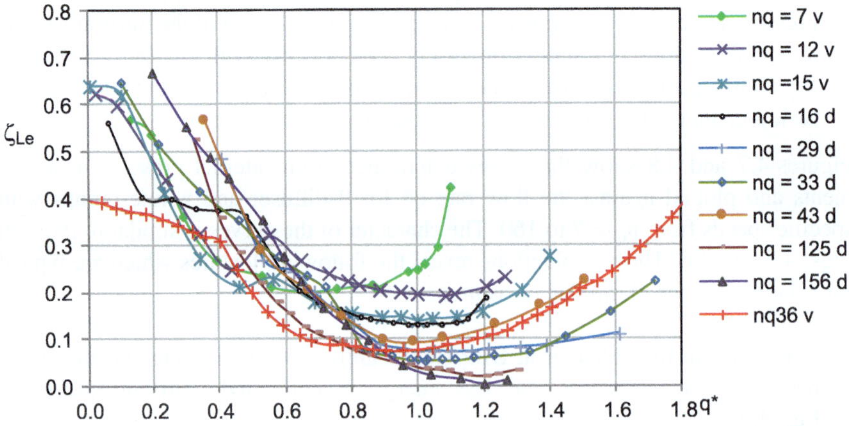

Fig. 4.8 Losses in vaned diffusers (*d*) or volutes (*v*) with various specific speeds

optimum deceleration in the triangular section is exceeded (refer to Figs. 4.5 and 4.6 $q^* = 0.4$ to 1). The optimum for this deceleration is estimated to be around $c_{3q}/c_2 = 0.8$. The strong pressure increase immediately downstream of the diffuser vane leading edge contradicts the hypothesis that shock losses due to incorrect approach flow to the vanes are significant for the features observed.

- The shock and deceleration losses ($\zeta_{Le} - \zeta_{Le,min}$) are plotted in Figs. 4.10 and 4.11 as a function of the deceleration ratio c_{3q}/c_2 where $\zeta_{Le,min}$ represents the minimum in the curve $\zeta_{Le,min} = f(q^*)$. With the exception of the semi-axial pumps with $n_q = 125$ to 160 the minimum diffuser loss closely equals the loss at the BEP, i.e. $\zeta_{Le,min} = \zeta_{Le,opt}$. The trend of the diffuser losses is well captured by

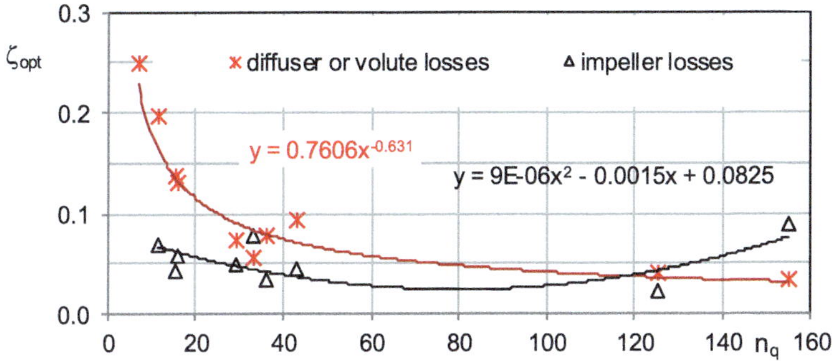

Fig. 4.9 Impeller and diffuser losses at BEP as function of the specific speed

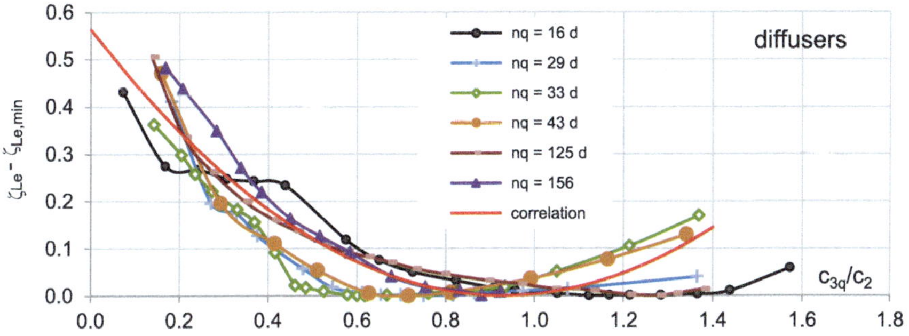

Fig. 4.10 Deceleration and shock losses in diffusers as function of the deceleration ratio c_{3q}/c_2. The correlation is given by (Eq. T3.8.25)

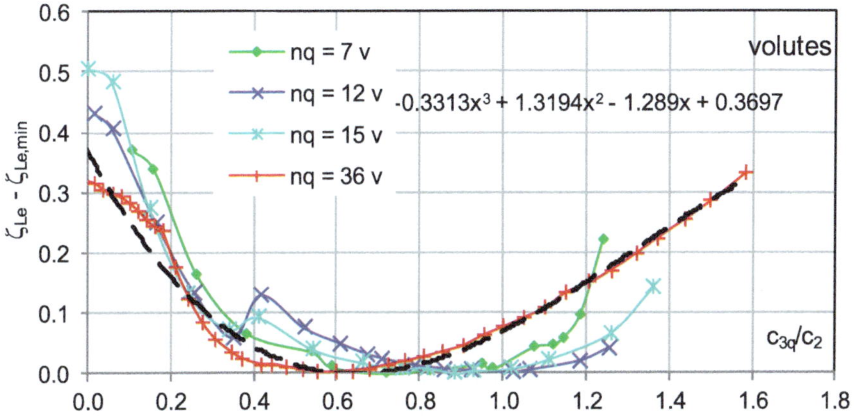

Fig. 4.11 Deceleration and shock losses in volutes as function of the deceleration ratio c_{3q}/c_2. The correlation is given by (Eq. T3.8.25)

the deceleration ratio which also seems to work for volutes. Minimum losses occur over a wide range of $c_{3q}/c_2 = 0.6$ to 1.2 with a focus between 0.8 and 1.0. The stator losses rise at high flow due to separation in the diffuser (which impairs the pressure recovery) and possibly due to cavitation. The shock and deceleration losses can be estimated by the correlation given as (Eq. T3.8.25) in Table 3.8.

- The non-uniformity of the impeller outlet flow contributes significantly to these losses.
- The diffuser losses reach their maximum value at Q = 0. A great portion of the high kinetic energy at the impeller outlet is dissipated through the exchange of momentum between backflow and through-flow. The exchange of momentum intensifies if structures prevent the fluid from rotating freely with the impeller. A smooth cylinder around the impeller would generate only minor friction losses, while an annular collector produces a greater exchange of momentum because of the larger fluid volume present. The power dissipated by recirculating fluid increases further when the impeller pumps into a volute; it reaches the maximum value in the combination with a vaned diffuser, since the diffuser vanes (which are located close to the impeller outlet) severely hamper the rotation of the fluid. In this way they cause the greatest velocity gradients and the most intensive exchange of momentum. This largely determines the power consumption at shut-off.
- Throughout a wide range, the impeller losses depend only to a moderate degree on the flow rate. They tend to reach a minimum at Q = 0 (refer to [5] for instance). Weakly pronounced shock losses can be detected only at high specific speeds (refer to $n_q = 156$ in Fig. 4.7). When recirculation starts at the impeller inlet the shock losses drop again with falling flow rate. Thus it appears that major shock losses do not occur since the effects of incidence and the deceleration from w_1 to w_{1q} are mitigated by the onset of recirculation. This is a kind of "self-healing effect" which may be explained by the principle of least resistance.

Let us consider the differences between operation with positive and negative incidence with reference to Fig. 4.12. With positive incidence a local separation zone forms on the blade suction surface just downstream of the leading edge. The throat area between the blades, however, is too large because $q^* \ll 1$ so that the flow has to be decelerated. This deceleration can take place downstream of the stalled fluid since sufficient length is available before the throat is reached. The stalled fluid acts almost like a change in blade profile to which the streamlines can often adapt quite well. Even when the stalled fluid blocks a part of the throat area, the flow is still decelerated. The situation is entirely different with negative incidence. Then the separation zone immediately blocks part of the throat area which is anyway too small for flow rates above $q^* > 1$. Consequently, the fluid is accelerated and the static pressure drops by virtue of the Bernoulli equation. Even if the flow downstream of the stalled fluid is reasonably well decelerated, the losses in the diffusing channel will grow at least with the square of the increase of the effective

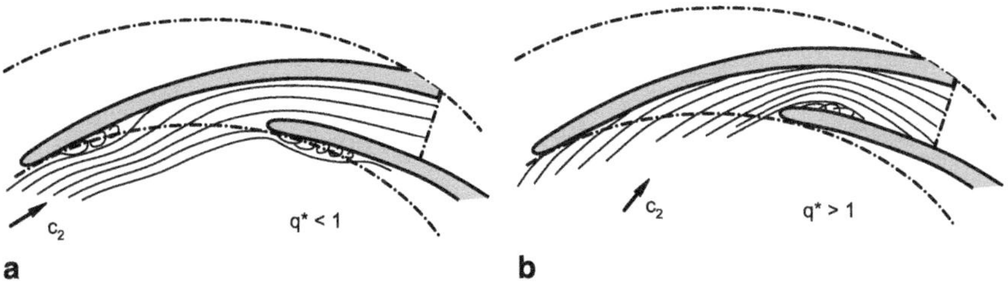

Fig. 4.12 Effect of local flow separation at impeller or diffuser inlet. **a** Positive incidence: deceleration downstream of stalled fluid (*little influence on flow through throat area*). **b** Negative incidence: stalled fluid blocks part of throat area, fluid is accelerated

velocity $c_{3q,eff}$ caused by the blockage (compared with the nominal value c_{3q} without cross sectional blocking). According to Chap. 1.6 the following equation qualitatively describes this situation (Fig. 4.12):

$$\frac{Z_{Le,eff}}{Z_{Le}} = \left(\frac{c_{3q,eff}}{c_{3q}}\right)^2 \frac{(1 - c_{p,eff})}{(1 - c_p)} \tag{4.8a}$$

These considerations apply to the impeller inlet as well as to diffusers or volutes. As will be discussed in Chap. 6 (and confirmed through measurements), the conditions are quite similar if the separation zone caused by the incorrect approach flow is not filled with stalled fluid but instead with vapor when cavitation occurs.

In summary, it should be noted that the sudden deceleration of c_2 to c_{3q} in the collector has a significant effect on the losses at partload and consequently the shape of the characteristic. Except near the best efficiency point, the losses in the collector at partload are significantly higher than the impeller losses ($Z_{Le} \gg Z_{La}$). In contrast, the impeller *losses* have a rather small impact on the partload characteristic (disregarding exceptional cases). However, the impeller *shape* has a significant effect on the recirculation and the Q-H-curve, Chap. 5.

4.1.4 Head and Power at Operation Against Closed Discharge Valve

The "shut-off head" (operation at Q = 0) determines the design pressure of the discharge pipe and, consequently, constitutes an important parameter which frequently has to be guaranteed by the pump manufacturer. The shut-off head can be estimated by means of the head coefficients ψ_o in Fig. 3.24. Diffuser pumps tend to have higher shut-off heads than machines with volute casings because the exchange of momentum (and consequently the energy transfer during operation at shut-off) between impeller and collector is enhanced by the diffuser vanes. By the same token pumps with vaneless diffusers or annular collectors tend to give lower shut-off heads, because in the absence of diffuser vanes the exchange of momentum between rotor and collector is minimized. The shut-off head increases (with otherwise identical parameters) with growing impeller outlet width b_2. For additional influence parameters see Chap. 5.

As will be extensively discussed in Chap. 5, recirculations at impeller inlet and outlet occur at low partload. They reach their maximum intensity when the pump is operated against a closed discharge valve, i.e. at Q = 0. Typical velocity profiles at the impeller inlet are shown in Fig. 4.13. Qualitatively these profiles are largely independent of the specific speeds and the impeller parameters. Recirculation is recognized by the negative meridional velocities near the outer streamline. With growing backflow from the impeller, the circumferential velocity components measured upstream of the blade leading edge increase since an angular momentum is transmitted by the blades to the recirculating fluid. The rotation of the recirculating fluid is partially dissipated in the suction chamber, further increasing the losses in the pump (turbulence losses, power dissipated by recirculating fluid).

The flow rate of the recirculating fluid Q_{Rec} and its angular momentum M_{Rec} can be obtained by integration over the recirculation zone:

$$Q_{Rec} = 2\pi \int c_{1m} r \, dr \tag{4.9}$$

$$M_{Rec} = 2\pi\rho \int c_{1m} c_{1u} r^2 \, dr \tag{4.10}$$

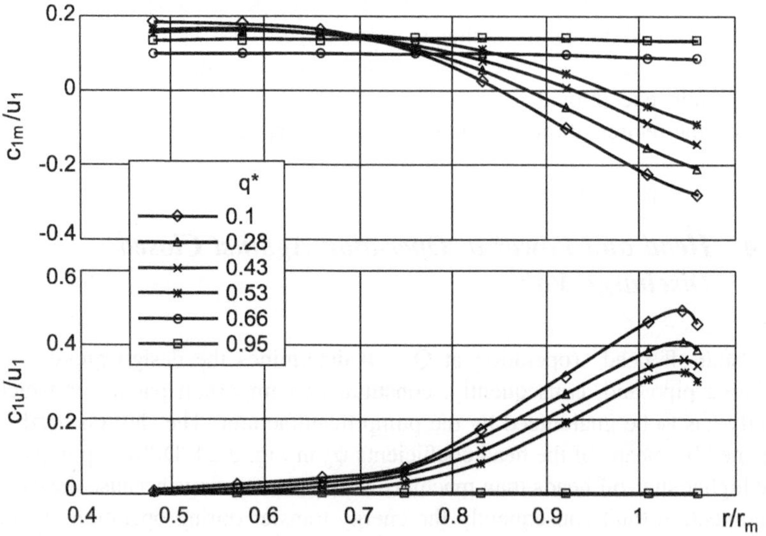

Fig. 4.13 Velocity profile at the impeller inlet of an axial pump, $n_q = 213$, [3]

The power dissipated by recirculation $P_{Rec} = \omega \times M_{Rec}$ is evaluated as power coefficient $\lambda_{1,Rec}$:

$$\lambda_{1,Rec} = \frac{2\,P_{Rec}}{\rho\,u_1{}^3 r_1{}^2} \tag{4.11}$$

The flow rate of the recirculating fluid is defined by a flow coefficient:

$$\varphi_{1,Rec} = \frac{Q_{Rec}}{A_1 u_1} \tag{4.12}$$

The evaluation of various measurements on radial and axial impellers in the range $20 < n_q < 200$ shows:

- The recirculation power at the impeller inlet according to Eq. (4.11) increases linearly from the onset of recirculation with falling flow rate until the amount $\lambda_{1,Rec,o} = 0.21 \pm 0.01$ is reached at Q = 0. This value is practically independent of the impeller or the specific speed. Consequently Eq. (4.11) can be used with $\lambda_{1,Rec,o}$ to estimate the recirculation power at the impeller inlet.
- The flow coefficient $\varphi_{1,Rec}$ of the recirculating fluid according to Eq. (4.12) increases about linearly from the onset of recirculation towards a maximum value at Q = 0. This value depends on the impeller geometry or the specific speed. In Fig. 4.13 it amounts to approximately $\varphi_{1,Rec} = 0.08$ at Q = 0. The evaluated tests (n_q = 20 to 220) were in the range of $\varphi_{1,Rec}$ = 0.05 to 0.1.

The exchange of momentum at the impeller outlet, the secondary losses described in Chap. 3.6 and the recirculation power at the impeller inlet add up to the power consumption at shut-off. This is plotted as power coefficient λ_o in Fig. 4.14, where λ_o is defined by Eq. (T3.4.9). With $d_{2b} = ½(d_{2a} + d_{2i})$ and $b_2 = ½(d_{2a} - d_{2i})$ this power coefficient can also be applied to axial impellers. The portion of the power dissipated by recirculating fluid in the total power consumption at Q = 0 grows from approximately 50% at n_q = 10 to over 95% at n_q = 250.

Diffuser pumps with $n_q > 20$ tend to absorb more power at shut-off than pumps with volutes or annular casings because a high number of diffuser vanes are arranged at a short distance from the impeller blades. Fluid rotation in the collector is thus reduced and the exchange of momentum is intensified as discussed above. When trimming the impeller, the distance between impeller blades and diffuser vanes (or cutwaters) increases and λ_o drops accordingly. With unusually large impeller outlet widths (such as are selected for sewage or dredge pumps) λ_o drops as evidenced by the test data in Fig. 4.14. The shut-off power P_o thus grows less than proportionally to the outlet width.

The dimensionless static head at the impeller outlet at shut-off is derived from (Eq. T3.3.12): $\psi_{p,o} = 1 - (w_2/u_2)^2 - \zeta_{La}$ with $w_1{}^2 = u_1{}^2 + c_1{}^2$ and $c_1 = 0$.

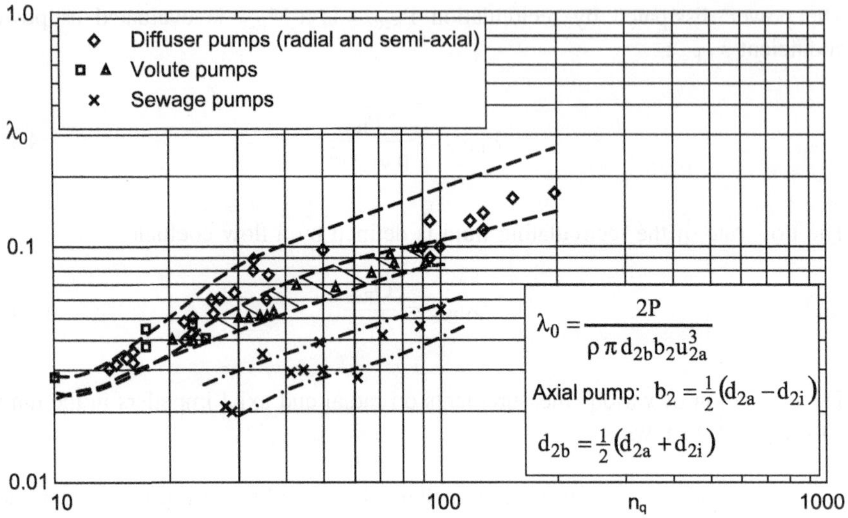

Fig. 4.14 Power consumption at operation against closed discharge valve

According to Chap. 5 and Fig. 5.15 w_2/u_2 is to be expected in the range of 0.3 to 0.5. Consequently $\psi_{p,o} = (0.75\ \text{to}\ 0.91) - \zeta_{La}$ is anticipated. With deceleration to $w_2 = 0$ (albeit physically impossible) and flow without any losses, $\psi_{p,th,o} = 1$ would be obtained as the upper limit.

Values measured on diffuser pumps ($n_q < 35$) are in the range of $\psi_{po,} = 0.8$ to 1. They point to low impeller losses at Q = 0. The theoretical head coefficient at shut-off is obtained from Eq. (T3.3.10) with $\gamma = 1$: $\psi_{th,o,max} = 2$. The minimum hydraulic efficiency consequently becomes: $\eta_{h,o,min} = \psi_o/\psi_{th,o,max} = {}^1\!/_2\psi_o$. If the slip factor γ is below 1, $\eta_{h,o}$ is correspondingly higher.

The following values were measured for the pressure recovery in the diffuser at shut-off for $n_q < 35$: $(H_3 - H_2)/H_o = 0.08$ to 0.29 and $(H_6 - H_2)/H_o = 0.12$ to 0.31 [B.20]. This implies that an increase of the static pressure occurs due to the exchange of momentum at Q = 0, but almost exclusively in the triangular section upstream of the diffuser throat.

4.1.5 Influence of Pump Size and Speed

The similarity laws for *geometrically similar* pumps derived in Chap. 3.4 apply to any flow rate (i.e. any incidence). If all velocity components are referred to the circumferential velocity u_2, the inlet and outlet triangles are independent of speed and pump size (Fig. 4.1). Any flow coefficient φ is linked to a specific head coefficient ψ_{th}. Since the Reynolds and roughness dependent friction losses make up only a small fraction of the hydraulic losses, the hydraulic efficiency (within a few percent) is independent of size and speed so that $\psi = f(\varphi)$ constitutes a first approximation of a universal characteristic for geometrically similar pumps. With regard to the secondary losses greater deviations may result when converting to another speed or size, but these rarely reach 5%. Consequently, the similarity laws according to Table 3.4 can be used with sufficient accuracy to scale the characteristics of a pump to any speed or impeller size. When doing so, a formula for the efficiency majoration is generally used in order to get the correct power (see Table 3.9). However, for pumping applications involving media with a high viscosity (such as heavy oil) or two-phase mixtures, corrections of the characteristics are necessary. These are discussed in Chap. 13.

4.1.6 Influence of Specific Speed on the Shape of the Characteristics

When calculating the ratio of the shut-off head coefficients to the best efficiency point in Fig. 3.24, it can be seen that $\psi_o/\psi_{opt} = H_o/H_{opt}$ increases with growing specific speed, Fig. 4.15. The reason for this is that the achievable head coefficient at the best efficiency point decreases more rapidly with the specific speed than the

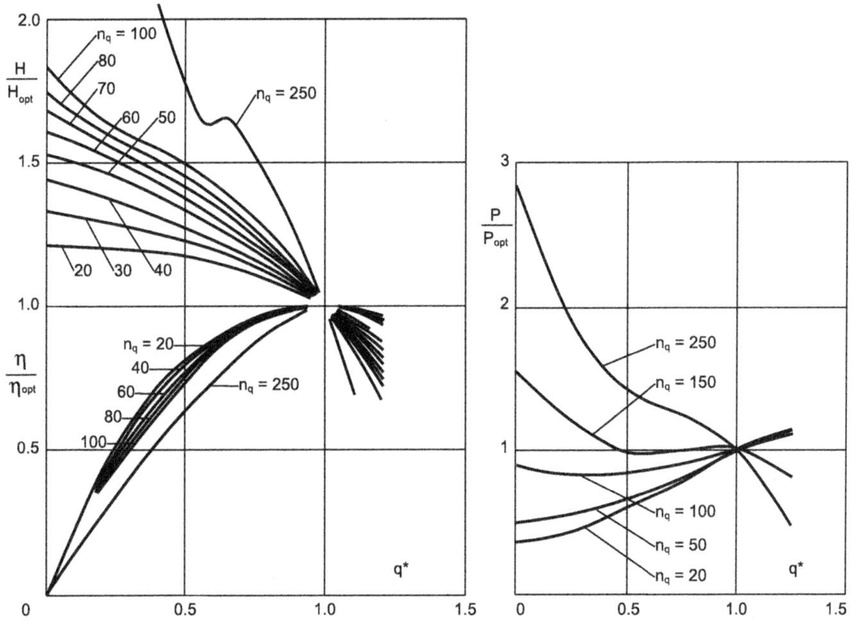

Fig. 4.15 Influence of specific speed on the shape of the pump characteristics

shut-off pressure which is largely determined by recirculations and exchange of momentum (Chap. 5).

The ratio P_o/P_{opt} of the power consumption at shut-off to the power at the design point increases from values around 0.5 at low specific speed to values above 2 to 4 in axial pumps. The efficiency curves in Fig. 4.15 become peakier with increasing specific speed since the losses caused by non-uniform velocity distributions over the blade and channel heights gain in significance.

4.2 Best Efficiency Point

According to Chap. 3.7 the fluid tends to retain its angular momentum (its "swirl") $\rho \times Q \times r_2 \times c_{2u}$ downstream of the impeller. Experience shows that the losses reach a minimum when the volute casing is designed so that the fluid at the design point flows according to the conservation of angular momentum $c_u \times r = c_{2u} \times r_2$. This means that the volute cross sections, especially the end cross section or "throat", have to be selected according to Eq. (3.15). If c_{2u} from Eq. (T3.2.8) is substituted in Eq. (3.15), the collector characteristic is obtained as a linear relationship $H_{Le} = f(Q_{Le})$ according to Table 4.1, Eq. (T4.1.1). The intersection of this straight line with the characteristic of the impeller according to Eq. (T3.3.7) or (T3.3.11) supplies the flow rate at which angular momentum conservation is satisfied in the volute throat area. Theoretically, the maximum efficiency should be expected at the resulting flow rate. Equation (T4.1.3) supplies this best efficiency point flow rate (in short, the "best efficiency point" or BEP), Fig. 4.16. Equation (T4.1.3) can be derived from Eq. (3.15): $Q = c_{2u} \times r_2 \times J_{sp}$ with c_{2u} from Eq. (T3.2.7) which yields Eq. (T4.1.3).

The volute/diffuser characteristic cannot be measured. It has meaning only near the best efficiency point. The concept is very useful for evaluating the shift of the best efficiency point due to a change of the volute or the diffuser throat area (Fig. 4.16).

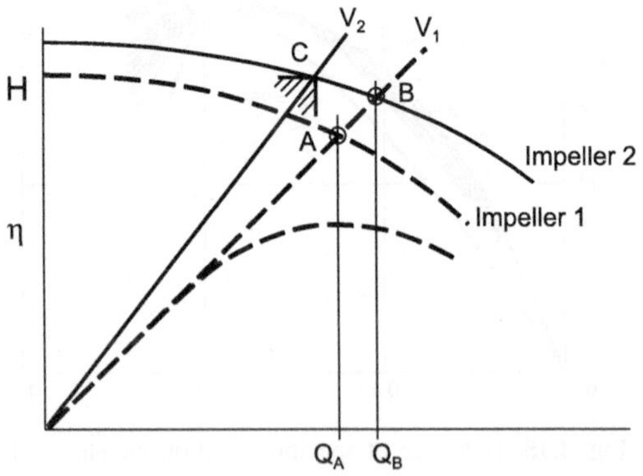

Fig. 4.16 Location of the best efficiency point

According to Fig. 4.16, a reduction of the best efficiency point flow rate from Q_B to point C (with impeller 2) is achieved by a reduction of the volute throat area, which causes the volute characteristic to change from V_1 to V_2. Conversely, two different impellers (impellers 1 and 2 in Fig. 4.16) in a given casing would have their best efficiency points Q_A and Q_B on the same volute characteristic V_1. If such calculations are performed for a given pump, it is recommended to substitute the actual slip factor calculated from the experiment according to Eq. (T3.2.9). Table 4.1 and Eq. (3.15) apply to both diffusers and volute casings. Diffusers and volutes with *rectangular* cross sections can be calculated using Eq. (4.13):

Table 4.1 Diffuser/volute characteristic and best efficiency point

Dimensional representation	**Dimensionless representation**	Eq.
$Q_{La} = Q + Q_{s1} + Q_{s2} + Q_E$ $Q_{Le} = Q + Q_{s3} + Q_E$		
$H_{Le} = \eta_h \frac{u_2^2}{g}\left\{\frac{Q_{Le}}{u_2\, r_2\, J_{sp}} - \frac{c_{1u}\, d_{1m}^*}{u_2}\right\}$	$\psi_{Le} = 2\eta_h \left\{\frac{2\pi\, b_2\, f_q\, \varphi_2\, Q_{Le}}{J_{sp}\; Q} - \frac{c_{1u}\, d_{1m}^*}{u_2}\right\}$	4.1.1
$H = \eta_h \frac{u_2^2}{g}\left\{\gamma - \frac{Q_{La}\,\tau_2}{f_q\, A_2\, u_2 \tan\beta_{2B}} - \frac{c_{1u}\, d_{1m}^*}{u_2}\right\}$	$\psi = 2\eta_h\left\{\gamma - \frac{\varphi_{La}\,\tau_2}{\tan\beta_{2B}} - \frac{c_{1u}\, d_{1m}^*}{u_2}\right\}$	4.1.2
$Q_{opt,th} = \dfrac{\eta_v\, f_q\, A_2\, u_2\, \gamma}{\frac{\tau_2}{\tan\beta_{2B}} + \frac{2\pi\, b_2\, f_q}{J_{sp}} \frac{Q_{Le}}{Q_{La}}}$	$\varphi_{opt,th} = \dfrac{f_q\, \eta_v\, \gamma}{\frac{\tau_2}{\tan\beta_{2B}} + \frac{2\pi\, b_2\, f_q}{J_{sp}} \frac{Q_{Le}}{Q_{La}}}$	4.1.3

$$J_{sp} = z_{Le}\, b_3 \ln\left(1 + \frac{2a_3}{d_3 + 2e_3}\right) \tag{4.13}$$

Experience shows that the measured best efficiency point is near the theoretical value calculated from Eq. (T4.1.3). For a volute with a *circular* cross section of radius r_{3q} at the throat the integral is:

$$J_{sp} = 2\pi\, z_{Le}\, r_z\left\{1 + \frac{r_{3q}}{r_z} - \sqrt{1 + 2\frac{r_{3q}}{r_z}}\right\} \tag{4.13a}$$

According to the definition $\eta = P_u/P = P_u/(P_u + \Sigma P_v) = 1/(1 + \Sigma P_v/P_u)$ the efficiency reaches its maximum when the ratio of all losses to the useful power $\Sigma P_v/P_u$ becomes minimal. In view of the different types of losses (Chaps. 3.6 and 3.7) and the interaction of the main flow with the flow in the impeller sidewall gaps (which influences disk friction and leakage losses) it is a "coincidence" rather than a strictly derivable physical law that the best efficiency point can be roughly described by the intersection of the diffuser and the impeller characteristics. The finding that the diffuser characteristic describes the conditions quite well is also explained by the fact that the diffuser losses according to Fig. 4.8 have a distinct minimum near the BEP flow while the impeller losses exhibit an entirely different character (also refer to Figs. 4.5 to 4.7 in this regard).

With increasing hydraulic losses the best efficiency point shifts towards smaller flow rates. This finding is *not* reflected by the concept of the diffuser characteristic since the hydraulic efficiency does not enter in Eq. (T4.1.3). Equation (T4.1.3) merely supplies the flow rate where the angular momentum conservation is satisfied in the throat of the collector at a given circumferential velocity c_{2u} (i.e. c_{2u} at $H_{optth,}$). Equation (T.4.1.3) does not imply any assumptions on hydraulic losses. With excessive hydraulic losses, such as occur when pumping highly viscous media, the calculation according to Table 4.1 is unable to predict the best efficiency point. Notwithstanding this, the best efficiency points are still approximately situated on the collector characteristic as shown in Chap. 13.1.

The position of the efficiency optimum depends on the following parameters:

- With low specific speeds the BEP flow rate is largely determined by the collector throat since the losses in the volute or diffuser dominate in comparison to the impeller losses.
- Above approximately $n_q > 75$ the best efficiency point is increasingly influenced by the flow rate at shockless impeller entry since the impeller losses then gain in significance. But the deceleration losses in the diffuser have also a strong impact at $q^* < 1$ as demonstrated by Fig. 4.8.
- Strictly speaking, the calculation applies only to the actual volute up to the throat since the hydraulic losses in the downstream diffuser/discharge nozzle are not taken into account. The greater these losses are, the more the actual best efficiency point shifts to the left of the theoretical value from Table 4.1. The same applies to diffusers: the calculation is valid only up to the diffuser throat. Hence: if the volute or diffuser is followed by components causing major losses, a slightly over-dimensioned throat area is recommended so that the kinetic energy at the diffuser inlet and the associated losses are reduced.
- Because of the short distance between the impeller outlet and the diffuser throat, the flow is scarcely able to develop according to the angular momentum conservation. Greater deviations between theory and experiment must therefore be expected when calculating diffusers rather than volutes. The losses in the return channels contribute to this. It should be remembered in this context that the conservation of angular momentum describes a flow *without external forces*. Such are however exerted by the presence of vanes and, in general, by any structures causing friction and flow deflection.
- A *local* expansion of the cross section near the volute throat has only a small effect on the position of the best efficiency point if the remainder of the volute is too narrow. Similarly it is not possible to shift the best efficiency point towards a smaller flow rate using a local constriction without adapting the remainder of the volute accordingly. (A local constriction would be comparable to a venturi

nozzle which has only a minor effect on the throughput of a long pipe.) With low specific speeds, however, it is possible to shift the BEP by modifying the throat area and the subsequent diffuser.

Sometimes (a) *different impellers are employed in a given casing* or (b) *a given impeller is used in modified or different casings.* Both cases shall be explained with reference to the experiments depicted in Figs. 4.17 and 4.18. Curves 1 to 3 in Fig. 4.17 show measurements with three different impellers in the same volute casing which was designed for $n_q = 16$. Curve 1 represents the baseline test with the impeller calculated for $n_q = 16$. The impeller tested with Curve 2 was designed for $n_q = 21$, while Curve 3 was measured with an $n_q = 13$ impeller. The best efficiency points of the three different impellers are situated close to the intersection of the three Q-H-curves with the volute characteristic of $n_q = 16$. The resulting best efficiency points are all very close to $n_q = 16$, although the impellers were dimensioned for $n_q = 13$, 16 and 21. These tests confirm that the casing largely determines the flow rate at BEP with *small* n_q and that the impeller losses are of minor influence - even though the best efficiency point of the oversized impeller ($n_q = 21$) is situated a few percent to the right of the intersection with the volute characteristic, while the undersized impeller is found slightly to the left.

Figure 4.18 shows how a *given impeller* behaves in *different casings*. Both tests were done with the impeller designed for $n_q = 21$. Curve 2 represents its performance in a casing designed for $n_q = 16$, while curve 4 was tested with the volute throat area increased by 37%. Because of differing hydraulic losses, the

Fig. 4.17 Three different impellers in the same casing (casing designed for $n_q = 16$). All Q-H-curves are normalized to the baseline test with $Q_{opt} = 280\,m^3/h$, $H_{opt} = 194\,m$

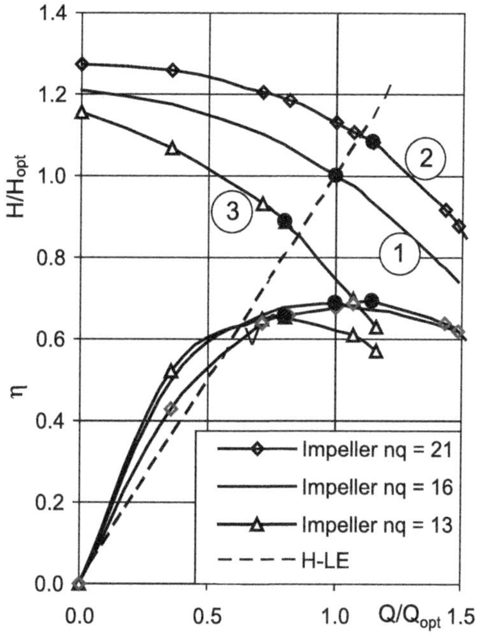

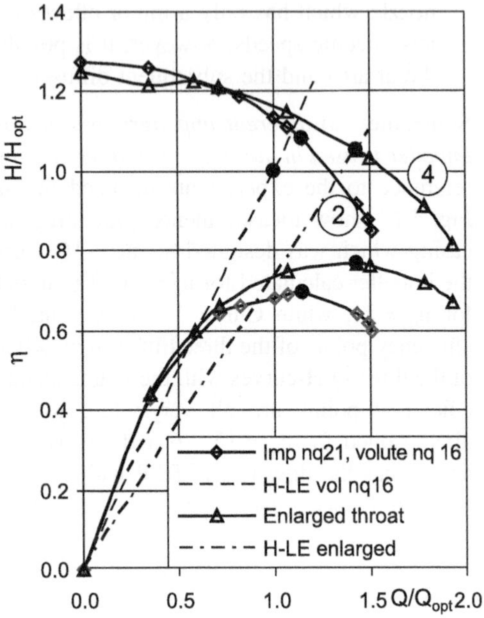

Fig. 4.18 Volute throat area increased by 37% (impeller designed for $n_q = 21$). All Q-H-curves are normalized to the baseline test with $Q_{opt} = 280\,m^3/h$, $H_{opt} = 194\,m$

characteristics 2 and 4 increasingly diverge at high flows - contrary to the idealized depiction in Fig. 4.16. This behavior cannot be expressed by the equations in Table 4.1 unless the various combinations are calculated with different hydraulic efficiencies.

Except for very low partload operation, the power consumption is basically determined by the impeller alone. This follows from Euler's equation. The maximum efficiency obtained increases with the flow rate or the useful power P_u since the secondary losses (disk friction, mechanical and leakage losses) are largely independent of the impeller and casing. Consequently, the ratio of the secondary losses to the power consumption diminishes with increasing P_u.

In order to expand the operation range of a given pump, attempts are sometimes made to shift the best efficiency point by modifying the volute or diffuser throat area as shown by Fig. 4.18.

By reducing the throat area the efficiency optimum is shifted towards smaller flow rates. However, the actual efficiency reached at the new BEP is reduced since the specific speed drops and the portion of the secondary losses in the power consumption increases. At partload the efficiency increases since the deceleration losses in the casing diminish. At $q^* > 1$ the efficiency drops steeply, since the acceleration of the fluid from the impeller outlet to the throat area causes major losses in the diffuser.

Finally, it must be emphasized that the relationships according to Figs. 4.17 and 4.18 are representative for low specific speeds of up to approximately $n_q = 35$. With growing n_q the influence of the impeller inlet increases and becomes dominant above $n_q = 100$. If the impeller blade inlet angles at the design flow rate are selected close to the flow angle (incidence near zero), the best efficiency point is situated close to the intersection of the impeller and collector characteristics even at high specific speeds, but with increasing incidence this is no longer true.

4.3 Prediction of Pump Characteristics

Since neither the hydraulic losses nor the effect of recirculation can be accurately calculated in advance, a prediction of the characteristics is only possible using empirical (or numerical) methods. Empirical methods are based on statistical evaluations of tested pumps. The more the features of a new pump to be calculated differ from those of the database, the less secure the prediction becomes. A possible approach is described in the following:

4.3.1 The data at the best efficiency point are known from the design calculation: Q_{opt}, H_{opt}, η_{opt}, $\eta_{h,opt}$, γ_{opt}. The secondary losses according to Chap. 3.6 can also be calculated.

4.3.2 Slip factor γ and hydraulic efficiency η_h depend on the flow rate ratio q*. An evaluation of many tests yielded the relations shown in Figs. 4.19 and 4.20. The hydraulic efficiency can be expressed by Eq. (4.14). The slip factor is given by (Eq. 4.14a) but the extrapolation to low (or even zero) flow is uncertain because the slip factor concept is not really applicable to operation with extensive recirculation.

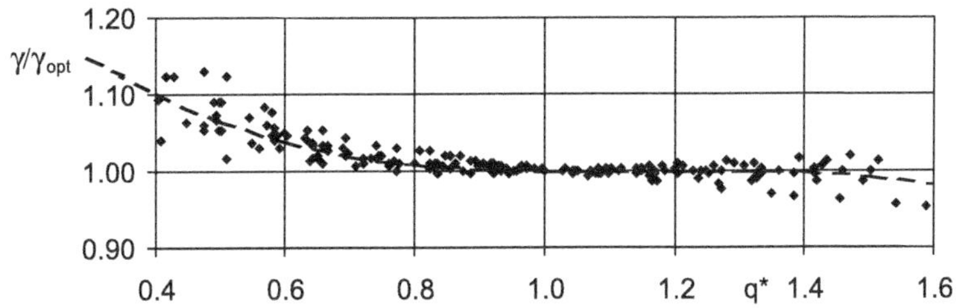

Fig. 4.19 Dependence of slip factor γ on the flow rate ratio

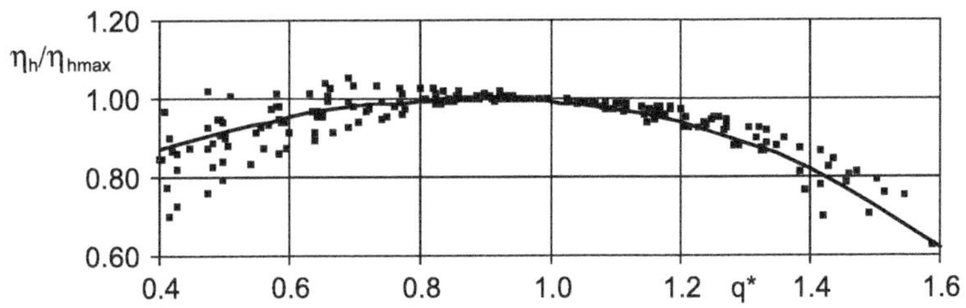

Fig. 4.20 Dependence of hydraulic efficiency η_h on the flow rate ratio

$$\frac{\eta_h}{\eta_{h,max}} = 1 - 0.6\,(q^* - 0.9)^2 - 0.25\,(q^* - 0.9)^3 \tag{4.14}$$

$$\frac{\gamma}{\gamma_{opt}} = 1.38 - q^* + 0.87\,q^{*2} - 0.25\,q^{*3} \tag{4.14a}$$

According to this evaluation the hydraulic efficiency reaches its maximum value at $q^* = 0.9$ (not at $q^* = 1$). As explained in Chap. 4.1.3, the calculation applies only to the flow range where no major recirculations occur. For the analysis of the data in Figs. 4.19 and 4.20 at $\beta_{2B} \leq 90°$ recirculation was assumed to be present when $\gamma \geq 0.95$.

4.3.3 Set up a table and enter the values η_h/η_{hmax}, and γ/γ_{opt} taken from Fig. 4.19 for Q = 0 to Q_{max}. Subsequently calculate η_h and γ. The head is then obtained from Eq. (T3.3.7).

4.3.4 The power consumption follows from Eq. (T3.5.1) where the recirculation power is set to $P_{Rez} = 0$ and the secondary losses are determined from Table 3.5 or Chap. 3.6.

4.3.5 These calculation steps are performed only in the load range where $\gamma < 0.95$ (the limit can also be set slightly lower). Head coefficient and power coefficient at Q = 0 can be estimated from Figs. 3.24 and 4.14. The curves for head and power between Q = 0 and the range without recirculation calculated as per step 4.3.2 must be interpolated based on experience. In doing so it is possible to utilize the empirical observation that the characteristic at the best efficiency point is tangent to the straight line through the points $Q(H_{th} = 0)$ and Q_{opt}, H_{opt}.

The weaknesses of the above approach are obvious. They include the large scatter of ψ_o and λ_o and the uncertainty of the statistical data from Figs. 4.19 and 4.20. In particular, a possible instability of the Q-H-curve is not really discernable (Figs. 4.19 and 4.20).

4.4 Range Charts

As pointed out in Chap. 2, centrifugal pumps are employed over very wide ranges of flow rates and heads in many different types and designs to suit the particular applications. A certain pump type, for example single-stage process pumps, covers a specific performance field to suit the relevant market requirements, e.g. Q = 3 to 800 m^3/h and H = 20 to 250 m. For economic reasons - costs of design, testing, patterns, manufacturing, spare parts and stocking - such a range should be covered in an optimum way with the fewest pump sizes possible.

Figure 4.21 shows a range chart for single-stage volute pumps operating at n = 2950 rpm. It contains 27 pump sizes which are described by the nominal diameter of the discharge nozzle (first figure) and the nominal impeller outer diameter. In the USA the pump size is defined by (for example) "8 × 6 × 14

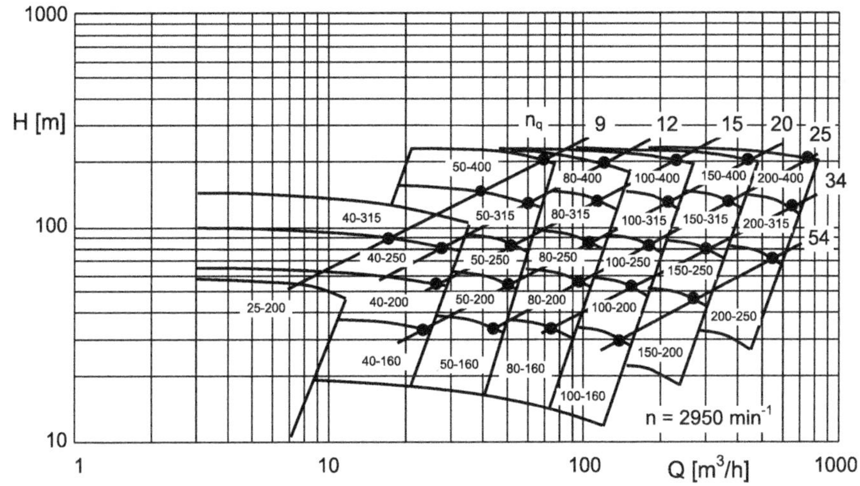

Fig. 4.21 Range chart of process pumps, Sulzer Pumps

MSD", where the first figure represents the size of the suction nozzle, the second the discharge nozzle and the third the impeller diameter (all given in inches); MSD in this example indicates the pump type of a specific pump manufacturer. Sizes arranged in the chart in a nearly vertical column deliver similar volumetric flow rates and therefore have the same nominal discharge nozzle diameter. Sizes arranged in horizontal rows have identical impeller diameters and produce a similar head. The specific speeds of the individual sizes increase from top down and from left to right. Pumps with the same specific speed are found, in a double-logarithmic graph, on inclined straight lines drawn in Fig. 4.21.

In theory, the impellers and collectors of all sizes with the same specific speeds are geometrically similar (which allows cost savings for design and testing), but in practice different sizes of the same n_q often deviate in some design details from geometric similarity (e.g. shaft and hub diameters). The impeller outer diameters are stepped with the standard number $10^{0.1} \approx 1.26$. All pumps in this range chart can also be used with the lower standard speed so that the application range of the series is expanded accordingly. However, unless the available $NPSH_A$ demands a pump with a low speed, the selection of such a pump can be uneconomical.

Range charts of the type shown in Fig. 4.21 allow a rapid selection of possible pumps which can be considered for a specific application. For multistage pumps the range chart can show the heads for different stage counts. In the case of very large pumps running with lower speeds, the range chart can also be drawn up with the speed as a parameter. In general, all parameters (impeller diameter, specific speed, number of stages and standard speeds) must be used to optimally cover the planned performance range. To define the steps in the flow rates, different criteria can be considered: (1) Most important is the complete coverage of the field. (2) Number of sizes required. (3) Specification of the major customers: for instance that the rated performance of the pump is to be in the range $0.8 < q^* < 1.1$. (4) Avoidance of impermissible partload or overload operation. (5) Efficiency losses when the pump operates near the border of its individual field. For example the flow rates can be graduated according to the criterion that an efficiency drop of $\Delta\eta = 3\%$ defines the limit of the field (Fig. 4.21).

When defining the head steps, the permissible amount of impeller trimming must be taken into account (Chap. 4.5.1). In general the steps in flow rate and head should be refined with increasing pump size, power consumption and impeller tip speed, since the losses in terms of investment, energy and maintenance costs would become onerous with steps selected too coarse.

When drawing up a range chart, stepping factors for flow rate and impeller size are selected based on the above considerations. The factor for the flow rates is typically between 1.5 and 2. The factor for the impeller diameter is typically between 1.12 and 1.32 (it is usually a standard number $10^{\times}$). The best efficiency point should be near the right border of the performance field of each size to avoid selecting a pump in the overload range. With the speed selected and the specific speeds used it is then possible to calculate the best efficiency points for all sizes.

Constant-efficiency curves according to Fig. 4.22 are frequently used to present the characteristics of a pump size with different impeller trim diameters, Chap. 4.5.1. To create such curves, horizontal lines (e.g. at 50, 60, 70, 75, 80 and 82%) are placed through the efficiency curves of the different impeller diameters. The intersections are then transferred to the Q-H curves at selected flow rates. The points obtained there yield curves of constant efficiency. The procedure for $\eta = 0.75$ is shown in Fig. 4.22.

4.5 Modification of the Pump Characteristics

Adjusting the characteristic to various operating points is necessary in many applications. To do so, there is a wide range of possibilities and combinations thereof (too many to be discussed exhaustively). Power and/or head reductions as well as increasing the power or head are common. According to Table 4.2 the measures range from simple trimming or filing to the retrofit of new impellers (e.g. with changed blade angles) or to the replacement of an entire cartridge of a multistage barrel casing pump. Table 4.2 lists possibilities for adapting the characteristics to changing operation requirements. Which of these options is feasible in the individual case depends on the design details of the pump under consideration.

4.5.1 Impeller Trimming

For economic reasons, a given pump size is used to cover a certain performance range as discussed in Chap. 4.4. This can be accomplished by a reduction of the

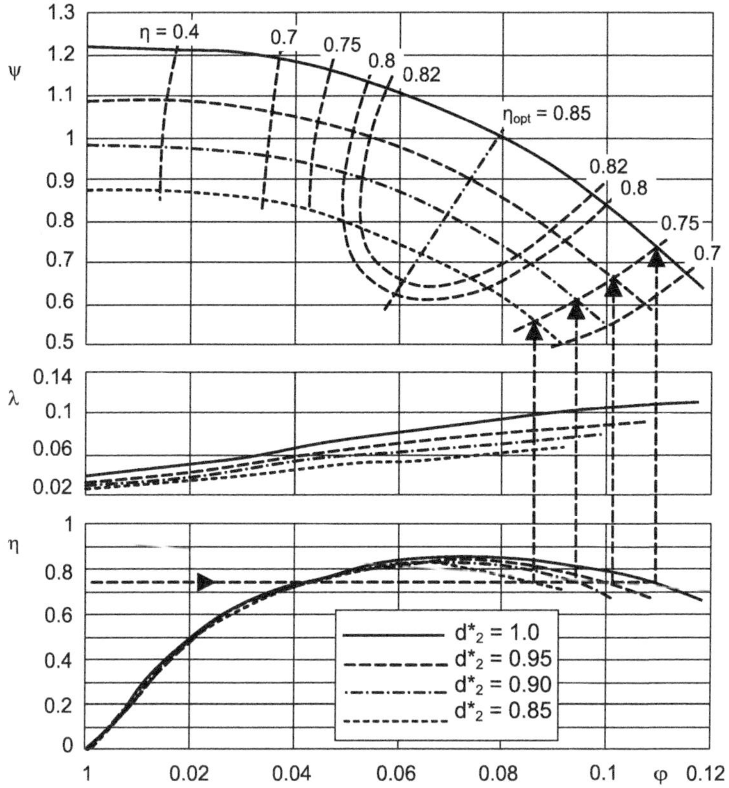

Fig. 4.22 Determination of constant-efficiency curves (note that all pressure coefficients are referred to the *full* impeller diameter); $d_2^* = d_2'/d_2$

impeller outer diameter. By means of this "trimming" the head of a given pump is reduced to the value required by the plant without the need to unnecessarily waste energy by throttling.

The characteristics of a pump with a trimmed impeller cannot be calculated according to the similarity laws since a trimmed impeller is not geometrically similar to the impeller with the full diameter. The impeller outlet width remains identical or increases slightly, the blade outlet angle may change and the blades become shorter. Blade shortening leads to higher blade loading and lower flow deflection resulting in a drop in the slip factor γ. These conditions are illustrated in Fig. 4.23a by means of the outlet triangle. At a given flow rate c_{2m} increases roughly proportional to the reduction of d_2 causing the head to be reduced according to Eq. (T3.3.7). The discharge angle in the absolute reference frame increases from α_2 to α_2'. Since the collector remains unchanged, the ratio c_{3q}/c_2 grows at a given flow, see Fig. 4.23b (refer to Chap. 5.3.1 for a detailed discussion). The best efficiency point is consequently shifted towards lower flow than with full diameter. This is also reflected in Chap. 4.2 and Fig. 4.16. The volute characteristic remains unchanged while the head is reduced. However, the BEP shift is smaller than in the case of a geometrically similar reduction of impeller and collector (Fig. 4.23) because the impeller is oversized for the reduced flow rate.

Table 4.2 Modification of pump characteristics

Task	Possibilities	Remarks
1. Increase head at Q = const.	1.1 Under-file impeller blades at outlet on suction side 1.2 Weld-up blades and under-file as 1.1 1.3 Reduce hydraulic losses 1.4 New impeller: greater d_2, z_{La}, b_2, β_{2B} 1.5 Add stages (if multistage) 1.6 Fluid counter-rotation at impeller inlet	Q-H-curve gets flatter, increasing risk of instability Check shaft and casing stresses
2. Reduce head and power at Q = const.	2.1 Reduce impeller outer diameter (trimming) or smaller impeller 2.2 Reduce number of stages (if multistage) 2.3 In multistage pumps: increase pre-rotation at impeller inlet: a) shorten the return vanes b) add a gap between return vanes and stage casing elements (see sketch) 2.4 New impeller with smaller b_2, β_{2B} or less blades 2.5 Add a throttle in discharge nozzle (this results in a high loss in efficiency since the power stays the same!)	Q-H-curve gets steeper! 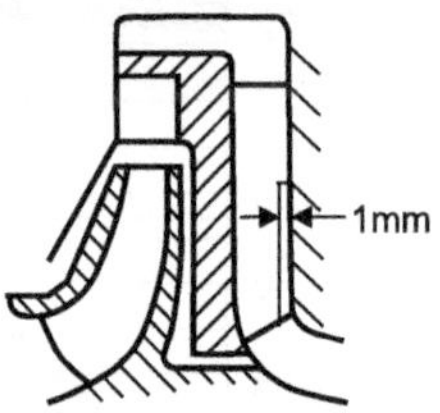
3.Increase head and shift BEP to higher flow	3.1 Enlarge diffuser or volute throat area; achievable shift of BEP up to about $\Delta Q_{opt} \approx 15\%$ 3.2 In addition to 3.1 new impeller with greater d_2, z_{La}, b_2, or β_{2B}. With high n_q also greater β_{1B} und d_1	Q-H-curve gets flatter, possibly instable (saddle). Effect increases for small n_q. Check $NPSH_A$ requirements! Risk of pressure side cavitation: check the incidence and w_{1q}/w_1, unless β_{1B} and/or d_1 are adapted.
4. Decrease head and shift BEP to lower flow	4.1 Reduce diffuser or volute throat area; achievable shift of BEP: $\Delta Q_{opt} \approx 25\%$ 4.2 In addition to 4.1, new impeller with smaller z_{La}, b_2, or β_{2B} and d_1; β_{1B}	Effect increases at small n_q Q-H-curve gets steeper Check the risk of diffuser or volute cavitation for $n_q < 25$.

Figure 4.24 shows the characteristics of a pump with a double-entry impeller ($n_q = 25$) for the full impeller diameter and with a trim of 10%. The characteristics (H_{gs}, η_{gs}) of a 10% smaller, *geometrically similar,* pump are shown for comparison. The magnitude of the deviations from the similarity law caused by the trimming depends on the geometry of the impeller and the pressure distribution on the blades. The shorter the blade length (i.e. the higher the specific speed), the greater is the

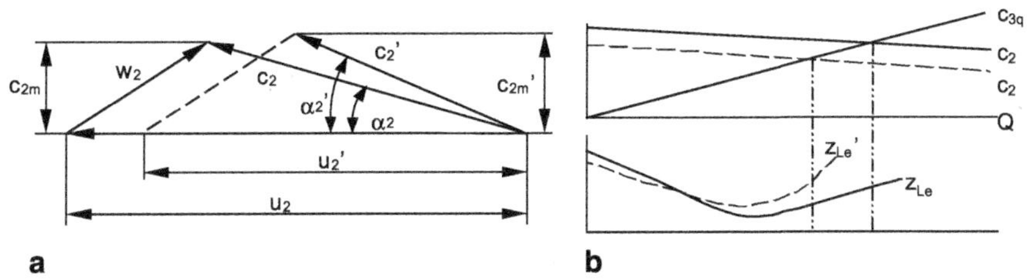

Fig. 4.23 Impeller trimming

head reduction for a given trim ratio. Since these relationships cannot be calculated accurately enough from theory, the calculation of the characteristic for a trimmed impeller is performed according to empirical relationships. In principle these relationships, like the impeller geometry, are specific to each pump type. The procedure according to [B.17] is shown in Fig. 4.24. To derive the characteristic $H' = f(Q')$ with trimmed diameter d_2' (from the characteristic $H = f(Q)$ with full impeller diameter d_2), random rays (1 to 3 in Fig. 4.24) are drawn from the origin to the characteristic $H = f(Q)$. For every pair of Q and H so obtained, Q′ and H′ are calculated according to Eq. (4.15).

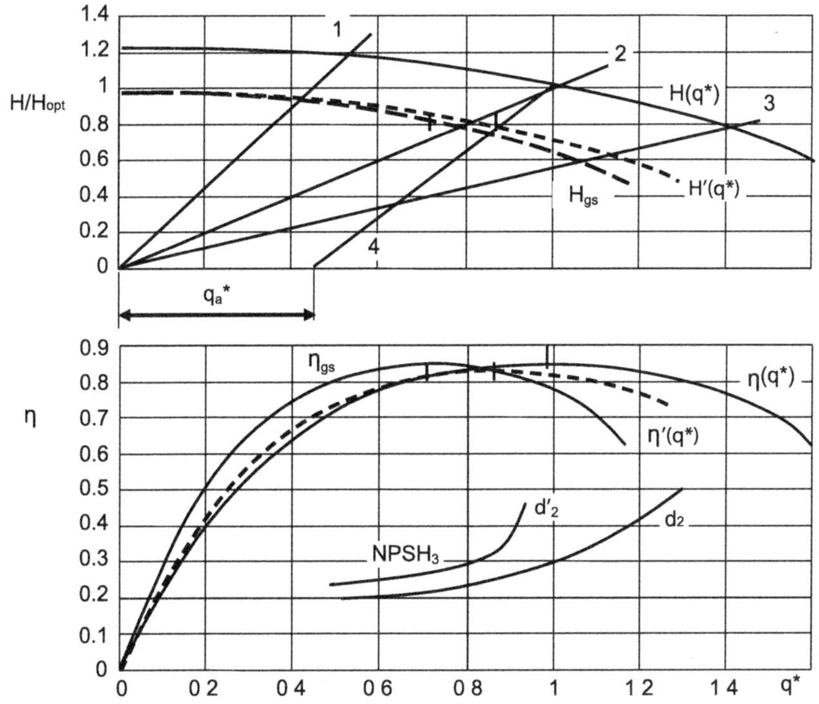

Fig. 4.24 Modification of pump characteristics by trimming the impeller

$$\frac{Q'}{Q} = \left(\frac{d_2'}{d_2}\right)^m \quad \text{and} \quad \frac{H'}{H} = \left(\frac{d_2'}{d_2}\right)^m = \frac{Q'}{Q} \tag{4.15}$$

The exponent m lies between 2 and 3; it can be calculated from Eq. (4.15a) for a given trim diameter ratio and from Eq. (4.17a) for a specified head reduction. If the blades are under-filed at the outlet, m is closer to 3 for trims smaller than the profile length. If trimming exceeds 5% of the impeller diameter, m is near 2. If β_{2B} and b_2 do not change with the radius, in theory m = 2 applies.

Exponent m for a given trim diameter ratio;

$$m = 2.08 + 1.1\,\mathrm{EXP}\left(15\left\{\frac{d_2'}{d_2} - 1\right\}\right) \tag{4.15a}$$

The exponent m can be calculated from trimming tests.

$$m = \frac{\ln\frac{H'}{H}}{\ln\frac{d_2'}{d_2}} \tag{4.16}$$

The efficiency optimum Q'_{opt} for a pump with a trimmed impeller is to the *right* of the ray running from the origin through the best efficiency point of the full diameter impeller (ray 2 in Fig. 4.24). The best efficiency points of all trim curves are located on a line (# 4 in Fig. 4.24) through the best efficiency point measured with the full d_2, which intersects the abscissa at a value $q^*_a \approx (0.005 \text{ to } 0.01)\, n_q$ (Fig. 4.24).

Different trimming laws have been published: According to [B.5] m = 2 should be used and the best efficiency points of the trimmed impellers should be assumed on a straight line through the origin.

The trim required to achieve a certain head H′ is calculated from:

$$\frac{d_2'}{d_2} = \left(\frac{H'}{H}\right)^{\frac{1}{m}} \tag{4.17}$$

Exponent m for a specified head reduction:

$$m = 2.08 + 1.2\,\mathrm{EXP}\left(7.5\left\{\frac{H'}{H} - 1\right\}\right) \tag{4.17a}$$

In order to apply Eqs. (4.17) and (4.17a) draw a straight line from the origin of the Q-H-curve through the specified (rated) trim point performance $Q_R, H_R (= H')$. This line intersects the Q-H-curve given for the full impeller diameter at the head "H" which is to be used in Eqs. (4.17) and (4.17a).

In [N.2] the following formulas are given for the calculation of the trim curves:

$$\frac{Q'}{Q} = K \text{ and } \frac{H'}{H} = K^2 \text{ with } K = \sqrt{\frac{d'^2_{2b} - d^2_{1b}}{d^2_{2b} - d^2_{1b}}} \tag{4.18}$$

Due to trimming the efficiency at the BEP approximately drops according to:

$$\Delta\eta = \varepsilon(1 - d_2'/d_2) \tag{4.19}$$

Set $\varepsilon = 0.15$ to 0.25 for volute pumps and $\varepsilon = 0.4$ to 0.5 for diffuser pumps. With small n_q the efficiency sometimes increases through a slight trimming since the impeller disk friction losses decrease with the 5th power of the diameter according to Eq. (T3.6.2).

By virtue of Eq. (4.19), the power consumption of the trimmed impeller increases ($\eta' = \eta_o - \Delta\eta$) as compared to an untrimmed impeller (η_o). The additional power is approximately:

$$\Delta P = \frac{P'}{\eta_o}\varepsilon\,(1 - d_2{}^*) \tag{4.20}$$

($d_2{}^* = d_2'/d_2$ and d_2 = full impeller diameter). If k_{kW} is the capitalized energy cost per kW of installed power, the power increase according to Eq. (4.20) requires an additional energy cost ΔK_E of:

$$\Delta K_E = z_{pp}\,k_{kW}\frac{P'}{\eta_o}\varepsilon\,(1 - d_2{}^*) \tag{4.21}$$

Here, z_{pp} stands for the number of parallel operating pumps affected by the trim. If the additional energy costs according to Eq. (4.21) exceed the expenditure for a new pattern (and design costs), a volute casing or a diffuser adapted to the reduced impeller diameter would be economically (and ecologically) justified.

For volute pumps (with $\varepsilon = 0.2$) Eq. (4.21) was evaluated for a capitalized energy cost of $k_{kW} = 2000$ \$/kW, Fig. 4.25. This figure gives an indication of how the trim diameter of pumps with high power consumption should be limited. If more than one pump is affected ($z_{pp} > 1$), the sum of the additional power must be considered according to Eq. (4.21) since the pattern costs occur only once but the total energy loss amounts to $z_{pp} \times \Delta P$. An example is included in the legend to Fig. 4.25. If a limit for the energy cost ΔK_E is selected, the applicable trim diameter can be calculated from:

$$d_2{}^* = 1 - \frac{\eta_o\,\Delta K_E}{z_{pp}\;\varepsilon\,k_{kW}\,P'} \tag{4.22}$$

In addition to the economical limitation of the trim ratio of large pumps, the operational behavior should be considered, since it may deteriorate if blade shortening through trimming is excessive. To visualize this issue, consider

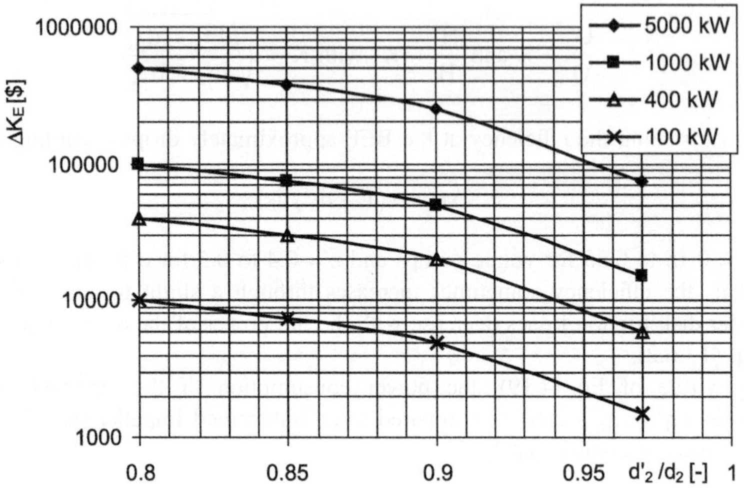

Fig. 4.25 Additional energy costs caused by efficiency impairment due to trimming; valid for volute pumps, calculated with k_{kW} = 2000 \$/kW, $\eta_o = 0.8$ und $\varepsilon = 0.2$. Enter the graph with the combined power of all trimmed pumps $(z_{pp} \times P)$, where P is the power of *one* pump

Fig. 4.26 where the characteristics for the full impeller diameter ($d_2^* = 1.0$) and for the trimmed diameter $d_2^* = 0.8$ of a pump with $n_q = 25$ are plotted. The operation (or guarantee) point shown is assumed at 80% of the best efficiency point of the impeller trimmed to $d_2^* = 0.8$ (which, according to [N.6], would still appear permissible). Since the flow rate of shockless impeller entry is not changed by the trimming, the impeller *inlet* runs at approximately 57% of the BEP flow rate delivered by the impeller with full diameter. A pump selected in this way will practically always operate with inlet recirculation. This may be acceptable with small pumps of low power. However in high-power pumps, vibrations and noise are likely to become excessive through the dissipation of the recirculation power. Due to the trim the blades are shortened and the blade loading increases. This can cause higher cavitation induced noise. Large low-pressure pumps selected in this way may become very noisy because the large surfaces of rather thin-walled casings radiate noise very effectively. Specifically, this problem is often encountered with single-stage pumps with double-entry impellers where the noise problem is compounded by the non-uniform approach flow observed in this type of inlet casing, see Chap. 7.13.

Example: The impellers of four pumps in parallel, each with a power of 250 kW, kW are to be trimmed by 10%; the total power consumption is thus z_{pp} P = 1000 kW. With $d_2' = 0.9$ read from curve "1000 kW" the additional capitalized energy costs of 50'000 \$. If the costs of a new pattern or a pattern modification are below this amount, it is economical to invest in new patterns. This would also make sense with respect to material costs and ecology (Fig. 4.25).

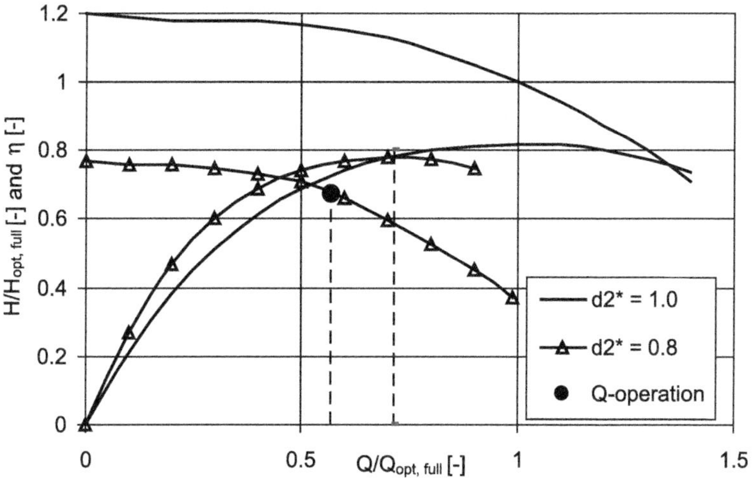

Fig. 4.26 Characteristics for impeller trimming of 80% of the full diameter. The operation point shown is at 80% of the BEP of the curve for $d_2^* = 0.8$, but at 57% of the BEP with the full impeller diameter

To avoid excessive shortening of the blades, it is advisable to limit the diameter reduction of *volute casing* pumps as follows:

$$\begin{aligned} &\frac{d_2'}{d_2} \geq 0.8 \text{ to } 0.85 \quad \text{for} \quad n_q < 40 \\ &\frac{d_2'}{d_2} \geq (0.8 \text{ to } 0.85) + 0.0025(n_q - 40) \quad \text{for} \quad 40 < n_q < 100 \end{aligned} \tag{4.23}$$

The limits defined by (Eq. 4.23) apply to medium-sized pumps. The impellers of small pumps (where noise and vibrations are not considered an issue) are sometimes trimmed to diameters smaller than stipulated by Eq. (4.23). However, the greater the power consumption of the pump, the less trimming should be employed in order to reduce the risk of vibrations, noise and cavitation. The economic limit for $d_2^* = d_2'/d_2$ defined by Eq. (4.21) or Fig. 4.25 may then be considered as a selection criterion.

If little margin over $NPSH_R$ is available, if the inflow conditions are unfavorable or if stringent limits on noise and vibrations are specified, the value 0.85 should be used in Eq. (4.23) or trimming limited even further.

Trimming Recommendations

- In the case of *diffuser* pumps, the characteristic may become unstable if impeller trimming is excessive. The limit must be determined experimentally. In the absence of tests, it is recommended not to trim by more than 5%. If greater trimming is required, a diffuser with a smaller inlet diameter can be designed. Since the diffuser pattern costs are much lower than the costs of a casing pattern, a new diffuser, adapted to the trimmed impeller, is economically justified at power ratings smaller than would be the case with a volute pump.

- In the case of diffuser pumps, the impeller shrouds are left at full diameters (only the blades are trimmed according to Fig. 4.27b) to forestall instabilities of the characteristic and axial thrust excursions at partload and to keep the exchange of momentum between main flow and the liquid rotating in the impeller sidewall gaps as low as possible (Chaps. 5 and 9.1).
- Since the impeller shrouds of diffuser pumps are not trimmed, the efficiency loss due to trimming is greater than with volute pumps (the impeller disk friction losses are proportional to d^5).
- In the case of volute pumps, the impeller shrouds are trimmed to the same diameter as the blades to reduce disk friction losses, Fig. 4.27a.
- In general, an oblique trim according to Fig. 4.27c increases the shut-off pressure compared with parallel trimming (as discussed in Chap. 5) and guards against a characteristic drooping towards $Q = 0$. The trim angle is mostly selected between 5° and 15°.
- An oblique trim is recommended for specific speeds above $n_q > 40$ so that the outer streamline does not become too short.
- With semi-axial impellers the position of the blade trailing edges has a major influence on power and head at shut-off (Chap. 5). The smaller the ratio d_{2a}/d_{2i}, the smaller P_o and H_o become.
- Pumps with double-entry impellers of low n_q are mostly trimmed so that the blade trailing edges remain parallel to the rotor axis, Fig. 4.27e.

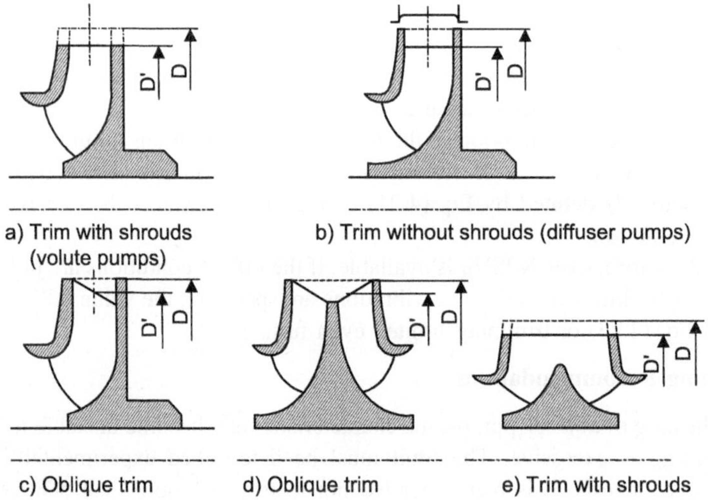

Fig. 4.27 Impeller trimming

- As a rule, noise and vibrations diminish with slight trimming since the distance between the impeller blades and the collector increases. However, with major trimming the hydraulic excitation forces may rise again since the blade loading increases.
- Since the head reduction caused by trimming depends on parameters that cannot be accurately predicted, an impeller is preferably trimmed in steps. The risk to miss the contractual/desired operating point by an excessive trim is thus reduced.
- Severe trimming (more than 10%) also causes the suction characteristics to deteriorate: $NPSH_3$ increases as hinted by the curves in Fig. 4.24. One of the reasons for this is the changed pressure distribution due to blade shortening and increased blade loading, (Chap. 6). In addition, a 3% drop in a severely reduced head constitutes an absolutely smaller change in head and thus less cavitation. Therefore, in multistage pumps the suction impeller should preferably not be trimmed. Or the suction impeller is trimmed less than the series stages if the pump has only a few stages.

Inlet rings and low-flow impeller design

As discussed in the context with Fig. 4.26, a large impeller trim is likely to cause the pump to operate with inlet recirculation even at rated flow. Outlet recirculation may occur too because the ratio b_2/d_2' of the trimmed impeller can become quite large. Even though the distance between impeller and volute cutwater is enlarged, noise and vibrations often cause problems; a case history of a single-stage pump with double-entry impeller and double-volute is reported in Chap. 10.7.3.

Onset and intensity of inlet recirculation and associated cavitation, vibrations and noise can be reduced by designing an inlet ring according to the recommendations given in Chap. 6.8.3. Different shapes of such rings are depicted in Fig. 6.39. The throat diameter d_R of the inlet ring can be calculated for a selected incidence angle with the rated flow Q_R (or the targeted BEP) of the *trimmed* impeller. Depending on the maximum flow rate specified for the trimmed impeller and the particular requirements and circumstances of the project, the incidence (including the blade blockage effect τ_{1a}) could be selected to $i_{1a}' = 1$ to 3°. The diameter d_R of the ring throat area can then be calculated for the rated flow Q_R of the *trimmed* impeller with $\alpha_1 = 90°$ from Eq. (4.24):

$$d_R = 2.9\sqrt[3]{\frac{Q_R \tau_{1a}}{f_q\, n\, k_n \tan(\beta_{1B,a} - i_{1a}')}} \quad \text{valid for } \alpha_1 = 90° \tag{4.24}$$

with $\beta_{1B,a}$ blade angle of the existing impeller. Since the shaft or hub diameter d_n is given, iteration over k_n is required. This calculation is done for the outer streamline. Due to the ring, the NPSH-curve is shifted to lower flow rates. The resulting curve $NPSH_R = f(Q)$ can be estimated from Table 4.4, Eqs. (4.4.26) to (4.4.29). Since the inlet rings reduce the intensity of the impeller inlet recirculation, head and power at shut-off are reduced by a few per cent when installing inlet rings.

If warranted for the project, a low-flow impeller can be developed with a smaller inlet diameter and smaller outlet width. In this way, inlet and outlet recirculation of the trimmed impeller can be reduced or eliminated. A low-flow impeller may be considered when: (1) noise, vibrations or cavitation problems were encountered in a plant with an excessive impeller trim - or if the impeller inlet has been over-sized in the first place; (2) the trim diameter required to reach a specified operation point is lower than the limits given by Eq. (4.23).

The optimum design parameters to be chosen for the low-flow impeller depend on the rated point Q_R and H_R (or BEP) of the *trimmed* impeller and on the geometry of the original impeller. The procedure outlined below can be followed when designing the low-flow impeller:

1. Select design point (rated conditions or BEP) Q_R and H_R and $NPSH_3$ for the specified $NPSH_A$. Apply appropriate NPSH-margins as per Table 15.1.
2. Calculate suction specific speed n_{ss} for new BEP (or Q_R).
3. Determine impeller inlet diameter from Eqs. (15.1.12) and (15.1.13) or with the help of Fig. 6.21.
4. Select incidence $i_{1a}' = 1$ to 3° and determine impeller blade inlet angles from Eq. (T7.1.7)
5. Calculate trim diameter d_2' required from Eq. (4.17)
6. Calculate the performance curves for the trimmed impeller. The trim diameter must be varied iteratively to match the specified rated conditions.
7. Select impeller outer diameter $d_{2,LF}$. The impeller diameter depends on the specified project data and the choice of other design parameters. As a starting point $d_{2,LF}/d_{2nom} = (d_2'/d_{2nom})^{0.5}$ could be tried. However, if the maximum shut-off pressure is specified, $d_{2,LF}$ must be selected to ensure that the shut-off head is below that limit.
8. Select number of impeller blades z_{La}: If the baseline impeller has 6 or 7 blades, $z_{La} = 5$ could be selected if a steeper Q-H-curve is needed. The impeller outlet width b_2 should be reduced as compared to the baseline design in order to reduce impeller outlet recirculation; Eq. (7.1) can be used to determine b_2. Once z_{La} and b_2 have been defined, the blade outlet angle β_{2B} is determined that way that the rated point (Q_R and H_R) is met. The hydraulic efficiency used in this calculation is expected to be a little lower than the baseline with $d_{2,nom}$.
9. Calculate performance curves including $NPSH_R = f(Q)$ for low-flow impeller according to Table 4.4 using the performance curves with the nominal (full) impeller diameter as baseline. Note that the volute characteristic does not change (unless the volute throat area would be modified).
10. Develop inlet ring which matches the new impeller inlet diameter determined from step 3 above, see also Chap. 6.8.3 for details of ring dimensions.

4.5.2 *Under-Filing and Over-Filing of the Blades at the Trailing Edge*

By profiling the impeller blade trailing edge *on the suction side* ("under-filing") the distance between the blades is increased from a_2 to a_2' (see Table 2 and Fig. 3.6). Consequently, the fluid is decelerated more effectively and the head increases. The magnitude of the gain in head depends on the pressure distribution near the blade trailing edge. If the blade loading is high, or the flow is close to stall, the fluid is unable to follow the contour generated by under-filing and the gain in head is minor. The smaller the outlet angle β_{2B}, the greater is generally the relative gain $\Delta H/H$. Figure 4.28 shows as an example the under-filing of an impeller in a volute pump of $n_q = 28$. In general under-filing results in the following:

- The head increases at the best efficiency point by the ratio $H'/H = (a_2'/a_2)^x$ with the exponent x = 0.4 to 0.8; x diminishes with increasing blade outlet angle.
- At shut-off the gain $\Delta H/H$ is approximately half as high as at BEP.
- $\Delta H/H$ grows with increasing flow rate and is greatest at overload $q^* > 1$.
- The BEP shifts on the diffuser/volute characteristic according to Chap. 4.2.
- The efficiency at BEP mostly increases by 0.5 to 1% (rarely more) since the wake flowing off the trailing edge becomes narrower, unless the flow separates.
- The power consumption increases in accordance with the increase in head but corrected by the efficiency gain if applicable.

Obviously the blades must only be profiled to the extent that their mechanical strength is not jeopardized. Tip speed u_2, blade width (or n_q) and the distance between the impeller blades and diffuser vanes must be considered when evaluating the mechanical characteristics (Chap. 14.1, Chap. 10 and Table 10.1).

If qualified (and approved) welding procedures are available, it is possible to increase the head (in an extreme case) by filler-welding the blades on the pressure side followed by under-filing on the suction side. In this way the blade outlet angle and a_2 can be increased.

Contrary to some opinions, the head *cannot* generally be reduced or the characteristic made steeper by profiling on the *pressure* side ("over-filing"). This

Fig. 4.28 Under-filing of the impeller blade trailing edges: a_2 has been increased by 6.5%

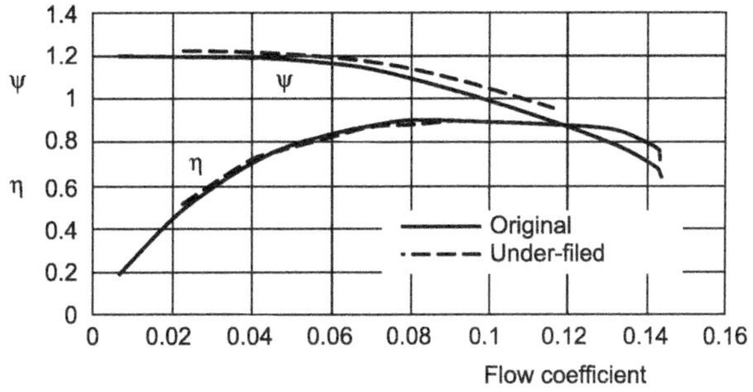

because the distance between the blades (a_2) remains unchanged by such a modification. Over-filing merely narrows the wake flowing off the trailing edge (Chap. 10.1). This reduces the losses due to wake vortices and even can produce a slight *increase* in head as sometimes measured.

4.5.3 Collector Modifications

By increasing or reducing the volute or diffuser throat it is possible to shift the flow rate at the best efficiency point. Figure 4.18 shows an example of a pump with $n_q = 16$ whose volute throat was enlarged by 37%. This increased the BEP flow rate by 25% and the specific speed rose from $n_q = 15.9$ to 18.2. The BEP flow rate roughly corresponds with the volute characteristic (the large gain in efficiency points to a non-optimum design of the diffuser/discharge nozzle prior to the change).

The following are typical effects of an increase of the diffuser or volute throat (a throat area reduction shows the inverse tendencies):

- The head at $q^* > 1$ increases since the losses in the collector and the deceleration ratio c_{3q}/c_2 diminish (the flow rate where $c_{3q} = c_2$ is reached increases). With growing specific speed this effect becomes weaker.
- Shut-off head and partload head drop slightly.
- The power consumption when operating without recirculation is (theoretically) determined only by the impeller and hardly changes. At shut-off the power consumption rises due to an increased exchange of momentum.
- With low specific speeds the efficiency increases when opening up the volute/diffuser throat area since the losses in the collector drop and the best efficiency flow rate grows. The secondary losses (which remain constant) are thus referred to a greater useful power P_u.
- At $n_q < 12$ all the efficiency curves measured with different volute/diffuser throat areas are essentially enveloped by one curve as demonstrated by Fig. 7.15. In contrast, at higher specific speeds the partload efficiency improves if the casing cross section is reduced. The maximum improvement is found in the vicinity of $q^* = 0.5$.
- As the specific speed increases the reaction of the best efficiency point location to casing modifications diminishes. With volute casing pumps this statement roughly applies to $n_q > 60$ to 80. In the case of diffuser pumps the influence of the area A_{3q} is more profound than with volutes; however, the sensitivity decreases here too with increasing n_q.
- Apart from differences in sensitivity, diffuser and volute pumps react similarly to throat area modifications.

4.6 Analysis of Performance Deviations

It may happen on the test stand or in the plant that the measured performance is below the expected values and outside of the tolerances. The allowed deviations are given by specifications issued by the client. Depending on the quality class (Chap. 15) of the pump and the test configuration, performance tolerances typically are in the range of ±2 to ±5%. If the actual deviations exceed the usual measurement and manufacturing tolerances, it is necessary to determine the causes of the deviations and to define corrective action. This task is often difficult and Table 4.3 may provide some help for a focused analysis (note that the high number of possible errors and error combinations precludes an exhaustive discussion). Deviations and corrective measures for solving problems with NPSH-curves are discussed in Chap. 6.9 and Table 6.4.

Before dealing with the detailed analysis according to steps 3 to 9 below, it is advisable to first perform the checks according to steps 1 and 2:

1. Verify the expected values: Are the design calculations and the data base used correct?
2. Verify measurements:

   - Direction of rotation
   - Check measurements using redundant instruments
   - Any air in the measurement lines?
   - Is the measured inlet pressure distorted through partload recirculation (see Fig. 5.16)?
   - Potential flow rate measurement problems: (1) ensure that there are adequate straight pipe lengths for velocity equalization upstream and downstream of the instrument; (2) air separation at low pressures; (3) cavitation in an orifice or a nozzle; (4) bypass in the test loop (for instance leaking valves); (5) pre-rotation; (6) correct installation of the orifice or the instrument.
   - Electric power measurements are frequently the cause of considerable uncertainties (accuracy, motor efficiency, power factor).
   - Gear box efficiency, if pump is driven via a gear
   - Check the test set-up, notably inlet flow conditions. Particular attention is needed with high specific speed pumps and wet pit installations (pre-rotation, effects of partload recirculation)
   - Measurements in plants are often problematic and demand extreme care to prevent incorrect conclusions.

3. If the expected values and the measurements are correct, geometrical checks of impeller, diffuser and volute casing have to be performed and compared with the desired dimensions on the drawings. As a first step the following dimensions must be measured on the impeller: d_2, b_2, a_2, a_1, d_1, d_n, blade thickness and profiling at blade inlet and outlet if applicable. The dimensions d_2, b_2, a_2, are essential for achieving the head; a_2 and a_1 provide an indication whether the blade position is roughly correct. On the diffuser, d_3, a_3, b_3 must be measured

Table 4.3 Analysis of performance deficits

Problem/Findings	Possible causes	Possible remedies
1. $H < H_{required}$ and $P < P_{required}$	1.1 b_2, a_2, β_{2B} too small. Blades too thick. If the efficiency is correct, b_2, a_2, or β_{2B} are too small 1.2 Excessive pre-rotation	1.1 Under-file impeller blades at outlet (sharpen on suction side) according to Sect. 4.5.2 1.2 Check approach flow
2. $H < H_{required}$ and $P \geq P_{required}$	2.1 Hydraulic losses too high 2.2 Throttling effects 2.3 Excessive clearances (Q_{sp}, Q_E). During operation clearances may increase due to wear, abrasion or corrosion 2.4 Flow rate measuring error	2.1 Calculate η_h for test and required geometry according to Eq. (T3.5.8). If $(H/\eta_h)_{test} = (H/\eta_h)_{required}$, the hydraulic losses are too high. Reduce roughness. Possibly increase diffuser/volute throat area. 2.2 Check A_{1q}, a_3, b_3, A_{3q}, A_5, A_6 2.3 Check the clearances (reduce if possible)
3. $H = H_{required}$ and $P > P_{required}$ $\eta_{opt} < \eta_{opt,\ required}$	3.1 Incorrect power measurement 3.2 P_m, P_{RR}, P_{s3}, P_{ER} too large 3.3 b_2, a_2, β_{2B} too large and η_h too low	3.1 Check measurement method and instruments 3.2 Check the secondary losses 3.3 Check the impeller geometry and surface finish; correct as necessary
4. $H > H_{required}$ and $P > P_{required}$	b_2, a_2, β_{2B} too large	If the required efficiency is achieved, trim impellers (Sect. 4.5.1). If η is too low, see items 2 and 3 of Table 4.3
5. $Q_{opt} < Q_{opt,required}$	$n_q < 35$: 5.1 Diffuser/volute throat area too small 5.2 Hydraulic losses too high	5.1 Diffuser: increase a_3. Volute: increase A_{3q} 5.2 See items 2 and 3.
6. $Q_{opt} > Q_{opt,required}$	$n_q < 35$: Diffuser/volute throat area too large $n_q < 60$: Impeller vane inlet angles and/or throat too large	Decrease throat area (insert, weld-up, re-design)
7. Head only about 50% of required	Wrong sense of impeller rotation Double-entry impeller incorrectly mounted on shaft	• Check motor sense of rotation • Wrong sense of impeller rotation (pattern error) • Turn by 180°
8. Q-H-curve unstable		see Chap. 5
9. H_o or P_o too high		see Chap. 5
10. Pump runs at full speed, produces head but does not deliver any fluid	Pump is unable to fill siphon in discharge pipe (Chap. 11.8, Fig. 11.28)	Add structures to prevent rotation of fluid due to impeller inlet recirculation

and the return vane outlet has to be checked. Do these vanes generate unintentional pre- or counter rotation? The dimensions a_5, b_5, a_6 and b_6 (Table 2) can be measured easily. If these are too small due to casting tolerances, a head loss is generated; furthermore, the approach flow to the subsequent impeller is modified. On the volute casing only the throat area and the width can be measured easily. As discussed above, the throat area has a crucial influence on the losses and on the position of the best efficiency point. Unfortunately, the shape of the meridional section of a closed impeller can hardly be measured; the same applies to the blade shape and to complicated diffuser channels. The shapes of these 3D-channels can have a strong impact on the stability of the Q-H-curve. This issue is difficult and costly to resolve.

4. Determine the clearances of the annular seals on impellers and balance device. The lower the specific speed, the higher is their impact on performance.
5. Measure and evaluate the surface quality and roughness of the hydraulic channels. With low and moderate specific speeds, the collector roughness is particularly important, Chap. 3.10.
6. The loss analysis according to Tables 3.5 to 3.8 (and Fig. 3.33) is recommended for determining which effects are important and which parameters are less significant. For instance, the surface quality of the collector rather than that of the impeller should be improved when dealing with pumps of low specific speeds, Chap. 3.10.
7. If major geometrical deviations are discovered, their effects and the measures considered can be evaluated quantitatively according to Chap. 4.7.
8. The dimensionless characteristics ψ_{opt}, ψ_o, λ_{opt}, λ_o, σ often tell more than absolute values and allow an assessment of whether the results are plausible.
9. The cause for a performance deficit is frequently not found in a single striking defect but in a combination of different (sometimes minor) deviations which all act in the same sense. For example: If it is found that b_2 and a_2 are both at the lower tolerance limit, the blades are too thick, the annular seal clearances are at the upper tolerance limit and the roughness of the flow channels excessive, a large performance deficit may result due to the combined effect of these deviations.

During years of operation in the plant, the annular seal clearances at the impellers (and, if applicable, at the axial thrust balance device) gradually increase. Such wear can be caused by occasional contact, erosion corrosion due to high flow velocities in the seal and/or abrasion due to solid particles present in the fluid. The increase in leakage becomes particularly evident in pumps with low specific speeds. This is demonstrated by the example provided in Fig. 4.29 of a 3-stage pump ($n_q = 14$) with balance piston. Head and power consumption are shifted by the increase in leakage rate to the left; the efficiency drops and the best efficiency point shifts towards a slightly higher flow rate. The leakage flow also generates a certain pre-rotation upstream of the impeller since it leaves the seal with a circumferential velocity which is approximately half of u_1.

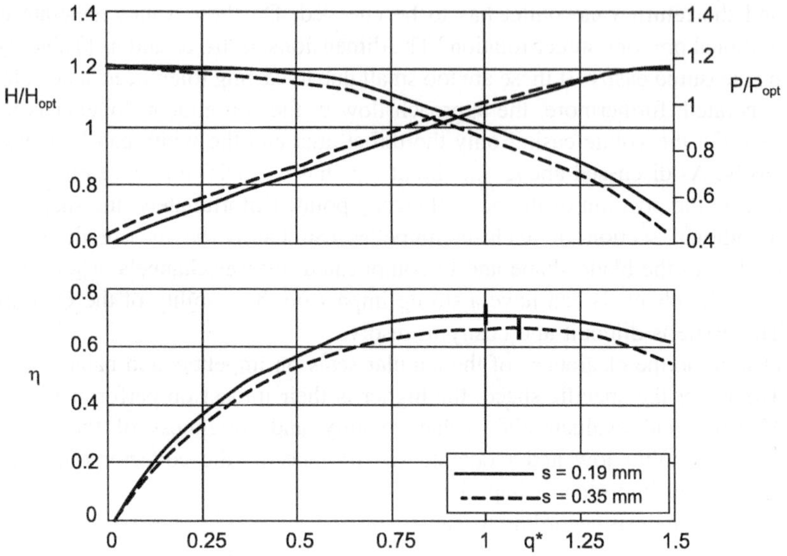

Fig. 4.29 Influence of annular seal clearance on pump characteristics (n_q = 14)

4.7 Calculation of Modifications of the Pump Characteristics

Sometimes it happens that a given (measured) characteristic has to be changed to adapt a pump to new conditions or that the influence of intentional or unintentional geometrical changes needs to be investigated. Application examples include:

- The analysis of performance deficits and of the effects of possible remedies according to Chap. 4.6 and Table 4.3.
- Performance upgrade of a power plant.
- Retrofit of existing plants for different operation conditions, including the effects of high viscosity when the product changes.
- Achieving a steeper or flatter characteristic by installing a different impeller in a given casing. To this end, the blade number, the impeller outlet width or outlet angle and/or pre-rotation can be modified.
- Calculation of the influence of blade thickness or blade distance a_2 at the impeller outlet.
- The shift of the best efficiency point brought about by changing the volute or diffuser throat area.
- The influence of the annular seal clearances or the hydraulic efficiency on the characteristic curves.
- Calculate the impact of pre-swirl or counter-rotation on performance.

The detailed calculation procedure is shown in Table 4.4. To begin with, the secondary losses for the *given* characteristic (in Table 4.4 called "Baseline") are

Table 4.4 Modification of pump characteristics

Given: Characteristic of a pump, "**Baseline**":
$H = f(Q)$; $\eta = f(Q)$; $P = f(Q)$ and geometrical data

Required: Characteristic of a modified pump; in the following termed "**modified**"

1	Leakage losses at BEP	$(Q_{s1}, Q_{s2}, Q_{s3}, Q_E)_{opt}$ according to pump type and axial thrust balance device	Table 3.7(1)
2	Disk friction losses Piston or disk friction Interstage seal Mechanical losses	ΣP_{RR} sum of all stages ΣP_{er} sum of all pistons or disks $\Sigma P_{s3} = (z_{st} - 1)\, P_{s3,st}$ P_m according to bearings and shaft seals	Table 3.6 Table 3.5

As function of flow rate, to be calculated for the **Baseline**; step 3 to 13

3	Leakage losses as a function of flow rate	$Q_{sx} = Q_{sx,opt}\sqrt{\dfrac{H}{H_{opt}}}$	4.4.1
4	Volumetric flow rate through the impeller	$Q_{La} = Q + Q_{s1} + Q_{s2} + Q_E$	4.4.2
5	Volume flow through the diffuser	$Q_{Le} = Q + Q_{s3} + Q_E$	4.4.3
6	Hydraulic efficiency	$\eta_h = \dfrac{\rho\, g\, H_{tot} Q_{La}}{P - \sum P_{RR} - \sum P_{s3} - P_m - P_{er}}$	4.4.4
7	Pre-rotation due to α_1 or α_6 and induced by leakage flow $c_{u,sp}/u_{sp}$ from Eq. (T3.11a)	$\dfrac{c_{1u}}{u_2} = \dfrac{c_{u,sp}}{u_{sp}} d^*_{sp} \dfrac{(Q_{s1} + Q_{s2})}{Q_{La}} + \dfrac{(Q+Q_E)^2}{f_q\, A_1\, u_2\, Q_{La} \tan\alpha_6}$	4.4.5
8	Slip factor $H = H_{tot}/z_{st}$	$\gamma = \dfrac{g\, H}{\eta_h\, u_2^2} + \dfrac{Q_{La}\, \tau_2}{f_q\, A_2\, u_2 \tan\beta_{2B}} + \dfrac{d^*_{1m}\, c_{1u}}{u_2}$	4.4.6
9	Volume flow through the impeller referred to BEP	$q^*_{La} = \dfrac{Q_{La}}{Q_{La,opt}}$	4.4.7
10	Slip factor and hydraulic efficiency as function of q*.	$\dfrac{\eta_h}{\eta_{h,opt}} = f(q_{La}{}^*)$	4.4.8
11	These data are necessary for the calculation of the modified characteristics.	$\dfrac{\gamma}{\gamma_{opt}} = f(q_{La}{}^*)$	4.4.9
12	Deceleration from impeller outlet to diffuser throat	$\dfrac{c_{3q}}{c_2} = \dfrac{Q + Q_E + Q_{s3}}{z_{Le} A_{3q} \sqrt{\left(\dfrac{Q_{La}}{f_q A_2}\right)^2 + u_2^2 \left(\gamma - \dfrac{Q_{La}\tau_2}{f_q\, A_2 u_2 \tan\beta_{2B}}\right)^2}}$	4.4.10
13	For volutes of arbitrary cross sections the factor J_{sp} can be calculated from the baseline test	$J_{sp} = \dfrac{2\pi\, b_2\, f_q}{\dfrac{f_q\, A_2\, u_2\, \gamma}{Q_{Le,opt}} - \dfrac{\tau_2 Q_{La,opt}}{\tan\beta_{2B} Q_{Le,opt}}}$	4.4.11

Best efficiency point „ **modified**"; step 14 to 18			
14	Definition of **modified** quantities, *marked by* (′)	e.g. A'_{3q}, b'_2, d'_2, z'_{La}, β'_{2B}, e', α'_1, α'_6, η'_h, Q'_E, Q'_{s1} or impeller blade trailing edge profile (a'_2)	
15	Slip factor at BEP γ_{opt} from Eq. (4.4.6)	$\gamma'_{opt} = \gamma_{opt}\dfrac{\left(1-\dfrac{\sqrt{\sin\beta'_{2B}}}{z'^{0.7}_{La}}\right)}{\left(1-\dfrac{\sqrt{\sin\beta_{2B}}}{z^{0.7}_{La}}\right)}$	4.4.12
16	a_3 from J_{sp}	$a_3 = \dfrac{d_z + 2e_3}{2}\left\{\exp\left(\dfrac{J_{sp}}{z_{Le}\, b_3}\right)-1\right\}$	4.4.13
17	Theoretical BEP for **modified** *impeller* flow rate	$Q'_{La,opt} = \dfrac{f_q\, A'_2\, u'_2\, \gamma'}{\dfrac{\tau'_2}{\tan\beta'_{2B}} + \dfrac{2\pi\, b'_2\, f_q}{J'_{sp}}\dfrac{Q'_{Le}}{Q'_{La}}}$	4.4.14
18	Hydraulic efficiency	$\eta'_{h,opt} = \eta_{h,opt}$ from Eq. (4.4.4) or estimate from Table 3.8, or 3.9	4.4.15
As function of flow rate, to be calculated for the **modified hydraulic**; step 19 to 28			
19	Volume flow through the impeller	$Q'_{La} = q^*_{La}\, Q'_{La,opt}$	4.4.16
20	Leakage losses Q'_E, Q'_{s1} Q'_{s3}, as functions of flow rate	$Q'_{sx} = Q'_{sx,opt}\sqrt{\dfrac{H}{H_{opt}}}$	4.4.17
21	Hydraulic efficiency $\eta'_h = f(Q'_{La})$	$\eta'_h = \eta'_{h,opt}\dfrac{\eta_h}{\eta_{h,opt}}$; $\dfrac{\eta_h}{\eta_{h,opt}}$ from Eq. (4.4.8)	4.4.18
22	Slip factor $\gamma' = f(Q'_{La})$	$\gamma' = \gamma'_{opt}\dfrac{\gamma}{\gamma_{opt}}$; $\dfrac{\gamma}{\gamma_{opt}}$ from Eq. (4.4.9)	4.4.19
23	Useful flow rate	$Q' = Q'_{La} - Q'_{s1} - Q'_{s2} - Q'_E$	4.4.20
24	Pre-rotation due to α'_1 or α'_6 and induced by leakage flow	$\dfrac{c'_{1u}}{u'_2} = \dfrac{c'_{u,sp}}{u'_{sp}} d'^{*}_{sp}\dfrac{(Q'_{s1}+Q'_{s2})}{Q'_{La}} + \dfrac{(Q'+Q'_E)^2}{f_q\, A'_1\, u'_2\, Q'_{La}\tan\alpha'_6}$	4.4.21
25	Deceleration from impeller outlet to diffuser throat	$\dfrac{c'_{3q}}{c'_2} = \dfrac{Q'+Q'_E+Q'_{s3}}{z'_{Le}A'_{3q}\sqrt{\left(\dfrac{Q'_{La}}{f_q A'_2}\right)^2 + u'^2_2\left(\gamma' - \dfrac{Q'_{La}\tau'_2}{f_q A'_2 u'_2 \tan\beta'_{2B}}\right)^2}}$	4.4.22
26	Head	$H'_{tot} = z_{st}\dfrac{\eta'_h\, u'^2_2}{g}\left\{\gamma' - \dfrac{Q'_{La}\tau'_2}{f_q\, A'_2\, u'_2 \tan\beta'_{2B}} - \dfrac{d'^{*}_{1m} c'_{1u}}{u'_2}\right\}$	4.4.23
27	Power consumption	$P' = \dfrac{\rho\, g\, H'_{tot}\, Q'_{La}}{\eta_h} + \sum_{st} P'_{RR} + \sum P'_{S3} + P'_m + P'_{er}$	4.4.24
28	Efficiency	$\eta' = \dfrac{\rho\, g\, H'_{tot}\, Q'}{P'}$	4.4.25

NPSH calculation			
Given: Characteristic of a pump, „Basis“ (baseline), $NPSH_3 = f(Q)$ and geometrical data Required: NPSH-curve of a modified pump „modified“			
29	Determine cavitation coefficient as function of incidence for **baseline**	$\sigma = \frac{2g\,NPSH}{u_1^2}$	4.4.26
		$i_1' = \beta_{1B} - \beta_1' = \beta_{1B} - \arctan\frac{c_{1m}\,\tau_1}{u_1 - c_{1u}}$	4.4.27
30	It is assumed that the cavitation coefficient $\sigma = f(i_{1a}')$ is the same for the modified and the baseline. The calculation is done for the outer streamline		
31	Calculate flow rates at given values of incidence as determined in step 29	$Q' = \frac{f_q\,A_1\,u_1}{\frac{1}{\tan\alpha_1} + \frac{\tau_1}{\tan(\beta_{1B} - i_1')}}$	4.4.28
32	The NPSH values linked to the flow rates determined in step 31 follow from:	$NPSH' = \sigma(i_1')\frac{u_1^2}{2g}$	4.4.29

calculated according to the information given in Tables 3.5 to 3.7. This loss analysis supplies the hydraulic efficiency from Eq. (T4.4.4) and the slip factor from Eq. (T4.4.6). Because of its circumferential velocity c_{usp}, the leakage flow mixing with the main flow induces a pre-rotation. This effect can be taken into account by Eq. (T4.4.5). This equation is obtained from the balance of the angular momentums of the main and the leakage flows.

If quantities that have an effect on the slip factor are changed for the modified pump a corrected value γ' is calculated according to Eq. (T4.4.12) from the baseline slip factor γ (which has been derived from the loss analysis above). This is to avoid the inaccuracies involved when calculating the slip factor directly from a correlation.

In order to obtain the correct volute characteristic matching the measured BEP with volutes of arbitrary cross sections, the factor J_{sp} (Table 4.1) can be calculated from (Eq. T4.4.11). Modifications to the volute or diffuser throat area are best taken into account by multiplying this value of J_{sp} by an appropriate correction factor.

If the hydraulic efficiency of the modified hydraulics is expected to differ from the baseline, the new value should be entered; otherwise η_h from Eq. (T4.4.4) is also used for the new characteristic. The effects of the roughness on the hydraulic efficiency can be calculated from Table 3.10 and Chap. 3.10.3. The effects of geometric modifications made to the hydraulic components can be estimated from Table 3.8.

According to the procedure in Table 4.4 it is possible to combine any number of modifications, for instance, in order to evaluate the effects of various casting tolerances.

Figure 4.30 shows a calculation of an increase in the flow capacity of a pump with a double-entry impeller. The impeller outlet width is increased by 10% and the volute throat area is enlarged by 15%. In this way, the flow rate at the BEP is increased by 10%.

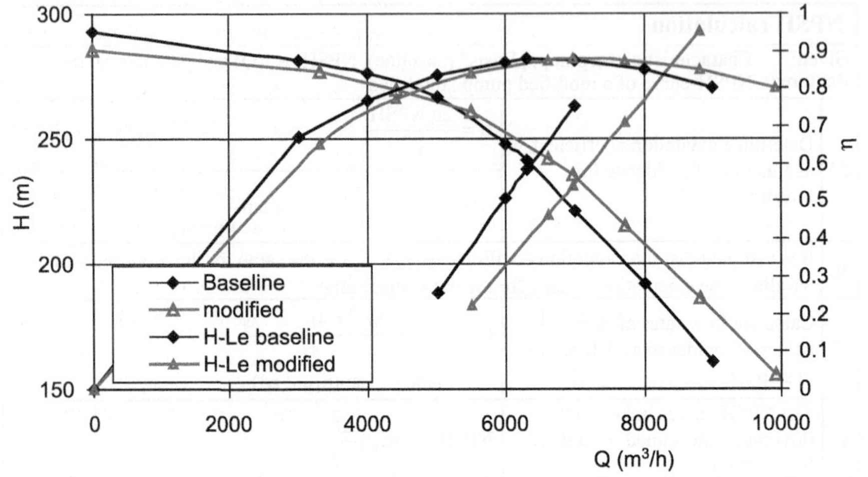

Fig. 4.30 Calculation of the effects of an increase of the impeller outlet width b_2 (by 10%) and a volute cross section enlargement A_{3q} (by 15%); n = 1490; rpm double-entry impeller $n_q = 23$. Performance curves prior to modification are termed "baseline"

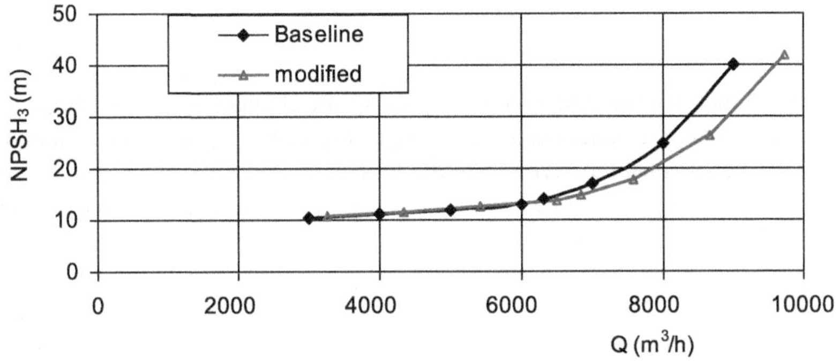

Fig. 4.31 Increasing the flow capacity by increasing the impeller inlet diameter by 2.3%. Same pump as Fig. 4.30. The NPSH-curve prior to modification is termed "baseline"

The effect of design modifications on the $NPSH_3$ can be evaluated according to Eqs. (T4.4.26) to (T4.4.29). In doing so, it is assumed that the correlation between cavitation coefficient and incidence $\sigma = f(i_{1a}')$ is the same for the modified and the baseline impeller. When increasing the blade inlet angle, this assumption may be somewhat optimistic. Figure 4.31 shows the effect of increasing the impeller inlet diameter by 2.3%; pump data are the same as in Fig. 4.30. The effects which can be handled by Table 4.4 (4) include: modifications of the impeller inlet diameter, the hub diameter, the blade inlet angle and the angle of pre-rotation.

References

1. Gülich, J.F.: Bemerkungen zur Kennlinienstabilität von Kreiselpumpen. Pumpentagung Karlsruhe, **B3** (1988)
2. Hunziker, E.: Einfluß der Diffusorgeometrie auf die Instabilitätsgrenze eines Radialverdichters. Dissertation, ETH Zürich (1993)
3. Toyokura, T.: Studies on the characteristics of axial-flow pumps. Bull JSME **4**(14), 287–293 (1961)
4. Ubaldi, M., Zunino, P.: Experimental investigation of the stalled flow in a centrifugal pump-turbine with vaned diffuser. In: ASME Paper 90-GT-216 (1990)
5. Wesche, W.: Experimentelle Untersuchungen am Leitrad einer radialen Kreiselpumpe. Dissertation, TU Braunschweig (1989)

Chapter 5
Partload Operation, Impact of 3-D Flow Phenomena Performance

A pump working significantly below the best efficiency flow rate is said to operate at partload. At low specific speeds this can be roughly assumed at $q^* < 0.8$, at high n_q below $q^* < 0.9$. Since blade inlet angles and channel cross sections are too large for the reduced flow rate, flow patterns during partload operation fundamentally change compared with the design point. The flow becomes highly 3-dimensional since it separates in the impeller and the collector. Finally, recirculations are observed at impeller inlet and outlet at sufficiently low flow. An easy means to obtain information on the impeller flow are stroboscopic observations of tufts. Flow patterns in a radial impeller of $n_q = 22$ gained in this way are shown in Fig. 5.1 [B. 20]. It can be seen that the flow is attached at $q^* > 0.8$ while increasingly large zones with separation and recirculation are observed at a lower flow rates. Similar flow patterns were found on impellers of $n_q = 26$ and 33 and in a CFD study in [64].

The interaction between impeller and casing at inlet and outlet is highly significant for the partload behavior of a pump. It largely determines the Q-H-curve, radial and axial forces, hydraulic excitation forces, noise and cavitation. Such complex flows are difficult to understand. Designing pumps with good partload behavior, therefore, is largely based on experience. When doing so, it is helpful to classify the experimental observations according to the physical mechanisms involved. This is the objective of the following discussion.

Although local stall in impeller and collector also occur at operation significantly above the best efficiency point ($q^* > 1$), no flow recirculation is observed. Steady flow phenomena are discussed in Chap. 5; for unsteady flow see Chap. 10.

5.1 Basic Considerations

Let us first analyze which physical characteristics of the partload behavior can be deduced from basic principles, observation and experience.

J. F. Gülich, *Centrifugal Pumps*,
https://doi.org/10.1007/978-3-030-14788-4_5

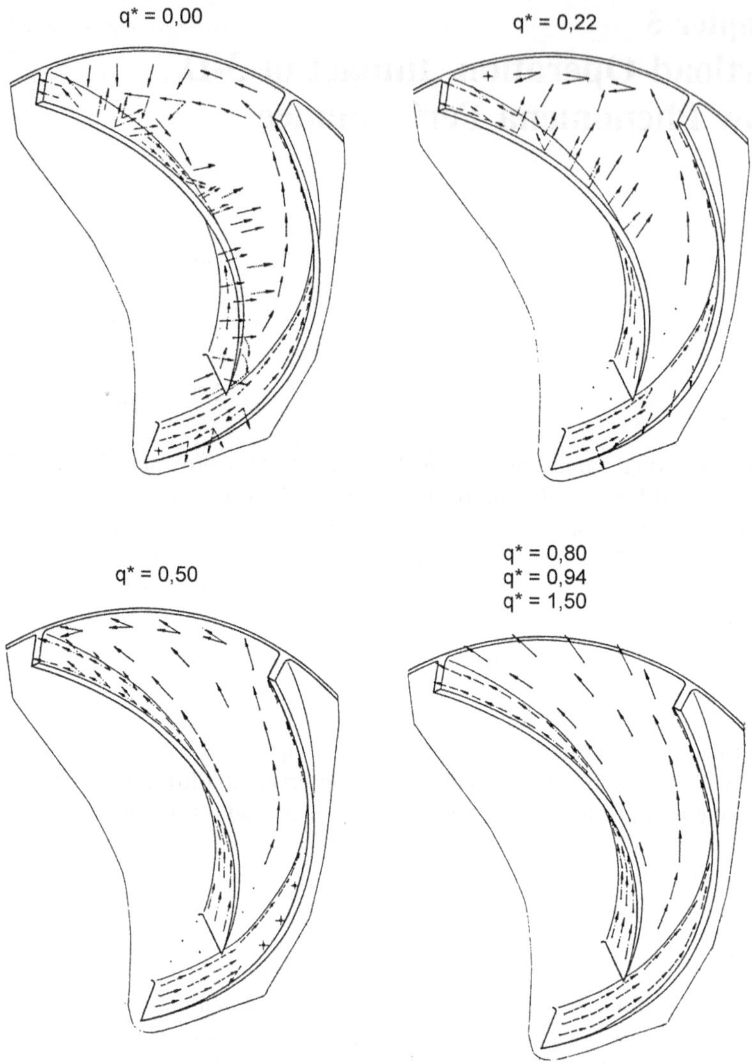

Fig. 5.1 Stroboscopic flow observations in an impeller (tufts) [B.20]

Well-designed pumps achieve hydraulic efficiencies of 85 to 95% in the best efficiency point - depending on the specific speed, size and surface quality. It can therefore be assumed that the flow near the best efficiency point is “attached”. This implies that no significant flow separations occur since these would be associated with greater losses. Measurements and observations of tufts in impellers and collectors confirm this statement.

Conversely, full recirculation is bound to occur both at the impeller inlet and outlet when operating against a closed valve. This is shown by the following

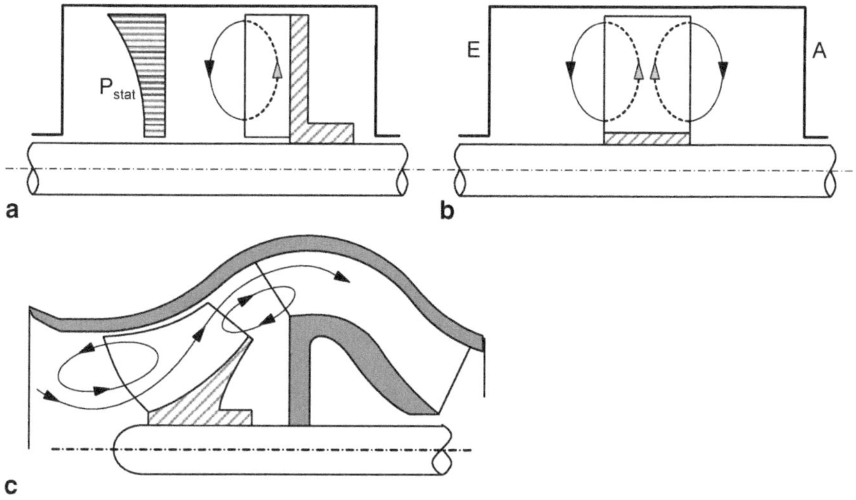

Fig. 5.2 Generation of recirculation. **a** Radial blades, open on one side; **b** radial blades, open on both sides; **c** semi-axial pump

consideration. Imagine a fluid-filled vessel in which a disk with radial ribs ("blades") is arranged on a shaft as shown in Fig. 5.2a. Obviously the fluid will be entrained in the circumferential direction when the shaft is rotated. Since centrifugal forces act on the rotating fluid, a pressure gradient perpendicular to the rotor axis is created. A parabolic pressure distribution is thus generated in the vessel according to Eq. (1.27). The highest pressure develops at the outer radius of the rotor. This causes the fluid in the casing to move radially inwards and, in the bladed rotor, radially outwards. In this manner, a "recirculation" is established. For reasons of continuity, the flow rate entering the impeller at the hub must be equal to the outflow in the outer region.

Considering a bladed rotor according to Fig. 5.2b without a rear shroud, the flow conditions described above develop in the same way on sides A and E. Making abstraction of the casing walls at A and E, Fig. 5.2b represents an axial pump operating against a closed valve. *The recirculating fluid leaves the impeller at the outer streamline* and flows back into the suction pipe. It is also observed that a recirculation develops at the impeller outlet which *re-enters the impeller at the hub.*

These considerations obviously apply in all cases where noticeable differences in radii exist between the outer and inner streamline; in other words, with all impellers not having axis-parallel blade leading or trailing edges. This is true for semi-axial and axial impellers at the inlet and outlet. The same situation is found at the inlet of radial impellers if (as is common practice) the blades are advanced into the impeller eye. Radial impellers with axis-parallel trailing edges do not have this type of recirculation driven by centrifugal forces.

The pattern of recirculation at $Q = 0$ determined by the centrifugal forces is not changed in essence either by the type of impeller blading or the design of the casing. As is shown by experience and a large number of measurements, this basic recirculation pattern is found at the impeller inlet of multistage radial pumps with return vanes, in axial pumps with suction bells as well as at the impeller outlets of semi-axial and axial impellers, irrespective of whether a volute, an annular collector or a diffuser is arranged downstream.

Figure 5.2c shows the flow patterns resulting from Fig. 5.2b in a semi-axial pump (described by Pfleiderer in 1938, [37]). As deduced above and confirmed by experience, these conditions develop at $Q = 0$ as a matter of course. Since this recirculation pattern cannot arise all of a sudden, already at higher flow rates conditions must develop at which some fluid exits from the impeller near the outer streamline and re-enters near the hub, Fig. 5.2c.

The recirculation described above develops due to centrifugal forces. Qualitatively it is independent of the type of pump and its design parameters. If the flow is attached during operation near the best efficiency point, although we observe fully developed recirculation shut-off, we are bound to conclude that the first separations occur in the impeller and in the collector somewhere between best efficiency point and operation at $Q = 0$. It is fundamental to realize that this behavior is inevitably found in all pumps - regardless of the Q-H-curve being stable or unstable, whether or not excessive hydraulic excitation forces and pressure pulsations occur or cavitation problems due to partload are observed. Consequently, *the occurrence of separation and recirculation is a necessary but not sufficient condition for that kind of partload problems.* During the design process, it can be attempted to shift separation and onset of recirculation towards the smallest possible q^*, but separation and recirculation cannot be totally avoided. The aim should rather be to avoid *damaging* recirculations. Unfortunately, this limit is not clearly defined since the (possibly detrimental) effects of a recirculation with a given intensity depend on the design of the pump. For example, the vibration amplitudes measured on a pump with given hydraulic excitation forces are influenced by the rotor damping in annular seals and bearings. Or, the occurrence or absence of cavitation erosion is not only determined by the hydraulic cavitation intensity but also by the impeller material and the fluid characteristics (Chap. 6.6).

The finding that the flow is attached at the best efficiency point and that fully developed recirculation occurs at $Q = 0$, allows yet another conclusion. The physical mechanisms of energy transfer from the impeller to the fluid are likely to be fundamentally different in attached flows and fully separated flows. According to Chap. 4.1.6, the ratio of shut-off head to rated head increases strongly with increasing specific speed, reaching values around 3 to 5 in axial pumps. If such high pressures are generated during recirculation, these mechanisms must be highly effective (irrespective of the fact that the absolute value of the head coefficient falls with increasing n_q).

5.2 The Flow Through the Impeller

5.2.1 *Overview*

The 3-dimensional flow distribution in the impeller determines not only the energy transfer from the blades to the fluid (including the slip) and the hydraulic losses but also the pressure recovery and possible separations in the collector. This is because the velocity distribution at the impeller outlet has a considerable impact on the flow development in the diffusing elements. To ignore this fact is one of the main weaknesses of the one-dimensional theory covered in Chaps. 3 and 4. Numerous measurements in impellers have shown that there is an infinite variety of velocity profiles which depend on many geometrical and hydraulic parameters. To describe the flow patterns by simple means is impossible. The models developed in the following are intended to help understand the complex relationships and to draw conclusions for the evaluation of numerical flow calculations and the design of impellers using such insight.

The impeller flow follows curved paths in the absolute reference frame. As discussed in Chap. 1.4.1, this is possible only if a pressure gradient is acting in the direction of the momentary center of streamline curvature. This pressure gradient generates the centripetal force required to keep the fluid element on its curved path. The pressure gradient caused by the streamline curvature constitutes the sole mechanism by means of which an impeller is able to increase the static pressure. This fact explains why the fluid is able to flow *from the lower suction pressure to the higher outlet pressure* and why the static pressure in the impeller increases although the flow in the absolute reference frame is accelerated.

Consequently, the pressure rise in the impeller is exclusively generated by the blades imparting an absolute velocity with tangential components c_u to the fluid. The pressure field develops in a way that the equilibrium with the body forces is maintained. If the impeller flow is treated from the view of an observer in the rotating system where centrifugal and Coriolis acceleration have to be considered, these body forces are generated through: (1) the centrifugal acceleration due to rotation $b_z = \omega^2 \times r$; (2) the Coriolis acceleration $b_c = 2\omega \times w$; (3) the centrifugal acceleration due to the streamline curvature $b_z = w^2/R_{sl}$ (R_{sl} being the momentary radius of the streamline). Neglecting shear stresses due to wall friction and exchange of momentum between streamlines, it is possible to write the equilibrium of forces in steady flow conditions as follows:

$$\frac{1}{\rho}\frac{\partial p}{\partial n} = 2\omega \times w + r\omega^2 \frac{\partial r}{\partial n} - \frac{w^2}{R_{sl}} \qquad (5.1)$$

Because of the boundary layers, the flow turning in the meridional section and flow deflection by the blades, the velocity distribution in an impeller is always non-uniform. The flow kinematics determined by the impeller geometry creates a pressure field which is imposed on these non-uniform velocity distributions. To maintain the equilibrium of forces according to Eq. (5.1) at any point, the streamline curvature adapts and compensatory flows perpendicular to the main flow develop. These "secondary flows" and their effects on the velocity profiles in the impeller will be discussed in the following sections. They have considerable influence on the losses and the pressure recovery in the collector and the stability of the head-capacity characteristic ("Q-H-curve").

Generally, secondary flows are induced when the fluid is subjected to forces acting perpendicular to the main flow direction. These forces generate corresponding pressure gradients which are determined by the resultant of the centrifugal and Coriolis accelerations according to Eq. (5.1). Consequently, the ratio of these accelerations determines the direction into which the flow will be deflected. The "Rossby number" defines this ratio:

$$\mathrm{Ro} = \frac{b_z}{b_c} \tag{5.2}$$

Since the Rossby number so defined relates two opposing forces acting perpendicular to the main flow it can be aptly described as "secondary flow parameter". As is shown in the following, at least three Rossby numbers must be used to describe the secondary flows between the blades, in axial impeller flow and in the curved meridional section. Different definitions of the Rossby number are used in the literature, which should be considered for comparisons.

The flow distribution in the impeller essentially depends on the interaction of the following effects:

- Blade forces (circulation or lift), caused by the flow around the blades.[1]
- Centrifugal forces. Experience shows that the circumferential velocity imparted by an impeller to the fluid increases with the ratio of outlet to inlet radius. The explanatory term "centrifugal effect" is used for this phenomenon even though centrifugal forces do not exert any direct pumping effect.
- Coriolis forces
- Velocity distribution at the impeller inlet
- Boundary layers
- Annular seal leakages
- Interaction between impeller and stator during recirculation

Experience shows that shape and *stability* of the Q-H-curve are hardly affected by the rotor speed (in the range of practical interest) [15, 31, 51, 53]. Experiments on a three-stage pump in [19] yielded very similar φ-ψ-curves in a speed range of 1:4 and in a range of Reynolds numbers of almost 1:20 (tests with cold and hot water). This finding allows the conclusion that the *global* partload behavior is not sensitive to the Reynolds number or to boundary layer effects, even though these should be

[1]The term "blade forces", introduced to ease understanding, is not quite exact: the force acting on the blade is the integral of the pressure distribution over the blade; this pressure distribution solely depends on the velocity distribution in the impeller.

responsible for the first local separation. Consequently, *the partload characteristic is essentially determined by the equilibrium of Reynolds-independent body forces according to* Eq. (5.1). Fully developed recirculation - but not the onset of recirculation - is therefore modeled quite well by 3D-Euler (i.e. non-viscous) calculations.

5.2.2 Physical Mechanisms

5.2.2.1 The Effect of the Impeller Rotation

Consider a radial impeller with back-swept blades, Fig. 5.3. The velocity component perpendicular to the drawing plane is considered zero. A fluid element at a given point on a curved path with radius R_{s1} at the relative velocity w is exposed to the following body forces (in addition to the pressure gradients):

- the component of centrifugal acceleration $b_{z1} = w^2 \times R \times \sin\beta$ acting in the direction of w,
- the component of the centrifugal acceleration $b_{z2} = w^2 \times R \times \cos\beta$ normal to w,
- the acceleration due to streamline curvature $b_{z3} = w^2/R_{s1}$ perpendicular to w,
- the Coriolis acceleration $b_c = 2w \times \omega$ normal to w, but opposite b_{z2} and b_{z3}.

The accelerations perpendicular to the momentary flow direction are primarily responsible for the secondary flow and, consequently, for the velocity distribution in the impeller channel. Obviously, the ratio $(b_{z2} + b_{z3})/b_c$ determines whether the

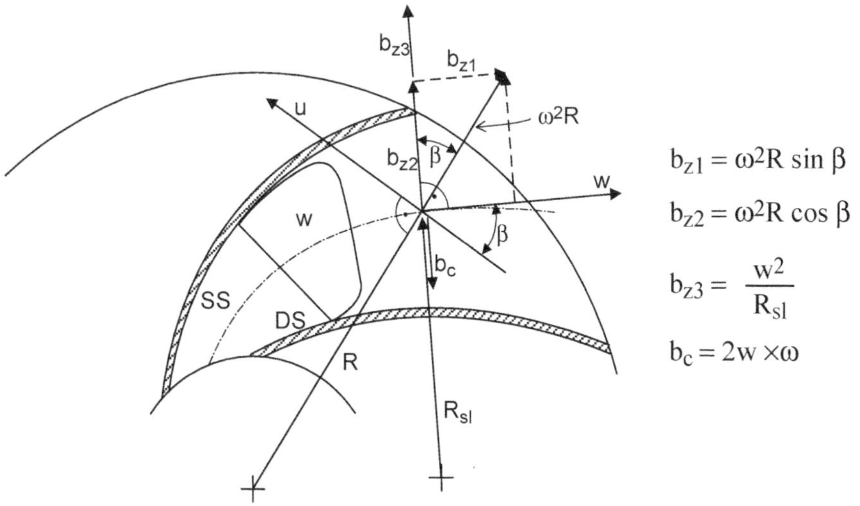

Fig. 5.3 Accelerations acting on a fluid element in a radial impeller

fluid element is deflected towards the suction or the pressure surface of the blade. This ratio is the Rossby number (with subscript B for "Blade") defined in Eq. (5.2):

$$\mathrm{Ro_B} = \frac{\mathrm{u}\,\cos\beta}{2\,\mathrm{w}} + \frac{\mathrm{w}\,\mathrm{R}}{2\,\mathrm{u}\,\mathrm{R_{sl}}} \tag{5.3}$$

If the Rossby number is near 1.0, no noticeable secondary flow would be expected; if it is smaller than 1.0, the Coriolis force predominates and the flow is deflected in the direction of the blade pressure surface; if it exceeds 1.0, the secondary flow transports fluid to the suction surface of the blades. Since the flow distribution is always non-uniform due to boundary layers, meridional curvature and blade forces, secondary flows also occur at $\mathrm{Ro_B} = 1$ in reality.

Setting $w = w_u/\cos\beta$, $w_u = u - c_u$ and $\psi_{th} = 2c_u/u$, the following expressions for the Rossby number are obtained:

$$\mathrm{Ro_B} = \frac{\cos^2\beta}{2\left(1 - \frac{c_u}{u}\right)} + \left(1 - \frac{c_u}{u}\right)\frac{\mathrm{R}}{2\cos\beta\,\mathrm{R_{sl}}} \tag{5.4}$$

$$\mathrm{Ro_B} = \frac{\cos^2\beta}{2 - \psi_{th}} + \left(1 - \frac{\psi_{th}}{2}\right)\frac{\mathrm{R}}{2\cos\beta\,\mathrm{R_{sl}}} \tag{5.5}$$

Since the local relative velocity in the impeller channel is not constant, the local accelerations acting on the individual fluid elements vary as well. In a zone filled with stalled fluid, the relative velocity w is low, $\mathrm{Ro_B}$ tends toward infinity, and the centrifugal acceleration prevails. It transports stalled fluid towards the blade suction surface as shown by the direction of the tufts in the stalled fluid zones on the blade pressure surface in Fig. 5.1.

These relationships show qualitatively the effect of the impeller rotation on the secondary flow and the pitch-wise velocity distribution. The influence can be seen most clearly in a rotating pipe or channel. In this case cosβ tends towards zero and $\mathrm{R_{sl}}$ to infinity. Then $\mathrm{Ro_B}$ is close to zero and a secondary flow towards the pressure surface develops. Measurements on a rotating channel of square cross section [8] in Fig. 5.4 and on a rotating diffuser [35] confirm this statement.

The influence of the Coriolis acceleration disappears in the boundary layers, because the relative velocity close to the wall tends toward zero and $\mathrm{Ro_B}$ tends toward infinity. The fluid transported by the secondary flow to the pressure surface can flow back to the suction surface in the zones near the wall (satisfying continuity). This is evident from Figs. 5.4 and 5.10. Although the conditions with impellers are significantly more complicated because of the additional influences to be discussed below, it is nevertheless possible to evaluate approximately the influence of the rotation:

- With essentially radial blades ($\beta_{2B} \approx 90°$), the flow conditions are similar to those in a rotating channel. The Rossby number tends toward zero and a strong secondary flow in the direction of the pressure surface must be expected. This

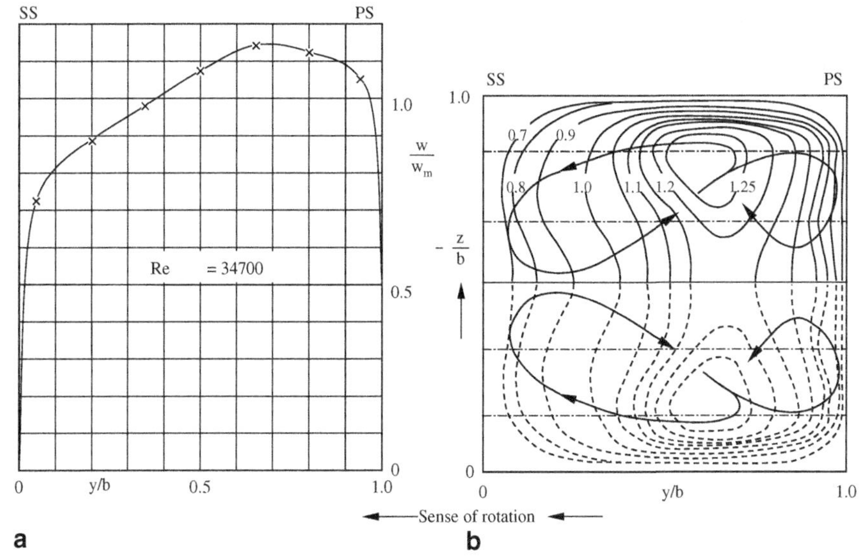

Fig. 5.4 Rotating channel of quadratic section, *SS* Suction side, *PS* pressure side, [8] **a** velocity distribution in the direction of the channel axis. **b** Constant-velocity curves and secondary flow

implies a pitch-wise velocity profile having a maximum ("jet") near the pressure surface and, consequently, a deficit in the suction surface ("wake"), Figs. 5.4 and 5.11. The jet-wake flow pattern has been observed frequently in radial compressor impellers.

- In the case of back-swept blades, the Rossby number according to Eq. (5.5) is close to 1.0. The flow is in a "delicate equilibrium". The velocity distribution depends on many influences and is difficult to predict. Small changes in the geometry or in the flow rate can cause the flow pattern to switch between different states.
- The Rossby number increases slightly with falling flow rate as per Eq. (5.3) since the relative velocity decreases at partload and the first term gains in significance. The flow tends to shift towards the pressure surfaces of the blades.

Within the impeller channels the Rossby number Ro_B introduced above is presumably below 1.0 in most cases (with the exception of zones with stalled fluid) so that the flow is deflected in the direction of the blade pressure surface. Equations (5.2–5.4) show that the secondary flow between the blades depends on several parameters which have an opposing effect on the two terms on the right-hand side of the equations. The flow equilibrium in radial impellers with strongly back-swept blades can consequently develop in various ways which are difficult to predict. Therefore, the analysis of a large number of measured velocity profiles at the outlet of such impellers did not yield any general tendencies.

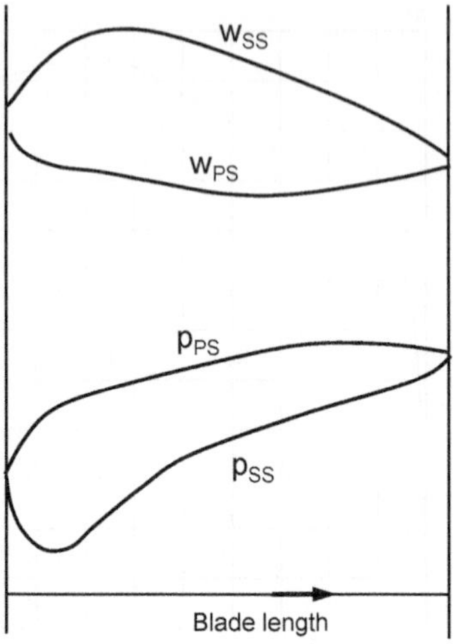

Fig. 5.5 Pressure and velocity distribution on an impeller blade (schematically)

5.2.2.2 The Effect of Blade Forces

An impeller transfers energy to the fluid when a lower pressure prevails on the suction surface than on the pressure surface. This is the case when the flow around the blades generates higher relative velocities on the suction surface than on the pressure surface, as sketched in Fig. 3.3. Consequently, the distributions of the static pressure and the relative velocity over the blade length develop as shown schematically in Fig. 5.5. The theoretical blade work is obtained from the integral over the pressure distribution; the proportionalities $Y_{th} \sim (p_{PS} - p_{SS})$ and $Y_{th} \sim (w_{SS}{}^2 - w_{PS}{}^2)$ apply. As the flow rate is reduced, Y_{th} increases; this would also be true for $(w_{SS} - w_{PS})$ if the secondary flow brought about by the Coriolis force did not partly or completely equalize this velocity difference.

At the blade trailing edge, the pressure differential tends toward zero since a discontinuity in the static pressure distribution cannot exist in the free flow downstream of the impeller. The velocity distribution in the impeller adapts itself towards the outlet so that this discharge condition is fulfilled (Fig. 3.3). This process results in the reduced flow deflection discussed in Chap. 3 which mainly develops in the last part of the blade downstream of a_2 (see figure in Table 0.2).

To understand the impeller flow, it is important to remember that the effects of the blade forces and the Coriolis force are opposed to each other. The flow around the blade tends to produce a velocity maximum near the suction surface (Fig. 3.3a, Profile k), while the Coriolis force according to Fig. 5.4 would rather shift the maximum of the relative velocity to the pressure surface. Measurements frequently show that the velocity distribution at the impeller outlet is more uniform in the range of $q^* = 0.8$ to 0.9 than at the best efficiency point. The mixing losses are correspondingly lower here. This feature is partly responsible for the maximum of the hydraulic efficiency being frequently found in this domain too. It does not coincide with the maximum of the coupling efficiency. This statement is also borne out by Fig. 4.20.

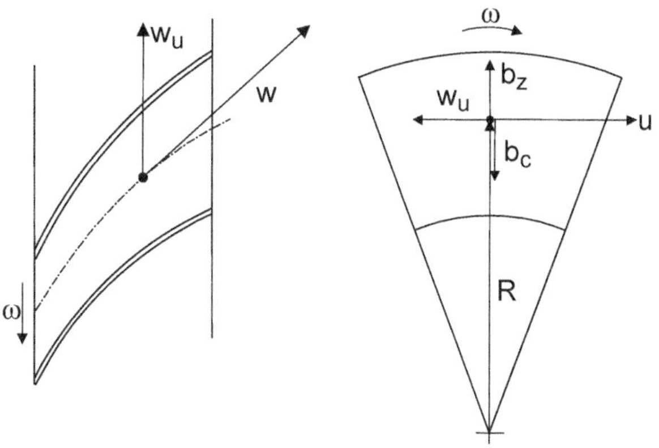

Fig. 5.6 Accelerations acting on a fluid element in axial flow

5.2.2.3 The Effect of Axial Flow

In axial and semi-axial impellers as well as in the axial inlet of a radial impeller, the centrifugal acceleration $b_z = u^2/R$ and the Coriolis acceleration $b_c = 2w_u \times \omega$ act on a fluid element according to Fig. 5.6. This produces a Rossby number which has been given the subscript "ax":

$$Ro_{ax} = \frac{1}{2\frac{w_u}{u}} = \frac{1}{2\left(1 - \frac{c_u}{u}\right)} = \frac{1}{2 - \psi_{th}} \tag{5.6}$$

Equation (5.6) shows that the Rossby number is below 1.0 in non-separated flow so that the fluid tends to be deflected towards the hub. With increasing flow rate, Ro_{ax} drops and the secondary flow in the direction of the hub intensifies.

If Ro_{ax} exceeds 1.0, the secondary flow is directed outwards. In a stalled fluid zone (w_u approaches zero) the Rossby number tends toward infinity and the flow is accelerated radially outwards. Ro_{ax} becomes very large also in the boundary layers on the blades (because of the liquid adhering to the walls $w \rightarrow 0$). The boundary layer flow is therefore accelerated outwards through the centrifugal forces. The secondary flows in an inducer derived from measurements confirm these considerations; see Fig. 5.7 according to [32].

5.2.2.4 The Effect of Meridional Curvature

In a radial impeller the flow is deflected by 90° from the axial entry to the radial outlet. The meridional velocity component follows the behavior of the flow through a bend according to Fig. 1.10. At the entry to the bend, the velocity profile exhibits the highest velocity on the inner streamline due to the conservation of angular momentum ($c \times r = \text{constant}$). As the fluid progresses through the bend, a secondary flow builds up owing to centrifugal forces which shift the maximum of the velocity distribution near the outlet of the bend towards the outer wall. In Fig. 5.8 these mechanisms are applied to the meridional flow of a radial impeller.

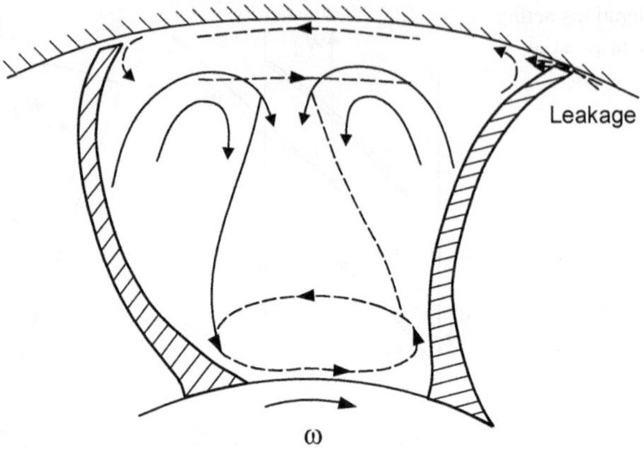

Fig. 5.7 Secondary flow in an axial inducer [32]

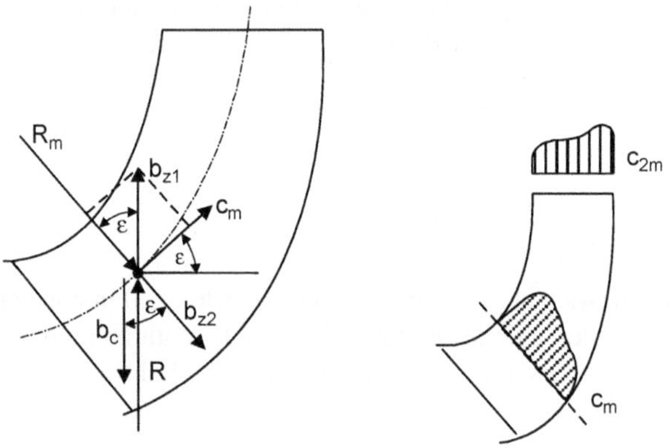

Fig. 5.8 Accelerations and velocities acting in the meridional section

The pressure difference between rear and front shrouds created through the centrifugal acceleration $b_{z2} = c_m^2/R_m$ amounts to $\Delta p = \rho \times c_m^2 \times B/R_m$ according to Eq. (1.29). The centrifugal acceleration generated by the bend and a component of the Coriolis acceleration $b_c = 2w \times \omega$ both contribute to divert the flow towards the hub, Fig. 5.8. This tendency is opposed by the component of the centrifugal acceleration $b_{z1} = u^2/R$ which is caused by the rotation (see Figs. 5.6 and 5.8). The secondary flow can be described through the Rossby number Ro_m:

$$Ro_m = \frac{b_{z1}\cos\varepsilon}{b_{z2} + b_c\cos\,\varepsilon} = \frac{1}{\left(\frac{c_m}{u}\right)^2 \frac{R}{R_{m\cos\varepsilon}} + 2\frac{w_u}{u}} \tag{5.7}$$

The effect of the blade forces and the blade Rossby number are superimposed on this process. The flow becomes fully 3-dimensional and unpredictable because one

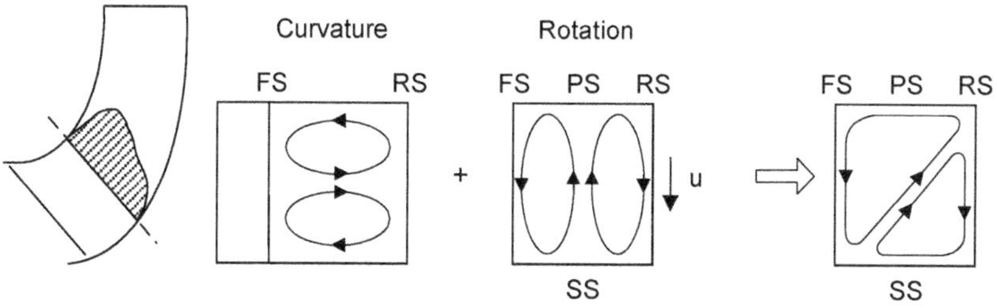

Fig. 5.9 Combination of the secondary flows caused by curvature and rotation; *FS* front shroud, *RS* rear shroud, *SS* suction side, *PS* pressure side

or the other effect is predominating at times. Added to this is the backflow from the collector at partload. Thus the maximum in the velocity distribution can be found either near the rear shroud, when the influence of the meridional curvature is controlling the flow, or near the front shroud, when the centrifugal force predominates in those parts of the blade channels, where the flow has an axial component.

It should well be noted that the meridional curvature exerts a crucial influence on the *span-wise* velocity distribution, while the *pitch-wise* distribution is determined by the blade forces and rotation.

5.2.3 The Combined Effect of Different Mechanisms

The discussion of the various forces acting on the flow has shown that the velocity profile in the impeller (and at the impeller outlet) can have its maximum near the rear or front shroud as well as on the pressure or suction surface of the blade. What the velocity distribution of a given pump really looks like at a given flow can be neither easily predicted nor readily explained by measurement or numerical calculation, unless a distinct mechanism can be found to predominate. Particularly difficult to assess is the flow in radial impellers of specific speeds from $n_q = 50$ to 100 since the blade trailing edges are near the zone where the curvatures of the meridional section are severe. If the secondary flow is asymmetrical a *spiral-shaped* flow pattern develops in the impeller channel. In this case the velocity distribution at the impeller outlet depends on the point at which the impeller trailing edge "cuts through the spiraling flow". This model explains why apparently totally inconsistent velocity distributions are observed at the impeller outlet as functions of the specific speed as well as the flow rate.

Asymmetrical spiral-shaped flow patterns can also be obtained from the superimposition of the curved flow through the meridional section (Fig. 5.8) with the secondary flow (Fig. 5.4) caused by the rotation. This generates a flow pattern according to Fig. 5.9, see also [21].

In general it can be said: *the secondary flows caused by body forces direct fluid in the core flow from low-pressure zones to zones of higher pressure.* In the

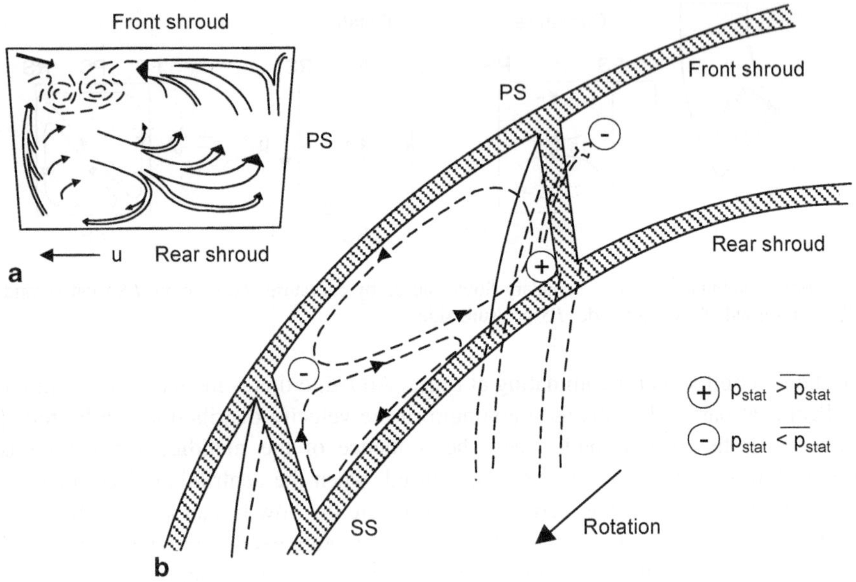

Fig. 5.10 Secondary flow in a radial impeller. **a** from [9]; **b** according to the models developed in this chapter; *SS* suction surface, *PS* pressure surface

boundary layers the fluid then flows back from the higher to the lower pressure for reasons of continuity, Figs. 1.12, 5.9 and 5.10.

Where the effects of the meridional curvature (Chap. 5.2.2.4) and the blade forces (Chap. 5.2.2.2) are superimposed, a zone with minimum energy transfer (called "wake") is generated. It is located in the corner between the front shroud (c_{2m}-minimum according to Fig. 5.8) and the blade suction surface (w-maximum according to Fig. 3.3). Added to this is the effect of the Coriolis acceleration which transports fluid in the direction of the blade pressure surface. For reasons of continuity, fluid is transported back through the boundary layers on the shrouds to the blade suction surface.

Low-energy fluid from the boundary layers thus accumulates in the front shroud suction surface corner. The wake flow is enhanced in this way. Figure 5.10 shows which type of secondary flow is established.

This type of flow pattern has frequently been observed in measurements, especially in highly loaded radial compressors but also in pumps. Figure 5.11 shows the c_m-distribution at the outlet of a compressor impeller with back-swept blades. Another example is given by the tests on semi-axial impellers reported in [13], which clearly show the energy deficit in the front shroud suction surface corner.

The zone of low energy transfer in an impeller is frequently located on the front shroud covering the entire pitch rather than being concentrated in the corner next to the blade suction surface. The wake has an effect on the pressure recovery in the collector, the recirculation at the impeller outlet and the stability of the Q-H-curve.

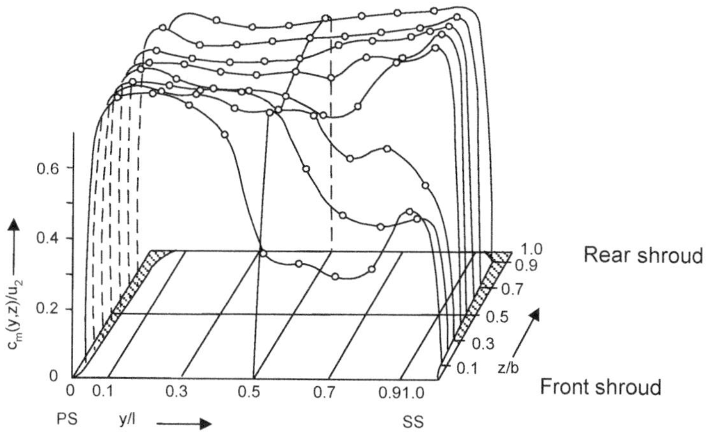

Fig. 5.11 Meridional velocity at the outlet of a radial compressor impeller [10]

The theoretical blade work is obtained from the integral over u × c_u at the impeller outlet, while the useful energy transferred to the fluid follows from the integral of thc total pressure at the impeller outlet (in each case after deduction of the corresponding integral values at the impeller inlet). The quotient of both quantities yields the hydraulic impeller efficiency. Due to the diffuser losses, it is higher than the hydraulic efficiency of the pump. A *specific* value of the integral can be generated by *an infinite variety* of different velocity profiles at the impeller outlet. Consequently, the shape of the impeller outlet velocity profile has no direct impact on the transferred energy at a given value of the integral. This influence becomes really effective in the collector.

5.2.4 Recirculation at the Impeller Inlet

During operation at shut-off, recirculation is bound to develop inevitably at the impeller inlet as discussed in Chap. 5.1.2. The fluid flows back on the outer streamline from the impeller into the suction chamber, from where it re-enters the impeller near the hub. Due to the energy received by the blades, the fluid recirculating from the impeller has a circumferential velocity. The recirculated fluid mixes with the liquid in the suction chamber and induces a pre-swirl to the net flow through the pump by exchange of momentum. The pre-rotation at low flow decreases from the outer to the inner streamline as shown in Fig. 4.13.

The rotation of the fluid generates a parabolic pressure distribution in front of the impeller. If no ribs or structures upstream of the impeller hamper the rotation, the swirl progressively enters the suction pipe when the flow rate is reduced. The induced swirl can be detected in the suction pipe as far as L/D > 10. If no structures are available for preventing fluid rotation, the measurements are falsified by the suction pressure being recorded too high. Consequently, the head measured is too

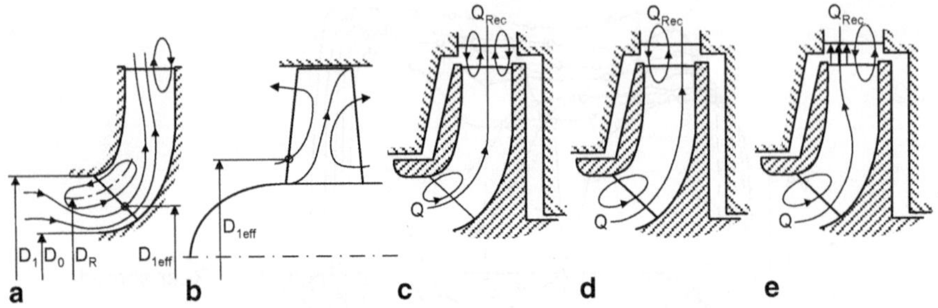

Fig. 5.12 Recirculation flow patterns (components in the meridional section)

low, whereas the NPSH values are too high. When testing axial suction pumps a flow straightener or a cross in the suction pipe is therefore absolutely required. At low load the inlet pressure must be measured upstream of such elements; near best efficiency point and beyond, the inlet pressure is measured downstream, Chap. 5.2.6 and Fig. 5.16.

The features of recirculating flow described above are confirmed without exception by countless tests on radial, semi-axial and axial impellers. The recirculation patterns developing in the meridional sections of radial and axial impellers are schematically shown in Fig. 5.12 (for semi-axial impellers see Fig. 5.2c). Figure 4.13 shows the velocity distributions measured at the inlet of an axial impeller. In that example the recirculation starts on the front shroud at about $q^* = 0.6$. The intensity of recirculation rises with decreasing flow rate; thus the extent of the recirculation zone, the negative c_m and the pre-rotation c_u all increase.

The flow patterns induced by recirculation (Figs. 5.2, 5.12 and 5.14) suggest that strong vortices are generated. These can form "cross-channel vortices" which are perpendicular to the vanes and reach from the suction to the pressure surface of the blades. The vortices are perpendicular to the suction surface of the blade and they rotate with the impeller. The presence of such vortices is demonstrated by Figs. 5.46, 5.47, 5.48, 5.49 and 11.21 for axial pumps and by Figs. 6.43 to 6.45 for vaned diffusers.

Even if the conditions described above are qualitatively the same with all impellers, there are considerable quantitative differences in terms of onset and intensity of recirculation. The onset of inlet recirculation was the topic of many investigations, e.g. [11, 43, 44, 50, 51], but a generally applicable method for prediction has not been found to date (nor is it likely to be developed).

Two prerequisites must be fulfilled for triggering impeller inlet recirculation in the form observed:

1. The flow must separate locally.
2. Strong pressure gradients must develop perpendicular to the direction of the main flow.

One condition alone is not sufficient: neither separation over the entire blade height in a plane cascade nor pressure gradients across the flow (e.g. in a sharp bend) will result in a recirculation of the type observed at the impeller inlet. The local flow

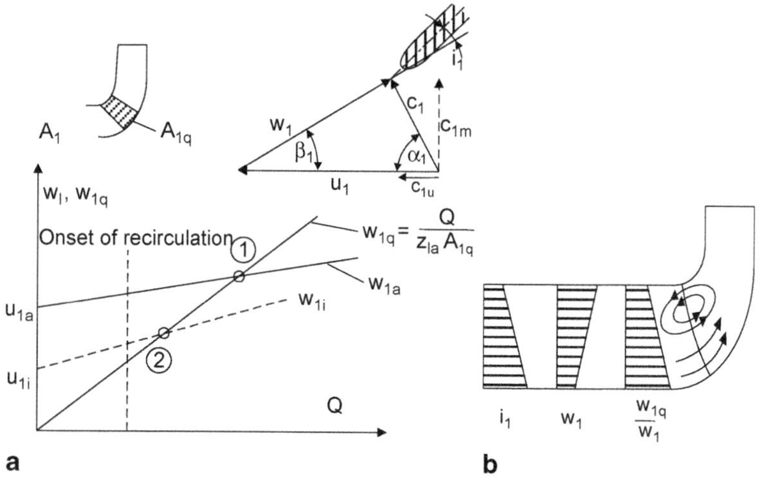

Fig. 5.13 Flow deceleration at the impeller inlet, [B.20]

separation not only depends on boundary layer effects but also on the equilibrium of forces. As discussed above, this equilibrium is determined by many geometrical parameters as well as the flow rate. These effects can be calculated by numerical methods at best. Any attempt to capture them by a few simple parameters in a generally valid way in order to predict the onset of recirculation has little chance of success, given the complex geometries and flow conditions. Simple correlations may be successful for a specific group of pumps of a similar design, [11], but fail to yield generic results [47, B.20].

The incidence, seen as responsible for the onset of recirculation by many authors, has an effect on local flow separation near the leading edge, but it neglects the effects of the pressure gradients across the flow and the deceleration in the impeller throat. A correlation of the onset of recirculation with the incidence is therefore not possible, as was demonstrated in [B.20] utilizing data from [51][2]. During the tests in [51] semi-axial impellers (n_q = 180) with five different bladings were investigated: the recirculation at the inlet started at incidences between *zero* and 11° or deceleration ratios from w_{1q}/w_{1a} = 0.42 to 0.61. Recirculation can thus start *even at zero incidence* if the flow is decelerated strongly downstream of the leading edge. The flow phenomena involved are explained in Fig. 5.13; there are plotted against the flow rate: (1) the relative velocity w_1, obtained from the velocity triangles for the outer and the inner streamlines, (2) the mean velocity $w_{1q} = Q_{La}/(z_{La} \times A_{1q})$ in the throat area at the impeller inlet, obtained from the continuity equation. To the right of point 1 the flow is accelerated from w_1 to w_{1q}; to the left of point 1 it is decelerated on the outer streamline. Since the velocity at Q = 0 cannot

[2]This was also the conclusion of Stoffel and Weiß based on LDV-measurements in three radial impellers, [48].

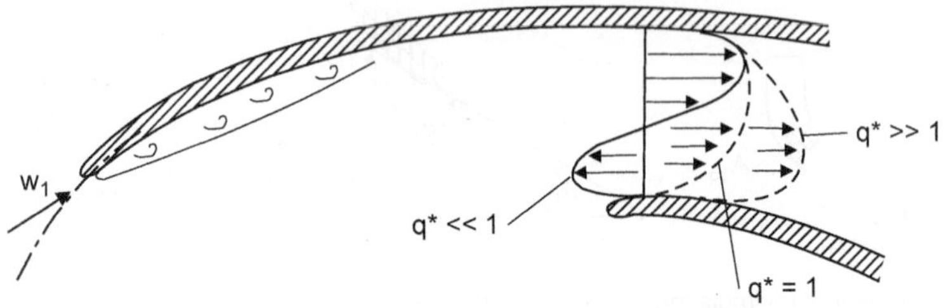

Fig. 5.14 Distribution of relative velocity in the impeller inlet throat area

be decelerated to zero, the flow is bound to separate at some stage. Essentially independent of the incidence, this separation is determined by the throat area (in other words, the deceleration ratio w_{1q}/w_{1a}). Figure 5.13a also shows how pressure differentials can develop perpendicular to the main flow: the deceleration near the inner streamline starts at a lower flow rate, since the relative velocity (because of the smaller radius) is considerably lower than on the outer streamline. Up to the point of stall, higher pressures develop on the outer streamline than on the hub. As soon as a separation zone develops in which the relative velocity drops, the stalled fluid is transported by centrifugal forces towards the outer streamline (see Chap. 5.2.2.3). This process ultimately causes back-flow into the suction plenum when the zone of stalled fluid has sufficiently grown in size. From this follows: The greater the ratio of the outer to the inner streamline radii r_{1a}/r_{1i}, the greater are the pressure differentials perpendicular to the flow direction and the higher are the onset and intensity of the recirculation. This fact also explains why the onset of recirculation $q^*_{RB} = Q_{RB}/Q_{opt}$ generally increases with growing specific speed, and why recirculation starts at relatively low flow rates in impellers with axis-parallel (or slightly oblique) leading edge. Conversely, the recirculation starts at identical flow rates, if a set of impellers of various specific speeds is designed with the same impeller inlet parameters - in particular with identical values for r_{1a}/r_{1i} and w_{1q}/w_{1a} (this situation is depicted in Fig. 3.8). The impeller outlet exerts little influence on the onset of recirculation as confirmed by measurements reported in [48].

The hypothesis that the incidence is crucial for the onset of recirculation, is also contradicted by the following consideration (refer to Fig. 5.13b): Since the blade angles near the inner streamline are usually larger than at the outer streamline, the incidence near the hub increases faster than on the shroud when the flow rate is reduced. If recirculation primarily depended on the incidence, it would occur near the hub rather than at the outer streamline. However, if the deceleration w_{1q}/w_1 over the blade height is considered, the greatest deceleration actually occurs near the outer streamline, since this is where the relative velocity is greatest. As discussed above, this deceleration also generates a pressure gradient across the main flow, which is a prerequisite for the onset of recirculation.

Strong deceleration of the relative velocity from w_1 to w_{1q} generates a non-uniform velocity distribution on the first portion of the blade between leading edge and throat area. This is depicted schematically in Fig. 5.14 for various q*.

Recirculation develops on the pressure surface of the blades from a velocity deficit (which is already evident at $q^* = 1$) because the relative velocity is decelerated more strongly on the blade pressure surface than on the suction surface. This pitch-wise non-uniformity is induced by the work transfer (see Fig. 3.3), except for high flow rates with $q^* \gg 1$ and $w_{PS} > w_{SS}$. As has been explained in Chap. 5.2.2.1, the effect of the Coriolis forces diminishes with respect to centrifugal forces in zones with low relative velocity. Stalled or low-velocity fluid on the pressure surface is thus transported outwards by centrifugal forces. This process triggers a flow recirculation of the type shown in Fig. 5.2.

Since the absolute blade inlet angles in a pump are usually small, the incidence at partload amounts to a few degrees only. Moreover, with falling flow rate a pre-rotation is increasingly induced by the blades acting on the approaching flow. Consequently, flow separation caused by incidence downstream of the leading edge is local and of limited extent as sketched in Fig. 5.14. Local separation as such does not necessarily induce recirculation.

Further investigations may be quoted in this context: (1) In experiments on a radial impeller with different approach flow conditions, the velocity distribution upstream of the impeller was modified by screens in order to change the incidence over the blade height, [50]. No correlation could be found between the onset of recirculation and the incidence. However, if these experiments are evaluated in terms of the deceleration ratio w_{1q}/w_1, the onset of recirculation is shifted to higher flow rates when w_{1q}/w_1 decreases. (2) A radial compressor was examined in [29]; in both impellers investigated recirculation was found to occur at the same flow rate despite different blading. This observation can be interpreted in the sense that the pressure difference across the main flow was instrumental for triggering recirculation since the ratio d_{1a}/d_{1i} was almost identical in both impellers.

Typically, recirculation starts at $Q_{RB}/Q_{opt} = (0.4 - 0.75) \pm 0.1$ (ascending from $n_q = 10$ to 300). According to the evaluations in [B.20], recirculation started when the deceleration ratio w_{1q}/w_{1a} was in the range of 0.4 to 0.65. Below $w_{1q}/w_{1a} = 0.4$, recirculation was observed in all of the pumps analyzed. However, an attempt to correlate the critical deceleration with geometrical dimensions was unsuccessful. The overall deceleration can be expressed through:

$$\frac{w_{1q}}{w_1} \approx \frac{Q}{f_q\, A_{1q}\, z_{La}\, u_1} = \frac{\varphi_1\, A_1}{z_{La}\, A_{1q}} \tag{5.7a}$$

An empirical correlation of the inlet flow coefficient for the onset of recirculation has been determined from the data reported in [B.20]:

$$\varphi_{1,RB} = 0.03 + 0.16(1 + b_1/R_{DS})(d_{1a}/d_{1i})(z_{La}A_{1q}/A_1) \tag{5.7b}$$

This correlation exhibits only a relatively low scatter of about ±10% although it comprises radial and axial impellers. A broader data base would be desirable to check whether this correlation may be applied in general. At any rate, Eq. (5.7b) contains three essential elements: (1) The term $z_{La} \times A_{1q}/A_1$ quantifies the deceleration of the relative velocity according to Eq. (5.7a) - the higher A_{1q}/A_1, the greater is the deceleration and the earlier is the recirculation onset. (2) The ratio d_{1a}/d_{1i} is a measure for the pressure differentials over the blade height - the greater d_{1a}/d_{1i}, the higher are the pressure gradients and the earlier the recirculation onset. (3) The ratio b_1/R_{DS} describes the curvature in the meridional section which is significant for separations in the impeller channel: severe curvature results in premature separation and recirculation. See also Fig. 7.4 for geometric dimensions.

Apart from the large recirculation zone at the leading edge, additional zones of more localized stall and recirculation are observed within the impeller channel at low load, see Fig. 5.1. Pronounced separation is noticeable on the blade pressure surface, roughly in the middle between inlet and outlet. Another recirculation is found on the rear shroud near the outlet. If stalled fluid from within the channels joins the recirculation at the inlet, fluid entering the suction plenum can attain considerable circumferential velocities. Thus $c_{1u} > u_1$ can be measured in the recirculation zone immediately upstream of the blade leading edge. Similar phenomena were reported in [48].

The following parameters determine the pressure gradient perpendicular to the flow direction and thus influence the onset and intensity of recirculation:

- The ratio of the diameters at the outer and inner streamlines d_{1a}/d_{1i} at the blade leading edge is presumably the most significant parameter. This ratio increases with the specific speed. The onset q_{RB}* and intensity of recirculation generally increase with n_q. Large-eye suction impellers with large d_{1a}/d_{1i} exhibit the same tendency. Impellers with axis-parallel leading edges ($d_{1a} = d_{1i}$) generate little recirculation (*therefore* tending to instability, as explained below).
- The smaller the throat area at the impeller inlet, the lower is the tendency towards recirculation since w_{1q}/w_1 increases.
- Incidence or blade inlet angle and the angle distributions along the streamlines exert rather a secondary influence.
- The position of the leading edge in the meridional section and in the ground view (forward or backward sweep).
- Curvature of the front shroud in the meridional section.
- Number of impeller blades.
- Flow rate and velocity distribution upstream of the impeller.
- Boundary layer thickness on the front shroud, which is also influenced by the amount of the annular seal leakage and the geometry introducing the leakage.

In summary, it should be noted:

1. Recirculation at the impeller inlet always occurs at the outer streamline. It starts when the deceleration of the relative velocity at the throat w_{1q}/w_{1a} exceeds a certain threshold, which depends on the shape of the impeller.
2. Even pumps with stable characteristics that do not present the slightest problems during partload operation *inevitably* are subject to partload recirculation.
3. Recirculation does not originate immediately downstream of the impeller blade leading edges but rather near the throat area or near a sharp curvature where the greatest pressure differentials over the blade height occur. *The equilibrium of forces between the streamlines is crucial for recirculation.*

4. Pressure gradients perpendicular to the main flow direction can be explained by centrifugal forces and by different decelerations of the relative velocity on the outer and inner streamline (w_{1q}/w_{1a} smaller than w_{1q}/w_{1i}).
5. Although boundary layer effects may influence the first local separation, onset and intensity of recirculation are virtually *independent of the Reynolds number*. This statement is confirmed by the experiments in [45] in the range of interest for water pumping. Many tests done with varying speeds confirm that instabilities in the Q-H-curves do not depend on the speed - were it otherwise, the similarity laws would not be applicable.
6. A correlation of the onset of recirculation with incidence is not possible.
7. Recirculation is preceded by separation on the pressure surfaces of the blades near the outer streamline.
8. Stall and recirculation depend on the 3-dimensional velocity distribution in the impeller and the approach flow; such effects cannot be fully described or predicted by simple methods.
9. Several zones with separation and recirculation can occur in the impeller channel. This is shown in particular by the observations of tufts in Fig. 5.1 or the measurements in [31, 48, 55].
10. A collector which guarantees largely uniform pressure distribution over the impeller circumference has only little influence on the inlet recirculation. In contrast, a volute casing generates a non-uniform pressure over the impeller circumference at partload. Then, the individual impeller channels operate at different operating points on the Q-H-curve, as imposed by the local pressure in the volute, Chap. 9.3. The inlet recirculation is affected by these asymmetries. Presumably back-flow is strongest at the circumferential position with the highest volute pressure, since the impeller channels there operate at the lowest flow, see Table D9.1.
11. Trimming of the impeller outer diameter changes the inlet recirculation only if the blades are shortened severely.

For measuring the onset and strength of impeller inlet recirculation, see also Chap. 5.2.6.

5.2.5 Flow at the Impeller Outlet

As shown in Fig. 5.12, recirculation also occurs at the impeller outlet. This topic is discussed in context with the diffuser flow in Chap. 5.3, since the recirculation occurs from the stator into the impeller. As a consequence, there is a strong interaction between both components. At low partload, due to excessive deceleration, flow separations also occur within the impeller channels. This is shown by Fig. 5.1 and corroborated by [55]. Extensive stall is in particular observed in impellers with large outlet angles or large relative outlet widths b_2/d_2 (as shown by measurements on a dredge pump in [24]).

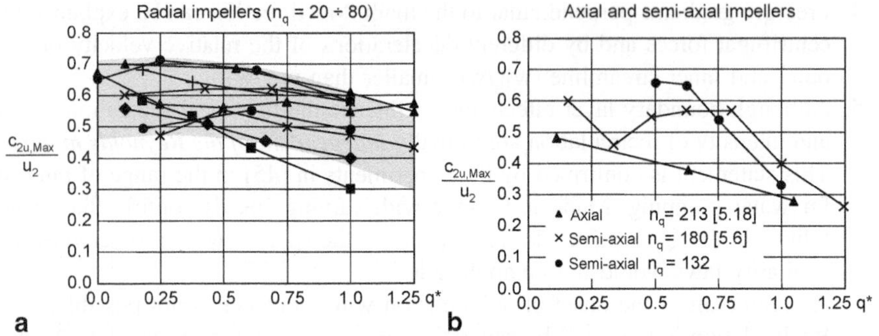

Fig. 5.15 Maximum local values of c_{2u}/u_2 for various impellers

According to Chap. 5.2.2, the velocity distribution at the impeller outlet depends on a sensitive equilibrium of various mechanisms. Therefore, it cannot easily be predicted whether the velocity maxima occur near the rear or front shroud or in the middle of the passage. As shown in Fig. 5.12, this also applies to the position of the recirculation zone: all cases shown there were observed. Nevertheless, the analysis of a large number of tests allows some general statements on the discharge flow distributions of radial, semi-axial and axial impellers:

- At any load point - even at Q = 0 - there is at least one domain of the impeller or "one streamline" where a "quasi-normal" energy transfer from the blades to the fluid can be assumed. This is demonstrated by Fig. 5.15 where the maximum circumferential component of the absolute velocity was plotted as $c_{2u,max}/u_2$ against q* ($c_{2u,max}$ represents the local maximum measured in the outlet profiles). Between q* = 0 and 0.5, the ratio $c_{2u,max}/u_2$ reaches values from 0.5 to 0.7 in all of the experiments. The recirculation and zones of stalled fluid therefore appear to develop according to the principle of the least resistance in a way that a part of the impeller, or a streamline, operates with that deceleration which is just tolerated without flow separation.
- The circumference-averaged c_{2m} and the c_{2u} distributions are strongly non-uniform at partload. Where c_{2m} becomes negative (that is, where recirculation occurs) c_{2u} becomes significantly smaller than with positive c_{2m}. This applies to radial, semi-axial and axial impellers. The finding is in contrast to what would be expected when drawing a velocity triangle for the considered "streamline" because a small meridional component would be connected to a large circumferential component. *The simultaneous occurrence of small c_m and c_u components is an indication of flow separation.*
- With semi-axial and axial impellers as well as radial impellers with an oblique trailing edge ("oblique trimming"), recirculation always occurs on the hub (rear shroud). This because the outer streamline generates more pressure at partload than the inner streamline due to the action of centrifugal forces, Chap. 5.1.3.

5.2.6 Experimental Detection of the Onset of Recirculation

The onset of recirculation at the impeller inlet can be determined experimentally by various methods. The easiest way is to measure the pressure increase generated by the fluid rotation according to Fig. 5.16a. To do so, one pressure tapping is placed as close as possible to the impeller inlet while a second pressure tapping is installed at some distance upstream in the suction pipe. Between both tappings a flow straightener, a cross or ribs prevent fluid rotation at the upstream measuring location. The pressure difference ΔH_s between both pressure tappings is recorded and plotted against the flow rate: At flows without recirculation, $\Delta H_s = f(Q)$ yields a parabola corresponding to the pressure loss between the two measuring points. Where noticeable recirculation is present, the curve $\Delta H_s = f(Q)$ clearly deviates from the parabola and ΔH_s changes the sign (as well demonstrated by Fig. 5.16). The pressure loss caused by the flow straightener is very small at partload and does not significantly influence the results, but its impact can be easily taken into account by calculation.

In a number of tests reviewed, the onset of inlet recirculation (Q_{RB}) was in a close range of $q^*_{RB} = Q_{RB}/Q_{opt} = 0.55$ to 0.65; Table D 5.1 gives some examples. The dimensionless measured head difference $2\ g\Delta H_s/u_1{}^2$ - plotted in Fig. 5.16b - exhibits similar trends for $n_q = 26$ to 90. It is also possible to approximately determine the circumferential component $c_{1u,a}$ on the outer streamline from the measured ΔH_s if it is assumed that the fluid rotates like a solid body: according to Eq. (1.27) the pre-swirl becomes $\omega_{swirl}/\omega = c_{1u,a}/u_1 = (2\ g\ |\ \Delta H_s\ |\ /u_1{}^2)^{0.5}$. The results are depicted in Fig. 5.16c. This calculation makes sense only if there are no swirl-breaking structures (ribs) present between the impeller and the pressure tapping closest to it (h_2 in Fig. 5.16a). All tests in Fig. 5.16 were done with single-stage volute pumps with axial inlet and radial impellers. The tests in [55] were done with air as operating fluid (for using hot wire anemometry).

At the onset of recirculation, the back-flow is limited to a small zone near the blade tip at d_{1a}. As the flow rate is reduced, the swirl progressively extends into the suction pipe (in the absence of ribs or structures which would prevent fluid rotation). The effect can be appreciated from the test data in Fig. 5.16d. The data can be correlated by Eq. (5.7c). If the pressure tapping is placed according to ISO 9906 at a distance of $L = 2d_s$ from the inlet flange, the actual distance from the impeller blade leading edge would be typically at $L = 3d_s$. Then the swirl in the suction pipe would be felt at the pressure tapping at $q^*_{RB} = 0.34$ rather than at 0.62:

$$\frac{Q_{swirl}}{Q_{opt}} = \frac{Q_{RB}}{Q_{opt}}\, e^{-0.2\frac{L_{ps}}{d_s}} \tag{5.7c}$$

In Fig. 5.16d, the onset of inlet recirculation is set as $q^*_{RB} = Q_{RB}/Q_{opt} = 0.62$, but if the onset is expected at other flow rates, these can be inserted in Eq. (5.7c) to estimate the length of the swirl in the suction pipe.

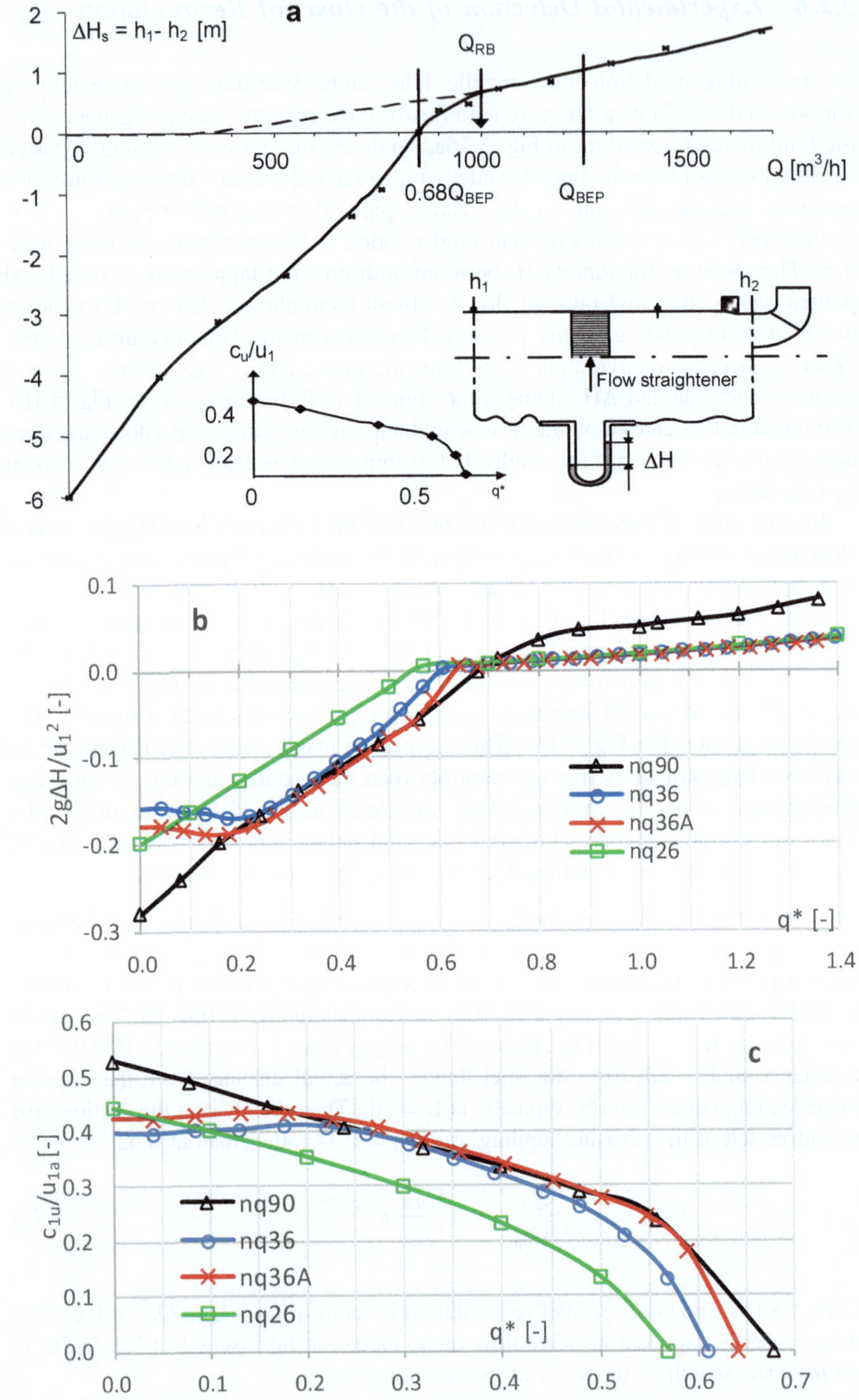
a
ΔHs = h1- h2 [m]
QRB
0.68QBEP
QBEP
Q [m3/h]
cu/u1
q*
h1
h2
Flow straightener
ΔH
b
2gΔH/u1² [-]
nq90
nq36
nq36A
nq26
q* [-]
c
c1u/u1a [-]
nq90
nq36
nq36A
nq26
q* [-]

◂**Fig. 5.16** Determination of the onset of recirculation by means of pressure measurements in the suction pipe. **a** Test set-up and absolute values, $n_q = 90$; **b** dimensionless pressure difference, various n_q; **c** tangential velocities induced by recirculation, various n_q; **d** Length L_{ps} of inlet recirculation in suction pipe of pump with axial inlet; no ribs in front of impeller, [55]

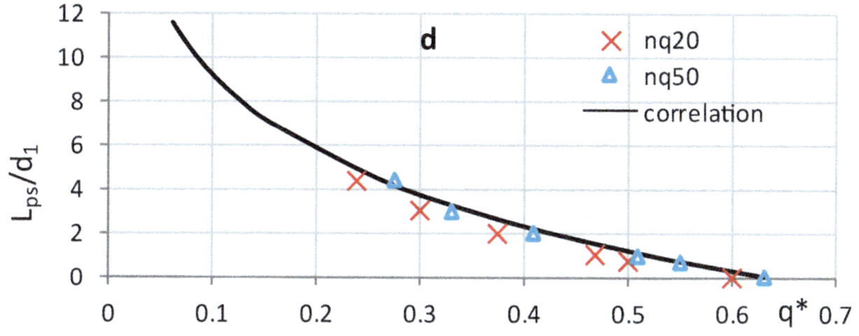

Fig. 5.16 (continued)

Table D5.1 Test data for impeller inlet recirculation

n_q	q^*_{RB}	$(c_{1u}/u_1)_{Q=0}$	Fig.	Remarks
26	0.57	0.45	5.16	estimated [34]
28	0.6		10.35	double-entry
29	0.62 to 0.67		[63]	4 impellers, data in Table D7.1-1
90	0.8	0.53	5.16	q^*_{RB} very high due to extreme shape of radial impeller with $n_q = 90$
150	0.55		5.29	
213	0.6	0.5	4.13	

Other methods for determining the onset of inlet recirculation include:

- Observation of tufts or similar indicators of the flow direction.
- Measurement of the velocity distribution using various velocity probes, hot wire instruments (when testing with air), or laser velocimetry
- Measurement of pressure fluctuations in the suction pipe. At the onset of recirculation the pressure fluctuations increase suddenly. At a lower flow rate, i.e. in fully developed recirculation, they fall again - albeit not to the same level as near the best efficiency point.

The experimental determination of the onset of the impeller outlet recirculation is not as easy as at the impeller inlet, because access is more difficult. The following methods are available: measurement of pressure fluctuations (the stochastic components increase at the onset of recirculation, [11, 31, 55]); velocity measurements by means of the methods mentioned above; observations of tufts by endoscope; determining the fluid rotation in the impeller sidewall gaps through measurement of pressure differences (described in Chap. 5.4.3 and Fig. 5.30).

5.3 The Flow in the Collector

5.3.1 Flow Separation in the Diffuser

The measurements depicted in Fig. 5.15 show that the absolute velocity at the impeller outlet at low partload reaches 50 to 70% of u_2. A flow with this velocity cannot be decelerated in the diffuser to zero at Q = 0 without a separation occurring. Observations of tufts in the diffuser of a pump with $n_q = 22$ in Fig. 5.17 confirm this, [B.20]. While the flow is attached at q* = 1.5, initial separations appear at q* = 0.94 near the diffuser outlet where the boundary layer is thickest. At q* = 0.8, the stall spreads over the greatest part of the vane suction surface. At q* = 0.6, additional separations occur on the vane pressure surface, and recirculation in the direction of the impeller is visible on the rear shroud just in front of the diffuser vane leading edge. The recirculation zone grows with falling flow. Unsteady cavitation is observed at q* = 1.5 on the pressure surface of the leading edge. The cavity appears each time an impeller blade passes a diffuser vane. Shape and location of the cavity suggest that it is caused by a negative incidence $i_3 = \alpha_{3B} - \alpha_3$ (flow angle exceeding vane angle), see also Chaps. 6.8.2.3 and 10.7.1 (rotor-stator interaction). Depending on the approach flow and the choice of the diffuser opening angle, the phenomena described occur with different pumps at different q*; but qualitatively they are typical for the flow through radial diffusers. In particular, recirculation in all experiments occurred upstream of the diffuser vane leading edges or just before the volute cutwater (never in the middle between two vanes or near the vane suction surfaces). This behavior was also observed in radial compressors [30].

Flow separation at partload does not occur simultaneously in all diffuser channels. When reducing the flow rate from BEP, stall is first observed in one channel. Upon further reduction in flow, stall occurs in 2, 3 or more channels. This phenomenon is termed "alternate stall", [42]. The stall cells may be unsteady. In this case the stall propagates from one diffuser channel to the next. This is known as "rotating stall" discussed in Chap. 10.7.2. Several reasons are responsible for the stall not occurring simultaneously in all diffuser channels: (1) flow at the impeller outlet non-uniform over circumference - for example due to non-uniform inlet flow; (2) geometric tolerances; (3) non-uniform pressure distribution downstream of the diffuser due to a volute or the discharge nozzle. In this case each diffuser channel can be considered as working against a slightly different system curve (see Chap. 10.7.1, Fig. 10.25). The

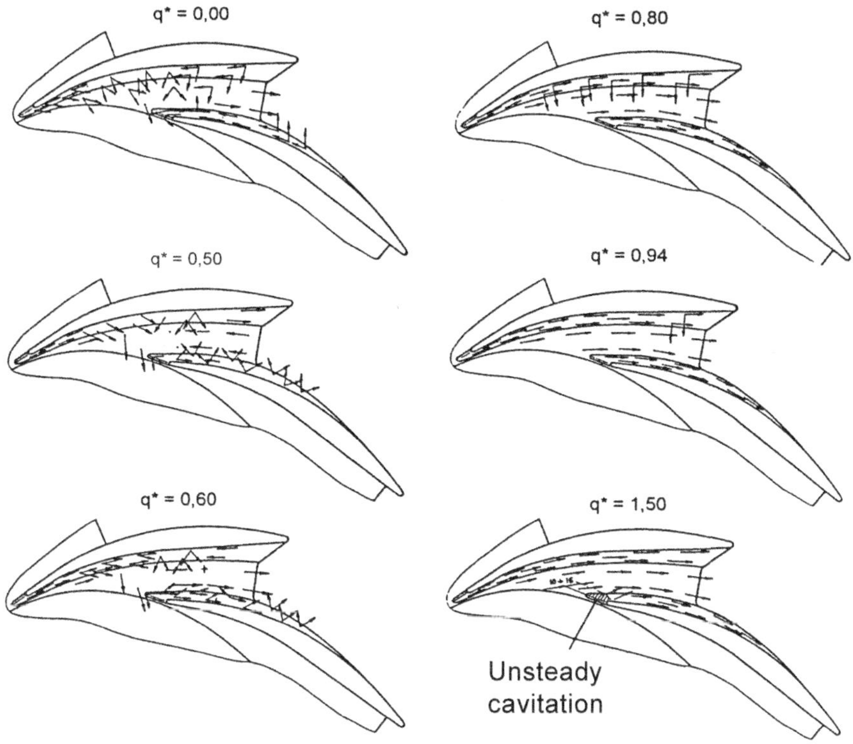

Fig. 5.17 Flow observation in a diffuser (nq = 22), same test as in Fig. 4.5 [B.20]

last stage of a multistage pump can thus behave differently from the series stages where the flow is more or less rotationally symmetric.

The flow conditions at the diffuser inlet can be assessed in a similar way as described for the impeller inlet in Fig. 5.13: while the absolute velocity c_2 at the impeller outlet falls with increasing flow rate, the velocity c_{3q} in the throat area of the diffuser calculated according to the continuity equation increases from the theoretical value zero at $Q = 0$ proportionally to the flow, Fig. 5.18. Both curves intersect at Q_1 (where $c_2 = c_{3q}$). At $Q > Q_1$ the flow in the triangular section at the diffuser inlet is accelerated, i.e. the static pressure in the throat area drops in accordance with Bernoulli's equation: at high flow rates cavitation is even possible in the diffuser, Chap. 6.2.5. If the pressure is measured in the throat area of the diffuser (subscript 3), H_3-H_2 becomes zero at $Q = Q_1$. Somewhere below $Q < Q_1$, the deceleration attains a level where the flow separates. At this point the curve (H_3-H_2) clearly flattens out, or, as in Fig. 5.18, has a local maximum.

Tests on two diffusers with different throat areas show the phenomena explained above, Fig. 5.19. It is clearly visible that the curve $\psi_{3\text{-}2}$ becomes zero at $c_2 = c_{3q}$ and that the separation point (maximum of $\psi_{3\text{-}2}$) shifts to the left when the diffuser throat area is reduced. The efficiency at partload increases accordingly. The flattening of

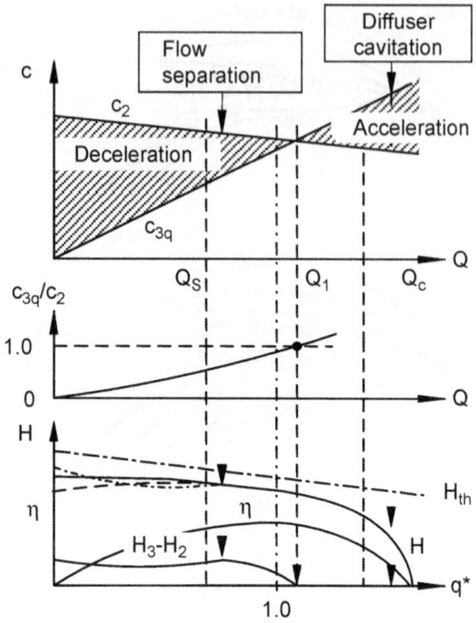

Fig. 5.18 Flow deceleration and acceleration in the collector throat area [22]

the Q-H-curve is evident where the separation in the diffuser occurs. The static pressure increase in the impeller ψ_p is hardly affected by the diffuser modification.

As demonstrated by the experiments in Fig. 4.6a, volutes behave in this respect very similarly to diffuser pumps so that the considerations made using Fig. 5.18 apply equally to volute pumps. In particular, a critical deceleration from c_2 to c_{3q} is also reached in volutes below a specific flow rate at partload where the flow separates (c_{3q} is then calculated with the local volute cross section or the volute throat area). Cavitation in the volute occurs where acceleration becomes excessive.

Also the incidence has an effect - at least with regard to local flow separation. If the first part of the diffuser vane is designed so that the incidence in the critical flow rate is near zero, separation still occurs as soon as the permissible deceleration is exceeded. This largely depends on the ratio c_{3q}/c_2 and the velocity distribution at the diffuser inlet. Measurements on diffuser pumps in the range of $15 < n_q < 35$ showed that the curve $\psi_{3\text{-}2} = f(Q)$ became flat or unstable when c_{3q}/c_2 reached values between 0.3 and 0.6, but a correlation with geometric data or the specific speed was not possible - most likely, because of the impact of the impeller outlet flow distribution on the flow in the diffuser.

5.3.2 Pressure Recovery in the Diffuser

The pressure recovery in diffusers largely decides on whether the Q-H-curve becomes stable or unstable. This is true for both low-specific speeds as well as

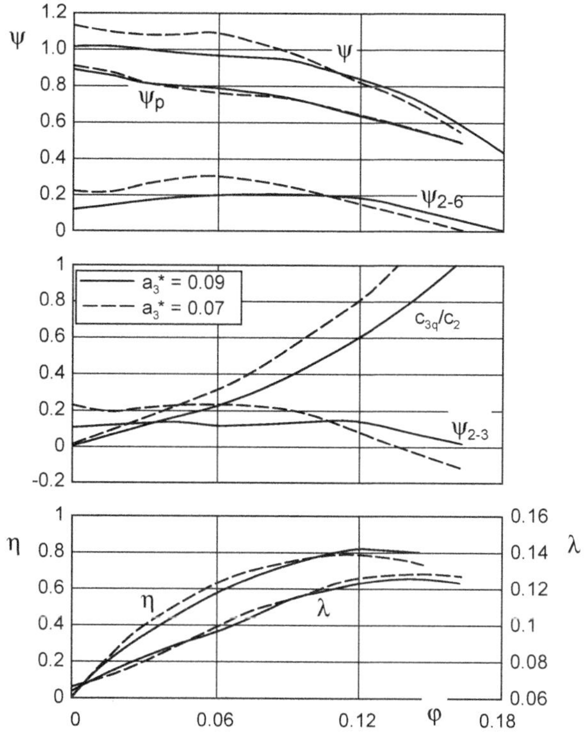

Fig. 5.19 Influence of the diffuser throat area on head capacity curve [22]

semi-axial pumps. Two zones must be distinguished for the pressure rise in the diffuser: (1) the triangular section from H_2 to H_3 in which the pressure rise (H_3-H_2) is measured; (2) the actual diffusing channel, where the pressure increases from H_3 to H_4, see Fig. 4.4 or sketches in Table 0.2. Up to the throat area, the flow in a volute behaves in a similar way as in the triangular section upstream of the diffuser throat, while the flow in the discharge nozzle is comparable to the flow in the diffusing channel from H_3 to H_4 of a vaned diffuser, see also tests in [56].

The measurements of (H_3-H_2) shown in Figs. 5.19, 4.5 and 4.6 reveal an important pressure recovery in the triangular section at the diffuser inlet (or in the volute). It *increases* with *falling* flow rate up to a maximum value at which the flow separates. The deceleration ratio c_{3q}/c_2 is crucial for this pressure rise. In contrast, the pressure recovery in the diffusing channel from H_3 to H_4 rises with increasing flow rate in a manner which is similar to a conical diffuser in a pipe. This is true as well for a diffusing discharge nozzle of a volute casing. Near the best efficiency point and above the pressure recovery (H_4 − H_3) essentially increases with the square of the flow. The pressure recovery coefficients c_p given in Figs. 1.14 or 1.15 can be applied for an approximate assessment. In contrast, at partload the c_p-values fail to predict the pressure rise since a highly non-uniform velocity distribution prevails in the throat area. The actual kinetic energy at the throat is considerably higher than the value that would be obtained by calculating c_{3q} from the continuity equation. This is why the maximum throat velocity (rather than the average) is significant for the pressure increase ($H_4 - H_3$) in the diffusing channel.

Below the flow rate at which the fluid stalls in the collector, especially at Q = 0, the pressure increase in the diffuser presumably follows other mechanisms than during normal diffuser flow. The Carnot shock could constitute such a mechanism because it describes the pressure increase following a sudden deceleration (expansion of the cross section or fully separated flow). Applied to the diffuser inlet, the head increase would then become:

$$(H_3 - H_2)_{stat} = \frac{c_{3q}(c_2 - c_{3q})}{g} \tag{5.8}$$

However, an evaluation of Eq. (5.8) with mean velocities shows that no pressure increase would occur at Q = 0 in the diffuser through this mechanism because of $c_{3q} = 0$. Even at higher flow rates, the pressure recoveries calculated in the partload range according to Eq. (5.8) turn out considerably lower than those actually measured. To correctly perform such calculations, the effect of recirculation and stalled fluid on cross sectional blockage would have to be considered.

Another possible mechanism for the pressure increase with fully separated flow in the diffuser - especially in the triangular section - is the exchange of momentum (compare to fluid couplings or regenerative pumps). This would mean that the head increases (like the power consumption) with the intensity of recirculation in fully separated flow.

5.3.3 *Influence of Approach Flow on Pressure Recovery and Stall*

If the flow near the wall is already strongly decelerated and thick boundary layers prevail at the diffuser inlet, the pressure recovery deteriorates and the risk of stall increases (Chap. 1.6). This is equally true when the velocity profile at the diffuser entry is asymmetrical, for instance because of an upstream bend. These considerations apply to vaned diffusers as well: if thick boundary layers or non-uniform velocity distributions are present at the inlet, the pressure recovery is impaired and the tendency to separation increases. This fact has a major effect on the hydraulic losses and the stability of the Q-H-curve.

The strong impact of non-uniform approach flows on diffuser pressure recovery is demonstrated by the tests in [1, 57] from which the relationship:

$$c_p/c_{p,o} = 1 - 2.25(U_{fl} - 1) \quad \text{with} \quad U_{fl} = \int \left(\frac{c_{2m}}{c_{2m,av}}\right)^2 \frac{dA}{A} \tag{5.8a}$$

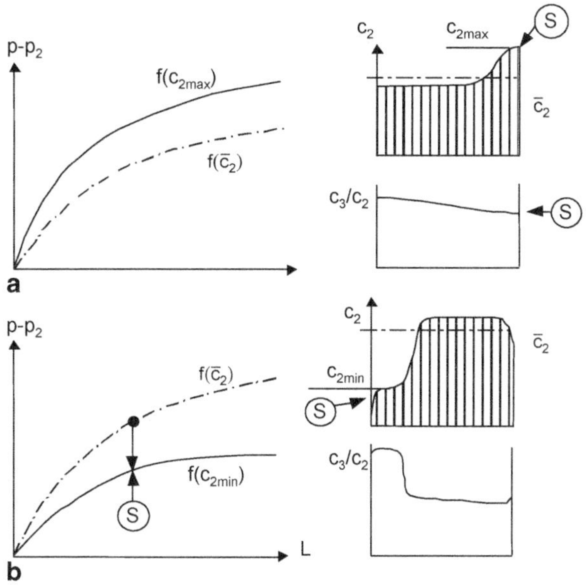

Fig. 5.20 Flow separation in the diffuser, S = flow separation, **a** separation at local excess velocity, **b** separation at local velocity deficit

can be derived as a very rough estimation of this effect (valid for $1 < U_{fl} < 1.4$). In Eq. (5.8a) $c_{p,o}$ is the coefficient with uniform and c_p the pressure recovery with non-uniform approach flow velocity distribution. The non-uniformity is defined according to Eq. (8.19) with the *meridional* component of the velocity. If c_{2m} near an impeller sidewall tends towards zero, the c_p value drops to a fraction of its value with uniform approach flow, since hardly any pressure is recovered in the diffuser in the near-wall fluid due to insufficient kinetic energy. Obviously, a highly non-uniform approach flow with an energy deficit close to a wall can be expected to induce early flow separation in the diffuser. As discussed below, this feature needs not be a disadvantage with respect to the stability of the Q-H-curve. However, the efficiency deteriorates through non-uniform approach flow.

As stated in Chap. 5.2.4, two conditions have to be satisfied for recirculation to occur: the flow must separate *and* sufficiently high pressure-gradients perpendicular to the main flow must be present. This also applies to the recirculation from the collector. Such pressure gradients can primarily be generated when the approach flow is non-uniform, causing the pressure build-up to vary across the diffuser width. To visualize this, the pressure rise in the diffuser could be considered for various streamlines of a non-uniform velocity profile according to Eq. (1.7).

Whether the flow in the diffuser separates on the side where c_2 is at a maximum or where c_2 is at a minimum depends on the other characteristics of the velocity profile. Consider a profile as sketched in Fig. 5.20a: due to deceleration the pressure rise is greatest where c_2 is at a maximum. This is where the flow will separate if either the pressure recovery in the diffuser is determined by the mean value of c_2, or the maximum of c_2 constitutes a local peak. In contrast, the maximum and average values of c_2 differ only to a minor extent in Fig. 5.20b; the average c_2 will therefore determine the pressure rise in the diffuser. On the side of the velocity deficit, the pressure build-up is smaller than average. Thus a secondary flow is generated and the fluid stalls where c_2 has a minimum near the wall.

The effect of the non-uniform approach flow is superimposed on the general deceleration c_{3q}/c_2, which increases as the flow rate is reduced, refer to Fig. 5.18. The ratio c_{3q}/c_2 is critical for the onset of Q-H-curve instabilities. The investigations described in [25] can also be interpreted in this manner: The instabilities in the characteristics occur if an ever-increasing non-uniformity of the diffuser approach flow - together with the growing general deceleration in the diffuser - leads to flow separation. Thereby the pressure recovery in the diffuser is greatly affected, while the static pressure rise in the impeller is hardly impaired.

According to the above considerations, the *span-wise* non-uniformity of the (pitch-wise averaged) velocity components is significant for hydraulic losses and the stability of the Q-H-curve. Conversely, the *pitch-wise* non-uniformity has an effect on the unsteady forces and pressure pulsations. It causes an unsteady approach flow to the diffuser, increases the turbulence and consequently can have a favorable effect on the pressure recovery in the diffuser.

5.3.4 *Flow in the Volute Casing*

A fluid element downstream of the impeller outlet moves in accordance with the conservation of angular momentum $c_u \times r = c_{2u} \times r_2 = \text{constant}$, Chap. 3.7. Consequently, the circumferential velocity in the volute drops with growing radius, while the static pressure increases from the inside to the outside according to Eq. (1.28). These pressure and velocity distributions, shown qualitatively in Fig. 5.21, are confirmed by experiments near the best efficiency point, e.g. [31]. At overload and partload operation, the volute cross sections do not match the impeller discharge velocities and the flow patterns in the volute deviate from the behavior shown in Fig. 5.21 (see also Chap. 9.3.3). The pressure gradients over the radius become flatter at partload, while they increase at overload.

Similar to the flow through a bend or any curved channel, secondary flows are generated in a volute according to the mechanisms discussed in Chap. 1.4. The flow pattern has the shape of a double vortex according to Fig. 5.21 (refer also to

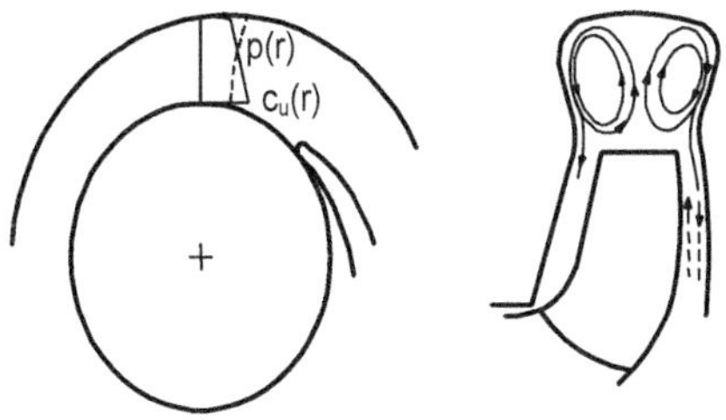

Fig. 5.21 Flow patterns in volute casings

Fig. 1.12); it becomes increasingly asymmetrical with growing non-uniformity of the impeller outflow. In flat, rectangular or trapeze-shaped volute cross sections a less intensive secondary flow is expected than in a volute of circular or square cross section (because of the smaller radius ratio).

The deceleration of the flow in the volute is governed by similar mechanisms as discussed in context with Fig. 5.18 for diffusers. This follows from the experiments shown in Fig. 4.6. However, the interaction between volute and impeller is weaker than with a diffuser: shut-off power and head (P_o, H_o and $H_{p,o}$) are therefore smaller. Volute casings react in a less sensitive manner to a non-uniform impeller outflow than diffusers. Saddle-type instabilities of the Q-H-curve are therefore rarely found with volute casing pumps below $n_q = 70$ (unless the characteristic $H_p = f(Q)$ is unstable).

5.3.5 Flow in Annular Casings and Vaneless Diffusers

In a vaneless diffuser, the flow pattern is largely in accordance with the conservation of angular momentum $c_u \times r = c_{2u} \times r_2$. The pressure increases from the inside to the outside following Eq. (1.28). Since the velocity decreases in the boundary layers, the flow path curvature increases in order to keep the equilibrium with the radial pressure gradients imposed by the main flow (see velocity c_{BL} in Fig. 1.12). As a consequence, a spiraling secondary flow is established where the fluid in the boundary layer flows back from the outside to the inside.

Since the effects of incidence are absent in a vaneless diffuser, the flow instabilities are primarily caused by non-uniform velocity distributions at the impeller outlet. If c_{2u} changes over the impeller width, the pressure recovery on different streamlines varies according to Eq. (1.28). Secondary flows are thus amplified and flow separations may be induced.

As explained in Chap. 3.7, the length of the flow path which a fluid element follows in the vaneless diffuser, and consequently the friction losses, depend on the flow angle α_3. At low flow rates the outflow angle decreases and the path length covered by a fluid element is extended. The growing boundary layer thickness will ultimately lead to stall. For this reason, a critical flow angle was sometimes observed, below which the boundary layer separated and recirculations occurred. The critical flow angle decreases if low-energy zones prevail at the side walls of the vaneless diffuser. The energy distribution over the width therefore plays a significant role. The flow tends to separate where the local kinetic energy at the impeller outlet is significantly below the average. The distribution of α_3 over the width of the vaneless diffuser is determined by the c_{2m} and c_{2u} profiles since $\tan \alpha = c_m/c_u$. Because of the strong impact of the local flow distribution on the secondary flows and local stall, it appears difficult to define a generally applicable critical outflow angle. The tendency to separations - or the critical outflow angle - increases with the radius ratio r_4/r_2. The separation tendency falls with increasing Reynolds number, turbulence and a growing ratio of width to radius b/r_2.

With unfavorable impeller designs, flow separation can occur even near the best efficiency point. These effects can be analyzed and avoided using numerical flow calculations. However, flow separations are unavoidable at low partload.

During operation against a closed valve, the fluid circulates in the vaneless diffuser far more intensively than in vaned diffusers. An exchange of momentum is effective only between the zones of stalled fluid in the vaneless diffuser and the impeller. Consequently, power consumption and shut-off head are lower than in vaned diffusers or volutes since the recirculating fluid has a higher circumferential velocity (refer to Fig. 4.6b).

5.4 The Effects of Flow Recirculation

5.4.1 Effects of Flow Recirculation at the Impeller Inlet

According to Table 3.3 the theoretical head can be written as:

$$H_{th} = \frac{u_2^2 - u_1^2}{2\,g} + \frac{w_1^2 - w_2^2}{2\,g} + \frac{c_2^2 - c_1^2}{2\,g} \tag{5.9}$$

The three terms on the right side of Eq. (5.9) have the following meaning:

- $H_z = (u_2^2 - u_1^2)/(2g)$ is the head component generated by the centrifugal forces which merely depends on the inlet to outlet diameter ratio d_1/d_2 and is not affected by the blade angles.
- $H_w = (w_1^2 - w_2^2)/(2g)$ is the head component generated by the deceleration of the relative velocity in the impeller. It depends on the blade angles, the flow rate and possible recirculation zones having an effect on the streamline pattern.
- $H_{p,th} = H_z + H_w$ constitutes the theoretical increase of the static pressure to be expected in the impeller if there were no losses. While H_p can be measured at the impeller outlet, H_z and H_w cannot be recorded separately.

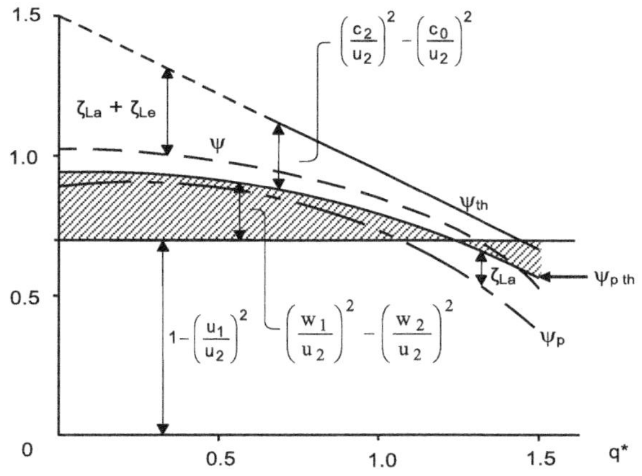

Fig. 5.22 Static and dynamic head components as a function of flow rate ratio q*

- $H_a = (c_2^2 - c_1^2)/(2g)$ is the head component created through acceleration of the absolute flow in the impeller. It is largely converted into static pressure in the collector through deceleration.

Figure 5.22 shows how these individual head components depend on the flow rate. In the domain where recirculation occurs, the presentation has a somewhat hypothetical character and applies at best to a streamline that works "more or less normally". The centrifugal head component has been assumed in the graph as independent of the flow rate. This is only correct in the domain where no recirculation at the impeller inlet is present, since the onset of recirculation will block the outer region of the impeller inlet cross section so that the flow is diverted to the hub (Fig. 5.12). The effective streamlines are shifted and enter the blading on a smaller radius than without recirculation. As a result, the centrifugal head component increases. This head component can be estimated from Eq. (5.10) where d_{1eff} describes the effective diameter on which the net through-flow enters the impeller:

$$H_Z = \frac{(u_2^2 - u_{1eff}^2)}{2\,g} \tag{5.10}$$

The theoretical head increases due to the recirculation by the amount:

$$H_{Rec} = \frac{u_2^2}{2\,g}\left(\frac{d_1^2}{d_2^2} - \frac{d_{1eff}^2}{d_2^2}\right) \tag{5.11}$$

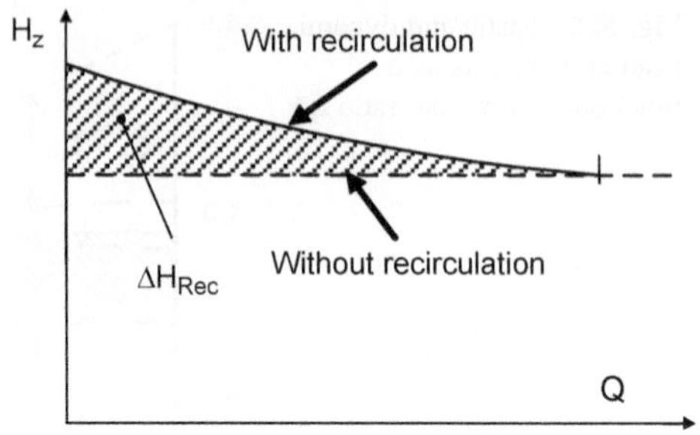

Fig. 5.23 Influence of inlet recirculation on centrifugal head component

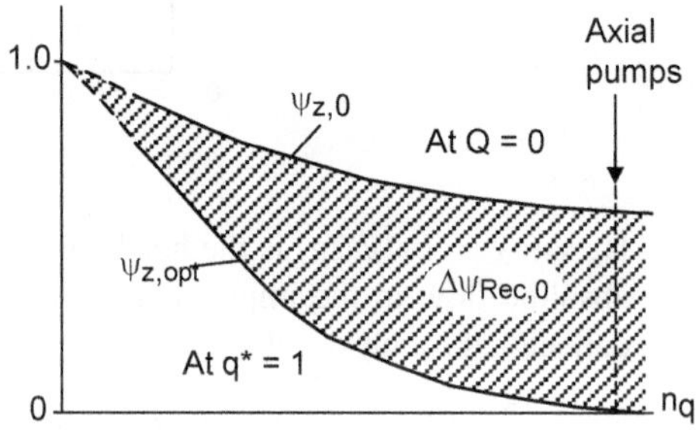

Fig. 5.24 Influence of specific speed on centrifugal head component

The head gained due to recirculation is determined by the extent to which the streamlines are shifted towards the hub. At the onset of recirculation, the effect is small but it grows with decreasing flow rate until the recirculation reaches its maximum intensity at Q = 0 (Fig. 5.23). According to Eq. (5.11), the head component caused by recirculation increases with the square of the ratio d_1/d_2.

Since d_1/d_2 rises with growing specific speed, the gain in head due to recirculation increases significantly with growing n_q. This is shown in Fig. 5.24 where the centrifugal head component at the best efficiency point $\psi_{z,opt}$ (without recirculation) and the component $\psi_{z,o}$ at shut-off are plotted against n_q. The difference between both curves is the head increase $\psi_{Rec,o}$ caused by the recirculation at shut-off. In axial pumps $\psi_{z,opt} = 0$ applies at the best efficiency point (because $d_1 = d_2$); at shut-off the streamlines shift strongly and $\psi_{Rec,o}$ reaches a maximum. In contrast, the head component caused by recirculation is small at low specific speeds, since the potential for streamline shifting diminishes because of the low blade height. As discussed in Chap. 5.1.2 and confirmed by experience, the intensity of recirculation increases with growing n_q.

The recirculation induces a pre-rotation in the incoming fluid (Fig. 4.13). Consequently, the theoretical head decreases according to Eq. (T3.3.1) (or Eq. 5.9: w_1 falls, c_1 increases). Experience confirms this: if the recirculation-induced pre-rotation is slowed down by structures, the head at partload and especially at

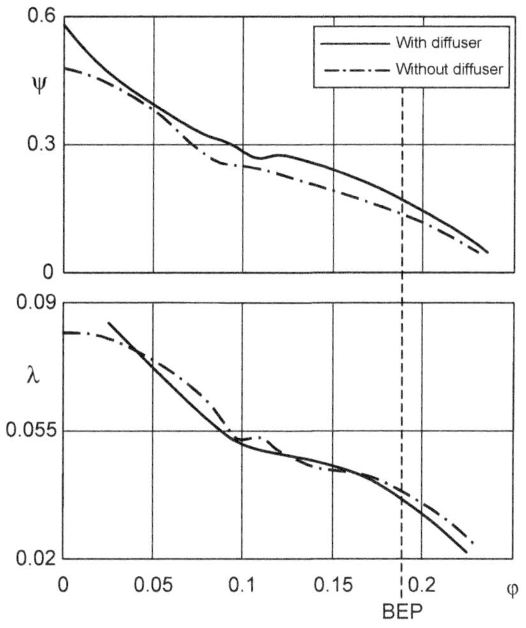

Fig. 5.25 Characteristics of an axial pump with and without diffuser [20]

shut-off increases. Examples for this are the return vanes of multistage pumps and ribs in the suction nozzle or inlet casing. The theoretical aspects of the induced pre-rotation are further discussed in Chap. 8.4.

The influence of the inlet recirculation on the head according to Fig. 5.24 is most clearly observed in axial pumps. Figure 5.25 shows the characteristics of an axial pump with a hub ratio of 0.4 where the same impeller was measured both with and without diffuser. At $q^* = 0.65$ the characteristic has a saddle which is due to flow separation. Below $q^* = 0.55$, the characteristic increases steeply, since the streamlines with increasing recirculation are shifted at the inlet towards smaller and, at the outlet, towards larger radii (Figs. 5.12 and 5.2). According to Eq. (5.11), $\psi_{Rec,o} = 0.84$ should be expected for $d_n/d_2 = 0.4$, while 0.49 to 0.59 was measured. The diffuser generates only a minor increase in head and power at shut-off, as demonstrated by a comparison of both characteristics.

Most axial and semi-axial impellers have characteristics that resemble those shown in Fig. 5.25. Consequently, the gain in head observed when the pump operates with fully separated flow can be explained by the action of centrifugal forces. These cause a greater deflection of the fluid than the streamlined flow around the blade profiles near the design point.

At low specific speeds, the effect of recirculation on the head predicted by Eq. (5.11) is far weaker than in the above example (Fig. 5.24). Figure 5.26 shows tests performed on a pump with $n_q = 16$, where the head at partload increased noticeably with an enlarged impeller inlet diameter (blade shape and angles

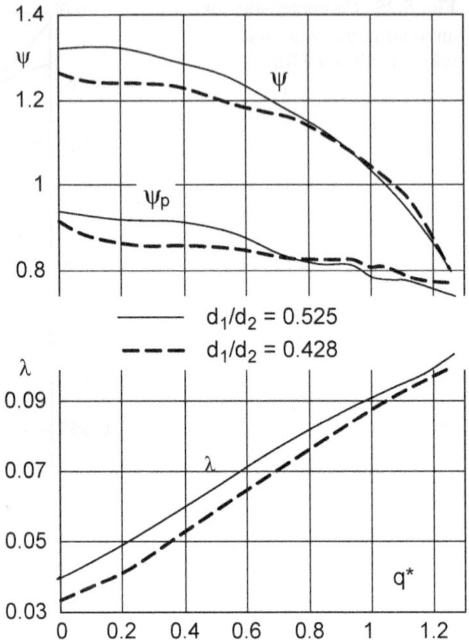

Fig. 5.26 Influence of impeller inlet diameter on performance curves, $n_q = 16$

remained the same during this modification). According to Eq. (5.11), the theoretical gain in head coefficient at Q = 0 is approximately 0.043; the measured value is 0.04. With the increased impeller eye diameter, the static pressure at the impeller outlet near the BEP is smaller than in the baseline test; it increases when operating with recirculation. Both observations are in agreement with Eq. (5.9) and Fig. 5.22. With the enlarged impeller inlet diameter, the power consumption increased due to the growing recirculation (this increase is unexpectedly high).

If the hub diameter (at otherwise identical geometry) is increased, the amount by which the streamlines can shift towards a smaller radius with recirculation is reduced; this causes the shut-off head to drop, [20].

According to the model discussed above, the head developed when operating with extensive inlet recirculation is significantly influenced by the effective diameters at which the still "intact" streamline enters and leaves the blading, Fig. 5.12. Consequently, the position of the blade leading edge must exercise a considerable influence on the partload characteristic. This is indeed confirmed by experience and numerous publications. Figure 5.27 shows measurements on four impellers with different blade leading edge diameters at the outer and inner streamlines (blade numbers and the impeller outlet width were changed as well). In these tests, shut-off head and power rise with increasing impeller inlet diameter and growing ratio d_{1a}/d_{1i} in an altogether consistent manner. In summary:

- If recirculation is fully developed at the impeller inlet, head, power consumption and the static pressure rise in the impeller increase with growing recirculation intensity. These effects are enhanced if the pre-rotation induced by the

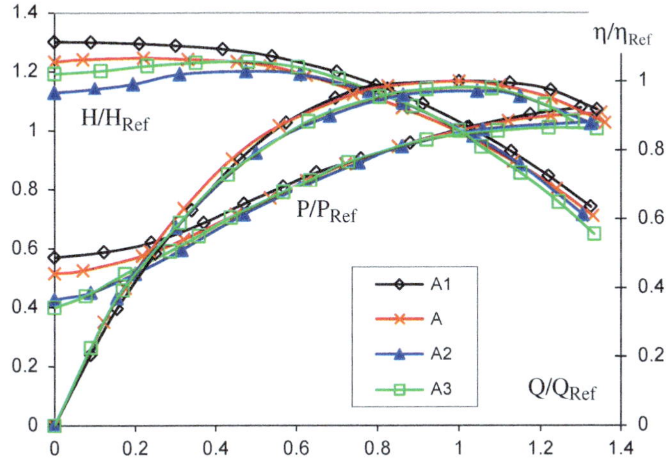

	d_1^*	$\frac{d_{1a}}{d_{1i}}$
A	0.61	1.55
A1	0.67	1.71
A2	0.6	1.1
A3	0.6	1.29

Fig. 5.27 Influence of impeller inlet on performance curves [44]

recirculation on the net through-flow is slowed down or eliminated through structures upstream of the impeller.

- With rising specific speed (or with growing ratio d_{1a}/d_{1i}) these effects increase strongly. This explains the steep Q-H-curves and the high shut-off power of pumps with high n_q and the influence of the blade leading edge on the partload characteristic.

5.4.2 *Effect of Flow Recirculation at the Impeller Outlet*

When a fluid element flows back from the collector into the impeller, its velocity has a circumferential component near zero. In the impeller, the fluid is re-accelerated to a higher velocity. If the conservation of angular momentum is applied to this process, the result is an increase of the power consumption. This is well confirmed by experience.

The effect of recirculation on the head is far less obvious. Generally it is assumed that the recirculation will impair the head and be responsible for possible instabilities of the Q-H-curve. However, a closer examination of numerous experimental observations reveals that this assumption is, at best, part of the truth. Consider three stages with reference to the characteristics in Fig. 4.5 and the flow observations in the diffuser shown in Fig. 5.17: (1) local separation in the diffuser, (2) fully developed recirculation, (3) a "transition zone" between these two states. Local separation, as occurring in Fig. 5.17 at $q^* = 0.94$, does neither significantly affect the pressure recovery in the diffuser nor the Q-H-curve. The pressure increase in the collector becomes unstable only with severe separation at $q^* = 0.4$. The Q-H-curve

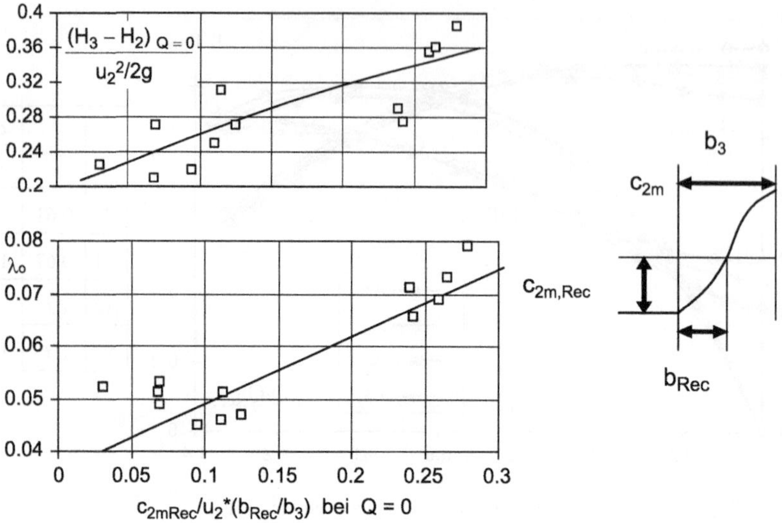

Fig. 5.28 Influence of outlet recirculation on pressure recovery in the diffuser and power consumption at shut-off, [B.20]

remains stable since the impeller at $q^* < 0.4$ generates significantly more static pressure due to recirculation at the impeller inlet (ψ_p increases markedly towards $Q = 0$). At shut-off the recirculation is fully developed. In the transition zone left to the maximum diffuser pressure at $q^* = 0.4$, the recirculation is not yet developed to the extent that it is able to make a contribution to the stabilization of the characteristic.

Numerous test results and observations seem to prove that fully developed recirculation at the impeller outlet increases the head:

1. As demonstrated by Fig. 5.28, the pressure recovery ($H_3 - H_2$) in the triangular section of the diffuser unmistakably increases with the intensity of recirculation at shut-off (as defined in Fig. 5.28). As expected, the power coefficient follows the same trend. Figure 5.28 covers 13 tests on radial pumps with $n_q = 15$ to 35.
2. Measurements on a semi-axial pump ($n_q = 150$) are shown in Fig. 5.29. Here, the static pressure $\rho \times g \times H_p$ at the impeller outlet was measured on the outer streamline. Its curve $H_p = f(Q)$ is stable. The slight saddle in the characteristic is caused by a deficit in the diffuser pressure recovery since the flow approaching the diffuser is non-uniform over the blade height. In contrast to the measurements on radial pumps discussed above, the diffuser does not contribute to the pressure rise below $q^* = 0.67$ (the contribution is even negative in a certain range). At the best efficiency point the pressure recovery in the diffuser is 15% of the head, at $q^* = 1.27$ even 30%. As can be derived from the measurements of the pressure differentials ΔH_s upstream of the impeller (installed and conducted according to Chap. 5.2.6, Fig. 5.16), the inlet recirculation started around $q^* = 0.55$. Below $q^* = 0.5$, head and static pressure rise

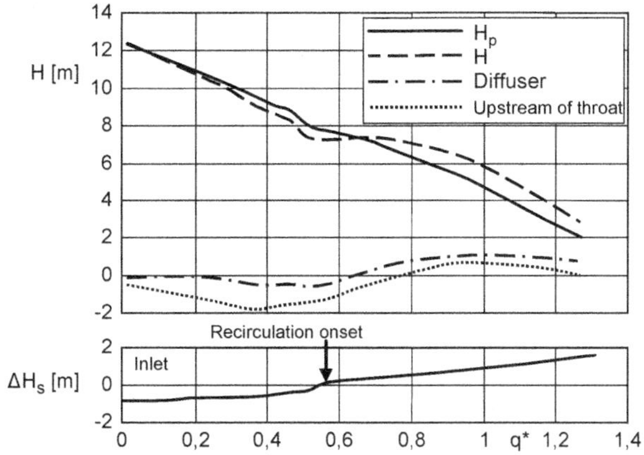

Fig. 5.29 Component characteristics of a semi-axial pump n_q = 150 H_p = static pressure rise in the impeller

in the impeller increase strongly due to fully-developed recirculation at the impeller inlet. Often the saddle-type instability is triggered near the onset of inlet recirculation. The reason is that the energy transfer at the outer streamline is strongly diminished when the flow rate approaches the recirculation onset. Then, the diffuser approach flow becomes skewed which impairs the pressure recovery in the diffuser at the same time.

3. Figure 5.30 shows measurements on two impellers (combined with the same diffuser): impeller A has a stable, impeller B an unstable Q-H-curve. Impeller A develops a recirculation on the front shroud which grows continuously from q* = 0.5 to q* = 0. In contrast, the recirculation in impeller B is lower throughout the flow range than in impeller A; the same applies to the head. In particular, the minimum head of impeller B occurs at q* = 0.25 where the recirculation is zero! This finding is confirmed by the power consumption which also has its minimum at q* = 0.25. The static pressure rise of impeller A at Q = 0 is higher than that of impeller B. Obviously this was also caused by the stronger recirculation.

   If zones of recirculation occur at the same time near the rear and front shrouds (Fig. 5.12c), mixing losses are bound to increase because there are two shear layers between forward and backward flow as compared to one shear layer if the backflow is either on the front or the rear shroud (Fig. 5.12d or e).
4. The interaction between collector and impeller caused by recirculation is well demonstrated by Fig. 4.6b: below the onset of outlet recirculation at q* = 0.6, the static pressure rise in the impeller is significantly higher in the test with a vaned diffuser than with the vaneless design. A quite similar behavior is found when comparing volute and diffuser tests in Fig. 4.6a.
5. The interaction between the flow in the impeller sidewall gaps E and F and the main flow is influenced by gap A between the front shrouds of diffuser and impeller and by the overlap or the length of this gap. If gap A is large, the interaction due to exchange of momentum between the flow in the impeller sidewall gap and the main flow increases. If gap A is small, interaction only

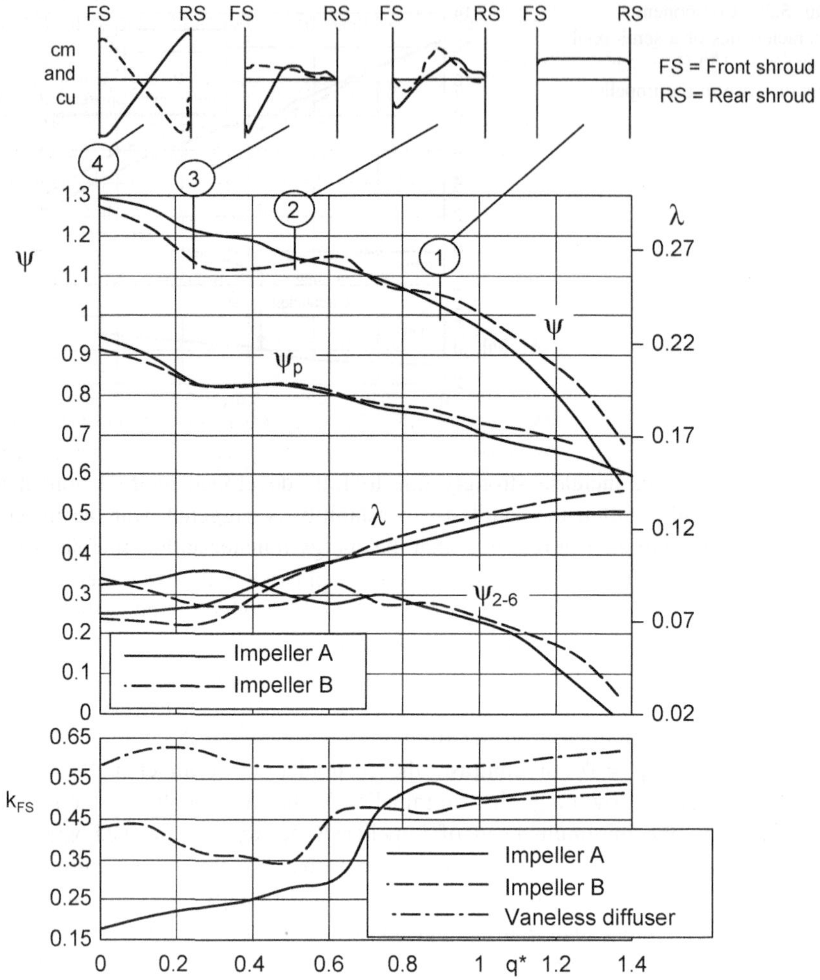

Fig. 5.30 Influence of impeller outlet flow distribution on stability, $n_q = 32$

occurs by way of the energy transport through the leakage flow (see also Chap. 9.1 and Fig. 9.1). Many experiments have confirmed without exception that the pressure against a closed valve of a multistage pump increases if gap A is narrow [B.20 and 18]. As shown in Fig. 5.31 the power consumption changes only slightly in the process. The gain in head at partload can therefore be explained by the enhanced exchange of momentum between diffuser and impeller. Consequently, the static pressure rise in the impeller and the pressure recovery in the diffuser increase if gap A is narrowed (Fig. 5.31).

6. Figure 5.32 shows the strong influence of the trailing edge position on the head of semi-axial pumps. Shape 1 produced significantly higher values for power consumption and head than shape 2 where d_{2a} was slightly reduced. Imagining the

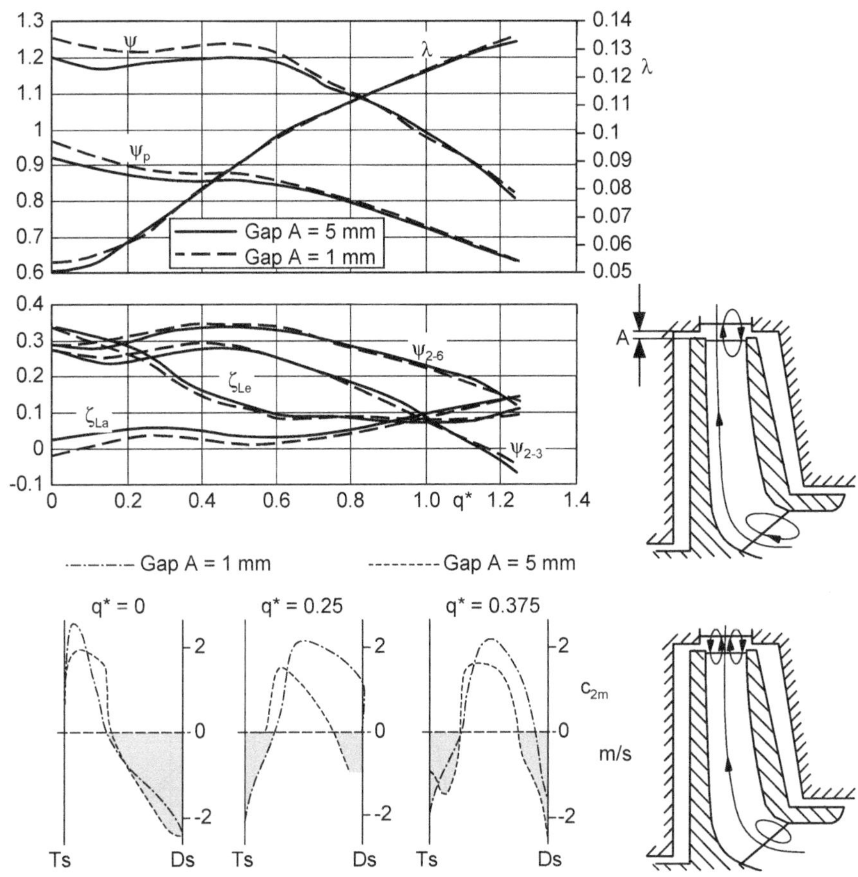

Fig. 5.31 Effect of gap A on the characteristics, $n_q = 26$ [B.20]

two extreme cases of an axis-parallel trailing edge (as with a radial impeller) and an essentially radial trailing edge (as with an axial impeller), the influence of the trailing edge on the recirculation is easily understood; see sketch in Fig. 5.32.

7. Even with radial impellers, an oblique trim of the impeller blade trailing edges causes an increase of the shut-off head since the difference in radii generates a more intensive recirculation. Figure 5.33 illustrates this: the impeller with oblique blade trailing edges (with the same average d_{2m}) generated at low flow rates higher head and power consumption than the impeller with axis-parallel trailing edges - an indication that the exchange of momentum was increased due to recirculation.
8. A large ratio of diffuser inlet width b_3 to impeller outlet width b_2 promotes flow separation in the diffuser since the meridional component is abruptly decelerated. Downstream of the sudden expansion, flow recirculation is enhanced, thus increasing power consumption and head at low flow rates. This is revealed by

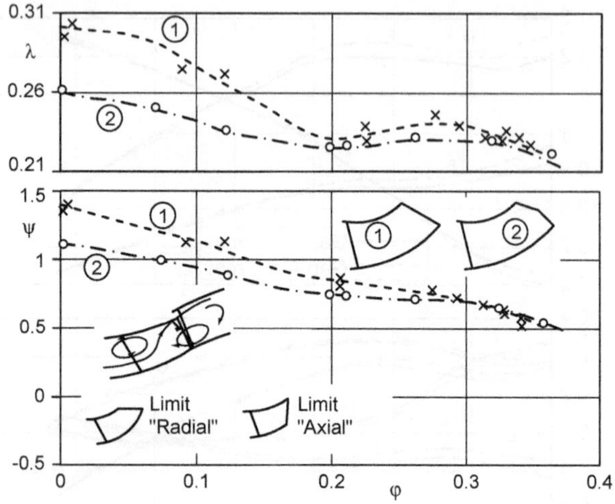

Fig. 5.32 Influence of impeller blade trailing edge on the shape of the characteristics [20]

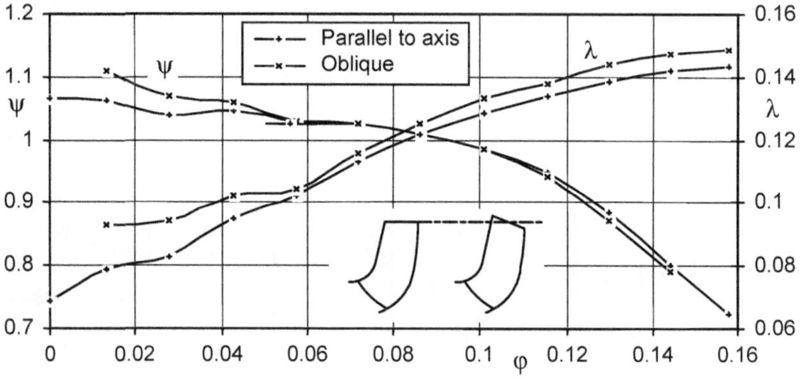

Fig. 5.33 Oblique versus axis-parallel impeller blade trailing edge at the same average impeller outlet diameter, n_q = 33, [20]

the tests in Fig. 5.34 where a given impeller was tested with two different diffusers which had almost the same throat area. On the wider diffuser the flow already separates at = 0.08. However, head and power consumption at <0.04 are higher than with the narrower diffuser which hardly shows any separation.

9. With multistage pumps, the axial rotor position can have a considerable influence on the Q-H-curve because the location of the recirculation zone can be affected by the rotor position. This can be inferred from Fig. 5.12. Sometimes the rotor is deliberately shifted from the position, where impeller mid and diffuser mid are aligned, in order to modify the Q-H-curve and/or the axial thrust. If for instance, the central rotor position yields a poorly developed recirculation at the front shroud, the recirculation may be intensified by shifting the rotor in the direction of the pressure side because the meridional velocity on the front shroud side is suddenly decelerated.

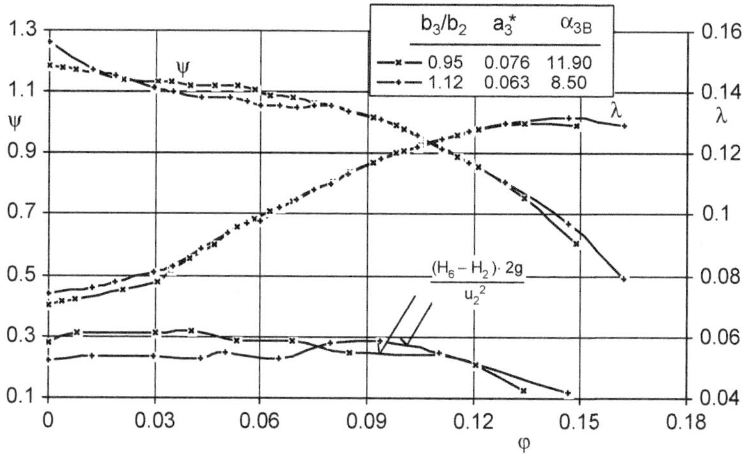

Fig. 5.34 Typical influence of diffuser parameters on stability and power [20]

10. With pumps of low specific speeds, the relative impeller outlet width b_2/d_2 must be selected sufficiently large to obtain a stable Q-H curve. According to Chap. 5.2.2 the velocity distribution at the impeller outlet becomes increasingly non-uniform with growing b_2/d_2. However, this promotes recirculation at the impeller outlet as discussed in Chap. 5.3.3.
11. The exchange of momentum, hence the shut-off head, increase if the distance between the impeller blades and diffuser vanes (gap B) or the diffuser throat a_3 is reduced. A smaller vane distance a_3 possibly reduces the circumferential component c_u of the recirculating fluid.
12. An increase of the impeller blade number yields a slightly higher shut-off head since the exchange of momentum is intensified.

The effect of recirculation could also be interpreted by another physical model: The recirculation zone blocks a part of the impeller outlet section. In the diffuser, too, recirculation blocks a part of the cross section, which might improve the pressure recovery in the healthy flow. According to this model, a stream tube would operate more or less normally in the impeller and diffuser as in a kind of "partial admission", albeit with reduced hydraulic efficiency. The stalled fluid zones "guide" the flow like solid walls, despite a certain exchange of momentum which is bound to occur in the shear layers between recirculation and main flow. According to this model, the partload head also increases with expanding recirculation zones. The experimental findings described above could also be interpreted according to this model.

With fully developed recirculation the *three-dimensional* flow distribution at the impeller outlet is crucial. This fact limits the validity of the simplified models discussed above.

It must be assumed that the shut-off head H_o increases with the kinetic energy at the collector inlet according to $H_o \sim c_3^2$ and $\psi_0 \sim (c_3/u_2)^2$. When low-energy liquid flows back from the collector into the impeller, it is re-accelerated and c_3 increases. In this case *both* shut-off pressure *and* power consumption increase. This becomes evident from Fig. 5.28 and the tests in Figs. 5.30, 5.31, 5.32, 5.33 and 5.34 as well as from Fig. 4.6a, b. If the exchange of momentum between impeller and collector is increased deliberately (as in Chaps. 7.3.2 and 7.3.3 or by selecting a large ratio of b_2/d_2 and/or d_{2a}/d_{2i}, d_{1a}/d_{1i} etc.), power consumption *and* shut-off pressure generally increase. The mechanisms involved may be comparable to a regenerative pump. Which portion of the kinetic energy of c_3^2 becomes effective in increasing the shut-off pressure (at a given recirculation) depends on the configuration of the collector. If much energy is dissipated, the head does not increase despite intensive recirculation. This fact is illustrated by the tests in Fig. 5.19: when the diffuser throat area was reduced by modifying a_3, the shut-off head increased considerably while the power consumption remained almost identical. These tests show that shut-off head and power do not always change in the same sense.

5.4.3 *Effect of Outlet Recirculation on the Flow in the Impeller Sidewall Gaps and on Axial Thrust*

While the recirculation at the impeller inlet does not noticeably affect the flow in the impeller sidewall gaps (gaps E and F), the impact of the outlet recirculation can be considerable. Experience shows this to be indeed the case when recirculation occurs near the front shroud; then recirculating fluid of low circumferential velocity c_u is transported into gap F by the annular seal leakage. As discussed in Chap. 9.1, fluid with low c_u slows down the rotation in the impeller sidewall gap. The same applies to the rear shroud (gap E) of impellers with balancing holes. In contrast, this effect is hardly noticeable in multistage pumps where the leakage on the rear shroud is directed radially outwards.

The strong impact of the recirculation on fluid rotation in the impeller sidewall gap F is well illustrated by tests shown in Fig. 5.30 where the measured rotation ratio k_{FS} is plotted against the flow rate (as defined in Chap. 9.1, k_{FS} is the ratio of the mean angular fluid velocity in gap F to the impeller tip speed). With impeller A, the recirculation near the front shroud expands continuously as the flow rate is reduced below the onset of outlet recirculation (occurring at $q^* \approx 0.75$). At the same time the rotation factor k_{FS} in gap F drops steadily from about 0.5 to 0.18 because fluid with low c_{2u} enters the impeller sidewall gap. The behavior of impeller B is totally different: At $Q = 0$, the recirculation appears near the rear shroud; this is why k_{FS} is not affected (k_{FS} is 3-times higher than with impeller A). The reduction of the rotation factor in the range of $q^* = 0.15$ to 0.65 is caused by the low c_{2u} on the front shroud. This is evident from a comparison of the curves $k_{FS} = f(q^*)$ with the velocity profiles. The behavior of k_{FS} consistently follows the pattern expected from the c_{2m}- and c_{2u}-distributions shown on top of Fig. 5.30. In the vaneless diffuser the circumferential velocity of the recirculating fluid is higher than in the vaned diffuser since it is not impeded by the presence of vanes.

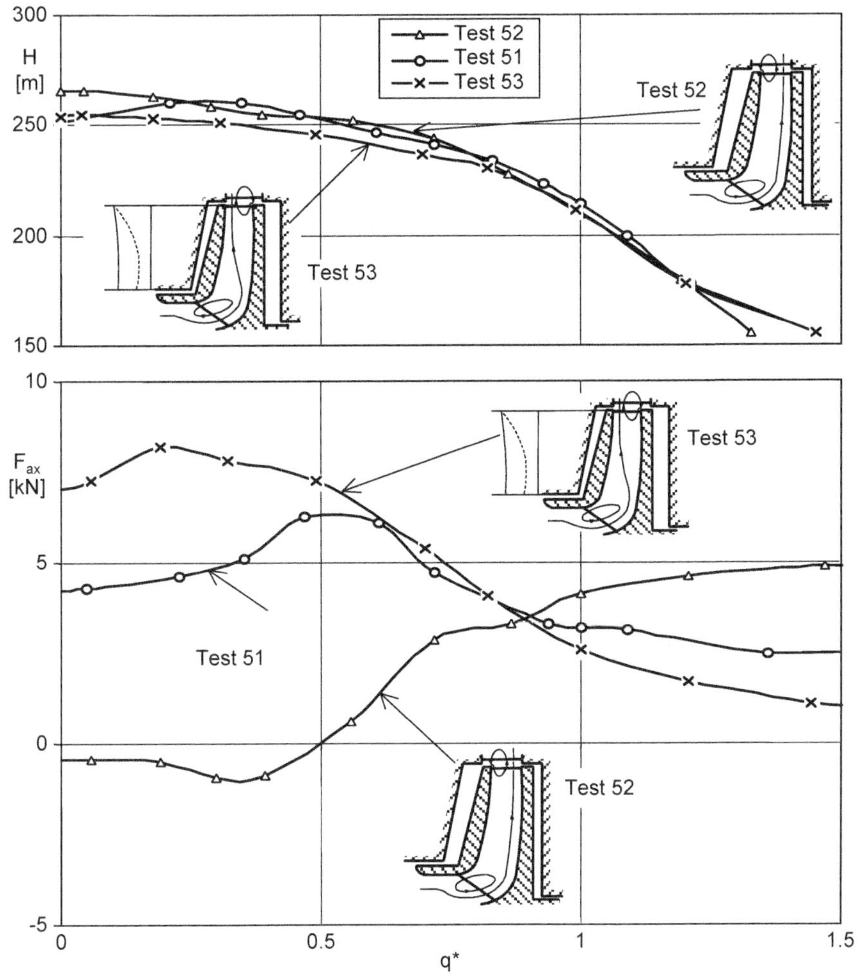

Fig. 5.35 Influence of axial rotor position on Q-H-curve and axial thrust, $n_q = 22$, Sulzer Pumps; Test 51: Rotor position centered; Test 52: Rotor shifted to discharge side; Test 53: Rotor shifted to suction side

Consequently, fluid rotation in the impeller sidewall gap becomes more intense; k_{FS} constant as shown by Fig. 5.30.

Investigations on a pump with $n_q = 22$ show the effect of flow recirculation on fluid rotation in the impeller sidewall gap E in a similar manner [26]: with a wide gap E, the recirculation occurred near the rear shroud throughout the flow rate range and fluid rotation was slowed down accordingly. With a very narrow gap E, the recirculation zone jumped from the rear shroud to the front shroud and the rotation factor increased. The change in location of recirculation manifests itself as a slight saddle in the Q-H-curve, Chap. 5.5.2.

Such changes of fluid rotation in the impeller sidewall gaps are bound to exert a major influence on the axial thrust (Chap. 9.2) [6, 18]. Since the axial forces in multistage pumps are balanced out to a large extent, changes of the rotation in the impeller sidewall gaps can result in thrust reversals as happened in the tests shown in Fig. 5.35. In these tests, the influence of the axial rotor position on the Q-H-curve and the axial thrust was investigated. Although velocity measurements are not available, the results can well be interpreted in the light of the above explanations. If the rotor is shifted towards the discharge side, the recirculation forms on the front shroud because of the back-step between the impeller front shroud and the diffuser side wall. This causes the fluid rotation in gap F to slow down. The integration of the pressure distribution on the front shroud yields a greater force than at the best efficiency point and the axial thrust acts towards the pressure side. The opposite is observed when the rotor is shifted towards the suction side: a back-step is created on the rear shroud and recirculation is provoked there. The rotation in the impeller sidewall gap on the front shroud remains unaffected and the axial thrust is directed towards the suction side as corresponds to design practice. With both rotor positions the *Q-H-curve* is *stable*; the *recirculation zone remains stable*, in both cases, on the side where the back-step is formed.

In contrast, if the impellers are centrally aligned relative to the diffusers, the axial thrust curve initially follows the case where the recirculation develops on the rear shroud. The axial thrust suddenly drops below $q^* < 0.6$. This suggests that the recirculation zone has jumped to the front shroud.

Above the best efficiency point, the axial thrust also depends on the rotor position, since the annular seal length and, consequently, the leakage and pressure distribution in the impeller sidewall gaps are affected (Chap. 9.1).

Axial thrust curves strikingly similar to Fig. 5.35 were published in [6]. By tightening gap A to about $A/r_2 = 0.005$ the axial thrust reversal was removed and flat axial thrust curves (close to theoretical prediction) were obtained over the whole flow range.

Conclusions:

- The partload characteristic can (but does not necessarily) depend on the rotor position. This is due to the impact that the velocity distribution at the diffuser inlet has on the pressure recovery.
- Some pumps are sensitive to the axial rotor position while others are not. This is explained by *differences in the impeller discharge velocity distribution*. Since the latter is usually unknown, the effect of rotor shifting cannot easily be predicted.
- For the above reason, it is difficult to predict in a particular application whether the rotor should be shifted towards the suction or the discharge side in an attempt to stabilize the Q-H-curve. Shifting the rotor towards the discharge side is expected to increase the shut-off head and thus stabilize a drooping Q-H-curve as demonstrated by Fig. 5.35.
- Gap A should be sufficiently small, as stipulated in Table 7.7, Eq. (T7.7.16), to get the desired improvement of the Q-H-curve and to control the influence of the impeller discharge recirculation on axial thrust - notably to avoid axial thrust excursions such as shown in Fig. 5.35.

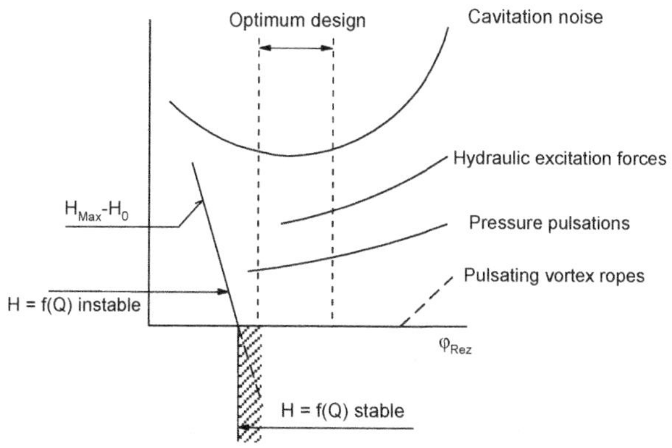

Fig. 5.36 Optimum intensity of recirculation

5.4.4 Damaging Effects of Partload Recirculation

As shown above, power consumption and head increase due to recirculations at impeller inlet and outlet when the pump is operated against a closed valve. In pumps with high specific speeds this increase is undesirable since the pipeline has to be calculated for the shut-off pressure and the motor must be selected for start-up against a closed valve. The additional power generated by the recirculation is dissipated in the pump and heats up the fluid (see Chap. 11.6, permissible operating range and minimum flow).

Extended operation in the range with strong recirculation often affects the service life of a pump owing to vibrations or cavitation, since the *entire* recirculation power is dissipated; refer to Chap. 10 and Eq. (10.17):

- Large eddies develop in the shear layers between recirculating fluid and the main flow (just as in stalled water downstream of the pier of a bridge). Especially at the impeller outlet, these eddies generate excitation forces and pressure pulsations due to the interaction between impeller and collector. As a consequence, increased wear on annular seals, damage to bearings and seals, fatigue fractures on various pump components, undesirable vibrations, pressure fluctuations in the system and noise can develop. Frequently observed problems of this kind, their causes and possible remedies are compiled in Table 10.1.
- The eddies generated in shear layers can result in cavitation damage at the impeller inlet (in special cases even at the impeller outlet or in the diffuser) if the pressure in the eddy core drops below the vapor pressure of the liquid. At the impeller inlet, the bubbles thus created usually implode on the pressure surface of the blades (Chap. 6.8 and Table 6.3).

Thus on the one hand recirculation can cause damage through excessive vibrations or cavitation, while, on the other hand, a certain amount of recirculation is necessary in order to obtain stable characteristics in the first place. Consequently, an optimum intensity of recirculation should exist as suggested in Fig. 5.36. Yet, such an optimum is not a universal quantity but depends considerably on the mechanical design of the pump. The extent of recirculation required for stabilization of the Q-H-curve cannot yet be quantified either.

5.5 Influence of Flow Separation and Recirculation on the Q-H-Curve

5.5.1 Types of Q-H-Curve Instability

In Chap. 4.1 has been discussed how the Q-H-curve is obtained from the theoretical blade work and the losses in the various parts of a pump, see also Fig. 5.22. According to the equations in Chap. 4.1, the Q-H-curves of impellers with outlet angles below 90° would fall steadily from q* = 0 with increasing flow rate (they would be "stable"). As discussed in Chap. 5.1, the flow is bound to separate in impeller and diffuser at some partload flow, since fully developed recirculation prevails at q* = 0 at inlet and outlet of the impeller. Flow separation can lead to instability of the Q-H-curve. We distinguish two types of instability:

Type F: The head-capacity characteristic can have a maximum at Q > 0 so that $H_o < H_{max}$. Examples are given by the impellers A, A2 or A3 in Fig. 5.27. This type of Q-H-curve falls towards shut-off; it will be called "type F instability" or "drooping curve" in the following. The lower the specific speed, the higher is the tendency of a pump towards this type of instability; above n_q = 25 to 30 such instabilities are rather an exception.

Type S: The Q-H-curve can have a saddle as shown by impeller B in Fig. 5.30. This form of instability will be called "Type S" or "saddle-type" [23]. From about $n_q > 30$, diffuser pumps are susceptible to this type of instability. However, it is also occasionally found in volute pumps of high specific speeds (e.g. [31] $n_q = 90$). With growing n_q the risk of a saddle-shaped instability increases and the flow rate where the Q-H-curve becomes instable moves ever closer to the design point. Typically this instability occurs at q* = 0.6 to 0.9; for that reason it is also sometimes called "full load instability" [27]. The Q-H-curves of axial pumps usually exhibit a saddle.

A saddle-type instability is sometimes connected with a hysteresis: with falling flow rate the sudden change of the head occurs at a lower flow rate than upon opening of the throttle valve. The Q-H-curves of pump turbines frequently have a hysteresis since the impeller is designed with large inlet cross sections (A_1 and A_{1q}) for optimization of the turbine operation. A hysteresis implies that a specific type of flow pattern locks-in over a certain range of flow rates; the flow pattern switches suddenly to a different type upon reaching a level of disequilibrium which cannot any more sustained by the lock-in effect. Examples for hysteresis are found in [31, 54] and in Fig. 6.17.

5.5.2 *Saddle-Type Instabilities*

Based on the analysis of numerous measurements, it is possible to develop simplified models for the flow processes causing instabilities in spite of the extraordinarily complex flow patterns. Some empirical facts and considerations from the preceding paragraphs form the basis for these models. These are reviewed in the following under items F1 to F8:

F1 At q* = 0 all pumps operate with fully developed recirculation at impeller inlet and outlet. The mechanisms of energy transfer are therefore different from non-separated flow processes near the best efficiency point.

F2 The high shut-off pressure of axial pumps can be interpreted as the effect of centrifugal forces. Another model considers the recirculations at inlet and outlet of an axial impeller as the result of a kind of "self-healing effect" of the flow which enables a "quasi-normal" work transfer to be effective along the streamlines evolving around the recirculation zones. Both models consider the recirculation as being responsible for the fact that the shut-off pressure can reach a multiple of the head at the best efficiency point. Centrifugal forces are a key factor in this process.

F3 Even at q* = 0, the pressure recovery in the collector reaches 10 to 30% of the shut-off pressure. Either the exchange of momentum or the maximum of the stagnation pressure of the local absolute velocity c_{2max} can be considered as the controlling mechanism. The pressure rise in the diffuser at q* = 0 increases with intensifying recirculation.

F4 At a certain deceleration ratio c_{3q}/c_2 the flow in the collector separates. The actual process depends on the velocity profile at the impeller outlet and on geometrical parameters.

F5 Recirculation occurs when both flow separation *and* a sufficiently high pressure gradient perpendicular to the main flow direction are present. This combination is a necessary and sufficient condition.

F6 At $n_q < 70$ the static pressure rise in the impeller usually contributes less to the instability than the pressure recovery in the diffuser. The curve of the static pressure rise is often stable when the overall characteristic has a saddle. In contrast, the pressure recovery in the diffuser is practically always unstable when the Q-H-curve has a saddle.

F7 The curve of the static pressure rise in the impeller is stabilized by the recirculation from the diffuser, Fig. 4.6b.

F8 A saddle, that is, a decrease in head when the flow is reduced, is a clear indication that the flow patterns in the impeller and/or diffuser have changed.

Let us now examine which kind of changes or which flow patterns can have an unfavorable effect on the stability. To this end, some hypotheses (H1 to H8) are formulated below and their plausibility is verified by analyzing various experiments.

H1 If the flow is attached during operation near the best efficiency point, but fully separates and recirculates at Q = 0, the following can be concluded: A pump should have a steadily falling characteristic, if the velocity distribution at the impeller outlet develops *continuously from the uniform profile at the best efficiency*

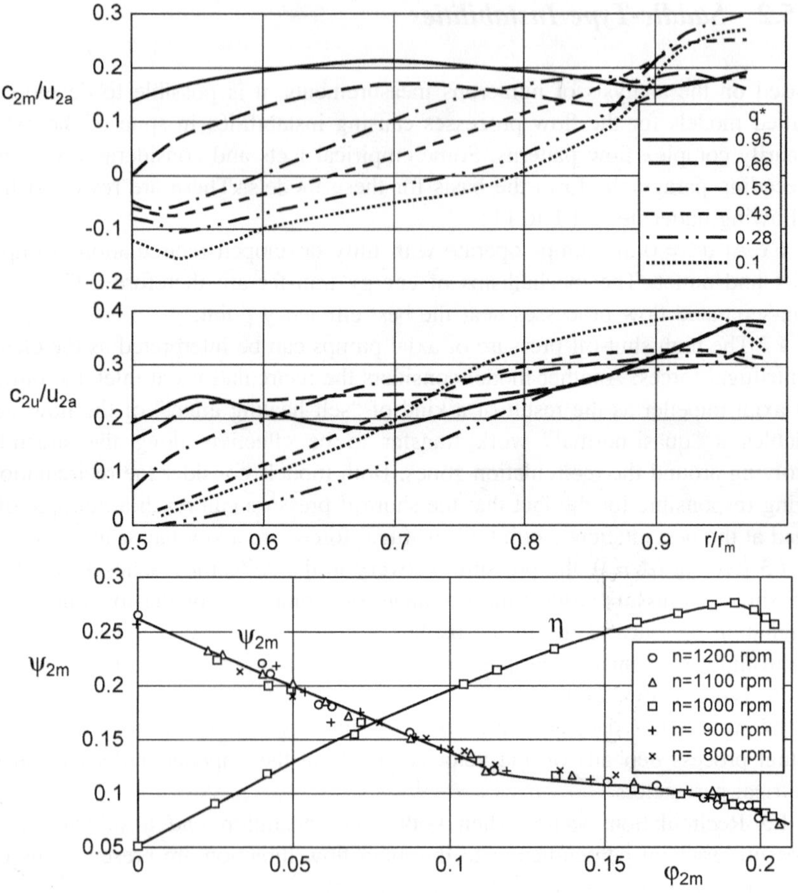

Fig. 5.37 Measurements on an axial pump $n_q = 213$ [52]

point to the asymmetrical distribution with fully developed recirculation. Sudden changes in the velocity distribution (or a "switching" of the flow pattern should neither occur at the inlet nor at the outlet nor within the impeller. *Separation and recirculation zones should increase continuously and remain at the same location at all flows.* Figure 5.37 shows an example: although the axial pump with $n_q = 213$ exhibits a distinct change in the gradient dH/dQ at the onset of inlet recirculation (shown in Fig. 4.13), there is no instability. The impeller has three blades with only about 70° wrap on the outer streamline, i.e. no blade overlap whatsoever. The recirculations at inlet and outlet are strong; once started, they increase continuously. Even above the recirculation onset, the velocity profiles at the outlet develop with a steadily increasing c_{2m}-deficit at the hub. In the zone without recirculation the flow is deflected in the direction of the hub according to Chap. 5.2.2.3; the maximum of c_{2m} is thus found in the inner half of the channel (according to Eq. 5.6 a Rossby number of 0.59 is obtained at q* = 1.04). The behavior of a semi-axial pump (n_q = 180) with a pronounced saddle in the characteristic which was reported on in [51] is quite different: the velocity profiles at the outlet develop non-uniformly and a c_{2u}-deficit on the *outer* streamline occurs in the unstable range.

H2 If the flow pattern suddenly changes upon a minor decrease in flow, a reduction of the head must be expected according to the principle of the least resistance. The more pronounced the sudden change in the flow pattern, the higher is the associated drop in head. It is frequently observed, for instance, that the recirculation zone at the impeller outlet jumps from the front shroud to the rear shroud (or vice versa). During such a change, a flow range is passed where the recirculation almost disappears. As per Figs. 5.28 and 5.30, this is associated with a drop in head. Sometimes a hysteresis is found in the Q-H-curve or in the axial thrust. The hysteresis is caused by the fact that the switch in the flow pattern occurs at different values of q* depending on whether the flow rate is reduced or increased (see also Fig. 6.17). Considering the strong influence that the location of the recirculation zone has on the fluid rotation in the impeller sidewall gap at the front shroud (see Fig. 5.30 and Chap. 5.4.3), it becomes evident that a hysteresis in switching flow patterns can exercise a major influence on the axial thrust. This type of flow pattern switching can have various causes, for instance:

(a) There can be a "separation delay": The separation was deliberately or inadvertently delayed for too long. When the separation is finally triggered, stalled fluid suddenly covers a large zone. Cavitation observations on impellers with flat pressure distributions on the suction surface show that such effects indeed occur. This is in contrasted with a "normal" flow separation which starts as a tiny separation bubble and grows continuously when the flow rate is reduced (refer to Figs. 5.1, 5.17 and 4.13).
(b) Upon the onset of the impeller *inlet* recirculation the flow distribution at the impeller *outlet* changes suddenly; for instance, because the recirculation zone jumps from the rear to the front shroud (or vice versa). This case presumably occurs quite frequently, leaving the observer with the impression that the inlet recirculation is the primary cause of the instability and therefore undesirable. The change in the flow pattern at the impeller outlet caused by inlet cavitation is an example for this mechanism; refer to the discussion of Fig. 6.17.

H3 The flow pattern switching gains major significance in semi-axial impellers and in radial impellers with high specific speeds, since the pressure recovery in the collector falls with increasing n_q and contributes little to the head. Moreover, the number of degrees of freedom for the flow grows, since the channel width increases and the guidance of the flow decreases. As mentioned above, flow pattern switching can occur especially upon the onset of impeller inlet recirculation. A possible mechanism can be postulated as follows: In Fig. 5.38 the incidences at the front shroud and on the hub are plotted against the flow rate. In the absence of recirculation, the incidence on the hub increases faster with decreasing flow than on the front shroud because the blade angles on the hub are greater (Fig. 5.14). At a critical incidence the flow *initially separates at the hub*. This can induce a

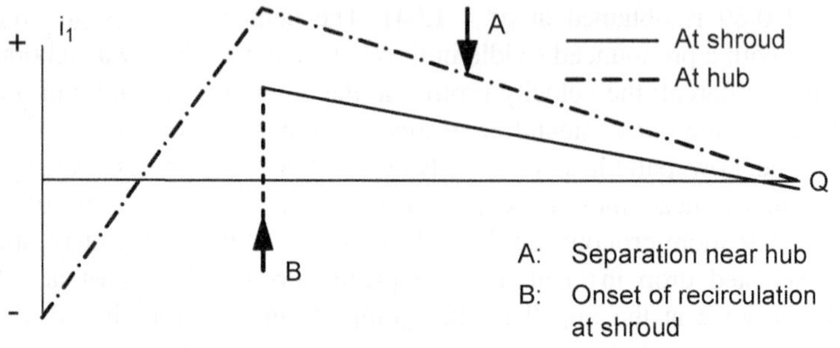

Fig. 5.38 Variation of incidence i_1 at impeller inlet

Fig. 5.39 Symmetric and asymmetrical velocity distribution (c_{2m}) at impeller outlet

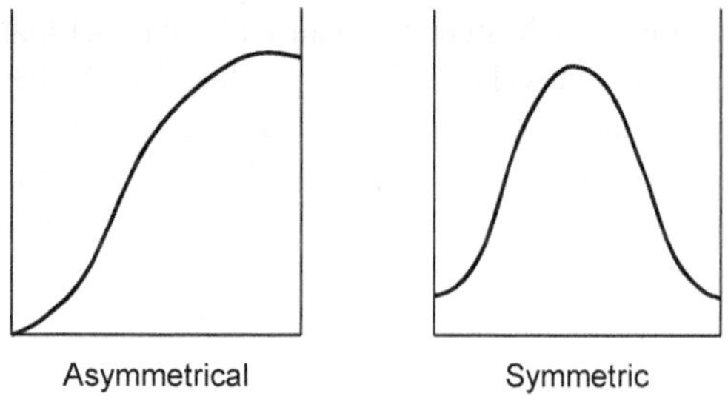

recirculation at the impeller outlet near the rear shroud because of the reduced energy transfer on the inner streamline. Recirculation has not yet developed at the impeller inlet, since significant pressure gradients perpendicular to the main flow direction have not yet build up. This happens only at a lower flow rate, when deceleration and the incidence at the outer streamline have reached critical values. Once the recirculation at the impeller inlet has sufficiently intensified, a pre-rotation is induced at the hub so that the incidence at the hub diminishes (it may even become negative at very low flow). Consequently, the separation at the hub, and hence the outlet recirculation at the rear shroud, disappear. The recirculation zone at the outlet subsequently jumps form the rear to the front shroud.

H4 If recirculation is thought to have a "self-healing effect" by blocking a part of the diffuser channel so that the remaining channel cross section operates quasi normally, asymmetric or one-sided velocity distributions at the impeller outlet are better than symmetrical distributions with zones of low energy on the rear and front shrouds, Fig. 5.39. This conclusion is also obtained from F5: the pressure gradients perpendicular to the main flow which are necessary for inducing recirculation develop more intensively when the velocity distribution is asymmetric. Reason: with symmetrical distributions only two weak recirculation zones develop and there is no well-defined pressure gradient over the channel width, Fig. 5.30.

H5 When the deceleration in the collector is not too large, symmetrical profiles are harmless (with regard to efficiency even favorable). The velocity distribution with low-energy zones on both the rear and front shroud becomes detrimental only when the general deceleration c_{3q}/c_2 has reached a threshold. This flow pattern

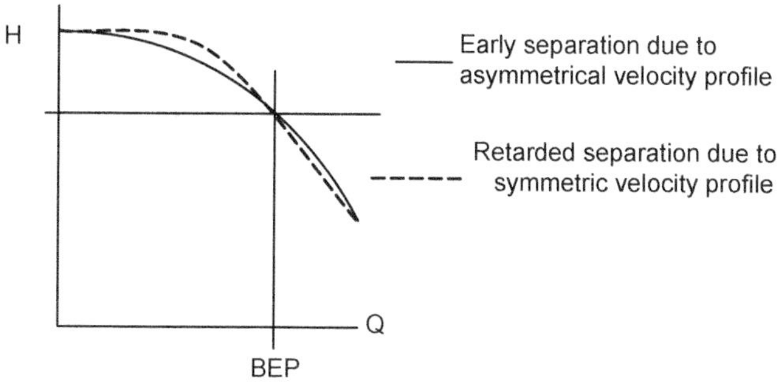

Fig. 5.40 Impact early flow separation on the shape of the Q-H-curve

leads to separation without (or with reduced) recirculation: "S*eparation without recirculation means a risk of instability*". This applies particularly to semi-axial and axial impellers whose Q-H-curve steepness is determined by the recirculation. Saddle-type instability is expected if the *recirculation* which immediately follows the separation in the impeller *is not sufficiently intense*.

H6 The flow in the collector should separate in the range of the Q-H-characteristic where the curve of the static pressure rise in the impeller is not yet flat or even unstable. Early separation in the diffuser has also the advantage that the head has not yet risen too much, the characteristic flattens at higher flow and there is still "potential" until the (limited) shut-off pressure is reached, Fig. 5.40. Retarded separation occurs more suddenly, earlier separation more gradually. As a rule, asymmetrical velocity profiles are expected to separate at higher flow than symmetrical ones.

H7 To avoid saddle formation with *semi-axial* and *axial* pumps, the impeller designed should avoid low-energy zones developing on the outer streamline near the impeller outlet. That way a switching of the flow pattern is prevented, because the recirculation occurs inevitably at the hub at low flow, Chap. 5.1. Early onset of hub recirculation helps with this endeavor since the through-flow is deflected towards the outer streamline. This increases the effect of the centrifugal forces, Figs. 5.2 and 5.12. The c_{2m}-distributions in Fig. 5.37 point in this direction - likewise the tests in [5, 16, 17] with open as well as closed semi-axial impellers. The measurements in [16, 17] show in particular that energy supply to the outer streamlines improves stability. In these tests this was accomplished by the flow through the gaps between the casing and the blades of the open impeller which, because of this effect, had a more stable characteristic than a closed impeller. The stability was also improved on a closed impeller through energy supply at the outer streamline. In a test series this was achieved by machining slots into the blades near the front shroud close to the impeller outlet. The slot length was approximately 50% of the blade length. Furthermore, the work transfer on the outer streamline of a closed impeller was improved in that the annular seal leakage was deflected axially instead of introducing it radially into the main stream (this produced thinner boundary layers on the outer streamline at the impeller inlet). Tests with a

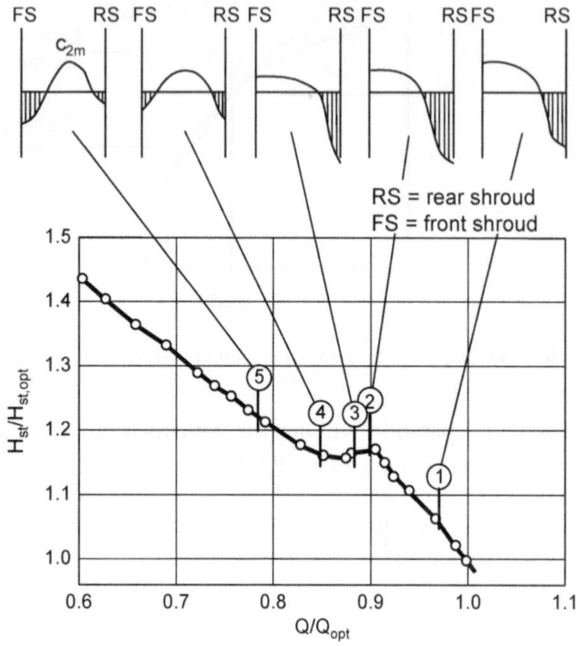

Fig. 5.41 Q-H-curve of a multistage pump with diffuser [25]

semi-axial pump (n_q = 160) with semi-open impellers confirmed the above findings: The pronounced saddle-type instability exhibited a head maximum at q* = 0.75. The dip of 10% in the Q-H-curve was strongly reduced when increasing the clearance from s_0/d_{1a} = 0.0015 to 0.0033, [58]. The effects on rotating stall and excitation forces are discussed in Chap. 10.7.2.

Tests and CFD studies of a volute pump with axial inlet (n_q = 28) were reported in [64]. The instability of the drooping Q-H-curve was much reduced by machining slots into the blades near the *hub*. The location of the slots was critical: with slots at r/r_2 = 0.75 the shut-off head was increased by about 5% and the instability almost vanished, while slots at r/r_2 = 0.86 even increased the instability with a reduction in H_0 (note that no rib in the suction nozzle was shown; thus the instability could have been induced by measuring the effect of prerotation as explained in context with Fig. 5.16).

H8 Inducers mostly have a steep, stable characteristic, Chap. 7.7. There are several reasons for this: (1) The ratio d_2/d_n is large, which means that a high shut-off pressure is obtained according to Eq. (5.11). (2) There is no collector in which the flow could separate. Often only the static pressure behind the inducer is measured; the curve of static pressure rise, however, is quite stable with most impellers according to F6. Even the integration of the total pressure usually yields a stable characteristic. With unstable Q-H-curves the deficit in the diffuser pressure recovery which occurs after separation always makes a significant contribution to the instability. (3) The blade channels of inducers are narrow, the flow is well guided; thus there is no sudden switch of the flow patterns.

The velocity profiles at the outlet of an inducer shown in [28] consequently exhibit a continuous increase of c_{2u} at the outer streamline with decreasing flow rate - quite similar as in Fig. 5.37. Towards shut-off, major pressure differentials between the outer and the inner streamlines occur downstream of the inducer. The pressure gradient across the main flow keeps the balance to possible pressure gradients perpendicular to the flow in the impeller. Consequently, the inducer suppresses any recirculation in the impeller which obviously would occur without inducer. These findings confirm indirectly that separation and pressure gradient across the flow have to act together for recirculation to occur.

As a further example to illustrate the above, Fig. 5.41 shows measurements from [25] with two features identified above as detrimental to stability: (1) the switching of the recirculation zone from the rear shroud to the front shroud and (2) symmetrical velocity profiles with low-energy zones on rear and front shrouds in the unstable range. Even at $q^* = 0.97$ a quarter of the outlet area is affected by recirculation (in other words the impeller is too wide). At $q^* = 0.9$ this zone has grown further. In the domain of the saddle at $q^* = 0.88$, it is significantly smaller than at $q^* = 0.9$. A low-energy zone has formed each on the rear and the front shroud at $q^* = 0.84$, forming a symmetrical velocity distribution with less recirculation than at $q^* = 0.9$. Finally at $q^* = 0.78$ recirculation dominates near the front shroud and the head starts to rise steeply towards shut-off.

A switch of the recirculation zone from the front to the rear shroud occurred also in the tests on a diffuser pump with unstable characteristic, [53]. Measurements on a pump turbine with unstable characteristic can also be found in [46]; prior to the onset of stall the low-energy zone was located on the front shroud and jumped to the rear shroud following flow separation.

Double-entry impellers are - apart from the manufacturing tolerances - symmetrical and tend to generate symmetrical outflow velocity profiles as long as there is no stall in the diffuser. It can be imagined that the unstable flow in this case has even more degrees of freedom than in single-entry impellers. Owing to the risk of flow pattern switching and insufficient recirculation, it is therefore difficult to achieve stable characteristics with double-entry diffuser pumps. Experience teaches that problems of unstable characteristics with this type of design are frequently quite difficult to solve [33]. The tests described in [33] become clear in the light of the models developed above: the saddle-shaped instability was reduced by: (1) narrowing the gap "A" between the front shrouds of impeller and diffuser; and (2) an oblique correction in a way that the blade diameters at the front shrouds were smaller than at the center rib of the impeller. Flow recirculation was intensified in this way. When flow separation and/or recirculation occur at the diffuser inlet, the pressure distribution at the impeller outlet changes. As a consequence, the flow through the impeller channels is affected. The resulting flow patterns cannot be symmetric anymore and the flow becomes highly unsteady. Both halves of a double-entry impeller are then likely to deliver different flow rates. This has been shown by CFD calculations in the unstable domain of the Q-H-curve [2], although experimental confirmation is not yet available.

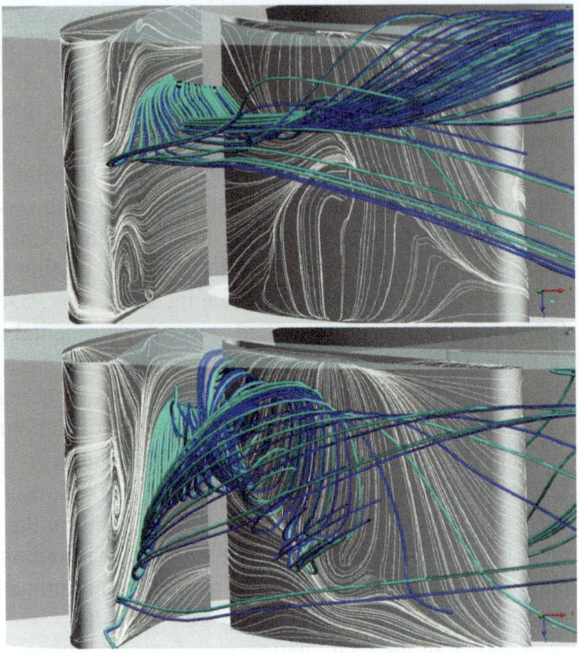

Fig. 5.42 Flow patterns in a diffuser of a pump with double-entry impeller, diffuser and volute, q* = 0.8, CFD calculations [2]

Figure 5.42 shows calculated flow patterns in the diffuser of a pump with double-entry impeller, diffuser and volute, [2]. The Q-H-curve had a hysteresis.

The view is from the volute into the diffuser channels. The flow rate ratio is q* = 0.8. The upper figure is for the upper branch of the Q-H-curve which is obtained when *decreasing* the flow rate. The lower figure is for the lower branch of the Q-H-curve which is obtained when *increasing* the flow rate (Fig. 8.20 shows the Q-H-curve). In the lower branch the flow is highly skewed leading to a strong vortex and increased losses in the diffuser. The higher losses account for the drop in head (leading to a positive slope or instability), see also Fig. 5.20. Since the vortex origins downstream of the throat area within the diffuser channel, recirculation into the impeller has not yet started at q* = 0.8. Recirculation to the impeller would tend to increase the head. In the present case, this effect becomes only into play below q* = 0.6 where the head starts to rise towards shut-off.

The stability of the Q-H-curve often reacts with surprising sensitivity to apparently minor differences in the inflow to the impeller - even with radial impellers. This shows that the velocity distribution upstream of the impeller can severely influence its outflow (if the flow is unstable), which in turn has an impact on the pressure recovery in the diffuser. *The impeller inlet and the deflection in the meridional section largely determine the* c_m*-distribution at the impeller outlet* (see Chap. 5.2).

Consequently, tests with only one stage are of little relevance for the partload behavior of multistage pumps [22] because the approach-flow from the return vanes is not modeled correctly. Multistage pumps sometimes exhibit better stability characteristics than single-stage pumps, possibly because the manufacturing tolerances in the individual stages result in slightly different flow patterns (influencing flow separation and recirculation).

Pumps with volute casings are usually less susceptible to saddle-type instabilities. This observation confirms the statement that the pressure recovery in the collector has a major influence on the stability of the Q-H-curve. During investigations into the stability of Q-H-curves it is therefore useful to measure the static pressure rise of the impeller, plotting it against the flow rate in addition to the pressure increase in the collector. Unfortunately, this fact was not taken into account in many investigations which makes their interpretation very difficult, for instance [5, 17, 25, 51]. In volute casings the pressure over the impeller circumference varies strongly at partload which creates major asymmetries in the flow. This, in accordance with the models developed above, could be responsible for the improved stability of volute casing pumps. The variation of the pressure over the impeller circumference appears to be responsible for the absence of flow pattern switching.

The above analysis shows that circumference-averaged absolute velocities at the impeller outlet are quite relevant for assessing the risk of instability. In particular the c_m-distribution plays a truly crucial role since it allows the detection of recirculation and low-energy zones. This becausc the locally transmitted energy is proportional to the product of $c_m \times c_u$, so that c_m points to such a zone even if c_u exhibits little or no deficit compared with the average velocity. Yet, as shown by the quoted measurements, c_m-deficits at partload are usually coupled with c_u-deficits.

It should well be noted that the models discussed above are simplifications which cannot take into account all details of the 3D-flow, in particular in radial impellers of high specific speeds. However, these simplified models allow to define measures of stabilization and to draw conclusions for the design (Chap. 5.6).

5.5.3 Type F Instabilities

Pumps with low specific speeds are at risk to produce Q-H-curves which exhibit a head deficit near shut-off (instability of type F according to Chap. 5.5.1). As discussed extensively above and backed by numerous measurements, recirculation at impeller inlet and outlet determines the head at low flow rates. If the shut-off head is too low, we are bound to conclude that the recirculation at the impeller inlet and/or outlet is insufficiently developed and has to be intensified in order to stabilize the Q-H-curve. The pre-swirl at the impeller inlet induced by the recirculation must also be suppressed as much as practical, Chap. 5.6.4.

Considerations similar to those in Chap. 5.5.2 apply to the separation point in the diffuser where the characteristic becomes flat. The outflow from the impeller has an effect on the recirculation also in this case. Low-energy zones prevailing simultaneously on rear and front shrouds are unfavorable while asymmetric velocity distributions generate more intensive recirculation, in this way enhancing the chances to get a stable Q-H-curve.

5.6 Means to Influence the Shape of the Q-H-Curve

5.6.1 Introduction

In principle, any characteristic can be stabilized if it is rendered sufficiently steep. This can be achieved by designing the impeller with a small relative outlet width, a small blade outlet angle, a low number of blades, large pre-rotation and/or with an adequately narrow throat area in the collector. However, with respect to the size of the pump (hence costs) it is usually necessary to select the head coefficient close to its upper limit - and this is given precisely by the Q-H-curve becoming unstable. Unfortunately, the exact location of this limit cannot be predicted. Even the qualitative understanding of the relevant mechanisms is incomplete. Ultimately, experiments are required to furnish the proof that the Q-H-curve is stable.

When a characteristic is unstable, usually a head deficit of only a few percent is involved. Therefore, any theoretical prediction must be highly accurate to be sufficiently reliable. With the complexity of separated unsteady flow and the large number of geometrical parameters that accuracy can hardly be achieved today. It is not possible either, to provide truly reliable criteria for achieving stable Q-H-curves based on a few main dimensions of the hydraulic components, even though such is sometimes suggested. As discussed above, the variety of flow patterns in the impeller depends on the shape of the meridional section and the blades as well as the velocity distribution upstream of the impeller. Furthermore, the combination of the various parameters is often more important than an individual parameter itself. Consequently, it may happen that a measure which has a stabilizing effect on pump "A" proves to be of no consequence or even destabilizing with pump "B".

Nevertheless there are a number of measures by means of which it is possible to deliberately influence the stability of the Q-H-curve. These measures agree with the above developed models and have proved themselves many times over in practice (refer also to Table 5.1). The mentioned restrictions show clearly that there cannot be any "magic formulas" to make all characteristics stable under all conditions. The higher the head coefficient at the best efficiency point is selected (this applies to all specific speeds), the more difficult it is to stabilize the characteristic.

Measures to stabilize the Q-H-curve frequently may have also undesirable side effects such as an unacceptable loss of efficiency. If the operator of the pump accepts instability because the plant operation never requires the pump to enter the unstable range, the head coefficient can be selected higher. By appropriate shaping of the hydraulic components it is then possible to achieve efficiencies higher than in cases where a Q-H-curve steadily rising to shut-off has been specified. This situation is often encountered in storage pump applications (which have to be designed for peak efficiencies owing to their high power consumption), if the operating range is narrow because of the water levels created by the topography.

5.6.2 Influencing the Onset of Recirculation at the Impeller Inlet

The onset of recirculation at the impeller inlet primarily depends on the deceleration of w_1 to w_{1q}, the pressure gradients perpendicular to the main flow and, secondarily, on the incidence. The onset of recirculation can consequently be shifted towards a smaller flow rate by:

1. Reducing the impeller inlet diameter,
2. Reducing the throat area at the impeller inlet,
3. Creating a pre-rotation, e.g. by inlet guide vanes or return vanes,
4. Increasing the blade inlet diameter d_{1i} at the hub or reducing the ratio d_{1a}/d_{1i},
5. Reducing the impeller blade inlet angle if the throat area is reduced too.

Measures 1 to 3 will reduce the deceleration at the impeller inlet while possible pressure gradients perpendicular to the main flow are reduced with measure 4. These pressure gradients are mainly governed by the distribution of the flow and geometry parameters over the blade height. However, it is difficult to specify quantitative rules of a generally applicable nature.

If the onset of recirculation is excessively shifted towards low flow there is a risk of an unstable characteristic since the intensity of recirculation at $Q = 0$ is reduced with the above measures. Given that the vast majority of pumps operate without problems even at low partload, it would be uneconomic to try to avoid operation with recirculation at any cost (see also Chap. 5.1.7).

5.6.3 Influencing the Onset of Recirculation at the Impeller Outlet

The onset of recirculation at the impeller outlet primarily depends on the deceleration c_{3q}/c_2 and on the velocity distribution at the impeller outlet. The latter is responsible for the pressure gradients perpendicular to the main flow at the diffuser inlet. Such pressure gradients must be present to initiate flow recirculation.

The separation in the collector can be shifted towards a lower flow rate using the following measures:

1. Reduction of the ratio b3/b2 (keeping the throat area constant by adapting a3).
2. Reduction of the throat area of the collector. With low specific speeds this route is only open insofar as the best efficiency point does not shift excessively to the left. With increasing specific speeds (from approximately nq > 80) the collector has less influence on the best efficiency point; a reduction in throat area would then be possible.

Table 5.1(1) Interpretation and modification of pump characteristics ψ_P = static pressure rise in impeller; H_{Le} = pressure rise in diffuser/volute

Problem/findings			**Possible cause or mechanism**	**Possible remedies**
Q-H-curve	**Internal pressures**	**Axial thrust, $\beta/\omega = f(Q)$**		Note: All modifications can produce unfavorable side effects
① Q-H-curve drooping towards Q =0 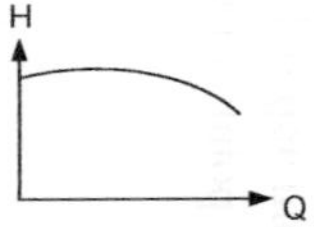 Typical for $n_q < 30$	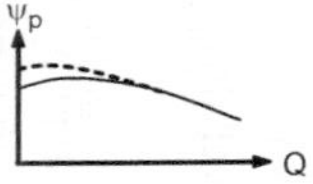 H_P, ψ_P flat towards Q = 0		Insufficient recirculation at impeller inlet (centrifugal effect too small)	Advance impeller vane leading edge near hub (reduce d_{1eff}) Advance return vane trailing edge (reduce c_{1u}) Increase impeller eye diameter Reduce hub diameter 
	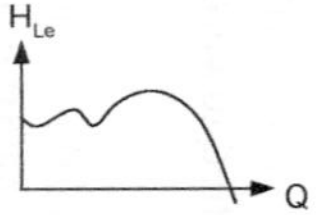 H_{Le}, flat or drooping towards Q = 0	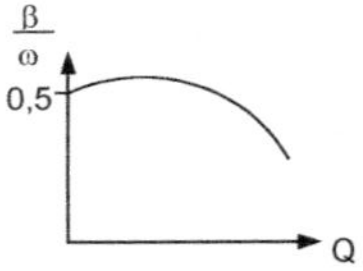	Insufficient recirculation at impeller outlet: exchange of momentum between impeller and diffuser or volute too small	Diffuser: reduce a_3 Volute: reduce A_{3q} Reduce gap A Increase δ_{TE} Increase b_2 reduce d_3* (attention: pressure pulsations) 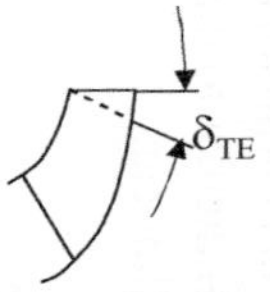
② Shut-off head and/or shut-off power too high 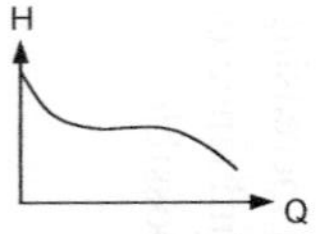 Typical for $n_q > 35$	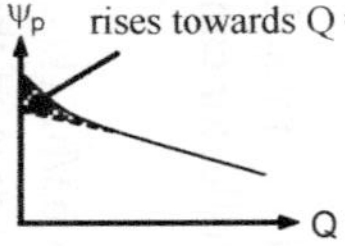 ψ_P rises towards Q = 0		Recirculation at impeller inlet too strong	Shorten return vanes Add inlet ring Reduce d_1 Increase d_{1i} or d_n 
	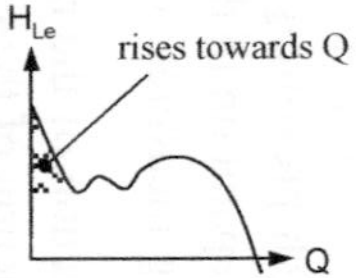 H_{Le} rises towards Q = 0	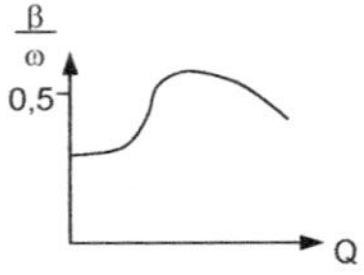 At low flow drop in fluid rotation due to recirculation	Recirculation at impeller outlet too strong	Diffuser: increase a_3 Volute: increase A_{3q} Decrease b_2 Decrease δ_{TE} Decrease b_3/b_2 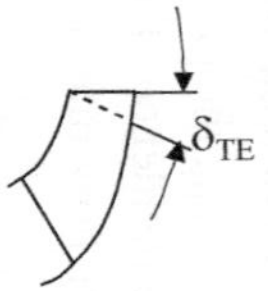

Table 5.1(2) Interpretation and modification of pump characteristics ψ_P = static pressure rise in impeller; H_{Le} = pressure rise in diffuser/volute

Problem/findings			Possible cause or mechanism	Possible remedies
Q-H-curve	**Internal pressures**	**Axial thrust, NPSH = f(Q)**		Note: All modifications can produce unfavourable side effects
③ Q-H-curve flat or with saddle 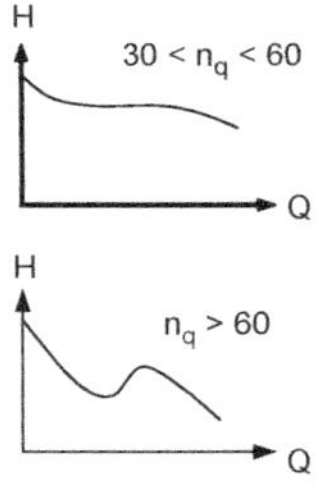	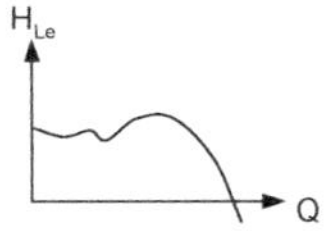	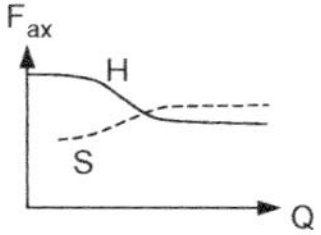 Axial thrust excursions 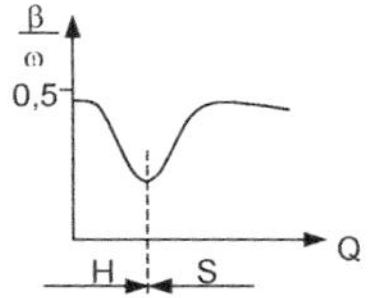 Outlet recirculation: S: at shroud; H: at hub	• Flow separation in volute or diffuser, but not yet fully developed recirculation • Back-flow zone shifts from hub to shroud or vice versa • Flow separation sensitive to manufacturing tolerances or impeller inlet conditions • Flow separation sensitive to rotor position	• Decrease b_3/b_2 • Modify meridional section • Analysis of flow pattern switching by CFD • Install different hydraulic components in different stages • Reduce gap A and increase overlap
④ Q-H-curve too steep at q* > 1 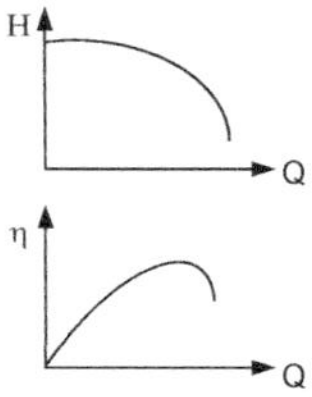	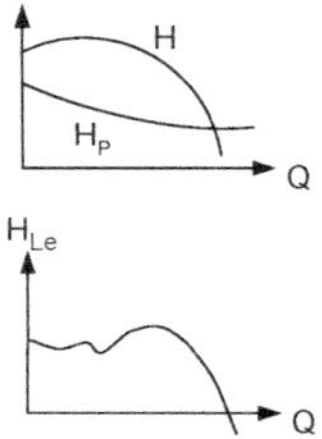	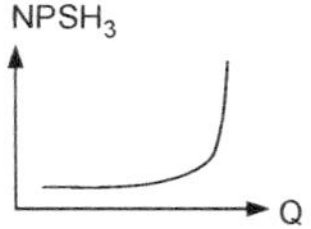 In case of diffuser or volute cavitation, the NPSH-steep rise is scarcely influenced by the impeller inlet geometry	• Cavitation in diffuser or volute • Strong acceleration of fluid from impeller outlet to diffuser or volute throat area • Throttling loss; blockage by foreign matter	• Increase casing throat area (Attention: BEP shifts to higher flow) • Remove blockage; possibly enhance surface finish

3. Reduction of the diffuser inlet angles. This measure is effective only to a limited extent if the throat area is not reduced at the same time. In the case of volute casings the influence of the cutwater angle is even less than with diffusers.

Since flow separation and recirculation at the diffuser inlet are largely controlled by the velocity distribution at the impeller outlet, the key for improving diffuser performance is often a re-design of the impeller and/or the components upstream of the impeller.

5.6.4 Eliminating a Type F Instability

To stabilize a Q-H-curve drooping towards shut-off, it is primarily necessary to intensify the effect of recirculation at the impeller inlet and outlet. In general this will also increase the power consumption at $Q = 0$. The effect of recirculation at the impeller outlet (exchange of momentum) can be intensified by:

1. Increasing the relative impeller outlet width $b_2^* = b_2/d_2$
2. Increasing the ratio b_3/b_2
3. Reducing gap A, Fig. 5.31
4. Reducing $a_3^* = a_3/d_2$ (Fig. 5.19) or reducing the volute throat area. However, this measure implies a shift of the best efficiency point and is therefore applicable only to a limited extent. To a certain degree, a reduction of a_3^* can be compensated for by an increase in b_3.
5. Reduction of the impeller/diffuser distance d_3/d_2; however, there are limits to this because of pressure pulsations and dynamic stresses (Table 10.2, Chap. 10.1.9). This measure is hardly effective with volute casings since the exchange of momentum is more influenced by the secondary flow in the volute than by the cutwater clearance.
6. Oblique impeller trailing edge, Fig. 5.33 (the effectiveness increases with b_2^*)
7. Asymmetrical velocity profile at the impeller outlet in the range where the characteristic flattens out (Chap. 5.5.2).
8. Possibly axial shift of the rotor, Fig. 5.35
9. Double volutes: design cutwaters with negative incidence in order to reduce flow separation at low partload: very low angles on low-pressure side of cutwater (side facing the impeller).

The effect of the impeller inlet recirculation can be influenced by:

1. Suppressing the pre-rotation caused by the recirculation utilizing ribs or similar structures. In multistage pumps the return vanes of the preceding stage must be arranged as close as possible to the subsequent impeller. For mechanical reasons it is necessary to maintain a minimum distance which can be selected roughly equal to the blade height at the impeller inlet.

2. In a large liquid volume upstream of the impeller the circumferential component of the recirculating fluid is largely dissipated; the effect is similar to that of the structures mentioned under item 9.
3. A diffuser installed upstream of the impeller (as in Fig. 6.39c) slows down the pre-rotation induced by recirculation less strongly than an axial approach flow. The shut-off head drops with increasing diffuser opening ratio. In the extreme one could imagine a smooth plate at a short distance upstream of the impeller: the small enclosed fluid volume is able to rotate almost unimpeded at Q = 0, causing the shut-off head to drop drastically.
4. Advancing the impeller blade leading edges into the impeller eye, especially at the hub, so that d_{1i} becomes about minimal.
5. Enlargement of the impeller inlet diameter
6. Reduction of the hub diameter

Measures 9 and 12 as well as 1 to 4 are most effective for achieving a characteristic which steadily rises towards shut-off at low specific speeds. These measures are important up to approximately $n_q = 25$.

Even though the shut-off head increases slightly with the impeller blade outlet angle and the blade number, the Q-H-curve flattens out and the risk of instability increases. The radial bladed impellers discussed in Chap. 7.3.3 are a certain exception, but their characteristics are very flat.

5.6.5 Influencing the Saddle-Type Instability of Radial Impellers with $n_q < 50$

The Q-H-curve stability of radial impellers with $n_q < 50$ is largely determined by the pressure recovery in the diffuser which, in turn, depends on the velocity distribution at the impeller outlet. Considering firstly diffuser pumps, the saddle formation of the Q-H-curve can be influenced by the following measures:

Flow separation in the diffuser can be shifted towards lower flow rates by reducing the diffuser inlet width b_3. Experience shows that this reduces the risk of saddle-type instabilities. At the same time the intensity of recirculation at Q = 0 decreases, causing the shut-off head to fall. This may be desirable in some cases but can also cause instability of type F, a waviness or flatness of the Q-H-curve or a second saddle. The recommended b_3-reduction apparently contradicts Chap. 5.5.2, item H6, but all of the measures discussed are really part of an optimization process.

1. An a_3-reduction can have different effects on the saddle; an optimum will have to be found for each pump. However, an a_3-reduction usually produces an increase of the shut-off head, a shift of the best efficiency point towards lower flow rates and a head reduction at $q^* > 1.0$, since the point with $c_{3q} = c_2$ shifts towards reduced flow (Figs. 5.18 and 5.19). The lower the specific speed, the more intensive is the reaction of the optimum efficiency position to any change

in a_3 (Chap. 4.2). In contrast, an a_3-increase shifts the separation in the diffuser towards a higher flow rate. As mentioned before, this can be favorable because the characteristic flattens out at an early stage. However, the stability of the Q-H-curve is improved only if the curve of static pressure rise produced by the impeller is sufficiently steep in this domain and if the pressure recovery in the diffuser is not unduly impaired by the early separation.

2. Shape and position of the impeller blade leading edge have a major effect on the flow in the impeller and, consequently, also on the velocity distribution at the impeller outlet. However, it is hardly possible to give general rules. In contrast to type F instabilities, it is not always favorable to advance the impeller blades on the hub as far into the eye as possible. This would produce very large inlet angles at the inner streamline and correspondingly severe changes in the incidents. These can provoke a switching of the flow patterns at the impeller outlet in accordance with the mechanism postulated when discussing Fig. 5.38.
3. In multistage pumps, geometrically different components could be designed for the individual stages. Flow separation in the stages would then occur at different q*. This design philosophy may be followed in pumps with two or three stages, if the pump concept anyway requires different patterns for the individual stages. For cost reasons this concept is scarcely viable for pumps with many stages, but it can be applied easily to the diffuser of the last stage.

Generally speaking, the measures mentioned in Chap. 5.6.4 are by no means suitable for rectifying the saddle-type instability; their effect on the saddle is varied or even counter-productive. To really understand and rectify the saddle, it is necessary to determine the velocity distribution at the impeller outlet by means of numerical calculations or even measurements. If that information is available, the impeller - and its approach flow passages - can be designed so as to avoid sudden switches in flow patterns within the impeller, at the impeller outlet or in the recirculation zones. Impeller outlet velocity profiles with low-energy zones simultaneously at rear and front shrouds are also unfavorable.

4. A sudden switch in the flow pattern can be prevented by creating a "deliberate separation point" in the impeller. This ensures that the low-energy zone and the recirculation at the impeller outlet always occur at the same location (usually on the front shroud). Since the meridional curvature tends to produce a low energy zone near the front shroud (Chap. 5.2.2.4), this location is the natural choice for a "deliberate separation point". Most of the available measurements revealed that the recirculation at Q = 0 occurs near the front shroud. For this reason, *radial* impellers should be designed so as not to have a low-energy zone or recirculation on the rear shroud at *any* flow rate (this, in contrast to semi-axial impellers, Chap. 5.6.6).
5. Following this argument a large front shroud radius, which might be expected to benefit the efficiency, may not always present the optimum shape in terms of stability. This could be one of the reasons why saddle-type instabilities are rarely found below $n_q = 25$, where the front shroud radius is quite small.

6. Symmetrical impeller outlet velocity profiles should be avoided at partload, since they imply low-energy zones simultaneously at the rear and front shroud. In this case the recirculation cannot develop sufficiently and there is a risk of a switch in the flow pattern: If the recirculation zone shifts from hub to shroud (or vice versa) upon a reduction of the flow rate, zero or very little recirculation will occur in a certain flow range; the lack of recirculation will cause a drop in head, as demonstrated by Figs. 5.28, 5.29, 5.30, 5.31 and 5.41.

Systematic investigations into these phenomena are not currently available since detailed measurements are highly elaborate and suitable numerical methods are involved and need substantial validation. When attempting to determine the true partload head, it is imperative that the numerical calculation captures correctly the interaction between impeller and diffuser (this follows from Fig. 4.6a, b). To assess the impeller design in terms of the outlet velocity profile, it may often be sufficient to calculate the impeller alone - provided the process has been validated properly. Measurements in [49] demonstrated that the outflow from an impeller of $n_q = 20$ yielded qualitatively similar velocity profiles when the pump was tested with a vaned or a vaneless diffuser. Several parameters allow to deliberately design the impeller so that asymmetric velocity profiles are obtained which continuously develop from the best efficiency point to shut-off. These include:

- Velocity profile at impeller inlet
- Optimization of the impeller blade leading edge according to position, twist, incidence distribution and shape, e.g. forward or backward sweep
- Meridional section curvature
- Boundary layer influenced by: annular seal flow introduction, openings in the front shroud at a suitable point, slots in the blade
- "Auxiliary blades" on rear or front shrouds (boundary layer fences)
- Unilaterally more intensified deceleration in the impeller (e.g. increasing a_2 at the front shroud side)
- Unilaterally intensified deceleration in the diffuser (e.g. increasing a_3 on the front shroud side)
- Rear shroud not rotational symmetric.

In pump design such types of three-dimensional flow effects have not been systematically examined nor employed to any great extent, but the exploitation of 3D-effects bears a potential for future optimization by means of numerical calculations. If such avenues are to be successful it is important to optimize the parameters in a way that the efficiency at BEP is not unduly impaired. This becomes increasingly difficult with radial impellers of high specific speeds since the hydraulic losses due to non-uniformity are dominating.

5.6.6 *Influencing the Saddle-Type Instability of Radial Impellers with $n_q > 50$*

Radial impellers with axis-parallel blade trailing edges are employed up to about $n_q = 100$. The impeller channels are very short relative to their width, separation and recirculation start at high values of q*, the flow has many degrees of freedom and is therefore in a kind of "quasi-stable equilibrium". Impeller flow and Q-H-curve can react with great sensitivity to changes in the approach flow. Furthermore, the annular seal flow injection affects the boundary layer on the outer streamline. Thus it is also possible to influence the Q-H-curve stability by choosing an open or closed impeller design.

The measures described in Chaps. 5.6.5 and 5.6.7 can be applied by analogy also for stabilizing the Q-H-curves of radial impellers with high specific speeds. The blade twist at the outlet, the distribution of the blade angles over the outlet width and the trailing edge shape need to be optimized. At high n_q the trailing edge is not necessarily executed parallel to the rotor axis.

5.6.7 *Influencing the Instability of Semi-axial and Axial Impellers*

As explained above, the outlet recirculation of axial impellers and of semi-axial impellers with sufficiently large ratio d_{2a}/d_{2i} is always located near the hub. The following measures are available for stabilizing the Q-H-curve:

1. Design the impeller so that the velocity profiles develop continuously from the best efficiency point to shut-off.
2. Avoid low-energy zones at the outer streamline as this would result in a flow pattern switch after the onset of recirculation: the recirculation displaces the through-flow to the outer streamline and the low-energy zone jumps to the hub Chap. 5.5.2, item H7, provides additional information.
3. According to Chap. 5.5.2, item H7, the recirculation should be triggered as soon as possible after the flow separates, because stalled zones displace the through-flow at the inlet and outlet so that its energy transfer is enhanced by centrifugal forces. This can be achieved by attempting to provoke a unilateral separation that does not extend over the entire channel width simultaneously. This statement is true for both impeller and diffuser.
4. As d_{2a}/d_{2i} is increased, the characteristic becomes steeper after the onset of recirculation. Although this is mostly undesirable with pumps of this type, it is necessary to optimize the ratio d_{2a}/d_{2i} in order to minimize or eliminate the formation of a saddle-type instability.
5. The ratios d_{1a}/d_{2a} and d_{1a}/d_{1i} are instrumental for optimizing the impact of flow recirculation on the risk of saddle-type instabilities.

6. Shape and position of the impeller blade leading edge have a major influence on the flow in the impeller and, consequently, on its outflow velocity profile. These parameters are very important for achieving stable characteristics for semi-axial impellers (refer also to Chap. 5.5.2, H3). The measures include sweep-back and sweep-forward.
7. The diffuser vane leading edge can have a major effect on the stability. According to investigations in [12], Q-H-curve stability, head coefficient and efficiency were improved by a design with back-swept leading edges as compared to the original diffuser whose vane leading edges were perpendicular to the rotor axis. With reference to the radial direction the sweep was 10° at the outer streamline and 30° at the hub according to Fig. 5.43 (the bow of the blade was also changed during these tests).
8. Presumably it is favorable to design the impeller so that the recirculation onset occurs at roughly the same flow rate on inlet and outlet. If the recirculation onset at the inlet occurs at a considerably lower flow rate than at the outlet, there is a risk of flow pattern switching as described in Chap. 5.5.3, H3.
9. From items 2 and 3 it can be concluded that a strong twist of the blades at the impeller outlet may be helpful for stabilizing the characteristic. It also appears to be favorable to select the solidity on the outer streamline, compared to the inner streamline, as large as possible in order to avoid a lower-energy zone at the outer streamline (the solidity is defined as the ratio of blade length to the pitch L_{sch}/t_2).
10. As demonstrated by Fig. 5.29 and other tests, the pressure recovery in the diffuser of a semi-axial pump strongly diminishes as the flow is reduced below BEP: it becomes zero or even negative at low flow. This is thought to be a generic problem in the design of diffusers for semi-axial pumps because the flow separates near the hub streamline. The flow separation is caused by the *combined effects* of: (1) incidence at part load, (2) flow deceleration, (3) and flow deflection at the inner streamline. The stalled fluid is thus found *on the suction side near the hub* where a low-pressure zone is created that way. A high-pressure zone on the pressure side of the vane near the diffuser outlet generates a secondary flow from the pressure to the suction surface. This secondary flow triggers a back-flow towards the stalled fluid in the suction side/hub corner. Detailed investigations on this flow structure were reported in [59, 60] but the phenomenon has been revealed by many CFD studies. Increasing the ratio of c_{3m}/c_{2m} by increasing the diameter of the inner streamline d_{3i} and reducing the diffuser inlet width b_3 can be attempted in order to mitigate the incidence. Applying a well-designed profile to the suction side of the diffuser vanes will help to reduce the flow separation below the BEP.

"Casing treatment" is used in the compressor industry to extend the stable operation range by shifting the onset of stall and surge to lower flow rates. In axial compressors slots of various geometries or honey combs are used for that purpose. These artifacts are machined into the casing in a way that the blades rotate *within* the treated part of the casing. One disadvantage associated with casing treatment is

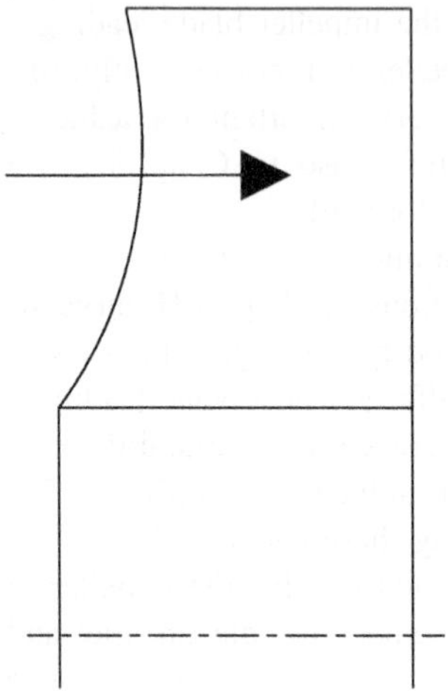

Fig. 5.43 Sweep back at the diffuser inlet of an axial compressor, [12]

a drop in efficiency. Investigations on an axial pump in [15] avoided this drawback by machining axial grooves into the ca*sing just upstream of the impell*er. In the pump tested the grooves resulted in a completely stable Q-H-curve and an improvement of the required $NPSH_3$ at partload without efficiency impairment. Details of the geometry tested are given in Chap. 5.7; they may help to design a casing treatment in new applications. Whether this type of casing treatment is effective, depends on the nature of the flow in the impeller; for example: flow separations near the hub are unlikely to be influenced by the grooves tested.

Longitudinal grooves or ribs in the suction pipe – even high roughness – reduce the pre-rotation induced by impeller inlet recirculation. Hence, the theoretical head increases according to Eq. (3.4). The tangential and axial velocity components at the impeller inlet are modified and, as a consequence, the velocity profile at the diffuser inlet changes too. In turn, this affects the pressure recovery in the diffuser which strongly depends on the approach flow. By means of optimizing the geometry of grooves ("J-grooves") the instability in the Q-H-curve of a semi-axial pump (n_q = 107) could be removed without loss in efficiency, [61]. The grooves are mainly upstream of the impeller leading edge, but about 20% of the groove length is arranged over the blades. If the grooves cover some portion of the blades, the work done by the blades tends to suffer (the effect is similar to increasing the clearance of semi-open impellers). To keep performance impairment to a minimum, the grooves must be shallow and the length $L_{g,Bl}$ over the blades must be kept short.

For the particular pump of [61], the optimum groove-geometry was converted to a dimensionless representation which better allows adaptation to different pumps, see Table 5.2 and Fig. 5.43-1.

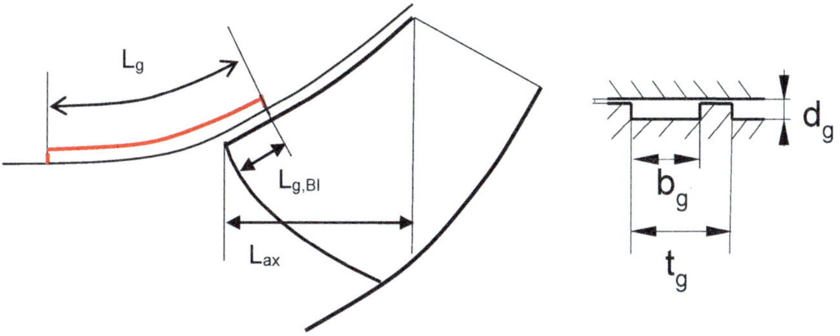

Fig. 5.43-1 J-grooves at the inlet of semi-open impellers [61]

Table 5.2 Optimum groove dimensions according to [61]

Number of grooves	z	28	Groove factor: $J_E = \frac{4\ z^2}{\pi^2}\left(\frac{b_g}{d_{1a}}\right)^3\left(\frac{L_{g;Bl}}{L_{ax}}\right)\left(\frac{L_g}{d_g}\right)$ (5.12) Optimum $J_E = 0.17$
Groove length	L_g/d_{1a}	0.256	
Groove length over blade	$L_{g,Bl}/L_g$	0.22	
Groove width	b_g/d_{1a}	0.051	
Groove depth	d_g/d_{1a}	0.01	Efficiency loss: $\Delta\eta = 3.6(J_E - 0.17)$ [%] (5.13)

According to [61], the factor J_E - defined by Eq. (5.12) - should not be lower than 0.17; otherwise, the effect of the grooves was not sufficiently strong to completely suppress the instability of the Q-H-curve. If J_E exceeds 0.17, an efficiency loss is expected from Eq. (5.13). These recommendations can scarcely be expected to be generic, but they may help in instability problems.

Longitudinal grooves have also been applied on inducers to suppress back-flow-induced cavitation which rotated at $f/f_n = 0.037$, [62]. Applying the grooves also reduced the required $NPSH_3$, particularly so, at low flow. Swirling in the draft tube of a Francis turbine (i.e. the effect of the vortex rope) can also be reduced by longitudinal grooves.

5.6.8 Reduction of Head and Power at Shut-off

Although a steep characteristic and a high shut-off pressure may be demanded in special cases, a low shut-off head with a flat, steadily rising characteristic is desirable in most applications in order to reduce the costs of the discharge pipe. This requirement is highly relevant for semi-axial and axial pumps. The shut-off power of these machines should rise as little as possible above the power at the best efficiency point in order to lessen the motor costs. In pumps with high specific speeds, it is therefore important to reduce the effect of recirculation at the impeller inlet and outlet. This can be performed using the following means:

1. Ribs and structures upstream of the impeller, which slow down the pre-rotation induced by the recirculation, can be provided at a greater distance upstream of the impeller inlet. This applies especially to high specific speeds ($n_q > 50$) because shut-off and power fall strongly when the distance between the impeller blade leading edges and the rotation-braking structures is increased. However, such structures must not be eliminated altogether - otherwise the characteristic will become unstable due to excessive pre-swirl. If the pump takes suction from an open sump, rotation-braking structures are also required to reduce the risk of air-drawing vortices as well as wall and floor vortices, Chap. 11.7.3. A detailed analysis of tests with various devices installed upstream of the impeller is reported in [4].
2. A large volume of fluid upstream of the impeller has a similar effect as ribs since the swirl velocity of the recirculating fluid is effectively dissipated in a large plenum. To reduce head and power consumption at shut-off, the fluid volume upstream of the impeller must therefore be kept as small as practicable.
3. Shut-off head and power can be reduced by designing a flatter impeller blade trailing edge leading to a smaller ratio d_{2a}/d_{2i} (see test results in Fig. 5.32).
4. The ratio d_{1a}/d_{1i} at the impeller inlet must not be chosen too large if a low shut-off pressure is to be achieved. Shut-off head and power of pumps with high specific speeds can be effectively reduced by increasing d_{1i} (possibly by cutting back the blades).

The diffuser parameters have little influence on the shut-off head of axial pumps as is evident from Fig. 5.25. As for the rest, and at low specific speeds, the measures mentioned in Chap. 5.6.4 can be applied in the inverted sense. This is a compromise between the demands for a stable characteristic, low shut-off head and power as well as a high efficiency.

5.7 Flow Phenomena in Open Axial Impellers

The flow patterns and mechanisms discussed above are present in both open and closed impellers. However, the leakage around the tip of the impeller blades exerts a significant influence on the flow patterns in open impellers and, consequently, on the part-load behavior of the pump.

Fluid flows from the blade pressure surface to the suction surface through the gaps between casing and impeller blades. The associated flow phenomena near the casing wall were visualized in tests with axial compressors and in an axial pump by oil coating images (in the case of the pump also by cavitation). The leakage through the gaps has an effect on the Q-H-curve and the generation of rotating stall. Compressors were discussed in [7, 38–41] while the axial pump was dealt with in [3, 14].

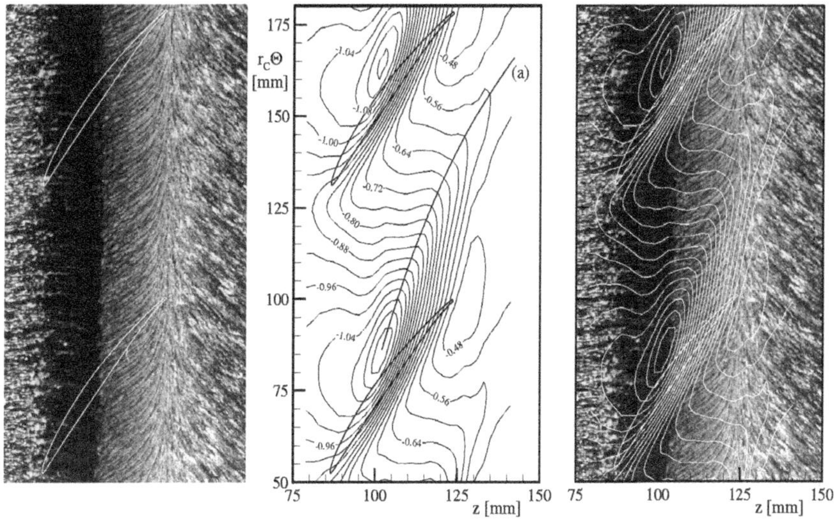

Fig. 5.44 Structures of absolute flow near the casing at q* = 1 in an axial compressor; left: oil coating image; center: measured isobars: line (a) direction of gap vortex axis; right: isobars overlaid on oil coating image, [40, 41]. Flow is from left to right, rotor moves from top to bottom of the figure

The tip leakage flow follows the path of least resistance; the velocity in the relative reference frame is therefore directed essentially perpendicular to the blade. As a consequence, the axial component of the velocity in the tip clearance is (during pumping) directed *against the main flow*. The lower the blade inlet and outlet angles, the larger is this axial velocity component.[3] In addition, the rotation of the blades creates a velocity component in circumferential direction due to shear stresses on the face end of the blades so that the flow near the casing wall follows a curved path in the absolute reference frame. By means of oil coating images of the casing wall it is possible to render the resulting streamlines visible. Oil coating images represent time-averaged streamlines of the (unsteady) absolute flow. Figure 5.44 shows such a picture; the streamlines sketched in Fig. 5.45 facilitate interpretation.

Where the main axial forward flow (left domain in Fig. 5.45) meets the tip leakage flow directed axially backwards, the casing boundary layer separates. A *separation line* is created over the circumference of the casing. In Fig. 5.44 it appears as a narrow black band which is situated nearly above the middle of the rotor at the operating point shown (refer also to Fig. 5.45). The wall shear stresses reach their minimum in this band.

[3]Inducers and helico-axial impellers for 2-phase mixtures (Fig. 13.24) have very flat angles so that the leakage flow is in essence directed axially backwards. In contrast, in radial-bladed impellers according to Chap. 7.3.3 the leakage flows in circumferential direction (if the gaps are not too large).

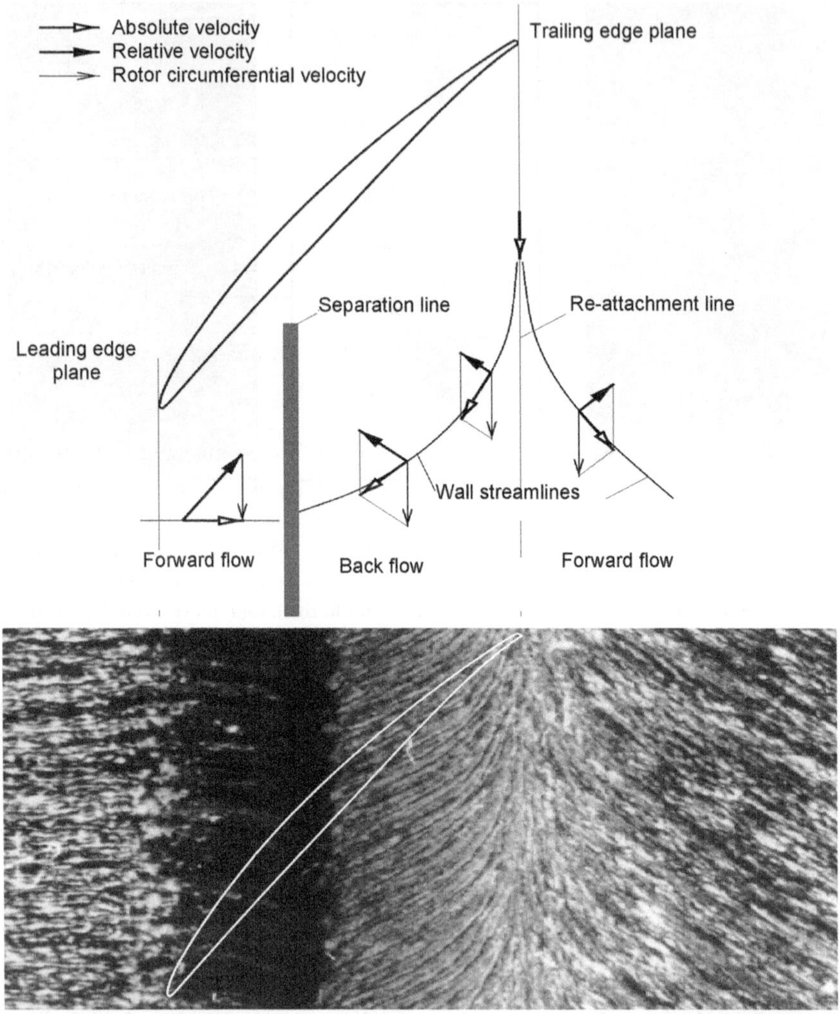

Fig. 5.45 Flow pattern on casing wall; left region: forward flow upstream and near impeller inlet; center region: back-flow within the impeller; right region: forward flow towards diffuser, [41]. Flow is from left to right, rotor moves from top to bottom of the figure

Immediately downstream of the trailing edge of the blade the flow re-attaches to the casing wall. The *re-attachment line* is noticeable in Fig. 5.44 as a white line over the circumference of the casing. Since the effects of tip leakage vanish downstream of the blading, the re-attachment line is frequently observed immediately behind the rotor. Depending on blade tip geometry, blade outlet angle and the load distribution on the outer streamline, the re-attachment line can also form upstream of the rotor outlet plane, [14, 38].

The separation line changes its position with the flow rate: to the right of the best efficiency point ($q^* > 1$) it is found near the impeller outlet and shifts *upstream* with diminishing flow rate. As the flow is reduced, the incidence at the blade leading edge increases; this causes the flow deflection and the pressure differential between pressure and suction surfaces to increase on the first portion of the blade. Hence, the tip leakage velocity grows and shifts the separation line.

In summary: the higher the blade loading near the inlet, the closer the *separation line* moves towards the inlet plane of the rotor. The greater the load near the outlet, the closer the *re-attachment line* moves to the rotor outlet plane. The blade loading near inlet and outlet increases with falling flow rate.

Figure 5.44 also shows the pressure distribution measured on the casing wall. Line "a" indicates a pressure trough along which the vortex induced by the tip leakage is moving. The tip clearance vortex starts at the point of minimum pressure on the separation line, developing downstream along a slightly curved axis. At low partload, just before the onset of stall, the vortex changes in appearance; it forms a spiraling pattern called "vortex breakdown", [14, 15]. In the process the tip leakage flow becomes increasingly unsteady. Figure 5.46 shows the cavitating vortex core in the impeller of an axial pump before and after the vortex breakdown for various flow rates.

Just prior to the onset of inlet recirculation the separation line is found at the blade leading edge. If the flow rate is reduced slightly more, rotating stall occurs in an axial compressor, while inlet recirculation starts in an axial pump. The Q-H-curve usually becomes unstable at that flow rate. The separation line jumps upstream (Fig. 5.47c) and the zone of the inlet recirculation expands in radial and axial direction. As a result of the high circumferential velocity of the recirculating fluid the head drops, mostly causing a saddle in the Q-H-curve, (see also Chap. 8.4).

Figure 5.47 shows oil coating images of the impeller blade suction surface of an axial pump shortly after the onset of inlet recirculation. At the inlet, a recirculation zone "B" can be recognized. A larger recirculation zone "A" is seen where the streamlines are directed radially outward (see Sketch c). These pictures agree well with the flow structures sketched in Fig. 5.12.

As discussed in Chaps. 5.1 and 5.2, recirculation at the impeller inlet is primarily determined by the equilibrium of forces over the blade height. Despite the influence of the tip leakage flow, this is also true for open axial impellers. Therefore it must *not* be concluded from the tests described above that the Q-H-curve could be stabilized by applying a cover band to the rotor.

In the domain with extended recirculation at the impeller inlet and outlet a vortex is formed which runs *across the blade channel*. The cross-channel vortex rotates with the blades; its axis is perpendicular to the blade. The cross-channel vortex originates on the blade suction surface near the outlet on the outer streamline and extends to just before the leading edge of the following blade. The vortex origin is clearly visible on the oil coating image in Fig. 5.48 ($q^* = 0.33$); when the $NPSH_A$ is lowered, the vortex cavitates, Fig. 5.48b.

Partload flow becomes highly three-dimensional in the diffuser too. This is well demonstrated by oil coating images of the casing wall and a diffuser vane suction

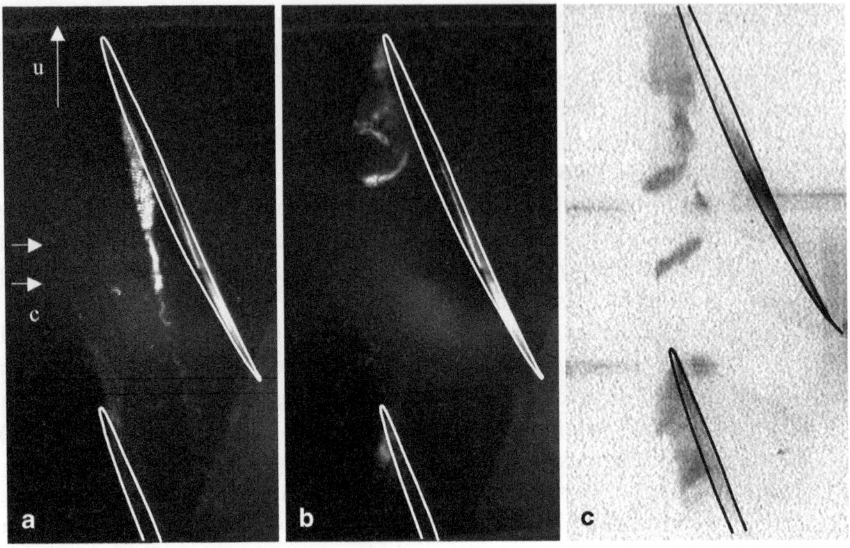

Fig. 5.46 Cavitating tip clearance vortex in an axial pump; **a** prior to vortex breakdown at q* = 1.3; **b** vortex breakdown at q* = 0.8; **c** vortex breakdown at q* = 0.75 (instantaneous, photo during transient operation), [14]. Flow is from left to right, rotor moves from bottom to top of the figure

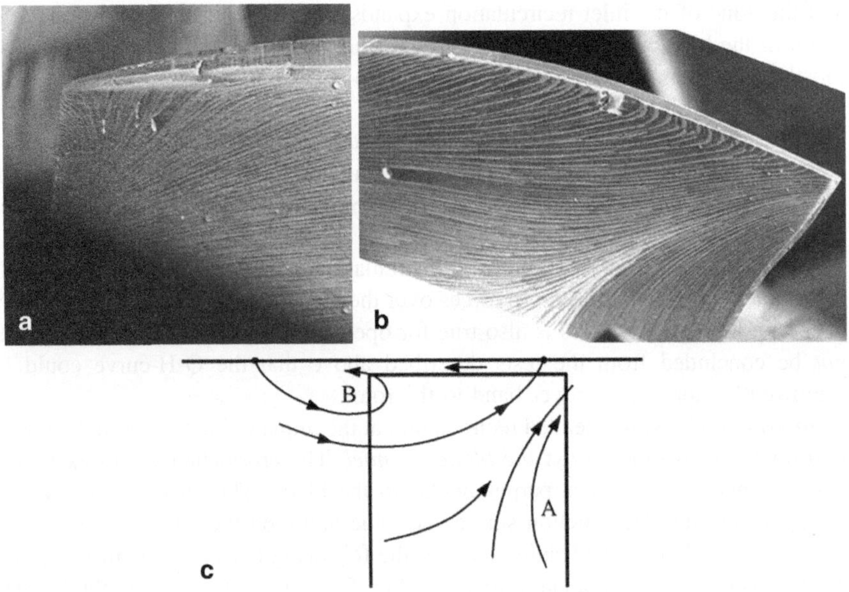

Fig. 5.47 Streamlines at the blade suction side of an axial pump at q* = 0.7; **a** at impeller inlet, **b** at impeller outlet, **c** sketch of flow pattern, [14]. Flow is from left to right of the figure

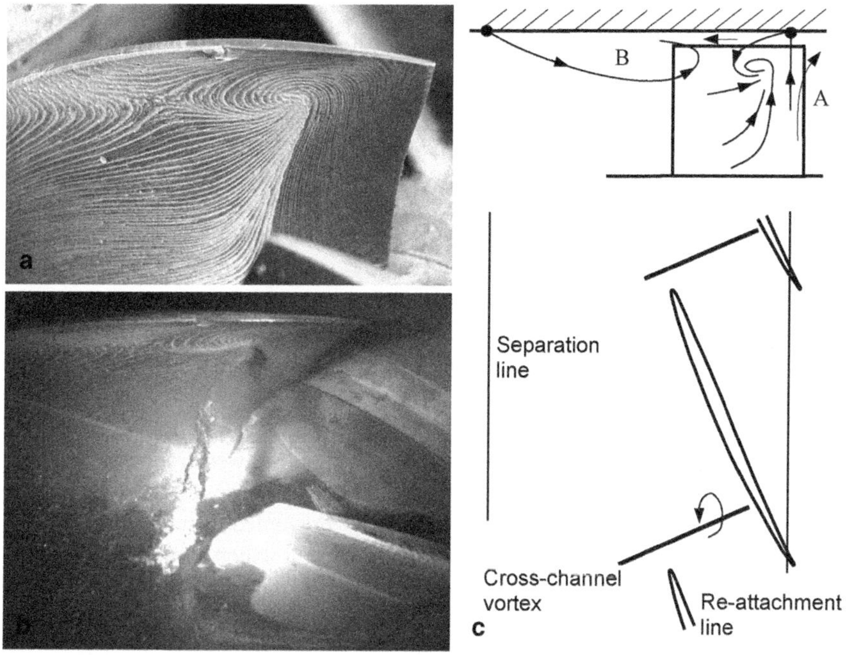

Fig. 5.48 Flow near the impeller outlet of an axial pump at q* = 0.33; **a** streamlines on the blade suction side; **b** cavitating vortex across the channel; **c** sketch of flow pattern, [14]. Flow is from left to right of the figure

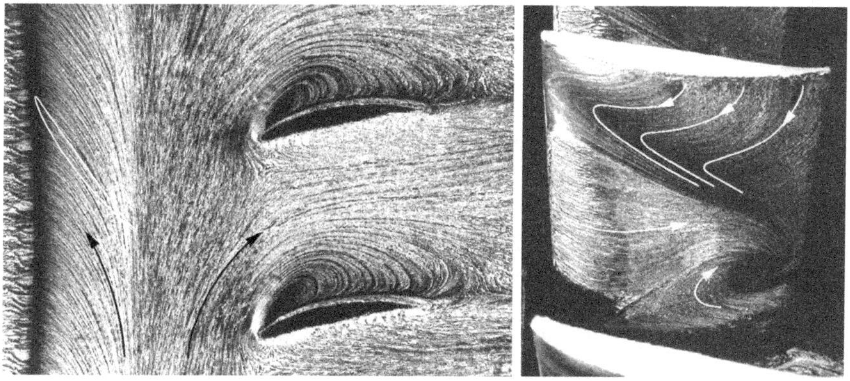

Fig. 5.49 Flow patterns in the diffuser of an axial compressor just prior to the onset of rotating stall; left: flow on the casing wall; right: vane suction side (the hub is at the bottom), [39]. Flow is from left to right, rotor moves from bottom to top of the figure

surface taken just before the onset of rotating stall in a single-stage compressor, Fig. 5.49. Because of the large incidence caused by the strongly developed tip leakage flow on the rotor, the flow in the diffuser separates near the casing on the suction surface. The outer channel half is largely blocked by fluid of low velocity which is absorbed at the outlet of the diffuser channel by a corner stall created near the hub.

The investigations in [3, 7, 14, 38–41] were made on an axial pump and various compressor bladings. The latter were tested both in a machine and in a wind tunnel on stationary vane cascades. The flow patterns observed were similar in all of the investigations. They are thus characteristic for axial pumps (and compressors). Similar flow patterns must be expected in semi-axial pumps of high specific speeds with open impellers. This was confirmed by unsteady CFD-calculations of a semi-axial pump (n_q = 155) examined in [36].

The highly three-dimensional flow evident from the oil coating images in Figs. 5.46, 5.47, 5.48 and 5.49 gives an impression of the mechanisms that generate vibrations and noise in pumps.

Further details to the above tests with the axial pump can be found in [15]. The test pump had an unstable Q-H-curve where head and power dropped suddenly at q* = 0.73 upon the onset of inlet recirculation, Fig. 5.50. The curve $NPSH_3$ = f(q*) had a peak at q* = 0.75, where high vibrations prevented normal operation. The available $NPSH_A$ had an influence on the onset of stall which shifted to higher flow rates when the $NPSH_A$ was lowered, Fig. 5.51.

The Q-H-curve became completely stable and the NPSH-peak as well as the excessive vibrations at partload vanished when grooves were machined into the casing upstream of the impeller. The effect of the grooves is to reduce the tangential velocity of the recirculating fluid by exchange of momentum between the swirling flow upstream of the impeller and the liquid in the grooves. The effect is similar to that of the back-flow catcher shown in Fig. 6.39d.

Because of this remarkable improvement, details of the pump and the casing treatment are given below, [15]:

- Pump data: n_q = 150, d_2 = 350 mm, Q_{opt} = 1780 m^3/h at 1485 rpm, η_{opt} = 0.78 (the efficiency is quite low when compared to Fig. 3.27, possibly due to a large distance between impeller and diffuser vanes), ψ_{opt} = 0.35, z_{La} = 6, z_{Le} = 11, d_n/d_2 = 0.535.
- Casing treatment: axial grooves machined into the casing just *upstream* of the impeller (no grooves overlapping the blades).
- Casing diameter in the grooved part: d_s/d_2 = 0.973. It is important that the casing diameter in the grooved part is *smaller* than the impeller diameter. Otherwise the grooves are not sufficiently effective to reduce the pre-swirl caused by the recirculating fluid.
- Groove length L_g/d_2 = 0.515. The minimum length investigated was $L_{g,min}/d_2$ = 0.15, but the Q-H-curve had a small instability close to shut-off.
- The groove dimensions were: pitch p_g/d_2 = 0.051, depth t_g/d_2 = 0.0114, width b_g/d_2 = 0.0286, width to pitch ratio b_g/p_g = 0.562, depth to width ratio t_g/b_g = 0.4.

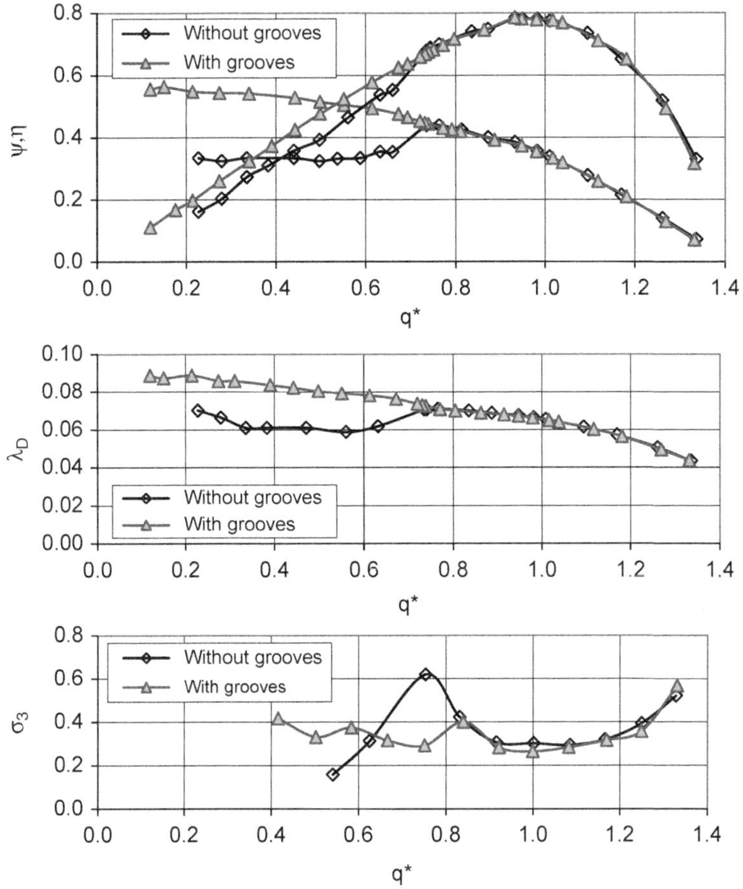

Fig. 5.50 Effect of grooves upstream of impeller on characteristics, [15]

Fig. 5.51 Onset of stall as function of available $NPSH_A$, [15]

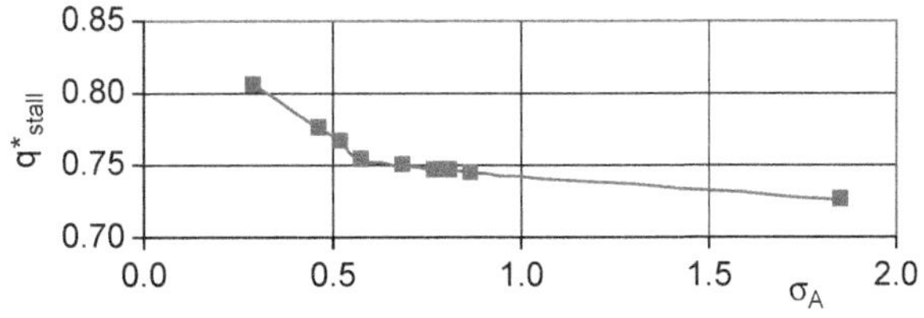

- At flows above the onset of inlet recirculation the grooves have virtually no influence on head, power, efficiency and $NPSH_3$. Partload efficiency is improved by the grooves.
- When the pre-swirl is reduced by the grooves, the velocity distribution at the impeller outlet is modified too. This change in flow pattern is instrumental in

making the Q-H-curve stable. Refer also to Chap. 6.2.5 and the discussion of Fig. 6.17.

- With grooves no cross-channel vortex is observed.
- The high tangential velocity of the recirculating fluid induces cavitation at the grooves. There is thus a risk of cavitation erosion.
- The grooves reduced vibrations near the peak in the $NPSH_3$, although not at all flow rates. However, noise increased at partload by roughly 10 dBA.

The grooves can be expected to improve the characteristics and vibration behavior if the instability is caused by inlet recirculation. If the instabilities occur at the impeller outlet, the effect of the grooves (if any) cannot be predicted.

5.8 Flow Instabilities in Double-Entry Impellers and Double Volutes

Double-entry impellers: Experience suggests that it is sometimes particularly difficult to achieve stable Q-H-curves for single-stage pumps with double-entry impellers and vaned diffusers - and even with double-volutes. The flow mechanisms responsible for the flow instabilities leading to this problem have not yet been fully demonstrated. Therefore, the goal of the following discussion is to provide some working hypotheses which seem able to explain some of the observed test results.

Double-entry impellers are cast in a horizontal position; this may lead to geometric differences between both of the impeller sides. As found when checking the dimensions of double-entry impellers, the inlet angles β_{1B} and the throat areas A_{1q} are often quite different on both sides of the impeller due to core shift caused by buoyancy. Consequently, both impeller sides provide a slightly different pressure rise. This is depicted in Fig. 5.52 where the static pressure rise H_p is plotted against the flow rate. Uneven or unsteady approach flow provides another source for creating small differences in the characteristics of both impeller sides.

Due to geometric tolerances and fluctuations in the approach flow, parallel channels will not stall simultaneously at exactly the same flow rate. By the same token, this applies to both halves of a double-entry impeller. During CFD studies in [2] it was found that the flow rate to both impellers halves differed by about $\pm 10\%$ of the net flow through the pump when the pump operated in the unstable range of the Q-H-curve, which had a hysteresis (Chap. 8, Fig. 8.20).

In Fig. 5.52 are plotted: (1) the nominal or average rise in static head H_p; (2) the curve LHH of the impeller side which provides a head lower than average (e.g. smaller b_2); (3) the curve HHH of the side with the higher H_p. The counter pressure at the impeller outlet, which is imposed by the system characteristic, intersects the curve LHH at a flow rate lower than average and the curve HHH at a higher flow. The static pressure might be expected to be fairly constant over the impeller outlet width because there is little flow-path curvature in the *axial* direction. Hence the "system characteristic" seen by both of the impeller halves would be more or less

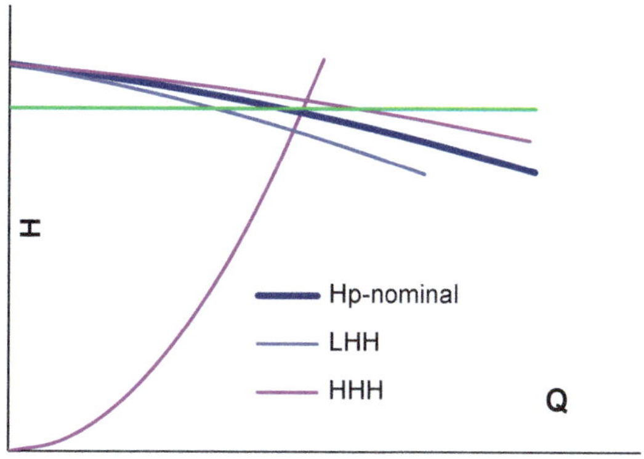

Fig. 5.52 Head curves of a double-entry impeller; LHH is for the impeller side with lower inlet angle; HHH corresponds to the side with inlet angle higher than nominal (or average)

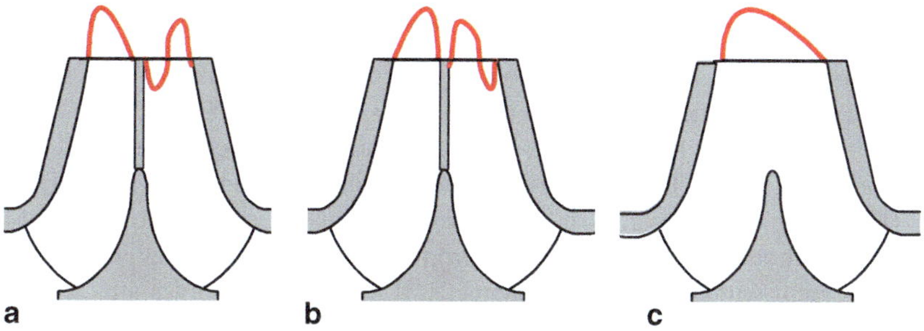

Fig. 5.53 Different velocity profiles at the exit of double-entry impellers

horizontal. The largest differences in flow rate occur where the Q-H_p-curve is flat. If the curve is slightly unstable, the operation becomes *dynamically instable* as explained in Chap. 11.3. Increased pressure pulsations, vibrations and axial thrust reversal are likely consequences. The flatter the Q-H_p-curve or the higher the instability, the higher flow fluctuations and excitation forces are to be expected.

Flow recirculations at the impeller inlet and outlet on the flatter curve HHH are likely to occur at higher q* than on the steeper curve LHH. When flow recirculation occurs on the impeller side with the higher flow, this side provides less flow to the discharge nozzle. In order to maintain the flow required by the system, the side LHH must compensate this deficit. This process is probable unstable in that through-flow fluctuations between both impeller halves are induced.

The differences in flow patterns between the two impeller sides make the approach flow to the diffuser or the volute more non-uniform. This will impair the pressure recovery in the stator. On the side with the lower pressure recovery the tendency to stall and flow recirculation will increase and provide a kind of positive feedback which amplifies the original perturbation, compare also to Fig. 5.20.

The effects of the geometric differences between the two impeller sides on the stability of the Q-H-curve and on hydraulic excitation forces are expected to depend

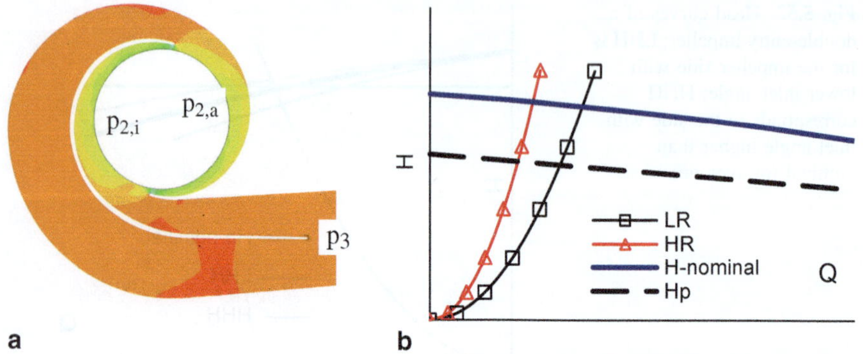

Fig. 5.54 Flow through a double-volute; HR is the channel with the higher resistance, channel LR has a resistance lower than average

on the flow patterns at the impeller outlet, because the recirculation may be either near one of the shrouds or near the centre rib as sketched in Fig. 5.53.

Double-entry impellers for high-head applications are mostly designed with a center rib, Fig. 5.53a. Pumps with lower heads can be found often without center rib (Fig. 5.53c) - depending on the specific speed i.e. the outlet width of the impeller. Without the center rib the flow from both sides of the impeller can mix out to some extent within the impeller channels. If the velocity profiles are of the type shown in Fig. 5.53a, the mixing helps to stabilize the flow. In this case the impeller without center rib would tend to yield the more stable Q-H-curve. However, if the velocity profile is of the type in Fig. 5.53b, the center rib would have little or no effect on the stability. This mechanism explains why it is sometimes observed that the center rib impairs the stability of the Q-H-curve, while it has no effect in other cases.

The flow patterns at part load are prone to being unstable which contributes to the hydraulic excitation forces. Depending on the actual flow pattern, it is sometimes possible to stabilize the flow by trimming. The impeller trailing edge can be trimmed in various ways: (1) both sides symmetric; (2) one impeller side only; (3) an oblique cut can be made either with the largest diameter near the shroud or near the centre rib. How the Q-H-curve reacts to the trimming depends largely on the velocity distribution at the impeller outlet as can be inferred from Fig. 5.53. The one-sided cavitation erosion on the impeller trailing edge of a double-entry impeller shown in Fig. 6.40 testifies to the uneven flow distribution at the impeller outlet.

Double volutes: Since the geometries of the inner and outer channel of a double-volute are different, the flow patterns through both channels are bound to be different too. The flow rates through these channels are given by their individual "system" characteristics, Fig. 5.54b. Both channels form parallel resistances (see Table 1.5) between the static pressure p_2 (H_p) at the impeller outlet and a given pressure p_3 in the discharge nozzles where both channels meet. However, the pressure p_2 is not constant around the impeller circumference: $p_{2a} \neq p_{2i}$; it adapts to the flow rates which establish themselves according to the resistance of the channels, see Chap. 10.7.1.

Since the inner and outer channels have different shapes, flow separation cannot be expected to start at exactly the same flow rate. When the flow stalls in one channel, the resistance of said channel increases. As a consequence, some fluid is displaced into the second channel (similar to rotating stall according to Chap. 10.7.2) in order to maintain the total flow through the pump as given by the system characteristic. This process may be instable and generate hydraulic excitation forces by alternating stall, Chap. 10.7.3. At operation near shut-off there could be forward flow through one channel and back-flow through the other. That the flow patterns in the two channels of a double volute are not identical at all flows, is well demonstrated by the distinctly different erosion patterns in the impeller side rooms of a double volute pump reproduced in Fig. 14.35.

The combination of double volutes with double-entry impellers compounds the problem because there are 4 intersection of the two Q-H-curves HHH and LHH with the two system characteristics.

It may be hypothesized that double-volutes with wrap angles less than 180° may create less excitation forces than with a wrap of 180° since some additional asymmetry is introduced. A wrap angle of 165° may more suitable than 180°.

As shown in Chap. 9.3.7, size and direction of the radial forces acting on an impeller depend on the circumferential asymmetry of the approach flow through the inlet casing. This applies to all pumps with between-bearings impellers. In such cases size and direction of the radial forces depend on the relative position of the discharge to the suction nozzle as well as on the flow rate.

5.9 Concluding Remarks

The 3D-flow in a pump is very complex. It is, therefore, not yet possible to accurately predict the efficiency, the stability of the Q-H-curve, shut-off head and power, the required NPSH and the hydraulic excitation forces. There is even no clue as to a universal method how to design a diffuser pump with a stable Q-H-curve - were it otherwise, the costly test beds of the pump manufactures would largely have vanished. This assessment of the state of the art may sound disappointing in view of the vast data bases and sound experience of major pump manufactures and institutions and the progress made in CFD analysis. Reasons and comments:

- The 3D-flow (hence performance) depends on the detailed geometric features of the pump from the suction to the discharge flange.
- There is a strong interaction of the main flow with the flow in the impeller side rooms including the leakages.
- The pressure recovery in the diffuser strongly depends on the velocity distribution at the impeller outlet which, in turn, is influenced by the approach flow (inlet casing).

- The combined effect of incidence at the leading edge and flow deceleration/acceleration in the throat of impeller, diffuser or volute is responsible for flow separation and recirculation. The incidence effect can be effectively mitigated by profiling the impeller blades, the volute cutwater or the diffuser vanes on the suction side.
- Therefore, performance (notably stability) cannot be described by any single geometric feature or group of parameters although many attempts to do so have been made (without generic success).
- While the details of the flow, and their impact on performance, cannot be exactly calculated, often their tendency can be predicted well enough to define design strategies and corrective action where required.
- There are features which are common to "all" pumps: recirculation and associated vortices, choking of the flow in the diffuser or volute throat at high flow, reducing rotor/stator interaction by increasing the distance between impeller blade and diffuser vanes or volute cutwater.
- While the onset of impeller inlet recirculation often is related to an instability of the Q-H-cure, fully developed recirculation at shut-off and low flow increases the head and the steepness of the curve at low flow.
- Ribs or grooves reduce the pre-rotation at the inlet at low flow and thus increase the theoretical head. Again, the entire flow history, including diffuser flow) is influenced by such structures.

References

1. Agrawal, D.P., et al.: Effect of inlet velocity distribution on the vaned radial diffuser performance, pp. 71–75. ASME Fluid Mach. Forum, Portland (1991)
2. Braun, O.: Part load flow in radial centrifugal pumps. Dissertation, EPF Lausanne (2009)
3. Bross, S., Brodersen, S., Saathoff, H., Stark, U.: Experimental and theoretical investigation of the tip clearance flow in an axial flow pump. In: 2nd European Conference on Turbomachinery, Antwerpen, pp. 357–364 (1997)
4. Canavelis, R., Lapray, J.F.: Effect of suction duct design on the performance of mixed flow pump. IMechE Paper, C333/88 (1988)
5. Carey, C., et al.: Studies of the flow of air in a rotor model mixed-flow pump by Laser/Doppler anemometry. NEL-Reports 698 (1985), 699 (1985), 707 (1987)
6. Cooper, P., et al.: Minimum continuous stable flow in centrifugal pumps. In: Proceedings of Symposium on Power Plant Pumps, New Orleans, 1987, EPRI CS-5857 (1988)
7. Dobat, A., Saathoff, H., Wulff, D.: Experimentelle Untersuchungen zur Entstehung von rotating stall in Axialventilatoren. VDI-Bericht. **1591**, 345–360 (2001)
8. Dobener, E.: Über den Strömungswiderstand in einem rotierenden Kanal. Dissertation, TH Darmstadt (1959)
9. Eckardt, D.: Detailed flow investigations within a high-speed centrifugal compressor impeller. ASME J. Fluids. Eng. **98**, 390–402 (1976)
10. Eckardt, D.: Flow field analysis of radial and backswept centrifugal compressor impellers. In: 25th International Gas Turbine Conference on ASME, New Orleans, pp. 77–86 (1980)
11. Fraser, W.H.: Recirculation in centrifugal pumps. In: ASME Winter Annual Meeting, Washington DC, pp. 65–86 (1981)

12. Friedrichs, J., et al.: Effect of stator design on stator boundary layer flow in a highly loaded single-stage axial-flow low-speed compressor. ASME J. Turbomach. **123**, 483–489 (2001)
13. Friedrichs, J.: Auswirkungen instationärer Kavitationsformen auf Förderhöhenabfall und Kennlinieninstabilität von Kreiselpumpen. Diss TU Braunschweig, Mitt des Pfleiderer-Instituts für Strömungsmaschinen, Heft 9. Verlag Faragallah (2003)
14. Goltz, I., Kosyna, G., Stark, U., Saathoff, H., Bross, S.: Stall inception phenomena in a single-stage axial pump. In: 5th European Conference on Turbomachinery, Prague (2003)
15. Goltz, I.: Entstehung und Unterdrückung der Kennlinieninstabilität einer Axialpumpe. Diss TU Braunschweig, Mitt des Pfleiderer-Instituts für Strömungsmaschinen, Heft 10. Verlag Faragallah (2006)
16. Goto, A.: Study of internal flows in a mixed-flow pump impeller at various tip clearances using 3D viscous flow computations. ASME Paper 90-GT-36 (1990)
17. Goto, A.: The Effect of tip leakage flow on partload performance of a mixed-flow pump impeller. ASME Paper 91-GT-84 (1991)
18. Gülich, J.F., et al.: Influence of flow between impeller and casing on partload performance of centrifugal pumps. ASME FED. **81**, 227–235 (1989)
19. Gülich, J.F., et al.: Rotor dynamic and thermal deformation tests of high-speed boiler feedpumps. EPRI Report GS-7405 (1991, July)
20. Gülich, J.F.: Bemerkungen zur Kennlinienstabilität von Kreiselpumpen. Pumpentagung Karlsruhe, B3 (1988)
21. Gülich, J.F.: Impact of 3D-Flow Phenomena on the Design of rotodynamic Pumps. IMechE. **213**(C1), 59–70 (1999)
22. Gülich, J.F.: Influence of interaction of different components on hydraulic pump performance and cavitation. In: Proceedings of Symposium on Power Plant Pumps, New Orleans, 1987, EPRI CS-5857 2.75–2.96 (1988)
23. Gülich, J.F.: Untersuchungen zur sattelförmigen Kennlinien-Instabilität von Kreiselpumpen. Forsch. Ing. Wes. **61**(4), 93–105 (1995)
24. Hergt, P., et al.: Fluid dynamics of slurry pump impellers. In: 8th International Conference on Transport and Sedimentation of Solids, Prague, D2–1 (1995)
25. Hergt, P., Jaberg, H.: Die Abströmung von Radiallaufrädern bei Teillast und ihr Zusammenhang mit der Volllastinstabilität. KSB Techn. Ber. **26**, 29–38 (1990)
26. Hergt, P., Prager, S.: Influence of different parameters on the disc friction losses of a centrifugal pump. In: Conference on Hydraulic Machinery, Budapest, pp. 172–179 (1991)
27. Hergt, P., Starke, J.: Flow patterns causing instabilities in the performance curves of centrifugal pumps with vaned diffusers. In: 2th International Pump Symposia, Houston, pp. 67–75 (1985)
28. Hergt, P.: Ergebnisse von experimentellen Untersuchungen des Förderverhaltens eines Inducers. Pumpentagung Karlsruhe, B 5–01 (1992)
29. Hunziker, E.: Einfluß der Diffusorgeometrie auf die Instabilitätsgrenze eines Radialverdichters. Dissertation, ETH Zürich (1993)
30. Inoue, M., Cumpsty, N.A.: Experimental study of centrifugal impeller discharge flow in vaneless and vaned diffusers. ASME J. Eng. Gas Turbines Power. **106**, 455–467 (1984)
31. Kaupert, K.A.: Unsteady flow fields in a high specific speed centrifugal impeller. Dissertation, ETH Zürich (1997)
32. Lakshminarayana, B.: Fluid dynamics of inducers—a review. ASME J. Fluids Eng. **104**, 411–427 (1982)
33. Martin, R., et al.: Partload operation of the boiler feedpumps for the new French PWR 1400 MW nuclear plants. ImechE Paper C344/88 (1988)
34. Meschkat, S.: Experimentelle Untersuchung der Auswirkung instationärer Rotor-Stator-Wechselwirkungen auf das Betriebsverhalten einer Spiralgehäusepumpe. Dissertation, TU Darmstadt (2004)
35. Moore, J.: A wake and an Eddy in a rotating radial flow passage. ASME J. Eng. Power **95**, 205–219 (1973)

36. Muggli, F., Holbein, P., Dupont, P.: CFD calculation of a mixed flow pump characteristic from shut-off to maximum flow: ASME FEDSM2001-18072 (2001)
37. Pfleiderer, C.: Vorausbestimmung der Kennlinien schnellläufiger Kreiselpumpen. VDI, Düsseldorf (1938)
38. Rohkamm, H., Wulff, D., Kosyna, G., Saathoff, H., Stark, U., Gümmer, V., Swoboda, M., Goller, M.: The impact of rotor tip sweep on the three-dimensional flow in a highly-loaded single-stage low-speed axial compressor: part II—test facility and experimental results. In: 5th European Conference on Turbomachinery—Fluid Dynamics and Thermodynamics, Prague, pp. 175–185 (2003)
39. Saathoff, H., Deppe, A., Stark, U., Rohdenburg, M., Rohkamm, H., Wulff, D., Kosyna, G.: Steady and unsteady casing wall flow phenomena in a single-stage compressor at partload conditions. Int. J. Rotating Mach. **9**, 327–335 (2003)
40. Saathoff, H., Stark, U.: Tip clearance flow in a low-speed compressor and cascade. In: 4th European Conference on turbomachinery, Florenz, pp. 81–91 (2001)
41. Saathoff, H.: Rotor-Spaltströmungen in Axialverdichtern. Dissertation, TU Branuschweig, ZLR-Forschungsbericht 2001–2005 (2001)
42. Sano, T., et al.: Alternate blade stall and rotating stall in a vaned diffuser. JSME Int. Ser. B **45** (4), 810–819 (2002)
43. Schiavello, B., Sen, M.: On the prediction of the reverse flow onset at the centrifugal pump inlet. In: ASME 22nd Annual Fluids Engineering Conf, New Orleans (1980, March), Performance Prediction of Centrifugal Pumps and Compressors
44. Sen, M., Breugelmans, F.: Reverse flow, prerotation and unsteady flow in centrifugal pumps. In: NEL Fluid Mechanics Silver Jubilee Conference, Glasgow (1979, November)
45. Stachnik, P.: Experimentelle Untersuchungen zur Rezirkulation am Ein- und Austritt eines radialen Kreiselpumpenlaufrades im Teillastbetrieb. Dissertation, TH Darmstadt (1991)
46. Stepanik, H., Brekke, H.: Off-design behavior of two pump-turbine model impellers. In: 3rd International Symposium on Transport Phenomena and Dynamics of Rotating Machinery, Honolulu, pp. 477–492 (1990)
47. Stoffel, B., Hergt, P.: Zur Problematik der spezifischen Saugzahl als Beurteilungsmaßstab für die Betriebssicherheit einer Kreiselpumpe. Pumpentagung Karlsruhe, B8 (1988)
48. Stoffel, B., Weiß, K.: Different types and locations of partload recirculations in centrifugal pumps found from LDV measurements. In: IAHR Symposium Valencia (1996)
49. Stoffel, B., Weiß, K.: Experimental investigations on part load flow phenomena in centrifugal pumps. World Pumps **337**, 46–50 (1994)
50. Stoffel, B.: Experimentelle Untersuchungen zur räumlichen und zeitlichen Struktur der Teillast-Rezirkulation bei Kreiselpumpen. Forsch. Ing. Wes. **55**, 149–152 (1989)
51. Tanaka, T.: An Experimental study of backflow phenomena in a high specific speed impeller pump. ASME Paper 80-FE–6
52. Toyokura, T.: Studies on the characteristics of axial-flow pumps. Bull. JSME. **4**(14), 287–293 (1961)
53. Ubaldi, M., Zunino, P.: Experimental investigation of the stalled flow in a centrifugal pump-turbine with vaned diffuser. ASME Paper 90-GT-216 (1990)
54. Weinerth, J.: Kennlinienverhalten und Rotorbelastung von axialen Kühlwasserpumpen unter Betriebsbedingungen. Dissertation, TU Kaiserslautern. SAM Forschungsbericht Bd 9 (2004)
55. Weiß, K.: Experimentelle Untersuchungen zur Teillastströmung bei Kreiselpumpen. Dissertation, TH Darmstadt (1995)
56. Wesche, W.: Experimentelle Untersuchungen am Leitrad einer radialen Kreiselpumpe. Dissertation, TU Braunschweig (1989)
57. Yoshinaga, Y., et al.: Study of performance improvement for high specific speed centrifugal compressors by using diffusers with half guide vanes. ASME J. Fluids Eng. **109**, 259–367 (1987)
58. Van Esch, B., Cheng, L.: Unstable operation of a mixed-flow pump and the influence of tip clearance. ASME AJK2011-06016

59. Miyabe, M., et al.: Rotating stall behavior in a diffuser of a mixed-flow pump and its suppression. ASME FEDSM 2008-55132
60. Miyabe, M., et al.: On improvement of characteristic instability and internal flow in mixed-flow pumps. J. Fluid Sci. Technol. **3**(6), 732–743 (2008)
61. Saha, S., et al.: Suppression of performance curve instability of a mixed-flow pump by use of J-groove. ASME JFE **122**, 592–597 (2000)
62. Choi, Y., et al.: Suppression of cavitation in inducers by J-grooves. ASME JFE **129**, 15–22 (2007)
63. Cowan, D., Liebner, T., Bradshaw, S.: Influence of impeller suction specific speed on vibration performance. In: 29th International Pump Users Symposium, Houston, pp. 18–47 (2013)
64. Li, X., et al.: Experimental and numerical investigations of head-flow curve instability of a single-stage centrifugal pump with volute casing. Proc IMechE Part A J Power Energy (2016)

Chapter 6
Suction Capability and Cavitation

Definition of terms: "Cavitation" means the partial evaporation of liquid in a flow system. A cavity filled with vapor is created when the static pressure in a flow locally drops to the vapor pressure of the liquid due to excess velocities, so that some fluid evaporates and a two-phase flow is created in a small domain of the flow field. The vapor condenses suddenly ("implodes") as soon as it is transported downstream into zones where the static pressure again exceeds the vapor pressure. With increasing extension of the cavitating zone with two-phase flow, the head and efficiency of the pump may be impaired, noise and vibrations excited and components damaged through cavitation erosion under certain conditions. When using the term "cavitation" the "cavitating flow" - i.e. the occurrence of local zones with two-phase flow - and "cavitation erosion" or cavitation damage must be well distinguished.

The term "*hydrodynamic cavitation intensity*" is used for the totality of the implosion energy of all bubbles. If this exceeds the "*cavitation resistance*", the material will be damaged by "cavitation erosion" if it is exposed sufficiently long to the attack. The cavitation resistance is a *material property* which is independent of the flow system. The possibility of employing cavitation processes for separating dissolved gases from liquids was investigated in [37].

6.1 Cavitation Physics

6.1.1 Growth and Implosion of Vapor Bubbles in a Flowing Liquid

Any substance can exist in the phases solid, liquid and gaseous. The transition from one phase to another (e.g. "formation of ice" or "evaporation") is described through phase equilibriums in a p-T-diagram. Accordingly, the vapor pressure curve $p_v(T)$

J. F. Gülich, *Centrifugal Pumps*,
https://doi.org/10.1007/978-3-030-14788-4_6

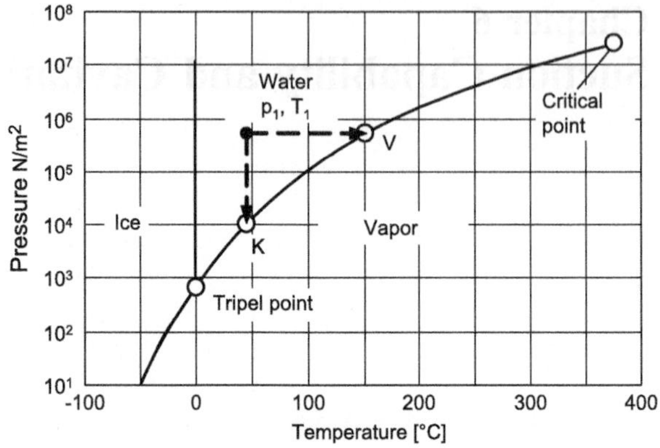

Fig. 6.1 Phase equilibrium of water

characterizes the equilibrium (the "saturation status") between liquid and vapor from the triple point to the critical point (Fig. 6.1). A liquid in condition p_1, T_1 with $p_1 > p_v(T_1)$ can be evaporated, if it is heated at constant pressure to the saturation temperature $T_v(p_1)$ (point V), or expanded to the vapor pressure $p_v(T_1)$ at constant temperature (point K in Fig. 6.1). During phase conversion from liquid to vapor the evaporation enthalpy must be added. In the reverse process, i.e. condensation, the evaporation enthalpy is liberated.

During cavitation a small portion of the liquid is expanded at constant temperature (in Fig. 6.1 from p_1 to K), to the extent that some liquid evaporates. Downstream of the low-pressure zone, the pressure rises again so that the saturation pressure is exceeded and the vapor condenses. This distinguishes cavitation from evaporation by "flashing". The nature of cavitating flows therefore implies that: (1) the static pressure in a system initially drops locally to the vapor pressure as a result of flow acceleration; (2) subsequently the flow is decelerated so that the pressure increases and the vapor bubbles implode. To illustrate the cavitation process, Fig. 6.2 shows the development of the static pressure in a Venturi nozzle. The minimum pressure p_{min} in the throat area can be calculated from Eq. (1.7), the pressure recovery in the diffuser from A_1 to A_2 according to Chap. 1.6.

Let the temperature of the fluid in otherwise identical conditions vary in four stages from T_1 to T_4 with the relevant vapor pressures p_{v1} to p_{v4}: in condition 1 with $p_{min} > p_{v1}$ (T_1) the vapor pressure is not reached and consequently no evaporation occurs. With $p_{min} \leq p_{v2}$ (T_2) saturation is reached and initial vapor bubbles develop. At p_{v3} (T_3) the zone with $p_{min} \leq p_{v3}$ takes up roughly one half of the diffuser length; this domain is covered with a large cavity which influences the flow in the channel. In condition 4 the pressure recovery in the diffuser is smaller than $p_{v4} - p_{min}$, so that the vapor created no longer condenses; the two-phase flow is maintained and covers the entire channel downstream of the diffuser. In accordance with the pressure loss and the thermodynamic equilibrium some fluid evaporates with increasing pipe length (only a fraction of the flowing fluid mass evaporates). Such a process is called expansion evaporation or "flashing".

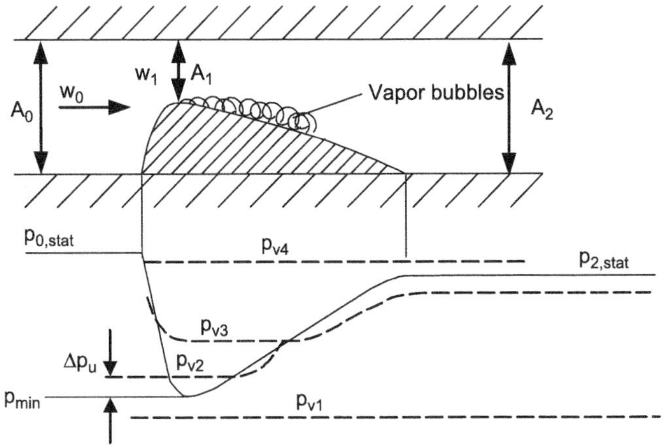

$$p_{0,stat} - p_{min} = \frac{\rho}{2}\left(w_1^2 - w_0^2\right) + \Delta_{pv} \qquad c_p \equiv \frac{p_{0,stat} - p_{min}}{\frac{\rho}{2} w_0^2} = \left(\frac{A_0}{A_1}\right)^2 - 1 + \zeta_v$$

Fig. 6.2 Cavitation in a nozzle

Because of the shear stresses between the flowing liquid and the cavity, the latter cannot be at rest. Bubbles develop continuously and are transported downstream to the end of the cavity where they implode when $p > p_v$ applies again. For evaporating the corresponding amount of liquid, the required evaporation enthalpy must be transported by heat conduction and convection from the surrounding liquid to the boundary of the vapor bubble. The heat transport requires a finite temperature difference ΔT_u. This temperature difference corresponds to a pressure difference Δp_u between the pressure in the bubble and the vapor pressure as shown in Fig. 6.2 (see also Chap. 6.4.1). Let us assume that the water flowing through the nozzle in Fig. 6.2 is saturated with dissolved air corresponding to the inlet pressure. As soon as the local pressure in the nozzle falls noticeably below the vapor pressure, air is liberated in accordance with Henry's law, Appendix A.3. This produces finely distributed individual bubbles when the pressure throughout the channel is sufficiently lowered or a cavity is formed as in Fig. 6.2. Consequently, gas separation can also create cavitation without the pressure dropping below the vapor pressure. However, a vapor component is always present in any gas-filled volume having an interface with a liquid. The *partial pressure* of the vapor is equal to the vapor pressure resulting from the fluid temperature. If the effect of gas separation is significant, the phenomenon is also called "gaseous cavitation". Non-condensable gases weaken the intensity of the implosion, thereby reducing the potential for noise, vibrations and material erosion.

6.1.2 Bubble Dynamics

Cavitation nuclei: Cavitation bubbles are generated only when the fluid contains "nuclei". These are accumulations of gas or vapor molecules which are present as microscopically small bubbles with diameters in the range of 10^{-3} to 10^{-1} mm. Free gas bubbles in a non-saturated fluid would gradually dissolve in the fluid through diffusion. The fact that nuclei are nevertheless stable in unsaturated fluid is explained by the hypothesis that the gas molecules are adsorbed on non-wetting particles (e.g. corrosion products or dirt). Experience has shown that sufficient nuclei are present in most technical systems where liquids are transported. This also applies to very pure de-aerated feed water (otherwise the boiling process in the steam generator would not be possible).

The interior of a nucleus contains gas and vapor. The pressure p_B in the nucleus corresponds to the sum of the partial pressures of the gas p_g and the liquid p_v. Because of the surface tension S_T it is greater than the pressure p of the surrounding liquid: $p_B - p = p_g + p_v - p = 2 \times S_T/R$ (R = radius of the nucleus). A given nucleus contains a certain mass of gas. If the pressure in the bubble changes, the bubble volume follows the law of a perfect gas $p \times V = m \times R \times T$. From this and the equilibrium of the forces on the bubble it follows that the radius of the nucleus varies with the surrounding pressure. If a nucleus enters a zone of low pressure - for instance in the throat area of a nozzle according to Fig. 6.2 - its radius grows. Since vapor pressure and vapor density are given by the liquid temperature (which remains constant), some liquid evaporates. In the process, only nuclei over a certain size are stimulated into growth. The lower the local pressure drops, the more nuclei are activated. *The number of nuclei that can be activated therefore grows with the square of the flow velocity.* The longer the low pressure zone and the lower the local pressure drops, the more cavitation bubbles are generated.

Water absolutely free of nuclei would be able to withstand high tensile stresses. In practice, the tensile stresses in the water are near zero because of the presence of nuclei, Chap. 6.4.3. If insufficient nuclei are available, the generation of cavitation bubbles is restricted. Above a threshold of the nuclei content, saturation occurs and the cavitating flow cannot be influenced through the supply of additional nuclei. This threshold falls with increasing flow velocity. From the above it follows that various forms of cavity flows (brought about by different pressure distributions) react differently to the nuclei spectrum. The form of pressure distribution - flat or peaky - consequently has an impact on the growth of bubbles.

Bubble implosion: When a vapor bubble is transported by the flow into zones where the local pressure exceeds the vapor pressure, the phase equilibrium according to Fig. 6.1 is disturbed and the vapor contained in the bubble condenses suddenly. Imagine a sphere-shaped bubble according to Fig. 6.3a whose vapor content suddenly condenses: since the pressure within the bubble collapses in the process, the bubble wall is radial-concentrically accelerated inwards by the higher

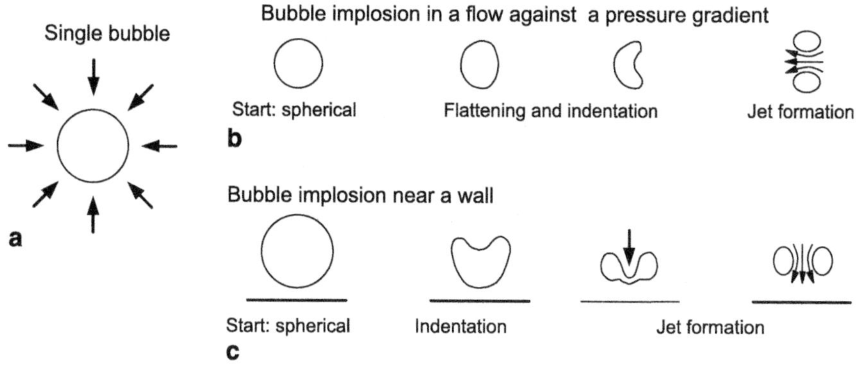

Fig. 6.3 Bubble implosion

surrounding pressure. Towards the end of the implosion the bubble wall reaches a very high velocity c_i, which can be calculated from "Rayleigh's equation":

$$c_i = \sqrt{\frac{2}{3} \cdot \frac{p - p_B}{\rho} \left(\frac{R_o^{\,3}}{R_e^{\,3}} - 1 \right)} \tag{6.1}$$

In Eq. (6.1) p is the pressure in the surrounding liquid, p_B the pressure in the bubble ($p_B \geq p_v$), R_o the bubble radius at the start and R_e at the end of the implosion. Extremely sharp pressure peaks are created during this process according to the ("water hammer equation")[1] $p_i = \rho \times a_o \times c_i$. These peaks are locally and temporally very limited. They can exceed 1000 bar and may consequently surpass the strength of the materials used in the pump.

The calculation of a concentric bubble implosion according Eq. (6.1) assumes idealized conditions, but it is a useful physical concept for understanding some essential aspects of cavitation damage. It is important to recognize that the implosion pressure and the damage potential increase with the bubble size R_o at the *start* of the implosion and the driving pressure differential, while it decreases with growing bubble radius R_e at the end of the implosion. The local pressure difference triggering the implosion is not only determined by the flow conditions but also by adjacent cavitation bubbles which can have an intensifying or diminishing effect.

In technical flows, bubbles implode asymmetrically as depicted in Fig. 6.3 and demonstrated by the photos in Fig. 6.4. In the process, a sharp micro-jet is formed which constitutes a shock-load according to $p_i = \rho \times a_o \times c_{jet}$ for the material.

The photos of a bubble implosion in Fig. 6.4 show very well the formation of the micro-jet. The bubble was generated through electric spark discharge (the two

[1]The water hammer equation is based on Joukowski; a_o is the speed of sound in the liquid.

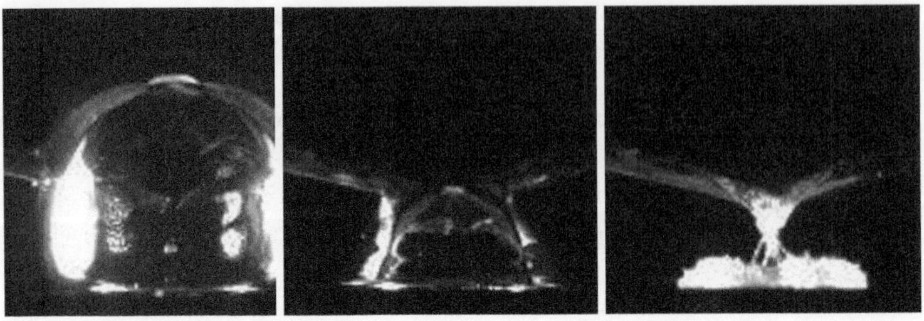

Fig. 6.4 Implosion of a bubble near a wall; Photos: Laboratory for hydraulic machinery, Ecole Polytechnique Fédérale, Lausanne; 24,000 frames per second; the scale is about 2:1

electrodes are faintly visible; their distance from the wall at the bottom of the picture was 6 mm). The bubble diameter in the photograph on the left is approximately 15 mm (the electrodes are also visible in the center of the bubble).

The asymmetry of the implosion is either caused by the proximity of a solid wall (Fig. 6.3c) or through pressure gradients (Fig. 6.3b), as can be expected wherever bubbles are transported by the flow into zones of higher pressure. This applies especially to the end of a cavity where the flow is decelerated (for instance in a nozzle according to Fig. 6.2 or in a pump).

At the start of the implosion typical bubble diameters in flow systems are in the range of 1 to 5 mm, towards the end of the implosion they are in the order of tenths of millimeters. The diameter of the micro-jet is consequently similar in size. High pressure peaks must be expected only in the immediate vicinity of the bubble at the final stage of the implosion. The bubbles must therefore implode very close to the solid surface to be able to attack the material.

The liquid imploding into the cavity during the implosion acts like a piston on the content of the bubble so that non-condensable gases are compressed adiabatically. The bubble content is heated accordingly so that a residual vapor component is also compressed in the final stage of the implosion (R_e in Eq. 6.1 remains finite). As a result of the energy stored in the compressed gas and vapor a bubble is formed again after the implosion ("rebound phase"). This process is repeated until the energy is dissipated.

At the final stage of the bubble collapse the compressibility of the surrounding liquid also plays a role. The shock waves associated with compressible flows contribute to the possible destruction of the material in addition to the micro-jet. Three additional effects limit the implosion pressure to finite values: (1) If the vapor density is high, the dissipation of the condensation heat has a limiting effect. (2) When the bubble-wall velocity c_i reaches the speed of sound, the physical model of Eq. (6.1) fails. (3) The implosion energy cannot be greater than the (finite) potential energy at the start of the bubble implosion, which is given by:

$$E_{pot} = \frac{4}{3}\pi\left(R_o{}^3 - R_e{}^3\right)\ (p - p_v) \tag{6.2}$$

E_{pot} is the maximum work that can be performed by the surrounding liquid during the bubble implosion to $R = R_e$.

6.2 Cavitation in Impeller or Diffuser

6.2.1 Pressure Distribution and Cavity Length

Cavitation zones at the impeller leading edge are created in a way similar to that in the nozzle sketched in Fig. 6.2. Consider the pressure conditions at the impeller inlet according to Fig. 6.5 which shows the pressure distribution on the blade suction surface *without cavitation.* The static pressure p_s and the velocity c_s prevail at the inlet nozzle. In the inlet casing the static pressure drops to the value p_1 immediately upstream of the impeller due to flow losses and acceleration (if any). Suppose the blade with the inlet angle β_{1B} receives a fluid approaching with the flow angle β_1 or incidence i_1. As the fluid flows around the profile, the static pressure is reduced by $\Delta p = 1/2 \times \rho \times \lambda_{w,i} \times w_1^2$ to the value p_{min} as the result of local excess velocities. As with an air foil, this reduction in pressure can be described by a profile coefficient $\lambda_{w,i}$ and the relative velocity w_1. $\lambda_{w,i}$ depends on the shape of the profile and the approach flow, especially on the incidence i_1. If the minimum local pressure p_{min} is above the vapor pressure, no cavitation occurs. If p_{min} reaches (or falls below) the vapor pressure, a cavity is formed.

As can be seen from the pressure distribution in Fig. 6.5, the cavity length increases with the length of the zone in which we find $p \ll p_v$. The appearance of the first bubbles at $p_{min} = p_v$ is termed "visual cavitation inception".

The longer and thicker the cavity, the greater is its retroactive effect on the flow. The pressure distribution then increasingly deviates from the form without cavitation. Since the same vapor pressure prevails throughout the cavity, the pressure distribution on the front part of the blade changes accordingly. In addition, the cavity displaces liquid due to the cross section blockage. Downstream of the cavity this displacement effect disappears and the liquid deceleration increases with the thickness of the cavity (see Chap. 4.1.3 in this regard). Figure 6.6 illustrates these relationships by means of the calculated pressure distributions on a radial impeller [31]. Without cavitation (full lines) the pressure drops sharply on the suction surface at the leading edge. The resulting "low-pressure peak" disappears when cavitation occurs (broken lines) and is replaced by a horizontal portion of constant pressure p_v. In this example, the cavity extends over approximately a third of the blade length. The steep rise in pressure downstream of the cavity due to deceleration is clearly visible.

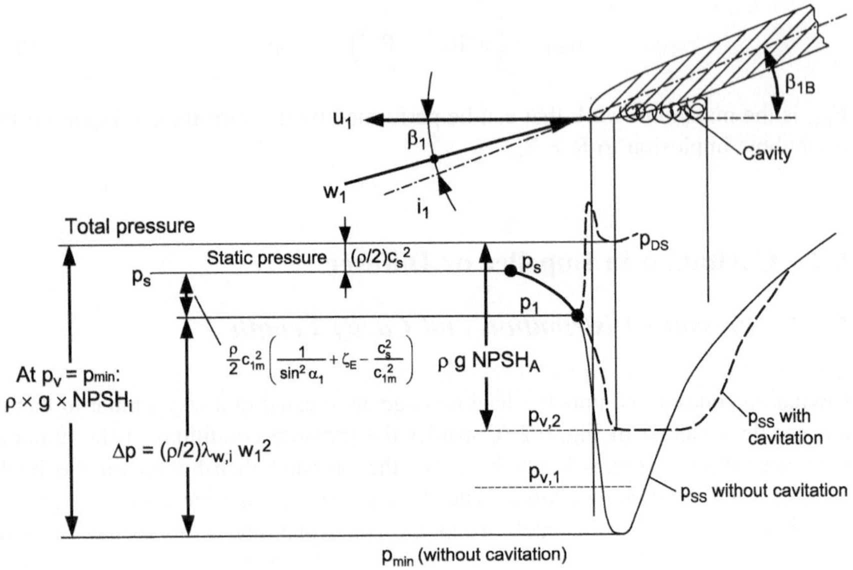

Fig. 6.5 Cavitation at the impeller blade leading edge

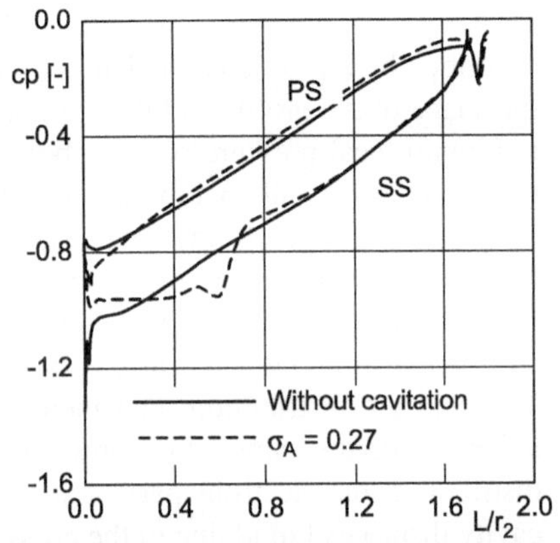

Fig. 6.6 Pressure distribution with positive incidence: i = 5°; $w_{1q}/w_1 = 0.71$ Calculation for outer streamline; *SS* suction surface, *PS* pressure surface, [31]

The integration of the pressure distribution essentially supplies the specific work performed by the impeller. It can be seen in Fig. 6.6 that (in the case shown) the head is not yet affected by cavitation. Although the low-pressure peak is "cut off", this deficit is compensated for towards the end of the cavity. As the length of the cavity progresses, eventually a condition is reached in which the blade work is impaired. As a rough approximation, this is the case when the cavity length on the blade suction surface exceeds the blade pitch $t_1 = \pi \times d_1/z_{La}$ and blocks a part of the throat area at the impeller inlet (the blockage leads to a flow acceleration).

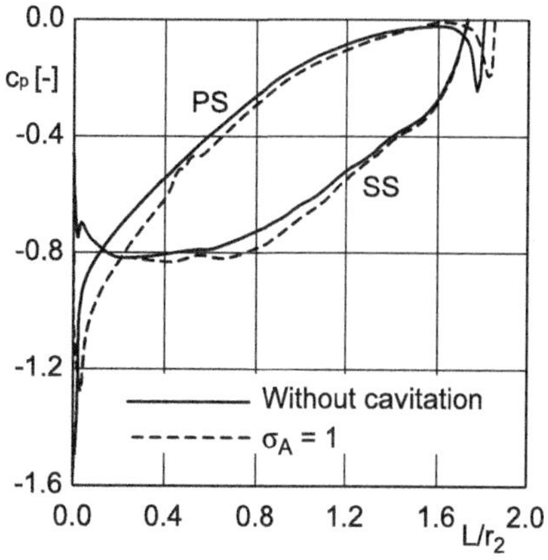

Fig. 6.7 Pressure distribution with negative incidence: i = −12°; w_{1q}/w_1 = 1.34 Calculation for outer streamline; *SS* suction surface, *PS* pressure surface, [31]

A low-pressure peak on the blade suction surface is obtained with positive incidence $i_1 \geq 0$ according to Fig. 6.6 and a stagnation point occurs on the pressure surface. With negative incidence, the stagnation point moves to the suction surface. Then the low-pressure peak develops on the pressure surface, where the first cavitation bubbles will subsequently occur (Fig. 6.7). The first part of the blade, where $p_{SS} > p_{PS}$ applies, "works as a turbine"; consequently, it reduces rather than generates head. Even minor bubble development on the blade pressure surface causes a drop in head, because even short cavities noticeably block part of the throat area and modify the pressure distribution. The effect is well illustrated by Fig. 4.12b.

6.2.2 *Required NPSH, Extent of Cavitation, Cavitation Criteria*

The length of the low-pressure zone, where $p \leq p_v$ applies, determines the extent of the cavitation zone. The cavity can be negligibly small (initial bubbles which cannot be detected on the pump from the outside) or involve most of the impeller channels ("full cavitation" and ultimately collapse of the flow and head). Criteria are therefore required to describe the cavitation in a pump quantitatively.

As discussed above, the *minimum local static pressure* at the impeller inlet determines the occurrence and the extent of cavitation. This pressure cannot be measured by any simple means. At best it could be calculated with numerical methods or determined in sophisticated laboratory tests. Industrial practice requires quantities which can be measured by simple means. For this reason, all cavitation processes are generally quantified by the NPSH value introduced in Chap. 2, which defines NPSH, or "net positive suction head", as the difference between the *total* head in the inlet nozzle and the vapor head.

$$\mathrm{NPSH} = \frac{p_s - p_v}{\rho g} + \frac{c_s^2}{2g} \tag{6.3}$$

If the pressure p_s is present in the inlet nozzle, the net positive suction head *available* in the plant ($NPSH_A$) is obtained for given values of vapor pressure and flow rate according to Eq. (6.3) and Fig. 6.5.

"Cavitation inception" is observed at $p_{min} = p_v$ where the first vapor bubbles are generated. The relevant net positive suction head is the $NPSH_i$. It constitutes the minimum NPSH *required* by the pump to operate without cavitation bubbles. If the inlet pressure is gradually reduced to value below that of $NPSH_i$, the cavity length increases until the cavitation zones finally become sufficiently large to impair the work transfer. In other words, *a pump requires different values of NPSH depending on the extent of cavitation considered or permitted.*

The *required* NPSH of the pump, $NPSH_R$, therefore constitutes that net positive suction head at which a *specific cavitation criterion* is maintained. Without specifying a cavitation criterion (i.e. the permitted extent of cavitation), a statement on the required NPSH of a pump is (in principle) meaningless. Popular cavitation criteria are:

- $NPSH_i$: visual cavitation inception: initial vapor bubbles are visible
- $NPSH_o$: starting of head drop
- $NPSH_1$: head of suction impeller drops by 1%
- $NPSH_3$: head of suction impeller drops by 3%
- $NPSH_{FC}$: "Full cavitation" or "head breakdown": head of impeller is severely reduced (it "collapses"). The impeller largely runs in a 2-phase flow.
- $NPSH_x$ with x percent of drop in head
- A defined drop of efficiency (e.g. beginning or 1% drop in efficiency)
- A defined loss of material or cavitation erosion
- A defined increase of noise due to cavitation
- A specified service life of the suction impeller (e.g. 50,000 h).

The most frequently used cavitation criterion is the $NPSH_3$; not because it is particularly relevant from a technical point, but because it is easy to measure. The symbol "NPSH" can therefore be found in the pump characteristics of many manufacturers without an explicit specification that it is actually the $NPSH_3$.

6.2.3 *Scaling Laws for Cavitating Flows*

As discussed above, origin, number and size of the cavitation bubbles depend on the *blade pressure distribution and on the nuclei spectrum.* As the low-pressure peak gets deeper (i.e. at higher circumferential velocity), increasingly smaller nuclei are activated. The local pressure distribution is affected by the Reynolds number. Even individual roughness elevations can lower the local pressure to the extent that a bubble is created and $NPSH_i$ is reached (these effects depend also on Re). Strictly applicable similarity laws which are able to cover such fine details are not known for centrifugal pumps. In practice, simplified model laws are therefore used for transferring measurements on a given pump to other speeds or other sizes of *geometrically similar* machines.

The flow condition at the impeller inlet is described by the flow coefficient φ_1, which is independent of size and speed as follows from the inlet velocity triangle.

$$\varphi_1 = \frac{c_{1m}}{u_1} \tag{6.4}$$

All pressure differences and pressure distributions in a pump are proportional to the square of the speed. Consequently, in first approximation, an Euler number must be used as a similarity characteristic for the cavitation criteria or the extent of the cavitation. The cavitation coefficients σ_i, σ_3, σ_x are defined according to:

$$\sigma_x = \frac{2g \; NPSH_x}{{u_1}^2} \tag{6.5}$$

From this it follows that NPSH values are scaled to the square of the speed and the diameter (subscript M applies to the known "model"):

$$NPSH = NPSH_M \left(\frac{n}{n_M}\frac{D}{D_M}\right)^2 \tag{6.6}$$

Additionally, the following model law applies to the cavity length L_{cav}:

$$\frac{L_{cav}}{d_1} = f(\sigma_A, \varphi_1) \tag{6.6a}$$

With given values of σ_A and φ_1 the ratio L_{cav}/d_1 with *geometrically similar impellers* is (in first approximation) independent of the size and the speed.

Prerequisites for applying the above model laws are:

- The pumps have to be geometrically similar. This requirement must be strictly satisfied for the impeller and the inlet. The collector and impeller outlet (impeller outer diameter) can also have an effect on cavitation - especially at partload recirculation. It is therefore not advisable to transfer the partload cavitation behavior of diffuser pumps to volute casing pumps (or vice versa). Since an impeller/collector interaction is not always present or perceptible in the cavitation behavior, it is sometimes argued that geometric similarity at the impeller outlet is not required when transferring cavitation experiments from the model to the prototype. This may be true in many instances but can also result in incorrect conclusions with serious consequences.

- Identical liquid of the same temperature, gas content and nucleus spectrum. Frequently, this requirement cannot be strictly satisfied in practice; refer also to Chap. 6.4.

During the measurements in [22] the speed of a test pump was varied in the range from 1000 to 3500 rpm. The model laws according to Eqs. (6.5 to 6.6a) were fulfilled within minor scatter. This finding was confirmed by the measurements in [25] during which not only the circumferential velocity but also the water temperature was varied between 27 and 81 m/s and 30 and 160 °C respectively. The test results were affected by the usual scatter but did not reveal any systematic deviations from which scaling laws for speed or temperature could be derived.

Model laws deviating from Eqs. (6.5 to 6.6a) were not yet introduced in the pump industry. If the scaling laws derived from the investigations in [33, 34, 57] are applied to pumps, unrealistic results are obtained, which grossly contradict decades of testing and plant experience.

When applying the model laws and when evaluating cavitation tests, nevertheless significant deviations occur at times. These can be caused by the following uncertainties:

1. Even small tolerances in the blade inlet profile or angle (even the roughness) can cause considerable deviations in the cavitation inception $NPSH_i$, but have little effect on $NPSH_3$.
2. Tolerances of the throat area and the inlet angle have an effect on the steep rise of NPSH for $Q > Q_{opt}$ and on the $NPSH_3$ values at the best efficiency point.
3. Because of the sensitivity to geometrical deviations, scaling laws for the effect of the impeller size could only be determined with very precise NC-milled impellers (systematic tests are not known to date).
4. With small impellers (inlet diameters below 100 to 140 mm) the required $NPSH_x$ increases with decreasing impeller size because of the boundary layer blockage and because the cavity has a greater blockage effect on the flow in narrow channels.
5. The Reynolds number affects the boundary layer thickness. At low Reynolds numbers, some roughness elevations and irregularities are located within the boundary layer. At high Reynolds numbers, the roughness generates tiny vortices which can increase the $NPSH_i$.
6. As long as no saturation with nuclei is reached, the $NPSH_R$ values depend on the nuclei spectrum. Even if this were known, its effect could not be predicted.
7. Air separation - especially with NPSH values below atmospheric pressure - can falsify the measurements.
8. Because of air separation and tolerances, measurements of the NPSH at low absolute inlet pressures are difficult and uncertain. For this reason, it is sometimes suggested that the exponent in Eq. (6.6) should be selected smaller than 2.0 when scaling down the NPSH from a higher to a lower speed. There are even measurements which seem to point to exponents greater than 2.0, [57], refer also to Chap. 6.4.

In summary: There is, as yet, no universally accepted alternative to the scaling laws given by Eqs. (6.4 to 6.6a) that would be suitable for application in industry. Therefore, Eq. (6.6) is also given in standards [N.2] and in [12]. Nevertheless, it should be noted that the scaling laws apply to the *flow* but not to the *activation* of cavitation nuclei. The nuclei activation process cannot be scaled, because it depends on the *absolute* pressure difference $\Delta p_u = p_v - p_{min}$ (Fig. 6.2). Nuclei activation cannot be expressed through relative quantities such as $c_{p,min}$ or σ.

When testing at low $NPSH_A$ (typically $NPSH_A < 20$ m) in test loops with air-saturated water, it is prudent to use exponents inferior to 2.0 when *scaling down* to lower speeds or sizes. In this way, overly optimistic statements about the required NPSH can be avoided. To this end, the speed exponents of tests reported in [42] were evaluated and plotted against the $NPSH_3$ in Fig. 6.8. The exponents derived are between 1.3 and 1.8.

In practice there is no uniform approach as revealed by an analysis of the sales documentation of 8 pump manufacturers. The apparent exponents derived from the sales curves range from 0.2 to 2.2, [59]. It would make little sense to apply an exponent of (say) 1.5 at "low" speeds and an exponent of 2.0 at "high" speeds. Rather a continuous function should be used. On the basis of the scarce test data available, the following tentative formula for scaling the $NPSH_3$ down to lower speeds or sizes is suggested:

$$NPSH_a = NPSH_M \left(\frac{n_a}{n_M}\frac{D_a}{D_M}\right)^x \quad \text{with}$$
$$x = 2\left(\frac{NPSH_3}{NPSH_{Ref}}\right)^{0.3} \quad NPSH_{Ref} = 20\ \text{m} \tag{6.6b}$$

The curve represented by Eq. (6.6b) is shown in Fig. 6.8. It is seen that Eq. (6.6b) was established in a way that conservative NPSH predictions are obtained. The following example illustrates the application of Eq. (6.6b): Suppose a $NPSH_3 = 6$ m is measured at a speed of 3000 rpm on a given pump. If the same pump is operated at 1500 rpm, Eq. (6.6b) gives x = 1.39 and $NPSH_3 = 2.3$ m.

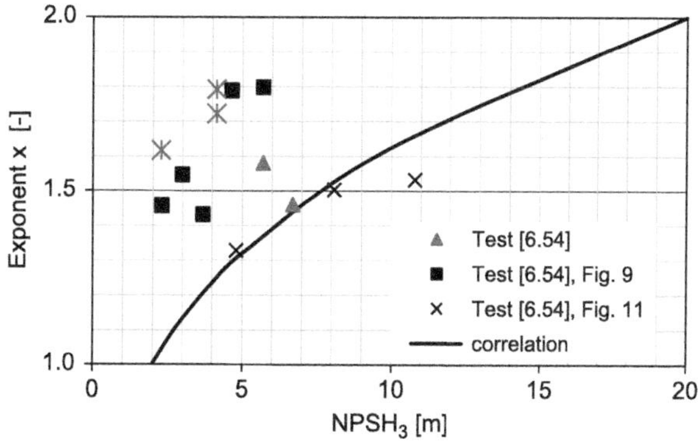

Fig. 6.8 Exponent x for scaling $NPSH_3$ at low speeds (tentatively) Test [6.54] is from [42]

It is recommended to use Eq. (6.6b) only for scaling *down*. To be on the safe side, the exponent 2.0, that is Eq. (6.6), should be used for *scaling up*. Note that the procedure is entirely empirical; its only objective is to avoid predicting overly optimistic $NPSH_3$ when scaling down.

6.2.4 The Suction Specific Speed

To evaluate the suction capability of a pump or the quality of a suction impeller with regard to $NPSH_3$ the suction specific speed n_{ss} has become the accepted parameter. It is defined at the best efficiency point of the pump to enable comparison of pumps that are *not* geometrically similar. It would be physically more meaningful to define the suction specific speed with the flow rate of the shockless entry since this corresponds to the design point of the impeller *inlet*. But doing this would make the comparison of pumps of different manufacturers more difficult since the impeller parameters are generally unknown.

$$n_{ss} = n\frac{\sqrt{Q_{opt}/f_q}}{{NPSH_3}^{0.75}} \tag{6.7}$$

In the case of double entry impellers the flow per impeller side must be considered. The suction specific speed has a form which is analogous to the specific speed and is formed with the same dimensions. It can be derived by eliminating the diameter from the two relationships $Q = k_1 \times n \times D^3$ and $NPSH = k_2 \times (n \times D)^2$. With the dimensionless coefficients defined by Eqs. (6.4 and 6.5) the suction specific speed takes the form of Eq. (6.8); $k_n = 1 - (d_n/d_1)^2$ is the hub blockage.

Table D6.1 Typical values of the suction specific speed, n_{ss} [min^{-1}, m^3/s, m]

Application	u_1 (m/s)	n_{ss}
Standard impellers for axial inlet or between-bearing pumps	<50	160 to 220
Suction impellers for axial inlet	<35	220 to 280
Suction impellers of between-bearing pumps, single or double-entry	<50	180 to 240
High pressure pumps designed for short cavity length	>50	160 to 190
Industrial inducers (Sect. 7.7.4)	<35 (45)	400 to 700
Inducers for rocket pumps		>>1000

$$n_{ss} = \frac{30}{\sqrt{\pi}}(2g)^{0.75}\frac{\sqrt{\varphi_1 k_n}}{{\sigma_3}^{0.75}} = 158\frac{\sqrt{\varphi_1 k_n}}{{\sigma_3}^{0.75}} \tag{6.8}$$

Table D6.1 provides the range of suction specific speeds for different impellers. The suction specific speed to be selected for a particular design depends on the application - especially on the circumferential velocity u_1 and the medium to be pumped.

Small pumps often have suction specific speeds lower than those given in Table D6.1 if sufficient $NPSH_A$ is available.

Note that the (very useful) concept of suction specific speed would break down, if the scaling laws of Eqs. (6.5 and 6.6) were not applicable.

Suction specific speed increase: To reduce hydraulic and leakage losses, the impeller inlet diameter should be selected so that the relative velocity at the impeller inlet is roughly minimum (see Table 7.1). Such a design does not always achieve a sufficiently low $NPSH_3$ and the suction capability has to be improved compared with this design concept. Different routes are available to do so:

- By enlarging the impeller inlet diameter d_1, suction specific speeds of up to roughly 300 are achieved. As explained in Chap. 5, large inlet diameters produce correspondingly intensive partload recirculations since the pressure gradients perpendicular to the flow direction increase (a greater ratio d_1/d_{1i}). The increase of the suction specific speed through larger d_1 is therefore limited by considerations of the behavior at partload.
- Increasing the blade inlet angle or the throat area at the impeller inlet corresponds to a design for a larger flow rate, which means that the point of shockless entry is shifted towards $Q_{SF} \gg Q_{opt}$. If the inlet angles and/or throat areas are increased at a given inlet diameter, $NPSH_3$ and $NPSH_{FC}$ are reduced to some extent, since the "vapor handling capability" increases. Thus the flow up to the throat area is more strongly decelerated and the main flow less affected by the blockage caused by the cavity. However, the impeller inlet operates at partload. Consequently, incidence and cavity volume increase in the usual operation range and the recirculation at the impeller inlet starts at higher flow rates; noise, vibrations and erosion can be the result. This measure for increasing the suction specific speed was sometimes employed in the past but is no longer recommended today: a moderate enlargement of the inlet diameter is preferable.
- Decreasing the number of blades reduces the blockage at the impeller inlet and improves the suction specific speed by several percent, but the head coefficient suffers. This deficiency could be compensated by enlarging the width or the blade angle at the impeller outlet, although with an increased risk of hydraulic vibration excitation due to uneven impeller outlet flow. A compromise is possible in the case of impellers with an even blade number by cutting back every second blade at the inlet ("splitter blades").
- Double-entry impellers divide the flow rate per impeller side by two. If each half of the double-entry impeller were to achieve the same suction specific speed as a single-entry impeller, a reduction of $NPSH_3$ by 37% would be achieved (multiply original NPSH by the factor of $(1/2)^{2/3} = 0.63$ according to Eq. (6.7)). This gain is rarely achieved since the double-entry impeller has a greater hub ratio, which causes the suction capability to deteriorate according to Chap. 6.3.2. Additionally, the inlet casing of a double-entry pump generates a considerably less uniform flow distribution than an axial inlet. This further impairs the achievable suction specific speed since the blades are subjected to different incidences in different sectors, Chap. 7.13.

- An inducer is a kind of axial impeller with 2 to 4 long blades upstream of the actual impeller, which produce little inlet blockage. The inducer works with developed cavitation. It increases the pressure upstream of the actual impeller to the extent that the latter operates without a cavitation-induced head drop, Chap. 7.7.

Suction specific speed limitation: it has been variously suggested to limit the suction specific speed since - *meanwhile obsolete* - operating statistics[2] hinted to an accumulation of cavitation and vibration damage on pumps whose suction specific speeds exceeded the value 213 (N_{ss} = 11,000 in US units)[3]. The explanation for this observation was that high suction specific speeds would require large impeller inlet diameters and large inlet angles which, due to excessive partload recirculation, could cause cavitation erosion, vibrations and pressure pulsations. This reasoning is correct tendency-wise at best since, in addition to the impeller inlet diameter, a number of other parameters are responsible for the intensity and damage potential of partload recirculation. It is therefore *incorrect* to derive the onset of recirculation from the suction specific speed as was done in the endeavor to facilitate pump selection. Such a suction specific speed limitation constitutes an impermissible simplification of the three-dimensional flow processes leading to partload recirculation. Moreover, as explained in Chap. 7, high suction specific speeds can be achieved not only through the enlargement of the impeller inlet, but far better through the careful design of the blades.

The suction specific speed issue (including case histories) is comprehensively discussed in [28]. As proven by many years of operation of countless pumps with suction specific speeds far above n_{ss} = 213 without damage, the suction specific speed limitation often means an unnecessary restriction. It would thwart many economical plant concepts since it would lead to unnecessarily large pumps or greater civil engineering costs to build installations with greater $NPSH_A$.

This view is well supported by tests as to the "limits" of the suction specific speed reported in [70]: volute pump with axial inlet (OH2), n_q = 29; n = 3560 rpm. The data of the 4 test impellers can be found in Chap. 7, Table D7.1-1. Based on the test results, a relation between the maximum n_{ss} and the specific speed was proposed which can be expressed by Eq. (6.8a). Below this limit vibrations would not exceed the criteria spelled out in ISO 13709 (API 610). It is not suggested in [70] that Eq. (6.8a) should be considered as a sharp limit; rather, there is room for further development up to roughly 15% of the n_{ss} proposed by Eq. (6.8a). One merit of this study is highlighting the fact that it will be progressively difficult to achieve good performance, low vibrations and high n_{ss} as the specific speed increases. Of course, Eq. (6.8a) cannot be considered as generic because the mechanical vibrations depend on many factors including the rotor speed, the

[2]The statistics originate from the period 1975 to 1981 on the operation of pumps developed in the sixties; details in [28].

[3]Different limits were propagated depending on the type of pump and author.

mechanical design and the excitation forces which depend on the detailed geometry of the hydraulic components.

$$n_{ss,max} = 290 - 40\left(\frac{n_q}{n_{q,Ref}} - 1\right) \text{with } n_{q,Ref} = 20 \quad (6.8a)$$

valid for $n_q > 20$; below $n_q = 20$ assume $n_{ss,max} = 290$.

6.2.5 Experimental Determination of the Required $NPSH_R$

The required NPSH of a pump is determined in a "suction test" . To this end, the head of the *suction stage* is measured at constant speed and constant flow rate while gradually lowering the inlet pressure. The head is plotted as a function of $NPSH_A$ or against σ_A as in Fig. 6.9. With sufficiently high $NPSH_A$ a horizontal part of the suction curve is obtained where the head is not yet affected by cavitation. If the inlet pressure drops below a specific value the head starts to fall. This drop is hardly perceptible at first but increasingly develops further with decreasing net positive suction head. On falling below a sort of threshold the head finally decreases steeply. In the end the head drops almost vertically. During this condition, called "full cavitation" or "head breakdown", two-phase flow prevails in large parts of the impeller channels. Due to the vapor content, the density of the 2-phase mixture in the hydraulic channels is reduced and the pressure build-up is impaired (apart from the additional losses).

Once the suction curves have been plotted with several flow rates the σ_R (as in Fig. 6.9) or the $NPSH_R$ values, which correspond to the various cavitation criteria $NPSH_o$, $NPSH_1$, $NPSH_3$, $NPSH_{FC}$ can be determined. They are subsequently plotted against the flow rate according to Fig. 6.10.

When testing multistage pumps, the criteria for the head drop must always be based on the head of the *suction* stage only: if a 3%-decrease of the *total* head of a multistage pump were to be incorrectly permitted, the suction stage would at the measured $NPSH_3$ already operate in full cavitation. The suction stage head of multistage pumps can be directly measured using an additional pressure tapping. If this is not practical, it can be determined through calculation. If all stages have the same impeller diameter, it is assumed that $H_{1.st} \approx H_{tot}/z_{st}$. But with a large number of stages, determining the NPSH values in this way becomes quite uncertain due to unavoidable measuring tolerances. This, because a drop of the overall head by a few tenths of a percent cannot be measured accurately. For example: in a 10-stage pump the $NPSH_3$ of the first stage occurs at a drop of 0.3% of the total head of the pump.

The inlet pressure can be varied during the test by: (1) throttling in the suction pipe; (2) changing the suction water level; (3) varying the gas pressure above the water level in the suction tank of a closed loop. Throttling using a suction valve is easily accomplished in the test, but frequently results in problems with gas separation when working with NPSH values that are below atmospheric pressure. This

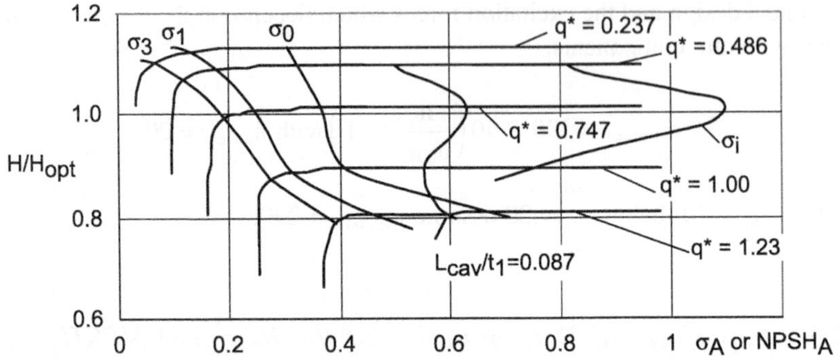

Fig. 6.9 Determination of the different cavitation criteria in a suction test with constant speed by successive reduction of the inlet pressure (each curve is measured at Q = constant)

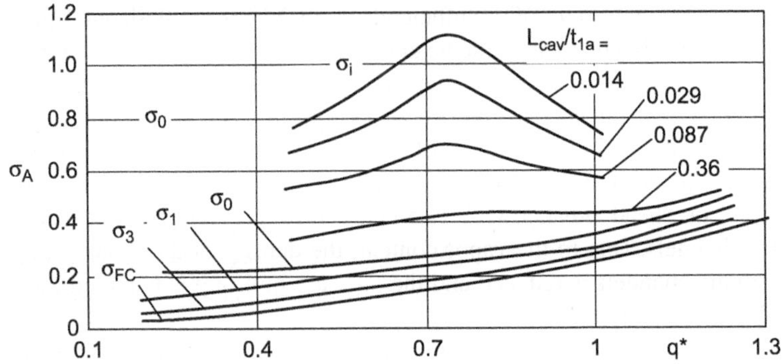

Fig. 6.10 Cavity length and head reduction as function of the cavitation coefficient and the flow rate

problem is aggravated by the minimum pressure in the throttling device resulting from excessive flow velocities. To reduce gas separation, the suction throttle valve should be installed as low as possible. When testing in open test loops, underwater valves installed in the basin immediately downstream of the entry into the suction pipe (mostly suction bell) proved to be suitable. When varying the suction water level, gas separates as soon as noticeable vacuums are reached. When testing in a closed loop, an efficient de-aeration of the liquid must be ensured since (contrary to open circuits) free gas cannot escape from the circuit.

Model pumps for suction impeller development (especially for high-pressure pumps) are frequently designed with a window through which cavitation at the impeller inlet can be observed by means of a stroboscope. In this way it is possible to determine the cavitation inception $NPSH_i$, and the length of the cavity can be visually estimated and recorded as a function of $NPSH_A$ (sketches, photos and video). The results of such a test are depicted in Figs. 6.9 and 6.10, while Fig. 6.11 shows a photo of a cavity. Cloud-type accumulations of bubbles in different forms

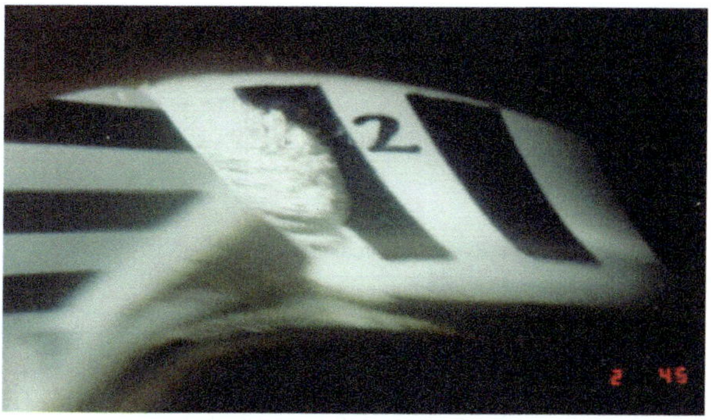

Fig. 6.11 Cavity at the impeller blade leading edge

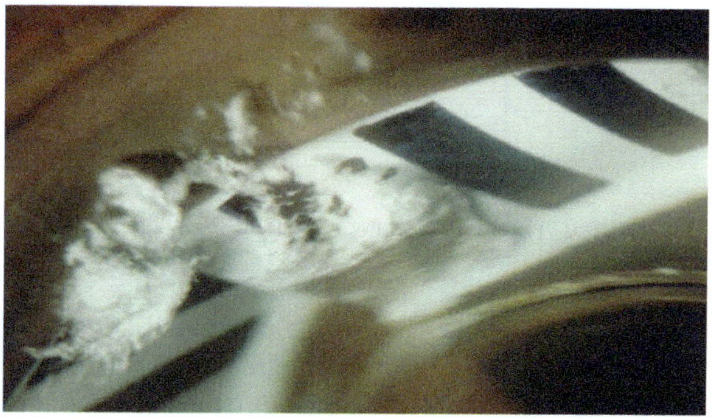

Fig. 6.12 Cavitation clouds at partload recirculation

and at different locations frequently occur at partload recirculation, Fig. 6.12 shows an example (photos Sulzer Pumps).

Without special provisions (transparent blades or reflecting surfaces on the hub or blades) only the cavitation bubbles on the blade suction surface can be recorded.

The extent of cavitation as defined by the different cavitation criteria depends on the flow rate. Figure 6.13 qualitatively shows this dependence in terms of the ratio $NPSH/NPSH_{i,SF}$ (that is, all NPSH-values are referred to the cavitation inception at shockless flow).

- The visual cavitation inception $NPSH_i$ is at a minimum at the flow rate of shockless entry Q_{SF}, which is calculated from Eq. (T3.1.10). With $Q < Q_{SF}$, bubbles occur on the (visible) suction surface, with $Q > Q_{SF}$ on the pressure surface of the blades.
- $NPSH_i$ and the cavity length increase with increasing incidence - i.e. decreasing flow rate - up to the start of recirculation from the impeller into the suction chamber. With increasing recirculation the incidence (and consequently the cavity length) decreases again. Three effects contribute to this: (a) the flow through the impeller increases by Q_{Rec} (Fig. 5.12). (b) The outer ring of the impeller inlet is blocked by recirculating liquid reducing the effective flow cross section with the effect that c_{1m} and β_1 grow. (c) The fluid Q_{Rec} recirculating

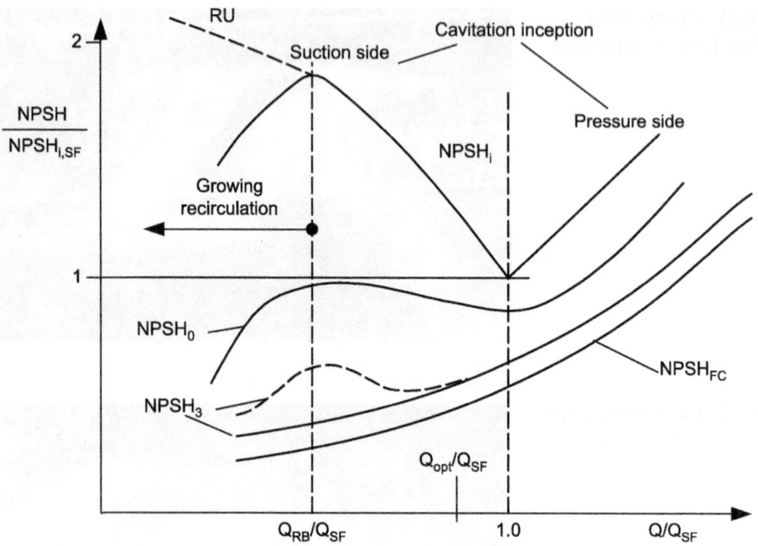

Fig. 6.13 Typical NPSH-curves of radial impellers (schematic)

from the impeller has a circumferential component in the direction of impeller rotation. The swirl of the recirculating fluid is transmitted to the entering flow through exchange of momentum, thus generating a pre-rotation. All three effects add up in the sense that the flow angles grow (incidence is reduced), leading thus to shorter cavities. If the pre-rotation is effectively suppressed through ribs or other structures, only the mechanisms (a) and (b) remain effective. Depending on the impact of the individual effects on the flow angle, the $NPSH_i$ curve decreases, flattens or even increases continually towards Q = 0 according to the curve "RU".

- Near the shockless flow rate, cavitation bubbles occur simultaneously on the suction and pressure surfaces if $NPSH_A$ is sufficiently low.
- The curve for full cavitation $NPSH_{FC}$ typically runs roughly towards the origin of the coordinates. At large flow rates the curve rises steeply. If one measures at a sufficiently large flow, the NPSH-curve rises almost vertically. The steep rise constitutes a kind of pumping limit similar to the choking of a compressor.
- Also the curve for $NPSH_3$ frequently tends towards the origin of the coordinates and increases steadily with increasing flow rate. Exceptions are suction impellers with very high suction specific speed, inducers and impellers with $n_q > 70$. Then the $NPSH_3$ often follows the broken curve in Fig. 6.13, which is characterized by intensive recirculation and changes in the flow patterns. The shape of the curves $NPSH_x = f(Q)$ is determined by: (1) the throat area blockage due to cavities; (2) the deceleration of the flow downstream of the cavity; (3) the deceleration or acceleration of the relative velocity from the vector w_1 to the velocity in the impeller throat area w_{1q}. Note that $NPSH_x$ would *increase* with

diminishing flow (same as $NPSH_i$) if it were determined by the incidence i_1. At low partload the throat area at the impeller inlet is much too large; hence some blockage by a cavity tends to improve the flow (compare to Fig. 4.12a). If the fluid is accelerated in the throat area, that is $w_{1q}/w_1 > 1$, the static pressure drops and the bubble development increases accordingly (see Fig. 5.13).

- Like $NPSH_i$, the curve for $NPSH_o$ frequently has a maximum near the onset of recirculation Q_{RB}, but it is considerably flatter than in the curve $NPSH_i(Q)$. The larger the suction specific speed, the more pronounced this maximum becomes. If the impeller inlet diameter is relatively narrow, no maximum is found in the curve $NPSH_o(Q)$; such is the case in Fig. 6.10.

Table D6.2 Typical NPSH-relationships at shockless entry

Type of impeller	$NPSH_i/NPSH_3$	$NPSH_0/NPSH_3$	$NPSH_{FC}/NPSH_3$
Normal design	4 to 6		
Leading edge profile for low $NPSH_i$	2 to 3	1.1 to 1.3	0.8 to 0.9
Low blade numbers $z_{La} = 3$	8 to 10		

Depending on the suction impeller design, the approach flow and the specific speed, deviations from the above tendencies are possible. The magnitude of the various NPSH values and their interrelationships depend on the impeller design and the pump type. Typical ratios at the point of shockless entry according to [30] are shown in Table D6.2.

A high blade loading at the impeller inlet produces pronounced low-pressure peaks and correspondingly elevated ratios of $NPSH_i/NPSH_3$. This explains the high values of $NPSH_i/NPSH_3$ given in Table D6.2 for small blade numbers[4]. These are also influenced by the fact that $NPSH_3$ drops with low blade numbers due to decreasing blade blockage.

For a given "NPSH margin" as defined by the ratio $F_{NPSH} = NPSH_A/NPSH_3$, very different cavity lengths are obtained depending on the blade design (this also follows from the data in Table D6.2). Figure 6.14 shows some examples of measured cavity length. Case 5.1 refers to a heat extraction pump with $n_{ss} = 278$, case 5.3 refers to a cooling water pump with $n_{ss} = 275$ (concept "L").

Shortening the impeller blades impairs their ability to compensate at the end of the cavity for the work deficit induced by the bubbles. This is one reason why the $NPSH_x$-values increase when the impellers are significantly trimmed at the outlet.

The suction curves $H = f(NPSH_A)$ at constant flow rate can assume different characteristic shapes:

1. If the flow is decelerated effectively downstream of the cavity, the impeller is able to handle relatively long cavities before a drop in head occurs. The more this is the case, the closer $NPSH_{FC}$, $NPSH_3$, $NPSH_1$ and $NPSH_0$, move together:

[4]As explained in Chap. 7.2.4, it is possible through suitable blade design to achieve a low blade loading even with small blade numbers.

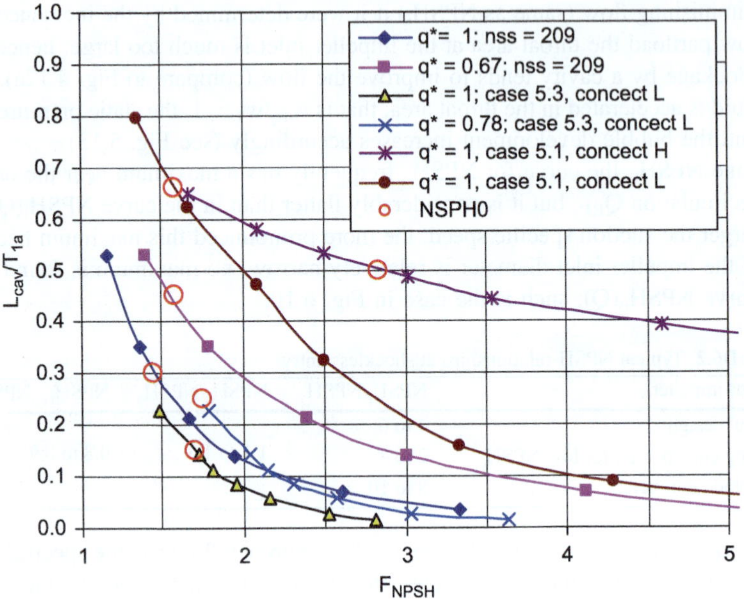

Fig. 6.14 Influence of the NPSH-margin on the cavity length (T_{1a} = pitch); concept "H": sharp profile, high incidence; concept "L": blade designed for flat pressure distribution, see Chap. 7.2.4, Fig. 7.10; $F_{NPSH} = NPSH_A/NPSH_3$, case histories described in [28]

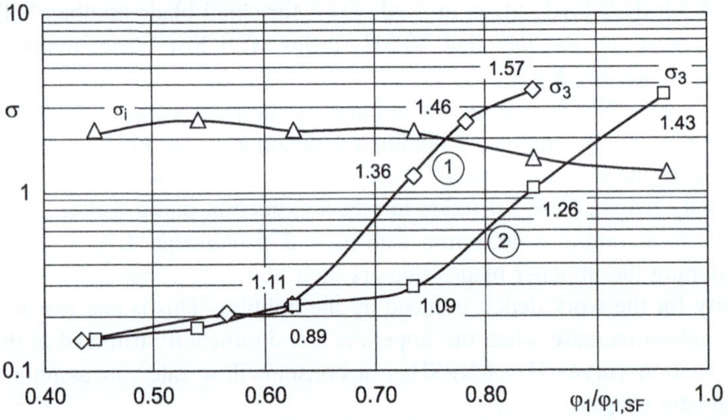

Fig. 6.15 Influence of the diffuser on cavitation with $n_q = 15$. With curve 2 the diffuser throat area was 25% larger than with curve 1. The figures given in the graph are the accelerations c_{3q}/c_2 from impeller outlet to the diffuser throat

a sharp knee occurs in the curve $H = f(NPSH_A)$. Such curves are typical for impellers with small to medium specific speeds having peaky pressure distributions. Observations and measurements in [9, 15, 47] confirm that sheet

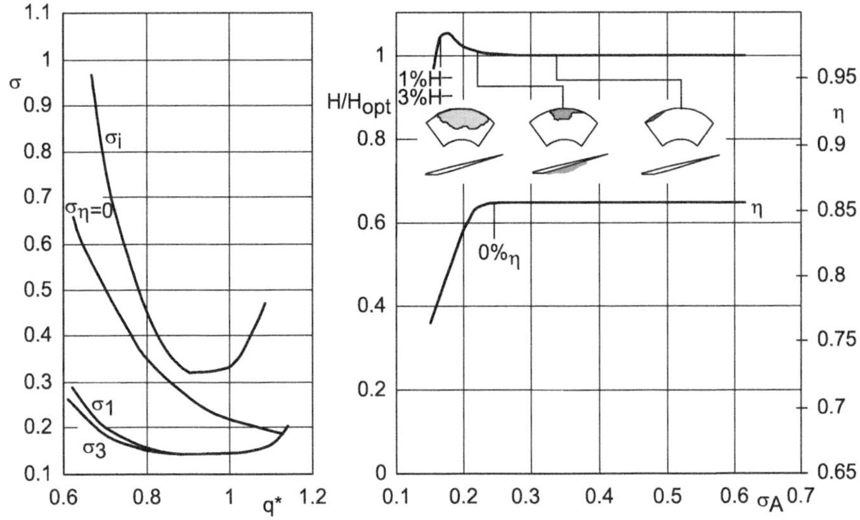

Fig. 6.16 Cavitation measurements on an axial pump, Sulzer Pumps

cavitation does not cause any loss of head even with quite long cavities since the flow behind the cavity is decelerated according to Fig. 6.6. When the cavity finally becomes so long that the energy transfer is affected, the head drops steeply. Conversely, unsteady cavities, as manifested through separation of bubble clouds, generate no additional lift. They induce a gradual loss in head as soon as bubbles occur since there is no deceleration downstream of the cavity. When bubble clouds travel from the suction surface to the pressure surface of the next blade the head is impaired according to observations in [15].

2. In the case of impellers with a flat pressure distribution $NPSH_{FC}$, $NPSH_3$, $NPSH_1$ and $NPSH_0$ are far apart as for instance in Fig. 6.9.
3. At $Q > Q_{SF}$ the loss in head is caused by bubbles on the blade pressure surface. The head drop starts at a high σ_A since even small cavities noticeably block the throat area. Consequently, the fluid is accelerated, the static pressure drops and more vapor is generated. In addition, no pressure recovery is present downstream of the cavity (Fig. 6.6).
4. With low specific speeds (approximately $n_q < 20$) the head is impaired at large flow rates $Q > Q_{opt}$ through cavitation in the diffuser or volute. This happens when the fluid is accelerated from the impeller outlet to the throat area of the collector - that is when c_{3q} is significantly above c_2 (Chap. 5.3 and Fig. 5.18). The lower the specific speed, the closer this point moves to the best efficiency point. (During the tests in Fig. 6.17 cavitation bubbles on the diffuser vane leading edges were observed only at $q^* > 1.3$.)
5. Figure 6.15 shows the results $\sigma = f(\varphi_1/\varphi_{1,SF})$ of tests with two diffusers with different inlet widths a_3 and an *identical* impeller (designed for $n_q = 15$). The throat area of diffuser 2 was 25% larger than that of diffuser 1. The cavitation

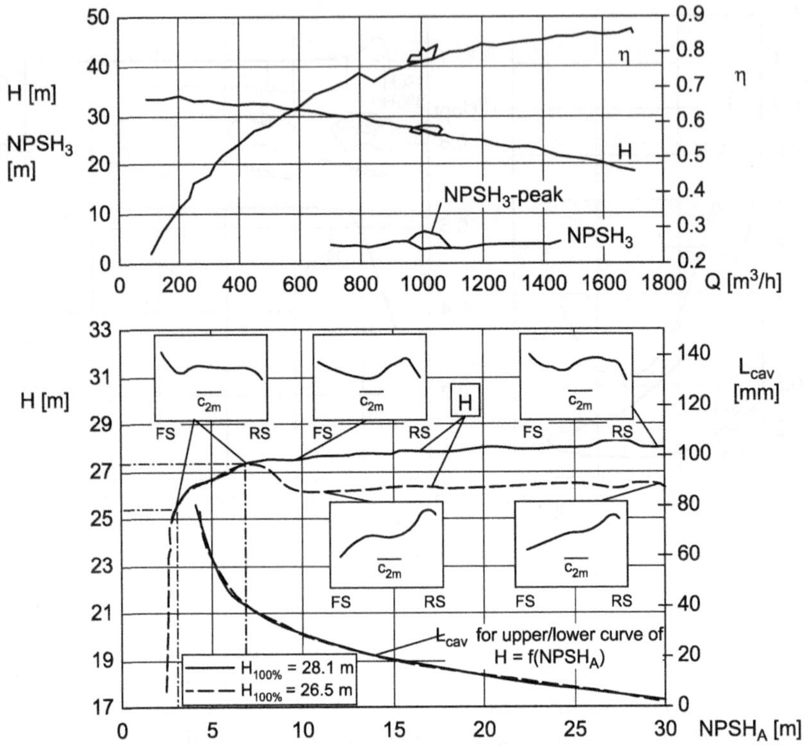

Fig. 6.17 Influence of cavities on the head, n = 1200 rpm, [46]

inception σ_i is identical in both tests, but the σ_3-curve from test 2 is shifted by about 17% towards larger flows in the domain of $\varphi_1/\varphi_{1,SF} > 0.63$ as compared with the curve from test 1.

The figures given at the measuring points stand for the acceleration ratios c_{3q}/c_2 of the fluid from the impeller outlet to the diffuser throat area. Up to $c_{3q}/c_2 \approx 1.1$, $NPSH_3$ is determined by the impeller. For accelerations significantly above 1.1, $NPSH_3$ increases according to $(c_{3q}/c_2)^x$ with x = 8 to 9 (the exponents apply only to these tests; they cannot be transferred to other conditions). The steep rise of the $NPSH_3$-curve and the application limit of the pump are clearly determined by the diffuser. This statement is generic to pumps of low specific speed.

6. Using the static pressure p_2 at the impeller outlet, it is possible to define a cavitation coefficient for the collector.

$$\sigma_{A,LE} \equiv \frac{p_2 - p_v}{\frac{\rho}{2}\, c_{3q}^2} = \frac{NPSH_A - \frac{c_s^2}{2g} + H_p}{\frac{c_{3q}^2}{2g}} = \sigma_A \left(\frac{u_1}{c_{3q}}\right)^2 - \left(\frac{c_S}{c_{3q}}\right)^2 + \psi\, R_G \left(\frac{u_2}{c_{3q}}\right)^2 \tag{6.9}$$

7. Whether the drop in head is induced by the diffuser or the impeller can also be determined by measuring the static pressure H_p at the impeller outlet and plotting it against $NPSH_A$: As long as the diffuser does not exert any influence, the loss in head and the drop of static pressure rise occur at the same $NPSH_A$. In contrast, if the diffuser is responsible for the drop in head, the curve H = f ($NPSH_A$) drops earlier than $H_p = f(NPSH_A)$.
8. It is sometimes observed that the head increases slightly, as in Fig. 6.16, before the head starts to drop. This feature confirms that the flow is affected by the cavity. The cavity causes a higher flow deflection and lift or a redistribution of the flow over the blade height. Axial pumps or impellers with thick profiles (such as used in sewage pumps) sometimes have this peculiar feature. Figure 6.16 shows measurements on an axial pump at $q^* = 1$ where this increase in head amounts to 5% while $\eta = f(NPSH_A)$ is not affected. Other pumps may show a slight improvement in efficiency. With increasing incidence the bubble development in axial pumps increases strongly; this has a particular influence on σ_i and, to a lesser extent, on σ_x, Fig. 6.16.
9. With semi-axial impellers a peak is sometimes observed in the $NPSH_3$-curve at partload as illustrated by Fig. 6.17. The peak is due to the effect of the cavity on the energy transfer in the impeller. As in Fig. 6.17, this can be combined with a hysteresis. Two stable flow conditions were therefore found during the tests reported in [46]. On the upper branch of the suction curves the c_{2m}-distribution at the impeller outlet is relatively uniform across the blade width. On the lower branch of the curve a velocity deficit was observed on the outer streamline with short cavities. With sufficiently long cavity lengths, however, the c_{2m}-distribution changed in such a way that the energy transfer at the outer streamline increased. Due to this, the head rose to values corresponding to the upper branch of the curve. Hence the deficit in head was caused by insufficient energy transfer at the outer streamline which becomes visible in the c_{2m}-distribution. Noteworthy is the finding that the cavity in both cases was identical in length, i.e. the NPSH-peak was not caused by extremely long cavities but by a change of the energy transfer in the impeller due to cavitation (which manifested itself as a change in the outlet velocity profile). These tests confirm the influence of flow pattern changes on the head and the stability of the Q-H-curve as discussed in Chap. 5.5. The importance of a preferably undisturbed energy transfer at the outer streamline is also demonstrated (Chap. 5.6.7). The tests moreover explain that the head can sometimes increase with cavitation as shown by Fig. 6.16. On all three impellers examined in [46] it was observed that the energy transfer in the impeller evened-out during cavitation (this may be interpreted as a sort of "self-healing effect" according to the principle of the least resistance). Refer also to the tests discussed in Chap. 5.7 where the peak in the curve $NPSH_3 = f(Q)$ and the instability of the Q-H-curve could be removed by machining grooves into the casing upstream of the impeller.

With strong recirculation sheet cavitation is reduced while bubble clouds appear in the impeller channels and in front of the impeller inlet. The vapor bubbles develop in the center of vortices caused by shear layers between the through-flow and recirculating fluid. The bubbles can implode on the pressure surface of the blades and on structures upstream of the impeller and cause material erosion.

While the cavity length of sheet cavitation can be determined easily during visual observation, the bubble clouds caused through the partload recirculation are difficult to quantify using stroboscopic light. Cavitation noise measurements (Chap. 6.5.2) or paint erosion tests under controlled conditions (Chap. 6.6.6) can help in this case.

Instead of the suction tests at constant flow rate described above, the $NPSH_R$ can also be determined by varying the flow rate according to Fig. 6.18. To do so, the position of the suction valve is kept constant and the discharge valve is successively opened. As a result the flow rate increases along the characteristic for cavitation-free operation, (curve "a" in Fig. 6.18), until the head starts to fall due to cavitation (points KB). With further opening of the discharge valve, the flow rate increases and the head drops due to cavitation. The gradient dH/dQ grows with increasing cavitation until the curves becomes almost vertical. As a result several curves as shown by b and c are obtained, from which $NPSH_o$, $NPSH_3$ and $NPSH_{FC}$ can be determined as sketched in Fig. 6.18. This procedure is less accurate than the suction tests according to Fig. 6.9 with constant flow rate but it is useful when, due to a lack of flexibility of the test loop, the horizontal part of the suction curve cannot be covered with a sufficient number of measuring points.

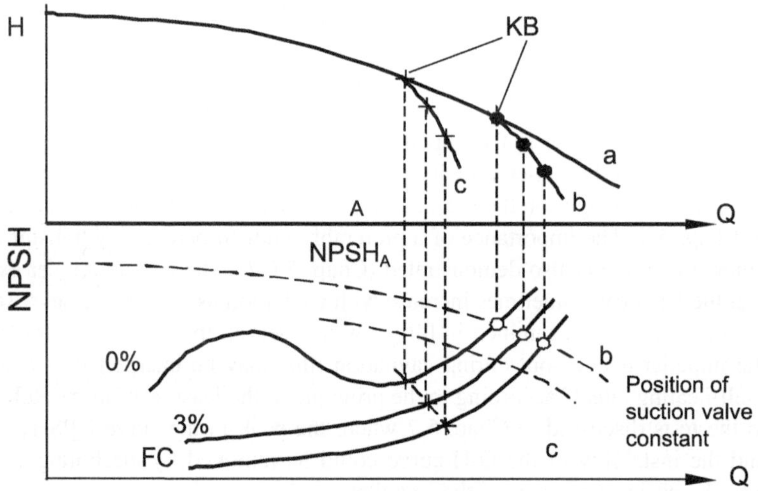

Fig. 6.18 Suction test at constant suction valve position

6.2.6 *Cavitation in Annular Seals*

Cavitation can occur in the annular seals of closed impellers and with open impellers in the gaps between the impeller blades and the casing. Here, too, vapor bubbles develop as soon as the local pressure reaches vapor pressure. Material erosion on the casing or on the impeller blades due to gap cavitation with open impellers has been observed when unsuitable materials are employed, Chap. 6.8.

Annular seal cavitation in closed impellers was investigated on a pump and a rotating cylinder in [36]. During these tests, cavitation bubbles were generated in the annular seal (close to its outlet) and in the zone where leakage and main flow mix. Cavitation increased when the suction nozzle had a smaller diameter than the impeller inlet (i.e. $d_s < d_1$). Attempting to generalize the test data from [36] the cavitation inception in an annular seal can be described by Eq. (6.9a):

$$\sigma_{sp,i} \equiv \frac{p_s - p_v}{\frac{\rho}{2}\, c_{sp}^2} = 1.2\left(\frac{u_{sp}}{c_{sp}}\right)^{0.8} \tag{6.9a}$$

This equation is based on the idea that the velocity in the seal is the parameter of primary importance. The flow rate in the suction nozzle plays a comparatively minor role. Any relationship based on the leakage flow fraction Q_{sp}/Q would be less meaningful since it would be difficult to transfer to conditions other than tested. The flow through the suction nozzle can be responsible for the generation of cavitation bubbles upstream of the impeller, on ribs or on a recess as might occur with $d_s < d_1$ (because of standardization, $d_s < d_1$ is frequently encountered).

6.3 Determination of the Required NPSH

6.3.1 *Parameters Influencing NPSH$_R$*

The formation of cavitation zones and, consequently, all values of the $NPSH_R$ according to the cavitation criteria discussed in Chap. 6.2.2 depend on many parameters:

The **inlet casing** determines the velocity profile at the impeller inlet, hence the approach flow angle of the blades. Additionally, it influences the partload recirculation. As a result, the shapes of the inlet casing as well as ribs or other structures have an effect on cavitation. An axial inlet without upstream disturbances such as bends, valves or other fittings supplies a rotationally symmetric flow to the impeller. However, it permits an intensive partload recirculation which reaches far back into the suction pipe if no ribs are present. Radial inlet chambers for between-bearing pumps generate zones with counter-rotation and pre-rotation which are caused by the flow around the shaft (Chap. 7.13, Fig. 7.50). This leads to significant variations of the cavity length over the circumference since the approach flow angle depends on the swirl upstream of the impeller inlet. Swirl is also created by piping bends arranged close to an axial nozzle.

The influencing parameters of the inlet are:

- Inlet type:
  - axial
  - radial, symmetrical (counter-rotation and pre-rotation zones over 180° each according to Fig. 7.50)
  - radial, asymmetrical (part-volute with counter rotation zone)
- Acceleration ratio in the inlet A_s/A_1
- Ribs, structures
- Hydraulic losses
- Formation of vortex ropes
- 3-dimensional velocity distribution due to disturbances upstream of the inlet nozzle
- The turbulence generated in the inlet casing has an effect on incipient and developed cavitation, [34], see also Chap. 6.4.3.

Impeller: Obviously, the impeller geometry has a decisive influence on cavitation. Experience shows that even small tolerances can have a huge effect on the cavitation inception and the erosion. Some of the parameters listed below interact; their effects cannot be determined separately - even by experiment. The impeller geometry (and consequently its influence on the cavity length) cannot be sufficiently described by the main dimensions and blade angles. For this purpose, the coordinates of the blade surfaces and front shrouds are also required. The impact of the actual impeller shape on the 3D-flow is responsible for the fact that no simple procedures for predicting the cavity length are available (apart from numerical methods). The following impeller parameters affect cavitation bubble formation and head drop (this applies mainly to radial-type impellers):

- Impeller inlet diameter and hub diameter
- Blade inlet angles on the various streamlines
- Blade number: Small blade numbers mean less blockage due to the blade thickness. Impellers with small z_{La} also have relatively long blades on a large pitch so that the cavity reaches the throat area only when the cavitation coefficients are low; σ_{FC} and σ_3 consequently fall with diminishing blade number (with otherwise identical parameters). As discussed above with reference to Table D6.2, this does not apply to the cavitation inception σ_i.
- The throat area at the impeller inlet determines the deceleration or acceleration w_{1q}/w_1 of the relative velocity. The throat area depends on the shape of the entire blade - even on the blade outlet angle, which exerts a significant influence with small blade numbers.
- Profiling of the blade leading edges, see Chap. 7.2.1. Cavitation inception is sensitive to tiny differences in geometry, hence also to manufacturing tolerances. Unfavorable leading edge profiles - such as semi-circular and most of all blunt

shapes - increase the cavity volume and length and thus impair performance. The $NPSH_3$ and $NPSH_{FC}$, rise accordingly as shown by the test data in [2] mentioned in Chap. 7.2.1.

- Position of the blade leading edges in the meridional section
- Position of the blade leading edges relative to the flow (swept forward or backward, see Chap. 7, Fig. 7.8)
- Blade angle development between inlet edge and throat area
- Shape and curvature of front shroud and hub/rear shroud
- Fillet radii between the blades and the impeller side walls
- Roughness of blades (influence on the first bubbles)
- Partload recirculation: σ_i and σ_o usually fall once recirculation sets in.
- If an impeller is trimmed significantly (Chap. 4.5.1), the $NPSH_R$ values increase since the blade loading (hence the local excess velocities) increase. Moreover, 3% of head of a severely trimmed impeller means a lower absolute drop than with full diameter so that a higher $NPSH_3$ value is measured after trimming. Example: if an impeller with full diameter produces a head of H_{opt} = 200 m, the $NPSH_3$ is given by a head drop of ΔH = 6 m; if the impeller is trimmed and only has H_{opt} = 140 m, $NPSH_3$ is obtained at a ΔH of only 4.2 m.

Seal clearances: The leakages through annular (or radial) seals have a dual effect: (a) They increase the flow rate Q_{La} through the impeller: increasing the seal clearance, means shifting the curve $NPSH_3$ = (Q) to lower flow by the amount ΔQ_{sp}. (b) The leakage increases the pre-swirl of the fluid entering the impeller since it has a circumferential component of approximately 50% of the circumferential speed at the annular seal $c_{u,sp} \approx ½u_{sp}$, Both of these effects increase the flow angle at the impeller inlet. This has three consequences: (1) Deterioration of the cavitation characteristics in the overload range; (2) Cavitation inception at lower σ_i for flows below shockless entry, but the effect may be offset by vortices created by the leakage entering the main flow; (3) Shorter cavity length at partload.

The leakage exits the seal at a high velocity like a jet which is dissipated in the main flow. This process is linked to vortices which may generate cavitation in the vortex core if the absolute pressure is sufficiently low. Balance holes for axial thrust reduction have similar effects as the seal leakage: the smooth flow is disturbed by additional turbulence and vortices. Overall, the sensitivity of the suction performance to the leakage depends on: (1) the clearances and the associated increase in the approach flow angle, (2) the blade angles and profile since these determine the incidence and pressure distribution, (3) the absolute $NPSH_A$ which has an impact on vortex cavitation.

The effect of seal clearances on $NPSH_3$ has been measured on an OH2-process pump (impeller with balance holes), [67]. Data are given in Table D7.1-1. Two impellers have been tested: impeller "n_{ss}186" with incidence $i_{1a,opt}$ = 4.8° and impeller "n_{ss}346" with $i_{1a,opt}$ = −3.4°. Tests were done with clearances s/s_{API} = 0.75; 1.0 and 2.0 with s_{API} from [N.6], Eq. (3.12). Upon increasing the clearances, the $NPSH_3$-curves were shifted left and upwards. At BEP this shift corresponds to a factor f = 1.04 for impeller "n_{ss}186", and f = 1.2 for impeller "n_{ss}346" when doubling the clearances i.e. at $s/s_{API} = 2$. With seal clearances at $s/s_{API} = 0.75$, the $NPSH_{3,opt}$ was reduced by about 4% as compared to $s/s_{API} = 1$ for both impellers. The higher sensitivity of impeller "n_{ss}346" to an increase in seal clearances could be caused by the strong negative incidence. Since the leakage increases the flow angle (as explained above), an impeller with positive incidence should be more forgiving for an increase in leakage than an impeller with negative incidence. Furthermore, the absolute $NPSH_A$ may have an impact on vaporization in front of the impeller where the leakage mixes with the main flow (vortices). The geometry at the exit of annular seal (axial gap between impeller neck-ring and casing) which influences the mixing of the leakage with the main stream had no measurable influence on $NPSH_3$. It is difficult to generalize these measurements to predict the effect of the leakage on $NPSH_3$. In order to estimate the effect of seal wear on $NPSH_3$, one can follow these steps: (1) Calculate leakage increase ΔQ_{sp} according to Table 3.7; (2) Shift curve $NPSH_3 = f(Q)$ by ΔQ_{sp} to the left; (3) Add some margin for the effect of increased pre-rotation. That margin could be 50 to 80% of the correction calculated from step 2. The effect of the increase in leakage on the Q-H-curve can be estimated from Table 4.4; a procedure for estimating the pre-swirl and shifting the NPSH-curve due to increased leakage is also given there. When defining NPSH-margins, the effect of leakage increase due to wear of the annular seals should be taken into account since many specifications require pumps to be able to operate at double the nominal seal clearances.

Collector: If intensive enough, partload recirculation from the collector can influence the recirculation at the impeller inlet which has an effect on the cavity shape and its length. The cavitation patterns at partload therefore depend on whether the impeller pumps into a volute or into a diffuser. The variation of pressure in the volute over the circumference causes the individual impeller channels to operate on different points of the Q-H-curve, i.e. at different flow rates. As a consequence, the individual blades operate with different incidence, which has an effect on the blade pressure distribution and the cavity (Chap. 9.3.3).

The essential parameters in this context are:

- Type of collector (volute casing, diffuser, annular casing)
- Throat area in the diffuser or volute casing (Fig. 6.15).

Since the flow separation in the collector depends on the inflow conditions, the geometry of the impeller outlet (outlet width, outlet angle etc.) has some effect on the cavitation at partload. During partload recirculation there is a strong interaction between the inlet casing, impeller and collector. A separate consideration of individual components is therefore not always able to supply adequate information on the flow behavior.

Operating parameters: The cavitation coefficient σ_A determines the extent of the cavitation in a given geometry and at a given flow coefficient. The flow coefficient (defined by the ratio c_{1m}/u_1) determines the incidence which has an impact on the pressure distribution on the blades. Since the cavitation erosion according to Chap. 6.6.3 increases with the 6th power of the circumferential speed, even a seemingly minor extrapolation of the available operating experience can lead to problems with high-speed pumps. The influence of fluid properties and gas content is discussed in 6.4.

6.3.2 *Calculation of the $NPSH_R$*

As explained in context with Fig. 6.5, the minimum local pressure at the impeller inlet results from: (1) the acceleration of the main flow and the losses in the inlet; (2) local excess velocities generated by the flow around the blade leading edge. The following relationship is therefore frequently applied:

$$\mathrm{NPSH} = \lambda_c \frac{c_{1m}^2}{2g} + \lambda_w \frac{w_1^2}{2g} \tag{6.10}$$

or expressed as the cavitation coefficient according to Eq. (6.5):

$$\sigma = (\lambda_c + \lambda_w)\,\varphi_1^2 + \lambda_w \left(1 - \frac{\varphi_1}{\tan\alpha_1}\right)^2 \tag{6.11}$$

The coefficient λ_c covers the acceleration and losses at the inlet; λ_w accounts for the low-pressure peak at the blades. The coefficient λ_c can be determined from: $\lambda_c = 1 + \zeta_E$ (ζ_E = loss coefficient of the inlet). For axial inflow, $\lambda_c = 1.1$ can be used and $\lambda_c = 1.2$ to 1.35 for pumps with radial inlet casings (depending on design). Since $c_1 \ll w_1$, any uncertainty with λ_c is not as significant as with λ_w.

Conversely, λ_w does not only depend on the cavitation criterion considered (e.g. $NPSH_i$ or $NPSH_3$), but also on all the geometrical and operational parameters listed in Chap. 6.3.1. Table 6.3 supplies typical ranges for λ_w in operation at shockless entry (or with $q^* = 1$):

Some impellers still lie outside these ranges. If Eq. (6.11) is substituted in Eq. (6.8) the following is obtained for the suction specific speed:

$$n_{ss} = \frac{158\sqrt{\varphi_1\, k_n}}{\left\{(\lambda_c + \lambda_w)\,\varphi_1^2 + \lambda_w\left(1 - \frac{\varphi_1}{\tan\alpha_1}\right)^2\right\}^{0.75}} \tag{6.12}$$

Equations (6.8) and (6.12) show that the achievable suction specific speed diminishes with increasing hub ratio $\nu = d_n/d_1$ since $k_n = 1 - \nu^2$.

Table D6.3 Coefficients λ_c and λ_w at shockless flow

		$\lambda_{w,3}$		$\lambda_{w,i}$
Impeller	$NPSH_3$	0.1 to 0.3	$NPSH_i$	0.4 to 1.5 (to 2.5)
Inducer		0.03 to 0.06		0.2 to 0.3

Although λ_w depends on the inlet and impeller geometry (and this dependency is largely unknown), a defined value of λ_w is frequently assumed for optimizing the impeller inlet diameter. To do so, the following is substituted in Eq. (6.10): $c_{1m} = 4 \times Q/\left[\pi \times (d_1{}^2 - d_n{}^2)\right]$; $w_1{}^2 = c_{1m}{}^2 + (u_1 - c_{1m}/\tan\alpha_1)^2$ and $u_1 = \pi \times d_1 \times n/60$. The resulting equation is differentiated $\partial NPSH/\partial d_1$. Finally, the expression obtained is set equal to zero and solved for the optimum inlet diameter $d_{1,opt}$:

$$d_{1,opt} = \left\{d_n^2 + 10.6\left(\frac{Q_{La}}{f_q\, n}\right)^{\frac{2}{3}}\left(\frac{\lambda_c + \lambda_w}{\lambda_w}\right)^{\frac{1}{3}}\right\}^{\frac{1}{2}} = \left\{d_n^2 + d_2^2\left(\varphi_{2_{La}} b_2^*\right)^{\frac{2}{3}}\left(32\frac{\lambda_c + \lambda_w}{\lambda_w}\right)^{\frac{1}{3}}\right\}^{\frac{1}{2}}$$

$$d_{1,opt} = \left\{d_n^2 + 1.48 \times 10^{-3}\,\psi\, n_q^{1.33} d_2^2\left(\frac{\lambda_c + \lambda_w}{\lambda_w}\right)^{\frac{1}{3}}\right\}^{\frac{1}{2}} \tag{6.13}$$

The optimum flow coefficient (calculated with this $d_{1,opt}$) is obtained as follows:

$$\varphi_{1,opt} = \left(\frac{Q_{La}\,\lambda_w}{f_q\, n(\lambda_c + \lambda_w)}\right)^{\frac{1}{3}} \frac{2.3}{\left\{d_n^2 + 10.6\left(\frac{Q_{La}}{f_q\, n}\right)^{\frac{2}{3}}\left(\frac{\lambda_c + \lambda_w}{\lambda_w}\right)^{\frac{1}{3}}\right\}^{\frac{1}{2}}} \tag{6.14}$$

In Eqs. (6.13) and (6.14) it was assumed that the absolute hub diameter d_n is determined by the mechanical design. Alternatively, the hub ratio $k_n = 1 - d_n{}^2/d_1{}^2$ can be assumed as constant, resulting in simpler equations:

$$d_{1,opt} = 3.25\left(\frac{Q_{La}}{f_q\, n\, k_n}\right)^{\frac{1}{3}}\left(\frac{\lambda_c + \lambda_w}{\lambda_w}\right)^{\frac{1}{6}} \tag{6.14a}$$

$$\varphi_{1,opt} = \sqrt{\frac{\lambda_w}{2(\lambda_c + \lambda_w)}} \tag{6.14b}$$

$$n_{ss} = \frac{98}{(\lambda_c + \lambda_w)^{0.25}}\left(\frac{k_n}{\lambda_w}\right)^{0.5} \tag{6.14c}$$

The Eqs. (6.13 to 6.14c) apply to inflow without pre-rotation. They can be used to optimize the impeller inlet for a selected suction specific speed ($NPSH_3$) or for low cavitation inception ($NPSH_i$). Merely the appropriate value for λ_w must be entered in each case. However, the resulting inlet diameter is only optimal *if the* λ_w *selected for the calculation is indeed achieved in practice*. Whether this is the case ultimately depends on all the parameters discussed in Chap. 6.3.1.

One could now assume that λ_w - as with an air foil - primarily depends on the incidence $i_1 = \beta_{1B} - \beta'_1$. Experience shows, and specific tests confirm, that $\lambda_{w.3}$ depends on both the *incidence* (i.e. on the flow rate with given impeller and inlet casing) and on the *absolute blade angle* (i.e. the cascade stagger angle). The minimum pressure coefficients $c_{p,min}$ increase with a growing flow angle β_1 or with the blade inlet angle β_{1B}. Figures 6.19 and 6.20 show an evaluation of various tests. The values of λ_w thus obtained show this dependency, albeit with a large scatter.

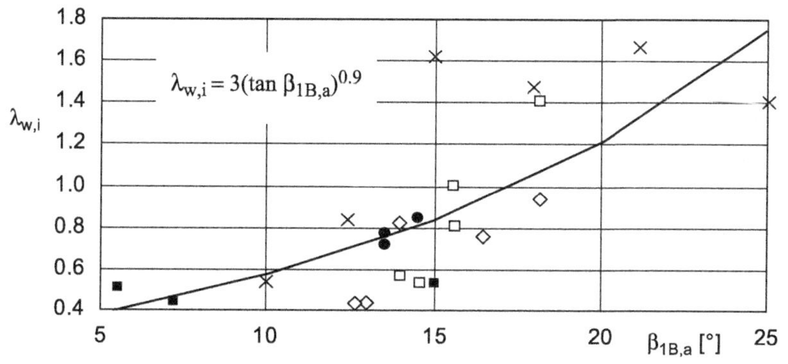

Fig. 6.19 Coefficients for cavitation inception as function of the impeller *blade* inlet angle at the outer streamline

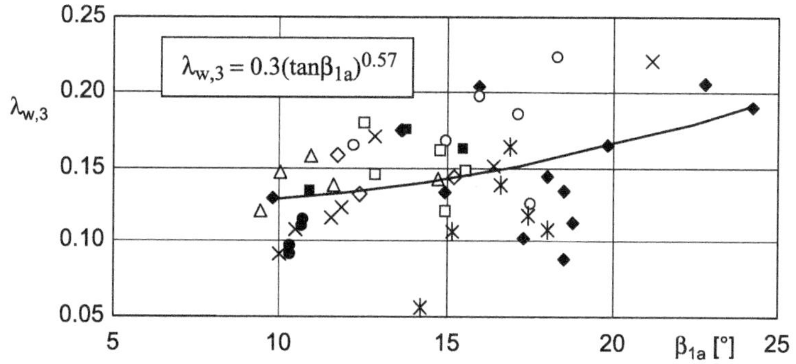

Fig. 6.20 Coefficients for 3% head drop as function of the *flow* angle at $q^* = 1$ at the outer streamline

Figures 6.19 and 6.20 apply to the operation at best efficiency point when it can be assumed that the flow rate of shockless entry is near Q_{opt}. Figures 6.19 and 6.20 show that $\lambda_{w,i}$ depends more on the blade angle than $\lambda_{w.3}$ does on the flow angle. Note that the data is better correlated by the *blade* angle in Fig. 6.19 and the *inflow* angle in Fig. 6.20. The tests in Fig. 6.19 were conducted on radial inlets and evaluated with $\lambda_c = 1.35$; the standard deviation amounts to ±35%. The data in Fig. 6.20 originates from measurements with radial inlets ($\lambda_c = 1.35$) and with axial suction ($\lambda_c = 1.1$). The standard deviation is ±25%. When applying blade coefficients $\lambda_{w,}$ it must always be borne in mind that these coefficients are very sensitive to blade design, flow rate and approach flow profile.

If $\lambda_{w,3}$ depends on the approach flow angle, it should also be possible to establish a correlation between the suction specific speed and the approach flow angle. If this is performed for a wide range of applications it must be taken into account that the suction specific speed according to Eq. (6.12) drops with increasing hub ratio and that the flow coefficient φ_1 should grow with increasing specific speed in order to obtain optimum impellers. With the three parameters of approach flow angle (or φ_1), specific speed and hub ratio it is possible to form a normalized suction specific speed $n_{ss}{}^{**}$:

$$n_{ss}{}^{**} = \frac{n_{ss}}{\sqrt{k_n}}\left(\frac{n_{q,Ref}}{n_q}\right)^{0.19} \quad \text{for } n_q < 170 \text{ with } n_{q,Ref} = 27 \tag{6.15}$$

The quantity $n_{ss}{}^{**}$ was plotted against the approach flow angle β_{1a} in Fig. 6.21. The test data comprises pumps with radial and axial approach flow and radial as well as semi-axial impellers in the range $n_q = 10$ to 160. The standard deviation amounts to ±14%. The solid curve given by Eq. (6.16a) can be used for design; the dashed curve represents the data in Table D7.1-1.

To determine the inlet diameter of an impeller for selected values of n_q, n_{ss} and d_n/d_1 (or k_n) it is possible to determine $n_{ss}{}^{**}$ from Eq. (6.15): read the approach flow angle to be designed from Fig. 6.21 and to calculate the inlet diameter d_1 from Eq. (6.16). If d_n is specified, iteration by way of k_n is required.

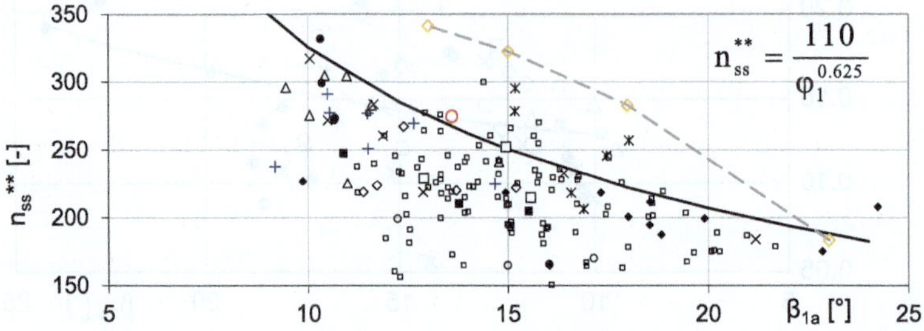

Fig. 6.21 Normalized suction specific speed as function of the *flow* angle at q* = 1 at the outer streamline; $n_{q,Ref} = 27$

$$d_1 = 2.9\sqrt[3]{\frac{Q_{La}}{f_q\, n\, k_n \tan\beta_1}\left(1 + \frac{\tan\beta_1}{\tan\alpha_1}\right)} \tag{6.16}$$

Or expressed in a dimensionless form valid for $\alpha_1 = 90°$:

$$d_1^* \equiv \frac{d_1}{d_2} = \frac{0.483\ \omega_s^{0.667}\ \psi^{0.5}}{(k_n \tan\beta_1)^{0.333}} \tag{6.16a}$$

The relationship between the approach flow angle and the normalized suction specific speed according to Fig. 6.21 can also be expressed by Eq. (6.16b); the suction specific speed for a given approach flow angle or flow coefficient $\varphi_1 = \tan\beta_{1a}$ can then be determined from:

$$n_{ss}^{**} = \frac{110}{\varphi_1^{0.625}} \tag{6.16b}$$

Conversely, it is possible to calculate the design flow coefficient for a desired suction specific speed:

$$\varphi_1 = k_n^{0.8}\left\{\frac{110}{n_{ss}}\right\}^{1.6}\left\{\frac{n_q}{n_{q,Ref}}\right\}^{0.3} \tag{6.16c}$$

These formulae would be applicable to a state-of-the-art design with low incidence and low blade loading near the leading edge, see Chap. 7.2.4.

6.3.3 *Estimation of the $NPSH_3$ as Function of the Flow Rate*

As seen from the test data in Fig. 6.20, the blade coefficients $\lambda_{w,3}$ at BEP are subject to a great scatter. If one wants to estimate the shape of the $NPSH_3$-curve over the flow rate, major uncertainties must be expected too. To determine $NPSH_3 = f(Q)$, the following consideration is applied: In the domain where the relative velocity w_1 up to the impeller throat area A_{1q} is decelerated to w_{1q} and while the incidence is still positive, the $NPSH_3$-curve is rather flat. In contrast, it rises steeply when a cavity forms on the pressure surface of the blade and when the flow is accelerated from w_1 to w_{1q}. The boundary Q_{sa} between these two domains therefore has to be defined. Furthermore, the diffuser influence must be evaluated. It can cause a premature steep rise of the $NPSH_3$-curves, especially with low specific speeds. Three flow rate criteria can be utilized to evaluate the steep rise of the NPSH-curve:

1. If the flow rate of shockless entry Q_{SF} is exceeded, the incidence becomes negative and cavitation bubbles tend to appear on the pressure surface of the blades:

$$Q_{SF} = \frac{f_q\, A_1 \tan\beta_{1B}\, u_{1a}}{\tau_1 \left(1 + \frac{\tan\beta_{1B}}{\tau_1 \tan\alpha_1}\right)} \tag{6.17}$$

2. The flow rate Q_w, where $w_1 = w_{1q}$, is calculated from:

$$Q_W = \frac{f_q\, A_1\, u_{1m}}{\sqrt{\left(\frac{A_1}{z_{La}\, A_{1q}}\right)^2 - 1} + \frac{1}{\tan\alpha_1}} \tag{6.18}$$

3. The flow rate Q_c, where $c_{2u} = c_{3q}$, is obtained from:[5]

$$Q_c = \frac{f_q\, u_2\, A_2\, \gamma}{\frac{f_q\, A_2}{z_{Le}\, A_{3q}} + \frac{\tau_2}{\eta_v \tan\beta_{2B}}} \tag{6.19}$$

To estimate the NPSH$_3$-curve it is possible to utilize the flow rates calculated according to Eqs. (6.17 to 6.19) as per Fig. 6.22 as follows:

1. It is assumed that the pressure surface cavitation becomes effective at $Q > Q_{PS}$, Q_{PS} being defined as the average value of Q_w and Q_{SF}:

$$Q_{PS} = \frac{1}{2}(Q_{SF} + Q_W) \tag{6.20}$$

2. If Q_c is smaller than Q_{PS}, it must be assumed that the collector has an effect on the NPSH value in the range $Q > Q_c$ and significantly determines the steep rise of the NPSH$_3$-curve.
3. The flow rate Q_{sa}, which marks the transition from the flat to the steep part of the NPSH-curve, is chosen as the smaller of the two values Q_c and Q_{PS}; hence $Q_{sa} = Q_{PS}$ if $Q_c > Q_{PS}$ and $Q_{sa} = Q_c$ if $Q_c < Q_{PS}$.
4. The NPSH$_3$ is now calculated for the design flow rate Q_{opt} according to Eq. (6.10). $\lambda_{w,3}$ at the best efficiency point is determined as a function of the approach flow angle $\beta_{1,a}$ from Fig. 6.20.

[5] Since α_2 of pumps with low specific speeds is very small, $c_2 \approx c_{2u}$ supplies a sufficient approximation that allows the simplified relationship according to Eq. (6.19).

5. In the domain where $Q < Q_{sa}$ applies, $NPSH_3 = f(Q)$ is determined from Eq. (6.10). To this end, the approach flow angle β_1 is calculated as a function of the flow rate and the applicable blade coefficient λ_w is determined from Eq. (6.21) so that an NPSH-curve is obtained that rises with the flow rate.

$$\lambda_w = \lambda_{w,opt}\left(\frac{\tan\beta_1}{\tan\beta_{1,opt}}\right)^{0.57} \tag{6.21}$$

   This calculation is pursued up to the flow rate Q_{sa} defined above, for which we obtain the value $NPSH_{sa}$.
6. In the domain with pressure surface cavitation defined by $Q > Q_{sa}$ one assumes that $NPSH_3$ increases with the square (or a higher power) of the flow rate:

$$NPSH = NPSH_{sa}\left(\frac{Q}{Q_{sa}}\right)^{x} \tag{6.22}$$

The exponent selected is x = 2 to 3, but it should be borne in mind that the NPSH-curve rises very steeply when a certain flow rate is exceeded. Then the flow rate hits a limit which is sometimes referred to as “choking” in analogy to compressor technology. In principle, the exponent increases with the flow rate and becomes very large at choking conditions.

The above described method can be easily adapted to test results to improve the accuracy of the NPSH-prediction. For $Q > Q_{opt}$ the curve for full cavitation generally lies only slightly below the values of the $NPSH_3$. In the domain of the steep rise the curve for $NPSH_o$ is frequently just a little above the $NPSH_3$-curve.

The cavitation inception $NPSH_i$ at the impeller inlet is best calculated by way of the pressure distribution using numerical methods according to Chap. 8. Since the cavitation inception is very sensitive to the approach flow distribution and the

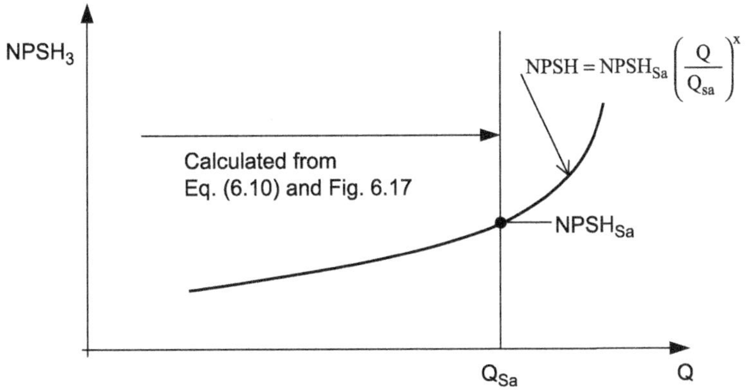

Fig. 6.22 Estimation of the curve $NPSH_3 = f(Q)$

geometrical characteristics of the impeller, a generally applicable empirical method is not available. Numerical methods can be used successfully since the pressure distribution at the blade leading edge depends little on viscous effects.

Figure 6.23 depicts typical curves indicating how the $NPSH_3$ depends on the flow rate. For low specific speeds $NPSH_3$ falls steadily with the flow as indicated by the curve for $n_q = 20$; at higher specific speeds the $NPSH_3$-curves increasingly run flatter at partload, ultimately rising again towards $Q = 0$. With axial impellers the $NPSH_3$-value increases sharply as soon as the flow rate deviates significantly from the shockless entry. Consequently, the operating range becomes relatively narrow (see Fig. 6.16).

Diffuser cavitation: As already mentioned, the influence of cavitation in the collector at $q^* > 1.1$ to 1.3 must not be neglected with specific speeds below $n_q = 20$ to 30. During this evaluation it is also possible to take into account the meridional component c_{2m} neglected in Eq. (6.19). The flow rate of the impeller, where $c_{3q} = c_2$, is then obtained from:

$$Q_{La,c} = u_2\, f_q A_2\, \gamma \frac{\sqrt{a^2 - 1} - \frac{\tau_2}{\tan \beta_{2B}}}{a^2 - 1 - \left\{\frac{\tau_2}{\tan \beta_{2B}}\right\}^2} \quad \text{with } a \equiv \frac{f_q\, A_2\, Q_{Le}}{z_{Le}\, A_{3q}\, Q_{La}} \qquad (6.22a)$$

This equation, in essence, agrees with the investigations in [48]. The pumping limit through full cavitation in the collector is reached when the static pressure p_3 in the throat area drops to the vapor pressure p_v. To determine the velocity $c_{3q,FC}$ at which this condition is reached, Bernoulli's equation is applied between the suction nozzle and the diffuser throat area and solved subsequently for $c_{3q,FC}$ or the volumetric flow $Q_{3q,FC} = c_{3q,FC} \times z_{Le} \times A_{3q}$:

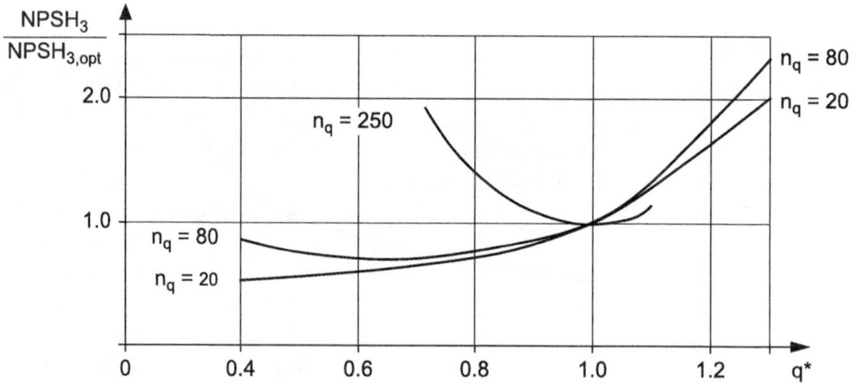

Fig. 6.23 Influence of the specific speed on the shape of the curve $NPSH_3 = f(q^*)$

$$\frac{p_s}{\rho} + \frac{c_s^2}{2} + g(H + H_{v,3-6}) = \frac{p_3}{\rho} + \frac{c_{3q,FC}^2}{2}$$
$$Q_{FC,Le} = u_2\, z_{Le} A_{3q} \sqrt{\sigma_A \left(\frac{d_1}{d_2}\right)^2 + \psi + \zeta_{3-6}} \tag{6.22b}$$

Investigations in [48] found the visual cavitation inception in the diffuser at approximately $c_{3q} = c_2$ according to Eq. (6.22a) while the pumping limit was reached at a volumetric flow that was slightly smaller than calculated according to Eq. (6.22b). The influence of the pressure loss in the diffuser and in the return channels in Eq. (6.22b) is covered by $\zeta_{3-6.}$ In view of the findings in [48] it may be neglected. Measurements with regard to the pumping limit according to Eq. (6.22b) have been reported on in [16].

6.4 Influence of the Fluid Properties

Development and disappearance of cavitation bubbles depend on the gas content and the physical properties of the fluid. These include the surface tension between vapor/gas and liquid as well as the diffusion processes during the dissolution of the gases. In the following we distinguish between:

- Thermodynamic parameters which determine the energy transport during evaporation and condensation of the bubbles
- The influence of dissolved and free gases
- The nuclei content.

On these complex processes little quantitative information is available that could be used in practice. The influence of the liquid properties must be clearly distinguished from the model laws or "scale effects" treated in Chap. 6.2.3. These laws cover the parameters of pump size, speed and viscosity (i.e. Reynolds number) - irrespective of the fact that the behavior of the cavitation nuclei depends on the flow velocity.

6.4.1 Thermodynamic Effects

To generate a vapor bubble of volume V_b the corresponding evaporation enthalpy must be made available. It corresponds to the energy $E = \rho'' \times V_b \times h_v$ (wherein ρ'' is the density of saturated vapor in the bubble; h_v is the evaporation enthalpy). The energy E is transported through thermal conduction and convection from the surrounding liquid to the boundary of the bubble. This process requires a finite temperature difference ΔT_u. According to the equation by Clausius-Clapeyron, ΔT_u corresponds to the pressure difference Δp_u by which the pressure in the bubble is below the vapor pressure:

$$\Delta p_u = \frac{\rho' \rho''}{\rho' - \rho''} \cdot \frac{h_v \Delta T_u}{T} \tag{6.23}$$

Because of the shear stresses between the flowing liquid and the cavity, a flow prevails in the cavity that can be sustained only if the vapor transported downstream is continually produced through evaporation in the front part of the cavity. Based on the measurements in [14], the generated vapor flow can be estimated from Q $'' = c_Q \times L_{cav} \times B_{cav} \times w_1$ with $c_Q = 0.0052$ as average value from measurements on a profile (the range measured is $c_Q = 0.004$ to 0.007). L_{cav} and B_{cav} are length and width of the cavity respectively. From the energy necessary for the evaporation of Q″ and the heat transport it is possible to derive a relationship to estimate the temperature difference between a bubble and the surrounding liquid [14]. In simplified form it is:

$$\Delta T_u = 2\,\mathrm{Pr}^{0.67} \frac{h_v\, \rho''}{c'_p\, \rho'} \tag{6.24}$$

Herein, $\mathrm{Pr} = \rho' \times \nu \times c_p'/\lambda'$ is the Prandtl number, c_p' is the specific heat and λ' is the thermal conductivity of the liquid. From the value calculated for ΔT_u according to Eq. (6.24) the pressure difference Δp_u is obtained with Eq. (6.23) as follows:

$$\Delta p_u = 2\,\mathrm{Pr}^{0.67} \frac{(\rho'' h_v)^2}{(\rho' - \rho'')\,T\, c'_p} \tag{6.25}$$

For water at 20 °C one obtains negligibly small values: $\Delta T_u = 0.07$ °C and $\Delta p_u = 10\ \mathrm{N/m^2}$. In practice the significance of the thermodynamic effects is felt above roughly 150 °C. Hydrodynamic cavitation intensity and damage risk are reduced accordingly.

Stepanoff used a factor $B = \rho' \times c'_p \times \Delta T_u/(\rho' \times h_v)$ to correlate the influence of the liquid properties while $\Delta T_u = f(\Delta p_u)$ was employed for an assumed pressure reduction according to Eq. (6.23), [B.2]. A comparison with Eq. (6.24) yields the relation $B = 2\ \mathrm{Pr}^{0.67}$.

The energy supplied during the bubble creation must be removed by heat convection and conduction during the implosion. In the initial phase the excess temperatures developing during the implosion may correspond to the values from Eq. (6.24). In the final phase of the implosion, the gas and a residual vapor component contained in the bubble are compressed almost adiabatically. Very high temperatures are then generated in a tiny volume. At times these have been observed as luminescent phenomena.

At the critical point, liquid and gaseous phase cannot be distinguished since $\rho' = \rho''$ and $h_v = 0$. On approaching the critical point any cavitation therefore disappears. The greater the value for ΔT_u calculated from Eq. (6.24), the more bubble growth and implosion are retarded by the required heat transport. As a rule of thumb, it can be assumed that thermodynamic effects at $\Delta T_u < 1$ °C can be neglected but that they increasingly reduce the hydrodynamic cavitation intensity at $\Delta T_u > 2$ to 5 °C until cavitation disappears at the critical point. The influence of the fluid properties on the cavitation intensity and the risk of damage must be assumed as significant when exceeding $\Delta T_u > 5$ °C.

As has been shown by tests with various liquids, the $NPSH_3$ required is reduced if noticeable thermodynamic effects are to be expected or ΔT_u becomes large. This improvement can be estimated by calculating Δp_u from Eq. (6.25) and the NPSH improvement $\Delta NPSH_3$ from Eq. (6.26) (the values obtained for water are shown in Fig. 6.24).

$$\Delta NPSH = a \left(\frac{\Delta\, p_u}{\rho' \, g \, H_{Ref}}\right)^{0.58} \quad \text{with } H_{Ref} = 1.0\,\text{m} \tag{6.26}$$

The empirical constant for water is a = 0.43 m and for hydrocarbons approximately a = 0.25 m. The reduction of the $NPSH_3$ according to Eq. (6.26) provides only a rough estimation. Similar values are obtained as from the graph in [N.4]. The tests used there cover the range up to about $\Delta T_u < 7$ °C and $\Delta p_u < 2.4$ bar. The improvement of the $NPSH_3$ expected from thermodynamic effects for a given pump application should be limited to a maximum of 50%, even if a greater improvement is calculated according to Eq. (6.26). Also the absolute value of the correction should be limited to $\Delta NPSH_{3,max} = 2$ to 4 m.

Stroboscopic cavitation observations at water temperatures of 60 to 180 °C in [25] demonstrated that the cavity length (as measured from the impeller blade leading edge) is independent of the water temperature - with only minor scatter. Experience confirms that the cavities observed on model pumps with cold water correspond well to the erosion patterns on geometrically similar impellers in power station pumps (temperatures up to roughly 200 °C). These observations do not

Fig. 6.24 Influence of the water temperature on cavitation

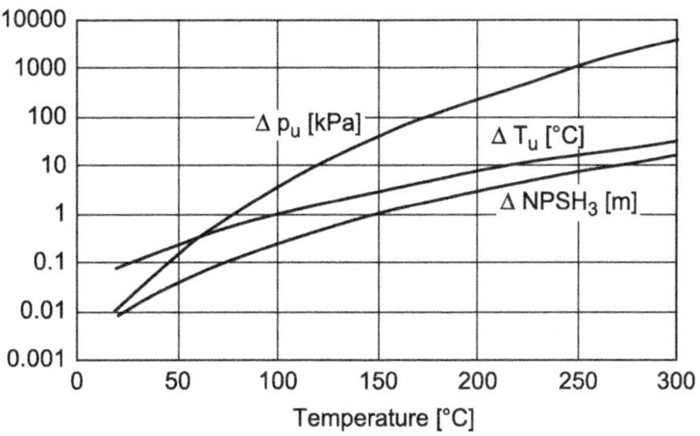

contradict the statement that the hydrodynamic cavitation intensity with water at temperatures above 120 to 140 °C significantly diminishes compared with cold water. This is because the damage potential is affected not only by the length of the cavity, but also by the de-aeration of the feed water, the cavity thickness or vapor production Q'' and the thermodynamics of the implosion process.

Equations (6.24 to 6.26) do not only apply to water but also to any other liquids such as hydrocarbons or refrigerants. If hydrocarbon mixtures contain components with different vapor pressures, ΔT_u must be calculated separately for each component. With mixtures of this type the individual components evaporate and condense to different degrees depending on local pressure. Thus the cavitation process takes place gradually and only minor cavitation intensities are generated.

Of the quantities used in Eq. (6.24) the vapor density has by far the strongest dependence on the water temperature. Therefore the vapor density has the greatest effect on the hydrodynamic cavitation intensity, since it directly determines the mass of liquid to be evaporated. Secondary influences are added to the parameters in Eq. (6.24) when considering cavitation erosion:

- Density of the liquid as a measure for the "strength of water hammer" during the implosion
- Speed of sound in the liquid (effect similar to density)
- The surface tension is significant for nuclei spectrum and bubble growth.
- The viscosity only has an effect on the cavitation if it reaches the values of heavy oil. Cavitation inception $NPSH_i$ depends on the Reynolds number.

6.4.2 *Non-condensable Gases*

If a medium of temperature T_f contains gases whose critical temperature $T_{crit} < T_f$, no conversion from the gaseous to the liquid phase occurs upon pressure increase: In the prevailing state such gases are "non-condensable". If a cavitation bubble contains non-condensable gases (e.g. air dissolved in water) these are adiabatically compressed during the implosion. The energy necessary for gas compression reduces the implosion pressures or the hydraulic cavitation intensity at a given potential energy as defined by Eq. (6.2). As a consequence, dissolved and free gases reduce cavitation noise, vibrations and damage. In special cases, even some air can be introduced into the suction pipe to alleviate cavitation problems.

If a liquid contains high amounts of dissolved gases which separate in the low pressure zones, large gas-filled cavitation zones can be created at the impeller inlet. Work transfer is consequently impaired. Pumping can even be interrupted in extreme cases ("gas-locking").

If x_D kg of gas per kg of liquid is dissolved in the suction reservoir at pressure p_e, the volumetric gas component Q_{gas}/Q' at pressure p is calculated as follows:

$$\frac{Q_{gas}}{Q'} \approx \frac{p_e - p_v}{p - p_v} GLR_{in} + \frac{\rho\, x_D\, R\, T}{p - p_v}\left(1 - \frac{p - p_v}{p_e - p_v}\right) \tag{6.27}$$

Q′ is the volumetric flow of the liquid; GLR_{in} is the volumetric gas/liquid ratio of the free gases which might already be present at the inlet of the suction pipe.

The calculation can be made for the pressure in the inlet nozzle ($p = p_s$) or an assumed minimal pressure at the blade (note that the pressure in the cavity $p_B = p_v + p_g$).

Equation (6.27) is based on Henry's law according to which the component of dissolved gases is proportional to the *partial pressure* of the gas above the liquid level in the suction reservoir. Gas is soluble in a liquid only up to a certain extent. The solubility described by this saturation limit depends on the type of the liquid, the type of the gas and the temperature. An air mass fraction of 24 ppm is soluble in water at a temperature of 20 °C and a partial pressure of 1 bar (which, in Eq. (6.27) would have to be substituted as $x_D = 24 \times 10^{-6}$). The solubility usually drops with increasing temperature; it is zero at the boiling point. In Appendix A3 these relationships are explained in more detail. Equations and diagrams to determine the solubility of various gases in water can also be found there.

Normal pumps, which are not specifically designed to transport gas-liquid mixtures, can generally accommodate (at inlet pressures near 1 bar) up to approximately 2% gas volume in the inlet nozzle without appreciable effect on the head. At gas contents above 5 to 10%, the pumps usually become gas-locked (i.e. the flow stops), Chap. 13.2.

6.4.3 Nuclei Content and Tensile Stresses in the Liquid

If a liquid contains few cavitation nuclei, the vapor generation is delayed and only occurs at lower pressures than in a liquid rich in nuclei. "The water is able to sustain tensile stresses" or a "boiling delay" occurs. Cavitation inception occurs at a lower $NPSH_A$ if the tensile strength Z increases:

$$NPSH_i = \frac{p_s - p_v - |Z|}{\rho\, g} + \frac{{c_s}^2}{2g} \tag{6.27a}$$

To exclude uncertainties with regard to the sign of the tensile strength, the absolute value of Z was substituted in Eq. (6.27a). The *gaseous* cavitation discussed in Chap. 6.1.1 is described by the local pressure and the solubility of the gas in a physically meaningful way and is therefore distinguished from the term "tensile stresses"; hence solubility effects are not covered by Eq. (6.27a). The effect of tensile strength and gas content on cavitation inception is shown schematically in Fig. 6.25.

The tensile strength depends on the gas and nuclei content which is affected by the water treatment. Water with dissolved gases in *highly turbulent flow* practically

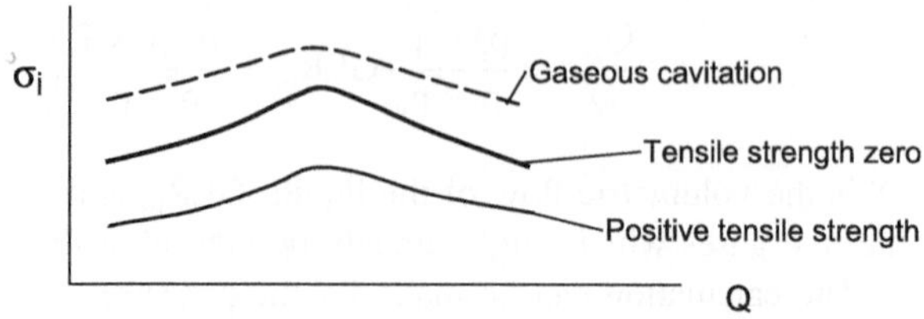

Fig. 6.25 Influence of tensile stresses and gas content on cavitation inception

sustains no tensile stresses. Such conditions are encountered in plants and test circuits with valves, branches, fittings, heat exchangers or pumps. Experience shows this to be true even for the piping systems of power stations, which operate with highly de-aerated water under high pressures (the gas content is in the order of 1 to 3‰ of saturation at 1 bar). This observation is in apparent contradiction to investigations such as [24, 33, 57]. A possible explanation is the high turbulence level prevailing in such plants. The gas or vapor molecules adsorbed on solid particles (e.g. metal oxides) also contribute to the generation of nuclei.

In industrial practice open systems or test stands often do not basically differ in their behavior from closed circuits with de-aeration, since many nuclei are generated in closed circuits by vortex shedding, cavitation in the impeller and/or in valves. This effect counteracts the influence of de-aeration. Efforts to reduce the nuclei content involve complicated systems to smooth the flow and to partially absorb the gas present in the nuclei (calming sections and large tanks). The following can be concluded from the work quoted above:

- As mentioned in Chap. 6.1.2, the number of nuclei activated for bubble growth increases with the square of the flow velocity. The influence of the tensile stresses on cavitation therefore depends on the test parameters. During the transposition from model tests at low speed with cold air-saturated water to the operating conditions in a power station with high speed, high temperature and extremely pure de-aerated water, several parameters change simultaneously. Their effects appear to largely compensate for each other, (see Chap. 6.4.1).
- Which nuclei are activated to growth at any specific location in the test circuit depends on the local velocity, the turbulence and the pressure profile (i.e. the history of the fluid). For this reason one tends to measure *a different tensile strength at any point in the circuit.* Attempts have been made to predict the "cavitation-effective" tensile strength in the impeller from the tensile strength measured at the sampling point by using correction factors, [57]. Unfortunately, such factors cannot be generic since they depend on the plant.
- Because of the influences that the pressure history in a circuit, the velocity and the turbulence have on the nucleus growth, the effects of the tensile strength and the flow features cannot be separated.
- The close interaction between nucleus growth/tensile strength and flow quantities leaves little prospect for finding generally valid relationships suitable for the industrial practice. The pressure profiles in circuits, the pressure distribution at the impeller blades, and consequently the residence time of the nuclei in the various pressure zones, are not subject to any generally applicable laws.

- If it is possible to produce water with a tensile strength greater than zero, σ_i tends to diminish with increasing tensile strength according to Eq. (6.27a).
- Although gas content and nuclei number do not directly correlate, it is possible to reduce the nuclei content through de-aeration so that Z could be up to 0.5 bar in de-aerated, *calmed* water. Tensile stresses over 1.0 bar are obtained by pressurizing filtered de-aerated water for several hours, resulting in an increasing (but by no means complete) dissolution of microscopic gas accumulations in the water.
- The tensile strength measured diminishes with increasing gas content.
- No information is available on tensile stresses in hot water (thermal power stations) or in the various fluids pumped in process engineering, petrochemical industry or refrigeration engineering.

In summary the following can be noted: Although the effect of the nuclei spectrum and the tensile strength on the cavitation bubble creation has undoubtedly been demonstrated, no generally valid methods are known by means of which these effects could be reliably captured in practical pump design. The worldwide application of the model laws according to Chap. 6.2.3 in the pump and water turbine industry over decades allows the conclusion - despite the scatter present in cavitation measurements - that the tensile stresses are (as a rule) of subordinate importance in pump operation.

The effects of water quality, gas and nuclei content are responsible for a part of the (at times quite significant) scatter which frequently afflicts the cavitation measurements. The nuclei content is not generally measured in the pump industry during model or acceptance tests. Partial degassing of the water is frequently performed in closed test circuits, and the air or oxygen content (dissolved components) is monitored. In this way the nuclei spectrum is indirectly affected via the reduction of the dissolved gases. Problems with air separation upstream of the impeller are thus avoided or reduced.

6.5 Cavitation-Induced Noise and Vibrations

6.5.1 Excitation Mechanisms

The implosion of vapor-filled zones (bubbles) generates pressure pulsations which excite vibrations and noise; various mechanisms are responsible for this:

- The implosion (of small) individual bubbles creates high-frequency fluid-borne noise in the kilohertz range that can be utilized to diagnose cavitation, [26].
- Fluctuations of the cavity are generated by unsteady impeller approach flows caused by circumferentially non-uniform velocities in radial pump inlets, turbulence, unfavorable approach flow conditions and vortex shedding on ribs.

- Intense pressure pulsations with frequencies below 10 Hz can occur on suction impellers or inducers with strong partload recirculation at low $NSPH_A$. The mechanism of their creation can be explained according to [64] in simplified manner as follows: Intense recirculation induces a pre-rotation, generating a roughly parabolic pressure distribution upstream of the impeller (refer to Fig. 5.16). If the pressure drops below the vapor pressure, a vapor core forms in the center of the suction pipe, which blocks a part of the approach flow cross section. Because of the blockage, the axial velocity in the outer domain of the inlet pipe increases. As a consequence the approach flow angle increases. This causes the recirculation to collapse and the vapor core to implode. Thus the blockage in the suction pipe center disappears, the approach flow angle diminishes and the cycle is repeated.
- Low-frequency pulsations of large amplitudes are created through large fluctuations of the cavitation zones. The compressibility of the cavities may result in cavitation surges. These are not related to instabilities of the Q-H-curve and can thus occur even with steadily falling Q-H-characteristics.
- Various forms of pulsating cavitation in inducers were discussed in [60]. Even rotating cavitation zones with super-synchronous frequency were observed near the best efficiency point. Measurements for rotating cavitation in a radial impeller (a design with rear and front shrouds in parallel) were discussed in [15]. Problems with rotating cavitation in industrial pump applications have not yet been reported; this has been found also in tests with three impellers with 3D-blades (n_q = 32, 52 and 82) in [65].
- Cavitation in the inlet casing can lead to pressure pulsations, vibrations and damage; a case history is discussed in Chap. 10.7.3, [66].

The pressure waves generated during bubble implosion create fluid-, solid- and air-borne noise. The air-borne noise is occasionally noticed as an irritating rattling, rumbling "cavitation noise" since it occurs in the frequency range of several hundred hertz where the human ear reacts with particular sensitivity. This noise is caused by large clouds of bubbles created through partload recirculation or through long cavities of the size of the blade pitch. The noise is particularly intensive in the operating range around $NPSH_A = NPSH_0$ to $NPSH_3$. As discussed below, the noise is *reduced* when $NPSH_A$ is reduced towards $NPSH_{FC}$. The noise increases with growing circumferential velocity u_1 while it diminishes with higher gas contents. By injecting or drawing in some air into the suction pipe the cavitation noise is effectively combated (obviously, this measure is only acceptable in very special cases).

6.5.2 Cavitation Noise Measurements for Quantifying the Hydrodynamic Cavitation Intensity

The fluid-borne noise caused by cavitation - in the following called "cavitation noise" - can be easily measured with a suitable pressure transducer in the pump inlet (for instance piezoelectric quartz pressure sensors). Since the intensity of the recorded pressure signal increases with the number of bubbles, bubble volume and implosion pressure, the sound pressure is a measure for the hydrodynamic cavitation intensity. This is why numerous works deal with cavitation noise. Emphasis is on three practical applications:

- Determining the cavitation inception by capturing the condition at which the sound pressure increases due to cavitation over the background noise (e.g. cavitation on the non-visible pressure surfaces of the blades).
- Determining a relationship between cavitation noise and erosion.
- Diagnosis of plant problems.

A proven methodology consists of registering the sound pressure NL during a suction test (Fig. 6.9) at each measuring point as a RMS-value in the frequency range from for instance 1 to 180 kHz. A curve NL = f(σ) is then obtained for each flow rate which constitutes the sound pressure as a function of the $NPSH_A$-value. The result of such a test is schematically shown in Fig. 6.26: At high inlet pressures, where no cavitation takes place as yet, the background noise NL_o is measured. It is generated by turbulence, unsteady blade forces and mechanical machine noises and does not depend on the suction pressure. The appearance of the first cavitation bubbles is manifested through an increase in sound pressure which marks the "acoustic cavitation inception". It often occurs at a slightly higher NPSH value than the visual cavitation inception since micro-bubbles can develop in the circuit before cavitation patterns are observed on the blades. Following cavitation inception the sound pressure increases with decreasing σ_A, slowly at first but more intensely later on up to a maximum which is expected around $NPSH_{o.}$ To the left of the maximum the sound drops steeply, mostly to levels below the background noise.

Such a maximum is observed in air-, fluid- and solid-borne noise as well as with erosion. It can be explained by means of Fig. 6.26. The potential implosion energy according to Eq. (6.2) depends on the driving pressure differential and the bubble volume. The number and the volume of the bubbles increase with falling σ_A. Conversely, the driving pressure differential (in theory, linearly) diminishes with σ_A. In reality the driving pressure differential may be initially amplified by the imploding neighboring bubbles. When the compressibility of the medium rises with growing vapor content, the pressure difference drops more strongly. These counter-acting effects produce the curve for $(p - p_v)$ shown in Fig. 6.26.

The character of the curves of $L_{cav} = f(\sigma_A)$ and $(p - p_v) = f(\sigma_A)$ is responsible for the maximum of the potential implosion energy and consequently for the peaks in cavitation noise and erosion.

Two additional reasons can be responsible for the drop of the sound pressure on reaching the maximum (Zone B in Fig. 6.26): (a) If large cavities are present at the impeller inlet, a part of the sound is absorbed in these zones with two-phase flow. The bubbles can implode for instance in the interior of the impeller channels whereby vapor bubbles shield the pressure transducer from the implosion zone. (b) At low inlet pressures air can separate in the pump inlet (or in an upstream

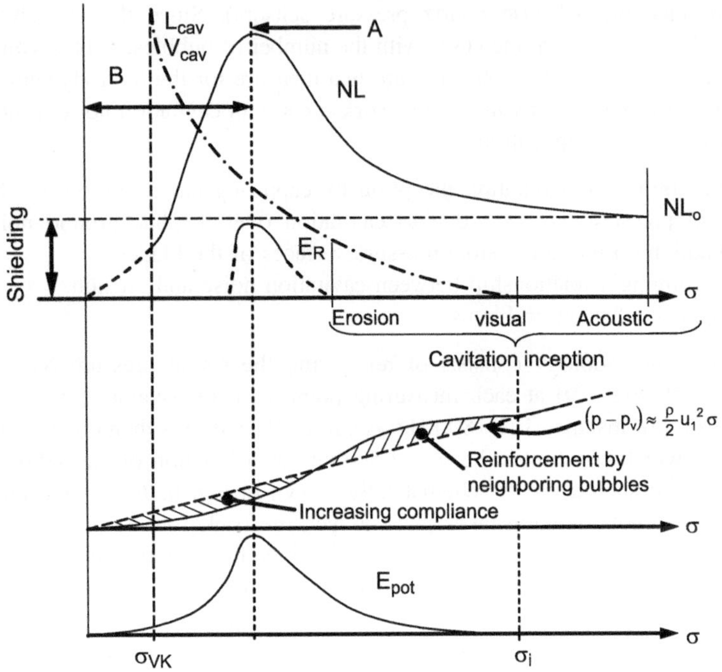

Fig. 6.26 Influence of the cavitation coefficient on cavity volume, cavitation noise, erosion and potential implosion energy

throttle valve), producing a flow containing bubbles which impair sound transmission. A two-phase flow changes the acoustic characteristics of the medium, notably the speed of sound. Compressibility and impedance jumps on the bubble walls are responsible for this. The speed of sound falls significantly below the value in water depending on the gas content and the 2-phase flow patterns, Chap. 13.2.

From the fact that the noise in Zone B falls below the background noise it may be concluded that the main cause of the background noise is not found in the inlet but in the impeller, i.e. in the unsteady blade forces. If the background noise were created by flow processes in the inlet, vapor zones in the impeller would not be able to shield the pressure sensors from the background noise.

When measuring cavitation noise, Zones A and B in Fig. 6.26 must be distinguished. In Zone A, cavitation is locally restricted and the cavitation noise recorded can serve as a measure for the cavitation intensity. In Zone B, large zones with two-phase flow absorb a high portion of the cavitation noise. Therefore, the sound pressure does not constitute a reliable measure for the cavitation intensity.

Cavitation noise measurements supply an indirect measure for the cavitation intensity which, in addition, depends on the instrumentation used - in particular on the recorded frequency range. This should be considered when comparing various publications. In order to facilitate comparison of the measurements with different pumps and operating conditions, a dimensionless presentation of a reference cavitation noise is introduced. To this end, the background noise NL_o must first be subtracted from the total signal NL to capture the effect of the cavitation alone. Since the addition of sound sources follows a quadratic relationship, the cavitation noise CNL is calculated from:

$$\mathrm{CNL} = \sqrt{\mathrm{NL}^2 - \mathrm{NL_0}^2} \tag{6.28}$$

Since every blade has a cavitation zone (hence presents a sound source) CNL is normalized to a reference blade number $z_{La,Ref} = 7$. It is thus possible to compare impellers with different blade numbers:

$$\mathrm{CNL_{Ref}} = \mathrm{CNL}\sqrt{\frac{z_{La,Ref}}{z_{La}}} \tag{6.29}$$

These sound pressures are then presented in dimensionless form:

$$\mathrm{NL}^* = \frac{2\,\mathrm{NL}}{\rho\ u_1^2} \tag{6.30}$$

Similarly NL_o, CNL and CNL_{Ref} are rendered dimensionless. As proven in Fig. 6.27a, Eq. (6.30) constitutes a useful scaling law for the conversion of the cavitation noise to different speeds: The dimensionless sound curves for the speeds 5000, 7000 and 8920 rpm tally well. Measurements reported in [54] gave cavitation noise curves very similar to Fig. 6.27a.

Cavitation noise measurements can not only be performed easily, they also offer the advantage that all forms of cavitation are registered - in particular bubble clouds and pressure surface cavitation, which are difficult to quantify visually. The cavitation noise CNL covers all parameters of the 2-phase flow and the bubble implosion for the considered operating condition, i.e.: flow velocity, cavitation coefficient, cavity length, incidence, profile shape, temperature, gas content,

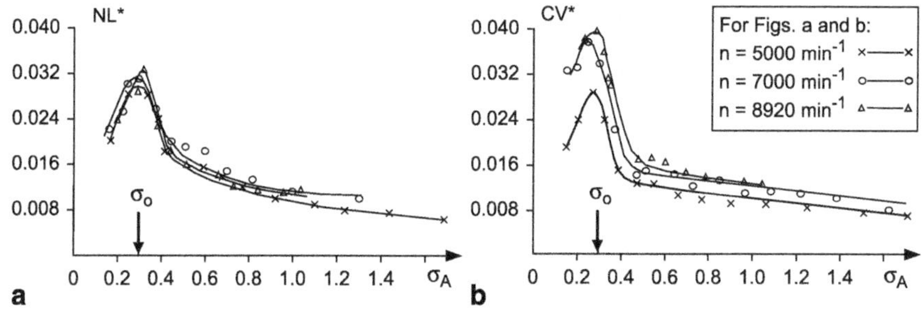

Fig. 6.27 Cavitation noise measurements on a high pressure stage (water temperature 60 °C). **a** Fluid-borne noise NL*; **b** solid-borne noise CV*, [25]

effective local pressure difference and bubble size. Conversely, the location where the bubbles implode cannot be determined. If cavitation noise measurements are conducted under controlled conditions, notably with the same instrumentation and test setup, the fluid-borne noise measured constitutes a measure for the cavitation intensity (but must not be confused with the latter).

It is also possible to measure the solid-borne noise CV instead of the fluid-borne noise. Figure 6.27b shows the solid-borne noise signals that were measured on the outside of the inlet casing using an acceleration sensor. These curves $CV^* = CV/(u_1{}^2/d_1)$ exhibit quite similar characteristics as the fluid-borne noise.

In the case of solid-borne noise measurements the exponent 2.5 would perhaps be more suitable, but the available data is insufficient to confirm this assumption.

The following can be considered as sources for the background noise: solid-borne noise caused by mechanical excitations (bearings, gear box, friction in seals or labyrinths), gas bubbles carried along in the flow, unsteady blade forces (impeller), turbulence of the flow and pressure fluctuations in the circuit (valves).

The application of noise measurements is discussed in Chaps. 6.6.4–6.6.6.

6.5.3 *Frequency Characteristics of Cavitation Noise*

Cavitation creates noise through *stochastic* impulses during the implosion of individual bubbles. Duration, magnitude and time sequences of the impulses are irregularly distributed (compare to turbulent fluctuations). The consequence is a continuous broad-band spectrum. Its energy distribution has a wide maximum which is brought about by the most frequently occurring implosion time. If the cavitation noise is measured as a broad-band RMS-value, this maximum can be expected in the range from 20 to 400 kHz. It strongly depends on the type and the operating conditions of the pump, because the frequencies where the energy density is greatest increase with the cavitation coefficient and the flow velocity. Individual bubbles with a small initial radius can have implosion times in the range of micro-seconds and thus generate frequencies in the MHz-domain.

As shown by stroboscopic observations of the bubble field on impellers, the cavities fluctuate with increasing length - i.e. decreasing σ_A. In the case of cavitation on individual profiles such fluctuations occur at discrete frequencies according to a Strouhal number $S_{Str} = f \times L_{cav}/w_1$. A typical range is $S_{Str} = 0.3 \pm 0.04$. The length of the fluctuations ΔL_{ER} grows with increasing approach flow velocity according to $\Delta L_{ER}/L_{cav} = (0.01 \text{ to } 0.015) \times w_1/w_{Ref}$ ($w_{Ref} = 1$ m/s), [10, 11].

In pumps, broad-band cavitation noise was observed in the range of several hundred Hertz to several kHz. The reason for this might be that the cavity fluctuations at the various blades are not in phase. According to $f = S_{Str} \times w_1/L_{cav}$ the frequency increases with the circumferential speed while it decreases as cavities get longer. In high-head pumps operating with long cavities, pressure pulsations generated in this way can cause operational problems, for example damage to mechanical seals as reported in [17]. In summary:

1. Cavitation noise and spectrum depend on: (a) the implosions of individual bubbles causing noise in the range from 10 kHz to 1 MHz; (b) the fluctuations of the cavity which occur with frequencies below 1000 to 2000 Hz. Presumably the cavity fluctuations modulate the implosion events since they influence the transport of the bubbles to the location of implosion.
2. With increasing speed and $NPSH_A$ the maximum energy densities of the cavitation noise spectrum shift towards higher frequencies.
3. If one wishes to capture the *hydrodynamic cavitation intensity* using cavitation noise measurements, it is necessary to measure in the widest possible frequency range. The low-frequency components must be measured as well since the cavity oscillations significantly contribute to the cavitation damage. These statements are confirmed by measurements of the unsteady pressures in a rotating impeller: At the end of the cavity, frequencies below 200 Hz were recorded while the dominating frequencies decreased with increasing cavitation, [15]. If the cavitation noise were only measured in a narrow frequency band, the range of maximum energy density would shift relative to the measuring band when changing the operating parameters. Misinterpretation of the results (e.g. relation between noise and speed) would be the consequence.[6]
4. If it is attempted to determine the *cavitation inception* by means of cavitation noise measurements, the high-pass filter must be set to a value which is above the frequency range of the background noise (e.g. 10 kHz). Only in this way will this method be sensitive enough to sufficiently distinguish the minor noise increase generated by the first bubble implosions from the background noise. An interfering factor in this context is annular seal cavitation or cavitation in the circuit, for example in valves, since these effects are likewise dependent on the inlet pressure.

6.6 Cavitation Erosion

Material erosion due to cavitation only occurs when the hydrodynamic cavitation intensity HCI exceeds the cavitation resistance CR of the material and the cavitation acts on the material for a sufficient period of time. HCI *exclusively* depends on the flow and liquid properties while CR is *purely a material quantity*: HCI and CR are *totally independent of each other*. The erosion rate E_R increases with rising HCI and falls with growing CR since $E_R \sim HCI^x/CR^y$. If the HCI is significantly greater

[6]Measurements reported in the literature were frequently performed only in narrow frequency bands. The choice of the frequency band must be considered as arbitrary. It renders the interpretation difficult. It also limits the comparability with other measurements and tends to falsify any correlations between noise and erosion or operating parameters.

than the CR, the material damage occurs after a short period of exposure (or a short "incubation time"). If the HCI hardly exceeds the CR, the incubation period becomes very long and tends towards infinity for HCI < CR.

To date, the hydrodynamic cavitation intensity can neither be measured directly nor calculated theoretically. Therefore empirical correlations are needed. Two approaches are discussed: (A) Estimation of the erosion rate based on the cavity length, the $NPSH_A$ and the fluid properties in Chap. 6.6.3. (B) Estimation of the erosion rate based on the cavitation noise in Chap. 6.6.4. Some applications of these methods were reported on in [27].

Different flow conditions lead to different cavitation zones and erosion patterns which allow conclusions on the causes of the damage, Chap. 6.8.2.

6.6.1 Testing Methods

Different test devices were developed to measure the erosion rate of various materials under cavitation in the laboratory:

- In the magneto-strictive vibrator a material specimen is moved in the test medium with a high frequency and low amplitude (usually 20 kHz and ±25 μm). Because of the strong acceleration (almost 4×10^5 m/s^2) the fluid is not able to follow the movements of the specimen and vapor bubbles are generated which violently implode on the sample causing a very high HCI. Unfortunately, vibration-induced cavitation is not representative for flow cavitation. At times, the material sample is arranged opposite the vibrator to run tests with a lower HCI.
- Flow channels with Venturi nozzles similar to Fig. 6.2 or bodies of various shapes placed in a flow are frequently employed to measure the influence of the velocity and the material on the erosion, [11].
- Rotating discs with holes (pikes or other bodies) which generate cavitation zones.
- Jet apparatus where the test medium is expanded with a high pressure difference (e.g. 250 bar) through a fine orifice (0.5 mm diameter). The vapor bubbles generated in the process implode on the material specimen to be examined.
- Droplet erosion causes similar damage as cavitation (think of the micro-jet); therefore tests with droplet erosion devices can also be employed, [41].

All these laboratory tests have two disadvantages: (1) The cavitation conditions are fundamentally different from the flow in centrifugal pumps. The hydrodynamic cavitation intensity is unknown. Transferring the results to pumps harbors large uncertainties. (2) To achieve short testing times cavitation intensities are used which are usually significantly greater than in pumps. The tests therefore apply to "developed erosion" and not for operation near the onset of erosion where the damage mechanism resembles a fatigue process.

However, if it is desired to determine the influence of the material or the liquid properties on developed cavitation in the form of material or corrosion factors, such test devices represent the only viable option because of the prohibitive costs of performing such tests on pumps.

Attempts have also been made to detect and quantify cavitation erosion by means of electrochemical measurements. The process relies on the fact that the passivation layers on steels are destroyed by the bubble implosion, creating increased localized corrosion. The potential difference between the eroding sample and a reference electrode is recorded. It serves as a measure for the erosion intensity, [14].

If bubbles implode with sufficient intensity on a ductile material, pitting due to plastic deformation is observed during the initial phase of material exposure. Depending on the size of the bubbles and the material, pitting diameters are in the range from 10 to 50 μm. The number and size of the pits are considered as a measure for the cavitation intensity. Attempts have been made to develop methods for the prediction of cavitation damage on this basis, [52]. The considerable testing costs that would be required in each individual case prevent its wide industrial application.

Despite the considerable efforts invested in tests with the devices described above, no methods have been developed for impeller damage prediction, which would be sufficiently general, accurate and simple in the industrial application. Reason: Cavitation erosion is a complex process with 3-dimensional two-phase flow, thermodynamic effects and material reactions in the micro-range, [27]. With theoretical approaches and erosion tests in simple test devices it is not possible to obtain generally valid solutions that could be transferred to pumps or other machines. This is because the flow and pressure distribution in the rotating impeller cannot be reproduced by rotating discs or stationary channels (Venturi) or air foils - much less by the magneto-strictive vibrator. This is why erosion rates in Venturi nozzles [11] or on air foils plotted over the cavity length yield entirely different tendencies than are observed in centrifugal pumps, [23].

Attempting to describe cavitation erosion in centrifugal pumps for the purpose of damage prediction in a strictly scientific way, the following effects must be captured analytically, numerically or experimentally in a generally applicable form:

- *Inlet flow*: Suction casings of between-bearing pumps produce strongly non-uniform fully 3-dimensional flows to the impeller. Consequently incidence, cavity length and bubble volume vary over the circumference.
- *Nuclei spectra* (number and size of nuclei for various industrial applications and water qualities).
- *Impeller flow*, taking into account partload recirculation (if applicable), characterized by the local velocity and pressure distribution on the blades. Development and growth of vapor bubbles are determined by the dynamic characteristics of the nuclei in the zone where the local static pressure drops to the vapor pressure. Gas content and thermodynamic fluid properties influence

the bubble growth. The cavity in turn interacts with the flow around the blade. Entirely different forms of cavitation can be found (see Chap. 6.8.2).

- *Implosion* of the bubbles upon entry into zones where the static pressure exceeds the vapor pressure. This process generates the hydrodynamic cavitation intensity which represents the load spectrum for the material. In fact, the growth and implosion zones cannot be separated since the cavity oscillates. The content of non-condensable gases and thermodynamic processes influence the implosion event and consequently the hydraulic cavitation intensity.
- *Reaction of the material* to the implosions. The resulting load spectrum is made up by an "infinite" number of individual localized shocks of extremely short duration.

There is little prospect of describing these complex physical relationships analytically, numerically and/or experimentally in a sufficiently general manner. Empirical correlations are (as yet) the only viable alternative for cavitation damage prediction in the industrial practice. Consider that even comparatively simpler processes such as pressure loss and heat transfer in turbulent single-phase or two-phase flows are also calculated almost exclusively utilizing empirical correlations.

When resorting to empirical methods, the greatest possible number of test results is correlated with physically relevant similarity parameters through power functions. In this way, the available tests with defined geometries and flow conditions can be transferred to other configurations. The scatter of empirical correlations grows with the number of physical parameters relevant to the process. Since empirical correlations of simple two-phase flows can be affected by uncertainties of more than ±100%, even greater scatter has to be accepted with the extremely complex phenomenon of cavitation erosion.

The methods for the prediction of erosion rates in centrifugal pumps discussed in the following are based on measured, calculated or estimated cavity lengths, or on fluid- and solid-borne noise measurements. Details on the development of the correlations for the erosion calculation are given in the literature, [19, 23, 25, 26].

6.6.2 Cavitation Resistance

The imploding vapor bubbles or the micro-jets attack the material by impacts with an extremely small diameter and a short duration but with a high energy and frequency. The impacts manifest themselves as local plastic deformations of the material as demonstrated by Fig. 6.28 (left photograph) which shows the damage created in satellite by a single cavitation vortex. The right photograph in Fig. 6.28 shows the cavitation erosion on a boiler feed pump impeller of $n_q = 26$ running at a tip speed of about $u_1 = 92$ m/s at $\sigma_A = 0.52$; the material was 1.4317 (see Table 14.12).

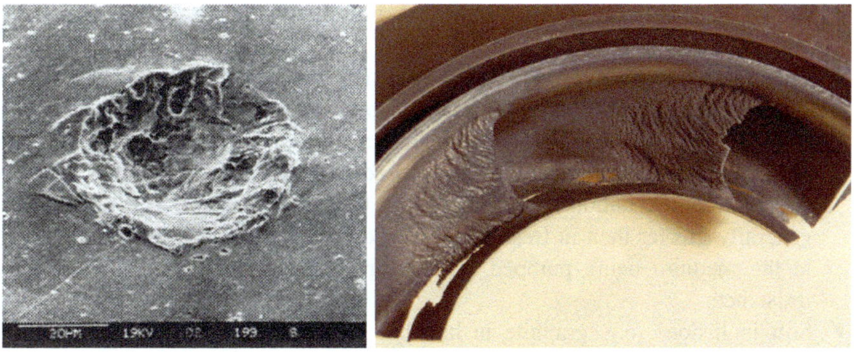

Fig. 6.28 Cavitation erosion. *Left*: Crater generated on stellite by the implosion of a cavitating vortex; the crater diameter is about 50 μm. Photo: Ecole Polytechnique Fédérale, Lausanne, [13]. *Right*: Cavitation damage on a boiler feed pump impeller

The shock-like loading by a micro-jet hitting a structure suggests that hardness can serve as a first approximation for the cavitation resistance. The extremely small local extension of the loading also suggests that the material structure is highly significant for the cavitation resistance. Micro-structural and other uncontrollable influences thus result in variations of the erosion rate in a ratio of up to 1:5, [40].

The state of knowledge for predicting the cavitation erosion with regard to cavitation resistance can be outlined as follows, see [14, 61]:

- None of the conventional material characteristics (such as hardness, tensile strength) are sufficient to accurately describe the cavitation resistance since the loading under cavitation differs too much from the type of load applied during a tensile test, hardness measurement etc.
- Within groups of *similar* materials the cavitation resistance rises with increasing hardness or tensile strength R_m. The erosion rate E_R then roughly follows the proportionality $E_R \sim 1/R_m^2$ or $E_R \sim 1/H_{Mat}^2$.
- With *identical* tensile strength the *tougher* material has the greater cavitation resistance. Extremely brittle materials have a low cavitation resistance despite great hardness.
- With identical hardness and identical toughness the material with the finer structure is the more resistant.
- The material attack preferentially occurs on grain and phase boundaries or pores on the surface. The presence of hard particles in the material structure diminishes the resistance to cavitation erosion while the resistance to abrasion increases (Chap. 14.5).
- The erosion resistance can be improved if one succeeds in increasing the tensile strength through suitable metallurgical measures (such as precipitation or dispersion hardening, martensite formation or strain hardening) without excessively decreasing the ductility [61]. A decisive factor for a high cavitation

resistance is the conversion of austenite into martensite through strain hardening, [50].

- Like most metals, steel is chemically resistant in water only if a passivation layer, or a protective layer of corrosion products, is formed on the boundary surface between metal and water. If this layer is destroyed by imploding bubbles, unprotected base material is exposed and a corrosive attack is superimposed on the cavitation load. This is why the cavitation erosion in sea water is generally greater than in fresh water. Good corrosion resistance of the material in the medium being pumped is therefore a prerequisite for high cavitation resistance.
- Soft inclusions (e.g. graphite in lamellar cast iron), soft or very brittle grain boundaries, pores, cracks and structural flaws reduce the cavitation resistance. This must also be considered in the development of coatings.
- Coatings must not only be free of defects but also possess sufficient hardness without being brittle. The layer must be thick enough to ensure that the base material is not plastically deformed under the implosion pressure peaks.
- Bubble implosions create a complex, multi-axis stress condition. Compressive residual stresses increase the cavitation resistance, tensile residual stresses reduce it.
- If a material is exposed to a cavitating flow, it takes some time before damage (if any) becomes visible. During this period (the so called "incubation time") the material reaction is determined by strain hardening. When metal loss starts, the surface is progressively roughened as exposure time increases (growing notch effects). Later on, cracks may be observed in the zone where the maximum erosion rate is reached and larger chunks of material are ripped out.
- Materials and alloys for overlay welding especially developed for high cavitation resistance have an extremely high strain hardening coefficient.
- Test methods suitable for directly measuring the cavitation resistance are not known. The measurement of the cavitation erosion cannot be utilized as an indirect measure for the cavitation resistance as long as the absolute hydraulic cavitation intensity cannot be determined.
- **Conclusion**: *high tensile strength, high strain hardening capacity, a homogenous structure and surfaces without pores or cracks are prerequisites for high cavitation resistance.*

When material erosion occurs, work is performed on the material. The energy per surface unit corresponds to a stress σ multiplied by a distance ε, i.e. the integral of the stress-strain diagram $\int \sigma \times d\varepsilon$. An idealized stress-strain curve is utilized which assumes that the elastic range is retained up to the ultimate tensile strength. Under this assumption the integral $U_R = \int \sigma \times d\varepsilon$ yields the material work per volume unit in the case of brittle fracture. This quantity is called the "ultimate resilience" $U_R = R_m^2/(2 \times E)$. It is used to evaluate the cavitation resistance. The product of the material work per unit volume and the erosion rate yields the specific erosion power $P_{ER} = U_R \times E_R$.

The use of U_R is a rough approximation since the influencing parameters on the cavitation resistance listed above cannot be described with by a single characteristic quantity. In fact, the mechanism of the material damage changes as a function of the hydrodynamic cavitation intensity: (1) if HCI $\approx$ FS, the fatigue strength FS *under cavitation* is relevant for the cavitation resistance; (2) if HCI $\approx$ HV, the hardness HV is the relevant parameter; (3) if HCI > R_m (i.e. in the presence of "developed erosion"), the tensile strength R_m or U_R describes the material behavior quite well.

For the various types of steel the use of hardness, tensile strength and "ultimate resilience" is equivalent, since the E-modulus and the ratio of hardness to tensile strength are roughly constant (Brinell hardness/tensile strength $\approx$0.293).

The effects of the metallurgical structure and the corrosion in the case of sea water are described through material and corrosion factors F_{mat} and F_{cor} which are determined from tests with a jet device, a rotating disk or the magneto-strictive vibrator, [18, 19]. For more information on material characteristics see Chap. 6.6.7.

6.6.3 Prediction of Cavitation Damage Based on Cavity Length

The risk and extent of cavitation damage can be estimated using empirical correlations; the equations required for this purpose are compiled in Table 6.1 (at the end of Chap. 6.8.8).

Damage correlation: As discussed in the preceding paragraphs, the hydraulic cavitation intensity increases with the total volume of all vapor bubbles created in the flow and the pressure difference driving the implosion. In a first approximation it can be assumed that the bubble volume grows with the length of the cavity L_{cav} and that the driving pressure differentials increase with the $NPSH_A$. One can then attempt to correlate the erosion rate E_R with the material and fluid properties, the $NPSH_A$ and the cavity length L_{cav}. Using the cavity length measured in a pump, has the advantage that it is not necessary to separately capture all the parameters which determine the cavitation flow (see Chap. 6.3.1) and whose prediction would be very complicated, difficult and uncertain.

Incidents of cavitation damage recorded on impellers from various plants and applications were evaluated to develop a correlation for the prediction of cavitation damage. To this end, the *damage lengths measured from the leading edge* (instead of the cavity length L_{cav} which was not measured in all cases) and the *maximum local depth of the material erosion* were used to describe the damage. For a given cavity the erosion depends on the $NPSH_A$, the cavitation resistance (captured by U_R) and the liquid properties which are characterized by vapor density, speed of sound and gas content. To find a correlation between all these parameters and the erosion rate, a similarity parameter Θ_u, Eq. (T6.1.1), was derived and plotted against the *damage* length measured from the blade inlet edge.

Pressure surface and vortex cavitation according to this evaluation proved to be approximately 50-times more aggressive with a given damage length than cavitation on the suction surface of the blades since the driving pressure differentials in the implosion zone or the bubble volume are greater. The data for the suction and pressure surface cavitation can be described as power functions $\Theta_u = f(L_{cav})$ and solved for the specific erosion power . In this way Eq. (T6.1.2) was obtained for the prediction of cavitation erosion for a measured or calculated cavity length. In this procedure it is implicitly assumed that damage length and cavity length are roughly equal. As was shown by the comparison of laboratory tests with plant measurements this is frequently the case - albeit with considerable data scatter. The cavity length can be estimated from Eq. (T6.1.3) if no measurements are available, [8]. Figure 6.29 contains about 70 data points gathered from plant operation and taken on a test pump with a head per stage of 800 m, [20, 21, 23, 25].

Consequently the damage correlation covers data from a wide parameter range:

- Demineralized and fresh water: 10 to 190 °C, gas content $\alpha = 0.03$ to 24 ppm
- $NPSH_A = 4$ to 230 m; $u_1 = 20$ to 90 m/s; $\sigma_A = 0.11$ to 0.95
- $q^* = 0.25$ to 1.3; $n_q = 22$ to 55; $d_2 = 270$ to 2400 mm
- Damage (or cavity) length: $L_{cav} = 10$ to 300 mm
- $R_m = 460$ to 1000 N/mm²; $E_R = 0.02$ to 80 μm/h.

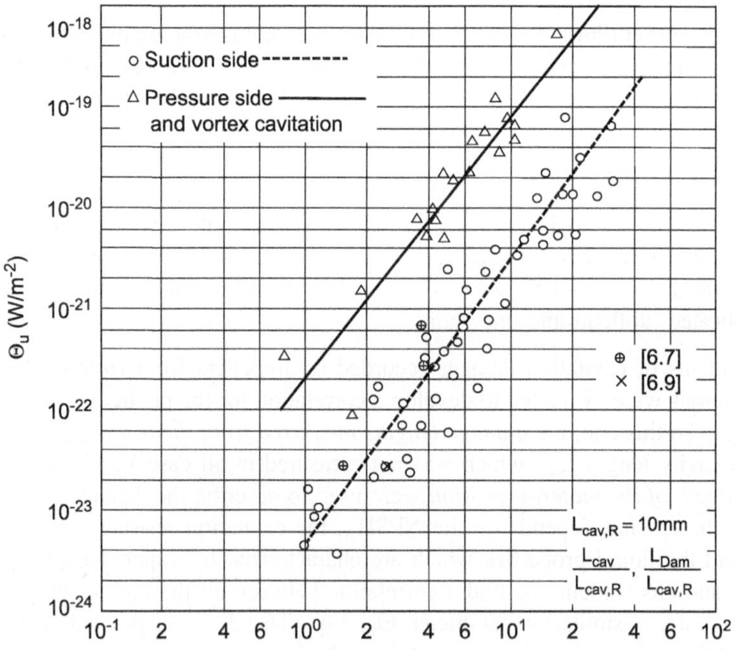

Fig. 6.29 Correlation between erosion and damage length; abscissa $L_{cav}/L_{cav,R}$

As proven by this broad spectrum of parameters the correlation comprises totally different pump types with correspondingly different impeller designs. Figure 6.29 contains measurements from [25] and data from literature [32, 43, 63]; consequently Eq. (T6.1.2) is not specific to the pumps of only one manufacturer. Cavitation damage in semi-axial pumps is also correlated by Eq. (T6.1.2) as demonstrated by 5 examples analyzed in [32]. Applications of the calculation methods listed in Table 6.1 and comparisons with experiences in various plants were discussed in detail in [6, 27, 32, 43, 44, 54, 62].

The data in Fig. 6.29 is also remarkable for the fact that the erosion data correlate with the *absolute* cavity length in the range of examined impeller diameters (d_2 = 270 to 2400 mm, range 1:9). This means that there is *no* noticeable influence of the impeller size. An attempt to plot the erosion data from Fig. 6.29 against the ratio cavity length to blade pitch (L_{cav}/t_{1a}), yielded a poorer correlation of the measurements than according to the equations in Table 6.1. This behavior can possibly be explained by the fact that the flow around the rear of the cavity is more influenced by the absolute cavity length than by the blade pitch. A longer cavity generally involves also a greater cavity thickness. The local deceleration downstream of the cavity, and consequently the driving pressure differential and the implosion energy, grow with increasing cavity thickness (see Figs. 6.5, 6.6 and 6.7).

Impeller life L_I: From a statistical analysis of the data in Fig. 6.29 it is possible to estimate by means of Eq. (T6.1.10) the probability with which a required impeller life $L_{I,req}$ is achieved in a specific application. The expected life $L_{I,exp}$ in this context is calculated from Eqs. (T6.1.2) and (6.1.9). Accordingly, the life is considered exhausted when the erosion depth has reached 75% of the blade thickness. (Obviously it would also be possible to insert a different permissible erosion depth into Eq. (T6.1.9) to determine the corresponding probability).

Uncertainties and limits of the correlation:

1. Equation (T6.1.2) largely applies to *developed* erosion. The erosion threshold cannot be determined from Eq. (T6.1.2) (refer to Chap. 6.6.7 in this respect).
2. The cavity length can only serve as a rough measure for the volume of the imploding vapor bubbles. For a more accurate estimation, the cavity thickness (which is very difficult to measure) would have to be considered. In general, the erosion cannot be solely a function of the cavity length. Cavity volume and erosion rate (at a given cavity length) will depend on the incidence for example. Furthermore, the different forms of cavitation discussed in Chap. 6.8.2.1 suggest that there may be differences in the local hydrodynamic cavitation intensity that cannot be captured by the cavity length.

Table 6.1 (1) Assessment of the risk of cavitation damage

Specific erosion power	$P_{ER} = E_R\, U_R$	**Material**	$\mathbf{F_{Mat}}$	$\mathbf{F_{cor}}$ Fresh water	$\mathbf{F_{cor}}$ Sea water
Ultimate resilience	$U_R = \frac{R_m^2}{2E}$	Ferritic steels	Table D6.4	1.0	1.5
		Austenitic steels		1.0	1.3
		Aluminum bronzes		1.0	1.1

			Eq.
Erosion rate from cavity length	$\Delta p = p_1 - p_v = \rho\, g\, NPSH_A - \frac{\rho}{2} c_1^2$		
	$\theta_u = P_{ER} \left(\frac{p_{Ref}}{p_1 - p_v}\right)^3 \frac{a_{Ref}}{a} \left(\frac{\rho''}{\rho''_{Ref}}\right)^{0.44} \frac{F_{Mat}}{F_{cor}} \left(\frac{\alpha}{\alpha_{Ref}}\right)^{0.36}$		6.1.1
	$P_{ER} = C_1 \left(\frac{\Delta p}{p_{Ref}}\right)^3 \frac{F_{cor}}{F_{Mat}} \left(\frac{L_{cav}}{L_{Ref}}\right)^{x_2} \frac{a}{a_{Ref}} \left(\frac{\alpha_{Ref}}{\alpha}\right)^{0.36} \left(\frac{\rho''_{Ref}}{\rho''}\right)^{0.44}$		6.1.2
$p_{Ref} = 1\ N/m^2$ $L_{Ref} = 10\ mm$ $\alpha_{Ref} = 24\ ppm$	$a_{Ref} = 1490\ m/s$ $\rho''_{Ref} = 0.0173\ kg/m^3$	Type of erosion | $C_1 (W/m^2)$ | x_2 Suction surface | 5.4×10^{-24} | 2.83 Pressure surface/vortex | 2.5×10^{-22} | 2.6	–
Estimation of the cavity length	$L_{cav} = \frac{\pi d_1}{z_{La}} \left\{1 - \left[\frac{\sigma_A - \sigma_3}{\sigma_i - \sigma_3}\right]^{0.33}\right\}$	for $\sigma_A < \sigma_i$	6.1.3
Erosion rate from fluid-borne noise (Measurement of NL or CV and Eq. 6.1.7)	$CNL_{Ref} = NL \sqrt{\frac{z_{La,Ref}}{z_{La}} \left\{1 - \frac{NL_0^2}{NL^2}\right\}}$	$I_{ac} = \frac{CNL_{Ref}^{\ 2}}{\rho\, a}$	6.1.4 6.1.5
	$P_{ER} = C_2 \frac{F_{cor}}{F_{Mat}} \left(\frac{I_{ac}}{I_{Ref}}\right)^{1.463}$	$C_2 = 8.8\times10^{-8} \frac{W}{m^2}$	6.1.6
Fluid-borne noise NL from solid-borne noise CV $[m/s^2]$	$NL = CV\, \rho\, h \sqrt{\frac{\rho_p\, a_L}{\pi\, \rho\, a \sqrt{3}}} \left(1 + \frac{R}{L} \frac{h_D}{h}\right)$	Casing considered as cylinder with radius R, wall thickness h, length L, density ρ_p, cover wall thickness h_D	6.1.7
	NL is RMS in range 1 to 180 kHz	$z_{La,Ref} = 7$ $I_{Ref} = 1\ W/m^2$	
Paint erosion test	$t_v = f\,(CNL_{Ref})$ from Fig. 6.31 for zinc paint		6.1.8
Expected service life of impeller in hours	$L_{I,exp} = \frac{0.75\, e}{3600\, \Sigma\, (\tau\, E_R)}$	$E_R = P_{ER}/U_R$ is the erosion rate in m/s e = blade thickness in m τ = duration of service at particular load considered; τ is referred to the impeller life load spectrum	6.1.9
Probability to reach required impeller life	$W = 1.17 - 0.53 \frac{L_{I,req}}{L_{I,exp}}$	for $0.4 \le \frac{L_{I,req}}{L_{I,exp}} \le 1.5$; $L_{I,req}$ = required service life	6.1.10

- W is the probability to reach $L_{I,req}$. W can be used for Eqs. (6.1.2), (6.1.6) and (6.1.13)
- All units are in the SI-System
- For double entry impellers set $2\times z_{La}$ (with z_{La} per impeller side)

Table 6.1 (2): Assessment of the risk of cavitation damage		Eq.
Erosion rate from implosion pressure p_i	$r_H = \sqrt{\frac{S\,\alpha_T}{8\pi(1-\alpha_T)}}$; $p_i = \frac{C_5\, r_H\, a\, CNL_{Ref}}{u_1 \sqrt{1+\frac{r_H{}^2}{r_x{}^2}}}$; Material of cover / α_T: Perspex 0.23; Steel 0.16	6.1.11 6.1.12
	$C_5 = 2500\ m^{-1}$; S = sound absorbing surfaces (pump inlet casing) r_x = distance between impeller and sensor, r_H = reverberation radius	
	$P_{ER} = C_3 \frac{F_{cor}}{F_{Mat}} \left(\frac{p_i}{p_{Ref}}\right)^3$; $C_3 = 8.5\times10^{-28}\ \frac{W}{m^2}$; For $p_i > 70\ N/mm^2$ with stainless steel	6.1.13
Suction surface: cavity length calculated from cavitation noise	$\frac{L_{cav}}{L_{Ref}} = 5.4\times10^5 \left(\frac{I_{ac}}{I_{Ref}}\right)^{0.52} \left(\frac{\Delta p_{Ref}}{\Delta p}\right)^{1.06} \left(\frac{a_{Ref}}{a}\right)^{0.35} \left(\frac{\alpha}{\alpha_{Ref}}\right)^{0.13} \left(\frac{\rho''}{\rho''_{Ref}}\right)^{0.15}$	6.1.14a
Pressure surface:	$\frac{L_{cav}}{L_{Ref}} = 3.9\times10^5 \left(\frac{I_{ac}}{I_{Ref}}\right)^{0.56} \left(\frac{\Delta p_{Ref}}{\Delta p}\right)^{1.15} \left(\frac{a_{Ref}}{a}\right)^{0.38} \left(\frac{\alpha}{\alpha_{Ref}}\right)^{0.14} \left(\frac{\rho''}{\rho''_{Ref}}\right)^{0.17}$	6.1.14b
Implosion pressure calculated from cavity length	$p_i = C_4 (p_1 - p_v) \left(\frac{L_{cav}}{L_{Ref}}\right)^{0.91} \left(\frac{a}{a_{Ref}}\right)^{0.33} \left(\frac{\rho''_{Ref}}{\rho''}\right)^{0.15} \left(\frac{\alpha_{Ref}}{\alpha}\right)^{0.12}$	6.1.15
Implosion pressure calculated from erosion rate	$p_i = 1.1\times10^9\, p_{Ref} \left(\frac{P_{ER}}{I_{Ref}} \frac{F_{Mat}}{F_{cor}}\right)^{0.333}$ $p_{Ref} = 1\ N/m^2$; Erosion / C_4: Suction surface 18; Pressure surface 67	6.1.16
Onset of erosion	$p_{i,ES} = F_S = (0.054\ \text{to}\ 0.15)\ R_m$; For ferritic steels in fresh water or boiler feed water	6.1.17
Relative erosion rate	$\varepsilon = \frac{E_{R,2}}{E_{R,1}} = \left(\frac{NL_2}{NL_1}\right)^{2.92} = \left(\frac{CV_2}{CV_1}\right)^{2.92}$	6.1.18

3. The pressure differential in the implosion zone is proportional to the $NPSH_A$ only in a first approximation. To capture this pressure difference more accurately, one would have to calculate the 3D-flow downstream of the cavity where the fluid is strongly decelerated and the local pressure is correspondingly increased. The imploding bubbles also have an effect on each other. As a result of increasing blade work the driving pressure differential increases with the cavity length; this might be a reason for the high exponent of L_{cav}.
4. It is difficult to observe quantitatively how close the bubbles implode to the blade surface since only a few tenths of millimeters are involved. However, the damage potential critically depends on this distance. For example, cases are known where no erosion was observed although considerable damage would have been expected based on the correlation. This finding indicates that the bubbles imploded in the liquid far enough away from the blade to preclude damage. In such cases the prediction according to Eq. (T6.1.2) is pessimistic.
5. The correlation for erosion on the suction surface largely applies to attached cavities. Cavitation forms caused by vortices (high local concentration of imploding bubbles) are better captured by the damage correlation for the blade

pressure surface. Since both cavitation types can occur simultaneously and merge gradually their distinction is sometimes difficult.

6. In calculating the similarity parameter according to Eq. (T6.1.1) some assumptions in terms of the exponents of $(p_1 - p_v)$, of the thermodynamic liquid properties and of the gas content had to be made. Although these are based on tests, they are subject to uncertainties. Such exponents can be roughly constant only in the domain of developed erosion. When approaching the erosion threshold, the exponents increase strongly as discussed in Chap. 6.6.7.
7. The reaction of the material to a given hydrodynamic cavitation intensity cannot be perfectly quantified by U_R.

Owing to the above uncertainties the standard deviation of Eq. (T6.1.1) is approximately ±120%. The scatter is due to various uncertainties: (1) the depth of the attack is difficult to measure due to the rough, rugged surface (it has been estimated visually in most cases); (2) the operation time at various working points, while the material was being eroded, was often not exactly available from the plant; (3) the real local material characteristics (structure of the casting); (4) the gas content in many cases had to be assumed from the type of water pumped. Even under laboratory conditions, the scatter of cavitation erosion tests is very large. Examples: (a) during vibration tests with identical material and identical test parameters the scatter was ±90% in [56]; (b) during the Venturi tests in [11] the scatter was ±50%; (c) Piltz reports of variations of 1:5 in his tests, [40].

6.6.4 Prediction of Cavitation Damage Based on Cavitation Noise

Damage correlation: As discussed in Chap. 6.5, sound waves are generated by the bubble implosions. The associated fluid-borne noise can be recorded and used as a measure for the hydrodynamic cavitation intensity. If the specific erosion power P_{ER} is plotted against the acoustic intensity, Fig. 6.30, a correlation is obtained that can be used for prediction of cavitation damage, Eqs. (T6.1.4 to T6.1.6). The fluid-borne noise used for this correlation corresponds to the root mean square RMS in the range from 1 to 180 kHz. The difference between the signal at a given suction pressure and the background noise corresponds to the component of the measured noise caused by the cavitation, which is determined from Eq. (T6.1.4). At the same time, a standard reference blade number is used since the number of noise sources - i.e. the impeller blade number - has an effect on the measured signal (in the case of double-entry impellers twice the blade number must be used since there are cavities in both entries). In Fig. 6.30 suction and pressure surface cavitation can be described by the same correlation. It is therefore concluded that the noise measurement correctly captures the fact that the implosion pressure on the pressure surface of the blades is higher than on the suction surface (due to the greater driving pressure differential).

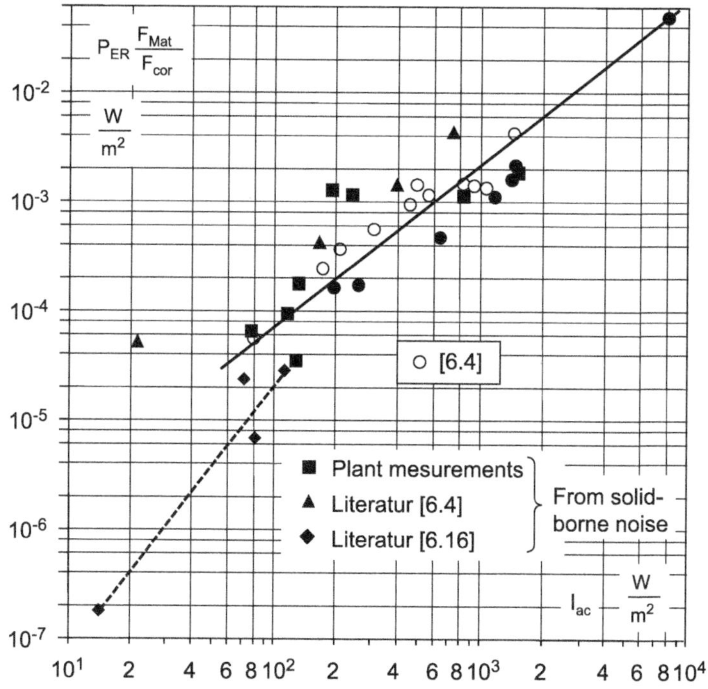

Fig. 6.30 Estimation of erosion rate based on cavitation noise [26, 27] Lit. [6.4] from [25]; Lit. [6.16] from [4]

The correlation according to Eq. (T6.1.6) yields a velocity exponent of 5.85; it thus confirms well the exponent of 6, which was implicitly used in Eqs. (T6.1.1 and 6.1.2), since $(p_1 - p_v)^3 \sim u_1^6$. By equating Eqs. (T6.1.2 and 6.1.6), it is also possible to estimate the cavity length from the measured cavitation noise, Eq. (T6.1.14).

Limits of the correlation: Like Eq. (T6.1.2), Eq. (T6.1.6) applies only to developed erosion. The onset of erosion cannot be extrapolated on the basis of the available data. Since the background noise (e.g. machine and test loop noise) is subtracted, the measured signal is a measure for the hydrodynamic cavitation intensity. The effective vapor volume, the driving pressure differential in the implosion zone, the liquid properties and gas contents are captured in this measurement in first approximation. The estimation of the cavitation risk based on Eq. (T6.1.6) supplements the evaluation according to Eq. (T6.1.2) since the uncertainties 2, 3 and 6 partly cease to apply. The acoustic properties of various test arrangements are additional parameters whose influences are difficult to assess. Based on the noise signals it cannot be determined where the bubbles implode. If this occurs in the free liquid, the bubbles are harmless and the correlation produces pessimistic predictions. Likewise, it cannot be evaluated if the bubbles are locally concentrated or distributed over larger areas of the blade inlet, which can mean major differences in local erosion (e.g. vortex cavitation).

The calculations are also influenced when cavitation occurs at different locations, e.g. simultaneously on the suction and pressure surfaces of the blades, or in annular seals, in the inlet casing or in the circuit in which the pump is operating. If the measured noise differs only little from the background noise, the erosion prediction becomes uncertain (no developed erosion is present either). Equation (T6.1.6) is preferably used only for $NL > 1.25\ NL_o$.

The effect of the temperature on the fluid-borne noise is not sufficiently known. The noise tends to decline with increasing temperature and increasing gas content. Generally valid statements for predictions cannot yet be made.

Even noise measurements at $u_1 < 20$ m/s appear problematic (because of the low HCI) with regard to the transfer to high circumferential speeds.

6.6.5 Solid-Borne Noise Measurements for Cavitation Diagnosis

While fluid-borne noise can be recorded during model tests by simple means, solid-borne noise measurements on the pump casing are often the only practical way to obtain information on the hydrodynamic cavitation intensity in plants. If the solid-borne noise is measured on the outer casing wall with an accelerometer, the signal of a given fluid-borne noise depends on the casing geometry and the material. Based on the statistical energy analysis, these influences can be captured in an approximate way. Even the fluid-borne noise can be calculated from the measured solid-borne noise according to Eq. (T6.1.7), [26]. The available data was also entered in Fig. 6.30. The points in the graph are characterized as solid-borne noise data from which the fluid-borne noise was calculated. Both groups of data can be presented through the same correlation. The solid-borne noise employed also constitutes an RMS-value. To avoid resonance with the transducer, the measurements were made in the range from 1 to 47 kHz. Figure 6.30 also contains tests on air foils from [4].

The relationship between erosion and solid-borne noise offers a very simple technique for cavitation diagnosis, but the method is subject to the same limitations as the fluid-borne noise measurements. In plants it is often difficult to determine the *load-dependent* background noise which must be known to make a reliable assessment - unless the cavitation noise clearly dominates.

As discussed earlier, fluid-borne noise measurements capture any bubble implosions in the pump even if these occur at a distance from the blades and do not contribute to erosion. To eliminate this influence, it has been suggested to measure solid-borne noise on the pump shaft with the object to record only those implosions which occur on the blades, [54]. When considering this option, it should be borne in mind that even implosions within the liquid cause shock waves and pressure pulsations which travel through the liquid till hitting a solid boundary. Therefore, even bubbles imploding in the free stream excite solid-borne noise in the same way as pressure pulsations traveling through a pipeline cause solid-borne noise in the pipe wall (part of which is radiated as air-borne noise to the environment). This argument is supported to some extent by a comparison of the measured solid- with the liquid-borne noise data in Fig. 6.27a, b.

6.6.6 Paint Erosion Tests to Determine the Location of Bubble Implosion

As shown above, the cavitation evaluation is pessimistic insofar as it cannot be determined (either based on cavity observations or on noise measurements) whether the bubbles implode harmlessly in the liquid or near a wall, thereby posing the risk of material erosion. Paint erosion tests can help in this regard. Such tests are only meaningful if the test duration is selected such that an essentially time-independent paint erosion pattern is obtained at the hydrodynamic cavitation intensity prevailing under the specific test conditions. To achieve this, the process must be calibrated as described in [24]. Calibration tests with various cavitation conditions (cavity lengths) and speeds were conducted with a given paint (zinc dust paint). During these tests the erosion pattern was recorded after defined operating intervals. After a certain test period the eroded area asymptotically reached a final form, meaning that the eroded area would not grow any more if the testing time were prolonged. The required test duration t_v thus determined was plotted against the fluid-borne noise CNL_{Ref} according to Eq. (T6.1.4) in Fig. 6.31. Because of differing acoustic properties different curves apply to pumps with steel cover and model pumps with a Perspex cover. When applying Fig. 6.31, it is also possible to measure the

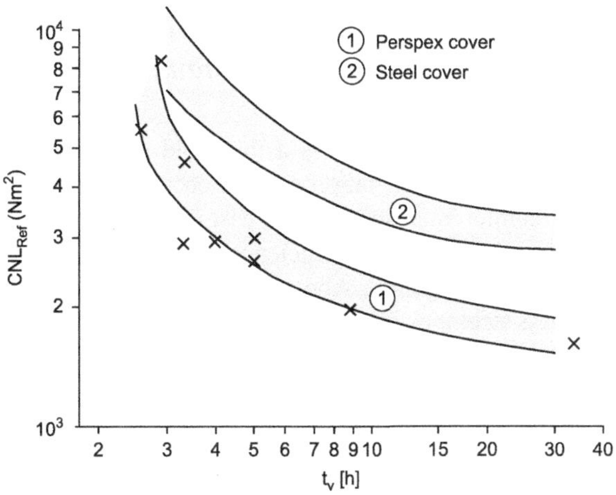

Fig. 6.31 Required testing time for paint erosion tests (zinc dust paint) [25]

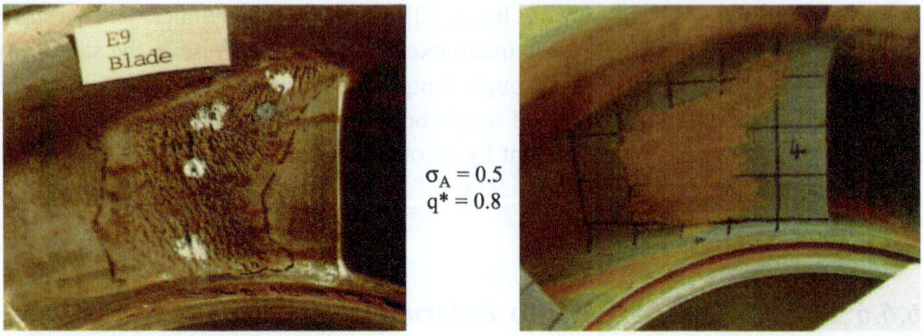

Steel impeller, 9000 min^{-1}/160 h   Paint erosion, 3000 min^{-1}/8 h

Fig. 6.32 Comparison of erosion patterns with steel impeller and paint erosion [25]

solid-borne noise and to calculate by means of Eq. (T6.1.7) the fluid-borne noise prior to reading the required test duration from the graph.

Figure 6.32 shows the paint erosion pattern from a test run for 8 h at 3000 rpm. The result is compared with the erosion pattern of a steel impeller with identical cavitation and flow coefficients σ_A and φ_1 after 160 h at 9000 rpm.

The paint erosion pattern observed is indeed representative for the damage pattern on the steel impeller - provided that the test duration has been set to an adequate period. However, this will only be successful if it is possible to quantify the hydrodynamic cavitation intensity using sound measurements. If the paint erosion test in Fig. 6.32 had been run for a period significantly shorter than 8 h, the eroded area would have been smaller. After too short a testing time possibly no paint erosion would have been detected at all and the damage potential would have been evaluated too optimistically. The calibrated paint erosion test supplies the true damage length, even on the non-visible pressure surface of the blades.

6.6.7 *Onset of Erosion and Behavior of Material Subject to Different Hydrodynamic Cavitation Intensities*

For every material there is a threshold of the hydrodynamic cavitation intensity below which no cavitation erosion occurs. Limits were proposed, for instance, for the circumferential speed u_1 for cast iron at 12 m/s and for austenitic steel at 22 m/s for operation at the best efficiency point, [30]. Such figures can only be valid for a specific water quality. Above figures apply to cold air-saturated water; in case of stainless steel they also can be used for de-aerated feed water at temperatures above about 140 °C, as follows from the Eq. (T6.1.2).

If the erosion rate from any test is plotted on a log-log graph against the hydraulic cavitation intensity HCI (or, as in Fig. 6.33, over σ), the specific erosion power P_{ER} asymptotically approaches a threshold HCI where P_{ER} becomes zero.

While it is possible in the range of "developed erosion" to calculate with a roughly constant exponent x in the relation $E_R \sim \sigma^x$, the exponent tends towards infinity upon approaching this asymptote. Consequently, almost any relationship between erosion rate and the independent test variables such as speed, noise, gas content, temperature, cavitation coefficient or cavity length can be measured in the domain close to the onset of erosion. This explains why values between 1 and 20 were published (somewhat uncritically) for the speed exponent. *Outside* of the range of *developed* erosion it is meaningless to use a material ranking based on measured erosion rates since the ratio of the erosion rates of two materials near the erosion onset can assume any value, becoming infinite at the threshold, Fig. 6.33. This fact has not been adequately considered in many publications.

Material rankings are only meaningful if they originate from tests done with developed erosion and if they are applied solely for this condition. The ratio of the erosion rates of two materials with highly differing cavitation resistance is always subject to considerable uncertainty in this context. When changing the material to improve impeller life, it was thus observed in practice that the hoped-for reduction of the cavitation damage was not achieved although it would have been expected when applying the material ranking measured in laboratory tests.

Ranking for any materials and cavitation intensities can be calculated from Eq. (T6.1.2). Figure 6.34 shows an example that was calculated for air-saturated water at 20 °C with $NPSH_A$ = 37 m and a cavity length of 50 mm.

These parameters were selected in a way that a cavitation erosion of 1 mm is obtained for ferritic chromium steel with 13% Cr and 4% Ni after 40,000 h of operation. The erosion rates written in Fig. 6.34 therefore simultaneously constitute the ratio of each material to this chromium steel. The ranking thus obtained produces similar relations as published by others; e.g. [50]. The calculation was done with the material factors F_{Mat} listed in Table D6.4.

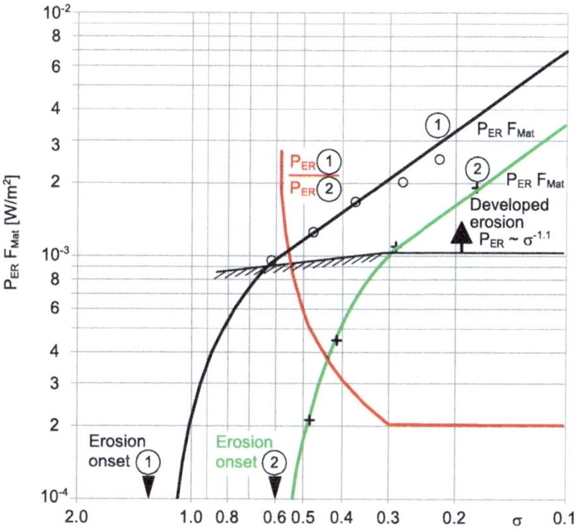

Fig. 6.33 Specific erosion power P_{ER} of two different steels subject to cavitation, [27]

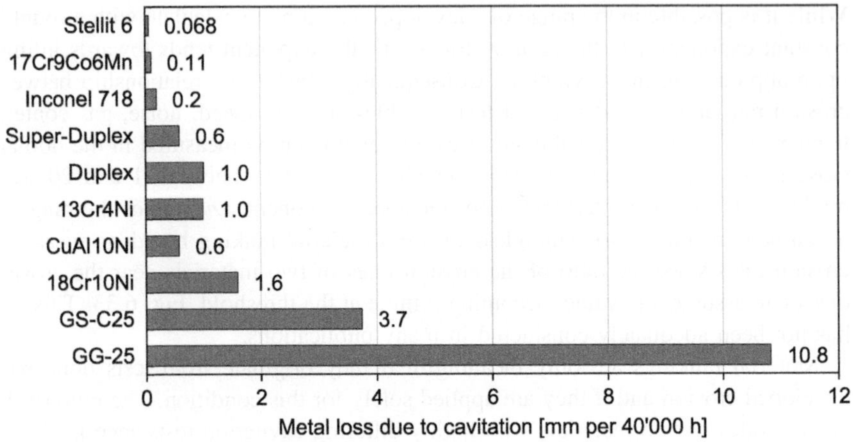

Fig. 6.34 Metal loss due to cavitation of different materials

While cast iron and carbon steel proved to exhibit little cavitation resistance, special alloys containing cobalt and manganese (such as 17Cr9Co6Mn) showed the best (currently known) cavitation resistance because of their high strain hardening, [38, 45, 49–51].

Table D6.4 Material factors for cavitation damage prediction

Material	F_{Mat}
Cast iron	0.6
Cast carbon steel	0.75
Ferritic steels	1.0
Duplex	1.3
Austenitic steels	1.6
Super Duplex	2.1
Aluminum bronzes	3
Inconel 718, [5]	5
Stellit 6	6
17Cr9Co6Mn	10

To determine the onset of erosion, a relationship is required between the *absolute* hydrodynamic cavitation intensity (e.g. the "implosion pressure") and the *fatigue strength* under cavitation. Since the implosion pressure cannot be measured, it has to be determined indirectly. To this end, first the intensity of the sound source is calculated from the measured fluid-borne noise with the help of the equations governing room acoustics. Using an acoustic efficiency, it is subsequently possible to estimate the mechanical power which has emitted the sound, [23]. These considerations lead to Eqs. (T6.1.11, T6.1.12).

The test data from Fig. 6.30 were evaluated in this way and shown in Fig. 6.35 thus providing a correlation between the specific erosion power and the implosion pressure, Eq. (T6.1.13). Due to the uncertainty of the necessary assumptions, this

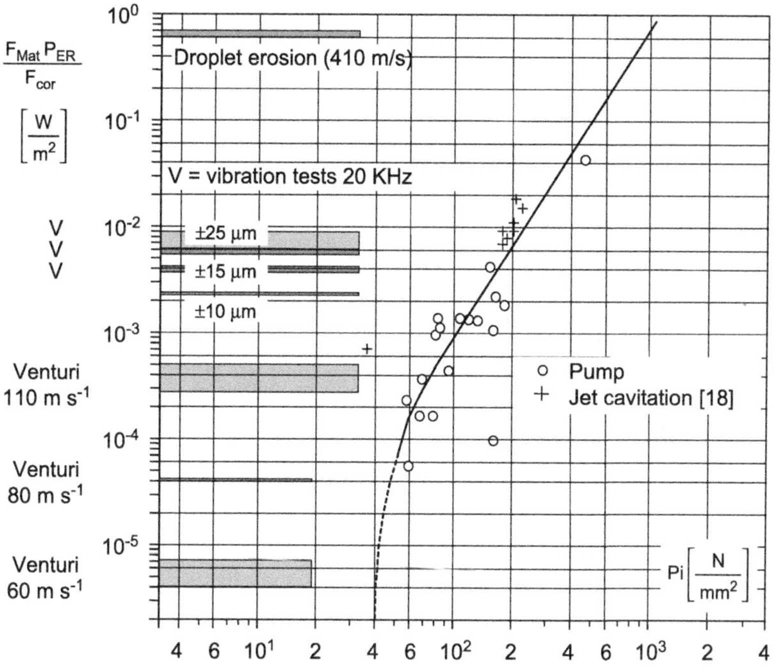

Fig. 6.35 Specific erosion power as function of the implosion pressure [18, 23]

equation has a highly empirical character. However, the wide range of the available data and the number of test points from various pumps give credence to the statistical relevance of the correlation.

Erosion measurements in various cavitation devices and machines can only be meaningfully compared if an absolute measure for the hydrodynamic cavitation intensity has been defined. The implosion pressure calculated here is indeed such a measure. Therefore, it was possible in Fig. 6.35 to plot together test results from a droplet erosion apparatus, magneto-strictive vibrators, Venturi nozzles and a jet cavitation device (data from [11]). With the droplet impact test device the implosion pressure can be estimated according to $p_i = \rho \times a \times w$. For the data in Fig. 6.35 this produces p_i = 610 N/mm^2. The comparison with the trend curve according to Eq. (T6.1.16) yields a value of similar magnitude (930 N/mm^2).

Fluid-borne noise and cavitation erosion were measured on various steels in a jet cavitation apparatus [18]. The implosion pressure was calculated using Eqs. (T6.1.4, 6.1.11 and 6.1.12). The results, included in Fig. 6.35, agree well with the erosion rates in pumps although the geometrical arrangement and the flow conditions are totally different. The estimations of implosion pressures found in the literature (quoted in [23]) are in the range of the data plotted in Fig. 6.35.

The correlation of test data from totally different devices and machines allows the conclusion that Eq. (T6.1.12) supplies a useful reference for the absolute hydrodynamic cavitation intensity. The value of p_i estimated in this way can be compared with the fatigue strength *under cavitation*, which for ferritic steels in fresh or demineralized water is likely to be of the order of $F_s = (0.054 \text{ to } 0.15) \times R_m$, [23]. With $p_i < F_s$ the erosion threshold is not reached; no damage occurs. If p_i is clearly above the Brinell hardness, developed erosion must be expected. Equating Eqs. (T6.1.2 and 6.1.13), it is possible to estimate the implosion pressure even if only cavity length or damage length are known, Eq. (T6.1.15). By comparison with Eq. (T6.1.17) it is even possible to evaluate the erosion threshold in this way.

6.6.8 Summarizing Assessment

The stroboscopic observation of the cavitation propagation at the impeller inlet is not only an indispensable tool in the development of cavitation-exposed impellers, but it also allows a quantitative evaluation of the erosion rate and even the erosion threshold, Eqs. (T6.1.15 and T6.1.17). The comparison of model tests with plant experiences and tests with water temperatures up to 180 °C [25] has shown that roughly the same cavity lengths must be expected in cold and hot water. The cavities observed in the test implicitly comprise the effects of all the geometrical parameters of suction chamber and impeller as well as the flow rate and $NPSH_A$. The relevance of stroboscopic bubble observations can be considerably enhanced and complemented by cavitation noise measurements and paint erosion tests.

Fluid- or solid-borne noise measurements are a measure for the hydrodynamic cavitation intensity. They capture both the local pressure differentials in the implosion zone as well as the effects of the liquid properties and the gas content. The sound pressures and the accelerations (for an initial estimation) can be scaled to other operating conditions in proportion to u_1^2.

To establish whether the bubbles implode near the blade, paint erosion tests are revealing if conducted under controlled conditions. The test period required to obtain a roughly stationary erosion pattern can be determined from Fig. 6.31 based on the fluid-borne noise measured in the test pump. Non-cavitation induced sounds are subtracted as background noise, but cavitation noise generated in other components of the circuit (e.g. in valves) can falsify the measurements.

The fluid-borne noise can be calculated from a solid-borne noise measurement using the methods of statistical energy analysis. Solid-borne noise signals can therefore be utilized for cavitation diagnosis.

The methods of room acoustics allow an estimation of the implosion pressure from the fluid- or solid-borne noise. Consequently, it is also possible to evaluate the erosion threshold and correlate erosion measurements from various test devices and machines, Fig. 6.35.

Each of the methods discussed above has its weaknesses and limits. However, if cavity length, paint erosion patterns, fluid- or solid-borne noise are measured for a given application, the damage risk can be evaluated quite comprehensively with relatively little effort. With flat pressure distributions at the impeller inlet one obtains (for instance), thin, translucent cavity fields and accordingly low sound signals. In this case Eq. (T6.1.6) may predict smaller erosion rates from the fluid-borne noise than Eq. (T6.1.2) based on the cavity length. In such a case, Eq. (T6.1.2) may yield a pessimistic prediction - provided the noise measurement is relevant. This is sometimes doubtful with low circumferential speeds and/or water containing gas. Conversely, if the cavity length, noise measurement and paint erosion all predict considerable erosion, future damage is highly probable. If both the fluid- and the solid-borne noises are measured, the uncertainties and the damage risk can be better evaluated.

For developed erosion (above the threshold value) the relationships between erosion rate and cavity length, velocity and tensile strength or hardness have been confirmed through many tests, even if considerable scatter must be expected. According to [52] the pitting volume created by the implosions in a material (for instance) grows according to $E_R \sim w^x \times R_m^{-y}$ with x = 5.2 to 6.7 and y = 2.2; this agrees well with the exponents found in the above correlations.

However, the traditional notion that cavitation erosion should depend on the flow velocity, is misleading. As Eqs. (6.1) and 6.2) demonstrate, the HCI depends on the pressure difference acting on an imploding bubble. While it is true that the pressure distribution depends on the square of the velocity, very different pressure differentials can be encountered at a given velocity. Examples are the differences between suction and pressure surface erosion (Fig. 6.29) or between erosion in an impeller and a Venturi nozzle, [23].

Cavitation erosion only occurs if three conditions are fulfilled:

1. Cavitation bubbles develop, which can be confirmed through visual or acoustic observation (fluid- or solid-borne noise).
2. The bubbles must implode *near a wall,* which can be verified through paint erosion tests under controlled conditions.
3. The absolute hydrodynamic cavitation intensity must exceed the cavitation resistance: the implosion pressure pi must be greater than the fatigue strength Fs *under cavitation.* Equations (T6.1.12) and (T6.1.15 to 6.1.17) allow a quantitative evaluation of the conditions:

   * $p_i \ll F_S$: no risk is assumed.
   * $p_i > R_m$: a very intensive attack must be expected.
   * $p_i >$ BHN (Brinell hardness): considerable damage must be expected.

Finally it is stressed once more that all equations provided are only able to provide rough estimates which are subject to major uncertainties. Because of the extraordinary complexity of the processes involved in cavitation erosion, anything different could hardly be expected. However, for the engineer's evaluation of cavitation problems, the equations in Table 6.1 constitute the only method known today to consistently describe the relationships between the various parameters involved. The equations cover a wide range and are suitable for practical application with little effort. As with all complex system problems, a careful analysis and interpretation of the observations and measurements in the plant are indispensable to prevent erroneous conclusions.

6.7 Selection of the Inlet Pressure in a Plant

As discussed with reference to Figs. 6.9, 6.10 and 6.14, *extended cavities appear well before a measurable drop in head,* such as $NPSH_3$, is noticeable. To prevent noise, vibrations and erosion damage, the suction pressure in the plant (the $NPSH_A$) for most applications must be set clearly above the $NPSH_3$ (an exception is the cavitation control discussed in Chap. 11.2). Essential aspects for determining the $NPSH_A$ are; see also [1]:

- the circumferential speed u_1 at the impeller inlet
- the impeller material
- the medium pumped, its temperature and gas content
- the required operation range (e.g. at partload recirculation or at $Q > Q_{SF}$)
- the characteristics of the pump, the hydraulic quality of the impeller and the approach flow conditions, all of which have an impact on the cavity volume.

Considering a *given pump*, whose required NPSH according to the various cavitation criteria is known as a function of the speed and the flow rate, a sufficient $NPSH_A$ must be provided in the plant to allow safe operation of the pump (see Table 2.2). To increase the $NPSH_A$ there are various possibilities:

1. Installing the pump on a lower level or the suction vessel on a higher location
2. Reducing the pressure losses in the suction line
3. Installing a booster pump
4. Increasing the gas pressure above the liquid level in a closed suction vessel.

Conversely, if a *given* $NPSH_A$ is assumed, the type, speed and size of the pump to be utilized must be selected so that trouble-free operation throughout the load range is ensured.

The criteria for trouble-free operation without damage are:

C1: Covering the required operating range, e.g.: Is it possible to achieve the maximum flow rate demanded? (Chap. 11).
C2: Avoidance of impermissible vibrations, pulsations and noise.
C3: Limitation of cavitation erosion or achieving the specified impeller service life.

Table 6.2 shows how the $NPSH_A$ can be established for various applications. It is stressed, however, that such data can only constitute rough recommendations that may frequently be conservative but cannot guarantee freedom of damage in problem cases.

Explanations to Table 6.2:

- Criterion C1 obviously must be satisfied in all applications.
- An unsuitable impeller, unfavorable suction conditions or inlet casings, extended partload or overload operation require a higher $NPSH_A$.
- Depending on the type of fluid, a lower $NPSH_A$ may be sufficient.

Table 6.2 Selection of available $NPSH_A$ to ensure safe operation

Purpose: this table allows to determine the available $NPSH_A$ necessary to ensure trouble-free operation. The following criteria are used:

- C1: Ensure that pump can run in the entire range of operation, in particular at maximum flow
- C2: Limit cavitation induced noise and vibration
- C3: Avoid (excessive) cavitation erosion

u_1 (m/s)	Fluid	Criteria	Determination of the necessary $NPSH_A$	
< 10	All	C1 (C2)	$NPSH_A \geq 1.25 \times NPSH_3$ Minimum: $NPSH_A = NPSH_3 + 0.6$ m	
< 20	Hydrocarbons (except kerosene and similar)		Objective: ensure operation range; cove tolerances and uncertainties	
10 to 50	Water	C2, C3	$\sigma_A \geq \sigma_3 + F_R\, F_F \{0.05 + 2\,(u_1/u_{Ref})^3\}$ u_{Ref} = 100 m/s. Apply only for $\sigma_A < \sigma_i$	
			Fluid factors	F_F
			Water with $\alpha > 5$ ppm; t < 200 °C	1.0
			Sea water or other corrosive media	≥ 1.15
			Hydrocarbons with $\Delta T_u > 5$ °C	0.75
20 to 60	Hydrocarbons		**Risk factors**	F_R
			Operation at $0.8 < q^* < 1.1$	1.0
			Operation at partload recirculation or pressure surface cavitation at $Q > Q_{SF}$	1.2
			Non-uniform approach flow	≥ 1.1
50 to 75	Water Cr-Ni-steel	C3 C1	Cavity length, cavitation noise or operation experience under the *same condition,* must be available for checking the impeller life according to Table 6.1	
> 60	Hydrocarbons Cr-Ni-steel	C1 C2	Cavity volume must be limited in order to avoid excessive vibrations, noise and pulsations	
> 75	Water at T < 200 °C Cr-Ni-steel	C1 C3	$NPSH_A \geq NPSH_i$ Practically bubble-free operation is necessary; otherwise damage after short periods of operation.	
10 to 25	De-aerated water $\alpha < 5$ ppm, T < 100 °C Cr-Ni-steel	C2	$\sigma_A \geq \sigma_3 + F_R\, F_F \{0.05 + 2(u_1/u_{Ref})^3\}$ u_{Ref} = 100 m/s	
> 25		C3	Very erosive; cavity volume or cavitation noise must be limited, Table 6.1	

Minimum: $NPSH_A \geq 1.25 \times NPSH_3$ or $NPSH_A = NPSH_3 + 0.6$ m whatever is greater

The safety factor on $NPSH_3$ is defined as: $F_{NPSH} = \dfrac{NPSH_A}{NPSH_3} = \dfrac{\sigma_A}{\sigma_3}$

The available NPSH determined in this way is only sufficient if the following is observed:

1. The impeller is designed for an incidence at BEP i < 3° (to 5°).
2. The approach flow is reasonably uniform (criteria are given in Chap. 11.7.3)
3. There is no excessive vortex formation.
4. The selected $NPSH_A$ is also sufficient for the maximum continuous flow.

Caution: The above procedure is empirical and can give no guarantee in extreme cases. It may be conservative in other instances.

- "Hydrocarbons" stands for hydrocarbon mixtures with various boiling curves or other fluids that liberate weak implosion energies, where $\Delta T_u > 5$ °C roughly applies according to Eq. (6.24).
- A prerequisite for the application of Table 6.2 is the use of a suitable material. The information "Cr-Ni-steel" means that the material employed must at least be an equivalent to 1.4317 (G-X5 Cr Ni 13 4) or (ASTM A743 Gr CA 6MN).
- Margins to the calculated $NPSH_A$ are necessary in the case of corrosive media.
- At circumferential speeds above $u_1 = 50$ m/s the risk of cavitation damage increases rapidly, even if high-strength stainless steels are used. Only limited cavity lengths can then be permitted. Above tip speeds of $u_1 = 75$ to 80 m/s the pump must be operated practically free of bubbles. The erosion risk can be evaluated using the methods discussed in Chaps. 6.6.3–6.6.8. This requires stroboscopic bubble observation, cavitation noise measurements or a reliable prediction of the cavity length or the $NPSH_i$ value.
- For the ferritic steel 1.4317 (ASTM A743 Gr CA 6MN) the maximum permissible cavity length required to reach an impeller service life of 40,000 h was determined according to Table 6.1. Metal erosion of 6 mm in 40,000 h was assumed for this calculation. The result is shown in Fig. 6.36. For other conditions it is possible to scale according to Eq. (6.31) unless the calculation according to Table 6.1 is preferred.
- In particularly critical cases more accurate analyses and additional margins are recommended.

$$L_{ca,al} = L_{cav,graph}\left\{\frac{R_m}{R_{m,Ref}}\right\}^{0.7}\left\{\frac{\Delta e\, L_{I,Ref}}{\Delta e_{Ref}\, L_I}\right\}^{0.35} \qquad \begin{aligned} R_{m,Ref} &= 800\,\mathrm{N/mm^2} \\ L_{I,Ref} &= 40{,}000\,\mathrm{h} \\ \Delta e_{Ref} &= 6\,\mathrm{mm} \end{aligned} \tag{6.31}$$

- In pumps with unfavorable suction conditions and large sound-radiating casing surfaces (possibly with low wall thickness) the noise level can increase

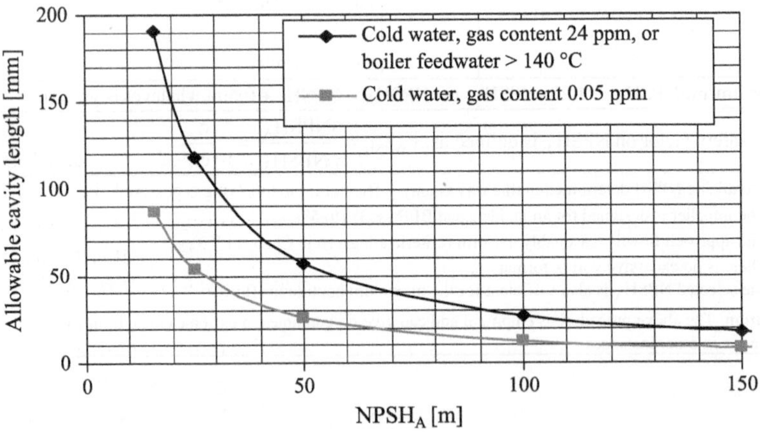

Fig. 6.36 Allowable cavity length as function of the $NPSH_A$ for steel 1.4317 (ASTM A743 Gr CA 6MN) for an impeller life of 40,000 h

considerably due to cavitation. If noise guarantees must be met, the $NPSH_A$ should be selected so that the pump operates far above $NPSH_o$ since, according to Chap. 6.5.2, the maximum cavitation noise must be expected near $NPSH_o$.

- Favorable operating experiences with a given pump in an identical application (u_1, load range, fluid) may justify a lower $NPSH_A$.

Often the ratio of the $NPSH_A$ to the $NPSH_3$ is used to define a safety margin F_{NPSH}. This value can be easily determined when establishing $NPSH_A$ based on Table 6.2.

Depending on the design of a specific impeller, *entirely different cavity lengths* must be expected for a given NPSH-margin F_{NPSH}, see Fig. 6.14. Consequently, the NPSH-margin is not a universal measure for risks in terms of cavitation-induced noise, vibrations and erosion.

6.8 Cavitation Damage: Analysis and Remedies

6.8.1 Record Damage and Operation Parameters

To define suitable measures for solving cavitation problems, a careful analysis of the operating conditions is indicated. It comprises the following elements (for use as a checklist):

- Record damage pattern and location: impeller inlet, suction surface, pressure surface, inlet, collector.
- Intensity of the attack: maximum depth ΔE_{Max}, L_{Damage} measured from leading edge, L_{start}, sketch, photo.
- Operating parameters: Q, H, n, $NPSH_A = f(Q)$.
- System characteristic and maximum flow rate.
- Operation time at various operating conditions (histogram if required).
- Pump design data: Q_{opt}, H_{opt} with applicable speed; characteristics and NPSH-curves (acceptance test). Does $NPSH_3$ increase at partload? (This could point to an excessive impeller inlet diameter).
- Material of component: tensile strength; material designation (chemical analysis if required).
- Fluid pumped: type, water analysis; corrosive agents, temperature, gas content
- Geometrical data (here for impeller): $d_{2,eff}$, d_1, d_n, β_{1B} (outside, center, hub), a_{1a}, a_{1m}, a_{1i}, A_{1q} (cardboard template), blade inlet profile (cardboard template, caliper).
- Deceleration ratios: w_{1q}/w_{1m} and c_{3q}/c_2 at Q_{opt}, Q_{max} and at the operating point.
- If the blade inlet angles are not known, these can be estimated from $\beta_{A1} = \arcsin(a_1/t_1)$.
- Suction nozzle or diameter of suction bell.
- Impeller trim $d_2^* = d_{2,eff}/d_{2,nom}$ (effective to nominal impeller diameter).
- Pump control, transient and upset conditions, (type, frequency, duration).

- Increase of water temperature at impeller inlet through the mixing with leakage flow and balance water upstream of the impeller; corresponding increase of vapor pressure and drop in $NPSH_A$.
- Cavitation diagnosis with solid- or fluid-borne noise according to Chap. 6.6.5.

Inlet conditions:

- *Dry pit installation*: Bends, branches and valves generate non-uniform velocity profiles unless a pipe section of sufficient length is available to ensure reasonably uniform flow. Two and more bends arranged in different planes create a swirl. (Remedy: turning vanes in the bends or installing a flow straightener upstream of the pump inlet nozzle).
- *Wet pit installation*: Visual observation of the suction bay for surface vortices, swirl, ground vortices, vortices drawing air.
- *Inlet casing* of between-bearing pumps: rib installed at the location where the two part-flows around the shaft meet? Ribs too thick or with unfavorable trailing edge profile?

Flow perturbations generated by the approach flow increase with rising flow rate. An unfavorable design of the inlet casing can result in vortex ropes or create impermissible distortions of the impeller approach flow. To analyze the inlet casing, criteria such as stream-wise area distribution and hub blockage must be utilized, Chap. 7.13.

6.8.2 Forms of Cavitation and Typical Cavitation Damage Patterns

Cavitation can occur at the impeller blade inlet, in the impeller channels, on ribs in the inlet casing, in the collector and in the annular seals of the impeller or the balance device. Depending on the geometrical parameters of impeller, inlet and collector, the pump flow rate and the prevailing suction pressure, different forms of cavitation can develop.

6.8.2.1 Cavitation Damage at the Impeller Inlet

Cavitation damage is mostly observed at the impeller inlet since this is where the lowest pressures occur in the pump. The erosion pattern frequently allows conclusions as to the cause of the damage. The flow rate of shockless entry Q_{SF} and the cavitation coefficient $\sigma_A = 2gNPSH_A/u_1^2$ should be calculated according to the data from Chap. 6.8.1 since these quantities may be significant for the evaluation. Typical patterns of cavitation damage at the impeller inlet are shown in Table 6.3.

- Frequently, attached cavities are observed on the blades ("sheet cavitation"), see Figs. 6.5 and 6.37 (the bubbles implode near the wall). Attached cavities can cause damage on the pressure surface at $Q > Q_{SF}$ and on the suction surface when $Q < Q_{SF}$. In the range $Q \approx Q_{SF}$ (or with variable operation conditions), damage can occur on the blade suction and pressure surfaces simultaneously.

Table 6.3 Cavitation damage

Damage pattern	Mechanism	Possible cause	Possible remedy
① Blade suction surface, start *near leading edge* 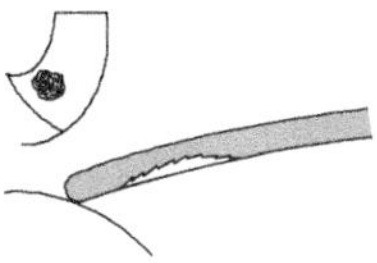	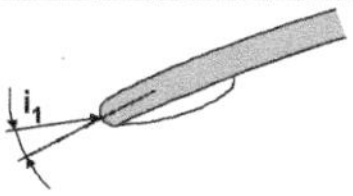 Attached cavity on suction surface ("sheet cavitation") $Q < Q_{SF}$ • Damage near shroud • Damage near hub	Incidence too large. Unfavorable leading edge profile. Unfavorable flow distribution upstream of impeller. • $\beta_{1B,a}$ too large • $\beta_{1B,i}$ too large	Operation at larger flow rate. Reduce blade inlet angle.Improve leading edge profile on suction surface. Reduce incidence by means of inlet ring. Reduce impeller inlet diameter. Increase pre-rotation.
② Within impeller channel on suction surface 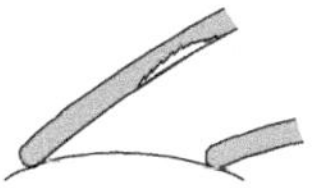 possibly also on pressure surface, hub or shroud	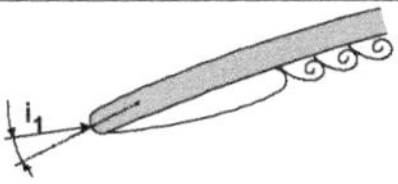 Cavitating vortices shed by long, thick cavities at $Q << Q_{SF}$	Large incidence combined with low σ_A. $\sigma_A < 0.15$ to 0.3 $NPSH_A$ too low. Unsuitable material.	Reduce impeller inlet diameter. Reduce inlet angle. Increase $NPSH_A$. Better material. Possibly: drill holes into blade to allow flow from pressure to suction surface.
③ Blade pressure surface, outer half of blade, start near leading edge 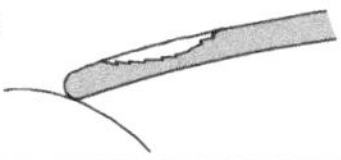	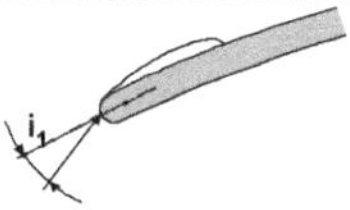 Attached cavity on pressure surface, $Q > Q_{SF}$	Negative incidence. Unfavorable leading edge profile. Excessive flow rate.	Reduction of flow rate. Cut back blade and profile on pressure surface. Improve leading edge profile.
④ Blade pressure surface, outer half of blade, start near leading edge	Cavity clouds in channel due to recirculation 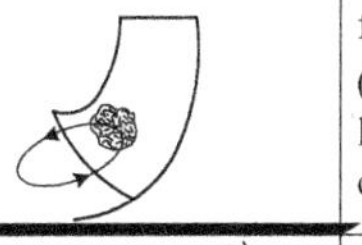	Intensive recirculation due to excessive flow deceleration (d_1, A_{1q}, β_{1B} too large). q^* too small.	Increase flow rate. New impeller with reduced d_1, A_{1q}, β_{1B}. Inlet ring: Fig. 6.39. Reduce recirculation.
⑤ Blade pressure surface *near hub*	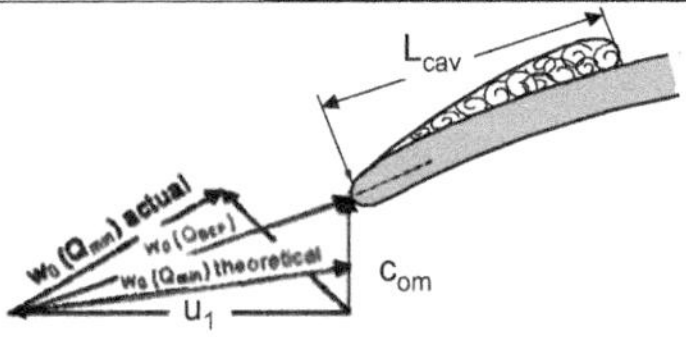	Negative incidence on hub due to recirculation.	Reduce recirculation. Improve leading edge profile (by grinding). Reduce pre-rotation.
⑥ Metal loss on hub, shroud or in fillet radii	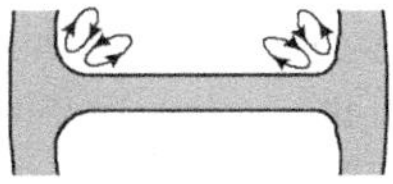 corner vortex due to large incidence	Inlet angle not correct. Fillet radii too large. Inlet flow perturbation.	Match inlet angle, possibly cut back blade and grind profile. Grind fillet radii.
⑦ Gap cavitation in open impellers	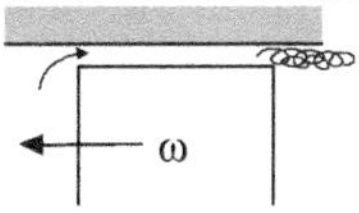	High blade loading.	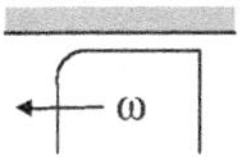 Rounding of blade edge on pressure surface

Fig. 6.37 Attached cavities, Sulzer Pumps

- The damage zone frequently starts close to the blade leading edge and roughly corresponds to the visually observed cavity length. The damage length will be longer if bubbles separate from the rear of the cavity and are transported downstream. Depending on the blade twist, incidence and approach flow, attached cavities can create damage over the entire width of the blade, or only near the outer or inner streamline.
- With greater incidence angles, the cavities increasingly become detached from the blades, larger vortex-type bubble accumulations are torn off the rear of the cavity and swept downstream. Because of increasing pressure gradients, they have a high damage potential. The deceleration at the rear of the cavity increases with the cavity thickness.
- Flat pressure distributions result in translucent cavities of low thickness; hence the hydraulic cavitation intensity is low because the bubble volumes are small and the pressure gradients at the cavity rear are low.
- The bubbles can implode on the blades or shrouds, in the free stream or on the opposite blade. If this occurs downstream of an air foil (e.g. in the case of axial pumps), the term "super-cavitation" is used.
- Vortex cavitation: The pressure in the center of a vortex drops as the result of centrifugal forces (Chap. 1.4). When the vapor pressure is reached, bubbles form which are entrained by the flow. Vortices occur in shear layers, as corner vortices or as the result of excessive incidence near the hub, but also downstream of attached or detached cavities. The 3-dimensional effects in the fillet radii at the rear or front shrouds are difficult to estimate. Damage is frequently observed in this zone; it can be particularly severe near the hub (including pitting on the shaft). Sometimes casting defaults occur near the fillet radii between blade and hub and pores are exposed as cavitation erosion progresses.

Large pores give rise to additional vortices and thus increase cavitation bubble formation resulting in accelerated damage.

- Long vortex ropes with a vapor core can be generated in the inlet casing of a pump. The vortex is cut by the blades when entrained by the flow into the impeller, Figs. 6.38 and 11.21. Due to the large cavity volumes involved, vortex ropes have a high potential for locally concentrated material erosion on the pressure surface of the blades. This problem must be rectified primarily through modifying the inlet casing.
- Flow separation at the trailing edge of ribs in the inlet casing can generate vortex streets which likewise may create cavitation bubbles once the vapor pressure is reached. The vortex intensity increases with the flow rate.

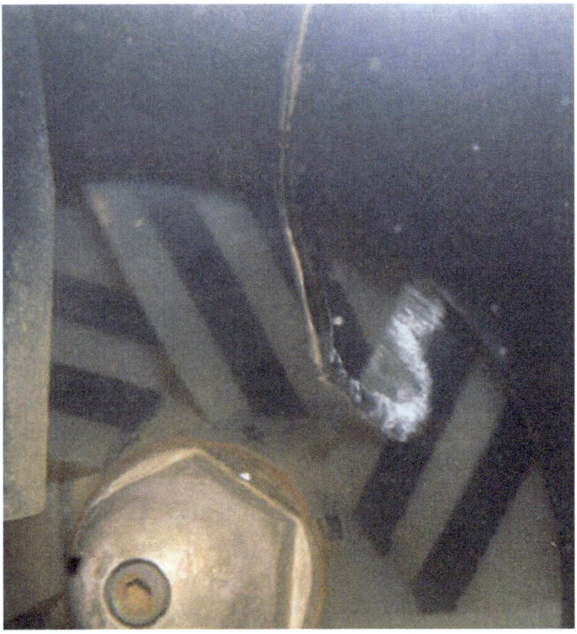

Fig. 6.38 Vortex rope generated in the inlet; the vortex is cut by the blades

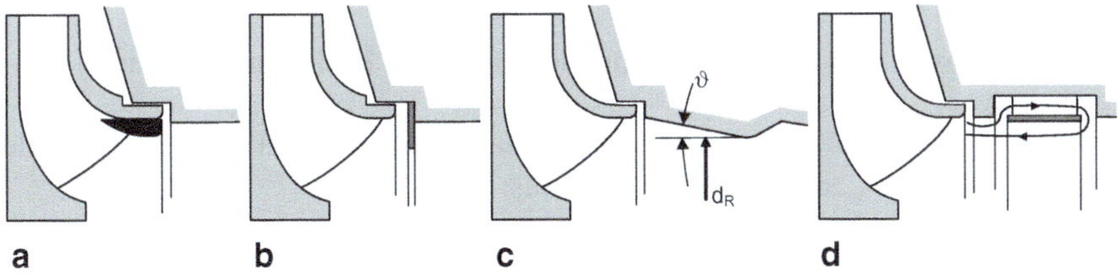

Fig. 6.39 Improvement of partload behavior if impeller inlet is excessively large, [30]. **a** Ring rotating with impeller; **b** orifice; **c** inlet diffuser; **d** recirculation brake, [53]

- Individual bubbles, which appear like pearls on the profile ("bubbly cavitation"), are rarely observed in pumps. They present no risk of damage because of their small vapor volume.
- The spiraling tip vortexes of propellers can cavitate as shown by Fig. 5.46. The vortex may be stable over one revolution.
- Gap cavitation in open impellers (e.g. axial pumps or semi-axial impellers) or the adjoining casing part which is frequently - and wrongly - made of a less cavitation-resistant material than the impeller.
- Partload recirculation mostly creates damage on the blade pressure surface. Frequently the erosion produces sharply defined zones in the outer half of the blade span. Sometimes it is difficult to determine from the damage pattern whether the cavitation is due to partload or overload operation. Since the local static pressure on the blade leading face is greater than on the suction surface, and since large bubble formations are often generated during recirculation, recirculation-induced cavitation damage can be very serious indeed.
- Extensive recirculation with small q^* can also cause damage on the blade pressure surface near the hub. The recirculating fluid creates a strong pre-swirl near the hub which results in large flow angles and creates flow separation on the pressure surface, Table 6.3, item 5.
- When operating with very low cavitation coefficients σ_A (e.g. condensate pumps or inducers) very long cavities develop. Downstream of these, vortex-type bubble formations separate which are transported far into the impeller channels. There they can lead to damage on the suction or pressure surface and on the rear or front shrouds. Sometimes this type of damage can be alleviated by drilling small holes in the zone between blade leading edge and the throat. Some water flows from the pressure to the suction surface through the holes with the result that the bubbles no longer implode near the blade. (Balance slots near the leading edge are discussed in [47].)

As discussed in Chap. 6.1.2, only the bubbles that implode close to the material surface are effective in causing damage. For this reason, even minor changes in the geometric shape can have a major effect on the extent of the damage if the distance of the bubble implosion from the material is modified as a result. Experience confirms that the erosion pattern can vary significantly from blade to blade even if the geometrical deviations of the blade geometry are hardly measurable. This is particularly true for operation near cavitation inception.

6.8.2.2 Cavitation Damage at Impeller Outlet or in the Inlet Casing

Although cavitation damage is most frequently observed at the impeller inlet, erosion sometimes occurs at other locations:

- In rare instances, localized erosion is observed at the impeller outlet in high-pressure pumps on the blade trailing edges and/or on the shrouds, Fig. 6.40. The damage in Fig. 6.40 is created through unsteady excess velocities

Fig. 6.40 Cavitation erosion at the blade trailing edge of a double-entry impeller; fluid: boiler feed water, material: 1.4317, $u_2 = 120$ m/s, diffuser $d_3/d_2 \approx 1.05$, operation at partload, $\sigma_{A,Le,c3} = 2.5$, Eq. (6.32)

which occur when the impeller blades pass the diffuser vanes, Chap. 10.1. The velocity distribution near the impeller outlet and thickness and shape of the blade trailing edge exert an influence on the damage which may be observed at high tip speeds. The one-sided cavitation erosion on the impeller trailing edge of this double-entry impeller testifies to the uneven flow distribution at the impeller outlet, Chap. 5.8.

- Profiling on the pressure surface or increasing the distance between impeller blades and collector are possible remedies. Care must be taken when increasing this distance since this can shift the best efficiency point or render the Q-H-curve unstable. If the problems cannot be solved in this way, a redesign of the impeller with lower blade loading at the outlet should be considered.
- Extensive damage far down in the impeller channels, near the outlet and/or in the diffuser point to operation with full cavitation - usually at flows well above BEP. At high circumferential speeds, material pitting during full cavitation can occur even after a short time (for instance due to incorrect operation of the pump or transient conditions). This mechanism is well understood and damage is prevented by avoiding to operate at excessive flow with insufficient $NPSH_A$.
- A speed-controlled pump can get into full cavitation if it is to operate with reduced head at the design flow rate. This can happen if the pump is over-sized or the plant requirements were overestimated or changed. When the speed is reduced in accordance with the required discharge pressure, the $NPSH_3$-curve shifts towards smaller flow (see Fig. 11.5), Q/Q_{SF} increases and the NPSHA and $NPSH_{FC}$ curves intersect at the operating point. With an over-sized multistage pump the associated loss in head of the first stage is hardly noticeable in operation. As a remedy, operating mode and control of the pump, including transients, should be analyzed in order to decide whether the operating parameters can be changed or a new impeller has to be designed for a suitably larger Q_{SF}. In case of multistage pumps, de-staging may be considered in order to keep the speed in the range where cavitation remains within acceptable limits.
- Cavitation damage in the inlet casing is usually caused by partload recirculation; sometimes also at excessive flow well above BEP. Gap cavitation can occur near the annular seal. The bubbles created through the recirculation can cause material erosion on ribs or the casing upstream of the impeller inlet. If the damaged parts are made of gray cast iron or nodular cast iron, more cavitation

resistant materials will provide a remedy. Sometimes it is sufficient to cut back and profile the ribs. If the impeller is incorrectly designed or unsuitable for the effective operating conditions, the measures discussed in Chap. 6.8.3 can be considered (refer also to Table 6.3). In addition, improvement can sometimes be achieved through an inlet ring or inlet diffuser; various options are shown in Fig. 6.39.

- With open impellers, gap cavitation can cause damage to the casing especially when cast iron or nodular cast iron is used. A better material should be employed as a remedy (replace the entire part, add a ring or apply a weld overlay, if a weldable material was used).
- Gap cavitation in annular seals on the impeller or balance devices.
- Cavitation damage can also be observed in vortex pumps and single-channel impellers of sewage pumps since these impellers are often made of cast iron (0.0625, GJL-250) which has a very low cavitation resistance. Furthermore, sewage pumps are frequently required to operate at flows much above BEP when the water level in the pit rises to its maximum (intermittent pumping between maximum and minimum water level). Operation at excessive flow leads to cavitation from the inlet right to the blade trailing edge as can be seen on the photographs in Fig. 6.41. Cavitation erosion can be detected in front of the leading edge and over a large area on the pressure surface of the blade extending almost up to the trailing edge with erosion through the blade near the front shroud. Cavitation erosion, vibrations and noise were significantly reduced by applying relatively large-sized bores through the front shroud (near the leading edge) and rear shroud, Fig. 6.42 from [58]. The bores lead some water from the ISR to the zones where cavitation is observed either visually or where damage did occur. The effect of these bores is to kind-of blow away the cavitation bubbles from the blade and the shrouds. However, the drop in efficiency

Fig. 6.41 Cavitation damage of a single-channel impeller from cast iron, $u_2 = 15$ m/s

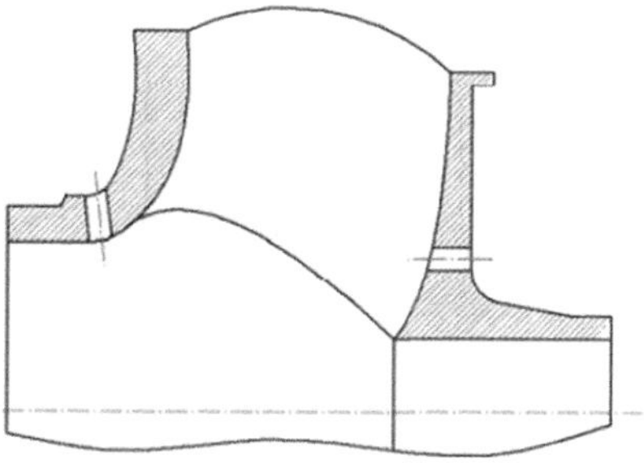

Fig. 6.42 Bores through the shrouds into the cavitation zones near the leading of a single-channel impeller, [58]

caused by the bores for the pump tested in [58] was 2% points ($n_q = 50$, $d_2 = 630$ mm). It is not known whether this remedy was applied in actual sewage plants.

6.8.2.3 Cavitation Damage in Diffusers or Volutes

In some instances cavitation erosion has been observed at diffuser vanes and volute cutwaters. Different mechanisms can be responsible for this type of damage:

1. Unsteady part load cavitation in diffusers: local cavitation can occur on the suction side of the leading edge, Figs. 6.43 and 6.44. The cause is unsteady sheet cavitation due to the pitch-wise variation of flow velocities and incidence as discussed in Chap. 10.1, Figs. 10.3–10.5. The extent of the damage on the diffuser vanes can vary around the circumference because the flow through the individual impeller channels depends on the pressure distribution in the volute, which becomes increasingly non-uniform as the flow rate is reduced, see Chap. 9.3.3.
2. Vortex cavitation in the diffuser channels near the diffuser throat area, Figs. 6.43 and 6.44. The vortexes may be created by shear layers caused by partload flow recirculation, see sketch on the right in Fig. 6.44. On the left, Fig. 6.44 shows a photograph of the cavitation erosion on the diffuser of a single-stage boiler feed pump with a head of typically 700 m. The diffuser shows both damage due to unsteady sheet cavitation near the leading edge and erosion in the channel due to vortex cavitation as sketched in Fig. 6.43. A sequence of photos taken with a high-speed camera is shown in Fig. 6.45. The pump has a 5-bladed double-entry

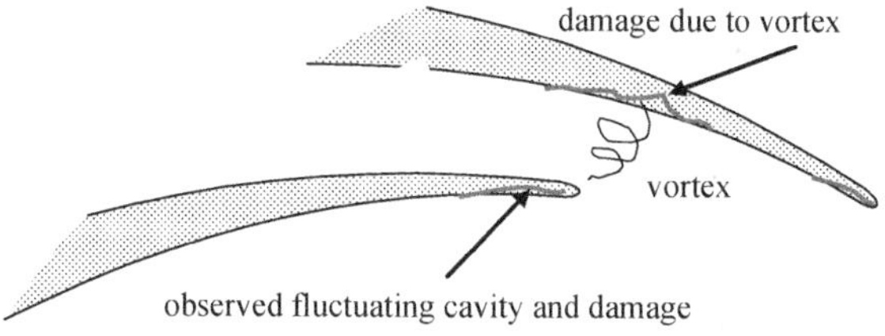

Fig. 6.43 Diffuser cavitation due to fluctuating cavities at the leading edge and due to vortex cavitation in the channel

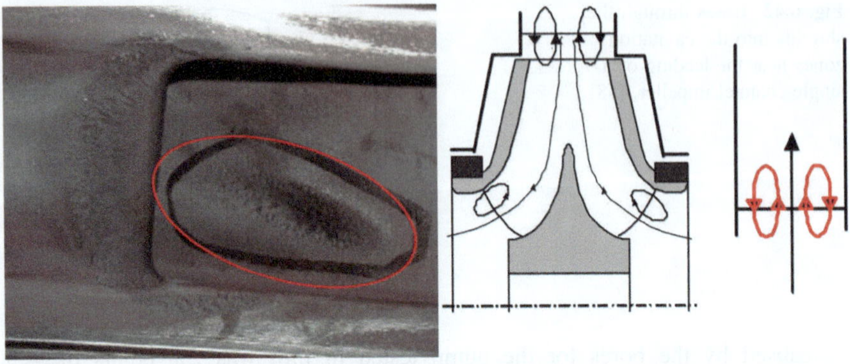

Fig. 6.44 Sheet cavitation at the diffuser vane leading edge and vortex cavitation in a diffuser channel

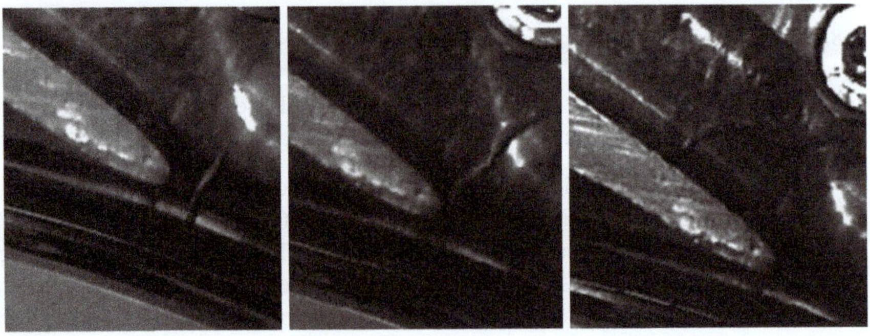

Fig. 6.45 Vortex formation at the impeller outlet and in the diffuser channels at $q^* = 0.4$, [3]

Fig. 6.46 Cavitation erosion at the cutwater of a double volute at $q^* = 1$

impeller ($n_q = 22$) and a 12-vaned diffuser followed by a single volute [3]. The distance between impeller blade trailing edge and the diffuser vane leading edge was about $d_3/d_2 = 1.05$.

3. As discussed in Chap. 5.3.1, unsteady diffuser cavitation can also be observed at high flow due to excess velocities induced by each passage of a blade under a diffuser vane, Fig. 5.17. Unsteady cavitation at the volute cutwater (CW) at high flow occurs in a way similar to diffuser cavitation, shown in Fig. 5.17 at $q^* = 1.5$. Photographs with a high-speed camera were published in [68] as shown in Fig. 6.47. Cavitation starts when a blade passes under the cutwater and vanishes at about mid-pitch of the impeller movement. The cavitation cloud then detaches and travels into the discharge nozzle. Laser particle image velocity measurements showed that the approach flow angle at the cutwater leading edge varied between about 25° to 75° as the blades moved over one pitch. It may be noted that the cutwater leading edge had a semi-circle shape and that a design with an elliptical profile could bring improvement. The available cavitation coefficient as defined by Eq. (6.32) is estimated to about 3.5 in Fig. 6.47. The σA at the impeller inlet corresponded to 3%-head drop, i.e. $\sigma A = \sigma 3 = 0.5$. Virtually no cavitation was observed at the impeller inlet, but the pressure surface of the blades could not be observed. Upon decreasing the NPSHA during these tests at constant flow rate, the head drop was continuous (no horizontal part of the curve H = f(NPSHA)) which suggests that the head drop was caused by volute rather than impeller cavitation.
4. Part load cavitation at volute cutwaters: local cavitation on the low-pressure side near the leading edge is shown in Fig. 6.46. The single-stage pump had a double-entry impeller ($n_q = 44$) and a double volute made from nodular cast iron 0.7042 (GGG42) with $d_z/d_2 = 1.16$; the impeller tip speed was $u_2 = 44$ m/s

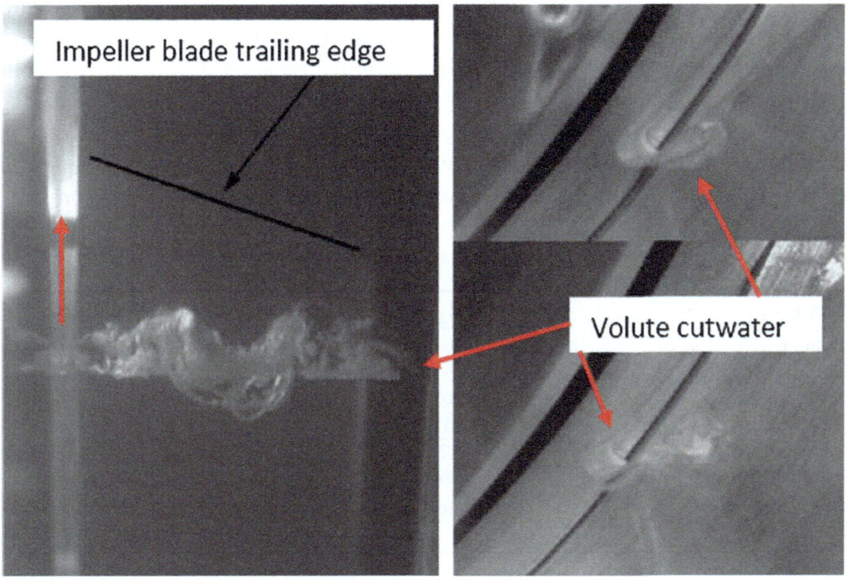

Fig. 6. 47 Unsteady cavitation at the volute cutwater, $q^* = 1.17$, [68]

and the head at BEP about 90 m. Similar damage was observed on single-stage boiler feed pumps with double-entry impeller ($u_2 > 130$ m/s) and double volutes at partload when the cutwater distance was insufficient. Increasing the distance between the impeller blade trailing edge and the volute cutwater and applying an elliptical profile to the low-pressure side removed this problem.

5. Cavitation damage throughout the impeller channel and deeply into the diffuser channels when operating the pump at excessive flow in full cavitation. This type of damage has been observed in multistage pumps with single-entry impellers, e.g. [7].

Cavitation occurs when the *static* pressure p_3 at the diffuser inlet *locally* drops down to the vapor pressure p_v. The available cavitation coefficient with this static pressure can be estimated from Eq. (6.32) which is analogous to Eq. (6.9) but uses c_3 rather than c_{3q}. It is referred to the average approach flow velocity c_3 to the diffuser vane or volute cutwater. The velocity at the impeller outlet c_{2u} can be estimated from the Table 3.2 as long as there is no recirculation from the diffuser to the impeller. At partload it can be estimated from Fig. 5.15. The velocity c_{3u} is subsequently obtained from the conservation of angular momentum $c_{3u} = c_{2u}\, d_2/d_3$ (the radial velocity component c_m can be neglected).

$$\sigma_{A,LE,c3} \equiv \frac{p_3 - p_v}{\frac{\rho}{2}\, c_3^2} = \frac{NPSH_A - \frac{c_s^2}{2g} + H_p}{\frac{c_3^2}{2g}} = \sigma_A \left(\frac{u_1}{c_3}\right)^2 - \left(\frac{c_S}{c_3}\right)^2 + \psi\, R_G \left(\frac{u_2}{c_3}\right)^2 \tag{6.32}$$

This calculation has been done for some pumps where diffuser or volute cavitation damage has been observed. The results are presented in Table D6.5. The cavitation coefficients where damage was observed are in the range of 2 to 2.5 for the boiler feed pumps, while they are roughly 6 for the water transport pumps. The damage observed at a $NPSH_A$ that high may be explained by the casing material, the non-uniformity of the velocity distributions and the geometric shape of the leading edge (cast volute cutwater versus diffuser vane leading edge with better finish).

Table D6.5 Cavitation erosion in diffusers or volutes with double-entry impellers

Stator		Diffuser			Double-volute			
Service		Boiler feed			Cold water		Boiler feed	
Column		1	2	3	4	5	6	7
Specific speed	n_q	22	22	28	44	44	22	28
Stator distance	d_3/d_2	1.07	1.054	1.047	1.16	1.16	1.1	1.14
Relative flow	q^*		<0.5	<0.7	1		<0.7	<0.7
Impeller tip speed	u_2 [m/s]	138	112	119	44	29	121	132
At impeller inlet	σ_A	0.22	0.84	0.90	0.95	0.60	0.60	0.80
At diffuser inlet	c_3/u_2	0.65	0.66	0.67	0.43	0.43	0.64	0.61
	$\sigma_{A,Le,c3}$	1.9	2.5	2.5	6.6	5.8	2.3	2.5

In Table D6.5 the case in column 1 deals with a 48-MW boiler feed pump with 4 stages producing a head per stage of 860 m, [39]. The first stage had a double-entry impeller. Cavitation damage occurred at the diffuser vane leading edges in the same way as shown in Figs. 6.43 and 6.44 with the original distance of $d_3/d_2 = 1.07$; reportedly, the problem was solved by increasing gap B to achieve $d_3/d_2 = 1.1$. Columns 2 to 7 in Table D6.5 refer to single-stage pumps with double-entry impellers as shown in Figs. 2.6 and 7.48. The cavitation coefficient defined by Eq. (6.32) is calculated with the average (steady) velocity; unsteady local velocity peaks are higher than that. Therefore, Eq. (6.32) represents an upper limit; the actual pressures are lower and closer to the vapor pressure.

While diffuser cavitation at *excessive* flow above BEP and/or low $NPSH_A$ has been observed in multistage diffuser pumps with single-entry impellers, *part load* diffuser cavitation has been observed only with *double-entry* impellers. This type of damage has not yet been noticed in multistage pumps with single-entry impellers - even when operating at heads per stage H_{st} up to 1200 m. Also single-entry volute pumps have not yet been reported to have been damaged by cavitation. Whether this is a coincidence cannot yet be ascertained. One hypothesis could be that double-entry impellers have more non-uniform partload velocity profiles at the impeller outlet - possibly due to uneven flow through both impeller sides, Chap. 5.8 . This may be compounded by the strongly distorted flow at the impeller inlet which is likely to cause higher excess velocities at the impeller outlet. Partload recirculation at the impeller outlet can occur near both of the shrouds which leads to excess velocities in the center of the impeller (erosion on the cutwater is strongest in the center as demonstrated by Fig. 6.46).

6.8.3 Reduction or Elimination of Cavitation Damage

Table 6.3 provides an overview of damage patterns, possible damage mechanisms, causes and remedies. In addition to the measures shown all kinds of cavitation damage can be reduced by:

1. Increasing the $NPSH_A$ in order to reduce the bubble volume. This measure is always favorable although the driving pressure differential increases with rising $NPSH_A$.
2. Using a more cavitation resistant material. If the damaged component was made of grey cast iron, nodular cast iron or bronze, it is indicated in most cases to employ a better material since it is rarely possible to sufficiently reduce the bubble volume by modifying the components.
3. If high-strength nickel chromium steels were used, the potential for improvement through another material is limited. Through tempering to higher tensile

strength, choice of another alloy or special materials the cavitation resistance can be improved to some extent. The corresponding improvements can be quantitatively evaluated using Table 6.1.

4. Analyzing and optimizing the impeller design, i.e. optimizing inlet diameter, inlet angles and angle development, meridional section, blade profile and deceleration ratio.
5. If the inlet provides an approach flow which varies strongly around the circumference, an optimization of the impeller does not necessarily solve the problem. Rather, the approach flow should be analyzed in order to devise ways for improvement.
6. When operating with large bubble volumes, i.e. $\sigma_A < 0.3$ to 0.35, the remedy frequently consists of a more cavitation-resistant material since the bubble volume cannot be sufficiently reduced, unless the blades operate at excessive incidence. The blade inlet profile has only a minor effect in the case of large cavity lengths or at low σ_A.
7. In the case of gap cavitation on open impellers the excess velocity in the gap (i.e. the jet contraction) can be reduced by rounding the blade edge on the pressure surface according to Table 6.3, [35].
8. If partload recirculation was established as the cause, the following measures should be considered:

   A. Reduction of impeller inlet diameter (paying attention to the maximum specified flow rate)
   B. Reduction of the impeller inlet diameter by installing a rotating ring, an orifice or an inlet ring with diffuser (Fig. 6.39). The following recommendations can be derived for dimensioning from [29 and 30]:

      $d_R/d_1 = 0.93$ to 0.96 for $\nu = d_n/d_1 > 0.35$
      $d_R/d_1 = 0.90$ to 0.95 for axial suction with $\nu = 0$
      diffuser opening angle $\vartheta = 5 - 10°$
      d_R is the smallest diameter of the inlet ring, ϑ is half the opening angle of the diffuser, Fig. 6.39.

Inlet rings have been applied to: (1) diminish the effects of cavitation, (2) reduce noise and vibrations, (3) improve partload performance. Recommendations for inlet-ring design are given in Chap. 4.5.1. Different types of rings have been applied as shown in Figs. 6.39 and 6.48; the choice often depends on the ease of installation and costs. The orifice -type ring in Fig. 6.39b is the least costly but severely jeopardizes the operation at high flow as demonstrated by the case history of a BB1-type pump discussed in Chap. 10.7.3. As a remedy, a slotted orifice-ring, chamfered by 30° at the inlet, can be installed, Fig. 6.49, [69]. The slotting reduced the blockage of the ring by 43%.

Table 6.4 Analysis of NPSH-Problems

Problem/findings	Possible causes	Possible remedies
1. NPSH at $q^* > 1$ too high (at $q^* < 1$ correct)	1. Air separation, if excessively throttled in the suction pipe 2. A_{1q} too small, β_{1B} too small 3. Cavitation in diffuser or volute 4. Pre-rotation 5. Leakage too high (Q_{sp}, Q_E)	1. Increase test speed, under-water valve, de-gassing of water 2. Cut back blade at leading edge inlet and profile on pressure surface 3. Increase A_{3q} (BEP moves to larger flow!) 4. Check inlet flow conditions
2. $NPSH_3$ increase at partload ($NPSH_3$ at $q^* >> 1$ too low)	Inlet diameter d_1 or A_{1q} too large	Possibly inlet ring, Fig. 6.39 Analyse approach flow angle β_1; see Table 7.1 Possibly new impeller
3. $NPSH_3$ too high at all flow ratios q^*	1. A_{1q} too small, β_{1B} too small (blades too long) 2. Blade manufacturing allowance not removed correctly, or applied on pressure surface 3. Inlet losses too high	1. Cut blades back at leading edge; profile on pressure surface 2. Check impeller, correct as necessary
4. Cavitation noise at partload	Inlet recirculation too strong: impeller inlet diameter d_1 or A_{1q} too large	Possibly inlet ring according to Fig. 6.39
5. Cavitation noise at $q^* > 1$	1. Pump runs at excessive flow (compared to BEP or shockless) 2. Unfavourable inlet with poor velocity distribution around circumference. Ribs or flow splitters in inlet too thick (vortex shedding) 3. Vortex formation in sump	1. Measure flow rate in plant to ascertain need for retrofit impller, if any. Profile rib according to Table10.13 2. Analyse contours of inlet casing or sump.
6. $NPSH_3$ at partload too high (axial inlet)	1. Error in suction pressure measurement due to pre-rotation 2. Impeller inlet diameter excessively large Excessive vane loading near inlet with high n_q impellers	1. Measure suction pressure upstream of flow straightener, flow splitter etc. Chap. 5.2.4 and. Fig. 5.16 2. Inlet ring according to Fig. 6.39 3. Impeller re-design
7. $NPSH_3$-curve with sharp peak at partload	Switch of impelle flow pattern triggered by cavityr	Chap. 5 Lit. [55] Possibly new impeller with lower blade loading on outer streamline

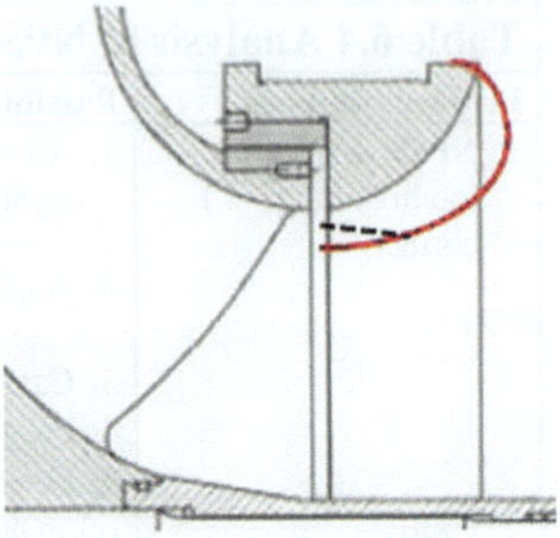

Fig. 6.48 Inlet ring without and with diffuser (dashed line)

Fig. 6.49 Orifice-type inlet ring slotted and chamfered to improve $NPSH_3$ at high flow [69], Courtesy by ITT Goulds Pumps and the authors

6.9 Insufficient Suction Capacity: Analysis and Remedies

If the measured $NPSH_3 = f(Q)$ does not correspond to design or attempts are made to improve the suction capability, the problem can be analyzed using Table 6.4. Before carrying out corrections on the impeller, the following checks should be made:

1. As with the analysis of performance deficits, the measurements must be checked and the impeller inlet measured. The relevant information is provided in Chap. 4.6.
2. During all NPSH-tests, care should be exercised to avoid falsifying the measurements by free gases separating in the suction line.
3. Check the quality of the approach flow through the suction piping and intake bay according to Chap. 11.7.
4. Machining allowances made at the blade leading edges lead time and again to cavitation problems due to manufacturing errors. Any machining allowance

should be clearly indicated on the drawings; it should be applied to the suction surfaces since the pressure surfaces of the blades are difficult or impossible to access for machining operations. Templates are required for grinding, Chap. 15.3.

References

1. Arn, C.: Analyse et prediction de la baisse de rendement des turbines Francis par cavitation à bulles. Diss. EPF Lausanne (1998)
2. Balasubramanian, R., Bradshaw, S., Sabine, E.: Influence of impeller leading edge profiles on cavitation and suction performance. In: Proceedings of 27th International Pump Users Symposium, Texas A & M, pp. 34–44 (2011)
3. Berten, S., et al.: Investigation of cavitating flow phenomena in a high-energy pump diffuser at partload operation. In: International Rotating Equipment Conference. Düsseldorf session 8–3 (2012)
4. Bourdon, P., et al.: Vibratory characteristics of erosive cavitation vortices downstream of a fixed leading edge cavity. In: IAHR Symposium Belgrade, Paper H3 (1990)
5. Cooper, P., et al.: Reduction of cavitation damage in a high-energy water injection pump. ASME AJK2011-06092
6. Cooper, P., et al.: Elimination of cavitation-related instabilities and damage in high-energy pump impellers. In: 8th International Pump Users Symposium, Houston (1991)
7. Cooper, P., Antunes, F.: Cavitation damage in boiler feed pumps. EPRI CS-3158 (1983)
8. Cooper, P.: Pump Hydraulics—advanced short course 8. In: 13th Intl Pump Users Symposium, Houston (1996)
9. Dreiß, A.: Untersuchung der Laufradkavitation einer radialen Kreiselpumpe durch instationäre Druckmessungen im rotierenden System. Diss TU Braunschweig, Mitt des Pfleiderer-Instituts für Strömungsmaschinen, Heft 5. Verlag Faragallah (1997)
10. Dupont, P.: Etude de la dynamique d'une poche de cavitation partielle en vue de la prédiction de l'érosion dans les turbomachines hydrauliques. Diss. EPF Lausanne (1993)
11. Durrer, H.: Kavitationserosion und Strömungsmechanik. Techn. Rundschau. Sulzer. **3**, 55–61 (1986)
12. EUROPUMP-brochure: NPSH for rotodynamic pumps: a reference guide. Elsevier (1999)
13. Farhat, M.: Contribution à l'étude de l'érosion de cavitation: mécanismes hydrodynamiques et prediction. Diss. EPF Lausanne (1994)
14. Franc, J.P., et al.: La Cavitation. Mechanismes physiques et aspects industriels. Presses Universitaires Grenoble (1995)
15. Friedrichs, J.: Auswirkungen instationärer Kavitationsformen auf Förderhöhenabfall und Kennlinieninstabilität von Kreiselpumpen. Diss TU Braunschweig, Mitt des Pfleiderer-Instituts für Strömungsmaschinen, Heft 9. Verlag Faragallah (2003)
16. Gantar, M.: The influence of cross section size of the diffuser channel on the hydraulic and cavitation characteristics of multi-stage radial pumps. In: Turboinstitut Conference on Fluid Flow Machinery. Ljubljana p. 469 (1984)
17. Gülich, J.F., et al.: Pump vibrations excited by cavitation. In: IMechE Conference on Fluid Machinery. The Hague (1990)
18. Gülich, J.F., Clother, A., Martens, H.J.: Cavitation noise and erosion in jet cavitation test devices and pumps. In: 2nd ASME Pumping Machinery Symposium. Washington (1993)
19. Gülich, J.F., Pace, S.: Quantitative prediction of cavitation erosion in centrifugal pumps. In: IAHR Symposium Montreal, Paper 42 (1986)
20. Gülich, J.F., Pace, S.E.: Solving pump problems related to hydraulic instabilities and cavitation. In: EPRI Power Plant Pumps Symposium, Tampa (1991)

21. Gülich, J.F., Rösch, A.: Kavitationserosion in Kreiselpumpen. Techn. Rundschau. Sulzer. **1**, 28–32 (1988)
22. Gülich, J.F.: Ähnlichkeitskenngrößen für Saugfähigkeit und Blasenausbreitung bei Pumpen. Techn. Rundschau. Sulzer. **2**, 66–69 (1980)
23. Gülich, J.F.: Beitrag zur Bestimmung der Kavitationserosion in Kreiselpumpen auf Grund der Blasenfeldlänge und des Kavitationsschalls. Diss. TH Darmstadt (1989)
24. Gülich, J.F.: Calculation of metal loss under attack of erosion-corrosion or cavitation erosion. In: International Conference on Advances in Material Technology Fossil Power Plants. Chicago (1987)
25. Gülich, J.F.: Guidelines for prevention of cavitation in centrifugal feedpumps. EPRI Report GS-6398, (1989)
26. Gülich, J.F.: Kavitationsdiagnose an Kreiselpumpen. Techn. Rundschau. Sulzer. **1**, 29–35 (1992)
27. Gülich, J.F.: Möglichkeiten und Grenzen der Vorausberechnung von Kavitationsschäden in Kreiselpumpen. Forsch. Ing. Wes. **63**(1/2), 27–39 (1997)
28. Gülich, J.F.: Selection criteria for suction impellers of centrifugal pumps. World Pumps, Parts 1 to 3, January, March, April, (2001)
29. Hergt, P., et al.: Influence of a diffuser in front of a centrifugal impeller. In: 8th Conference Fluid Machinery, Budapest, pp. 333–340 (1987)
30. Hergt, P., et al.: The suction performance of centrifugal pumps—possibilities and limits of improvements. In: Proceedings 13th International Pump Users Symposium, Houston pp. 13–25 (1996)
31. Hirschi, R.: Prédiction par modélisation numerique tridimensionelle des effects de la cavitation à poche dans les turbomachines hydrauliques. Diss. EPF Lausanne (1998)
32. Ido, A., Uranishi, K.: Tip clearance cavitation and erosion in mixed-flow pumps. ASME Fluid Mach. Forum FED **119**, 27–29 (1991)
33. Keller, A., et al.: Maßstabseffekte bei der Strömungskavitation. Forsch. Ing. Wes. **65**, 48–57 (1999)
34. Keller, A.: Einfluß der Turbulenz der Anströmung auf den Kavitationsbeginn. Pumpentagung Karlsruhe, C-4 (1996)
35. Laborde, R., et al.: Tip clearance and tip vortex cavitation in an axial flow pump. ASME J. Fluids. Eng. **119**, 680–685 (1997)
36. Ludwig, G.: Experimentelle Untersuchungen zur Kavitation am saugseitigen Dichtspalt von Kreiselpumpen sowie zu sekundären Auswirkungen des Spaltstromes. Diss TH Darmstadt (1992)
37. Marks, J.: Experimentelle Untersuchung der Stofftrennung mittels Kavitation am Beispiel von Ammoniak-Wasser-Gemischen. Diss. TU Berlin. Mensch & Buch Verlag, Berlin (2005)
38. McCaul, C., et al.: A new highly cavitation resistant casting alloy and its application in pumps. NACE-Corrosion, New Orleans (1993)
39. Michell, F.L., et al.: Twenty-three years of operating experience with the world's largest boiler feedwater pump. In: Proceedings of 14th International Pump Users Symposium, Texas A & M, pp. 75–83 (1998)
40. Piltz, H.H.: Werkstoffzerstörung durch Kavitation. Kavitationsuntersuchungen an einem Magnetostriktions-Schwinggerät. Diss. TH Darmstadt (1963)
41. Rieger, H.: Kavitation und erosion. VDI Ber. **354**, 139–148 (1979)
42. Rütschi, K.: Messung und Drehzahlumrechnung des NPSH-Wertes bei Kreiselpumpen. Schweiz. Ing. u. Arch. **98**(39), 971–974 (1980)
43. Schiavello, B., et al.: Improvement of cavitation performance and impeller life in high-energy boiler feedpumps. In: IAHR Symposium Trondheim (1988)
44. Schiavello, B.: Prescott M: Field cases due to various cavitation damage mechanisms: analysis and solutions. EPRI Power Plant Pumps Symp, Tampa (1991)
45. Schiavello, B.: Cavitation and recirculation troubleshooting methodology. In: 10th International Pump Users Symposium, Houston (1993)

46. Spohnholtz, H.H.: NPSH-Verhalten von Halbaxialpumpen bei Teillast. Diss TU Braunschweig, Mitt des Pfleiderer-Instituts für Strömungsmaschinen, Heft 4. Verlag Faragallah (1997)
47. Schmidt, T.: Experimentelle Untersuchungen zum Saugverhalten von Kreiselpumpe mittlerer spezifischer Drehzahl bei Teillast. Diss TU Braunschweig, Mitt des Pfleiderer-Instituts für Strömungsmaschinen, Heft 5. Verlag Faragallah (1997)
48. Scott, C., Ward, T.: Cavitation in centrifugal pump diffusers. Proc. ImechE. C. **452/042** (1992)
49. Simoneau, R., Mossoba, Y.: Field experience with ultra-high cavitation resistance alloys in Francis turbines. In: IAHR Symposium Trondheim, Paper K1 (1988)
50. Simoneau, R.: A new class of high strain hardening austenitic stainless steels to fight cavitation erosion. In: IAHR Symposium Montreal, Paper 83 (1986)
51. Simoneau, R.: Cobalt containing austenitic stainless steels with high cavitation erosion resistance. US Patent 4588440 (1986)
52. Simoneau, R.: Transposition of cavitation marks on different hardness metals. ASME FEDSM97-3300 (1997)
53. Sloteman, D.P., et al.: Control of back-flow at the inlets of centrifugal pumps and inducers. In: 1st International Pump Symposium, Houston (1984)
54. Sloteman, D.P.: Cavitation in high-energy pumps—detection and damage potential. In: Proceedings of 23rd Interntational Pump Users Symposium, Texas A & M, pp. 29–38 (2007)
55. Schmidt, T., et al.: NPSH-Verhalten von Halbaxialpumpen bei Teillast. Pumpentagung Karlsruhe, C-5 (1996)
56. Steller, K. et al.: Comments on erosion tests conducted in an ASTM interlaboratory test program. J. Test. Eval. 103–110 (1979)
57. Striedinger, R.: Beitrag zur Bedeutung der Wasserqualität und von Maßstabseffekten in Kreiselpumpen bei beginnender Kavitation. Diss TU Darmstadt, Shaker (2002)
58. Thamsen, P.U., et al.: Cavitation in single-vane sewage pumps. ISROMAC 12-2008–20196
59. Timcke, J.H.: NPSH-Umrechnung quadratisch oder nicht? ingenieur verlag nagel, Δp Das moderne Pumpenmagazin 7, Teil 1: Nr 3, 54–56 + 58–60, Teil 2: Nr 2, pp. 50–53 (2001)
60. Tsujimoto, Y., et al.: Observation of oscillating cavitation in an inducer. ASME J. Fluids. Eng. **119**, 775–781 (1997)
61. Uetz, H.: Abrasion und Erosion. Hanser, München (1986)
62. Visser, C.F.: Pump impeller lifetime improvement through visual study of leading-edge cavitation. In: 15th International Pump Users Symposium, Houston (1998)
63. Worster, D.M., Worster, C.: Calculation of 3D-flows in impellers and its use in improving cavitation performance in centrifugal pumps. In: 2nd Conference on Cavitation, Paper IMechE C203/83 (1983)
64. Yedidiah, S.: Oscillations at low NPSH caused by flow conditions in the suction pipe. ASME Cavitation and Multiphase Flow Forum (1974)
65. Skara, V.: Expermiental observation of cavitation phenomena in centrifugal pump impellers at part load. Diss. TU Braunschweig (2015)
66. Bolleter, U, Carney, B.: Solution to cavitation-induced vibration problems in crude-oil pipeline pumps. In: 8th Pump Users Symposium Texas A&M (1991)
67. Liebner, T., Cowan, D., Bradshaw, S.: Influence of wear ring geometry on suction performance. In: 32nd Pump Users Symposium Texas A&M (2016)
68. Bachert, R.: Dreidimensionale, instationäre Effekte kavitierender Strömungen – Analysen an Einzelprofilen und in einer Radialpumpe. Diss. TU Darmstadt (2004)
69. Bradshaw, S., Sabini, E.: Modification of BB1 vibration characteristics to meet ISO 13709 limits. In: Texas A&M Pump Symposium (2011)
70. Cowan, D., Liebner, T., Bradshaw, S.: Influence of impeller suction specific speed on vibration performance. In: 29th International Pump Users Symposium. Houston, pp. 18–47 (2013)

Chapter 7
Design of the Hydraulic Components

As per the state of the art, the hydraulic design of the various types of impellers, diffusers, volutes and inlet casing is based on empirical methods such as described in this chapter. The first design created that way is then subject to analysis and optimization by CFD. Even then the accuracy of performance prediction is not always satisfactory. This, because the 3D-flow through the pump depends on the complex shapes of the flow paths given by the inlet casing, impeller and collector. To compound the issue, the interaction of the main flow with the flow in the impeller side rooms can have an unexpectedly large influence on the Q-H-curve and efficiency (an example can be found in Chap. 9.1). Always remember that it is the *combination of all parameters and shapes* of the hydraulic channels which determines the flow patterns - hence performance. In order to reduce the uncertainties of performance prediction, a systematic approach to hydraulic design is advocated. To this end, Chap. 7.14 introduces a novel concept for a fully analytical description of the impeller geometry. Prior to starting the hydraulic design all requirements the pump has to fulfill and the boundary conditions imposed should be thoroughly reviewed and documented (see "hydraulic specification" in Chap. 17).

7.1 Methods and Boundary Conditions

7.1.1 Methods for the Development of Hydraulic Components

This chapter deals with one-dimensional calculation procedures and design methods for impellers, volute casings, diffusers and inlet casings. For developing these components, the main dimensions and blade angles are calculated in a first

J. F. Gülich, *Centrifugal Pumps*,
https://doi.org/10.1007/978-3-030-14788-4_7

step. Subsequently, the hydraulic contours are designed based on certain rules and methods. Many pump manufacturers employ computer programs for this work, and the drawings are generated on 2D-CAD systems. However, these methods are also increasingly replaced by 3D-CAD systems with which fully three-dimensional geometry models of a component can be created, Figs. 2.3 and 7.45. The complex hydraulic channels can be evaluated better with such models than with the conventional two-dimensional representations in various sections and views. Even more importantly, 3D-CAD systems provide the capability to directly manufacture the hydraulic components (or the casting patterns) by NC milling, stereo lithography or other fast-prototyping processes, [16]. The advantages of such processes are evident in terms of geometrical accuracy and lead times (not in the least also for model tests with milled or stereo-lithographed components). Since manual designs on the drawing board are rather the exception, the subsequent discussion of the design methods emphasizes the fundamental aspects of the design processes rather than a very detailed description of geometrical operations.

The development of hydraulic components often comprises the following steps:

1. Calculation of the main dimensions and blade angles by one-dimensional methods based on empirical correlations for slip factors and hydraulic efficiencies (based on databases and experience), Chap. 3.
2. Generating an initial design.
3. Optionally, quasi-3D methods can be applied to impellers for a pre-optimization of the design ("Q-3D", see Chap. 8.2).
4. Verification or optimization using 3D-Navier-Stokes programs where the geometry is changed, if required, until a satisfactory result is obtained.
5. Works or model tests for verification (and further optimization if necessary).

Frequently step (3) is skipped since Q-3D methods operate with velocity distributions that may considerably deviate from the reality (Chap. 8.2).

Efficiency, head, partload characteristics and cavitation depend on the local velocity and pressure distribution which cannot be captured in simple calculation and design processes. For this reason, all one-dimensional design methods are affected by uncertainties - even if derived from a large number of tests. Ultimately, it is the combination of the various geometry parameters and the shaping of the channels and blades that produce good hydraulics; these effects cannot be forced into unequivocal design rules. Experience still plays a crucial role in the hydraulic design, since empirical correlations, statistical data or "design methodologies" form the basis of the majority of pumps available on the market. Published design recommendations differ considerably; hydraulic components designed by different methods can nevertheless produce equivalent results. For instance, impellers can be designed for a specific pumping task with very different geometric parameters (e.g. blade angle, blade number, outlet width, blade twist etc.) and may achieve essentially the same efficiency and the same head coefficient.

7.1.2 The Hydraulic Specification

Depending on the pump type and application, the hydraulic design must meet a number of requirements and boundary conditions. Prior to starting the actual design, a hydraulic specification should be established thoroughly (obviously it is not necessary to specifically consider the following points in every application). A more comprehensive format for the hydraulic specification is proposed in Chap. 17.7.

1. The calculation is invariably based on the best efficiency point, defined by: Q_{opt}, H_{opt} at a specified speed n. If the planned main operating point, or the rated flow Q_R, in the plant does not coincide with the best efficiency point, the design flow rate Q_{opt} should be selected so that the condition $0.8 < Q_R/Q_{opt} < 1.1$ is satisfied.
2. The maximum flow rate Q_{max} should be defined to check the cavitation characteristics. To do so, the planned mode of operation must be known (important, for instance, in parallel operation, Chap. 11.1).
3. The suction conditions ($NPSH_A$) have a large impact on the design in many cases, refer also to Chap. 15.
4. In most applications a "stable" Q-H-curve is demanded where the head rises continuously when the flow rate is reduced.
5. The shut-off head H_o is often limited because of its impact on pipeline design pressure and costs. At $n_q < 40$, a head rise of $1.2 < H_o/H_{opt} < 1.25$ is desired (as close as possible to, but not lower than, 1.2).
6. With semi-axial and axial pumps $n_q > 100$, both shut-off head H_o and power P_o must be kept as low as possible because P_o determines the size of the motor when $P_o > P_{opt}$.
7. The cavitation characteristics must satisfy three criteria: (1) the pump must be able to operate in the planned range; (2) cavitation noise and vibrations must be limited; (3) no cavitation damage, Chap. 6.7.
8. To reduce pump size and costs, the head coefficient is frequently selected rather near the upper limit - except in cases where a steep Q-H-curve is explicitly demanded.
9. With respect to energy costs, and for ecological reasons, the highest possible efficiency must be aimed at (the energy costs are often capitalized and rated at 2000 to 5000 $ per kW of pumping power).
10. The hydraulic configuration is - sometimes profoundly - affected by the requirements of the mechanical design such as: the pump type, the longitudinal section of the pump, the required shaft diameter (criteria: torque, sag due to gravity or radial thrust, critical speed), impeller mounting (slide-fit, shrink-fit, shaft protection sleeves), annular seal clearances, axial thrust balancing (leakage) and the manufacturing costs.
11. The manufacturing process envisaged for the hydraulic components and/or the patterns may have an influence on the hydraulic design as well, e.g. on the draft to be applied on the blades for cast components or if sheet metal is used for

making the impellers and diffusers. Likewise, the materials must be considered e.g. metal versus plastics or ceramics.

12. The hydraulic excitation forces and pressure pulsations must be limited to an allowable magnitude.

Consequently, a pump must by not be optimized only for the best efficiency point but the entire operating range of $Q = 0$ to Q_{max} should be implicitly considered in the design. Since there is no comprehensive theory satisfying all of the listed requirements, usually the hydraulic design is based explicitly or implicitly on experience, statistical data, empirical approaches or proven similar hydraulics. Even if numerical methods are employed for further optimization, it is expedient to perform the initial design on such bases, since partload and cavitation characteristics cannot be controlled sufficiently in a wide operating range utilizing numerical methods.

7.1.3 Calculation Models

The main dimensions and blade angles of impeller and collector are selected applying one-dimensional methods so that the requirements listed above are satisfied as much as possible. During this calculation, impeller and diffuser can either be idealized as "cascades" or as "channels". Both models have some justification and are therefore briefly outlined:

Cascade model: Impeller and diffuser are considered as cascades. Consequently, the calculation focuses on matching the (geometrical) vane angles to the flow angles. Accordingly, the design is based on the velocity vectors, incidences and the slip factor. At partload and overload, losses due to excessive incidence are considered to play a critical role. Calculating cavitation according to Eq. (6.10) is also based on the idea of a flow around a profile. C. Pfleiderer can be considered as a proponent of the cascade concept, [B.1].

Channel model: Impeller and diffuser consist of channels whose cross sections change along the flow path. Deceleration or acceleration conditions are therefore the relevant physical quantities (instead of the angles of incidence and deviation). The channel model does not exclusively utilize velocity vectors but rather considers the velocities calculated from the continuity equation as $w = Q/A$ with the local cross section of the channel. A complete calculation method according to the channel model is not known - it would also be hardly sensible to ignore the incidence at the impeller inlet - but H. H. Andersons "area ratio method" goes in that direction, [B.4, B.11]. Instead of determining head and impeller outlet angle via slip coefficients, the design is based on the area ratio A_v, which relates the cross section $A_{2q} = a_2 \times b_2$ between the blades at the impeller outlet to the throat area of the collector $A_{3q} = a_3 \times b_3$. The following relationships apply:

- The area ratio is defined by: $A_v = z_{La} \times A_{2q}/(z_{Le} \times A_{3q})$
- Cross section at the impeller outlet: $z_{La} \times a_2 \times b_2 \approx 0.95 \times \pi \times d_2 \times b_2 \times \sin \beta_{2B}$
- Cross section at diffuser inlet: $z_{Le} \times a_3 \times b_3$

The main dimensions are then determined in the following way:

1. Select a value for A_V based on statistical data of executed pumps from the correlation: $A_v = (42 \text{ to } 53)/n_q$.
2. Obtain ψ_{th} and c_{3q}/u_2 from empirical functions that can be approximated by: $\psi_{th} = A_v^{0.23}$ and $:c_{3q}/u_2 = 0.31 \times A_v^{0.45}$.
3. Estimate the hydraulic efficiency (Table 3.9), determine $\psi = \eta_h \times \psi_{th}$, calculate d_2 from Eq. (T7.1.3). Calculate u_2 and $c_{3q} = u_2 \times (c_{3q}/u_2)$.
4. Select the number of diffuser vanes or volute channels and determine the throat area of the collector from $A_{3q} = Q_{opt}/(z_{Le} \times c_{3q})$.
5. Select b_2^* from Fig. 7.2 and the number of impeller blades. Calculate b_2 and A_{2q} from $A_{2q} = A_v \times z_{Le} \times A_{3q}/z_{La}$.
6. The distance between the impeller blades a_2 is determined from $a_2 = A_{2q}/b_2$. The outlet angle β_{2B} is finally determined so that the required a_2 is obtained.

Both models are incomplete idealizations of the complex flow processes in a pump. Determining the impeller outlet angle - as outlined above - according to the channel model appears neither expedient nor does it afford any physical insight since the approach is purely empirical. However, the channel model can provide a meaningful addition to the cascade model since it better explains some experimental facts and allows checks of the design according to additional criteria:

- The Q-H-curve falls steeply when the fluid at $q^* \gg 1$ is accelerated from the impeller outlet to the collector throat area. Then some of the static pressure generated in the impeller is re-converted into kinetic energy which is only recuperated downstream of the throat area with increased losses, Fig. 5.18. In the case of a diffuser, the "shock losses" generated by increased incidence could be considered as the cause of the steeper gradient of the Q-H-curve. However, in the case of a volute or annular collector the effects of incidence are negligible.
- Reducing the throat area of the collector increases the head at partload since c_{3q}/c_2 increases and the flow is better able to follow the imposed deceleration. That way the onset of stall is retarded and zones of stalled fluid are reduced, Fig. 5.18 . This phenomenon is hardly affected by the cutwater or diffuser vane inlet angle and poorly explained by the cascade model.
- The $NPSH_3$ curve increases steeply if the flow vector w_1 at the impeller inlet is greatly accelerated towards the throat area, Fig. 5.13 ($w_{1q} \gg w_1$).
- As explained in Chaps. 4.1.3, 5.2.4 and 5.3.1, the flow in the impeller and collector separates mainly when an allowable deceleration has been exceeded and less because of excessive incidence.

- If angles and channel cross sections are not well matched it is necessary to use the blade distances a_1 and a_2 (or A_{1q}) as design criteria in addition to the blade angles.

Since blade angles *and* channel cross sections have to match the intended flow, cascade and channel models are jointly used in the design procedures discussed in this chapter.

Airfoil model: If the blades are widely-spaced (low L/t), it is also possible to apply the airfoil theory. The lift coefficients of the individual airfoil may then be corrected to account for the influence of neighboring blades using suitable corrections, (Chap. 7.6).

7.2 Radial Impellers

The methods for impeller design discussed in this chapter are essentially generic. Based on these methods, radial impellers for pure fluids can be developed without additional information. Semi-axial impellers are treated in a similar way; however, some particular aspects must be considered which are explained in Chap. 7.5. Methods for axial impellers are presented in Chap. 7.6.

The procedure is the same for open and closed impellers; in the case of large clearances the performance loss must be considered as per Chap. 3.6.5.

7.2.1 Determination of Main Dimensions

Prior to actually drawing an impeller, the main dimensions and blade angles must be determined. The relevant calculation is explained step by step in the following; the required equations are listed in Table 7.1.

1. Given quantities: n, Q_{opt}, H_{opt} and the boundary conditions as per Chap. 7.1.2. As a result, the specific speed n_q can be calculated without which a meaningful design would scarcely be possible. Additionally, the approach flow conditions must be defined. In many applications the approach flow angle is $\alpha_1 = 90°$, and the c_m-distribution over the approach flow cross section is assumed constant.

2. Efficiencies: To calculate the head, the hydraulic efficiency has to be assumed; the equations in Table 3.9, or Figs. 3.30 to 3.32, can be used for this purpose. In general the hydraulic efficiency should be assumed conservatively in order to avoid the risk of a deficit in head.

The volumetric flow through the impeller exceeds the useful flow rate by the annular seal leakage losses and the balance water flow, Eq. (T7.1.1). These volumetric losses must initially be estimated for the impeller calculation, Chap. 3.6, Eqs. (T3.5.9 and 3.5.10). The balance water flow can be calculated when the axial thrust balance device (if any) has been dimensioned.

Table 7.1 Impeller calculation

Given quantities	n, Q_{opt}, H_{opt}, α_1; hence the specific speed is known		Eq.
Impeller flow rate	$Q_{La} = Q_{opt} + Q_{sp} + Q_E = Q_{opt}/\eta_v$		7.1.1
Shaft diameter	$d_w = \left(\frac{16\,P_{max}}{\pi\,\omega\,\tau_{al}}\right)^{\frac{1}{3}} = 3.65\left(\frac{P_{max}}{n\,\tau_{al}}\right)^{\frac{1}{3}}$	P_{max} in W n in rpm τ_{al} in N/m²	7.1.2
Impeller outlet diameter	$d_2 = \frac{60}{\pi\,n}\sqrt{\frac{2g\,H_{opt}}{\psi_{opt}}} = \frac{84.6}{n}\sqrt{\frac{H_{opt}}{\psi_{opt}}}$	ψ_{opt} from Eq. (3.26) or Table 7.2	7.1.3
Impeller inlet diameter for minimum relative velocity	$d_1{}^* = f_{d_1}\sqrt{d_n^{*2} + 1.48\times10^{-3}\,\psi_{opt}\,\frac{n_q{}^{1.33}}{(\eta_v\delta_r)^{0.67}}}$	$\delta_r = 1 - \frac{c_{1m}}{u_{1m}\tan\alpha_1}$	7.1.4
	Application	f_{d1}	
	Normal impeller	1.15 to 1.05 falling from n_q = 15 to 40	
	Suction impeller	1.25 to 1.15	
Impeller inlet diameter for selected $\beta_1 = \beta_{1a}$	$d_1 = 2.9\sqrt[3]{\frac{Q_{La}}{f_q\,n\,k_n\tan\beta_1}\left(1 + \frac{\tan\beta_1}{\tan\alpha_1}\right)}$		7.1.5
Impeller inlet diameter for selected coefficients λ_c,λ_w	$d_{1,opt} = \sqrt{d_n^2 + 10.6\left(\frac{Q_{La}}{f_q\,n}\right)^{\frac{2}{3}}\left(\frac{\lambda_c + \lambda_w}{\lambda_w}\right)^{\frac{1}{3}}}$		7.1.6
Blade angle for selected incidence i_1 = 0 to 4°	$\beta_{1B} = \beta'_1 + i_1 = \text{arc tan}\,\frac{c_{1m}\,\tau_1}{u_1 - c_{1u}} + i_1$	With constant c_{1m} und τ_1: $\tan\beta_{1B}(r) = \frac{r_a}{r}\tan\beta_{1B,a}$	7.1.7
Slip factor	$\varepsilon_{lim} = \exp\left\{-\frac{8.16\sin\beta_{2B}}{z_{La}}\right\}$	$k_w = 1 - \left(\frac{d^*_{1m} - \varepsilon_{Lim}}{1 - \varepsilon_{Lim}}\right)^3$	7.1.8
	$\gamma = f_1\left(1 - \frac{\sqrt{\sin\beta_{2B}}}{z_{La}{}^{0.7}}\right)k_w$	$k_w = 1$ for $d^*_{1m} \le \varepsilon_{Lim}$ Radial: $f_1 = 0.98$ Semi-axial: $f_1 = 1.02 + 1.2\times10^{-3}(n_q - 50)$	
Head ($A_2 = \pi\,d_{2b}\,b_2$)	$H = \frac{\eta_h\,u_2{}^2}{g}\left\{\gamma - \frac{Q_{La}}{f_q\,A_2\,u_2\tan\beta_{2B}}\left[\tau_2 + \frac{A_2\,d^*{}_{1m}\tan\beta_{2B}}{A_1\tan\alpha_1}\right]\right\}$		7.1.9
Effective blade loading, according to [70]	$\xi_{eff} = \frac{2\pi\,\psi_{opt}}{\eta_h\,z_{La}\,L^*_{sch}\,(w^*_1 + w^*_2)}$		7.1.10
Allowable blade loading [70]	$\xi_{al} = \left(\frac{n_{q,Ref}}{n_q}\right)^{0.77}$ $n_{q,Ref} = 40$ Band width ±15%		7.1.11
Lift coefficient according to [26]	$\zeta_a = \frac{\pi\,\psi_{th}}{z_{La}\,L^*_{sch}\sqrt{\varphi_2^2 + \left(1 - \frac{\psi_{th}}{4}\left\{1 - \frac{\sin\varepsilon_{MS}}{z_{La}}\right\}\right)^2}}$	allowable $\zeta_a < 0.9$	7.1.12

3. Shaft diameter d_w: Apart from the criteria mentioned in item 7.1.2.10, the torque has to be transmitted. When the material and the allowable sheer stress τ_{al} have been selected, the minimum required shaft diameter can be calculated from Eq. (T7.1.2). To satisfy all criteria given in item 7.1.2.10, it is frequently necessary to increase the shaft diameter above the minimum that has been calculated based on the torque only.

4. Impeller outer diameter d_2: The head coefficient is selected from Eq. (3.26), Fig. 3.24 or from Table 7.2 for sewage pumps. Subsequently, the impeller outer diameter d_2 is calculated from Eq. (T7.1.3). In order to achieve a stable Q-H-curve, good partload behavior and acceptably low hydraulic excitation forces, the head coefficient ψ_{opt} must be limited. The range of typical head coefficients is shown in Fig. 3.24 as a function of the specific speed. If ψ_{opt} is selected in the upper range of the band, flatter Q-H-curves and smaller impeller diameters are obtained, but the risk of a Q-H-curve instability increases. Head coefficients near the lower limit of the band lead to steeper Q-H-curves with less risk of instability. If extremely steep Q-H-curves are demanded, it is obviously possible to select ψ_{opt} below the range in Fig. 3.24 - but the pump becomes correspondingly larger and more expensive.

With low specific speeds (below $n_q = 20$ to 25) one tends to obtain better efficiencies with high head coefficients since the impeller disk friction loss falls with the 5th power of the impeller diameter, Eq. (T3.6.2). The smaller n_q, the greater is this influence. With $n_q > 50$ the impeller disk friction effects are insignificant and turbulent dissipation due to uneven velocity distributions largely determines the efficiency; in this case a high head coefficient tends to be less favorable.

5. Impeller blade number z_{La}: The selection of the blade number z_{La} depends on various criteria:

A. To reduce pressure pulsations and hydraulic excitation forces, impeller and diffuser blade/vane numbers must be matched according to Chap. 10.7.1: z_{La} and z_{Le} must be selected so that the values m = 0 and m = 1 for the parameter $m = |\nu_2 \times z_{La} - \nu_3 \times z_{Le}|$ are avoided up to $\nu_2 = 3$ and $\nu_3 = 3$.

B. The hydrodynamic blade loading should be in an optimum range: If the loading is too low, unnecessarily high friction losses must be expected. If the loading is too high, the turbulent dissipation losses increase due to uneven flow distribution. The blade loading can only be verified after completing the impeller design. This can be done using empirical values for the blade loading in step 14.

C. Eight and more blades are not normally recommended out of consideration of the stability of the Q-H-curve.

D. Less than 5 blades are unfavorable for high heads per stage since the impeller outlet flow becomes very non-uniform over the circumference due to the large blade spacing. The consequence would be unnecessarily high pressure pulsations, noise and vibrations. Therefore avoid $z_{La} < 5$ for $H_{st} > 100$ m.

Based on the above criteria most radial and semi-axial impellers are designed in a range of approximately $10 < n_q < 120$ with 5 to 7 blades. If no stable Q-H-curve is required since the plant only demands a narrow operating range near the best efficiency point, 9 blades are also designed (impellers with straight radial blades are discussed in Chap. 7.3.3).

At high specific speeds the ratio of blade length to blade pitch ("solidity") gains significance as a design criterion. The criteria A to D are sufficient for selecting blade numbers if used together with correlations for suitable hydrodynamic blade loading. Empirical formulae for blade number selection (as sometimes published) are not helpful because they are likely to violate the various criteria discussed above.

E. For suction impellers with high n_{ss} and flat curves $NPSH_3 = f(Q)$ one should select $z_{La} = 5$ (or possibly 6); since it would be difficult to get flat NPSH-curves with 7 blades.

6. Impeller inlet diameter d_1: The choice of the inlet diameter depends on the requirements made on the cavitation characteristics. Three cases are distinguished:

A. Dimensioning for minimum relative velocity at the impeller inlet: This design tends to minimize leakage, friction and shock losses. It is recommended when the available $NPSH_A$ is sufficiently high so that cavitation is avoided. This is true, for example, for the second (and following) stages of multistage pumps ("series stages"). The minimum of the relative velocity w_1 is obtained from $w_1^2 = c_{1m}^2 + (u_1 - c_{1m}/\tan\alpha_1)^2$, with $c_{1m} = 4 \times Q/\{\pi(d_1^2 - d_n^2)\}$ and $u_1 = \pi \times d_1 \times n/60$, by differentiation of $\partial w_1/\partial d_1$. Setting the expression obtained to zero, one solves for the inlet diameter $d_{1,min}$ and obtains Eq. (T7.1.4), where pre-rotation (if any) is covered iteratively by the swirl number δ_r. The inlet diameter is selected several percent above the minimum thus calculated to allow for secondary effects such as boundary layer blockage and uneven velocity distributions. Thus $d_1 = f_{d1} \times d_{1,min}$ is obtained with recommendations for the factor f_{d1} given in Table 7.1.

B. Design for a selected suction specific speed: For suction impellers a suction specific speed n_{ss} suitable for the intended application is chosen from Table D6.1. With the selected n_{ss} the approach flow angle β_{1a} is read from Fig. 6.21 and the inlet diameter is calculated from Eq. (T7.1.5). The inlet diameter can also be determined somewhat summarily from Eq. (T7.1.4) with a correspondingly larger factor f_{d1} (Table 7.1).

C. Design for selected λ-coefficients according to Chap. 6.3.2: Equation (T7.1.6) offers an alternative route to calculate d_1 when the coefficients λ_c and λ_w are chosen. The recommendations given in Table D6.1 should also be considered in this context: If the circumferential speed u_1 is high, the inlet diameter and the suction specific speed should not be selected too large.

According to Chap. 6.3.2, Eq. (T7.1.6) yields the inlet diameter at which the required $NPSH_R$ reaches its minimum for specified values of flow rate and rotor speed with *given* λ-coefficients. As discussed in Chap. 6.3.2, these coefficients are not well known since they depend on the impeller geometry and the approach flow. Consequently, the optimum inlet diameter calculated from Eq. (T7.1.6) is uncertain as well. In order to be able to evaluate the sensitivity of the required $NPSH_R$ to changes of the λ-coefficients and the impeller inlet diameter, it is convenient to calculate $NPSH_R = f(d_1)$ for various values of λ_w from Eq. (6.10) and to plot it in a graph. Figure 7.1 shows an example of such a calculation. The smaller the value of λ_w, the greater is the inlet diameter at which $NPSH_R$ reaches its minimum. At the same time, this minimum becomes flatter; hence the inlet diameter can be selected somewhat below the mathematical minimum without significant loss in suction capacity.

D. Dimensioning for minimum $NPSH_i$: Suction impellers of high-pressure pumps with $u_1 > 50$ m/s must be designed for short cavity lengths to avoid cavitation damage. Above $u_1 \approx 75$ m/s the impeller must be operated virtually without any cavitation. The inlet diameter can be calculated from Eq. (T7.1.6), into which the coefficient for cavitation inception $\lambda_{w,i}$ must be entered, Fig. 6.2 and Table D6.3. Since these values depend sensitively on blade profile, incidence and non-uniformities of the approach flow, CFD-calculations of the pressure distribution are useful to verify the design, Chap. 8.

As is evident from Figs. 6.19 and 6.20 and discussed in Chap. 6.3, the λ-coefficients and suction specific speeds are subject to a large scatter since many influence parameters are ignored in this simple approach. Sizing the impeller inlet on the basis of above concepts therefore implies some uncertainty. The risk of cavitation problems (vibrations, noise and erosion) should be evaluated according to Chap. 6 and Tables 6.1 and 6.2 when determining the impeller inlet geometry.

7. Blade inlet diameter at the inner streamline d_{1i}: The lower limit for d_{1i} is usually given by the method of mounting the impeller on the shaft. For between-bearing impellers and $n_q < 25$ to 30, the diameter d_{1i} should be selected as close as possible to the lower limit in order to improve the stability of the Q-H-curve. With overhung impellers and high specific speeds, d_{1i} should not be

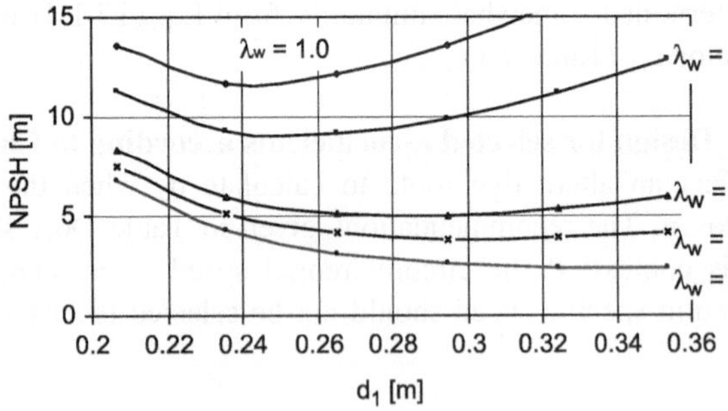

Fig. 7.1 Optimization of the impeller inlet diameter d_1

selected too small since very large blade inlet angles would be obtained. These cause excessive flow separations near the hub during partload operation.

8. Impeller blade inlet angles: Once the impeller inlet diameter has been determined from step (6), it is possible to calculate all quantities of the inlet velocity triangle according to Table 3.1. The blade inlet angle is obtained by adding the incidence i_1 to the flow angle β'_1 (with blade blockage) from Eq. (T7.1.7): $\beta_{1B} = \beta'_1 + i_1$. If cavitation is not an issue (e.g. series stages of multistage pumps or low-head pumps with ample $NPSH_A$), the incidence should be selected between 0 and 4°. The flow rate of shockless entry according to Eq. (T.3.1.10) is then close to, or slightly above, the best efficiency point. In contrast, suction impellers should be designed with low blade loadings - i.e. flat pressure distributions - at the impeller inlet. With increasing tip speed u_1 at the impeller inlet, lower incidences should be chosen in order to reduce the thickness of the cavities and hence the risk of cavitation damage, see also [10]. The blades can then be designed with very low (even strongly negative) incidence. When this is done, care must be taken that the minimum throat area A_{1q} is sufficiently large (see item 15 below).

Instead of using the incidence, it is also possible to apply an "angle exaggeration". This constitutes the ratio of the blade angle tangent to the tangent of the flow angle with blockage and is defined as $a_{ex} = \tan\beta_{1B}/\tan\beta'_1$. For radial impellers a_{ex} = 1.1 to 1.2 may be assumed and for semi-axial impellers a_{ex} = 1.05 to 1.1. If the impeller is designed for low $NPSH_i$, a_{ex} = 0.95 to 1.05 can be selected.

The above calculation of the flow and blade angles is performed for the outer, mean and inner streamlines (with high n_q use 5 or more streamlines). The meridional velocity c_m *upstream* of the leading edge is mostly assumed constant over the inflow cross section. Depending on the local hub blockage, a different value of c_m should be used for the calculation of each streamline if applicable, Table 3.1.

If the pump has to be operated far to the right of the best efficiency point ($Q_{max} \gg Q_{opt}$), inlet diameter and incidence must be enlarged accordingly. However, this causes the cavitation characteristics in the best efficiency point and at partload to deteriorate; therefore, the design options and risks should be carefully evaluated in such cases.

9. Outlet width b_2: Blade number, blade outlet angle and outlet width cannot be selected independently of one another. They must be matched so that the demanded head coefficient is achieved with a stable Q-H-curve. As shown by the outlet velocity triangle in Table 3.2, a large outlet width reduces the meridional velocity c_{2m} thereby increasing c_{2u} (and the head) - *unless the flow separates.* With given values of outlet angle and blade number the head consequently increases with the outlet width; the Q-H-curve becomes correspondingly flatter. As discussed in Chap. 5, the outlet recirculation intensifies with increasing b_2/d_2 and, consequently, shut-off pressure and power consumption rise as well.

A sufficiently large outlet width is required to achieve a stable Q-H-curve (in particular at low n_q). Conversely, the non-uniformity of the flow at the impeller

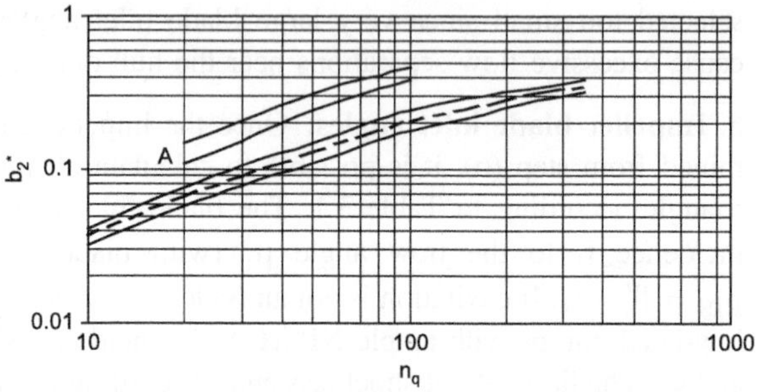

Fig. 7.2 Impeller outlet width. Radial and semi-axial impellers: $b_2{}^* = b_2/d_{2a}$; axial impellers: $b_2{}^* = 1/2(1-\nu)$ *Curves of range A*: sewage pumps with $z_{La} \geq 2$

outlet grows with the width of the impeller, increasing the turbulent dissipation losses in the collector as well as pressure pulsations and excitation forces. Finally, it should be ensured to maintain $b_2 < b_1$.

Since the above effects cannot be calculated theoretically, the relative outlet width $b_2{}^* = b_2/d_{2a}$ is commonly selected from empirical data. To render the impeller discharge flow near the best efficiency point as uniform as possible and to avoid unnecessary turbulent dissipation losses, it should be attempted to select $b_2{}^*$ as low as is permissible with respect to the stability of the Q-H-curve. Figure 7.2 gives a band for $b_2{}^* = f(n_q)$ which approximately covers the data provided in [B.8, B.18, 25]. Set $b_2{}^* = 1/2(1 - \nu)$ for axial impellers. The mean curve for the outlet width in Fig. 7.2 can be approximated by Eq. (7.1). For sewage pump impellers with 2 or 3 blades Eq. (7.1a) can be used to determine the impeller outlet width.

$$b_2{}^* = 0.017 + 0.262\frac{n_q}{n_{q,Ref}} - 0.08\left(\frac{n_q}{n_{q,Ref}}\right)^2 + 0.0093\left(\frac{n_q}{n_{q,Ref}}\right)^3 \quad (7.1)$$

$$b_2{}^* = 0.02 + 0.5\frac{n_q}{n_{q,Ref}} - 0.03\left(\frac{n_q}{n_{q,Ref}}\right)^2 - 0.04\left(\frac{n_q}{n_{q,Ref}}\right)^3 \quad n_{q,Ref} = 100 \quad (7.1a)$$

10. Outlet angle β_{2B}: The blade outlet angle must be determined so that the specified head, or the head coefficient selected in step (4), is achieved with the values of d_2, z_{La} and b_2 determined previously. The outlet angle β_{2B} which fulfills this requirement can be determined by trial and error when calculating the respective slip factor and the head from Eq. (T7.1.8 and T7.1.9), see also Tables 3.2 and 3.3. The hydraulic efficiency must be assumed according to Chap. 3.9.

Outlet angles of radial impellers with 5 to 7 blades are commonly in the range of 15 to 45°; in many cases angles are selected between 20 and 27°. Matching outlet angles and outlet width is an optimization task during which the requirements in terms of efficiency and stability of the Q-H-curve (shut-off pressure and saddle) have to be considered. A possible criterion is to minimize the deviation angle $\delta_2 = \beta_{2B} - \beta_2$ which should not exceed 10 to 14° in order to limit the turbulent dissipation losses caused by non-uniform flow distributions.

A specified head can be achieved according to Eq. (T7.1.9) with very different combinations of b_2, z_{La} and β_{2B}. Therefore no strict rules can be given for selecting the outlet angle. The blade length, obtained from the design, depends on the inlet and outlet angles; consequently this applies also to the resulting hydraulic blade loading. All these parameters must be collectively optimized and cannot be determined independently of one another (see also step 4).

With specific speeds above $n_q \approx 40$ to 60 the blade outlet angle should not be made constant over the impeller outlet width. The outer streamline is unloaded so that $\beta_{2B,a} < \beta_{2B,m} < \beta_{2B,i}$. The head is calculated on the basis of the mean angle $\beta_{2B,m}$. In this way, it is possible to match the lengths of all streamlines; in particular it can be ensured that the outer streamline is sufficiently long and, therefore, lowly loaded.

After the outlet angle has been determined, all parameters of the outlet triangle according to Table 3.2 can be calculated. To avoid premature stall and efficiency losses the deceleration ratio w_2/w_{1a} should not fall below the value of 0.7 ("de Haller criterion").

11. Blade thickness e: Requirements in terms of castability and mechanical strength determine the blade thickness which depends on the head per stage or tip speed u_2, the specific speed (impeller outlet width), the number of blades and the material and corrosion effects which limit the allowable alternating stress. Experience shows that these requirements are usually met when $e/d_2 = 0.016$ to 0.022 is selected for multi-vane impellers. The upper range applies to high-pressure impellers with more than 600 m of head per stage. The lower range applies to low heads and low specific speeds. With increasing blade width b_2 (i.e. also with growing specific speed) the blade stresses rise at a given head; thicker blades must be selected accordingly. A rational for checking the blade thickness with respect to the risk of impeller fatigue fractures is provided in Chap. 14.1. Equation (T14.3.1) in Table 14.3 allows to estimate the blade thickness as a function of the number of blades, the impeller tip speed and the specific speed; the correlation works even for single-channel impellers. The minimum thickness required to achieve good casting qualities must be observed in the case of cast impellers; it depends on the casting process and is 3 to 5 mm.

12. Blade leading edge profile (Fig. 7.3): Unfavorable leading edge profiling generates local excess velocities and a correspondingly intense low-pressure peak which impairs the cavitation behavior and may even affect the efficiency. Designing the leading edge as a semi-circle is very unfavorable in this regard and only acceptable for small pumps or applications with low requirements. Elliptical inlet profiles provide favorable pressure distributions. If the elliptical profiling extends only over a short distance, the blades react less sensitively to (excessive) incidence. In contrast, long wedge-like profiles perform well only at shockless entry ($i_1 = 0$) since high excess velocities are generated in incident flow. Long, thin (wedge-like) profiles are also unfavorable with respect to the casting quality (blade cracks caused by rapid cooling of thin profiles) and in terms of strength and risk of blade cracking.

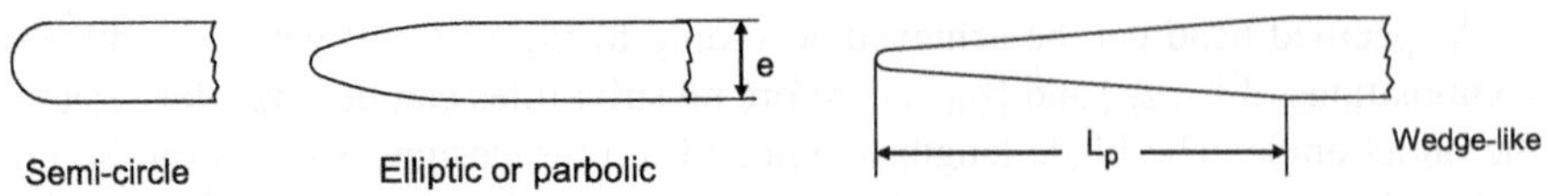

Fig. 7.3 Impeller blade leading edge profiles

According to calculations on individual profiles [65] quoted in [B.1], the minimum pressure coefficient $c_{p,min,SF}$ at zero incidence (shockless entry) is:

$$\Delta p = c_{p,min} \frac{\rho}{2} w_1^2 \quad \text{with} \quad c_{p,min,SF} = 0.373 \frac{e}{L_p}\left(2 + 0.373 \frac{e}{L_p}\right) \qquad (7.1a)$$

L_p is the length of the profile measured from the leading edge to the point of the transition to the full blade thickness e. Equation (7.1a) yields $c_{p,min,SF} = 2.05$ for a semi-circle ($e/L_p = 2.0$), while an elliptical profile with $e/L_p = 0.2$ yields $c_{p,min,SF} = 0.155$. As per Chap. 6.3.2 (Fig. 6.19), these results cannot be transferred directly to impellers since λ_w depends on the absolute value of the blade angle and other geometrical parameters of the impeller and the inlet. However, Eq. (7.1a) can be used for selecting and assessing a leading edge profile.

The effects of parabolic, elliptical, semi-circular and blunt leading edge profiles were investigated by testing and by CFD in [4], see also Chap. 8.7. The particular parabolic profile investigated was slightly better than the specific elliptical profile. This finding applied to all cavitation criteria (inception, 1 and 3% head drop and head breakdown). In this particular test the σ_3-values at BEP were 0.18 for the parabolic profile, 5% higher for the elliptical, 17% higher for semi-circular and 25% higher for the blunt leading edge profile. Evidently, the semi-circular and the blunt leading edge perform less well than the parabolic and elliptical profiles, see also Eq. (7.1a). The suspected reason for the parabolic profile to be slightly superior to the elliptical is that the parabolic profile was thinner which results in lower excess velocities as long as the incidence is low. It should be noted that the thickness distribution of parabolic and elliptical profiles can be made nearly identical by a suitable selection of the profile parameters.

The profiling and shaping of the blades are discussed further in Chap. 7.2.4

13. Blade trailing edge profiles: Two design philosophies are used in practice:

A. The blades according to Fig. 3.6 are tapered towards the trailing edge to roughly half of the blade thickness according to $e_2 = 1/2\,e$ in order to reduce the width of the wake, turbulent dissipation losses and pressure pulsations.
B. The full blade thickness is configured up to the trailing edge to preserve the possibility for under-filing in case of a deficit in head, Chap. 4.5.2. Outlet profiling is sometimes waived also for cost reasons.

After performing step 13, all parameters required for a first impeller design according to Chap. 7.2.2 are established. When the first design is available, blade lengths and channel cross sections are known so that blade loading and deceleration ratios or cross sections can be verified.

14. Blade loading: Hydrodynamic blade loading can vary in a wide range without quantifiable adverse effects on efficiency and hydraulic excitation forces. Truly reliable criteria for the optimum blade length are hence not available. However, various limits for the hydrodynamic blade loading have been suggested. Using these blade-loading coefficients, it is possible to verify the selected blade length and blade number. Applying such criteria allows a more consistent approach to impeller design.

A. The evaluation of a number of proven radial pumps produced a correlation for the blade loading that can be used to verify the blade length, [70]. According to this, the blade number and length are selected appropriately if the effective blade load ξ_{eff} calculated from Eq. (T7.1.10) is roughly within the range of the allowable loading ξ_{al} given by Eq. (T7.1.11). In general, it is recommended to select the blade loading typically 10% below the value given by Eq. (T7.1.11) in order to reduce pressure pulsations and hydraulic excitation forces.
B. Another possibility of checking the blade loading was published in [26]: the lift coefficient ξ_a calculated from Eq. (T7.1.12) should not exceed the value 0.9. Whether this criterion has proved itself in practice is not known. For axial impellers with $\varepsilon_{MS} = 0$, Eq. (T7.1.12) leads to the relationship Eq. (T7.5.8b) valid for propeller pumps.

15. Throat area A_{1q}: As discussed in Chap. 7.1.3, blade angles and channel cross sections must be matched to the intended flow. In the design point the deceleration of the relative velocity vector w_1 to the mean flow velocity w_{1q} in the throat area must not be too high in order to avoid premature inlet recirculation. However, with respect to cavitation at high flows, excessive acceleration must not be allowed either. According to evaluations on operating pumps the throat area should be selected so that the ratio w_{1q}/w_{1m} in the best efficiency point with axial approach flow is in the range of 0.75 to 0.85 and with radial inlet casings (between-bearing pumps) between 0.65 and 0.75. At the maximum flow rate w_{1q}/w_{1m} should not appreciably exceed the value 1.0. The vector of the relative velocity w_{1m} is calculated on the mean streamline d_{1m} (not d_{1b}).

16. Blade distance a_2 at the outlet: the distance between the blades at the outlet a_2 must match the outlet angle β_{2B} in order to ensure that the slip factor (hence the head) are obtained. To this end, an angle $\beta_{a2} = \text{arc sin } a_2/t_2$ can be defined and compared with β_{2B}. Experience shows that the ratio sin β_{a2}/sin β_{2B} should be in the range of 0.7 to 0.9, Chap. 3.3 and Eq. (3.8).

Since the impeller design is to a large extent empirical, it is recommended to systematically apply consistent design methods and rules such as outlined above. This approach fosters a better predictability of hydraulic performance and allows one to faster build a data base and gain experience.

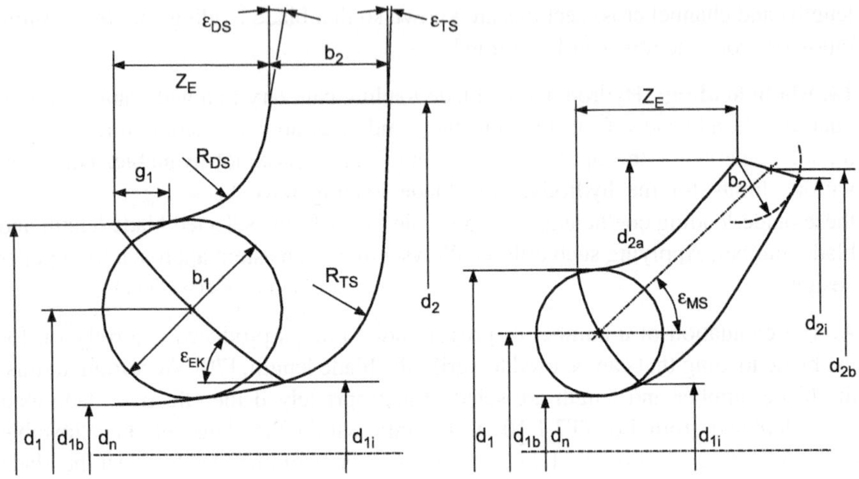

Fig. 7.4 Design parameters for the meridional section of the impeller. *Left*: Radial impeller, *right*: semi-axial impeller

7.2.2 *Impeller Design*

7.2.2.1 Design of the Meridional Section

A section through the impeller axis is called "meridional section". In this presentation the blade leading and trailing edges are projected into the drawing plane through "circular projection", Fig. 7.4.

To be able to design the meridional section, details for the position of the leading edge are required next to the dimensions (d_2, b_2, d_1, d_{1i}, d_n) determined above. Recommendations for the axial extension z_E and the radius of curvature R_{Ds} of the front shroud are given by Eq. (7.2):

$$z_E = a\frac{d_{1a} - d_n}{2}\left(\frac{n_q}{n_{q,Ref}}\right)^x \qquad \begin{array}{l} R_{Ds} = (0.6 \text{ to } 0.8)b_1 \\ b_1 = 1/2(d_1 - d_n) \\ n_{q,Ref} = 1 \end{array} \qquad \begin{array}{l} \text{for example:} \\ a = 0.75 \\ x = -0.05 \\ \text{see also Table 7.8} \end{array} \tag{7.2}$$

Favorable flow conditions are achieved by moving the leading edge forward into the impeller eye. In this way low blade loadings and correspondingly moderate low-pressure peaks are obtained and cavitation is reduced. The leading edge at the outer streamline should not be located in the region of high curvature but preferably before the start of the front shroud curvature. Therefore, the radius R_{Ds} should *not* be tangent to the point defined by z_E, but a short section g_1 = (0.2 to 0.3) × b_1 should be introduced with only a minor increase in radius in order to achieve flatter pressure distributions. For small pumps, or when a short axial extension of the impeller is a priority, smaller values are selected for z_E and R_{Ds} than calculated from Eq. (7.2).

The angle ε_{Ds} which defines the shape of the front shroud near the impeller outlet is determined according to the specific speed. It can be used to *influence the velocity profile at the impeller outlet.* For $n_q < 20$, $\varepsilon_{Ds} = 0$ is often selected. For radial impellers with higher specific speeds ε_{Ds} increases to approximately 15 to 20°. Selecting $\varepsilon_{Ds} > 0$, increases the flow deceleration in the impeller for given b_2 and β_{2B} and thus the head produced by the impeller; this applies also for low n_q.

With d_2, b_2, z_E, d_1, g_1, ε_{Ds} and R_{Ds} it is now possible to draw the outer streamline which can be defined by a free curve or assembled from straight lines and circular arcs. When using a design program, Bezier functions are often used. Unfortunately, general rules are not known which would ensure favorable hydraulic features. Meridional section and blade development must rather be optimized together.

The angle ε_{Ts} which defines the shape of the rear shroud near the impeller outlet can be chosen positive or negative. At $n_q < 30$, $\varepsilon_{Ts} = 0$ is often designed (some manufacturers also use negative values). At high specific speeds ε_{Ts} is always positive with $\varepsilon_{Ts} < \varepsilon_{Ds}$.

The inner streamline must be shaped in a way that the cross sections $A = 2 \times \pi \times r \times b$ vary continuously along the mean streamline. To this end, a mathematical rule may be selected which describes the cross section A or the channel width b in the meridional section from b_1 to b_2 as a function of the developed length of the outer streamline, Fig. 7.5. In addition to a linear or parabolic function, a cubic law could also be considered to obtain smaller changes in cross section near the inlet and outlet of the impeller. After the desired rule $b = f(L)$ has been defined, the width is marked off with dividers at a number of checkpoints on the outer streamline. Subsequently, the inner streamline is drawn as an envelope curve. The same construction can also be performed in sub-operations with several streamlines. Finally, the streamlines can be constructed from the theory of potential flows.

The blade leading edge is initially obtained from the chosen parameters d_1, z_E, d_{1i} and the designed inner streamline. The position of the leading edge has a major effect on the partload behavior according to Chap. 5. The following criteria serve for shaping the leading edge: (1) In multistage pumps with $n_q < 25$ the angle ε_{EK}

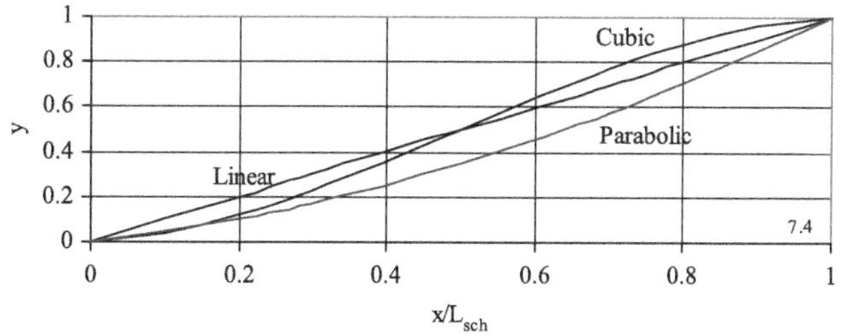

Fig. 7.5 Options for the development of the flow area in the impeller; L_{sch} = blade length

must be designed as large as possible to prevent the Q-H-curve from drooping towards Q = 0. (2) For impellers with an axial inlet ε_{EK} = 30 to 40° is about the optimum according to [25]. (3) If ε_{EK} is selected too small, problems with hub cavitation and instabilities of the Q-H-curve may be encountered. (4) Choosing ε_{EK} too large can result in too low values of d_{1i} and too large blade angles $\beta_{1B,i}$ on the hub.

As discussed in Chap. 5, the meridional section largely determines the flow distribution over the blade width and, consequently, the turbulent dissipation losses, the partload behavior and the stability of the Q-H-curve.

Criteria for the meridional section to be checked:

- The leading edge should join the outer streamline with the largest possible angle to reduce the local blockage in the case of twisted blades.
- The curvatures of the outer and inner streamlines and the cross sections $A = 2 \times \pi \times r \times b$ should vary continuously along the mean streamline.

7.2.2.2 Blade Design

The blade design serves to define the shape of the blades along the outer, inner and mean streamlines so that the inlet and outlet angles established in Chap. 7.2.1 are obtained. The blade describes a three-dimensional curve along each surface of revolution; for instance, along the outer streamline defined by the meridional section (or front shroud). This curve can be described through its projections into the meridional section and the plan view. Every point of this curve is defined in space by the coordinates r, z and ε.

With reference to Fig. 7.7a consider an element of the outer streamline which is positioned between the points 5 and 6 on the radii r_5 and r_6. In the meridional section it has the (approximate) length:

$$\Delta m = \sqrt{\Delta r^2 + \Delta z^2} \tag{7.3}$$

In the plan view, Fig. 7.7b, e, the position of the points 5 and 6 is defined by these radii and the angle Δε. The distance between points 5 and 6 is:

$$\Delta g = \sqrt{\Delta r^2 + \Delta u^2} \tag{7.4}$$

The length of the streamline element in space is obtained from the length in the plan view Δg and the extension in axial direction Δz (normal to the plan view):

$$\Delta L = \sqrt{\Delta g^2 + \Delta z^2} \tag{7.5}$$

By means of Eqs. (7.3 and 7.4) the relationship between the true length ΔL of the streamline element and its projection Δm in the meridional section and the projection Δu in the plan view is obtained:

$$\Delta L = \sqrt{\Delta g^2 + \Delta z^2} = \sqrt{\Delta r^2 + \Delta u^2 + \Delta z^2} = \sqrt{\Delta m^2 + \Delta u^2} \tag{7.6}$$

The distances ΔL, Δu and Δm according to Fig. 7.7c constitute a right-angled triangle which includes the true angle β between the streamline element ΔL and the tangential direction Δu.

If a streamline is developed step by step into the drawing plane according to this procedure, its true length and the true angles relative to the circumferential direction are obtained. This procedure is used to design the blade coordinates in the plan view from a defined meridional section and a blade development to be specified. In the above method, the curvilinear triangles have been replaced by rectilinear ones. The smaller the elements are selected, the better this approximation becomes. Of the different methods for designing the blades, the Kaplan method is described [B.2]; it comprises the following steps:

1. Depending on the width of the impeller (i.e. n_q) one to five additional surfaces of revolution are drawn into the meridional section in addition to the outer and inner streamlines. These surfaces can be designed as streamlines of identical part flows. To this end, (estimated) normals to the streamlines are drawn. In doing so, the width between two streamlines is determined in a way that the same flow $\Delta Q = 2 \times \pi \times r \times \Delta b \times c_m$ travels through each partial channel (frequently with the assumption c_m = constant).
2. All surfaces of revolution or streamlines are sub-divided into n elements of identical length Δm (e.g. Δm = 8 mm) according to Fig. 7.7a so that the points a_1 to a_n, b_1 to b_n etc. are obtained.
3. In the blade development according to Fig. 7.7d, n parallel straight lines are drawn at the distance Δm. Lengths $L_{a,m}$, $L_{b,m}$, $L_{c,m}$ etc. of the developed meridional section are obtained (which constitute the projection of the streamlines into the meridional section).
4. The length $L_{a,u}$ of the streamlines in circumferential direction is still unknown; within certain limits it can be freely selected. A possible approach consists in defining a development law according to Fig. 7.6. Starting with $\beta_{(j=0)} = \beta_{2B}$ at point A in Fig. 7.7d, the circumferential length Δu of the element Δm is calculated from Eq. (7.7) for each step j to obtain the position of the next point.

$$\Delta u_j = \frac{\Delta m_j}{\tan \beta_j} \tag{7.7}$$

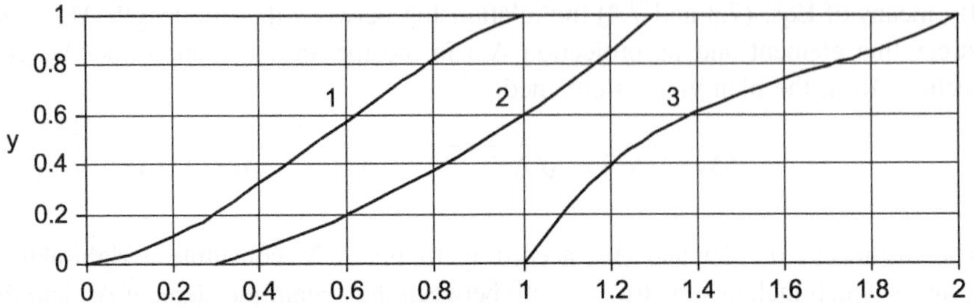

Fig. 7.6 Blade development options (the figures on the abscissa have no meaning)

The angles for each step are obtained from the selected law according to which the angle is to develop as a function of the blade length from β_{2B} to β_{1B}. In general it is possible to write:

$$\beta_j = \beta_{2B} - y(x)(\beta_{2B} - \beta_{1B}) \tag{7.8}$$

Any random function which satisfies the conditions y(0) = 0 and y(1) = 1 can be substituted for y(x). If a linear angle development from β_{2B} to β_{1B} is desired, the function y(x) has to be formulated as:

$$y(x) = \frac{1}{L_{a,m}} \sum_{0}^{j} \Delta m_j \tag{7.9}$$

This method can also be used in a way that the outlet angle is kept constant over n_a sections ($\beta_j = \beta_{2B}$ for $j < n_a$) and/or that the inlet angle is kept constant over n_e sections ($\beta_j = \beta_{1B}$ for $j > n - n_e$). Accordingly, Eq. (7.19) is only formulated for the middle part of the blade in which the angle changes from β_{2B} to β_{1B}.

5. An alternative way of determining the blade development is as follows: a line S1 is drawn from point A with an angle β_{2B}, Fig. 7.7d. A point E at distance $L_{a,m}$ is then selected such that a line S2 started with the angle β_{1B} intersects the line S1 in a point P. The ratio $L_p/L_{a,m}$ has an effect on the blade length and the channel cross sections. It must be selected so that the intended blade distances a_1 and a_2 as well as acceptable blade loadings according to the criteria in Eqs. (T7.1.10 to 7.1.13) are obtained. A smooth curve is then nestled into the geometry defined by the points A and E and the lines S1 and S2. This curve forms the chosen blade development.
6. The blade geometry must be represented in the plan view so as to allow the manufacturing of the impeller or pattern. The blade points 1 to 13 in the development (Fig. 7.7d) define triangles with the sides ΔL, Δu and Δm. These points are situated on the radii r_1 to r_{13} (see the meridional section in Fig. 7.7a). Starting on the outer radius the points are transferred to the radii r_1, r_2 etc. with

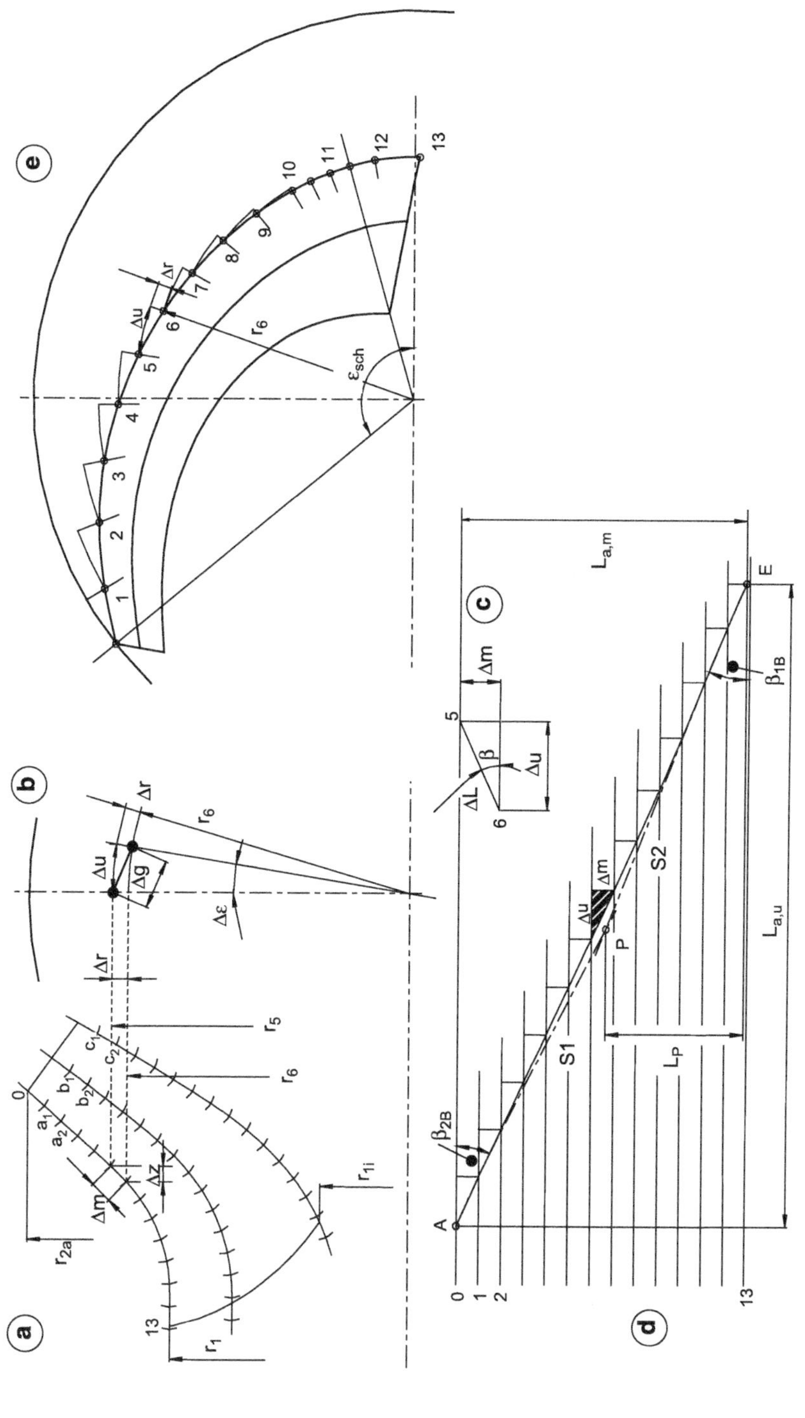

Fig. 7.7 Blade design according to Kaplan's method

Δu and Δr according to Fig. 7.7e. This yields the points 1, 2, 3, etc. in the plan view. Joining these points produces the streamline projected into the plan view.

7. The wrap angle ε_{sch}, formed by the blade in the plan view depends on the blade angles β_{1B} and β_{2B}, the radii ratio d_1/d_2 (hence n_q) and the streamline length in the meridional section. Wrap angles of radial impellers usually are within the following ranges: with $z_{La} = 5$: $\varepsilon_{sch} = 130$ to 160°; with $z_{La} = 6$: $\varepsilon_{sch} = 120$ to 140° and with $z_{La} = 7$: $\varepsilon_{sch} = 100$ to 130°.
8. Steps 4 to 6 are performed for all surfaces of revolution (streamlines).
9. To generate sufficient checkpoints for the manufacturing process, radial sections (A to Q in Fig. 7.8) are drawn in the plan view and their intersections with the surfaces of revolution or streamlines transferred into the meridional section (shown in Fig. 7.8 by way of Section K). The radial sections should yield smooth curves in the meridional section in order to avoid waviness in the blade surface. These radial sections need not always be curves; they can also be straight lines.
10. For manufacturing it is also possible to use contour lines ("board lines"). These are equidistant sections in the meridional section normal to the axis, No. 0 to 12 in Fig. 7.8. Their intersections with the streamlines and radial sections are transferred to the plan view and joined to form the board lines. These must present smooth curves in the plan view in order to prevent wavy blade surfaces.
11. The blade surface designed in this way constitutes the camber, the suction or the pressure surface - depending on the angles (camber, suction or pressure surface) that were used for the development. The selected blade profile can be drawn into the development as section normal to the blade surface in order to better appreciate its effect on the approach flow.
12. The blade profile is represented as developed onto the camber line where it is dimensioned as thickness = f(L) (dimensioning not shown in Fig. 7.8).

7.2.3 *Criteria for Shaping the Blades*

When constructing the blade development, the designer enjoys a certain freedom which he may utilize to obtain specific hydraulic features. Impellers designed for high $NPSH_A$, i.e. when cavitation is not an issue, will be primarily optimized for efficiency and the stability of the Q-H-curve. In order to minimize the losses, a smooth variation of geometric parameters should be targeted (for example by increasing the blade angles to a linear function from inlet to outlet, i.e. $\partial\beta/\partial L =$ constant). On the outer streamline the outlet angle β_{2B} is usually several degrees larger than the inlet angle $\beta_{1B,a}$ and a development similar to curve 2 in Fig. 7.6 is obtained. On the inner streamline of an impeller with axial inlet (end suction pump) $\beta_{1B,i}$ is generally much larger than β_{2B} so that a development similar to curve 3 in Fig. 7.6 is obtained.

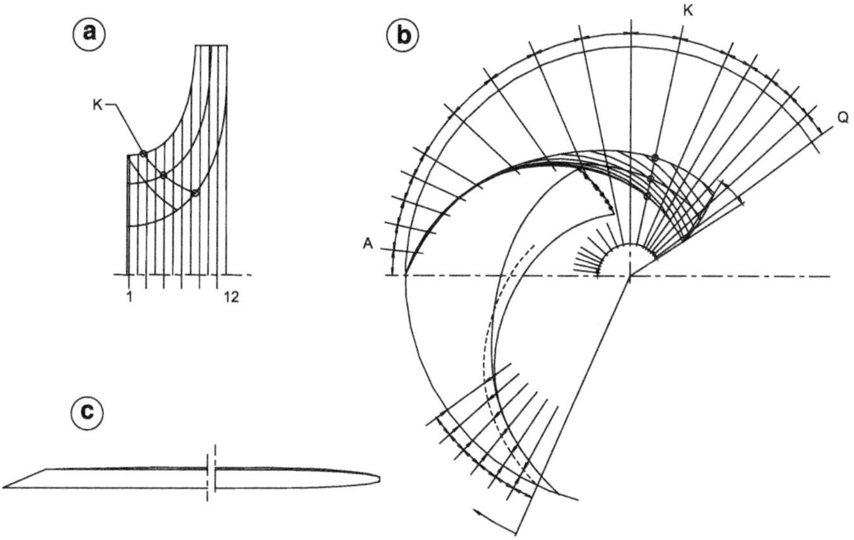

Fig. 7.8 Representation of the impeller coordinates by board lines 0 to 12 and radial sections A to Q. **a** Meridional section, **b** plan view, **c** blade profile (developed)

It is also possible to specify a law for u × c_u as a function of the blade length. For example, the design could be done with low blade loading at the inlet because of cavitation, with the strongest blade load in the middle part where the blades overlap and moderate blade loading towards the outlet with the object to achieve as uniform a flow as possible. It is thus attempted to minimize turbulent dissipation losses and pressure pulsations.

The developments of all streamlines must be matched to each other so that the desired leading edge is obtained. Since the inflow angles are usually small (between 12 and 18°), the approach flow of the blade leading edge can be assessed qualitatively by way of its position in the plan view. If the leading edge is radial (No. 1 in Fig. 7.9), it is hit by an almost perpendicular approach flow. If the leading edge is inclined by the angle ε against the radial direction, it meets the approach flow at an angle which tends to cause lower velocity peaks and losses (compare to a cylinder exposed to an inclined versus a perpendicular approach flow). A leading edge shape according to Nos. 2, 3 or 4 in Fig. 7.9 corresponds to a "sweep back" which is used in axial pumps, compressors and inducers. It tends to displace fluid away from the hub. Special shapes such as No. 5 in Fig. 7.9 ("sweep forward") are sometimes used to extend and unload the outer streamline (see Chap. 5). As discussed in Chap. 5, the leading edge also influences the partload behavior - especially shapes such as 4 and 5 in Fig. 7.9.

The blade twist at the inlet is obtained from Eq. (T7.1.7) from the calculated flow angles and the selected incidences. This involves major angle variations over the blade height if $d_{1i} \ll d_{1a}$ and $\beta_{1B,I} \gg \beta_{1B,a}$. In contrast, the blade twist at the

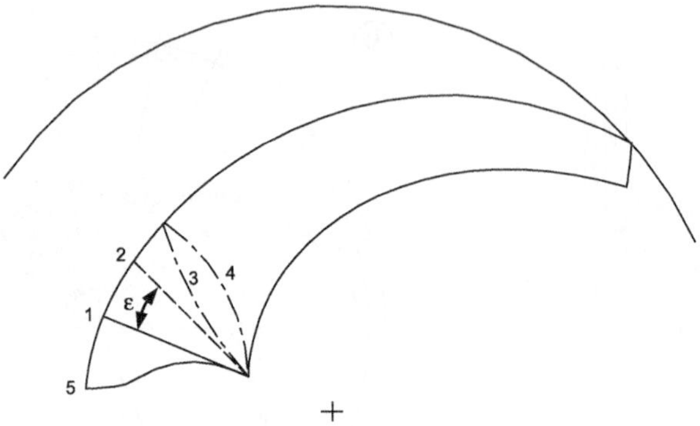

Fig. 7.9 Different shapes of the impeller blade leading edge

impeller outlet can be selected freely. The blades of radial impellers of low specific speeds are frequently not twisted at the outlet. The higher the specific speed, the more an outlet twist will be required to optimally match the streamline lengths. In order to reduce pressure pulsations, the blades can be twisted at the impeller outlet. The effectiveness of this measure increases with the ratio of the outlet width b_2 to the pitch of the blades t_2 and with the specific speed.

The blades can be arched deliberately in order to increase the throat area A_{1q}. The radial sections will then appear as pronounced curves in the meridional section. This design is used especially with radial impellers of high specific speeds (compare to the impellers of Francis turbines).

In general, all streamlines, blade developments, radial sections, board lines and channel cross sections should vary continuously along the flow path in order to prevent unnecessary accelerations, decelerations and velocity differences over the channel. Turbulent dissipation losses caused by uneven velocity distributions can thus be minimized.

7.2.4 Criteria for Suction Impeller Design

From the time when boiler feed pumps were built with heads up to 600 to 800 m per stage - or with u_1 up to 80 m/s - (so called “advanced class feed pumps” are operating even with 1200 m per stage), the design philosophy of the suction impellers has changed. In the past - with low circumferential speeds at the impeller inlet - large inlet angles and high incidences were selected and blades with long wedge-like inlet profiles were designed to obtain sufficiently low values of $NPSH_3$, concept “H” in Fig. 7.10c. Even when operating with low incidence, such impellers generate pronounced low-pressure peaks and large detached cavities. At $u_1 > 50$ m/s these can result in severe cavitation erosion.

In view of such damage, the manufacturers of high-pressure pumps were faced with the need to develop new blade shapes by means of stroboscopic observation of the cavities generated at the impeller inlet. These new designs feature: (1) Small incidences. (2) Relatively small absolute blade and approach flow angles which subsequently produce lower incidences at partload (example: with $\beta_{1B} = 20°$ an incidence of 10° is obtained at $Q/Q_{SF} = 0.5$, while the incidence is only $i_1 = 5°$ at $\beta_{1B} = 10°$). (3) Thick, short profiles which are shaped to be less sensitive to incident flow. (4) Low blade loadings in the inlet region up to roughly the throat area, concept "L" in Fig. 7.10d.

In order to achieve flat pressure distributions, the blades can also be designed with an asymmetric leading edge profile where the pressure surface is arranged according to the calculated/selected blade angle $\beta_{1B,PS} = \beta_{1B}$ and the thickness distribution is applied to the pressure rather than on the camber surface, Fig. 7.10e. The maximum blade thickness can be selected equal to the nominal thickness e_{nom} (left sketch in Fig. 7.10e) or greater than the nominal thickness (right sketch in Fig. 7.10e). In [10] $e_{max}/e_{nom} = 1.37$ had been successfully applied to reduce cavitation damage. In doing so, the blade leading edge was swept forward near the hub (compare to shape 4 in Fig. 7.9); this blading is called "biased wedge design", [57].

Based on these characteristics, flat pressure distributions are obtained. In addition the pressure increase over the first part of the blade should be kept small by limiting the increase in radius and blade angles at the outer streamline over a length of approximately $t_1 = \pi \times d_1/z_{La}$. This limits the increase in $u \times c_u$ in that region. This design achieves low coefficients σ_i for cavitation inception and thin translucent cavities with low vapor volume and correspondingly low hydraulic cavitation intensity. Figure 7.10a schematically shows how the specific work should develop over the blade length to obtain flat pressure distributions according to Fig. 7.10b. Despite the flat blade angles in the inlet region over a length of t_1 the throat area between the blades must be designed large enough so that the maximum specified flow rate can be achieved, Chap. 7.2.1, item 15. This can result in S-shaped developments of the camber line according to Fig. 7.6, curve 1.

Before designing suction impellers, the risk of cavitation damage should be assessed according to Chap. 6 and especially Chap. 6.7. In many applications the risk of damage is low and the aim is to achieve the lowest possible $NPSH_3$ and high n_{ss}. In contrast, the suction impellers of high-pressure pumps with considerable damage potential must be designed to achieve a low $NPSH_i$ and low cavity volumes. Typical parameters of these two extreme cases are compared in Table D7.1 (applications in-between can be interpolated). However, high suction specific speeds can also be achieved by designing impellers with flat pressure distributions - implying negative incidence and suction-side profiling. This is increasingly done and will be more so in the future. Some examples were discussed in [81]. Impellers with flat pressure distributions, designed according to the principles discussed above, have been tested in a single-stage volute pump with axial inlet, [80]. The data of interest are listed in Table D7.1-1. Three impellers with suction specific speeds have been designed with *negative incidence*. Interestingly, the onset of inlet recirculation $q^*_{RB} = Q_{RB}/Q_{opt}$ of all impellers was in a narrow range. Reason: with

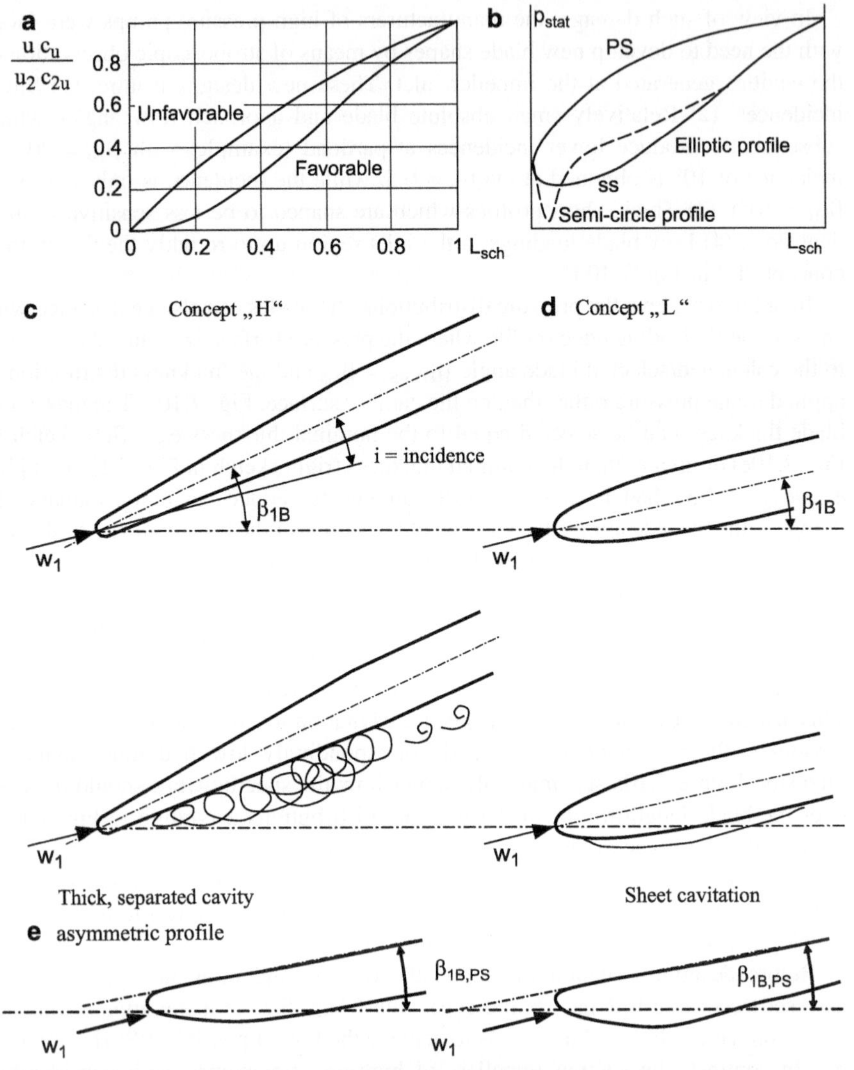

Fig. 7.10 Suction impeller design. **a** Work transfer; **b** pressure distribution; **c** design with high incidence and sharp profile; **d** design for a flat pressure distribution; **e** asymmetric profiling on the suction surface

increasing inlet diameter d_1^*, the onset of recirculation q^*_{RB} would be expected to rise, while q^*_{RB} drops as the incidence is reduced: in these tests, the effect of increasing the impeller inlet diameter apparently was compensated by reducing the incidence. Q_{RB} was predicted by CFD: incidentally, 2 cases were close to test, 2 cases were off (see Chap. 8.8). As in Fig. 6.21, the n_{ss} increases with decreasing approach flow angle β_{1a}; the data in Table D7.1-1 are situated above the upper limit of the scatter in Fig. 6.21.

Table D7.1 Suction impeller design		
Risk of erosion	**Low**	**High**
Impeller inlet tip speed	$u_1 < 35$ m/s	$u_1 > 50$ m/s
Fluid	Water above 220 °C and hydrocarbons	Water below 220 °C
Suction specific speed	$n_{ss} = 220$ to 260	$n_{ss} = 160$ to 180
Approach flow angle	$\beta_{1a} = 10$ to 15°	$\beta_{1a} = 14$ to 18°
Cavitation inception at q* = 1	$\sigma_i = 1.2$ to 2	$\sigma_i = 0.5$ to 0.8
Incidence at q* = 1	$i_1 = 3$ to 5° trend: smaller or negative	$i_1 = < 2°$ or negative
Shockless entry at q* = 1, but depending on Q_{max}	$Q_{SF}/Q_{opt} = 1.1$ to 1.2 trend: lower than 1.0	$Q_{SF}/Q_{opt} = 0.95$ to 1.05 or even lower
Resulting pressure distribution	Peaky	Flat

Figure 7.10.1 shows the cavitation coefficients $\sigma_3 = f(\varphi_1)$ calculated from the test curves in [80]. Interestingly, the three curves of the impellers with $n_{ss} > 280$ coincide over their respective test range while the curve with $n_{ss} = 186$ seems shifted upwards. This observation may show the influence of the meridional section: the high n_{ss}-impellers had similar meridional sections with generous radii at the outer streamline (OSL). In contrast, the impeller with $n_{ss} = 186$ had a strong curvature at the OSL and the axial extension z_E of the blades is much shorter (z_E: rough estimates in Table D7.1-1, definition in Fig. 7.4). The sketches in Fig. 7.10.1 show, very approximately, the impeller shapes. As often observed, the plot against the flow coefficient looks more telling than the plot against φ_1/φ_{1SF}. This finding seems to emphasize the dominant impact of the stagger angle on cavitation phenomena. With 3D-leading edges it is also difficult to define a meaningful blade blockage. Figure 7.10.1 shows also the blade coefficients λ_{w3} calculated from Eq. (6.11).

Good cavitation characteristics of high-pressure pumps can only be achieved if the inlet casing is optimized so that no vortices (vortex ropes) are created. The inflow to the impeller must be as uniform as possible so that the incidence does not vary over the circumference by more than ±1°. The partload behavior can be improved with a diffuser upstream of the impeller, [24].

The design criteria for suction impellers discussed above were predominantly developed empirically based on stroboscopic cavitation observations, partly supported by numerical calculations of inviscid flow. This follows from publications such as [11, 23–25, 34, 57, 58, 71]. Today efficient numerical methods can be employed for the prediction of cavitation inception and cavity length, [14, 27], provided the approach flow variations are kept under control.

Table D7.1-1. Suction impellers tested in [80], n_q = 29, type OH2

d_1^*	β_{1a}	$i'_{1,a}$	n_{ss}	$\varphi_{1,opt}$	Q_{SF}/Q_{opt}	$\varphi_{1,SF}$	q^*_{RB}	$q^*_{RB,CFD}$	$\varphi_1/\varphi_{1,SF}$	b_2^*	β_{2B}	z_E/r_{1a}
	°	°	–	–	–	–	–	–	–	–	°	–
0.445	22.6	4.8	186	0.402	1.30	0.521	0.62	0.48	0.771	0.091	24.0	0.42
0.482	17.7	−7.0	287	0.308	0.67	0.207	0.65	0.63	1.490	0.082	26.3	0.83
0.500	15.3	−2.3	327	0.263	0.90	0.235	0.67	0.63	1.116	0.077	29.0	0.82
0.527	13.3	−3.4	346	0.227	0.80	0.182	0.64	0.74	1.248	0.086	27.5	0.60

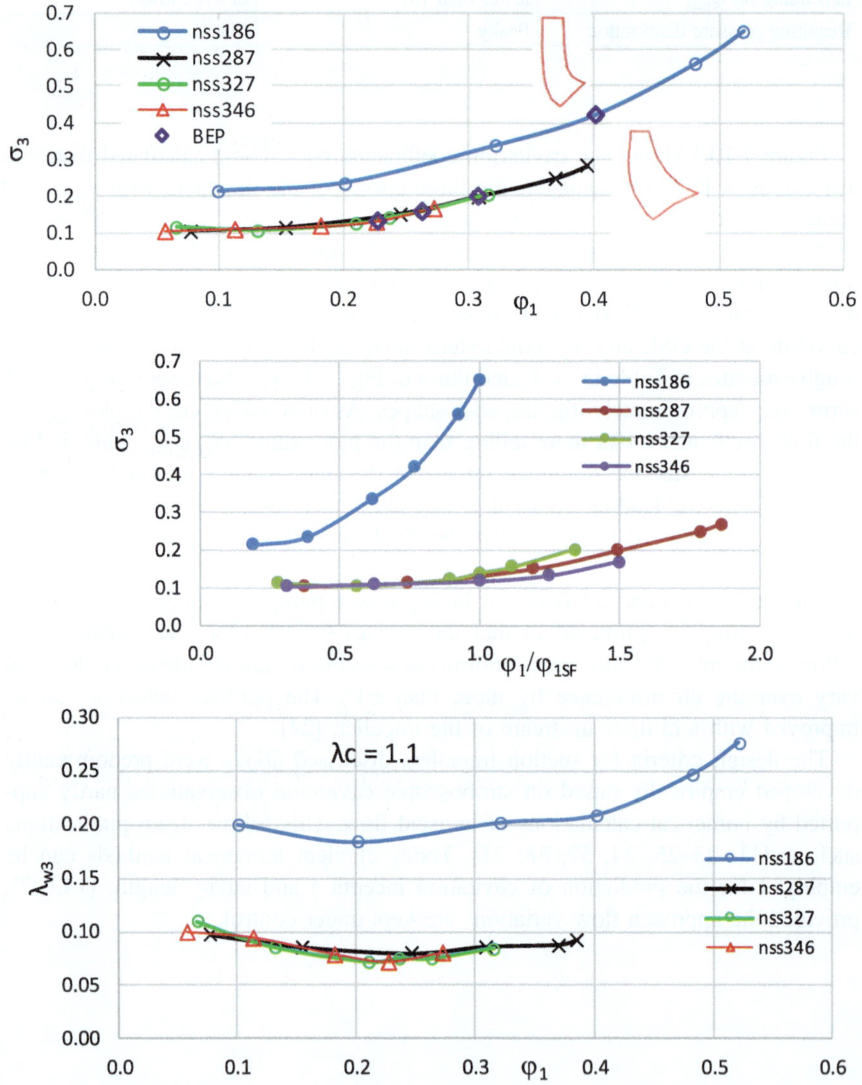

Fig. 7.10.1 Cavitation performance of impellers specified in Table D7.1-1; [80]

7.2.5 Exploiting Three-Dimensional Effects in Design

As explained in Chap. 5, the flow deflection (or the slip factor) and the losses are affected by secondary flows which, in the one-dimensional theory, can only be taken into account by empirical coefficients. The statistical data used do not do justice to the details of the geometry of a specific impeller. Applying numerical methods it is possible, however, to examine the impact of all geometrical characteristics on the flow. This opens up new possibilities for optimization by attempting to influence the flow by means of new kinds of geometrical effects and shapes. This does not only concern the minimization of the hydraulic losses in the design point but, especially, also the stability of the Q-H-curve and cavitation. 3D-effects increase with the blade height, i.e. with the specific speed. Examples for 3D-effects are:

- Sweep-forward: a leading edge pulled forward according to Fig. 7.9, shape 5, tends to improve the stability of the Q-H-curve of semi-axial pumps, Chaps. 5.5 and 5.6.
- Sweep-back reduces the extent of the cavitation zones in inducers and axial impellers.
- A leading edge pulled forward at the hub according to shape 4 in Fig. 7.9 helps to reduce the cavitation near the hub at partload according to [57].
- Axial blades can also be designed with a lean of their entire surface against a radial line. Numerical analyses of this effect were reported in [40].
- By means of blade twisting it is possible to affect the energy distribution over the blade height.
- The common practice to vary the impeller outlet angle over the blade height at high specific speeds can be employed more purposefully with the help of numerical calculations.
- The rear shroud need not necessarily be designed as a surface of revolution if the flow can be improved by a 3D-shape, [47]. A freely shaped rear shroud would not imply any major increase in expenditure for the pattern manufacture if NC-milling or stereo-lithography were employed for the purpose. However, no specific information is available on the benefit of such measures.

7.3 Radial Impellers for Small Specific Speeds

7.3.1 Two-Dimensional Blades

The outer and inner streamlines of cylindrical (untwisted) blades coincide in the plan view. At specific speeds below $n_q = 16$ to 18 untwisted blades can be designed, for example, as series stages of multistage pumps. Since the impeller losses at low specific speeds are small (see Fig. 3.33) and the shock losses at the

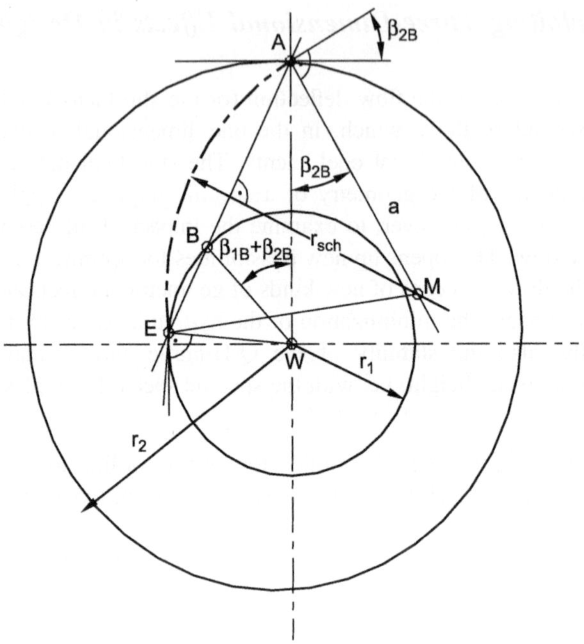

Fig. 7.11 Construction of circular blades

impeller inlet are insignificant because of the low ratio d_1/d_2, the cylindrical blades imply virtually no loss in efficiency. However, if good cavitation characteristics are required, twisted blades must be employed.

Impellers with cylindrical blades are calculated and designed in a similar way as per Chap. 7.2.1. Impellers for undemanding services can be made with circular arc blades instead of designing the blade development according to Chap. 7.2.2.2. In the plan view the blades can be designed so that inlet and outlet angles are achieved by *a single* circular arc. The design is explained by means of Fig. 7.11: (1) Circles are drawn with the inlet and outlet radii of the impeller. (2) At any point A on the outer radius r_2, the blade angle β_{2B} is applied and the normal A-a is erected in A. (Note direction of rotation.) (3) From the shaft center W the angle ($\beta_{1B} + \beta_{2B}$) is applied in the direction of rotation on the radius W-A. The resulting line intersects the circle with r_1 in a point B. (4) A line through A and B produces a second intersection E with the circle r_1, marking the blade leading edge. (5) The normal in the middle of the distance A-E intersects the normal A-a in a point M constituting the center of the circular arc. The latter is drawn with the radius r_{sch} = MA = ME. (6) The circular arc obtained constitutes the camber line of the blade. Suction and pressure surfaces of the blades are obtained from circular arcs with $r = r_{sch} \pm 0.5$ e around point M.

The radius r_{sch} of the circular arc blade can be calculated from Eq. (7.10):

$$r_{sch} = \frac{r_2{}^2 - r_1{}^2}{2(r_2 \cos \beta_{2B} - r_1 \cos \beta_{1B})} \tag{7.10}$$

Circular arc blades of this type result in an unfavorable development of the relative velocity which is decelerated to a value $w < w_2$ within the impeller channel. So as not to jeopardize the stability of the Q-H-curve, the blade leading edges must not be parallel to the rotor axis; they must indeed be brought forward to as small as possible a diameter d_{1i} (which is limited by the hub diameter).

Radial impellers with cylindrical blades are mainly employed in the range $8 < n_q < 18$ for small pumps and for less demanding service up to about $n_q = 25$.

The small outlet width ratio b_2^* of impellers with very low specific speeds can cause casting problems because of the narrow channels. These difficulties are avoided by resorting to pumping disks or semi-open impellers with radial blades discussed in the following.

7.3.2 Pumping Disks with Channels of Circular Section

The work transfer in radial impellers of low specific speeds (i.e. with a small ratio d_1^*) is largely accomplished by centrifugal forces. This method of energy transfer implies very low losses in the impeller (Fig. 3.33). Hence the hydraulic efficiency of the pump is essentially determined by the losses in the collector. When centrifugal effects prevail, the impeller channels can be designed as very simple shapes without jeopardizing efficiency. A rotating disk with cylindrical holes according to Fig. 7.12 can be employed as an impeller, as described for instance in [46]. The holes can be drilled in radial direction with $\beta_{2B} = \beta_{1B} = 90°$. Alternatively, the holes can be leaning back in circumferential direction with $\beta_{2B} = \beta_{1B} < 90°$ to produce less flow deflection and correspondingly lower head coefficients but steeper Q-H-curves.

The collector is designed as a volute or an annular casing which is often preferred because of its simplicity. The fluid leaving the collector is decelerated further in a diffuser. The throat area of the diffuser mainly determines the best efficiency flow rate since the losses in the diffuser by far outweigh the impeller losses. When the fluid from the impeller outlet is accelerated into the diffuser throat (i.e. at $c_{3q} > c_2$), the diffuser losses increase. Consequently, head and efficiency drop. With sufficient acceleration, the flow cavitates in the throat area and head and efficiency drop almost vertically ("choking"). The design of the annular collector follows Chap. 7.12, the diffuser is calculated according to Chap. 1.6, and the best efficiency point flow rate is obtained from Chap. 4.2.

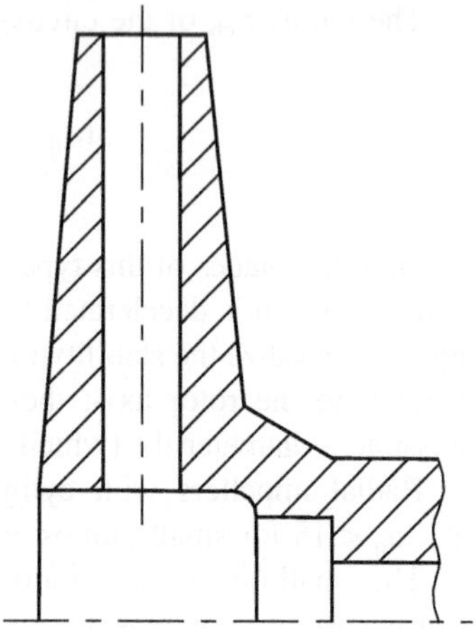

Fig. 7.12 Using a disk with drilled holes as an impeller

To obtain high head coefficients and low impeller losses, the velocity c_B in the holes must be selected low. It largely corresponds to the relative velocity in the holes with diameter d_B:

$$c_B = \frac{4\,Q}{\pi\, d_B^2\, z_{La}} \quad \text{and} \quad \varphi_B = \frac{c_B}{u_2} \tag{7.11}$$

According to Chap. 5.2 the Coriolis acceleration is $b_c = 2 \times \omega \times w$. In radial holes b_c is largely responsible for the slip factor which increases with growing flow rate because of $w \approx c_B$. From measurements it can be deduced that the flow coefficient in the design point should not exceed $\varphi_B = 0.06$ to 0.08 if high head coefficients are desired. The impeller losses are then in the range of a few percent of the head. If high head coefficients are targeted, number and size of the holes must be selected such that flow coefficients of $\varphi_B < 0.08$ are obtained at the design point. It is then that head coefficients in the range of 1.1 to 1.25 are achieved. However, the Q-H-curves become flatter and the head drops slightly towards Q = 0. If a steeper characteristic is required, a higher flow coefficient must be selected for the design point; even backward leaning holes may be required. Conversely, the impeller losses increase and the efficiency drops.

Since the annular casing according to Chap. 7.12 has a larger cross section than corresponds to the impeller outlet velocity c_{2u}, the flow is decelerated abruptly (similar to a sudden expansion according to Chap. 1.2.3) and an intense secondary flow is generated. According to Chap. 5, the head can be increased if the exchange of momentum between the impeller and the flow in the collector is intensified (similar to a regenerative pump, Chap. 2). This increases the kinetic energy at the inlet to the diffuser whose pressure recovery rises accordingly. The exchange of momentum grows with decreasing φ_B; it can be deliberately promoted by providing material recesses on the impeller outer diameter which accelerate the fluid

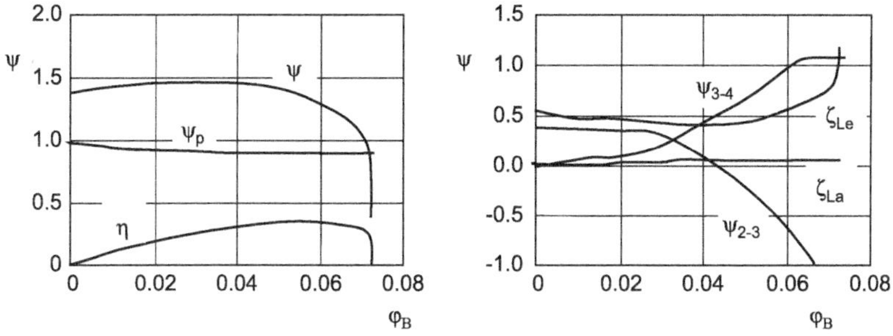

Fig. 7.13 Tests with a rotating disk with 12 drilled channels, Sulzer Pumps

circulating in the annular casing by exchange of momentum. In [46] for instance, blind holes were used for this purpose. In this way it is possible to considerably increase head coefficient and efficiency: at BEP ψ_{opt} = 1.3 to 1.4 is obtained while the maximum values at partload just reach ψ_{max} = 1.5. The Q-H-curve on the left of the best efficiency point becomes very flat and droops slightly towards shut-off.

Between impeller and casing, relatively large axial gaps (up to approximately $s_{ax} = 0.01 \times d_2$) can be permitted since the impeller sidewall operates like a friction pump. The head can even be improved slightly since the impeller sidewall boundary layer enters the collector with high circumferential speed - in this way increasing the kinetic energy of the fluid circulating there. However, a slight loss in efficiency is incurred since the increase of the power consumption exceeds that of the head.

Figure 7.13 shows the dimensionless characteristics of a rotating disk with 12 holes and recesses on the impeller outlet to increase the exchange of momentum between the annular casing and the impeller. The flow coefficient was calculated from Eq. (7.11). All head coefficients correspond to the definitions used in Chaps. 3 to 5. The Q-H-curve is very flat at flows below the best efficiency point; it drops vertically at $\varphi_B \approx 0.073$, which is due to cavitation in the diffuser.

Noteworthy is also the flat curve of the static pressure rise ψ_p in the impeller which shows that slip factor and impeller losses change only moderately with increasing flow rate. At Q = 0, ψ_p almost reaches the theoretical value of the flow without losses. According to Chap. 4.1.4 this corresponds to a hydraulic impeller efficiency of 0.96. The pressure recovery in the diffuser $\psi_{3\text{-}4}$ roughly increases with the square of the flow rate. The pressure increase $\psi_{2\text{-}3}$ from the impeller outlet to the diffuser throat is significant below $\varphi_B \approx 0.03$. Above $\varphi_B \approx 0.044$ the static pressure is reduced since the flow is accelerated from the impeller outlet to the diffuser throat area (i.e. $c_{3q} > c_2$). The measured impeller losses are insignificant compared with the losses in the collector.

7.3.3 Impellers with Straight Radial Blades

Impellers with straight radial blades with $\beta_{2B} = \beta_{1B} = 90°$ according to Fig. 7.14 are sometimes employed in small pumps with speeds up to 25,000 rpm[1]. The range of specific speeds is about $n_q = 6$ to 12. This type of pump is often equipped with an annular casing from which the fluid is discharged through a diffuser. Alternatively, vaned diffusers or double volutes are in use. The impeller is semi-open or open to reduce disk friction and annular seal losses. Depending on the number of blades, this design demands tight axial clearances of $s_{ax}/b_2 = 0.01$ to 0.02 between blades and casing walls to obtain acceptable efficiencies.

The impeller outlet width is selected relatively large, similarly to the curve shown for sewage pumps in Fig. 7.2. The impeller usually has 16 to 24 (32) blades. Typically 3 to 6 blades have the full blade length from d_1 to d_2 while the rest are splitter blades - otherwise blade blockage at the inlet would become excessive. The large impeller outlet width and large number of blades produce slip factors close to 1.0 and foster a strong exchange of momentum with the flow in the annular collector. As with drilled disks, head coefficients of $\psi_{opt} = 1.2$ to 1.5 are achieved. The Q-H-curve becomes correspondingly flat and often slightly unstable, similar to that in Fig. 7.13. Blade numbers over 16 increase the head coefficient in the best efficiency point only to an insignificant degree but yield a higher shut-off head due to improved exchange of momentum, thereby increasing the stability of the Q-H-curve.

The hydraulic characteristics of radial bladed impellers are similar to those of the drilled disks discussed in Chap. 7.3.2. According to Fig. 3.26, efficiencies up to approximately 45% can be achieved–depending on specific speed, pump size and mechanical losses. High-speed pumps are often equipped with an inducer to lower $NPSH_3$; the suction specific speeds are typically between $n_{ss} = 350$ and 450.

Figure 7.15 shows the dimensionless characteristics of a radial bladed impeller measured with two different diffuser inserts the throat diameters of which amounted to 71 and 39% of the impeller outlet width. With the large diffuser insert a specific speed of $n_q = 11$, a head coefficient of $\psi_{opt} = 1.48$ and an efficiency of 43% were obtained. By reducing the diffuser throat area, the specific speed was reduced to $n_q = 7$ while the efficiency dropped to 23%. The power consumption was practically identical in both tests so that the best efficiency point shift was only achieved through throttling in the diffuser. The fact that cavitation in the diffuser plays a significant role is borne out by the curves for the cavitation coefficient which have changed drastically even though the impeller was identical.

[1]These impellers are sometimes called "Barske impeller", [5, 13, 43].

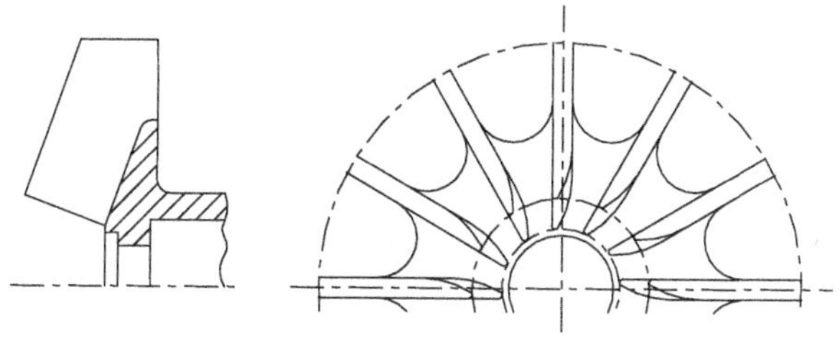

Fig. 7.14 Impeller with straight radial blades

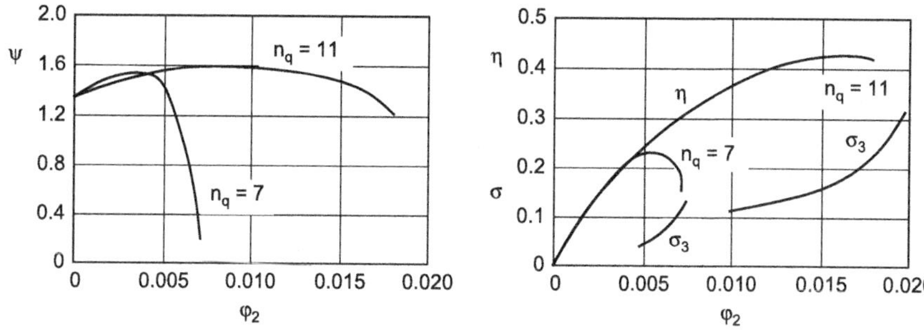

Fig. 7.15 Tests with an impeller with straight radial blades and two different diffusers, Sulzer Pumps

7.3.4 *Double-Acting Impeller with Straight Radial Blades*

A double-acting overhung impeller with straight radial blades is shown in Fig. 7.16 in a view from the inlet and in the meridional section. The impeller is used in a single-stage, single-entry end-suction pump of the type shown in Fig. 2.1. The heads achieved are well above 300 m, [12, 22]. Nearly identical blades are arranged on both sides of the center rib. Five large openings in the center rib near the hub allow half of the fluid to flow to the blades on the rear (that is why the impeller is labeled "double-acting"). On each side of the center rib there are 5 long and 20 short blades. The collector is built as a concentric casing followed by a partial volute which discharges into a diffuser.

In the present application the impeller is used for pumps in the range of roughly $1 < n_q < 10$. Figure 7.16c shows the dimensionless characteristics for $n_q = 2$, $n_q = 5$ and $n_q = 9$. While efficiencies above 50% are achieved with $n_q = 9$, the efficiency drops sharply for lower specific speeds.[2]

[2] In order to plot the curves for $n_q = 2$ to $n_q = 9$ on the same graph, a logarithmic scale was chosen for the flow. The shape of the efficiency curves therefore looks unusual.

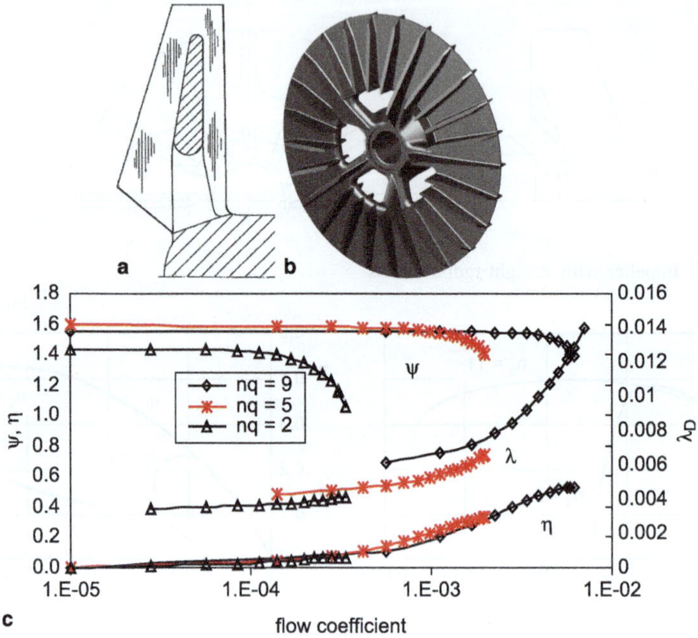

Fig. 7.16 End-suction pump with double-acting impeller; **a** meridional section; **b** front view; **c** performance characteristics, λ_D according to Eq. (T3.4.10)

The head coefficients at BEP are above 1.4; they drop at very low specific speeds due to increased losses in the collector. The Q-H-curves are stable but flat because of the high head coefficients. Since flow conditions on both sides of the impeller are very similar, the axial thrust is balanced to a large extent. There are no disk friction losses, which accounts for the relatively high efficiency. The gaps between the casing and the impeller blades can be made rather large without unduly impairing the efficiency. Operation on the right of the best efficiency point is limited by cavitation in the collector. Due to the low blade loading vibrations are very low.

"Double-acting" impellers with specific speeds of $n_q = 4$, 8 and 10 have been tested in [86]. In essence, they used grooves rather than blades and 6 large holes in the hub to allow flow to the rear side of the impeller; the concept is very similar to Fig. 7.16. Per side of the impeller, there were 180 grooves (long and short) with a depth of 3 mm (with $d_2 = 202$ mm). Groove width was 2 mm; the "land" between two grooves - which in fact corresponds to the blade thickness - was 1.5 mm. Performance was similar to Fig. 7.16, but the head coefficients were about 10% lower. Axial clearances as large as 5 mm (with groove depth 3 mm and total $b_2 = 12$ mm) had little effect on the head, while the efficiency suffered moderately (because of the scatter difficult to quantify). The suction specific speed was $n_{ss} = 194 \times (1 - 0.6 n_q/n_{q,Ref})$ with $n_{q,Ref} = 10$ (tested between $n_q = 4$ and 9). As with the other pumps of low specific speed, the high head coefficients are achieved by an intense exchange of momentum between the impeller channels and the casing near the impeller outlet. For this reason, performance is little affected by increased axial clearances between casing and impeller.

Another type of a "double-acting" impeller designed for $n_q = 6$ has been described in [87]. The meridional section of the *open* impeller with axial inflow resembles the shape of a conventional impeller for low n_q. The blades are arranged on both sides of a support plate which is attached to the hub by the blades themselves. Since the blade inlet angles can be selected to match the axial approach flow, the $NPSH_R$ is expected to be lower than with radial blades as applied in Fig. 7.16.

7.4 Radial Impellers for Non-clogging Pumps

In various applications coarse solids or fibrous matters are transported by the fluid pumped; this requires special designs to reduce the risk of clogging and damage. Typical applications are:

- Sewage pumps the operation of which is mainly threatened by clogging through textiles, plastic parts and foils.
- Dredge pumps where the excavated material can include coarse gravel and all kinds of waste.
- Turnip or fish pumps which are equipped with single-channel impellers in order to avoid damaging the goods pumped.
- Paper pulp pumps (Chap. 13.5).

Sewage pumps usually are single-stage pumps with radial, semi-axial or axial impellers. Radial and semi-axial pumps use volutes because diffusers would be prone to clogging. Axial pumps are built with diffusers with thick vane leading edges to reduce the risk of clogging. Two types of clogging should be distinguished for sewage pump design:

1. Clogging by large solids such as plastic parts or wood. To reduce the risk that foreign matter gets stuck in the impeller or volute, one specifies the diameter d_K of a ball which must be able to pass through the pump.
2. Clogging by fibrous or stringy matter such as textile waste or plastic bags. In essence, this issue is independent of the ball passage diameter and more difficult to handle than the passage of large solids. The reason is that fibrous matter can be caught at the impeller blade leading edge independently of the ball passage diameter or channel width. Fibrous matter can accumulate in the impeller eye and in the axial impeller sidewall clearances (gaps E and F). Clogging can also be caused by spinning fibrous matter into a long agglomeration in front of the impeller. Partload flow recirculation from the impeller could be instrumental in this process. In extreme cases, fibrous matter entering the impeller side rooms can lead to a blockage of the rotor. Expeller vanes can be applied and/or the impeller side rooms can be closed by introducing a gap A (Table 0.2) in order to reduce the ingress of fibrous matter into the side rooms. Various devices and

arrangements which are able to chop long fibers are applied to reduce the risk of clogging. In small pumps a chopper wheel with (for example) 3 knives can be installed in front of the impeller. In other designs cutting edges are integrated into the casing and impeller.

Figure 7.17 shows a selection of special impeller designs which were developed for pumping liquids containing solids with a risk of clogging or which are laden with gas.

Published design information on sewage pumps is scarce. The design data given in the following were derived from the sales curves of four major sewage pump manufacturers. The results of this study are presented in Table 7.2. Figure 7.18 shows the pressure coefficients thus derived for pumps with 1- to 3-bladed impellers and for vortex pumps. The curves shown represent in essence the upper limit, even though some individual data points can be found above these curves. However, many pumps offered on the market are designed for lower pressure coefficients - presumably with the object to increase the ball passage.

The efficiencies documented in the sales curves have been compared to those expected for normal water pumps from Table 3.9. To do so, for any given sewage pump with a BEP flow Q_{opt} and a specific speed n_q the reference efficiency $\eta_{opt,R}$ of a clear-water pump has been calculated from Eq. (T3.9.1). The ratio of the actual sewage pump efficiency to $\eta_{opt,R}$ yields an efficiency multiplier $f_\eta = \eta_{opt,sewage}/\eta_{opt,R}$. The multipliers resulting for the different pump types are given in Table 7.2. While the *absolute* efficiencies depend on the pump size and specific speed, no clear dependency of the multiplier f_η on the specific speed or the pump size could be established (with the exception of vortex pumps). Hence the concept of normalizing the sewage pump efficiency with Eq. (T3.9.1) provides a meaningful method for efficiency prediction and assessment of sewage pumps. Table 7.2 gives the average efficiency multipliers as well as the range in which most of the pumps analyzed were found.

The design of sewage pumps is largely determined by the size of foreign matter that must pass through the pump without clogging. The impeller inlet diameter d_1, the impeller throat width a_1 or area A_{1q}, the impeller outlet width b_2 and the volute throat diameter d_{3q} or area A_{3q} all must be chosen to be larger than the ball passage diameter d_k. To select these parameters significantly larger than for clear-water pumps (which are optimized for efficiency) inevitably will impair the efficiency.

A sufficiently large blade distance is achieved by selecting the blade number between 1 and 4. The blade development is designed to yield the required a_1 as a minimum. The ball passage increases with the specific speed. Typical values for the ball passage are given in Table 7.2 and Fig. 7.19. In vortex pumps, the ball passage usually equals the diameter of the discharge nozzle. The same should be attempted for single-channel impellers, but in practice this goal is by far not always achieved as demonstrated by Fig. 7.20.

To obtain the necessary ball passage, the outlet width of impellers with low and medium specific speeds is designed considerably larger than for pumps intended for transporting pure liquids. The upper curve in Fig. 7.2 gives an idea of the relative

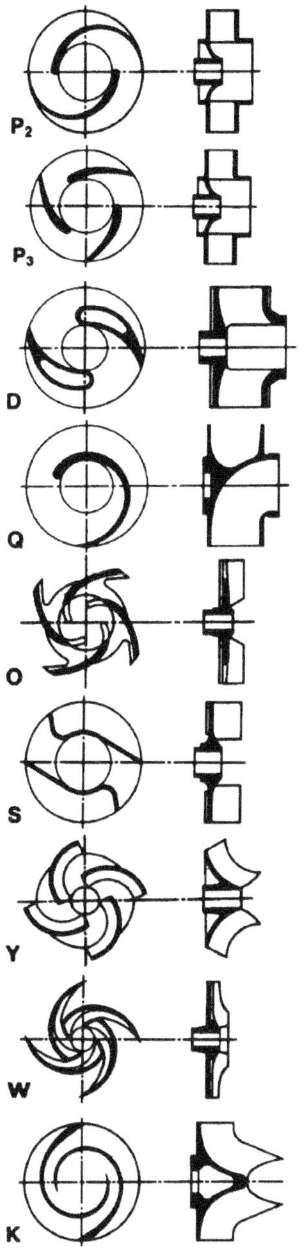

P_2: Two-channel impeller with very large passages for handling charged liquids.

P_3: Three-channel impeller with large passages for handling charged or viscous liquids.

D: Two-channel impeller for handling liquids with stringy matter or large solids without screening. Free passage with same diameter as suction nozzle.

Q: Single-channel impeller for handling liquids with stringy matter or large solids without screening. Free passage with same diameter as suction nozzle

O: Open radial impeller with 3 or 4 blades for handling gaseous and pulpy fluids.

S: Special open impeller with S-shaped blades, for handling liquids charged with fibers or small solid particles or stringy matter.

Y: Open, semi-axial screw-type impeller for handling large flow rates of clean or slightly contaminated fluids.

W: Open radial impeller with multiple blades of varying channel widths for handling liquids with high gas content, in particular *paper stock*

K: Open two-channel impeller with projecting blades for handling strongly contaminated viscous and gaseous liquids, high-consistency liquids, juices in particular paper pulp up to highest concentrations.

Fig. 7.17 Impellers for handling liquids charged with solids or gases, Sulzer Pumps

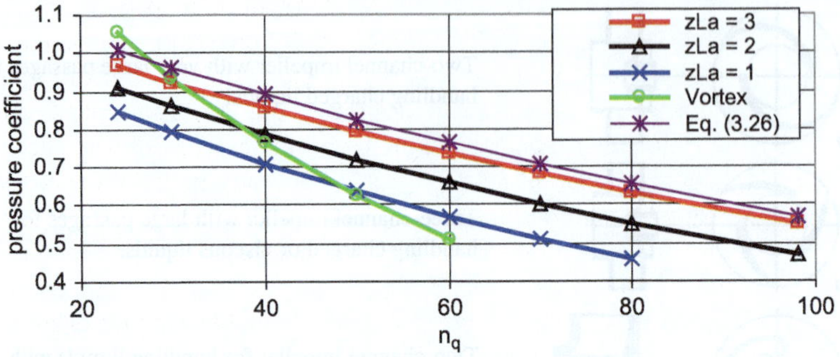

Fig. 7.18 Pressure coefficients for sewage pumps

outlet width of sewage impellers with 2 to 4 blades; single-channel impellers of small pump sizes are often larger. The data in Fig. 7.19 are representative of b_2/d_2 because of the condition $b_2 \geq d_k$. The large relative outlet width tends to yield high head coefficients, but because of the low blade numbers the slip factor γ drops, reducing the head (Chap. 3.3, Tables 3.2 and 3.3). Typical slip factors are given in Table 7.2; they depend on b_2/d_2, d_1/d_2, ε_{sch} and β_{2B}. The BEP head coefficients which can be achieved are shown in Fig. 7.18 and Table 7.2 which also provides head coefficients at shut-off.

The flow through pumps with single-channel impellers which are designed for very large ball passage is highly 3-dimensional and unsteady, see Chap. 9.3.9. Flow rate, pressure rise and torque are fluctuating strongly over one impeller revolution. Rough running and impairment in efficiency are the consequence. To appreciate the impact of the ball passage on efficiency, suppose that the ball passage is selected to the criterion $d_k = d_d$ (d_d = diameter of the discharge nozzle). As a consequence, the impeller inlet diameter d_1, the outlet width b_2 and the volute or casing throat area d_{3q} must at least be equal to the discharge nozzle diameter: $d_{1a} = b_2 = d_{3q} = d_d$. The impact of selecting $d_k = d_d$ on the efficiency is threefold: (1) The large impeller outlet width causes a strongly non-uniform flow distribution and mixing losses in the casing; (2) The axial velocity at the impeller inlet $c_{ax} = 4Q/(\pi d_{1a}^2)$ is larger than the radial velocity at the blade leading edge $c_{rad} = Q/(\pi d_{1m} b_1)$. With $d_{1m} = (d_{1a} + d_{1i})/2$ the deceleration of the flow from c_{ax} to c_{rad} becomes: $c_{rad}/c_{ax} = d_{1a}/\{2b_1(1 + d_{1i}/d_{1a})\}$. With $d_{1a} = b_1 = b_2 = d_k = d_d$ we get $c_{rad}/c_{ax} = 0.5/(1 + d_{1i}/d_{1a})$ which yields $c_{rad}/c_{ax} = 0.25$ for $d_{1i}/d_{1a} = 1$ or $c_{rad}/c_{ax} = 0.33$ for $d_{1i}/d_{1a} = 0.5$. This strong deceleration at the impeller inlet, together with the sudden flow turn from the axial to the radial direction leads to flow separation and losses. (3) The requirement $d_{3q} = d_d$ implies that there is no flow deceleration in the diffuser; the associated sudden deceleration from the impeller outlet velocity c_2 to the throat velocity $c_{3q} = c_d$ could cause additional losses depending on the casing design. On the other hand, the flow deceleration in the volute is quite efficient, but the BEP may be shifted to higher flows due to the large throat area.

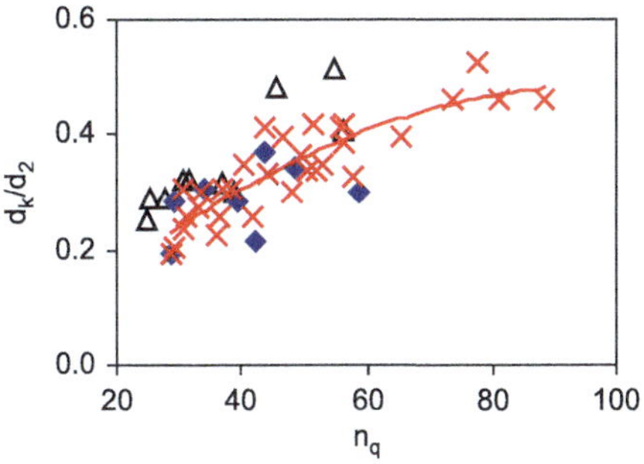

Fig. 7.19 Ball passage diameter d_k referred to impeller diameter d_2 for single-channel impellers from three manufacturers

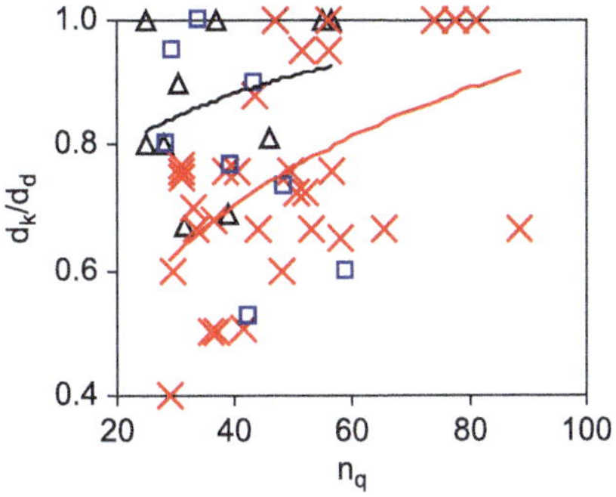

Fig. 7.20 Ball passage diameter d_k referred to discharge nozzle diameter d_d for pumps with single-channel impellers

Efficiency impairment with 2- and 3-bladed impellers is less severe than in single-channel impellers, see data in Table 7.2. Essentially it is due to the non-uniform flow in the impeller which is induced by the large impeller outlet width and the large pitch of the blades.

Studies on optimizing single-channel impellers were reported in [60, 64]. In [60] the impeller blade wrap angle was varied between 289° and 420°; the optimum efficiency (64%) was found at 370° while it dropped to 50% at a wrap angle of 289° and was $\eta_{opt} = 61\%$ at 420°.

Open versus closed impellers: A parametric study of a pump with open single-channel impellers has been done in [8]; the main data were $n_q = 45$, $\beta_{1B} = 5°$, $\beta_{2B} = 30°$, $\varepsilon_{sch} = 325°$. When increasing the axial gap width s between the casing and the blade of the open impeller from $s/b_2 = 0.005$ to 0.053, the BEP flow rate shifted to lower flow by about 30%, the head coefficient dropped by 12% and the efficiency diminished from 52 to 39%.

A closed impeller performed better than an open impeller with a gap width of $s/b_2 = 0.0057$: the BEP shifted slightly to higher flow; the shutoff head was not affected, but the difference in head increased linearly with the flow, attaining 35% at $q^* = 1.5$. The efficiency at BEP increased by 1%, but at $q^* = 1.5$ the efficiency of the closed impeller was 44% as opposed to the open impeller with only 35%.

According to a comparison of efficiency curves published in [49] closed and open impellers achieved the same efficiency at BEP, but the open impellers had

Table 7.2 Sewage pump design data		$n_q^* = n_q/n_{qRef}$ with $n_{qRef} = 1$			
Efficiency of water pumps according to Table 3.9 $Q_{Ref} = 1\ m^3/s$		$m = 0.1\,a\left(\frac{Q_{Ref}}{Q}\right)^{0.15}\left(\frac{45}{n_q}\right)^{0.06}$		$Q \le 1\ m^3/s$	$a = 1$
				$Q > 1\ m^3/s$	$a = 0.5$
		$\eta_{opt,R} = 1 - 0.095\left(\frac{Q_{Ref}}{Q}\right)^m - 0.3\left\{0.35 - \log\frac{n_q}{23}\right\}^2\left(\frac{Q_{Ref}}{Q}\right)^{0.05}$			
Sewage pump efficiency		$\eta_{opt} = \eta_{opt,R} f_\eta$			
Pump type	Efficiency multiplier f_η		Blade wrap	Ball passage	
	Average	Range	ε_{sch} [°]	d_k/d_2	
$z_{La} = 1$	0.82	0.65 to 0.95	280 to 350	0.2 to 0.5 (Fig. 7.19)	
$z_{La} = 2$	0.91	0.8 to 0.98	220 to 270	$d_k/d_2 = 0.13\text{Ln}(n_q^*) - 0.26$	
$z_{La} = 3$	0.94	0.9 to 0.99	170 to 220	$d_k/d_2 = 0.08\text{Ln}(n_q^*) - 0.11$	
Vortex	0.6	0.45 to 0.8	$f_\eta = (0.7\text{ to }0.8)\left(\frac{Q_{opt}}{Q_{Ref}}\right)^{0.06}$ $Q_{Ref} = 1\ m^3/s$		
Pressure coefficient at shut-off		Vortex	$\psi_0 = 1.38 - 0.0112\ n_q/n_{qRef}$		
		$z_{La} = 1$	$\psi_0 = 1.45 - 0.0046\ n_q/n_{qRef}$		
		$z_{La} = 2$	$\psi_0 = 1.36 - 0.0035\ n_q/n_{qRef}$		
		$z_{La} = 3$	$\psi_0 = 1.39 - 0.002\ n_q/n_{qRef}$		
Pressure coefficient at BEP		Vortex	$\psi_{opt} = 1.7\exp(-0.02\ n_q/n_{qRef})$		
		$z_{La} = 1$	$\psi_{opt} = 1.03\exp(-0.0095\ n_q/n_{qRef})$		
		$z_{La} = 2$	$\psi_{opt} = 1.1\exp(-0.0087\ n_q/n_{qRef})$		
		$z_{La} = 3$	$\psi_{opt} = 1.17\exp(-0.008\ n_q/n_{qRef})$		
Typical slip factors for $n_q < 60$ from Eq. (T3.2.9)		$z_{La} = 1$	$\gamma_{opt} = 0.48$ to 0.6		
		$z_{La} = 2$	$\gamma_{opt} = 0.53$ to 0.65		
		$z_{La} = 3$	$\gamma_{opt} = 0.67$ to 0.75		
Impeller tip speed		Vortex	$u_{2,max}$ [m/s] $= 45 - 0.42\ n_q/n_{qRef}$		
		$z_{La} = 1$	$u_{2,max}$ [m/s] $= 42 - 0.32\ n_q/n_{qRef}$ up to $n_q = 60$		
Minimum to reduce risk of clogging: $u_{2,min} = 12$ m/s		$z_{La} = 2$	$u_{2,max}$ [m/s] $= 38 - 0.15\ n_q/n_{qRef}$		
		$z_{La} = 3$	$u_{2,max}$ [m/s] $= 42 - 0.15\ n_q/n_{qRef}$		

much lower efficiencies than the closed designs at off-design conditions. This finding applies to single-channel and two-channel impellers. This difference in efficiency reached roughly 10% at q* = 0.5 and at q* = 1.5.

The ratio of the axial thrust of the closed impeller to that of the open impeller was about 0.5 at shutoff and 0.3 near BEP. Overall, closed impellers give better performance and lower axial thrust than open impellers.

Impeller tip speed: The impeller tip speed of single-channel impellers must be limited in order to keep the excitation forces due to the hydraulic unbalance

(Chap. 9.3.9) within manageable limits. Values for the maximum tip speeds are given in Table 7.2. A minimum tip speed of about 12 m/s should be considered to avoid deposits in the volute.

Since the blades are not well adapted to the flow, vortex pumps are prone to cavitation noise which increases with the flow rate. Therefore, the tip speed of vortex pumps must also be limited as stipulated in Table 7.2.

Blade design: The blade thickness for impellers with 1, 2 or 3 channels can be estimated from Eq. (T14.3.1) in Chap. 14.1. At the inlet, the blades of impellers with 2 or 3 blades must be thickened to approximately $e_{max}/d_2 = 0.05$ and well profiled to prevent textiles or plastic foils from getting caught on the leading edges. For single-channel impellers the leading edge thickness is sometimes higher: in [60] $e_{max}/d_2 = 0.1$ was used. An even thicker leading edge profiled on the pressure side was found to reduce the risk of clogging by long fibers (up to 300 mm), [64]. The blade angles at the impeller inlet are chosen relatively small to improve the partload and noise characteristics. The range of wrap angles of the blades is given in Table 7.2.

Impact of hydraulic design on clogging: Open impellers are proposed by some manufactures as being less prone to clogging then closed impellers. This may be true if the blade is designed that way that stringy matter tends to move towards the outer streamline where it is caught in the clearance between wear plate and blades. There, the textile matter can be chopped into smaller pieces if a cutting groove is provided in the wear plate.

How the blades should be designed in order to reduce the risk of clogging is somewhat speculative. Injecting a sufficient load of rags, every pump will clog. On the other hand, unfavorable (highly loaded) blade designs will attract any kind of stringy matter, while lightly loaded blades are likely to remain clean if the concentration of rags is moderate.

An attempt to calculate the forces acting on a rag appears to be questionable since these forces cannot be quantified with sufficient confidence. Thinking of a piece of a plastic bag sticking on the leading edge of the blade, one would have to integrate the pressure distribution on the blade pressing the rag onto the blade surface. Together with the surface properties of blade and rag, the shear stresses between rag and blade and between the fluid and the rag must be determined. A piece of plastic bag on a smooth surface tends to stick there also because of capillary forces.

Qualitatively, the following tendencies could be postulated:

1. Blades which are highly loaded near the leading edge (i.e. large inlet angles) are apt to attract rags.
2. Large inlet angles with the stagnation point on the pressure surface lead to flow separation on the suction surface (as well as cavitation). Trash or rags are likely to be caught in a zone of stalled fluid (think of a dead-water zone in a river).

3. In general terms, the flow in an impeller is decelerated on the pressure surface of the blade. The resulting reduced velocities imply lower shear stresses acting on any rag wrapped around the leading edge. As a consequence, the flow exerts lower forces on the rag.
4. In contrast, the flow is accelerated on the suction surface. Higher velocities, creating higher drag, are likely to better be able to carry away a rag. However, one can always imagine a situation where the forces acting on the rag are the same on the pressure and the suction surface that way that equilibrium is established. In this sense, a probabilistic approach would be needed. The rag may eventually be swept away be fluctuations in the approach flow.
5. Low inlet angles, i.e. blades lightly loaded near the leading edge with the stagnation point on the suction surface, tend to eliminate the flow separation and cavitation on the suction surface.
6. A strong sweep of the blade introduces radial flow components which might be able to shift rags along the leading edge. Once the rag is set in motion, it is more likely to be swept away from the leading edge. A high ratio d_{1a}/d_{1i} at the radial leading edge is apt to induce stronger secondary flows which transport rags radially outwards.
7. A thick leading edge with a short elliptical profile should be used. The profile length may be L_p = (1 to 1.5) e_{max} if e_{max} is the maximum blade thickness.
8. An oval-shaped leading edge would have the effect of a sweep back on the outer streamline as well as on the hub. Both outer and inner streamline are unloaded that way. Fluids (hence rags) are displaced towards the mid streamline. The mid streamline is then to be designed with a low inlet angle - hence a stagnation point on the suction surface - in order to sweep the rags away.

In summary, it is suggested tentatively that the blade should be designed according to the features listed below in order to reduce the risk of clogging by fibrous matter or rags - even though hard experimental proof is difficult to obtain.

- Low inlet angles, stagnation point on the suction surface. The blade development becomes S-shaped in order to achieve the required ball passage.
- Introduce radial movement of the flow by sweeping the leading edge
- Secondary flow is apt to set rags radially in motion. A lean of the blade may have similar effects as the sweep.
- Small blade diameters d_{1i} at the inner streamline or a high ratio of d_{1a}/d_{1i} are apt to create radial flow components which can set a rag in motion.
- An oval-shaped leading edge appears worth a try. Some manufacturers seem to have this feature. Near the mid-streamline very low inlet angles may be developed to prevent rags from getting caught there.
- The leading edge profile should be thick and short.

Figure 7.21 shows a sewage pump designed with a 3-bladed impeller. The rear shroud includes expeller blades in order to reduce the axial thrust and to keep the impeller sidewall gap free of coarse solids. Wear plates are installed adjacent to the front and rear shrouds to prevent casing wear.

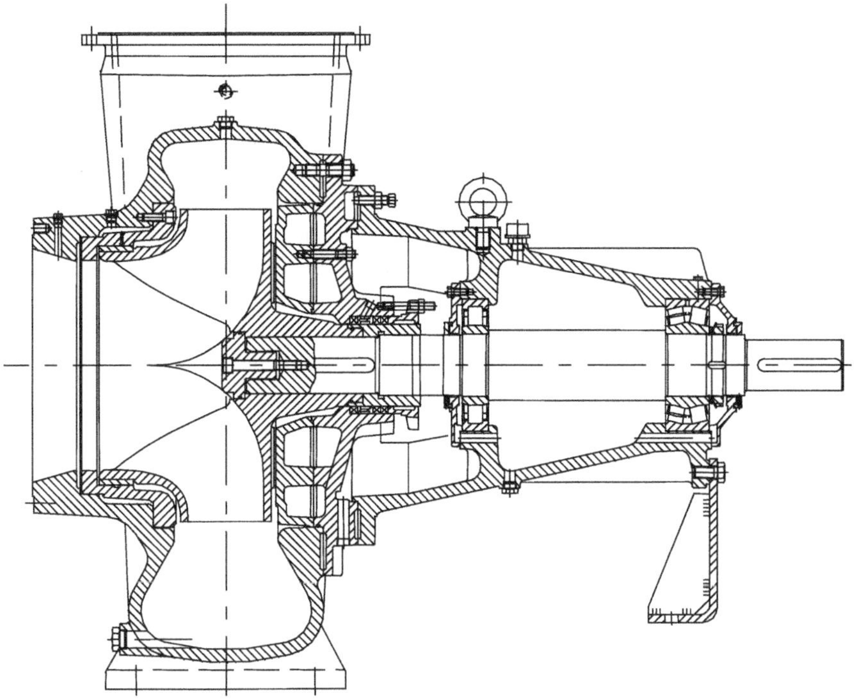

Fig. 7.21 Sewage pump $n_q = 71$; $Q_{opt} = 1.1\ m^3/s$, $H_{opt} = 18\ m$ at n = 590 rpm, Sulzer Pumps

7.5 Semi-axial Impellers

As discussed in Chap. 2, semi-axial pumps are designed with specific speeds of about $n_q = 30$ to 200. This wide range is given rather by design than by hydraulic requirements. Hydraulically, the semi-axial design is at an optimum between about $n_q = 40$ and $n_q = 160$. With regard to the shapes of the impeller, the transition from radial to semi-axial and axial pumps is continuous (Fig. 3.8). Semi-axial impellers with specific speeds below $n_q = 50$ differ little from radial pumps in their hydraulic characteristics; both are therefore designed according to the same principles using slip factors.

Assuming an optimum design, semi-axial pumps are superior to pumps with radial impellers above $n_q \approx 60$ since the discharge flow from radial impellers becomes more non-uniform with increasing specific speed due to the flow deflection in the meridional section and growing blade span. At specific speeds above $n_q \approx 150$, semi-axial pumps have characteristics similar to propeller pumps and the calculation according to the slip factor concept becomes uncertain due to the widely spaced blades.

Large cooling water pumps (CWP) in power stations or chemical process plants are one important application of semi-axial hydraulics. The rotor speed of CWPs must be selected on the available NPSH and on mechanical considerations. The $NPSH_A$ of CWPs is given by the atmospheric pressure and the submergence, resulting typically in $NPSH_A$ = 12 to 15 m (unless the pumps are installed at higher elevations). The required $NPSH_3$ at maximum flow must be sufficiently below the $NPSH_A$. Considering this requirement and suction specific speeds in the range of n_{ss} = 220 to 260 at BEP, the maximum rotational speed of the pump can be estimated from Eq. (7.11a). This gives a first estimate; the actual speed selected then depends on the standard motor speed, head, specific speed, available models and the specific requirements of the project. The maximum head can be checked against Eq. (7.11b). Equations (7.11a) and (7.11b) represent an upper band of CWPs in service, but no absolute limit. These limits can be exceeded by providing a very robust design with suitable blade thicknesses, adequate separation margins to eigen frequencies, suitable line-shaft bearings and mechanical components.

$$n = n_{Ref}\left(\frac{Q_{Ref}}{Q_{opt}}\right)^{0.45} \quad \text{with } Q_{Ref} = 10{,}000\,\text{m}^3/\text{h and } n_{Ref} = 650\,\text{rpm} \qquad (7.11a)$$

The maximum head for this type of application can be assessed from:

$$H_{opt} = H_{Ref}\left(\frac{n_{q,Ref}}{n_q}\right)^{1.1} \quad \text{with } H_{Ref} = 27\,\text{m and } n_{q,Ref} = 100 \qquad (7.11b)$$

Semi-axial impellers are basically designed according to Chap. 7.2; however, a number of special features must be considered which result from the radius ratio d_{2a}/d_{2i} at the impeller outlet. The geometrical features of a stage are shown in Figs. 7.22 and 7.4.

1. The head coefficient $\psi = \psi_{2a}$ is selected from Fig. 3.24 and the blade diameter d_{2a} on the outer streamline calculated from Eq. (T7.1.3).
2. The shape of the impeller in the meridional section is strongly influenced by the angle ε_s formed by the outer streamline with a normal to the rotor axis. This angle increases with the specific speed approximately according to Eq. (7.12); for axial impellers it becomes 90° by definition.

$$\varepsilon_s = 90^\circ\left(\frac{n_q}{n_{q,Ref}}\right)^{0.74} \quad n_{q,Ref} = 200 \qquad (7.12)$$

Executed pumps are in the range of ±5° of the angles calculated from Eq. (7.12), [B.2, B.18, 25].

3. The mean impeller outlet diameter can be written as:

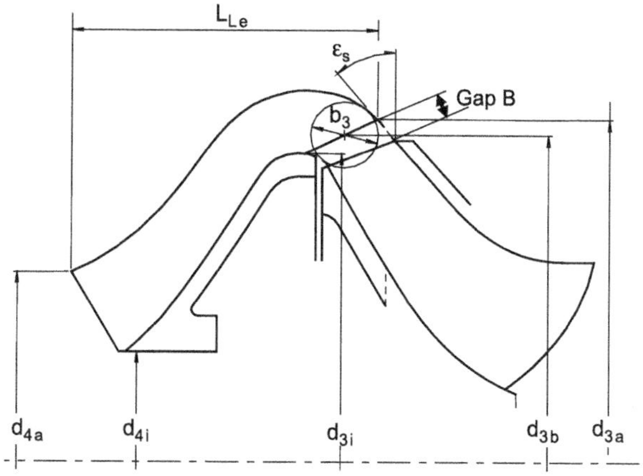

Fig. 7.22 Some design parameters of a semi-axial stage

$$\frac{d_{2m}}{d_{2a}} = e^{-0.04\left(\frac{n_q}{n_{q,Ref}}-1\right)} \quad n_{q,Ref} = 30 \tag{7.13}$$

The diameter of the inner streamline is obtained from:

$$\frac{d_{2i}}{d_{2a}} = \sqrt{2\left(\frac{d_{2m}}{d_{2a}}\right)^2 - 1} \tag{7.14}$$

Equations (7.13 and 7.14) produce a ratio $d_{2i}/d_{2a} = 0.52$ for $n_q = 200$, which, according to Fig. 7.24, quite accurately corresponds to the hub ratio of an axial pump of this specific speed. The empirical correlation provided by Eq. (7.13) consequently produces a smooth transition from semi-axial to axial impellers.

4. As in Sect. 7.2.1 and Table 7.1 the impeller inlet is calculated according to the required suction specific speed. The blade inlet angles are generally in a range of 12 to 18°. With specific speeds above $n_q = 150$ the explanations in Chap. 7.6 concerning axial pumps must also be observed. According to [B.18] the flow coefficient in the design point can be selected as follows:

$$\varphi_1 = (0.18\,\text{to}\,0.27)\left(\frac{n_{q,Ref}}{n_q}\right)^{0.3} \quad n_{q,Ref} = 200 \tag{7.15}$$

The upper limit must be selected for optimum efficiency and the lower limit for high suction specific speeds.

5. Number of blades: For specific speeds below $n_q \approx 140$ the criteria given in Chap. 7.2 apply. In the lower n_q-range, 7 blades are a good choice; nonetheless, 5 to 7 blades can be selected throughout the range $20 < n_q < 140$. Table D7.3

(Chap. 7.6.2) gives an indication for the blade number for $n_q > 140$. However, the solidity L/t is the more important criterion. The solidity required at the outer streamline can be estimated from Eq. (7.16) according to [B.18].

$$\left(\frac{L}{t}\right)_a = 0.64\left(\frac{n_{q,Ref}}{n_q}\right)^{0.74} \quad n_{q,Ref} = 200 \tag{7.16}$$

As further indication of the required blade length on the outer streamline the following relationship can be derived from data given in [B.18]:

$$\left(\frac{L}{d_2}\right)_a \geq 1.1\left(\frac{\beta_{2a}{}^{\circ}}{25^{\circ}}\right)\left(\frac{n_q}{n_{q,Ref}}\right)^{0.4} \quad n_{q,Ref} = 200 \tag{7.17}$$

The blade length resulting from Eq. (7.17) is considered the minimum necessary to forestall an instability of the Q-H-curve.

6. The impeller outlet width can be determined according to Fig. 7.2.
7. The impeller outlet angle on the outer streamline $\beta_{2B,a}$ is determined as in Chap. 7.2. The following information for the selection of the blade angles is given in [B.18]:

   - The highest efficiencies are obtained when the blade outlet angle on the mean streamline is selected as $\beta_{2B,m} = 20$ to 26°.
   - On the streamline with the diameters d_{1b} and d_{2b}, inlet and outlet angles should not deviate by more than ±2°. In the blade development the mean streamline will then almost result in a straight line.
   - In addition, $\beta_{1B,a} < \beta_{2B,a}$ should be executed at the outer streamline and $\beta_{1B,i} > \beta_{2B,i}$ at the inner streamline.

To determine the distribution of the impeller blade outlet angle over the blade height an assumption has to be made with regard to the energy transfer on the various streamlines. In principle, this energy distribution can be selected arbitrarily. However, to obtain reproducible results the designer usually selects a specific distribution of the angular momentum on the impeller outlet according to: $u \times c_u = f(r)$ or $c_u = c_{u,a}(r_a/r)^m$.

As per Table D7.2 three possibilities are compared in the following:

A. In the free vortex design with m = 1 (that is $u \times c_u = u_a \times c_{u,a}$ = constant) the same blade work is transmitted theoretically on each streamline. The circumferential component of the absolute velocity grows from the blade tip to the hub according to: $c_u(r) = c_{u,a} \times r_a/r$.
B. When selecting the exponent m = 0, the absolute velocity over the blade height becomes constant: $c_u(r) = c_{u,a}$.
C. The fluid rotates with constant angular velocity like a solid body if m = −1 is chosen: $c_u(r) = c_{u,a} \times r/r_a$.

Table D7.2 Swirl distribution at impeller outlet

		c_{2m} = constant $\alpha_1 = 90^{\circ}$		The figures at the sketches are valid for ν = 0.5			
		Angular momentum constant: m = 1.0		**Constant swirl velocity: m = 0**		**Solid body rotation m = -1.0**	
Circumferential component of absolute velocity	$\frac{c_u}{c_{u\,a}} = \left(\frac{r_a}{r}\right)^m$	$\frac{r_a}{r}$	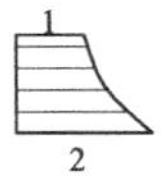	1.0	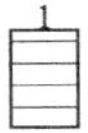	$\frac{r}{r_a}$	
Theoretical head	$\frac{H_{th}}{H_{th,a}} = \frac{r\,c_u}{r_a c_{ua}}$	1.0		$\frac{r}{r_a}$		$\frac{r^2}{r_a^2}$	
Static pressure	$\frac{p_a - p}{\frac{\rho}{2} c_{u,a}^2}$	$\frac{r_a^2}{r^2} - 1$	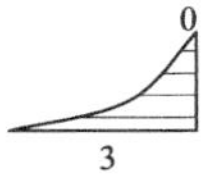	$2 \ln \frac{r_a}{r}$		$1 - \frac{r^2}{r_a^2}$	
Total pressure	$\frac{p_{tot,a} - p_{tot}}{\frac{\rho}{2} c_{u,a}^2} = \frac{p_a - p}{\frac{\rho}{2} c_{u,a}^2} + 1 - \frac{c_u^2}{c_{u,a}^2}$	0				$2\left(1 - \frac{r^2}{r_a^2}\right)$	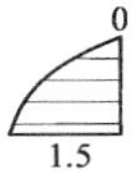
Absolute flow angle at impeller outlet	$\frac{\tan\alpha}{\tan\alpha_a} = \frac{c_{u,a}}{c_u}$	$\frac{r}{r_a}$		1.0	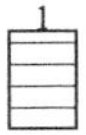	$\frac{r_a}{r}$	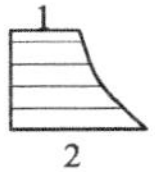
Relative flow angle at impeller outlet	$\frac{\tan\beta}{\tan\beta_a} = \frac{1 - \frac{\psi_{th}}{2}}{\frac{r}{r_a} - \frac{c_u}{c_{u,a}} \frac{\psi_{th}}{2}}$	$\frac{1 - \frac{\psi_{th}}{2}}{\frac{r}{r_a} - \frac{r_a}{r} \frac{\psi_{th}}{2}}$	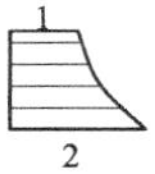	$\frac{1 - \frac{\psi_{th}}{2}}{\frac{r}{r_a} - \frac{\psi_{th}}{2}}$			

These three options for allocating the angular momentum over the blade height at the impeller outlet yield different distributions of the specific work, the static pressure, the total pressure and the flow angles in the relative and absolute reference frames. The corresponding equations, which were derived assuming constant meridional velocity and swirl-free approach flow, can be found in Table D7.2.

If the blading is designed according to the law of the free vortex, the radial equilibrium is inherently satisfied. This implies that the axial velocity (apart from boundary layer effects) remains roughly uniform since no forces normal to the main flow direction cause a re-distribution of the fluid. However, the strong flow deflection on the hub requires a correspondingly high blade twisting. The hub diameter must be selected sufficiently large in order to limit the twist and avoid flow separation, see Chap. 7.6. The static pressure drops significantly from the tip to the hub while the total pressure over the blade height remains constant. At the hub, kinetic energy is at its maximum and exceeds the value on the outer streamline by the factor $(r_a/r_i)^2$. The diffuser approach flow angle is smaller on the hub than on the outer streamline. Consequently, the flow deflection to be achieved in the diffuser is greater on the hub than on the tip, which is likely to cause flow separation near the hub. Furthermore, a large amount of kinetic energy has to be converted into static pressure on the hub to avoid losses due to secondary flows and turbulence. Due to these unfavorable flow distributions, the diffuser is an altogether critical element in free vortex designs since the diffuser vanes are subjected to extreme loading at the hub. The recoverable kinetic energy at the diffuser inlet diminishes with rising specific speed because the ratio of the circumferential velocity c_{2u} to the axial velocity c_{2m} decreases.

If the impeller is designed for constant absolute outlet velocity, the blade work diminishes from the outer to the inner streamline in proportion to the radius. Radial equilibrium is no longer ensured so that radial flows are induced and a constant meridional velocity cannot be expected. Static pressure and total pressure decrease slightly towards the hub, but to a significantly lesser degree than with free vortex blading.

When the blading is designed according to the forced vortex, the radial equilibrium is disturbed most severely. The twist of the impeller blades is minimized in this design. Energy transfer and total pressure rise on the hub are lowest. According to [B.18], the blading designed for the forced vortex has advantages when trimming the impeller, since the loss in efficiency is reduced.

As extensively discussed in Chap. 5, stability of the Q-H-curve, power consumption and head at shut-off operation greatly depend on the meridional section and the position of the leading and trailing edges. The explanations in Chap. 5.6 should therefore be observed when selecting these design parameters. It is important to manufacture the impellers of high specific speeds with good surface finish and to profile the trailing edges on the suction surface, [B.18].

7.6 Axial Impellers and Diffusers

7.6.1 Features

Axial (or "propeller") pumps are used for large flow rates with low heads. Typical applications are irrigation, sewage treatment plants, cooling water pumps and drainage systems. As a rule, these pumps draw the water from an open sump. Vertical pumps are used in wet pit and dry pit installations. Smaller units are also equipped with submersible motors.

Since suction is taken from surface waters, the atmospheric pressure and the necessary submergence (Chap. 11.7.3) determine the available $NPSH_A$ which amounts to 10 to 14 m in the majority of applications. This circumstance has a significant effect on pump selection and impeller design.

The blades of high-quality propeller pumps are designed as slim airfoils. They are individually attached to the hub unless a cast impeller is used which is the case for pumps of small or moderate size. For mechanical reasons and because of the given $NPSH_A$, the impeller tip speeds are limited to about u_2 = 25 to 28 m/s (apart from special designs).

Axial pumps are mainly employed in the range above $n_q \approx 150$. Although axial pumps can be designed for lower specific speeds, semi-axial pumps are superior to axial pumps in the range up to approximately $n_q = 170$ unless the plant layout or other design reasons call for a purely axial design. The majority of axial pump applications is in the range $180 < n_q < 300$. These are single-stage machines with axial approach flow to the suction bell.

As a result of flow separation due to excessive deceleration and incidence, the Q-H-curves of propeller pumps, as a rule, have a pronounced saddle-shaped instability which restricts the allowable operating range (see Figs. 4.10 and 5.25, also Chap. 11).

Since the impeller inlet and outlet diameters of an axial pump are identical by definition, the energy is only transmitted through deflection of the absolute flow and deceleration of the relative velocity, see definition of Y_{sch} in Eq. (T3.3.2). During operation near the best efficiency point, there is no centrifugal effect. As discussed in Chap. 5, centrifugal forces become effective only with pronounced partload recirculation. The achievable head coefficients are therefore low, Fig. 3.24. Accordingly, the deflection of the flow is minor. Consider a propeller pump of the specific speed $n_q = 230$ with axial approach flow, a head coefficient of $\psi_{opt} = 0.22$ and a hydraulic efficiency of 88% as an example. The required flow deflection is calculated from Eqs. (T3.3.9), (T3.2.11 and T3.2.15) as follows:

$$\Delta\beta = \beta_2 - \beta_1 = \arctan\frac{\tan\beta_1}{1 - \frac{\psi}{2\eta_h}} - \beta_1 \tag{7.18}$$

Assuming $\beta_1 = 12°$, flow turning is only $\Delta\beta = 1.7°$ on the outer streamline in the present example. As a consequence, axial pumps react very sensitively to minor errors in the blade angles which would hardly be noticed with a radial impeller. It is therefore useful to design the blade fixation on the hub so that the blade position can be modified if the required performance data are not achieved. This design also allows performance adjustments when operating requirements change.

Because of the sensitivity to minor geometrical deviations and since the blades do not form actual channels due to a lack of overlap, calculation methods according to the slip factor concept are unsuitable for axial impellers. For this reason, propellers are designed according to the airfoil theory unless numerical methods are employed. CFD is particularly suitable for this application insofar as strong centrifugal and Coriolis forces are acting normal to the main flow direction. The blade height is large compared with the flow path length through the impeller; hence streamline theory has its limitations. Nevertheless, the streamline theory is useful to make the first design before a numerical optimization can be started.

As mentioned above, cavitation limits the impeller tip speed, hence the rotor speed for a given set of performance data. In order to reliably predict the cavitation characteristics in the design stage, the pressure distributions on the blades should be calculated. This calls for numerical methods if high-quality pumps are to be designed. Not only Navier-Stokes programs but also simpler - and considerably faster - methods (e.g. singularity methods) can be employed for this type of calculation. With the frequently used free vortex blading, simple methods may be used with success since the radial equilibrium is satisfied. If the blades are designed for different vortex distributions, 3D-programs should be considered, Chap. 7.5.

7.6.2 Calculation and Selection of Main Dimensions

In contrast to radial or semi-axial impellers, the conditions at the impeller inlet and outlet of propeller pumps cannot be treated independently. Even the speed is not necessarily needed when specifying the pumping task. In the following we consider therefore the general case where only flow rate Q_{opt}, head H_{opt} and the suction conditions in the system (i.e. $NPSH_A$) are specified. Since the head is low, losses in inlet, column pipe, outlet elbow and valves can take a considerable fraction of the head. Such additional losses should therefore be considered when establishing the design head to be developed by the impeller since they have an effect on the optimization of the hydraulics, [25].

The first task of the design is to determine the speed n, the impeller outer diameter d_2, the hub ratio $\nu = d_n/d_{1a}$ (or hub diameter d_n), the specific speed n_q, the head coefficient ψ, the flow coefficient φ and the cavitation coefficient σ. These seven unknowns must be selected such that the specified head is achieved and the pump can be safely operated with the specified $NPSH_A$ value. Seven equations according to Table 7.3 are available to determine these quantities.[3]

Explanations to Table 7.3

- Axial approach flow with $\alpha_1 = 90°$ and blading according to $c_u \times r$ = constant are assumed (Chap. 7.5).
- The hydraulic efficiency (e.g. from Table 3.9) must be assumed for the initial calculation and verified after designing the blading.
- Corresponding to the specified $NPSH_A$ a cavitation criterion or the required $NPSH_R$ of the pump must be chosen. To this end, a safety margin F_{NPSH} according to Table 6.2 can be determined to calculate $NPSH_R = NPSH_A/F_{NPSH}$. An alternative criterion would be to assume $NPSH_A = NPSH_{(\Delta\eta=0)}$. The pump would then operate with that extent of cavitation which is just below the level where efficiency would begin to drop. This assumption allows to calculate $NPSH_{(\Delta\eta=0)}$ from Eq. (T7.3.6) which originates from [28]. However, no information is available about the scatter of the test data around this correlation.
- Equation (T7.3.2) gives an indication of the maximum head coefficients to be selected in practice. With $f_\psi = 1$ it approximately constitutes the upper limit of the curve given for ψ_{opt} in Fig. 3.24 which reflects the data of various publications (among these [B.22, 25]). If cavitation must be optimized, larger impeller diameters are required; the head coefficients and the constant f_ψ must be reduced accordingly (e.g. $f_\psi = 0.9$). The head coefficients given by Eq. (T7.3.2) can be exceeded in cases where a targeted high head takes precedence over cavitation and partload characteristics.
- Hub ratio: Like the relative outlet width b_2^* of radial impellers, the ratio $\nu = d_n/d_2$ is a most important design parameter which has an effect on the efficiency and the blade twist at the impeller inlet and outlet. Various criteria can be utilized here: (1) The blade outlet angle on the hub $\beta_{2B,i}$ should be less than 90°. (2) The deceleration ratio of the relative velocity on the hub should not be below an allowable value, i.e. $w_{2i}/w_{1i} \geq 0.6$. (3) In a swirling flow, a core of stalled fluid is formed when the diameter ratio drops below a specific limit. The criterion for core stall is $c_m/c_{2u,i} \geq 1$. For blading according to $c_u \times r$ = constant this leads to the condition stated by Eq. (T7.3.5) which is used subsequently. (4) The hub diameter can also be calculated according to the values for ψ_i given in Fig. 3.24. For the blading according to $c_u \times r$ = constant this produces a hub ratio of $\nu = (\psi_a/\psi_i)^{0.5}$. The attainable head coefficient is limited by the hub ratio. Higher head coefficients can also be achieved with axial impellers when the hub ratio ν is increased.

---

[3]One could also formulate an equation $d_2 = f(n)$ by inserting Eqs. (T7.3.1 and T7.3.2) into Eq. (T7.3.3). Further equations for $\nu = f(n)$ and $NPSH = f(n, d_2)$ could be derived to obtain three more complicated equations to determine the three quantities n, d_2 and ν.

- The flow coefficient is calculated from Eq. (T7.3.4), if Eq. (T7.3.5) is substituted for ν in Eq. (T7.3.1) and the resulting quadratic equation is solved for φ.
- Instead of calculating $NPSH_{(\Delta\eta=0)}$ from Eq. (T7.3.6a), it is possible to use the coefficients λ_c and λ_w according to Eq. (6.10) and Eq. (T7.3.6.b). Following this approach, the λ-values have to be selected in accordance with the selected cavitation criterion. The dependency of the λ-values on the flow coefficient and the geometry should also be considered. It may be derived from test data gathered on similar pumps and impeller designs.

The equations of Table 7.3 supply a single solution for given values of Q_{opt}, H_{opt} and $NPSH_A$, which can be found iteratively. However, a more direct and obvious method than the iteration is to treat the speed as an independent variable and to plot the $NPSH_R$, the circumferential speed and the approach flow angle against the rotor speed. If the speed range is selected sufficiently large, a minimum value in the curve $NPSH_R = f(n)$ will be found as shown in Fig. 7.23. Entering the diagram at the given $NPSH_R$ value (e.g. at $NPSH_A/F_{NPSH}$), one obtains exactly the speed which is required to accomplish the specified pumping task. At the same time, the seven equations are satisfied and the selected NPSH-margin is maintained. When the speed has been determined in this manner, all other quantities (d_2, ν, φ, ψ) can be calculated from Table 7.3.

If an electric motor is employed as a direct drive, the standard speeds must be used according to the frequency f of the electrical power supply. The synchronous speed is calculated with the number of pole pairs p = 1, 2, 3, from Eq. (7.19).

$$n_{syn} = \frac{60\,f}{p} \qquad (7.19)$$

The slip of asynchronous motors is 1 to 1.5% of the synchronous speed. The speed of asynchronous motors is thus obtained by multiplying n_{syn} from Eq. (7.19) with a factor between 0.985 and 0.99.

However, additional criteria must be observed when selecting the speed:

- The approach flow angle β_1 must not be selected too large. According to [25] the best cavitation characteristics are obtained at $\beta_1 = 10$ to 12°. The larger the approach flow angle, the higher will be the flow rate at which the Q-H-curve becomes instable. For this reason, a small approach flow angle is also propitious for achieving a wide operating range. In general, the approach flow angle should hardly exceed $\beta_1 = 15°$, but $\beta_1 = 10$ to 12° is recommended when good cavitation characteristics are important. At $n_q < 250$, this requires smaller head coefficients than given by Eq. (T7.3.2) and correspondingly larger pumps. A factor $f_\psi < 1$ must then be employed in optimization calculations according to Table 7.3. It is also possible to select the approach flow angle (or the flow coefficient) and determine the head coefficient iteratively from Eq. (T7.3.8), while maintaining the hub ratio as given by Eq. (T7.3.5).

Table 7.3 Main dimensions of axial impellers. Note: n_q with rpm, m^3/s, m

			Eq.
Assumptions	axial approach flow: $\alpha_1 = 90°$ Constant angular momentum: $c_u \times r = c_{2u,a} \times r_2$		
Given/selected quantities	Q_{opt}, H_{opt}, $NPSH_A$, F_{NPSH}, η_h		
Required $NPSH_R$	$NPSH_R = NPSH_A/F_{NPSH}$	F_{NPSH} acc. Table 6.2	
Desired quantities	n, d_2, ν, n_q, ψ, φ, σ		
Specific speed	$n_q = n\frac{\sqrt{Q_{opt}}}{H_{opt}^{0.75}}$	$n_q = 158\frac{\sqrt{\varphi(1-\nu^2)}}{\psi^{0.75}}$	7.3.1
Pressure coefficient ($f_\psi = 1$ gives about maximum achievable values)	$\psi \equiv \frac{2\,g\,H}{u_{2a}^2} = 0.29\,f_\psi\left(\frac{n_{q,Ref}}{n_q}\right)^{1.44}$	$n_{q,Ref} = 180$	7.3.2
Impeller outer diameter	$d_2 = \frac{60}{\pi\,n}\sqrt{\frac{2\,g\,H}{\psi}} = \frac{84.6}{n}\sqrt{\frac{H}{\psi}}$		7.3.3
Flow coefficient	$\varphi \equiv \frac{c_m}{u_{2a}} = \frac{a}{2} + \sqrt{\frac{a^2}{4} + \left(\frac{\psi}{2\,\eta_h}\right)^2}$	$a = \left(\frac{n_q}{158}\right)^2 \psi^{1.5}$	7.3.4
Hub ratio	$\nu = \frac{\psi}{2\,\eta_h\varphi}$		7.3.5
Cavitation coefficient Note: $\lambda_w = f(\varphi)$	$\sigma_{(\Delta\eta=0)} = 0.14 + 1.14\,\varphi^2 + 2\varphi^3$		7.3.6a
	$\sigma = (\lambda_c + \lambda_w)\,\varphi^2 + \lambda_w$		7.3.6b
Required $NPSH_R$	$NPSH_R = \sigma\frac{u_2^2}{2g}$ (note: $u_2 = u_1$)		7.3.7
Calculation of pressure coefficient for selected flow coefficient	$\psi = \varphi^{\frac{4}{3}}\left\{\varphi\left(\frac{n_q}{158}\right)^2 + \frac{\sqrt{\psi}}{4\,\eta_h^2}\right\}^{-\frac{2}{3}}$		7.3.8
Plot in a graph: $NPSH_R$, u_2 and β_1 = arc tan φ as a function of rotor speed. If pump is driven directly by electro motor: use standardized speeds			

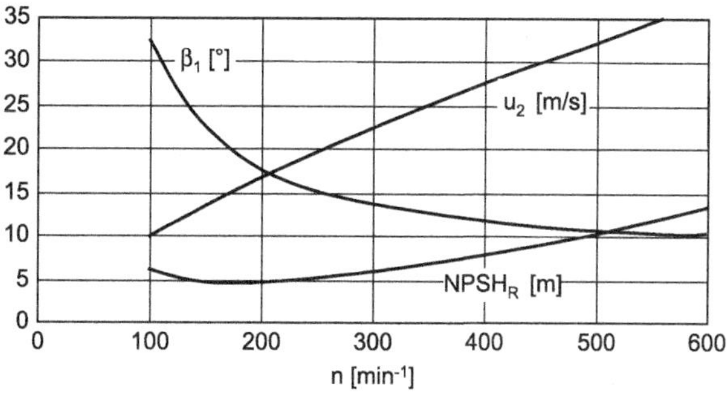

Fig. 7.23 Selection of the speed of an axial pump for a given NPSH

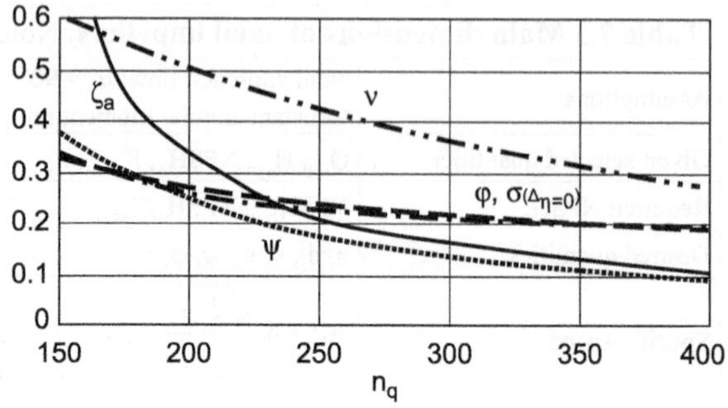

Fig. 7.24 Dimensionless coefficients of axial pumps. (The curves for φ and $\sigma_{(\Delta\eta = 0)}$ virtually coincide)

Table D7.3 Impeller blade numbers for $n_q > 140$

n_q	140 to 170			160 to 230	220 to 290	>290
z_{La}	5	6	7	4	3	2
z_{Le}	7	8	9	7	5	5
	8	11	10	9	7	7
	9	13	11,12	11	10	9

- The hydraulic efficiency drops with increasing specific speed. If a high efficiency is desired, the speed must be selected lower than would be permissible with respect to cavitation in order to avoid too high a specific speed.
- As mentioned above, the impeller tip speed should also be limited for mechanical reasons.

By means of the equations in Table 7.3, the dimensionless parameters ψ, φ, ν and σ can be calculated and plotted as functions of the specific speed. Figure 7.24 shows the characteristics obtained, which are based on the head coefficients given by Eq. (T7.3.2). Figure 7.25 supplies the parameters when a flow coefficient of $\varphi = 0.22$ (or $\beta_1 = 12.4°$) is specified. Optimization calculations in [62] produced similar (at high n_q up to 10% higher) hub ratios as Figs. 7.24 and 7.25.

If a maximum impeller tip speed is assumed, the attainable head and the necessary $NPSH_R$ can also be plotted against n_q as done in Fig. 7.26. These diagrams provide a quick overview of the design options for a specific pumping task (subject to the assumptions made for the calculations). The curves depend especially on the head coefficients selected according to Eq. (T7.3.2).

Impeller blade number z_{La}: The optimum number of blades z_{La} decreases with increasing specific speed as shown by Table D7.3. The optimum blade number is determined by the blade loading so that the final selection has to be verified during the blade design process.

7.6.3 Basic Properties of Airfoils

Airfoil geometry: As mentioned above, the blades of axial pumps (with the possible exception of small inexpensive machines for uncritical service) are

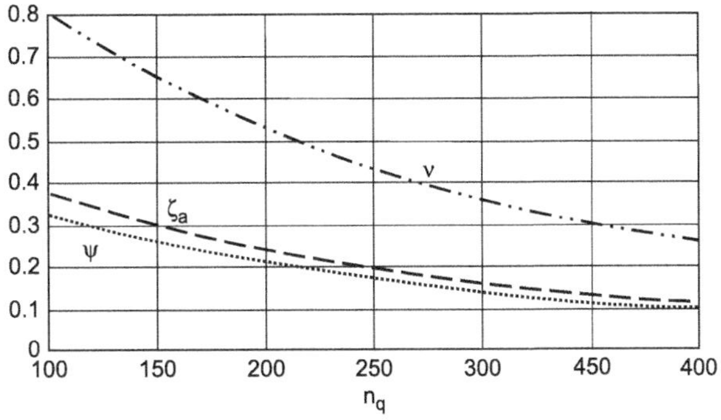

Fig. 7.25 Hub ratio, lift and pressure coefficients of axial pumps calculated for $\varphi = 0.22$

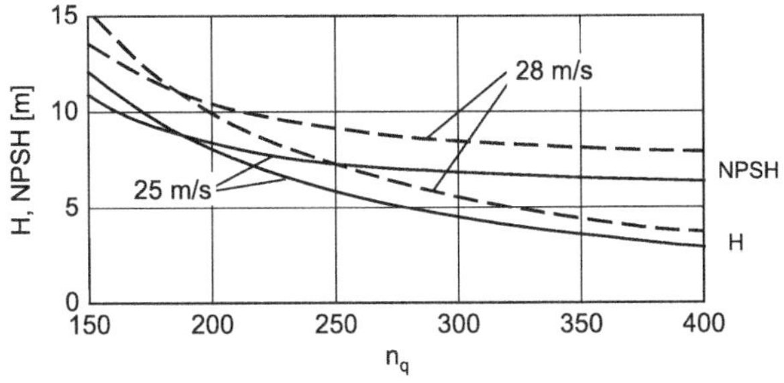

Fig. 7.26 Approximate head and $NPSH_{(\Delta\eta=0)}$ of axial pumps at $u_{2,max}$ = 28 and 25 m/s

designed as airfoils whose geometrical characteristics are explained with reference to Table 7.4. An airfoil can be imagined as a camber line (with maximum camber f) on which a symmetrical profile is superimposed according to a specific thickness distribution. The resulting coordinates of the profile, upper (or suction) surface $y_o(x)$ and the lower (or pressure) surface $y_u(x)$, completely describe the airfoil. The chord on the camber line (NACA airfoils) or the profile tangent (Göttinger airfoils) is used as reference axis or x-axis. All dimensions are based on the profile length L. The following geometric parameters have an effect on the flow characteristic (see figure in Table 7.4):

- maximum thickness ratio d/L
- relative thickness position x_d/L
- camber f/L and its position x_f/L on the profile chord
- nose radius r_k/L
- shape of camber line (e.g. S-shaped)

Airfoil flow: If an airfoil is exposed to the undisturbed approach flow velocity w under the angle of attack δ_A, a stagnation point is formed near the leading edge, Fig. 7.27. At δ_A = 0 the stagnation point is right on the leading edge, but it moves away from the edge with increasing incidence (hence there is a flow around the leading edge at $\delta_A \neq 0$).

Table 7.4 Airfoils

Geometry	A_{fl} area: $A_{fl} = L \times b$; b = const. b span (width) β_s stagger angle d maximum thickness f camber L length, chord r_k leading edge (nose) radius x_d location of maximum thickness x_f location of maximum camber y_o coordinates: suction surface y_u coordinates: pressure surface	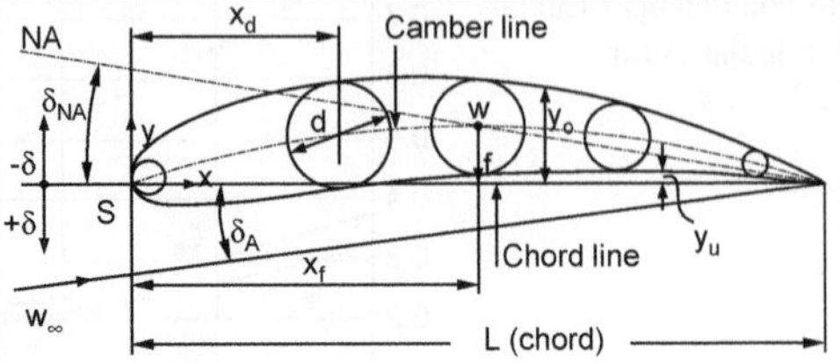 reference line: chord S NA = direction of zero lift Note sign of angles	
Aerodynamic quantities	Lift coefficient	$\zeta_a = \frac{F_a}{\frac{\rho}{2} w_\infty^2 A_{fl}}$	7.4.1
	Drag coefficient	$\zeta_w = \frac{F_w}{\frac{\rho}{2} w_\infty^2 A_{fl}}$	7.4.2
	Sliding coefficient	$\varepsilon = \frac{F_w}{F_a} = \tan\lambda$	7.4.3
	Pressure coefficient	$c_p = \frac{p - p_\infty}{\frac{\rho}{2} w_\infty^2} = 1 - \left(\frac{w}{w_\infty}\right)^2$	7.4.4
	Reynolds number	$Re = \frac{w_\infty L}{\nu}$	
	F_a lift F_w drag β_∞ angle of approach flow δ_A angle of attack (chord) $\delta_{A,o}$ angle of attack referred to zero lift direction δ_{NA} angle of zero lift direction λ sliding angle, $\lambda = \arctan \varepsilon$ w_∞ approach flow velocity	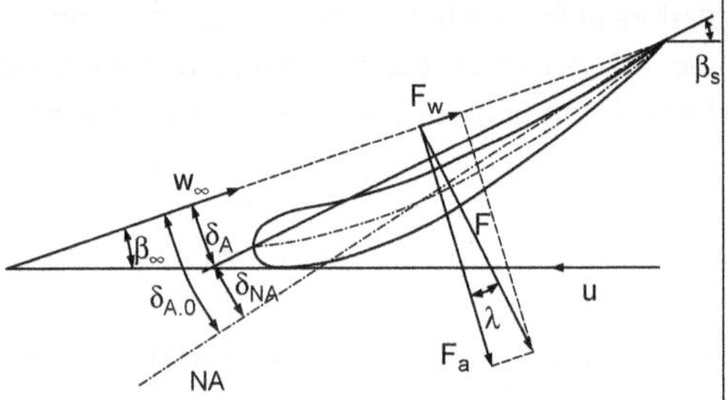 	
Direction of zero lift in degrees referred to chord Note sign of δ_{NA} !		$\delta_{NA} = -100 \frac{f}{L} \left\{ 0.82 + \frac{\left(\frac{x_f}{L}\right)^2}{1 + 5\frac{d}{L}} \right\}$	7.4.5
Lift coefficient Profile efficiency: η_P = 0.85 to 0.92		$\zeta_a = 2\pi\, \eta_P \sin(\delta_A - \delta_{NA})$ $\zeta_a = \frac{d\zeta_a}{d\delta_A}(\delta_A - \delta_{NA})$	7.4.6
Maximum of sliding coefficient		$\varepsilon_{min} \approx 0.012 + 0.02\frac{d}{L} + 0.08\frac{f}{L}$	7.4.7

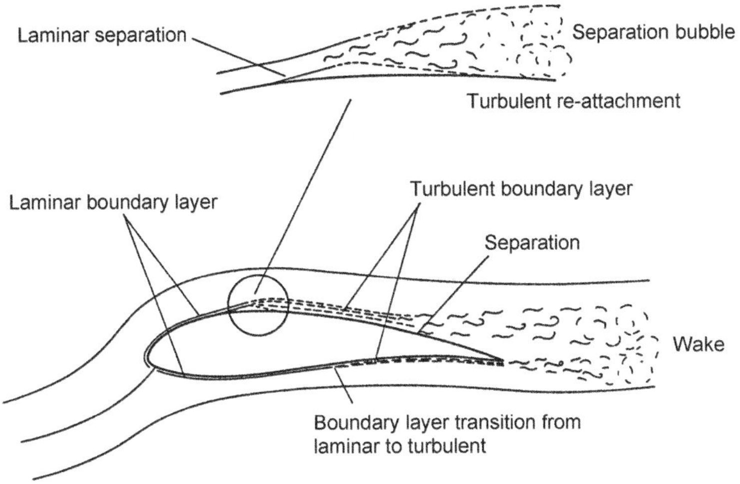

Fig. 7.27 Flow around an airfoil according to [15]

On the suction surface of the airfoil the flow is accelerated and the static pressure drops accordingly. Downstream of the maximum airfoil thickness, where the profile tapers off, the flow is decelerated and the static pressure rises up to the surrounding pressure p_{∞}.

Near the nose of the airfoil the boundary layer is laminar, but it becomes turbulent after a specific flow path length. This transition depends on the Reynolds number, the turbulence of the approach flow and the roughness. Flow separation occurs where the boundary layer becomes too thick (or the deceleration too high). The wake is formed by the boundary layers and stalled fluid.

The local velocities determine the pressure distribution which, when integrated, yields the forces acting on the airfoil. The pressure distribution also determines the onset and extent of cavitation, Chap. 6. According to Bernoulli's equation the local pressure coefficient is obtained from $c_p = 1 - (w/w_{\infty})^2$. As per Eq. (T7.4.4), c_p represents the distribution of the local static pressure referred to the stagnation pressure. With increasing incidence the low-pressure peak grows, the flow around the profile nose increases and the location of the pressure minimum moves closer to the leading edge, Fig. 7.28. The magnitude of $c_{p,min}$ grows with increasing thickness and curvature ratio of the profile. The pressure remains constant in the separated flow.

Lift and drag: The fluid movement exerts a force F on the airfoil. This force is obtained by integrating the pressure and shear stress distribution on the airfoil.

The resultant force is made up of a component F_a normal to the approach flow direction, the "lift" and a force F_w in the direction of the approach flow, the "drag" or "resistance". The resistance consists of the "skin friction drag" and the "pressure drag" (or "form drag"). The friction drag practically does not change with the

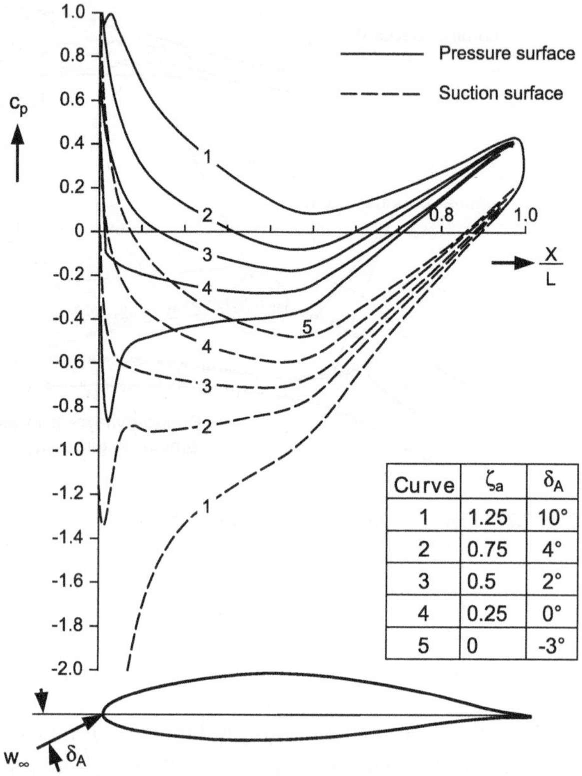

Fig. 7.28 Pressure distribution on airfoil NACA 65_2-415 according to [50]

incidence while the form drag increases with rising angle of attack and grows severely when the flow separates from the airfoil.

The flow characteristics of an airfoil are described by the lift and drag coefficients, the pressure distribution and the minimum pressure coefficient $c_{p,min}$. With a given profile these values primarily depend on the angle of attack δ_A, which constitutes the angle between the direction of the undisturbed approach velocity w and the profile chord (or the profile tangent). The *local* incidence $i_1 = \beta_{1B} - \beta_1$ between the velocity vector at the leading edge and the camber angle must be distinguished from the angle of attack δ_A between w_∞ and the profile chord. The incidence i_1 determines the pressure distribution near the leading edge; thus it is relevant for cavitation. The incidence of shockless entry ($i_1 = 0$) is described as δ_{SF}; it corresponds to the "design angle" of the profile. The angle between the lift and the resulting force F is called the "gliding angle" λ. The ratio $\varepsilon = F_w/F_a = \tan \lambda$ is termed "drag-to-lift ratio". The drag-to-lift ratio of commonly used airfoils is in the range of 0.01 to 0.04 depending on the angle of attack. It can be seen as a measure for the hydraulic quality of the profile. At a specific approach flow angle - the "zero lift direction" - the airfoil is not subjected to any lift.

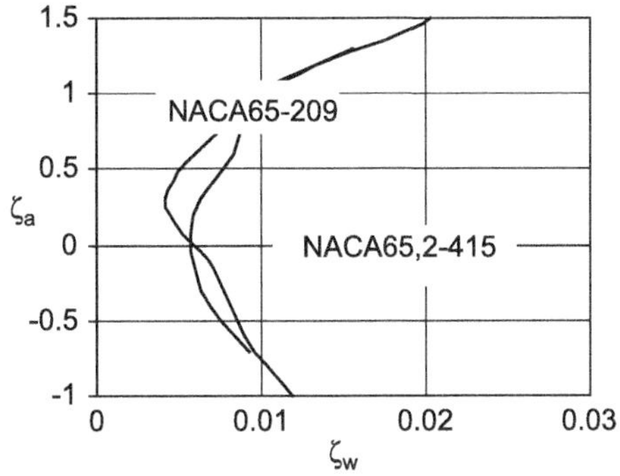

Fig. 7.29 Polar diagram for NACA65$_2$-415 and NACA65–209, Re = 6×10^6

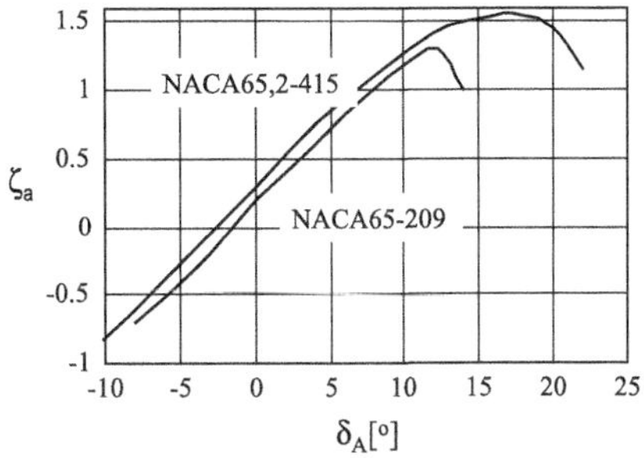

Fig. 7.30 Lift coefficients as function of the angle of attack δ_A, Re = 6×10^6

Profile coordinates and flow characteristics of airfoils are published in airfoil catalogues and manuals based on measurements and calculations, e.g. [1, 15, 50]. The lift coefficients are often presented in "profile polars" according to Fig. 7.29 as a function of the drag coefficient. A line drawn from the coordinate origin to any point on the curve represents the dimensionless resultant force acting on the profile with the components of lift and drag. The tangent from the origin to the polars defines the operation point with the optimum gliding angle or minimum loss. The design point of airfoils is preferably selected close to this optimum.

For application in pump design (where the resistance is less significant than with aircrafts) the presentation of the lift coefficients as a function of the angle of attack according to Fig. 7.30 is more convenient. From this plot, it is possible to read the value of δ_A required for a specific lift coefficient. In the domain of small angles of attack, the lift coefficient grows virtually in proportion to δ_A. The maximum lift is obtained just prior to the onset of stall. The gradient is mostly in the range of $d\zeta_a/d\delta_A = 0.09$ to 0.11. If δ_{NA} stands for the zero-lift angle, the lift coefficient can be calculated from Eq. (7.22):

$$\zeta_a = \frac{d\zeta_a}{d\delta_A}(\delta_A - \delta_{NA}) \tag{7.20}$$

Influence of the airfoil parameters on the aerodynamic coefficients

1. The maximum local velocity increases with the thickness ratio d/L roughly according to:

$$w_{max}/w_{\infty} = 1 + f_p d/L \quad \text{with}\, f_p = 1.3\, \text{to}\, 1.8 \tag{7.21}$$

2. With increasing thickness ratio d/L the magnitude of $c_{p,min}$ consequently increases according to:

$$c_{p,min} = 1 - \left(\frac{w_{max}}{w_{\infty}}\right)^2 = -f_p \frac{d}{L}\left(2 + f_p \frac{d}{L}\right) \tag{7.21a}$$

   The pressure minimum remains at the same location. Tables for the velocity ratios $w(x)/w_{\infty}$ were established for many airfoils so that the pressure distribution can be calculated for a selected profile, [1]. The lift coefficient of a profile in incident flow increases slightly with the growing thickness ratio since the streamline curvature on the suction surface is amplified. The thickness ratio has only little effect on the gradient $d\zeta_a/d\delta_A$.
3. When the relative thickness position x_d/L is increased, the location of the pressure minimum and the transition laminar/turbulent are shifted further downstream.
4. If the curvature gets stronger, the lift and the magnitude of $c_{p,min}$ increase.
5. The minimum drag coefficient is reached when the flow approaches in the zero-lift direction. The drag increases with the thickness ratio.
6. The point of attack of the resulting force (the "neutral point") is located in the range of $x_N/L = 0.25$ to 0.28.

The linear relationship between lift coefficient and angle of attack allows an approximate formula as follows:

$$\zeta_a = a\frac{d}{L} + b\,\delta_A \tag{7.22}$$

The gradients b of many profiles are within the range of b = 0.09 to 0.11. The property of the airfoils described by Eq. (7.22) allows to multiply the y-coordinates of the profile geometry with a constant factor and to calculate the lift coefficient of the thickened (or slimmed down) profile from Eq. (7.22). In this way it is possible to thicken blades from the outer streamline to the hub without the need to select different profiles. After a profile has been selected, the applicable constant "a" can be easily calculated with Eq. (7.22) from the curve $\zeta_a = f(\delta_A)$, since the gradient b is known as well.

The zero-lift direction δ_{NA} depends on the camber and the thickness of the profile. For any profile δ_{NA} can be estimated either from Eq. (T7.4.5) or in graphic form by drawing a line from the trailing edge through the point of maximum camber (see upper figure in Table 7.4: line NA through point W).

If the angle of attack $\delta_{A,o}$ is referred to the zero lift direction, the theoretical lift coefficient of a flat plate in incident flow is $\zeta_a = 2\pi \times \sin \delta_{A,o}$. According to the figure in Table 7.4, $\delta_{A,o} = \delta_A - \delta_{NA}$ applies (take note of the sign of the angles). Experience shows that the lift coefficient of any profile can be estimated according to the above theoretical relationship if a profile efficiency η_P is introduced which results in the approximation formula Eq. (T7.4.6). The drag coefficients, too, can be determined from an empirical relationship, Eq. (T7.4.7).[4]

The lift and drag coefficients depend on the Reynolds number. This is true especially in the range below the critical Reynolds number $Re = w_\infty \times L/\nu$, which is around 10^5. The lift coefficient increases with the Reynolds number. The transition laminar/turbulent is shifted upstream with increasing Re while the resistance drops. Except for oils and similar viscous media, pumps operate above the critical range. Rough surfaces reduce the lift coefficient, especially with high Reynolds numbers.

7.6.4 Blade Design

It is possible to design an axial pump for any swirl distribution at the impeller outlet (three possible distributions have been discussed in Chap. 7.5). The following discussion focuses on the design of the blading according to the free vortex concept with $c_u \times r = c_{2u,a} \times r_2$. Here, the same blade work $Y_{sch} = g \times H_{th}$ is transmitted on all streamlines and the axial velocity over the radius is constant. If other swirl distributions are used, the c_m-distribution should be calculated iteratively from the radial equilibrium. Apart from the variation of the axial velocity, the equations in Table 7.5 are written so that they can be used for any swirl distribution.

The blade design is described in Table 7.5 which includes the velocity triangles for axial approach flow. The step-by-step procedure is described below:

1. Given are the performance data. The hydraulic efficiency η_h can be estimated from Table 3.9.
2. Several streamlines (cylinder sections) are defined on which the calculation according to Eqs. (T7.5.4 to 7.5.8) is performed. The meridional velocity is assumed to be identical on all sections.
3. The approach flow angle in the relative system increases from the tip to the hub according to the ratio of the radii, Eq. (T7.5.2).
4. Define shape of the hub (axial or slightly semi-axial). The hub diameter at the outlet can be calculated from Eq. (T7.3.5).

[4]These approximation formulae originate from [B.1].

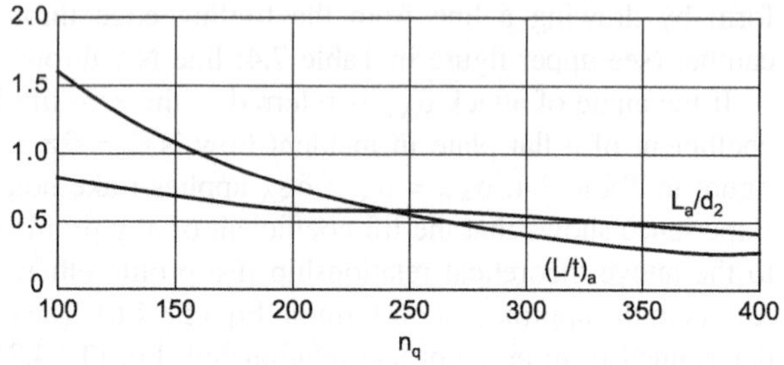

Fig. 7.31 Selection of the blade length at the outer streamlines of axial pumps

5. The circumferential velocity component at the outer streamline which is required to achieve the specified head is calculated from Eq. (T7.5.3).
6. At the outlet, the circumferential velocity component c_{2u} and flow outlet angle α_2 increase from the tip to the hub, Eqs. (T7.5.4 and 7.5.5). *Equation (T7.5.5) supplies the flow outlet angle for which the blading has to be designed.*
7. When designing the blades as airfoils, the approach flow velocity w_∞ and angle β_∞ are calculated from Eqs. (T7.5.6 and 7.5.7).
8. According to Eqs. (T7.5.8a or 7.5.8b) it is now possible to determine the flow deflection coefficient $\zeta_a L/t$ for each of the streamlines. In this formula ζ_a stands for the lift coefficient, L for the length of the chord and $t = 2\pi \times r/z_{La}$ is the blade pitch on the cylinder section considered (definitions for airfoils see Table 7.4). Equation (T7.5.8) can be derived from the conservation of angular momentum, Chap. 3.2, if the external torque on the rotor is calculated from the forces acting on the airfoil according to Table 7.4. The term $\cos \lambda$ in the numerator of Eq. (T7.5.8a) is frequently not written explicitly but set at $\cos \lambda \approx 1.0$ because the gliding angle λ of relevant airfoils is below 5°. This was done in the expression on the right-hand side of Eq. (T7.5.8a). Sometimes, $\sin(\beta_\infty + \lambda) \approx \sin \beta_\infty$ is assumed as an approximation resulting in Eq. (T7.5.8b). In this case the design implies some head margin since a higher deflection coefficient is specified.
9. The velocity triangles at inlet and outlet can now be calculated from Tables 3.1 and 3.2. As a result we get the velocities and the *flow* angles which are required to achieve the specified head.
10. Blade length and pitch (solidity) must now be selected so that the deflection coefficient calculated in the respective cylinder section is obtained. Initially, only the product of ζ_a and L/t is known, which constitutes the required deflection coefficient. One of these two quantities must be selected so that the second one can be calculated. One option is to estimate $(L/t)_a$ on the outer streamline from Fig. 7.31 as a function of the specific speed.[5] Alternatively, the

[5]Figure 7.31 was derived from information provided in [B.22].

Table 7.5 (1) Blade design of axial impellers A. Calculation of velocities required to achieve the specified head		
Assumptions and definitions * c_m = constant * $x \equiv (r_2/r)^m$	Constant angular momentum: $m = 1$, $c_u = c_{2u,a} \times r_2/r$ Solid body rotation: $m = -1$, $c_u = c_{2u,a} \times r/r_2$ Constant whirl velocity: $m = 0$, c_u = constant	
Given or selected quantities	$Q_{opt}, H_{opt}, \eta_h, n, d_2, \nu, n_q, \psi, \varphi, r_i = \nu \times r_2$	Eq.
Meridional velocity	$c_m = \dfrac{4Q}{\pi d_2^2 (1-\nu^2)}$	7.5.1
Flow angle at impeller inlet as function of radius	$\tan\beta_1 = \dfrac{r_2}{r}\tan\beta_{1a}$	7.5.2
Circumferential component of absolute velocity at outer streamline	$c_{2u,a} = \dfrac{g\,H}{\eta_h\, u_{2,a}} + c_{1u}$	7.5.3
Circumferential component of absolute velocity	$c_{2u} = c_{2u,a}\, x$	7.5.4
Impeller flow outlet angle of relative velocity	$\tan\beta_2 = \dfrac{c_m}{u - x\, c_{2u,a}}$	7.5.5
Approach flow angle	$\beta_\infty = \arctan \dfrac{c_m}{u - \frac{1}{2}(c_{2u} + c_{1u})}$	7.5.6
Approach flow velocity	$w_\infty^2 = c_m^2 + \left[u - \frac{1}{2}(c_{2u} + c_{1u})\right]^2$	7.5.7
Flow deflection coefficient	$\zeta_a \dfrac{L}{t} = \dfrac{2g\, H_{th}\, c_m \cos\lambda}{u\, w_\infty^2 \sin(\beta_\infty + \lambda)} = \dfrac{2\,\Delta c_u \sin\beta_\infty}{w_\infty \sin(\beta_\infty + \lambda)}$	7.5.8a
	$\zeta_a \dfrac{L}{t} \approx \dfrac{\psi_{th}}{\sqrt{\varphi^2 + (1 - 0.25\psi_{th})^2}} \approx \dfrac{2\,\Delta c_u}{w_\infty}$	7.5.8b
Hydraulic efficiency	$\eta_{sch} \equiv 1 - \dfrac{Z_{sch}}{H_{th}} = 1 - \dfrac{w_\infty\, \varepsilon}{u \sin(\beta_\infty + \lambda)}$	7.5.9
Sliding coefficient relevant for pump	$\varepsilon_{res} = 0.02 + \dfrac{0.008}{1-\nu}$	7.5.10
	$\varepsilon_{res} = \varepsilon + \dfrac{0.02\, t}{\zeta_a\, r_a (1-\nu)} + 0.018\, \zeta_a$	7.5.11
Diffusion factor Outer streamline max. $D_{fz,a} = 0.45$ Hub maximum: $D_{fz,i} = 0.6$	$D_{fz} = 1 - \dfrac{w_2}{w_1} + \dfrac{w_{1u} - w_{2u}}{2 w_1}\left(\dfrac{t}{L}\right)$	7.5.12
Length to pitch ratio (solidity) [41][B.22]	$\left(\dfrac{L}{t}\right)_a = 3.2\, \psi_a$	7.5.13

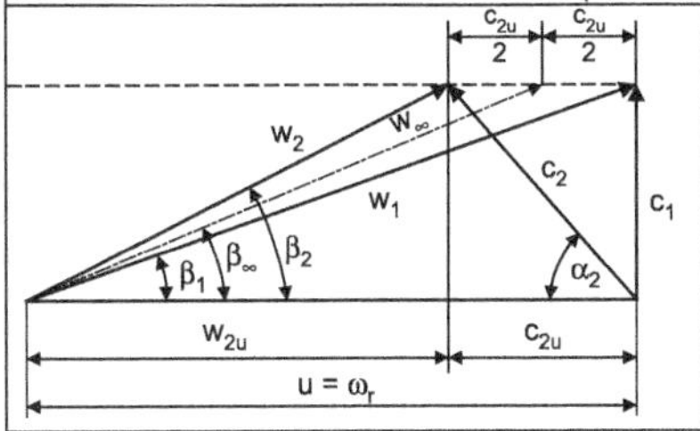

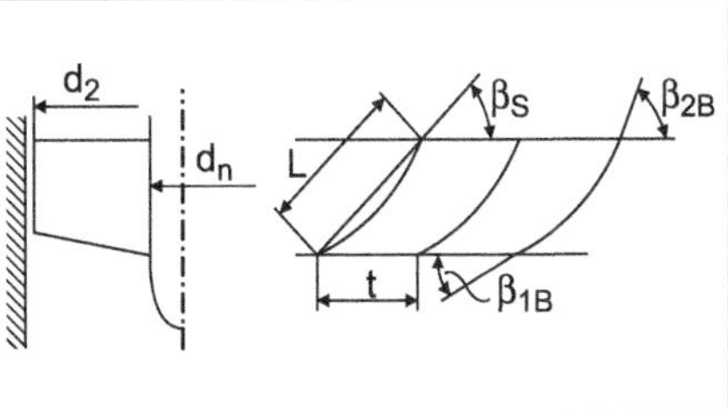

Table 7.5 (2) Blade design of axial impellers according to [32]
B. Determine incidence, deviation angle and blade angles

The complete velocity triangles and w_∞, β_∞ are known from Table 7.5 (1)			
Select	z_{La}, z_{Le} from Table D7.3		Eq.
Blade tip clearance (tentative)	$\Delta D/d_2 = 0.004$		7.5.14
Select solidity at outer streamline from Fig. 7.31. Eq. (7.5.15) gives higher length in order to reduce blade loading	$\sigma_a = \left(\frac{L}{t}\right)_a = 0.7\left(\frac{n_{q,Ref}}{n_q}\right)^{0.46}$	$n_{q,Ref} = 200$	7.5.15
Select solidity or blade length at inner streamline $L_i/L_a = 0.7$ to 0.9	$\sigma_i = \left(\frac{L}{t}\right)_i$		7.5.16
Relative blade height, measured from hub	$h = \frac{r - r_i}{r_a - r_i}$		7.5.17
Design blade length in a way that smooth leading and trailing edges are achieved. The factors f_2 and f_3 can be suitably selected; f_1 follows from Eq. (7.5.19)	$\frac{L}{L_a} = \frac{L_i}{L_a} + f_1 h + f_2 h^2 + f_3 h^3$		7.5.18
	$f_1 = 1 - \frac{L_i}{L_a} - f_2 - f_3$		7.5.19
Select airfoil type and thickness ratio d/L at hub and tip. Thickness distribution over relative blade height from Eq. (7.5.20) if linear. Tangent on blade thickness at tip is zero. Alternatively a_1 and b_1 can be selected	$\frac{d}{L} = \left(\frac{d}{L}\right)_i + a_1 h + b_1 h^2$ $b_1 = \left(\frac{d}{L}\right)_i - \left(\frac{d}{L}\right)_a$ $a_1 = -2\left[\left(\frac{d}{L}\right)_i - \left(\frac{d}{L}\right)_a\right]$		7.5.20
Incidence factor taking into account the thickness ratio d/L, [32], Fig. 186, p. 236 ($K_i = 1.0$ for $d/L = 0.1$)	$K_i = 18.88\left(\frac{d}{L}\right) - 118\left(\frac{d}{L}\right)^2 + 292\left(\frac{d}{L}\right)^3$		7.5.21
Design incidence for zero-cambered profile with $d/L = 0.1$. Simplified correlation, [32], Fig. 187, p. 237	$i_o = 5\,\sigma\left(1 - \frac{\beta_1 - 20°}{70°}\right)$		7.5.22
Effect of 3D-flow in impeller on incidence, compared to 2D-cascade data; [32], Fig. 201, p. 247	$(i_c - i_{2D}) = 2.16 - 7.24h + 2.277h^2$		7.5.23
Effect of 3D-flow in impeller on deviation angle compared to 2D-cascade data; [55], Fig. 202, p. 247 According to tests in [55] and [56] typically 2° should be added to the values given by Eq. (7.5.24)	For $h < 0.355$: $(\delta_c - \delta_{2D}) = 1.55 - 5.5h$ For $h > 0.355$: $(\delta_c - \delta_{2D}) = -0.4$		7.5.24
[32], Fig. 195, p. 242	$m = 3.488 \times 10^{-5}\,\beta_1^2 - 5.777 \times 10^{-3}\,\beta_1 + 0.4107$		7.5.25
[32], Fig. 196, p. 243	$b = -9.435 \times 10^{-5}\,\beta_1^2 + 0.01594\,\beta_1 + 0.2812$		7.5.26
Correction factor for deviation angle derived from experiments with axial pumps	$K_{exp} = 1.25$ tentatively		

Table 7.5 (3) Blade design of axial impellers according to [32]

Factor for deviation angle taking into account the thickness ratio d/L, [32], Fig. 193, p. 241 ($K_\delta = 1.0$ for d/L = 0.1)	$K_\delta = 6.6\left(\frac{d}{L}\right) + 34\left(\frac{d}{L}\right)^2$	7.5.27
Deviation angle for zero-cambered profile with d/L = 0.1. Simplified correlation [32], Fig. 194, p. 242	$\delta_o = a_\delta\, \sigma^y$ $a_\delta = 4.4103 - 0.1075\, \beta_1 + 0.000706\, \beta_1^2$ $y = 0.0000006258\, \beta_1^4 + 0.0001034\, \beta_1^3 - 0.006011\, \beta_1^2 + 0.137\, \beta_1 - 0.1438$	7.5.28
$n = -0.68 + 0.007\beta_1 - \sigma^2(-0.0000191\, \beta_1^2 + 0.00152\, \beta_1^2 + 0.0179) - \sigma\,(-0.1305 - 0.0103\, \beta_1 + 0.0002124\, \beta_1^2 - 0.0000009543\, \beta_1^3)$		7.5.29
$\frac{d_\delta}{di} = a_i\ e^{b_i \beta_1}\left\{1 - 1.4075\sigma + 0.6419\sigma^2 - 0.0885\sigma^3\right\}$ $a_i = 1 + 0.58\,\sigma + 0.1273\,\sigma^2$ and $b_i = 0.0047\,\sigma^2 - 0.0322\,\sigma$		7.5.30
Required camber angle (m from Eq. 7.5.25): $\gamma = \dfrac{\beta_2 - \beta_1 + K_{exp} K_\delta \delta_o - K_i i_o + (\delta_c - \delta_{2D}) - (i_c - i_{2D})\left\{1 - \left(\frac{d\delta}{di}\right)_{2D}\right\}}{1 - \frac{m}{\sigma^b} + n}$		7.5.31
Incidence	$i_c = K_i i_o + n\,\gamma + (i_c - i_{2D})$	7.5.32
Deviation angle $\delta_c = \beta_{2B} - \beta_2$	$\delta_c = K_{exp} K_\delta \delta_o + \frac{m}{\sigma^b}\gamma + (\delta_c - \delta_{2D}) + (i_c - i_{2D})\left(\frac{d\delta}{di}\right)_{2D}$	7.5.33
Blade inlet angle	$\beta_{1B} = \beta_1 + i_c$	7.5.34
Blade outlet angle	$\beta_{2B} = \beta_2 + \delta_c$	7.5.35
The stagger angle is determined after selection of the profiles:		
Direction of zero lift in degrees referred to chord	calculate δ_{NA} from Eq. (T7.4.5)	7.5.36
	for comparison calculate δ_{NA} from: $\delta_{NA} = -\operatorname{atan}\left(\frac{f}{1 - x_f}\right)$	7.5.37
Angle of attack from required lift and selected gradient $d\delta_a/d\delta_A$	$\delta_A = \delta_{NA} + \left(\frac{\zeta_a}{d\zeta_a/d\delta_A}\right)$	7.5.38
Stagger angle, use average δ_A from Eqs. (T7.4.5) and (T7.5.38)	$\beta_s = \beta_\infty + \delta_A$	7.5.39
Camber line (mean line), [1], p. 74, Eq. (4.26) for uniform blade loading	$y_c = -\frac{\zeta_a}{4\pi}\left\{(1 - x)\ln(1 - x) + x \ln x\right\}$	7.5.40
Calculate the camber required for a specific lift ζ_a from a known profile designed for $\zeta_{a,base}$; applicable to any blade loading	$y_c = y_{c,base}\,\frac{\zeta_a}{\zeta_{a,base}}$	7.5.41
Calculate slope dy_c/dx of actual camber line from tabled slope for base	$\left(\frac{dy_c}{dx}\right)_{actual} = \left(\frac{dy_{c,base}}{dx}\right)_{base}\frac{\zeta_a}{\zeta_{a,base}}$	7.5.42

Table 7.5 (4) Blade design of axial impellers		
Blending function for final trailing edge thickness, $x_{BL} = 1 - x_{TE}$	$\Delta y = \frac{e_{TE}}{2\,L}\frac{x - x_{BL}}{1 - x_{BL}}$	7.5.43
Thickness distribution of *symmetrical* profile without TE correction	$y_{sp} = \frac{(d/L)_{actual}}{(d/L)_{base}} y_{sp,base}$	7.5.44
Thickness distribution of *symmetrical* profile with TE correction	$y_{sp} = \frac{(d/L)_{actual}}{(d/L)_{base}} y_{sp,base} + \frac{e_{TE}}{2L}\frac{x - x_{BL}}{1 - x_{BL}}$	7.5.45
Blade coordinates, *suction* surface	$x_{SS} = x - y_{sp}\sin\Theta$ $y_{SS} = y_c + y_{sp}\cos\Theta$ $\Theta = \operatorname{atan}\left(\frac{dy_c}{dx}\right)$	7.5.46
Blade coordinates, *pressure* surface	$x_{PS} = x + y_{sp}\sin\Theta$ $y_{PS} = y_c - y_{sp}\cos\Theta$ from Eq. (7.5.42)	7.5.47
$\sigma = L/t$ solidity Dimensionless representation $x = X/L$, $y = Y/L$ X and Y are the absolute coordinates in mm or m		
Forces, stresses, natural frequency		
Forces acting on a section of the blade		
average height of blade section	$\Delta r = 0.5(\Delta r_a + \Delta r_i)$	7.5.48
average radius	$r = 0.5(r_a + r_i) - 0.5\Delta r$	7.5.49
lift acting on blade section	$\Delta F_a = 0.5\rho\, w_\infty^2\, \zeta_a\, L\, \Delta r$	7.5.50
axial force	$\Delta F_{ax} = \Delta F_a \cos\beta_\infty$	7.5.51
tangential force	$\Delta F_u = \Delta F_a \sin\beta_\infty$	7.5.52
bending moment on root of blade	$\Delta M_B = \Delta F_a (r - r_{i,m})$	7.5.53
Total forces and bending moment are obtained as the sum of all blade sections: $F_a = \Sigma\Delta F_a$ etc.		
Mean bending stress at root of blade W = section modulus at the hub	$\sigma = \frac{M_B}{W}$	7.5.54
Mass per unit length A = area of blade section $f_{am} = 1.5$ accounts for the effect of the added mass of the surrounding water	$\mu = \rho_{mat}A + f_{am}\,\rho\,A$	7.5.55
First natural frequency of blade I = area moment of inertia at the hub	$f_n = \frac{3.52}{2\pi}\sqrt{\frac{E\,I}{\mu\,L^4}}$	7.5.56

lift coefficient ζ_a can be determined from Figs. 7.24 and 7.25. Executable values of ζ_a are slightly above the head coefficient so that $\zeta_a = (1.2$ to $1.35) \times \psi_{a,opt}$ can serve as a basis for the outer streamline. The blade length at the outer streamline is believed to be critical for achieving good performance in terms of the allowable operation range. Low blade loading is beneficial with respect to cavitation and the point of stall.

11. An airfoil profile is selected according to the criteria given in Chap. 7.6.5.
12. In order to determine the *blade* angles necessary to accomplish the flow deflection calculated from step 8, the camber angle $\gamma = \beta_{2B} - \beta_{1B}$ must be determined. This can be done according to the procedure given in [32], Table 7.5(2). Incidence, deviation angle and blade angles follow from Eqs.

(T7.5.21 to T7.5.35). The procedure yields incidences which become *negative* at the outer streamline and positive at the inner streamline. However, it is necessary to design the blades for zero-incidence at the outer streamline in order to realize the specified BEP flow rate. With negative incidence the optimum flow rate would be below the design, as found for example in [35].

13. The deviation angle predicted from [32] may be too low. This can be derived from test data in [55, 56] where the pressure rise predicted was not achieved. To compensate this deficiency, a correction factor K_{exp} has been introduced in Eq. (T7.5.31). This factor is to be derived from test data with axial pumps; tentatively it can be set to $K_{exp} = 1.25$.
14. As a result we have now the blade inlet and outlet angles, Eqs. (T7.5.34 and 7.5.35).
15. The camber determines the lift of the blade section. Consequently a suitable camber line (called "mean line" in [1]) is selected according to the lift coefficient required for the considered blade section.
16. The required angle of attack δ_A and the stagger angle follow from the lift coefficient ζ_a as given by Eqs. (T7.5.37 to T7.5.39).
17. A symmetrical profile close to the required thickness (e.g. $d/L = 0.06$) is selected; e.g. NACA 65-006. The tabled coordinates of the symmetrical profile are thus known, e.g. [1].
18. The actual thickness required for any given blade section is obtained by correcting the blade thickness coordinates by the factor $f_d = (d/L)_{actual}/(d/L)_{base}$. This slightly modifies the profile characteristics, but the influence is neglected.
19. For manufacturing reasons a final trailing edge thickness e_{TE} is required; e.g. $e_{TE}/d_{2a} = 0.001$. The corresponding blade thickness correction is added to the blade coordinates. The trailing edge thickness is blended into the profile coordinates over a selected length $\Delta x_{TE}/L$; for example over 20% of the chord: $\Delta x_{TE}/L = 0.2$. Using a linear blending function gives Eq. (T7.5.43).
20. The coordinates of the symmetrical profile are calculated from Eq. (T7.5.45).
21. Finally, the coordinates of the suction and the pressure surface are obtained by superposing the thickness distribution of the symmetrical profile on the camber line, Eqs. (T7.5.46 and T7.5.47).
22. All geometric parameters must vary smoothly over the blade height. The present procedure ensures that blade length, thickness and camber fulfill this requirement, Eqs. (T7.5.18 to T7.5.20).

Various recommendations for selecting these parameters can be found in the literature. Unfortunately, most data originate from tests on axial compressors (notably [32, 38, 39]) where the blading is significantly different from that of propeller pumps. The extent to which these recommendations can be applied to pumps is therefore uncertain. Nevertheless, the methods given in [32] are often applied for designing axial pumps.

It is possible to derive the relationships for ζ_a and L/t given as Eq. (T7.5.13) from the literature. Summarizing the various data the following is obtained:

- The lift coefficient in the design point ought to be selected in the range of the lowest gliding angle: $\zeta_a = 0.2$ to 0.8 but not greater than: $\zeta_a = (1.2$ to $1.35) \times \psi_a$
- Maximum lift coefficient (at the inner streamline): $\zeta_{a,max} = 1.25$
- Solidity L/t = 0.4 to 1.2
- Maximum deflection coefficient (at the inner streamline): $(\zeta_a L/t)_{max} \approx 1.5$
- On the hub, ψ_i must also be verified according to Fig. 3.24.

These recommendations must be observed primarily for the outer section of the blades. Compromises may be required near the hub. The greater the required flow deflection $\Delta\beta = \beta_2 - \beta_1$, the greater L/t must be selected.

The lift and drag coefficients given in the literature apply to individual airfoils of infinite span b (frequently measured for L/b = 5). The flow from the pressure to the suction surface around the tip reduces the lift of an airfoil of finite span. In an axial pump this flow is restricted by the narrow gap between the casing and the tip of the blade so that the profile in the pump tends to supply a slightly greater lift. In a cascade there is some interaction between the blades. In principle, some correction for the effects of span and cascade would therefore have to be made when applying data from individual profiles to the impeller of an axial pump. Since generally applicable rules for the cascade correction are not known, it seems to be common practice to ignore these corrections when $t/L \geq 1$, [B.1, 25]. Span correction is generally ignored as well since its effect is minor and to some extent compensates for the effect of the neighboring blades.

The hydraulic losses consist of the actual blade losses which can be determined from the airfoil data plus friction losses on casing, hub, inlet and diffuser. The actual blade losses are obtained from Eq. (T7.5.9) when the drag-to-lift ratio of the profile is inserted. The hydraulic efficiency of the entire pump can also be estimated by means of Eq. (T7.5.9) if the resulting drag-to-lift ratio ε_{res} is employed according to the correlations (T7.5.10 or 7.5.11), [B.1]. This calculation is performed for the mean diameter d_m of the blading:

$$d_{1m} = d_2\sqrt{0.5(1 - \nu^2)} \tag{7.22a}$$

The allowable deceleration of the relative velocity on the various cylinder sections can be verified by means of the diffusion factor D_{fz} suggested by Lieblein [37], Eq. (T7.5.12). According to [7, 9] a value of $D_{fz,a} = 0.45$ on the outer streamline and $D_{fz,i} = 0.6$ on the hub should not be exceeded. In [B.3] it is recommended to limit the value on the outer streamline to $D_{fz,a} = 0.35$.

Forces and moments acting on the blade can be calculated from the lift and drag forces according to Table 7.5 for various cylinder sections. Integrating these over the blade height allows to determine the bending stresses at the root of the blade. In addition to the hydraulic forces, centrifugal forces are acting on the blade. The lowest natural frequency of the blade must be calculated to ensure that it is sufficiently above the excitation frequency $z_{Le} \times n/60$, Chap. 10.7.1. The axial force acting on the impeller can be determined by integrating the axial force components over the blade height (unless the approximation formulae according to Chap. 9.2 are preferred). Comments to calculation of forces and stresses:

- The maximum head at partload should be used for checking the blade strength. To this end the ratio H_{max}/H_{opt} is applied to the stresses calculated for BEP conditions. This factor may be set to $H_{max}/H_{opt} = 1.6$ to 2.0.
- The mean (steady) bending stress and alternating stresses are considered. At partload close to stall and flow recirculation the alternating lift forces are expected to be high. The dynamic lift components can be assumed to amount to 35% of the steady forces.
- The section modulus W is calculated for the blade section at the hub. In doing so, the camber is conservatively neglected. The area moment of inertia is calculated for the blade section at the hub. Again the camber is neglected making the calculation conservative with respect to the natural frequency of the blade.
- The tapering of the blade thickness from the hub to the outer streamline is neglected. This makes the calculating conservative by typically 2 to 4%.
- For geometrically similar impellers the natural frequency is proportional to $f \sim 1/d_2$.

7.6.5 *Profile Selection*

The camber determines the lift of the blade section. According to the theory of thin blade sections, *the lift is directly proportional to the camber*. Consequently, the camber y_c required to achieve a specific lift ζ_a at a specific blade section can be calculated by means of Eq. (T7.5.41) from any known (NACA or other) baseline profile having the lift coefficient $\zeta_{a,base}$. The camber line of the selected baseline profile is given by $y_{c,base} = f(x)$ as tabled in [1]. $\zeta_{a,base}$ is the design lift coefficient of the tabled profile. Criteria for the selection of the profiles:

1. *The thickness distribution and the camber line both determine the pressure distribution*. In order to achieve good cavitation characteristics, a profile with a flat pressure distribution and a low pressure coefficient $c_{p,min}$ should be selected. With such profiles, the maximum thickness is sufficiently far away from the leading edge; $x_d = 0.5$ appears to be close to optimum (e.g. the NACA65 series). Higher values of x_d/L are not recommended since this would increase the risk of flow separations towards the rear of the profile. Lift coefficient (blade loading), low-pressure peak $c_{p,min}$ and sensitivity to cavitation increase with growing profile thickness and camber.
2. Shaping the camber line as a circular arc is unfavorable because such profiles are sensitive to excessive incidence. A parabolic camber line is superior in terms of

cavitation since the first section of the blades can be configured so that a flatter pressure distribution is obtained.

3. Although sharp profiles promise minimal losses in shockless approach flow (which would be favorable at BEP), they are sensitive to incidence and unsuitable in terms of partload and cavitation characteristics.
4. The low-pressure peak $c_{p,min}$, and consequently the cavitation sensitivity, increase with growing profile thickness d/L. For this reason, the profile thickness is usually not selected larger than d/L = 0.15 to 0.18.
5. A pillar subjected to inclined approach flow produces smaller low-pressure peaks than a structure in perpendicular approach flow. Leading edges that appear radial in the plan view are thus not optimal. Consequently, sweep-back similar to Fig. 7.32 (see for instance [61]) is frequently used in order to reduce blade loading on the hub streamline and secondary flows. Sweep-back can significantly improve the stability of the Q-H-curve, the cavitation characteristics (reducing the $NPSH_R$ by typically 10 to 30%) and even the efficiency, see Fig. 5.43 and [21]. The sweep-back also prevents textiles, plastic foils and all kinds of fibers from being caught on the blade leading edges. Sweep-back is indispensable for sewage applications, but such contaminations also occur in irrigation and drainage systems. The effects of sweep-back have been investigated in [18, 20, 35, 48].
6. For mechanical and hydraulic reasons, the profile thickness has to increase from the tip to the hub. The thickness coordinates of a profile may be multiplied by a constant factor to achieve this.

Based on the above criteria, the profiles of the NACA 6 series can be selected for axial pumps since the camber lines are designed for uniform hydraulic loading from the leading edge up to a point a/L. The designation of the profiles indicates their flow characteristics. The profile "NACA 65_2-415, a = 0.7" may serve as an example: (a) The first Fig. (6) stands for the profile series and its design principles. (b) The second Fig. (5) indicates the position of the pressure minimum for the

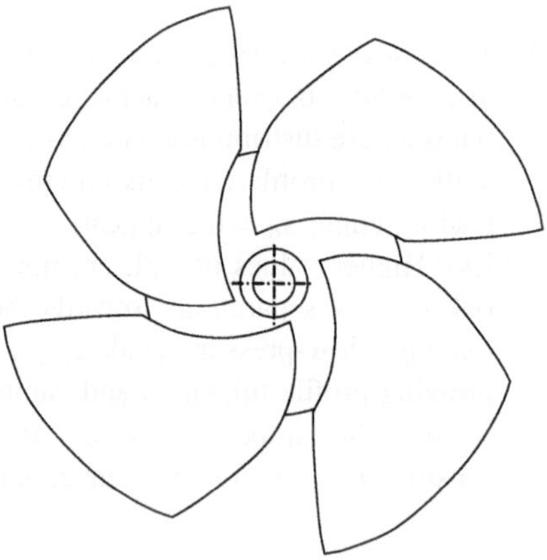

Fig. 7.32 Axial impeller with sweep-back, similar to [61]

symmetrical basic profile with zero lift in terms of $10 \times x_{cp,min}/L$. In the present case this means that the point of the lowest pressure under the above conditions is situated at 50% of the profile length. (c) The third figure indicates the domain of the lift coefficients around the design value in which favorable pressure gradients prevail on the upper and lower surface of the profile. (d) The figure after the dash signifies the design lift coefficient in terms of $10 \times \zeta_a$; in the present example this means that the profile is designed for $\zeta_a = 0.4$. (e) The last two figures signify the maximum profile thickness in percent, i.e. d/L = 0.15. (f) The additional quantity "a" describes the hydraulic load of the camber line; in the present case the hydraulic load is uniform from the leading edge up to 70% of the profile length; downstream it decreases to zero in a straight line up to the trailing edge.

The characteristics of the profiles NACA 65-209 and NACA 65_2-415 are shown in Figs. 7.28, 7.29 and 7.30. The pressure distribution (according to [50]) for various lift coefficients is shown in Fig. 7.28, where the approximate angle of attack required to achieve the corresponding lift coefficient has been entered for illustration. The pressure minimum on the suction surface (according to the design) is located at 50% of the profile length. The pressure distribution remains flat up to an angle of attack of 2° but a clear low-pressure peak appears on the suction surface when the angle of attack exceeds 4°. Figure 7.29 shows the profile polars $\zeta_a = f(\zeta_w)$ for both profiles which differ in the thickness ratios and the design lift coefficients. However, both polars are very similar. The minimum of the flow resistance is situated roughly at the design lift. Figure 7.30 shows the lift coefficients as function of the angle of attack according to the data in [1]. Both curves confirm the mentioned linear increase of the lift with the angle of attack up to a maximum lift which, with the thicker, more cambered profile "425", is higher than with profile "209". In the linear range both profiles reach the same lift when the angle of attack of the thinner profile is increased by about one degree.

7.6.6 *Design of Axial Diffusers*

The absolute impeller outlet velocity is $(c_2/u_2)^2 = (c_{2u}/u_2)^2 + (c_m/u_2)^2 = 1/4\,\psi_{th}^2 + \varphi^2$. If a diffuser is arranged downstream of the impeller, the flow is decelerated therein to the outlet velocity $c_4 = c_m/A_R$ (A_R being the ratio of the annular cross sections of the diffuser outlet to the diffuser inlet). The diffuser also serves to reduce the circumferential velocity component to a value as close as possible to zero because any swirl at the pump outlet would imply additional losses.

If ζ_{Le} is the pressure loss coefficient of the diffuser, the pressure recovery in the diffuser is obtained from the Bernoulli equation:

$$\Delta\psi_{Le} = \left(\frac{c_2}{u_2}\right)^2 (1-\zeta_{Le}) - \left(\frac{c_4}{u_2}\right)^2 = \frac{\psi_{th}^2}{4}(1-\zeta_{Le}) + \varphi^2\left(1-\zeta_{Le}-\frac{1}{A_R^2}\right) \quad (7.23)$$

For cost reasons an expansion of the channel is frequently avoided so that $A_R = 1.0$. The lower the head coefficient, the lower will be the possible pressure recovery because of unavoidable hydraulic losses. If, for instance, $\beta_1 = 12.4°$ (i.e. $\varphi = 0.22$), the pipe friction coefficient $\lambda = 0.04$ and the channel length $L/D_h = 3$ ($\zeta_{Le} = \lambda \times L/D_h = 0.12$), no pressure recovery can be achieved in the diffuser with $A_R = 1$ if the head coefficient is $\psi_{th} = 0.15$, Eq. (7.23). At specific speeds above approximately 270 it must therefore be carefully examined if a diffuser is able to produce any noticeable pressure recovery at all - or if it would even cause a loss in efficiency. If a vaned diffuser brings no advantages, a conical diffuser can be employed which opens up in the direction of the hub and/or to the outside (design according to Chap. 1.6).

If a diffuser is required to achieve a good efficiency, the vanes can be designed as airfoils. The equations in Table 7.5 can be converted accordingly. Whether the cost for the manufacture of profiled vanes is justified or not, has to be examined in each individual case. Velocity and flow angle distribution at the diffuser inlet are also afflicted with some uncertainties. This is true especially at the hub where complicated flow conditions prevail already in the impeller. Flow separation is frequently observed near the diffuser hub and careful analyses and optimization steps are required to avoid or limit stall in this region. On the outer streamline the flow distribution also deviates from the inviscid calculation due to boundary layer effects and the tip clearance flow (see Chap. 5.7 for details). If the diffuser is designed with unprofiled vanes, the calculation is performed according to the channel model:

1. The absolute velocities and flow angles at the impeller outlet are known: $c_{3m} = c_m$, $c_{3u} = c_{2u}$, $\alpha_3 = \alpha_2$. These correspond to the values at the diffuser inlet, since radii as well as cross sections are constant. All of these quantities are functions of the radius.
2. The number of vanes is often selected between 5 and 8 while the criteria according to Chaps. 7.2.1 and 10.7.1 have to be taken into account. Axial length and vane number of the diffuser must be matched so that the ratio of vane length L to pitch ("solidity") is sufficiently large in order to achieve the intended flow deflection. This means approximately $L/t = 1$ to 1.5.
3. To limit dynamic vane loading and pressure pulsations a distance of $a/L = 0.05$ to 0.15 is required between impeller blades and diffuser vanes (L is the chord length of the impeller blades). With regard to the efficiency and head coefficient there is presumably an optimum distance even though generally applicable design rules are lacking. Excessive distances produce unnecessary friction losses and impair efficiency and head coefficient. At distances $a/L < 0.05$ the efficiency drops again according to the tests in [2], presumably because the non-uniformities of the impeller outlet flow are not sufficiently smoothed out at

the diffuser inlet. Greater losses generated by fluctuations of the approach flow velocity at the diffuser inlet are the consequence.

4. The vane angles at the diffuser inlet can be calculated according to Eqs. (3.18 and 3.19). The vane blockage is generally low but can be considered in c_{3m}'.
5. To limit the pressure losses in the column pipe and the outlet bend, the flow velocity in the column pipe should not exceed 5 m/s. The annular cross section at the diffuser outlet must be dimensioned accordingly.
6. Since the design aims at a swirl-free discharge flow with $\alpha_4 = 90°$ and the flow in the diffuser cannot be vane-congruent, the vane angle should be selected at $\alpha_{4B} = 94$ to 96°.
7. The vanes can be designed as arcs of circles. The leading edges can be rounded or shaped as elliptical profiles. The trailing edge is designed blunt (rounding on the outlet would be unfavorable because of the formation of vortex streets, Table 10.13).

7.7 Inducers

To reduce the required $NPSH_R$ of a pump it is possible to install an inducer upstream of the actual impeller. As an example, Fig. 2.1 shows the section of a single-stage process pump with inducer. An inducer typically allows a reduction in the value of the required $NPSH_R$ of a pump to half of the value without inducer.

The inducer increases the static pressure upstream of the impeller. Cavitation bubble development on the impeller blades is reduced or suppressed accordingly. As a consequence, the pump can be operated at a higher speed or a lower $NPSH_A$ than without an inducer, provided that the inducer does require significantly less $NPSH_R$ than the impeller.

Inducers for industrial application typically reach suction specific speeds of $n_{ss} = 400$ to 700, Chap. 6.2.4. These high suction specific speeds are achieved by small approach flow angles β_{1a} (refer to Fig. 6.21), thin leading edge profiles with low blade blockage and long channels in which the cavitation bubbles generated in the inducer implode at least partially before the fluid enters the actual impeller. The low approach flow angle and the small flow coefficient φ_1 at the inducer inlet require a considerably enlarged inlet area compared with the impeller. This is achieved by reducing the hub ratio and by enlarging the inlet diameter.

The recommendations and data for the inducer design discussed in the following largely originate from work done by the NASA published in [19, 29, 30, 45], and in [B.16, B.24]. The information is tailored for industrial applications with suction specific speeds of up to $n_{ss} \approx 600$. Inducers with far higher suction specific speeds are employed in the aerospace technology where even two-stage inducers are used at times. The design methods discussed in the following do not fundamentally differ from those of the high-performance inducers.

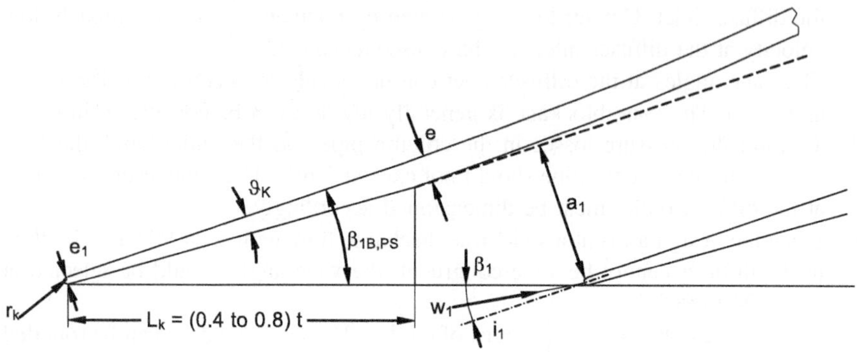

Fig. 7.33 Selection of the blade geometry at the inducer inlet

As is evident from the example in Fig. 2.1, the flow in the inducer is essentially axial. Often the inlet diameter is slightly larger than that of the downstream impeller. The hub diameter at the inducer inlet is designed as small as possible so that the hub contour is generally conical with the hub radius growing towards the impeller inlet. As with axial impellers, the static pressure in the inducer is increased in that the relative velocity is decelerated, Eq. (T3.3.8). This must be considered when deciding on one of two possible blading concepts:

- Blades with constant pitch designed as screw surfaces with blade angles $\beta_{1B} = \beta(L) = \beta_{2B}$ = constant on cylindrical sections. These blades only generate pressure when the incidence $i_1 = \beta_{1B} - \beta_1$ is greater than zero or for as long as the flow in the blade channel is decelerated according to $w_{1q} < w_1$. The advantage of this blading is its simple manufacture.
- Blades with variable pitch (Fig. 7.34, right) where the angle on cylinder sections grows from the inlet to the outlet as with a propeller pump: $\beta_{2B} > \beta(L) > \beta_{1B}$.

7.7.1 Calculation of Inducer Parameters

Determination of main dimensions and inlet conditions, Table 7.6

1. **Basis:** Known parameters are the pump selected for the relevant application, its speed n, flow rate Q_{opt}, $NPSH_R$ curve and the dimensions of the impeller. In addition, the suction specific speed to be achieved by the inducer or its $NPSH_R$ is specified. Furthermore, the boundary conditions for the mechanical design must be ascertained. Usually the approach flow angle is $\alpha_1 = 90°$ and the axial velocity is assumed as constant over the inlet cross section.
2. **Design flow rate:** Like axial impellers, inducers have a steep Q-H-curve. The pressure rise therefore drops rapidly to zero above the shockless entry (depending on the type). For this reason, the maximum required flow rate of the

pump must be taken into account when determining the design flow rate of the inducer by selecting $Q_{ind} = (1.1 \text{ to } 1.15) \times Q_{opt}$ or by making use of Eq. (7.6.7).

3. **Blade number z_{VLa}:** Inducers usually have 2 to 4 blades.
4. **Inducer inlet diameter d_1:** The inlet is designed for the selected suction specific speed. The optimum flow coefficient for the required suction specific speed can be selected from Eq. (T7.6.3). The appropriate inlet diameter can then be determined. Alternatively, the optimum inlet diameter can be calculated for selected coefficients λ_w and λ_c from Eq. (T7.6.1). The expected suction specific speed is obtained from Eq. (T7.6.4). Finally, it is also possible to calculate $\varphi_{1,opt}$ from Eq. (T7.6.2). All of these options yield the same result. For inducers with axial inlet $\lambda_c = 1$ to 1.1 and $\lambda_w = 0.02$ to 0.08 can be assumed. The inlet diameter must also be matched to the inlet nozzle and the downstream impeller, which can imply considerable restrictions on the design options in case of standard pumps.

Iteration may be required with respect to the hub ratio. Two uncertainties must be considered in this calculation: (a) As discussed in Chap. 6.3.2, the coefficients λ_c and λ_w depend on the geometry; the values assumed in the calculation will be attained only if the blade design is as good as in the measurements from which the coefficients had been derived. (b) The $NPSH_R$ of the pump is not only governed by the inducer but also by the downstream impeller. If the impeller has not been specifically designed for operation with the inducer, the design suction specific speed of the inducer may not be reached under certain conditions, Chap. 7.7.3. This risk is particularly high in the range of large flow rates ($q^* > 1$); it grows with increasing suction specific speed of the inducer.

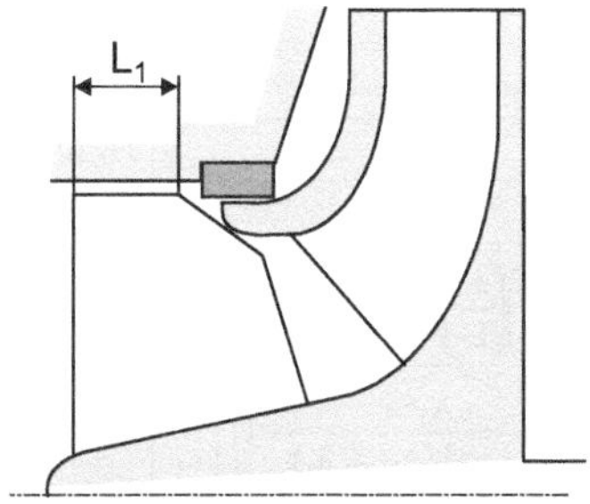

Arranging an inducer in the meridional section

Fig. 7.34 Impeller with variable-pitch inducer

Table 7.6 Inducer design, $\alpha_1 = 90°$ (Fig. 7.33)			Eq.
Optimum inlet diameter for given λ_w	$\lambda_c = 1.1$ $\lambda_w = 0.02$ to 0.08	$d_{1,opt} = 3.25\left(\frac{Q_{La}}{n\,k_n}\right)^{\frac{1}{3}}\left(\frac{\lambda_c+\lambda_w}{\lambda_w}\right)^{\frac{1}{6}}$	7.6.1
Optimum flow coefficient for given λ_w	$\varphi_{1,opt} = 0.1$ to 0.18	$\varphi_{1,opt} = \sqrt{\frac{\lambda_w}{2\,(\lambda_c+\lambda_w)}}$	7.6.2
Optimum flow coefficient for given n_{ss} $n_{ss,Ref} = 400$	$\varphi_{1,opt} = 0.15\left(\frac{n_{ss,Ref}}{n_{ss}}\right)^{0.93}$	$d_{1,opt} = 2.9\left(\frac{Q_{opt}}{k_n n \tan\beta_1}\right)^{\frac{1}{3}}$	7.6.3
Suction specific speed $n_{ss} = 350$ to 700	$\varphi_1 = \left\{\frac{n_{ss,Ref}}{n_{ss}}\right\}^{0.93}$ $n_{ss,Ref} = 52$	$n_{ss} = \frac{98}{(\lambda_c+\lambda_w)^{0.25}}\left(\frac{k_n}{\lambda_w}\right)^{0.5}$	7.6.4
Flow inlet angle at outer streamline	$\beta_{1,a,opt} = \arctan\varphi_{1,opt}$	$\beta_{1,a,opt} = \arctan\frac{240\,Q_{opt}}{\pi^2\,d_{1,opt}^3\,n\,k_n}$	7.6.5
Pressure surface blade inlet angle at outer streamline $i_1 = 2$ to $4°$	Criterion: design incidence: $i_{1,opt} = (0.35$ to $0.45) \times \beta_{1B,PS}$	$\beta_{1B,a,DS} = (1.5$ to $1.8) \times \beta_{1,a,opt}$	7.6.6
	Criterion: incidence at maximum flow rate: $i_{1,Qmax} = 1°$	$\beta_{1B,a,DS} - 1° \geq \beta_{1,max} = \arctan\frac{240\,Q_{max}}{\pi^2\,d_1^3\,n\,k_n}$	7.6.7
Blade inlet angles over blade height; same incidence for all streamlines		$\tan\beta_{1B,DS}(r) = \frac{r_a}{r}\tan\beta_{1B,a,DS}$	7.6.8
Blade leading edge profile (s. Fig. 7.33)		$\vartheta_k = (0.35$ to $0.5) \times \beta_{1B,a,DS}$	7.6.9
		$L_k = (0.4$ to $0.8) \times t$	
Check also: $i_1 \geq \vartheta_k + 0{,}5°$		$r_k = (0.03$ to $0.05) \times e$	
Cavitation coefficient only for $i_1 > 0$ $k_3 = 0.22$ to 0.65	$\sigma_3 = k_3\,k_n\left\{\frac{\varphi_1}{\sqrt{k_n}} - \frac{\varphi_1^2}{\tan\beta_{1B}}\left(1 - 0.36\frac{k_n}{\tau_1}\right) + \frac{3.18\times10^{-3}}{\tau_1\left(\tan\beta_{1B} - \varphi_1\,\tau_1\sqrt{k_n}\right)}\right\}$		7.6.10
Influence of tip clearance on n_{ss}	$\frac{n_{ss}}{n_{ss,s=0}} = 1 - k_4\sqrt{\frac{s}{b_1}}$	$k_4 = 0.5$ to 0.65	7.6.11
Blade length		$(L/t)_a = 1$ to 2.5	7.6.12
Hydraulic efficiency		$\eta_h = 1 - 0.11 \times L/t$	7.6.13
Deviation angle δ_2 Flow angle $\beta_2 = \beta_{2B} - \delta_2$		$\delta_2 = \left(2 + \frac{\beta_{2B}-\beta_{1B}}{3}\right)\left(\frac{t}{L}\right)^{1/3}$	7.6.14
Static pressure rise in inducer. The calculation is done for the mean streamline		$\psi_p = \eta_h\left[1 - \varphi_1^2\left\{\frac{A_1^2}{A_2^2\sin^2(\beta_{2B}-\delta_2)} - 1\right\}\right]$	7.6.15
Note: Use *pressure surface* blade angle $\beta_{1B,PS}$ rather than camber angle.			
Number of inducer blades: $z_{Ind} = 2$ to 4. Sweep-back $\varepsilon_{sb} = 65$ to $90°$ according to Fig. 7.37			
Hub diameter (given by mechanical design)	**Inlet:** Select hub diameter close to minimum: $\nu_{1,min} \approx 0.15$ **Outlet:** Match to impeller hub; a large hub diameter helps to increase the head developed by the inducer (centrifugal forces)		

5. **Inlet diameter at the inner streamline** d_{1i}: The lower limit for d_{1i} is obtained from a minimum hub ratio of $v_{1,min} \approx 0.15$. Smaller values of d_{1i} are unfavorable since excessive blade inlet angles would result at the inner streamline.
6. **Blade inlet angle** β_{1B}: When the inducer inlet diameter has been determined according to step (4), all quantities of the inlet velocity triangle according to Table 3.1 can be calculated. The first section of the inducer blades is designed according to Fig. 7.33. The calculation is based on the angle on the pressure surface of the blade $\beta_{1B} = \beta_{1B,PS}$ which is obtained from the flow angle plus an incidence of $i_1 = 2$ to $4°$ from Eqs. (T7.6.5 to 7.6.7). In this way, the flow and blade angles are calculated for outer, mean and inner streamline. When determining the inlet angles, it must be observed that the steep rise of the $NPSH_R$-curve (depending on the blade configuration) occurs near the flow rate of shockless entry. In addition to the design point it is therefore necessary to consider the maximum required flow rate at which the incidence should not drop below one degree to avoid cavitation on the pressure surface of the blades, Eq. (T7.6.7).
7. **Inlet throat area** A_{1q}: As with impellers (see Chap. 7.2.1, item 15), the deceleration of the relative velocity vector w_1 to the mean flow velocity w_{1q} in the throat area must be limited in order to avoid premature inlet recirculation. Conversely, the fluid must not be accelerated at the maximum flow rate with respect to overload cavitation. The throat area at the inducer inlet must therefore be selected so that the ratio w_{1q}/w_1 in the design point is in the range of 0.6 to 0.75 but does not exceed the value $w_{1q}/w_1 = 1$ at the intended maximum flow rate.
8. **Inducer outlet diameter** $d_{2,a}$ **and** $d_{2,i}$: The diameters on the outer streamline and on the hub are obtained from the dimensions of the downstream impeller. A conical hub of the inducer with $d_{2i} > d_{1i}$ is favorable since the head is increased through centrifugal forces and since the flow on the hub requires less severe deflection with increasing d_{2i} (lower tendency to stall). If the inducer diameter at the outlet is smaller than at the inlet, $d_{2,a} < d_1$, the first cylindrical part (i.e. dimension L_1 in Fig. 7.34) must be designed sufficiently long as to cover at least the throat area of the inducer.

Relationships at the outlet of the inducer: The blade outlet angle must be chosen so that the static pressure in front of the impeller increases sufficiently in the specified operating range to cover the NPSH requirement of the impeller. The effect of the static pressure rise in the inducer on the NPSH requirement of the system inducer/impeller is explained in Fig. 7.35 which was calculated on the basis of measurements on a process pump. The measured $NPSH_{3,imp}$ of the impeller alone was plotted against the flow rate, curve 1.

The static pressure rise H_p in the inducer is plotted as curve 3. Curve 3 passes through zero at Q_B. The inducer operates as a pump at $Q < Q_B$, but as a brake at $Q > Q_B$, (Chap. 12). If the static pressure rise in the inducer (curve 3) is subtracted from the $NPSH_{3,imp}$ of the impeller (curve 1: without inducer), curve 5 is obtained which passes through zero at Q_u, In the domain with $Q < Q_u$, (that is at

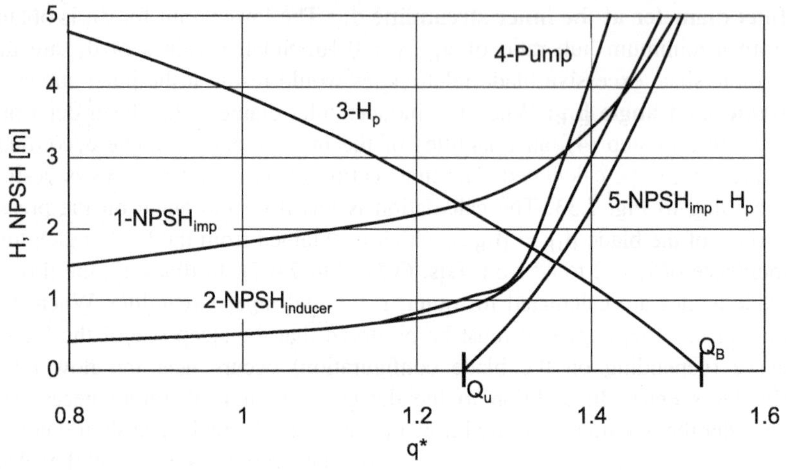

Fig. 7.35 Inducer characteristics

$H_p > NPSH_{3,imp}$) curve 5 is situated below the flow rate axis. In this domain the $NPSH_R$ of the system inducer/impeller is largely determined by the $NPSH_{3,inducer}$ of the inducer. In the region $Q_u < Q < Q_B$ the control of the $NPSH_R$ of the pump passes from the inducer to the impeller which now obtains insufficient pressure. The operating limit is located to the left of curve 5 and left of the point in which curves 1 and 2 interChap. However, it is not easy to accurately predict the steep rise of curves 2 and 5. In brake operation $Q > Q_B$, curve 5 is located above curve 1. Here the pump cannot be operated under any circumstances because the impeller receives less inlet pressure than without an inducer. The calculated NPSH requirement $NPSH_{3,inducer}$ of the inducer is plotted as curve 2 while curve 4 constitutes the measured $NPSH_3$ of the system inducer plus impeller.

These relationships determine the procedure for calculating the inducer outlet as follows:

9. The measured or calculated $NPSH_{3,imp}$ of the impeller (without inducer) is plotted as a function of the flow rate, curve 1.
10. The curve of the required $NPSH_{3,inducer} = f(Q)$ of the inducer is estimated according to Eq. (T7.6.10) [30]. Compared with other correlations this relationship has the advantage that it produces a steep rise of the curve with increasing flow rate when the incidence tends to zero (the nominator of the term on the right side of this equation then tends to zero). Since the constant k_3 varies over a wide range, it is recommended to estimate the suction specific speed at the design point from Eq. (7.23a) and use Eq. (7.6.10) only for the shape of the curve $NPSH_3 = f(Q)$.
11. A blade outlet angle β_{2B} on the outer streamline is selected and the static pressure rise in the inducer calculated. This is done using Eq. (T7.6.15) which can be derived from Table 3.3. The head coefficient is calculated on the mean

streamline. The flow deflection is taken into account by way of the deviation angle using Eq. (T7.6.14). The hydraulic efficiency of the inducer can be estimated from Eq. (T7.6.13). Depending on the flow rate ratio q* and the geometry, it is in the range of $\eta_h = 0.7$ to 0.9. For this calculation it would also be possible to use correlations for the head coefficient given in the literature. When applying correlations, it must be considered that individual geometrical parameters such as the hub diameters on the inlet and outlet of the inducer cannot be assessed in their effect on the pressure increase. Furthermore, the geometries of the inducers examined in the literature are not adequately known. In addition, the *static* pressure rise (not the head rise or the total pressure rise) must be used as a criterion since it is solely responsible for the NPSH improvement.

12. The rise in the static pressure caused by the inducer must be calculated for the flow rate range between shockless entry or $H_p = 0$ and the flow rate Q_u where H_p is greater than $NPSH_{3,imp}$ of the impeller alone. At partload the calculation becomes uncertain due to growing inlet recirculation. H_p increases due to the centrifugal effect (streamline displacement) so that roughly linear characteristics $H_p = f(Q)$ are obtained which tend towards the point $\psi_{p(Q=0)} = 1.0$.
13. The difference between $NPSH_{3,La}$ (curve 1) and the static pressure rise H_P (curve 3) gives an indication of the maximum flow rate which is definitely to the left of the resulting curve 5.
14. To facilitate assessing this limit, the curve $NPSH_{3,inducer} = f(Q)$ of the inducer (curve 2) is also taken into account.
15. If the maximum flow rate does not meet the requirements, the calculation according to steps (10) to (13) must repeated with another outlet angle. With respect to partload operation, over-sizing should be avoided.
16. When the outlet angle has been established, all parameters of the outlet velocity triangle according to Table 3.2 can be determined. To verify the configuration, w_2/w_{1a} should be calculated too. To avoid premature flow separation and efficiency loss, the deceleration ratio w_2/w_{1a} should not fall below the value of 0.7 ("de Haller criterion"). The lift coefficient according to Eq. (T7.5.8) should not exceed the value $\zeta_{a,a} = 0.5$ on the outer streamline.
17. The blade angles on the mean and inner streamlines are selected according to $u \times c_u$ = constant while a compromise is made on the hub (if required) to avoid excessive blade angles, Chaps. 7.5 and 7.6. The larger the hub diameter, the easier it is to configure this free-vortex blading.

The calculation of the pressure rise in the inducer is quite uncertain since flow deflection and hydraulic losses are affected by intense secondary flows which occur as the result of thick boundary layers and centrifugal forces, Fig. 5.7.

7.7.2 Design and Shaping of an Inducer

After determining the inlet and outlet diameters as well as the blade angles, the inducer can be designed. The design follows Chap. 7.2.2. The following information can help to select appropriate blade shapes and hub contours:

- Blade profile at inlet: The first section of the blade is usually designed wedge-shaped with a leading edge as thin as possible. The minimum required blade thickness is given by blade stresses and natural frequencies. The length L_k of the wedge-shaped portion is between 50 and 80% of the pitch, Fig. 7.33. The wedge angle ϑ_k, which makes up the profiling on the suction surface, is selected as $\vartheta_k = (0.3 \text{ to } 0.5) \times \beta_{1B,a,PS}$. The thin inlet profile is important to obtain good suction capability. It also means low cavitation inception coefficients ($\sigma_i \approx 0.2$ to 0.3) in shockless approach flow. However, σ_i increases rapidly even with a minor incidence, as shown in Fig. 7.36.
- The blade angle β_{1B} is best kept constant over the first part of the blade length, approximately up to x/L = 0.25, in order to produce less deflection of the flow and consequently reduce the low-pressure peak and the generation of cavitation bubbles.
- If the pressure distribution is calculated, it should be checked that the blade design ensures that the static pressure in the throat area on the blade suction surface exceeds the vapor pressure, that is $p_{ss(x=t)} > p_v$.
- A conical hub fosters the pressure build-up in the inducer due to centrifugal forces. At the same time the tendency towards flow separation decreases with the amount of hub taper that can be designed.
- The blade lengths are selected in a relatively large range L/t = 1.5 to 2.5. If the overhung of the inducer permits, values in the range of 1.5 to 1.8 should be the target. If the blades are made too short, the pressure rise in the inducer is impaired since the deviation angle increases. Moreover, the suction capability suffers and there is a growing tendency for pressure pulsations at partload. Blades with L/t > 2.5 do not seem to bring about any advantages.
- The blade outlet angles on the outer streamline generally are below 20°, but this depends on the inducer and impeller inlet diameters. The blades are slightly tapered towards the outlet.

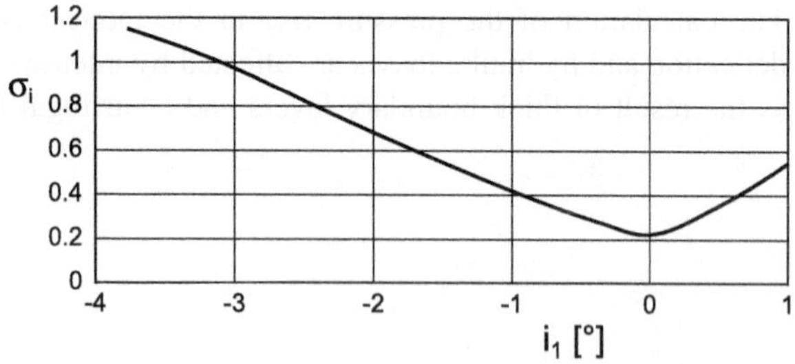

Fig. 7.36 Visual cavitation inception of an inducer as a function of the incidence

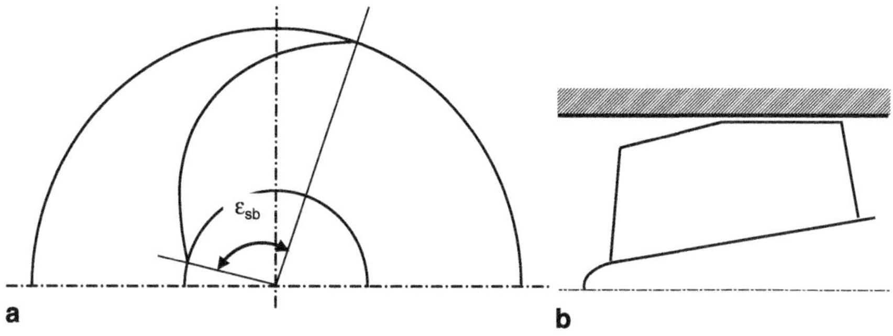

Fig. 7.37 Sweep-back of the inducer blade leading edge

- Efficient inducers are nearly always designed with a "sweep-back" according to Fig. 7.37a. A sweep-back does not only increase the suction specific speed but also moderates the tendency to pressure pulsations. During the tests in [3] the optimum wrap angle ε_{sb} (defined according to Fig. 7.37) lay between 65 to 90° on the outer streamline. At half the blade height ε_{sb} should be selected roughly half as large as on the outer diameter. Compared with a basic test with ε_{sb} = 29°, a suction specific speed increase from n_{ss} = 280 to n_{ss} = 510 was obtained with optimum sweep-back. The cavity extensions observed were correspondingly smaller and the pressure pulsations were also lower. It should be noted that the blade leading edge in Fig. 7.37 is already strongly inclined towards a radial line on the hub, i.e. it does not join the hub radially.
- The outer contour of the inducer is often tapered at the inlet, Fig. 7.37b. At the inducer inlet there is therefore a large gap between the blades and the casing. This design also reduces the pulsations at low partload and low $NPSH_A$.

7.7.3 Inducer Performance Prediction

If no test data for the blade parameters (λ_w and λ_c) are available for the type of design intended, a statistical correlation can be used to determine the expected n_{ss} and $NPSH_R$ from selected geometric parameters. In order to establish such a correlation, inducer tests from various sources (including [82–85]) were analyzed. The results are shown in Fig. 7.37-1 were the cavitation coefficient is plotted as function of the quantity x_σ which contains the flow coefficient and geometric data. The correlation, Eq. (7.23a) is based on the data near design flow (symbol x), while the graph also shows data at off-design conditions (symbol Δ). Most of the data points are within ±15% of the correlation, while the scatter is larger at off-design operation (not surprisingly). In view of the low absolute $NPSH_R$ the uncertainty of the test data is presumably even greater than ±15%. Furthermore, the data from literature are often incomplete and not always easy to interpret. The plot demonstrates

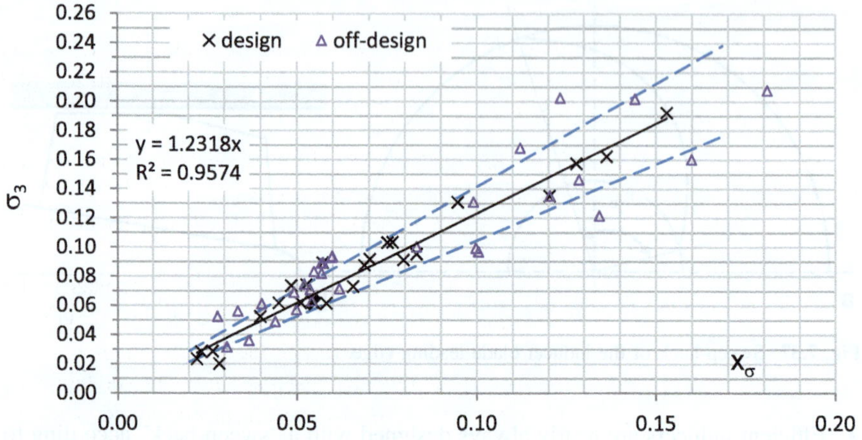

Fig. 7.37-1 Cavitation coefficient for inducers correlated by Eq. (7.23a). The dashed lines mark ±15% of the correlation

that the correlation can also be used to predict the $NPSH_R$ as a function of the flow rate, which is important because the BEP of an inducer is not well defined (the BEP of the pump is given by the volute and the impeller - and with low n_q mostly by the volute). An evaluation of the factor k_3 in Eq. (7.6.10) from the same test data gave no useful correlation.

$$x_\sigma = \varphi_1^{0.43}\varphi_{1,SF}^{0.75}\left(\frac{L}{t}\right)_a^{-0.2}\left(\frac{d_{1i}}{d_{1a}}\right)^{0.3}\left(\frac{d_{2a}}{d_{2i}}\right)^{-0.1}\left(1+\frac{\varepsilon_{sb}}{100}\right)^{-0.47}\left(\frac{z_v - La}{3}\right)^{0.24}$$

$$\sigma_3 = 1.26 x_\sigma \tag{7.23a}$$

For $L/t > 2$, set $(L/t)_{max} = 2$ in the correlation since blades longer than that are not expected to bring much further improvement. The correlation covers the following range of design data (n_{ss} = 280 to 800).

Table D7.3-1 Inducer data to Fig. 7.37-1

	z_{La}	d_{1a}	d_{1i}/d_{1a}	d_{2i}/d_{1i}	φ_1	i_{1a}	$\beta_{1B,a}$	$\beta_{2B,a}$	$(L/t)_a$	ε_{SB}	$\varphi_{1,SF}$	$\varphi_1/\varphi_{1,SF}$
		mm	–	–	–	°	°	°		°	–	
Average			0.27	1.4	0.15	2.8	11	14	2.0	32	0.20	0.78
Maximum	4	300	0.53	2.9	0.30	7.5	19	22	3.3	93	0.34	1.20
Minimum	2	50	0.13	1.0	0.057	−1.4	6.5	8	1.0	0	0.11	0.42

The main parameters according to the correlation are:

(1) The inlet flow coefficient φ_1: Low $NPSH_R$ require low φ_1.

(2) The coefficient of shockless entry flow $\varphi_{1,SF} = \tan\beta_{1B,a}/\tau_{1a}$ accounts for the effect of the blade inlet angle at the outer streamline: low $NPSH_R$ requires low $\beta_{1B,a}$ in order to reduce blade loading and cavitation.
(3) Length to pitch ratio$(L/t)_a$: Longer blades reduce blade loading and tend to reduce cavitation. However, $(L/t)_a > 2.5$ to 3 must not be surpassed. Above $(L/t)_a > 1.5$ to 2 little or no improvement is expected. Even though extra-long blades have little effect on the head produced by the inducer, cavitation bubbles may diminish in long channels.
(4) Number of blades: reducing the blade number increases the triangular section up to the throat area and reduces blade blockage; both effects reduce the $NPSH_R$. However, the axial length of the inducer increases for a selected ratio $(L/t)_a$. For this reason, 3 blades are often selected even though two blades would promise lower $NPSH_R$.
(5) Reducing the hub ratio at the inlet, d_{1i}/d_{1a} increases the flow area and reduces $NPSH_R$.
(6) Reducing the hub diameter d_{2i} at the inducer outlet (increasing the ratio d_{2a}/d_{2i}) seems to be beneficial - maybe due to larger channels.
(7) Sweep back ε_{SB}: improves $NPSH_R$ and reduces pressure pulsations and instabilities of the cavity flow.

Overall, these parameters and their trends are plausible (in fact most act similar to impellers). Attempts to make a correlation with the ratio $\varphi_1/\varphi_{1,SF}$ failed because the absolute blade angle or stagger angle has a dominating influence on the blade loading, hence the pressure minimum at the leading edge and cavitation.

7.7.4 Inducer Design Systematic

Similar to the concept introduced in Chap. 7.14 for impellers, a design systematic can be developed for a range of inducers. The idea is to use a set of rules and equations by which an inducer can be designed for any given flow rate, speed and inlet diameter. The latter is given by the pump into which the inducer shall be installed. The objectives and concepts are:

- "Design systematic" means that the inducer *geometries are fully described by analytical equations.* This approach will reduce the uncertainty of $NPSH_R$-prediction although the endeavor is limited by the effect of the impeller downstream of the inducer
- The performance curves are given by: $\sigma_R = f(\varphi_1)$and $\psi_p = f(\varphi_1)$
- The inducer geometry is fully described in dimensionless terms. Therefore, the inducers can be applied *to any pump (independent of n_q)* by:
  - Scaling of any inducer selected from the range to any size and speed
  - Develop inducer geometry for any new φ_1 not yet designed

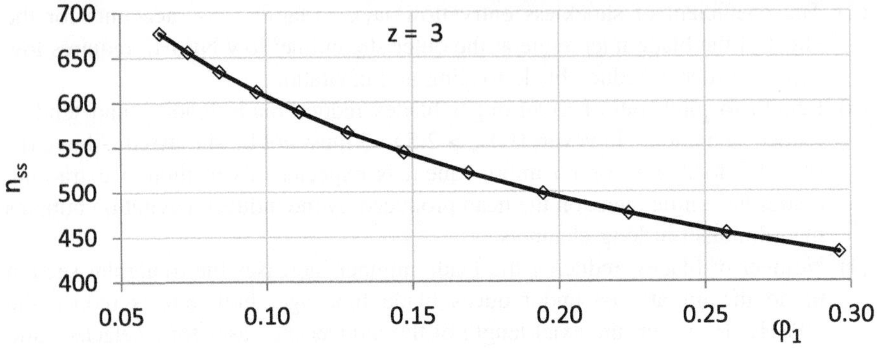

Fig. 7.37-2 Suction specific speed for a series of inducers with 3 blades

  - Since the geometry follows from analytical equations, the new geometry is consistent with the design of the adjacent geometries (designed for lower resp. higher flow coefficients)

- The range covers the flow coefficients used for industrial inducers. As an example, Fig. 7.37-2 shows the suction specific speeds as function of the inlet flow coefficient.

For any given application n, Q_{opt} and impeller inlet diameter d_{1a} are specified. Selecting the inducer inlet diameter (equal to d_{1a} or d_{sp} for example) and the hub diameter at the inlet, φ_1 and n_{ss} can be calculated. If the n_{ss} is not sufficient, the inducer inlet diameter, must be increased.

7.7.5 *Matching the Inducer to the Impeller*

Frequently the inducer is mounted upstream of the impeller of a standard pump whose $NPSH_R$ is unable to satisfy the specific requirements of the application. If an impeller of moderate suction capability is combined with an inducer of high suction specific speed, there is, however, no guarantee that the expected reduction of the $NPSH_R$ will actually be achieved. Especially the maximum flow rate can be limited through excessive incidence of the downstream impeller if cavitation occurs on the blade pressure surfaces. To check this, the flow angle β_2 at the inducer outlet is calculated from $\beta_2 = \beta_{2B} - \delta_2$ with δ_2 from Eq. (T7.6.4) (or according to Table 3.2). If β_2 is larger than the blade inlet angle β_{1B} of the downstream impeller, the blade leading edges of the latter can be adjusted to a certain extent by cutting back the blades and profiling them on the pressure surface. To keep the discharge angle as low as possible, the pressure increase in the inducer should not be chosen unnecessarily large.

Figure 7.38 shows measured flow velocities and angles at the outlet of an inducer. Characteristically the meridional velocity near the outer streamline is far

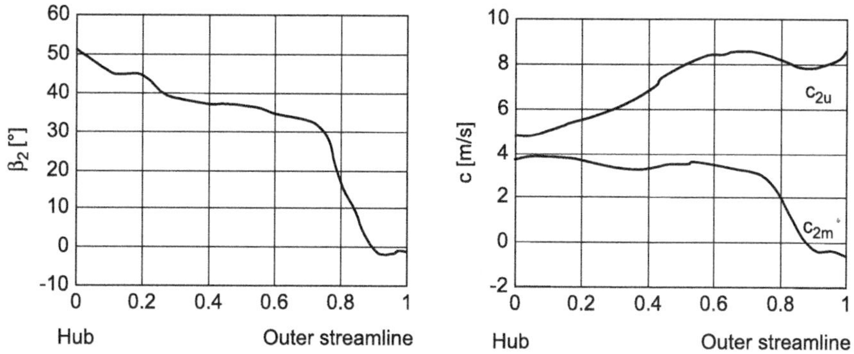

Fig. 7.38 Outlet flow angle and absolute velocity at the outlet of an inducer, Sulzer Pumps

below the mean value while the circumferential velocity is above average. Consequently, the discharge angles are considerably smaller on the outer streamline and far greater near the hub than would be expected from the calculation. These three-dimensional effects are already pronounced in the design point and are reinforced at partload (see also Chap. 5.2 and Fig. 5.7). The deficit in the meridional velocity near the outer streamline is also caused by the tip clearance flow and the boundary layer thickness. With growing tip clearance the deviation angle or the difference between blade and flow angles increases. Owing to the distorted flow conditions the calculation of the flow angles at the inducer outlet becomes quite uncertain.

An acceleration of the fluid from the inducer outlet to the impeller inlet causes a corresponding drop in static pressure and, consequently, an increased tendency to cavitation. The ratio of the velocity w_{1q} in the throat area of the impeller to the approach flow velocity w_1 (which is identical to the relative velocity w_2 at the inducer outlet) should therefore also be verified, Chap. 7.2.1 item (15): at the maximum flow rate $w_{1q}/w_2 = 1$ should not be significantly exceeded.

7.7.6 Recommendations for Inducer Application

The large diameter ratio d_1/d_{1i} causes partload recirculation to start near the design point. At low flow rates recirculation then becomes very intense inducing a strong swirl in the inlet pipes of end-suction pumps. As a consequence of the rotation a parabolic distribution of the static pressure p(r) builds up at the inducer inlet. At low suction pressures a vapor-filled vortex can be generated in the center of the suction pipe. As discussed in Chap. 6.5.1, such vortexes can cause severe pulsations. By a suitable design of the inducer it is possible to largely avoid such kinds of pulsations. Favorable in this respect seem to be: long blades (high solidity L/t), variable pitch design, sweep-back and a tapered outer contour of the inducer. However, it is not yet possible to provide generally valid design criteria for mastering this highly three-dimensional two-phase flow. For this reason it is necessary to prove by way of experiment that no impermissible pulsations and excitation forces occur in the demanded operating range. To this end, the water in the test circuit has to be largely de-aerated since the pulsations are greatly attenuated in air-saturated water. If at all possible, the test should be done with the same circumferential velocity as in the planned operation.

As a rule, pumps for industrial application are required to operate also at low partload. To be able to satisfy this requirement without excessive recirculation, the suction specific speeds of industrial inducers are limited to approximately $n_{ss} = 500$ to 700. If this range is markedly exceeded, the $NPSH_3$ at partload increases so that the inducer may no longer satisfy its actual purpose of lowering $NPSH_3$ throughout the operating range.

Obviously the target should be to restrict the partload operation as far as possible and to entirely avoid operation at $q^* < 0.3$ (except during start-up).

Major cavitation zones are created in the inducer in most applications. Despite the low $NPSH_A$ - i.e. low implosion pressure - the implosion energy can therefore be considerable. While hardly any cavitation damage is to be expected when pumping hydrocarbons, the circumferential velocities of the inducers when pumping water have to be limited in order to reduce the risk of cavitation damage (especially with de-aerated cold water, refer also to Table 6.2). For pumping water it is therefore advisable to use high-alloy steels with suitable cavitation resistance. The limit for the circumferential velocity cannot be determined generally; it may be within a range of 25 to 30 m/s when pumping water and 35 to 40 m/s with hydro carbons. These limits can be inferred from published applications, e.g. [33]. Initial cavitation bubbles (corresponding to $NPSH_i$) are created in the gap between casing and inducer.

Inducers are able to some extent to transport gas-laden liquids. Inducers combined with closed impellers are claimed to handle gas fractions of up to 25% of the flow rate at the inducer inlet; combined with semi-open impellers 35% and with open impellers even 40%. The gas-handling capabilities depend on the geometry and should be confirmed by testing, see also Chap. 13.2.

The Q-H-curve and the efficiency virtually do not change near the best efficiency point when installing an inducer. Due to the energy transfer in the inducer the circumferential component of the absolute velocity at the inducer outlet is increased, but the downstream impeller has to perform correspondingly less work in accordance with the Euler equation, Eq. (T3.3.1). The efficiency of the pump tends to fall since the hydraulic losses in the inducer are proportionally slightly higher than in the impeller, but this effect is sometimes hardly measurable.

However, an inducer increases the shut-off head by several percent due to the intense inlet recirculation and the associated rise of the centrifugal head component, Chap. 5.4.1. In the overload range the head drops slightly compared with the pump without inducer but this is likely to become noticeable only when the inducer operates in the range $H_p < 0$ ($Q > Q_B$), Fig. 7.35.

Inducers have relatively thin blades with a large blade height. When the inducers are cast possible blade deformations must be monitored. In particular it should be

Table 7.7 Diffuser and volute design calculations		Eq.
Given	n, Q_{opt}, H_{opt}	
Diffuser or volute design flow rate	$Q_{Le} = Q_{opt} + Q_{s3} + Q_E$	7.7.1
Circumferential component of absolute velocity at impeller outlet	$c_{2u} = \frac{g\,H}{\eta_h\,u_2} + \frac{u_{1m}\,c_{1u}}{u_2}$	7.7.2
Circumferential component of absolute velocity at diffuser inlet	$c_{3u} = \frac{d_2\,c_{2u}}{d_3}$	7.7.3
Meridional component of absolute velocity at diffuser inlet. For volutes set: $\tau_3 = 1.0$	$c_{3m}' = \frac{Q_{Le}\,\tau_3}{\pi\,d_3\,b_3}$	7.7.4
Flow angle at diffuser inlet with blockage	$\tan\alpha_3' = \frac{c_{3m}'}{c_{3u}}$	7.7.5
Diffuser vane inlet angle	$\alpha_{3B} = \alpha_3' + i_3$; incidence $i_3 = \pm 3°$	7.7.6
Throat area of a volute of circular section or equivalent circle of a volute with an arbitary section	$X_{Sp} = \frac{Q_{Le}}{\pi c_{2u} r_2}\frac{\varepsilon_{sp}}{2\pi}$ $d_{3q} = X_{Sp} + \sqrt{2\,d_z' X_{Sp}}$; $A_{3q} = \frac{\pi\,d_{3q}^2}{4}$	7.7.7
Factor for diffusers or volutes with rectangular cross section	$X_{Le} = \exp\left\{\frac{Q_{Le}}{b_3\,c_{2u}\,r_2\,z_{Le}}\right\}$	7.7.8a
Outer radius r_a of the end section (throat area) of a rectangular volute	$r_a = r_z'\,X_{Le}$	7.7.8
Height of the end section (throat area) of a rectangular volute	$a_3 = r_a - r_z' = r_z' \exp\left\{\frac{Q_{Le}}{b_3 c_{2u} r_2}\frac{\varepsilon_{sp}}{2\pi} - 1\right\}$	7.7.9
Diffuser inlet width $f_{a3} = 1.1$ to 1.3	$a_3 = f_{a3}\,\frac{d_3}{2}\{X_{Le} - 1\}$	7.7.10
Diffuser inlet width calculated for a selected deceleration ratio c_{3q}/c_2	$a_3 = \frac{Q_{Le}}{z_{Le}\,b_3\sqrt{c_{2m}^2 + c_{2u}^2}}\left(\frac{c_2}{c_{3q}}\right)$	7.7.11
Return channel outlet width	$b_6 = \frac{Q_{Le}}{\pi\,d_6\,c_{6m}}$; $c_{6m} = (0.85 \text{ to } 0.9)\,c_{1m}$	7.7.12
Meridional component of absolute velocity at return channel inlet	$c_{5m} = \frac{Q_{Le}}{\pi\,d_5\,b_5}$	7.7.13
Circumferential component of absolute velocity at return channel inlet	$c_{5u} = \frac{c_{4u}\,r_{4m}}{r_{5m}}$	7.7.14
Flow angle at return channel inlet	$\alpha_5 = \arctan\frac{c_{5m}}{c_{5u}}$	7.7.15
Gap A and overlap x_{ov} (Fig. 9.1)	$\frac{gapA}{r_2} = 0.007 \text{ to } 0.01$; $\frac{x_{ov}}{gapA} = 2 \text{ to } 4$	7.7.16
Gap B	s. Table 10.2	
ε_{Sp} is given in radians; for volutes with equal spacing: $\varepsilon_{Sp} = 2\,\pi/z_{Le}$		

checked that the desired blade outlet angles are achieved: if β_{2B} is too small, the necessary pressure increase is not attained and the steep rise of the NPSH occurs at too low a flow rate. The thin blades also require a verification of the natural frequencies and stresses in order to prevent blade fractures.

When installing an inducer in an end-suction pump, the impeller overhung increases. This has to be considered when evaluating the shaft deflection and the critical speed. Also some additional excitation forces in radial direction must be expected. Radial excitation forces can (for instance) be created by the tip clearance flow or through periodically variable flow conditions that might develop in the channels of the inducer as the result of varying cavity volumes, [63].

7.8 Volute Casings

7.8.1 Calculation and Selection of Main Dimensions

In the volute casing the kinetic energy present at the impeller outlet is converted into static pressure with as few losses as possible. The volute casing then directs the fluid into the discharge nozzle or, in multistage pumps, into the following stage. Prior to calculating a volute, the boundary conditions for the casing design must be established. In particular it must be decided whether a single or double volute is required. The selection criteria are:

- **Single volutes** are the least expensive solution in terms of manufacturing costs; they are best accessible for the dressing of the cast channels. Their disadvantage consists in considerable radial forces which develop during operation at off-design conditions according to Chap. 9.3 through disturbances of the circumferential symmetry of the flow in the casing. The radial forces generate bearing loads, bending stresses in the shaft and shaft deflection which can threaten the reliability of the machine. The heads up to which single volutes can be sensibly used depends on the specific speed, the design of the pump (especially of the bearing housing), the shaft thickness and the bearing span or the overhung of the impeller. When pumping water, the limit is about $H_{opt} = 80$ to 120 m at $n_q < 40$. At high specific speeds a double volute may be indicated already for heads above $H_{opt} = 60$ to 80 m. When pumping liquids with densities considerably lower than water, the limit is correspondingly higher since the radial forces are proportional to $\rho \times g \times H$.
- **Double volutes** are employed when bearing loads, shaft stresses and shaft deflection become impermissibly high without measures for radial thrust reduction so that their control would require an excessively high design effort and cost. The rib between the inner volute and the outer channel also reduces the casing deformation under internal pressure, which facilitates the casing design at high specific speeds. The pressure to be applied during the hydrostatic test of the casing must also be considered.

- **Twin volutes** differ from double volutes mainly in that both partial volutes end in separate channels - and not in a common discharge nozzle as is the case with double volutes. Apart from special designs they are found in multistage volute pumps or vertical pumps where the partial volutes end in a central column pipe.
- **Three or four volutes** are sometimes used in borehole pumps (for example) instead of semi-axial diffusers. Their application allows smaller casing diameters to be used.

When volute type and design boundary conditions have been established, the main dimensions must be selected or calculated according to Table 7.7 in Chap. 7.9.2:

1. Wrap angle of partial volutes: In the case of double or twin volutes the wrap angle of the partial volutes generally amounts to $\varepsilon_{sp} = 180°$. In this case an even impeller blade number must be avoided in order to reduce pressure pulsations, Chap. 10.7.1. If nevertheless a 6-bladed impeller (for example) is to be employed it is advisable to reduce the wrap angle of the inner volute ε_{sp} to 165° or 170° so that two blades never pass the cutwaters simultaneously. A similar procedure should be employed in the case of multiple volutes. The selection of a wrap angle of 165° may reduce the risk of flow instabilities and alternating stall as discussed in Chaps. 5.8 and 10.7.3. This design option could be recommended for general use since the increase in static radial thrust as compared to 180° is moderate and a defined radial force might be advantageous with respect to bearing design and vibrations.

In the case of axial-split pumps wrap angles of less than 180° are sometimes selected to avoid that the center rib passes through the split flange of the casing (matching difficult because of casting inaccuracies). The radial thrust increases according to Chap. 9.3 and Table 9.4.

Double volutes with wrap angles under 90° must not be used since they are unable to reduce the radial forces. Wrap angles over 180° should be avoided as a matter of principle since the long outer channel causes additional losses, and high radial thrusts must be expected at $q^* > 1$ (see Chap. 9.3.4).

2. Casing design flow rate Q_{Le}: To ensure that the actual best efficiency flow coincides with the design flow rate, the volute must be designed for Q_{opt} (if dimensioning were done for another flow rate, the best efficiency point would be

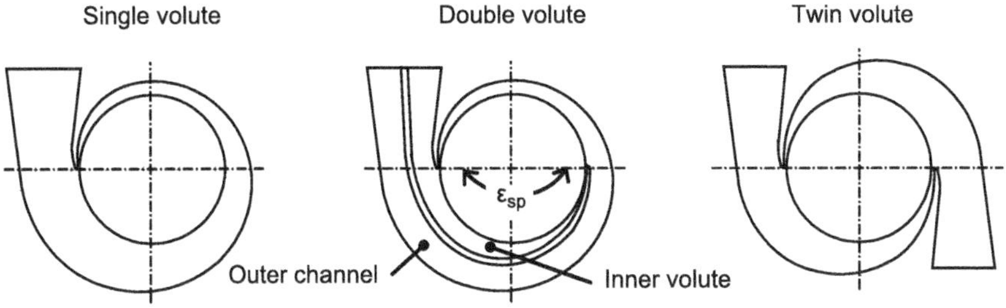

Fig. 7.39 Volute casings

shifted according to Chaps. 3.7 and 4.2). The design flow rate must be augmented by possible leakages (Q_E and Q_{s3}) which may flow *through the volute*, Eq. (T7.7.1). Note that the leakage losses at the impeller inlet do not flow through the volute.

3. Inlet velocity: The circumferential component of the absolute velocity at the impeller outlet is calculated from Eq. (T7.7.2) or Table 3.2. Downstream of the impeller it develops in accordance with the conservation of angular momentum as per $c_{3u} = c_{2u} \times r_2/r_3$, Chap. 3.7. With some types of pumps a diffuser or stay vanes according to Chap. 7.11 are arranged between the impeller and the volute. In this case the circumferential velocity c_{4u} at the outlet of these components must be used as inlet velocity to the volute.

4. Cutwater diameter d_z*: Between impeller and cutwater a minimum clearance (gap B) must be maintained to limit pressure pulsations and hydraulic excitation forces to allowable levels, Chap. 10. The ratio of the cutwater diameter $d_z^* = d_z/d_2$ is calculated from the formula given in Table 10.2 as a function of n_q and H_{st}.

5. Shape of the volute cross sections: The shapes must be selected so as to suit the pump type, taking into account casing stresses and deformations as applicable. Figure 7.40 shows some design options for the volute cross section. When designing the casing, also the requirements of economic manufacture of patterns and castings must be taken into account. With the rectangular and trapezoidal basic shapes, for instance, all corners must be well rounded for casting reasons and flat surfaces designed with a slope ("draft angle") in accordance with the pattern split plane. Concrete volutes have specific shapes to facilitate shuttering, Fig. 7.40f to h.

Rectangular and trapezoidal shapes offer the advantage to be developed on surfaces of revolution, something that facilitates design and manufacture. Volute casings with circular cross sections that cannot be arranged on surfaces of revolution are employed, for instance, when the volute is welded together from segments. Semi-axial impellers may have asymmetric cross sections as per Fig. 7.40d.

In double volutes the cross sections of the inner volute and the outer channel must be matched so that the casing outer wall is given an easily producible shape, and a simple transition to the circular discharge nozzle is achieved, Fig. 7.40e.

Basically, the designer has considerable freedom to configure the cross section without risking major losses in efficiency. It is expected that flat cross sections (similar to flat elbows) result in less intense secondary flow than circular cross sections, thereby generating fewer losses. For this reason, flat cross sections with a ratio of width B to height H in the range of B/H = 2 to 3 may be considered the optimum.

6. Inlet width b_3: The inlet width is obtained from the impeller outlet width b_2, the necessary width of the impeller sidewall gaps and requirements of the casing design. In particular the transition from the volute to the discharge nozzle should be smooth. This requirement means a relatively large b_3 in single volutes so that the ratio h/b at the volute throat area is close to 1.0. In a double volute the ratio h/b is rather near 0.5. The ratio b_3/b_2 depends on the specific speed; it can be selected in wide limits without major effect on the efficiency. Open impeller sidewall gaps

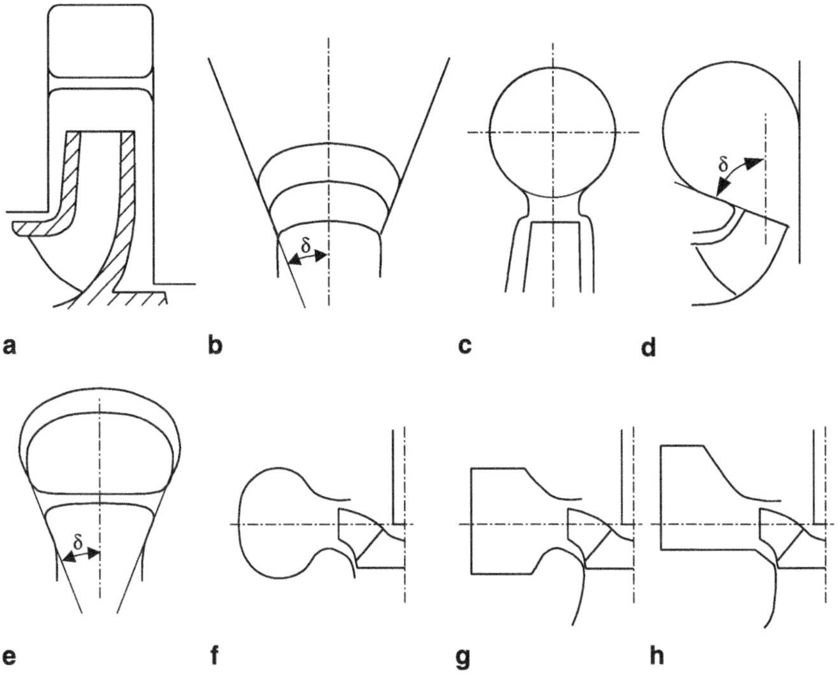

Fig. 7.40 Cross section shapes of volute casings. **a** Rectangular volute, **b** trapezoidal, **c** circular, **d** asymmetric (semi-axial impeller), **e** double volute, **f** concrete volute 1, **g** concrete volute 2, **h** concrete volute 3

according to Fig. 7.40a are favorable with respect to efficiency and radial thrust. At small specific speeds they result in values of $b_3/b_2 = 2.0$ to 4.0. In some applications vibration problems were reported on pumps with impeller sidewall gaps wide open to the volute, see Chaps. 10.7.3 and 9.1 for design guidelines. However, large values of b_3/b_2 are prohibited for design reasons at high specific speeds and relatively wide impellers; $b_3/b_2 = 1.05$ to 1.2 may then be selected. At high n_q large ratios of b_3/b_2 are unfavorable because strong secondary flows and turbulent dissipation losses would result.

7. Cutwater profile: The cutwater should be profiled elliptically in order to render it as insensitive as possible to changes of the approach flow direction with changing operating conditions. In practice such profiling is feasible for large pumps. The cutwater is highly stressed and any profile must be short and thick (except the nose radius) to reduce the risk of cracks forming in the cutwater. The leading edge thickness e_3 of the cutwater can be defined by a circle; its diameter should be approximately $0.02 \times d_2$. In the case of sewage pumps or abrasive wear at least twice that cutwater thickness should be selected. For double volutes the required thickness of the center rib is obtained from stress calculations (and the minimum casting wall thickness required).

The cutwater has a camber angle α_{3B} with the circumferential direction which has to be designed to match the approach flow angle according to Eqs. (T7.7.3 to 7.7.6). As discussed in Chap. 3.7, it is recommended to design the cutwaters with low camber angles and incidences.

It is even advantageous to shape the cutwater with an asymmetric profile on the low-pressure side (i.e. the side facing the impeller). Then the incidence becomes negative, the pressure distribution is flatter and the tendency to flow separation at partload is reduced. This design feature can help with the Q-H-curve stability of double volutes. The resulting thick cutwater profiles are also beneficial with respect to mechanical stresses.

In order to reduce noise and vibrations an oval-shaped leading edge of the cutwater may be selected (Table 10.2). This design could also help to reduce erosion near the cutwater caused by horse shoe vortexes (slurry, dredge or waste water pumps).

8. Volute throat area A_{3q}: As explained in Chaps. 3.7 and 4.2, volute casings are mainly designed in accordance with the theorem of angular momentum conservation. Different approaches to the volute design did not produce any measurable improvements in efficiency, [17]. A partial volute with the wrap angle ε_{sp} must be designed for the flow $Q_{Le} \times \varepsilon_{sp}/2\pi$ where $\varepsilon_{sp} = 2\pi$ applies to single volutes ($z_{Le} = 1$). $\varepsilon_{sp} = \pi$ is used for twin volutes or double volutes ($z_{Le} = 2$) with $2 \times 180°$ and $\varepsilon_{sp} = 2/3\pi$ for a triple volute ($z_{Le} = 3$). If all partial volutes have the same wrap angle, $\varepsilon_{sp}/2\pi = 1/z_{Le}$ applies. The throat area of each partial volute must satisfy Eq. (3.14); this is obtained from:

$$\int_{r_z'}^{r_A} \frac{b}{r} dr = \frac{\varepsilon_{sp} Q_{Le}}{2\pi c_{2u} r_2} \tag{7.24}$$

Although the thickness of the volute cutwater causes local accelerations and excess velocities, it has only a minor impact on the best efficiency point flow rate (this may be visualized by comparison to a Venturi nozzle which has scarcely any effect on the flow rate through a long pipe at a given driving pressure differential). Integration therefore proceeds from the stagnation point radius $r_z' \approx r_z + e_3/2$ to the outer limitation of the volute cross section at the radius r_a, without taking into account the cutwater thickness, [68]. The integral of Eq. (7.24) can be solved analytically for rectangular cross sections of constant width $b = b_3$. The outer volute radius r_a is obtained from Eq. (T7.7.8) and the height of the throat area $a_3 = r_a - r_z'$ from Eq. (T7.7.9).

An analytical solution can also be given for volutes with circular cross sections. The diameter of a circular throat area d_{3q} is obtained from Eq. (T7.7.7). Also, d_{3q} can serve as an approximation for the rapid estimation and assessment of cross sections of any shape if considered as an equivalent cross section giving the actual throat area of the casing. As discussed in Chap. 4.2, large hydraulic losses which occur downstream of the volute in the diffuser shift the best efficiency point to smaller flow rates than would be obtained according to the conservation of angular momentum. In the case of double and twin volutes with low specific speeds the volute must therefore be designed for a slightly higher flow rate by substituting, for instance, $Q_{Le} = (1.05 \text{ to } 1.25) \times Q_{opt}$ in Eq. (T7.7.1).

7.8.2 *Design and Shaping of Volute Casings*

After the main dimensions have been established and the decisions in terms of type and basic shape of the volute have been made as per Chap. 7.8.1, the volute and subsequent diffuser can be designed:

1. Volute cross sections designed according to the conservation of angular momentum: The area distribution of the partial volutes is best developed on surfaces of revolution which are inclined at an angle δ relative to a normal to the rotor axis, Figs. 7.40b and 7.41. As mentioned in Chap. 7.8.1 item (5), the shape of the cross section is defined by straight lines and arcs of circles for casting reasons. The shape cannot be described analytically by a single expression. It is therefore generally expedient to design the cross sections so that they can be optimally integrated into the desired casing shape and to finally calculate the circumferential angle at which the cross section thus designed satisfies the angular momentum conservation, Fig. 7.41. If this design is done manually, the integral according to Eq. (7.24) is replaced by a summation of finite surface elements $\Delta A = b \times \Delta r$, where b and Δr have to be taken from the drawing. The circumferential angle on which the cross section under consideration has to be arranged is obtained according to Eq. (3.14) from:

$$\varepsilon = 360° \frac{c_{2u} r_2}{Q_{Le}} \int_{r_z'}^{r_A} \frac{b}{r} dr = 360° \frac{c_{2u} r_2}{Q_{Le}} \sum_{r_z'}^{r_A} \frac{b}{r} \Delta r \tag{7.25}$$

Step-by-step procedure (Fig. 7.41):

1. Use b_3, r_z' and δ to design the two surfaces of revolution on which the volute cross sections are to be developed.
2. Draw the outer limitation for various cross sections: AF1 to AF3 in Fig. 7.41. The height to which these cross sections are required can be estimated from the throat area of an equivalent volute with circular cross section from Eq. (T7.7.7).
3. Divide the cross sections into elements of height Δr located on the radii r. The width b of these surface elements can then be measured.
4. Enter the values b, r and Δr of all surface elements located in the cross section considered into a table.
5. Calculate the sum $\Sigma(r \times \Delta r/b)$ of all elements.

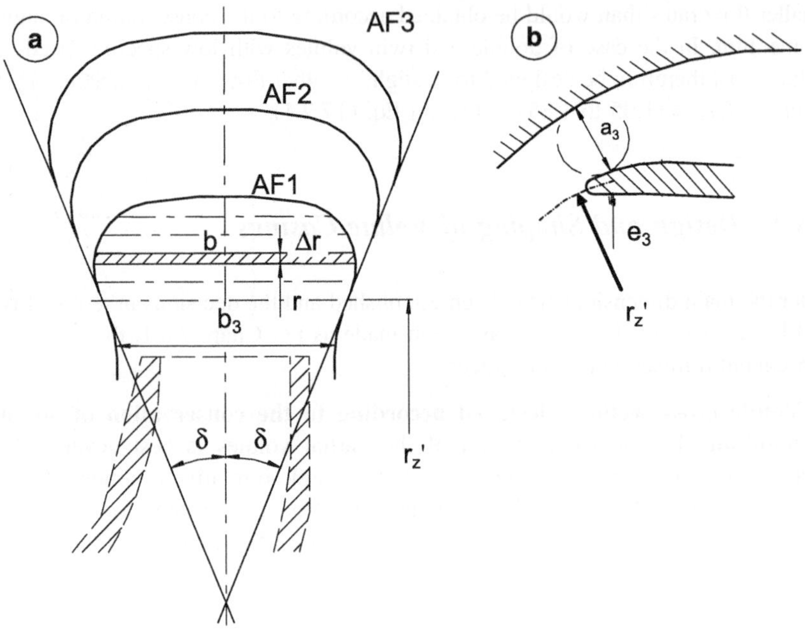

Fig. 7.41 Design of the volute sections. **a** Meridional section; **b** cutwater details

6. Enter this sum into Eq. (7.25) and calculate circumferential angle ε on which the designed cross section has to be arranged.
7. The outermost point of the cross section on radius r_a is obtained at the same time so that the layout of the outer volute shape $r_a(\varepsilon)$ can be drawn.

As a rule, this task will be completely or at least partly performed on the computer describing the cross sectional shapes by a set of equations.

2. Volute cross sections designed for constant velocity: Alternatively, volute casings are frequently designed by assuming a constant velocity in all cross sections over the circumference, [B.2]. The volute throat area A_{3q} is again calculated in accordance with the conservation of angular momentum from Eq. (7.24) or Eqs. (T7.7.7 to 7.7.9) - otherwise the specified best efficiency point would be missed. The velocity in the throat area then amounts to:

$$c_{3q} = \frac{Q_{Le}}{A_{3q}} \quad \text{with} \quad A_{3q} = \int_{r_z'}^{r_A} b\, dr \tag{7.26}$$

All the other cross sections of the volute $A(\varepsilon)$ for any wrap angles ε are now calculated from this velocity:

$$A(\varepsilon) = \frac{Q(\varepsilon)}{c_{3q}} = \frac{Q_{Le}}{c_{3q}}\frac{\varepsilon}{\varepsilon_{sp}} = A_{3q}\frac{\varepsilon}{\varepsilon_{sp}} \tag{7.27}$$

Thus the cross sections are directly proportional to the circumferential angle. The efficiencies of the volutes designed for constant velocity are at least on a par with the ones designed according to the conservation of angular momentum. At low specific speeds even a slight advantage might be expected since the flow velocities in the first part of the volute are reduced and lower friction losses can thus be expected. In contrast, volutes designed according to the conservation of angular momentum might have an advantage at high specific speeds because of reduced secondary flow in cross sections which match perfectly. Furthermore, the casing dimensions become noticeably smaller at high n_q if the volute is designed to the theorem of angular momentum conservation. Following these arguments it appears that an optimum philosophy may be:

- Design volutes for $n_q < 25$ to 35 for constant velocity as per Eqs. (7.26 to 7.27)
- Design volutes for $n_q > 25$ to 35 according to the conservation of angular momentum as per Table 7.7.

3. Cutwater correction: Because of its finite thickness, the cutwater interferes with the flow, generating local excess velocities and deviations of the streamlines from the pattern given by the conservation of angular momentum. The interference increases with the thickness of the cutwater. When thick cutwaters are required, this effect cannot be neglected (e.g. sewage, dredge or slurry pumps). The relative blockage increases as the volute cross sections get narrower with falling specific speed. Consequently, an allowance for the cutwater thickness must be applied to the

Fig. 7.42 Volute casing with radial discharge nozzle; R_{sp} represents a mean streamline

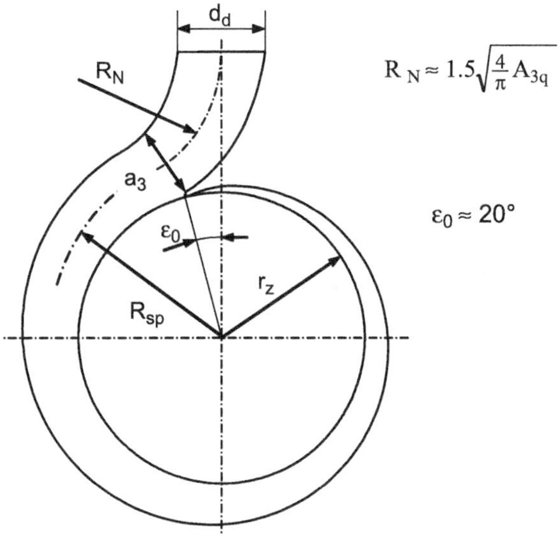

throat area of all volute casings; it can be estimated from Eq. (7.28), suggested in [67], where e_3 and a_3 are measured at the throat, see Fig. 7.41b:

$$\frac{\Delta a_3}{a_3} = 0.2\left(\frac{e_3}{a_3}\right)^2 \tag{7.28}$$

The cross sections upstream and downstream of the throat area must be adjusted in accordance with the cutwater correction so that continuous shapes are obtained. Inevitably this implies local deviations from the cross sections calculated from the conservation of angular momentum.

4. Discharge nozzle/diffuser: As a rule, a diffuser follows the volute throat area. In the case of single-stage pumps the diffuser simultaneously forms the discharge nozzle. At high specific speeds (above $n_q \approx 80$) the fluid should not be further decelerated downstream of the volute throat area - otherwise the discharge nozzles would become uneconomically large. In contrast, at low specific speeds the deceleration in the diffuser is significant and every care should be applied to design smooth hydraulic contours which minimize the losses. Maximum allowable diffuser opening angle, pressure recovery and losses are calculated according to Chap. 1.6. The diffuser length and the area ratio have to be matched so that the pressure recovery (or the c_p-value) is maximized. In the case of non-circular cross sections the optimization is done using the concept of an equivalent diffuser as per Eq. (1.45). The diffuser is designed for standardized discharge nozzle sizes. The length of the diffuser/discharge nozzle is limited by cost considerations and by the moments and stresses induced by pipeline forces (since these moments are proportional to the nozzle length). Experience shows that the data of straight diffusers according to Chap. 1.6 can be applied to curved discharge nozzles without correction, presumably since the secondary flow in the volute has a favorable effect on the boundary layers. In this way, the negative effect of a slight to moderate curvature is largely compensated for. A more prudent approach would be to apply the curvature correction according Eq. (1.54).

Fig. 7.42-1 Short and long cross-over for multistage pumps

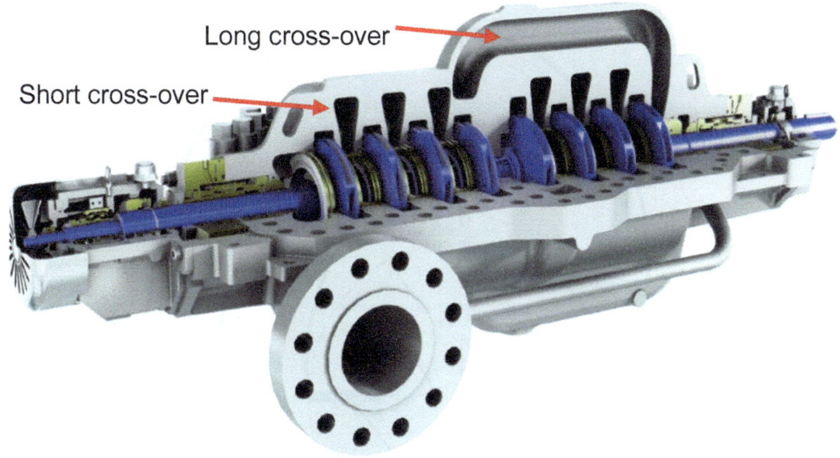

Fig. 7.42-2 8-stage pump with double volutes, Sulzer

In addition to the tangential discharge nozzle according to Fig. 7.39, radial discharge nozzles as per Fig. 7.42 and intermediate shapes with a slight offset are frequently found. According to [25], the volute should start roughly at an angle of $\varepsilon_o = 60°$ for tangential nozzles and at about $\varepsilon_o = 20°$ for radial nozzles. The mean radius of radial discharge nozzles can be estimated from Eq. (7.29), [25]:

$$R_N \approx 1.5\sqrt{\frac{4}{\pi}A_{3q}} \tag{7.29}$$

These recommendations can only give some hints; actually the angle ε_o and the radius R_N result from the design process.

Excessive local curvatures often result at the transition from the volute to the discharge nozzle if the calculated cross sections are strictly adhered to. In order to prevent distorted velocity profiles and local flow separation, the volute contour should be adjusted so that a smooth transition between volute and diffuser is obtained. The resultant local increase in flow area near the throat usually has only little effect on the best efficiency flow rate - provided the remainder of the volute is dimensioned correctly. Neither would it be possible to shift the best efficiency point towards lower flow rates by way of a local area reduction, without suitably adjusting the remainder of the volute. A local area reduction would be comparable to the effect of the cutwater thickness as discussed in Chap. 7.8.1 item 8.

5. Double volutes: At small and medium specific speeds the fluid is decelerated in the outer channel downstream of the throat area. Again this diffuser is dimensioned according to Chap. 1.6. Since the flow resistances of the two partial volutes with their unlike outlet channels generally differ, different flow rates are obtained in both channels. The individual flow rates in both volutes/channels can be estimated

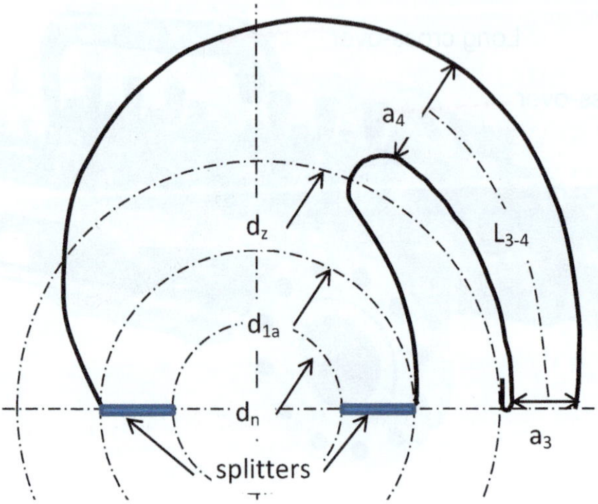

Fig. 7.42-3 Short cross-over design

according to Table 1.5. The pressure loss coefficients of the partial volutes and the various diffusers are calculated from Eqs. (T3.8.21) and (T3.8.22) to this end. The friction and deflection losses in the outer channel are calculated likewise from Table 1.4 and Chap. 1.5.1. Outer and inner volutes constitute two parallel flow resistances. After all resistances of the two channels have been combined into a total resistance according to Table 1.5, the two individual flows can be calculated. In this calculation it is assumed that both channels are subject to the same pressure difference, as given by the pressure at the discharge flange minus the pressure at the impeller outlet. In this way it is possible to evaluate the flow unbalance between both partial volutes and to optimize the geometry of the outer channel so that both partial volutes receive the same flow rates as far as possible. Since the impeller, in both partial volutes, works at different operating points (Chap. 9.3.3), the flow rates through the volutes adjust rather more favorably in reality than is borne out by the above estimate.

6. Multistage pumps with double volutes: In multistage pumps with double (or "twin") volutes (as per Figs. 2.9 and 7.42-2), "short crossovers" are used to guide the fluid from one stage to the next, Fig. 7.42-1 (left). Downstream of the volute throat area, the fluid is decelerated in a diffuser from the throat velocity c_{3q} to the velocity c_4 at the inlet to the cross-over which has the same function as the return channels in a vaned diffuser pump. With the back-to-back impeller arrangement, the "long cross-over" is employed to guide the fluid from the last stage of the 1st group of stages to the 2nd stage group, Fig. 7.42-1 (right).

Recommendations for the design of twin volutes for multistage pumps, Fig. 7.42-3:

a. The volutes and throat areas (A_{3q}, a_3) are designed according the equation for conservation of angular momentum as for any other type of volute, Table 7.7.
b. However, it may be useful to increase the design flow rate Q_{Le} of the volute by an amount similar to the factor f_{a3} in Table 7.7. In this way, high losses occurring downstream of the throat area could be compensated for to some extent. Such losses tend to move the BEP to lower flow and increasing the factor f_{a3} may be necessary to offset this effect. In this way, the throat area can be optimized to minimize the overall losses in the casing and to achieve the specified BEP flow rate.
c. Downstream of the throat area at a_3 the flow is decelerated in a diffuser to a_4, see Fig. 7.42-3.
d. It is most important to optimize this diffuser according to the rules set forth in Chap. 1.6: The allowable opening angle as per Eq. (1.44) must not be exceeded and the pressure recovery c_p should be maximized with the aid of Fig. 1.19 (or equivalent data) by properly selecting the area ratio A_R and the length ratio $L_{3\text{-}4}/R_{1,eq}$.
e. Since the diffuser is curved, Eq. (1.54) can be applied to estimate the pressure recovery and losses.
f. The area should increase in a linear fashion from a_3 to a_4 in Fig. 7.42-3. Specifically, flow deceleration should start right from the diffuser inlet in order to reduce friction losses in the region where velocities are highest.
g. After the diffuser outlet at a_4, the channel turns sharply and the area expands. Flow distortion and separation downstream of a_4 is the almost unavoidable consequence. In order to keep the associated losses as small as feasible, the velocity head in the area a_4 should be a low fraction of the head per stage: $c_4^2/(2gH_{opt,st}) = 0.02$ to 0.04. This, because the losses are directly proportional to this fraction. Optimization of the return channel has been attempted. An additional guide vane near the inner streamline has been studied in [73]. The effect would be to reduce the non-uniformity of the flow, losses and pressure pulsations. The splitter must not be too long within the diffuser in order to reduce friction and shock losses in the region with high flow velocities. Other studies attempted to reduce the curvature of the inner streamline by an increased wrap angle, [74, 75].
h. Effective flow splitters in the axial split-plane of the pump casing are essential to reduce the pre-rotation to the subsequent impeller, to stabilize the flow and help to prevent the Q-H-curve from drooping towards shut-off.
i. In multistage pumps, the cutwater distance ratio d_z/d_2 is usually made smaller than for single-stage pumps (see Table 10.2), but it should not be made smaller than for diffuser pumps.
j. If the impellers are arranged back-to-back as in Fig. 2.9, the head loss in the long cross-over has to be taken into account. In order to limit these losses, the velocity head should be made as low as compatible with the compactness of the design; again a range of $c_4^2/(2gH_{opt,st}) = 0.02$ to 0.04 could be envisaged. Typically, the long-cross-over (Fig. 7.42-1, right) consists of a flow turn at the inlet with a loss coefficient of $\zeta_{in} = 0.5$, a straight channel with $\zeta_{ch} = \lambda_R\, L/d_h = 0.2$, another flow

turn radially inwards $\zeta_b = 0.3$ and the inlet losses to the next stage $\zeta_E = 0.2$. The total loss coefficient is the sum of these individual loss components: $\zeta_{L\text{-}XO} = 1.2$. The head loss in the long cross-over $\Delta H_{L\text{-}XO}/H_{opt,st} = 0.024$ to 0.048 referred to one stage which would result in a head and efficiency loss of a 10-stage pump 0.24 to 0.48% assuming $c_4^2/(2gH_{opt,st}) = 0.02$ to 0.04. These figures can just give a rough indication; for any given design, the losses must be estimated from the actual geometry

7. Concrete volutes: are sometimes used for pumps with flow rates in excess of about 10 m^3/s and for heads up to 30 m. If the volutes are constructed without steel lining, the maximum local flow velocity must be limited to 10 m/s to avoid erosion of the concrete. For radial impellers, cross sectional shapes according to Fig. 7.40f, g are preferred, while shapes with a flat base according to Fig. 7.40h are used for semi-axial pumps. Shapes g and h are less costly to build than shape f. However, they involve a loss of efficiency of approximately 1% at $n_q = 70$ to 100 (at still higher specific speeds the differences in efficiency are insignificant), [36, 59].

7.8.3 *Influence of the Volute Shape on Hydraulic Performance*

When shaping volute and discharge nozzle, design requirements derived from the overall machine concept must be taken into account. Even the manufacturing process of the volute casing has an impact on the design: Consider for example the difficulties of core support for long and narrow channels of double volutes for low specific speeds, or the special requirements of welded designs or concrete volutes.

The effects of various geometrical parameters of the volute casing on the hydraulic behavior of pumps with specific speeds of up to approximately $n_q = 70$ are discussed in the following. The tendencies revealed originate from a number of experimental investigations such as [6, 17, 31, 44, 72]. The numerical values indicated illustrate the approximate magnitude of the effects as found in the mentioned tests; these figures may not be considered as generic.

- Wide impeller sidewall gaps according to Fig. 7.40a (or large b_3/b_2) foster pressure equalization over the impeller circumference, in this way reducing the radial thrust by up to 50% (Chap. 9.3). Wide impeller sidewall gaps yield clearly better efficiencies at partload than narrow gaps between impeller and casing (typically up to 3%). At Q_{opt} the efficiency is also slightly better (approximately 1%) even though shock losses due to sudden deceleration of the meridional velocity c_{2m} develop according to Eq. (T3.8.24). Wide impeller sidewall gaps lead to lower hydraulic losses since the low-energy boundary layer is able to flow from the volute into the impeller sidewall gap on the outer wall, as shown in Fig. 9.3. In as much as this fluid does not flow back to the

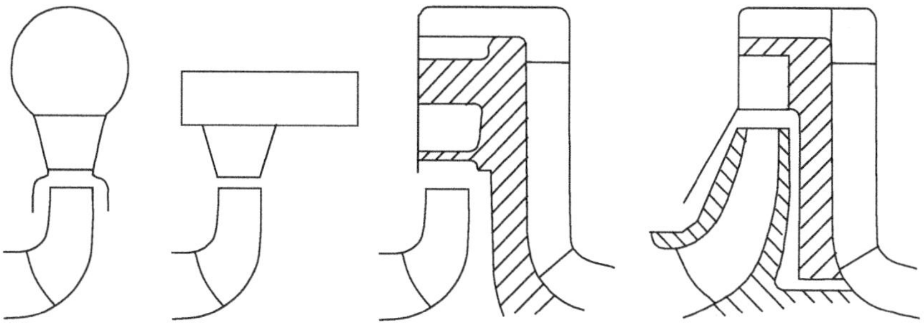

Fig. 7.43 Diffuser arrangements. **a** Diffuser with volute, **b** diffuser and annular casing as last stage, **c** overflow from diffuser to return vanes through vaneless annulus, **d** lateral overflow from diffuser to return vanes

suction casing in the form of annular seal leakage, it is re-accelerated on the rotating shroud. In this way the fluid enters the volute from the impeller shroud boundary layers with a high circumferential velocity contributing to the energy transfer.[6]

- The opening angle δ of the surfaces of revolution (Fig. 7.40b) has an influence on the height to width ratio H/B of the volute; it should be selected so as to minimize secondary flow, i.e. to get an optimum H/B ratio under the given design constraints.
- Cutwater distances of up to 10% of r_2 hardly have an effect on the efficiency in the case of thin cutwaters. If the distance is increased from 10 to 20%, the efficiency drops by approximately 1%. With thick cutwaters the efficiency falls rather continuously when increasing cutwater distance. If the impeller sidewall gaps are large, the effect of the cutwater distance on the radial force is small.
- A cutwater thickness of up to $e_3/d_3 = 0.04$ hardly has any effect on the efficiency at the best efficiency point. With thick cutwaters the partload efficiency increases since the cutwater becomes less sensitive to excessive incidence (provided the cutwater is properly profiled).
- A reduction of the cross sections in the outer channel of a double volute (or an increase of the resistance) yields greater radial forces - especially at $q^* > 1$.
- If the cutwater distances of partial volutes are not the same, the resulting asymmetries of the pressure distribution may lead to higher radial forces (the effect of casting tolerances must be considered).
- In radial discharge nozzles generally higher losses would be expected than in tangential nozzles because of the additional flow deflection. Tests reported in [31], for instance, yielded a loss in efficiency of 4% with $n_q = 23$ and measurements in [18] with $n_q = 45$ a loss of 1%. According to other authors, measurable differences should hardly be expected with optimally designed

[6]See also Chap. 3.10.3 for the effect of roughness on the pumping action of shrouds.

diffusers. The hydraulic conditions can be visualized as follows: The radial discharge nozzle produces a deflection in the opposite direction relative to the flow in the volute. This results in a certain equalization of the velocity profile at the volute outlet. The equalization is only efficient if both curvatures are correctly matched. At low specific speeds R_{sp}/a_3 is large while the curvature in the discharge nozzle is much more severe; the additional losses therefore predominate (R_{sp} is the radius of the mean streamline in the volute). At high specific speeds the conditions are much better: R_{sp}/a_3 falls with increasing n_q and moves closer to the curvature ratio R_N/a_3 in the nozzle; the counter curvature has a balancing effect in this case. In the light of this consideration, the above quoted test results on $n_q = 23$ and 45 are quite plausible. It can therefore be assumed that radial and tangential discharge nozzles roughly produce identical efficiencies at high specific speeds. Conversely, the losses in radial nozzles are expected to increase with falling specific speed implying a drop in efficiency and head.

7.9 Radial Diffusers with or Without Return Channels

7.9.1 Calculation and Selection of Main Dimensions

The diffuser serves to convert the kinetic energy present at the impeller outlet into static pressure with as little losses as possible. Various diffuser configurations are in use as shown in Fig. 7.43. In the case of single-stage pumps the diffuser discharges into a downstream volute casing or into an annular collector; the same applies to the last stage of multistage pumps. The diffusers of multistage pumps form a unit with the return channels which direct the fluid to the next stage (with the exception of the last stage).

The diffusers and return channels of multistage pumps are built in various types. While the diffuser inlet and the subsequent diffusing channel of all types are basically very similar in design, the overflow geometry from the diffuser outlet to the return vanes can be configured in quite different ways: (1) Diffuser and return vanes can be designed to form a single continuous channel according to Fig. 7.44 which is similar to a three-dimensional bend. This type of design tends to give the lowest hydraulic losses, but is more expensive in terms of design and manufacture. However, modern methods of 3D-CAD and numerically controlled manufacturing mitigate this disadvantage. (2) Diffuser and return channels can be separated as per Fig. 7.43c by a vaneless annulus. The fluid exiting radially from the diffuser is deflected in this annulus by 180° to enter the return channels radially. (3) The liquid flows laterally from the diffuser channels so that it enters the return channels with a 90° deflection (Fig. 7.43d). These basic shapes can be configured in quite different ways as shown by the examples in Fig. 7.45.

The diffuser designs discussed above feature an outer diffuser diameter d_{Le} which is typically 30 to 50% larger than the impeller diameter d_2. More compact

Fig. 7.44 Continuous diffuser and return vanes, Sulzer Pumps

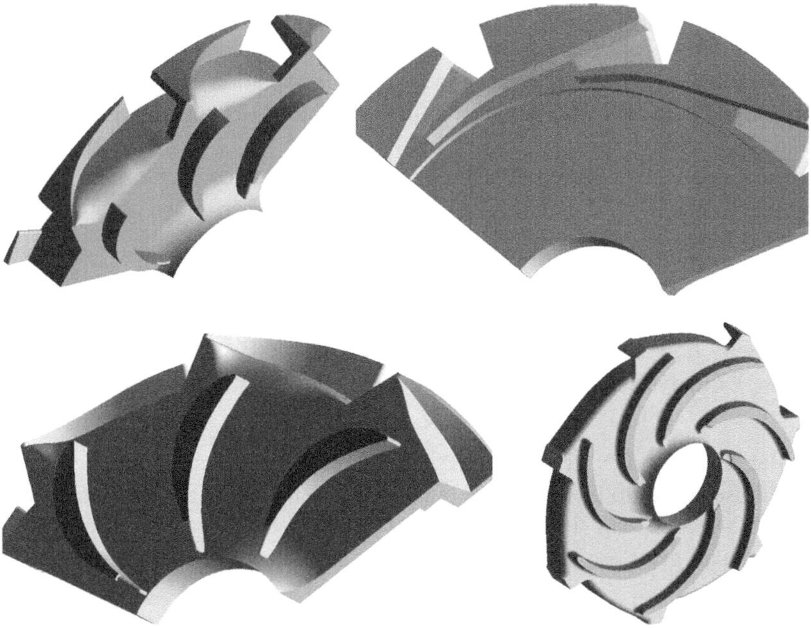

Fig. 7.45 Diffuser and return vanes as 3D-models, Sulzer Pumps

diffusers would be desirable in various applications, for example in submersible pumps. Figure 7.46 shows a most compact design where the radial impeller discharges axially into a diffuser whose outer diameter equals the impeller diameter, i.e. $d_{Le} = d_2$. The development of such diffusers is described in [51–54, 66]. After

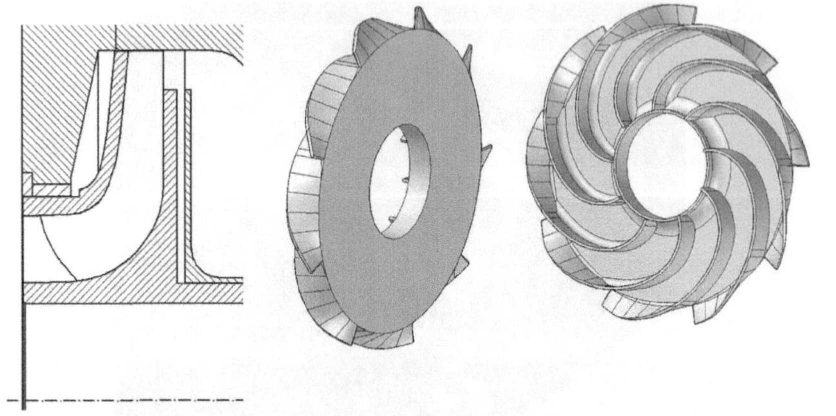

Fig. 7.46 Compact diffusers, University Kaiserslautern, [51–53]

extensive optimization of this type of diffuser it was found that the loss in efficiency was only about 2% compared to diffusers according to Fig. 7.43.

Compared to pumps with diffusers as shown in Fig. 7.43, the impellers used for compact diffusers as per Fig. 7.46 create a smaller axial thrust (reducing the balance piston diameter and the balance flow rate). The vibration behavior was not significantly influenced by the compact diffusers, but the Q-H-curves exhibited a slight saddle-type instability.

The design of diffusers with return vanes is discussed in the following. The configuration of the diffusers for single-stage pumps is performed accordingly. The required formulae are listed in Table 7.7 (Chap. 7.9.2).

1. Calculation of the flow rate Q_{Le}: In order to achieve the specified best efficiency point, the diffuser must be designed for Q_{opt} (if dimensioning were based on a different flow rate, the best efficiency point would be shifted according to Chaps. 3.7 and 4.2). The diffuser must therefore be calculated (just as with volutes) for the flow rate $Q_{Le} = Q_{opt} + Q_E + Q_{s3}$.

2. Inlet diameter d_3*: A clearance ("gap B") must be maintained between impeller and diffuser vanes to limit pressure pulsations and hydraulic excitation forces to an allowable level, Chap. 10. The diameter ratio $d_3^* = d_3/d_2$ can be calculated from the formulae given in Table 10.2. The minimum diametrical clearance for small

Table D7.4 Impeller-diffuser vane combinations

<table>
<tr><td>z_{La}</td><td colspan="3">5</td><td>6</td><td colspan="4">7</td></tr>
<tr><td>z_{Le}</td><td>7</td><td>*8*</td><td>12</td><td>10</td><td>9</td><td>*10*</td><td>*11*</td><td>*12*</td></tr>
<tr><td colspan="9">Vane numbers printed italic are considered best with respect to low vibrations</td></tr>
</table>

impellers is 2 mm giving $d_3 = d_2 + 2$ mm, but only if this does not violate the criteria given in Table 10.2.

3. Inlet velocity: The circumferential component of the absolute velocity at the impeller outlet c_2 is calculated from Eq. (T7.7.2) or Table 3.2. The diffuser inlet flow parameters follow from Eqs. (T7.7.3 to 7.7.5).

4. Inlet angle α_{3B} and leading edge profile: The diffuser vanes are to be set at a camber angle α_{3B} with respect to the circumferential direction. α_{3B} is calculated from the approach flow angle as per Eq. (7.7.6). The incidence is selected in the range of $i_3 = \pm 3°$.

As discussed in Chap. 3.7, it is recommended to design the diffuser vanes with low camber angles and incidences. The leading edge can be designed with an asymmetric profile on the low-pressure side (i.e. the side facing the impeller). Then the incidence becomes negative, the pressure distribution is flatter and the tendency to flow separation at partload is reduced. The asymmetric profile avoids accelerating the flow in the throat as might be the case with a symmetric profile.

5. Diffuser vane number z_{Le}: To reduce hydraulic excitation forces and pressure pulsations, it is necessary to carefully match diffuser vane and impeller blade numbers. When selecting the diffuser vane number, it is therefore imperative to consider the criteria given in Chap. 10.7.1. Together with the outer diameter d_{Le} and the throat width a_3, the number of diffuser vanes has an influence on the shape and the length of the actual diffusing channel whose deceleration is critical for the hydraulic losses in the diffuser. Taking into account the castability (or the machining costs, if the diffuser is milled), excessively narrow channels are also unfavorable. Therefore, the parameters z_{Le}, a_3 and d_{Le} cannot be selected independently of each other. Commonly used blade/vane number combinations are: $z_{Le} = 12$ with $z_{La} = 7$ for specific speeds from approximately 20 to 35 and $z_{Le} = 8$ with $z_{La} = 5$ for $n_q < 20$. Circumferential velocity u_2 and impeller size d_2 should also be considered in this context. With high specific speeds and/or very large pumps 15 diffuser vanes are also sometimes employed. Further combinations are obtained from Table D7.4. When there is a risk of resonance between an impeller eigen frequency of the two-diameter mode and the excitation frequency $\nu_3 \times z_{Le} \times f_n$, all combinations resulting in $|\nu_3 \times z_{Le} - \nu_2 \times z_{La}| = 2$ should be avoided. Including this criterion leads to the preferred vane combinations (printed in italic face) in Table D7.4.

6. Inlet width b_3: The inlet width b_3 is selected larger than the impeller outlet width; the ratio is in the range of $b_3/b_2 = 1.05$ to 1.3. Shock losses due to a mismatch of the impeller and diffuser channels should be avoided. Mismatch can be caused by manufacturing tolerances. Thermal expansion of multistage pumps must also be considered in this context. If the ratio selected for b_3/b_2 is small, a chamfer should be introduced to avoid the shock losses and associated hydraulic excitation forces, see figure in Table 0.2 (2). According to Chap. 5, the ratio b_3/b_2 has an effect on the partload characteristics: if b_3/b_2 is large, recirculation and shut-off head increase; if b_3/b_2 is reduced to values near 1.0, the risk of a saddle-type Q-H-curve is reduced. The choice of the inlet width also has an effect on the overlap of the shrouds of impeller and diffuser. As discussed in Chaps. 5.4.2 and 9.1, it is recommended to decouple the impeller sidewall gaps from the main flow to avoid axial thrust excursions when recirculating fluid enters the impeller sidewall gap. According to [B.20], gap A between the shrouds should be (0.007 to 0.01) $\times$ r_2 and the overlap x_{ov} = (2 to 4) $\times$ A (where A is the width of gap A), Fig. 9.1 and Table 0.2.

7. Throat width a_3: With a given diffuser inlet width b_3 and vane number, the throat width a_3 largely determines the best efficiency point, Chap. 4.2. It also has an effect on the shut-off head and the shape of the Q-H-curve (saddle formation), Chap. 5. Equation (3.20) supplies a throat width $a_{3,th}$ which corresponds to the dimensioning in accordance with the conservation of angular momentum. The final throat width is designed according to Eq. (T7.7.10) as $a_3 = f_{a3} \times a_{3,th.}$. Generally, f_{a3} = 1.1 to 1.3 should be assumed; in other words, the executed flow deceleration is greater than that corresponding to the conservation of angular momentum. Values $f_{a3} > 1$ also foster good overload characteristics at $q^* > 1$. With very wide impellers and low specific speeds it is sometimes necessary to design $f_{a3} < 1$ to achieve the specified best efficiency point.

Instead of using the conservation of angular momentum design as a base, it is also possible to calculate the throat width for a chosen deceleration ratio c_{3q}/c_2 from Eq. (T7.7.11). The deceleration ratio is in the range of c_{3q}/c_2 = 0.7 to 0.85.

Because of the effect of the throat width on the partload and overload characteristics, the choice of the ratio c_{3q}/c_2 or of the factor f_{a3} is an optimization task for which no generally applicable rules are available. If the flow at the diffuser inlet is decelerated excessively, the stall occurs at too high flow and the Q-H-curve can become unstable. Insufficient deceleration may mean a loss of efficiency. It is also possible that the flow separation at partload occurs too *late* which can produce a saddle in the Q-H-curve, Chap. 5.5. The effect of the deceleration ratio c_{3q}/c_2 on the shape of the Q-H-curve is evident from the tests in Figs. 4.5, 4.6 and 5.19.

8. Diffuser outer diameter d_{Le}: With respect to size and costs one would try and make the outer diameter as small as possible. But low hydraulic losses require optimum diffuser length and sufficient space for flow turning; when choosing d_{Le} a compromise must be sought between efficiency and costs. The required outer diameter increases with growing design flow rate, i.e. with the specific speed. For multistage pumps it is in the range of d_{Le}/d_2 = (1.05 to 1.15) + 0.01 n_q. The upper limit applies when the efficiency has priority over the pump size. If the diffuser discharges into a volute its outer diameter can be kept smaller. The diffuser channels require a sufficient overlap if it is intended to reduce the radial thrust.

9. Diffusing channel: Downstream of the throat area A_{3q} the channel opens similar to a plane or pyramidal diffuser in a pipe or channel (this section may be called "diffusing channel", see Table 0.2). Maximum allowable diffuser opening angle, pressure recovery and losses are calculated according to Chap. 1.6. Diffuser length and area ratio are matched in such a way that the pressure recovery (or the c_p value) is maximized. The diffuser losses can be estimated from Eqs. (T3.8.16 to 3.8.19). Since the cross sections are not circular, the calculation is done using the concept of an equivalent diffuser as per Eq. (1.45) unless the channel is designed as a plane diffuser. If the diffuser opens in the axial direction, $\vartheta_b = \vartheta_a$ may be selected with tan $\vartheta_a = 0.5(a_4 - a_3)/L_{3-4}$. The diffusing channels are often designed with a slight curvature since this helps to achieve a greater diffuser length and, consequently, a higher pressure recovery or lower losses with a given outer diameter d_{Le}. Experience shows that the data of straight diffusers according to Chap. 1.6 can be applied to slightly curved diffusers (with up to 25° deflection) without correction - presumably since the unsteady approach flow of the diffuser fosters thinner boundary layers, in this way compensating for the potentially negative effect of the channel curvature. The final diffuser length and outlet width a_4 are obtained only when drawing the diffuser; iteration may be necessary.

10. Diffuser criteria: Once all parameters discussed above have been determined and a diffuser drawing has been made, the quality of the diffuser design should be verified according to the following criteria:

A. The optimum pressure recovery in a diffuser of given length L_{3-4} according to Fig. 1.19 is achieved when the area ratio $A_{R,opt}$ is approximately:

$$A_{R,opt} = 1.05 + 0.184 \frac{L_{3-4}}{R_{eq}} \quad \text{with } R_{eq} = \sqrt{\frac{a_3\, b_3}{\pi}} \tag{7.30}$$

The pressure recovery coefficient will then be about:

$$c_{p,opt} = 0.36 \left(\frac{L_{3-4}}{R_{eq}}\right)^{0.26} \tag{7.31}$$

B. Deceleration ratio: $c_{3q}/c_2 = 0.7$ to 0.85 (at $n_q < 12$ to 15, c_{3q}/c_2 may be higher).
C. Ratio length to throat width: $2.5 < L_{3-4}/a_3 < 6$.
D. The ratio of inlet width b_3 to throat width a_3 should preferably be in the range of $0.8 < b_3/a_3 < 2$. Cross sections which are too flat are unfavorable but can sometimes not be avoided with very wide impellers and low specific speeds.
E. The outlet cross section A_4 of the diffusing channels of multistage pumps should be dimensioned so that the kinetic energy at the diffuser outlet is in the range of $c_4^2/(2 \times g \times H_{opt}) = 0.02$ to 0.04 in order to keep the losses in the overflow channels low. However, the outlet velocity c_4 must also be matched to the inlet velocity c_{1m} into the downstream impeller: $0.85 < c_4/c_{1m} < 1.25$.
F. The equivalent opening angle must be checked according to Eq. (1.45); it must not exceed the allowable opening angle given by Eq. (1.44) or a more stringent criterion based on test results should such be available. Exceeding the allowable diffuser angle would lead to premature stall the consequence of which could be increased pressure pulsations and dynamic stresses. This is particularly true for the last stage diffuser of a multistage pump or when the diffuser is followed by a volute

where the discharge pressure is non-uniform around the circumference, Chap. 10.7. Obviously this criterion gathers significance as the head per stage is increased.

11. Return vanes: The outlet velocity c_{6m} from the return vanes should be lower than the inlet velocity c_{1m} into the following impeller in order to smooth the impeller approach flow through a slight acceleration: $c_{6m} = (0.85 \text{ to } 0.9) \times c_{1m}$. Accordingly, the width b_6 is obtained from Eq. (T7.7.12).

The course of the flow areas between the return channel and the impeller inlet should be designed so that the flow is accelerated continuously. The width b_5 at the inlet of the return vanes can be selected equal to the outlet width (or different, depending on the configuration of the channel). The meridional component of the inlet velocity is obtained from Eq. (T7.7.13).

The velocity c_4 is obtained from continuity with dimensions a_4 and b_4 from the drawing. The circumferential component c_{4u} and the radial component of the vector c_4 are subsequently determined according to Fig. 7.47. The circumferential component of the velocity at the inlet to the return channels is calculated from the conservation of angular momentum, Eq. (T7.7.14). From c_{5u} and c_{5m} it is possible to determine the inlet angle of the return vanes as per Eq. (T7.7.15). The camber angle of the return vanes can be selected equal to this flow angle (plus blockage effect).

To obtain the desired flow angle at the outlet of the return channels, a slight angle exaggeration is required. For example, if an outflow angle of $\alpha_6 = 90°$ is desired, the camber angle of the vane should be selected 4 to 6° larger, that is $\alpha_{6B} = 94°$ to 96°.

If the diffuser is designed to give some pre-swirl to the subsequent impeller, the outflow angle can be estimated from $\alpha_6 \approx \arcsin a_6/t_6$.

Frequently return vane and diffuser vane numbers are selected identical. Sometimes fewer return vanes are selected (i.e. $z_R < z_{Le}$). In this case the solidity of the return vanes should be $L/t > 2$ so that the flow is deflected sufficiently.

7.9.2 *Design and Shaping of Radial Diffusers*

The design of radial diffusers according to Fig. 7.47 comprises the following steps:

1. Circles with r_3 and r_4 or r_{Le} are drawn in the plan view.
2. Based on the vane pitch the start of two vanes is constructed. The vane thickness at the inlet is $e_3 = (0.01 \text{ to } 0.015) \times d_2$. The camber angle α_{3B} is drawn and the profile defined at the inlet.
3. An arc of a circle with radius a_3 defines a point on the opposite vane surface which determines the throat area. The concave vane surface can now be provisionally drawn as a smooth curve (or a combination of circular arcs) up to the outer diameter d_4. From the intersection of this curve with the diameter d_4 it is now possible to mark off a_4 as an arc of a circle. This produces a point on the

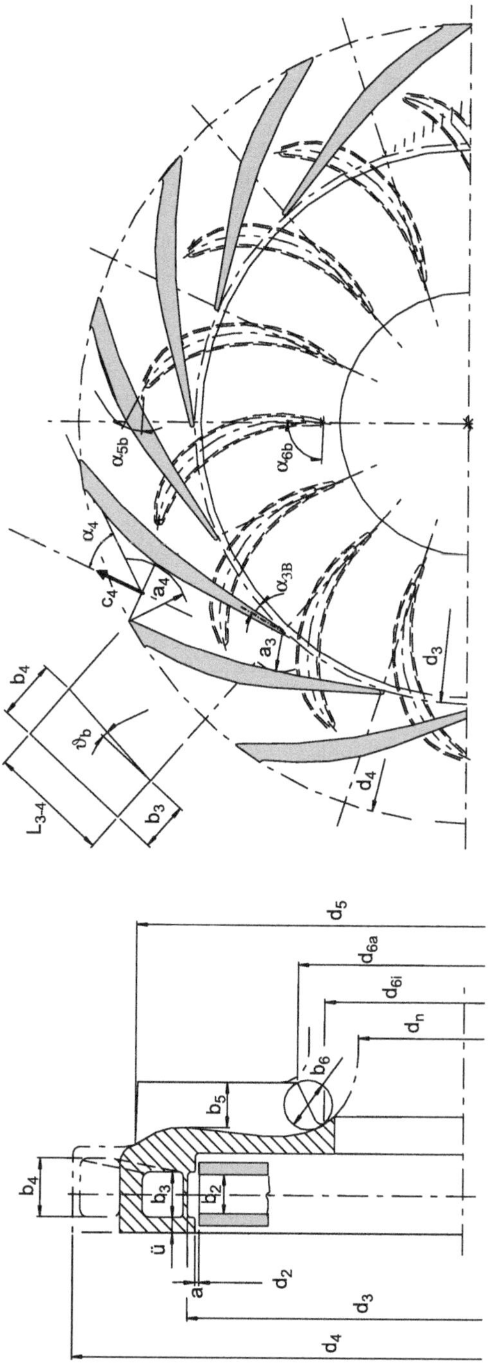

Fig. 7.47 Construction of a radial diffuser

convex vane surface which - together with the concave vane surface - supplies the vane thickness at the end of the channel. The resulting vane thickness must satisfy the requirements in terms of strength and castability. If this is not the case, the selected values of d_4 and a_4 cannot be maintained simultaneously but must be modified iteratively.

4. By changing the assumed curve of the concave vane surface (if adjusting d_4 is not desired) the channel geometry can be optimized so that the desired values for a_4, L_{3-4} and the pressure recovery coefficient c_p are achieved as closely as possible. Flow areas and vane curvatures should vary smoothly in order to reduce the risk of premature stall. Alternatively, the first section of the vane can also be configured as a logarithmic spiral or with special profiles, such as were analyzed in [69].
5. The meridional section is drawn and, taking into account d_1 and d_n of the subsequent impeller, the inlet and outlet widths b_5 and b_6 of the return channels and the position of the leading and trailing edges of the return vanes on the diameters d_5, d_{6a} and d_{6i} can be defined. At the same time, the section through the stage must be drafted.
6. The camber angle α_{5B} and d_5 is now drawn in the plan view. The camber line of the return vanes can, for instance, be constructed as a circular arc of the radius r_{sk}:

$$r_{sk} = \frac{r_5^2 - r_6^2}{2(r_5 \cos \alpha_{5B} - r_6 \cos \alpha_{6B})} \tag{7.32}$$

and subsequently given a thickness distribution and with leading and trailing edge profiles. This construction is performed in a way similar to Chap. 7.3.1 and Fig. 7.11. An elliptical profile at the inlet makes the vane less sensitive to incident flow. This is important since the approach flow direction is only a rough estimation in view of the complex geometry between diffuser outlet and return vane inlet. The return vanes should taper off to a thin trailing edge in order to reduce the blockage, but sufficient vane strength must be ensured.
7. With regard to position and effect of the trailing edges of the return vanes on the stability of the characteristics, reference is made to Chap. 5.6.

7.10 Semi-axial Diffusers

When designing a semi-axial diffuser, it is first necessary to determine the inlet and outlet diameters for the outer and inner streamline. Drafting therefore starts with the meridional section the shape of which has to be matched to the impeller (especially d_{2a} and d_{2i}). The diffuser outlet of single-stage pumps must be matched to the column pipe and the shaft protection pipe if applicable. In multistage pumps the diffuser outlet is matched to the impeller inlet of the subsequent stage. Design procedure (see also Fig. 7.22):

1. The inlet and outlet diameters d_{3a}, d_{3i}, d_{4a}, d_{4i} are largely determined by aspects of the mechanical design. The distance between impeller blades and diffuser vanes can be selected according to Table 10.2 to avoid excessive pressure pulsations and hydraulic excitation forces. According to [B.18] it can also be selected as Gap $B = 0.3 \times b_2$.
2. The axial length of the diffuser must be sufficiently large to keep the losses due to flow deflection and deceleration within acceptable limits. Iteration is usually necessary since the allowable opening angle of the diffuser can only be verified when the length of the channel has been established from a drawing. A reference value for the minimum axial length L_{Le} can be determined from [B.18]:

$$\frac{L_{Le}}{D_{2m}} = 0.72\left(\frac{n_q}{n_{q,Ref}}\right)^{0.19} \quad n_{q,Ref} = 200 \tag{7.33}$$

3. The diffuser inlet width is obtained with the impeller outlet width from $b_3 = (1.02 \text{ to } 1.05) \times b_2$. The smaller values apply to higher specific speeds.
4. After the main dimensions have been determined in this way, the course of the flow areas in the meridional section can be established so that a steady deceleration and an even curvature are obtained. The procedure is similar to Chap. 7.2.2.1.
5. When drawing the meridional section, various requirements of the mechanical design need to be taken into account, for example flange and stud dimensions, intermediate bearings, shaft protection sleeves and annular seal configurations. With high specific speeds the meridional section may be made of cylinders and conical sections, in particular when welded components are used. This type of design can yield very good efficiencies because flow turning can be minimized.
6. After the outer and inner streamlines have been drawn, the mean streamline can be constructed. Alternatively, the design can start from a mean streamline. The channel width from b_3 to b_4 is then defined by drawing circles which yield a smooth course of the flow area from inlet to outlet. Outer and inner streamlines are subsequently found from the envelopes to said circles. The procedure is similar to that described for impellers, Chap. 7.2.2.1.
7. The vane angles at the diffuser inlet are calculated for each streamline from Eqs. (T7.7.2 to 7.7.6).
8. The vane angles at the diffuser outlet are usually designed for swirl-free outflow. A slight angle exaggeration of 4 to 6° can be employed since the flow is not able to completely follow the vanes.
9. The number of diffuser vanes is usually selected greater than the number of impeller blades. The rules of Chap. 7.9.1 are applicable. In particular, diffuser and impeller should not have the same number of vanes nor vane number

combinations with common multipliers in order to avoid excessive pressure pulsations.

10. The diffuser throat area can be calculated according to Chap. 7.9.1 and Eq. (T7.7.10) which take into account the fact that the fluid in the vaneless space between impeller and diffuser flows in accordance with the conservation of angular momentum. The calculation therefore applies also to large distances between impeller and diffuser. The factor f_{a3} is close to 1.0; it has to be optimized in terms of hydraulic losses, stability and steepness of the Q-H-curve in the overload range. Generally applicable guidelines for controlling these three-dimensional effects are not available. CFD may be used for optimization.
11. The development of the vanes is drafted with the selected inlet and outlet angles α_{3B} and α_{4B} for various streamlines. The vane design procedure is similar to Chap. 7.2.2.2. In the inlet region of the vanes it is advisable to keep α_{3B} roughly constant in order to produce little flow deflection in the region with strong deceleration.
12. After the vanes have been drafted, the cross sections must be verified to ensure as steady as possible a flow deceleration. Because of the complex geometry this work is highly involved; it is best carried out using a 3D-CAD system.

7.11 Volutes Combined with a Diffuser or Stay Vanes

In some types of pumps a vaned diffuser is arranged between the impeller and a single volute in order to reduce the radial thrust, Fig. 7.48. The diffuser and the volute are configured according to Chaps. 7.9 and 7.8, respectively.

At high specific speeds, stay vanes between impeller and volute are often necessary for very large pumps to reduce the deformation of the volute casing under the internal pressure and to reduce casing stresses. The length and thickness of the stay vanes must be determined on the basis of strength and requirements of mechanical design. The stay vanes are usually shaped in a way that they have little effect on the flow in the design point. They are not designed to serve as a diffuser but to allow the fluid to flow according to the conservation of angular momentum with the least impediment possible. To this end, the vane inlet angle is dimensioned according to Eqs. (T7.7.2 to 7.7.6) with zero incidence. The camber line is drafted as a logarithmic spiral with $b_3 = b_4$ (or approximately as an arc of a circle). Such types of stay vanes hardly reduce the radial thrust.

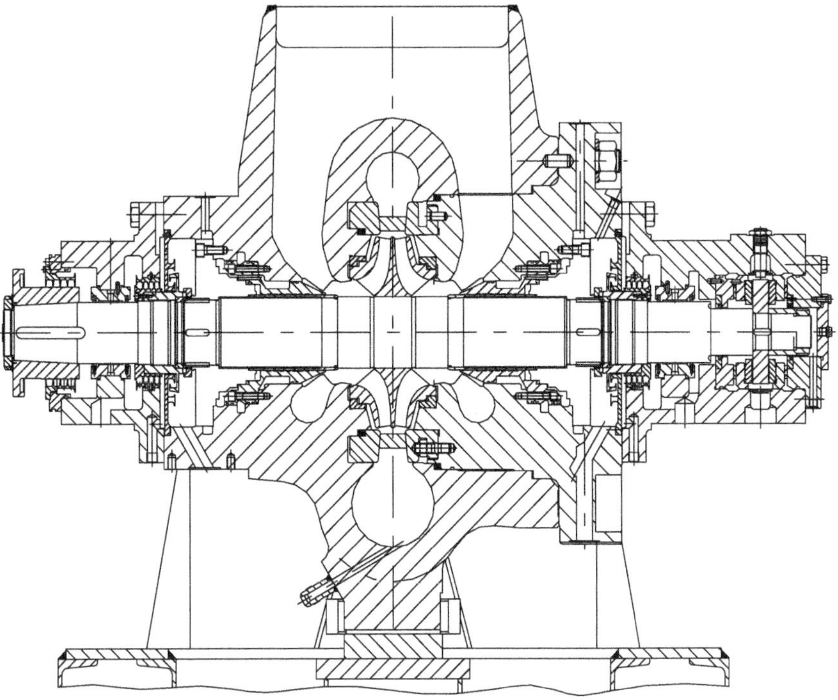

Fig. 7.48 Boiler feed pump with double-entry impeller, volute and diffuser, Sulzer Pumps

7.12 Annular Casings and Vaneless Diffusers

Because of their lower manufacturing costs, annular casings are sometimes preferred to volute casings for small pumps. Annular collectors also offer an alternative for pumps with low specific speeds since the efficiency impairment caused by the annular casing diminishes with falling n_q. For $n_q < 12$ even higher efficiencies are expected than with volutes according to [B.9]. The ratio of the pump efficiency expected with an annular collector η_{RR} to that with a volute η_{Sp} amounts to approximately $\eta_{RR}/\eta_{Sp} = (n_{q,Ref}/n_q)^{0.45}$ (with $n_{q,Ref} = 12$). This relation is valid for $n_q < 14$; it was derived from a diagram in [B.9].

To reduce the losses in the annular collector, the exchange of momentum between the impeller and the flow in the annulus should be intensified and smooth surfaces should be produced (see Chap. 7.3.2). The ratio b_3/b_2 must therefore not be selected too small; it amounts to approximately $b_3/b_2 = 1.2$ to 1.4. A low radial extension of the annular collector is favorable to reduce friction losses and secondary flows. The ratio width B to height H is likely to be at an optimum in the range of B/H = 1.5 to 2.5. The cross section $A_{RR} = B \times H$ of the annular collector must be considerably larger than corresponds to the circumferential component c_{2u} as stipulated by Eq. (7.34).

$$A_{RR} = (1.5\,\text{to}\,2)\frac{Q_{opt}}{c_{2u}} \tag{7.34}$$

The fluid is discharged from the annular collector through a diffuser the throat area of which is calculated from Eq. (T7.7.9). At low specific speeds the diffuser throat area determines the best efficiency flow rate, as corroborated by Fig. 7.15. Optimization of the diffuser is carried out in accordance with Chap. 1.6 or Eqs. (7.30 and 7.31). The diffuser is arranged tangentially because high deflection losses would result at the diffuser inlet if it were configured radially.

It is also possible to combine an annular collector with a volute by designing an annular collector over (for example) 180° followed by a volute leading the fluid into the discharge nozzle/diffuser. This combination improves the efficiency since the transition to the discharge nozzle can be shaped much more favorably. In addition, the radial thrust at Q = 0 and low partload is reduced compared with the volute, Chap. 9.3.

Vaneless diffusers are rarely used in pumps. Their hydraulic characteristics have been discussed in Chap. 3.7. The pressure increase in the vaneless diffuser is calculated from Eq. (7.32). The radius ratio is determined on the basis of the desired pressure increase and design requirements. In general the inlet width b_3 should be selected not much larger than the impeller outlet width b_2 in order to reduce the tendency to flow separation.

7.13 Inlet Casings for Between-Bearing Pumps

For between-bearing pumps an inlet casing is required to deflect the fluid from the inlet nozzle - mainly arranged normally to the axis - into the axial direction. Generally a small inlet casing is desired with respect to costs. The most important hydraulic requirement is to design the inlet casing so as to achieve as uniform a velocity distribution as possible at the impeller inlet. Quantitative criteria for the quality of the approach flow are given in Chap. 8.8.2, item 17. These criteria could be applied when optimizing inlet casings by CFD or testing.

Different pump types require various types of inlet casings according to Fig. 7.49. With reference to a section through the pump axis, symmetrical inlet casings, such as used for barrel pumps, can be constructed as annular collectors with surfaces of revolution (Fig. 7.49a). Alternatively, symmetrical inlet casings can be designed similar to volutes with cross sections decreasing in circumferential direction (Fig. 7.49b). In order to reduce the tangential velocity components near the rib R1, a compromise between the angular shape and the volute-type area distribution can be sought.

Axial-split pump casings require asymmetric inlet casings where the inlet nozzle is located below the split flange. Asymmetric casings can either be designed so that the flow rate is divided "quasi-symmetrically" into two identical flows around the

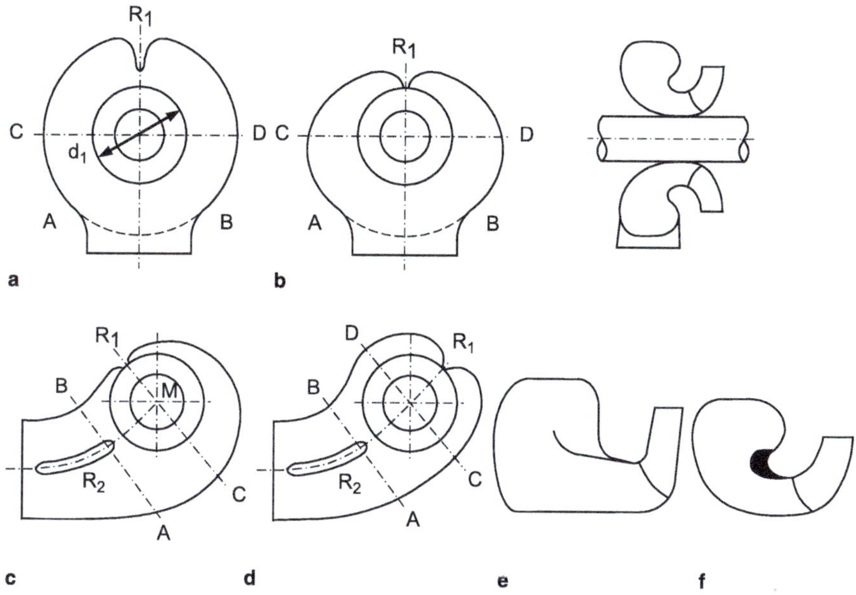

Fig. 7.49 Types of inlet casings. **a** Annulus, **b** spiral-shaped symmetrical, **c** asymmetric, **d** quasisymmetrical, **e** inlet cone, **f** inlet ring

hub (Fig. 7.49d) or that the lower half handles a significantly larger flow rate according to Fig. 7.49c.

All casing types have in common that the radial suction nozzle and the flow around the hub generate *circumferential* components in the velocity distribution at the impeller inlet. As a result, pre-rotation is generated over roughly one half of the impeller circumference and counter-rotation in the second half, Fig. 7.50. However, the flow through both halves is not uniform either since the liquid impinges on the rib which is absolutely indispensable in each and every inlet. As a consequence,

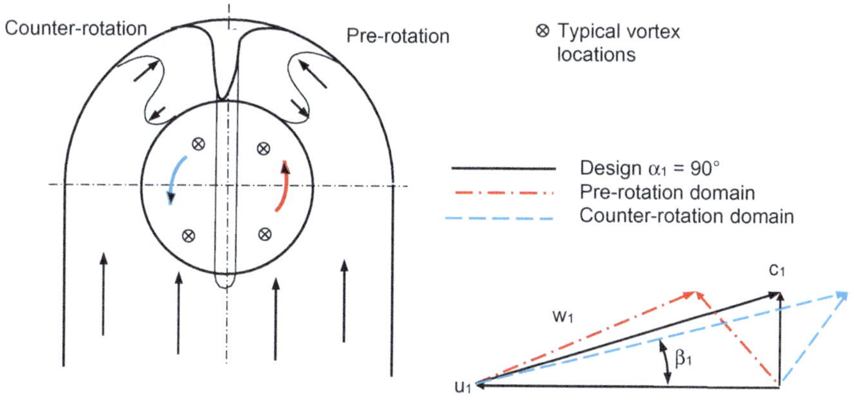

Fig. 7.50 Flow patterns in inlet casings

counter-currents near the hub occur in the region of the rib as indicated in Fig. 7.50. In the half with pre-rotation the approach flow angle increases compared to an approach flow with $\alpha_1 = 90°$. Conversely, the flow angle decreases in the half with counter-rotation, as shown by the velocity triangles in Fig. 7.50 (refer also to Fig. 3.1).

The uneven velocity distribution at the impeller inlet has the following disadvantages:

- The angle of the approach flow to the impeller blades varies over the circumference and the radius, resulting in correspondingly different pressure distributions and cavity lengths on the blades.
- Noise and vibrations are excited.
- Loss in efficiency
- According to Chap. 9.3.7, Fig. 9.32, radial forces are generated.

Inflow disturbances of this kind always grow with the specific speed, with the circumferential velocity u_1 and (due to increasing inertia) with the flow rate. For a given pump, they are highest at overload $q^* > 1$.

The flow through inlet casings generates 3-dimensional boundary layers the separation of which can cause vortex ropes. The pressure in the vortex core is lower than in the surrounding fluid due to the centrifugal forces caused by the rotation. If the pressure in the core drops below the vapor pressure, the vortex cavitates as discussed in Chap. 6.8.2. If the liquid contains appreciable amounts of dissolved gas, a gaseous vortex core appears even at local pressures above vapor pressure when gas separates from the liquid.

The 3-dimensional flow in the inlet casing cannot be calculated by simple means, but some general rules for favorable flow conditions can be given:

1. The meridional component of the approach flow velocity profile upstream of the impeller inlet can be smoothed out by flow acceleration and an extended axial flow path upstream of the impeller. The acceleration should preferably take place in the region with the strongest flow turning (or downstream of it).
2. To foster a more uniform flow over the circumference of the impeller, the fluid has to be accelerated towards the impeller inlet. The cross section A-B in Fig. 7.49 must therefore be selected so that an acceleration ratio $c_{1m}/c_{A\text{-}B} = 1.5$ to 2.2 is obtained. The lower value applies to $n_q = 15$ and the upper value to $n_q > 60$. Half of the flow rate in cross section A-B flows through cross section C-D (or C-M in Fig. 7.49c). The cross sections C-D or C-M must be designed so that approximately the same velocity as in A-B prevails there or that the same acceleration ratio $c_{1m}/c_{C\text{-}D} = 1.5$ to 2.2 is achieved from C-D to the impeller inlet.
3. The circumferential component of the approach flow velocity can be decreased by reducing the blockage caused by the hub. If the inlet casing is very large, an inlet cone (or bellmouth) according to Fig. 7.49e ensures a favorable flow distribution upstream of the impeller. This type of bellmouth can be considered for a retrofit of an over-sized problem casing.

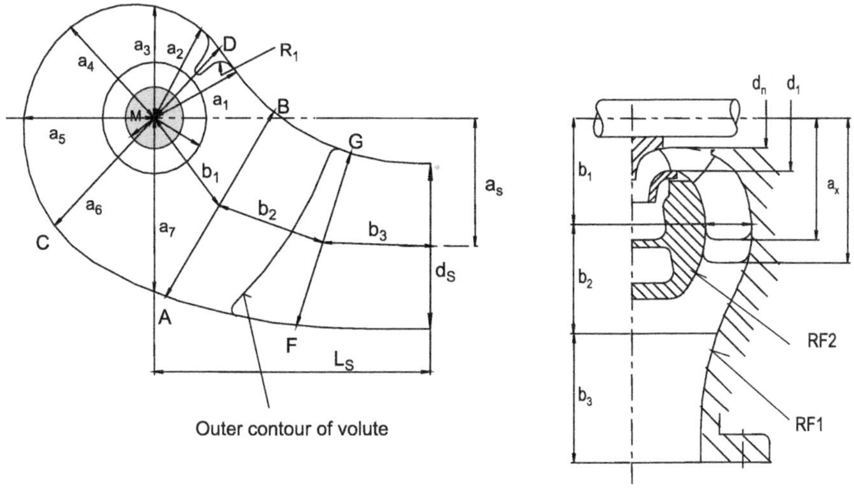

Fig. 7.51 Inlet casing design

4. If the inlet casing is over-sized, the flow can be smoothed out using a ring as shown in Fig. 7.49f. The effect of an *asymmetric* ring of this type is reported on in [42]. The asymmetric ring is better able to re-distribute the fluid to both halves of the inlet casing. Inlet rings according to Fig. 6.39 can also help. If the radial expansion of the inlet casing is large, it must be noted that the fluid tends to maintain its angular momentum according to $c_u \times r$ = constant when moving radially inward, which causes the circumferential velocities to be intensified and increases the risk of generating vortices. An inlet casing should therefore be expanded as little as possible in radial direction but rather axially.
5. Severe local decelerations from the suction nozzle to the inlet casing cause flow separation and vortices. Over-sized annular casing cross sections (which per se would favor the flow distribution over the circumference) should therefore be avoided.
6. Every inlet inevitably requires a rib R1 (Fig. 7.49) at the location where the two partial flows around the hub meet. If this rib is omitted, cavitation characteristics and work transfer (head coefficient and efficiency) are seriously affected. A periodic pre-rotation can also be induced which would result in unsteady operating.
7. The rib R2 is not absolutely necessary with symmetrical inlets but desirable for reasons of mechanical design in order to limit the casing deformation under the internal pressure (note pressure of hydrostatic test, if applicable). With asymmetric inlets a rib R2 can be expected to foster a more uniform flow.
8. All ribs must be rounded at the leading edge and profiled asymmetrically at the trailing edge according to Table 10.13 in order to reduce the formation of vortices and prevent vortex streets.

The step-wise procedure for drafting an inlet casing for a single-stage double-entry pump is explained below with reference to Fig. 7.51. This design of an asymmetric inlet casing is particularly complex because the inlet casing has to be developed around the volute casing. The design methodology essentially follows [B.2, B.9, B.21, B.22, 41]. The cross sections develop on the surfaces of revolution RF1 and RF2.

1. Known parameters are the inlet type, the inlet nozzle diameter d_s and the impeller inlet: d_1, d_n and the impeller inlet area $A_1 = (\pi/4) \times (d_1^2 - d_n^2)$. The position of the inlet nozzle is given by the dimensions L_s and a_s.
2. Impeller inlet and hub diameter are drawn in the plan view. In an inlet according to Fig. 7.49c, the rib R1 is positioned approximately 45° over the partition plane of the axially split casing.
3. The outer contour from the rib to point C is defined by the radial dimensions a_1 to a_6:

$$\begin{aligned} a_1 &= (1.5 \text{ to } 1.8) \times r_1 \\ a_6 &= (2.5 \text{ to } 2.8) \times r_1 \end{aligned} \qquad (7.39)$$

   The dimensions a_2 to a_5 increase with the circumferential angle ε according to:

$$\begin{aligned} a_i &= a_1 + (a_6 - a_1) \times \varepsilon^\circ/180. \\ a_7 &= (3 \text{ to } 3.5) \times r_1 \\ (A - B) &= (1.7 \text{ to } 2.2) \times d_1 \end{aligned} \qquad (7.40)$$

   The distance A-B is positioned (approximately at a distance of $b_1 \approx d_1$ from the shaft axis) normally to the flow direction. With the details for the inlet nozzle (d_s, a_s, L_s) it is now possible to draw the inlet contour in the plan view as a connection of the checkpoints defined above. With regard to the split flange of axial-split pumps (near point B), a steep contour facilitates the positioning of the studs as closely as possible to the area subjected to pressure.

4. The outline of the volute is drawn. Another section F-G is positioned between the suction nozzle and A-B. The distances b_1 to b_3 between the centers of the sections A-B and F-G and the suction nozzle are defined.

5. The meridional section of the impeller and the volute casing are now drawn and the surfaces of revolution RF1 and RF2 are designed taking into account the hub contour which usually houses part of the shaft seal.

6. It is now possible to draft the flow areas ensuring a continuous velocity variation along the flow path. As an important element the cross section A-B must be selected so that the acceleration ratio $c_{1m}/c_{A\text{-}B}$ defined under item (2) is achieved. The cross sections with the radial dimensions a_6 to a_1 are determined such that the velocity is approximately identical to that in cross section A-B. In doing so, it is assumed that the flow rate in the cross section M-C (or D-C) is

half that of the cross section A-B and that the flow rate subsequently decreases in proportion to the circumferential angle.

7. The areas of the cross sections between the inlet nozzle A_s and section A-B should vary in a continuous fashion. If the fluid is decelerated in this region, flow separation must be prevented under any circumstances (to be verified according to the rules for the diffuser design in Chap. 1.6). Any diffusion in the inlet casing should be designed with opening angles lower than those given by Eq. (1.44) or Fig. 1.20 in order to take into account possible flow distortions in the suction pipe.
8. The rib R2 in the inlet is arranged essentially along the mean streamline.

Vaned inlet casings: While all of the inlet casings depicted in Fig. 7.49 invariably generate a non-axisymmetric velocity distribution in front of the impeller blades, the tangential velocity components - and the resulting variation of incidence over the circumference - can be almost eliminated by a vaned inlet casing. An earlier version developed in the 1970s is presented in Fig. 7.51-1. A more compact design is shown in Figs. 7.51-2, 2.8 and 2.10 and discussed in [77, 78]. The tangential velocity components are virtually removed by several radial vanes with a length-to-pitch ratio of typically L/t > 2. This can be verified by visual observation

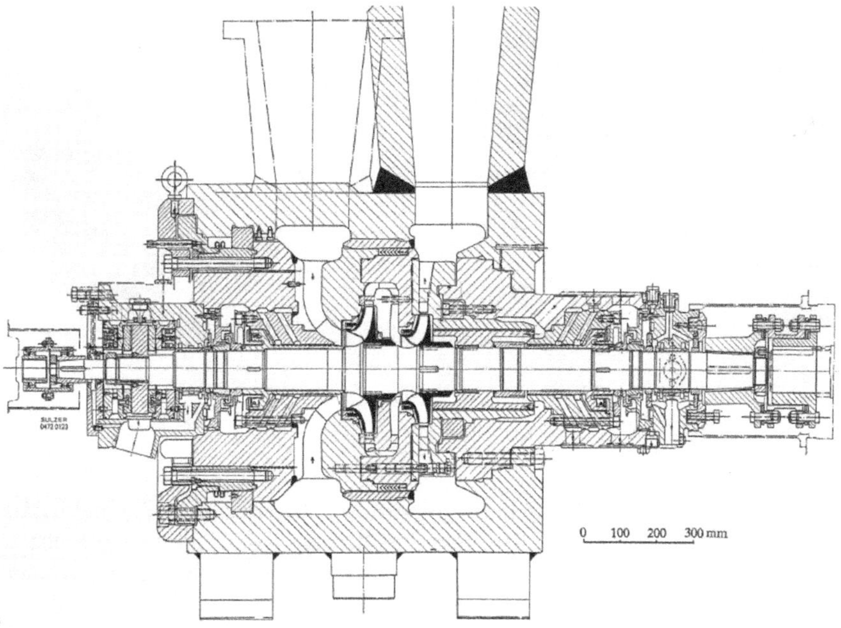

Fig. 7.50-1 "Advanced-class" boiler feed pump $n_q = 26$ with $H_{opt,\ st} = 1200$ m, [76]

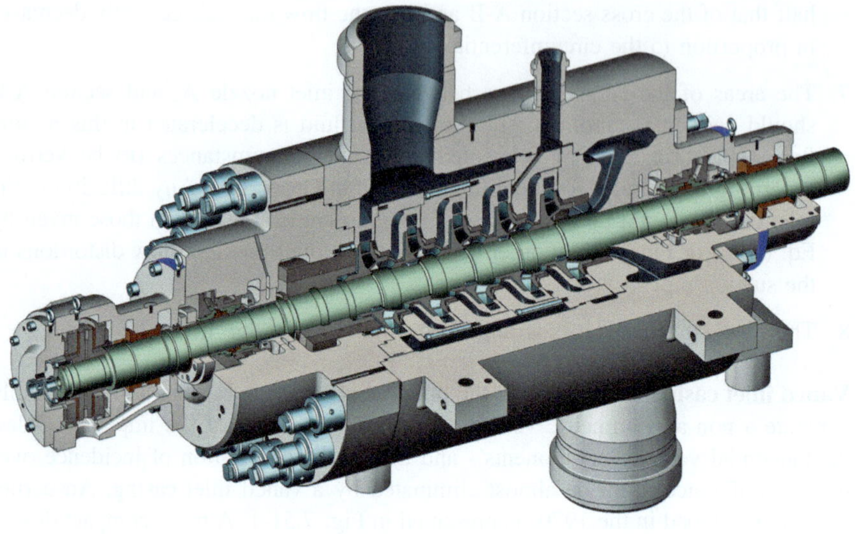

Fig. 7.50-2 5-stage boiler feed pump with $H_{opt,st}$ = 800 m, [77]

Fig. 7.51-3 Model test for flow observation in a vaned inlet casing, [77]

of the cavity length (which is then uniform over the circumference) and by CFD analysis as shown in Figs. 7.51-3 and 7.51-4 respectively. Furthermore, the ribs provide an even force transmission from the stage casings to the bearing assembly which greatly benefits the mechanical design. Slight disadvantages of the vaned inlet casing are increased hydraulic losses caused by higher velocities and friction losses on the ribs. Therefore, this design is specific to high-pressure pumps where these losses and the associated increase in $NPSH_R$ are insignificant.

Fig. 7.51-4 CFD study of vaned inlet casing, [77]

7.14 Analytical Method for Impeller Design

7.14.1 Motivation, Scope and Objectives

For a specified performance given by n, Q_{opt}, H_{opt} and $NPSH_{R,opt}$ and for given boundary conditions, an impeller can be designed in quite different ways. For the following argument, let us assume that the pressure coefficient ψ_{opt} is selected. In this case, the impeller size is fixed. Still, the main parameters z_{La}, b_2, d_{1a}, d_{1i}, β_{2B}, incidence, rake and others can be selected in a certain range. Even if the main parameters would be selected according to a design systematic, the meridional section of the impeller is usually drawn as free-style curves; formerly this was done by hand, nowadays Bezier coefficients are often used. The same applies to the blade development which determines the blade load distribution over the blade length.

The consequences are:

- For a given task every designer gets a different geometry since each designer makes his individual (arbitrary) choices: every impeller is a *unique* design. The complex shapes have an impact on performance, which cannot be quantified. There is a virtually infinite number of combinations of selected parameters and streamline and blade shapes. It is the *combination of these parameters and shapes* rather than any single parameter which determines the efficiency, the stability of the Q-H-curve, the hydraulic excitation forces and the cavitation characteristics $NPSH_R = f(Q)$.
- Since every designer produces a different geometry for a given set of specified conditions, each design will generate different velocity distributions and performance data. The differences in performance are admittedly often within the more or less accepted range of tolerances. However, the arbitrariness inherent to the generation of the impeller geometry is one of the reasons why the accuracy of the prediction of efficiency, head, Q-H-stability, $NPSH_R$ and excitation forces is not always entirely satisfactory - and once in a while downright disappointing.

Since the flow undoubtedly depends on the geometry of the flow passages, it may be concluded that we must attempt to control the geometry in order to be able to control the flow patterns in the pump. In other words: consistent geometry is a *necessary* (but unfortunately *not sufficient*) condition for *consistent* performance.

In the following, design procedures are developed, which allow designing any range of pumps in a mathematically consistent way. The objectives and considerations are:

- Fully analytical description of impeller geometry: every designer produces exactly the same impeller geometry for a given set of specified data and boundary conditions.
- Develop and optimize a design system ("design systematic") rather than individual hydraulics.
- Improve accuracy of performance prediction and reduce the costs for corrective action in case of performance deficits.
- Reduction of development effort: the impeller geometry can be created within minutes.
- A design systematic is equally useful for the design of: (1) standardized pump ranges, (2) engineered pumps, (3) pumps tailored to any client's specifications.
- Even prior to developing a single pump it is advisable to establish a design systematic in order to have a basis for future pumps of the same type.
- The design systematic is not meant to restrict the freedom of the innovative designer in any way. If better rules are known or new ideas become available, these should be implemented in the systematic.
- The design systematic produces consistent designs with a minimum of effort. It attempts to achieve more consistent performance through a tight control of the geometry. It does not, as such, guarantee better performance. As with all other hydraulic designs this can be achieved only, if the design systematic has been tested with good results.

Limitations: (1) Due to secondary flows the fluid moves on a kind of spiraling streamlines through a given impeller. Hence, the velocity profiles at the diffuser/volute inlet depend on the specific speed. The velocity profile is critical for the losses in diffuser or volute and the stability of the Q-H-curve which is always very difficult to predict. (2) Mechanical design constraints which have an impact on the flow path but do not correlate with the specific speed. Depending on the priorities defined for a development project, a pump range can be designed in two ways:

1. Performance priority: The dimensions of the hydraulic components are given from a design systematic, previous projects or a model test. The mechanical design is made to integrate these components without modifications to the flow paths. If a tested hydraulic is used, the performance uncertainty is essentially determined by manufacturing tolerances only.
2. Cost priority: The mechanical design is largely influenced by considerations of product standardization and costs. The hydraulic components are designed to fit into the mechanical design. Scaling of hydraulic components with a given

specific speed to different pump sizes is no more possible. The mechanical constraints may not allow to optimize efficiency and required NPSH. Performance prediction is subject to higher uncertainties.

7.14.2 Meridional Section

Obviously, the shape of the outer streamline has a profound impact on the flow through the impeller *and* the diffuser - hence on efficiency, $NPSH_R$, Q-H-curve stability and hydraulic excitation forces. Second to this is the inner streamline. As a consequence, the key issue for developing a systematic impeller design procedure without any arbitrariness of the individual designer is a *fully analytical f*ormulation of the outer and inner streamlines of the impeller meridional section. This description must be independent of the specific speed, the suction specific speed (or impeller inlet diameter), the pump size and the hub diameter which is given by mechanical design constraints.

Over the years, quite a number of attempts have been made by the author to develop equations which would fulfill the above requirements. The difficulty was that the tangents to the streamline vary between zero at the inlet and infinite at the outlet of radial impellers (obviously, there is no such problem with semi-axial impellers). The only satisfactory general solution found so far is presented in the following.

Both, the outer and inner streamlines are calculated from "normalized streamlines" (NSL) by the procedure described in Table 7.8. The NSL are given by a number of discrete points in z*, r*-coordinates. The parameters used for developing the meridional section from normalized streamlines are shown in Fig. 7.52. The curved portion of the outer and inner streamlines can be followed by a straight section between r_{ga} (and r_{gi}) and r_2. The correlations (T7.8.5 and T7.8.6) define these straight sections as a function of the specific speed. The constants in these equations could be modified to get different meridional shapes.

One possible set of streamlines is defined by the dimensionless coordinates $r^* = f(z^*)$ given in Table 7.9, but *any other shape* of streamline can be normalized by Eqs. (T7.8.1 and T7.8.2). The number of points might be increased by a Nurbs function or other forms of interpolation in order to get a more accurate fit of the curves (say 100 points per streamline).

The selected type of normalization ensures that the method can be used for:

- any specific speed
- shapes where ε_{DS} significantly increases with the specific speed
- the method is equally suitable for designing semi-axial impellers
- any impeller inlet diameter (e.g. large-eye suction impeller, small-eye series stages impeller)
- any number of different streamline definitions could be implemented

Table 7.8 Normalized stream line equations

OHH = overhung impellers; BB5 = impellers for multistage pumps; $n_{q,Ref} = 1.0$

Normalization	axial coordinate	$z^* = \frac{z}{z_{ga}}$	7.8.1
	radial coordinate	$r^* = \frac{r - r_1}{r_{ga} - r_1}$	7.8.2
Streamline The outer streamline starts at $z = z_{in}$	axial coordinate	$z = z_{in} + z_{ga} z^*$	7.8.3
	radial coordinate	$r = r_1 + (r_{ga} - r_1)\, r^*$	7.8.4
| OHH and sewage | BB5 a | 2.75 | 2.0	r_{ga}/r_{1a}	$\frac{r_{ga}}{r_{1a}} = a\left(\frac{n_{q,Ref}}{n_q}\right)^x$	7.8.5
x | 0.16 | 0.1	r_{gi}/r_{ga}	$\frac{r_{gi}}{r_{ga}} = 0.83\left(\frac{n_q}{n_{q,Ref}}\right)^{0.021}$	7.8.6
Straight exit, r is the independent variable	between r_g and r_2	$z = z_2 - (r_2 - r)\tan\varepsilon_{DS}$	7.8.7
Inlet length, relevant for annular seal length	OHH: $a = 0.12$, $d_n = 0$ BB5: $a = 0.25$	$z_{in} = a\,(d_{1a} - d_n)$	7.8.8
Axial length of outer streamline (projection)	| OHH | BB5 a | 0.75 | 0.33 x | –0.05 | 0.227	$z_E = \frac{d_{1a} - d_n}{2}\left(\frac{n_q}{n_{q,Ref}}\right)^x a$	7.8.9
Diameter of mid stream-line at leading edge	| OHH | BB5 a | 0.04 | 0.05 x | 0.05 | 0.00		7.8.10
	$d_{1m,LE} = d_{1m,z=0} + a\left(\frac{n_q}{n_{q,Ref}}\right)^x\left(\frac{r_{ga}^* + r_{gi}^*}{2}\,d_2 - d_{1m,z=0}\right)$		7.8.11
Diameter of mid stream-line at $z = 0$	$d_{1m,z=0} = \left(\frac{d_{1a}^2 + d_n^2}{2}\right)^{0.5}$		7.8.12
Diameter of inner stream-line at leading edge (OHH)	if $d_n < 0.35 d_{1a}$, $d_{1i} = 0.35\, d_{1a}$; else $d_{1i} = 1.25\, d_n$		7.8.13
Start of inner streamline	$\frac{z_{iSL}}{z_E + b_2} = a + b\left(\frac{n_q}{n_{q,Ref}}\right)$ sewage: $a = 0.2$; $b = 0.002$; other: $a = b = 0$		7.8.14

Table 7.9 Normalized stream line coordinates

Outer streamline		Inner streamline	
z_a*	r_a*	z_i*	r_i*
1.0000	1.0000	1.0000	1.0000
0.9986	0.9335	0.9911	0.8068
0.9945	0.8692	0.9735	0.6959
0.9878	0.8072	0.9526	0.6195
0.9784	0.7475	0.9302	0.5610
0.9664	0.6901	0.9070	0.5134
0.9519	0.6351	0.8834	0.4729
0.9349	0.5826	0.8595	0.4374
0.9155	0.5325	0.8353	0.4056
0.8938	0.4849	0.8110	0.3767
0.8698	0.4397	0.7863	0.3500
0.8437	0.3971	0.7614	0.3253
0.8156	0.3569	0.7362	0.3021
0.7855	0.3192	0.7106	0.2803
0.7537	0.2839	0.6846	0.2597
0.7201	0.2511	0.6581	0.2401
0.6850	0.2206	0.6310	0.2214
0.6484	0.1925	0.6033	0.2036
0.6106	0.1667	0.5749	0.1865
0.5716	0.1431	0.5458	0.1701
0.5317	0.1218	0.5158	0.1543
0.4910	0.1025	0.4849	0.1392
0.4496	0.0852	0.4531	0.1246
0.4079	0.0700	0.4202	0.1106
0.3658	0.0565	0.3861	0.0972
0.3237	0.0449	0.3508	0.0843
0.2818	0.0348	0.3143	0.0719
0.2402	0.0264	0.2763	0.0602
0.1992	0.0193	0.2370	0.0489
0.1723	0.0153	0.2099	0.0418
0.1458	0.0119	0.1821	0.0349
0.1199	0.0089	0.1536	0.0283
0.0944	0.0064	0.1244	0.0220
0.0820	0.0053	0.1095	0.0190
0.0697	0.0043	0.0944	0.0160
0.0576	0.0034	0.0792	0.0131
0.0456	0.0026	0.0637	0.0103
0.0339	0.0018	0.0481	0.0076
0.0224	0.0012	0.0323	0.0050
0.0111	0.0005	0.0162	0.0024
0.0000	0.0000	0.0000	0.0000

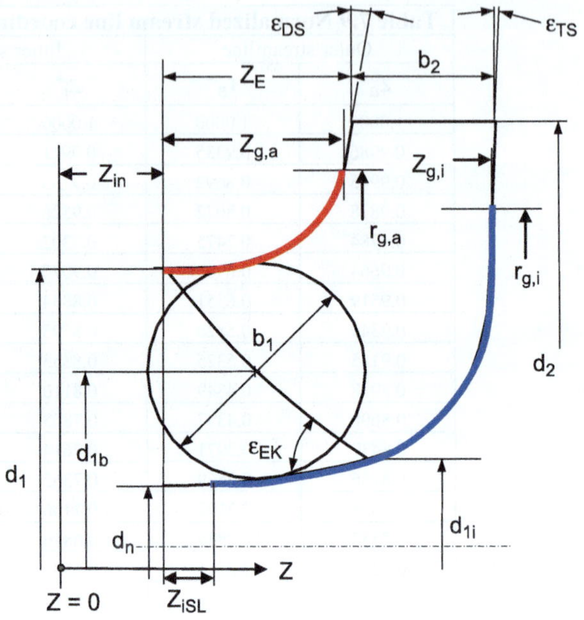

Fig. 7.52 Meridional section with outer and inner streamline. The colored streamline sections are normalized

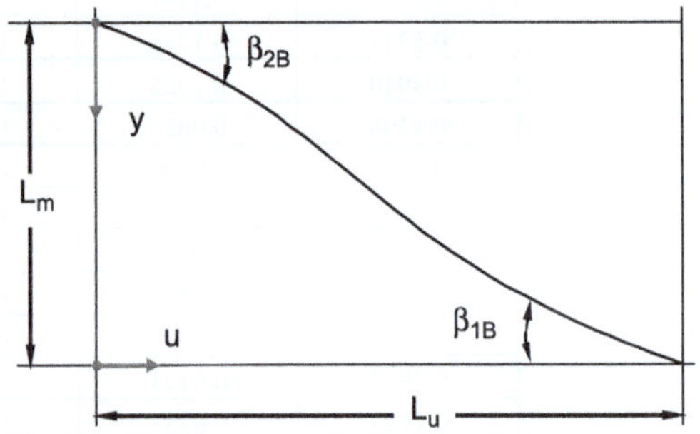

Fig. 7.53 Blade development

- Overhung impellers or between bearing impellers (single- or multi-stage pumps) with any hub diameter for high-pressure as well as low-pressure applications.

Once the outer and hub streamlines have been calculated, additional streamlines can be developed from geometric operations on the basis that the same amount of fluid flows between any two adjacent streamlines. For conventional designs the mid streamline can be calculated by averaging the dimensionless coordinates (z*, r*) of the outer and inner streamlines. The analogous procedure can be applied for developing additional streamlines between outer and mid as well as inner and mid streamlines.

The streamline defined in Table 7.9 yields a radial impeller exit. Some arguments might be advanced in favor of this design with $\varepsilon_{DS} = \varepsilon_{TS} = 0$:

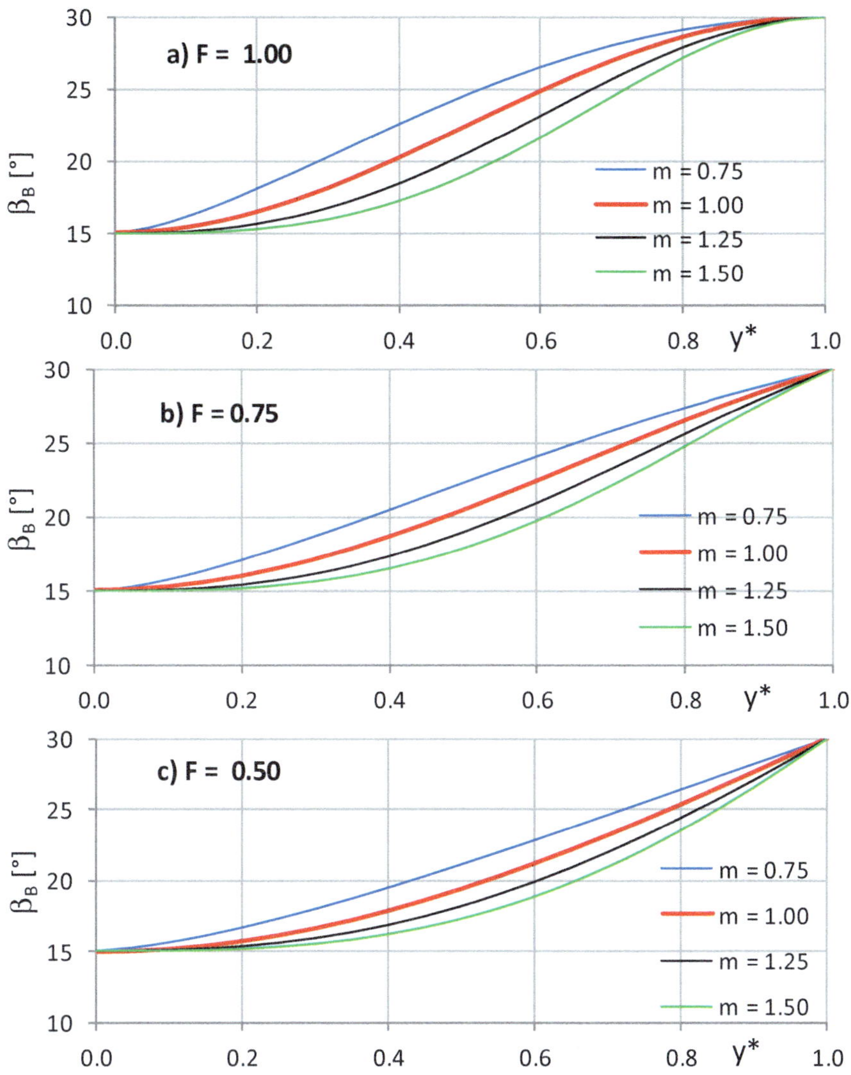

Fig. 7.53-1 Blade development according to cosine functions: effect of exponent m and factor F on blade loading distribution

- Smallest axial velocity component at the impeller outlet (the goal is $c_{2,ax} = 0$); hence less mixing losses in diffuser or casing
- Strongest decrease in width "b" in the region of strongest curvature (compare to Kaplan bend design)
- Radial forces are reduced since the projected area of the front shroud is smaller than in a design with $\varepsilon_{DS} > 0$. This because, according to Chap. 9.3.3, the contribution of the front shroud to the radial force is large if not dominating.

- Lower hydraulic impeller interaction improves rotor dynamic stability; this would be notably relevant for multistage pumps with a high head per stage, Chap. 10.6.3

Table 7.10 Blade design

Blade development on 5 streamlines			
Developed length of streamline in meridional section	length element	$\Delta m = \sqrt{\Delta z^2 + \Delta r^2}$	7.10.1
	coordinate	$y_j = y_i + \Delta m$	7.10.2
	total length of streamline	$L_m = \Sigma\, \Delta m$	7.10.3
	normalized streamline developed in meridional section	$y^* = \frac{y}{L_m}$	7.10.4
Normalized blade angle **Linear option:** a = 1.0, b = c = 0 For single-blade impellers a 6th order polynomial is used		$\beta_B^* \equiv \frac{\beta_B - \beta_{1B}}{\beta_{2B} - \beta_{1B}} = a y^* + b y^{*2} + c y^{*3}$	7.10.5
Blade development according to cosine function		$\beta_B^* \equiv \frac{\beta_B - \beta_{1B}}{\beta_{2B} - \beta_{1B}} = 0.5 - 0.5\cos(\pi F y^{*m})$ $\beta_B = \beta_{1B} + \frac{(\beta_{2B} - \beta_{1B})\beta_B^*}{0.5 - 0.5\cos(\pi F)}$ exponent m = 1.0 and F = 1.0 yields true cosine function	7.10.6 7.10.6.a
Blade angle: *inlet to outlet* (modified)		$\beta_B = \beta_{1B} + (\beta_{2B} - \beta_{1B})\beta_B^*$	7.10.7
		$\Delta u = \frac{\Delta y}{\tan \beta_B} = \frac{\Delta m}{\tan \beta_B}$	7.10.8
		$y = y^* L_m$	7.10.9
		$u_j = u_i + \Delta u$	7.10.10
Length of blade element		$\Delta L = \sqrt{\Delta u^2 + \Delta y^2}$	7.10.11
Wrap angle	of one element r is the mean radius under the element Δu	$\Delta \varepsilon = \frac{\Delta u}{r}$ [radian]	7.10.12
	of entire blade on the individual streamline	$\varepsilon_{sch} = \Sigma\, \Delta\varepsilon$	7.10.13
Total blade length		$L_{sch} = \Sigma\, \Delta L$	7.10.14
Elliptical blade profile e_{nom} = nominal blade thickness, according to Table 14.3, Eq. (T14.3.1) L_E = length of leading edge profile		$\frac{y}{e_{nom}} = \sqrt{1 - \frac{x^2}{L_E^2}\left(1 - \frac{e_1^2}{e_{nom}^2}\right)}$	7.10.15

- A stronger shroud curvature might improve the Q-H-stability of diffuser pumps since it provides a defined location of flow separation at partload - thus reducing the risk of flow pattern switching, Chap. 5.6.5.
- Suction impellers can be designed with little increase in radius near the inlet for a given axial length z_E.

However, radial impellers with $\varepsilon_{DS} > 0$, typically 3 to 10°, are frequently used. Flow deceleration within the impeller channels, head coefficient and efficiency may be improved. A systematic experimental study should show which design is optimum.

7.14.3 Blade Design

The blades are developed on 5 streamlines. Starting points are the coordinates z, r at the impeller blade trailing edge of these streamlines in the meridional section. The parameters used are shown in Fig. 7.53. The design procedure has been discussed already in Chap. 7.2.2.2; the formulae used are given in Table 7.10.

During the layout process, the blade inlet and outlet angles β_{1B} and β_{2B} are calculated for each streamline in order to get the specified head and desired incidence at the impeller inlet. As discussed in Chap. 7.2.2.2 the blade angles between inlet and outlet can be determined by any functions which yield these values of β_{1B} and β_{2B} as boundary conditions at the leading and trailing edge. The normalized blade development given by Eqs. (T7.10.5 to 7) ensures just this. Therefore, a function can be chosen for defining $\beta_B{}^* = f(y^*)$, which allows to optimize the hydraulic features for any specific application. Three examples may be quoted: (1) a linear function for the series-stages of multistage pumps or low-pressure pumps with ample $NPSH_A$; (2) if cavitation criteria significant for the design, a cosine function as per Eq. (T7.10.6) or a 3rd order polynomial may be adequate for achieving a low blade loading near the impeller inlet; (3) a single-blade sewage impeller will need a 6th order polynomial.

The cosine function offers a simple, but very efficient means to control the blade loading. Cosine-blades are unloaded at the inlet and at the outlet. Unloading at the inlet is beneficial to reduce cavitation and incidence effects, while unloading at the outlet is expected to reduce the rotor/stator interaction, noise and vibrations. Unloading inlet and outlet implies that the blade angles increase more strongly in the mid-portion of the blades (roughly between A_{1q} and A_{2q}). As a consequence, more work is transferred in the domain where the blades overlap; the slip may decrease slightly in this way. In addition, the blade angle distribution according to Eq. (T7.10.6) can be modified: by applying an exponent m on y*, and/or by truncating the range of $\beta_B{}^*$ which is used: truncation factor F. If F is not equal to 1.0, Eq. (T7.10.6a) must be used for calculating the blade angles in order to obtain the specified blade outlet angle β_{2B}.

Figure 7.53.1a demonstrates how the blade loading can be controlled by the exponent m in Eq. (7.10.6). Increasing exponent to $m > 1$ unloads the blade at the inlet and puts more load towards the rear. In contrast, $m < 1$ unloads the blades at the outlet and reduces the rotor/stator interaction but rendering the impeller more susceptible to cavitation. Figure 7.53-1b, c show the effect of the truncation factor F. In summary, the cosine blade development provides much freedom to define an optimized blade loading in a fully analytical way.

7.14.4 Procedure for Developing a Design Systematic

As an integral part of a comprehensive product planning a hydraulic specification should be established the outset of the development of a pump range (or an individual pump). A sample format is given in Chap. 17.7; it may also serve as a check list.

From the hydraulic specification are given:

- Range of pumps: range chart, for each pump size: n, Q_{opt}, H_{opt}
- Required suction specific speed or $NPSH_{3,opt}$
- Boundary conditions for mechanical design

Procedure:

1. As far as possible and expedient all parameters and criteria are described by analytical functions of the specific speed, suction specific speed or geometric quantities.
2. Use smooth correlation for pressure coefficient, Eq. (3.26) with appropriate factor for f_T, impeller outlet width, Eq. (7.1), impeller inlet diameter as function of suction specific speed, Eqs. (T15.1.12 and T15.1.13) and other geometric parameters as required.
3. If necessary, adapt correlations for hydraulic shapes as per Table 7.8 and 7.10; notably Eqs. (T7.8.5 to T7.8.13) and Eq. (T7.10.5).
4. Adapt conditions and correlations for the mechanical design; e.g. shaft/hub diameter, impeller fixation, wall thicknesses, axial length.
5. Determine main dimensions of all pump sizes.

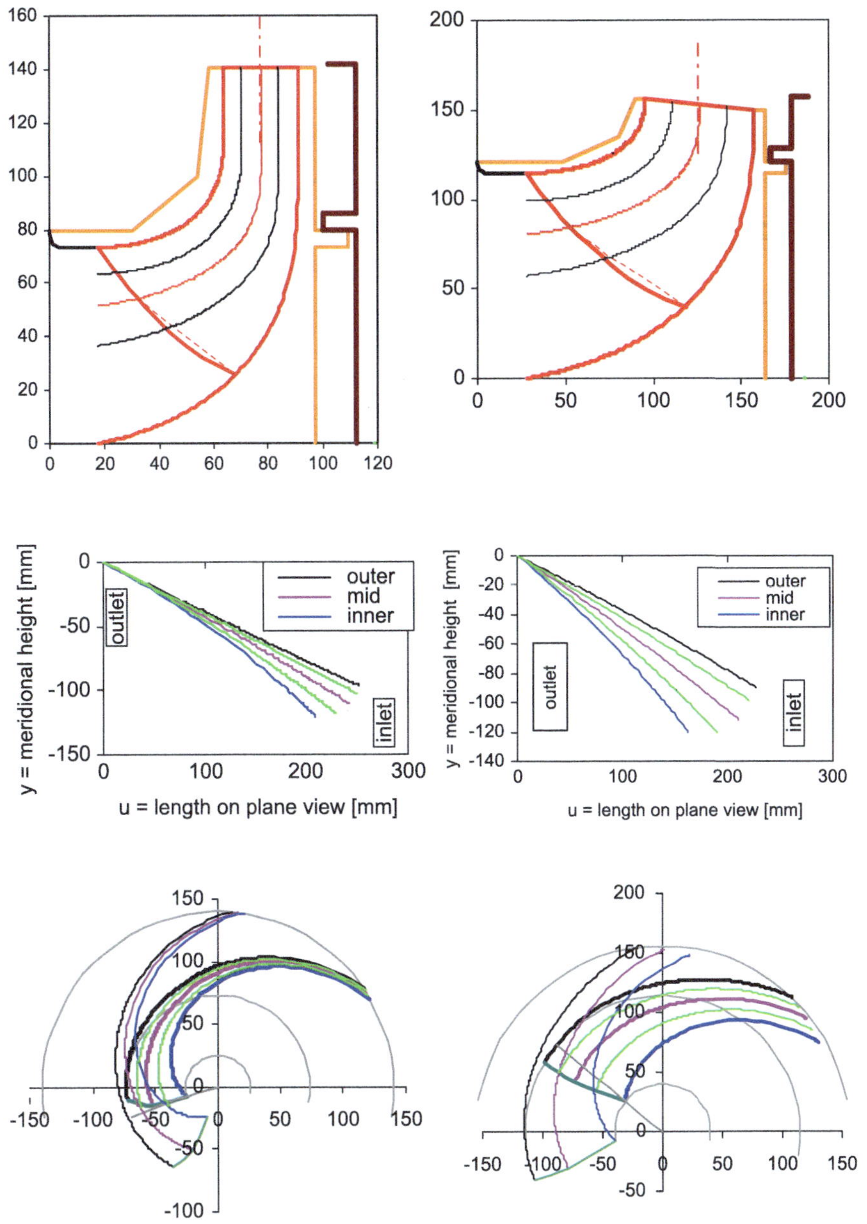

Fig. 7.54 Impellers with axial inflow from analytical design procedure; *left*: $n_q = 36$, *right*: $n_q = 100$

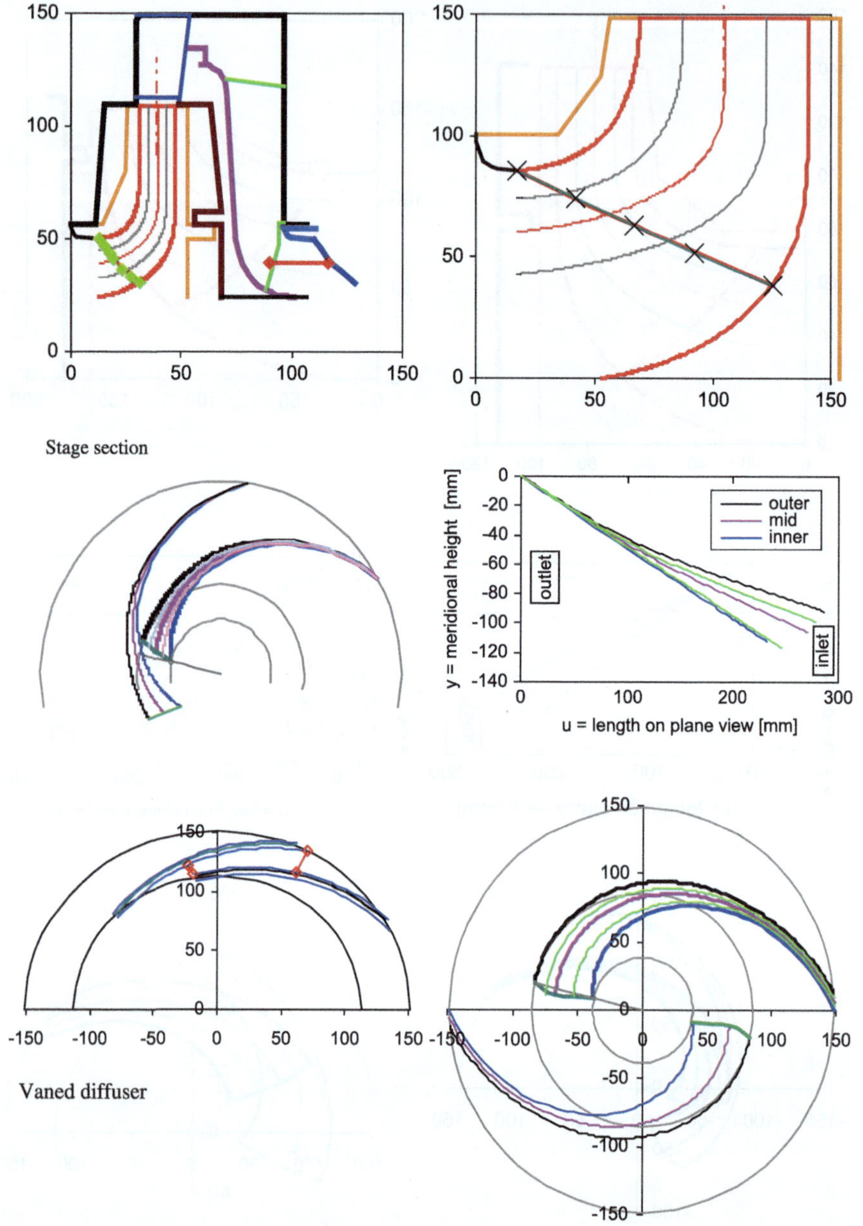

Fig. 7.55 Components from analytical design procedure; *left*: multistage pump $n_q = 24$, *right*: sewage pump with 2 blades, $n_q = 46$

Table D7.5 Sample calculations

Case		Single-stage pump with over-hung impeller		Multistage pump	Sewage pump
n_q		36	100	24	46
n	rpm	1450	1450	2950	1450
Q_{opt}	m^3/h	200	954	105	290
H_{opt}	m	20	15	58	18
d_{2a}/d_{2i}	mm	281	312/300	217	297
d_1	mm	147	229	101	171
d_n	mm	0	0	48	0
n_{ss}		220	220	180	–

6. Check the meridional section of impeller and diffuser (volute), the stage section, the blade angle development and blade plane view.

   * Adapt dimensions in stage section as required
   * Adapt length of streamlines as required to get the desired shape of the leading edge.

7. Step 6 is to be done for (say) 3 to 5 specific speeds which cover the entire range of n_q required for the project. The correlations can then be adapted to get smooth geometry variations throughout the range. This step should ensure that all remaining specific speeds can well be interpolated.
8. After completing step 7, all parameters are fixed in a way which ensures that the impeller geometries vary smoothly over the entire n_q-range required. Once this is achieved, *no more changes are allowed to individual impellers*. Otherwise the systematic approach will be lost: geometry, hence performance, are no longer under control.
9. Subsequently, all impellers can be designed in a very short time.
10. The coordinates of the meridional section and of the blade camber surfaces are transferred to a program which creates the 3D-model of the impeller. The blade thickness with proper profiling is applied in this program. The leading edge profile can be calculated from Eq. (T7.10.15).
11. Check impeller throat area to achieve the desired deceleration at the BEP:

    - axial approach flow: $w_{1q}/w_{1m} = 0.75$ to 0.85 at Q_{opt}
    - radial approach flow: $w_{1q}/w_{1m} = 0.65$ to 0.75 at Q_{opt}

    Note: at specified maximum flow keep $w_{1q}/w_{1m} < 1.0$
    w_{1m} is determined at the geometric mean diameter of the inlet $d_{1m} = \{(d_{1a}^2 + d_n^2)/2\}^{0.5}$

12. Check smoothness of blade surface in the 3D-model.
13. CFD calculations at BEP with one impeller channel and axial approach flow with the objective to check:

    * theoretical head

* pressure distribution at impeller inlet at BEP and maximum flow ($q^* = 1.3$ to 1.5) in order to asses cavitation inception. Aim: get flat pressure distribution at the inlet of suction impellers
* quality of flow, non-uniformity factors (span-wise and pitch-wise) according to Table 8.2.

14. If the results of steps 11 to 13 reveal a potential for optimizing the geometry, the design systematic is to be adapted accordingly.

7.14.5 Some Results

In the following some examples of impellers or stages designed according to the above procedures are presented. Figure 7.54 shows impellers designed for a range of single-stage pumps with overhung impellers and axial approach flow; depicted are the designs for $n_q = 36$ and $n_q = 100$. Figure 7.55 shows the impeller and diffuser of a low-pressure multistage pump and the stage section in the left column. The right column shows a sewage pump impeller with two blades for $n_q = 46$.

All of these designs are examples from ranges of pump sizes which roughly cover $n_q = 7$ to 100 for the OHH pumps, $n_q = 10$ to 50 for the multistage pumps and $n_q = 20$ to 100 for the sewage pumps with 2 or 3 blades. The layout and design strictly followed the rules given in this book. The main data of the sample designs are listed in Table D7.5.

References

1. Abbot, I.H., Doenhoff, A.E.: Theory of wing sections. Dover, Mineola (1959)
2. Aschenbrenner, A.: Untersuchungen über den Einfluß des Abstandes zwischen Lauf- und Leitrad auf das Betriebsverhalten einstufiger Axialpumpenbeschaufelungen. Dissertation, TU Braunschweig (1965)
3. Bakir, F., et al.: Experimental analysis of an axial inducer influence of the shape of the blade leading edge on the performances in cavitating regime. ASME J. Fluid. Mech. **125**, 293–301 (2003)
4. Balasubramanian, R., Bradshaw, S., Sabine, E.: Influence of impeller leading edge profiles on cavitation and suction performance. In: Proceedings of 27th International Pump Users Symposium, pp. 34–44. Texas A & M (2011)
5. Barske, U.M.: Development of some unconventional centrifugal pumps. Proc. IMechE. **174**, 2 (1960)
6. Bergen, J.U.: Untersuchungen an einer Kreiselpumpe mit verstellbarem Spiralgehäuse. Dissertation, TU Braunschweig (1969)
7. Bernauer, J., et al.: Technik und Anwendung moderner Propellerpumpen. KSB Techn. Ber. **19** (1985)
8. Böke, J.: Experimentelle und theoretische Untersuchungen der hydraulischen Kräfte an einschaufeligen Laufrädern von Abwasserpumpen unter Berücksichtigung der Änderung geometrischer Parameter. Dissertation, TU Kaiserslautern, Schaker Aachen (2001)
9. Conrad, O.: Belastungskriterien von Verzögerungsgittern. MTZ. **26**(8), 343–348 (1965)

10. Cooper, P., et al.: Reduction of cavitation damage in a high-energy water injection pump. ASME AJK2011–06092
11. Cooper, P.: Pump hydraulics—advanced short course 8. In: 13th International Pump Users Symposium, Houston (1996)
12. Cropper, M., Dupont, P., Parker, J.: Low flow—high pressure. Sulzer. Tech. Rev. 3 + 4, 15–17 (2005)
13. Dahl, T.: Centrifugal pump hydraulics for low specific speed application. In: 6th International Pump Users Symposium, Houston (1989)
14. Dupont, P., Casartelli, E.: Numerical prediction of the cavitation in pumps. ASME FEDSM2002–31189
15. Eppler, R.: Airfoil design and data. Springer, Berlin (1990)
16. Favre, J.N.: Development of a tool to reduce the design time and to improve radial or mixed-flow impeller performance. ASME Fluid Mach. FED. **222**, 1–9 (1995)
17. Flörkemeier, K.H.: Experimentelle Untersuchungen zur Optimierung von Spiralgehäusen für Kreiselpumpen mit tangentialen und radialen Druckstutzen. Dissertation, TU Braunschweig (1976)
18. Forstner, M.: Experimentelle Untersuchungen an vor- und rückwärts gepfleilten Axialpumpenschaufeln. Dissertation, TU Graz (2002)
19. Furst, R., Desclaux, J.: A simple procedure for prediction of NPSH required by inducers. ASME FED. **81**, 1–9 (1989)
20. Glas, W.: Optimierung gepfleilter Pumpenschaufeln mit evolutionären Algorithmen. Dissertation, TU Graz (2001)
21. Goltz, I., Kosyna, G., Stark, U., Saathoff, H., Bross, S.: Stall inception phenomena in a single-stage axial pump. In: 5th European Conference on Turbomachinery, Prag (2003)
22. Gülich, J.F.: Blade wheel for a pump. US patent US 8,444,370 B2 (2013)
23. Hergt, P., et al.: The suction performance of centrifugal pumps—possibilities and limits of improvements. In: Proceedings of 13th International Pump Users Symposium, pp. 13–25. Houston (1996)
24. Hergt, P.: Design approach for feedpump suction impellers. In: EPRI Power Plant Pumps Symposium Tampa (1991)
25. Hergt, P.: Hydraulic design of rotodynamic pumps. In: Krishna, R. (ed.) Hydraulic design of hydraulic machinery. Avebury, Aldershot (1997)
26. Hergt, P.: Lift and drag coefficients of rotating radial and semi-axial cascades. In: 7th Conference on Fluid Machinery, Budapest (1983)
27. Hirschi, R.: Prédiction par modélisation numerique tridimensionelle des effects de la cavitation à poche dans les turbomachines hydrauliques. Dissertation, EPF Lausanne (1998)
28. Holzhüter, E.: Einfluß der Kavitation auf den erreichbaren Wirkungsgrad bei der Berechnung des Gitters einer Axialkreiselpumpe. Pumpentagung Karlsruhe), K12 (1978)
29. Jacobsen, J.K.: NASA space vehicle criteria for liquid rocket engine turbopump inducers. NASA SP-8052 (1971)
30. Janigro, A., Ferrini, F.: Inducer pumps. Von Karman Inst. LS. **61** (1973)
31. Jensen, R.: Experimentelle Untersuchungen an Einfach- und Doppelspiralen für Kreiselpumpen. Dissertation, TU Braunschweig (1984)
32. Johnsen, I.A., Bullock, R.O.: Aerodynamic design of axial-flow compressors. NASA-SP36 (1965)
33. Kowalik, M.: Inducers-state of the art. World Pumps 32–35 (1993)
34. Krieger, P.: Spezielle Profilierung an Laufrädern von Kreiselpumpen zur Senkung von NPSH. VGB Kraftwerkstechnik. **72**, Nr 5 (1992)
35. Kuhn, K.: Experimentelle Untersuchung einer Axialpumpe und Rohrturbine mit gepfleilten Schaufeln. Dissertation, TU Graz (2000)
36. Lapray, J.F.: Seventy-five years of experience in concrete volute pumps. IMechE Paper C439/026 (1992)
37. Lieblein, S., et al.: Diffusion factor for estimating losses and limiting blade loadings in axial-flow-compressor blade elements. NACA RM 252, (1963)

38. Lieblein, S.: Incidence and deviation angle correlations for compressor cascades. ASME J. Basic. Eng. **82**, 575–587 (1960)
39. Lieblein, S.: Loss and stall analysis of compressor cascades. ASME J. Basic. Eng. **81**, 387–400 (1959)
40. Lohmberg, A.: Strömungsbeeinflussung in Laufrädern von Radialverdichtern durch Neigung det Schaufeln in Umfangsrichtung. Dissertation, Ruhr-Universität Bochum (2000)
41. Lomakin, A.A.: Zentrifugal- und Axialpumpen, 2nd edn. Maschinostrojenje, Moskau (1966)
42. Lottermoser, H.: Anforderungen an die Sicherheitseinspeisepumpen eines Kernkraftwerkes. Pumpentagung Karlsruhe, A2 (1984)
43. Maceyka, T.D.: New two-stage concept optimizes high-speed pump performance. IMechE Paper C110/87 (1987)
44. Meier-Grotian, J.: Untersuchung der Radialkraft auf das Laufrad bei verschiedenen Spiralgehäuseformen, Dissertation, TU Braunschweig (1972)
45. NASA (ed.): Liquid rocket engine axial-flow turbopumps. NASA SP-8125 (1978)
46. Nicklas, A., Scianna, S.: Kreiselpumpe an der Grenze zur Verdrängerpumpe. Pumpentagung Karlsruhe, A 5-03 (1992)
47. NREC-Bulletin
48. Penninger, G.: Schwingungen und mechanische Belastungen von Axialpumpenschaufeln mit und ohne Pfleilung im kavitierenden off-design Betrieb. Dissertation, TU Graz (2004)
49. Radke, M., et al.: Einfluß der Laufradgeometrie auf Betriebssicherheit und Lebenszykluskosten von Abwasserpumpen. KSB Technik kompakt. Ausgabe 4 Juni (2001)
50. Riegels, F.W.: Aerodynamische Profile. Oldenbourg, München (1958)
51. Roclawski, H., Hellmann, D.H.: Numerical simulation of a radial multistage centrifugal pump. AIAA 2006-1428, 44th AIAA Aerospace Sciences Meeting (2006)
52. Roclawski, H., Hellmann, D.H.: Rotor-Stator interaction of a radial centrifugal pump stage with minimum stage diameter. WSEAS Trans. Fluid Mech. **1**(5) (2006)
53. Roclawski, H., Weiten, A., Hellmann, D.H.: Numerical investigation and optimization of a stator for a radial submersible pump stage with minimum stage diameter ASME FEDSM2006-98181 (2006)
54. Roclawski, H.: Numerische und experimentelle Untersuchungen an einer radialen Kreiselpumpenstufe mit minimalem Stufendurchmesser. Dissertation, TU Kaiserslautern (2008)
55. Schiller, F.: Theoretische und experimentelle Untersuchungen zur Bestimmung der Belastungsgrenze bei hochbelasteten Axialventilatoren. Dissertation, TU Braunschweig (1984)
56. Schroeder, C.: Experimentelle Untersuchungen zur Auslegung hochbelasteter Axialventilatoren. Dissertation, TU Braunschweig (1982)
57. Sloteman, D.P. et al.: Design of high-energy pump impellers to avoid cavitation instabilities and damage. In: EPRI Power Plant Pumps Symposium, Tampa (1991)
58. Spring, H.: Critique of three boiler feedpump suction impellers. ASME Pump. Mach. Symp. FED. **81**, 31–39 (1989)
59. Srivastava, J.: Large vertical concrete sea water pumps. Indian Pump Manufacture Conference (1991)
60. Stark, M.: Auslegungskriterien für radiale Abwasserpumpenlaufräder mit einer Schaufel und unterschiedlichem Energieverlauf. VDI Forschungsheft. **57**, Nr. 664 (1991)
61. Strinning, P. et al.: Strömungstechnischer Vergleich zweier Auslegungskonzepte für Axialpumpen in Tauchmotorausführung. Pumpentagung Karlsruhe, B 4–08 (1992)
62. Tsugava, T.: Influence of hub-tip ratio on pump performance. ASME FEDSM97-3712 (1997)
63. Tsujimoto, Y., et al.: Observation of oscillating cavitation in an inducer. ASME J. Fluids. Eng. **119**, 775–781 (1997)
64. Ulbrich, C.: Experimentelle Untersuchungen der Pumpencharakteristiken und Geschwindigkeitsfelder einer Einschaufel-Kreiselpumpe. Dissertation, TU Berlin (1997)
65. Weinig, F.: ZAMM **13**, 224 ff (1933)

66. Weiten, A.: Vergleich der strömungsmechanischen und rotordynamischen Eigenschaft von Gliederpumpenstufen mit radialen Leiträdern und mit minimalem Stufendurchmesser. Dissertation, TU Kaiserslautern (2006)
67. Wesche, W.: Auslegung von Pumpenspiralen mit dicken Gehäusezunge. Techn. Rundschau. Sulzer. **4**, 157–161 (1980)
68. Wesche, W.: Beitrag zur Auslegung von Pumpenspiralen. VDI Ber. **424** (1981)
69. Wesche, W.: Experimentelle Untersuchungen am Leitrad einer radialen Kreiselpumpe. Dissertation, TU Braunschweig (1989)
70. Wesche, W.: Method for calculating the number of vanes at centrifugal pumps. In: Proceedings of 6th Conference on Fluid Machinery, pp. 1285–1293. Budapest (1969)
71. Worster, D.M., Worster, C.: Calculation of 3D-flows in impellers and its use in improving cavitation performance in centrifugal pumps. In: 2nd Conference on Cavitation, Paper IMechE C203/83 (1983)
72. Worster, R.C.: The flow in volutes and its effect on centrifugal pump performance. Proc. ImechE. **177**(31), 843–875 (1963)
73. Cooper, P., Graf, E.: Computational fluid dynamical analysis of complex internal flows in centrifugal pumps. In: Proceedings of 11th International Pump Users Symposium, Houston, pp. 83–93 (1994)
74. Teesink, P., Visser, F., Jochems, J.: Efficiency optimization of a 30-MW boiler feed pump. ASME FEDSM 2013-16090
75. Van der Schoot, M., Visser, F.: Efficiency upgrade of a double-case pump using CFD and model testing. AJKFluids 2015-33379
76. Weldon, R.: A new type of boiler feed pump for 660-MW electricity generating sets. Tech. Rev. Sulzer **3**, 189–198 (1973)
77. Gerdes, R., Dupont, P., Meuter, P.: A long history of high pressure. Sulzer Tech. Rev. **2&3**, 9–15 (2009)
78. Germaine, B., Fitch, G., Harris, C.: State-of-the-art boiler feedpump upgrade for Ratcliffe power station. IMech S966/003/2004. In: 2nd International Symposium on Centrifugal Pumps
79. Elsässer, T., et al.: Hydraulische und konstruktive Ausführung und erste Betriebserfahrungen mit der Turbospeisepumpe für den Block K des Braunkohlekraftwerks Niederaußem. Pump Users Intl Forum 2004
80. Cowan, D., Liebner, T., Bradshaw, S.: Influence of impeller suction specific speed on vibration performance. In: 29th International Pump Users Symposium, Houston, pp. 18–47 (2013)
81. Gülich, J.F.: Selection criteria for suction impellers of centrifugal pumps. World Pumps, Parts 1 to 3, January, March, April (2001)
82. Mejri, I., et al.: Comparison of computational results obtained from a VOF cavitation model with experimental investigations of three inducers. ASME FEDSM2005-77084. Parts 1 and 2, 1-44
83. Ashihara, K., Goto, A.: Effect of blade loading on pump inducer performance and flow fields. ASMEJFE2002-31200 (2002)
84. Kang, D., et al.: Suppression of cavitation instabilities in an inducer by circumferential groove and explanation of higher frequency components. Int. J. Fluid Mach. **3**(2), 137–149 (2010)
85. Bakir, F., et al.: Design and analysis of axial inducer performance. ASME FEDSM 98-5118
86. Kagawa, S., Kurokawa, J.: New centrifugal pump in very low specific speed range. ASME AJK2011-06018
87. Begnini, H., Jaberg, H.: Development of a novel centrifugal pump with lowest specific speed. In: 11th Pump User's International Forum (2016)

Chapter 8
Numerical Flow Calculations

Real flows are described by partial differential equations which cannot be solved analytically in the general case. By dividing a complex flow domain into a multitude of small cells, these equations can be solved in an *approximate manner* by numerical methods. Because of their wide range of application, numerical flow calculations ("computational fluid dynamics" or "CFD" for short) have become a special discipline of fluid dynamics.

The information given in this chapter is intended to help the understanding and interpretation of numerical flow calculations of centrifugal pumps. The focus is on viscous methods since these provide the best chance of describing realistically the boundary layers and secondary flows in decelerated flow on curved paths as encountered in radial and semi-axial pumps. This statement is not meant to preclude that simpler methods may be applied meaningfully to make a first design.

The following discussion of CFD methods and possibilities focuses also on limits and uncertainties of CFD modeling as well as on quality criteria and issues in CFD applications, Chaps. 8.3.2, 8.3.3, 8.8 and 8.10.

8.1 Overview

Because of the complex flow phenomena in centrifugal pumps, the design of impellers, diffusers, volutes and inlet casings is frequently based on empirical data for determining the flow deflection in the impeller and estimating performance and losses. The design of flow channels and blades then relies on experience and coefficients which originate from test data, Chaps. 3 and 7. The availability of relatively inexpensive computers with high computing powers has fostered the development of numerical methods which are able to solve the 3-dimensional Navier-Stokes-equations in complex components with reasonable effort. Therefore, numerical methods are used also in the pump industry with the object of optimizing

J. F. Gülich, *Centrifugal Pumps*,
https://doi.org/10.1007/978-3-030-14788-4_8

the hydraulic components, to increase the reliability of performance prediction and thus to reduce testing costs.

Depending on the task at hand and the available resources, different types of CFD programs can be applied (see also Chap. 8.10 in this context):

1. Design programs for impellers or diffusers where the blades and the meridional section are modified based on simplified flow calculations until the designer considers the component as acceptable or optimum. Quasi-3D-calculations with computation times of a few minutes are often used for this purpose.
2. Separate calculations of impellers, diffusers, volutes and inlet casings using 3D-Navier-Stokes-methods ("3D-NS"). If the results of such an analysis are unsatisfactory, the components are modified until an optimum design is reached. The success depends on the experience of the designer in using the programs and in assessing the results. However, the effort necessary to modify the geometry and the considerable amount of computation times limit the number of optimization cycles which are feasible in practice.
3. Stage calculations which model inlet casing, impeller and collector as an entity, e.g. [17]. Stage computations can be conducted as unsteady calculations by the "sliding mesh" method (long computer times, high memory requirements), or as steady calculation by the "frozen rotor" method or by introducing "mixing planes" between stationary and rotating component, [33, 60].
4. Calculation of hydraulic forces.
5. Inverse methods which determine that geometry which is able to generate a *prescribed* pressure distribution. Such methods may be used in special cases, for example for designing blades with favorable cavitation characteristics in two-dimensional flows, [22]. Other procedures are developed for optimizing blade shapes which produce less secondary flows, [90].
6. Automatic optimization of impellers where the program varies the geometry until the losses are minimized and/or a prescribed pressure distribution or any other target values or criteria are reached.
7. Expert systems for design and analysis of the components and for interpretation of the results. They may be based on the methods according to items 1) and 2) (possibly item 3) and comprise programs for design and geometry generation in 3D. Furthermore, programs can be integrated into the system for generating the data for numerically controlled machining (e.g. for producing patterns and components) and for stress analysis.
8. Calculation of two-phase flows.
9. Modeling cavitating flows which capture the impact of the cavities on the flow field in the impeller.

The methods mentioned are available in many variations. In addition, methods for non-viscous flows are in use, such as 3D-Euler calculations or singularity methods for 2-dimensional flows - for example for developing blades of axial pumps with favorable cavitation characteristics. 3D-Euler programs can be supplemented by boundary layer models in order to include friction losses and flow separation. However, the losses caused by turbulent exchange of momentum (dominating in turbulent flows) cannot be captured that way. These methods will not be discussed in detail because they are being replaced more and more by 3D-Navier-Stokes programs which receive most of the development efforts.

Three mathematical theories for the numerical calculation of incompressible flow shall be mentioned:

1. Irrotational non-viscous (potential) flows are handled by the Laplace equation.
2. Rotational non-viscous flows are handled by the Euler equations.
3. Rotational viscous flows are handled by the Navier-Stokes equations: laminar flows by means of Prandtl's equation; turbulent flows by means of the Reynolds equations and turbulence models; modeling of large vortices (LES: "large-eddy-simulation"); direct numerical simulation of the turbulence (DNS: "direct-numerical-simulation").

There is a wealth of literature on numerical flow calculations, only a fraction of which can be quoted. An overview of newer developments and basic concepts, theories and procedures can be found in [7, 23, 24, 29, 43, 45, 49, 53, 56, 61, 64, 71, 72, 74, 79]. Details on turbulence modeling are given in [9, 26, 57, 66, 67, 88].

The performance of some commercial 3D-Navier-Stokes programs was compared using the test case of a turbulent flow through a 180°-bend, [28]. It was found that the patterns of secondary flow in the bend strongly depend on the turbulence model - even qualitatively. The velocity profiles in the bend were not at all well predicted. The evaluation of three 3D-Navier-Stokes programs and the comparison between CFD calculation and velocity measurements in a pump by laser anemometry was reported in [69, 70]. The attainable accuracy of 3D-Navier-Stokes calculations and their sensitivity to the influences of the calculation grid and numerical parameters was investigated in [41].

Technically speaking, 3D-Navier-Stokes calculations are basically appropriate for all applications because they offer the chance - if used correctly - of getting the most accurate solution. This does not imply that their use would be in every case economically justifiable or technically sensible. Many a case attacked by CFD could presumably be handled with a fraction of the effort by an engineering assessment and engineering common sense.

8.2 Quasi-3D-Procedures and 3D-Euler-Calculations

8.2.1 Quasi-3D-Procedures

The quasi-3D-method ("Q-3D") was developed by Wu [89] at a time when high-performance computers were not yet available. In the Q-3D method the 2-D Euler-equations are solved in cylindrical coordinates in that the flow between the front and rear shrouds ("S2-surfaces") and between the blades ("S1-surfaces") of an impeller is iteratively superimposed. In doing so, rotational symmetry is assumed for the flow on the S2-surfaces in the meridional section. At the impeller blade trailing edge equal static pressures are imposed on the suction and pressure surfaces. The continuity equation is solved via a stream function. Frequently only one average stream surface is used in the meridional section in order to accelerate convergence. In this way efficient design programs may be developed, for example the "real-time-procedure" described in [86]. In this program system the first draft of an impeller can be optimized by interactive modification of the blades, while the resulting pressure and velocity distributions are instantly ("in real time") available. The flow calculations can be done either by the Q-3D-method or by 3D-Euler calculations.

Quasi-3D-procedures are unable to predict losses and secondary flows. Within limits they may be used for:

- Impeller calculations at the best efficiency point (BEP) if the influence of secondary flows is deemed small. Q-3D is therefore less appropriate for radial impellers of high specific speeds and/or large outlet widths.
- Impeller design systems within the limits stipulated above.
- Determination of cavitation inception based on the calculated pressure distributions on the blades - provided the approach flow conditions can be specified correctly and the grid resolution is made sufficiently high.
- Q-3D-methods (in combination with empirical 1D-design procedures) were used extensively for the first draft of an impeller and its pre-optimization, [6]. The procedure needs extensive calibration by comparisons between calculation and test data in order to adapt the calculation parameters and to ease interpretation of the calculations. Essential geometric parameters must be defined, based on tested impellers with good performance, prior to being able to make efficient use of such programs.
- Q-3D-methods can be supplemented by a boundary layer model in order to estimate friction losses.

Quasi-3D-methods are not suitable for the calculation of diffusers, volutes, inlet casings and curved channels. They are of little use for impeller calculations at partload. As discussed extensively in Chap. 5.2, secondary flows in curved and/or rotating channels are influenced by non-uniform velocity distributions generated by boundary layers. Therefore, velocity distributions calculated for non-viscous flow always deviate (more or less) markedly from test data. Consequently, non-viscous calculations of an inlet casing or a return channel yield incorrect incidences which falsify the pressure distributions. With *axial* approach flow or when losses are deemed insignificant, non-viscous calculations can predict the theoretical head (i.e. the flow deflection in the impeller) reasonably well at the best efficiency point. However, the velocity distribution over the impeller outlet width usually exhibits a trend which is contradicted by measurements; examples may be found in [73]. It may therefore appear difficult to draw relevant conclusions for improving the design from such velocity distributions (which does not preclude achieving good designs, as is indeed also possible without resorting to CFD).

8.2.2 Three-Dimensional Euler-Procedures

The Euler equations represent the non-viscous terms of the Navier-Stokes equations discussed in Chap. 8.3. Consequently, 3D-Euler methods do not predict losses, but the balance of centrifugal, Coriolis and pressure forces is captured correctly with the exception of the impact of shear stresses and boundary layers. 3D-Euler methods fail in applications where boundary layers play an essential role, such as local flow separation and secondary flows which are dominated by boundary layer flow. The theoretical head is predicted correctly only if the velocity distribution at the impeller outlet is not affected by boundary layer effects and secondary flow to the extent that the integral of $u \times c_u$ over the exit is falsified.

3D-Euler procedures are better suited for accelerated flows (turbines) and for the same applications as listed above for Quasi-3D. Accuracy and relevance are significantly superior to with Quasi-3D. In fully separated flow, the recirculation in the impeller is determined to a large extent by centrifugal and Coriolis forces. Fully developed recirculation (but not the onset of recirculation) can therefore be modeled rather well by 3D-Euler calculations.

The application of Euler methods is not recommended for diffusers and volutes. With respect to velocity distributions the same limits apply as for Quasi-3D.

8.3 Basics of Navier-Stokes Calculations

8.3.1 The Navier-Stokes Equations

Because of the widespread use of 3D-Navier-Stokes programs, this topic is discussed here in detail in order to show the possibilities and limits of numerical calculations and to give recommendations for assessing CFD results.

Consider a 3-dimensional, incompressible flow with the relative velocities w_x, w_y, w_z in a Cartesian x, y, z-coordinate system; the rotation is around the z-axis. Since the present discussion focuses on the basic aspects, conservation of momentum is written only for the x-direction (complete equations and derivations can be found in many text books, e.g. [1, 101]:

$$\begin{aligned}&\frac{\partial w_x}{\partial t}+w_x\frac{\partial w_x}{\partial x}+w_y\frac{\partial w_x}{\partial y}+w_z\frac{\partial w_x}{\partial z}+\frac{1}{\rho}\frac{\partial p}{\partial x}-\omega^2 x+2\omega w_y\\&=\nu\left(\frac{\partial^2 w_x}{\partial x^2}+\frac{\partial^2 w_x}{\partial y^2}+\frac{\partial^2 w_x}{\partial z^2}\right)+\left(\frac{\partial \sigma'_x}{\partial x}+\frac{\partial \tau'_{xy}}{\partial y}+\frac{\partial \tau'_{xz}}{\partial z}\right)\frac{1}{\rho}\end{aligned}\qquad(8.1)$$

On the left side of the equation are the substantial acceleration terms (Chap. 1.4.1), the effect of the pressure and the body forces in the rotating system (i.e. centrifugal and Coriolis acceleration, gravity being negligible in comparison). On the right-hand side are the loss terms; the first loss term describes the effects of the molecular viscosity while the second term covers the losses due to turbulent exchange of momentum.

Equation (8.1) represents a very general form of the conservation of momentum which comprises some technically significant special cases:

1. Equation (8.1) can be employed for investigating the flow in a rotating impeller in the relative system or, with $\omega = 0$, for calculations of stationary components (diffusers, volutes, inlet casings).
2. By setting the right-hand side to zero, the Euler equation for non-viscous flow is obtained, which is solved in Euler-programs.
3. If the second term on the right-hand side is set to zero, the Navier-Stokes equations for laminar flows are obtained. The remaining term covers the shear stresses due to the molecular viscosity ν. As long as constant temperature is assumed, ν is a property of the fluid independent of the flow characteristics.

Together with the continuity Eq. (8.2) the three component equations, exemplified by Eq. (8.1), yield a system of four partial differential equations for the four unknown functions p, w_x, w_y and w_z.

$$\frac{\partial w_x}{\partial x} + \frac{\partial w_y}{\partial y} + \frac{\partial w_z}{\partial z} = 0 \tag{8.2}$$

If the velocities in Eq. (8.1) are considered as unsteady, the Navier-Stokes equations would be sufficient to calculate turbulent flows. This direct numerical simulation ("DNS") of turbulent flows in pump components is, however, beyond present computing capabilities. In order to exactly describe all turbulent fluctuations, extremely fine calculation grids and excessive computing times are required. The number N of elements needed in a grid for DNS can be estimated from $N \approx Re^{9/4}$; in the order of 10^{13} elements would be required for $Re = 10^6$ [43].

Reynolds therefore replaced the unsteady velocities by $w + w'(t)$ with w representing the time-averaged velocity and $w'(t)$ its turbulent fluctuation. Equation (8.1) then gives the "Reynolds-averaged" Navier-Stokes-equations (referred to as "RANS"). Present-day Navier-Stokes-Programs work on this basis. The stresses caused by turbulent exchange of momentum (the "Reynolds-stresses"), are given by Eq. (8.3) (the over-bars signifying time-averaged quantities):

$$\sigma'_x = -\rho\overline{w_x'^2} \quad \tau'_{xy} = -\rho\overline{w'_x w'_y} \quad \tau'_{xz} = -\rho\overline{w'_x w'_z} \tag{8.3}$$

Since the fluctuating velocities w_x', w_y' and w_z' are unknown, the system of the four equations cannot be solved ("closure problem"). Therefore, by means of a "turbulence model", additional empirical equations must be devised for the velocity fluctuations. This can be done for example by establishing a relation between the Reynolds stresses and the average velocity components.

8.3.2 *Turbulence Models*

In general terms the turbulence model describes the distribution of the Reynolds stresses in the flow domain. All turbulence models in use are of an *empirical* nature. They contain constants and concepts which were selected so that CFD calculations agreed as closely as possible with the test results in *the considered particular geometry and flow regime* (for example a one-sided plane diffuser). If a turbulence model contains five (or even more) empirical constants, a given set of test data gathered on a specific flow can be represented by different combinations of said constants which are more or less equivalent for the specific case. It is however difficult to decide which of the combinations would be physically the most relevant and the best suited for extrapolation to different flow situations. A broader validation would require investigating a statistically relevant number of measurements, which appears scarcely practicable.

The above discussion leads to the conclusion that *there is no universally valid turbulence model which will yield optimum results for all applications.* Instead it is necessary to select the turbulence model most suitable for the components to be calculated and to carefully validate it by comparing the CFD results to test data, Chap. 8.8.2.

The choice of the turbulence model and the pertinent turbulence parameters thus harbors one of the main uncertainties of 3D-Navier-Stokes calculations of turbo machines. Turbulence models have therefore received considerable attention in a great number of publications, e.g. [9, 10, 20, 26, 53, 57, 66, 67, 72].

From the multitude of turbulence models only a few are briefly discussed below. Each of the models presented is available in different versions. This situation unmistakably testifies to the fact that the problem of turbulence modeling is not yet solved satisfactorily.

Most turbulence models are based on the concept of the eddy viscosity ν_t (introduced in Chap. 1.5.1) which is determined *solely by flow features* (in contrast to the molecular viscosity which is a property of the fluid). The eddy viscosity is frequently expressed by a velocity scale V_t and a length scale L_t of the turbulence, through $\nu_t = V_t \times L_t$.

Turbulence models can be classified by the number of transport equations employed. These are differential equations which attempt to describe the transport of the turbulence quantities in the flow domain.

Zero-equation models (or "algebraic" turbulence models) assume that the eddy viscosity depends essentially on local flow quantities (such as velocity gradients) and on a given length scale for the energy carrying vortices. This dependence is expressed by algebraic equations. Zero-equation models are applicable to attached flows as well as to fluid jets and wake flows; their advantage is the comparatively low calculation time required.

Single-equation turbulence models take into account (in addition to the local flow quantities) the "history", i.e. the phenomena upstream of the fluid element considered. This is done by means of a transport equation for the velocity scale of the turbulence, for example in form of a transport equation for the kinetic energy k of the turbulence.

Two-equation turbulence models use two transport equations for the turbulence parameters V_t and L_t (or k and ε). The **k-ε-model** belongs to this group. Usually it is combined with a wall function. The model is based on the specific kinetic energy k and the dissipation rate ε of the turbulent fluctuations. The production of turbulence is calculated from the local velocity gradients. The turbulence T_u and length scale of the vortices L_t are employed; these parameters are related by Eqs. (8.4) and (8.4a):

$$k = \frac{1}{2}\left(\overline{w_x'^2} + \overline{w_y'^2} + \overline{w_z'^2}\right) \approx \frac{3}{2}\overline{w_x'^2}; \quad T_u = \frac{\sqrt{\overline{w_x'^2}}}{w_{Ref}} = \frac{1}{w_{Ref}}\sqrt{\frac{2}{3}k}; \quad \varepsilon = \frac{k^{3/2}}{L_t} \tag{8.4}$$

The eddy viscosity is calculated from:

$$\nu_t = c_\mu\, L_t \sqrt{k} = c_\mu\, \frac{k^2}{\varepsilon} \tag{8.4a}$$

The "standard-k-ε-model" uses the constant $c_\mu = 0.09$. In addition, there are four more empirical constants in the transport equations for the turbulent kinetic energy k and the dissipation rate ε, which (among others) control the production of the kinetic energy of the turbulence.

In two-dimensional attached boundary layers, the production and dissipation rates of the turbulent kinetic energy are in local equilibrium. In this case the relation of Eq. (8.4b) holds between the wall shear stress τ_w and the turbulence parameters (see also Table 8.1):

$$w_\tau \equiv \sqrt{\frac{\tau_w}{\rho}} = c_\mu{}^{1/4}\sqrt{k} \tag{8.4b}$$

The standard k-ε-model shows weaknesses when modeling the following types of flow:

- Flows on curved paths
- Decelerated flows
- 3-dimensional boundary layers
- Rotating components, since the body forces influence the boundary layers
- Swirling flows
- Strong secondary flows
- Secondary flows which are solely induced by turbulence cannot be captured. Such secondary flows are encountered in channels with non-circular cross sections, see for example Fig. 1.6.
- The production of turbulent kinetic energy is over-predicted in locations with strong velocity gradients (for example near a stagnation point). Therefore flow separations are not (or insufficiently) recognized.

Almost all of the phenomena listed above are encountered in impellers, diffusers, volutes and inlet casings. As a consequence of the above deficiencies, the calculation of losses becomes unreliable and zones with flow separation are either predicted too small or they are not recognized at all, [8]. Furthermore, the velocity distributions in decelerated flow or curved streamlines are not calculated correctly, as was demonstrated with a diffuser in [47] and with bends in [28, 54]. The k-ε-model failed equally when calculating the swirling flow in a pipe (as induced for example by a vortex rope): the decay of the swirl due to friction was over-predicted and the velocity distribution was not captured correctly, [54].

It should be noted that the *standard k-ε-model with the logarithmic wall function fails in apparently simple geometries such as diffusers or bends*. It can therefore be concluded that the k-ε-model is in principle unsuitable for the calculation of pumps: after all, the flow through impellers, diffusers and volutes is decelerated, follows curved paths and is frequently subject to separation. The inability of the standard k-ε-model to capture these phenomena is not only due to the equations and parameters describing the turbulence. The situation is aggravated by the use of the logarithmic wall function which imposes rather than predicts the flow near the solid boundaries, Chap. 8.3.3.

In spite of its severe shortcomings, the k-ε-model has been widely used because convergence was better than with other turbulence models. Furthermore, as it is one of the first industrially applicable two-equation turbulence models, there is a broad validation basis. Some programs employ a modified k-ε-model which attempts to capture the influence of curved flow paths.

Realizable k-ε-model: This modification of the standard k-ε-model prevents non-physical solutions such as negative values of k and ε as well as negative normal turbulent stresses. Unrealistically high production of turbulent kinetic energy is thus avoided.

k-ω-model: This model was developed specifically for flows against strong pressure gradients (such as encountered in a diffuser). It solves a transport equation for the frequency ω of "large" vortices (with $L_t = k^{1/2}/\omega$). The k-ω-model captures the flow near the walls more accurately than the k-ε-model, while the latter better describes the processes in the core flow. Therefore, both models were combined providing a smooth transition from near-wall to core-flow, [57]. Close to flow separation there is a mismatch between turbulence production and dissipation. This situation is not well handled by the eddy viscosity concept employed in the k-ω-and k-ε-models.

Shear stress transport model (SST): This model is designed to mitigate this shortcoming, [57]. It employs the k-ε-model in the core flow, the k-ω-model near solid surfaces and a modified relation for the eddy viscosity which limits the shear stresses due to pressure gradients. The k-ω-model uses five empirical constants. Additionally, the SST-model includes empirical blending functions for the transition between k-ω- and k-ε-model. Flows on curved paths are still not well simulated; to remedy this deficiency "curvature corrections" are provided. In the pump industry, the k-ω SST model is used by many authors, e.g. [5, 39, 91–100]; at the time of writing this may be considered as a kind of standard.

SAS-SST model: In the "scale-adaptive simulation (SAS)" the turbulence spectrum is resolved in unsteady flow regions. This is accomplished by means of the von-Karman length-scale which is dynamically adjusted to calculate turbulence in a similar way as in the large-eddy simulation (LES), [102]. The SAS-SST model has been used in [95, 100 and 109] and found to be closer to test data than the SST-model. The improvemnts are most remarkable at flows with large zones of stalled fluid (that's what the SAS is designed for). The URANS cannot well calculate vortex shedding from a cylinder because the vortex decay into turbulence is not well captured. However, the SAS-SST model can better solve this task, [111].

Kato-Launder k-ε-model: This model is a modification of the standard k-ε-model. It avoids an over-production of turbulence in domains with high velocity gradients (e.g. near the stagnation point at the vane leading edge) by an alternative formulation of the production term, [51]. Since the model over-predicts turbulence production in rotating channels its application to pumps is not really attractive, [39].

Low-Reynolds k-ε-model: This model is able to better resolve the flow closer to solid surfaces. To this end, the constants of the standard k-ε-model are multiplied by functions which depend on a Reynolds number defined with turbulence quantities. The maximum distance of the cells to the wall must be limited to $y^+ \leq 2$ (y^+ is defined in Table 8.1). The first cells near a wall are to be located in the viscous sublayer. Since the sublayer in highly turbulent flow is very thin, a sufficient resolution of near-wall flow requires a high number of nodes.

Two-layer model: This model equally attempts to more accurately describe near-wall flows including roughness effects. A comparison between calculations and test data has been reported in [10, 20]. This procedure also relies heavily on empirical data. The limit between the near-wall layers and the core flow is often selected at $\nu_t/\nu \approx 20$.

A comparative study of 5 turbulence models has been reported in [106] where the performance of a single-stage volute pump (n_q = 14) has been calculated between q* = 0.3 to 1.7. Findings: Below q* = 1.3, all turbulence models gave very similar head and efficiency; the head was up to 5% higher than test. The efficiency was over-predicted by up to 3% at BEP, rising to 20% at q* = 0.3. At q* = 1.7, four models over-predict head and efficiency by 10 to 15% but k-ε model was by 50% too high. Overall, k-ω SST model comes out best. However, it should be noted that the number of elements was only 310,000 which is far too few for reliable loss prediction in a low-n_q pump, [92] and [94]. Of course, if different models give similar results, this does not necessarily mean that those results are correct.

Reynolds stress transport model: All components of the Reynolds stresses, Eq. (8.3), are calculated by transport equations without resorting to the eddy viscosity concept. The transport equations are derived by inserting w + w′(t) in the Navier-Stokes equation and subtracting the Reynolds-averaged equation. The resulting additional 7 coupled equations (6 Reynolds stress components and 1 length scale) require a considerable calculation effort and severely impair convergence. It is also necessary to define boundary conditions for the additional parameters. Therefore, this model is not popular in the turbo machinery industry.

Large eddy simulation (LES): Consider a local flow separation where vortices are generated in the shear layer between main flow and stalled fluid. At their origin the vortices are large. When traveling downstream of the stalled zone, the vortices decay through mixing. Large vortices have a high energy content and hence an impact on the flow distribution. When vortices break up into smaller sizes, energy is dissipated. In large eddy simulation, the movements of the individual large-scale vortices are followed by unsteady calculations, while the small eddies are handled by a statistical turbulence model, [75, 76]. In LES calculations the wall treatment is difficult and wall functions are required to determine the unsteady wall shear stresses. LES calculations of a sharp 90°-bend with R/D = 1.0 in [68] and of a diffuser in [47] agreed well with test data in terms of flow separation, velocity distributions and friction coefficients. Also, a diffuser opening to one side could well be calculated by LES, while parallel calculations with the k-ε-model yielded disappointing results because velocity profiles and friction losses deviated strongly from the measurements. However, because of contradictory investigations such results should not be considered as generic.

LES calculations require very fine grids which should be as isotropic as possible. The number of elements needed increases with the square of the Reynolds number. Therefore, LES applications in pumps are scarcely feasible at present because of limited computation power. Thus, experience with pump calculations is lacking. Flows with Reynolds numbers of roughly 5,000, possibly up to 20,000, can be handled (depending on computer and geometry).

In [47] a "v^2-f" turbulence model (comparable with the SST model) was presented which (when used with a linear wall function) yielded results on the flow in a diffuser very similar to the LES calculations.

Detached eddy simulation (DES): Attached turbulent boundary layers are solved with RANS turbulence models (k-ω SST) while separated flows away from the walls are solved by LES-like models. The switching between RANS and LES depends on the local grid topology; this may lead to grid-induced flow separation in attached boundary layers, [109]. Another drawback may be that the onset of flow separation which can be critical for instabilities of the Q-H-curve cannot be predicted any better than with the RANS model. The DES may be most appropriate for calculating vortex shedding from bluff bodies where flow detachment is - independent from Re - dictated by sharp corners.

8.3.3 Treatment of Near-Wall Flows

All velocity components, including fluctuations, vanish close to solid walls. The necessary boundary conditions are usually provided by "wall functions" according to Table 8.1. Two concepts may be distinguished:

- A fine resolution of the calculation domain down to the viscous sublayer, implying cells as close to the wall as a dimensionless wall distance of $y^+ = 1$ (y^+

is defined in Table 8.1). A linear velocity distribution is assumed in the viscous sublayer according to Eq. (T8.1.8), called linear wall function. This type of modeling requires an accordingly high number of cells. It is employed with: the k-ε-model for *low Reynolds numbers*, the two-layer model, the v^2-f turbulence model, SST model, Reynolds stress transport model and LES.

- A coarser resolution of the calculation domain near the wall is sufficient when a logarithmic wall function according to Eqs. (T8.1.8 to 8.1.10) is employed, Table 8.1. The logarithmic wall function is used in the standard k-ε-model because the transport equations in this model do not yield meaningful values for k and ε near a solid surface. When designing the grid, the cell sizes near the walls must be selected in such a way that y^+ is in the range of 30 to 100.

The wall functions listed in Table 8.1 are valid for flows without significant pressure gradients in the flow direction, for example for flat plates, pipes and channels with hydraulically smooth or rough surfaces. Wall functions describe the effects of roughness in a statistical manner, since an exact geometric and hydraulic modeling of individual roughnesses would be altogether impracticable.

Comments to Table 8.1:

- All equations are only applicable to $y > \varepsilon$, because it is impossible to define a velocity distribution within the height of the roughness.
- Equations (T8.1.9 and T8.1.10) can be used approximately almost to the center of the pipe. Equation 8.1.11 is also applicable for $y^+ > 70$, but for rough pipes larger deviations are to be expected in comparison to Eq. (T8.1.9 or 8.1.10).
- At $y^+ = 11.6$ Eqs. (T8.1.8 and T8.1.9) yield the same value $w^+ = 11.6$ for *hydraulically smooth* ($\varepsilon^+ = 0$) surfaces.
- The limit $y^+ = 11.6$ in Eq. (T8.1.9) is an approximation; fully turbulent friction is only encountered at $y^+ > 70$. The transition in the range of $5 < y^+ < 70$ is often described by a 3rd order power function.
- The Eqs. (T8.1.1 to 8.1.10) are valid for pipes, channels and flat plates, while Eqs. (T8.1.11 and 8.1.12) are applicable only to pipes. They may serve for the purpose of comparison.

Applying wall functions implies some basic problems:

1. If a wall function is applied, the result desired by performing the calculation is anticipated - at least to some extent. In finite-volume-procedures the wall shear stress in the cell adjacent to the wall is determined from the wall function. The velocity distribution in the control volume is then derived from the momentum balance. If the other momentum fluxes or the source term dominate over the wall shear stress, the error induced by applying a wall function may be moderate. For this reason logarithmic wall functions sometimes yield reasonable results in flow situations where they should theoretically not be employed.
2. The wall function imposes the wall shear stress and has thus a direct impact on the hydraulic losses. This applies to the friction on the wall as well as to the exchange of turbulent momentum between near-wall fluid layers (or cells) and the main flow.

Table 8.1 Law of the wall and velocity distributions				Eq.
Valid only for *fully developed* flow over flat plates, in pipes (channels); not valid for diffusers, bends, impellers and volutes. **Attention**: In this table ε is the surface roughness, which must not be confused with the turbulence dissipation in the k-ε-model!				
Definitions	Shear stress velocity	$w_\tau = \sqrt{\frac{\tau_w}{\rho}}$	$\frac{w_\tau}{w_m} = \sqrt{\frac{c_f}{2}} = \sqrt{\frac{\lambda_R}{8}}$	8.1.1
	Dimensionless wall distance	$y^+ = \frac{y\, w_\tau}{\nu}$		8.1.2
	Dimensionless velocity	$w^+ = \frac{w}{w_\tau}$		8.1.3
	Roughness parameter	$\varepsilon^+ = \frac{\varepsilon\, w_\tau}{\nu}$	ε = equivalent sand roughness	8.1.4
	Thickness of viscous sublayer *only for hydraulically smooth*	$\delta_1 = 5\frac{\nu}{w_\tau} = 5\frac{\nu}{w_m}\sqrt{\frac{2}{c_f}}$		8.1.5
	Average velocity	$w_m = \frac{Q}{A}$	w_{max} = maximum velocity in pipe center	8.1.6
Type of friction	**$y^+ < 5$**	**$5 < y^+ < 70$**	**$y^+ > 70$**	
	Viscous	Laminar – turbulent	Fully turbulent	
	Hydraulically smooth	**Transition domain**	**Hydraulically rough**	
	$0 < \frac{\varepsilon\, w_\tau}{\nu} < 5$	$5 < \frac{\varepsilon\, w_\tau}{\nu} < 70$	$\frac{\varepsilon\, w_\tau}{\nu} > 70$	8.1.7
	All roughness peaks remain within viscous sublayer	Some roughness peaks reach out of the viscous sublayer	All roughness peaks reach out of the viscous sublayer	
	c_f and $\lambda_R = f(Re)$	c_f and $\lambda_R = f(Re, \varepsilon/D_h)$	c_f and $\lambda_R = f(\varepsilon/D_h)$	
Velocity profiles				
Hydraulically smooth	For $y^+ < 5$ it is $w^+ = y^+$	For $y^+ > 11.6$ it is $\frac{w}{w_\tau} = 2.5 \ln y^+ + 5.5$		8.1.8
Transition domain For hydraulically smooth set $\varepsilon^+ = 0$		$\frac{w}{w_\tau} = 2.5 \ln \frac{y^+}{1 + 0.3\, \varepsilon^+} + 5.5$ Valid for $y > \varepsilon$ and $11.6 < y^+ < 300$ to 500		8.1.9
Hydraulically rough		–	$\frac{w}{w_\tau} = 2.5 \ln \frac{y}{\varepsilon} + 8.5$ Valid for $y > \varepsilon$ and $11.6 < y^+ < 1500$ to 3000	8.1.10
Center domain of a pipe ("law of the wake"), valid for $y^+ > 1500$ to 3000		$\frac{w}{w_{max}} = 1 + 2.5 \ln \frac{y}{R} \sqrt{\frac{c_f}{2}} \frac{w_m}{w_{max}}$		8.1.11
Ratio between average and maximum velocity; theoretically the factor would be 3.75 instead of 4.07		$\frac{w_m}{w_{max}} = \frac{1}{1 + 4.07 \frac{w_\tau}{w_m}}$		8.1.12

3. In decelerated flows with $dp/dx > 0$ (e.g. in a diffuser) the wall shear stress decreases with growing roughness and drops to $\tau_w \approx 0$ at the point of flow separation. Imposing the wall shear stress by a wall function therefore suppresses local stall: *when applying a logarithmic wall function in decelerated flow, separations cannot be recognized correctly* or they are predicted at considerably lower flow rates than measured. Similar arguments apply to the flow on curved paths (unless the flow is strongly accelerated) because zones with deceleration are found in sharp bends, Chap. 1.4.3 (Fig. 1.12). In calculations of a pump with vaneless diffuser the wall treatment had a strong impact on the velocity profiles and stall; applying a logarithmic wall function suppressed local stall at the blade leading edge and in the annular diffuser, [39].
4. Measurements and calculations in [20] confirm that an increased roughness leads to premature stall in a flow against a pressure gradient (decelerated flow).
5. As discussed in detail in Chap. 3.10.3, the losses due to flow through a component strongly depend on the interaction between roughness, near-wall flow, turbulence and velocity distribution. Handling roughness effects in CFD using the equations in Table 8.1 is therefore not satisfactory. *Large discrepancies between calculated and real hydraulic losses are frequently to be expected.*

In summary it can be stated:

- Flow separation and hydraulic losses can be influenced strongly when employing wall functions - this applies particularly to flows on curved paths and against a pressure gradient (decelerated flows).
- Roughness effects are not well captured by CFD calculations. The problem of correctly assessing the effects of roughness on losses and flow distribution discussed in Chap. 3.10 apply equally to numerical flow calculations.
- Consequently, the calculation of centrifugal pumps should always employ *linear wall functions or a combination of logarithmic and linear wall functions* which allow modeling the flow close to solid surfaces down to $y^+ < 2$. This recommendation is borne out by investigations in [92] and [94] were calculations were done firstly with wall functions ($y^+ = 100$) and, secondly, with the low-Re model ($y^+ = 1$): at $q^* = 1.2$ the wall function gave 6% higher head and efficiency than the test and the low-Re (between $q^* = 0.65$ to 1.0 differences were small). Flow separation at $q^* = 1.2$ in the discharge nozzle (and the associated losses) could only be captured with high resolution of the boundary layers.

8.3.4 Grid Generation

In pump components the Navier-Stokes equations (supplemented by a turbulence model) can only be solved numerically. To this end, the calculation domain must be subdivided into a multitude of small cells. Prior to tackling this task, a 3-dimensional geometric model of the component must be created which completely and unequivocally describes the calculation domain by coordinates. Powerful programs for generating the geometry of impellers, diffusers, volutes and casings are therefore needed which allow interactive modifications of the hydraulic contours for efficient optimization. Commercial 3D-CAD-programs are increasingly used for that purpose. If the calculation domain has been described completely by coordinates, the geometric data can be transferred to a grid generator.

Sufficiently fine high-quality grids are among the most important prerequisites for obtaining reliable CFD solutions. Coarse grids can severely falsify the results; the same is true for grids where the elements have unfavorable shapes in terms of angles or proportions. The grid generation thus is of pre-eminent importance; it is often the most difficult task requiring most of the effort for a CFD application.

Structured and unstructured grids are in use, Figs. 8.1 and 8.2. Structured grids usually are made up from rectangular elements which cover the calculation domain in a uniform way. They offer little flexibility but have a simple data structure; their integration into 3D-CAD-systems is more difficult. Unstructured grids fill the calculation domain by tetrahedrons, without generating continuous grid lines. They are therefore flexible and easily integrated into 3D-CAD-systems; their data structure is more complex and the calculation time increases significantly as compared to structured grids.

Block-structured grids combine the advantages of both systems. Different parts of the calculation domain are filled with structured blocks, but the blocks are linked in an unstructured way.

One of the difficulties is to design the grids in a way that gridlines are as perpendicular to the walls as possible. The problem occurs for example with low

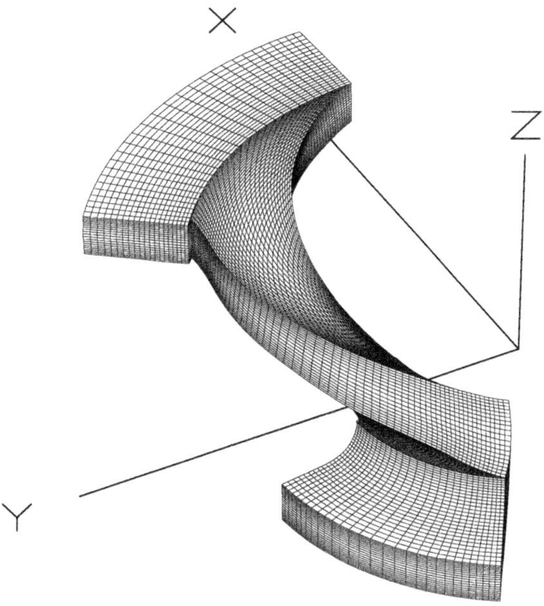

Fig. 8.1 Structured grid of an impeller channel with annular inlet and outlet section (75,000 nodes)

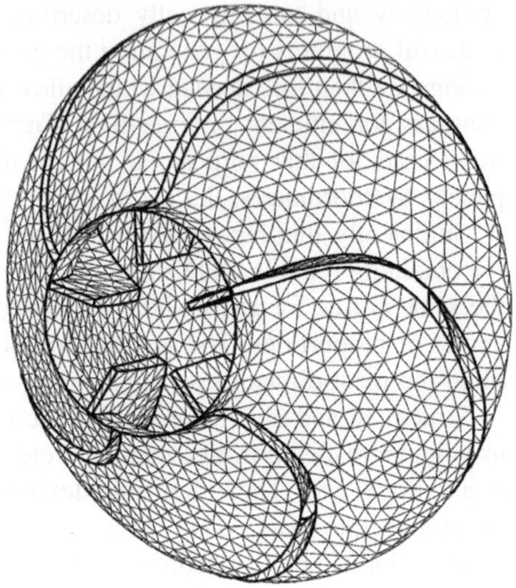

Fig. 8.2 Unstructured grid of a diffuser in a semi-axial pump

vane angles as encountered at the inlet to pump impellers and diffusers (in an inducer these would be as low as 4 to 8°).

The calculation domain must extend sufficiently upstream and downstream of the component to be investigated, as shown by the impeller grid in Fig. 8.1. It is recommended in [43] that the length of these extensions is one half of the blade length, while [39] uses a length equivalent to one impeller inlet diameter.

Guidelines for grid generation:

1. Gridlines should be as orthogonal as possible to the walls. Likewise, grid lines should be orthogonal within the calculation domain. The elements should have no angles lower than 20° or above 160°. Angles below 20° not only impair the accuracy but also the convergence. Block-structured grids are a possible remedy in which an O-grid is put around the blades, [43]. However, with highly twisted vanes a limited grid distortion is unavoidable.
2. To avoid low gridline angles, close to the wall, circular cross sections are meshed as shown in Fig. 8.4. The same technique applies to near-circular sections as may be found in volutes.
3. The gridlines should join the inflow and outflow boundaries with angles of close to 90°.
4. No intersections of grid lines (no negative volumes) are to be allowed.
5. As far as possible gridlines should roughly follow streamlines (difficult for 3D-flows and stalled zones).
6. The size of the elements should not change sharply. "Expansion rate": The dimensions of the cells should not change from one cell to the next by more than a factor of 1.1 near walls and 1.5 to 2 in the bulk flow domain; or the cell volume should not increase more than by a factor of 1.5. The aspect ratio of the elements should not exceed the factor 10 near walls and 5 in the bulk flow

regions. It is particularly important to fulfill these requirements in zones with strong velocity gradients.

7. The grid must be refined in zones with high velocity gradients (near the vane leading and trailing edges, in narrow gaps and near walls). The boundary between refined and coarser grid zones should not be located in zones with strong velocity gradients either. Some CFD-programs automatically refine the grids in zones with strong velocity gradients.
8. If the numerical solution exhibits zones with high residuals, the grids can be refined there; some programs are able to do this automatically.
9. A particularly high grid quality must be implemented at periodic boundary conditions.
10. The impeller blade trailing edges deserve particular attention because the calculated results can be sensitive to their modeling. In one study it was concluded that more than 20 elements should be used to resolve the wake.

The general tendency is: the finer the grid, the higher is the accuracy to be expected; conversely, calculation times and convergence difficulties increase. For pumps the minimum number of nodes is in the range of 70,000 to 100,000 per impeller or diffuser channel. If required by the geometry and the type of flow, more nodes are used. In principle the accuracy increases with the number of nodes; if that is seemingly not the case when comparing CFD results with test data, the numerical and modeling errors may have compensated for each other (provided the measurements are correct), Chap. 8.8.

Impellers are calculated in the rotating system, diffusers and casings in the absolute reference frame. When calculating impellers or diffusers separately from other components, only one channel is usually modeled by applying periodic boundary conditions. When calculating the impeller in the relative reference frame, inlet and outlet parts of the calculation domain and stator components in the absolute system are rotated with $-\omega$; this also applies, of course, to the casing wall when dealing with semi-open impellers.

8.3.5 Numerical Procedures and Control Parameters

In numerical flow calculations the partial differential equations are transformed into algebraic equations applying finite element techniques (in CFD usually finite volumes rather than finite differences). Different algorithms are available for solving the system of algebraic equations. Much in use is the "SIMPLE" algorithm (Semi-Implicit Method for Pressure Linked Equations) and its derivatives. The various algorithms differ in calculation times, convergence and accuracy. Details to numerical procedures can be found, for example, in [24, 61, 64, 71, 79].

Algebraic difference equations determine how the physical quantities - velocity and pressure - behave *within* the cells. To achieve correct results, the program user must specify the order of the corresponding polynomials. Usually at least three options are provided:

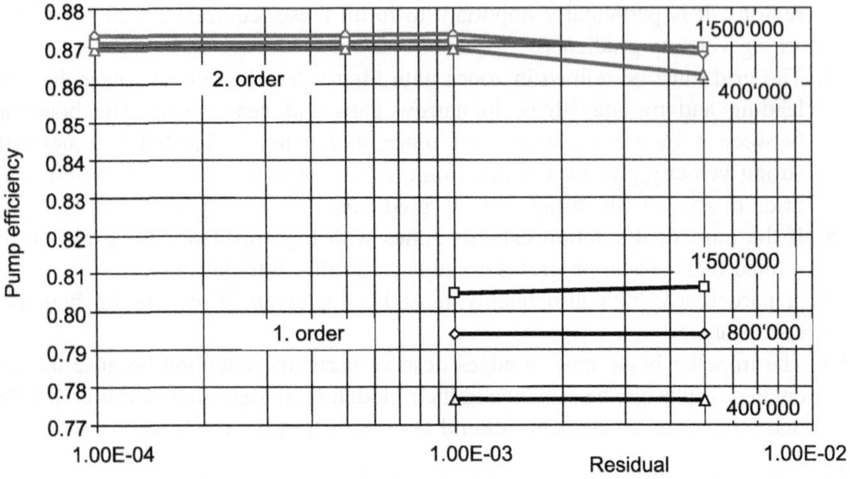

Fig. 8.3 Influence of grid resolution and order on the calculated efficiency, [42]

1. First order truncation error: the physical quantities are constant within a cell,
2. Second order truncation error: the physical quantities vary linearly,
3. Third order truncation error: the physical quantities are square functions within a cell.

Calculations with first order truncation errors are not considered sufficiently accurate in turbo machinery applications. At least the second order is stipulated by [21, 27]. "Numerical diffusion" (viscosity) in first order calculations severely limits the accuracy, [71]. An example in Fig. 8.3 illustrates this by means of calculations with a single-stage volute pump ($n_q = 35$, reported in [84]). In this case high differences are observed between calculations with first and second order schemes. It is further demonstrated that the sensitivity to the fineness of the grid is much higher with the first than with the second order scheme, but such results should not be considered as generic.[1]

The higher the order of the polynomial, the better is the numerical accuracy, but the convergence of the iteration suffers. In case of convergence problems the calculations can be started with the first order scheme; the converged solution can subsequently be re-started with a higher order.

Each program has various numerical parameters which allow the optimization of accuracy and convergence: for example damping factors, relaxation factors and time steps. These parameters may improve convergence, but damping factors can influence the results. Only by comparison with measurements and by a consistent

[1]The order of the truncation error alone does not determine the accuracy; it essentially implies that the exact solution is reached faster with increasing grid refinement.

application of the same parameters for similar cases can the reliability of the calculations be improved and the risk of fortuitous results reduced.

As a rule, an insufficiently converged solution should be discarded. The convergence can be assessed by the "residuals" and the course of the convergence. The residuals indicate how far an obtained solution is from exactly fulfilling the *discretized* equations of conservation of mass and momentum. Commonly the residuals are dimensionless quantities (for example the residual for the conservation of the mass is referred to the total mass), Chap. 8.8.

8.3.6 Boundary Conditions

For all variables in the transport equations "boundary conditions" must be prescribed at the borders of the calculation domain. At solid walls defined by the user, the boundary conditions are set automatically by the CFD-program: (1) no-slip condition for the relative velocity components parallel to the surface; (2) velocity components perpendicular to the wall are set zero. The turbulent kinetic energy and its gradient at the wall are set to zero while the dissipation rate assumes finite values above zero.

Meaningful boundary conditions for pump components are discussed below.

Inlet and outlet boundary conditions: Option (1): The velocity distribution at the inlet to the calculation domain is specified by the components in x-, y-, z-direction. As a consequence, the inlet-mass flow rate is prescribed. The pressure is then given at the outlet; either in one cell or as an average static pressure. Option (2): When calculating cavitation phenomena, the static pressure at the inlet and the mass flow rate at the outlet are specified. These conditions should also be used when unsteady pressures are to be calculated because specifying the outlet pressure could then falsify the pressure fluctuations calculated. In general, option (2) seems to be indicated if it converge equally well as option (1).

If the pressure difference between inlet and outlet would be imposed instead of the mass flow rate, convergence problems can be encountered at partload with flat or unstable Q-H-curves. This because the mass flow can oscillate when trying to establish itself as imposed by the pressure difference.

Turbulence parameters, e.g. the turbulence level and the length scale (alternatively the ratio of the eddy to the molecular viscosity) must be defined at the inlet boundary as well. The length scale of the turbulence can be selected, for example, as 1 to 10% of the hydraulic diameter, the blade height or the impeller inlet diameter. The higher the length scale, the more intense is the exchange of momentum (or the mixing) perpendicular to the main flow direction. Differences in velocity equalize then on a shorter flow path length. The turbulence level in pumps is relatively high, for example 5% at the impeller inlet and 10% at the diffuser inlet[2].

[2]In the investigations reported in [41] the CFD results were not very sensitive to the turbulence parameters. This finding was confirmed by the CFD studies reported in [5].

By means of a sensitivity study it is possible to assess the influence of the turbulence parameters in a particular application.

The flow in a diffuser critically depends on the velocity distribution and the turbulence at the diffuser inlet, Chaps. 1.6, 1.3, 5.3.3. The same applies to all flows with local deceleration - consequently also to bends. To calculate such components without knowing in detail, and specifying correctly, the inlet conditions can only produce fortuitous results. The turbulence parameters have a strong influence on diffuser calculations.

The velocity distribution at the inlet is often given by the outlet conditions of a preceding complex component. Examples are the inlet chambers of between-bearing pumps feeding an impeller and the impeller outlet flow distribution serving as inlet boundary condition for a diffuser calculation. In such cases coupled calculations are recommended. Alternatively, one component is calculated after the other. The outlet velocity and pressure distributions from the preceding part are transferred as the inlet conditions to the downstream component. However, the interaction of the different components gets lost in this procedure (which can scarcely be considered as state of the art any more).

When calculating closed impellers, the leakage through the annular seal must, in principle, be modeled since it influences the velocity distribution at the impeller inlet - in particular at the outer streamline.

Periodic boundary conditions: In order to save computing time, often only one impeller or diffuser channel is calculated. To this end, "periodic boundary conditions" are introduced mid-pitch between two vanes. The calculation is then controlled in a way that identical pressures and velocities result at corresponding cells at these borders.

According to Chap. 1, a fully developed pipe flow needs a considerable length to establish (asymptotically). Instead of calculating a long length of pipe to get a fully developed velocity profile, the concept of periodic boundary conditions can be used to advantage: According to Fig. 8.4 *only one cell* in flow direction is needed if inlet and outlet to that cell are defined as periodic boundary conditions. This implies that the program iterates until identical - hence fully developed - velocity distributions are obtained at inlet and outlet. During the iteration the pressure gradient in flow direction is varied until the specified flow rate is reached.

The results of this calculation, in the form of a fully developed velocity and turbulence field, can subsequently be used as inlet condition to an impeller, a diffuser, a bend, or any other device. An added benefit is that the usually unknown turbulence parameters are fully consistent with the resultant velocity profile.

Periodic boundary conditions can also be applied if the series stages of a multistage pump are to be analyzed. At the inlet and outlet to each stage identical pressure and velocity distributions are achieved iteratively wherein the average pressure at the outlet exceeds the inlet pressure by the pressure rise in the stage. This is the easiest way to get the correct inlet conditions to the stage.

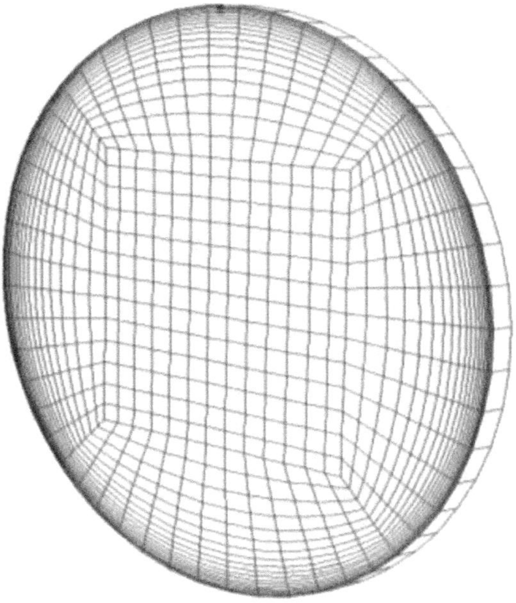

Fig. 8.4 Grid for calculating fully developed pipe flow by means of periodic boundary conditions, ANSYS Germany, Otterfing

Symmetry condition: In order to save computer capacity only one half of a symmetric component may be analyzed - in principle. A symmetry condition bounding a calculation domain is treated as a wall *without friction or any other effect*. Velocity components perpendicular to the surface are set zero. The gradients of the velocity components parallel to the surface are equally zero. However it should well be noted that geometric symmetry of a component is not sufficient but that the inlet flow must also be truly symmetric. This requirement is obviously not fulfilled if a volute is fed by a single-entry impeller. A pipe bend should not be calculated using a symmetry condition, refer to Chap. 10.12. Following the same logic, a symmetry condition should not be applied to a volute either, even if it is fed by a double-entry impeller (that is with nominally symmetric flow).

In [68] a LES calculation of a 90°-bend has been performed using "non-reflecting" outlet boundary conditions. The objective is to prevent wave reflections which could lead to non-physical solutions.

8.3.7 Initial Conditions

To start a calculation, initial values are required in all cells. These can be set automatically by the program (possibly at zero) if not given by the user.

In general it will be assumed that the converged solution is independent of the initial conditions. However, this is by no means always true. If there is a hysteresis in the Q-H-curve, the axial forces or any other quantity, the converged solution will indeed depend on the initial conditions. To analyze hysteresis effects, it is necessary to use the converged solution from the preceding operation point as initial condition for the next point to be analyzed. This must first be done with increasing flow rate and in a second series of calculations with decreasing flow rate (or vice versa). In this way it is possible to reproduce a measured Q-H-curve hysteresis by means of CFD, [30].

When analyzing *unsteady* flows, the initial conditions must be a solution of the differential equation. The system status at the time $t = 0$ is used to calculate transients. If it is not possible to define a solution of the differential equation, the trivial solution "zero" is used.

8.3.8 Possibilities of 3D-Navier-Stokes-Calculations

In the pump industry there is a broad range of actual and potential CFD applications and possible objectives:

1. Calculation of the head at the best efficiency point. Minimizing hydraulic losses.
2. Attaining specific impeller outlet velocity profiles which are considered beneficial with respect to diffuser performance and/or the stability of the Q-H-curve, Chap. 5.6.
3. Prediction of the partload characteristic. Stage calculations of rotor and stator promise the only chance for a realistic solution, Chaps. 8.6.2 and 8.6.3.
4. Calculation of hydraulic forces, in particular axial and radial thrust.
5. Calculation of flows through annular seals, impeller sidewall gaps including rotor-dynamic forces or coefficients, [3, 78, 91, 92, 94].
6. Visualizing flows in confined spaces which are hardly accessible to experimental equipment; for example gap vortices as in [43].
7. Unsteady calculations of the Q-H-curves and the unsteady pressure and velocity distributions at the impeller outlet as well as rotor/stator interaction forces and phenomena, Chap. 8.6.3.
8. Calculation of cavity flows to determine cavitation inception, the cavity length as a function of the $NPSH_A$ and of the cavitation induced drop in head or efficiency, Chap. 8.7.
9. Calculation of the flow paths of entrained solid particles for the assessment of hydro-abrasive wear, [55, 115]. The momentum equations of the solid particles are solved in addition to the Navier-Stokes equations yielding the impingement angles and velocities of the solids on the structural surfaces which bound the flow domain. This information can be used in a wear model to estimate the metal loss due to abrasion.
10. Calculation of two-phase flows.
11. Calculation of performance with highly viscous fluids (Chap. 13.1). Such calculations should model the heating of near-wall fluid due to wall shear

stresses, turbulent dissipation and heat transport phenomena. The calculation domain therefore must include the impeller sidewall gaps and the annular seals. As in supersonic flows, the dissipation terms must be activated in the energy equation and temperature-dependent fluid properties must be implemented. This type of application covers laminar and turbulent flows as well as the transition between the two conditions. Since measurements are scarce and there is a multitude of different fluids to be handled in practice, the potential application of CFD is promising (testing such fluids would be very expensive).

12. Theoretically, it should be possible to calculate the complete pump characteristics (as described in Chap. 12.2) by CFD stage calculations. While turbine modes would be easy to handle, the calculation of brake operation modes may suffer from convergence problems because of extensive flow recirculations and unsteadiness, Chap. 8.8.4. Conversely, the required accuracies are much less than in the usual pump applications. This CFD application looks quite attractive because measured data and prediction methods are scarce and testing is very expensive.
13. Calculation of intake structures with free liquid level with the object to optimize the intake for obtaining a uniform approach flow free from swirl and vortices; see Chaps. 10.14 and 11.7.3 where literature with CFD applications can be found.

Two approaches may be followed when tackling these tasks: (A) Ideally, absolute values are considered (e.g. heads and efficiency). However, high precision is required in the pump industry, since minus tolerances on efficiency and head are rarely accepted and generous margins for covering large uncertainties are precluded for economic reasons. (B) Knowing or assuming that absolute values are unreliable, the assessment is rather based on differences between various calculations. In doing so, it is supposed that the impeller which looks best in the calculations is also the best in reality when combined with the inlet casing and collector. However, this frequent conception should be reviewed critically, Chap. 8.8.

Either the pump can be analyzed as a unit or the components can be calculated individually which implies the following possibilities:

Inlet casings: The goal of the calculation is to determine the velocity distribution at the impeller inlet and possibly the losses in the casing. While the losses usually are of secondary importance (except for very high specific speeds), the optimization of complex inlet casings for achieving a uniform approach flow to the impeller can be highly relevant with respect to cavitation, noise, vibration excitation and efficiency. The significance of such an optimization increases with the specific speed and the power transferred by the impeller. Even if the accuracy of the calculation is moderate, 3D-Navier-Stokes methods appear well suited for this purpose, [15] - all the more so because experimental investigations are very expensive. It is necessary to employ a turbulence model and wall treatment which allows the prediction of local stall and of velocity profiles which are at least qualitatively correct.

Impeller: The impeller calculation yields the static pressure rise $\rho \times g \times H_p$, the theoretical head H_{th}, the rise in total pressure $\rho \times g \times H_{La}$, the losses expressed, for example, by the hydraulic efficiency of the impeller $\eta_{h,La} = H_{La}/H_{th}$ and the outlet velocity profile. Furthermore, hydraulic forces and moments can be obtained by integrating over the pressure distributions.

The calculation of an individual impeller can neither give the head nor the hydraulic efficiency of the pump - not even as an approximation - since the losses in the collector are ignored. The strong rotor/stator interaction during partload recirculation precludes individual calculations of impeller and collector. This is demonstrated by Fig. 4.6 where the tests of a given impeller with a vaned and a vaneless diffuser are compared. At flows below the onset of outlet recirculation, the static pressure rise of the impeller is considerably higher with a vaned diffuser than that with a vaneless annulus. A separate calculation is unable to predict this effect. The frequently observed agreement between measured and calculated characteristics near half load is therefore fortuitous since measured and calculated characteristics often intersect in this range.

In order to appreciate the limits of an individual impeller calculation, consider Fig. 5.30 which demonstrates how sensitively the head at partload can respond to the interaction of impeller and collector.

Diffuser: The diffuser calculation yields the static pressure rise and the losses. As stressed above, the correct impeller outlet velocity profile is required as inlet boundary condition.

Volutes: The volute calculation gives the pressure recovery, the losses and the pressure distribution. Because of the lack of rotational symmetry, the entire volute must be calculated over 360°. If the volute is fed by a double-entry impeller, a symmetry condition in mid-plane could be considered (in principle) in order to cut the number of nodes by half. However, any unsteady behavior of the secondary flow cannot be captured that way, Chap. 10.10. If done at all, individual volute calculations should be restricted to the best efficiency point. At partload the pressure distribution in the volute varies around the circumference (Chap. 9.3.3). Consequently, the impeller outflow varies as well: *the impeller operates in every circumferential position at a different flow rate*, Chap. 10.7.1, Fig. 10.23. At flows of $q^* \neq 1$, it is therefore preferable to perform unsteady calculations combining impeller and volute, Chap. 8.6.3, but often the frozen rotor model is used for lack of resources.

When optimizing components separately, it is tacitly assumed that the individually optimized components will well work together. However, pressure recovery and losses in the diffuser depend strongly on the impeller discharge flow distribution. Mixing losses in the diffuser due to uneven approach flow must be charged to the impeller. The non-uniformity of the impeller outlet velocity is therefore an important criterion when assessing the merits or deficiencies of an impeller design, refer also to Chap. 5.3.3.

8.4 Averaging and Post-processing

Numerical flow calculations produce a vast quantity of data which is meaningless unless it is reduced to a small number of comprehensible physical parameters and flow pictures which lend themselves to easy interpretation. Only a well-designed "post-processing" allows drawing conclusions from the calculated data for the optimization of the hydraulic contours. The post-processing covers the steps below:

1. Integration and averaging of velocities, pressures, forces and momentums at various control surfaces. These averages represent the global performance parameters of head, torques and efficiency as well as the hydraulic losses which are an essential optimization criterion.
2. Velocity profiles at specific control surfaces, in particular at inlet and outlet (examples in Chap. 8.5.4, Figs. 8.11 and 8.10).
3. Averages of angular momentum, total pressure and static pressure along the flow path (example in Chap. 8.5.4, Fig. 8.10).
4. Pressure distribution on various streamlines along the blades or channels for assessing the blade loading, incidence and risk of stall (example: Chap. 8.5.4, Fig. 8.9).
5. Distribution of loss parameters.
6. The residuals of mass, momentum and energy in various control surfaces are needed for checking convergence and assessing the quality of the numerical solution.
7. Graphical representation of flow patterns, velocities and streamlines (possibly by means of special graphics programs).

Calculating integrals and averages is necessary to check that the specified performance is reached. To this end, control surfaces are defined at the inlet and the outlet of the calculated component or pump. The averages of velocities and pressures are determined by summation over all calculation elements in said control surfaces.

Different averaging procedures are available; for example, mass-averages and area-averages. Physically meaningful averaging is done in a way that the *conservation equations of mass, angular momentum and energy (enthalpy or total pressure) are fulfilled.* In order to determine which type of averaging fulfills this requirement, consider two different impellers mounted on the same shaft each of which having uniform (but different) velocities and pressures in their respective control surfaces. In this case the conservation equations for mass, angular momentum and energy can be written separately for both impellers (Chap. 1.2). Obviously, the sum over both impellers fulfills the conservation equations of the *rotor*. This consideration proves that mass-averaging is required to fulfill the conservation equations for incompressible flow. Consequently, the global flow quantities are to be determined by summation over all elements in a control surface as stipulated by Eqs. (8.5) to (8.7) (c_n is the velocity component normal to the respective control surface of area A):

$$\text{volumetric flow rate: } Q_{La} = \sum_A c_n\, dA \tag{8.5}$$

$$\text{angular momentum: } M_j = \rho \sum_A r\, c_u\, c_n\, dA + \sum_A r\, \tau_u\, dA \tag{8.6}$$

$$\text{energy flow (useful power): } P_j = \sum_A p_{Stat}\, c_n\, dA + \frac{\rho}{2} \sum_A c^2\, c_n\, dA \tag{8.7}$$

Angular momentum and power are thus *mass*-averaged while the *area*-average yields the correct flow rate. These summations are done over the outlet control surface (j = 2) and the inlet control surface (j = 1). The differences between outlet and inlet yield the torque $M_{th} = M_2 - M_1$ transferred by the impeller and the useful energy of the impeller $P_{u,La} = P_2 - P_1$.

Equations (8.5 to 8.7) are equally applicable to diffusers, volutes and inlet casings. Provided that $c_u \neq 0$, a hydraulic moment is obtained from Eq. (8.6) also for these components; it causes a corresponding reaction force in the casing support.

In Eq. (8.6), τ_u is the tangential component of the shear stresses in the control surface; τ_u vanishes if the discharge flow is truly radial (i.e. $\alpha_2 = 90°$). Such shear stresses are essentially due to turbulent exchange of momentum, which increases with growing gradients in the outlet velocity profile. These shear stresses therefore increase with the non-uniformity of the velocity profile at the impeller outlet and are maximum during flow recirculation.

If there is no recirculation at the control surfaces, i.e. the normal velocity in each cell is $c_n \geq 0$, Eq. (8.6) yields the theoretical head (with $M_\tau = 0$):

$$Y_{sch} = g\, H_{th} = \frac{M_{th}\, \omega}{\rho\, Q_{La}} = \overline{u\, c_{u2}} - \overline{u\, c_{u1}} = \frac{\omega}{Q_{La}} \sum_A r\, c_u\, c_n\, dA \tag{8.8}$$

and the total pressure is obtained from:

$$\overline{p_{tot}} = \frac{1}{Q_{La}} \left(\sum_A p_{stat}\, c_n\, dA + \frac{\rho}{2} \sum_A c^2\, c_n\, dA \right) \tag{8.9}$$

Thus, the total pressure must be *mass*-averaged in order to fulfill the energy equation. To calculate the radial forces acting on the impeller, the conservation of momentum must be applied in the *radial* direction; the static pressure is to be *area*-averaged when performing this evaluation. In this way, the radial force components acting on the shrouds can be determined. The resultant radial force acting on the blades should be - at least in theory - determined by integrating the static pressure and shear stresses on the blade pressure and suction surfaces. This is done in the same way as calculating the torque as per Eq. (8.25). All radial forces components acting on an impeller are shown in Fig. 9.24.

Note that the impeller calculation must always be done with the actual flow through the impeller i.e. with $Q_{La} = Q + Q_E + Q_{sp}$. If the leakages were ignored, the calculation would obviously apply to the wrong operation point.

The CFD results of an impeller calculation in dimensionless form are expressed by

$$\text{static pressure rise in impeller: } \psi_p = \frac{2(\overline{p_{2stat}} - \overline{p_{1stat}})}{\rho u_2^2} \tag{8.10}$$

$$\text{theoretical head: } \psi_{th} = \frac{\psi}{\eta_h} = 2\left(\frac{\overline{(u\,c_u)_2}}{u_2^2} - \frac{\overline{(u\,c_u)_1}}{u_2^2}\right) \tag{8.11}$$

$$\text{increase in total pressure: } \psi_{La} = \frac{2(\overline{p_{2tot}} - \overline{p_{1tot}})}{\rho\, u_2{}^2} \tag{8.12}$$

$$\text{hydraulic losses in the impeller: } \zeta_{La}\frac{2g\,Z_{LA}}{u_2{}^2} = \psi_{th} - \psi_{La} \tag{8.13}$$

$$\text{hydraulic efficiency of the impeller: } \eta_{h,La} = \frac{\psi_{La}}{\psi_{th}} \tag{8.13a}$$

The Eqs. (8.5) to (8.13a) are applicable to radial, semi-axial and axial impellers, since they capture the integral values of the blade torque and the enthalpy flux.

The impeller losses defined by Eq. (8.13) are given by the difference between the theoretical head and the rise in total pressure which exceeds the useful head of the pump by the losses in the collector. If the impeller outlet velocity is nearly uniform, the definition of impeller losses according to Eq. (8.13) is meaningful. However, if the impeller discharge is strongly non-uniform, mixing losses due to exchange of momentum are created in the collector; these mixing losses should be charged to the impeller but do not appear in ζ_{La}. Furthermore, at a given flow rate a non-uniform velocity profile usually contains more kinetic energy than a uniform discharge. This becomes evident if the kinetic energy is written for a given channel and constant flow according to Eqs. (8.9) and (8.5) for $c_n = c$, yielding:

$$\frac{2\,p_{dyn}}{\rho\,c_{av}^2} = \frac{\int_A (c/c_{av})^3\,dA}{\int_A (c/c_{av})\,dA} \tag{8.14}$$

The mean velocity $c_{av} = Q/A$ in the channel remains constant for a given flow rate and area (independently of the actual velocity distribution as distorted it may be). In contrast, the integral over c^3 rises with increasing non-uniformity.

Turning now to the discharge flow of an impeller, a non-uniform outlet velocity profile does not necessarily contain more kinetic energy than a uniform one, because the relative distribution of c_{2m} and c_{2u} is relevant. However, it can frequently be observed that the rise in total pressure in the impeller increases with growing non-uniformity of the flow (while the static pressure rise ψ_p is unaffected). The impeller losses are then under-predicted because the excess kinetic energy cannot - or only incompletely - be recovered in the collector. In case of low specific speeds the calculation may produce, in the extreme, hydraulic impeller efficiencies of 1.0 (or even slightly higher). When comparing two calculations "A" and "B" it thus can happen that losses *and* static pressure rise in the impeller are predicted lower for case "A" than for case "B". Such a result is a contradiction in itself, because *smaller* losses fundamentally should imply a *higher rise in static pressure*. A separate optimization of impeller and collector can thus be misleading if the impeller outlet flow is non-uniform and collector losses make up a considerable portion of the total hydraulic losses of the pump (as is the case at low and moderate specific speeds).

According to the above consideration, the losses calculated from Eqs. (8.12) and (8.13) represent a sort of lower limit. In order to avoid underrating the impeller losses, the losses in a mixed-out condition can be analyzed by re-writing Eqs. (8.12 and 8.13) with average velocities:

$$\psi_{La,min} = \psi_p + \left(\frac{\overline{c_2}}{u_2}\right)^2 - \left(\frac{\overline{c_1}}{u_2}\right)^2 \tag{8.15}$$

The average velocities at the outlet are calculated from the angular momentum, Eq. (8.11), and the continuity equation according to $c_{2u}/u_2 = \psi_{th}/2 + d^*_{1m} \times c_{1u}/u_2$ and $c_{2m}/u_2 = \varphi_{2La}$ and $c_2{}^2 = c_{2u}{}^2 + c_{2m}{}^2$. The result is:

$$\zeta_{La,max} = \psi_{th} - \psi_p - \left(\frac{\psi_{th}}{2} + \frac{d^*_{1m} c_{1u}}{u_2}\right)^2 - \varphi_{2,La}{}^2 + \left(\frac{\overline{c_1}}{u_2}\right)^2 \tag{8.16}$$

Applying the mixed-out condition, the impeller losses are overrated; this value may therefore be called $\zeta_{La,max}$. Assuming that half of the excess kinetic energy can be recovered in the collector, an average may be used as a plausible estimate:

$$\zeta_{La} = 0.5\left(\zeta_{La,min} + \zeta_{La,max}\right) \tag{8.17}$$

Obviously, the values $\zeta_{La,min}$ and $\zeta_{La,max}$ cannot represent absolute maxima or minima which would also cover numerical and modeling errors; they solely describe the effect of a non-uniform discharge flow. The post-processing may include a non-uniformity coefficient to help assessing the quality of the impeller outlet flow. Two possible definitions (U_{F1} and U_{F2}) are provided by Eqs. (8.18 and 8.19) where "X" stands for the velocity components or pressures investigated:

$$U_{F1} = \frac{\int X^2 \, dA}{\overline{X}^2 A} \tag{8.18}$$

$$U_{F2} = \frac{1}{\overline{X}}\sqrt{\frac{1}{A}\int \left(\overline{X} - x\right)^2 dA} \tag{8.19}$$

In order to assess the effect of a non-uniform outlet flow quantitatively, consider a linear velocity distribution of the form $c = c_{min} + \Delta c x^*$, with $\Delta c = c_{max} - c_{min}$ and a mean velocity of $c_{av} = c_{min} + \Delta c/2$. Integrating the Eqs. (1.37 and 1.38) for these distributions, the following relations are obtained:

$$U_{F1} = \left(\frac{c_{min}}{c_{av}}\right)^2 + \frac{c_{min}\,\Delta c}{c_{av}^2} + \frac{1}{3}\left(\frac{\Delta c}{c_{av}}\right)^2 \tag{8.20}$$

$$U_{F2} = \sqrt{\left(1 - \frac{c_{min}}{c_{av}}\right)^2 + \frac{1}{3}\left(\frac{\Delta c}{c_{av}}\right)^2 - \left(1 - \frac{c_{min}}{c_{av}}\right)\frac{\Delta c}{c_{av}}} \tag{8.21}$$

$$c_p = 2\left(\left(\frac{c_{min}}{c_{av}}\right)^2 + \frac{c_{min}\,\Delta c}{c_{av}^2} + \frac{1}{3}\left(\frac{\Delta c}{c_{av}}\right)^2 - 1\right) = 2\left(U_{F_1} - 1\right) \tag{8.22}$$

$$c_{pth} = \left(\frac{c_{min}}{c_{av}}\right)^2 + \frac{c_{min}\,\Delta c}{c_{av}^2} + \frac{1}{2}\left(\frac{\Delta c}{c_{av}}\right)^2 - 1 \tag{8.23}$$

$$\zeta_{mix} = c_{p_{th}} - c_p = U_{F2}^2 \tag{8.24}$$

Evaluating these equations with arbitrary figures we find that for *linear* velocity distributions 2/3 of the kinetic energy is converted by exchange of momentum into static pressure while 1/3 is dissipated. The pressure loss coefficient equals U_{F2}^2, while the pressure recovery equals $2 \times (U_{F1} - 1)$. Even with non-linear velocity distributions, these relations help to assess the effect of non-uniform outlet velocity profiles.

Another way to account for the mixing losses downstream of the impeller is to integrate the outlet velocity profile according to Eqs. (1.37) and (1.38). The resulting mixing loss coefficient ζ_{mix} represents the lower limit of the mixing losses which is to be added to the loss coefficient according to Eq. (8.13).

Alternatively to Eq. (8.6), the torque acting on the impeller can be determined through the integration of the pressure and shear stress distributions on the blades.

$$M = z_{La}\int_{r_1}^{r_2}\int\left(\left(p_{PS} - p_{SS}\right)dBrdr + \left(\tau_{PS} + \tau_{SS}\right)\frac{dBrdr}{\tan\beta_B}\right) \tag{8.25}$$

The integral over the tangential components of the shear stresses acting on front and rear shroud within the impeller channels must be added to Eq. (8.25). Integrating only over the pressure distribution, a slightly smaller torque value than the true torque on the impeller would be obtained.

If the shear stresses τ_u in Eq. (8.6) are neglected, the torque calculated is too low and the hydraulic impeller efficiency is predicted too high. In general, a higher torque and lower hydraulic impeller efficiency are expected from Eq. (8.25) as compared to Eq. (8.6). The difference in efficiency is typically in the order of 1 to 2%; this implies a difference of 10 to 20% in terms of losses. A separate evaluation of the integrals in Eq. (8.25) over the pressure and the shear stresses allows one to differentiate the friction losses.

Partload operation with recirculation: Consider an impeller with flow recirculation according to Fig. 8.5 with a net through-flow of $Q_{La} = Q + Q_E + Q_{sp}$. The flow rates of recirculating fluids can be calculated from:

$$Q_1 = Q_{La} + |Q_{R1}| \quad Q_2 = Q_{La} + |Q_{R2}| \tag{8.26}$$

$$Q_{R1} = \frac{1}{2}\left(\sum_{A_1} |c_{1n}|\, dA - Q_{La}\right) \quad Q_{R2} = \frac{1}{2}\left(\sum_{A_2} |c_{2n}| dA - Q_{La}\right) \tag{8.27}$$

The conservation of angular momentum according to Eq. (8.6) is fulfilled even with recirculation, when the signs of c_u and c_n are defined as per Fig. 8.5. Equation (8.25) is equally valid with recirculation if the shear stresses are applied with their actual direction (locally positive or negative).

For assessing the tangential velocities of the recirculating fluid Q_R, when it re-enters the impeller, the following considerations should be noted:

- If a *friction-free* control surface is placed around the impeller, all components of the angular momentum which are caused by recirculation are canceled. In this case the tangential components of the fluid re-entering the impeller have no

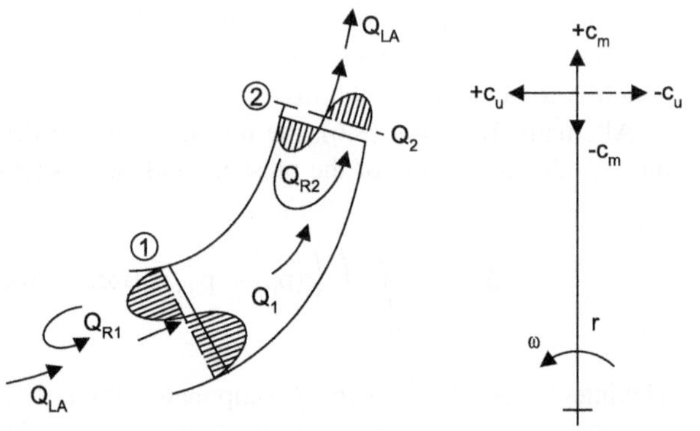

Fig. 8.5 Treatment of flow recirculation

effect on the head. The same applies if the control surface is placed in a way that all streamlines of the recirculating fluid remain within the control volume. In other words, the pre-rotation of the fluid upstream of the impeller is induced by the impeller itself and therefore has no impact on the balance of angular momentum.

- In view of countless tests and experimental observations which demonstrate that the inlet geometry upstream of an impeller has a considerable impact on the head at partload, it becomes obvious that reality is far from the frictionless theory. Nevertheless, the conservation of angular momentum applies to operation with flow recirculation as well. When calculating the conservation of angular momentum for recirculating flows, it must be taken into account that the tangential velocity of the fluid re-entering the impeller is *always smaller* than when it exits from the impeller into the suction chamber. The reasons for this are: (a) wall friction in the suction plenum; (b) the transfer of angular momentum from the recirculating fluid to the net through-flow is subject to (mixing) losses: large water volumes in the inlet bay show this mixing effect clearly; (c) ribs, return vanes and any structures in the inlet chamber have a strong impact on the Q-H-curve at partload because they reduce the swirl upstream of the impeller, Chap. 5.6.4, items 9 and 10.

Consequently, the tangential velocity c_{1u} when operating with recirculation can only be calculated if the effect of the *stator elements on the swirl of the recirculating fluid is correctly modeled.* At the impeller inlet these are the inlet casing, ribs or return vanes in case of multistage pumps, while the diffuser or volute must be included in the calculation domain at the impeller outlet. If the impeller sidewall gaps are more or less open to the main flow, these have a strong impact too, Chap. 9.1. In a pump with an axial inlet, without ribs, the pipe length upstream of the impeller must be modeled sufficiently long (L/D = 15 to 20), since the inlet swirl at low load may reach up to a length of L/D = 15 into the suction pipe.

As mentioned above, the tangential velocity $c_{u,Rec}$ of the fluid re-entering the impeller is always smaller than that of the fluid exiting from the impeller; the high power at shut-off testifies to this. It follows that the numerator in Eqs. (8.8 and 8.9) remains finite even when the net flow Q_{La} through the impeller is zero. Consequently, Eqs. (8.7 to 8.9) *do not produce physically meaningful results* when applied to operation with recirculation: at low flow - in particular at Q_{La} close to zero - theoretical head and total pressure rise to values above $u_2^2/(2g)$, which is impossible with blade angles below $\beta_{2B} < 90°$. This explains the frequent observation that Q-H-curves calculated by CFD rise at partload above experimental data. If such is the case, either post-processing fails to properly take into account the above considerations or the physical modeling is not correct (e.g. disregarding the effect of the impeller sidewall gaps, simplifications of the stator geometry).

Partload operation *with considerable recirculation* can be handled with consideration to the following:

1. Only unsteady or steady stage calculations allow the correct evaluation of the angular momentum of the recirculating fluid because its tangential velocity $c_{u,Rec}$ can solely be determined from the rotor/stator interaction.
2. The hydraulic torque acting on impeller and collector can be calculated from Eqs. (8.6) or (8.25) (it is best to calculate both values for the sake of comparison and for checking plausibility).
3. The head H of the pump is determined by the *area*-averaged *static* pressure at the stage outlet or at the discharge nozzle. Even if the calculation shows some back-flow through the outlet control surface, the impact on the *area*-averaged values is considered small. The difference in kinetic energy between discharge and suction nozzle is added to the static pressure rise in order to get the total head (Table 2.2). Since the dynamic head at partload is small (often negligible), it is calculated with the average velocities from c = Q/A. The relevant outlet control surfaces can be selected in the following manner: In the case of volute pumps it is the discharge nozzle (diffuser outlet); in the case of multistage pump it is the return vane outlet. In case of a pump with diffuser and subsequent volute (as in Fig. 7.48) it is the discharge nozzle, because of the non-uniform pressure distribution in the volute.
4. The partload head is of interest primarily for the assessment of the stability of the Q-H-curve.
5. At partload recirculation the hydraulic efficiency and the theoretical head are of no practical relevance and need not be calculated. An attempt to determine the hydraulic efficiency from $\eta_h = \rho \times g \times H \times Q_{La}/(M \times \omega)$ with M from Eq. (8.6 or 8.25) and the theoretical head from $H_{th} = H/\eta_h$, at low flow (Q_{La} tending to zero) would give too high values of H_{th}. This has often been observed in CFD calculations done by uncritical users.
6. As already mentioned, the integration of Eqs. (8.7 to 8.9) yields no meaningful results when operating with recirculation. Total pressures at the impeller outlet and theoretical heads have no physical meaning during recirculation. In the frame of scientific investigations (e.g. for comparison with measurements) the recirculating flows Q_{R1} and Q_{R2} and the pertinent fluxes of angular momentum could be integrated separately. In practice such data are scarcely of interest. Primarily the torque taken by the impeller (e.g. at Q = 0) is required; it results from Eqs. (8.6 and 8.25).
7. The area-averaged static pressure at the impeller outlet can be used to determine the radial force and to assess the contribution of the impeller to the stability of the Q-H-curve.

Table 8.2 gives recommendations for a comprehensive post processing for different applications. In addition an evaluation of the y^+-values is warranted. An y^+-statistic may be made for that purpose against which the efficiency or losses could be correlated.

Table 8.2 (1) CFD-post processing

- Table 8.2 refers to a calculation of the entire pump between the suction and discharge nozzles or the impeller only.
- Most quantities are to be understood as difference between outlet and inlet of the component considered or of the pump
- The first operation point calculated is the BEP
- The flow rate through the impeller Q_{La} and diffuser or volute Q_{Le} must include the respective leakage losses (if any).
- Column "C": p = calculation of entire pump; io = impeller only; sc = single-channel impeller

Note N1: With *open* ISR (large gap A) use only above a tentative limit of $Re > 10^5$, Eq. (T3.6.1), since otherwise pumping action of shrouds cannot be neglected when calculating torque and H_{th}.

Quantity			Eq.	Objectives and comments	C
Torque from integration of pressures and shear stresses (N1)	M	impeller outlet	8.25	Check power absorbed and efficiencies	p io
Torque from integration of angular momentum (N1)	M_{th}		8.6	Check M_{th} against M from Eq. (8.25) ratio: $0.98 < M/M_{th} = 1.01$	
Theoretical head; valid only for flow rates where no recirculation occurs. (N1)	H_{th} ψ_{th}		8.8 8.11	$H_{th} = \frac{M\ \omega}{\rho g Q_{La}}$ use M rather than M_{th} Impeller only: check head according to $H_{opt} = \eta_{h,opt}\, H_{th}$ with $\eta_{h,opt}$ as used for design	
Total (useful) head of pump	H ψ_{opt}	nozzles	2.1		p
Hydraulic efficiency (N1)	$\eta_{h,opt}$			$\eta_{h,opt} = H_{opt}/H_{th}$	
Total pressure rise	p_{tot} ψ_{La}	impeller outlet	8.9 8.12	Assessment of hydraulic losses	p io
Static pressure rise in impeller	ψ_p		8.10	For comparing different impeller designs the ratio ψ_p/ψ_{th} is an additional criterion	
Hydraulic losses in impeller	ζ_{La}		8.13		
Diffuser or volute losses	ζ_{Le}		4.8		p
Total non-uniformity of flow	U_{F1} U_{F2}	impeller outlet	8.18 8.19	Evaluate for: $p_{2,stat}$, c_2, c_{2u}, c_{2m}, $p_{2,tot}$ Significance: mixing losses	p io
Axial velocity component $c_{2,ax}$ (target: zero for radial impellers)	$\zeta_{La,ax}$			$\zeta_{La,ax} \equiv \frac{c_{ax}^2}{\psi u_2^2} = \frac{\int c_{ax}^2\, dA}{\psi u_2^2 A_2}$ This is largely a mixing loss.	

Table 8.2 (2) CFD-post processing

Quantity		Location	Eq.	Objectives and comments	Calculation
Impeller inlet recirculation	Q_{R1}	inlet	8.27	Relevant for partload and oversized impeller inlet.	p
Impeller outlet recirculation	Q_{R2}	impeller outlet	8.27	With very large impellers (dredge, sewage) strong recirculation is expected even at BEP.	p
Pitch-wise non-uniformity $c_2 = f(\varepsilon)$ with c_2 massaveraged *over span*	$U_{F1,p}$		8.18	Relevant for hydraulic excitation forces caused by rotor/stator interaction (the span-wise non-uniformity has no influence since mixed out)	p io
Span-wise non-uniformity $c_{2m} = f(z)$ with c_{2m} area-averaged *over pitch*	$U_{F1,s}$		8.18	Relevant for Q-H-curve stability (flow pattern switching) and mixing losses.	
Radial forces, static, on blades and shrouds	F_R k_{Ru}		9.7	Polar plots for F_R and/or k_{Ru}	
Radial forces, dynamic, on blades and shrouds determined from integration of pressure and shear stress distribution over blade pressure and suction surfaces.	$F_{R,dyn}$ $k_{Ru,dyn}$			• pressure distribution $p_2 = f(\varepsilon)$ *area-averaged over span* • k_{Ru} as orbit in relative system • k_{Ru} as spectrum in absolute system (peaks over one revolution) • sc: F_R depends on circumferential position of blade trailing edge versus cutwater ISR flow has a large impact. Annular seals best to be neglected since eccentricity not known (seal stiffness)	
Axial forces, static and dynamic	F_{ax} f_{ax}		9.2.15	Interaction between flow in ISR and diffusor or volute	p io
Pressure distribution at outer, mid and inner streamline				Assess blade loading: no sharp peaks near leading edge; lowly loaded towards impeller outlet.	
Cavitation at impeller inlet	σ_i, σ_0 L_{cav} σ_3				
Cavitation at volute cutwater or in diffuser throat				This is an unsteady phenomenon with minimum pressure when blade(s) or velocity peaks pass the cutwater or diffuser vanes. Criterion: fluid volume where $p_{stat} < p_v$	p

Table 8.2 (3) CFD-post processing					
Quantity		**Location**	**Eq.**	**Objectives and comments**	**Calculation**
Impact of cavitation at impeller inlet and diffuser/volute on flow pattern switching, see Figs. 6.17 (6.16)				Assess stability and excitation forces	p
Torque fluctuations as function of circumference: M = f(ε)				There is a coupling with the system? Numerical stability?	sc
Pressure fluctuations at impeller outlet and in discharge nozzle				Noise and vibrations	p
Flow rate fluctuation through impeller as function of circumference: Q = f(ε)				Depends on the pressure distribution in the (volute) casing	sc
Temperature increase in impeller side wall clearances and annular seals				Highly viscous flow, typically for Re < 10^5, Eq. (T3.6.1). Evaluate pumping acting of shrouds.	p

8.5 Impeller Calculations

Impeller-alone calculations are only appropriate for axial approach flow in a range of 0.7 < q* < 1.1 - that is in zones with little interaction between rotor and stator. Possible goals: (1) qualitatively check flow patterns in impeller (incidence, secondary flows, flow separation). (2) Compare impeller losses near BEP of different impeller designs. Limitation: mixing losses (due to uneven outlet velocity profiles downstream of the impeller) cannot be evaluated. (3) Cavitation inception and cavity length.

8.5.1 Impeller Performance at Best Efficiency Flow Rate

In order to assess possibilities, limits and accuracy of Navier-Stokes calculations, 38 radial and semi-axial impellers in the range of n_q = 12 to 160 were analyzed by comparing calculated performance at the best efficiency point to test data from diffuser pumps, [41]. In the course of this investigation it became evident that consistent, reproducible results and good accuracy could be achieved only if strict rules were followed for grid generation, calculation parameters and post-processing procedures:

1. Boundary conditions: the static pressure in one cell was specified at the impeller outlet; the mass flow rate and the direction of the velocity vector were given at the inlet.
2. Figure 8.1 shows a typical grid. The number of nodes was between 48,000 and 125,000. In flow direction 80 to 120 nodes, over the blade pitch 20 to 50 nodes were selected as linear function of the number of impeller blades. Over the blade span 12 to 30 nodes were selected as linear function of b_2/d_2.
3. The standard k-ε-turbulence model was used (at that time) with a turbulence level of 5% and a turbulence scale of 1 to 1.5-times the blade thickness.
4. Relative residuals: 10^{-4}; convergence was achieved after 70 to 100 iterations.
5. The leakage through the annular seals and the balance piston was taken into account when defining the flow through the impeller. The inflow geometry of the leakage at the impeller inlet was not modeled.
6. The control surface at the inlet was implemented at the first node at the entrance to the calculation domain. The control surface at the impeller outlet was at one node after the blade trailing edge. The position of the control surfaces had virtually no influence on the results with the selected type of averaging.
7. The size of the cells near the walls is an important parameter. The grids were designed according to Chap. 8.3.4 so that the criterion $y^+ = 25$ to 300 was largely fulfilled.

When observing the above rules, the theoretical head ψ_{th} could be predicted with an accuracy of about ±3% as shown by Fig. 8.6. The dashed curves in Fig. 8.6 reflect the uncertainty of the test data as per Eq. (T3.5.8), Chap. 8.8.3.

The hydraulic losses in the impeller are depicted in Fig. 8.7. They were referred to the total of hydraulic losses in the pump determined from Eq. (T3.5.8). The curves shown correspond to the expected trends. All calculated absolute impeller loss coefficients ζ_{La} are within a band of about ±0.02.

According to Eq. (8.15) the hydraulic loss in the impeller is calculated as a small difference of two nearly equal figures. Small tolerances on ψ_{th} and ψ_{La} cause large *relative* errors in the impeller loss. For example: at $\psi_{th} = 1.1 \pm 0.01$ and $\psi_{La} = 1.06 \pm 0.01$ the loss coefficient is in the range of $\zeta_{La} = 0.04 \pm 0.02$ (hence ±50%).

As discussed in Chap. 8.3.4, the grid must be designed so that realistic velocity distributions (corresponding to turbulent flow) are calculated near the walls. Otherwise the loss calculation is falsified. Loss coefficients of various impellers are plotted in Fig. 8.8 against the width of the near-wall cells. The calculated impeller losses decrease strongly if larger wall cells are used. This result shows: In order to achieve consistent results which can be used for performance prediction, grid generation rules should be developed, validated and followed in practice.

The uncertainties inherent to loss calculations should be appreciated when optimizing impellers since the calculated trends are likely to be within the band of

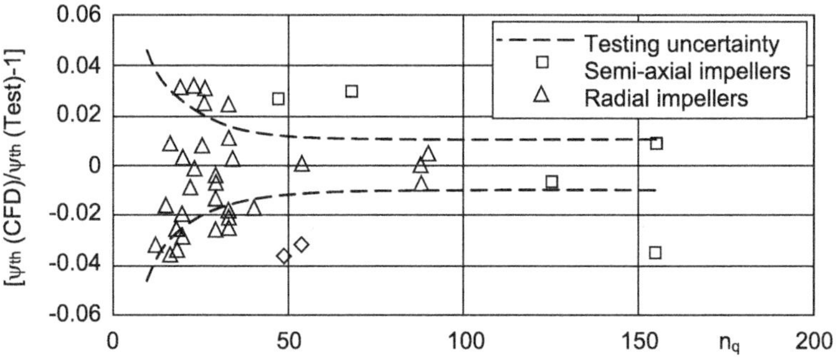

Fig. 8.6 Specific work at Q_{opt}: comparison between calculation and measurement

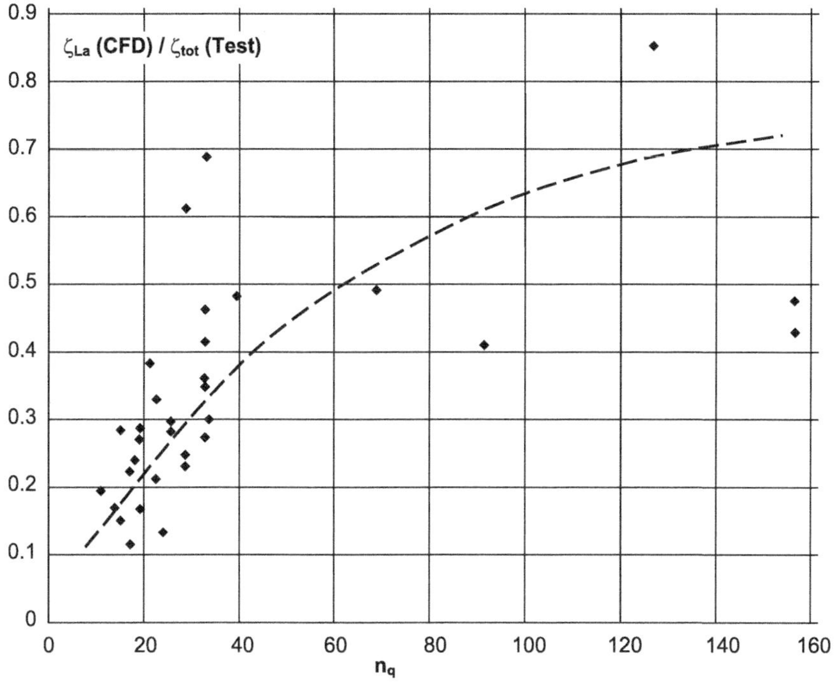

Fig. 8.7 Hydraulic losses in the impeller

uncertainties caused by variations in wall cell size. If such is the case, it is difficult to decide which of the calculated options promises the best test results. As demonstrated by Fig. 8.8, the grid must be changed as little as possible from one step of an optimization process to the next in order to minimize grid influences on the losses. The calculated velocity profiles are also sensitive to the grid.

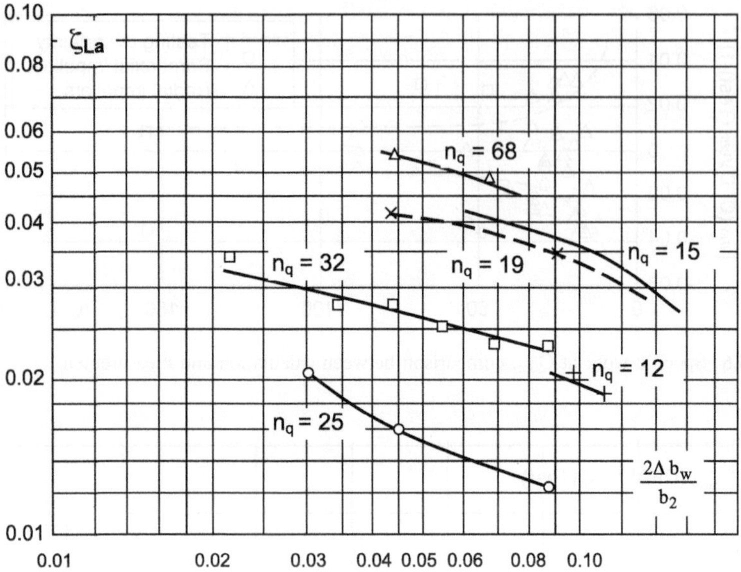

Fig. 8.8 Influence of near-wall node size Δb_w on impeller losses

In a qualitative way, impeller calculations can reveal weaknesses of a design, such as non-uniform discharge flow (e.g. if the impeller outlet width is too large) or shock losses at the inlet due to non-uniform approach flow or inadequate choice of the blade angles (incidence effects).

8.5.2 *Velocity Profiles*

Calculated and measured velocity profiles at the impeller inlet usually agree reasonably well near the best efficiency point and above. An exception is the zone near the outer streamline where the axial velocity tends to be under-predicted. It appears that the calculation over-estimates the effects of friction, Chap. 8.3.3. Velocity profiles at partload and near the onset of recirculation appear to be predicted correctly only if the influence of the leakage is considered, [40].

The type of the grid (C- or H-grid) can have an influence on the velocity profiles near the onset of recirculation or in unstable flow conditions. In contrast, fully developed recirculation at the impeller inlet is quite stable since it is mainly determined by the equilibrium of forces acting on the streamlines at different radii. Fully developed inlet recirculation is therefore rather well predicted by 3D-Navier calculations and even 3D-Euler programs.

An analysis of numerous measurements of the velocity profiles at the impeller outlet gave no general tendencies how these profiles develop as functions of flow rate and specific speed. Examples can be found where measurements and calculations agree quite well, while other cases show large discrepancies even qualitatively. Differences may be expected primarily in unstable flow conditions; but the term "unstable" must not be confused with "non-uniform", because non-uniform velocity profiles can occur in very stable flow situations such as recirculation.

8.5.3 *Influencing Parameters*

As discussed above, the global performance parameters result from the integration of the velocity profiles at impeller inlet and outlet. Thus, all quantities having an impact on the velocity profiles potentially influence the theoretical head and the losses. Sensitivity studies in this respect have shown:

- The grid can have a large influence on global performance and velocity profiles. In this respect not only the absolute number of nodes is relevant, but also their distribution in flow direction as well as span-wise and pitch-wise. Selecting unsuitable grid parameters can easily change global performance by ±5%. In other cases, the differences were below 0.5% when varying the number of nodes between 45,000 and 110,000.
- In principle 3D-Navier-Stokes calculations model the influences of speed, viscosity and impeller size (i.e. the Reynolds number) in a way similar to efficiency scaling according to Chap. 3.10, [41, 73]. In sample CFD calculations at BEP with water between 10 to 180 °C and speeds between 3000 to 6000 rpm the inner and the hydraulic efficiency scaled according to $(1 - \eta_2) = (1 - \eta_1)\,(n_1/n_2)^{0.075}(\nu_2/\nu_1)^{0.075}$ this exponent is close to the exponents found in some efficiency-scaling formulae published in the past.
- Impeller calculations with increased surface roughness resulted (as expected) in higher losses but also in a higher flow turning, i.e. in a higher theoretical head. The reason for this may be that the program calculates with thicker boundary layers where c_u approaches u_2. When calculating the angular momentum according to Eq. (8.8) a higher H_{th} is obtained. As discussed in Chap. 3.10, tests with rough impellers sometimes showed slightly increased heads too.
- If the turbulence scale or the turbulence level is increased, the hydraulic losses rise because more turbulent energy is dissipated. Higher turbulence implies a more intense exchange of momentum across the flow. Therefore, the turbulence parameters have an impact on the velocity distributions and hence (however slightly) on the theoretical head, [41, 73].

8.5.4 *Sample Calculation*

An example of an impeller calculation is discussed in the following. It is the impeller of a multistage pump $n_q = 31$ with a grid similar to Fig. 8.1. Figure 8.9

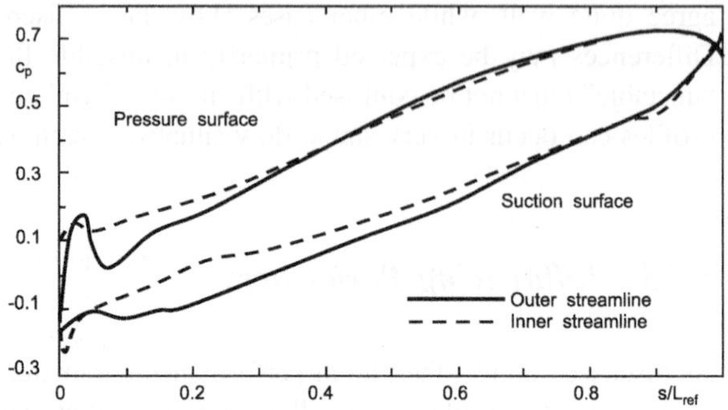

Fig. 8.9 Pressure distribution over blade length in a radial impeller at $q^* = 1$

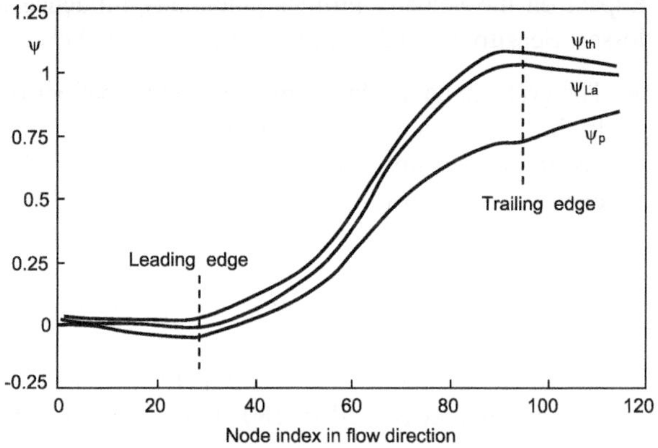

Fig. 8.10 Angular momentum, total pressure and static pressure in the calculation domain of a radial impeller at $q^* = 1$

shows the distribution of the static pressure over the blade length on the outer and inner streamlines. At the outer streamline an unfavorable acceleration is noticed on the pressure side shortly after the leading edge; it impairs the work transfer. This flow may be caused by an uneven approach flow due to the sharp flow turn at the outlet of the return channels. Apart from this deficiency at the inlet, blade loading is quite uniform on outer and inner streamlines.

The dimensionless values of theoretical head ψ_{th}, the rise in total pressure ψ_{La} and rise in static pressure ψ_p are plotted in Fig. 8.10 as per Eqs. (8.10 to 8.12).

In the inlet domain from node 1 to 26 (which corresponds to the contour of the return channel) the total pressure decreases slightly due to losses, while the static pressure decreases more strongly because of flow acceleration.

In the first part of the impeller channel (nodes 26 to 45) the work transfer is moderate, in the center part it is strongest and towards the outlet downstream of a_2 it decreases again due to slip effects, Chap. 3.3, Fig. 3.4. Downstream of the blade trailing edge theoretical head and total pressure decrease due to losses in the annulus while the static pressure rises approximately according to Eq. (1.28).

Figure 8.11 depicts the circumferentially averaged velocity distributions at the impeller inlet. The meridional component in Fig. 8.11 suggests that the onset of recirculation should occur at about $q^* = 0.3$ to 0.35 (in the test the onset was observed at $q^* = 0.5$ to 0.6). Recirculation is fully developed at $q^* = 0.12$,

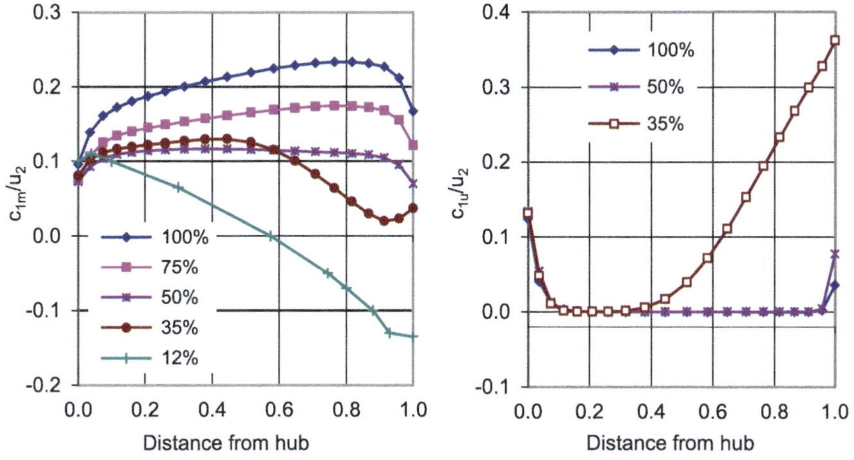

Fig. 8.11 Velocity distribution at the impeller inlet; the abscissa is the relative distance from the hub

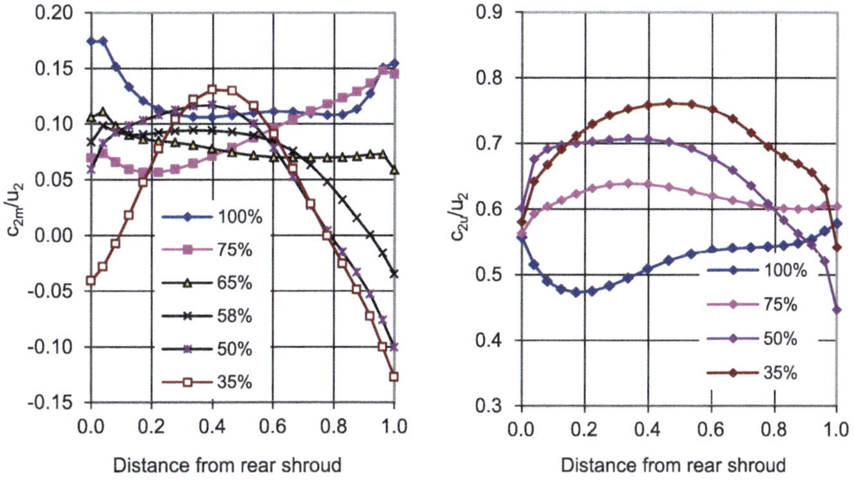

Fig. 8.12 Velocity distribution at the impeller outlet; the abscissa is the relative distance z/b_2 from the rear shroud

covering almost half the height of the blade. As soon as the meridional velocity near the outer streamline drops significantly below the average velocity, high tangential velocities are observed near the outer streamline even before recirculation sets in, Fig. 8.11.

Figure 8.12 shows for different flow rates the circumferentially averaged velocity profiles at the impeller outlet. At partload the profiles are kind of symmetric. This feature would suggest a tendency to unstable flow according to Chap. 5.5.2 (the measured stage characteristic is slightly unstable).

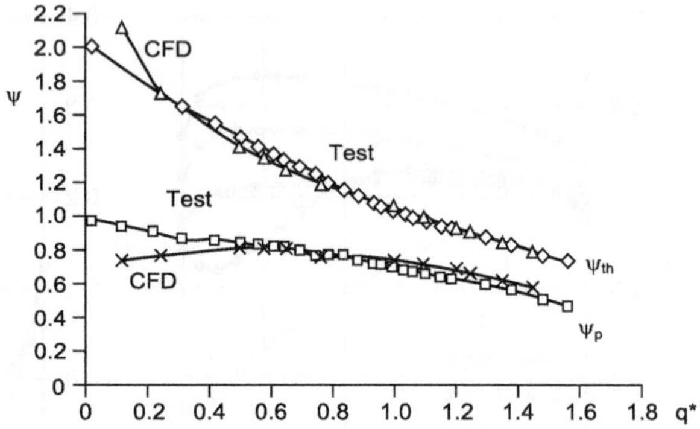

Fig. 8.13 Theoretical work impeller ψ_{th} and static pressure rise in impeller ψ_p; comparison between calculation and measurement

Measured and calculated theoretical head agree well in the flow range without recirculation, Fig. 8.13. The calculated head is too high below q* = 0.2, apparently because the recirculation effects are not correctly taken into account. The curve of the rise in total pressure produced by the impeller is roughly parallel to ψ_{th} (it is not shown in Fig. 8.13 since it has no practical relevance and is unable to indicate the risk of an instability).

At high flow the calculated static pressure rise of the impeller ψ_p exceeds the measured value. This may be a measurement problem or an under-prediction of the losses. Below q* < 0.5 the static pressure rise is predicted too low presumably because the recirculation is not modeled correctly (refer also to Fig. 4.6b).

8.6 Calculation of Collectors and Stages

8.6.1 Separate Calculation of the Collector

Pumps of any specific speeds can achieve competitive efficiencies only if an optimum pressure recovery is achieved in the collector. As mentioned above, the velocity distribution at the inlet is crucial for diffuser performance. Ignoring this influence is one main weakness of one-dimensional theories and of CFD calculations performed separately for impeller and collector. If separate calculations are performed in spite of this severe restriction, the impeller is calculated first. Using the impeller outlet velocity distribution thus obtained, the collector is calculated in a second step as in Chap. 8.6.2.

The sample computation below used the calculated circumferentially averaged distributions of velocity and total pressure at the diffuser inlet and a constant pressure at the diffuser outlet as boundary conditions, [8]. The H-grid employed had 62,000 nodes which were arranged in a way to coincide with the pressure measurements indicated in Fig. 8.14. Figure 8.14 shows the calculated and measured pressure recovery in the diffuser made dimensionless by $\frac{1}{2}\rho \times u_2^2$. The pressure distribution at q* = 1.0 on the front shroud is plotted for three different velocity

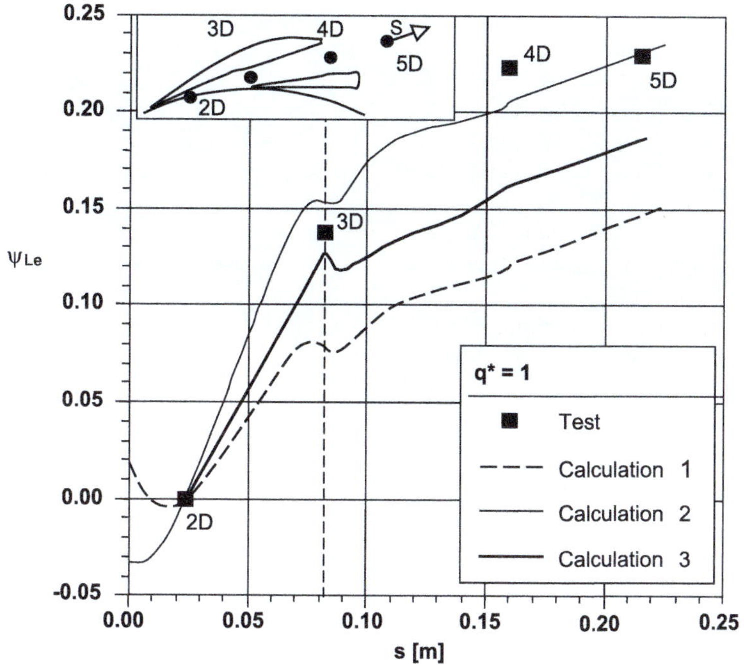

Fig. 8.14 Pressure rise in the diffuser near the front shroud at q* = 1. Velocity distribution at the diffuser inlet: *calculation 1*: constant; *calculation 2*: from impeller calculation; *calculation 3*: measured. Sulzer Innotec AG [8]

profiles at the impeller outlet: curve 1 for a constant velocity, curve 2 for the calculated impeller outlet flow and curve 3 for the measured velocity distribution (laser measurement). The calculations confirm the (repeatedly stressed) strong influence of the velocity distribution at the diffuser inlet on the pressure recovery. The calculations also capture the pressure rise in the triangular inlet section up to the throat area which has been evidenced by many measurements, Figs. 4.5, 4.6, Chap. 5.3. Figure 8.15 depicts the calculated streamlines in the return channels of a multistage pump.

8.6.2 Steady Calculations of Stages or Complete Pumps

Given sufficient computing power, the combined calculation of inlet casing, impeller and collector represents the state of the art. Such "stage calculations" allow to model the interaction between the different components, notably the exchange of momentum between the net through-flow and recirculating fluid at partload.

The head at low partload (in particular at shut-off) depends on the pre-swirl induced by the recirculating fluid which is slowed down by ribs, return vanes or a

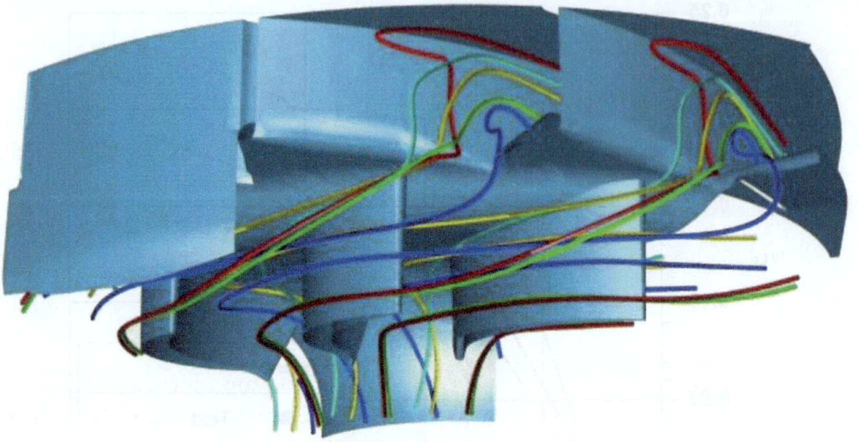

Fig. 8.15 Streamlines calculated in the return vanes of a diffuser

large fluid volume. These effects must be modeled truthfully when calculating partload characteristics. *Ribs in the suction nozzle or the inlet casing must not be omitted*. Numerical flow calculations of a semi-axial pump (n_q = 155) very well demonstrated the strong impact of ribs at the impeller inlet on the head at low flow, [60].

Frequently, convergence represents a problem at low flow and shut-off because the flow is highly unsteady, see Chap. 8.8.4 for a detailed discussion. The flows in impeller sidewall gaps must also be modeled, because of their interaction with recirculating fluid and the main flow, Chaps. 9.1 and 5.4.3.

When calculating multistage pumps, the velocity distribution at the outlet of the return vanes has to be modeled correctly in order to give the relevant boundary conditions for the impeller inlet. As discussed in Chap. 8.3.4, this can be achieved by setting periodic boundary conditions (alternatively two stages must be calculated where only the second stage is considered relevant for the performance).

The small distance between the impeller blade trailing edges and the diffuser vane leading edges requires very fine grids in this region since velocity gradients are high and the curvatures at impeller blade trailing edge and diffuser vane leading edge are high.

Because of the high computational effort needed for a complete unsteady calculation of inlet, impeller and collector, stationary stage calculations are (still) employed in many applications. Two ways are available to handle the rotor/stator interactions in this case:

1. A "*mixing plane*" is introduced between impeller and collector where *circumferentially averaged* velocity and pressure profiles are used at the diffuser inlet. The non-uniformities between front and rear shroud are thus taken into account by the calculation; this is essential for investigations into the stability of the Q-H-curve according to Chap. 5. The pitch-wise non-uniformity of the impeller

discharge flow is considered less relevant for steady performance than the span-wise non-uniformity. Calculations with a mixing plane would therefore be preferred for diffuser pumps. Investigations into the unsteady effects caused by pitch-wise non-uniformity (i.e. the rotor/stator interaction) need fully unsteady calculations as per Chap. 8.6.3.

2. When applying the "*frozen rotor concept*", the 3-dimensional velocity distribution at the impeller outlet is given at an arbitrarily selected fixed circumferential rotor position as inlet boundary condition to the collector. At the interface between rotor and stator, the velocity vectors are transformed from the relative to the absolute system. *The calculated performance depends on the rotor position*. Several rotor positions must therefore be calculated and the results averaged for performance prediction.

Sometimes, not all of the impeller and diffuser channels are modeled in order to reduce the computation time. Since vane number combinations with common multipliers should not be used, an error in the vane pitch is incurred.

In general terms, the mixing-plane approach may be preferable for diffuser pumps where the pressures and velocities change strongly over one diffuser vane pitch. The head calculated would then depend strongly on the (arbitrary) rotor position if using the frozen rotor method. Volute pumps are better calculated with the frozen rotor because the rotor position versus the cutwater has little effect on the overall performance. Frozen-rotor calculations often converge better than the mixing-plane analysis. Convergence of the latter is difficult because of the small rotor/stator clearance (gap B) used in diffuser pumps.

Because of the strong pressure variation over the circumference, the results of off-design calculations of volute pumps are doubtful and convergence may be difficult for both the mixing plane and the frozen rotor approach.

An application of frozen rotor calculations was reported in [16]: inlet, impeller and double-volute of a feed pump with n_q = 34 were modeled with 280,000 nodes and the Q-H-curve calculated from q^* = 0 to 1.4. While calculation and test agreed well near BEP, deviations reached up to 10% at partload and high flow. Similar deviations were found in stage calculations published in [60]. When the uncertainties reach ±10%, an investigation of the stability of the Q-H-curve becomes questionable. The secondary flows in the volute and flow separation at the cutwater were well modeled by CFD and it was possible to improve the cutwater geometry based on the calculated flow patterns.

8.6.3 Unsteady Calculations

Since the flow in a pump is fundamentally unsteady because of the rotor/stator interaction, the final goal of CFD is an unsteady calculation of the complete machine. The sole impediment for a wide application of such calculations is computing time. Unsteady calculations using "sliding meshes" are done over a few revolutions of the impeller (usually not exceeding 5 to 8 near BEP) until a converged solution is obtained which is *periodic with the blade passing frequency*. The number of revolutions needed to reach convergence increases with the unsteadiness of the flow. Overall performance parameters have to be determined by averaging over one impeller revolution of a *fully converged (periodic)* solution.

Unsteady calculations to determine pressure and velocity fluctuations as well as rotor/stator interaction forces were compared to test data and reported in [13, 29, 82]. In [60] one partload operation point of a semi-axial pump with n_q = 155 was analyzed with respect to head, pressure recovery in the diffuser and pressure pulsations. As to be expected, pressure pulsations at blade passing frequency are dominant. A comparison with test data can be misleading since pressure pulsations are influenced by the system (Chap. 10). In Fig. 8.15-1, the flow structures in the diffuser of a semi-axial pump were gleaned from unsteady CFD calculations. Note the cross-channel vortex near the throat area of the diffuser. Compare also to the cross-channel vortexes shown in the impellers of axial pumps, Figs. 5.48 and 11.21.

Two volute pumps (n_q = 35 and n_q = 18) were investigated in [85]. Steady frozen rotor calculations gave excessively high head variations over the circumference and convergence was poor. In contrast, unsteady calculations of impeller, volute and sidewall gaps (with 160,000 and 900,000 nodes) converged well and produced plausible results. Initial oscillations decayed after about two revolutions. Subsequently head, efficiency and other flow quantities were periodic with the blade passing frequency. It is interesting to note that the program was unable to find a steady solution because the flow in the volute is unsteady in reality, Chap. 8.8.2.

Fig. 8.15-1 Vortex structures forming in the diffuser of a semi-axial pump n_q = 150. Sulzer

In a further investigation in [85], the complete characteristic was calculated in one run by continuously varying the flow rate. This attempt failed because of inertia effects: when reducing the flow rate (i.e. during deceleration) higher heads were calculated than with increasing flow rate (acceleration). Extensive unsteady calculations were also done in [5].

Unsteady calculations - even though expensive - have meantime become the state of the art when reliable results are sought. Fully unsteady (transient) calculations of volute pumps from suction to discharge nozzle were reported in [91] to [110], extensive calculations of the flow in side-channel pumps in [114]. The following conclusions can be drawn from these studies (the figures given below are typical values which vary over a wide range):

- Between 1.5 and 24 million nodes were used for this type of calculation. The lower number leads to $y^+ = 60$ typically; the higher number can resolve the flow down to $y^+ = 1$ for small n_q. For larger n_q it may be impracticable to achieve $y^+ = 1$ because of limited computing resources (meshes with more than 25 million nodes can hardly be managed).
- Time step equivalent to 1°, possibly starting with larger time step and narrowing down as the solution converges.
- The ISR needs to be modelled because of strong interaction with volute, Figs. 5.30, 5.31, 5.35, 9.6, and 9.7. ISR studies are reported in [91, 92, 108]; these agree well with the concepts discussed in Chap. 9.1.
- The low Reynolds number k-ω SST model was used in these investigations. Some authors apply the scale-adaptive simulation SAS.

As discussed in Chap. 9.3.9, the flow in pumps with single-blade impellers is extremely unsteady. An unsteady CFD-analysis of this type of pump was reported in [100]; data of this study: $n_q = 41$; $d_2 = 220$ mm; $b_2/d_2 = d_1/d_2 = 0.36$; number of nodes: 3.6×10^6; $y^+ = 40$ in impeller and volute; test results $\psi_{opt} = 0.74$, $\eta_{opt} = 0.66$. The unsteady calculations were compared to test data and steady calculations with (a) mixing plane ("stage interface") and (b) frozen rotor. Not surprisingly, the stage interface and frozen rotor calculations were not able to produce physically meaningful results for highly unsteady flow: at BEP stage interface yielded only 60% of measured head, while frozen rotor gave 70 to 80% depending on rotor position. At $q^* = 0.8$, frozen rotor gave efficiencies of 0.42 to 0.5 as opposed to the test 0.64. An unsteady calculation using the scale-adaptive (SAS), high-resolution SST model over-predicted the efficiency by up to 20% at partload. Trying different numerical schemes, the discrepancies could be reduced: head was about 6% below test, efficiency 3% above and power 9% below test. At $q^* = 0.8$, the calculated fluctuations were about ±25% of the average head and ±30% of the power. However, matching CFD-results to test data in this way by trial and error, risks to produce coincidental results as discussed in detail in Chap. 8.8.4: whether the numerical scheme would also work successfully for other pumps with highly unsteady flow at very low q^* and with unstable Q-H-curves - or whether it was just a convenient fix in the present case - remains to be demonstrated.

Pressure pulsations calculated by CFD combined with structural analysis by FEM were reported in [104]. Air-borne noise radiation from the volute casing was calculated by the "boundary element method" which calculates the casing surface vibration velocity generating airborne noise ($n_q = 27$, $d_2 = 168$, $d_3^* = 1.1$). A variation of the impeller outlet width b_2 suggested a noise-minimum at about $b_2/d_2 = 0.063$ but this observation cannot be considered as generic because the noise generated by a pump does not only depend on the pressure fluctuations but also on the casing material and wall thickness (influencing the many structural eigen modes of the casing hence the reaction to a given spectrum of pressure pulsations). This study shows the broad range of applications but the predicted noise level was 22 dB below expectation.

8.7 Two-Phase and Cavitating Flows

The objectives of calculating cavity flows are to predict cavitation inception, cavity length and cavitation induced head drop (Chaps. 6.2 and 6.3).

Cavitation inception and flows with short cavity length (when σ_A is only a little smaller than σ_i) react sensitively to minor deviations in geometry (even to roughness elevations). Therefore, the geometry of the blade leading edge must be described very accurately by the program employed for geometry generation and a very fine grid needs to be used near the leading edge. Even then, the pressure distributions calculated with single-phase flow tend to show sharp local peaks since the leading-edge tip radius is usually in the range of 1 to 3 mm only. Probably such peaks are not really relevant for the assessment of the cavitation because the time available for bubble growth in such peaks is extremely short (mostly below 10^{-4} s).

The local velocity distribution upstream of the impeller has an impact on cavitation inception; inlet boundary conditions therefore must be carefully defined or calculated. Inlet casings of between-bearing pumps generate a strongly three-dimensional velocity distribution. Cavitation inception and cavity lengths therefore vary around the circumference requiring an unsteady calculation. The influence of the leakage entering upstream of the impeller needs to be considered since the incidence is influenced by that leakage.

All programs used for analyzing cavitation are based on models which strongly simplify the complex interaction between the liquid and the vapor phase, Chaps. 6 and 13.2.

Calculation of the cavity length without coupling to the main flow: The phase boundary between the liquid and the cavity is determined from the pressure distribution calculated for purely liquid flow. In doing so, it is assumed that there is no interaction between the cavity and the liquid flow and that there is no flow in the cavity. According to [44] the calculation comprises the following steps:

1. 3D-Navier-Stokes calculation of the inlet casing in order to determine the velocity distribution upstream of the impeller.

2. With the inlet boundary conditions thus defined, cavitation inception ($NPSH_i$) is derived from a single-phase liquid 3D-Navier-Stokes calculation assuming $\sigma_i = c_{p,min}$ (Chaps. 6.1.1 and 6.2.1).
3. Bubble growth and implosion are calculated from the Rayleigh-Plesset equation which describes the dynamic behavior of a spherical bubble, filled with vapor and gas, as a function of the local pressure on the blade, [25]. In the low-pressure zone the bubble radius grows till reaching a maximum at the lowest local pressure p_{min}. Subsequently the bubble radius decreases according to the pressure rise to vanish at $p > p_v$. The envelope of these bubble radii yields an initial cavity length, Fig. 8.16.
4. The boundary of the cavity is then iteratively modified according to the pressure distribution until the cavity contour corresponds to the isobar $p = p_v$. The grid is adapted to the cavity contour during each iteration step.
5. The theoretical head and the rise in total pressure created by the impeller are determined according to Chap. 8.4. The performance drops when the cavity extension is sufficiently large. The high effort required for this calculation can be avoided if the cavitation-induced head drop is estimated from an empirical criterion, e.g. assuming that $NPSH_3$ is reached when the cavity length equals the pitch at the impeller inlet ($t_1 = \pi \times d_1/z_{La}$) or the cavity reaches the throat.

The pressure distributions shown in Figs. 6.6 and 6.7 were calculated according to the above procedure. Comparing predicted and measured cavitation inception and cavity length, it is concluded that the calculation without coupling between cavity and main flow is well appropriate for predicting the cavity length at least up to $L_{cav} = ½t_1$ [19] and [18]. The prediction of $NPSH_3$ using the empirical criterion is less reliable (in particular with pressure side cavitation), because the effect of the cavity extension on the main flow, work transfer and losses cannot be described by simple criteria.

Evaporation at constant enthalpy: If the pressure of a fluid of given temperature T and enthalpy h(T) is reduced, a fraction x of the liquid evaporates as soon as the saturation pressure is reached, Fig. 6.1 and Chap. 6.1.1. This fraction is calculated from $x = f_g \times \{h - h'(p)\}/h_v$ (h = enthalpy of the liquid upstream of the impeller, h_v = evaporation enthalpy, $h'(p)$ = enthalpy of the saturated liquid at the local pressure p). Thus the vapor volume can be calculated in zones in which the pressure falls to the vapor pressure. An empirical factor f_g is introduced to describe to what

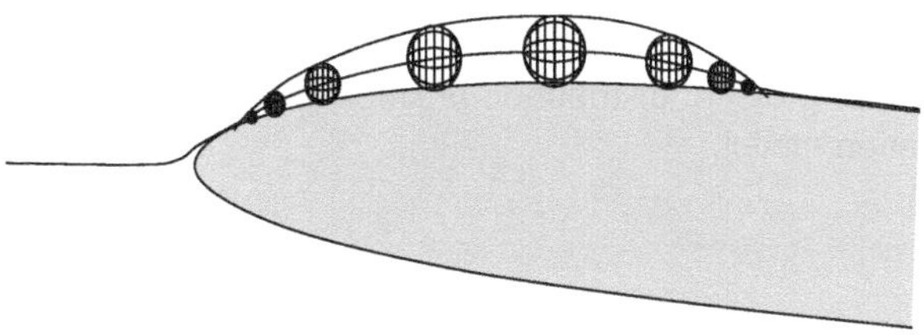

Fig. 8.16 Start solution for the calculation of the cavity length [44]

extent thermodynamic equilibrium is established during the expansion: at $f_g = 1$ vapor is produced according to this equilibrium while any evaporation is suppressed at $f_g = 0$. An application of this procedure is reported in [85].

Mixing of gas and liquid phase: The fluid is treated as a homogenous mixture of two substances having different densities. The density in each calculation element is determined from $\rho_{hom} = (1 - \alpha) \times \rho' + \alpha \times \rho''$, Eq. (T13.3.5). α is the volumetric gas content or vapor volume fraction, ρ' the density of the liquid and ρ'' that of the gaseous phase, see Chap. 13.2. Evaporation and condensation when reaching or exceeding the vapor pressure are evaluated in each element as by source terms and by a simplified Rayleigh-Plesset equation where surface tension and 2nd-order terms are neglected. These processes can be controlled by empirical factors which determine the rate of evaporation and condensation. Furthermore, nuclei size and population have to be defined in the source terms. Overall, four factors can thus be chosen freely to adapt the calculated result to test data.

Even though no free surfaces or individual bubbles are considered, attached cavities or clouds of bubbles in the free stream can be captured in this method. For example the cavitation vortices created at the rear from an attached cavity could be calculated.[3] Calculations with this model were reported in [1, 4, 19, 52, 93, 103]. Cavitation (at $q^* > 1$) on the volute cutwater of a pump with $n_q = 12$ was calculated with the low-Reynolds model using a fine grid giving $y^+ < 1$ in the volute, [93]. Since the head was over-predicted at $q^* > 1$, the surface roughness of the volute was increased to 140 μm (N12!) in order to match H_{opt} to the test data.

Different shapes of impeller leading edge profiles were investigated in [1] by using this homogenous two-phase mixture model. The CFD results were compared to measurements on a test pump ($n_q = 29$) which had a radial inlet as used for multistage machines. In spite of the distorted approach flow inherent to this type of inlet casing, the CFD analysis was made with one impeller channel only (assuming uniform inlet velocity profiles). The calculated pressure distributions and vapor volume fractions α are plotted as a function of the normalized blade length in Fig. 8.17 for 3% head drop and in Fig. 8.18 for head breakdown. The calculated results apply to the BEP at mid-span of the blade with the parabolic leading-edge profile. The CFD analysis shows:

- The low-pressure peak on the suction side, which would be observed in single-phase flow at $\sigma_A > \sigma_i$, is cut off at the vapor pressure. This zone occupies about 30% of the blade length at 3% head drop and well above 50% at head breakdown. It is obvious that the energy transfer of the blade in this zone is impaired. This behavior could not be inferred by way of approximation from an analysis of the pressure distribution in single-phase flow, i.e. without resorting to a cavitation model.

[3] Such vortices are particularly aggressive with respect to cavitation erosion, Chap. 6.

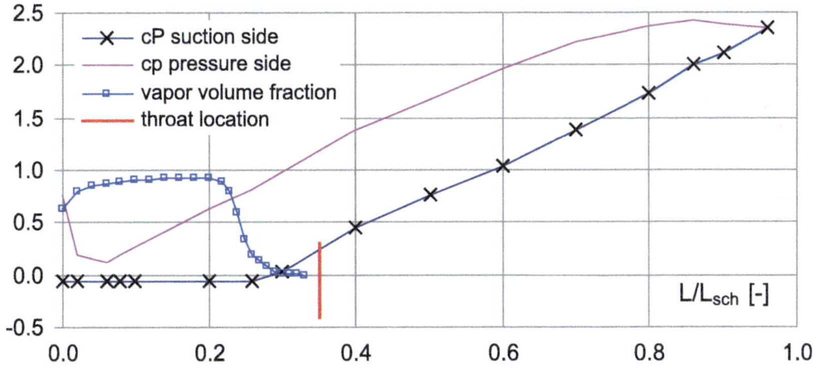

Fig. 8.17 Pressure distribution and vapor volume fraction at mid span, $q^* = 1$, $\sigma_A = 0.18$, at 3% head drop [1]

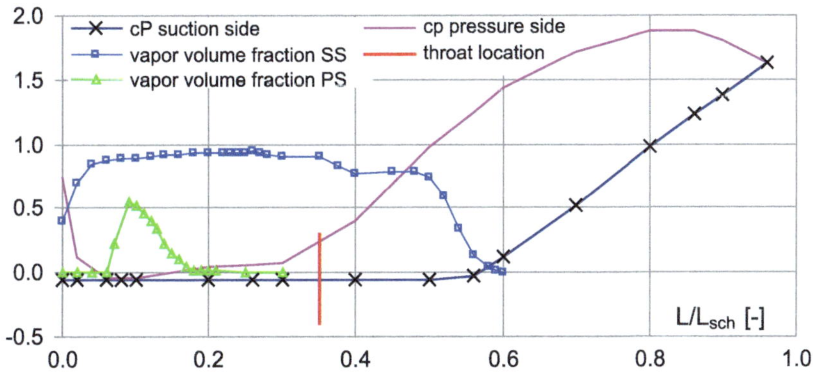

Fig. 8.18 Pressure distribution and vapor volume fraction at mid span, $q^* = 1$, $\sigma_A = 0.17$, at head break down [1]

- The flat portion of the suction-side pressure distribution is observed in the domain where the vapor volume fraction exceeds about $\alpha = 0.1$ at 3% head drop and about $\alpha = 0.18$ at head breakdown.
- At 3% head drop the cavity has not yet reached the impeller throat at mid-span (but it may have done so near the outer or inner streamline)
- Between 3% head drop at $\sigma_A = 0.18$ and head breakdown at $\sigma_A = 0.17$ an almost explosive increase in vapor volume fraction is observed. As a consequence, the curve $H = f(\sigma_A)$ drops almost vertically. The vapor volume fraction in the throat area region attains 0.8 to 0.9.
- At head breakdown cavities were generated on the suction side as well as on the pressure side of the blades.
- The head drop due to cavitation was well predicted by CFD even at 3% head loss and full cavitation.

The two-phase flow through a helico-axial impeller (as shown in Fig. 13.24) was calculated in [83] for a volumetric gas content of 20%. At the best efficiency point a thin gas layer was found on the pressure surface of the blades; this agreed with experimental observations.

Two-phase models are under development, which abandon the simplification of the homogenous density model but attempt to model the interaction between liquid and gas phase.

Judging from the published information, all of the methods described above produce similar results when compared to measurements. This is no coincidence, since all procedures determine the gas-filled zones essentially from the pressure distribution while phase interaction and thermodynamic effects are completely ignored. However, the pressure distribution is largely determined by the main flow away from the walls (in homogenous mixtures it is furthermore independent of the density) - and this calculation scarcely differs in the various procedures (refer also to Chap. 8.10).

Summary: All methods employ empirical factors by which the calculation is adapted to test data. In this way, the CFD results are manipulated to come closer to the test data of the individual authors. However, there is a considerable risk that the factors optimized for the investigated case fail in new applications which are different from the test set-up. Specifically, it should well be borne in mind that cavitation is extremly sensitive to the pressure distribution on the blade, to local pressure in the core of vortices, gas content, thermodynamic properties, the residence time of nuclei in the low-pressure zone and the nuclei spectrum. At present, no convincing generic approach can be offered.

Another application in the domain of two-phase flows is "particle tracking". The behavior of solid particles in a fluid, droplets in a gas stream or bubbles in a liquid flow can be investigated. The streamlines and the local particle concentration are calculated by this method. Fundamentals may be found in [79], applications in [55] and [115].

8.8 Calculation Strategies, Uncertainties, Quality Issues

The commercial Navier-Stokes programs employed in industry are extremely complex. The available program options for physical modeling, grid generation, numerical solution and post-processing leave the program user ample liberty to select suitable models and parameters in order to adapt the CFD results to specific measurements. Little is to be said against such an optimization of calculation parameters - which is even inevitable at the present state of the art - if a large number of similar applications (e.g. a range of pumps covering a wide domain of specific speeds) can be handled with the selected set of parameters and assumptions *in a consistent way*. However, if adaptations need to be made in each case, the parameters are obviously unsuitable for general use. CFD calculations are subject to considerable uncertainties, A number of publications dedicated specifically to quality issues testify to this statement. The pertinent publications cover uncertainties and quality of CFD calculations, [12, 27, 38, 62, 63, 65, 80], and the development of CFD quality assurance and guidelines by various organizations, [21, 38].

8.8.1 Uncertainties, Sources and Reduction of Errors

"Uncertainties" result from a lack of knowledge; therefore they cannot be quantified. In contrast, "errors" are made due to avoidable simplifications or negligence. Applying the suitable effort and diligence, errors can be quantified or avoided. The various uncertainties and errors of a numerical flow calculation are partly caused by the program and partly by the user. These comprise:

1. **Modeling errors** due to uncertainties in the physical models, assumptions and data. Unknown are (for example) the real turbulence characteristics, the true geometry of a cast impeller and the roughness. The Reynolds equations of turbulent flow themselves are a "model" of an unknown physical process. The models used at present for describing two-phase flow, cavitation and roughness effects are rudimentary and based on gross simplifications.
2. **Simplifications** belong to the group of modeling errors. Important uncertainties are frequently generated by simplifications of the geometry which are made because of limited computer resources. Again, such uncertainties cannot be quantified. Consider for example a feed pump according to Fig. 7.48 with $u_1 \approx 80$ m/s and $P \approx 10{,}000$ kW in one stage. Its high performance and relevance for the reliable operation of the power plant would certainly justify an in-depth CFD analysis. The inlet casing (which is similar to Fig. 7.50) generates at the impeller inlet a distorted 3D-velocity distribution varying over 360°; the single volute following the diffuser equally requires modeling around the entire circumference at off-design flow. In order to model this type of pump correctly, the entire fluid volume between the inlet and the discharge nozzle needs to be meshed including the impeller sidewall gaps and the annular seals; the instabilities discussed in context with Figs. 10.35 and 10.36 demonstrate the necessity to do so. Many a user may be tempted to simplify this complex geometry because of a lack of resources.
3. **Numerical errors** [71]: The *numerical error* of a CFD calculation is the difference between the exact solution of the differential equation and the actual solution obtained. It is composed of the discretization error and the solution error. The *discretization error* is the difference between the exact solution of the *differential* equation and the exact solution of the *discretized* equation; it is determined by the order of the selected solution, Chap. 8.3.5. In contrast, the *solution error* is the difference between the exact solution of the discretized equation and the actual solution obtained; it comprises convergence errors (residuals) and rounding errors.

4. **Rounding errors** may often by negligible. Be careful however when calculating losses which represent a small difference of two comparatively large figures; rounding errors may then be more significant. Rounding errors increase with: (1) the number of nodes; (2) high aspect ratios of cell height to width; (3) when the variables (pressures and velocities) cover wide ranges - which indeed they do in a pump. Comparing the CFD results obtained with "single-precision" and "double-precision" allows an assessment of possible rounding errors.
5. **User errors** are caused by negligence, lack of knowledge of the complex programs and of physical phenomena in the pump itself as well as inadmissible simplifications made in an attempt to perform a CFD analysis with a minimum of effort and resources.

Discretization errors: Errors caused by the grid are *fortuitous*. It is therefore not possible to predict in a specific case whether a coarse grid will produce higher or lower heads and efficiencies than a fine grid. Discretization errors caused by the grid can be severe. By successive grid refinement, using different types of grids and searching for grid-independent solutions this error can be quantified and reduced considerably.

To illustrate this, Fig. 8.19 (derived from [84]) shows the hydraulic efficiency of an impeller as a function of the flow rate ratio q* from calculations with 11,000 and 32,000 elements: in two domains the finer grid yields higher efficiencies than the coarse grid; in the range $0.86 < q^* < 1.05$ the reverse is true. At $q^* = 0.86$ and 1.05 a grid-independent solution would be assumed because *coincidentally* the curves intersect at these points. Considering, at $q^* = 1.4$, the impeller loss coefficient $\zeta_{La} = 1 - \eta_{h,La}$, we find $\zeta_{La} = 0.04$ with the finer grid and $\zeta_{La} = 0.06$ with the coarse grid; in this case the discrepancy amounts to as much as 50%.

The finding of Fig. 8.19 is typical for CFD calculations of pumps. In spite of a multitude of investigations, no general tendencies whatsoever can be given which would allow one to predict the deviations between a CFD calculation and a test qualitatively and quantitatively, see Chap. 8.8.4 for a detailed analysis.

Numerical errors can be quantified, [71]. To this end, the problem at hand is calculated with different cell widths h, $2 \times h$, $4 \times h$, which differ by the factor of 2; the resulting solutions (for example heads, powers, efficiencies) are H_h, H_{2h}, H_{4h}.

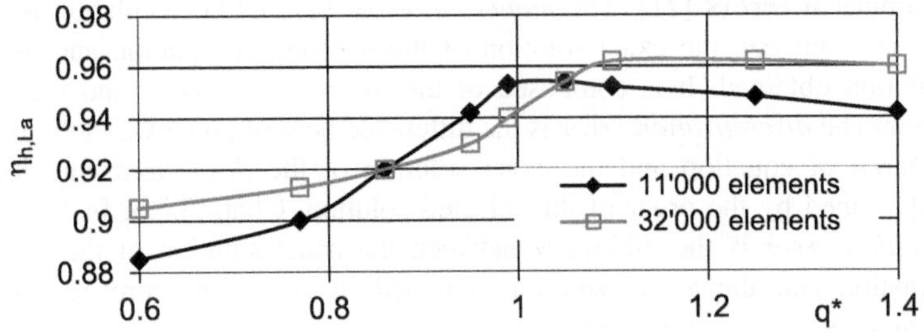

Fig. 8.19 Influence of number of elements on the hydraulic impeller efficiency [84]

The order "p" of the solution is then calculated from Eq. (8.28). Even when the discretization scheme used was of the 2nd order, there is no guarantee that the solution is really 2nd order if the number of cells selected was too small.

$$p \approx \frac{1}{\log 2} \log\left(\frac{H_{2h} - H_{4h}}{H_h - H_{2h}}\right) \tag{8.28}$$

Equation (8.29) yields estimations of the discretization error e_h and the grid-independent solution H_{nu}

$$e_h \approx \frac{H_h - H_{2h}}{2^p - 1} \qquad H_{nu} \approx H_h + \frac{H_h - H_{2h}}{2^p - 1} \tag{8.29}$$

As demonstrated by Fig. 8.19, this method (called "Richardson extrapolation") is not always meaningful since it would fail at the points q* = 0.86 and 1.05.

However, for complex problems, such as calculating an entire pump, the effort for this type of error estimation would appear prohibitive. In order to assess grid influences on the solution, the error analysis may then be restricted to calculations with different grids and discretization schemes (1st to 3rd order) as in Fig. 8.3.

8.8.2 *CFD Quality Assurance*

In order to build confidence in the results of numerical flow calculations, a considerable effort for quality assurance is required. It includes:

1. The *numerical verification*: "Are the equations solved correctly?" This is primarily the task of the developer of the CFD program. No test data are required to do this job.
2. The *validation* of the physical models and assumptions: "Are the correct equations solved?" This validation is done by comparing CFD results to test data. Since the measurements are taken as "reality", they must be sufficiently reliable and testing tolerances should be known, Chap. 8.8.3. Validation is done on the one hand using published general test cases ("benchmark tests"), e.g. [28, 47], on the other hand by the program user with specific machines and devices (pumps, turbines, compressors). At the present state of the art a comprehensive validation by comparing CFD to measurements on similar pumps is imperative. Loss calculations, roughness effects, cavitation and two-phase flows need extensive validation.
3. *Prior to any validation* effort numerical errors and grid problems must be appreciated and eliminated as required, Chap. 8.8.1, Eq. (8.29).
4. Calculations of a given geometry with different types of grids and variable grid resolution allow to assess grid influences and to obtain grid-independent

solutions (as far as practicable). Note that a "grid-independent solution" is not necessarily a "correct" solution.

5. Calculations with different turbulence models and comparison with well documented measurements such as pipe bends, diffusers, rough pipes, two-phase flow in pipes (Fig. 13.12) are part of a comprehensive validation.
6. Sensitivity studies of the effect of the simplifications made, selected control surfaces for averaging and numerical parameters.

When systematically applying CFD on an industrial scale, validation is an ongoing task as long as programs and applications evolve. The validation is not restricted to selected comparisons of CFD and tests but includes extensive sensitivity studies with respect to grid resolution, grid structure, the local grid refinement required, numerical parameters, turbulence modeling and wall treatment. A comprehensive validation for a broad industrial application could be approached in the following way:

1. Define the range of Reynolds numbers and pump sizes for the planned application.
2. With the selected wall treatment and turbulence model the wall cell size is obtained from Table 8.1: for a linear wall function (for example) from the requirement $y^+ = 1$, for a logarithmic wall function (for example) from $y^+ = 30$.
3. The grid resolution required near the wall can now be determined from the maximum grid size variation factor as per Chap. 8.3.4 item 6.
4. With the grid resolution defined in this manner, losses and velocity distributions are calculated for simple test cases such as smooth and rough pipes, 90°-bends, straight and curved diffusers. Compare these CFD results to measurements.
5. If step 4 is successfully concluded with a set of consistent assumptions and numerical parameters, the near-wall grid topology and numerical settings validated in this way are a sound basis for calculating pumps.
6. The calculations of pump components are validated with measured global performance data and velocity distributions (in particular at the impeller inlet and outlet).

A basic problem of validation - as of all CFD calculations - resides in the fact that numerical errors and deficiencies in the physical models can *compensate for or add* to each other according to statistical laws, since both categories of errors *are completely independent of each other and are thus statistically distributed.* In an individual case, a first-order solution with a coarse grid may happen to yield a head which agrees well with test data thus suggesting a success-full validation. However, in other applications the selected settings would produce erroneous results.[4]

Having developed in an individual case a good impeller by means of CFD does not justify the unqualified conclusion that the applied procedures and program settings should be correct and appropriate for future applications.

[4]Compare to CFD programs used in meteorology, which frequently give the correct forecast but fail without recognizable reason in other cases.

The risk of producing fortuitous results can be reduced by strict adherence to some basic CFD quality criteria such as listed below:

1. **Exact specification of the actual task**:

   - What problem shall be calculated?
   - What purposes shall be served by the results?
   - What type of conclusions can be drawn from the results?
   - Which are the criteria to assess the results? Example: Upstream of the pump, the suction pipe in a plant generates a non-uniform flow distribution due to various bends or fittings. With considerable effort and care (bends are quite difficult to model correctly!) the velocity profile at the impeller inlet may be calculated. However, the sole question of interest remains open, namely whether the resulting velocity profiles will cause excessive noise, vibrations and cavitation or impair head or efficiency in a measurable way. Unless acceptance criteria are defined how to assess the non-uniformity of the approach velocity, the exercise may be futile. If the task is restricted to improving the velocity profiles by modifications of the geometry, the selected turbulence model and wall treatment must be confirmed to be able to achieve this goal, which is by no means always guaranteed, Chap. 8.3.2.
   - What is the accuracy required to achieve the goals set for the CFD analysis? Is CFD really able to produce this accuracy? Example: Pumps with low specific speeds have flat Q-H-curves; the difference which decides between a stable and an unstable characteristic is a few per cent of the head. A difference in head of only $\pm 2\%$ can decide whether the pump characteristic is acceptable or not.

2. **Choice of the calculation domain**: Frequently, considerable simplifications are made because of limited computing capabilities. In order to avoid false conclusions, it should be analyzed which effects are ignored, what impact this may have on the results and what type of information the simplified calculations may be able/unable to provide. Special problems are:

   - The flow distribution at the impeller inlet if the inlet casing or the approach flow is not modeled.
   - Neglecting the flow in the impeller sidewall gaps: (1) at low specific speeds front and rear shroud provide a considerable pumping action, Chap. 3.10; (2) the inflow of the seal leakage has an impact on the incidence at the outer streamline and thus on cavitation inception, losses and the stability of the Q-H-curve; (3) the exchange of momentum at the impeller outlet between the main flow and the sidewall gaps influences the Q-H-curve, losses and axial forces, Chaps. 5 and 9.1.
   - If stage calculations are made without modeling all channels, there is an error in the vane pitches if $\nu_2 \times z_{La} \neq \nu_3 \times z_{Le}$ (which usually should be ensured to avoid vibration problems).

A sufficient number of nodes must be placed upstream of the domain of interest in order not to falsify the results by the inlet boundary conditions, Chap. 8.3.4. The outlet of the calculation domain must be placed away from zones where geometry and/or flow are subject to strong variations or gradients.

3. **Grid generation**: As repeatedly stressed above, the reliability of a CFD calculation critically depends on the quality of the grid; guidelines are given in Chap. 8.3.4. Faultless grids also ease the convergence. The effort for designing good grids thus pays double dividends.
The 3D-geometry model to be meshed must be free of gaps, errors and inaccuracies. Fillet radii between vanes and side walls are critical areas to be watched. Obviously, the local cell sizes must be much smaller than the smallest geometric details to be resolved by the CFD calculation (e.g. gaps between open impellers and casing). This is particularly true for impeller blade leading edges with small radii when cavitation inception is to be calculated.
After a first calculation it must be checked that the grid is sufficiently fine in regions with strong velocity gradients. Grid parameters such as y^+-values and angles between the elements can be checked by means of histograms.
4. **Boundary conditions**: The turbulence parameters at the inlet to the calculation domain are usually unknown and assumptions must be made, Chap. 8.3.6. The influence of the turbulence parameters can be assessed by means of a sensitivity study. If more cells are introduced between the entrance to the calculation domain and the component to be investigated, the influence of the turbulence parameters at the inlet boundary decays, but the control surface must be placed sufficiently away from said boundary.
No back-flow should be allowed at the outlet boundary.
5. **Initial conditions** can influence the solution - for example in flow conditions which are subject to a hysteresis. If unsteady processes are investigated, the initial conditions must represent a physically possible state of the system. When investigating transients this usually is the steady condition at time $t = 0$.
6. **Fluid properties** must be correctly specified. This requires due care if the program has a very general data structure. When calculating with high-viscosity fluids, the temperature-dependence of the viscosity must be implemented: Since the fluid heats up in the impeller side rooms and annular clearances, disk friction and volumetric losses are affected by local decrease in viscosity, Figs. 13.6, 13.7 and Eq. (13.6a).
7. **Turbulence modeling**: From the ample literature it is to be concluded, that there is as yet no universally applicable turbulence model and that the standard k-ε-model with logarithmic wall function is not really satisfactory for decelerated flows and curved flow paths. Therefore, it should not be applied to impellers, diffusers, volutes, inlet casings and bends. In the pump industry, turbulence models should be preferred which use linear or linear/logarithmic wall functions. The options include: (1) SST k-ω-model (now kind of standard); (2) Two-layer model; (3) other models with linear or linear/logarithmic wall functions as for example in [47, 54]. When employing linear wall functions,

sufficiently fine grids must be used so that the near-wall grid points are located within the viscous sublayer. Otherwise, errors similar to those expected with logarithmic wall functions may occur.

When calculating pumps with the standard k-ε-model, an apparently reasonable agreement is frequently found between calculation and test data, while other cases exhibit unexplainable discrepancies between CFD and test. This observation may be explained at times by the fact that the various errors in physical modeling, simplifications and numerical solutions sometimes *compensate* and sometimes *add*, as is the case with statistically distributed quantities. Frequently the Reynolds stresses are small compared to other terms in the transport equations (pressure gradients, Coriolis and centrifugal forces). In such cases the deficiencies of the turbulence models have less impact - even Euler programs may then produce reasonable solutions.

Sometimes it is argued that CFD results (in particular losses) can be relevant for comparing different design options for the purpose of component optimization, even though they may not be absolutely correct. However, theif the turbulence model is not able to reliably recognize flow separations in their initial stage and since wall functions impose wall shear stresses and influence the momentum exchange between near-wall flow and core flow, such comparisons and optimizations are not above all doubt.

Even the purely qualitative assessment of calculated flow patterns in impellers, diffusers or volutes remains questionable if the models used are unable to correctly predict the velocity distributions in simple components such as a pipe bend.

If the flow is laminar in some locations, the standard turbulence models fail since they are designed for high Reynolds numbers. Instead, turbulence models for low Reynolds numbers must be validated.

8. **Wall function**: As mentioned above, reliable results are not to be expected when applying logarithmic wall functions for curved flow paths and decelerated flows. The wall treatment has a considerable impact on losses. In those cases where the use of logarithmic wall functions is acceptable, the cell sizes near the walls must be selected so that $30 < y^+ < 100$ is achieved; in no case $y^+ < 11$ should then be admitted. For rough surfaces the wall function must be adapted according to Table 8.1. These recommendations for y^+ apply to calculations with logarithmic wall functions only. When using a two-layer model, the SST-model or other low-Re turbulence models, $y^+ < 2$ should be maintained in near-wall cells. At least 10 cells should then be located between the wall and $y^+ = 20$. These criteria can be checked by means of a histogram of the y^+-values.
9. **Modeling of the blade trailing edges** has an impact on the pressure distribution, flow deflection and losses, which can be quite significant and therefore needs optimization for the particular conditions.
10. **Transition between impeller and stator**: The impeller calculation is done in the relative system. Consequently, stator parts have the angular velocity of -ω.

Due to the no-slip condition, a fluid particle on the impeller shrouds has the tangential component $w_u = 0$ in the relative system and, correspondingly, $c_u = u_2$ in the absolute system. Immediately downstream of the impeller outlet a fluid particle at the casing wall has the velocity $c_u = 0$ due to the no-slip condition in the absolute system. This sudden jump from $c_u = u_2$ to $c_u = 0$ can cause problems - in particular with low impeller outflow angles α_2 (low specific speeds), because there is apparently a high friction. The problem can be avoided if the casing walls are defined as symmetry conditions (instead of a solid wall rotating with $-\omega$), but this procedure is unphysical and thus not really satisfactory. Since the sharp transition from rotor to casing exists in reality, the CFD program should be able to handle it. To mitigate the problem, a turbulence model must be used which limits the production of turbulence in strong velocity gradients (SST-model).

11. **Performance averaging**: The position of the control surfaces where head, power and losses are determined (Chap. 8.4) can have an influence on the results. Control surfaces in the suction and discharge nozzles harbor little uncertainties since velocities and velocity gradients are low (with the possible exception of axial pumps). However, in the range with strong gradients, in particular at the impeller outlet, head, torque and losses depend on the position of the control surface. Example $n_q = 35$: impeller loss coefficient $\zeta_{La} = 0.035$ at $r/r_2 = 1.01$ and 0.039 at $r/r_2 = 1.1$. As discussed in Chap. 8.4, partload operation with flow recirculations needs special procedures.
12. **Checking convergence**: A calculation which did not converge properly is to be rejected. The residuals of local quantities should be below 10^{-4} and residuals of RMS-values not exceed 10^{-5}. However, these values depend on the program used and the case studied. The particular definition and normalization of these residuals should be appreciated according to the particular program used.

- Target values such as head, torque and efficiency plotted against their residuals should asymptotically approach constant values in steady and periodic values in unsteady calculations. The course of this plot should be monotonic; initial oscillations must decay prior to dropping to the final level. This requirement equally applies to residuals plotted against the iteration steps or time.
- Mass flow residuals should be checked at different control surfaces (inlet and outlet of calculation domain and of the various components).

In case of convergence problems:

- Check quality of grid, improve as necessary, Chap. 8.3.4.
- Stepwise increase order of solution as per Chap. 8.3.5 (start with 1st order).
- The boundary conditions must be physically meaningful in order to allow the program to find a solution at all.
- Analyze domains with high residuals.

- Attempt unsteady calculation: *a flow which is unsteady in reality may not converge as a steady calculation!* Chap. 8.6.3 and [84].
- Better initial conditions, e.g. data from a converged solution with a different flow rate, Chap. 8.3.7.
- Modify relaxation parameters or time step according to user manual.

13. **Numerical errors**:

- According to Chap. 8.5.3, the scheme for interpolating the variation of the physical quantities within the cells should be at least 2nd order.
- If nevertheless a 1st order scheme is used, the grid must be extremely fine, Fig. 8.3.
- The magnitude of the numerical errors can be estimated as per Chap. 8.8.1.
- The selected time step must not exert any impact on the solution. Determine time steps as fractions or multiples of a typical physical time scale. A typical time scale can be calculated by dividing a length scale by a velocity scale. In general the same scales can be used as for calculating the Reynolds number. For impeller calculations the impeller outer radius r_2 would be a typical length scale and the tip speed $\omega \times r_2$ a typical velocity scale. The time scale then results from "factor"/ω. The magnitude of the optimum "factor" depends on the solver. Sequential solvers such as the SIMPLE-algorithm require smaller values than coupled solver algorithms which use factor $\approx$1. The optimum factor can be found by a few calculations with relatively coarse grids.

14. **Loss calculations**:

- The hydraulic losses in large pumps are only a fraction of the work transferred by the impeller. If it is attempted to apply CFD for the optimization, the required accuracy of the loss calculation must be better than $\pm$10%. As an illustration, consider a pump with a hydraulic efficiency of $\eta_h = 0.9$ and assume that the uncertainty of the losses calculated by CFD is $\pm$10%. Consequently, a CFD analysis of this pump would yield a hydraulic efficiency in the range of $\eta_h = 0.89$ to 0.91. This band width covers all the difference between a competitive and an unacceptable hydraulic design.
- When using the standard k-ε-model with the logarithmic wall function, the friction losses in decelerated flow (impeller, diffuser and volute) are over-estimated, while the mixing losses are presumably under-predicted. If the designer is guided by these results, he will select shorter blades and channels in order to reduce friction losses. The consequences would be: high blade loadings, non-uniform impeller outlet velocity profiles, high hydraulic excitation forces (noise and vibrations), enhanced mixing losses and possibly even instabilities of the Q-H-curve due to sudden changes in flow patterns.
- In general loss calculations by present-day CFD capabilities must be considered as quite uncertain. This applies to the absolute values as well as to the comparison between different design options. Think of the effect of the

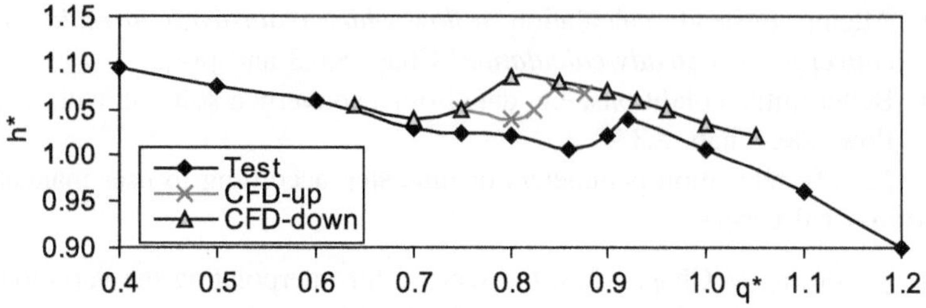

Fig. 8.20 CFD-study of a pump with instable Q-H-curve, h* = H/H_{opt} [5]

statistical distribution of errors from physical modeling and numerical procedures.[5]

- The influence of roughness can scarcely be evaluated because the interactions between roughness, turbulence and velocity profiles are largely unknown and difficult to quantify, Chap. 3.10.

15. **Instabilities in the Q-H-curve**: Instabilities in the Q-H-curve can often be predicted by CFD, but the onset of flow separation may be found to occur in the test at higher flow rates than calculated. This observation is borne out by investigations in [5] on a pump with double-entry impeller, diffuser and volute. Figure 8.20 compares the test to CFD calculations with increasing flow (curve "CFD-up") and decreasing flow rate (curve "CFD-down"). It is the same pump with a hysteresis in the Q-H-curve as in Fig. 5.42. While the onset of the instability is not well captured by CFD, the dip in the head of about 4% is well predicted. However, the impeller side rooms were not modelled in CFD. Energy dissipation into the ISR may be responsible for the fact that the instability in the test was observed at higher flow than in CFD. Also: using log-wall functions with $y^+ = 100$ may not be appropriate to aptly capture the onset of instability (the unsteady calculation of the entire pump was done with 2.5×10^6 nodes).

    Studies on a pump turbine with axial inlet, guide vanes, stay vanes and a volute gave similar results: the instability occurred in the test at q* = 0.95, while CFD predicted the onset at q* = 0.86 with a coarse grid, at q* = 0.84 with a medium grid and at q* = 0.83 with a fine grid, [5].

16. **Velocity distributions**: As demonstrated by investigations of pipe bends and diffusers, calculations with the standard k-ε-model with logarithmic wall functions produce quantitatively and even *qualitatively* wrong velocity distributions in flows with adverse pressure gradients. No better results can be expected for pump diffusers.

[5]For example, the disk friction losses calculated by CFD in [33] were only 10% of the values expected according to Eq. (T3.6.11) which is confirmed by various measurements.

The velocity distribution in impellers is determined by two basically different effects: (1) centrifugal, Coriolis and blade forces and (2) boundary layer effects. These forces act equally in viscous and non-viscous flow; they are therefore captured even by a Euler calculation. In as much as the flow distribution is determined by forces, the choice of the turbulence model and wall function has an indirect influence only. In contrast, boundary layer effects, such as prevailing in a diffuser, are largely determined by wall functions and turbulence models.

In those cases where the velocity distribution in an impeller is determined by forces a reasonable agreement between calculation and test can be expected. When the flow is controlled by boundary layer effects, the choice of the wall function and turbulence model is important. These relationships may explain why calculated and measured velocity profiles sometimes agree quite well while they are even qualitatively wrong in other cases (as can happen with a given impeller at different flow rates).

Before it is attempted to design an impeller so that no sudden changes in flow patterns occur at partload (Chap. 5), it must be ascertained by means of measurements that the selected turbulence model and wall function allow velocity distributions to be obtained which are at least qualitatively correct.

These limitations basically also apply to the assessment of flow patterns based on which it is attempted to improve a design by reducing the non-uniformities of velocity and pressure distributions.

17. **Inlet casings**: When optimizing the approach flow, the uniformity factor according to Eqs. (8.18) and (8.19) can be evaluated at the reference plane at the impeller in order to have a single quantity which allows comparing different designs. For assessing the quality of the approach flow, criteria similar to those in Chap. 11.7.3 can be defined. Depending on the application, the specific speed and the tip speed u_1, the maximum local deviation from the average axial approach flow velocity could be limited to $(c - c_{av})/c_{av} = \pm 0.1$ to ± 0.15 and the approach flow angle deviation in the relative system to $\Delta\beta = \pm 1$ to $3°$ for example. The more stringent criteria apply to impellers with a high risk of cavitation damage and high specific speeds.
18. **Sensitivity analyses** are appropriate to investigate how the target values (head, efficiency, stability of the Q-H-curve) depend on the simplifications and assumptions made for the CFD calculation.

8.8.3 Comparison Between Calculation and Experiment

A thorough comparison of CFD results with test data is necessary for building confidence in the calculations and assessing the possibilities and limits of numerical flow calculations. To do such a validation correctly, it is necessary to also appreciate the uncertainties of the measurements.

8.8.3.1 Evaluation of Measured Velocity Profiles

As discussed above, H_p, H_{th}, H_{La} and ζ_{La} can be determined from an impeller calculation by integration over the impeller outlet control surface according to Eqs. (8.10) to (8.13). This applies not only to the CFD calculation but also to measurements by probes or laser velocimetry, since these are done also on a grid of measuring points. The global performance data can thus be evaluated by means of the same formulae. All quantities in these equations represent spatial and temporal averages derived from a finite number of measuring points for which the following measuring uncertainties may be assumed: (a) static pressure at the impeller outlet as an average of a number of measuring points: ±5%. While the individual pressures are more precise than ±5%, the calculation of a representative average is subject to a greater uncertainty; (b) velocity at the impeller outlet: ±5%. From these individual errors of averaged velocities and pressures an uncertainty of ±5% would also result for H_p, H_{th} and H_{La}. Since H_{La} and H_{th} are affected by the same error (resulting from the velocity measurement), these errors can largely cancel in the calculation of the impeller's hydraulic efficiency $\eta_{h,La} = H_{La}/H_{th}$. If the velocity profiles at the impeller outlet are non-uniform, the interpretation of the losses encounters the difficulties discussed in Chap. 8.4.

As demonstrated by experience, predicted global performance can agree quite well with measured average performance even if calculated and measured velocity profiles match neither qualitatively nor quantitatively. This is no contradiction, because *a given average value can result from an infinite number of different velocity profiles*. Furthermore, a converged solution implies that the *average* meridional velocities are correct since mass flow is conserved.

It was also observed that calculated and measured velocity profiles at the impeller inlet usually agree quite well throughout the entire range of flow rates - even with partload recirculation. In contrast, large discrepancies are often observed between calculated and measured pressure and velocity profiles at the impeller outlet (e.g. [50]), because the selected turbulence model and wall function were unsuitable for decelerated flow on a curved path and because the interaction between impeller and collector flow was not modeled. Another reason for such deviations is the sensitive equilibrium of the flow in a radial impeller with back-swept blades (Chap. 5.2).

As shown in Chap. 5.3, the stability of the Q-H-curve depends significantly on the pressure recovery in the collector. Therefore, the calculated or measured total pressure at the impeller outlet often does not tell much about the shape of the Q-H-curve of the pump. The rise in total pressure provided by the impeller can represent a stable curve $H_{tot,imp} = f(Q)$ even if the Q-H-curve of the pump is quite unstable. Only if the curve $H_{tot,imp} = f(Q)$ of the impeller is unstable, will this also apply to the Q-H-curve of the pump. Investigations focusing only on the total pressure at the impeller outlet are therefore open to considerable doubt.

8.8.3.2 Evaluation of Industrial Performance Tests

If only head and power measurements are available, the hydraulic efficiency of the tested pump can be estimated from Table 3.5, Eq. (T3.5.8) and compared to the CFD results. The theoretical head coefficient is given by $\psi_{th} = \psi/\eta_h$ and Eq. (T3.2.8) yields an average tangential velocity c_{2u}. All *average* velocity components at the impeller outlet can be calculated from Table 3.2. The overall uncertainty of the hydraulic efficiency can be determined from the individual errors according to the law of error propagation. To this purpose, the following tolerances may be assumed typically:

- mechanical losses ±20%
- disk friction ±25%
- annular seal leakage ±30%
- measuring tolerance of efficiency at coupling ±1%
- measuring tolerance of head ±0.5%
- measuring tolerance of balance fluid ±2%

The uncertainty of the hydraulic efficiency determined from Eq. (T3.5.8) increases with falling specific speed, because disk friction, leakage losses and mechanical losses absorb a larger fraction of the power at the coupling. Depending on type and execution of the tested pump, the uncertainty of the hydraulic efficiency amounts to typically ±3.5% at $n_q = 15$; to about ±2.4% at $n_q = 20$; and roughly to ±1.2% at $n_q > 40$. The same uncertainties can be assumed for H_{th} and c_2.

The uncertainty of the leakage flow rates at the impeller and a possible balance device imply also an uncertainty as to the exact flow to be used in the CFD calculation, which always must be done with the total flow through the impeller rather than the useful flow through the discharge nozzle.

If the static pressure at the impeller outlet has been measured, the losses in impeller and collector can be estimated according to Chap. 4.1.3. However, the uncertainty of these estimates can be quite high depending on the position of the pressure tappings used and the variation of the static pressure around the impeller circumference (a measuring error of a few per cent may even lead to negative losses in impeller or collector).

If the entire pump (including impeller side rooms and annular seals) is calculated by CFD, the head is determined between discharge and suction nozzle. The calculated input power corresponds to the inner power P_i since the mechanical losses are not modelled. P_i is also determined from the test from $P_i = P - P_m$ (Table 3.5). The comparisons between measured and calculated head and inner efficiency η_i are straightforward, but the comparison of the detailed losses is subject to considerable uncertainty. In order to determine the hydraulic efficiency from CFD, the following steps may be taken: (1) Calculate torque M on impeller from Eq. (8.25); (2) determine the flow through the impeller $Q_{La} = Q +$ leakages; (3) calculate theoretical head from $H_{th} = M\omega/(\rho\ g\ Q_{La})$; (4) hydraulic efficiency $\eta_h = H/H_{th}$. While this procedure works well for single-stage pumps, multi-stage pumps are more difficult to handle - in particular if the first and last stages differ from the series stages because the input power can only be measured for the ensemble of all stages.

8.8.4 Accuracy of Numerical Performance Prediction

When comparing CFD with test results, the uncertainty of the test data needs to be considered as well: (1) Measuring tolerances are expected to be about ±1% (maybe a little less for the head). (2) The calculated (theoretical) geometry deviates from tested geometry due to manufacturing tolerances - notably castings and annular seal clearances. If the hydraulic components are machined, the effect of geometric tolerances may be in the order of ±1% and the combined effect of geometry and testing tolerances about ±1.5%. Therefore, deviations below ±1.5% might be ambiguous. Casting tolerances can be much larger; they must be checked carefully if a meaningful comparison between test and CFD is aimed at.

Head (and thus efficiency) are subject to a *systematic* error between CFD and test: In CFD the head is calculated from the static pressures p_{stat} in the suction and discharge nozzles and the velocity heads p_{dyn} are integrated over the nozzle areas from the actual, nonuniform velocity distribution. In contrast, the head in the test is determined from $p_{stat} + \rho/2\ (Q/A)^2$. As discussed in Chap. 8.4 in context with Eq. (8.14) a non-uniform velocity profile contains more kinetic energy than the uniform profile defined by $c = Q/A$. In a specific case ($n_q = 36$), the resulting errors were: CFD *over*-predicted the head by 0.1% at BEP, but by 0.6% at $q^* = 1.3$. These figures are not generic because the errors will increase with the ratio p_{dyn}/p_{stat} and will thus reach their maximum with axial and semi-axial pumps. In general terms, the error increases with n_q and q^*. When comparing CFD with test results, one may be advised to evaluate head and efficiency in CFD the same way as in the test by calculating the head from $p_{stat} + \rho/2\ (Q/A)^2$ - in addition to the integration of the velocity profiles.

Countless authors compare CFD results to test data for specific pumps. Mostly, so it seems, the tacit goal was to prove that CFD can predict the test results accurately. Few, if any, publications attempt a broader assessment of the accuracy which can be achieved by CFD, [113]. In many publications, the discrepancies between test and CFD are not spelled out in precise percentages; instead, they must be inferred from performance curves presented in graphs with insufficient resolution. Often, published data are limited to the q*-range where the discrepancies are not too appalling.

CFD is a numerical *approximation* of the physical processes. Therefore, and given unavoidable testing tolerances, even an exact match of CFD and test would be coincidental. To get a broader view on the accuracy attained by CFD for pump performance prediction, 30 different pumps from 7 renowned institutions have been analyzed by comparing CFD results to test data. Of course, this arbitrary sample can reflect only *what these CFD users have accomplished in the recent past*. Indeed, some users may have lately become more attentive to quality issues.

In the following, the deviations between test and CFD are given, in per cent, as per Eq. (8.30):

$$\delta_H = 100 \times \frac{H_{CFD} - H_{Test}}{H_{Test}} \quad \delta_P = 100 \times \frac{P_{CFD} - P_{Test}}{P_{Test}} \quad \delta_\eta = 100 \times \frac{\eta_{CFD} - \eta_{Test}}{\eta_{Test}} \tag{8.30}$$

The results of the analysis are presented in Table 8.3 in terms of:

1) the average deviations at BEP, which would tend towards zero for a large pump population if the errors are statistically distributed;
2) the standard deviation which is about ±4%;
3) the 95%-confidence limit amounting to ±6 to 8%;
4) the maximum and minimum deviations found in the ensemble of the 30 pumps investigated (one pump with 17% efficiency error was removed from the lot since considered atypical).

Five types of errors have been discussed in Chap. 8.8.1; none of these falls into the category of "systematic errors" which could be corrected. Instead, it may be assumed that the errors are *statistically distributed*. Hence error theory was applied to estimate the probability of reaching any specified level of accuracy for predicting efficiency, head or power. Figure 8.22 shows the result: For example, if a band width of ±2% would be considered acceptable for the efficiency calculated by CFD, the probability of reaching that accuracy was about 40%. The 95%-confidence limit was about ±7%. As with any statistic: (1) an individual can be dead-on the prediction; (2) but also outside the ensemble of individuals used for establishing the statistic; (3) the data reflects *observations in the past* based on an arbitrary sample; (4) extrapolation of the past to the future is invalid if conditions change, e.g. users become more critical, diligent or develop better methods, or more stringent rules and quality standards.

Deviations at off-design conditions - ±15 to 20% are not uncommon - were much larger than at BEP, but no figures can be presented because the pumps had been calculated over different q*-ranges. Many publications show discrepancies of a similar magnitude as found in the present study. Other authors published higher accuracies, but it is usually not possible to decide from the published information whether this is due to a more diligent use of CFD or due to coincidence. It is also common practice to try different set-up parameters, numerical schemes and meshes to reduce the discrepancies between calculation and test, e.g. [100]. Such adaptations can be successful for an individual case but may fail when applied to different pumps.

Table 8.3 CFD error statistic

	δH_{opt}	$\delta \eta_{opt}$	δP_{opt}
	%	%	%
Average	1.1	2.0	−1.1
Std-dev	±4.0	±3.7	±3.1
95% confidence	±7.8	±7.3	±6.1
Maximum	8.8	8.8	6.6
Minimum	−5.2	−3.8	−5.3

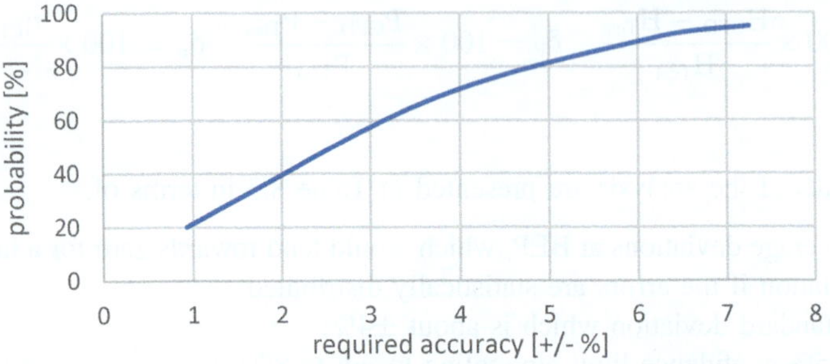

Fig. 8.21 Probability to achieve a specified accuracy of efficiency prediction by CFD-based on past observation

Figure 8.21 can likewise be used for head and power with sufficient confidence. Three examples of the comparisons between CFD and test are shown in Fig. 8.22; another sample (in a different format) is depicted in Fig. 8.23. These graphs illustrate the broad range of patterns the deviations between CFD and test can follow.

The following can be concluded from the above analysis:

1. On average, head is over-predicted by 1.1%. This bias could be due to surface roughness effects which usually are not modeled in CFD but are present in the tested pumps. The calculated power tends to be on the low side (by -1.1% on average). This can be due to the mechanical losses not considered in CFD and neglecting surface roughness (roughness increases the power consumption). Both effects add to overpredicting the efficiency by about 2%, and that is what Table 8.3 shows.
2. Notoriously, the agreement between CFD and test is usually (but not always) best in the range $0.8 < q^* < 1.1$. The discrepancies increase *erratically* as the flow is reduced towards shut-off.
3. Typically, at $q^* > 1.2$, CFD tends to diverge strongly from the test data (that's why some users do not calculate in this domain at all). This is especially true for pumps with low specific speeds (say $n_q < 25$) because volute losses are dominating. Possible reasons:
   (a) Flow separation occurs in the discharge nozzle due to negative incidence at the cutwater, see Fig. 9.28. This is particularly harmful since part of the volute throat area is then blocked by stalled fluid; see Fig. 4.12.
   (b) Volute cavitation is triggered by flow acceleration ($c_{3q}/c_2 > 1$) and incidence. It should be checked in CFD that the minimum absolute static pressure at the cutwater or the diffuser vane leading edge is above the vapor pressure (no cavitation model needed to do this simple check).
   (c) The volute losses are under-predicted if roughness is ignored; friction losses increase with Q^2 in this domain. As discussed in Chap. 8.3.3, it is

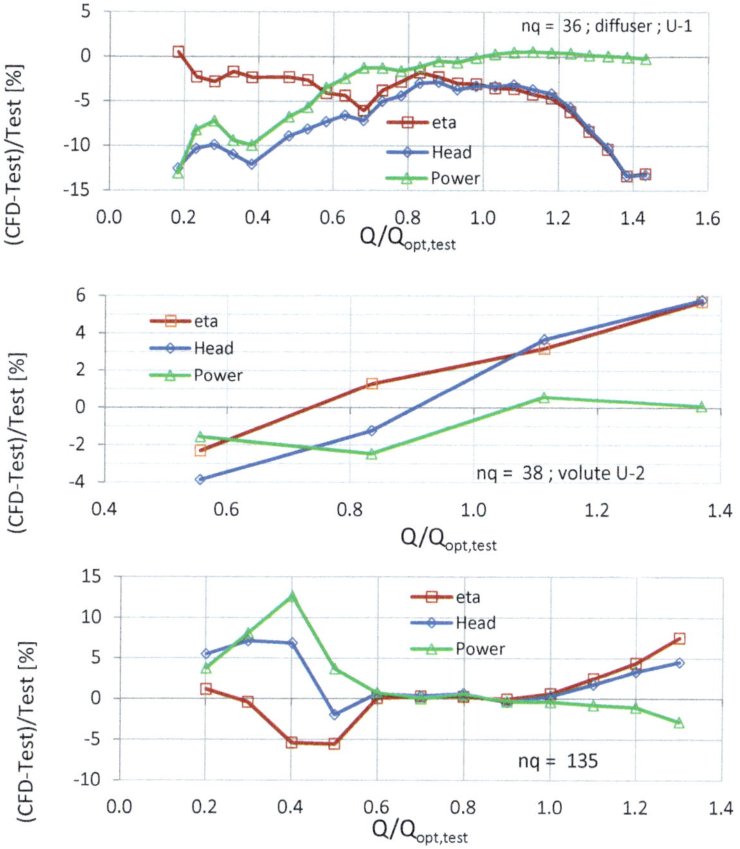

Fig. 8.22 Examples for the deviations between CFD and test

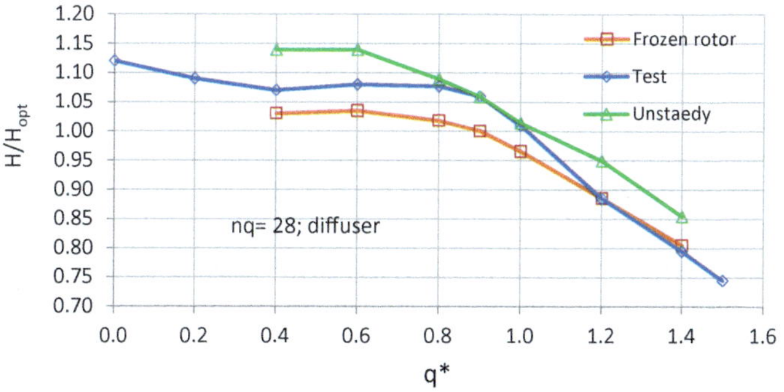

Fig. 8.23 Comparison between test data, unsteady CFD and frozen rotor CFD

necessary to resolve the wall-near flow down to $y^+ = 1$ to 2 to well capture flow separation because logarithmic wall functions cannot do so. This is well confirmed by investigations in [92] and [94].

(d) Convergence problems due to high unsteadiness of the flow (as discussed below).

4. The deviations $\delta = f(q^*)$ between CFD and test show no general pattern; rather they seem to be quite *erratic*. Apart from the slight bias discussed above (item 1), no *systematic* error, common to all pumps investigated, could be detected: Hence the reason of the scatter is unlikely to be found in the basic equations.
5. No dependence on specific speed, type of pump or group of users could be detected even though the different users had applied different meshing, set-up parameters and CFD-methods.
6. The wavy or "erratic" nature of $\delta = f(q^*)$ suggests that the effect of the parameter "flow rate" on the CFD-solution is not well captured. Possible causes: (a) numerical; (b) local flow pattern changes which cannot be calculated with the turbulence models and wall treatment; (c) solution not properly converging in highly unsteady flows.
7. Often, the predicted shapes, or gradients, of the Q-H-curves significantly differ from the test (again we may say that CFD fails to appreciate the parameter "flow rate" correctly). Then, the predicted and measured curves intersect at some q^*. Near the intersection, CFD seems to predict performance correctly; see Figs. 8.19, 8.20 and 8.23.
8. In Fig. 8.23, the frozen rotor calculation coincides with the test curve at $q^* = 1.2$ to 1.4 (kind of unusual), while sliding mesh results agree with the test near $q^* = 0.9$. The sliding mesh and the frozen rotor curves exhibit an off-set of roughly 10%, but both curves are "correct" in specific flow rate ranges. It appears difficult to explain the peculiar results in Fig. 8.23 and the erratic behavior of the discrepancies between CFD and test by physical phenomena or application errors. This type of coincidental agreement at arbitrary q^* between different numerical schemes or test data can be observed frequently; Fig. 8.19 is another example and additional cases can be found in [108]. From these findings, one might be tempted to conclude that CFD results contain a fortuitous component which may be in the order of 5 to 10% of the real value near BEP and considerably larger at off-design conditions. This uncertainty seems not to depend much on the quality of the calculation.
9. CFD results depend on the *combination* of all set-up parameters, grid topologies and modelling assumptions. Since there is an infinite number of combinations, an infinite number of solutions can be found in a band of (say) ±5% of the test data. This hypothesis may account for postulated fortuitous components of the CFD solutions.
10. When doing an optimization study, one compares design "A" with design "B" by modifying a specific design parameter (e.g. b_2). In view of the above

observations: Does the erratic behavior of the deviations really always justify the assumption that a meaningful tendency can be derived? What would be the expected error-band below which any calculated efficiency differences become meaningless? When comparing impeller "A" with impeller "B" in an optimization study, most important for the subsequent optimization steps is the question: has "A" a higher or lower efficiency than "B". If CFD is unable to reveal this trend correctly, then the optimization process will be questionable. Example: in an impeller optimization study (at BEP) using single-channel calculations, 4 design options were compared to a baseline design. The result was disappointing: only in one case the efficiency trend was predicted correctly (i.e. efficiency decreased both in CFD and test) while CFD predicted the wrong trends in 3 other cases. The transient calculations predicted the current trend only in one case. A none-technical person would have a 50%-chance to bet correctly on a plus or minus trend. At least in the present case, even a pre-optimization using single-channel and transient calculations would lead the designer into the wrong direction. In addition, this finding indicates that the mixing losses at the impeller outlet need to be included in impeller optimization studies. However, the efficiencies of all tested design options were in a narrow range of −0.3 to +0.7% with the testing tolerance of ±0.5%. Therefore, it is difficult to draw final conclusions.

11. Deviations of H and P with the same sign (plus or minus): errors compensate to some extent: η is seemingly quite accurate - this is frequently observed. Deviations of H and P with opposing signs (one positive, the other negative): errors add: η can be grossly off.
12. From the *present* study, the probability that the CFD performance predictions are grossly misleading (useless) was about 40%.
13. Flow separation in the diffuser or volute is anticipated where the Q-H-curve flattens. This feature seems to be often captured by CFD - at least qualitatively as in Fig. 8.23. However, the flow rate where the curve flattens or becomes unstable is mostly lower in CFD than in the test, see Figs. 8.23, 8.20. In other words: flow separation is predicted too late by CFD.
14. At flows above BEP, the power is *under-predicted* by typically 3 to 12% in many instances. It appears that this has little to do with losses; rather the flow deflection seems to be calculated too low. Since little impact of the diffuser or volute on the impeller flow is anticipated due to the absence of recirculation, the reason for the problem may be in the impeller. Cavitation should not be the reason since any cavitation (in the casing or the impeller) could reduce the power in the test, but not in single-phase CFD. In one extensive study, the power prediction for single-channel impellers was by 20 to 30% too low. The mechanism responsible for predicting too low power and the problems at $q^* > 1.2$ are not yet fully clarified.

Possible reasons for the discrepancies between CFD and test are:

- Geometry not adequately modelled: manufacturing tolerances, inlet and outlet or other simplifications; insufficient resolution of annular seals and impeller side rooms: disk friction, leakage losses and interaction of ISR flow with main flow.
- Leakage losses are very difficult to model: (1) the small gap width (typically 0.2 to 0.3 mm) requires very fine meshing; (2) the relative roughness in the seal is high (typically $\varepsilon_{max}/s_0 = 0.02$ to 0.05).
- Numerical errors are minimized where the grid lines follow the stream lines. Such is not possible for regions of stalled fluid and the mesh elements around leading and trailing edges of impellers, vaned diffusers or volute cutwaters. It is in these locations with high velocity gradients and skewed grid elements, that numerical errors are highest. Deviations of the calculated from the real flow in these locations influence the entire downstream-flow, [113].
- The transition between the rotating and stationary reference frames are not well modelled: high velocity gradients, interaction of main flow with impeller side rooms, recirculation across the interfaces, Figs. 9.6, 9.7.
- *Turbulence models and wall treatment are unable to well capture losses and flow separation*: The flow through a pump has little in common with the flow over a flat plate, which is assumed in CFD for modelling friction and flow separation. The boundary layers have a strong influence on the flow distribution and losses. The transition between laminar and turbulent boundary layers depend on the Reynolds number, surface roughness, velocity gradients, turbulence and curvature - all of which effects cannot well be captured by CFD. Often CFD results could be interpreted as *over-predicting the effect of friction* but *under-predicting the losses caused by non-uniform flow distributions.* CFD-results then suggests that shorter blades, possibly reduced width (b_2), would yield higher efficiencies, while the test data may show that the lower losses were obtained with longer blades (due to more uniform flow). This observation from a specific study, of course, cannot be generic.
- The CFD solution is not converged properly. Convergence at off-design condition is delayed by large-scale unsteadiness of the flow. During intense partload recirculation, size and shape of the zones with stalled fluid change with time; stochastic pressure and force fluctuations are present in addition to periodic phenomena. The stochastic components may be expected to hamper - or even prevent - convergence because there may be no unique flow pattern to converge to. More revolutions must be calculated in unsteady off-design calculations than near BEP with no flow separation. It may also be necessary to average the performance over more than one revolution.
- Roughness effects neglected (mostly).
- *Even when staying within recommended parameters and best-practice guidelines,* the CFD results can be adapted by trial and error to better match the test results by selecting grid size, modifying grid topology, y^+-values, resolution of

flow behind trailing edge, location r/r_2 where impeller outlet parameters are computed, angular position in frozen-rotor calculations, interface between rotating and stationary domain and other set-up parameters. Presumably, most authors had adapted their set-up in this way to reduce the differences between CFD and test, [113]. However, this procedure cannot lead to generic results - unless the same set of parameters is used successfully for a *statistically-relevant* number of pumps over a wide range of geometries and specific speeds. Otherwise, the set-up might fail when applied to new projects. Manipulating the results in this way may be particularly dangerous if computation resources are moderate: the errors due to poor resolution are then camouflaged by unwittingly introducing other modeling and numerical errors, see also Fig. 8.19. Insight as to why the "reasonable" agreement has been found may then be mostly lacking. When comparing CFD studies from different publications, it is mostly impossible to assess the influence of the above effects on the results because of insufficient detailed information.

- Although one would generally expect that the accuracy of loss prediction increases as y^+ is reduced, no strict proportionality can be expected between y^+ and accuracy. If good agreement between CFD and test data is occasionally observed with coarse grids, this should be attributed to coincidence.

Conclusions:

- CFD is like a toolbox: the quality of the work done with a toolbox largely depends on the worker: it is almost impossible to make an *objective* assessment of CFD capabilities separately from the user influence. Only some numerical errors can be quantified by objective methods (Chap. 8.8.1).
- The above investigation of uncertainties covers volute and diffuser pumps over a wide range of specific speeds. Different calculation methods, meshes and set-up parameters were applied.
- The statistical evaluation is meant to: (1) show the CFD accuracy obtained in the past on an arbitrary sample; (2) demonstrate the necessity for improvement and for a more critical assessment of CFD results.
- The almost universal belief that CFD is able to correctly predict the *differences* in efficiency when comparing design "A" and "B" during an optimization process, needs to be assessed more critically. In view of the above analysis one might be tempted to conclude:

  - Large changes in efficiency, say above $\Delta\eta_{opt} > \pm 2\%$, have a good chance of being captured by CFD.
  - Fine-tuning a design, based on predicted efficiency differences below $\Delta\eta_{opt} = \pm 0{,}2\%$, is likely to be meaningless.
  - Further investigations are needed to define which accuracy can be expected in predicting $\Delta\eta_{opt}$.

  - The accuracy at off-design conditions is considerably less than at BEP. Optimization studies for improving the stability of the Q-H-curve may rely more on the analysis of flow patterns than on quantitative head prediction.
- Possible improvements may include:
  - Systematic approach to grid generation and set-up
  - Sufficiently fine grids (y^+ control)
  - Reject non-converged solutions: unsteady calculation over more revolutions with averaging over longer periods.

8.8.5 CFD Set-Up Parameters

Commonly used CFD tools are multi-purpose programs with comprehensive user manuals covering a variety of calculation options over several hundred pages, e.g. [102]. While it is necessary for the user to study the relevant manuals, it seems appropriate to discuss set-up parameters and methods which have been successfully used for pumps. Based on a number of publications and reviews with experienced CFD users, Table 8.4 attempts to present recommendations for state-of-the-art calculations.

The methods and set-up parameters depend - to some extent - on the accuracy aimed at by the user. For the case of argument, three levels may be defined:

1. Professional performance prediction (PPP) with considerable commercial risk if the specified performance is not met: The targeted tolerances on head and power may be maximal ±3 to 4% corresponding to quality class G1 in Table 15.2. Efficiency tolerances are usually less: if a large pump is short in efficiency by 2%, this can be a serious technical and commercial issue. Impeller side rooms (ISR) and annular seals must be included in the CFD model because of the associated losses and their interaction with the main flow.
2. Academic studies (AS) with no commercial risk. The targeted tolerances on head and power possibly will be limited to ±10% - otherwise the calculated performance might be almost meaningless.
3. CFD could be used as a theoretical "flow visualization tool" with no specific requirements on accuracy. However, correct velocity distributions are even more difficult to achieve than global performance. This mode is widely used - with good reason - for detecting likely locations of flow separation or non-uniform flow with the object of optimizing the design of hydraulic components. Indeed, this application may be the most valuable benefit which can be drawn from CFD, [113].

Of course, the highest possible accuracy should always be aimed at to prevent erroneous conclusions, but many users are unable or unwilling to commit the required computing resources.

Table 8.4 CFD Set-up					
		Steady (RANS)		Unsteady (URANS)	
	Rotor/stator interface	Volutes: Frozen rotor. Vaned diffusers: mixing plane		Transient (sliding mesh)	
	Accuracy level	AS	PPP	AS	PPP
1.	Flow rate range: q*	0.5 to 1.2	0.8 to 1.1	0 to max	
2.	Include ISR and seals (in the rotating reference frame)	recommended	yes	recommended	yes
		Leakage flows must be included to ensure that impeller is calculated with actual Q_{La} and diffuser with Q_{Le}			
3.	Boundary condition *inlet*	Total pressure, direction of velocity: normal to boundary; turbulence level: 5%; eddy viscosity ratio: 10			
		Location: axial inlet: $L/d_s > 10$, radial inlet: $L/d_s > 5$. No flow recirculation through the inlet cross section			
4.	Boundary condition *outlet*	Mass flow rate; location: $L/d_d > 5$ No flow recirculation through the outlet cross section			
5.	Impeller outlet rotor-stator numerical interface location	Location: $r_{interface}/r_2 = (d_3/d_2)^{0.5}$ or $(d_z/d_2)^{0.5}$			
6.	Equivalent time step	$0.3/\omega$ to $1/\omega$		1° to 3°	
7.	Turbulence model	k-ω SST	k-ω SST	k-ω SST	k-ω SST
	Scale-adaptive SAS	no	no	no	yes
	Curvature correction	no	no	no	yes
8.	Wall treatment				
	Wall function	$y^+ < 50$	–	$y^+ < 50$	–
	Low-Re model		$y^+ < 2$		$y^+ < 2$
9.	Surface roughness	most users do not use roughness models because of their uncertainty			
10.	Number of elements	$> 5 \times 10^6$	$> 10 \times 10^6$	$> 5 \times 10^6$	$> 10 \times 10^6$
11.	Time discretization			2nd order	
12.	Accuracy	double-precision floating point recommend only for 2-phase flows and cavitation			
13.	Turbulence advection term spatial discretization	Turbulence must be modelled by "bounded discretization schemes to avoid values below zero; example: "high resolution" scheme [102].			
14.	Diffuser or volute cavitation	Monitor c_{pmin} (check for cavitation at $q^* > 1$): minimum local pressure at cutwater or vane leading edge must exceed vapor pressure (single-phase, not cavitation model).			
15.	Maximum residuals	Monitor mass flow rate, head, efficiency, power			

Comments to Table 8.4

1. Performance calculations are done for the entire pump. Head and $NPSH_R$ are evaluated in the suction and discharge nozzle at a distance of $L/d_{Nozzle} = 2$ (ISO 9906) from the respective flange.
2. The table gives *typical* values (which can vary over some range) with a bias to high-quality calculation.
3. Because of the high unsteadiness of the flow, off-design calculations outside the range $q^* = 0.8$ to 1.1 should be done unsteady. With strong flow recirculation (say typically at $q^* < 0.25$ to 0.3) the flow patterns become chaotic and a valid solution may not be obtained. Some authors claim that solutions down to $Q = 0$ can be obtained, others disagree.
4. *Volutes:* Steady-state calculations may best be done with frozen rotor because the rotor position has less influence on overall performance (only close to the cutwater). However, several rotor positions should be calculated and averaged to increase the accuracy of the performance prediction. In one example, the head varied by about ±1.5% when the impeller was rotated versus the cutwater by one blade pitch.
   Vaned diffusers: Steady-state calculations may best be done with the mixing plane because pressure and velocities vary strongly over one pitch t_3 of the diffuser. However, authors differ about the best choice of frozen rotor versus mixing plane.
5. Mesh generation is mostly done by commercial programs; see Chap. 8.3.4 for mesh quality requirements.
6. The number of elements or nodes depends to some extent on the absolute size of the pump. No absolute figures can be given. The number of elements for a given geometry rather follows from the rules: (1) resolution of wall-near flow (e.g. target $y^+ < 2$); (2) expansion rate near wall <1.1, in the bulk-flow domain <1.5 to 2; (3) aspect ratio of elements. In practice, the number of nodes is limited by the available computer capabilities; but the compromises taken in this respect are often questionable.
7. *Boundary conditions*: when specifying the mass flow rate at the inlet and pressure at the outlet (as done by many authors) pressure fluctuations in unsteady calculations may be influenced by the imposed pressure. That's why Table 8.4 stipulated mass flow rate at the outlet and total pressure at the inlet. In steady-state calculations, both options are considered equivalent, but convergence may be different. Specifying the pressures at inlet *and* outlet impairs the convergence - mainly so where the Q-H-curve is flat or even unstable.
8. The pressure given as boundary condition should not influence the results because only the calculated pressure differences are relevant in incompressible flow.
9. The SAS model is recommended for globally unstable flows, [111]. Since the SAS model requires finer grids and smaller time steps, calculation times increase significantly. If the flow instability is not strong or if grid and time steps are too coarse, the URANS model is used per default. For calculating "normal" pumps near BEP the SAS brings, therefore, no advantages. For

calculating single-channel impellers, unstable head curves or sewage and dredge pumps with very large ratios of b_2/d_2 the SAS model is worth a try.

10. Impeller outlet rotor-stator numerical interface location: in order to use a systematic approach, it is suggested to use $r_{interface}/r_2 = (d_3/d_2)^{0.5}$ or $(d_z/d_2)^{0.5}$. This would work for volutes with large gap B. However, if a low-head diffuser pump with, e.g. $d_3/d_2 = 1.025$, is calculated, any interface would cut through the trailing edge vortex. Moreover, the extremely high velocity gradients and low outlet flow angles α_2 impair convergence as well as accuracy.
11. *Mesh size:* Although many publications claim, that "good" agreement between CFD and experiment has been obtained with smaller mesh size, larger y^+ and less stringent set-up parameters, it is suggested that following the requirements stipulated in Table 8.4 should reduce the risk of obtaining fortuitous or manipulated results when calculating new pump applications. The meshing is to be optimized to the specific requirements of the calculation: (1) When calculating the performance of a pump with low n_q, the volute or diffuser losses are dominating, and boundary layers in the stator should be resolved down to $y^+ = 1$. In contrast, the impellers losses are small; and $y^+ = 30$ to 50 may be acceptable in the impeller. (2) For large pumps and large n_q computer limitations will not allow to resolve down to $y^+ < 2$ and larger values will be adopted. (3) When leading edge cavitation is to be resolved, fine grids on the blades are necessary, while coarser meshing may be acceptable in the stator. (4) Highly viscous fluids can use meshes coarser than in water to achieve $y^+ < 2$.
12. Impeller and impeller side rooms (ISR) are calculated in the rotating frame of reference. The relative velocity vector is then following the gridlines to some extent if there is no appreciable flow separation. Inlet and outlet casing are calculated in the stationary domain. When calculating an axial approach flow to the impeller in the rotating frame of reference (RFR), high numerical errors can be introduced. The "alternate rotation model" (ARM) should be used in this case according to [102]. However, when using the ARM, the impeller is calculated in the absolute reference frame: the absolute velocity vectors then deviate strongly from the gridlines which can increase the errors in the impeller calculation. As an alternative, the suction pipe can be linked to the rotating domain by a frozen rotor interface. This may be most appropriate when calculating pumps with radial inlet casings. Many users apply the ARM as standard in spite of said errors which can be introduced when calculation the impeller in the absolute system.
13. Time step: some users start with larger time steps (e.g. 12°) and refine subsequently to about 1° or less.
14. In unsteady calculations, typically 5 to 10 revolutions must be calculated to obtain a *periodic* solution. As the unsteadiness increases at off-design conditions, the required number of revolutions may increase to 20 or more. Once a periodic solution is obtained, the head, power, efficiency and forces are evaluated by averaging these parameters over one or several revolutions. Pressure pulsations and unsteady forces can be evaluated from the periodic solution in the time domain and as spectra.

15. Surface roughness: should be - in principle - accounted for to be able to make meaningful comparison with test data. However, the roughness-models are uncertain. Therefore, most calculations are done assuming hydraulically smooth surfaces.
16. As discussed above, the head is often over-predicted by CFD at q* > 1. The reason could be cavitation occurring in the test in the volute or the diffuser. If c_{pmin} is monitored at these locations, it could be checked that the minimum local pressure at cutwater or vane leading edge does not drop below the vapor pressure. This is done in the single-phase calculation without using any cavitation model.

It should be noted that defining set-up parameters and rules as in Table 8.4 will, most likely, not guarantee a *unique* CFD-solution. It must be suspected that also the *combination* of parameters as well as different mesh topologies will have an influence. In other words: two independent users calculating the same geometry and flow conditions are likely to get different performance predictions.

8.9 Criteria for Assessment of Numerical Calculations

8.9.1 General Remarks

The frequently observed tendency to uncritically believing in computer results increases with the complexity of the program used and the amount of data produced. However, at the present state of the art of CFD applications in turbomachinery design, a critical analysis of CFD results is essential in preventing failure. Only from careful assessment of the calculations and from comparison with test data and experience is it possible to derive criteria and define strategies for the optimization of hydraulic components. The program user may consider the comparison of CFD with test data as her or his permanent task. In order to prevent conclusions from fortuitous results, such comparisons should be made in a statistically relevant number on a broad range of geometries and applications.

3D-Navier-Stokes programs are primarily tools for analyzing an available design. Therefore, the first draft of a new design should be of a reasonable quality. If there are fundamental errors in the first draft, the CFD analysis may not be able to guide the program user to a satisfactory solution. Prior to the calculation a careful check of the grid should be performed according to the criteria given in Chap. 8.3.4.

8.9.2 Consistence and Plausibility of the Calculation

Prior to a detailed assessment of a calculation or even geometry modifications, check the results for consistency and plausibility. The actual criteria for doing so may depend on the program used and the application, but some general hints can be given:

- Did the calculation converge properly? Usually a non-converged solution should be discarded since continuity may be violated - thus falsifying the global performance parameters.
- Is the mass flow rate conserved at the relevant control surfaces?
- Are changes in angular momentum sufficiently close to zero in zones without energy transfer from the blades?
- Are global performance parameters such as ψ_{th} or $\eta_{h,La}$, plausible? Their magnitude can be predicted quite well from conventional performance estimations according to Chap. 3.
- Are the average values of velocities and pressures plausible when compared to streamline performance prediction methods according to Chaps. 3 and 4?
- Is the pre-swirl induced by the impeller excessive?
- Backflow through the control surface at the outlet of the calculation domain? Even though such backflow may be prevented by a conical reduction of the flow area towards the outlet, the reasons for such anomalies should be investigated.

8.9.3 *Will the Specified Performance Be Reached?*

For designing an impeller, the head is usually specified. Estimating the hydraulic efficiency from Chap. 3.9 the theoretical work Y_{th} can be calculated. If this target is not reached in the CFD calculation, blade angles or the impeller outlet width can be modified until the impeller design is able to produce the specified head. If the calculated losses suggest that the estimated hydraulic efficiency may not be obtained, the target for Y_{th} can be adapted accordingly.

8.9.4 *Maximization of the Hydraulic Efficiency*

Minimizing the hydraulic losses in impeller and collector is one of the main goals of applying CFD. When developing appropriate criteria for assessing the losses predicted by CFD, it should be recognized that the hydraulic losses in many cases account for 7 to 15% of the power absorbed at the coupling. Possible improvements are likely to be in the order of 1 to 3% of said power. The potential for improving the impeller is only a fraction of this. If CFD is to capture efficiency improvements of typically below one per cent, extreme care and accuracy are required when performing and analyzing CFD calculations.

Consider Eq. (8.1) for assessing the losses: Since the velocity magnitudes are given by continuity and the kinematics of the flow around the blades, losses fundamentally manifest themselves in a reduction of the static pressure. Furthermore, viscous losses due to skin friction in fully turbulent flow are small compared to the losses caused by turbulent exchange of momentum. Consequently, it should be expected that the losses depend strongly on the selected turbulence model and wall treatment and that the static pressure is falsified if the losses are not predicted correctly. Navier-Stokes calculations therefore often predict the theoretical head quite well, while the static pressure rise (and thus total head and losses) deviates much more from test data.

In radial and semi-axial impellers a low-energy zone („wake“) forms in the front shroud suction side corner due to the curvature of the meridional section and the pressure distribution on the blades. The c_m- and c_u- or w-distribution as well as a graphic representation of the local loss coefficient analogous to Eq. (8.13) can reveal zones of high losses. Some authors analysed the local generation of entropy for showing locations with high losses, e.g. [112] and [114]. The following criteria may be of help for optimizing an impeller design and assessing losses:

1. Geometrical parameters and flow quantities should vary as smoothly as possible over the width and the length of the blades, [7]. Strong gradients cause secondary flows and additional losses.
2. Velocity and pressure distributions along the blades are plotted for different streamlines (e.g. outer, mean and inner streamline). Either the static pressure or the "reduced pressure" $p_{red} = p - ½\rho \times u^2$ is used for the purpose. The following criteria are relevant:

   - Sharp low-pressure peaks may be due to high incidence, unsuitable leading-edge profiles or blade angle developments.
   - Local excess velocities, unnecessary acceleration or deceleration, high gradients induce additional losses. Such deficiencies should be removed by modifying the blade development and/or the meridional section.
   - Well-known blade loading criteria such as "de Haller", "Lieblein", or similar should not be exceeded in order to reduce the risk of flow separation and additional losses [7] - unless reliable tests with similar impellers are available which prove that higher blade loadings can be tolerated.
3. Additional losses are created by excessive secondary flows which can be visualized by specific graphic programs. Appreciable pressure gradients perpendicular to the main flow direction also indicate strong secondary flows and may be used for quantifying such effects, [70].
4. Non-uniformities in the velocity distribution which are generated at the impeller inlet are *reinforced* within the impeller since the flow is turned and decelerated.
5. At a first glance, the c_{2m}-distribution may appear as insignificant because c_{2m} amounts to only a fraction of the velocity vector and the kinetic energy. However, the c_{2m}-distribution is actually the most sensitive indicator for the risk of flow separation and recirculation, as discussed in Chap. 5 and confirmed by the c_{2m} and c_{2u}-distributions in Fig. 5.30. Also, the low-energy zone ("wake") is easier to recognize by a c_{2m}-deficit than by considering the c_{2u}-profile, because the flow deflection in the wake can be similar to the rest of the

channel. Since the energy transfer is proportional to the product $c_m \times c_u$, a deficit remains though. A non-uniform c_{2m}-distribution usually implies a variation of the discharge flow angle α_2 and associated shock losses over the width of the channel. The impeller design should therefore aim at a uniform c_{2m}-distribution.

6. A uniform c_{2u} or w_{2u}-distribution is required in order to transfer the same energy on all streamlines. If this criterion is violated, the static pressure varies over the impeller width. Since high differences in static pressure cannot be maintained when the trailing edge is parallel to the rotor axis, secondary flow is enhanced and losses increase. Curved streamlines at the impeller outlet are an indication of secondary flows and should therefore be avoided by modifying the design. Conversely, secondary flow reduces the pressure differences between pressure and suction surface of the blade. Presumably there is an optimum amount of secondary flow in an impeller, albeit no criteria are available for quantifying said amount.
7. Flow separation generates losses, pressure pulsations and hydraulic excitation forces and should therefore be avoided at the design flow rate.
8. Recirculation or strongly distorted velocity profiles at the impeller outlet should be avoided at the design flow rate; they may indicate that the impeller outlet width was selected too large and/or the curvature of the meridional section is too strong. As a rule, flow recirculation at the impeller inlet at BEP should not be allowed either.
9. Obviously, friction losses on excessively large wetted surfaces are to be avoided; but the margin for optimization is quite small if the main parameters for the first design are selected properly based on previous tests and experience. The impact of friction losses in impellers is expected in general to be quite small: at low specific speed the losses in the diffuser or volute dominate, while mixing losses prevail at high n_q due to the non-uniform velocity profiles.
10. Compare calculated velocity and pressure distributions with those of tested impellers (examples with high and low efficiency, stable and unstable Q-H-curves are equally revealing).

Even though the pressure rise in the diffuser or volute of a radial pump only amounts to about 25% of the total head, roughly 40 to 70% of the hydraulic losses are generated in the collector. As stressed above, the velocity distribution at the impeller outlet has a strong impact on these losses. This effect can be taken into account in the impeller design if the numerical procedure used allows a reliable calculation of the impeller discharge flow distribution.

The design criteria given in the following for the diffuser can be applied to volutes as well - recognizing however that volutes are less sensitive than vaned diffusers because of the larger cutwater distance and because incidence effects are weaker (only one or two cutwaters at a large gap B).

1. Non-uniformity in the impeller outlet flow generates mixing losses in the collector. The pressure recovery is impaired due to variations in local incidence

(possibly inducing stall) and excess velocities inducing losses due to exchange of momentum, Chap. 5.3.3.
2. Non-uniformities in the velocity and pressure distribution at the diffuser inlet are *reinforced* within the diffuser. They cannot equalize because the flow is decelerated and possibly turned: locally thick boundary layers at the inlet will grow excessively as the flow is further decelerated.
3. The general criteria given above for the impeller are largely applicable to diffusers too.

8.9.5 Stability of the Head-Capacity Curve

As discussed extensively in Chap. 5.3, the pressure rise in a diffuser is accomplished at partload mainly in the triangular inlet section (this phenomenon was measured in radial pumps as well as in a semi-axial pump with $n_q = 160$). It was further demonstrated that the inlet flow boundary conditions exert an essential influence on the pressure recovery. Both of these effects are crucially important for the head generated at partload and for the stability of the Q-H-curve. Since flow recirculation can occur in an unsuitable design even near the best efficiency point, the impeller discharge velocity profile must be analyzed over the whole operation range. In order to achieve a stable Q-H-curve, the velocity distribution at the impeller inlet (and within the impeller) must be "designed" so as to develop continuously over the flow rate and, in particular, to avoid sudden switching of the flow patterns, Chap. 5.2. This task can be accomplished only by means of numerical flow calculations which have been validated as to predict reliably the flow distributions in impellers. Velocity profiles which are favorable for achieving maximum efficiencies may not necessarily be best for the stability of the Q-H-curve. Stage calculations are needed for investigating the partload characteristics, but present methods need improvement and extensive validation.

8.9.6 Unsteady Forces

While the *circumferentially* (pitch-wise) averaged c_{2m} and c_{2u}-distribution has been found to be highly relevant for steady performance and in particular the stability of the Q-H-curve, the *span-wise* averaged distribution of the impeller outlet velocity vector determines the rotor/stator interaction, i.e. pressure pulsations, hydraulic excitation forces and noise, Chaps. 10.1 and 10.7.1. Span-wise averaging allows a better judgment and quantification of these effects than just the presentation of the 3D velocity distribution, which is difficult to interpret. The span-wise averaged c_2 may be referred to the mean outlet velocity determined from conservation of angular momentum; $c_{2,max,min}/c_{2,average}$ can then be used as a measure for comparing different design options.

In order to reduce hydraulic excitation forces at the source, the following design options may be considered (thereby keeping the design pressure coefficient constant); refer also to Chap. 10.11.3:

- lower blade loading towards the impeller trailing edge
- profiling of the impeller blade trailing edge on the pressure surface
- blade twist ("rake")
- optimize relative impeller outlet width b_2/d_2
- optimize shape of meridional impeller section
- improve impeller approach flow conditions.

In a first step the geometric modifications may be tried for the design point calculating the impeller only. If the modifications appear to be successful, unsteady stage calculations over the entire operation range follow in a second step.

8.10 Fundamental Considerations on CFD-Calculations

As discussed above, numerical flow calculations pose a number of difficult questions which are not yet fully answered. Some fundamental issues are emphasized therefore in the summary below.

In spite of the broad application of CFD programs, the development of methods is by no means completed. The deficiencies of the prediction of losses (problems of turbulence modeling, wall treatment and roughness) represent, at the present, the most severe restrictions since the correct loss prediction is a prerequisite of a true component optimization.

There is not even consensus as to the methods which are optimum for the different applications. Apart from a wealth of publications which propagate the use of 3D-NS calculations, other contributions show that non-viscous programs - at least in the considered application - can yield similar results. This apparent contradiction has physical reasons:

1. According to Chap. 1.4 pressure gradients are generated in flows on curved paths. The pressure increase in an impeller is thus primarily determined by kinematic conditions (i.e. geometry). Friction effects are of secondary importance.
2. As discussed in Chap. 5.2, the equilibrium of forces is largely determined by body forces (centrifugal and Coriolis forces). Again, friction effects exert an indirect (rather moderate) impact in causing boundary layers and influencing secondary flow.
3. Body forces and the kinematics of the flow over a blade are captured *in the same way* by non-viscous flow as by Navier-Stokes calculations. In all applications where these effects largely determine the flow behavior both methods (if

applied correctly) will produce similar pressure distributions on the blades, global flow deflection and blade work Y_{th}.

4. The hydraulic impeller efficiencies at the design point are in the range of $\eta_{h,La} = 0.97 \pm 0.02$ at $n_q = 10$ and with an axial pump of $n_q = 250$ in the range of $\eta_{h,La} = 0.95 \pm 0.02$. Even a correct non-viscous impeller calculation is within a few per cent close to a correctly evaluated test. The same applies, of course, to a Navier-Stokes calculation - quite independently of the selected turbulence model and wall treatment (unless the calculation is affected by blatant errors).
5. Non-viscous methods are therefore a valid option for the *design* of an impeller according to the criteria "equilibrium of forces and pressure distribution"; especially since the hydraulic impeller efficiency can be estimated with acceptable accuracy as per item 4 above. All the more so, if the first draft of the impeller is free from flow separation and serious shortcomings.
6. Considering the dominance of body forces and blade kinematics as well as the level of the hydraulic impeller efficiencies, and considering the statistical distribution of the various numerical and modeling errors, it becomes clear why calculated CFD results and test data quite often can be found within a band width of about ±5% (possibly up to 10%), so that pertinent publications accordingly claim "good agreement".
7. While the flow through the impeller always follows a strongly curved flow path in the absolute system (even with radial blades as per Fig. 7.14), flow conditions are fundamentally different in the diffuser: body forces are absent and streamline curvature is low (or zero) in many cases. Two extremes may be distinguished in diffuser design: (a) in straight diffusers or diffusers with low streamline curvature the flow is controlled mainly by boundary layer effects. Non-viscous methods are bound to fail. (b) In strongly curved diffuser channels (such as used in semi-axial or axial pumps for turning the flow into axial direction) the pressure distribution on the vanes is to a large extent given by kinematic conditions (geometry). Non-viscous methods can therefore be considered for the design of the vanes - within the limitations discussed below.
8. In principle, diffuser calculations without correctly specifying the inlet velocity distribution and without loss model are subject to severe restrictions which should be fully appreciated when using non-viscous programs.
9. By definition, hydraulic losses can be captured only by viscous programs. As discussed extensively above, loss predictions by Navier-Stokes programs are subject to considerable uncertainty. Therefore, Euler methods combined with a boundary layer model which makes use of the extensive boundary layer research offer an alternative for diffuser design and analysis near BEP flow. These methods are also able to recognize flow separation. They may be used for diffusers with more than 30° flow turning (and impellers); they may not be used for diffusers with straight or slightly curved streamlines.
10. The velocity distribution near walls, stall and secondary flow cannot be handled by non-viscous methods. As discussed in Chap. 5, saddle-type instabilities of Q-H-curves are influenced by unstable velocity distributions. Non-viscous

methods are therefore unsuitable for partload calculations and investigations into the stability of flow patterns.

11. Since the pressure distribution on a blade, global flow deflection and specific work done by the blades are independent of the density, the above discussion applies equally to cavitation and 2-phase flow, as far as the methods rely in essence on calculations of the pressure distribution and neglect the interaction between the phases.

References

1. Balasubramanian, R., Bradshaw, S., Sabine, E.: Influence of impeller leading edge profiles on cavitation and suction performance. In: Proceedings of 27th International Pump Users Symposium, Texas A&M, pp. 34–44 (2011)
2. Bartsch, P.: Numerische Untersuchung der Leitrad-Laufrad-Wechselwirkungen in axialen Kreiselpumpen. Diss., TU Berlin (1994)
3. Bissig, M., Staubli, T.: Numerische Berechnung der Fluid-Rotorinteraktion im Radseitenraum von Hydromaschinen. In: VDI-Tagung Fluid-Struktur-Wechselwirkung, Heidelberg (2002)
4. Bouziad, A.Y., Farhat, M., Guennoun, F., Kueny, J.L., Avellan, F.: Physical modeling and simulation of leading edge cavitation, application to an industrial inducer. In: 5th International Symposium on Cavitation, Osaka, Cav03-Os-6-014 (2003)
5. Braun, O.: Part load flow in radial centrifugal pumps. Diss., EPF Lausanne (2009)
6. Casey, V.M.: Computational methods for preliminary design and geometry definition in turbomachinery. In: Nato AGARD Lecture Series 195 (1994)
7. Casey, V.M.: The industrial use of CFD in the design of turbomachinery. In: Nato AGARD Lecture Series 195 (1994)
8. Casey, V.M. et al.: Flow analysis in a pump diffuser. Part 2: Validation of a CFD code for steady flow. ASME FED **227**, 135–143 (1995)
9. Cebici, T.: Turbulence Models and their Application. Springer, Berlin (2004). ISBN 3-540-40288-8
10. Chen, C.J., Patel, V.C.: Near-wall turbulence models for complex flows including separation. AIAA J. **26**, 641–648 (1988)
11. Cheng-I, Y.: A simulation of viscous incompressible flow through a multiple-blade-row turbomachinery. ASME FED **227**, 11–18 (1995)
12. Coleman, H.W.: Some observations on uncertainties and the verification and validation of simulations. ASME J. Fluids Eng. **125**, 733–735 (2003)
13. Combes, J.F., et al.: Numerical investigation of the rotor-stator interaction in a centrifugal pump using a finite element method. ASME FEDSM97-3454 (1997)
14. Combes, J.F., et al.: Numerical and experimental analysis of the flow in a centrifugal pump at nominal and partial flow rate. ASME Paper 92-GT–284
15. Cooper, P., Graf, E.: Computational fluid dynamical analysis of complex internal flows in centrifugal pumps. In: Proceedings of the 11th International Pump Users Symposium, Houston, pp. 83–93 (1994)
16. Cugal, M., Baché, G.: Performance prediction from shutoff to runout flows for a complete stage of a boiler feedpump using CFD. ASME FEDSM97-3334 (1997)
17. Dawes, W.N.: A simulation of the unsteady interaction of a centrifugal impeller with its vaned diffuser: flow analysis. ASME J. Turbomach. **117**, 213–221 (1995)
18. Dupont, P.: Numerical prediction of cavitation in pumps. In: Proceedings of 18th International Pump Users Symposium, Texas A&M, pp. 59–65 (2001)

19. Dupont, P., Casartelli, E.: Numerical prediction of the cavitation in pumps. ASME FEDSM2002–31189
20. Durbin, P.A., et al.: Rough wall modification of two-layer k-ε. ASME JFE **123**, 16–21 (2001)
21. ERCOFTAC Special interest group on "Quality and trust in industrial CFD". Best practice guidelines. http://www.ercoftac.org
22. Favre, J.N.: Resolution du problème inverse par petites perturbations d'un écoulement potentiél incompressible. Diss., EPF Lausanne (1988)
23. Ferziger, J.H.: Review: simulation of incompressible turbulent flows. J. Compo Phys. **69**(1), 1–48 (1987)
24. Ferzinger, J.H., Peric, M.: Computational Methods for Fluid Dynamics. Springer, Berlin (1997)
25. Franc, J.P., et al.: La Cavitation. Mechanismes physiques et aspects industriels. Presses Universitaires Grenoble (1995)
26. Fraser, S.M., et al.: Improved k-ε-modeling of impeller flow performance of a mixed-flow pump under off-design operating states. Proc. IMechE. **207**, 219–229 (1993)
27. Freitas, C.J.: Engineering editorial policy statement on the control of numerical accuracy. ASME. J. Fluids Eng. **115**, 339–340 (1993)
28. Freitas, C.J.: Perspective: selected benchmarks from commercial CFD codes. ASME. J. Fluids Eng. **117**, 208–218 (1995)
29. Ginter, F.: Berechnung der instationären, turbulenten Strömung in hydraulischen Strömungsmaschien. Diss TU Stuttgart, 1997, Mitteilung Nr 12 Inst für Strömungsmechanik und hydraulische Maschinen
30. Ginter, F., Staubli, T.: Performance discontinuity of a shrouded centrifugal pump impeller. In: IMech Conference, pp. 1027–1049 (1999)
31. Ginter, F., et al.: Entwicklung eines Pumpenzulaufkrümmers mit Hilfe der Strömungsberechnung. Pumpentagung Karlsruhe B5 (1992)
32. Goto, A.: Study of internal flows in a mixed-flow pump impeller at various tip clearances using 3D viscous flow computations. ASME Paper 90-GT–36
33. Goto, A.: Numerical and experimental study of 3D flow fields within a diffuser pump stage at off-design condition. ASME FED **227**, 1–9 (1995)
34. Göde, E.: 3-dimensional flow simulation in a pump-turbine. ASME FED **86**, 29–34 (1989)
35. Graf, E.: Analysis of centrifugal impeller BEP and recirculating flows: comparison of quasi-3D and Navier-Stokes solutions. ASME Pumping Machinery Symposium, pp. 235–245 (1993)
36. Graf, E. et al.: Three-dimensional analysis in a multi-stage pump crossover diffuser. ASME Winter Annual Meeting, pp. 22-29 (1990)
37. Greim, R., et al.: Berechnung dreidimensionaler Strömung in Pumpenlanfrädem. Pumpentagung Karlsruhe B6 (1992)
38. Guide for verification and validation of computational fluid dynamics solutions. AIAA Guide G-077-l998, http://www.aiaa.org
39. Gugau, M.: Beitrag zur Validierung der numerischen Berechnung von Kreiselpumpen. Diss., TU Darmstadt (2004)
40. Gülich, J.F.: Berechnung von Kreiselpumpen mit Navier-Stokes-Verfahren - aus der Sicht des Anwenders. Forsch. Ing. Wes. **60**, 307–316 (1994)
41. Gülich, J.F., Favre, J.N., Denus, K.: An assessment of pump impeller performance predictions by 3D-Navier-Stokes calculations. ASME FEDSM97-334l (1997)
42. Hansen, T.: Comparison of steady-state and transient rotor-stator interaction of an industrial centrifugal pump. In: CFX Users Conference (2001)
43. Hildebrandt, T.: Weiterentwicklung von 3D Navier-Stokes-Strömungsrechenverfahren zur Anwendung in hochbelasteten Verdichter- und Turbinengittern. Diss., Universität der Bundeswehr, München (1998)
44. Hirschi, R.: Prédiction par modélisation numerique tridimensionelle des effects de la cavitation à poche dans les turbomachines hydrauliques. Diss., EPF Lausanne (1998)

45. Holbein, P.: Berechnung dreidimensionaler reibungsbehafteter inkompressibler Innenströmungen. Diss., TU Hannover (1993)
46. Howard, J.H.G. et al.: Flow analysis in a spiral inducer impeller. ASME Paper 93-GT–227
47. Iaccarino, G.: Predictions of a turbulent separated flow using commercial CFD codes. ASME. J. Fluids Eng. **123**, 819–828 (2001)
48. Kamemoto, K., et al.: Analysis of unsteady characteristics of flows through a centrifugal pump impeller by an advanced vortex method. IAHR Symposium, Valencia, pp. 729–738 (1996)
49. Kaps, A.: Numerische Untersuchung der Strömung in einer radialen Kreiselpumpe mit dem Ziel einer wirkungsgrad- und lagerkraftoptimierten Gehäusegestaltung, Diss., TU Berlin (1996)
50. Kaupert, K.A.: Unsteady flow fields in a high specific speed centrifugal impeller. Diss., ETH Zürich (1997)
51. Kato, M., Launder, B.E.: The modeling of turbulent flows around stationary and vibrating square cylinders. In: 9th Symposium on Turbulent Shear Flows, Kyoto, paper 10-4 (1993)
52. Kubota, A., Kato, H., Yamaguchi, H.: Finite difference analysis of unsteady cavitation on a two-dimensional hydrofoil. In: 5th International Conference on Numerical Ship Hydrodynamics, Hiroshima (1989)
53. Lakslnninarayana, B.: An assessment of computational fluid dynamic techniques in the analysis of turbomachinery. ASME. J. Fluids Eng. **113**, 315–352 (1991)
54. Liu, W.: Modeling of swirling turbulent flows. Diss., TU Stuttgart (2001)
55. Mack, R., Drtina, P., Lang, E.: Numerical prediction of erosion on guide vanes and in labyrinth seals in hydraulic turbines. Wear, pp. 233–235, 685–691 (1999)
56. Majidi, K.: Numerische Berechnung der Sekundärströmung in radialen Kreiselpumpen zur Feststoffförderung. Diss., TU Berlin (1997)
57. Menter, F.R.: A comparison of some recent eddy-viscosity turbulence models. Transactions ASME **118**, 514–519 (1996)
58. Meier-Grotian, J.: Untersuchungen der Radialkrft auf das Laufrad einer Kreiselpumpe bei verschiedenen Spiralgehäuseformen. Diss., TU Braunschweig (1972)
59. Meschkat, S., Stoffel, B.: The local impeller head at different circumferential positions in a volute casing of a centrifugal pump in comparison to the characteristic of the impeller alone. In: 21st IAHR Symposium on Hydraulic Machinery and Systems, Lausanne (2002)
60. Muggli, F., Holbein, P., Dupont. P.: CFD calculation of a mixed flow pump characteristic from shut-off to maximum flow: ASME FEDSM2001-18072 (2001)
61. Noll, B.: Numerische Stromungsmechanik. Springer, Berlin (1993)
62. Oberkampf, W.L., Trucano, T.G.: Validation methodology in computational fluid dynamics. AIAA paper 2000-2549 (2000)
63. Oberkampf, W.L., Trucano, T.G.: Verification and validation in computational fluid dynamics. Sandia National Laboratories report 2002-0529 (2002)
64. Oertel, J.R.H., Laurien, E.: Numerische Strömungsmechanik. Springer, Berlin (1995)
65. Roache, P.J.: Verification and validation in computational science and engineering. Hermosa, Albuquerque, http://www.hermosa-pub/hermosa (1998)
66. Rodi, W.: Turbulence modelling for incompressible flows. Phys. Chem. Hydrodyn. **7**(5/6), 297–324 (1986)
67. Rodi, W.: Turbulence Models and Their Application in Hydraulics, 3rd edn. Balkema, Rotterdam (1993)
68. Rütten, F.: Large eddy simulation in 90°-pipe bend flows. J. Turbul. **2**, 003 (2001)
69. Schachenmann, A., Gülich, J.F.: Vergleich von drei Navier-Stokes Berechnungsverfahren mit LDA-Messungen an einem radialen Pumpenlaufrad. Pumpentagung Karlsruhe B7 (1992)
70. Schachenmann, A., et al.: Comparison of 3 Navier-Stokes codes with LDA-measurements on an industrial radial pump impeller. In: ASME Fluids Engineering Conference, Los Angeles (1992)
71. Schäfer, M.: Numerik im Maschinenbau. Springer, Berlin (1999)

72. Schilling, R.: A critical review of numerical models predicting the flow through hydraulic machinery bladings. In: 17th IAHR Symposium, Beijing, GL2 (1994)
73. Schilling, R.: Stand der numerischen Strömungssimulation bei hydraulischen Turbomaschinen. Festschrift zum Jubiläum 100 Jahre Turbomaschinen und 50 Jahre Fluidantriebstechnik an der TU Darmstadt (1997)
74. Schönung, B.E.: Numerische Strömungsmechanik. Springer, Berlin (1990)
75. Song, C.C.S. et al.: Simulation of flow through Francis turbine by LES method. In: IAHR Symposium, Valencia. pp. 267–276 (1996)
76. Song, C.C.S., et al.: Simulation of flow through pump-turbine. In: IAHR Symposium, Valencia (1996)
77. Staubli, T., et al.: Verification of computed flow fields in a pump of high specific speed. ASME FED **227**, 75–82 (1995)
78. Staubli, T., Bissig, M.: Numerical parameter study of rotor side spaces. In: 21st IAHR Symposium Hydraulic Machinery and systems, Lausanne (2002)
79. Steinmann, A.: Numerische und experimentelle Untersuchung der ein- und zweiphasigen Strömung in einem technisch belüfteten Abwasserteich. Diss TU Berlin (2002)
80. Stem, F., et al.: Comprehensive approach to verification and validation of CFD calculations Part 1: Methodology and procedures. ASME. J. Fluids Eng. **123**, 793–802 (2001)
81. Tanabe, S., et al.: Turbulent flow analysis in a pump impeller. ASME Fluid Mach. Forum FED **119** 1–6 (1991)
82. Torbergsen, E., White, M.F.: Transient simulation of impeller/diffuser interactions. ASME FEDSM97-3453 (1997)
83. Tremante, A., et al.: Numerical turbulent simulation of the two-phase flow (liquid/gas) through a cascade of an axial pump. ASME FEDSM2001–18086
84. Treutz, G.: Numerische Simulation der instationären Strömung in einer Kreiselpumpe. Diss., TU Darmstadt (2002)
85. Visser, F.C.: Some user experience demonstrating the use of CFD for cavitation analysis and head prediction of centrifugal pumps. ASME FEDSM2001–18087
86. Watzelt, C., et al.: Real-time design of hydraulic machinery bladings on a parallel environment system. ASME FED **227**, 45–51 (1995)
87. Chen, W.-C., et al.: CFD as a turbomachinery design tool: code validation. ASME FED **227**, 67–74 (1995)
88. Wilcox, D.C.: Turbulence Modeling for CFD. DCW Industries, La Canada, California (1998)
89. Wu, C.H.: A general theory of three-dimensional flow in subsonic and supersonic turbomachines of axial, radial and mixed-flow types. Trans. ASME **74**, 1363–1380 (1952)
90. Zangeneh, M., Goto, A.: Turbodesign: next generation design software for pumps. World Pumps, pp 32–36 (2003 Feb)
91. Will, B.C., Benra, F.K., Dohmen, H.J.: Investigation of the flow in the impeller side clearances of a centrifugal pump with volute casing. J. Therm. Sci. **21**(3), 197–208 (2012)
92. Juckelandt, K., Wurm, F.H.: Applicability of wall-function approach in simulations of turbomachinery. GT 2015-42014
93. Limbach, P., Skoda, R.: Numerical simulation and experimental validation for NPSH prediction of a low-specific speed centrifugal pump with a variation of cavitation model mass transfer parameters. In: 11th Pump User's International Forum Düsseldorf (2016)
94. Juckelandt, K., Wurm, F.H.: Experimental and numerical analysis of a low-specific speed pump. In: 11th Pump User's International Forum (2016)
95. Höller, S., et al: Low-specific speed mixed-flow API pump for single- and multistage usage. In: 11th Pump User's International Forum (2016)
96. Nishi, Y., et al.: Effect of blade outlet angle on radial thrust of single-blade centrifugal pump. In: 26th IAHR Symposium Hydraulic Machinery and Systems (2012)
97. Berten, S., et al.: Experimental and numerical analysis of pressure pulsations and mechanical deformations in a centrifugal pump impeller. ASME AJK2011-06057

98. Feng, J., Benra, F.K., Dohmen, H.J.: Numerical investigation on pressure fluctuations for different configurations of vaned-diffuser pumps. Int. J. Rotating Mach. (2007)
99. Miyabe M et al: Rotating stall behavior in a diffuser of a mixed-flow pump and its suppression. ASME FEDSM-2008-55132
100. Schiffer, J., et al.: Performance analysis of a single-blade impeller based on unsteady 3D numerical simulation. In: 11th Pump User's International Forum Düsseldorf (2016)
101. Lechler, S.: Numerische Strömungsberechnung. Springer, Berlin (2018)
102. ANSYS CFX Theory Guide (internet) www.sharcnet.ca/Software/Ansys/16.2.3
103. Chatagny, L., Berten, S.: Challenges and open questions in cavitation simulations on centrifugal pump applications. In: 11th Pump User's International Forum Düsseldorf (2016), L01-046
104. Liu, H., Ding, J., et al.: Numerical research on hydraulically generated vibration and noise of a centrifugal pump volute with impeller outlet width variation. Math. Probl. Eng., article ID 620389 (2014). Hindawi
105. Liu, H., et al.: Investigation into transient flow in a centrifugal pump with wear ring clearance variation. Adv. Mech. Eng., article ID 693097 (2014). Hindawi
106. Wang, Y., et al.: Applicability of turbulence models to low-specific speed centrifugal pump. In: 26th IAHR Symposium (2012)
107. Spence, R., Amaral-Teixeira, J.: Investigation into pressure pulsations in a centrifugal pump using numerical methods supported by industrial tests. Computers & Fluids **37**, 690–704 (2008)
108. Roclawski, H.: Numerische und experimentelle Untersuchungen an einer radialen Kreiselpumpenstufe mit minimalem Stufendurchmesser. Diss., TU Kaiserslautern (2008)
109. Si, Q., et al.: Numerical investigation of pressure fluctuations in a centrifugal pump volute based on SAS model an experimental validation. Adv. Mech. Eng. (2014 Feb)
110. Roclawski, H., Hellmann, D.H.: Numerical simulation of a radial multistage centrifugal pump. AIAA 2006-1428
111. Menter, F.R.: Best practice: scale-resolving simulations in ANSYS CFD. Version 2, 2015
112. Li, X., et al.: Experimental and numerical investigations of head-flow curve instability of a single-stage centrifugal pump with volute casing. Proc. IMechE Part A: J. Power Energy (2016)
113. Denton, J.D.: Some limitations of turbomachinery CFD. ASME GT2010-22540
114. Fleder, A.: Numerische und experimentelle Untersuchung des Einflusses wichtiger Geometrieparameter auf die Performance von Seitenkanalpumpen unter Berücksichtigung der Strömungsverluste. Thesis, TU Kaiserslautern (2015)
115. Fecarotta, O., et al.: Preliminary development of a method for impact erosion prediction in pumps running as turbines. MDPI Proc. **2**, 680 (2018)

Chapter 9
Hydraulic Forces

Abstract The pressure rise in the impeller generates hydraulic forces and moments which act on the rotor. In particular forces in axial and radial direction are significant for appropriate sizing of shaft and bearings. While the radial force is determined by the pressure distribution *around the impeller circumference*, the axial force is governed by the flow through the impeller sidewall gaps and the resulting pressure distributions on the shrouds.

Because of its high relevance in the determination of the axial forces acting on turbo machine rotors there is a wealth of publications dealing with the subject of flow through the impeller sidewall gaps. A comprehensive collection of relevant quotations can be found in [19].

Note: The (classical) discussion of the flow phenomena in the impeller sidewall gaps (or the "impeller side rooms" ISR) in Chaps. 9.1 and 9.2 is based on the assumption that the flow in the ISR is *rotationally symmetric*. This is largely true for diffuser pumps and the near the BEP of volute pumps. It is not true for volute pump operating at off-design conditions as discussed in Chap. 9.3.3.

Chapter 9 deals mostly with steady forces, while unsteady flow phenomena, excitation forces and their impact on noise and vibrations are the topic of Chap. 10. Synchronous radial forces created in single-channel impellers are discussed in Chap. 9.3.9.

9.1 Flow Phenomena in the Impeller Sidewall Gaps

For reasons of mechanical design, axial clearances are required between the shrouds of a closed impeller and the casing (called "impeller sidewall gaps" or "impeller side rooms" ISR). Width and shape of the resulting liquid-filled spaces between impeller and casing are essentially determined by aspects of the mechanical design as influenced by the standardization of components and manufacturing and material costs.

J. F. Gülich, *Centrifugal Pumps*,
https://doi.org/10.1007/978-3-030-14788-4_9

The fluid contained in the impeller sidewall gaps cannot be at rest when the impeller rotates: immediately at the shroud the fluid adheres to the solid wall and has thus the velocity $c_u = \omega \times r$. A boundary layer is formed in which the tangential velocity drops with increasing distance from the shroud. The fluid also adheres to the casing wall where the velocity is zero $c_u = 0$. The velocity increases in the casing wall boundary layer, Fig. 9.1; definitions in Fig. 9.2.

With tight sidewall gaps or low Reynolds numbers both boundary layers are merged; with wide gaps they are separated and a *core flow* is formed. The flow in the impeller sidewall gap can be laminar or turbulent. In most applications with water pumping the flow is turbulent and the boundary layers are separated as depicted in Fig. 9.1; only this case is treated in the following. In highly viscous flow the fluid in the ISR is heated by the disk friction as discussed in Chap. 13.1.

The centrifugal forces in the rotating boundary layer generate a fluid transport radially outwards. Thus the shroud acts in a similar way as a friction pump (Chap. 3.10.3). For reasons of continuity, fluid flows back radially inwards along the casing wall. In the meridional section a circulating flow is thus established in the impeller sidewall gaps. Figure 9.1 shows the corresponding velocity distributions in radial and circumferential direction. Usually the radial velocities are much smaller than the circumferential components prevailing in the core flow.

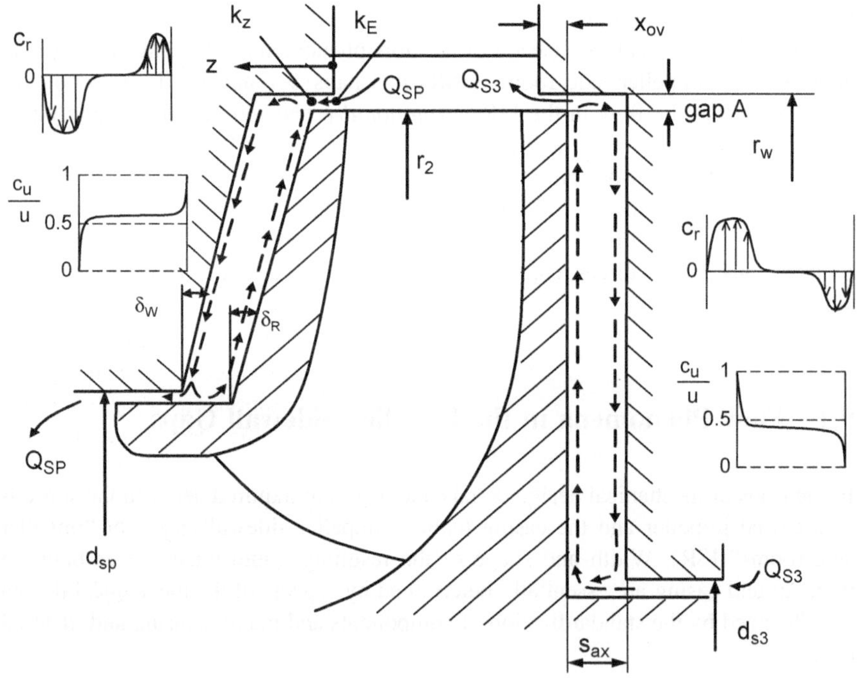

Fig. 9.1 Velocity distribution in the impeller sidewall gaps

Due to seal leakages, a net flow through the impeller sidewall gap can be superimposed on the circulating flow induced by the impeller rotation. Such a leakage is virtually always present on the *front shroud*; it flows along the casing wall *radially inwards from the outer radius to the seal ring*. This leakage carries the angular momentum $\rho \times Q_{sp} \times c_{2u,FS} \times r_2$ into the impeller sidewall gap and thus enhances the fluid rotation ($c_{2u,FS}$ is the *local* tangential velocity near the front shroud at the impeller outlet). If no friction were present, the leakage would conserve its angular momentum $c_u \times r$ = constant when flowing radially inwards. In viscous flow the rotation is smaller, but it will still grow when flowing inwards if $k_E = c_{2u,FS}/u_2 > 0.5$ (Fig. 9.1).

On the *rear shroud* three cases must be distinguished:

1. If the pump is designed with an annular seal and balance holes (Figs. 2.1 and 9.15 in Table 9.1), flow conditions are very similar as on the front shroud. The same is true for the last stage of multistage pumps with central axial thrust balancing (Fig. 9.18).
2. In pumps without balance holes (Fig. 9.3) the shaft seal prevents any net flow through the impeller sidewall gap; only the circulating flow described above is present on the rear shroud.
3. In multistage pumps with a central axial force balance device, a leakage Q_{s3} flows through the inter-stage seal. It runs through gap E along the rear shroud *radially outwards from the seal ring to the diffuser inlet*, Fig. 9.1. This leakage carries only little angular momentum into the sidewall gap. The leakage flow rate is *accelerated* in tangential direction by the shear stresses on the rotating shroud. Interstage seal leakage therefore *slows down* the fluid rotation in gap E. The leakage flow rate establishes itself in a way that the pressure decay in the impeller sidewall gap plus the pressure difference across the interstage seal equals the pressure recovery in the diffuser, fulfilling Eqs. (T3.7.4 and 3.7.6).

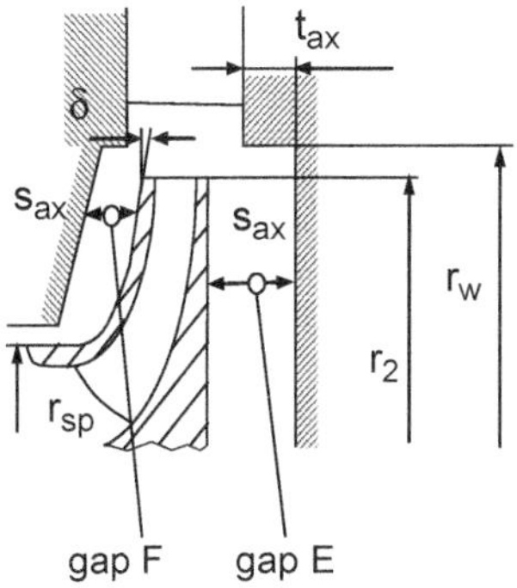

Fig. 9.2 Definitions

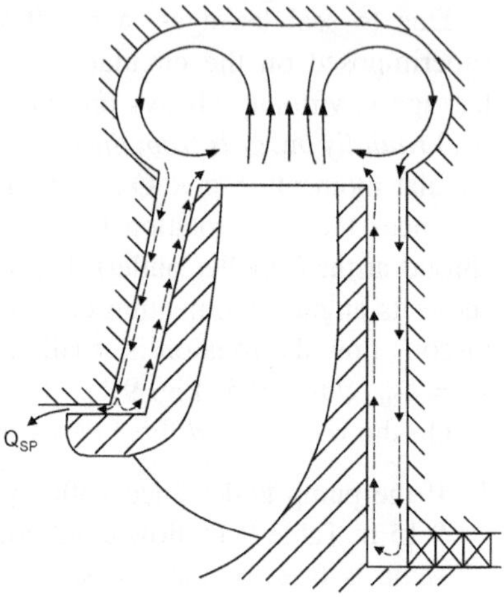

Fig. 9.3 Influence of secondary flows in the volute on the impeller sidewall gaps

Depending on the design, the main flow at the impeller outlet and the flow in the impeller sidewall gap are more or less coupled. If gap A (Fig. 9.1) is tight and the overlap x_{ov} is large, both flows are largely decoupled. This is particularly true for radially outward leakage flow. If the leakage is radially inwards, some angular momentum is carried into the sidewall gap (as given by $c_{2u,FS}$), even if gap A is relatively small; the rotation of the fluid is thus enhanced. The tangential velocity changes in magnitude from k_E to k_z due to shear stresses on the stationary diffuser sidewall and the rotating shroud, Fig. 9.1. These friction effects can be estimated by means of Eq. (T9.1.17) or Eq. (3.11a); the relevant length is x_{ov}.

In volute pumps the leakage flow is fed from the boundary layer flow on the volute casing wall, Fig. 9.3. The circumferential velocity components are therefore smaller than in diffusers. During partload recirculation in the collector, the circumferential velocity components are much reduced or close to zero. Any recirculating fluid entering the impeller sidewall gaps will slow down the rotation (Chap. 5.4.3). Figure 9.4 illustrates the impact of the velocity distribution at the

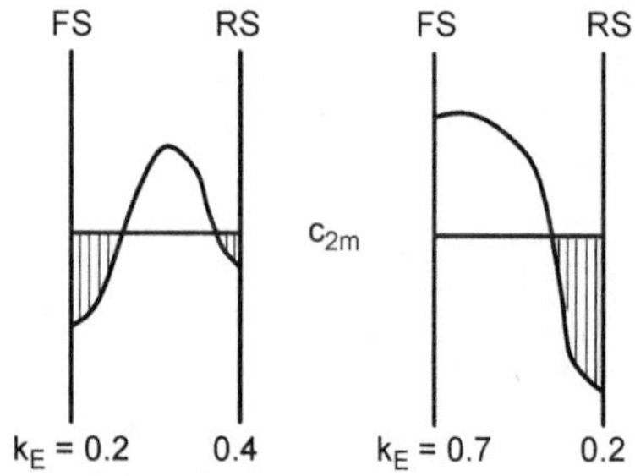

Fig. 9.4 Influence of the velocity distribution at the impeller outlet on the boundary condition to the impeller sidewall gap

impeller outlet on the inlet boundary condition $k_E = c_{2u,local}/u_2$ for the flow through the impeller sidewall gap.

If the impeller sidewall gaps are largely open to the main flow (large gap A, or volutes as shown in Fig. 9.3), the flow in the impeller sidewall gap is intensely coupled to the main flow by exchange of momentum which accelerates or retards the fluid rotation - depending on the magnitude of $c_{2u,FS}/u_2$.

The tangential velocity c_u results from a balance of all moments acting on the fluid contained in the impeller sidewall gap. These moments and their influence parameters are listed below:

- Fluid friction (shear stresses) on the rotating shroud acts as a *driving* moment; the friction effects fall with rising Reynolds number but increase with the roughness.
- Fluid friction (shear stresses) on the casing wall generates a *retarding* moment, which increases with the size of the wetted casing surface adjacent to the impeller sidewall gap. Again, the shear stresses fall with increasing Reynolds number and rise with the surface roughness.
- Turbulent dissipation in the impeller sidewall gap retards fluid rotation. Dissipation increases with enlarged gap width or the fluid volume set into motion.
- The exchange of momentum grows with increasing difference between the velocities in the main flow and the flow in the sidewall gap. During partload recirculation, the velocity gradients attain a maximum since the fluid recirculating from the collector has a low tangential velocity. As mentioned above, the exchange of momentum can be minimized by a design with a tight gap A and a generous overlap x_{ov}.
- Depending on magnitude, radial direction and inlet swirl $c_{2u,FS}/u_2$ of the leakage flowing through the impeller sidewall gap, fluid rotation is either accelerated or slowed down. If the leakage is directed radially inwards, the inlet swirl $c_{2u,FS}/u_2$ depends on the flow rate (q*), the impeller design (outlet velocity profile) and the type of collector.

The "disk friction loss" is also a result of this balance of moments; it is thus no universal quantity - not even for a given pump, since it depends on the leakage and on c_{2u} (hence on q*). An exact analytical calculation of the fluid rotation from the balance of moments is not possible for the general case of an impeller sidewall gap with net through-flow. When numerical procedures are used, the sidewall gap, the impeller and the collector must be modeled in a coupled calculation in order to capture correctly the inlet boundary conditions to the sidewall gap. This is particularly true for partload recirculation and for volutes, Fig. 9.3. In practice empirical coefficients and procedures are employed which describe the core flow by rotation factors k. Equation (9.1) defines k as the ratio of the tangential fluid velocity $c_u = \beta \times r$ to the circumferential velocity $u = \omega \times r$ (β is the angular velocity of the fluids in the core flow):

$$k = \frac{c_u}{u} = \frac{\beta}{\omega} \tag{9.1}$$

With net flow through the impeller sidewall gap, k(r) is a function of the radius. When there is no net through-flow, a constant value of k_o is applied, which can be calculated from Eq. (T9.1.3). This is a simplified version of Eq. (9.1a) derived in [50]. The flow with k_o = constant corresponds to a forced vortex (or a "solid-body rotation") with a constant angular velocity β of the fluid, Eq. (1.27).

$$k_o = \cfrac{1}{1 + \sqrt{\cfrac{\frac{1}{\cos\delta_W}\left\{\left(\frac{r_w}{r_2}\right)^5 - \left(\frac{r_i}{r_2}\right)^5\right\} + 5\frac{t_{ax}}{r_2}\left(\frac{r_w}{r_2}\right)^4}{\frac{1}{\cos\delta_R}\left\{1 - \left(\frac{r_i}{r_2}\right)^5\right\} + 5\frac{t_{ax}}{r_2}\left(\frac{r_i}{r_2}\right)^4\left\{1 + \frac{r_w - r_i}{t_{ax}}\tan\delta_W - \frac{r_2 - r_i}{t_{ax}}\tan\delta_R\right\}}\left(\frac{c_{f,w}}{c_{f,R}}\right)}} \tag{9.1a}$$

In Eq. (9.1a) δ_R is the inclination angle of the impeller side wall, δ_w is the inclination angle of the casing wall (Fig. 9.1) and r_i is the inner radius of the surface (e.g. r_1, r_{sp} or r_n). The tests in [50] were done with diffuser pumps, hence closed ISR where $t_{ax} = s_{ax}$, applies. Therefore, Eq. (9.1a) is only valid for closed ISR.

With open impeller side rooms the exchange of angular momentum between main flow and ISR cannot be neglected. For axial force calculations it may be better to estimate the rotation factors from experiments as shown in Fig. 9.8.

It happens that the impeller side rooms (ISR) are large and of complex shapes for reasons of mechanical design and production costs. Figure 2.5 shows an example. The large wetted surface of the ISR slows down the rotation of the liquid. As a consequence, the axial force on the respective shroud increases and the efficiency diminishes because of the increased disk friction.

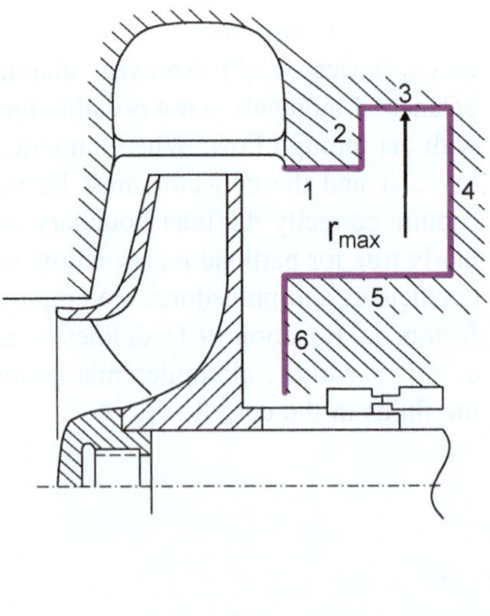

Fig. 9.5 Large impeller side room ISR on the rear shroud slowing down the rotation of the fluid

The rotation factor k_o for an ISR of any size and shape can be estimated from the balance of momentum discussed above which leads to Eq. (9.1b).

$$k_0 = \frac{1}{1+\sqrt{B}} \tag{9.1b}$$

The factor B describes the ratio of the effect of the surfaces (area and roughness) of the casing to that of the impeller as defined by Eq. (9.1c)

$$B = \frac{A_C c_{f,casing}}{A_R c_{f,impeller}} \tag{9.1c}$$

The ratio of the friction factors in Eq. (9.1c) takes into account any differences in the roughness of the casing and the impeller shrouds. A_R and A_C are calculated from Eqs. (9.1d) and (9.1e). A_R represents the impeller shroud and hub. A_C represents all wetted casing surfaces in the ISR. These surfaces are highlighted in Fig. 9.5. In Eq. (9.1e) r_y stands for the outer radius and r_x for the inner radius of any surface element which is perpendicular to the rotor axis (surfaces 2, 4 and 6 in Fig. 9.5), while r_z and b_z stand for the radius and width of any cylindrical surface element (surfaces 1, 3 and 5 in Fig. 9.5).

$$A_R = \frac{1}{5\cos\delta_R}\left(r_2{}^5 - r_n{}^5\right) + r_n{}^4 b_n \tag{9.1d}$$

$$A_C = \frac{1}{5}\sum\left(r_y{}^5 - r_x{}^5\right) + \sum r_z{}^4 b_z \tag{9.1e}$$

In the case of an ISR open to the volute with $t_{ax} = 0$, $r_w = r_2$ and equal roughness on casing and impeller A_c equals A_R, B becomes unity and $k_o = 0.5$, same as with Eq. (9.1a) or Eq. (T9.1.3).

Side room designs with $r_{max} > r_2$ are particularly obnoxious to the efficiency since the factor B in Eq. (9.1.b) increases with the 5th power of r_{max}/r_2. Incidentally, in a series of tests with small and large ISR it was found that the loss in BEP efficiency $\Delta\eta_{opt}$ was roughly proportional to $(r_{max}/r_2)^5$.

The effect of large and complex ISR on the disk friction can be estimated from Eq. (T3.10.9). With an ISR open to the volute or a large gap A, the main flow exiting from the impeller interacts with the flow in the ISR through an exchange of turbulent momentum. When k_o is much lower than c_{2u}/u_2, i.e. if there are large velocity gradients, part of the energy from the main flow is dissipated in the ISR. As a consequence, the head is reduced. The efficiency impairment due to an ISR of complex shape is thus caused not only by an increased disk friction but also by a drop in the hydraulic efficiency.

An example is provided by the tests shown in Fig. 9.6. One test had a large ISR similar to Fig. 9.5 with $r_{max}/r_2 = 1.35$ leading to $k_o = 0.17$ while the second tests had a small ISR with $k_o = 0.5$. Compared to the small ISR, the head dropped over

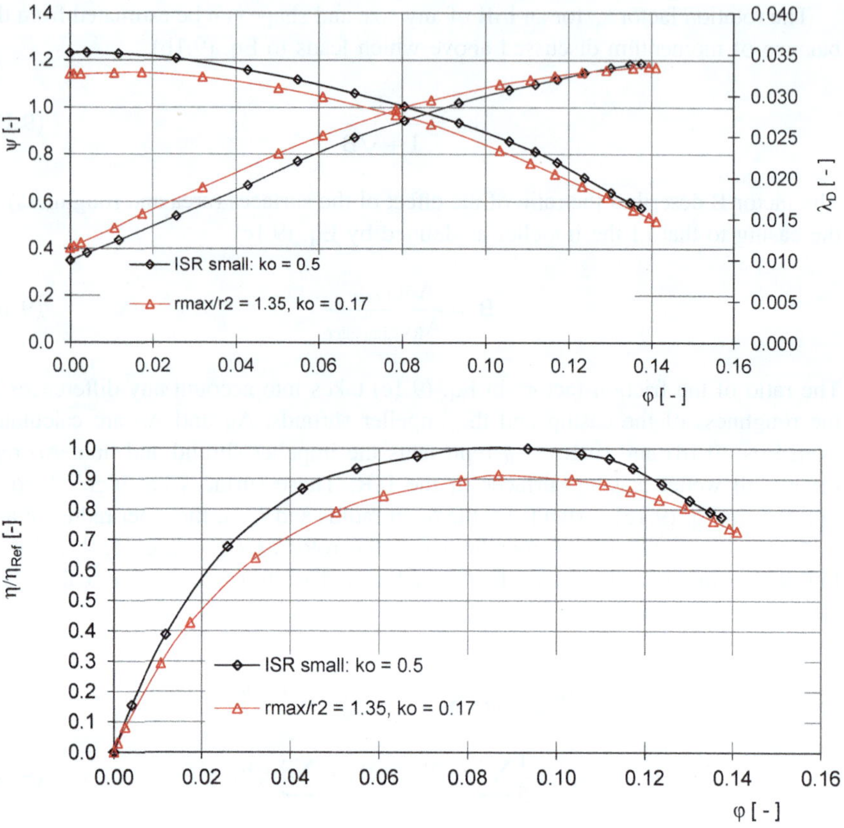

Fig. 9.6 Effect of ISR on performance, $n_q = 26$, $Re_{u2} = 1.6 \times 10^6$, WILO SE

the entire flow range, but most at low flow. According to Eq. (T3.10.9) the gain in disk friction power when increasing k_o from 0.17 to 0.5 amounts to $\Delta P/P_{opt} = 0.05$. The efficiencies in Fig. 9.6 are referred to the BEP of the pump with the small ISR.

The velocity distribution at the impeller outlet depends on the impeller geometry, the flow rate and on the approach flow distribution at the impeller inlet. It has an impact on the pressure recovery in the volute, but also on the interaction between the main flow and the flow in the ISR. Figure 9.7 illustrates the relevant mechanisms: If the circumferential velocity c_{2u} at the rear shroud (RS) is higher than the velocity in the ISR, some energy of the main flow is transferred to the ISR by turbulent exchange of momentum. This effect is felt by a loss in useful head; the liquid rotation in the ISR grows. In contrast, if the main flow velocity is lower than the velocity in the ISR, energy is transferred from the ISR to the main flow. Furthermore, the pumping action of the impeller shroud contributes to the head. If δ is the width of the opening between main flow and ISR flow, the gain or loss in head may be expected to follow the proportionality of Eq. (9.1f).

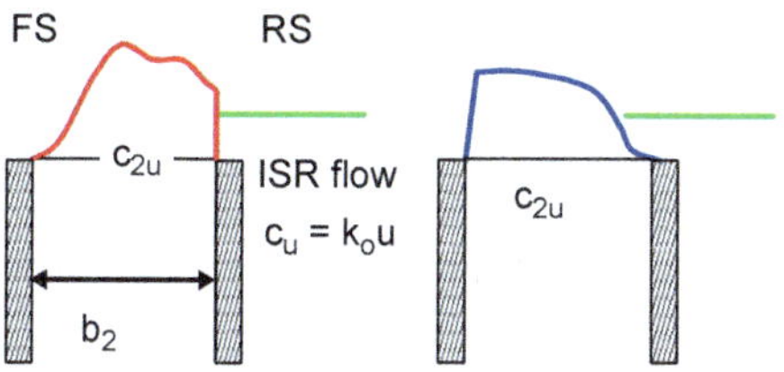

Fig. 9.7 Influence of the velocity distribution at the impeller outlet on the interaction between the main flow and the ISR flow. Left: energy from main flow dissipated to ISR; right: pumping action of shroud accelerates main flow

$$\Delta H \approx \delta(k_0 u_2 - c_{2u}) \quad (9.1f)$$

Since the velocity distribution at the impeller outlet depends strongly on the flow rate through the pump, ΔH may be positive over a certain flow range and negative over another (as borne out by some tests).

The effect described by Eq. (9.1f) may have the same tendency on both the rear and the front shroud (reinforcing each other) or it may be positive on one shroud and negative on the other (compensating to some extent).

Equation (9.1f) also shows that the interaction of main flow and ISR flow is controlled by both the gap width and the velocity gradient. It is possible to observe a strong interaction even though the gap is relatively small; likewise little interaction may be noticed with large gaps if the velocities match well. The effects described above are well illustrated by Fig. 5.30 which shows the influence of the velocity distribution on the liquid rotation in the ISR on the front shroud of a diffuser pump.

According to Fig. 9.1 the turbulent *core flow* is subject neither to circumferential nor radial velocity gradients; the core flow is thus *free of shear stresses*. The radial velocity in the core flow is zero ($c_r = 0$), even when a leakage flows through the sidewall gap. If the leakage is directed radially inwards, the radial velocity c_r on the casing is higher than on the shroud. In contrast, if the leakage is directed radially outwards, the radial velocity c_r on the rotating shroud is greater than on the casing wall.

If the leakage is directed radially inwards, the tangential velocity in the core flow increases with the inlet swirl $Q_{sp} \times c_{2u}$ so that $k > 0.5$ is often observed. In contrast, if the leakage is directed radially outwards, the tangential velocity in the core flow decreases with growing Q_{s3} resulting in $k < 0.5$. These features, which are qualitatively depicted in Fig. 9.1, were also found in numerical flow calculations, e.g. in [43]. The axial sidewall clearance s_{ax} and the Reynolds number have virtually no influence on the core flow. For this reason, correlations using a Reynolds number calculated with s_{ax} (rather than r_2) are of little relevance. Only the *ratio* of the casing roughness to the shroud roughness - not their absolute size - has an impact on the core flow. This is also demonstrated by Eq. (T9.1.3).

Spiral flow patterns are observed in the boundary layer on a rotating disk. The flow angles on the disk depend on the local Reynolds number. They attain about

40° to 45° at $Re < 2 \times 10^5$; above this limit they drop continuously to about 10° to 15° at $Re \approx 10^7$, [19]. An attempt to establish a unique relation between the flow angle and the rotation in the core flow failed, [19].

If, as in Fig. 9.3, a cylindrical casing wall is missing, $r_w = r_2$ and $t_{ax} = 0$ must be set in Eq. (T9.1.3). An inspection of Eq. (T9.1.3) reveals that k_o can attain (with $c_{f,w} = c_{f,R}$) a maximum value $k_o = 0.5$ which was frequently assumed as a first approximation. The rotation factor k_o diminishes if the length of the cylindrical portion t_{ax} is increased. If the friction coefficient (or roughness) on the casing ($c_{f,w}$) exceeds the friction on the shroud ($c_{f,R}$), the rotation factor k_o drops.

As mentioned above, k depends on the radius if there is a net flow through the impeller sidewall gap. Since direct measurements of the velocities are quite involved, the rotation factor $\bar{k}$ has often been derived as an average between two pressure tappings applied in the sidewall gap. To this end Eq. (1.26) is integrated with $c = \beta \times r = k \times \omega \times r$ yielding the pressure distribution for $\bar{k}$ = constant:

$$p = p_2 - \frac{\rho}{2} u_2^2 \bar{k}^2 \left(1 - \frac{r^2}{r_2^2}\right) \tag{9.2}$$

Based on Eq. (9.2), the average rotation factor between the radii r and r_2 can be calculated from the measured pressure difference $\Delta p = |p - p_2|$ according to Eq. (T9.1.8). An average rotation factor determined in this manner predicts the expected pressure decay in the impeller sidewall gap, but it does *not* yield the correct axial force acting on the shroud which results from the integration of the pressure distribution (refer to Chap. 9.2.1).

The dimensionless coefficient c_p used in the following describes the pressure decay in the impeller sidewall gap against the static pressure p_2 at the impeller outlet; c_p is defined by Eq. (T9.1.5) (note the negative sign).

The average rotation factors $\bar{k}$ have been determined from measuring the pressure difference between the impeller outlet and the inlet to the annular seals in multistage pumps, [14]. The results are plotted in Fig. 9.8 against the flow rate ratio q*. Due to the leakage flowing radially inwards the average rotation factors on the front shroud are in the range of $\bar{k} = 0.55$ to 0.72 for diffuser pumps. On the rear shroud the rotation is slowed down due to the leakage flowing radially outwards; consequently we only find $\bar{k} = 0.4$ to 0.45. When the leakage is flowing radially inwards, the rotation in volute casings is significantly lower than in diffuser pumps, because the swirl at the inlet to the sidewall gap is lower, as has been explained above with reference to Fig. 9.3.

When reducing the flow rate, the rotation initially increases because c_{2u} rises. Below $q^* < 0.5$, the rotation factor decreases slightly due to recirculation at the impeller outlet. Rotation factors down to k = 0.2 were measured when extensive recirculation was present. The tests in Fig. 5.30 demonstrate the strong impact of the main flow (in terms of c_{2u}) on the rotation factor, [18, B.20].

Based on specific tests, numerous methods have been proposed for calculating disk friction and fluid rotation in impeller sidewall gaps. A comparison of various

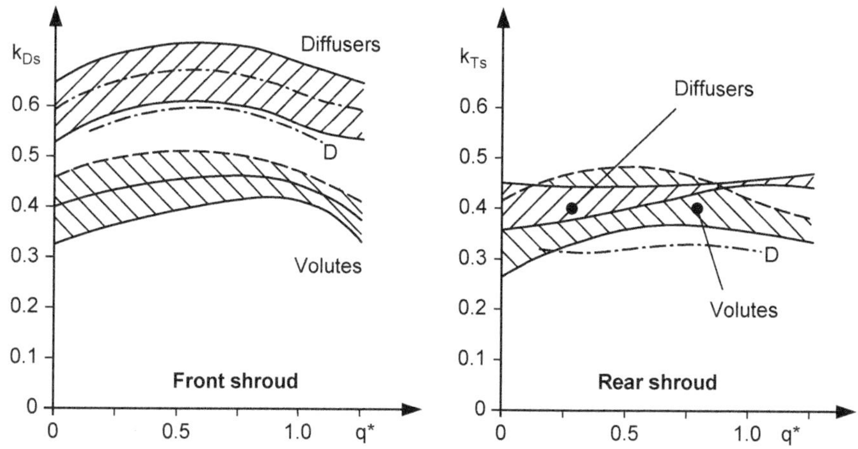

Fig. 9.8 Fluid rotation in the impeller sidewall gaps [14]. Curve D from tests with multistage diffuser pumps [48]

methods has been published in [27] and more recently in [53]; it identifies considerable uncertainties and deficiencies in the analyzed calculation procedures which are due to the close interaction of disk friction, rotation and roughness effects. Former methods were unable to cover these phenomena by a common concept. Some measurements in the IRS are also published in [53, 54]; these studies confirm the phenomena described above.

The new calculation procedure presented in Table 9.1 allows determining (as a function of the radius) the disk friction, the rotation factor and the axial force on the shroud, even with excessive leakage flowing through the impeller sidewall gap. The step-wise procedure takes into account the effects of the absolute roughness as well as the difference in the roughness of the rotating shroud and the stationary casing. The procedure is simple and easy to program; it predicts the published test results (e.g. of [44]) on average with the least amount of scatter, [15]. The inherent uncertainties reside in the definition of the boundary condition k_E at the entry to the impeller sidewall gap and in the roughness.

A simple formula for estimating the average rotation factor has been derived from the data in Figs. 9.11 and 9.13; it is given as Eq. (T9.1.4) in Table 9.1[1]. If the leakage flows radially inwards, Q_{sp} is positive, but if the leakage is directed radially outwards Q_{sp} is negative.

The calculation procedure in Table 9.1 comprises the following steps[2]:

[1] A graph in [20] yields values similar to those calculated from Eq. (T9.1.4).

[2] The calculation of the fluid rotation according to Table 9.1 is based originally on [36], but there are a number of improvements covering the calculation of disk friction losses, roughness effects, handling of gap A, simplifications and convergence.

1. After calculating the leakage flow rate through the impeller sidewall gap according to Table 3.7, the flow coefficient φ_{sp} and the Reynolds number are determined from Eq. (T9.1.1).
2. The rotation factor k_o for $\varphi_{sp} = 0$ results from the geometric dimensions, Eq. (T9.1.3).
3. With leakage $Q_{sp} > 0$, the rotation factor k(x) is calculated step-wise from Eqs. (T9.1.9, 9.1.10); $x = r/r_2$ is the radius ratio. Depending on the direction of the leakage through the impeller sidewall gap two cases are to be distinguished:
   A. When the leakage is directed radially **inwards**, the calculation follows the flow direction *from outside to inside*, because the tangential velocity c_{2u} (or its local value near the shroud) is needed as boundary condition at the entry to the impeller sidewall gap. The leakage flow (φ_{sp}) is counted as *positive*.
   B. When the leakage is directed radially **outwards**, the calculation follows the flow direction *from inside to outside*, because the tangential velocity $c_u(r_{s3})$ is needed as boundary condition at the entry to the impeller sidewall gap. The leakage flow (φ_{sp}) is counted as *negative*. At this stage, only k(x) *from inside to outside* has to be calculated.
4. As discussed above the tangential velocity $c_{u,E}$ at the entry to the impeller sidewall gap has a strong impact on fluid rotation. An appropriate choice of $k_E = c_{u,E}/u_2$ is therefore important:
   A. When the leakage flow is directed radially **inwards**, two cases must be distinguished: (1) If the impeller sidewall gap is *open* towards the casing (similar to Fig. 9.3) usually $k_{E(x\,=\,1)} = c_{2u}/u_2$ may be assumed unless it is intended to estimate the effect of recirculation (or the impact of the local velocity distribution at the impeller outlet) on the axial force. (2) There is a *defined gap A* with an overlap x_{ov} according to Fig. 9.1. This situation is often found in diffuser pumps (which indeed should be designed that way). The tangential velocity at the outlet of gap A is k_z. This value should be used as boundary condition for the step-wise calculation of the flow through the impeller sidewall gap. The larger the overlap x_{ov} and the smaller the width of gap A (Fig. 9.1), the more the tangential velocity at the outlet of gap A approaches the asymptotic value of $c_u = ½u_2$. The fluid rotation in gap A is accelerated by shear stresses on the rotating impeller shroud and slowed down by friction on the opposite cylindrical part of the casing. Any difference between the tangential velocity c_{2u} of the main flow and the fluid rotation in the impeller sidewall gap gives rise to an exchange of momentum by turbulent shear stresses which can be described by the eddy viscosity ε_τ. The change in the tangential velocity due to gap A can be calculated from the balance of moments generated by friction and the effect of the eddy viscosity in a step-wise procedure using Eq. (T9.1.17). The eddy viscosity is difficult to capture; a value of $\varepsilon_\tau = 0.01\ m^2/s$ can be derived from [29], which may be used in the absence of more precise data. Equation (T9.1.17) covers only the effect of gap A. The cylindrical part of the casing t_{ax} has not been included in this calculation, since it is taken into account by Eq. (T9.1.3)

which captures the effect of the friction on the surfaces $2\pi \times r_w \times t_{ax}$ and $\pi(r_w^2 - r_2^2)$ on k_o.

B. Any leakage directed radially **outwards** exits from the inter-stage seal into gap E. The *inlet* swirl to this seal roughly corresponds to c_{1u} (according to α_6). The tangential velocity at the seal *outlet* can be calculated from Eq. (3.11a). Numerical flow calculations reported in [43] yielded results similar to Eq. (3.11a).

5. The pressure decay $c_p(x)$ in the impeller sidewall gap is obtained for every calculation step from Eq. (T9.1.11), while Eq. (T9.1.12) yields the axial force reduction $c_A(x)$, introduced in Chap. 9.2 that is caused by the rotation. This calculation is carried out proceeding always *from outside to inside* (even when the leakage flow is directed radially outwards), i.e. the calculation starts at $x = 1$, where $c_p = c_A = 0$ is known as initial condition.

Equation (T9.1.9) has been derived from the balance of moments acting on a fluid element in the impeller sidewall gap. The term 2 k/x represents the non-viscous flow according to the conservation of angular momentum $r \times c_u$ = constant. The first term on the right side of the equation empirically describes the effect of friction: if $k < k_o$, shear stresses on the rotating shroud enhance the fluid rotation; in contrast, shear stresses on the casing slow down the rotation if $k > k_o$. If the leakage flow is very high, the friction term of Eq. (T9.1.9) becomes small compared to 2 k/x. Thus the formula yields the physically correct answer if Q_{sp} or Re_u tend to infinity. If the leakage flow becomes very small, the rotation factor k tends to k_o while dk/dx tends towards zero. Under this condition Eq. (T9.1.9) is undetermined. Very high gradients dk/dx result in the case of $k_E > k_o$; the solution becomes numerically unstable. Equation (T9.1.9) should therefore be applied only above $\left|\varphi_{sp} \times Re_u^{0.2}\right| > 0.002$ below this limit set $k = k_o$.

The calculation is quite sensitive to the value used for k_o which depends among other parameters on the ratio of the roughnesses of shroud and casing. *The rotation in the impeller sidewall gap strongly depends on boundary conditions at the entry to the gap: if leakage flow Q_{sp} and tangential velocity c_u at the entry are not captured correctly, neither the calculation according to Table 9.1 nor numerical flow calculations can yield reliable results.*

Figure 9.9 shows a comparison between test results from [51] performed to determine the influence of the leakage on the fluid rotation and calculations according to Eq. (T9.1.9). The local rotation factor k is plotted against the radius ratio x. One can see that the local tangential velocity in the impeller sidewall gap may exceed the circumferential speed $\omega \times r$ of the impeller, if the leakage is high and the radius ratio low. This implies that the local disk friction loss becomes "negative" at $k > 1$. This means that the fluid "drives the impeller" as it does in a friction turbine. The radius where $c_u = \omega \times r$ (i.e. $k = 1.0$) is reached can be estimated from data given in [19] from the correlation:

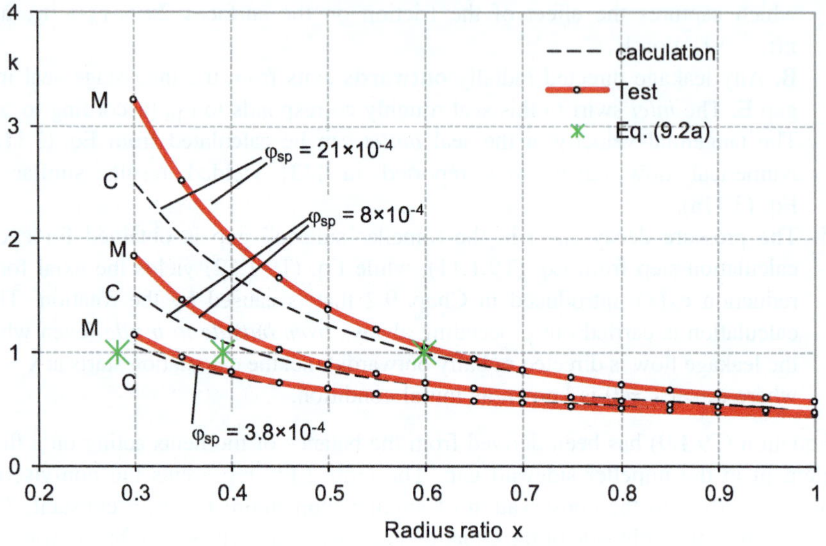

Fig. 9.9 Fluid rotation in impeller sidewall gaps. Comparison of calculation according to Table 9.1 with measurements from [51]. *C* = calculation, *M* = measurement

$$\frac{r_{(k=1)}}{r_2} = 9\,\varphi_{sp}^{0.44} \quad \text{with} \quad \varphi_{sp} \text{ from Eq. (T9.1.1)} \tag{9.2a}$$

Equation (9.2a) yields similar radius ratios as read from Fig. 9.9 for k = 1.0.

The large influence of the leakage flow φ_{sp} through the impeller sidewall gap and of the inlet swirl k_E is demonstrated by Figs. 9.10, 9.11, 9.12 and 9.13. These graphs were calculated according to Table 9.1 (2) with $k_o = 0.45$, $d_2 = 400$ mm, n = 3000 rpm, roughness $\varepsilon_R = \varepsilon_w = 4$ µm, $s_{ax}/r_2 = 0.065$ and $\nu = 10^{-6}$ m²/s.

Figures 9.10, 9.11, 9.12 and 9.13 show:

Flow Radially *Inwards*:

- **$k_E > k_o$**: If the leakage flow enters the impeller sidewall gap with a high tangential velocity, fluid rotation is *accelerated* and disk friction losses diminish with growing leakage flow rate, Fig. 9.10. As demonstrated by the calculation and the tests in [51], the disk f*riction loss* can drop to a fraction of its value without net through-flow if leakage and inlet swirl are high. If the annular seal clearances are executed according to Eq. (3.12), the seal flow coefficient φ_{sp} is in the order of 5×10^{-4}. In this case the differences in disk friction calculated with and without net through-flow remain below 25%. If annular seal clearances double due to wear during service, the seal flow coefficient attains the level of about $\varphi_{sp} \approx 15 \times 10^{-4}$.
- **$k_E < k_o$**: However, if the leakage flow enters the impeller sidewall gap with a low tangential velocity, fluid rotation is *slowed down* and disk friction losses increase with growing leakage flow rate, because the entering fluid mass must be accelerated to that tangential velocity which corresponds to k_o. Such can be the

Table 9.1 (1) Rotation of the fluid in the impeller sidewall gaps			Eq.
Definitions Subscripts: w: casing; R: impeller	$\varphi_{Sp} = \frac{Q_{Sp}}{\pi r_2^2 u_2}$ $\quad x = \frac{r}{r_2}$ $\quad Re_u = \frac{u_2 r_2}{\nu}$		9.1.1
Rotation of fluid	$k = \frac{c_u}{u} = \frac{c_u}{\omega r} = \frac{\beta}{\omega}$		9.1.2
Estimation of leakage flow for normal clearances, Eq. (3.12)	$\varphi_{sp} = 5.5\times10^{-4}\,\psi_{opt}^{1.5}\left\{\frac{n_q}{n_{q,Ref}}\right\}^{0.4}$ $\quad n_{q,Ref} = 20$		9.1.2a
Rotation of fluid without flow through impeller sidewall gaps ($\varphi_{sp} = 0$). Open sidewall gaps: set $r_w = r_2$ and $t_{ax} = 0$! t_{ax} is the *cylindrical* part of the casing with r_w	$k_o = \frac{1}{1+\left(\frac{r_w}{r_2}\right)^2\sqrt{\left(\frac{r_w}{r_2}+5\frac{t_{ax}}{r_2}\right)\frac{c_{f,w}}{c_{f,R}}}}$		9.1.3
Estimation of average rotation factor for $r_{sp}/r_2 > 0.3$ and $k_E \approx 0.5$ Leakage: inwards: φ_{sp} positiv; b = 1.0; outwards: φ_{sp} negativ; b = 0.65	$\frac{k_{cp}}{k_o} = \exp\left\{300\,\varphi_{sp}\left(\left[\frac{r_2}{r_{sp}}\right]^b - 1\right)\right\}$		9.1.4
Pressure reduction coefficient $\Delta p \equiv p - p_2$ c_p ist negative	$c_p \equiv \frac{\Delta p}{\frac{\rho}{2}u_2^2} = -2\int_1^x k^2 x\,dx$	$c_p = -k_o^2(1 - x^2)$	9.1.5
Axial thrust coefficient giving the reduction of the axial force on impeller shroud due to fluid rotation; c_A ist positiv	$c_A \equiv \frac{8\int_{r_2}^{r}\Delta p\,r\,dr}{\rho\,u_2^2 r_2^2} = 4\int_1^x c_p\,x\,dx$	$c_A = k_o^2\left(1-x^2\right)^2$	9.1.6
Pressure distribution in the impeller sidewall gap	$p = p_2 - \frac{\rho}{2}u_2^2\left(\bar{k}\right)^2\left(1-\frac{r^2}{r_2^2}\right) = p_2 + \frac{\rho}{2}u_2^2 c_p$		9.1.7
Calculation of average rotation factor from the measurement of a pressure difference in the impeller sidewall gaps	$\bar{k} = \sqrt{\frac{2(p_2 - p)}{\rho\,u_2^2\left(1-\frac{r^2}{r_2^2}\right)}}$		9.1.8
Calculation of average rotation factor from axial thrust coefficient	$\bar{k}_A = \frac{\sqrt{c_A}}{1-x^2}$		9.1.8a

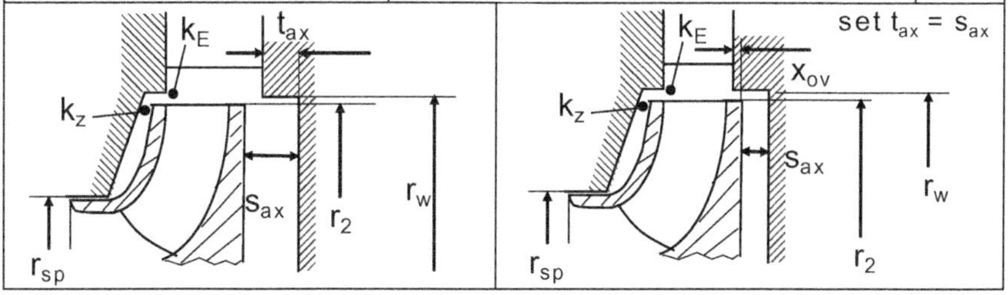

Table 9.1 (2) Rotation of the fluid in the impeller sidewall gaps		Eq.
Step-wise calculation of rotation, pressure coefficient and axial thrust coefficient: Only for $\lvert \varphi_{sp}\, Re_u^{0.2} \rvert > 0.002$ **1.** Leakage radially inwards: φ_{sp} is positive. Calculation of k from outer to inner radius. The calculation starts with $k_E = c_{2u}/u_2$ or k_z from Eq. (T9.1.17) if $x_{ov} > 0$. **2.** Leakage radially outwards: φ_{sp} is negative. Calculation of k from inner to outer radius. The calculation starts with $k_{out} = c_u/u_{sp}$ from Eq. (3.11a)		
$\frac{dk}{dx} = \frac{0.079 x^{1.6}}{\varphi_{Sp}\, Re_u^{0.2}} \left\{ \left(\frac{1-k_o}{k_o} \cdot k \right)^{1.75} \frac{1}{\cos\delta_W} - \lvert 1-k \rvert^{1.75} \frac{1}{\cos\delta_R} \right\} - 2\frac{k}{x}$		9.1.9
$k_{n+1} = k_n + \frac{dk}{dx}(x_{n+1} - x_n)$		9.1.10
$c_{p,n+1} = c_{p,n} + \left(x_n k_n^2 + x_{n+1} k_{n+1}^2 \right)(x_{n+1} - x_n)$		9.1.11
$c_{A,n+1} = c_{A,n} + 2\left(x_n c_{p,n} + x_{n+1} c_{p,n+1} \right)(x_{n+1} - x_n)$		9.1.12
Torque due to friction on an annular element of the shroud (dimensionless)	$dM^* \equiv \frac{dM}{\rho\, \omega^2\, r_2^5} = \frac{0.287 x^{3.6} \lvert 1-k \rvert^{1.75} dx}{Re_u^{0.2}} \frac{c_{f,R}}{c_{f,glatt}} \operatorname{sign}(1\text{-}k)$	9.1.13
Disk friction coefficient from summation of all elements	$k_{RR} = \frac{\sum_n dM^*}{1 - d_{sp}^{*\,4.6}}$	9.1.14
Disk friction coefficient from integral of Eq. (T9.1.13) with laminar term added; for $k = k_o$ = constant	$k_{RR} = \frac{\pi\, R}{2 Re\, s_{ax}} + \frac{0.0625}{Re^{0.2}} \cdot \left(1 - k_o\right)^{1.75} f_R\, f_L$	9.1.15
Rotation factor k_z at outlet from gap A, [29]	$k_z = 0.33 + 0.28\, k_E + 126\, \varphi_{sp}\, (k_E - 0.31)$	9.1.16
Sidewall gap inlet: development of rotation factor in overlap x_{ov} (Fig. 9.1); $z^* = z/r_2$ is the coordinate in axial direction. Stepwise calculation from $z^* = 0$ to $z^*_{max} = x_{ov}/r_2$; Fig. 9.1	$\frac{dk}{dz^*} = \frac{c_{f,R}\,(1-k)^2 - r_w^{*2}\, c_{f,W}\, F_{Form}\, k^2}{\frac{\varphi_{Sp}}{2} + \frac{(r_w^* - 1)\, \varepsilon_\tau}{\omega r_2^2}}$ F_{Form} is a form factor (usually 1.0). For $x_{ov} = 0$ no change in rotation is expected; set $k_z = k_E$	9.1.17

case during partload recirculation, as indicated in Fig. 9.10 and confirmed by tests in [51]. The tests reported in [11, 36] were executed on rotating bodies at $k_E < k_o$; that is why the disk friction slightly increased with rising flow through the impeller sidewall gap - quite contrary to pumps, [15].

Flow Radially Outwards:

- If the leakage flows radially outwards through the impeller sidewall gap, disk friction losses always rise with increasing leakage. If the leakage is high (enlarged annular seal clearances), the increase in disk friction is significant. In multistage pumps according to Fig. 9.1, however, the effects of the leakage *on disk friction* largely compensate on the front and rear shroud if seal clearances

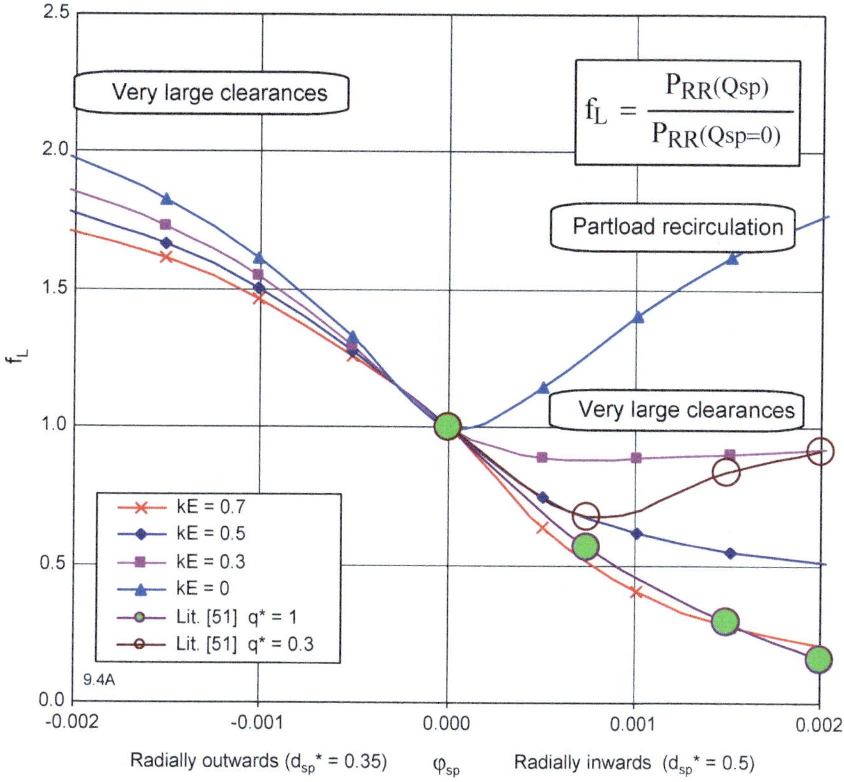

Fig. 9.10 Influence of leakage and inlet swirl on disk friction losses

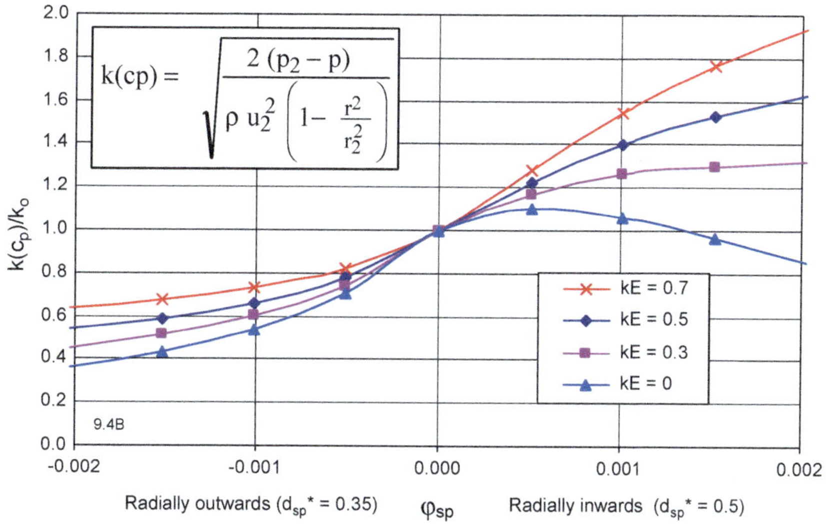

Fig. 9.11 Influence of leakage and inlet swirl on the rotation in the impeller sidewall gaps; average rotation factor without leakage $k_o = 0.45$

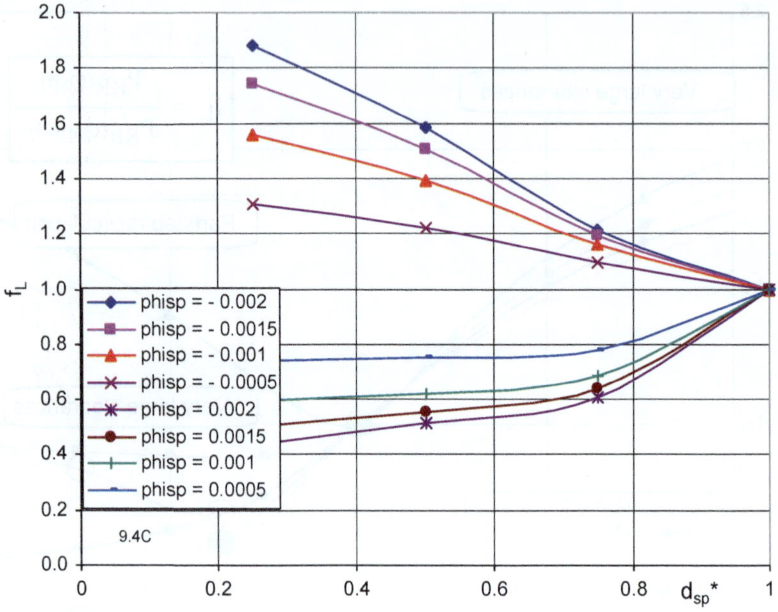

Fig. 9.12 Influence of leakage and seal diameter on disk friction losses, $k_E = c_{2u}/u_2 = 0.5$

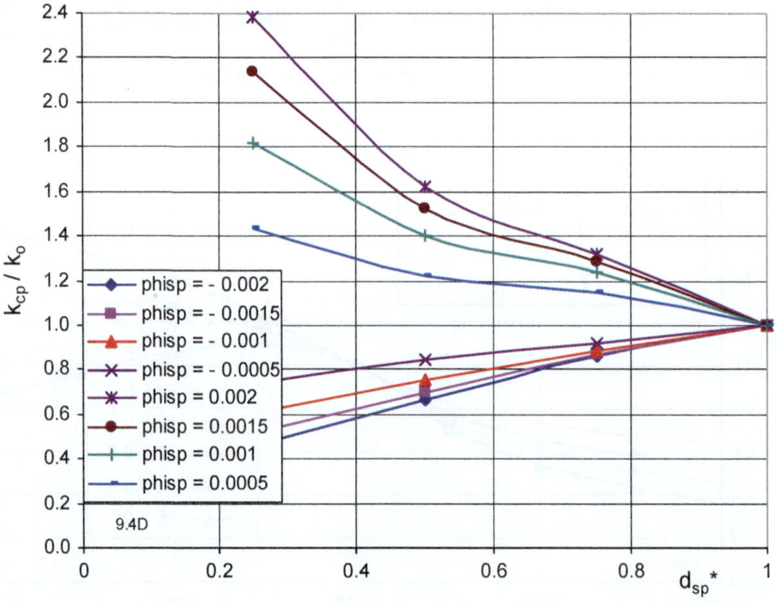

Fig. 9.13 Influence of leakage and seal diameter on fluid rotation in impeller sidewall gap, $k_E = c_{2u}/u_2 = 0.5$

are enlarged at the impeller inlet *and* at the interstage seal. (As discussed below the effects of increased leakage add with respect to the axial force!) Since the leakage flow enters the impeller sidewall gap at a small radius, the impact of the inlet swirl k_E is much weaker than with radially inward flow.

Figure 9.11 shows the influence of φ_{sp} and k_E on the fluid rotation in the impeller sidewall gap, where k_{cp} represents that average of k which yields the correct pressure decay according to Eq. (T.9.1.5). As demonstrated by this graph again, the calculations of the flow in the impeller sidewall gap can (in the general case) promise reliable results only if leakage flow rate and inlet swirl k_E are specified correctly.

The factors f_L and k_{cp}/k_o represented in Figs. 9.10 and 9.11 depend on the ratio of the annular seal diameter to the impeller outer diameter: $d_{sp}^* = d_{sp}/d_2$. These factors become 1.00 at $d_{sp}^* = 1.0$. The influence of the diameter ratio d_{sp}^* is shown in Figs. 9.12 and 9.13 which were calculated for $k_E = 0.5$ (corresponding to $\psi_{th} = 1.0$).

The impeller sidewall gap flow and the disk friction losses as well as axial forces are functions according to $(k, P_{RR}, F_{ax}) = f(k_E, \varphi_{sp}, d_{sp}^*, Re, \varepsilon_R, \varepsilon_w)$, which cannot be represented by simple analytical formulae. It should be noted that k_E has to be determined from the *local* circumferential velocity component at the entry to the impeller sidewall gap; $k_E = c_{2u}/u_2 = ½\ \psi/\eta_h$ is only an estimation which fails completely at partload recirculation. The calculation according to Table 9.1 (2) gives a good approximation. When using CFD, setting the correct boundary conditions (in particular k_E) is critical. Figures 9.10, 9.11, 9.12 and 9.13 allow a good qualitative and quantitative assessment of how important these effects might be in a particular application.

Because of the close interaction between the main flow at the impeller outlet and the fluid in the impeller sidewall gap, it is difficult to measure the true disk friction losses in a pump. In an attempt to capture the effect of the volute on disk friction losses, certain tests were performed with disks or bodies rotating in a volute casing. Such tests are doomed to failure because the boundary layers thrown off the rotating disk are strongly slowed down at the large wetted casing surface. Thus the fluid enters the impeller sidewall gap with a very low tangential velocity (low k_E), Fig. 9.3. As a consequence, the disk friction losses measured exceeded those determined in tight casings by roughly 40 to 70%. In the tested volute casings the ratio r_w/r_2 was about 1.3 to 1.4; using these data, Eqs. (T9.1.3 and 9.1.15) yield indeed friction coefficients of the magnitudes measured. Applying the measured data to the loss analysis of pumps would however be a mistake, because the flow conditions in the volute, when operating with an impeller, are completely different due to the high tangential velocity (high k_E).

Tests in [55] used a rotating disk in a volute casing, whereby the interaction between the flows in the impeller sidewall gaps and the volute could be partly decoupled by a cylindrical cover. The stronger the decoupling, the lower were the measured disk friction losses, which finding confirms the statements made above.

Other investigators used stationary vanes to induce a pre-rotation at the entry to the sidewall gaps of their rotating disk test rig. These attempts failed because it is very difficult to induce in that way a swirl of the magnitude prevailing at the impeller outlet of a real pump. As a consequence, the measured disk friction coefficients exceeded those to be used for pump operation by 50 to 80%, [11]. Only the test rigs used in [46, 51], where the disk and the impeller are driven by *independent motors* are really representative for the flow conditions in pumps and appropriate for measuring the true power absorbed by disk friction.

Guidelines for the Design of the Impeller Sidewall Gaps

1. Axial clearance: $s_{ax}/d_2 = 0.015$ to 0.04. The absolute minimum is given by the mechanical design. Note that small axial clearances reduce the structural impeller eigen frequencies and risk to increase fluid-structure interactions, Chap. 10.7.3.
2. Avoid excessive fluid volumes in the impeller sidewall gaps in order to prevent unnecessarily high disk friction losses due to turbulent dissipation. In high-energy pumps large fluid volumes in the impeller sidewall gaps are particularly harmful if gap A is large and the overlap is small, Chap. 10.7.3.
3. Avoid ribs on, or complex contours of, the casing wall which would slow down fluid rotation, increase disk friction and reduce efficiency. Impeller side rooms with $r_{max}/r_2 > 1.0$ (Fig. 9.5) are particularly harmful to the efficiency; they even can cause a loss in head. The loss in efficiency is expected to be roughly proportional to $(r_{max}/r_2)^5$.
4. Volute pumps: If the impeller sidewall gaps are open to the volute, the casing should be shaped in a way that low-velocity fluid flowing from the volute into the impeller sidewall gaps does not interfere with the pumping action of the shrouds. The boundary layer fluid thrown off the shrouds must be allowed to transfer its energy to the main flow (see Chap. 3.10.3 for test data). The efficiency is not impaired by this type of design, refer also to Fig. 9.3.
5. Wide impeller sidewall gaps open to the volute may not be acceptable in high-energy pumps because recirculating fluid can modify the axial force and excite vibrations. In this case the casing can be designed to incorporate a sufficiently tight gap A with an adequate overlap as stipulated under item 6 below, [31]. However, gap A is usually not controlled in volute pumps. With a small gap A head and efficiency may slightly suffer at BEP since the pumping action of the shrouds is more likely to be dissipated in the impeller sidewall clearances.
6. In diffuser pumps, the flow in the sidewall gaps should be effectively decoupled from the main flow in order to avoid drastic flow pattern changes in the impeller sidewall gap due to partload recirculation (axial force excursions), [18, B.20]. Gap A between the impeller shrouds and the diffuser side plates (or the casing) should be selected in the range of: $A/r_2 = 0.007$ to 0.01 with an overlap of $x_{ov}/A = 2$ to 4 (Fig. 9.1).

Fluid rotation in the impeller sidewall gaps can be increased by auxiliary vanes on the shrouds; details are given in Chap. 9.2.7.

9.2 Axial Forces

9.2.1 General Procedure for Calculating Axial Forces

For sizing the axial bearing and a possible device for balancing the axial thrust it is necessary to know the axial force acting on the pump rotor. The resultant force comprises as per Fig. 9.14: the forces F_{RS} and F_{FS} caused by the pressure distributions acting on the impeller shrouds, the momentum F_I, unbalanced axial forces F_w acting on the shaft and the rotor weight in the case of vertical pumps.

As a boundary condition for calculating the pressure distribution, the static pressure p_2 at the impeller outlet is required. It is calculated from the static pressure rise generated by the impeller, which is given by Eq. (T9.2.1) for zero pre-swirl, Table 3.7(1). For the calculation of the axial forces acting on the shrouds, only pressure *differences* above the impeller inlet pressure p_1 are needed. The pressure at the impeller outlet can also be estimated from Eq. (T9.2.2) without the need to calculate the velocities used in Eq. (T9.2.1). Simplifying Eq. (T9.2.2) we may set $\eta_{h,La}/\eta_h = 1$ taking into consideration that the pressure rise at the impeller outlet is under-predicted by a few percent.

In Eq. (T9.2.1), the quantity $R_G = H_p/H$ is called "degree of reaction" or "reaction ratio". As a general tendency, R_G increases with the specific speed as demonstrated by Fig. 9.14-1 which was calculated from Eq. (T9.2.2). Test data are shown in Fig. 9.14-2 in terms of the ratio R_G and as coefficient of the static pressure rise ψ_p in the impeller in Fig. 9.14-3. The curves $R_G = H_p/H = f(q^*)$ exhibit a minimum near or below BEP. R_G rises steeply above BEP because the total head drops due to the weak pressure recovery in the diffuser while the pressure rise in the impeller decreases linearly with the flow rate. The lower the specific speed, the

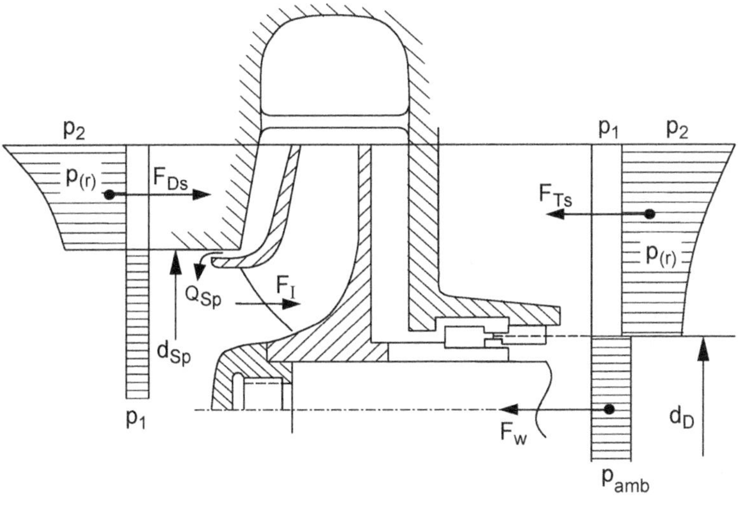

Fig. 9.14 Pressure distribution and axial forces on an impeller of a single-stage pump

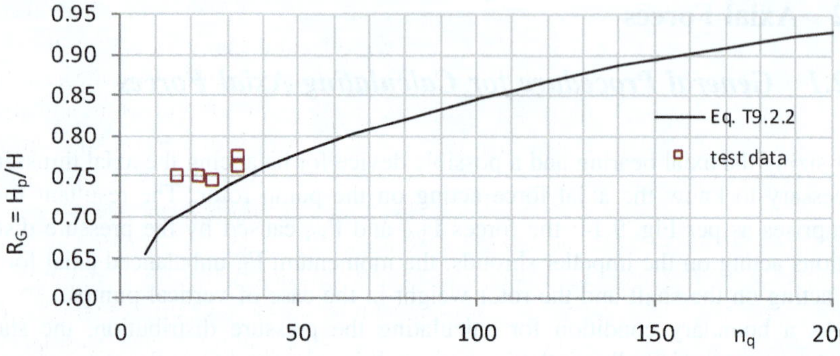

Fig. 9.14-1 Degree of reaction RG at BEP as function of specific speed, calculated from Eq. (T9.2.2)

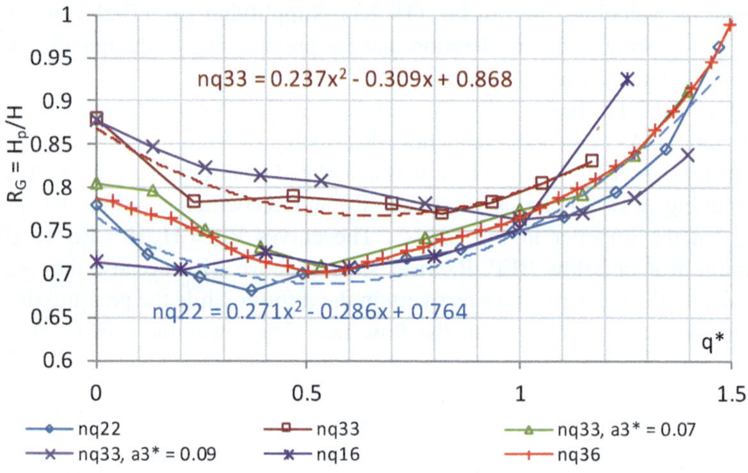

Fig. 9.14-2 Degree of reaction RG as function of q* derived from test data with diffuser pumps

steeper is the curve $R_G = f(q^*)$ above BEP because of the flow acceleration in the casing throat area. At part load, the static pressure at the impeller outlet (or R_G) rises as the exchange of momentum between rotor and stator increases.

The static pressure rise $\psi_p = f(q^*)$ exhibits essentially linear curves which start at $\psi_{p,0} = 0.9$ to 0.94 at Q = 0; the theoretical value is $\psi_{p,0} = 1.0$. For the axial force calculation, it may be more appropriate to estimate the curve $\psi_p = f(q^*)$ from Fig. 9.14-3 rather than using the data in Fig. 9.14-2. The curve for $n_q = 36$ is quite flat (and slightly unstable) because the impeller has a high blade outlet angle reaching $\psi_{opt} = 1.0$.

The axial forces acting on the shrouds result from the integral $F = 2\pi \times \int p \times r \times dr$ over the pressure distribution given by Eq. (9.2). The integration is done

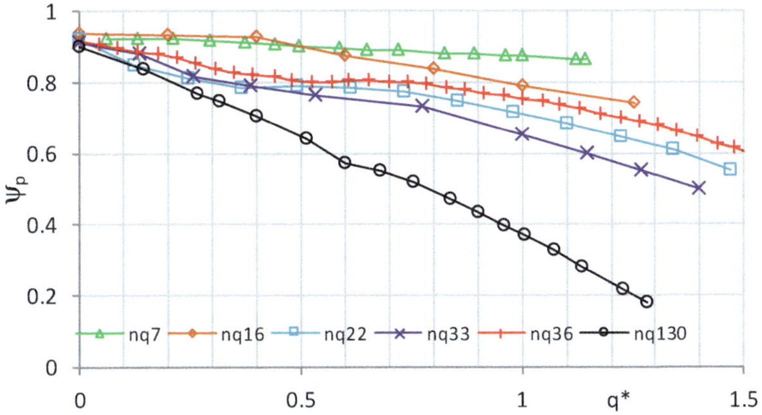

Data to figures 9.14-2 and 9.14-3					
n_q	7	16	22	33	130
Fig.	13.10	5.26	4.5	5.19	5.29

Fig. 9.14-3 Static pressure rise in impeller, derived from test data; nq36 with $\psi_{opt} = 1.0$ (volute)

from r_{sp} or r_D to r_2; in the general case it is performed separately for rear and front shroud. There are two options to do this:

A. An average is used for the rotation factor which can be derived from Eqs. (T9.1.3 and 9.1.4) or read from Fig. 9.8. The force acting on *one* shroud is:

$$F = \pi r_2^2 \left\{ \left(1 - x^2\right) \Delta p_{La} - \frac{\rho}{4} u_2^2 \left(\bar{k}\right)^2 \left(1 - x^2\right)^2 \right\} \tag{9.3}$$

Set $x = x_{sp} = d_{sp}/d_2$ for the front shroud and $x = x_D = d_D/d_2$ for the rear shroud. The calculation with an average rotation factor is appropriate if the leakage is small or zero.

B. If it is sought to determine the influence of the leakage, Eq. (T9.1.9) is evaluated in a first step as described above. Subsequently, the axial force reduction coefficient c_A is calculated according to Table 9.1. This coefficient captures the reduction ΔF of the axial force caused by the rotation of the fluid as compared to the case with $p = p_2 = \text{constant}$ (equivalent to $k = 0$). The reduction is referred to a force F_{Ref} which corresponds to a pressure distribution calculated with $k = 1$ and $\pi \times r_2^2$ The coefficient c_A is thus defined by:

$$c_A = \frac{\Delta F}{F_{Ref}} = \frac{4\,\Delta F}{\pi\,\rho\,u_2^2 r_2^2} = \frac{8\int_{r2}^{r} \Delta p r\,dr}{\rho u_2^2\,r_2^2} = 4\int_1^x c_p x\,dx \tag{9.4}$$

By means of coefficient c_A the force acting on one shroud becomes:

$$F = \pi r_2^2 \left\{ (1 - x^2)\Delta p_{La} - \frac{\rho}{4} u_2^2 c_A \right\} \tag{9.5}$$

It is possible to determine from c_A an average rotation factor $\bar{k}_A$ which gives exactly the same axial force as Eq. (9.5) when inserted into Eq. (9.3). $\bar{k}_A$ can be calculated from Eq. (T9.1.8a).

In view of the uncertainties involved and the margins taken on the capacity of the axial bearing, the calculation with average rotation factors according to Eq. (9.3) may be considered adequate in many applications. If the impact of the leakage and/or worn annular seal clearances is investigated, the calculation has to be performed according to Eq. (T9.1.9) and option B above.

The net axial force acting on the impeller is given by the difference between the forces acting on rear and front shrouds $F_{HY} = F_{RS} - F_{FS}$; in the general case with $k_{RS} \neq k_{FS}$ it is calculated from Eq. (T9.2.8). If the rotation factors on front and rear shrouds can be assumed to be equal, the simplified Eq. (T9.2.7) may be used.

In principle the averages of k_A resulting form Eq. (T9.1.8a) should be inserted into Eqs. (T9.2.3, 9.2.7 and 9.2.8). If the forces are calculated using average rotation factors $\bar{k}$ from Eq. (T9.1.8) rather than $\bar{k}_A$ from Eq. (T9.1.8a), the axial force is slightly over-predicted and the calculation is conservative.

According to the conservation of momentum (Chap. 1.2.3) an axial force $F_I = \rho \times Q(c_{1m} - c_{2m} \times \cos\varepsilon_2)$ is acting on the impeller; ε_2 is the angle between the mean streamline at the impeller outlet and the rotor axis (hence $\varepsilon_2 = 90°$ with radial impellers), Fig. 9.15.

Unbalanced axial forces F_w acting on the shaft have to be analyzed separately for each type of pump. In single-stage pumps with overhung impellers according to Fig. 9.14 it is: $F_w = \frac{1}{4}\pi d_D^2(p_{amb} - p_1)$.

The axial thrust acting on the rotor is the resultant of all forces discussed above: $F_{ax} = F_{HY} - F_I + F_w + F_{coupl}$. In the case of multistage pumps the forces must be determined for all stages and summed up to get the net axial force on the rotor. If the coupling selected is able to transmit axial forces, the coupling thrust F_{coupl} must be taken into account. A force acting towards the suction side is counted positive.

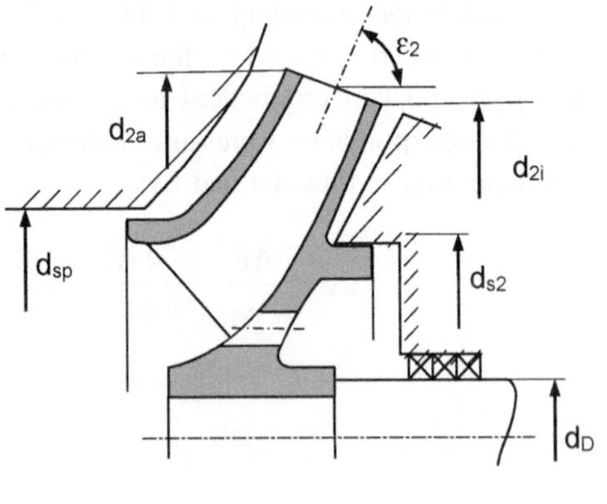

Fig. 9.15 Impeller with balance holes

If the suction pressure exceeds the atmospheric pressure, the force component F_w is negative; it unloads the bearings in single-stage pumps with overhung impellers.

The above discussion applies to closed radial and semi-axial impellers. The static pressure rise created by a semi-axial impeller will not be identical on the outer and inner streamline. If such is the case, Eqs. (T9.2.3 and 9.2.4) must be evaluated separately: for the rear shroud use $\Delta p_{La,RS}$ and k_{RS} while $\Delta p_{La,FS}$ and k_{FS} apply to the front shroud.

The calculation of the axial forces is subject to uncertainties because the following quantities cannot be determined accurately: (a) the inlet swirl k_E to the impeller sidewall gap (this is the main uncertainty since the velocity distribution at the impeller outlet is not known; the impact of partload recirculation is very strong); (b) impeller losses and consequently p_2; (c) the rotation factors k_{RS} and k_{FS} and the magnitude of the leakages; (d) possible differences between $p_{2,RS}$ and $p_{2,FS}$ (in particular at partload) in the case of radial impellers of high specific speeds and semi-axial pumps; (e) geometric tolerances such as annular seal clearances, axial rotor position and casting tolerances of the impeller. Because of these uncertainties, safety margins on the capacity of the axial bearing are indicated. In order to define appropriate margins, the formulae given in Tables 9.1 and 9.2 may be used for a sensitivity study investigating the impact of various assumptions and parameters (e.g. annular seal clearances) on the calculated axial thrust.

In view of the uncertainties of the various assumptions, simple formulae are alternatively in use for providing a rough estimate of the axial thrust, for closed radial and semi-axial impellers for example Eq. (T9.2.12). The use of such formulae may be justified if the axial bearing is bound to be oversized for reasons of mechanical design or if specific experience is available proving that the capacity of the axial bearings selected that way is adequate.

9.2.2 Single-Stage Pumps with Single-Entry Overhung Impellers

If special devices for reducing the axial thrust are not provided (as shown in Fig. 9.14), the calculations follow Chap. 9.2.1. While the effect of the momentum F_I is usually small, magnitude and direction (sign) of the unbalanced axial forces acting on the shaft F_w can cause problems in special cases. If inlet or system pressures are high (as in boiler circulation pumps, for example) the direction of the axial thrust is reversed and acts towards the pressure side; this needs to be taken into account when selecting the bearing.

In applications with uncontaminated liquids, the pumps are frequently designed with balance holes and an annular seal on the rear shroud according to Fig. 9.15 (in Table 9.1). Supposing equal annular seal diameters, constant pressure p_2, identical leakages and equal geometry of the impeller sidewall gaps, the forces on both shrouds are theoretically balanced between d_{sp} and d_2. If the balance holes are

Table 9.2 Calculation of axial thrust on impeller

Table 9.2 Calculation of axial thrust on impeller		Eq.
Static pressure rise in impeller (above impeller inlet) to $\alpha_1 = 90°$	$\Delta p_{La} = p_2 - p_1 = \eta_{h,La}\frac{\rho}{2}\left(u_2^2 - w_2^2 + c_1^2\right) = \rho\, g\, H_P = \rho\, g\, H\, R_G$	9.2.1
	$\Delta p_{La} \approx \rho\, g\, H\left(1 - \frac{\psi}{4\eta_h}\right)\frac{\eta_{h,La}}{\eta_h}$	9.2.2
Axial force on *one* impeller shroud F_{RS} und F_{FS} [(a)]	Front shroud $x = x_{sp} = d_{sp}/d_2$: $F = \pi\, r_2^2\left\{\left(1-x^2\right)\Delta p_{La} - \frac{\rho}{4}u_2^2\bar{k}^2\left(1-x^2\right)^2\right\}$	9.2.3
	Rear shroud: $x = x_D = d_D/d_2$: $F = \pi\, r_2^2\left\{\left(1-x^2\right)\Delta p_{La} - \frac{\rho}{4}u_2^2 c_A\right\}$	9.2.4
Dimensionless force on *one* shroud	$f = \psi_p\,(1-x^2) - \frac{\bar{k}^2}{2}(1-x^2)^2 = \psi_p\,(1-x^2) - \frac{c_A}{2}$	9.2.5
Axial force on impeller	$F_{Hy} = F_{RS} - F_{FS}$ — F_{RS} = axial force on rear shroud; F_{FS} = axial force on front shroud	9.2.6
Axial force on impeller when $\mathbf{k} = \mathbf{k}_{FS} = \mathbf{k}_{RS}$	$F_{Hy} = \frac{\pi}{4}\left(d_{sp}^2 - d_D^2\right)\left\{\Delta p_{La} - \frac{\rho}{2}\bar{k}^2 u_2^2\left(1 - \frac{d_{sp}^2 + d_D^2}{2\, d_2^2}\right)\right\}$	9.2.7
Axial force on impeller when $\mathbf{k}_{FS} \neq \mathbf{k}_{RS}$	$F_{Hy} = \pi r_2^2\left\{\Delta p_{La}\left(x_{sp}^2 - x_D^2\right) - \frac{\rho}{4}u_2^2\left(\bar{k}_{RS}^2(1-x_D^2)^2 - \bar{k}_{FS}^2(1-x_{sp}^2)^2\right)\right\}$	9.2.8
Axial force on closed semi-axial impeller	$F_{d2a-d2i} = \frac{\pi}{4}\left(d_{2a}^2 - d_{2i}^2\right)\Delta p_{La}$ when $\mathbf{d}_{2i} < \mathbf{d}_{2a}$	9.2.8a
Momentum force [(b)]	$F_I = \rho\; Q\,(c_{1m} - c_{2m}\cos\varepsilon_2)$	9.2.9
Unbalanced axial force on shaft	$F_w = 1/4\,\pi\, d_D^2\,(p_{amb} - p_1)$	9.2.10
Force on impeller hub in multistage pumps if $d_{s3} > d_n$	$F_{hub} = \frac{\pi}{4}\left(d_{s3}^2 - d_n^2\right)\rho g H_{st}$ use $\mathbf{d}_n$ **or** $\mathbf{d}_{shaft}$	9.2.10a
Resultant force on rotor	$F_{ax} = F_{HY} - F_I + F_w + F_{d2a-d2i} + F_{hub} + F_{coupl}$ — F_{coupl} = axial force of coupling	9.2.11
Estimate for radial and semi-axial pumps	$F_{ax} = (0.7 \text{ to } 0.9)\,\rho\, g\, H_{tot}\frac{\pi}{4}\left(d_{sp}^2 - d_D^2\right)$	9.2.12
Axial thrust with different seal diameters, Figs. 9.15 and 9.16	$F_{Hy} = \frac{\pi}{4}\left(d_{sp}^2 - d_{s2}^2\right)\left\{\Delta p_{La} - \frac{\rho}{2}\bar{k}^2 u_2^2\left(1 - \frac{d_{sp}^2 + d_{s2}^2}{2d_2^2}\right)\right\}$	9.2.13
Resultant force on impeller *with balance holes*	$F_{ax} = (0{,}1 \text{ to } 0{,}2)\, F_{Hy,Eq.9.2.7} + F_{Hy,Eq.9.2.13} - F_I + F_w + F_{coupl}$	9.2.14
Double-entry impeller	$F_{ax} = f_{ax}\;\frac{\rho}{2}u_2^2\left(d_2^2 - d_{sp}^2\right)$ — Steady forces $f_{ax,stat} = 0.01$ to 0.02; Unsteady forces $f_{ax,dyn} = 0.02$ to 0.06	9.2.15
Semi-axial impellers, *closed*	$F_{ax} = \rho\, g\, H\, f_{ha}\frac{\pi}{4}\left(d_{sp}^2 - d_D^2\right)$ — $f_{ha} = \left(\frac{n_q}{n_{q,Ref}}\right)^{0.17}$ $n_{q,Ref} = 220$	9.2.16
Semi-axial impellers, *open*	$F_{ax} = \rho\, g\, H\, f_{ha}\frac{\pi}{4}\left(d_1^2 - d_D^2\right)$ — $f_{ha} = \left(\frac{n_{q,Ref}}{n_q}\right)^{0.28}$ $n_{q,Ref} = 200$ for $n_q < 200$	9.2.17
Axial impellers	$F_{ax} = (1 \text{ to } 1.1)\frac{\pi}{4}\left(d_2^2 - d_n^2\right)\rho\, g\, H$	9.2.18

(a) Semi-axial impellers produce different pressure rise at outer and inner streamline: $\Delta p_{La,FS} \neq \Delta p_{La,RS}$
(b) ε_2 = angle between mean streamline and rotor axis at impeller outlet (90° for radial impellers)
Subscripts: RS = rear shroud; FS = front shroud; D = shaft seal diameter

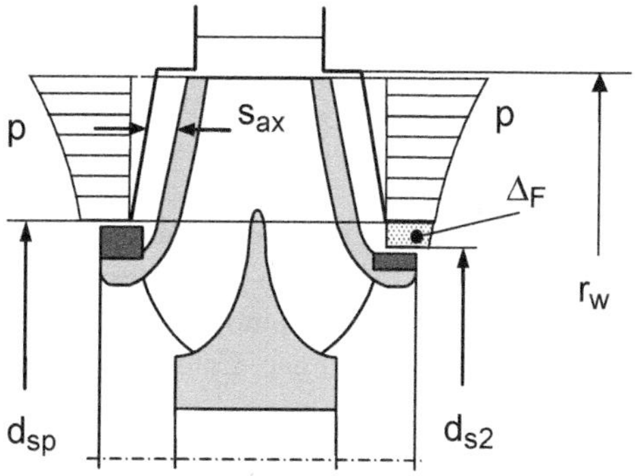

Fig. 9.16 Double-entry impeller

correctly sized, the suction pressure essentially prevails on the rear shroud at the diameter where the balance holes are drilled. The fluid rotation in the unloaded space on the rear shroud causes a pressure distribution which can be calculated using rotation factors according to Table 9.1.

As discussed in Chap. 9.1 the velocity distribution at the impeller outlet determines the boundary condition for the leakage flow into the ISR (k_E in Fig. 9.4).The variation of the swirl k_E as a function of the flow rate is one main uncertainty of the axial force prediction for impellers with balance holes and annular seal on the rear shroud. The effect of k_E on the axial thrust can be calculated in the same way as for double-entry impellers according to Fig. 9.20 and Eq. (9.5e).

Since a complete symmetry cannot be achieved, it is recommended to first calculate the axial thrust of an impeller *without balance holes* from Eq. (T9.2.7) and to assume a residual thrust of 10 to 20% of the unbalanced force. As an alternative or supplement a calculation can be done according to Eq. (9.5e) with (for example) $(k_{E,DE} - k_{E,NDE}) = 0.5$.

To this are added the unbalanced axial forces acting on the shaft F_w and the momentum F_I. Another force is created when different seal diameters d_{sp} and d_{s2} are selected. This finally leads to Eq. (T9.2.14) which gives the resulting axial force acting on the rotor. The seal diameter d_{s2} on the rear shroud can be made different from the ring diameter d_{sp} at the impeller inlet; the objective is to create a defined thrust or to balance a high axial force F_w acting on the shaft to the pressure side. The axial thrust of this type of design is calculated from Eq. (T9.2.13) derived from Eq. (T9.2.7) when replacing d_D by d_{s2}.

Usually one balance hole is drilled per impeller channel. In order to prevent a pressure build-up on the rear shroud due to the throttling effect of the balance holes, the sum of the areas of the holes should be at least 4- to 5-times the flow area in the annular seal clearance, i.e. $\frac{1}{4} \times z \times d^2 > 4 \times d_{sp} \times s$ (z = number of holes, s = radial seal clearance). When selecting the hole diameter according to the above criterion, wear in the annular seal must be taken into account; s may thus be set as $s = 2 \times s_o$ (i.e. twice the nominal clearance).

Balance holes double the leakage losses, Table 3.7(1), and impair the efficiency of the pump. Furthermore, the ingress of the balance fluid perturbs the flow at the impeller inlet. Above $n_q = 50$ an efficiency drop of about 1% (point) can be expected.

Expeller vanes on the rear shroud can be employed instead of balance holes to balance part of the axial thrust. This type of thrust balancing is less expensive than an annular seal on the rear shroud; it is, for example, customary in pumps for acids. Expeller vanes are much in use for pumping liquids which contain solid particles (e.g. sewage or slurry pumps), because they reduce the amount of solids entering the impeller sidewall gap. Calculation and design of expeller vanes are discussed in Chap. 9.2.7.

9.2.3 Multistage Pumps

Axial forces in high-pressure multistage pumps reach many tons so that a hydraulic axial thrust balancing is mandatory. Even then an axial bearing is required; its size depends largely on the accuracy of the axial thrust prediction and on the type of the balance device. An assessment of the uncertainties of the axial force prediction (discussed in Chap. 9.2.1) shows that the band width of the expected residual thrust is significant as compared to an economic bearing size. Therefore, bearing sizes and axial thrust balance systems of multistage pumps are frequently determined on the basis of axial force measurements carried out on each particular type of pump and hydraulic.

Any calculations must take into account the significant differences in fluid rotation on front and rear shrouds as per Table 9.1 caused by the different directions of the leakage flow (except in the last stage). Once the rotation factors are determined, Eqs. (T9.2.3–5 or 8) can be used. As discussed above, the radially outward leakage on the rear shroud *reduces* the rotation factor to $k < 0.5$ and increases the axial force. In contrast, the radially inward leakage on the front shroud enhances rotation to values $k > 0.5$ and reduces the force acting on the front shroud, Fig. 9.17. As a consequence, *both effects add* and significantly increase the axial thrust compared to earlier publications where $k \approx 0.5$ has been proposed to calculate the axial forces. If the clearances wear during service, the axial forces acting on the rotor increase. Accordingly, balance device and bearings must be dimensioned for new as well as for the maximum allowed clearances (often the maximum allowable clearances are defined as two-times the new clearances).

As discussed in Chap. 5.4.3, the axial rotor position can have an influence on the axial thrust. This is particularly true if gap A is too large and/or the overlap x_{ov} is insufficient, i.e. if main flow and impeller sidewall gap are not sufficiently decoupled and the partload recirculation can affect the flow in the sidewall gaps. An example of a thrust reversal induced in this way is shown in Fig. 5.35.

It is important to note that the *residual thrust is a fraction only of the hydraulic forces acting on the rotor.* Consequently, *even small tolerances in the calculation of the hydraulic thrust imply large changes in the residual thrust* which is a small difference of two large figures. This fact should well be considered when calculating the axial thrust of a multistage pump.

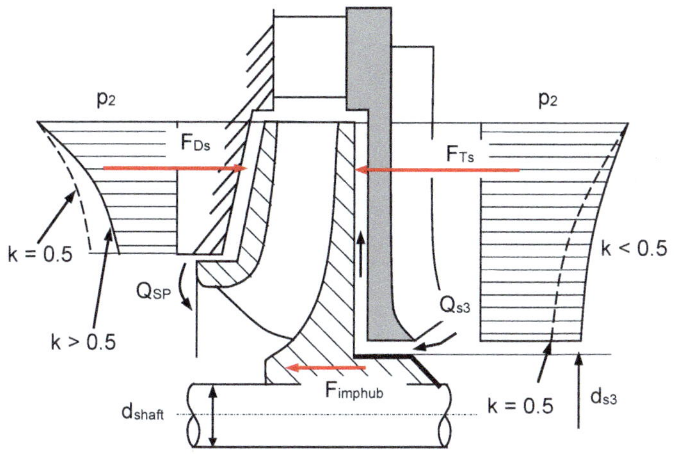

Fig. 9.17 Pressure distribution on an impeller of a multistage pump

Unbalanced axial forces acting on the shaft of a multistage pump merit a careful analysis since they can have a considerable influence on the residual thrust on the rotor due to the high pressures involved. Because of the complicated rotor design and the different pressures prevailing in the individual stages, there is a risk to overlook these force components or to assess their effect erroneously. In a design like Fig. 2.8 with shrink-fitted impellers, where the interstage seal diameter d_{s3} is greater than the shaft diameter d_{shaft}, an axial force acts on the impeller hub according to Eq. (T9.2.10a).

Various options are available for balancing the axial thrust of multistage pumps, Fig. 9.18:

Balance disks (Fig. 9.18a) create forces which oppose the axial thrust generated by the impellers. When dimensioned correctly, balance disks are able to completely balance the axial thrust as the required pressure automatically builds up in the radial gap due to an axial displacement of the rotor. The axial clearance s_{ax} at the disk thus depends on the axial force acting on the rotor, hence on the operation point of the pump.

Since the axial thrust is compensated automatically, the uncertainties of the calculation are less serious than with a balance piston. However, problems may be encountered due to transients during which water can evaporate in the radial gap. As a remedy, another throttling gap downstream of the disk can be included in the design. Balance disks are usually preceded by a cylindrical gap ("pre-throttle"), as shown in Fig. 9.18a, which is necessary for the stable operation of the balance disk. The throttling effect in the cylindrical gap allows larger widths s_{ax} of the radial gap because the pressure is reduced from p_1 to p_z.

If the axial force acting on the impellers increases, the rotor is shifted to the left in Fig. 9.18a, the axial clearance s_{ax} at the disk diminishes and thus throttles the balance flow rate. Due to the reduced flow rate the pressure loss in the pre-throttle

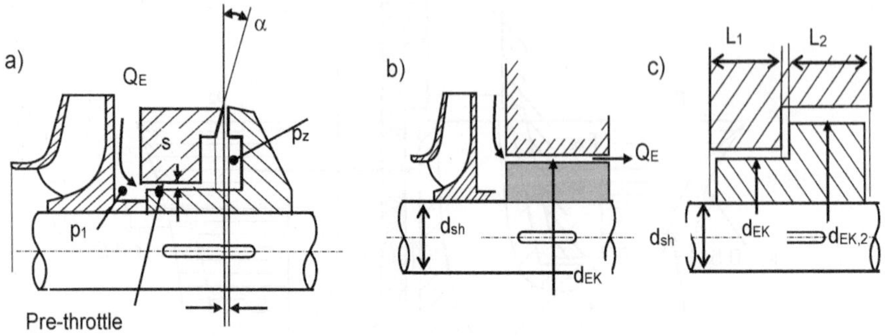

Fig. 9.18 Axial thrust balancing of multistage pumps. **a** Balance disk; **b** balance piston; **c** stepped balance piston

decreases. Therefore, the pressure p_z rises generating a higher balancing force acting on the disk. Equilibrium between the forces acting on the rotor and on the disk is thus established in every operation point. The gap widths s and s_{ax}, the length of the pre-throttle and the inner and outer radius of the disk must be tuned carefully. When doing so, it must be ensured that the counter forces created by the disk are at all times greater than the resulting axial forces which act on the rotor at varying operation conditions including transients. This requirement must be equally fulfilled when the clearances widen due to wear.

The greatest force on the disk is created when the axial clearance approaches zero, because then the pressure drop in the pre-throttle is small (i.e. p_z is maximum). This maximum force can be significantly increased if the gap is made slightly conical, as depicted in Fig. 9.18a, by the angle α, [32].

Mechanically the disk must be made very stiff in order to limit deformation under load to the very minimum. Details for the design of balance disks and considerations on the stability of operation can be found in [26, 30, 32, 41]. Due to shaft deflection, the balance disk will be slightly tilted against its the stationary counterpart. This deviation from parallelism generates a tilting moment which is stabilizing since it tends to restore the disk to a parallel position. The tilting has virtually no influence on the resulting axial force but it increases the risk of cavitation in the gap, [26, 41].

Balance pistons (Fig. 9.18b) generate a force opposing the axial thrust. As an approximation, the force created by the piston can be calculated from the area of the piston and the pressure at the entrance to the piston:

$$F_{EK} = p_{EK}\frac{\pi}{4}\left\{d_{EK}{}^2 - d_{sh}{}^2\right\} - \Delta F_{EK} \tag{9.5a}$$

$$\Delta F_{EK} = \frac{\pi}{4}\left(d_{EK}^2 - d_{sh}^2\right)^2\frac{\rho}{4}\omega^2\left(k_{in}{}^2 - k_{out}{}^2\right) \tag{9.5b}$$

The term ΔF_{EK} accounts for the effect of fluid rotation: at the inlet to the piston set $k_{in} = 1.0$ and at the outlet of the piston set $k_{out} = 0.2$ (r_{sh} = radius of shaft at piston).

The piston diameter is typically of a similar magnitude as the annular seal diameter $d_{EK} \approx d_{sp}$ at the impeller inlet; and the piston length is roughly $L_{EK} = 0.8 \times d_{EK}$. Since the piston is unable to automatically adapt to the hydraulic forces on the rotor, the axial thrust should be known as accurately as possible for a correct dimensioning of the piston. A relatively large axial bearing is required to take up the uncompensated thrust which is given by the uncertainties of the force calculation, changes in thrust as function of the flow rate through the pump and increase in thrust caused by wear in the clearances of the annular seals. The advantage of the piston is its robust design and high reliability in operation. The disadvantages are a higher leakage and a larger axial bearing, both of which impair the efficiency of the pump. According to operation and damage statistics of large boiler feed pumps, 310 out of 533 pumps with balance disks experienced damage, while only 27 incidents were found on 511 pumps with balance pistons, [32].[3]

Stepped balance piston (Fig. 9.18c**):** With this design it is attempted to benefit from the advantages of balance disks without incurring the disadvantages of the piston. But this works only if the radial gap between the two pistons can vary with the load. The pressure at the step can be calculated from:

$$p_z = \frac{p_{EK}}{1 + \dfrac{L_1 d_{EK,2}^2}{L_2 d_{EK}^2}} \tag{9.5c}$$

If no throttling is assumed at the step and the effect of rotation on the step is neglected, the balancing force can be estimated from:

$$F_{EK} = p_{EK}\left\{d_{EK}^2 - d_{sh}^2\right\}\frac{\pi}{4} + p_z\left\{d_{EK,2}^2 - d_{EK}^2\right\}\frac{\pi}{4} - \Delta F_{EK} \tag{9.5d}$$

For an exact calculation, the axial gap width must be known. This depends on the design of the axial bearing and the elasticity of its support.

The fluid leaking through the balance device (piston or disk) is led via a pipe back to the suction nozzle of the pump. Therefore, the suction pressure (i.e. $NPSH_A$) prevails at the low-pressure side of the balance device. In addition, there is a small pressure loss in the balance pipe. In the absence of a booster pump the balance fluid can also be returned to the suction tank. To prevent evaporation ("flashing") in the balance line, a throttle orifice can be installed at the inlet to the suction tank in order to create a sufficiently high counter-pressure at the balance device. It is recommended to monitor the proper functioning of the balance device by measuring the leakage flow rate during service in order to early recognize an

[3]Whether these figures are still representative for the present, is unknown since no recent statistics of this amplitude are available.

increase in the flow due to enlarged clearances, as may be caused by abrasive wear. **Back-to-back design:** The axial thrust of multistage, double-entry pumps according to Fig. 2.14 is almost perfectly balanced without any sacrifice in efficiency (Chap. 9.2.4). The axial thrust of single-entry pumps with several stages in back-to-back design according to Fig. 2.10 is balanced to a large extent. Disadvantages of this design are the long cross-over channels which involve complicated castings. Conversely, the piston placed in the middle of the rotor greatly improves the rotor dynamics (Chap. 10.6).

A two-stage back-to-back design according to Fig. 9.19 needs particular attention because the axial thrust is not well balanced, since the pressure distributions on the rear shrouds of stages 1 and 2 are *not at all equal*. The leakage on the rear shroud of the second stage flows radially *inwards*. There the fluid rotation is higher than on the rear shroud of the first stage, where the leakage flows radially *outwards*. This effect can well be calculated by the method given in Table 9.1 applying different rotation factors to rear and front shrouds of both stages, Eqs. (T9.2.8) or (T9.2.3 or 9.2.4). If the clearances of the center bushing are enlarged due to wear during service the axial thrust of the 2-stage back-to-back design increases strongly. Erroneously considering this design to have well balanced axial thrust, is likely to result in bearing failures when clearances increase due to wear during service. Axial force predictions according to the procedures in Table 9.1 are quite accurate near BEP and at high flow. However, the axial force calculation at partload can substantially deviate from the measured forces because the tangential velocity at the inlet to the ISR can be quite low due to recirculation.

Individual stage balancing: Balancing the individual stages as in Fig. 2.15 and 9.15 in Table 9.1 is common practice in vertical pumps with semi-axial impellers

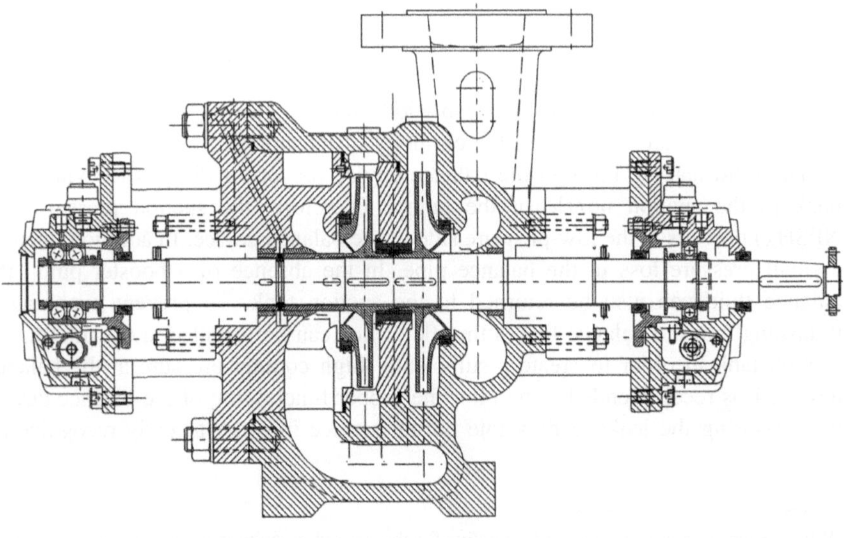

Fig. 9.19 Two-stage back-to-back process pump, Sulzer

and diffusers. Individual balancing is not an economic solution for high pressure pumps (lower efficiency and higher costs), since a throttling device is needed anyway to reduce the pressure at the shaft seal. However, low-pressure pumps are sometimes designed with individual stage balancing in order to avoid a balance piston. The main uncertainty of the axial thrust prediction is given by the variation of the inflow conditions (factor k_E) to the ISR. The expected variation of the axial force can be calculated in the same way as for double-entry impellers, Chap. 9.2.4, Eq. (9.5e).

Expeller vanes (as per Chap. 9.2.7) could also be considered for individual stage balancing. However, with respect to efficiency this design is less favorable than the balance devices discussed above because the axial clearances between expeller vanes and casing would become too large in multistage pumps due to the requirements of manufacturing tolerances and thermal expansion. Furthermore, a throttling device is needed anyway to reduce the pressure at the shaft seal.

9.2.4 Double-Entry Impellers

Theoretically the axial thrust of double-entry impellers should be perfectly balanced, but in reality steady as well as unsteady axial forces are created by various asymmetries as there are: (1) unavoidable geometric tolerances of the impeller and inlet casing; (2) differences in the annular seal clearances generating different rotation factors on both sides; (3) asymmetric partload recirculation. Frequently the unsteady axial forces are greater than the steady component so that low-frequency thrust reversal may occur. Sometimes axial rotor movement can be visually observed on low-speed pumps operating at partload. The period of such axial rotor movement may be in the order of several seconds. This type of thrust reversal can be acceptable with low-speed pumps unless there are problems with the mechanical seals, but damage to the shaft seals and excessive vibrations may be encountered in high-energy pumps.

Sometimes it is attempted to create a definite axial thrust on the rotor by designing either side of the impeller with different annular seal diameters. However, it is hardly practical to make the diameter difference sufficiently large to create a resultant static axial force which exceeds the unsteady axial force components. The calculation (including the influence of different seal leakages on both sides of the impeller) can be performed using Eq. (T9.2.8), where x_D is set to $x_D = d_{s2}/d_2$. According to Fig. 9.16 the smaller seal diameter is d_{s2} (Table 9.1). If $d_{s2} \neq d_{sp}$ is executed and the same rotation factor is assumed for both sides, also Eq. (T9.2.13) can be used. The resulting force F_I generated by the momentum is zero.

Pumps with double-entry impellers need axial bearings which can be dimensioned according to Eq. (T9.2.15), adding the coupling thrust, the rotor weight in the case of vertical pumps and appropriate margins. Equation (T9.2.15) uses axial force coefficients f_{ax} derived from measurements on pumps with double-entry impellers. Ranges of f_{ax} are given for static and unsteady force components in

Table 9.2. The lower limits apply to operation near the best efficiency point, while the upper limit is relevant for partload operation below $q^* \approx 0.40$.

In partload operation the flow distribution at the impeller outlet cannot be expected to be uniform (see Chaps. 5.8 and 10.7.3). As a consequence, an axial force is created by differences in the circumferential velocity c_{2u} on both sides of the impeller. The boundary conditions for the ISR integration according to Table 9.1 change accordingly. The results of a sample calculation for an impeller with a specific speed of $n_q = 27$ are presented in Fig. 9.20. The rotation factors at the inlet to the ISR - $k_{E,DE}$ and $k_{E,NDE}$ - were treated as independent variables. The subscript DE stands for the drive end and NDE for the non-drive end. In an actual pump it is not known on which side the highest value k_E will occur; hence the direction of the calculated axial forces could act either towards DE or NDE. In this calculation $k_{E,DE}$ was varied between 0.2 and 0.7 and $k_{E,NDE}$ varied between 0.2 and 0.5. When plotting f_{ax} against the difference of rotation factors ($k_{E,DE} - k_{E,NDE}$), all resulting f_{ax} fall in a narrow range which can be approximated by a linear correlation as given by Eq. (9.5e):

$$f_{ax} \equiv \frac{F_{ax}}{\frac{\rho}{2} u_2^2 (d_2^2 - d_{sp}^2)} = 0.14(k_{E,DE} - k_{E,NDE}) \tag{9.5e}$$

In the extreme case one could expect $k_{E,DE} = 0.7$ and $k_{E,NDE} = 0.2$; the resulting axial force factor would be 0.07.

The axial force measurements shown in Chap. 10.7.3, Fig. 10.35 show variations from +10,000 to −25,000 N; this corresponds to $f_{ax,stat} = 0.048$. The dynamic force fluctuation were $f_{ax,dyn} = \pm 0.007$.

Figure 9.20 and Eq. (9.5e) can be used for an estimate of the axial forces on a double-entry impeller, which are created by geometric tolerances and non-uniformities of the flow at partload. Also, the effect of the ISR inflow conditions on the axial forces on an impeller with balance holes and annular seal on the rear shroud (Fig. 9.15) can be estimated from Eq. (9.5e).

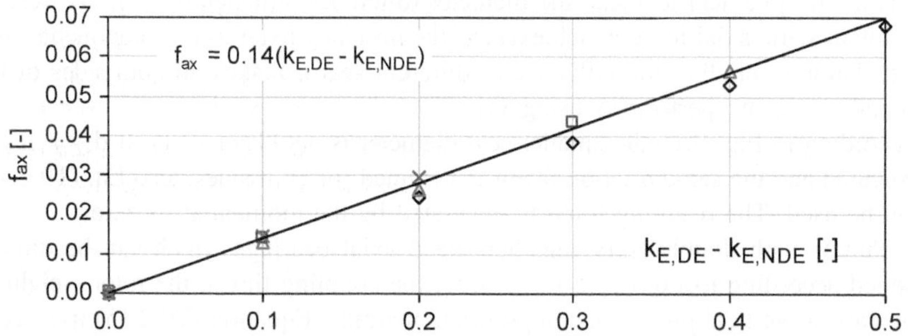

Fig. 9.20 Axial force factor for double-entry impeller

In addition to the variation of the velocity distribution at the impeller outlet, some variation in the static pressure may be expected too.

9.2.5 *Semi-axial Impellers*

Closed semi-axial impellers can be calculated according to Chap. 9.2.1 and Tables 9.1 and 9.2. Since the front shroud diameter is larger than the rear shroud diameter ($d_{2a} > d_{2i}$), the calculation follows Eqs. (T9.2.3) and (9.2.4) separately for front and rear shroud setting r_{2a} and r_{2i} accordingly. At the impeller outlet, the force $F_{d2a\text{-}d2i}$ according to Eq. (T9.8a) is to be added. The uncertainties of the calculation are enhanced by the circumstance that the static pressures at the outer and inner streamlines are different. These differences are not very well known - in particular at partload. Therefore, measurements or approximation formulae are alternatively employed for dimensioning the capacity of the axial bearing. Equations (T9.2.12 or 9.2.16) can be applied to closed impellers. In these formulae, set the shaft diameter for impellers without balance holes (Fig. 9.14), but use the annular seal diameter d_{s2} for the rear shroud if balance holes are provided, (Fig. 9.15 in Table 9.1). The empirical factor f_{ha} (from test data given in [B.18]) is calculated from Eq. (T9.2.16) for closed impellers. For semi-open impellers f_{ha} is determined from Eq. (T9.2.17) setting $d_{sp} = d_1$, [B 18] and [B 2].

9.2.6 *Axial Pumps*

In order to calculate the axial thrust of axial pumps, the axial component of the lift forces acting on a blade can be integrated over the blade height. However, the approximation formula according to Eq. (T9.2.18) may be often adequate. Added to this force are the unbalanced axial forces acting on the shaft which can be calculated as per Chap. 9.2.1 similar to Fig. 9.14. The momentums at the inlet and outlet cancel. Since the Q-H-curves of axial pumps are very steep, the increase of the axial thrust at partload needs attention.

9.2.7 *Expeller Vanes*

By means of expeller vanes on the rear shroud (also called "rudimentary" or "pump-out" vanes) the fluid rotation in the sidewall gap E can be intensified and the axial thrust reduced. Frequently expeller vanes are employed in order to reduce the pressure at the shaft seal and/or to impede the ingress of solid particles into gap E (e.g. in sewage, dredge and slurry pumps).

As far as possible, the axial clearance between the casing (wear plate) and the expeller vanes is selected as the minimum required by considerations of mechanical design. This minimum depends on the machining tolerances, thermal expansion and casing deformation under load. However, when designing a range of pumps with standardized bearing frames, it is necessary to vary the expeller vane height h and the axial clearance s to the rear cover in order to achieve a constant distance between the impeller center plane and the rear cover for a given bearing frame and given impeller diameters.

Expeller vanes are often executed as radial ribs with rectangular cross sections. Sometimes they are built as back-swept auxiliary vanes following (near the outer impeller diameter) the contour of the impeller blades. This design is also chosen when balance holes are used in addition to the expeller vanes. Little or no efficiency impairment is incurred with such an arrangement since the expeller vanes essentially create the same pressure rise as the main blades (they can be considered as working in a similar way as a semi-open impeller with the same blade shape); compare also to the discussion of Fig. 7.16.

The fluid rotation grows with the number of expeller vanes and their height h; it decreases if the width s of the axial gap is enlarged (Fig. 9.21 in Table 9.3). Enhanced fluid rotation entails an additional pumping power compared to a plain rear shroud- at least if *radial* expeller vanes are used. The efficiency of the pump is then impaired. The loss increases with decreasing specific speed and becomes significant below $n_q = 20$. Some tests seem to indicate that back-swept expeller vanes would cause little additional losses, but systematic investigations which would confirm this finding are not yet known.

The power increase caused by the expeller vanes grows with the 5th power of their outer diameter (as with disk friction losses or, indeed, the power absorbed by an impeller). Since pressure differences are proportional to the square of the diameter, power can be saved by not extending the expeller vanes up to the impeller outer diameter: by selecting $d_{Rs} < d_2$, power losses can be significantly reduced without much sacrifice in axial thrust balancing or pressure relieve at the shaft seal (this is the most common design).

Fig. 9.21 Closed impeller with expeller vanes

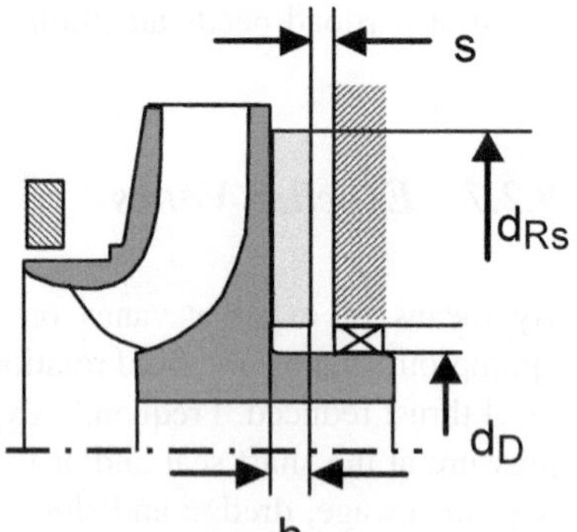

Table 9.3 Semi-open impellers; expeller vanes		Eq.
Semi-open impellers with or without expeller vanes	$F_{ax} = \frac{\pi}{4} d_2^2 \left[\Delta p_{La}\left(1-\frac{d_D^2}{d_2^2}\right) - \frac{\rho}{4}\bar{k}_{RS}^2 u_2^2 \left(1-\frac{d_D^2}{d_2^2}\right)^2 - \frac{\Delta p_{La}}{2}\left(1-\frac{d_1^2}{d_2^2}\right)^2\right]$	9.3.1
Semi-open impellers *without* expeller vanes	$F_{ax} = \frac{3\pi}{16}\rho r_2^4 \omega^2 \left(1-d_1^{*2}\right)^2$ $f_{ax} \equiv \frac{8F_{ax}}{\pi d_2^2 \rho u_2^2} = \frac{3}{8}\left(1-d_1^{*2}\right)^2$	9.3.1a
Expeller vanes Rotation factors k_{Rs} and k_{av}	$k_{Rs} = \frac{1}{1+0.13\frac{s}{s+h}\sqrt{\frac{r_2}{h\,z}}}$	9.3.2
k_o from Eq. (T9.1.3)	$\bar{k} = \sqrt{\left(\frac{d_{Rs}}{d_2}\right)^{\left(2-0.9\frac{d_{Rs}}{d_2}\right)}\{k_{Rs}^2 - k_o^2\}+k_o^2}$	9.3.3
	Geometry as per [36]: Number of vanes z: $d_2 < 200$ mm: z = 4 $d_2 > 200$ mm: z = 6 to 8 $d_{RS}/d_2 = 0.75$ to 0.85 $h/r_2 = 0.03$ to 0.05 $s/r_2 = 0.008$ to 0.015 s/h = 0.1 to 0.2. vane width = 2 h	
Power loss of *radial* expeller vanes $\zeta_{Rs} \equiv \frac{P_{RS}}{\rho\,\omega^3 r_2^5}$	$\zeta_{Rs} = \frac{0.1}{Re_u^{0.2}}\left[\left(\frac{d_{Rs}}{d_2}\right)^4\left(\frac{h+s}{r_2}+0.24\right)+\frac{1}{4}\left(\frac{h+s}{r_2}\right)^{0.1}\left\{1-\left(\frac{d_{Rs}}{d_2}\right)^5\right\}\right]$	9.3.4

Impellers with *radial* expeller vanes are usually employed without balance holes. Furthermore the influence of any leakage on the flow in gap E is small if expeller vanes are present. The effect of expeller vanes is therefore calculated without leakage. The rotation factor k_{Rs} in the section equipped with expeller vanes is then essentially independent of the radius; it is calculated from Eq. (T9.3.2).

If the expeller vanes are trimmed (i.e. $d_{Rs} < d_2$), the fluid rotation in the plain outer part of the rear shroud (between d_{Rs} and d_2) is higher than k_o from Eq. (T9.1.3) but lower than k_{Rs} from Eq. (T9.3.2). The resulting fluid rotation is determined by an exchange of momentum between fluid in the vaned and plain parts of the shroud. The average rotation factor k_{av} on the entire rear shroud is given by Eq. (T9.3.3). The pressure prevailing at the shaft seal can be calculated by inserting the value k_{av} into Eq. (T9.1.7). When defining the impeller outlet pressure p_2, the suction pressure has to be taken into account. The axial thrust can be calculated from Eqs. (T9.2.3 or 9.2.8) using said value of k_{av}.

The power absorbed by *radial* expeller vanes can be estimated from Eq. (T9.3.4) which can be used for trimmed ($d_{Rs} < d_2$) as well as full ($d_{Rs} = d_2$) expeller vanes. The formulae originate from the tests reported in [36] but have been simplified to ease their application in practice. Table 9.3 provides recommendations for suitable geometric dimensions of expeller vanes. The following recommendations for the design of expeller vanes can be given:

- number of expeller vanes equal to number of impeller blades: $z = z_{La}$
- shape of expeller vanes same as of impeller blades (mostly back-swept)
- outer diameter: $d_{Rs} = 0.85d_2$
- vane height $h/r_2 = 0.04$ to 0.05 (adapt as required for rear cover standardization)
- axial clearance $s/r_2 = 0.01$, or $s/h = 0.2$ (adapt as required for rear cover standardization)
- width of expeller vanes equal to height h or equal to nominal blade thickness e_{nom} (whatever is greater)

When using expeller vanes, the supporting effect (stiffness) of an annular seal on the rear shroud is absent. The shaft deflection caused by the radial thrust and the alternating shaft stresses increase. Therefore, high-head single-stage pumps can suffer shaft failures when the axial thrust balancing is modified from a design with an annular seal and balance holes on the rear shroud to a design with expeller vanes.

In slurry pumps, expeller vanes are sometimes employed on both the rear and the front shrouds in an attempt to reduce abrasive wear. The axial thrust increases accordingly. The calculation follows the same logic as discussed above treating front and rear shrouds separately and determining the rotation factors from Table 9.3.

Apparently, the majority of manufacturers of single-stage pumps with single-entry impellers elect balance holes and annular seal on the rear shroud rather than expeller vanes. The reasons for this choice are not quite evident. Table 9.5 in Chap. 9.2.10 provides a comparison of these two methods of thrust balancing. This evaluation seems to come out rather in favor of expeller vanes. A systematic test series to elucidate this question is lacking.

9.2.8 Semi-open and Open Impellers

Impellers without a front shroud are termed "semi-open". The axial thrust is higher than with closed impellers, because the pressure in the *vaneless* impeller sidewall gap always exceeds the pressure within the *bladed* impeller. Assuming that the pressure within the impeller increases linearly with the radius, the axial thrust of a semi-open impeller can be calculated from Eq. (T9.3.1), see Fig. 9.22 in Table 9.3. Added to this are unbalanced axial forces F_w acting on the shaft and the momentum F_I. For k_{RS} set k_o from Eq. (T9.1.3) if the rear shroud is plain; or set $k_{RS} = k_{av}$ from Eqs. (T9.3.2 and 9.3.3) if there are expeller vanes. The axial thrust of semi-open impellers is frequently reduced by employing expeller vanes.

The axial thrust can be reduced further by cut-outs in the rear shroud as shown in Fig. 9.23. The term "open impeller" is applied to this type of design. Possible expeller vanes follow the contour of the blades. When calculating the axial thrust of an open impeller, it is assumed that the axial forces are balanced down to the minimum diameter d_{co} at the cut-outs (defined in Fig. 9.23).

Table 9.4. Overview on axial thrust calculations

	Axial force balancing	Rear shroud		Equation for axial force F_{Hy}:	Sensitivity to increase in clearances	Comments
		Q_{sp}	k_{RS}			
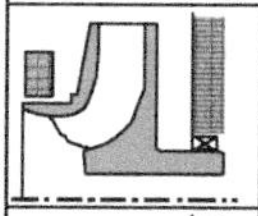	none	0	< 0.5	9.2.8	moderate increase of F_{Hy}	$k_{RS} = k_o$
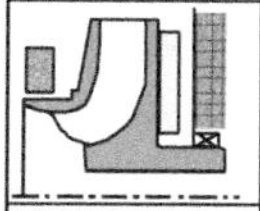	expeller vanes	0	0.7 – 0.9	9.2.8	moderate	k_{RS} from Table 9.3
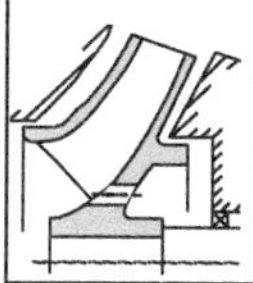	rear shroud seal and balance holes	> 0	0.55 – 0.65	9.2.14 9.2.13 9.2.8 9.5a	weak, if the same clearance on front and rear shroud	axial force determined by tolerances and partload recirculation entering the ISR
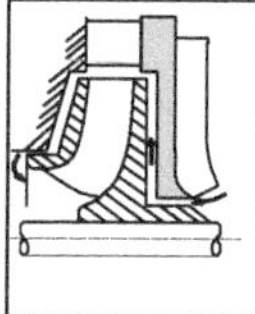	multistage (with piston or disk)	< 0	0.35 – 0.45	9.2.8	*strong:* effects on front and rear shroud add	main issues: (1) partload recirculation entering the ISR on front shroud; (2) increased clearances
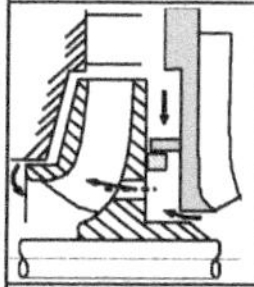	multistage: rear shroud seal and balance holes	> 0	0.55–0.65	9.2.8 9.2.14 9.2.13 9.5a	weak	main issue is *partload recirculation* entering the ISR
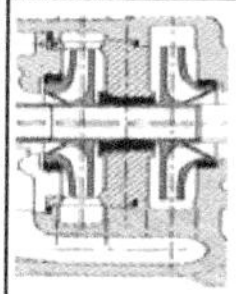	back-to-back arrangement	1st: < 0 2nd: > 0	1st Stage: 0.35 – 0.45 2nd Stage: 0.55 – 0.65	9.2.8	*Very strong*, effects of both stages add	careful analysis needed because forces on rear shrouds are not balanced
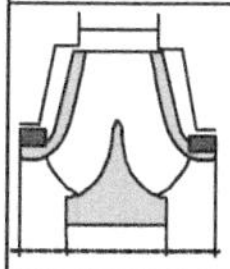	double-entry impeller	> 0	0.55 – 0.65	9.2.15 9.2.8 9.2.13 9.5a	weak, if the same clearance on both sides	nominally axial forces are well balanced; main issue is partload recir-culation entering the ISR, chap. 10.7.3

In all cases: on front shroud: $Q_{sp} > 0$; $k_{Ds} > 0.5$ (typically: 0.55 to 0.65)
$Q_{sp} < 0$: leakage flow is radially outwards

For impellers with *multiple radial* blades ($\beta_{1B} = \beta_{2B} = 90°$) another formula can be used for axial thrust prediction. It is given by Eq. (9.3.1a) which was derived from Eq. (T9.2.8) by assuming a forced vortex with ω between the radial impeller blades and a forced vortex with ½ω between rear shroud and casing. This equation agrees with the measurements given in [8] for the case without balance holes. With multiple balance holes, well distributed over the rear shroud, the axial thrust can be reduced to about 1/4 of the force predicted by Eq. (9.3.1a).

Table 9.5 Balance holes versus expeller vanes

	Rear shroud seal and balance holes 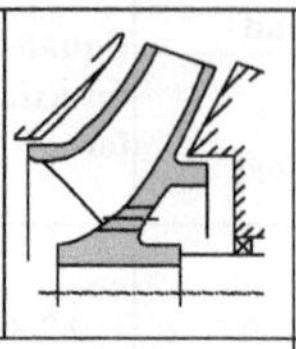	**Expeller vanes** 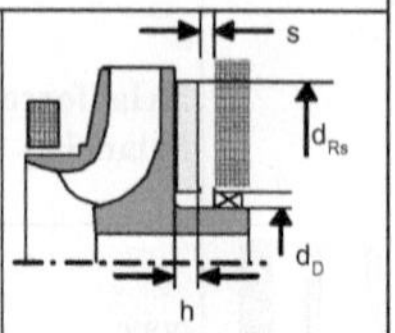
design requirements	annular seal on rear shroud may increase the length of the casing (e.g. in multistage pumps)	• axial gap needs to be controlled for effective pumping action • for radial vanes: expeller vane diameter $d_{Rs}/d_2 < 0.85$ to reduce losses • backward swept expeller vanes to reduce losses
impeller side room design on rear shroud	more freedom to design rear cover since side room can be large	rear cover design: freedom restricted because of controlled axial gap
Impacts on costs	• additional annular seal • machining more costly • rear cover more expensive depending on design • higher spare part inventory	• as such, expeller vanes are the least costly design: much less machining • easier standardization of rear cover since no seal • differences in axial position can be compensated by expeller vane height
efficiency	*additional* volumetric losses through rear shroud seal	• Virtually no additional losses with backward swept expeller? Sufficient reliable data lacking, comparison tests required!
hydraulic performance	rear shroud seal leakage causes flow perturbation at the impeller inlet, additional mixing losses and cavitation	pumping action of expeller vanes adds energy to main flow but also causes mixing losses
axial forces	• if seal diameters equal, force is well balanced near BEP • partload recirculation on one of the shrouds will increase axial thrust; estimation acc. Eq. (9.5e) • at partload the direction of the force can be either towards suction or towards drive end	• the axial force on the rear shroud depends little on the operation point • depending on the velocity profile at the impeller outlet, the axial thrust at partload could increase or decrease
axial force reversal	axial force is sensitive to differences in flow conditions, see Fig. 9.20 with comments	little impact of partload recirculation on force on *rear* shroud, but force on front shroud is affected
sensitivity to wear in annular seal	Possibly less sensitive if both seals wear by the same amount. However, at partload the reverse may be true.	axial forces towards suction increases with wear of seal on front shroud
radial forces	annular seal on rear shroud provides additional stiffness which reduces shaft deflection and increases rotor damping	little effect expected

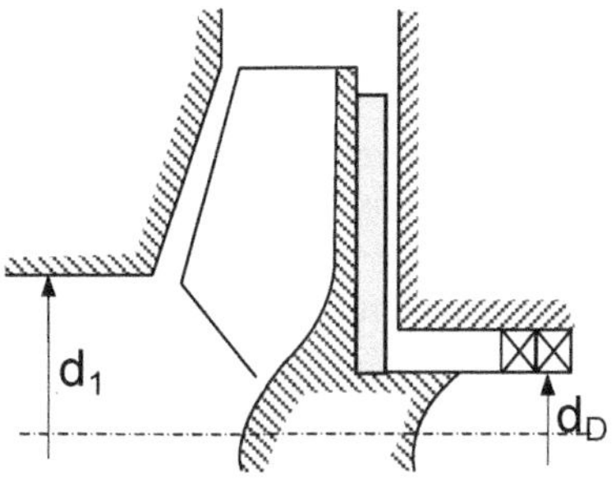

Fig. 9.22 Semi-open impeller with expeller vanes

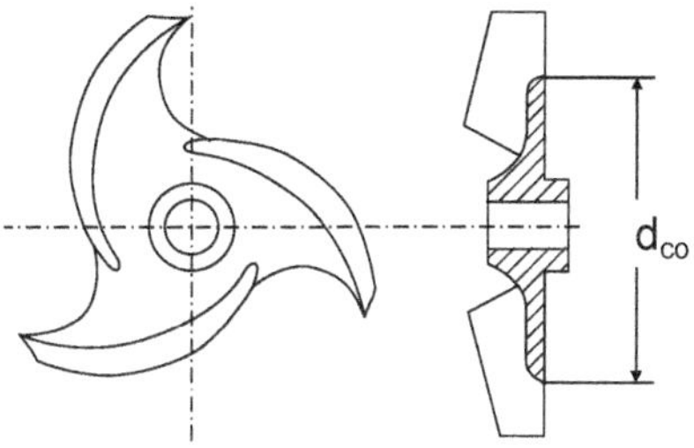

Fig. 9.23 Open impeller

9.2.9 Unsteady Axial Thrust

During start-up of a pump, transient axial forces exceed the forces in normal operation because it takes a few seconds till fluid rotation in the impeller sidewall gaps is fully developed, while the pressure rise in the impeller is proportional to the square of the speed with virtually no inertia effects. When starting a vertical pump, the axial thrust can thus be acting upwards during a short period. This fact needs to be taken into account when selecting the axial bearing.

If a vertical pump is started with an open discharge valve, the axial thrust F_I due to the momentum at very high flow can also be acting upwards - in particular with high specific speeds.

The spectrum of unsteady axial forces exhibits low- and high-frequency components similar to the spectrum of pressure pulsations. In practice such axial force fluctuations rarely cause problems, with the possible exceptions of high-energy pumps operating at partload if gap A was selected too large.

Measurements on multistage pumps in [B.20] gave $k_{ax} = 0.005$ at $q^* = 0$ and $k_{ax} = 0.0025$ at $q^* = 1$, where k_{ax} is the coefficient of the unsteady axial thrust per stage according to Eq. (T9.2.15). It is defined as RMS-value in the frequency range $f = (0.2–1.25)\ n/60$. About half of these values were measured in the range $f < 0.2$ n/60.

9.2.10 Axial Thrust Calculation Overview

Table 9.5 gives an overview of commonly used impeller configurations, the relevant equations to be applied, the sensitivity to annular seal clearances and typical rotation factors to be expected. Table 9.5 provides a comparison between axial force balancing by expeller vanes versus balancing holes and annular seal on the rear shroud.

Important notice: As extensively discussed above, the interaction between the main flow and the fluid in the ISR has an altogether dominating effect on the rotation factor and the axial forces when the ISR is open to the volute or diffuser. The associated energy transfer can scarcely be quantified. This should be borne in mind when calculating rotation factors and axial forces. In critical cases it is recommended to study the sensitivity of the results to the circumferential velocity at the impeller outlet which defines the boundary condition k_E for the ISR integration (Figs. 9.1, 9.2, 9.3 and 9.4). In a sensitivity study the inlet conditions to front and rear shroud ISR could be varied in a range of from $k_E = 0.2$ to 0.7 (for example). This range would cover rather extreme cases; see also Chap. 9.2.4. At partload the value $k_E = 0.7$ approximately represents the maximum as can be inferred from Fig. 5.15, while the value $k_E = 0.2$ is about the minimum to be expected with recirculation at the impeller outlet (Fig. 5.30). The effect of inlet conditions to the ISR on the axial forces can be estimated from Fig. 9.20 and Eq. (9.5e) for double-entry impellers or single-entry impellers with balance holes and annular seal on the rear shroud.

The two single factors which represent the largest uncertainty of axial force calculations are: (1) partload recirculation and (2) influence of the leakage flow. While reasonable estimates of the impact of the leakage can be made by employing the procedure in Table 9.1, the estimation of the influence of partload recirculation on the axial force is beyond the present state of the art. As discussed in Chap. 9.1 recirculation at the diffuser or volute inlet slows down the rotation in the ISR and thus increases the force on the shroud affected. But it usually is impossible to predict whether the recirculation occurs on the front or the rear shroud.

By designing small gaps A with sufficient overlap x_{ov} (Table 0.2) the effect of the recirculation can be mitigated when there is a radially inward leakage through the ISR or largely eliminated when there is no radially inward leakage to the ISR.

It is not desirable to design a pump rotor for zero axial thrust. Rather, the axial forces should be selected such that the rotor is under tension and that no thrust reversal occurs in the allowable operation range. This way the rotor eigen frequencies are improved and rough running due to thrust reversal (unloading the bearing) is avoided.

9.3 Radial Forces

9.3.1 Definition and Scope

The radial force acting on an impeller (the "radial thrust") must be known in order to calculate bearing loads, shaft stresses and shaft deflection. Fundamentally, radial forces are generated when the circumferential distribution of the static pressure p_2 at the impeller outlet is non-uniform. Flow asymmetries in the collector as well as a rotationally non-symmetric impeller inflow can thus create radial forces.

Since the pressure distribution at the impeller outlet is unsteady, its integration over the circumference yields a time average, the *static* radial thrust, as well as a spectrum of unsteady radial force components, which are called "*dynamic* radial thrust". Thus the dynamic radial thrust represents hydraulic excitation forces leading to forced vibrations, which are discussed in Chap. 10.7.

The radial forces created by time-dependent pressure distributions at the outlet of a closed impeller are influenced by various physical effects:

Excitation forces with static and dynamic components which are independent of rotor vibrations:

1. Non-uniform flow in the collector acting on the impeller outlet width *and on the projection of the shrouds* unless these are perpendicular to the rotor axis.
2. Non-uniformities of the flow in the impeller sidewall gaps, which can be created by the pressure distribution in the collector, as well as asymmetries in the leakage flow if the impeller runs with some eccentricity in the annular seal (as is usually to be expected). Perturbations of rotational symmetry are also created if the impeller shrouds and casing walls are not machined.
3. If the impeller runs with some eccentricity in the annular seal due to shaft deflection (for example, sag caused by the rotor weight) the pressure distribution in the seal is non-uniform; hence a static force in the annular seal is created as discussed in Chap. 10.6.2.

Unsteady reaction forces created by rotor vibration:

- Hydraulic impeller interaction, Chap. 10.6.3
- Forces in annular seals, Chap. 10.6
- Forces in open (axial and semi axial) impellers created by clearance variations due to orbiting of the rotor ("Alford effect", Chap. 10.7.3).

Radial forces act on all surfaces of the impeller which are exposed to the non-uniform pressure distribution in the casing (Fig. 9.24): resultant of blade forces $F_{R,blade}$, both of the cylindrical surfaces of the impeller side plates $F_{R,c}$, front shroud $F_{R,FS}$, rear shroud $F_{R,RS}$ and the annular seals. If the front shroud is inclined, $F_{R,FS}$ can well be the *dominant* radial force component. If the shroud surface is perpendicular to the axis (as the rear shroud in Fig. 9.24) the radial force is zero; in this case the force $F_{R,RS}$ is only acting on the hub or shaft.

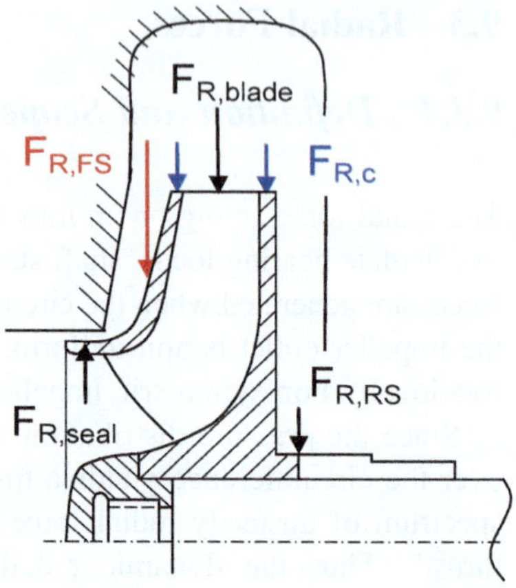

Fig. 9.24 Radial forces acting on impeller

All of these effects cannot be easily separated when measuring the radial forces, nor are they amenable to an exact theoretical prediction which would need modeling the three-dimensional flows in impeller, the collector, the impeller side rooms and the annular seals. To exactly calculate the variation of the annular seal clearance around the circumference would need to take into account the fluid-structure interaction between shaft deflection and hydraulic seal forces. Therefore, empirical coefficients are commonly used for estimating the radial thrust; these were derived from measurements and thus correspond to statistical data. Mostly the published radial thrust coefficients are valid for impellers with annular seals built with common clearances; they represent a combination of all of the effects mentioned above. The radial thrust coefficients can be defined with the head or the impeller tip speed u_2 as well as with the outlet width or the outlet diameter d_2. Below table shows the resulting formulae

	Referred to head	Referred to u_2
Calculated with outlet width	$k_R = \frac{F_R}{\rho g H d_2 b_{2tot}}$ (9.6)	$k_{Ru} = \frac{2F_R}{\rho u_2^2 d_2 b_{2tot}}$ (9.6a)
Calculated with outlet diameter	$k_{R,D} = \frac{F_R}{\rho g H d_2^2}$ (9.7)	$k_{Ru,D} = \frac{2F_R}{\rho u_2^2 d_2^2}$ (9.7a)

The actual figures of k_R and k_{Ru} differ by the factor of the pressure coefficient; that is, $k_{Ru} = \psi \times k_R$ applies. F_R is the radial force and b_{2tot} is the impeller outlet width including the wall thickness of rear and front shrouds. The coefficient k_R can also be interpreted as $k_R = \Delta p_{La}/(\rho \times g \times H)$ where Δp_{La} represents the average pressure difference which acts on the projected surface $d_2 \times b_{2tot}$.

Unless qualified by an additional subscript ("dyn" for unsteady thrust components, or "tot" for the sum of static and unsteady components), all subsequent references to k_R only imply the static (i.e. steady) radial thrust coefficient. As demonstrated by various tests, the radial thrust coefficients are virtually independent of rotor speed and Reynolds number in the range of practical interest; for geometrically similar pumps they are independent of the impeller size, e.g. [6, 17]. For a given pump the radial thrust depends primarily on the flow rate ratio q*.

Traditionally, the radial force coefficients are referred to the surface $\pi \times d_2 \times b_{2tot}$ even though the pressure in a fluid can create a force only if acting on a solid surface. This procedure is convenient, but in actual fact, the resultant force component $F_{R,blade}$ must be calculated by integrating the pressure and shear stress distributions on the blade surfaces and on the surfaces of front and rear shroud which form the impeller channel - and not by integrating the pressure distribution $p_2(z, \varepsilon)$ over the surface $\pi \times d_2 \times b_2$. The procedure follows the same logic as used for calculating the torque acting on the impeller, Eq. (8.25). Incidentally, some tests, as well as CFD studies, suggest that the forces determined by integrating the pressure over the surface $\pi \times d_2 \times b_2$ are quite similar to those obtained from integration over the blade and shroud surfaces. This finding applies to impellers with 5 or more back-swept blades where the blade overlap is large. Then the outlet pressure p_2 acts on the portion of the blade $b \times t_2 \times \cos \beta_{a2}$ between the throat at a_2 and the trailing edge where $\beta_{a2} = \operatorname{asin}(a_2/t_2)$. The blade portion exposed to p_2 is similar to the surface $b_2 \times t_2$ at the outer impeller diameter. It is also to be noted that only pressure variations over the circumference (rather than absolute pressures) are relevant for the radial thrust. The situation will be different for impellers like the one shown in Fig. 9.23 where the integration over the blade surfaces should give a force different from $p_2 \times \pi \times d_2 \times b_2$.

9.3.2 Measurement of Radial Forces

Several methods have been developed for measuring the radial forces acting on an impeller. Their principal characteristics are reviewed below since testing method and conditions should be appreciated in order to be able to correctly assess the significance of experimental radial thrust data. The tests are always done on single-stage pumps. Details on instrumentation and test data evaluation can be found in the quoted literature as well as in [2, 28].

Integration of the pressure distribution: The radial force can be derived by integrating (over the circumference) the static pressure distribution measured at the impeller outlet by means of probes and/or pressure tappings in the casing, e.g. [22, 33]. If only pressure tappings in the casing walls are used, this method is quite simple since no special instrumentation is required. The accuracy is moderate depending on the number of pressure tappings installed. If the static pressure is recorded in a tight grid of measuring points covering the entire impeller circumference, the effective radial thrust can be determined *without falsification through the forces generated in the annular seals*. However, if the pressure is integrated only over the impeller outlet width $b_{2,tot}$ (including shrouds thicknesses) the radial force would be under-estimated because the force $F_{R,FS}$ on the projection of the shroud surface would be missing.

Measurement of bearing forces: In most of the investigations, the bearing forces were measured by means of strain gages or force transmitters. Strain gages are applied to brackets which hold the bearing. The brackets must be flexible enough to allow the radial forces to produce sufficient elastic strain in order to get accurate measurements. Such flexible elements reduce the eigen frequencies of the test rig; the test speeds must therefore be selected such that the measurements are not falsified by resonances. This can be verified by mounting a known mechanical imbalance in lieu of the impeller: in the range where the forces recorded increase proportional to the square of the rotor speed, there is no falsification due to resonances. The mechanical imbalance serves also for calibrating the test rig.

The measurement of the bearing forces yields the resultant of all forces acting on the rotor. It is not possible to separate the radial thrust from forces created by the seal - unless a radial seal is used instead of an annular seal. The radial thrust determined in this way depends on the characteristics of the annular seal, in particular on the clearance and type of surface of the seal (plain or serrated).

Measurement of the shaft deflection: If the shaft deflection is measured by proximity probes, the radial forces acting on the impeller can be determined after proper calibration of the test rig and taking into account the shaft deflection under the rotor weight. The calibration can be done by means of a mechanical imbalance or static forces (e.g. dead weights). As when measuring bearing forces, a dynamic calibration is important in order to avoid falsification of the test results by resonance effects. Due attention must be paid to the bearing clearances because these impair the accuracy of the measurements. This testing method is relatively simple but not very accurate (bearing clearances, run-out and dynamic characteristics of the test rig). Only the resultant of radial thrust and annular seal forces can be measured (as with bearing force measurements).

Measurement of shaft stresses: All forces and moments resultant from impeller and seals can be measured by means of strain gages applied to the shaft close to the overhung impeller, [5, 6]. This complex and expensive method may be used when the interest is focused on unsteady forces. The calibration is performed by a mechanical imbalance. The dynamic characteristics of the test rig must be checked as described above to avoid resonances.

Force measurements by means of magnetic bearings: Active magnetic bearings can be used to measure the forces acting on a rotor. The rotor of the test rig is centered by two magnetic fields which are generated by electromagnets. The electrical current in the magnets is controlled by proximity probes and electronics so that the rotor is held in its centric position. The bearing forces can be determined from the measured current and the air gap between rotor and magnet (i.e. from the signal of the proximity probes), [13]. An advantage of this method is that the test pump can be built with sufficient stiffness to avoid resonances.

In all of the force measurements discussed above, the influence of the seal forces could be eliminated by using a radial seal (Fig. 3.15) or employing annular seals with deep serrations which have a low stiffness, Chap. 10.6.2.

9.3.3 Pumps with Single Volutes

The physical mechanisms which generate radial thrust shall be discussed in detail using the example of a single volute dimensioned according to the conservation of angular momentum as per Chaps. 3.7 and 4.2. In this context consider Fig. 9.25, where flow conditions and pressure distributions in a developed volute are sketched for three flow rates:

1. Near the best efficiency point, the flow angle at the impeller outlet matches the cutwater camber angle α_z. The flow deceleration largely follows the conservation of angular momentum and the pressure distribution is nearly uniform except

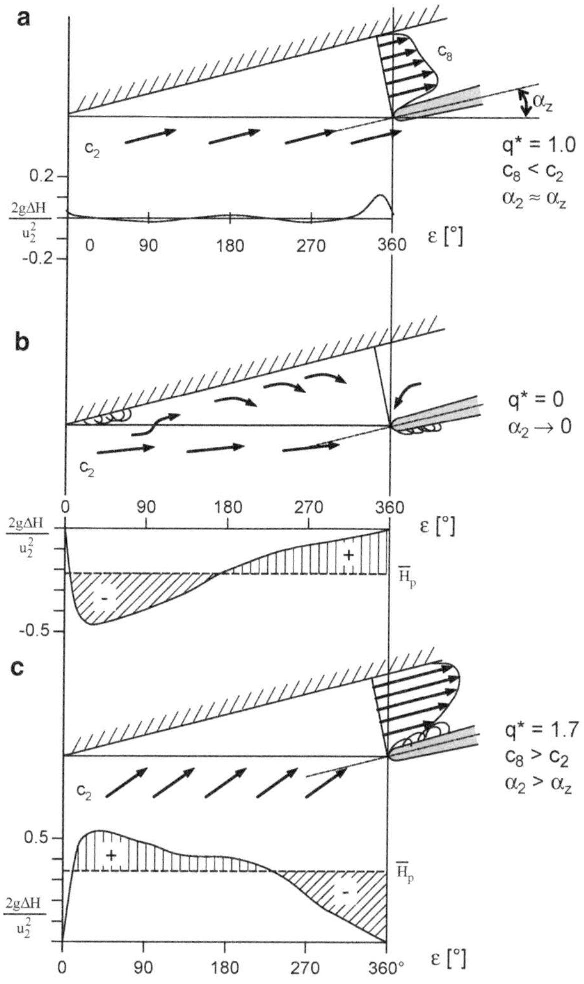

Fig. 9.25 Distribution of the static pressure and flow conditions in volutes

for a local perturbation induced by the flow around the cutwater. In order to keep this perturbation low, the camber angle α_z should be selected to match the approach flow angle with an incidence of i_3 zero to $-2°$. The cutwater should be profiled on the suction side. The profile should be little sensitive to incident flow; apply an elliptical profile or similar shape (do not use semi-circle!).

2. At partload ($q^* \ll 1$) the flow area in the volute is too large at every point on the circumference. The fluid exiting the impeller with the absolute velocity c_2 is decelerated to an average velocity in the volute given by $c_{sp} = Q/A$ (Q and A being the local values of flow rate and area). The fluid approaches the cutwater with a positive incidence, causing a local depression in the static pressure. The static pressure rises from a minimum downstream of the cutwater to a maximum which is reached near the throat area according to the flow deceleration from c_2 to c_{sp}. At low load - in particular at Q = 0 - there is a strong mismatch between c_2 and c_{sp} and the fluid enters the volute similar to a jet entering a plenum. Consequently, an increasing fluid volume circulates in the casing since the fluid cannot be at rest even at Q = 0 due to the rotation of the impeller. The circulation is impeded by the cutwater where the flow separates reinforcing the pressure minimum, Fig. 9.25b.
   Since the pressure distribution at the BEP varies little over the impeller circumference, its integration only yields a small resultant radial force (with an infinitely thin cutwater the force would be theoretically zero). In contrast, the non-uniform partload flow induces a strong variation of the static pressure over the circumference which results in a radial force usually reaching its maximum at Q = 0. Due to flow separation, the pressure recovery immediately downstream of the cutwater is lower than further on in the volute. As a consequence, the radial force is directed towards this pressure minimum, Fig. 9.27a.
3. At high flow rates ($q^* > 1$) the volute cross sections are too small. The flow is thus accelerated downstream of the impeller. Correspondingly, the static pressure decreases in circumferential direction from a maximum (stagnation pressure) at the cutwater. The approach flow angle to the cutwater is too large (negative incidence) generating a flow separation in the discharge nozzle. Downstream of the cutwater there is a stagnation point and a pressure maximum. The vector of the resultant radial force is directed away from that maximum, Figs. 9.25c and 9.27b.

Pressure measurements reported in [34] well confirm the mechanisms and tendencies described above. The static pressures measured at $r/r_2 = 1.03$ in a single volute ($n_q = 25$) are shown in Fig. 9.26. Interestingly, the curves for all flow rates intersect near the circumferential position of about 110° downstream of the cutwater. Solely at this location, the pressure is independent of the flow rate. Pressures as low as $\psi = 0.24$ were found just upstream of the throat area at $q^* = 1.63$ where the flow is accelerated by an estimated ratio of 1.3 into the discharge nozzle.

The static pressure per se can create a force only if it acts on a solid structure; thus it produces a resultant radial force on front and rear shrouds which is added to the resultant of the blade forces discussed above.

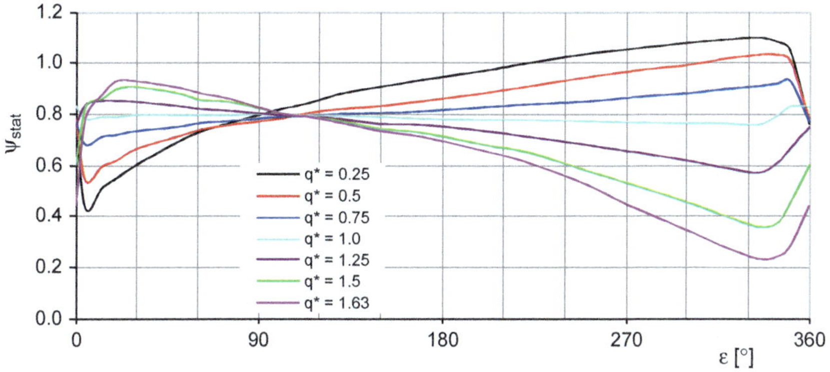

Fig. 9.26 Wall pressures measured at different flow rates in a single volute at $r/r_2 = 1.03$. The cutwater position is at $\varepsilon = 0$; [34]; $n_q = 35$

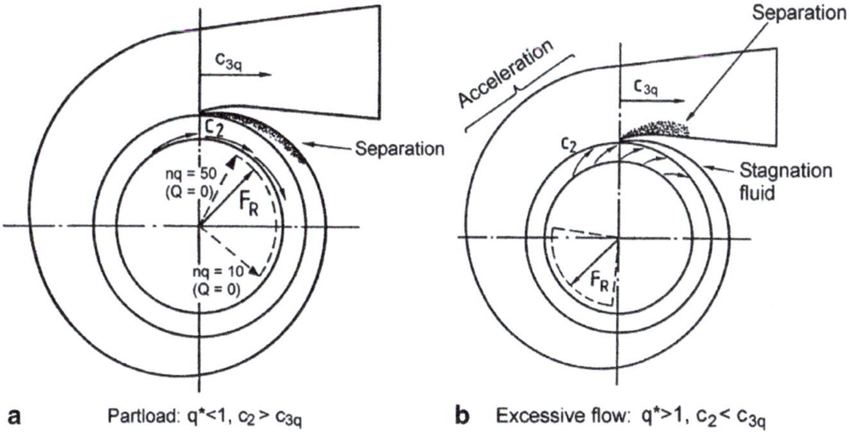

Fig. 9.27 Radial thrust in single volutes

From the flow behavior described above and from Figs. 9.25 and 9.27 result the following relationships which have been confirmed by numerous investigations:

- The radial force attains its minimum near the design point *of the volute*. The (relatively low) radial thrust in the design point is caused by: (1) asymmetries induced by the flow around the cutwater (in particular with thick cutwaters), (2) geometrical tolerances, (3) friction losses are not constant over the circumference; they influence the pressure build-up in the volute. As discussed in Chap. 7.8.2, not all volutes are designed strictly according to the conservation of angular momentum. The effective position of the BEP results from the dependence of the various losses on the flow rate. Therefore the minimum of the radial force does not necessarily coincide with the BEP-flow or the design point.

- The flow rate at which the radial forces have their minimum is essentially determined by the design point of the *volute*, because the most uniform volute flow is expected at that flow rate. If the design point of the impeller differs from that of the volute, the effect on the radial forces is only of secondary influence. If two different impellers are tested in a given volute casing, the minimum of the radial thrust is observed at about the same flow rate, i.e. it is found on the volute characteristic, [10]. In contrast, if a given impeller is tested in two different volute casings, the minimum of the radial thrust shifts to a higher flow rate with increasing volute throat area.
- The radial force rises at partload and at high flow; it attains a relative maximum at partload which is usually found at Q = 0 (see curve for single volute in Fig. 9.28).

  Figure 9.28 Radial thrust in various types of collectors, nq = 19, [42]
- The larger the volute cross sections for a given cutwater diameter, the less uniform is the flow in the volute at partload. Therefore, the radial thrust coefficients of single volutes increase with growing specific speed until a maximum is reached in the range of n_q = 50 to 60, refer to Fig. 9.46 in Table 9.7. Presumably this maximum is given by a maximum blade loading: if the non-uniformity of the partload pressure distribution in the volute exceeds some specific threshold, back-flow from the volute to the impeller inlet limits the local pressures which the impeller is able to sustain.

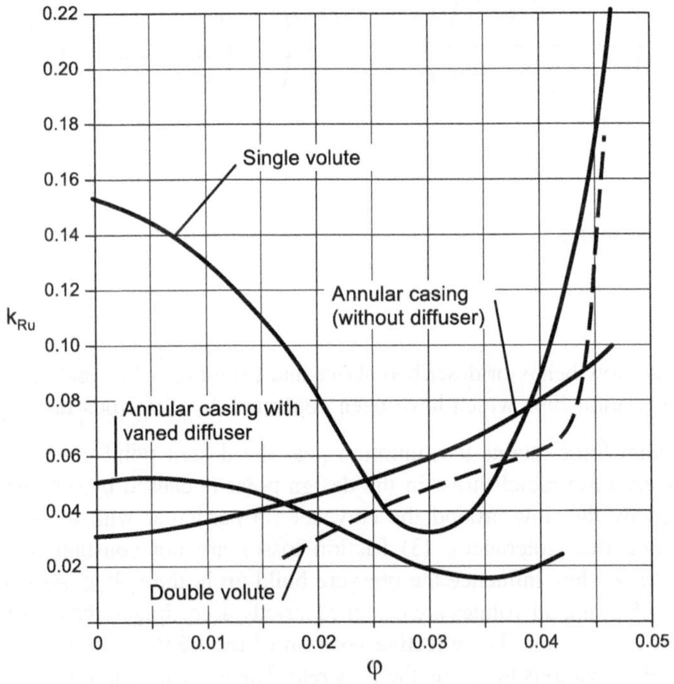

Fig. 9.28 Radial thrust in various types of collectors, n_q = 19, [42]

- At low partload the radial force acting on the impeller is directed to a point downstream of the cutwater (Fig. 9.27a), while at high flow rates the radial force is directed to a point shifted by roughly 180° as compared to partload (Fig. 9.27b). Consequently, the radial force changes its direction by roughly 180° in the range around the best efficiency point. Near the BEP the force direction is thus uncertain and difficult to predict since it depends on tolerances and the effects mentioned above.
- With a given pump, the direction of the radial force vector varies with the flow rate q*. Considering different pumps, the force direction depends on the form of the casings and thus on the specific speeds. This dependence is depicted in Fig. 9.27a for Q = 0 based on measurements in [1].
- All data concerning the magnitude and the direction of the radial forces are subject to uncertainties, because the area distributions in the volute, the impeller sidewall gaps and the *shapes offront and rear shrouds* all have an impact on the pressure distribution, hence the radial thrust.
- As discussed above, it is not possible to separate the forces acting in annular seals from the radial thrust proper without special devices. With annular seals the radial forces measured depend on the seal clearances. That is why Table 9.7 provides radial thrust coefficients for "normal" and double clearances; normal clearances can be considered to be roughly corresponding to Eq. (3.12).
- If the distance between impeller and cutwater is increased strongly, the circulation of the fluid at Q = 0 is less impeded by the presence of the cutwater; the radial thrust drops by a few percent as compared to a small cutwater distance. In an infinitely large casing there are no radial forces.

If the pressure distribution in the volute is non-uniform, the same applies *by necessity* to the impeller side wall gaps. Pressure differences in the ISR induce fluid flows from the volute into the ISR (and vice versa) from regions of high pressure to low-pressure domains. These flows tend to limit the pressure differences. Consequently, at *off-design operation, the flow in the ISR is by no means rotationally symmetric.*

Figure 9.29 illustrates this by CFD calculations of a single volute (n_q = 26) at partload (left side) and overload. In regions of low pressure we find high velocities while velocities are low in high-pressure zones (as expected from the Bernoulli equation). At partload, with high pressures upstream of the throat, fluid flows around the cutwater from the volute into the ISR; the stagnation point is on the pressure side of the cutwater and the flow separates on the suction side. At overload, with low pressures upstream of the throat, fluid flows from the ISR into the volute; the stagnation point is on the suction side of the cutwater. The fluid stalls on the pressure side in the discharge nozzle and pressure recovery in the diffuser is accordingly impaired.

The flow patterns discussed in context with Fig. 9.29 may be compared to the erosion patterns in the IRS of volute pumps depicted in Figs. 14.33, 14.34, 14.35 and 14.36.

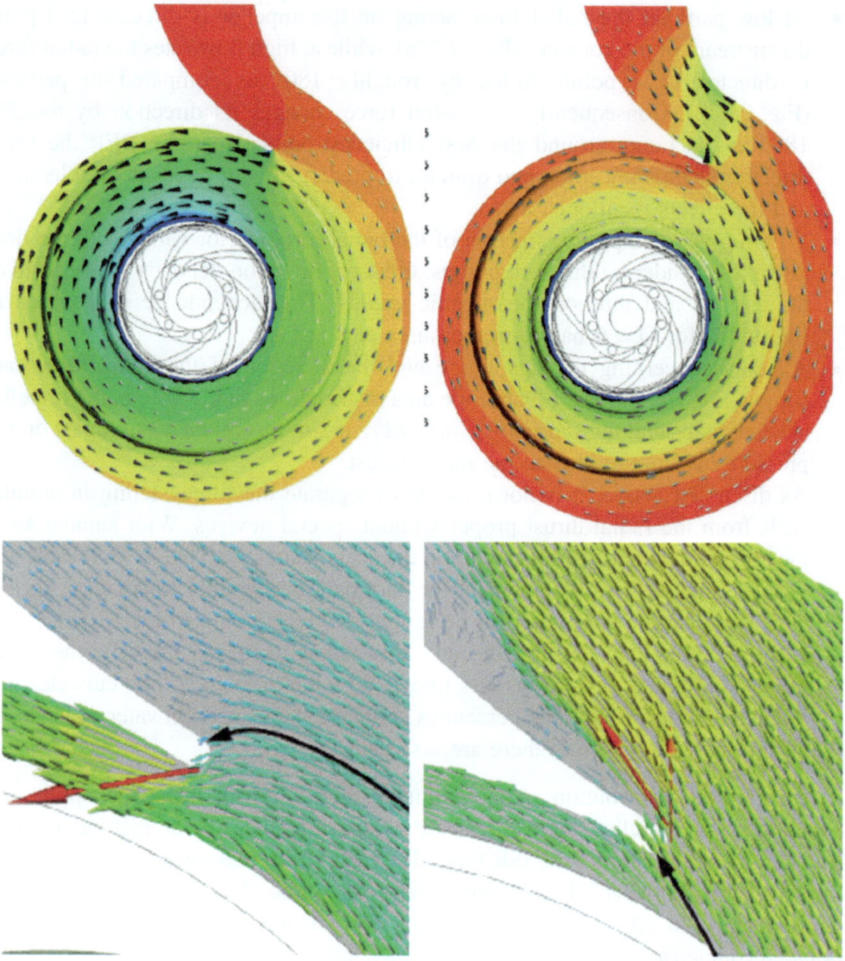

Fig. 9.29 Pressures and velocity vectors in the ISR, calculation Lucerne University of Applied Sciences and Arts

The calculated flow and pressure distributions qualitatively agree with those depicted in Fig. 9.25: at partload flow separation downstream of the cutwater, at overload in the discharge nozzle.

The calculated pressure distribution in the ISR is shown in Fig. 9.30. Interestingly, the amplitude of the pressure variation over the circumference *increases* with decreasing radius. Near the inlet of the annular seal at $r/r_2 = 0.6$, the pressure varies between $c_p = 0.46$ and 0.78, while the variation at the impeller outer diameter is between $c_p = 0.74$ and 0.93. As a consequence, the pressure difference across the annular seal, the leakage and the seal forces vary strongly over the circumference. The flow rates Q_i through the individual impeller channels have

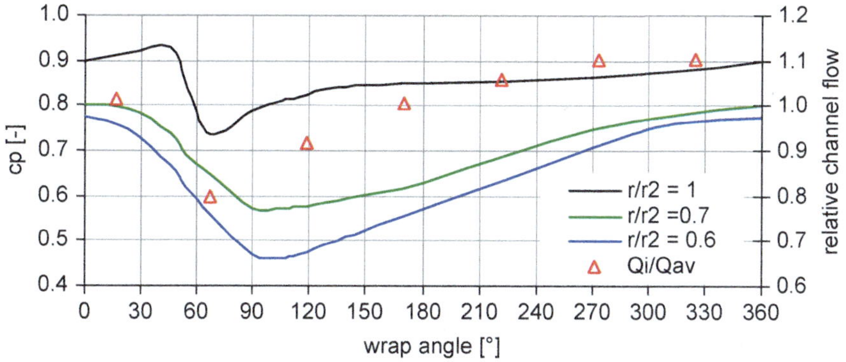

Fig. 9.30 Pressure distribution in the ISR and ratio of flow rate through individual impeller channels Q_i to average flow rate $Q_{av} = Q_{total}/z_{La}$; $q^* = 0.5$, unsteady calculation with 7×10^6 nodes, Lucerne University of Applied Sciences and Arts

been plotted in Fig. 9.30. They are referred to the average flow Q_{av} measured in the discharge nozzle. The impeller sidewall gaps were fully open to the rectangular volute; the width was $s_{ax} = 0.05 \times d_2$.

The variation of the pressure distribution around the circumference could also give rise to a tilting moment on the impeller.

The asymmetries of the flow through the ISR explain the non-uniform erosion patterns observed on the casings of volute pumps in abrasive service; some pictures can be found in Chap. 14.5.6.

In contrast to the results in Fig. 9.30, the pressure distributions in the ISR measured in [34] were fairly constant (deviation from average pressure less than 5%) because gap A was quite small; in addition the ISR was very large (roughly $s_{ax} = 0.07 \times d_2$). This test confirms that a small gap A with sufficient overlap and a wide ISR is apt to equalize the pressures in the ISR and reduce the radial force on the impeller front shroud.

9.3.4 Pumps with Double Volutes

Double volutes are employed in order to reduce the radial thrust. If a second cutwater is introduced at a spacing of 180° from the first cutwater, the rotational symmetry is improved as per Fig. 9.31. The pressure distribution in both of the partial volutes (over a wrap angle of 180°) is analogous to Figs. 9.25 and 9.26, as shown by measurements in [24]. As demonstrated by Fig. 9.31, even a rather short rib is able to drastically reduce the radial thrust at shut-off and below $q^* < 0.5$, (test 4). Increasing the length of the rib to obtain a full double volute (test 2) yields an almost constant radial force over the entire flow range. In general, a flat and

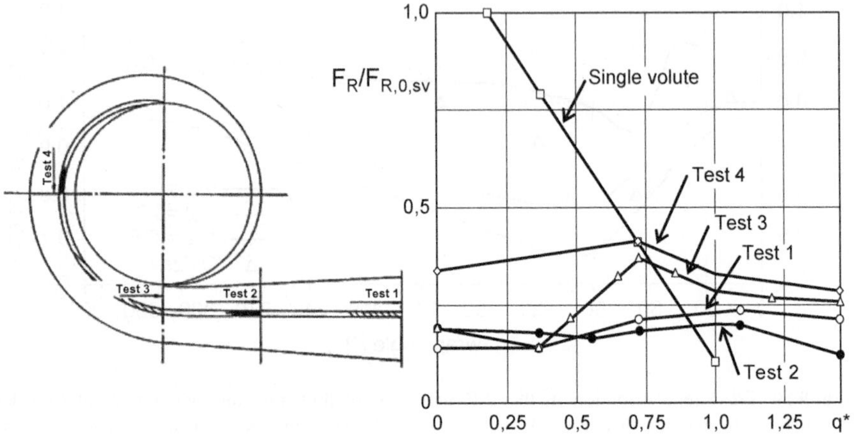

Fig. 9.31 Radial thrust balancing by means of double volutes; all forces F_R are referred to the radial thrust of the single volute at shut-off, $F_{R,0,sv}$, [B.9]

unsystematic distribution of the radial thrust can be expected with double volutes in the flow range of q* = 0 to 1.1, e.g. [1]. The directions of the radial forces are uncertain too.

As demonstrated by Fig. 9.28, the radial thrust in double volutes can rise strongly at flow rates well above the best efficiency point. This increase is caused by differences in the flow resistances of the outer and inner channels. Consequently, the flow rates through both channels will not be the same and the impeller operates in both partial volutes at different points on the Q-H-curve (refer to the discussion in Chap. 10.7.1). The difference in the static pressure at the impeller outlet prevailing in both of the partial volutes generates a radial force. Since the resistances rise with the square of the flow rate, the radial force rises steeply at q* > 1, while the impact at partload is small. However, designers make different assumptions for designing double-volutes; and this has an impact on the radial forces.

At high specific speeds, double-volutes may give rise to higher variations $k_R = f(q^*)$ than at moderate n_q because the large areas of the volute sections cause stronger non-uniformities in the flow distribution. Radial forces in a process pump ($n_q = 90$) with a double-volute were determined by integrating the static pressures measured around the circumference of the impeller on a radius of $r/r_2 = 1.05$. The results are shown in Fig. 9.30-1.

If both cutwaters are not spaced by 180° apart, the circumferential symmetry is destroyed and the radial thrust increases in comparison to a 180°-volute. Thus a cutwater spaced by less than 90° brings no radial thrust reduction at all. This is proven by the tests shown in Fig. 9.48 (in Table 9.7) where the radial thrust reduction factor F_{Dsp} achieved by a double volute is plotted against the volute wrap angle, [17].

Double volutes with wrap angles of less than 180° are built for design reasons, e.g. in axial split pumps or for allowing complete draining of the casing (pumps

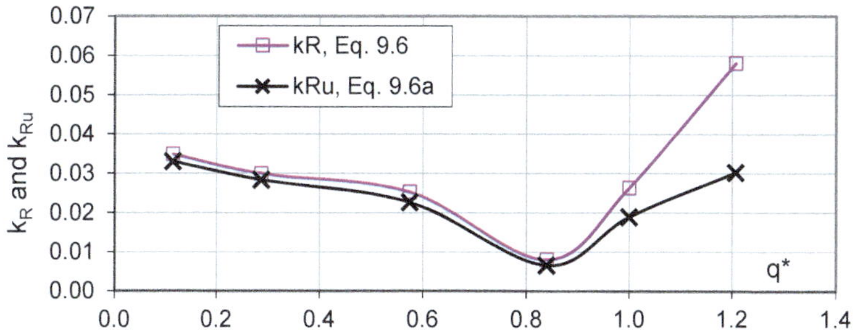

Fig. 9.30-1 Radial force coefficients of double-volute, n_q = 90, [24]

with the discharge nozzle vertically upwards). If the wrap angle is near 120°, as may be the case in axial split pumps, a rib in the upper half of the casing can reduce the radial thrust.

With reference to Fig. 9.31 there is an optimum length of the rib in the discharge nozzle: while the rib in test 3 was too short (leading to a force maximum at q* = 0.75) the rib of test 1 appears to be too long generating higher flow resistance and higher radial thrust at high flow.

9.3.5 Pumps with Annular Casings

As discussed above, the high radial thrust created by single volutes at shut-off is generated by the cutwater impeding the free rotation of the fluid in the casing and leading thus to flow separation and a pressure minimum downstream of the cut-water. In contrast, the fluid rotates freely at Q = 0 in a vaneless annular casing. Consequently, the radial thrust of pumps with annular casings reaches its mini-mum at Q = 0. The radial force rises in a slightly upward curve with increasing flow rate until the best efficiency point, Fig. 9.28. At overload, when the annular cross section is much too small for the flow, a pronounced pressure minimum is generated in the domain upstream of the discharge nozzle. The radial thrust rises accordingly; the discharge nozzle acts like a strong depression.

In Table 9.7 a formula is given for estimating the radial thrust in annular casings. The available data do not allow to discern an influence of the specific speed. The relation given for the function k_R = f(q*) is quadratic, predicting a strong increase at q* > 1.

9.3.6 Diffuser Pumps

The radial thrust in diffuser pumps is created by geometric tolerances of the diffuser and by asymmetries in the flow downstream of the diffuser, as caused by the discharge nozzle, for example. The more so, if the diffuser channels are short and have a small overlap. Therefore a ring of stay vanes with little or no flow deflection is scarcely able to reduce the radial thrust.

From the available radial thrust measurements no clear dependence of the radial thrust coefficient from the specific speed or geometric parameters could be found. Only the eccentricity of the impeller with respect to the diffuser generates defined radial forces which are approximately proportional to the eccentricity. These forces are *de-centering*, [21]. Since only very small eccentricities are encountered in practice, these radial force components are usually insignificant; they are implicitly covered by the statistical test data in Table 9.7.

9.3.7 Radial Forces Created by Non-uniform Approach Flows

Inlet chambers of between-bearing pumps (as used in multistage or double-entry single-stage pumps) generate an impeller approach flow with a circumferentially non-uniform velocity distribution, Chap. 7.13. One half of the impeller is approached with a pre-rotation while the other half is subject to counter-rotation, Fig. 7.50 . This variation of the circumferential velocity component c_{1u} leads, according to the Euler-equation, to a variable work transfer in different sectors of the impeller. Thus a steady radial force is created whose magnitude and direction depend on the flow rate. Figure 9.32 shows measurements from [B.20] with an impeller of n_q = 33: test 1 was made with an inlet chamber common to multistage pumps, while in test 2 an insert with radial ribs was mounted in order to reduce the circumferential velocity components and to smooth out the impeller approach flow. According to these tests, the radial forces created by non-uniform approach flow increase strongly at high flow ($q^* \gg 1$), because the perturbations of the flow induced by the inlet casing increase with growing inertia (i.e. high flow velocities).

Tests with non-uniform approach flow to a semi-axial pump (estimated n_q = 160) reported in [52] gave radial forces with similar trends and magnitudes as the tests with a radial impeller from [B.20], Fig. 9.32. The maximum velocity of the approach flow was roughly 25% higher than the average (c_{max}/c_{av} = 1.25). While Fig. 9.32 presents force coefficients referred to head and b_2, as per Eq. (9.6), Fig. 10.68 gives the same data referred to u_2 and d_{2a}, as per Eq. (9.7a), which is more appropriate for semi-axial and axial pumps.

Note the strong dependence of the direction of the radial force on the flow rate. Such force variations have an influence on the bearing load and thus on the vibration behavior of the pump, see Chap. 10. These tests confirm that different sectors of the impeller operate (even under steady operation) at different flow conditions, which implies their working on different points of the Q-H-curve. Such asymmetries - and the resulting radial forces - can be caused either by the geometry *upstream or downstream* of the impeller. The directions of the force vectors in Fig. 9.32 depend on the geometry of the test set-up and the coordinate system

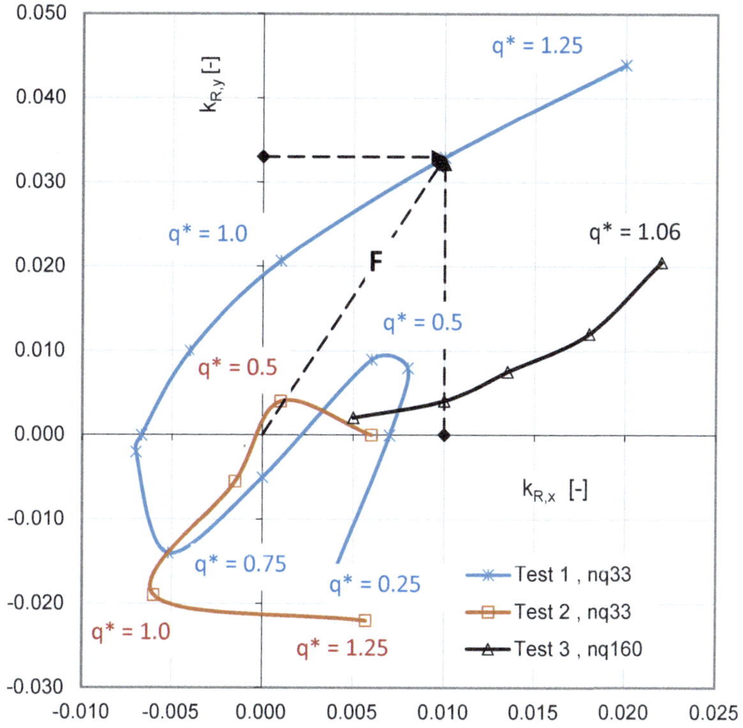

Fig. 9.32 Influence of the approach flow on the static radial thrust coefficients. Plotted are the x- and y-components of the radial force coefficients k_R defined by Eq. (9.6). *Test 1*: with inlet chamber of a multistage pump; *Test 2*: with ribs, which provide essentially a rotational symmetric approach flow, [B.20]. *Test 3* [52]. A line from the origin to any point on the curves represents the force vector (direction and size). Example: the vector F represents the radial force coefficient at the flow rate ratio $q^* = 1.1$ with the dimensionless force components in y-direction $F_y = 0.033$ and in x-direction $F_x = 0.01$

selected; they cannot be compared between different tests without knowing the required details.

In single-stage pumps with double-entry impellers and double volutes the radial forces created by non-uniform approach flow (Fig. 9.32) may be of similar magnitude as the forces generated by an uneven pressure distribution in the volute. Direction and size of the radial thrust in these pumps may, consequently, be expected to depend strongly on the relative position of the suction and discharge nozzles.

The forces created by a non-uniform approach flow to an axial or a semi-axial impeller of high specific speed can be estimated tentatively from Eq. (9.8).

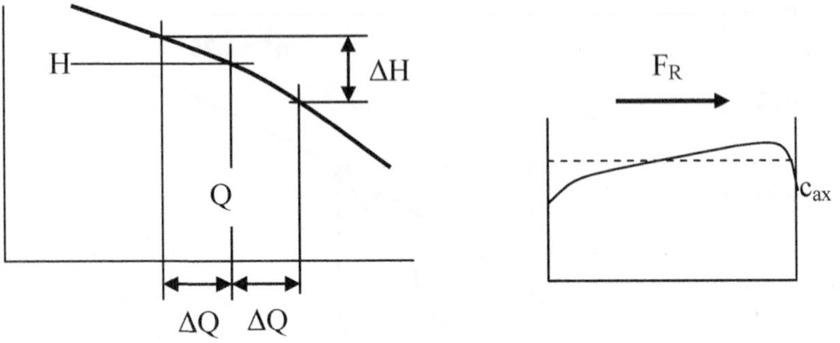

Fig. 9.33 Effect of asymmetric approach flow on the radial force created by an impeller of high specific speed

$$F_R = k_{R,D}\,\rho\,g\,H\,d_2^{\,2} \quad \text{with} \quad k_{R,D} = f \sin\bar{\beta}\,\cos\bar{\beta}\,\frac{\Delta H}{H}\left(\frac{L_{sch}}{d_2}\right)^2 \tag{9.8}$$

H is the head of the considered operation point and ΔH is the head difference read from the Q-H-curve for the difference in flow rate ΔQ caused by the non-uniform approach. For example: if the axial flow velocity is asymmetrical by ±5% of the average as shown in Fig. 9.33, ΔH is read for ΔQ = ±0.05 × Q. L_{sch} is the blade length and $\bar{\beta} = 0.5(\beta_{1B} + \beta_{2B})$ is the average blade angle.

The factor "f" in Eq. (9.8) needs experimental verification. Setting f = 1.0 gives factors $k_{R,D}$ of similar magnitude as given in Table 9.7 for axial pumps.

CFD studies of unsteady hydraulic forces created by highly-distorted approach flow to a semi-axial pump were reported in [9]. The impact of the approach flow on the vibrations of vertical pumps is discussed extensively in Chap. 10.14.

9.3.8 Axial Pumps

Radial forces in axial pumps are essentially caused by non-uniform approach flows induced for example by inlet bends or vortexes. Perturbations of rotational symmetry downstream of the impeller equally contribute to the generation of radial forces. The mechanisms were discussed in Chap. 9.3.7; further insight can be gained from Chap. 10.14.

The experimental radial force coefficients given in Table 9.7 are defined with the impeller outer diameter. According to the tests in [42] the coefficients for steady radial forces are $k_{R,D} = 0.02$ at $q^* < 1.2$. At even larger flow rates, the radial forces rise strongly because of the increasing non-uniformity of the flows at the impeller inlet and outlet. The coefficients for unsteady radial forces are expected in the order of $k_{R,D} = 0.01$.

If the rotor of an axial pump runs in an eccentric position, radial forces are created by the impact of the flow through the gaps between the blade tips and the casing on the work transferred by the individual blades, Chap. 10.7.3.

9.3.9 *Radial Forces in Pumps with Single-Channel Impellers*

Single-channel impellers are used in sewage pumps and other applications where solid matter, typically as large as the nozzle size, is required to pass through the pump, Chap. 7.4. Since there is only one blade, the mass distribution of the blade is not rotationally symmetric. Therefore, there is a resultant mechanical imbalance. To compensate for this, material is removed from the shrouds (mainly opposite the blade trailing edge) during a shop balancing process. Thickening the blade near the leading edge and opposite the trailing edge helps compensating the mechanical imbalance. The size of the *mechanical* imbalance is independent of the angular position of the impeller; it induces a purely sinusoidal radial vibration (on an isotropic structure) and a *circular* orbit of the force or the displacement.

The pressure distribution on a single blade cannot be rotationally symmetric either. Consequently, a radial hydraulic force is created on the impeller which rotates with the frequency f_n given by the rotor speed. It is called the "hydraulic imbalance". The pressure distribution on the blade depends on the flow rate - and so does the hydraulic imbalance. The flow in the pump becomes highly unsteady and 3-dimensional. Two main effects provoke the hydraulic imbalance:

a. The shape of the blade determines the blade loading. The pressure distribution on the blade results from the blade angle as function of the radius $\beta_B = f(r)$ or as function of the wrap angle $\beta_B = f(\varepsilon)$. In a rotationally symmetric (or infinitely large) casing this pressure distribution would be independent of the angular position of the impeller. As with the mechanical imbalance, the force orbit would have a circular shape.
b. As discussed in Chap. 10.7.1 the flow through the individual channels $Q_i(\varepsilon)$ of a multi-bladed impeller varies around the impeller circumference if the pressure distribution $p_2(\varepsilon)$ in the casing is not constant. The same applies to single-channel impellers: as dictated by the local pressure in the casing, the velocity distribution (or the blade loading) in the impeller depends on its angular position - in particular on the position of the blade trailing edge versus the discharge nozzle or cutwater. Near the BEP (of the *casing*) this variation is small since the pressure distribution is fairly constant if the volute is designed according to the conservation of angular momentum; the orbits are then nearly circular. At off-design conditions when $p_2(\varepsilon)$ becomes increasingly non-uniform the orbits are not any more circular, but oval or distorted.

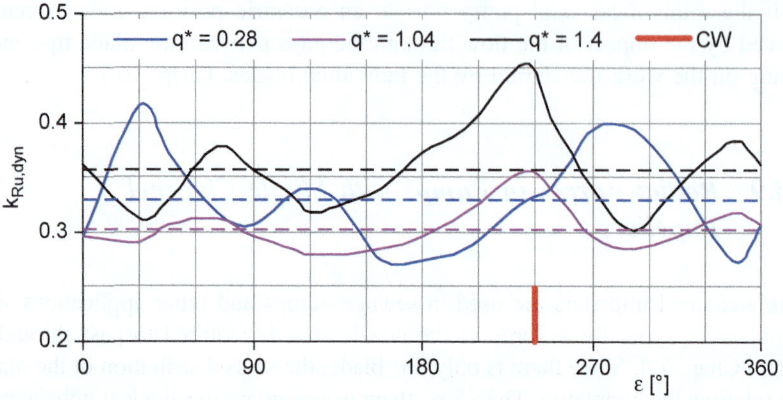

Fig. 9.34 Variation of radial force over one impeller revolution, test T2. The volute cutwater position CW is at 240°, [7]

From the above observation we can postulate: *The deviation from the circular orbit reflects the influence of the casing*. Abundant experimental evidence confirms that the hydraulic imbalance depends on both the blade shape $\beta_B = f(\varepsilon)$ and on the pressure distribution in the casing $p_2(\varepsilon)$. The pressure $p(\varepsilon)$ is determined by the area distribution $A = f(\varepsilon)$ in the casing. Which effect is dominating is not easy to decide because there is a strong interaction between the flow in the impeller and the flow in the casing. In order to get a better appreciation of both effects, published test data are analyzed subsequently by grouping the tests into two categories: (1) Tests done with different impellers in a given casing demonstrating the effects of the blade shape; (2) Tests done with a specific impeller in different casings revealing the impact of the casing.

When measuring the radial forces, the signal of the strain gages is recorded over one impeller revolution (time signal). This signal represents the total force $k_{Ru,tot}$ acting on the device[4]. The *static* force is the arithmetic average of the time signal. The unsteady or "*dynamic*" force is then calculated from $k_{Ru,dyn} = k_{Ru,tot} - k_{Ru,stat}$.

As an example, Fig. 9.34 shows the variation of the force over one impeller revolution for different flow rates, while Fig. 9.35 exhibits the force orbits for the same test. The test corresponds to "T2" in Table D9.1. Taking for example the flow $q^* = 1.4$, the maximum amplitude is $k_{Ru,dyn,peak} = 0.45$. This peak (i.e. the highest force during one revolution) would be relevant for the excitation of vibrations and mechanical loading. The average force over one impeller revolution is $k_{Ru,dyn,av} = 0.36$ (black dotted line in Fig. 9.34). The ratio peak to average is 1.25.

In Fig. 9.34 three distinct peaks can be recognized at high flow. The highest amplitude occurs when the blade trailing edge nears the cutwater because then the highest energy fluid hits the casing. When the impeller inlet opening A_{1q} is turned

[4] For easier reading, the term "force" is used for the force coefficient k_{Ru} in the following.

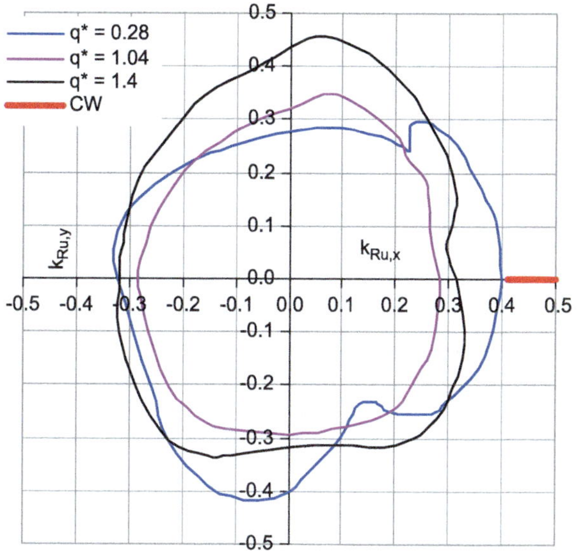

Fig. 9.35 Radial force in orbit representation, test T2. The discharge nozzle axis (cutwater CW) is on the positive x-axis, [7]

to a position close to the discharge nozzle, there is kind of a "direct communication" between the discharge and suction nozzles. As a consequence, the local head is much reduced and a minimum of the radial force is observed (at about 285° in Fig. 9.34). At low flow with very strong recirculation, there are also three peaks, but these are shifted versus the peaks at high flow (the flow patterns responsible for this being as yet unknown).

The test data reported in the literature quoted were converted to k_{Ru} as defined by Eq. (9.7). The coefficients of the static (or steady) force $k_{Ru,stat}$ are plotted in Fig. 9.36; the unsteady (dynamic) forces $k_{Ru,dyn}$ are shown in Fig. 9.37 for volutes and in Fig. 9.38 for annular casings. Relevant data of the tested pumps are listed in Table D9.1.

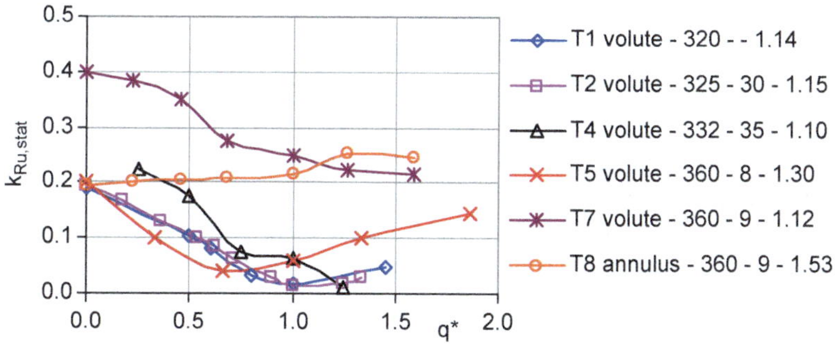

Fig. 9.36 Static radial forces

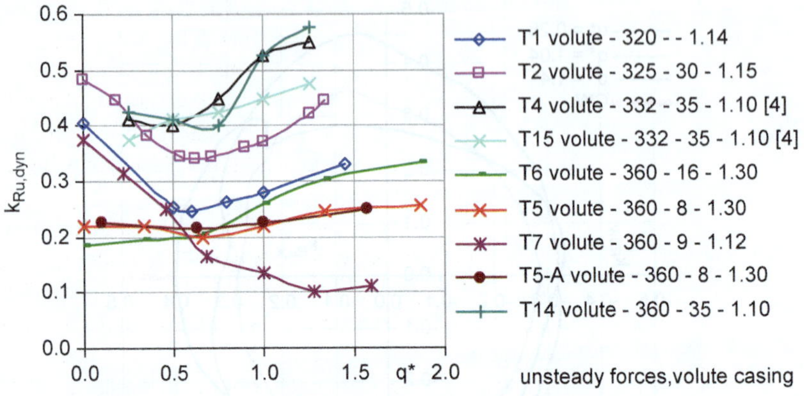

Fig. 9.37 Dynamic radial forces in volutes, maxima over one impeller revolution

Table D9.1 Radial force measurements with single-blade impellers

Test	Casing	n_q	ε_{sch}	d_1^*	β_{2B}	b_2^*	d_z^*	Lit.	Data
		–	°	–	°	–	–		
T1	Volute	50	320	0.5			1.14	[12]	Average
T2	Volute	45	325	0.48	30	0.49	1.15	[7]	Average
T3	Annulus	44	332	0.42	35	0.33	1.58	[4]	Peak
T12	Annulus	44	332	0.42	35	0.33	1.58	[4]	Peak
T13	Annulus	44	360	0.42	35	0.33	1.58	[4]	Peak
T4	Volute	44	332	0.42	35	0.33	1.10	[4]	Peak
T15	Volute	44	332	0.42	35	0.33	1.10	[4]	Peak
T14	Volute	45	360	0.42	35	0.33	1.10	[4]	Peak
T5-A	Volute	43	360	0.35	8	0.29	1.30	[38]	Peak
T5	Volute	43	360	0.35	8	0.29	1.30	[39]	Peak
T6	Volute	40	360	0.35	16	0.29	1.30	[39]	Peak
T7	Volute	32	360	0.48	9	0.24	1.12	[40]	Peak
T8	Annulus	27	360	0.48	9	0.24	1.53	[40]	Peak
T9	Annulus	43	332	0.42	35	0.33	1.58	[45]	
T10	Annulus	43	420	0.42	35	0.33	1.58	[45]	
T11	Annulus	43	332	0.42	35	0.33	1.58	[49]	

For easy reference the tests have been given arbitrary numbers T1, T2, etc. The labels of the test curves have the format defined by way of the example of test 2: T2 volute-325-30; $d_z^* = 1.15$ [7]. The first figure stands for the wrap angle of the blade $\varepsilon_{sch} = 325°$, the second figure defines the blade outlet angle $\beta_{2B} = 30°$ and the relative cutwater distance is $d_z^* = d_z/d_2$.

The following should be appreciated when using the data:

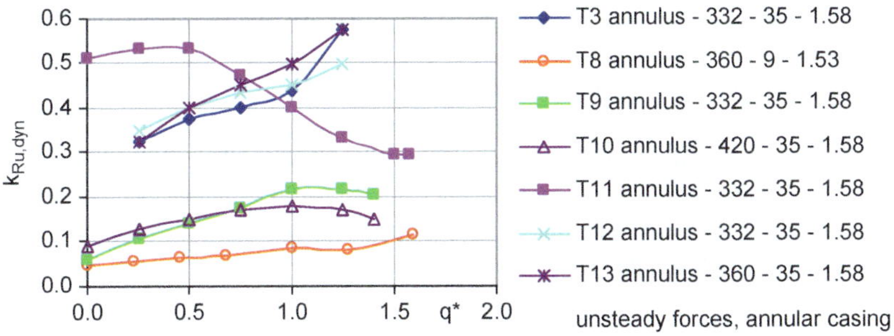

Fig. 9.38 Dynamic radial forces in annular casings, maxima over one impeller revolution

- Forces from [40] to [4] are *maximum* amplitudes recorded during one impeller revolution. Some of these values were read from the orbits.
- Data from [12] and [7] were published as forces *averaged* over one impeller revolution. The ratio between peak amplitudes and average can be expected to be in the range of 1.2 to 1.3 as derived from tests in [7] and [4]. In order to ease the comparison with the other data in Fig. 9.37, the average forces have been multiplied by a factor of 1.25 giving thus an estimate of the maximum force amplitudes (this implies some scatter).
- The static forces measured in [40] are very high: The shape of the curves suggests that the casing was very large and that the BEP of the casing is not to be found within the flow range tested. In contrast, the dynamic forces at high flow are much below the data published by the other sources. Possibly, the lower dynamic forces are due to the lower specific speed and a large casing.
- The leading edges of the impellers of [38, 39] were designed with $d_{1i}/d_{1a} = 0.87$. The leading edge of the impeller of [40, 39] was designed with $d_{1i}/d_{1a} = 0.73$. All other publications quoted used $d_{1i}/d_{1a} = 1.0$. The reduced d_{1i}/d_{1a} could also have contributed to the lower forces reported in [38, 39].
- All of the data were read from published graphs. The accuracy is not much better than ±10% when the static component was not given (in particular when reading the forces from published orbits).
- The force measurements in [45, 49] were done by using an in-service balancing procedure. As compared to the strain gage measurements used by the other sources, there is some doubt as to the validity of the results. This because the same pump was used in [4, 49] which was a scaled model of the pump in [45]. Consequently, tests T3, T9 and T11 should be quite close, while in fact they differ widely. Because of this uncertainty, some doubts remain as to the tendencies found in these tests.
- Cavitation is another factor which is likely to contribute to the differences between the hydraulic imbalances measured by different authors.

The above analysis allows the following conclusions:

1. The highest forces in Fig. 9.37 were recorded for outlet angles of $\beta_{2B} = 35°$. The force coefficients k_{Ru} in terms of Eq. (9.7) diminish to less than half if β_{2B} is reduced from 35° to below 10°.
2. With the exception of the data in [40], the *static* force coefficients in Fig. 9.36 follow the typical trend for single volutes, as shown in Fig. 9.28. However, the force coefficients are smaller than in Fig. 9.46 because front and rear shroud of the impeller are perpendicular to the pump axis; hence no radial forces act on the shrouds in the impeller side rooms.
3. With the exception of the data in [40], the hydraulic imbalance exhibits a minimum near $q^* = 0.6$ to 0.7 when the impeller is mounted in a volute, see Fig. 9.37. But unsteady forces rise steadily with q^* when the impeller is mounted in an annular casing. The behavior is similar to the static radial forces in pumps with multi-bladed impellers, Fig. 9.28. This observation confirms that the pressure distribution $p(\varepsilon)$ in the volute exerts a large influence on the hydraulic imbalance.

Effect of the blade shape $\beta_B(\varepsilon)$: The blade loading near the exit is determined by the blade outlet angle β_{2B} and the wrap angle ε_{sch} of the blade. The blade imposes on the fluid a tangential velocity which hits the cutwater. The larger the blade outlet angle, the higher the velocity and the associated shock. Low outlet angles (possibly combined with large wrap angles) reduce the blade loading and the hydraulic imbalance *in a given casing*. Tests with different impellers in a given casing show the following (refer to Figs. 9.36, 9.37 and 9.38 and Table D9.1):

- Two impellers with the same wrap angle (360°) were tested in [39] with blade outlet angles of $\beta_{2B} = 8$ and 16° (test T5 and T6). At high flow the hydraulic imbalance increased significantly with the higher outlet angle, while there was virtually no influence on the static radial force.
- The impellers tested in [40] had very low outlet angles ($\beta_{2B} = 8°$ to 9.5°) with wrap angles from 320° to 400°. Blade loading was low; the hydraulic imbalance was comparatively small, Fig. 9.37. However, the static radial forces are unusually high, Fig. 9.36. This could be due to a grossly oversized volute as the character of the curve of the static force suggests. The wrap angle (at almost equal β_{2B}) had little effect on the head coefficients and on the hydraulic imbalance.
- A parametric study of a pump with an open single-channel impeller in a volute has been performed in [7]; the main data were $n_q = 45$, $\beta_{1B} = 5°$, $\beta_{2B} = 30°$, $\varepsilon_{sch} = 325°$. Keeping exactly the same blade shape, the impeller outlet width was varied from $b_2/d_2 = 0.29$ to 0.69. The width b_3 of the casing was adapted to the impeller width, while the impeller inlet diameter was kept constant at $d_1/d_2 = 0.48$. Varying the impeller outlet width had little effect on the coefficients of the radial force k_{Ru} as defined by Eq. (9.7). This finding confirms that the absolute dynamic force is proportional to the impeller outlet width. At the same time this implies that Eq. (9.7) represents a reasonable approach. The blade loading (or the velocity distribution in the impeller) apparently was little affected by the impeller outlet width. This may be so because the casing width had been

adapted when varying the width of impeller. If the casing width would be maintained constant, the flow recirculation and the interaction between impeller and casing flow might be quite different.

- In [4] three impellers have been tested in a given annular casing with an annulus diameter of $d_z^* = 1.58$, Fig. 9.39, and in a given volute with $d_z^* = 1.1$, Fig. 9.40. All of the impellers had the same inlet and outlet angles ($\beta_{1B} = 9°$, $\beta_{2B} = 35°$). Compared to the baseline tests T3 and T4, the impeller in tests T12 and T15 had a lower blade loading near the exit, hence a higher loading over the first half of the blade length. The impeller of tests T13 and T14 had the same $\beta(r)$ as the baseline but a higher wrap angle (360° instead of 332°). These figures show: (1) With given exit blade loading, the larger wrap angle brings no advantage; (2) The lower blade loading near the exit brings a small force reduction in the volute, but brings scarcely an advantage when an annular casing is used; (3) The circumferential locations of the force maxima and minima depend on the interaction between the flow in the impeller and casing - hence on the impeller (and casing) design.

Effect of the casing area distribution A(ε): The flow in the casing determines the pressure distribution p(ε) around the impeller circumference. The higher the difference between the tangential velocities in impeller and casing, the higher is the interaction between both components. In order to reduce the hydraulic imbalance at high flow, the casing areas upstream of the discharge nozzle must be sufficiently large: the flow must not be accelerated at high flow into the discharge nozzle;

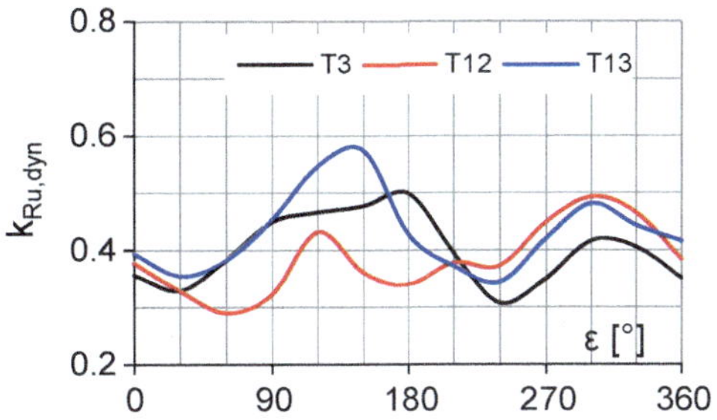

Fig. 9.39 Radial force of different impellers in the same *annular* casing at q* = 1. *T3* baseline; *T12* lower exit blade loading; *T13* higher wrap angle [4]

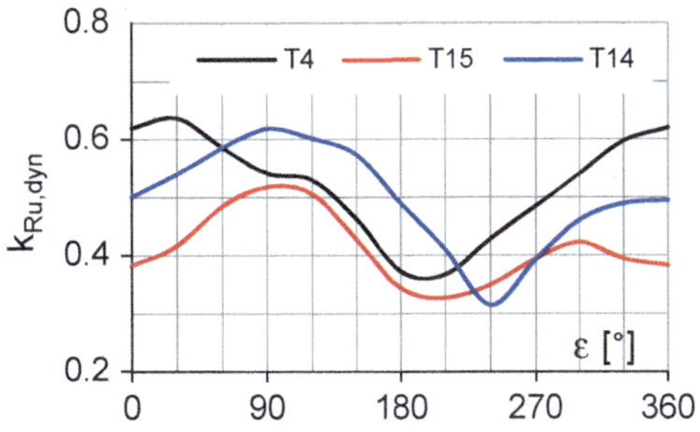

Fig. 9.40 Radial force of different impellers in the same *volute* at q* = 1. *T4* baseline; *T15* lower exit blade loading; *T14* higher wrap abgle [4]

otherwise a low-pressure zone would be created and the radial force which is directed into this zone would increase.

The findings from the available tests are summarized below:

- The influence of the volute cutwater distance was investigated in [12] by inserting different cutwaters in a given casing. In the range $d_z^* = d_z/d_2 = 1.07$ to 1.24, the volute cutwater distance had little influence on the hydraulic imbalance at $q^* > 0.8$. At shut-off $k_{Ru,dyn}$ increased from 0.3 to 0.35 when the cutwater distance was reduced from $d_z^* = 1.24$ to 1.07. Note that these findings apply to the unsteady radial force averaged over one impeller revolution. The static radial force behaved in the same way. The head was scarcely affected by the variation in cutwater distance. The efficiency diminished when increasing d_z^* - in particular at $q^* > 1$; the BEP was shifted to higher flow (presumably A_{3q} increased).
- Because of the interaction between the flows in the casing and the impeller, a given blade performs differently in different casings; Fig. 9.41 shows an example. In these tests the force peak in the annular casing was at $q^* = 0.5$ by 40% lower than with the volute. The reduction is due to the fact that the flow in annular casings becomes increasingly more uniform as the flow rate is reduced. At $q^* = 1.25$ the force maximum in the annular casing is about 5% higher than in the volute because the variation in pressure $p(\varepsilon)$ increases with increasing flow in annular casings. These observations confirm the tendency found in pumps with multi-bladed impellers that annular casings have lower forces than volutes in partload operation, while the forces increase at flows above BEP, Fig. 9.28.

As put forward at the beginning, the deviation from the circular orbit should reflect the impact of the casing on the hydraulic imbalance. To demonstrate this influence, the orbits of the radial forces of three different impellers tested in a volute and in an annular casing have been analyzed. The ratios of the major to the minor orbit axis were plotted against the flow rate in Figs. 9.42 and 9.43. The ratio of the major to

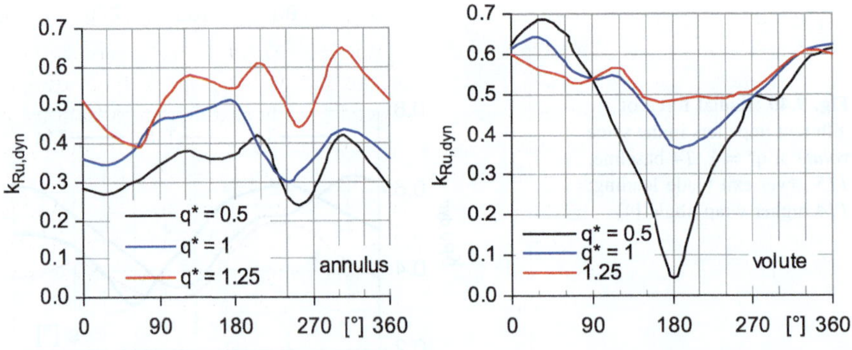

Fig. 9.41 Radial forces of a given impeller in different casings T3 (annulus), T4 (volute), [4]

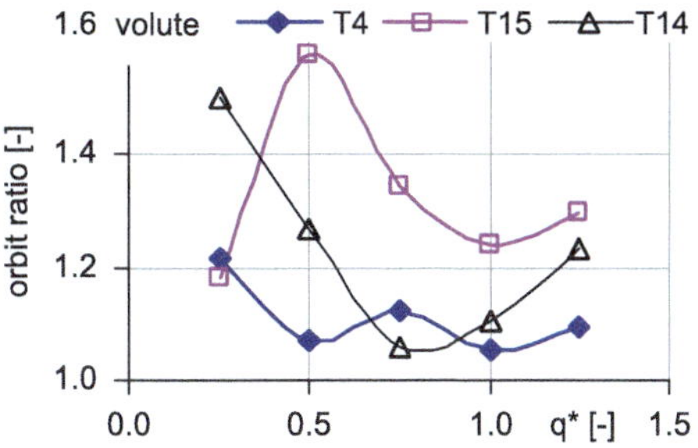

Fig. 9.42 Ratio of major to minor orbit axis of radial forces. Test of different impellers in the same *volute*. *T4* baseline; *T15* lower exit balde loading; *T14* higher wrap angle [4]

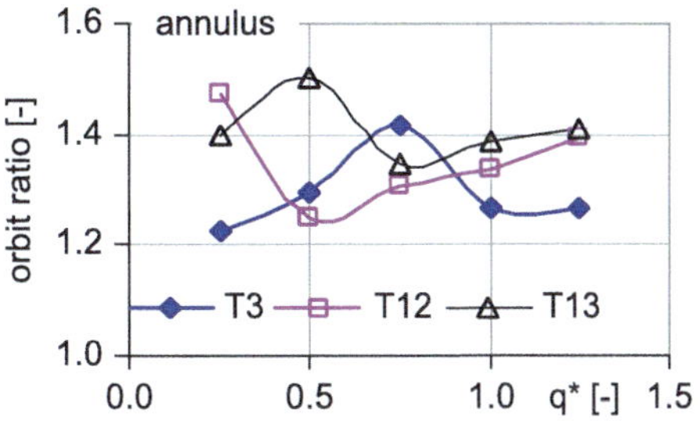

Fig. 9.43 Ratio of major to minor orbit axis of radial forces. Test of different impellers in the same *annular casing*. *T3* baseline; *T12* lower exit blade loading; *T13* higher wrap angle [4]

the minor orbit axis becomes 1.0 with a circular orbit. Hence the square root of the orbit ratio approximately corresponds to the influence of the casing on the hydraulic imbalance. For $q^* > 0.7$ this influence is below 20% in the test data in Figs. 9.42 and 9.43. These graphs also illustrate that the casing influence depends on the blade loading of the impeller. The impeller with reduced exit loading (consequently higher front loading) has the most distorted orbits in the volute, even though the peak forces were not higher than with the other impellers. It follows that (in principle) the casing areas $A(\varepsilon)$ and blade angles $\beta(\varepsilon)$ should be optimized together in order to minimize the hydraulic forces. Vice versa, the wrong combination of a "good" impeller and a "good" casing could produce unexpectedly high hydraulic imbalance if casing and impeller characteristics are not well matched.

Torque fluctuations: Investigations in [7]: The torque fluctuations measured were: at $q^* = 1 : \pm 3\%$; at $q^* = 0.28 : \pm 20\%$; and at $q^* = 1.4 : \pm 8\%$. The largest variation occurs when the impeller blade trailing edge passes the volute cutwater. An impeller in a rotationally symmetric (or in an infinitely large) casing would generate no torque fluctuations. Torque fluctuations thus reflect the influence of the casing on the flow, hence to some extent on the radial forces. Depending on q*, 70 to 90% of the torque fluctuation amplitude occurs at the rotational frequency; the rest mainly at $2f_n$ and $3f_n$. The unsteady radial forces also show peaks at $2f_n$ and $3f_n$ which are directly related to the peaks in the graphs $k_{Ru,dyn} = f(\varepsilon)$ in Figs. 9.34 and 9.39, 9.40 and 9.41: two peaks giving force components at $2f_n$ and three peaks yielding components at $3f_n$.

Axial force fluctuations were of similar magnitude as the torque fluctuations. When increasing the axial gap width s between the casing and the blade of the open impeller from $s/b_2 = 0.005$ to 0.053, the *static* radial force coefficient $k_{Ru,stat}$ was little affected; the hydraulic unbalance at low flow decreased (at shutoff by 20%) but increased at high flow by up to 20%. At a similar seal clearance, the hydraulic imbalance of the closed impeller was 7% higher than with the open impeller - presumably because of the force on the added shroud.

Scaling laws: Overall, speed and size have no systematic influence on the radial force coefficients as long as there are no resonances between the hydraulic imbalance and eigen frequencies of the test set-up, [7]. The tested range was n = 600 to 1100 rpm, $Re = u_2 r_2/\nu = (0.9 \text{ to } 1.6) \times 10^6$, $d_2 = 156$ to 240 mm. At the highest speed the hydraulic imbalance increased by about 5% at high flow; cavitation is the suspected reason.

The system and the approach flow can have a high impact on the hydraulic imbalance as demonstrated by tests in [49], where a resonance with flow fluctuations in the suction bay doubled the hydraulic imbalance force coefficient k_R when the speed was reduced from 1450 to 1000 rpm. This finding is due to system influences rather than a putative violation of the scaling laws.

CFD-studies: Tests and CFD studies of a single-channel impeller in a volute pump were reported in [38]. Data of interest are: $n_q = 42, \varphi_{2,opt} = 0.03, \psi_{opt} = 0.62, \eta_{opt} = 0.63, b_2{}^* = 0.29, d_1{}^* = 0.35, \beta_{1B} = 13°, \beta_{2B} = 8°$, blade wrap angle $\varepsilon_{sch} = 360°$. Unsteady CFD calculations were done of the pump from suction to discharge nozzle with a total of about 900,000 elements (finite element code).

The head coefficient was well predicted by this calculation over a range of $q^* = 0.25$ to 1.4; but at $q^* = 1.75$ the head was under-predicted by about 25%. Considering the strongly non-uniform unsteady flow, the good agreement up to $q^* = 1.4$ is noteworthy.

The radial forces were generally under-predicted by CFD; at $q^* = 1.57$ the ratio of the test data to CFD data was roughly $F_{R,test}/F_{R,CFD} = 1.45$. Some of the discrepancy might be due to neglecting the forces on the shrouds and on the annular seal in the calculation. The radial forces calculated showed in essence circular polar force plots, while the polar plots of the measured forces were progressively more oval with increasing flow. It can therefore be concluded that the interaction between the flows in the casing and the impeller was not well captured by CFD. This deficiency could be due to the turbulence modeling not well capturing energy dissipation in recirculating flows. CFD-calculations reported in [3] gave very similar results: calculated orbits close to circular but measured orbits oval with discrepancies up to $F_{R,test}/F_{R,CFD} = 1.5$.

The maximum force amplitude was directed to locations a little upstream of the cutwater or discharge nozzle as would be expected from Figs. 9.25 and 9.27 for the static forces. The distribution of the static pressure measured and calculated around the impeller circumference confirmed this finding. The static pressure ψ_p in the volute varied around the circumference as measured at $d/d_2 = 1.1$ and calculated by CFD. At $q^* = 1.57$ the range was $\psi_p = 0.3$ to 0.41 (or ±15% around the average static pressure at the impeller outlet). The variation was minimum at $q^* = 0.63$. The pressure variation in the casing is responsible for the static radial force which, accordingly, has its minimum at $q^* = 0.63$, Fig. 9.36.

The forces calculated by CFD in the relative reference frame must be converted into the absolute system by Eq. (9.8a):

$$\begin{aligned} F_{x,abs} &= F_{x,rel} \cos \varphi - F_{y,rel} \sin \varphi \\ F_{y,abs} &= F_{x,rel} \sin \varphi + F_{y,rel} \cos \varphi \end{aligned} \tag{9.8a}$$

Cavitation: Often the audible noise, the hydraulic imbalance and the synchronous vibrations increase strongly at flows much above BEP - even though the head strongly decreases with q^*. Cavitation is one possible reason, but it has not yet been demonstrated that cavitation would be the dominating mechanism. Undoubtedly, cavitation contributes to the noise and vibrations, [47]. The effect of cavitation on excitation forces can contribute directly through the collapse of large clouds of cavitation bubbles. This effect certainly contributes much to the air-borne noise at high flow. But cavitation could also have an indirect effect through changes in flow pattern (hence energy transfer) in a way similar to the effects discussed in context with Fig. 6.17. It can also be observed, that the vibrations rise steeply at flows above BEP, but drop again at very high flow. This behavior is thought to be due to the increased compliance of the medium in the presence of large zones of 2-phase flow (cushioning effect as discussed in Chap. 6.5.2).

Flow patterns: Because of the required ball passage, the impellers are designed with excessively large outlet width b_2. Consequently, strong flow recirculation at the impeller outlet is already present at the BEP. The flow leaving the impeller at one location (say near the rear shroud), escapes laterally and re-enters the impeller at another location (e.g. near the front shroud). In this way, much energy is dissipated by vortices and large-scale turbulence during flow recirculation, see sketch in Fig. 9.44. In the presence of flow recirculation, pressure pulsations and vibrations exhibit broadband character rather than discrete frequencies. With growing flow rate, however, the recirculation diminishes. As a result, the flow becomes *more*

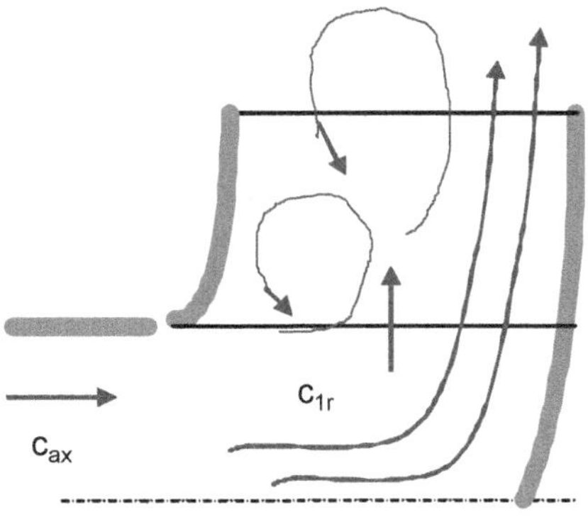

Fig. 9.44 Flow pattern in single-channel impeller

structured and the energy may be concentrated in a smaller frequency band. In other words, excitation at discrete frequencies (in this case the synchronous component) could increase. The flow distortion is compounded by the sharp flow turning at the impeller inlet when the blade is in essence a 2-D design.

A hydraulic radial force F_R acting on the impeller can be expressed as an equivalent pressure difference Δp acting on the projected area $d_2 \times b_2$: $\Delta p = F_R/(d_2 \times b_2)$ or $\Delta p/(\rho g H) = k_R = k_{Ru}/\psi$. Taking, for example, test T2 at maximum flow $q^* = 1.34$ with a force peak of $k_{Ru} = 0.44$ and $\psi = 0.55$ we get $k_R = 0.8$. In other words the head fluctuations amount to 80% of the total dynamic head.

Overall, it is the 3D-flow which determines the radial forces - as well as the efficiency. The 3D-flow patterns are determined by all of the design parameters (d_1^*, b_2^*, d_z^*, d_{1i}/d_{1a}, β_{2B}, β_{1B} etc.) and the shapes of the meridional section, the 2D- or 3D-blade $\beta_B(\varepsilon)$ and the casing $A(\varepsilon)$. The *combination* of these features, rather than any single parameter or shape, establishes the flow field.

Large fillet radii between the blade and the shrouds block a portion of the channel. Therefore, some impellers are designed as kind of spiral channels with more or less constant sections, which lead from the axial inlet to the radial outlet. This design might have beneficial effects on the hydraulic imbalance as well as on the efficiency and cavitation.

Optimizing the flow with a 3D-blade is another option. Simple rules (if such are possible) to take care of all these parameters and features have, as yet, not been published. Measuring the displacement orbit in the annular seal can reveal the direction of the radial force. If the imbalance force is directed at high flow towards the region upstream of the volute cutwater seen in Fig. 9.27, it may be expedient to enlarge the volute area in the region in an attempt to more decelerate the flow and buildup a higher pressure.

Prediction of radial forces: The *static* radial force in volutes can be estimated from the curves (T1 and T2) in Fig. 9.36 since these curves are reasonably close. Anyway, the dynamic forces are the dominating factor.

Based on the data in Fig. 9.37, it is possible to develop a tentative correlation for predicting the hydraulic imbalance of single-blade impellers in volute pumps (there are not sufficient reliable data for annular casings). The formulae derived are given in Table 9.6.

Incidentally, the minima of the curves $k_{Ru,dyn}$ in Fig. 9.37 are close to $q^* = 0.6$. Therefore, as a first step, a correlation for $k_{Ru,dyn,min}$ is developed, Eqs. (T9.5.1) and (T9.5.2). It depends definitely on the blade outlet angle and weakly on the wrap angle, the relative impeller inlet diameter and the volute cutwater distance. The tendencies found are plausible, but more test data (in a broader range) would be required to confirm the exponents. Above $q^* = 0.6$ the forces increase according to a square law with the flow rate as described by Eq. (T9.5.3).

The correlations in Table 9.6 hold down to about $q^* = 0.5$. No attempt has been made for correlating forces below that flow because the values at partload do not exceed those at high flow which can be used for design calculations. Furthermore, operation at $q^* < 0.5$ is restricted in order to avoid deposition of solids in the piping and the casing. The data from [40] have not been included since too little is known about the possible reasons why these results differ widely from the other tests.

A comparison between the test data and the correlations from Table 9.6 is shown in Fig. 9.45. The standard deviation is only ±2.5%, but it must be borne in mind that the data base is quite slender and that the correlations can only cover the geometries which have been tested as listed in Table D9.1. The correlation Eq. (T9.5.2) describing $k_{Ru,dyn,min}$ as function of geometry is quite accurate. Obviously, the correlation cannot cover 3D-effects such as blade angle distribution, casing area distribution, fillet radii and 3D-blade designs.

The correlations in Table 9.6 could be used to select design parameters with the object of minimizing hydraulic excitation forces. In addition, it could be attempted to develop a simplified mechanical model for calculating the forced response of the mechanical design to the predicted hydraulic imbalance. The model could involve a flexible beam, a mass and a torsional spring describing the flexibility of the support or, alternatively, two flexible beams one for the pump/motor assembly, the other for the support.

While reducing the blade outlet angle diminishes the hydraulic imbalance, it also reduces the pressure coefficient. Consequently, the specified head of a given pump can be achieved only if the impeller outer diameter is increased to make up for the loss in ψ_{opt}. Since the radial force follows the proportionality $F_{R,opt} \sim k_{Ru,dyn,opt}/\psi_{opt}^{1.5}$, there should be an optimum blade outlet angle at which thc hydraulic imbalance becomes minimum. This proportionality is derived from Eqs. (9.7) and (T7.1.3).

Table 9.6 Hydraulic imbalance of single-blade impellers in volutes

		Eq.
Correlation covering geometric parameters $\beta_{Ref} = 10°$; $\varepsilon_{Ref} = 330°$	$x = \left(\frac{\beta_{2B}}{\beta_{Ref}}\right)^{0.4} \left(\frac{\varepsilon_{Ref}}{\varepsilon_{sch}}\right)^{0.2} \left(\frac{d_1^*}{0.5}\right)^{-0.1} d_z^{*\,-0.15}$	9.5.1
Minimum $k_{Ru,dyn}$ at $q^* = 0.6$	$k_{Ru,dyn,min} = 0.695 - 0.99x + 0.5x^2$	9.5.2
Hydraulic imbalance for $q^* > 0.5$	$k_{Ru,dyn} = k_{Ru,dyn,min} + 0.053 \left(\frac{\beta_{2B}}{\beta_{Ref}}\right)^{1.29} (q^* - 0.6)^2$	9.5.3

Tentative design recommendations for low hydraulic imbalance

Parameter	**Effect on $k_{Ru,dyn}$**	**Side effects**
Reduce blade loading near exit: low β_{2B}, S-shaped blade	strong reduction of excitation force demonstrated	• reduced pressure coefficient • impairment of η_{opt} expected
Large fillet radii between blade and shrouds	needs optimization as to the most effective shapes and locations	• slightly reduced pressure coefficient? • posiitive effect on η_{opt}?
$d_{1i} < d_{ia}$	likely	none expected
3D-blade	needs optimization	positive effect on η_{opt} expected
Casing shape Tentative recommendation: do not undersize volute	area $A(\varepsilon)$ requires optimization	little effect expected
Cutwater distance d_z/d_2	slight improvement expected	little effect expected

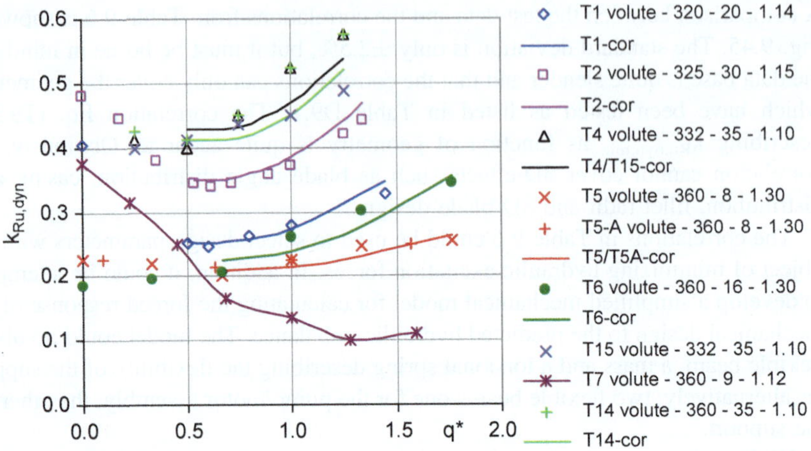

Fig. 9.45 Hydraulic imbalance correlations for single-channel impellers in volute pumps

9.3.10 Radial Thrust Balancing

If the mechanical components of a pump are not dimensioned and designed correctly for the true steady radial thrust, operational problems may be encountered:

- Excessive shaft deflection and wear in the annular seals
- Radial forces generate alternating shaft stresses which can lead to shaft breakage due to fatigue
- Bearing overload and bearing damage
- Damage to the shaft seals (in particular to mechanical seals) due to excessive shaft deflection. For pumps built according to [N.6], the shaft deflection at the mechanical seal must be limited to 50 μm.

The steady radial thrust can be reduced by the following design measures:

1. Double volutes or a rib spaced at 180° from the cutwater, refer to Fig. 9.31. The double volute casings tested in [1] produced very flat curves $F_R = f(q^*)$. The radial thrust coefficients (over the entire flow range) amounted to 10 to 20% of the thrust measured with a single volute at shut-off.
2. Multiple volutes can be considered in special cases, for example for multistage pumps (depending on manufacturing costs).
3. With reference to Fig. 9.31 multiple ribs similar to test 4 may be considered. For example two ribs each with a wrap of about 90° spaced at 120° and 240° from the cutwater.
4. A combination of diffuser and volute. The diffuser must have sufficient vane overlap in order to achieve the desired pressure equalization.
5. A combination of vaned diffuser and annular casing, see curve in Fig. 9.28.

6. At partload pumps with annular casings experience much smaller radial forces than single-volute pumps, Fig. 9.28. Since the annular casing at low specific speeds implies only a small penalty in efficiency, this option may have advantages when the pump operates mainly below the best efficiency point (as is frequently the case with process pumps). However, note that the radial thrust in annular casings increases strongly at $q^* > 1$.
7. As mentioned in Chap. 7.12, also a combination of an annular casing with a volute can be beneficial to reduce the radial thrust. Measurements were published in [1]. The tested casings consisted of a concentric annulus wrapped over 270° of the circumference followed by a single volute. The specific speeds were $n_q = 23$, 41 and 68, the cutwater distances were $d_z^* = 1.33 - 1.84$ and the casing width ratios were $b_3/b_2 = 1.39 - 1.75$. The radial forces at shut-off amounted to between 20 and 50% of the values measured with a single volute, but (probably) the reduction of the radial force is achieved only with wide impeller sidewall gaps. However, at the BEP (and above) the radial thrust in the combination of annular casing and volute is higher than in a single volute.
8. If the flow in the impeller sidewall gap is effectively decoupled from the flow in the volute casing by minimizing gap A, the non-uniform pressure distribution does not act on the projected area of the shroud. The radial thrust has been reduced that way in pumps with double volutes and double-entry impellers. If the decoupling is effective, any circumferential pressure variations in the impeller sidewall gaps tend to diminish unless they are induced by uneven flow through the annular seals (e.g. due to eccentricity of the rotor).
9. Wide impeller sidewall gaps foster the pressure equalization over the circumference; the radial thrust - in particular at partload - is reduced; the effect is maximum at $Q = 0$. This effect is important for the combination of an annular casing with a volute

The effects of items 8 and 9 are to some extent opposite; which effect is dominating in a specific case depends on the design but is difficult to predict.

9.3.11 Radial Thrust Prediction

For dimensioning the bearings of single-stage pumps and for calculating the shaft and its deflection, the radial forces acting on the impeller must be known for the entire flow range. These forces are commonly determined from experimental radial force coefficients according to Eqs. 9.6 or 9.7. Table 9.7 provides all data necessary for that purpose for single and double volutes, annular casings and diffuser pumps, [17]. Explanations to Table 9.7:

1. In Table 9.7, q^* always refers to the *design point of the collector* which is usually identical to the best efficiency point of the pump. The difference only becomes relevant if various impellers and collectors are combined.

Table 9.7 (1) Radial thrust calculation

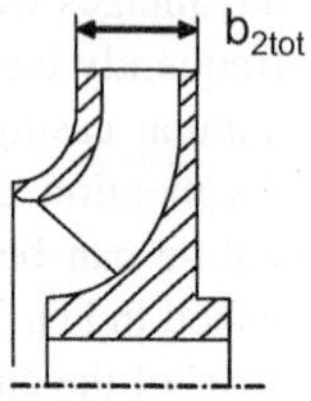

q^* refers to the layout point of the *casing* (volute or diffuser)			
Radial force	$F_R = k_R\, \rho\, g\, H\, d_2\, b_{2tot}$		
1. Steady radial forces			
Single volute	$q^* = 0$	k_{R0} from Fig. 9.46	
	$q^* = 0.5$	from Fig. 9.47	
	$0 < q^* < 1$	$k_R = (k_{R0} - k_{R,opt})\,(1 - q^{*2}) + k_{R,opt}$	
	$q^* = 1$	from Fig. 9.47 or $k_{R,opt} = 0.03$ to 0.08	
	$q^* > 1$	$k_R = 0.09\, q^{*2}$ (or use $k_{R,opt}$ instead of factor 0.09)	
Double volute	$k_{R,Dsp} = F_{Dsp}\, k_{R0}$ with $F_{DSp} = (1.75 - 0.0083\, \varepsilon_{sp}°)$ from Fig. 9.48		
	Below $q^* \approx 1.1$, F_{DSp} and k_R depend little on flow rate; above $q^* >> 1$ the radial thrust can increase sharply. If wrap angle ε_{sp} is smaller than 180°, the radial force coefficient for $q^* = 0.5$ or $q^* = 1.0$ can be estimated from the following interpolation formula: $f = \frac{k_{Rx}}{k_{R0,SV}} + \left(1 - \frac{k_{Rx}}{k_{R0,SV}}\right)\frac{\varepsilon_{sp}° - 90}{90}$ valid for $90 < \varepsilon < 180°$ k_{Rx} stands for the k_R-values of single volute pumps for either $q^* = 0.5$ or $q^* = 1.0$. $k_R = k_{R0,SV}\, F_{dsp}\, f$ k_{R50} and $k_{R,opt}$ are taken from Fig. 9.47		
Twin volute	$k_{R,Zsp} = (0.3$ to $0.5)\, k_{R,Dsp}$		
Annular casings	$k_{R0} = 0.03$ to 0.1 $k_{R,opt} = 0.1$ to 0.2	$k_R = k_{R0}\,(1 + q^* + a\, q^{*2})$ "a" depends on the geometry; tentatively a = 0,18	
Diffuser	$k_{R0} = 0.02$ to 0.09	$q^* = 1$: $k_{R,opt} = 0.01$ to 0.06	
2. Unsteady radial forces		$q^* < 0.5$	$q^* = 1$
All types of volutes and annular casings, *except single-channel impellers*		$k_{R,dyn} = 0.07$ to 0.12	$k_{R,dyn} = 0.01$ to 0.05
Diffusers		$k_{R,dyn} = 0.04$ to 0.16	$k_{R,dyn} = 0.01$ to 0.09
3. Axial pumps			
$F_R = k_{R,D}\, \rho\, g\, H\, d_2^2$	steady force: $k_{R,D} = 0.02$ for $q^* < 1.2$ unsteady forces: $k_{R,D} = 0.01$; see also Sect. 10.14.		

Table 9.7 (2) Radial thrust calculation

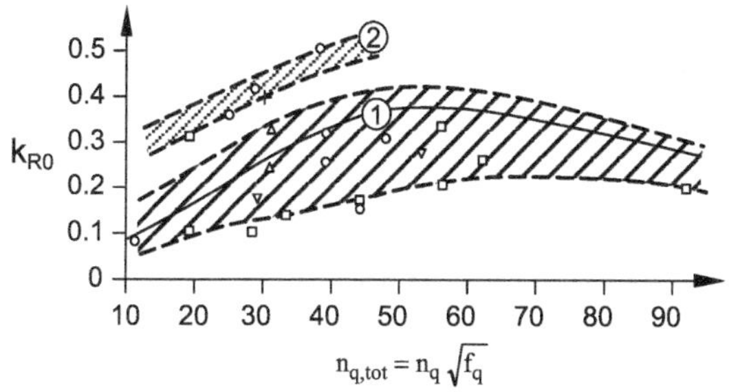

Curve 1: Seal clearance according to Eq. (3.12)
Curve 2: Double seal clearances

single-entry impellers: $f_q = 1$
double-entry impellers: $f_q = 2$

Approximation formula for curve 1:
$k_{R0} = 3.730\text{E-}08x^4 - 7.274\text{E-}06x^3 + 3.610\text{E-}04x^2 + 2.041\text{E-}03x + 3.944\text{E-}02$
Approximation formula for curve 2 (average):
$k_{R0} = -5.000\text{E-}05x^2 + 9.200\text{E-}03x + 1.910\text{E-}01$

Fig. 9.46 Static radial force coefficients for single volutes at shut-off (Q = 0)

2. Figure 9.46 in Table 9.7 (2) gives the radial thrust coefficients k_{R0} of *single volute pumps* operating against closed discharge valve as a function of the specific speed. Note that the specific speed must be calculated with the *design flow of the volute*. Attention is required when dealing with double-entry impellers. Under this provision, Fig. 9.46 is valid for single- as well as double-entry impellers. The abscissa in Fig. 9.46 is therefore labeled $n_{q,tot} = n_q \times f_q^{0.5}$.
3. A finite radial thrust is also to be expected in the best efficiency point. The dependence on the specific speed is weak, Fig. 9.47.
4. The data provided in Fig. 9.46 originate from [17] and [1]; they compare reasonably well with the published data. When the annular seal clearances increase, higher radial force coefficients must be applied as stipulated by curve 2. These must also be used for radial or diagonal seals as applied in pumps which are exposed to abrasive wear, Fig. 3.18. Note: any radial forces calculated by CFD must be compared to curve 2, unless the forces created in annular seals are taken into account in the CFD computations.
5. When calculating the radial thrust in *double volutes*, the coefficient k_{R0} read from Fig. 9.46 must be multiplied by the correction factor F_{Dsp} from Fig. 9.48 (in Table 9.7) which depends on the wrap angle of the inner volute. The value thus obtained may be applied in the range of $q^* = 0$ to 1.1. As discussed above, the radial thrust can increase strongly at $q^* \gg 1$ due to the pressure difference caused by different flow resistances in the outer and inner channels.

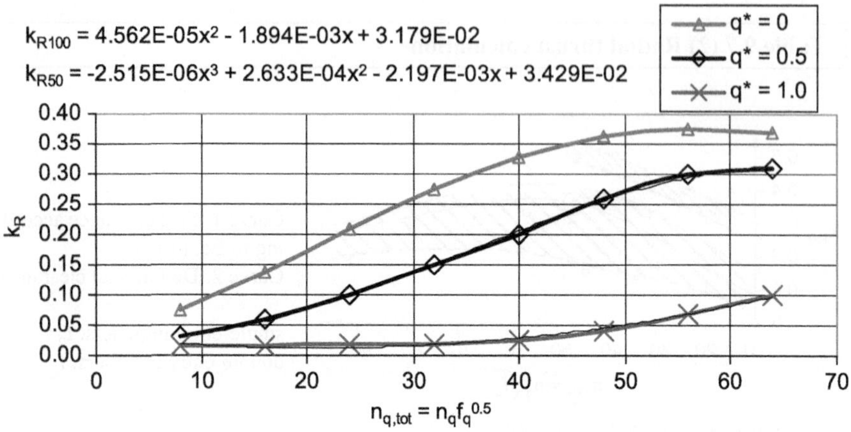

Fig. 9.47 Static radial force coefficients for single volutes, data from [1]

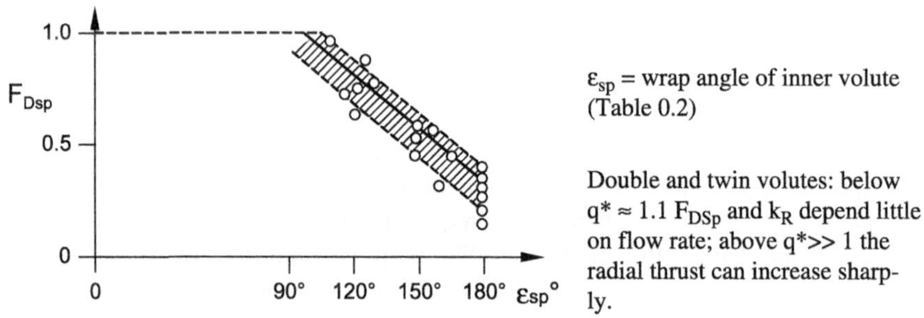

Fig. 9.48 Factors for radial force coefficients of double volutes

6. If the two channels of a double volute are exactly symmetrical ("twin volutes") according to Fig. 7.39, the radial forces amount to only 30 to 50% of the forces created in double volutes with a common discharge nozzle. Twin volutes are found in the series stages of multistage pumps according to Fig. 2.9.
7. Data for annular casings are scarce in the literature; great uncertainties in the thrust prediction are therefore to be expected. All the more so, because there are no generally accepted rules for dimensioning annular casings. The function $k_R(q^*)$ depends on the design of the annular casing. In the range where the ratio c_R/c_2 is small, the radial thrust is expected to increase moderately with the flow rate. Where the fluid is strongly accelerated near the discharge nozzle, radial forces presumably rise with the square of the flow.
8. The dependency (if any) of the radial force coefficients of diffuser pumps on geometric parameters or the specific speed is not documented.
9. The unsteady radial forces are quite similar for all types of collectors. The data given in Table 9.7 may be considered as broadband RMS values for the entire

load spectrum of practical interest. Figure 10.42 provides broadband values for specific frequency ranges, refer also to Chap. 10.7.

10. Open and semi-open impellers can experience slightly higher radial thrust coefficients than closed impellers, because they lack the effect of some pressure equalization in the impeller sidewall gaps. On the other hand, there are no radial forces acting on the front shroud.
11. Cavitation has a significant impact on steady radial forces only when the pump runs with full cavitation. However, extensive cavitation can generate unsteady radial forces which increase the level of vibrations, [17].
12. It is prudent to use the sum of static and dynamic radial thrust for the calculation of the shaft and the bearings, i.e. $k_{R,tot} = k_R + k_{R,dyn}$. For an in-depth analysis of the bearings the dynamic loading due to unsteady radial forces may be considered separately. In the absence of specific measurements the data given Fig. 10.42 can be used for that purpose.
13. In general it is recommended to apply at least a coefficient of $k_{R,tot} = 0.15$ for the mechanical design.
14. As discussed in Chap. 9.3.3 the flow in the impeller side wall gaps is extremely complex. It interacts with the flow through the annular seals. Therefore, the radial forces acting on the shrouds and in the annular seals represent a large uncertainty in radial thrust prediction because their effects were ignored in the published radial force coefficients.

References

1. Agostinelli, A., et al.: An experimental investigation of radial thrust in centrifugal pumps. ASME. J. Eng. Power **82**, 120–126 (1960)
2. Bachmann, P.: Fortschritte im Erfassen und Auswerten von Kräften und Momenten an Rotoren hydraulischer Modell-Turbomaschinen. Escher. Wyss. Mitt. 69–81 (1980)
3. Benra, F.K., Sommer, M.: Comparison of calculated and measured hydrodynamic forces of a centrifugal sewage water pump. 1. In: Conference on Experiments/Process/System Modelling IC-EpsMsO Athens (2005)
4. Benra, F.K., et al.: Experimental investigation hydrodynamic forces for different configurations of single-blade centrifugal pump. In: ISROMAC, pp. 2006–2011 (2011)
5. Bolleter, U., et al.: Measurement of hydrodynamic interaction matrices of boiler feedpump impellers. ASME. J. Vib. **109**, 144–151 (1987)
6. Bolleter, U., et al.: Rotordynamic modeling and testing of boiler feedpumps. EPRI Report TR-100980 (September 1992)
7. Böke, J.: Experimentelle und theoretische Untersuchungen der hydraulischen Kräfte an einschaufeligen Laufrädern von Abwasserpumpen unter Berücksichtigung der Änderung geometrischer Parameter. Diss TU Kaiserslautern, Schaker Aachen (2001)
8. Dahl, T.: Centrifugal pump hydraulics for low specific speed application. In: 6th International on Pump Users Symposium, Houston (1989)
9. Dupont, P., et al.: CFD analysis of sump flow and its impact on the hydraulic forces acting on the impeller of a vertical pump. In: International Rotating Equipment Conference 2008, Düsseldorf (2008)

10. Faschallegg, E.: Radialkräfte und Rückenschaufeln in Spiralgehäusepumpen. Europa. Ind. Rev. **1** (1969)
11. Geis, H.: Experimentelle Untersuchungen der Radseitenverluste von Hochdruck-Wasserturbinen radialer Bauart. Diss. TH Darmstadt (1985)
12. Gikadi, T., et al.: Untersuchung der hydraulischen Kräfte bei Abwasserpumpen. 3Rinternational **34**(8), 420–425 (1995)
13. Guinzburg, A., Buse, F.: Magnetic bearings as an impeller force measurement technique. In: 12th International Pump Users Symposium, Houston (March 1995)
14. Gülich, J.F.: Influence of interaction of different components on hydraulic pump performance and cavitation. In: Proceedings of Symposium Power Plant Pumps, New Orleans, 1987, EPRI CS-5857, pp. 2.75–2.96 (1988)
15. Gülich, J.F.: Disk friction losses of closed turbomachine impellers. Forsch. Ing. Wes. **68**, 87–97 (2003)
16. Gülich, J.F.: Effect of Reynolds-number and surface roughness on the efficiency of centrifugal pumps. ASME. J. Fluids Eng. **125**(4), 670–679 (2003)
17. Gülich, J.F., et al.: Review of parameters influencing hydraulic forces on centrifugal impellers. Proc. IMechE. **201**(A3), 163–174 (1987)
18. Gülich, J.F., et al.: Influence of flow between impeller and casing on partload performance of centrifugal pumps. ASME. FED. **81**, 227–235 (1989)
19. Hamkins, C.P.: The surface flow angle in rotating flow: Application to the centrifugal impeller side gap. Diss. TU Kaiserslautern, Shaker, Aachen (2000)
20. Hergt, P.: Hydraulic design of rotodynamic pumps. In: Rada Krishna, H. C. (ed.) Hydraulic Design of Hydraulic Machinery. Avebury, Aldershot (1997)
21. Hergt, P., Krieger, P.: Radialkräfte in Leitradpumpen. KSB. Techn. Ber. 32–39 (1973)
22. Iversen, H., et al.: Volute pressure distribution and radial force on the impeller. ASME. J. Eng. Power **82**, 136–144 (1960)
23. Jensen, R.: Experimentelle Untersuchungen an Einfach- und Doppelspiralen für Kreiselpumpen. Diss. TU Braunschweig (1984)
24. Kaupert, K.A.: Unsteady flow fields in a high specific speed centrifugal impeller. Diss. ETH Zürich (1997)
25. Kleinert, H.J.: Beitrag zur verbesserten Berechnung hydraulischer Ausgleichsvorrichtungen. Diss TU Dresden (1971)
26. Kosyna, G.: Untersuchungen an radial durchströmten Spaltdichtungen unter Berücksichtigung von Parallelitätsfehlern. Diss. TU Braunschweig (1976)
27. Lingelbach, T., Wiederuh, E.: Die Axialschubberechnung radialer Turbomaschinen. Fortschrittber VDI Reihe **7**, 154 (1989)
28. Liess, C.: Die Ermittlung dynamischer Radialkräfte in hydraulischen Maschinen. Voith. Forsch. Konstruktion. 3.1–3.9 (1982)
29. Lauer, J.: Einfluß der Eintrittsbedingung und der Geometrie auf die Strömung in Radseitenräumen von Kreiselpumpen. Diss. TU Darmstadt (1999)
30. Lees, A.W.: The performance of balance disk in boiler feed pumps. I Mech. E, C **56**(79), 29–36 (1979)
31. Makay, E., Barret, J.A.: Changes in hydraulic component geometries greatly increased power plant availability and reduced maintenance cost: Case histories. In: 1st International Pump Symposium, Houston (1984)
32. Makay, E., Szamody, O.: Survey of feedpump outages. EPRI FP-754, Final report RP 641 (1978)
33. Meier-Grotian, J.: Untersuchungen der Radialkrft auf das Laufrad einer Kreiselpumpe bei verschiedenen Spiralgehäuseformen. Diss. TU Braunschweig (1972)
34. Meschkat, S.: Experimentelle Untersuchung der Auswirkung instationärer Rotor-Stator-Wechselwirkungen auf das Betriebsverhalten einer Spiralgehäusepumpe. Diss. TU Darmstadt (2004)

35. Meschkat, S., Stoffel, B.: The local impeller head at different circumferential positions in a volute casing of a centrifugal pump in comparison to the characteristic of the impeller alone. In: 21st IAHR Symposium on Hydraulic Machinery and Systems, Lausanne (2002)
36. Möhring, U.K.: Untersuchung des radialen Druckverlaufes und des Drehmomentes im Radseitenraum von Kreiselpumpen bei glatter, ebener Radseitenwand und bei Anwendung von Rückenschaufeln. Diss. TU Braunschweig (1976)
37. Ni, L.: Modellierung der Spaltverluste bei halboffenen Pumpenlaufrädern. Fortschrittber VDI Reihe **7**, 269 (1995)
38. Nishi, Y., et al.: Radial thrust of single-blade centrifugal pump. ASME-JSME-KSME AJK 2011-06011 (2011)
39. Nishi, Y., et al.: Effect of blade outlet angle on radial thrust of single-blade centrifugal pump. In: 26th IAHR Symposium Hydraulic Machinery and Systems (2012)
40. Okamura, T.: Radial thrust in centrifugal pumps with a single-vane impeller. JSME **23**(180), 895–901 (1980)
41. Petermann, H., Kosyna, G.: Spaltstrom und Kräfte radial durchströmter Dichtspalte in Kreiselpumpen. Pumpentagung Karlsruhe. K6 (1978)
42. Rebernik, B.: Radialkräfte von Kreiselpumpen mit unterschiedlichen Gehäuseformen. „25 Jahre ASTRÖ", Aströ, Graz, pp. 55–60 (1979)
43. Schenkel, S.: Modellierung und numerische Simulation der Strömungsvorgänge am Laufradeintritt von Turbomaschinen. Diss. TU Darmstadt (1999)
44. Schubert, F.: Untersuchungen der Druck- und Geschwindigkeitsverteilung in Radseitenräumen radialer Strömungsmaschinen. Diss. TU Braunschweig (1988)
45. Stark, M.: Auslegungskriterien für radiale Abwasserpumpenlaufräder mit einer Schaufel und unterschiedlichem Energieverlauf. VDI Forschungsheft. **57**(664) (1991)
46. Tamm, A., Stoffel, B.: The influence of gap clearance and surface roughness on leakage loss and disk friction of centrifugal pumps. ASME. Fluids Engng. Meeting Montreal, FEDSM 2002-31324 (2002)
47. Thamsen, P.U., et al.: Cavitation in single-vane sewage pumps. ISROMAC 12-2008–20196
48. Thomae, H., Stucki, R.: Axialschub bei mehrstufigen Pumpen. Techn Rundschau Sulzer **3**, 185–190 (1970)
49. Ulbrich, C.: Experimentelle Untersuchungen der Pumpencharakteristiken und Geschwindigkeitsfelder einer Einschaufel-Kreiselpumpe. Diss. TU Berlin (1997)
50. Zilling, H.: Untersuchung des Axialschubes und der Strömungsvorgänge im Radseitenraum einer einstufigen Radialpumpe. Strömungsmech Strömungsmasch. **15** (1973)
51. Hergt, P., Prager, S.: Influence of different parameters on the disc friction losses of a centrifugal pump. In: Conference on Hydraulic Machinery, Budapest, pp. 172–179 (1991)
52. Van Esch, B.P.M.: Performance and radial loading of a mixed-flow pump under non-uniform approach flow. In: ASME JFE, vol. 131 (May 2009)
53. Will, B.C., Benra, F.K., Dohmen, H.J.: Investigation of the flow in the impeller side clearances of a centrifugal pump with volute casing. J. Therm. Sci. **21**(3), 197–208 (2012)
54. Roclawski, H.: Numerische und experimentelle Untersuchungen an einer radialen Kreiselpumpenstufe mit minimalem Stufendurchmesser. Diss. TU Kaiserslautern (2008)
55. Nemdili, A.: Einzelverluste von Kreiselpumpen mit spezifischen Drehzahlen von nq = 15 bis 35 rpm. Diss. Uni Kaiserslautern, SAM Forschungsberichte, Bd 1 (2000)

Chapter 10
Noise and Vibrations

As explained in Chap. 5, the flow at the impeller outlet is non-uniform. The diffuser vanes or volute cutwaters are thus approached by an unsteady flow. The flow in the stator acts back on the velocity field in the impeller. The related phenomena are called "rotor/stator interaction" (RSI). As a consequence of the RSI, hydraulic excitation forces are generated. These give rise to pressure pulsations, mechanical vibrations and alternating stresses in various pump components. The vibrations transmitted to the foundations spread as solid-borne noise throughout the building. The pressure pulsations excite the pump casing to vibrations. They travel as fluid-borne noise through the piping system, where they generate vibrations of the pipe walls. The vibrating walls and structures radiate air-borne noise.

The interaction of fluid forces with the pump and the piping is very complex; it is scarcely amenable to analytic solutions. Figure 10.1 depicts the interrelations between various components and physical mechanisms which are treated in this chapter.

As the importance of all these effects increases with the circumferential speed of the impeller, vibration problems are encountered particularly in high-pressure pump applications. Whether the vibrations and the noise of a pump are felt as disturbing, largely depends on its design and application. Even very small pumps, for example, central heating circulation pumps, may be subject to stringent requirements concerning the limitation of noise.

In many applications the pumps are built and tested to standards and specifications defining the maximum allowable vibration amplitudes (process pumps and almost all large pumps). This chapter focuses in the first place on the hydrodynamic aspects of vibration excitation, as well as on the basic physical mechanisms involved, in order to help with the analysis of vibration problems. The interaction between hydraulic excitation and mechanical vibrations is discussed too, but the theory of mechanical vibrations is beyond the scope of this text. There is a wealth of specific literature on vibrations which should be consulted for calculation and analysis methods, mechanical vibrations, testing methods and data treatment, e.g. [1, 8, 21, 30, 137]. Some basic notions of mechanical vibrations are given in Chap. 17.6.

J. F. Gülich, *Centrifugal Pumps*,
https://doi.org/10.1007/978-3-030-14788-4_10

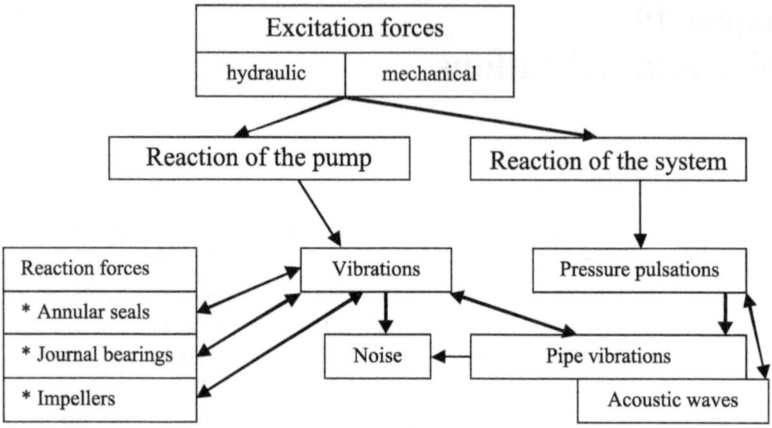

Fig. 10.1 Interaction of hydraulic and mechanical effects

10.1 Unsteady Flow at the Impeller Outlet

The velocity distribution at the impeller outlet is pitch-wise non-uniform as a result of the work transmitted by the blade to the fluid and due to secondary flows. According to Chap. 5.2, the qualitative distribution of the relative velocity shows a maximum near the suction surface of the blades, Fig. 10.2. The finite blade thickness, the blockage caused by the boundary layers and possible flow separation add to the distortion of the velocity profile immediately downstream of the impeller outlet.

Considering the fluid flow in the stationary system, a velocity *minimum* in the *relative* system corresponds to a *maximum* in the *absolute* reference frame. A fluid particle carried out of the boundary layer into the collector approaches the velocity of $c_{2,max} \approx u_2$ in the absolute reference frame. The average absolute velocity at the impeller outlet is roughly $c \approx 0.5\ u_2$. The pitch-wise variation in stagnation pressure caused by the wake is therefore in the order of:

$$\Delta p_s = p_{dmax} - \bar{p}_d = \frac{\rho}{2} u_2^2 - \frac{\rho}{2}\left\{\frac{u_2}{2}\right\}^2 = 0.75 \frac{\rho}{2} u_2^2 \quad \text{or} \quad \Delta p^* = \frac{2\Delta p_d}{\rho u_2^2} = 0.75 \tag{10.A}$$

The instantaneous stagnation pressures measured downstream of an impeller well confirm the estimation given by Eq. (10.A): at a radius of $r/r_2 = 1.015$ the peak measured was $\Delta p^* = 0.9$, curve 1 in Fig. 10.6, [4]. Extrapolating this curve to a location still closer to r_2 yields values near 1.0.

When the local velocity maximum (or the blade) is somewhat upstream of the diffuser vane or volute cutwater, a pressure maximum is observed which is related to the stagnation pressure. When the blade passes under the diffuser vane or volute

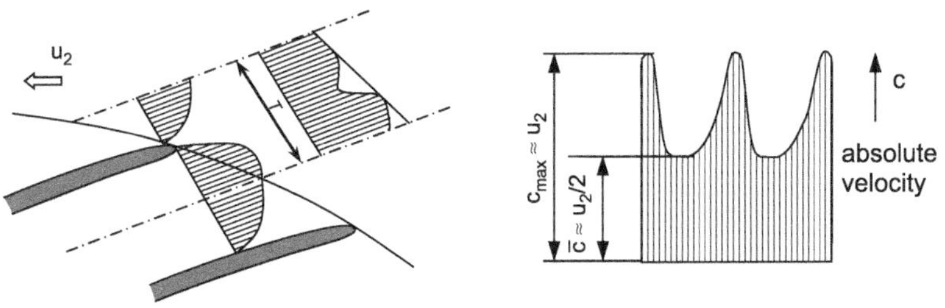

Fig. 10.2 Wake flow on impeller outlet in the rotating (left) and absolute (right) reference frame

cutwater, there is a pressure minimum due to the effect that the fluid is kind of "squeezed" through gap B, Fig. 10.3, see also Chap. 10.7.1.

The unsteady velocity variations at the impeller outlet caused by the rotor-stator interaction can be measured by means of fast velocity probes or laser velocimetry. As an example, Fig. 10.4 shows the absolute velocity and the static pressure downstream of a radial impeller. The measuring point A1 is at a distance of 4% of the impeller radius r_2, the measuring point D3 is placed at the diffuser throat and F3 is just upstream of the diffuser outlet. The figure shows one revolution of an impeller with 7 blades at n = 1000 rpm. Clearly discernible are 7 velocity maxima. The velocity maximum corresponds (according to the Bernoulli equation) to a minimum of the static pressure occurring at the same phase. In contrast, there is a small phase shift between the pressure maximum and the velocity minimum. This may be due to the fact that that the pressure recovery is subject to considerable losses, while the acceleration is not. The pressure and velocity variations decay significantly as the flow progresses through the diffuser channel. The velocity peaks at the locations A1, D3 and F3 shifted since convected with the flow velocity in the diffuser channel.

Figure 10.5 shows the variation of the vectors of the relative and absolute velocities during the passage of one blade pitch in the period $\Delta t = 60/(z_{La} \times n)$, corresponding to 0.0086 s in this test. During this short time, the velocity vectors vary in size and direction. The fluctuations increase at partload, as can be seen when comparing q* = 1 and q* = 0.5 in Fig. 10.5. The incidence at the diffuser vane or volute cutwater varies over one pitch which can cause unsteady cavitation as observed in Fig. 5.17 and Fig. 6.46-1 and explains some features of the interaction forces as shown by Figs. 10.24-1. The variations of the vectors of the relative velocity imply that the lift acting on the impeller blades is unsteady. Consequently

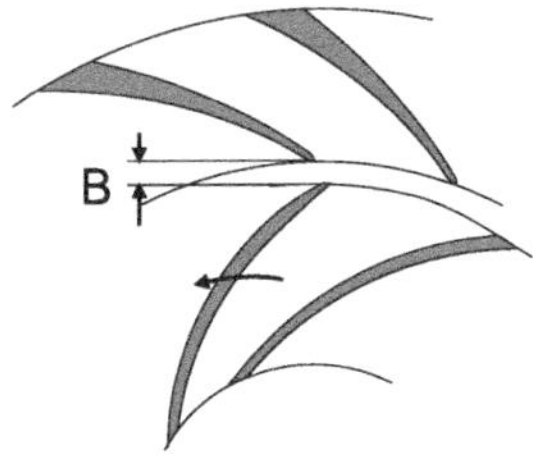

- Minimum pressure where vanes are directly opposite; here the velocity is maximum.
- Applies also to volute cutwaters.
- Measured with single-channel impellers too.

Fig. 10.3 Rotor-stator interaction

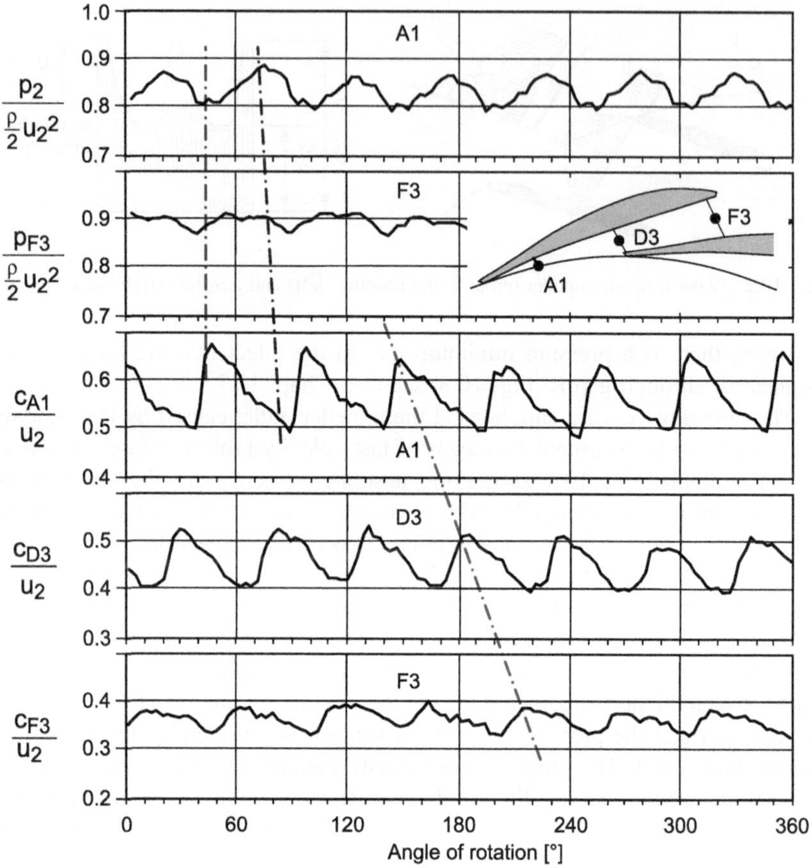

Fig. 10.4 Unsteady velocities and pressures in a diffuser, operation at q* = 1; Measurements Sulzer Innotec, [16]

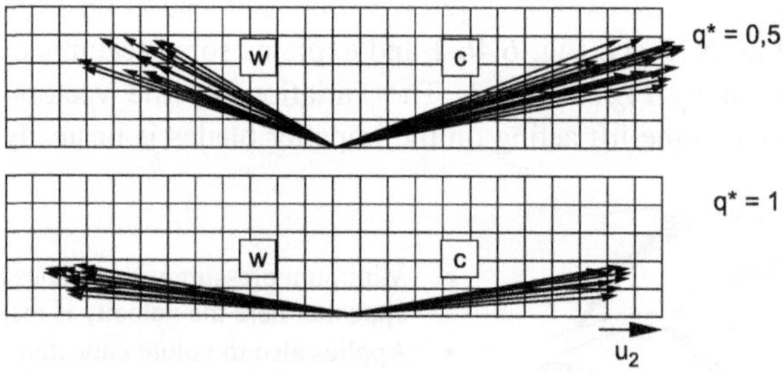

Fig. 10.5 Velocity vectors at the diffuser inlet (A1) within one pitch of the impeller blades, *w* relative velocity, *c* absolute velocity. Sulzer Innotec, [16]

the impeller blades experience alternating forces and stresses. In a similar way the approach flow to the collector is unsteady inducing alternating hydrodynamic forces on the diffuser vanes or the volute cutwater. Thus the local flow field is subject to significant variations each time an impeller blade passes a diffuser vane or a volute cutwater. This rotor/stator interaction creates perturbations with the frequency $f = z_{La} \times n/60$ called "blade passing frequency". The unsteady flow around the diffuser vanes interacts with the flow in the impeller. While this causes only a local disturbance of the velocity field, the impact on the pressure field extends over the whole of the impeller channel, [82].

The interaction between the impeller blades and the collector discussed above is not the only phenomenon to cause an unsteady flow around the impeller or diffuser vanes. *Any disturbance in the circumferential symmetry* of the flow field will have that effect. Circumferential disturbances are created by: (1) Non-uniform velocity distributions of the impeller approach flow as generated by inlet chambers, bends upstream of the suction nozzle, structural elements (ribs) and vortices; (2) Flow deflection, secondary flow, flow separation and vortices inside the impeller forming a complex non-uniform flow field at the impeller outlet (Chap. 5); (3) Unsymmetries in the discharge flow path as given by volutes, annular chambers and the discharge nozzle; (4) The individual impeller channels are not completely identical due to casting tolerances. This gives rise to periodical changes of the impeller blade lift forces and incidence variations in the diffuser approach flow.

Owing to the 3-dimensional structure of the flow at the impeller outlet and to vortices, discrete frequencies ("sounds") as well as stochastic fluctuations are generated. The velocity fluctuations result from the geometry and from boundary layer effects. They likewise produce fluctuations in the pressure field, the integration of which yields the momentary blade forces. A fraction of the mechanical energy contained in the unsteady pressure fields is transmitted as liquid-borne noise ("pressure pulsations"). Apart from vibration measurements at the shaft or the casing, pressure pulsations in practice are often the only practicable way of getting information, by an indirect approach, about the frequencies and the intensity of the unsteady hydrodynamic effects generated in a pump. That is why pressure pulsations are often considered to be an indirect measure for the alternating forces acting on vanes and structural components since such forces cannot easily measured. Note that alternating forces and stresses are *not* directly proportional to the pressure pulsations, [67].

Whereas the time-average of the interaction forces between rotor and stator is zero, this does not apply to the momentary forces causing excitation in the baseplate or the foundation.

10.2 Pressure Pulsations

Unsteady fluid flow in a *compressible* medium generates pressure variations that travel through the system at the speed of sound that is characteristic to the medium (refer to Chap. 17.5). The compressibility of water, though low, is sufficient to allow this. Pressure pulsations can give rise to acoustic system resonances and fatigue fractures of pump components. They are also responsible for the radiation of air-borne and solid-borne noise.

10.2.1 Generation of Pressure Pulsations

Pressure pulsations are generated not only by the impingement of the wake flow on the diffuser vanes, but also by vortices forming downstream of the trailing edge due to shear layers and flow separation. As a rule, the wake flow is an important source of pressure pulsations in the pump. It creates frequencies corresponding to the "blade passing frequency" $z_{La} \times n/60$ and its harmonics. Due to asymmetries in the impeller channels, pulsations with the rotational frequency and its multiples are generated as well, [10]. Pressure amplitudes of the effects discussed here manifest in the spectrum as rather sharp peaks ("discrete frequencies"). With certain combinations of impeller and diffuser blade numbers, the pressure pulsations of two or more blades may be in phase and amplify each other through "phase resonance", Chap. 10.7.1.

In contrast, the character of vortices in separated flow and turbulence is chiefly stochastic. Their effects do not show up as discrete frequencies, but produce a more continuous spectrum over a certain frequency domain. These are termed "broadband" pressure pulsations (or "white noise"). Under attached flow conditions, turbulent fluctuations are a weak source of pressure pulsations. However, with flow separation at partload and subsequent recirculation at the impeller outlet, strong vortices are created in the shear layer between the main flow and the recirculating fluid. This increases broadband pressure pulsations below the blade passing frequency $z_{La} \times n/60$.

Likewise, partload recirculation at the impeller inlet induces strong interactions between the impeller and components of the inlet casing, particularly ribs, or returns vanes in the case of multistage pumps. Due to shear layers between the main flow and the recirculating fluid, vortices are generated which give rise to broadband pressure pulsations mainly in the domain below the rotational frequency.

Vortex shedding downstream of ribs or similar structures causes pressure pulsations whose frequencies can be described by Strouhal numbers, Chap. 10.12.4. If an interaction occurs between the sequence of vortices thus created ("vortex streets") and the subsequent impeller, periodic lift forces are generated.

The range of maximum pressure recovery in a diffuser is achieved under flow conditions where alternating flow separation/re-attachment occurs. Such separations are expected to have frequencies of 2 to 12 Hz (up to 25 Hz) and show a broad peak in the spectrum. The frequency depends on the upstream and downstream piping and the diffuser geometry. It *drops* with increasing flow velocity, [58]. Also cavitation generates pressure pulsations and noise, Chap. 6.5.

10.2.2 Noise Generation in a Fluid

If the local pressure in a compressible medium is subject to a sudden variation due to fluctuations of the flow regime, sound waves are generated. For illustration, consider a sphere moving periodically to and fro in a fluid. The movements cause periodic variations in local velocity and pressure. Most of the energy input necessary to sustain the movement of the sphere is dissipated into friction and vortex decay. A small fraction of the energy is used to compress the fluid and this part is radiated as sound waves. The zone of liquid around the sphere where the displacement of the fluid is noticeable is termed "hydrodynamic near field" (measurable, e.g., by means of laser velocimetry). The radiated acoustic energy can be measured far from its source (in the "far field") by means of pressure transducers.

A stationary object placed into the hydrodynamic near field acts as an additional source of noise since it causes strong fluctuations in the velocity field and disturbs the hydrodynamic short circuit around the primary source. The acoustic energy radiated from this secondary source depends on the intensity of the velocity and pressure variations it produces. Therefore the acoustic radiation of the secondary source can exceed that of the primary source, [47]. This phenomenon is thought to explain very well the behavior of centrifugal pumps: The impeller blade wake flow is a relatively weak primary source whereas the diffuser vanes (or the volute cutwater) in the near field of the impeller act as a strong secondary source. This can be inferred from the important impact that the distance between the impeller and diffuser vanes has on the pressure pulsations.

The fluid dynamic aspects of the near field are described by well-known hydrodynamic laws, e.g. Bernoulli's equation. Consequently, the pressure variations in the near field are proportional to the square of the velocities. With regard to the far field, this is not generally true since the acoustic efficiency increases with increasing speed. As the Mach number of pumps is nearly always far below 0.1 this influence is probably small. For a first approximation the square law can be applied.

10.2.3 Influence Parameters of the Pump

The pressure pulsations experienced in a plant depend on a number of pump as well as system parameters. Basically, the primary causes of the pressure pulsations are the velocity distribution over the blade pitch at the impeller outlet and unsteady approach flows upstream of the impeller. They generate the hydrodynamic near field. Any structural element situated in the near field at the impeller inlet or outlet will act as a secondary source of sound generation. The pressure pulsations at the impeller outlet reach their minimum at that flow rate where the velocity distribution is the most uniform. This flow rate is often situated just a little below the BEP, which is the reason why the hydraulic efficiency has its maximum there too, Eq. (4.14) and Fig. 4.17. Recirculations at partload and flow separation in the high-flow range usually cause the pressure pulsations to increase considerably. The following parameters have an influence on the generation of pressure pulsations:

Parameters affecting the non-uniformity of the wake flow

- Thickness and shape of the impeller blade trailing edge
- Hydrodynamic vane loading at the impeller outlet and the blade pitch or number of blades; measurements relating to these parameters are found in [67]
- Velocity distribution over the pitch and, consequently, the whole geometry of the impeller (especially the blade outlet angle)
- Approach flow to the impeller
- Flow separation in the impeller
- Interaction between the impeller and the collector during partload recirculation
- There is a tendency for the pressure pulsations to rise with the specific speed, i.e. with increasing impeller outlet width and impeller inlet diameter.
- Reynolds number. In theory the wake should be expected to get smaller with an increasing Reynolds number due to thinner boundary layers on the impeller blades. This effect, however, seems to be slight and hardly measurable because of the inevitable influence of the system on such measurements, [44].

For keeping the pressure pulsations low, it should be attempted in the first place to obtain as uniform a flow as possible at the impeller outlet by designing impellers free from flow separation with low blade loading and weak secondary flow.

Geometric parameters: The velocity and pressure fluctuations diminish strongly with increasing distance from the impeller outlet. Numerous measurements in the literature testify to this fact. Some results are shown in Fig. 10.6 where unsteady stagnation pressures, pressure pulsations and dynamic diffuser vane stresses reported in the literature are plotted against the distance between impeller outlet and stator. The definitions of the dimensionless quantities plotted in Fig. 10.6 are given in said figure.

The findings measured by different authors on different pumps are quite consistent. Notably, the gradients by which the unsteady effects decay are very similar.

The effects of the RSI decay downstream of the impeller according to a power function $(d_3/d_2 - 1)^x$. The exponents x resulting from the various tests are listed in the table below Fig. 10.6; the references are listed there too. These exponents can be used for estimating the effects of increasing gap B on the excitation forces, dynamic stresses and noise caused by the RSI.

For volute pumps, the gradients flatten when gap B is increased above 10%. This trend would be expected.

The most important design parameters to control dynamic effects due to the RSI are:

- The distance between the impeller blade trailing edges and the diffuser vane leading edges or the volute cutwater ("gap B") is the single most important design parameter for limiting the pressure pulsations generated at high circumferential speeds to an acceptable level and for avoiding fatigue fractures of diffusers, impellers, or other components. According to the data in Fig. 10.6, pressure pulsations and unsteady stresses in the diffuser vanes are proportional to

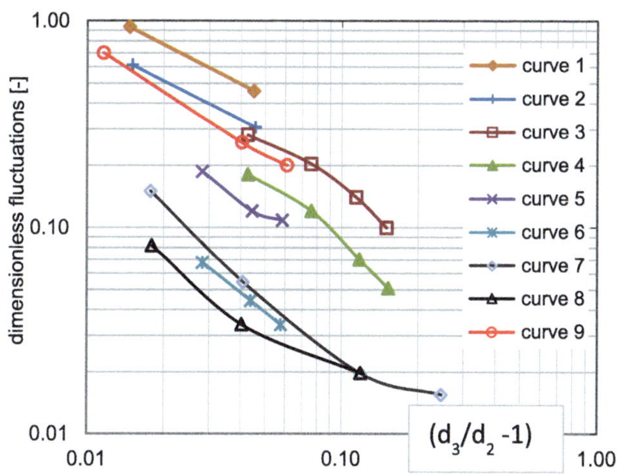

curves 1 to 4:

$$\Delta p^* = \frac{c_2^2}{u_2^2}$$

curves 5 to 8:

$$\Delta p^* = \frac{2\,\Delta p_d}{\rho\, u_2^2}$$

curve 9:

$$\Delta p^* = \frac{2\,\sigma}{\rho\, u_2^2}$$

general trend:

$$\Delta p^* \propto \frac{1}{\left(\frac{d_3}{d_2} - 1\right)^x}$$

	Measured quantity	Location		lxl	Lit.
1	Unsteady stagnation pressure	downstream of trailing edge	local	0.63	[4]
2			average	0.63	
3		vaneless diffuser	low flow	0.8	[88]
4			high flow	0.92	
5	Pressure pulsations	diffuser inlet	q* = 0.6	0.77	[67]
6			q* = 1.0	0.95	
7		volute	q* = 0.5	0.88	[79]
8			q* = 1.0	0.74	
9	Unsteady stress in diffuser vane	diffuser leading edge	1/10 of measured stresses are plotted	0.77	[33]

Fig. 10.6 Influence of gap between impeller and diffuser/volute on pressure pulsations and stresses

$(d_3/d_2 - 1)^{-0.77}$. Doubling gap B consequently reduces excitation forces by about 40%. Yet this is strictly true only if a *given* impeller is combined with different diffusers having different inlet diameters. Trimming the impeller for increasing gap B, increases the blade loading to the effect that the benefits of an enlarged gap B can be less than stipulated above. As a rule however, it can be justly assumed that impeller trimming will reduce pressure pulsations, especially in diffuser pumps. In rare cases an increase of gap B can even lead to higher pulsations, as demonstrated by measurements taken on a volute pump, [46]. This finding is explained by a more non-uniform flow distribution at the impeller outlet caused by higher vane loading and possibly by thicker blade trailing edges. Volute pumps are more prone to such abnormal behavior than diffuser pumps: at large d_3/d_2 the exponent flattens and the *relative* gain through trimming becomes small. In this case, the effect of increased blade loading may prevail.

- Combination of the number of impeller and diffuser vanes (Chap. 10.7.1).
- Staggering of the impellers on the shaft of multistage pumps.
- Staggering of the two halves of double-entry impellers.

10.2.4 Influence of the System

Pressure pulsations are affected to a large extent by the system in which the pump is operating and the location where the measurements are taken. Variations in cross section (e.g. valves) cause reflections; forward traveling waves can be in phase with reflected waves and thus give rise to standing waves. In this way acoustic resonances and interferences are created which cause the pressure pulsations to increase or decrease (or vanish) at certain frequencies. For this reason the results of the pulsation measurements depend strongly on the location of the measuring points.

Figure 10.7 shows an example from [11]: The measurements in figure a and b were made with rubber bellows installed several meters downstream of the pump. Due to their compliance, elastic bellows act like an open end of a pipe and largely reflect sound waves. At a distance of 355 mm downstream of the pump (figure a) the blade passing frequency ("5 $\times$ n") and its 1st harmonic ("10 $\times$ n") are just recognizable in the spectrum. In contrast, owing to standing waves, these components are several times higher when recorded at a distance of 755 mm (figure b). Higher pressure pulsations are thus sometimes recorded at a certain distance from the pump than next to the discharge or suction nozzle. Without bellows (figure c) the blade passing frequency is measured as a strong peak at a distance of 355 mm (in contrast to the test with bellows). In addition, strong broadband pressure pulsations are measured which originate from the throttle valve.[1] Other system parameters are:

- Fluid properties, including the content of free gases. Since the fluid properties, in particular the velocity of sound, depend on the temperature, the pressure pulsations may also vary with temperature, particularly if resonances occur, [46]. The influence of free gas is discussed in Chaps. 6.5 and 13.2.4.
- Approach flow conditions as given by the piping arrangement, valves and fittings, or intake vortices generated at a free fluid surface
- Cavitation phenomena in the plant (valves, other pumps), Chap. 6.5.

The impact of the system on the spectra at a given q*, as shown in Fig. 10.7, is kind of straightforward: the excitation is constant, only the acoustic system response varies. More difficult to interpret are the tests reproduced in Fig. 10.8 where the overall pressure pulsations are plotted against the flow for three different piping arrangements. The test speed was 1450 rpm, the impeller had 7 blades and the

[1]Examples for the influence of the system on pressure pulsations can be found in [B.20]. The topic is discussed in more detail also in Chap. 10.12.3.

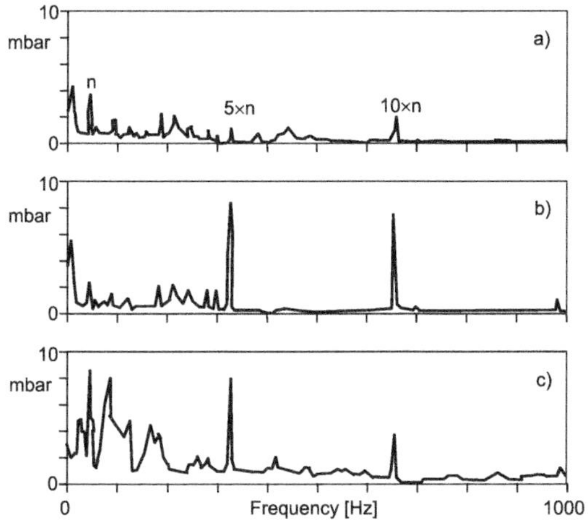

Figure	Bellows	A (mm)
a	yes	355
b	yes	755
c	no	355

Fig. 10.7 Pressure pulsations in the discharge nozzle, $z_{La} = 5$ [11]

diffuser 12 vanes ($n_q = 37$), [50]. The three test configurations have different lengths L between reflecting components (orifice, butterfly valve and hand valve). The corresponding fundamental acoustic frequencies f, given below Fig. 10.8, were estimated from f = a/(2 L) assuming a sound velocity of a = 1200 m/s. Thereby it should be noted that the locations of wave reflection at the pump are not clearly defined; they may also depend on the frequency. At a flow rate of about q* = 1.2, configuration 1 yields about twice the pulsations of the other pipe layouts suggesting a sharp resonance in a small range of flow rates, which is totally absent in the other configurations. Possible explanations:

- A given pump produces a specific flow pattern (hence a spectrum of excitation forces and pressure pulsations) at any given flow. Apparently, pulsations with 67 Hz are strongly excited at q* = 1.2, while there is no particular excitation at 55 and 167 Hz. In contrast, pulsations of 167 Hz (blade passing frequency) appear to be excited at q* = 1.13. If this hypothesis would be true, even a change in flow rate as small as 6% would give a drastic change in excitation forces and system response.
- There could be an interaction of the acoustic waves in the system with the flow in the pump. In other words: the system characteristic is unsteady due to the pressure variations caused by standing waves. As a consequence, fluctuations of the flow rate through the impeller channels are generated even at these high frequencies. The mechanism is similar to the process described in Fig. 10.25 when the head variation ΔH is thought to be caused by the standing waves. In view of the tests discussed in context with Fig. 10.23, the interaction of the flow through the impeller with acoustic waves is the more likely explanation.
- Pressure pulsations in the valves and the orifice have an influence on the measurement, but it is unlikely that their characteristics should change when the flow rate is modified by a few per cent.

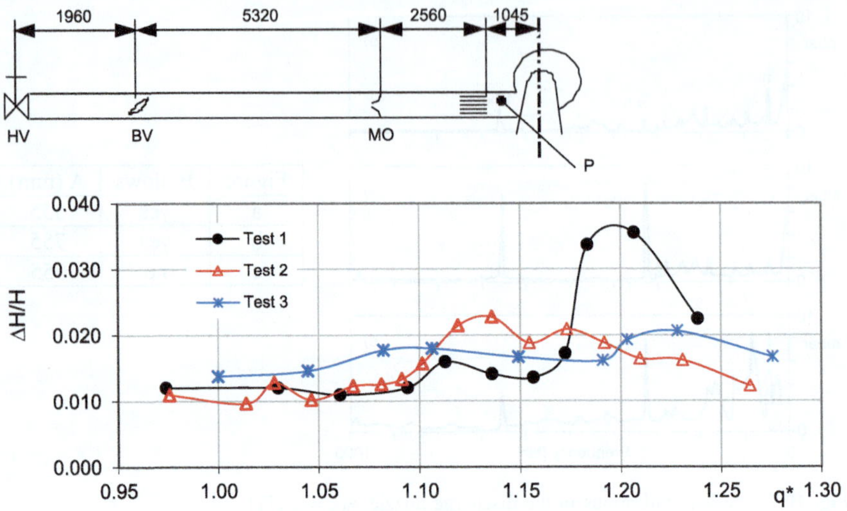

Test	Configuration	L [m]	f [Hz]
1	without MO, flow control by BV	8.9	67
2	with MO, flow control by BV	3.6	167
3	without MO, flow control by HV	10.9	55
P pressure transducer; MO measuring orifice; HV hand valve; BV butterfly valve			

Fig. 10.8 Influence of system on measured pressure pulsations, [50]

10.2.5 Scaling Laws

No exact methods for scaling pressure pulsations have been established which could be easily applied in practice because the acoustic efficiency is unknown. It represents that part of the mechanical energy that is converted into sound. However, experience shows that pressure pulsations in geometrically similar pumps often increase roughly with the square of the circumferential speed. Therefore, Eq. (10.1) seems to be a meaningful non-dimensional coefficient:

$$\Delta p^* = \frac{2\Delta p_d}{\rho\, u_2^2} \tag{10.1}$$

Of course, Eq. (10.1) *holds only in the absence of acoustical resonances.*

Noise generated by hydrodynamic phenomena can be described by various models according to the dominant physical mechanisms, [44, 70]: Net volume fluctuations generate noise as "monopoles". These pulsations increase with the square of the speed. Fluctuating lift forces radiate noise as "dipoles". Their amplitudes increase with the third power of the speed. Shear layers and vortices radiate noise as "quadrupoles", whose signals are proportional to the fourth power of the speed. All of these three mechanisms contribute in pumps to generate pressure pulsations the amplitudes of which consequently increase with an exponent between 2 and 4 of the circumferential speed. Presumably, the diffuser vanes (or the volute cutwater) in unsteady flow do act as monopoles due to their displacement effect. Moreover, the hydrodynamic phenomena in the near field follow the square law. At any rate, experience shows that pressure pulsations in centrifugal pumps are best described by the square law of Eq. (10.1). This is borne out, for example, by the measurements in [67] at rotor speeds between 1500 and 3300 rpm for the overall level as well as blade passing frequency. Pressure pulsation measurements in [90] covered a speed range from 1020 to 2520 rpm. The similarity law, Eq. (10.1), was well fulfilled - except where a resonance in the suction pipe occurred. The resonance occurred in a dead-end pipe; it vanished when a valve in the dead-end pipe was opened (the flow through the pipe disrupting the standing waves as discussed in Chap. 10.12.5).

Though deviations from the square law have often been noticed, they can be explained to a large extent as being caused by the influence of system parameters of the kind discussed in Chap. 10.2.4. For practical purposes, the best estimate will often result from using $\Delta p^* = f(f/f_n)$ according to Eq. (10.1), which gives a first approximation for the pressure pulsation spectrum of geometrically similar pumps ($f_n = n/60$ is the rotational frequency).

10.2.6 Measurement and Evaluation of Pressure Pulsations

Measurements of pressure pulsations are subject to considerable uncertainties due to the influence of the system and of the location of the pressure transducers (as shown in Chap. 10.2.4). Thus the values measured may be distorted typically by a factor of up to 10. To obtain accurate measurements, anechoic elements should be placed on the pipes upstream and downstream of the pump. However, in a liquid-filled system such is usually impracticable.

If it is possible to vary the pump speed (or the water temperature) over a large range, the accuracy of the measurements can be improved by taking the root mean square of the measured pressure pulsations which must be normalized according to Eq. (10.1) for the purpose. Yet system influences can only be mitigated by this procedure; it is not possible to eliminate them completely. System influences can also be reduced by placing pressure transducers at three different locations on the pipe and taking the root mean square of the measured values.

Pressure pulsations are measured usually by piezoelectric pressure transducers in the time domain. However, the "time signal" (pressure as a function of time, measured similarly to Fig. 10.40) does not allow drawing conclusions as to the likely excitation mechanisms. In order to interpret the measurements and to develop remedies for reducing detrimental pulsations, the time signal is treated by a frequency analysis. This yields a frequency spectrum according to Fig. 10.7. As mentioned before, that spectrum contains discrete peaks, such as the blade passing frequency and broadband components generated by vortices.

For analyzing non-harmonic pressure pulsations (purely sinusoidal vibrations are rarely observed in machines) the Fourier analysis is the accepted method. It allows any periodic non-harmonic signal to be described as a sum of sinusoidal partial components, Eq. (10.2):

$$p(t) = p_o + \sum_{\nu=1}^{\infty} a_\nu \cos \nu\omega_o t + \sum_{\nu=1}^{\infty} b_\nu \sin \nu\omega_o t \qquad (10.2)$$

The average pressure p_o (static pressure at the measuring location) and the amplitudes of the partial pressure components a_ν and b_ν result from Eq. (10.3) with the integers $\nu = 1, 2, 3\ldots$

$$p_o = \frac{1}{T}\int_0^T p(t)dt; \quad a_\nu = \frac{2}{T}\int_0^T p(t)\cos \nu\omega_o t\, dt; \quad b_\nu = \frac{2}{T}\int_0^T p(t)\sin \nu\omega_o t\, dt \qquad (10.3)$$

Pressure pulsations can be described in various ways:

- peak-to-peak: Δp_{p-p}
- amplitude or peak: $\Delta p_a = ½\Delta p_{p-p}$
- energetic average called "root mean square" (RMS) Δp_{RMS}:

$$\Delta p_{RMS} = \sqrt{\frac{1}{N}\sum_{i=1}^{N} p_i^2} \qquad (10.4)$$

$$\Delta p_{RMS} = \frac{\Delta p_a}{\sqrt{2}} = \frac{\Delta p_{p-p}}{2\sqrt{2}} \qquad (10.4a)$$

Equation (10.4a) is strictly valid only for sinusoidal signals, but it can be used for estimation purposes. It should be remembered, however, that peaks of a broadband signal can be several times higher than the RMS value. Whether such peaks are detrimental or not depends on their energy content. Peak values are therefore not more significant than RMS values, even though these disguise the peaks. In actual fact, the RMS values are representative for the energy content. This is a strong argument for using RMS values when evaluating the effect of pressure pulsations on the pump and the system.

RMS values can be represented either as spectra (e.g. 1 Hz bandwidth) or as broadband RMS values, e.g. for the range of 1 to 2000 Hz, if *n* is the number of narrow band values whose sum is taken, Eq. (10.5):

$$\Delta p_{RMS,BB} = \sqrt{\sum_n (\Delta p_{RMS})^2} \qquad (10.5)$$

amplitude definition

10.2.7 Pressure Pulsations of Pumps in Operation

Statistical data of pressure pulsations measured in the discharge nozzles are plotted as RMS values for three frequency ranges in Fig. 10.9, [B.20, 44]. They have been made dimensionless according to Eq. (10.1). The curves AA represent the arithmetic average, the curves SD the standard deviation and CL the 95% confidence limit.

The data result from measurements in 36 single and multistage pumps. They largely correspond with the findings in [26] and can be used for estimates and comparisons. With special pump designs, such as used for marine pumps, substantially lower levels are obtained.

The pressure pulsations shown in Fig. 10.9 apply to diffuser as well as volute pumps. The tested diffuser pumps were designed according to $d_3/d_2 = 1.04$. If the distance between impeller blades and diffuser vanes is smaller than that ratio, the pressure pulsations increase in accordance with $(d_3/d_2 - 1)^{-0.77}$ as per Chap. 10.2.3. Figure 10.9 applies to $z_{La} \geq 5$ only. With 3- or 4-bladed impellers distinctly higher pressure pulsations are to be expected because of the non-uniform flow at the impeller outlet resulting from the large blade pitch. The test data in [B.20] showed the following trends:

- In the range $0 < f < 0.2 \times f_n$, the 3-stage pumps generated roughly twice as much pressure pulsations as the single stage pumps.
- For $0 < f < 0.2 \times f_n$ the pressure pulsations in the suction nozzle were about half those in the discharge nozzle.
- In the range above $f > 0.2 \times f_n$ there are no general tendencies. The level of pressure pulsations in the suction nozzle was on average as high as in the discharge nozzle. It was not significantly different in single stage and 3-stage pumps.
- At $q^* = 0.25$ the pressure pulsations are roughly twice as high as near the best efficiency point.

Since the hydraulics tested have been used in plants in trouble-free pump operation, the levels of pressure pulsations measured may be considered as acceptable.

10.2.8 Damaging Effects of Pressure Pulsations

A certain level of pressure fluctuations is unavoidable and has no detrimental effects. Excessive pressure pulsations, however, can excite pump and pipe vibrations and might even cause damage. Some typical problems are (see also Table 10.1 based on [B.20]):

- Bearing housing vibrations (possibly resonances), sometimes causing fatigue fracture of instrumentation or auxiliary pipes.
- Baseplate vibrations. Multiplying the pressure fluctuations (read from Fig. 10.9) by the cross sections of the suction or the discharge pipes provides a rough estimate of the excitation forces at work.

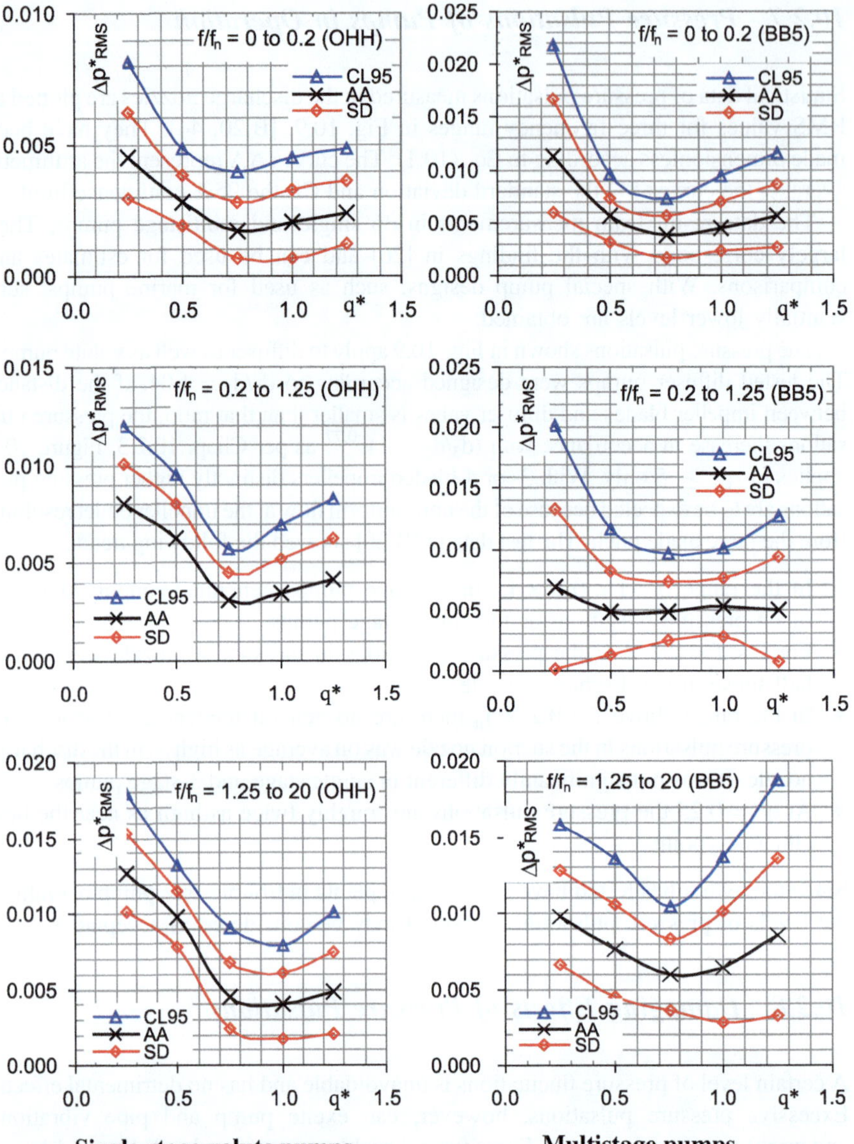

Fig. 10.9 Statistical data of pressure pulsations in different frequency ranges, measured in discharge nozzle [B.20]. AA = average, SD = standard deviation; CL95 = 95% confidence limit Definition: $\Delta p^*_{RMS} = \frac{2\Delta p_{RMS}}{\rho u_2^2}$

- Fatigue fracture of tie bolts in multistage segmental pumps or joining elements in the stage casings of barrel pumps.
- The pressure pulsations generated by the pump can give rise to standing waves (acoustic resonances) in the system. These may upset the control system, induce pipe vibrations, cause breakage of instrument lines, or get into resonances with other components - for example with the fuel rods of a nuclear reactor as reported in [46] or the tubes of a heat exchanger.
- Finally, pressure pulsations are one of the most important sources for noise emissions from the pump: air-borne noise is radiated from the casing, the piping and the baseplates, and solid-borne noise is radiated into the piping and the foundation.

10.2.9 Design Guidelines

Table 10.2 lists various design features recommended for reducing pressure pulsations, vibrations and noise at the source, [B.20, 52]. The most important steps to reduce the rotor/stator interaction are:

1. The choice of appropriate combinations of impeller blade and diffuser vane numbers, Chap. 10.7.1, Eq. (10.13).
2. The distance (gap B) between the impeller blades and the diffuser vanes or volute cutwater according to the formulae given in Table 10.2. Excitation forces decrease with $(d_3/d_2 - 1)^{0.77}$.
3. Low blade loading at the impeller outlet. The pressure pulsations decrease with the number of blades as demonstrated by tests with 7, 9 and 11 blades reported in [67].
4. The rotor/stator interaction can be reduced by an oblique or an oval shape of the volute cutwater, see sketches in Table 10.2.
5. Stagger impeller on the rotor of multistage pumps.
6. In multistage pumps with back-to-back impeller design (Fig. 2.9) a resonance condition can occur between standing waves in the long cross-over and the blade passing frequency, [74]; for analysis refer to Chap. 10.12.3.

If vane fractures or other damage, excessive vibrations, pressure pulsations or noise are encountered on a given pump, oblique trimming of the impeller blade trailing edge, increasing of the distance between the impeller and the collector and profiling of the impeller blades are relatively simple remedies. All of these, however, have an impact on the pump characteristics.

Table 10.1 Hydraulically induced damage to pump components

Damage	Possible causes	Possible remedies	Remarks
1. Blade breakage at impeller outlet	**1.1** High blade interaction forces cause excessive alternating stresses	**1.1** Increase distance between impeller and diffuser vanes **1.2** Oblique trim of impeller trailing edge	• Check for resonance of excitation with impeller eigen frequencies • Further remedies according to Table 10.2 • Insufficient casting quality often contributes • Notch effect at transition between blades and shrouds • Pulsations and alternating stresses decrease with $(d_3^*-1)^{-0.77}$
2. Vane fracture at diffuser inlet	**2.1** As 1.1 **2.2** Open diffuser, where closed diffuser would be necessary, Chap. 14.1	**2.1** As 1.1, 1.2 **2.2** Install closed diffuser	
3. Impeller shroud fracture	**3.1** as 1.1 **3.2** Distance between impeller and diffuser vanes too small **3.3.**Insufficient thickness of impeller shrouds **3.4** Resonance	**3.1** As 1.1, 1.2 **3.2** Increase d_3/d_2 **3.3** Increase shroud thickness **3.4** Change z_{La} or z_{Le}	
4. Damage to mechanical shaft seal	**4.1** High pressure pulsations **4.2** Pulsations due to cavitation **4.3** Radial shaft vibrations **4.4** Axial shaft vibrations	**4.1** see Table 10.2 **4.2** Reduction of cavity volume: new impeller, higher $NPSH_A$ **4.3** see Table 10.9	Mechanical seals are sensitive elements. Further causes: mechanical problems, wear, cavitation at sliding surfaces, corrosion, deformation, electrostatic charging inducing sparks
5. Shaft breakage	**5.1** Radial thrust too high **5.2** Alternating stresses due to forces acting on balance piston (rotor setting!)	**5.1** Possibly double volute	Further causes: notch effect, insufficient shaft sizing, corrosion fatigue
6. Labyrinth wear	**6.1** Radial thrust/shaft deflection too high **6.2** Thermal deformation of shaft and/or casing **6.3** Excessive vibrations	**6.1** Double volute, stronger shaft **6.3** see Table 10.9	Further mechanical causes: unsuitable materials, shaft deflection
7. Damage to radial bearings	**7.1** Radial thrust too high **7.2** High dynamic loads (e.g. hydraulic imbalance) **7.3** Rotor instability **7.4** Bearing load too low	**7.1** Possibly double volute **7.2/7.3** see Table 10.9	Many mechanical damage mechanisms: shaft deflection causing high local contact loads, lubrication, material
8. Damage to axial bearing	**8.1** Axial thrust too high, e.g. because of effect of partload recirculation **8.2** Labyrinth wear	**8.1** Measure axial thrust, modify balance piston or hydraulic components accordingly **8.2** Replace labyrinth rings	Mechanical problems or transient operation

Table 10.2 Design guidelines for low pressure pulsations [B.20]

Parameter	Recommendation		Remarks
Distance between impeller and diffuser vanes Gap B	minimum $d_3/d_2 = 1.015$ for $H_{St} < 100$m	$n_q < 40$	n_q [rpm, m³/s, m] $n_{q,Ref} = 40$ $H_{Ref} = 1000$ m d_3 applies to diffusers $\rho_{Ref} = 1000$ kg/m³
	$\frac{d_3}{d_2} \ge 1.015 + 0.08\left(\frac{\rho\ H_{st}}{\rho_{Ref}\ H_{Ref}} - 0.1\right)^{0.8}$	$n_q < 40$	
	$d_3/d_2 = 1.04 + 0.001\,(n_q - n_{q,Ref})$	$n_q > 40$	
Distance between impeller and volute cutwaters	$\frac{d_z}{d_2} \ge 1.03 + 0.1\frac{n_q}{n_{q,Ref}} + 0.07\frac{\rho\ H_{st}}{\rho_{Ref}\ H_{Ref}}$		d_z applies to volutes, see Table 0.2
Vane number combinations	$m = \nu_3\, z_{Le} - \nu_2\, z_{La}$ $m = 0$ must be avoided for $\nu < 4$		Chapter 10.7.1
Impeller blade number	Select z_{La} so that $z_{La} \times f_n$ does not coincide with acoustical natural frequency in cross-over channel.		Only relevant in special cases (e.g. long cross-over channel)
	For pumps with $H_{st} > 100$ m avoid $z_{La} < 5$		Blade loading (!)
Impeller blade outlet profile	Design for low blade loading near outlet; profile on pressure side of blades.		Trailing edge not too thin (stresses!)
Twisted blades at impeller outlet (rake)	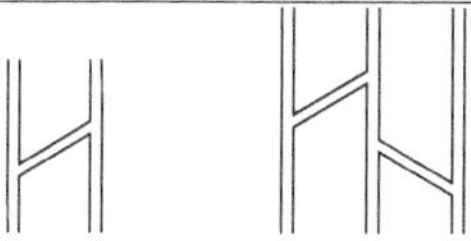		Staggering blades with double-entry impellers (may not be always effective)
Oblique blade cut at impeller outlet	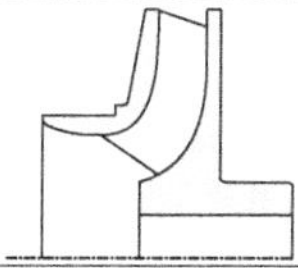		Effective retrofit correction since d_3/d_2 is increased (but there is a head reduction)
Shape of volute cutwater	Oblique 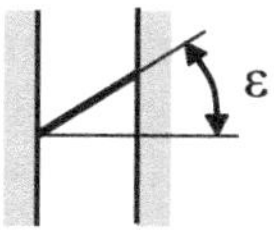	Oval 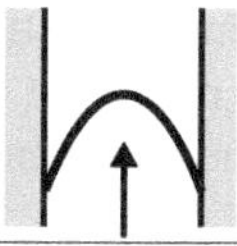	Angle $\varepsilon > 35°$
Partload pressure pulsations	Avoid excessive flow deceleration at BEP at impeller inlet and in volute or diffuser throat area. c_{3q}/c_2 not much below 0.8 (influence on BEP!)		Chap. 5
Inlet casing	➢ Profile ribs according to Table 10.13 in order to avoid vortex shedding. ➢ Optimize area distribution to avoid vortices.		Chap. 10.12.4 Chap. 7.13
Suction piping	➢ Design for uniform velocity profile at impeller inlet. ➢ Avoid successive bends in different planes. ➢ Install flow straightener if necessary.		Chap. 11.7
Pump sump	➢ Avoid air drawing vortices ("vortex ropes"). ➢ Avoid bottom vortices. ➢ Structures for suppressing effects of partload recirculation.		

10.3 Component Loading by Transient Flow Conditions

Strong pressure waves generated in the system during transient or upset conditions can damage pump components by creating momentarily extreme peak loads or unusual load directions. In this way not only tie bolts and screws but also the axial thrust balance disks or pistons may be overloaded. Under the impact of water hammer, for example, the pump shaft can be subject to excessive stress transmitted via the split ring which axially positions the balance piston, Fig. 10.10a.

Excessive pressure pulsations in the system are likely to increase the alternating stresses in the impeller or diffuser vanes and in the impeller shrouds. This phenomenon can be explained as follows: When a wave reaches the gap between impeller and diffuser, it enters the impeller sidewall gap and is reflected at the annular seal while it travels through the impeller channels. Thus the impeller shrouds are subject to fluctuations of the pressure difference between the impeller channels and the sidewall gaps. Alternating stresses increase. This interpretation suggests that fatigue fractures can be caused or accelerated by system-induced pressure fluctuations due to differences in the reflection times. If the discussed mechanism is active, it would appear that gap A should not be made too small in order to reduce the (unavoidable) wave reflection caused by the change in impedance at gap A. Standing waves in the discharge pipe and the exit chamber of a pump can thus increase alternating stresses in impellers and diffusers. Examples are standing waves generated in the annular discharge chamber of a barrel casing pump, Fig. 2.8, or in the long crossovers of back-to-back pumps in Figs. 2.9, 2.10 or 2.11.

A transient reversal of the force direction can lead to damage in a multistage pump, if the impellers are axially positioned on the shaft by a split ring or a shoulder. The axial thrust, normally acting towards the impeller inlet, pushes the impeller against the split ring. Sometimes the impellers are then held on the shaft by means of a slight shrink fit. If the pump experiences high pressure fluctuations generated in the system, the direction of the axial force is reversed when the waves

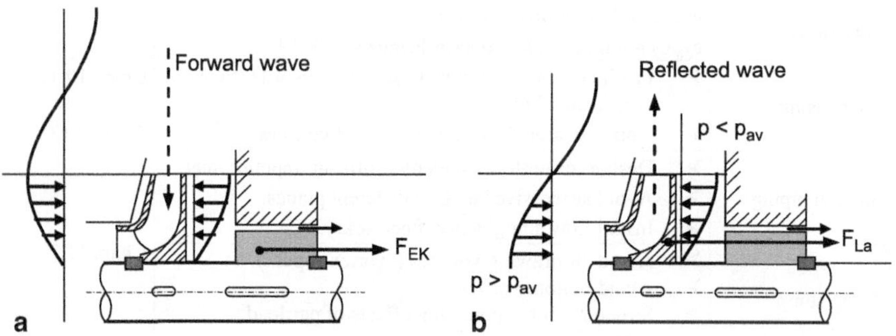

Fig. 10.10 Influence of pressure waves on pressure distribution and forces. **a** Excessive loading of piston split ring and shaft with forward wave; **b** axial thrust reversal with wave reflection

are traveling back through the pump after reflection. As a consequence, the impeller of the last stage may be axially displaced and rub against the casing. For illustration, Fig. 10.10b shows the effect of pressure waves on the pressure distribution on the impeller shrouds which determine the axial thrust: when the wave is running forward (from the discharge pipeline to the impeller), the axial force does not change much because the rear and front shrouds are subject to roughly the same load, Fig. 10.10a. However, the reflected wave shown in Fig. 10.10b causes a momentary pressure peak at the impeller inlet, whereas the impeller outlet and the impeller sidewall gaps are still in the trough of the wave.

The described axial displacement and subsequent rubbing of the last stage impeller has been actually observed in plants when strong transient pressure pulsations are generated during filling of a discharge pipe containing gas inclusions. The impeller of the first stage can be axially displaced by a similar mechanism during extreme pressure fluctuations in the suction pipe if a high pressure peak is generated at the impeller inlet by a forward wave, while the pressure downstream of the impeller is still low.

The differences between the reflection times of the waves entering the impeller sidewall gap and the waves traveling through the pump can generate the forces described above only in shock-like transients. Significant pressure differences at the impeller are created only if the pressure surge in the impeller sidewall gap occurs in a time Δt which is smaller than (or of similar magnitude as) the reflection time. With the sound velocity a, the reflection time is $\Delta t = (d_2 - d_n)/(2 \times a)$. The corresponding length of the $\lambda/4$-wave is: $\lambda/4 = \frac{1}{2}(d_2 - d_n)$ and the corresponding frequency results from $f = a/\lambda = \frac{1}{2}a/(d_2 - d_n)$. If water hammer phenomena are measured in a plant, the corresponding frequency is calculated from the recorded pressure increase dp/dt by means of $f = (dp/dt)/(p_{max} \times \pi)$; p_{max} is the amplitude of the pressure surge, [73].

Water hammer, inclusions of air or vapor in the suction or discharge pipelines, as well as damaged or cavitating valves, are likely to cause strong pressure fluctuations leading to the dynamic loading and possible damage. Such events are seldom recorded accurately in plant operation and are thus difficult to prove. The discussion of the above (somewhat hypothetical) mechanisms may help to analyze such failures.

10.4 Radiation of Noise

10.4.1 Solid-Borne Noise

Liquid-borne noise (pressure pulsations, cavitation noise), radial as well as axial hydraulic excitation forces and mechanical imbalance all generate vibrations in the suction and discharge pipelines and in the pump foundation that are radiated as solid-borne noise, which is extensively discussed in [25]. This process is governed by complicated transfer functions. Solid-borne noise generally consists of an infinite number of vibration modes in the form of longitudinal and transverse waves. The eigen frequencies of these vibrations can be excited by broadband (turbulence, cavitation) or by narrowband forces (e.g. rotational frequency, blade passing frequency). Acoustic resonances, too, would generate solid-borne noise. That is why the spectra of solid-borne noise (just like the spectra of pressure pulsations) are quite complex. Solid-borne noise is usually measured by means of accelerometers or, in special cases, by means of load cells mounted between the pump and the baseplate.

The mode density of both solid and liquid-borne noise rises rapidly with increasing frequency. In the high-frequency domain (above roughly 5 to 10 kHz) the solid-borne noise b_{ks} induced by excitation from pressure pulsations Δp can be estimated by means of the statistical energy analysis method. To do so, the component (e.g. a pump casing) is idealized as a cylinder with the radius R, wall thickness h, length L, density ρ_p and cover wall thickness h_D, [47].

$$b_{ks} = \frac{\Delta p}{\rho h \sqrt{\frac{\rho_p}{\sqrt{3}\pi\, \rho\, a}\left(1 + \frac{R}{L}\frac{h_D}{h}\right)\sqrt{\frac{E}{\rho_p}}}} \tag{10.6}$$

In this equation b_{ks} is the root mean square of the acceleration of the pump casing or of the piping and Δp is the root mean square of the pressure pulsations. The statistical energy analysis method is applicable to problems characterized by a high density of vibration modes in the solid component as well as in the liquid-filled space, and if the acoustic field in the liquid is of a diffuse and broadband nature.

Solid-borne noise can become a nuisance due to mechanical vibrations or through the radiation of secondary air-borne noise. Since the structural damping in metallic pipelines, steel frames and concrete structures is low, solid-borne noise is transmitted efficiently. For this reason, it is most effective to reduce solid-borne noise at the source (pressure pulsations, cavitation, mechanical and hydraulic imbalances). In addition, secondary measures, such as acoustic attenuation, are often required.

Noise control can be accomplished either by *damping* or *insulation*. Noise insulation is based on the fact that sound waves are reflected whenever they encounter a sudden change of the impedance $\rho \times a$ in the sound path. In the case of solid- and liquid-borne noise, only soft lightweight materials such as cork, foamed plastics and foamed rubber are able to provide that sort of impedance jump. In order to prevent the radiation of solid-borne noise into pipelines, elastic bellows can be used, if such is feasible with respect to system pressure. For the sound properties of plane waves see Chap. 10.12.3.

In order to avoid the excitation of vibrations in the pump foundation, frequently the baseplate is mounted on vibration insulators such as soft springs or rubber-metal components. A pump installed in that way represents a mass-spring-system that can be treated according to the methods established for vibration calculations. For vibration control to be efficient, the springs used have to be soft enough that the eigen frequency of the system is not greater than half the lowest excitation frequency.

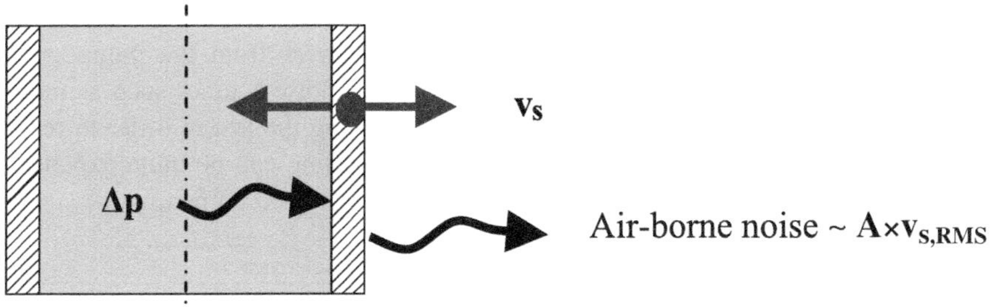

Fig. 10.11 Pressure pulsations Δp cause vibration velocity v_s of pipe which radiates air-borne noise

This because the unsteady forces transmitted to the pump foundation will get lower than the excitation forces only for frequencies above f = 1.4 × f_{eigen}, [57]; see Chap. 17.6, Table 17.6. Gaps and interstices in a structure (for example between the pump foundation and the building) will effectively reduce or prevent the radiation of solid-borne noise. However, noise radiated through "noise-bridges", as formed by components indispensable for force transmission, is very difficult to control.

10.4.2 Air-Borne Noise

Part of the energy contained in solid-borne noise is radiated from the pipelines, the pump casing and the baseplate as air-borne noise. The level of air-borne noise radiated is roughly proportional to the product of the effective (RMS) vibration velocity of the vibrating structure multiplied by its surface area, Fig. 10.11.

The noise radiated from machines is measured or specified as sound power or sound pressure. It is given as "noise level" in decibel (dB).[2] As a rule, the A-weighted sound pressure level, dB(A), is used. The A-weighting (DIN 45635) takes into consideration the annoyance caused by various frequencies as perceived by the human ear. For this reason, the greatest emphasis is laid on the frequency domain roughly between 800 and 5000 Hz, whereas frequencies outside this range contribute less to the A-weighted level.

For assessing the effect of various noise sources - e.g. pump, motor, gear box, throttle valve - the *sound power* levels of these components are added. Accordingly the total sound power level results from the logarithm of this sum.

Table 10.3 gives the different noise level definitions and provides formulae for adding noise levels. Table D10.1 shows some examples of the addition of two or more noise sources.

[2]The measurement is done according to DIN 45635, [N.19] and ISO standards.

When determining the noise according to the "prescribed surface method" the sound pressure is measured at a distance of *one meter* from the pump on an imaginary cuboidal measuring surface S. Extraneous noise sources such as motor, gear box, valves and piping have to be shielded during the test in order to record only the noise radiated from the pump. Such a shielding can be quite expensive. Moreover all values measured in a reverberating room need correcting.

Table 10.3 Noise emission of pumps		Reference
Sound power	$L_W = 10\ \lg \frac{W}{W_o}$	$W_o = 10^{-12}$ W
Sound pressure (p is the RMS)	$L_p = 20\ \lg \frac{p}{p_o}$	$p_o = 2\times10^{-5}$ N/m²
Measuring surface S = measuring surface in a distance of 1 m	$L_s = 10\ \lg \frac{S}{S_o}$	$S_o = 1$ m²
Calculate sound power from sound pressure	$L_W = L_p + L_s$	
Addition of n noise sources: Sound *power* level	$L_{W,tot} = 10\ \lg\left(\sum_i^n \frac{W_i}{W_o}\right) = 10\ \lg\left(\sum_i^n 10^{0.1\,L_{W,i}}\right)$	
Addition of n noise sources: sound *pressure* level	The individual sound power levels $L_{W,i}$ are added as above and the total sound power level is determined on the chosen measuring surface	
Addition of n *equal* noise sources	$L_{W,tot} = L_{W.1} + 10\ \lg n$	$L_{p,tot} = L_{p.1} + 10\ \lg n$
A-weighted levels in dB A		
Pump type	**Sound power**	**Sound pressure**
Side channel pumps [B.28]	$L_{WA} = 67 + 12.5\lg \frac{P_{opt}}{P_o}$	$L_{pA} = 44 + 11.5\lg \frac{P_{opt}}{P_o} + 3\lg \frac{n}{n_o}$
Applicable to **all** pump types according to [32] in the range of: $10 < P_{opt} < 10^4$ kW $200 < n < 6000$ rpm	Sound power	$L_{WA} = 72 + 10\ \lg \frac{P_{opt}}{P_o} - \lg \frac{P_{opt}}{10\,P_o}$ ± 4 dB
	Measuring surface	$L_s = 10\lg \frac{S}{S_o} = 12 + \lg \frac{P_{opt}}{P_o}$
	Sound pressure	$L_{pA} = 60 + 9\ \lg \frac{P_{opt}}{P_o} - \lg \frac{P_{opt}}{10\,P_o}$ ± 4 dB
All formulae are valid for operation at BEP with the power P_{opt} (without driver) Reference quantities: $P_o = 1$ kW $n_o = 1$ rpm $S_o = 1$ m²		

Table D10.1 Addition of noise sources

Addition of n equal noise sources							
Number of equal noise sources	2	3	4	5	6	7	8
Increase in level ΔL [dB]	3	4.8	6	7	7.8	8.5	9
Addition of two different noise sources							
Level difference [dB]	0	4	8	12	16	> 24	
Increase in level ΔL [dB]	3	1.5	0.6	0.3	0.1	0	
Total level: $L_{tot}=L_1+\Delta L$ if L_1 is the greater of the two noise levels							

The (preferable) "intensity measuring method" to a large degree avoids the draw-backs inherent to the prescribed surface method. It is based on measuring the sound intensity at various locations around the pump (ISO 9614-1) or by scanning the complete measuring surface (ISO 9614-2). The sound intensity (in W/m^2) is defined as the sound power per surface unit. Frequently pump operators specify the maximum values for air-borne noise which the pump must not exceed. Table 10.3 gives formulae for this purpose which are applicable for operation at the best efficiency point without extensive cavitation. The noise levels calculated according to Table 10.3 may be considered as the values to be statistically expected for industrial pumps. A poor pump design will lead to higher noise levels.

During partload operation and at $q^* > 1$ the noise levels increase. A quantitative assessment of this effect can be made by using Fig. 10.9. The equations given in Table 10.3 take into account neither the influence of the motor nor of the gear box, pipes and valves which increase the noise of a single stage volute pump by typically 5 dB.

Noise pollution in a pumping system can be reduced by a number of remedies:

1. Correct pump selection: Avoid partload operation resulting from over-sizing.
2. Make available sufficient $NPSH_A$ or, to put it another way, select for a given $NPSH_A$ a pump that will operate without extensive cavitation. To obtain this it is necessary that $NPSH_A > NPSH_0$.
3. Low pump speed.
4. Well balanced pump rotor. Careful alignment of the drive train.
5. Unsuitable valves used for throttling high pressure differences can be important sources of noise. Therefore quiet valves should be chosen. This also implies providing a sufficiently high back pressure in order to avoid cavitation in the valve which would generate considerable noise.
6. Low flow velocities help to reduce the noise radiated from the pipes.
7. Branches, orifices, elbows, sharp bends and sudden expansions cause flow separations resulting in noise and should therefore be designed for smooth flow or avoided as far as possible.
8. Arrange all fittings at a distance of at least $L > 10 \times D$ upstream of the pump in order to avoid increasing noise generation by an inadequate impeller approach flow. Install flow straighteners if the approach flow is distorted, for example by multiple bends closely upstream of the pump.
9. Choose quiet motors, couplings and gear boxes, if any. Select fans for cooling of electric motors for one sense of rotation only.
10. Control solid-borne noise by the choice of appropriate pipe supports.
11. Solid-borne noise radiation in pipelines can be suppressed to a large extent by the use of elastic bellows provided there are no noise bridges.
12. Install small units on elastic supports if that does not cause vibration problems.
13. Thick-walled casings and pipes reduce noise. Pumps with large, thin-walled casings are prone to noise radiation.
14. The design of the building has an important impact on noise pollution. In a room with hard, smooth (tiled) walls, floor and ceiling, the sound is efficiently

reflected and the noise level increases due to the absence of absorbing surfaces. Noise absorbing walls, ceilings and structural elements reduce the noise.

15. Noise and vibration insulation help to reduce the amount of noise radiated to the surroundings. Thermal insulating material consisting of mineral wool covered by sheet metal (thickness 1 to 2 mm) reduces the noise level by 1 to 4 dB. The thicker the sheet metal, the better is the noise insulation. Mineral wool without a metal cover is ineffective, aluminum is less effective than steel because of its lower density.
16. By means of a hood, noise can be reduced by 10 to 30 dB depending on hood design. The insulation achieved by a hood is more effective in the high-frequency domain than at low frequencies.
17. In order to fight air-borne noise at its source, the measures to reduce pressure pulsations as per Table 10.2 can be applied.
18. Inlet casings which generate a non-uniform impeller approach flow should be avoided.
19. Air-borne noise can be reduced by increasing the wall thickness of sound radiating surfaces (pump casing, pipes and baseplates). Doubling the wall thickness reduces the noise by about 4.5 dB for a given strength of the noise source.
20. For non-grouted installations the baseplate is often a major source of noise. Damping treatment (e.g. by damping cassettes) can be applied to the baseplate panels to reduce their noise radiation.

When attempting to reduce the noise in a pumping station it should be borne in mind that the pump drive train (e.g. electric motor with fan, gear box), piping, auxiliaries, valves or other extraneous noise sources have a considerable (if not dominating) impact on the overall noise level. More information on noise can be found in [47, 54, 57].

10.5 Overview of Mechanical Vibrations of Centrifugal Pumps

The various types of vibrations count among the most frequent sources of operational problems with centrifugal pumps. In order to solve vibration problems, it is often vital to attempt to understand the interaction of hydraulic and mechanical phenomena. Basically three types of vibrations should be distinguished:

Free vibrations occur if a system consisting of mass, spring and damping is excited by a single event (e.g. a shock or a blow) and afterwards left to itself. The system then vibrates at its natural (or "eigen") frequencies. The resulting amplitudes decay faster with increased damping. Free vibrations, for example, are created by an earthquake or by an impact test to determine eigen frequencies of bearing housings, base plates or other structures.

Forced vibrations are generated if a system consisting of mass, spring and damping is excited by a periodic force. In this way, an imbalance excites rotor vibrations which are transmitted by the bearings and the casing to the foundation. The reaction of the system depends on the ratio of the excitation frequency ω to the eigen frequency ω_E and the damping. When resonance occurs at a ratio of $\omega/\omega_E = 1$ the amplitudes can become very high if damping is low (resonance amplification).

The time relation between excitation and system response is described by the phase angle. With low damping and $\omega << \omega_E$ the system follows the excitation virtually without a time lag. When the system goes through a resonance the phase angle jumps from below 90° to over 90°. If the system is highly damped, the amplitudes scarcely increase in a resonance; in such cases the phase change provides a better indication of a resonance condition than the amplitude.

An eigen frequency is not only excited to resonance by a close-by discrete excitation frequency. It can also be excited selectively by broadband mechanisms such as turbulence. For example: if broadband pressure pulsations in the range of 5 to 40 Hz are observed at partload, a baseplate eigen frequency of 25 Hz could show up in the spectrum as a resonance peak, even though the pressure pulsation spectrum exhibits no peak at 25 Hz. Further examples of resonance problems with pumps are the excitation of bearing housing vibrations by impeller/diffuser vane interactions or shaft vibrations caused by rotating stall.

Self-excited vibrations occur if there is a feedback from the vibrating structure to the excitation mechanisms. A vibrating pump rotor creates reaction forces in the annular seals. If the flow in the seal generates forces which excite the rotor orbit in the direction of the rotor speed and damping is sufficiently low, a self-excited vibration is created. The rotor becomes instable and its amplitudes are limited only by non-linear effects such as rubbing in the annular seals (Chap. 10.6.6).

While free vibrations are rarely significant in pump operation, forced and self-excited vibrations frequently cause problems. Typical categories are:

1. **Lateral shaft vibrations** are measurable as forced vibrations in every pump because of an unavoidable residual imbalance of mechanical and hydraulic origin. Shaft vibrations are measured by proximity probes installed close to the bearings. This type of vibration is discussed in the section "Rotor dynamics".
2. **Torsional vibrations** of a pump rotor can be significant if the driver provides an unsteady torque. Such is the case with electric motors which are fed from a frequency converter or pumps driven by a reciprocating engine, Chap. 10.13.
3. **Bearing housing vibrations** are generated as reactions of the bearing housing to lateral shaft vibrations and to unsteady deformations in the pump casing. They are measured by accelerometers installed in vertical, horizontal and axial direction. Resonances of the bearing housing with the blade passing frequency are not uncommon, in particular with semi-open bearing housing designs. Closed bearing housing designs are stiffer, but excessive vibrations still cause sometimes problems. Bearing housing vibrations are discussed in Chap. 10.11.

4. **Vibrations of the pump/baseplate system** are excited by rotor vibrations and pressure pulsations and transferred to the pump foundation. Baseplates are considered as continuous structures with many degrees of freedom. Baseplates of high-pressure pumps must be analyzed for the risk of resonances with the excitation frequencies. Excitation forces can also originate from the driver or the piping.
5. **Axial rotor or bearing housing vibrations** are frequently observed on single stage, double-entry pumps during partload operation. They are caused by unsteady impeller inlet and discharge flows. Axial rotor shuttling is particularly pronounced when flow recirculation at the impeller outlet influences in an unsteady way the flow in the impeller sidewall gaps. Such effects usually occur at low frequencies, so that axial rotor movements may be visually observed.
6. **Vibrations of vertical pumps** (see Chap. 10.14): Vibrations of the rotor and the column pipe, as well as the coupling between both phenomena, are to be analyzed. Furthermore, motor and motor pedestal vibrations may be an issue. Depending on the length of the column pipe, its eigen frequencies may be near the rotational frequency or in the range of a few Hertz only. Non-uniform approach flow, in particular vortices, frequently lead to vibration problems, Chaps. 10.14 and 11.7.3. Since the journal bearings of vertical pumps have no defined loading, stiffness and damping provided by the lubrication films are low. Therefore, such bearings are prone to instabilities which may be recognized by chaotic vibration orbits.

Excessive vibrations can create a number of problems such as rotor rubbing and premature wear of annular seals, damage to bearings, couplings and shaft seals as well as baseplate vibrations and noise (see also Table 10.1).

10.6 Rotor Dynamics

10.6.1 Overview

The pump rotor as movable element and origin of energy transfer is the source of all vibrations which are excited by the pump. The reliable prediction of shaft vibrations and the limitation of hydraulic excitation forces are therefore important for the safe operation of high-energy pumps.[3] The task of the rotor dynamic analysis is to determine eigen frequencies and eigen modes, to calculate forced response and to assess excitation forces. Although vibration problems can be encountered with any

[3]As per [N.25] high-energy pumps are defined as pumps whose heads per stage exceeds. $\frac{H_{st,opt}}{H_{Ref}} > 275\left(\frac{n_{q,Ref}}{n_q}\right)^{1.85}\left(\frac{\rho_{Ref}}{\rho}\right)$ with $n_{q,Ref} = 25, H_{Ref} = 1\,\text{m}$ and $\rho_{Ref} = 1000\,\text{kg/m}^3$. The formula is valid for $25 < n_q < 67$. Below $n_q = 25$ the limit is $H_{st,opt} = 275$ m; no limits are defined for the range above $n_q = 67$. The topic is discussed in more detail in Chap. 15.4 where three quality classes are defined.

type of pump and though the physical mechanisms discussed in the following are generic in nature, rotor dynamic investigations are most significant for multistage pumps with high impeller tip speeds, [1, 21, 30].

A pump rotor is essentially subject to the following forces:

1. *Steady forces*, such as the rotor weight as well as hydraulic radial and axial forces, determine the reaction forces in the bearings. The bearing forces determine stiffness and damping coefficients and have a great impact on the dynamic behavior of the rotor.
2. *Excitation forces* are generated by the mechanical imbalance and by periodic and stochastic hydraulic forces in radial (and axial) direction whose synchronous components are frequently the most significant. Other excitation forces are due to rotating stall, unbalanced blade forces and stochastic forces caused by flow separation, recirculation and turbulence. *The excitation forces are always present, independently of the rotor vibration.*
3. Hydraulic *reaction forces* are generated *by the orbital movement* of the rotor. Due to the radial displacement, the pressure distribution around the impeller changes. This is true for both closed impellers and open impellers ("Alford effect"). Considerable reaction forces can thus be created in high-pressure pumps, even though the vibration amplitudes usually are below 0.1 mm. Rotor orbiting also creates non-uniform pressure distributions in annular seals leading to high reaction forces. *Reaction forces of this kind are acting only if the rotor vibrates.*
4. Likewise reaction forces are created in journal bearings, since the pressure distribution in the lubricating film depends on the rotor movement.
5. The shaft coupling can excite rotor vibrations through alignment errors (e.g. created by cold/hot operation), through coupling deficiencies and through transmission of vibration from the driver or a gear, [12, B.19].

10.6.2 Forces in Annular Seals

Forces created in the annular seals at the impeller or at an axial thrust balance device have an altogether dominating impact on pump rotor vibrations. To a large extent they determine the eigen frequencies, eigen modes, amplitudes of forced response and the stability limit of multistage pumps. The "dry" critical speed (in air) of a multistage pump calculated without the effect of annular seals is typically only 30 to 50% of the effective ("wet") critical speed during pump operation. Rotor dynamic calculations of a multistage pump in air cannot be used even as a first approximation (except, of course, for dry running, which is sometimes required).

The fluid forces created in an annular seal are usually calculated according to a linearized model, e.g. [13, 21]. The radial and tangential force components are described by spring (k), damping (c) and mass coefficients (m) which are proportional to the displacement, the velocity and the acceleration of the rotor movement:

$$-\begin{pmatrix} F_x \\ F_y \end{pmatrix} = \begin{pmatrix} k & k_c \\ -k_c & k \end{pmatrix}\begin{pmatrix} x \\ y \end{pmatrix} + \begin{pmatrix} c & c_c \\ -c_c & c \end{pmatrix}\begin{pmatrix} \dot{x} \\ \dot{y} \end{pmatrix} + \begin{pmatrix} m & m_c \\ -m_c & m \end{pmatrix}\begin{pmatrix} \ddot{x} \\ \ddot{y} \end{pmatrix} \quad (10.7)$$

The coefficients without subscript describe those reaction forces which act in the direction of the displacement, the vibration velocity or acceleration, while the terms with subscript c ("cross-coupled terms") mark those reaction forces which act perpendicular to said vectors. The resulting forces can be interpreted as integrals over the pressure distribution and wall shear stresses in the seal. Equation (10.7) was derived for oscillations around a central rotor position, but Eq. (10.7) can be used approximately for static eccentricities up to 50% of the seal clearance. In Eq. (10.7) rotational symmetry was assumed (the diagonal terms in the matrices are equal). This assumption is essentially fulfilled for short seals with $L_{sp}/d_{sp} < 0.5$. For long seals with $L_{sp}/d_{sp} > 0.75$, such as a balance piston, a fourth matrix must be added to Eq. (10.7) in order to account for the tilting of the rotor axis in the seal. The fourth matrix is not needed if the long seal is interrupted by deep annular grooves which allow a pressure equalization around the circumference (such grooves can be recognized in Fig. 2.8).

The coefficients in Eq. (10.7) depend on the geometry of the seal and on the flow conditions; they are determined by experiments or calculated by means of various computer programs, [3, 13, 34, 38, 56, 66, 75, 77, 146]. Figure 10.12 provides the range of coefficients expected for short annular seals, as encountered at the impeller inlet or as interstage seals. These data are applicable to smooth seals or seals with shallow serrations (groove depth up to 0.5 mm). Figure 10.12 provides all equations defining the seal coefficients and the range of parameters which have been covered by the tests. Figure 10.12 demonstrates that the cross-coupled stiffness k_c* grows with increasing pre-rotation U_t* at the seal entrance and with an increasing ratio of the circumferential speed $\omega \times R$ to the axial flow velocity c_{ax}. In contrast, the direct damping c* diminishes, if the ratio of the axial flow velocity to the circumferential speed increases. All other coefficients scarcely depend on flow conditions; the ranges measured for k*, c_c*, m* and m_c* are listed in Fig. 10.12. The domains which apply to typical impeller inlet seals and interstage seals are highlighted in Fig. 10.12. Essential for the assessment of rotor behavior are the resulting radial and tangential forces which depend on the amplitude A and the vibration frequency Ω. With the coefficients from Fig. 10.12, the radial and the tangential forces F_r and F_t are given by Eq. (10.8):

$$\begin{aligned} F_r &= LR\,\Delta p\frac{A}{s}\left\{-k^* - c_c^*\frac{\Omega}{\omega} + m^*\left(\frac{\Omega}{\omega}\right)^2\right\} \\ F_t &= LR\,\Delta p\frac{A}{s}\left\{+k_c^* - c^*\frac{\Omega}{\omega} - m_c^*\left(\frac{\Omega}{\omega}\right)^2\right\} \end{aligned} \quad (10.8)$$

A negative *radial* force opposing the momentary rotor displacement has a centering effect and increases the rotor eigen frequencies. A negative *tangential* force

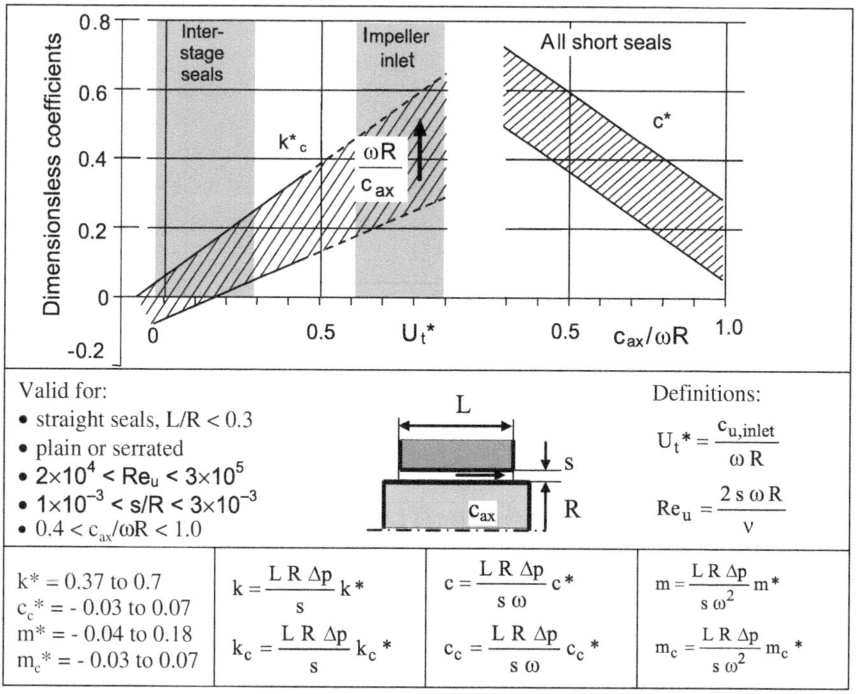

Fig. 10.12 Dimensionless force coefficients for short annular seals, [13]

opposing the momentary circumferential orbit movement has a damping effect. If the orbit moves against the direction of shaft rotation and if $F_t > 0$, the rotor runs stably. Figure 10.13 represents these conditions. Neglecting the small influence of the cross-coupled mass, the stability limit is derived from Eq. (10.8) for the condition $F_t = 0$ yielding $k_c^* = c^*\, \Omega/\omega$. From this the limiting frequency of the orbit movement follows as:

$$\Omega_{\text{limit}} = \frac{\omega k_c^{*}}{c^*} \tag{10.9}$$

Above Ω_{limit} a positive damping is achieved and the rotor operation is stable. Below this limit self-excited vibrations render operation impossible. The frequency limit rises in proportion to the operation speed ω and the cross-coupled stiffness k_c^*; it decreases with increased direct damping c^*. If the rotor vibrates at its lowest (speed-dependent) eigen frequency f_{e1}, the stability limit is reached at the following operation speed n_{e1} (in rpm):

$$n_{e1} = 60\, f_{e1} \frac{c^*}{k_c^{*}} \tag{10.10}$$

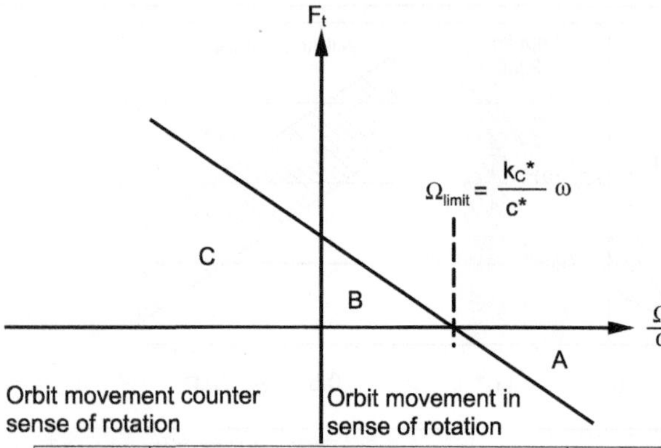

Domain	Orbit movement	k_c^*	F_t	Rotor
A	In the sense of rotation	$\frac{k_c^*}{c^*} < \frac{\Omega}{\omega}$	$F_t < 0$	Stable
B	$\Omega > 0$	$\frac{k_c^*}{c^*} > \frac{\Omega}{\omega}$	$F_t > 0$	Unstable
C	Against the sense of rotation $\Omega < 0$			Stable

Fig. 10.13 Rotor instabilities

When the rotor is subject to imbalance-induced forced vibrations, the orbit frequency equals the rotational frequency. From Eq. (10.9) follows:

$$1 \le c^*/k_c^* \tag{10.11}$$

This means that the rotor runs stably at all speeds if the ratio c^*/k_c^* is sufficiently greater than 1.0. This condition must be fulfilled to enable pump operation in the first place.

The effect of annular seals is governed by the following principles:

- All coefficients are proportional to the pressure difference across the seal as well as to its diameter.
- All coefficients increase in inverse proportion to the radial seal clearance.
- Since the pressure differences across the seal are proportional to the fluid density, seal forces decrease with falling density. As compared to operation with cold water, seal stiffness and damping decrease when light hydrocarbons are pumped. Consequently, eigen frequencies and vibration measurements on the test bed with cold water cannot be transferred directly to plant operation with light hydrocarbons or other low-density fluids.
- The radial force is determined by the direct stiffness, because it dominates over the mass term and the cross-coupled damping. The direct stiffness of a short annular seal (say $L_{sp}/r_{sp} < 0.3$) is caused by the following mechanism: The flow velocity on the side with the smaller gap (due to the higher flow resistance given by $\lambda \times L/s$) is lower than on the side with the larger clearance, Fig. 10.14a. Due to the greater dynamic pressure and inlet loss $\frac{1}{2}\rho c_{ax}^2\,(1 + \zeta_E)$ on the side with

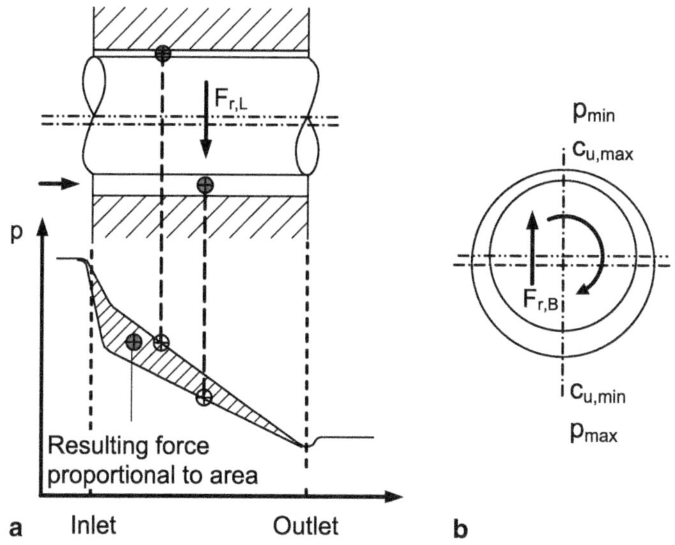

Fig. 10.14 Radial forces in annular seal. **a** Effect of axial flow (Lomakin effect). **b** Effect of circumferential flow (Bernoulli effect)

the higher velocity, a pressure difference between the narrow and the wide gap is created. This pressure difference generates a force which is opposed to the rotor displacement and thus has a centering effect on the rotor. The mechanism is frequently referred to as the "Lomakin effect". The stiffness k of a short plain annular seal with diameter d_{sp}, radial clearance s, friction coefficient λ and length L_{sp} can be estimated from [36]:

$$k = \frac{1.4\, a\, L_{sp}\, d_{sp}\, \Delta p}{(1+a)^2\, s} \quad \text{with} \quad a = \frac{\lambda\, L_{sp}}{2\, s} \tag{10.12}$$

$$k_{Ber} = -\frac{\pi \rho a_{SL} L R\, c_u^2}{s} \quad \text{with} \quad c_u = \gamma_{av} \omega R;\ a_{SL} \text{ from Table10.4(1)} \tag{10.12a}$$

- A positive stiffness augments the eigen frequencies of the rotor. Consequently, the eigen frequencies rise with growing pressure differences Δp across the seal and with the rotor speed since $\Delta p \sim n^2$.
- The centering "Lomakin effect" is opposed by a "Bernoulli effect", Fig. 10.14b. For reasons of continuity the circumferential flow, induced by shear stresses on the rotating shaft, has its velocity maximum on the side with the narrow gap. According to the Bernoulli equation the highest velocity corresponds to the lowest pressure. As a consequence, a de-centering force component is acting from the wider to the narrower gap, [15]. Since it takes some distance to

generate an appreciable circumferential flow by wall shear stresses, the Bernoulli effect is weak in short seals with low pre-swirl, but it becomes dominating in long annular seals. The counter-acting Lomakin and Bernoulli effects lead to a maximum of the direct stiffness which can be expected around L/D = 0.5. The foregoing statement is true only if the fluid entering the seal does not have a strong circumferential velocity. Very long seals which are not interrupted by deep annular grooves are de-centering, [21].

- The de-centering Bernoulli stiffness term k_{BER} is proportional to the square of the average circumferential velocity c_u in the seal; it can be calculated from Eq. (10.12a). This same expression represents the negative term in Eq. (T10.4.10).
- The smaller the gap width, the higher are the centering forces. If the seals wear during service (for example due to abrasion), the eigen values and critical speeds of a rotor drop.
- While deep grooves are apt to reduce the leakage, they impair the stiffness of the seal and are therefore detrimental to rotor dynamics. At given conditions smooth annular seals yield the highest stiffness but they are sensitive to wear and increase the risk of rotor seizing. Seals with shallow grooves provide a good compromise between these extremes. Grooves on the stator, due to the increased friction surface, reduce the circumferential velocity component in the seal and, consequently, the cross-coupled stiffness.
- The tangential force is that component which acts in (or against) the direction of the vibration orbit. It determines the damping of the rotor and its stability. The *effective* damping results from two opposing effects: (1) The *direct damping* which is proportional to the orbit frequency stabilizes the rotor because it opposes the movement of the orbit. Direct damping decreases with a growing ratio $c_{ax}/(\omega R)$, Fig. 10.12. (2) If the cross-coupled stiffness acts in the direction of the orbit movement, it feeds energy to the rotor and is destabilizing. A strong pre-rotation at the seal inlet and a large ratio of the circumferential to the axial velocity increase the cross-coupled stiffness, reduce the effective damping and are thus destabilizing. If the tangential force becomes zero or positive, self-excited vibrations occur unless the rotor is sufficiently damped by the bearings or other features.
- The influence of the mass term is considerable with long seals (high L_{sp}/d_{sp}), such as the balance piston.

Smooth gaps have a strong impact, seals with deep grooves exert only a minor influence on the eigen frequencies. Deep grooves also impair rotor damping as compared to smooth seals. The choice of the groove depth is therefore a compromise between leakage (efficiency) and rotor dynamic criteria. While the coefficients from Fig. 10.12 are applicable only to short seals, the physical mechanisms are relevant also for long seals such as balance pistons or throttling bushes used as shaft seals.

Seal calculations: Equations for calculating forces in plain annular seals were published in [3]. The procedure can be compressed into the set of equations reproduced in Table 10.4(1). It can be tentatively applied also to serrated seals by using an equivalent seal clearance s_{eq} defined by Eq. (T10.4.2). This concept is based on the idea that the stiffness decreases with the fluid volume contained in the seal because the liquid in the seal provides less stiffness (due to its compressibility) than the solid metal.

Based on some comparisons between calculations according to Table 10.4(1) and test data, the following tentative conclusions may be drawn:

- The procedure can be applied to short plain seals with $L/D < 0.5$ and shallow serrations with $g/s < 1$ or about 0.5 mm groove depth. These parameters cover most pump applications.
- The merits of the procedures include: (1) the effect of different parameters and assumptions can be much more easily assessed than with more complex computer programs and CFD; (2) the seal coefficients calculated from other programs can be checked for plausibility and gross errors (e.g. due to erroneous input). Note that such programs have a considerable range of uncertainty as well.
- When calculating seals with deep serrations, the circumferential velocity in the seal cannot well be captured by the method.
- Serrations on the rotor, which should be avoided, cannot well be calculated.

For all seal calculations it should be noted:

- All seal coefficients are proportional to the pressure difference Δp_{sp} across the seal. Δp_{sp} depends on the ISR integration and on the leakage flow which cannot be predicted accurately. Wear increases seal clearances during service; even in new conditions manufacturing tolerances have an impact.
- The cross-coupled stiffness critically hinges on on the rotation at the seal inlet $U_t^* = c_{u,in}/(R\omega)$; this depends on: (1) the local $c_{2u,local}(k_E)$ at the ISR inlet which is largely unknown during partload recirculation (see Figs. 10.15, 10.35/10.36, 5.35, 9.6/9.7); (2) ISR integration according to Table 9.1; (3) swirl brake effectiveness e_{SB} which is largely unknown: tentatively one may assume $U_{t,SB}^* = (1 - e_{SB})U_t^*$; with $e_{SB} = 0.7$.
- There are insufficient test data on long seals with $L/D > 0.5$. If long seals are interrupted by deep grooves, which allow some pressure equalization around the circumference, the method may be adapted.
- The relative roughness: $\varepsilon/(2s_0)$ is in the order of 0.008. In high-head pumps, the seals are "hydraulically rough".
- The uncertainty of the coefficients is expected to be considerable; but this applies to all methods as well as to test data.
- Outlet loss coefficient ζ_A has a big influence. $\zeta_A = 0.85$ may be assumed as typical, but $\zeta_A = 0.85$ gives higher stiffness than $\zeta_A = 1.0$ and is not conservative.
- ζ_E depends on inlet swirl U_t^* and on sharpness of edges at seal inlet: tiny radius or chamfer drastically reduce ζ_E.

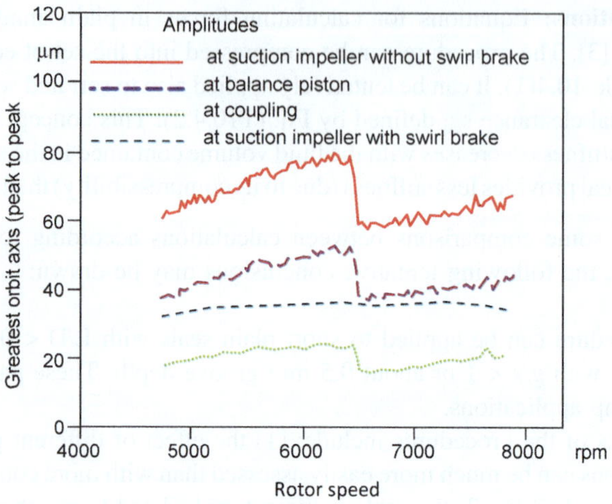

Fig. 10.15 Influence of part-load flow on rotor vibrations at q* = 0.25, [43]

- Geometrical tolerances, notably of the gap width. Assuming manufacturing tolerances H7/h7 typically giving $\Delta D_{max} = 2s_0 + 0.092$ mm for (H7/h7) and $\Delta D_{min} = 2s_0$.
- Gap width variation over length (manufacturing tolerances).

For hole pattern seals, the stiffness can be estimated from $k_{HP} = 0.75 \times k_{Plain}$ [21] or according to Table 10.4(1) with the concept of the equivalent seal clearance. The hole patterns tested in [21] had approximately the following dimensions referred to the radial seal clearance s: hole diameter d: d/s = 10; hole depth h: h/s = 3 to 4; hole area/total area: $\gamma = 0.4$ to 0.5; d/h = 3.3 to 2.5.

Numerical calculations of the rotor dynamic coefficients of a balance piston (L/D = 1.2) with serrations have been reported in [101]. The impeller side room was included in the model in order to get the correct inlet swirl to the annular seal (CFD: $c_u/u_{sp} = 0.96$ while Table 9.1 yielded 0.9). CFD set-up: 8×10^6 grid elements; SST k-ω; $y^+ = 1.1$; five whirling frequencies Ω were evaluated, seal coefficients were then determined by curve fit. CFD results were compared to finite difference and bulk-flow codes. While the three methods gave similar values for the direct stiffness, large discrepancies were found for all other force coefficients. The deviations were extreme for the direct mass.

Balancing pistons: Usually, the length of the piston is in the range of L/D = 0.7 to 1.0. With long seals, tilting of the rotor versus the bushing must be considered because it impairs the stiffness of the seal. Therefore, deep annular grooves are machined in the balance piston bushing at two or three locations along its length in order to foster pressure equalization around the circumference of the piston. The force coefficients should then be calculated for the 2 or 3 sections separately. For any section following one of the deep grooves assume $c_{u,in}/(\omega R) = 0.5$; $\zeta_E = 0.15$;

Table 10.4 (1) Force coefficients for annular seals			
Total loss coefficient of seal	$\zeta_{tot} = \zeta_E + \zeta_A + \frac{\lambda L}{2s}$	[21]: ζ_E = 0 to 0.3 ζ_A = 0.7 to 1.0	10.4.1
Serrated seals calculated with equivalent seal clearance s_{eq}	$\frac{s_{eq}}{s_0} = 1 + \frac{b\,g}{t\,s}$; $\frac{A_{st}}{A_r} = 1 + \frac{2g}{t}$		10.4.2
	shallow serrations from [13]: g/s = 1.0; b/s = 2.7; t/b = 2.0	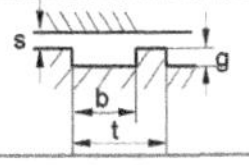	
In order to avoid confusion γ is used instead of k for c_u/u_{sp} Determine γ_{in} from Table 9.1.	$a_\gamma = \frac{\lambda L}{4s}\left(1 + \frac{0.75}{1 + 4\left[\frac{Re}{Re_u}\right]^2}\right)$		10.4.3
Tangential velocity in seal as function of $c_u = f(z)$	$\gamma(z) \equiv \frac{c_{u,z}(z)}{u_{sp}} = 0.5 + (\gamma_{in} - 0.5)\exp\left(-\frac{a_\gamma z}{L}\right)$		10.4.4
Average tangential velocity c_u in seal	$\gamma_{av} \equiv \frac{c_u}{u_{sp}} = 0.5 + \frac{(1 - e^{-a_\gamma})}{a_\gamma}(\gamma_{in} - 0.5)$		10.4.5
Tangential velocity (average)	$c_u = \gamma_{av} u_{sp} = \gamma_{av}\omega R$		10.4.6
Force coefficient	$X_{SF} = \frac{L R \Delta p}{s}$		10.4.7
Axial velocity	$c_{ax} = \left(\frac{2\Delta p}{\rho\zeta_{tot}}\right)^{0.5}$		10.4.8
Seal length influence	$a_{SL} = 1 - \frac{\tanh(L/D)}{L/D}$		10.4.9
Direct stiffness	$k = \frac{2\pi a_{SL} L R \Delta p}{s\,\zeta_{tot}}\left\{\frac{1.95\lambda\,(2 + \zeta_E - \zeta_A)R^2}{L\,s\,\zeta_{tot}} - \frac{c_u^2}{c_{ax}^2}\right\}$ $k^* = \frac{k\,s}{L\,R\Delta p} = \frac{k}{X_{SF}}$ $k^* = \frac{2\pi a_{SL}}{\zeta_{tot}}\left\{\frac{1.95\lambda(2 + \zeta_E - \zeta_A)R^2}{L\,s\,\zeta_{tot}} - \frac{c_u^2}{c_{ax}^2}\right\}$		10.4.10
Angle between axial and tangential velocity	$\varphi_c = a\tan\frac{c_u}{c_{ax}}$		10.4.11
	$B = 1 + 0.75\cos^2\varphi_c$		10.4.12
	$E = \frac{0.5(2 + \zeta_E - \zeta_A)}{\zeta_E + \zeta_A + \frac{\lambda L}{4s}B}$		10.4.31
Cross-coupled stiffness	$k_c = \frac{3\pi a_{SL} R^2 \rho c_{ax} c_u}{16s}\left\{(1 + 6E)B\lambda\frac{L}{s} + 48E\right\}$ $k_c{}^* = \frac{k_c}{X_{SF}}$		10.4.14
Direct damping	$c = \frac{R}{c_u}k_c$	$c^* = \frac{c\omega}{X_{SF}} = \frac{\omega R}{c_u}k_c{}^* = \frac{u_{sp}}{c_u}k_c{}^*$	10.4.15
Cross-coupled damping	$c_c = \frac{2\pi a_{SL} L R^2 \rho c_u}{s}$	$c_c{}^* = \frac{c_c\omega}{X_{SF}}$	10.4.16
Direct mass	$m = \frac{\pi a_{SL} L R^3 \rho}{s}$	$m^* = \frac{m\omega^2}{X_{SF}}$	10.4.17

$\zeta_A = 0.85$. Experimental data on the coefficients of long seals are difficult to obtain in the range of pressures relevant for high-pressure pumps, because the pressure differences of several 100 bar generate extremely high forces which are difficult to harness in a test rig. The coefficients for long seals are therefore derived from theoretical calculations, [34, 66].

The balance piston of a high-pressure pump, which breaks down for example 200 bar, has a strong centering effect on the rotor. The radial rotor position near the piston is therefore essentially dictated by that force. If the relative position of the piston and the bearings is not well adjusted, alternating shaft stresses and even shaft failures may result.

A high level of damping is the most important criterion for the operation of a multistage pump with low vibrations. Two ways are open to increase the rotor damping and boost the stability margin:

1. Reduction of the circumferential flow velocity c_u in the seal. The benefits of this measure are two-fold: firstly, the direct damping is enhanced and secondly, the cross-coupled stiffness (as the driving force) is reduced. The cross-coupled stiffness is in fact approximately proportional to the average circumferential velocity in the seal. A reduction of c_u can only be achieved by either a rough stator surface or by structures impeding the flow in circumferential direction, e.g. honeycomb seals as per [22, 37]. If serrated seals are applied to reduce the leakage and/or the risk of rotor seizure, the serrations should be machined into the stator bushing in order to decrease c_u by an increased friction surface on the stator. Also spiral grooves in the stator bushing reduce the cross-coupled stiffness if they are designed to achieve a pumping action against the sense of rotor rotation, [21]. However, the shorter the seal, the less efficient will be such measures, because the flow path is too short to effectively reduce the inlet swirl within the short seal length. A rough stator is efficient if the seal is long ($L_{sp}/d_{sp} > 0.5$) and/or the pre-rotation at the seal inlet is strong ($c_u/u > 0.5$).
2. Reduction of the pre-rotation at the seal inlet by a swirl brake, [41]. Tests in [13, 43] showed a strong increase in damping by introducing a swirl brake at the balance piston. When operating with twice the design clearances, vibration amplitudes dropped by more than 50% when the swirl brake was introduced. A swirl brake is also effective at the seal at the impeller inlet. In contrast, there is virtually no pre-rotation at the interstage seal and a swirl brake placed there would bring no improvement. In very short seals damping can only be improved by means of a swirl brake.

If the leakage flows radially inwards, the pre-rotation at the seal inlet increases with the leakage flow rate, the tangential velocity c_{2u} at the impeller outlet and the ratio d_2/d_{sp} of the impeller to the seal diameter, refer to Chap. 9.1. Therefore, a particularly high pre-rotation is present in pumps with low specific speeds (Fig. 9.9). If the seal clearances grow during service due to wear, the pre-rotation strongly increases and rotor eigen frequencies as well as damping drop considerably.

If, in contrast, fluid recirculating from the collector enters the impeller sidewall gap with low tangential velocity, the pre-rotation at the seal is reduced, rotor damping increases and rotor amplitudes drop. This can provoke a sudden change in vibration amplitudes, as demonstrated by tests on a boiler feed pump at $q^* = 0.25$ shown in Fig. 10.15, [43]. For this test the clearances at the balance piston and at the impellers were opened up to twice their design value. When the pump was run up on a speed ramp, the amplitudes dropped sharply as the speed exceeded about 6600 rpm. This remarkable behavior is explained by the following mechanism: At $q^* = 0.25$ recirculation at the impeller outlet is well developed. As the balance water flow rate grows with increasing speed, more fluid with low tangential velocity is drawn into the impeller sidewall gap and the pre-rotation at the balance piston is reduced. Apparently this reduction was as effective as a swirl brake. This is concluded from the level of the amplitudes attained at $n > 6600$ rpm.

10.6.3 Hydraulic Impeller Interaction

If an impeller moves around its centric position due to vibrations of the rotor, the pressure distribution over the impeller circumference changes. Reaction forces are thus generated in a similar way as with annular seals. These forces can be described by the linearized model of Eq. (10.7) in terms of coefficients for stiffness, damping and mass. The interaction forces do not only act on the impeller outlet width but also on the shroud. If the shroud is not perpendicular to the rotor axis, this effect can be considerable. Figure 10.16 shows the coefficients of stiffness, damping and mass measured on a diffuser pump, [13]. The equations defining the force coefficients and the test parameters are given in Fig. 10.16. The coefficients quoted in [15] are of similar size as those in Fig. 10.16.

The test data show the following trends: the radial force resulting from Eq. (10.8) has little impact on the eigen frequencies because the direct stiffness is low and the effects of the mass and cross-coupled damping (both are significant) compensate to a large extent. Overall, the effect of the hydraulic impeller interaction on the eigen frequencies is slightly negative. The tangential force is destabilizing because the cross-coupled stiffness is high while the direct damping is weak.

The limiting speed at which the impeller contributes to rotor instability follows from Eq. (10.9) and Fig. 10.13. Typically the hydraulic impeller interaction is one order of magnitude smaller than the effect of the annular seals.

Measuring a given impeller once in combination with a volute and then with a diffuser revealed little differences in the impeller interaction coefficients. Likewise, differences between vaned and vaneless diffusers were insignificant. Very low interaction forces were measured on impellers whose shrouds were perpendicular to the rotor axis. From this finding it is concluded that the hydraulic impeller interaction is mainly influenced by the front shroud, [21]. In order to minimize its destabilizing effect, the radial projection of the impeller front shroud should expose as small an area to radial forces as possible.

Tests carried out with rotating conical bodies for modeling the impeller sidewall gaps demonstrated that the coefficients grow with decreasing sidewall clearance.

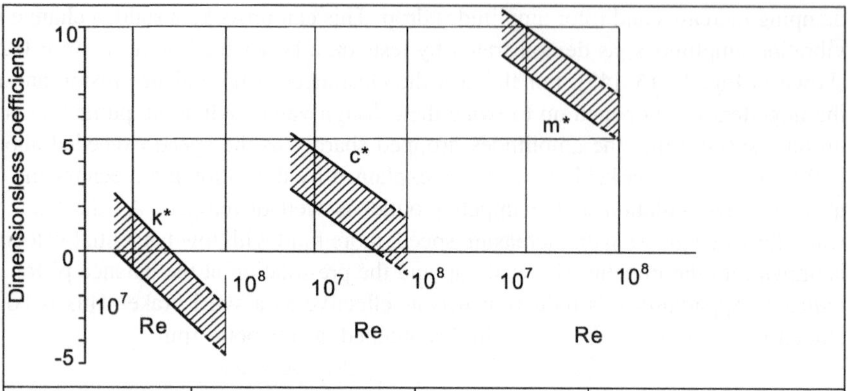

$L_S/r_2 = 0.125$ $5\times10^6 < Re < 10^8$ $0.8 < q^* < 1.2$ $20 < n_q < 35$	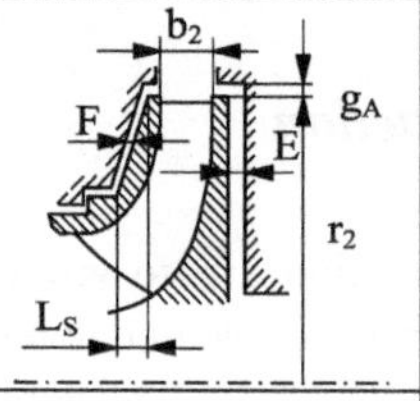	$k_c^* = 7$ to 10 $c_c^* = 7$ to 9.5 $m_c^* = 0.3$ to 3	$Re = \frac{\omega r_2^2}{\nu}$ A = 1 mm F = 5.7 mm E = 6.3 mm
$k = X_F \omega^2 k^*$ $k_c = X_F \omega^2 k_c^*$	$c = X_F \omega c^*$ $c_c = X_F \omega c_c^*$	$m = X_F m^*$ $m_c = X_F m_c^*$	$X_F = \pi \rho r_2^2 b_2$

Fig. 10.16 Hydraulic impeller interaction coefficients, [13]. Note: the Re-dependence of k* is questionable, see discussion in the text

The cross-coupled stiffness, and thus the destabilizing effect of the front shroud, increases with growing pre-rotation. The leakage flowing through the impeller sidewall gap affects primarily the direct stiffness, which has a de-centering effect and thus reduces the eigen frequency, [40].

The physical mechanism underlying the hydraulic impeller interaction can be thought similar to the "Bernoulli effect" discussed in context with Fig. 10.14 since tangential velocities are dominating in the ISR and on the bands of the shrouds. The flow is not governed by an imposed pressure differential and gap widths are large. Therefore, no "Lomakin effect" can be expected on the impeller shrouds including the effect of gap A.

Based on the test data published in [13, 99, 143–145], correlations have been developed which allow to estimate the impeller interaction forces as function of geometry, Reynolds number and flow rate ratio q*, see Table 10.4(2).

For high-head centrifugal pumps, the tests in [13] appear to be the most relevant in terms of geometry, Reynolds number and inlet swirl to the ISR. The correlations and trends are discussed in the following:

1. The correlations are based on 29 test points covering the range of: $Re = 7 \times 10^5$ to 7×10^7; $q^* = 0.25$ to 1.4; $L_s/r_2 = 0$ to 0.25; $s_{ax}/r_2 = 0.016$ to

0.18; g_A/r_2 = 0.0045 to 0.06; d_2 = 160 to 370 mm. The effects of s_{ax}/r_2 and g_A/r_2 are expected to be strongly non-linear. When using the correlations, it is advisable not to extrapolate beyond above limits but use the cut-off values instead as given by above maxima and minima. The term used in the correlations for quantifying the effect of the projected shroud length yields data measured on a 2D-impeller when $L_s = 0$, see [99].

2. There is conflicting evidence on the influence of the Reynolds number. Some tests in [13] showed a distinct Re-dependence of direct coefficients but none on the cross-coupled coefficients. In contrast to this, the speed (or Re) had no systematic effect on the coefficients in the range of n = 500 to 2000 rpm in the tests reported in [99, 143–145]. For the tangential (cross-coupled) forces a Reynolds-dependence seems plausible; the Re-exponents which best fit the data are higher than for the direct force terms. The apparent contradiction in the Reynolds influence between [13] and the other data may be due to the fact that the impeller forces were obtained by subtracting the seal forces from the overall measurement in [13].
3. The Re-exponents in Table 10.4(2) were determined by minimizing the scatter around the correlations. But the result is hampered by a lack of data. The high Re-exponent of the cross-coupled mass term may be doubtful; anyway, the physical meaning of this term seems somewhat elusive.
4. The flow rate ratio q* has in impact on c_{2u}, hence the swirl at the ISR inlet, Fig. 5.15. This effect is particularly important when recirculation occurs at the impeller outlet, Figs. 5.30, 5.35, 10.15, 10.35, 10.36. Since the 3D-flow depends on the design of impeller and diffuser, it is not possible to give general rules. The test data in [13] show some general tendencies (as captured by the correlations) but with strong variations around the correlations.
5. While the ratio d_3/d_2 (gap B) has a dominating impact on the RSI, it has virtually no influence on the hydraulic impeller interaction. One reason may be that the whirling amplitudes are small compared to gap B.
6. The number of impeller blades, blade outlet angles or diffuser hydraulic design have little influence on the coefficients.
7. The direct stiffness is always de-centering; i.e. it tends to reduce the eigen frequencies of the rotor. In the tests with annular seals in [13], some small centering direct stiffness was found, but it is suspected that this result is due to an incorrect separation of the effects of the seal and the impeller.
8. The main geometric parameter gleaned so far from the tests are: gap A, front shroud projection L_s and s_{ax} (gap F). Large L_s and small gaps A und F increase the interaction forces. Specifically, the de-centering effect of the impeller increases; *a negative value of k* yields a positive force which is de-centering,* Eq. (10.8).
9. L_s depends on the shape of the shroud, and its determination with curved front shrouds is to some extent arbitrary. The cylindrical extension to the seal has an influence which cannot be separated. An integral over the entire shroud would be necessary to get the true interaction forces.

Table 10.4 (2) Hydraulic impeller interaction coefficients		
Definitions, see also Figure 10.16	$s^*_{ax} = \frac{s_{ax}}{r_2}$; $L^*_s = \frac{L_s}{r_2}$; $g^*_A = \frac{g_A}{r_2}$ (g_A = Gap A)	10.4.21
Direct stiffness	$k^* = \frac{-3.4}{0.53+0.47q^*}(0.25+6L^*_s)\left(\frac{0.033}{s^*_{ax}}\right)^{0.32}\left(\frac{0.02}{g^*_a}\right)^{0.55}$	10.4.22
Cross-coupled stiffness	$k^*_c = 4.1\left(\frac{10^7}{Re}\right)^{-0.25}(0.67+2.64L^*_s)\left(\frac{0.033}{s^*_{ax}}\right)^{0.4}\left(\frac{0.02}{g^*_a}\right)^{0.2}$	10.4.23
Direct damping	$c^* = 3.8\left(\frac{10^7}{Re}\right)^{0.1}(0.67+2.64L^*_s)\left(\frac{0.033}{s^*_{ax}}\right)^{0.5}\left(\frac{0.02}{g^*_a}\right)^{0.55}(1.67-0.67q^*)$	10.4.24
Cross-coupled damping	$c^*_c = 10.4\left(\frac{10^7}{Re}\right)^{0.2}(0.67+2.64L^*_s)\left(\frac{0.033}{s^*_{ax}}\right)^{0.7}\left(\frac{0.02}{g^*_a}\right)^{0.25}(1.67-0.67q^*)$	10.4.25
Direct mass	$m^* = 8.3\left(\frac{10^7}{Re}\right)^{0.1}(0.25+6L^*_s)\left(\frac{0.033}{s^*_{ax}}\right)^{0.65}\left(\frac{0.02}{g^*_a}\right)^{0.17}(1.33-0.33q^*)$	10.4.26
Cross-coupled mass	$m^*_c = 3\left(\frac{10^7}{Re}\right)^{-0.55}(0.25+6L^*_s)\left(\frac{0.033}{s^*_{ax}}\right)^{0.9}\left(\frac{0.02}{g^*_a}\right)^{-0.75}$ Limit m_c* to a maximum of 3	10.4.27
Resulting radial and tangential forces Ω = whirl frequency A = amplitude	$F_r = \pi\rho\omega^2 r_2^2 b_2 A\left\{-k^* - c^*_c\frac{\Omega}{\omega} + m^*\left(\frac{\Omega}{\omega}\right)^2\right\}$ $F_t = \pi\rho\omega^2 r_2^2 b_2 A\left\{+k^*_c - c^*\frac{\Omega}{\omega} - m^*_c\left(\frac{\Omega}{\omega}\right)^2\right\}$	10.4.28

10. There appear to be some fundamental problems with the tests published so far: (1) When face seals were used as in [13, 99, 143–145], seal forces are largely eliminated, but all terms where tangential velocities are involved - i.e. direct damping and all cross-coupled terms - will be present. (2) Using face seals, the cylindrical extension of the shroud to the face seal will affect direct stiffness and mass, but also all other coefficients. (3) When testing the impeller with an annular seal (i.e. configuration as in the real pump), the forces of the annular seals must be subtracted by calculation. Since seal forces are an order of magnitude higher than impeller interaction, impeller coefficients are found as the result of subtracting two large figures to obtain a tiny rest with large uncertainties. As a consequence, the results obtained in [13] by this method failed to yield data similar to the face seal tests and other publications.
11. A large uncertainty is inherent to the test procedure. Standard deviations quoted in [13] were given as follows in per cent; the first figure is for the BEP, the second figure for $q^* = 0.25$: k^*: ±25(±100); k_c^*: ±25(±110); c^*: ±50 (±200); c_c^*: ±70(±260); m^*: ±35(±120); m_c^*: ±30(±150).
12. The correlations given in Table 10.4(2) are also subject to large uncertainties. Only the correlation for the direct stiffness looks quite good; see Fig. 10.16-1 where the measured stiffness is plotted against the stiffness predicted from the correlation. The correlation coefficient of c^* was $R^2 = 0.89$ while all other coefficients correlated poorly ($R^2 = 0.63$ to 0.7).

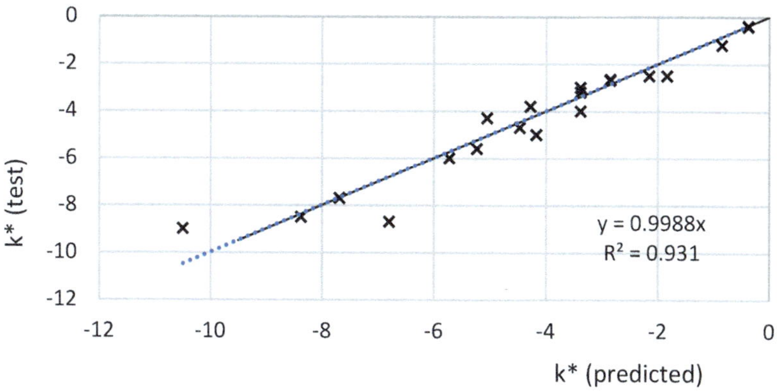

Fig. 10.16-1 Hydraulic impeller interaction: Correlation of measured direct stiffness versus prediction from Eq. (10.4.22)

10.6.4 Bearing Reaction Forces

The pressure distribution in the lubricating film of a journal bearing provides the reaction force which balances the external forces transmitted from the shaft. The higher the bearing load, the thinner is the film. Since the film thickness variations are within a few hundredths of a millimeter, radial journal bearings have considerable stiffness. Journal bearings provide damping in a wide range of operation conditions. Setting the mass terms zero, the dynamic behavior of journal bearings can be described by Eq. (10.7), if vibration-induced movements are small. The coefficients necessary for evaluating Eq. (10.7) can be obtained from tests carried out by the bearing manufacturer or from computer programs. The controlling parameters are:

- Most important is the type of bearing: cylindrical bearings are unsuitable for high-speed pumps, since they are prone to instability ("oil whip"). The threshold of instability is reached when the tangential force becomes positive at $\omega/\Omega > 2$, Eq. (10.8). Since the rotor vibrates at its eigen frequency, a bearing instability arises when the operation speed reaches twice the critical speed. For high-speed pumps multi-lobe bearings are used primarily because of their good damping. Roller bearings have a very high stiffness but provide little damping; they are unsuitable for high speeds at high bearing loads. Instabilities of the kind discussed in Chap. 10.6.6 cannot occur with roller bearings.
- Viscosity (hence temperature) of the lubricating oil. Temperature effects should also be borne in mind when comparing measured with predicted vibrations.
- Bearing clearance and width to diameter ratio B/D
- Bearing load: highly loaded bearings are stabilizing, lightly loaded bearings are de-stabilizing.

Often the calculation of the bearing load which directly determines the dynamic bearing coefficients harbors great uncertainties. While the rotor weight can be determined with sufficient accuracy, the size of radial hydraulic forces generated in diffuser or volute pumps is determined from statistical data only (such as Table 9.4). The direction of the radial hydraulic forces is even more difficult to predict. Furthermore, size and direction of the radial hydraulic forces depend on the flow rate as is well demonstrated by the test in Fig. 9.32. Thus it may happen that rotor weight and radial forces act in the same direction at a specific flow rate (hence heavily loading the bearing), while they are opposed to each other at another flow and unload the bearing. Such effects may be responsible for many apparent paradoxes observed in the vibration behavior of pumps.

A strongly loaded tilting-pad *axial* thrust bearing generates a restoring moment which acts against the shaft deflection and provides additional damping. This effect is significant when the eigen mode of the shaft is dominated by the overhung of the axial bearing; that is, if the highest vibration amplitudes occur near the axial bearing.

10.6.5 Eigen Values and Critical Speeds

The rotor dynamic analysis of a multistage pump is invariably performed by a computer program designed to model the rotor, the dynamic reactions of the bearings, the annular seals and the hydraulic impeller interaction. The objective of such an analysis is to determine the damped eigen values; these are the speed-dependent rotor eigen frequencies, the eigen modes and the damping. The results are presented preferably in a Campbell-diagram according to Fig. 10.17. There the eigen values are plotted against the speed for design and twice the design clearances of the annular seals. The straight line for synchronous vibration through the origin of the graph intersects these curves at the "critical speeds" f_{kr} which strongly depend on the restoring (stiffness) forces in the annular seals. The *critical speed* should be clearly distinguished from the *eigen frequency at the operation speed* f_{EB}. The latter is the frequency at which the shaft would vibrate if excited by a blow during operation. As demonstrated by Fig. 10.17, the eigen frequencies strongly depend on the restoring forces in the annular seals. In a *rotor instability*, the shaft vibrates at its *eigen frequency*. In an *imbalance-excited resonance* the rotor vibrates at a *critical speed*. As can be concluded from Fig. 10.17, a continuous change of the seal clearances would lead to an infinite number of eigen frequencies and critical speeds.

Some damping coefficients D are inserted in Fig. 10.17, where D is Lehr's damping; D = 1 corresponds to aperiodic damping or "critical damping". Resonance amplification is approximately given by $Q_A = 1/(2 \times D)$. In the example shown in Fig. 10.17 the damping at a given speed is reduced by about 40% when seal clearances are doubled. A swirl brake at the balance piston increases damping considerably but has virtually no influence on the eigen frequencies.

In general the shaft of a multistage pump does not vibrate in a plane as does a simple beam supported at two points, but each eigen value is associated with a specific "eigen mode". The eigen mode represents the 3-D bending line of the shaft with which the rotor vibrates at that particular eigen frequency. Figure 10.18 shows as an example the two least damped eigen modes of a boiler feed pump, which were calculated for an operation speed of 7900 rpm (132 Hz). Mode A shows the highest amplitudes mid-span of the rotor; the eigen frequency is lower than the rotational frequency. In contrast, mode B experiences vibration maxima at the overhung coupling and axial bearing, but the eigen frequency is slightly above the rotational frequency. Mode A is better damped than mode B, because the vibration amplitudes in the annular seals and the bearings are larger.

Each eigen mode is associated with a specific damping. Since damping is proportional to the vibration velocity, the following mechanisms apply: If a bearing or an annular seal is located at or near a maximum of the amplitudes, rotor damping is increased. If a bearing or a seal is located close to a vibration node (amplitude minimum) it does not contribute to the damping. Vice versa, the detrimental effects of de-stabilizing elements are least felt in a vibration node, but are most harmful when placed in a vibration maximum. These mechanisms should be considered when designing a rotor in order to maximize overall damping.

Usually the clearances of annular seals widen slowly during service due to wear so that their stiffening and damping effects diminish. This fact must be taken into account when designing a pump, by defining a maximum allowable clearance (frequently twice the design clearance) and analyzing the rotor for new and worn conditions as done in Fig. 10.17. This analysis frequently reveals that the critical speeds under worn conditions fall into the range of operation speeds. Such is acceptable if the damping is sufficiently high, because a resonance is scarcely detectable during service in that case. This is in sharp contrast to turbo compressors,

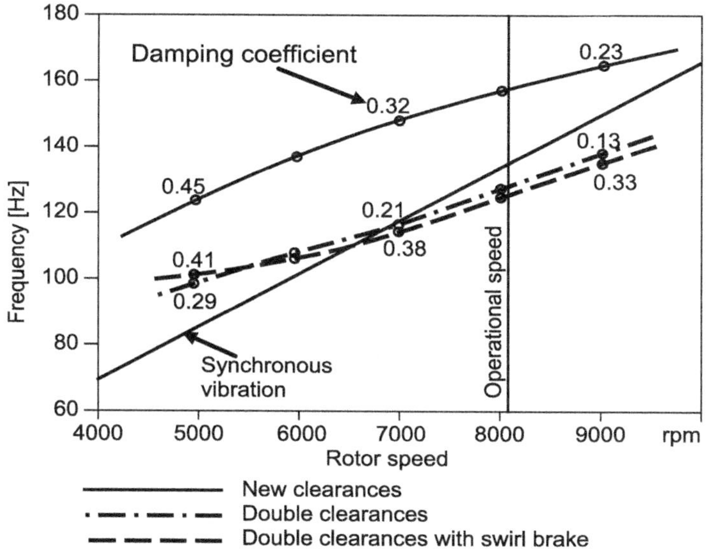

Fig. 10.17 Campbell diagram

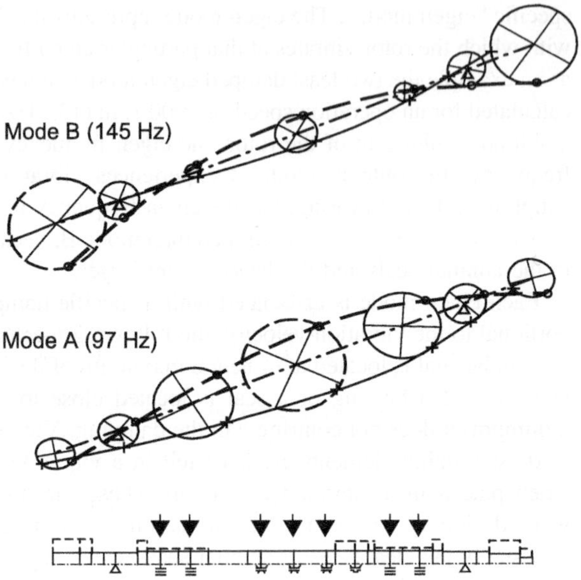

Fig. 10.18 Calculated eigen values and eigen modes, $n = 7900$ rpm

which cannot be operated at a critical speed due to the lack of damping. Experience with high-pressure pumps shows that operation near or at a critical speed is perfectly acceptable with D > 0.2. With worn clearances, when running at the maximum operational speed, a minimum damping of D > 0.05 should be maintained in order to prevent rotor instabilities, [13].

If the amplitudes of shaft vibrations are recorded as functions of the speed, the rotor damping can well be assessed qualitatively from the behavior of the amplitudes and the phase angle: (1) If damping is strong, the amplitudes rise as a function of the speed with exponents smaller than 2. (2) In the extreme they scarcely grow with the speed as demonstrated by the test with a swirl brake shown in Fig. 10.15. (3) If damping is weak, the amplitudes grow in the sub-critical range with the square of the speed. (4) Near a resonance, exponents greater than 2 are observed. With high damping, the phase shift is smooth when going through resonance; with low damping the phase angles vary strongly near a resonance.

Apart from considering the damped eigen values with the help of a Campbell-diagram, the rotor dynamic analysis includes calculations of forced shaft vibrations. To this end, synchronous excitation forces are applied. The calculated amplitudes are then compared to allowable limits and/or factors are defined which quantify the sensitivity of the pump to excitation by imbalance. The sensitivity factors are calibrated by comparison with pumps in service with known operation characteristics, [13].

In summary, the following parameters have an impact on rotor vibrations and need to be considered in an analysis:

- Rotor geometry: Bearing span, shaft diameters, distribution of masses on the rotor, in particular overhung components (coupling and axial bearing).

- Radial bearings: type, relative clearance and width-to-diameter ratio b/d.
- Bearing load: rotor weight, radial hydraulic forces as a function of q*.
- Viscosity of lubricating film (oil temperature).
- Axial bearing: mass and overhung, loading (stiffness and damping), restoring moment against bending in lateral vibrations.
- Coupling: type, mass and overhung, alignment errors.
- Annular seals at impeller and balance device; surface geometry (plain, serrations or honeycomb), clearance, length-to-diameter ratio, pressure difference, pre-rotation, Reynolds numbers (in axial- and tangential direction), tilting of long annular seals, leakage through the impeller sidewall gap.
- Hydraulic impeller interaction: impeller geometry (b_2, d_2), angle between front shroud and rotor axis, leakage through the impeller sidewall gap, swirl of the leakage flow entering the impeller sidewall gap.

10.6.6 Rotor Instabilities

Self-excited vibrations ("rotor instabilities") can be caused by radial journal bearings, annular seals or by the hydraulic interaction of both closed and open impellers ("Alford effect"). Such instabilities are created by the following mechanism: The pressure distribution generated by the vibrating body creates a force component which acts in the direction of the movement of the vibration orbit. Thus energy is fed from the circumferential fluid motion to the rotor; consequently, the orbital movement is accelerated. For instability to occur the energy fed to the rotor must exceed the direct damping which opposes the movement of the orbit. Two effects combine to create such a pressure distribution: Firstly, fluid is displaced by the orbit (similar to a positive displacement pump). Secondly, there is a journal bearing effect through which fluid is transported by shear stresses in the boundary layer of the rotating shaft.

Rotor instabilities can be generated by all of the mechanisms which cause the orbit to move in the direction of rotation, [31]:

1. Turbulent flow through the annular seals at the impeller and the balance piston
2. Laminar flow in a journal bearing yields a ratio of vibration to rotational frequency $\Omega/\omega = 0.45$ to 0.48
3. Hydraulic impeller interaction
4. Excitation forces created by open impellers ("Alford effect")
5. Inner friction in a rotor, e.g. due to microscopic movements in a shrink fit that is too long and/or too weak: $\Omega/\omega = 0.2$ to 1.0
6. Liquid accumulation in the rotor, e.g. in a hollow shaft or in a coupling: $\Omega/\omega = 0.5$ to 1.0

The effects 5 and 6 can be easily avoided by an appropriate design. In mechanisms 1 to 4, the flow conditions are such that the vibrating body generates a tangential force which acts in the direction of rotation and opposes the direct damping. The damping increases with the ratio Ω/ω, so that instabilities occur *below* a specific limit of the orbit frequency.

Characteristically for most of the instabilities, the rotor vibrates at its lowest eigen frequency below the actual operation speed of the shaft. Therefore, *an instability is usually recognized as sub-synchronous vibration.* If the speed is increased after the onset of instability, the frequency of the vibrations remains nearly constant, while the amplitudes increase sharply until limited by non-linear effects such as rubbing in the annular seals.

Figure 10.19 very clearly shows the described behavior: the shaft vibrations recorded were measured on a multistage pump during run-up from 1500 to 5500 rpm (pump operating in air). Such a graph is called "Bode plot". Peak-to-peak amplitudes in μm are plotted on the ordinate. The spectrum (top figure) exhibits an eigen frequency at 35 Hz. The cascade diagram (bottom) shows the synchronous amplitudes (due to imbalance) increasing with the speed. At about 5000 rpm, all of a sudden, there appear sub-synchronous vibration peaks which rise sharply. This is a rotor instability which locks in at the rotor eigen frequency of 35 Hz.

On rare occasions super-synchronous instabilities have been observed, which may be explained by fluid entering the seals with tangential velocities higher than $c_u > \omega \times R$. If that is the case, swirl brakes appear to be the obvious remedy. Two cases of super-synchronous rotor instabilities have been reported in [129].

Case 1: A two-stage pump, $n_q = 11$, with opposed semi-open impellers (pump similar to Fig. 9.19) exhibited rotor instability at $1.34 \times f_n$ when the center shaft sleeve was designed with grooves on the *rotating* part (the stator bush had a plain surface). Note that no instabilities were experienced with *plain* rotor and stator. The instability occurred at partload and overload - but not near BEP. The grooves (depth 1.3 mm, width 1.6 mm, land 1.6 mm) led to high cross-coupled stiffness and negative damping because the fluid rotation in the gap is enhanced by friction on the large grooved rotor surface. In addition, the inlet swirl to the center bush is very high because of the high ratio d_2/d_{sp} inherent to $n_q = 11$ giving $c_{u,in}/(\omega R) > 1$ (refer to Fig. 10.12).

The instability vanished after abolishing the grooves of on the shaft sleeve. Expeller vanes on the rear shrouds were also able to remove the instability; reason: the flow in the impeller side room (ISR) changes to the forced-vortex type with low $c_{u,in}$ at the inlet to the bushing. Pressure pulsations (measured in the center bush flush line) showed exactly the same type of instability: *they were caused by the high rotor amplitudes* at off-design conditions.

Case 2: A large sewage pump with $n_q = 86$ was subject to super-synchronous rotor vibrations. The rotor instability was caused by *hydraulic impeller interaction* (see Chap. 10.6.3) at the long cylindrical front shroud with a shape similar to Fig. 7.21. Depending on flow rate and speed, different instabilities with forward and backward rotor orbiting were recorded. Three identical pumps in identical systems behaved quite differently, which testifies to the sensitivity of the rotor to non-definable differences in flow pattern in the ISR. While backward rotor whirl is often due to rubbing of the rotor on the stator, this was not so in the present case. The backward

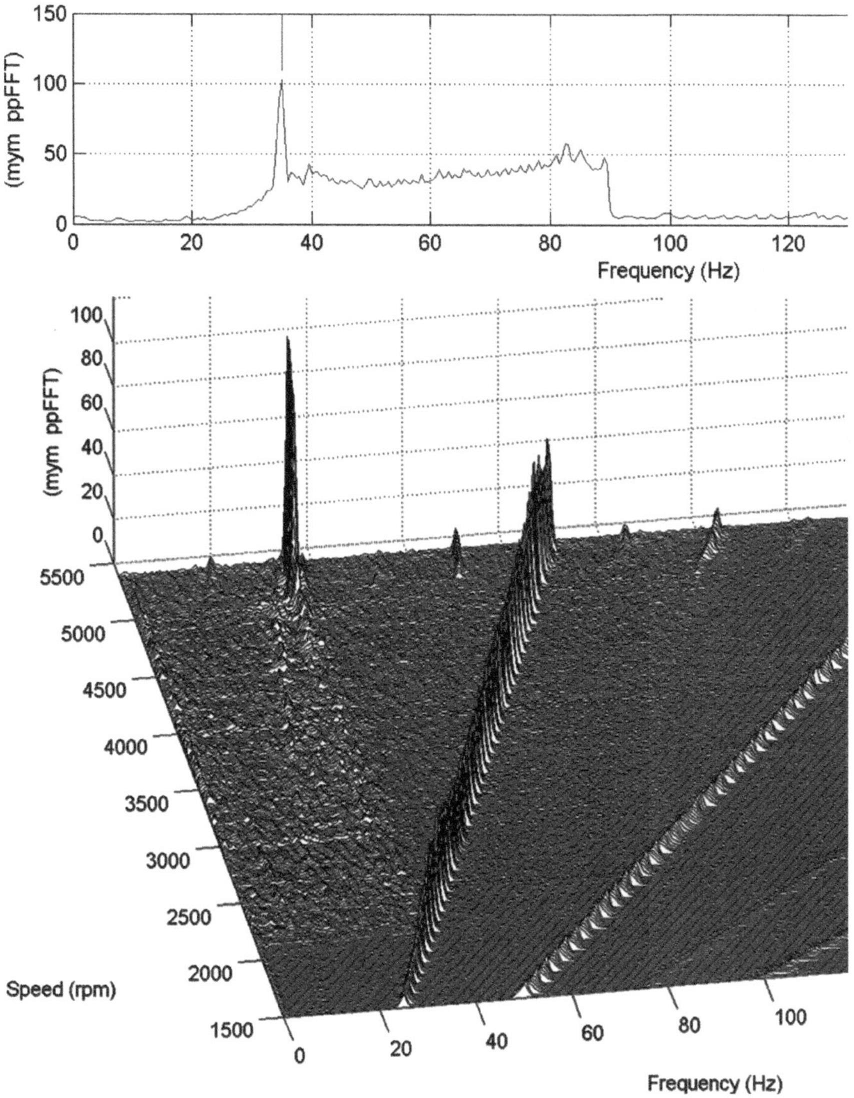

Fig. 10.19 Rotor instability of a multistage pump recorded during run-up

whirl may have been caused by flow separation in the volute. Since a pump with $n_q = 86$ has a very large volute casing, backflow from the volute may have entered the ISR on the front shroud and caused the backward whirl. Pressure pulsations were measured in the ISR; these were phase-locked with rotor instability vibrations; these *pulsations were caused by the rotor movement* (and not vice versa). The rotor instability could be eliminated by air injection in the ISR, by expeller vanes or by ribs ("spoilers") on the casing. All of these measures are apt to disrupt the instable flow patterns in the ISR.

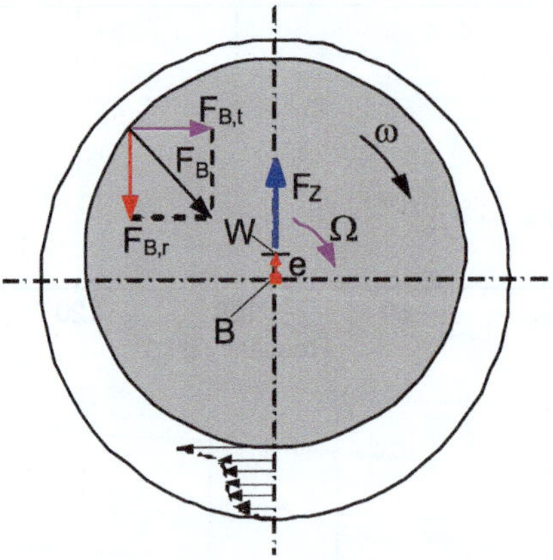

W = shaft center

B = center of bore

e = eccentricity

ω = angular speed of shaft

Ω = angular orbit speed

F_B = force due to hydrodynamic bearing effect

F_Z = centrifugal force due to orbit movement

Fig. 10.20 Instability mechanism

The unsteady 3-dimensional flow leading to instability in a seal or bearing is difficult to visualize, but a simplified explanation may be offered with reference to Fig. 10.20: The fluid circulates in the gap of an annular seal with about half the circumferential velocity of the rotor. Fluid circulation is caused by wall shear stresses and by a fluid displacement induced by the vibration orbit $e \times \Omega$. The pressure distribution in the gap is not symmetrical to the direction of the rotor displacement, but has its maximum upstream of the narrowest gap (as given by journal-bearing theory). The force F_B created by the pressure distribution thus has not only a radial component $F_{B,r}$ opposed to the shaft displacement, but also a tangential component $F_{B,t}$ acting in the direction of the shaft rotation. If this tangential force exceeds the damping, the gap is reduced further by the centrifugal force F_z which is caused by the movement of the orbit. Thus a self-excited vibration (or instability) is generated. The stability limit is reached at twice the speed that corresponds to the lowest rotor eigen frequency. There are three reasons for this: (1) The driving tangential force increases with the average circumferential fluid velocity c_u in the gap; (2) an unstable rotor vibrates at its eigen frequency; (3) the average tangential velocity is about half the circumferential rotor speed. In other words: an instability in a bearing or an annular seal occurs *when the tangential velocity* c_u *in the gap equals the circumferential rotor speed at the lowest rotor eigen frequency* ω_E, i.e. if $c_u = r \times \omega_E$.

As mentioned in Chap. 10.6.2, the rotor stability can be considerably increased by a reduction of the mean tangential velocities in annular seals. This can be accomplished by: (1) rough stator surfaces, (2) swirl brakes, (3) honeycombs, (4) hole patterns. Likewise, the stable operation range can be enlarged by increasing the (lowest) eigen frequency of the rotor, because it is the ratio ω/ω_E of operation speed to the lowest eigen frequency that determines the stability limit.

The mechanisms leading to instability in an annular seal or in the impeller sidewall gap (which is an important aspect of the hydraulic impeller interaction) can be thought to be qualitatively similar to those discussed for journal bearings, even though the flow is usually turbulent. The differences between forced and self-excited vibrations are summarized in Table 10.5.

Table 10.5 Forced versus self-excited vibrations

Feature	Forced vibrations	Self-excited vibrations
Frequency	Vibration frequency equals *excitation* frequency: usually frequency of rotation, rotating stall, or blade passing frequency and its harmonics.	Vibration frequency equals *natural* frequency of rotor; practically constant and little dependent on speed; mostly sub-synchronous.
Amplitude	Amplitude appears in spectrum as peak which increases with speed. With resonance and weak damping strong amplification of amplitudes.	Vibration appears suddenly above a speed (stability) limit and grows strongly when speed is increased further.
Orbit movement	Always in direction of rotation.	Mostly in direction of rotation.
Rotor stresses	If excited with the frequency of rotation, the rotor moves in its deformed shape (bending line); hence no alternating stresses are generated.	Rotor is subject to alternating stresses the frequencies of which are equal to the difference between the frequencies of rotation and vibration.
Influence of damping	Additional damping reduces the amplitudes, but has little impact on the frequency of the vibration.	Additional damping increases the speed limit, but has no influence on the amplitudes.
Influence of geometry	Mechanical and hydraulic imbalances determine the excitation.	Amplitudes are independent of rotational symmetry: self-excitation above stability limit.
Measures	• Mechanical balancing • Reduce hydraulic imbalance (precision casting) • Increase damping in order to limit amplitudes • Shift critical speed in order to avoid resonance	• Increase damping in order to get higher stability limit • Reduce cross coupling stiffness (swirl brake, rough stator) • Increase natural frequency

10.7 Hydraulic Excitation of Vibrations

10.7.1 Interaction Between Flows Through Rotor and Stator (RSI)

Velocity variation over the pitch of the blade: Size and angles of the absolute velocity at the impeller outlet vary over the blade pitch as shown in Fig. 10.5. The unsteady impeller outlet flow, interacting with the flow in the stator, is an important source of excitation forces (rotor/stator interaction "RSI"). The unsteady impeller outlet flow generates a pressure field that rotates with a frequency $z_{La} \times n/60$. The equally time-dependent approach flow to the diffuser, in turn, induces complex pressure fields in the stator. Both pressure fields move relative to each other; they contain the impeller periodicity $p_2 = \nu_2 \times z_{La}$ and the diffuser periodicity $p_3 = \nu_3 \times z_{Le}$. In the *absolute* system the pressure field rotates with $p_2 \times f_n$. Casing and bearings are subjected to excitation forces with that frequency. In the *relative* system (or rotating reference frame) the pressure field moves with $p_3 \times f_n$; impellers and rotor are excited with that frequency.

These relations are well illustrated by the measurements shown in Fig. 10.21: Figure (a) shows the unsteady pressures measured in the volute casing (i.e. in the absolute system) over one impeller revolution. The 7-bladed impeller produces a clear periodicity of the pressure fluctuations which decay only slowly with increasing distance from the impeller outlet. The pressure signal in the relative system was plotted over two revolutions by a transducer mounted in the impeller, Fig. 10.21b. Since the pump has a double volute the pressure field measured in the relative system has a basic periodicity of 2.

Although the interference of the pressure fields of impeller and diffuser cannot be calculated exactly, it is of great importance for the selection of suitable combinations of impeller and diffuser vane numbers, [10, 17, 27, 29]. Without knowing the complex structure of the pressure fields in the impeller and diffuser in detail, it is possible to apply a Fourier series to each of both fields and determine the resulting pressure field as the product of both series; see Eq. (10.12b) where φ_2 and φ_3 are the angular coordinates of the rotor and the stator.

$$p_2 = \sum_{\nu_2=1}^{\infty} a_{2,\nu} \cos \nu_2 z_{La}\, \varphi_2 \qquad p_3 = \sum_{\nu_3=1}^{\infty} a_{3,\nu} \cos \nu_3 z_{Le}\, \varphi_3 \tag{10.12b}$$

The new pressure field obtained in this way contains two periodicities which correspond to the difference $p_2 - p_3$ and the sum $p_2 + p_3$.

In a single-volute casing ($z_{Le} = 1$) the impeller moves through one pressure cycle per revolution as can be seen in Fig. 9.26. But the same is true for any other disruption of the rotational symmetry of the flow in the stator - for example if there are one or more stalled channels in a diffuser.

In diffuser pumps, the sum $p_2 + p_3$ mostly has no practical relevance, since the corresponding pressure pulses have very high frequencies and therefore contain little energy. However, high frequency noise corresponding to the sum $p_2 + p_3$ can

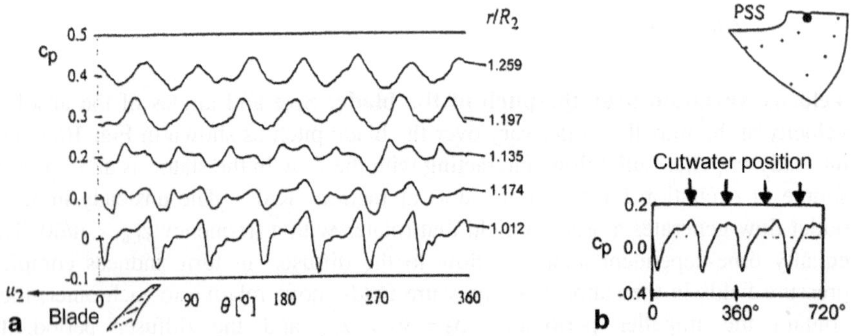

Fig. 10.21 Unsteady pressure curves. **a** Rotating with the impeller, measured in the absolute system at q* = 1; **b** measured in the relative system at q* = 0.7; $d_z/d_2 = 1.22$, [53]

become a problem if the pulsation amplitudes are reinforced by phase resonance, if casing or piping are particularly prone to noise radiation in this frequency range or if the hydraulic design has deficiencies such as a small distance between impeller and diffuser or a pulsating impeller discharge flow.

Most importantly, the difference:

$$m = |p_2 - p_3| = |\nu_2 \times z_{La} - \nu_3 \times z_{Le}| \tag{10.13}$$

is significant for the selection of suitable impeller and diffuser vane numbers. Three conditions have high practical relevance:

Case m = 0: Strong pressure pulsations are generated at m = 0, that is when impeller and diffuser vane numbers have common integer multipliers. In this case the rotor/stator interactions at two or more blades occur simultaneously (i.e. they are directly in phase) and reinforce each other. Such combinations should be avoided altogether for pumps with diffusers and double or multiple volutes. The pressure pulses excite the impeller shrouds with zero diameter nodes (like an umbrella). This type of excitation can also lead to axial thrust fluctuations and is linked to unsteady torque which may produce torsional vibrations.

Case m = 1: At m = 1 the blade forces of the impeller have a non-zero resultant. Therefore lateral vibrations are excited at the blade passing frequency (and higher orders). The impeller shroud is excited with a one-diameter node. To avoid difficulties with shaft vibrations at $\nu_2 \times z_{La} \times f_n$, m = 1 should never be allowed in the first and second order and should be avoided up to the third order if possible. Numerous field problems have been caused by vane number combinations which led to m = 1 in the 2nd order; notably 5&9 or 5&11 (Fig. 10.49), 6&11, 4&5.

Case m $\geq$ 2: The impeller is excited to vibrations with a frequency of $\nu_3 \times z_{Le} \times f_n$. If a structural eigen frequency of the impeller gets into resonance with this excitation frequency, fatigue fractures in the front or rear shroud could occur. The strain imposed on the blades by shroud vibrations can also lead to blade fractures. The number m corresponds to the number of diameter nodes with which the impeller is excited, Fig. 10.22: At m = 1 the impeller oscillates around one, at m = 2 around two diameter nodes etc.

Note: Lateral excitation forces are NOT necessarily proportional to pressure pulsations - even though that may be observed in some instances. This fact follows from theory, e.g. [17, 27]. It has been confirmed by tests in [67] and by [126] where

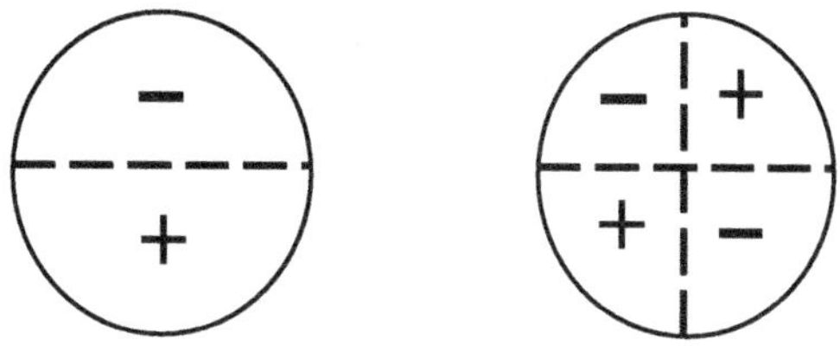

Fig. 10.22 Mode shapes of impeller vibrations with one and two diametrical modes

an impeller with 6 blades was tested with diffusers with 5 and 6 vanes: the combination 6-6 (yielding m = 0) gave high pulsations and low lateral forces, while the combination 6-5 (hence m = 1) had low pulsations but high lateral forces. There are two reasons why there is no direct proportionality between forces and pulsations: (1) The lateral forces follow from the integration of the pressure field around the blades: this integral cannot be measured by just one transducer in an arbitrary location; (2) Consider vane combinations with even number of blades and diffuser vanes (or common multiples); i.e. cases with m = 0: these combinations generate high pressure pulsations but low forces because RSI-forces of opposite blades compensate whereas pressure fluctuations do not compensate.

The relation between the rotating (subscript 2) and the stationary reference frame (3) is given by: $\varphi_3 = \varphi_2 - \omega \times t$. The pressure nodes "m" given by Eq. (10.13) rotate in the *stationary* system with the spin speed ω_S and in the *rotating* reference frame with ω_R [122, 107]:

$$\omega_S = \frac{\nu_2 z_{La}}{m}\omega \quad \omega_R = \frac{\nu_3 z_{Le}}{m}\omega \tag{10.13a}$$

If the mode "m" from Eq. (10.13) is positive, the pressure field spins in the same direction as the rotor; when m is negative, it rotates against the direction of the rotor movement. The frequencies measured in the stationary and the rotating reference frames are:

$$\frac{f_S}{f_n} = m\frac{\omega_S}{\omega} = \nu_2 z_{La} \quad \frac{f_R}{f_n} = m\frac{\omega_R}{\omega} = \nu_3 z_{Le} \tag{10.13b}$$

For illustration of Eq. (10.13b), Table D10.1-1 shows modes for the combination $z_{La} = 7$ with $z_{Le} = 12$ and Fig. 10.22-1 illustrates their graphical representation in the stationary system. At blade passing frequency, a backwards spinning mode with 5 diametrical pressure nodes is observed. At 2-times BPF, the pressure field has two nodes and spins in the same sense as the rotor.

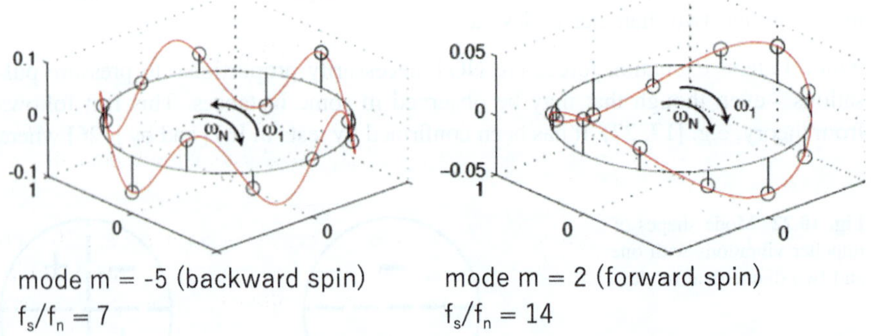

Fig. 10.22-1 Mode shapes of RSI pressure field in stationary system. Y-axis: Δp^* as per Eq. (10.1), $\omega_1 = \omega_S$; $\omega_N = \omega$; $z_{La} = 7$; $z_{Le} = 12$, [6]

Table D10.1-1 Pressure field spinning modes

			Stationary system		Rotating system	
		Nodes	Spin speed	Frequency	Spin speed	Frequency
ν_3	ν_2	m	ω_S/ω	f_S/f_n	ω_R/ω_n	f_R/f_n
1	1	−5	−1.4	7	−2.4	12
1	2	2	7.0	14	6.0	12
2	3	−3	−7.0	21	−8.0	24
3	5	−1	−35	35	−36	36

Resonance condition: To excite mechanical resonances with impeller or casing structural eigen frequencies, or acoustical resonances, the *spinning frequency* must coincide with an eigen frequency.

An application of the theory to an incident with costly consequences can be found in [107] which reports on damage to primary coolant circulation pumps for boiling water reactors. The impeller side room at the rear shroud was made up by a disk of 1000 mm diameter with a thickness of 25 mm. This annular disk was welded to an inner cylinder; the welds cracked to due fatigue during service. Parts of the broken disk fell on the impeller rear shroud which was destroyed subsequently by debris rubbing on the rotating impeller (thereby 30 kg's of metal powder were dispersed throughout the reactor and cooling water loop). The impeller head 5 blades working in a double volute. The cause of breaking the welds of said disk was a resonance between the eigen frequency of the disk and the spinning pressure fields linked to the RSI. The critical mode of the disk was that with two diametrical nodes. In the stationary system, the 2nd order spinning pressure fields ($\nu_2 = 2$) occurred at $\nu_3 = 6$ with m = −2 (backward) and at $\nu_3 = 4$ giving m = 2 (forward spinning). Both modes spin with about 224 Hz (for n = 1344) rpm in opposite directions. If there are a forward and a backward wave with the same diametrical mode and frequency, the superposition of the two waves gives a standing wave. It might be for this reason that the strain measured on the disk for the mode m = 2 was double the values which were recorded for m = 0 and m = 3 which was a resonance condition at 1292 rpm. The pressure modes with m = 2 were in resonance with the eigen frequency of the disk, which was 353 Hz in air but only about 224 Hz in water. The effect of the added mass on the eigen frequency was thus k = 224/353 = 0.63 which is quite close to the estimate given by Eq. (10.15a) below. A given mode shape (e.g. m = 2 with two diameter nodes) can be achieved by many combinations of ν_2 and ν_3; therefore, it is necessary to investigate these up to sufficiently high orders - depending on the circumstances of the specific project. Unfortunately, there is no way to estimate the amplitude of the pressure pulsations. It is therefore, difficult to assess and control the risk of failure other than through avoiding resonances. In view of many possible modes it will not always be easy to avoid resonances in case of a variable speed drive. A contributing factor to the failure (if not the most significant) could be too a small cutwater distance d_z/d_2.

Experimental evidence of the spinning pressure fields with several diametrical modes according to Eq. (10.13) can be found in [122] where two impellers ($z_{La} = 6$ and 9) were tested in a stator with 20 vanes. In these tests, the time-dependent pressure distributions (mode shapes) at the impeller outlet were recorded around the circumference by 10 pressure transducers. The modes shapes measured corresponded to the nodes expected from Eq. (10.13) for m = − 2 with the frequency $f_R/f_n = \nu_2 \times z_{La}$ at $\nu_2 = 3$ and $\nu_3 = 1$; peaks at $\nu_2 = 1$ and 2 with $\nu_3 = 0$ were also observed. Casing deformations due to the rotating pressure fields were derived from a large number of accelerometers installed on the cover of the same pump turbine in [123]. The time-dependent deformations well agreed with the modes expected from Eq. (10.13). The spinning of the vibration modes (pressure, acoustic, deformations or stresses) can be measured as differences in phase over the circumferential locations in the rotating or the stationary system.

Pressure pulsations and lateral excitation forces were measured on the rotating impeller in [125] with $z_{La} = 6$ combined with three diffusers with $z_{LE} = 5$, 7 and 11. Note that these combinations should not be applied in practice. Forward and backward whirl of pressure pulsations and of lateral forces were well predicted by the theory, Eq. (10.13b). The shaft whirled in the same direction as the forces. Whirling directions are the same in the relative and absolute system.

Staggered double-entry impellers: Double-entry impellers are often designed with a center rib extending to the impeller outer diameter; this allows to stagger the blades on both impeller halves, see sketch in Table 10.2. Correctly done, staggering tends to reduce pressure pulsations and radial excitation forces at blade passing frequency. Since staggering is of most interest for pumps with high heads, staggered impellers are dominantly used in pumps with double volutes. If the volute wrap angle is $\varepsilon_{sp} = 180°$, staggering with half the impeller pitch, yields a combination where two blades simultaneously pass both cutwaters. This would increase the pulsations experienced by pump and system. In order to avoid this situation, the following criteria are suggested for selecting the stagger angle δ_{st}:

- Double-volutes with wrap angle $\varepsilon_{sp} = 180°$: $\delta_{st} = ¼\ 360/z_{La}$; i.e. $\delta_{st} = 13°$ for 7 blades and $\delta_{st} = 18°$ for 5 blades.
- Double-volute with wrap angle $\varepsilon_{sp} = 165°$: $\delta_{st} = ½\ 360/z_{La}$ can be used for any number of blades.
- An impeller with 6 blades should not be combined with $\varepsilon_{sp} = 180°$; but $\varepsilon_{sp} = 165°$ should be selected instead
- For single-volutes, the stagger angle could be half the blade pitch $\delta_{st} = ½\ 360/z_{La}$

With staggered impellers, pressure pulsations and radial forces may have the dominant peaks at the BPF = $2 \times z_{La} \times f_n$ (rather than at $z_{La} \times f_n$) - although these peaks would be expected to be lower than with a non-staggered impeller because the pulses have a width of b_2 rather than $2 \times b_2$ (b_2 being defined as the outlet width *per impeller side*).

In general terms, the stagger affects the velocity distribution at the impeller outlet which depends on q*. The optimum stagger would then depend on the velocity distribution at the impeller outlet,

Tests with different stagger angles were reported in [147] with a single-volute pump $n_q = 37$. The effect of the stagger on efficiency was negligible. Staggering of about ½ blade pitch gave a head reduction of about 3% (scarcely generic). Staggering had virtually no effect on the pulsations in the inlet casing. Pulsations were reduced by roughly 50% with stagger angles of 24° and 36° (i.e. near half the blade pitch of 30° with $z_{La} = 6$). While pulsations at BPF are reduced, broad band pulsations may increase due mixing-out of the velocity distribution.

Flow variations over the circumference of the impeller: As discussed in Chap. 1.4.1, the pressure distribution in a centrifugal pump is determined solely by the velocity distribution. In the absolute system, the flow follows curved paths and the local pressure distribution establishes itself in a way that it balances the centrifugal forces induced by the flow path curvature, Eq. (1.26). Therefore, if variable pressures are measured in the volute over the impeller circumference, the flow around the impeller blades, *by necessity*, depends on the momentary circumferential position of the blades. The blade forces thus change over the circumference; their resultant yields the radial thrust created by the blades. From the very existence of a radial force acting on the impeller it is to be concluded that the impeller channels operate at every circumferential position in a different operation point. The experimental prove of this conclusion has been provided by [63, 64]: In a volute pump ($n_q = 26$), the local values of the flow rate and the total head were determined in 12 circumferential positions. Thereby the flow rate was determined by measuring the radial discharge velocity over one blade pitch by means of hot wire anemometry (the tests were done with air as the fluid). The head coefficient ψ_{tot} was measured by a total pressure probe integrating the signal over one pitch.[4]

The results are plotted in Fig. 10.23 where q* equals the *local* flow rate through one impeller channel divided by the nominal flow rate through one impeller channel at BEP (Q_{opt}/z_{La}). Each of the closed curves in Fig. 10.23 represents the head-flow characteristic of an individual impeller channel as it rotates by 360° at a given average discharge flow in the nozzle (e.g. at $q^*_{tot} = 0.75$).

In addition to the characteristics of the individual impeller channels, Fig. 10.23 shows the average head-capacity curve measured between suction and discharge nozzle as well as the ψ_{tot}-q*-curve of the impeller discharging freely into atmosphere without any casing.

The angles shown in the graph give the position of the impeller channel from the cutwater which is at $\varepsilon = 0°$. Looking at the minimum and maximum flows measured, we find:

[4]The definition of the total head ψ_{tot} is analogous to Eq. (8.9).

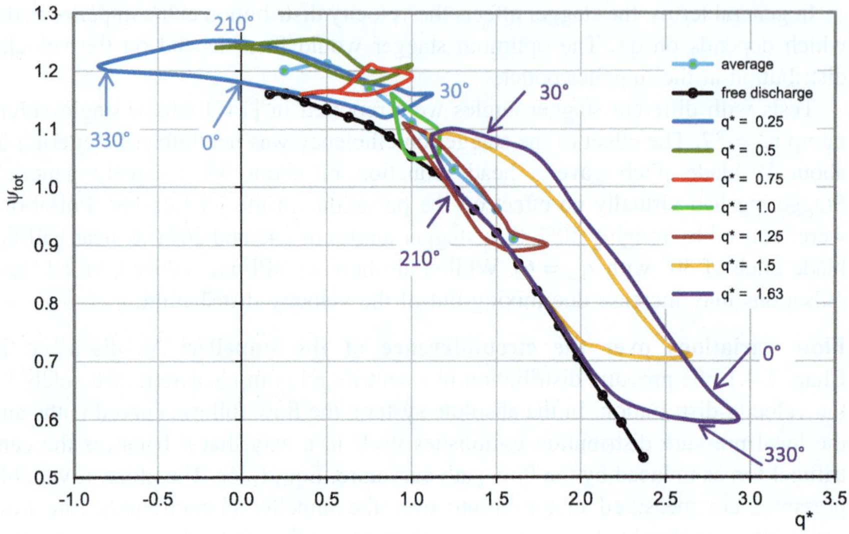

Fig. 10.23 Transient head-capacity curves of an individual impeller channel as it rotates over 360° in a single-volute, adapted from, [63]

- When the pump operates at $q^*_{tot} = 0.25$ (as measured at the discharge nozzle) with an average head coefficient of $\psi_{tot} = 1.2$, the flow rate of an individual impeller channel varies between the extremes of $q^* = -0.8$ (backflow) to $q^* = 1.1$ (forward flow) with heads from $\psi = 1.13$ to 1.25. The static wall pressures in the volute (shown in Fig. 9.26) vary between the extremes of $\psi_{stat} = 0.42$ and 1.1. The strongest backflow occurs at about 330° where the *static* pressure in the volute is highest as demonstrated by Fig. 9.26.
- When the pump operates at $q^* = 1.63$ (as measured at the discharge nozzle) with an average head coefficient of $\psi_{tot} = 0.91$, the flow rate of an individual impeller channel varies between the extremes of $q^* = 1.1$ to $q^* = 2.9$ with heads from $\psi = 0.6$ to 1.1. The static wall pressures in the volute vary between the extremes $\psi_{stat} = 0.23$ and 0.94, Fig. 9.26. The strongest forward flow occurs at about 340° where the *static* pressure in the volute has its minimum, see Fig. 9.26.
- At BEP the variation of flow and head around the circumference is small since the pressure distribution is nearly uniform, see Fig. 9.26.

Superimposed on the ψ-q*-curves of the individual impeller channels in Fig. 10.23 (but not shown for simplicity) were pressure peaks which traveled with the sound velocity through the volute. These peaks are created by the passage of each of the blades under the cutwater. They thus represent pressure pulsations at blade passing frequency *in the acoustic far field*. This most interesting finding (if confirmed) suggests that acoustic waves do interact with the flow through the impeller. In other words, the counter-pressure determining the flow through an impeller channel can be influenced by acoustic waves from the system (similar to the system characteristic): a momentary increase of the local pressure in the volute would reduce the momentary flow through the affected impeller channel - at least up to certain frequencies. By the same token, a momentary drop in local pressure would increase the flow through the channel. Hence the mechanisms described in Fig. 10.25 are effective also when the local pressure in the stator varies because of acoustic waves.

With increasing frequency of the acoustic waves, the inertia of the fluid in the channel will reduce the reaction of the flow to the acoustic pressure variation. It can be surmised that inertia effects increase when the acoustic frequency f_{ac} exceeds the frequency $f_T = 1/T_T$. This frequency is given by the transit time $T_T = L_c/w$ of the fluid through the impeller channel. To estimate the order of magnitude of the "transit frequency" f_T, let us assume the channel length $L_c = (r_2 - r_1)/\sin\beta_{av}$ and $\beta_{av} = ½(\beta_{1B} + \beta_{2B})$. With $r_1 = r_2/2$ and $\beta_{av} = 30°$ we get $L_c = r_2$; and assuming the average relative velocity through the channel to $w = u_2/2 = 0.5\omega \times r_2$, the frequency defined by the transit time becomes $f_T = \omega/2$ or $f_T/f_n = \pi$. According to this criterion, acoustic waves would have an influence on the flow through the impeller channels at least up to 3-times the rotational frequency. In the tests in [63] the influence was measured even at blade-passing frequency i.e.7-times the rotational frequency.

Pressure pulsations: Pressure measurements on the rotating impeller of a pump turbine model are shown in Fig. 10.24a (top), [14]. The pump had 20 adjustable guide vanes and 20 diffuser vanes followed by a volute. Dominant peaks are seen at the rotational frequency and at $f = z_{Le} \times f_n$, but there were also peaks at $\nu \times z_{Le} \times f_n$ the amplitudes of which decrease with increasing ν. The peaks at $\nu = 2$ to 19 are attributed to the uneven pressure distribution in the volute. According to Figs. 9.25 to 9.27 the non-uniformity of the pressure in the volute increases as the flow is reduced - and so do the pressure pulses measured in Fig. 10.24 which shows the amplitudes for two different flow rates. Exactly the same behavior was found in tests with the last impeller of a multistage pump where the diffuser was followed by an annular casing and a discharge nozzle, [6, 7]. The diffuser had 12 vanes; hence the highest peak in the spectrum is found at $f = 12 \times f_n$, Fig. 10.24b (bottom).

A possible mechanism causing the variations of the pressure fluctuations over the circumference is sketched in Fig. 10.25: the uneven discharge pressure distribution in the annular discharge chamber (or volute) causes the flow rate through the individual impeller channels to vary around the circumference. These flow rate variations generate the pressures pulsations.

Spectra very similar to those in Fig. 10.24a, b were published in [45] and in [125], Fig. 10.24c. The tests were done with a single-stage pump with vaned diffusers ($d_3/d_2 = 1.01$) followed by a volute. Again, the pulsations were measured on the blade pressure surface near the trailing edge. The static pressure distribution in the volute followed the same pattern as in Fig. 9.27. As a consequence of the variable counter-pressure in the volute, the flow rates through the individual impeller channels vary around the circumference - and so do the pressure pulsations: *they are maximum where the volute pressure is minimum.* The variation of the pressure pulsations around the circumference is weak near the BEP, but it increases as the operation moves further away from the BEP. At $q^* > 1.0$, the minimum pressure is upstream of the cutwater. There, the local flow rate through the impeller

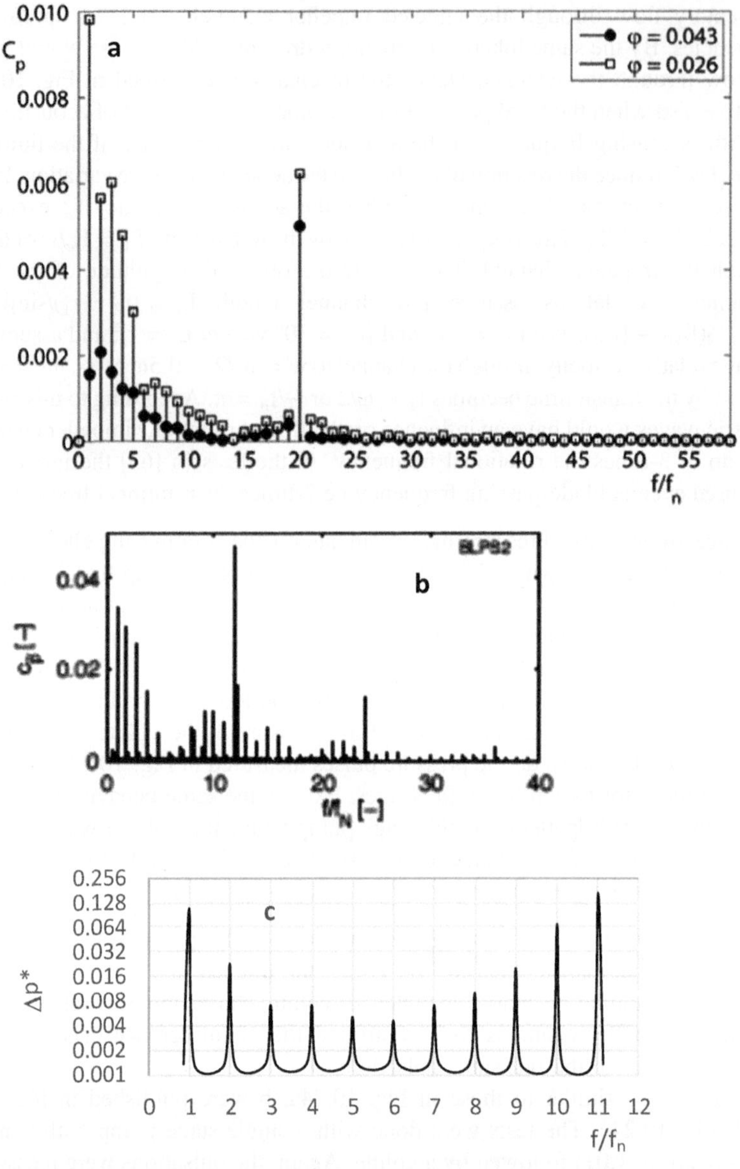

Fig. 10.24 Measured pressure pulsations on a rotating impeller, $c_p = \Delta p_d^*$, Eq. (10.1) **a** Pump turbine ($n_q = 30$) with vaned diffuser and volute, [14]; **b** pump with vaned diffuser and annular casing at $q^* = 0.8$, $n_q = 22$, [6]; **c** $z_{Le} = 11$ (schematic) similar to [45] and [125]

channels reaches its maximum, Fig. 10.23. Strong variations of the pressure pulsations over the circumference were also measured in tests with a single-volute pump reported in [90]. At $q^* = 0.2$ the peaks vary from $\Delta p^* = 0.05$ at $\varepsilon = 30°$ down to $\Delta p^* = 0.015$ at $\varepsilon = 270°$ ($\varepsilon = 0$ at cutwater).

According to the tests in [45], an uneven pressure distribution in the volute (likewise to be expected in an annular casing) causes pressure fluctuations at frequencies with the orders $\nu = 2$ to $(z_{La} - 1) \times f_n$. The pressure pulses were measured on the rotating impeller (as in Fig. 10.21). At high flows ($q^* > 1.5$), the highest pulsations were found when the blade moved into the low-pressure zone *upstream* of the volute cutwater. At partload, the highest pulses were recorded in the low-pressure zone *downstream* of the cutwater (see Figs. 9.25 and 9.27). Pressure fluctuations which vary over the circumference generate side bands with frequencies of $\nu \times z_{Le} \times f_n$ as measured by strain gages mounted on the impeller shrouds. In this way the sidebands can excite natural frequencies of the impeller.

Some general trends of RSI - applicable to volute and diffuser pumps - emerge from many tests and numerical investigations, e.g. [6, 7, 45, 53, 90, 125, 130]:

- When analyzing pressure pulsations in the time domain, the blade passing causes pressure *dips or minima* at partload ($q^* < 0.8$), Fig. 10.25-1(b). In contrast, pressure *peaks or maxima* are observed above BEP (say $q^* > 1.1$ to 1.2), Fig. 10.25-1(a).
- Referring to Fig. 10.25-1, the difference in the shape of the curves p(t) between $q^* > 1$ (pressure peaks) and $q^* < 1$ (pressure minima) may be explained as follows: Low local counter pressure is found at high flow rates, hence at high radial velocity c_{3m} and high *negative incidence*. The stagnation point is on the suction surface of the diffuser vane leading edge, while a low-pressure zone is observed on the pressure surface, i.e. within the diffuser channel. As a consequence, cavitation can occur at this location as demonstrated by Fig. 5.17 where an unsteady cavity was observed at $q^* = 1.5$ whenever a blade passes under the

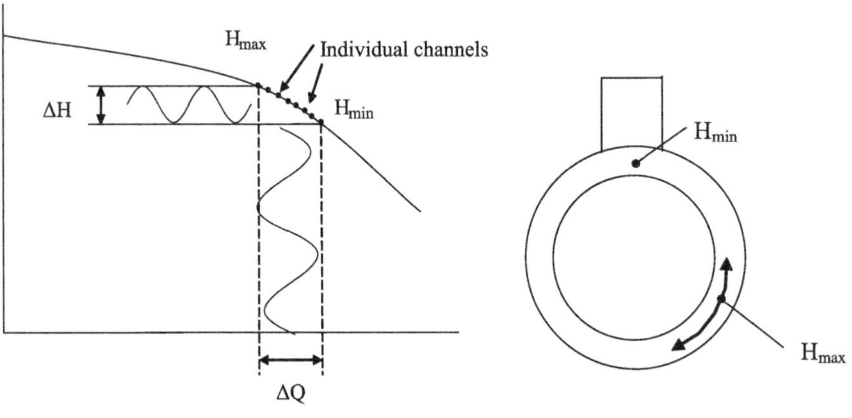

Fig. 10.25 Flow rate variations induced in individual impeller channels by an uneven pressure distribution in an annular casing

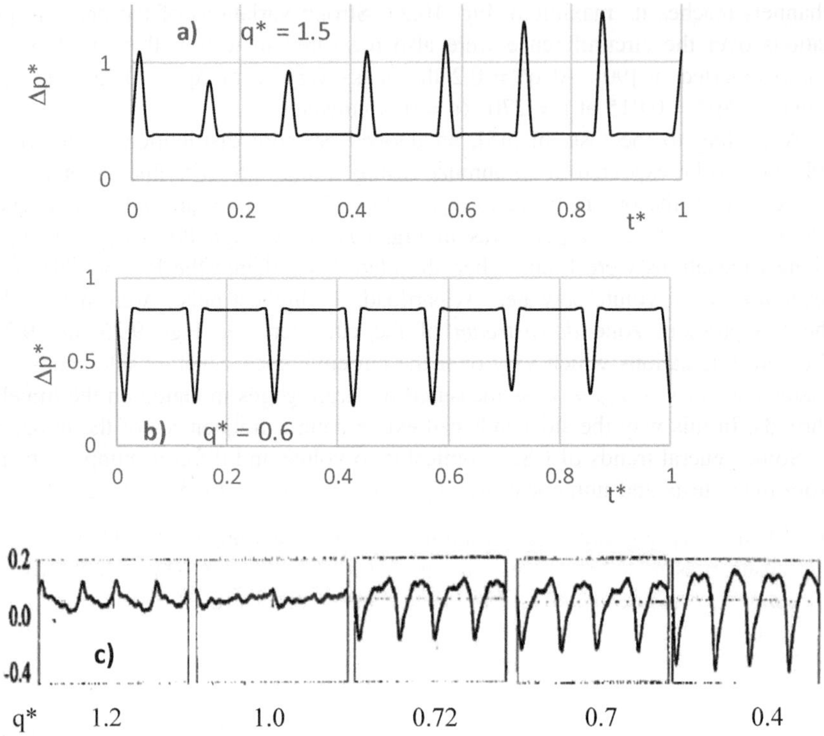

Fig. 10.25-1 Pressure pulsations measured in relative system on the impeller blade trailing edge: a) and b): pump with 7 diffuser vanes (one revolution, schematic). Bottom: pump with double volute, two revolutions, [53]). Note the pressure peaks at q* > 1 (negative incidence) and minima at q* < 1 with positive incidence

diffuser vane leading edge. Exactly the same phenomena were observed by high-speed camera in a volute pump: at q* = 1.17 unsteady cavitation occurred at the pressure side of the cutwater (i.e. within the discharge nozzle) at every passage of an impeller blade, see photo Fig. 6.45-1. Thereby the angle α_3 of the flow approaching the cutwater leading edge varied from about 20° to 75° [148]. In contrast to the negative incidence at high flows, *incidence is positive* at part load and the stagnation point is on the pressure surface of the diffuser vanes causing a low-pressure minimum on the suction surface, which is recorded when the pressure transducer mounted on the impeller blade passes the diffuser vane. Cavitation, if any, occurs then on the suction surface as evidenced by the erosion depicted in Figs. 6.43 and 6.44.

- Pressure fluctuations near BEP are low; the pressure minima are slightly more pronounced than the maxima, suggesting a small positive incidence - as one might expect for a design at BEP.

- At very low q^*, outlet recirculation and flow separation give rise to more chaotic flow patterns with more broadband pulsations. The effect of the incidence is less clearly seen.
- The pressure pulsations measured close to the impeller blade trailing edge are much higher than those measured in the discharge pipe presented in Fig. 10.9 (or in the suction nozzle).
- The highest amplitudes are expected at the impeller blade trailing edge on the *pressure side* because this point is closest to the leading edge of the diffuser vanes or volute cutwater. Hence unsteady velocities are highest at this location. Pulsations are lower on the suction side of the blade trailing edge.
- The pressure peaks tend to become sharper when the ratio d_3/d_2 is decreased.[5]
- Within the impeller channels, pulsations are lower than at the impeller outlet: in mid-channel pulsations are roughly half, and near the impeller throat about 1/4 to 1/5, of the values at the outlet.
- The difference in pressure pulsations between the impeller outlet and the impeller side rooms (ISR) grows when gap A is reduced and the overlap x_{ov} is enlarged. In [6] the pulsations in the ISR were typically 1/3 of those at the impeller trailing edge with gap $A/r_2 = 0.0085$.
- The highest pressure pulsations in the diffuser are observed on the leading edge of the vanes.
- Side bands with $\Delta f = f_n$ are caused by the non-uniform pressure in the volute or annular casing. Side band frequencies occur at $\nu \times z_{Le} \times f_n$ but most of the side bands are much lower than the main peaks; they often remain unnoticed.
- Any circumferential non-uniformity modulates the blade passing frequency. The non-uniformity of the pressure distribution in the volute or casing can be viewed as a wave with a wave-length corresponding to 360° (or 180° in a double-volute). Likewise, side bands at frequencies $\Delta f = f_n$ can also be caused by some perturbation which rotates with the impeller such as hydraulic imbalance due to casting tolerances or the impeller bore being slightly off-center.

Figure 10.25-2 shows measured pressure amplitudes (zero to peak) at blade passing frequency from various sources. Table D10.1-2 gives the details of the different tests. In order to allow a better comparison of tests at different ratios of d_3/d_2, all data are referred to $d_3/d_2 = 1.06$ according to Eq. (10.13c). This applies also to the tests in [90] with a cutwater distance $d_3/d_2 = 1.1$ (even though the pulsations were measured at $d_x/d_2 = 1.025$). The same normalization was used for radial force coefficient in Fig. 10.28-1.

[5]It may be noted that the tests in [125] were done under conditions which are not representative for high-head pumps which should not be designed with $d_3^* = 1.01$, vane combinations leading to $m = 1$ and usually are not required to operate at flows as high as $q^* > 1.5$.

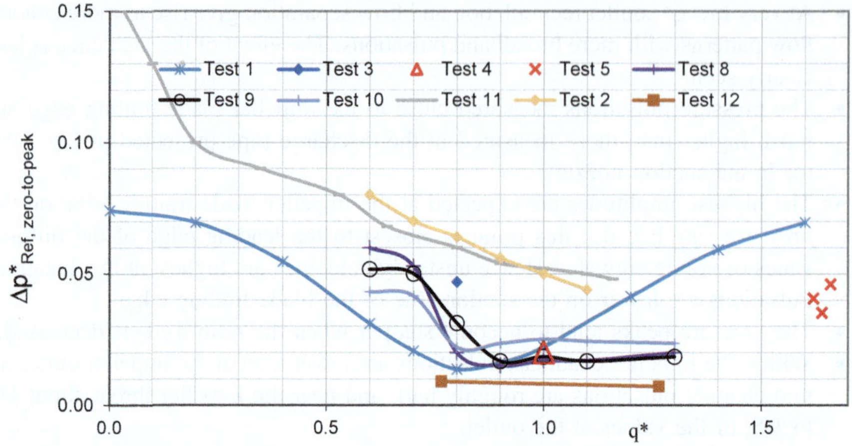

Fig. 10.25-2 Measured pressure pulsation amplitudes (zero-to-peak) at blade passing frequency, test conditions according to Table D10.1-2

Table D10.1-2 Test data to Fig. 10.25-2

	Stator	System	z_{La}	z_{Le}	d_3/d_2	n_q	Frequency	Lit.
Test 1	volute	stationary	7	1	1.1	25	$z_{La}*f_n$	90
Test 2	diffuser	rotating	7	12	1.058	22	$z_{Le}*f_n$	7
Test 3	diffuser	rotating	7	12	1.058	22	$z_{Le}*f_n$	6
Test 4	diffuser	stationary	7	12	1.058	22	$z_{La}*f_n$	6
Test 5	diff+vol	rotating	6	5	1.011		$z_{Le}*f_n$	125
Test 6	diff+vol	rotating	6	7	1.011		$z_{Le}*f_n$	125
Test 7	diff+vol	rotating	6	11	1.011		$z_{Le}*f_n$	125
Test 8	diffuser	stationary			1.058		$z_{La}*f_n$	67
Test 9	diffuser	stationary			1.028		$z_{La}*f_n$	67
Test 10	diffuser	stationary			1.044		$z_{La}*f_n$	67
Test 11	diffuser	stationary	7				$z_{La}*f_n$	67
Test 12	diff+vol	rotating	9	20	1.095		$z_{Le}*f_n$	14

$$\Delta p_{Ref} = \Delta p_{test}\left(\frac{d^*_{3,test}-1}{d^*_{3,Ref}-1}\right)^{0.77};\ k_{Ref} = k_{test}\left(\frac{d^*_{3,test}-1}{d^*_{3,Ref}-1}\right)^{0.77} \quad \text{with} \quad d_{3,Ref}* = 1.06 \tag{10.13c}$$

In roughly half of the tests, pulsations were measured at the impeller trailing edge, while the rest of the data were recorded in the stationary system (volute or diffuser). Figure 10.25-2 is intended to show trends and give an order of magnitude. The actual values depend on the specific geometry of the pumps tested as well as on the methods of data treatment by the individual authors.

The data in Fig. 10.25-2 can be compared to the measurements in Figs. 10.4, 10.21, 10.24 and 10.26, 10.27, 10.28, but the influence of the flow rate and the diffuser or cutwater distance must be taken into account when comparing different tests.

Influence of diffuser stall: The rotor/stator interaction increases as the flow at the impeller outlet becomes less uniform. That is why pressure pulsations (Fig. 10.9) and excitation forces (Figs. 10.40, 10.41, 10.42) strongly increase at partload. The same is true if one or more diffuser channels stall, which is the case in alternate stall or rotating stall. Measurements reported in [6, 7] well demonstrate the effect of stalled diffuser channels on the pressure pulsations - and the alternating stresses in the impeller. The pressure measured over one revolution in the rotating system near the impeller trailing edge is shown on Fig. 10.26, while Fig. 10.27 depicts the pressure recorded in the stationary system near the diffuser throat over 300 revolutions. On the left in Fig. 10.26, one diffuser channel is stalled as a consequence of which the peak-to-peak amplitude increases by a factor of about 2 in one of the 12 cycles given by the 12-vaned diffuser. On the right in Fig. 10.26, there are two stalled channels inducing two pressure fluctuations twice as large as on the un-stalled channels. As the pressure at the inlet to the stalled channel rises, the flow through the diffuser channel is reduced; the channel is partially or totally blocked by stalled fluid. *The flow through the impeller channel passing the stalled diffuser channel is reduced too* (see Figs. 10.23 and 10.25). The fluid displaced by the stalled diffuser channel is diverted to the neighboring channel where the velocity rises accordingly as given by continuity. Consequently, the local pressure drops at the inlet to the channel with increased through-flow. As demonstrated by Fig. 10.5, the incidence at the diffuser vane leading edge varies over the pitch inducing low-pressure peaks as shown in Figs. 10.26 and 10.27. It is also seen in Fig. 10.4 that a minimum of the instantaneous pressure coincides with a maximum of the velocity, and vice versa (compare curves for p_2 and c_{A1}). Flow observations using tufts (see also tests in Fig. 5.17) showed that the flow patterns vary over time from channel to channel. Figure 10.26-1 depicts two instantaneous flow patterns: in the left photo there is through-flow in the center channel, while the fluid stagnates in the right photo taken an instant later (unsteady stall).

Depending on the number of stalled diffuser channels z_s, additional pressure peaks are observed at $f = z_s \times (f_n - f_{RS})$ in the rotating system; i.e. each time when an impeller blade passes a stalled diffuser channel. Measuring the pressure fluctuations at any position in the stationary system, the peaks occur with a frequency f_{RS} if one stall cell is involved.

Figure 10.27 shows pressure measurements in the diffuser throat over 300 revolutions, while Fig. 10.28 depicts, as a function of the dimensionless time, the pressure in the throat, at the diffuser outlet and in the mid between those points.[6] These figures show:

---

[6]The dimensionless time is defined by $t^* = t \times f_n$.

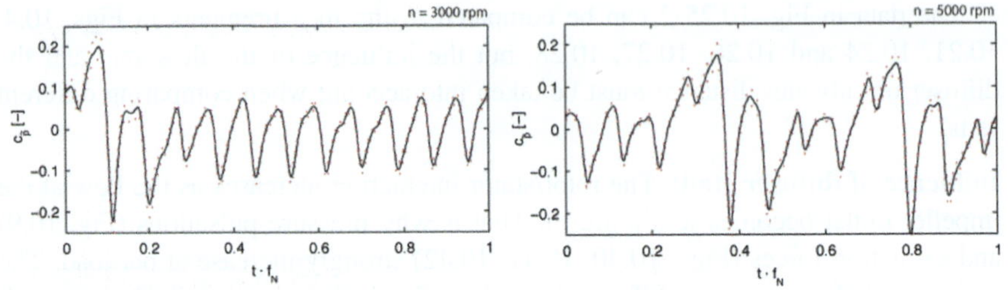

Fig. 10.26 Measured pressure pulsations on a rotating impeller, $z_{Le} = 12$, $q^* = 0.9$; $n_q = 22$; left: $n = 3000$ rpm; right: $n = 5000$ rpm, [6]

Fig. 10.26-1 Tufts in diffuser channel showing different flow patterns in adjacent channels, [138]

- Normal, less disturbed, flow is observed in domain "A" of Fig. 10.27, up to t* = 30 in Fig. 10.28 and for t* > 90.
- The high-pressure peak (domain "C" in Fig. 10.27 and t* = 70 in Fig. 10.28) corresponds to the stalled diffuser channel: there is virtually no through-flow (the channel is blocked) because the pressure difference between mid-channel and outlet is zero.
- The low-pressure peak (domain "B" in Fig. 10.27 and t* = 45 in Fig. 10.28) is an indication of a high flow rate through the diffuser channel because a high flow engenders a high pressure recovery.
- Since the diffuser outlet pressure is imposed by the annular exit chamber, the pressure at the channel outlet remains fairly constant; already in mid-channel the pressure fluctuations are only a fraction of those at the diffuser inlet. These findings agree well with the tests shown in Fig. 10.4.

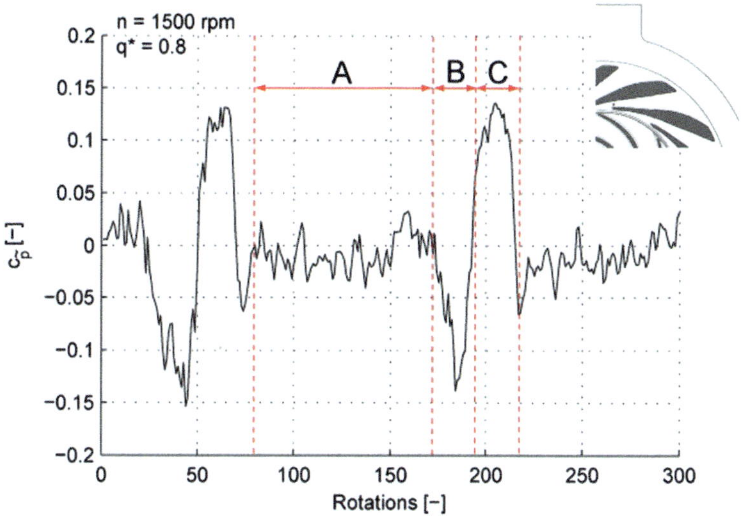

Fig. 10.27 Measured pressure pulsations in the diffuser throat, $z_{La} = 7$, $q^* = 0.8$; $n_q = 22$, [6]

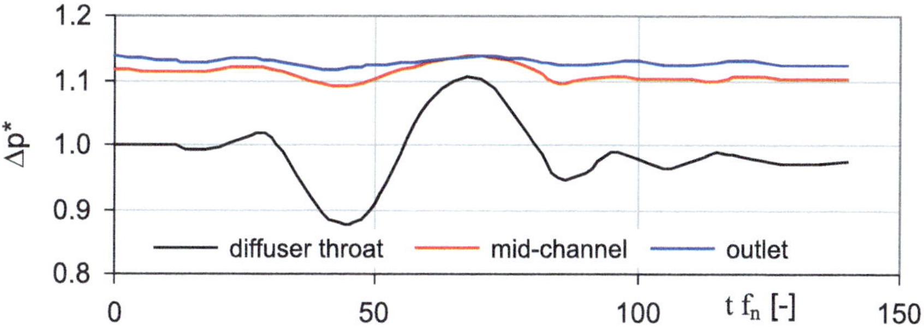

Fig. 10.28 Unsteady pressures in the diffuser during stall passage, $q^* = 0.8$; $n_q = 22$, [6]

- It is significant to note that these tests had been carried out on a single-stage pump with a diffuser discharging into an annular chamber followed by the discharge nozzle. In high-pressure pumps of this type, the annulus is quite narrow. As a consequence, the pressure distribution around the circumference varies significantly. As discussed in context with Fig. 10.23, the flow through the individual impeller channels is expected to vary around the circumference according to the local (counter-) pressure at the diffuser inlet. While this feedback is likely to reinforce the flow instabilities as well as the impeller stresses, the dominating effect is believed to be the diffuser stall. However, the observation that impeller fatigue failure is more likely in the last stage rather than in the series stages could be explained by the impact of the uneven pressure distribution in the annular casing on the momentary flow through the impeller channels and the unsteady flow phenomena thus reinforced.

Interestingly, in these tests, the number of stall cells depended on the rotor speed, Fig. 10.26. This could be an effect of the Reynolds number on the flow separation (somewhat less likely at $Re > 10^7$). As an alternative explanation, the flow through the impeller channels could have been influenced by pressure pulsations in the system (acoustic effects) as discussed above in context with Fig. 10.23: for example standing waves getting into resonance with the blade passing frequency. Still another hypothesis: even though the relative pressure differences in the pump are independent of the speed, the *absolute* pressure variation $p(\varepsilon)$ around the circumference of the annular casing increases with the square of the speed; this could induce more than one diffuser channel to stall as per the mechanism discussed in context with Fig. 10.25.

That stall occurs in one or two (rather than in all) diffuser channels may have at least two reasons: (1) The pressure distribution around the circumference of the annular casing is not constant; stall is likely to occur in the channel which experiences the highest counter pressure; (2) Because of fluctuations and geometric tolerance the flow conditions are not identical in all of the channels; hence stall is unlikely to occur in all channels at exactly the same flow rate. Once stall occurs in one channel, fluid is displaced to neighboring channels where the tendency to flow separation is consequently diminished.

The pressure peaks created by the presence of blocked diffuser channels induce alternating stresses in the impeller. In order to reduce the risk of early stall, the flow deceleration in the diffuser channels must not exceed (possibly be lower than) the limit stipulated by Eq. (1.44).

Radial forces due to RSI: Radial forces acting on the impeller at diffuser vane passing frequency have been measured in [125] with a radial impeller ($n_q \approx 30$) and with a semi-axial impeller ($n_q \approx 160$) in [131]. Strain gages on the shaft were used for these tests. The radial pump had 6 impeller blades combined with 5, 7 and 11 diffuser vanes. Yielding $m = 1$ according to Eq. (10.13), these vane number combinations generate high RSI-forces at vane passing frequency. Furthermore, the gap between impeller and diffuser was very small ($d_3/d_2 = 1.01$). For both of these reasons, high RSI-forces are to be expected. The measured forces were normalized to $d_3/d_2 = 1.06$ according to Eq. (10.13c); the results are presented in Fig. 10.28-1. These forces decrease when increasing the number of diffuser vanes - presumably, because the flow becomes more uniform as the pitch of the diffuser vanes gets smaller. This observation is confirmed by CFD studies in [130], where pressure pulsations in the combination $z_{La} = 6$; $z_{Le} = 9$ were lower than with $z_{La} = 6$; $z_{Le} = 7$. The RSI-forces of the radial stage given in Fig, 10.28-1 are rough estimates from the cascade diagrams reproduced in [125].

As cautioned above, these forces are high due to the unfavorable vane number combination and the small gap B. The curves for the $z_{La} = 6/z_{Le} = 11$ were $m = 1$ is reached in the 2nd order are below the curves where $m = 1$ is fulfilled in the 1st order. This type of stage would be totally unacceptable for a high-pressure pump, as may be inferred from a sample calculation: for a stage with $H_{opt,st} = 600$ m, $d_2 = 400$ mm, $n = 5200$ rpm the combination $z_{La} = 6$ with $z_{Le} = 11$ would yield at

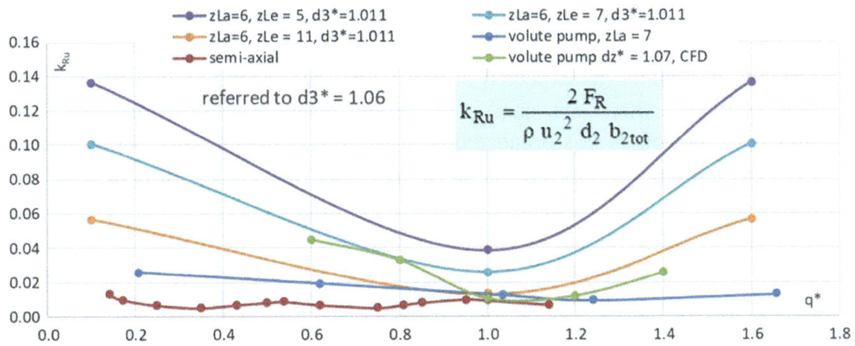

Fig. 10.28-1 Unsteady radial forces zero-to-peak at blade passing frequency, [125, 131, 135]. k_{Ru} according to Eq. (9.6a), referred to $d_3^* = 1.06$. Note: the 3 upper curves must only be used for vane combinations where m = 1 according to Eq. (10.13) is fulfilled

BEP an excitation force at blade passing frequency of roughly ±5000 N (zero to peak) and about double that value at q* = 0.6. Designing this stage with $d_3/d_2 = 1.06$ would reduce the force at BEP to about ±1300 N according to Fig. 10.6.

The RSI-forces of the semi-axial pump ($z_{La} = 6$; $z_{Le} = 7$) are an order of magnitude lower than those of the radial pump, [131]. The vane number combination is unfavorable, but the gap between impeller and diffuser was presumably large. The RSI-forces as a function of the flow rate exhibit maxima and minima which depended on the instability of the Q-H-curve: the *maximum head* of the unstable curve coincided with the *force minimum* at q* = 0.75 in Fig. 10.28-1. And the force maximum near q* = 0.5 matched with the head minimum. These extrema and their variations are caused by sudden changes in the velocity profile at the impeller outlet. These profiles trigger flow separation and back-flow in the diffuser. It can be surmised that the force minima are found where the velocity profile at the impeller outlet is more uniform than at the force maxima. Increasing the clearances between the impeller blades and the casing reduced the instability of the Q-H-curve and the RSI-forces, (similar to the test reported in Chap. 5.5.2). Impeller blade trailing edge and diffuser vane leading edge were presented as radial in the meridional section in [131]. It is noteworthy that the criterion of Eq. (10.13) seems applicable also to semi-axial and axial pumps. Figure 10.28-1 shows also tests with a volute pump [135]; it is the same pump as in [90].

The hydraulic excitation forces and the vibrations measured at the blade passing frequency are caused by different mechanisms whose presence and dominance depends on various design parameters:

a. Force F_w due to the wakes of the blades, which are always present due to the finite trailing edge thickness and the boundary layer blockage. The boundary layer thickness is expected to decrease with increasing flow rate (higher through-flow Reynolds number). F_w may decrease with Q.

b. Force F_{nu} due to the non-uniformity of the velocity distribution over the pitch, which increases at off-design conditions. If the flow distribution varies from channel to channel (as expected in part load), frequency components with $\nu_2 f_n$ with $\nu_2 = 1, 2, 3, \ldots$ etc. can occur. At off-design conditions the flow in the impeller channels is less organized or structured than near BEP. Consequently, F_{nu} is minimum near (or a little below) BEP and rises as the flow is reduced or increased.
c. Force $F_{m=1}$: With m = 1 according to Eq. (10.13) the individual blade forces have a non-zero resultant, this mechanism would be present even with infinitely thin blades and zero boundary layer thickness and even with perfectly uniform velocity distribution. The excitation forces are expected to increase with the blade loading - hence with the work transfer provided by the blades. If the power curve P = f(Q) rises with the flow rate (typically the case for pumps with $n_q < 80$), the blade loading and associated excitation forces at BPF will rise with the flow rate (notably at $q^* > 1$). $F_{m=1}$ may increase with Q.

Depending on the relative strength of these forces (F_w, F_{nu}, $F_{m=1}$) the amplitudes will vary in different ways with the flow rate. As shown in Chap. 9.3.9, the synchronous excitation forces in single-channel impellers increase with the square of the flow rate (above a minimum at $q^* = 0.6$). This observation can be explained by the blade loading increasing with the flow rate.

The pressure pulsations (Fig. 10.25-1) and the radial excitation forces (Fig. 10.28-1) at BPF were taken from measured spectra (as opposed to data from the time domain). Peaks in the time domain at BPF can be much higher than in the spectra. The reason for this difference is that the signal in time domain contains not only the BPF but also the harmonics of the BPF. This can be visualized by superposing three sine waves with, for example; 1 × BPF, 2 × BPF and 3 × BPF (in general terms superimposing different frequencies being integer multiples of a base frequency). The pressure fluctuations at the blade trailing edge calculated from measured impeller blade stresses in Fig. 14.7-2 are therefore much higher than those in Fig. 10.25-2. This should also be borne in mind when using excitation forces at BPF from Fig. 10.28-1 to estimate the forced response of bearing housing vibrations.

Phase resonance: If the phase velocity of the impeller/diffuser vane interference reaches or exceeds the speed of sound, intense pressure waves are radiated into the fluid. The phase velocity of the impeller pressure field is given by $c_{ph} = u_2 \times p_2/m$. Consequently, large values of p_2/m as per Eq. (10.14) should be avoided so as not to risk intense pressure pulsations due to phase resonance [10].

$$\frac{p_2}{m} = \frac{\nu_2 z_{La}}{|\nu_2 z_{La} - \nu_3 z_{Le}|} \tag{10.14}$$

Another criterion for phase resonance according to [17] is:

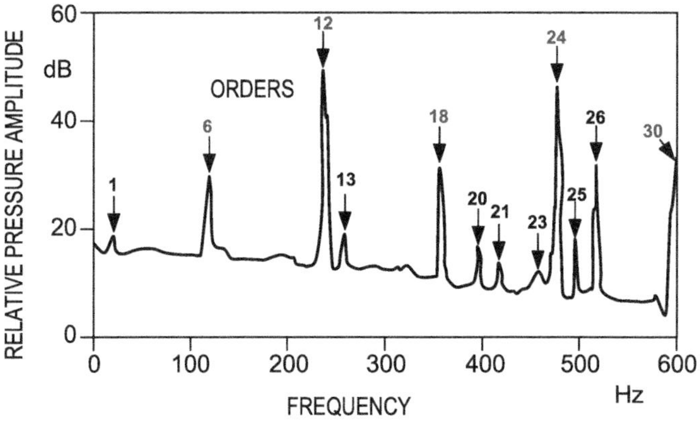

Order		ν_2	1	2	3	4	5	6	7
ν_3		$\nu_2{}^*z_{La}$	6	12	18	24	30	36	42
1	$\nu_3{}^*z_{Le}$	11	1.2	**12**	2.6	1.8	1.6	1.4	1.4
2		22	0.4	1.2	**4.5**	**12**	3.8	2.6	2.1
3		33	0.2	0.6	1.2	2.7	**10**	**12**	**4.7**
4		44	0.2	0.4	0.7	1.2	2.1	**4.5**	**21**

Fig. 10.29 Pressure pulsations measured in the discharge nozzle, $z_{La} = 6$, $z_{Le} = 11$. The higher the figure resulting from Eq. (10.14) the stronger is the phase resonance, [11]

$$\frac{z_{La}}{z_{Le}}\left\{\frac{z_{Le} - z_{La}}{z_{La}} \pm \frac{u_2 d^*_{vol}}{a \pm c_{vol}}\right\} = \frac{m}{\nu_2} \tag{10.15}$$

Phase resonance occurs if the pressure waves generated on two diffuser vanes by the passing impeller blades simultaneously arrive at the volute cutwater, in the discharge nozzle or at the measuring position. Increased pressure fluctuations will then occur at these locations. These pulsations have the frequency $\nu_2 \times z_{La} \times f_n$; i.e. they correspond to the order ν_2 of the blade passing frequency. The plus sign in Eq. (10.15) stands for phase resonance in the discharge nozzle, the minus sign for resonance at the volute cutwater. These relationships were derived for large storage pumps where a diffuser is arranged between impeller and volute. The quantity $d^*_{vol} = d_{vol}/d_2$ is representative of the distance traveled by a wave from the diffuser vane leading edge to the discharge nozzle or to the cutwater, d_{vol} is the mean volute diameter and c_{vol} the mean velocity (mostly negligible as compared to the speed of sound a).

Phase resonance is present when Eq. (10.15) is approximately satisfied for $m = 0$ or $m = \pm 1$; the integers $\nu_2 = 1, 2, 3$, describe the vibration order. Since the distances traveled by the waves and the speed of sound cannot be determined very accurately, Eq. (10.15) has to be evaluated for a certain range of uncertainty. This means that the left and right sides have to be situated in the band of ±0.02, for example. There may also be a kind of "lock-in" effect since the points of wave reflection in a complex geometry may vary according to the modes excited.

The excessive pressure pulsations measured in [17, 78, 80] could be explained by Eq. (10.15). However, phase resonance is likely to occur quite often without being noted if the amplitudes are of no consequence. An example is provided in Fig. 10.29 which shows pressure pulsations measured in the discharge nozzle of a pump with 6 impeller blades and 11 diffuser vanes. The tabled data below the figure were calculated according to the phase resonance criterion given by Eq. (10.14) which indeed predicts high amplitudes for the orders 12, 24 and 30. The 18th order is measured to be significantly lower (note the dB-scale!) which might be expected according to the data tabled below Fig. 10.29. Obviously, this quantitative tendency is to some extent coincidental because the actual amplitudes depend on the system response to each individual excitation frequency.

Phase resonance may sometimes be overlooked as a source of high pressure pulsations and noise because it occurs at the same frequencies as the rotor/stator interaction i.e. at $\nu_2 \times z_{La} \times f_n$. If phase resonance occurs, it should first be tried to reduce the rotor/stator interaction by the remedies given in Table 10.2 (notably the distance between rotor and stator) prior to changing the vane combination which would need new components and would thus be more costly.

Airborne noise of a small multistage pump ($H_{st} < 10$ m, $d_2 < 100$ mm) with a plastic casing and extremely thick blade trailing edges showed peaks in the spectrum at frequencies $\nu_2 \times z_{La} \times f_n$ at all of the orders $\nu_2 = 1$ to 12. The impeller had 6 blades combined with 7 diffuser vanes. The phase resonance criterion of Eq. (10.14) predicted the peaks at the orders $\nu_2 = 1$ to 7, while peaks at the orders $\nu_2 = 8$ to 12 would be expected according to Eq. (10.15).

In the pump industry, phase resonance has rarely been identified as the sole cause of high vibrations, component damage or noise. However, phase resonance may often explain some of the peaks in a spectrum. Unacceptably high excitation may be observed if other factors are contributing such as unusually high non-uniform flow at the impeller exit, acoustic resonances in the system or in the hydraulic channels of the pump, weak structures of the components (mechanical resonances) or high noise radiation capabilities.

In order to avoid or mitigate the effects of phase resonance, common multiples of the blade/vane combination as well as $z_{Le} - z_{La} = 1$ should be avoided when designing the impeller and diffuser. By a suitable selection of the vane count, the frequencies of phase resonance can be shifted to as high orders as possible with the goal to reduce the energy content at those frequencies.

Phase resonance has more often been reported as a source of high pressure pulsations and noise in water turbines and large storage pumps, [28]. As possible remedies (though not always effective) are quoted: (1) a hole in the volute cutwater with the object to mitigate the pulsations by wave interference; (2) increase compliance of fluid in volute by a hose filled with pressurized air wrapped over the circumference in the volute, [28]. However, a hole in the cutwater would imply some loss in efficiency.

Impeller vibrations: Forms of vibration with m > 2 are of little practical importance since the structural eigen frequencies are usually sufficiently high so that no resonances occur. An exception are the impellers of very large pumps or pump turbines (with typically d_2 > 2000 mm) and high heads, especially light-weight constructions, where even modes with more than two diameter nodes (m > 2) can cause problems [80]. To verify the risk of resonance between an impeller eigen frequency and the excitation frequency $\nu_3 \times z_{Le} \times f_n$, the natural frequencies must be calculated or measured. When doing so, the movement of the water induced through the impeller vibration - the "added fluid mass" - must not be neglected since it considerably reduces the natural frequencies. From a diagram in [80] it is possible to derive the relationship, Eq. (10.15a), for the ratio κ of the impeller eigen frequency in the pump to that in air:

$$\kappa \equiv \frac{f_{water}}{f_{air}} = 0.38 + 2.14\frac{s_{ax}}{d_2} \qquad (10.15a)$$

Equation (10.15a) applies to the vibration with two diameter nodes; with one diameter node κ is still about 35 % lower. The narrower the impeller sidewall gap s_{ax}, the greater is the added mass of water (and the lower is the natural frequency) since greater amplitudes are imposed on the fluid in the narrow gap through the deformation of the impeller. Strictly speaking, Eq. (10.15a) applies only to the geometries examined in [80], but it can be used for rough estimations in the absence of more precise information. From measurements in [45] the following ratio of the impeller natural frequencies κ in the pump and in air can be derived:

- without a diffuser (this corresponds to open impeller sidewall gaps) κ = 0.72; 0.67; 0.64 for m = 1; 2; 3 respectively (one to three diametrical nodes)
- with a diffuser (this corresponds to closed impeller sidewall gaps) κ = 0.58; 0.59; 0.56 for m = 1; 2; 3 respectively (one to three diametrical nodes)

Closing the impeller sidewall gaps by the diffuser side plates increased the added mass by restricting the liquid movement in the sidewall gaps. Physically the effect is similar to reducing the axial sidewall clearance. The model tested in [45] was a single stage pump with a diffuser and a volute (estimated n_q = 30). The impeller had 6 blades with β_{2B} = 32°; the distance between impeller and diffuser vanes was d_3^* = 1.01 only.

Alternating forces on the impeller and diffuser vanes can cause fatigue fractures of the blades or the impeller sidewalls. Inadequate impeller sidewall or blade thickness and a lack of casting quality aggravate this type of problem. The main cause of such fractures is frequently given by an insufficient distance between impeller and diffuser vanes (see Chap. 14.1 and Table 10.2) and/or resonances between an excitation frequency and a structural eigen frequency. The selection of a material with insufficient fatigue strength is another possible cause as discussed in Chap. 14.

10.7.2 Rotating Stall (RS)

Rotating stall in the impeller or diffuser excites vibrations at discrete frequencies below the rotational frequency. The mechanism leading to rotating stall shall be explained with reference to Fig. 10.30: Suppose the cascade presented there operates with large incidence (or great deceleration) at the limit where local separations occur. If the flow first separates on blade "A" because of the local incidence exceeding the mean value due to irregularities in the flow or the geometry, the stalled fluid will block a part of the channel cross section. This blockage displaces fluid to the two adjacent channels. On blade "B" this enlarges the incidence so that the flow will now separate on blade "B". On blade "C", however, the incidence is reduced and the tendency for stall drops accordingly. In this way the stalling zone shifts against the direction of rotation at a fraction of the approach flow velocity (w_1 at the impeller, c_3 at the diffuser). Consequently, the perturbation rotates in the direction of rotation in the absolute system. The frequency ratios of rotating stall f_{RS} generated by impeller stall are in the range of $f_{RS}/f_n = 0.5$ to 0.9, those induced by diffuser stall are $f_{RS}/f_n = 0.1$ to 0.25 of the rotor rotational frequency f_n, [48].

Recent experiments show that rotating stall does also occur at low frequencies and that the number of stall cells and propagation frequencies depend on the flow rate ratio q*. Some examples are given below; they include closed radial and semi-axial impellers as well as open impellers:

1. In a pump turbine ($n_q = 30$) with $z_{La} = 9$, $z_{Le} = 20$ measured rotating stall occurred at $q^* = 0.42$ to 0.78 with frequency ratios of $f_{RS}/f_n = 0.014$ to 0.028 and the number of stall cells varying between 3 and 5, [14].
2. In a pump with $z_{La} = 7$, $z_{Le} = 12$ measured rotating stall occurred between $q^* = 0.6$ to 0.8 with $f_{RS}/f_n = 0.007$ and the number of stall cells varied between 1 and 3, [6]. Figure 10.31 shows the pressure pulsations measured on the rotating impeller. At $q^* = 1$, the 12 pressure peaks (of uniform amplitude) caused by the 12 diffuser vanes are found in every rotation measured at the same circumferential position. In contrast, one stall cell with a higher peak is detected at $q^* = 0.8$. This peak shifts by about 180° during 70 rotations giving a frequency of the rotating stall of $f_{RS}/f_n = 0.5/70 = 0.007$. The passage of the stall cell through the diffuser can be seen in Fig. 10.32 where the pressure was

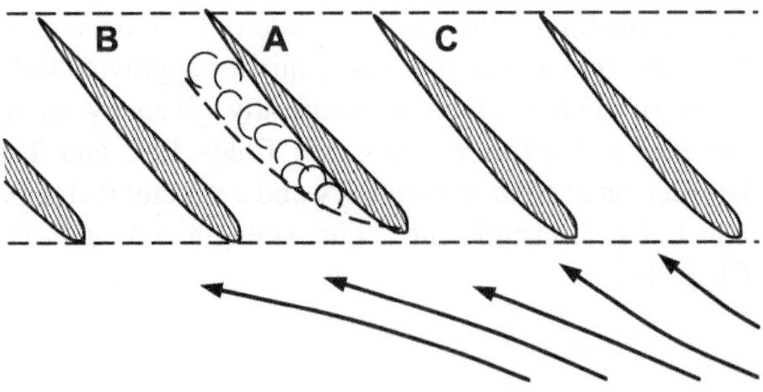

Fig. 10.30 Rotating stall

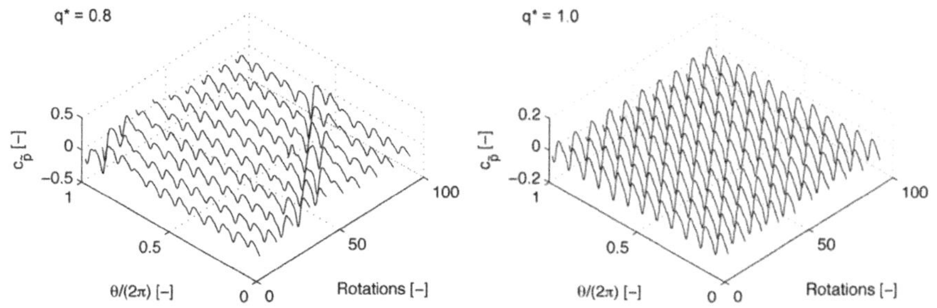

Fig. 10.31 Propagation of a stall cell through the diffuser measured in the rotating impeller; left: $q^* = 0.8$, one stall cell; right: $q^* = 1$, no stalled diffuser channel; $z_{Le} = 12$, [6]

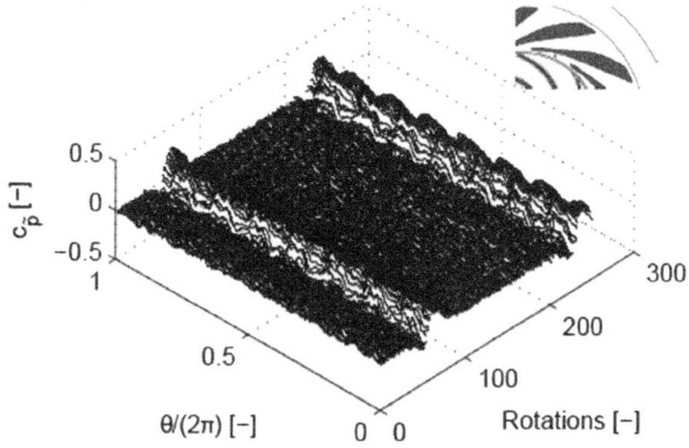

Fig. 10.32 Propagation of a stall cell through the diffuser; measured at the diffuser inlet; $q^* = 0.8$, [6]

recorded at the diffuser inlet. A pressure peak with a width of about 20 revolutions moves around the circumference in about 140 revolutions. In these investigations, the number of stall cells was found to depend also on the speed.

3. In a pump with $z_{La} = 7$, $z_{Le} = 8$ measured rotating stall occurred with 2 and 3 stall cells at $f_{RS}/f_n = -0.008$ to 0.02, [72]. The onset and type of stall depended on the flow rate and the ratio d_3/d_2 which was varied between 1.035 and 1.1.
4. In a semi-axial pump $n_q = 160$ with $z_{La} = 6$, $z_{Le} = 7$, rotating stall was observed in the unstable region of the Q-H-curve, [131]. Three types of stall occurred: (1) Slightly below head maximum at $q^* = 0.7$, rotating stall started at a frequency slightly below f_n. This type of stall vanished near the head minimum at $q^* = 0.5$. (2) Below this flow rate, another stall rotated near $2 \times f_n$; it vanished at $q^* = 0.3$. (3) There, one more stall occurs at $3 \times f_n$. All of these frequencies were not at constant ratios to the rotational speed but decreased with the flow rate.
5. In a semi-axial pump with open impeller, $n_q = 90$ with $z_{La} = 5$, $z_{Le} = 9$, rotating stall was observed in the diffuser, [132]. The stall occurred in the unstable region of the Q-H-curve ($q^* = 0.68$ to 0.82). There were two stall cells; propagation frequency was $f_{RS} = 0.42$ Hz or $f_{RS}/f_n = 0.042$.

6. In a semi-axial pump with closed impeller ($n_q = 43$, $z_{La} = 6$, $z_{Le} = 13$ and $z_{Le} = 7$) rotating stall occurred in the diffuser. The Q-H-curve was unstable with a hysteresis. The pressure fluctuation reached amplitudes of $\psi_s = \pm 0.2$ at $q^* = 0.5$. At $q^* = 0.62$ there was one stall cell with a stall progression speed of $f_{RS}/f_n = 0.083$. At $q^* = 0.3$ two stall cells were observed with $f_{RS}/f_n = 0.133$. There were small side bands around the BPF, [133, 134].
7. Rotating stall (RS) in the turbine brake mode was analyzed by measuring pressure fluctuations between impeller and guide wheel and by flow visualization done by tufts of thin fluorescent wires (0.14 mm dia.), [136]. A mathematical model was then developed to detect RS by analyzing the signals of accelerometers which were mounted on the protuding shaft of the adjustable guide vanes. This technique can be used for diagnosis during operation without intrusion into the pressure boundary. The rotating stall in the guide wheel occurred at $f_{RS}/f_n = 0.7$. The pressure fluctuations caused by the rotating stall allowed to observe four phases during one RS-cyle: (1) normal through-flow through the guide wheel was observed during pressure increase; (2) flow separation and blockage of the channel occurred near the pressure maximum - as described in context with Figs. 10.26, 10.27, 10.28; (3) during pressure decrease back-flow develops; (4) the channel blocks again when backflow ceases near the pressure minimum. Subsequently, the cycle starts with normal through-flow as pressure rises again. The stall-caused pressure fluctuations modulate the blade passing frequency causing side bands at $\pm 0.7 \times f_n$ at $1 \times z_{La} \times f_n$ and at $2 \times z_{La} \times f_n$. Interestingly, the vibrations were more severe during inception of rotating stall than with fully developed (RS), while the *opposite* is true for the pressure fluctuations. Again we see that vibrations do not necessarily increase with the pressure pulsations (there is no proportionality): the excitation forces result from the integration of the pressure distribution over the vane, while the pressure pulsations are a local phenomenon.

Rotating stall is often linked to flat or unstable Q-H-curves which should be precluded by design; operation in this range must be avoided. Incidents where rotating stall at the low frequencies described above manifestly lead to problems have not been found. The stall as such may present a problem if causing excessive pressure pulsations, but the fact that it is slowly moving around is not. The rotating stall may then remain undetected.

While incidence is the classical explanation for the origin of rotating stall, it is suggested that excessive deceleration in the diffuser channels can induce the same phenomena.

As discussed extensively in Chap. 5, partload flow separation does occur by necessity in every pump somewhere between BEP and shut-off. This is equally true for pumps with or without vibration problems and with or without unstable Q-H-curves. Furthermore, due to geometric tolerances, flow fluctuations and flow asymmetries, flow separation is unlikely to start *simultaneously* at the same flow rate in all of the parallel channels of a component. It is also likely that flow patterns are not exactly the same in all parallel channels in fully stalled conditions.

Therefore, it may be postulated *that alternating or rotating stall is the rule rather than the exception in vaned diffuser pumps.*

Obviously, pumps with vaned diffusers are more sensitive to rotating and alternate stall than pumps with single volutes. Vibration spectra of any kind show many peaks which depend on the flow rate and vary with the measuring location in a largely unexplained way. Unless any of these peaks exceed allowable limits they are usually overlooked since they do no harm.

The flow patterns in the impeller channels are strongly influenced by centrifugal and Coriolis forces, while flow separation in a diffuser is governed by boundary layer effects and the upstream velocity profile (which amounts to the same). Therefore, rotating and alternate stall are more likely to be observed in the diffuser than in the impeller.

Rotating stall is not necessarily linked to large flow rate fluctuations, but these can be generated if the stall is in resonance with acoustic waves in the system (which frequently happens in compressors). On compressors, with their large number of blades, rotating stall nearly always occurs close to the maximum of the Q-H curve. This represents the "surge limit" of the compressor. Most pump impellers have fewer blades and incidents of rotating stall were rarely reported. Sometimes they may be confused with rotor instabilities, when observed as sub-synchronous vibration near the operating speed. An example is reported in [62]: resonance between broad-band excitation forces and the rotor eigen frequency was observed on a feed pump in slightly over-critical operation. The excitation, which occurred even close to the best efficiency point, was caused by the recirculation at the impeller inlet. The vibration frequency was at a fixed ratio with the rotational frequency. To rectify this situation, a "recirculation brake" was installed at the impeller inlet. To this end, the case ring was radially extended inwards and furnished with axial slots. In this way, the circumferential component of the recirculating fluid was reduced and the vibration excitation was largely suppressed.

Incidents of rotating stall similar to the type described in Fig. 10.30 presumably occur more frequently in cavitating between-bearing pumps where a surge-like noise is sometimes observed. If the impeller is exposed to non-uniform inlet flow, the size of the cavitation zones varies over the circumference. The periodic fluctuation of the cavities causes a low-frequency modulation of loud cavitation noise.

10.7.3 Various Hydraulic Excitation Mechanisms

Apart from the mechanism discussed above there are several other phenomena which generate hydraulic excitation forces. These are discussed subsequently by way of case histories if available.

Cavitation in impeller and inlet casing: Extensive cavitation generates (broadband) pressure pulsations which can excite vibrations and noise, as discussed under item 1, Chap. 6.5.1. All types of pumps can be affected. The range where broadband pulsations and noise can occur varies widely; incidents observed include: 600 to 800 Hz; 1000 to 2000 Hz; 3000 to 4000 Hz; 6000 to 8000 Hz. The frequency range may depend on: (1) hydraulic design (vapor generation and local pressures driving bubble implosion); (2) water temperature and gas content; (3) structural eigen frequencies (i.e. response of the casing to the excitation).

Strong cavitation-induced pressure pulsations of a few Hertz can be generated by flow recirculation at the impeller inlet. Structural vibrations can be excited.

Rotating cavitation zones can cause *super-synchronous* radial excitation forces at $f/f_n = 1.1$ to 1.2, which are to be distinguished from rotating stall, [15, 81]. The phenomenon was observed with inducers when operating at cavitation coefficients where some head drop is induced by extended cavities.

Case history semi-axial pump: Impeller cavitation, at high flow in a choking situation, caused high peaks in the bearing housing vibrations at BPF and 2xBPF in a large semi-axial pump (2000 kW). Presumably the cavities caused large wakes which increased the rotor-stator interaction and may have caused the 2xBPF peaks. When the problem of the impeller cavitation was resolved, these peaks went down to normal/acceptable levels.

Case history pipeline pump: single-stage, double-entry crude oil pump, [42, 91]. The mechanical seals suffered endemic damage due to pressure pulsations in the range of 1200 to 1800 Hz caused by extensive cavitation. The 11-MW pump conveyed at 3600 rpm $Q_{opt} = 1.85\ m^3/s$ at $H_{opt} = 616$ m ($n_q = 28$). Stroboscopic observations revealed on the impeller blades large cavity volumes which would have caused heavy cavitation erosion in water, but produced no damage in crude oil. At flows above the BEP, cavitation vortexes and bubbles were observed at the trailing edge of the rib in the inlet casing and at the impeller neck ring, see sketch in Fig. 10.32-1. Sharp corners were responsible for these bubbles. Removing these sharp edges and profiling the formerly blunt rib, as depicted in Fig. 10.32-2, brought about a drastic reduction in vibration levels. This improvement was verified by testing the pumps on site prior and after the casing modification, Fig. 10.32-3. A new impeller, designed according to "concept L" replaced the baseline impeller (concept "H") as per Fig. 7.10; the original and re-designed blade leading edges can also be seen in Fig. 10.32-2. Pressure pulsations, fluid-borne cavitation noise, solid-borne cavitation noise and radial excitation forces of a model pump, all demonstrated a consistent reduction of the signals when comparing the modified to the baseline hydraulic. The radial excitation forces measured on the bearing brackets increase with growing bubble volume (i.e. a reduction in the cavitation coefficient), Fig. 10.32-4.

Cavitation in diffusers or volutes: Extensive cavitation in the volute or diffuser can also cause excessive vibrations and abnormal noise. Such problems were encountered in a multistage pump where cavitation in the double-volute of the 1st stage occurred (pump design similar to Fig. 2.11). Due to casting problems, the throat areas were appreciably below design. Even at BEP, an inlet pressure

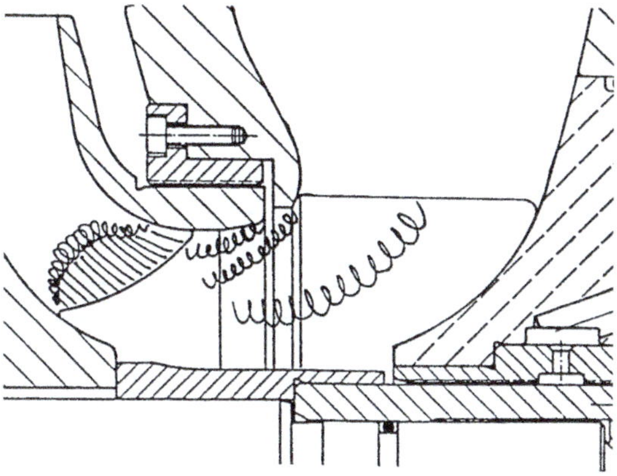

Fig. 10.32-1 Vortexes and bubble extension observed on baseline hydraulic, [91]

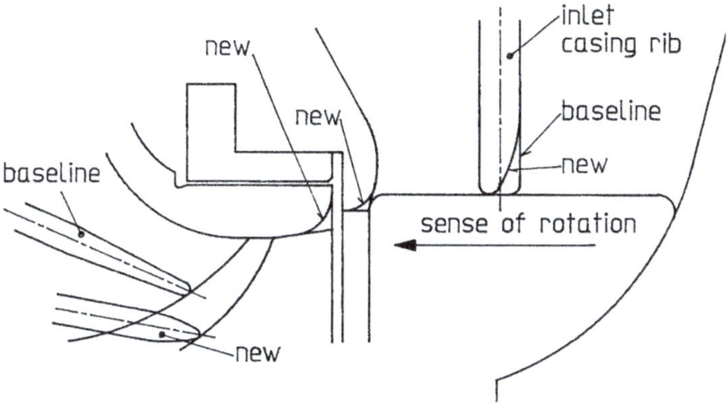

Fig. 10.32-2 Casing and impeller modifications compared with baseline hydraulic, [91]

Fig. 10.32-3 Influence of casing modifications measured on site with new impeller at 3600 rpm and $\sigma_A = 0.45$, [91]

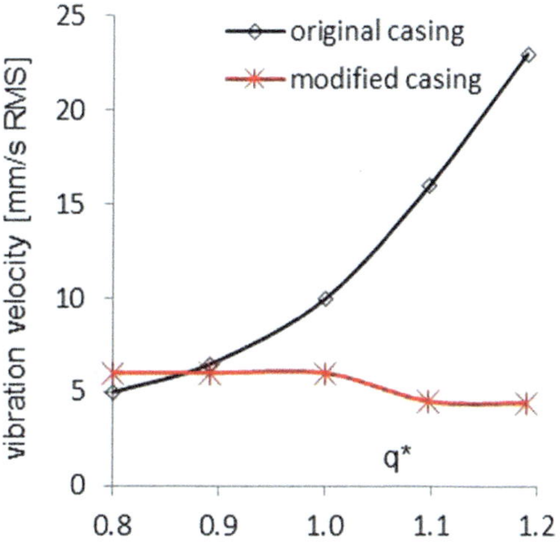

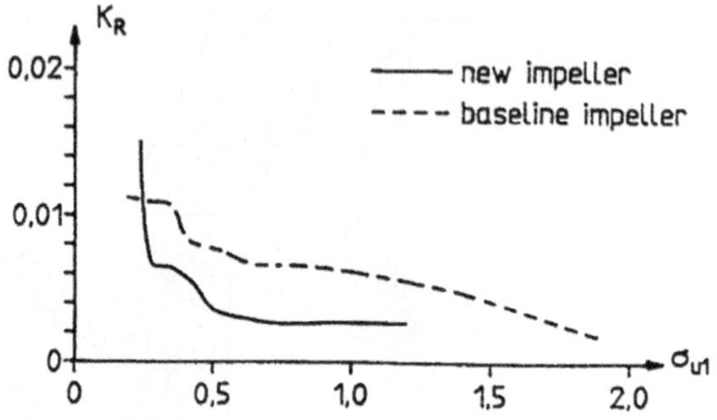

Fig. 10.32-4 Influence of cavitation on the radial excitation forces, [91]

equivalent to $\sigma_A = 3.5$ was necessary to suppress the abnormal noise and reduce vibrations below the specified limits. The cavitation noise was measured with an accelerometer as solid-borne noise on the casing; the records showed broad band noise in the range of 2500 to 6000 Hz. Bearing housing vibration velocity at blade passing frequency was also increased when cavitation was present even though the cavitation was in the volute throats. The severe cavitation in the volute was explained by a strong acceleration of the fluid, which was estimated to $c_{3q}/c_2 = 1.8$ to 2 at $q^* = 1.3$; compare to Fig. 6.15 where strong cavitation in the diffuser was present when $c_{3q}/c_2 > 1.1$ was exceeded.

Non-uniform approach flow: Non-uniform approach flow and vortex ropes are prone to excite vibrations including column pipe vibrations of vertical pumps. The associated flow phenomena are discussed in Chap. 11.7.3. Vortex ropes induce vibrations at blade passing frequency since each cutting of a blade through a vortex induces a shock loading on the blade; Fig. 11.21 may visualize this. As observed in some pumping stations with large intake structures, the generation of vortexes may occur in *irregular intervals* which may take several minutes (up to one hour!). Vibrations than may increase drastically when the vortex passes the impeller, while amplitudes are on an acceptable level after the event. Air accumulation in the pump or the system may produce similar behavior which is difficult to detect since it may take sometimes hours or days for the perturbation to build up. Vertical-pump vibrations are discussed in Chap. 10.14.

Partload recirculation: Partload recirculation at the impeller inlet and outlet is a frequent cause of high vibrations due to the strong interactions between rotor and stator. The excitation is to a large extent broadband/stochastic and vibrations are mostly below the rotational frequency. But also vibrations at blade passing frequency use to increase at partload. At low partload, broadband excitation in the low-frequency range grows strongly, but the increase is of a gradual rather than a sudden nature. As an example, Fig. 10.33 shows the overall RMS vibration velocities measured with two impellers. One impeller had a very large inlet diameter with an approach flow angle as low as $\beta_1 = 7°$, while the second impeller had a smaller d_1 yielding $\beta_1 = 13°$. Installing the impeller with the smaller inlet diameter reduced the vibrations at $q^* < 0.5$ by roughly 50%. The level of excitation

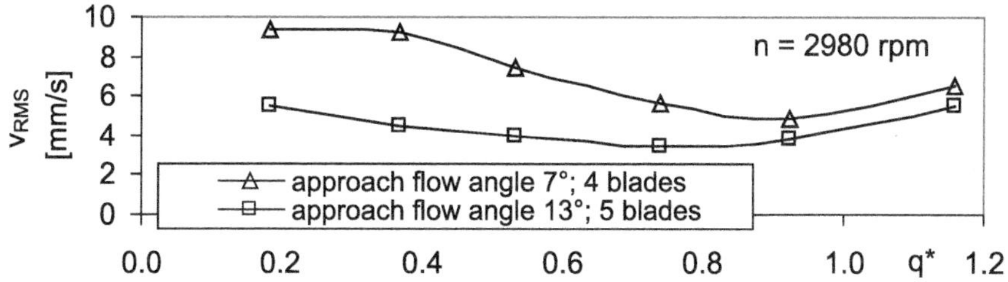

Fig. 10.33 Effect of impeller inlet recirculation on the vibration velocities of a single stage volute pump with double-entry impeller, $n_q = 20$, $P_{opt} \approx 300$ kW, $H_{opt} \approx 150$ m

forces created by partload recirculation can be very high indeed. This because the whole recirculation power P_{Rec} is dissipated into turbulence (and finally into heat). The recirculation power of a large pump can reach several megawatt near minimum flow; refer to Eq. (10.17) and Fig. 10.50 for estimations.

The strong rotor/stator interaction induced by partload recirculation can cause excessive discharge pressure pulsations, axial vibrations and subsequent damage to shaft seals and bearings if there is a strong coupling between the main flow and the flow in the impeller sidewall gaps. Case histories to this type of problem are documented in [23, 61]. In one case, a single stage volute pump with double-entry impeller ($n_q = 31$, $H_{opt} = 300$ m) suffered high axial vibrations and damage to the mechanical seals and bearings, [23]. Pressure pulsations below $q^* = 0.6$ exceeded 10 bar peak-to-peak. When gap A was reduced to $A/r_2 = 0.006$, the pressure pulsations dropped to 1.6 bar peak-to-peak and shaft seals and bearings performed properly. Pulsating and cavitating vortices due to partlod recirculation are discussed in Chap. 6.5.1, [87].

Vibration problems with a single-stage pump ($n_q = 35$) with double-entry impeller were reported in [140]. Measured bearing housing vibrations exceeded the specified limits below $q^* = 0.9$. Vibrations were at blade passing frequency and broad-band at low flow due to inlet recirculation. Three reasons contributed: (1) The impeller had 6 unstaggered blades and the double volute had a spacing of 180° between cutwater and rib. High pressure pulsations and vibrations at BPF occurred as a consequence of this design, Eq. (10.13). (2) The impeller was trimmed to 88% of the nominal impeller diameter. According to Fig. 4.26 the onset of inlet recirculation is expected close to the BEP of the trimmed pump: it shifted from about to $Q_{RB}/Q_{opt} = 0.65$ with full d_2 to $Q_{RB}/Q_{opt} = 0.8$ with the trimmed impeller. (3) The impeller inlet diameter was oversized in order to achieve a high suction specific speed ($n_{ss} = 235$). Two remedies were implemented: (1) The cutwater leading edge was cut oblique by 30° as per Table 10.2 and the rib of the double volute was cut back to reduce the wrap angle from 180° to 168° (cut also oblique by 30°). In this way the vibrations at BPF were effectively reduced. (2) "Bull rings" were mounted at the impeller inlet to reduce the effect of inlet recirculation. Initially, these rings were designed as sharp-edged orifices according to Fig. 6.39b. While these rings helped at partload below q* = 0.75, vibrations increased at high flow due to strong cavitation: at q* = 1.2 the $NPSH_3$ increased by a factor of 1.75 (i.e. from 4 to 7 m). As a remedy the bull rings were slotted in order to reduce by 43% the blockage effected by the rings, and the leading edge was chamfered by 30°, see photo Fig. 6.49. As a consequence, vibrations were down below the allowable and $NPSH_3$ at q* = 1.2 was back to almost the original value while NPSH was reduced by 40% at q* = 0.5.

Alternating stall: Alternating stall in the channels of double-volutes causes unsteady radial forces on the impeller, see also Chap. 5.8. Such forces can induce damage to annular seals, mechanical seals or bearings. As an example, Fig. 10.34 shows the damage to a bearing and the shaft orbits measured on a pump with double-entry impeller and a 180°-double-volute. The peak-to-peak amplitudes attain 130 μm at the rotational frequency. This relatively high frequency should be appreciated in context with the flow variations in the impeller channels as they rotate through a non-uniform pressure field in the casing, see discussion of Fig. 10.23.

Unstable Q-H-curves and axial force reversal: Increased vibrations are to be expected in the unstable region of a Q-H-curve. Unstable Q-H-curves can cause high axial rotor and bearing housing vibrations on single-stage pumps with double-entry impeller - in particular when axial thrust reversal is observed. Figure 10.35 shows the unstable Q-H-curve of a high-head pump, the axial thrust and shaft as well as lateral and axial bearing housing vibrations. Above q* = 0.8 the axial force is close to zero (as expected in a pump with double-entry impeller) and vibrations are low. Most likely, flow separation in the diffuser starts near q* = 0.79 causing the vibrations to rise significantly. The Q-H-curve becomes unstable at

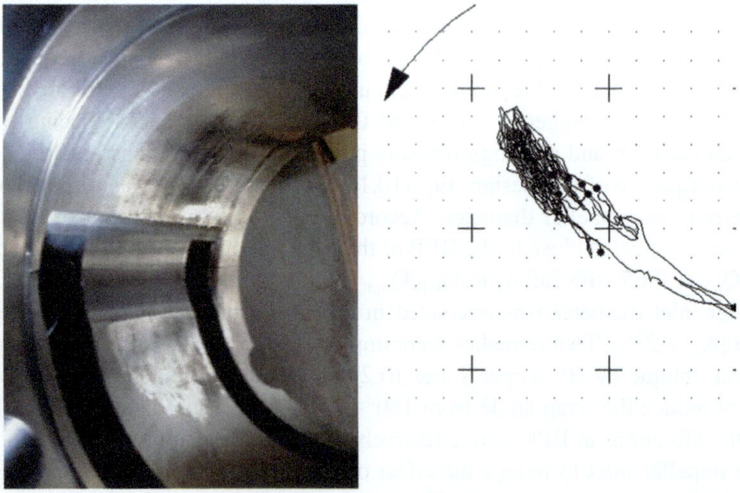

Fig. 10.34 Bearing damage and shaft orbits at q* = 0.5; speed 1780 rpm

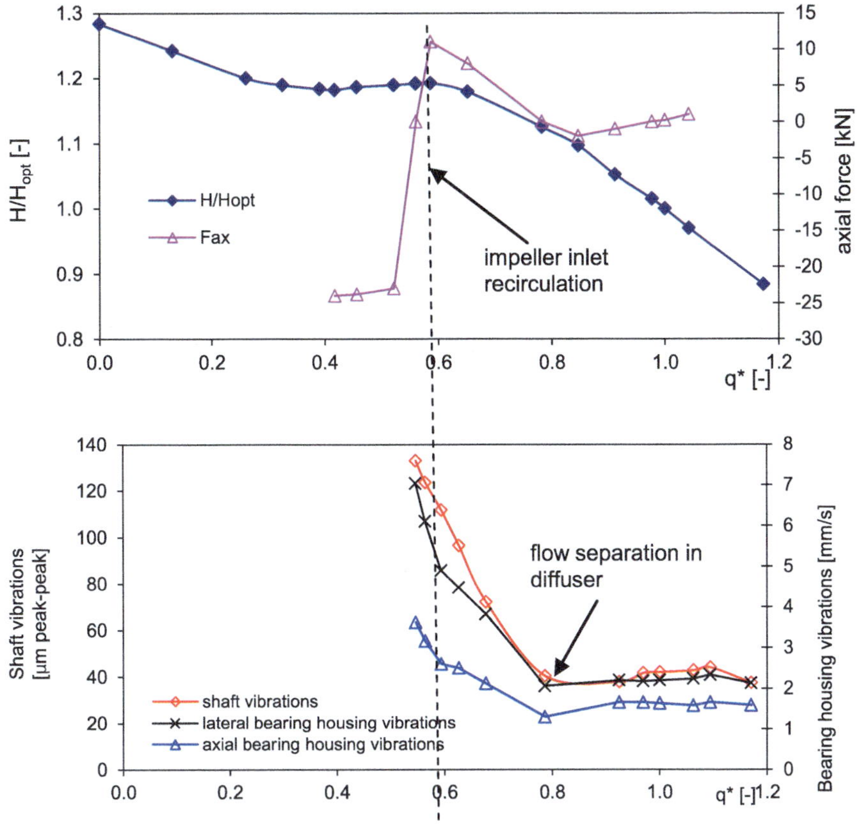

Fig. 10.35 Single-stage pump with double-entry impeller, vaned diffuser and volute; n = 4900 rpm. Vibrations amplitudes are referred to BEP averages of DE and NDE horizontal and vertical

q* = 0.59 - presumably at the onset of impeller inlet recirculation. This induces a reversal of the axial force and vibrations rise more steeply in the unstable range.

The flow mechanisms likely to be responsible for the axial thrust reversal are depicted in Fig. 10.36. At high flow (Fig. HF) the velocity distribution at the impeller outlet is quite uniform. This implies that the circumferential velocity component at the inlet to the ISR $k_E = c_{2u}/u_2$ is nearly the same on both sides of the impeller. The axial force is small; it is due also to a slight difference in annular seal diameters on both sides of the impeller. When flow separation is present in the diffuser, the velocity distribution at the impeller outlet becomes skewed, Fig. DS. The inlet conditions to the ISR (as given by k_E) are now different on both sides. As a consequence, the pressure distributions in both of the ISR are modified: on the shroud with the lower c_{2u} (side NDE in Fig. 10.36) the pressure distribution becomes flatter, while it becomes steeper on side DE with the higher k_E. The axial force acting in the direction of DE increases. The impeller side with the higher c_{2u} effects the higher deceleration of the relative flow velocity; this side can be said "to

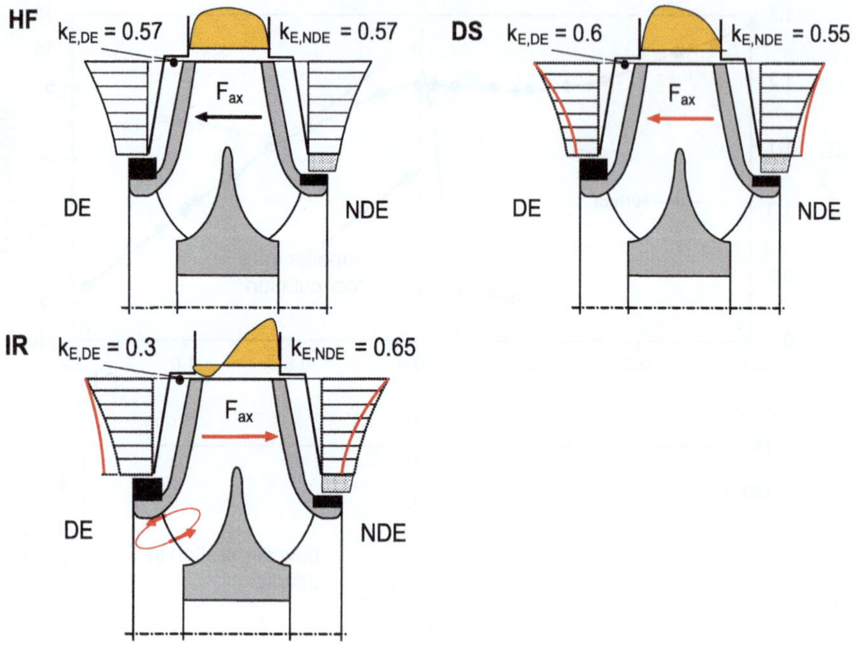

Fig. 10.36 Flow mechanisms causing axial thrust reversal in Fig. 10.35. *HF*: high flow, *DS*: diffuser separation; *IR*: impeller inlet recirculation

work more at partload". Hence inlet flow recirculation may be expected to first start on this same side.

When flow separation starts in the diffuser, it first appears in one or two channels, Fig. 10.27: for reasons of asymmetries in the flow as well as of geometric deviations, stall does not appear in all channels simultaneously. Likewise impeller inlet recirculation is not expected to occur simultaneously on both sides of a double-entry impeller; it is likely to start on the side with the higher flow deceleration and/or larger inlet angles and throat areas A_{1q}. The onset of impeller inlet recirculation changes the velocity profile at the impeller outlet, Fig. 10.36 (IR). Back-flow from the diffuser is induced and k_E drops to very low values. The pressure distribution on the side with recirculation becomes quite flat. The axial force is then reversed to the opposite side as shown in Fig. 10.36 (IR).

Note that the described mechanism of axial thrust reversal is not necessarily linked to an unstable Q-H-curve since flow separation and recirculation occur by necessity in all pumps (Chap. 5). The static pressure at the impeller outlet may also vary slightly over the outlet width. Any such differences will influence the resulting axial force in addition to the effect of the velocity distribution.

Figure 10.37 shows spectra of shaft and bearing housing vibrations measured on NDE at $q^* = 0.56$ (test in Fig. 10.35). An analysis of the frequency behavior of this pump reveals:

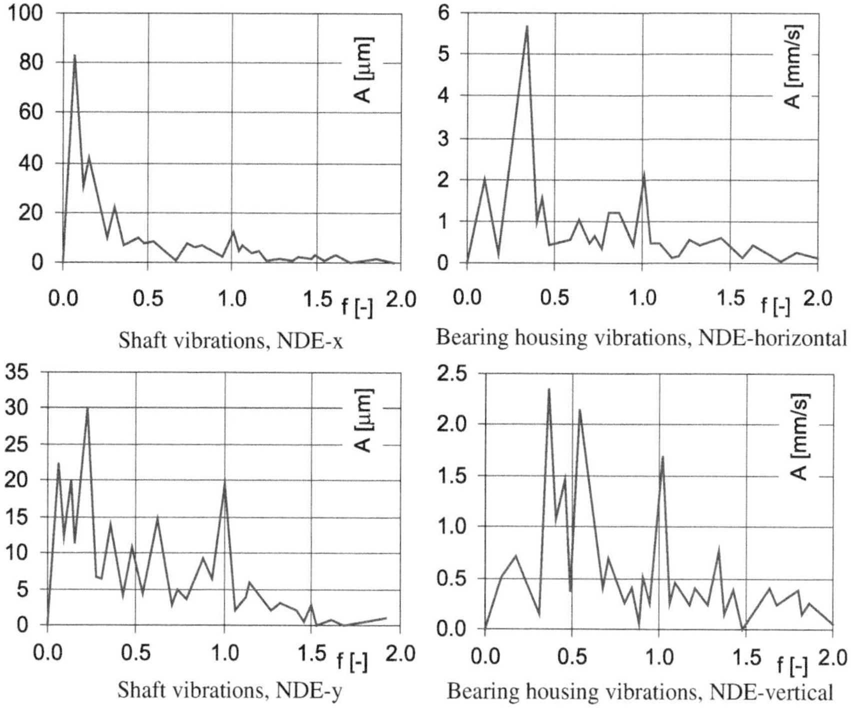

Fig. 10.37 Idealized order vibration spectra at q* = 0.56 (same test in Fig. 10.35)

1. At partload there are several peaks at varying frequencies, typically in the range of $f/f_n = 0.03$ to 0.6. As discussed above, this is the range with impeller inlet recirculation.
2. Bearing housing vibrations: peaks near $f/f_n = 0.35$ and around $f/f_n = 0.20$ are present at all flows (they were weak at $q^* > 0.79$).
3. Shaft vibrations: above $q^* > 0.79$ only synchronous vibrations are observed.
4. From items 2 and 3 it might be postulated that system properties which are independent of the flow rate have an impact on the peaks near $f/f_n = 0.35$ and $f/f_n = 0.20$.
5. Peaks on DE and NDE are different in magnitude and frequency.
6. Peaks of horizontal and vertical vibrations different in magnitude and frequency. For the shaft vibrations this can be due to differences in the local bearing characteristics which depend on the load direction. For the bearing housing vibrations this can be due to anisotropic stiffness of the bearing housings and on the influence of the system (pressure pulsations in suction and discharge piping).
7. Peaks of shaft and bearing housings can occur at different frequencies.

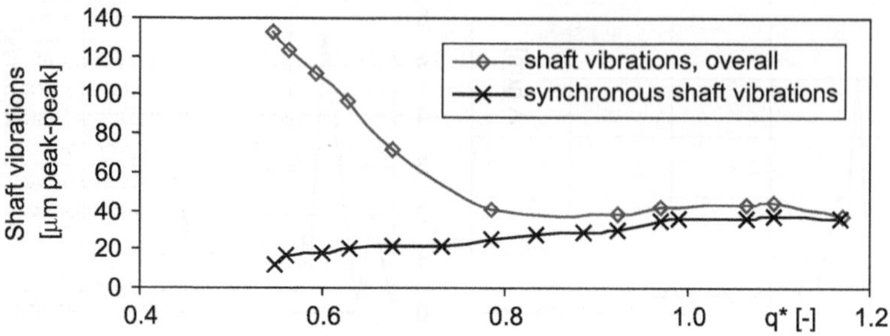

Fig. 10.38 Shaft vibrations (same test as Fig. 10.35)

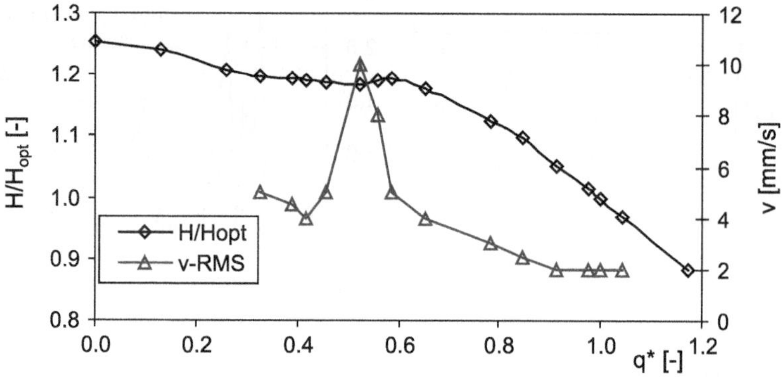

Fig. 10.39 Bearing housing resonance with blade passing frequency in the unstable range of the Q-H-curve

8. Some frequency peaks depend on the flow rate in a stochastic manner.
9. The synchronous peak (hydraulic imbalance) is small at partload. It significantly increases with the flow rate, but levels off above BEP where the hydraulic imbalance diminishes - presumably because the head decreases, see Fig. 10.38.
10. System pressure pulsations and piping vibrations are likely to influence the reaction of the pump to the various hydraulic excitations.

Unstable Q-H-curves; bearing housing vibrations: In a 6-stage pump a bearing housing resonance with blade passing frequency was observed only in a narrow operation range where the Q-H-curve was slightly unstable. Outside the unstable range excitation was not sufficiently high to detect the resonance, Fig. 10.39. The resonance was observed only in a small speed range: even when increasing the speed above the range with unacceptable vibrations, vibration amplitudes drop to normal. This is a clear indication that a mechanical resonance was the problem and not “excessive” hydraulic excitation.

Hydraulic imbalance: Geometric deviations between the individual impeller channels create a non-uniform pressure distribution at the impeller outlet, which rotates with the rotor speed and causes synchronous vibrations. The resulting radial hydraulic force has the same effect as a mechanical imbalance and is therefore called "hydraulic imbalance". If impeller casting quality is inadequate, the hydraulic imbalance can be excessive. Typical casting errors are: variations in the outlet width between or within the channels, non-uniform blade pitches, deviations in blade angles at the outlet or inlet and deviation of the true hydraulic center of the impeller from the center of the shaft bore. Measured force coefficients for the hydraulic imbalance can be found in Fig. 10.42.

The hydraulic imbalance usually exceeds the mechanical imbalance when the head per stage is larger than roughly 400 m; it becomes an important factor for the reliable operation of the pump if the head per stage is above 500 to 600 m. The synchronous hydraulic imbalance grows with increasing flow rate; it is very small near shut-off. The reason for this behavior is that the hydraulic imbalance is caused by differences in lift on the individual blades. Lift differences diminish in separated flow, where the shape of the blades has little influence, Chap. 5.

Impellers with one blade only (as used for sewage pumps) generate a high hydraulic imbalance which must be taken into account when designing such pumps, [55]. This topic is discussed in Chap. 9.3.9.

Impeller-diffuser alignment: Vibrations can be excited if impellers and diffusers are not in the correct axial position in that their centerlines do not match. To smooth the flow when the ratio b_3/b_2 is small, a chamfer at the diffuser inlet is recommended, [61]; refer to the figure on the last page of the symbols.

Vaneless diffusers: Pressure pulsations with frequencies of 60 to 90 % of the blade passing frequency can be generated in vaneless diffusers or if the clearance between impeller and volute cutwater is large (say, at $d_3 > 1.25$), [60]. The fluctuations depend on the flow rate, since they are caused by the non-uniform velocity distribution at the impeller outlet, Chap. 5.3.4.

Vortex streets: Excitation of vibrations by vortex streets can be a problem in large pump turbines but also with ribs or other structures, e.g. with velocity probes or similar devices, Chap. 10.12.4. No incidences are known where vortex shedding from the pump impeller blade trailing edges has been proven as a source for excessive vibrations, presumably because the flow is too irregular to form organized vortices.

Open axial impellers: The orbital rotor movement causes the gap width of open impellers to vary around the circumference (similar to annular seals). The resulting differences in the flow through the gap between the impeller blades and the casing have an impact on the work transferred by the blades. When operating in the position with the smallest gap the energy transfer reaches its maximum according to [76]. The differences in head thus created around the circumference induce radial as well as tangential forces on the rotor (called also "Alford-effect"), [2, 15].

Measurements of the cross-coupled stiffness k_c with open pump impellers are not available; for a rough estimation tests with axial compressors from [76] may be used in the absence of more relevant data:

$$k_c = \frac{M\beta}{D_m h_B} \quad \text{with} \quad \beta = -254 + 466\,q^* - 213\,q^{*2} \qquad (10.16)$$

In Eq. (10.16) M is the torque acting on the impeller, D_m is the mean blade diameter and h_B is the blade height; $h_B = ½(D_a - D_i)$. The tests were carried out only in the range of $0.92 < q^* < 1.08$; an extrapolation of Eq. (10.16) is therefore difficult. For turbines, β = 2 to 5 applies and the orbit moves in the direction of rotation. Hence the tangential force is destabilizing. Negative values of β imply a backward moving orbit.

The Alford effect is a *reaction force* (or a fluid-structure interaction between the flow around the blades and the pump rotor); it is not an excitation force because an infinitely stiff (centric) rotor does not produce this type of forces. Since the effect is induced by the rotor orbit, the vibration frequency observed should be an *eigen frequency of the rotor*, or there must be an excitation mechanism inducing *forced* vibrations (e.g. imbalance or rotating stall). Severe vibrations would be expected if an eigen frequency locks in with the Alford forces; this would amount to a self-excited vibration or an instability . With a *forward* whirl, these forces are *stabilizing* in pumps since opposed to the whirl movement. With a *backward* whirl these forces are de-stabilizing since acting in the same direction as the whirl movement. In turbine operation the forward whirl is de-stabilizing while the backward whirl is stabilizing. The effect increases with A/b_h if A is the vibration amplitude and b_h is the blade height.

Measured hydraulic excitation forces: Figures 10.40 and 10.41 show measured radial forces acting on an impeller in the range of $q^* = 0.25$ to 1.25. The data are represented in the time domain (Fig. 10.40) and as RMS-spectra (Fig. 10.41). The forces are normalized according to Eq. (9.6). At partload a continuous increase of broadband excitation is observed in the low-frequency domain. It should be noted that there is no sudden increase of excitation force at the onset of recirculation. The hydraulic excitation forces depend on the design of the impeller and diffuser or volute (in particular so at partload), but they essentially reach the same level, independently of the type of collector (be it volute, diffuser or annulus), refer to Table 9.4.

Figure 10.42 provides statistical data for the hydraulic excitation forces for four different frequency ranges. When using this data it should be noted that peak values can be 2- to 3-times the RMS-values because of the stochastic nature of the signals measured, Figs. 10.40 and 10.41. The synchronous radial force is of particular practical relevance; it adds as an additional vector to the mechanical imbalance. The data range marked “1” in Fig. 10.42 applies to impellers with small casting tolerances (such as precision or ceramic core castings). The extremely high excitation forces from [83] may either be due to very poor casting quality or to errors in the

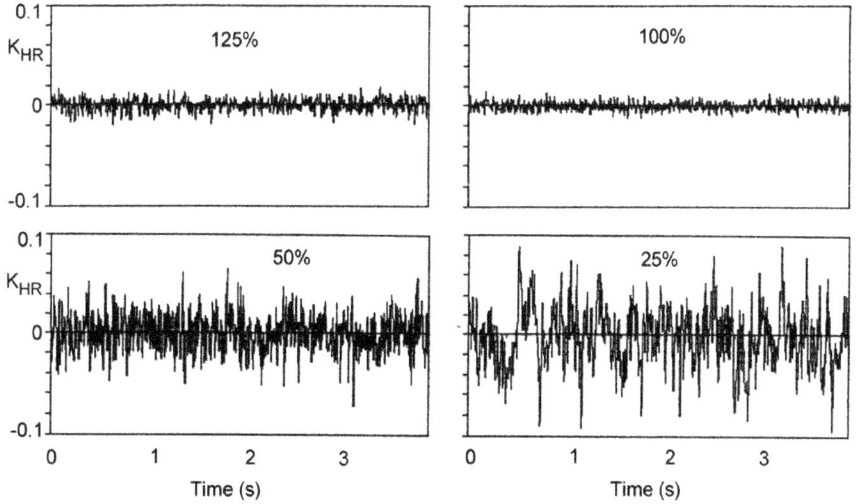

Fig. 10.40 Hydraulic radial excitation forces in the time domain (broadband), q* = 0.25 to 1.25, frequency band 0.01 to 125 Hz, n = 4000 rpm. $K_{HR} \equiv k_R$ as per Eq. (9.6), [13]

Fig. 10.41 Spectra of hydraulic radial excitation forces (broadband) at different flow rates, n = 4000 rpm [13]. $K_{HR} \equiv k_R$ according to Eq. (9.6)

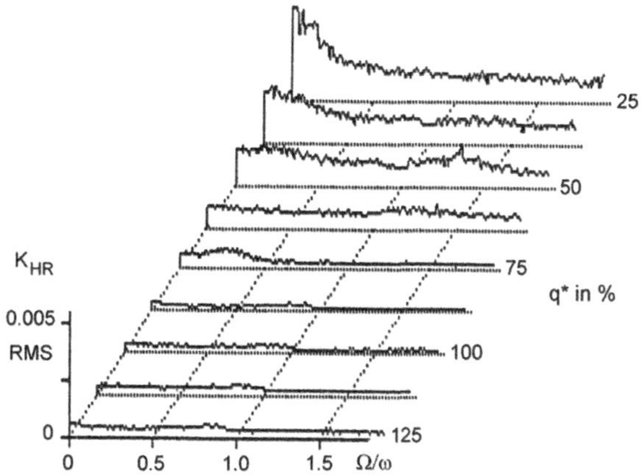

test evaluation (the forces were not derived from force measurements, but from rotor modeling and transfer functions).

Figure 10.42 can be used to estimate the radial hydraulic excitation forces in various frequency ranges. The directions of the excitation forces acting on the different impellers of a multistage pump are unknown. In order to estimate the resulting radial force on the rotor, the force acting on one impeller may be multiplied by the square root of the number of stages in order to get a kind of statistical RMS-value. This approach can also serve to estimate the excitation forces transmitted to base plate and foundation.

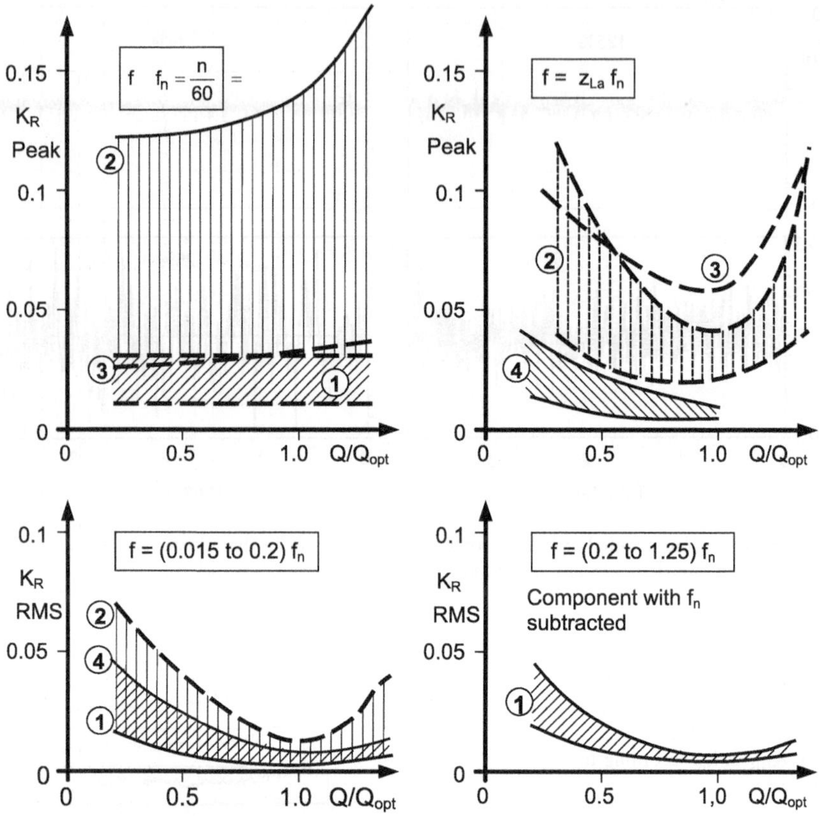

Fig. 10.42 Hydraulic radial excitation forces, (1) [B.19]; (2) [83]; (3) [128]; (4) [52]

Summary of basic principles, Fig. 10.43: Closing this section, the chain of events shall be emphasized once more: The unsteady flow as given by velocity differences between rotor and stator, recirculation and large-scale turbulence creates pressure fluctuations (called "near field" in acoustic terms). The resulting unsteady pressure distributions on the blades, diffuser vanes, casing and piping cause alternating stresses in these components. A tiny fraction of the pressure fluctuations is used to compress the fluid giving rise to liquid-borne noise which travels throughout the system (the acoustic "far field").

Pressure fluctuations and mechanical vibrations: Due to differences in reflection time and non-uniform pressure distributions all types of pressure fluctuations have a potential for exciting vibrations. This applies to pressure pulsations created by the pump as well as to those generated in the system by valves, fittings, or other pumps. The excitation of a structural resonance can be broad-band excitation or by discrete frequencies. Acoustic resonances with standing waves can reinforce the excitation considerably.

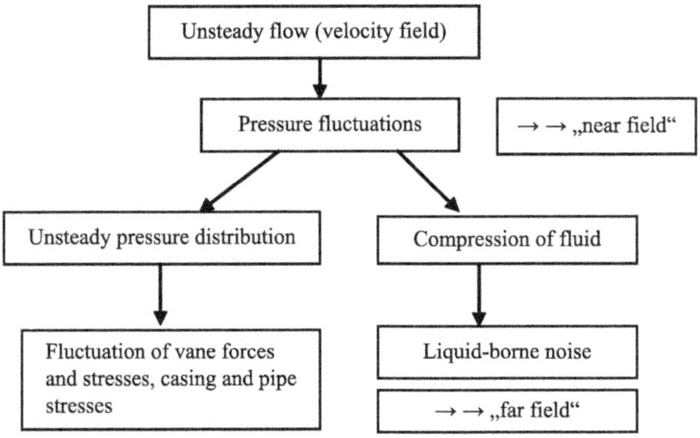

Fig. 10.43 Hydraulic excitation of vibrations, basic principles

- In general, the relation between pressure pulsations and mechanical vibrations cannot be quantified: there is no direct proportionality. When considering discrete frequencies, such as the blade passing frequency, a strong coherence between pressure pulsations and stresses can be expected.
- All pressure pulsations cause mechanical stresses in various components and thus reactions of the pump and the piping. For illustration, Fig. 10.44 shows the unsteady pressures (c_p) and the unsteady stresses (σ), both measured in the front shroud of the rotating impeller. The fluctuations of pressures and stresses are very much coherent.
- As a consequence, the complexity of unsteady flow phenomena manifests itself also in the mechanical vibrations of the rotor, the bearing housings and the piping. For this reason, the spectra of said vibrations mostly show various peaks the true origins of which remain unknown because the details of the unsteady flow phenomena and their effect on the components can hardly be captured.
- Specifically, the uneven pressure distributions on the rotor generate unsteady forces acting on the rotor which potentially induce shaft vibrations . This is borne out by the measured shaft displacements δ shown in Figs. 10.45 and 10.46. These tests from [6] have been done at $q^* = 0.8$ where a slowly rotating stall was present in the diffuser. The stall cell propagated at a frequency of $f_{RS}/f_n = 0.005$ (or about $f_{RS} = 0.43$ Hz). The response of the rotor to the hydraulic radial force created by the stalled diffuser channels is seen in Fig. 10.45: the location of the orbits follows the progression of the stall cell. The pressure distributions measured in the impeller at the same conditions are shown in Fig. 10.46. We see 12 peaks created by the RSI with the 12 diffuser vanes. Two peaks are larger than average; they are caused by the stalled cells (see also Fig. 10.26). Point "a" in the graph corresponds to the location of the hydraulic radial force resulting from integration of the pressure distribution, while point "b" shows the measured shaft displacement. The displacement is on the opposite side of the resulting force showing thus the direction of the force. The right graph in Fig. 10.46 is 70 revolutions later than the left graph.

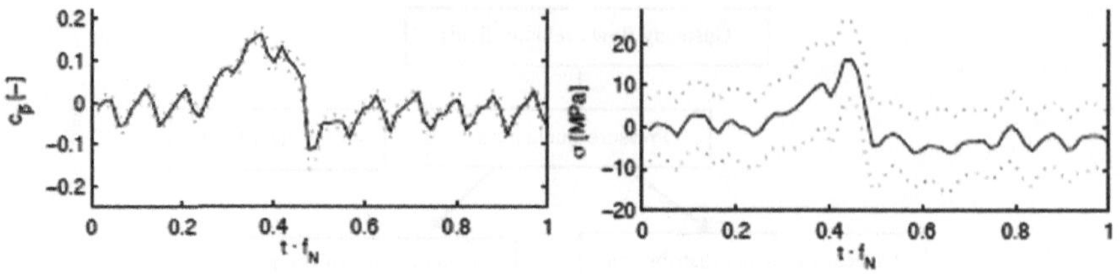

Fig. 10.44 Unsteady pressures (*left*) and unsteady stresses (*right*) measured in the rotating impeller, [6]

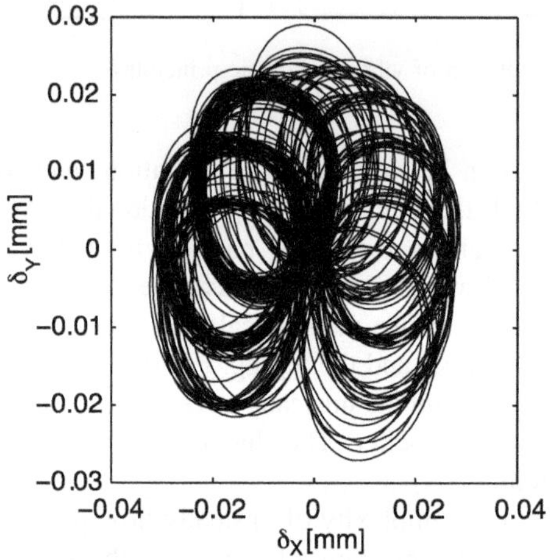

Fig. 10.45 Rotor orbits as function of time due to slowly rotating diffuser stall, [6]

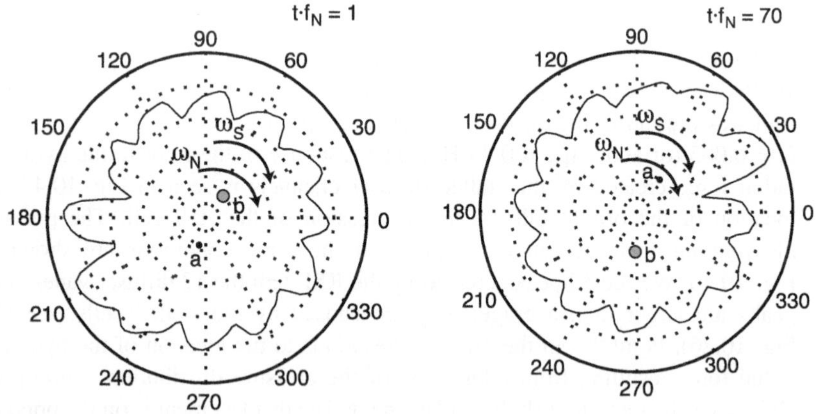

Fig. 10.46 Pressure distribution measured in the rotating impeller; point **a** location of resultant radial force, point **b** shaft displacement, [6]

10.8 Guidelines for the Design of Pumps with Low Sensitivity to Vibrations

Even though the large variety of pump types, designs and applications precludes the formulation of exhaustive design guidelines, some basic rules can be given. The violation of these rules has proven time and again to lead to costly vibration problems. The necessity of following the recommendations given below increases with the sensitivity of the pump to vibrations, i.e. with the impeller tip speed u_2 and the specific speed, but also with the width of the operation range and the number of stages.

1. Impeller/diffuser vane number combinations should be selected such that m = 0 and m = 1 as per Eq. (10.13) are avoided for the lowest three harmonics, i.e. for ν_2, ν_3 = 1, 2, 3. High fluctuations of pressure, torque and axial loads are expected at m = 0. Unbalanced blade forces excite lateral rotor vibrations at m = 1. Vane number combinations implying m = 2 should be avoided up to the third order for pumps with a high head per stage. This, to reduce the risk of resonances between the impeller eigen frequency with two diameter nodes and the excitation by the *diffuser vane passing frequency* $z_{Le} \times n/60$ as seen by the impeller.
2. The distance gap B between impeller and diffuser or volute should not be smaller than stipulated in Table 10.2. Since the last stage of a multistage pump radiates pressure pulsations into the discharge pipe and since pressure pulsations in the system and the exit chamber add to the excitation of the last-stage-impeller (impact on the risk of impeller fractures), gap B in the last stage may be made larger than in the preceding stages.
3. When trimming the impeller, the blade loading increases. Noise and vibrations will increase when blade loading becomes excessive. The trim of impellers of large pumps should therefore be limited, Chap. 4.5.1. The risk of operational problems increases with the specific speed.
4. In order to reduce the excitation at blade passing frequency $z_{La} \times n/60$, *the impellers should be mounted on the shaft in such a way that the blades in the different stages are staggered in the circumferential direction.*
5. The blades on both sides of a double-entry impeller should be staggered if the center rib extends to the impeller outer diameter (Table 10.2).
6. Blade numbers below $z_{La} = 5$ should be avoided if the head per stage exceeds 80 to 100 m. With low blade numbers, impeller discharge flow becomes strongly non-uniform over the blade pitch and impeller/stator interaction increases accordingly. Further, the blade passing frequency is lower and the energy content of the pulses increases.
7. If the head per stage exceeds about 500 m, the casting tolerances of the impellers should be limited as per Table 15.2 to reduce the hydraulic imbalance.

8. The impeller shrouds should be machined (at least near the periphery), since as-cast shrouds cause flow and pressure fluctuations in the impeller sidewall gaps.
9. Diffusers and impellers should axially be correctly centered to a perfect match. For small ratios of b_3/b_2 a chamfer as per Table 0.2(2) should be introduced.
10. Excessive partload recirculation should be avoided. To do this, it must be ensured that flow deceleration at the impeller inlet and in the diffuser throat is within the limits stipulated in Chaps. 5 and 7.
11. The non-uniformities of cast casings can also induce fluctuations in the impeller sidewall gap flow, for example by the possible offset between the upper and lower casings of axial-split pumps.
12. Non-uniform velocity profiles at the impeller inlet increase noise and vibrations, the detrimental effects increasing with the specific speed and the head per stage. The inlet casings of between-bearing pumps or inline pumps and the suction piping or intake structures should be designed to minimize this source of excitation.
13. The clearance gap A between the impeller and diffuser shrouds should not be made too large. There should be a sufficient overlap in order to decouple the flow in the impeller sidewall gaps from the main flow, Chap. 9.1, Table 7.7. The larger the impeller sidewall gap, the more important is this decoupling. If Gap A is too small, the eigen frequencies of the impeller will be reduced; the dynamic loading of the impeller will increase too.
14. The impeller sidewall gaps should not be too narrow if there is a risk of a resonance between the impeller eigen frequency and the excitation by $\nu_3 \times z_{Le} \times f_n$. In general, narrow impeller sidewall gaps are presumably unfavorable with respect to vibrations, since fluid/structure interactions are expected to increase. In the absence of more specific information, the sidewall gaps should not be made smaller than $s_{ax}/d_2 = 0.02$ with an absolute minimum of about 4 mm for reasons of mechanical design and tolerances.
15. Check the impeller/diffuser vane number combination also with respect to phase resonance according to Eqs. (10.14) and (10.15).
16. The annular seals at impellers and balance devices of multistage pumps must be designed to provide sufficient rotor damping, Chaps. 10.6.2, 10.6.5. Damping can be improved by swirl brakes and rough stator designs such as hole patterns or honeycombs, which reduce the tangential velocity in the seals. Hole pattern seals should be combined with swirl brakes.
17. Long seals (as with a balance piston) can lead to instabilities if the gap is divergent (for example due to tilting of the rotor or due to manufacturing tolerances). The risk of instability can be reduced by machining one or two deep grooves in the sleeve of the balance piston (grooves equally spaced over the length of the piston), see Fig. 2.8.

18. The rotor dynamic analysis should comprise the calculation of the damped eigen values for new and worn conditions, taking into account the dynamic characteristics of the journal bearings, annular seals, axial bearing and the hydraulic impeller interaction. The rotor damping of pumps with new clearances should be $D > 0.2$. In worn conditions (double seal clearances) the damping should not be below $D = 0.05$. If calculations are uncertain a minimum damping of $D > 0.08$ should be aimed at.
19. The rotor should be designed so that coupling-dominated eigen modes are avoided, since these are weakly damped. This implies the selection of an appropriate coupling. A well-loaded axial bearing provides some damping but modes which are dominated by the axial bearing require attention so that their impact is minimized.
20. Damping elements should be located near an amplitude maximum, while destabilizing elements do the least harm when placed at a vibration node.
21. In critical applications, acoustic resonances of standing waves within the pump or in the adjacent piping with hydraulic excitation mechanisms should be analyzed.
22. Structural resonances of hydraulic excitation mechanisms with the eigen frequencies of bearing housings and baseplates should be prevented by analyzing the design.
23. The critical speeds - and consequently the stability limit - can be increased by the following measures:

    - Thicker shaft, shorter bearing span.
    - Mitigate impact of overhung of coupling and axial bearing: reduce mass, increase shaft thickness, shorten span.
    - Increase bearing stiffness.
    - Increase stiffness of annular seals at impellers and axial thrust balance device: smaller clearance, no - or shallow - serrations, optimize L_{sp}/d_{sp}, increase pressure difference across seal.
    - By injecting fluid at 50% of the length of the annular seal of the balance piston, the stiffness can be increased according to investigations in [84]. Practical applications of this measure have not been reported.
    - Reduce de-centering effect of the hydraulic impeller interaction by maximizing the angle between front shroud and rotor axis.

24. Measures to increase the damping and the stability limit:

    - Swirl brakes at the inlet to annular seals at impellers and balance piston.
    - Rough stator in annular seals, hole patterns, honeycomb.
    - Smaller clearances of annular seals.
    - Increased seal length.
    - Reduction of destabilizing effects of the hydraulic impeller interaction: larger impeller sidewall gaps: minimize ratio L_s/R_2 as per sketch in Fig. 10.16.
    - Multi-lobe bearings.

10.9 Allowable Vibrations

The allowable vibration amplitudes are defined in standards such as [N.6] or [N.16]. The standards define different vibration limits for a "preferred operation range" POR and an "allowed operation range AOR". These operation ranges may be defined in the contract, for example the POR by $0.85 < q^* < 1.1$, while the "allowed operation range" could be limited by the minimum and maximum flows q^*_{min} and q^*_{max} or by the minimum and maximum *continuous* flows with reference to Fig. 11.15.

Shaft vibrations are judged on the basis of the unfiltered peak-to-peak shaft displacement $S_{(p-p)}$ measured on the major orbit axis relative to the bearing housing. The "unfiltered" amplitudes comprise all frequencies within the measuring range used. Amplitudes of discrete ("filtered") frequencies are judged in relation to the overall level as stipulated in the various standards.

According to Table 10.6, shaft vibrations are assessed by comparing the unfiltered peak-to-peak amplitudes $S_{(p-p)}$ with the diametrical bearing clearance ΔD, [B.19, N.16]. This method is based on a kind of plausible physical concept.

The ISO-guidelines [N.9] for assessing shaft vibrations are applicable to coupled machines with journal bearings operating near the BEP at design speed. The speed range covered by the standard is 1,000 to 30,000 rpm.

Table 10.6 Assessment of shaft vibrations based on the ratio of peak-to-peak amplitudes to bearing clearance in new conditions

$S_{(p-p)}$ = major peak-to-peak shaft displacement amplitude measured in two orthogonal directions
ΔD = diametrical bearing clearance in new conditions

$S_{(p-p)}/\Delta D$	Assessment	Description	
< 0.33	Good	New conditions in the preferred operation range (near BEP)	
< 0.5	Acceptable	Allowed for unrestricted long-term operation in the allowable operation range	
< 0.7	-	Short-term operation	
> 0.7	Alarm	Risk of damage	Trip after time delay of 10 s
0.9	Trip	-	

Table 10.7 Assessment of shaft vibrations according to ISO 7919-3 and ISO 13709 (API 610)

Simplified representation; for information only; always refer to newest edition of respective standard!

<table>
<tr><td colspan="4">Applicable to pumps with journal bearings.</td></tr>
<tr><td colspan="4">Upper limit for unfiltered peak-to-peak vibrations measured during stationary operation relative to the bearing housing.
S(p-p) = major peak-to-peak shaft displacement amplitude measured in two orthogonal directions.</td></tr>
<tr><td rowspan="3">ISO 7919-3
[N.9]
Rotor speed:
1000 to 30’000 rpm</td><td>New conditions, good</td><td>A</td><td>$S_{(p-p)} = \frac{4800}{\sqrt{n}}$ µm</td></tr>
<tr><td>Allowed for unrestricted long-term operation in the allowable operation range; alarm</td><td>B</td><td>$S_{(p-p)} = \frac{9000}{\sqrt{n}}$ µm</td></tr>
<tr><td>Not allowed for long-term operation; short-term operation possible; trip</td><td>C</td><td>$S_{(p-p)} = \frac{13200}{\sqrt{n}}$ µm</td></tr>
<tr><td rowspan="7">ISO 13709
(API 610) [N.6]
Domain:
5 to 1000 Hz</td><td rowspan="3">Pumps with overhung impeller or between-bearing pumps</td><td rowspan="2">Unfiltered</td><td>$A_u \equiv S_{(p-p)} = \frac{2280}{\sqrt{n}}$ µm</td></tr>
<tr><td>Upper limit: 50 µm</td></tr>
<tr><td>Filtered for $f < f_n$</td><td>$A_f \equiv S_f < 0.33\ S_{(p-p)}$</td></tr>
<tr><td rowspan="3">Vertical pumps</td><td rowspan="2">Unfiltered</td><td>$A_u \equiv S_{(p-p)} = \frac{2490}{\sqrt{n}}$ µm</td></tr>
<tr><td>Upper limit: 100 µm</td></tr>
<tr><td>Filtered</td><td>$S_f < 0.75\ S_{(p-p)}$</td></tr>
<tr><td colspan="3">Outside of the preferred operation range 30% higher amplitudes are allowed.</td></tr>
</table>

Table 10.7 and Fig. 10.47 compare the allowable shaft vibrations measured relative to the bearing housing according to API 610 [N.6] and ISO 7919-3 [N.9].

The allowable bearing housing vibrations according to [N.6] and [N.16] are compared in Table 10.8. The ISO-standard [N.16] distinguishes two categories of pumps; the applicable categories is to be defined in the contract.

10.10 General Vibration Diagnostics

10.10.1 Overview

Since vibration problems are among the most common causes of operational problems with rotodynamic pumps, efficient vibration diagnostics are of high

Table 10.8 (1) Assessment of bearing housing vibrations according to ISO 13709 (API 610) and ISO 10816-7 (2009): Vibration velocities

Simplified representation; for information only; always refer to newest edition of respective standard!

RMS of vibration velocities in mm/s for lateral and axial vibrations measured during stationary operation. Applicable to all pumps with $z_{La} \geq 3$ and $P > 1$ kW

Unfiltered vibration velocity v_u			**v_u [mm/s] RMS**			
ISO 10816-7 [N.16]	Category		I		II	
	P_{opt} [kW]		< 200	> 200	< 200	> 200
	New conditions in the preferred operation range (near BEP)		2.5	3.5	3.2	4.2
	Allowed for unrestricted long-term operation in the allowable operation range		4.0	5.0	5.1	6.1
	Short-term operation		6.6	7.6	8.5	9.5
	Risk of damage		> 6.6	> 7.6	> 8.5	> 9.5
	Alarm	Trip after time delay of 10 s	5.0	6.3	6.4	7.6
	Trip		8.3	9.5	10.6	11.9
	Plant acceptance test	Preferred operation range	2.5	3.5	3.2	4.2
		Allowed operation range	3.4	4.4	4.2	5.2
	Factory acceptance test	Preferred operation range	3.3	4.3	4.2	5.2
		Allowed operation range	4.0	5.0	5.1	6.1
ISO 13709 API610 [N.6] Range: 5 to 1000 Hz	Pumps with overhung impeller or between-bearing pumps P_{st} in kW is power consumption per stage n in rpm	n < 3600 rpm P_{st} < 300 kW	3			
		n > 3600 rpm P_{st} > 300 kW	$v_u = 3\left(\frac{n}{3600}\right)^{0.3}\left(\frac{P_{st}}{300}\right)^{0.21}$			
			Valid for $3 < v_u < 4.5$ [mm/s]			
	Vertical pumps		5 mm/s			
Discrete frequencies	ISO 10816-7	f_n and $z_{La} \times f_n$ in preferred operation range, all acceptance tests Allowable operation range: 1.3 v_f	Class I: v_f < 2 mm/s			
			Class II: v_f < 3 mm/s			
	API 610		$v_f < 0.67\, v_u$			
	Outside of the preferred operation range (but within the allowed operation range) 30% higher amplitudes are allowed for both ISO 10816-7 and ISO 13709					

Table 10.8 (2) Vibration assessment of non-rotating parts according to ISO 10816-7 (2009): Vibration displacements for pumps running at n < 600 rpm (Simplified representation; for information only; always refer to newest edition of respective standard!)		
Peak-to-peak vibration displacement in μm for filtered vibrations. The limits are applicable for each of the frequencies: $0.5f_n$; $1f_n$ and $2f_n$ Measured during stationary operation. Applicable to all pumps with $z_{La} \geq 3$ and P > 1 kW		
Filtered vibration displacement		**μm peak-to-peak**
New conditions in the preferred operation range (near BEP)		50
Allowed for unrestricted long-term operation in the allowable operation range		80
Short-term operation		130
Risk of damage		> 130
Alarm	Trip after time delay of 10 s	100
Trip		160
Plant acceptance test	Preferred operation range	50
	Allowed operation range	65
Works acceptance test	Preferred operation range	65
	Allowed operation range	80

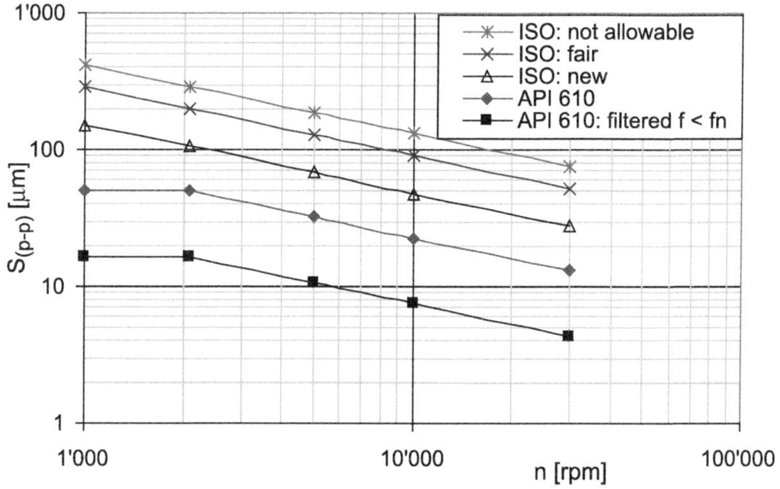

Fig. 10.47 Shaft vibration limits, unfiltered peak-to-peak $S_{(p\text{-}p)}$ measured on the major orbit axis; ISO 7919-3 and ISO 13709 (API 610 unfiltered and filtered)

practical relevance. This chapter provides methods for vibration diagnostics in general with some bias to shaft vibrations. Bearing housing vibrations are treated more specifically in Chap. 10.11. The following measurements can serve as a basis for vibration diagnostics:

1. For diagnosing rotor dynamic problems, the direct measurement of shaft vibrations by proximity probes is often considered the most relevant. Usually, the probes are mounted in the bearing housings close to the journal bearings. The probes thus record the *relative movement* between bearing housing and shaft. In doing so, it is tacitly assumed that the bearing housing is stiff in comparison to the rotor and that coupled vibrations between rotor and stator need not be considered. When interpreting measurements, it should be recognized that the displacement x decreases with increasing frequency for a *given* vibration velocity v according to $x = v/\omega$.
2. A simpler alternative to (1) is the measurement of the bearing housing vibrations by means of accelerometers whose signals are integrated to give the vibration velocities. Accelerometers are often installed in addition to proximity probes, thus enhancing the chance for an efficient and correct vibration diagnosis. To diagnose rotor dynamic problems, bearing housing vibration measurements may be considered less relevant than proximity probes, because the vibration characteristics of the bearing housings and the system have an (unknown) impact on the measured accelerations. System influences can arise from structural resonances, the drive train or the piping.
3. Pressure pulsations can be measured at various locations by means of piezoelectric transducers in order to diagnose problems linked to cavitation, noise, pipe vibrations, acoustic resonances, flow or pressure fluctuations in the system, structural resonances and fatigue damage to components.
4. It can be attempted to apply solid-borne noise measurements to diagnose problems as listed under item (3) above. In contrast to measuring pressure pulsations, solid-borne noise can be measured by accelerometers mounted on the outside of the casing or piping without harming pressure-bearing components. With the exception of thin-walled components (not relevant for pump applications), only the high-frequency vibrations can be detected. This, because the dynamic forces are too weak to excite low-frequency vibration modes of thick-walled components. Solid-borne noise measurements are thus, in principle, appropriate for diagnosing high frequency phenomena such as cavitation or rubbing in seals. The difficulty is to distinguish between different noise sources unless one particular phenomenon is really dominating.
5. Impact tests are used to determine the structural eigen frequencies of bearing housings, base plates and piping. Such tests are standard practice to analyze resonance problems.
6. Measuring the air-borne noise of a pump can sometimes give a first hint as to the possible source of a vibration problem, for example, if the blade passing frequency produces a high peak in the noise spectrum.

10.10.2 Vibration Measurements

General information for the measurement and interpretation of vibrations is given in [N.17], basic methods and problems which can be encountered are discussed below.

Frequency range: Define range of frequencies to be measured. If nothing else has been specified, the measuring range should be $5 < f < 2.5 \times z_{La} \times n/60$ Hz. It is thus suggested to include at least twice the blade passing frequency with a margin of about 15 %.

Bearing housing vibrations are often measured with piezoelectric accelerometers, which can be glued, screwed, or attached by a magnet to the bearing housing. The probes are mounted in vertical, horizontal and (possibly) axial direction on the DE and NDE bearing housings. The instructions of the instrument manufacturer and the imposed vibration standards should be followed closely to reduce the risk of errors resulting from insufficient contact between accelerometer and bearing housing. The resonance frequency of piezoelectric transducers is 10 to 50 kHz. Measurements below 5 Hz are uncertain because of the unfavorable signal-to-noise ratio in this range. These transducers are sensitive to insulation defects of the connectors and leads (moisture). Possible problems are:

- The resonance frequency of the accelerometer must be sufficiently above the highest frequency to be measured; the usable frequency range is about 1/5 to 1/3 of the accelerometer resonance frequency.
- Insufficient contact between accelerometer and bearing housing (surface not plane, too rough, poor gluing, e.g. due to high metal temperature in service).

Shaft vibrations are usually measured with proximity probes mounted in the bearing housing close to the journal bearing. Two probes are mounted under 90° in one measuring plane on both the drive end (DE) and non-drive end (NDE) bearing housings. Often axial shaft movement is also recorded by another proximity probe. The axial signal can be used to indicate the direction of the axial thrust. The signals measured represent the *relative movement of the shaft versus the bearing housing*. The probes generate a high frequency magnetic field, which induces an eddy current in the moving shaft. An inhomogeneous shaft material therefore creates a perturbation signal, the "electrical run-out". The sum of electrical and mechanical run-out must not exceed a moderate fraction of the allowable vibrations in order not to falsify the measurements. Some shaft materials are unsuitable for shaft vibration measurements. In this case the shaft may need to be fitted with a sleeve of an appropriate material.

Absolute shaft movement can be measured by optical methods based on laser technology.

Pressure pulsations are usually measured by piezoelectric (quartz) pressure transducers . Considerations:

- The natural frequency of the transducer must be above the measuring range to avoid resonance.
- As discussed in Chap. 10.2.4, the pressure pulsations recorded depend on the position of the transducer with respect to the pressure nodes of standing waves. For example: A transducer placed in a pressure node of the waves caused by blade passing is unable to capture that frequency.
- The transducer should be mounted essentially flush with the inner surface of the pipe or pump casing (and not in a cavity) in order to avoid perturbations induced by the flow over the cavity or (worse) trapping gas bubbles.
- Sometimes it is only practical to install the transducer at the end of an instrument pipe. If this is done, the length of the instrument pipe must be short enough to avoid resonance with standing waves. The length of the instrument pipe should not exceed 80% of one quarter wavelength ($\lambda/4$ waves as per Table 10.12) for the highest frequency of interest in the measuring range.
- Gas bubbles trapped in the instrument piping would greatly falsify the measurements.

Recording of vibration data and signal analysis: The acceleration, pressure pulsations or shaft amplitudes are measured as a function of time. Vibration spectra are then extracted from the data from the "time domain" by fast Fourier transformation "FFT". The process is outlined below:

- Each signal is recorded in a number N of data blocks (N = 64 to 100).
- Each data block contains m data points, where $m = 2^k$ (k is an integer). A "standard" FFT sample block size is m = 1024, but larger block sizes may be required to get the desired resolution.
- To record or analyze the data, a suitable sampling interval ΔT is selected. The sampling rate f_s is related to the sampling interval by $f_s = 1/\Delta T$. The sampling rate determines the upper frequency $f_{Ny} = f_s/2$ (the "Nyquist frequency") which can be resolved by an FFT. For example: in order to resolve frequencies up to 1000 Hz, the sampling frequency must be larger than 2000 Hz. To include a sufficient margin, the sampling rate should be $f_s = (2.2\text{–}3) \times f_{max}$, where f_{max} is the maximum frequency to be resolved.
- If the sampling rate is too low, the spectral lines above the Nyquist frequency fold back into the domain below f_{Ny}. This phenomenon is known as "aliasing".
- If the sampling rate is too high, the resolution in the high frequency range is impaired, since resolution Δf and sampling interval are related through $\Delta f \times \Delta T = 1/m$. The resolution is thus determined by the sampling rate f_s and the number of data points m in a sampling block: $\Delta f = f_s/m$. In [69] a resolution of 1 Hz is recommended; this would often require block sizes larger than m = 1024.

- Number of FFT spectrum lines: 1 + m/2.
- The length of the time record follows from $T = N \times m \times \Delta T = N \times m/f_s$.
- The error of the analysis decreases with increasing number N of data blocks; the error is $1/N^{0.5}$.
- To increase the statistical accuracy, the number N of data blocks must be increased. For a better resolution, the block size m must be increased. Both measures increase the time T needed to record the data.
- The frequency above the Nyquist frequency f_{Ny} must be filtered out before digitization by an analogue low pass filter.

Most of the above parameters are determined by the design of the FFT-analyzer. The user has to select the input range to avoid overloads, frequency range, resolution (i.e. number of spectral lines) and averaging.

Possible checks:

- It is good practice to monitor the signal in the time domain and check for overloads to ensure that both transducer and measurement system are operating within their permissible dynamic range.
- Measure with a second set of instruments and/or data acquisition method.
- If the overall vibration is much larger than any visible peaks in the spectrum, the range should be increased to find additional peaks or zones with broadband excitation.

10.10.3 Vibration Diagnostics

In order to diagnose the cause of unacceptable vibrations, it is usually necessary to analyze the spectrum of the vibration signals recorded, because the various vibration mechanisms discussed above can be recognized by their characteristic frequencies. When measuring shaft vibrations, the orbits should be plotted too, since the orbit shape is specific to some types of vibrations.

Table 10.9 shows how different types of vibrations can be recognized from the characteristic features of the spectra and shaft orbits. The information given in Table 10.9 can help with a *methodical failure analysis*. In doing so, not only spectra and orbits are considered. It can also be most revealing to analyze how the vibration frequencies and amplitudes depend on speed, flow rate ratio q*, inlet pressure or $NPSH_A$ and, possibly, temperature. Columns 3 and 4 of Table 10.9 describe some characteristic features.

In many cases only the combination of the various features gives a clue as to the cause of the problem. The following information may help to focus the vibration diagnostic:

Table 10.9 (1) Vibration Diagnostics				[1] With bearing problems orbit is nearly always chaotic	
Spectrum	**Orbit** [1]	**Speed dependence**	**Flow dependence**	**Possible causes**	**Remarks/remedies**
1 Synchronous vibration (peak at f_n) 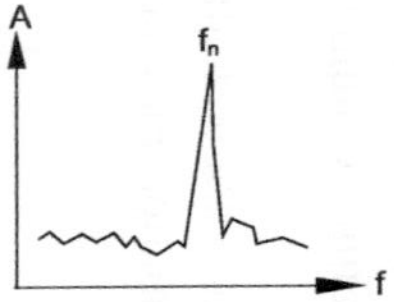	Oval to circular, only small vibrations above f_n 	Peak follows the speed	Weak or absent However, hydraulic unbalance increases with q*	1.1 Mechanical unbalance, e.g. lost parts, coupling, foreign matter caught in impeller 1.2 Hydraulic unbalance caused by impeller tolerances (dominant if $u_2 > 100$ m/s) or by uneven flow into impeller. 1.3 Operation at critical speed (rotor, coupling overhung) 1.4 Extremely low damping (e.g. product lubricated bearings, antifriction bearings) 1.5 Excessive run out (mechanical or electrical) 1.6 Bowed shaft (thermal deformation) 1.7 Resonance with bearing housing or foundation 1.8 Alignment error (piping forces) 1.9 Casing distortion [B.15] 1.10 Upper/lower casing halves offset [B.15] 1.11 Bearing eccentricity or damage [B.15]	1.1 Improve balancing 1.2 Precision casting, more stringent quality standards 1.3 Review rotor dynamics, bearings (oil viscosity), seals 1.4 Swirl brakes; better damping seals 1.5 Electrical run out: more appropriate material for shaft, sleeve below proximity probes 1.7 Detune bearing housing by mass or stiffening 1.11 Check bearing temperature/repair
2 Sub-synchronous vibration at $(0.45{\div}0.95)f_n$ 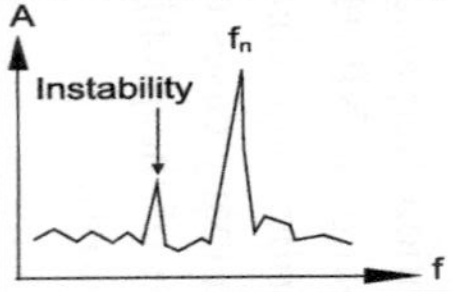	Tendency to chaotic orbit at location of instability 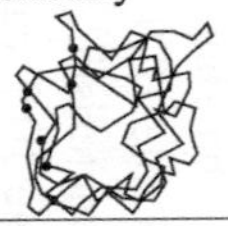	The amplitudes jump to excessive levels above a certain speed. The frequency increases with the speed.	Weak except with item 2.4	**2 Rotor instability:** 2.1 Increase of seal clearance due to wear 2.2 Unsuitable seals (damping too low) 2.3 Radial bearings unstable (type, clearance) 2.4 Unloading of radial bearings with variation of flow rate q* (radial thrust varies) 2.5 Inner friction (e.g. shrink fit too weak) 2.6 Open impellers: Alford effect	2.1 new seal rings 2.2 Swirl brakes; better damping seals (honeycomb, shallow serrations only) 2.3 Install multi-lobe bearings

Table 10.9 (2) Vibration Diagnostics

1) With bearing problems orbit is nearly always chaotic

Spectrum	Orbit 1)	Speed dependence	Flow dependence	Possible causes	Remarks/remedies
3 Sub-synchronous vibration **3.1** at $(0.5 \div 0.95)f_n$ **3.2** at $(0.01 \div 0.3)f_n$ at various q^*	Irregular, shaky	Peak proportional to speed (at given q^*). Amplitudes increase slightly with speed.	At specific values of q^* depending on vane loading. Number of stall cells can vary.	**3 Rotating stall** or similar periodic excitation 3.1 Recirculation at impeller inlet 3.2 Rotating stall in diffuser 3.3 Resonance with natural frequency of shaft at operation above critical speed	3.1 Recirculation brake at impeller inlet 3.2 Increase damping (swirl brake, annular seal) 3.3 Replace impeller and/or diffuser
3a Super-synchronous vibration at $(1.1 \div 1.2)f_n$		proportional to speed		Rotating cavitation observed in some inducers when operating with cavitation-induced head drop.	Geometric features of sensitive inducers not known
4 Peaks at $\nu \times z_{La} \times f_n$ 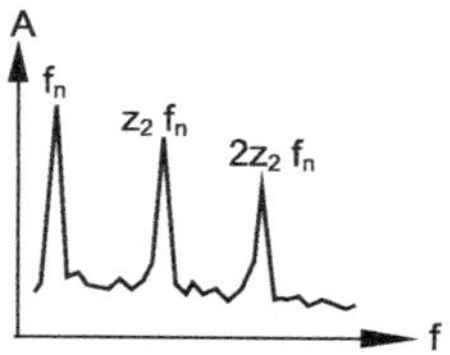 	Oval to circular with regular higher order vibrations 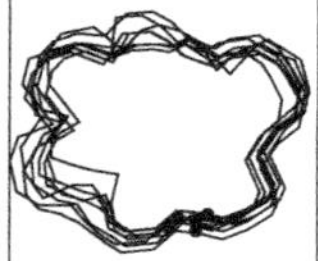	f strictly proportional to speed, if not influenced by acoustic resonances (4.3), except 4.4	Amplitudes are smallest in the range of $q^* = 0.8$ to 1.0. Increasing at partload and $q^* > 1$	4.1 Resultant of blade forces greater than zero due to unfavorable combination of impeller and diffuser vane count i.e. $\lvert \nu_3 \times z_{Le} - \nu_2 \times z_{La} \rvert = 1$ 4.2 Vane interaction forces at $z_{La} \times f_n$ 4.3 Phase resonance 4.4 Acoustic resonances excited by $z_{La} \times f_n$ 4.5 Resonance with bearing housing natural frequency 4.6 Peaks at $\nu \times f_n$ ($\nu = 1, 2, 3, \ldots$) can be caused by geometric tolerances of the individual impeller channels (deviation in blade pitch, blade angles, exit width, blade thickness, blade distance, e.g. a2) 4.7 Uneven flow into the impeller 4.8 Vortices (vortex ropes)	4.1 Change z_{Le} (or z_{La}) 4.2 Reduce vane interaction forces according to Table 10.2. 4.4 Change z_{La}; detune 4.5 Detune (mass and/or stiffness) 4.6 Improve casting quality (entire channel including inlet is important)

Table 10.9 (3) Vibration Diagnostics

[1] With bearing problems orbit is nearly always chaotic

Spectrum	Orbit [1]	Speed dependence	Flow dependence	Possible causes	Remarks/remedies
5 Peaks at low multiples or fractions of f_n ($2f_n, \frac{f_n}{2}$ etc.)	Alignment error: 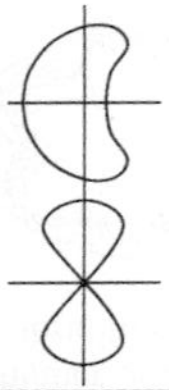	f proportional to f_n	Weak	5.1 Alignment error (sometimes also measurable at f_n only). Alignment can deteriorate during operation: (A) Thermal deformation due to hot fluids or the sun shining on the pump. (B) Deformation due to piping forces. Alignment errors are often accompanied by large axial vibrations. 5.2 Non-linearities caused by loose parts (bearings, sleeves etc.) 5.3 Rotor rubbing with $f = ½f_n$	5.1 Improve alignment under hot conditions; better piping support. 5.2 Thorough design analysis
6 Broadband vibrations at $f < 0.2 \times f_n$ increasing towards $f = 0$ 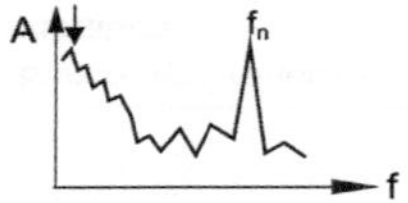	Irregular, oval to chaotic	Weak	Typically found at $q^* < 0.5$. Amplitudes growing with falling q^*	6.1 Recirculation at impeller inlet and/or outlet 6.2 Damping too low: unsuitable type of bearings (clearance), unsuitable type of seals or increased clearance (wear) 6.3 Fluctuating cavities (in this case amplitudes depend on $NPSH_A$)	6.1 Check impeller and diffuser (volute) design: do not select the parameters w_{1q}/w_1 and c_{3q}/c_2 at BEP too small
7 Relatively broad peaks, which are unrelated to rotational frequency 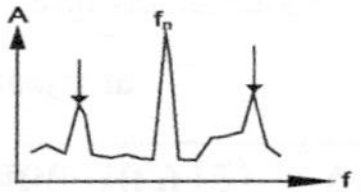		**Not** dependent on speed. Most important distinction to **3**, **4** and **5**	At $q^* < 0.5$ increasing due to recirculation	7.1 Structural natural frequency excited by broadband excitation forces such as partload recirculation and turbulence 7.2 Cavitation	Natural frequency can be determined by impact testing 7.1 Improve partload behavior (Chap. 5) 7.2 New impeller with less cavitation; increase $NPSH_A$

Table 10.9 (4) Vibration Diagnostics

Spectrum	Speed dependence	Flow dependence	Possible causes	Remarks/remedies
8 Broadband vibrations above 500 Hz, dependent on $NPSH_A$	Amount of cavitation depends on speed and flow rate		Pressure pulsations and vibrations caused by cavitation	as 7.2
9 Relatively significant peaks at frequencies between $0.5 < f < 16$ Hz	Weak	9.1 to 9.3 increasing with falling q* 9.4 to 9.6 in range of unstable or flat Q-H-curve	9.1 If dependent on $NPSH_A$: pulsations due to water hammer caused by condensation of vaporous vortex core, generated by recirculation 9.2 Broadband excitation of low structural natural frequencies (base plate, piping) 9.3 Periodic axial thrust reversal 9.4 Acoustic modes with little damping due to flat or unstable Q-H-curve 9.5 Fluctuations caused by parallel operation of pumps with flat or unstable Q-H-curve 9.6 Instabilities of control system 9.7 Air drawing vortices, bottom vortices in sump	9.1 Improve impeller or inducer. If possible, reduce fluid rotation by ribs or structures upstream of the impeller. 9.7 Remove vortices as per Chap. 11
10 Peaks at grid frequency	None	None	Electrical perturbation of instrumentation	Shield instruments
11 Peaks at motor speed or multiples thereof	Yes	None	Motor vibrations are transferred via base plate to pump	
12 Peaks at multiples of rotational frequency with antifriction bearings	Yes	None	12.1 Bearing damage, peaks possibly proportional to the number of rotating elements in bearing 12.2 Gear problems, peaks proportional to number of teeth	
13 Excitation of vertical pump by approach flow normal to pump body/column pipe, Chap. 11.7.3	None	Increases with Q^2	Excessive velocity in intake causing vortex shedding and excitation according to Table 10.13.	

- Usually different vibration mechanisms act at the same time and various system influences obscure the measured signals so that a correct diagnosis becomes very difficult.
- The chain of events as to cause and effect is often not evident. For example: High amplitudes lead to wear of the annular seal, increased seal clearance reduces rotor damping, vibrations increase and foster wear of the seals. If a pump in new conditions exhibits acceptable vibrations which rapidly deteriorate during service, at least three scenarios can be considered: (a) The wear was caused by a bowed shaft (bow due to thermal or mechanical effects) or an unsuitable rotor setting. (b) The wear was caused by high vibrations, e.g. due to excessive hydraulic imbalance, rotating stall or operation with excessive part-load recirculation. (c) Unsuitable rotor design and/or selection of unsuitable annular seals with marginal damping in new conditions but rapid deterioration during service. Which scenario applies, may not be easy to decide from the spectra and orbits, but obviously each of the scenarios would require quite different remedies.
- Very irregular or chaotic orbits point to bearing problems: excessive clearance, low damping, bearing damage, loose sleeve, or instability. Again the bearing problems can be the *root cause* of the difficulties or the *symptom* in that the bearing damage is caused by excessive vibrations induced, for example, by a resonance. Rubbing in annular seals can also generate chaotic orbits.
- Usually the measured spectra contain various peaks which cannot be explained readily by any known vibration mechanism. Furthermore, various peaks depend on the flow rate in a stochastic way which is difficult (if not impossible) to explain. These observations apply to the spectra of shaft and bearing housing vibrations as well as to pressure pulsations. This behavior may be due to subtle changes in the flow pattern within the pump or within the system and to acoustic phenomena. Unless such vibration components contribute significantly to the overall level, it may be a futile effort to try and understand their origin. It is therefore recommended to: (1) Analyze the spectrum (quantitatively) in order to pinpoint those discrete frequency peaks and broadband regions which determine to a large extent the overall level and exceed allowable limits; (2) Focus the effort to find root cause and remedy for the largest contributors.
- When measuring vibrations on the test bed, the influences of the foundation and the piping can falsify the results, because the temporary arrangement on the test bed usually is less rigid than in the plant.

Table 10.9 (5) Shaft versus bearing housing vibrations

Bearing housing vibrations	Shaft vibrations relative to bearing	Interpretation	Probable cause
insignificant	high	• bearing housing behaves as rigid compared to shaft • shaft leads life of its own	• rotor dynamic problems: rotor instability (seals, bearings) or resonance with rotor eigen frequency
high	insignificant	• hydraulic excitation forces are transmitted via pump casing or rotor to bearing housing • rotor follows or dictates the movement of the bearing housings	• resonance with eigen frequency of bearing housing (most frequent cause) • forced vibration of bearing housing through hydraulic excitation

For monitoring large pumps it is quite common to measure both the vibration velocities at the bearing housings and shaft displacement by proximity probes mounted close to the bearings. The latter yield the *relative* movement between shaft and bearing housings. Often it is observed that shaft vibrations are quite low whereas bearing housing vibrations exceed the allowable limits considerably. If that is the case, it may be surmised that rotor dynamic behavior is good and that possible vibration problems are caused by other effects, for example by a resonance of the bearing housing with some hydraulic excitation mechanism. However, the behavior may just be the opposite: shaft vibrations exceed the allowable limits while bearing housing vibrations are quite low. In this case rotor dynamic problems are suspected to be the root cause; Table 10.9(5) may help with the diagnosis.

A non-uniform pressure distribution in the stator (volute, annular casing or diffuser) causes a static bending line of the shaft in the same way as the dead weight of the rotor. As a consequence, the shaft is subject to alternating stresses with a risk of shaft breakage due to fatigue. Bearing loads are static forces. A static radial force does not induce vibrations. In the *absolute* system the "orbit" reduces to a radial displacement. In the *relative* reference frame the orbit is circular.

In contrast, with a non-uniform pressure distribution in the impeller (hydraulic imbalance) the shaft rotates in its deformed state with static bending stresses in the shaft. Bearing loads are dynamic forces. Synchronous lateral vibrations are excited. In the *absolute* system the orbit is circular. In the *relative* reference frame the orbit reduces to a radial displacement.

The various mechanism and damage patterns associated with bent shafts are described in Table 10.9(6).

Table 10.9 (6) Damage patterns caused by shaft deflection
Ω = vibration orbit measured in absolute reference frame

Origin of radial forces	Non-uniform pressure distribution in stator (volute or diffuser)	Non-uniform pressure distribution in impeller channels
Nature of radial forces	• Static radial thrust F_R (Table 9.6), rotor weight: $\Omega = 0$, • Forces from misalignment of balance piston, or line shaft bearings in vertical pumps: $\Omega = 0$ • Alternating stall in diffuser or double-volutes $\Omega > 0$ • Rotating stall in diffuser: $\Omega > 0$	• Hydraulic and mechanical imbalance F_u: $\Omega = \omega$ (synchronous excitation) • Rotating stall in impeller: $\Omega > 0$ • Bent shaft, run-out: $\Omega = \omega$
Shaft movement	Shaft rotates around bending line	Shaft rotates in its deformed state
Bearing load	static	dynamic
Rotor bending line	Static deflection: 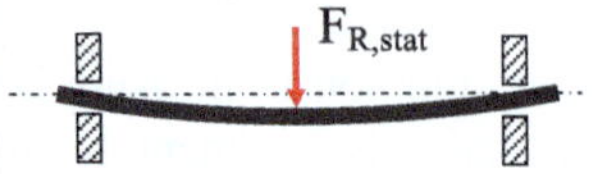	Dynamic deflection: 
Orbit in *absolute* reference frame, f = deflection	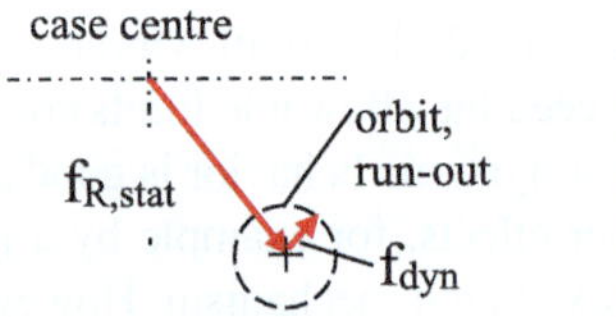	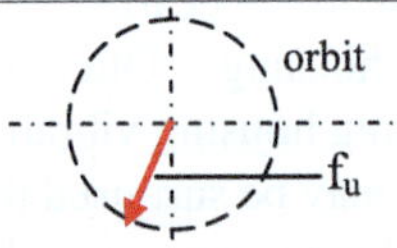
Orbit in *relative* reference frame	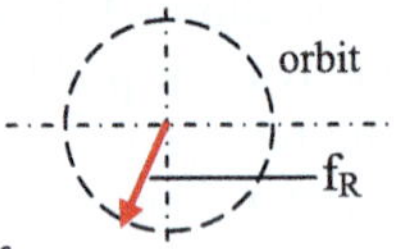 f_R = radial displacement (due to static radial force F_R)	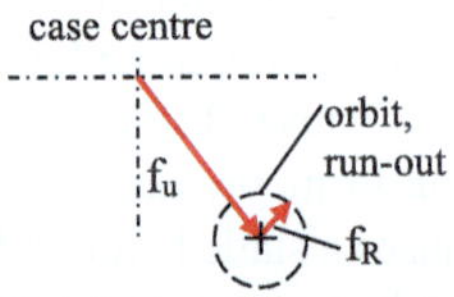 f_u radial displacement (imbalance)
Rotor stresses	*Alternating* bending stresses	*Static* bending stresses
Rotor wear	uniform	local
Stator wear	Case ring: local wear 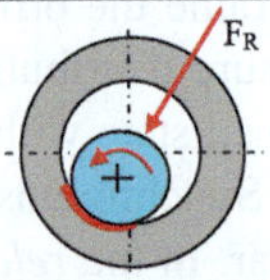	Case ring: uniform wear 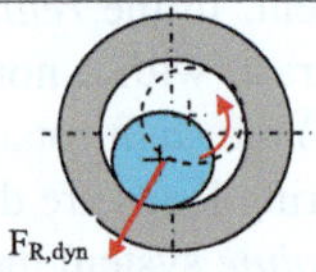
Failure modes	• Risk of shaft breakage due to fatigue at stress concentration • Bearing wear due to shaft deflection	• High vibrations (rotor, bearing housing) • Shaft seals, bearings • Rotor rubbing

10.11 Bearing Housing Vibrations: Mechanism, Diagnostics, Remedies

Impeller/diffuser interaction forces (Chap. 10.7) can be transmitted via *the rotor and the stator* to the bearings. In reaction to this, the bearing housings are subject to vibrations which are easily detected by means of accelerometers or vibration velocity transducers. In both methods vibration velocities are used as acceptance criterion, since these are limited by pump specifications and standards, Table 10.8. The vibration velocity is a measure of the energy of vibration and is considered the best indicator of machinery condition. This because the vibration velocity is *independent* of the frequency (the vibration displacement decreases with the frequency while the acceleration increases with the frequency).

Excessive bearing housing vibrations can lead to the damage of seals, bearings, auxiliary piping, or structural components. An assessment of bearing housing vibration severity based on operation experience is proposed in [B.15], where the zero-to-peak values are given (these have been divided by $2^{0.5}$ to compare with the RMS values specified in Table 10.8):

- up to 3.8 mm/s *zero-to-peak* (up to 2.7 mm/s RMS): good mechanical performance
- 13 mm/s *zero-to-peak* (9 mm/s RMS): bearing and seal failures likely
- 25 mm/s *zero-to-peak* (18 mm/s RMS): catastrophic failure likely

With high-head pumps (specific speed below $n_q = 50$) bearing housing vibrations can exceed the allowable limits specified in Table 10.8, when operating at low flow rates. Then recirculation is expected at the impeller inlet and outlet. This occurs typically below 40 to 70 % of BEP flow (the onset increasing with n_q). In contrast, vibrations above the onset of recirculation are by and large well within specification. The vibration spectra usually show peaks at rotational frequency f_n, blade passing frequency (BPF) $z_{La} \times f_n$ and broadband components as well as peaks at odd frequencies of various origins. This is illustrated by Fig. 10.48, which shows bearing housing vibrations at DE and NDE. In this case, both spectra have almost exactly the same shape (suggesting the same excitation mechanisms on both sides). However, the amplitudes on the NDE are roughly half because of its heavy design. Peaks are found at (or close to) f_n, $5 \times f_n$ (BPF suction impeller), $7 \times f_n$ (BPF series stages), $10 \times f_n$ and $14 \times f_n$ (2nd order BPF of suction and series stages impeller). The excitation grows with increasing flow recirculation, because the flow at the impeller outlet becomes less uniform, thus increasing impeller/stator interaction. In many of the problems encountered, bearing housing resonance occurs with either the blade passing frequency or broadband excitation.

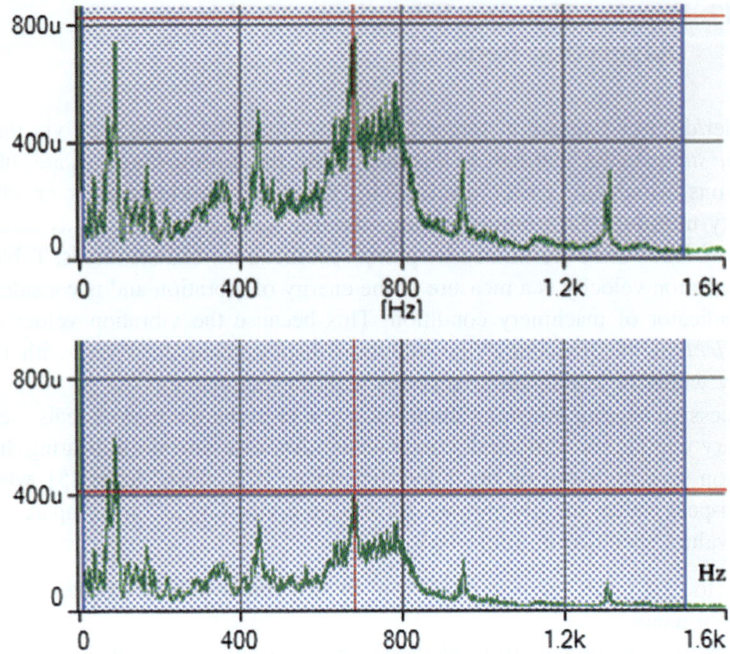

Fig. 10.48 Spectra of bearing housing vibrations of a 5-stage pumps at q* = 0.25; vibration velocities in μ m/s; *top*: DE (overall 6.8 mm/s RMS); *bottom*: NDE (overall 3.9 mm/s RMS); rotor frequency f_n = 94 Hz

10.11.1 Hydraulic Excitation Mechanisms

The most important excitation mechanisms relevant for bearing housing vibrations are summarized below (refer to Chap. 10.7 for other effects and more details):

1. Hydraulic imbalance generates vibrations at synchronous frequency f_n which are caused by geometric differences in the individual impeller channels. *Amplitudes usually increase with the flow rate and are minimum near shut-off.* If such a behavior is observed, geometrical checks of the impeller can reveal differences in the impeller channels (e.g. b_2, a_2 or β_{2B}).
2. If the vane number combination as per Eq. (10.13) is not correct in that it gives m = 1, radial forces are created on the rotor which are transmitted to the bearing housings. First order excitation (e.g. z_{La} = 7 combined with z_{Le} = 8) as well as 2nd order excitation (e.g. z_{La} = 5 with z_{Le} = 9) have been observed to cause excessive vibrations, which were eliminated by changing the number of diffuser (or impeller) vanes in order to avoid m = 1. As an example Fig. 10.49 shows bearing housing vibrations of a 6-stage pump at minimum flow, q* = 0.21. The number of diffuser vanes is z_{Le} = 9. On the NDE blade passing vibrations of the series stages (z_{La} = 7) were dominant, Fig. 10.49a. The

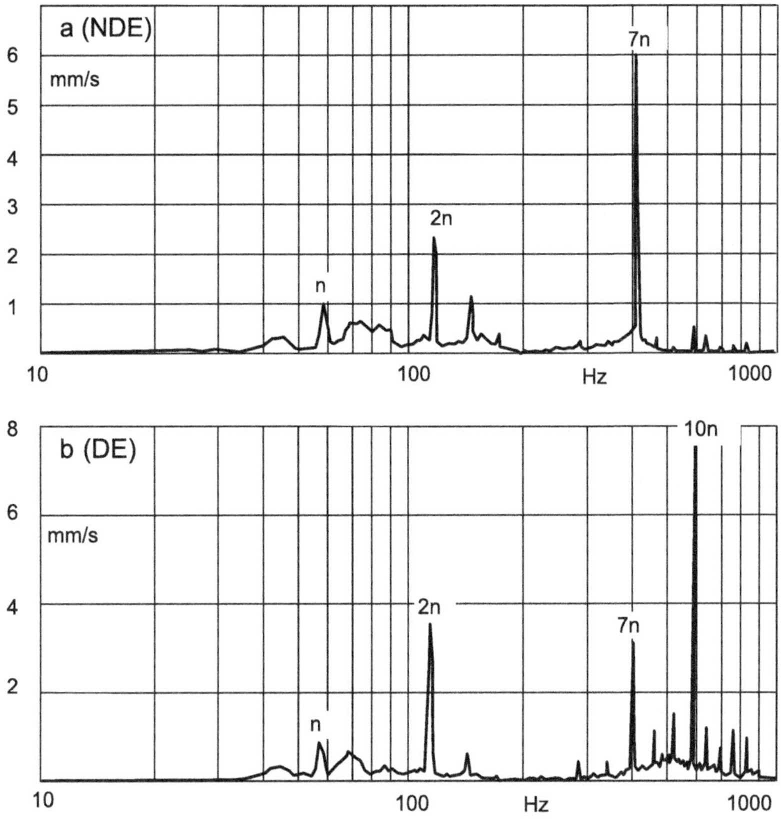

Fig. 10.49 Vertical bearing housing vibrations of a 6-stage pump; $f_n = 59$ Hz, $u_2 = 56$ m/s, $q^* = 0.21$. **a** NDE: series stages with $z_{La} = 7$, overall 8.3 mm/s; **b** DE: 1st stage with $z_{La} = 5$, overall 12.5 mm/s

suction impeller (near DE) has 5 blades which give high second order blade passing excitation at $2 \times z_{La} \times f_n = 10 \times f_n$ (Fig. 10.49b). Misalignment due to thermal deformation shows up at $2 \times f_n$; it is more pronounced at the DE. Imbalance effects are small. All peaks are very sharp, testifying to the low structural damping of the bearing housings.

3. Vane number combinations with m = 0 must also be avoided, since they can cause high axial vibrations of the bearing housing (high pressure pulsations). Furthermore, pressure distributions which are not symmetric around the impeller shrouds can generate a tilting moment on the impeller. The bending moments in the shaft created in reaction to this have a similar effect on bearing housing vibrations as a lateral excitation of the rotor. Impeller shrouds and casing walls should therefore be machined in high-head pumps or pumps which are known to be sensitive to this type of excitation and which are built to stringent specifications.

4. Vibrations at $z_{La} \times f_n$ are excited by the *pitch-wise* non-uniform velocity profile at the impeller outlet. This velocity profile depends on: (1) The impeller inlet flow distribution (in particular for higher n_q); (2) the impeller geometry in its broadest terms: "blade loading" (pressure coefficient) and secondary flows; (3) the flow rate; (4) flow recirculation at the impeller inlet and outlet. The pitch-wise non-uniformity interacts with the stator vanes by two mechanisms, (a) stator vane incidence variations (lift fluctuations) Fig. 10.5; (b) varying impact velocity on stator vanes as demonstrated by Fig. 10.2.
5. The rotor/stator interaction decreases with the distance between impeller and diffuser vanes according to $(d_3^* - 1)^{-0.77}$ (with $d_3^* = d_3/d_2$). An insufficient distance between the impeller and diffuser vanes can lead to excessive vibrations and even impeller shroud fractures; see Table 10.2 for minimum recommended distance d_3^* or d_z^*. Depending on hydraulic design, n_q, u_2 and allowable vibrations, the distance d_3^* in Table 10.2 may need to be increased.
6. Impellers with low blade numbers ($z_{La} = 3$ or 4) have large pitches at the outlet ($t_2 = \pi d_2/z_{La}$) with strong pitch-wise non-uniformity of the flow. Impellers with 3 or 4 blades are typically used for specific speeds below $n_q = 12$ to 14. Examples are process pumps with single or double-entry impellers. Pumps susceptible to giving vibration problems should therefore have at least 5 impeller blades.
7. Broadband pressure fluctuations can selectively excite structural resonances. Broadband excitation is caused by vortices and large-scale turbulence, which increases strongly as the flow is reduced from BEP to shut-off, since essentially the difference between power input and useful power is dissipated. Impeller inlet and outlet recirculation at low flow accounts for this.
8. Instabilities in the Q-H-curve increase hydraulic excitation (see Chap. 10.12). The same is true for sudden changes in flow pattern, which can cause a hysteresis in the Q-H-curve or the axial thrust curve.
9. Pressure pulsations caused by cavitation can excite vibrations. The effect increases with the cavity volume, which is proportional to cavity length, impeller blade inlet width b_1 (thus increasing with the specific speed) and incidence.
10. High pressure pulsations at blade passing frequency may be reduced by *over-filing* the impeller blades at the trailing edge (Fig. 3.6). In contrast, *under-filing* the blades can increase the excitation because of the higher blade distance a_2 and its impact on momentum exchange during recirculation; also blade loading and non-uniformity of impeller outlet flow are increased by under-filing. However, if gap B is increased by trimming the impeller outer diameter, some head can be recovered by under-filing while vibrations still may be reduced.
11. A resonance between blade passing frequency and standing waves in the long crossover aggravates bearing housing vibrations on pumps of the type shown in Fig. 2.9
12. Rotating stall in the impeller is expected to excite vibrations at frequencies in the range of $f = (0.5–0.95) \times f_n$, while rotating stall in the diffuser excites $f = (0.05–0.3) \times f_n$.

With the exception of hydraulic imbalance, hydraulic excitation is lowest in the range of $q^* = 0.9$ to 1.0. Excitation increases at partload and at very high flow ($q^* \gg 1$) similar to the tendencies shown in Fig. 10.9 (pressure pulsations). In practice, excessive hydraulically induced vibrations are therefore mostly partload problems (except hydraulic imbalance). This implies that excessive vibrations occur during operation with *fully separated, recirculating flows*. However, in separated recirculating flows the blade shape is expected to exert little influence on the flow pattern. This is borne out by the observation that the shut-off power depends little on the blade shape or the blade inlet angles. Further, the hydraulic imbalance, which is caused by differences in blade shapes or impeller channels, *increases rather than decreases with the flow rate*. However, the impeller blade outlet angles β_{2B} have some influence on shut-off head and power, since the blade distance a_2 increases with β_{2B}. This fosters the exchange of momentum during recirculation at the impeller outlet. Blade loading optimization at BEP may therefore not necessarily help to reduce *broadband* excitation at low flow. In contrast, impeller outlet width b_2 and inlet diameters d_1, d_{1i} have a strong effect on flow recirculation. These parameters are possibly subject to optimization. In particular, too large an impeller outlet width creates strongly non-uniform flows possibly leading to bearing housing vibrations at partload (flow phenomena described in [49]).

The power dissipated during partload recirculation may be considered as an indirect measure of the broadband excitation, even though it cannot be expected that vibrations would be strictly proportional to the recirculation power. The mechanical reaction to broadband excitation is more like a *filter*: natural frequencies, in the range considered, are excited selectively by the broadband activity, whereas the forced response of the heavy structures involved is much weaker - if not altogether negligible. The recirculation power can be estimated in the following way: When the power at the coupling P is measured (or known), the recirculation power P_{Rec} can be calculated from Eq. (10.17):

$$P_{Rec} = P - \frac{\rho\, g\, H(Q + Q_{s1} + Q_E)}{\eta_h} - \sum_{st} P_{RR} - \sum P_{S3} - P_m - P_{er} \qquad (10.17)$$

The hydraulic efficiencies (including that at $q^* = 0$) can be estimated from Eq. (4.14). Equations (10.17) and (4.14) hold for all flow rates below the onset of recirculation including shut-off. Equation (10.17) can be evaluated over the entire flow range. At some flow the recirculation power becomes zero. The corresponding flow rate is an estimate for the (overall) onset of recirculation.

P_{Rec} is the sum of the impeller inlet and outlet recirculation power. In order to assess the relative contribution of both inlet and outlet recirculation to the broadband excitation, the recirculation power dissipated at the impeller inlet $P_{Rec,in}$ can be calculated from the information given in Chap. 4.1.4 using Eq. (4.11). At shut-off the inlet recirculation power is then given by Eq. (10.18):

$$P_{Rec,in,o} = \frac{\rho}{2}\lambda_{1,Rec,o} u_1^3 r_1^2 \quad \text{with} \quad \lambda_{1,Rec,o} = 0.21 \pm 0.01 \tag{10.18}$$

The onset of inlet recirculation is calculated from Eq. (5.7b). For want of more accurate information, it can be assumed that the power dissipated by inlet recirculation decreases linearly with the flow rate from the value $P_{Rec,in,o}$ calculated from Eq. (10.18) to zero at the onset of inlet recirculation. The recirculation power at the outlet is given by the difference $P_{Rec} - P_{Rec,in,o}$. Figure 10.50 shows the example of a 5-stage pump with a specific speed of $n_q = 30$ and a power consumption at BEP of about 5200 kW. At minimum flow $q^* = 0.25$ the total recirculation power is roughly 900 kW (250 kW due to inlet, 650 kW due to outlet recirculation). Even though this type of estimation can provide very rough figures only, it gives a clue as to the energy available for broadband excitation.

10.11.2 Mechanical Reaction to Hydraulic Excitation

The mechanisms by which the hydraulic excitation forces are transmitted from the impeller outlet to the *bearing housing* are not well understood quantitatively. The interaction forces between rotor and stator vanes act equally on the stator and rotor (Newton's law: "action = reaction"). Consequently, vane interaction forces are transmitted through both the shaft and the stator to the bearing housing. Note that measurements of the shaft vibrations cannot answer the question of the transmission path, since they (usually) record only the *relative* movement between bearing housing and shaft. Hence it is often observed that shaft vibrations are within allowable limits whereas bearing housing vibrations are excessive. In this case, the movements of the bearing housing determine the vibration behavior of the rotor. The opposite behavior can be found in a pump with rotor dynamic problems: in this case bearing housing vibrations are low while shaft vibrations are high.

The transmission of forces through the shaft to the bearing housing is influenced by the oil film in the journal bearings or by the roller bearings, but also by the rotor

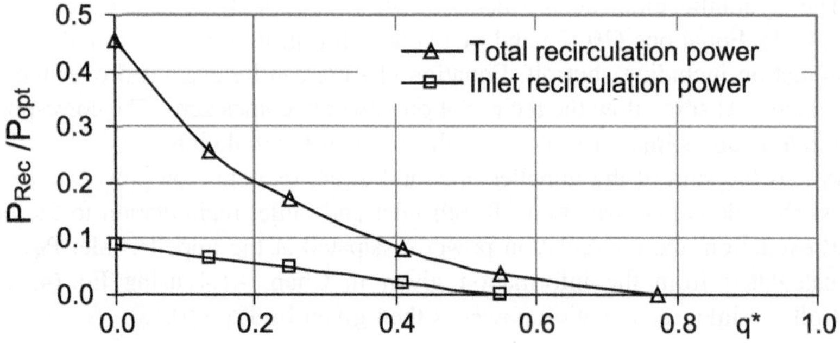

Fig. 10.50 Estimated recirculation power of a 5-stage pump; n = 5600 rpm; $n_q = 30$; $P_{opt} = 5200$ kW; $H_{opt} = 2200$ m

damping in the close running clearances. That forces are transmitted through the shaft to the bearing housing is proven by tests with high mechanical imbalance (which virtually has no impact on casing deformation). Vibration transmission through the casing depends on the design of the pump. Various mechanisms can contribute:

- Pressure pulsations generate unsteady casing deformations, as sketched in Fig. 10.51 (see also Chap. 10.12.6 and Fig. 10.63). Small deformations near the center of the casing are magnified towards the bearing housing - the more so, the longer the bearing span. This mechanism might explain the effect which acoustic resonances in long crossovers can have on bearing housing vibrations of pumps of the type shown in Fig. 2.9. Likewise, pressure pulsations in the system are prone to create casing deformations rather than lateral forces on the rotor. Apart from casing deformations, differences in reflection time of pressure waves within the pump may create non-uniform pressure distributions around the impeller and thus generate lateral forces on the rotor which are transmitted to the bearings.
- In a similar way, tie bolt deformation can contribute to the bearing housing vibrations in segmental type pumps shown in Fig. 2.7. Tie bolts may deform in longitudinal and bending modes.
- With barrel pumps (Fig. 2.8), deformations of both the barrel casing and the stage elements and associated components can occur. The force transmission paths are further influenced by bolts, seats, clearances and gaskets.

Frequently, large differences in vibrations are observed on the drive end (DE) and non-drive end (NDE) of the pump. They may be caused by: (1) differences in the force transmission path; (2) differences in mass or in stiffness of the bearing housing and casing; (3) the balance device (piston or disk); (4) differences in the hydraulic design of the first and last stages; (5) the damping provided by other close-running clearances or swirl brakes .

Numerous mechanical design features can strongly influence pump vibrations. Some are quite obvious, others more difficult to ascertain:

1. Bearing housing resonances. This item is one of the first points to be checked out thoroughly when high blade passing frequencies are detected, because eigen frequencies of the bearing housings are often in the vicinity of the blade passing frequency. The eigen frequencies are calculated by FE analysis and/or measured by impact testing. This can be done (a) *without* the rotor being mounted, (b) *with* the rotor mounted, but at stand-still and (c) *during operation*. While the

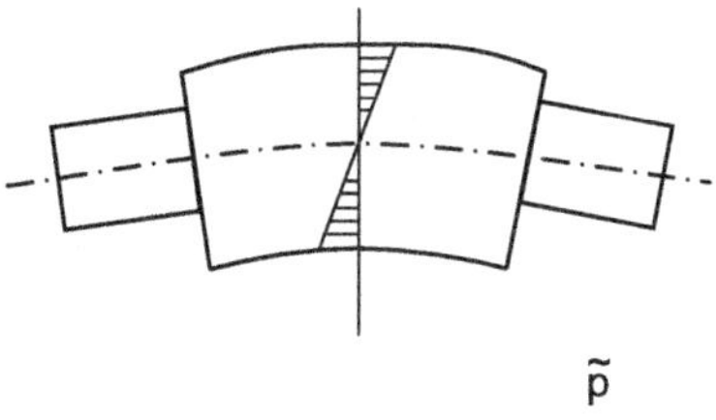

Fig. 10.51 Casing deformation due to pressure pulsations

resulting three frequencies are likely to be different, the frequency measured during operation would appear to be the truly relevant one. However, impact testing during service requires sufficiently high impact energy to get clear signals in the spectrum of an operating pump (whose energy content is very high).
2. In multistage pumps the impellers of the different stages must be staggered. Staggering causes a phase shift between the different impellers, but has no influence on the (blade passing) frequency: *only the amplitudes are affected.*
3. Vibrations can be excited if impellers and diffusers are not well axially positioned so that the middles of the impeller and diffuser channels do not match. To smooth the flow if the ratio b_3/b_2 is small, a chamfer at the diffuser inlet is recommended, Table 0.2(2); refer to the figure on the last page of the symbols. Exceptions are cases where an axial offset of the diffuser versus the impeller is deliberately introduced to improve the stability of the Q-H-curve and/or reduce the axial thrust.
4. Operation near an eigen frequency of the rotor. For remedies see Chap. 10.8.
5. If the bearing housing design is strongly non-isotropic, *vertical* vibration amplitudes may be different from *horizontal* vibrations.
6. Differences between vertical and horizontal vibrations may also be influenced by the baseplate and/or the foundation, since their vibrations are felt by the accelerometers (as during an earthquake). The orientation of the suction and discharge nozzles may also have an impact because of the forces created by pressure pulsations in the system since these forces act in the direction of the nozzles.
7. Multistage pumps with balance pistons or disks may behave differently because of the impact of the balance device on rotor damping and eigen frequencies.

Measured drive-end (DE) and non-drive end (NDE) vibrations can be quite different. Apart from coupling or driver problems (discussed below), several mechanisms can be responsible for this:

- The bearing housings are bolted to the casing, to the cover or to the cartridge (e.g. Figs. 2.7 to 2.13, 7.48); the measured eigen frequencies therefore depend not only on the bearing housing itself but also on the component to which it is bolted and which experiences deformation as well.
- In a multistage pump with a balance piston the forces transmitted through the rotor to the bearing housing may be damped by the piston and could thus be lower on the NDE (if the piston is on that side) than on the DE.
- Forces transmitted through the casing to the bearing housing depend very much on the design: wall thicknesses or material sections, seats, gaskets or bolting between different elements in the path of force transmission all exert an influence. Force transmission paths to DE and NDE are usually quite different.
- For any *given* spectrum of excitation forces the forced response of a structure decreases with increasing mass, since acceleration is proportional to force divided by mass. The accelerated mass on DE and NDE might be quite different.

Frequently, the vibrations exceed allowable limits on the DE, while they are within specification on the NDE (or vice versa). Even though the excitation forces may differ on DE and NDE (e.g. due to particular design features of a suction stage, as per Fig. 10.49, or a last stage diffuser), such cases suggest that the differences between DE and NDE vibrations largely result from differences in the mechanical design as suggested by Fig. 10.48. In that case hydraulic excitation is excessive only with respect to the particular DE or NDE design. In other words: to a vibration problem there can be a mechanical as well as a hydraulic solution.

10.11.3 Hydraulic Versus Mechanical Remedies

The hydraulic excitation forces discussed above are always present. With a given pump they increase with the square of the tip speed and the density, i.e. $\rho \times u_2^2$. The bearing housing and shaft vibrations of the pump would also increase with $\rho \times u_2^2$ if it were not for resonance phenomena. The dynamic response to a high-frequency solicitation is usually not identical to deformation under a static load of the same magnitude. If (as a mental experiment) the speed of a given pump were allowed to increase continually, at some point a tip speed u_2 would be reached above which the allowable vibrations (as per Table 10.8) would be exceeded. Consequently, hydraulic excitation forces are "excessive" only for a *given mechanical design* above a *specific limit of the impeller tip speed.*

The hydraulic excitation forces can be measured by the methods described in Chap. 9.3.2 (which is seldom done) but they cannot as yet be predicted theoretically with reasonable precision. Therefore, it is difficult to state whether or not any given hydraulic components create "excessive" excitation forces. Neither is it practicable to predict the amplitudes of the bearing housing vibrations generated by known hydraulic excitation forces.

Modifications to the hydraulic components may be indicated when known or proven criteria are violated, such as: (1) wrong vane combination; (2) high synchronous vibrations due to hydraulic imbalance; (3) insufficient distance between rotor and stator vanes; (4) unstable Q-H-curve; (5) low number of blades (z_{La} = 3 or 4) in high-head pumps; (6) as-cast (not machined) impeller shrouds in high-head applications; (7) acoustic resonance of blade passing frequency with standing waves in long crossovers. Further hydraulic design features, which might possibly contribute to excessive hydraulic excitation, can only be assessed by comparison to hydraulics with a better vibration record, or by checking against general hydraulic design guidelines. Such factors include: relative impeller outlet width $b_2^* = (n_q)$, deceleration ratios c_{3q}/c_2, w_{1q}/w_{1a}, w_2/w_{1a}, b_3/b_2, blade loading limits (Table 7.1), pressure coefficient ψ_{BEP}, shut-off power coefficient λ_o or the approach flow angle β_{1a} which allows to assess the appropriate choice of the impeller inlet diameter.

Modifications to the mechanical design are warranted if there are manifest resonances between a bearing housing natural frequency and the blade passing frequency, or if there are resonances with the baseplate or the foundation. When attempting to detune the bearing housing against the blade passing frequency either the bearing housing should be *stiffened* or mass should be *added* since, in doing so, the forced response to broadband excitation is reduced at the same time. Detuning should *not* be attempted by reducing the stiffness or the mass. Any reduction would, firstly, increase forced response. Secondly, resonance vibrations can occur due to broadband excitation which is equally present at the detuned eigen frequency. Increasing the stiffness to raise the eigen frequency usually requires the addition of mass, which in itself tends to lower the eigen frequencies. Hence, major changes in stiffness may be required to have any significant effect.

To tackle the root cause of vibrations at blade passing frequency requires modifications to the impeller, diffuser and/or approach flow. Changes to approach flow and impeller aim at achieving a more uniform velocity distribution at the impeller outlet in order to reduce the rotor/stator interaction. Changes to the diffuser or volute intend to mitigate the effects of non-uniform impeller outflow. Since the problems mainly occur at partload, such modifications must be effective with separated and recirculating flows, which are difficult to analyze. Table 10.10 lists the parameters which are thought to have an impact on bearing housing vibrations. These parameters are quite similar to Table 10.2, which should be referred to for further details.

10.11.4 Bearing Housing Vibration Diagnostics

If the measured vibrations exceed the allowable levels as defined by Tables 10.6, 10.7 and 10.8, [N.6] or [N.16], it is necessary to first find the root cause of the vibrations and then to devise remedies. This process is often quite involved and can become very expensive, since the cause of apparently exceeding the allowable vibrations may be found in at least six different areas:

1. Measurements and/or data treatment can be misleading or erroneous.
2. Excessive hydraulic excitation forces.
3. Mechanical design of the pump: bearing housing resonance, operation near an eigen frequency of the rotor, too low mass or stiffness, loose seats or gaskets in force transmission path.
4. Excessive vibrations may result from mechanical interaction with the system (baseplate, foundation and piping).
5. Excessive vibrations may be influenced by hydraulic/acoustic interaction with the piping system and its components.
6. Excessive vibrations may be due to the drive train (motor, gear box, coupling and misalignment).

Once it is ascertained that measurements and data evaluation are correct, but prior to embarking on an expensive investigation of the above problem areas, it is worthwhile to make a survey of likely contributors from items 2 to 6 to the problem at hand. It would make little sense to perform a lengthy hydraulic analysis or to modify, or even replace, hydraulic components to find out in the end that the problem is not caused by an unfavorable hydraulic design, but that there is a bearing housing resonance, a resonance with a natural frequency of the baseplate of the pump, acoustic resonances in the piping or even that the measurements or data treatment were not correct.

Table 10.10 Impact of hydraulic design on bearing housing vibrations

Parameter	Effect	Remarks
Distance between impeller blade TE and collector LE	Increase of d_3* always helps	Vane interaction decreases with $(d_3/d_2-1)^{-0,77}$
	d_2- trim increases blade loading, which may be dominating in rare cases	
Diffuser to impeller width b_3/b_2	Recirculation is expected to increase with the ratio b_3/b_2	
Oblique impeller trim	Smoother flow to diffuser	Quite effective since d_3/d_2 increases, but head reduction
Vane count	$m = \nu_3 \times z_{Le} - \nu_2 \times z_{La}$; avoid m = 1 for $\nu \leq 3$	proven, Chap. 10.7.1
Impeller blade number	Select z_{La} to avoid acoustic resonance in long crossover, column pipe or system.	
	If H_{st} > 100 m, avoid z_{La} < 5	High unsupported span
Impeller trailing edge profile	*Pressure side* profiling (Fig. 3.6) *reduces* excitation; flow follows pressure side even during recirculation, Fig. 3.4	Avoid too thin blade trailing edge (cracking!)
	Suction side profiling (Fig. 3.6) *can increase* excitation; vane distance a_2 increases; higher momentum exchange during recirculation	
Twisted blades at the impeller outlet, "rake". Staggered blades with double-entry impellers	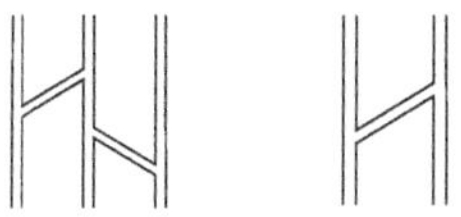	Not always effective
Impeller outlet width $b_2^* = b_2/d_2$	Large b_2* fosters non-uniform velocity distribution at the impeller outlet and increases the momentum exchange during recirculation	For a given ψ_{opt} there may be an optimum b_2*
Pressure coefficient ψ_{opt} and outlet angle β_{2B}	High ψ_{opt} and high β_{2B} (high blade loading) is expected to increase non-uniformity of impeller outflow	High ψ_{opt} also gives flatter Q-H-curves, hence less system damping, Chap. 10.12.
Oblique or oval volute cutwater leading edge	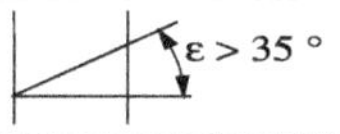	
Gap A and overlap x	Interaction of main flow with sidewall gap flow may increase instabilities and excitation	Influence possible
Meridional shape of impeller	Impact on impeller discharge flow non-uniformity	Influence expected but not demonstrated
Impeller inlet flow	Non-uniform inlet flow induces vibrations; the effect increases with the specific speed	More details in Table 10.2
Impeller inlet diameter	Minimize inlet recirculation	Select close to $d_1^*{}_{min}$ as per Eq. (T7.1.4)
Deceleration ratios	Limit w_{1q}/w_1 and c_{3q}/c_2 to reduce recirculation	

Of the six problem areas listed above, item 1 has been discussed in Chap. 10.10.2, item 2 in Chap. 10.11.1 and item 3 in Chap. 10.11.2. Items 4 to 6 will be discussed in the following in terms of vibration mechanisms, things to check, sources of error and methods of assessment. In doing so - explicitly or implicitly - bearing housing vibration is in the foreground, but the methods apply to shaft vibration or excessive pressure pulsations as well.

Impact tests to determine eigen frequencies: Figure 10.52 shows the results of an impact test on a bearing housing; plotted is the mobility (mobility = velocity/ force) against the frequency. The test was done with the rotor at standstill. Dominant eigen frequencies appear at 660 and 730 Hz. The broadband behavior above 550 Hz may be due to the damping provided by the oil film in the journal bearings. Tests without rotor have sharper peaks because of less damping.

Excessive vibrations due to mechanical system interaction: The accelerometers mounted on the bearing housings measure absolute accelerations (such as may be induced also by an earthquake). Therefore vibration amplitudes measured on the bearing housings can be increased by (resonant) vibrations of foundation, baseplate and piping:

1. Pipe vibrations can be induced by the throttle valve or dead-end pipes and other components in the test loop - in particular, if mechanical eigen frequencies of the piping are excited by standing waves. Remedy: tightly bolt down the piping, additional supports, piping dampers (in plants). When applying such measures, thermal expansion and allowable loads on suction and discharge nozzles must be taken into account.

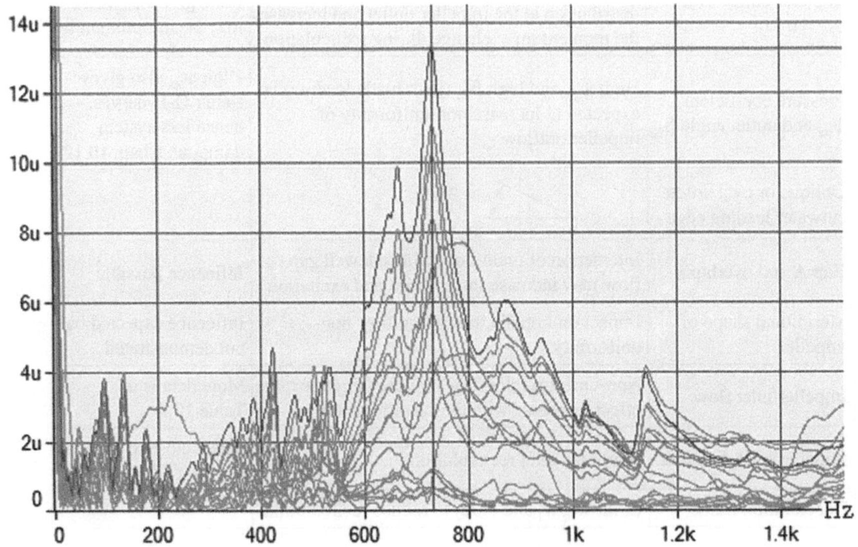

Fig. 10.52 Spectrum of an impact test with the rotor mounted at standstill

2. Vibrations of the pump foundation. This problem may, for example, occur on the test bed (works tests) if the pump is mounted on beams. The best way to avoid this problem is to bolt the baseplate of the pump to a solid concrete slab (thickness typically 1000 mm).
3. Resonant vibrations of baseplates, which are often complicated cast or welded structures with many eigen modes. Baseplate eigen modes may be detected by impact testing or calculated by finite element analysis. Blade pass vibrations usually are expected to excite modes of higher order. As the vibration order increases, eigen frequencies of different modes come increasingly closer, which makes it difficult to clearly detect and avoid resonances.

The items 1 and 2 above are typical for a works test. There, the piping and mounting of the pump is of temporary nature because the experimental set-up needs to be modified for a variety of pumps to be tested in such installations.

Vibration measurements on the baseplate enable a judgment of the likely contribution of the baseplate to the bearing housing vibrations. The coherence between baseplate and bearing housing vibrations can be determined by means of a multi-channel analyzer.

Excessive vibrations due to hydraulic/acoustic system interaction: Pressure pulsations in the test loop are expected to have an influence on bearing housing vibrations through a number of possible mechanisms (see also Chap. 10.12):

1. The pressure pulsations in the inlet and discharge nozzles exert unsteady forces on the pump which are roughly related to the nozzle area times the amplitudes of the pressure pulsations.
2. On a test bed (usually) the useful energy of the pump ($P_u = \rho \times g \times H \times Q$) is dissipated in a throttle valve as large-scale turbulence giving rise to high broadband pressure pulsations. The resulting excitation (manifesting itself also in high audible noise) is often aggravated by cavitation in the valve.
3. Standing waves and their possible interaction with test loop components are discussed in Chap. 10.12. Such effects are not easy to detect, since the reflection conditions at changes in cross sections are difficult to assess; in particular so at the impeller and diffuser throats, at the transitions from the suction pipe to the inlet chamber and from the outlet chamber to the discharge nozzle. At all of these locations the portion of acoustic energy reflected depends on the frequency, but no quantitative information is available.

Acoustic waves generated by impeller/diffuser vane interaction in the different stages of a multistage pump travel different path lengths to any given point of reflection or measurement in the loop. Amplitudes of all these waves are superimposed (they add or cancel partially).

Excessive vibrations due to drive train problems: The motor, gear box or coupling can have an impact on pump vibrations through various mechanisms, which depend on the variety of drivers and gears available. Some of the problems are:

1. Misalignment usually shows up in the spectra as peaks at $2 \times f_n$ and/or $1 \times f_n$. It is often linked to high axial vibrations. Misalignment can be due to: (a) assembly errors; (b) thermal deformation (high temperature fluids or day/night cycle in out-door pump stations); (c) excessive piping forces.
2. Problems with the coupling. Too heavy a coupling lowers the rotor eigen frequencies at the DE overhung. Coupling-induced vibrations have also an impact on the DE bearing housing vibrations.
3. Torsional vibrations induced by frequency controlled electric motors.
4. Motor vibrations can be transmitted via the foundation to the pump. Such vibrations show up in the spectrum at the rotational frequency of the motor rotor or multiples thereof.
5. The impact of a gear box or a fluid coupling is to be assessed by examining the characteristics of the particular components selected.

Vibration analysis methodology: If a specific pump exceeds allowable vibration limits, the procedure described below may be followed.

Step 1: Prior to starting a vibration analysis a review of operation or testing observations could give clues to possible sources of the problem; such a review may include:

- Operation history
- Actual operation conditions (speed, flow rate, suction and discharge pressure, power)
- Unusual noise from pump, bearings, valves, piping, gear box or motor?
- Cavitation noise?
- Surging type of noise?
- Large fluctuations of pressure or flow rate readings?
- Type of baseplate, foundation and bolting: weak or rigid?
- High pipe vibrations?
- Review of approach flow (velocity, perturbations such as bends, branches etc. refer to Chap. 11.7)
- Liquid level in suction tank (if any), required submergence, gas drawing vortices?
- Free gas in suction or discharge piping: insufficient venting (in particular closed circuits) or gas coming out of solution in zones of low pressure?
- Is there a suction screen or a filter? If so, is it intact and effective to eliminate the risk of foreign matter caught in the impeller (imbalance)?
- Bearing temperatures normal or excessive?
- Integrity of coupling?
- Are all auxiliaries working properly?
- Pumps operating in plants: modifications to pump, system or operation? Have there been large transients in the past with the potential of damaging the pump?

Step 2: Determine fundamental frequencies and review data treatment conditions. Check suitability of sampling rate, selected measuring range with respect to natural frequencies of vibration captors and frequency resolution as per Table 10.11A.

Table 10.11 Bearing housing vibration diagnostics
A Fundamental frequencies and data treatment conditions

The figures given in the table are an example for n = 6000 rpm, $z_{La,1} = 5$, $z_{La} = 7$ and $z_{Le} = 12$

1.	Test speed of driver if $n_{driver} \neq n$		n_{driver} = rpm	Does this speed show a peak in the spectrum?
2.	Test speed of pump rotor		n = 6000 rpm	
3.	Rotational frequency of pump rotor		$f_n = n/60 = 100$ Hz	
4.	Number of impeller blades:	first stage	$z_{La,1} = 5$	$f_n \times z_{La,1} = 500$ Hz
5.		series stages	$z_{La} = 7$	$f_n \times z_{La} = 700$ Hz
6.	number of diffuser vanes:	first stage	$z_{Le,1} = 12$	
7.		series stages	$z_{Le} = 12$	
8.	Measured frequency range		$5 < f < f_{max} = 2000$ Hz	At least $f_{max} = 2.5 \times f_n \times z_{La}$
9.	Accelerometer natural frequency		$f_{ac} > 3\ f_{max} > 6000$ Hz	At least $f_{ac} = (3 \text{ to } 5) \times f_{max}$
10.	Sampling rate $f_s = (2.2 \text{ to } 3) \times f_{max}$		$f_s = 4400$ Hz	
11.	FFT sample block size		m = 1024	
12.	Frequency resolution $\Delta f = f_s/m$		$\Delta f = 4400/1024 = 4.3$ Hz	
13.	Number of data blocks		N = 64	
14.	Number of FFT spectrum lines: 1 + m/2		513	
15.	Time required for one record $T = N \times m/f_s$		$T = 64 \times 1024/4400 = 15$ s	

Step 3: Gather and review main pump data as per Table 10.11B. These data relate to the test conditions. If the test conditions differ from the plant operation, the consequences for plant operation need to be assessed separately.
Step 4: Check suitability of vane number combination. This must be done for all stages, in so far as vane numbers differ. The assessment can be done as per Table 10.11C: calculate parameter m according to Eq. (10.13) for the selected combination of z_{La} and z_{Le} (shaded area in Table 10.11C). Example: with the combination $z_{La} = 5$ with $z_{Le} = 9$ lateral vibrations are excited at 10-times the rotational frequency.

Having completed steps 3 and 4, the most obvious hydraulic parameters have been checked:

a. The impeller/diffuser vane number combination.
b. The impeller/diffuser (volute) relative vane distance $d_3{}^*$ ($d_z{}^*$), Gap B.
c. Machining of the impeller shrouds.
d. The extent of cavitation in terms of the available cavitation coefficient σ_A.
e. Gap A and overlap x.

Deficiencies in items A to C are notorious for creating vibration problems and are therefore candidates for possible correction. Vibrations induced by cavitation (with low NPSH-margin or low σ_A) are particularly relevant for single stage pumps and gain importance with increasing specific speeds.

Table 10.11 Bearing housing vibration diagnostics

B Pump main data

No.	Item		Symbol	Test conditions		Plant operation		Unit
				Test conditions		Plant operation		-
16.	Type of fluid							-
17.	Fluid temperature		T					°C
17.	Fluid density		ρ					kg/m³
18.	Speed range min/max		n					rpm
19.	Type of pump							-
20.	Number of stages		z_{st}					-
21.	Type of bearings							-
22.	Specific speed		n_q					-
23.	Impeller outer diameter		d_2					mm
24.	Impeller inlet diameter		d_1	1st stage				mm
25.	Impeller circumferential speed	at d_2	u_2					m/s
26.	Impeller circumferential speed	at d_1	u_1	1st stage				m/s
27.	n, Q, H, P, Q_E, $NPSH_A$, $NPSH_3$ = f(Q) *as tested*			Q-H-curve unstable or flat?				
28.	Available NPSH		$NPSH_A$		m	(as function of flow, if necessary)		
29.	Cavitation coefficient and NPSH-margin		σ_A =			-	allows judgment of effect of cavitation	
29.			$NPSH_A/NPSH_3$ =			-		
30.	Diffuser inlet (or volute cutwater) diameter		d_3 (d_z)		mm	**Important:** compare to minimum values from Table 10.2		
30.			d_3* (d_z*)					
30.	Gap A and overlap x		Gap A		mm	Recommendations:		
30.			$2A/d_2$			$2A/d_2$ = 0.007 to 0.01	refer to Chap. 9.1 and Fig. 9.1	
30.			x		mm			
30.			x/A		mm	x/A = 2 to 4		
31.	Impeller shrouds		Check/confirm that impeller shrouds are machined					
32.	Impeller sidewall gaps		Circumferential variations in impeller sidewall gaps (not machined or with an offset between casing halves of axial split pumps) may have an impact on vibrations					

More subtle hydraulic design parameters are analyzed in step 9.

Table 10.11 Bearing housing vibration diagnostics
C Vane number combinations; stage number: ....

33.		v_2	1	2	3
	$z_{La} = 5$		$v_2 \times z_{La}$		
	$z_{Le} = 9$		5	10	15
	v_3	$v_3 \times z_{Le}$	$m = \lvert v_2 z_2 - v_3 z_3 \rvert$		
	1	9	4	**1**	6
	2	18	13	8	3
	3	27	22	17	12
	4	36	31	26	21

Significance of parameter m

	Excitation Mechanisms	Recommendations
m = 0	Strong pressure pulsations Axial thrust fluctuations Torsional vibrations	Absolutely avoid vane combinations with common multipliers
m = 1	Unbalanced radial blade forces excite lateral shaft/bearing housing vibrations with $\mathbf{f = v_2 \times z_{La} \times n/60}$	Diffuser pumps: avoid m = 1 up to 3rd order. Not always possible with double volute pumps.
m = 2	Impeller vibrations with two diameter nodes are excited by $\mathbf{f = v_3 \times z_{Le} \times n/60}$	Check resonance with impeller eigen frequency

Step 5: Organize measured vibration data; for example in the format presented by Table 10.11D:

- What exactly has been measured (item 35)?
- Compare overall RMS values as functions of flow rate with allowable vibrations as per specification or Table 10.8 (item 36).
- Analyze spectra in order to find relevant peaks as per item 37. This has to be done for various flow rates.
- Tables 10.9 and 10.10 can help to find possible causes and remedies for the various peaks.

Table 10.11D is set up for vibration velocities measured at the bearing housings, but a similar format can be used for measured shaft displacements or even pressure pulsations.

Step 6: Determine (assess) the natural frequencies of the bearing housings DE and NDE, the baseplate and the foundation (if not rigid), Table 10.11E. Calculated natural frequencies possibly need checking by impact testing. Review spectra of impact testing to reveal peaks and broadband regions (if any). Impact tests can be done with or without the rotor at standstill. In a particular example the measured natural frequencies were only 2 to 5 % higher with the rotor than without. With the rotor mounted, the peaks were broader than without. This was due to the oil film damping effect mentioned above. Impact tests can even be done during operation, if a good coherence of the signals can be obtained and if it is possible to achieve high enough impact energies.

In order to evaluate possible acoustic resonances, the frequencies of standing waves in the pump or test loop must be calculated. This is basically done as outlined in Table 10.12. Standing waves could occur in the long crossover of multistage back-to-back pumps, in the column pipe of vertical pumps or in the test loop. The wave length of these components is not always easy to define because of two reasons: (1) It appears that the waves can to a certain extent adapt to the components (lock-in effect); for example, different wave lengths can occur in diffusers or volutes since the point of reflection can be anywhere in the conical portion. (2) It is not always clear whether a component reflects as a closed or an open end. Frequently wave reflections are not total, but only a portion of the sound waves is reflected (for example, at a change in cross section).

Table 10.11 Bearing housing vibration diagnostics

D Measured vibrations

34.	What has been measured?		overall RMS					
			zero-to-peak					
			peak-to-peak					
			discrete frequencies					
			spectra					
35.	Overall RMS = f(Q) or f(q*) At which flow rates are allowable vibration levels exceeded?	Q	q*	overall RMS velocity (mm/s)				
		m^3/h		DE-H	DE-V	NDE-H	NDE-V	allowable
36.	Q =	m^3/h	q* =	DE-H	DE-V	NDE-H	NDE-V	allowable
	Overall RMS velocity (mm/s)							
	discrete peaks in mm/s	peaks at f_n =	Hz					
		at sub-synchronous $f < 0.5 \times f_n$						
		at sub-synchronous $f < 0.95 \times f_n$						
		at $2 \times f_n$						
		at $z_{La} \times f_n$						
		others at f =	Hz					
		others at f =	Hz					

The accuracy of one-dimensional wave calculations in a complex system is questionable and even misleading, but may serve as a first assessment. The separation margins needed depend on the uncertainty of the measurement or prediction. For variable speed pumps, the entire operation range must be considered.

Table 10.11 Bearing housing vibration diagnostics **E Natural frequencies**					Separation margin
37.	Natural frequencies f_{eigen} and spectra of DE bearing housing	Impact test with rotor at standstill	f_{e1} =	Hz	$f_{e1}/f_n \times z_{La}$ =
		Impact test without rotor	f_{e2} =	Hz	$f_{e2}/f_n \times z_{La}$ =
		Impact test during operation	f_{e3} =	Hz	$f_{e3}/f_n \times z_{La}$ =
		Calculated	f_{e4} =	Hz	$f_{e4}/f_n \times z_{La}$ =
38.	Natural frequencies f_{eigen} and spectra of NDE bearing housing	Impact test with rotor at standstill	f_{e1} =	Hz	$f_{e1}/f_n \times z_{La}$ =
		Impact test without rotor	f_{e2} =	Hz	$f_{e2}/f_n \times z_{La}$ =
		Impact test during operation	f_{e3} =	Hz	$f_{e3}/f_n \times z_{La}$ =
		Calculated	f_{e4} =	Hz	$f_{e4}/f_n \times z_{La}$ =
39.	Acoustic resonance between $f_n \times z_{La}$ and standing waves in long crossover, column pipe, etc.		f_{e1} =	Hz	$f_{e1}/f_n \times z_{La}$ =
			f_{e2} =	Hz	$f_{e2}/f_n \times z_{La}$ =
			f_{e3} =	Hz	$f_{e3}/f_n \times z_{La}$ =
			f_{e4} =	Hz	$f_{e4}/f_n \times z_{La}$ =
40.	Natural frequencies and spectra of baseplate or foundation from impact test	There are a number of vibration modes and natural frequencies which must be checked against f_n and $f_n \times z_{La}$. Using a multi-channel analyzer, the coherence between bearing housing and baseplate can be determined.			

Step 7: Diagnostics and remedies. Natural frequencies can be excited by discrete peaks, such as f_n or $z_{La} \times f_n$ or by broadband turbulence.
Step 8: Review data relevant for rotor dynamics, Table 10.11G. If the rotor speed is close to an eigen value, high vibrations are expected at the bearing housings too. Remedies include modifications to the bearings or the annular seals.

Note: Often shaft vibrations are acceptable while bearing housing vibrations exceed the allowable limits. Since the shaft displacement is (usually) measured relative to the bearing housing, such an observation allows no conclusions as to the cause of the vibrations. In addition, shaft displacement diminishes with increasing frequency. Therefore, the higher frequencies (e.g. blade passing frequency) contribute little to the shaft vibrations.

Step 9: A non-optimum hydraulic design can cause excessive vibrations (for a given mechanical design), but without measuring the excitation forces and comparing them to some standard it is difficult to decide whether hydraulic excitation is unusually high, the circumferential speed is too high or the mechanical design is too weak. The hydraulic design can focus on two issues:

- Reduce blade loading in non-separated flow (near BEP).
- Reduce the excitation induced by flow recirculation at partload.

Table 10.11 Bearing housing vibration diagnostics

F Diagnostics and remedies

	Observations	Possible mechanisms	Possible remedies
41.	Bearing housing in resonance with discrete excitation frequency	• Vane combination m = 1 • Excitation at $f_n \times z_{La}$ • Excitation at f_n • Acoustic resonance	• Increase stiffness or increase mass of bearing housing and support system (detuning) • Increase mass • Replace impeller with different z_{La} • Over-file impeller blade trailing edge • Remove acoustic resonance if possible (e.g. length of a dead-end pipe)
42.	Excessive bearing housing vibrations without a manifest cause or resonance	• Excessive hydraulic excitation, possibly broadband or cavitation • Mechanical design too weak for selected hydraulic components and speed • System influences	• Over-file impeller blade trailing edge • See Table 10.2 for various hydraulic effects and modifications • Increase stiffness **or** increase mass of bearing housing and support system (reduce forced response) • Increase rotor damping: e.g. swirl brakes at impeller inlet rings or different wear ring geometry
43.	Vertical amplitudes higher than horizontal	• Vertical bearing housing stiffness lower than horizontal • Baseplate vibrations • Vertical nozzles • Acoustical resonances in long cross-over (item 46)	• Re-design of bearing housing and support system • Check coherence between bearing housing and baseplate vibrations • Grouting, re-design
44.	DE and NDE quite different vibration levels	• Misalignment • Coupling imbalance • Damping by balance piston • Different force transmission paths • Different mass and/or stiffness	Check alignment and balance of coupling; balance adjustment on site. Possibly: redesign. If one end has acceptable vibrations, while the other exceeds limits, one conclusion may be that a mechanical design can be found which is able to cope with the given hydraulic excitation
45.	Acoustic resonance between $f_n \times z_{La}$ and standing waves in long crossover	Pressure pulsations in the cross-over/cross-under cause vertical vibrations to be higher than horizontal vibrations due to the casing deformations induced by the arrangement of these channels.	Replace impeller with different z_{La} and/or reduce excitation as in item 43
46.	Weak foundation		Hard rubber pads (25 mm thick) to isolate foundation from pump
47.	Interaction with piping (mechanical)	Pipe vibrations shake the pump	Better piping support and/or damping

Table 10.11 Bearing housing vibration diagnostics **G Data relevant for interaction with rotor dynamics**			
48.	Eigen frequencies of rotor close to test speed		
49.	Bearings: Type: Oil viscosity (oil temperature) Bearing load as function of flow		Remove possible instability Improve damping
50.	Roller bearings	Peaks at frequencies corresponding to the number of roller elements times f_n:	
51.	Annular seal clearances		
52.	Type of annular seals, swirl brakes		
53.	Balance piston or disk		
54.	Orbit characteristics	Refer to Table 10.9	

Table 10.11 Bearing housing vibration diagnostics **H Hydraulic design**				
55.	Review features as per Tables 10.10 and 10.2			
56.	Review impeller inlet recirculation approach flow angle		Calculate β_{1a} from Table 3.1 f_{d1} =	A low β_{1a} or a high f_{d1} may indicate excessive inlet recirculation.
57.	Incidence at impeller inlet		i_{1a} =	Calculate from Table 3.1
58.	Excessive inlet recirculation		$P_{Rec,in}$	Calculate from Eq. (10.18)
59.	Excessive shut-off power			May indicate non-optimum design.
60.	Excessive impeller outlet width leading to non-uniform discharge flow		$b_2^* = b_2/d_2$ =	Compare with data base of successful designs or Fig. 7.2
61.	Deceleration ratios	Diffuser	c_{3q}/c_2 =	Excessive deceleration may indicate non-optimum design and high partload recirculation.
62.		Impeller inlet	w_{1q}/w_{1a} =	
63.		Impeller outlet	w_2/w_{1a} =	

10.12 Hydraulic and Acoustic Excitation of Pipe Vibrations

Pipelines, in particular those for hot fluids, are mounted on elastic supports in order to absorb thermal expansion. The more elastic the support, the higher is the risk that the piping is excited to vibrations. Excessive pipe vibrations are sometimes encountered in the complex piping systems of thermal power stations or in other plants. All the more so, because high excitation forces are generated in large power station pumps.

When pipe vibrations in a plant exceed allowable limits, attention readily turns to the pumps operating in the system as the probable cause of excessive hydraulic excitation, but a careful analysis of pipe vibrations extends to several phenomena:

1. The possible excitation mechanisms generated by the pumps
2. The acoustic eigen frequencies (standing waves) in the piping system
3. Excitation by vortex shedding in valves, T-branches, orifices, filters; that is to say, in all components with strong flow separation
4. The interaction or coupling of the phenomena 1 to 3
5. The mechanical eigen frequencies and their possible resonances with standing waves
6. Vapor or gas inclusions in suction pipes due to evaporation, ingress of gas by large vortices (Chap. 11.7.3) or gas coming out of solution; gas inclusions in the discharge pipe generated when the system is filled by the pump.

The interaction of the various hydraulic, acoustic and mechanical phenomena in a pipe system with branches is very complex indeed. The discussion of the features of these mechanisms can only provide working hypothesizes and general trends for analyzing pipe vibration problems. In a complex pipe system with many branches and different types of components, there is an almost infinite number of acoustic and mechanical eigen frequencies, which sometimes render unequivocal conclusions as to the root cause excitation mechanism difficult, if not impossible.

10.12.1 Excitation of Pipe Vibrations by Pumps

Unstable characteristics: As discussed in Chap. 11.3, a pump operating in an unstable range (hence with a positive gradient dH/dQ) feeds energy into the system during each cycle of a flow fluctuation. Self-excited vibrations are thus generated, Fig. 11.12. The energy fed into the system, hence the excitation, is proportional to the gradient +dH/dQ. Pumps with an unstable or flat Q-H-curve can excite severe pipe vibrations. In contrast, system damping is enhanced with increasing steepness of a Q-H-curve which rises continuously towards shut-off.[7]

For analyzing pipe vibrations in a system with unstable pumps the following may be noted:

- Self-excited vibrations are expected in the entire range where the Q-H-curve is flat or unstable.
- The amplitudes rise with the gradient +dH/dQ: the greater the discontinuities in the characteristic, the higher are the amplitudes to be expected in a given system (this follows immediately from Chap. 11.3 and Fig. 11.12).

[7] An example for this type of damping is given in [86]: The investigated system of a burner and fan was susceptible to self-excited vibrations below a specific flow rate. When testing fans with differently steep characteristics, the stable operation range increased with the steepness of the Q-H-curve.

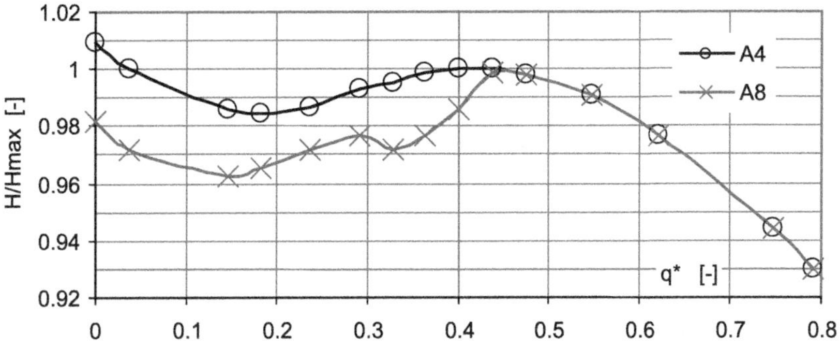

Fig. 10.53 Unstable pump characteristic. Curve A4: pump operation was possible in the specific plant; curve A8: unacceptably high pipe vibrations

- The threshold of the gradient +dH/dQ susceptible of inducing excessive pipe vibrations depends on the sensitivity of the system. As an example, Fig. 10.53 shows two Q-H-curves of a single-stage feed pump with double-entry impeller and a double-volute for a large thermal power station. The pump could be operated with curve A4 (head deficit 1.5 %), whereas operation with curve A8 (head deficit 3.8 %) produced excessive pipe vibrations which inevitably required a modification of the pump. Note that this specific example does by no means allow the conclusion that instabilities according to curve A4 would be acceptable in all plants. That is definitely not the case: a Q-H-curve continuously rising towards shut-off in the planned operation range remains an indispensable requirement for the safe operation of a plant. Interestingly, shaft vibrations and bearing housing vibrations were within the allowable limits even with the configuration causing the highly unstable curve A8.
- In the unstable range of a Q-H-curve, vibrations do not necessarily rise when the flow is reduced. Amplitudes are largest where the gradient +dH/dQ is maximum; vibrations may decrease at lower flow when operating in a flatter portion of the Q-H-curve. This behavior is different from broadband excitations as caused by partload recirculation. Broadband excitation tends to rise continuously with falling flow (as demonstrated by Figs. 10.9, 10.41 and 10.42).
- At the discontinuity in the Q-H-curve (highest local gradient +dH/dQ), the highest vibration amplitudes are expected (e.g. in the range of q* = 0.3 to 0.45 in Fig. 10.53). When increasing the flow rate above the instability threshold, vibration amplitudes drop abruptly (usually to acceptable levels).
- As far as known, unstable characteristics do not generate typical frequencies. Depending on the operation point (valve position, flow rates through different branches), different frequencies may result as a system response to the (stochastic) excitation by the pump.

Pressure pulsations at discrete frequencies: Rotational frequency f_n, blade passing frequency $z_{La} \times f_n$ (and higher orders), as well as rotating stall represent periodic excitations which can get into resonance with acoustic and/or mechanical eigen frequencies of the system. If inadmissible vibrations are observed at these frequencies, either the eigen frequencies must be shifted by changing the pipe length or distance between pipe supports or the excitation frequency must be modified (e.g. by selecting a different number of impeller blades).

Broadband excitation by pressure pulsations: During partload recirculation shear layers between recirculating fluid and net through-flow create vortices and large-scale turbulence. The resulting broadband pressure pulsations can excite pipe vibrations. Cavitation also generates broadband pressure pulsations. All acoustic and mechanical eigen frequencies which lie in that frequency band can be excited selectively by broadband pressure pulsations, if the energy density of the excitation is sufficiently high. As discussed with reference to Eq. (10.17) and Fig. 10.50, broadband excitation at partload can be very high indeed.

Cavitating or air-drawing vortices which are generated by recirculation at the impeller inlet (Chap. 6.5.1) can induce strong low-frequency pulsations - in particular with degassed water (implosion of the cavitating vortex) when inducers or large-eye impellers are employed. Air-drawing vortices can also cause vibrations if the pump takes suction from a vessel with a free fluid level, Chaps. 11.7.3 and 10.14.

Excessive cavitation in a pump operating at $NPSH_A \ll NPSH_o$ may be able to cause kinds of self-excited vibrations, since energy is stored in large vapor volumes which can couple to standing waves in the system.

Double-entry impellers are susceptible to partload instabilities because geometric tolerances can create flow pattern fluctuations between both sides of the impeller. These may couple to standing waves in the system.

10.12.2 Excitation of Pipe Vibrations by Components

If high pressure differentials are throttled in a valve, considerable pressure pulsations are created. Usually these generate broadband spectra, but in addition vortex streets with discrete frequencies can be present in some types of valves (e.g. check valves), Chap. 10.12.4. When the flow rate is high, a considerable power (i.e. $P_v = Q \times \Delta p$) is converted from mechanical to thermal energy. If an unsuitable valve has been selected and/or the back pressure is too low, the throttling may be accompanied by violent cavitation. Broadband pressure pulsations can then selectively excite acoustic and mechanical eigen frequencies of the piping.

In order to reduce pressure pulsations and hydraulic excitation due to valves with high pressure differentials, the throttling should be done in two (or more) stages rather than in one stage. Furthermore, many small throttle openings generate less strong pulsations than one large opening. Reference [5] reports on a case of steam pipe cracking due to standing waves downstream of a control valve which was designed to break down the pressure differential in a single large opening. By modifying the valve to two stages and many small openings in the second stage, the dynamic piping stresses were reduced by a factor of 4.

It is scarcely possible to give general rules as to which pressure pulsations are acceptable. The allowable limit in an actual plant depends on many parameters, in particular, on whether mechanical eigen frequencies of the piping are excited. Plant problems resulting from excessive pressure pulsations reported in the literature cover at least the range of ±5 to ±20 bar. A sufficient counter pressure must be provided to avoid cavitation in valves or other components.

Unsuitable check valves can vibrate at their eigen frequency. If the flow rate is small, the check valve opens only a little and the damping is low. Vortex streets can be generated in the wake of a butterfly or a check valve, Chap. 10.12.4.

Diffusers working in parallel can generate instabilities due to stall fluctuating between both units, [39]. Vibration excitation by pipe bends and T-branches is discussed in Chap. 10.12.4.

10.12.3 Acoustic Waves in Pumping Systems

Due to wave reflections, standing waves are created at discrete frequencies in pipelines (similar to organ pipes). Standing waves are the acoustic eigen frequencies in a pipe or a piping system. Every system has an infinite number of this kind of eigen frequencies. If one of these eigen frequencies coincides with an excitation frequency (e.g. pump speed, blade passing, vortex street), resonance occurs with an amplitude amplification which depends on the damping of the system at that particular frequency. Acoustic resonances between the eigen frequencies and the excitation frequencies of the pump sometimes induce pipe vibrations or upset the control system of the pump.

In the frequency range of interest to pumping applications, longitudinal waves in a pipe can be considered as a *plane* waves. In a plane wave the fluid particles are displaced by the amplitude x in the direction of the pipe axis. The particle velocity $w_p = dx/dt$ is the alternating velocity of the fluid particles. It must not be confused with the sound velocity which is the speed by which a wave travels through the fluid. In a (free-progressive) plane wave:

- The particle velocity is given by $w_p = p/(\rho \times a)$ where p is the sound pressure.
- The impedance is $Z = p/w_p = \rho \times a$
- The sound intensity is $J = p \times w_p = p^2/(\rho \times a)$
- The sound power is given by the integral of the sound intensity over the pipe cross section $W = \int J dA = \int p^2/(\rho \times a) dA$

A sound wave is described by the time-dependent sound pressure and particle velocity. At a wall (of a sufficiently high mass), the particle velocity is bound to be zero. Therefore, we find a velocity node and a maximum of the sound pressure at a closed pipe-end. At an open pipe-end, the pressure is imposed by the environment (e.g. by the pressure in a tank); it cannot be changed by the wave. In contrast, the movement of the fluid particles is not impeded. At an open pipe-end, we therefore find a pressure node and a maximum of the particle velocity.

Sound waves are reflected at closed as well as at open pipe-ends. Consequently, standing longitudinal waves are established between two reflecting components if the vibration mode fulfills the boundary conditions mentioned above. The ratios of the wave length λ to the length L between the reflecting components are integer numbers. The frequencies of the standing waves follow from $f = a/\lambda$.

The sound velocity "a" in a pipe differs from the sound velocity a_o in the liquid because of the flexibility of the pipe wall. It can be calculated from Eq. (10.19), where ρ is the liquid density, D_m is the mean pipe diameter, h is the wall thickness and E is Young's modulus of elasticity of the pipe material. Chapter 17.5 gives sound velocities in liquids.

$$a = \frac{a_o}{\sqrt{1 + \frac{D_m}{h}\frac{\rho\, a_o^2}{E}}} \quad \text{with} \quad D_m = 0.5(D_a + D_i) \tag{10.19}$$

Based on the above principles, Table 10.12 shows the acoustical properties of various system components. *Note that pipe bends do not reflect sound waves.*

At every change in cross section (change in impedance) a portion of the sound pressure is reflected, while another portion is transmitted to the downstream pipe or component. Table 10.12 gives the basic relationships for the reflected and transmitted pressure waves, [24]. Figure 10.54 shows the reflection and transmission coefficients for a variation in cross section. If the cross section changes from A_1 to A_2, the following characteristics can be found when $\varepsilon = A_2/A_1$ is the area ratio:

- The reflection coefficient R is always less than (or equal to) unity; i.e. the pressure in the reflected wave is lower than in the incident wave.
- For an area contraction ($\varepsilon < 1$; $R > 0$) the reflected wave is in phase with the incident wave. An incident compression wave is reflected also as a compression wave. The transmission coefficient is greater than unity; i.e. the pressure in the transmitted wave is higher than in the incident wave. This fact may be interpreted as a concentration of the acoustic energy on a smaller area or volume.
- For an area expansion ($\varepsilon > 1$; $R < 0$) the reflected wave is 180° out of phase with the incident wave. An incident compression wave is reflected as an expansion wave. The transmission coefficient is smaller than unity. Hence the pressure in the transmitted wave is lower than in the incident wave: the acoustic energy distributed on a larger area or volume.
- A large area expansion acts like an acoustic filter which protects the downstream piping or components from pressure pulsations created in the upstream piping.

In a complex system, there are many locations where the impedance changes and where a fraction of the sound energy is reflected. These multiple partial reflections depend also on the frequencies. This makes the analysis of a piping system difficult and not really amenable to one-dimensional wave theory.

Table 10.12 Acoustical properties of system components

Component	Particle velocity	Sound pressure	
Open pipe-end • tank • rubber bellows	Maximum	Node	
Closed pipe-end Change in cross section > 50%	Node	Maximum	
Pipe open on both ends 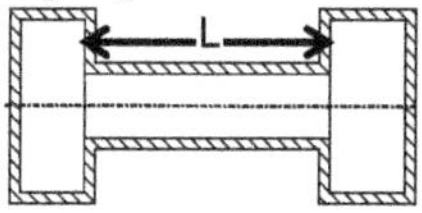	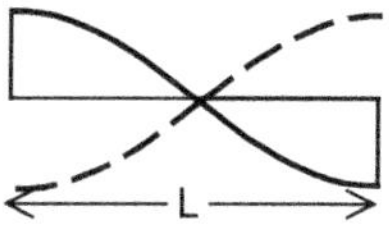	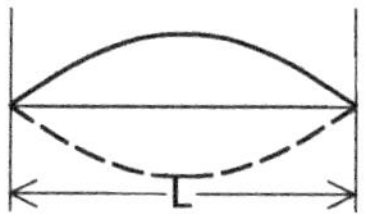	$f_\nu = \nu \frac{a}{2L}$ $\lambda_\nu = \frac{2L}{\nu}$ $\nu = 1, 2, 3 \ldots$
Pipe closed on both ends 	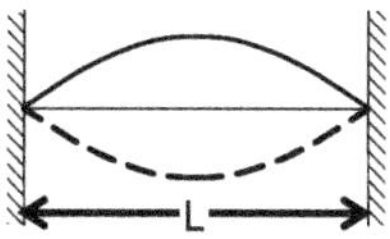	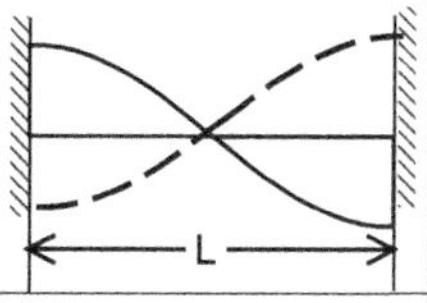	
Pipe open on one end 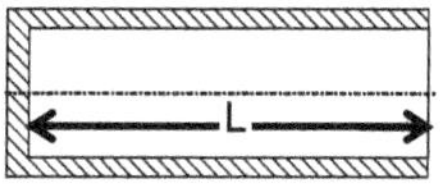	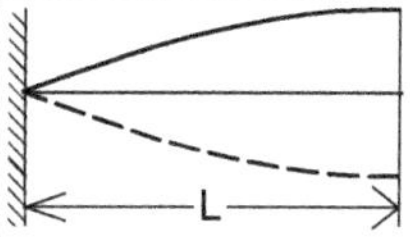	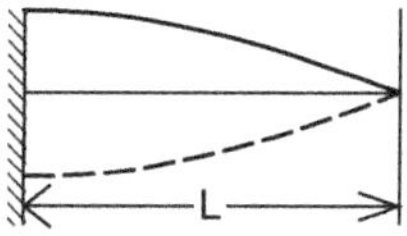	$f_\nu = (2\nu - 1)\frac{a}{4L}$ $\lambda_\nu = \frac{4L}{2\nu - 1}$
Change in cross section 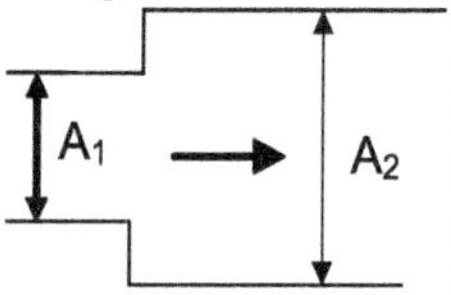	Reflected wave	$R = \frac{1-\varepsilon}{1+\varepsilon}$	Area ratio $\varepsilon = \frac{A_2}{A_1}$
	Transmitted wave	$T = \frac{2}{1+\varepsilon}$	
Pressure reflection coefficient	$R = \frac{p_R}{p_I} = \frac{\text{pressure amplitude of reflected wave}}{\text{pressure amplitude of incident wave}}$		
Pressure transmission coefficient	$T = \frac{p_T}{p_I} = \frac{\text{pressure amplitude of transmitted wave}}{\text{pressure amplitude of incident wave}}$		
Piping bend	Waves travel virtually without impediment through bends.		
Steam generator	Nearly reflection free.		

A pump connected to a pipe represents a change in impedance. Some sound energy travels through the pump, but a portion is reflected. The following tendencies have been found:

- Waves traveling through the suction pipe encounter a drastic change in cross section at the impeller throat, where high-frequency waves are largely reflected like in a closed pipe. Here a pressure maximum and a velocity node may be found.

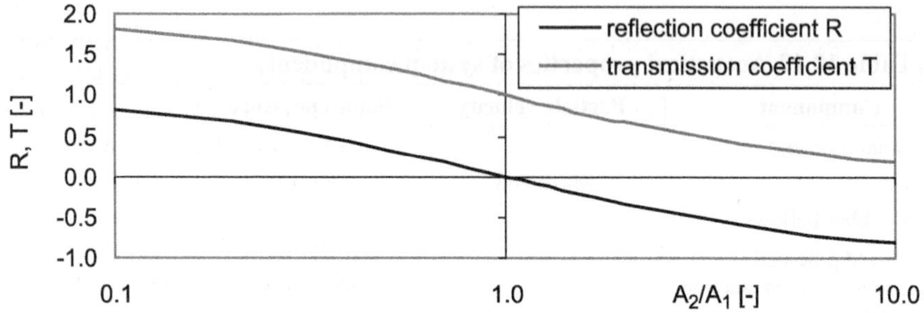

Fig. 10.54 Reflection and transmission coefficients due to a change in cross section

- The same applies to a discharge pipe which connects to diffuser or a volute if the area ratio of discharge pipe to the throat is large.
- If the discharge pipe connects to an annular exit chamber (Figs. 2.8, 2.10 and 2.11), wave reflections are expected to be similar to an open pipe-end. In that case a pressure node and a velocity maximum would be expected at the pump. It can also be argued that a pressure node at the discharge nozzle opening would not allow pressure pulsations to be radiated into the discharge pipe, because a particle velocity greater than zero is required to transmit energy into the pipeline.
- In diffusers or volutes there may be no sharp points of reflection. Standing waves of variable wave length or frequency possibly occur in such geometries within a certain range.

Apart from the longitudinal waves discussed above, transverse waves occur above a limiting frequency $f_{limit} = 0.586 \times a/D$ (such waves are less significant in pumping applications).

The effect of a centrifugal pump on sound waves can be described by the following, [11, 12]:

- Up to 5 Hz the waves travel rather unimpeded through the pump, while waves above roughly 80 Hz are largely reflected at the pump. The amount of reflected energy depends on the ratio of the impedances of the pipe to the pump and on the change in cross sections in the impeller and diffuser.
- With f < 20 Hz the system damping is given approximately by R = −dH/dQ. If the Q-H-curve is continuously falling with increasing flow (i.e. dH/dQ is negative), the damping is positive and the wave energy decreases when traveling through the pump. If the Q-H-curve is instable (dH/dQ positive), the damping is negative and self-excited vibrations are generated .

- With no or little cavitation the pump is passive. If strong cavitation and correspondingly large vapor volumes are present in the pump, pressure pulsations can be amplified because of the energy stored in the compressible vapor volumes (the pump is "active").

Reflection and transmission of acoustic waves through pumps were studied extensively in [149] (from which all raw test data presented below were derived). The test pumps were at *standstill* in a closed loop while an auxiliary pump was run at different speeds to generate pressure pulsations traveling to either the suction or the discharge nozzles of the test pumps. Four pressure transducers were installed each in the suction pipe and the discharge pipe to record acoustic waves entering and exiting the nozzles of the test pumps. The sampling frequency was 10 kHz; the tested frequency range was 150 to 290 Hz.

Three single-stage process pumps with axial inlet and single volute of the KSB-range "Etanorm" were tested; their data are given in Table D10.1-3; the cross section is similar to Fig. 2.1. Test loop piping was of steel; the pump casings were made of grey cast iron. The tests were done without cavitation with $\sigma_A = 6$ ($\sigma_{3,opt} = 0.53$). With extensive cavitation, the compliance increases and amplitudes measured in the suction nozzle diminished.

Wave reflection and transmission at the nozzles of the pump are governed by the *change in impedance* from the pipe to the pump. It is therefore critical to get the best estimate of the effective sound velocity (a_{case}) in the casing. The reflection and transmission coefficients discussed in the following are valid for the impedances of the piping and the pumps tested. Connecting the same pumps to piping made of plastics, or if the casing would be of cast steel instead of cast iron, one would get completely different results.

The sound velocity a_{case} in the casing depends on compliance of the casing; i.e. on the geometry and the material, the casing cover and its bolting, static and shaft seals might have an influence too. It can be attempted to estimate a_{case} from Eq. (10.19a) where V is the fluid volume in the casing (between suction and discharge flange) and ΔV is the volume increase by applying a pressure difference Δp on the casing. $\Delta V/\Delta p$ can be determined by finite element models of the pump. For a long pipe Eq. (10.19) can be derived from Eq. (10.19a) by considering the circumferential strain only. For the present test pumps from cast iron, the effective sound velocity can be correlated by Eq. (10.19b) while Eq. (10.19c) provides an estimate for cast steel. These formulae can be used only in the range n_q = 20 to 80, but they cannot be used for different casing geometries.

$$a_{case} = \frac{a_o}{\sqrt{1 + \frac{\Delta V \rho a_o^2}{V \Delta p}}} \tag{10.19a}$$

for gray cast iron $(n_{q,Ref} = 10)$:

$$a_{case,CI} = 165 + 520 \ln \left[\frac{n_q}{n_{q,Ref}} \right] \quad (10.19b)$$

for cast steel $(n_{q,Ref} = 10)$:

$$a_{case,steel} = 385 + 490 \ln \left[\frac{n_q}{n_{q,Ref}} \right] \quad (10.19c)$$

The measured reflection and transmission coefficients can be correlated by the Helmholtz number (He) defined by Eq. (10.19d) where h_2 is the distance between the pump axis and the discharge flange.

$$He = \frac{L_{Ref}}{\lambda} = \frac{L_{Ref}}{a_{eff}} f \quad \text{with} \quad L_{Ref} = 0.5(d_z = d_s) + h_2 \quad (10.19d)$$

The resulting correlations are given in Table D10.1-3. As a sample, Fig. 10.54-1 shows a graph of the reflection coefficient measured at the discharge nozzle.

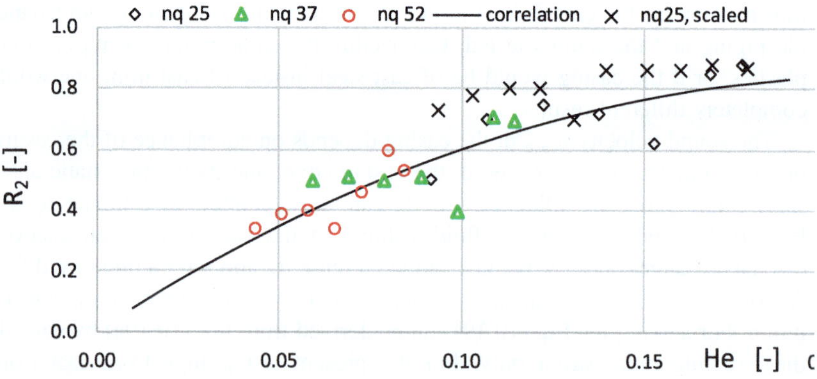

Fig. 10.54-1 Reflection of acoustic waves at the discharge nozzle correlated to Helmholtz number

Table D10.1-3 Test pump data and correlations

Pump	Etanorm	80-65-200	80-65-160	80-65-125
n_q		25	37	53
n	rpm	2900		
Q_{opt}	m^3/h	124	115	114
H_{opt}	m	62	34	21
ψ_{opt}	-	1.1	0.96	0.9
η	-	0.83	0.83	0.8
z_{La}	-	6		
d_{2a}	mm	219	174	141
d_s	mm	80		
d_d	mm	65		
d_z	mm	235	193	158
d_z/d_2	-	1.07	1.11	1.12
a_{pipe}	m/s	1350		
a_{case}	m/s	625	850	1025
h_2	mm	225	200	180
L_{Ref}	mm	383	337	299

Reflection at discharge nozzle	$R_2 = 7.8He - 18He^2$
Reflection at suction nozzle	$R_1 = 7.8He - 16He^2$
Transmission at discharge nozzle	$T_{21} = 0.92 - 3.38He$ or: $T_{21} = 1.16 - R_2$
Transmission at suction nozzle	$T_{12} = 1.3 - 4.85He$

The test data revealed the following:

1. Reflection increases and transmission diminishes as the frequency is increased—as has been found in the earlier studies mentioned above.
2. Plotted against the frequency, the reflection coefficients of the test pumps result in three distinct curves, Fig. 10.54-2. In contrast, all data can be represented by a single curve and correlation when plotting the data against the Helmholtz number defined by Eq. (10.19d), Fig. 10.54-1.
3. The test data, as presented in Fig. 10.54-2, can be correlated by Eq. (10.19e), where R_{vt} is the reflection coefficient for a change in cross section with the ratio $\varepsilon = A_{3q}/A_{nozzle}$ (calculated from the formula given in Table 10.12).
4. Eq. (10.19e) shows that the contribution of R_{vt} is small. Thus, the impedance of the pump seems to be dominantly determined by the effective sound velocity a_{case} in the casing rather than by changes in cross section (as from the discharge nozzle to the volute throat).
5. Trimming impeller from 219 mm to 165 mm ($d_z/d_2 = 235/165 = 1.43$) gave the same reflection and transmission coefficients as the test with $d_2 = 219$ mm. This finding confirms the dominance of a_{case}; the impeller influence seems weak.
6. Influence of pump size: The pump with $n_q = 25$ was scaled by a factor of 1.54 (to $d_2 = 336$ mm). The reflection and transmission coefficients of both pump sizes were, within reasonable scatter, equivalent when plotted against the Helmholtz number; see Fig. 10.54-1.
7. The test data in Fig. 10.54-2 (same as in 10.54-1) exhibit relative minima and maxima (which were ignored when developing the correlations). These extrema may be caused by wave interference but no detailed investigations into their origin has been reported.

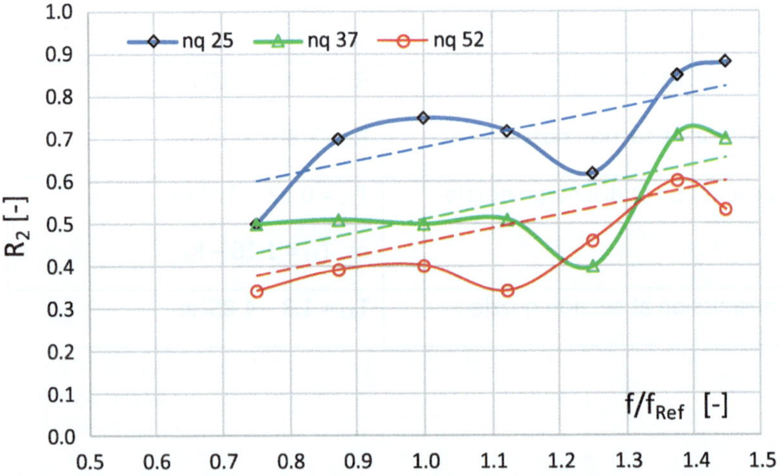

Fig. 10.54-2 Reflection of acoustic waves at the discharge nozzle as function of frequency, f_{Ref} = 200 Hz. Dashed lines are calculated from Eq. (10.19e)

$$R_2 = 0.06 + R_{vt} - 0.32\left(1.45 - \frac{f}{f_{Ref}}\right) \quad \text{with} \quad f_{Ref} = 200\,\text{Hz} \qquad (10.19e)$$

Assuming that the blade passing frequency is the strongest potential source for acoustic resonances with standing waves, the pressure sensors in suction and discharge pipe should be placed according the following rules given in [149]: $f = f_{BPF}$

- First sensors: distance from the flanges: $L_1 = 5D$
- The sensors must not be equidistant
- Wave length to be considered: $\lambda_{BPF} = a_{pipe}/f_{BPF}$
- The distance ΔL between sensors must not be equal to multiples of half the wave length $\nu \times \lambda_{BPF}/2$ with $\nu = 1, 2, 3, \ldots$
- Put some margin between the pressure nodes and the sensor positions according to $\lambda/2 + \lambda/10 > \Delta L < \lambda/2 - \lambda/10$
- The minimum distance between sensors should exceed the double wave length of sampling frequency: $\Delta L_{min} = 2\,\lambda_{samp} = 2a_{pipe}/f_{samp}$

Fundamentally, the damping or excitation effect of a component is strongest if it is placed in a velocity maximum, respectively in a pressure node. The following recommendations can be given for the arrangement of components in a piping system and for analyzing pipe vibration problems:

1. The risk of resonances of the pump speed, the blade passing frequencies and their harmonics with standing waves in dead-end pipes should be checked. A typical arrangement is the minimum flow pipe which is closed during normal operation. The length of the dead-end pipe should be selected so as to avoid this type of resonance. In order to exclude resonances with high-order standing waves, *dead-end pipes should be selected as short as possible.*
2. In pipe systems with many branches, frequencies below about 30 Hz are of particular relevance since they are the least damped.
3. All damping components, such as pumps with a stable Q-H-curve, valves, orifices and filters (i.e. any local resistance) should be placed in velocity maxima (respectively in pressure nodes) in order to maximize system damping, Fig. 10.55. *The damping provided by a component is directly proportional to its pressure loss.*
4. Pumps with unstable or flat Q-H-curves should not be placed in or near a velocity maximum (pressure node), in order to reduce the risk of self-excited vibrations. The detrimental effects of an unstable or flat Q-H-curve are minimized if the pump is placed in a velocity node.
5. Of primary concern are system frequencies of up to 30 Hz and discrete frequencies where excitation by the pump is high; such are the rotational and the blade passing frequencies and their harmonics. Actually, problems with excessive pipe vibrations have been observed at frequencies well above 30 Hz without evidence of excitation by discrete frequencies generated by the pump.

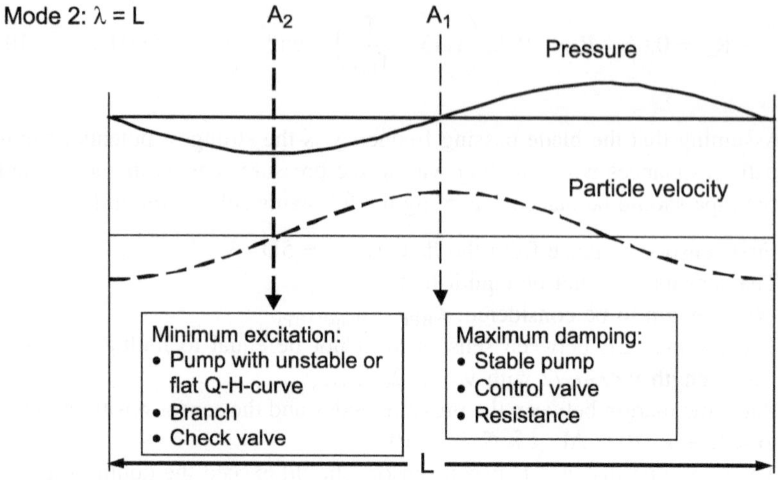

Fig. 10.55 Optimum placement of components with respect to standing waves

6. Acoustic resonances can be diagnosed by measuring vibrations at various speeds or sound velocities (fluid temperatures). Since system damping is usually low, resonance peaks are expected to be sharp so that relatively small variations of the above parameters are sufficient to find out whether an acoustic resonance is present - unless there are lock-in effects.

In order to analyze or prevent operational problems, the pipe system can be analyzed by means of simplified acoustic models, [12, 127]. In doing so, the primary objectives are: (1) To maximize system damping by selecting the pipe length between the different components (pumps, flow resistances and variations in cross section) according to the rules given above; (2) To avoid resonances of standing waves with the harmonics of the rotational frequency, blade passing and possible vortex streets.

If the pump analyzed operates at variable rotor speeds, with different fluids or temperatures (hence with different sound velocities), the investigation has to cover the entire operation range required. An example may illustrate how critical acoustic resonances can be, [46]: when using an impeller with 5 blades, an amplification factor of nearly 5 was observed in the pressure pulsations at a water temperature of 190 °C (leading to resonance with a structural eigen frequency of fuel rods). Upon changing the impeller to 7 blades, much weaker resonance amplification occurred at 240 °C.

Details of acoustic problems in closed systems can be found in [59].

10.12.4 Hydraulic Excitation by Vortex Streets

Vortex streets caused by alternating flow separation and vortex detachment are encountered in the wake flow of cylinders or bluff structures in cross flow but also on thick plates in parallel flow.

Each individual vortex originates at the separation point between the attached boundary layer and the stalled flow zone downstream of the body. A vortex forming on one side of the body momentarily influences (retards) the flow separation on the second side of the body. In this way the vortex shedding on both sides does not occur simultaneously but with a time interval Δt_{VS}.

As a consequence, the vortex shedding is periodic with a frequency f which can be described by a Strouhal number $S_{Str} = f \times \delta_w/w$ according to Table 10.13. Here, w is the approach flow velocity; δ_w is a characteristic length which can be approximated by the diameter of a cylinder, the thickness of a plate or a profile, or in general by the width of the wake caused by the body. The frequency f is the number of vortex pairs shed per second from the structure: $f = 1/(2\Delta t_{VS})$.

The lift of a profile fluctuates with the frequency f as defined above. In contrast, a body placed in the wake flow is excited by the total number of vortices swept by; i.e. by the frequency $f_w = 2 \times f = 1/\Delta t_{VS}$). The drag of the profile also fluctuates with the frequency $2 \times f$. Since alternating vortices can have various effects, both frequencies f and $2 \times f$ should be considered when a vibration problem is analyzed

. The location of flow separation on a bluff body (with sharp edges) is independent of the Reynolds number; consequently, the Strouhal number is also independent of the Reynolds number. If, in contrast, the location of flow separation varies with the Reynolds number (as, for example, on a cylinder) the Strouhal number does depend on the Reynolds number.

While vortex streets are quite stable in laminar flow, vortices decay quickly in highly turbulent flow so that little influence of the vortex street has to be assumed after a length of $x > 10 \times \delta_w$ downstream of the body.

Apart from the vortex frequency, the amplitudes of pressure pulsations or unsteady lift forces are of interest. These depend on the geometry of the body exposed to the flow. The maximum amplitudes are found for bodies with $x/\delta_w = 0.2$ to 1.2 when δ_w is the thickness of a bluff body and x is the length of the downstream extension after δ_w (see sketch in Table 10.13). This range should definitely be avoided when designing trailing edges of any kind, even trash racks, [28]. The trailing edges of profiles or ribs should be optimized so as to minimize the amplitudes of the perturbations created by the vortex street. Table 10.13 provides the necessary information and lists typical Strouhal numbers for a wide range of applications.[8] Vortex streets shed from ribs upstream of an impeller can interact with the flow around the blades. Vortex shedding in various components can also excite standing waves in the system and induce pipe vibrations.

---

[8]From the wealth of literature on flow-induced vibrations, [9, 65] are quoted.

Table 10.13 Excitation of vibrations by vortex streets

Component	Values from [17] and [18]				S_{Str}	$\hat{\zeta}_a$
Cylinder $S_{Str} = f(Re)$		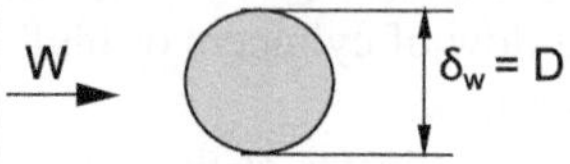			0.2 to 0.3	0.3 to 0.7
Quadratic profile		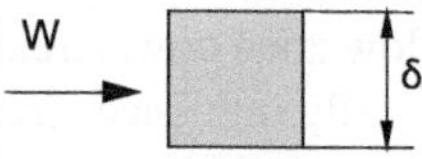			0.125	0.55
Rectangular profile	$\frac{L}{\delta_w} = 0.5$	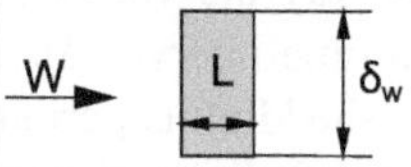			0.17	0.45
	$\frac{L}{\delta_w} = 2$	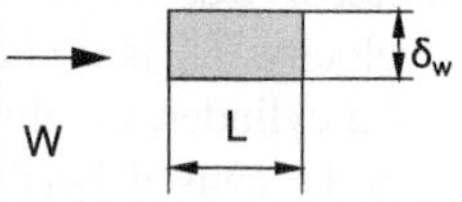			0.068	0.3
Flat plate in parallel flow width B	**Trailing edge**	**Geometry**			0.2 to 0.24	**RA**
	Blunt	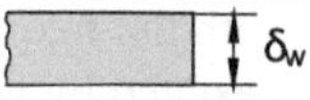				1
	Semi-circle					2.6
	Symmetrically sharpened		α	30°		≈ 0
				45°		0.45
				60°		4
				90°		3
	Sharpened on one side	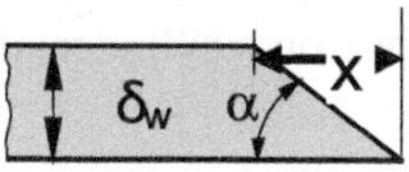	α	<10°		> 2
				30°		0.1
				45°		0.4
				60°		0.5
	Sharpened on one side and rounded	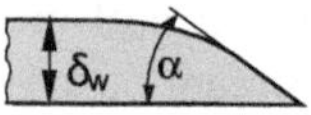	α	30°		< 0.2
				45°		< 0.5
	Hollow					< 0.5
Check valve [20]	δ_w = width of separated flow				0.08 to 0.32	
Flow divider (T-branch) [89]	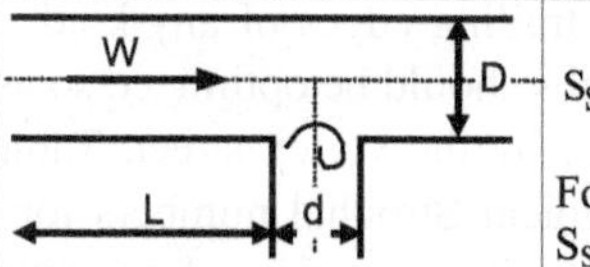	$S_{Str} \equiv \frac{f\,d}{w} = 0.8\left(\frac{d}{D}\right)^{0.27}\left(\frac{L}{D}\right)^{-0.08}$ Formula valid for L/D < 15; for L/D > 15 S_{Str} does not decrease further				
Dynamic lift coefficient (amplitude)	$\hat{\zeta}_a = \frac{2F}{\rho w^2 \delta_w B}$	$S_{Str} = \frac{f \delta_w}{w}$	RA = relative amplitude Reference: amplitude with blunt trailing edge			

The design of the trailing edge (TE) of a structure is central for the excitation forces generated by vortex shedding, which depend critically on how the vortexes detaching from both sides of the structure interact. The physical mechanisms controlling this process have been investigated on a hydrofoil (NACA 0009) with (a) a truncated and (b) an oblique trailing edge bevelled at 30° which resulted roughly in $x/\delta_w = 1.7$, [142]. The Strouhal number of the hydrofoil was about $S_{Str} = 0.22$. The vortex shedding locked in with an eigen frequency of the hydrofoil in a range of about ±20% of the resonant frequency. Vibration amplitudes were negligible outside the resonance and lock-in. The resonant vibration amplitudes of the bevelled profile were about 40% lower than those of the truncated foil. Flow observation showed that the vortexes detached from the sharp corners of the profile (boundary layers were fully attached on the upper and lower profile surfaces). With the bevelled TE, the vortex separation at the lower surface is shifted with respect to the detachment at the upper surface that way that the upper and lower vortexes collide. Part of their energy is thus broken up into turbulence and the excitation forces and vibration amplitudes are reduced.

Damage can occur if vortex shedding frequency gets into a resonance with a structural eigen frequency. A simple example is a thermocouple protection tube or a velocity probe mounted in a pipe. A resonance between the vortex frequency and the first bending eigen frequency of said tube could lead to breakage in a short time.

The excitation forces, hence the movements of the vibrating body, are perpendicular to the approach flow velocity. The lateral amplitudes have a strong influence on the flow around the body hence on the vortex shedding. As a consequence, the vortices become more organized and the dynamic lift or excitation forces increase. Over a certain range of the approach flow velocity, the vortex frequency is then controlled by the eigen frequency of the body. This phenomenon is known as "lock-in" effect. An example for the lock-in effect is provided in [28] where the oscillating torque acting on a stay vane was measured. The vibration frequency locked in at the eigen frequency of 300 Hz over a flow range of about ±16 %, while the torque amplitude was virtually zero outside the range of resonance and lock-in. Large-amplitude vibrations (say $y/D = 0.4$ with y = vibration amplitude and D = cylinder diameter) can lock-in up to ±40 % of the eigen frequency, [9]. In order to avoid the vortex shedding frequency f locking in on the eigen frequency f_e of a structure the range $0.7 < f/f_e < 1.3$ should be avoided, [9].

According to [28] stay vane failures due to vortex shedding were frequent until the rules similar to Table 10.13 for designing the trailing edges were developed. Another issue is high tonal noise generated by vortex shedding in hydraulic turbines. High noise up to 108 dB-A was observed at 400 Hz on a Francis turbine were the runner blade trailing edges had too short a profile with about $x/\delta_w = 0.5$. After a minor correction of the trailing edge to $x/\delta_w = 1.5$ the noise level dropped by 15 dB-A, [28]. The noise depended strongly on the load (flow rate).

With the possible exception of large storage pumps, noise and vibration problems with pumps are rarely reported. The reasons might include: (1) structural eigen frequencies are sufficiently above the vortex shedding frequencies so that resonance is avoided; (2) the close distance between impeller and diffuser vanes prevents the formation of organized vortex streets; (3) the flow patterns in decelerated flow are more irregular than in accelerated flow through turbines; flow patterns in pumps are less prone to fostering the formation of vortex streets.

Table D10.2 Excitation forces in 90°-bend at R/D = 1.0 according to [71] $S_{Str} = f\,D/c_{ax}$

	Axial excitation		Lateral excitation		Force definition
Re	S_{Str}	F_{ax}* [RMS]	S_{Str}	Fy* [RMS]	
5000	0.3	0.043			
10,000	0.3	0.028			$F^* = \dfrac{F}{\frac{\rho}{2} c_{ax}^2 D^2}$
27,000	0.2	0.019	0.014 – 0.0055	0.0082 – 0.0063	

Vortex streets can also be generated in T-branches, check valves, orifices and butterfly valves. T-branches produce vortex streets when the flow passes through the main pipe. The vortex street can excite standing waves in the branched-off pipe (compare to the sound generated when blowing over the end of a pipe). The frequencies of the vortices are described by a Strouhal number $S_{Str} = f \times d/w$ which is calculated with the diameter d of the *branched-off pipe* and the flow velocity w in the *main pipe*. The Strouhal number depends on the ratio d/D and on the distance L from a possible bend upstream of the T-branch, Table 10.13. If there is no bend or if the distance is $L/D > 15$, the formula has to be evaluated with $L/D = 15$.

According to investigations in [71], excitation forces are induced by a flow through a bend. These forces act axially and laterally to the pipe axis (problems occurring in practice due to this effect were not reported). The *axial* forces are created by a periodic roll-up of the shear layer between stalled fluid and the main flow. The highest force amplitudes in the spectrum were found at Strouhal numbers of 0.2 to 0.3 (like in other components with separated flow as shown in Table 10.13).

The vortices of secondary flow in a bend (Fig. 1.12) are unsteady in that the vortices change in size and position. Low-frequency forces acting *laterally to the pipe axis* are the consequence. Frequencies and RMS force amplitudes can be estimated from Table D10.2; when doing so, it should be noted that the forces decrease with increasing ratio R/D.

In summary, unacceptable vibrations, damage or excessive noise caused by vortex streets are only likely if resonances between the vortex shedding frequencies and structural or acoustic eigen frequencies occur. The large lock-in range of up to ± 30 % must be taken into account in the risk assessment.

10.12.5 Coupling of Flow Phenomena with Acoustics[9]

Free shear layers generated by flow separation are *inherently unstable, because the size of a vortex increases with the distance traveled downstream* (i.e. time). Vortices can interact with acoustic waves or with downstream structures. Such a coupling creates order out of chaos in that the vortices arrange in a more regular pattern. As a consequence the vortices gain energy in a smaller frequency band, [85].

Hydrodynamic coupling: When vortices impinge on a solid structure, the fluctuations can grow by the following feedback mechanism: flow processes at the location of vortex impingement modify the vortex pattern; the altered flow conditions have an impact on vortex shedding at the location where the flow separates. Figure 10.56 depicts the hydrodynamic coupling between a shear layer created by mixing two air streams of different velocities and a wedge placed downstream in the vortex street.

As an example, Fig. 10.57 shows a test where two air streams with different velocities are employed to create a shear layer. A wedge was placed in the vortex street generated in this way at a distance L from the trailing edge. The velocity fluctuations (measured at a distance x/L = 0.78) demonstrate the effect of the hydrodynamic coupling on the flow pattern and on the potential for the hydraulic excitation forces. On the right picture *without* the wedge the velocity fluctuations are only 0.45 % of the velocity v_1. When the vortices impinge on a wedge (left picture), the fluctuations increase by a factor of 18. As evidenced by the flow patterns in Fig. 10.57, the vortices become more structured in the presence of the wedge and the energy is concentrated in a smaller band. In this way the potential for exciting vibrations can rise considerably.

This coupling between separation and impingement is of a purely hydrodynamic nature since acoustic waves are not involved. The vortex frequency is directly proportional to the flow velocity. In order to avoid this type of excitation, a sufficiently long distance must be provided between the locations of separation and impingement.

Acoustic coupling: Vortex structures can interact with acoustic waves if the frequencies of the vortex street and the acoustic wave are sufficiently close to each other. The mechanism is shown in Fig. 10.58 and explained in context with the T-branch shown in Fig. 10.59, where a fluid flows in a main pipe D with a velocity w over a T-branch with diameter d. At the trailing edge TE of the branch, vortices are generated which can excite a standing wave in the branched-off pipe. At the open end, the particle velocity has a maximum and the oscillating fluid particles (w_p) have an influence on the vortex shedding, Fig. 10.59.

Resonance between vortex shedding and standing waves occurs if the vortex frequency defined by the Strouhal number of the T-branch (Table 10.13) equals the

[9]This section is largely based on information given in [89].

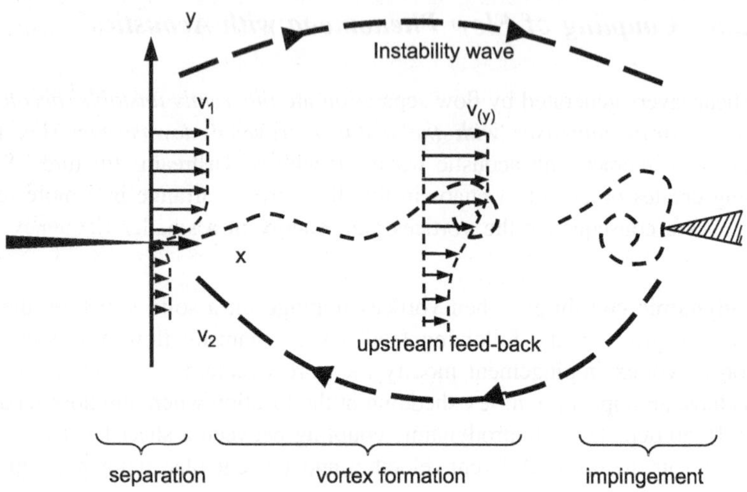

Fig. 10.56 Hydrodynamic coupling, [89]

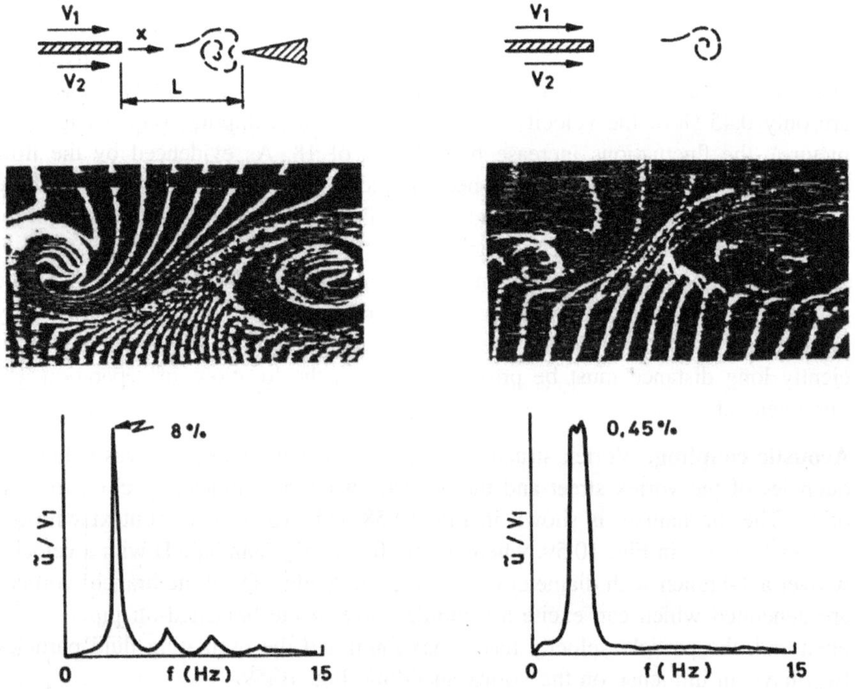

Fig. 10.57 Velocity fluctuations increased by hydrodynamic coupling, [89]

frequency of the standing wave in the branch according to Table 10.12. In the case of a dead-end pipe, this condition is fulfilled at the critical velocity according to Eq. (10.20):

$$w_{krit} = \frac{fd}{S_{Str}} = (2\nu - 1)\frac{a\,d}{4L\,S_{Str}} \tag{10.20}$$

$$w_{krit} = \frac{f(d+r)}{S_{Str}} = (2\nu - 1)\frac{a(d+r)}{4L\,S_{Str}} \tag{10.20a}$$

Resonance occurs not only when the frequencies match exactly but in a range of roughly ±10 to 20 % of the acoustic eigen frequency ("lock-in"). A corresponding range applies to the critical velocity w_{krit} given by Eq. (10.20).

In contrast to hydrodynamic coupling, the vibration frequency is largely independent of the flow velocity, because it is given by the acoustic eigen frequency of the dead-end pipe. If the flow velocity varies in a wide range, the vibration frequency may jump to the next acoustic eigen frequency.

A coupling between standing waves and vortex layers is created only if the vortex-shedding component is located in a zone where the standing wave has a sufficiently high particle velocity; that way the particle movement can have an impact on the movement of the vortices.

A possible coupling is therefore strongest if the source of the vortices - as in Fig. 10.59 - is located in a pressure node (i.e. a maximum of the particle velocity). Coupling is minimal (theoretically non-existent), if the source of the vortices is located in a pressure maximum, respectively in a node of the particle velocity. Consequently, vortex-shedding components should be arranged always as near as feasible to a *closed* pipe-end.

When rounding the corners of a T-branch by radius r, the critical velocity increases by the factor of (d + r)/d according to Eq. (10.20a). Rounding the downstream corner of the T-branch is most effective since the interaction between the acoustic wave and the vortices is maximum at this location. In contrast, the rounding of the upstream corner has little effect. In case of a simple T-ranch the amplitudes of the pressure pulsations are reduced. In tandem T-branches rounding the corners can reduce or even increase the pressure pulsations. The effect of the rounding depends on the form of the acoustic waves forming in the branches and the spacer (length L_B in Fig. 10.61).

The coherent two-dimensional vortex street can be destroyed by spoilers or inserts which create an unstructured turbulence not susceptible of exciting standing waves. The same effect is achieved by increased turbulence in the main pipe such as created, for example, by an orifice. An orifice with d/D = 0.7 in a distance of L/D = 5.5 upstream of the T-branch cut the amplitudes down to 25 % of the level without orifice, [89].

In case of resonance the pressure pulsations can attain 10-times the stagnation pressure in the main pipe. The amplitudes decrease with growing d/D. In general, only weak pulsations are expected for branches with d/D > 0.5. However, if the

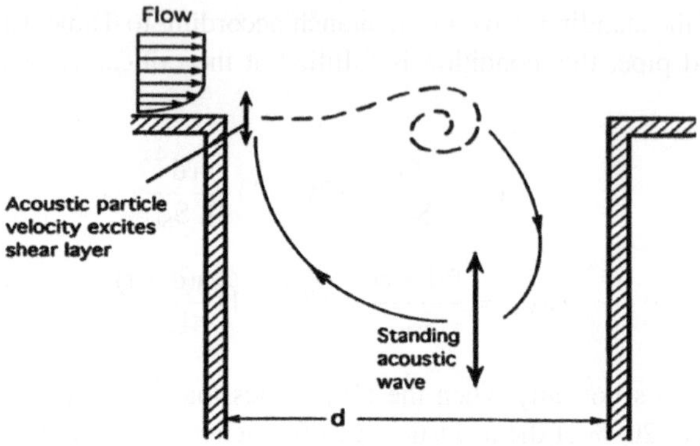

Fig. 10.58 Acoustic coupling, [89]

wave lengths in the main pipe and the branch are similar and the branch is located in a pressure node, strong pulsations are expected even with d/D > 0.5.

The ratio of the amplitudes to the stagnation pressure in the main pipe depends on the acoustic properties of the pipe system. While the amplitudes created by a single T-branch decrease with an increasing ratio d/D, the pulsations rise strongly if T-branches are co-axial or arranged in series (in "tandem") because the waves can reinforce each other, as shown by the measurements in Fig. 10.60. In order to avoid this type of resonance amplification, the piping arrangement can be detuned by executing the branches with different lengths $L_1 \neq L_{2.3}$. With co-axial branches a difference in length as small as $(L_1 - L_3)/D = 1$ is sufficient to cut the amplitudes to a fraction of the level with $L_1 = L_3$. In case of a tandem design, effective detuning is

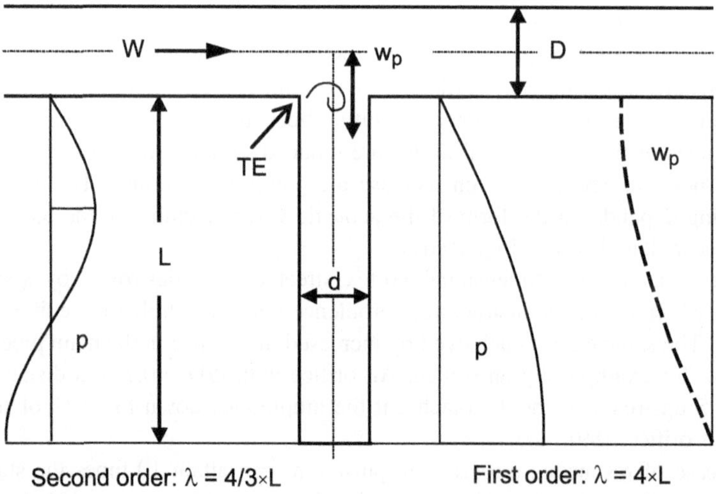

Fig. 10.59 Interaction of a standing wave in a dead-end pipe and vortex shedding in a T-branch; w_p = velocity (particle velocity); p = pressure wave, [89]

accomplished only if the lengths are selected according to the condition $(L_1 - L_2)/L_B \geq 1$, Fig. 10.61.

A coupling between vortex streets and acoustic waves can be induced by various components and thus lead to high pressure pulsations and pipe vibrations. In one example reported in [20], vortices shed by a check valve coupled to the second order of standing waves in the pipe between the feed water tank and the control valve of a thermal power station. By displacing the control valve, the distance between check and control valve was reduced. The vortex shedding then occurred in a location where the particle velocity was small and the coupling (and the vibrations) vanished.

If sound waves are able to interact with vortex streets, a coupling of acoustic waves in the system with unsteady flow phenomena in the diffuser or impeller is equally conceivable. In particular a coupling would be expected when operating a pump with an unstable Q-H-curve. Influenced by acoustic waves, the flow in the pump could switch between (two) quasi-steady patterns as discussed in Chap. 5.4.2 . Up to which frequencies a possible coupling could be active is open to debate.[10] In the tests discussed in Chap. 10.7.1 (Fig. 10.23) such coupling had been observed even at blade-passing frequency. Due to the postulated interaction, strong waves generated in the system (for example, by an unsuitable throttle valve) could reinforce the pressure pulsations created by the impeller and cause damage to the impeller shrouds, tie bolts, or other components.

The coupling between flow and acoustic waves is governed by the particle velocity w_p (and not by the much higher sound velocity). The order of magnitude can be estimated from $w_p = p_{dyn}/(\rho \times a)$; thus a pressure amplitude of $p_{dyn} = 5$ bar in cold water is linked to a particle velocity of about 0.3 m/s.

Reference [5] reports on high noise and severe wear in safety relief valves caused by λ/4 waves (valves are closed). The excitation mechanism was found to be vortex shedding in the dead-end pipe, as shown in Fig. 10.59. Shortening the dead-end pipe on which the valves were mounted increased the frequency of the noise but did not stop wear. When the system was modified to a conical branch with rounded entrance (Fig. 10.62), the loud noise stopped and there was no more wear on the valves. The cone had an opening angle of about 18° per side, the inlet to outlet area ratio was 2.0 and the rounding r/d = 0.2. The mechanism causing this improvement may be due to the modified Strouhal number (Table 10.13 and Chap. 10.12.4) or to the changed reflection conditions in the cone.

---

[10] The possible interaction between acoustic waves and unsteady flow in a pump could be verified experimentally by generating pressure pulsations in a system by means of a device (e.g. piston or spark) and measuring their impact on the flow patterns in the diffuser and impeller (e.g. by laser velocimetry).

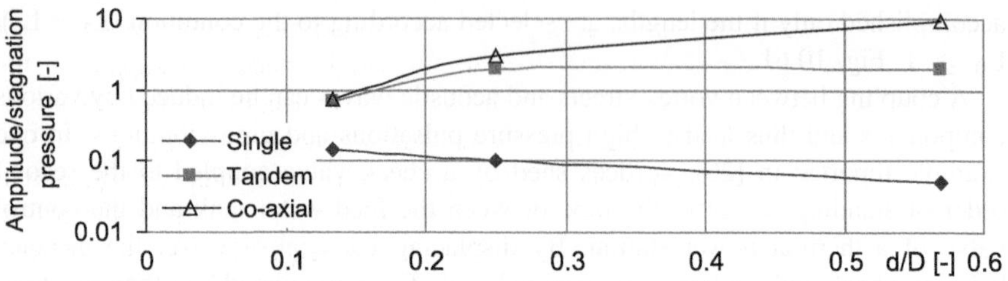

Fig. 10.60 Pressure pulsations created by a T-branch, [89]

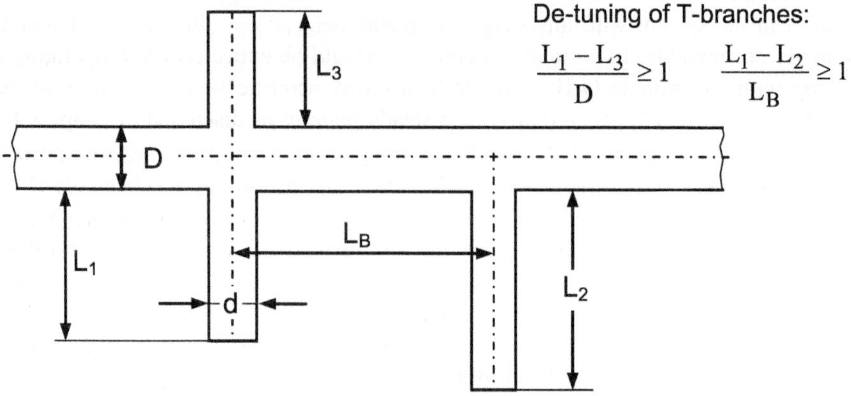

Fig. 10.61 T-branches in co-axial or tandem arrangement

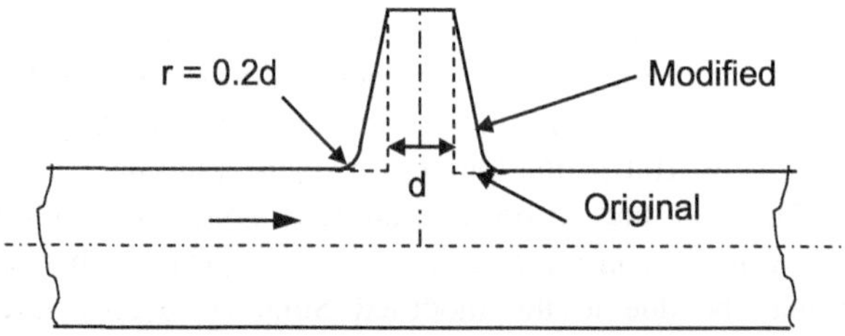

Fig. 10.62 Modified T-branch to eliminate excessive noise and valve wear [5]

10.12.6 Pipe Vibration Mechanisms

Pipe vibrations piping vibrations" are generated by volumetric flow fluctuations in the system, which are inherently linked to pressure pulsations. The oscillation of flow and pressure induces elastic deformations of the piping. Two mechanisms could be responsible for this (1) Flow rate oscillations represent non-balanced mass forces which provoke reaction forces according to the conservation of momentum.

(2) Time-dependent pressures in a pipe induce corresponding dynamic stresses in the pipe wall resulting in elastic deformations which create the vibration.

The dynamic stress in the direction of the pipe axis is $\sigma_{dyn} = p_{dyn} \times D/(4 \times h)$ where p_{dyn} is the pressure amplitude and h is the wall thickness of a pipe with diameter D. Because of the large mass which would have to be accelerated in a longitudinal vibration and because of asymmetries, the pipe deforms laterally and primarily bending vibrations are generated. According to this hypothesis, the stresses would be in phase with the pressures. The *mechanical vibration modes would then follow the acoustic modes*[11]. Consequently, the greatest pressure amplitudes would be measured at the location of the largest pipe deflection.

In this context, consider a pipe with three legs which connect two components of the system, Fig. 10.63. This type of subsystem, which can be found in many plants is characterized by: (1) The components (due to their mass) are fixed points which form nodes of the mechanical vibration; (2) The components represent a strong change of impedance for acoustic waves. At least some portion of the wave energy is reflected at the components which embody nodes of the standing waves.

Since waves are not reflected by bends, pressure distributions as sketched in Fig. 10.63 are expected (plotted against the developed pipe length). In Fig. 10.63a the solid line represents the pipe under the static operation pressure p_{stat}, while the dotted lines indicate the deformations created by a change in pressure (say by $\pm 10\%$ of p_{stat}). If the pressure variations are unsteady (as created by waves), the pipe experiences complicated unsteady deformations. When comparing the modes 1 and 3 shown in Fig. 10.63, it becomes evident that various mechanical vibrations modes can be excited depending on the acoustic mode.

The following conclusions can be drawn from the mechanisms discussed in Chap. 10.12:

- A pipe is excited by acoustic waves primarily at its eigen modes. Pipe vibrations exceeding allowable limits therefore occur at *resonances between mechanical and acoustic modes.*
- Elements for pipe support and vibration damping are not effective at reducing vibrations if they are located near a node of the mechanical mode (i.e. where pipe displacement is small or even zero). For an optimum arrangement of such devices, the *reflection characteristics* of the components connected by the piping and the *acoustic and mechanical eigen modes* must be known. To appreciate this, consider Fig. 10.63: If the components reflect as open pipe-ends (Fig. 10.63b), a damper or a support placed in the center of the pipe (line A1) would influence only the 1st and the 3rd order, but the pipe could vibrate unimpeded in the 2nd and 4th order. If, in contrast, the components reflect as closed pipe-ends (Fig. 10.63c), a damper in the center of the piping would be altogether ineffective with respect to mode 1 (and all odd modes), while it would

---

[11]If the excitation mechanism were caused by unbalanced mass forces, the mechanical vibration would be out of phase with the excitation.

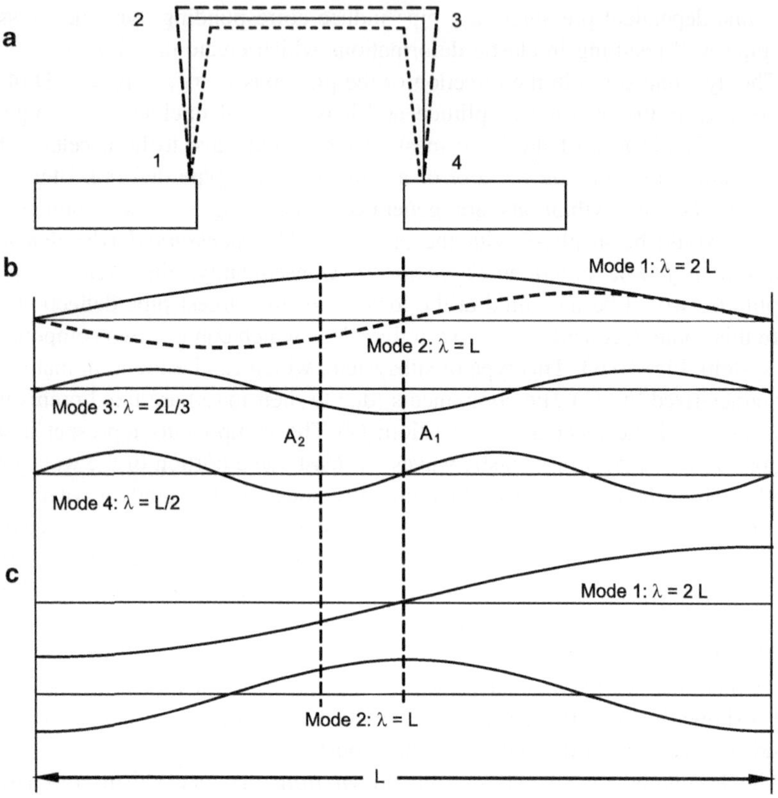

Fig. 10.63 Pressure waves in a pipe; **a** static deformation of the pipe under internal pressure; **b** eigen modes, if the components reflect as open ends; **c** eigen modes, if the components reflect as closed ends

be in the optimum position for reducing vibrations with mode 2 (and all other even modes).

- In case of doubt, it may be considered to arrange pipe supports between two reflecting components at irregular intervals, line A_2 in Fig. 10.63.
- The preceding argument applies equally to placing pressure transducers for measurements: the level of pressure pulsations recorded at a specific location depends on the wave form as demonstrated by Fig. 10.63. In order to define appropriate locations of pressure transducers with the objective of measuring and analyzing problems of pressure pulsations or pipe vibrations, an assessment of the most probable excitation mechanisms should be made.
- Instrument lines should be as short as possible, otherwise $\lambda/4$-waves according to Table 10.12 (pipe open at one end) could strongly falsify the measurements.
- Flow resistances should be placed in pressure nodes (maxima of particle velocity) in order to maximize system damping, Figs. 10.55 and 10.63.
- It should be attempted to create predictable locations of wave reflection.

- Dead pipe-ends should be as short as feasible in order to push the frequencies of standing waves as much as possible above the excitation frequencies.
- T-branches should be placed in pressure maxima (velocity nodes) of the main pipe in order to avoid an acoustic coupling of the waves in the main pipe and in the branch.
- Tandem or co-axial branches should be made of different lengths in order to detune the piping, Fig. 10.61.
- Vortex-generating components should be placed near a closed pipe-end (velocity node), in order to avoid a coupling between the waves and the vortices (by no means should they be located near an open pipe-end, i.e. in a maximum of the particle velocity).
- Components with strong vortex shedding should be placed at a distance sufficiently long to avoid a hydraulic coupling between the components; conversely, the distance should not be too large in order to prevent a coupling of the vortex streets with low-frequency acoustic waves. This means that components with strong vortex shedding should be installed as close together as possible, but with a minimum distance of $L/\delta_w = 10$ to 20, where δ_w is the width of the wake of the body shedding the vortices. An example is a check valve installed upstream of a control valve [20]: If the distance between both valves is short, the vortices shed from the check valve remain near a velocity node. Excitation is correspondingly low (the control valve reflects in a similar way as a closed pipe-end because of its drastic change in cross section).
- If standing waves are excited in a dead-end pipe by a vortex street, a small flow through said pipe can break up the vortices and thus reduce the vibrations, [19].

In practice, optimizing the piping design for minimum vibrations is not easy, since there are many acoustic and mechanical eigen frequencies to be found in a complex system. Much of the acoustic energy is broadband; therefore, standing waves at any discrete frequency can be excited. Furthermore, the determination of the eigen frequencies is uncertain, because the frequency-dependent reflection characteristics of the components are not well known. Last but not least, there are usually a lot of boundary conditions and plant design criteria which restrict the options for designing the piping solely on vibration criteria.

Pipelines must be installed in a way to allow free expansion under operational temperatures and stresses. The resulting flexible supports often are unable to prevent the pipe from vibrating. In power plants and in the process industry viscous damping devices are used in critical cases in order to keep pipe vibrations within acceptable limits. A viscous damper consists of a cylinder filled with a viscous liquid and a piston immersed in said liquid, Fig. 10.64. The piston is rigidly fixed to the pipe, while the cylinder is rigidly fixed to the pipe support structure (e.g. a wall) - or vice versa. The piston can move easily in 6 degrees of freedom (displacement and rotation) within the cylinder. The damping provided by such movement is ideally proportional to the velocity of the relative movement between piston and cylinder. In practice, the damping decreases with the frequency of the pipe vibrations. The damping effect is due to viscous shear and liquid displacement.

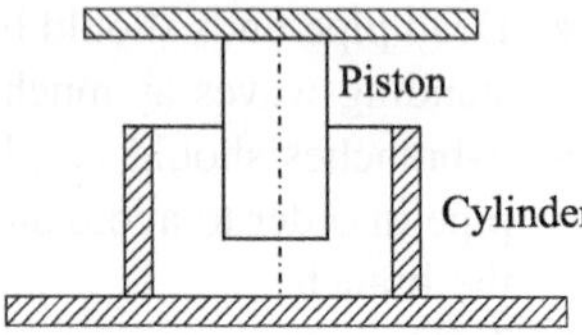

Fig. 10.64 Viscous pipe vibration damper, [68]

The viscous fluid is selected according to the operation temperature of the fluid. The fluid temperature is only 20 to 30 % of the liquid temperature in the pipe. The dampers can carry no static loads; these must be taken by pipe elastic hangers or similar devices. The damper allows free thermal expansion/contraction of the pipe with virtually no forces because such expansion is a very slow process. These dampers are also effective when shock loading occurs on the pipe (water hammer, earthquake). The dampers must be installed at locations where pipe vibrations are close to maximum; they are ineffective at vibration nodes. Several small dampers are more effective than one large damper, [68].

Table D10.3 provides criteria according to ISO 10816-1 for the assessment of pipe vibration velocities (RMS-values). Accordingly, the allowable vibration velocity is about 10 mm/s RMS (this follows also from other standards). Note that peak values are considerably higher than RMS-values (in measurements in a specific plant peak values were up to 5-times higher than RMS).

Table D10.3 Allowable pipe vibration velocities according to ISO 10816-1. (Simplified representation; for information only; refer to newest edition of the standard!)

Vibration velocity (mm/s) RMS	Assessment
<3.5	Good
3.5 – 7	Acceptable
7 – 16	Improvement recommended
>16	Not acceptable

10.13 Torsional Vibrations

Under steady torque an elastic shaft is twisted by the angle $\vartheta = 32 \times M \times L/(\pi \times d^4 \times G)$ where M is the torque, L is the length of the shaft, d is the shaft diameter and G is the shear modulus; $G = E/(2 + 2\nu)$ (for steel: $G \approx 8 \times 10^{10}$ N/m^2 with Poisson ratio $\nu = 0.3$). If a fluctuating torque is applied to an elastic shaft, the twist angle becomes time-dependent; in other words: torsional vibrations are generated.

If the driver provides a periodic torque, forced vibrations are created which can get into resonance with a torsional eigen frequency of the drive train. Shock-like events or transient excitations which apply tangential forces on the rotor generate free vibrations at the torsional eigen frequencies of the rotor. Transient loads occur during start-up of an electric motor or during accidental events such as short-circuits. Then the load-factor of the torque can be a multiple of rated torque and this event should be evaluated with a low-cycle fatigue analysis in order to ensure the integrity of the shaft. This stress analysis must include all stress concentration factors (e.g. at key ways). A minimum safety factor of 2 against the corrosion fatigue strength is recommended in [B.15]. A forced response stress analysis with respect to high-cycle fatigue is also warranted if a resonance cannot be avoided. The highest shear stress τ occurs at the location with the highest gradient $d\vartheta/dx$ because of $\tau = \frac{1}{2} \times G \times d \times d\vartheta/dx$ which leads to the well-known relation $\tau = 16\ M/(\pi d^3)$. Transient loading exciting a torsional eigen frequency also occurs during a reversal of the sense of rotation in reverse flow through a vertical pump after shut-down in the absence of a check valve.

While lateral vibrations and pressure fluctuations can be a cause of trouble in almost any pump application, torsional vibrations are rarely reported as a problem. This observation is borne out by: (1) a wealth of literature on lateral vibrations and the comparatively few publications on torsional vibrations; (2) monitoring equipment for lateral vibrations is almost a standard for large pumps, while torsional monitoring is the exception rather than the rule; (3) incidences of excessive torsional vibrations in pump applications are quite uncommon.

Since severe torsional vibrations could lead to motor damage (loosening of windings), coupling wear, fretting on components fastened by a shrink fit to the shaft, gear failures or shaft failures due to fatigue, a torsional analysis is required in some pump applications (see [51] for a detailed discussion). These include:

- The driver provides a periodic torque. Examples are electric motors with variable frequency or reciprocating combustion engines. If a variable speed drive is employed, operation is within a certain range of speeds. Consequently, the design should avoid any torsional eigen frequency in the specified speed range. If such is impracticable, the rotor should be designed so that any resonance should occur at as low a speed as possible in order to minimize excitation forces.
- A right-angle gear is mounted between a horizontal driver and a vertical pump. In this case there is also a coupling between lateral and torsional vibrations.
- The excitation frequency of the driver coincides with, or is sufficiently close to, a torsional eigen frequency of the rotor train
- Large pumps driven by steam or gas turbine may need to be analyzed because of the wide speed range and the use of flexible couplings.
- The specified level of excitation forces which are transmitted to the pump foundation is very low.
- Vertical pumps with long shafts

Vertical pumps with long column pipes (Figs. 2.14 and 2.15) will have low torsional eigen frequencies. The need for a torsional analysis of a vertical pump can be inferred from estimating the lowest eigen frequency of a simple rotor consisting of the impeller with mass moment of inertia J_{imp}, a shaft with spring constant k_{res} and a motor with mass moment of inertia J_{mot}. The shaft may be built from n sections with diameter d_i and length L_i. The spring constant of every shaft section is given by Eq. (10.21). The resulting spring constant is calculated for n springs in series from Eq. (10.22). The first natural frequency $f_{1,f\text{-}f}$ of the rotor with *both ends free* follows from Eq. (10.23), [8]. In contrast to lateral vibrations, the bearings have virtually no influence on torsional eigen frequencies. For this reason the estimate provided by Eq. (10.23) can be quite close to the eigen frequency calculated by a finite element analysis which is indicated if Eq. (10.23) predicts an eigen frequency close to the speed of the rotor.

$$k_i = \frac{\pi d_i^4 G}{32 L_i} \quad i = 1 \text{ to } n \tag{10.21}$$

$$k_{res} = \frac{1}{\frac{1}{k_1} + \frac{1}{k_2} + \cdots \frac{1}{k_n}} \tag{10.22}$$

$$f_{1,f-f} = \frac{1}{2\pi}\sqrt{\frac{k_{res}\,(J_{imp} + J_{mot})}{J_{imp} J_{mot}}} \tag{10.23}$$

Equation (10.23) can also be helpful when doing a sensitivity study or for assessing the possible benefits of corrective action such as detuning the rotor.

When performing an impact test to determine the eigen frequencies of a rotor of a vertical pump at stand-still, the motor rotor is not able to move because of the high resistance of the thrust bearing which is subject to the whole weight of the rotor. In this case the eigen frequencies measured correspond to the boundary condition "clamped-free". The first eigen frequency $f_{1,c\text{-}f}$ of the rotor can then be estimated with the mass moment of inertia of the impeller from Eq. (10.24). The mass moment of inertia of the shaft J_{shaft} can be taken into account, but the effect is usually negligible. It should be noted that, under the described conditions, the impact test is unable to yield the true eigen frequencies of the rotor in operation.

$$f_{1,c-f} = \frac{1}{2\pi}\sqrt{\frac{k_{res}}{J_{imp} + J_{shaft}/3}} \tag{10.24}$$

The mass moment of inertia for a cylindrical body of diameter d and length L (shaft, disk, motor rotor) is calculated from Eq. (10.25):

$$J = \rho_{mat}\frac{\pi d^4 L}{32} \tag{10.25}$$

The mass moment of an electric motor with rated torque M_R can be estimated from Eq. (10.26) according to [B.15], but the data scatter is ±35% (and even higher). Presumably, for low-speed motors the upper range is more relevant.

$$\frac{J}{J_{ref}} = 3.5 \times 10^{-4} \left(\frac{M_R}{M_{ref}} \right)^{1.38} \quad J_{ref} = 1\,kgm^2 \quad M_{ref} = 1\,Nm \tag{10.26}$$

The system to be analyzed includes all rotating components on the drive train: motors, gears, couplings, pump and auxiliary equipment (e.g. a booster pump driven by the same motor). When doing a torsional rotor dynamic analysis, eigen frequencies, eigen modes, damping and forced response are addressed in a similar manner as in a lateral analysis, Chap. 10.6.5. The added mass of the water acting on the impeller and coupling stiffness need particular attention.

There is also a coupling between torsional vibrations of the shaft and lateral vibrations of the column pipe. Gyroscopic effects may be thought to be involved in this type of coupled vibrations.

Constant speed asynchronous and synchronous motors provide the following excitation frequencies, [B.15]: (1) synchronous $f = f_n$, (2) number of poles times slip frequency $f = (f_{grid} - f_n) \times p$, (3) $f = p \times f_n$ (where p is the number of poles). Variable frequency drives of former technology provided a significant excitation through the "control pulse" frequencies, [B.15]. State-of-the-art variable frequency drives (VFD) trigger the sinusoidal electric current quite accurately, so that nearly no torsional excitations occur, [51]. Steam and gas turbine drives give an excitation frequency of driver speed times the turbine blade number.

Typical excitation torque is 1 to 5 % of rated torque; the higher values are for partload operation and impellers with less than 5 blades, [B.15]. Higher torque fluctuations may be expected when operating with vortexes or high air content (expulsion of air from the discharge pipe after start-up with an incompletely vented discharge pipe). In a test of a semi-axial vertical cooling water pump (with about 3000 kW, length 10 m, $n_q = 85$) the following torque fluctuations have been measured: near BEP ± 2 %, partload ± 10 % and during air expulsion from discharge pipe up to ±18 % of average torque. Torque fluctuations for single-channel impellers can be considerable; some test data are given in Chap. 7.4.

Although there are hydraulic phenomena which induce torque fluctuations, these are apparently sufficiently damped. It is not known that they could lead to self-excited vibrations. A possible explanation for this is the following: Suppose a mechanism tends to accelerate the rotor in circumferential direction leading to an increase in speed by Δn. As a result, the power (resisting the speed increase) grows by the factor $(1 + \Delta n/n)^3$. After reaching the maximum speed amplitude, the rotor is retarded to the speed $(n - \Delta n)$. During this part of the cycle the impeller operates in the brake (or turbine) mode. Hence the energy fed back from the liquid to the rotor is small or even nil.

Another argument may be as follows: If there were hydraulic phenomena which excite torsional vibrations, this should affect all pumps and motors. The phenomenon should have surfaced at least in some of the millions of pumps in service.

Small torque fluctuations at blade passing frequency and at $f = z_{La} \times z_{Le} \times f_n$ are present in all pumps due to the rotor/stator interaction manifest through pressure pulsations. Torque fluctuations at lower frequencies can be caused for example by:

- uneven approach flow velocity
- entrained air
- surface, bottom and wall vortices as can be present in intake structures, Chap. 11.7.3. Since these vortices are not steady, the rotor frequencies observed shift around the rotational frequency depending on the momentary location of the vortex. Hence the excitation frequency is $\nu \times f_n$ where ν is *no* integer.

In case operation is close to a resonance, the following may be expected:

- Structural damping is low and amplification factors are consequently high. However, excitation forces may be low and stresses might still be acceptable. Damping determined from the torsional resonance curves of a large vertical semi axial pump was about D = 0.05.
- Low damping implies sharp peaks; therefore, a separation margin of 10 % may be sufficient depending on the accuracy of prediction, [51].
- Potential damage includes shaft ruptures due to fatigue at locations of stress raisers (e.g. key ways).
- The reaction of the driver on load rejection upon sudden shaft rupture should be assessed in the risk analysis.

Unfortunately, torsional vibrations are not detected by the usual vibration monitoring instruments, be it shaft displacement and/or accelerometers on the bearing housing or motor of a closed coupled vertical pump. For this reason, problems due to torsional vibrations tend to surface only when some damage becomes manifest. To detect torsional vibrations, proximity probes with appropriate markers on the shaft can be used. Possibly, monitoring the motor current fluctuations may be another way to detect torsional vibrations in a plant or on the test bed.

10.14 Vertical-Pump Vibrations

10.14.1 Scope

Because of the close interaction between approach flow effects and vibrations, this sub-chapter is dedicated specifically to the topic of vertical-pump vibrations even though the basics of intake structure hydraulic design are discussed in Chap. 11.7.3. The present text focuses on vibrations of single-stage vertical cooling water pumps (CWP) with powers of typically 1000 to 3500 kW as installed in large thermal power stations and other services. With the object of minimizing costs, the CWP intake structures are built as compact as appears acceptable. As a consequence of such endeavors, non-uniform approach flow to the CWP generates, time and again, hydraulic excitation forces causing pressure pulsations, structural vibrations, bearing damage or even cracks in the column pipe (CP) or discharge bend. The design of the CP, motor support structures and discharge head are tailored to the individual plants. An accurate prediction of the structural eigen frequencies is as vital to reliable operation as are the careful hydraulic design of the intake structures and the pump itself. The diversity of plant layouts and CWPs makes of each plant a prototype design which harbors the potential for disagreeable surprises during plant operation. In view of the above it is instructive to review the problems in terms of *case histories* (referred to subsequently as CH1, CH2, etc.) from a variety of sources. The overall goals are: (1) increase understanding of damage mechanisms, (2) identify common failure modes, (3) quantify excitation forces in uniform and non-uniform approach flow, which can be used for the mechanical design of the bearings and the CP, (4) demonstrate limits of Froude model testing of intakes, (5) review methods for troubleshooting.

Figure 11.23 shows a pump sump with standard dimensions, while Fig. 2.15 shows a typical semi-axial cooling water pump. The cooling water requirements for thermal power stations are in the range of 80 to 140 m^3/h per MWe with the higher values for nuclear plants (MWe represents the electrical power of the plant). Typical range of pump data: $n = f(Q_{opt})$ = 200 to 700 rpm. Q_{opt} = 6,000 to 60,000 m^3/h; H_{opt} = 10 to 50 m; P_{opt} = 1000 to 3500 kW; n_q = 40 to 200; n_{ss} = 220 to 260; $NPSH_A$ = 12 to 15 m; length of column pipe (subscript CP) L_{cp} = 6 to 12 m (see Fig. 2.15).

10.14.2 Vibrations Related to Non-uniform Approach Flow and Inlet Recirculation

Design guidelines for intake structures can be found in Chap. 11.7.3 and numerous publications, e.g. [92, 93]. Even though model tests are often done for the individual plant layout, these tests - and adherence to the guidelines - are no guarantee for the absence of vortices and sufficiently low excitation forces which can be caused by non-uniform and unsteady approach flow. Vortices with a gas/vapor core can form on the free water surface or on the floor or the side walls of the pump sump. The impeller blades passing through a vortex experience a shock-like loading. This is readily perceived by observing a blade passing through a vortex as demonstrated by Fig. 11.21.

- *Steady* vortexes give rise to frequencies at the blade passing frequency (BPF): $f = \nu_2\ z_{La}\ f_n$ where ν_2 is an *integer*. Mostly, the first order with $\nu_2 = 1$ is dominant. Experimental evidence: Case histories CH1and [120]
- *Unsteady* vortexes move slowly in the sump in a stochastic way, or possibly with a vortex precession speed. Then the excitation frequency becomes $f = a_x\ z_{La}\ f_n$ where a_x is *not* integer. If a constant precession speed is postulated, the vibration spectrum could show a peak in the vicinity of the BFP (say 10% lower) which may be difficult to explain. Unsteady vortices have been observed visually in [93, 94] and calculated by CFD in [96, 110]. As the unsteadiness of the flow increases (e.g. in compact intakes), the vortices will become more unsteady too.

- If the column pipe has an eigen frequency near the BPF, it is conceivable that the vortex precession speed should lock-in with this eigen frequency.

10.14.2.1 Slow Vortex Formation (CH1)

Vortex formation can take a surprisingly long time which is demonstrated by vibration problems in plant CH1.

Pump data: n = 470 rpm, Q = 24,000 m^3/h, H = 22 m, $n_q = 120$, P = 1600 kW, $z_{La} = 4$, BPF = 31 Hz.

Intake structure: Water is taken from a large surface water with constant level: The intake channel (L > 60 m) is fully submerged (closed channel flow). This duct feeds several parallel pump sumps where rear wall and floor splitters were installed. A scaled model test had been done, but some site constraints made it highly uneconomic to stay within the design recommendations given in the literature. The approach flow velocity to the sump was on the high side ($c_A = 0.6$ m/s), and the rear wall of the sump had a semi-circular shape. This design fosters swirl at the rear wall as suggested in Figs. 7.50 and 10.70.

Problems: vibration velocities fluctuated over time exceeding allowable levels, but no significant component damage.

Root cause: Stochastic increase in vibrations due to bottom or side-wall vortices.

Remedy: none implemented because modification of the huge underground intake would have been too expensive.

Figure 10.65 shows vibration velocities near the axial bearing as function of time. Vibrations increase sharply from a level of 6 to 8 mm/s to 20 mm/s or sometimes above; the periods between these bursts vary between 10 to 50 min. The long build-up time suggests that large zones of swirling flow are generated in the intake channel or that an external unsteadiness triggered the vortex build-up. The spectrum (Fig. 10.66) shows peaks at 1st, 2nd and 3rd harmonics of BPF. The presence of bottom or wall vortices was confirmed by underwater video camera.

One hypothesis for the long periods of vortex formation could be strong waves on the vast water surface (e.g. due to wakes from ships or boats), but no attempt was undertaken to check this hypothesis.

10.14.2.2 Fast Vortex Formation (CH2)

A similar phenomenon was observed in plant CH2, but the vortex formation takes much less time than in plant CH1.

Pump data: n = 490 rpm, Q = 13,000 m^3/h, H = 40 m, $n_q = 60$, P = 1600 kW, $z_{La} = 4$, BPF = 33 Hz.

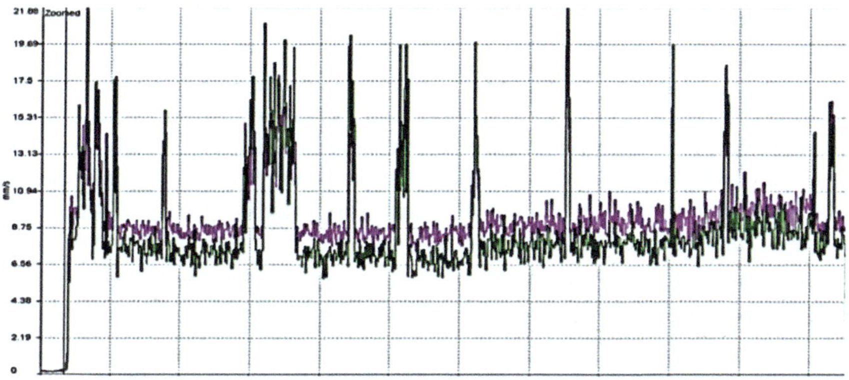

Fig. 10.65 Vibration velocitiy in mm/s-RMS as funtion of time. Period 10 to 50 min. Scale: x-axis about 33 min per divide, y-axis 0 to 22 mm/s

Fig. 10.66 Vibration spectrum, x-axis in Hz

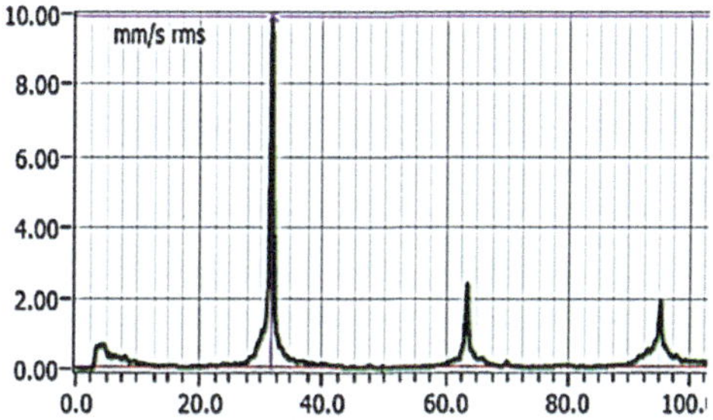

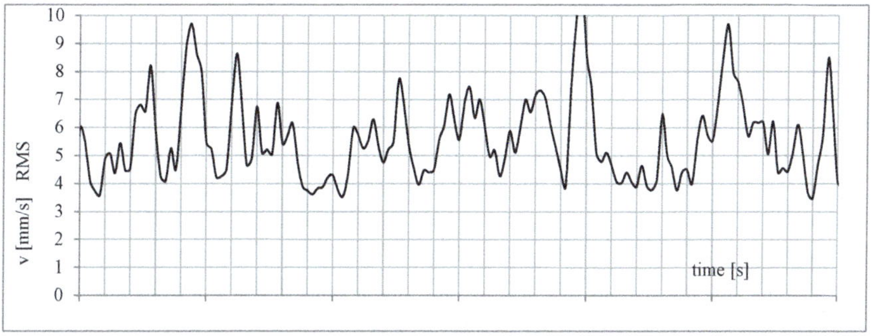

Fig. 10.67 Bearing housing vibration velocities in mm/s as funtion of time. Fluctuation periods 20 to 70 s. Scale: x-axis 10 s per divide

Intake structure: The water comes from an open channel which feeds a number of parallel pumps by a sharp 90°-turn. No flow-smoothing devices were installed in the sumps. A scaled model test had been done in which a *cross beneath the suction bell* was employed to eliminate sub-surface vortices. The approach flow velocity in the sump was below 0.2 m/s.

Problem: fluctuating vibrations exceeding allowable limits. Expected cause: A-periodic increase in vibrations due to bottom and side-wall vortices caused by the 90°-flow turn from the main channel to the sump of individual pumps.

Figure 10.67 shows vibration velocities as function of time. Vibrations increase from a level of 4 mm/s up to about 10 mm/s. The periods (20 to 70 s) between the peaks are much shorter than in plant CH1. Building-up higher peaks seems to take more time, while lower peaks follow in a more rapid succession. It can be surmised that the higher peaks stem from larger vortices and that the generation of such takes more time than for smaller vortices. This is in line with the observations in plant CH1. Free-surface air-drawing vortices in the Froude model tests took typically 6 s for developing as reported in [93].

Lesson: An unfavorable flow feature - the sharp 90°-turn - was seemingly remedied in the model test by another unfavorable feature: the cross under the suction bell with a structure *perpendicular* to the flow direction.

10.14.2.3 Radial Forces Generated by the Approach Flow

Radial forces on the impeller: Non-uniform approach flow causes *steady* lateral forces on radial, semi-axial and axial impellers as discussed in Chap. 9.3.7. A *steady* radial force on the impeller causes alternating bending stresses in the shaft, Table 10.9(6). If the force is sufficiently large, shaft deflection leads to rubbing of the rotor on the stator. This type of rubbing will cause a uniform wear on the rotating part and local wear on the stator. If the non-uniform approach flow fluctuates, *unsteady* radial forces are generated; the wear pattern can then be more difficult to interpret.

The test data on the effect of non-uniform approach flow on radial forces are presented in terms of force coefficients defined by Eq. (9.7a). Figure 10.68 shows *steady* forces in terms of the components F_x and F_y of the force vector. For dimensioning the bearings or other components, the resultant force needs to be calculated. Expressed as force coefficients, the resultant becomes $k_{Ru,D} = (k_{Ru,Dx}{}^2 + k_{Ru,Dy}{}^2)^{0.5}$. For example, the semi-axial pump $n_q = 160$ yields $k_{Ru,D} = 0.006$ at $q^* = 1.06$ (Fig. 10.68). The tests in [100] were done with a skewed approach flow velocity profile with $c_{max}/c_{av} = 1.25$ and $c_{min}/c_{av} = 0.75$ (roughly). Note that the data in Fig. 10.68 are the same as in Fig. 9.32, but different definitions of the force coefficient have been used. Reason: when applying the test data to other pumps, using either head or tip speed may yield different forces; the more conservative estimate may be used for design.

In tests with *uniform* approach flow, unsteady forces were measured in [93]; steady components were then negligible. The tests were done with an axial pump $n_q = 150$. Forces were derived from strain gages on the shaft. Figure 10.69 shows the unsteady force amplitudes zero to peak for different sump designs. In test 1, the pump took suction from a large basin with virtually zero approach flow velocity.

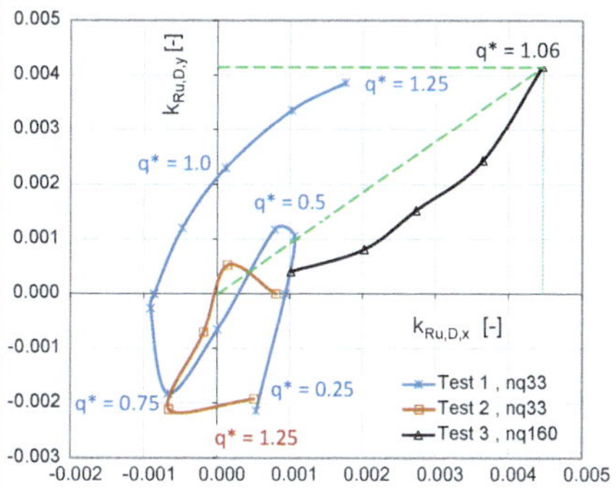

Data from [111]	
q*	1
$k_{Ru,D,stat}$	0.005 to 0.008[b]
$k_{Ru,D,dyn}$	0.0025[bb]
$k_{R,D,stat}$[a]	0.02 to 0.03
$k_{R,D,dyn}$[a]	0.01
[a] $k_{R,D}$ defined with head $k_{R,D} = k_{Ru,D}/\psi$ [b] estimate since ψ is unknown	
$k_{Ru,D} = \frac{2F_R}{\rho u_2^2 d_{2a}^2}$ $k_{R,D} = \frac{F_R}{\rho g H d_{2a}^2}$	

Fig. 10.68 *Steady* radial forces caused by non-uniform approach flow. Tests 1 and 2 from Fig. 9.22, Test 3 from [100]

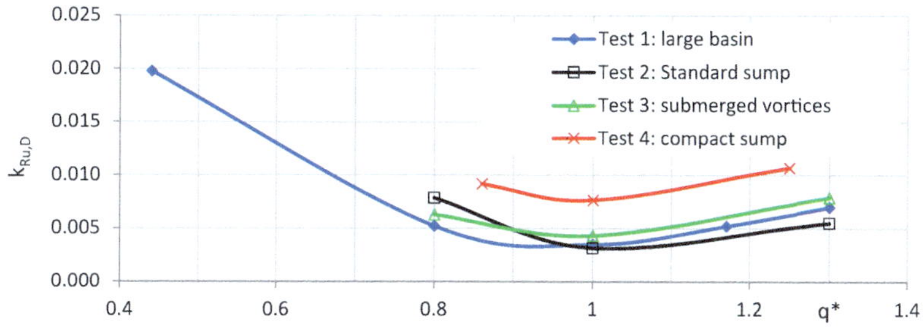

Fig. 10.69 *Unsteady* radial forces on impeller; amplitudes zero to peak, calculated from data in [93]

Tests 2 and 3 were done with a standard sump according to [92] with $W/D_T = 2$, $X/D_T = 0.75$ and $C/D_T = 0.5$. Submerged vortices were present in test 3; these increased the forces at high flow by a factor of 1.4 compared to test 2. Test 4 is with the compact sump from case history CH3 below. *The dominating component is synchronous excitation.* The Q-H-curve was unstable below $q^* = 0.8$ which caused the high force fluctuations seen at low flow. Note: the force coefficients in Fig. 10.69 are rough estimates derived from graphs in [93].

Measurements on an axial pump in [111] provided steady and unsteady forces caused by non-uniform inflow and the effect of the discharge bend which was close to the impeller and thus influenced the pressure distribution at the diffuser outlet; see detailed discussion of rotationally-asymmetric boundary conditions on the flow through an impeller in context with Fig. 10.23.

From above test data, the radial force coefficients listed in Table D10.4 can be assumed tentatively near BEP for design purposes. Attempting such estimates is subject to considerable uncertainty because the force coefficients depend on the approach flow velocity.

Table D10.4 Radial forces created by the approach flow near BEP

Approach flow in sump	Steady radial forces $k_{Ru,D,stat}$	Unsteady radial forces $k_{Ru,D,dyn}$
Uniform	Negligible	Figure 10.69
Non-uniform (±25%)	0.004 to 0.008	

Radial forces caused by non-uniform approach flow increase with the flow rate because the kinetic energy of the flow in the sump increases with the square of the flow velocity. Reducing the size of the sump, i.e. increasing the approach flow velocity c_A, will increase the steady and unsteady radial excitation forces. As the water level h_{WL} decreases in a given open sump at a specified flow rate Q, the approach flow velocity ($c_A = Q/(W \times h_{WL})$ increases and so does the flow reflection on the rear wall. This has also been observed in tests reported in [113].

Radial forces on the pump body: The water entering the sump exerts forces on the *outside* of the assembly of suction bell, bowl and that part of the column pipe which is exposed over length L_{exp} to the approach flow velocity c_A. Looking at the pump body as a cylinder in cross flow with exposed frontal area $A_A = D_{cp} \times L_{exp}$, these forces can be estimated from:

$$F = \zeta \frac{\rho}{2} c_A^2 A_A \tag{10.27}$$

The force acting in the flow direction is estimated from the drag coefficient ζ_D; it has steady and unsteady components. An unsteady force acting perpendicular to the flow direction is described by the dynamic lift coefficient ζ_L. Drag and lift coefficient depend on the geometry and the Reynolds number. Theoretically, the lift force F_L fluctuates at a frequency $f = S_{Str} \times c_A/D_{cp}$ where S_{Str} is the Strouhal number of a cylinder in cross flow.

Even test data for circular cylinders in a free stream vary in a wide range as shown by Table D10.5, data from [35, 119, 101]. On a cylinder in a free stream, the transition from laminar to turbulent occurs at $Re = 2 \times 10^5$. Up to this Reynolds number vortex shedding is regular with the dominant frequency given by the Strouhal number; there is also broad band excitation by turbulence with much less energy. The same applies to the trans-critical range $Re > 5 \times 10^6$. In the range of $Re = 2 \times 10^5$ to 5×10^6 vortex shedding is stochastic, and force coefficients depend strongly on Re with minima occurring near $Re = 10^6$.

In addition to the forces acting on the *outside* of the suction bell and pump body, the momentum of the water entering the suction bell creates forces on the *inside* of the suction bell if the incoming flow is not rotationally symmetric. The mechanism is similar to the fluid-momentum acting on a pipe bend, but the in-flow is entering the suction bell unevenly and unsteadily from all sides.

Table D10.5 Strouhal numbers and force coefficients for cylinders in cross flow

Reynolds number: Re	500 to 2×10^5	2×10^5 to 5×10^6	$> 5 \times 10^6$
Vortex shedding	With S_{Str}	Stochastic	With S_{Str}
Strouhal number: S_{Str}	0.2	–	0.3
Drag coefficient: ζ_D	1.0 to 1.1	0.4 to 0.7	0.7
Lift coefficient: ζ_L	0.35 to 0.7	0.15 to 0.2	0.2 to 0.3

The fluctuating lift force is present in uniform as well as in non-uniform approach flow. When a cylinder is placed closed to a wall at a distance e, vortex shedding is suppressed for $e/D < 0.2$. Clear vortex shedding has been observed down to $e/D = 0.3$ as demonstrated by test data in [106]. The wall distance in a standard sump is $e/D_T = 0.5$ - or larger depending on the ratio of the column pipe to the suction bell diameter, but the rear wall of the sump might have an influence on vortex shedding. Therefore, the frequencies and forces estimated in this way can provide only an order of magnitude.

Damage induced by strongly non-uniform approach flow (CH3):

Data: $Q = 52{,}000\ m^3/h$, $H = 21$ m, $n_q = 138$, $P = 3300$ kW, $z_{La} = 4$, $z_{Le} = 7$.

Problems: damage occurred to the lower bearing, wear due to rubbing of impeller on the casing, cracks in discharge bend splitter vane, cracks in the column pipe *parallel* to the pump axis caused by local stresses at the bearing support spider; in addition, pressure and power readings were fluctuating. The pump (357 rpm) was driven by a motor (1493 rpm) via a gear. Measured vibrations at the gear reached up to 11 mm/s RMS. Motor vibrations were between 5 and 8 mm/s RMS with synchronous excitation being dominant. The rotational frequency of the motor $f_n = 24.9$ Hz and the BPF $f_{BPF} = 23.8$ Hz caused a "beating" of the vibration signals with $f = f_n - f_{BPF} = 1.1$ Hz. There were daily fluctuations of the vibrations signals due to solar heating of the motor stool, which caused misalignment. Maximum vibrations at the suction bell reached almost 10 mm/s RMS at blade passing frequency.

Intake structure (Fig. 10.70): The water entered the compact sump through a small opening like a jet with $c_A = 2.2$ m/s (inlet to suction bell $c_T = 2.1$ m/s). Due to the sharp edges at the entrance to the opening, the velocity in the vena contracta was considerably higher. The sudden expansion of the jet downstream of the opening is bound to create flow separation, unsteady flow and vortices. The flow hits the rear wall with high velocities that way that it is turned back and creates, on the impeller, a radial force which is *opposed* to the inlet velocity, Fig. 10.71. A cross-type anti-vortex device (AVD) was installed underneath the suction bell. Even though the cross is designed to deflect some fluid upwards to the suction bell inlet, flow separation and additional unsteadiness is caused by the water flowing over that part of the AVD which is *perpendicular* to the flow direction.

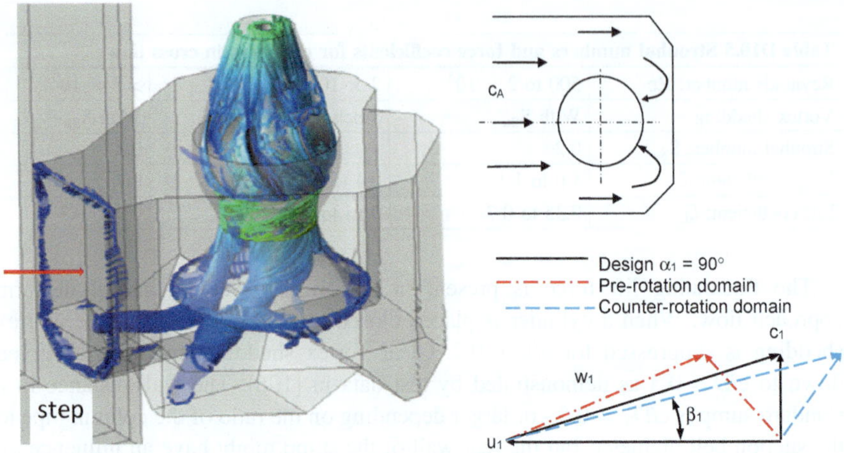

Fig. 10.70 Vortex formation in pump sump, plant CH3; CFD from [96]

The reasons for damage to the lower bearing were investigated by means of unsteady CFD calculations of the flow in the intake structure, [95, 96]. As reported in [93, 114], comprehensive model testing had been done for developing this type of intake structure, which is much more compact than standard intakes designed according to [92]. Velocity profiles, measured at the location of the impeller inlet, were within the allowable deviations of ±5% and the swirl angle was below 5°. Thus, the traditional criteria with respect to uniformity of the flow and absence of swirl and vortices had been plainly fulfilled. In spite of this, the flow in this compact sump was shown by CFD to be quite non-uniform with strong vortices developing at the abrupt flow expansion at the inlet to the sump and in the rear of the CP, Fig. 10.70. The presence of a strong vortex in the plant was confirmed by some erosion observed on the concrete walls of the sump, [117]. The strong secondary flows in the suction bell and at the impeller inlet (Fig. 10.71) clearly demonstrate that the average swirl is by no means representative of the local swirl patterns. The swirl angle measured in a model test by a vorticity meter is an *integral* over the cross section where counter- and pre-swirl largely compensate.

The unsteady local swirl leads to varying incidence angles as per Fig. 10.70. As a consequence, the work transferred by the blades varies over space and time creating lateral forces on the impeller. Steady and unsteady radial forces calculated by CFD are shown in Fig. 10.71. The forces on the impeller were directed *against* the inlet flow velocity. This result is well borne out by the one-sided local wear on the bearing support which occurred in the plant after the stationary bearing bush had been completely destroyed. The wear pattern suggests that a static radial force was dominant.

As demonstrated by Fig. 10.69, the radial forces in the compact sump are 2-times higher than in the standard sump. In [93] it was noted in passing that fluctuating lateral forces in the compact sump had increased, but the possible

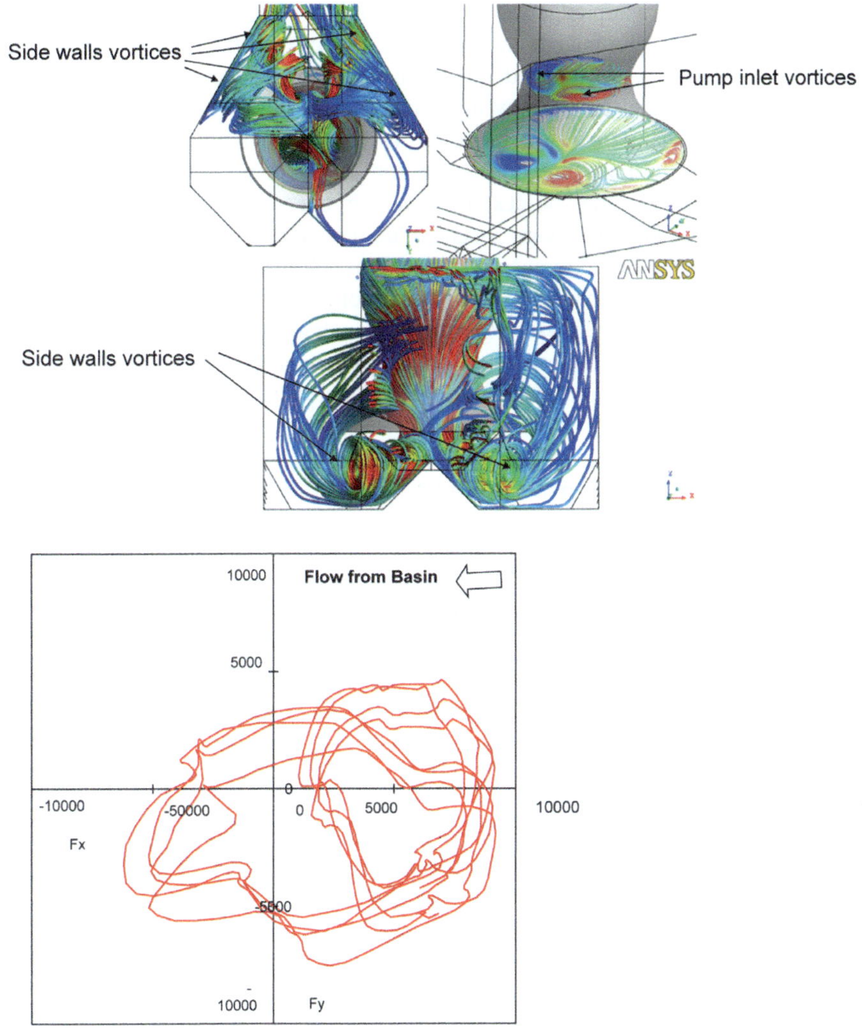

Fig. 10.71 Unsteady radial forces on shaft created by highly distorted approach flow to a semi-axial pump, CFD analysis [96]

implications were apparently not followed up. As reported in [117] several installations, built with this compact intake, experienced damage to the pumps. It can be inferred that the severity of these problems increased with velocity c_A through the inlet opening of the sump as discussed above.

Extensive experimental and unsteady numerical studies on this compact intake have been carried out in order to develop remedies, [117]. The most effective modifications were: (1) The step on the floor was removed. In this way the velocity of the jet entering the sump was reduced and flow separation on said step eliminated. (2) The sharp edges at the sump entrance were rounded. Jet contraction was thus virtually eliminated; the actual jet velocity in the vena contracta and associated vortices were reduced. (3) Horizontal ribs on the diffusing side walls were installed parallel to the flow direction whereby the intensity of the vortices upstream of the suction bell was reduced. (4) Ribs of varying height were installed perpendicular to the flow direction on the roof of the sump in the rear part of the sump. The intensity of the vortices in the wake of the suction bell was reduced that way. These modifications were installed in the plant. In addition, a cage-like strainer at the outer circumference of the suction bell was tested. It consisted of 48 radial ribs ($r_i/r_a = 0.91$) extending from the rim of the suction bell down to the floor of the sump. This strainer was effective in reducing the intensity of the vortices passing into the pump, but it was not installed in the plant. These studies were done by CFD, model testing and plant observation (underwater video). In [117] it was concluded that the CFD procedures applied were able to detect strong vortices, but failed to predict unsteady vortices of low or moderate intensity. Grid refinement or using different turbulence models did not improve the ability to predict unsteady vortices.

The CFD-calculated steady and unsteady forces acting on the impeller and on the pump body (suction bell, bowl, CP) are represented in Table D10.6 in absolute terms and by dimensionless force coefficients. The unsteady forces are much below the test data with the compact sump shown in Fig. 10.69, but both test and CFD are subject to large uncertainties.

Table D10.6 CFD results CH3

Force on shaft	$F_{R,stat}$	N	5,000	$k_{Ru,D} = \frac{2F_R}{\rho u_2^2 d_{2a}^2}$
	$F_{R,dyn}$	N	4,000	
	$k_{Ru,D,stat}$	–	0.0033	
	$k_{Ru,D,dyn}$	–	0.0027	
Inner force on CP due to momentum	$F_{CP,in,stat}$	N	12,000	$f_{CP} = \frac{F_{CP}}{\rho Q c_T}$
	$f_{CP,in,stat}$	–	0.38	
	$F_{CP,in,dyn}$	N	2,100	
	$f_{CP,in,dyn}$	–	0.067	
Outer force on CP due to approach flow	$F_{CP,out,stat}$	N	7,500	$\zeta_{CP} = \frac{2F_{CP}}{\rho c_A^2 A_A}$
	$\zeta_{CP,out,stat}$	–	0.52	
	$F_{CP,out,dyn}$	N	1,000	
	$\zeta_{CP,out,dyn}$	–	0.069	

The momentum of the fluid entering the suction bell creates forces inside the bell which are larger than the forces acting on the outside. The momentum of the water $F_I = \rho \times Q \times c_T$ entering the suction bell exceeds 31,000 N; that fraction of this momentum which can be described by the non-uniform inflow accounts for the resultant force acting on the inside of the suction bell. A force coefficient can be defined by $f_{CP} = F_{CP}/F_I = F_{CP}/(\rho \times Q \times c_T)$. The force created by the approach flow on the outside of the pump can be described by a coefficient $\zeta = 0.52$

according to Eq. (10.27); this coefficient includes the resultant of drag and lift. Inside and outside forces add since they act into the same direction: the flow hits the CP from the rear and the lateral momentum of the water entering the suction bell is highest on the rear. The forces acting on the outside of the CP are unusually large because of the high approach flow velocity $c_A = 2.2$ m/s.

Vibration velocities measured by accelerometers at the suction bell showed dominant peaks at BPF with fluctuating signals up to 10 mm/s. The frequencies calculated by CFD or Eq. (10.13) were not present in the measurements. This is not really surprising since the amplitude actually measured on a pump at any given frequency depends on: (1) the strength of the excitation at said frequency, and (2) the response of the structure at that same frequency. Both of these effects cannot be accounted for by the criterion Eq. (10.13).

10.14.2.4 Asymmetric Approach Flow Causing Premature Inlet Recirculation (CH4)

A thorough investigation of vibration problems in cooling water pumps has been reported in [95]; the contribution contains a wealth of information on the tests performed and on trouble shooting methodology. Pump data: n = 585 rpm, Q = 11,900 m^3/h, H = 46 m, $n_q = 61$, P = 1700 kW, $n_{ss} = 206$; $z_{La} = 7$, $L_{cp} \approx 9000$ mm, $D_{cp} \approx 1400$ mm

Problems: high motor vibrations, damage to the lower bearing, impeller rub in suction casing and fluctuating vibrations exceeding allowable limits, excitation of standing waves in the discharge pipe, condenser vibrations and damage. Root cause: Non-uniform approach flow causing impeller inlet recirculation - well above BEP at $q^* = 1.1$ *to* 1.2! Remedies: Adding structures below the curtain wall in order to smooth the flow in the sump, new suction bell design with 10 ribs which provided 20° pre-rotation.

The water is distributed from a header through short circular pipes to the 3 parallel pumps. The sharp 90°-turn (as in a T-branch) leads to flow separation in the short pipe and an asymmetric jet entering the open sump, Fig. 10.72. There is a curtain wall, but it leaves too high a distance to the floor and does not provide a good dissipation of the jet. The flow reaching the pump is highly skewed and a

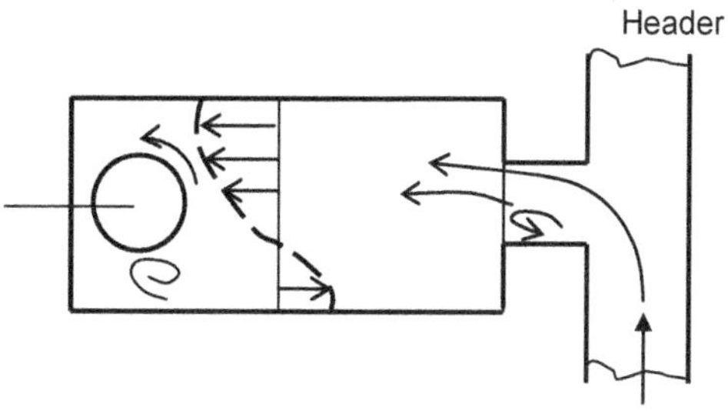

Fig. 10.72 Asymmetric approach flow caused by a sharp 90°-turn from the header to the sump

vortex is forming. As a consequence, part of the impeller inlet is subject to pre-rotation, while another part is exposed to counter-rotation. The pumps operated at $q^* = 1.2$ where the average nominal incidence calculated with constant inlet velocity was *minus* 2.6° (i.e. above shock free inlet $Q/Q_{SF} = 1.07$). This condition would normally be expected to absolutely preclude inlet recirculation. However, in the zone with counter-rotation the flow angle decreases and the incidence increases. This mechanism manifestly triggered a strong recirculation at the impeller inlet which caused high excitation forces responsible for the problems mentioned above. When increasing the flow rate, the vibrations and pressure pulsations abruptly dropped to acceptable levels.

The recirculation caused broad band excitation in the range of 1 to 2000 Hz which excited eigen frequencies of the motor as well as standing waves in the discharge pipe. The onset of recirculation was detected by a sudden rise in pressure pulsations, pressure measurements in the suction bell (static pressure varying around the circumference), increase of motor vibrations and fluctuations of the motor current. A scaled model test of the intake had been done, but failed to give any clues to potential problems.

10.14.2.5 Excessive Inlet Recirculation (CH5)

Problems: Severe column pipe cracking and vibrations on a pump with n = 590 rpm, $n_q = 60$, P = 2000 kW. Root cause: operation with excessive inlet recirculation.

The impeller had an over-sized inlet diameter and large blade inlet angles. As a consequence, the pumps operated at about 50% of the flow of shockless entry with an incidence angle above 10°. The resulting recirculation at the impeller inlet induced a strong swirl in the sump. This swirling flow excited strong column pipe vibrations leading to severe cracks at the upper end of the column pipe in a short time, Fig. 10.73. The sump had no rear wall splitter which could have mitigated the swirling. In addition, the distance between the bottom floor splitter and the suction bell looked large. Due to the absence of a rear wall splitter, the swirling flow can pass between the pump and the rear wall. With reference to Fig. 10.74, the following mechanism could be hypothesized: the flow is accelerated in the small gap. According to the Bernoulli equation the pressure in the gap at the rear wall drops (the mechanism can be compared to the destabilizing "Bernoulli effect" in an annular seal, Chap. 10.6.2), Fig. 10.14. Due to vortex shedding this process is unsteady - similar to the Karman vortices shed from a cylinder in cross flow. If the vibration amplitudes of the pump body are large, the vortex shedding could couple with the mechanical vibration. The low eigen frequency of the column pipe (about 3 Hz) makes it likely that the excitation locks into this frequency. Indeed, the short time to failure justifies the hypothesis that fluid-structure coupling was instrumental in producing the cracks.

Vibrations measured on the discharge bend (which were within the allowable limits of 7 mm/s) showed fluctuations quite similar to Fig. 10.67 with periods of 15

Fig. 10.73 Cracks in column pipe

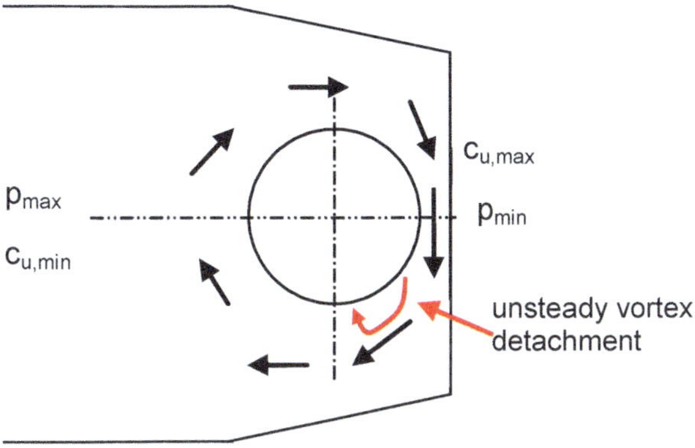

Fig. 10.74 Swirl in pump sump induced by impeller inlet recirculation (without rear wall splitter)

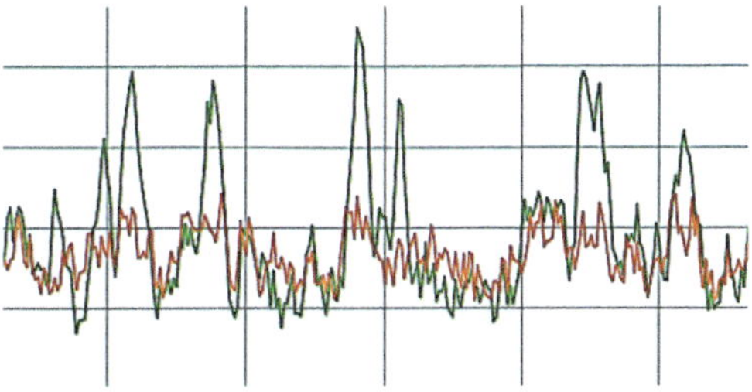

Fig. 10.75 Vibration velocity near discharge bend as funtion of time. Fluctuation periods 15 to 80 s. Scale: x-axis 50 s per divide; scale: y-axis 1 mm/s per divide)

to 80 s, Fig. 10.75. This behavior could be due to build-up of swirl and subsequent swirl break-down. The mechanism could be the following: (1) Flow recirculation from the impeller induces a swirl in the sump as described above; (2) If the water in the sump surrounding the pump body is strongly swirling, the flow angle of the fluid approaching the impeller increases; (3) As a consequence, the incidence is reduced: this way the cause of the flow recirculation vanishes and the recirculation diminishes; (4) But then the approach flow angle decreases and incidence increases again: flow recirculation is induced again and a new cycle begins. The mechanism is similar to that described in Chap. 6.5.1 in context with low-frequency pulsations which can be caused by part load operation of inducers.

10.14.2.6 Model Testing Limitations

Careful model testing of the intake structures had been done in all of the case histories presented above, but the sump-related flow problems were not detected in those tests. As per Chap. 11.7.3, model testing of the intake structures focuses on three quality criteria for the approach flow: (1) absence of significant free-surface or submerged vortices, (2) limits on the non-uniformity of the axial velocity profile and (3) limits on swirl as measured by a swirl-meter. The swirl is measured by counting the revolutions of a propeller which is made of 4 radial blades. It is noteworthy that this swirl meter determines the *overall swirl integrated over the pipe section*, but is unable to detect local zones of pre-swirl and counter-swirl (see Figs. 10.70 and 10.71). The secondary flow in a pipe bend (Fig. 1.12) would produce virtually zero-swirl. The *overall* swirl influences the pump performance (head and power). However, the *local swirl*, which is not considered in the model tests, induces variations of the incidence angle at the blades and thus has an impact on excitations forces.

Model tests on intake structures are done according to the Froude similarity; i.e. the Froude numbers Fr of the scaled model (M) and the plant (P) are equal: $Fr_M = Fr_P$

$$Fr = \frac{\text{inertia force}}{\text{gravity force}} = \frac{c_T}{\sqrt{g\,D_T}} \quad \text{with} \quad c_T = \frac{4Q}{\pi\,D_T^2} \tag{10.28}$$

At $Fr_M = Fr_P$ the velocity in the model is $c_M = c_P/\lambda^{0.5}$ when $\lambda = D_P/D_M$ is the scaling ratio. The Reynolds number ratio is $Re_M = Re_P/\lambda^{1.5}$ and the flow rate ratio is $Q_M = Q_P/\lambda^{2.5}$. Occasionally, another test is done at the same velocity as in the plant: $c_M = c_P$ yielding $Re_M = Re_P/\lambda$, but it recognized that it is impractical to carry out model tests at the same Reynolds number as in the plant.

The Froude number represents the ratio of the inertia to gravity force. Therefore, it is relevant for flow phenomena with a free liquid surface, which are *governed by gravity* (for example ocean waves). The Froude number is relevant only for detecting the generation of vortices with an air/vapor core: (1) the local liquid level depression in the vortex core generated at the liquid/gas interface is directly proportional to the acceleration due to gravity (Table 1.1); (2) the gas/vapor-filled core of submerged vortices is influenced by the static pressure of the surrounding liquid, hence is proportional to $\rho \times g \times h$ (if h is the water depth at the considered location). The Froude number is meaningless for flow phenomena like flow separation, velocity distributions, vorticity or swirl, secondary flows and hydraulic excitation forces.

Considering a T-branch like in Fig. 10.72, it is evident that flow separation, swirl and vortices would be much the same in an open channel as in a closed (covered) duct. Hence the flow in a T-branch cannot be modeled in a Froude test. By definition, turbulence (local swirl) is governed by the Reynolds number, i.e. by the ratio of inertia to viscous forces: friction factors decrease with increasing Re. In the plant, the Reynolds numbers are typically 25- to 60-times higher than in the model. As a consequence, vortex decay in the plant is slower than in the model. For supporting the argument: the inlet length in turbulent pipe flow (required to achieve a fully-established velocity profile) increases with the Reynolds number, Eq. (1.19d). By the same token, a non-uniform velocity profile at the sump entrance will take in the plant a longer ratio L/d_h to smooth out than in the model. While subsurface vortices are often present in the initial configuration tested in the model, and measures are taken to make them disappear in the model, they may still occur in the plant.

The effects of non-uniform and unsteady flow on excitation forces are likely to increase with the ratio of inertia to dissipative forces - hence with the Reynolds number. They can be expected to be higher in a plant than in a scaled model. Hydraulic forces which are proportional to the square of the speed are negligible in the model when compared to the plant: $F_M/F_P = 1/\lambda^3$ (if the forces F are proportional to a dynamic pressure acting on an area).

Sometimes anti-vortex devices (AVD) in sumps are installed on the floor (as done in plants CH2 and CH3), which are *perpendicular* to the flow direction. Such AVDs may break up and suppress vortices with an air/vapor core because of increased flow unsteadiness and turbulence, but the unsteadiness in itself is a potential excitation mechanism. Therefore, this type of stratagem to get rid of vortices in a scaled model is not recommended. Likewise, steps and sudden area changes in pump sumps should be avoided because they are bound to induce flow separation, swirl and unsteadiness - even if no vortices are observed in the Froude model test.

Since Reynolds number effects are not properly accounted for in Froude models, numerical flow calculations of the flow in the sump can be used to get additional insight. CFD-studies of intakes (too numerous to list here) have been performed with Navier Stokes codes for examples in [96, 97, 117]. A new approach using the so-called "Lattice Boltzmann method" (LBM) has been reported in [110], where the flow field in a sump was calculated with unsteady large eddy simulation (LES) and compared to test data (velocity distributions and vortex structures). The LBM was said to be particularly suitable for analyzing intakes because of: (1) the ability to describe complex geometries without manual meshing (resulting in very low pre-processing effort); (2) fast solver technique for unsteady flows; (3) applying LES for dealing with turbulence. In this way, meandering and unsteady vortices were well captured in [110].

In summary, the model test failure to reveal the approach flow problems in the case histories discussed above can be explained by:

- the Reynolds number which is much lower in the model than in the plant
- the disregard of local swirl (secondary flow) at the impeller inlet, Fig. 10.71 (the measured swirl is an integral)

- disregard of flow unsteadiness since the measured velocity profiles are time-averaged
- the disregard of vorticity other than that leading to the generation of air/vapor-filled vortex cores
- the disregard of radial forces induced by the approach flow on the impeller as well as on the column pipe and pump body as per Chap. 10.14.2.3

Vortex suppression in Froude model tests by selecting deliberately high velocities, by introducing flow unsteadiness and turbulence as intended in [117] can backfire through high excitation forces acting on the pump. While it is impractical to carry out model tests at the same Reynolds number as in the plant, improvements could be considered by:

- CFD studies of the full-size intake of the plant if properly validated by test results. However, a quantitative criterion for the intensity of the vortices is needed because CFD could show any number of small vortices which would do no harm.
- Steady and unsteady forces acting on the suction bell/column pipe could be measured in Froude model test by supporting the CP by a frame with 4 brackets equipped with strain gages.
- Measure pressure pulsations between suction bell and impeller inlet.
- If the approach flow velocity significantly exceeds the limit of $c_A = 0.5$ m/s, unsteady forces acting on the CP and the impeller need to be considered in the pump design.

10.14.3 Hydraulic Design Issues

Hydraulic design issues are addressed here only in as much as they have a significant impact on vibrations. In the past, designers often used large impeller inlet diameters and high incidence angles in order to obtain high suction specific speeds. Excessive incidence induces flow separation near the leading edge and large zones of cavitation. Hence this type of design tends to be counterproductive since inducing strong part load recirculation. Advanced designs focus on flat pressure distributions on the first part of the blades. This is achieved with low incidence, suction side profiling of the blade at the leading edge and a slow increase of the blade angle over the first part of the blade, Chap. 7.2.4. These design features allow achieving high suction specific speeds with smaller impeller inlet diameters, [102]. Depending on the maximum flow, the incidence at the outer streamline should be selected in the range of 0° to 3° (calculated with leading edge blade blockage).

In order to reduce the rotor/stator interaction, the following design features should be adopted:

- The combination of impeller and diffuser vane numbers should be selected to avoid common multipliers and radial forces on the impeller (avoid m = 0 and m = 1 in Eq. 10.13).
- The distance between the impeller trailing edge and the diffuser leading edge, measured on the mean streamline, in flow direction (often referred to as "gap B") should be sufficiently large. Table 10.2 gives gap $B/r_2 = n_q/1000$ for $n_q >$ 40, but larger gaps are often selected and may indeed be beneficial in reducing hydraulic excitation forces.
- The impeller blade trailing edge should be profiled on the pressure side and truncated by a small radius, Fig. 14.7-5. Semi-axial pumps are quite sensitive to the shape of the trailing edge. The optimum shape depends on the pressure distribution on the blade. The design can be optimized by model testing.

The hydraulic imbalance which excites synchronous vibrations must be kept under control by adequate casting quality. This is no easy matter because casting large semi-open impeller to close tolerances requires considerable experience.

Rotating stall could have an influence on vibrations, but no vibration incident due to this effect has been documented for semi-axial pumps. Diffuser rotating stall linked to instability of the Q-H-curve has been investigated in [132]; data: $n_q = 90$; $z_{La} = 5$, $z_{Le} = 9$. It occurred between $q^* = 0.68$ and 0.82 with a propagation frequency $f_{RS} = 0.42$ Hz or $f_{RS}/f_n = 0.042$. The cause of stall was identified as flow separation near the diffuser inlet at the hub near the suction side of the vane. A second instability was observed at $q^* = 0.6$ due to onset of impeller inlet recirculation.

Sometimes suction bells are designed with 2 to 4 ribs. The orientation of these ribs with respect to the floor splitter may have an influence on the flow into the suction bell (lateral forces, local flow separation).

In general, it may be questioned whether such ribs are desirable in applications where an appropriate floor splitter is installed in the sump. The objective of such ribs would be to reduce the effect of part load recirculation on the flow in the sump. However, operation at very low flow with excessive recirculation is probably only an issue if the pump design is not optimum (excessive inlet diameter and incidence). In usual applications, the floor splitter seems to be sufficient.

To reduce the effect of vortex shedding, the trailing edge of any such ribs should *not* be of semi-circular shape. The trailing edge can be designed according to Table 10.13. While such effects cannot be quantified and may be insignificant, a design which minimizes the potential for vortex shedding can be introduced with virtually no extra cost.

10.14.4 Influence of Discharge Pipe on Vibrations. Pressure Pulsations

Pressure fluctuations in the discharge flange represent excitation forces which could be estimated roughly by multiplying the peaks in the pressure pulsation spectrum with the cross section of the discharge nozzle. If of sufficiently high energy contents, broadband pressure pulsations can excite mechanical as well as acoustic eigen frequencies as reported in [95]. To lessen the effect of such forces on the lateral vibrations of the pump body (i.e. the structures usually above, but occasionally below, the foundation) the discharge nozzle should be as close as possible to the foundation (reducing the lever effect). Standing waves created by periodic excitation such as f_n and $z_{La} \times f_n$ can be evaluated in simple piping systems by assuming 1-D waves from basic theory according to Chap. 10.12.3, but the results can be misleading in complex systems, [24].

High pressure pulsations are generated when the pump is operating in a flat or instable range of the Q-H-curve. Stochachstic peaks and broadband excitation would then be strong.

Works tests may be more affected by pressure pulsations than plant operation - or vice versa. In works tests the total head and power is throttled in a valve and dissipated into broadband pulsations which is not the case in plant operation. Furthermore the discharge pipe is different from the plant (support, stiffness, standing waves). Systems with several pumps feeding into a common header can be of a problem because of flow separation and vortex shedding at the 90° - flow turn and multiple reflections points.

10.14.5 Eigen Frequency Analysis

Mechanically, motor, motor stool, column pipe and rotor form an entity subject to vibrations which are excited by the imbalances of the rotors of pump and motor, misalignment, hydraulic forces created by the approach flow, the flow through the pump and pressure pulsations in the discharge pipe. Finite element analysis (FEA) allows calculating - *in one complex model* - this entity which is bolted to the foundation and connected to the discharge pipe. To do so, corresponds to the state of the art; the analysis should comprise all eigen frequencies and mode shapes up to $f = 1.25 \times z_{La} \times f_n$ at least, [104].

10.14.5.1 Shaft Eigen Frequencies and Vibrations

The unsupported span between line-shaft bearings can be selected from experience and checked with rotor dynamic calculations. The ISO 13709 (API610) standard [109] can be used to check conformity (up to 100 mm shaft diameter).

Since the radial bearings in vertical pumps are not loaded by the rotor weight, no unequivocal bearing load can be defined. Hence the stiffness of the line-shaft bearings cannot be calculated. Likewise, it is not possible to quantify the bearing loads caused by the impeller imbalance which mainly acts on the lower bearing. In addition, shaft imbalance, bent shafts and bearing misalignment provide unknown loadings and bearing stiffness. All of these effects correspond to tolerances and, as such, are statistically distributed. As a consequence, the bearings have an unpredictable influence on the eigen frequencies. The *flow-dependent* axial force on the impeller causes tensional shaft stresses which increase the lateral eigen frequencies by typically 10%, [104].

When the eccentricity of the shaft in the bearing becomes close to the bearing clearances, the stiffness gets highly non-linear. Weakly-loaded cylindrical bearings are prone to rotor dynamic instabilities with sub-synchronous vibrations close to $\frac{1}{2} \times f_n$ ("half-speed whirl").

10.14.5.2 Eigen Frequencies of Motor and Structures Above the Foundation

The eigen frequencies of the structures above the foundation, i.e. motor, motor stool and possibly bearing housing can be predicted by finite element analysis (FEA). Uncertainties inherent to such calculations are mainly the influence of: (1) the stiffness of the foundation; (2) the quality of the grouting; (3) the bolting which tie the pump to the foundation; (4) the discharge pipe and its supports. The mode shapes of the motor/motor stool resemble a clamped cantilever beam (so-called "reed mode"). It is excited usually by the motor imbalance [105], but can be also influenced by alignment errors, impeller imbalance, hydraulic excitation forces and pressure fluctuations in the discharge pipe.

The piping, fittings, expansion joints, the water mass contained therein and piping supports need to be modeled up to a boundary which can be selected within an imagined sphere corresponding to a radius $R = 2 \times H_{motor}$ (motor height above foundation), [104].

10.14.5.3 Eigen Frequencies of Column Pipes

One eigen frequency of the column pipe is quite often close to the rotational frequency of the shaft. An accurate prediction of the column pipe eigen frequencies is, therefore, vital for the reliable operation of the pump. When calculating the eigen frequencies of the column pipe (CP), the water mass contained *within* pump and column pipe is added to the metal mass. Three other influences more are difficult to quantify: (1) The stiffness of the connection to foundation, (2) The added mass of water surrounding the column pipe; (3) The effect of the rotor mass and bearing stiffness on the CP eigen frequency.

Stiffness of connection to foundation (usually a concrete floor): the steel flange and bolting are modeled by FEA, while information on the foundation must be obtained from civil engineering. One could include part of the concrete foundation in the FEA model in the form of a annular plate with the inner diameter given by the hole through which the pump is installed, while selecting the outer diameter to $D_{boundary} = (2 \text{ to } 3) \times D_{flange}$ or $R_{boundary} = 2 \times H_{motor}$, [104]. The boundary is chosen that way that the deformation at the edge is negligible and the plate can be assumed to be clamped at the outer edge.

The coupling between the structures above and below the foundation depends on the stiffness's of: (1) the foundation; (2) the connection flange and bolting; (3) the stiffness of the pump body in this region. If the resultant stiffness is very strong, then motor and column pipe vibrations are little related. With infinite stiffness, motor and column pipe could be considered as two independent clamped cantilever beams. In contrast, if the stiffness of the connection between the pump/motor unit and the foundation is weak, the vibrations of motor and column pipe are intimately related. With very low stiffness, the pump/motor unit would "pivot" around said connection. In this case, the discharge pipe and support would also have more influence on the CP eigen frequencies.

Added mass of water surrounding the column pipe: In an infinite water volume, the added mass μ_{ad} can be set equal to the water displaced by the CP; i.e. per unit length: $\mu_{ad} = \frac{1}{4}\pi \times \rho \times D_{cp}^2$, [9]. In a confined space, the added mass increases because the free movement of the fluid displaced by the vibrating CP is impeded by the sump walls. A rough estimate may be made from analytical formulae, using e.g. the model for the added mass μ_{ad} contained between two concentric cylinders (the inner cylinder is considered vibrating, the outer as stiff), [9].

$$\mu_{ad} = Y\rho\frac{\pi}{4}D_{cp}^2 \quad \text{with} \quad Y = \frac{(W/D_{cp})^2 + 1}{(W/D_{cp})^2 - 1} \tag{10.29}$$

μ_{ad} = added mass (in kg/m) of the water surrounding the CP (in addition to the water mass within the CP)
D_{cp} = column pipe outer diameter
W = width of the sump; standard sump: $W = 2 \times D_T$; for $D_T/D_{cp} = 2^{0.5}$ we get Y = 1.29. The rear wall splitter (if any) impedes the movement of the water displaced by the vibrating column pipe. This tends to increase the added mass effect. Overall, the value of Y = 1.3 to 1.5 is considered reasonable.

The added mass depends on the water level. If the separation margin of the column pipe eigen frequency to the synchronous excitation is small, vibration amplitudes may be sensitive to the water levels (as in CH7 and CH8).

Coupled vibrations: Effect of the rotor on the eigen frequencies of the column pipe: If the shaft vibrates with amplitudes which are only a tiny fraction of the clearances of the water-lubricated bearings, the rotor has no appreciable effect on the eigen frequencies of the column pipe. If a column pipe eigen frequency is close to the rotational frequency, vibration amplitudes will be high - even with little imbalance. In this case, the bearing stiffness will be considerable and the rotor needs to be taken into account when calculating the eigen frequencies of the CP. State-of-the-art FEA programs allow calculating coupled vibrations with spring-elements coupling the column pipe with the shaft. Specifically, this should be done for slender column pipes with $L/D_{cp} > 4$, tight vibration specifications, risk of standing waves and weak foundations or weak piping supports. The coupling is expected to be strong if the masses of rotor and CP, or their respective eigen frequencies, are of similar magnitude. Then the rotor/stator system needs to be analyzed in one model because the coupled eigen frequency will differ considerably from the values calculated separately for rotor and CP. Since the goal of the eigen frequency ananlysis is just this: to determine the risk of resonances, it is recommended to employ coupled-vibration models in order to thoroughly check the design of large vertical pumps. The coupling of shaft and column pipe through appropriate bearing stiffness is vital for getting meanigful results [108].

Apart from the lateral eigen values a forced response calculation is to be done with the maximum expected hydraulic and mechanical imbalance. The modeling should include torsional and combined lateral/torsional modes.

A Sensitivity study should be made in order to get an idea about the range of uncertainties of the predicted eigen frequencies of shaft, column pipe and motor: the effects to be investigated include: added mass, foundation stiffness, bearing stiffness, the influence of discharge piping and other parameters as required.

10.14.6 Case Histories "Eigen Frequencies"

CH6. Bearing instability. A case with high vibrations and recurrent failures of bearings and mechanical seals was reported in [108]. The vibration spectra of the column pipe were dominated by peaks at $0.49 \times f_n$ which shot up when reaching a threshold speed. The column pipe ($L_{CP} = 13.7$ m) was quite slim with an estimated $L_{CP}/D_{CP} = 37$; there were 8 plain ungrooved cylindrical bearings. Diagnosis testing methods used: proximity probes near the impeller for shaft vibrations, several accelerometers along the column pipe. A two-level *rotor dynamic* FEA model comprising column pipe, shaft and motor stool was used for analyzing the behavior of the pump as built and for developing tilting pad bearings which were applied to solve the problem. The calculations revealed several *casing* modes and *rotor* modes below the maximum operation speed. In "casing modes" the shaft follows the movement of the column pipe, while in "rotor modes" the rotor vibrations are dominating and the CP vibrations are minimal. Eigen frequencies, measured by impact tests at standstill, were found near 5, 15, and 30 Hz. Calculations with $n = 0$ and with $n = 1600$ rpm gave the results in Table D10.7:

Table D10.7 Calculated eigen frequencies versus test (*casing* modes)

Eigen frequency from test (n = 0) in Hz	5	15	30
FEA versus test at n = 0: f_{FEA}/f_{test}	1.12	1.04	0.96
FEA versus test at n = 1600 rpm: f_{FEA}/f_{test}	1.27	1.17	1.1
Influence of speed from FEA: $f_{n=1600}/f_{n=0}$	1.14	1.12	1.14
Damped versus undamped from FEA at 1600 rpm: $f_{dampe}/f_{undamped}$	0.91	0.92	0.92

Conclusions from Table D10.7:

- The lower modes were less well predicted than those at high frequency
- The accuracy of test versus prediction (at n = 0) was 4 to 12%
- Influence of speed according to FEA: eigen frequencies at 1600 rpm were about 12% higher than at n = 0.
- Damped eigen frequencies were about 8% lower than undamped
- Using the test data at n = 0 for determining the required separation margins would be too uncertain because the influence of the speed on the eigen frequency is too strong to be neglected - at least in this particular case.

The above results were derived for a slim column pipe. Their transposition to large cooling water pumps remains uncertain. The damped analysis revealed a rotor-mode eigen frequency with negative damping (i.e. rotor instability) close to half the operating speed. The half-speed-whirl instability thus explained the observed damages quite well. Subsequent analysis with preloaded of 4-pad tilting-pad bearings showed that positive damping could be achieved (confirmed by testing).

Vibrations depending on water level

For a given pumping station, the water level can have an impact on the vibrations for three reasons:

- The eigen frequencies of the column pipe increase with sinking water level (the added mass effect is reduced).
- The tendency for generating vortices increases with lower submergence of the suction bell.
- The approach flow velocity $c_A = Q/(W \times h_{wl})$ increases when the water level h_{wl} falls. The flow becomes less uniform and radial excitation forces grow.

CH7: Pump data: n = 740 rpm, Q = 8,250 m^3/h, H = 40 m, $n_q = 70$, P = 1070 kW. Reference [98] reports on vibration problems with pumps where the separation margin SM between column pipe and rotational frequency were so narrow that the tidal variation of the water level was responsible for significant fluctuations of the vibrations measured on top of the of the motor (up to 17 mm/s RMS at low tide). The separation margin (ratio) between motor eigen frequency was 0.93 (at low tide). With rising water level, the added mass of the CP increased, the eigen frequency decreased and the SM improved. Thus, vibration amplitudes were minimum at high tide and reached their maximum at low tide.

The problem was solved by 3 passive vibration absorbers screwed to the motor, Fig. 10.76. These absorbers (designed as a single-mass oscillators) consisted of a mass and a spring in the shape of a rod. As suggested by the photo, the design allowed tuning the eigen frequency in the field by changing the vertical position of the mass on the rod. Column pipe vibrations have not been mentioned in [98]. It might well be that the case was solely a question of resonance with weak imbalance. However, it must be realized that the symptoms have been cured, but the excitation forces had not been reduced.

CH8: Vibrations fluctuating with the tide were also reported in [103] on a vertical 300-kW pump running at 1180 rpm. At high tide, vibrations were 11 mm/s with a CP eigen frequency of 19.3 Hz (SM = 0.98) and at low tide 6 mm/s with an eigen frequency of 20.8 Hz (SM = 1.06). To diagnose the problems an accelerometer was mounted on the CP somewhat above the bowls of the 2-stage pump. This case demonstrates again how sensitive the separation margin can be to the added water mass and how accurate the prediction must be to prevent such problems - unless the separation margin is sufficiently large to cover the uncertainties of the prediction. As a remedy, the pumps were shortened in order to keep the CP eigen frequency sufficiently above the running speed.

Vibrations at different water levels were measured by accelerometers on the suction bell in [113]. The main response frequency increased as the water level was lowered because the eigen frequency of the column pipe increased at lower submergence of the suction bell.

CH9: Bearing damage due to column pipe resonance. Recurrent damage to the line-shaft bearings troubled the operator of semi-axial pumps which had very

Fig. 10.76 Passive vibration absorber, [98]

slender column pipes; that way that a 3rd order eigen frequency was excited by the rotor speed (894 rpm, 14.9 Hz). Vibration velocities at the motor reached 10 mm/s at f_n. However, since the separation margin was 23%, the problem was not a motor resonance but an eigen frequency of the column pipe. This has been verified by installing an accelerometer on the suction bell where 19 mm/s were measured at f_n. The eigen frequencies of the lowest three column pipe modes were measured as: 1, 5.5 and 15.9 Hz. Thus the 3rd mode had a separation margin of only 7% away from imbalance excitation. All CP eigen frequencies showed as sharp peaks in the spectra measured at top of the motor as well as at the suction bell. But vibrations at the suction bell were much higher than on the motor proving that the problems are below ground rather than above. As a remedy, the column pipe was shortened in order to increase the separation margin to about 16%.

CH10: Column pipe resonance causing shaft ruptures and cracks in the CP, [112]. CWP data: $Q = 15{,}000\ m^3/h$, $H = 21$ m, $P = 1000$ kW, $n_q = 106$, $n = 510$ rpm ($f_n = 8.5$ Hz). Initially, motor vibrations were 2.5 mm/s, but they increased over time to 5 mm/s when a shaft fracture due to high-cycle fatigue occurred near the coupling. Slight rubbing of the impeller was also observed, and a number of bolts holding the pump to the foundation ruptured. After the first repair (with new shaft), motor vibrations rapidly increased to 10 mm/s zero to peak. Since the separation margins (at f_n) to the rotational speed were only 4%, the motor stool was reinforced by welding braces to the discharge head/motor stool. Experimental modal analysis revealed an eigen frequency of 8.19 Hz (SM = −3.7%) when the bolts connecting the discharge head to the foundation were loosened - as opposed to 8.44 Hz (SM = −0.7%) with tightened bolts. In the latter case, vibrations at the top of the motor went up to 29 mm/s zero to peak. After welding braces to the discharge head, motor vibrations were reduced by 50% but remained still unacceptable. Furthermore, cracks in the column pipe were detected. Accelerometers were then installed at the suction bell where vibration velocities at rotational speed reached 90 mm/s zero to peak (displacement 1.7 mm). The calculated eigen frequency of the column pipe was 8.1 Hz (SM = −4.5%). The root cause of all the problems encountered were thus an insufficient separation between column pipe eigen frequency and rotational frequency. To improve the separation margin, the mass of the suction bell was increased to 2000 kg; thereby decreasing the eigen frequency of the CP to 7.3 Hz (SM = −15%). After this modification, vibrations of the suction bell went down to 3 mm/s, while only 0.5 mm/s were recorded on the top of the motor. Lessons:

- Resonances of the column pipe cannot be detected sufficiently well by measurements on structures above the foundation.
- Stiffening of the structures above the foundation and tightening the connection to the foundation can aggravate vibration problems related to the column pipe.

CH11: Column pipe resonance causing bearing damage. CWP with 2000 kW. While vibrations of motor and structures above the foundation were within acceptable limits, cracks were found on the column pipe and the lowest bearing was subject to heavy wear. In addition, the suction bell/casing wall showed strong marks of impeller rubbing. Hence high displacement amplitudes were manifestly present. Vibration peaks measured above the foundation were dominantly synchronous because the calculated eigen frequency of the column pipe was too close to the running speed. Vibration measurements on the column pipe had not been made. But arguing that the actual stress must have reached the corrosion fatigue strength in order to initiate cracks, the calculated forces and deflections required to create such stresses were far beyond any possible imbalance. It was therefore concluded that coupled vibrations near resonance were responsible for the high amplification factors requisite to explain the cracks. In summary, strong shaking of the column pipe is believed to be at the origin of the cracks and the bearing wear.

Initially, also vibrations on the motor and the structures above ground were too high. Only after proper grouting and adding supports, they were brought down to acceptable levels. Thus, the symptoms disappeared, but the excitation forces were not reduced. This case, which is quite similar to cases CH9 and CH10, demonstrates again that acceptable vibration levels measured on the structures above the foundation do not always allow the conclusion that the problems have been solved.

CH12: Cooling water pump failures were reported in [115]; Q = 4,500 m^3/h, H = 53 m, n = 885 rpm. After less than 2 months of operation severe damage occurred to the impeller wear rings, impellers and bearings. Impeller and shaft were worn on one side, while wear on the stator parts was uniform around the circumference. Hence high synchronous rotor vibrations must have been present. The vibrations fluctuated with periods of a few minutes. This "beating" was found to be caused by the fact that 6 parallel pumps were mounted on a common soft concrete structure with the discharge pipes connected to a common header. This way, the vibrations of the individual pumps influenced each other - leading to the beating when the vibrations were in phase and out of phase.

A variable-speed shaker mounted on the motor was used to determine the eigen frequencies from Bode plots. Two velocity transducers were attached to the bowl. Both the motor and the column pipe eigen frequencies were very close to the running speed (SM less than 2%). The low separation margin was responsible for the excessive vibrations (at the motor top 46 mm/s or 0.5 mm displacement) and the resulting component damage. It is interesting to note that the resonance with the column pipe caused the same wear patterns as an imbalance. Vibrations were very sensitive to the stiffness of the connection of the pump/motor unit to the foundation. Eventually, the problems were solved through stiffening the structures above ground by adding braces and anchor bolts.

Beating of vibrations due to the interaction of parallel pumps mounted on a common foundation was also reported in [116]: the vibrations varied with a period of 18 min between 1.6 and 6.3 mm/s.

10.14.7 Amplification Factors and Separation Margins

The separation margin is calculated as the ratio of the eigen frequency to the excitation frequency SM = f_{eigen}/f_{ex}; it can be expressed as ratio or in per cent: SM = $100(f_{eigen}/f_{ex} - 1)$. Selecting a suitable separation margin serves two purposes:

1. Ensure that vibration amplitudes remain within acceptable limits. This implies that the amplifications factors must be sufficiently low. In practice, the damping factor can at best be roughly estimated. It is expected in the range of D = 0.02 to 0.05.
2. Cover the uncertainties of the eigen frequency calculation: These are expected in the range of 5 to 10% if the modeling is done with utmost diligence.

When dealing with variable-speed pumps, the entire speed range needs to be considered, but the highest speed is the most critical because the excitation forces increase with the square of the speed. Amplification factors of forced vibrations of a single-mass vibrator model under periodic excitation forces may be used to get an idea what separation margins should be selected - even though the pump assembly is not an ideal single-mass vibrator. Nevertheless, this simple mechanical model can be used to illustrate the behavior of the pump under periodic excitation. Figure 10.77 suggests that separation margins should not be selected below ±15% in order to keep the amplification factors below 3 to 3.5.

It is possible to estimate the amplification factors and damping coefficients by an impact test or by measuring the transfer functions when exciting the bowl by a shaker at various frequencies.

In summary: *Separation margins below 10% must be regarded as critical*; SM below 15% should be avoided; SM = 20% is recommended in ISO 13709 [109] but is often difficult to achieve. Recommendations given in [124]: minimum SM below running speed for reed analysis: ±20% for n > 1200 rpm; +25/−20% for n < 900 rpm.

10.14.8 Line-Shaft Bearings

Bearing selection and design is of utmost importance for the reliable operation of large vertical pumps: bearing failures occurred in 7 of the 13 case histories discussed above; 6 bearing failures were caused by high vibrations (one primarily by sand erosion). But the reverse can be true as well: for example, if bearing wear is caused by sand, vibrations will increase and may exceed the allowable levels as in CH13 below.

CH13. Problems with line-shaft bearings of river water pumps were reported in [121]. Pump data: variable pitch axial pump, 495 rpm, 11,500 m^3/h, 7.5 m, n_q = 195, d_2 = 1000 mm. Erosion of the original grease-lubricated bronze bearings due

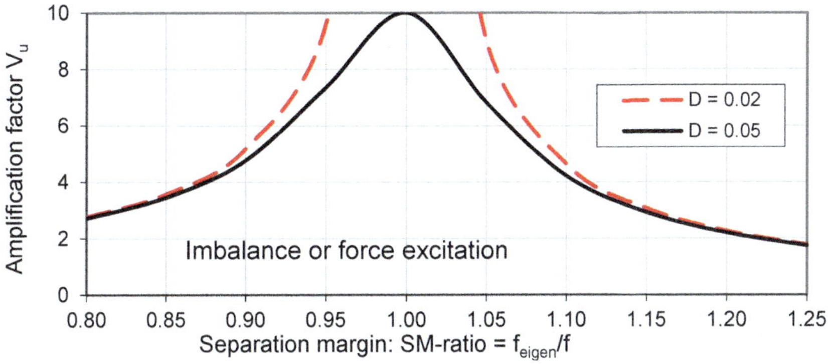

Fig. 10.77 Amplification factors V_u for excitation by imbalance or force as function of $SM = f_{eigen}/f$

to sand-laden water led to a first retrofit with monolithic SIC bearings. Even though SIC components were resistant to sand erosion, the bearings failed because monolithic SIC is extremely brittle and sensitive to shock loading. Due the occasional formation of "grease ice" in the impeller (or foreign matter) vibrations at the bowl went up to 15 mm/s which apparently contributed to the bearing failures. In a second retrofit fiber-reinforced SIC (which is more elastic) had been used for the shrunk-on shaft sleeves while the monolithic SIC bushings were backed up by a soft material (rubber).

In essence, the line-shaft bearings consist of a plain sleeve fixed on the shaft and a bushing fixed on the stator (bearing support structure). Mostly, polygon bearings with longitudinal grooves are used. Bearing requirements include: high stiffness; high margin against rotor instability; high damping; dry running capability during startup of pump; high resistance against wear due to sand. Typical bearing clearances $\Delta d/d = 0.0015$ to 0.002, length $L/d = 1.5$ to 2.5. If no stiffness data are available, $k_{bearing} = 7 \times 10^6 \times (d/d_{Ref})$ [N/m] can provide a rough estimate, [104]; d = shaft sleeve diameter, $d_{Ref} = 100$ mm. The loading of the lower bearing (installed as close as possible to the impeller) is given by: (1) mechanical imbalance; (2) hydraulic imbalance; and (3) steady as well as unsteady radial forces acting on the impeller. The mechanical imbalance can be estimated from the balancing of the impeller which usually is done according to class G6.3 which means G = 6.3 mm/s. The corresponding eccentricity (e in mm) of the rotor mass m_R is $e = G/\omega$ with $\omega = \pi \times n/30$. The imbalance is $U = m_R \times e = G/\omega \times m_R$ in kg-mm and the imbalance excitation fore becomes:

$$F_{imb} = m_R \times e \times \omega^2 \times 10^{-3} = U \times \omega^2 \times 10^{-3} \qquad (10.30)$$

For class G = 6.3 this gives: $F_{imb} = 0.66 \times m_R \times (n/n_{Ref})$ with F_{imb} in N, m_R in kg and $n_{Ref} = 1000$ rpm.

The radial hydraulic excitation forces caused by the approach flow can be estimated from the force coefficients provided in Table D10.4. High dynamic loading (unsteady forces) may disrupt the liquid film and induce bearing wear.

10.14.9 Trouble Shooting and Diagnostic Tools

Typically, a large power plant may have 3 to 10 CWP-units each weighing several tons with a total length from suction bell to motor top of 8 to 20 m. It is obvious that any serious trouble with the intake structure, pump or motor is likely to cause considerable costs not only to the pump manufacturer but also to the operator/owner of the plant. Every handling of this huge equipment is a major operation involving significant manpower and costly down-time. Finding the root cause and implementing remedies can take several months - if not years.

Measuring vibrations on the motor, motor stool or bearing housing tells little of the behavior of the column pipe: if these structures above the foundation are stiff - which is the design goal - the column pipe can be represented by a cantilever beam clamped to the base plate with the maximum stress at the clamped end. The stiffer the baseplate and foundation, the less the vibrations measured above ground tell about the behavior of the column pipe.

If high vibrations are measured on motor, bearing housing and/or discharge nozzle, these structures are sometimes stiffened by adding supports or by modifying the motor stool. This only can help if the high vibrations are caused by low separation margins between eigen and excitation frequencies because *the excitation forces cannot be reduced* by adding stiffness. *The stresses in the column pipe are likely to even increase*.

Works tests can tell little on the behavior of the pump in the plant because the systems are totally different with respect to:

- the sump geometry and approach flow,
- the stiffness of the foundation and discharge pipe supports,
- standing waves and pressure pulsations in the discharge pipe where the total head is throttled in the works test.

To quote a pessimistic view from [116]: "A shop test will only offer reasonable assurance that a unit is balanced well enough to run sufficiently long to solve any problems in the field".

As a consequence of the above, it is expedient to think about trouble shooting and diagnostic methods. Table 10.14 gives an overview of the tools available - most of which have been used in practice. Table 10.15 may help with finding the root cause of a problem and define remedies. Naturally, troubleshooting starts with the items which can be checked with less efforts and progresses, as need be, to more complex and expensive investigations. Depending on the type of problem, the first steps include checking alignment and mechanical components such as couplings.

Comments to Table 10.14:

- Time signals: non-periodic peaks (bursts) occurring at intervals from a few seconds to minutes (or even hours) hint to vortex formation in the sump, Figs. 10.65 and 10.67.
- Analyze long-term vibration trend to detect degradation of mechanical components.
- Spectra give indications as to the nature of the excitation mechanism and mechanical or acoustic eigen frequencies.
- If pressure pulsation spectra show the same peaks as mechanical vibrations, the cause is more likely to be hydraulic or acoustic. If a peak in the mechanical spectrum is not present in pressure pulsation spectrum, the likely causes are mechanical (eigen frequency, imbalance of motor or impeller, misalignment). However, broad band energy (from turbulence, air ingress or flow recirculation) and sump flow fluctuations can excite eigen frequencies so that the decision as to the root cause remains to some extent ambiguous.
- Column pipe vibrations can be measured by waterproof or protected accelerometers attached to the CP. An alternative was reported in [104], where an accelerometer, leak-tight mounted in a nylon tube, was introduced through a small hole in the foundation and attached by a magnet near the bowl. Measurements with an accelerometer at the lower end of the column pipe are an expedient - if not indispensable - way to diagnose problems related to CP vibrations and bearing wear.
- Column pipe vibrations can also be measured by vibration velocity transducers. According to [115] these have an advantage over accelerometers or proximity probes in that the velocity transducers produce an electric signal without a need for a power supply and signal conditioner. Velocity transducers operated for several months in river water without special waterproofing [115].
- It may be considered to equip large pumps with a small tube (diameter 20 to 30 mm) reaching down from the mounting flange to the bowl. This tube would be sealed on top and bottom. In case of need, an accelerometer or a velocity probe could be let down from above the foundation to the bowl to measure CP vibrations. Once a proper design has been done for this type of instrumentation, the equipment could be used in any plant for diagnosing trouble.

Cracks in the column pipe definitely prove that the fatigue strength has been exceeded. Hence high amplitudes of the CP vibrations must have been present - even if low amplitudes were measured on the structures above ground. Since the fatigue strength can be expected to be in the order of $0.3 \times R_m$ (and roughly $0.15 \times R_m$ for sea water), the resulting stresses are considerable. Therefore, it can help to diagnose the root cause of the problem by estimating the forces and deflections required to induce such stresses and comparing these forces to any plausible excitation mechanism - e.g. to impeller imbalance. To do so, the column pipe can be modeled as cantilever beam clamped at the top. The deflection y_{ci} or the force F_{ci} at the suction bell required to initiate cracking near the top of the CP can be estimated from:

Table 10.14 Diagnostic tools

Item	Measured quantity	Instrument	Output or goal, comments
Performance	n, Q, H, P, T, water level	As available in plant; Q and P can be inferred from measured head	Determine operation point (q*) and $NPSH_A$
Motor, motor stool, bearing housing, discharge pipe	Vibration velocity or acceleration	Velocity probe or accelerometer	• Vibration in the time domain • Spectra • Compare to allowable limits • Diagnosis First step if excessive vibrations
Impact test	Response of component to an impact	Instrumented hammer and accelerometers; This test is standard in case of problems.	• Determine eigen frequencies and mode shapes • Response may allow to determine the damping
Power fluctuations	Motor current [95]	Current probe clamped around one motor power line	Power fluctuations may indicate: • Flow fluctuations in the sump • Strong vortices • Air entrainment • Unusual rotor/interaction • Flow fluctuations in the discharge pipe
Shaft vibrations near coupling	shaft displacement	Proximity probes	• Vibration in the time domain • Spectra • Orbits Shaft vibrations are measured *relative* to column pipe. May help to detect bearing instability or alignment errors
Shaft vibrations in column pipe		Underwater proximity probes [104]	
Pressure pulsations in discharge pipe	Fluctuating pressure	Quartz pressure transducer	• Pulsations in the time domain • Spectra • Compare to spectra of mechanical vibrations • Detect standing waves in the system Easy to do, but influenced by system (standing waves)
Column pipe vibrations and cracks	Acceleration or vibration velocity near suction bell or bowl	Waterproof or capsuled accelerometer, [95, 103] velocity probe [115]	• Vibration in the time domain • Spectra
Column pipe stresses and cracks	Unsteady stresses	Strain gages applied near top of the column pipe (below the foundation and below gussets used for stiffening the CP)	• Unsteady stresses can be compared to fatigue strength • Spectra can be indicative of eigen frequencies and excitation mechanisms • The vibration amplitudes (deflection) near the suction bell can be estimated from the clamped cantilever-beam model
Visual flow observation	Flow patterns	Underwater video camera	Detect strong impeller inlet recirculation, vortices, air bubbles or debris

$$\frac{y_{ci}}{L_{cp}} = \frac{2\sigma_{ci}}{3\alpha E}\left(\frac{L_{cp}}{D_{cp}}\right) \quad F_{ci} = \frac{\sigma_{ci}\pi}{32\alpha L_{cp} D_{cp}}\left(D_{cp}^4 - d^4\right) \tag{10.31}$$

Subscript ci = crack initiation; σ_{ci} = alternating bending stress required to initiate cracks; this is the local peak stress amplitude (zero to peak).

With $\sigma_{ci} = R_m/3$ and Young's modulus $E = 2 \times 10^{11}$ N/m^2 the strain amplitude required to initiate cracks for R_m = 600 N/mm^2 becomes ε_{ci} = 0.001. Assuming a stress concentration factor of α = 4 Eq. (10.31) becomes:

$$\frac{y_{ci}}{L_{cp}} \approx \frac{10^{-3}}{6}\left(\frac{L_{cp}}{D_{cp}}\right)\left(\frac{R_m}{R_{m,Ref}}\right) \quad \text{with } R_{m,Ref} = 600\,\text{N/mm}^2 \tag{10.32}$$

For the cracks shown in Fig. 10.73, Eq. (10.32) would yield an amplitude at the suction bell in the order of 10 mm; and for the cracks in Fig. 10.78 an amplitude in the order of 1 to 2 mm would be expected.

- Step 1: Estimate the corrosion fatigue strength of the column pipe material. Note that fatigue strength could be impaired in weld-influenced zones due to increase in residual tensile stresses, increase in grain size or precipitations. Due allowance must be made for the mean stress in the CP caused by the weight and internal pressure. If the CP cannot be considered as a slender beam, the 3D-stress distribution will differ from the beam model.
- Step 2: Assume stress concentration factor; for example, α = 4 if the cracks originate at a weld.
- Step 3: Calculate the force required to induce that amplitude of the peak stress, which equals the corrosion fatigue strength. The beam model predicts bending stresses which lead to CP cracks in the circumferential direction. A bearing support spider would induce local stresses in the CP which would require much lower crack-initiation forces.
- Step 4: Calculate the imbalance of the impeller which would be required to generate this force. If the resultant imbalance is beyond any plausible defect on the impeller, excessive imbalance cannot be considered the root cause. Resonance with a column pipe eigen frequency or high excitation forces from the sump may then be investigated as possible culprits. This analysis is likely to often point to an eigen frequency problem.
- Step 5: Calculate the deflection (amplitude) of the column pipe at the suction bell. Comparing this amplitude to the clearance between impeller and stator gives another indication as to the severity of the vibrations. Rubbing of the impeller on the stator can be demonstrated by this analysis.

The above estimate cannot be applied if the cracks occur at a local stress riser such as a bearing support spider unless one has an idea about the stress distribution near the support spider. Figure shows 10.78 shows a crack in the CP originating at a spider bracket directly welded to the column pipe. This type of design causes high local stresses if CP vibrations are high. To reduce to local stresses, the brackets of the bearing support spider can be welded to a ring rather than directly to the CP. Also, the spider can be designed to have increased flexibility as compared to radial brackets. In addition, holes can be drilled into the brackets to increase flexibility; there a number of design options used by different pump manufacturers.

Troubleshooting

- Determine operation point q*
- Operation in unstable range of Q-H-curve?
- Review pump selection: actual operation versus specification
- Review operation procedures
- On-site inspection:
  - observe flow in sump for vortices, turbulence or air bubbles in the water
  - cracks in column pipe, support structures, concrete floor, discharge elbow
  - grouting
  - bearing damage
  - coupling damage
  - rubbing of impeller in casing
- Review problems related to quality, e.g. casting tolerances (hydraulic imbalance), other issues related to design or workmanship

10.14.10 Conclusions and Recommendations

Incidentally, the 13 case histories discussed above fall into two categories: (1) Non-uniform approach flow was the root cause of 4 of the plant problems, (2) while insufficient separation margins of the column pipe were responsible for the damage in the 8 incidents. The following conclusions may be drawn from the analysis:

Fig. 10.78 Cracks in CP originating at a radial bracket of a bearing support spider

1. Apparently-successful model tests according to Froude similarity do not guarantee absence of trouble due to unsteady and non-uniform approach flow to the pump. Reasons:

   (1) Reynolds number in the plant is an order of magnitude higher than in the model. Consequently, unsteadiness, swirl and vortices decay more slowly in the plant than in the model.
   (2) The swirl angle measured in model test is an integral - not relevant for local incidence variation
   (3) The radial forces generated by the approach flow on the pump is disregarded in the model tests

2. Non-uniform approach flow causes steady and unsteady radial forces at low frequencies. These forces are expected to increase with the square of the flow velocity at the inlet to the sump ($F_R \sim c_A^2$) due to: (1) increasing energy in the unsteadiness of the approach flow, and (2) due to the reflection of the water hitting the rear wall of the sump, Fig. 10.70. Therefore, in general, the approach flow velocity to the sump should not exceed the value of $c_A = 0.5$ m/s recommended in [92]. If the velocity is appreciably higher than this recommendation, an in-depth analysis of the lateral forces acting on the pump is warranted. In addition, there is a risk of vortex formation even if the model test does not show unacceptable vortices.
3. In two of the case studies, a *cross beneath the suction bell* was employed in the model test to get rid of submerged vortices. This subterfuge worked in the model but not in real plant operation. Flow separation in the sump greatly increases the velocity fluctuations and the unsteady radial forces. Anti-vortex devices *perpendicular* to the flow direction cause such flow separation; they should be avoided by all means.
4. Horizontal fillets between sump side walls and floor and rear wall and floor can be recommended in the general case (see Fig. 11.23). Vertical fillets between the rear wall and the side walls are beneficial for low approach flow velocities ($c_A < 0.5$ m/s). At high flow velocities, they may lead to strong flow reflection at the rear wall, Fig. 10.70. In such cases, the usefulness of the fillets needs careful investigation.
5. A floor splitter of triangular shape below the pump center-line leads to less radial excitation forces than a rib of rectangular cross section according to tests in [120].
6. Splitters and fillets should be designed with oblique shapes facing the incoming flow - blunt shapes would induce flow separation and swirl, Fig. 11.23.
7. Covered sumps without rear-wall splitter create strong vortices because of the increased velocities below the cover [94]. A rear wall splitter should always be installed to reduce vortices in the wake of the CP (see also Fig. 10.70). The higher the approach flow velocity, the more a rear wall splitter is needed - particularly so in covered sumps.

Table 10.15 Diagnosing the root cause and finding remedies

Observation	f	Possible cause	Measures, Steps		Conclusion
High motor/motor stool vibrations	1×	Motor imbalance	**Step 1:** Decouple motor from pump and test again	Vibration acceptable	Motor imbalance acceptable; go to step 2
				Vibration still high	Motor is possible cause: balance motor
	1× 2×	Misalignment	**Step 2:** Check alignment		If okay, step 3
	1× 2× z_{La}×	Eigen frequencies too close to excitation frequencies	**Step 3:** • Impact test • Determine SM • Estimate amplification factors, Fig. 10.77	SM > 1.25	Eigen frequencies of motor is NOT a likely cause → Steps 4, 5
				SM < 1.15	Resonance problem likely • Check grouting • Check options for stiffening
	1× z_{La}×	Excitation comes from below ground • High hydraulic excitation forces • Mechanical or acoustic resonance	**Step 4:** • review hydraulic design • review approach flow		
High discharge pipe vibrations		• Pipe support too weak • Standing waves	**Step 5:** • Review pipe supports • Measure pressure pulsations; check for standing waves		
High column pipe vibrations	1× z_{La}×	• **Eigen frequencies too close to excitation frequencies** • Imbalance	**Step 6:** • Measure CP vibrations, see Table 10.14 • Determine eigen frequencies and SM		Eigen frequency problem is the most likely scenario for high CP vibrations

(continued)

(continued)

Table 10.15 Diagnosing the root cause and finding remedies

	< 1×	• **sub-synchronous** (0.4 to 0.9) f_n bearing whirl	• Review bearing design • Pre-load bearing through intentional mis-alignment	
	$\neq \nu \times f_n$	• Impeller inlet recirculation • Vortices • Unsteady approach flow	• Visual flow observation in sump • Review hydraulic design • Review intake structure design	
Vibrations at multiples of f_n $\nu = 2, 3, 4 \ldots$	$\nu \times f_n$	• blade pitch error of impeller	• Check blade pitch	
Vibrations NOT related to f_n	$\neq \nu \times f_n$	• Unsteady vortices in sump • Air ingress • Mechanical or acoustic eigen frequencies excited by broad band pressure pulsations • alternate or rotating stall	• Visual flow observation • Determine eigen frequencies • Review part load recirculation • Measure pressure pulsations • Check for standing waves • Check impeller and diffuser incidence and deceleration in throat areas	Review options for modification of sump, e.g. installing a screen
Vibrations at f_n and $z_{La} \times f_n$ varying with time; periods 10 s to 1 h				
Bearing wear: *shaft sleeve uniform*, bushing local, over circumference		• Steady and unsteady radial force due to uneven approach or discharge flow		Wear could also be caused by sand, but is then likely to be more uniform.
Bearing wear: *shaft sleeve local*, bushing uniform, over circumference		• Excessive imbalance • Bent shaft • High CP vibrations (small SM)	• Check imbalance on balancing machine • Check impeller tolerances (hydraulic imbalance)	If balancing is okay, check separation margins step 3 & 6. Wear could also be caused by sand.

(continued)

(continued)

Table 10.15 Diagnosing the root cause and finding remedies

Rubbing of semi-open impeller in casing; stator wear uniform, over circumference	• High CP vibrations (small SM) • Bearing wear • High imbalance	• Check bearing wear, clearances • Check imbalance including corrosion and cavitation effects	Rubbing implies large amplitudes either due to shaft or high CP vibrations
Cracks in discharge bend	• Pipe support too weak • Standing waves • High CP vibrations (small SM)	Measure pressure pulsations	
Cracks in column pipe	• Resonance with CP eigen frequency • Recirculation at the impeller inlet causing swirl in the sump	• Analyze force or deflection required to exceed fatigue strength • Review hydraulic design (incidence, approach flow angle) • Check for asymmetric approach flow in sump • Review swirl breaking devices in sump: *rear wall* and floor splitter	Fatigue strength has been exceeded: • High amplification factor (SM too low) • Excessive swirl in sump due to impeller inlet recirculation • Imbalance NOT likely to be the primary cause
Vibrations varying with time	• Tidal water level variations (impact on CP eigen frequencies) • Misalignment caused by solar heating of motor stool	• Analyze vibrations as function of water level • Shield motor stool from sun	
Vibrations fluctuating with time. "Beating" of vibrations	• Parallel pumps influencing each other via foundation or discharge piping. Vibrations in phase: maximum amplitude; out of phase: minimum	• Do vibrations depend on the number of pumps in operation? • Loose anchor bolts to check influence on vibrations • Check eigen frequencies (motor and CP) • Check influence of foundation stiffness • Check influence of discharge pipe (mechanical and acoustic)	

(continued)

(continued)

Table 10.15 Diagnosing the root cause and finding remedies

	• Two excitation frequencies are close together. Example: CH3	
Flow disturbances in intake structure	• Review model test results if any • Visual observation of water surface • Under-water video camera* • CFD of sump flow* • Cross flow at inlet to individual pump sump? • Do vibrations depend on number of pumps in service? • Effective rear wall splitter installed? • Floor splitter design effective? • Check dimensions against standards • Excessive velocities: $c_A > 0.5$ m/s? • Steps, sudden enlargement, structures perpendicular to flow direction in sump? • Air ingress (e.g. from wet cooling towers)?	

*) Note: vortex formation can take a long time, see CH1. Observation time has to cover this effect; CFD calculations too

8. Every pump sump should have *effective* anti-swirl devices. Otherwise pre-swirl would induce head loss during part load recirculation.
9. Vortices are mostly unsteady; they can cause non-periodic fluctuations of the vibration signals. According to the case histories discussed above, their formation can take between a few seconds up to one hour.
10. Vibrations measured above the foundation are not always sufficiently representative for the column pipe vibrations.
11. If column pipe vibrations are dominating (operation close to a CP eigen frequency), stiffening the structures above ground may give a bogus solution since the root cause of the problem had not been addressed.
12. Generally, it can be expected that mechanical and hydraulic imbalance present a problem only if the separation margins are too low. In other words: If excitation forces are not manifestly (beyond any doubt) out of control, the resonance question needs to be seriously addressed.
13. For predicting the eigen frequencies, a complete model of motor, discharge head, column pipe and rotor is required. The model is to include part of the foundation and discharge pipe, added water mass effects as well as bearing stiffness. This type of analysis of the complete motor/pump unit is considered vital for analyzing and preventing vibration problems, [104–106, 108, 112, 115, 116, 118].
14. Separation margins below 10% are considered insufficient in many cases; the target for design is SM = 20%, but SM = 15% may be considered a reasonable compromise. The actual minimum required SM cannot be predicted accurately because it depends on the excitation forces, the damping and the mechanical design.
15. Vibrations can also fluctuate ("beating") due to the interaction of parallel operating pumps mounted on the same foundation and/or discharging into a common header. Beating is a modulation or superposition of slightly different frequencies. The smaller the difference between the frequencies of two superimposed vibrations, the longer is the period of the resulting vibration. For detecting beating phenomena, vibration records in the plant must be taken over sufficiently long intervals covering the beating periods.
16. To avoid interaction of the vibrations of parallel pumps ("beating") it would be beneficial to mount the pumps on individual foundations.

References

1. Adams, M.L.: Rotating Machinery Vibration. Marcel Dekker Inc (2001)
2. Alford, J.S.: Protecting turbomachinery from self-excited rotor whirl. ASME J. Eng. Power **87**, 333–344 (1965)
3. Amoser, M.: Strömungsfelder und Radialkräfte an Labyrinthdichtungen hydraulischer Strömungsmaschinen. Diss. ETH. Nr. 11150 (1995)

4. Arndt, N., et al.: Unsteady diffuser vane pressure and impeller wake measurements in a centrifugal pump. In: Proceedings of 8th Conference on Turbomachinery, Budapest, pp. 49–56 (1987)
5. Au-Yang, M. K.: Flow-induced Vibrations of Power and Process Plant Components. Professional Engineering Publishing Ltd (2001)
6. Berten, S.: Hydrodynamics of high specific power pumps for off-design operating conditions. Dissertation, EPF, Lausanne (2010)
7. Berten, S., et al.: Experimental Investigation of Flow Instabilities and Rotating Stall in a High-Energy Centrifugal Pump Stage. ASME. FEDSM (2009)
8. Blevins, R.D.: Formulas for Natural Frequency and Mode Shape (Reissue). Krieger, Malabar (1995)
9. Blevins, R.D.: Flow-Induced Vibrations (Reprinted 2nd edn.). Krieger Publishing Company, Malabar (2001)
10. Bolleter, U.: On blade passage tones of centrifugal pumps. Vibrations **4**(3), 8–13 (1988)
11. Bolleter, U.: Generation and propagation of pressure pulsations in centrifugal pump systems. In: AECL Seminar on Acoustic Pulsations in Rotating Machinery, Toronto (1993)
12. Bolleter, U., et al.: Hydraulic and mechanical interactions of feedpump systems. EPRI Report TR-100990 (Sept 1992)
13. Bolleter, U., et al.: Rotordynamic modeling and testing of boiler feedpumps. EPRI Report. TR-100980 (Sept 1992)
14. Braun, O.: Part load flow in radial centrifugal pumps. Dissertation, EPF Lausanne (2009)
15. Brennen, C.E.: Hydrodynamics of Pumps. Concepts ETI, Norwich (1994)
16. Casey, V.M., et al.: Flow analysis in a pump diffuser. Part 2: validation of a CFD code for steady flow. ASME FED **227**, 135–143 (1995)
17. Chen, Y.N.: Wasserdruckschwingungen in Spiralgehäusen von Speicherpumpen. Techn Rundschau Sulzer. Forschungsheft, 21–34 (1961)
18. Chen, Y.N., Beurer, P.: Strömungserregte Schwingungen an Platten infolge Karman'scher Wirbelstraßen. Pumpentagung, Karlsruhe (K6) (1973)
19. Chen, Y.N., Florjancic, D.: Vortex-induced resonance in a pipe system due to branching. IMech C109/75 (1975)
20. Chen, Y.N., et al.: Reduction of vibrations in a centrifugal pump hydraulic system, pp. 78–84. IAHR Karlsruhe (1979)
21. Childs, D.: Turbomachinery Rotordynamics. Wiley, New York (1993)
22. Childs, D.W., et al.: Annular honeycomb seal test results for leakage and rotordynamic coefficients. ASME Paper. 88-Trib-35
23. Cooper, P., et al.: Minimum continuous stable flow in centrifugal pumps. In: Proceedings of Symposium Power Plant Pumps, New Orleans, 1987 EPRI CS-5857 (1988)
24. Corbo, M.A., Stearns, C.F.: Practical design against pump pulsations. In: Proceedings of 22nd International Pump Users Symposium, pp. 137–177. Texas A&M (2005)
25. Cremer, R., Heckl, M.: Körperschall, 2nd edn. Springer, Berlin (1995)
26. Deeprose, W.M., et al.: Current industrial pump and fan fluid-borne noise level prediction. IMechE Paper. C251/77 43–50 (1977)
27. Domm, U., Dernedde, R.: Über eine Auswahlregel für die Lauf- und Leitschaufelzahl von Kreiselpumpen. KSB. Techn. Ber. **9** (1964)
28. Dörfler, P., Sick, M., Coutu, A.: Flow-Induced Pulsation and Vibrations in Hydroelectric Machinery. Springer, London (2013)
29. Dubas, M.: Über die Erregung infolge der Periodizität von Turbomaschinen. Ing. Archiv. **54**, 413–426 (1984)
30. Ehrich, F.F.: Handbook of Rotordynamics. McGraw Hill, New York (1992)
31. Ehrich, F.F., Childs, D.: Self-excited vibration in high-performance turbomachinery. Mech. Eng. **106**, 66–79 (1984)
32. Europump Leitfaden: Geräuschemission bei Kreiselpumpen (2002)
33. Florjancic, D.: Entwicklung der Speisepumpen und grossen mehrstufigen Pumpen für die Wasserversorgung. Tech. Rev. Sulzer. **4**, 241–254 (1973)

34. Florjancic, S.: Annular seals of high energy centrifugal pumps: a new theory and full scale measurement of rotordynamic coefficients and hydraulic friction factors. Dissertation, ETH Zürich (1990)
35. Försching, H.W.: Grundlagen der Aeroelastik. Springer, Berlin (1974)
36. Freese, H.D.: Querkräfte in axial durchströmten Drosselspalten. Pumpentagung, Karlsruhe (K6) (1978)
37. Gaffal, K.: Innovatives, umweltfreundliches und wirtschaftliches Speisepumpenkonzept erprobt. VGB Kraftwerkstech. **73**, 223–230 (1993)
38. Graf, K.: Spaltströmungsbedingte Kräfte an berührungslosen Dichtungen von hydraulischen und thermischen Turbomaschinen. Dissertation, ETH Nr. 9319 (1991)
39. Greitzer, E.M.: The stability of pumping systems. ASME J. Fluids Eng. **103**, 193–242 (1981)
40. Guinzburg, A.: Rotordynamic forces generated by discharge to suction leakage flows in centrifugal pumps. California Institute of Technology Report E249.14 (1992)
41. Gülich, J.F.: European Patent EP 0224764 B1 (1989)
42. Gülich, J.F., et al.: Pump vibrations excited by cavitation. In: IMechE Conference on Fluid Machinery, The Hague (1990)
43. Gülich, J.F., et al.: Rotor dynamic and thermal deformation tests of high-speed boiler feedpumps. EPRI Report GS-7405 (July 1991)
44. Gülich, J.F., Bolleter, U.: Pressure pulsations in centrifugal pumps. ASME J. Vibr. Acoust. **114**, 272–279 (1992)
45. Guo, S., Maruta, Y.: Experimental investigation on pressure fluctuations and vibration of the impeller in a centrifugal pump with vaned diffusers. JSME In. J. **48**(1), 136–143 (2005)
46. Hartlen, R.T., et al.: Dynamic interaction between pump and piping system. In: AECL Seminar on Acoustic Pulsations in Rotating Machinery, Toronto (1993)
47. Heckl, M., Müller, H.A.: Taschenbuch der Technischen Akustik. Springer, Berlin (1975)
48. Hergt, P., Krieger, P.: Radialkräfte in Leitradpumpen. KSB. Techn. Ber. 32–39 (1973)
49. Hergt, P., et al.: Fluid dynamics of slurry pump impellers. In: 8th International Conference Transport and Sedimentation of Solids, Prague, D2-1 (1995)
50. Höller, K.: In "25 Jahre ASTRÖ". Aströ, Graz (1979)
51. Kaiser, T., Osman, R., Dickau, R.: Analysis guide for variable frequency drives operated centrifugal pumps. Proceedings of the 24th International Pump Users Symposium, Texas A&M, pp. 81–106 (2008)
52. Kanki, H., et al.: Experimental research on the hydraulic excitation force on the pump shaft. ASME Paper 81-DET-71
53. Kaupert, K.A.: Unsteady Flow Fields in a High Specific Speed Centrifugal Impeller. Dissertation, ETH, Zürich (1997)
54. Kollmann, F.G.: Maschinenakustik. Grundlagen, Meßtechnik, Beeinflussung. 2. Aufl. Springer, Berlin (2000)
55. Krieger, P.: Wechselwirkungen von Laufrad und Gehäuse einer Einschaufelpumpe am Modell der instationären Strömung. Forsch. Ing. Wes. **54**(6), 169–180 (1988)
56. Kündig, P.: Gestufte Labyrinthdichtungen hydraulischer Maschinen. Experimentelle Untersuchung der Leckage, der Reibung und der stationären Kräfte. Dissertation, ETH. Nr. 10366 (1993)
57. Kurtze, G.: Physik und Technik der Lärmbekämpfung. Braun, Karlsruhe (1964)
58. Kwong, A.H.M., Dowling, A.P.: Unsteady flow in diffusers. ASME J. Fluids Eng. **116**, 843–847 (1994)
59. Lucas, M.J., et al.: Handbook of the Acoustic Characteristics of Turbomachinery Cavities. ASME Press, New York (1997)
60. Luce, T.W., et al.: A numerical and LDV investigation of unsteady pressure fields in the vaneless space downstream of a centrifugal impeller. ASME FEDSM97-3327 (1997)
61. Makay, E., Barret, J.A.: Changes in hydraulic component geometries greatly increased power plant availability and reduced maintenance cost: case histories. In: Proceedings of the 1st International Pump Symposium, Houston (1984)

62. Marscher, W.D.: Subsynchronous vibration in boiler feedpumps due to stable response to hydraulic forces at part-load. In: Proceedings of IMechE, vol. 202, pp. 167–175 (1988)
63. Meschkat, S.: Experimentelle Untersuchung der Auswirkung instationärer Rotor-Stator-Wechselwirkungen auf das Betriebsverhalten einer Spiralgehäusepumpe. Dissertation, TU, Darm-stadt (2004)
64. Meschkat, S., Stoffel, B.: The local impeller head at different circumferential positions in a volute casing of a centrifugal pump in comparison to the characteristic of the impeller alone. In: 21st IAHR Symposium on Hydraulic Machinery and Systems, Lausanne (2002)
65. Naudascher, E., Rockwell, D.: Flow-Induced Vibrations. An Engineering Guide. Balkema, Rotterdam (1994)
66. Nordmann, R., et al.: Rotordynamic coefficients and leakage flow for smooth and grooved seals in turbopumps. In: Proceedings IFToMM Meeting, Tokyo (Sept 1986)
67. Offenhäuser, H.: Druckschwankungsmesssungen an Kreiselpumpen mit Leitrad. VDI. Ber. **193**, 211–218 (1973)
68. Reinsch, K.H., Barutzki, F.: Erhöhung der Lebensdauer von Rohrleitungssystemen durch den Einsatz viskoser Dämpfer. Rohrleitungstechnik, 7. Auf., Vulkan-Verlag, Essen
69. Robinet, F., Gülich, J.F., Kaiser, T.: Vane pass vibrations—source, assessment and correction—a practical guide for centrifugal pumps. In: 16th International Pump Users Symposium, Houston, pp. 121–137 (1999)
70. Ross, D.: Mechanics of Underwater Noise. Pergamon Press (1976)
71. Rütten, F.: Large eddy simulation in 90°-pipe bend flows. J. Turbul. **2**, 003 (2001)
72. Sano, T., et al.: Alternate blade stall and rotating stall in a vaned diffuser. JSME Int. Ser. B. **45**(4), 810–819 (2002)
73. Schneider, K.: Das Verhalten von Kreiselpumpen beim Auftreten von Druckwellen. Dissertation, TU, Stuttgart (1986)
74. Schwartz, R., Nelson, R.: Acoustic resonance phenomena in high energy variable speed centrifugal pumps. In: 1st International Pump Symposium, Houston, pp. 23–28 (1984)
75. Spirig, M.: Einfluß der Kammerströmung auf die strömungsbedingten Kräfte im endlich langen Spalt einer hydraulischen Labyrinthdichtung. Dissertation, ETH. Nr. 13288 (1999)
76. Storace, A.F., et al.: Unsteady flow and whirl-inducing forces in axial-flow compressors. ASME J. Turbomach. **123**, 433–445 (2001)
77. Storteig, E.: Dynamic characteristics and leakage performance of liquid annular seals in centrifugal pumps. Dissertation, MTA-00-137 TU, Trondheim (2000)
78. Strub, R.A.: Pressure fluctuations and fatigue stresses in storage pumps and pump turbines. ASME Paper No. 63-AHGT-11 (1963)
79. Sudo, S.: Pumping plant noise reduction. Hitachi Rev. **29**(5), 217–222 (1980)
80. Tanaka, H.: Vibration behavior and dynamic stress of runners of very high head reversible pump-turbines. In: IAHR. Symposium, Belgrade, Beitrag U2 (1990)
81. Tsujimoto, Y., et al.: Observation of oscillating cavitation in an inducer. ASME J. Fluids. Eng. **119**, 775–781 (1997)
82. Ubaldi, M., et al.: An experimental investigation of stator induced unsteadiness on centrifugal impeller outflow. ASME J. Turbomach. **118**, 41–51 (1996)
83. Verhoeven, J.: Unsteady hydraulic forces in centrifugal pumps. IMechE Paper C348/88 (1988)
84. Warth, H.: Experimentelle Untersuchungen axial durchströmter Ringspalte von Hybridentlastungseinrichtungen. Dissertation, TU Kaiserslautern, SAM Forschungsbericht Bd. 2 (2000)
85. Weaver, D.S.: Interaction of fluid flow and acoustic fields. In: AECL Seminar on Acoustic Pulsations in Rotating Machinery, Toronto (1993)
86. Weber, M.: Geräusch- und pulsationsarme Verbrennungsluftgebläse und deren Einfluß auf selbsterregte Brennkammerschwingungen. Dissertation, TU Kaiserslautern, SAM Forschungsbericht Bd 7 (2002)
87. Yedidiah, S.: Oscillations at low NPSH caused by flow conditions in the suction pipe. ASME Cavitation and Multiphase Flow Forum (1974)

88. Yuasa, T., Hinata, T.: Fluctuating flow behind the impeller of a centrifugal pump. Bull. JSME **22**(174), 1746–1753 (1979)
89. Ziada, S.: Flow-excited resonances of piping systems containing side-branches: excitation mechanism, counter-measures and design guidelines. In: AECL Seminar on acoustic pulsations in rotating machinery, Toronto (1993)
90. Parrondo, J., et al.: The effect of the operating point on the pressure fluctuations at BPF in the volute of a centrifugal pump. ASME JFE **124**, 784–790 (2002)
91. Bolleter, U. et al.: Solution to cavitation-induced vibration problems in crude-oil pipeline pumps. In: 8th Pump Users Symposium, Texas A&M (1991)
92. ANSI/HI 9.8 Standard: Pump Intake Design (2012)
93. Rosenberger, H.: Experimental determination of the rotor impacts of axial pumps in intake structures under distorted approach flow. Thesis TU Kaiserslautern. SAM Forschungsbericht Bd 5 (2001)
94. Weinerth, J.: Kennlinienverhalten und Rotorbelastung von axialen Kühlwasserpumpen unter Betriebsbedingungen. Diss TU Kaiserslautern. SAM Forschungsbericht Bd 9 (2004)
95. Schiavello, B., Smith, D.R., Price, S.M.: Abnormal vertical pump suction recirculation problems due to pump-system interaction. In: 21st Pump Users Symposium, Houston, pp. 18–47 (2004)
96. Dupont, P. et al.: CFD analysis of sump flow and its impact on the hydraulic forces acting on the impeller of a vertical pump. In: Rotating Equipment Conference 2008, Düsseldorf
97. Krueger, S., et al.: Pump sump CFD for vertical pump. ASME FEDSM2009-78162
98. May, F.: Sulzer Technical Review 3/2015
99. Ohashi, H.: Influence of impeller and diffuser geometries on lateral fluid forces of whirling centrifugal pump impeller. NASA CP **3026**, 285–322 (1988)
100. Van Esch, B.P.M.: Performance and radial loading of a mixed-flow pump under non-uniform approach flow. ASME JFE **131** (2009)
101. Untaroiu, A., et al: On the dynamic properties of pump liquid seals. ASME JFE **135** (2013)
102. Gülich, J.F.: Selection criteria for suction impellers of centrifugal pumps. World Pumps, Parts 1 to 3, January, March, April, 2001
103. Manning, T., Serge, D.: River water pump cyclic vibration. In: 29th Pump Users Symposium, Houston (2013)
104. Marscher, W.D.: End users guide to centrifugal pump rotor dynamics. In: 30th Pump Users Symposium, Houston (2014)
105. Marscher, W.D.: The effect of fluid forces at various operation conditions on the vibrations of vertical turbine pumps. In: IMechE Seminar Radial Loads and Axial Thrust, pp. 17–38 (1986)
106. Sumer, B., Fredsoe, J.: Hydrodynamics Around Cylindrical Structures, Revised Edition (2006)
107. Ohashi, H.: Case study of pump failure due to rotor-stator interaction. Intl J of Rotating Machinery **1**, 53–60 (1994)
108. Corbo, M.A., et al.: Practical use of rotor dynamic analysis to correct a vertical long-shaft pump's whirl problem. In: 19th Pump Users Symposium, Houston, pp. 107–120 (2002)
109. ISO 13709: Centrifugal Pumps for the Petrochemical Industry, 2nd edn (API 610) (2009)
110. Schneider, A., Conrad, D., Böhle, M.: Lattice Boltzmann simulation of the flow field in pump intakes—a new approach. ASME JFE **137**, 031105 (2015)
111. Rebernik, B.: Radialkräfte von Kreiselpumpen mit unterschiedlichen Gehäuseformen. "25 Jahre ASTRÖ", Aströ, Graz, pp. 55–60 (1979)
112. Boyadjis, P., Onari, M.: Diagnosing and correcting a damaging below-ground column natural frequency in a vertical pump using field testing and FEA. In: 28th Pump Users Symposium, Houston (2012)
113. Nakato, T.: Field-tested solutions to pump vibrations. In: 1st International Symposium on Noise and vibrations, Paris (1993)
114. Schubert, F., Rosenberger, H.: Development of a compact intake chamber for vertical tubular pumps. In: ASCE Joint Conference Water Resources, Minneapolis (2000)

115. Smith, D., Woodward, G.: Vibration analysis of vertical pumps. In: 15th Turbomachinery Symposium Texas A&M, pp. 61–68 (1986)
116. Corley, J.E.: Vibration problems of large vertical pumps and motors. In: 9th Turbomachinery Symposium Texas A&M, pp. 75–82 (1980)
117. Kirst, K.: Experimentelle und numerische Untersuchungen von Zulaufbedingungen vertikaler Pumpsysteme. Dissertation TU Kaisers-lau-tern. SAM Forschungsbericht Bd 22 (2012)
118. Kaiser, T., et al.: Analysis guide for variable-frequency drive operated centrifugal pumps. In: 24th International Pump Users Symposium Texas A&M, pp. 81–106 (2008)
119. Blevins, R.D.: Applied Fluid Dynamics Handbook. Van Nostrand Reinhold, New York (1984)
120. Schäfer, F.: Untersuchung des Einflusses hydraulischer und mechanischer Anregungen auf das Betriebsverhalten einer axialen Rohrgehäusepumpe. Dissertation TU Kaiserslautern. SAM Forschungsbericht Bd 18 (2008)
121. Peters, M.: Fiber-reinforced ceramic bearings for cooling water pump applications. In: IMech 2014 Turbomachinery Symposium
122. Franke, G., et al.: On pressure mode shapes arising from rotor-stator interactions. IAHR WG1-2003 Meeting
123. Fischer, R.K., et al.: Contribution to improved understanding of the dynamic behaviour of pump turbines. In: 22nd IAHR Symposium on Hydraulic Machinery and Systems (2004)
124. Buckler, M.: Tutorial on vertical pumps. In: Calgary Pump Symposium (2013)
125. Guo, S., Okamoto, H.: An experimental study on the fluid forces induced by rotor-stator interaction in a centrifugal pump. Int. J. Rotating Mach. **9**(2), 135–144 (2003)
126. Zhang, M., Tsukamoto, H.: Unsteady hydrodynamic forces due to rotor-stator interaction on a diffuser pump with identical number of vanes on the impeller and diffuser. J. Fluid Eng. **127**, 743–751 (2005)
127. Nicolet, C.: Hydro-acoustic modelling and numerical simulation of unsteady operation of hydro-electric systems. Ph.D. thesis EPFL (2007)
128. Gülich, J.F., et al.: Review of parameters influencing hydraulic forces on centrifugal impellers. Proc. IMechE **201**(A3), 163–174 (1987)
129. Smith, D.R., et al.: Centrifugal pump vibration caused by super-synchronous shaft instability. In: 13th International Pump Users Symposium, Houston, pp. 47–69 (1996)
130. Feng, J., Benra, F.K., Dohmen, H.J.: Numerical investigation on pressure fluctuations for different configurations of vaned-diffuser pumps. Int. J. Rotating Mach. (2007)
131. Van Esch, B., Cheng, L.: Unstable operation of a mixed-flow pump and the influence of tip clearance. ASME AJK2011-06016
132. Miyabe, M., et al.: Rotating stall behavior in a diffuser of a mixed-flow pump and its suppression. ASME FEDSM 2008-55132
133. Miyabe, M., et al.: On improvement of characteristic instability and internal flow in mixed-flow pumps. J. Fluid Sci. Technol. **3**(6), 732–743 (2008)
134. Miyabe, M., et al.: Unstable head-flow characteristic generation of a low-specific speed mixed-flow pump. J. Therm. Sci. **15**(2), 115-ff (2006)
135. Barrio, E., et al.: The effect of impeller cutback on the fluid-dynamic pulsations and load at blade-passing frequency in a centrifugal pump. ASME JFE **130**, 111102 (2008)
136. Botero F et al: Non-intrusive detection of rotating stall in pump-turbine. Mech. syst. signal process. (2014)
137. Figliola, R.S., Beasley, D.E.: Theory and Design for Mechanical Measurement, 5th edn. Wiley, Hoboken (2011)
138. Berten, S.: R&D pump investigations. Technical Review Sulzer, No. 1 (2017)
139. Berten, S., et al.: Experimental and numerical analysis of pressure pulsations and mechanical deformations in a centrifugal pump impeller. ASME AJK2011-06057 (2011)
140. Bradshaw, S., Sabini, E.: Modification of BB1 vibration characteristics to meet ISO 13709 limits. In: Texas A&M Pump Symposium (2011)

141. Berten, S., et al.: Experimental investigation of pressure fluctuations in a high-energy centrifugal pump at off-design conditions. In: IMechE Conference (2014)
142. Zobeiri, A., et al.: How oblique trailing edge of a hydrofoil reduces the vortex-induced vibration. J. Fluids Struct. **32**, 78–89 (2012)
143. Jery, B., et al.: Forces on centrifugal pump impellers. In: Texas A&M Pump Symposium, pp. 21–32 (1985)
144. Franz, R., et al.: The hydrodynamic forces on a centrifugal pump impeller in the presence of cavitation. ASME JFE **112**, 264–271 (1990)
145. Jery, B.: Experimental study of unsteady hydrodynamic force matrices on whirling centrifugal pump impellers. Ph.D. thesis, Caltech (1987)
146. Dietzen, F., Nordmann, R.: Calculating rotordynamic coefficients of seals by finite-difference techniques. NASA CP 2443, pp. 77–96 (1986)
147. Fu, D.C., et al.: Impact of impeller stagger angle on pressure fluctuation of double-suction centrifugal pump. Chin J Mech Eng (2018)
148. Bachert, R.: Dreidimensionale, instationäre Effekte kavitierender Strömungen – Analysen an Einzelprofilen und in einer Radialpumpe. Diss. TU Darmstadt, (2004)
149. Lehr, C., Linkamp, A., Brümmer, A.: Abschlussbericht zum Verbundprojekt: Entwicklung von Grundlagen für instationär betriebene hydraulische Pumpsysteme in flexiblen Kraftwerken. TU Dortmund Februar (2019)

Chapter 11
Operation of Centrifugal Pumps

Reliable pump performance can be achieved only if the pump is well integrated into the system. This implies the implementation of adequate controls which ensure that the pump operates in the range between the allowable minimum and maximum flow - in particular if pumps are operated in parallel. Proper procedures must be implemented for start and stop. Transient operation and unsteady phenomena which need attention include: dynamic instabilities caused by an unstable Q-H-curve, water hammer and air-drawing vortices in wet pump installations. Suction pressure transients and distorted approach flow to the pump harbor also potential for trouble. Operation in various brake and turbine modes is discussed in Chap. 12. The effects of high viscosity, free gas or large amounts of solids on the performance curves are the topic of Chap. 13.

11.1 System Characteristics, Operation in Parallel or in Series

A positive displacement pump delivers at a fixed speed a nearly constant flow rate independently of the back pressure. In contrast, the flow rate of a centrifugal pump depends on the pressure difference $\Delta p = \rho \times g \times H_A$ imposed by the system on the pump. The pressure rise Δp generally depends on the flow rate because of hydraulic losses. Thus the system characteristic $H_A = f(Q)$ is understood as the difference in total pressure which must be supplied by the pump to maintain a specific flow rate through the system, Eq. (T2.2.6). The operation point of a centrifugal pump is given by the intersection of the characteristics of system and pump, Fig. 11.1.

The head H_A required by a plant consists of a constant (static) component H_{stat} (which may be zero) and a flow-dependent component H_{dyn}.

J. F. Gülich, *Centrifugal Pumps*,
https://doi.org/10.1007/978-3-030-14788-4_11

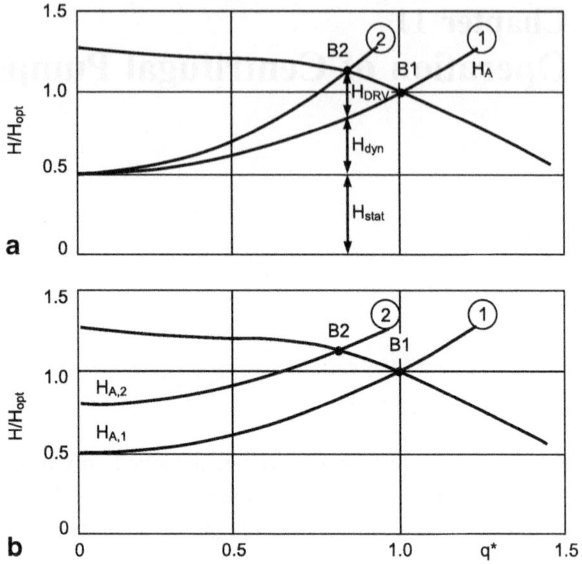

Fig. 11.1 System characteristics. **a** With throttle valve, **b** with variable static head

$$H_A = H_{stat} + H_{dyn} \tag{11.1}$$

The static component includes geodetic head differences H_{geo} and differences in system pressure (expressed as head), see Table 2.2:

$$H_{stat} = H_{geo} + (p_a - p_e)/(g\,\rho) \tag{11.2}$$

The dynamic head includes all head losses H_v in the suction pipe, the discharge pipeline including components (e.g. heat exchangers), the adjustable head drop in the throttle valve H_{DRV} and the increase in kinetic energy:

$$H_{dyn} = \frac{c_a^2 - c_e^2}{2g} + H_v + H_{DRV} = H_{dyn,r}\frac{Q^2}{Q_r^2} = R\,Q^2 \quad \text{with} \quad R \equiv \frac{H_{dyn,r}}{Q_r^2} \tag{11.3}$$

$H_{dyn,r}$ is the dynamic head which is calculated for a specific flow rate Q_r (usually the design flow rate of the system). The system characteristic defined according to Eqs. (11.1 to 11.3) is a parabola whose apex is at H_{stat} on the ordinate. This behavior is very often seen in practice; some typical applications are discussed in Chap. 11.2. In principle, however, the system characteristics can have any shape, depending on the process. For example, sliding-pressure operation in a power plant yields essentially linear system characteristics. There are also instances where the operation point is not determined by means of the usual pump characteristic at n = constant, but the combined characteristic of pump and driver is used instead. Examples are combustion engines without speed control and variable speed electric motors (of small power) whose speed largely depends on the power.

If the dynamic head changes (e.g. due to throttling) and/or the static head varies (e.g. owing to water level or boiler pressure variations), the operation point will shift accordingly (curves 2 in Fig. 11.1a, b). The analysis of various operation modes and upset conditions is an essential element to enable the project engineer to make a correct pump selection. Lack in diligence and foresight in doing this job leads time and again to costly problems. Systems with several pumps operating in parallel or in series call for particular attention in this respect.

Parallel operation: Many plants are designed for parallel operation, for example, because of strongly varying flow requirements, or in order to provide standby capacity, or because the required flow rate cannot be attained with a single pump for technical or economic reasons.

The operation point of each pump is derived from the intersection of the system characteristic with the combined characteristics of all pumps in operation. In parallel operation each unit has to pump against the same pressure difference as imposed by the system. Hence the combined characteristic of the pumps is obtained by *adding the flow rates* of all operating pumps *at constant head.* Figure 11.2 demonstrates this procedure for a system with two identical pumps operating in parallel. Those parts of the pipeline which are common to both pumps are characterized by the flow resistance R_{com}, while the resistance elements (suction and discharge pipelines combined) which are individual to each pump are described by the resistance R_{single}. If Q_{tot} is the total flow rate delivered through the system and z_{pp} is the number of pumps operating in parallel, the system characteristic is calculated from:

$$H_A = H_{stat} + \left(\frac{R_{single}}{z_{pp}^2} + R_{com}\right) Q_{tot}^2 \qquad (11.4)$$

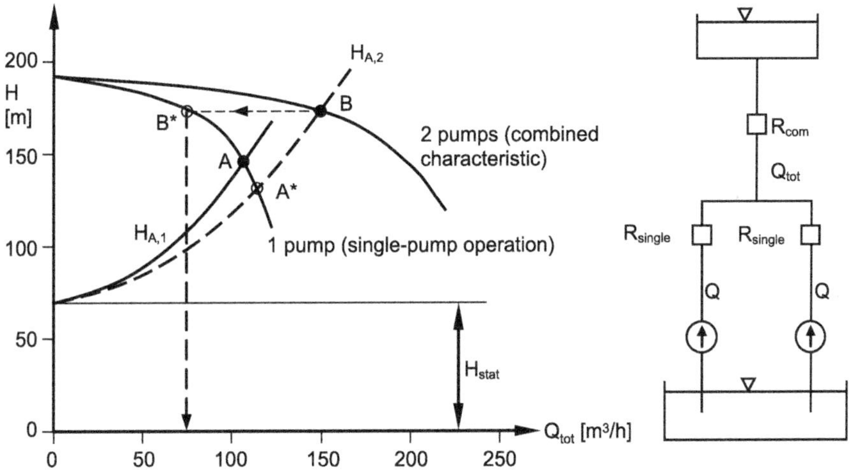

Fig. 11.2 System characteristics for parallel operation

Equation (11.4) gives the system characteristic for operation with a single pump as well as with any number of pumps in parallel. If two pumps are operating, the operation point B represents the flow rate Q_{tot} delivered through the system while point B* gives the flow rate of each individual pump.

If a single pump is operating, its flow rate is given by the point A (Fig. 11.2). The difference between the system characteristics for operation with a single pump $H_{A,1}$ and operation with two pumps will increase if the contribution of the individual resistances R_{single} becomes more important. Often R_{single} is negligible, with the result that $H_{A.1}$ and $H_{A.2}$ coincide (single-pump operation in point A*).

In the example shown in Fig. 11.2 the system flow rate increases by just about 35% when the second pump is added (it is not possible to obtain twice the flow by adding a pump). The greater the dynamic component of the system characteristic and the flatter the pump characteristic, the smaller is the gain in flow resulting from the addition of parallel pumps. *Parallel operation is mainly appropriate for applications with a high static head* H_{stat}, *that is, with a flat system characteristic.*

Instead of adding the pump characteristics to obtain the combined curve (as has been done in Fig. 11.2), it is often easier and clearer to have the system characteristics for various cases of parallel operation plotted into the characteristic of one pump, Fig. 11.3. With $Q_{tot} = z_{pp} \times Q$ inserted in Eq. (11.4), the system characteristic results from Eq. (11.4) as:

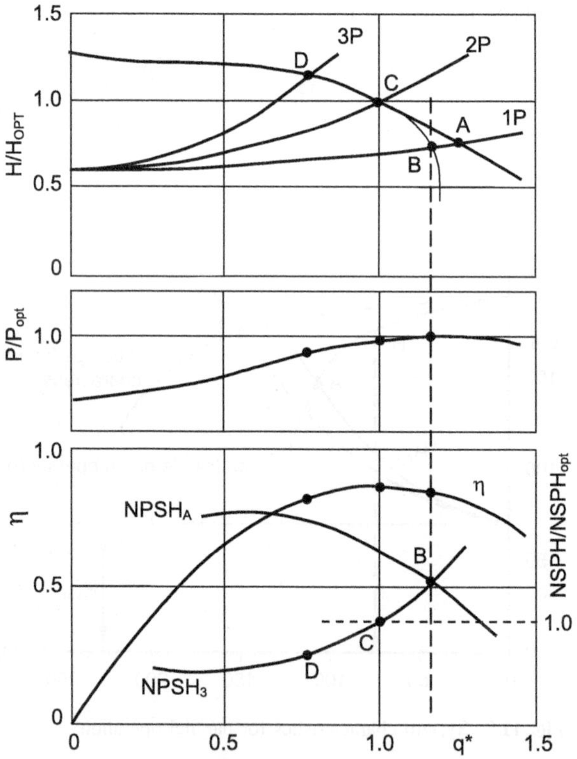

Fig. 11.3 Parallel operation of centrifugal pumps

$$H_A = H_{stat} + (R_{single} + z_{pp}^2 R_{com})Q^2 \tag{11.4a}$$

The advantage of plotting the characteristics according to Fig. 11.3 rather than Fig. 11.2 is that only the system characteristics have to be calculated from Eq. (11.4a). It is not necessary to add up all the head curves, which can be quite involved if the Q-H-curve has a saddle or several pumps are installed in parallel.

Figure 11.3 shows a system with a relatively high dynamic head. Putting an additional pump into service (or taking one out) gives rise to great shifts of the operation point which can cause a pump in single operation to operate at excessive flow rates and cavitation. This is demonstrated by the curves for $NPSH_3$ and $NPSH_A$ in Fig. 11.3. If two or three pumps are working (operation points C and D), normal operation is possible because the $NPSH_A$ is sufficiently above the $NPSH_3$. A pump in single-operation would work in point A if the $NPSH_A$ is sufficient. However, in Fig. 11.3 the $NPSH_A$ at Q_A is clearly below the $NPSH_3$, so that the pump would run in full cavitation. If the system were left to itself, the flow rate would be limited to point B by cavitation. In that case the ensuing flow rate and extent of cavitation are such that the produced head is in balance with the head required by the system (see Chap. 11.2 "cavitation control").

For most applications this type of operation mode is inadmissible (lower efficiency, cavitation damage, noise, vibrations). Only in exceptional cases a limitation of the flow rate by means of cavitation is tolerated, for instance, if this operation mode occurs only for a very short time during the automatic switch-over of a pump or in the case of cavitation control. Any system with pumps operating in parallel therefore must be carefully analyzed in order to check: (1) whether the maximum allowable flow rate could be exceeded in the case of a single-pump operating (or under any other conditions); (2) what possible remedies are required keep the flow within the allowable operation range.

If the $NPSH_A$ is low, pumps operating in parallel should be equipped with individual suction pipes in order to avoid a drop in the $NPSH_A$ caused by increased pressure losses when more pumps are switched on.

For parallel operation stable, steadily falling Q-H-curves are required, because unambiguous intersection points between the combined pump characteristics and the system curve must be obtained. If the pump Q-H-curve is flat in the partload range, one pump can displace another during parallel operation, since the Q-H-curves of the individual pumps are not exactly identical due to manufacturing tolerances and wear. Hence the impeller diameters of two identical pumps operating in parallel should be trimmed to the same size when adaptations in the plant are necessary. Problems of that kind only arise if the pumps operate in the flat or unstable portion of the Q-H-curves. With storage pumps, which often operate near the best efficiency point because of a fixed water level, unstable Q-H-curves during parallel operation would not cause problems of this kind.

Even pumps whose Q-H-curves are clearly different can be operated in parallel. In that case, the pump with the lower shut-off head should be started up only when

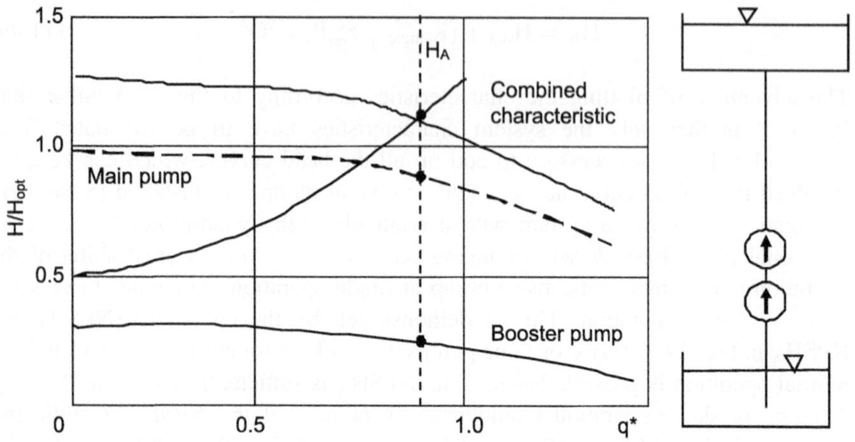

Fig. 11.4 Operation of two pumps in series

the required head is lower than the shut-off pressure of the pump to be added. Similarly, the pump with the lower shut-off pressure has to be shut down before the head required by the system exceeds the shut-off pressure of the smaller pump.

Operation in series: Pumps are also installed in series. Applications include: (1) pipelines transporting water or oil over large distances; (2) installation of a booster pump with the object to provide sufficient $NPSH_A$ to the main pump. A typical example is a single-stage, double-entry, low-speed booster pump with a head of 150 to 200 m feeding a high-speed boiler feed pump generating a head of 4000 m (Fig. 2.8).

When operating in series (Fig. 11.4), all pumps deliver the same flow rate. Hence the curve representing the combined characteristics of all pumps is obtained by *adding the heads* of the individual pumps at *constant flow*. In series operation the individual pumps can be quite different from each other but they must be designed to cover the range of flow rates required. Again, the operation point of the pumps results from the intersection between the combined characteristics and the system characteristic. All pumps operate at the same flow rate (unless part of the pumpage is taken away or added in between). The casings of all pumps and the piping must be designed for the maximum pressure which occurs when the discharge valve of the last pump is closed.

11.2 Pump Control

Variations in the process parameters of a plant make it necessary for the pumps to be operated within wide ranges. Parameters susceptible to variations are: the required flow rate (e.g. boiler feed pumps and condensate pumps following the load, water supply following demand or drainage pumps), liquid levels, process pressures, flow resistances (which may change due to corrosion, sediments, filter loading) and fluid properties (e.g. viscosity and density).

The pumps have to be controlled in such a way that all of the operation requirements of the specific process will be met. This can be done by manual control or by means of automatic control circuits. Apart from putting into service additional pumps installed in parallel or in series (as discussed in Chap. 11.1), a variety of methods and devices for pump control are available: by throttling or bypass control the *system* characteristics are modified, whereas the *pump* characteristic is changed by means of speed control, pre-rotation control or blade adjustment. In cavitation control the pump characteristic automatically adjusts to the prevailing operation condition. Overall the purpose of the control is to supply the system with the necessary flow rate at the required pressure, to optimize the operation with respect to energy and maintenance costs and to prevent the pump from working in inadmissible operation ranges.

The choice of the suitable control system is chiefly determined by the plant configuration, operation requirements and economic considerations. It also depends on the type of pump, since pre-rotation control and blade adjustment, for instance, only make sense at high specific speeds. The optimization of energy consumption has to be carried out, in principle, for each individual application, since not only the pump and system characteristics need to be considered, but also the duration over which the pump is to be operated at the various conditions.

Frequently the pump is controlled from the liquid level. The measured actual level supplies the control signal for the speed, the throttling position of the control valve, inlet guide vane control or adding or removing of pumps from service.

Throttle control: By adjusting a valve in the discharge pipe, the system characteristic is modified so that the required flow rate is attained, Fig. 11.1. Since a throttling device in the suction pipe would diminish the $NPSH_A$, this option must be ruled out. Throttle control requires relatively low capital investment. For this reason it is often applied to pumps of low and medium power (where it is almost the rule). Its disadvantage is the loss of mechanical energy and the resulting impact on operation costs. The higher the dynamic head component H_{dyn} in the system characteristic, the higher are these losses. The power loss caused by throttling amounts to $P_v = \rho \times g \times Q \times H_{DRV}/\eta$ (Fig. 11.1); it is smallest when the pump and the system characteristic are flat. Throttle control is therefore best suited for systems where a high static pressure is required (e.g. boiler feed pumps) and for pumps with low specific speeds, because their power consumption decreases at lower flow rates (Fig. 4.15). It is unsuitable for pumps with high specific speeds (especially axial pumps) whose power consumption grows with falling flow rate. Applying throttle control in this case over a wide flow range would mean a waste of energy. Another disadvantage of throttle control is that the pumps may be operated far from the best efficiency point, which entails the risk of increased wear, cavitation damage, noise and vibrations. Finally, throttling high pressure differences frequently implies wear and damage to the valve, noise and pipe vibrations.

Speed control: Speed control is employed to mitigate the disadvantages of throttle control, especially unnecessary energy consumption. The characteristics and the $NPSH_R$ required at various speeds are calculated from the similarity laws (Table 3.4). The best efficiency points (like any other values of q* = constant) are located on parabolas through the origin of the coordinate system, Fig. 11.5. If the system characteristic consists only of dynamic components ($H_{stat} = 0$), the pump operates at all speeds at the same q*, i.e. with the same incidence. For reasons of energy conservation, operation should be as close to the BEP as possible. In this case the advantages of speed control can be fully exploited. Conversely, the greater the static component of the system characteristic, the less energy can be saved by means of speed control. If the pump characteristic is very steep (axial and semi-axial impellers), speed control is yet again advantageous: *the steeper the pump and system characteristics, the greater are the savings on energy* as compared to throttle control.

If a speed-controlled pump (given an appropriate system characteristic) is running at partload at a reduced speed, the hydraulic excitation forces and the risk of cavitation decrease. Speed control is therefore also apt to reduce maintenance costs. Only the high capital investment for the driver unit stands in the way of a wide application of speed control. Variable speed drivers are: frequency-controlled electric motors, hydraulic couplings, combustion engines, steam or gas turbines for high power and adjustable-speed belt drives.

Fig. 11.5 Speed control

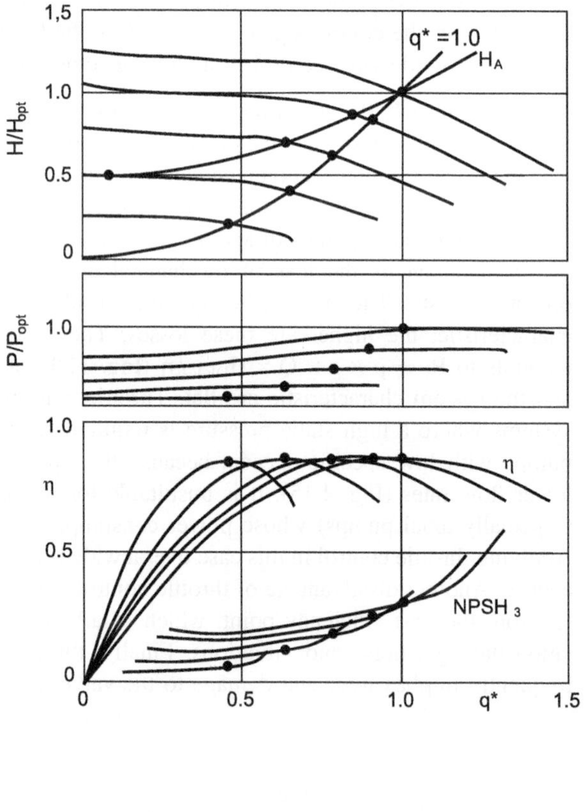

Bypass control: A fraction of the flow rate is returned from the discharge pipe to the suction side of the pump through a bypass line in order to avoid operation at low load, Fig. 11.6. The system characteristic is modified by the bypass valve in such a way that the flow rate required at any point in time is delivered through the discharge pipeline to the process. Special care has to be taken to prevent the pump from running into excessive cavitation at high flow rates. The combined system characteristic H_A of discharge pipe $H_{A,V}$ and bypass $H_{A,By}$ is determined by adding the flow rates that are delivered through both pipelines at a constant head as shown in Fig. 11.6 (the procedure is analogous to that used for parallel operation, see Fig. 11.2). The pump operates at the intersection (point B) of the pump characteristic with the combined system characteristics. The flow rates Q_v delivered to the process and through the bypass Q_{By} (Fig. 11.6) are derived from the head H_B. The throttling in the bypass implies the power loss $P_v = \rho \times g \times Q_{By} \times H_B/\eta$ which is dissipated into heat. As a consequence the temperature of the fluid at the pump inlet increases.

In the case of pumps whose power consumption increases with falling flow rates (e.g. axial pumps) bypass control is energy-wise more efficient than throttle control. Bypass control also allows to avoid pump operation in the unstable range (if any). At high flow rates, however, a bypass is likely to require considerable expense and space for piping and fittings. If the power consumption of the pump decreases when reducing the flow rate, the bypass control is energy-wise worse than the throttle control. For this reason it is scarcely used for controlling pumps with low specific speeds.

Nevertheless, high-pressure pumps are usually equipped with a "minimum-flow bypass" leading fluid back to the suction reservoir in order to protect the pump against overheating and excessive excitation forces when operating at low flow (see Chap. 11.6).

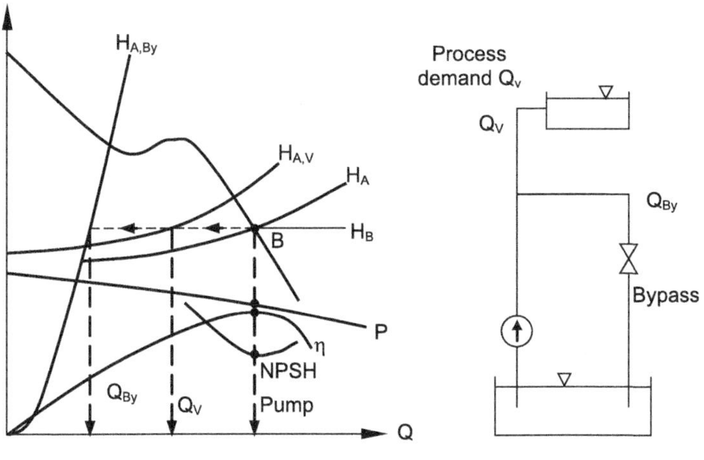

Fig. 11.6 Bypass control, the same arrangement is used for minimum flow systems

Cavitation control: Consider a pump with a characteristic according to Fig. 11.7. Given a sufficiently high suction pressure (so that $NPSH_A > NPSH_o$), it generates the head H_B at a given flow rate Q_B. If the suction pressure is lowered to $NPSH_A < NPSH_o$, the head will drop below H_B because of cavitation. From any value of $NPSH_A$ thus results a "cavitation characteristic" which diverges from the non-cavitating characteristic until full cavitation is reached and a stable delivery may become impossible. The points of intersection of the system characteristic with these cavitation characteristics form stable operation points provided that they are placed above a certain limit represented by curve A. Below this limit the flow is interrupted or the operation becomes unstable with periodical large-amplitude oscillations (these must not be confused with cavitation-induced high-frequency pressure pulsations, which usually decrease at full cavitation). The stable range between cavitation-free operation and the limiting curve A can be used for auto-control of the flow rate. This type of control may be employed in systems where the $NPSH_A$ directly depends on the amount of liquid supplied to the suction tank. The scheme works in the following way: If, at a given time, the rate of fluid fed into the tank exceeds the momentary flow rate drawn by the pump, the liquid level and the $NPSH_A$ rise; as a result the operation point is shifted towards higher flows. Conversely, if the ingress of liquid falls momentarily below the flow rate of the pump, the liquid level and the $NPSH_A$ will drop. As a consequence, cavitation at the impeller inlet increases, the pump is unable to maintain the required head and the operation point is shifted to lower flows. In the case of a multi-stage pump, cavitation may extend to the second stage.

Prerequisites for cavitation control are: (1) low circumferential speeds; (2) a choice of suitable materials for keeping cavitation damage within reasonable limits. These conditions are occasionally given with condensate pumps on ships or in small power plants. Apart from special cases, cavitation control is rarely used.

Impeller blade adjustment: Semi-axial and axial pumps above $n_q \approx 150$ can be controlled with very good efficiency by means of adjustable impeller blades in a

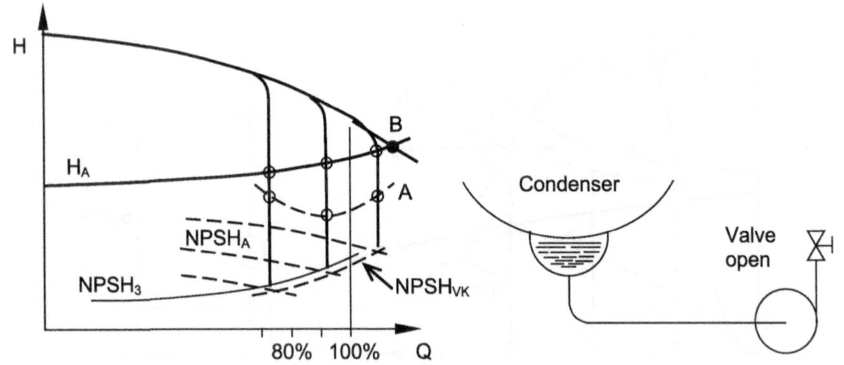

Fig. 11.7 Cavitation control

wide range, [7]. Unfortunately the mechanism for continuous blade adjustment during operation requires a rather expensive design.

Each blade position is associated with an individual characteristic as depicted in Fig. 11.8 for three different angle settings. By adjusting the blades, the incidence corresponding to the required flow rate can be matched almost at an optimum. Hence the efficiency loss with regard to the design point is small. This kind of control is particularly well suited to axial pumps since the flow rate at best efficiency point can be shifted by means of the blade setting angle without considerably modifying the head at the BEP. Given a flat system characteristic (constant liquid levels with low dynamic head), an efficient control over a wide range of flows is thus possible.

Semi-axial pumps can be controlled well by blade adjustment in operation against steep or flat system characteristics including water level fluctuations. Because of hydrodynamic and centrifugal forces acting on the blades the torque required for blade adjustment needs careful analysis [21].

In radial pumps with low to medium specific speeds, losses incurred in the diffuser largely outweigh the losses in the impeller when operating at partload. Impeller blade adjustment would neither be feasible nor bring any technical benefits.

Pre-rotation control: According to Euler's equation, Eq. (T3.3.1), the head can be modified by changing the swirl at the impeller inlet. Pre-rotation reduces the head while counter-rotation increases the head as demonstrated by Fig. 3.1. The greater the ratio d_1/d_2, the more a pre-rotation control is effective. The possible range of control consequently increases with growing specific speed. Hence pre-rotation control is only employed for semi-axial and axial impellers.

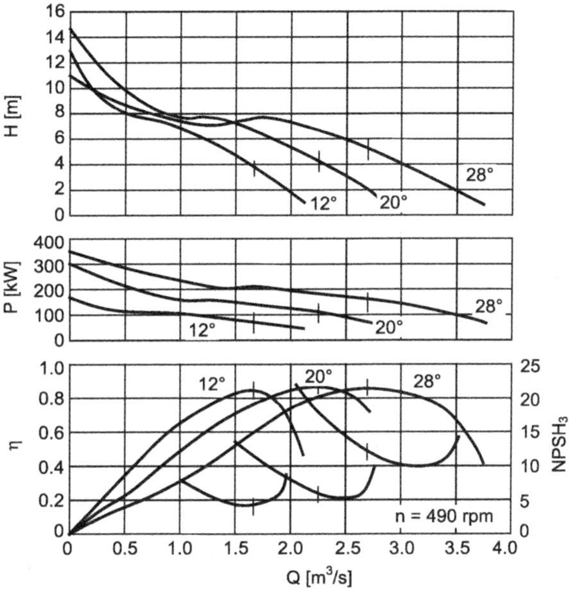

Fig. 11.8 Propeller pump with variable pitch impeller

Figure 11.9 shows the characteristics of a semi-axial pump with pre-rotation control. With increasing pre-rotation the efficiency decreases considerably (much more than it does with impeller blade adjustment). This restricts the useful range of control. The head at best efficiency point is modified in the same sense as the BEP flow rate: when increasing the approach flow angle, both H_{opt} and Q_{opt} rise.

The values of $NPSH_R$ change little but the flow rates where $NPSH_R$ is minimum shift as much as the flow rates at BEP. As demonstrated by Fig. 11.9, the range available for controlling the flow rate is wider if the system characteristic is flat ($H_{A,1}$) than with a steep system curve ($H_{A,2}$). Consequently, pre-rotation control is suitable for applications where a wide range of flow rates is required at essentially constant water level (or head). When a fixed flow rate is to be pumped at variable heads (or water levels), the range which can be covered by pre-rotation control is limited by cavitation. This can be inferred from Fig. 11.9 when considering the curves $NPSH_3 = f(Q)$ for different inlet guide vane angles.

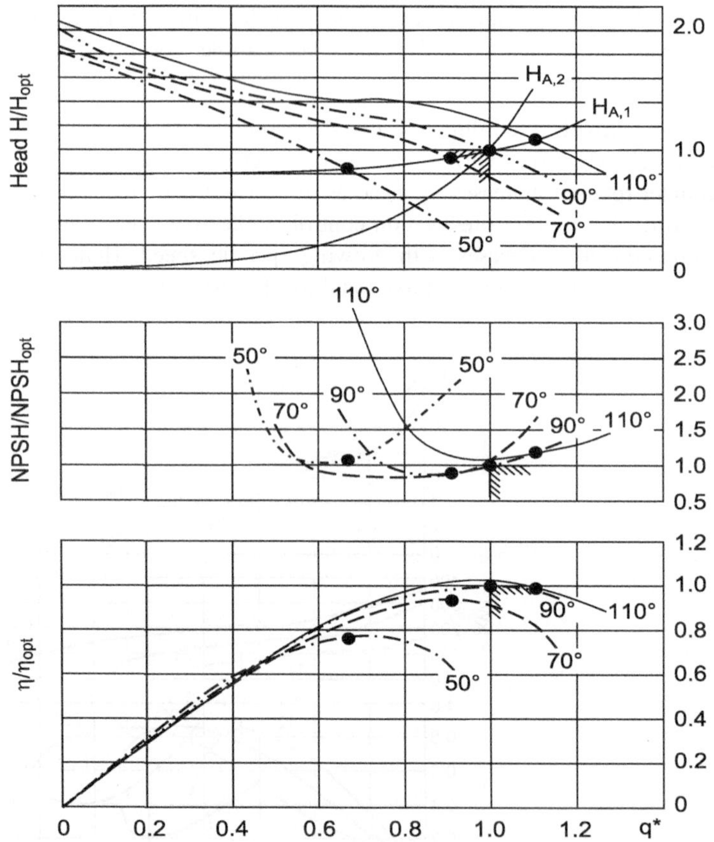

Fig. 11.9 Characteristics of a semi-axial cooling water pump with pre-rotation control; all data are referred to the BEP with no pre-rotation ($\alpha_1 = 90°$)

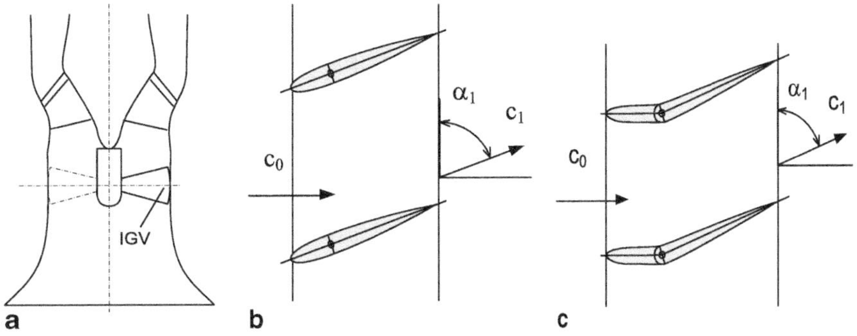

Fig. 11.10 Semi-axial pump with inlet guide vanes (IGV), **a** general arrangement, **b** IGV without flap, **c** IGV with flap

The pre-swirl is created by means of a set of adjustable inlet guide vanes (IGV) with an appropriate profile placed at a short distance upstream of the impeller, Fig. 11.10. If the swirl controlling device is built without flaps, only the stagger angles of the cascade are adapted, Fig. 11.10b. When changing the stagger angle the incidence rapidly increases and flow separation will occur on these vanes leading to noise and vibrations and a loss in efficiency.

The pre-rotation control device can be made of a fixed cascade with 90° vane angles and flaps, Fig. 11.10c. Thus only the rear-portion (or flap) is adjusted for modifying the pre-rotation. In this way the camber of the inlet guide vanes is modified rather than the incidence and the effects of flow separation are mitigated, [2, 10, 19]. Therefore this design allows a wider range of flow control than the device without flaps. The wake flowing off the inlet guide vanes increases the unsteadiness of the flow at the impeller inlet, in particular if flow separation occurs in the IGV. The unsteadiness of the flow increases noise, vibrations and impeller blade stresses and impairs the efficiency, [6].

Diffuser vane control: By adjusting the diffuser vanes of a radial pump with low or medium specific speed, the best efficiency point can be shifted within relatively wide limits. The deceleration ratio c_{3q}/c_2 and the shock losses generated at the diffuser inlet at the flow rate required by the system can be reduced that way, Fig. 5.18. This type of control allows operation at high efficiencies, but implies a complex mechanical design. It is used in pump turbines and storage pumps.

11.3 Static and Dynamic Stability

A *pump characteristic* is regarded as stable if dH/dQ is negative, i.e. if the head drops when the flow rate is increased. If the *system curve* has a positive gradient dH_A/dQ, the pump operation remains stable because the pump will run back to its original operation point whenever there is a minor deviation of ±dQ. The

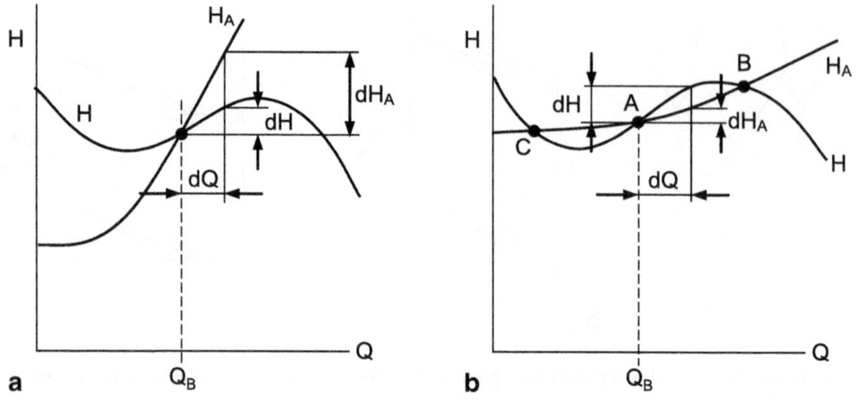

Fig. 11.11 Static stability. **a** Stable operation, **b** unstable operation

mechanism can be visualized with reference to Fig. 11.11 as follows: When the flow is momentarily increased by the amount dQ due to a perturbation, the head generated by the pump drops whereas the system requires a higher head $dH_A > dH$. Consequently, the criterion for *static* stability, where the time-average operation point with a given system characteristic remains the same, is $\mathbf{dH/dQ < dH_A/dQ}$. In order to assess the static stability it is therefore sufficient to know the pump and system characteristics.

The criterion of static stability may be fulfilled even when dH/dQ is positive (and consequently the characteristic is "unstable"). This situation is encountered if the system characteristic is steeper than the pump characteristic so that both characteristics intersect in only *one* point, Fig. 11.11a. Given a momentary deviation of dQ, the system then requires dH_A whereas the pump only delivers dH.

Conditions are quite different in Fig. 11.11b where the system characteristic exhibits a smaller gradient than the pump. To a momentary deviation of dQ, the pump reacts with a higher head rise than required by the system. The system can absorb the additional flow; hence the deviation increases and the operation point shifts from point A to the higher flow rate at point B. With a momentary deficit of -dQ, the operation point would shift from point A to point C. The operation points B and C are stable (they meet the stability criterion). Point A is unstable if the system characteristic is sufficiently flat, i.e. when pumping against a high static pressure with low head losses in the system.

A dynamic instability is a self-excited vibration where periodic fluctuations of flow rate and pressure occur around a given operation point. There are two requirements for the excitation of such vibrations: (1) the pump characteristic has to be unstable (hence, it must exhibit a positive gradient dH/dQ); (2) the system must have sufficient compressibility for storing energy during the vibration cycle. Compressible volumes in this sense are: vapor-filled spaces such as found in de-aerators or boilers, gas volumes in a tank for suction pressure control, cavitation

Fig. 11.12 Dynamic instability

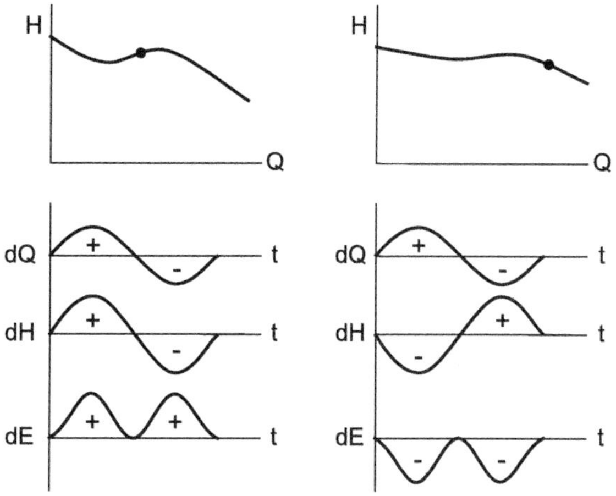

zones or the elasticity and compressibility of hot water in large-volume piping systems. Dynamic instabilities are hence an issue in boiler feed pump service.

The mechanism of a dynamic instability can be explained with reference to Fig. 11.12, [8]. Consider the operation in the unstable range of the characteristic represented in the left graph and in the stable range (right graph) where flow rates and pressures are plotted against the time. In the unstable range dH/dQ is positive which means that H oscillates in phase with Q. The energy fed into the system per cycle is $dE(t) = \rho \times g \times dH(t) \times dQ(t)$. The energy dE put into the system is positive; it enhances the original perturbation and the oscillation increases (refer to explanations in Chap. 10: self-excited vibrations arise if more energy is fed into a vibratory system during a cycle than is dissipated through damping.)

In contrast, dH/dQ is negative on the stable branch of the characteristic. Fluctuations of pressure and flow rate are in opposite phase, dE becomes negative. Consequently, the energy added during the perturbation is dissipated and the amplitudes decline (the damping is higher than the excitation).

Note: In the above argument and in Fig. 11.12 dH(t) and dQ(t) are *fluctuations in time* which can be quantified only by a dynamic model of the system. These terms cannot be derived by differentiating the useful energy dE/dQ with $E = \rho \times g \times H \times Q$ which would just give the gradient of the steady power curve P(Q) and has nothing to do with dynamic instability.

The stability analysis is performed using the methods of control engineering. In addition to the pump and system characteristics, the oscillating fluid mass and the compressibility of the system must be known. The risk of self-excited vibrations increases with growing instability of the pump characteristic (refer also to the discussion of Fig. 10.53). Together with inadequate system damping even slight instabilities of the characteristic (which normally would barely be detected) can cause self-excited vibrations. The actual occurrence of vibrations depends on the degree of instability of the Q-H-curve and on the system properties (particularly damping). Dynamic instability can occur even with one pump in single operation. It manifests as low-frequency fluctuations of pressure and flow rate that can cause pipe vibrations and disturb the functioning of the control system.

As discussed in Chap. 10.7.1 the tests reported in [15] suggest that the pressure fluctuations in the system (including the effect of standing waves) can couple to the flow through the impeller channels (Figs. 10.23 and 10.25). Such coupling explains the high impact of a flat or unstable Q-H-curve on the vibrations of the piping or the pump.

11.4 Start-Up and Shut-Down

The pump must be filled for start-up with liquid up to the level where the impeller and casing are flooded since the pressure rise created by the impeller operating in air is virtually zero owing to the low density, Chap. 13.2.

The torque $M_A(n)$ generated by the motor during the starting process must sufficiently exceed the torque M(n) required by the pump at any time in order to be able to accelerate the rotor of the unit. The pump torque must therefore be evaluated as a function of the speed (or time) for analyzing the start-up process. According to the similarity laws, $M/M_N = (n/n_N)^2$ is considered a sufficiently accurate approximation for this purpose.[1]

For starting a pump, several processes are common which shall be discussed with reference to Fig. 11.13. The torque characteristics are plotted against the speed in Fig. 11.13a, the pump and system characteristics are shown in Fig. 11.13b, while the torque characteristics are plotted versus q* in Fig. 11.13c. Curve 1 in Fig. 11.13a shows the torque characteristics according to the relation $M/M_N = (n/n_N)^2$ which, in theory, would be zero at n = 0. However, the bearing and shaft seal require a break-away torque at n = 0 (point L) which the driver has to overcome. The break-away torque is only a few per cent of the nominal torque. As a rule, it does not require a detailed analysis. The torque required by bearings and seals drops with growing speed so that point A on curve 1 is reached at approximately 10 to 20% of the nominal speed.

Start-up against closed valve: Pumps whose power consumption is lower at Q = 0 than at the BEP (i.e. pumps of low to medium specific speeds) are often started against a closed valve because the starting torque is lowest at Q = 0. At $n = n_N$ the relation $M_{GS}/M_N = P_o/P_N$ applies (Fig. 11.13c). Hence the torque characteristics result from $M/M_N = P_o/P_N \times (n/n_N)^2$: curve L-A-GS in Fig. 11.13a. As soon as the nominal speed is reached, the discharge valve must be opened. This causes the torque in Fig. 11.13a to shift from GS to N.

[1] For a discussion of extremely fast starting transients as encountered in rocket pump applications, refer to [1].

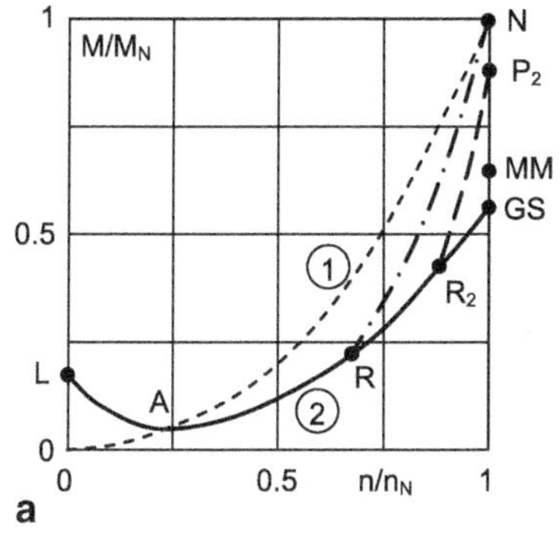

Point	
N	Operation at nominal flow (here $Q_N = Q_{opt}$)
L	Start-up torque at n = 0
GS	Closed valve $\frac{M_{GS}}{M_N} = \frac{P_0}{P_N}$
R	Opening of check valve $\frac{M_R}{M_N} = \frac{P_0}{P_N}\left(\frac{n_R}{n_N}\right)^2 = \frac{P_0}{P_N}\cdot\frac{H_0}{H_N}$

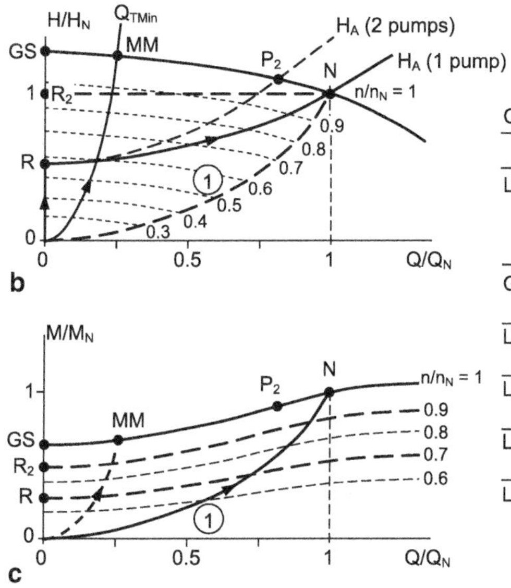

Curve	
(1)	$M/M_N = (n/n_N)^2$
L-A-GS	Start-up against closed valve $\frac{M}{M_N} = \frac{P_0}{P_N}\left(\frac{n_R}{n_N}\right)^2$
GS-N	Opening of throttle valve till Q_N is reached.
L-A-R-N	Start-up against check valve which is loaded by $H_{A,Stat}$
L-A-R_2-P_2	Start-up against check valve which is loaded by H_N
L-A-MM	Start-up with open minimum flow bypass
L-A-N	Start-up with open valve $H_{stat} = 0$, $H_{Dyn} = H_N$

Fig. 11.13 Start-up of a centrifugal pump. **a** Torque as function of speed M(n), **b** characteristics, **c** torque as function of flow rate M(Q)

Start-up with liquid-filled discharge pipe: Starting a pump with a long liquid-filled pipeline can be done in the same way as starting against a closed valve - even with the valve open - because a great mass of fluid has to be accelerated generating a counter-pressure by inertia effects.

If the fluid is accelerated in the time Δt from the velocity c_1 to c_2, the system reacts with a head difference ΔH_b which must be added to the steady-state system head H_A, [B.15]:

$$\Delta H_b \approx \frac{L(c_2 - c_1)}{g\,\Delta t} \tag{11.5}$$

Equation (11.15) is obtained by integrating the unsteady term in Eq. (1.7) for $\partial c/\partial t = (c_2 - c_1)/\Delta t$. By means of Eq. (11.15) it is easy to assess whether ΔH_b has to be taken into account in the analysis of the start-up process or may be ignored because of $\Delta H_b \ll H_A$.

If H is the head at the operation point and H_o is the shut-off head, the time needed for accelerating the water column to c_{max} is approximately [N.3]:

$$\Delta t \approx \frac{2\,L c_{max}}{g(H_0 - H)} \tag{11.6}$$

Start-up with minimum flow valve open: High pressure pumps must not be operated against a closed valve in order to prevent excessive heating of the fluid (Chap. 11.6). For this reason a minimum flow pipeline is required and the pump is started with the minimum flow valve open, Fig. 11.6. The torque characteristics are similar to the start-up against a closed valve since the power consumption is only slightly higher than at Q = 0.

Start-up with discharge valve open against a closed check valve: If the static head of the system prevailing at the check valve is H_{stat} (point R in Fig. 11.13b), the pump will start against the closed valve as discussed above until H_{stat} is reached at $n_R/n_N = (H_{stat}/H_o)^{0.5}$ and the check valve opens. Subsequently operation shifts towards point N on the system curve. The torque characteristic is represented by the curve L-A-R-N in Fig. 11.13a.

If the pump is to be started while another pump is already operating at the nominal point (or any other operation point), the pressure imposed on the check valve corresponds to the momentary operation point (in Fig. 11.13b this is H_N). Thus the second pump will operate against closed valve until that pressure is reached at R2 and the check valve opens. Subsequently, the pump operates on the system characteristic until the parallel operation point P2 of both pumps is reached at $n = n_N$. The torque characteristic follows L-A-R2-P2.

Start-up with open discharge valve: If there is no appreciable static back pressure, the operation point will run up along the system characteristic. This situation encountered if the system characteristic has only dynamic components and if the pipe is short (hence the inertia of the liquid in the pipe is negligible). The torque follows a square law through the operation point. If this point coincides with the best efficiency point, the start-up proceeds along curve 1 (L-A-N in Fig. 11.13a). Any other throttle valve position is feasible as well - for instance start-up at minimum flow, which would be preferable with regard to the starting torque of pumps with low specific speeds. All of the possible torque characteristics are between the curves 1 and 2 in Fig. 11.13a (as long as $Q < Q_N$).

Pumps with high specific speeds whose power consumption grows with falling flow rates are preferably started up with the discharge valve open (or partially open). The throttle valve position to choose must allow the operation point to be more or less positioned near to the best efficiency point in order to prevent the starting torque getting too high. If a pump with a high shut-off power ratio P_o/P_N has to be started against a closed check valve, it is advisable to open a bypass during the start-up.

Start-up with the discharge pipeline drained: At the beginning of the start-up process the pump will run at H = 0. Consequently it operates far to the right of the BEP with strong cavitation (similar to Fig. 11.7). As the pipeline is progressively filling and back pressure is building up, the operation point shifts back to lower flow rates until the pipeline is full and steady operation is reached. This kind of start-up is mainly performed with semi-axial and axial pumps. In this process, the operation period with extensive cavitation has to be as short as possible. Pumps built with balance disks are at risk during start-up without back pressure in that the disks may rub due to unbalanced axial thrust and cavitation.

Start-up of a pump with a long suction pipe: A pump installed in a pipeline may have a suction pipe several kilometers long. In this case the inertia of the liquid column in the suction pipe must be taken into account in order to ensure that the pump has ample NPSH available at all times during the start-up process. Otherwise, the liquid column could separate with water hammer as a consequence. With reference to Table 2.2, consider a pump drawing liquid from a tank with the head H_e through a suction pipe of length L. The run-up time of the pump is Δt_{start}. An ample margin on the $NPSH_3$ of the pump should be selected for the transient; say $F_{NPSH} > 2.5$ (see Table 6.2). The column of water in the suction pipe is accelerated by the pressure difference between the tank and the suction nozzle of the pump (minus the pressure drop). From Eq. (11.15) we can calculate the maximum allowable velocity and flow rate which maintains the minimum NPSH available with the stipulated margin: with $c_1 = 0$; $c_2 = c_{max}$; $\Delta H_b = H_e - F_{NPSH} NPSH_3 - H_{v,s}$ we get:

$$c_{max} \approx \frac{g\,\Delta t_{start}\left(H_e - F_{NPSH}\,NPSH_3 - H_{v,s}\right)}{L} \tag{11.6a}$$

The corresponding maximum start-up flow rate can then be calculated from c_{max} and the pipe diameter. If the pressure drop in the suction pipe $H_{v,s}$ at the calculated c_{max} is high an iteration may be required. If it is sufficiently covered by the margin F_{NPSH}, it can be neglected. The pump must not be started without taking into account the maximum flow derived above against an open discharge valve or against a low-resistance discharge pipe. In summary: a pump with a long suction pipe must be started at controlled flow against the minimum flow bypass or a slightly opened discharge before taking the pump in slow steps to full flow. If the pump is driven by a variable-speed motor, the speed can be increased slowly during start-up to maintain ample $NPSH_A$ during the process.

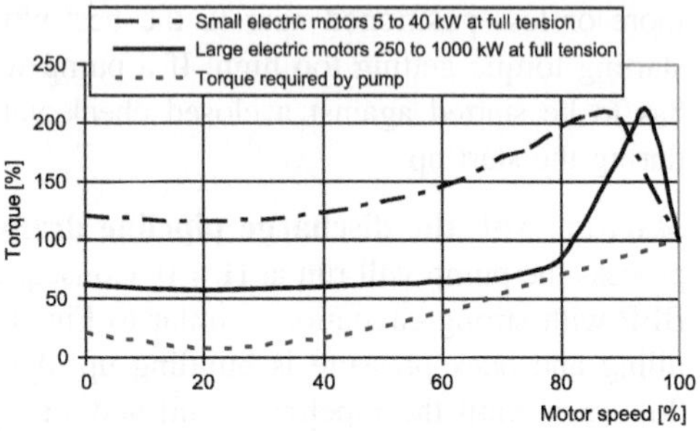

Fig. 11.14 Starting-torque curve of an electric motor and pump

Start-up time: Figure 11.14 shows the respective characteristics for an electric motor of a low and a high power class together with the torque characteristics of a pump. For the acceleration of the unit the torque $M_B = M_A - M$ is available (M_A is the torque provided by the driver); M_B must definitely be positive during the entire start-up to prevent the group from "hanging up". With the mass moment of inertia J of all the rotating parts (driver, pump rotor, coupling and gear, if any) it follows from Newton's law: $M_B = J\,d\omega/dt$, from which the start-up time can be calculated:

$$t = \int_0^{\omega} \frac{J}{M_B} d\omega = \frac{\pi J}{30} \int_0^{n} \frac{dn}{M_B} = \frac{\pi J}{30} \Sigma \frac{\Delta n}{M_B} \tag{11.7}$$

Instead of solving the integral the sum over finite steps can be calculated. As a rule, the start-up process should be as fast as possible. Consequently M(n) should be kept as small as possible.

Electric motors can be switched on directly - provided that the power supply system does allow doing so. However, the direct switch-on implies high starting torques and high currents which are often not acceptable with respect to the available power supply and installations. By means of a start-up transformer or a star-delta connection the power consumption can be reduced, but then the starting torque may prove inadequate for accelerating the unit so that switching from star to delta again gives rise to high peak loads. On account of the heating of the motor an analysis of these processes must also take into consideration how often the motor is switched on.

Coast-down time: After switching off the driver, the rotor speed decreases in accordance with the mass moment of inertia of the drive train and the retarding torque M_R generated by the rotor. M_R consists of the pump torque $M = M_N \times (n/n_N)^2$ and the power losses of the driver, including gears. The coast-down time is calculated from Eq. (11.7) with M_R substituted for M_B. The speed decreases as a function of time similar to a hyperbolic curve, because the pump torque acts in the beginning of the coast-down as a strong brake. Typically the coast-down time is between 10 and 60 s, depending on the pump size; for small pumps it is even less.

11.5 Power Failure, Water Hammer

Sudden flow variations in a pipe give rise to pressure surges which travel at the velocity of sound through the system (the phenomenon is known as "water hammer"). The order of magnitude of a pressure peak Δp generated by a velocity variation Δv can be estimated from $\Delta p = \rho \times a \times \Delta v$, where "a" is the speed of propagation of a disturbance in the pipeline according to Eq. (10.19).

The magnitude of the pressure peaks and the potential damage to be expected from water hammer increases with the length of the pipeline and the flow velocity in the pipe. Water hammer is caused, for example, by the rapid closing of a valve or by a pump driver failure with subsequent abrupt closure of the check valve, by operator error, by trapped air and also when pumps are switched over or started up. The basic theory concerning water hammer, calculation methods and remedies can be found in [4, 11, 25].

The highest pressure peaks are caused by the abrupt closure of a check valve, since this induces a sudden deceleration of the fluid to zero. A wave travels from the check valve through the system until it is reflected at a major change in impedance (Table 10.12). The pressure drops after each wave front so that even pressures below atmosphere may result.

Water hammer can cause considerable damage: pipe ruptures because of too high pressures or vacuum, separation of the liquid column when the vapor pressure is reached and subsequent thermal water hammer. Low pressure systems are especially at risk because of their low design pressure and correspondingly thin pipe walls. The analysis of water hammer problems and the choice and optimization of means for water hammer protection are indispensable when dealing with large piping systems. The graphical methods formerly used for analysis have been replaced largely by computer programs.

There are various protective measures to prevent damage: (1) The velocity variation causing the water hammer can be reduced by means of a flywheel which extends the coast-down time of the pump following a driver failure. However, this remedy is only sufficient for pipes up to a length of about 2000 m. (2) The lowest pressure behind the wave front can be raised by means of water fed into the pipeline from an air vessel installed downstream of the check valve. (3) A stand pipe or a surge tank placed at a prominent high point of the system can shorten the active length of the pipeline, since the pressure is imposed on the pipe at this location. (4) Blow valves are suitable for breaking vacuum. (5) Pressure relief valves or rupture disks avoid damage due the pressure peaks. (6) Optimizing the valve closing time helps to reduce the velocity difference Δv.

If liquid runs back from the system through the pump after a shut-down, the pump is accelerated in the opposite sense of rotation like in turbine operation. As a rule, this should be prevented by the installation of a check valve. However, some systems allow short reverse running which necessitates a suitable mechanical design of pump and driver. During reverse running there is no external torque acting on the pump and driver. The run-away speed is only limited by the effects of disk friction, mechanical losses, hydraulic losses due to friction, excessive incidence and the residual swirl of the liquid flowing off the impeller inlet into the suction pipe. The resulting reverse running speed increases with the specific speed; it can be estimated from Eq. (T12.1.8).

11.6 Allowable Operation Range

Obviously a pump should be selected so that it operates predominantly close to the best efficiency point in the so called "preferred operation range". This mode of operation is apt to bring about the lowest energy and maintenance cost and to reduce the risk of system problems since hydraulic excitation forces and cavitation risk, as a rule, attain a minimum near the BEP. However, off-design operation for limited periods cannot be avoided. Rules are needed for defining the allowable ranges and modes of operation in order to reduce the risk of damage and excessive wear. To this effect, limits must be defined for continuous operation and for short-term operation at maximum and minimum flow. Such limits cannot be calculated theoretically; they are derived from experience and should not be considered as sharp delimitations. The following criteria may be used for defining allowable operation ranges:

- Energy costs
- Heating of the fluid at low load
- Type of pump and service
- Fluid to be pumped and its temperature
- Power class of the pump, head per stage or impeller tip speed u_2
- Risk of cavitation; criteria: $NPSH_A$, u_1, suction specific speed or inducer
- Partload recirculation: hydraulic excitation forces, noise, cavitation
- Vibration behavior of the pump and the system
- Stability of the Q-H curve
- Power consumption of the motor

The range of the allowable **continuous operation** may be defined as those conditions at which the pump can be operated for many thousand hours without damage or excessive wear. This range could be defined, for instance, by the requirement that the efficiency must not fall below 80 to 85% of the maximum efficiency of the pump in question. Such a criterion is not only sensible with respect to the energy consumption but it also ensures that the pump does not operate with excessive partload recirculation (below $q^* < 0.5$) or with flow separation at high flows above BEP. In practice many pumps are oversized due to margins applied on head and/or flow. As a consequence, these pumps often work with moderate partload recirculation. At low or medium circumferential speeds and with a robust design this usually does not lead to noticeable damage if the pumps are working within in the range of allowed continuous operation.

Figure 11.15 shows the allowed continuous operation range as a function of the specific speed, which is obtained by applying the efficiency criterion. This recommendation is valid primarily for large pumps from 500 to 1000 kW. The higher the head per stage and the power, the greater is the importance of chiefly operating

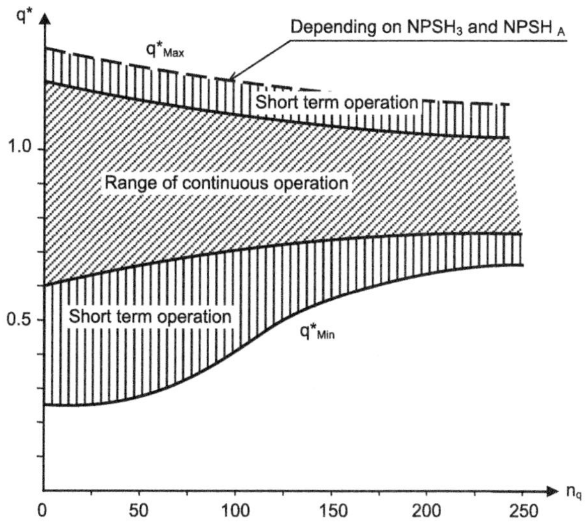

Fig. 11.15 Recommended ranges of minimum and maximum flow for continuous and short-term operation

near the best efficiency point. In this regard also the pumped fluid has to be taken into consideration. The range recommended in Fig. 11.15 may be modified by an in-depth analysis of the actual pump and system according to the criteria listed above. Small pumps, especially those with low specific speeds, are sometimes even permanently operated at low partload (below $q^* \approx 0.5$). Moreover, the pump manufacturer may recommend other ranges, for instance because of an instability in the characteristic of an axial pump.

The maximum flow rate has to be restricted, particularly with regard to the risk of cavitation in the high flow range. Generally applicable rules can scarcely be established, since the cavitation risk depends on the suction impeller design, the flow rate of shockless entry, the $NPSH_A$, the circumferential speed at the impeller inlet and on material and fluid properties. With respect to cavitation, the upper limit given in Fig. 11.15 for continuous operation is valid rather for applications with a low cavitation risk. The dependence of the axial thrust on the flow rate is another criterion to be checked at high flows, in particular for multistage pumps.

Whereas pumps with low specific speeds are relatively robust, the allowable operation range of axial pumps is often restricted by: (1) the stable region of the Q-H-curve; (2) by the maximum power consumption of the selected motor; (3) the strong impact of cavitation effects observed at partload and overload as demonstrated by the NPSH-curve in Fig. 6.16. This should be given careful consideration in the planning stage of a pumping station.

Short-term operation may be defined as covering abnormal or upset conditions which result as a rule in premature wear of the pump and whose cumulative duration per year does not exceed typically 100 h. To this category belong: (1) operation in the high-flow range with increased cavitation (e.g. when pumps operating in parallel are switched over); (2) partload operation with strong recirculation, especially working on the minimum flow bypass. The limits shown in Fig. 11.15 provide a rough guideline. Depending on the pump type and manufacturer, great differences of vibration and cavitation behavior may be encountered, which may call for wider or justify narrower limits.

Minimum flow operation: A high-pressure pump with a high power cannot be operated against a closed valve since the temperature in the pump would rise by several degrees per second because of the high ratio of the power input to the mass of the pump. Excessive cavitation, vapor locking and damage would occur before any control mechanism could be activated. Such pumps are therefore provided with a minimum flow bypass leading back to the suction reservoir, similar to Fig. 11.6. The minimum flow valve opens automatically when the discharge valve is closed. With high-pressure pumps roughly up to $n_q = 35$, the minimum flow devices are typically sized as follows: (1) up to a power of roughly 2000 kW for $q^*_{min} = 0.1$ to 0.15; (2) at considerably higher powers frequently for $q^*_{min} = 0.25$ to 0.35. Details concerning the layout of minimum flow systems, their control and the associated components (such as valves and throttling devices) can be found in several papers published in [9].

The minimum flow is mainly determined by three criteria: (1) the *thermal* minimum flow limits the temperature rise of the fluid; (2) the *hydraulic* minimum flow attempts to limit the hydraulic excitation forces and vibrations; (3) finally excessive cavitation induced by partload recirculation is to be considered.

Since the intensity of the partload recirculation increases with rising specific speed, the lower operation limit q^*_{min} defined in Fig. 11.15 increases with n_q. A possible instability of the head-capacity curve below the minimum flow usually should be considered as irrelevant since the pump is not operated in that range. Pumps with inducers possibly necessitate a limitation of the operation range because of pulsations at partload, Chap. 7.7.4.

Temperature rise in the pump: The enthalpy rise in the pump is given by $\Delta h_{tot} = P_i/\dot{m}$ (with the inner power $P_i = P \times \eta_m$). Since the enthalpy remains constant during throttling, the temperature of the fluid (with specific heat c_p) rises when the *flow rate is throttled down to the pump inlet pressure* by:

$$\Delta T_{Dr} = \frac{\Delta h_{tot}}{c_p} = \frac{P_i}{c_p \dot{m}} = \frac{g\,H}{c_p \eta_i} \qquad (11.8)$$

The temperature of a fluid passing through the minimum flow line thus increases by the amount given by Eq. (11.8). The same temperature rise is generated in the axial thrust balance device or during operation in a closed circuit.

Even in the discharge nozzle, the temperature is slightly higher than in the suction tank. In the vast majority of applications this temperature rise is scarcely perceptible and of no consequence. Conversely, for high-pressure pumps with heads above 2000 m (especially with hot water) the temperature rise must be taken into account when the density is determined for exact calculations and test data evaluation. The temperature rise in the pump consists of two components: $\Delta T = \Delta T_v + \Delta T_{is}$. The portion ΔT_v results from the losses in the pump, while ΔT_{is} is caused by the isentropic compression of the fluid. The temperature rise from the inner losses (i.e. disk friction, leakages and head losses in the hydraulic channels) is calculated from Eq. (11.9):

$$\Delta T_v = (1 - \eta_i)\frac{P_i}{c_p\,\dot{m}} = \left(\frac{1}{\eta_i} - 1\right)\frac{\Delta p}{\rho\,c_p} = \left(\frac{1}{\eta_i} - 1\right)\frac{g\,H}{c_p} \tag{11.9}$$

The isentropic compression is obtained from the water/steam tables or from an enthalpy/entropy diagram. A rough estimation for water can be made according to:

$$\Delta T_{is} = 0.7\frac{T_s}{T_{Ref}}\cdot\frac{H}{H_{Ref}} \tag{11.10}$$

T_s = temperature of the approach flow in °C, T_{Ref} = 100 °C and H_{Ref} = 1000 m.

If the temperature difference between the suction and the discharge nozzle $\Delta T = T_D - T_S$ is measured and ΔT_{is} is determined from the water property tables [N.7], $\Delta T_v = \Delta T - \Delta T_{is}$ can be calculated. In this way the inner efficiency of a pump can be determined from Eq. (11.9) as $\eta_i = \Delta h_{is}/\Delta h_{tot}$ (with $\Delta h_{is} = g \times H$). This is the basis of the thermometric efficiency measurement, [N.2]. As the temperature rise usually lies within the range of 1 to 4 °C, great diligence is called for during such measurements and their evaluation.

The effects discussed above are moderate for water; they can be quite important when pumping liquid natural gas, hydrocarbons or other liquids. A detailed analysis of the heating of the fluid is then mandatory.

11.7 The Approach Flow to the Pump

Centrifugal pumps are fed from intake bays in wet-pit or dry-pit arrangements or via a suction pipe from tanks. An approach flow as uniform as possible and free of swirl and vortices is important for a reliable operation of the pump. Any pre-rotation will reduce the head, while any counter-rotation will lead to an increase in power and head, possibly over-loading the motor. Non-uniform approach flow, vortices and air entrainment generate noise and vibrations and can impair the efficiency. With growing specific speed the pump becomes increasingly sensitive to approach flow perturbations, because the inlet diameter ratio d_1/d_2 rises with the specific speed. By virtue of Eq. (T3.3.1) the non-uniformities of the approach flow gather influence on the work transfer in the impeller with rising d_1/d_2.

If the pump takes suction from a reservoir with a free liquid level, air drawing vortices are created when the submergence of the suction pipe inlet is not sufficient. Air drawing vortices and their avoidance are discussed in Chap. 11.7.3.

11.7.1 Suction Piping Layout

A number of criteria should be considered for the layout of the suction piping and for the analysis of possible problems:

1. Principle: The suction pipe should be as short and as straight as practicable.
2. Valves (if any) should open to the full pipe cross section with as little restrictions as possible.
3. If bends in the suction pipe cannot be avoided, the bends should be in one plane only, because bends in the 3D-space create defined swirls due to superimposing secondary flows created in the individual bends. A straight pipe section of L/D > 5 to 8 should be provided between the pump inlet nozzle and any upstream pipe fitting or valve. A longer distance may be required in the case of bends in the 3D-space in order to dissipate the swirl generated in the bends. If the plant layout does not allow to provide sufficient straight pipe lengths, the approach flow can be improved by a flow straightener and/or deflection baffles in the bends. A non-uniform axial approach flow can be smoothed out to some extent by flow acceleration (Chap. 1.8), but possible tangential velocity components c_u *increase* when reducing the pipe diameter because of the conservation of angular momentum according to $c_u \times R = \text{constant}$. Tests for optimizing a suction pipe of an arrangement similar to Fig. 11.16 were reported in [26]. In this case, the best results were achieved by a combination of acceleration and deflection baffles since the latter reduce tangential velocity components.
4. Pipes branching off a main pipe at 90° and dead-end pipes generate vortices and often require measures for smoothing the flow, e.g. as in Fig. 11.16.
5. The transition from a tank to the suction pipe should be rounded, chamfered or made with a conical section in order to reduce jet contraction and associated vortices. Frequently a cross is installed to prevent ("bath-tub") vortex generation. The submergence of the suction pipe inlet should be at least 2 to 4 × D, for details see Chap. 11.7.3. If saturated liquids are to be pumped, a smooth

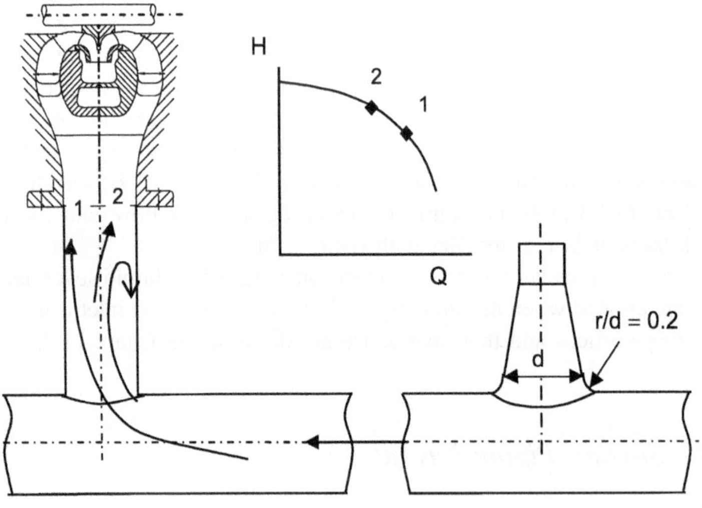

Fig. 11.16 Asymmetric approach flow to double-entry impeller and vortex generation at branches or at dead-end pipes

flow from the tank into the suction pipe is particularly important in order to avoid cavitation due to flow acceleration, jet contraction and vortices.

6. If a saturated liquid is pumped, the suction pipe connected to the tank should have a vertical downward section as long as possible prior to continuing with horizontal or sloped pipe legs (see Fig. 11.17). That way, the risk of liquid flashing into vapor can be reduced because the static pressure rises in the vertical down-flow pipe section. If the pumpage contains a large amount of dissolved gas, this arrangement helps to prevent or reduce gas separation.
7. The pressure drop must be minimized to get a maximum $NPSH_A$. The flow velocities are therefore often limited to 1 to 2 m/s. In general it is recommended not to exceed 4 m/s.
8. Gas separation must be minimized to avoid the risk of de-priming. Suction pipes should never be installed with high points where gas could accumulate. If the liquid is drawn from a tank installed at a level lower than the pump, the pressure in the suction pipe is below atmospheric. A portion of the dissolved gases can separate from the liquid in this case. In order to avoid gas accumulations, the suction pipe should rise to the pump with a slope of at least 10°.
9. If the tank is located above the pump and if there is a risk of flashing of vapor and/or gas (process engineering, feedwater and hot-water pumps), the suction pipe should be installed with a downward slope of at least 10°. Gas and vapor accumulations in horizontal pipes can lead to rough running or even water hammer if the vapor suddenly condenses.
10. If the pump inlet pressure is below atmospheric, the ingress of air must be prevented by selecting vacuum-prove seals (shaft seal as well as static seals in the piping, if any).
11. Valves, fittings and the suction pipe upstream of a double-entry pump must be installed in a symmetry plane in a way that both sides of the impeller receive the same flow. This problem is discussed with reference to Fig. 11.16 in more detail below.

Figure 11.16 shows an arrangement which is often used when several pumps are installed in parallel. The pumps are fed from a common pipeline or header via (frequently too short) suction pipes which are branched off under 90°. The pumps discharge into a common header arranged in a similar manner. This concept allows a compact layout of the pumping station. The disadvantage of this arrangement can be seen from Fig. 11.16. The 90°-branch causes flow separation and vortices at the inlet to the suction pipe. As a consequence, noise and vibrations increase and the efficiency is impaired. Vortex streets may be generated which could lead to acoustic resonances, Chap. 10.12.

If this arrangement is employed for a pump with a double-entry impeller, both sides of the impeller are subject to different inflow conditions, Fig. 11.16. They may work at slightly different points on the Q-H-curve with an increased axial thrust.

Several remedies may be applied (singly or in combination) to mitigate or avoid such problems: (1) increase the length of the branched-off suction pipe to $L/D > 8$ to 10 in order to allow the flow to smooth out; (2) no sharp-edged inlet to the suction pipe, provide a generous radius or chamfer instead; (3) provide a cone at the

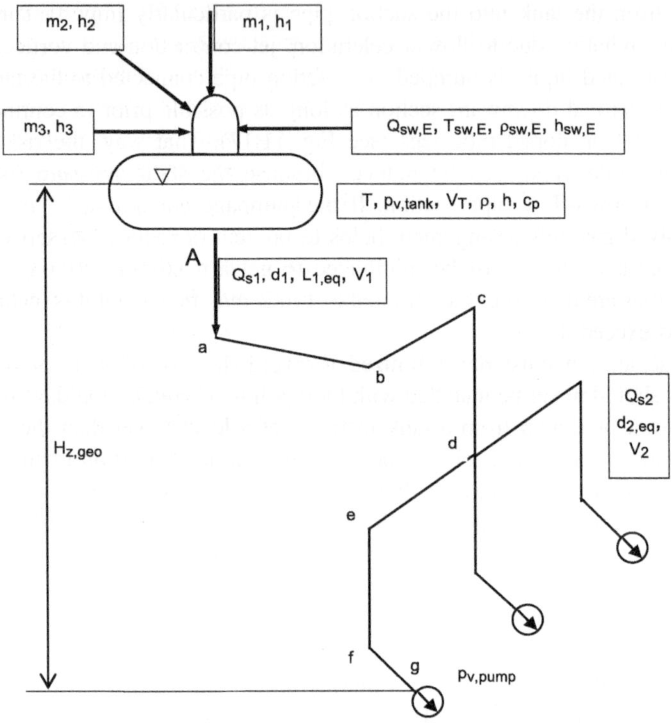

Fig. 11.17 De-aerator and suction piping

inlet to the branched-off suction pipe which accelerates the flow, Chap. 1.8; (4) install a flow straightener in the suction pipe downstream of the branching; (5) branch off the suction pipe under 45° or less.

11.7.2 Transient Suction Pressure Decay

When operating with a saturated liquid, the pump gets the fluid from a tank in which a free liquid level is established (examples are boiling liquids in the process industry or the de-aerator in a power plant). The vapor pressure prevailing in the plenum above the liquid level is given by the temperature of the liquid, $p_v = f(T)$. The pump requires a positive geodetic suction head, and the tank must be installed sufficiently above the pump. The pipe between the tank and the pump should not have high points nor long horizontal sections in order to avoid vapor accumulations in case of some liquid flashing into vapor. If vapor is trapped in the pipe, water hammer can be induced by a sudden condensation due to a pressure surge.

During a load reduction, the energy input into the tank may decrease, for example due to the ingress of colder water and/or a drop in the enthalpy of the heating steam drawn from a turbine. As a consequence, the temperature of the liquid inventory in the tank decreases and the vapor pressure drops. Load transients in this type of plant require a margin on the available $NPSH_A$ in order to avoid unacceptable cavitation.

If the pressure above the liquid level drops from p_1 to p_2, a fraction of the liquid stored in the tank evaporates till the saturation temperature $T_v = f(p_2)$ is reached. Since the liquid in the tank cannot be superheated, this thermodynamic equilibrium is established quickly. However, some liquid with higher temperature still travels through the suction pipe for a short time. If the geodetic head (or the $NPSH_A$) is too low, some of the liquid in the pipe will flash into vapor. Relevant for this process is the *local* state of saturation which depends primarily on the geodetic head.

As compared to steady state operation, the geodetic suction head must be increased so that no flashing occurs in the suction pipe during the pressure decay in the tank. The available geodetic head $H_{z,geo}$ yields $NPSH_A = H_{z,geo} - \sum H_v$ since the vapor pressure $p_{v,tank}$ prevails above the liquid level in the tank. In order to avoid evaporation upon a pressure decay in the tank, the available NPSH must exceed the required NPSH, i.e. $NPSH_A > NPSH_R + H_{trans}$; H_{trans} represents a margin added to cover the transients assumed for the plant.

The transit time of a fluid particle through a suction pipe of length L_s amounts to $t_s = L_s/c_s = V_s/Q$ (c_s = flow velocity in the suction pipe, V_s = volume of liquid contained in the pipe). The allowable pressure decay velocity in the tank is then given by:

$$\left(\frac{d_p}{d_t}\right)_{al} = \frac{\rho\, g\, H_{trans}}{t_s} = \frac{\rho\, g\, H_{trans}}{L_s} c_s = \rho\, g\, H_{trans} \frac{Q}{V_s} \tag{11.11}$$

The sum of the hydraulic losses in the suction pipe can be expressed according to Table 1.5 by a total resistance coefficient ζ_{tot} defined by $\sum H_v = \zeta_{tot} \times c_s^2/2g$.

From $NPSH_A = NPSH_R + H_{trans} = H_{z,geo} - \zeta_{tot} c_s^2/2\,g$ we can derive:

$$H_{trans} = H_{z,geo} - NPSH_R - \zeta_{tot} \frac{c_s^2}{2g} \tag{11.12}$$

$$\left(\frac{d_p}{d_t}\right)_{al} = \frac{\rho\, g\, c_s}{L_s}\left(H_{z,geo} - NPSH_R - \Sigma\zeta_s \frac{c_s^2}{2g}\right) \tag{11.13}$$

Differentiating Eq. (11.13) and setting the obtained expression equal to zero, we obtain the optimum velocity in the suction pipe which yields the highest margin against flashing:

$$c_s = \sqrt{\frac{2g}{3\Sigma\zeta_s}\left(H_{z,geo} - NPSH_R\right)} \tag{11.14}$$

Thus it is possible to calculate the required geodetic head and an optimum suction pipe diameter for a specified pressure decay velocity and a given transit time of a fluid particle through the suction pipe.

The expected pressure decay velocity in a plant results from process calculations which establish an energy balance of the tank (e.g. a de-aerator) as a function of time during transient conditions (e.g. the load rejection of a steam turbine). The energy balance accounts for all liquid and vapor mass flow rates entering and leaving the tank during a time step as well as for the changes in the energy stored in the tank. This type of investigation is highly relevant for the de-aerator of a thermal power plant, [5, 22, 23]. Process plants can be studied in the same way.

The calculation procedure is described in the following with reference to Table 11.1 and Fig. 11.17 which represents the de-aerator/feed water tank of a thermal power station or any suction tank from which a pump draws a liquid at saturation temperature. The water flows from the de-aerator to the feed pump (or its booster pump). Preferentially each pump is fed by a separate suction pipe, but the general case is we considered in the following where the water leaves the tank through a common pipe branching off to feed z_{PP} parallel pumps. Frequently three 50-%-pumps are installed out of which one is at standby while two units are operating in parallel (as shown in Fig. 11.17).

The individual steps of the calculation are:

1. Analysis of the piping system. In the system depicted in Fig. 11.17 we distinguish: (1) pipe run #1 from point A to d; and (2) pipe run #2 from point d to g.

   - **Pipe run #1** consists of "i" straight pipe sections which may have different diameters d_i and "j" components such as bends branches, valves or filters. The flow resistances of all these pipe sections and fittings are described by one equivalent total resistance coefficient $\Sigma\zeta_1$ which can be calculated with the help of Table 1.5. According to Eq. (T11.1.2) this total resistance is referred to the pipe diameter d_1 or the flow velocity $c_1 = 4 \times Q_{s1}/(\pi \times d_1^2)$. The *total flow rate* Q_{s1} of all operating pumps flows through pipe run #1.
   - **Pipe run #2** too consists of "i" straight pipe sections (possibly with different diameters d_i) and "j" components. The resistances of these elements are described by one equivalent total resistance coefficient $\Sigma\zeta_2$ as per Table 1.5, Eq. (T11.1.3), which is referred to the diameter d_2 and the flow velocity $c_2 = 4 \times Q_{s2}/(\pi \times d_2^2)$. Pipe run #2 handles the flow rate of *one* of the z_{PP} pumps which are in operation, i.e. $Q_{s2} = Q_{s1}/z_{PP}$.

2. A common resistance coefficient $\Sigma\zeta_{tot}$ combining both of the pipe runs can now be calculated from Eq. (T11.1.4). It accounts for the fact that different flow rates are processed by both pipe runs. If each pump is fed by an individual suction pipe, $\Sigma\zeta_{tot} = \Sigma\zeta_1$ applies while $\Sigma\zeta_2$ is zero.
3. The total transit time from the tank to the pump inlet nozzle results from the sum of the individual transit times needed by the fluid to travel through pipe runs #1 and #2, Eq. (T11.1.7). When calculating the fluid volumes in the pipe runs from Eqs. (T11.1.5 or 11.1.6), all components must be taken into account (for example also the fluid accumulated in a filter). Each component, as well as each change in pipe diameter, is expressed by equivalent pipe lengths $L_{1,eq}$ and $L_{2,eq}$ as defined by Eqs. (T11.1.5 and 11.1.6) which yield the actual fluid volume when using the reference diameter.

Table 11.1 Calculation of suction pressure decay transients			Eq.
Required NPSH	$NPSH_R \geq 1.3\ NPSH_3$	*minimum* 30% margin on $NPSH_3$	11.1.1
Total pressure coefficient of *common* suction piping with reference diameter d_1	$\Sigma\zeta_1 = \sum_{i_1} \lambda_i \frac{L_i}{d_i}\left(\frac{d_1}{d_i}\right)^4 + \sum_{j_1} \zeta_j \left(\frac{d_1}{d_j}\right)^4$	Combine all resistances as per Table 1.5. Each pipe run consists of i_1 (or i_2) straight pipes and j_1 (or j_2) fittings (bends, branches, valves etc.)	11.1.2
Total pressure coefficient of the *individual* suction pipe with reference diameter d_2	$\Sigma\zeta_2 = \sum_{i_2} \lambda_i \frac{L_i}{d_i}\left(\frac{d_2}{d_i}\right)^4 + \sum_{j_2} \zeta_j \left(\frac{d_2}{d_j}\right)^4$		11.1.3
Total pressure coefficient of the inlet piping referred to c_1 respectively d_1	$\Sigma\zeta_{tot} = \Sigma\zeta_1 + \Sigma\zeta_2 \frac{Q_{s2}^2\, d_1^4}{Q_{s1}^2\, d_2^4}$		11.1.4
Fluid volumes and equivalent lengths of pipe runs 1 and 2. Each pipe run consists of i components	$V_2 = \frac{\pi}{4}\sum_{i_1} d_i^2 L_i = \frac{\pi}{4} d_1^2 L_{1,eq}$	$L_{1,eq} = \sum_{i_1} \frac{d_i^2}{d_1^2} L_i$	11.1.5
	$V_2 = \frac{\pi}{4}\sum_{i_2} d_i^2 L_i = \frac{\pi}{4} d_2^2 L_{2,eq}$	$L_{2,eq} = \sum_{i_2} \frac{d_i^2}{d_2^2} L_i$	11.1.6
Transit time of a fluid particle from tank to pump inlet	$t_s = \frac{V_1}{Q_{s1}} + \frac{V_2}{Q_{s2}} = \frac{V_1}{Q_{s1}}(1+a)$	$a \equiv \frac{Q_{s1}\, d_2^2\, L_{2,eq}}{Q_{s2}\, d_1^2\, L_{1,eq}}$	11.1.7
Optimum velocity in pipe run 1	$c_{1,opt} = \sqrt{\frac{2g}{3\Sigma\zeta_{tot}}\left(H_{z,geo} - NPSH_R\right)}$		11.1.8
Optimum diameter of pipe run 1	$d_{1,opt} = \left\{\frac{24\, Q_{s1}^2\, \Sigma\zeta_{tot}}{\pi^2\, g\left(H_{z,geo} - NPSH_R\right)}\right\}^{0.25}$		11.1.9
Allowable pressure decay velocity	$\left(\frac{d_p}{d_t}\right)_{al} = \frac{\rho g Q_{s1}}{V_1(1+a)}\left(H_{z,geo} - NPSH_R - \frac{8\, Q_{s1}^2}{\pi^2\, g\, d_1^4}\Sigma\zeta_{tot}\right)$		11.1.10
Change of mass in tank during time step Δt	$\Delta m = \left(\dot{m}_1 + \dot{m}_2 + \dot{m}_3 + Q_{sw,E}\, \rho_{sw,E} - Q_{s1}\, \rho\right)\Delta t$		11.1.11
Enthalpy *decrease* in tank during time step Δt (m_T = liquid mass in tank): $\Delta h = \frac{\Delta t\left(\rho\, Q_{s1}\, h - \dot{m}_1\, h_1 - \dot{m}_2\, h_2 - \dot{m}_3\, h_3 - \rho_{sw,E}\, Q_{sw,E}\, h_{sw,E}\right) + \Delta m\, h}{m_T + \Delta m + 0.5\, \rho\, Q_{s1}\, \Delta t}$			11.1.12
Temperature in tank after time step Δt	$T_{(n+1)} = T_{(n)} - \Delta T$ with $\Delta T = \frac{\Delta h}{c_p}$		11.1.13
From $T_{(n+1)}$ or $h_{(n+1)} = h_{(n)} - \Delta h$ the vapor pressure in the tank $p_{v,tank}$ and all other fluid properties can be determined from property tables or equations.			
Vapor pressure decay at the pump inlet	for $t < t_s$: $\Delta p_v(t) = p_{v,pump}(t=0) - p_{v,tank}(t)$ for $t > t_s$: $\Delta p_v(t) = p_{v,pump}(t - t_s) - p_{v,tank}(t)$		11.1.14
Reduction of $NPSH_A$	$\Delta NPSH_A = \frac{\Delta p_v}{\rho g}$		11.1.15

4. Once the flow resistances of the suction pipe have been defined, the optimum flow velocity and the optimum pipe diameter can be determined. Using the above definitions, these quantities are calculated from Eqs. (T11.1.8 or 11.1.9) either for the branched system (Fig. 11.17) or for the case of individual suction pipes. The diameters of the pipe runs #1 and #2 are determined so that the fluid travels with the same velocity through both runs.
5. The resistances $\Sigma\zeta_{tot}$, $\Sigma\zeta_1$ and $\Sigma\zeta_2$ according to steps 1 to 3 must now be checked using the selected pipe diameters; repeat step 4 as necessary.
6. The allowable pressure decay velocity can now be calculated from Eq. (T11.1.10) using the final diameters d_1 and d_2 selected. In order to provide a sufficient margin against evaporation, the required NPSH used in the calculation should be at least 30% higher than $NPSH_3$, [B.26]. Therefore set $NPSHR = 1.3 \times NPSH_3$ in the calculation as per Table 11.1. Obviously, any other safety factor (or $NPSH_o$) could be used instead. The allowable pressure decay velocity $(dp/dt)_{al}$ as per Eq. (T11.1.10) represents a quadratic equation with respect to the flow rate Q_{s1}. For any given system, $(dp/dt)_{al}$ exhibits a minimum when plotted against the flow. Usually several operational transients must be analyzed.
7. The effective pressure decay velocity resulting from plant transients (such as load reduction or steam turbine load rejection) should not exceed the allowable value calculated from step 6. All mass flow rates entering and leaving the de-aerator and their associated enthalpies must be defined as functions of the time for calculating such transients. For the example shown in Fig. 11.17 these are: (1) the feedwater flow $Q_{sw,E}$ entering the tank and its temperature $T_{sw,E}$; (2) further mass flow rates of liquid or vapor m_1, m_2, m_3 etc. with the enthalpies h_1, h_2, h_3 (mass flows entering the tank are positive; mass flow rates leaving the tank are negative). The liquid stored in the tank (V_T or m_T) at the start of the transient (i.e. at time $t = 0$) must be known. If it is desired to take into account the heat capacity of the walls and internal structures of the tank, the corresponding "water values" may be calculated and added to the heat capacity of the liquid.
8. The calculation is done in time steps which may be selected as a fraction of the transit time t_s, for example $\Delta t = t_s/10$. The mass balance is calculated from Eq. (T11.1.11) for each time step. The drop in enthalpy during the time step is obtained from the energy balance according to Eq. (T11.1.12). When deriving Eq. (T11.1.12), it was assumed that the enthalpy during a time step decreases linearly with the time; in other words the fluid drawn from the tank has an average enthalpy of $h_n - \Delta h/2$ during the step. (The energy balance results from Eq. (1.1) and Eq. (1.3) since the process is unsteady.) Equation (T11.1.12) implies ideal mixing of the liquid inventory with all fluids entering the tank. Heat loss to the environment is neglected.
9. After calculating the enthalpy drop during the time step, the decrease in temperature, the vapor pressure in the tank $p_{v,tank}$ and the fluid properties can be derived from the water/steam tables (Table 17.2).

10. Finally, the drop in the available NPSH can be calculated from Eqs. (T11.1.14 and 11.1.15). This drop is caused by the fact that the vapor pressure in the tank $p_{v,tank}$ is lower than the vapor pressure at the pump inlet $p_{v,pump}$ where the water temperature is still higher than in the tank. Thus we have: $\Delta p_v = p_{v,tank} - p_{v,pump}$ and $\Delta NPSH_A = \Delta p_v/(\rho \times g)$.
    The effective pressure decay velocity becomes $(dp/dt)_{eff} = \Delta p_v/\Delta t$. It should be noted that the fluid temperature at the pump inlet remains equal to the temperature at $t = 0$ for all times $t < t_s$. Only for times $t > t_s$ successively colder liquid enters the pump. Therefore $\Delta p_v(t) = p_{v,tank}(t) - p_{v,pump}(t = 0)$ applies to $t < t_s$, while for $t > t_s$ we have: $\Delta p_v(t) = p_{v,tank}(t) - p_{v,pump}(t - t_s)$. The effective pressure decay velocity attains its maximum at $t = t_s$; it drops for $t > t_s$ because colder fluid then enters the pump. Figure 11.18 shows these relationships with an example where the transit time is $t_s = 80$ s.
11. In order to prevent evaporation at the time of the highest pressure decay velocity (i.e. for $t = t_s$), it must be checked that the local *static* pressure in the suction pipe remains at all locations sufficiently above the vapor pressure. The static pressure at the outlet of each pipe section ("a" to "g" in Fig. 11.17) and at each component is obtained from Bernoulli's equation, i.e. Eq. (1.7). Particular attention should be focused on the locations in the suction pipe situated downstream of a component with a high flow resistance. Likewise critical are locations after a horizontal pipe section where the pressure decreases due to friction losses. Horizontal pipe sections should be as short as possible and the pressure drop of fittings as low as feasible.

The above investigation must be performed for different transients and plant conditions whenever it is not easy to clearly ascertain the worst case scenario. The turbine trip after load rejection is presumably the steepest transient in a power station. If the effective pressure decay velocity exceeds the allowable limit, the following remedies may be considered: (1) Install tank at a higher level above the pump; (2) increase liquid level; (3) change process conditions; (4) heat input into tank (e.g. by steam injection); (5) shorter suction pipe; (6) pump with higher suction specific speed; (7) reduce flow resistances.

11.7.3 Pump Intakes and Suction from Tanks with Free Liquid Level

Large flow rates drawn from surface waters usually are brought to the pumps through special intake structures in which screens may be integrated as necessary for taking solid matter out of the water. Recommendations for the layout of intakes with static or traveling screens are given in [12, 18].

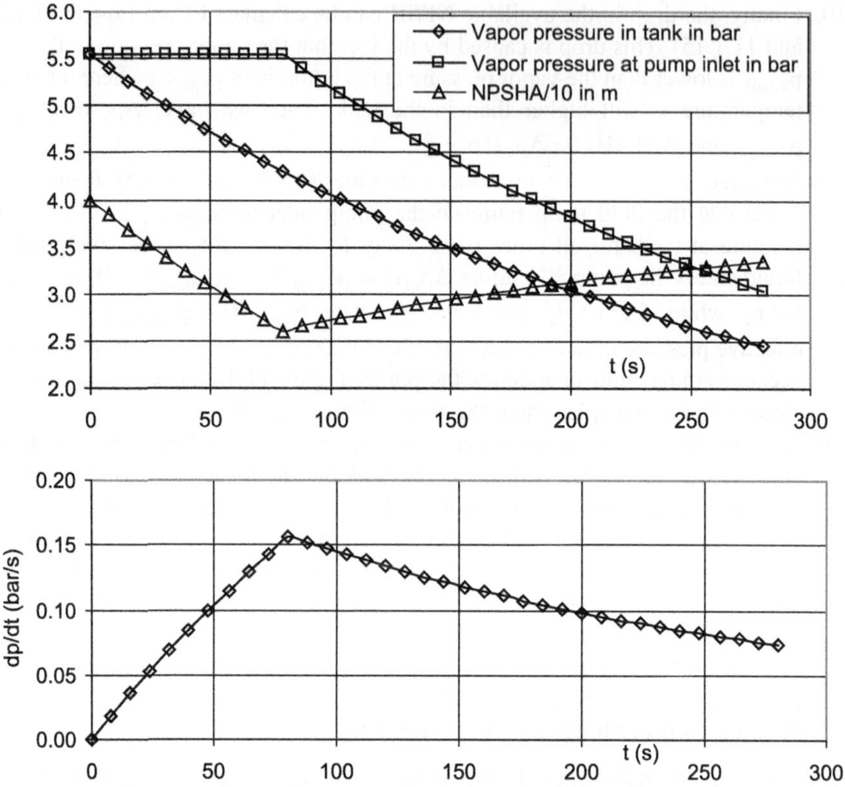

Fig. 11.18 Transient suction pressure decay, example with a transit time of 80 s

Applications are pumping stations for irrigation, cooling water or sewage plants for example. The task of the intake is to lead the water to the suction bell of the pump with as uniform an approach flow as possible.

If more than one pump is installed, each pump as rule draws water from a separate pit in order to reduce interference between parallel operating pumps. The lower end of the pump with the suction bell can either be submerged in the pit ("wet pit design") or the pump can draw the water from the pit via a short suction pipe ("dry pit design"). Depending on the layout of the plant, many different types of intakes and pit arrangements can be found. Model tests are frequently done for optimizing the intake. Information and design recommendations on intake structures are given for example in [N.4, N.10, 5, 12, 13, 20, 24, B.5, B.15, B.17]. An overview is provided in [20].

Large flow rates require pumps of high specific speed. Because of their large impeller inlet diameters, pumps with high n_q are particularly sensitive to non-uniform velocity profiles at the impeller inlet. Non-uniform approach flows in intakes with a free liquid level generate vortices which should be avoided by an appropriate design. Since intake structures of large plants imply a significant investment and since operational problems with the pumps may be very costly, this topic has been investigated extensively.

The basic flow features discussed in the following can be found not only in large intake bays but also in any vessel with a free liquid level from which a pump takes suction. The recommendations given in this chapter apply by analogy to all situations where a pump is fed from a reservoir, a pipeline or a channel with free liquid levels. Examples are de-aerators in a thermal power station, all sorts of suction tanks in the process industry or incompletely filled pipes from which a pump takes suction via a branched-off pipe.

Generation of vortices: *Small* vortices (turbulences) are characteristic for any smooth turbulent flow; such vortices do not jeopardize the safe operation. *Large* vortices form in shear flows, that is, in locations where strong velocity gradients prevail. Large vortices are inherent to stalled or strongly decelerated flow. According to Chap. 1.4.2, the pressure in a vortex drops from the outer zones to the core. A *vortex rope* is created if the fluid rotation is sufficiently strong to generate a gas- or vapor-filled core. On a free liquid surface the liquid levels in the vortex core drops because a constant gas pressure is imposed on the liquid level.

Vortex ropes lead to problems if they reach the suction bell or the outlet opening of the tank; they are called “air-drawing vortices”. The risk of such vortices increases with decreasing liquid level or submergence S of the outlet opening, Fig. 11.23. In the following we distinguish:

- The submergence S is the difference in height between the liquid level and the outlet opening or suction bell.
- The “*critical submergence*” S_{cr} is that liquid level at which the vortex just reaches the outlet opening.

Therefore $S > S_{cr}$ is a necessary condition for the safe operation of a pump. There are *surface vortices* which originate at the water level and *bottom vortices* or *wall vortices* (“submerged vortices”) which are completely submerged and originate on a structure, Fig. 11.19. The core of submerged vortices is filled with gas and vapor (according to the respective partial pressures). Vortices in intakes are generated by the following mechanisms:

- non-uniform approach flow
- asymmetrical arrangement of the inflow or outflow openings in a channel, a pit or an intake
- flow turning, for example when the water flows from a channel to the pit
- deceleration or acceleration of the flow
- flow around obstacles (screens, supporting structures, pumps installed in a row)
- flow over a recess.

The crucial factor is the fluid rotation (or swirl) generated by the above mechanisms. The risk of vortices thus increases with growing swirl. The critical submergence which is required to avoid vortices is directly proportional to the rotation, Eqs. (T11.2.6 to 11.2.8).

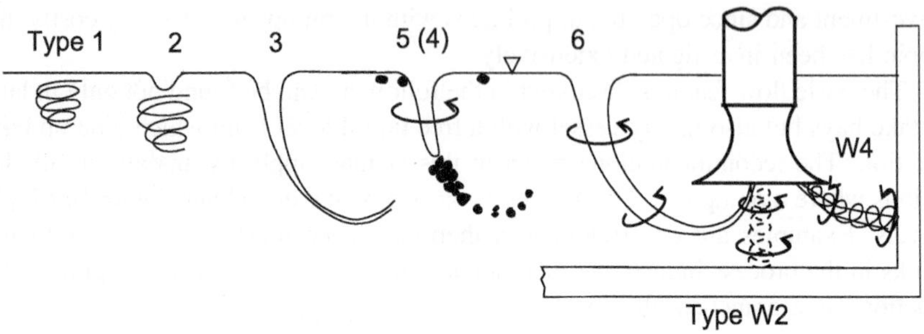

Fig. 11.19 Classification of free surface vortices and submerged vortices

Commonly the strength of a vortex is classified by the following criteria, refer to Fig. 11.19, [13, 20, 24]:

Free surface vortices:
Type 1 Swirl visible on liquid level
Type 2 Surface dimple
Type 3 Dye core to intake
Type 4 Vortex pulling trash (but not air) to intake
Type 5 Vortex pulling air bubbles to intake
Type 6 Vortex pulling air continuously to intake.
Submerged vortices:
Type W1 Weak swirl (no coherent core)
Type W2 Coherent dye core
Type W3 Organized air core (air separation)
Type W4 Vapor core (strong cavitation noise); rotation is sufficiently strong to reach the vapor pressure in the core.

Figure 11.19 schematically shows the vortices to be observed according to the above classification. Figure 11.20 shows the photograph of a surface vortex of type 6 as well as a submerged vortex of the type W4. Such vortices must not be allowed since they are likely to impair the head and/or reduce the flow rate. In addition, they generate unsteady forces which act on the impeller, the shaft and the bearings and cause vibrations. Cavitation and noise are another consequence.

As shown by photographs in [27] taken with a high-speed camera, the vortices are not necessarily destroyed by the impeller blades even though the vortex is cut by the blade. In the tested axial pump (n_q = 150) the vortex passed through the 6-bladed impeller into the diffuser. The vortex cut by the blade is just slightly shifted downstream on the suction surface of the blade as shown by Fig. 11.21. In the wake of the blade both parts of the cut vortex re-unite, albeit slightly disturbed. As can well be seen in Fig. 11.21 the vortex expands in the low-pressure region of the suction surface into a large cavitation zone.

Fig. 11.20 Surface vortex type 6 and bottom vortex type W4. Photo: Hydraulic Machinery Institute, Technical University Kaiserslautern

Fig. 11.21 Vortex passing through a propeller pump. Photo: Hydraulic Machinery Institute, Technical University Kaiserslautern, [27]

Presumably the blade overlap has an influence on the vortex passing through the pump but no information is available on the behavior of other (notably radial) impellers.

Effects of vortices on pump operation: Vortices reaching the suction bell or the suction pipe impair the safe operation of the pump through various mechanisms:[2]

1. Change of power, head or flow rate due to rotation induced at the impeller inlet (counter-rotation or pre-rotation).
2. Reduction in head, flow rate and efficiency due to ingress of air. Up to 2% of air has little impact on performance, but the pump is likely to de-prime at gas contents in the range of 5 to 10% (Chap. 13.2.4).
3. Air-drawing vortices are unsteady in nature even if the water level and the flow rate are constant, because the rotation in the vortices carried by the approach flow fluctuates. Vortices thereby excite vibrations and noise and induce head and power fluctuations.
4. Each cutting of a blade through a vortex implies a shock-like loading which can cause fractures of the blades or damage to the bearings.
5. Submerged bottom or wall vortices change their position periodically or stochastically. Since every blade cuts through the vortices, vibration frequencies are observed with $\nu \times z_{La} \times f_n$ of the blade passing frequency where ν *is no integer*. More or less steady surface vortices produce an excitation at $z_{La} \times f_n$ (Chap. 10.7.1). If submerged vortices are linked to cavitation, concrete structures can be damaged.
6. A one-sided approach flow to the impeller generates steady radial forces which induce alternating bending stresses in the shaft and may lead to shaft failure, [16], refer also to Fig. 9.32, Chap. 9.3.7. The radial force created by a given velocity profile on a high-specific speed impeller can be estimated from Chap. 9.3.7, Eq. (9.8).
7. Ingress of air into the pump causes noise and vibrations and increases the hydraulic excitation forces. When operating an axial test pump ($n_q = 150$) with vortices of type 6, the hydraulic forces on the impeller increased by a factor of 5 and the vibration velocity by a factor of 2 in comparison to operation without vortices. These tests were done at the best efficiency flow rate with a submergence of $S/D_T = 1.04$ [20]. The pressure pulsations linked to operation with air-drawing vortices can cause damage when exciting system components to vibrate, Chap. 10.12.
8. The hydraulic excitation forces induced by pre-rotation and vortices can lead to damage of various components in the pump or the system, including cracks in the column pipe and impeller blade fractures, [17].
9. Air inclusions in the discharge pipe can lead to operational problems and damage to the piping due to water hammer.

[2]The actual figures quoted in the following originate from tests reported in [12, 13, 20]. This information serves for illustration only and is scarcely generic.

10. If the approach flow is reasonably uniform, surface vortices of type 6 usually draw an amount of air which makes up less than 2% of the total flow. However, with extremely non-uniform approach flow, up to 18% air has been drawn into the pump as observed in [13]. If surface vortices draw an air flow of 3 to 6% of the total flow, the efficiency of the pump may be impaired by 7 to 20% according to [13].
11. If a rotation is induced in the pipe, the measurement of the flow rate can be impaired. A rotation of $\Delta\alpha = \arctan c_u/c_{ax} = 20°$ caused an error of 2% when measuring the flow rate with a nozzle of d/D = 0.6 as reported in [13]).

Empirical vortex data: No simple methods are available for predicting the circulation and vortices generated by non-uniform approach flow. It is also scarcely possible to extrapolate test data gathered on a specific intake geometry to configurations other than tested. Various experimental data can be found in [12, 13]. From these data some correlations have been derived as given in Tables 11.2 and 11.3

Table 11.2 Critical submergence of suction bells

					Eq.
Air drawing vortices are expected if the submergence is $S < S_{cr}$ D corresponds to the suction bell diameter D_T or to the inner pipe diameter if no suction bell is installed.			$Fr = \frac{c}{\sqrt{g\,D}}$		Eq.
Intake with submerged suction bell. Criterion: vortex type 4	$\left(\frac{S}{D}\right)_{cr} = 0.8 + 1.7\,Fr$	Different distances to rear wall or bottom [24] $0.3 < Fr < 0.5$			11.2.1
	$\left(\frac{S}{D}\right)_{cr} = 1.35\,Fr$	Different distances to rear wall or bottom [12] $0.1 < Fr < 1.2$			11.2.2
Symmetric approach flow	$\left(\frac{S}{D}\right)_{cr} = 1.7\,Fr$	Observed in executed intake plants with horizontal pipe; quoted in [12]	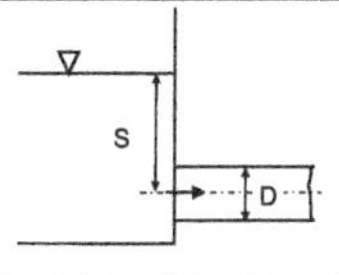		11.2.3
Asymmetrical approach flow	$\left(\frac{S}{D}\right)_{cr} = 2.3\,Fr$				11.2.4
90°-branching a suction pipe off a header with a free liquid level	$\left(\frac{S}{D}\right)_{cr} = 1.29\sqrt{Fr}$	Diameter ratio $D_T/D_1 = 0.3$ $0.4 < Fr < 3$ [3]	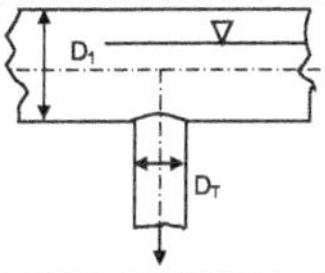		11.2.5
	$\left(\frac{S}{D}\right)_{cr} = 0.8 + 0.5\,Fr$				
Critical submergence, if swirl is known	$C_D = c_{u,k}\,r_k = \frac{\Gamma}{2\,\pi}$	Γ = circulation C_D = swirl constant r_k, $c_{u,k}$ = referred to vortex core (Chap. 1.4.3)			11.2.6
	$\left(\frac{S}{D}\right)_{cr} = k_{sc}\frac{C_D}{D\sqrt{g}}$	Outflow direction of pipe	φ	k_{sc}	11.2.7
		Downwards	0	110	
	$k_{sc} = \frac{110}{1+\frac{\varphi}{2\pi}}$	Horizontal	$\pi/2$	90	11.2.8
		Upwards	π	75	

Table 11.3 Empirical data for air-drawing surface vortices[a]			Eq.
Gas volume fraction of entrained air (pump intake), [13]	$\alpha = \frac{Q''}{Q' + Q''} = a_1(Fr - 0.4)$	Upper envelope of data: $a_1 = 0.1$ approx. average: $a_1 = 0.03$	11.3.1
Swirl (angular momentum)	$K = \rho\, \pi\, R^3\, c_{ax}\, c_u$	$R = D/2$ = pipe radius	11.3.2
Momentum of the axial flow	$I_{ax} = \rho\, Q\, c_{ax} = \rho\, \pi\, R^2\, c_{ax}{}^2$	$c_{ax} = 4\, Q/(\pi\, D^2)$	11.3.3
Dissipation of swirl in a pipe section of length Δx, [13]	$\beta = 0.15 \exp\left(-7.43 \frac{K_o}{D\, I_{ax}}\right)$	At $Re = (1.4 \text{ to } 4) \times 10^5$	11.3.4
	$\frac{K}{K_o} = \frac{c_u}{c_{u,o}} = \exp\left(-\beta \frac{\Delta x}{D}\right)$	$\frac{K}{K_o} = \frac{c_u}{c_{u,o}}$ with D = constant	11.3.5
Pressure drop increase caused by swirl, [13]	$\frac{\Delta H_v(c_u > 0)}{\Delta H_v(c_u = 0)} = 4.3 \frac{c_{u,av}}{c_{ax}}$	At $Re = 2 \times 10^5$ depends on Re and roughness	11.3.6
Relation between core radius and vortex depth h, [13]	$\frac{r_k}{D} = \frac{1}{k_{sc}} \sqrt{\frac{h}{D}}$	At the critical submergence. When vortex reaches the suction bell, it is $h = S_{cr}$	11.3.7
k_{sc} from Eq. (T11.2.8)	$\frac{r_k}{D} = 0.0109 \sqrt{\frac{h}{D}} + 1.45$	From tests with outflow in horizontal direction	11.3.8

[a]The correlations in Table 11.3 originate from measurements in specific test stands. Information on the relevance of these correlations in practice is not is available.

which can be helpful for a quantitative assessment of situations where vortices may present a problem. Notably Table 11.2 provides equations for the critical submergence in various geometries. If the circulation is known, its influence can be estimated from Eqs. (T11.2.6 to 11.2.8).

Table 11.3 provides empirical data on air-drawing vortices which can jeopardize the safe operation of the pump as discussed above. In [5] the volumetric air content was plotted against the Froude number; these statistical data can be described by Eq. (T11.3.1). If it is assumed that these data apply to the critical submergence, Eq. (T11.3.1) can be used for a rough estimation of the amount of gas entrained at the critical submergence. Observing or estimating the diameter of the vortex in the outlet opening, the air flow rate can be estimated on the assumption that the air has roughly the same axial velocity as the water. Thus the volume fraction α of the gas becomes $\alpha = (d_k/D)^2$ if d_k is the vortex diameter and D the pipe diameter.

A swirl induced in a long pipe slowly decays due to friction and turbulent dissipation. This effect can be estimated from Eqs. (T11.3.2 to 11.3.5). The pressure drop in the pipe rises when the ratio of the swirl momentum to the axial momentum (respectively the ratio c_u/c_{ax}) is increased; this effect can be estimated from Eq. (T11.3.6). The analysis of test data made in [13] revealed that the ratio of the radius of the vortex core to the length of the vortex is constant. This observation allows estimating the radius of the vortex core for the critical submergence at which the vortex length equals by definition the water depth above the outlet opening.

Thus h = S_{cr} is determined from Table 11.2 and the radius of the vortex core is obtained from Eqs. (T11.3.7 or 11.3.8) using k_{sc} from Eq. (T11.2.8). An increase of the vortex core radius is to be expected if the submergence drops below the critical value (i.e. $S < S_{cr}$).

Criteria for the quality of the approach flow: Not only vortices but any kind of non-uniform flow can impair performance and induce vibrations and unsteady forces. The approach flow to a large pump of high specific speed should therefore fulfill the following quality criteria:

- Non-uniformities in the axial approach velocity profile at the impeller inlet should be limited to a maximum of ±10% of the average flow velocity. On an annular section, it should even be limited to ±5%. Moreover, the non-uniformity should not exceed ±5% for applications with $n_q > 200$, [B.5].
- The maximum deviation in the approach flow angle should be limited to $\Delta\alpha = \pm 5°$; with $\Delta\alpha = \arctan c_u/c_{ax}$.
- The strength of surface vortices should be limited to type 2 (surface dimple), since type 3 vortices induce a defined swirl into the suction bell or suction pipe.
- The strength of bottom and wall vortices should be limited to type W1 since vortices of type W2 carry a defined swirl into the outlet opening.
- No discernable unsteadiness and waves in the approach flow
- No vortices which reach the suction bell
- No air entrainment.

Model tests: Because of the high investment costs of intake structures, the risks presented by design deficiencies and the large variety of plant concepts, model tests are often performed for the optimization of intakes. Since flows with free liquid levels are involved, the influence of gravity must be modeled correctly. It is therefore imperative to maintain the same Froude number $Fr = c_T/(g \times D)^{0.5}$ in the model and in the plant. It is safe to enlarge the test range to cover the condition $c_{model} = c_{plant}$. Since the Reynolds number in the plant usually is much higher than in the model it is common practice to observe only the Froude similarity. According to [N.10] viscous effects are not expected to falsify the test results if the conditions $Re > 6 \times 10^4$ and $We > 240$ are respected. Thereby, the influence of the surface tension is assessed by means of the Weber number defined by $We = c_T(\rho \times D_T/S_T)^{0.5}$. It represents the ratio of inertia effects to surface tension forces. However, severe damage to pumps can occur due to vortices and non-uniform approach flow even if model tests have been performed successfully according to the above recommendations. For a detailed discussion of vibrations of vertical pumps see Chap. 10.14.

The geometrical scaling ratio between plant and model should *not* be larger than 15.

Guidelines for the design of intake structures: From the above discussion, notably Table 11.2 and 11.3, it follows that problems linked to vortices increase with the Froude number or the flow velocity. Problems caused by the approach flow are therefore mostly observed at high flow rates $q^* > 0.8$ while partload operation is rather dominated by the effects of recirculation at the impeller inlet. The interaction between the recirculating fluid and the structures and the liquid volume in the intake is strong.

The basic features may be discussed with reference to the wet-pit installation shown in Fig. 11.23, [B.17]. The vertical pump draws the water through a suction bell of diameter D_T. In order to achieve a uniform approach flow free of vortices, the following guidelines and criteria should be considered:

1. Flow velocity c_T at the suction bell inlet: For flow rates above 1 m^3/s a flow velocity of $c_T = 1.7$ m/s (range: 1.2 to 2.1 m/s) is recommended at the inlet to the suction bell. Below 0.3 m^3/s a range of $c_T = 0.6$ to 2.7 m/s is allowed, [N.10]. If the suction bell diameter is calculated for the BEP with $c_T = 1.7$ m/s the upper limit of $c_T = 2.1$ m/s is reached at about $q^* = 1.25$. Thus reasonably good flow conditions may be expected even at maximum flow. Typically, the ratio between suction bell inlet and impeller inlet diameter is in the range of $D_T/d_1 = 1.75 \pm 15\,\%$ ([17] gives 1.75).

2. Submergence: The suction bell must be covered by a sufficient amount of liquid, the "submergence" S, in order to avoid *air-drawing vortices* which start from the water surface. In addition, the submergence must be sufficient to guarantee the needed $NPSH_A$. The risk of wall vortices or bottom vortices equally increases when the liquid level is lowered because the static pressure, as given by the water depth, decreases. Less rotation is then sufficient to reduce the pressure in the vortex core to a level where gas separates or vapor is created. The required submergence depends on the design and the quality of the approach flow. If the guidelines for intake design given below are observed, a reasonably uniform approach flow can be expected and the submergence for uncovered pits *without curtain wall* (Fig. 11.23) can be calculated from Eq. (11.15), [N.10]:

$$\frac{S}{D_T} \geq 1 + 2.3\,Fr \quad \text{with Froude number } Fr = \frac{c_T}{\sqrt{g\,D_T}} \quad \text{and} \quad c_T = \frac{4Q}{\pi D_T^2} \tag{11.15}$$

$$\text{or:} \quad S = D_T + \frac{9.2Q}{\pi {D_T}^{1.5}\sqrt{g}} \tag{11.15a}$$

Inspection of Eq. (11.15a) shows that there must be an optimum suction bell diameter at which the submergence - hence the costs for civil engineering - become minimum. Differentiating Eq. (11.15a) with respect to D_T and setting $\partial S/\partial D_T = 0$ yields the optimum suction bell diameter $D_{T,opt}$ which gives the minimum submergence:

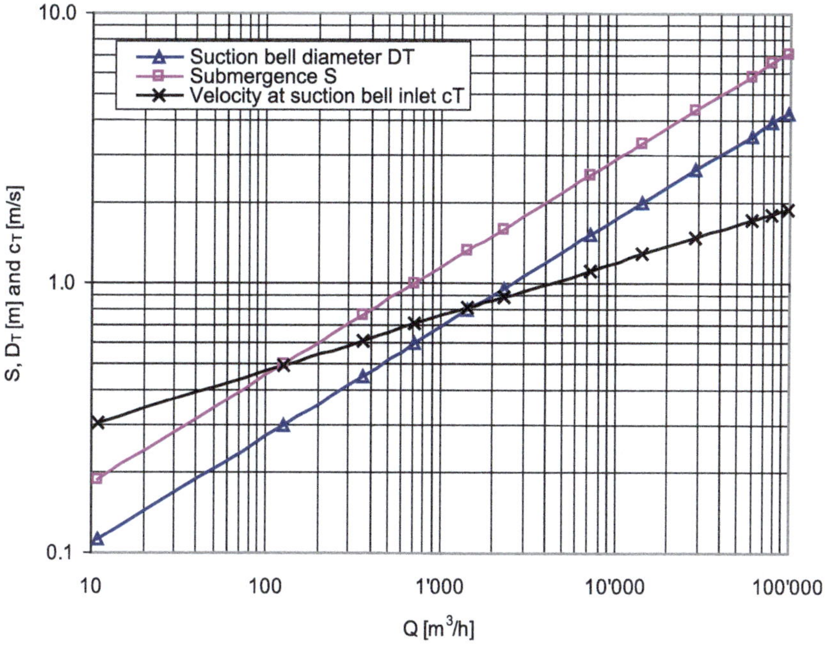

Fig. 11.22 Minimum submergence and optimum suction bell diameter D_T for a layout according to Eqs. (11.15, 11.15b)

$$D_{T,opt} = 1.81\left\{\frac{Q}{\sqrt{g}}\right\}^{0.4} \quad \text{and} \quad c_{T,opt} = 0.39\,Q^{0.2}\,g^{0.4} \qquad (11.15b)$$

The optimum suction bell diameter is linked to the constant values of $Fr_{opt} = 0.29$ and $S/D_{T,opt} = 1.67$.[3] The optimum values resulting from Eqs. (11.15 and 11.15b) for submergence, suction bell diameter and flow velocity c_T are plotted in Fig. 11.22 against the flow rate.

Measurements with intakes reported in [24] demonstrate that submergences according to Eq. (11.15) are plainly sufficient to suppress severe vortices. Only vortices of type 2 - surface dimples - were observed. These did not impair pump operation. An analysis of plants with vortex problems showed that the submergence was below $S_T/D_T = 1.5$ in nearly all of the problem cases, [14].

The submergence according to Eq. (11.15) is applicable to intakes of the type shown in Fig. 11.23, but *without* splitters on the floor or on the rear wall and without a curtain wall or other devices to suppress swirl. Equation (11.15) further represents the upper envelope of the data available in [13]. Intakes designed

[3]At extremely high flow rates in excess of $Q = 16.7\ m^3/s = 60{,}000\ m^3/h$ the limit of the recommended velocity $c_T = 1.7$ m/s is exceeded.

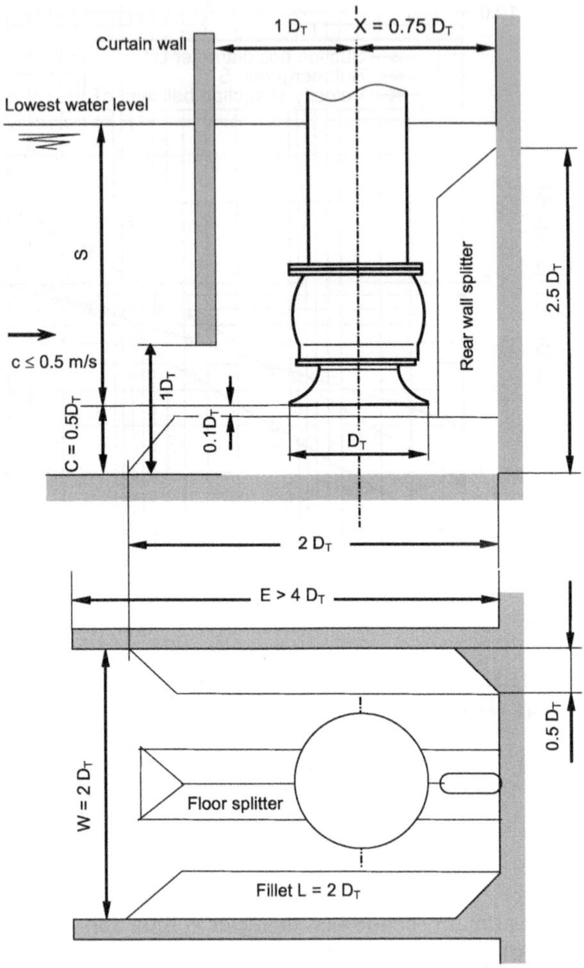

Fig. 11.23 Intake chamber for wet pit pump; very often without curtain wall (preferred)

according to Eq. (11.15) therefore contain a certain margin if the approach flow is reasonably uniform. Consequently, the submergence can be selected 10 to 25% lower than stipulated by Eq. (11.15) and Fig. 11.22 if the intake is designed carefully with splitters on the floor and the rear wall according to Fig. 11.23.

A number of measures are available for reducing the required submergence and the costs for civil engineering:

- A covered intake according to Fig. 11.26 or a curtain wall as per Fig. 11.23 reduces the required submergence by roughly 30% compared to Eq. (11.15). However, the approach flow velocity in the sump increases in this type of design. Severe damage to the pumps can be the consequence as shown in chapter 10.14 which reports on several case histories.

- An annular curtain wall (i.e. a cylinder around the pump) reduces the submergence S by about 50%. However a minimum submergence of $S/D_T = 1.0$ must always be ensured.
- Also very effective for reducing the required submergence is a plate, a grid or a perforated sheet. These horizontal baffles should be arranged a little below the minimum liquid level (100 to 200 mm) since they are ineffective if covered by too much water.
- Floating baffles (rafts) are able to suppress air-drawing vortices. Such devices may be considered for retrofitting an intake structure which present vortex problems.

The design submergence according to Eq. (11.15) must be greater than the *critical* submergence at which the vortices reach the outlet opening.

3. Approach flow velocity in the intake: Former design rules given in [N.4] and [B.5] limited the approach flow velocity to 0.3 m/s. These conservative designs often led to excessively large and expensive intake structures. Therefore, the rules were amended by the Hydraulic Institute [N.10] and the recommended limit for the approach flow velocity is now 0.5 m/s. This value is used in the following. If the limit of 0.5 m/s is exceeded the stagnation pressure on the rear wall induces unfavorable lateral velocity components and vortices.

Higher approach flow velocities are required in sewage plants in order to avoid deposits of solids; c = 0.7 to 2 m/s is mentioned in [20].

The static and dynamic forces created by high approach flow velocities may lead to vibration problems and component failure as discussed under item 14 and in chapter 10.14.

4. Chamber dimensions (Fig. 11.23): In order to prevent vortex formation certain proportions of the pump chamber must be selected. The following recommendations can be derived from the quoted literature (preferred dimensions are printed in bold characters):

- Chamber length: $\mathbf{E/D_T > 4}$
- Chamber width: $\mathbf{W/D_T = 2}$ (up to 2.5); widths up to $W/D_T = 3$ are often considered acceptable but such designs are clearly inferior to $W/D_T = 2$; this value should be exceeded as little as possible.
- Bottom clearance: $\mathbf{C/D_T = 0.4}$ **to 0.5** (0.3 to 0.75). Too low a bottom clearance leads to bottom vortices.
- Rear wall clearance: $\mathbf{X/D_T = 0.75}$ **to 1.0**. Too low clearances harbor the risk of bottom or wall vortices; too large clearances induce unfavorable flows around the pump body and thus lead to vortex streets and vortices.
- The corners between side walls and rear walls should have a 45°-fillet of the size $0.2 \times W$ as depicted in Fig. 11.23, or fillet radii with $r = 0.2 \times W$.

During the tests reported in [24] the combinations $C/D_T = 0.4$ with $X/D_T = 0.75$ and $C/D_T = 0.5$ with $X/D_T = 0.75$ proved best (lowest vortex formation) and the combination $C/D_T = 0.4$ with $X/D_T = 1.0$ was the most unfavorable.

5. Swirl breaking structures serve for reducing the risk of bottom vortices. During partload recirculation they prevent excessive rotation in the sump. This can be achieved by a splitter (with fillets) as in Fig. 11.23 or a cone with ribs as in [B.5]. These elements are installed below the suction bell. If the bottom clearance is too large, ribs or a cross can be cast or welded in the suction bell. Without swirl-breaking structures there is a risk of excessive vibrations. In addition, the shut-off head is reduced by excessive pre-rotation which can at times pose problems during start-up of the pump, Chap. 11.8.

If a cone is used, only one splitter oriented in the flow direction is installed below the suction bell. Any rib or splitter perpendicular to the flow would cause flow separation and vortices, [27]. With large flow rates ($q^* > 1$) a vortex can form at the apex of the cone if the longitudinal splitter is omitted, as observed in [27].

6. Bottom vortices are generated by flow separation or strong velocity gradients in the approach flow. Screens, flow deflections, support structures, sudden changes in cross section or diffusers with excessive opening angles can generate bottom vortices in this way.

7. Flow straightening by screens (Chap. 1.8): If the flow in the intake is non-uniform because of design constraints, the flow can be smoothed out by screens installed upstream of the pump chamber or even within the chamber itself, Fig. 11.24. Such flow straightening can be accomplished for example by: (1) perforated walls (pigeon-holed brick work); (2) perforated sheets or grids; (3) stakes or profiles. If the approach flow is strongly non-uniform, two or more screens in a row may be installed. The blockage ratio of the screen (or the sum of the areas of the openings to the frontal surface of the screen) can be selected around 50 to 70%. The screen should have a sufficient distance L from the pump in order to allow the vortices created by the screen to decay, Fig. 11.24. This length should be $L/D_h > 20$ if Dh is the hydraulic diameter of the openings in the screen, refer to Tables 1.2 and 1.3.

8. Flow deflections: An approach flow perpendicular to the pump chamber as shown in Fig. 11.24 generates vortices due to the 90°-flow turning und must therefore be avoided. The pump chambers should be arranged at an angle of 45° to the approach flow channel. The lengths of the chambers should be at least $L = 8 \times D_T$. Literature [16] reports on model tests carried out to solve vibration and cavitation problems caused by an unfavorable intake design with a 90°-flow turning. During these tests the flow could be drastically improved by bottom and rear wall splitters and perforated walls even though the submergence was below $S/D_T = 1.0$.

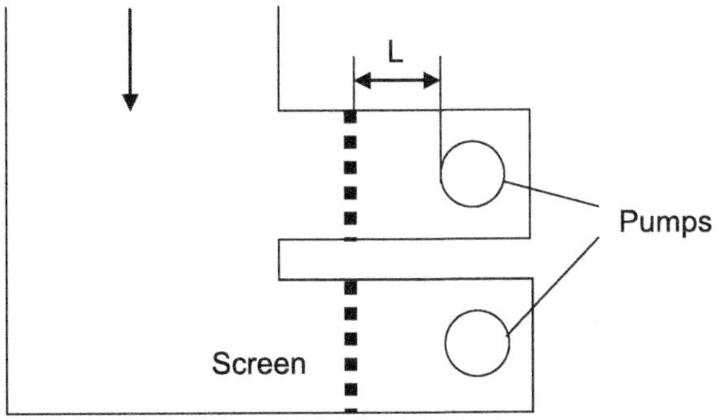

Fig. 11.24 Unfavorable approach flow conditions due to 90°-turn are improved by screens which provide a suitable head loss

9. Sloped walls which decelerate or accelerate the flow should not exceed a maximum angle of 10° (15°) in order to prevent high velocity gradients which would induce vortices.

10. Pumps operating in parallel must not interfere with each other. This can be avoided by: (1) Installing the pumps in separate chambers if they are arranged side-by-side in a plane perpendicular to the direction of the approach flow (in Fig. 11.24). (2) If the pumps are arranged in a row in the main stream, by providing large distances between the pumps and by keeping flow velocities low as given in item 11 below. (3) Installing annular curtains around the pump units.

11. Pumps installed in a row: If several pumps are arranged in flow direction in a channel, the flow around the pump bodies is likely to generate vortex streets, Chap. 10.12.4. To reduce the interaction of vortices and distorted flow with downstream pumps, the distance x_a between the pump rotor axis' should be at least $x_a = (2 + 20 \times c/c_{ref}) \times D_T$, where c is the flow velocity in the channel and $c_{ref} = 1$ m/s. To reduce excitation by vortex streets shed from the pump bodies the flow velocity should not exceed 0.3 m/s.

12. Air entrainment: The injection of the fluid into the suction reservoir or inlet bay must not be arranged above the liquid level in order to prevent air entrainment. Feeding an intake over a weir is unsuitable for the same reason. This rule also applies to all sorts of tanks with a free liquid level in order to prevent entrainment of gas or vapor into the suction pipe. Any reservoir should be designed so as to allow the entering fluid to tranquillize and mix well with the stored liquid. A short-cut between inlet pipe and outlet opening may create a swirl which could enter the suction pipe.

13. Dry pit installations: Large pumps are usually fed via a bend which provides an appropriate acceleration to the fluid. The approach flow velocity in the chamber should not exceed 0.5 m/s and the velocity at the inlet to the bend should be below 1.7 m/s. To avoid air entrainment, a sufficient submergence must be provided. It can be determined from Eq. (11.15c) giving results similar to [B.17].

$$\frac{S}{D_T} \geq 0.2 + 1.5\frac{c_T - c}{\sqrt{g\, D_T}} \quad \text{with} \quad c_T = \frac{4Q}{\pi\, D_T^2} \quad \text{or} \quad c_T = \frac{Q}{A_T} \tag{11.15c}$$

c = flow velocity in intake chamber; A_T = inlet cross section of (rectangular) bend. The inlet to the bend or suction pipe should be rounded with R/D > 0.1 to 0.2 in order to avoid jet contraction, vortices and air entrainment, see also Table 1.4.

14. Covered intake chambers (Fig. 11.25): The height of the chamber should be about $H \approx 1 \times D_T$ and/or dimensioned so that the flow velocity at the inlet to the chamber does not exceed 0.7 m/s. If the height is too small, fluid acceleration is too strong, and unfavorable cross flows are created. These would reinforce any swirl or asymmetries in the approach flow due to the conservation of angular momentum. Covered chambers therefore should always be equipped with floor and rear splitters to prevent cross flows which could induce vortices at the bottom or the walls. If the fluid acceleration in the covered chamber is too strong, the water hits the rear wall in a way that generates vortices. Fillets in the corners according to Fig. 11.23 reduce or prevent vortex generation.

In some compact intake designs the traveling screen discharges the water with high velocities into the intake chamber. The approach velocity to the pump can reach c = 2.5 m/s. The water hitting the pump body creates static and dynamic forces which are proportional to the stagnation pressure of the approach velocity, see Chap. 10.14.2.3.

15. Traveling screens and trash racks: Devices to prevent debris from entering the pump will increase the flow velocities due to the reduction in flow area. A distance of $L = 6 \times D_T$ between the outlet of screens or racks and the pump is recommended in [N.10] to allow the jet created by such devices to dissipate, see also Table 1.2. By a careful intake design with floor and rear wall splitters as well as fillets in the corners the required distance can be reduced to $L = 3 \times D_T$. This can significantly reduce the costs of civil engineering, [27]. The flow velocity at the outlet of screens or racks should not exceed 0.7 m/s.

The most important measures for a good intake design (and for retrofitting in case of trouble) are summarized below:

- Uniform symmetric approach flow
- Adequate submergence
- Chamber dimensions according to the guidelines given above
- Floor and rear wall splitters
- Corner fillets.

16. Suction bell design: The height h of the suction bell should be selected to about h/d_{1a} = 0.7. The lower rim of the suction bell should be horizontal, and the transition from horizontal to vertical should be made with a low curvature in order to avoid flow separation. Figure 11.25-1 shows a shape which is thought to give good flow conditions. The shape in Fig. 11.25-1 is calculated as clothoide which results in a change in curvature proportional to the length of the flow path. An elliptical

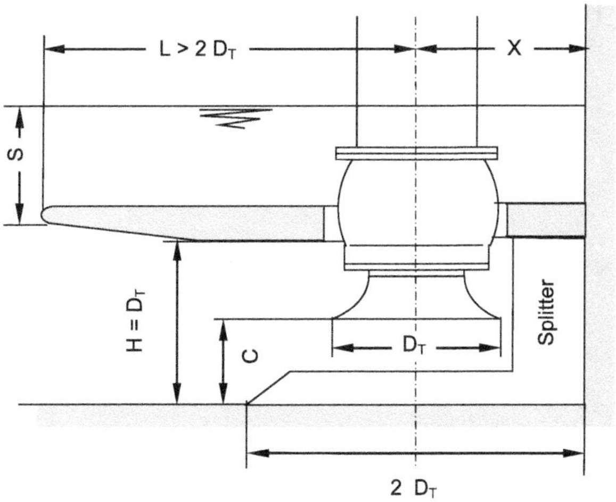

Fig. 11.25 Covered intake chamber

Fig. 11.25-1 Suction bell as clothoide with $D_T/d_{1a} = 1.8$ and height $h/d_{1a} = 0.72$

curve would have a similar shape. In contrast, the suction bell in Fig. 2.15 seems not optimum because the horizontal rim is too small and the transition from said rim to vertical is too sharp. Welded suction bells should have a generous horizontal rim plus a minimum of 3 conical sections which follow the shape in Fig. 11.25-1 as closely as possible.

11.7.4 Can Pumps

Vertical pumps in a suction can (or barrel) according to Fig. 11.26 are employed for example as condensate pumps in power stations or as process pumps in refinery service. To avoid unfavorable approach flow to the impeller and associated operational trouble the following recommendations can be given:

- Flashing of gas or vapor at the highest point above the inlet nozzle should be avoided. A free liquid level in the can should not be allowed. Implement proper procedures for filling and venting.
- The axial flow velocity c_{ax} between the suction can and the pump body should be as a rule below 1 to 1.5 m/s.
- The flow velocity in the inlet nozzle should be small ($c_s < 2$ m/s) because the non-uniformity of the approach flow to the impeller increases with the dynamic

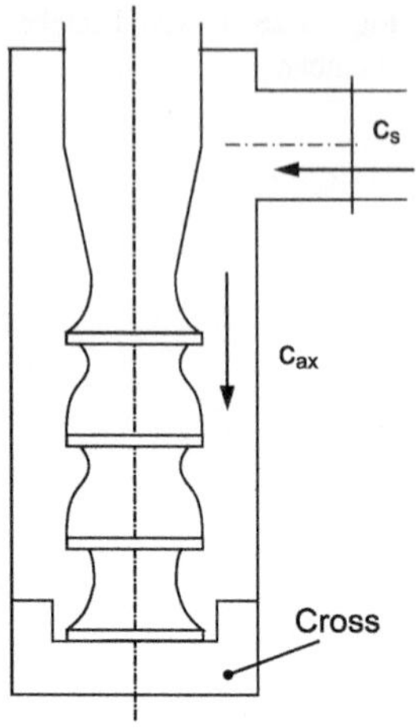

Fig. 11.26 Approach flow to can pumps

head in the inlet nozzle. A large distance between suction bell and inlet nozzle helps to smooth out the flow.

- Distance between suction bell and can bottom equals $^1/_2 D_T$.
- A cross at the bottom of the tank (as in Fig. 11.26), ribs in the suction bell, a suction strainer or other structures apt to prevent swirl are required. Otherwise, partload recirculation would induce a strong fluid rotation in the tank and generate excessive vibrations and alternating stresses (leading, for example, to fatigue cracking in the column pipe). Also the shut-off head and partload characteristics would be affected by the omission of swirl-breaking structures. These are therefore mandatory - in particular at high specific speeds.

11.8 Discharge Piping

One aspect of discharge piping design is a trade-off between investment and energy costs - provided that the pressure losses in the discharge pipe make up a significant portion of the head. If this is true, the energy costs diminish with increasing pipe diameter due to the lower head required. Additionally, pump costs may decrease due to the lower head. Investment costs for the piping, of course, increase with larger pipe diameters (the more so, the higher the design pressure).

The flow velocities can vary in a wide range: In long pipelines of several kilometers length they may be limited to 1.5 to 3 m/s with respect to the pressure drop; in short discharge pipes a maximum of 5 m/s may be used, but in very short high-pressure piping 10 to 20 m/s may be chosen in view of the investment cost (provided that an appropriate pipe material is selected, Chap. 14). The medium pumped has an influence too: for example it may be necessary to limit the velocity if there is a risk of erosion corrosion or if solid particles are present in the fluid.

Conversely, a minimum velocity must be maintained in slurry pumping if there is a risk of solids settling in the pipe, Chap. 13.4.

Siphons are discharge pipes in which energy is recovered in a pipe section with a down-slope. A siphon can also be installed in situations where it is absolutely mandatory to prevent back-flow after pump shut-down, for example when pumping over a dike where a failure of the check valve would have disastrous consequences. Typical applications with siphons are irrigation plants, de-watering (drainage) schemes with pumping over a dam, or cooling water systems in thermal power stations. Figure 11.27 depicts schematically a system with a siphon in which water is transported to a high point (defined by $H_{stat,F}$) in the discharge pipe. From the high point the water flows down to an upper basin which is by $H_{stat,D}$ at a higher elevation than the reservoir from which the pump takes suction. The recovered head is given by the difference between the high point and the discharge water level; i.e. by $z_s = H_{stat,F} - H_{stat,D}$. Figure 11.27 also shows the profile of the static pressure in the piping system. The head H_A required by the system can be calculated by applying the Bernoulli equation between the points #0 and #2, Eq. (1.7). Likewise, the minimum pressure p_{min} at the high point is obtained by writing Eq. (1.7) for the points #1 and #2:

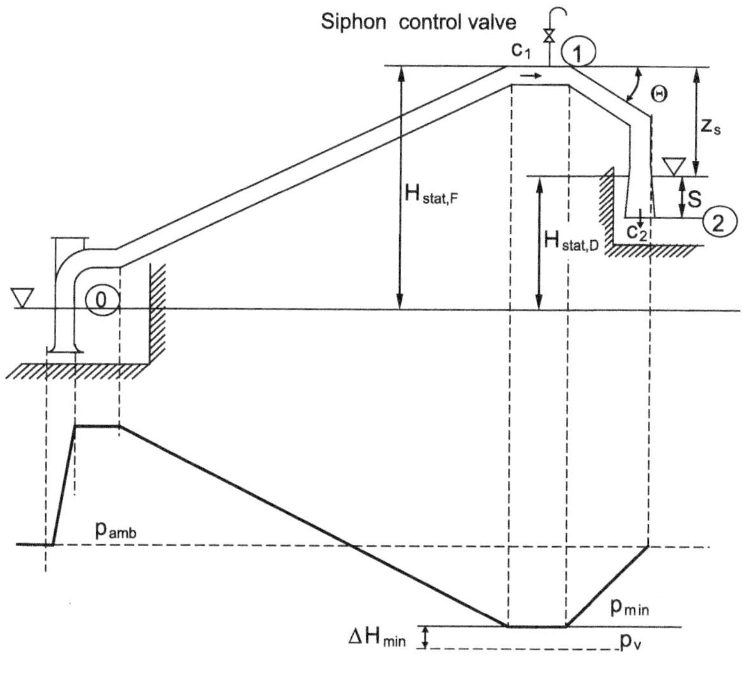

Fig. 11.27 Pumping station with a siphon

$$H_A = H_{stat,D} + \frac{c_2^2}{2g} + \Sigma H_{v,0\text{-}2} \tag{11.16}$$

$$p_{min} = p_{amb} - \rho\, g\, z_s - \frac{\rho}{2}(c_1^2 - c_2^2) + \Delta p_{v,1\text{-}2} \tag{11.17}$$

In order to prevent separation of the liquid column, the pressure p_{min} at the high point must at all times remain sufficiently above the saturation pressure p_v. A head margin of $\Delta H_M = (p_{min} - p_v)/(\rho \times g) = 1.5$ to 2.5 m is recommended.

The upper boundary should be used if the water is saturated with air in order to limit air separation in the zone where the static pressure is below atmospheric. If the pressure losses $\sum H_{v.1\text{-}2}$ between points #1 and #2 are high and not well known, a higher margin is required. The maximum allowable siphon height can be obtained from Eq. (11.17) for a selected margin ΔH_M:

$$z_{s,max} = \frac{p_{amb} - p_v}{\rho\, g} - \Delta H_M + \Sigma H_{v,1-2} - \frac{c_1^2 - c_2^2}{2g} \tag{11.18}$$

If the head difference given by the system exceeds the maximum allowable siphon height $z_{s,max}$, the excess head can be destroyed by throttling, i.e. by increasing $\sum H_{v.1\text{-}2}$. An alternative (as used for example in cooling water systems in power stations) is to install an open intermediate basin at a level which corresponds to the maximum siphon height. The siphon discharges into this basin with the pipe outlet well submerged. Subsequently the water flows over a weir out of the intermediate basin to level #2 (which could be a river or a lake for example).

The siphon can function only if: (1) its outlet opening is well submerged at all times, (2) the pipe is properly vented, (3) the system is vacuum-tight. The siphon could be started with the help of a vacuum pump, but this option is expensive. Starting with an empty pipe is possible if the flow velocities are sufficiently high to be able to sweep out the air. According to [B.18] the minimum flow velocity to accomplish this is given by Eq. (11.19):

$$c_{min} = 1.2\sqrt{g\, D \sin \Theta} \tag{11.19}$$

If the pump is started with an empty pipe, the trapped air is first compressed until the pressure is increased by $\Delta p = \rho \times g \times S$ (S = submergence according to Fig. 11.27). Subsequently the air is pushed out. The start-up procedure begins at point A in Fig. 11.28a (the pump starts running in full cavitation). As the pipe is being filled, the water level rises and the operation point shifts accordingly to lower flows via point B to C till the value $H_{stat,F}$ is reached at the high point. The flow rate corresponding to $H_{stat,F}$ in point C (in Fig. 11.28 labeled Q_c) is relevant for the calculation of c_{min} from Eq. (11.19).

If the power consumption rises towards shut-off, the motor must be selected for the highest power required during start-up. As soon as the air has been swept out of the siphon and a closed water column is established in the pipe section with the down-slope, the siphon becomes effective and the operation point shifts from C to B.

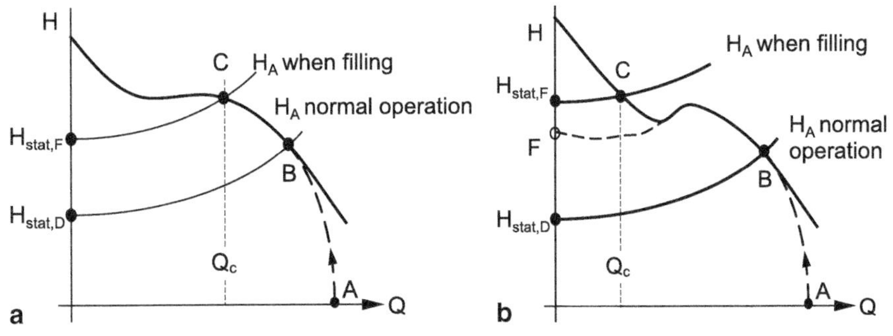

Fig. 11.28 Start-up of a pump in a siphon system. **a** High point C is reached, **b** with unstable Q-H-curve the pump hangs up at point F (Q = 0), only part of the pipe is filled

The pump must be selected so as to ensure that the high point C can be reached and that the flow rate is sufficient to sweep out the air. Preferably the high point should be outside any unstable portion of the Q-H-curve. If this is not the case (as in Fig. 11.28b), the pump is only able to traverse the unstable region if the Q-H-curve is sufficiently steep. The high point C is reached in this case. In contrast, if the Q-H-curve is flat (as depicted by the broken line to point F in Fig. 11.28b), it is impossible to reach point C. The water level stops at level F at Q = 0 below the high point. Then it is impossible to start the flow through the siphon. This problem can occur in the absence of swirl-breaking structures in front of the impeller inlet. The differences between the arrangement during the works test and the installation of the pump in the plant need to be considered; the parameters influencing the shut-off head are discussed in Chap. 5.6.8.

An automatic valve is installed at the high point which lets air into the pipe when the pump is shut down. Thus a vacuum is avoided and the water is prevented from flowing back to the pump from the upper basin. Only the water contained in the piping upstream of the high point flows back through the pump which may then for a short time operate as a turbine. Pump and driver must be selected accordingly and it must be ensured that the motor cannot be started prior to a complete stop of the rotor. Because of possible air accumulations in a system with a siphon, there is a risk of separation of the liquid column and subsequent water hammer. Siphons thus need to be analyzed carefully in this respect.

References

1. Barrand, J.P., Picavet, A.: Qualitative flow visualizations during fast start-up of centrifugal pumps. In: IAHR Symposium Valencia, pp. 671–680 (1996)
2. Bross, S.: Entwicklung neuer Schaufelgitter aus Profilen variabler Geometrie zum Einsatz in Leiträdern drallgeregelter Turbomaschinen. Dissertation TU Braunschweig, ZLR-Forschungsbericht 93–10 (1993)

3. Chang, K.S., Lee, D.J.: Experimental investigation of the air entrainment in the shut-down cooling system during mid-loop operation. Ann. Nucl. Energy **22**(9), 611–619 (1995)
4. Chaudhry, M.H.: Applied Hydraulic Transients, 2nd edn. Van Nostrand Reinhold, New York (1987)
5. De Vries, M., Simon, A.: Suctions effects on feedpump performance; a literature survey. EPRI Report CS-4204, Aug 1985
6. Dues, M.: Experimentelle Untersuchung der Interferenz zwischen Leitrad und Laufrad einer axialen Kreiselpumpenstufe. Dissertation TU Berlin (1994)
7. Fickelscher, K.: Theoretischer Vergleich der Verstellpropeller- und der Drallregelung bei Kühlwasserpumpen. VDI-Z. **108**, 785–789 (1966)
8. Greitzer, E.M.: The stability of pumping systems. ASME J. Fluids Eng. **103**, 193–242 (1981)
9. IMechE Conference on Centrifugal pump low-flow protection (1991)
10. Jarius, M.: Untersuchung einer Axialgitterschaufel mit Höchstumlenkung durch struktur- und niederfrequente Wölbungsvariation. Dissertation TU Berlin (2000)
11. Jaeger, C.: Fluid Transients. Blackie, Glasgow (1977)
12. Knauss, J.: Wirbelbildung in Einlaufbauwerken - Luft- und Dralleintrag. DVWK Schrift 63, Paul Parey, ISBN 3-490-06397-X (1983)
13. Knauss, J. (Hrsg): Swirling flow problems at intakes. In: IAHR Hydraulic Structures Design Manual. AA Balkema, Rotterdam (1987). ISBN 90 6191 643 7
14. Melville, B.W., Ettema, R., Nakato, T.: Review of flow problems at water intake sumps. Iowa Institute of Hydraulic Research, University of Iowa. EPRI Report RP-3456-01 (1994)
15. Meschkat, S.: Experimentelle Untersuchung der Auswirkung instationärer Rotor-Stator-Wechselwirkungen auf das Betriebsverhalten einer Spiralgehäusepumpe. Diss. TU Darmstadt (2004)
16. Nakato, T., et al.: Field-tested solutions to pump vibrations. In: SHF Symposium, pp. 435–442 (1993)
17. Paterson, I.S., Adam, B.R.: Installation effects on wet pump performance. IMechE **C180**(77), 63–68 (1977)
18. Prosser, J.M.: The hydraulic design of pump sumps and intakes. BHRA, Bedford/CIRIA, London (1977)
19. Radke, M.: Strömungstechnische Untersuchung des Einflusses von Vorleiträdern variabler Geometrie auf das Betriebsverhalten axialer Kreiselpumpen. Fortschrittber VDI Reihe **7**, 210 (1992)
20. Rosenberger, H.: Experimental determination of the rotor impacts of axial pumps in intake structures under distorted approach flow. Thesis TU Kaiserslautern, SAM Forschungsbericht Bd. 5 (2001)
21. Saalfeld, K.: Vergleichende Darstellung der Regelung von Pumpen durch Vordrall und durch Laufschaufelverstellung. KSB Techn Ber. **7**, 22–31 (1963)
22. Stoll, A.: Speisewasserentgasung beim gleitendem Druck. Siemens Z. **36**(8), 608–618 (1962)
23. Strub, R.A.: Abfall des Saugdruckes von Speisewasserpumpen bei starken Lastschwankungen. Techn Rundschau Sulzer. **3**, 41–44 (1960)
24. Tillak, P., Hellmann, D.H., Rüth, A.: Description of surface vortices with regard to common design criteria of intake chambers. In: 2nd International Conference on Pumps and Fans, Beijing, pp. 863–874 (1995)
25. Thorley, A.R.D.: Fluid Transients in Pipeline Systems, 2nd edn. Wiley, New York (2004)
26. Weinerth, J., Rosenberger, H., Hellmann, D.H., Hausen, W.: Optimierung der Betriebsbedingungen von Wassertransportpumpen mit Hilfe von Modellversuchen. Pump Users Intl Forum Karlsruhe (2000)
27. Weinerth, J.: Kennlinienverhalten und Rotorbelastung von axialen Kühlwasserpumpen unter Betriebsbedingungen. Dissertation TU Kaiserslautern, SAM Forschungsbericht Bd. 9 (2004)

Chapter 12
Turbine Operation, General Characteristics

If the system head imposed on the pump exceeds the shut-off head, liquid flows back through the pump from the discharge to the suction nozzle in the "pumping-brake mode". There are a number of other abnormal modes of operation which are described by the "general characteristics". These include a turbine mode where the liquid transfers power to the rotor, which can be used to drive an electrical generator or any other machine.

12.1 Reverse Running Centrifugal Pumps Used as Turbines

12.1.1 Theoretical and Actual Characteristics

Reverse-running centrifugal pumps can be used as turbines for energy recuperation (in [N.6] termed "Hydraulic Power Recovery Turbines", HPRT). Applications are processes where a large amount of fluid energy is dissipated in valves or other throttling devices. In some processes dissolved gases separate from the fluid or liquid is flashed into steam during the expansion. Higher energy differences are then available for power recovery than in pure liquid flow, Chap. 13.3.

The preceding chapters dealt with centrifugal pumps which transfer energy to the liquid. The direction of rotation and flow for this type of operation are defined as *positive*. But a centrifugal pump can also transform liquid energy into mechanical power as a prime mover if the directions of rotation and flow are reversed i.e. both become negative (for abnormal turbine operation refer to Chap. 12.2). In turbine operation the pressure in the inlet nozzle is higher than in the outlet nozzle. The

J. F. Gülich, *Centrifugal Pumps*,
https://doi.org/10.1007/978-3-030-14788-4_12

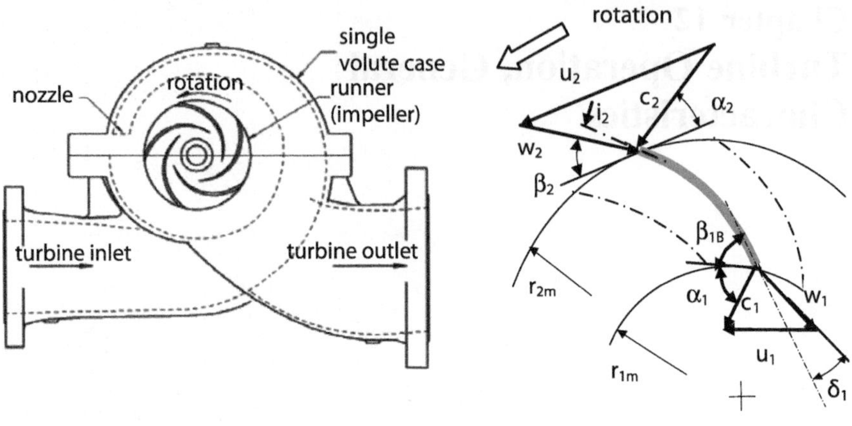

Fig. 12.1 Direction of rotation and flow in a pump running as turbine

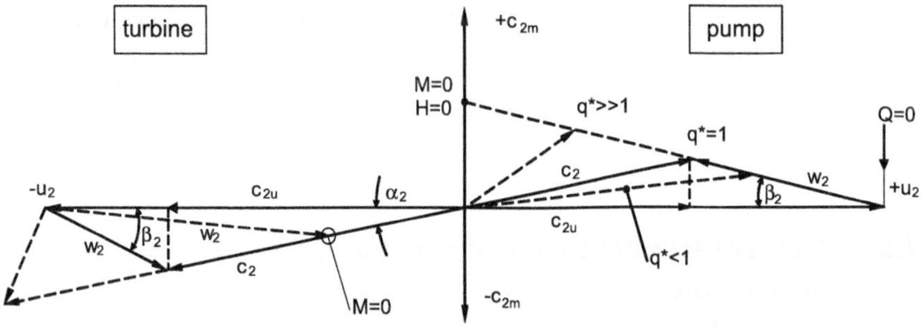

Fig. 12.2 Velocity trianges in pumps (*right*) and in turbine operation (*left*)

diffuser ("guide wheel") or volute feeds the liquid to the outer diameter of the runner.[1] The discharge nozzle of the pump is thus an inlet nozzle to the turbine, while the pump suction nozzle becomes the turbine outlet (or "exhaust") nozzle, Fig. 12.1.

The velocity vectors at the inlet and outlet of a runner blade are depicted in Fig. 12.1 on the right. The relative velocity w_2 at the inlet hits the blade with an incidence i_2 while the liquid leaves the blade with the relative velocity w_1 at a deviation angle δ_1. With reference to Fig. 12.2, let us consider the resultant flow conditions by comparing the velocity triangles for pump and turbine operation for an impeller with back-swept blades.[2]

[1]In turbine technology the term "runner" is commonly used instead of impeller and the term "guide wheel" is used instead of diffuser.

[2]The same subscripts are used for the calculation stations of turbine and pump operation (Chap. 3.1). The fluid in the turbine flows from high to low pressure in accordance with the subscripts 6, 5, 4, 3, 2, 1.

In *pump operation* the inflow angle to the impeller α_1 and the outflow angle β_2 from the impeller are largely *independent* of the flow. The reasons are: (1) angle α_1 is determined by the inlet geometry upstream from the impeller; (2) β_2 is essentially given by the impeller blade outlet angle β_{2B}. This is borne out by the observation that the slip factor γ depends little on the flow rate for $q^* > 0.7$ (see Fig. 4.19). As a result, the absolute velocity c_2 at the impeller outlet - and therefore the specific work of the pump according to Eq. (3.4) - drop with increasing flow (Tables 3.3 and 4.1).

In *turbine operation* the guide wheel or volute determines the runner inflow angle α_2 which is largely *independent* of the flow q^* when fixed guide vanes are used. The fluid leaves the runner with the angle β_1, which likewise depends little on the flow rate. Consequently, the inflow velocity c_2 to the runner increases with the flow rate, Fig. 12.2. The specific work increases proportionally to the flow through the turbine according to Eq. (3.4). This is demonstrated by Eq. (12.1) when we replace in Eq. (3.4) the circumferential components of the absolute velocity by the meridional components and the corresponding flow angles according to Fig. 12.2. In this way $c_{2u} = c_{2m} \times \cot \alpha_2$ and $c_{1u} = u_1 - w_{1u} = u_1 - c_{1m} \times \cot \beta_1$ are inserted into Eq. (3.4). The specific work of a turbine runner thus becomes:

$$Y_{sch} = Y_{th} = \frac{P_{sch}}{\rho\, Q_{La}} = u_2\, c_{2u} - u_1\, c_{1u} = u_2\, c_{2m} \cot \alpha_2 - u_1^2 + u_1\, c_{1m} \cot \beta_1 \quad (12.1)$$

The angles α_2 and β_1 in Eq. (12.1) are *flow* angles. The inflow angle α_2 to the runner can be determined from the guide wheel or volute geometry. An approximation of the inflow angle α_2 can be calculated from the cross section A_{3q} of the throat as demonstrated with reference to Fig. 12.3. In the case of a guide wheel, Fig. 12.3 represents one vane pitch; in the case of a single volute set $z_{Le} = 1$ and with a double volute set $z_{Le} = 2$. The fluid enters through the throat area A_{3q} with the velocity $c_{3q} = Q/(z_{Le} \times A_{3q})$; its circumferential component is $c_{3u} = c_{3q} \times \cos \alpha_{3B}$. Since α_{3B} is a relatively small angle, errors in assuming $\alpha_3 \approx \alpha_{3B}$ have a negligible effect on $\cos \alpha_{3B}$. Thus the blade angle can be estimated from $\alpha_{3B} = \arcsin\ a_3/t_3$.

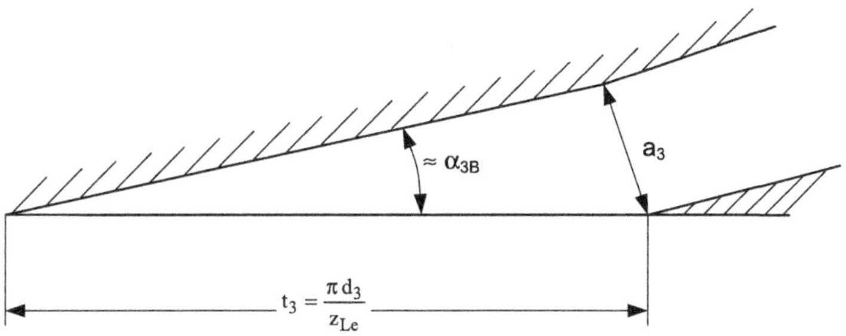

Fig. 12.3 Determination of the outflow angle from the throat area and pitch; applicable to a guide wheel, a volute or a runner

While the total flow rate Q entering the turbine inlet nozzle flows through the guide wheel, the actual flow through the runner is reduced by the volumetric efficiency $Q_{La} = Q \times \eta_v$ due to leakage losses. The approach flow angles result from:

$$c_{2m} = \frac{Q\,\eta_v}{\pi\, f_q\, d_{2b}\, b_2} \qquad c_{2u} = c_{3u}\frac{r_{3,eff}}{r_2} = \frac{r_{3,eff}\, Q \cos\alpha_{3B}}{r_2\, z_{Le}\, A_{3q}}$$
$$\tan\alpha_2 = \frac{r_2\, z_{Le}\, A_{3q}\, \eta_v}{r_{3,eff}\, f_q\, A_2 \cos\alpha_{3B}} \tag{12.2}$$

$$\tan\beta_2 = \frac{c_{2m}}{u_2 - c_{2u}} = \frac{Q\,\eta_v}{f_q A_2\left(u_2 - \frac{r_{3,eff}\, Q \cos\alpha_{3B}}{r_2\, z_{Le}\, A_{3q}}\right)} \tag{12.3}$$

The approach flow angles in the absolute and relative system depend mainly on the volumetric flow rate and the cross-sections $z_{Le} \times A_{3q}$ and A_2. For double-exit impellers A_2 and b_2 are defined per side of the impeller. The radius r_3 is defined by $r_{3,eff} = r_3 + e_3 + k_3 \times a_3$ where e_3 is the thickness of the diffuser vane leading edge or the volute cutwater and k_3 is an empirical factor derived from experiment (tentatively $k_3 = 0.25$). The condition for shock-free entry is $\tau_2 \times \tan\beta_2 = \tan\beta_{2B}$. Equation (12.3) then yields the flow rate for shock-free entry as per Eq. (12.4):

$$\frac{Q_{SF}}{u_2\, A_2\, f_q} = \frac{\tan\beta_{2B}}{\tau_2\,\eta_v + \frac{r_{3,eff}\, A_2\, f_q \tan\beta_{2B} \cos\alpha_{3B}}{r_2\, z_{Le}\, A_{3q}}} \tag{12.4}$$

The flow rate for shock-free entry increases with the cross-section A_{3q} of the guide wheel and the blade angle β_{2B}. The best efficiency point in *turbine* operation is close to the flow rate of shock-free entry. In contrast, the best efficiency point in *pump* operation is found at a discharge flow angle which is considerably smaller than the impeller blade outlet angle because of the effect of the slip factor.

The angle β_1 of the fluid exiting from the runner differs from the blade angle β_{1B} because a vane-congruent flow cannot be expected in turbine operation either. In analogy to Fig. 12.3, β_1 can be estimated from the throat area A_{1q}. To this end, A_{1q} replaces A_{3q} and c_{1q} replaces c_{3q}. The velocity in the throat area A_{1q} is $w_{1q} = Q \times \eta_v/(f_q \times A_{1q} \times z_{La})$. The circumferential component is $w_{1u} = w_{1q} \times \cos\beta_{A1}$. The following relations are obtained:

$$w_{1u} = \frac{\eta_v\, Q \cos\beta_{A1}}{z_{La}\, f_q\, A_{1q}} \quad \text{and} \quad c_{1u} = u_1 - \frac{\eta_v\, Q \cos\beta_{A1}}{z_{La}\, f_q\, A_{1q}} \tag{12.5}$$

$$\tan\beta_1 = \frac{z_{La}\, A_{1q}}{A_1 \cos\beta_{A1}} \quad \text{with} \quad \beta_{A1} = \arcsin\frac{A_{1q}}{b_1\, t_1} \tag{12.6}$$

Inserting Eqs. (12.2 and 12.5), into Eq. (12.1) the specific work results in:

$$Y_{sch} = u_2^2 \left\{ \frac{Q}{u_2\, z_{Le}\, A_{3q}} \left(\frac{r_{3,eff}}{r_2} \cos\alpha_{3B} + \frac{d_1^*\, \eta_v\, z_{Le}\, A_{3q}}{z_{La}\, f_q\, A_{1q}} \cos\beta_{A1} \right) - d_1^{*2} \right\} \tag{12.7}$$

According to Eqs. (12.7 and 12.1) the specific work of a turbine increases linearly with the flow rate. At Q = 0, the straight line $Y_{sch} = f(Q)$ intersects the ordinate at $Y_{sch} = -u_1^2$, see Fig. 12.4. The gradient of this straight line increases with a decreasing throat area $z_{Le} \times A_{3q}$ of the guide wheel or volute. The power transferred from the fluid to the runner blades is in accordance with Eqs. (12.1 and 12.7):

$$P_{sch} = \rho\, Q_{La}\, Y_{sch} = \rho\, Q_{La}^2 \left(\frac{u_2}{f_q\, A_2} \cot\alpha_2 + \frac{u_1}{f_q\, A_1} \cot\beta_1 \right) - u_1^2\, \rho\, Q_{La}$$

$$P_{sch} = u_2^2\, \rho\, \eta_v Q \left\{ \frac{Q}{u_2\, z_{Le}\, A_{3q}} \left(\frac{r_{3,eff}}{r_2} \cos\alpha_{3B} + \frac{d_1^*\, \eta_v\, z_{Le}\, A_{3q}}{z_{La}\, f_q\, A_{1q}} \cos\beta_{A1} \right) - d_1^{*2} \right\} \tag{12.8}$$

Therefore, $P_{sch} = f(Q)$ represents a parabola through the origin of the coordinate system. According to Eq. (12.8), the runner is able to deliver power above a flow rate of $Q_{L,th}$ which is obtained by setting $P_{sch} = 0$ or $Y_{sch} = 0$:

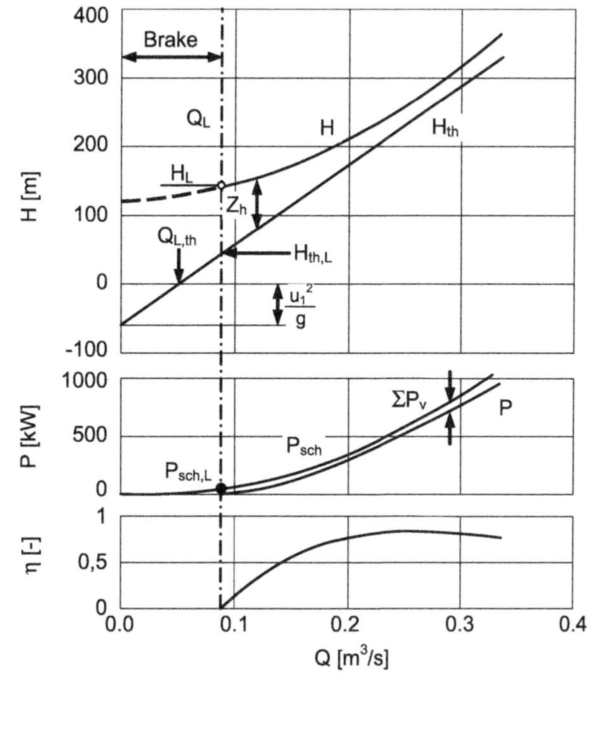

Fig. 12.4 Theoretical and actual turbine characteristics

$$Q_{L,th} = \frac{u_1^2}{\frac{u_2}{f_q\, A_2}\cot\alpha_2 + \frac{u_1}{f_q\, A_1}\cot\beta_1} = \frac{u_2\, z_{Le}\, A_{3q}\, d_1^{*2}}{\frac{r_{3,eff}}{r_2}\cos\alpha_{3B} + \frac{d_1^*\,\eta_v\, z_{Le}\, A_{3q}}{z_{La}\, A_{1q}\, f_q}\cos\beta_{A1}} \tag{12.9}$$

The parabola $P_{sch} = f(Q)$ has a second intersection with the abscissa at $Q_{L,th}$ and its minimum value at $Q = 0.5 \times Q_{L,th}$ (not seen in Fig. 12.4 because of the scale).

The energy difference between the inlet and exhaust nozzles is $Y = g \times H$. The head H is defined as in pump operation by the formulae given in Table 2.2.

The specific work $Y_{sch} = g \times H_{th}$ transferred from the fluid to the runner is smaller than the energy difference $Y = g \times H$ between inlet and exhaust nozzles due to hydraulic losses Z_h. The following relation applies: $H_{th} = \eta_h \times H = H - Z_h$.

The useful power P available at the coupling of the turbine is smaller than the supplied energy ($\rho \times g \times H \times Q$) because of secondary losses. In analogy to Table 3.5, the power balance of a turbine (Fig. 12.5) is given by:

$$\rho\, g\, H\, Q = P + \rho\, g\, H(1 - \eta_h)\, Q_{La} + \rho\, g\, H\,(Q_{sp} + Q_E) + \sum_{st} P_{RR} + \sum P_{s3} + P_m + P_{er} \tag{12.10}$$

On the left of Eq. (12.10) is the energy input and on the right the energy output as the sum of the useful power P and all losses. These include the hydraulic losses, leakages, disk friction, interstage seals, mechanical losses and friction losses

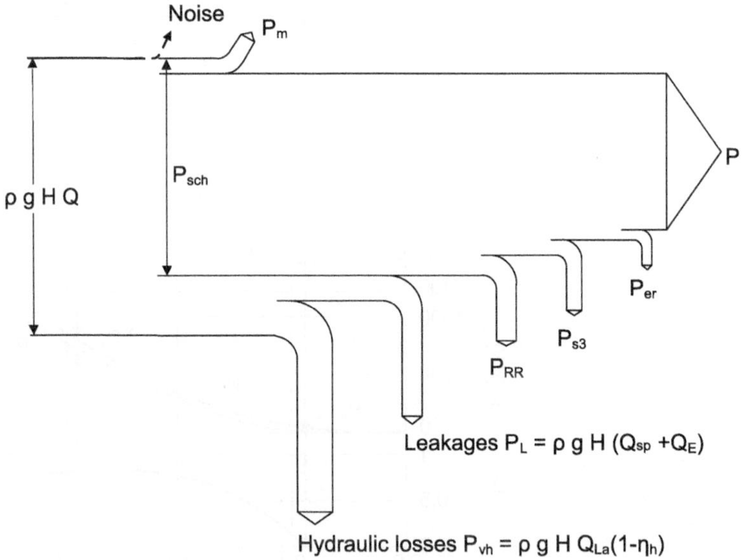

Fig. 12.5 Energy balance of a turbine

induced by the axial thrust balance device. The secondary losses can be calculated from Tables 3.5 to 3.7. The hydraulic losses and leakages reduce the power P_{sch} transmitted to the runner blades, Eq. (12.11):

$$\rho\, g\, H\, \eta_h\, (Q - Q_{sp} - Q_E) = P_{sch} = P + \sum_{st} P_{RR} + \sum P_{s3} + P_m + P_{er} \quad (12.11)$$

In order to obtain the power P available at the coupling, disk friction and mechanical losses must be subtracted from P_{sch} as per Eq. (12.11). The efficiency η at the coupling is then given by:

$$\eta = \frac{P}{\rho\, g\, H\, Q} \quad (12.12)$$

As shown by Eq. (12.1), the specific work of a turbine runner drops with increasing circumferential component c_{1u} at the outlet, i.e. with the residual swirl (sometimes called "vortex effect"). As a rule, the corresponding kinetic energy cannot be recovered downstream of the runner. Residual swirl reduces the turbine efficiency, since it is dissipated through turbulence between the runner outlet and the exhaust nozzle.

Equations (12.7 and 12.8) represent the theoretical characteristics of a turbine (assuming a flow without losses). The actual characteristics are determined according to Fig. 12.4 by adding the hydraulic losses to the theoretical head ($H = H_{th} + Z_h$) and by subtracting the secondary losses from the power transferred to the blades according to Eq. (12.11). Due to these losses the minimum flow rate necessary to deliver power to the coupling also shifts from $Q_{L,th}$ to a higher value Q_L. In order to pass the flow Q_L through the turbine, the head H_L is required; H_L comprises the losses Z_h and the quantity $H_{th,L}$ which serves to overcome the disk friction and the mechanical losses.

12.1.2 Runaway and Resistance Characteristics

The operating condition with M = 0 and P = 0 is to be understood as the idling speed of the turbine operating with no load ("runaway speed"). This condition is met upon start-up of the turbine when the operating speed is attained (start-up without load when opening the throttle valve). The runaway characteristic connects all the points H(Q) which occur for M = 0 at various speeds.

The runaway speed depends strongly on the test or plant conditions which fact may be illustrated by means of three cases: (1) If the coupling fails in a plant thereby disconnecting the turbine from the driven machine, the highest runaway speed is reached. (2) If the turbine drives an electric generator, the runaway speed upon power failure will be much less than in case of a coupling failure because of the higher rotor mass and resistance provided by the generator. (3) Finally the runaway speed will be quite low if the turbine would drive a pump (unless the pump would suddenly de-prime and run in air).

The time to reach the runaway speed may be in the order of 1 to 2 s only. A power failure will also be linked to sudden large changes in flow and thus generate pressure surges, Chap. 11.5.

According to the similarity laws (Table 3.4) the runaway characteristic curve is obtained, with $Q_L \sim n$ and $H_L \sim n^2 \sim Q^2$, as a parabola through the origin of the coordinate system, Fig. 12.6. The rotor also accelerates to this runaway characteristic if reverse flow is established in a pump upon driver failure, Chap. 11.5.

If the static head $H_{A,stat}$ is available in the plant, the turbine runs at a head of H_B which is lower than $H_{A,stat}$ because of the flow losses H_v created in the inlet and outlet piping systems of the turbine, Fig. 12.6. Upon load rejection, the turbine will run at point D_1 on the runaway characteristic if H_B remained constant. As a result of the reduced hydraulic losses, a somewhat higher runaway speed occurs corresponding to point D in Fig. 12.6 (in the plant the turbine runs directly from B to D on the plant characteristic curve).

With a locked rotor (n = 0) and at a given head, a certain flow passes through the turbine which depends on the flow resistance of the machine. The correlation $H_w \sim Q^2$ yields the "resistance curve" (turbine characteristic at n = 0) which is represented by a parabola through the origin of the coordinate system. The stationary rotor is subject to a hydraulic moment ("torque at n = 0") which is proportional for a given machine to the available head: $M_w \sim H_w \sim Q^2$. Flow rate

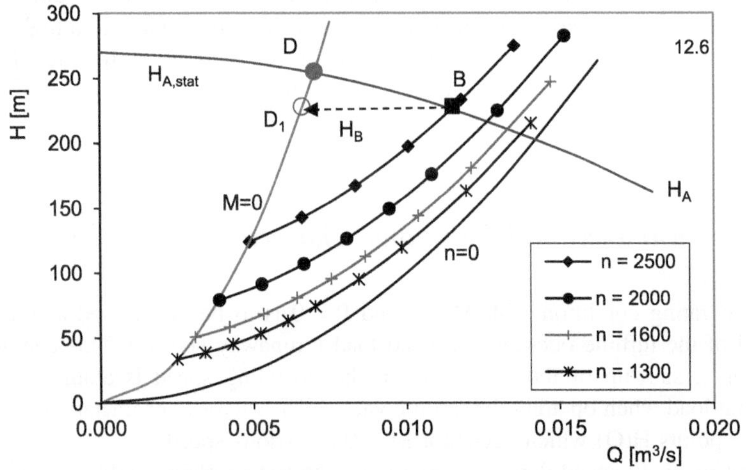

Fig. 12.6 Turbine characteristics at various speeds. Runaway H_L at M = 0, resistance H_w at n = 0 and plant characteristics H_A

and torque depend slightly on the position of the runner versus the casing. Since the speed n is zero, the similarity laws according to Table 3.4 obviously do not apply. The machine represents a pure flow resistance following a square law.

Whereas with radial machines the runaway characteristic (M = 0) for a given flow rate Q is above the resistance curve (n = 0) as in Fig. 12.6, the reverse is true with axial machines; here the runaway characteristic lies below the characteristic at n = 0, [12, 15][3]. Consequently, the flow rate of a radial machine decreases with load rejection, whereas it increases in the case of an axial turbine.

12.1.3 Estimation of Turbine Characteristics from Statistical Correlations

Since neither the hydraulic losses in the turbine nor the runaway nor resistance characteristics can be predicted from basic principles, the turbine characteristics are often estimated from statistical correlations, if no test data are available. To this end, the performance data at maximum efficiency in turbine operation are related to the corresponding data at the best efficiency point of pump operation (Table 12.1). The ratios $H_{opt,T}/H_{opt,P}$ and $Q_{opt,T}/Q_{opt,P}$ are either related to the total efficiency or to the hydraulic efficiency, or represented as a function of the specific speed. In [17] eight correlations of this kind were compared with one another on the basis of 35 pumps in the range of $12 < n_q < 190$. Of these, the formulae in [14] were found to be the most accurate. The relevant relations are listed in Table 12.1 as Eqs. (T12.1.1 to 12.1.3). Unfortunately, a considerable scatter has to be expected when applying such correlations. Thus the data from [4] are about 20% higher than the predictions according to Eqs. (T12.1 to 12.1.2). It could hardly be expected that the same conversion factors apply to a radial pump with $n_q = 25$ and a semi-axial pump with $n_q = 150$, though both machines happen to have the same efficiencies.

Table 12.1 also contains conversion factors according to Eqs. (T12.1.4 to 12.1.7), which were correlated with the specific speed and which originate from tests on single stage standard pumps with volute casings, [13]. The measurements from [4] are rendered by this correlation with deviations of up to 13%. As far as larger pumps with greater efficiencies are concerned, Eqs. (T12.1.6 and 12.1.7) should be used with caution.

A given machine achieves essentially the same efficiencies in the turbine and pumping modes; the deviations are generally in the range of ±2%. At low specific

[3]It may be concluded from this observation that runaway and resistance characteristics with increasing specific speed come closer, restricting thus the turbine operating range, and that (in theory) they would coincide at a certain specific speed.

Table 12.1 Turbine characteristics. NOTE: n_q with [rpm], [m³/s], [m]			Eq.
Flow rate at BEP in turbine mode to flow at pump BEP	$\frac{Q_{opt,T}}{Q_{opt,P}} = \frac{1}{\eta_{opt,P}^{0.8}}$	$12 < n_q < 190$ [17]	12.1.1
Head at BEP in turbine mode to head at pump BEP	$\frac{H_{opt,T}}{H_{opt,P}} = \frac{1}{\eta_{opt,P}^{1.2}}$		12.1.2
Specific speed in turbine mode to pump specific speed	$\frac{n_{q,T}}{n_{q,P}} = 0.95\sqrt{\eta_{opt,P}}$		12.1.3
Flow rate at BEP in turbine mode to flow at pump BEP	$\frac{Q_{opt,T}}{Q_{opt,P}} = \frac{2.5}{\eta_{h,opt,P}} - 1.4 \quad \pm 7\,\%$	volute pumps $8 < n_q < 70$ The scatter approximately represents the standard deviation [13]	12.1.4
Head at BEP in turbine mode to head at pump BEP	$\frac{H_{opt,T}}{H_{opt,P}} = \frac{2.4}{\eta_{h,opt,P}^{2}} - 1.5 \quad \pm 14\,\%$		12.1.5
Specific speed in turbine mode to pump specific speed	$\frac{n_{q,T}}{n_{q,P}} = 1.3\eta_{opt,P} - 0.3 \quad \pm 10\,\%$		12.1.6
Efficiency at BEP in turbine mode to efficiency at pump BEP	$\frac{\eta_{opt,T}}{\eta_{opt,P}} = 1.16 - \frac{n_{q,P}}{200} \quad \pm 5\,\%$		12.1.7
Runaway speed (M = 0) referred to *pump* speed	$\frac{n_L}{n_N} = \left(\frac{n_{q,P}}{12}\right)^{0.19}\left(\frac{H}{H_{opt,P}}\right)^{0.5}$	from data in [4] and [B.18]	12.1.8
Flow rate at runaway (M = 0); referred to *turbine*	$\frac{Q_L}{Q_{opt,T}} = 0.3 + \frac{n_{q,P}}{400}$	acc. to [5]	12.1.9
Head at runaway (M = 0); referred to *turbine*	$\frac{H_L}{H_{opt,T}} = 0.55 - 0.002\, n_{q,P}$	from data in [18] and [4]	12.1.10
Flow rate at runaway (M = 0); referred to *pump*	$\frac{Q_L}{Q_{opt,P}} = 0.45 + \frac{n_{q,P}}{150}$	from data in [10] and [4] $n_q < 180$	12.1.11
Head at runaway (M = 0); referred to *pump*	$\frac{H_L}{H_{opt,P}} = \left(\frac{12}{n_{q,P}}\right)^{0.38}\left(\frac{n}{n_N}\right)^{2}$		12.1.12
Flow rate at locked rotor (n = 0) referred to *pump*	$\frac{Q_W}{Q_{opt,P}} = \left(\frac{41}{n_{q,P}}\right)^{0.28}\left(\frac{H}{H_{opt,P}}\right)^{0.5}$		12.1.13
Torque at locked rotor (n = 0) referred to *pump*	$\frac{M_W}{M_{opt,P}} = \left(\frac{120}{n_{q,P}}\right)^{0.22}\frac{H}{H_{opt,P}}$		12.1.14
Turbine characteristic	$H_T = H_{opt,T} - \frac{H_{opt,T} - H_{L,N}}{Q_{opt,T}^2 - Q_{L,N}^2}\left(Q_{opt,T}^2 - Q_T^2\right)$		12.1.15
Required $NPSH_R$ of a turbine	$NPSH_R = 0.1\, H_{T,st}\left(\frac{n_{q,T}}{45}\right)^{1.5}$		12.1.16

speeds, turbines often have slightly higher efficiencies than the corresponding pump, because the turbine power is greater than that of the pump. This implies that the secondary losses take a lower fraction of the useful power. In the case of semi-axial and axial pumps the efficiency of the turbine can be expected to be somewhat less than that of the pump because the trailing edges of the blades are not designed for turbine flow and hydraulic losses are accordingly higher.

The efficiency curve $\eta = f(Q)$ can be estimated from Fig. 12.7, where the turbine efficiency related to the best point $\eta_T/\eta_{opt,T}$ was plotted against the ratio $(Q - Q_L)/(Q_{opt,T} - Q_L)$; these data originate from [4, 9] and from [7].

The specific speed of the turbine is lower than that of the pump (n_q is always formed with the data at the BEP), because $(H_{opt,T}/H_{opt,P})^{0.75}$ is always greater than $(Q_{opt,T}/Q_{opt,P})^{0.5}$, Eqs. (T12.1.3 and 12.1.6).

The runaway speed can be calculated from Eq. (T12.1.8) by inserting the head which corresponds to the operating point to be investigated, for example the load rejection of a turbine or the failure of a pump drive. The flow rate Q_L and the corresponding head H_L on the runaway characteristic can be estimated from Eqs. (T12.1.9 and 12.1.10) where the data are related to the best point of the *turbine*. The data from [10] allow correlations to be established for Q_L and H_L, which are listed as Eqs. (T12.1.11 to 12.1.12) in Table 12.1; these correlations refer to the best point in *pump operation*. Equations (T12.1.8 and 12.1.12) were laid through the range of the measured data points in a way that the similarity law $H_L \sim n^2$ is fulfilled.

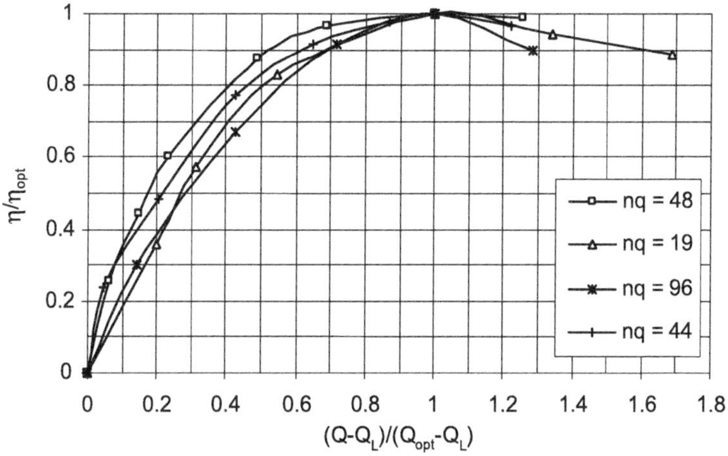

Fig. 12.7 Turbine efficiency curve versus flow rate

The turbine characteristic and the torque at n = 0 can be estimated from Eqs. (T12.1.13 and 12.1.14); these correlations were derived from data in [10]. All the correlations mentioned are based on some 25 to 35 measurements in the quoted publications. The tests cover different pump types of various manufacturers in the range of $10 < n_q < 220$. The scatter is typically ±20%; individual values may show far greater deviations the cause of which cannot be investigated due to lack of information in the cited publications. Figure 12.8 shows the data used. Some data

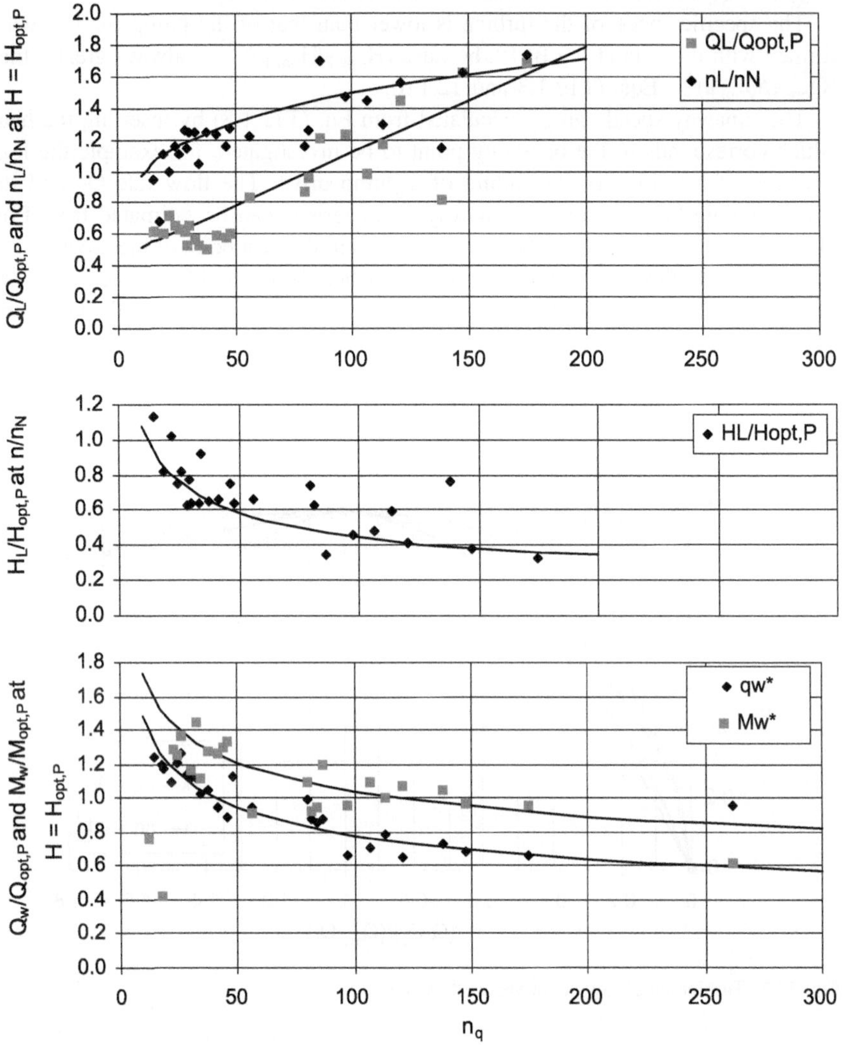

Fig. 12.8 Correlations for calculating runaway (subscript L) and resistance *curves* (subscript w)

fall way outside the spread, particularly those for the torque at locked rotor (n = 0). Equation (T12.1.8) also renders quite well a graph in [B.18] in which the reverse running speeds of centrifugal pumps are plotted as a function of n_q.

The turbine characteristic for a particular centrifugal pump can be determined as follows:

1. Given are the BEP performance data of the pump at the nominal speed n_N: $H_{opt,P}$, $Q_{opt,P}$ and $\eta_{opt,P}$. The turbine characteristic is calculated for the nominal speed too. If necessary, the characteristic is subsequently converted to other speeds according to the similarity laws in Table 3.4.
2. The data at the turbine best efficiency point are calculated from Eqs. (T12.1.1 to 12.1.3) or Eqs. (T12.1.4 to 12.1.7) from the BEP of the pump.
3. The runaway point $Q_{L,N}$, $H_{L,N}$ for the nominal speed n_N is obtained from the turbine best efficiency point by means of Eqs. (T12.1.9 and 12.1.10); or from Eqs. (T12.1.11 and 12.1.12) with the performance data at the pump BEP. It is also possible to take the average from both procedures.
4. The runaway characteristic then is obtained from $H_L = H_{L,N}(Q_{L,x}/Q_{L,N})^2$, if $Q_{L,x}$ signifies any flow rate on the runaway characteristic.
5. The characteristic $H_T = f(Q_T)$ can be approximated as a parabola through the best efficiency point (Step 2) and the runaway point $Q_{L,N}$, $H_{L,N}$ (Step 3), Eq. (T12.1.15).
6. The turbine efficiency results from the pump efficiency according to Eq. (T12.1.7), or it is assumed as $\eta_{opt,T} = \eta_{opt,P} \pm 0.02$. The efficiency curve is calculated by means of Fig. 12.7 with $\eta_T = 0$ at $Q = Q_{L,N}$.
7. The power curve is calculated from $P_T = \eta_T \times \rho \times g \times H_T \times Q_T$.
8. The locked rotor characteristic at n = 0 and the locked rotor torque can be calculated from Eqs. (T12.1.13 and 12.1.14). In doing so, the head H is treated as an independent variable: for selected values of H the flow rate is calculated from Eq. (T12.1.13) and the torque from Eq. (T12.1.14).

It is emphasized again that calculating the turbine characteristics on the basis of these statistical correlations is subject to considerable uncertainties, because such statistics cannot take into account the actual geometric properties of the machines under consideration. Thus the runner eye exerts a significant influence on the head-capacity curve in turbine operation, whereas it only slightly changes the pump characteristics at low or moderate specific speeds. A suction impeller with good NPSH properties in the high flow range in pump operation thus shifts the turbine characteristic to a considerably higher flow rate. This can be seen from the influence of A_{1q} in Eq. (12.7) and from the measurements in [9]. The inlet casings of between-bearing pumps can have an effect as well.

An attempt at developing a method for the estimation of turbine characteristics based on the geometric data of the machine is reported in [3]. However, a reliable and complete procedure for the calculation of turbine characteristics has not been published. In an actual application measured turbine characteristics are therefore indispensable.

12.1.4 Estimation of Turbine Characteristics from Loss Models

An alternative to the performance prediction from statistical data is based on a loss analysis. This procedure is based on two steps:

Step 1: Determine correlations for BEP flow and head from turbine test data. The calculations required for step 1 are given in Table 12.2 and described below:

- Test data are available on a range of pumps running as turbines.
- The leakage losses, disk friction and mechanical losses can be determined from Tables 3.5 to 3.7.
- The flow at shock-free entrance can be calculated from the geometric parameters by using Eq. (T12.2.1). The turbine BEP is expected to be close to the flow at shock-free entrance. The ratio $Q_{opt,T}/Q_{SF}$ can be correlated as a function of specific speed for different types of pumps.
- The theoretical power is calculated from the individual losses, Eq. (T12.2.3). From this follows the theoretical head and the hydraulic efficiency.
- The theoretical head can also be determined from the geometry of the machine by using Eq. (12.7). It depends on the effective radius $r_{3,eff}$ which is representative of the swirl entering the impeller. This radius is derived from the test data by varying the factor k_3 in Eq. (T12.2.8) till the theoretical heads calculated from Eqs. (T12.2.5) and (12.7) do match exactly. k_3 can subsequently be correlated as a function of specific speed and type of pump.

Table 12.1 (1) Turbine test data evaluation			Eq.
Available: measured turbine performance curves. Goal: determine dimensionless parameters required for future performance predictions			
Flow rate at shock-free entrance		$Q_{SF} = \dfrac{u_2 A_2 f_q \tan\beta_{2B}}{\tau_2\,\eta_v + \dfrac{r_{3,eff}\, A_2 f_q \tan\beta_{2B} \cos\alpha_{3B}}{r_2\, z_{Le}\, A_{3q}}}$	12.2.1
Correlate correction factor for flow rate at BEP in turbine mode		$k_f = \dfrac{Q_{opt,T}}{Q_{SF}} = f(n_q, \text{type})$	12.2.2
Theoretical power		$P_{th} = P + \Sigma P_{RR} + P_m + P_{er} + P_{s3}$	12.2.3
Volumetric efficiency		$\eta_{v,opt} = \dfrac{Q_{opt,T} - Q_{s1} - Q_{s2} - Q_E}{Q_{opt,T}}$	12.2.4
Theoretical head		$H_{th} = \dfrac{P_{th}}{g\,\rho\,Q\,\eta_v}$	12.2.5
Hydraulic efficiency		$\eta_{h,opt,T} = \dfrac{H_{th}}{H}$	12.2.6
Correlate		$\dfrac{\eta_{h,opt,T}}{\eta_{h,opt,P}} = f(n_q, \text{type})$	12.2.7
Effective radius which gives the actual swirl acting on the impeller		$r_{3,eff} = r_3 + e_3 + k_3 a_3$ — d_3 applies to diffusers, d_z applies to volutes	12.2.8
Determine k_3 so that it yields exactly the same theoretical heads from Eqs. (12.7) and (T12.2.5)		Correlate $k_3 = f(n_q, \text{type})$ expected: $k_3 = 0.2$ to 0.5	12.2.9
Friction loss in vaneless space		$\zeta_{LR} = \dfrac{2\, c_f\, r_2}{b_3 \sin\alpha_3 \cos^2\alpha_3}\left(\dfrac{c_{3u}}{u_2}\right)^2\left(1 - \dfrac{r_2}{r_3}\right)$	12.2.10
Volutes	Wetted surface of volute	$A_{wet} = \pi^2 \dfrac{d_{3q}}{2}\left(d_z + d_{3q}\right)$	12.2.11
	Volute loss coefficient	$\zeta_{vol} \equiv \dfrac{2g\, Z_{vol}}{u_2^2} = \dfrac{c_f\, c_{vol}^3\, A_{wet}}{Q\, u_2^2}$	12.2.12
	Loss in inlet nozzle, tentatively $\zeta_c = 0.04$	$\zeta_d \equiv \dfrac{2g\, Z_d}{u_2^2} = \left(\dfrac{c_{3q}}{u_2}\right)^2\left(\zeta_c + \lambda\dfrac{L}{d_h}\right)$	12.2.13
	Total loss in volute casing	$Z_{vol,tot} = \left(\zeta_{LR} + \zeta_{vol} + \zeta_d\right)\dfrac{u_2^2}{2g}$	12.2.14
Guide wheels	Loss in annular chamber	$\zeta_d = \zeta_d'\left(\dfrac{c_d}{u_2}\right)^2$ tentatively $\zeta_d' = 1.0$	12.2.15
	Loss in return channels	$\zeta_{RV} = \zeta_{RV}'\left(\dfrac{c_{1m}}{u_2}\right)^2$ tentatively $\zeta_{RV}' = 1.5$	12.2.16
	Loss in diffuser channels	$\zeta_{dif} = \zeta_{dif}'\left(\dfrac{c_{3q}}{u_2}\right)^2$ tentatively $\zeta_{dif}' = 0.1$	12.2.17

Table 12.2 (2) Turbine test data evaluation

Total loss in guide wheel (one stage)	$Z_{Le,tot} = \left(\zeta_{LR} + \frac{\zeta_d}{z_{st}} + \zeta_{RV} + \zeta_{diff}\right)\frac{u_2^2}{2g}$	12.2.18
Losses in impeller and exhaust casing	$Z_{La} = H - H_{th} - Z_{vol,tot}$	12.2.19
Impeller loss coefficient is correlated against Q/Q_{opt} or Q/Q_{SF}	$\zeta_{La} - \zeta_{La,opt} = f\left(\frac{Q}{Q_{SF}}\right)$	12.2.20
Correlate impeller loss coefficient at BEP	$\zeta_{La,opt} \equiv \frac{2g\, Z_{La}}{u_2^2} = f(n_q, \text{type})$	12.2.21

- In turbine operation there is accelerated flow in the volute casing (or diffuser). There is little or no flow separation. Consequently, the hydraulic losses ζ_{Le} in the casing can be estimated from simplified models as shown in Table 3.8. This leads to Eqs. (T12.2.10) to (T12.2.18). Typical results are shown in Fig. 12.9; the data cover volute and diffuser pumps with $n_q = 21$ to 32 and single and double-entry impellers.
- The impeller losses ζ_{La} can than be calculated from Eq. (T12.2.19): They are correlated in the form of ($\zeta_{La} - \zeta_{La,opt}$) against Q/Q_{opt} or Q/Q_{SF} as given by Eq. (T12.2.20). The impeller losses at the BEP in the turbine mode were as an average of 11 turbines $\zeta_{La,opt} = 0.042$ with a range of $\zeta_{La,opt} = 0.02$ to 0.08. Figure 12.10 shows a typical example of such a plot. The data comprise volute and diffuser pumps with single- and double-entry impellers with specific speeds in the range of $n_q = 14$ to 52. From the shape of the curve in Fig. 12.10 it is concluded that high shock looses occur at the impeller inlet. The behavior of the *impeller* losses of a *turbine* exhibits a noteworthy resemblance to the *diffuser* losses in *pump* operation (compare Figs. 4.8 and 12.10). It is also observed that the impeller losses of both diffuser and volute pumps behave in the same way.

Step 2: Use the correlations determined in step 1 for turbine performance prediction. The calculations required for step 2 are given in Table 12.3 and described below:

- Ascertain geometric data of pump and rated speed of turbine.
- Calculate flow rate of shock-free entry, Eq. (T12.3.1). The ratio $Q_{opt,T}/Q_{SF}$ is expected for volute pumps in the range of 0.75 to 0.9 and for diffusers between 0.9 and 1.0, decreasing with growing specific speed.
- Determine best efficiency flow rate in turbine mode, Eq. (T12.3.2) from correlation developed in Table 12.2, Eq. (T12.2.2)
- Calculate theoretical head of turbine, Eq. (T12.3.3)

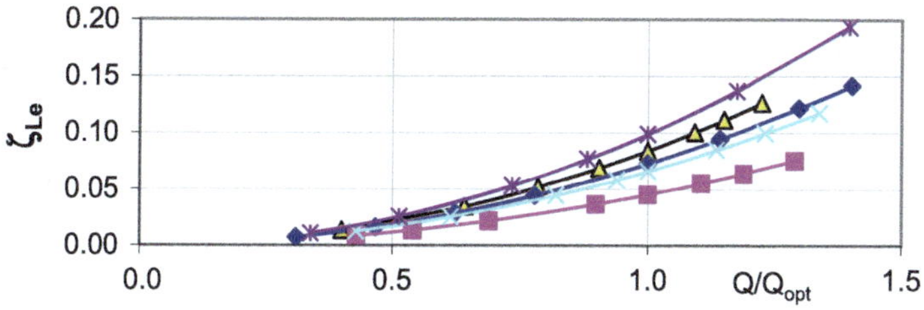

Fig. 12.9 Diffuser and volute loss coefficients of various pumps operating as turbines

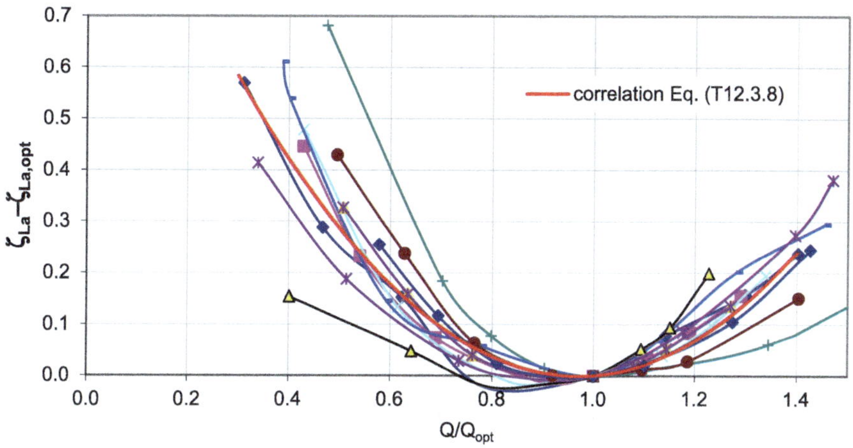

Fig. 12.10 Impeller loss coefficients of various pumps operating as turbines

- Calculate hydraulic losses in collector (volute or diffuser) as a function of the flow rate, step (T12.3.4)
- Determine hydraulic losses in impeller at BEP according to option A or B, Eqs. (T12.3.5) to (T12.3.7)
- Determine hydraulic losses in impeller as a function of the flow rate, Eqs. (T12.3.8) and (T12.3.9)
- Calculate head as a function of flow rate, Eq. (T12.3.10)
- Determine secondary losses P_{RR}, P_m, P_{s3}, P_{er}, Q_{sp}, Q_E according to Chap. 3, step (T12.3.11)
- Calculate power and efficiency, Eqs. (T12.3.12) and (T12.3.13).

12.1.5 Behavior of Turbines in Plants

Control: As mentioned when discussing Fig. 12.6, a turbine works at the intersection of the turbine and the system characteristics (the same as in pump operation, Chap. 11.1). Varying operation conditions can be achieved by means of throttle, bypass or speed control. Radial, semi-axial or axial machines can be controlled also by adjusting the guide vanes.

Table 12.3 Turbine performance prediction		Eq.
Available: pump geometry. Goal: determine turbine performance		
Flow rate at shock-free entrance	$Q_{SF}=\dfrac{u_2 A_2 f_q \tan\beta_{2B}}{\tau_2\,\eta_v+\dfrac{r_{3,eff}\,A_2 f_q \tan\beta_{2B}\cos\alpha_{3B}}{r_2\,z_{Le}\,A_{3q}}}$	12.3.1
Flow rate at BEP in turbine mode	$Q_{opt}=k_f\,Q_{SF}$ $\qquad k_f=\dfrac{Q_{opt,T}}{Q_{SF}}=f(n_q,\text{type})$	12.3.2
Theoretical head $H_{th}=f(Q)$	$H_{th}=\dfrac{u_2^2}{g}\left\{\dfrac{Q}{u_2\,z_{Le}\,A_{3q}}\left(\dfrac{r_{3,eff}}{r_2}\cos\alpha_{3B}+\dfrac{d_1^*\,\eta_v\,z_{Le}\,A_{3q}}{z_{La}\,f_q\,A_{1q}}\cos\beta_{A1}\right)-d_1^{*2}\right\}$ with $r_{3,eff}$ from Eqs. (T12.2.8) and (T12.2.9)	12.3.3
Diffuser or volute losses as function of flow: $\zeta_{Le}=f(Q)$	calculated from Table 12.2 or estimated from Fig. 12.9: $\zeta_{Le}=\zeta_{Le,opt}\,q^{*2}$ and $Z_{Le}=\zeta_{Le}\dfrac{u_2^2}{2g}$	12.3.4
Impeller losses at BEP	Option A: at BEP: $\zeta_{La,opt}=f(n_q,\text{type})$	12.3.5
	Option B: If option A is used, check that $\zeta_{La,opt}$ from Eq. (12.3.7) is positive $\dfrac{\eta_{h,opt,T}}{\eta_{h,opt,P}}=f(n_q,\text{type})$	12.3.6
	$\zeta_{La,opt}=\psi(1-\eta_{h,opt,T})-\zeta_{Le,opt}$	12.3.7
Impeller losses as function of flow: $\zeta_{La}=f(Q)$ $Z_{La}=f(Q)$	$\zeta_{La}=\zeta_{La,opt}+a\,q^*+b\,q^{*2}+c\,q^{*3}$ $\qquad c=1.3;\ x_o=0.97$; $a=c\,x_o^2$; $b=-2\,x_o\,c$	12.3.8
	$Z_{La}=\zeta_{La}\dfrac{u_2^2}{2g}$	12.3.9
Head: $H=f(Q)$	$H=H_{th}+Z_L+Z_{casing,tot}$	12.3.10
Determine secondary losses from Chap.3	$P_{RR},\ P_m,\ P_{s3},\ P_{er},\ Q_{sp},\ Q_E$	12.3.11
Power: $P=f(Q)$	$P=g\,H_{th}\,(Q-Q_{sp}-Q_E)-P_{RR}-P_m-P_{er}-P_{s3}$	12.3.12
Turbine efficiency $\eta=f(Q)$	$\eta=\dfrac{P}{\rho\,g\,H\,Q}$	12.3.13

Throttle control should be used if the operation point (D in Fig. 12.11) is above the turbine characteristic, that is to say if the flow rate Q_1 fed to the turbine is smaller than the value corresponding to the point of intersection A of the system and the turbine characteristics: $Q_1 < Q_A$ according to Fig. 12.11. By this process

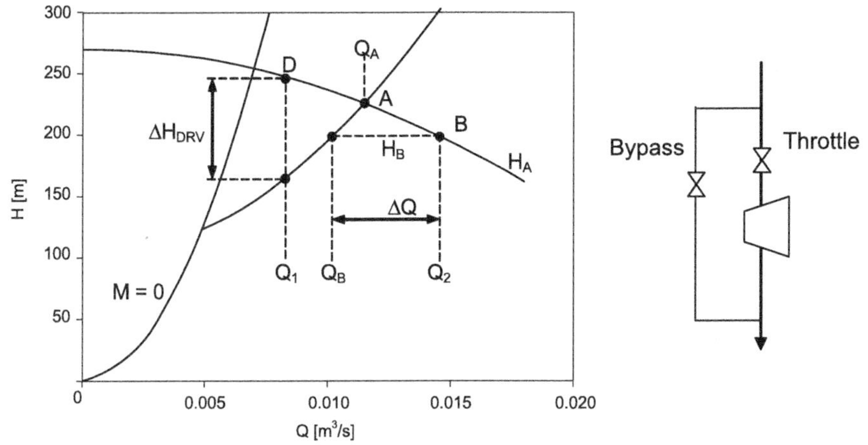

Fig. 12.11 Turbine control by throttle or bypass

the head difference ΔH_{DRV} is throttled and the power $P_v = \rho \times g \times \Delta H_{DRV} \times Q_1$ is dissipated.

A turbine of constant speed and fixed blade position cannot process a flow rate which is greater than the value at the point A given by the intersection of the system characteristic with the turbine curve. If the process requires expanding a flow rate Q_2 (point B in Fig. 12.11) which exceeds Q_A, the excess fluid $\Delta Q = Q_2 - Q_B$ must be led through a bypass in which all of the head H_B is throttled. The power $P_v = \rho \times g \times H_B \times (Q_2 - Q_B)$ is dissipated. Hence bypass control should be applied whenever the required operating point B of the system is below the turbine characteristic.

Regarding speed control, it should be noted that a *higher* flow rate requires a *lower* speed at a given head, as can be seen from Fig. 12.6. Note that the torque increases according to the ratio P/ω. Speed control can be advantageous in cases where flow rate and head of the system behave in a way that allows the turbine always to be run close to the best point. Whether or not speed control is economically advantageous depends on how close together the runaway and resistance characteristics are (here the specific speed has an effect) and which control range has to be used. In many cases runaway and resistance characteristics are too close together to allow a speed control over a sufficiently large range.

With turbine characteristics as shown in Fig. 12.6, speed control at constant head covers an acceptable flow range. Likewise, it can be seen from Fig. 12.6 that with constant flow rate and variable head a relatively large control range could be covered. With a low flow rate and a high head a speed controlled turbine operates at low efficiencies at partload (low flow rates requires *high* speeds). In that case speed control offers only slight advantages (or none at all) over throttle control, (refer also to [16]).

The control strategy to be selected for a given application is largely determined by the process. Frequently the exhaust pressures will be monitored (kept constant), the inlet pressure may be controlled by throttling and any excess liquid is expanded through a bypass.

In processes with large variations in flow rate adjustable guide vanes are suitable for control since they allow the operation of a turbine over a wide range at essentially the same head with only slightly impaired efficiency at low flow. Figure 12.12 demonstrates this by comparing the flat envelope of the BEP efficiencies which are obtained by adjustable guide vane control to the efficiency curve with fixed guide vanes. Figure 12.13 depicts the head and efficiency characteristics

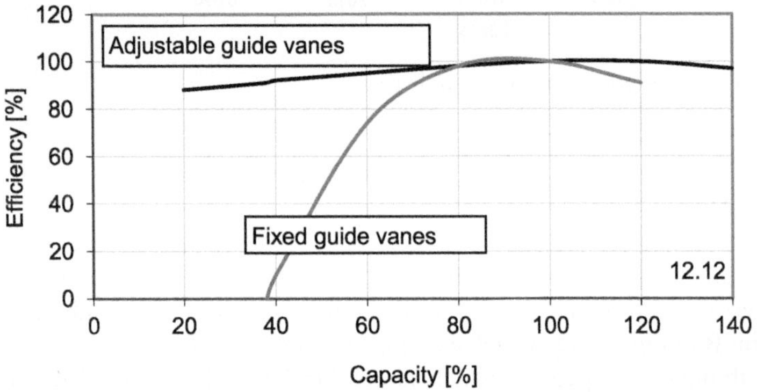

Fig. 12.12 Turbine control by adjustable versus fixed guide vanes; all efficiencies are referred to the design point of the turbine

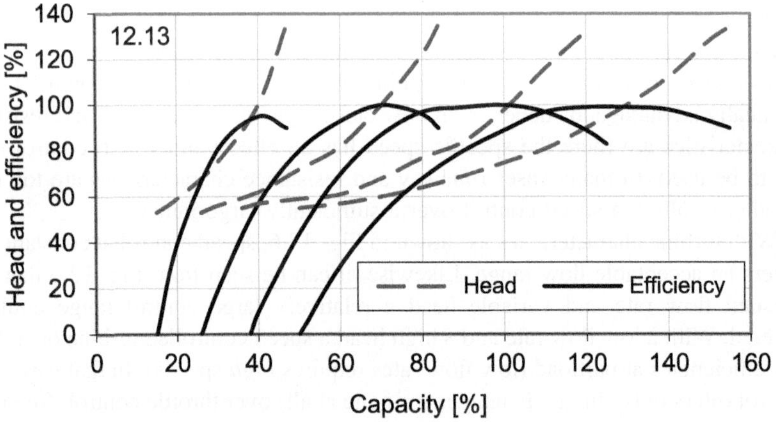

Fig. 12.13 Turbine head and efficiency with adjustable guide vane control; all efficiencies and heads are referred to the design point of the turbine

at various guide vane positions. Both figures confirm that a turbine with adjustable guide vanes can well be operated in ranges which would be impossible to reach with fixed guide vanes.

Modification of turbine characteristics: During *pump* operation, a reduction in the impeller diameter ("impeller trimming") allows a very effective reduction of the power consumption and the head. In contrast, the *turbine* characteristic near the best efficiency point is little affected by impeller trimming. Equation (12.7) supplies the explanation for this experimental finding: if the impeller diameter is reduced from d_2 to d_2', the tip speed u_2' decreases, but c_{2u} *increases* according to Eq. (12.2) by the factor d_2/d_2' due to the conservation of angular momentum. Therefore we find $u_2' \times c_{2u}'/(u_2 \times c_{2u}) \approx 1$. Hence the theoretical specific work virtually does not change upon impeller trimming.

As a consequence of impeller trimming, the runaway head drops to $H_L'/H_L \approx (d_2'/d_2)^2$ and the efficiency at partload improves, (tests reported in [4, 9]). However, the shape of the runaway characteristic does not change considerably. The turbine characteristic at locked rotor (n = 0) upon impeller trimming shifts to lower flow rates (at H = constant), [4]. So the resistance at n = 0 increases with impeller trimming. The turbine characteristic at locked rotor and the overload characteristic therefore become steeper.

It follows from Eq. (12.7) that the specific work at a given flow rate is reduced by opening the collector throat area (cross-section A_{3q}) and the characteristic becomes flatter. If the required turbine flow rate cannot be achieved (or if the flow should be increased in order to increase the power), the collector throat area must consequently be enlarged. If a steeper characteristic or a lower flow rate is required, A_{3q} has to be reduced accordingly. The extent of the required modification can be estimated from Eq. (12.7). Changing the throat area (or the "nozzle") is a very effective way to modify the turbine characteristics and to adapt a given machine to the required service. The effect of modifying the throat area on the characteristics can be appreciated by considering Fig. 12.13.

When applying multistage pumps as turbines, it is recommended to design the machine for easy removal or addition of one stage. It is then possible to adapt the turbine to the plant requirements with little cost. By adding a stage the operation point of the turbine is shifted to lower flows (at constant head). Vice versa the flow through the turbine and the power generated can be increased by removing one stage. At constant flow the power would be proportional to the number of stages. The output of a turbine can be increased by designing a special turbine impeller with more blades and higher blade outlet angles, Fig. 12.21.

By cutting back the runner vanes at the turbine outlet, A_{1q} can be increased and a flatter characteristic can be achieved. This remedy is supposed to be helpful especially at higher specific speeds; its effect can be assessed from Eq. (12.7).

By sharpening the blades at the impeller outer diameter, the angle β_{2B} can possibly be increased and the point of shock-free entry and the BEP be shifted towards a slightly higher flow rate. H_{opt} increases but H(Q) remains roughly the same.

Application guidelines:

1. In turbine operation a minimum back pressure must be available at the exhaust nozzle in order to avoid cavitation problems such as noise, vibrations or material erosion. According to a diagram in [12], the required NPSH of a turbine can be estimated from Eq. (T12.1.16). Another rule of thumb states that the $NPSH_R$ in turbine operation is 35 to 50% of the $NPSH_R$ of the pump [1]. The NPSH of the plant must be above the required $NPSH_R$ by an appropriate safety margin. For example, 50% of the margin recommended in Table 6.2 could be used for turbines.
2. Considerably smaller radial forces can be expected in turbine operation than in the pump, because the fluid passes through the volute without flow separation and, as a result, important disturbances in the circumferential symmetry do not occur.
3. The axial forces in turbine operation hardly depend on the flow rate according to measurements in [18]; they were 40 to 70% of the axial forces at the best efficiency point of the pump in the machines investigated in [18].
4. Impeller design: the impeller blades and shrouds should be rounded according to Fig. 12.14. This reduces the losses created when the fluid enters the impeller and consequently improves the efficiency at BEP by 1 to 2%. In partload operation the efficiency gain is even greater owing to reduced shock losses, [B.9]. The blade leading edge may become less sensitive to incidence when it is shaped as an elliptic profile which can be symmetric or asymmetric either on the pressure or the suction side - depending whether the BEP flow is to be increased or reduced.
5. As the efficiency decreases rapidly at partload but remains flat for $Q > Q_{opt,T}$, partload operation is disadvantageous; therefore do not oversize turbines!
6. Turbine performance can be improved by employing special impellers which have more blades and higher blade outlet angles than commonly selected for pumps, Fig. 12.21.

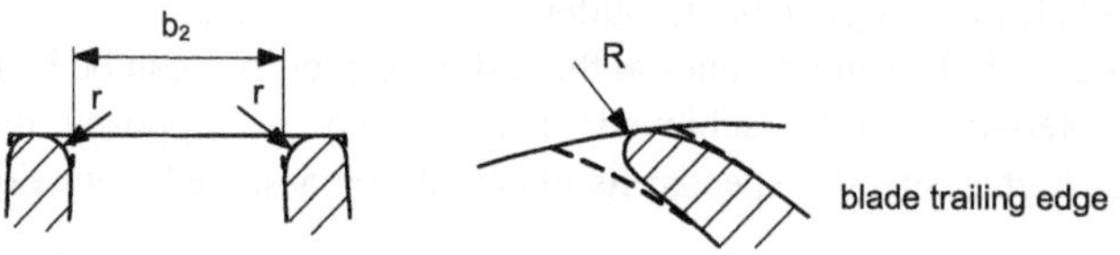

Fig. 12.14 Rounding of blades and shroud edges on the runner outer diameter for turbine operation

12.2 General Characteristics

According to the definitions given above, a centrifugal pump works with positive direction of flow and rotation when operating as a pump, whereas it runs in the negative direction of flow and rotation as a turbine. In both cases the pressure in the discharge nozzle is higher than in the suction nozzle. This condition will be termed as "positive head". The torque is counted as positive as well.

In theory it is possible for n, Q, H and M to assume any one of 16 combinations of positive and negative signs. Eight of them are significant in practice, because they can occur in transients and in upset or abnormal operation conditions, [8, 10]. Some examples of such conditions are: (1) reverse running of a pump caused by back-flow from the discharge pipe following a power failure; (2) start-up of a pump against a back-flow from the discharge pipe; (3) wrong sense of rotation; (4) flow, without power supply from the driver under an imposed positive or negative pressure difference.

The analysis of water hammer following a power failure (Chap. 11.5) or of upset conditions such as the flow through a pump after a pipe rupture has to rely on quantitative data concerning the pump behavior under all possible rotation and flow directions (termed "general characteristics").

According to [B.2] the mentioned eight operation modes of practical interest, A to H, can be represented in the four quadrants of a coordinate system with the flow rate as abscissa and the speed as ordinate, Fig. 12.16. All values are in relation to the performance data at the BEP during operation as a pump. The same conditions are represented also in Fig. 12.15 for the positive and the negative sense of rotation. The eight operation modes of practical interest are shown in Fig. 12.16; they are discussed in the following.

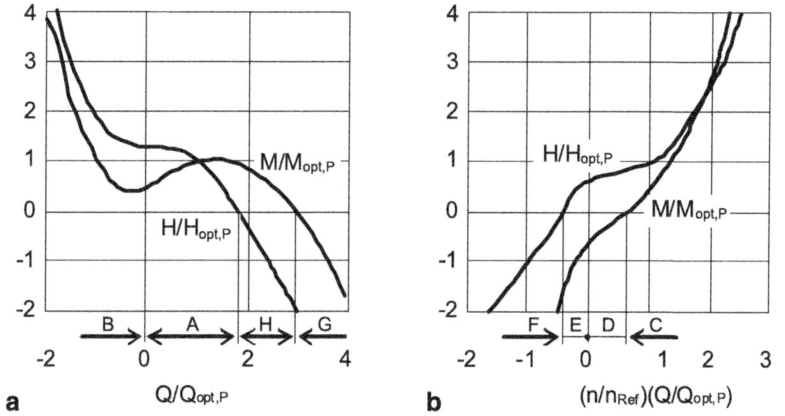

Fig. 12.15 General characteristics presenting head and torque as a function of the flow rate; **a** positive sense of rotation; **b** negative sense of rotation; M = torque

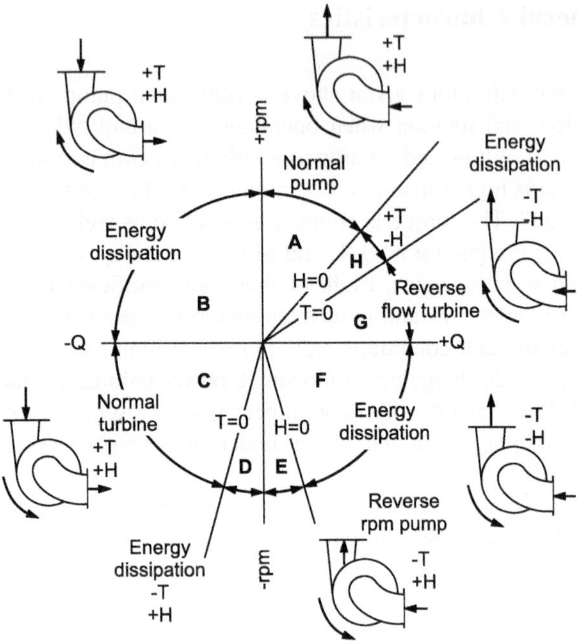

Fig. 12.16 Operation modes, [B.2]; T = torque

A. Normal operation as a pump between maximum flow rate Q_{max} at H = 0 and operation against closed discharge valve with H_o. If the pressure against which the pump delivers increases, the operation point runs back on the characteristic until at $H = H_o$ the point Q = 0 is reached and the flow is stopped.

B. If the discharge pressure exceeds H_o, the fluid flows back through the pump, even though the rotor is driven in the positive direction because the impeller is unable to sustain any externally imposed head $H > H_o$. Head and torque are positive. All of the power supplied at the coupling is dissipated. The hydraulic energy resulting from the through-flow Q_R and the head difference between the discharge and suction nozzles ($\rho \times g \times H \times Q_R$) is dissipated too. The pump is running in the "brake mode". This condition occurs for a short time if there is a power failure and the discharge pipe is not equipped with a check valve (or the valve fails to close).

C. Normal operation as a turbine in the range between runaway and locked rotor characteristics, Chap. 12.1. In the representation shown in Figs. 12.11 and 12.17 the operation points of the locked rotor characteristic are placed on the abscissa (n = 0); the runaway speeds are to be found on the straight line with M = 0.

D. If the flow rate is too low ($Q < Q_L$) for overcoming the idling or runaway torque of the machine, external torque must be applied to the rotor in order to process the flow rate $0 < Q < Q_L$. The turbine works as a brake and dissipates energy. Consequently this operation range is between the runaway characteristic of the turbine and the ordinate, Fig. 12.4.

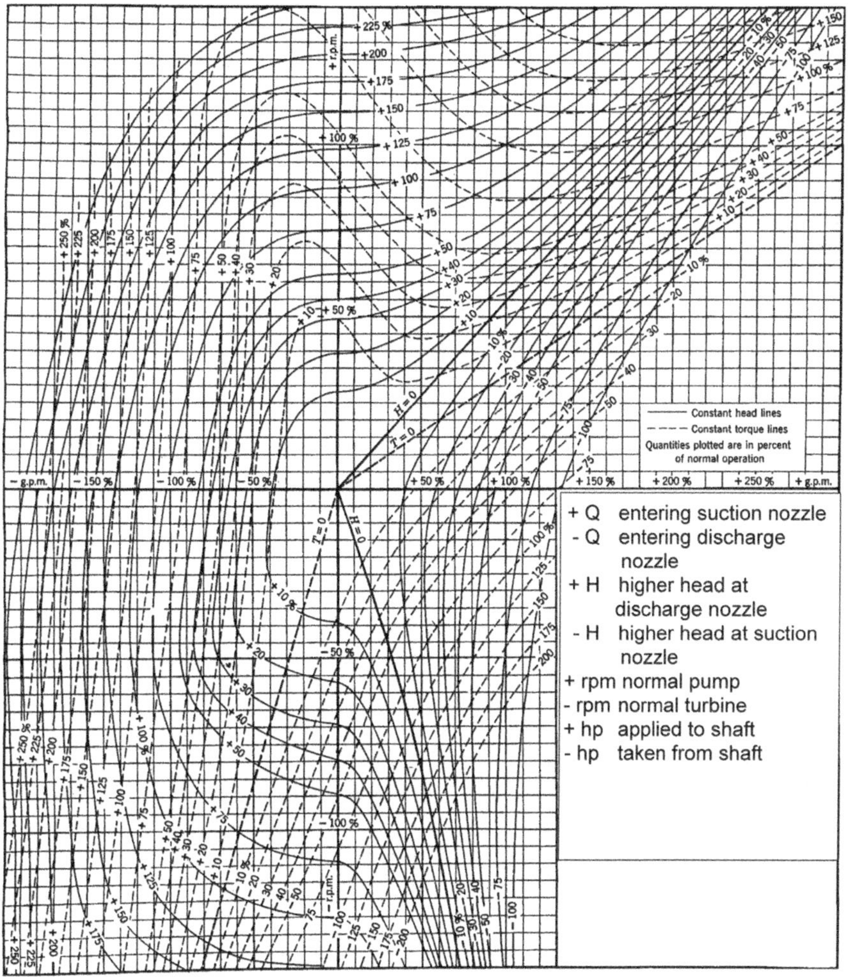

Fig. 12.17 General characteristic of a double-entry pump n_q = 36, [B.2]; T = torque

E. If the pressure in the discharge nozzle drops further, the machine pumps from the suction to the discharge nozzle (Q > 0) with an abnormal sense of rotation. This occurs, for example, if the poles of an electric motor are connected wrongly or a double-entry impeller is mounted on the shaft in the wrong way. Because of important shock losses caused by excessive incidence, head and efficiency are greatly reduced in that case.

F. If the pressure in the discharge nozzle drops below the pressure prevalent in the suction nozzle (i.e. H becomes negative), the pump runs with an abnormal sense of rotation and works as a brake.

G. If the pressure in the suction nozzle is sufficiently high above the pressure in the discharge nozzle (i.e. if a sufficiently high negative head is available), the machine works with a positive sense of rotation as a turbine (with an abnormal direction of *turbine* rotation). This occurs for instance with two (or more) pumps connected in series, if the drive of the second pump fails and the upstream pump forces fluid through the machine left without drive. Another example of an upset condition is a rupture of the discharge pipeline when the suction or system pressure is sufficiently high (e.g. in primary coolant system of a reactor).

H. Between that range where the rotor can deliver power as a turbine and the limit H = 0 of normal pump operation, both the energy difference prevalent between the suction and the discharge nozzle and the power at the coupling are dissipated (Fig. 12.18).

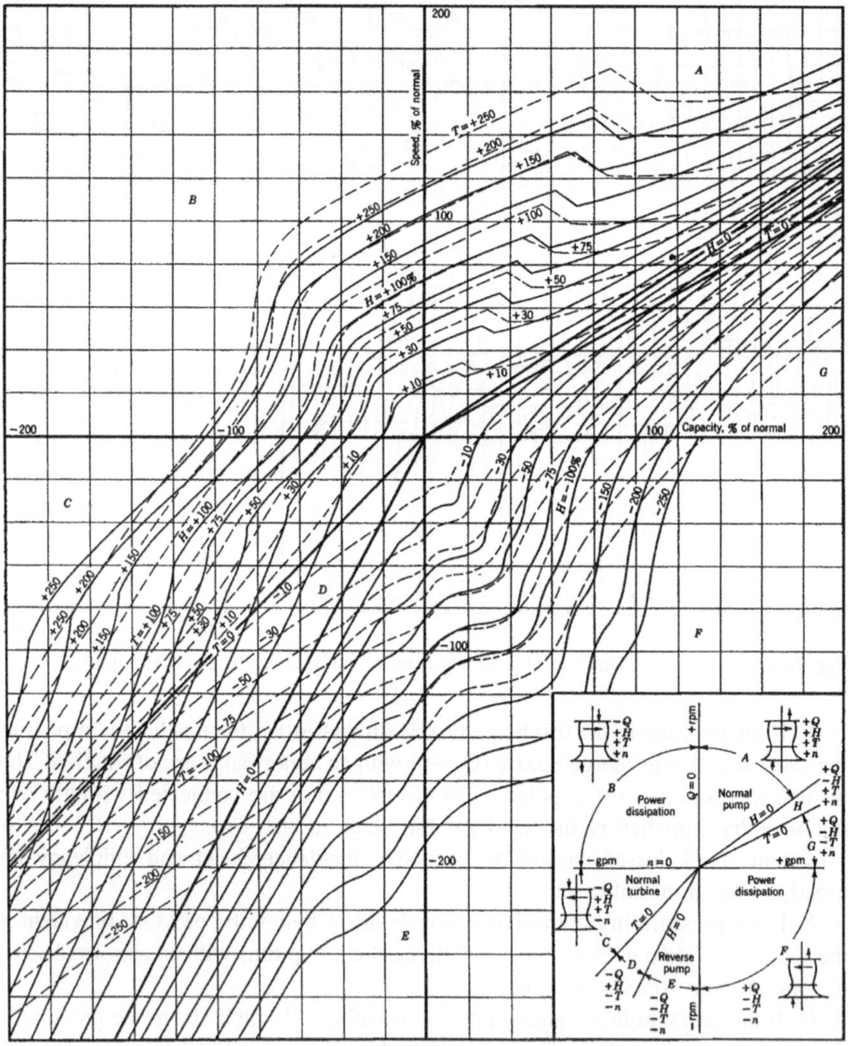

Fig. 12.18 General characteristics of a semi-axial pump n_q = 145, [B.2]; T = torque

In Figs. 12.17, 12.18 and 12.19 from [B.2], the general characteristics are plotted in accordance with the four-quadrant representation for the specific speeds $n_q = 36$, 145 and 260. In the absence of specific test results these diagrams facilitate a rough estimation of the operation conditions to be expected during upset and transient conditions. Measurements at abnormal operation conditions are found in [6, 10, 11]. For the detailed analysis of water hammer problems and upset conditions computer programs are used which require the general characteristics to be represented in appropriate form, [2, 10, 11].

An example of an operational transient is shown in Fig. 12.20 where the rotor speed, flow rate and head experienced upon a failure of the pump drive are plotted

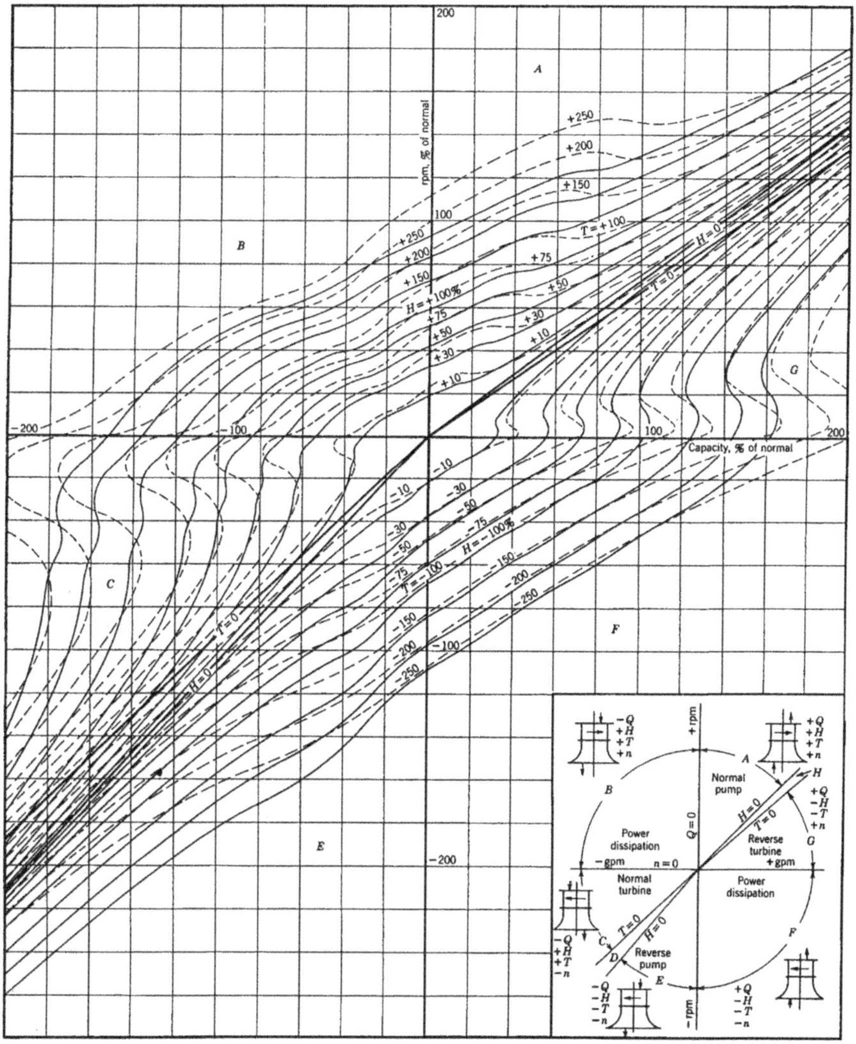

Fig. 12.19 General characteristic of an axial pump $n_q = 260$, [B.2]

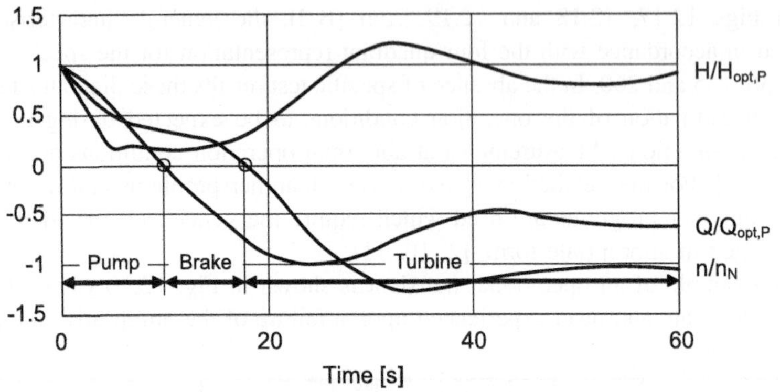

Fig. 12.20 Time-dependent speed, head and flow rate after failure of the pump drive in a system without check valve or flow restriction devices

Fig. 12.21 Double-outlet turbine runner with 14 blades

against the time. Assume that with t < 0 the pump operates at the design point and that at the time t = 0 there is a drive failure and that there is no check valve. Owing to the inertia of the rotating masses and the flowing liquid, a positive flow is initially being maintained upon the drive failure.

The machine in this range is still working as a pump. When the rotor has decelerated to the extent that it is unable to produce sufficient head, the flow direction is reversed (in Fig. 12.20 for t > 10 s) and the machine will work as a brake. As soon as the kinetic energy of the rotor is exhausted (i.e. when the point n = 0 is reached; in Fig. 12.20 at t ≈ 18 s), the rotor is driven by the fluid and the machine will run as a turbine (though without delivering useful power).

The pressures experienced at the nozzles of the pump are largely determined by pressure waves ("water hammer"). During the transient the relation between speed, flow rate and head can therefore not be derived any more from the similarity laws (Fig. 12.21).

References

1. Buse, F.: Using centrifugal pumps as hydraulic turbines. Chem. Eng. **88**, 113–117 (1981)
2. Chaudhry, M.H.: Applied Hydraulic Transients, 2nd edn. Van Nostrand Reinhold, New York (1987)
3. Cohrs, D.: Kennlinienvorausbestimmung bei Kreiselpumpen im Turbinenbetrieb. Pumpentagung, Karlsruhe (A3) (1996)
4. Diederich, H.: Verwendung von Kreiselpumpen als Turbinen. KSB Techn Ber **12** (1966)
5. Engeda, A., et al.: Auswahl von Kreiselpumpen als Turbinen. Pumpentagung, Karlsruhe (A6) (1988)
6. Florjancic, D.: Neue Entwicklungen auf dem Gebiet der Umkehrmaschine für Pumpspeicherwerke. Techn Rundschau Sulzer Forschungsheft (1961)
7. Gülich, J.F.: Energierückgewinnung mit Pumpen im Turbinenbetrieb bei Expansion von Zweiphasengemischen. Techn Rundschau Sulzer **3**, 87–91 (1981)
8. Hergt, P., et al.: Die strömungstechnischen Eigenschaften von Kreiselpumpen im Turbinenbetrieb. Pumpentagung, Karlsruhe (C1) (1984)
9. Hirschberger, M., Kuhlmann, J.: Entwicklung und Einsatz doppelströmiger Kreiselpumpen als Entspannungsturbinen. In: Pumpen als Turbinen. Faragallah, Sulzbach (1993)
10. Martin C.S.: Representation of pump characteristics for transient analysis. ASME Winter Annual Meeting, Boston, FED, vol. 6, pp. 1–13 (1983)
11. Patterson, I.S., Martin C.S.: Effect of specific speed on pump characteristics and hydraulic transients in abnormal zones of operation. In: IAHR Symp Stirling, pp. 151–172 (1984)
12. Priesnitz, C.: Einsatzmöglichkeiten von rückwärtslaufenden Standardkreiselpumpen als Turbinen zur Energierückgewinnung. In: Pumpen und Verdichter Informationen, pp. 3–12 (1987)
13. Schmiedl, E.: Pumpen als Turbinen. Pumpentagung, Karlsruhe (A6) (1988)
14. Sharma, K.R.: Small hydroelectric projects—use of centrifugal pumps as turbines. Kirloskar Electric Co., Bangalore (1985)
15. Surek, D.: Axialpumpen im Turbinenbetrieb. In: Pumpen als Turbinen. Faragallah, Sulzbach (1993)
16. Wesche, W.: Vergleichende Betrachtung von Kreiselpumpen im Turbinenbetrieb. In: Pumpen als Turbinen. Faragallah, Sulzbach (1993)
17. Williams, A.A.: The turbine performance of centrifugal pumps: a comparison of prediction methods. Proc. IMechE. **208**, 59–66 (1994)
18. Yang, C.S.: Performance of the vertical turbine pumps as hydraulic turbines. ASME Winter Annual Meeting Boston, pp. 97–102 (1983)

Chapter 13
Influence of the Medium on Performance

High viscosities (low Reynolds numbers) impair the pump performance. Liquids with up to $\nu = 3000\ mm^2/s$ (3000 cSt) can be pumped with centrifugal pumps, but efficiencies drop to very low levels which make the operation highly uneconomic. Large amounts of free gas entrained in the liquid seriously deteriorate performance and restrict the range where pumping is possible at all. Centrifugal pumps can handle large amounts of solid matter even though the efficiency suffers because the flow paths of the solids deviate from those of the liquid due to differences in density. Empirical methods are used to estimate the drop in performance in applications which involve the fluids mentioned above.

13.1 Pumping Highly Viscous Fluids

13.1.1 Effect of Viscosity on Losses and Performance Characteristics

When a centrifugal pump is used for transporting a fluid with a viscosity much higher than cold water, additional losses impair the performance. Therefore, the pump characteristics determined for water must not be applied without correction to pumping highly viscous fluids as encountered for example in the oil and process industry.

At a sufficiently high viscosity ν the flow regime becomes laminar. Depending on the size and the speed of the pump, the transition from turbulent to laminar flow occurs at about $\nu = 10^{-4}\ m^2/s$. For $\nu < 10^{-5}\ m^2/s$ the viscosity influence is small; therefore, the methods of efficiency scaling explained in Chap. 3.10 may be applied in this range.

How the performance data and pump characteristics are changed from service with water (subscript w) to operation with a viscous fluid (subscript v) is calculated by empirical methods. To this end, Eq. (13.1) defines correction factors for flow rate, head and efficiency (similar to Chap. 3.10.3):

J. F. Gülich, *Centrifugal Pumps*,
https://doi.org/10.1007/978-3-030-14788-4_13

$$f_Q = \frac{Q_v}{Q_w} \quad f_H = \frac{H_v}{H_w} \quad f_\eta = \frac{\eta_v}{\eta_w} \qquad (13.1)$$

Prior to discussing the empirical methods in detail, let us examine some general principles concerning the physical phenomena involved and the magnitude of the losses to be expected. Consider the impact of high viscosities on the power balance of a pump and the secondary losses according to Chaps. 3.5 and 3.6:

1. The mechanical losses are independent of the properties of the medium delivered; for pumping a viscous fluid they are the same as for pumping water.
2. Leakage losses through the annular seals diminish with decreasing Reynolds number or growing viscosity. However, this drop is less important than would be expected from calculating with the viscosity prevalent at the temperature in the suction nozzle: As discussed in Chap. 3.6.2 high shear stresses are generated in the narrow gaps of annular seals. Per surface unit the mechanical energy $P_d/A = \tau_o \times w$ is dissipated into heat (τ_o is the wall shear stress). Consequently, extreme shear stresses in an annular seal inevitably result in heating up of the medium flowing through the gap. The more so, because the flow through the gap is small and the thermal transport properties of oil are not as good as those of water.[1] Since the viscosity of oil decreases strongly with rising temperature, the viscosity in the seal drops to values lower than at the temperature in the suction nozzle. Therefore, it may be assumed that the leakage losses slightly diminish with increasing viscosity, but that their overall impact on the efficiency is small when changing from operation with water to a viscous fluid. Changes in volumetric efficiency may thus be conservatively neglected.
3. Disk friction losses grow with decreasing Reynolds number or increasing viscosity. Their influence on the efficiency is very important, as will be shown later on, especially with low specific speeds.
4. The hydraulic losses in inlet, impeller and volute or diffuser according to Chap. 3.8 are made up by friction losses which depend on the Reynolds number and by losses due to turbulent dissipation which are virtually independent of the Reynolds number.
5. As mentioned above, the flow regime of highly viscous fluids tends to be laminar. That is why the roughness of the impeller sidewalls and of the hydraulic channels is of little importance for operation with highly viscous fluids.

Disk friction losses and friction losses in the hydraulic channels are the controlling factors in viscous pumping. They can be estimated in accordance with Chap. 3 as follows:

Disk friction losses: The impact of disk friction can be determined from Table 3.6, calculating the friction factor k_{RR} from Eq. (T3.6.3)[2]. The ratio P_{RR}/P_u of the disk

[1]The term "oil" in the following text is meant to include all highly viscous fluids.

[2]This equation covers the whole range from laminar to fully turbulent flow.

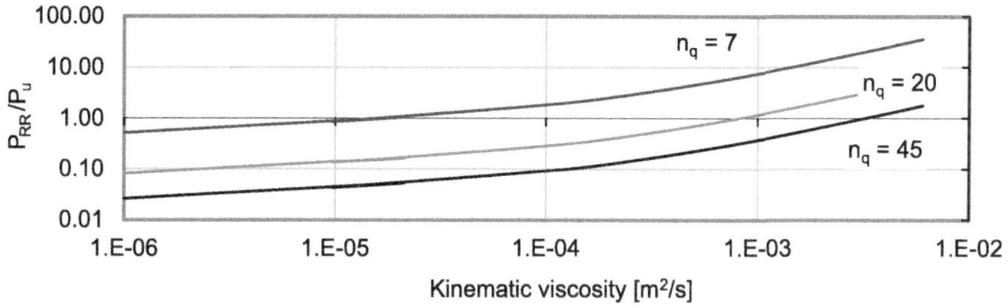

Fig. 13.1 Influence of the viscosity on the disk friction losses P_{RR}/P_u; calculated with $n = 1450$ rpm, $d_2 = 350$ mm and $s_{ax}/r_2 = 0.035$

friction losses P_{RR} to the useful power P_u of an impeller is derived from Eq. (T3.5.13).

Figure 13.1 shows this ratio as a function of the viscosity with the specific speed as a parameter.

The following data have been assumed for this calculation: head coefficient ψ_{opt} according to Eq. (3.26), $n = 1450$ rpm, $d_2 = 350$ mm and $s_{ax}/r_2 = 0.035$. The character of the curves in Fig. 13.1 depends little on these assumptions. If the viscosity increases from 10^{-6} to 3×10^{-3} m²/s, the ratio of the disk friction to the useful power grows by a factor of 35. At a specific speed of $n_q = 7$, disk friction losses are about 18-times larger than the useful power. Even at $n_q = 45$, the disk friction is almost as high as the useful power if the viscosity reaches 3×10^{-3} m²/s. The power consumption of the pump increases over the whole flow rate range by the difference in disk friction losses $\Delta P_{RR} = (P_{RR,v} - P_{RR,w})$. The power curve thus shifts upward nearly parallel to the curve for water.

Disregarding any other influences, let us now consider how disk friction losses *alone* affect the efficiency when the pump operation is changed from water to oil. The efficiency of a single-stage pump can be defined according to Table 3.5 (with $P_{Rec} = P_{s3} = P_{er} = 0$) by the approximate expression:

$$\eta \approx \frac{\eta_m}{\frac{1}{\eta_{vol}\,\eta_h} + \frac{P_{RR}}{P_u}} \tag{13.2}$$

Assuming $\eta_{vol}\,\eta_h = 0.86$, the calculation with the values for P_{RR}/P_u from Fig. 13.1 yields the efficiency correction factors shown in Fig. 13.2. These factors include *solely* the effect of the higher disk friction resulting from growing viscosity (the mechanical efficiency cancels if the factor f_η is formed).

From these considerations as well as from Fig. 13.2, it is evident that the increase in power consumption and the drop in efficiency of a pump in viscous service depend strongly on the specific speed - *and this alone due to the effect of disk friction*.

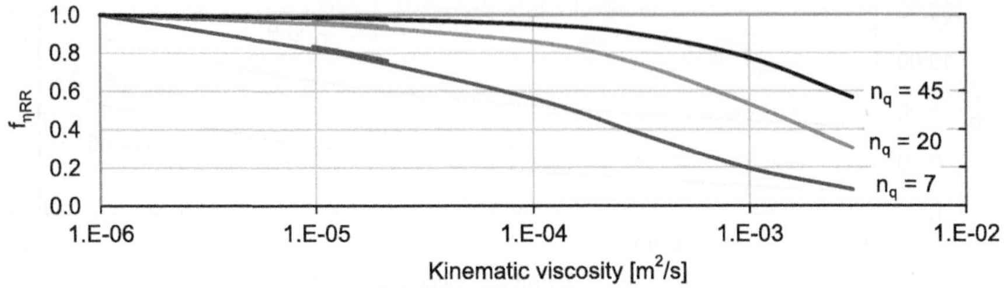

Fig. 13.2 Influence of the disk friction losses on the efficiency; calculated with, n = 1450 rpm, d_2 = 350 mm and $s_{ax}/r_2 = 0.035$

Hydraulic losses: The influence of the viscosity on the hydraulic losses is discussed with reference to Chap. 3.10.3, where these losses are considered as the sum of Reynolds-dependent friction losses Z_R and mixing losses Z_M which do not depend on the viscosity. Consequently, the theoretical head for operation with water and a highly viscous fluid can be expressed as (see also Table 3.8):

$$H_{th} = H_w + Z_{R,w} + Z_{M,w} = H_v + Z_{R,v} + Z_{M,v} \tag{13.3}$$

In Eq. (13.3) it is assumed that the viscosity has no influence on the slip factor, hence none on the theoretical head either. According to tests in [1] on a pump with $n_q = 30$, this assumption seems to be justifiable even at $v = 1200 \times 10^{-6}\ m^2/s$ (at least as a first approximation).

Since the mixing losses are considered to be independent of the Reynolds number, $Z_{M,v} = Z_{M,w}$ in Eq. (13.3) cancels. Hence it is possible to relate H_v to H_w in terms of the multiplier f_H defined in Eq. (13.1):

$$f_H = \frac{H_v}{H_w} = 1 - \frac{Z_{R,w}}{H_w}\left(\frac{Z_{R,v}}{Z_{R,w}} - 1\right) = 1 - \frac{Z_{R,w}}{H_w}\left(\frac{c_{f,v}}{c_{f,w}} - 1\right) \tag{13.4}$$

Equations (13.3) and (13.4) apply at any specific flow rate Q_x, i.e. for a given theoretical head. However, at the reduced flow Q_v the theoretical head is higher than at Q_w as demonstrated by Fig. 13.3. This implies that the hydraulic efficiency is lower than f_H if the slip is not affected by the viscosity, Eq. (13.4a). In contrast, if the slip factor diminishes ($\gamma_v < \gamma_w$), the factors $f_{\eta h}$ and f_H come closer together.

$$f_{\eta h} \equiv \frac{\eta_{h,v}}{\eta_{h,w}} = \frac{H_v\, H_{th,w}}{H_w\, H_{th,v}} = f_H \frac{H_{th,w}}{H_{th,v}} = f_H \frac{\gamma_w - \dfrac{\varphi_{2,La}\tau_2}{\tan\beta_{2B}}}{\gamma_v - \dfrac{\varphi_{2,La}\tau_2}{\tan\beta_{2B}} f_Q} \tag{13.4a}$$

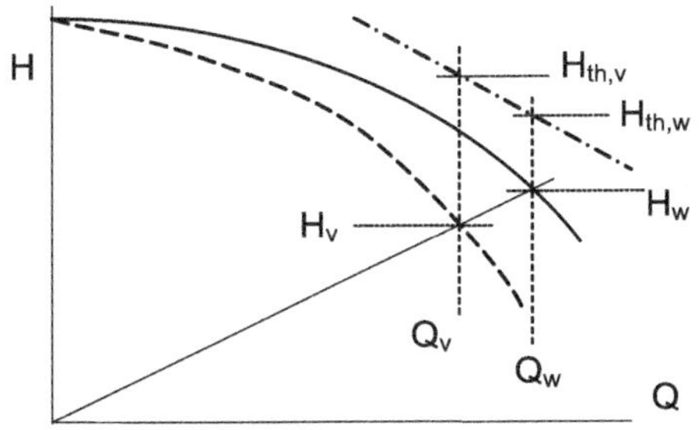

Fig. 13.3 Derivation of hydraulic efficiency factor

If the ratio of the friction losses to the head in operation with water $Z_{R,w}/H_w$ is known, the head loss in pumping viscous fluids can be estimated from Eq. (13.4). Head correction factors have been calculated in this way for the above example. The result is shown in Fig. 13.4 where the ratio of the friction losses to the head $Z_{R,w}/H_w$ was used as a parameter.

With reference to Chap. 3.10.3, the efficiency correction factor given by Eq. (13.5) can be derived from Eqs. (13.2 and 13.4):

$$f_\eta = f_{\eta h} \frac{1 + \left(\frac{P_{RR}}{P_u}\right)_w \eta_{vol}\, \eta_{h,w}}{1 + \left(\frac{P_{RR}}{P_u}\right)_w \frac{k_{RR,v}}{f_Q k_{RR,W}} \eta_{vol}\, \eta_{h,w}} \qquad (13.5)$$

The efficiency correction factor given by Fig. 13.5 was derived for $f_Q \approx f_H$ by means of Eq. (13.5). Figure 13.5 shows once more the great impact of the specific speed. Moreover, the efficiency impairments calculated from this theoretical treatment of losses are similar to those resulting from the tests in Fig. 13.9.

Thermal effects: The efficiency correction factor includes the effects of disk friction, hydraulic losses and volumetric losses. The ratio of the disk friction coefficients in viscous pumping to those in water is calculated from Eq. (T3.10.9) which includes the factor f_{therm} accounting for the viscosity decrease in the impeller sidewall gaps. This drop is caused by high shear stresses which raise the temperature in the boundary layers due to dissipation, as has been explained above when discussing the volumetric losses [2]. Consequently, the wall friction at high viscosities is smaller than would be predicted by calculating with the nominal temperature prevailing in the suction nozzle. From tests reported in [2], it may be concluded that at viscosities above about 400×10^{-6} m^2/s the fluid is heated appreciably due to the dissipation caused by shear stresses. Based on these tests and an analysis of data from tests with pumps operating with highly viscous fluids, an empirical factor f_{therm} has been derived in [3] that allows estimating how much the disk friction is reduced because of thermal effects:

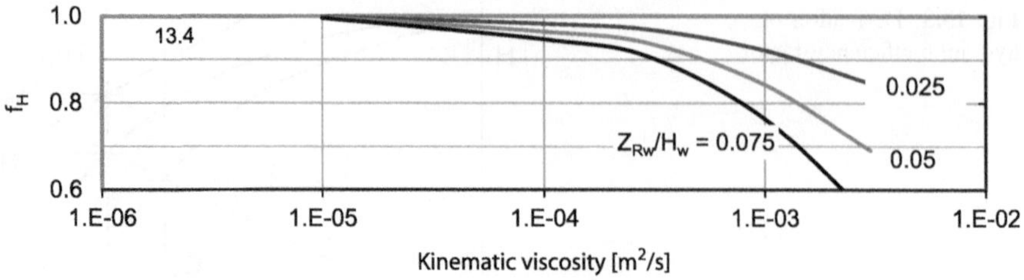

Fig. 13.4 Influence of the viscosity on the hydraulic losses; calculated with, n = 1450 rpm, d_2 = 350 mm

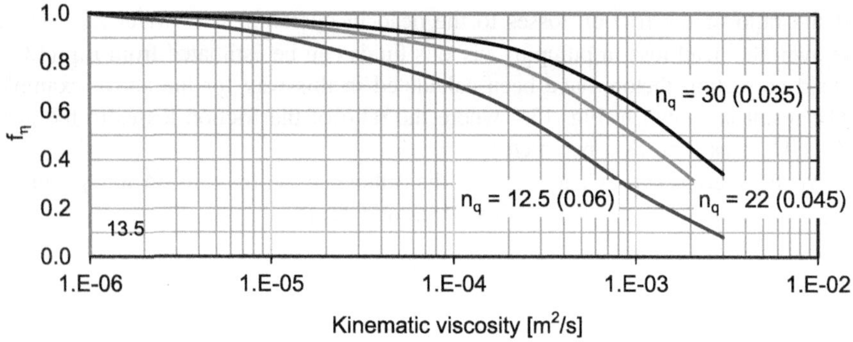

Fig. 13.5 Influence of the viscosity on the efficiency; calculated with ψ_{opt} from Eq. (3.26), n = 1450 rpm, d_2 = 400 mm and s_{ax}/r_2 = 0.035. The figures in parentheses represent the fraction of friction losses in water $Z_{R,w}/H_w$

$$f_{therm} = \exp\left\{-2 \times 10^{-5}\left(\frac{\nu}{\nu_{Ref}}\right)^{1.34}\right\} \quad \text{with} \quad \nu_{Ref} = 10^{-6}\ m^2/s \tag{13.6}$$

Temperature measurements within a single-stage process pump confirm that the fluid temperature significantly increases in the impeller side rooms and the annular seals. Typical results are shown in Fig. 13.6 which gives the suction temperature T_s, the temperature T_{2a} in the impeller side room (ISR) at the diameter d_2 and the temperatures near the inlet ($T_{sp,1}$) and at the outlet ($T_{sp,2}$) of the annular seal. In addition, the viscosities at these local temperatures are given.

Two tests are shown in Fig. 13.6: (1) with a viscosity at suction of 450 mm²/s the oil is heated in the impeller sidewall clearance (ISR) locally by up to 15 °C and the local viscosity drops from 450 to 240 mm²/s. In the annular seal the oil is heated by another 2 °C; (2) with a viscosity at suction of 205 mm²/s the oil is heated in the impeller sidewall clearance by 7 °C and the local viscosity drops from 205 to 150 mm²/s. In the annular seal the oil is heated by 1 °C.

The temperature rise from the suction to the discharge nozzle was 5.6 °C in the test with a viscosity of 450 mm²/s and 3.3 °C in the test 2 with 205 mm²/s. The temperature rise in the impeller is only about 20% of these values because the

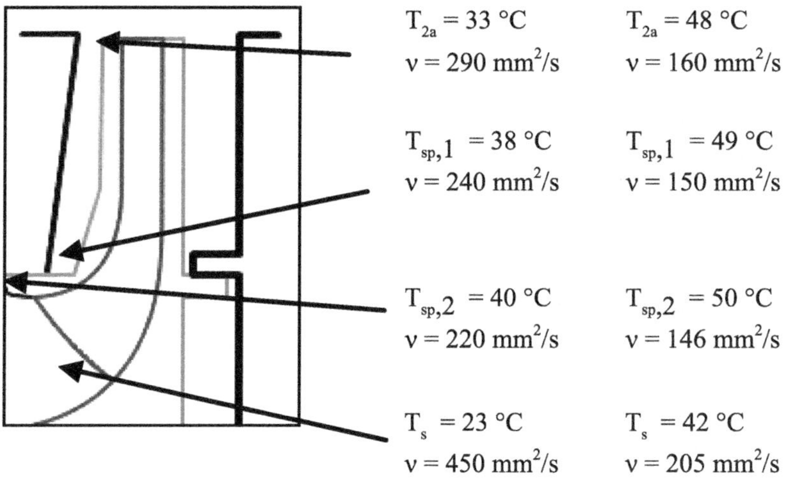

Fig. 13.6 Temperatures in an oil pump $n_q = 7$, $n = 3000$ rpm, $q^* = 1$: left column: $T_s = 23$ °C and $\nu = 450$ mm^2/s; right column: $T_s = 42$ °C and $\nu = 205$ mm^2/s, $A_{gap}/r_2 = 0.04$

hydraulic losses in the impeller are much smaller than in the volute (as discussed in detail below). The main flow temperature rise at the impeller outlet is thus about 1.1 °C at 450 mm^2/s and 0.7 °C at 205 mm^2/s. Overall it is recognized that there are large temperature gradients between the main flow and the ISR and within the ISR. Such gradients, which depend on the flow conditions (geometry), make a theoretical analysis of the disk friction losses very difficult because the local temperatures and viscosities cannot be described by analytical methods.

The heating of the fluid within the pump increases with the shear stress, hence with the impeller tip speed and the viscosity. This is confirmed by the sample test shown in Fig. 13.7 which shows the temperature rise in the ISR due to disk friction as a function of the tip speed u_2 and the viscosity. Consequently, the factor f_{therm} defined by Eq. (13.6) should include the impeller tip speed. But other features of the pump (for example the impeller side wall clearance and gap A) have an influence too.

From Figs. 13.6 and 13.7 it becomes clear that the disk friction losses are over-predicted if calculated with the viscosity at the inlet to the pump. As a consequence, it is *not possible* to determine the hydraulic efficiency from the loss analysis and from Eq. (T3.5.8) as done in tests with water if the actual temperatures and viscosities in the ISR and annular seal cannot be predicted. Hence it is equally impossible to determine the slip factor from Eq. (T3.2.9).

For the tests presented in Fig. 13.7 the thermal effects can be described by Eq. (13.6a), where $k_{RR}(T)$ is the disk friction coefficient calculated from Eq. (T3.6.3) with the viscosity at the average temperature in the impeller sidewall clearance and $k_{RR}(T_s)$ is the friction coefficient calculated with the viscosity at suction.

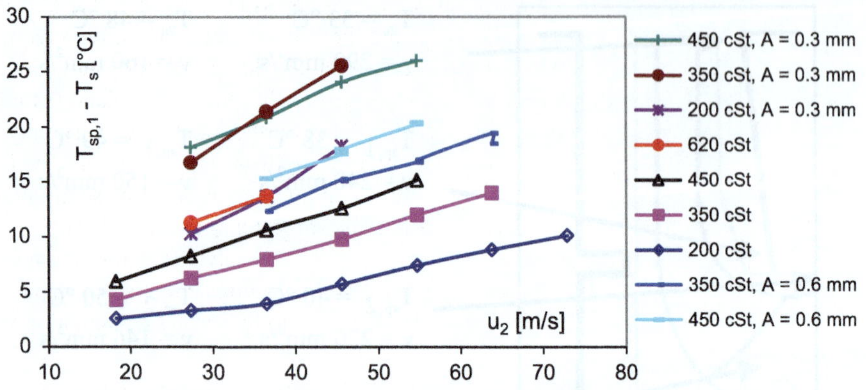

Fig. 13.7 Sample test: temperature rise in impeller sidewall clearance; the data are specific to the particular test set up and the test fluid

$$f_{therm} \equiv \frac{k_{RR}(T)}{k_{RR}(T_s)} = \left(1 - 1.45 \times 10^{-5}\left(\frac{A_{gap}}{r_2}\right)^{-1.54}\right)\left(\frac{u_{Ref}}{u_2}\right)^x \exp\left\{-3.2 \times 10^{-4}\frac{\nu}{\nu_{Ref}}\right\}$$
$$x = 1 - \exp\left\{-2.2 \times 10^{-4}\frac{\nu}{\nu_{Ref}}\right\} \tag{13.6a}$$

with $u_{Ref} = 27.3$ m/s and $\nu_{Ref} = 10^{-6}$ m²/s.

The correlation given by Eq. (13.6a) covers a range of $A_{gap}/r_2 = 0.0019$ to 0.04 (A_{gap} is the radial clearance at gap A according to Table 0.2).

The factor f_{therm} from Eq. (13.6a) is specific to the pump and oil tested since thermal effects depend on several parameters:

1. Heat removal via the leakages through the impeller sidewall gap, hence annular seal geometry.
2. An important effect is the exchange of fluid between the impeller sidewall clearances and the main flow. It is brought about by the pumping action of the shrouds (similar to a friction pump) and by exchange of momentum through gap A.
3. Heat removal from the fluid owing to convection and conduction through the impeller shrouds and the casing.
4. Dependency of the viscosity on temperature: $\nu = f(T)$. The steeper the curve $\nu = f(T)$ the stronger will be the reduction of the losses in a given pump. In other words: a given pump with a given viscosity at the suction nozzle will have quite different losses and efficiency multipliers if the fluids have different curves $\nu = f(T)$. This can be the case with different oils, but the effect can be even stronger when pumping different fluids, e.g. oil versus molasses.
5. Heat transport properties of the pumped fluid (thermal conductivity, specific heat, Prandtl number). Because of different transport properties, some substances - e.g. oil and molasses - behave in different ways even though they may have the same viscosity.

Given all of these influences and the lack of precise data, Eq. (13.6a) has to be considered as a rough approximation only. The strong impact of the heating of the fluid on the efficiency - even at moderate viscosities - contributes to the scatter and uncertainties of statistical viscosity correction factors which are discussed in Chap. 13.1.2.

Characteristics: Figure 13.8a shows how the characteristics are changed when highly viscous fluids are pumped as compared with water transport. If the characteristics for viscous fluids are referred to the data at the best efficiency point (BEP) with water pumping, the multipliers according to Eq. (13.1) can be read directly from this plot. Suppose that curve 1 in Fig. 13.6a shows the characteristics measured with water that correspond to the a hydraulic efficiency of $\eta_{h,w} = 0.9$. Imagine that the same pump operates with appreciably higher hydraulic losses, such as could be caused by excessive surface roughness, throttling or indeed by a higher viscosity. As a consequence, the Q-H-curve may degrade down to curve 2. Higher hydraulic losses cause the best efficiency point to shift to lower flow rates. According to Chap. 4.2 it might be expected (as a first approximation) that the BEP moves along line 3 which represents the volute/diffuser characteristics.

Figure 13.9 shows test results from [1] performed on a single-stage volute pump $n_q = 30$ which confirm this assumption quite well (as demonstrated by the volute characteristic and its intersection with the curve for 1200 mm²/s).

Likewise, tests with viscosities up to 3000 mm²/s on three single-stage volute pumps of $n_q = 12$, 22 and 45 yielded best efficiency points that were close to the values resulting from the volute characteristics [5]. If such is the case, it follows that $f_H = f_Q$ applies. With low Reynolds numbers (high viscosities, low speeds, small impeller diameters) the efficiency curves become very flat, see for example Fig. 13.10. It becomes than difficult to define the BEP. The volute characteristic drawn to the water BEP provides the best guess in this case.

Theoretically, the formulae given in Table 4.1 for calculating the best efficiency flow do not allow a prediction of the BEP in viscous fluids from the collector characteristic, because the hydraulic efficiency cancels in the equations.

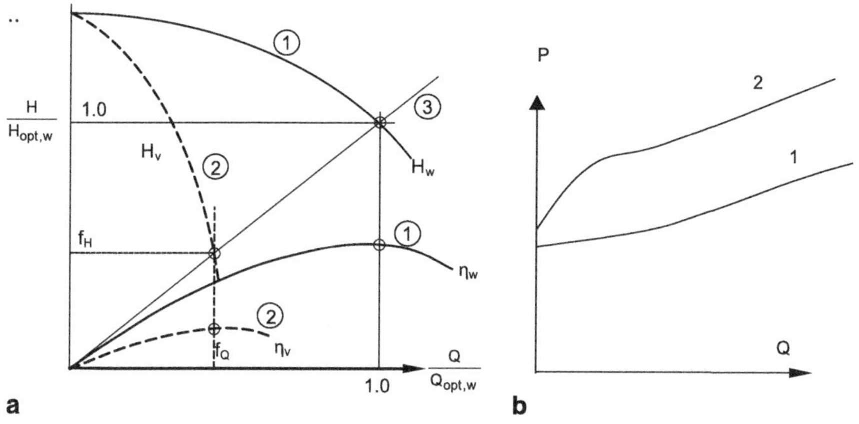

Fig. 13.8 Change of performance curves when pumping highly viscous fluids

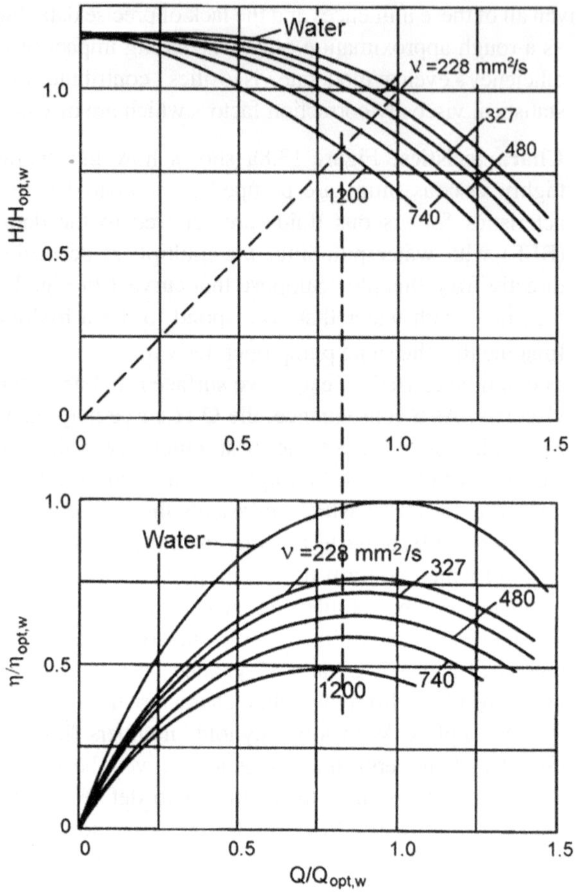

Fig. 13.9 Characteristics of a single-stage volute pump measured with different viscosities, $n_q = 30$ [1]

Empirically, an estimation of the BEP flow rate is possible by drawing a straight line, representing the volute/diffuser characteristic, from the origin to the BEP with water and then calculating the head for pumping viscous fluids from Eq. (13.7):

$$H_v = H_w f_H \tag{13.7}$$

With viscous fluids the power consumption shifts upwards, roughly parallel to the curve for pumping water, as shown in Fig. 13.8b. At flow rates near Q = 0 the power increase is lower because the fluid is heated in the pump. The power increase caused by viscous fluids is almost *entirely* due to greater disk friction losses on the impeller shrouds and annular seals, because the theoretical head H_{th} (and consequently P_{th}) remain essentially constant.

The effect of viscosity on the individual losses and the efficiency as well as on the BEP can be estimated by means of the formulae from Chaps. 3.6, 3.8, 3.10.3 and 4.2 as discussed above. This procedure allows the specific design parameters of any pump to be taken into account. Conversely, only empirical methods (or

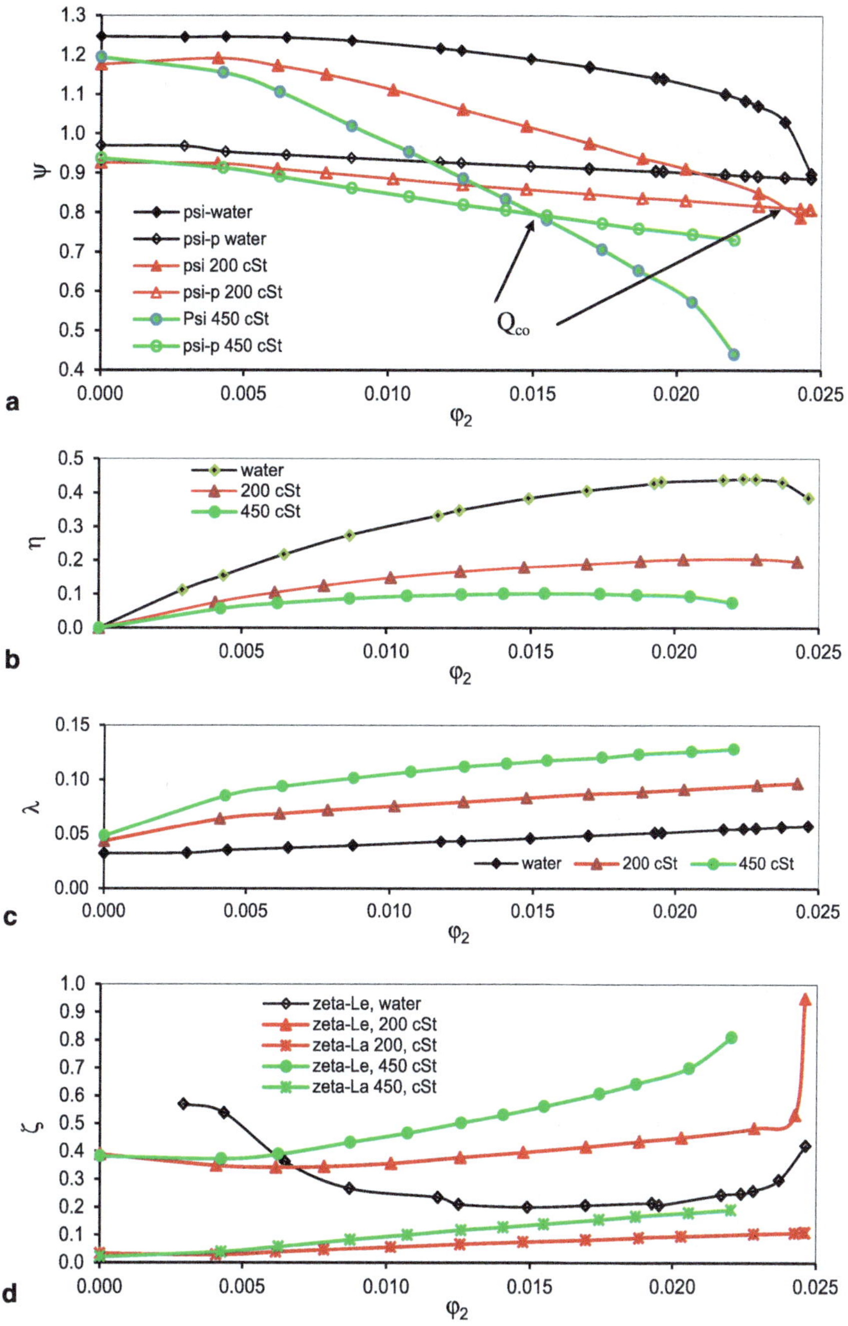

Fig. 13.10 Pressure coefficient, static pressure rise in impeller (**a**), efficiency (**b**), power coefficient (**c**), and impeller and volute losses (**d**); n = 3000 rpm; $n_q = 7$

possibly CFD) can be relied upon for investigating the drop in viscosity caused by the local heating of the boundary layers.

Dimensionless performance curves measured on a single-stage volute pump $n_q = 7$ with viscosities of 450 and 200 mm^2/s are plotted in Fig. 13.10. Figure a shows the pressure coefficients $\psi(\varphi)$ and the static pressure rise created by the impeller in terms of $\psi_p(\varphi)$ determined as per Chap. 4.1.3. Interestingly, the static pressure rise $\psi_p(\varphi)$ crosses the curve $\psi(\varphi)$ at a flow rate Q_{co} which drops with increasing viscosity. At flow rates higher than this cross-over, the *static head at the impeller outlet exceeds the total head in the discharge nozzle*. Accordingly, energy is dissipated in the volute above the cross-over flow rate. Only below the cross-over flow rate Q_{co}, volute or diffuser are able to fulfill their function of pressure recovery. Additional tests with higher viscosity and lower speed showed that the cross-over flow rate drops also when decreasing the speed at a given viscosity. Consequently, the Reynolds-number is the controlling factor.

The shift of the cross-over flow rate to lower flows when decreasing the Reynolds number can be explained by the boundary layer blockage in the volute and diffuser (discharge nozzle). The smaller the hydraulic channels (the lower the specific speed), the higher is the relative boundary layer blockage. As a consequence of the reduced flow deceleration (c_{3q}/c_2) the onset of stall and recirculation is shifted to lower flow rates. For this reason, pumps in viscous operation work smoothly without undue vibrations at flows with low ratios of $Q/Q_{opt,water}$ where water pumps would be affected strongly by flow recirculation. The steep drop of the head coefficient in Fig. 13.10a at the largest flow rate measured is due to cavitation in the volute throat area.

In Fig. 13.10 the BEPs are found at flows close to the cross-over of the curves $\psi_p(\varphi)$ with $\psi(\varphi)$. From this observation it can be concluded that boundary layer blockage is also a main responsible for the shift of the BEP to lower flows when the viscosity is increased. This because higher disk friction and lower leakage flows would tend to shift the BEP rather to higher flow rates.

In order to increase the capacity of a given machine for pumping high-viscosity fluids, an opening of the volute (or diffuser) throat area would therefore be worth considering.

As discussed above, it is *not possible* to determine the hydraulic efficiency from the loss analysis as per Eq. (T3.5.8). Therefore, the slip factor has been assumed to be equal to that derived from the tests with water. Using this slip factor, the theoretical head can be calculated from Eqs. (T3.2.7) and (T3.3.1). The hydraulic efficiency then follows from $\eta_h = H/H_{th}$. The hydraulic losses in impeller and volute can subsequently be derived as per Chap. 4.1.3. The losses derived that way are plotted in Fig. 13.10d. These results suggest: (1) There are virtually no shock losses in the impeller. This observation may be explained by the hypothesis that the fluid motion is able to adapt to the blades which act in the *upstream* direction due to the high viscosity. (2) The losses in the volute and the diffusing discharge nozzle make up about 80% of the hydraulic losses (as expected at low specific speeds); (3) character and magnitude of the impeller losses are quite similar to the water test data shown in Fig. 4.7; (4) in contrast, the dependence of the volute losses on the flow rate is completely different from Fig. 4.8 in that there is scarcely a minimum in the function $\zeta_{Le} = f(\varphi)$. With a viscosity of 200 cSt the casing losses depend little on the flow rate, but with increasing viscosity the losses rise steadily from shut-off.

As mentioned above the power curves are shifted up by a constant amount due to the increased disk friction losses when the viscosity is increased - except near shutoff where the viscosity drops because the oil is heated due to the large power dissipation at small or no through-flow. The test in Fig. 13.10c well confirms this observation.

13.1.2 Estimation of Viscous Performance from the Characteristics Measured with Water

The discussion in Chap. 13.1.1 demonstrates (as could be expected a priori) that the hydraulic losses and the efficiency of a pump operating with viscous fluids depend on the *Reynolds number* and the *specific speed.* The latter reflects, as a first approximation, the geometry of the hydraulic channels and the distribution of losses. In addition, the flow rate ratio q* may be expected to have an influence. Therefore, the correction factors defined by Eq. (13.1) are in general terms described by the expression:

$$f_x = f(Re, n_q, q^*) = f(n, Q, H, d_2, q^*, \nu)$$

For estimating the performance with highly viscous fluids from test data measured with water, two methods are available: (A) Loss analysis; (B) Various empirical procedures.

13.1.2.1 Loss Analysis

If sufficient geometrical data of the pump to be investigated are available, a loss analysis as described in Chap. 3.10.3 and Table 3.10 promises to yield the most accurate results. Nevertheless, it is always advisable to compare the results with empirical methods since the local variations of viscosity resulting from thermal effects are difficult to capture.

The test data reported in [1, 5–8] have been compared to the predictions of the loss analysis method. This comparison has been discussed in detail in [3, 9]. The results are given in Figs. 13.11 and 13.12 which show the correction factors for efficiency and head as measured and as predicted from the loss analysis. In order to determine the influences of Reynolds number, pump type (single- or double-entry) and specific speed, these correction factors are plotted against a modified Reynolds number which is defined by Eq. (13.8):

$$Re_{mod} = Re\, \omega_s^{1.5}\, f_q^{0.75} \qquad (13.8)$$

The test data plotted in Figs. 13.11 and 13.12 cover the following range: $250 < Re_{mod} < 10^7$ or $1500 < Re < 10^8$; $140 < d_2 < 510$; $1 < \nu < 3000 \times 10^{-6}$ m^2/s; $0.28 < \eta_{opt} < 0.86$; $6 < n_q < 45$. The standard deviation between calculated

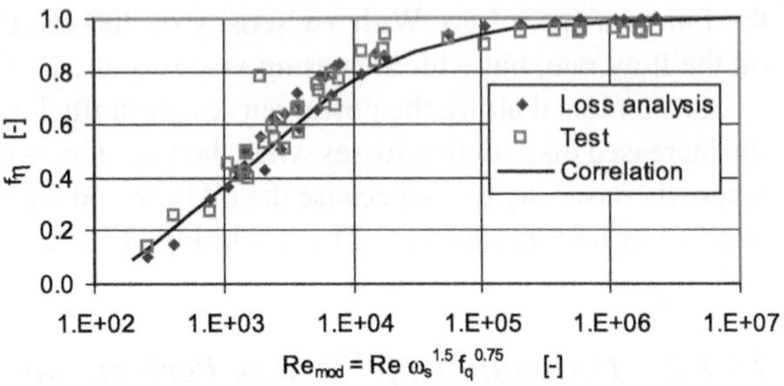

Fig. 13.11 Efficiency correction factor, comparison of loss analysis with experiment

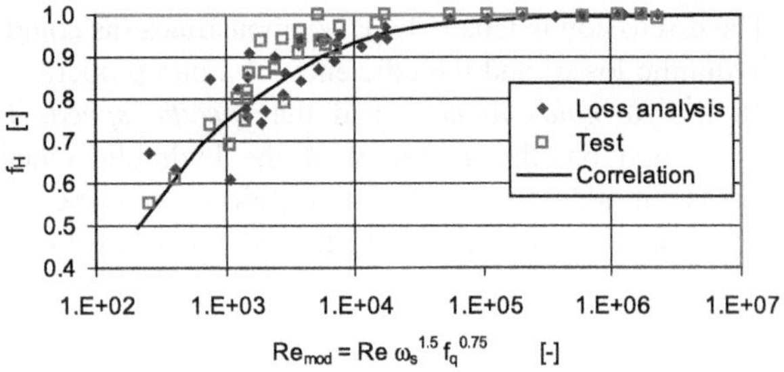

Fig. 13.12 Head correction factor, comparison of loss analysis with experiment

and measured data is about ±15%. A more accurate prediction method has not yet been published. Figures 13.11 and 13.12 can be used for performance prediction as well as Eqs. (T13.1.3 and 13.1.4).

For carrying out the loss analysis, Eqs. (T3.10.1 to 3.10.20) are solved step by step. In this way, the factors f_H and $f\eta$ for the BEP are obtained. Subsequently, the pump characteristics are determined as described in Chap. 13.1.2.3.

Since the loss analysis method usually relies on various assumptions (e.g. actual roughness), it is recommended to apply appropriate margins on head and power. In an actual application, a sensitivity study may be made to investigate how the performance of the selected pump responds to the various assumptions.

13.1.2.2 Empirical Correlation of Data Gained from a Loss Analysis

The average curves in Figs. 13.11 and 13.12 can be represented by Eqs. (13.9 and 13.10). The upper and lower envelopes of the data are described by Eqs. (T13.1.3 to T13.1.4) with the appropriate exponents x and y as given in Table 13.1. These envelopes can help to assess the uncertainty of viscous performance predictions and to define suitable margins. In that regard, this methods is

$$f_\eta = \mathrm{Re}_{mod}{}^{-y} \quad \text{with} \quad y = \frac{19}{\mathrm{Re}_{mod}{}^{0.705}} \tag{13.9}$$

$$f_{H,opt} = \mathrm{Re}_{mod}{}^{-x} \quad \text{with} \quad x = \frac{6.7}{\mathrm{Re}_{mod}{}^{0.735}} \tag{13.10}$$

superior to others. If an exact loss analysis is not required, the performance characteristics measured with water can be converted into those relevant for viscous fluids by calculating the correction factors from Eqs. (13.9 and 13.10) or reading them from Figs. 13.11 and 13.12.

13.1.2.3 Calculation of Viscous Performance Characteristics

The method for predicting the performance characteristics for viscous pumping is presented in Table 13.1 which comprises the necessary formulae, a tabular format for doing the calculation and a sketch that shows how the points shift from water to viscous performance. The procedure is as follows:

1. Take the performance curves measured (or calculated) for cold water.
2. The correction factors f_H, f_η for the best efficiency point Q_{opt} are determined either by loss analysis, according to Eqs. (T13.1.3 to 13.1.4), or by one of the empirical calculation methods presented in Tables 13.2(1) or 13.2(2).
3. The flow rate at the best efficiency point for viscous flow results from $f_Q = f_H$. The available tests show this to be *on average* the best assumption, although deviations from this average are observed.
4. As schematically shown by the figure in Table 13.1, the data with water (Q_w, H_w, and η_w) are read for various flow rates (e.g. for q* = 1.0/0.8/0.6 and 1.2) from the water characteristics and multiplied by said correction factors. Whereas the factors f_Q, and f_η are independent of q*, $f_H(q^*)$ must be calculated according to according to Eq. (T13.1.6).
5. The shut-off head with oil does not differ much from that with water; sometimes it is slightly lower.
6. For multi-stage pumps the *head per stage* must be used.
7. For calculating the power, the actual density of the pumped fluid (not the density of water) has to be inserted into Eq. (T13.1.7). From Eq. (T13.1.7) it is evident that a factor for the power can be derived too. This factor is defined as $f_P = f_Q \times f_H/f_\eta$.
8. Assess uncertainties and define adequate margins by means of Eqs. (T13.1.3) and (T13.1.4). Using the different exponents for x and y, the expected (average) value and the upper and lower limits of the correction factors can be calculated for this purpose.

Table 13.1 Estimation of characteristics for pumping viscous fluids

			Eq.
Calculation options: 1. Loss analysis according to Table 3.10 2. Correlations according to Figs. 13.11 and 13.12 and Eqs. (T13.1.1 to 13.1.4) 3. Empirical correction factors according to Table 13.2			
Reynolds number	$Re = \frac{u\, r_2}{\nu} = \frac{\omega\, r_2^2}{\nu}$		13.1.1
Modified Reynolds number	$Re_{mod} = Re\ \omega_s^{1.5}\ f_q^{0.75}$		13.1.2
Correction factor for head H_{opt} according to Fig. 13.12	$f_{H,opt} = [Re_{mod}]^{-\left\{\frac{6.7}{Re_{mod}^{x}}\right\}}$	Exponent x **Mean** 0.735 · Max. 0.81 · Min. 0.68	13.1.3
Correction factor for efficiency according to Fig. 13.11	$f_\eta = [Re_{mod}]^{-\left\{\frac{19}{(Re_{mod})^{y}}\right\}}$	Exponent y **Mean** 0.705 · Max. 0.77 · Min. 0.65	13.1.4
Correction factor for flow rate	$f_Q = f_{H,opt}$		13.1.5
Correction factor for head for $q^* \neq 1$	$f_H(q^*) = 1 - (1 - f_{H,opt})(q^*)^{0.75}$	$q^* = \frac{Q}{Q_{BEP}}$	13.1.6
Power consumption	$P_v = \frac{\rho_v\ g\ Q_v\ H_v}{\eta_v}$	$f_P = \frac{f_Q\ f_H}{f_\eta}$	13.1.7
Correction factor for the required $NPSH_3$ (to be applied at Q = constant)	$f_{NPSH} = 1 + \zeta_E \frac{c_{f,v}}{c_{f,w}} \frac{c_{1m}^2}{2\ g\ NPSH_3}$	Axial inlet: $\zeta_E = 0.1$ to 0.15 Radial inlet: $\zeta_E = 0.25$ to 0.5	13.1.8

Given: n, $Q_{opt,w}$, $H_{opt,w}$, $\eta_{opt,w}$, ν, n_q, ρ. Subscript: w = water, v = viscous fluid

	Water characteristics			Correction factors			Characteristics for viscous fluid			
Eq.				13.1.5	13.1.6	13.1.4	$Q_v = f_Q\ Q_w$	$H_v = f_H\ H_w$	$\eta_v = f_\eta\ \eta_w$	13.1.7
q_w^*	Q_w	H_w	η_w	f_Q	f_H	f_η	Q_v	H_v	η_v	P_v
—	m^3/s	m	—	—	—	—	m^3/s	m	—	kW
1.2										
1.0										
0.8										
0.6										
0	0		0	—	1.0	—	0		0	

Range of validity:
- Kinematic viscosity up to $4000 \times 10^{-6}\ m^2/s$
- $7 < n_q < 50$
- Open and closed impellers
- Tests with single-stage volute pumps; for multistage pumps calculate with head per stage

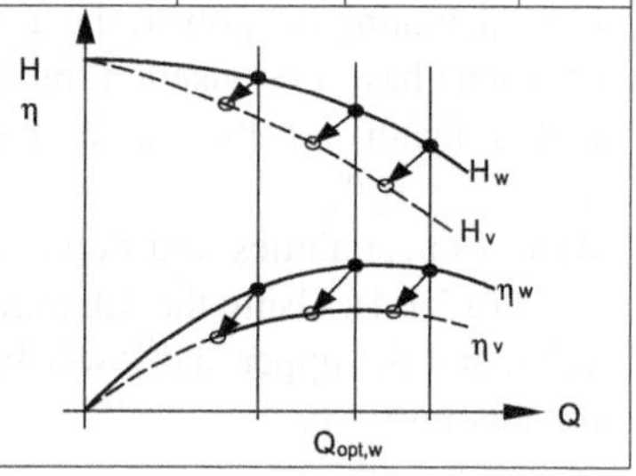

If only the viscous performance data Q_v and H_v are specified, the pump design data Q_w and H_w for pumping water must be calculated by iteration according to Eq. (13.11):

$$Q_w = \frac{Q_v}{f_Q} \quad \text{and} \quad H_w = \frac{H_v}{f_H} \tag{13.11}$$

The easiest way to get a first estimate for $f_Q = f_H$ is from Eq. (T13.2.2) by inserting $Q = Q_v$ and $H = H_v$ into Eq. (T13.2.1); this, because neither the impeller diameter nor the specific speed are known at the outset. The factors f_Q and f_H are corrected afterwards by iteration.

13.1.2.4 Empirical Procedures

The empirical procedure according to [N.4] has been replaced by [N.11]. Since [N.4] has been in use for decades, it is presented in order to allow comparisons. The correction factors used in [N.4] were by and large derived from tests with single-stage pumps. The results had been presented in a generic form in graphs.

The procedure published in [N.4] yields the correction factors as a function of Q, H, ν and q*. Q, H and ν are implicitly included in a parameter B_{HI} defined by Eq. (T13.2.1). If all values are inserted in this equation in consistent units, the result is a dimensionless number that corresponds to $B_{HI} \sim 1/Re^{0.5}$. B_{HI} is thus a physically meaningful parameter for laminar flow.[3] The specific speed whose impact on the losses of a pump is very important (as explained in Chap. 13.1) does not explicitly enter in the calculation. This is a disadvantage of the procedure.

The procedure described in [B.5] is quite similar, but it includes the influence of the specific speed in the parameter B, see Eq. (T13.2.1 right column) [4]. For the specific speed of $n_q = 15$ the parameters B and B_{HI} are equal. However, the graph published in [B.5] does not take into account the influence of the ratio q* of the actual flow rate to the flow at the BEP.

By using mathematical expressions (instead of graphs), the influence of the ratio q* can be taken into account. The representation by formulae is preferable to a graph with regard to using computers or pocket calculators. To this end, Table 13.2 gives all the necessary equations for calculating the correction factors. Once these factors have been determined, the rest of the calculation can be done in the way shown in Table 13.1.

Using the procedure described in [B.5] attention has to be paid to the following points: For impellers with expeller vanes f_η is derived from Eq. (T13.2.5). For calculating impellers without expeller vanes the expression $f_{\eta,o} = 0.4 + 0.6 \times f_\eta$ is

---

[3]This can be verified by inserting into Eq. (T13.2.1) $Q \sim n \times d^3$ and $H \sim n^2 \times d^2$ and $Re \sim n \times d^2/\nu$ according to the similarity laws listed in Table 3.4.

Table 13.2 (1) Estimation of pump characteristics when pumping viscous fluids according to [N.4] and [B.5] SI-units; Q [m³/s]; H [m]				Eq.
Procedure	**[N.4]**	**[B.5]**		
Parameter	$B_{HI} = \frac{480\sqrt{\nu}}{Q^{0.25}\,(g\,H)^{0.125}}$	$B = 100\sqrt{\frac{\nu}{Q\,n^{(s)}}}\,(g\,H)^{0.25} = B_{HI}\sqrt{\frac{15}{n_q}}$		13.2.1
Correction factor for flow rate	$f_Q = e^{-0.11\,(\log B_{HI})^{5.5}}$	$f_Q = \left(\frac{15}{n_q}\right)^{0.013\,B} e^{-0.165\,(\log B)^4}$		13.2.2
Correction factor for flow rate ratio	$f_{q*} = 1 - 0.014(B_{HI}-1)(q*-1)$			13.2.3
Correction factor for head	$f_H = (0.25 + 0.75\,f_Q)\,f_{q*}$			13.2.4
Correction factor for efficiency	$\alpha = 0.05\,e^{0.04\sqrt{B_{HI}}-0.5}$	$\beta = 0.083\,B^{\,0.59}$	$f_\eta = B^{-\beta} - \Delta n_q$	13.2.5
With expeller vanes	thought valid for all configurations	$n_q < 25$: $\Delta n_q = 0.005\,(25 - n_q)$ $n_q > 30$: $\Delta n_q = 0.005\,(n_q - 30)$		
Without expeller vanes	$f_\eta = e^{-\alpha(B_{HI}-0.5)^{1.08}}$	$f_{\eta,o} = 0.4 + 0.6\,f_\eta$		13.2.6
Power	$P_v = \frac{\rho_v\,g\,Q_v\,H_v}{\eta_v}$			13.2.7

suggested in [B.5] (according to that expression f_η cannot be less than 0.4, which sets a limit to the validity of the formula).

The influence of the specific speed is manifest, according to [B.5], in each of the three factors. The reason for f_η dropping with decreasing specific speed can be seen in the disk friction losses (see Chap. 13.1.1). But it is hard to explain why f_η should drop between $n_q = 30$ and $n_q = 45$ as much as between $n_q = 25$ and $n_q = 6$. The portion of the friction losses decreases with rising specific speed, whereas the losses caused by non-uniform velocity distributions increase. Mixing losses due to non-uniform flow are not supposed to be significantly influenced by the Reynoldsnumber. The procedure [B.5] relies on a single test at $n_q = 46$ reported in [5]. Whether the result is generally applicable, is not easy to decide.

The methods [N.4] and [B.5] arrive at quite different results for some ranges: for $n_q > 20$, for example, according to [4] the procedure [N.4] risks to greatly over-predict power consumption with viscous fluids, whereas the power predicted for $n_q < 15$ risks to be too low. The tests shown in Fig. 13.9 with a pump $n_q = 30$ give an example for this discrepancy: the efficiency measured at 1200×10^{-6} m²/s corresponds to a factor $f_\eta = 0.49$, whereas method [N.4] predicts $f_\eta = 0.2$.

Applying either of the statistical methods to various pump types harbors large uncertainties because pump design, hydraulic layout and surface finish have an

Table 13.2 (2) Estimation of pump characteristics when pumping viscous fluids according to [N.11] SI-units; Q [m³/s]; H [m] Simplified representation; for information only; always refer to newest edition of respective standard!			Eq.
Parameter	$B = \frac{480\sqrt{v}}{Q^{0.25}\,(g\,H)^{0.125}}\left\{\frac{n_{q,Ref}}{n_q}\right\}^{0.25}$	$n_{q,Ref} = 20$	13.2.8
Correction factor for flow rate	$f_Q = e^{-0.165\,(\log B)^{3.15}}$		13.2.9
Correction factor for head at BEP	$f_{H,BEP} = f_Q$		13.2.10
Correction factor for head at $q^* \neq 1$	$f_H(q^*) = 1-(1-f_{H,BEP})(q^*)^{0.75}$	$q^* = \frac{Q}{Q_{BEP}}$	13.2.11
Correction factor for efficiency	$f_\eta = B^{-\beta}$ with $\beta = 0.0547\,B^{0.69}$		13.2.12
Power at coupling	$P_v = \frac{\rho_v\,g\,Q_v\,H_v}{\eta_v}$		13.2.13
Correction factor for $NPSH_3$ $n_{ss,Ref} = 200$	$f_{NPSH} = 1 + A_1\left\{\frac{1}{f_{H,BEP}} - 1\right\}\left\{\frac{n_{ss,Ref}}{n_{ss}}\right\}^{1.33}$ End suction pumps: $A_1 = 0.1$; pumps with radial inlet: $A_1 = 0.5$ Apply f_{NPSH} at constant flow.		13.2.14

effect on the magnitude of the various losses. In view of these uncertainties, an exact description of the procedures reported in [N.4] and [B.5] has not been attempted when deriving the equations in Table 13.1(1). The formulae in Table 13.2(1) agree well with the general characteristics of both of these methods, but they do *not* always yield identical values as can be read from the graphs.

The method described in [N.11] is empirical too. It yields values quite similar to those resulting from the calculation according to Eqs. (T13.1.1 to 13.1.4). Table 13.2(2) gives a representation which is equivalent to [N.11].

All of the empirical procedures for predicting pump characteristics as reported in [N.4], [N.11] and [B.5] are based on tests with relatively small, single-stage, single-entry process pumps with Reynolds numbers of approximately 4×10^6 when operating with water. Therefore, the empirical methods can be expected to produce reasonably accurate predictions for this type of pump only. For large pumps with Reynolds numbers well above 10^7 in water service, these methods are not accurate enough. For such pumps the absolute Reynolds number (and not just the viscosity) has to be taken into consideration. The performance estimation of a large pump for oil transport starting from this data base of relatively small model pumps is likely to be too conservative and a loss analysis is recommended instead.

13.1.3 Influence of Viscosity on the Suction Capacity

The procedures in [N.4] and [B.5] give no information about cavitation behavior. Due to additional losses at the inlet of the pump, a certain increase of the required $NPSH_3$ should be expected. Hence a safety margin is recommended. The inlet losses according to Chap. 6.3.2 are: $H_{v,E} = \zeta_E \times c_{1m}^2/(2\ g)$. Assuming that these losses increase with the ratio of the friction coefficients $c_{f,v}/c_{f,w}$, the correction factor for NPSH can be calculated from:

$$f_{NPSH} = 1 + \zeta_E \frac{c_{f,v}}{c_{f,w}} \frac{{c_{1m}}^2}{2\ g\, NPSH_3} \tag{13.12}$$

The friction coefficients $c_{f,v}$ and $c_{f,w}$ are calculated from Table 3.10. According to Chap. 6.3.2 the loss coefficient of the inlet follows from $\zeta_E = \lambda_c - 1$. For pumps with an axial inlet $\zeta_E = 0.1$ to 0.15 can be expected, for between-bearing pumps (radial inlet casings) $\zeta_E = 0.25$ to 0.5.

The $NPSH_3$ for water pumping has to be multiplied by the factor f_{NPSH} so that at least part of the additional losses occurring in a viscous fluid is taken into account. This correction is applied at constant flow rates (i.e. without shifting the flow by the factor f_Q).

Test results on the suction behavior with highly viscous fluids are not available. Due to thermodynamic effects, a slight reduction of the NPSH required is expected for various substances in comparison with water, see Chap. 6.4.1. Thermodynamic effects can thus compensate for the impact of higher viscosities on the $NPSH_3$ to some extent.

13.1.4 Start-Up of Pumps in Viscous Service

When operating multistage pumps with high-viscosity fluids, the heating of the medium from stage to stage may be appreciable at low flow rates. The temperature rise in each stage ΔT_{st} can be derived from Eq. (11.9). With the power and efficiency of the stage represented by P_{st} and η_{st} we get:

$$\Delta T_{st} = (1 - \eta_{st}) \frac{P_{st}}{c_p \rho\, Q} = \left(\frac{1}{\eta_{st}} - 1\right) \frac{g H_{st}}{c_p} \tag{13.12a}$$

A stage-by-stage calculation can be done using the temperature and viscosity at every stage inlet as calculated from Eq. (13.12a). Also the axial thrust balance device receives fluid heated by the previous stages. These effects increase as the flow rate is reduced. The lower the specific speed, the higher is the temperature rise from stage to stage.

During start-up at low flow, heating effects become most pronounced. When pumping a viscous fluid in a closed system (e.g. a pump for transformer oil cooling), the system can be warmed up during start-up by operating the pump at low flow (in the case of a variable speed drive also at reduced speed). Thus it is possible to start the pump without needing to especially size the motor power for cold start-up. Likewise, processes can be started by recycling fluid over a minimum flow bypass if temperatures of the product are higher during normal operation than under start-up conditions (an example are oil wells).

13.1.5 Viscous Pumping Applications-Recommendations and Comments

1. According to the available test data, pumps with specific speeds in the range of $n_q = 20$ to about 40 will give the best efficiencies when pumping viscous fluids.
2. With a given head and speed, high head coefficients yield smaller impeller diameters, and consequently lower disk friction losses and better efficiencies. That is why high head coefficients should preferably be chosen for pumping highly viscous fluids. Since the Q-H-curve becomes very steep when viscous fluids are pumped, impellers with higher than normal pressure coefficients can be used without the risk of an unstable Q-H-curve (some tests are presented in [6]). However, for water pumping such impellers would not be acceptable.
3. As the disk friction is much higher when pumping fluids more viscous than water, expeller vanes do not seem to be a good choice with regard to efficiency. Rather, the obvious solution for balancing the axial thrust are balance holes combined with an annular seal on the rear shroud, because the annular seal losses become less important with increasing viscosity. Disk friction losses will be reduced by the heating of the fluid in the impeller sidewall gaps due to dissipation because the local viscosity decreases at higher temperatures.
4. In turbulent flow the disk friction losses depend very little on the width of the impeller sidewall gap, in laminar flow these losses grow in inverse proportion to the axial clearance between the casing and the impeller, Eq. (T3.6.3). Therefore, narrow impeller sidewall gaps should be avoided when pumping highly viscous fluids. The width of the impeller sidewall gap is also an important factor of uncertainty in performance prediction, since it does not enter into the calculation as per [N.11] and [B.5]. This is especially true at low specific speeds where the viscous power increase is governed by extremely high disk friction losses. It is thus to be expected that two otherwise identical pumps with different axial clearances of the impeller sidewall gaps attain different efficiencies when pumping high-viscosity fluids. The high losses are to some extent offset by thermal effects resulting from dissipation in the narrow sidewall gaps.

5. Predictions of the pump characteristics based on the correction factors given in [N.11] and [B.5] must be considered as relatively rough approximations, because the dependence of the losses on the Reynolds number can vary appreciably in different pumps. The absolute values of efficiency and Reynolds number, the pump design, as well as the quality of execution, all have an impact on viscous performance predictions.
6. Two otherwise identical pumps with different surface roughness have different efficiencies in turbulent flow, but equal efficiencies in laminar flow. Consequently, different factors f_H und f_η should be used for estimating the performance in viscous pumping from data measured with water. This effect can only be accounted for by a detailed loss analysis based on known surface roughness.
7. Due to the high losses when pumping highly viscous fluids the thermometric method can be easily applied in a plant to determine the efficiency from Eq. (11.9) as described in Chap. 11.6. The temperatures mentioned in context with Fig. 13.6 are an example.
8. The suction pipe should be as short as possible; hence the pump should take suction immediately from the tank. If the suction pipe has some length, its diameter should be one nominal size large than the suction nozzle of the pump.
9. Pumping viscous fluids requires more motor power. The starting torque and the start-up current are higher as well. Therefore, it has to be checked whether the pump shaft, coupling and driver are suitable for these conditions.
10. Given the uncertainties of the calculation methods and the possible variations of the physical properties of the fluids to be pumped, adequate margins are necessary when selecting the pump and the motor.
11. Viscous fluids risk disturbing the adequate functioning of auxiliary equipment (such as shaft seals, barrier fluid injection or cooling circuits, etc.). Therefore care should be taken when specifying these systems.
12. The calculation methods discussed apply to Newtonian fluids; the behavior of Non-Newtonian fluids can be different, as discussed in Chap. 13.5.
13. Special care is required when emulsions are to be pumped.

Emulsions are mixtures of immiscible liquids such as oil and water. For example, this situation is encountered in oil production since water is often injected into oil wells in order to increase oil production by boosting the pressure in the well. As the oil field is depleted, more water is injected and the "water cut", defined by $WC = Q_w/(Q_w + Q_{oil})$, increases. Two types of emulsions must be distinguished: (1) *water* droplets are dispersed in oil up to a limiting WC which depends on the oil properties and other components such as asphaltenes; (2) when the limiting WC is exceeded, *oil* droplets are dispersed in water. The behavior below and above this limit is quite different. This is exemplified by the test data in Fig. 13.12-1 where the critical WC is at 0.62. When water is added to oil, the viscosity rises exponentially to WC = 0.62 up to about 47-times the viscosity of the water; for WC > 0.63 the viscosity drops sharply. As the viscosity increases, head and efficiency decrease and the BEP is shifted to lower flow rates. The data in Fig. 13.12-1 were derived from

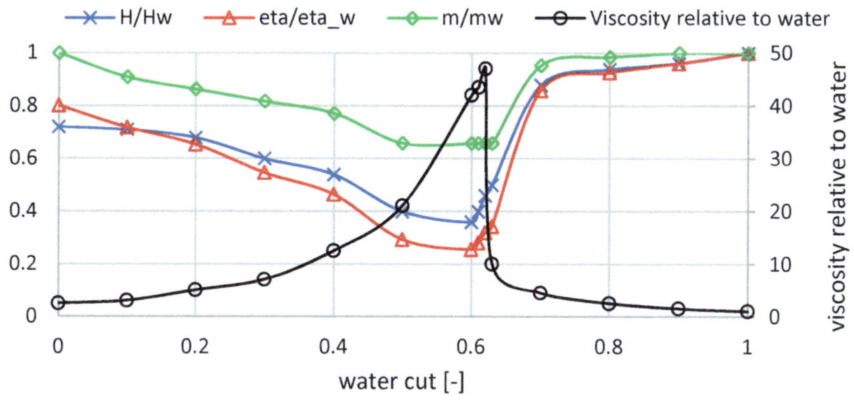

Fig. 13.12-1 Effect of water-cut on emulsion viscosity and pump performance. WC = 0: oil; WC = 1.0: pure water; m is the mass flow rate. Curves calculated from data in [61]

tests in [61] carried out with a small process pump and light crude oil with a viscosity of 2.5×10^{-3} Pas; head, efficiency and mass low rate are referred to performance with cold water. The factors in Fig. 13.12-1 - although not generic - demonstrate that the effect of emulsions can be very strong and that pump operation in the range of WC = 0.4 to 0.7 should be avoided. The drop in head and efficiency at WC = 0 is due to the oil viscosity.

13.2 Pumping of Gas-Liquid Mixtures

Gas-liquid mixtures are encountered in the process industry as well as in the production and transport of oil, which often is accompanied by natural gas. An important application of high economical significance is the transport of oil-gas-mixtures coming out of oil wells. By means of two-phase pumps the pressure at the well-head can be decreased, thus boosting oil production. However, handling mixtures with high fractions of gas by centrifugal pumps is a difficult task since gas and liquid tend to separate because of their large density differences. Pumping can then become very inefficient or even impossible. As discussed in detail below, this phase separation is caused by body forces and buoyancy which is due to pressure gradients perpendicular to the main flow direction.

13.2.1 Two-Phase Flow Patterns in Straight Pipe Flow

Prior to the discussion of the complex flow phenomena in pump impellers and collectors, it is appropriate to review the different flow patterns in a horizontal pipe or channel. All the more so, since this type of flow is relevant for the inlet to a two-phase pump and can have an impact on performance.

Figure 13.13 shows sketches of the different phase distributions which can be encountered. The graph delineates the approximate boundaries between these flow patterns which were correlated with the mass velocity [10]. Definitions are given in Table 13.3: one apostrophe refers to the liquid, two apostrophes to the gas phase. The mass velocities of the gas fraction on the ordinate and of the liquid fraction on the abscissa are multiplied by fluid properties as indicated in Fig. 13.13. These corrections are required when media other than water and air at atmospheric pressure flow through the pipe.

Discussion of flow patterns in Fig. 13.13: If small gas bubbles are finely dispersed in a liquid flowing turbulently through a pipe, a "bubbly flow" regime is observed. The higher the flow velocity and the turbulence and the smaller the bubbles, the more homogenous is the mixture. Turbulence and surface tension

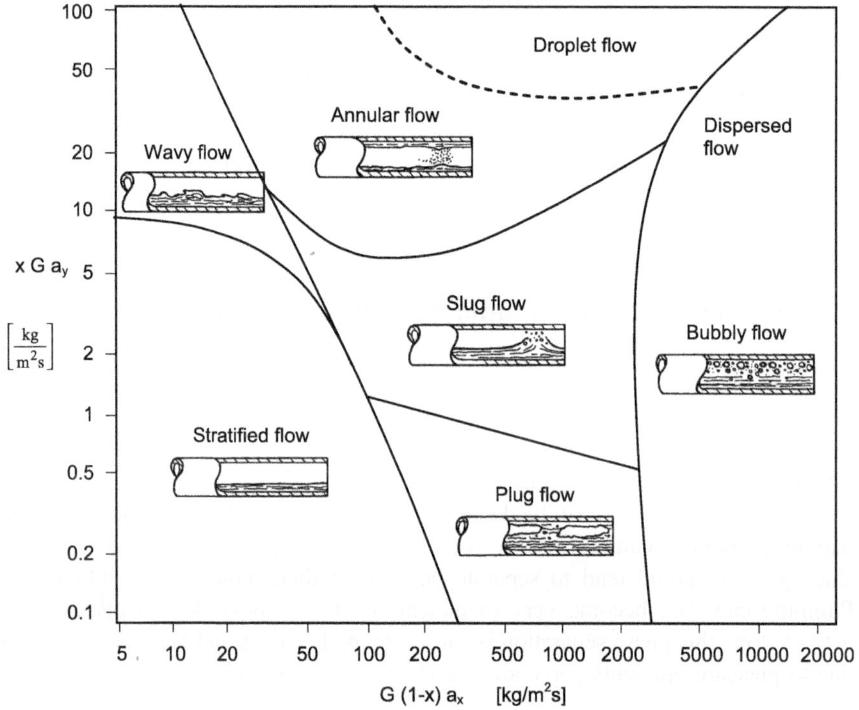

$$a_y = \sqrt{\frac{\rho_a \rho_w}{\rho' \rho''}}\ ;\ a_x = \frac{S_{T,w}}{S_T}\left(\frac{\nu' \rho_w}{\nu_w \rho'}\right)^{0.333}$$

Fig. 13.13 Flow patterns in horizontal pipes, according to [10]. S_T = surface tension, $G = m_{tot}/A$ = mass velocity. Subscripts: w = waster, a = air at 20 °C and 1 bar

acting on small bubbles tend to create a relatively uniform distribution of bubbles. In a horizontal pipe, however, the bubble concentration increases from bottom to top due to buoyancy. As the average pipe flow velocity is reduced, more bubbles travel to the upper domain in the pipe. At very high flow velocities the bubbly flow becomes frothier or dispersed due to increasing shear forces.

If the flow velocity is reduced, bubbles tend to coalesce until larger gas accumulations ("plugs") are created. These can block an appreciable portion of the upper half of the pipe section. The effect grows with decreasing mixture velocity. In upward flow through a vertical pipe the plugs flow faster than the liquid due to buoyancy. There is thus a "slip" between the liquid and gas phase.

When the fluid stagnates, the gas rises towards the top of a horizontal pipe filled with a liquid/gas mixture. Such a phase separation also occurs at low flow velocities. Then the liquid fills only a part of the pipe section. Gas and liquid flow are fully separated (almost comparable to open-channel flow). Such a "stratified flow" is created when both the velocities of liquid and gas phase are small, Fig. 13.13. If the gas flow rate is increased and the gas velocity appreciably exceeds that of the liquid, surface waves are created due to shear stresses (similar to wind creating waves on a lake). The stratified flow then becomes "wavy".

When the wave crests reach the upper wall of the channel, a "slug flow" is created, in which gas and liquid distributions vary strongly over the pipe length. Liquid slugs can also alternate with gas pockets. The length of the slugs may attain several pipe diameters. Slug flow is therefore prone to pulsations which can create operational problems such as vibrations of piping and pumps. Vibrations can be excited by slugs in the *discharge pipe* as well as in the inlet pipe.

If a high gas flow is combined with a low liquid flow rate, an "annular flow" is created, since high shear stresses distribute the liquid in the form of a thin layer over the entire circumference of the pipe wall. The phases are separated to a large extent. In horizontal flow the thickness of the liquid film varies over the circumference due to gravity (the highest thickness is observed at the pipe bottom). If the gas velocity is increased further, the liquid film is disrupted and droplets are entrained. A droplet or "mist flow" is created and the mixture becomes more homogenous. This process is unsteady because of continuous liquid entrainment and deposition effects.

Figure 13.13 is valid for horizontal pipes. Different flow patterns are observed in vertical channels where the flow direction has an impact on the phase distribution. In vertical upward flow the gas flows faster than the liquid due to buoyancy; in downward flow the liquid is accelerated more strongly than the gas due to gravity.

The transitions between different flow patterns are smooth. They depend on operation conditions and liquid/gas properties. Therefore, various flow pattern maps have been published; some of these may be found in [11]. A comprehensive calculation of two-phase flow patterns in horizontal or inclined pipes was developed by [12]; the procedure can also be found in [11, 13].

As the discussion of the two-phase flow patterns in a pipe flow shows, we can distinguish three categories of typical behavior:

A. Homogenous phase distribution: If flow velocities are high *and at the same time* one phase fraction is dominating, a relatively homogenous flow with gas bubbles or liquid droplets is created. Under these conditions the kinetic energy of the turbulence is sufficiently able to mix the phase of the smaller volume fraction with the dominating phase. The effect of body forces is of secondary importance. The dispersed phase (droplets in a gas stream or bubbles in a liquid flow) is transported by the dominating phase in a process similar to pneumatic or hydraulic transport of particles by a "carrier fluid".

B. Separated phases in co-current flow: If flow velocities are *low* and turbulent mixing action is correspondingly weak, body forces dominate and the phases are essentially separated, giving stratified or wavy flow. If flow velocities are *high* so that shear stresses dominate over body forces, annular flow is created when the liquid fraction is low.

C. Separated phases in intermittent flow: If turbulent mixing action, body forces or shear stresses are dominating, the phase distribution *along the pipe axis* scarcely varies (in such cases, phase concentration is more likely to be non-uniform over the pipe section). If none of the above mechanisms is strong, the effects of bubble coalescence and surface tension take precedence and larger gas accumulations are generated. Phase slip is also important. As a consequence, intermittent slug and plug flow patterns are observed.

In summary, the tendency for phase separation is influenced by the following physical mechanisms:

- The risk of phase separation grows with increasing gas volume fraction α. This is true for a volumetric gas content of up to about 80 to 90%.
- The risk of phase separation grows strongly with increasing body forces. Gravity is relevant in pipe flow. In rotating impellers, centrifugal and Coriolis forces are very strong body forces. Centrifugal forces also induce phase separation in stationary curved channels (volutes, diffusers or pipe bends).
- Differences in the phase velocities (slip): Mostly the gas velocity exceeds that of the liquid phase. If the liquid volume fraction is small, the faster gas flow tends to entrain the liquid as observed in droplet, annular, wavy or stratified flow. In bubbly, slug or plug flow, the pressure gradient in the flow direction (dp/dx) accelerates the gas phase, because the gas expands. This is due to the "buoyancy effect" discussed in detail below.
- The tendency for phase separation diminishes with decreasing liquid to gas density ratio $DR \equiv \rho^* = \rho'/\rho''$. In the limiting case, when the gas density approaches the liquid density, no phase separation due to body forces or buoyancy is possible.
- Viscous and turbulent shear stresses at phase boundaries foster phase mixing.
- The surface tension tends to keep the form of the bubbles or droplets. Bubble coalescence creates large bubbles and slugs.

- In highly viscous liquids, bubble movement is severely hampered as can be seen by inspecting Eq. (13.15). It is therefore difficult to separate small gas bubbles from oil in a gravity separator.
- The ratio of the body forces to inertia has an important effect: Dominating body forces separate, while high inertia in straight channels (high velocities) mixes the phases.
- The exchange of momentum between the phases and turbulent velocity fluctuations foster phase mixing.

The pressure drop in a channel with two-phase flow likewise depends on all of the above parameters as well as on the flow patterns. Prediction methods can be found in various books, e.g. [11, 13, 56].

13.2.2 Two-Phase Flow in Pumps. Physical Mechanisms

The capability of a centrifugal pump to convey a two-phase mixture depends in the first place on whether gas and liquid form a homogenous mixture or to what extent the two phases separate. Fine gas bubbles dispersed in a liquid can be considered as a quasi-homogenous mixture. The bubbles are entrained by the liquid flow, but there is some slip between the phases which causes additional losses. Likewise, droplets of liquid carried along by a gas stream can be represented by the homogenous flow model. The transport mechanism is similar to pneumatic conveying or to the hydraulic transport of solids (discussed in Chap. 13.4). The droplets tend to be deposited on the blades or in the collector channels as liquid films; but the process is unsteady since the films may be disrupted by shear forces between the gas stream and the liquid. Slip and losses due to exchange of momentum are much higher than in bubbly flow. With larger gas volume fractions bubbly flow is no longer possible, since small bubbles tend to coalesce to form larger gas accumulations (as in slug or plug flow in a pipe). When gas accumulates in the pump the delivery can be disrupted; the pump becomes "gas-locked".

Even though geometry and body forces in a pump are completely different from those of a flow through a straight channel, the same physical mechanisms (discussed in Chap. 13.2.1) are effective in determining the flow patterns which, in turn, have an impact on the energy transfer. While in a straight channel pressure gradients in the flow direction are usually moderate, the pressure field in a rotating impeller strongly influences the phase distribution. Apart from this "buoyancy effect", body forces in rotating impellers have a great impact.

Body forces: To illustrate the influence of body forces, consider a rotating disk with straight radial channels according to Fig. 13.14 (see Chap. 7.3.2 for pumping with such a device). The fluid in the channel is subject to a strong Coriolis acceleration $b_c = 2 \times \omega \times w$. To give an example: at $n = 3000$ rpm and $w = 10$ m/s this amounts to $b_c = 6280$ m/s^2. A fluid particle thus experiences

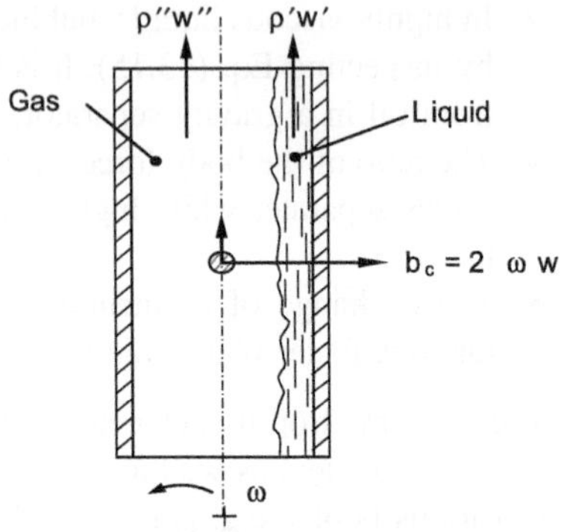

Fig. 13.14 Phase separation in a rotating channel or an impeller with radial blades

640-times the acceleration due to gravity. The Coriolis force acts perpendicularly to the direction of the relative velocity (which follows essentially the channel axis). Consequently, the liquid is transported to the leading face ("pressure surface") of the channel and a stratified flow develops if a sufficient amount of gas is present. This occurs even if the channel diameter is small. The Coriolis acceleration is equally strong in radial impellers with back-swept blades. The effect of impeller rotation is therefore similar to Fig. 13.14; but additional forces need to be taken into account as described in Chap. 5.1.

Since the static pressure at the impeller outer radius r_2 is constant around the circumference, gas and liquid phases have the same static pressure at the impeller outlet. If both phases are *completely separated,* the pressure rise in the impeller is limited by the gas density, refer also to [14]. In *separated* flow the pressure rise in the impeller would thus be only the fraction $\rho''/\rho' = 1/\rho^*$ of the pressure rise in single-phase liquid flow (SPL). In reality there is some interaction between the phases, and the pressure rise may be somewhat higher.

Gas and liquid are accelerated to a high impeller outlet velocity c_2. Because of its high density the liquid phase, therefore, contains a high kinetic energy $E' = \frac{1}{2} \times \rho' \times Q' \times c_2^2$. This energy cannot be well recovered in the collector since, with essentially separated phases the pressure recovery in a diffuser is again limited by the gas density. In addition, both phases have different velocity vectors, which generate shock losses in the diffuser or volute.

In *separated* two-phase flow the static pressure rise in the impeller is essentially limited by the gas density:

$$\Delta p_{La} = \frac{\rho''}{2}(u_2{}^2 - u_1{}^2 + w_1{}^2 - w_2{}^2) - \rho'' g Z_{La}$$

Buoyancy effects: A gas bubble moves in a pressure field to the location of the lowest pressure. This effect may be called "buoyancy". Consider a gas bubble of density ρ'' in a liquid with a density ρ', as in Fig. 13.15. A volume element of this bubble may have the area dA and the height Δz. At one boundary of the element the

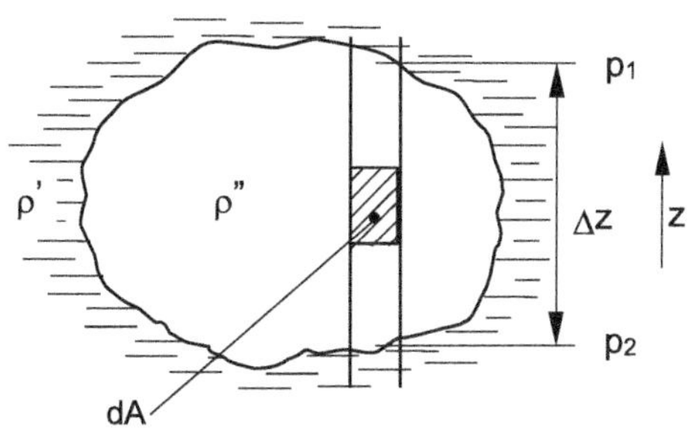

Fig. 13.15 Buoyancy of a gas bubble in a liquid with a pressure gradient

pressure is p_1 and at the opposite boundary p_2. The volume element then experiences a force: $dF_A = (p_2 - p_1) \times dA$. The pressures p_2 and p_1 result from the pressure field in the surrounding liquid: $p_2 = p_1 + (\partial p/\partial z \Delta z$. The force exerted by the pressure field on the bubble then becomes:

$$dF_A = \frac{\partial p}{\partial z}\Delta z\, dA = \frac{\partial p}{\partial z} dV \tag{13.13}$$

In the gravity field the pressure gradient is $\partial p/\partial z = \rho' \times g$. The buoyancy accordingly is $F_A = \rho' \times g \times V$. If the gravity force acting on the bubble is also taken into account, the resulting force is:

$$F_{res} = (\rho' - \rho'')gV \tag{13.14}$$

Equation (13.13), which was derived for an arbitrary pressure field, demonstrates that a gas bubble always tends to move towards the location of the lowest pressure. Such movement is opposed by inertia and by the fluid dynamic drag F_w, which the liquid exerts on the bubble due to the viscosity. If inertia is neglected, the bubble rise velocity c_A in a liquid at rest in the gravity field can be calculated from the equilibrium of buoyancy and drag $F_A = F_w$.

$$c_A = \sqrt{\frac{4\, g\, d_B}{3\, \zeta_w}\left(1 - \frac{\rho''}{\rho'}\right)} \tag{13.15}$$

In Eq. (13.15) d_B is the diameter of a spherical gas bubble and ζ_w is the Reynolds-dependent drag coefficient as per Eq. (T13.5.6). In a similar way, the velocity of a gas bubble relative to the liquid in an arbitrary pressure field with the gradient $\partial p/\partial z$ can be expressed by the proportionality, [15]:

$$W_{rel} \sim \sqrt{\frac{2\, d_B}{\zeta_w\, \rho'}\left(-\frac{\partial p}{\partial z}\right)} \tag{13.16}$$

The work transfer from the impeller blades to the fluid induces a pressure gradient from the pressure to the suction surface of a blade. This is given qualitatively by $\partial p/\partial z \approx (p_{PS} - p_{SS})/t$, where t is the pitch of the blades. Gas bubbles consequently have the tendency to accumulate near the suction surface of the blades. The complex forces discussed in Chap. 5.2 generate secondary flows which have an impact on the phase distribution in the impeller channels. The tendency described above can thus be reversed to the effect that gas accumulates near the pressure rather than the suction surface. The actual phase distribution depends on the equilibrium of forces acting on a fluid element in the impeller. Secondary flow and recirculation can foster phase mixing and thus improve the gas handling capacity of a pump.

The influence of geometry and flow rate on the balance of forces cannot be calculated by simple means. Flow pattern prediction and interpretation in an impeller with two-phase flow is far more difficult than with single-phase flow. Published flow pattern observations in impellers report gas accumulations mostly near the suction surface at partload, while gas pockets near the pressure surface seem to occur at high flows at $q^* > 1$. The reports are sometimes contradictory (probably due to the different geometries investigated). Therefore, it is difficult to establish general rules.

The effect of body forces on a liquid particle is greater than the force on the gas phase by the factor $\rho^* = \rho'/\rho''$. The behavior of the liquid is hence dominated by body forces and inertia, while the distribution of the gas is primarily determined by the pressure field (buoyancy) and secondary flows. Turbulence, surface tension and shear stresses are thought to play a secondary role. These factors foster phase interaction and mixing with increasing flow velocity.

As demonstrated by Fig. 13.14, Coriolis forces transport liquid to the pressure surface, while buoyancy effects move gas to the suction surface of the blades of radial impellers. *Both effects add;* that is why the gas handling capabilities of radial impellers are very limited. Phase separating effects grow with increasing flow rate because the pressure gradient from the pressure to the suction surface (buoyancy) as well as the Coriolis acceleration ($b_c = 2 \times \omega \times w$) become stronger with increasing flow rate ratio q^*. In addition to this, in a radial impeller gas bubbles travel *against a strong centrifugal pressure gradient* towards the outlet (a buoyancy effect opposing outward flow).

Gas thus tends to accumulate in locations with strong deceleration. According to [16] gas bubbles are found near the impeller inlet on the pressure surface of the blades, where they coalesce due to deceleration. Further downstream in the impeller channel, buoyancy effects move the gas from the pressure towards the suction surface. If a critical gas volume fraction is exceeded, this cross-movement of the gas blocks the channel and the head breaks down ("gas-locking").

In Chap. 5.2 the Rossby number has been introduced to qualitatively explain impeller flow patterns. This concept can be applied to two-phase flow as well. In the axial inlet of a radial impeller the Rossby number according to Eq. (5.6) is close to $Ro_{ax} = 0.5$ (since $w_{1u} = u_1$ at $\alpha_1 = 90°$). Consequently, liquid is forced towards the hub and gas gathers near the outer radius of the impeller eye. That is why it has been often observed that gas accumulates near the annular seal of radial impellers, e.g. [14]. When the gas pocket covers a sufficient portion of the blade length, the head developed by the pump drops sharply and the flow stops.

In semi-open impellers, liquid can flow from the pressure surface of the blades through the gap between blades and casing into the wake region (the shroud suction surface corner, Chap. 5.2); here gas tends to accumulate due to the buoyancy effect discussed above. The leakage disturbs and re-mixes the gas accumulation, thus counteracting to some extent the buoyancy effect. Furthermore, the detrimental effects of gas accumulation in the sidewall gap on the front shroud are absent. In comparison to closed radial impellers, the gas handling capabilities of semi-open impellers are improved appreciably. The phase mixing effect of the flow through the gaps increases when the gap width is enlarged. With extremely large gaps, the flow patterns will resemble those encountered in vortex pumps which can handle two-phase flow quite well (depending on design). Due to the strong recirculation in vortex pumps, gas is well dispersed in the liquid.

According to [17], impellers with radial blades ($\beta_{2B} = 90°$) perform better in two-phase flow than impellers with back-swept blades. This finding seems to contradict tests in [18] which attributed better two-phase performance to blade outlet angles as low as 10°. As discussed in Chap. 5, it is difficult to explain the 3-dimensional flow patterns in impellers (be it in single or two-phase flow) by simple geometric parameters. Blade numbers, meridional shape, blade length and channel width all have an influence on flow patterns and gas accumulations.

Semi-axial, axial and helico-axial impellers: Semi-axial impellers work better in two-phase flow than radial impellers, because centrifugal and Coriolis accelerations have opposing components. Consequently, the resulting forces and the tendency to phase separation are reduced. Secondary flows contribute to better phase mixing. This can be visualized by Fig. 5.9 which shows secondary flow patterns developing in inducers. These are reported to be able to pump higher gas volume fractions than radial impellers. Even radial impellers with an oblique trailing edge performed better in two-phase flow than normal radial impellers according to the tests in [19] see Chap. 13.2.4.

Helico-axial impellers (Fig. 13.25) can be viewed as axial impellers with a (more or less) conical hub. In essence they can handle any gas volume fraction. This performance is achieved (to some extent) by balancing centrifugal with Coriolis forces and by cross-channel mixing induced by the shape of the hub as well as secondary flows similar to those in Fig. 5.7. The small blade height evident from Fig. 13.25 may also help to limit separation effects.

As with radial impellers, the Rossby number in the impeller inlet is close to $Ro_{ax} \approx 0.5$ because $w_{1u} = u_1$ (when $\alpha_1 = 90°$). In the inlet region, liquid is thus expected to migrate towards the hub. At the impeller outlet, where the relative velocity is decelerated, $Ro_{ax} \approx 1$ is expected and liquid movement is now rather away from the hub. This tendency is fostered by increasing the hub diameter towards the outlet. With growing flow rate the relative velocity at the outlet increases. The Rossby number drops accordingly and the liquid tends to migrate

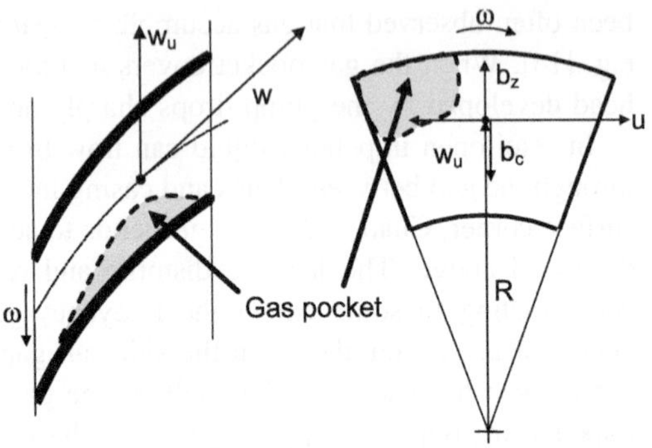

Fig. 13.16 Centrifugal and Coriolis forces in axial or semi-axial impellers and likely locations of gas accumulation

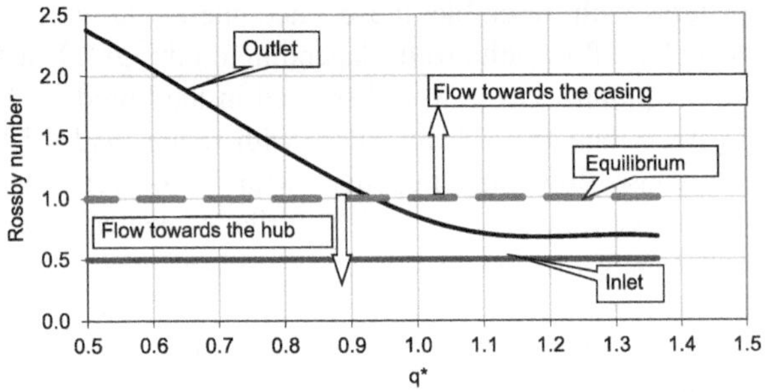

Fig. 13.17 Rossby number indicating the tendency of liquid migration

towards the hub. This may explain why two-phase performance impairs with increasing flow rate notably at $q^* > 1$. At partload the opposite effects take place, the Rossby number at the outlet increases, centrifugal forces take precedence and the liquid migrates away from the hub. Two-phase performance consequently improves. At low flow, with recirculation, mixing improves even more. Figures 13.16 and 13.17 illustrate these effects, which are present in axial and semi-axial pumps, and also at the inlet of radial impellers.

When liquid and gas phases separate at high gas content (typically above 80%) liquid films are expected to form on solid walls in a similar way as in annular flow in a pipe.[4] Pump performance then depends on the magnitude of shear stresses and their ability to disrupt the liquid films and create droplet or mist flow patterns. The blade tip leakage of open impellers fosters this process. Liquid flowing through the boundary layers on the blades is transported by centrifugal forces towards the outer streamline (see also Fig. 5.7).

Pressure recovery in the diffuser, Fig. 13.18**:** The pressure recovery in a diffuser depends on the liquid/gas phase distribution. In a homogenous mixture it is largely determined by the density of the "carrier fluid". When a liquid stream carries gas bubbles, the pressure rise is dominated by the *liquid density*. But when a gas carries droplets of liquid, the pressure rise is to a large extent determined by the *gas*

[4]Compare this process to water films forming on the blades of wet steam turbines.

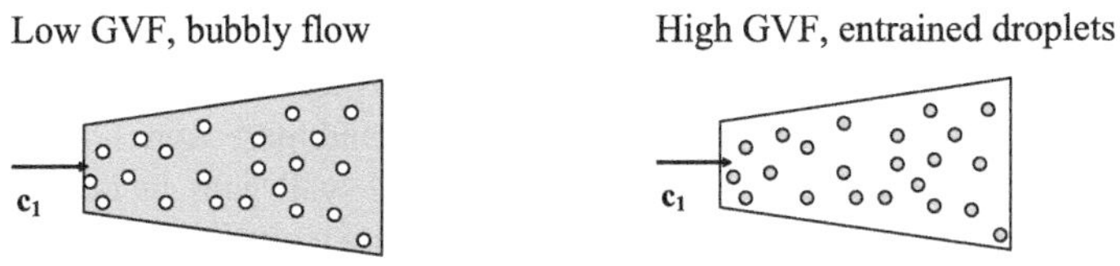

$$\Delta p = c_p \frac{\rho_{mix}}{2} c_1^2 \qquad \Delta p = c_p \frac{\rho''}{2} c_1^2$$

Fig. 13.18 Pressure recovery in a diffuser in bubbly or droplet flow

density. In homogenous flow the diffuser performance may be estimated by using the density of the mixture in calculations according to Chap. 1.6.

In contrast, when the phases are separated (stratified, wavy or annular flow) the diffuser works very inefficiently because the pressure recovery is limited by the gas density if there is little interaction between the phases. Then the kinetic energy of the liquid is to a large extent dissipated. With increasing phase interaction diffuser performance is between the extremes of homogenous and separated flow. High losses must also be expected in intermittent (plug or slug) flow, but phase interaction in such flows is quite strong.

Two-phase performance of plane diffusers has been measured in [20]. The tests showed that the onset of stall was not greatly affected by the gas content. Below a gas volume fraction of $\alpha < 0.4$ there was bubbly flow with bubble diameters of about 3 mm; above $\alpha > 0.4$ there was churn flow.

Up to $\alpha = 0.72$ the peak pressure recovery occurred at the same geometry $c_p^* = f(A_R, L/h_1)$ as would be predicted from Fig. 1.18. As the phase slip increased, pressure recovery dropped sharply. Annular flow had weak phase interaction, high slip and low pressure recovery.

Little quantitative data is available on the pressure recovery in pump diffusers or volutes. The strong secondary flow is expected to carry the liquid to the outer wall of the volute. Gas accumulates near both sides of the impeller outlet, where the gas interacts with the flow in the impeller sidewall gaps.

Effect of the impeller sidewall gaps: In some tests the impeller sidewall gaps were thought to have an influence on the gas handling capability of a pump [17]. Gas tends to accumulate in the impeller sidewall gaps, because there is little radial flow which could carry gas bubbles away. Due to the rotation of the fluid and the induced centrifugal forces, gas gathers near the annular seal at the impeller inlet and the hub of the rear shroud. With increasing rotation and gas volume fraction the gas pocket grows towards larger radii. When the gas ring thus formed reaches the impeller outer radius, an efficient pressure build-up in the volute becomes impossible. Gas flows back into the impeller channels and the pump cannot maintain the flow. Radial ribs on the casing walls in the impeller sidewall gaps were therefore proposed to reduce fluid rotation and increase the gas handling capacity. Bleeding gas from the impeller sidewall gaps has been suggested to improve gas handling and avoid interruption of the flow when pumping gas-laden liquids, [17].

The above observations may be explained by the fact that there was no leakage flow through the impeller sidewall gap in the pumps tested (a mechanical seal was used at the impeller inlet instead of an annular seal, and there were no balance holes in the rear shroud). The impact of the impeller sidewall gaps as described above should therefore not be generalized. It may be advantageous for two-phase pumping to decouple the flow in the volute from the flow in the impeller sidewall gaps by designing the diffuser or volute with a small gap A and a good overlap x_{ov} (refer to Fig. 9.1), but test data are not available.

Gas accumulations near the shaft can impair shaft seal cooling. A single rib or a similar feature on the casing wall is able to disrupt the rotation of the water in the impeller sidewall gap in such a way as to sweep out the gas.

Unsteady phenomena: A gas pocket may form somewhere in the impeller channel (or anywhere in the pump) and grow with time. Consequently, the liquid flow is accelerated due to the blockage effect caused by the gas. As the liquid velocity rises, the gas may eventually be swept away. Subsequently the gas pocket forms again. A periodic or stochastic variation of head and flow rate would be the consequence. Surge-like phenomena are therefore observed in two-phase pumping. Tests reported in [16, 18] showed this kind of unstable behavior. The head was observed to jump from about 90% down to 40% of water head when the gas volume fraction was increased from typically 5 to 6%. Two stable states of operation were thus observed at 90 and 40% of water head, *but a stable operation between these two levels was impossible*. In other words, a stable regime with (say) 65% of water head could not be established by adjusting the gas flow rate.

Influence of viscosity: When pumping oil/gas mixtures, an increase in the oil viscosity is expected to mitigate phase separation effects since the drag of a bubble rises with the viscosity (this follows from Eqs. 13.15 and 13.16). Tests with oil viscosities of 10 and 18 mm^2/s showed an improvement of two-phase performance at $q^* > 1$, but little effect at partload and near BEP, [18].

Influence of circumferential velocity: If a pump is operated with a given two-phase mixture at a specific flow rate ratio q* with different speeds, the velocity triangles remain similar, Chap. 3.4. According to Chap. 5.2 the ratios between centrifugal and Coriolis acceleration remain unchanged. All pressure differences and forces are proportional to the square of the tip speed. This is actually the basis of dynamic similarity for geometrically similar hydraulics, Chap. 3.4.[5] Since phase separation is largely determined by body forces and pressure gradients, the above relations apply to two-phase flow as well. Therefore, the two-phase multipliers, *in a first approximation*, are independent of impeller tip speed and consequently independent of size and speed of the pump. This statement is confirmed by tests at various speeds.

[5]In single-phase flow these relations are modified by the Reynolds-dependent losses. In turbulent flow the effect is small (Chap. 3.10); in laminar flow the losses increase considerably as discussed in Chap. 13.1.

Nevertheless, mixing of the phases is somewhat improved with increasing tip speed and turbulence. But it is often difficult to discern a slight improvement clearly from the data scatter inherent to two-phase testing.

Increasing the rotor speed of a given pump has two beneficial effects: (1) Higher velocities in the suction pipe can provide a more homogenous flow to the impeller. (2) Disruption of large gas bubbles. Tests in [18] demonstrate a higher performance with increased speed when operating with two-phase flow. In that particular case, near the best efficiency point, 4 to 5% gas could be handled with a speed of 1450 rpm while 6 to 7% were pumped at 2000 rpm.

The similarity laws and the two-phase multipliers can be applied as a first approximation to predict the performance of one stage of a two-phase pump. But the similarity laws *cannot be applied to the entirety of a multistage pump.*

Influence of specific speed: Since the shape of an impeller depends on the specific speed, there is an apparent effect of n_q on the gas handling capacity of a pump. During two-phase tests in [18] an impeller with $n_q = 15$ performed better than a pump with $n_q = 23$. But it would be wrong to conclude that the gas handling capability should increase with falling n_q. As explained above phase distribution in the hydraulic channels depends on many factors which cannot be described by the single parameter n_q. One effect may be the blade length which increases with falling n_q. Longer blades mean lower blade loading and consequently reduced buoyancy effects; they also give the flow a chance to recover downstream of a local gas accumulation, refer also to Fig. 4.12.

Influence of approach flow: The phase distribution in the suction pipe could be expected to have an impact on performance. A homogenous bubbly flow tends to improve performance while stratified flow in a horizontal pipe fosters phase separation in the impeller. Intermittent (slug) flow amplifies unsteady effects, thus increasing the risk to encounter problems with surge and vibrations due to variations in mixture density and blade loading. During tests in [18] gas-locking was observed when the flow pattern in the suction pipe changed from bubbly to stratified or slug flow. But since the gas content had been increased at the same time, this observation may be a coincidence. In contrast, tests with the same pump reported in [16] scarcely showed any difference in performance when changing from bubbly to stratified flow in the suction pipe. These observations leave some uncertainties as to the real causes since the flow patterns in the suction pipe were changed by variations of the rotor speed. When changing the suction pipe diameter at constant rotor speed, virtually no effect on gas handling capability was found. The homogenizing effect of the rotor speed may therefore be considered as dominating. Likewise, tests in [17] showed little influence of the approach flow. Even with stratified flow in the suction pipe, the gas phase was well dispersed just downstream of the blade leading edge. Gas accumulations occurred further downstream in the channels. Extensive testing with various mixing devices in the suction pipe was reported in [21]. Again, it was found that none of the mixers brought about a significant improvement in 2-phase pumping performance. Solely an inducer was able to boost 2-phase performance, but this effect is explained by the compression of the gas accomplished by the inducer rather than through a more homogenous approach flow to the impeller.

Tests with wet gas compressors were performed by injecting liquid into the gas flow in the inlet pipe. Some tests were done with annular film flow while in others the droplets were finely dispersed by nozzles. No impact of the suction pipe flow patterns on performance was detected, [22]. The range of test parameters was $\rho^* = 12$ to 30, GVF = 0.97 to 0.995 (x = 0.52 to 0.9).

Overall it is concluded that the two-phase distribution in the suction pipe has little impact on the performance. Any attempt to significantly improve two-phase pumping by installing mixing devices at the pump inlet is likely to fail. However, slug flow at the pump inlet should be avoided for operational reasons.

Mach number: If a *compressible* fluid is expanded through an accelerating nozzle from the pressure p_1 prevailing in a reservoir to the pressure p_2, the flow velocity in the throat of the nozzle at first rises with the square root of the pressure difference $(p_1 - p_2)^{0.5}$. In contrast to incompressible flow, the throat velocity cannot be increased infinitely through lowering the downstream pressure p_2. Maximum flow in the nozzle is reached when the local flow velocity attains the local speed of sound. The reason for this is: A pressure perturbation in a fluid travels through the system at the sound velocity. If the pressure downstream of the nozzle is decreased but the sound velocity prevails in the nozzle, the change in pressure cannot be felt upstream of the nozzle and is thus unable to accelerate the fluid. The Mach number is defined by $M = w/a$ (w = local relative velocity, a = local sound velocity); its effect is important in the design of thermal turbo machines.

Since gas-liquid mixtures are compressible, Mach number limits may be expected also in two-phase pumping. However, it has not yet been reported that such limits have actually been hit in pump operation or testing. The sound velocity in two-phase flow depends on the gas volume fraction α, the density ratio ρ'/ρ'', the phase distribution (or flow patterns) and on the sound velocities in the pure gas and liquid phases (a'' resp. a'). When the phase distribution is unknown, it is difficult to reliably predict the sound velocity. Therefore, depending on the various test conditions, different prediction methods were published.

Table 13.3(1) gives equations for calculating the sound velocity in two-phase flow for different flow patterns according to [23]. These equations are consistent in that they yield the physically correct sound velocities for $\alpha = 0$ and $\alpha = 1$, i.e. for a pure liquid or gas. The sound velocity is usually lower in two-phase flow than the sound velocities in the pure phases because wave reflections occur at the gas/liquid phase boundaries. Thereby the sound energy is attenuated. Reflection and absorption increase with the change in impedance $\rho' \times a'/(\rho'' \times a'')$ at the phase boundaries. Therefore, the sound velocity decreases strongly when density ratios ρ'/ρ'' are high. For given gas and liquid properties the function $a = f(\alpha)$ exhibits a minimum. Finely dispersed gas bubbles, even with low gas content, drastically reduce the sound velocity if the density ratio is high. Further studies can be found in [55] and [57].

13.2.3 Calculation of Two-Phase Pump Performance

The complex physical processes discussed above are not amenable to sufficiently accurate theoretical calculations: empirical correlations are employed in practice.

Definition of basic two-phase flow parameters (Table 13.3):

- In the following text, quantities without superscript refer to the mixture. The liquid phase is designated by an apostrophe (e.g. Q′) and the gas phase by two apostrophes (e.g. Q″).
- The volumetric flow rate of a mixture results from adding liquid and gas flow rates: $Q = Q' + Q''$. The volumetric flow rate determines the velocities and is relevant for the flow through the pump. In particular the velocity triangles and the flow coefficient must be calculated from the volumetric flow of the mixture. Due to the compression, the volumetric gas flow Q″ is reduced when progressing through the pump. The same applies to the volumetric mixture flow while the liquid flow rate Q′ remains constant.
- The fraction of the gas phase *of the mixture* (or its concentration) is described by the "gas volume fraction" GVF or α, also termed "void fraction", which is defined by Eq. (T13.3.3). As the mixture is compressed the GVF is reduced.
- In contrast, the *mass flow rate* through the pump is not affected by the compression. By the same token the "gas mass fraction" x of a mixture remains constant throughout the pump; it is defined by Eq. (T13.3.4).
- Alternatively, the volumetric gas content is often defined by the "gas/liquid ratio" $GLR = Q''/Q'$. At constant temperature the GLR is directly proportional to the absolute pressure.
- The density of a homogenous mixture results from the volume fractions of liquid and gas; it is given by Eq. (T13.3.5).

Neither the gas volume fraction α nor the mass fraction x nor the homogenous density are able to describe the conditions in the pump because of uneven phase distributions and phase slip.

Quasi-isothermal compression: In order to calculate pump performance in two-phase flow the following assumptions are made:

1. Independently of the phase distribution (be it homogenous or separated) the conservation of mass, momentum and energy is applied separately to both the liquid and the gas phase.
2. Up to a gas volume fraction of about 80%, the compression is quasi-isothermal because the heat capacity of the liquid phase is much larger than that of the gas and both phases are sufficiently mixed with respect to heat transfer.
3. Effects of evaporation or condensation of vapor as well as the dissolution of gases in liquids are neglected.

In order to describe the useful work transfer in a pump, liquid and gas phase are considered separately. The useful work transferred to the liquid is thus:

$$Y' = \frac{p_2 - p_1}{\rho'} + \frac{c_2'^2 - c_1'^2}{2} \tag{13.17}$$

Y′ corresponds to the head ($g \times H = Y'$) in incompressible flow.

If a gas is compressed isothermally from p_1 to p_2, the useful specific work transferred per kg gas is given by:

$$Y'' = \int_{p_1}^{p_2} \frac{dp}{\rho} + \int c\,dc = R\,T \int_{p_1}^{p_2} \frac{dp}{p} + \int c\,dc = R\,T\,\ln\frac{p_2}{p_1} + \frac{c_2''^2 - c_1''}{2} \tag{13.18}$$

With the gas mass content x, the specific useful work transferred to the mixture is:

$$Y = Y'(1 - x) + Y''x \tag{13.19}$$

The useful power of the pump is: $P_u = \dot{m} \times Y = (\dot{m}' + \dot{m}'') \times Y$. The stage efficiency in two-phase pumping is defined accordingly by $\eta_{TP} = P_u/P$, Eq. (T13.3.9).

If a possible phase slip is neglected, the useful energy transferred to an isothermal gas/liquid mixture can be calculated from Eqs. (13.17 to 13.19), yielding Eq. (T13.3.7) in Table 13.3. This equation is valid also for the limiting cases of single-phase liquid (x = 0) and gas (x = 1), in as much as the compression is isothermal (which is not the case in usual thermal turbo machines at x = 1).

The coefficient of useful work for two-phase flow is obtained if Y is divided by the number of stages and $\frac{1}{2}u_2^2$ as per Eq. (T13.3.10). The work transfer depends primarily on the gas content, the density ratio and the phase distribution, which is influenced by the velocities or the flow rate, Chap. 13.2.2.

Polytropic compression: For high GVF, say above 0.8, the heating of the mixture during the compression is not to be neglected and the polytropic model should be applied. The necessary formulae are given in Table 13.3(2). Various quantities are defined below:

- The polytropic efficiency η_{pol} includes all losses which heat up the mixture. These are the hydraulic losses characterized by friction, mixing losses which *include losses caused by the interaction between the phases*, as well as disk friction and leakage losses. If the balance fluid is introduced into the inlet nozzle, the polytropic efficiency includes the leakage through the device for axial thrust balancing. The polytropic efficiency is determined by multiplying the efficiency measured in single-phase flow by the two-phase multiplier f_η; hence $\eta_{pol,Tp} = \eta_{i,SP} \times f_\eta$. Figure 13.19 shows the process in the enthalpy-entropy diagram.
- The choice of the mixture properties is important because the liquid content increases the apparent specific heat of the mixture and reduces the temperature rise caused by the compression. The specific heat $c_{p,mix}$ is calculated from

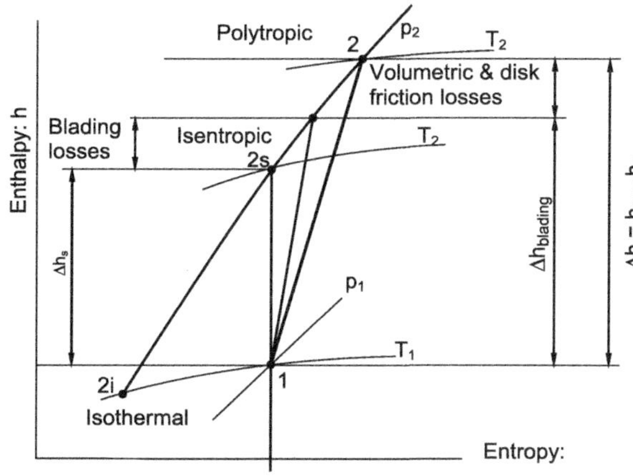

Fig. 13.19 Isentropic, polytropic and isothermal compression in enthalpy-entropy diagram

Eq. (T13.3.17). The exponents for isentropic and polytropic compression are then determined from $c_{p,mix}$ and the gas constant R from Eqs. (T.13.3.18 to 13.3.20). It is absolutely necessary that these relations are fulfilled; otherwise, the equations for useful work and work input will yield inconsistent results.

- Useful power and required work input can be calculated from Table 13.3(2) for the gas and for the mixture.
- The efficiency of the pump may be either defined as polytropic efficiency as per Eq. (T13.3.26) or as isentropic efficiency according to Eq. (T13.3.28).
- The 2-phase multipliers are determined the same way as in Eqs. (T13.3.11 and 13.3.12). They can either be based on polytropic or isentropic compression.
- The temperature of the two-phase mixture increases from stage to stage as the fluid is heated due to compression and dissipation of the various losses. Consequently, some liquid is evaporated. The vapor pressure is determined from the temperature at the stage inlet. The mass of liquid evaporating can be calculated from the enthalpies. Evaporation is neglected in the equations of Table 13.3(2).
- The temperature rise must be calculated from Eq. (T13.3.30) because the relation $T_2/T_1 = (p_2/p_1)^m$ *is not applicable to two-phase mixtures* (it works only for calculating the compression of gases).
- For test data evaluation the polytropic efficiency is determined from the measured internal power and pressure rise by solving iteratively Eq. (T13.3.26). It is not correct to use the expression $n = \ln(p_2/p_1)/\ln(\rho_{mix,2}/\rho_{mix,1})$ which is valid for gases only (this is in contrast to statements in [22]).
- As far as pure gas compression is concerned the formulae in Table 13.3 (2) agree with [N.22].

Theoretically the polytropic model can be used for the whole range from low gas contents up the compression of pure gas (i.e. for $0 < \alpha = 1$).

Derivation of empirical coefficients: The relations Y = f(α, ρ*, q*, geometry) and η = f(α, ρ*, q*, geometry) are not amenable to theoretical treatment, but empirical coefficients ("two-phase multipliers") can be derived from experiments. Such coefficients are valid only for the tested geometry and within the range of parameters investigated. Probably the easiest way to define two-phase multipliers is to refer the measured two-phase work coefficient ψ_{TP} and the efficiency η_{TP} to the equivalent data (ψ_{SPL} and η_{SPL}) measured in single-phase liquid (SPL) flow [24]. The corresponding two-phase multipliers are defined by Eqs. (T13.3.11 and 13.3.12). They can be derived from test results in the following way:

Step 1 A baseline test with single-phase liquid flow (subscript SPL) gives the usual dimensionless characteristics: $\psi_{SPL} = f(\varphi)$ and $\eta_{SPL} = f(\varphi)$.

Step 2 Tests with two-phase flow are carried out with various gas volume fractions GVF, liquid/gas density ratio DR (or ρ^*) and flow rate ratio q^*. The test matrix is designed to fully cover the range of interest in terms of GVF, DR and q^*.

Step 3 If the isothermal model is used, Eqs. (T13.3.7 to 13.3.10) are evaluated for each measuring point. When the polytropic model is applied, the equations given in Table 13.3(2) are evaluated for each data point. As a result, ψ_{TP} (α, ρ^*, q^*) and η_{TP} (α, ρ^*, q^*) are available.

Step 4 For each data point, the flow coefficient φ is calculated with the mixture flow rate. Using this value of φ, the corresponding single-phase data ψ_{SPL} and η_{SPL} can be determined from the SPL-test under step 1.

Step 5 By dividing ψ_{TP} (α, ρ^*, q^*) and η_{TP} (α, ρ^*, q^*) resulting from step 3 by the corresponding SPL data from step 4, the two-phase multipliers defined by Eqs. (T13.3.11 and 13.3.12) are obtained for isothermal or polytropic compression.

Step 6 The two-phase multipliers may be presented in the format of Figs. 13.23, 13.27 and 13.28 or by appropriate correlations.

Performance prediction: The performance prediction follows the formulae given in Tables 13.3(1) and 13.2(30) for isothermal or polytropic compression. In order to avoid inconsistencies, it is imperative to apply exactly the same logic and model (isothermal or polytropic) to the evaluation of the two-phase multipliers and to performance prediction.

The performance of multistage pumps is calculated stage by stage. In this process, the outlet conditions of the previous stage are entered as inlet conditions to the following stage. Gas density, flow rate and velocity triangles change from stage to stage due to the compression of the gas phase. The procedure is described for a polytropic compression in a multistage pump; it can be applied by analogy to an isothermal compression or a single-stage machine:

1. Determine liquid and gas properties, flow rates, gas fraction and density ratio.
2. Select hydraulics and single-phase liquid performance curves. In general, several hydraulics will be required to match the impeller/diffuser geometry to the flow as the gas is successively compressed.

Table 13.3 (1) Gas-Liquid-Mixtures		Eq.
Volumetric flow rate of mixture	$Q = Q' + Q''$	13.3.1
Mass flow rate of mixture	$\dot{m} = \dot{m}' + \dot{m}'' = \rho' Q' + \rho'' Q''$	13.3.2
Gas volume fraction or void fraction: GVF ≡ α	$\alpha = \frac{Q''}{Q'+Q''} = \left\{1+\frac{1-x}{x}\frac{\rho''}{\rho'}\right\}^{-1} = \frac{x\,\rho'}{x\,\rho'+(1-x)\,\rho''} = \frac{GLR}{1+GLR}$	13.3.3
Gas mass fraction	$x = \frac{\dot{m}''}{\dot{m}'+\dot{m}''} = \frac{\rho''}{\rho'}\frac{\alpha}{1-\alpha\left(1-\frac{\rho''}{\rho'}\right)} = \alpha\frac{\rho''}{\rho_{hom}} = \frac{GLR}{GLR+\frac{\rho'}{\rho''}}$	13.3.4
Gas/liquid ratio GLR	$GLR = \frac{Q''}{Q'} = \frac{\alpha}{1-\alpha} = \frac{x\,\rho'}{(1-x)\,\rho''}$	13.3.4a
Density of homogeneous mixture	$\rho_{hom} = (1-\alpha)\,\rho' + \alpha\,\rho''$	13.3.5
Dynamic viscosity of mixture	$\frac{1}{\mu} = \frac{x}{\mu''} + \frac{1-x}{\mu'}$	13.3.6
Specific energy for pressure rise from p_1 to p_2 (isothermal flow)	$Y_{isot,TP} = (1-x)\frac{p_2-p_1}{\rho'} + x\,R\,T\ln\frac{p_2}{p_1} = f_\psi\,\psi_{SPL}\,z_{st}\frac{u_2^2}{2}$	13.3.7
Useful power	$P_u = \dot{m}\,Y_{isot,TP} = (\dot{m}'+\dot{m}'')\,Y_{isot,TP}$	13.3.8
Efficiency with two-phase flow	$\eta_{TP} = \frac{P_u}{P} = \frac{\dot{m}\,Y_{isot,TP}}{P} = \frac{(\dot{m}'+\dot{m}'')\,Y_{isot,TP}}{P}$	13.3.9
Work coefficient with two-phase flow	$\psi_{TP} = \frac{2\,Y_{isot,TP}}{z_{st}\,u_2^2}$	13.3.10
Two-phase multiplier for work coefficient	$f_\psi = \frac{\psi_{TP}}{\psi_{SPL}} = \frac{\text{work coefficient with two-phase flow }(x>0)}{\text{work coefficient with single phase flow }(x=0)}$	13.3.11
Two-phase multiplier for efficiency	$f_\eta = \frac{\eta_{TP}}{\eta_{SPL}} = \frac{\text{efficiency with two-phase flow }(x>0)}{\text{efficiency with single-phase flow }(x=0)}$	13.3.12
Sound velocity in homogenous flow	$a_H = \frac{a''}{(1-\alpha)\sqrt{(1-\alpha)\left(\frac{a''}{a'}\right)^2+\alpha\frac{\rho'}{\rho''}}+\alpha\sqrt{\alpha+(1-\alpha)\frac{\rho''}{\rho'}\left(\frac{a''}{a'}\right)^2}}$	13.3.13
Sound velocity in slug flow	$a_s = \frac{a''}{\alpha+(1-\alpha)\frac{a''}{a'}}$	13.3.14
Sound velocity in stratified flow	In the gas phase: $a_{EG} = \frac{a''}{\sqrt{1+\left(\frac{a''}{a'}\right)^2\frac{\rho''}{\rho'}\cdot\frac{1-\alpha}{\alpha}}}$ In the liquid phase: $a_{EL} = \frac{a'}{\sqrt{1+\left(\frac{a'}{a''}\right)^2\frac{\rho'}{\rho''}\cdot\frac{\alpha}{1-\alpha}}}$	13.3.15 13.3.16
Superscripts: ′ liquid phase; ″ gas phase	Sound velocity in gases: $a'' = \sqrt{\kappa R T}$; R = gas constant, κ = exponent of isentropic expansion	
Note: For the sake of clarity, the equations are written for static conditions.		

Table 13.3 (2) Polytropic compression of Gas-Liquid-Mixtures		Eq.
Specific heat of mixture	$c_{p,mix} = \frac{\rho' Q' c'_p + \rho'' Q'' c''_p}{\rho' Q' + \rho'' Q''} = \frac{c'_p + \frac{GLR}{DR} c''_p}{1 + \frac{GLR}{DR}}$	13.3.17
Exponent of isentropic compression	$\kappa = \frac{c_{p,mix}}{c_{p,mix} - R\,Z}$; $\frac{\kappa}{\kappa - 1} = \frac{c_{p,mix}}{R\,Z}$	13.3.18
Polytropic stage efficiency in 2-phase flow	$\eta_{pol,TP} = \eta_{st,SPL}\, f_\eta$; defined by: $\eta_{pol,TP} = \frac{\dot{m}\, Y_{pol,TP}}{P} = \frac{Y_{pol,TP}}{h_2 - h_1}$	13.3.19
Exponent n of polytropic compression	$m \equiv \frac{n-1}{n} = \frac{\kappa - 1}{\kappa \eta_{pol}} = \frac{R\,Z}{c_{p,mix}\, \eta_{pol}}$	13.3.20
Useful polytropic work per kg gas	$Y''_{pol} = \frac{p_1}{m\, \rho''_1} \left\{ \left(\frac{p_2}{p_1} \right)^m - 1 \right\}$	13.3.21
Polytropic *work input* required per kg gas	$h''_2 - h''_1 = \frac{\kappa}{\kappa - 1} \cdot \frac{p_1}{\rho''_1} \left\{ \left(\frac{p_2}{p_1} \right)^m - 1 \right\}$	13.3.22
Useful polytropic work per kg mixture	$Y_{pol,Tp} = (1 - x) \frac{p_2 - p_1}{\rho'} + \frac{x}{m} \cdot \frac{p_1}{\rho''_1} \left\{ \left(\frac{p_2}{p_1} \right)^m - 1 \right\}$	13.3.23
Work input required per kg mixture	$h_2 - h_1 = \frac{p_2 - p_1}{\rho' \eta_{pol}} (1 - x) + \frac{x\, \kappa}{\kappa - 1} \cdot \frac{p_1}{\rho''_1} \left\{ \left(\frac{p_2}{p_1} \right)^m - 1 \right\}$	13.3.24
Inner power	$P_i = \dot{m}\, (h_2 - h_1)$	13.3.25
Derive polytropic efficiency from tests	$\frac{p_2 - p_1}{\rho' \eta_{pol}} (1 - x) + \frac{x\, \kappa}{\kappa - 1} \cdot \frac{p_1}{\rho''_1} \left\{ \left(\frac{p_2}{p_1} \right)^{\frac{\kappa - 1}{\kappa \eta_{pol}}} - 1 \right\} = \frac{Pi}{\dot{m}}$	13.3.26
Useful isentropic work	$Y_{s,TP} = (1 - x) \frac{p_2 - p_1}{\rho'} + x \frac{\kappa}{\kappa - 1} \cdot \frac{p_1}{\rho''_1} \left\{ \left(\frac{p_2}{p_1} \right)^{\frac{\kappa - 1}{\kappa}} - 1 \right\}$	13.3.27
Isentropic efficiency	$\eta_{s,TP} = \frac{\dot{m}\, Y_{s,TP}}{P} = \frac{Y_{s,TP}}{h_2 - h_1}$	13.3.28
Determine tip speed, number of stages or pressure rise	$f_\psi\, \psi_{SPL}\, z_{st} \frac{u_2^2}{2} = Y_{pol,Tp} = (1 - x) \frac{p_2 - p_1}{\rho'} + \frac{x}{m} \cdot \frac{p_1}{\rho''_1} \left\{ \left(\frac{p_2}{p_1} \right)^m - 1 \right\}$	13.3.29
Temperature rise of mixture	$T_2 - T_1 = \frac{h_2 - h_1}{c_{p,mix}}$	13.3.30

Note: For the sake of clarity, the equations in Table 13.3 are written for static conditions. If total conditions are considered, the differences in kinetic energy ½ $(c_d^2 - c_s^2)$ must be added to all equations concerning the specific work.
R = gas constant; $R = R_u/M$; R_u = 8315 J/kg K; M = molecular mass of gas
Z = real gas factor; gas state equation: $p/\rho = Z\,R\,T$

3. Select size and speed.
4. Calculate flow coefficient φ at the inlet of the first stage and determine the pressure coefficient ψ_{SPL} and efficiency η_{SPL} from the SPL-curves.
5. Determine two-phase multipliers f_ψ and f_η from GVF, DR, and q* from test data (such as Figs. 13.27 and 13.28).
6. Calculate polytropic efficiency from Eq. (T13.3.19).
7. Calculate pressure rise in stage from Eq. (T13.3.29).
8. Calculate power absorbed by stage from Eqs. (T13.3.24 to 13.3.25).
9. In polytropic flow calculate the stage outlet temperature from Eq. (T13.3.30).
10. In quasi-isothermal flow the temperature rise due to the losses in the stage can be estimated in a similar way as from Eq. (13.12a) from the difference between the power absorbed by the stage P_{st} and the useful power in the stage $P_{u,st}$:

$$\Delta T_{st} = \frac{P_{st} - P_{u,st}}{c_{p,mix}\, \rho_{mix}\, Q_{mix}} \tag{13.19a}$$

11. Calculate the gas density at the outlet of the stage.
12. Use the stage outlet data as inlet conditions for the following stage.
13. When the performance of all stages has been determined, the total pressure rise and power of the pump are known. Speed, stage geometries and number of stages may subsequently be varied and optimized to achieve the specified pressure rise with the minimum power. Surge limits and maximum allowable flow must be observed during this selection process.
14. Check mechanical constraints in terms of allowable rotor length, stresses and tip speeds.

13.2.4 Radial Pumps Operating with Two-Phase Flow

As mentioned above, the gas handling capability of radial pumps is poor unless special design features for improving that capacity are implemented. Phase separation leads to gas accumulations in which the pressure build-up is limited by the gas density, because the achievable pressure rise in a closed gas volume (or "gas pocket") is much lower than in the liquid (refer to Fig. 13.14). The part of a blade that is covered by a gas pocket therefore transfers little energy. Gas accumulations also block part of the cross sections of the impeller channels. The relative velocity increases and work transfer is reduced accordingly, as follows from the velocity triangles. The liquid phase is accelerated, but the energy transferred does not contribute to increase the static pressure. These mechanisms may be responsible for the experimental finding that the performance of radial pumps deteriorates (often, but not always) more quickly at $q^* > 1$ than at partload. Even with fairly homogenous flow-frequency surging becomesow there is an exchange of momentum between gas und liquid phase, generating additional losses which are absent in single-phase flow.

Figure 13.20 shows tests performed on a single-stage radial pump ($n_q = 26$), which demonstrate the drastic performance impairment even with small gas volume fractions α, [25]. The tests were done with a ring section pump similar to Fig. 2.6 but with only one stage. The pump had a radial inlet chamber and a diffuser. The impeller had no balance holes.

All measuring points in the graphs are referred to the flow and head at BEP when pumping pure water. While head and efficiency are little influenced near BEP when 2% gas is present ($\alpha = 0.02$), the characteristics with $\alpha = 0.04$ deviate strongly from single-phase performance at $q^* > 0.8$. The performance deficit grows with increasing flow and amounts to more than 50% at $q^* > 1.1$. In contrast, partload operation below $q^* < 0.8$ is little affected. This situation changes with $\alpha = 0.05$,

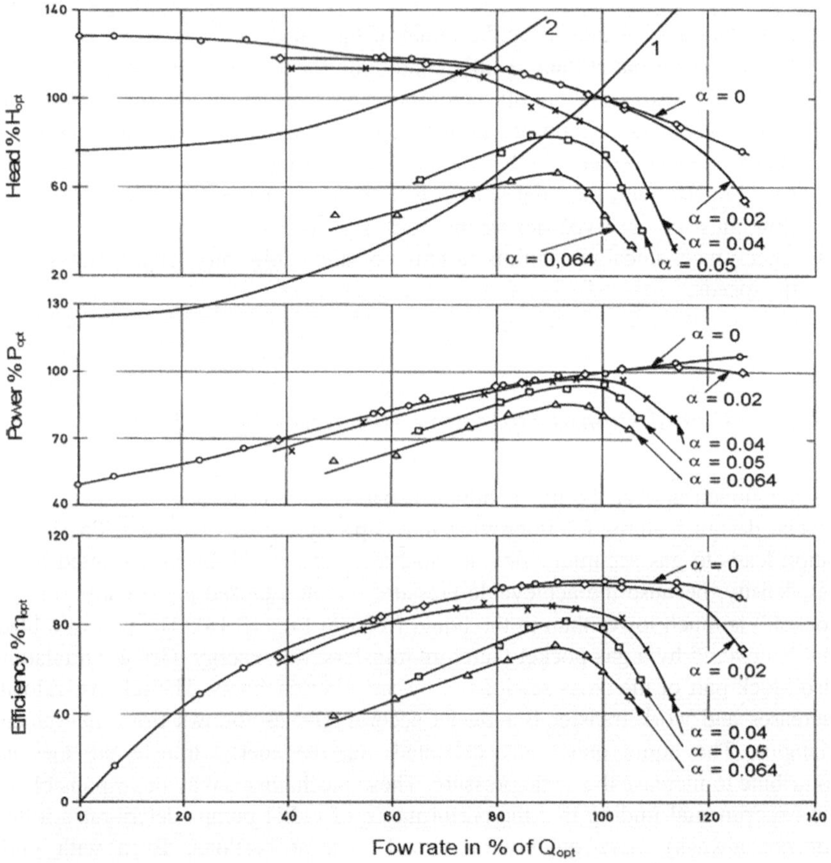

Fig. 13.20 Influence of air content on the characteristics of a single-stage pump ($n_q = 26$) at an inlet pressure of 2.5 bar absolute, $\rho'/\rho'' = 334$, [25]

where operation is only possible in a window which, in this specific test, is defined roughly by $0.6 < q^* < 1.1$. Increasing the gas volume fraction from 4 to 5%, changes partload performance completely. Gas accumulations in the impeller are expected to be responsible for this.

The tests in Fig. 13.20 were done at a density ratio of DR = $\rho'/\rho'' = 334$. In another test with DR = 186 the gas handling capability improved in that performance with $\alpha = 0.06$ was similar to the curve for $\alpha = 0.04$ in Fig. 13.20. It is therefore important to note that any two-phase performances measured are valid *only at the density ratio* tested; an application at higher or lower density ratios would be altogether meaningless. The lower the liquid to gas density ratio, the higher is the pump performance in two-phase flow. In the extreme, at $\rho'/\rho'' = 1$ there would be no performance impairment at all. Naturally these arguments also apply to the operation limits evident from the curves at $\alpha = 0.05$ and $\alpha = 0.064$ in Fig. 13.20. For the influence of the density ratio see also Fig. 13.27.

Results similar to those in Fig. 13.20 were obtained in tests reported in [16, 18]. A single-stage volute pump ($n_q = 23$) of the type shown in Fig. 2.1 was operated with oil/air mixtures (oil viscosity 18 mm^2/s, density ratio about 700). Figure 13.21 shows the characteristics with gas contents of 2 and 5%.

With a gas content of 6% the drop in head was roughly 60% at BEP. Trimming the rear shroud to 70% of the impeller diameter reduced gas the accumulations in the impeller sidewall gap and improved two-phase performance. The mechanism is explained as follows: Removing part of the rear shroud induces a recirculation at the impeller outlet; it breaks up the gas accumulations into finer bubbles which are carried away by the liquid. The effect is similar to the oblique trailing edge used in [19] for improving the gas handling capability.

Pumping limit: The limit of head breakdown (or gas-locking) shown in Fig. 13.21 delineates the range where no operation is possible. A peculiarity of these tests are the Z-shaped curves H = f(Q) measured for specific gas volume fractions.

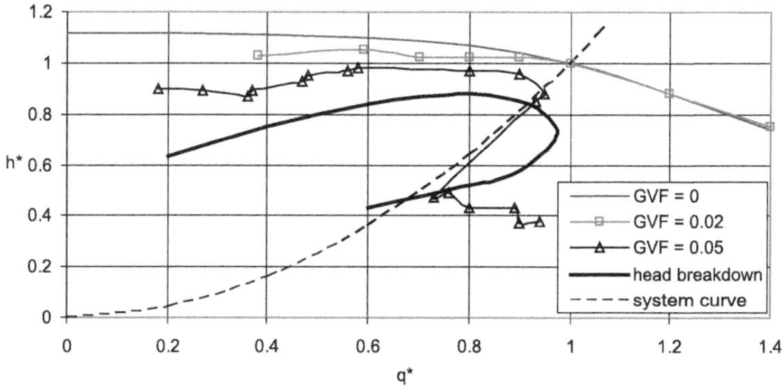

Fig. 13.21 Influence of air content on the characteristics of a single-stage volute pump $n_q = 23$; $\rho'/\rho'' = 700$; oil viscosity 18 mm^2/s; $h^* = H/H_{opt}$ [16]

According to the curve for GVF = 0.05 the operation jumps from the point q* = 0.93, h* = 0.85 down to the point q* = 0.73, h* = 0.47. In between these two points stable operation is impossible. This behavior can be explained by a sudden change between two stable flow patterns and phase distributions. The mechanisms are similar to the switch in flow patterns described in Chap. 5.5.2 (saddle-type instability) and the impact of cavitation induced instabilities of the Q-H-curve described in context with Fig. 6.17. Hysteresis effects can thus also be expected in two-phase pumping. Such unstable behavior depends on the details of the geometric features of the actual impeller; it is not a general characteristic. With a different impeller having an outlet angle of 10°, Z-shaped Q-H-curves were not measured, [18].

When the gas content at given operation conditions is continuously increased, the pumping limit or gas-locking announces itself by growing fluctuations of power, pressure and flow. Low-frequency surging becomes audible when the gas content is raised until large fluctuations of the flow rate occur. Observing the flow at the pump inlet, momentary stagnation of the fluid is recorded. When finally gas-locking occurs, the flow in the suction pipe stops and the impeller operates as against a closed discharge valve with liquid gathering near its outlet (similar to the operation of a dynamic seal).

Figure 13.22 shows two-phase multipliers for work transfer (or head) calculated as per Eq. (T13.3.11) for single-stage pumps with radial impellers (the test data are from various sources). The strong influence of the density ratio DR is clearly demonstrated. The scatter is (among other effects) due to different operation points and different geometries. In a similar way Fig. 13.23 gives two-phase multipliers for work transfer for multi-stage pumps with radial impellers. The data were derived from [25] for a 3-stage pump with air/water mixtures and from [19] for an 8-stage pump with fuel-oil/CO_2 and air/water mixtures (both pumps had specific speeds near n_q = 26). Multi-stage pumps perform better than single stage pumps, since the gas is compressed successively as it progresses through the stages (thereby reducing the void fraction and the density ratio).

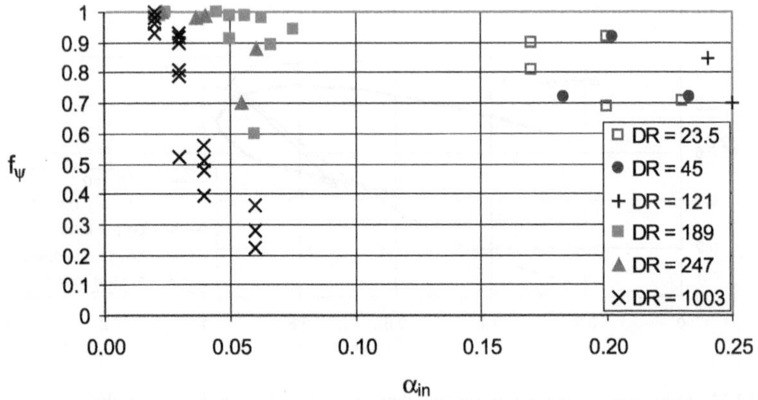

Fig. 13.22 Two-phase multipliers for head of single-stage pumps with radial impellers

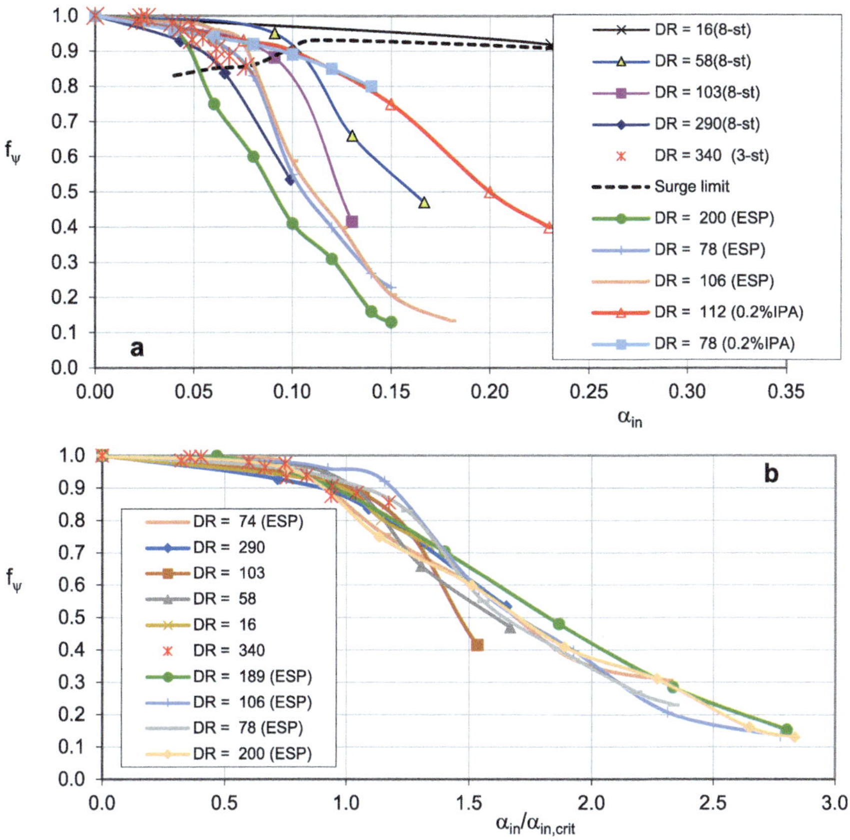

Fig. 13.23 Two-phase multipliers for head of multi-stage pumps with radial impellers; ESP-data from [60, 62]. Chart a: f_ψ plotted against void fraction; Chart b: f_ψ plotted against ratio of actual to critical void fraction

Since flow patterns and energy transfer depend on the geometrical shape of the hydraulic components and the fluid properties, the data in Figs. 13.22, 13.23 and 13.24 can be used only for a rough assessment of two-phase performance. Note also that f_ψ is plotted against the *inlet* void fraction. Alternatively, the averages of void fraction, density ratio and q* could be used for correlating the 2-phase multipliers. This procedure would be more meaningful to capture the effects of the gas compression within the pump when the number of stages or the tip speed u_2 are varied.

For lack of more precise data the multiplier for efficiency is assumed to be the same as for the head, hence $f_\eta = f_\psi$.

At high gas contents the operation range becomes narrow and flow instabilities can jeopardize reliable operation. It is therefore necessary to limit the selection of pumps with conventional radial hydraulics to a maximum GVF which depends on the density ratio. Figure 13.24 proposes tentative limits for single- and multistage pumps. Note that the head multipliers in Figs. 13.22 and 13.23 were calculated

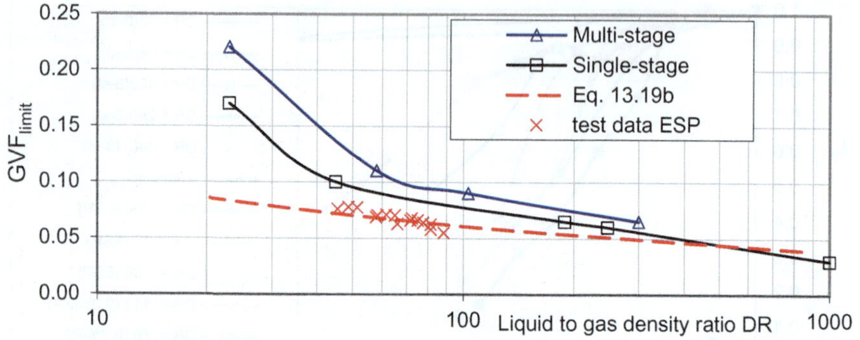

Fig. 13.24 Tentative limits for maximum allowable gas volume fraction for conventional radial impellers. Density ratio: $DR \equiv \rho'/\rho''_{in}$; gas volume fraction: $GVF \equiv \alpha$. Test data ESP from [58]

from the various test data by applying the formulae in Table 13.3. Therefore, the calculations include the isothermal compression of the gas.

Before pumping is altogether disrupted by gas locking, operation becomes unstable when the gas content exceeds some limit (the "surge limit") which depends on the specifics of each individual pump, the system and the fluid properties. Running in surge, will produce pressure and flow fluctuations; operation then becomes impossible due to excessive vibrations. Example: Surging tests in [62] showed pressure fluctuations of up to ±80% at 0.07 Hz and flow rate fluctuations of up to ±50% at 0.17 Hz (typical values). The surging mechanism may be imagined as follows: When gas pockets block a substantial portion of an impeller (say 20% of the channel), the pressure rise of the impeller is impaired due to the low gas density. As a consequence, the flow rate is momentarily reduced. Subsequently, when the gas pocket exits the channel, pressure rise and instantaneous flow rate are re-established till the next gas pocket enters the impeller.

In this way, the surge limit in radial impellers is reached when the phase distribution changes from quasi-homogenous flow to conditions with larger bubbles and gas accumulations (slug flow). Studies into these effects have been reported in [58] where three flow patterns in the impeller have been visually observed in a test pump: (1) homogeneous flow with tiny spherical bubbles dispersed inside the impeller channels with virtually no phase slip; (2) bubbles of uneven shape and size moving with noticeable phase slip; (3) large bubbles coalescing into gas pockets. Pressure surging indicates a flow pattern transition from dispersed bubbles to large non-spherical bubbles. This process is governed by an equilibrium between: (a) turbulence breaking up bubbles versus (b) surface tension and a growing bubble population fostering bubble coalescence.

Theoretical models for the critical bubble size have been developed, above which spherical bubbles get unstable and tend to coalesce. The theory, which can be found in [13, 58], assumes that the critical bubble size is reached when the resonance frequency of the bubble equals a characteristic turbulence frequency. From this bubble size, the critical void fraction α_{crit} has been derived as represented by Eq. (13.19b); α_{crit} was claimed to represent the limit for pressure surging, [58].

$$\alpha_{crit} = 0.13 \frac{We^{0.2}}{\left(1 - \frac{\rho''}{\rho'}\right)^{0.5} \left(\frac{\rho'}{\rho''}\right)^{0.2}} \left(\frac{\varphi_2 \psi}{f_{b2}}\right)^{0.4} \tag{13.19b}$$

$$We = u_2 \left(\frac{\rho' r_2}{S_T}\right)^{0.5} \tag{13.19c}$$

$$f_{b2} = \frac{V_{La}}{\pi {r_2}^2 b_2} \tag{13.19d}$$

The Weber number "We" is defined by Eq. (13.19c). The flow coefficient φ_2 is defined by Eq. (T3.4.6), the head coefficient by Eq. (3.4.8). V_{La} is the fluid volume contained in the impeller. Since this is often unknown, the factor f_{b2} has been introduced by Eq. (13.19c); f_{b2} can be estimated; it is close to 1 to 1.2 for radial impellers with moderate n_q.

The dimensionless representation of Eq. (13.19b) was derived from the original equation presented in [58] in order to arrive at a more generic formula which is made up of dimensionless terms only. In this way, the data can be compared meaningfully to other tests with different pumps and fluid properties, and Eq. (13.19b) can be used for predicting the behavior of other pumps.

Testing in [58–60] was done with a 14-stage pump with semi-axial, closed impellers ($n_q = 31$, $d_2 = 112$ mm). This type of pumps (known as "electrical submersible pumps" or "ESP") is widely used for artificial lift in oil wells for increasing oil production. The surge limits determined experimentally, were in the range of $\alpha = 0.05$ to 0.10 depending on the suction pressure, the flow rate and the stage considered (surge limit in stage 3 is lower than in stage 7 due to lower gas density in stage 3). Together with data from other sources, the measured head-multipliers f_ψ (Eq. T13.3.11) are plotted as function of the inlet void fraction α_{in} in Fig. 13.23a (data labelled "ESP"). The curves of Fig. 13.23a have been re-plotted in Fig. 13.23b as a function of the ratio α/α_{crit}. In this way, the curves come quite close together: apparently, the steep drop in performance, which occurs when gas bubbles coalesce into larger accumulations, does not too strongly depend on the actual geometry of the pump. Compared to other sources, the 2-phase performance and surge limit of this pump appear rather low, Fig. 13.23a. The approximate surge limit is shown there by the dashed curve; it occurs where the curves $f_\psi = f(\alpha_{in})$ start to drop steeply. From the character of these curves, it is quite obvious that growing gas accumulations are responsible for this.

The surge limits calculated from Eq. (13.19b) are shown in Fig. 13.24 together with the actual test data, marked "ESP". The agreement between test data and Eq. (13.19b) is fairly good. Compared to the tentative operation limits derived from the other tests in Fig. 13.23, the ESP surge limits seem on the conservative side. The criterion applied may be too restrictive; it may also be surmised that the exponent (0.2) on the density ratio is too low – or rather, that the exponent is not constant.

Equation (13.19b) has the merit that the influence of various parameters on the GVF limiting stable operation (α_{crit}) can be evaluated. Accordingly, the critical GVF:

- decreases with $DR^{0.2}$ for density ratios DR > 50
- increases with the rotational speed according to $n^{0.2}$
- increases with the impeller diameter according to $d_2^{0.3}$
- decreases with $S_T^{0.1}$; lower surface tension allows bubble break-up and improves 2-phase performance
- exhibits a flat optimum near the BEP with respect to the flow rate.

The tests covered the range of q* = 0.68 to 1.12. As demonstrated by Figs. 13.20 and 13.21, the flow range where stable operation is possible in 2-phase pumping is limited. This effect cannot be captured by Eq. (13.9b).

As with the flow in pipes (Fig. 13.13) - as in fact with any component - increasing the flow velocity tends to homogenize the flow and to disrupt larger gas bubbles. Therefore, the critical void fraction grows with increasing speed, flow and pressure coefficients and decreases with the fluid volume contained in the impeller. A larger fluid volume implies a larger transit time through the pump.

Significant improvements of α_{crit} and f_ψ have been found by injecting surfactants during the tests reported in [60, 62] where 0.2% isopropyl alcohol (IPA) was added to an air/water mixture. The main effect of surfactants is due to foam formation (i.e. more homogenous mixture and less slip between gas and liquid) since the reduction in surface tension was negligible in these tests. When 0.2% of IPA was added to the mixture, pressure fluctuations decreased from roughly 80 to 5% peak to peak. The curves f_ψ decreased continuously with GVF in the range tested (up to GVF = 0.25); no critical GVF could be detected, see Fig. 13.23a: at GVF = 0.15 and DR = 106/112, $f_\psi = 0.75$ with 0.2% IPA while $f_\psi = 0.2$ without IPA-injection. Further improvement when injecting 0.41% (rather than 0.2%) IPA was only minor.

The concept of the critical GVF and above test data can be applied to estimate 2-phase performance of radial and semi-axial pumps of moderate n_q:

- Step 1: calculate α_{crit} from Eq. (13.19b). This may help with selecting an appropriate pump.
- Step 2: calculate the head degradation $f_\psi = (\alpha_{in}/\alpha_{crit})$ due to 2-phase flow as function of GVF (α_{in}) from the data in Fig. 13.23b.
- Assume $f_\eta = f_\psi$ for estimating the efficiency.

The head breakdown at partload is caused by flow separation and recirculation where gas may be locked in a recirculation zone. Gas-filled zones are likely to lead to premature stall, possibly because the flow cannot follow the high deceleration downstream of a gas pocket. The head breakdown at high flow may be caused by gas accumulations near the impeller or diffuser throat causing acceleration of the liquid. Whether the Mach number has an influence on the head breakdown remains to be demonstrated. If this were the case, a limit should be reached when increasing the rotor speed; but doing so actually improved performance.

If the gas/liquid density difference is high (say $\rho'/\rho'' > 150$) or the inlet pressure is low, the following limits apply to radial pumps without special design features for improving two-phase performance:

- Up to 2% gas volume ($\alpha = 0.02$) virtually no performance impairment is expected for $q^* < 1.2$.
- Gas-locking is to be expected at $\alpha = 0.05$ to 0.08; rarely up to 10% gas can be pumped. With gas contents above 5%, the possible operation range is narrow and the drop in head and efficiency is quite high, Figs. 13.20 and 13.21.
- In multistage pumps the gas volume fraction decreases from stage to stage due to the successive compression of the gas. Unless the first stage becomes gas-locked, multistage pumps can tolerate more gas than single-stage machines, Fig. 13.24.
- Tests with an axial pump ($n_q = 157$) showed a similar performance impairment as that in Fig. 13.20 ($n_q = 26$), [26].
- The gas handling capabilities of semi-open impellers are better than those of closed radial impellers due to the mixing action of the gap flow (discussed in detail above). The clearance between impeller and casing has an influence on the gas handling capability, Table D13.1.

Table D13.1 Pumping limits of radial impellers at suction pressure of about 1 bara

	conservative	upper range
Closed impellers	0.01	0.02 to 0.03
Semi-open impellers: clearance $\Delta R/R_2 = 0.002$	0.02	0.04 to 0.05
Semi-open impellers: clearance $\Delta R/R_2 = 0.01$	0.04	0.06

With decreasing liquid to gas density ratio DR the gas handling capability of a given pump improves. As a very rough estimate it can be assumed that the limiting gas content follows the proportion $\alpha_{limit} \sim (\rho''/\rho')^{0.5}$. This assumption allows an assessment in the domain of $\rho'/\rho'' > 100$ of how a pump would behave with different gas densities. According to the above proportionality, the pump from Fig. 13.20 would have with $\rho'/\rho'' = 120$ and $\alpha = 0.067$ a characteristic similar to the curve for $\alpha = 0.04$ and the pumping limit might rise from $\alpha = 0.06$ to $\alpha = 0.1$.

By venting gas from the impeller sidewall gaps and from the impeller itself (through holes drilled into front and rear shrouds) gas volume fractions up to 65% could be pumped as claimed in [17].

During tests with air/water mixtures on a 22-stage semi-axial pump gas volume fractions above 50% could be pumped with a density ratio DR = 106, [27].

The gas-locking observed with radial impellers suggests that the two-phase multiplier for the specific work drops to zero when a critical GVF is exceeded. The data in Figs. 13.21 and 13.23 point into the same direction. At very high gas mass content, say at $x > 0.5$ (GVF close to unity) radial impellers are a again able to transfer energy when working in the range of droplet flow patterns. This is the domain of wet gas compressors, where the gas works as a carrier fluid that entrains the liquid in the form of droplets and films. Just like compressors, such machines need high impeller tip speeds in order to achieve an acceptable pressure rise. The allowable speeds are limited by droplet erosion (as observed in wet steam turbines).

13.2.5 Helico-axial Multiphase Pumps

Multistage "helico-axial" pumps were developed for pumping mixtures of oil, natural gas, water and sand with gas volume fractions from zero to 97% or even above [28–32]. The impellers have a very high hub ratio d_n*, which increases from the inlet to the outlet, Fig. 13.25. The blading resembles that of an inducer with very low blade height. Downstream of the impeller, a diffuser decelerates the flow and deflects it into an essentially axial direction. The diffuser hub ratio decreases from inlet to outlet in order to match the impeller inlet of the subsequent stage. Special shaping of the hub and the blades limits the detrimental effects of phase separation so that an acceptable pressure rise is achieved even with high gas content. There is a flow through the gap between the impeller blades and the casing. An annular seal at the hub limits the back-flow around the diffuser. Figure 13.26 shows an 11-stage pump of this type. The inlet nozzle is on the left and the discharge nozzle on the right in the figure. The large hub diameter ratio gives a stiff rotor. This is necessary for mastering the vibration behavior of the machine because the annular seals lose much of their damping in two-phase flow. A balance piston reduces the axial thrust and the pressure at the mechanical shaft seal. The residual thrust is taken by a high-capacity axial bearing which can be seen on the far right of the figure.

Just as for compressors, the stage geometry must be adapted to the actual mixture flow rate at the stage inlet. Different stage groups of impellers/diffusers are therefore mounted on the rotor as can be seen in Fig. 13.26.

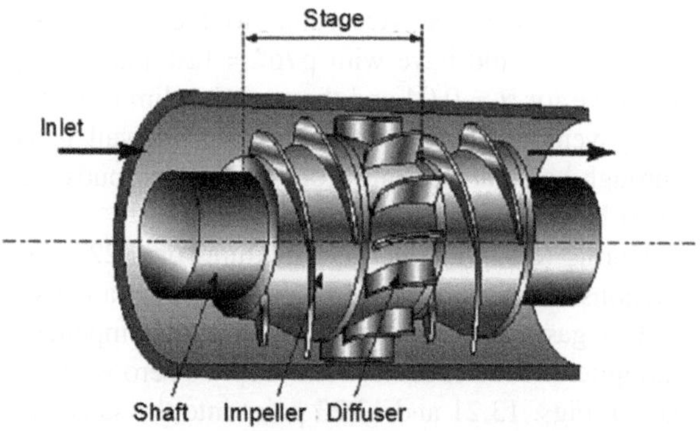

Fig. 13.25 Semi-axial stage of a multi-phase pump [32]

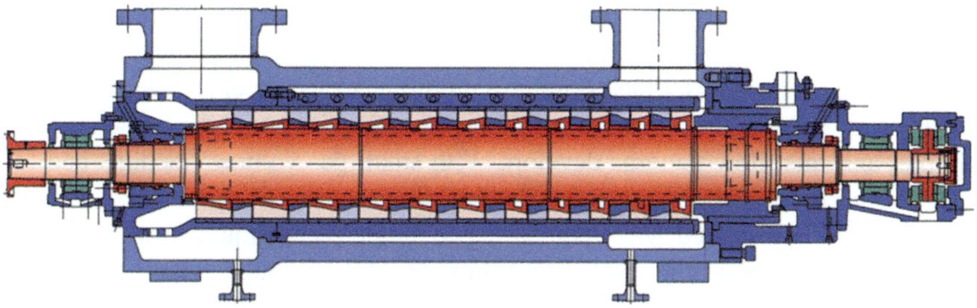

Fig. 13.26 11-stage multi-phase pump for oil-gas mixtures, motor 6000 kW, Sulzer Pumps

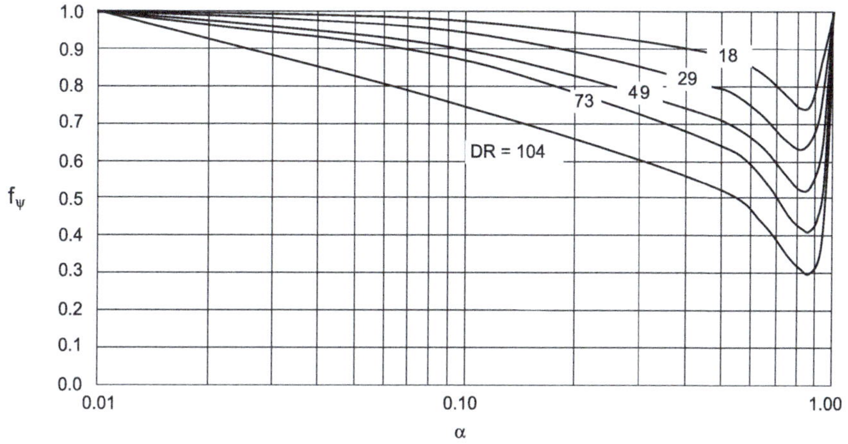

Fig. 13.27 Two-phase multipliers for the work coefficient measured on pumps according to Figs. 13.25 and 13.26. Parameter is the density ratio DR = ρ'/ρ'' [32]

Two-phase multipliers: Tests in [32] show the two-phase multipliers that can typically be achieved with helico-axial pumps. The measurements were done with a 13-stage machine with speeds up to 6600 rpm.[6]

Figure 13.27 shows the two-phase multipliers $f_\psi = f(\alpha, DR = \rho'/\rho'')$ of the work coefficient derived from tests with diesel oil and nitrogen. These multipliers were calculated for isothermal compression similarly to Table 13.3. The measuring points (omitted in Fig. 13.27) exhibit a scatter of roughly ±15% around the curves which were extrapolated to the theoretical limits $f_\psi = 1$ at $\alpha = 0$ and $\alpha = 1$.

In order to achieve a more general representation of the data, the curves in Fig. 13.27 were plotted with the density ratio DR as a parameter (rather than the inlet pressure used in [32]). The test results show that the highest impairment of the

[6]Maximum tip speed in these tests was 86 m/s. The first group of 8 stages had impeller diameters of 250 mm, the last 5 stages were 232 mm.

work coefficient is observed in the range of $\alpha = 0.83$ to 0.86, where phase separation effects are apparently most severe (this finding is specific to the pump investigated). For $\alpha > 0.9$ the two-phase multipliers raise steeply towards $f_\psi = 1.0$ at $\alpha = 1.0$, because the phase distribution becomes increasingly homogenous as droplet and mist flow patterns develop. With small gas volume fractions and high density ratios a rather homogenous bubbly flow can be expected, because work transfer is little impaired when comparing to single-phase liquid flow.

Figure 13.28 shows the same data as Fig. 13.27, but plotted against the *mass*gas content x instead of the gas *volume* content α. Comparing these two figures it is seen that phase separation (or the performance minimum) is controlled by the *volumetric* rather than the mass gas content. Conversely, the domain with droplets in a gas stream is best correlated by the gas *mass* content because the ability of the "carrier fluid" (the gas stream) to transport droplets is directly proportional to the gas mass flow $\rho'' \times Q''$.

For lack of more precise data the multiplier for efficiency is assumed to be the same as for the head, that is $f_\eta = f_\psi$.

Stability limit: Since the gas phase is compressible, all two-phase pumps experience a flow rate below which pumping tends to become unstable. As with compressors there is a "surge limit". Low-frequency fluctuations of power; pressures and flow as well as increased vibrations are the consequences. The level of vibrations and fluctuations depends on the pressure rise provided by the pump and the dynamic properties of the system in which the pump is operating. Although the multiplier for useful specific work f_ψ and the pressure rise may become quite small at high density ratios, gas-locking has not been observed in testing or operation of helico-axial multiphase pumps.

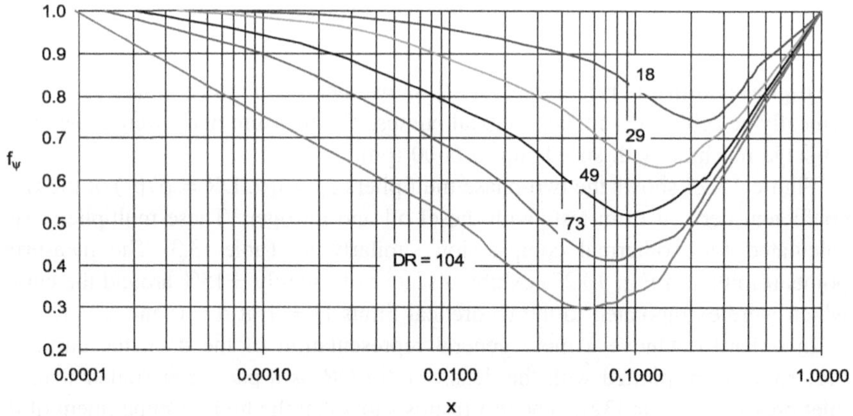

Fig. 13.28 Two-phase multipliers for the work coefficient. Same data as in Fig. 13.27 but plotted against gas mass content x. Parameter is the density ratio DR = ρ'/ρ''

Fig. 13.29 Characteristics of a two-phase pump at constant GVF and inlet pressure

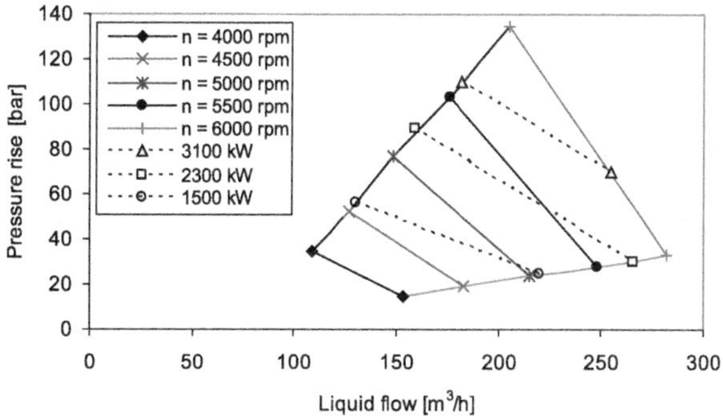

Two-phase pump characteristics: Figure 13.29 shows the characteristics of a multistage helico-axial two-phase pump at speeds between 4000 and 6000 rpm. Pressure and gas volume fraction *at the inlet* are constant. The upper boundary of the operational field is given by the surge limit, the lower boundary by the maximum flow. The steepness of the curves at constant rotor speeds grows strongly with the absolute value of the speed because the void fraction and density ratio throughout the pump decrease with rising compression (i.e. growing speed). The graph demonstrates that the similarity laws cannot be applied to the whole pump. The apparent deviation from similarity increases with the compression of the mixture. At the upper boundary of the operation range where compression is highest, $\Delta p \sim n^{3.3}$, at the lower boundary only $\Delta p \sim n^{1.9}$. With a growing number of stages and other factors apt to increase the compression the speed exponents get higher.

Figure 13.30 shows the characteristics of the same pump as in Fig. 13.29 at constant speed and inlet pressure. The gas/liquid ratio is varied between GLR = 2 to 15. With increasing gas content the achievable pressure rise drops drastically and the allowable operation range narrows because the stages are not adapted to the low compression achieved at high GLR.

Fig. 13.30 Characteristics of a two-phase pump at constant *speed* and inlet pressure

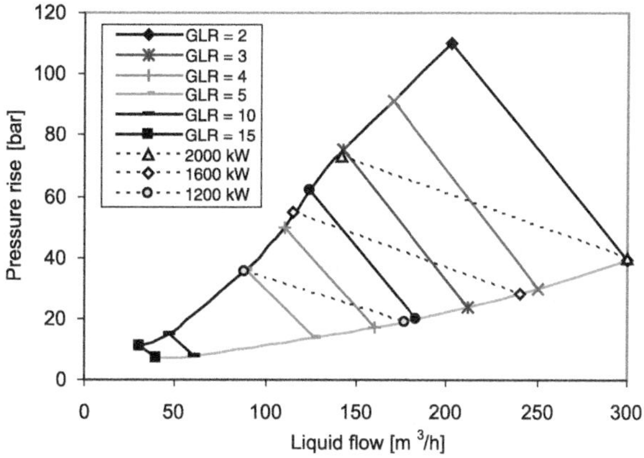

13.2.6 System Curves

As with single-phase flow, the operation point of a two-phase pump is given by the intersection of the system characteristic with the Q-H-curve at constant speed. In two-phase flow both the system and the Q-H-curves depend on the gas content (GVF) and the gas and liquid properties. For a given pump in a given system we therefore find an "infinite number" of pump and system characteristics as illustrated by Fig. 13.31. Two systems are here considered:

a. A recirculation mode, where the pump returns the mixture to the suction reservoir, system curve 1 in Fig. 13.32. In this system the pump has to overcome only the pressure drop in the suction and discharge piping, i.e. the system characteristic has no static head. Deviations occur when the average densities in the suction and discharge pipes are not identical due to differences in the gas contents in any vertical sections of the piping.
b. The system curve has a static component (curve 2 in Fig. 13.32) when the pump injects the mixture into a discharge plenum at a higher geodetic location and/or at a higher pressure.

Consider the behavior of the pump with reference to Fig. 13.32: If the system curve has no static head (curve 1), it is more likely to find intersections with the two-phase

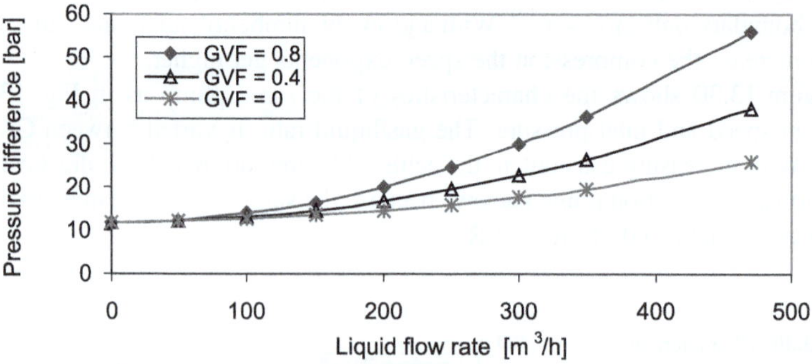

Fig. 13.31 System curves with different gas contents

Fig. 13.32 Impact of system curve on performance in two-phase flow

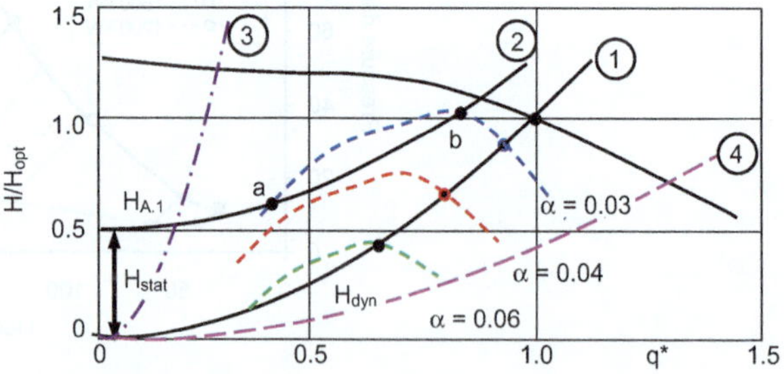

performance curves of the pump for the various GVFs. However, the higher the static head of the system curve, the less will the pump be able to meet the system demand, as demonstrated by curve 2. There is no intersection between system and two-phase performance curves at higher GVF. At very low flow, system curve 3, or very high flow, curve 4, there is no intersection between any stable performance curve and the system characteristics either.

A system curve may have two intersections with the performance curve of a given GVF as shown by points "a" and "b" on curve 2. This operation is unstable: either the pump works at point "a" or "b", or the GVF in the impeller fluctuates (somewhere between 0.03 and 0.04 in the example of Fig. 13.32). Unstable operation would also occur if the system curve (with or without static head) twice intersects the curve which delineates the head breakdown. This situation is shown by the system characteristic plotted in Fig. 13.21.

Two-phase pump testing is usually done in closed loops with system curves made up only by the pressure drop in the test loop as curve 1 in Fig. 13.32. This should be borne in mind when transferring test data to configurations with a static head as in curve 2.

The above discussion is mainly relevant for radial pumps. Multistage helico-axial pumps are not known to have a restricted stable flow rate, but the minimum flow or the "surge limit" mentioned above must be observed.

The system characteristics 1 to 4 in Fig. 13.32 are found in short pipes, e.g. on a test bed or in a process plant. Long pipelines behave differently because they can store liquid and gas due to the compressibility of the gas. If the flow rate of the pump changes momentarily due to some disturbance $\Delta V = \Delta Q \times \Delta t$, the pressure in a long pipeline changes by the ratio $(V + \Delta V)/V$, where $V = \frac{1}{4} \times \pi \times d^2 \times L$ is the volume in the pipe. In long pipelines this ratio is close to 1.0.

The system characteristics are then time-dependent: If flow deviations are short-term, the system curve is horizontal and the operation point shifts on a horizontal line when the flow through the pump changes (e.g. due to variations in the GVF). In other words: when the transit time $T_{Tr} = L/c_{mix}$ through a pipeline is long compared to the time of a perturbation of the flow through a multiphase pump, the momentary system curve is virtually horizontal due to the compressibility or inertia of the pipeline. In such a case it would be impossible to control the pump speed from a pressure rise set point. For a long-term change in the flow (with $t \gg T_{Tr}$) the system curve essentially follows the square law with possible deviations due to variations in flow patterns.

13.2.7 Slugs and Gas Pockets

Because of variations in the composition of the two-phase mixture over time, a variable speed drive is normally selected. If the gas content rises, the speed is increased in order to achieve the required pressure difference. If the liquid content increases, the rotor speed is reduced accordingly.

When the pump is operated with high gas content at high speed and the liquid flow rate increases suddenly due to ingress of a liquid slug, the pump may be at risk due to shock loading of the blades and a sudden increase in torque. Slugs alternating with gas pockets lead to cyclic loading of the pump and driver. Gas pockets of moderate size could lead to gas-locking which would trip the pump. To avoid such problems, "buffer tanks" of various designs are used. The principle is described in [33] and briefly discussed below with reference to Fig. 13.33.

The two-phase mixture enters the tank through a "slug breaker" in order to avoid damage to the internals of the buffer tank as could be caused by the kinetic energy of slugs entering the tank. The slug breaker may be a short cylinder with a number of holes which break up the incoming flow into small jets. The mixture enters a stilling chamber; gas and liquid separate to a large extent in the tank. One or more intake pipes draw liquid and gas out of the tank and lead the mixture to one or more pumps. The intake pipe has one or several rows of holes (or slots) and is open on the top or has large gas openings near the top of the tank. The liquid level forming in the tank depends on the gas and liquid flow rates.

The liquid enters the intake pipe through those holes which are below the variable liquid level while the gas enters the intake pipe through the openings at the top and through those holes which are above the liquid level. In this way, the liquid and gas streams fed to the pump are mixed in the intake pipe.

The liquid enters the intake pipe by gravity and by an ejector effect caused by the pressure drop created by the gas flow. The ejector effect increases with rising gas flow rate. An empirical correlation derived from tests is used for the intake pipe design and for predicting the liquid level under various operating conditions.

When a liquid slug enters the tank the liquid level rises and the liquid flow rate increases. When a gas pocket enters, the liquid level and the flow rate drop. The ability of the buffer tank to smooth out process fluctuation increases with the

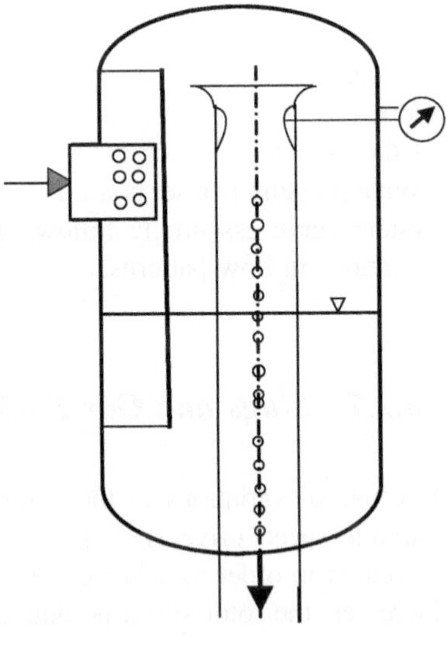

Fig. 13.33 Buffer tank, [33]

surface area of the liquid level A_L since the increase in liquid level Δz_L caused by a liquid slug of the volume V_{slug} is given by $\Delta z_L = V_{slug}/A_L$.

Properly calibrated, the buffer tank arrangement allows to determine the liquid and gas flow rates. An orifice or a Venturi tube is installed in the discharge pipe of the pump in order to accomplish this (for details see [33]).

Measuring the pressure rise generated by the pump, the power absorbed at the coupling and the rotor speed, provides another way to estimate the liquid and gas flow rates. The measured quantities allow an iterative performance calculation according to the equations given in Table 13.3, which yields the gas content and liquid flow rate, hence the actual operation point in the performance map as per Figs. 13.29 and 13.30.

13.2.8 Free Gas, Dissolved Gas and NPSH

The performance impairment induced by two-phase flow can be caused by three different effects: (1) free gas; (2) separation of dissolved gas from the liquid; (3) generation of vapor in zones where the local pressure drops to the vapor pressure (this process is known as "cavitation").

In a two-phase pump, free gas (as given by the gas volume fraction at the inlet nozzle) is dominating. The presence of free gas implies that the liquid is virtually saturated with dissolved gas. Part of the void is made up by vapor according to Dalton's law stating that the total pressure in the gas is the sum of the partial gas and vapor pressures $p = p_{gas} + p_v$, Chap. 17.3. The vapor pressure is determined from the liquid temperature. When the flow enters the impeller, the gas expands in the low-pressure zone (Fig. 6.5) according to the gas law $p \times V = m \times R \times T$, some gas comes out of solution according to Henry's law and a tiny quantum of liquid evaporates. When the two-phase multipliers are determined by experiment, all these effects are present in the pump; they are accounted for by the measurement and the multipliers derived thereof. Any gas separation due to pressure losses upstream of the impeller can be calculated according to Chap. 17.3 and added to the free gas content if deemed necessary.

When a NPSH-test is performed on a centrifugal pump with water, the suction pressure (or $NPSH_A$) is successively reduced, Chap. 6.2.5, Fig. 6.9. Vapor bubbles are generated at the impeller inlet below a specific inlet pressure (corresponding to the $NPSH_i$). As the inlet pressure is reduced further, the vapor volume (void fraction) increases till performance is impaired. Finally the head "breaks down" and the flow through the pump is disrupted when the impeller channels are largely filled with 2-phase flow; the pump becomes "vapor-locked". Note that the density ratio is very high in cavitation tests carried out with water (at 20 °C we find DR = 57,800 and at 180 °C DR = 172). This highlights the drastic loss in performance observed in extended cavitation.

In contrast, a two-phase pump is designed for continuous operation in two-phase flow - a situation which corresponds to a liquid pump working under full cavitation

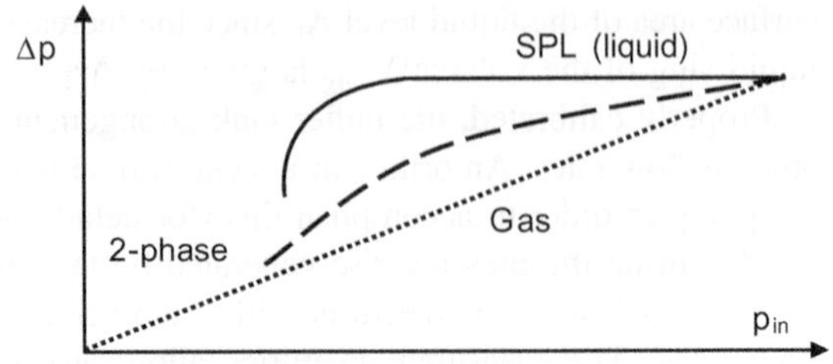

Fig. 13.34 "Suction test" with liquid pump versus two-phase pump and compressor

conditions. If a two-phase pump were subjected to the same type of testing as done to determine the NPSH-required of a liquid pump, a continuous drop of the pressure rise (head) would be expected, since the void fraction grows as the inlet pressure is reduced. Nevertheless, performance may be poor because of the low vapor density.

Figure 13.34 explains the behavior of a two-phase pump in such a test as compared to the NPSH-test of a liquid pump according to Fig. 6.9. In the tests depicted in Fig. 13.34 the speed and the total volumetric flow rate are kept constant while the inlet pressure is reduced. In contrast to liquid pumping, *the density does not remain constant* in two-phase pumping. Consequently, the pressure rise of a pump in two-phase flow *degenerates continuously as soon as the inlet pressure is reduced*, even when starting the tests at a very high pressure. The behavior described is similar to that of a compressor tested at different inlet pressures. According to Eq. (T13.3.29) the pressure ratio p_2/p_1 of a given compressor at a given speed is (theoretically) constant. The pressure difference $\Delta p = p_2 - p_1$ is then proportional to the inlet pressure as shown by Fig. 13.34.

In contrast to liquid pumping, the vapor (making up a fraction of the void as given by its partial pressure) does not condense or "collapse". Hence there is no possibility for bubble implosion and cavitation erosion. Actually the fluid temperature rises in gas-liquid pumping due to the compression of the gas and due to losses. This causes some liquid to evaporate. As a consequence, the terms of cavitation, NPSH or suction specific speed are *meaningless for gas/liquid pumping*.

13.3 Expansion of Two-Phase Mixtures in Turbines

13.3.1 Calculation of the Work Transfer

In various chemical engineering processes, fluids containing dissolved or undissolved gases are expanded from a higher to a lower pressure level (some applications are described in [34, 35, 51]). If the energy released by this process is sufficiently large, it may be worthwhile to expand the fluid in a pump operating as a turbine. In this way up to 80% of the mechanical energy can be put to use which otherwise would be dissipated into heat in a throttle valve. If standard pumps are employed, the investment costs can be relatively low. Therefore, energy recovery deserves consideration even when the available hydraulic power is moderate.

A two-phase flow is established in the turbine if the liquid contains free gas, if dissolved gases come out of solution or if part of the liquid evaporates during the expansion on reaching the vapour pressure. As the gas contained in the fluid is expanded, additional energy is released in comparison with incompressible flow. An exchange of momentum between the gas and the liquid causes higher losses. The higher the gas content, the more energy (or "head") is therefore required at the turbine inlet to pass a given volume flow rate through the machine. Thus the flow resistance grows and the turbine characteristic is altered with increasing gas volume fraction of the working fluid.

In a turbine a pressure difference *in the flow direction* is established between the inlet and the exhaust nozzle. Consequently, the buoancy effect discussed in Chap. 13.2 tends to accelerate the gas in the direction of the flow. In contrast to operation in the pumping mode, no gas-locking is possible - even with high gas content. Apart from a possible Mach number limit, phase separation problems do not restrict the stable operation of a turbine: "the gas is always blown through the machine", albeit with high losses and possibly little useful work.

Figure 13.35 shows the characteristics of a 3-stage pump with $n_q = 22$ operating as a turbine. The tests were executed with an air/water mixture and a back pressure of 1 to 8 bar. The curve $x = 0$ represents pure water flow; the curves with various gas mass contents ($x > 0$) apply to two-phase flow. At a constant volume flow rate (φ = constant) the work coefficient ψ rises with increasing gas content, while the efficiency drops owing to the additional losses. With growing gas content the efficiency optimum shifts towards higher volumetric flow rates, probably because phase mixing is better at higher flow velocities.

Similar characteristics were measured at higher pressure levels; but the curves for the same gas contents as in Fig. 13.35 come closer together. Estimating the hydraulic efficiency from $\eta_h = \eta^{0.5}$ at a given flow coefficient, it is found that the theoretical heads $\psi_{th} = \psi \times \eta_h$ are nearly the same at $x = 0$ and $x > 0$. This means that the velocity triangles are not changed in two-phase flow as compared to single-phase flow.

As in Chap. 13.2.3, the method to calculate the turbine characteristics with two-phase flow is based on the concept of an isothermal flow and on two-phase multipliers derived from the tests mentioned. The equations listed in Table 13.3 can fully be applied by analogy [24, 36]; they are expressed for turbine operation in Table 13.4.

As when operating in the pumping mode, head coefficient and efficiency in turbine operation depend on the gas content and the density ratio. To arrive at a generally valid presentation of the test results and to enable performance predictions at any other conditions, two-phase multipliers were calculated in [24] using Eqs. (T13.3.11 and 13.3.12). All measuring points can be represented as functions of the *aerage* gas volume fraction - regardless of the density ratio or the operating point on the characteristic. The results are given as $f_\psi = f(\alpha)$ by Eq. (T13.4.9) and $f_\eta = f(\alpha)$ by Eq. (T13.4.10). The scatter of the measured data points around these correlations was only ±4%. The domain of validity of Eqs. (T13.4.9 and 13.4.10) is given by the range covered by the test data listed in Table 13.4. It is interesting to note that

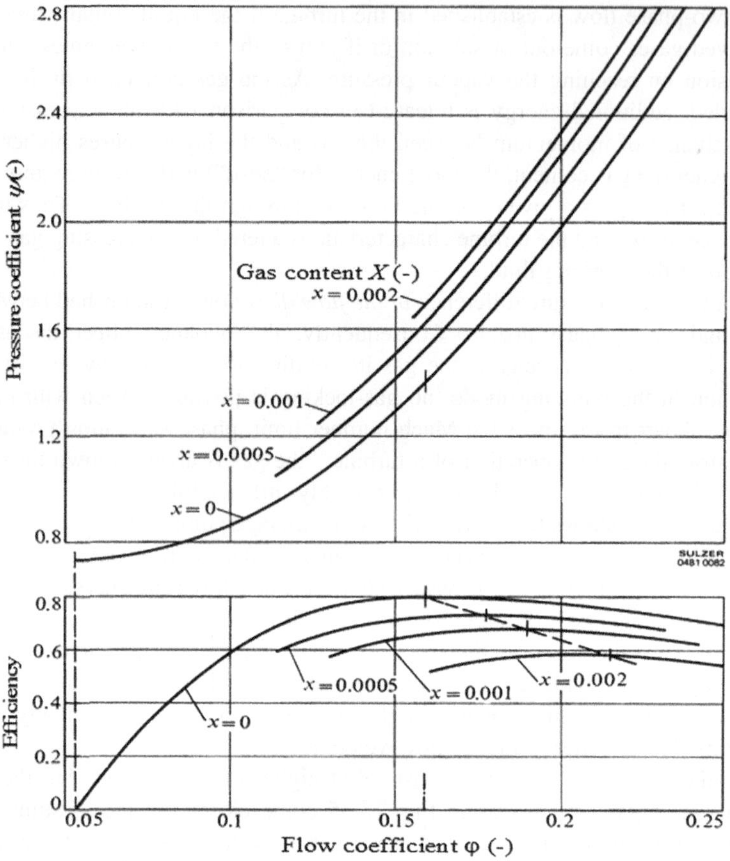

Fig. 13.35 Dimensionless turbine characteristics with a counter pressure of 1 bar, [24]

the additional losses caused by two-phase flow are virtually independent of the turbine operating point - quite unlike what is seen in pumps.

13.3.2 Prediction of Turbine Characteristics for Two-Phase Flow

Starting from a measured or calculated turbine characteristic for single-phase flow, the turbine characteristics can be predicted for any liquid/gas mixture with gas contents up to $\alpha_1 = 0.75$ at the turbine exhaust by means of the equations given in Table 13.4. If gas separates from the liquid during the expansion, it is advisable to calculate the expansion stage by stage. This may be done by iteration because the amount of gas after each stage is unknown. A similar procedure has to be applied if part of the liquid evaporates in the turbine. However, with highly developed

Table 13.4 Expansion of gas-liquid mixtures			Eq.
Density of gas phase	Inlet: $\rho''_2 = \frac{p_2}{R\,T}$	Outlet: $\rho''_1 = \frac{p_1}{R\,T}$	13.4.1
	Average density: $\rho'' = \frac{1}{2}(\rho''_1 + \rho''_2)$		
Average volume flow of gas phase	$Q'' = \frac{\dot{m}''}{\rho''}$		13.4.2
Specific energy available with isothermal expansion from p_2 to p_1	$Y_{isot} = (1-x)\frac{p_2 - p_1}{\rho'} + x\,R\,T \ln\frac{p_2}{p_1} + \frac{c_2^2 - c_1^2}{2}$		13.4.3
Equivalent head per stage with single-phase liquid flow	$H_{st} = \frac{Y_{isot}}{g\, z_{st}(1+0.45\alpha)}$	$\psi_{SPL} = \frac{2\,Y_{isot}}{z_{st}\, u_2^2 (1+0.45\alpha)}$	13.4.4
Specific speed calculated for design point with mixture flow rate	$n_{q,a} = \frac{n\sqrt{Q_{mix}/f_q}}{H_{st}^{0.75}}$		13.4.4a
Required impeller diameter for design point (subscript a)	$u_2 = \sqrt{\frac{2g\,H_{st}}{\psi_a}}$ $\quad \psi_a = \psi_{SPL}$	$d_2 = \frac{60\,u_2}{\pi\,n}$	13.4.5
Efficiency	$\eta_{TP} = \eta_a(1 - 0.55\alpha - \alpha^3)$ $\quad \eta_a = \eta_{SPL}$		13.4.6
Useful power at coupling	$P = \dot{m}\, Y_{isot}\, \eta_{TP}$		13.4.7
Efficiency during two-phase expansion	$\eta_{TP} = \frac{P}{\dot{m}\, Y_{isot}} = \frac{P}{(\dot{m}' + \dot{m}'')\, Y_{isot}}$		13.4.8
Two-phase multiplier for work coefficient	$f_\psi = \frac{\psi_{TP}}{\psi_{SPL}} = 1 + 0.45\alpha$		13.4.9
Two-phase multiplier for efficiency	$f_\eta = \frac{\eta_{TP}}{\eta_{SPL}} = 1 - 0.55\alpha - \alpha^3$		13.4.10
Range of validity for Eqs. (13.4.9) and (13.4.10):			
Gas volume fraction at inlet	$0 \le \alpha_2 \le 0.29$	Expansion ratio: $1.3 \le \frac{p_2}{p_1} \le 9.3$	
Average gas volume fraction	$0 \le \alpha \le 0.4$		
Gas volume fraction at outlet	$0 \le \alpha_1 \le 0.65$	Density ratio: $80 \le \frac{\rho'}{\rho''} \le 400$	
Average gas mass fraction	$0 \le x \le 0.0032$		

evaporation the expansion ceases to be isothermal. In that case sufficiently small calculation steps must be chosen for the assumption of quasi-isothermal flow per step to be justifiable. The drop in temperature can also be estimated from Eq. (13.19a).

The process of gas desorption is controlled by diffusion. During the very short transit time of a fluid particle through the turbine, solution equilibrium may usually not be expected. Considerable uncertainties of performance prediction can result from not knowing the magnitude of this delay. For the design of a two-phase expansion turbine to be reliable, it must be ascertained how quickly the solution equilibrium establishes itself when the gas-saturated fluid is exposed to a pressure reduction. Because of this uncertainty, any guarantee of the performance data of an expansion turbine should be given for operation with pure water. In that way the performance test will be unambiguous.

Formulae and solubility data for determining the amount of gas desorbing are given in Chap. 17.3. A factor ε has been introduced into Eq. (17.2) to account for a time lag in gas desorption; $\varepsilon = 1.0$ applies to equilibrium (no time lag in the diffusion process) and $\varepsilon = 0$ if virtually no gas desorbs in the short transit time of the turbine. Further details of such calculations can be found in [36].

In general, the following data are necessary for calculating a two-phase characteristic: (1) Turbine characteristics of the selected machine for single-phase flow ($x = 0$); (2) At the design point of the turbine: the mass flows of the liquid and the gas, from which the gas mass fraction x ensues; (3) The temperature of the liquid; (4) Pressure p_2 and gas content x_2 at the inlet; (5) Pressure p_1 at the outlet; (6) Number of stages z_{st} and speed n.

Determination of the impeller diameter and/or the number of stages

1. Calculate the mixture properties (Table 13.3) and the average gas density, Eq. (T13.4.1). Using the average gas density, the average volume flow rate of the gas is determined from Eq. (T13.4.2) and the mean gas volume fraction α from Eq. (T13.3.3).
2. The specific work Y_{isot} of the mixture is calculated by means of Eq. (T13.4.3); it corresponds to the total head $g \times H$ in incompressible flow.
3. By dividing the available specific work Y_{isot} of the mixture by the factor $f\psi$ and the number of stages, the equivalent head per stage in single-phase flow is derived from Eq. (T13.4.4).
4. Next, a curve $n_q = f(\varphi)$ is plotted into the characteristic $\psi_{SPL} = f(\varphi)$.
5. By means of the average mixture flow rate and the equivalent head, the specific speed $n_{q,\,a}$ is calculated for the design point considered, Eq. (T13.4.4a).
6. The values φ_a, ψ_a and η_a can be read with $n_{q,\,a}$ for the considered design point from the curve $n_q = f(\varphi)$. By means of ψ_a the impeller diameter can be calculated from Eq. (T13.4.5).
7. Efficiency and useful power with two-phase flow are calculated from Eqs. (T13.4.6 and 13.4.7).
8. The number of stages and/or the speed may be changed in an iterative procedure to optimize the selection of the machine.

Determination of the turbine characteristics: When the size of the turbine has been selected in this way, all of its characteristics for two-phase expansion have to be determined. This calculation is simplified by treating the inlet pressure p_2 as an independent variable and calculating that volume flow rate which passes the turbine at given values of p_2, p_1, x_2, x_1. For this calculation the gas mass content is considered constant. If the gas content grows due to liquid evaporating or gas desorbing during the expansion, the average between inlet and outlet has to be inserted. The calculation can be done as follows:

1. x_2, x_1, p_1 are given for each point on the characteristic; p_2 has to be chosen.
2. Calculation of the mixture data and of the specific work Y_{isot} according to Eq. (T13.4.3).
3. Next, the equivalent head coefficient in single-phase flow, which is necessary for processing the mixture defined in step (1), is calculated from Eq. (T13.4.4).
4. By means of ψ_{SPL} the flow coefficient φ and the efficiency η_{SPL} can be read from the single-phase characteristic. Using these values, the volume flow $Q = A_2 \times u_2 \times \varphi$ and the mass flow rate as well as the efficiency and power are calculated from Eqs. (T13.4.6 and 13.4.7).
5. For turbines, Mach number limits similar to those for pumps are to be expected (refer to Chap. 13.2.2). The sound velocity can be estimated by means of the equations in Table 13.3. However, no publications have been found, which report on Mach number limits having been attained by a pump operating as a turbine with two-phase mixtures.

Note: Check gas separation in exhaust pipe of the turbine: if a large amount of gas separates there due to high pressure losses, the head available for the turbine is reduced and the expected power recovery may not be met.

13.4 Hydraulic Transport of Solids

Granular goods such as sand, gravel, coal, minerals and ashes can be transported as suspensions in a carrier fluid (mostly water) in pipelines [52, 54]. Economically this is particularly interesting when the material to be transported is processed as a suspension, owing to mining, extraction or production methods. Centrifugal pumps are very appropriate for transporting solids; they are used for example in the mining industry, for pumping ashes in power plants, as dredge pumps, and in flue gas desulphurization plants [50, 53].

Pumps for the hydraulic transport of solids (also known as "slurry pumps") are two-phase pumps since they transport the phases solid and liquid. In principle, the same basic considerations apply as in Chap. 13.2 - with the important difference that both phases are virtually incompressible. The solid phase (subscript s) can absorb energy only in the form of kinetic energy; it cannot store it as static pressure. In the hydraulic transport of solids, the density ratio ρ_s/ρ is considerably less than in pumping gas-liquid mixtures: with coal it is roughly 1.5, with sand about 2.7, and with minerals it can be as high as $\rho_s/\rho = 5$. Consequently, segregation phenomena are much weaker than in gas-liquid flows. However, segregation is governed by the laws discussed in Chap. 13.2.2, since the paths of the solid particles are determined by centrifugal, Coriolis, lift and drag forces. In radial impellers large particles, due to the Coriolis force, move towards the pressure surface of the blades (see Fig. 5.3), while smaller particles rather move towards the suction surface. This is also evident just from considering the inertia of coarse grains. Consequently, wear at the impeller outlet is especially severe on the pressure surface of the blades, Chap. 14.5.5.

Downstream of the impeller in the volute casing coarse grains move on tangential paths because of their inertia. Particle motion is also strongly influenced by the drag acting on particles that move relative to the carrier fluid. Such a movement can be characterized by the settling velocity $w_{s,o}$ of a particle in a stagnant liquid; $w_{s,o}$ can be calculated on the analogy of Eq. (13.15), it is given by Eq. (T13.5.7) [37].

Due to density differences between solid and liquid, the flow paths of the particles deviate from the streamlines of the liquid. The resultant cross movement leads to additional losses because of an exchange of momentum between the solid and liquid phases. Further losses are generated by shock and friction between the particles and the solid walls (inlet, impeller and volute). As a result, head and efficiency are impaired in comparison with water pumping. This loss can be described (as in Chaps. 13.1 to 13.3) by empirical factors which are applied to the performance of the pump with water. The additional losses grow with a rising concentration of solid particles, the density ratio ρ_s/ρ and the grain size because the segregation tendency increases with these parameters.

Many test results have been published from which correlations were derived for calculating the head and efficiency impairment in the hydraulic transport of solids [38–40]. Depending on the different solid-liquid mixtures, test conditions and specific speeds, the proposed correlations diverge considerably from one another. Theoretical calculation models have also been developed [41, 42]. In [41] for instance, the head reduction is calculated from an increased slip caused by the solid particles and from the additional losses induced by shock and friction of the particles in the impeller and the volute. This calculation yields results similar to the empirical correlations from [39, 40]; hence it seems appropriate to investigate the influence that various parameters exert on the head.

Solid particle transport is described quantitatively as per Table 13.5 by the following parameters:

- The concentration of solid particles is given as volume concentration c_v according to Eq. (T13.5.2) or as mass concentration x according to Eq. (T13.5.3); these definitions correspond exactly to α and x in Table 13.3. The slip between the phases is not taken into account. Transport and volume concentration are thus assumed to be equal. This assumption is to some extent valid for pseudo-homogeneous mixtures only; that is to say at high velocities or with fine-grained mixtures. The coarser the particles and the lower the velocity, the more important is the phase slip (refer also to [43]).
- The homogeneous mixture density is determined by Eq. (T13.5.4).
- The settling velocity $w_{s,o}$ of any single particle in a stagnant liquid is a measure of the segregation tendency. It depends as per Eq. (T13.5.7) on the grain size, the density ratio, and the Reynolds-dependent drag coefficient of the particle. Since the solids pumped are usually described by a grain size spectrum, an average grain diameter must be chosen. The settling velocity falls with high solids concentrations because the particles affect one another as the concentration increases, Eq. (T13.5.8). Equation (T13.5.6) is valid for spherical particles; grains with sharp edges have higher drag coefficients. According to [44], drag

Table 13.5 Hydraulic transport of solids		Eq.
Volume flow (mixture)	$Q_{mix} = Q + Q_s$	13.5.1
Transport concentration c_T Volume concentration c_v	$c_T = \frac{Q_s}{Q + Q_s}$ $c_v = \frac{\rho_{mix} - \rho}{\rho_s - \rho}$ $c_T \approx c_v$	13.5.2
Mass concentration x	$x = \frac{\dot{m}_s}{\dot{m} + \dot{m}_s} = \frac{\rho_s}{\rho} \frac{c_v}{1 + c_v\left(\frac{\rho_s}{\rho} - 1\right)} = \frac{\rho_s}{\rho_{mix}} c_v$	13.5.3
Mixture density	$\rho_{mix} = (1 - c_v)\,\rho + c_v\,\rho_s$	13.5.4
Reynolds number (solid) d_s = average particle diameter	$Re_s = \frac{d_s w_{s,o}}{\nu}$	13.5.5
Drag coefficient of spherical solid particles	$\zeta_w = \frac{24}{Re_s} + \frac{4}{\sqrt{Re_s}} + 0.4$ for $10^{-2} < Re_s < 2\times 10^5$	13.5.6
Settling velocity of a free particle (at $c_v = 0$)	$w_{s,o} = \sqrt{\frac{4\,g\,d_s}{3\,\zeta_w}\left(\frac{\rho_s}{\rho} - 1\right)}$	13.5.7
Settling velocity at $c_v > 0$	$\frac{w_s}{w_{s,o}} = (1 - c_v)^\beta$ $\beta = \frac{4.4}{Re_s^{0.095}}$	13.5.8
Correction factor for head $f_H = H_{mix}/H_w$ at $f_Q = 1$; $d_{Ref} = 0.023$ mm	$f_H = 1 - \frac{x}{26}\left(\frac{\rho_s}{\rho} - 1\right)\left(1 + 4\frac{\rho}{\rho_s}\right) \ln\frac{d_s}{d_{Ref}}\left(\frac{25}{n_q}\right)^{0.34}$	13.5.9
Correction factor for the efficiency	$f_\eta \equiv \frac{\eta_{mix}}{\eta_w} \geq f_H$	13.5.10
Pressure rise in pump	$\Delta p = \rho_{mix}\,g\,H_{mix} = \rho_{mix}\,g\,f_H\,H_w$	13.5.11
Power consumption	$P = \frac{\rho_{mix}\,g\,H_{mix}\,Q_{mix}}{\eta_{mix}} \approx \frac{\rho_{mix}}{\rho} P_w$	13.5.12
Increase of pressure loss in piping for $Re_s > 2$ D = pipe diameter	$\frac{\Delta p_{mix}}{\Delta p_w} = 1 + 83\,c_v\left\{\frac{g\,D}{c_{mix}^2\sqrt{\zeta_w}}\left(\frac{\rho_s}{\rho} - 1\right)\right\}^{1.5}$	13.5.13
Critical velocity in pipe for $d_s > 0.5$ mm	$c_{krit} = 1.3\,c_v^{0.13}\left\{2\,g\,D\left(\frac{\rho_s}{\rho} - 1\right)\right\}^{0.5}$	13.5.14

The correction factors f_H and f_η are applied at Q = constant, i.e. at $f_Q = 1$.
ρ = density of carrier fluid. Subscripts: s = solid w = water mix = mixture

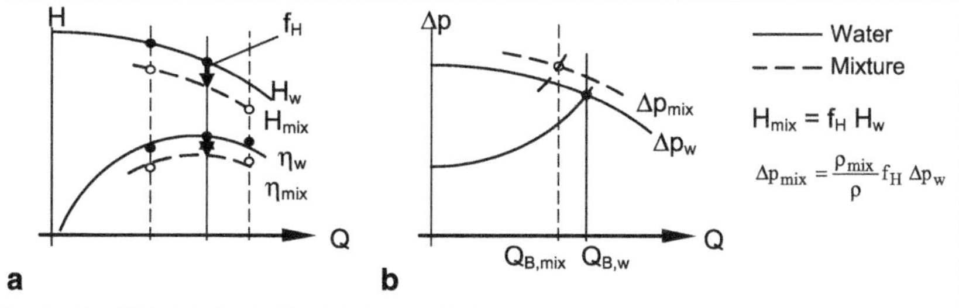

coefficients calculated for spheres (having the same volume as the considered particle) must be multiplied by the following factors: octahedron 2.4; cube 3.2; tetrahedron 4.7.

- Equation (T13.5.9) gives an empirical correlation for the head impairment which is described by the multiplier f_H, based on [39]. According to this correlation, the head impairment increases in proportion to the mass concentration of solids and, in a non-linear way, with the density ratio and the grain size. The correlation given in [39] was supplemented by adding the influence of the specific speed, which was derived from calculations published in [41].[7]
- The efficiency impairment is described by a factor f_η which is frequently assumed to be equal to the factor f_H, Eq. (T13.5.10). Other test results show f_η to be a few per cent higher than f_H [40]. The tests in [45] also record a distinctly higher value for f_η than for f_H. Thus the increase in power consumption was less significant than would be expected from the higher density of the mixture. This finding may be explained by the disk friction losses which do not increase in proportion to the mixture density but are mainly caused by the fluid alone. In addition, the theoretical head could be lower because the solid particles are diverted less effectively than the fluid, meaning $c_{2u,s} < c_{2u}$. To demonstrate this argument a term for the solids may be added to the momentum equation, Eq. (3.1), giving $M_{blade} = \rho \times Q_{Fluid} \times r_2 \times c_{2u} + \rho_s \times Q_s \times r_2 \times c_{2u,s}$ at $\alpha_1 = 90°$. A slip between solid and liquid would seem quite plausible.
- In contrast to Table 13.1, the correction factors f_H and f_η used for solids transport are applied at constant flow rates, as shown in Figure "a" of Table 13.5 (therefore $f_Q = 1$). This assumption is only valid in a range where the additional losses are not significant enough for shifting the best efficiency point to lower flow rates. At $f_H \ll 1$, it should indeed be expected that the BEP shifts to lower flow and thus $f_Q < 1$ would apply. Some mixtures behave like highly viscous fluids; Q-H-curve and BEP then shift in a similar way as described in Chap. 13.1.
- The factors f_H and f_η can be considered as more or less independent of the flow rate ratio q*.
- In spite of the head impairment, the pressure increase in the pump is higher during hydraulic transport of solids than with pure water because the effect of the mixture density outweighs that of the losses, Eq. (T13.5.11) and figure "b" in Table 13.5.

---

[7]In [38] a (quite unrealistic) exponent of 2.46 for the specific speed was derived from tests with $n_q = 27$ and 30. Considering the scatter inherent to this type of measurement, it appears unwise to attempt on so narrow a test range to determine an exponent for extrapolation to situations outside the tested range. The calculations in [41] produced a more plausible exponent for n_q.

- If a fluid with a viscosity higher than that of water is used as carrier liquid (e.g. oil/coal transport), Eq. (T13.5.9) has to be adapted in such a way that the Reynolds number Re_s resulting from a calculation with the settling velocity is used instead of the grain size d_s.
- A finely dispersed suspension with $Re_s < 0.02$ is calculated as a homogeneous flow with $f_H = f_\eta = 1$ with the homogeneous density as per Eq. (T13.5.4); an example for this are lime suspensions used for flue gas desulphurization.
- The uncertainty of head prediction for hydraulic transport with appreciable solids concentration is estimated as roughly $\pm 0.2 \times (1 - f_H)$.
- The similarity laws listed in Table 13.4 are applicable to the hydraulic transport of solids.

The correlation according to [39] was chosen because it represents a good average of the values given in various publications (comparisons are to be found in [38, 41]). Furthermore, this correlation is backed up by tests with relatively large pumps. The following parameters are covered by the test data: $c_{v,max} = 0.4$; $d_2 = 370$ to 710 mm; $\rho_s/\rho = 2.64$ to 4.6; $n_q = 25$; $n = 600$ to 1300 min^{-1} and $d_s = 0.17$ to 1.3 mm. This parameter range is enlarged to $n_q = 17$ and $\rho_s/\rho = 1.48$ when considering the equation in [40] which is very similar. The calculations in [41] seem to allow the extrapolation to grain sizes up to 25 mm.

In [46] a correlation was derived from the test data reported by various authors (in total 850 measuring points). Here, the head impairment is proportional to the mass concentration x, Eq. (13.20):

$$f_H = 1 - k_H \quad \text{with} \quad k_H = 2.705 \times \left(\frac{\rho_s}{\rho} - 1\right)^{0.64} \left(\frac{d_s}{d_2}\right)^{0.313} \tag{13.20}$$

All measuring points are within a range of ±15% of the correlation; the average error in k_H is 8% and the standard deviation is ±0.6%. Once again $f_\eta = f_H$ is the best approximation, even though the scatter of f_η is slightly higher than with f_H. The test data include the following parameters: $\rho_s/\rho = 1.5$ to 6.2; $d_2 = 210$ to 825 mm; $d_s = 0.03$ to 27 mm; maximum mass concentration $x = 0.66$; $n = 590$ to 1780 rpm; $q^* = 0.25$ to 1.4. Equation (13.20) has the advantage that the pump size enters into the calculation. It appears from tests in [47] that with very large pumps (impeller diameter 2667 mm) head and efficiency were impaired by only 5% at volume concentrations of as much as 50% (the grain size being 0.3 mm).

Another empirical correlation for the calculation of the head loss k_H caused by solid particles is Eq. (13.21) derived from [48] as quoted in [42]:

$$k_H = 0.664 c_v^{0.9} \left(\frac{\rho_s}{\rho} - 1\right)^{0.92} \left[\ln\left(1 + \frac{d_s}{d_{Ref}}\right)\right]^b (\tan \beta_{2B})^{0.2} z_{La}^{0.14} \left(1 + \frac{2.75}{e^{n_q/n_{q,Ref}}}\right) \tag{13.21}$$

with $f_H = 1 - k_H$ and $n_{q,Ref} = 11.8$

for $d_s < 8$ mm: $d_{s,Ref} = 1.43$ mm; $b = 0.8$

for $d_s > 8$ mm: $d_{s,Ref} = 0.2$ mm; $b = 0.4$

For a given application it is recommended to evaluate the head loss predicted from Eqs. (T13.5.9), (13.20) and (13.21); the resulting correction factors may be averaged and their range used for assessing the uncertainties. By means of Eq. (13.21) the effects of various design parameters on performance can be estimated.

The grain size *spectrum* has an important influence on the factors f_H and f_η. In [45] experiments were reported in which three different mixtures were tested.

The mixture with the *highest* particle density (yet with the same average grain size) caused the *slightest* head impairment, because the grain size spectrum contained a high fraction of fine particles. The fine grains help with the transport of the coarser particles. The quicker the particles settle, the higher is their tendency to separate in the pump and the more the head is impaired. Having the same concentration and the same average grain diameter, a mixture with a broad grain size spectrum is easier to handle than a mixture with largely uniform grain size. The rheological properties of a suspension depend on the density of the mixture and the grain size spectrum with the effect that the mixture can behave either like a Newtonian or like a Bingham-type fluid. The equations for calculating the head impairment do not cover these effects. All of the correlations discussed above for calculating the head impairment apply to heterogeneous mixtures ("settling slurries") that behave like Newtonian fluids. Larger uncertainties of head prediction are to be expected for homogeneous mixtures ("non-settling slurries"), such as clay or slick, which exhibit non-Newtonian behavior. In these cases a correction for increased viscosity, as discussed in Chap. 13.1, is possibly necessary.

The grain shape has an influence on slurry pumping as well: sharp-edged particles cause higher losses and, correspondingly, a higher impairment of the efficiency than spherical solids.

The more uniform the flow, the better the particles are able to follow the streamlines of the carrier fluid. Flow separations, abrupt changes of direction or strong secondary flows foster segregation. Hence a careful hydraulic design of impellers and volutes for slurry pumps brings two benefits: (1) The impairment of head and efficiency is reduced; (2) The effects of wear are mitigated since wear is very intense in zones with separated flow and vortices, Chap. 14.5. For these reasons double-curvature blades are preferred to cylindrical bladings; they also show less head impairment. This effect is not taken into account in the correlations given above.

Figure 13.36 shows a large dredge pump whose power exceeds 4000 kW when the mixture density is high. The double-walled casing of the pump makes it possible for the volute to be cast in a wear-resistant material that would be inappropriate for pressure-bearing parts because of its brittleness. Moreover, the *pressure-compensated* inner casing needs to be replaced only when it is worn down to a wall thickness of nearly zero. Adjacent to the impeller sidewall gaps the casing is protected by wear plates. As balance holes are not possible because of the risk of abrasion, the entire axial thrust is carried by a high-capacity taper roller bearing.

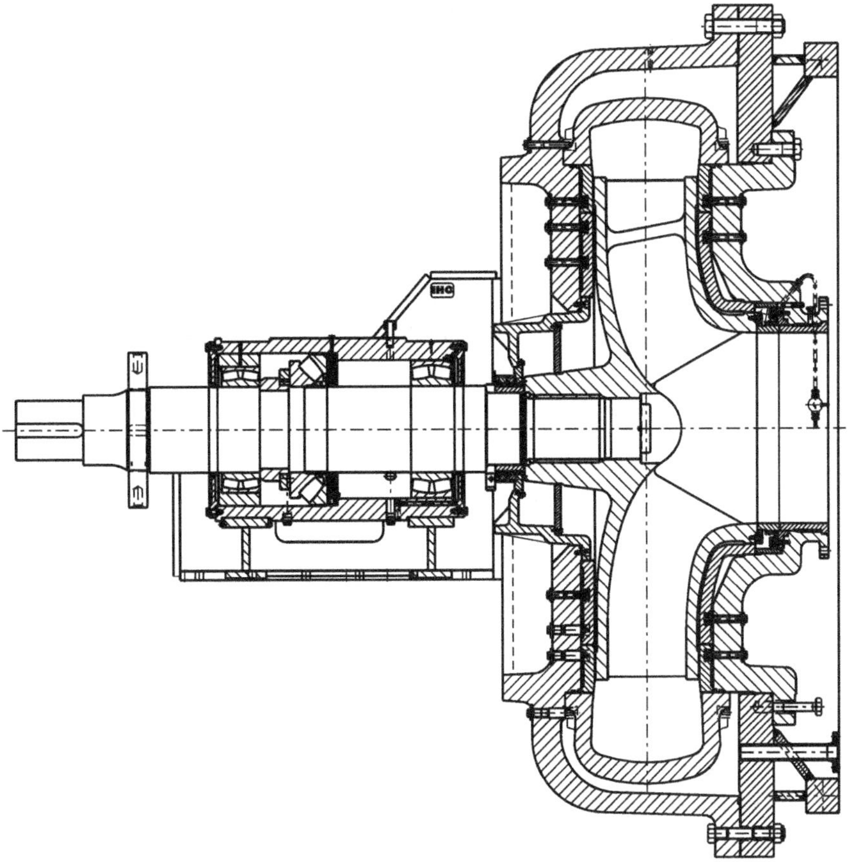

Fig. 13.36 Double-walled dredge pump, d_2 = 2600 mm, n = 240 rpm, Q_{opt} = 3.5 m^3/s, H_{opt} = 67 m, n_q = 19; IHC Holland

Application guidelines:

1. Solid particles are deposited in a pipeline whenever the mixture velocity falls below a critical velocity which can be estimated from Eq. (T13.5.14).[8] The critical velocity rises with increasing settling velocity of the particles, hence with the grain diameter and the density ratio. To avoid deposits, the flow velocity in the pipeline must be at least 0.3 to 0.5 m/s higher than the settling velocity. The critical velocity depends on the length of the pipe and its inclination angle; therefore, Eq. (T13.5.14) can only give a first indication.

[8]Equation (T13.5.14) was derived from a graph in [49]; it is valid for particle diameters greater than 0.5 mm. With smaller particles the critical velocity is lower than predicted from Eq. (T13.5.14).

2. In order to avoid solids settling in the pipe during operation with $c_{mix} < c_{crit}$, pumps for hydraulic transport must not be operated at low flow rates. Another reason to avoid partload operation is the excessive wear generated by recirculation. Since the system characteristic in hydraulic transport of solids shifts upwards, the operation point often shifts towards lower flow rates. However, this is not always the case: If the discharge pressure of the pump increases more than the resistance in the pipeline, the operation point will even shift to higher flow rates; such is the case when fine particles are pumped and f_H is close to 1.0. For controlling/limiting the shift of the operation point, it is advantageous to use steep Q-H-characteristics in hydraulic transport. Centrifugal pumps operating in series are favorable in this respect.
3. The rise in piping pressure loss caused by solids can be calculated according to [49] as follows: (1) The flow is considered homogeneous for $Re_s < 0.02$ and pseudo-homogeneous for up to $Re_s = 2$; in this range $\Delta p_{mix} = \Delta p_w \times \rho_{mix}/\rho$ applies, similar to Eq. (T13.5.11). (2) For heterogeneous flows with $Re_s > 2$ the pressure loss can be estimated from Eq. (13.5.13). However, for a pipe diameter of 1000 mm this equation yielded pressure losses about 50% higher than the test results [47]; so considerable uncertainties - depending on the plant configuration - are to be expected.
4. Abrasive wear by solid particles determines to a large extent the selection and design of the pump, first of all the allowable impeller tip speed u_2 and the choice of the material. Seals, replaceable wear plates and the shaping of the flow passages are further considerations. For fine grain sizes, rubber coated components or plastics (often polyurethane) are employed. For coarser particles, wear resistant alloyed cast iron, as listed in Table 14.7(4), is selected because rubber and plastics are no long-term match for the impingement energy of large grains. Abrasive wear is discussed in detail in Chap. 14.5. In that chapter further recommendations are given for the design of pumps for abrasive service and for limiting impeller tip speed, Chap. 14.5.5.
5. As a consequence of wear the seal clearances widen and the leakage losses correspondingly increase (the calculation can be done according to Table 3.7). This problem can be avoided by the design shown in Fig. 13.36: the annular seal at the impeller inlet is replaced by a special seal into which clean water is injected as a barrier fluid.
6. A widened annular seal that is several times the normal clearance not only requires appreciably more power for a given useful flow but also increases the required $NPSH_3$, because the impeller flow rate grows and the leakage through the gaps induces a higher pre-rotation. The NPSH-curve rises steeply and the possible operating range shifts towards lower useful flows as a consequence.
7. The impeller blades - especially at the outlet - can suffer important material loss due to abrasion. With increasing operation time, extensive wear can impair head and efficiency. Appropriate margins may then be indicated when selecting the pump.

8. Margins may also be required to cover variations of grain size and concentration during service.
9. Since both the composition of the fluid to be pumped and the prediction methods are subject to uncertainties, a generous sizing of the motor is recommended.
10. A motor with variable speed allows a wide operation range and covers uncertainties of production. V-belt drives are often used in the mining industry to adapt the speed to production requirements.
11. For the hydraulic transport of solids mainly single-stage radial pumps with volute casings are selected. If the required head cannot be attained with one single pump, several units are arranged in series. Multistage pumps are not popular in slurry service because of problems with abrasive wear.

The NPSH *required* is not appreciably affected by the presence of solid particles. However, when determining the $NPSH_A$, it must be taken into account that the mixture density prevails in any vertical pipe sections. As an example, consider a suction dredger pumping a sand-water mixture of density ρ_{mix}, the inlet of the suction pipe being placed at a depth z_{wl} below water level and the suction nozzle of the pump being placed at the level z_e above the water. According to Eq. (T2.2.8) the NPSH available is calculated as:

$$NPSH_A = \frac{p_{amb} - p_v}{\rho_{mix} g} - z_{wl}\left(1 - \frac{\rho}{\rho_{mix}}\right) - z_e - H_{v,s} \qquad (13.22)$$

Thus the height of the equivalent liquid column corresponding to the ambient pressure is reduced in proportion to the mixture density when compared to water pumping. In the considered case, moreover, the hydrostatic pressure gradient in the suction pipe is higher than in the surrounding liquid; this causes the $NPSH_A$ to decrease even more as given by the term with z_{wl} in Eq. (13.22). Thus the suction capacity is sometimes the limiting factor for the maximum solid particle concentration that can be pumped.

13.5 Non-Newtonian Liquids

In "Newtonian" liquids the shear stresses rise linearly with the velocity gradient according to $\tau = \rho \times \nu \times \partial w/\partial y$; they vanish at w = 0. All liquids not having this feature are termed "Non-Newtonian". Various liquids can, even at rest (i.e. w = 0), sustain finite shear stresses, which increase linearly with the velocity gradient $\partial w/\partial y$; when a flow is established (w > 0). Such fluids are called "Bingham liquids". Into this category fall paper stock and fine slurries with a high solids concentration whose rheology is different from that of heterogeneous mixtures, Chap. 13.4.

In paper production, "stock" designates a suspension of cellulose fibers in water. Since paper production involves a multitude of pumping tasks, "stock pumps" are a frequent application where Non-Newtonian liquids are transported. Stock is a 3-component, 3-phase mixture made up of water, solids and (unwanted) air. The solid substances, which eventually make up the paper, are cellulose fibers of 1 to 3 mm length and a diameter of roughly 25 μm; their density is slightly higher than the water density. The *mass* concentration x, called "consistency", defines the dry content of fibers in the mixture.

Bingham liquids move in a pipe like a piston with an almost rectangular velocity profile with weak shear stresses in the core flow, but high wall shear stress. Such flow patterns are created since the axial pressure gradient in the pipe, which is responsible for the flow, presses some carrier liquid out of the mixture (similar to squeezing a sponge). This liquid gathers near the pipe wall and acts like a lubricating film between the wall and the pumpage.

As mentioned above, Bingham liquids have a finite wall shear stress at w = 0. While the friction losses with water (or Newtonian fluids) in turbulent flow represent a parabola through the origin of a graph ΔH = f(w), paper stock exhibits a finite gradient ΔH/L at w = 0 which increases with the consistency, Fig. 13.37. With growing velocity, the head gradient ΔH/L of stock approaches that of water. With low consistencies friction losses of stock can even be slightly lower than with water, as sketched in Fig. 13.37. This improvement is explained by the formation of the lubricating film mentioned above.

The consistency varies during the different stages of the process. Up to consistencies of 1% the rheology of stock is very similar to that of pure water. Centrifugal pumps with closed impellers of normal design can therefore be selected for pumping stock with consistencies of up to 1.5%; no impairment of head or efficiency is expected in such applications. These conditions are found in those pumps which transport the stock to the paper machine (called "fan pumps"). For obtaining a perfect paper quality, the stock must be fed very uniformly to the paper machine. The allowable pressure pulsations are below 0.5% (peak-to peak) of the head, because otherwise the paper becomes cloudy. The head-capacity curves must be continuously rising towards shut-off without flat portions, in order to avoid low-frequency flow fluctuations. The measures listed in Table 10.2 are implemented to achieve low pressure pulsations: blade number $z_{La} \geq 7$, low blade

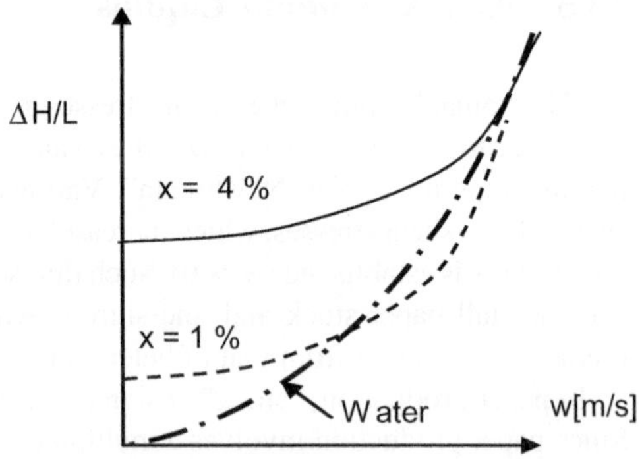

Fig. 13.37 Friction losses ΔH/L per meter pipe length for paper stock; x = consistency

loading, staggered blades to be applied when selecting double-entry impellers, and a very uniform approach flow to the impeller. The volutes are manufactured with high-quality surface finish in order to reduce turbulence.

In the processing train of a paper mill, the stock is pumped between various vessels ("chests") in order to accomplish the different steps of the process (e.g. de-inking). In this section the stock is maintained at high consistencies in order to reduce water and energy costs.

For pumping stock with up to 6% consistency, open impellers with 3 to 5 blades are used. Consistencies greater than 6% require special pump types. Figure 13.38 shows an example: the working principle of that pump is similar to that of a vortex pump. The ingestion of the stock into the pump is eased by a sort of inducer ("screw feeder") with three very long spiraling blades which reach into the chest. If air content and consistency are very high, air can be vented from the stuffing box chamber. The venting can be done by a pipe (as seen in Fig. 13.38) or through bores in the shaft.

Depending on the type of stock, the process stage, and the particulars of the plant, stock contains relatively large amounts of dissolved and free air. The free gas occurs in the form of very fine bubbles suspended in the water and adsorbed on the fibers. The gas concentration therefore increases with the consistency and other ingredients in the stock, such as glue. Proper piping layout, such as feeding the stock below the liquid level into a chest, helps to reduce the air content.

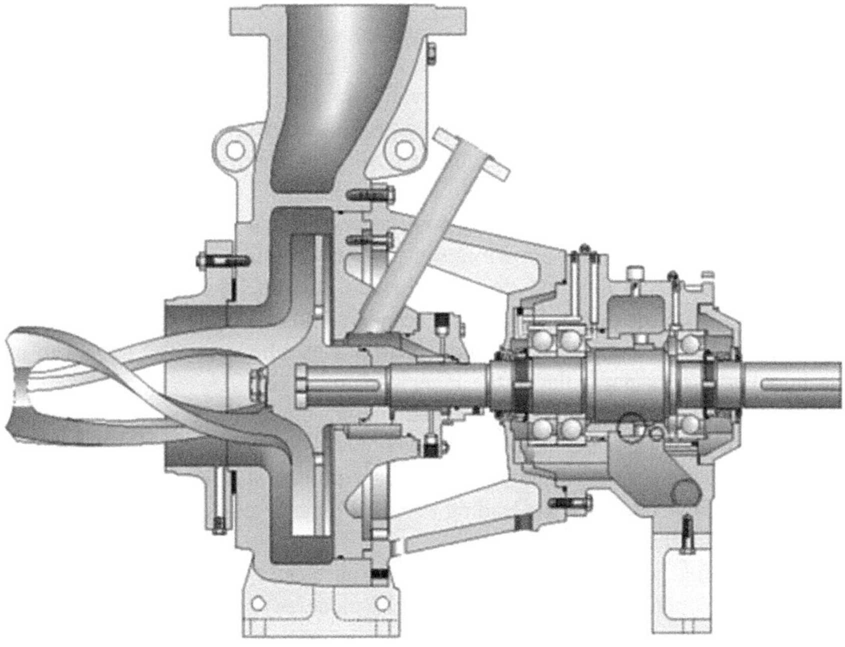

Fig. 13.38 Paper pulp pump for consistencies of 8 to 18%; Sulzer Pumps

The gas content impairs pump performance in the same way as discussed in Chap. 13.2, notably Fig. 13.20. Suction capability is also reduced. Presumably it is the air content which is the main cause of the head impairment sometimes observed in stock pumping. Therefore, the air content should be known when selecting a stock pump.

With excessive air content the pump can become gas locked. A higher suction pressure eases the pumping due to the lower GVF and density ratio, as discussed in detail in Chap. 13.2.

Influence of stock on pump performance: It can be considered as generally accepted that the power increases proportionally to the consistency. Pumping a stock of a consistency of 8% thus needs 8% more power than pumping water. The possible influence of the consistency on the head depends primarily on the air content. Since impeller and volute design have an impact on the gas handling capability and on the head impairment at a given gas content, it is difficult to give general rules. Furthermore, hydraulic losses increase with the length of the fibers. Since all of these parameters are usually not well known, the available information is sometimes contradictory.

A correction factor to account for the head impairment as compared to water pumping has been derived from a graph in [B.15], Eq. (13.23):

$$f_H \equiv \frac{H_{stock}}{H_w} = 1 - q^* e^{-\frac{0.00095}{x^{2.5}}} \tag{13.23}$$

This formula is valid for $q^* > 0.5$ and consistencies $x < 0.06$. The correction factor f_H is applied at constant flow rate (i.e. $f_Q = 1$), as illustrated by the figure in Table 13.5. Presumably, Eq. (13.23) is valid for cases where the gas content is approximately proportional to the consistency. Unfortunately, no details are known about the impeller design and the test conditions underlying the data on which Eq. (13.23) is based.

In contrast to Eq. (13.23), virtually no head impairment was measured at all with an impeller of the form "W" shown in Fig. 7.17 when pumping stock consistencies between 1 and 10%. Similar types of impellers gave the same result. Sometimes even a slight improvement in head as compared to water pumping is observed (compare to Fig. 13.37). Such impellers have blades which are well advanced into the impeller eye, in particular at the inner streamline (hub).

When appropriate impellers are selected and gas content is low, high margins on head are not necessary and should indeed be avoided since over-sizing of a pump would bring a number of disadvantages: larger pumps and motors, higher investment costs, part load operation leading to higher energy costs, higher maintenance and impaired reliability of the plant.

As mentioned above, high consistencies require open impellers with scalloped rear shrouds (openings in the rear shroud as close as feasible down to the hub, Fig. 9.23). If semi-open impellers were used, the stock would "burn" or "bake" in the impeller sidewall gap because of insufficient heat removal.

For pumping stock, single-stage volute pumps are almost exclusively selected. Large impeller trims should be avoided. The suction pipe should protrude into the chest; it must be as straight and as short as possible; its diameter should be one pipe size larger than the suction nozzle. If these rules are ignored, it will be difficult to bring stock with consistencies higher than 3% to the impeller. Complicated suction piping also presents the risk of dewatering the stock. If consistencies are high, the pump must take suction directly from the chest - as the case may be, using special devices as shown in Fig. 13.38.

Abrasive particles in the stock will cause erosion. Due to the lack of turbulence, the aspect of that kind of wear is unlike the wavy pattern shown in Figs. 14.19 and 14.20; the abrasion resembles rather a mechanical grinding process.

References

1. Hergt, P., et al.: Verlustanalyse an einer Kreiselpumpe auf der Basis von Messungen bei hoher Viskosität des Fördermediums. VDI Ber. 424 (1981)
2. Stoffel, B., et al.: Untersuchungen von Einzelverlusten in Kreiselpumpen bei viskosen Flüssigkeiten. Pumpentagung Karlsruhe, K10 (1978)
3. Gülich, J.F.: Pumping highly viscous fluids with centrifugal pumps. World Pumps. **395/396**, Aug/Sept (1999)
4. Holzenberger, K.: Vergleich von zwei Umrechnungsverfahren für die Kennlinien von Kreiselpumpen bei der Förderung zäher Flüssigkeiten. KSB. Techn. Ber. 25 (1988)
5. Mollenkopf, G.: Einfluß der Zähigkeit des Fördermediums auf das Betriebsverhalten von Kreiselpumpen unterschiedlicher spezifischer Schnelläufigkeit. Pumpentagung Karlsruhe, K10 (1978)
6. Hamkins, C.P., et al.: Prediction of viscosity effects in centrifugal pumps by consideration of individual losses. ImechE paper C112/87, 207–217 (1987)
7. Li Wen, G.: The "sudden-rising head" effect in centrifugal oil pumps. World Pumps **409**, 34–36 (2000)
8. Saxena, S.V., et al.: Ermittlung von Korrekturfaktoren für Hochleistungs-Pipeline-Kreiselpumpen beim Fördern von Mineralölen mit erhöhter Viskosität. Pumpentagung Karlsruhe C, 7–3 (1996)
9. Gülich, J.F.: Effect of Reynolds-number and surface roughness on the efficiency of centrifugal pumps. ASME. J. Fluids. Eng. **125**(4), 670–679 (2003)
10. Baker, O.: Design of pipe lines for simultaneous oil and gas flow. Oil. Gas. J. **26** (1954)
11. Collier, J., Thome, J.R.: Convective Boiling and Condensation, 3rd edn. Clarendon Press, Oxford (1996)
12. Taitel, Y., Dukler, A.E.: A model for predicting flow regime transitions in horizontal and near-horizontal gas-liquid flow. AIChE J. **22**, 47–55 (2003)
13. Bertola, V. (ed.): Modelling and Experimentation in Two-phase Flow. Springer, Wien (2004)
14. Patel, B.R., Runstadler, P.W.: Investigation into two-phase flow behavior of centrifugal pumps. Polyphase flow in turbomachinery. ASME. 79–100 (1978)
15. Kecke, H.J.: Zweiphasenströmung bei Radialkreiselpumpen. Pumpentagung Karlsruhe. **A**, 1–3 (1996)
16. Tillack, P.: Förderverhalten von Kreiselpumpen bei viskosen, gasbeladenen Flüssigkeiten. Diss. TU Kaiserslautern, (1998)
17. Kosmowski, I., Hergt, P.: Förderung gasbeladener Medien mit Hilfe von Normal- und Sonderausführungen von Kreiselpumpen. KSB. Techn. Ber. **26** (1990)

18. Sauer, M.: Einfluss der Zuströmung auf das Förderverhalten von Kreiselpumpen radialer Bauart bei Flüssigkeits-/Gasförderung. Diss. TU Kaiserslautern, (2002)
19. Turpin, J.L., Lee, J.F., Bearden, J.L.: Gas-liquid flow through centrifugal pump—correlation of data. In: Proceedings of 3rd International Pump Symposium, Texas A&M, 1986, pp. 13–20
20. Hench, J.E., Johnston, J.P.: Two-dimensional diffuser performance with subsonic, two-phase, air-water flow. ASME. J. Basic. Engng. **94**, 105–121 (1972 March)
21. Thum, D.: Untersuchung von Homogenisierungseinrichtugen auf das Förderverhalten radialer Kreiselpumpen bei gasbeladenen Strömungen. Diss. TU Kaiserslautern (2007)
22. Brenne, L., et al.: Performance evaluation of a centrifugal compressor operating under wet gas conditions. In: Proceedings of the 34th Turbomachinery Symposium, pp. 111–120 (2005)
23. Nguyen, D.L.: Sonic velocity in two-phase systems. Int. J. Multiphase. Flow. **7**, 331–320 (1981)
24. Gülich, J.F.: Energierückgewinnung mit Pumpen im Turbinenbetrieb bei Expansion von Zweiphasengemischen. Techn. Rundschau. Sulzer. **3**, 87–91 (1981)
25. Florjancic, D.: Einfluß von Gas- und Luftzuführung auf das Betriebsverhalten ein- und mehrstufiger Pumpen. Techn. Rundschau. Sulzer. Forschungsheft. 35–44 (1970)
26. Murakami, M., Minemura, K.: Effects of entrained air on the performance of centrifugal and axial pumps. Memoires Faculty Eng. Nagoya Univ. **124**, 23–1 (1971)
27. Pessoa, R., Prado, M.: Two-phase flow performance for electrical submersible pump stages. SPE Prod. Facil. **18**, 13–27 (2003)
28. Bratu, C.: Rotodynamic two-phase pump performance. Soc. Petroleum. Eng. SPE. **28516**, 555–567 (1994)
29. Bratu, C.: Multiphase production systems. OMAE 1996. In: 15th International Conference Offshore Mechanics, Florence (1996)
30. Arnaudau, P.: Development of a two-phase oil pumping system, Poseidon project. In: Offshore Technology Conference OTC 5648, Houston (1988)
31. Arnaudau, P., Bratu, C.: Transport of unprocessed oil and gas in multiphase pumps. BHRA Seminar on Multiphase Pumping Technology, Cranfield, June 16 (1988)
32. Gié, P., et al.: Poseidon multiphase pump: field test results. Offshore. Tech. Conf. OTC. **7037** (4), 489–501 (1992)
33. Gülich, J.F.: Apparatus and method for mixing, measuring and forwarding a multiphase gas mixture. US Patent 5,841,020, (1998)
34. Gopalakrishnan, S.: Power recovery turbines for the process industry. In: 3rd International Pumping Symposium, Houston (1986)
35. Hamkins, C.P., et al.: Pumps as energy recovery turbines with two-phase flow. In: ASME Pumping Machinery Symposium, San Diego (1989)
36. Gülich, J.F.: Energierückgewinnung bei der Expansion von Zweiphasengemischen. In: Pumpen als Turbinen. Faragallah, Sulzbach, (1993)
37. Weber, M.: Strömungsfördertechnik. Krauskopf, Mainz (1973)
38. Holzenberger, K.: Energiebedarf von Kreiselpumpen beim hydraulischen Feststofftransport. VDI. Ber. **424**, 89–98 (1981)
39. Cave, I.: Effects of suspended solids on the performance of centrifugal pumps. Hydrotransport 4, Paper H 3, BHRA Fluids Engineering, (1976)
40. Gahlot, et al.: Effect of density, size distribution and concentration of solid on the characteristics of centrifugal pumps. ASME. J. Fluids. Engng. **114**, 386–389 (1992)
41. Gneipel, G., et al.: Berechnung der Energiedifferenzzahlen von Kreiselpumpen bei der Förderung von heterogenen, grob-dispersen Flüssigkeits-Feststoff-Gemischen. Pumpentagung Karlsruhe. **A**, 1–1 (1996)
42. Kreuzfeld, G.: Berechnung der Zweiphasenströmung in Kreiselpumpenbauteilen. Diss. TU Dresden (1999)
43. MatouŠek, V.: Flow in Mechanism of Sand-water Mixtures Pipelines. University Press, Delft (1997)
44. Gneipel, G.: Berechnung der Partikelbahnen bei der Förderung von Fluid-Feststoffgemischen. Diss. B Bergakademie Freiberg (1990)

45. Gandhi, B.K., Singh, S.N., Seshadri, V.: Performance characteristics of centrifugal slurry pumps. ASME. J. Fluids. Eng. **123**, 271–280 (2001)
46. Engin, T., Gur, M.: Comparative evaluation of some existing correlations to predict head degradation of centrifugal slurry pumps. ASME. J. Fluids. Eng. **123**, 149–157 (2003)
47. van den Berg, C.H., Vercruijsse, P.M., Van den Broeck, M.: The hydraulic transport of highly concentrated sand-water mixtures using large pumps and pipeline diameters. Hydrotransport. **14**, 445–453 (1999)
48. Gneipel, G., Tuong, P.N.: Stoß- und Reibungsverluste beim hydraulischen Feststofftransport und deren Einfluß auf die Verminderung der Druckzahl der Pumpe. In: 10th Intnl Conf on Hydromechanisation, Zakopane (1998)
49. Weber, M.: Grundlagen der hydraulischen und pneumatischen Förderung. VDI Ber. **371**, 23–29 (1980)
50. Radke, M., et al.: Untersuchung kostenbestimmender Faktoren bei Kreiselpumpen in Rauchgasentschwefelungsanlagen. VGB Kraftwerkstech. **71**, 455–461 (1991)
51. Verhoeven, J.: Energy recovery in reverse running pumps. Pumpentagung Karlsruhe. B1 (1992)
52. Bischof, F.: Experimentelle Untersuchungen an einer Kreiselpumpe zur Feststofförderung. Diss. TU Braunschweig (1983)
53. Radke, M., et al.: Neue konstruktive Entwicklungen für Kreiselpumpen in Rauchgasentschwefelungsanlagen. Konstruktion **42**, 53–60 (1990)
54. Wilson, K.C., Addie, G.R., Sellgren, A., Clift, R.: Slurry transport using centrifugal pumps, 2nd ed. Blackie Academic and Professional, London. ISBN 0 7514 0408 X (1997)
55. Hellmann, D.H.: Pumps for multiphase boosting. In: 2nd International Conference on Pumps and Fans, Beijing, paper IL 4 (1995)
56. Govier, G.W., Aziz, K.: The Flow of Complex Mixtures in Pipes. Van Nostrand Reinhold, New York (1972)
57. Brennen, C.E.: Fundamentals of Multiphase Flow. Cambridge University Press, Cambridge (2005)
58. Zhu, J., et al.: Experimental study and mechanistic modeling of pressure surging in electrical submersible pump. J. Nat. Gas Sci. Eng. **45**, 625–636 (2017)
59. Zhu, J., Zhang, H.Q.: Numerical study on electrical submersible pump two-phase performance and bubble size model. SPE production & operations Aug. 2017, 267–278
60. Zhu, J.: Experiments, CFD simulation and modelling of ESP performance under gassy conditions. Ph.D. Thesis University of Tulsa (2017)
61. Bellary, S.A., et al.: Effects of crude oil-water emulsions at various water cuts on the performance of a centrifugal pump. Int. J. Oil Gas Coal Technol. **1**, 71–88 (2017)
62. Zhu, J., et al.: Surfactant effect on air/water flow in a multistage electrical submersible pump. Experimental Thermal and Fluid Science (2018)

Chapter 14
Selection of Materials Exposed to High Flow Velocities

Degradation or failures of materials due to fatigue, corrosion, abrasion and cavitation erosion time and again cause costly problems to pump operators. This could be avoided in most cases by a careful material selection. Frequently one of two causes can be made responsible for a wrong material selection: (1) the corrosive properties of the liquid pumped are not clearly specified (or are unknown), or (2) for cost reasons (competitive pressure) the least expensive material is chosen which appears just to be able to do the job.

The severity of attack of pump components by fatigue, abrasion, cavitation and erosion corrosion increases exponentially with the flow velocity, but the application limits of the various materials are not easily defined. They depend on the flow velocity as well as on the corrosive properties of the mediums pumped and the concentration of entrained solid particles, if any. Also, the alternating stresses induced by pressure pulsations and rotor/stator interaction forces (RSI) cannot really be quantified. That is why the thicknesses of blades, shrouds and vanes are usually chosen from experience and engineering judgment.

The present discussion of materials focuses on the interaction between flow phenomena and material behavior. To this end, some background information on corrosion and frequently used materials was deemed necessary, but a *comprehensive guide for material selection is clearly beyond the scope of this text.*

In this chapter methods are developed which foster a systematic and consistent approach for selecting materials and analyzing material problems in the field.

Four criteria are relevant for selecting materials which are exposed to high flow velocities:

1. The fatigue strength (usually in corrosive environment), since high velocities in pumps are inherently linked to high pressure pulsations, rotor/stator interaction forces and alternating stresses.
2. Corrosion induced by high velocities, in particular erosion corrosion.
3. Cavitation erosion, as has been extensively discussed in Chap. 6.
4. Abrasion as metal loss caused by solid particles entrained by the fluid.

J. F. Gülich, *Centrifugal Pumps*,
https://doi.org/10.1007/978-3-030-14788-4_14

Abrasion and cavitation erosion are primarily mechanical wear mechanisms which can, at times, be reinforced by corrosion. In contrast, corrosion is a chemical reaction between the metal, the medium pumped, oxygen and chemical substances. This reaction is always present - even if it is scarcely perceptible. Finally, the impeller tip speed can be limited by hydraulic forces or vibration and noise.

14.1 Impeller or Diffuser Fatigue Fractures

14.1.1 Allowable Head per Stage

Fatigue fractures of impeller blades, shrouds or diffuser vanes can be avoided by applying the state of the art; they are rarely observed. In highly-loaded pumps, when ignoring basic design rules or applying insufficient care in manufacturing, this type of damage is nevertheless sometimes encountered. The main causes of fractures in vanes or shrouds include:

- Too small a distance (gap B or ratio $d_3^* = d_3/d_2$) between impeller blades and diffuser vanes or volute cutwater (Table 10.2).
- Insufficient blade or shroud thickness.
- Deficiencies in quality: fillet radii between blades and shroud missing or too small, casting deficiencies, brittle material (insufficient toughness) caused by inadequate heat treatment.
- Insufficient corrosion fatigue strength (wrong material).
- Possibly, excessive pressure pulsations caused by the pump or the system, Chap. 10.3.
- Resonances between an eigen mode of the impeller with a hydraulic or an acoustic excitation. There may also be a fluid-structure interaction between the impeller side plates and the flow in the impeller sidewall gaps.

The rotor/stator interaction and pressure pulsations discussed in Chap. 10 generate alternating stresses in the impeller blades and the shrouds as well as in the diffuser vanes. An accurate analysis of these stresses is almost impossible (even though the components can well be analyzed by a finite-element program) because the hydraulic loading of the impeller by unsteady pressure distributions cannot be defined. It depends not only on the flow in the impeller, the collector and the sidewall gaps, but also on acoustic phenomena and possibly on pulsations in the system (refer also to Chap. 10.3).

In order to develop a consistent empirical procedure for assessing the loading of impellers and diffusers, for selecting vane and shroud thicknesses or for the analysis of damage, the model of a simple beam under a uniform load may be used as a starting point. Accordingly, the blades of closed impellers or diffusers are modeled by a beam clamped at both ends. Open impellers or diffusers are described by a beam clamped at one end, but free at the other.

The calculation according to Tables 14.1 and 14.2 is based on the following assumptions[1]:

1. Consider the last section of a blade at the impeller outlet as a beam clamped at both ends with a width of $x = 5 \times e$ and the span $L = b_2$ (e = nominal blade thickness without a possible profile). If the blade is profiled, an average blade thickness e_m is used to determine the section modulus; e_m is defined by Eq. (T14.3.3).
2. The impeller blades are subject to a steady uniform load which is composed of the centrifugal force given by Eq. (T14.1.2) and the pressure difference acting on the blade (ψ_{Load}) according to Eq. (T14.1.1). The hydraulic blade loading can be determined from the pressure distribution on the blade (e.g. by CFD). It is opposed to the centrifugal force, and ψ_{Load} is consequently negative, Eq. (T14.1.5). In diffusers there is no centrifugal force, and ψ_{Load} is positive. In the following $\psi_{Load} = 0.1$ is assumed.
3. The steady forces caused by the work transfer and centrifugal effects generate a mean stress σ_m according to Eq. (T14.1.6), which represents the maximum stress at the clamped ends.
4. The pressure pulsations are a measure for the unsteady, alternating pressures acting on the considered blade section. The smaller the distance between impeller and diffuser, the higher are the pressure pulsations. According to the data reported in [17], Eq. (T14.1.3) may be used. The pressure pulsations obtained correspond approximately to those measured in [41] at $q^* = 0.6$.
5. The fluctuating pressure creates alternating bending stresses σ_w according to Eq. (T14.1.7). The nominal stress obtained from Eq. (T14.1.4) must be multiplied by a notch factor α_k in order to determine the peak stresses which are relevant for generating fatigue cracking.
6. The notch factor is determined by two parameters:

   (1) The fillet radius r_f between the blades and the shrouds. Equation (T14.1.7) provides $\alpha_k = f(r_f/e)$. If the fillet radii are unknown, the notch factor should be selected to $\alpha_k = 2$ at least. If the corners are almost sharp, set $\alpha_k = 4$.

   (2) The casting quality, notably casting defects. The resulting notch factors depend on the size of the defect and its location relative to the casting surface. An accurate assessment of casting defects would require fracture-mechanical criteria and thus as a rule is not feasible. The following values may be assumed:
   poor casting quality: $\alpha_k = 4$
   average casting quality: $\alpha_k = 2$
   high-quality castings with $r_f/e > 0.5$: $\alpha_k = 1.5$.

7. The allowable alternating stress σ_w depends on the mean stress σ_m according to Fig. 14.1. As per Eq. (T14.1.8) and Fig. 14.1, a safety factor against fatigue $S_{bw} = 2$ and a safety factor against forced rupture S_z have been introduced.

[1]Most of the formulae used for the stress analysis can be found in [48].

Table 14.1 Impeller or diffuser vane loading

						Eq.
Bending of a beam by an equivalent uniform loading	Static or dynamic pressure difference		diffuser 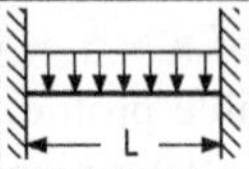		Open impeller or diffuser 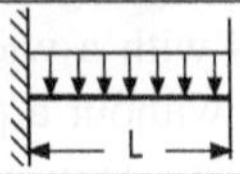	---
Static vane loading	$\Delta p_{stat} = \frac{\rho}{2} u_2^2 \psi_{load}$		$\psi_{Load} = \frac{2\,(p_{DS} - p_{SS})}{\rho\, u_2^2}$			14.1.1
Static load due to centrifugal force	$\Delta p_z = \rho_{Mat}\, \omega^2\, r\, e \cos\beta_{2B}$					14.1.2
Dynamic vane loading	$\Delta p_{dyn} = \frac{\rho}{2} u_2^2\, \Delta p^*$		$\Delta p^* = \frac{0.008}{(d_3^* - 1)^{0.9}}$		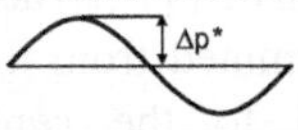	14.1.3
Bending moment M, section modulus W, bending stress σ	$M = F_T \frac{x\, \Delta p\, L^2}{12}$		$W = \frac{x}{6} e_m^2$		$\sigma = \frac{M}{W} = F_T \frac{\Delta p}{2}\left(\frac{L}{e_m}\right)^2$	14.1.4
	F_T	L	Y		ψ_{Load}	---
Impeller: closed	1	b_2	4	Centrifugal force acting	Set negative	14.1.5
Impeller: open	6					
Diffuser: closed	1	b_3	0	No centrifugal force	Set positive	
Diffuser: open	6					
Impeller blade mean stress: σ_m	$\sigma_m = F_T \frac{\rho}{4} u_2^2 \left(\frac{b_2^*}{e_m^*}\right)^2 \left\{\psi_{Load} + Y e_m^* \frac{\rho_{mat}}{\rho} \cos\beta_{2B}\right\}$					14.1.6
Impeller blade: alternating bending stress σ_w	$\sigma_w = F_T\, \alpha_k\, \Delta p^* \frac{\rho}{4} u_2^2 \left(\frac{b_2^*}{e_m^*}\right)^2$		$\alpha_k = 1.2\left(\frac{e}{r_f}\right)^{0.15}$		r_f = fillet radius between blade and Shroud	14.1.7
Allowable alternating bending stress $\sigma_{bw,al}$	$\sigma_{bw,al} = \frac{\sigma_{bw}}{S_{bw}}\left(1 - \frac{\sigma_m}{R_m} S_z\right)$					14.1.8
Allowable impeller circumferential speed	$u_{2,al} = 2\frac{e_m^*}{b_2^*}\sqrt{\frac{\sigma_{bw}}{F_T \rho \left\{S_{bw}\alpha_k \Delta p^* + S_z \frac{\sigma_{bw}}{R_m}\left(\psi_{load} + Y e_m^* \frac{\rho_{mat}}{\rho}\right)\right\}}}$					14.1.9
Tangential stress at r_i due to centrifugal force in a disk with a central bore	$\sigma_t = \frac{3+\nu}{4} \rho_{mat}\, u_2^2 \left\{1 + \frac{1-\nu}{3+\nu}\left(\frac{r_i}{r_2}\right)^2\right\}$				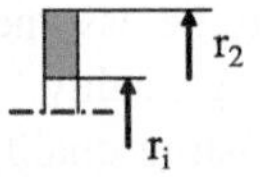	14.1.10

x in Eq (14.1.4) is the width of the "beam" considered; x cancels when calculating the stress.
S_{bw} = safety factor against fatigue fracture; S_Z = safety factor against forced rupture; α_k = notch Factor, ν = Poisson's ratio (mostly 0.3); ρ_{mat} = density of impeller material;
σ_{bw} = fatigue strength (endurance limit); * = quantity is referred to d_2

Table 14.2 Impeller shroud loading			Eq.
Alternating pressures acting on impeller shroud	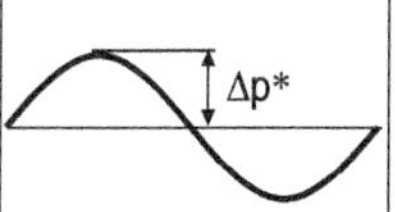	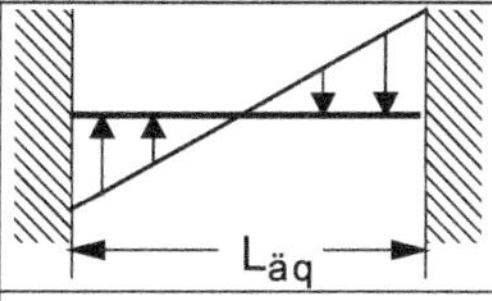	---
Static load on impeller shroud	$\Delta p_{stat} = \frac{\rho}{2} u_2^2 \psi_{Rs}$	$\psi_{Rs} = \frac{2\,(p_{Rs} - p_{La})}{\rho\, u_2^2}$	14.2.1
Dynamic load	$\Delta p_{dyn} = \frac{\rho}{2} u_2^2 \Delta p^* \left(2\frac{x}{t_2} - 1\right)$	$\Delta p^* = \frac{0.008}{(d_3^* - 1)^{0,9}}$	14.2.2
Bending moment M, section modulus W, bending stress σ	$M = \frac{x\, \Delta p\, L_{äq}^2}{60}$	$W = \frac{x}{6} e_{Rs}^2$; $\sigma = \frac{M}{W} = \frac{\Delta p}{10}\left(\frac{L_{äq}}{e_{Rs}}\right)^2$	14.2.3
Impeller shroud thickness e_{Rs}	$e_{Rs} = 1.25\, e$	e = blade thickness (without profile) For e_{Rs} set an average value which is representative for $\Delta r = 5\, e_{Rs}$	14.2.4
Static load on impeller shroud	$\psi_{Rs} = R_G \psi \left\{1 - \frac{r^* - r_1^*}{1 - r_1^*}\right\} - k^2\left(1 - r^{*2}\right) \approx 0.1$		14.2.5
Effective radius (load center)	$r^* = 1 - \frac{2\,\pi}{3\, z_{La}} \sin\frac{a_2}{t_2} \cos\frac{a_2}{t_2} \approx 0.9$		14.2.6
Equivalent span between two blades	$L^*_{äq} = \frac{L_{äq}}{d_2} = \frac{\pi}{2\, z_{La}}\left(1 + 0.8 \sin\beta_{2B}\right) \approx 0.67\frac{\pi}{z_{La}}$		14.2.7
Mean stress σ_m	$\sigma_m = \frac{\rho}{4} u_2^2 \left(\frac{L^*_{äq}}{e^*_{Rs}}\right)^2 \psi_{Rs}$	$t_2^* = \frac{t_2}{d_2} = \frac{\pi}{z_{La}}$	14.2.8
Alternating bending stress σ_w	$\sigma_w = \alpha_k \frac{\rho}{20} u_2^2 \left(\frac{L^*_{äq}}{e^*_{Rs}}\right)^2 \Delta p^*$		14.2.9
Allowable alternating bending stress $\sigma_{bw,al}$	$\sigma_{bw,al} = \frac{\sigma_{bw}}{S_{bw}}\left(1 - \frac{\sigma_m}{R_m} S_z\right)$		14.2.10
Allowable impeller circumferential speed	$u_{2,al} = 2\frac{e^*_{Rs}\, z_{La}}{\pi}\sqrt{\frac{\sigma_{bw}}{\rho\left\{S_{bw}\,\alpha_k \frac{\Delta p^*}{5} + S_z \frac{\sigma_{bw}}{R_m}\psi_{Rs}\right\}}}$		14.2.11

x in Eq (14.2.3) is the width of the "beam" considered; x cancels when calculating the stress.
S_{bw} = safety factor against fatigue fracture; S_Z = safety factor against forced rupture;
α_k = notch factor, ν = Poisson's ratio (mostly 0.3); ρ_{mat} = density of impeller material;
σ_{bw} = fatigue strength (endurance limit); * = quantity is referred to d_2*

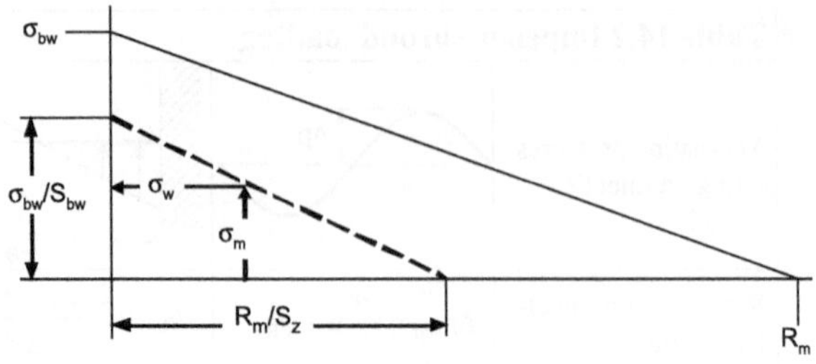

Fig. 14.1 Goodman diagram for determining the allowable stress amplitude σ_w as function of the mean stress σ_m

These were selected according to the ultimate elongation of the material. The values used for the different materials are listed in Tables 14.12 to 14.17

8. If the alternating bending stress from Eq. (T14.1.7) is set equal to the allowable stress given by Eq. (T14.1.8) and if the mean stress from Eq. (T14.1.6) is introduced, the resulting equation can be solved for the allowable impeller tip speed $u_{2,al}$. Using the pressure coefficients from Fig. 3.24, or Eq. (3.26), the allowable head per stage can be calculated.
9. The impeller shrouds are subject to centrifugal forces. A disk of constant thickness experiences the maximum tangential stress σ_t at the boundary of the center bore as given by Eq. (T14.1.10). Since the simple model of a disk of constant thickness cannot do justice to the complex stress distribution in an impeller, high safety factors S_{zz} are required.[2] These have been selected in view of the ultimate elongation, (see Tables 14.12 to 14.17). If Eq. (T14.1.10) is solved for u_2, another limit for the impeller tip speed is obtained.
10. If the impeller blades are subject to dynamic loading, the same must apply to the shrouds. Two kinds of dynamic loading are to be assumed:

    (1) When an impeller blade passes a diffuser vane or volute cutwater, the pressure distribution in the impeller changes. This implies alternating pressures over the pitch t_2 as shown by the figure in Table 14.2.

    (2) In addition, pressure pulsations can be generated by the flow in the impeller sidewall gaps. Such pulsations can be caused by seal clearances varying over the circumference (eccentric rotor position), or when the shrouds and/or the casing are not machined, or by offsets between upper and lower casing halves of an axial-split pump. The pressure fluctuations may be reinforced by system influences, Chap. 10.3.

    The dynamic loading of the shrouds is assumed according to Eq. (T14.2.2).
11. The static pressure in the impeller sidewall gaps p_{Rs} exceeds the pressure in the impeller channels p_{La}. The resulting pressure difference generates a steady bending stress σ_m in the shrouds. The difference between p_{Rs} and p_{La} is zero at the outer impeller radius and increases towards the center. It can be determined

[2]Note the difference between S_z and S_{zz} in the above discussion.

from Eq. (T14.2.5) which was derived from Eqs. (14.1 and 14.2). The pressure in the impeller sidewall gap is:

$$p_{Rs} - p_1 = \rho \frac{u_2^2}{2} \left\{ \psi_P - k^2 (1 - r^{*2}) \right\} \tag{14.1}$$

The pressure rise in the impeller channels is assumed to increase proportionally to the radius from the inlet to the outlet. One can write:

$$p_{La} - p_1 = \rho \frac{u_2^2}{2} \psi_p \left\{ 1 - \frac{r^* - r_1^*}{1 - r_1^*} \right\} \tag{14.2}$$

According to Eq. (T14.2.6), r* may be assumed as the center of gravity of the triangle defined by t_2 and a_2, see figure in Table 0.2 for defining a_2. Observed shroud fractures usually can be roughly described by this triangle.
The steady loading of the shroud results from the difference between Eq. (14.1 and 14.2) as given by Eq. (T14.2.5).

12. The pressure difference $p_{Rs} - p_{La}$ generates a steady mean bending stress σ_m according to Eq. (T14.2.8). The equivalent span $L_{äq}$ between two blades was chosen as the average of t_2 and a_2: $L_{äq} = ½ (t_2 + a_2)$. It can be expressed by Eq. (14.2.7).
13. The centrifugal forces generate stresses perpendicular to these bending stresses. Therefore they have been neglected. Alternatively, a combined stress could be used as σ_m in Eq. (T14.2.10).
14. Similar to the calculation of the blades in Table 14.1, the allowable impeller tip speed relevant for the integrity of the shrouds can be calculated as given by Eq. (T14.2.11). It is determined by the fatigue strength of the selected material.

The fatigue strength σ_{bw} (or endurance limit) depends on various parameters:

- Peak stresses at changes in cross section or shape of a component have a large impact on the fatigue strength. These effects are captured by stress concentration or notch factors.
- A specimen with a coarse surface has lower fatigue strength than a fine surface. The fatigue strength of an as-cast surface is about 30 to 50% lower than that measured on a polished specimen.
- Material: ductile materials are better than brittle ones because peak stresses at the root of the crack are reduced by local plastic deformation. Highly loaded impellers, diffusers or shafts should therefore have a minimum ultimate elongation of A > 18% (preferably A > 25%).
- Material defects (in particular in castings) impair the fatigue strength in a similar way as a notch. Cast materials often have lower fatigue strength than forged or wrought materials. The microstructure of a material should be homogenous with the finest possible grains in order to achieve optimum fatigue strength.

Precipitations at grain boundaries are unfavorable. The effect of the grain size is illustrated by the example of Inconel 718 (R_m = 1300 N/mm^2): the corrosion fatigue strength (CFS) at 10^8 cycles in sea water dropped from 400 to 240 N/mm^2 when the grain size was increased from 0.01 to 0.15 mm, [50]

- Corrosion drastically reduces the fatigue strength. When using metallic components in water, the fatigue strength is always impaired. Therefore the fatigue strengths measured in air must not be used without due correction for assessing stresses of components operating in water. In sea water the reduction of the fatigue strength is considerable indeed. *The fatigue strength increases with the resistance against local corrosion.*
- During fatigue tests in water or sea water, the endurance limit (the asymptotic region of the curve stress versus number of load cycles) is frequently not reached so that an extrapolation to an infinite number of cycles becomes very difficult. For example the tests reported in [50] did not allow to define the endurance limit of high-alloy steels and bronzes in sea water even after 10^8 load cycles. The polished specimens of the investigated materials failed at 10^8 cycles, at stress levels which for stainless steels depended on the pitting index. Figure 14.2a shows test data at 10^7 and 10^8 load cycles. The strong influence of the corrosion resistance and of the number of load cycles can be appreciated from this data. The CFS of some other materials tested in [50] are presented in Fig. 14.2b (including Ni-resist).
- The residual stresses in a component should be as low as possible. Residual *tensile* stresses (e.g. induced at the surface by grinding) lower the fatigue strength. Residual *compressive* stresses near the surface (e.g. from shot peening) increase the fatigue strength.

Because of the various influence parameters discussed above, very different values of fatigue strength can be found in the literature for a given material. The data spread is mainly due to the test conditions. Therefore it is very difficult (if not impossible) to define meaningful values for the fatigue strength of impellers and diffusers. The fatigue strengths measured under various test conditions are by no means comparable. A ranking of materials by measured fatigue strength is only meaningful within a specific test series with strictly controlled test conditions.

For calculating the allowable head shown in Figs. 14.3, 14.4, 14.5, 14.6 and 14.7, $\sigma_{bw}/R_m = 0.3$ was assumed. The corrosion fatigue strength (CFS) in sea water can be expected in the range of $\sigma_{bw}/R_m = 0.12$ to 0.15, but this level can only be attained with steels which are resistant to sea water, Chap. 14.4.2.

The impeller tip speeds allowable for some commonly used types of materials were calculated from Tables 14.1 and 14.2 assuming the impeller outlet width according to Eq. (7.1). With the pressure coefficients from Eq. (3.26), the allowable heads per stage at the BEP were then determined and plotted in Figs. 14.3, 14.4, 14.5, 14.6 and 14.7. The following assumptions were made for these calculations and figures; they should well be appreciated when applying this information:

1. The calculation applies to forced vibrations. They do not account for resonances between impeller eigen modes and hydraulic or acoustic excitation

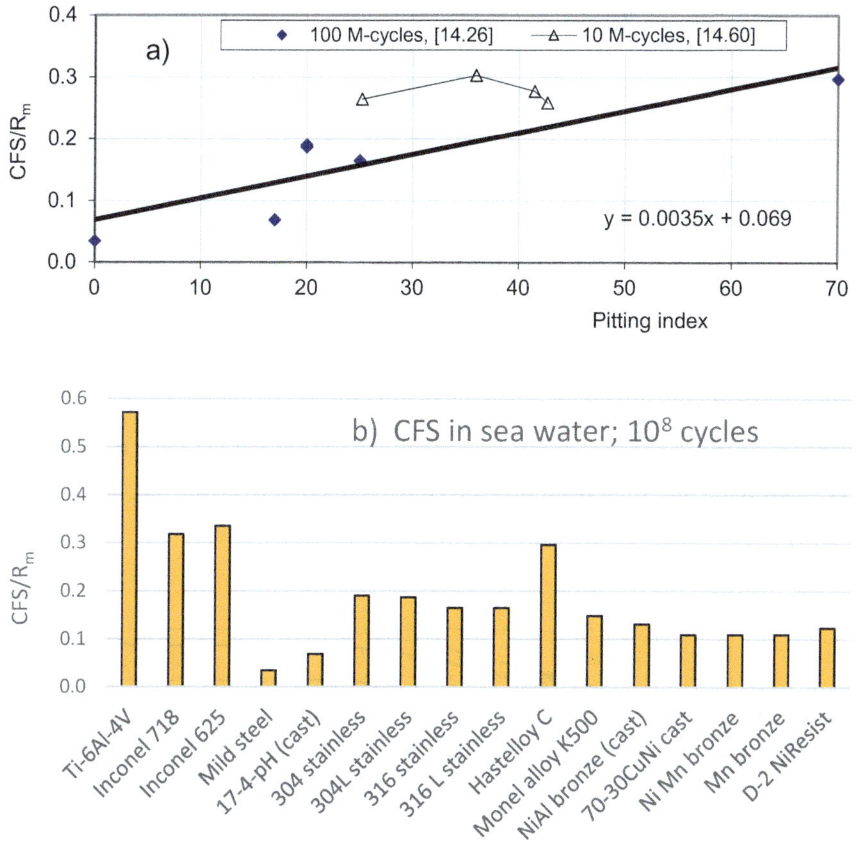

Fig. 14.2 Corrosion fatigue strength CFS of polished specimens in sea water. **a** Stainless steels; the pitting index is defined by Eq. (14.8). **b** Various materials [50, B.15]. R_m is the ultimate tensile strength

phenomena. In order to consider operation near a resonance, the amplification factors can be estimated from Table 17.5 using a damping factor of about D = 0.05 which has been derived from measurements in [54].

2. Impeller outlet width from Eq. (7.1); diffuser inlet width from: $b_3 = 1.15\ b_2$
3. Pressure coefficients from Eq. (3.26).
4. Tensile strengths R_m are minimum values from Tables 14.12 to 14.17.
5. Fatigue strength $\sigma_{bw}/R_m = 0.3$. This level of fatigue strength applies to fresh water and boiler feed water i.e. water specifications W1, W2, W5 and W6 as defined in Chap. 14.4.1. It does not apply to sea water (W3 and W4). If $\sigma_{bw}/R_m = 0.12$ to 0.15 is assumed for sea water, the allowable head per stage is distinctly lower than stipulated by Figs. 14.3, 14.4, 14.5, 14.6 and 14.7. The thicknesses of blades and shrouds must be increased accordingly for corrosive service.

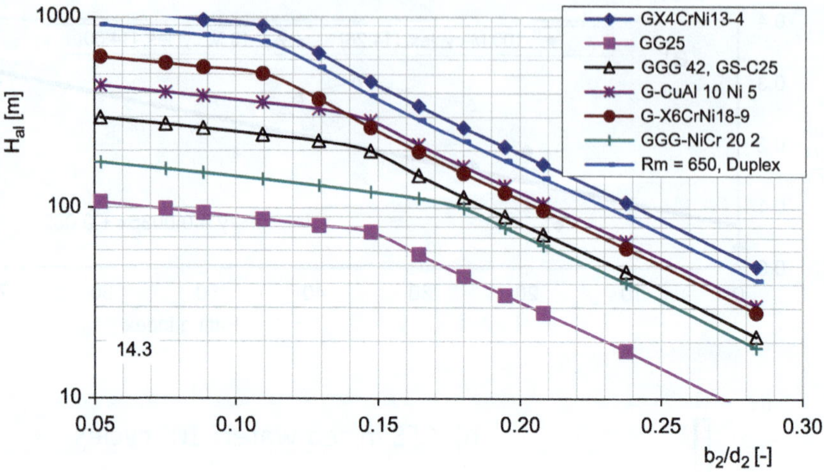

Fig. 14.3 Allowable head H_{al} limited by blade strength of *closed* impellers; applicable at $\rho = 1000$ kg/m^3 and for water specifications W1, W2, W5, W6 according to Chap. 14.4.1

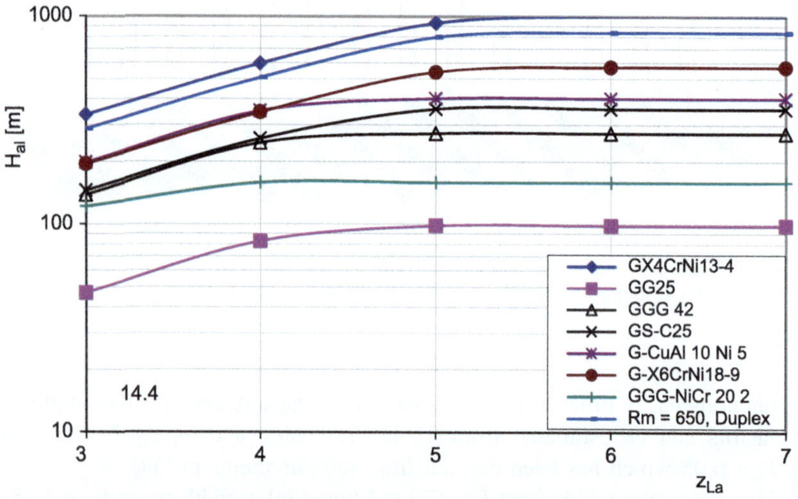

Fig. 14.4 Allowable head H_{al} limited by shroud loading of impellers; applicable at $\rho = 1000$ kg/m^3 and for water specifications W1, W2, W5, W6 according to Chap. 14.4.1

6. Safety factor against fatigue fractures $S_{bw} = 2$.
7. Safety factor for the mean stress S_z from Tables 14.12 to 14.17 (S_z increasing with falling ultimate elongation).

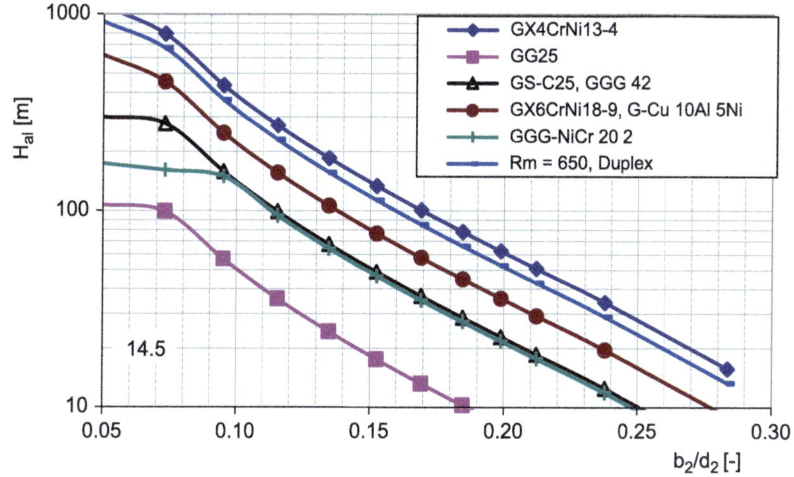

Fig. 14.5 Allowable head H_{al} limited by blade loading of *open* impellers; applicable at $\rho = 1000$ kg/m^3 and for water specifications W1, W2, W5, W6 according to Chap. 14.4.1

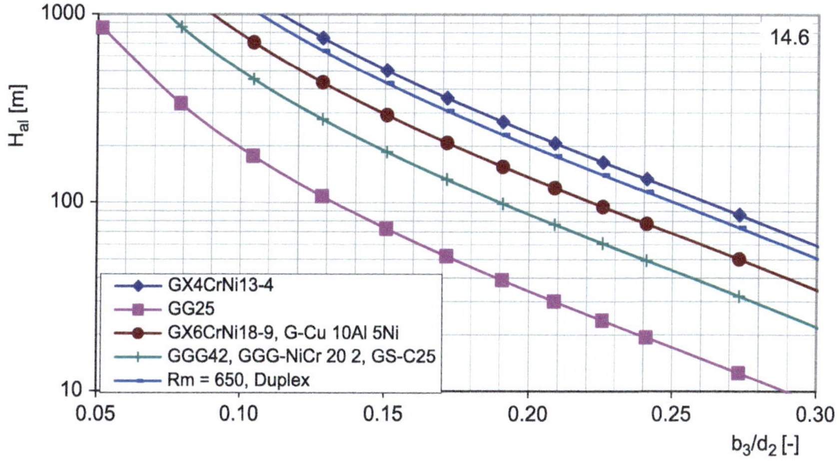

Fig. 14.6 Allowable head H_{al} limited by vane loading of *closed* diffusers; applicable at $\rho = 1000$ kg/m^3 and for water specifications W1, W2, W5, W6 according to Chap. 14.4.1

8. Safety factor against forced rupture under centrifugal forces in shroud S_{zz} from Tables 14.12 to 14.17 (S_{zz} increasing with falling ultimate elongation).
9. Notch factor (stress concentration) $\alpha_k = 2$.
10. Relevant average impeller blade thickness ratio $e_m^* = 0.015$; for diffuser vanes $e_m^* = 0.016$; for shrouds $e_{RS}^* = 0.0225$.

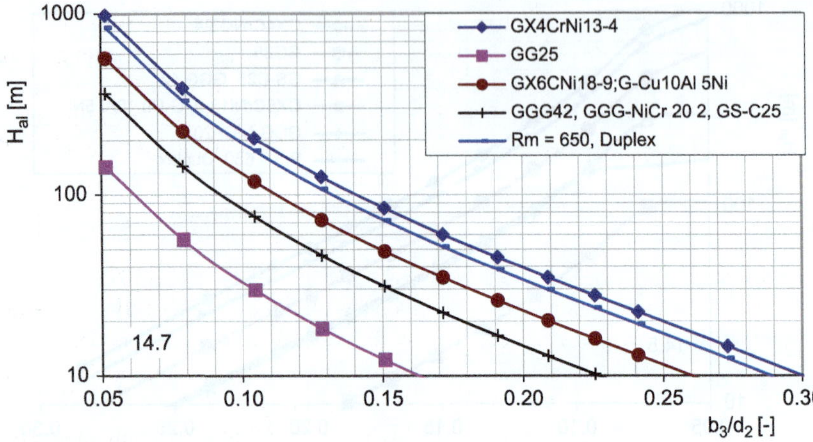

Fig. 14.7 Allowable head H_{al} limited by vane loading of *open* diffusers; applicable at SG = 1.0 and for water specifications W1, W2, W5, W6 according to Chap. 14.4.1

11. Distance between impeller and diffuser vanes: $d_3^* = 1.04$.
12. Average vane loading: impeller: $\psi_{Load} = 0.1$; diffuser $\psi_{Load} = 0.45$.
13. Average pressure difference loading on the shrouds: $\psi_{Rs} = 0.1$.
14. The allowable heads plotted in Figs. 14.3, 14.4, 14.5, 14.6 and 14.7 are valid for the best efficiency point. Higher loads at partload are covered by the safety factors.

The safety factors selected above may appear as rather high. They are justified because the assumed loading is uncertain and the stress models used are very simple. Furthermore, the safety factors must cover off-design operation as well as casting defects and the inherent uncertainties of the corrosion fatigue strength.

Figures 14.3, 14.4, 14.5, 14.6 and 14.7 demonstrate the following facts:

- The allowable head decreases strongly with growing impeller outlet width and thus with increasing specific speed.
- Up to about $n_q = 30$, impellers were built for heads per stage of 1200 m. Hence, operation conditions confirm the corresponding range in Fig. 14.3.
- Because of the increasing pitch t_2 at the impeller outlet, the allowable head drops with a decreasing number of impeller blades as shown by Fig. 14.4. Double-entry impellers with 5 blades (as in Fig. 7.48) achieve heads at BEP between 700 and 800 m. Operation experience thus confirms the corresponding range in Fig. 14.4.
- If several unfavorable factors combine, the allowable head can drop drastically below the values given by Figs. 14.3, 14.4, 14.5, 14.6 and 14.7. In a particular example, impeller shroud fractures occurred in a multistage pump with a head per stage below 200 m (with $d_3^* = 1.025$) because the fillets between blades

and shrouds were almost sharp, and the shroud thicknesses were much below design due to core shifting. With $\alpha_k = 4$ and $e_{Rs}{}^* = 0.012$, a calculation according to Table 14.2 yields an allowable head of 180 m. Thus the damage to these impellers is well predicted by the calculation.

- Alternating stresses σ_w in the impeller shrouds of a multistage diffuser pump ($n_q = 22$) have been measured in [3]. In addition, pressure pulsations were recorded in the rotating impeller. The pressures and stresses from the test at $q^* = 0.8$ had been reproduced in Fig. 10.44. The measured stresses can be compared to the prediction according to Eq. (T14.2.9). The results are presented in below table. These stress calculations have been done using the measured Δp^* listed in column 2.

q^*	test: Δp^*	$\sigma_{w,test}$	σ_w predicted from Eq. (T14.2.9)
–	–	MPa	MPa
1.0	0.05	3.5	3.3
0.8	0.13	10	8.6

The measuring uncertainty of the stresses in [3] may be in the order of ±15 to 20%. Predicted and measured stresses correspond surprisingly well. However, in view of the various assumptions made in deriving Eq. (T14.2.9) the agreement must be viewed to some extent as coincidental. The pressure pulsations recorded at partload were $\Delta p^* = 0.13$; this compares reasonably well to the value of $\Delta p^* = 0.11$ estimated for $d_3/d_2 = 1.055$ from Eq. (T14.2.2).

- Open impellers or diffusers are limited to lower heads than closed components.
- **Note**: Because of the lower endurance limits, the allowable heads per stage in sea water obviously are lower than given by Figs. 14.3, 14.4, 14.5, 14.6 and 14.7. At high heads larger shroud and blade thicknesses may be required in sea water.

The calculation according to Tables 14.1 and 14.2 and the data given in Figs. 14.3, 14.4, 14.5, 14.6 and 14.7 facilitate a consistent assessment of the risk of fatigue fractures of impellers and diffusers. However, when applying the method, the uncertainties and assumptions should well be appreciated.

When the limits given by the curves in Figs. 14.3, 14.4, 14.5, 14.6 and 14.7 for the various materials are exceeded in a specific application, a more detailed analysis is warranted. Conversely, by proper design, analysis and quality control, components can be built whose heads are above the allowable heads stipulated by Figs. 14.3, 14.4, 14.5, 14.6 and 14.7.

Table 14.3 gives some guidelines for selecting the thicknesses of the blades, the shrouds, diffuser vanes and leading as well as trailing edges. Equation (T14.3.1) can be used for multi-blade impellers as well as sewage pump impellers with 1, 2 or 3 blades. The required thickness increases with the impeller tip speed and the specific speed (since the impeller outlet width and pressure pulsations increase).

Table 14.3 Thickness of blades, vanes and shrouds

				Eq.
Closed impellers	Nominal blade thickness e (without profiling)	$u_{Ref} = 100$ m/s; $n_{q,Ref} = 100$; $z_{Ref} = 7$ applicable to $z_{La} = 1$ to 10		14.3.1
	$e^* \equiv \frac{e}{d_2} = 0.015\left(\frac{z_{Ref}}{z_{La}}\right)^{0.4}\left(1+0.25\left\{\frac{u_2}{u_{Ref}}\right\}^2\right)\left(1+0.5\left\{\frac{n_q}{n_{q,Ref}}\right\}^2\right)$			
	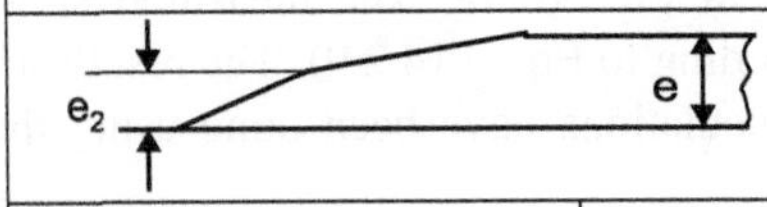			
	Trailing edge thickness	$e_2 = e/2$		14.3.2
	Mean blade thickness for calculation according to Table 14.1	$e_m = 0.5(e_2 + e) = 0.75 \times e$		14.3.3
	Shroud thickness	$e_{Rs} = 1.25 \times e$	e from Eq. (14.3.1)	14.3.4
Open impellers	Blade thickness at outer streamline	$e_a = e$		14.3.5
	Blade thickness at inner streamline (rear shroud or hub). Note: make e_i not much larger than e_{Rs} to avoid casting problems.	$e_i \leq 2 \times e_a$		14.3.6
	Mean blade thickness for calculation according to Table 14.1	$e_m = 1.125 \times e$		---
Diffusers	Leading edge thickness	$e_3 = 0.01 \times d_2$		14.3.7
	Mean vane thickness for calculation according to Table 14.1	$e_m = e_3 \times (1 + 5 \tan\delta)$		14.3.8
	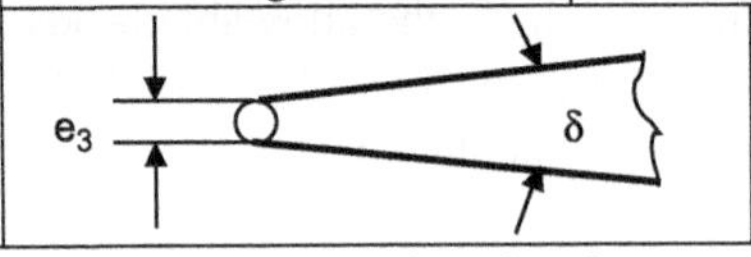	δ = angle between suction and pressure side of diffuser vanes: $\delta \geq 5°$		---

- With careful design and manufacturing these recommendations are thought to be on the safe side, but beware of resonances between impeller eigen frequencies and hydraulic or acoustic excitation.
- High-head sea water applications require higher thicknesses.
- The minimum thickness required for the actual casting process must be observed.
- Cast impellers minimum: $e = 0.015 \times d_2$

If the required head per stage exceeds the allowable value calculated from Tables 14.1 and 14.2, the following remedies may be considered:

1. Increase material thickness;
2. Stronger material: higher R_m, higher ductility (ultimate elongation);
3. Select a material with improved resistance to corrosion (higher endurance limit);
4. Attempt a more accurate calculation of the loading as input to a finite element stress analysis;

5. Increase distance between impeller and diffuser vanes (gap B) in order to reduce the excitation forces;
6. Improve casting quality and finish; reduce residual stresses; review heat treatment (grain size, homogeneity of material, precipitations);
7. Modify design, larger fillet radii to reduce notch factor.

14.1.2 Impeller Failures—Analysis and Prediction

In many instances, impeller failures are caused by the rotor/stator interaction (RSI) at the impeller outlet because the pressure pulsations are highest at this location. Figure 14.7-1 shows cracks in the blades originating at the blade trailing edge at the junction of the blade to the shroud (left photo). A front shroud failure is shown on the right photo. In this case, the cracks may have originated from the shroud or from the blade: when cracks are created in the blade, the unsupported span of the shroud increases which would lead to the type of failure observed in the right photo. The stress amplitudes (zero to peak) given in these figures are estimated from Eqs. (14.2a to e) by the procedure developed below. Obviously, these stress estimates are subject to large uncertainties (notch factors, scatter of Eq. (14.2e), actual hydraulic loading, simplifications of the model in Table 14.1-2); but they can give an order of magnitude.

Sometimes cracks in the blades originate at the leading edge near the hub, Fig. 14.7-2. This can have several reasons:

- The blades are too thin near the leading edge, e.g. if the leading-edge profile is too long as per concept "H" in Fig. 7.10. To avoid this kind of failure, short elliptical (or parabolic or similar) profiles (as per concept "L" in Fig. 7.10) should be employed.
- Ratio of blade span to blade thickness too large (specifically with semi-open impellers)
- High residual stresses (improper heat treatment). This problem is compounded if thin, long leading-edge profiles are used because the thin blades cool faster than the hub with its important material accumulation. Since the crack in Fig. 14.7-2 (left photo) opens up, tensile residual stresses may have been instrumental in causing this failure.
- Inadequate weld repairs done on the casting.
- Poor leading-edge finish and/or small fillet radii will favor the initiation of cracks.
- The cracks can occur during assembly/disassembly of the impeller when heating the hub using a torch.

There is no direct relation between the pressure pulsations and the alternating stresses in the impeller, diffuser or other pump components. In order to get data on the overall hydraulic vane loading, alternating bending stresses (simple amplitude)

cracks originating from trailing edge
$\sigma_w = \pm 12$ MPa with $\alpha_k = 2$

cracks originating in the shroud and/or from the trailing edge
$\sigma_w = \pm 17$ MPa (with $\alpha_k = 1$)

Fig. 14.7-1 Impeller failure in high-head seawater pump (super-duplex, PI = 39)

closed impeller; probably casting quality issue: residual tensile stress

semi-open impeller; $\sigma_w = \pm 28$ MPa with $\alpha_k = 2$; austenitc stainless steel

Fig. 14.7-2 Cracks at the impeller leading edge

were evaluated from tests with impellers, diffuser and one volute. The equivalent pressure fluctuations required to cause the measured impeller stresses were calculated from Eq. (14.2a) for stresses measured on the shroud and from Eq. (14.2b) for stresses measured on the blades or a diffuser vane. These formulae were derived from Tables 14.1 and 14.2.

The resulting "pressure pulsations" are a measure for the *total* vane loading; i.e. a summation of all oscillating forces experienced by the vane. These values were subsequently converted to the pulsations expected with $(d_3/d_2)_{Ref} = 1.06$ according to Eq. (14.2d) as per Fig. 10.6. This normalization allows to compare tests with different d_3/d_2 ratios. A "design curve" was estimated from the data; it is shown in Fig. 14.7-3 where test points from [3] were added. The design curve represents an upper envelope of the available tests.

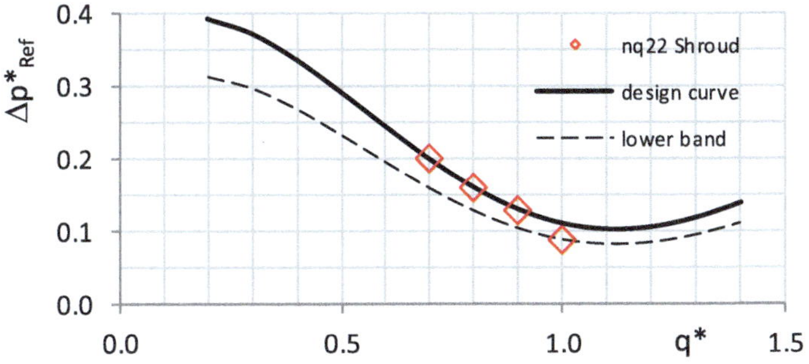

Fig. 14.7-3 Dynamic loading (pressure fluctuation) amplitude calculated from strain measurements; referred to $d_3/d_2 = 1.06$

$$\Delta p^*_{shroud} = 10\left(\frac{e_{shroud}}{L_{eq}}\right)^2 \sigma_w{}^* \tag{14.2a}$$

$$\Delta p^*_{blade} = \frac{2}{\alpha_k}\left(\frac{e}{b_2}\right)^2 \sigma_w{}^* \tag{14.2b}$$

$$\sigma_w{}^* = \frac{\sigma_w}{\frac{\rho}{2} u_2{}^2} \tag{14.2c}$$

$$\Delta p^* \propto \frac{1}{\left(\frac{d_3}{d_2} - 1\right)^{0.77}} \tag{14.2d}$$

$$\Delta p^*_{w,Ref} = 0.1 + 0.55(q^* - q_o)^2 - 0.18(q^* - q_o)^3 - 0.44(q^* - q_o)^4 \tag{14.2e}$$

Eq. (14.2e) can be used for design; $\Delta p^*_{w,Ref}$ at $(d_3/d_2)_{Ref} = 1.06$ with; $q_o = 1.12$

Comments:

- The data measured on the shroud are within the range of the other tests even though the other strain measurements were taken on the blades. This observation tends to support the approach taken to generalize these measurements. It also appears to confirm the simplified model for the shroud stresses used in Table 14.2.
- The hydraulic blade loading of all pumps shows a consistent trend as the function of the flow rate. The trend is similar to that of pressure pulsations and rotor/stator lateral interaction forces as per Chap. 10.7.1
- None of the test data are likely to have been influenced by resonances since most of the data were gathered at low speeds. During the test of [3] one had looked specifically for a resonance (because this was one of the failure hypotheses) but none was found.

- In summary, the data suggest a considerable hydraulic loading, but it must be borne in mind that the data can be applied only with the simplified mechanical models of Tables 14.1 and 14.2 because they were derived under these simplifications.
- The blade loading Δp^* derived from the strain measurements is higher than the Δp^* assumed in Table 14.1. The alternating stresses in the impeller are given by the sum of fluctuations over the entire frequency range.
- The range of specific speeds covered by the test data is $n_q = 20$ to 50 (radial impellers). Extrapolation to high specific speeds and semi-axial impellers is uncertain.

The "design curve" in Fig. 14.7-3 can be used to estimate the stresses in any specific application; it is given by Eq. (14.2e).

Commonly, high levels of alternating stress, insufficient corrosion fatigue strength and deficits in casting quality contribute to the failure. In most incidents of impeller failure the stress level cannot be considered as the *sole cause* because not all of the blades, and not all impellers of a multistage pump, are affected simultaneously. The reasons for this include:

(1) testing of corrosion fatigue strengths CFS is subject to a wide statistical scatter; the same applies to cast components
(2) casting defects as origin for crack initiation
(3) varying stress concentration factors, e.g. due to differences in fillet radii between blade and shroud
(4) varying blade and shroud thicknesses due to casting tolerances
(5) differences in residual stresses
(6) differences in material structure including those which may be induced by weld repairs of the casting.

Unless vane or shroud thicknesses are manifestly insufficient (e.g. core shifting), corrosion and/or quality issues are usually involved in impeller failures and need to be investigated. When using austenitic stainless steel, the risk of stress corrosion cracking in the presence of chlorides should be assessed, Fig. 14.9 (e.g. sewage pumps). Resonances of blade eigen frequencies with diffuser vane passing frequency $f_n \times z_{Le}$ may be an issue to be checked with semi-open impellers.

The following recommendations can help to reduce the risk of impeller failure:

- Most importantly: sufficient gap B: d_3/d_2 according to Table 10.2
- Adequate blade and shroud thicknesses according to Table 14.3
- The corrosion fatigue strength CFS must be realistically ascertained. In sea water, the CFS may drop according to a power law with the number (N) of load cycles CFS $\sim 1/N^x$ without ever reaching an endurance limit ($x = 0.05$ to 0.1)
- Ensure excellent casting quality because cracks originate at casting deficits

- The trailing edge should be profiled on the *pressure* surface in order to reduce the width of the wake hence RSI and pressure pulsations. The flow on the pressure surface is able to follow the blade shape
- In high-pressure pumps, the impeller blade trailing edge thickness should be $e_2 = (0.5 \text{ to } 0.6)e_{nom}$ and truncated with the radius. Profile length possibly $L = (3 \text{ to } 4)e_{nom}$
- Trailing edge profile and truncation radius should be optimized with respect to pressure pulsations during model tests
- The proposed trailing edge shape with the radius also reduces stress concentration (local peak stresses) at sharp corners
- Fillet radii $r/e_{nom} = 0.5$
- Fillet radii near the leading and trailing edges to be finished carefully
- Leading and trailing edge profiles to be ground and checked according to finger template or equivalent process as per the relevant quality class according to Chap. 15.3
- Figure 14.7-4 shows a typical profiling applied to the pressure surface
- The front and rear shrouds of the impeller should be designed to reduce stress concentration factors in sharp corners, Fig. 14.7-4 shows the principle. $d_{2D}/d_2 = 1.01$ to 1.015.

Material selection for sea water pumps with high heads per stage is a critical issue: tentative selection limits are given below:

Material selection for sea water pumps with high head per stage		
$H_{opt,st}$ (m)	Material	Pitting index
≤ 450	Super Duplex	PI $\geq$ 40
≤ 550	Super Duplex	PI $\geq$ 43
>550	Inconel 718	not defined

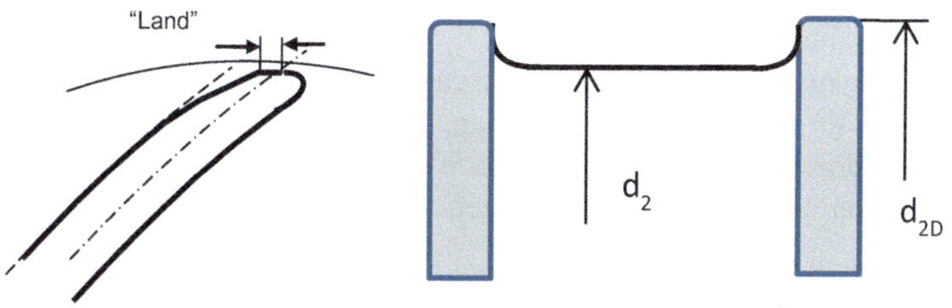

Fig. 14.7-4 Trailing edge profile on the pressure surface of the blade and impeller outlet design to reduce stress concentration, [B.20]

14.2 Corrosion

14.2.1 Corrosion Fundamentals

In principle, all metallic materials used in pumps (with the exception of noble metals) are unstable when exposed to water - at least if the water contains oxygen or traces of acids. However, if the material is selected correctly with respect to the *specific medium*, acceptable component life is achieved because protective films of corrosion products are formed on the metal surface. These films impede the diffusion and reaction of corrosive substances on the component surface and hamper further attack. Quality and thickness of the protective film determine the corrosion velocity. Four types of protective films may be distinguished:

1. **Passivation**: Usually the most effective protection is afforded by "passivation". During passivation a very resistant oxide film forms on the metal surface. Its thickness is in the order of only 3 nm (0.003 μm), but it virtually prevents a further oxidation. The passive film is rich with chromium. It is formed spontaneously when the metal is exposed to air, oxygen or water. Stainless steels with a minimum of 12% chromium content, nickel alloys, titanium and aluminum all are passivating in this way. Abrasion, cavitation erosion or corrosive substances (e.g. H_2S in sea water) can damage the passivation film and thus limit the use of a material. However, passivation does not afford a universal protection. The domain of passivation of any individual material depends on the pH-value, the temperature and on corrosive substances. It is usually determined by electro-chemical methods through measuring the corrosion current in an electrolyte.
2. **Calcareous rust scales**: In carbonated water a (relatively thick) protective layer is formed from deposits of rust and lime on unalloyed steels. Under specific conditions this film is able to sufficiently protect a component in that it impedes the diffusion of corrosive agents and oxygen from the water to the metal surface. However, because of its porosity, a deposit of rust *without lime* does not effectively impede corrosion.
3. **Magnetite or hematite films** form on carbon steels in fully demineralized water under certain conditions, refer to Chap. 14.3.
4. Tightly adhering protective films made up by various corrosion products, for example on copper alloys in sea water.

When protective films are formed (as well as in the corrosion process), a chemical reaction takes place between metal, water, oxygen and chemical substances. In order to become effective, the corrosive substances taking part in the reaction must move to the metal surface. This mass transfer is generated by diffusion and convection. The convective transport can be characterized by mass transfer coefficients which are determined by the boundary layer developing in the flow. Consequently, flow velocity and turbulence have a considerable influence on many corrosion phenomena. In situations where the mass transfer controls the speed of the chemical reaction, the corrosion rate will be proportional to $c_{cor} \sim w^x$, with x = 0.7 to 1.0 in turbulent flow depending on the geometry.

Thousands of publications are dedicated to the topic "corrosion". Chapter 14 mainly relies on [N.12 to N.15, N.19, B.5, B.17, B.28] and the individual publications quoted throughout the text. As far as possible, the material properties given have been taken from standards. Extensive tables providing the suitability of various materials in many liquids can be found in [B.5, B.8, B.17, B.28, 52], corrosion fundamentals in [1, 29].

14.2.2 Corrosion Mechanisms

Almost all corrosion mechanism are electro-chemical processes similar to "galvanic corrosion". Local corrosion elements, consisting of tiny anodes and cathodes, are formed on the metal surface. Considering iron in watery solutions, electrons travel from anodic to cathodic zones with iron-ions going into solution at the anodes while hydroxide or molecular hydrogen is generated at the cathodes. There are many forms of corrosion which are caused by different mechanisms. The types of corrosion most relevant for pump operation are reviewed in the following, [43, 52, 69, N.12].

Uniform corrosion (also termed "general corrosion") is characterized by a uniform metal loss which occurs in the absence of a protective film on the entire wetted surface. Examples are the rusting of carbon steels or the dissolution of metals in acids. The metal loss increases with the concentration of oxidizing substances in the water (i.e. oxygen or acids) and with the mass transfer coefficient (i.e. velocity and turbulence) which controls the transport of corrosive substances to the metal surface.

In order to avoid uniform corrosion, either the water chemistry must be established in a way that protective films can be formed or a more corrosion resistant (that is usually a passivating) material must be selected. Non-metallic or metallic coatings are also apt to prevent uniform corrosion.

Another form of general attack produces a wavy corrosion pattern which is uniform over the surface. It is similar to uniform corrosion and the same remedies can be applied.

Local corrosion: **pitting and crevice corrosion**: *Pitting* occurs at defects of the passivation film on the surface of the material. The defects are acting as anodes in local corrosion elements. These are localized (small-scale) differences in the electro-chemical potential which give rise to corrosion currents similar to galvanic corrosion. Cathodic protection is an effective method to prevent pitting.

Crevice corrosion: Corrosive substances can concentrate in gaps or crevices which are filled with stagnant liquid. This mechanism generates a strong local corrosion attack. The corrosion absorbs the oxygen required for passivation. Due to a drop of the pH-value (rising concentration of acids in the crevice) and due to the dearth in oxygen, crevice corrosion is more aggressive than pitting and therefore requires materials with a higher corrosion resistance. Due to these mechanisms, cathodic protection has only limited effect against crevice corrosion.

Exactly the same mechanisms are operative under *deposits of sludge.* In a similar way growth of algae, micro-organisms or mussel can cause a concentration of corrosive substances and lead to comparable damage. As an example, Fig. 14.8 shows local ("crevice"-type) corrosion on the casing of a submersible pump. Large corrosion pits developed under heavy deposits of sludge. The casing material was 1.4460 (X8CrNiMo27-5) with a pitting index PI = 36, Eq. (14.8). The pump operated in brackish water with a chloride content of 14,000 ppm. From Table 14.9 and Fig. 14.21 it is easily appreciated that a moderate increase in chloride concentration, acidity and impurities within the sludge deposit will bring the material to its limit.

Many components in a pump are sensitive to crevice corrosion which can occur near gaskets, screws, casing inserts or seats. The longer and narrower the gap, the higher is the risk of crevice corrosion.

Local corrosion is triggered by chlorides or sulfides and all halogens with the exception of fluorine which is more likely to cause uniform corrosion. The attack increases with the temperature and the chloride concentration. The larger the ratio of the cathodic to the anodic surface, the stronger is the local corrosion. Passivated stainless steels in stagnant fluids are susceptible to local corrosion which can occur in pumps filled with sea water during standby. The resistance of stainless steels against local corrosion depends on various parameters which are discussed in Chap. 14.4.2.

Fig. 14.8 Crevice corrosion of Duplex steel 1.4460 in brackish water under heavy sludge deposits (right picture), [10]

Aeration elements: Deposits, mussel growth and suspended substances can cause local corrosion on passivating metals. Remedies include: cleaning of the surface, filtering, chlorination and design measures for preventing deposits and growth of algae or mussels.

Galvanic corrosion (also "contact corrosion") can occur if two metals with different electrochemical potentials are exposed to the same electrolyte. If, in addition, the metals are connected by an electric conductor, a corrosion element is created. The anode (the material with the lower potential) is attacked more, and the cathode (the material with the higher potential) is dissolved less than would happen if the metals were not connected by a conductor. If differences in potential cannot be avoided, the surface with the lower potential should be large as compared to the surface with the higher potential. Only then will the corrosion current density (and hence the corrosion rate) be sufficiently low. As a consequence, (1) *the material of the impeller and the wear rings always must have a higher potential than the casing material*; (2) *the wear ring material should at least have the same potential as the impeller material*. These requirements gain importance with the strength of the electrolyte or the corrosivity of the pumped medium (for example in sea water pumps). The material with the lower potential is attacked more strongly at the transition to the material with the higher potential; for example at the border of a weld overlay.

Table 14.4 lists the potentials of materials exposed to *flowing* sea water according to [39]; similar values are given in [1, 8]. Materials, which are sensitive to local corrosion, exhibit lower potentials and are attacked more severely in *stagnating* water. Relevant for the material selection are only *potential differences* between two materials, because the absolute potentials may vary depending on the test conditions. The higher the potential difference, the more the material with the lower potential is attacked. But there is no quantitative relation between the corrosion rate and the potential difference.

High flow velocities can shift the potential and reinforce erosion corrosion in some cases, Chap. 14.3.

Galvanic corrosion can be prevented by cathodic protection of the material with the lower potential. A pump built from passivating materials (e.g. stainless steel) is protected cathodically by a pipeline made from carbon steel. Should the pipeline be replaced by stainless steel, corrosion damage can be experienced by the pump because the cathodic protection is removed.

If the anodic material (e.g. a carbon steel casing) is coated, a high corrosion rate is expected at locations where the coating begins to fail. This attack is very strong because of the unfavorable area ratio between anodic and cathodic material. Because of the risk of failure, it is not recommended to coat anodic components in sea water pumps.

Table 14.4 Electrochemical potential in flowing sea water

Material	Potential [V]
Magnesium and magnesium alloys	–1.6
Zinc and zinc alloys	–1
Aluminum alloys	–0.75 to –1
Mild steel, cast iron	–0.6 to –0.7
Low alloy steel	–0.6
Austenitic nickel cast iron GGG-NiCr 20-2 (EN 0.7660)	–0.45 to –0.55
Aluminum bronze	–0.3 to –0.42
Naval brass, yellow brass, red brass, aluminum brass	–0.3 to –0.4
Tin	–0.32
Copper	–0.32
Manganese bronze	–0.31
Silicon bronze	–0.28
Tin bronze	–0.3
Stainless steel types 410, 416 (13% Cr)	–0.28 to –0.34
90-10 copper-nickel alloys	–0.26
80-20 copper-nickel alloys	–0.23
Stainless steel types 430 (17% Cr)	–0.2 to –0.26
70-30 copper-nickel alloys	–0.2
Nickel aluminum bronze	–0.19
Nickel chromium alloys (Alloy 600, 76Ni16Cr8Fe)	–0.15
Nickel 200	–0.1 to –0.2
Silver	–0.12
Austenitic stainless steels such as X6CrNi 18-10	–0.07
Nickel copper alloys 400, Ni Cu 30 Al (EN 2.4374), Monel	–0.03 to –0.13
Austenitic stainless steels such as X2CrNiMo 19-11-2	0 to –0.1
Alloy 20 (EN 2.4660) 34Ni 20Cr 2Mo 3Cu	0
Alloy 825 (EN 2.4858) 42Ni 22Cr 3Mo 2Cu	0
Titanium	0
Platinum	+0.23
Graphite	+0.25

Intergranular (or inter-crystalline) corrosion: can occur at the grain boundaries due to chemical differences in the microstructure of the material. Examples are austenitic steels if chromium carbides are precipitated during heat treatment or welding. The chromium-depleted zones are attacked if there is not sufficient chromium left (at least 12% are required) to achieve passivation. To avoid such problems: (1) Select steels with less than 0.03% carbon ("L-grades"); (2) Add Tantal, Niobium or Titanium; (3) Annealing. Austenitic steels in the presence of chlorides are very sensitive to intergranular corrosion; and so are duplex steels.

Acid environments are particularly dangerous. Alloys with more than 25% chromium are not susceptible to intergranular corrosion. Damage is mostly caused by inadequate heat treatment.

Trans-granular (or trans-crystalline) corrosion: The cracks induced by corrosion are not along grain boundaries but go right through the crystals, Fig. 14.10.

Stress corrosion cracking (SCC) can occur in alloys which are subject to sustained tensile stresses in a corrosive environment; in austenitc steels stresses above 20 N/mm^2 may induce SCC. Stress corrosion cracking can also be triggered by pitting or crevice corrosion, Fig. 14.10 (right photo). Consequently, the resistance of stainless steels against SCC increases with the resistance against local corrosion. Typical for the aspect of SCC are branched cracks as seen in Fig. 14.10 (left).

The risk of SCC in austenitic steels increases with the stress level, the temperature, and the chloride concentration (even traces of Cl are dangerous). In the presence of additional substances, damage can occur even at room temperature. The material can be further sensitized by unfavorable heat treatment or cold working.

According to Fig. 14.9 the risk of SCC strongly depends on the nickel content. Austenitic steels with about 10% nickel exhibit a minimum of resistance; they are very sensitive to chloride-induced SCC. As demonstrated by Fig. 14.9, duplex steels or high-alloy chromium-nickel steels with more than 25% nickel can be employed to reduce the risk of SCC.

Ammonia (NH_3) and NH_4^+ induce SCC in copper alloys (in particular in brass); stresses as low as 5 N/mm^2 may induce SCC, Fig. 14.10 (left). As a remedy CuAlNi-bronzes or pure copper may be used. High-strength manganese bronzes are very sensitive to SCC and should not be used for pumps.

Ni-Resist (GGG-NiCr20 2, 0.7660) is subject to SCC if not furnace stress relieved.

SCC can lead to sudden catastrophic failure because SCC cannot be easily detected at an early stage.

Corrosion fatigue is characterized by a reduction of the endurance limit in a corrosive environment. Due to the cyclic mechanical loading, a passivation film cannot form at the end of the crack or on the location of stress concentration (e.g. at a notch). This applies to all metals in water. The cracks are mostly trans-cristalline. The topic is discussed extensively in Chap. 14.1

Fig. 14.9 Influence of nickel content on the risk of stress corrosion cracking (SCC)

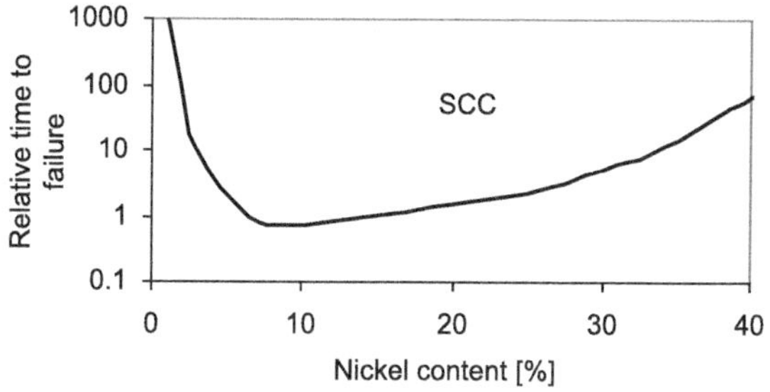

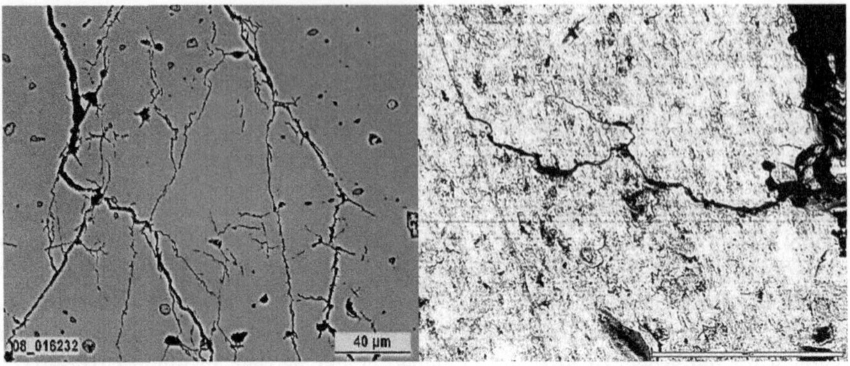

Fig. 14.10 Trans-crystalline stress corrosion cracking due to ammonia in brass, [10], *left picture*. Crack originating from a corrosion pit on a fractured surface of a highly stressed duplex component operating in sea water, *right picture*

Selective corrosion: In alloys with two or more phases, the less noble phase can be leached selectively from the material. Examples are: (1) Graphitic corrosion of grey cast iron where iron is leached from the material and a soft graphite matrix is left.[3] In this process the graphite flakes in the iron matrix of grey cast iron form local galvanic cells where iron is the anode. Anaerobic conditions with pH = 4 to 5.5 foster graphitic corrosion; could occur in sea water under mussles or other biologic growth. (2) Dezincification where zinc is selectively leached from copper-zinc alloys containing less than 85% copper. (3) Ni-Resist in sea water. (4) Leaching of aluminum out of certain aluminum bronzes: if the aluminum content exceeds 8% or if the nickel content is less than 4%. (5) The sliding rings of mechanical seals are sometimes sensitive to selective corrosion.

Austenitic steels can be sensitized to grain disintegration if the component is subject to improper welding or heat treatment procedures. Remedy: heat treatment adapted to the material and application of steels with low carbon content ($C < 0.03\%$).

Erosion corrosion is a metal loss caused by high flow velocities, vortices and turbulence (with or without abrasive particles) which dissolve or mechanically damage the protective film and prevent its re-building. Erosion corrosion is among the most-frequent cause for damage to process pumps in the chemical industry. The aspect of damaged parts shows a typical wavy pattern free from corrosion products, see Fig. 14.15. High-concentration acids are dangerous for stainless steels; sulfides and H_2S attack copper alloys.

[3]Graphitic corrosion is commonly called "graphitization", but this term rather describes the formation of graphite in iron or steel in a metallurgical process.

Fretting corrosion can be induced by corrosion and simultaneous microscopic oscillatory movements between two parts, e.g. between impellers, sleeves or other components mounted on a shaft. The passivation film is disrupted mechanically by tiny movements which are generated by shaft vibrations while corrosion evolves because of corrosive substances. Rests of grease hardened by oxidation can aggravete the damage. Molybdinum sulfite (MoS_2) and copper-additives to the grease inhibit the fretting. Micro-cracks can be generated which can lead to a fatigue fracture of the shaft. Remedy: tight shrink-fit, inert layer between the parts, spray coating or silver-plating of the seats (to be renewed after each disassembly).

Hard chroming is not recommended because the chromium layer is too brittle. The micro-movements due to alternating bending stresses in the shaft induce fine cracks in the chrome plating.

Biological corrosion is caused by micro-organisms (in particular on ferrous alloys). The organisms form most easily in stagnant water when the pump is shut down for an extended time. Remedies: (1) addition of chemicals; (2) keep components clean of deposits. Figure 14.11 shows an example of bio-fouling on brass caused by bacteria which had reduced nitrate present in drinking water to ammonia causing stress the corrosion cracking shown in Fig. 14.11.

The types and mechanisms of corrosion described above critically depend on the water analysis. Four groups of water quality are discussed in the following:

1. Fresh water, sewage with low concentration of corrosive substances, Chap. 14.2.3
2. Sea water, Chap. 14.2.4
3. Water used in flue gas desulfurization (FGD-water), Chap. 14.4.7
4. Demineralized water, Chap. 14.3 (erosion corrosion).

Fig. 14.11 Film of bio-fouling on brass formed by bacteria and causing SCC. [10]

14.2.3 Corrosion in Fresh Water, Cooling Water, Sewage

Rust, FeO(OH),[4] is generated when steel is exposed to water that contains plenty of oxygen. Due to its porosity, a layer of rust is unable to protect the steel from further attack. Even at moderate flow velocity, the rust is washed off the surface and deposited in the system as sludge. The actual water analysis of the individual application (pH, water hardness, salt content and oxygen concentration) has an important influence on the corrosion process and thus on the selection of the material. The parameters relevant for the corrosion are discussed in the following:

Oxygen content: The corrosion attack always increases with the oxygen concentration, but a minimum O_2-content is required for building protective films.

Temperature: In many situations the corrosion rate is doubled by a temperature rise of 10 to 30 °C.

The **pH** of the water indicates whether the reaction is acid or alkaline (base). At 25 °C and pH = 7, the water is neutral; at pH < 3 it is considered as strongly acid; at 4 < pH < 6 as weakly acid; at 8 < pH < 10 as weakly alkaline, and at pH > 11 as strongly alkaline. The pH of chemically pure water decreases with rising temperature. In general terms:

- neutral or weakly alkaline water is little aggressive
- weakly or strongly acid water must almost always be classified as strongly corrosive which requires a careful selection of the materials.

The **water hardness** (total hardness) is a measure for the concentration of alkali metal ions in the water, in particular calcium ions Ca^{++} and magnesium ions Mg^{++}. Most relevant for the corrosion is the carbonate hardness, i.e. the concentration of the Ca^{++}-ions. This concentration is specified as $c(Ca^{++})$ in mmol/l or in ppm. The water hardness in °dH ("deutsche Härte"="German hardness") is an obsolete unit, but it is still often encountered in practice. Meaning and conversion factors of hardness specifications are given in Table 14.5.

Table 14.5 Water hardness

1 mmol/l Ca^{++} =	100 ppm $CaCO_3$	5.6 °dH	56 mg CaO/l	40 mg Ca^{++}/l
1 °dH =	0.1786 mmol/l Ca^{++}	17.86 ppm $CaCO_3$	10 mg CaO/l	7.19 mg Ca^{++}/l
1mg/l = 1 ppm at density 1000 kg/m^3				

The water hardness has an influence on the corrosion and on the formation of "boiler scale" deposits. The protective calcareous rust scales can only form in sufficiently hard water. Soft water is often aggressive, for example in the presence

[4]Rust consists of oxides and hydrated oxides of iron. The process of rusting is complex and involves several stages which are not described in all publications in the same way.

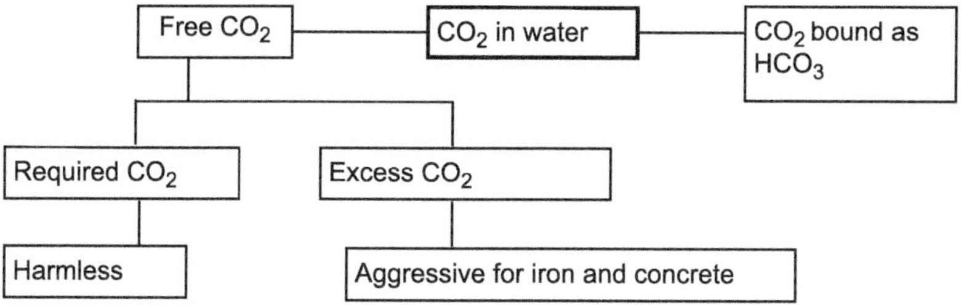

Fig. 14.12 Influence of carbon dioxide on the corrosion

of chlorides, Eq. (14.6) or at high flow velocities, Chap. 14.3. Water up to 8 °dH is considered as "soft", water with more than 18 °dH is said to be "hard".

Carbon dioxide (CO_2) is present in all natural waters. Part of the CO_2 (the "bound CO_2") is bound as calcium or magnesium carbonate and thus not available for corrosion. The rest of the CO_2 ("free CO_2") is dissolved in the water as a gas, Chap. 17.3. Part of the free CO_2 (the "required CO_2") is needed to keep the carbonates in solution (equilibrium between carbonates and CO_2); this portion is not effective for corrosion. The portion of free CO_2 which exceeds the required CO_2 (the "excess CO_2") acts aggressively on steel and concrete, Fig. 14.12.

If the concentration of free CO_2 in the water is equal to the content of the required CO_2, the protective calcareous rust scale is able to form and impede further attack - but only in low flow velocities and at temperatures below 30 °C. In contrast, no protective film can form if an excess of CO_2 is present. The corrosion rate rises with the concentration of the excess CO_2. The water becomes increasingly acid. At a water temperature of 17 °C, the equilibrium is described by Eqs. (14.3 or 14.4). If the actual pH exceeds the value calculated from Eqs. (14.3 or 14.4), a protective film is able to form; if the pH is lower, the water is aggressive.

$$\mathrm{pH} = 8.71 - 0.775\, c_{(Ca^{++})} + 0.0757\,\{c_{(Ca^{++})}\}^2 \quad \text{with} \quad c_{(Ca^{++})} \text{ in mmol/l} \qquad (14.3)$$

$$\mathrm{pH} = 8.71 - 0.135^\circ\mathrm{dH} + 0.00242\{^\circ\mathrm{dH}\}^2 \qquad (14.4)$$

The portion of bound CO_2 can be calculated from: 7.86 ppm bound CO_2 = 1°dH.

With rising temperature, more CO_2 is required for keeping the carbonates in solution. If the actual CO_2 concentration is lower than the required CO_2, a layer of hard deposits of $CaCO_3$ and $MgCO_3$ is precipitated on solid structures (boiler scale). The portion of the required free CO_2 can be calculated from Eq. (14.5):[5]

$$CO_{2,\text{free,required}}\,[\mathrm{ppm}] = 7.31 \times 10^{-3}(^\circ\mathrm{dH})^3 e^{2.9T^*} \qquad (14.5)$$

[5]Equations (14.5) and (14.6) were derived from data given in [69].

In Eq. (14.5) set $T^* = T/T_{Ref}$ and $T_{Ref} = 100$ °C.

Chlorides (respectively all halogens) are always corrosive; the more so, the softer the water. At a chloride concentration of more than 10 ppm, the corrosion rate of carbon steel drops with increasing HCO_3-concentration according to Eq. (14.6)[1]:

$$c_{Cor}\,[\text{mm/year}] = 60 - 0.45\,c_{(HCO_3)}\,[\text{ppm}] \tag{14.6}$$

Sulfates cause pitting on steel and graphitic corrosion on gray cast iron if the concentration exceeds about 250 ppm SO_4-ions. Sulfates also attack concrete. Frequently water contains 10 to 30 ppm sulfates, [B.28].

Hydrogen sulfide (H_2S) and **Fluorides** are aggressive to all metals.

Ammonia (NH_3) is not corrosive to carbon steels [52] (it is even used for boiler feed water conditioning). However, ammonia causes stress corrosion cracking in all types of brass and increased uniform corrosion on all copper and copper-nickel alloys.

Conductivity: The electrical conductivity is a measure for the amount of ions present in the water. It is relevant for boiler feed water.

Flow velocity: At flow velocities below about 1 m/s, deposits or aeration elements can be formed which lead to the local concentration of corrosive substances and thus induce local corrosion. *Above* a velocity threshold which depends on the medium and the material, either the corrosion rate increases due to accelerated mass transfer or erosion corrosion is induced.

Corrosion inhibitorscorrosion inhibitors: Some substances (e.g. phosphates, aluminum-compounds or SiO_2) which can be present in natural water inhibit corrosion and foster the formation of protective films. Such inhibitors can be responsible for the observation that a given steel exhibits different corrosion resistances in apparently identical waters. Corrosion inhibitors are often added to the water in closed circuits.

14.2.4 Corrosion in Sea Water and Produced Water

Sea water typically contains 3.5% (mass) dissolved salts, dominantly sodium and magnesium chlorides (fraction of NaCl 70 to 80%). The O_2 content of sea water at 10 to 15 °C is between 10 and 8 ppm (at saturation); the pH is 7.5 to 8.5. Close to the shore and in ports, the water is often contaminated by corrosive substances with hydrogen sulfide being particularly aggressive. Even H_2S-concentrations as low as 5 ppm can attack the passivation film and induce local corrosion. Apparently minor contamination of this type must be taken into account when selecting the materials for sea water applications. NACE MR0175/ISO 15156-3 provides requirements and recommendations for the selection of corrosion-resistant alloys for use in oil and natural gas production in H_2S environments.

Produced water from oil wells can have salt contents up to 30% with pH in the weakly acid domain. Frequently, the produced water contains dissolved H_2S, but no oxygen which reacts with H_2S.

Figure 14.13 shows the behavior of stainless steels in sea water. Depending on the flow velocity, three domains are to be distinguished: (1) At standby and at low flow velocities, there is a risk of local corrosion. The attack is strongest at w = 0 and decreases with growing velocity owing to an enhanced transport of oxygen to the metal surface. (2) Above a velocity of about 1 to 2 m/s, no differences in the local concentration are expected so that local elements are not present. The material is passive in this domain. (3) Above a threshold velocity, which depends on the material and the medium, the metal loss rises exponentially because the passivation film is attacked by erosion corrosion. Sensitive materials, such as austenitic steels, require measures against standby corrosion. These include: (1) Operation of the pump at regular intervals, e.g. once a week; (2) Filling the pump with fresh water if available; (3) Cathodic protection. However, if there are long narrow gaps (which allow only a very slow reduction of the concentration in the gaps by diffusion), these measures may not be sufficiently effective. Then materials are required which are more resistant to crevice corrosion.

If different materials are used in the same system, the effects of galvanic corrosion must be analyzed, Table 14.4.

The high flow velocities present in centrifugal pumps with heads above 20 to 30 m require passivating materials for impellers, diffusers and casings. Carbon steel or cast iron would attain only a short component life. This can be deduced from the corrosion rates given in Table 14.6 and Fig. 14.14.

The passivating materials include stainless steels which are resistant to uniform corrosion but sensitive to local corrosion attacking the passivation film. Pitting during standby, crevice corrosion and possibly deposits are the dominant corrosion mechanisms. Austenitic steels without molybdenum (1.4308, AISI type 304) are resistant against crevice corrosion in cold water (<25 °C) up to a chloride

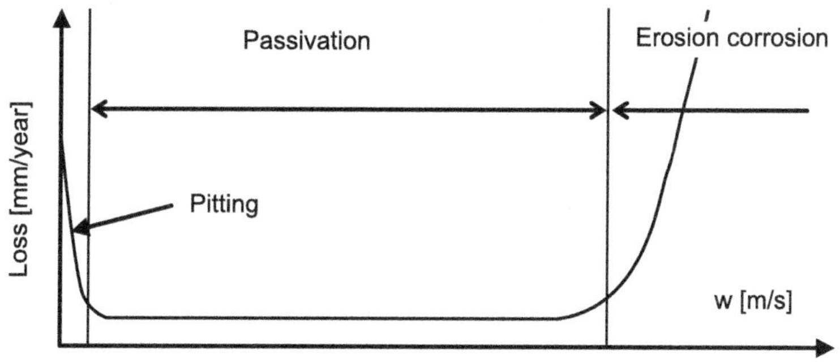

Fig. 14.13 Corrosion of stainless steels in sea water near the application limit

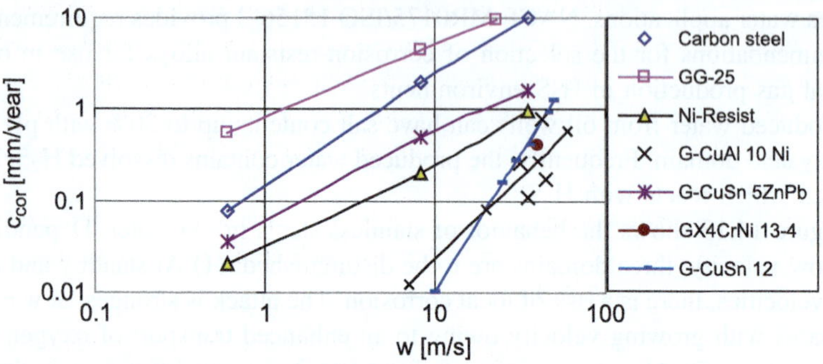

Fig. 14.14 Corrosion rate of different materials in sea water. Data according to [39, 43, 73]; w is the relative or absolute velocity (but not the circumferential speed). Above a loss of $c_{cor} > 0.1$ mm/year the material should be considered as unstable

concentration of 200 ppm, while austenitic steels alloyed with molybdenum (1.4409, AISI type 316) can be used up to about 1000 ppm chloride content, [39].

Table 14.6 gives an overview on the relevant corrosion mechanisms, the behavior of various materials, approximate limits for flow velocities and heads per stage, based on data in [39, 43, 45, 73]. If available the corrosion rates for uniform corrosion (c_{cor}) and for pitting (c_{LK}) are indicated as well. These corrosion velocities facilitate an estimation of the component life in sea water. The limits given in Table 14.6 apply to impellers, diffusers and casings but not to wear rings. If efficiency is an issue, the metal loss in annular seals should be limited to 0.05 to 0.1 mm/year (rotor dynamics may require also a limitation).

In summary, the following recommendations can be given for material selection for pumps[6] delivering sea water with pH > 5.5 which is free of *either* H_2S *or* O_2:

1. Carbon steels and cast irons (including Ni-resist) are unsuitable at high velocities because of severe uniform corrosion.
2. High-alloy cast iron of the type GGG-NiCr 20 2 (0.7660) is suitable for casings if flow velocities do not exceed 5 to 8 m/s. Velocities in volute casings, as a rule, are above that level and GGG-NiCr should not be used. GGG-NiCr 20 2 (0.7660) is unsuitable for impellers or for the suction bells of vertical pumps because the corrosion rate would be excessive at w > 10 m/s (roughly 0.3 mm/year). Due to the poor cavitation resistance, impeller blades and suction bells would suffer cavitation erosion if cavitation occurs on the blades or in the gaps between suction bell and blades. During partload recirculation, cavitation damage is also possible in the inlet casing. In spite of its good non-galling characteristics, GGG-NiCr 20 2 (0.7660) is unsuitable for wear rings because of a rapid increase in the seal clearances due to erosion corrosion.

[6]Since the flow velocities in pipelines are much lower than in pumps, these recommendations have only limited relevance for pipes.

Table 14.6 (1) Materials for sea water at T < 30 °C $H_{al} \approx \frac{1}{5} w_{al}^2$ at $\psi_{opt} \approx 1$

Type of material	Sea water	Natural		Polluted
	pH	> 5.5		4 to 5.5
	O_2-concentration	> 1 ppm	< 0.01 ppm	>1 ppm
	H_2S-concentration	≈ 0 ppm	< 10 ppm	Up to 500 ppm
Carbon steels, GS-C25, cast iron (GG-25, GGG-40)	UC (w < 0.5 m/s)	$c_{cor} \approx 0.1$ mm/year		Not applicable, heavy attack in a short time
	Erosion corrosion	$c_{cor} \approx 0.15\, w_r^{1.2}$ mm/year		
	Application limit	$w_{al} \approx 2$ m/s		
	LC at w < 1 m/s	$c_{LC} \approx 0.3$ to 0.8 mm/year		
	SC (GG, GGG)	Graphitization		
Ni-Resist Type GGG-NiCr 20 2	UC (w < 0.5 m/s)	$c_{cor} \approx 0.07$ mm/year		
	Erosion corrosion	$c_{cor} \approx 0.031 w_r^{0.95}$ mm/year		
	Application limit	$w_{al} \approx 20$ m/s	H = 80 m	
	LC at w < 1 m/s	$c_{LC} \approx 0.1$ mm/year		
	SC	Resistant		
Martensitic Cr-Ni-steels, Type 1.4317	UC (w < 0.5 m/s)	$c_{cor} \approx 0.1$ mm/year		
	Erosion corrosion	$c_{cor} \approx 10^{-6}\, w_r^{3.8}$ mm/year (?)		
	Application limit	*Not recommended*		
	LC at w < 1 m/s	$c_{LC} \approx 1.5$ mm/year		
Austenitic Cr-Ni-steels, Type 1.4409	UC (w < 0.5 m/s)	$c_{cor} \approx 0.01$ mm/year		
	Erosion corrosion	Fairly resistant		Limits application
	Application limit	$w_{al} \approx 40$ m/s	H = 300 m	$w_{al} \approx 25$ m/s; H = 120 m
	LC at w < 1 m/s *very sensitive to LC*	10%Ni: $c_{LC} \approx 1.8$ mm/year 29%Ni: $c_{LC} \approx 0.2$ mm/year *Not recommended.*		Application not recommended
	SC	To be avoided with appropriate alloy: Nb, Ti, C < 0.03		
Super-austenitic Ni-Cr-steels, Ni > 25% Type 1.4587	UC (w < 0.5 m/s)	$c_{cor} \approx 0.01$ mm/year		
	Erosion corrosion	Fairly resistant		
	Application limit	$w_{al} \approx 60$ m/s		$w_{al} \approx 55$ m/s
	LC at w < 1 m/s			
	SC	To be avoided with appropriate alloy: Nb, Ti, C < 0.03		
Duplex-steels	UC (w < 0.5 m/s)	$c_{cor} \approx 0.01$ mm/year		Suitable with PI > 40, for definition of PI see Eq. (14.8)
	Erosion corrosion	Fairly resistant		
	Application limit	$w_{al} \approx 55$ m/s	H = 600 m	
	LC at w < 1 m/s	$c_{LC} \approx 0$ mm/year		

UC: Uniform corrosion
LC: Local corrosion (crevices, pitting)
SC: Selective corrosion
c_{LC}: Pitting velocity

$w_r \equiv w/w_{Ref}$ with $w_{Ref} = 1$ m/s

w_{al} is the velocity relative to the component ($w \neq u$); w_1^*, c_{2u}^* can be estimated from Fig. 3.25

Table 14.6 (2) Materials for sea water at T < 30 °C		$H_{al} \approx \frac{1}{5} w_{al}^2$ at $\psi_{opt} \approx 1$		
Type of material	**Sea water**	**Natural**		**Polluted**
	pH	> 5.5		4 to 5.5
	O_2-Concentration	> 1 ppm	< 0.01 ppm	>1 ppm
	H_2S-Concentration	≈ 0 ppm	< 10 ppm	Up to 500 ppm
Tin bronze Copper-Tin-Alloys G-CuSn 12 2.1052.01	UC (w < 0.5 m/s)	$c_{cor} \approx 0.03$ mm/year		Not applicable, heavy attack in a short time
	Erosion corrosion	$c_{cor} \approx 10^{-5} w_r^3$ mm/year		
	Metal loss increases with ammonia content			
	Application limit	$w_{al} \approx 23$ m/s	H = 100 m	
	LC at w < 1 m/s	$c_{LC} \approx 0.25$ mm/year		
Copper-tin-zinc-alloy G-CuSn 5ZnPb 2.1096.01	UC (w < 0.5 m/s)	$c_{cor} \approx 0.03$ mm/year		
	Erosion corrosion	$c_{cor} \approx 0.06 w_r^{0.95}$ mm/year		
	Metal loss increases with ammonia content			
	Application limit	$w_{al} \approx 10$ m/s	H = 25 m	
Aluminum-bronze G-CuAl 10Ni 2.0975.01	UC (w < 0.5 m/s)	$c_{cor} \approx 0.05$ mm/year		
	Erosion corrosion	$c_{cor} \approx 3.6 \times 10^{-4} w_r^{1.8}$ mm/year		
	Metal loss increases with ammonia content			
	Application limit	$w_{al} \approx 30$ m/s	H = 180 m	
	LC at w < 1 m/s	$c_{LC} \approx 0.2$ mm/year		
	SC	With ≥ 4% Ni no SC		
Nickel base alloy NiCu30Al 2.4374	UC (w < 0.5 m/s)	$c_{cor} \approx 0.03$ mm/year		–
	Erosion corrosion	Fairly resistant		–
	Application limit	$w_{al} \approx 50$ m/s	H = 500 m	$w_{al} \approx 5$ m/s
	LC at w < 1 m/s	$c_{LC} \approx 0.4$ mm/year		–
Hastelloy C NiMo16Cr15W 2.4819	UC (w < 0.5 m/s)	$c_{cor} \approx 0.01$ mm/year		–
	Erosion corrosion	Resistant		–
	Application limit	$w_{al} \approx 60$ m/s	H = 700 m	$w_{al} \approx 60$ m/s H = 700 m
	LC at w < 1 m/s	$c_{LC} \approx 0$ mm/year		–
UC: Uniform corrosion LC: Local corrosion (crevices, pitting) SC: selective corrosion c_{LC}: Pitting velocity		$w_r \equiv w/w_{Ref}$ with $w_{Ref} = 1$ m/s		
w_{al} is the velocity relative to the component (w ≠ u); w_1^*, c_{2u}^* can be estimated from Fig. 3.25				

3. The corrosion attack increases with the oxygen content. In sea water which is virtually free of oxygen (O_2 < 10 to 20 ppb) with limited H_2S contamination, the same materials can be used as in sea water which is saturated with oxygen but contains no H_2S, Table 14.6.
4. High-alloy steels are essentially resistant against uniform corrosion, but sensitive to crevice corrosion and pitting. This risk increases strongly with growing chloride concentration, rising temperature, and falling pH. Crevice corrosion is more aggressive than pitting. In order to achieve a sufficient margin against crevice corrosion, steels with a pitting index PI > 40 should be selected, Chap. 14.4.2, Eqs. (14.8 and 14.9), Fig. 14.21.
5. Martensitic Cr-Ni steels of the type GX4CrNi 13-4 (1.4317) are unsuitable because of their high sensitivity to local corrosion.
6. Austenitic Cr-Ni steels of the type GX2CrNiMo 19-11-2 (1.4409) are essentially corrosion resistant in flowing sea water, but they are sensitive to standby corrosion which must be prevented by the measures discussed above.
7. Super-austenitic steels and duplex steels are essentially stable up to 40 °C. The same applies to nickel base alloys and special alloys.
8. Special alloys such as NiMo16Cr15W (Hastelloy C) must contain sufficiently high amounts of chromium and molybdenum in order to resist local corrosion.
9. Alloys with more than 70% copper are resistant against local corrosion, but can be employed only at moderate flow velocities because of erosion corrosion. These alloys are not recommended for applications with H_2S contamination.
10. Tin bronzes of the type G-CuSn 10 (2.1050.01) are suitable for impellers and casings with velocities up to about 10 m/s and heads up to 25 m. At these flow velocities, the resistance to cavitation is also sufficient.
11. Aluminum bronzes of the type G-CuAl 10Ni (2.0975.01) are well suited for impellers and casings up to velocities of 30 m/s and heads up to 200 m.
12. Titanium is stable even in boiling sea water.
13. In order to avoid galvanic corrosion, the materials used for impellers and wear rings must have higher potential then the casing material. The wear ring material should exhibit at least the same potential as the impeller, Table 14.4, Chap. 14.2.2.
14. Austenitic steels are very sensitive to stress corrosion cracking. The risk increases with the temperature. However, no incidents of damage in pumps were reported with a water temperature below 65 °C.
15. If chlorine is added to the water to prevent biological corrosion, no more than 2 ppm free chlorine must be dissolved in the water at the pump inlet; otherwise metal loss will be accelerated. The chlorine must be injected sufficiently far upstream of the pump to provide an adequate mixing length.
16. Acid sea water with pH < 4 or water with a noticeable H_2S contamination requires a molybdenum content of 5 to 7% to make the steel sufficiently resistant.
17. The materials discussed should be understood as representatives of their class of materials. The same application limits may be used for materials with similar chemical analysis. Low carbon content (C < 0.03%) is essential in this context.

18. Obviously, material selection is not only influenced by corrosion aspects, but also by many other criteria; some of them are discussed in Chap. 14.4.
19. If $H_{stage} > 400$, note that the endurance limit is drastically reduced in sea water.

14.3 Erosion Corrosion in Demineralized Water

In order to avoid corrosion and deposits, thermal power stations operate with partially demineralized water with life steam pressures up to about 60 bar. Nuclear power plants and boilers operating at pressures of more than 60 bar use fully demineralized de-aerated water which contains only traces of corrosive substances and thus has very low conductivities. In principle this type of water is marginally corrosive. In spite of this, considerable corrosion damage will be experienced if water conditioning or material selection is unsuitable. This is why operators, in practice, have the impression of demineralized water being particularly aggressive. This apparent contradiction is due to the inability of the materials to build protective films in *oxygen-free* water. Material selection, flow velocity and water conditioning must ensure that protective films can form. Three kinds of water conditioning are commonly employed for feeding *high-pressure steam generators*:[7]

- In *alkaline conditioning* pH > 9 is established by adding *volatile* alkaline substances (ammonia). The conductivity[8] is below 0.2 μS/cm and the oxygen content below 10 ppb (1 ppb = 10^{-9} kg/kg). Due to the extremely low oxygen content, the corrosion product on carbon steel is *magnetite* (Fe_3O_4) which grows in the form of tiny crystals on the steel surface. At *low flow velocities* magnetite forms a sufficiently resistant protective film. The color of magnetite is black. As hinted by its name, it is magnetic and therefore can be largely removed from the water by magnetic filters in order to avoid crud formation in steam generators. The large amount of magnetite generated and deposited in the system, and the poor resistance of the protective film to high flow velocities and turbulence are disadvantages of alkaline conditioning.
- In *neutral conditioning* pH = 7 is established. In order to enable the formation of protective films, *oxygen is added* to give a concentration of 50 to 200 ppb. The corrosion product is *hematite* (Fe_2O_3) which has a reddish-brown appearance and builds very resistant (crystalline) protective films. Since an oxidizer (O_2) is present, virtually no other corrosive substances (for example chlorides) are allowed in the feed water. Therefore, the conductivity must be limited to <0.2 μS/cm. It may be seen as a disadvantage of neutral conditioning that the water purity must be rigorously maintained and that the system is intolerant to the ingress of corrosive substances e.g. due to condenser leakage.

---

[7]Water conditioning should follow the actual recommendations in force, e.g. [68].

[8]Measured downstream of a strongly acid cation exchanger.

- In the *combined conditioning*, ammonia is used for alkalization at pH = 8 to 8.5 and *oxygen is added* to give a concentration of 50 to 100 ppb. Thus combined conditioning attempts to minimize the drawbacks of the neutral and the alkaline processes.

Due to its low oxygen content, alkaline conditioning is more tolerant against contaminants (condenser leakage) than neutral or combined conditioning. Low-pressure boilers employ also less expensive methods of water conditioning; frequently phosphate is added for alkalization.

Throughout the world, *alkaline conditioning* is the most frequent process. With this type of conditioning, extensive damage due to erosion corrosion has been experienced, because the magnetite serving as protection film is soluble. Excessive metal loss due to erosion corrosion did not only require the replacement of piping, pre-heaters and pump casings, but also led to severe damage through magnetite deposits in boilers and nuclear steam generators, [13].

In the following, erosion corrosion is considered as a chemical corrosion attack which is caused by *dissolving the protective magnetite film*. Erosion corrosion is thus understood as a mass transfer process (mechanical wear caused by wall shear stresses of the flowing liquid is not involved). The mass transfer coefficients, and consequently the risk of erosion corrosion, increase with the flow velocity and the turbulence. Figure 14.15 shows the wavy aspect typical for damage caused by erosion corrosion.

The metal loss depends in a characteristic way on the temperature and the pH. Its calculation as a mass transfer process is discussed in the following, refer to [14–16]. The calculation applies only to carbon steels and low-alloy steels which form magnetite in *alkaline demineralized water.* High-alloy passivating steels are not susceptible to erosion corrosion. An exception is the erosion caused by leakage through seats which break down high pressure differences.

Consider a flow through a pipe with its hydrodynamic boundary layer according to Fig. 14.16. If the surface consists of a soluble substance (in the present case magnetite), the concentration rises steeply near the wall ("concentration boundary layer").

At low flow velocities, the thickness δ of the magnetite film grows with time according to $\delta \sim t^m$ (m = 0.5 to 0.9). The corrosion rate diminishes with increasing operation time. Under these conditions, the magnetite *formation rate* m_{ox}

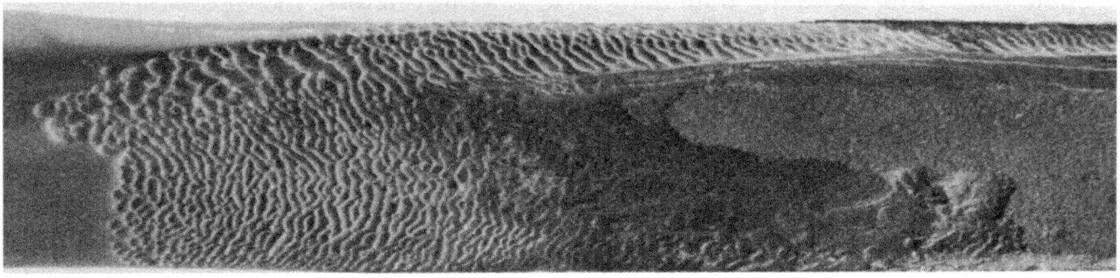

Fig. 14.15 Erosion corrosion on an electrode of an electric boiler, [14]

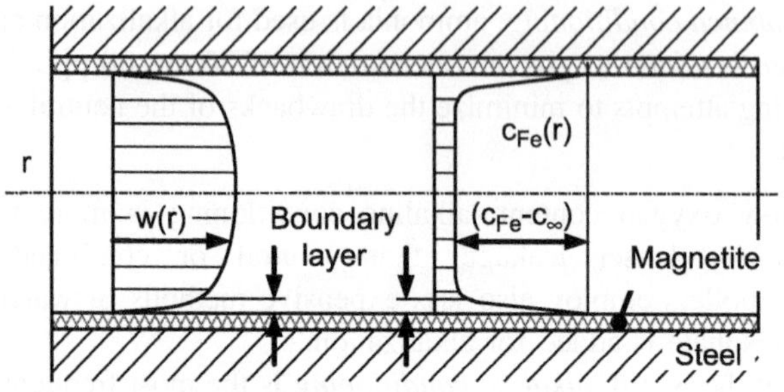

Fig. 14.16 Velocity profile w (r) and concentration profile $c_{Fe}(r)$

at the steel surface exceeds the magnetite *dissolution rate* m_s at the interface between magnetite and water, i.e. $m_{ox} > m_s$. Consequently, saturation in the boundary layer is exceeded and magnetite crystals grow on the surface building a protective film.

With increasing flow velocity (and, consequently, diminishing boundary layer thickness), the layer where magnetite crystals can form out of the over-saturated solution becomes thinner. When a state of equilibrium is reached, the magnetite formation rate (oxidation rate) equals the dissolution rate, i.e. $m_{ox} = m_s$. The specific iron loss rate in kg/(m^{2}s) can then be calculated from the mass transfer process according to Eq. (14.7):

$$m_s = \beta\rho(c_{Fe} - c_{\infty}) \tag{14.7}$$

The mass transfer coefficients β can be calculated for any geometry from published mass or heat transfer correlations of the form $Sh = a \times Re^b \times Sc^c$. If actual mass transfer correlations are missing, the analogy between heat and mass transfer can be used. Mass transfer correlations for various geometries are given in Table 14.9. For situations not listed in Table 14.9, the mass transfer coefficient can be estimated from known situations where similar turbulence may be expected.

Usually the metal loss rate is to be calculated for the location of strongest attack. To do so, the local maximum mass transfer coefficients are required. The maximum values given in Table 14.9 have been derived from heat transfer measurements. These maxima are expected to represent rather the lower limit since they correspond to an average of a certain area on the tested body.

The mass transfer in the boundary layer is controlled by the diffusion coefficient D. Together with other fluid properties needed for the calculation, the self-diffusion coefficient of water is plotted in Fig. 14.17 against the temperature, [38].[9]

The concentration c_{Fe} of the Fe-ions in the boundary layer can be read from Fig. 14.17. It is needed for the calculation of the metal loss rate according to

[9]In strongly diluted solutions (in the present case even applicable to the boundary layer) the diffusion coefficients of the corrosion partners equal the self-diffusion coefficient, [56].

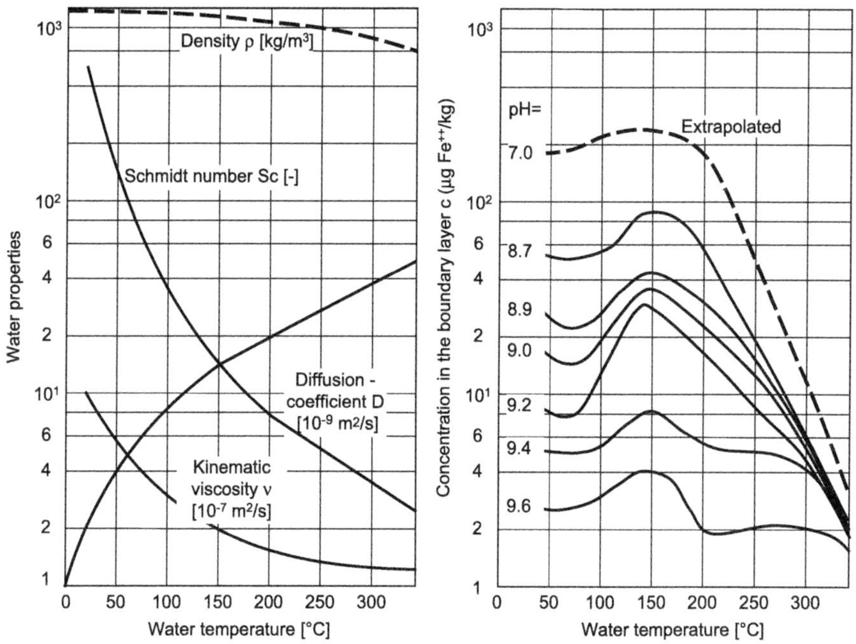

Fig. 14.17 Properties of water and solubility of magnetite, conditioned with ammonia, O_2-content less than 20 to 40 ppb, parameter: pH at 25 °C

Eq. (14.7). Figure 14.17 shows the solubility of magnetite as a function of temperature and pH as per [24] (the curve for pH = 7 is extrapolated). The model for estimating the metal loss implies that magnetite forms at the steel surface and that part of the magnetite is dissolved according to its solubility. In the laminar sublayer of the hydraulic boundary layer, the concentration of Fe^{++}-ions corresponds to the saturated solution.

Figure 14.17 is valid only for carbon steels in alkalinized water with oxygen contents below 20 to 40 ppb. The magnetite solubility with low-alloyed steels and/or oxygen contents above about 100 ppb is significantly lower than that given by Fig. 14.17. This can be derived from operation experience with erosion corrosion.

The Fe^{++}-ion concentration c_∞ in the bulk flow is assumed to be negligible in comparison to the concentration at the Fe_3O_4/H_2O interface, because a self-limitation of the corrosion process through an increased iron concentration in the bulk flow cannot be observed in practice.

As can be inferred from Fig. 14.17 and Eq. (14.7), the metal loss due to erosion corrosion drops strongly if the pH is increased. Accordingly, pH = 9.3 to 9.4 would be beneficial for suppressing erosion corrosion. However, establishing high pH has some disadvantages: (1) high expense for chemicals; (2) The corrosion of copper-containing alloys increases with the pH, so that operating at pH > 9.2 to 9.3 may harbor problems (depending on the materials used in the condenser and pre-heaters).

At any given pH the metal loss rate exhibits a maximum near 150 °C which is confirmed in a similar way from the data published in [32]. In contrast to the empirical model in [32], erosion corrosion can also occur at temperatures significantly above 200 °C as demonstrated by Fig. 14.17. This finding is confirmed by erosion corrosion damage actually observed in boilers.

The risk of damage due to erosion corrosion in a component (e.g. in a pump casing) can be assessed according to Table 14.8 for any set of conditions as specified by temperature, pH, geometry and flow velocity. Equations (T14.7.2 to 14.7.4) yield the metal loss.

The iron loss rates of carbon steel in demineralized water have been calculated according to Table 14.8 for various pH and temperatures and plotted in Fig. 14.18 as a function of the flow velocity. The mass transfer coefficient was calculated for the inlet to a pipe of 200 mm diameter, but the graphs can also be used for a rough assessment of other applications.

In Eq. (T14.7.2) the factor of 1.7 was introduced which results from the comparison of erosion rates calculated from Table 14.8 and those actually measured in power stations, [15, 16]. To this end, 21 erosion corrosion incidents in power plants and laboratory test data were evaluated according to Table 14.8 as reported in [16]. In [15] test results from [30] were analyzed according to the same procedure. Overall the investigated range covers the following parameters: mass transfer coefficients $\beta = 0.0013$ to 0.035 m/s, operation time 200 to 63,000 h, concentration difference $c_{Fe} - c_{\infty} = 5$ to 100 µg Fe^{++}/kg, pH = 8.5 to 10, temperature 50 to 287 °C, geometries: rotating disks, orifices, pipe bends, valves and pump casings. From these comparisons it can be concluded that the calculation would have been able to predict the damage in all cases where a significant metal loss was observed.

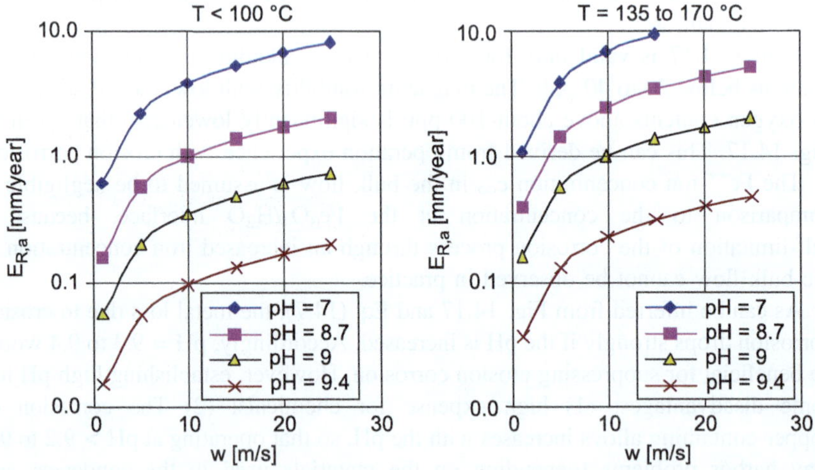

Fig. 14.18 Erosion corrosion of carbon steels in demineralized water, oxygen content <40 ppb

Table 14.7 Metal loss due to erosion corrosion in demineralized water			
Prerequisites	• Demineralized water • O_2-content smaller than 20 to 40 ppb • No corrosive substances (except traces) • Carbon steel or low-alloy steel		
Properties	Diffusion coefficient, Schmidt number, density, viscosity from Fig. 14.17		
	Solubility of magnetite $c_{Fe} = f(T, pH)$ from Fig. 14.17		
Geometry	Select mass transfer correlation from Table 14.9		
Calculate maximum local mass transfer coefficient	Reynolds number	$Re = \frac{wX}{\nu}$	
	Sherwood number from selected mass transfer correlation		Eq.
	$\beta_{max} = \left(\frac{\beta_{max}}{\beta}\right)\frac{Sh\, D}{X}$	$\left(\frac{\beta_{max}}{\beta}\right)$ local maximum	14.7.1
Metal loss in $kg/(m^2 s)$	$m_s = 1.7\, \beta_{max}\, \rho\, c_{Fe}\, f_{Cr}$	c_{Fe} in kg Fe^{++}/kg H_2O!; 1 µg Fe^{++}/kg = 10^{-9} kg Fe^{++}/kg	14.7.2
Erosion rate in m/s	$E_R = 1.7 \frac{\rho}{\rho_{Fe}} \beta_{max}\, c_{Fe}\, f_{Cr}$		14.7.3
Erosion rate in mm/year	$E_{R,a} = 3.15\times10^{10}\, E_R$		14.7.4
Material factor	$f_{Cr} = 0.17\left(\frac{Cr_{Ref}}{Cr}\right)^{0.77}$	$Cr_{Ref} = 1\%$	14.7.5

c_{Fe}	[kg Fe^{++}/kg H_2O]	Concentration in boundary layer
c_∞	[kg Fe^{++}/kg H_2O]	Concentration in bulk flow ($c_\infty = 0$)
D	[m^2/s]	Diffusion coefficient
f_{Cr}	[-]	Material factor for chromium
m_s	[kg/m^2s]	Metal loss due to mass transfer
Sc	[-]	Schmidt number
Sh	[-]	Sherwood number
X	[m]	Characteristic length
β	[m/s]	Mass transfer coefficient
ρ_{Fe}	[kg/m^3]	Density of iron
ρ	[kg/m^3]	Density of water

Material factor	
Cr (%)	f_{Cr}
< 0.1	1.0
0.1< Cr <12	Eq. 14.7.5
or from Fig. 14.19	
> 12	0

In Table 14.8 the factor f_{Cr} was introduced which covers the influence of the chromium content of the steel. This for the reason that low-alloy steels with 0.7 to 3% chromium experience significantly lower corrosion rates than carbon steel. The large influence of the chromium content on the metal loss is demonstrated by the test results plotted in Fig. 14.19. According to these measurements, very low additions of chromium are apt to drastically reduce the metal loss due to erosion corrosion. Figure 14.19 also shows that the band width of the erosion rates can cover a factor of 30. Building and dissolving the protective magnetite film consequently react very sensitively to nuances in material properties, manufacturing of the specimens as well

Table 14.8 Mass transfer correlations

Geometry	Correlation	Range of validity	X
Rotating disk 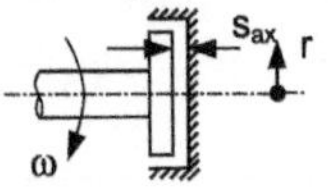	$Sh = 0.017\ Re^{0.8}\ Sc^{0.33}$	$Re > 2.5\times10^{5}$	Radius r $X = r$
	$Sh = 0.05\ Re^{0.78}\ Sc^{0.33}$	$Re > 6\times10^{5}$	
	Valid for separated boundary layers	$\delta = \dfrac{0.0174\,r}{Re^{0.139}} < \dfrac{s_{ax}}{2}$	$w = r\times\omega$
Flat plate in parallel flow 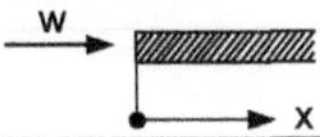	$Sh = a\ Re^{0.8}\ Sc^{0.33}$ Mass transfer: local: $a = 0.029$ average: $a = 0.037$	$Re > 5\times10^{5}$	$X = x$
Pipes, channels, gaps 	$Sh = 0.027\ Re^{0.8}\ Sc^{0.33}\left\{1+\left(\dfrac{D_t}{L}\right)^{\frac{2}{3}}\right\}$ At pipe inlet and for bends: $\dfrac{\beta_{max}}{\beta} = 2$ to 2.5	$Re > 10^{4}$ $\dfrac{L}{D_t} > 60$	pipes: D_t gaps: $X = 2s$ s = gap width
Sudden expansion, maximum local mass transfer 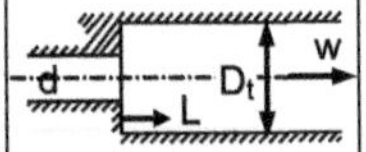	$Sh_{max} = 0.27\ Re^{0.67}\ Sc^{0.33}\left(\dfrac{D_t}{d}\right)^{0,67}$ Location of highest metal loss: $\dfrac{L}{D_t} = 4\left(1-\dfrac{d}{D_t}\right)$	$4000 < Re < 1.5\times10^{5}$ $0.7 < Sc < 6$ $0.06 < \dfrac{d}{D_t}\ 0.94$	$X = D_t$
Cylinder in cross flow 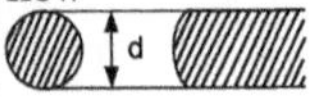	$Sh = (0.52\ Re^{0.5} + 0.00145\ Re)\ Sc^{0.33}$ Maximum local mass transfer: $\dfrac{\beta_{max}}{\beta} = 1.5$ to 2.5	$10^{3} < Re < 2\times10^{6}$	$X = d$
Plate in cross flow 	$Sh = 0.23\ Re^{0.731}\ Sc^{0.33}$ Also for a jet impinging on a surface (e.g. diffuser outlet flow impinging on casing)	$4000 < Re < 1.5\times10^{4}$	$X = \dfrac{2a}{\pi}$
Two-phase flow	$\dfrac{\beta_{TP}}{\beta_{SPL}} = 3.5\left(\dfrac{x}{1-x}\right)^{0.45}\left(\dfrac{\rho'}{\rho''}\right)^{0.25}\left(\dfrac{\mu''}{\mu'}\right)^{0.05}$	x = vapor content (mass fraction) defined by Eq. (T13.3.4)	
Definitions	$Re = \dfrac{w\,X}{\nu}$; $Sh = \dfrac{\beta\,X}{D}$; $Sc = \dfrac{\nu}{D}$	$Pr = \dfrac{c_p\,\rho\,\nu}{\lambda}$	$Nu = \dfrac{\alpha\,X}{\lambda}$
X = characteristic length The mass transfer correlations can also be used to calculate heat transfer coefficients. The Sherwood number is replaced by the Nusselt number and the Schmidt number by the Prandtl number.		Pr Prandtl number Nu Nusselt number α heat transfer coefficient c_p specific heat λ thermal conductivity	

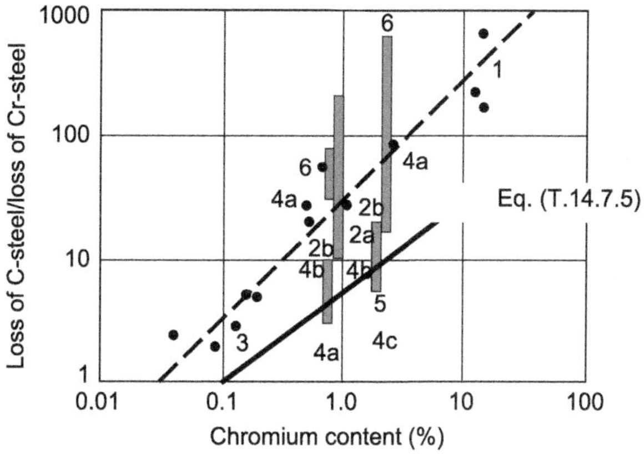

Fig. 14.19 Influence of chromium content on the metal loss in erosion corrosion; the graph corresponds to $1/f_{Cr}$

as chemical and hydraulic test or operation conditions. When selecting materials it is prudent to consider low-alloy steels with less than 0.1% chromium to be nearly as sensitive as carbon steel. As a rough guideline it can be assumed that the metal loss of steels with about 2% chromium content is only ¼ of the corrosion on carbon steel, [32].[10] In plant operation some margin is needed for pH-excursions and changing operation conditions. Equation (T14.7.5) provides that margin by using the lower range of improvement derived from the test data in Fig. 14.19.

The calculation according to Table 14.8 yields finite metal loss rates even at low flow velocities. Even during operation there is always a minimum corrosion rate which may be in the order of 0.01 to 0.05 mm/year depending on material, temperature and water conditioning. Only when the loss is appreciably above that level can we speak of erosion corrosion.

The risk of erosion corrosion can be assessed according to the following criteria:

- $E_{R,a} < 0.05$ mm/year: Erosion corrosion unlikely.
- $E_{R,a} = 0.1$ to 0.2 mm/year: Slight erosion corrosion likely; the plant becomes sensitive to pH-excursions.
- $E_{R,a} = 0.4$ to 0.8 mm/year: Probability of damage greater than 90%.
- $E_{R,a} > 1.2$ mm/year: Strong metal loss to be expected.
- As a general rule: The higher the predicted metal loss, the greater is the probability of experiencing erosion corrosion and the shorter is the incubation time.

Erosion corrosion is influenced by these additional parameters:

Potential differences: If more noble materials, which are resistant to erosion corrosion (e.g. stainless steels or copper alloys), are in contact with carbon steel, slight potential differences are created in an electrolyte (refer to galvanic corrosion

[10]This value appears more realistic for long-term operation than the reduction ratios between 1:10 and 1:100 measured in short-term tests reported in [22]. Results from short-term tests can be misleading because of the effects of incubation time and the accuracy of measuring small corrosion rates.

discussed above). As demonstrated by plant experience, an increase in metal loss due to erosion corrosion is expected on the carbon steel components if they are exposed to high flow velocities. This situation is encountered for example where a stainless steel bushing is mounted in a carbon steel casing or at the transition from a nobler weld-coating to a less noble base material. When weld-coating a carbon steel component with stainless steel, the transition must be placed in a location where flow velocities and turbulences are sufficiently low.

Oxygen content: The solubility given in Fig. 14.17 is applicable if the oxygen content is below 20 to 40 ppb. The calculation would over-predict the metal loss if applied to conditions with oxygen contents appreciably above that range, because the dissolution of the protective film can no longer be described by the concentrations differences given in Fig. 14.17. The addition of oxygen to water with low conductivity (less than 0.2 μS/cm) leads to more stable protective films according to various publications, e.g. [6]. The erosion rate is strongly reduced if the oxygen content exceeds 50 ppb in water with conductivities below 0.2 μS/cm, [31]. The influence of the pH on the corrosion rate is negligible at oxygen contents above 200 ppb, [6].

It should be noted that the oxygen content may vary throughout the plant because oxygen is consumed by corrosion processes in various components.

Alloy additions: The addition of molybdenum or manganese is claimed to reduce the risk of erosion corrosion. In [31] for example, the sum of chromium and molybdenum content is used to correlate the metal loss rates of low-alloyed steels. How-ever, it can be misleading to apply the results of short-term tests (gathered in test loops with small liquid inventories) to situations in real plant operation because the water chemistry is difficult to control correctly in a small test loop. Furthermore, the measuring of small metal loss rates is quite uncertain because of the influence of the incubation time. Marginally resistant materials have no margin to cope with changing operation parameters (pH, temperature, velocity and turbulence) and cannot be recommended.

The present procedure for estimating the metal loss rate is semi-empirical. The complex phenomena evolving on the interfaces metal/oxide and oxide/water are simplified by the model used. But it can hardly be considered as a coincidence that the model is able to describe very well the characteristic behavior of the metal loss as a function of temperature and pH, [16, 22, 30, 32]. The above ideas and concepts developed in [14–16] correspond to the discussion on magnetite films in steam pipes presented in [37].

14.4 Material Selection and Allowable Flow Velocities

The maximum allowable flow velocity of any given *material* in a specific *medium* at given *temperature* is determined by the criteria corrosion, stress levels, fatigue, cavitation and possibly abrasion if the liquid contains solid particles. From the wealth of metallic materials used in pumps, only a few examples can be discussed here. Each of the materials selected is representative for a specific class of materials with similar chemical composition. It has been attempted to select the most commonly used materials which are able to cover a wide spectrum of applications. Section 14.4 is structured in the following way:

- Definition of commonly encountered water chemistries
- Discussion of the properties of frequently used metallic materials
- Material selection for impellers, diffusers and casings
- Materials for wear rings
- Materials for shafts
- Materials for specific applications.

14.4.1 Definition of Frequently Encountered Fluids

As mentioned above the *exact water analysis of each individual application* needs to be known in order to enable a judicious choice of the materials to be used. In Tables 14.6 and 14.13, application limits of various materials are given for typical water qualities which are defined by "W1" to "W6":

W1 Weakly corrosive water comprising

- Water with pH above the limit given by Eqs. (14.3 and 14.4) where CO_2 is in equilibrium with carbonates.
- Water which is able to form protective films on carbon steel; according to DIN 50 930 this is water with:

  $O_2 > 3$ ppm (not below 2 ppm)
  $7 < pH < 8.5$ (but as high as possible)
  $c(Ca^{++}) > 0.5$ mol/m^3 (=mmol/l) respectively °dH > 2.8
  Chlorides: Cl < 2 ppm.

- Water conditioned with effective corrosion inhibitors.
- Corrosion-chemically neutral drinking water.
- No protective film can form if the water is virtually free from oxygen. The corrosion rate is controlled by the residual oxygen. The metal loss is negligible if $O_2 < 0.1$ ppm *and* pH > 8.5. This applies also to hot water (e.g. warm-water heating systems). Partially or fully demineralized water does not belong to this group but to W5 and W6.
- Presumably, resistant protective films can form only above a flow velocity of about 0.1 m/s. If the velocity exceeds roughly 4 m/s, protective films are increasingly destroyed, [N.12].

The temperature does not exert a significant influence. The oxygen present in the system during filling of closed circuits vanishes through corrosion unless there is ingress of oxygen during service.

W2 Surface water and sewage: Natural water such as fresh water from lakes, rivers or rain usually contains chloride ions in concentrations which can vary in a wide range (10 to 250 ppm). The metal loss depends on the water hardness according to Eq. (14.6). As a rule, these waters are nearly saturated with air. The corrosion attack increases with the availability of oxygen and excess free carbon dioxide dissolved in the water. The mass transfer of O_2 and CO_2 to the metal surface increases with the flow velocity (see also Chap. 14.3). Since contaminants can vary in a wide range, such water should be classified as weakly to moderately aggressive. Sometimes bio-fouling aggravates the situation.

W3 Sea water without H_2S *or* without O_2: All applications with chloride concentrations above 1000 ppm at temperatures below 40 °C fall into this category. This group also includes brackish water, mining water and brine. If the sea water is virtually free of oxygen ($O_2 < 0.01$ ppm), up to 10 ppm H_2S can be present without unduly impairing the corrosion resistance. Within this limit the same materials can be selected for H_2S-containing, oxygen-free sea water as for oxygen containing, H_2S-free water (H_2S and O_2 react and are therefore not simultaneously present in equilibrium water).

W4 Sea water with H_2S or similar contaminants as well as "produced water" coming out of oil reservoirs and brackish water are very aggressive. The attack grows with falling pH and increasing temperature, see Eq. (14.9) and Fig. 14.21.

W5 Partially demineralized water including treated water, e.g. in heating systems. In order to avoid lime deposits, the water circulating in systems at elevated temperature must be softened, Eq. (14.5). However, the protective calcareous rust scales scale is unable to form in softened water, Chap. 14.2.3. In the presence of oxygen and free carbon dioxide, the water becomes aggressive.

W6 Demineralized water conditioned according to Chap. 14.3, as well as distilled water and condensate.

Water in flue gas desulfurization plants (FGD): The special requirements of FGD-pumps will be discussed in Chap. 14.4.7.

Hydrocarbons as a rule are not corrosive at low temperatures, provided they are free of water containing corrosive substances. However, hot hydrocarbons can be corrosive. The same applies to hydrocarbons containing acids or phenol. In principle all materials listed in the tables can be used for pumping hydrocarbons. Nevertheless, it should be checked in each individual application whether the medium to be pumped is free of corrosive substances. The allowable tip speed is primarily limited by vibrations, stress levels, fatigue and temperature; it can be determined according to Tables 14.1 and 14.2. Gray cast iron (often for short "cast iron"), nodular cast iron, unalloyed cast steel and carbon steel can be selected for service temperatures of up to 230 °C. Above this temperature high-alloy cast steels such as GXCrNiMo 12–1 (1.4008) are employed.

14.4.2 Metallic Pump Materials

Metallic materials are employed for the majority of pumps. Plastics are used extensively for small pumps. Ceramics are selected for special applications. Mainly metallic materials are discussed in the following.

Cast iron of various types contains carbon of $\geq 3\%$ which leads to graphite precipitation at the grain boundaries. Therefore the microstructure of the casting is inhomogeneous. Consequently, *all* types of cast iron exhibit little resistance to cavitation and abrasion. Their mechanical properties severely limit the allowable tip speed.

In gray cast iron (GG-18–GG-50), the graphite precipitations are lamellar in nature which makes the casting very brittle (the elongation is nearly zero). Sudden catastrophic failure of pump casings cannot be excluded with brittle materials. Therefore, lamellar cast iron must not be used for pump casings where safety is an issue.

In contrast, graphite precipitation in nodular cast iron (GGG-40–GGG-50) generates spherical grains. The casting characteristics are similar to steel. Unalloyed grey and nodular cast iron behaves similarly to carbon steels as far as corrosion is concerned (an exception are special media, e.g. concentrated sulfuric acid).

Due to its low hardness nodular cast iron (GGG-40) is less resistant to erosion corrosion than gray cast iron (GG-25), [20]. Austenitic cast iron with about 15 to 20% nickel (trade name "Ni-Resist") is used in sea water applications.

All types of cast iron mentioned above are relatively inexpensive because they are easy to cast and to machine. Unfortunately, weld repairs are nearly impossible. This is a severe disadvantage in manufacturing or when local wear occurs during service, such as cavitation damage or erosion at the cutwater or cavitation in inlet casings due to partload recirculation.

Carbon steels (also called "mild steels" or "unalloyed steels") are stable only if a protective film is formed, Chap. 14.2.1, or if the water is free from oxygen and corrosive substances. However, these prerequisites are insufficient with *demineralized* water at high flow velocities, Chap. 14.3.

Low-alloyed steels contain for example chromium, molybdenum and/or manganese, in order to increase high-temperature strength or resistance to erosion corrosion. The contents of alloying elements are usually below 5%.

Stainless steels include martensitic, austenitic, duplex, fully-austenitic (or super-austenitic) steels. In all steels, ultimate tensile strength and yield strength increase with the carbon content owing to the formation of carbides, while the ductility decreases. Since the corrosion resistance drops with rising carbon content, stainless steels are manufactured with very little carbon, mostly below $C < 0.07\%$. If high corrosion resistance is required, carbon is even limited to $C < 0.03\%$. State-of-the-art foundry technology is able to produce steels with low carbon. The steels listed in Tables 14.12–14.17 have been preferred because of their low carbon content compared to otherwise similar alloys with higher carbon.

All stainless steels are passivating and not susceptible to uniform corrosion. Passivation requires a chromium content of at least 12 to 13%. However, stainless steels are sensitive to local corrosion (pitting and crevice corrosion) in the presence of chloride (generally all halogen) or sulfide ions in the water. The sensitivity to local corrosion is influenced by several parameters:

- The risk increases with the concentration of halogen ions (mostly Cl-ions), the sulfide concentration (H_2S) and the temperature. As the pH is reduced, the sensitivity rises strongly.
- Steels without molybdenum should be used only up to Cl $\approx$ 200 ppm in order to avoid crevice corrosion, [8] and [39].
- Stagnating water, gas bubbles and deposits enable a local concentration of corrosive substances and foster local corrosion since the passivation film may be locally destroyed.
- Local corrosion is often the starting point of stress corrosion cracking (SCC) in the presence of tensile stresses, which may be caused by the loading applied during service or by thermal and residual stresses.
- The resistance against chloride induced local corrosion increases with the content of chromium, molybdenum, copper, tungsten and nitrogen. The pitting index PI, defined by Eq. (14.8), is an often applied criterion to assess the risk of local corrosion.

$$PI = Cr - 14.5C + 3.3Mo + 2Cu + 2W + 16N \quad [Cr, C, Mo, etc.\ in\%] \qquad (14.8)$$

- The pitting index applies to high-alloy steels only; it must not be used for nickel base alloys. The resistance against local corrosion and SCC increases with the value of PI. When applying the pitting index as criterion for material selection, it should be noted that different definitions of PI are in use, which yield different numbers for a given material. The definition as per Eq. (14.8) from [N.6] is used in the present text, and the PI is calculated with the *average content* of the components of the alloy.[11]
- Since 1% carbon content can transform 16% chromium to Cr-carbide, a low carbon content (C < 0.03%) or stabilization of carbon with titanium or niobium is required in order to improve the resistance against intergranular corrosion as well as weldability. The required Cr-content can be estimated from: Cr [%] = 10 × C[%] + 12. With higher carbon content, there is a risk of precipitations in the heat-affected zone during welding, which make the component susceptible to corrosion.
- The nickel content fosters re-passivation when the passivation film has been damaged. Nickel also improves machinability and weldability. Manganese has a negative impact on the corrosion resistance.

[11]For example Table 14.12 specifies for 1.4409 Cr = 18 to 20%. Consequently PI is calculated with Cr = 19%.

- In highly corrosive environments the microstructure must by all means be free from precipitations, in particular at the wetted surfaces of a component. Since the skin of a casting is usually less resistant than the microstructure below, it must be removed if the corrosion is very aggressive. The quality of the material is then checked by metallographic sections with high resolution (scale 1000:1).
- Chloride-induced SCC can be avoided by selecting duplex steels or nickel alloys containing more than 25% nickel, refer to Fig. 14.9.
- All stainless steels are subject to local corrosion in chloride-containing electrolytes when a *critical temperature is exceeded.* Then local corrosion will be visible already after 24 h of exposure. The critical temperature T_{crit}, which increases with the pitting index, can be determined by measuring the electro-chemical corrosion potential. The criterion is the steep rise of the corrosion velocity when T_{crit} is exceeded. No general rules can be given for the domain where the material is stable. This domain depends on the specific alloy, any impurities in the alloy, the heat treatment of the individual casting, the chemical composition of the electrolyte, pH and the temperature. Therefore, every case must be assessed individually. The sensitivity to apparently secondary influences may explain why sometimes contradictory information is found in the literature or is obtained from experts. The reports are valid for the investigated conditions but cannot be considered as generally applicable.
- Standard tests are used to rank different materials in terms of their critical pitting temperature CPT or their critical crevice temperature CCT (wich is considerably lower than the CPT). While such tests are useful to compare different materials with respect to their resistance to local corrosion, they cannot say anything about the stability of any given material in a specific corrosive environment (e.g. in sea water with a large amount of H_2S). The ferric chloride solution test ASTM G-48 has been used for ranking austenitic, duplex, super-duplex and super-austentic steels with pitting index PI in the range of 27 to 56, test data in [75]. The results can be correlated by: CPT = 2.4(PI-23) with CPT in °C.
- Since molybdenum fosters the creation of ferrite, its content must be limited in austenitic and duplex steels. However, resistance against local corrosion increases with Mo-content. Therefore, sea water pumps operating at high temperatures and high chloride and H_2S contents require steels with Mo up to 6 to 8%.
- Nitrogen increases not only the resistance to local corrosion but also the tensile strength and ductility.
- Corrosion fatigue is fostered by chlorides, but it can also occur in water which is free of chlorides.

Martensitic steels with nickel contents up to 5% exhibit high strengths in the range of $700 < R_m < 1200$ N/mm^2, depending on the composition of the alloy and the heat treatment. Martensitic steels are unsuitable for sea water because of their sensitivity to local corrosion and erosion corrosion. In a corrosive environment martensitic steels are prone to stress corrosion cracking [23]. Under alternating stresses, there is a high risk of corrosion fatigue. Martensitic steels are also susceptible to hydrogen embrittlement. In the presence of boric acid, martensitic steels are sensitive to chloride corrosion, even at chloride concentrations as low as 1 ppm (depending on the fluid temperature), [23]. In cold water of the types "W1" and "W2", martensitic steels may be used up to between 20 and 50 ppm chlorides, depending on additional corrosive substances present.

Martensitic steels are very well suited for impellers, diffusers and casings of pumps operating with fresh water or demineralized water. In particular the steel 1.4317 (formerly 1.4313) GX4CrNi13 4 (ASTM A743 Grade CA 6NM) can almost be considered as standard material for highly loaded pumps and water turbines - not in the least because of its good cavitation resistance.

Austenitic Steels with nickel contents of at least 8% (type 1.4409) exhibit considerably lower tensile strength ($400 < R_m < 650$ N/mm^2) than martensitic or duplex steels. The low strength of austenitic steels often limits their use because the allowable head, fatigue strength and resistance to cavitation and abrasion are too low for high-head applications. While the resistance of austenitic steels against uniform corrosion and erosion corrosion is good, they are sensitive to local corrosion, intergranular corrosion and stress corrosion cracking (SCC). At high temperatures and in aggressive environments, stress levels as low 10% of the yield strength are able to initiate SCC. The sensitivity to local, intergranular and stress-induced types of corrosion depends on the type of steel and the parameters discussed above. It diminishes with decreasing carbon content, which should be limited to $C < 0.03\%$.

Austenitic steels with at least 4.5% molybdenum are resistant against pitting in cold water but remain susceptible to crevice corrosion in critical cases.

Duplex steels (with $600 < R_m < 800$ N/mm^2) have an austenitic-ferritic microstructure. They are more resistant to local corrosion than austenitic steels and nearly insensitive to SCC and intergranular corrosion. Duplex steels are selected when strength, abrasion and cavitation resistance of austenitic steels are not sufficient. In a highly corrosive environment, the resistance of duplex steels strongly depends on small differences in alloy composition, foundry processes and *heat treatment*. Improper heat treatment can lead to the precipitation of the brittle, intermetallic "sigma phase" which diminishes the local chromium content and thus fosters local corrosion, Fig. 14.20. The sigma phase clearly visible in Fig. 14.20 was one of the factors responsible for the local corrosion on 1.4460 shown in Fig. 14.8.

The sensitivity to heat treatment may be due to the fact that the austenitic and ferritic phases are chemically different. The austenitic phase contains less chromium and molybdenum, and might form a weak point in the corrosion resistance. In components susceptible to corrosion fatigue failure, local accumulations of the

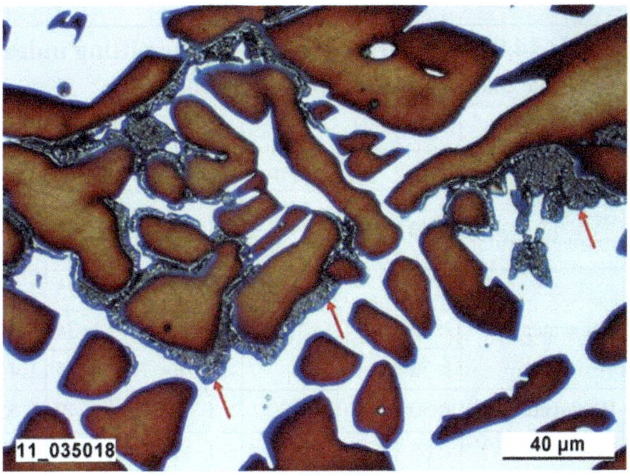

Fig. 14.20 Precipitation of sigma phase (*black-gray mottled areas*) in duplex steel 1.4460; brown areas are austenitic, white areas ferritic, [10]

austenitic phase might initiate cracks. Corrosion fatigue strength may also be impaired by residual stresses. These can be quite elevated in duplex steels because austenite and ferrite have different coefficients of thermal expansion.

At high temperatures duplex steels abruptly loose their corrosion resistance and the endurance limit drops drastically [71]. The application limit of duplex steels strongly depends on the material and the medium and is therefore debated. Therefore, applications above 100 to 150 °C need careful examination. At temperatures above 450 °C, phase transformations can impair the strength as well as the corrosion resistance.

Recommendations and criteria for the selection of duplex steels:

- Carbon content: C < 0.03%.
- Ultimate elongation: A > 20%.
- The ferritic component of duplex steels should amount to 40 to 60% (preferably 50% ferrite and 50% austenite). Mechanical and corrosion properties are optimum with this composition.
- The chromium content should be 24 to 27%, the molybdenum content at least 2.5% and the nitrogen content 0.1 to 0.2%. When pumping brackish water, in the presence of sulfides or in FGD applications, the steel should contain 3 to 4% copper.
- The pitting index as per Eq. (14.8) provides an essential (but not universal) criterion for material selection as demonstrated by Table 14.10 and Fig. 14.21. Duplex steels with a pitting index PI > 40, are called "super-duplex".
- Since cracks in the material can start from corrosion pitting (Fig. 14.10), the pitting index to be selected for a specific pump material should increase with the stress levels expected, hence with the head per stage.

Table 14.9 Steel selection based on the pitting index PI

Cl = ionic chloride concentration ≈ 0.55 × salt content

Fluid	T	pH	Cl	O_2	H_2S	PI	Suitable materials		Literature
	°C		ppm	ppm	ppm				
Sewage, untreated water	< 30	> 6	< 200	> 1	< 0.1	> 20	1.4309	without Mo	8, 39
			< 1000			> 25	1.4409	with Mo	39
Sea water	< 10	> 6	≈ 2×10^4	> 1	< 0.1	> 25	1.4409		8, 39
	< 25					> 34	1.4470	1.4458	8, 39
	< 50			**< 0.01**	< 10	> 35	1.4458	1.4568	
Produced water	< 50	> 4	≈ 10^5	> 1	< 50	> 40	1.4517	1.4587	4
FGD-water	< 65	> 2.5	≈ 5×10^4	> 1		> 45	1.4471	1.4573	

In order to preclude crevice corrosion, a material with PI > 40 is recommended for sea water applications. The same applies if there are uncertainties about contaminants of the water or if passivation is jeopardized by cavitation or abrasion

Fully austenitic steels ("super-austenitic") with nickel contents of at least 25% (Type 1.4458) exhibit a high resistance to SCC, local corrosion and erosion corrosion. They attain the same strength as austenitic steels and may be selected for applications where the corrosion resistance of duplex steels is insufficient. A steel with higher strength and higher corrosion resistance is the "654SMo", 1.4652, UNS

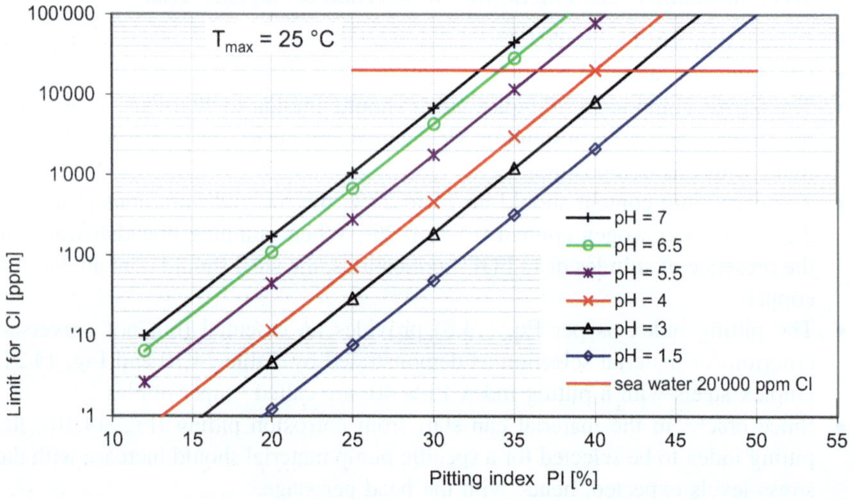

Fig. 14.21 Limits for the chloride content as a function of pitting index and pH; valid for T < 25 °C. The limits are valid for pitting but not for crevice corrosion Example: for a fluid with 10,000 ppm chlorides and pH = 5.5 the alloy should at least have a pitting index of PI ≈ 35

S32654, X0.03CrNiMoN24-22-7 with $R_m = 600$ N/mm^2 and $R_{p02} = 350$ N/mm^2 and A > 40%. With a pitting index of PI = 56 it offers extremely high resistance to local corrosion but is difficult to cast. The grain size is large which could impair the corrosion fatigue strength (no data on this are known). In a test with a rotating disk (42 m/s) metal loss due to cavitation erosion was only about 15% of that of 1.4317 (CA-6MN), [75]. The good cavitation resistance is probably due to the combination of high ductility with high tensile strength which suggests high work hardening capability, Chap. 6.6.2. For any given material the limiting chloride ion concentration can be estimated from Eq. (14.9) or from Figs. 14.21 and 14.22 as a function of the pitting index PI, the pH and the temperature.

$$\mathrm{Cl}\,[\mathrm{ppm}] = 4 \times 10^{-5} e^{(0.4\mathrm{PI} + 0.9\,\mathrm{pH})} \left(\frac{T_{Ref}}{T}\right)^{1.85e^{-0.04\mathrm{PI}}} \quad \text{with} \quad T_{Ref} = 80\,°\mathrm{C} \quad (14.9)$$

The chloride ion concentration of sea water can be estimated from the salt content: chloride-content ≈ 0.55 × salt-content. Sea water with 3.6% salt content contains about 2% chlorides or 20,000 ppm. Equation (14.9) and Fig. 14.21 have been derived from FGD corrosion test data given in [35]. The application limits according to Eq. (14.9) allow a consistent, quantitative assessment and material selection. The factor 4×10^{-5} or the other coefficients could be easily adapted to specific data from operation experience or measurements.

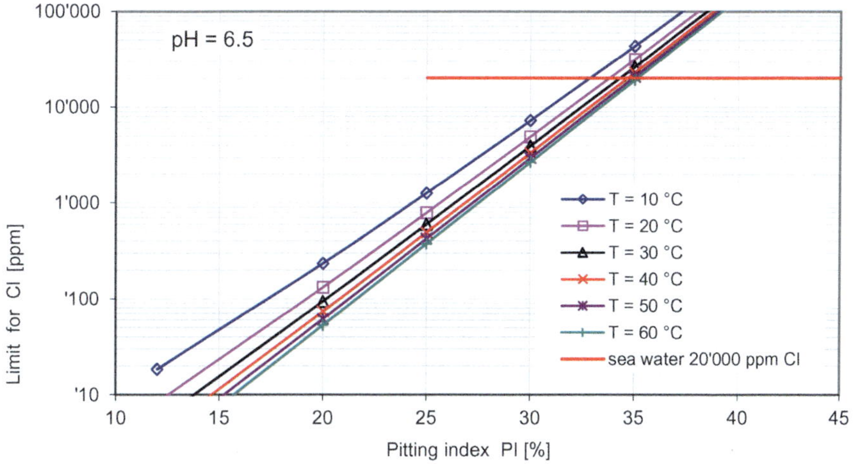

Fig. 14.22 Limits for the chloride content as a function of pitting index and temperature at pH = 6.5. The limits are valid for pitting but not for crevice corrosion

For any specific application as defined by the chloride content, pH and temperature, the required pitting index can be determined from Eq. (14.9) or Fig. 14.21 or Fig. 14.22. If the water contains additional corrosive substances - in particular sulfides, bromides, iodine and fluorides - or abrasive particles, a higher PI will be required.

Copper alloys are commonly designated as bronzes. Aluminum, tin and silicon bronzes (but not brass) are employed in pumps. Bronzes are resistant to uniform corrosion since a well-adhering protective film forms on them. Pitting is possible in cold and hot water. Low pH, gas bubbles, sulfide, chloride and sulfate ions, as well as oxygen increase the sensitivity to pitting. No pitting is to be expected if the oxygen content is below 0.1 ppm. Bronzes must *not* be used in the presence of ammonia, H_2S, inorganic or organic acids (causing uniform corrosion) or if the fluid contains abrasive particles.

Aluminum bronzes of the type G-CuAl-10Ni are marginally sensitive to pitting and are stable even in boiling sea water. The resistance to corrosion increases with the aluminum content. Tensile strengths of up to $R_m = 600\ N/mm^2$ are achieved.

Copper-tin alloys (tin-bronzes), such as G-CuSn10 and copper-tin-zinc alloys such as G-CuSn5ZnPb, are quite stable in sea water at low flow velocities. Due to their low strength, they can only be selected for moderate heads. They are also sensitive to erosion corrosion, cavitation and abrasion.

The corrosion rates of copper alloys increase with the oxygen available at the metal surface. Corrosion is thus determined by mass transfer coefficients (flow velocity and turbulence). The copper-nickel alloys CuNi10, CuNi20, CuNi30, which are used successfully for piping, are not suitable for the high flow velocities encountered in pumps.

Nickel base alloys or titanium may be employed in media where duplex steels or super-austenitic steels are not sufficiently resistant to corrosion. Nickel base alloys must contain chromium and molybdenum if they are to be used in sea water. Inconel 718 has been used in seawater injection pumps with a head per stage of 750 m at an inlet tip speed of $u_1 = 78$ m/s, [7]. The approximate composition is: Cr 19%, Ni 53%, Mo 3%, Co + Ta 5%, Fe balance. Inconel 718 has very good mechanical properties, excellent cavitation resistance (see Table D6.4 and Fig. 6.34) and is likely to have a fatigue resistance in sea water, which is higher than super-duplex; tensile strength in the order of $R_m = 860\ N/mm^2$, $E = 2.08 \times 10^5\ N/mm^2$ and elongation A = 20%.

Hastelloy C22 (alloy 22) with typically >50% Ni, 20% Cr and 14%Mo is used in pumps with different grades of similar composition.

Titanium is insensitive to pitting and very resistant to crevice corrosion. Nickel-chromium-iron alloys *without molybdenum* e.g. NiCr15 (Inconel 600, 2.4816) or X10NiCrAlTi32–21 (Inconel 800, 1.4876) are susceptible to pitting and can not be employed in sea water, [8].

14.4.3 Impellers, Diffusers and Casings

As compared to piping systems, impellers, diffusers and casings of centrifugal pumps are exposed to high flow velocities. The allowable velocities depend on the combination of several parameters so that each application should be assessed individually. This is particularly true when extrapolating operation experience to new conditions. Since the risk of cavitation increases proportionally to w^6 (and abrasion with w^3), an apparently insignificant increase in impeller tip speed can sometimes lead to unexpected problems.

Tables 14.12(1) to 14.11(6) list the properties of commonly used pump materials and Table 14.18 gives tentative limits for flow velocities. Concerning these tables the following should be noted:

- The allowable velocities stipulated in the tables should be considered as very rough, tentative guidelines, which are thought to account for the combined effects of possible damage due to corrosion fatigue, cavitation and corrosion.
- Whether damage occurs to a given material at a given velocity depends to a considerable degree on the design of the pump and the actual operation parameters. For example, the risk of cutwater erosion depends on the width of gap B between the cutwater and the impeller. Also, a well-profiled cutwater would be less prone to damage than a semi-circular shape. Finally, the surface finish of the casting has an impact.
- The recommendations are intended to be rather conservative, but they cannot guarantee freedom of damage. The price difference between a material "just acceptable" for a particular application and a well-resistant material usually is only a fraction of the expense needed to correct operational problems caused by material degradation in service.
- The material selection is done with the head or the velocities at the best efficiency point. Higher velocities at off-design operation are thought to be covered by these recommendations.
- Metal loss due to abrasion or erosion corrosion is determined by the (unknown) maximum local velocities, rather than by average values. The local excess velocities are supposed to be covered by the conservative assumptions of the allowable velocities.
- For assessing stresses, the tip speed u_2 is relevant, while the average velocity should be used for assessing corrosion and abrasion. Mostly an estimate of the average velocities will be sufficient: at the impeller inlet $w_1 \approx u_1$ and at the diffuser inlet $c_3 \approx u_2/2$ may be roughly assumed for $n_q < 50$. An exact calculation of these velocities can be done by means of Tables 3.1 and 3.2 or Fig. 3.25. In the tables "w" stands for the velocity relative to the component (i.e. the absolute velocity for the casing and the relative velocity for the impeller). Since the velocity near the cutwater is of similar magnitude as at the diffuser inlet, the same velocity limits can be used for diffusers and volute casings. Corrosion, abrasion or cavitation at volute cutwater is not really an exceptional occurrence.

- For annular seals the mean velocity vector is used: $w = (c_{ax}^2 + u_{sp}^2/4)^{0.5}$; it can be estimated from Fig. 3.25.
- If the allowable velocity w_{al} for a given material in a given medium has been selected from Tables 14.6 or 14.18, the allowable impeller tip speed and the allowable head can be determined from Fig. 3.25 and Table 14.11. For any given pump, the specific speed n_q can be calculated. Subsequently, the velocities referred to the tip speed u_2 and the pressure coefficient can be read from Fig. 3.25. Finally, the actual values of u_2, w_1, c_{2u} etc. can be calculated from the equations given in Table 14.11.
- In a given pump at a given rotor speed, impellers, diffusers, volutes, stage casings of multistage pumps as well as suction bells of vertical pumps with semi-open impellers are *locally* exposed to roughly the same maximum velocities. Therefore, they must be built from materials of similar resistance. The inlet and outlet casings of multistage pumps or the casings of barrel pumps are exposed to lower velocities. They could be manufactured from other materials if acceptable with respect to erosion corrosion and galvanic corrosion.
- The allowable velocities in annular seals of a specific material in a given environment are lower than for casings or impellers because even a small metal loss would entail appreciably larger clearances and a loss in efficiency (in particular with low n_q).
- The exact composition and properties of any given material depend on the applicable standard and the foundry. The figures given in the tables can only give an indication, since originating from a variety of sources with (at times) contradicting information.
- The application limits, stipulated in the tables for any specific material, are valid for the whole group of similar materials. For example, the limits given for GG-25 apply equally to GG-18 through GG-50. When considering stainless steels, the chemical composition, including the carbon content and the type of steel (i.e. ferrite, austenite, duplex) are relevant for comparing a particular material to the materials listed in the tables. Low concentrations of some elements in an alloy should not be overlooked, since they may have a significant impact on the corrosion resistance (e.g. N, Cu, Nb).
- In so far as quantitative information was available, any material in a specific medium has been termed as "stable" if the metal loss is less than $c_{cor} < 0.1$ mm/year. Materials with $c_{cor} > 1$ mm/year were classified as "instable".
- The hardness numbers should be understood in terms of Vickers-hardness HV30. The differences in hardness resulting from different test loads are irrelevant in the present context. By the same token, Brinell- and Vickers-hardness can be assumed to be equal, and the hardness can be estimated from the tensile strength R_m according to: HV ≈ HB ≈ (0.29 to 0.32) R_m.
- Mechanical properties such as tensile strength and elongation as well as coefficients of thermal expansion, thermal conductivity and specific heat depend on the temperature. The data listed in the tables are valid roughly for the range of 20 to 100 °C. For more accurate calculations, the material standards or the data of the foundry or manufacturer should be consulted.

Table 14.10 Determination of the allowable head per stage

	Read from Fig. 3.25	Calculate allowable tip speed $u_{2,al}$ from allowable velocities w_{al} determined from Tables 14.6 or 14.13
Impeller inlet	$w_1^*(n_q)$	$u_{2,al} = \frac{w_{al}}{w_1^*}$
Impeller outlet	$w_2^*(n_q)$	$u_{2,al} = \frac{w_{al}}{w_2^*}$
Diffuser inlet	$c_3^*(n_q)$	$u_{2,al} = \frac{w_{al}}{c_{2u}^*}$
Volute inlet	$c_3^*(n_q)$	$u_{2,al} = \frac{w_{al}}{c_{2u}^*} d_z^*$
Wear ring	$w_{sp}^*(n_q)$	$u_{2,al} = \frac{w_{al}}{w_{sp}^*}$
Allowable head per stage	$\psi_{opt}(n_q)$	$H_{opt,al} = \psi_{opt} \frac{u_{2,al}^2}{2\,g}$

Depending on each individual case, the application limit of a specific material may be given by fatigue, cavitation, erosion corrosion or abrasion. But there are a number of other aspects to consider when selecting materials. These criteria are discussed in the following:

1. **Corrosion fatigue**: The allowable tip speed or head per stage decreases with increasing corrosivity of the medium and rising specific speed (or blade span). It is much lower for open impellers or diffusers than with closed components. The impeller shroud thickness needs particular attention with blade numbers below $z_{La} < 5$. The analysis of corrosion fatigue problems can be done according to Chap. 14.1, Tables 14.1 to 14.3 and Figs. 14.2, 14.3, 14.4, 14.5, 14.6 and 14.7.
2. **Cavitation**: The erosion risk of impellers can be assessed according to Chaps. 6.6 and 6.7. The inlet casing may suffer cavitation erosion due to partload recirculation and/or cavitation at the outlet of the annular seal. Cavitation can also occur at the volute cutwater or in the diffuser. Although the available NPSH at the impeller outlet appears to be elevated, Eq. (6.9), a non-uniform impeller outlet velocity distribution can produce local excess velocities and cavitation. Due to the high pressure differentials, the cavitation intensity is significant according to Eq. (6.12). Therefore, the risk of local cavitation damage should be taken into account when selecting the casing material. For this reason the allowable velocities stipulated in Table 14.18 are quite low for all types of cast iron (GG-25, GGG-40 and GGG-NiCr 20–2). Due to the gap cavitation likely to be present in open impellers, all types of cast

iron are considered unsuitable for suction bells - at least in all applications where $u_1 > 10$ m/s.
Tables 14.12 to 14.17 give the relative cavitation rate referred to the steel 1.4317 (CA6NM). These figures were calculated according to Chap. 6.6.7 with the same assumptions as for Fig. 6.34, but they are applicable only in the absence of severe corrosion.

3. **Water analysis**: Oxygen content, chemical substances and impurities determine the corrosion risk. *Without knowing the exact properties of the water to be pumped, a sound material selection is risky.*
4. **Erosion corrosion** limits the application of carbon steels when pumping demineralized water. Steel 1.4317 (AISI410, ASTM A 743 Gr CA6NM) can almost be considered as the standard material for these applications.
   Erosion corrosion determines largely the material selection for sea water pumps - in particular in the presence of H_2S.
5. **Acid or base corrosion**: The information given in Chap. 14 is focused on centrifugal pumps for various types of water. Extensive tables giving the resistance of different materials in numerous fluids can be found in [B.5, B.8, B.28 and 52].
6. **Abrasion**: The expected metal loss can be estimated according to Chap. 14.5. The relative abrasion rates referred to steel 1.4317 (CA6NM) may serve as a first assessment. These figures were either derived from test data or calculated from Table 14.22. They apply only in the absence of severe corrosion.
7. **Temperature**: Applications at temperatures considerably below 0 °C require materials which do not loose their ductility. At temperatures above about 200 ° C, the high-temperature strength of the material must be taken into account. Both topics are beyond the scope of this text.
8. **Weldability**: It is highly desirable that the selected material should allow weld repair of casting defaults (or local wear suffered during operation) without loosing corrosion resistance in the weld-affected zone. Usually heat treatment is required after welding. Good weldability is an important criterion for selecting the casing material.
9. **Castability**: Casting defects can lead to damage and increase manufacturing costs if detected during machining or non-destructive material testing.
10. **Machinability**: Some special alloys are very difficult to machine which limits their application and causes extra costs in manufacturing.
11. **Costs**: Tables 14.12 to 14.17 provide the relative costs of the various materials referred to steel 1.4317 (ASTM A743 CA6NM). These figures allow a first assessment of the costs related to selecting a specific material. Obviously, the actual costs evolve with changing market conditions. The cost of the reference material (steel 1.4317) was assumed at 13 Euro/kg as per spring of the year 2007.

Table 14.11 Properties of cast irons and cast carbon steels

			Cast iron	Nodular cast iron	Ni-Resist	Cast steel	
DIN 17006			GG-25 GJL-250	GGG-40 GJS-400-15	GGG-NiCr 20-2	GS-C25 GP240GH	GX8 CrNi 12
Material-no. (DIN, EN)			0.6025	0.7040	0.7660	1.0619	1.4107
ASTM			A278 30	A536-60-40-18	A439 Type D2	A216 Gr WCB	A217 Gr CA-15
Unified numbering system UNS			F 12401			J 03002	J 91150
Chemical composition	C	%	3.4	3	3	0.21	< 0.1
	Cr		-	-	1-2.5	0.3	11.5-12.5
	Ni		-	-	18-22	-	0.8-1.5
	Mo		-	-	-	-	0.2-0.5
	Cu		-	-	<0.5	-	-
	Si		-	-	1.5-3	0.4	< 1
	Mn		-	-	0.5-1.5	0.5-0.8	< 1
Microstructure			Perlitic	Ferritic	Austenitic-Carbide	Bainite	Ferritic
Yield strength	$R_{p0.2}$	N/mm^2	-	250	210-250	240	> 500
Tensile strength	R_m	N/mm^2	245	400	370-480	420-600	> 590
Elongation	A	%	-	> 15	7-20	22	> 16
Young's modulus	E	N/mm^2	1.1×10^5	1.72×10^5	1.2×10^5	2.1×10^5	2.1×10^5
Vickers hardness HV30	HV		180-240	130-180	140-200	130-270	170-240
Density	ρ_{mat}	kg/m^3	7300	7100	7400	7800	7700
Specific heat capacity	c_p	J/kg K	540			460	460
Thermal expansion	α	$10^{-6}/K$	9	12.5	18.7	12.6	10.5
Thermal conductivity	λ	W/m K	48	36	12.6	45	26
Machinability			Very good			Good	Good
Weldability			Very poor	Poor	Poor	Good	Limited
Safety factors for chapter 14.1	S_z		10	4	4	4	2
	S_{zz}		18	10	15	8	5
Cavitation erosion	Relative to 1.4317 (CA6NM)		11	5.5	4.1	4.7	1.5
Loss due to abrasion			1.7	2	1.9	1.8	1.2
Relative costs			0.25	0.35	0.4	0.55	0.8

Table 14.12 Properties of martensitic and austenitic cast steel

Type			Martensitic		Austenitic		
DIN 17006			GX4CrNi 13-4	GX4 CrNiMo 16-5-2	GX2CrNi 19-11	GX2 CrNiMo19-11-2	GX2CrNi MoNb 17-13-4
Material-no. (DIN, EN), [N.14]			1.4317	1.4411	1.4309	1.4409	1.4446
ASTM: A743 Grade			CA 6NM	CB 7Cu2	CF-3	CF-3M	
AISI					304L	316L	
Unified numbering system UNS:			J 91540		J 92500	J 92800	
Chemical composition	C	%	< 0.06	< 0.06	< 0.03	< 0.03	< 0.03
	Cr		12-13.5	15-17	18-20	18-20	16.5-18.5
	Ni		3.5-5	4-6	9-12	9-12	12.5-14.5
	Mo		< 0.7	1.5-2	-	2-2.5	4-4.5
	Cu		-	-	-	-	
	N		-	-	< 0.2	< 0.2	0.12-0.22
	Nb		-	-	-	-	
Microstructure			Ferritic - Martensitic		Austenitic		
Pitting index	PI	Eq.(14.8)	14	21	20	25	34
Yield strength	$R_{p0.2}$	N/mm^2	> 550	> 540	> 185	> 195	> 210
Tensile strength	R_m	N/mm^2	760-960	760-960	440-640	440-640	440-640
Elongation	A	%	> 15	> 15	> 30	> 30	> 20
Young's modulus	E	N/mm^2	2.1×10^5	2.1×10^5	1.93×10^5	1.93×10^5	1.93×10^5
Vickers hardness HV30	HV		240-300	260-320	130-200	130-200	130-180
Density	ρ_{mat}	kg/m^3	7700	7800	7880	7900	7900
Specific heat capacity	c_p	J/kg K	460	460	530	530	530
Thermal expansion	α	$10^{-6}/K$	10.5	11	16.8	15.8	16
Thermal conductivity	λ	W/m K	26	17	15.2	14.5	13.5
Machinability			Good	Good	Good	Good	Good
Weldability			Good	Good	Good (with low C-content)		
Safety factors for Chapter 14.1	S_z		2	2	2	2	2
	S_{zz}		5	5	5	5	5
Cavitation erosion	Relative to 1.4317 (CA6NM)		1.0	0.8	1.5	1.5	1.5
Loss due to abrasion			1.0	1.0	1.3	1.3	1.4
Relative costs			1.0	1.2	1.1	1.5	1.4

Table 14.13 Properties of duplex and super-austenitic cast steel

Type			Duplex steels			Super-Austenitic	
DIN 17006			GX2CrNi MoN 22-5-3	GX2CrNi MoCuN 25-6-3-3	GX2CrNi MoN 25-6-3	GX2 NiCrMo 28-20-2	GX2NiCr MoCuN 29-25-5
Material-no. (DIN, EN) [N.14]			1.4470	1.4517	1.4468	1.4458	1.4587
ASTM: A890 Grade:			CD3MN	4B	3A		
AISI							
Unified numbering system UNS:			J 92205	J 92205	J 93371		
Chemical composition	C	%	< 0.03	< 0.03	<. 0.03	< 0.03	< 0.03
	Cr		21-23	24.5-26.5	24.5-26.5	19-22	24-26
	Ni		4.5-6.5	5-7	5.5-7	26-30	28-30
	Mo		2.5-3.5	2.5-3.5	2.5-3.5	2-2.5	4-5
	Cu		-	2.75-3.5	-	< 2	2-3
	N		0.12-0.2	0.12-0.22	0.12-0.25	< 0.2	0.15-0.25
			-	-		-	-
Microstructure			Ferritic-Austenitic			Super-Austenitic	
Pitting index	PI	Eq.(14.8)	34	44	38	35	48
Yield strength	$R_{p0.2}$	N/mm^2	> 420	> 480	> 480	> 165	> 220
Tensile strength	R_m	N/mm^2	600-800	650-850	650-850	430-630	> 480
Elongation	A	%	> 20	> 22	> 22	> 30	> 30
Young's modulus	E	N/mm^2	2.0×10^5	2.1×10^5	2.1×10^5	1.93×10^5	1.93×10^5
Vickers hardness HV30	HV		180-250	200-270	200-270	130-200	150-220
Density	ρ_{mat}	kg/m^3	7700	7700	7700	8000	8000
Specific heat capacity	c_p	J/kg K	450	450	450	500	500
Thermal expansion	α	$10^{-6}/K$	13	13	13	14.5	14.5
Thermal conductivity	λ	W/m K	18	17	17	16	17
Machinability			Fair	Fair	Fair	Fair	Fair
Weldability			Good	Good	Good	Good	Good
Safety factors for Chapter 14.1	S_z		2	2	2	2	2
	S_{zz}		5	5	5	5	5
Cavitation erosion	Relative to 1.4317 (CA6NM)		1.1	1.0	1.0	1.5	1.3
Loss due to abrasion			1.2	1.1	1.1	1.4	1.3
Relative costs			1.5	2	1.6	2.3	2.6

Table 14.14 Properties of super-duplex cast steel and wear ring material

			Super-duplex steels			Wear ring material
DIN 17006			GX3CrNiMo WCuN 27-6-3-1	GX3CrNiMo CuN 26-6-3-3	GX4CrNiMo CuN 24-6-5-3	GX120CrMo 29-2
Material-no. as per SEW 410 [N.19]			1.4471	1.4515mod	1.4573	1.4138
ASTM: A890 Grade:			6A CD3MWCuN		CE3MN	
Chemical composition	C	%	< 0.03	< 0.03	0.04	0.9-1.3
	Cr		25.5-28	25-26	22-25	27-30
	Ni		5.5-8	6-7.5	4.5-6.5	-
	Mo		3-4	3-3.5	4.5-6	2-2.5
	Cu		0.8-1.3	1	1.5-2.5	-
	N		0.15-0.28	0.17-0.25	0.15-0.25	-
			1 W	1 W		
Microstructure			Ferritic, Austenitic, each 50%	Ferritic, Austenitic	Ferritic, Austenitic	Ferritic, Carbides
Pitting index	PI	Eq. (14.8)	45	43	47	-
Yield strength	$R_{p0.2}$	N/mm^2	> 480	> 480	> 485	-
Tensile strength	R_m	N/mm^2	650-850	650-850	690-890	-
Elongation	A	%	> 22	22	> 22	-
Young's modulus	E	N/mm^2	2.1×10^5	2.1×10^5	2.1×10^5	2.1×10^5
Vickers hardness HV30	HV	Hardened				
		Annealed	200-260	200-260	200-320	260-330
Density	ρ_{mat}	kg/m^3	7700	7800	7800	7700
Specific heat capacity	c_p	J/kg K	450	450	450	500
Thermal expansion	α	10^{-6}/K	13	14	14	9.5
Thermal conductivity	λ	W/m K	17	15	15	19
Machinability			Fair			
Weldability			Good	Good	Limited	No
Cavitation erosion	Relative to 1.4317 (CA6NM)		0.9	1.0	0.9	1.0
Loss due to abrasion			1	1.1	0.9	0.9
Relative costs			2	2	2	
Application	pH		> 2.5	> 4	> 2.5	>7
			Sea water, FGD			Wear rings

Table 14.15 Properties of wear resistant cast iron

			GX300CrNiSi 9-5-2	GX300CrMo 15-3	GX300CrMo 27-1	GX150CrNi Mo 40-6
Material-no. (DIN, EN)			0.9630	0.9635	0.9655	1.4475
			DIN 1695			
ASTM			A532			
Chemical composition	C	DIN 1695	2.5-3.5	2.3-3.1	3-3.5	
	C	reduced	-	2.4-2.6	1.5-1.8	1.4-1.7
	Cr		8-10	14-17	23-28	39.5-42
	Ni		4.5-6.5	< 0.7	< 1.2	5-7
	Mo		< 0.5	1-3	1-2	2-3
	Cu		-	-	-	< 1.2
	N		-	-	-	0.1-0.2
	Si		1.5-3.2	0.2-0.8	0.2-1	-
Microstructure			Chromium carbide in martensitic, perlitic, austenitic structure			Ferritic, Austenitic, Carbide, each 33%
Pitting index	PI	Eq. (14.8)	-	-	-	31
Tensile strength	R_m	N/mm^2	500-600	450-1000	450-1000	
Elongation	A	%	-	-	-	
Young's modulus	E	N/mm^2	1.96×10^5	1.72×10^5	1.72×10^5	
Vickers hardness HV30	HV	Hardened	600-750	700-900	600-800	
		Annealed	Natural hard	< 400	< 400	
Density	ρ_{mat}	kg/m^3	7700	7700	7600	
Thermal expansion	α	$10^{-6}/K$	14.5	13		
Thermal conductivity	λ	W/m K	13.8	13.8		
Machinability			Very poor	With reduced C-content after annealing good		
Weldability						
Loss due to abrasion	Relative to 1.4317 (CA6NM)		0.08	0.04	0.07	0.1
Relative costs			1.1	1.1	1.25	1.65
Application	Fluid		Neutral with respect to corrosion		Weakly corrosive	Fairly corrosive, Eq. (14.9)
	Service		Slurry and dredge pumps	Bauxite, ore, or carbon slurries; sand laden sewage		

Table 14.16 Properties of copper and nickel alloys

Type				**Tin bronze**	**Inoxida**	**Monel K500**	**Hastelloy C**
DIN 17006			G-CuSn 5ZnPb	G-CuSn 10	G-CuAl 10Ni	NiCu30Al	NiMo16Cr 15W
Material-no. (DIN, EN)			2.1096.01	2.1050.01	2.0975.01	2.4374	2.4819
ASTM:			B584 C83600	B427 C90700	B148 958		A574 N10276
Unified numbering system UNS:			C83600	C90700	C95800	N05500	N10276
Pitting index	PI	Eq. (14.8)	-	-	-	-	70
Chemical composition	Cu	%	84-86	88-90	76	30	Cr 15-17
	Ni		2.5	2	4-6.5	Rest	51-64
	Al		-	-	8.5-11	2-4	Mo 15-17
	Sn		4-6	9-11	-	-	W 3-4.5
	Zn		4-6	-	-	-	-
	Fe		-	-	3.5-5.5	0.5-2	4-7
	Pb		4-6	-	-	-	C < 0.015
	Mn		3	3	3	-	1
Yield strength	$R_{p0.2}$	N/mm^2	90	130	270	590	280
Tensile strength	R_m	N/mm^2	220	270	600	600-880	700
Elongation	A	%	16	18	12	12	35
Young's modulus	E	N/mm^2	0.85×10^5	1.0×10^5	1.2×10^5	1.79×10^5	
Vickers hardness HV30	HV		60	80	140	170-230	
Density	ρ_{mat}	kg/m^3	8700	8700	7600	8500	8900
Specific heat capacity	c_p	J/kg K		400	≈400	≈525	
Thermal expansion	α	$10^{-6}/K$	17	20	≈17	≈15	
Thermal conductivity	λ	W/m K	58	45	≈50		
Weldability					Very good		
Safety factors for Chapter 14.1	S_z		4	4	4	4	2
	S_{zz}		10	10	10	10	5
Cavitation erosion	Relative to 1.4317 (CA6NM)		5 ?	4.5 ?	1 ?	1 ?	1 ?
Loss due to abrasion			2.5	2.2	1.6	1.3	1.2
Relative costs			1.0	0.9	1.2	5.0	1.8

Table 14.17 Velocity limits for cast materials $H_{al} \approx 0.2 \times w_{al}^2$ at $\psi_{opt} \approx 1$

DIN 17006			GG-25, EN-GJL-250	GGG-40, EN-GJS-400-15	GGG-NiCr 20 2	GS-C25, GP240GH	GX8CrNi12	GX4CrNi 13-4	GX4CrNiMo 16-5-2	GX2CrNi 19-11	GX2CrNiMo 19-11-2	GX2CrNiMoNb 17-13-4	GX2CrNiMoN 22-5-3	GX2CrNiMoCuN 25-6-3-3	GX2NiCrMoCuN 29-25-5	G-CuSn 10	NiCu30Al	G-CuAl 10Ni
Material-number			0.3025	0.7040	0.7660	1.0619	1.4107	1.4317	1.4411	1.4309	1.4409	1.4446	1.4470	1.4517	1.4587	2.1050.301	2.4374	2.0975.01
UNS number			F 12401			J 03002	J 91150	J 91540		J 92500	J 92800		J 92205	J92205				
ASTM			A 278 Class 30	A 536-60-40-18	A 439 Type D2	A216 Gr WCB	A 217 Gr CA-15	A 743 Gr CA6NM	A 743 Gr CB 7Cu2	A 743 Gr CF-3	A 743 Gr CF-3M		A 890 Gr CD3MN	A 890 Gr 4A				
Pitting index		PI	-	-	-	-	-	14	21	20	25	34	34	44	48	-	-	-
Impeller, diffuser, volute	W1 forming protective films	Qualific.	u	u	u	g	e	e	e	e	e	e	e	e	e	g	e	e
		w_{al} (m/s)	17	20	20	30	45	70	70	50	50	50	60	60	50	22	50	45
	W2 untreated water, sewage	Qualific.	u	u	u	u	e	e	e	e	e	e	e	e	e	g	e	e
		w_{al} (m/s)	15	17	17	25	45	70	70	50	50	50	60	60	50	22	50	45
	W5 & W6 Demineralized water	Qualific.	0	0	0	0	e	e	e	e	e	e	e	e	e	g	g	g
		w_{al} (m/s)	0	0	0	0	45	70	70	50	50	50	60	60	50	22	50	45
Wear rings	W1 forming protective films	Qualific.	u	u	u	u										g		
		w_{al} (m/s)	12	15	15	17										15		
	W2 untreated water, sewage	Qualific.	u	u	u	u										g		
		w_{al} (m/s)	10	12	12	15										15		
	W5 & W6 demineralized water	Qualific.	0	0	0	0										g		
		w_{al} (m/s)	0	0	0	0										15		

Qualification: 0 = not appropriate, not recommended; u = usable ; g = good; e = excellent

Note: w_{al} is the allowable velocity relative to the component (not the tip speed u_2, which is roughly twice as large for $n_q < 50$)

14.4.4 Wear Ring Materials

As explained in Chap. 3.6.2, the radial clearances in annular seals at the impeller or the balance piston are only a few tenths of a millimeter in order to limit the efficiency impairment due to leakages, Eq. (3.12). There is a considerable risk of the rotor touching the stator because of excessive vibrations or thermal deformations of the casing or the rotor during transient operation of pumps for hot liquids. Contact between the rotor and the stator can induce wear (increase in clearance and drop in efficiency). Contact can even lead to the destruction of the pump due to seizure. The correct choice of the material combination in the close running clearances is therefore altogether crucial for the reliable operation of the pump.

At high circumferential velocities and high pressure differences across the annular seal, the wear rings must also be resistant against erosion corrosion. Many materials with good galling characteristics (for example, cast iron) are susceptible at high velocities to metal loss due to erosion corrosion. Thus their use as wear ring material is very limited. Wear ring materials need to fulfill the following requirements:

1. Most importantly the wear ring material must exhibit little tendency to galling and seizure upon rotor/stator contact. Frequently a difference in the Vickers- or Brinell hardness of at least 50 HV (BHN) between rotor and stator is specified in order to reduce the risk of galling. In [B.15] a hardness difference of 10 HR_C (HV = 100) is recommended for materials with $HR_C < 45$ (HV < 450).
2. Whenever possible, different alloys should be used for rotor and case ring.
3. The metal loss resulting from corrosion or erosion corrosion should be very small during service in order to limit the increase in clearances and to avoid undue loss in efficiency and rotor damping.
4. The wear ring material must at least have the same potential as the casing and impeller materials in order to avoid galvanic corrosion, Chap. 14.2.
5. If abrasive solids are entrained by the liquid, the wear rings must be made from abrasion-resistant materials, since seals are subject to very high wear.

Table 14.19 lists suitable combinations of wear ring materials. The limit for the head H_{max} corresponds to the head per stage at the best efficiency point (calculated with $\psi_{opt} \approx 1$). When considering the balance piston or disk, H_{max} is the total head of the pump at BEP.

The risk of rotor-stator contact, as well as erosion corrosion, increases with the impeller tip speed. While suitable wear ring materials are readily available for pumps with low heads, affordable materials which fulfill the above requirements in an ideal way for high-head pumps are unknown. Depending on the pump manufacturer, a multitude of material combinations are offered. Some of these are discussed below:

Table 14.18 Wear ring materials

Rotor material	Stator material	Medium	H_{max} (m)	Remarks
All (as far as resistant in the medium pumped)	GG-30, GGG-40	W1, W2	100	Stator material with good galling resistance. Velocity limited by erosion corrosion
	GGG-40, GGG-NiCr 20-2		120	
	GX120CrMo 29-2 (1.4138) wnt (soft nitrated)		Limited by H_{stage}	
	GG-30, GGG-40	Hydro carbons	200	
	G-CuSn 12 2.1052.03	W1 to W3, W5, W6	80	
Ferritic, martensitic, austenitic steels, duplex, Stellite, Colmonoy,	X20Cr 13 (1.4021) HV > 250	W1, W2, W5, W6	Limited by H_{stage}	Also for balance pistons. Hardness difference between stator and rotor as high as possible, but at least HV50.
	GX120CrMo 29-2 (1.4138) wnt nitrated			
	X22Cr17 (1.4057) possibly nitrated			
	Stellite, Colmonoy			
All (as far as resistant in the medium pumped)	PEEK Simsite (Chap. 14.4.8)	W1 to W6	35 bar 140 bar	Relevant is pressure difference across labyrinth
All (as far as resistant in the medium pumped)	X5CrNiMo 18-10 (1.4401)	**W3** W2 W1	Limited by H_{stage}	Could be used in W5, W6 but uncommon
	X8CrNiMo 27-5 (1.4460)			
	Stellite, Colmonoy			
	Waukesha 88			
All (as far as resistant in the medium pumped)	X2CrNiMoCuN 25-6-3-3 (1.4517)	W4	Limited by H_{stage}	Resistant in W1 to W6
	GX2NiCrMoCuN 29-25-5 (1.4587)			
	Hastelloy C NiMo16Cr15W (2.4819)			
All types of bronze	NiCu30Al (2.4374) Monel K500	W3	400	Resistant in W1, W2, W5, W6 Impeller (rotor) not from duplex or austenitic steel
	G-CuAl 10 Ni (2.0975.03)		150	

Impeller: H_{max} is valid at $\psi_{opt} \approx 1$; balance piston: $H_{max} = H_{total, pump}$
Water specifications W1 to W6 are described in Chap. 14.4.1.

- In so far as cast iron is resistant against corrosion in the specified medium, it is a very suitable wear ring material because of its excellent galling characteristics brought about by the graphite in the microstructure.
- Martensitic chromium steels with 12 to 30% chromium can be used in cold and hot water of the types W1, W2, W5, and W6. Examples are X20Cr13 or X22Cr17 cast or wrought. Preferentially the case ring is heat treated to higher strength than the impeller or running ring. The case rings are sometimes surface-hardened, e.g. by soft nitrating or hard chroming. Even when rotor and stator rings have differences in hardness and/or alloy, there remains a considerable risk of galling and seizure, if rotor and stator are contacting under high forces.
- The plastic material PEEK (Polyetherketone, reinforced by carbon fibers) is very suitable for case rings. PEEK is resistant in the types of water W1 to W6. There is no risk of galling when contacting metallic impellers or running rings. Two types of PEEK are available: (1) PEEK with short fibers ("chopped carbon fiber") can be employed up to 20 bar pressure difference in the range of −30 to 135 °C; (2) PEEK with long wound fibers ("continuously wound carbon fiber") can be used up to 35 bar pressure difference in the range of −30 to 230 °C. If the PEEK bushing or ring is sufficiently supported by other structures, PEEK can be applied up to 140 bar, [N.6].
- Similar to PEEK, a composite material with the trade name Simsite is used in sea water pumps for wear rings. The allowable pressure differences are claimed to be equivalent to PEEK. More details are given in Chap. 14.4.8.
- In spite of their tendency to galling, austenitic steels of the type 1.4409 are used as wear rings for vertical pumps in sea water because of their high resistance to erosion corrosion [39]. In vertical pumps the forces leading to a contact between the rotor and stator are, as a rule, not excessive.
- In sea water pumps duplex-rings are also combined with austenitic steels.
- Tin bronzes, aluminum bronzes and nickel-aluminum bronzes are used as wear rings in sea water because of their low tendency to seizure. However, the flow velocities must be limited according to Table 14.6 (see also Fig. 14.14). These wear ring materials are combined with copper alloys as well as with austenitic steels, [39].
- In spite of their good galling characteristics, the various types of Ni-Resist (e.g. 0.7660) are unsuitable for wear rings in sea water applications, since the metal loss caused by erosion corrosion is too high, Fig. 14.14, [39].
- Martensitic/ferritic steels cannot be employed in sea water because of severe erosion corrosion.
- In sea water Monel wear rings must not be combined with duplex, austenitic or super-austenitic steels because of galvanic corrosion.
- Satellites are a family of cobalt base alloys. Colmonoy are nickel base alloys. Both groups of materials have a high hardness and relatively low tendency to seizure. In sea water there may be a risk of galvanic corrosion. In spite of this, these materials are used in sea water, [39].

- Hard spray coatings, e.g. from tungsten carbide, are sometimes employed in order to reduce the tendency to seizure and/or to enhance the resistance to abrasive wear by solid particles. In the finished condition, such coatings should have a thickness of at least 0.8 mm, [N.6]. Weld coatings, e.g. from Stellite, can also be used for the same purpose.
- Wear rings from SiSiC (reaction-bonded silicon carbide) are used in FGD-pumps, [46].
- Waukesha 88 is a nickel base alloy with 12Cr, 4Sn, 1Mn, 3Mo, 2Fe, 4Bi. It has low tendency to seizure and a high resistance against erosion corrosion in sea water, [39].
- Wear rings from tungsten carbide are employed in fluids containing abrasive particles, Chap. 14.5.
- In low-density fluids (say below 500 kg/m^3), there is an increased risk of galling because of the weak lubricating effect of the liquid. According to [B.15] it is common practice to make the case rings from metal-filled graphite. Such rings are also used for liquid CO_2 applications.

Sometimes the pump operator specifies running rings to be mounted on the impeller. These can be made for example from Stellite, Colmonoy or Waukesha 88.

14.4.5 Shaft Materials

Since the mechanical dimensioning of shafts and their manufacture are beyond the scope of this text, only aspects of corrosion are discussed. Important requirements on shaft materials include:

- Sufficient elongation, A > 18% (better > 20%), in order to render the shaft less sensitive to notch effects and fatigue damage. For highly loaded shafts A > 25% is recommended.
- High endurance limit in the specific water to be pumped in the individual application (water analysis!).
- Good *form stability* during manufacturing of the shaft and during operation.
- High yield strength in case of highly loaded shafts.
- Resistance against corrosion in the specified fluid, unless the shaft is completely and reliably protected from the liquid by properly sealed sleeves.

Some commonly used materials for shafts wetted by the pumped fluid are listed in Table 14.20. The approximate application limits for the heads focus primarily on aspects of corrosion. Total torque and number of stages are to be considered for the dimensioning of the shafts of multistage pumps.

Monel K-500 (2.4374) was frequently used as shaft material for sea water pumps. Since no endurance limit is reached at high load cycles, this material is not recommended. The development of duplex steels has made the application of 2.4374 as shaft material obsolete (not in the least for cost reasons).

Table 14.19 Materials for shafts exposed to the fluid pumped

Medium	Qualification	Head per stage	
		$H_{opt} < 300$ m	$H_{opt} > 300$ m
W1 forming protective films	Good	Ck45 (1.1191)	42CrMo 4 (1.7225)
	Excellent	X20Cr 13 (1.4021)	
W2 untreated water, sewage	Fair	Ck45 (1.1191)	42CrMo 4 (1.7225)
	Good	X5CrNiMo18-10 (1.4401)	X20Cr 13 (1.4021)
	Excellent	Duplex X2CrNiMoN 22-5-3 (1.4462)	
W3 sea water without H_2S	Good	X5CrNiMo18-10 (1.4401)	
	Excellent	Duplex X2CrNiMoN 22-5-3 (1.4462)	
W4 sea water with H_2S	Good	Super-Duplex X2CrNiMoCuN 25-6-3-3 (1.4515)	
	Excellent	X2CrNiMnMoN Nb 23-17-6-3 (1.3974) up to 60 °C	
W5 partly demineralized water	Good	X20Cr 13 (1.4021)	
W6 fully demineralized water (demineralized water)	Good	X20Cr 13 (1.4021)	
	Good	X22CrNi 17 (1.4057)	
	Excellent	X4CrNi 13-4 (1.4313)	
	Good	Duplex X2CrNiMoN 22-5-3 (1.4462)	

14.4.6 Materials for Feed Water and Condensate Pumps

As discussed in Chap. 14.3, neither cast irons nor carbon steels can be used for pumping feed water or condensate because of erosion corrosion in regions of high flow velocity. Common types of materials include:

Casings: In the inlet and outlet casings of ring section pumps (Fig. 2.7), as well as in the casings of barrel pumps (Fig. 2.8), moderate flow velocities prevail (up to 10 to 15 m/s). Depending on the water chemistry, low-alloy steels can be selected, for example for forged casings 10CrMo 9-10 (1.7380) and for cast components GS17CrMoV 5-11 (1.7706). High-alloy martensitic steels are also very suitable.

Impellers, diffusers, volutes and stage casings: Because of the high flow velocities, high-alloy martensitic steels are required: GX7CrNiMo 12-1 (1.4008) up to about 400 m head per stage and GX4CrNi 13-4 (1.4317) above H > 400 m. In addition, GX7CrNiMoNb 15-5 (ASTM A747 CB 7Cu-1, 17-4 pH) could be employed. In "advanced class" feed water pumps, impeller tip speeds up to about 150 m/s have been executed. These martensitic steels are not sufficiently stable in oxygen-free feed water if pH < 7. Under these conditions, austenitic or duplex steels should be used in the range of 6 < pH < 7.5.

Suction impellers: For applications with water temperatures above 140 °C, the same materials and limits apply as for impellers, but the cavity length must be limited in order to avoid cavitation damage. However, suction impellers pumping cold de-aerated water below 50 °C are exposed to a high risk of cavitation damage if the pump operates with extensive cavitation (i.e. with low $NPSH_A/NPSH_3$). In such cases, the peripheral velocity at the suction impeller inlet is to be limited to about u_1 = 27 m/s. As discussed in Chap. 6, the hydrodynamic cavitation intensity is very high in cold de-aerated water and large cavities are generated at the usually prevailing low cavitation coefficients of σ_A = 0.18 to 0.25. The risk of cavitation damage can be reduced to some extent by selecting higher strength materials such as GX4CrNi 13-4 (1.4317) or GX7CrNiMoNb 15-5 (17-4 pH) tempered to a tensile strength of about 1200 N/mm^2. Steel GX4CrNiMo 16-5-2 or special alloys such as 17Cr 9Co 6Mn (Chap. 6.6.7) can be considered. These materials can also be employed to reduce the risk of cavitation damage at elevated water temperatures.

At water temperatures above roughly 250 °C (for example in boiler circulation pumps), the risks of erosion corrosion and cavitation erosion diminish so that less resistant materials may be considered.

14.4.7 Materials for FGD-Pumps

As a rule, the water in flue gas desulfurization plants (FGD) is highly corrosive due to elevated chloride concentrations and, at the same time, very abrasive due to high contents of solid particles. The solids are mostly $CaCO_3$ (lime-stone) or $CaSO_4$ (plaster); their concentration is in the range of 10 to 60% (mass). Roughly 60% of the particles have grain sizes up to 50 μm, while the rest reaches 200 μm (maximum 300 μm). Mostly lime-stone is used which causes severe abrasive wear, Fig. 14.23. If $Ca(OH)_2$ is employed instead, the pumpage is much less abrasive [60]. Also, corrosion at standby is an issue.

The fluid properties can vary in a wide range, typically: pH = 3 to 8, chloride content 5000 to 80,000 ppm, temperature 40 to 65 °C (data according to [19, 21, 44, 60]). The high chloride content provokes strong corrosion which grows with falling pH and increasing temperature according to Eq. (14.9) and Fig. 14.21. Fluoride concentrations as low as 10 ppm significantly reinforce the corrosion in the presence of chlorides [19]. Bromide and iodine ions are also very aggressive. Therefore, an accurate water analysis is necessary for a correct material selection.

Because of the strong abrasion, the required component life (16,000 to 24,000 h) cannot be reached with austenitic steels. Materials with high carbon content, which would be preferred for an improved resistance to abrasion, are not sufficiently resistant to corrosion at pH < 4. Super-duplex steels with a pitting index PI > 42 can be used at pH > 2.5 (unless mineral castings or elastomeres are selected). With

Fig. 14.23 Impeller of a FGD pump after more than 40,000 h of service (virtually no wear in the volute casing)

all duplex steels, *a specific heat treatment adapted to the individual alloy is critical for achieving the desired corrosion resistance*. One objective of the heat treatment is to obtain a material with a microstructure free of precipitations. This can be checked by metallographic sections magnified 1000:1. Higher carbon contents in the skin of the casting are equally inadmissible, since it is particularly important to limit the carbon content at the wetted surface to $C < 0.03\%$.

Below $pH = 2.5$ these steels suffer acid corrosion. For $pH > 4$ some materials mentioned in Table 14.16 can be envisaged since they better resist abrasion due to their high hardness. If such materials are considered, the risk of corrosion damage must be assessed carefully on the basis of the correct water analysis. If the pH drops below the application limit of $pH = 4$ during commissioning or under upset conditions, these materials suffer severe wear. The ferritic matrix is selectively leached and the abrasion attack is intensified due to increased turbulence in the corrosion marks which are generated in the process.

The materials with high carbon contents are sensitive to local corrosion, and they are difficult to machine. Due to their low ductility, materials with high carbon contents are not recommended for casings (the risk of catastrophic failure presenting a safety issue). The different elements composing the microstructure are sensitive to selective corrosion (casings made from 1.4464 failed due to cracking during service), [19, 47].

Impellers, casings and wear plates of all-metal pumps are usually made from the same material. Sometimes plastics ("elastomeres") or rubber coatings are selected. Some types of rubber are largely immune against corrosion. With the prevailing small particle size, rubber exhibits good abrasion resistance. However, large foreign matter can destroy components clad with elastomeres or rubber. Water absorption and swelling of the materials as well as aging and temperature effects need to be considered too, [18, 34].

14.4.8 Composite Materials

PEEK was mentioned in Chap. 14.4.4 as a composite material used for wear rings. Under the trade name "Simsite®" another composite material is available which is used for complete pumps, impellers, wear rings, casings, bearings and other parts. Main applications are sea water pumps in marine service (Navy and merchant ships). Other applications include chemical process industry, pulp and paper, mining, waste water, and petrochemical plants.

Table 14.20 Properties of composite materials (as given by the manufacturer)

Type			Simsite 300	Simsite 375
Tensile strength	R_m	N/mm^2	138	375
Elongation	A	%	2	2
Young's modulus	E	N/mm^2	9000	22'000
Density	ρ_{mat}	kg/m^3	1450	1800
Thermal expansion	α	$10^{-6}/K$	18	10
Maximum service temperature	T_{max}	°C	154	175
Moisture absorption by mass in 48 hours		%	1.75	0.15

Simsite is a structural composite made of thermoset resins which are reinforced by continuously interwoven fibers in a tri-directional weave. The resin systems primarily used to make Simsite grades are phenolic, epoxy and polyimide. The reinforcing fibers used in the manufacture of Simsite are primarily, fiberglass, synthetic, and combination fibers. The type of fiber can be selected to suit the strength requirements of the particular application. Most grades of Simsite are enriched with graphite, which provides self-lubrication in the case of rotor/stator contact. The self-lubrication characteristics of the product make the parts less sensitive to galling and seizure; even dry running is possible for short periods of time.

The material is resistant to sea water and a number of chemicals. It can also be employed for pumping hydrocarbons. Since the material does not conduct electricity, galvanic corrosion does not occur. The resistance to cavitation erosion is expected to be typically 3-times higher than that of aluminum bronze.

All parts, including impellers and volute casings, are machined from solid blocks of composite. This applies even to closed impellers.

Some material properties are given in Table 14.21; note the low density as compared to most metallic materials used in pumps. Heads per stage and pressure ratings of the casings are limited by the strength. Also note the low elongation. Impellers with heads per stage of up to 150 m at the best efficiency point have been built at low specific speeds (possibly at n_q = 20 to 25). No precise data are available on the maximum allowable head per stage as a function of the impeller outlet width. Tentatively it can be assumed to follow similar trends as given by Figs. 14.3 and

14.4 for cast iron. The costs of a Simsite impeller are similar to those of a bronze impeller.

Simsite absorbs moisture; therefore the volumes of the parts slightly increase when exposed to water during service, [57, 58]. A pump built with Simsite impellers and wear rings must be operated for a few hours after filling the first time with water in order to prevent blocking in the closed running clearances.

14.5 Hydro-abrasive Wear

14.5.1 Influence Parameters

Hydro-abrasive wear (for short "abrasion") is a loss of material induced by solid particles entrained in the liquid. Frequently, the problem is caused by low concentrations of sand which unintentionally enters the pump (e.g. water drawn from wells, rivers, run-off from glaciers or oil wells). In contrast, large amounts of solids in high concentrations are pumped in hydraulic transport (Chap. 13.4) or in FGD plants. In these applications, abrasion is the dominating factor for pump design and operation.

Abrasion generates a wavy wear pattern with a characteristic aspect as shown by Fig. 14.24. The wavy structure of the attacked surface is caused by vortices which form at tiny surface irregularities (think of the effect of the form resistance created by roughness elevations on the pressure loss, Chap. 1). The particles are pressed against the metal surface by the centrifugal forces generated in these vortices. Therefore, the metal loss is strongest in region A, and the edge K is ground sharp by the vortex on the lee side, Fig. 14.25 [72]. This wear model originates from Ackeret and de Haller.

Every particle which comes into contact with a component, when flowing through the pump, contributes to the wear. A particle can hit a structure (e.g. an impeller leading edge) and cause a shock-like load (like an imploding cavitation bubble). Or it can slide along a wall or a blade and interact with surface irregularities or roughness elevations and cause wear by a friction-like process.

Depending on the type of flow, different abrasion mechanisms can be expected. The same applies to the influence of the grain size: in dredge pumps which transport coarse gravel ("stones" of 50 mm diameter), impeller and casing are subject to severe shock loading. In contrast, the wear process in a FDG pump with a lime-stone suspension of 50 μm grain size resembles a polishing or grinding process.

While the fine lime-stone particles largely follow the streamlines of the flow, coarse gravel trajectories strongly deviate from the streamlines of the liquid.

The abrasion is determined by the 3D-particle movement close to the walls or blades. The flow processes and the reactions of the material are almost as complex as with cavitation erosion. Exact generic methods for predicting abrasive wear in a

Fig. 14.24 Hydro-abrasive wear at the impeller inlet

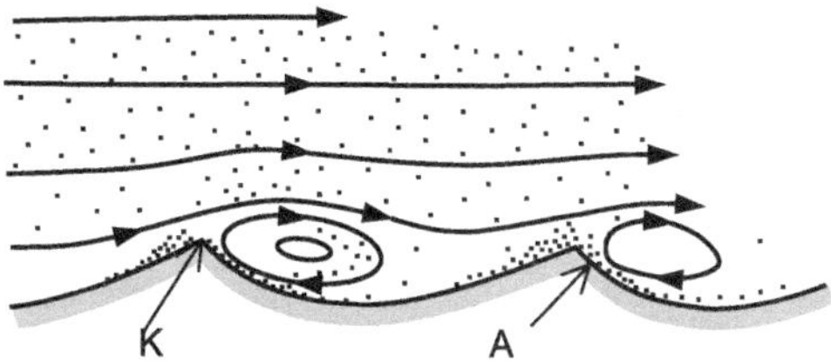

Fig. 14.25 Generation of wavy wear patterns by vortices

pump are therefore not available. Rough estimations and assessments of the severity of abrasive wear may be made on the basis of wear tests or operation experience. Prior to developing an empirical wear model, let us review the various parameters which determine the abrasive wear in a pump.

Concentration of solids: The metal loss (respectively the erosion rate E_R) increases with the number of particles which come into contact with a structure and their kinetic energy. Many experiments show that the metal loss is roughly proportional to the solids concentration c_s. Depending on the flow pattern, the curve of the loss rate $E_R = f(c_s)$ may level off at high concentrations if the particles interact with each other in a way that less particles contact the walls. The solids concentration can be defined as: (1) solids content c_s in kg solids/m^3 water; (2) mass concentration x; (3) volume concentration c_v according to Table 13.5; (4) solids content in ppm (1 ppm solids corresponds to $c_s = 0.001$ kg/m^3 = 1 g/m^3). The conversion of the different concentrations can be done using Eq. (T14.21.1).

Flow velocity: The higher the flow velocity, the more particles reach the walls at a given concentration. The kinetic energy of a particle rises with the square of the velocity w. As a consequence, the loss rate due to abrasion is theoretically proportional to the third power of the velocity. Many experiments confirm quite well the relation $E_R \sim w^3$ but the exponents found range between 0.9 and 5. This finding is possibly caused by testing difficulties and inaccuracies of a similar kind as discussed in Chap. 6.6.7.

Flow patterns: The local velocity distribution on a component, as given by incidence, work transfer, secondary flows, stall or partload recirculation, determines the local solids concentration near the walls as well as the kinetic energy with which the particles hit the walls. These processes, which cannot be quantified, often determine the abrasion far more than the average velocity.

Vortices generate high local flow velocities near the wall. In addition the centrifugal forces in the rotating fluid transport solids to the wall. Therefore vortices are very abrasive. Vortices are created by local flow separation, for example at a bore for venting the casing or by a flow deflection. Corner vortices between vanes and side plates, in particular in incident flow, can lead to severe abrasion damage, compare to Fig. 3.20.

Turbulence fosters the transport of fluid perpendicular to the direction of the main flow and also carries solids towards the walls. The loss of material thus rises with the turbulence intensity and the Reynolds number. The "mobility" of the particles, as can be characterized by the settling velocity according to Table 13.5, has an influence too. Thus a suspension of fine particles in viscous oil is expected to cause less abrasion than in water.

Impingement angle: If a fluid jet laden with solid particles impinges onto a wall at 90°, nearly all particles reach the surface with the full velocity and contribute to the abrasion. In contrast, in a flow parallel to a surface, the velocity in the boundary layer is smaller than the mean flow velocity. Consequently, the effect of the abrasive particles flowing through a channel is overrated when calculated with the nominal average velocity. Likewise, the solid concentration in the boundary layer will deviate from the average concentration due to gravity, centrifugal and Coriolis forces. The abrasion caused by a jet impinging under an angle ε on a wall is therefore usually higher than in parallel flow with $\varepsilon = 0$. The loss rate of a brittle material rises continuously to attain a maximum at $\varepsilon = 90°$. In contrast, the wear maximum of a ductile metal is attained at approximately $\varepsilon = 30°$ and that of elastomeres at about $\varepsilon = 15°$.

Grain size: The kinetic energy of a particle contacting the wall at a given velocity and angle increases with the particle mass. In principle, the wear rate rises with the grain size d_s, but not all tests exhibit this behavior.

As a rule, the grain sizes comprise a spectrum which is characterized by the average particle diameter d_s for which various definitions are in use. For example, d_s is defined as the diameter at which either 50% of the solids mass (or 50% of the solids volume) are below/above this average diameter.

Grain hardness, grain shape: The harder the abrasive, the higher is the wear. Angular grains cause higher wear than spherical particles (at otherwise equal conditions). Angular grains have a higher flow resistance coefficient than spheres; consequently, their flow paths are different (Chap. 13.4).

Corrosion and cavitation can increase the rate of abrasive wear. If abrasion and corrosion (or cavitation) are simultaneously present, the choice of a suitable material is difficult because high-strength steels (due to the required carbon content) usually do not resist corrosion very well.

Material properties: Abrasive wear always diminishes if the hardness H_{Mat} at the surface of the material is increased. The microstructure of the material has an impact too. Similar to cavitation erosion (Chap. 6.6.7), the material in abrasive environment is exposed to an "infinite" number of tiny shocks. Each individual load does not produce visible damage, but the sum of these loadings leads to destruction of the component. Depending on the ratio of the hardness of the abrasive H_s to the material hardness H_{Mat} and depending on the intensity of the solicitation (velocity and grain size), different damage mechanisms are active. For example, coarse gravel hitting the impeller blade leading edge may cause local plastic deformation. In contrast, a fine lime-stone suspension flowing along the blades interacts weakly with the irregularities of the surface. In principle, it can be assumed that the damage mechanism and the wear rate depend on the ratio $\Sigma E_{kin}/H_{Mat}$ of the specific kinetic energy of the solids to the material hardness. The hardness ratio H_s/H_{Mat} of the abrasive particles to the material has also an influence.

14.5.2 *Quantitative Estimation of Hydro-abrasive Wear*

14.5.2.1 Development of a Wear Model

The above discussion of the influencing parameters suggests that it is not possible to exactly quantify all flow, material and solids characteristics which are relevant for abrasive wear. Even for simple geometries, a prediction of the hydro-abrasive wear is subject to considerable uncertainties. Nevertheless, methods for a rough estimation of the abrasion rate are required for a quantitative assessment of the material in order to allow a meaningful choice.

For developing a model for hydro-abrasive wear, think of a short pipe section of length ΔL through which a sand-water mixture is flowing. Let's assume that the loss of metal due to abrasion is proportional to the number z_w of particles per second coming into contact with the material and to their kinetic energy. The "hydro-abrasive intensity" in case of spherical particles is thus given by:

$$\tfrac{1}{2}\rho_S\, w_{mix}^2 (\pi/6) d_S^3 z_w \tag{14.10}$$

If on the wetted surface $A_b = U \times \Delta L$ a layer of thickness ΔE is removed in the time Δt, the work done on the material is $V_E \times R_x$ (U is the perimeter of the wetted surface). The removed volume is $V_E = A_b \times \Delta E$. The quantity R_x stands for the resistance exerted by the material against the abrasion. R_x may be the tensile strength, the hardness or the endurance limit. The (surface-) hardness is a measure for the resistance which allows the material to oppose deformation on contact with a particle. The work done on the material may then be set equal to $V_E \times H_{Mat}$ (when the hardness is expressed in N/m^2).

The solid particles also suffer wear during the process so that some work E_s is done on the solids (angular particles become rounder, friable particles may disintegrate). The softer the particles in comparison to the component, the greater is the relative absorbed energy $E_s = f(H_{Mat}/H_s)$. The energy balance of the abrasion process thus becomes:

$$\tfrac{1}{2}\rho_S\, w_{mix}^2 (\pi/6) d_S^3 z_w\, \Delta t = V_E\, H_{Mat} + E_S(H_{Mat}/H_S) \qquad (14.11)$$

The number of spherical particles z per second results from the amount of solids carried by the liquid or its volumetric flow rate Q_s as: $z = 6 \times Q_s/(\pi \times d_s^3)$ or $z = 6 \times c_s \times Q/(\rho_s \times \pi \times d_s^3)$, because of $Q_s = c_s \times Q/\rho_s$ (Q is the liquid flow rate). Near the wall, in a fluid layer of thickness δ, the number of particles transported amounts to $z_w/z = U \times \delta/A$ (assuming the same solids concentration as in the bulk flow). Finally, it is assumed that the thickness δ of the fluid layer, where solids interact with the wall, is proportional to the grain size d_s. After some manipulations, the following proportionality for the abrasion process is obtained from the above arguments:

$$E_{R,a} \sim \frac{c_S\, d_S\, w_{mix}^3}{H_{Mat}\left(1+\frac{c_s}{\rho_s}\right)\left(1+f\left\{\frac{H_{Mat}}{H_s}\right\}\right)\Delta L} \qquad (14.12)$$

According to Eq. (14.12) the abrasive wear increases with the third power of the mixture velocity w_{mix}. The wear is proportional to the grain size and inversely proportional to the hardness. At low solids contents ($c_s \ll \rho_s$), the loss rises almost linearly with the concentration. This rise weakens with increasing concentration due to the term $(1 + c_s/\rho_s)$ in the denominator of Eq. (14.12). The abrasion diminishes with a growing ratio of H_{Mat}/H_s (i.e. the softer the abrasive is compared to the component). Equation (14.12) contains no geometric parameters (such as the ratio of the wetted surface to the flow area for example). If the solids/liquid mixture is assumed to be homogenous, geometric parameters cancel. In reality, the probability that a particle comes into contact with a wall depends on the velocity, the concentration distribution and on the turbulence. These parameters are determined by the geometry, but they cannot be quantified by simple algorithms.

As mentioned above, the abrasion process is too complex to be described by a simple model, but Eq. (14.12) can be used as a basis for correlating test data. To this purpose, various empirical factors have been introduced in Eq. (T14.21.2) in Table 14.22, which are discussed in the following.

Since abrasive wear in annular seals impairs the pump efficiency due to increased clearances, various tests were carried out on abrasion in annular seals, [11, 12, 28, 33, 34, 65, 66]. The published test data on seals and bores in [33] allow to develop correlations for the metal loss due to abrasion. The correlations derived will be made subsequently more generic for other geometries. The tests reported in [33] were carried out under the following conditions:

- Seal geometry: Inner diameter of the stator: D = 26 mm, length L = 20 mm, radial clearance s = 0.2 to 0.65 mm. The following dimensionless parameters can be derived: s/R = 0.16 to 0.053; L/d_h = 15 to 50 (d_h = 2 s). The ratios of clearance to radius in pumps are much lower; s/R = 0.002 to 0.004 would be expected.
- Tests with bores of 3 to 6 mm diameter (length 20 mm).
- Solids concentration up to x = 0.12; grain diameter d_s = 0.016 to 0.19 mm.
- The metal loss on the stator was on average about 40% lower than on the rotor, which suffered stronger wear at the inlet than near the outlet (possibly due to vortices created at the sharp-edged inlet). In the tests reported in [12] and [11] with s/R = 0.004 to 0.006, the metal loss on the stator was considerably more severe than on the rotor, which suffered little wear because it was protected by a spray coating. The tests in [12] were performed with low axial velocities c_{ax} u, while testing in [33] was predominantly done with $c_{ax} \geq u$.
- The average circumferential velocity in the annular seal is $w_u = ½\omega \times r$ according to Chap. 3.6.2 and Eq. (T3.7.10). This value was used for evaluating the test data because the axial velocity component was unknown (Table 14.22).[12]
- A particle with the circumferential velocity w_u is exposed to the centrifugal and the Coriolis force as per Eq. (5.6). At $w_u = ½\omega \times r$, the Rossby number becomes Ro = ¼. Consequently, the particles are deflected primarily towards the rotor. This could explain why higher wear was found on the rotor than on the stator in [33]. During the tests in [12] the residence time of the particles in the seal was high because of the low axial velocity $c_{ax} \ll u$. The distribution of solids, and consequently the wear, is sensitive to the flow pattern. Therefore, it is not easy to conduct really representative tests. When developing the correlations in Table 14.22, it was not attempted to capture specific details of the test setup which are not relevant to pumps.
- No information is available on the geometric tolerances of the tested components, but these can have a strong impact on annular seals. In addition, there is an effect of the inlet to the gap: How sharp were the edges at the start of the test? To what degree were the edges rounded by wear during the testing? The increase in seal clearance during the test may have an influence too. It is small when a hard material is tested, but high if the specimens are soft. Finally, uncertainties in the material properties and measuring tolerances (in particular with low wear on hard materials) contribute to the uncertainties. Overall, the scatter of this type of wear tests can be assumed to be in the order of 20 to 30% if not higher. This fact should be borne in mind when appreciating the test data evaluated below.

---

[12] In [33] the evaluations were based on $w_u = \omega \times r$.

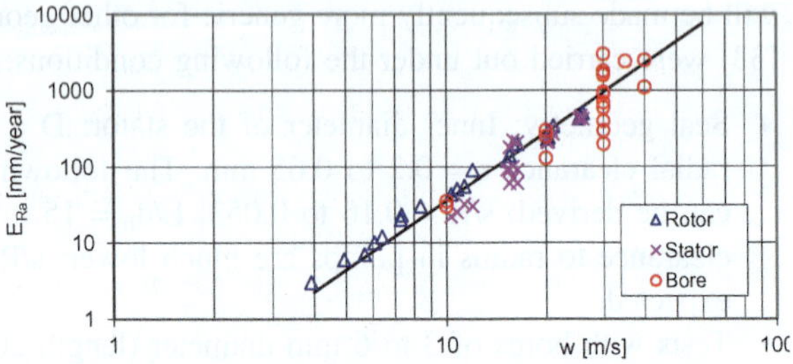

Fig. 14.26 Metal loss due to abrasion; $d_s = 83\ \mu m$, $x = 0.06$, $c_s = 64\ kg/m^3$; Tests with annular seal: rotor and stator from X105CrCoMo 18-2, $H_{Mat} = 700$ HV; Tests with cylindrical bores: duplex $H_{Mat} = 212$ HV

Figure 14.26 shows part of the test data used which were gathered on annular seals and bores. The trend-line corresponds to $E_{Ra} \sim w^3$. The scatter is caused by geometric variations and testing procedures. In particular it was found that the wear strongly increased with the diameter of the bores (d = 3 to 6 mm L/d = 3.3 to 6.7). This may be caused by vortices created at the inlet to the bore. Testing at L/d = constant may be more representative. Therefore the influence of the bore diameter has been ignored when correlating the test data.

14.5.2.2 Calculation of Wear Rates

Table 14.22 provides a method for estimating the hydro-abrasive wear. In developing this procedure, the following assumptions have been made:

- The **mixture velocity** is relevant for the wear rate (at low solids contents c_s it can be set equals to the fluid velocity).
- **Geometry**: The different geometries and flow patterns are described by empirical form factors F_{Form}. Some of these factors given in Table 14.22 had to be estimated on the basis of observations. A broad range of form factors is expected when dealing with vortices, recirculation and incident flow.
- **Annular seals**: The form factors F_{Form} for annular seals were derived from tests. The measurements in [33] yielded for the rotor $F_{Form} = 7.4$ and for the stator $F_{Form} = 4.4$. In general, it is recommended to use the average $F_{Form} = 5.9$ for stator and rotor, since it cannot be ascertained to which extent the differences between rotor and stator are caused by the specific test arrangement used in [33]. The solids concentration in the annular seal at the impeller inlet is lower than in the interstage seal because of the centrifugal forces counteracting the ingress of particles into the sidewall gap at the front shroud. For this reason, different form factors have been proposed in Table 14.22.
- For the same reason, the solids concentration in the impeller sidewall gaps is expected to be lower than in the impeller. The flow in the impeller sidewall gaps is smoother than in the impeller and wear is correspondingly less severe.
- The **impeller or diffuser vane leading edge** is exposed to impinging flow. The conditions were assumed to be similar to an orthogonal jet but less severe.

- Measurements of abrasive wear on **rotating disks** yielded $F_{Form} = 0.03$, if the peripheral velocity u_2 is used for the calculation. The actual conditions are not obvious because the velocity of the mixture rotating in the tank is not known. Furthermore, the local velocity and the wear vary over the radius.
- **Relevant velocities**: The velocities to be used for different geometries are defined in Table 14.22. The average circumferential velocity in an annular seal is $c_u = ½u_{sp} = ½\omega \times r_{sp}$. Therefore the average velocity vector can be calculated from Eq. (T14.21.3). The relative velocity in the impeller sidewall gaps may be assumed to $w = c_u = ½u_2$ as a first approximation. A more accurate calculation of $c_u(r)$ can be done according to Chap. 9.1. For the impeller inlet w_1 from Eq. (T14.21.5) can be assumed, for the impeller outlet w_2 from Eq. (T14.21.6), and for the diffuser or volute inlet c_3 from Eq. (T14.21.7) is relevant.
- **Solids concentration**: According to the available experiments, the abrasion rate can be assumed to increase proportionally to the solids content up to concentrations of about $c_v = 0.05$ or $x = 0.1$. At higher concentrations, the curve $E_R = f(c_s)$ presumably levels off. The wear in slurry pumps transporting mixtures with $x > 0.1$ is probably over-estimated when calculating according to Table 14.22. To avoid over-estimation of the wear, the abrasion could be calculated with $x = 0.1$ ignoring the effects of solids beyond that concentration. The wear rate thus estimated may be considered as a lower limit (as far as the influence of solids content is concerned).
- **Typical sand contents** : Well water up to 0.3 ppm; oil wells 6 to 600 ppm (peaks up to 3000 ppm); glacier run-off 500 ppm (peaks 2500 ppm).

  The **solids hardness** H_s is described by a factor F_{Hs} and an equivalent concentration $c_{s,eq}$ as defined by Eq. (T14.21.8) in which the fractions of a solids mixture are weighted according to their individual hardness. The reference hardness is that of quartz sand, since the correlations in Table 14.22 were derived from wear tests carried out with sand. If the solids consist solely of quartz sand, set $c_{s,eq} = c_s$. With a hardness of lime stone of 250 HV, the equivalent concentration would be $c_{s,eq} = (250/1150) \times c_s$ when the solids are made up entirely from lime. However, the calculation based on $c_{s,eq}$ is proven only when the solids pumped are similar to quartz. If, for example, only lime-stone is present, the calculation with $c_{s,eq}$ would overestimate the wear. This can be deduced from the measurements with lime stone reported in [72]. They produced 1/50 to 1/60 of the metal loss in quartz sand. From the data reported in [66], a ratio of 1/20 to 1/60 results (depending on the intensity of the attack). Therefore, an additional empirical factor F_{Hs} has been introduced in Table 14.22. The test data from [72] can be correlated by $F_{Hs} = 0.017$ with acceptable scatter as shown by Fig. 14.31.
- The information on the hardness of various solid particles exhibit considerable scatter: for quartz 1100 to 1450 HV, lime stone 100 to 150 HV as per [72] but 200 to 300 HV according to [67]. Other substances: feldspar 600 to 800 HV; limonite 270 to 490 HV; ground coke 580 to 640 HV; glass 580 to 640 HV; flint

930 to 1040 HV; garnet 1260 to 1560 HV; corundum 1800 to 2140 HV, [72]. Sometimes the hardness of minerals is specified by the hardness scale of Mohs. The Mohs hardness MH can be converted to Vickers hardness HV by the relation: $HV = (MH/0.7)^3$.

- **Grain size**: The influence of the grain size d_s is described by the factor F_{KG} according to Eq. (T14.21.9). The abrasive wear is proportional to the grain size d_s. (at least for small d_s). Considered over a wide range, the impact of the grain size on the wear becomes uncertain since the available test results are contradictory. During the tests reported in [62, Fig. 7.8], it was found that the metal loss was roughly proportional to the grain size in the range tested which covered $d_s = 0.6$ to 3.4 mm. The proportionality between grain size and wear in annular seals can only be assumed if d_s is sufficiently smaller than the seal clearance. If particles get stuck in the gap, the wear rises drastically (according to [67] by the factor of 10).
- **Grain shape**: The tests reported in [33] were executed with ground quartz sand; angular grains must be assumed for these data. Spherical sand grains, found in natural environments, are less abrasive than angular particles as demonstrated by the tests described in [72]. There the metal loss caused by rounded grains was only about 60% of the wear measured when using angular grains. However, it is difficult to decide to what extent this relation may be generic. Soft, friable particles (as lime stone or coke) will presumably wear off quickly into fairly round shapes when flowing through the system, even if they were initially angular. According to tests reported in [63], the relative metal loss (referred to round grains) increases according to the relation $1 + 2.1(\zeta/\zeta_{sphere} - 1)$, if ζ is the resistance coefficient of the considered particles and ζ_{sphere} that of a sphere (of equal diameter) calculated from Eq. (T13.5.6).
- **Material factors**: The reaction of the material to the abrasion depends on the "hydro-abrasive intensity". Therefore the material factors cannot be considered as universal quantities, Chap. 14.5.3.

The comparison between calculated and measured wear rates in Figs. 14.27 and 14.31 suggests that the estimation of erosion rates according to Table 14.22 can be of help for material selection. When doing this type of calculation, an uncertainty of at least ±50% should be expected because of the many influencing parameters which cannot be quantified accurately.

14.5.3 Material Behavior and Influence of Solids Properties

The material factors were determined from the measured data provided in [64]. The tests were done with bores and orthogonal jets, which exhibit quite different behavior since the material reaction to abrasion depends on the ratio of the hydro-abrasive intensity to the abrasion resistance (similar to cavitation as per Chap. 6.6.7). Figure 14.28 demonstrates this fact clearly. The relative metal loss of

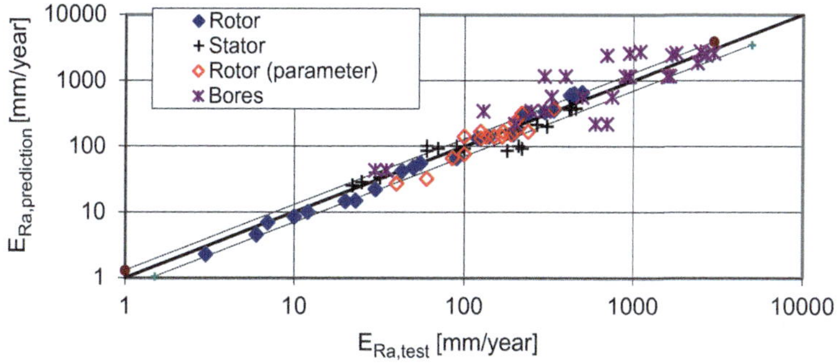

Fig. 14.27 Comparison of test data with prediction according to Table 14.22; rotor/stator: X105CrCoMo 18-2, H_{Mat} = 700 HV; bores in duplex steel H_{Mat} = 212 HV

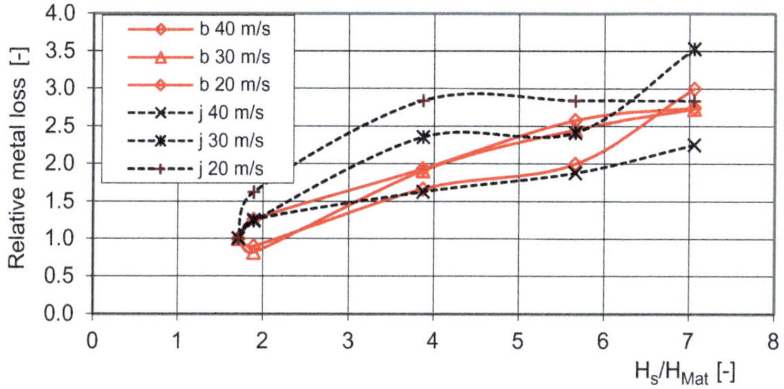

Fig. 14.28 Relative metal loss as function of the hardness; data from [64]; *b = bore, j = jet*

various materials (referred to the wear of the reference material 1.4528 with H_{Mat} = 700 N/mm^2) depends on the test conditions (i.e. the hydro-abrasive intensity). In order to determine the influence of the material on abrasion for purposes of wear prediction, appropriate averaging is necessary. To this end the wear tests carried out with bores and orthogonal jets were considered separately and plotted as material factors F_{Mat} against the hardness. According to Eq. (14.13), F_{Mat} is defined as the ratio of the wear of any material to the wear of a reference material with the hardness H_{Mat} = 700 HV *at identical test conditions*:

$$F_{Mat} \equiv \frac{\text{metal loss} \times \text{with} H_{Mat} = x}{\text{metal loss with } H_{Mat} = 700} \tag{14.13}$$

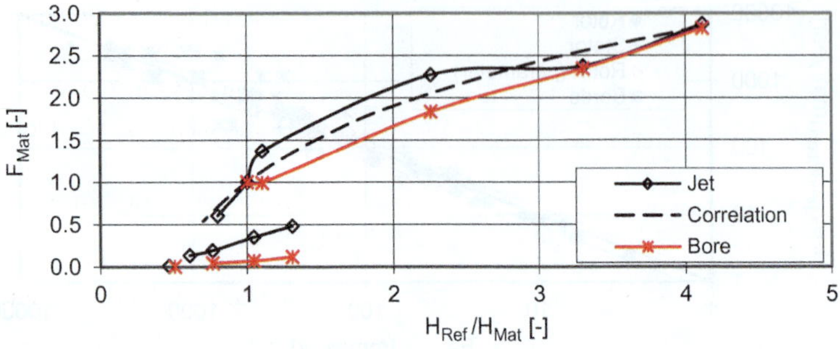

Fig. 14.29 Determination of material factors

The material factors determined are plotted in Fig. 14.29. For ductile materials, this graph allows a meaningful correlation, which is valid for orthogonal jets as well as for the cylindrical bores. The correlation is given by Eq. (T14.21.10) in Table 14.22 and shown in Fig. 14.29. Very hard materials yield factors F_{Mat} which are quite different for the jet and the bore. In Table 14.22 both correlations are given: from the tests with the bore follows Eq. (T14.21.11), while the measurements with the jet yield Eq. (T14.21.12).

In some of the data, the relation between wear and hardness is essentially linear;[13] see also the derivation of Eq. 14.12).

Wear tests carried out with a disk rotating in a sand-water mixture ($d_s \approx 1$ mm, x = 0.5, $u_2 \approx 9$ m/s) gave also an essentially linear relation between wear and hardness, albeit with a large scatter inherent to the testing techniques employed, see [5, 36, 59] and Fig. 14.30. Consequently, such test devices appear to be appropriate for determining the influence of the material on the wear (e.g. for material ranking).

Using the material factors thus determined, the available test data were compared to the wear prediction according to the procedure and equations given in Table 14.22, Fig. 14.31.

In addition, Fig. 14.31 comprises test data reported in [72]. These tests were done with a rotating pipe section at a speed of u = 8 m/s. The mixture had a volumetric concentration of 1:6 solids to water. The solids tested included: (1) sand (round) $d_s < 5$ mm, (2) quartz sand broken (angular) $d_s = 0.2$ to 1.5 mm, (3) ground coke (angular) $d_s < 5$ mm. The materials tested were carbon steel St 37 (190HV) and steel St70H (785HV). As for annular seals, a form factor $F_{Form} = 5.9$ was used. With these assumptions, the calculation according to Table 14.22 can well predict the test data.

Even though this calculation is uncertain as to the absolute wear rates, comparisons within the test series are of interest, because these reveal the influences of

---

[13]When comparing with [33, 34] it should be noted that a linear function represents a curve in the double-logarithmic plot.

Table 14.21 Estimation of metal loss due to hydro-abrasive wear

Solids content c_s in kg/m³	ρ = fluid density ρ_s = solids density	$c_s = \frac{x}{1-x}\rho = \frac{c_v}{1-c_v}\rho_s$	$x = \frac{c_s}{\rho + c_s}$	14.21.1
Metal loss rate in mm/year Valid for: $c_s < 150$ kg/m³	$\frac{E_{R,a}}{E_{R,Ref}} = \frac{F_{Form}\, F_{Mat}\, F_{KG}\, F_{KF}\, F_{Hs}}{1 + c_s/\rho_s}\left(\frac{c_{s,eq}}{c_{s,Ref}}\right)\left(\frac{w_{mix}}{w_{Ref}}\right)^3$			14.21.2
	$E_{R,Ref}$ = 1 mm/year; $c_{s,Ref}$ = 1 kg/m³; w_{Ref} = 10 m/s			
Geometry		**F_{Form}**	**Relevant velocity**	
Annular seals	Annular seals at impeller inlet, balance piston	3 to 5	$w = \sqrt{c_{ax}^2 + \left(\frac{u}{2}\right)^2}$	14.21.3
	Inter-stage seals	4 to 6		
Impeller sidewall gaps		3 to 5	$w \approx ½\, u_2$	14.21.4
Impeller inlet	Leading edge	20 to 100	$w_1 = \sqrt{c_{1m}^2 + (u_1 - c_{1u})^2}$	14.21.5
	Corner vortex	20 to 100		
	Blade surface	6 to 30		
Impeller outlet	Blade pressure surface	20 to 60	$w_{2u} = u_2 - c_{2u}$	14.21.6
Diffuser inlet, volute cutwater	Leading edge, corner vortex	10 to 30	$c_{3u} = \frac{d_2\, c_{2u}}{d_3}$	14.21.7
Cylindrical bore or channel [33]		3.3	$w = Q_{mix}/A$	
Jet impinging at 90° on a structure [33]		68	Velocity at jet orifice	
Rotating disk in wear device		0.03	$w = u_2$	
Equivalent solids concentration $c_{s,eq}$	$c_{s,eq} = \Sigma\left(c_s \frac{H_s}{H_{Quartz}}\right)$	The fractions of a solids mixture are weighted according to their hardness. H_{Quartz} = 1150 HV		14.21.8
Grain size	$F_{KG} = \frac{d_s}{d_{Ref}}$	d_{Ref} = 1 mm; for $d_s < 0.75 \times s$ (s = radial gap width)		14.21.9
Grain shape	F_{KF} = 1 for milled quartz sand; F_{KF} = 0.6 for round grains			
Grain hardness	F_{Hs} = 1 for quartz sand; F_{Hs} = 0.017 to 0.05 for lime stone			
Material hardness H_{Mat}	Ductile metals (A > 5%)		$F_{Mat} = 1 + 1.3\, \mathrm{Ln}\frac{H_{Ref}}{H_{Mat}}$	14.21.10
	Stellite 20 H_{Mat} = 670 HV Ferro-Titanite H_{Mat} = 535 to 1150 HV	gap or bore	$F_{Mat} = 0.14\frac{H_{Ref}}{H_{Mat}} - 0.063$	14.21.11
		jet	$F_{Mat} = 0.54\frac{H_{Ref}}{H_{Mat}} - 0.22$	14.21.12
H_{Ref} = 700 HV	Material	H_{Mat}	F_{Mat}	
Conversion:		HV	Cylindrical bore	Orthogonal jet
HV ≈ 0.29×R_m	GX250CrMo15-3	876	0.25	0.6
HV ≈ HB	Hard metal 82.5 WC	1380	0.004	0.01 (estimated)
	Silicon carbide SiC	1500	0.0035 (estimated)	0.008
	Tungsten carbide WC		0.0012	0.003 (estimated)
	WC-CoCr spray coatings		0.006 to 0.04	
Conversion of Rockwell to Vickers hardness			$HV \approx 125\exp(0.029 HR_C)$	

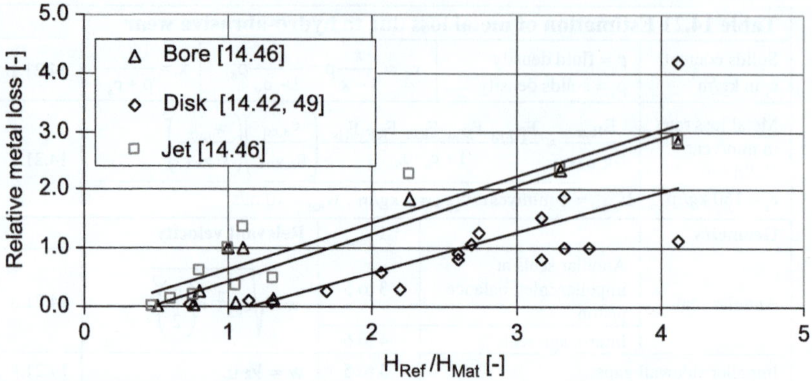

Fig. 14.30 Relative metal loss as function of the hardness in different test devices; data from [36, 59] and [64]

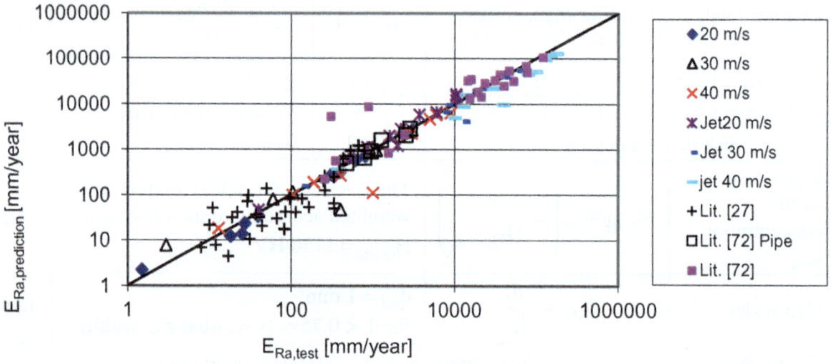

Fig. 14.31 Comparison of tests with different materials and test devices with wear prediction according to Table 14.22. Tests from [27] with coatings in the gap

the particle hardness and the grain shape. The tests further confirm the choice of the material factors.

Tests with rotating pins were reported in [72]. The pins rotated with u = 6.4 m/s in a mixture of one volume part solids to one part water. Different types of solids, grain sizes and steels were investigated. Since the flow is perpendicular to the pins, the form factor $F_{Form} = 68$ of the orthogonal jet was assumed for the calculation according to Table 14.22. With the exception of two measuring points with steel St60H and flint or glass, the data fit well into Fig. 14.31.

For the wide group of ductile metallic materials, the material factors from Eq. (T14.21.10) provide a good estimate. The assessment of hard materials and coatings is more difficult because the measured hardness alone is not necessarily sufficient to describe the process. It appears that the abrasion resistance may be improved by carbides without increasing the global hardness. In spite of this uncertainty, Eqs. (T14.21.11 and T14.21.12) can be used for a first evaluation as shown by Fig. 14.31 which contains test data on coatings reported in [72]. The measurements in [27] are better described by Eq. (T14.21.12) than by Eq. (T14.21.11). However, the scatter of test data is greater than with ductile materials. When dealing with hard materials or coatings, it is preferable to use measured values of F_{Mat}. The available data is given in Table 14.22. As can be seen, the factors depend strongly on the hydro-abrasive intensity, i.e. on the test device or machine, the flow velocity, grain size and hardness as well as solids concentration.

14.5.4 Material Selection

For an application with specified solids (type and concentration), Table 14.22 allows an assessment whether normal materials can be employed or a wear protection is required in order to achieve the desired component life.

In high-pressure pumps, sand contents as low as 5 ppm will cause unacceptable wear in the annular seals and, in particular, on the balance piston if duplex-steels or other steels are used. Even Stellite wear rings do not achieve the required component life. As a remedy spray coatings or parts made from tungsten or silicon carbide may be considered.

Parts made from SiC or WC should be manufactured with a metal-binder in order to achieve a sufficient ductility since fully ceramic components are extremely sensitive to shock-loading because of their brittleness. But also metal-bound carbides require a careful mechanical design in order to avoid fractures induced by the deformation of surrounding components. Such deformation can be due to mechanical loads as well as differences in thermal expansion. Silicon carbides have been selected for wear rings in FGD-pumps, [46] while WC was employed successfully in sea water injection pumps, [4].

Spray coatings are applied according to compositions and procedures specific to each manufacturer (few details are published). According to investigations reported in [27], high-velocity flame-sprayed coatings from tungsten carbides in a chromium-cobalt matrix performed best within the range of tested materials. A hardness of 1100 to 1300 HV was achieved. This type of coating was employed in sea water injection pumps to increase the component life by the factor of 13 compared to Stellite 12, [4]. The application of coatings is limited by the accessibility of the surfaces to be protected (e.g. impeller channels or narrow volute casings). Wear rings require expensive machining after the coating process.

The thickness of spray coatings needs optimization. In principle the thickness should be at least 0.8 mm, [N.6]. However, the adherence of the coating to the base material has the highest priority, since flaking-off of the coating may lead to seizure of the rotor if debris enters the annular seals. To prevent flaking-off, the coating must be sufficiently ductile.

Possibly, there is a limit to the maximum grain size. This limit may be attained when the kinetic energy of individual particles hitting the component is sufficiently high to plastically deform the base material. Since the coatings often are porous and unable to protect the base material from contact with the liquid, the base material must be resistant to local corrosion in the specified medium. The cavitation resistance of coatings is frequently inferior to that of steel 1.4317.[14]

Weld overlays (e.g. of Stellite), boronizing, nitriding and other types of surface hardening are also employed in order to increase the abrasion resistance. The improvement of the abrasion resistance by weld overlays is moderate according to [27]. Cracks, pores and deformation of the component caused by the heat input during welding are additional difficulties to be counted with. Boronized steels and nickel base alloys showed good behavior in the tests reported in [27]. It is claimed that machining of the parts after the boron process is not required, even though some deformation can occur. Coatings made from Al_2O_3 + $20ZrO_2$ did not show good results in the tests done in [27] because they are too brittle. The same applies to coatings made by hard chroming according to [40].

The impellers and casings of slurry pumps are frequently coated with rubber or plastics, e.g. polyurethane. The quality, in particular the Shore hardness, must be selected in accordance with the type of solids to be transported. The larger the particles, the more work is done on the coating due to elastic deformation caused by the impact of the solids. The absorbed energy leads to aging of the elastomeres which are hence best for small grain sizes. Their application limit is in the range of d_s = 1 to 2 mm.

Wear resistant materials according to DIN 1695 are selected for pumping coarse solids if the fluid is not (or weakly) corrosive. Table 14.16 provides the properties of some of these materials which can be used for impellers and casings. While the "naturally hard" materials (trade name "NiHard") can scarcely be machined, there are some types with a carbon content lower than DIN 1695 which can well be machined after annealing. After the machining the components are hardened up to HV1000, [5].

14.5.5 Abrasive Wear in Slurry Pumps

Because of the high hydro-abrasive wear, the velocities in slurry pumps must be limited. To define appropriate limits in a specific application, it can be attempted to estimate the wear rates by means of Table 14.22. This type of calculation can provide only a rough indication of the expected component life because the local abrasion rates vary strongly throughout the pump. Usually the most aggressive wear is observed at the impeller leading edge, in the corners between blades and shrouds (in particular with incident flow), on the pressure surface of blades near the impeller outlet, and at the volute cutwater.

---

[14]Efforts have been made to improve cavitation resistance, but details have not been published.

Recommendations for the maximum impeller tip speed are given in [74] as functions of grain size and mixture density. According to Eq. (T14.21.2), the erosion rate is proportional to $E_{R,a} \sim d_s \times c_s \times w^3$. For a selected allowable erosion rate, it follows that the velocity should be reduced with increasing grain size and mixture density according to $w \sim (d_s \times c_s)^{1/3}$. Using this proportionality, a relation for the maximum recommended impeller tip speed $u_{2,max}$ can be derived which corresponds roughly to the classes of slurry service defined in [74]. It is given by Eq. (14.14) which has the advantage of a continuous function:

$$u_{2,max} = u_{2,Ref}\left\{1 - 0.44\left(\frac{x}{(1-x)}\frac{d_s}{d_{Ref}}\right)^{0.33}\right\} \quad \text{with} \quad d_{Ref} = 1\,\text{mm} \qquad (14.14)$$

The reference speed in Eq. (14.14) for metallic materials is $u_{2,Ref} = 47$ m/s and for rubber $u_{2,Ref} = 31$ m/s. The velocities in the discharge nozzle should be limited to $c_d = (0.21 \text{ to } 0.28) \times u_2$ as derived from [74].

For the design and operation of pumps for highly abrasive service, some particularities need to be considered, [21, 26]:

- Annular seals must not be used because of severe wear caused by flow turning. Instead diagonal or radial gaps are recommended, see Fig. 3.15. The pump design should allow the clearance of the seal to be adjusted from the outside without opening the pump. Stepped gaps lead to high wear due to flow deflection and should be avoided. (The dredge pump depicted in Fig. 13.19 avoids flow through the seal altogether by flushing.)
- Expeller vanes on the rear shroud tend to keep large particles from entering the impeller sidewall gaps and reduce the pressure at the shaft seal. However, expeller vanes (also used sometimes on the front shroud) create strong vortices in the impeller sidewall gaps and generate higher wear than smooth shrouds. Therefore, expeller vanes should be avoided when pumping hard solids of high concentration.
- The shaft should be protected from the pumpage by appropriately sealed sleeves.
- Bores in the casing for draining or venting cause vortices and local wear. They should therefore be avoided.
- The impeller blades should be twisted (Francis vanes) and the blade inlet angles be well matched to the approaching flow. If feasible with respect to space, the blades should be well advanced into the impeller inlet. The leading edge should be thickened and designed with a profile that is little sensitive to variations in incidence (e.g. with an elliptical profile).
- Flow separation and local excess velocities lead to increased local wear. They should be avoided by a careful hydraulic design of impeller and casing, see also Chap. 13.4. Therefore, cylindrical and circular-arc blades are unsuitable for slurry pumps.
- Large fillet radii between blades and shrouds reduce the effect of corner vortices and the associated wear at the impeller inlet and outlet.

Fig. 14.32 Hydro-abrasive wear at the outlet of a double-entry impeller

- The smooth surfaces of impeller, diffuser and casing suffer less wear than the seals. The flow around the impeller trailing edge causes strong wear on the pressure surface as shown by Fig. 14.32 (see streamlines in Fig. 3.4). A low blade loading towards the exit is advisable. Sometimes the wear on the pressure surface is higher near the rear shroud than at the front shroud where a zone of reduced velocity (wake) is observed, Figs. 5.9 and 5.10. However, the wear pattern can be just the opposite if strong vortices are involved.
- Usually the number of blades should not be higher than $z_{La} = 5$ in order to reduce the effect of the inlet blockage caused by the thick leading edge profiles.
- Single volutes are preferable to double volutes if the material is difficult to cast. Welding and grinding are also easier to perform.
- Casing wear is highest at the volute cutwater. Large fillet radii between cutwater and casing walls, elliptical profiles and low incidence at the design point are important means to reduce wear. A large distance between impeller and cutwater allows some adaptation of the flow and reduces average as well as excess velocities.
- Use replaceable wear plates in the impeller sidewall gaps.
- The pump should be operated close to the best efficiency point or the flow rate of shockless entry in order to avoid wear caused by incidence and flow separation. Therefore, the flow rate of shockless entry should be the same for impeller and volute.
- Operation with partload recirculation causes wear in the suction nozzle.
- Diffusers are scarcely suitable for slurry service.

14.5.6 Erosion Patterns and Flow Mechanisms

Shape and structure of eroded surfaces give an impression of the flow mechanisms to which the components have been exposed. To recognize the underlying flow mechanisms can help to diagnose a problem and to develop remedies. With this purpose in view, various components of pumps operating in abrasive service are discussed in the following.

Figure 14.33 shows a view on a casing wall opposed to the impeller front shroud. At the volute cutwater there is wear from a horse shoe vortex (depicted in Fig. 3.20). The helical damage pattern in the suction nozzle suggests that the pump operated with extended recirculation at partload. The erosion around the suction port is not rotationally symmetric (the same applies to the flow through the ISR). As discussed in Chap. 9.3.3 (Fig. 9.29) fluid flows at partload from the volute around the cutwater into the impeller side rooms.

It may be possible to reduce the erosion due to the horse shoe vortexes by an oval-shaped leading edge (volute cutwater or impeller) as depicted in Table 10.2. This design unloads the streamlines near the walls; fluid, hence solid particles, are displaced away from the wall towards mid-channel.

Erosion due to horse shoe vortexes at the leading edge is observed frequently in abrasive service on impellers, diffusers and at volute cutwaters, but also on screws, studs or any type of protrusions.

Figure 14.34 shows a view on the rear casing wall (on the bearing frame) of a process pump operating with water with a high sand concentration. The wear on the rear shroud and the expeller vanes can be seen on the photo on the right of Fig. 14.34. In spite of the expeller vanes, the flow in the impeller side wall clearance is by no means rotationally symmetric (see also Chap. 9.3.3, Fig. 9.29).

Figure 14.35 shows a view on the casing wall of a double-volute pump. Spiraling flow paths can be recognized in the impeller sidewall clearance. The differences in flow patterns on the left and right halves of the eroded surface suggest that the flow rates (pressure distributions) through both channels of the double volute were quite different, see Chaps. 5.8 and 10.7.3.

Figure 14.36 shows severe erosion on the casing and the volute cutwater. Presumably the pump operated at partload since some flow enters from a

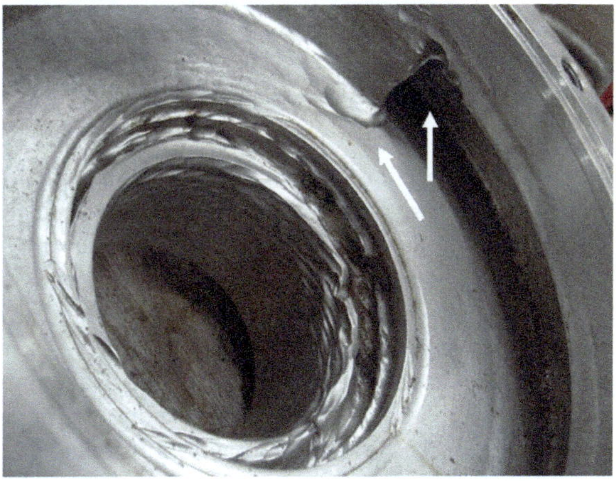

Fig. 14.33 Wear in suction nozzle and at volute cutwater of a process pump: horse shoe vortex (white arrows), [42]

Fig. 14.34 Casing side of ISR and view of expeller vanes on rear shroud of an open impeller, $q^* = 0.5$; sand $c_s = 150\ kg/m^3$, $u_2 = 28$ m/s, $d_s = 32$ μm, material: Duplex, HV260

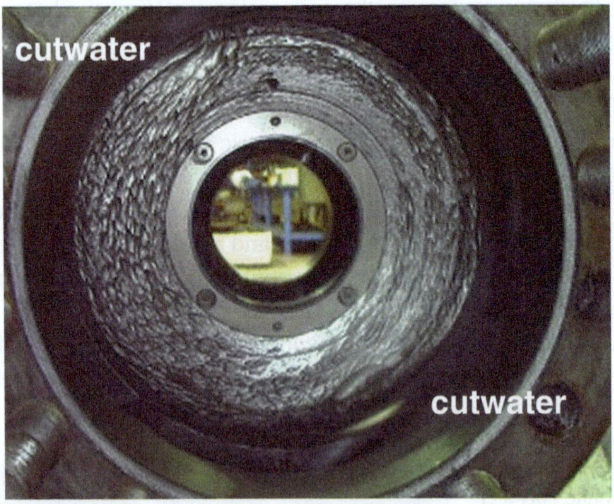

Fig. 14.35 Casing side of ISR of a double-volute pump, [42]

high-pressure zone upstream of the cutwater into the ISR. As in Fig. 14.33 note also the asymmetric wear around the suction nozzle; this confirms that the flow in the ISR was not rotationally symmetric. The relevant flow mechanisms have been discussed in context with Fig. 9.29.

Figure 14.37 depicts an impeller with cylindrical blades leading to flow separation and erosion downstream of the throat area. The rear shroud is eroded through its entire thickness starting from the inside of the impeller channels.

Figure 14.38 shows erosion on a semi-open impeller. At BEP erosion at the leading edge is high on the hub and mid streamlines. This would almost suggest

Fig. 14.36 Casing side of a process pump; fluid: catalyst fines

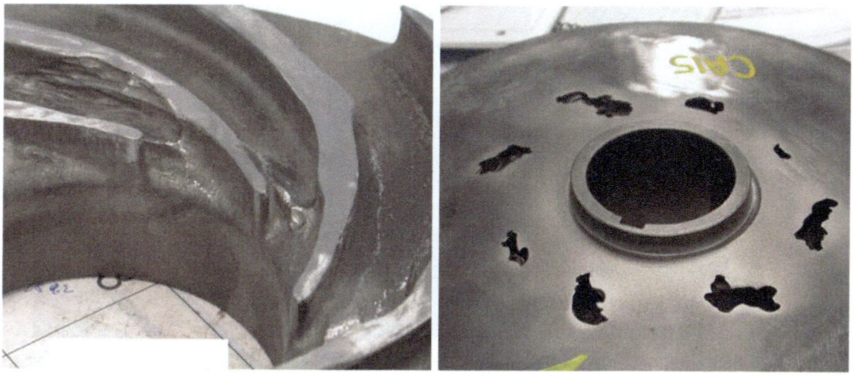

Fig. 14.37 Impeller of a process pump; fluid: catalyst fines

that flow recirculation was present even at the BEP. At overload ($q^* = 1.3$) erosion is most severe at the outer streamline where it would be expected because the highest velocity occurs at this location. At all flow rates there was a severe metal loss on the pressure side of the blades near the trailing edge, refer to Fig. 3.4.

Vortex erosion on the impeller blade trailing edge is frequently observed, Fig. 14.39 shows three examples. The impeller in the mid suggests that the rake of the blade unloaded the outer streamline (front shroud) in that the particles are displaced towards the hub where higher wear is observed than near the front shroud. The mechanisms creating the vortices include: (1) corner vortexes as discussed with Fig. 1.6; (2) horse shoe vortexes as discussed with Fig. 3.20; (3) secondary flows of

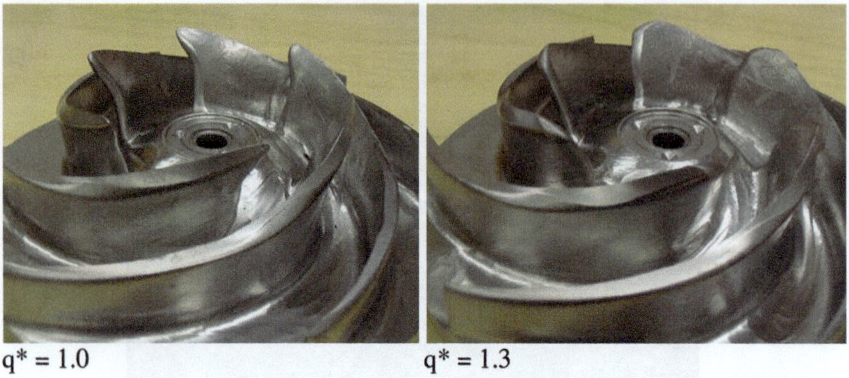

Fig. 14.38 Semi-open impeller of a process pump operating with a sand-water mixture of c_s = 150 kg/m^3, u_2 = 26 m/s, d_s = 32 μm, material: Duplex

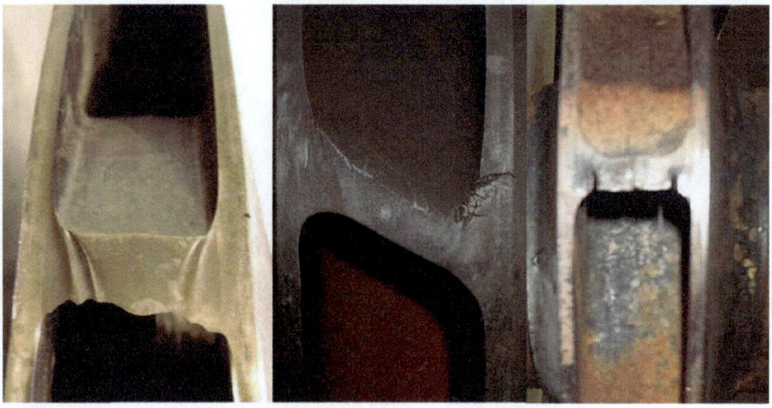

Fig. 14.39 Vortex erosion on the impeller blade trailing edge, [42]

various origins as discussed in Chap. 5.2; (4) there may also be a (presumably weak) interaction with the boundary layers thrown off the impeller shrouds.

Figure 14.40 shows exceptional erosion patterns at the leading edges of the impeller blades and on the blade suction surfaces. Apparently, the leading edges have been eroded to a kind of blunt shape which subsequently led to vortex shedding. The vortex streets created in this way exhibit the remarkable erosion pattern.

Figure14.41 shows the erosion at the leading edge of the vaned diffusers of a boiler feed pump after many years (more than 100,000 h) of service. It is interesting to note that, even in high-purity boiler feed water, the extremely low contents of solids can lead over the years to noticeable erosion (the impeller trailing edge erosion in the mid picture of Fig. 14.39 is also from a boiler feed pump). The solids

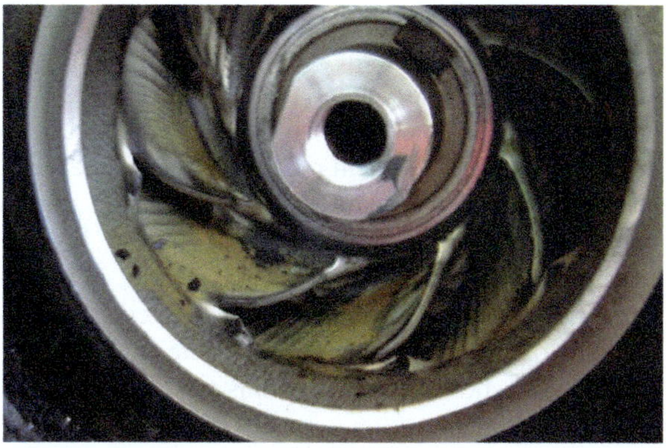

Fig. 14.40 Erosion at the impeller inlet of a water injection pump, [42]

Fig. 14.41 Erosion at the inlet to a vaned diffuser of a boiler feed pump, $u_2 > 100$ m/s

Fig. 14.42 Erosion at the diffuser outlet, [42]

stem from the corrosion products (hematite or magnetite) of the feed water system. Particle size is in the range of μm only, but the impingement velocities at the diffuser leading edge are high (impeller tip speed above $u_2 > 100$ m/s). Furthermore, the density of the solids is quite high (roughly 5000 kg/m^3). The impingement energy of the particles is accordingly increased. High density differences between particle and fluid imply dissimilarities in the fluid flow paths and the solids trajectories, which tends to increase wear.

All stages of this multistage pump were equally affected.

Figure 14.42 shows erosion at the outlet of a vaned diffuser. Presumably, the highest sand concentration was near the bottom of the channels. Flow separates at the outlet which leads to vortexes in the wakes causing the high local erosion.

Heavy sand erosion in a single-blade impeller (sewage pump) can be seen inside of the channel on the front shroud in Fig. 14.43. Note the variation in relative flow angle at the impeller outlet which varies around the circumference and locally attains or exceeds the blade exit angle of $\beta_{2B} = 45°$, see also Chap. 9.3.9. The flow angle in the relative system must have been quite steady (although varying over the circumference) - otherwise the erosion should not have left clearly defined flow paths.

Wear on the volute casing of a dredge pump can be seen in Fig. 14.44: a horseshoe vortex originating at the volute cutwater caused local wear right through the wall of the inner casing. Elsewhere, the wear is quite uniform with a wavy pattern. A sectional drawing of this type of pump is shown in Fig. 13.36.

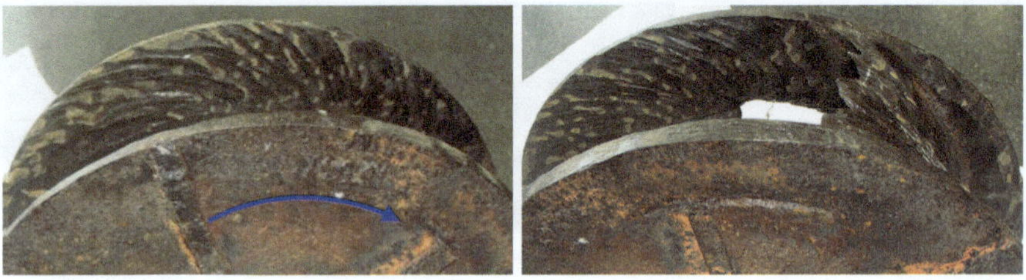

Fig. 14.43 Erosion on the front shroud and the blade pressure surface near the trailing edge of a single-channel impeller, u_2 = 15 m/s, material: cast iron

Fig. 14.44 Erosion through the volute casing of a large dredge pump

References

1. ASM Handbook volume 13A, Corrosion: Fundamentals, Testing, and Protection. ASM International (2003). ISBN 0-87170-705-5 (2003)
2. Berger, C., Schlücker, E.: Untersuchungen zur verbesserten Auslegung von Bauteilen aus Duplexstahle oder Superaustenit unter Schwingungsrisskorrosionsbelastung in wässrigen Medien. AiF report 13248 N (2005)
3. Berten, S.: Hydrodynamics of High Specific Power Pumps for Off-Design Operating Conditions. Diss, EPF Lausanne (2010)
4. Bolliger, W.: Advancements in Material Technologies For Longer Life When Pumping Difficult Liquids. Proc Indian Pump Manufacturers Association, Mumbai (2001)
5. Brendel, H., Dwars, A.: Neue Anwendungsgebiete für Umwälzpumpen in der Industrie- und Verfahrenstechnik. Industriepumpen und Kompressoren **1**, 20–26 (1999)
6. Bursik, A.: Vergleichende Untersuchungen zur Konditionierung von Kesselspeisewasser. VGB-Konferenz Chemie im Kraftwerk (1981)
7. Cooper, P., et al.: Reduction of cavitation damage in a high-energy water injection pump. ASME AJK2011–06092
8. Dechema Werkstofftabellen E 40 für Meerwasser (1998)
9. Effertz, P.H., Forchhammer, P., Hickling, J.: Spannungsrißkorrosionsschäden an Bauteilen in Kraftwerken – Mechanismen und Beispiele
10. Galsterer, D., Heimgartner, P.: Analyzing and effectively avoiding corrosion. Sulzer Tech. Rev. **1**, 22–25 (2013)
11. Grein, H., Krause, M.: Research and prevention of hydroabrasive wear. In: XVII IAHR Symposium, Beijing (1994)
12. Grein, H., Schachenmann, A.: Abrasion an Maschinen hydraulischer Kraftwerke. Sulzer Techn. Rundsch. 1/92
13. Gülich, J.F.: Bedeutung der thermohydraulischen Verhältnisse für die Korrosionssicherheit von DWR-Dampferzeugern. Atomwitschaft **2**, 82–87 (1975)
14. Gülich, J.F.: Abtragsraten bei der Erosionskorrosion unlegierter Stähle in Kesselspeisewasser. Techn. Rundsch. Sulzer. **4**, 19–22 (1986)
15. Gülich, J.F.: Calculation of metal loss under attack of erosion-corrosion or cavitation erosion. In: International Conference on Advances in Material Technology Fossil Power Plants. Chicago (1987)
16. Gülich, J.F., et al.: L'érosion-corrosion dans les pompes d'alimentation et d'extraction. Colloque EdF "Chimie de l'eau et corrosion, Seillac (1980)
17. Gülich, J.F., Bolleter, U.: Pressure pulsations in centrifugal pumps. ASME J. Vibr. Acoust. **114**, 272–279 (1992)
18. Günther, R., Knapp, P., Scherrer, G.: Erfahrungen über die Beständigkeit verschiedener Werkstoffe gegenüber hydroabrasivem Verschleiß in Kalksteinsuspensionen. Pumpentagung Karlsruhe, C8 (1988)
19. Hagen, M., Bolliger, W., Schöffler, W.: Influence of fluoride in chloride-containing FGD suspensions on the corrosion of duplex stainless steel. Vortrag DECHEMA (1997)
20. Heil, K.: Erosionskorrosion an unlegierten Fe-Werkstoffen in schnellströmenden Wässern. Dissertation, TH Darmstadt (1979)
21. Heimgartner, P., Schöffler, W., Bitterwolf, F.: Kreiselpumpen in Rauchgas-Entschwefelungsanlagen. Techn Rundsch. Sulzer. **3**, 13–19 (1987)
22. Heitmann, H.G., et al.: Erosionskorrosion in Wasser-Dampfkreisläufen. VGB-Konferenz Chemie im Kraftwerk (1981)
23. Heitz, E., Litzkendorf, M., Meysenburg, C.M., Weber, J.: Das Verhalten rostfreier Stahlgußlegierungen unter gleichzeitiger mechanischer und korrosiver Belastung. Z Werkstofftechnik. **11**, 244–258 (1980)
24. Held, H.D.: Zur Frage der Magnetitablagerungen in Zwangsdurchlaufkesseln. VGB Kraftwerkstech. **54**(6), 406–408 (1974)

25. Heubner, U., et al.: Nickelwerkstoffe und hochlegierte Sonderedelstähle, 2. Aufl., Bd. 153. Kontakt & Studium Werkstoffe, expert verlag. ISBN 3-8169-1011-4 (1993)
26. Heumann, A.: Konstruktive Gestaltung von Spiralgehäusepumpen sowie Werkstoffauswahl für schleißendes Fördergut. Pumpentagung Karlsruhe, A3-03 (1992)
27. Höppel, H.W.: Schädigungsmechanismen und hydroabrasives Verschleißverhalten unterschiedlicher Hartstoffsysteme. Dissertation Universität Erlangen-Nürnberg (1997)
28. Hutchings, L.M.: Tribology, Friction and Wear of Engineering Materials. Edward Arnold. ISBN 0-340-56184-x (1992)
29. Kaesche, H.: Die Korrosion der Metalle. Springer, Berlin (1990)
30. Kastner, W., Riedle, K., Tratz, H.: Untersuchungen zum Materialabtrag bei Erosionskorrosion. VGB Kraftwerkstech. **64**(5), 452–465 (1984)
31. Kastner, W., Hofmann, P., Nopper, H.: Erosionskorrosion in Kraftwerksanlagen – Entscheidungshilfe für Maßnahmen zur Schadensvermeidung. VGB Kraftwerkstech. **70** (11), 939–948 (1990)
32. Keller, H.: Erosionskorrosion an Naßdampfturbinen. VGB Kraftwerkstech. **54**(5), 292–295 (1974)
33. Kießling, R.: Zur Modellierung und Simulation des hydroabrasiven Verschleißes ringförmiger Strömungsspalte. Dissertation Universität Erlangen-Nürnberg (1994)
34. Knapp, P., et al.: Werkstoffe für Pumpen in der Rauchgasentschwefelung – Langzeiterfahrungen und neue Entwicklungen. Pumpentagung Karlsruhe A2-01 (1992)
35. Kovach, C.W.: High performance stainless steels. NiDI Reference book Series 11021, Nickel Development Institute
36. Kratzer, A., Tischner, H.: Noridur 9.4460, ein ferritisch-austenitischer CrNiMo-Stahlguß für Pumpen und Armaturen in Rauchgasentschwefelungsanlagen. KSB Techn Berichte **18**
37. Kunze, E. (Hrsg): Korrosion und Korrosionsschutz, 6 Bände. Wiley. ISBN 3-527-29994-7 (2001)
38. Landolt, B.: Zahlenwerte und Funktionen, 6. Auflage, Bd. 2, Teil 5a, S. 575 (1969)
39. Materials for saline water, desalination and oilfield brine pumps. NiDI Reference book No 11004, 2nd edn. Nickel Development Institute (1995)
40. Naidu, B.S.K.: Silt erosion problems in hydro power stations and their possible solutions. Workshop on silt damages in hydro power stations, New Delhi (1996)
41. Offenhäuser, H.: Druckschwankungsmesssungen an Kreiselpumpen mit Leitrad. VDI Ber. **193**, 211–218 (1973)
42. Palgrave, R.: Personal communication
43. Pini, G., Weber, J.: Werkstoffe für die Förderung von Meerwasser und hochchloridhaltigen Medien. Techn. Rundsch. Sulzer. **2**, 69–80 (1979)
44. Plant, W.H.D., Mathay, W.L.: Nickel-Containing Materials in Flue Gas Desulphurisation Equipment. Nickel Development Institute
45. Practical Guidelines for the Fabrication of Duplex Stainless Steels. Intl Molybdenum Association, Nickel Development Institute (2001)
46. Radke, M., et al.: Neue konstruktive Entwicklungen für Kreiselpumpen in Rauchgasentschwefelungsanlagen. Konstruktion **42**, 53–60 (1990)
47. Radke, M., et al.: Untersuchung kostenbestimmender Faktoren bei Kreiselpumpen in Rauchgasentschwefelungsanlagen. VGB Kraftwerkstech. **71**, 455–461 (1991)
48. Roark, R.J., Young, W.C.: Formulas for Stress and Strain, 5th edn. McGraw-Hill (1986)
49. Schläpfer, H.W., Weber, J.: Austenitic-Ferritic Duplex Steels. Material und Technik 14 H 2 (1986)
50. Sedricks, A.J., Money, K.L.: Corrosion fatigue properties of Nickel-containing materials in seawater. NiDI Druckschrift Nr 1258 (1977)
51. Stainless steels and specialty alloys for modern pulp and paper mills. NiDI Reference book Series 11025, Nickel Development Institute. (2002)
52. Stoffhütte. Taschenbuch der Werkstoffkunde. 4. Aufl. W Ernst & Sohn, Berlin (1967)
53. Syrett, B.C.: Erosion-corrosion of copper-nickel alloys in seawater and other aqueous environments—a literature review. Corros. NACE. **32**(6), 242–252 (1976)

54. Tanaka, H.: Vibration behavior and dynamic stress of runners of very high head reversible pump-turbines. In: IAHR Symposium Belgrade, Beitrag U2 (1990)
55. Taschenbuch der Chemie. 5. Aufl. Harri Deutsch, Zürich (1973)
56. Technik-Lexikon, Rowohlt Verlag, Bd 35, S. 94
57. Timcke, J.H.: Quellverhalten eines Schiffskreiselpumpen-Laufrads aus Gefüge-Verbundwerkstoff. Industriepumpen + Kompressoren **3**, pp. 191–196 (1999)
58. Timcke, J.H.: Swelling behaviour of pump impellers made from a structural composite. World Pumps, Dec 28–31 (1999)
59. Tischner, H.: Hochlegierte Gußwerkstoffe zur Förderung korrosiver und abrasiver Fluide. Pumpentagung Karlsruhe C7 (1988)
60. Tischner, H.: Werkstoffwahl für Pumpen in Rauchgasentschwefelungsanlagen. KSB Techn Berichte 25 (1988)
61. Tuthill, A.H., Lamb, S.: Guidelines for the use of stainless steel in municipal wastewater treatment plants. NiDI Technical series No 10076, Nickel Development Institute www.nidi.org (1998)
62. Uetz, H.: Abrasion und Erosion. Hanser, München (1986)
63. Van den Berg, C.H.: Grain shape effects on wear. Ports Dredging **152**, 11–15 (1999)
64. Vetter, G., Kießling, R.: Verschleißverhalten von Pumpenwerkstoffen bei hydroabrasiver Strahlbeanspruchung. Konstruktion **47**, 186–190 (1995)
65. Vetter, G., Kießling, R.: Zur Auslegung von Spaltdichtungen in Pumpen gegen hydroabrasiven Verschleiß. Konstruktion **48**, 167–173 (1996)
66. Vetter, G., Klotzbücher, G.: Einige tribologische Grundlagenuntersuchungen zum abrasiven Gleit- und Strahlverschleiß von Pumpenwerkstoffen. Konstruktion **45**, 371–378 (1993)
67. Vetter, G., et al.: Multiphase pumping with twin-screw pumps. In: 17th International Pumps Users Symposium, pp. 153–169. Texas A & M (2000)
68. VGB-Richtlinien für Kesselspeisewasser: VGB Kraftwerkstechnik **60**(10), 793–800 (1980)
69. Weber, J.: Korrosion und Ablagerungen in Kühlsystemen – Ursachen und Bekämpfung. Techn. Rundsch. Sulzer. **3**, 219–232 (1972)
70. Weber, J.: Materials for Seawater Pumps and Related Systems. Pumps Offshore/Onshore, GOL, Norway (1984)
71. Weber, J., Bolliger, W.: Werkstoffe für die Förderung korrosiver Medien. Dechema, 10–12.6 (1987)
72. Wellinger, K., Uetz, H.: Gleitverschleiß, Spülverschleiß, Strahlverschleiß unter der Wirkung von körnigen Feststoffen. VDI-Forschungsheft 449, Ausg. B, Bd. 21 (1955)
73. Werkstoffe für Meerwasseranlagen. International Nickel, Druckschrift 54 (1968)
74. Wilson, K.C., Addie, G.R., Sellgren, A., Clift, R.: Slurry Transport Using Centrifugal Pumps, 2nd edn. Blackie Academic and Professional, London. ISBN 0 7514 0408 X (1997)
75. Riihimäki, A.: Corrosion-resistant cast steels for process industry pumps Vuoriteollisuus - Bergshanteringen 4/2002 (in Finnish)

Chapter 15
Pump Selection and Quality Considerations

Operational problems can often be traced back to a non-optimum - or even downright wrong - pump selection. An unsuitable selection may result from insufficient knowledge of pump operation and installation conditions or from not giving them careful consideration and analysis. Common errors in pump selection are:

- The operation range between maximum and minimum flow of the pump has not been defined, its relevance has not been understood or it is different from what had been expected. If the pump is over-sized because too large a "safety margin" was added to the head and/or the flow rate, the pump will run at partload. This entails not only a reduced efficiency but also more vibrations and noise as well as higher wear and the possibility of cavitation damage.
- The maximum possible flow rate in the plant has not been specified or determined correctly. For its definition the minimum head that can result in the plant has to be determined by considering as applicable: (1) the minimum back pressure; (2) the maximum inlet pressure in the process; (3) the minimum discharge water level; (4) the maximum suction level; (5) the minimum pipe resistances. As explained in Chap. 11.1, plants with pumps operating in parallel have to be examined very carefully in this regard.
- For the sake of cost savings an existing pump size is used far outside of the range for which it was designed. This implies heavy trimming of the impeller in order to obtain the specified operation point. Even if the operation point is not too far off the BEP of the *trimmed* impeller, the pump still runs too far to the left of shockless impeller entry flow Q_{SF}, because Q_{SF} remains unchanged by the trimming. In that case the pump may permanently operate with recirculation at the impeller inlet, with excessive noise, vibrations, and cavitation damage as a possible consequence, Fig. 4.26.
- The installation conditions of the pump are not sufficiently taken into account. This is particularly relevant in the case of an inappropriate arrangement of the suction pipe or unfavorable approach flow conditions, Chap. 11.7.

J. F. Gülich, *Centrifugal Pumps*,
https://doi.org/10.1007/978-3-030-14788-4_15

- The special conditions required by the process in which the pump is to operate have not been analyzed adequately. This not only applies to the intended normal operation but also to load cycling, start-up and shut-down procedures and upset conditions.
- A pump with insufficient margin between $NPSH_A$ and $NPSH_3$ (or $NPSH_i$) is selected leading to problems with vibrations, cavitation noise and/or damage.
- Inappropriate materials are chosen (corrosion, abrasion, cavitation erosion).
- Inappropriate machine elements (bearings, seals, etc.) are used.

15.1 The Pump Specification

Every pump is part of a *system* working in an optimum way only if all components are well tuned to each other. Therefore it is necessary, for making a good pump selection, to know those plant features which have an impact on the operation of the pump. In addition to the required flow rate and the head, knowledge of the $NPSH_A$ is of particular relevance. In the absence of a clear specification, an idea of its order of magnitude has to be formed at least. Is it a process with a few meters of $NPSH_A$ or is it a high-pressure pump fed by a booster pump? If the pump is fed from an open pit, the minimum $NPSH_A$ of the plant will result from the atmospheric pressure (depending on the installation site above sea-level), the minimum submergence that must be determined according to Chap. 11.7.3, and the pressure losses in the suction pipe, if any.

The following conditions should (explicitly or implicitly) be known in view of correctly selecting a pump. In Chap. 17.7 a comprehensive hydraulic specification is proposed.

1. The task of the pump in the system.
2. The system pressure and temperature.
3. Data for rated performance: Q_R, $H_{R,tot}$. Often rated performance equals the guaranteed point Q_g, H_g. The rated and/or guaranteed performance may be identical to the BEP (but this is not necessarily so).
4. The $NPSH_A$ of the plant at rated, guaranteed or BEP conditions and, as necessary, at other operation conditions.
5. Performance data for other specific operation points (if necessary).
6. The maximum and minimum flow rates in the domain of *continuous operation.*
7. The maximum and minimum flow rates during *short-term operation* or in transient conditions, e.g. during a switch-over of parallel working pumps, at load rejection or other.
8. For pumps operating in parallel the maximum flow rate (run-out) is determined by the operation of a single pump. At run-out the available $NPSH_A$ must be sufficient to prevent excessive cavitation, see Fig. 11.3.

9. When pumps are installed in series, their interaction has to be analyzed with regard to control and upset conditions such as one pump falling out of service.
10. The type and the chemical composition of the medium to be pumped, in particular corrosive substances.
11. The physical properties of the pumpage if it is any other than water or a common, clearly defined medium. In this case the vapor pressure must be correctly specified in order to ensure that the effects and risks of cavitation can be assessed.
12. Viscosities appreciably above that of cold water need corrections for Q, H, P, η and NPSH according to Chap. 13.1.
13. Possible inclusions of free gas or dissolved gases that might separate from the liquid in the suction pipe. The free gas content at the impeller inlet can be estimated from Chap. 6.4.2 and Appendix A3. The available NPSHA must be selected so that the volume fraction of free gas at the impeller inlet is below typically 2 to 4% at low suction pressures, refer also to Chap. 13.2.
14. Possible inclusions of solids (abrasion).
15. The type of driver (electric motor, turbine, combustion engine).
16. Fixed or variable speed? Speed range, if applicable.
17. Is a gear box necessary?
18. What kind of control is intended?
19. How much standby capacity is required (e.g. 2 × 100% or 3 × 50% pumps)?
20. Operation mode: Continuous or short-term operation? Cyclic operation with frequent start-ups and shut-downs?
21. Installation conditions: Horizontal or vertical arrangement?
22. Approach flow or suction conditions: Open or closed circuit? Open pit?
23. Fluid level variations in the suction and discharge reservoirs or pressure variations on the suction and discharge side of the pumping system.
24. The system characteristic or at least its static part H_{stat} resulting from the geodetic head differences and/or the pressure differences between the suction tank and the discharge vessel (Chap. 11.1).
25. Are there any special requirements concerning the head-capacity characteristic (steepness, head rise, shut-off pressure)?
26. The maximum admissible shut-off pressure with the allowed tolerance, if applicable.
27. For correctly sizing the driver, the maximum power consumption must be determined; with a small specific speed it occurs at about the maximum flow rate, with a medium n_q near the BEP, and with very high specific speeds at shut-off, refer to Fig. 4.15.
28. Are there any special requirements regarding vibrations or noise? Have limits been specified for the sound level? (refer to Table 10.4).
29. What tolerances are permitted for manufacturing and measurements? Which standard is to be applied for the acceptance test?
30. The guarantee and acceptance conditions, including possible penalties on efficiency or power consumption.

31. The operation period per year and the energy costs (e.g. \$/kWh) or an assessment of the capitalized energy costs (e.g. \$/kW). Minimization of the energy costs per year according to the intended operation scenarios.
32. Safety considerations, explosion protection, zero-leakage to environment, ecological aspects.

Depending on the specific application various issues concerning the system should be analyzed:

- Process and equipment for start-up and shut-down of the pump. Viscosity at start-up. The start-up current. Is the electric supply sufficient for the start-up current?
- Transients and upset conditions.
- The minimum requirements in regards to the instrumentation necessary to ensure safe operation.
- Is a safety analysis in case of earthquakes required?
- Is a water hammer analysis necessary and, if so, are protection measures to be taken?
- Check valve for preventing backflow.
- Minimum flow operation: choice of appropriate valves; ensure sufficient back pressure to avoid cavitation (cavitation might damage the valve and is likely to excite vibrations and noise).
- Is reverse running of the pump possible?
- The filling of the suction and discharge pipes (pressure pulsations, cavitation).
- The method and, possibly, the equipment to limit the maximum flow rate.

Of course, the investment, operation and maintenance costs are also of eminent importance.

In order to find the best economical and technical solution that meets all the criteria listed above, the plant and the pump should be seen and optimized as an entity. For example, the $NPSH_A$ will influence: (1) the construction costs, (2) the speed and, consequently, type, size, and cost of the pump and of the driver, (3) the cost of a booster pump and driver, if any, and the accompanying piping and fittings.

Meeting the requirements for standby capacity may lead to quite different pump types and system concepts. The layout of the intakes or the suction piping has an effect on the selection of the pump type and on trouble-free operation. Control and safety issues and the optimization of the total energy consumption require close cooperation between the architect engineer and the pump manufacturer at the planning stage.

In addition to the general aspects listed above, specific requirements apply to the multitude of different pump services (some examples are listed in Table 2.4). Thus, some understanding of the processes and system is often indispensable to correctly select the pump.

15.2 Determination of Pump Type and Size

In most cases the pump is selected from catalogues or the corresponding computer programs supplied by the pump manufacturer. By using range charts as shown in Fig. 4.21, the pump selections possible for the specified flow and head can quickly be determined. This may be done for different pump types, e.g. single-entry or double-entry, single-stage or multi-stage. Individual characteristic curves of the pre-selected pump sizes can then be checked to ascertain: (1) whether the pump will operate safely with the $NPSH_A$ available in the plant, (2) to what extent the anticipated operation range can be covered, (3) which pump promises the best efficiency, (4) which pump selection is the most cost-efficient, (5) whether it is possible to meet all operational requirements. This evaluation includes numerous criteria from the properties of the fluid to the mechanical equipment, Chap. 15.1.

In the case of very expensive or large pumps (e.g. $P > 500$ kW), for difficult technical requirements or non-standardized equipment, the pump selection can be made in the way described below (some of the applicable formulae are listed in Table 15.1).

Initially the type of driver has to be chosen (fixed or variable-speed electric motor, gear box, turbine or combustion engine).

The pump selection as well as the optimization is usually an iterative process because any initial assumptions are to be verified at the end. Depending on which data have been specified, four different cases are presented here:

Case 1: Given: Q, H_{tot}, $NPSH_A$

Required: type, n, n_q, z_{St}, d_2, f_q, z_{pp}

If the $NPSH_A$ is specified, it often determines the pump type, the speed, the size and, consequently, also to a large extent the cost of the pump.

In order to get a first idea of the rotor speed possible for the specified flow rate Q_{opt} and $NPSH_A$, a suction specific speed may be selected from Table D6.1. According to Chap. 6.7 it should further be taken into account that $NPSH_A$ must be sufficiently above $NPSH_3$. To this end, safety margins F_{NPSH} according to Eq. (T15.1.5) may be used.[1] After selecting the pump and knowing the impeller inlet diameter, it is possible to check the $NPSH_A$ against Table 6.2.

The selection of the rotor speed and of the maximum head is discussed in Chap. 7.5 for semi-pumps and Chap. 7.6.2 for axial pumps.

Step 1: $NPSH_3$ is calculated using Eq. (T15.1.5). A suction specific speed is selected from Table D6.1 and the corresponding speed for the specified conditions is calculated from Eq. (T15.1.1). This calculation is possible for single- and for double-entry pumps. If the calculation is done for the range of suction specific speeds given in Table D6.1, a range of possible driver speeds is obtained.

---

[1]Equation (T15.1.3) was derived from a graph in [B.17], Eqs. (T15.1.4 to T15.1.6) follow from Eq. (T15.1.3).

Table 15.1 Pump selection			Eq.
Calculate speed from $NPSH_A$ and selected suction specific speed n_{ss}	$n = \frac{n_{ss}\,NPSH_{3,opt}^{0.75}}{\sqrt{Q_{opt}/f_q}}$		15.1.1
Calculate $NPSH_3$ from suction specific speed n_{ss}	$NPSH_{3,opt} = \left(\frac{n}{n_{ss}}\sqrt{\frac{Q_{opt}}{f_q}}\right)^{1.333}$		15.1.2
Required NPSH-safety margin calculated from $NPSH_3$	$F_{NPSH} = 1.16\left(\frac{NPSH_3}{NPSH_{Ref}}\right)^{0.14}$	$NPSH_{Ref} = 1\ m$	15.1.3
Required NPSH-safety margin calculated from $NPSH_A$	$F_{NPSH} = 1.14\left(\frac{NPSH_A}{NPSH_{Ref}}\right)^{0.123}$		15.1.4
$NPSH_3$ calculated from $NPSH_A$	$\frac{NPSH_3}{NPSH_{Ref}} = 0.878\left(\frac{NPSH_A}{NPSH_{Ref}}\right)^{0.877}$		15.1.5
$NPSH_A$ calculated from $NPSH_3$	$\frac{NPSH_A}{NPSH_{Ref}} = 1.16\left(\frac{NPSH_3}{NPSH_{Ref}}\right)^{1.14}$		15.1.6
Maximum head per stage; for $n_q < 32$ use $H_{st,max} = 800$ m	$H_{st,opt,max} = 60\left(\frac{n_{q,Ref}}{n_q}\right)^{2.27}$ $\quad n_{q,Ref} = 100$		15.1.7
Estimation of maximum number of stages	$z_{st,max} = \frac{H_{Ref}}{H_{st,max}}$ $\quad H_{Ref} = 4000$ m		15.1.8
	for $4 < z_{st} < 16$ and $n_q < 35$		
Pressure coefficient at BEP from Fig. 3.24	$\psi_{opt} = 1.21 e^{-0.77 n_q/n_{q,Ref}}$ $\quad n_{q,Ref} = 100$		15.1.9
Outer impeller diameter calculated with pressure coefficient from Eq. (15.1.9)	$d_2 = \frac{60}{\pi n}\sqrt{\frac{2g\,H_{opt}}{\psi_{opt}}} = \frac{84.6}{n}\sqrt{\frac{H_{opt}}{\psi_{opt}}}$		15.1.10
Normalized suction specific speed according to Fig. 6.21	$n_{ss}^{**} = \frac{110}{\varphi_1^{0.625}}$		15.1.11
Flow coefficient required to achieve selected suction specific speed	$\varphi_1 = k_n^{0.8}\left(\frac{n_{ss,Ref}}{n_{ss}}\right)^{1.6}\left(\frac{n_q}{n_{q,Ref}}\right)^{0.3}$ $n_{q,Ref} = 27 \quad n_{s,Ref} = 110$		15.1.12
Impeller inlet diameter with $\tan\beta_1 = \varphi_1$	$d_1 = 2.9\sqrt[3]{\frac{Q_{La}}{f_q\,n\,k_n\tan\beta_1}\left(1+\frac{\tan\beta_1}{\tan\alpha_1}\right)}$		15.1.13
Pressure coefficient at Q = 0 for diffuser pumps, Fig. 3.24	$\psi_o = 1.31\,e^{-0.3 n_q/n_{q,Ref}}$ $\quad n_{q,Ref} = 100$		15.1.14
Pressure coefficient at Q = 0 for volute pumps, Fig. 3.24	$\psi_o = 1.25\,e^{-0.3 n_q/n_{q,Ref}}$ $\quad n_{q,Ref} = 100$		15.1.15
Equations (15.1.3) to (15.1.6) yield the same result.			

Step 2: On the basis of these calculations, a speed has to be selected. In most cases the pump is driven by an electric motor; therefore, a standard speed Eq. (7.19) must be chosen. If the pump is driven via a gear, an electric motor with variable speed, a combustion engine or a turbine, any speed may be selected within certain limits.

Step 3: With the speed thus determined, it is possible to calculate with the *total* flow rate and the *total* head a specific speed $n_{q,tot}$ and to make a first assessment ($n_{q,tot}$ is thus independent of z_{st} or f_q):

- For $n_{q,tot} > 400$ to 450 either the speed has to be reduced or the flow rate has to be distributed over two or more parallel units. This also depends on the absolute value of the flow rate.
- For $180 < n_{q,tot} < 400$ a single-stage axial pump will result.
- In the range $100 < n_{q,tot} < 180$ single-stage semi-axial pumps are most often used.
- For $15 < n_{q,tot} < 140$ double-entry pumps are also an option (Chap. 2.3.3). In this case once again, the absolute value of the flow rate is another criterion to be considered.
- For $n_{q,tot} < 15$ and heads above 300 m multi-stage pumps will in most cases be the solution, provided the flow rate is not too small.

Step 4: With regard to the number of stages there are three selection criteria.

- *Efficiency:* As may be seen from Fig. 3.26, the efficiency of a centrifugal pump drops at $n_q < 30$ with falling specific speed so that centrifugal pumps below $n_q = 8$ (in principle) make economical and ecological sense only for small performances. The higher the power, the more it should be attempted to select the pump in the n_q-range of optimum efficiencies, which is roughly between $n_q = 30$ and 60. Consequently, for a first assessment of the reasonable number of stages, the useful power $P_u = \rho \times g \times H_{tot} \times Q$ is used. With $P_u < 30$ kW conventional centrifugal pumps are feasible until about $n_q = 5$ to 6 (special impellers even lower). With $n_{q,tot} < 10$ and a power of $P_u > 100$ kW, multi-stage pumps should be considered, but the head $H_{opt,tot}$ as well as the absolute value of the flow rate are to be taken into account. The planned running hours per year are another selection criterion, because energy costs are significant if the pump runs all year round, but are less relevant if the pump runs only a few hours per year. Ultimately the pump selection is determined by the sizes available on the market.
- *Head per stage:* Equation (T15.1.7) offers an approximate value for the maximum head per stage. It depends on the material choice and a number of other factors discussed in detail in Chap. 14.
  - *Maximum number of stages:* The number of stages and the resulting length of the pump are limited for reasons of mechanical design. The higher the impeller tip speed or the head per stage, the fewer stages are allowable with respect to rotor dynamics. In addition, in a pump with a horizontal rotor axis, the sagging of the shaft caused by gravity must be smaller than the annular seal clearances in order to avoid the impellers rubbing against the wear rings of the annular seals. Alternatively, the casing elements can be arranged along the bending line of the rotor. In this case it must be taken into consideration that the sagging may be smaller during operation owing to the axial load on the shaft. A rough estimate of the maximum feasible number of stages may be obtained from Eq. (T15.1.8). As a rule, the number of stages is limited to about 16, dependent on the impeller size. However, there are special designs

of vertical bore-hole pumps for oil pumping with 60 and more stages. Considerations of mechanical design and fabrication constraints are important factors in the choice of the number of stages. For a given shaft diameter and flow rate the number of stages is limited by the maximum admissible torque. Pumps with impellers arranged back-to-back as shown in Figs. 2.9 to 2.11 (with the piston in the middle of the rotor) may be equipped with more stages than pumps with impellers arranged in-line (as in Fig. 2.8). The back-to-back design is applied for low flow rates, resulting in low n_q where many stages are required. This may be the case where a direct-drive at 3000 or 3600 rpm is selected in order to avoid a booster pump (if necessary, with double-entry impeller in the first stage). The principal reason for selecting the back-to-back design is rotordynamics: The center bush provides a high stiffness which increases the eigen frequencies of the rotor. Rotor-dynamic considerations for switching from in-line to back-to-back include the rotor speed and the density of the fluid (annular seals provide low stiffness when fluid density is low, e.g. light hydrocarbons). Typical limits are given below:

Maximum number of stages for pumps with impellers arranged in-line		
n [rpm]	Fluid density [kg/m^3]	$z_{st,max,in\text{-}line}$
<3600	>850	12
	>450	10
3600 to 6000	>850	10
	>450	8
6000 to 7000	>850	6
up to 8500	>850	4
For larger number of stages select back-to-back arrangement with $z_{st,max}$ = 16		

– The annular seal clearances increase during service due to wear; often 2-times the new clearances are allowed or specified. The augmented leakage entails an increase in axial thrust (see Chap. 9.2.3). The balance piston of in-line pumps is often designed in a way that the axial-thrust direction reverses when the seal clearances increase from new to worn conditions. This then requires a high capacity of the axial bearing in pumps with in-line impeller arrangement. In contrast, the change in axial forces due to seal wear are balanced in the back-to-back design which requires a much smaller bearing. As compared to an in-line pump, the efficiency can be a little higher in the back-to back design: less piston leakage losses and smaller axial bearing.

Step 5: After the number of stages has been chosen, the true specific speed is calculated. Next, the efficiency to be expected can be determined from Table 3.9. The impeller diameter is calculated from Eq. (T15.1.10) using the head coefficient from Fig. 3.24 or Eq. (T15.1.9). In this way it is possible to get an idea of the pump size and to assess its feasibility.

Step 6: If so desired, the impeller inlet diameter can now be determined. To this end, the diameter of the shaft or of the impeller hub is estimated by means of Eq. (T7.1.2); but some aspects of mechanical design need to be considered as well. In Fig. 6.21a "normalized" suction specific speed n_{ss}** is plotted against the flow inlet angle. By means of Eq. (T15.1.12) this relation allows calculating the flow coefficient φ_1 that is necessary to obtain the required suction specific speed. By inserting φ_1 into Eq. (T15.1.13) the impeller inlet diameter is obtained.

Step 7: The shut-off pressure can be estimated by means of Eq. (T15.1.14 or 15.1.15).

Step 8: Fig. 15.4 allows a very rough estimation of the weight of the pump.

Case 2: Given: Q, H_{tot}, $NPSH_A$, n

Required: type, n_q, z_{St}, d_2, f_q, z_{pp}

Once the speed and the $NPSH_A$ are specified, $NPSH_3$ is calculated from Eq. (T15.1.5) and the necessary suction specific speed from Eq. (T3.4.17). Checking the result against Table D6.1 it can be seen whether (and, if so, in which way) it is possible to attain that suction specific speed. If the necessary suction specific speed cannot be achieved by a single-entry nor a double-entry impeller, the speed has to be reduced, the $NPSH_A$ of the plant to be raised, or using an inducer may be considered. In that case it is imperative to observe the restrictions and comments discussed in Chap. 7.7.4. If n, n_{ss} and $NPSH_A$ are feasible and it has been decided whether the single- or a double-entry pump is the better option, the procedure of Case 1, Steps 3 to 8 can be followed.

Case 3: Given: Q, H_{tot}, n

Required: type, $NPSH_A$, n_q, z_{St}, d_2, f_q, z_{pp}

A suction specific speed is selected from Table D6.1 and the required $NPSH_3$ is estimated from Eq. (T15.1.2). Equation (T15.1.6) then yields the corresponding $NPSH_A$. Next follows the assessment whether the $NPSH_A$ thus determined is feasible: does it correspond to the plant layout? Can it be attained by economically justifiable measures? If so, Case 1, Steps 3 to 8 can be followed for single- or double-entry pumps.

Case 4: Given: Q, H_{tot}

Required: type, n, n_q, $NPSH_A$, z_{St}, d_2, f_q, z_{pp}

In this case both n_q and n_{ss} may be chosen. It is possible to choose n_q and calculate the speed by means of Eq. (T3.4.15), then to choose n_{ss} and determine $NPSH_3$ and $NPSH_A$ from Eqs. (T15.1.2 and 15.1.6). This has to be repeated until feasible speeds (possibly the standard speed of an electric motor) as well as feasible and economically justifiable values for the $NPSH_A$ of the plant will result. Once this goal is achieved, Steps 3 to 8 of Case 1 can be followed.

Safety margins: The actual requirements of the plant may be different from the data used to draw up the pump specification. Also the actual pump performance may deviate from the predicted performance because of calculation and manufacturing tolerances. Therefore some margins on flow and/or head are often applied. When deciding upon the safety margins the following aspects should be considered:

1. The smallest tolerances are to be expected when the hydraulic components are cast using existing patterns and when test results are available.
2. Greater tolerances are likely when new patterns (for a new pump size) of a known (measured) hydraulic must be manufactured.
3. In the case of new hydraulics, when no measurements are available with the impellers and diffusing elements, there are in addition the uncertainties inherent in performance prediction.
4. Manufacturing tolerances of the annular seal clearances.
5. The labyrinth clearance may increase during operation due to wear.
6. Even the geometry of the impeller or the diffusing elements may change when pumping highly abrasive fluids.
7. Uncertainties of the required head or flow (e.g. uncertainty of the pressure loss calculation, of the expected water level fluctuations, etc.).
8. If highly viscous fluids are to be pumped or for the hydraulic transport of solids, larger margins may be added because the performance prediction is subject to higher uncertainties. Furthermore, product variations (temperature and composition) have also to be taken into account.
9. If negative tolerances on flow rate or head are not permitted by the specification, higher margins are necessary than with symmetrical tolerance fields.
10. Appropriate margins are added in order to ensure that the selected pump will meet the projected operation requirements. There are essentially four ways to build a safety margin into the pump selection process (items 11 to 14 below):
11. For a pump with variable speed it is recommended, as a rule, to provide a margin for the speed. Thus a range of ± 2 to $\pm 3\%$ of the design speed would cover any manufacturing tolerances.
12. If a pump is selected with a trimmed impeller diameter, a margin on the trim diameter can easily be provided.
13. A third option is to increase the impeller diameter. Thereby either the impeller outlet width is kept constant or the shapes of the front and the rear shrouds are extended by extrapolation. However, the minimum distances between the impeller and the collector (Table 10.2) must not be substantially reduced.
14. If the impeller blade trailing edges are sufficiently thick, under-filing might possibly serve as a margin for increasing the head, Chap. 4.5.2.

As has been pointed out before, an over-sized pump will run at partload, which means higher energy and maintenance costs. When choosing margins, the risk associated with partload operation must be carefully balanced against the operational requirements and possible consequences of the actual performance falling slightly short of the specification.

15.3 Technical Quality Criteria

A number of technical quality criteria can be defined for the selection of a pump. The observation of these criteria helps to minimize energy and maintenance costs and to prevent production losses due to pump failures. Achieving this objective assists also in improving the ecological balance.

In the following only hydraulic aspects are discussed. Mechanical pump design features or machine elements are not topics of this book. For the choice of appropriate materials, which is often decisive for trouble-free operation, refer to Chap. 14.

15.3.1 Hydraulic Criteria

For making the best choice and use of a pump the following hydraulic criteria should be considered (the specific figures given below should be understood as showing trends, not sharp limits).

1. **Rated flow Q_R:** The rated flow may be defined as the flow rate at which the pump is most often operated. Often Q_R is identical to the guarantee point Q_g. According to [N.6] the pump selection has to meet the condition that $0.8 < Q_R/Q_{opt} < 1.1$. The higher the power or the flow rate of the pump, the greater is the importance of running it near the BEP. This is true with regard to energy and maintenance costs as well as to smooth running and avoiding wear or cavitation damage.
2. **Continuous operation domain:** Continuous operation should be limited to about $0.6 < Q/Q_{opt} < 1.20$ for pumps with $n_q < 50$ (for large pumps or those with a risk of cavitation the upper limit is preferably set at $q^* < 1.1$). For $n_q > 50$ the continuous operation domain may be chosen according to Fig. 11.15 or according to the documentation of the pump manufacturer.
3. **Stability of the head-capacity curves:** Within the planned operation range the head should decrease *steadily* with increasing flow rate. Even a Q-H-curve which is flat over a relatively large domain can lead to excessive pump or pipe vibrations and should therefore not be accepted (at least not for high energy pumps). However, an unstable head-capacity curve is acceptable below the minimum flow rate. This applies also to the flow ranges where it is *impossible* to operate the pump because of the particular layout of the plant. Thus a storage pump whose maximum head H_{max} is given by the geodetic elevations of suction and discharge reservoirs may have an unstable Q-H-curve above H_{max}.

4. **Minimum flow rate:** According to Chap. 11.6 a thermal minimum flow rate is necessary in order to prevent overheating. Another criterion is the "hydraulic" minimum flow which can be read from Fig. 11.15.
5. **Maximum flow rate:** The maximum flow rate has to be determined on the basis of the system characteristics resulting from the planned operation modes. If the pump is to be operated at the maximum flow rate over a long period, there must be sufficient $NPSH_A$ available to avoid cavitation on the pressure surface (where the risk of damage according to Chap. 6.6.3 is usually highest). If the maximum flow rate occurs only during transient conditions (e.g. during the switch-over of pumps working in parallel) operating the pump *for a very short period* with cavitation is acceptable.
6. **Efficiency:** The efficiency has a strong impact on the energy costs (and ecological aspects) of a plant and is therefore an important criterion. In practice, however, its assessment is often less straightforward than expected. The offered efficiency may be compared with the data given in Table 3.9 and Figs. 3.26 to 3.29. Doing this, the scatter band according to Eq. (T3.9.10) has to be taken into consideration. If the offered efficiency is appreciably below this scatter band it is worthwhile to investigate the reasons for this shortcoming. Too low an efficiency may point to poor hydraulic design, e.g.: inadequate impeller design with very non-uniform velocity profiles, non-optimum design of the collector, extremely compact design, recirculation too close to the BEP, excessive incidence i_1 of the impeller or the diffuser, poor layout of the diffuser and the discharge nozzle downstream of the volute throat, inadequate surface quality, or excessive impeller trimming because an appropriate pump size has not been available. Efficiencies appreciably above the statistical data given in Chap. 3.9 call for a critical assessment too: Are the measurements reliable? Is the efficiency just the result of too small seal clearances that cannot be maintained during operation? For pumps of less than 10 kW with a low specific speed the overall efficiency is often not really relevant for assessing the hydraulic design, especially if the mechanical losses make up an important portion of the power at the coupling. The situation is different when a plant is equipped with a considerable number of pumps and the total energy costs of all pumps are evaluated accordingly. Obviously, an efficiency evaluation has also to take into account whether the pump will be operated continuously or only a few hours per year.
7. **Efficiency scaling:** If the works acceptance test of the pump is not executed at full speed and temperature (which is often impractical) there is an element of uncertainty with regard to the efficiency differences between the works test and the plant operation. As already discussed in Chap. 3.10, "efficiency majoration" are not really helpful. The loss analysis according to Table 3.10 is not free from uncertainties either, because an exact recording of the roughness geometry and its effect on the pump losses is impossible. Thus the plant operator is left with the following options: (1) requiring a guarantee for the efficiency at agreed conditions in the works acceptance test only (ignoring the exact efficiency under plant conditions), (2) requiring an additional acceptance test in the plant which means a

considerable amount of time and money spent for attaining an acceptable precision, (3) requiring rigorous control of the manufacturing quality and the measurement methods applied in so far as the efficiency is affected by them.

8. **Annular seals:** A high pump efficiency measured on the test stand is of no use to the plant operator if it deteriorates rapidly during operation due to wear in the annular seal clearances at the impeller and at a possible axial thrust balancing device. Annular seal clearances, annular seal length, surface structure and materials have all to be assessed in this respect. For narrow annular seals non-galling non-seizing materials combinations are of prime importance, Chap. 14.
9. **Over-sizing:** Over-sizing of the pump results in partload operation. This may lead to early wear, noise, vibrations and impaired efficiency. Supposed "safety margins", if chosen too large, may actually *jeopardize the safe operation.*
10. **$NPSH_A$:** According to Chap. 6 a margin must be provided for the NPSH of the plant that is adequate for avoiding cavitation induced vibrations, noise and damage. Table 6.2 provides criteria for selecting NPSH-margins. The higher the circumferential speed u_1 at the impeller inlet, the larger must be the distance between $NPSH_A$ and $NPSH_3$. For $u_1 > 50$ m/s only very limited bubble volumes are admissible during operation in water below 200 °C. In this case special precautions have to be taken in order to prevent damage, see Chap. 6. and especially Chap. 6.7.
11. **Approach flow:** Reasonably uniform velocity distributions at the impeller inlet are required to avoid efficiency impairment, noise, vibrations and possible cavitation damage. An adequate layout of the suction piping or the pump sump is necessary (Chap. 11.7). In addition, the inlet casing of between-bearing pumps has to ensure good approach flow conditions. To this end, a sufficient flow acceleration from the inlet nozzle to the impeller eye is necessary, Chap. 7.13. In their endeavor to keep costs low by using a very compact design, pump planners often risk developing system concepts prone to difficulties during operation (noise, vibrations, cavitation damage due to non-uniform approach flow). If these problems are understood in the planning stage, their consequences may often be reduced at moderate (or even negligible) cost by designing appropriate structures (e.g. flow straighteners, deflection baffles in bends, corner pieces and ribs in wet pit designs, etc.), Chap. 11.7.
12. **Incidence:** The incidence at the impeller or diffuser inlet should not exceed 4 to 6° (for $u_1 > 50$ m/s even less). The higher the tip speed u_1, the more important is it to limit the incidence.
13. **Trim on impeller outer diameter:** As discussed in Chap. 4.5.1, it is common practice to trim the impellers of standard pumps in order to cover a defined performance range with a given pump size. The lower the specific speed and the power of a pump, the larger is the range that can reasonably be covered in this way. With increasing specific speed the trim on the impeller should be limited according to Eq. (4.23), see Chap. 4.5.1. If large pumps are involved, the efficiency loss and the resulting higher energy cost may be compared, by means of Eq. (4.21), with the cost of making a new casing pattern (and, if need be, of the impeller) for an in-between size. Said comparison will show whether

an in-between casing size might be more economic than a heavily trimmed impeller. Smoother running and lower maintenance costs would signify an added advantage.

14. **Vibrations and noise:** Section 10.8 contains a list of measures and criteria for low-vibration pumps. Of particular importance are: (1) the distance between the impeller and the diffuser or the volute cutwater according to Table 10.2, (2) avoiding unsuitable combinations of the number of impeller blades with the diffuser vane count according to Chap. 10.7.1, Eq. (10.13), (3) avoiding at the BEP any excessive deceleration at the impeller inlet, keeping $w_{1q}/w_1 > 0.75$ for an axial inlet and $w_{1q}/w_1 > 0.65$ for a radial inlet Chap. 7.2, (4) restricting the deceleration at the diffuser inlet too, so that $c_{3q}/c_2 > 0.75$ Chap. 7.9.[2]
15. **Partload recirculation:** As discussed in Chap. 5, partload recirculation is present in every pump. Recirculation is even necessary, to some extent, for obtaining a stable pump characteristic. The energy contained in the recirculating fluid (the recirculation power P_{Rec}) is dissipated as a large-scale turbulence that contributes to higher levels of noise, vibrations, and wear. That is why excessive (or unnecessary) partload recirculation should be avoided. Indirect indicators for excessive recirculation are: (1) Too much power consumption at $Q = 0$ (Fig. 4.14 may serve as a reference); (2) A flat curve $NPSH_3 = f(Q)$ at $q^* > 1$; (3) An increase of, or a maximum in, the curve $NPSH_3 = f(Q)$ at partload indicates that the impeller eye diameter may be oversized or the impeller inlet angles were selected too large. Avoiding such NPSH-curves is possible without any loss of suction capacity. However, the suction specific speed as such, which is often considered as an indicator of excessive recirculation, is *not a useful criterion*, Chap. 6.2.4 and [2].
16. **Radial thrust balancing:** Whether or not a double volute has to be used depends on the pump type, the head and the specific speed. The decision will result from an assessment of the bearing load and the shaft deflection at the annular seals and the shaft seals. Caution is required with double volutes with a wrap angle significantly smaller than 180°, because in that case according to Table 9.4 the radial thrust increases considerably, Fig. 9.46.
17. **Pumping plant:** Selecting a "good" pump is not sufficient to guarantee reliable operation. The layout of the pumping plant has also to be planned carefully. Important items are: good approach flow conditions in accordance with Chap. 11.7, the correct choice of valves and fittings that will not generate inadmissibly high pressure fluctuations, adequate instrumentation, control and monitoring suitable for minimizing transients, upset conditions and water hammer. Just as with the pump, minor savings on plant investment costs can avenge themselves with costly production losses, retrofits and lengthy arguments between the parties concerned.

[2]Maximum values of w_{1q}/w_1 are close below 1.0. In a correctly designed hydraulic the ratio c_{3q}/c_2 follows from the collector calculation according to Chap. 7. Low values of c_{3q}/c_2 point to a non-optimum design and a design check would be indicated.

15.3.2 Manufacturing Quality

The vast majority of industrially used impellers, diffusers and volute casings, due to their complicated forms, are manufactured by various casting methods. Frequently cast components exhibit considerable dimensional deviations from the design according to the drawing. This results from non-uniform shrinking of the casting when cooling down, from core shift, and from pattern, mold and core tolerances. Furthermore, there is some loss of material because of scaling. This is taken into account in pattern making by adding a small allowance in order to make up for the assumed metal loss.

Head and efficiency of a pump according to Chap. 3.10 are considerably influenced by the surface quality which depends on the casting method and temperature, the material used, the molding method and material as well as on the cleaning and dressing of the castings.

The casting tolerances of the hydraulic components are a decisive factor in the effort of accurately meeting the projected pump characteristics. They have an influence also on efficiency, vibrations and noise, even impeller fractures and on the risk of cavitation damage. For the operation range outside of partload recirculation a quite good quantitative assessment of the effects of various tolerances on the pump characteristic is possible using the procedure given in Chap. 4.7 (Table 4.4).

The specification of a defined accuracy of the pump characteristic leads inevitably to the maximum allowable deviations for the impeller outlet width, the blade angles and the throat area of the collector (within the limits of uncertainty of such calculations). In contrast, an estimate of the effect of deviations in the shape of the meridional section or the blades is not easy to make.

Since better accuracy means, as a rule, more expensive casting methods and additional time for dressing, the tolerances are chosen with regard to the intended use of the pump. With this aim in view three quality classes (G1 to G3) are defined in Table 15.2 according to the criteria below. A more specific definition of the quality classes is given in Chap. 15.4.

1. The operator's requirements as to the allowable deviations of the actual from the projected pump characteristics. The accuracy may be specified for example according to ISO 9906, [N.2] (in Table 15.2 these tolerances are quoted).
2. The specific speed: at low n_q the diffuser/volute throat area must be executed within narrow tolerances, at high n_q the impeller tolerances are important.
3. The operator's requirements as to the maximum admissible vibrations according to Tables 10.6 to 10.8, [N.6] or [N.16].
4. The impeller tip speed u_2 or the head per stage as a measure for the steady loads and hydraulic excitation forces.
5. The circumferential speed u_1 at the impeller inlet as a measure for the risk of cavitation damage.
6. The power consumption is a measure for the energy and maintenance costs of the pump as well as the excitation of vibrations and noise. This because the recirculation power at partload is entirely dissipated as turbulence.

Table 15.2 Tolerances for impellers, diffusers and volutes

Quality class		G1	G2	G3		
Tolerances on Q_g, H_g, η according to ISO 9906	t_Q	± 5%	± 8%	± 9%	Tolerance on flow rate	
	t_H	± 3%	± 5%	± 7%	Tolerance on head	
	t_P	+ 4%	+ 8%	+ 9%	Tolerance on power consumption	
	t_η	0 (–3%)	– 5%	– 7%	Tolerance on efficiency	
API 610	v_u	3	-	-	mm/s RMS	Vibrations according to [N.6] or [N.16]
ISO 10816-7	v_u	3	3.7	5.6		
Outlet	u_2	> 90	40-90	< 40	m/s	Circumferential velocity
Inlet	u_1	> 50	15-50	< 15		
Head	H_{st}	> 400	80-400	< 80	m	Head per stage at BEP
Power	P	> 3000	300-3000	< 300	kW	Power consumption

- Figures 15.2, 15.3 and 15.5 provide more specific criteria for the head per stage, impeller tip speed and power per stage defining the quality class.
- **The most stringent criterion determines the quality class.**

Parameter		Tolerances			Check with	Significance
Impeller outlet width	b_2	± 2.5%	± 3.5%	± 5%	Slide gauge	Head Stability Excitation of vibrations
Vane distance at outlet	a_2	± 2.5%	± 3.5%	± 5%	Calipers, circular disks	
Outlet angle	β_{2B}	± 4%	± 7%	± 10%	Gauge, coordinate measuring machine	
Inlet angle	β_{1B} α_{3B}	± 1°	± 2°	± 3°	Gauge, coordinate measuring machine	Cavity length $NPSH_i$, $NPSH_3$
Vane distance at inlet	a_1	± 3%	± 4%	± 6%	Calipers, circular disks	$NPSH_3$ at $Q > Q_{SF}$
Inlet profile		± 4%	± 8%	-	Profile gauge	$NPSH_i$,
Outlet profile		± 5%	± 10%	-	Profile gauge	Head Pressure pulsations
Vane thickness	e	± 7%	± 10%	± 15%	Slide gauge, calipers	Head Stresses
Throat area of diffuser or volute	A_{3q}	± 5%	± 7%	± 10%	Calipers, gauges, circular disks	Flow rate at BEP Efficiency, Shut-off head, Stability
Vane distance[1)]	a_3	± 5%	± 7%	± 10%		
Width[1)]	b_3	± 5%	± 7%	± 10%		

[1)] The tolerances given are allowable only in as much as the tolerances of the diffuser/volute throat area are not exceeded

Which one of these criteria is deemed to be determinant for the quality class depends in each application upon the prevailing conditions and requirements. It is important to assess each criterion independently from the others. *The most stringent criterion or requirement determines the quality class*. For example, if great demands are made on the limitation of vibrations, a high quality has to be chosen even though the process might allow rather large tolerances on the performance curves and even if the power consumption of the pump is low.

In Table 15.2 geometric tolerances for the three quality classes are listed. The following observations should be kept in mind:

- The data given are applicable to radial and semi-axial pumps $n_q < 150$. They must not be used for the profile geometry of propeller pumps.
- Machined dimensions (such as the impeller outer diameter) are not taken into account, because the machining accuracy is within a few tenths of a millimeter.
- Since the impeller outlet angles β_{2B} may vary in a wide range, their tolerances are expressed as percentages (not as degrees). However the inlet angles β_{1B} do not exceed a relatively narrow range. That is why they are preferably expressed as angle differences.
- The tolerances for the impeller outlet width b_2 apply to the average of all channels as well as to the differences between individual channels which are important for the excitation of synchronous vibrations ("hydraulic unbalance"). These tolerances are meant furthermore for deviations in the same channel, because the outlet width frequently varies appreciably over one pitch (especially in impellers with a small number of blades).
- The effects of the casting tolerances may either *add up or compensate*. For that reason, the tolerances given should be considered as approximate values. If, for example, the outlet width of an impeller is too large but the outlet angles are too small (and perhaps the blades too thick, causing a higher blockage) it may still be possible to attain the performance data within the specified tolerances. The effect of various tolerances on the pump characteristic may be calculated by means of Table 4.4.
- For high values of H_{st} or P attention should always be paid to deviations of β_{2B} and b_2 from channel to channel and within a pitch. Deviations between the channels and errors in the pitch must be kept small in order to reduce the excitation of vibrations.
- If the parts are ordered as castings in accordance with a quality class defined in a standard, the corresponding tolerance ranges of that standard apply insofar as they are not wider than the values given in Table 15.2.
- Sometimes asymmetrical tolerance ranges are to be recommended, e.g. b_2 + 4%/ −0 in the case that no other margin has been provided for the head; or b_2 0/ −2% in order to avoid the risk of an unstable characteristic. Similar observations apply to the outlet angle β_{2B} and the blade distance a_2.
- The blade distance a_2 gives a clue to the accuracy of the blade positions within the casting. The width a_1 at the impeller throat has an important influence on the $NPSH_3$ curve in the high flow range.

- Measurements of the shapes of the rear and front shrouds and of the impeller blades and diffuser channels are rather complicated. If they are at all possible without destruction of the component, they have to be made by means of a coordinate measuring machine or templates. That is why deviations from the required shapes are often not recognized. Even if known, it is hardly possible to make a quantitative assessment of their effect on the pump characteristic. For this reason complying with the tolerances given in Table 15.2 is not a guarantee for success.
- An assessment of the allowable roughness for each pump application is possible according to Chap. 3.10. The allowable roughness depends on the Reynolds number and the pump size. For hot water it is lower than for oil. The figures given in Table 15.2 are only approximate values for average conditions.
- As discussed in Chap. 6, for circumferential speeds of $u_1 > 50$ m/s at the impeller inlet only very small cavitation bubble volumes are admissible. For $u_1 > 75$ m/s the pump should be operated virtually without cavitation bubbles, Chap. 6.7. Accurate compliance with the impeller blade inlet profile is crucial in this case because even small deviations from the required geometry have an influence on cavitation inception ($NPSH_i$). Special machining techniques are most often indispensable, such as: (1) careful grinding of the blade inlet using profile templates ("finger templates") and templates for the suction surface ("cone templates"); (2) electrical discharge machining; (3) NC milling operations with subsequent grinding to remove machining marks in inaccessible regions. The finished impeller undergoes a final check by means of said templates. As a proof that the pump will operate in the plant without bubbles, the operator may in critical cases even ask for separate acceptance tests of each impeller at the specified $NPSH_A$ with stroboscopic observations of the cavitation development at the impeller inlet. This requires a special test pump associated with additional costs.
- The tolerances for the impeller inlet profile given in Table 15.2 refer to local deviations of the blade thickness. They are only meant as approximate indications. It goes without saying that the shape of the profile has to be smooth and steady. A thin leading edge, followed by an almost concave blade surface, is likely to cause local flow separation and cavitation. The inlet profile rather should be "convex" on the suction and pressure surfaces in order to reduce the effect of incidence.

Figure 15.1 shows a "cone template" used for checking the blade shape at the inlet. It is designed as an intersection of a cone with the suction surface of the blade. The cone angle is selected so that the template touches the blade at an angle of about 90°. Figure 15.1 (on the right) shows a "finger template" for checking the leading edge profiles of the blades.

Table 15.3 contains supplementary quality requirements for cast components which are significant for the integrity of the component as well as for vibrations and efficiency of the pump. For high-duty impellers or inducers an adequate blade thickness and fillet radii between the blades and the impeller shrouds are important in order to avoid component failure due to fatigue, Chap. 14.1.

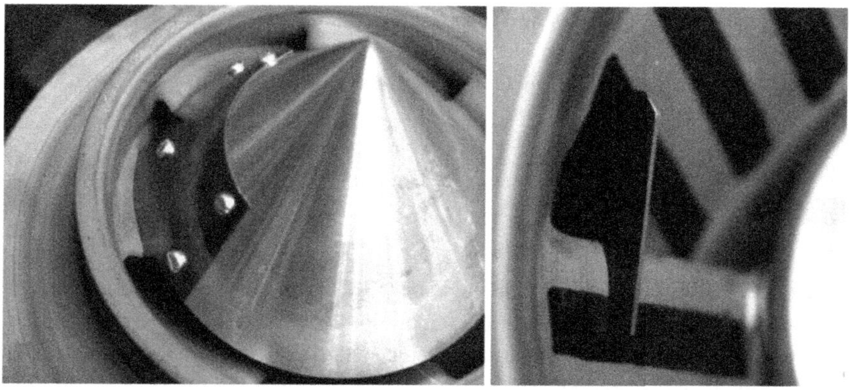

Fig. 15.1 Templates for checking the blades at the impeller inlet; left: cone template for checking the blade suction surface; right: finger template for checking the blade profiles, Sulzer Pumps

Table 15.3 Quality requirements for impellers and diffusers

Class	G1	G2	G3
Impeller shrouds	Machined	Machined	Cast[1)]
Roughness (Rugo test, visual check)	N8	N8 to N9	N9 to N10
Casting process for impeller and diffuser	Investment casting Ceramic core	Quality castings Ceramic core	Sand casting
Fillet radii, Sect. 14.1	Check very important: low notch effect	Check	Visual check
Variations of thickness of impeller shrouds and hydraulic channels	Check very important with respect to excitation of vibrations	Dimensional check	Visual check
Casting checks of impeller and diffuser	Surface crack checking[2)] Dye penetrant testing	Visual check of complete channels (endoscope if necessary); clean surfaces without rests of sand or slag	

The casting checks to be carried out primarily depend on the specification.
[1)] If no limits on vibrations are specified
[2)] Surface crack checking: magnetic test methods, if material is sufficiently magnetic; or dye penetrant testing (more expensive) with non-magnetic materials

In summary:

- *The quality class is a measure for the level of engineering analysis and manufacturing effort required for a specific application in order to fulfill the needs of the pump owner.*

- *The various criteria applied to the selection of the quality class allow taking into account the many aspects which should enter this decision.*
- *The most stringent criterion or requirement determines the quality class.*

15.4 High-Energy Pumps

As impeller tip speed, head per stage or power per stage increase, pumps are exposed to higher risk of vibration, cavitation and mechanical damage. By an appropriate design and by selecting suitable materials it must be ensured that the pump is fit for the intended service. A rationale for defining the severity of pump service may help to make appropriate choices as to materials, casting quality and the required efforts in engineering analysis and testing.

Cavitation damage can be prevented by providing sufficient $NPSH_A$, by selecting an appropriate material and by a careful hydraulic design which minimizes the creation of cavitation bubbles. Main selection criteria are the circumferential speed at the impeller inlet and the type of fluid and its properties. Table 6.2 gives guidelines for making the appropriate choices. Therefore cavitation criteria need not be explicitly included in the definition of high-energy pumps. The same applies to the effects of erosion-corrosion and abrasive wear discussed in Chap. 14.

The risk of impeller or diffuser damage due to alternating stresses can be evaluated according to Chap. 14.1. The relevant parameters are pressure rise per stage, impeller geometry (specific speed), type of fluid and materials. The risk of impeller blade cracking was used as criterion for defining high-energy pumps in [N.23], [B.15] and [1]. According to this definition the "limiting" curve between "low-energy" and "high-energy" pumps stipulates that the pressure rise per stage drops with increasing specific speed. This is true for blade stresses but does not cover the risk of impeller shroud fractures, which increases at low n_q where structural resonances are more likely to occur.

As per [N.23] high-energy pumps are defined as pumps whose head per stage exceeds the limit given by the respective curve shown in Fig. 15.2 and approximately by Eq. (15.1) with $H_{Ref} = 1$ m, $n_{q,Ref} = 25$ and $\rho_{Ref} = 1000$ kg/m^3:

$$\frac{H_{st,opt}}{H_{Ref}} > 275\left(\frac{n_{q,Ref}}{n_q}\right)^{1.85}\left(\frac{\rho_{Ref}}{\rho}\right) \qquad (15.1)$$

This formula is valid for $25 < n_q < 67$. Below $n_q = 25$ the limit is $H_{st,opt} = 275$ m; no limits are defined for the range above $n_q = 67$.

The weaknesses of Eq. (15.1) are: (1) The domain of specific speeds covered is quite narrow; (2) There is one single jump from "low-energy" to "high-energy" pumps; (3) The limiting pressure rise per stage is quite low; (4) The risk of impeller shroud fractures at low specific speeds due to structural resonances is not evidenced by the method.

Elaborating on the concept of quality classes G1 to G3 introduced in Chap. 15.3, the above shortcomings can be removed. To this end, the quality classes are defined by Figs. 15.2 and 15.3 where boundaries for the head per stage at BEP (or the impeller tip speed) are plotted against the specific speed. Curve 1 is roughly representative of the upper limits achieved by industrial applications (rocket pumps are excluded from the consideration). Curve 2 is calculated as $(1/\sqrt{3})$ of the heads given by curve 1; while curve 3 is calculated as 1/3 of the heads from curve 1. Comments to Figs. 15.2 and 15.3:

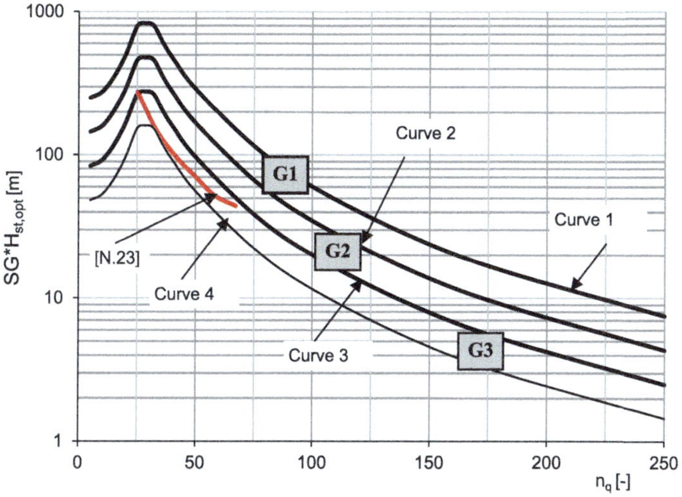

Fig. 15.2 Definition of quality classes in terms of head per stage: SG = specific gravity; SG = ρ/ρ_{Ref} with ρ_{Ref} = 1000 kg/m^3

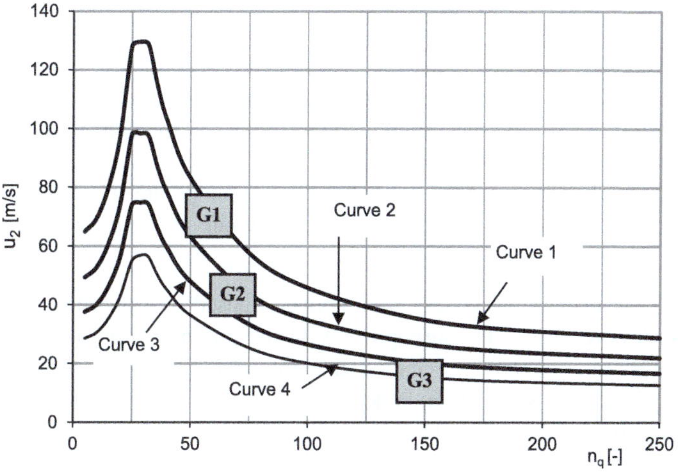

Fig. 15.3 Definition of quality classes in terms of impeller tip speed u_2; valid for liquid density of ρ = 1000 kg/m^3

Table 15.4 Energy level of pumps and quality classes, curves in Figs. 15.2, 15.3 and 15.5

Class	**G1**	**G2**	**G3**
Energy level	High between curves 1 and 2	Medium between curves 2 and 3	Low below curve 3
Energy level for **sea water** applications	High between curves 2 and 3	Medium between curves 3 and 4	Low below curve 4

- For defining the recommended quality class: (1) determine the head per stage at the best efficiency point $H_{st,opt}$; (2) multiply $H_{st,opt}$ by the specific gravity SG; (3) enter Fig. 15.2 with the specific speed n_q and SG × $H_{st,opt}$ in order to determine the quality class.
- The quality classes are defined by: Class G1 is between curves 1 and 2; it corresponds to "high-energy pumps". Class G2 is between curves 2 and 3. Class G3 is below curve 3; it corresponds to "low-energy pumps", Table 15.4.
- The curves exhibit maxima which are due to the fact that impellers with low specific speeds are likely to have low structural eigen frequencies—in particular for the mode with two diameter nodes (Chap. 10.7.1).
- Curve 1 does not represent an absolute limit, but impellers with heads per stage above curve 1 need careful design and structural analysis in terms of stresses (fatigue), natural impeller frequencies and possible resonances. Resonances with $\nu_3 \times z_{Le} \times n/60$ are more likely in diffuser pumps than with volute casings. Detailed structural analysis may not be required if the design is proven by truly representative long-term operation experience with the same impeller geometry, tip speed, material and fluid.
- In essence the curves are based on experience and engineering judgment. This is particularly true for the domain to the left of the maxima of the curves. The heads to the right of the maxima follow the rationale developed in Chap. 14.1.
- The impeller material must be selected in accordance with Chap. 14.1. Figures 15.2 and 15.3 are relevant for metallic pumps only.
- The requirements for the different quality classes are defined in Chap. 15.3, Tables 15.2 and 15.3.
- The limits for sea water (or produced water) pumps are to be downgraded by one class. Low-energy pumps are thus below curve 4, high-energy pumps are above curve 3.

The definition of quality classes given above is based on the criterion of impeller fractures. A criterion for the susceptibility of the pump to vibrations has not been applied even though it is understood that the vibration levels tend to rise with increasing head per stage or impeller tip speed. Another weakness for defining high-energy pumps from Fig. 15.2 is the absence of the influence of the pump size or the power. An alternative approach is discussed in the following.

The rotor/stator interaction (RSI) can be taken as representative of the unsteady hydraulic forces susceptible to excite vibrations. The RSI rises with the impeller tip speed u_2, the density of the liquid and the impeller outlet width which increases with the specific speed. The radial excitation forces can thus be assumed to be proportional to:

$$F_{dyn} \sim \frac{\rho}{2} d_2^2 \, b_2^* u_2^2 \Delta p^* \tag{15.2}$$

The reaction of the pump to a given level of excitation forces depends (among other factors) on the mass of the pump. The vibration velocity is assumed to be proportional to the excitation force divided by the mass of the pump m_{pump} and the angular velocity ω of the rotor. In this way a vibration velocity parameter v_v can be defined by Eq. (15.3):

$$v_v \sim \frac{F_{dyn}}{m_{pump}\,\omega} = \frac{\rho\, d_2^2 b_2^* u_2^2 \Delta p^*}{2\, m_{pump}\,\omega} = \frac{d_2^2 b_2^* u_2^2 \Delta p^*}{2\, m_{spu}\,\omega\, g\, H_{opt,tot}\, Q_{opt}} \tag{15.3}$$

The pump mass is expressed as the product of the specific mass m_{spu} of the pump (in terms of kg/kW) and the useful power at the best efficiency point. The specific mass is defined by Eq. (15.4):

$$m_{spu} = \frac{m_{pump}}{P_{u,opt}} = \frac{m_{pump}}{\rho\, g\, H_{opt,tot}\, Q_{opt}} \tag{15.4}$$

Figure 15.4 shows specific mass data of single-stage pumps with overhung impellers and between-bearing pumps; the latter include multistage pumps and single-stage pumps with double-entry impellers. The specific mass of a multistage pump diminishes with increasing number of stages for a given design. Depending on the design pressure and other design features the specific masses of various pump types are to be found in a wide band as suggested by Fig. 15.4.

The impeller outlet width can be determined from Eq. (7.1). Knowing the specific mass for a given type of pump and assuming a dimensionless pressure pulsation of Δp^*, defined by Eq. (T14.1.3), typical vibration velocities can be calculated from Eq. (15.3).

The data can be arranged in different groups of specific speeds and subsequently fitted by smooth curves or power functions. Finally, it is possible to calculate a graph of the type shown in Fig. 15.5 where curves with constant vibration velocity parameter are plotted as $P_{u,st,opt} = f(n_q)$.

Of course, it is *not possible* to calculate the actual vibration velocity of a pump from Eq. (15.3). Therefore, the absolute figures calculated are not relevant (they have been omitted from the graph); the curves can only give tendencies. Interestingly, the curves in Fig. 15.5 reflect the same trend as the curves in Figs. 15.2 and 15.3 to the left of the maximum; that is, the boundaries for the power per stage defining the classes G1 to G3 decrease as the specific speed is reduced.

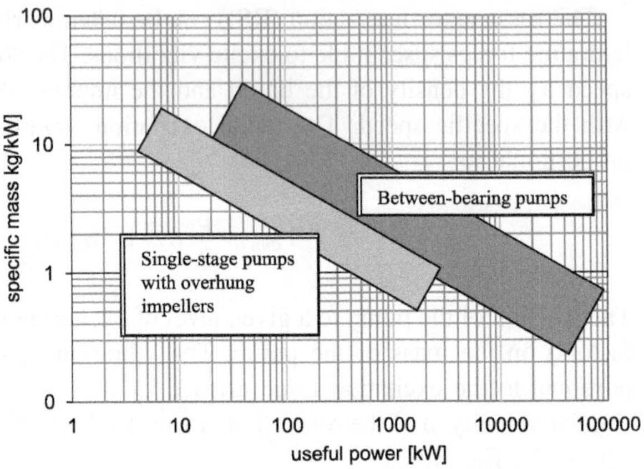

Fig. 15.4 Specific mass of bare-shaft metallic pumps referred to the useful power

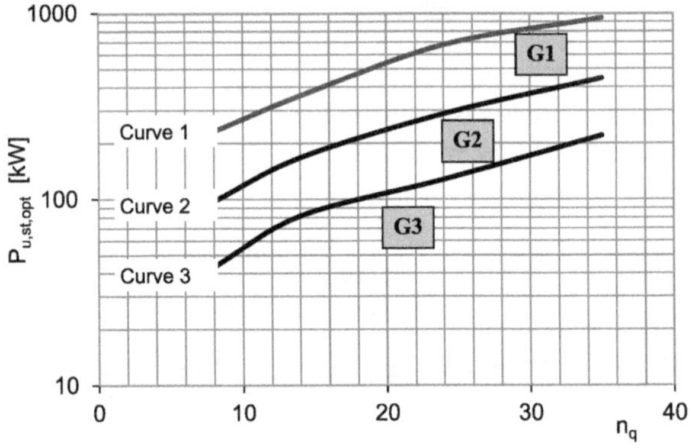

Fig. 15.5 Tentative definition of quality classes in terms of useful power per stage

The concept introduced by way of Fig. 15.5 relies on experience and engineering judgment. This is inevitable since an exact calculation in general terms is not possible. It may be attempted to validate the concept in a more quantitative manner by a comparison with test data.

References

1. Cooper, P., et al.: Minimum continuous stable flow in centrifugal pumps. In: Proc. Symp. Power Plant Pumps, New Orleans, 1987, EPRI CS-5857 (1988)
2. Gülich, J.F.: Selection criteria for suction impellers of centrifugal pumps. In: World Pumps, Parts 1 to 3, January, March, April (2001)

Chapter 16
Pump Testing

Tests on pumps are done with the object to verify performance and fitness for service, to find the reasons for malfunctions and to develop new or improved hydraulic components. Acceptance tests are usually done according to standards, notably ISO 9906, [N.2]. The details of these standards will not be repeated here since it is mandatory to always follow the latest edition in force when planning a test.

Some fundamental aspects of testing were treated in earlier chapters in order to ease the understanding and appreciation of the phenomena discussed. For example, this applies to the definition of head and NSPH (Chap. 2.2), cavitation (Chap. 6.2.5), radial forces (Chap. 9.3.2), vibrations (Chap. 10) and the quality classes defined in Chap. 15.3.2. These items will be referenced in the following as appropriate but not repeated.

As a consequence, the present chapter intends to give an overview on testing and instrumentation, to explain suitable test loop configurations and to discuss some of the many things which can go wrong when testing a pump.

16.1 Types of Tests and Measurements to Be Taken

Different types of tests are done on centrifugal pumps:

1. Factory acceptance tests (FAT) are performed to prove to the pump purchaser that the guaranteed performance is met and that the pump is adequate for service. The tests are done according to national or international standards and to the client's specification. Hydraulic performance is preferentially checked according to ISO 9906, [N.2]. For vibration limits and standards refer to Chap. 10.9.

J. F. Gülich, *Centrifugal Pumps*,
https://doi.org/10.1007/978-3-030-14788-4_16

2. Factory (or works) tests are done by the pump manufacturer in order to prove that the pump is fit for the intended service with respect to performance, vibrations and mechanical integrity. For large pumps, the works tests usually are witnessed by the client or an authority mandated by him. The works test is then identical to the FAT. For mass-produced small pumps the works test is done by the manufacturer on a selected number of specimens since it would be impractical and uneconomic to test thousands of pumps. Such tests may include endurance tests over months in order to ensure the reliability of the product.
3. If so agreed in the contract, acceptance or works tests can be done at reduced speed and with cold water even though the pump operates with higher speeds and/or hot water (or other liquids) in the plant.
4. Sometimes "string tests" are specified in order to test the proper functioning of pump, driver, gear box (if any) and auxiliaries. Typical applications are pumps with a power above several MW driven by gas turbines via a gear box or pumps with variable frequency drives. If an issue, torsional vibrations can be checked during a string test.
5. Field tests in the plant (e.g. in thermal power stations) are sometimes done to check performance - notably efficiency - at actual plant conditions, i.e. at full speed and temperature. But it is quite difficult to determine the efficiency with sufficient accuracy to warrant the high effort and cost involved. More often field tests are performed for finding the cause of malfunctions and for checking the suitability of remedies taken.
6. For very large pumps (e.g. storage pumps) works tests would become prohibitively expensive. Model acceptance tests can then be done in lieu of testing the full-size pump. The scaling of the model and its performance is done according to Table 3.4. The scaled-down model should have a minimum impeller diameter of typically 300 mm, while the scaling ratio should not be considerably above 10. Manufacturing of the model must be done to tight tolerance (e.g. numerically machined hydraulic components) in order to ensure that the test results are representative for the performance of the full-scale prototype. Sometimes model acceptance tests are mandated by the client to neutral institutes who are specialized for this kind of task and own test loops and instrumentation more accurate than the test equipment of the pump manufacturer. Cavitation tests often include the stroboscopic observation of the impeller inlet or even at the diffuser. While model acceptance tests - executed with the required care - give reliable results as to hydraulic performance, cavitation and hydraulic axial and radial forces (if measured), vibration levels recorded on the model pump would not be representative of the full-scale prototype since the mechanical response of the structures to the hydraulic excitation forces are totally different. The same is true for the airborne noise radiated from the model. However, by eliminating (as far as practical) system influences, it is possible to measure and scale the pressure pulsations generated by the pump as discussed in Chaps. 10.2.5 to 10.2.7. The pressure pulsations can also be measured as hydro-acoustic energy radiated into the discharge and suction pipe.

7. Model tests are common for the development of new hydraulic components. Depending on the type of pump and the objectives of the test, development tests can include measurements of: performance, cavitation (including stroboscopic observation), internal pressures as discussed in Chap. 4.1.3, radial and/or axial forces, pressure pulsations and even stresses in the diffuser vanes or in the rotating impeller, [B.19], [2, 3]. As with model acceptance tests discussed above, vibration measurements are not relevant unless the tests would be carried out with exactly the same mechanical design as the prototype.

The measurements to be taken mandatorily and optionally in the different types of tests are listed in Table 16.1. Testing for troubleshooting must always include measurements which allow determining the operation point in terms of $q^* = Q/Q_{opt}$ but the required accuracy is much lower than for performance tests.

Table 16.1 Measurements

Measurement	Acceptance or works tests	Field tests	Model acceptance tests	Model development tests
Performance mandatory	T, n, Q, H, M or P	T, n, Q, H	T, n, Q, H, M or P	T, n, Q, H, M or P
Performance optional		M or P		
Thermometric efficiency, chap. 11.6	possible	often the only practical means	possible	possible
$NPSH_R$	if specified	rarely possible	mandatory	mandatory
Bearing housing vibrations	if specified	almost mandatory	usually not representative of prototype vibrational behavior	
Shaft vibrations	if probes installed			
Stroboscopic cavitation observation	no	no	likely	"mandatory" for high inlet tip speeds u_1
Balance fluid flow Q_E and leakage through fixed throttle shaft seal if any	recommended	if installed	recommended	recommended
Internal pressures, chap. 4.1.3	no	no	possibly	recommended for detailed analysis
Radial and/or axial forces	no	no	possibly	recommended for volute pumps

T = water temperature at the inlet ; M = torque; P = power

16.2 Test Loop Configurations

The layout of the test loop has an important impact on the accuracy of the test results as well as on testing time and costs. The optimum layout depends on the type of pump and the kind and extent of the testing to be done. Various test loop options are discussed in the following.

General requirements:

- The test loop should be designed to allow measuring the true performance curves without cavitation-induced drop in head or efficiency - right up to maximum flow! To this end, sufficient $NPSH_A$ must be provided for the entire flow range to be tested as per the appropriate standard or the relevant specification of the client. In general testing should be performed at least up to $q^* = 1.3$ (preferably up to $q^* = 1.5$ or maximum possible). If the pump is tested at various speeds, it must be ensured that no cavitation occurs at the highest speed. Otherwise incorrect speed dependence might be erroneously derived from the tests.
- The test loop is to be designed to ensure smooth approach flow in order to avoid additional testing and correction costs when the specified vibration and noise limits are exceeded.
- Designing the piping as simple as feasible reduces: (1) investment costs, (2) operation costs for venting and trouble shooting
- Vibration testing is standard for many pumps. In order not to falsify the measurements, the test pumps must be rigidly mounted. A concrete slab of typically 1000 mm thickness is required as foundation for large pumps.
- When testing pumps with axial suction nozzle, partload flow recirculation can falsify the suction pressure measurement (affecting head and NPSH). The test loop layout should then be built as described in context with Fig. 5.16. Ribs in the suction nozzle are not always sufficiently effective.
- If a vacuum pump is used for $NPSH_A$ control, the seals used in the piping and valves must be designed for vacuum. Otherwise the ingress of air will falsify the measurements. Shaft seals must be flushed or otherwise be appropriate for operation under vacuum.
- Corrosion of carbon steel is a problem. Rust deposits in low-flow zones e.g. in heat exchangers.

Safety of personnel must be ensured by appropriate technical measures and procedures in accordance with legal rules. This topic needs high attention and strict guidelines.

- Test loops must be designed for the selected design pressure and temperature. If a vacuum pump is used for NPSH testing, the equipment must also be designed for an external pressure of 1 bar.
- Air vessels for NPSH control must be equipped with a safety valve.

- A pressure switch should be installed in the suction piping which trips the motor when the allowable pressure is exceeded as could be the case upon mal-operation of a valve (e.g. subjecting the suction pipe to the shut-off pressure).
- A pressure switch should also be installed in the discharge pipe which trips the motor when the design pressure is exceeded.
- Electrical equipment requires a special set of rules to ensure safety.

16.2.1 Types and Layout of Closed Test Loops

The features of three types of closed test loops are compared in Table 16.2. Air separation in the test loop limits the minimum $NPSH_A$ which can be achieved with each of these configurations. An estimate of the $NPSH_{A,min}$ is given in the table; it is meant as a rough guidance only.

16.2.2 Closed Test Loop with Pressurizer

The suction pressure control is effected by a tank ("pressurizer") installed in parallel to the main loop with the test pump.

Main Loop: Fig. 16.1 shows a closed test loop with no flow through the pressurizer. During the test, the water is circulated by the test pump through the main loop only (blue lines). The flow rate is controlled by the discharge throttle valve DTV.

In *low-pressure* test loops (up to a design pressure of 16 or 25 bar) the discharge throttle valve DTV is placed *downstream* of the magnetic-inductive flow measuring device MID because high velocities created in the DTV lead to low local pressures with the risk of air separation. Typically the local pressure in the DTV drops to the suction pressure if a valve with a single opening is installed. In *high-pressure* test loops (design pressure above 16 or 25 bar) the discharge throttle valve DTV is placed *upstream* of the magnetic-inductive flow measuring device MID. This, to reduce the cost of the MID, suction pipe and components placed between the DTV and the test pump. The high-pressure discharge pipe between pump discharge flange and throttle valve can be made quite short, but a minimum distance (say $L > 5D$) is required to avoid too strong interaction of flow and pressure fluctuations created by the valve with the pump.

Suction pressure control: In parallel to the main loop there is a tank with a free water level called "pressurizer". The pressurizer is linked to the main loop by a small (DN25) breathing pipe. The pressure maintained above the liquid level is

Table 16.2(1) Closed test loop layout options

	semi-open test loop	**closed test loop with pressurizer**	**closed test loop with flow through tank**
Layout	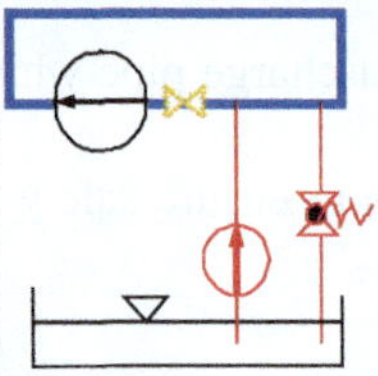	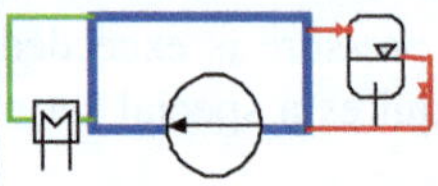	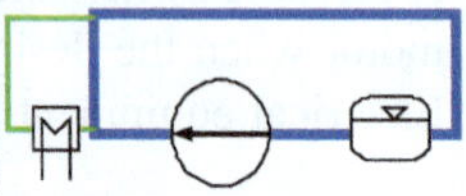
Details	Fig. 16.2	Fig. 16.1	Fig. 16.3
Principle	Pump operates in a closed loop; no free liquid level.	NO flow through pressurizer (tank)	Flow through tank with free water level.
Suction pressure control	By water injection and relief valve	By pressurizer, with compressed air and vacuum pump	By tank, with compressed air and vacuum pump
NPSH testing	➢ For $NPSH_A > 10$ m: no limitations ➢ $NPSH_A < 10$ m: throttle valve between injection and test pump ➢ Min-$NPSH_A$ = 5 m	➢ No limitations other than design pressure of pressurizer ➢ Min-$NPSH_A$ = 2 m	Min-$NPSH_A$ = 6 m unless vacuum is applied during testing; but testing times may become prohibitively long.
De-aeration of water	Not possible if air-saturated water is injected	➢ Very well possible ➢ De-aeration helps venting process ➢ No ingress of air through small breathing pipe	➢ Water is air-saturated ➢ Gas content is proportional to air pressure in the tank. ➢ De-aeration when applying vacuum.
Cooling	Water injection pump or city water	Heat exchanger	Heat exchanger
Venting	Venting less easy since no de-aeration possible.	Water is circulated through pressurizer prior to test to ease venting	➢ Layout of the piping should allow air to escape during the test to the tank. ➢ When testing with high $NPSH_A$, the water dissolves much air which separates in low-pressure zones. De-aerating during NPSH-test takes too long
Corrosion or dirt deposits	Least sensitive system	Critical component: cooler	Critical component: cooler

Table 16. 2 (2) Closed test loop layout options

	semi-open test loop	closed test loop with pressurizer	closed test loop with flow through tank
Cost considerations *for large pumps*	➢ Most simple system ➢ No cooler in loop ➢ Least expensive system ➢ Least space requirements	➢ Most flexible system ➢ Most suitable for accurate NPSH tests	➢ High-cost large tank ➢ Most expensive system ➢ Highest space requirements
Application	➢ Up to largest pump sizes ➢ High-pressure pumps ➢ Test with high suction pressure (no limitation) ➢ High-temperature tests ➢ Suitable also for temporary test loops adapted to the specific requirements of large projects	➢ Up to largest pump sizes ➢ High-pressure pumps ➢ Test with high suction pressure (no limitation) ➢ High-temperature tests ➢ Suitable for fluids other than water; notably oil with high gas solubility.	➢ least expensive system for small pumps if no NPSH-testing required ➢ Low-pressure pumps ➢ Tests with suction pressure up to 6 bar
Conclusion	Recommended for temporary test loops; high-temperature and/or high-pressure testing	Recommended for universal use	No real advantages over other systems.

imposed in this way on the suction pipe of the test pump. The small diameter breathing pipe ensures that virtually no gas is dissolved into the water in the main loop even when operating with de-aerated water (the diffusion of air into the water through the breathing pipe being very slow).

The pressure in the pressurizer (and thus the suction pressure) is controlled by compressed air for $NPSH_A > 10$ m, while a vacuum pump allows testing with $NPSH_A < 10$ m down to about 2 m or even lower depending on the vacuum which can be achieved. The purpose of the pressurizer is:

- Provide room for expansion and contraction of the water when the temperature increases due to dissipation of the power of the pump or when the loop cools off.
- Keep the loop completely flooded when water is lost through some leakage.
- Maintain and control the suction pressure upstream of the test pump. When measuring the Q-H-curves, the suction pressure is maintained high enough to ensure operation without cavitation-induced head drop or efficiency loss. For measuring the NPSH, the test starts with a high suction pressure (compressed air). Subsequently the pressure in the pressurizer is reduced and finally the vacuum pump is operated when suction pressures lower than atmospheric are required.
- Prior to the test, the water in the loop can be partially de-aerated. Even if no test with de-aerated water is required, de-aeration helps to remove gas from the main

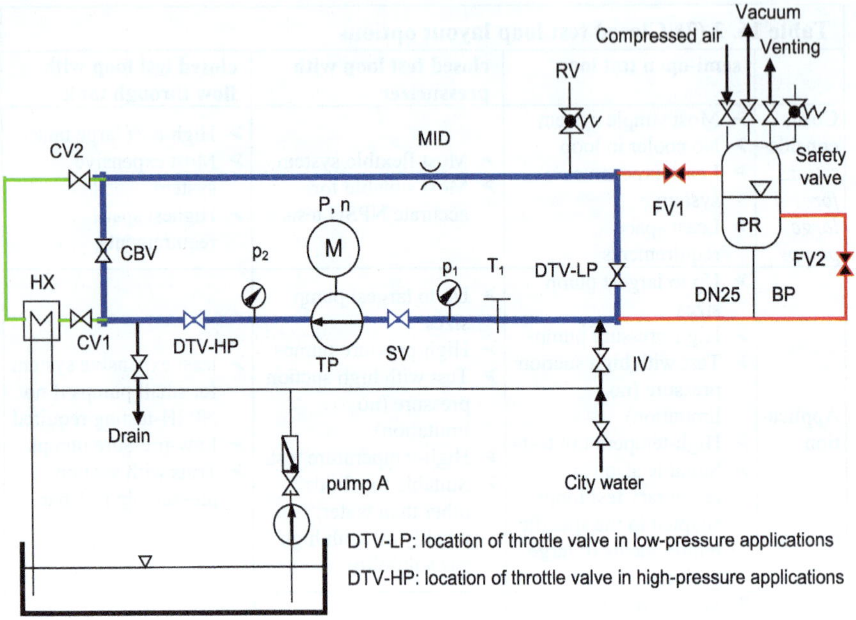

Fig. 16.1 Closed loop *without* flow through the pressurizer

loop. During the de-aeration the flushing valves FV1 and FV2 are open and water is circulated by the test pump through the pressurizer in which a vacuum is maintained by the vacuum pump. The de-aeration can be measured with an oxygen meter. The de-aeration process can thus be stopped when a selected oxygen content has been achieved, for example 1 ppm. At this level more than 80% of the air dissolved at saturation at 1 bar has been removed from the water.

- When the water is effectively de-aerated, the lowest $NPSH_A$ which can be achieved without air separation is likely to be near 2 m.
- During the test, the water level in the pressurizer should be maintained slightly above the highest point of the test loop piping.

The main components of the test loop are listed in Table 16.3. The main functions are described in the following:

Cooling: The power dissipated in the water when the test pump is operated is removed from the test loop via a cooler. Cooling water can be provided from: (1) via pump A from a wet pit (if available and sufficiently large); (2) via pump A from an external cooling device (cooling tower); hybrid cooling tower in summer with evaporation, in winter dry air cooling; (3) city water mains (if available and water costs acceptable); pump A is not required.

Table 16.3 Component list

Component		Function
TP	test pump	
SV	suction valve	keep water in loop when dismantling the test pump
DTV	discharge throttle valve	• control flow through test pump • keep water in loop when dismantling the test pump
CBV	cooler bypass valve	provide sufficient pressure drop to force water through the cooler
IV	injection valve	• filling of test loop from pit (or city water) • water injection for cooling if no cooler is provided
CV1, CV2	cooler valves	separate cooler from main loop
FV1, FV2	flush valves	open for flushing, venting, de-aeration of main loop
HX	cooler	heat removal; keep water temperature at selected level
pump A	auxiliary pump	• provide cooling water • filling of test loop
PR	pressurizer	suction pressure control, absorb water volume variations: leakage and density (temperature)
BP	breathing pipe	impose suction pressure; prevent diffusion of air into main loop
MID	magnetic inductive flow meter	flow rate measurement

Since a cooler designed for the full flow rate of the test pump could be prohibitive expensive, the cooler is installed in a bypass. The cooler has a higher pressure drop than the pipe section of the main pipe. To ensure a sufficient flow through the cooler, the cooler bypass valve CBV is throttled. It may be necessary to adjust the position of the valve CBV during the test in order to operate over the full flow rate range of the test pump.

Alternative: For low-pressure pumps the head at maximum flow may not be sufficient to allow creating a sufficiently large pressure drop across the cooler. In this case, an auxiliary cooling water pump could be installed at the location of valve CV1 in Fig. 16.1. This avoids the CBV and allows an easy operation of the test loop.

Pressurizer: There are no stringent criteria for sizing the pressurizer. A vertical tank (or a large-bore pipe section) could be used. For fairly large pumps the tank diameter may be about 1000 to 1200 mm, its height 1500 to 1800 mm. For small loops a smaller tank would do. The additional volume created by expansion of the water upon heating should be absorbed by the tank without too much of a level difference. Otherwise testing would be more complicated by monitoring and adjusting the water level.

This type of closed loop with a pressurizer has been used also for testing large vertical pumps with flow rates up to 26 m^3/s in a wet pit. The wet pit was a *closed* concrete basin. The $NPSH_A$ during the tests was controlled by a small pressurizer which allowed testing in the range of $NPSH_A = 6$ to 20 m, [7]. Apparently, the design followed the principle shown if Fig. 16.1, but de-aeration of this large installation was not possible.

Discharge throttle valve DTV: The DTV breaks down the total head of the pump minus the pressure drop in the loop. For a low-pressure loop the type of valve selected must be able to open to nearly the full pipe cross section. Otherwise, the maximum flow rate of the pump may not be obtained. If a gate valve or a butterfly valve is used, there is a single valve opening which causes a strong jet with a velocity of roughly $c = (2gH)^{0.5}$. This leads to noise and vibrations and possibly to air separation due to the low local pressures in the vena contracta. But there is little other choice for large pipe diameters since valves with smoother throttling characteristics are very expensive.

Shut-off valves: The suction valve SV and valve DTV-HP are used to keep the water in most of the loop when opening or dismantling the test pump. This not only saves time for filling and draining of large water quantities but also to reduces corrosion in the part of the loop kept filled.

In low-pressure test loops, shut-off valves should be able to open up to nearly the full cross section of the piping in order to make little pressure drop and enable testing to the highest flow rates specified.

Butterfly valves and gate valves could be used for the purpose. Butterfly valves may present some advantages over gate valves:

1. smaller dimensions
2. less costs
3. no risk of air accumulation (in gate valves air may be trapped)
4. less risk of problems due to accumulation of dirt

Quality of approach flow: If the piping upstream of the pump has a straight length of $L \geq (8 \text{ to } 10)D$ and there are no bends in different planes upstream of said straight section, the flow approaching the pump or the MID may be deemed sufficiently uniform. However, if the approach flow to the MID or the test pump is distorted, a flow straightener is required.

16.2.3 Semi-open Test Loop, Fig. 16.2

Unless noted otherwise, refer to the description of components and criteria given in Chap. 16.2.2.

Main Loop: Fig. 16.2 shows a semi-open test loop. It is called "semi-open" since air-saturated water is injected for cooling and controlling the suction pressure. The water is circulated through the loop by the test pump; the flow rate is controlled by the discharge throttle valve DTV. Careful venting is required since any air trapped in the system cannot escape.

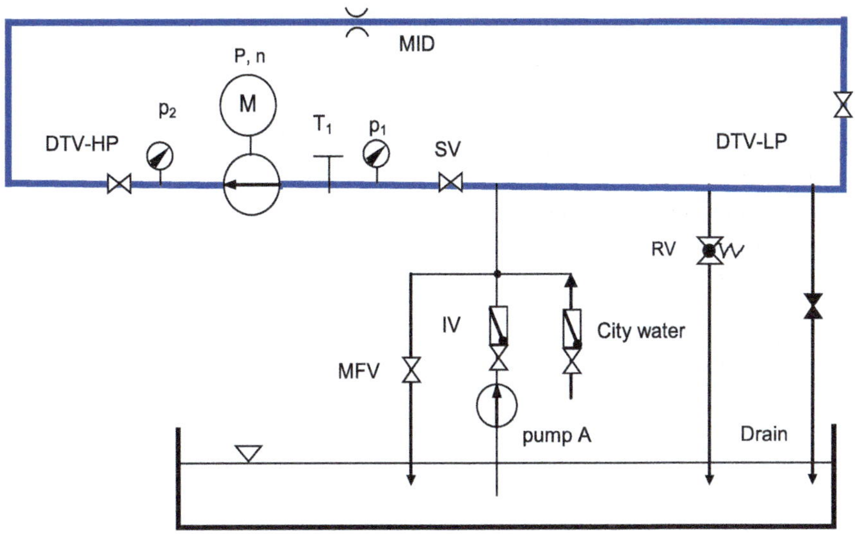

Fig. 16.2 Semi-open test loop without a tank; with injection cooling

Suction pressure control: The pressure in the loop is maintained by water injection through pump A which takes water from the pit or from the city mains. In order to avoid overheating of the injection pump which operates nearly at shut-off conditions, a minimum flow valve MFV is required.

The suction pressure is controlled down to atmospheric pressure by the relief valve RV. Pressures lower than atmospheric can be achieved by throttling suction valve SV. Air will be released in the process due to high velocities and low pressures in the valve. Therefore, the minimum achievable $NPSH_A$ is considered to be roughly $NPSH_A = 5$ m.

Cooling: The inner power P_i dissipated in the water when the test pump is operated is removed by drawing hot water from the loop through relief valve RV. The water is replaced by injecting cold water through pump A.

The flow rate of water Q_w to be injected can be calculated from Eq. (16.1):

$$Q_w = \frac{P_i}{\rho\, c_p\, \Delta T_{al}} \tag{16.1}$$

P_i = power dissipated in W; inner power of pump: $P_i = P - P_m$
ρ = density of water (1000 kg/m^3)
c_p = specific heat of water (4180 J/kg K)
ΔT_{al} = $T_1 - T_{injection}$ = allowable temperature rise in loop
T_1 = suction temperature during testing, e.g. 30 °C; the maximum for "cold water testing" is 40 °C
$T_{injection}$ = temperature of water injected; e.g. 15 °C

16.2.4 Closed Test Loop with Flow Through Tank with Free Water Level, Fig. 16.3

Unless noted otherwise, refer to the description of components and criteria given in Chap. 16.2.2.

Main Loop: Fig. 16.3 shows a closed test loop where the water is circulated through a large tank which is operated with a free water level. The flow rate through the test pump is controlled by the discharge throttle valve DTV.

Suction pressure control: The suction pressure can be controlled by the pressure above the liquid level and to some extent by the suction valve SV. When applying compressed air to the tank, more gas is dissolved in the water according to Henry's law (Chap. 17.3). This gas will separate on de-pressurizing the loop for NPSH-testing. Partial de-aeration by means of a vacuum pump is possible, but testing time will then become prohibitively long because it takes time to de-aerate the large water volume stored in the test loop.

Tank sizing: There are no definite rules for sizing the tank. In general terms a tranquil flow at low velocity should be achieved. The velocity in the tank (using the area below the liquid level) should probably not exceed about 0.7 m/s.

The jet formed by the water entering the tank should be allowed to dissipate prior to reaching the exit of the tank. Baffles may be used for that purpose. Applying these criteria, a very large tank volume may result. The exit of the tank must be designed to create a smooth swirl-free flow into the suction pipe.

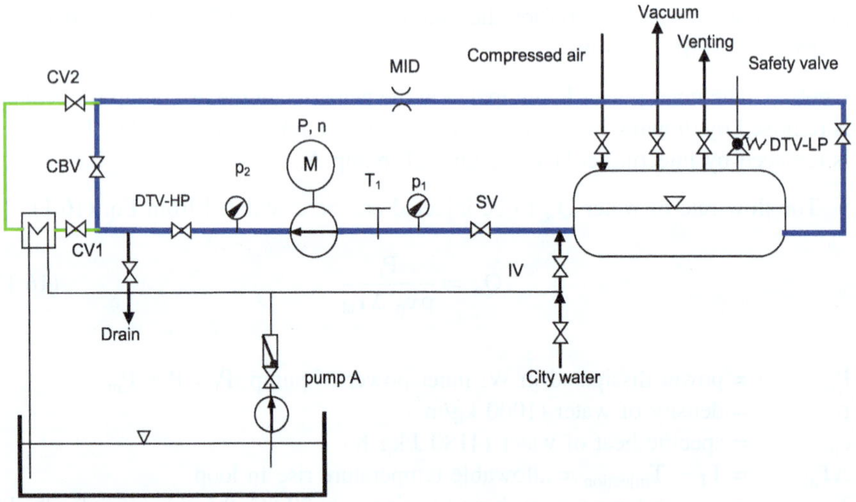

Fig. 16.3 Closed test loop with flow through a tank with a free water level

16.2.5 Open Test Loops

Open test loops are used for testing vertical pumps in wet pit design, but they could be applied as well to testing dry pit pumps of any configuration including very large pumps fed by a booster pump. Unless noted otherwise, refer to the description of components and criteria given in Chap. 16.2.2.

Figure 16.4 shows an arrangement for testing a vertical pump in a wet pit test loop. The flow rate through the test pump is controlled by the discharge throttle valve DTV.

Figure 16.5 shows a possible layout of the pit and the piping. The water delivered by the test pump is led back to the far end of the pit opposite to the pump. The end of the discharge pipe is perforated. This pipe is surrounded by a concentric perforated pipe ("flow distributor"). This arrangement breaks up the kinetic energy exiting the discharge pipe into a large number of small jets which dissipate quickly in the pit and allow a smooth approach flow to the test pump. The flow distributor takes less space and is less expensive than screens and baffles normally used in wet pits for smoothing the flow.

In order to ensure a smooth approach flow to the test pump, the wet pit should be designed according to the rules spelled out in Chap. 11.7.3; notably, the velocity in the pit should not exceed 0.5 m/s.

Suction pressure control: In theory, NPSH tests can be made by adjusting the water level in the pit. This procedure is possible with small pits and small pumps. However, performing NPSH tests with large pumps is altogether impractical. Rather the NPSH must be measured on a model pump; for example installed horizontally in a closed loop as per Chap. 16.2.2.

Sufficient submergence must be provided for the suction bell, see Chap. 11.7.3 The $NPSH_3$ of the test pump (possibly scaled from a model test) should be known in order to be able to judge the flow range where the test pump can be operated without cavitation-induced performance loss.

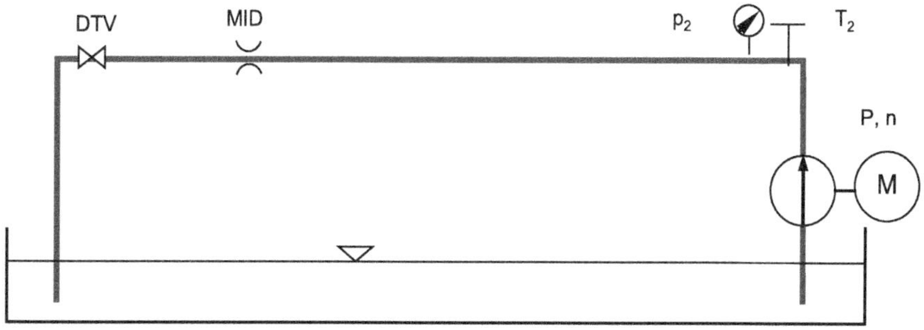

Fig. 16.4 Testing a vertical pump in a wet pit

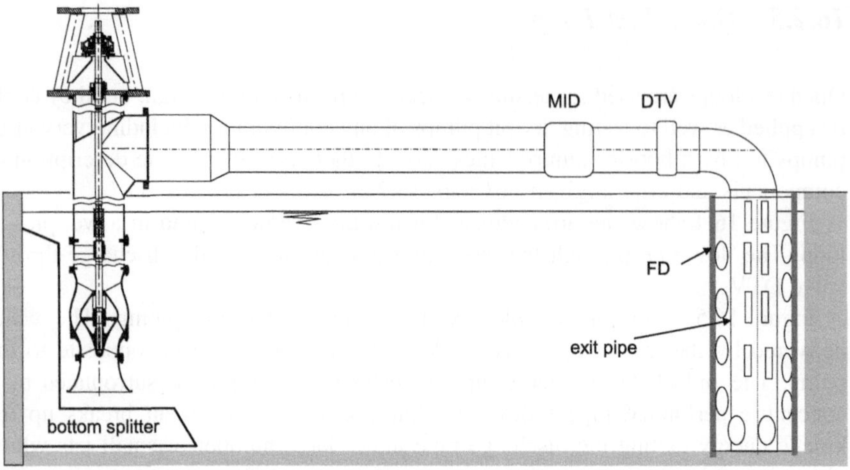

Fig. 16.5 Wet pit testing arrangement

During partload operation of large pumps the recirculating water would induce a strong swirl in the pit if no swirl breaking devices are installed. As a consequence, the water in the pit may rotate to a degree that shut-off head and power are measured too low.

If the suction bell of the test pump is close to the bottom splitter (Fig. 16.5) no swirl breaking structures in the suction bell are required. If the distance between suction bell and bottom splitter is large, swirl breaking structures (ribs) must be present in the suction bell.

When small pumps are tested in a large pit, the energy of the swirling flow recirculating from the impeller during partload operation is sufficiently dissipated so that the impact on the measurements can be neglected.

Cooling: The power of the test pump is dissipated in the water of the pit. When the allowable water temperature in the pit is exceeded, city water may be added while removing warmer water from the pit. If a cooling tower is available, the water in the pit may be cooled by that device.

Flow distributor: Exit pipe and flow distributor FD are designed with numerous openings to break up the kinetic energy of the exiting water and ensure a smooth approach flow.

The sum of the openings in the exit pipe could be dimensioned to give a velocity of about 1 m/s at the maximum flow rate for which the test loop is designed. The sum of the openings in the flow distributor could be dimensioned to give a velocity of about 0.5 m/s at maximum flow.

The openings in both the exit pipe and the flow distributor need to be neither precise nor carefully machined. They could be flame-cut in the pipes. The distribution of the openings over the cylindrical surface should be about uniform. The flow distributor is bolted to the bottom of the pit. The exit pipe is laterally supported in the flow distributor and closed at the bottom.

16.3 Instrumentation

A large variety of instruments is used for pump testing. The scope of the present discussion is therefore restricted to some of the basic principles. Automatic data acquisition can be almost considered as the standard. Therefore, non-electronic instrumentation which required manual handling and reading has lost some of its significance. However, dead weight manometers are still suitable for calibrating electronic pressure transducers and spring manometers may still be useful for operating the test loop or they may serve as a rough check of the electronic equipment during the testing.

All of the instrumentation used for measuring the performance (acceptance tests as well as model tests) must be calibrated on a regular basis as stipulated by the appropriate ISO standards.

The rotor speed can be measured with electronic counters with a precision of $\pm 0.1\%$. Other means include: stroboscopic devices or manual or automatic revolution counters.

The accuracies mentioned below for various instruments are those given by the vendors. These values may be attained under ideal conditions. However, the achievable accuracies in industrial pump testing are believed to be lower than those given in the vendor's documentation.

16.3.1 Pressure Measurement

Principle: Electronic pressure transducers measure the deformation of a membrane exposed to the pressure to be measured on one side and to atmospheric pressure on the second side of the membrane which is open to the environment. Electronic *differential pressure* transducers measure the difference in static pressure between two measuring points, for example $\Delta p_{stat} = p_d - p_s$ between discharge and suction nozzle. Calibration can be made quite easily with dead weight manometers. The accuracy is typically about $\pm 0.3\%$ of the span of the transmitter.

The pressure taps must be sharp-edged, flush with the inner pipe wall and free of a burr. If the flow is decelerated (tap in a recess) or accelerated (tap protruding in the main flow) errors of a few per cent of the stagnation pressure can occur. The bore may be 3 to 4 mm in order to avoid blockage by dirt or sludge. Suction and discharge pressure taps are installed at a distance of L = 2D of straight pipe from the respective flange.

The instrument lines between the location of the pressure tap and the manometer or transducer should be as short as practicable and of a simple layout avoiding high points. This eases venting and flushing of the instrument lines. The difference in elevation (if any) between pressure tap and instrument must be accounted for with the appropriate sign with respect to the reference datum used for the test in order to correctly assess differences in geodetic head. If possible, it is recommended to install all pressure reading instruments on the same elevation as the pump axis (for horizontal machines). This eliminates the need for accounting for the geodetic head and avoids possible falsification of the readings if some air bubbles are in the instrument lines.

16.3.2 Flow Rate Measurement

Electromagnetic flow measurements, ISO 9104: Magnetic-inductive flow measuring devices (MID) are used widely; they can almost be considered as a kind of standard for hydraulic testing. The following information is based largely on [5].

Principle: if an electrolytic liquid flows through a magnetic field, a current is induced in a coil (Faraday's law). The measuring device consists of a non-ferromagnetic tube with an electrically insulating inner surface and magnetic coils as well as two electrodes that are arranged diametrically opposite on the tube and are in contact with the process liquid. The electric current flowing through the coils generates a magnetic field with an induction B perpendicular to the tube axis. This magnetic field penetrates the tube walls and the electrolytic process fluid.

In accordance with the law of induction, a voltage is induced in the fluid. This voltage is proportional to the magnetic induction B, the velocity w of the fluid and the inner diameter D of the tube (i.e. the spacing of the electrodes). As a consequence, the MID provides a signal which is linear with the flow velocity over a wide range. The induced voltage of this signal is $U_i = k \times B \times w \times D$ (k is a dimensionless constant).

Properties and conditions:

- The fluid must be electrically conductive; the minimum conductivity is 50 μS/cm (MID hence not suitable for hydrocarbons). Normal drinking water has a conductivity in the order of 400 μS/cm. However, the signal does *not* depend on the actual conductivity. Below a conductivity of 5 μS/cm a capacitive signal pick-up can be used.
- The MID is fairly immune against electric and magnetic fields.
- The measuring tube must be completely filled.
- Measuring is equally possible with flow in both directions.
- The signal is largely independent of viscosity and density (transition laminar to turbulent with minor loss in accuracy).
- The measuring tube is made of an electrically insulating material (plastics, ceramics). When a metallic tube is used, its liner is of non-conductive material.

- High solids content (slurry, waste water) can be present without impairing the signal; example: dredging vessels pumping 50% solids, with diameter up to 500 mm at velocities of 15 m/s.
- Gas bubbles increase the volumetric flow rate. In the presence of gas bubbles, the liquid flow rate is lower than would be expected from the signal.
- In order not to falsify the measurement, the separating of gas from the liquid must be prevented by a correct installation of the device.
- The installation can be horizontal or vertical. The electrode axis must be always installed horizontally.
- Installation: in a pipe section with upward flow; not on suction side of the pump; not at high point; not in down-comer.
- The device causes no additional pressure drop if it has the same nominal diameter as the pipe.
- The measurement accuracy is little affected by the velocity profile as long as this is radially symmetric.
- Examples for the impact of disturbed velocity profiles: (1) one 90°-elbow at L = 5D upstream in a plane parallel to electrode axis caused a deviation of −0.2%; (2) swirl created by two 90°-bends arranged in different planes at L = 5D upstream of the MID caused a deviation of +0.1% above 2 m/s and up to 0.4% below 1 m/s.

According to ISO 9104 a straight pipe length of L = 10D is required upstream of the MID. However, according to the above data a straight length of L = 5D should be sufficient. Downstream of the MID a straight pipe length equals at least L = 2D is to be installed. The necessary straight pipe lengths upstream and downstream of the MID determine to a large extent the space needed for a test loop. The straight pipe length upstream of the MID can also be reduced by installing a flow straightener.

Since the MID provides a linear signal over a wide range of velocities, the measuring range is claimed to be up to 1:100. It is therefore possible to cover a wide range of test pumps with a single measuring device. Consequently, the MID can be selected that way that a high velocity is present at the highest flow rate to be tested in a given loop. Then a good accuracy is still achieved when testing pumps with lower flow rates. At the same time the costs of the MID and the associated piping are reduced. Most importantly, the total length of the test loop can be kept minimum.

Selecting the MID to the above criteria, the BEP flow rates for the pumps to be tested in a given test loop can cover a range of typically 1:12 with a nominal accuracy of ±0.3% in the range of 1 to 12 m/s and a range of 1:50 with an accuracy of ±0.6% at a velocity as low as 0.25 m/s.

Orifices, nozzles or Venturi nozzles, ISO 5167-1: Principle: the flow is accelerated from the approaching pipe with diameter D in the device with diameter d. The associated decrease in static pressure Δp in the vena contracta is measured in order to determine the volumetric flow rate. Sketches are shown in Fig. 16.6. These

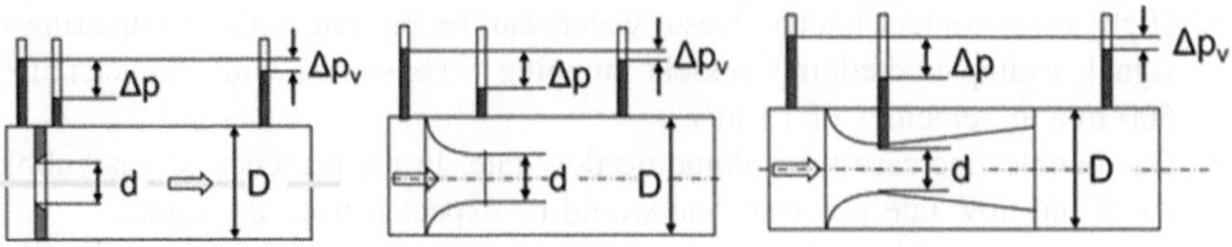

Fig. 16.6 Measuring orifice, nozzle, Venturi nozzle (from left to right)

devices create a permanent pressure loss Δp_V and there is a risk of air separation and even cavitation in the vena contracta if the counter-pressure is too low.

The flow rate can be calculated by means of Eq. (16.2) from the measured Δp.

$$Q = \alpha \frac{\pi}{4} d^2 \sqrt{\frac{2\Delta p}{\rho}} \quad d = \text{diameter of throat area} \tag{16.2}$$

Equation (16.2) applies to sharp-edged orifices, nozzles and Venturi nozzles, but the discharge coefficient α depends on the type of the device. The discharge coefficient is given by Eq. (16.3).

$$\alpha = \frac{C}{\sqrt{1-\beta^4}} \quad \text{with the diameter ratio} : \beta = \frac{d}{D} \tag{16.3}$$

The flow coefficients C depend weakly on the Reynolds number; typical values are given below:

- sharp-edged orifices (ISO 5167-2): C = 0.6 to 0.61 for Reynolds numbers Re > 50,000 and diameter ratios $\beta = d/D \leq 0.6$
- Venturi nozzles (ISO 5167-3): C = 0.924 to 0.985
- long-radius nozzles (ISO 5167–3): C = 0.944 to 0.995
- nozzles (ISA 1932): C = 0.929 to 0.938
- Venturi pipes (ISO 5167-4): C = 0.957 to 0.995.

The pressure difference Δp *measured* at the vena contracta is given by Eq. (16.4). The permanent pressure *loss* Δp_v of orifices can be calculated from Eq. (16.5). For Venturi nozzles this loss is much smaller: a rough estimate is given by the equation $\Delta p_v = 0.3(1 - \beta)\Delta p$.

$$\textit{measured}\ \text{pressure difference:}\ \Delta p = \frac{\rho}{2}\left(\frac{4Q}{\alpha\pi d^2}\right)^2 = \frac{\rho}{2}\left(\frac{4Q}{C\pi d^2}\right)^2 (1-\beta^4) \tag{16.4}$$

$$\textit{permanent}\ \text{pressure}\ \textit{loss}\ \text{of orifices:}\ \Delta p_v = \left(1-\beta^{1.9}\right)\Delta p \quad \text{with:} \quad \beta = \frac{d}{D} \tag{16.5}$$

A uniform approach flow is mandatory for this type of measurement. Depending on the type of fittings upstream of the device, the standards prescribe minimum required lengths of straight upstream pipe.

Accurate flow measurements with orifices require considerable care, [1, 4]:

- The approach flow must be quite uniform since orifices are sensitive to distorted velocity profiles and swirl (such as would be created by double-elbows out of plane). The measuring errors due to non-uniform approach flow increase with the diameter ratio β. Examples from data in [1] may serve for illustrating this effect: With a 90°-elbow L = 8D upstream of the orifice plate the flow coefficient C is −0.6% in error with an orifice of β = 0.5, but −2.1 % with β = 0.75. Even with L = 24D the error in C is still −0.25% with an orifice of β = 0.5 and −1% with β = 0.75.
- The roughness of the approach pipe must be below the limits given in the standards.
- Dulling of the sharp edge of the orifice by wear, nicks and deposition of dirt impair the accuracy significantly since the flow coefficient C will increase.
- The piping must be completely filled.
- Sufficient counter pressure must be provided in order to cavitation and air separation.
- The design must strictly follow the standards. If such is the case, no calibration of the orifice is required.
- Pressure taps must be drilled in the required location; they must be sharp-edged and without a burr.

Under ideal conditions the accuracy of the flow coefficient C of sharp-edged orifices is ±0.6% for $\beta \leq 0.6$ and ±β% for $\beta > 0.6$. Taking into account the above listed influences on the flow coefficient, the actual accuracy achievable in practice would be expected in the range of ±1 to 1.5%.

Ultrasonic flow meters: Mostly work on the transit time differential method, [6]. Two or up to six electro-acoustic converters (“piezos”) are arranged on opposite lines of a pipe with inner diameter D under an angle α. The piezos emit and receive short ultrasonic pulses in the MHz-range through the liquid (or gas) flowing through the pipe. A pulse travelling in the direction of the flow needs a shorter time to reach the opposite pipe wall than a pulse travelling against the current, as given by Eqs. (16.6) and (16.7).

$$\text{travelling time in flow direction: } T_{wf} = \frac{D}{(a + w\ \cos\alpha)\sin\alpha} \tag{16.6}$$

$$\text{travelling time counter flow direction: } T_{cf} = \frac{D}{(a - w\ \cos\alpha)\sin\alpha} \tag{16.7}$$

The difference ΔT of both travelling times is linear with the flow velocity which can be calculated from the measured ΔT from Eq. (16.8).

$$\Delta T = T_{cf} - T_{wf} = \frac{v \sin(2\alpha)}{D} T_{cf} T_{wf} \tag{16.8}$$

The sound velocity (a) in the fluid can be determined directly from the sum of both travelling times and the distance travelled by the pulse, Eq. (16.9).

$$a = \frac{2D}{\sin \alpha (T_{cf} + T_{wf})} \tag{16.9}$$

Properties and conditions:

- The limit for the dynamic viscosity is given by $\mu_{max} = 0.1 \times (D_{Ref}/D)$ Pa s with $D_{Ref} = 1000$ mm. At high viscosities the signal appears to be kind of damped.
- Solid particles reflect sound waves and falsify the measurement. The volume concentration of solids should not exceed 5%.
- Gas bubbles have a profound impact on the propagation of sound waves through reflection and damping, Chap. 13.2. Therefore, the volumetric gas content should not exceed 1%. In order not to falsify the measurement, the separation of gas from the liquid must be prevented by a correct installation of the device (refer to discussion of MID above).
- The impact of the velocity profile on the measurement accuracy depends on the number of acoustic beams employed. Single-beam devices must be used only for turbulent flow because the measuring error at the transition to laminar flow would be about 30%. Dual-beam devices achieve an accuracy of ±0.5% at the transition from laminar to turbulent flow. Ultrasonic flow meters with 5 beams achieve an accuracy of ±0.15% even at the laminar to turbulent transition.
- The device causes no additional pressure drop if it has the same diameter as the pipe.

Ultrasonic flow meters can be used for plant measurements where no flow meter was installed in the system because they can be clamped on a pipeline. The sound transmission properties of the pipe (wall thickness, material, and diameter) and the process fluid are taken into account. The straight pipe length upstream of the flow meter should be $L > 10D$ and the Reynolds number $Re > 10{,}000$. The accuracy achieved is ±2%.

16.3.3 Power, Torque and Efficiency Measurement

With the exception of thermometric measurement the pump efficiency can be determined only by measuring the power or the torque. These quantities are the most difficult to measure accurately with reasonable effort. The options include:

- There are various types of torque meters available. The accuracy usually is claimed to be about ±1%. A problem may be the electronic drift which requires that the "zero" signal (no load) must be checked prior and after the test, possibly also in between during the test.
- The motor power can be measured by different types of Watt meters.
- The energy fed to the motor can be measured by a kWh-counter over a specific time.
- If the motor power is determined that way, the motor efficiency must be known in order to calculate the power input to the pump at the coupling. Calibrated motors can be used, but the motor efficiency as a function of load harbors an uncertainty which may not be easy to assess.
- Swivel-bearing motors with force captor which can be calibrated with dead weights. The accuracy may be about ±0.5 to 1. This is a good choice for model tests, but the motor is expensive.
- If the pump is driven by the motor via a gear box, the gear losses must be known with sufficient accuracy. Since these losses depend on the load and the oil viscosity (temperature) the measuring uncertainty suffers significantly when using a gear box (unless a torque meter is used).
- For pumps with a head above about 250 m the thermometric measurement of the efficiency is an option which can be applied also for determining the efficiency under operation conditions in the plant. Extreme care must be taken with measuring the small differences involved. The equations for evaluating such tests are given in Chap. 11.6.

16.4 Test Preparation and Test Procedures

16.4.1 Test Preparation

The preparation of the test may include the following steps:

1. Prepare a test matrix which defines the quantities to be measured, the parameters to be varied and the number of test points (in terms of flow rates) to be taken for each test.
2. Prepare software for evaluating the test including measurements taken which are not part of standard procedures (e.g. for model testing).
3. Ensure that suction pressure reading is not falsified by pre-swirl induced by partload recirculation. This could lead to a seemingly unstable Q-H-curve and excessive NPSH at partload, see Chap. 5, Fig. 5.16.
4. Know what you will be testing: the actual main dimensions of the hydraulic components should be checked; the values should be recorded and compared to the design: are they within the tolerances stipulated on the hydraulic drawings?

5. Record roughness of the hydraulic channels (impeller, diffuser, volute, impeller side rooms).
6. Measure and record the clearances of the annular seals at impeller and axial thrust balance device if any.
7. Record rotor setting in radial direction with respect to pump casing centerline.
8. Record axial rotor setting with respect to axial clearances and impeller/diffuser alignment.
9. After assembly of the test pump: can the rotor be turned freely by hand without rubbing?
10. Ensure proper protection of coupling and other rotating parts.
11. Check sense of rotation.
12. Filling and careful venting of tests loop including instrument lines.
13. Check leak tightness.
14. Safety instruction: personnel, witnesses and visitors.
15. Safety check of equipment: electrical, high pressure components, accessibility, emergency exit routes.
16. Check instrumentation according to type and vendor instructions: venting, zero point of various transducers, validity of calibration, shielding of instrument cables as required.
17. Check data acquisition system.
18. Circulate water through test loop for de-aeration as described in Chap. 16.2.2.
19. When de-aeration has brought down the gas (oxygen) content to the desired level, close the flush valves FV1 and FV2 (Fig. 16.1). The loop is then ready for starting the test.

16.4.2 Procedure for Performance Test

The test can start at low flow or at the maximum flow rate. Prior to taking the first reading, pump operation should well be stabilized with respect to the test point (flow rate) but also as to the bearing temperatures because of the mechanical losses.

For each test point the quantities n, T_s, p_s, p_d, Q, P (or M) as per Table 2.2 are recorded simultaneously (as far as practical). When using an automatic data recording system, each quantity is recorded (for example) 5 times in a short interval. The readings are subsequently averaged and the scatter (standard deviation) is determined and compared to the allowable fluctuations.

After recording this data, the next test point (flow rate) is adjusted by means of the discharge throttle valve DTV. Again flow conditions must be stabilized prior to taking the next record. Depending on the test equipment, stabilizing may take typically 2 min for pumps up to 50 kW. With increasing power stabilizing the flow conditions will take progressively more time.

The number of test points to be taken to get the true Q-H-curve depends on the steepness of the curve. While steep curves require only a few points, flat or unstable curves require more flow rates to be measured in order to be sure to detect a possible instability. If a pronounced instability would be detected, one test may be done from high flow to low flow and a second test from low to high flow in order to detect a possible hysteresis, see Fig. 8.21 for an example.

It is desirable to measure shut-off head and power, but this becomes increasingly difficult (respectively impossible) with growing power input to the pump because of the heating of the water due to the dissipated power. For example: in a pump with a shut-off power of 1000 kW a water volume of 1 m^3 is heated by about 14 °C during 60 s of operating at $Q = 0$. If a reading would be taken under similar conditions, the associated decrease in density should be taken into account when calculating the head - otherwise a seemingly unstable Q-H-curve would be measured. If a variable-speed drive is used, shut-off head and power could be measured at reduced speed and scaled to the full speed according to the model laws.

After the test the pump is dismantled and checked for possible mechanical damage (e.g. rubbing in the annular seals, damage to bearings or shaft seals).

16.4.3 Procedures for Cavitation Testing

The recommended procedure to get the most accurate cavitation test results achievable under given circumstances with a given pump is maintaining speed and flow rate constant and successively decreasing the available NPSH in the test loop. This procedure has been described in Chap. 6.2.5 in context with Fig. 6.9. It is quite accurate if a sufficiently large portion of the recorded H($NPSH_A$)-curve is horizontal (as is the case in Fig. 6.9).

Another way to determine the effect of cavitation is to keep the suction valve at a constant position and successively open the discharge valve to increase the flow rate till the head deviates from the Q-H-curve measured *demonstrably* without cavitation, Fig. 6.18. The $NPSH_A$ (apart from the losses in the suction pipe which increase with the flow rate) remains essentially constant in this test, while the flow rate increases. At high flow rates it becomes increasingly difficult to be sure that cavitation has not yet started to impact performance prior to opening the DTV. For this reason this method is considered less accurate than measuring at constant flow and speed.

One can also keep the discharge valve at a constant position while closing successively the suction valve till the head deviates from the Q-H-curve measured *demonstrably* without cavitation. In this test the $NPSH_A$ is reduced, while the flow rate diminishes.

In all of these tests, the $NPSH_A$ can be controlled by varying the pressure above the liquid level, by throttling the flow in the suction pipe or by changing the liquid level. It is also possible to reduce the $NPSH_A$ by increasing the vapor pressure by heating the liquid to higher temperature, but this option is rarely chosen.

16.5 Test Evaluation and Accuracy

Head and NPSH are calculated from the measured pressures and flow rates according to Table 2.2, taking into account differences in geodetic height between the elevation of the pressure tap and the instrument as appropriate.

If the density varies between inlet and discharge nozzle (hot water, high pressure rise), the average density is to be used.

When testing vertical pumps in wet pit installation as per Fig. 16.5, the water level in the pit is used as inlet plane to calculate the inlet head (the velocity head there is zero).

The measurement uncertainties of the different quantities add up to the overall uncertainty. The individual uncertainties are: f_z for the elevations z (Table 2.2), f_{ps} for the suction pressure, f_{pd} for the discharge pressure; f_Q for the flow rate and f_P for the f_H of the head can be calculated from Eq. (16.10).

$$f_H = \pm\sqrt{\left(\frac{z_d - z_s}{H}\right)^2 f_z^2 + \left(\frac{p_d}{\rho g H}\right)^2 f_{pd}^2 + \left(\frac{p_s}{\rho g H}\right)^2 f_{ps}^2 + \frac{(c_d - c_s)^4}{(gH)^2} f_Q^2} \quad (16.10)$$

The uncertainty of the efficiency follows from Eq. (16.11).

$$f_\eta = \pm\sqrt{f_H^2 + f_P^2 + f_Q^2 + f_\rho^2} \quad (16.11)$$

16.6 Potential Testing Problems and Remedies

Problems during pump testing show up as (1) performance deviations (usually deficits), (2) unexpectedly high head or power at partload, (3) deviations from the expected NPSH-curves, (4) unusual noise, (5) vibrations exceeding the allowable limits. When trouble of that kind occurs, one of the first steps to take is checking the validity of the test results. If possible, the measurements could be checked by a redundant instrument (easily done with pressure readings).

Quite a number of problems were discussed in earlier chapters. For easy reference these are listed in Table 16.4.

Table 16.4. Potential testing problems and remedies

Problem	Possible causes	Actions, remedies, references
Doubtful pressure readings	• instrument bypass open • air in instrument lines • instrument line blocked by dirt	• close bypass • venting • flushing • re-calibration • redundant measurement with manometer
Doubtful volume flow reading	• air separation in flow measurement devices • disturbed approach flow	• re-design piping • de-aeration • venting • install flow straightener
Doubtful power reading	• unknown motor losses • unknown gear losses (if any) • rubbing in annular seals • mechanical problems	• motor too large (working at low load with low efficiency)? • for high heads: thermometric efficiency measurement as back-up • calibrate motor • inspect pump
Doubtful torque meter reading	• torque meter misalignment • wrong parameter settings	• check drift • correct alignment • calibration • check data sheet • redundant power reading
Performance deficits	• **geometric tolerances** • Q, H, P measurement incorrect • leakage through bypass (e.g. parallel flow rate measuring sections)	see chap. 4.6, Table 4.3
NPSH rises at partload towards Q = 0; Q-H-curve drooping towards shut-off	**Pre-rotation** (pumps with axial inlet or inline pumps without rib in suction nozzle)	flow straightener see chap. 5.2.6 and **Fig. 5.16**
Excessive vibrations and noise	• soft foundation • shaking piping • problem coupling • driver misalignment • distorted approach flow • standing waves in test loop • discharge valve too close to discharge flange causing interaction of flow fluctuations in valve and pump	• check stiffness of pipe work • re-design loop (T-piece) • check motor or/and coupling alignment • check type of coupling • displace valve downstream of the pump

References

1. Baker, R.C.: Flow Measurement Handbook. Cambridge University Press, Cambridge (2000)
2. Berten, S. et al.: Experimental investigation of flow instabilities and rotating stall in a high-energy centrifugal pump stage. ASME FEDSM (2009)
3. Bolleter, U. et al.: Rotordynamic modelling and testing of boiler feed pumps. EPRI Report TR-100980, Sept 1992
4. Fiedler, O.: Strömungs- und Durchflussmesstechnik. Oldenburg Verlag München (1992)
5. Hofmann, F.: Electromagnetic flow meters, 3rd edn. Krohne (2003)
6. Hofmann, F.: Fundamentals of Ultrasonic Flow Measurements in Industrial Applications. Krohne (2001)
7. Bobowski, L.: Neuer Prüfstand für Rohrgehäusepumpen. Festschrift zum Jubiläum 100 Jahre Turbomaschinen und 50 Jahre Fluidantriebstechnik an der TU Darmstadt (1997)

Chapter 17
Appendices

Contents

- Units and unit conversion
- Physical properties of water in tabular form and as simple polynoms which can be easily programmed in various applications
- Solubility of gases in water
- Atmospheric pressure as a function of the height above sea level
- Acceleration due to gravity as a function of latitude and height above sea level
- Sound velocities in liquids
- Basic notions of mechanical vibrations (spring-mass-damper model)
- Formulae for natural frequencies of simple structures

A comprehensive specification for hydraulic design projects.

17.1 Units and Unit Conversion

Acceleration			**Heat capacity, specific**		
1 ft/s^2	$= 3.0480 \times 10^{-1}$	m/s^2	1 kcal/kg °C	$= 4.1868 \times 10^3$	J/kgK
			1 cal/g °C	$= 4.1868 \times 10^3$	J/kgK
Area			1 Btu/lb °F	$= 4.1868 \times 10^3$	J/kgK
1 in^2	$= 6.4516 \times 10^{-4}$	m^2			
1 ft^2	$= 9.2903 \times 10^{-2}$	m^2	**Heat conductivity**		
1 yd^2	$= 8.3613 \times 10^{-1}$	m^2	1 kcal/m h °C	= 1.1630	W/m K
1 acre	$= 4.0469 \times 10^3$	m^2	1 cal/cm s °C	$= 4.1868 \times 10^2$	W/m K
1 $mile^2$	$= 2.5900 \times 10^6$	m^2	1 Btu/ft^2 hr(°F/in)	$= 1.4423 \times 10^{-1}$	W/m K
			1 Btu/ft hr °F	= 1.7308	W/m K
Caloric quantities per volume					
1 $kcal/m^2$	$= 4.1868 \times 10^3$	J/m^3	**Heat flux**		
1 Btu/ft^3	$= 3.7260 \times 10^4$	J/m^3	1 $kcal/m^2$ h	= 1.1630	W/m^2
			1 $kcal/ft^2$ hr	= 1.2518 x 10	W/m^2

J. F. Gülich, *Centrifugal Pumps*,
https://doi.org/10.1007/978-3-030-14788-4_17

Density

1 lb/ft^3	$= 1.6018 \times 10$	kg/m^3

Energy, heat, work

1 Nm	= 1.00	J
1 Ws	= 1.00	J
1 kpm	= 9.8067	J
1 kcal	$= 4.1868 \times 10^3$	J
1 kWh	$= 3.6000 \times 10^6$	J
1 Btu	$= 1.0551 \times 10^3$	J

Enthalpy, specific

1 kcal/kg	$= 4.1868 \times 10^3$	J/kg
1 Btu/lb	$= 2.3260 \times 10^3$	J/kg

Force

1 kp	= 9.8067	N
1 dyn	$= 1.00 \times 10^{-5}$	N
1 Dyn	= 1.00	N
1 lbf	= 4.4482	N

Mass

1 kps^2/m	= 9.8065	kg
1 lb	$= 4.5359 \times 10^{-1}$	kg
1 ton (long)	$= 1.0160 \times 10^3$	kg
1 ton (short)	$= 9.0714 \times 10^2$	kg

Mass flow rate

1 lb/hr	$= 1.2600 \times 10^{-4}$	kg/s
1 ton/hr short	$= 2.5200 \times 10^{-1}$	kg/s
1 ton/hr long	$= 2.8224 \times 10^{-1}$	kg/s

Mass velocity

1 lb/hr ft^2	$= 1.3562 \times 10^{-3}$	kg/m^2 s
1 kg/hr ft^2	$= 2.9900 \times 10^{-3}$	kg/m^2 s
1 lb/s ft^2	= 4.8824	kg/m^2 s

Power

1 mkp/s	= 9.8065	W
1 kcal/h	= 1.1630	W
1 PS	$= 7.3548 \times 10^2$	W
1 m^3 atm/h	$= 2.8150 \times 10$	W
1 ft lbf/min	$= 2.2597 \times 10^{-2}$	W
1 ft lbf/s	= 1.3558	W
1 Btu/hr	$= 2.9308 \times 10^{-1}$	W
1 hp (UK)	$= 7.4570 \times 10^2$	W

1 cal/cm^2 s	$= 4.1868 \times 10^4$	W/m^2
1 Btu/ ft^2 hr	= 3.1546	W/m^2

Heat transfer coefficients

1 kcal/m^{2}h °C	= 1.1630	W/m^{2}K
1 cal/m^{2}s °C	$= 4.1868 \times 10^4$	W/m^{2}K
1 kcal/ft^{2}hr °C	$= 1.2518 \times 10$	W/m^{2}K
1 Btu/ft^{2}hr °F	= 5.6785	W/m^{2}K

Length

1 Å	$= 1.00 \times 10^{-10}$	m
1 μ (micron)	$= 1.00 \times 10^{-6}$	m
1 in	$= 2.5400 \times 10^{-2}$	m
1 ft = 12 in	$= 3.0480 \times 10^{-1}$	m
1 yd = 3ft	$= 9.1440 \times 10^{-1}$	m
1 thou	$= 2.5400 \times 10^{-5}$	m
1 mile (st)	$= 1.6094 \times 10^3$	m
1 mile (n.)	$= 1.8532 \times 10^3$	m

Temperature difference

1°C	= 1	K
1°F	= 5/9	K

Viscosity, dynamic

Pa s ≡ N s/m^2 ≡ kg/ m s

1 kps/m^2	= 9.8065	Pa s
1cP	$= 10^{-3}$	Pa s
1 Poise =1 g/cm s	$= 1.00 \times 10^{-1}$	Pa s
1 lb/ft hr	$= 4.1338 \times 10^{-4}$	Pa s
1 kg/ft hr	$= 9.1134 \times 10^{-4}$	Pa s
1 lb/ft s	= 1.4882	Pa s

Viscosity, kinematic

1 c St = 1 mm^2/s	$= 10^{-6}$	m^2/s
1 Stoke	$= 1.0000 \times 10^{-4}$	m^2/s
1 ft^2/hr	$= 2.5806 \times 10^{-5}$	m^2/s
1 ft^2/s	$= 9.2903 \times 10^{-2}$	m^2/s

Volume

1 in^3	$= 1.6387 \times 10^{-5}$	m^3
1 ft^3	$= 2.8317 \times 10^{-2}$	m^3
1 yd^3	$= 7.6455 \times 10^{-1}$	m^3
1 US gal	$= 3.7853 \times 10^{-3}$	m^3
1 UK gal	$= 4.5460 \times 10^{-3}$	m^3

Pressure, stresses			1 barrel (US)	$= 1.5898 \times 10^{-1}$	m^3
1 bar	$= 1.0000 \times 10^{5}$	Pa			
1 at	$= 9.8067 \times 10^{4}$	Pa	**Volume, specific**		
1 kp/cm^2	$= 9.8067 \times 10^{4}$	Pa	1 ft^3/kg	$= 2.8317 \times 10^{-2}$	m^3/kg
1 atm	$= 1.0133 \times 10^{5}$	Pa	1 ft^3/lb	$= 6.2428 \times 10^{-2}$	m^3/kg
1 Torr	$= 1.3332 \times 10^{2}$	Pa			
1 mmHg	$= 1.3332 \times 10^{2}$	Pa	**Volume flow rate**		
1 mmWS	$= 9.8067$	Pa	1 ft^3hr	$= 7.8658 \times 10^{-6}$	m^3/s
1 lbf/in^2	$= 6.8948 \times 10^{3}$	Pa	1 ft^3/min =1cu min	$= 4.7195 \times 10^{-4}$	m^3/s
			1 ft^3/s = 1 cu sec	$= 2.8317 \times 10^{-2}$	m^3/s
Temperature			1 US gal/hr	$= 1.0515 \times 10^{-6}$	m^3/s
ϑ°C	$= (\vartheta + 273.15)$	K	1 UK gal/hr	$= 1.2628 \times 10^{-6}$	m^3/s
ϑ°F	$= 5/9(\vartheta - 32) + 273.15$	K	1 barrel/day (US)	$= 1.8401 \times 10^{-6}$	m^3/s
1°R	$= 5/9$	K	1 US gal/min	$= 6.3089 \times 10^{-5}$	m^3/s
			1 UK gal/min	$= 7.5766 \times 10^{-5}$	m^3/s

17.2 Properties of Saturated Water

Table 17.1 provides the properties of saturated water. They are rounded. In the liquid phase the properties depend little on pressure. The values from the table can therefore be used approximately also for pressures higher than the saturation pressure: at a pressure of 1000 bar the water density is about 4 to 5% higher than the saturation pressure.

Symbols used in the tables and the approximate equations:

$'$	liquid phase
$''$	vapor
p_v	saturation pressure
ρ'	density of water
ρ''	density of vapor
$c_{p,av}$	average specific heat capacity between 0 and T °C
c_p'	specific heat capacity at temperature T
h'	enthalpy
Δh_v	enthalpy of evaporation
λ'	thermal conductivity
ν'	kinematic viscosity
S_T	surface tension water/vapor
a_o	sound velocity in water

The critical point is at T = 374 °C and p = 221.2 bar. There water and vapor have the same properties; the specific heat tends to infinite; the enthalpy of evaporation and surface tension become zero.

Table 17.2 gives formulae for the calculation of the water properties, which are easy to program, but are approximate only. Column "F" gives typical errors, but some individual values may have deviations larger than shown in column F.

The formulae are sufficiently accurate for most calculations. But they must not be used for test data evaluation. In particular the accuracy is not sufficient for model or acceptance tests. For official measurements the actual water-vapor tables must be used.

Table 17.1 Properties of saturated water

T	p_v	ρ'	ρ"	h'	$c_{p'av}$	c_p'	λ'	ν'	S_T	Δh_v	a_o
°C	bar	kg/m^3		kJ/kg	kJ/kg K		W/mK	$10^6 m^2/s$	N/m	kJ/kg	m/s
0.01	0.0061	999.8	0.0049	-	-	4.217	0.569	1.75	0.076	2501	1405
10	0.0123	999.7	0.0094	42.03	4.203	4.193	0.587	1.3	0.074	2477.4	1445
20	0.0234	998.3	0.0173	83.86	4.193	4.182	0.603	1	0.073	2453.9	1480
30	0.0424	995.7	0.0304	125.61	4.187	4.179	0.618	0.8	0.071	2430.3	1505
40	0.0737	992.3	0.0512	167.34	4.184	4.179	0.632	0.656	0.070	2406.5	1520
50	0.1233	988	0.0830	209.11	4.182	4.181	0.643	0.551	0.068	2382.6	1540
60	0.1992	983.2	0.1302	250.91	4.182	4.185	0.654	0.471	0.066	2358.4	1550
70	0.3116	977.7	0.1981	292.78	4.183	4.19	0.662	0.409	0.064	2333.8	1560
80	0.4736	971.6	0.2932	334.72	4.184	4.197	0.67	0.361	0.063	2308.8	1560
90	0.7011	965.2	0.4233	376.75	4.186	4.205	0.676	0.322	0.061	2283.4	1560
100	1.013	958.1	0.5974	418.88	4.189	4.216	0.681	0.291	0.059	2257.3	1550
110	1.433	950.7	0.826	461.13	4.192	4.229	0.684	0.265	0.057	2230.5	1540
120	1.985	942.9	1.121	503.5	4.196	4.245	0.687	0.244	0.055	2202.9	1520
130	2.701	934.6	1.496	546.1	4.201	4.263	0.688	0.226	0.053	2174.4	1505
140	3.614	925.8	1.966	588.9	4.206	4.285	0.688	0.211	0.051	2144.9	1485
150	4.760	916.8	2.547	631.9	4.213	4.31	0.687	0.197	0.049	2114.2	1460
160	6.180	907.3	3.259	675.2	4.220	4.339	0.684	0.186	0.047	2082.2	1440
170	7.920	897.3	4.122	718.8	4.228	4.371	0.681	0.177	0.044	2048.8	1415
180	10.00	886.9	5.16	762.7	4.237	4.408	0.677	0.168	0.042	2014	1390
190	12.55	876	6.398	807	4.247	4.449	0.671	0.161	0.040	1977.4	1360
200	15.55	864.7	7.865	851.8	4.259	4.497	0.665	0.155	0.038	1939	1325
210	19.08	852.8	9.596	897.1	4.272	4.551	0.657	0.149	0.036	1898.7	1290
220	23.20	840.3	11.63	943	4.286	4.614	0.648	0.145	0.033	1856.2	1260
230	27.98	827.3	14.00	989.6	4.303	4.686	0.639	0.14	0.031	1811.4	1220
240	33.48	813.6	16.77	1036.9	4.320	4.77	0.628	0.136	0.029	1764	1185
250	39.78	799.2	19.99	1085.1	4.340	4.869	0.618	0.134	0.026	1713.7	1150
260	46.94	783.9	23.74	1134.3	4.363	4.986	0.603	0.131	0.024	1660.2	1105
270	55.05	767.8	28.11	1184.5	4.387	5.126	0.59	0.129	0.021	1603	1065
280	64.19	750.5	33.21	1236.1	4.415	5.296	0.575	0.128	0.019	1541.2	1020
290	74.45	732.1	39.2	1289.3	4.446	5.507	0.558	0.127	0.017	1475.2	975
300	85.92	712.2	46.25	1344.2	4.481	5.773	0.541	0.127	0.014	1403.1	925
310	98.70	690.6	54.64	1401.3	4.520	6.12	0.523	0.125	0.012	1324.1	875
320	112.90	666.9	64.75	1461.3	4.567	6.586	0.508	0.124	0.010	1236.5	820
330	128.65	640.5	77.15	1524.8	4.621	7.248	0.482	0.124	0.008	1138.1	-
340	146.08	610.3	92.76	1593.5	4.687	8.27	0.46	0.124	0.006	1025.6	-
350	165.37	574.5	113.4	1670.3	4.772	10.08	0.437	0.123	0.004	893.2	-
360	186.74	528.3	143.5	1762.2	4.895	14.99	0.399	0.124	0.002	722.6	-
374	221.20	315.5	315.5	2099.7	5.612	∞	0.238	-	0	0	-

Table 17.2: Approximate formulae for the properties of saturated water and vapor

Note dimensions!	These formulae are **NOT** sufficiently accurate for test data evaluation		
Approximation for T < 350 °C	$x \equiv \frac{T[°C]}{T_{Ref}}$ $T_{Ref} = 100$ °C	Error [%]	Valid to °C
Saturation pressure in bar	$p_v = 0.978x^4-0.345x^3+0.355x^2+0.0184x+0.0067$ for $T < 100$ °C	0.2	100
	$p_v = 1.4x^4-1.293x^3+0.666x^2+0.586x-0.346$ for $100 < T < 330$ °C	0.2	330
Density in kg/m³	$\rho' = 999.5+10.7x-94.6x^2+69.5x^3-35.5x^4+9.34x^5-1.01x^6$	0.13	350
Specific heat capacity in kJ/kg K	$c_p' = 4.225-0.3148x+0.5481x^2-0.3058x^3+0.0714x^4$	0.3	300
Average specific heat capacity in kJ/kg K	$c_{p,av}' = 4.1987-0.0284x+0.008x^2+0.0107x^3$ average from 0 to T	0.13	300
Enthalpy in kJ/kg	$h' = c_{p,av}T$ T in °C	0.13	310
	$h' = 423x-12x^2+6.8x^3$	0.35	
Enthalpy of evaporation in kJ/kg	$\Delta h_v = 2501-197x-80.8x^2+43.2x^3-11.8x^4$	0.3	280
Kinematic viscosity in m²/s	$\nu = 10^{-6}(0.3303x^6-2.3962x^5+7.0454x^4-10.902x^3+9.7548x^2-5.2896x+1.7473)$ for $T < 200$ °C	2.3	200
	$\nu = 10^{-6}(-0.012x^3+0.1172x^2-0.3877x+0.5575)$ for $T > 200$ °C	1.5	340
Thermal conductivity in W/m K	$\lambda = 0.569 + 0.191x - 0.0908x^2+0.0158x^3-0.00488x^4+0.00117x^5-0.000135x^6$	0.2	310
Gas constant of vapor in m²/s² K	$R = 461+6.2385x-14.577x^2+2.9915x^3-1.3068x^4$	-	340
Density of vapor in kg/m³	$\rho' = \frac{p_v}{R\ T}$ T in °K	0.5	340
Sound velocity in water in m/s	$a_o = 1416.2+356.8x-253x^2+26.6x^3$	-	320

17.3 Solution of Gases in Water

Water dissolves gases up to a maximum concentration, the solubility, which depends on the type of gas, the temperature and the partial pressure above the liquid level. In the process industry, frequent applications of the procedure given below are gases separating ("desorbing" or "flashing") in the suction pipe, Fig. 17.1. The mass fraction of selected gases dissolved in water can be calculated from Table 17.3 or taken from Figs. 17.2 and 17.3. Solubilities of other gas/liquid combinations can be found in [B.29].

According to Henry's law the solubility is proportional to the *partial pressure of the gas above the liquid level*, see Table 17.3, which provides formulae for the calculation of gas mixtures. Henry's law is an approximation for low pressures and gas concentrations. For example, it cannot be applied to determine the solubility of carbon dioxide (CO_2) at pressures above 25 bar, not even for an estimate.

With reference to Fig. 17.1 the total pressure in the tank above the liquid level is p_e and the fluid has a vapor pressure of $p_v(T)$. At the partial pressure of the gas, $p_{gas} = p_e - p_v$, the amount of gas dissolved is x_D kg gas/kg water. When the static in the suction pipe drops to a local value p, only the fraction $x = x_D \times (p - p_v)/(p_e - p_v)$ is still in solution. The difference $(x_D - x)$ separates from the water (assuming equilibrium) and forms free gas (bubbles). The free gas occupies the volume Q_{gas} (R is the gas constant):

$$Q_{gas} = \frac{\varepsilon R T}{p - p_v}(x_D - x)\rho Q \tag{17.1}$$

Referred to the liquid volume flow rate Q, the ratio of the free gas to the liquid is given by Eq. (17.2):

$$\frac{Q_{gas}}{Q} = \frac{\rho \varepsilon x_D R T}{p - p_v}\left(1 - \frac{p - p_v}{p_e - p_v}\right) \tag{17.2}$$

Gas dissolution is controlled by diffusion. The factor ε accounts for cases where the solution equilibrium is not attained: $\varepsilon = 1.0$ means that the equilibrium is established, while $\varepsilon = 0$ applies when no gas separates because the transit time is too short to allow for appreciable diffusion.

Not only water but all liquids dissolve gasses; an important application is the solution of gases in oil. Each individual combination liquid–gas has its individual solubility, which can be found in various manuals. No general rules are available to estimate the solubility.

For estimating the amount of air dissolved in oil, the Bunsen coefficient $\alpha_v = 0.09$ m^3 air/m^3 oil can be used. The Bunsen coefficient α_v is defined as the volume of air at 273.15 K (0 °C) and 1.013 bara, which is absorbed under equilibrium per unit volume of oil at a specified temperature, under the partial air pressure of 1.013 bar. The solubility α_v depends on the type of oil, the temperature and the viscosity; the value $\alpha_v = 0.09$ is for 40 °C and roughly 30 mm^2/s. The solubility increases with the temperature and decrease with the viscosity. These effects can be estimated

Table 17.3: Solubility of gases in water

Note: all pressures are *absolute!*

Correlations for calculating the molar solubility, [B.29]

Gas	M_g	The solubility is valid for a *partial pressure* of p_{Ref}= 1.013 bar. Temperature **T in K!**	Valid for
O_2	32	$\ln x_g = 18.554 \ln T + \frac{6889.6}{T} - 139.485$	0 to 350°C
N_2	28	$\ln x_g = 18.7292 \ln T + \frac{6921.99}{T} - 141.2677$	0 to 330°C
CO_2	44	$\ln x_g = 21.6694 \ln T + \frac{8741{,}68}{T} - 0.0011026\ T - 159.854$	0 to 80°C
Methane CH_4	16	$\ln x_g = 51.9144 \ln T + \frac{13282.1}{T} - 0.042583\ T - 338.217$	0 to 250°C
		$\ln x_g = 20.6794 \ln T + \frac{7478.8}{T} + 0.75316 \ln p_{ges} - 152.777$	0 to 350°C 6 to 2000 bar
		$p_{ges} = p_{CH4} + p_v =$ total pressure in bar !	

Definition of the molar solubility	$x_g = \frac{n_g}{n_g + n_w}$	n_g = number of moles of dissolved gases n_w = number of moles of water
Number of moles of dissolved gases calculated from x_g	$n_g = \frac{x_g\ n_w}{1 - x_g}$	
Mass fraction of dissolved gases per kg water	$x_D = \frac{x_g}{1 - x_g} \frac{M_g}{M_w}$	$x_D\,[\text{ppm}] = \frac{x_g}{1 - x_g} \frac{M_g}{M_w} 10^6$
M_g = molecular mass of gas; M_w = molecular mass of water; M_w = 18		
Fraction of dissolved gases at *partial* pressure p_{gas} (Henry's law)	$x = \frac{p_{gas}}{p_{Ref}} x_{Ref}$	x_{Ref} equal to n_g or x_D as calculated above; p_{Ref} = 1.013 bar

Calculation of gas mixtures using air as example:

Gas mixture above water level	$p_{ges} = p_v + p_{N2} + p_{O2}$	Total pressure = sum of partial pressures
Volume fraction r = mole fraction n	$\frac{p_v}{p_{ges}} + \frac{p_{N2}}{p_{ges}} + \frac{p_{O2}}{p_{ges}} = r_w + r_{N2} + r_{O2} = 1.00$	
The volume fraction of water vapor r_w is calculated from the vapor pressure $p_v = f(T)$	$r_w = \frac{p_v}{p_{ges}}$	
Ratio of volume fractions of oxygen to nitrogen	$\frac{r_{O2}}{r_{N2}} \quad a = \frac{0.21}{0.79} =$	Argon (1%) included in nitrogen
Volume fractions	$r_{N2} = \frac{1 - r_w}{1 + a}$	$r_{O2} = a\, r_{N2}$
Partial pressures	$p_{N2} = \frac{p_{ges} - p_v}{1 + a}$	$p_{O2} = \frac{a}{1 + a}(p_{ges} - p_v)$

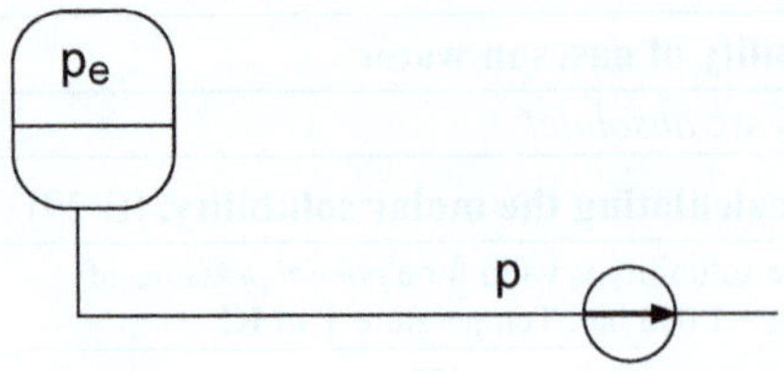

Fig. 17.1 Gas evolution in a suction pipe

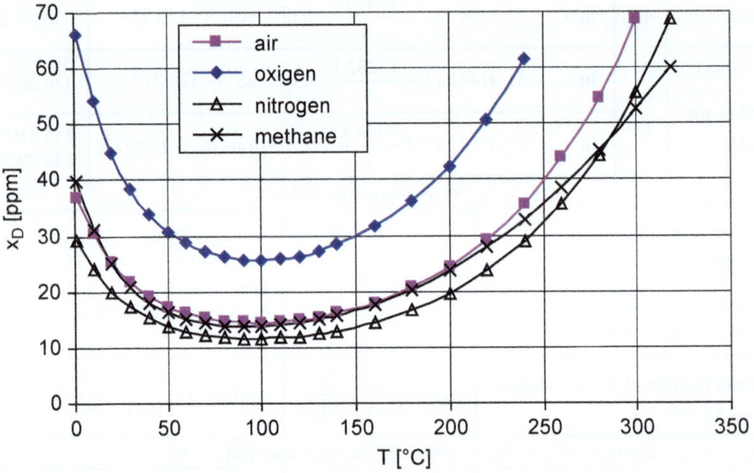

Fig. 17.2 Solubility of different gases in water at a partial pressure of 1.013 bar; solubility as mass fraction in ppm

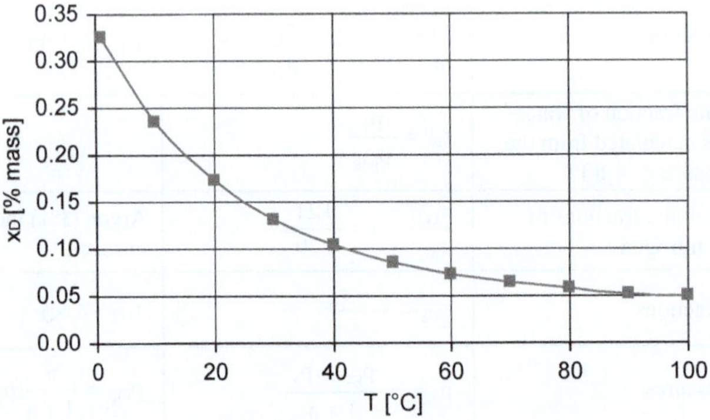

Fig. 17.3 Solubility in mass per cent of CO_2 in water at a partial pressure of 1.013 bar

from Eq. (17.3), but the relations may be different for various oils. The applicability of Eq. (17.3) is to be checked against the actual oil relevant for the project.

The volume of air at standard conditions (subscript std) dissolved in oil at pressure p is given by Eq. (17.3):

$$\frac{V_{air,std}}{V_{oil}} = \alpha_v \frac{p}{p_{Ref}} \left(\frac{T}{T_{Ref}}\right)^{0.56} \left(\frac{\nu_{Ref}}{\nu}\right)^{0.07} \tag{17.3}$$

with $\alpha_v = 0.09$, $p_{Ref} = 1.013$ bara, $T_{Ref} = 313$ K and $\nu_{Ref} = 30\ mm^2/s$

The mass fraction of dissolved air can be calculated from Eq. (17.4):

$$x_D = \alpha_v \frac{p}{p_{Ref}} \frac{\rho_{air,std}}{\rho_{oil}} \quad \text{with} \quad \rho_{air,std} = 1.29\,kg/m^3 \tag{17.4}$$

The amount of dissolved gas has virtually no influence on the density.

17.4 Physical Constants

17.4.1 Atmospheric Pressure

The atmospheric pressure p_{amb} decreases with increasing height H above sea level according to Eq. (17.5); $p_{amb,NN} = 1.013$:

$$\frac{p_{amb,H}}{p_{amb,NN}} = \left(\frac{288 - 6.5H}{288}\right)^{5.255} \quad \text{with H in [km]} \tag{17.5}$$

Due to weather and climatic conditions the atmospheric pressure varies on average by ±3% around the value given by Eq. (17.5) (higher deviations are possible).

17.4.2 Acceleration Due to Gravity

The acceleration due to gravity depends on the latitude φ and the height H above sea level, Eq. (17.6):

$$g = 9.7803(1 + 0.0053 \sin^2\varphi) - 0.003\,H \quad \text{with H in [km]} \tag{17.6}$$

17.5 Sound Velocity in Liquids

The speed of sound a_o in a liquid can be calculated by means of Eq. (17.7) where K_s are the isentropic bulk modulus and ρ the density of the liquid.

$$a_o = \sqrt{\frac{K_s}{\rho}} = \sqrt{\frac{c_p}{c_v}\frac{|p_2 - p_1|}{|\rho_2 - \rho_1|}} \tag{17.7}$$

The ratio of specific heats c_p/c_v can be assumed to be very close to 1.0. The sound velocity increases with the absolute pressure due to the compressibility of the liquid. At high pressures this influence is not negligible.

The sound velocity in sea water increases with the salinity S. The effect can be estimated from Eq. (17.8):

$$\frac{a_{seawater}}{a_{freshwater}} = 1 + 7 \times 10^{-4}\frac{S}{S_{ref}} \quad \text{with} \quad S_{ref} = 1000\,\text{ppm} \tag{17.8}$$

For many liquids it is difficult to find K_s. An estimate of the sound velocity can be made by means of Fig. 17.4 or Eq. (17.9) derived from [1]. The uncertainty of this estimate is about ±15%

$$a_0 = 1000\left(1.45 + 0.89\ln\left\{\frac{\rho}{\rho_{ref}}\right\}\right) \quad \text{with} \quad \rho_{ref} = 1000\,\text{kg/m}^3 \tag{17.9}$$

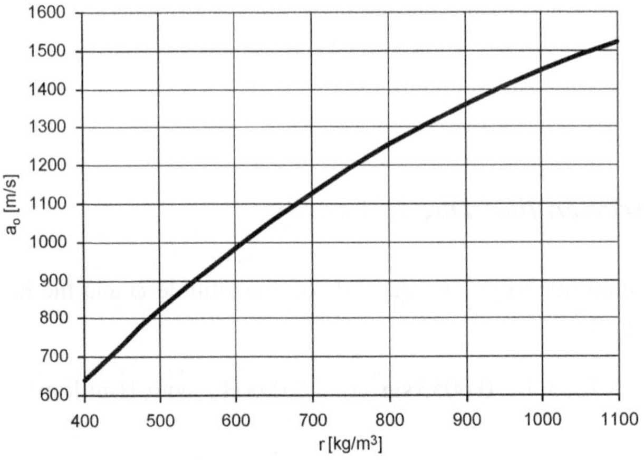

Fig. 17.4 Sound velocity a_o estimated from liquid density ρ. **Note:** small amounts of free gas in a liquid lower the speed of sound drastically

17.6 Mechanical Vibrations—Basic Notions

The simple system consisting of spring, mass and damping element is useful even for more complex situations because:

1. The model often helps the fundamental understanding of the basic behavior of complex systems.
2. Many real problems can be described sufficiently well by this model. An example would be the rotor of a pump with a double-entry impeller.

See Chap. 10.5 for an introduction to various types of vibrations.

1. Free vibrations with viscous damping: If a mass m supported by springs and damping elements is excited by a single blow, free vibrations occur.

Example of free vibrations: determining the natural frequency of a structure (e.g. the bearing housing of a pump) by an impact test. The structure is hit with an instrumented hammer and the resulting spectrum of free vibrations is recorded. Figure 10.52 gives an example (Table 17.4).

2. Forced vibrations: Forced vibrations are excited by periodic forces with an excitation frequency ω. Examples are a mechanical or a hydraulic unbalance. Depending on the ratio of excitation to eigen frequency $\eta = \omega/\omega_o$ the amplitudes of forced vibrations are amplified by the "amplification factor" V which is defined as the ratio of the amplitudes at any given frequency to the static deflection x_{stat} which would occur under a steady force of the same magnitude as the periodic excitation force considered.

When the excitation frequency equals the eigen frequency, "resonance" is said to occur. The amplitudes at resonance diminish with increasing damping. At low damping, the resonance amplitudes tend in theory to infinity; they are limited by non-linear effects leading often to damage.

The lag between the excitation force F and the amplitude x is given by the phase angle. At resonance the phase angle changes from below 90° to above 90°. If damping is high, the shift in phase angle is a better indication of resonance conditions than the behavior of the amplitudes (Tables 17.5, 17.6 and 17.7).

Table 17.4: Free vibrations		
If a mass m supported by springs and damping elements is excited by a single blow, free vibrations occur.		
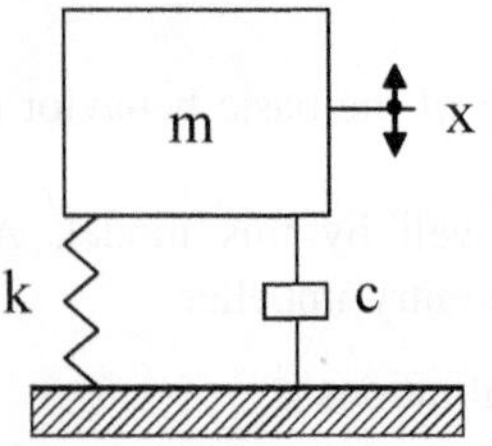	m = mass [kg] k = spring constant [N/m] c = damping [Ns/m] x = amplitude	
Differential equation of motion	m □+ c □+ k x = 0	
Undamped eigen frequency	$\omega_o = \sqrt{\frac{k}{m}}$	
Damped eigen frequency	$\omega_E = \sqrt{\omega_o{}^2 - \lambda^2} = \omega_o \sqrt{1 - D^2}$	$\lambda = \frac{c}{2\,m}$
The amplitudes of free vibrations decrease with time due to damping (i.e. losses caused by the movement of the body).	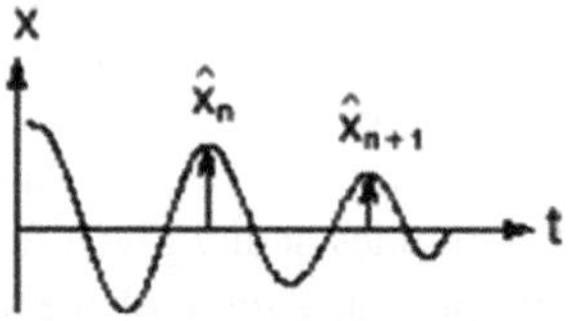	
Logarithmic decrement	$\delta = \ln \frac{\hat{x}_n}{\hat{x}_{n+1}}$	$\delta = -\frac{2\,\pi\,\lambda}{\omega_E}$
Damping ratio ζ (also called "Lehr's damping ratio)	$\zeta \equiv D = \frac{c}{c_c} = \frac{c}{2\,m\,\omega_o}$	
System aperiodically damped	$\lambda^2 = \omega_o{}^2 \qquad \omega_e = 0$	$c = c_c = 2\,m\,\omega_e$
Over-damped: $\zeta > 1$	Critically damped: $\zeta = 1$	Under-damped: $\zeta < 1$
Relations between ζ and δ	$\zeta \equiv D = \frac{\delta}{\sqrt{(2\pi)^2 + \delta^2}}$	$\delta = \frac{2\pi\zeta}{\sqrt{1 - \zeta^2}}$
Amplification factor Q	$Q \approx \frac{1}{2\,D}$ for $D < 0.25$	

Table 17.5: Forced vibrations with viscous damping (periodic force)	
Case A: Mass m is excited by a periodic force $F = F_0 \sin \omega t$	
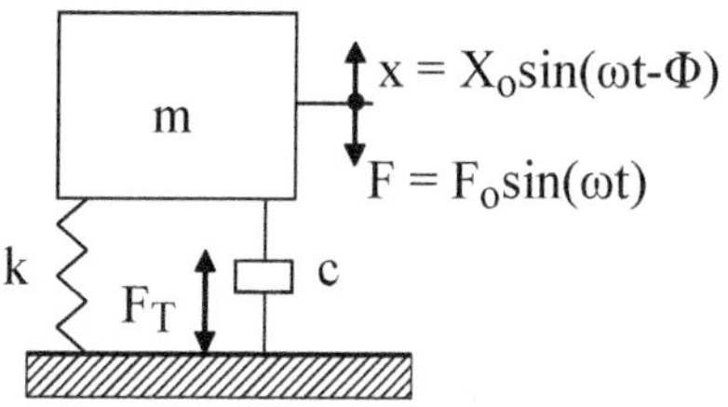	m = mass [kg] k = spring constant [N/m] c = damping [Ns/m] Φ = Phase angle between excitation force and displacement displacement: $x = X_0 \sin(\omega t - \Phi)$ The force F_T transmitted to the foundation can be determined from Table 17.6
Differential equation of motion	$m\ddot{x} + c\dot{x} + kx = F_0 \sin \omega t$
Particular solution	$x = X_0 \sin \omega t$
Frequency ratio	$\eta = \frac{\omega}{\omega_0}$
Undamped eigen frequency	$\omega_0 = \sqrt{\frac{k}{m}}$
Damped eigen frequency	$\omega_E = \sqrt{\omega_0^2 - \lambda^2} = \omega_0 \sqrt{1 - D^2}$ $\lambda = \frac{c}{2m}$
Damping ratio ζ (also called "Lehr's damping ratio" D)	$\zeta \equiv D = \frac{c}{c_c} = \frac{c}{2 m \omega_0}$
Amplification factor	$V_F = \frac{X_0 k}{F_0} = \frac{1}{\sqrt{(1-\eta^2)^2 + (2\zeta\eta)^2}}$
Phase angle between excitation force and displacement	$\tan \Phi = \frac{2\zeta\eta}{1-\eta^2}$
At resonance $\eta = 1$	$V_F = \frac{X_0 k}{F_0} = \frac{1}{2\zeta}$
Determination of damping from measured amplitudes: for relatively low damping: $\zeta \approx \frac{\Delta\omega}{2\omega_0}$	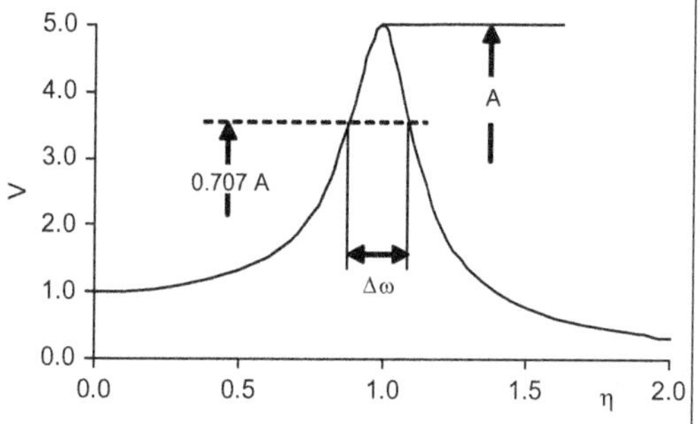

Case A: Mass m is excited by a force $F = F_0 \sin \omega t$

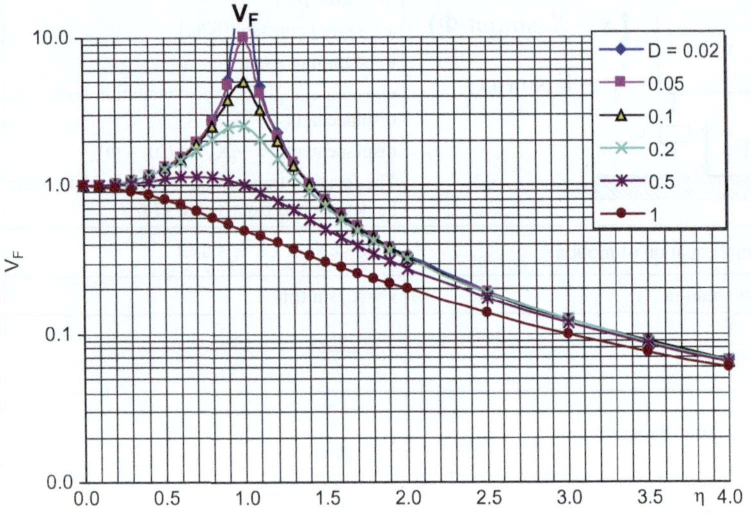

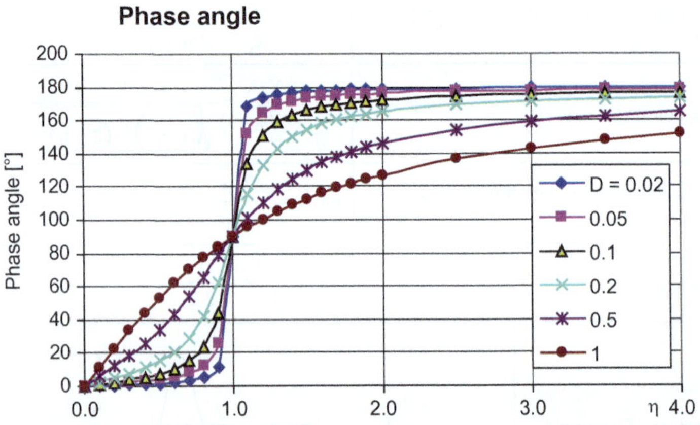

Table 17.6: Forced vibrations with viscous damping		
Case B: Baseplate excitation		
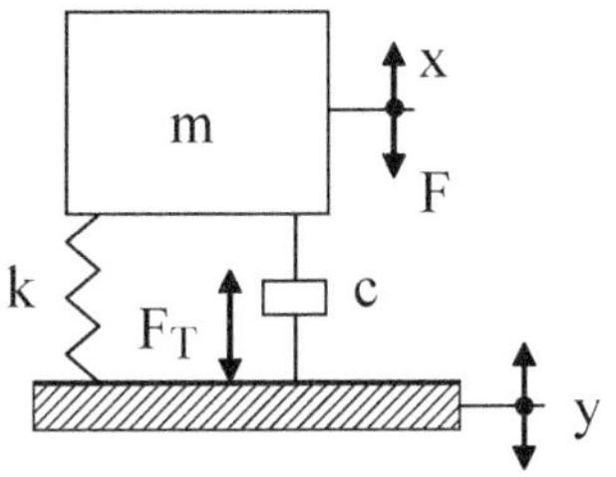	m = mass [kg] k = spring constant [N/m] c = damping [Ns/m] F_T = force transmitted from (or to) foundation by the springs and the damping elements displacement of mass m: $x = X_o \sin \omega t$ applied displacement: $y = Y_o \sin \omega t$	
Differential equation of motion	$m\,\ddot{x} + c\,\dot{x} + k\,x = m_u\ \omega^2 y \sin \omega t$	
Frequency ratio	$\eta = \dfrac{\omega}{\omega_o}$	
Undamped eigen frequency	$\omega_o = \sqrt{\dfrac{k}{m}}$	
Damped eigen frequency	$\omega_E = \sqrt{\omega_o^2 - \lambda^2} = \omega_o \sqrt{1 - D^2}$	$\lambda = \dfrac{c}{2\,m}$
Damping ratio ζ (also called "Lehr's damping ratio)	$\zeta \equiv D = \dfrac{c}{c_c} = \dfrac{c}{2\,m\,\omega_o}$	
Amplification factor (V_{BE} is also called transmissibility)	$V_{BE} = \dfrac{X_o}{Y_o} = \dfrac{F_T}{F} = \dfrac{\sqrt{1 + (2\,\zeta\,\eta)^2}}{\sqrt{\left(1 - \eta^2\right)^2 + (2\,\zeta\,\eta)^2}}$	
Phase angle between x and y or F_T and F	$\tan \Phi = \dfrac{2\,\zeta\,\eta^3}{1 - \eta^2 + (2\,\zeta\,\eta)^2}$	

Case B: Excitation of baseplate

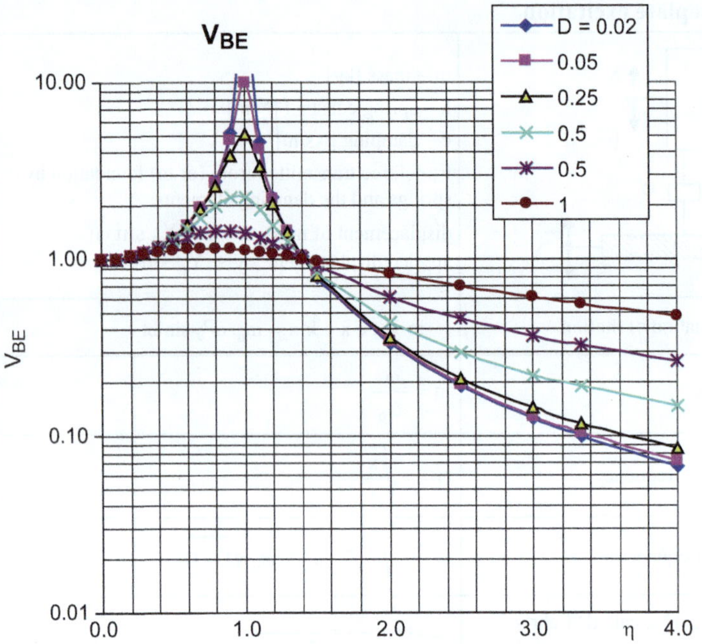

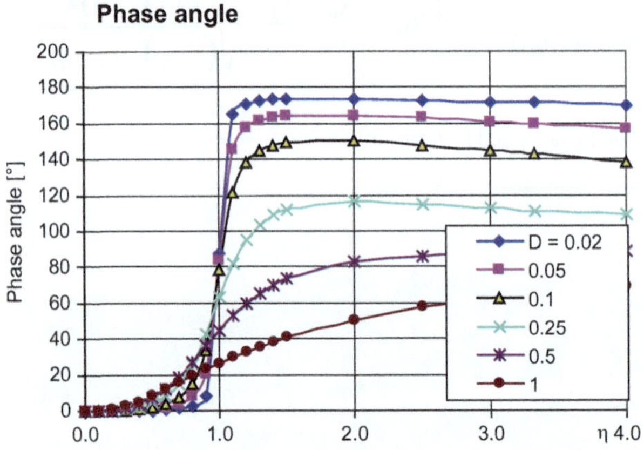

Table 17.7: Forced vibrations with viscous damping (imbalance excitation)	
Case C: Imbalance excitation	
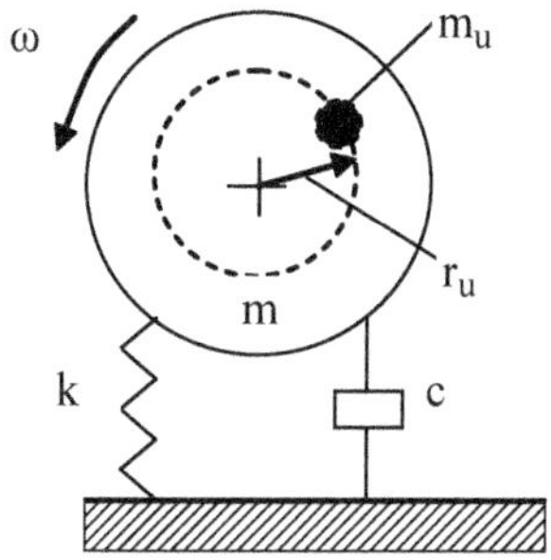	m = rotor mass [kg] e = eccentricity of rotor m_u = mass of an added imbalance [kg] r_u = radius of an added imbalance [m] k = spring constant [N/m] c = damping [Ns/m]
Differential equation of motion	$m\,\ddot{x} + c\,\dot{x} + k\,x = m_u\,r_u\,\omega^2 \sin\omega t$ or $m\,\ddot{x} + c\,\dot{x} + k\,x = m\,e\,\omega^2 \sin\omega t$
Frequency ratio	$\eta = \frac{\omega}{\omega_o}$
Undamped eigen frequency	$\omega_o = \sqrt{\frac{k}{m}}$
Damped eigen frequency	$\omega_E = \sqrt{\omega_o^2 - \lambda^2}$; $\lambda = \frac{c}{2\,(m + m_u)}$
Damping ratio ζ (also called "Lehr's damping ratio)	$\zeta \equiv D = \frac{c}{c_c} = \frac{c}{m\,\omega_o\,2}$
Amplification factor	$V_u = \frac{\eta^2}{\sqrt{\left(1-\eta^2\right)^2 + \left(2\,\zeta\,\eta\right)^2}}$
Amplitude if rotor operates with eccentricity e (added imbalance $m_u = 0$)	$X_o = e\,V_u$
Amplitude if rotor operates without eccentricity (e = 0) with added imbalance m_u auf Radius r_u	$X_o = \frac{m_u\,r_u}{m + m_u}\,V_u$
If the rotor operates with eccentricity e and with an imbalance $m_u > 0$ added on radius r_u, the imbalance forces due to e and r_u must be added as vectors.	
Phase angle between amplitude x and imbalance force vector	$\tan\Phi = \frac{2\,\zeta\,\eta}{1 - \eta^2}$

Case C: Imbalance excitation

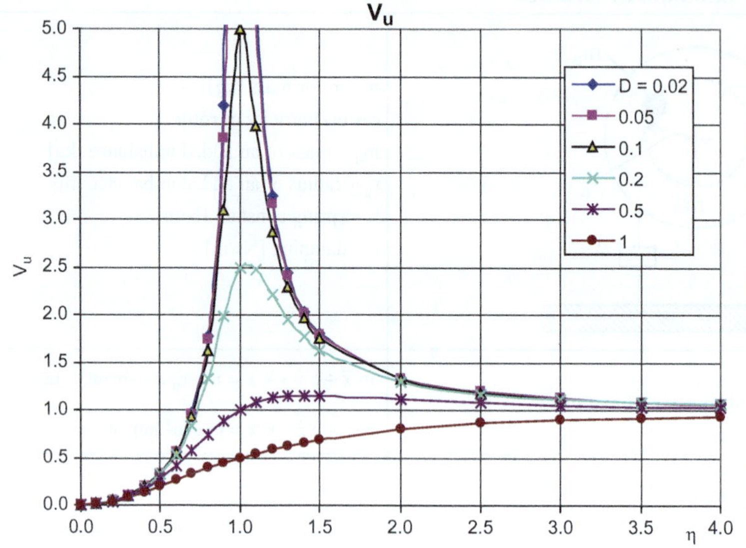

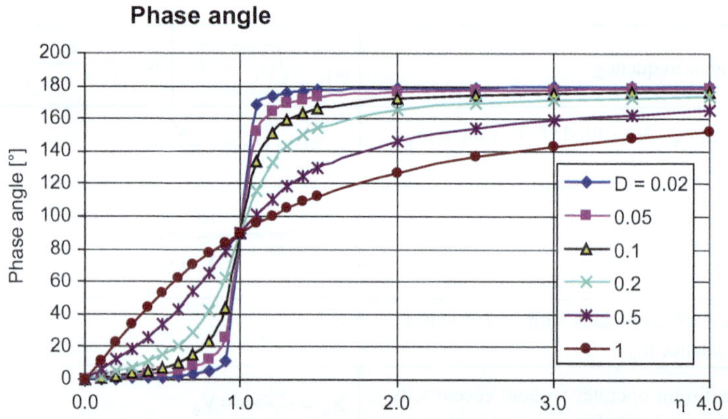

3. Eigen frequencies of simple structures: Table 17.8 gives eigen frequencies of some simple structures. If the structure vibrates in water (or any other liquid), the fluid lowers the eigen frequencies because of the "added mass" effect which accounts for the inertia of the liquid set into motion by the vibrating structure.

If a liquid-filled pipe vibrates in air, the added mass is given by the liquid contained in the pipe. The added mass of any structure submerged in a liquid can be estimated as 1.5-times the fluid volume displaced by the structure.

Table 17.8: Eigen frequencies of simple structures, from [2]

Mass-less beam with individual mass at free end (cantilever)	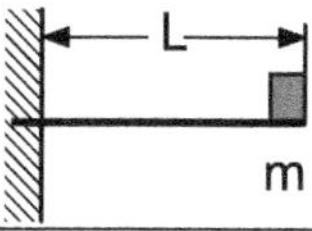	$f_1 = \frac{1.732}{2\pi}\sqrt{\frac{E\,I}{m\,L^3}}$	
Uniform beam; one end clamped, one end free (cantilever)	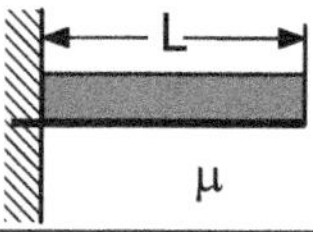	$f_n = \frac{k_n}{2\pi}\sqrt{\frac{E\,I}{\mu\,L^4}}$	$k_1 = 3.52$ $k_2 = 22$ $k_3 = 61.7$
Uniform beam; one end clamped, one end free; with individual mass at free end	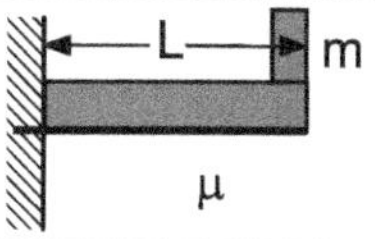	$f_1 \approx \frac{1.732}{2\pi}\sqrt{\frac{E\,I}{m\,L^3 + 0.236\,\mu\,L^4}}$	
Mass-less beam with individual mass in center; simply supported	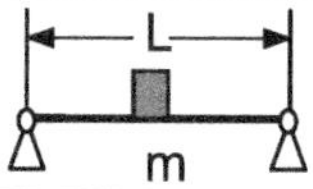	$f_1 = \frac{6.93}{2\pi}\sqrt{\frac{E\,I}{m\,L^3}}$	
Uniform beam; simply supported	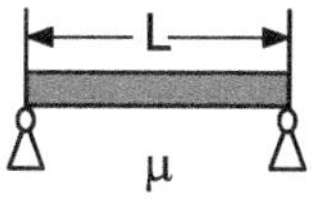	$f_n = \frac{k_n}{2\pi}\sqrt{\frac{E\,I}{\mu\,L^4}}$	$k_1 = 9.87$ $k_2 = 39.5$ $k_3 = 88.8$
Uniform beam with individual mass in center; simply supported	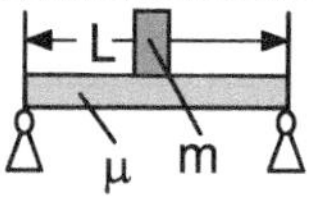	$f_1 \approx \frac{6.93}{2\pi}\sqrt{\frac{E\,I}{m\,L^3 + 0.496\,\mu\,L^4}}$	
Flat circular plate of thickness h and radius R vibrating at two nodal diameters	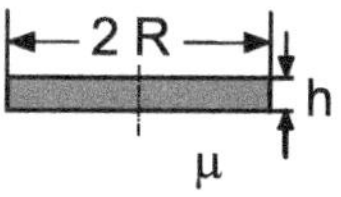	$f_1 \approx \frac{5.25}{2\pi}\sqrt{\frac{E\,h^3}{12(1-\nu^2)\,\mu\,R^4}}$	
E = Young's modulus of elasticity I = area moment of inertia f = eigen frequency [Hz] μ = load per unit length [kg/m] or unit surface [kg/m²] including added mass effect if applicable ν = Poisson's ratio (mostly 0.3)			

17.7 Hydraulic Specification

The hydraulic specification intends to gather all information which is needed for, or could have an impact on, the hydraulic design. The objectives are: (1) to help with a thorough product planning; (2) to serve as a checklist; (3) to carefully document the choices made for a hydraulic development project. Obviously, not every of the criteria listed below are relevant for every type of pump and application (to be marked accordingly). On the other hand, below proposal may not be complete. Only those aspects of mechanical design are considered, which have an impact on the hydraulic design.

Revision	Date	Author	Reviewed	Approved
1				
2				
3				
4				

1. General requirements

1. Pump type	
2. Installation	☐ horizontal; ☐ vertical
3. Pump service	
4. Medium pumped	type of liquid: temperature: T = °C density: ρ= kg/m³ viscosity: ν = mm²/s (cSt) vapor pressure p_v = bara (absolute)
	impurities apt to induce corrosion:
	sewage applications: type of sewage (risk of clogging): ball passage diameter: d_k = mm
	minimum and maximum temperature (e.g. during start-up or upset conditions)
5. Contamination with solids	sand content: c_s = kg/m³ grain size: d_s = mm other solids:
6. Free gas	type of gas: gas volume fraction GVF:
7. Dissolved gas	type of gas: dissolved mass content: ppm
8. Foaming risk	
9. Design pressure and temperature	p_d = bar; T_d = °C
10. Suction conditions	☐ from tank with free liquid level ☐ from wet pit; minimum submergence: m ☐ from booster pump ☐ other (specify):
11. Available $NPSH_A$:	$NPSH_A$ = m liquid level above (+) or below (-) pump axis centerline: z_E = m losses in suction pipe: $H_{v,s}$ = m transient conditions (e.g. load rejection):
12. Liquid level variations	☐ maximum liquid level: m ☐ minimum liquid level: m ☐ fast level fluctuations risk of ingress of gas (air):
13. NPSH safety margins	
14. Suction pipe layout	provide isometric drawing as required
15. Suction strainer	☐ yes ☐ no
16. Installation conditions	Location above sea level: m (effect on $NPSH_A$)

17. Type of driver	☐ electric motor; ☐ other (specify):
18. Nominal speed	n_{nom} = rpm
19. Variable speed	☐ yes ☐ no speed range: n_{max} = rpm; n_{min} = rpm
20. Direction of rotation viewed from driver	☐ clockwise ☐ anti-clockwise
21. Q_{max} and Q_{min} in *continuous* service	$Q_{max,cs}$ = m^3/h ; $Q_{min,cs}$ = m^3/h
22. Q_{max} and Q_{min} in *short-term* operation	Q_{max} = m^3/h ; Q_{min} = m^3/h
23. Number of parallel pumps	
24. Parallel operation	Q_{max} with single pump operation: $NPSH_A$ sufficient? Maximum flow limited by cavitation? („cavitation control")
25. Pump control	☐ throttle valve; ☐ speed control ☐ other (specify):
26. Operation conditions	☐ continuous ; ☐ intermittent
27. System characteristic	geodeticorstatichead:$H_{A,stat}$ = m
28. Q-H-curve stability	☐ mandatory ; ☐ stable above q* =
29. Maximum allowable shut-off head	$H_{o,max}$ = m Tolerance = Head rise: H_o/H_{opt} =
30. Maximum power:	at end of curve: P_{max} = kW at shut-off: $P_{o,max}$ = kW
31. Vibration and noise	Vibration limits: Noise level: Sound pressure L_p = dBA
32. Reverse flow due to geodetic head?	
33. Specific operation conditions which may cause problems	
34. Standby pump	
35. Start-up procedure	
36. Applicable standards	• for acceptance test: • for design: • for vibrations: • other (specify):
37. Guarantees ☐ at BEP ☐ at rated conditions	Efficiency: Head: Flow rate: Power:
38. Energy cost:	* operation hours per year: * Euro/kWh: * Euro/kW:
39. Model or prototype tests?	
40. Instrumentation	
41. Safety requirements, explosive protection, environmental aspects	

2. Hydraulic design

42. Performance			BEP	rated	max.	min.	zero
	n	rpm					
	Q	m^3/h					0
	H_{stage}	m					
	H_{total}	m					
	$NPSH_A$	m					
	$NPSH_3$	m					
	n_q	-					
	n_{ss}	-					

43. Head margin	
44. Number of stages	$z_{st} =$
45. Maximum number of stages	$z_{st,max} =$
46. Range chart	
47. Pump selection allowed in range of:	$q^* =$ to
48.Maximum impeller trim	$\frac{d_2'}{d_2} \geq$
49. Impeller trim diameter	
50. Cylindrical impeller extension?	up to: %
51. Component standardization requirements	
52. Sectional drawing	
53. Impeller fixation on shaft: >> provide design drawing	
54. Inducer required/allowed?	
55. Impeller type	☐ single entry ☐ double entry ☐ closed ☐ open
56. Impeller material	
57. Impeller manufacturing:	☐ cast ☐ other (specify):

58. Impeller design		suction stage	series stages
suction specific speed	n_{ss}		
number of blades	z_{La}		
hub diameter	d_n [mm]		
59. Blade design:			
60. Pressure coefficients:			
61. Applicable cavitation criteria	☐ $NPSH_3$ ☐ $NPSH_i$ ☐ other (specify):		
62. Impeller surface roughness:	channels: µm outer side of shrouds: µm		
63. Casing material			

64. Casing manufacturing:	☐ cast; ☐ other (specify):
65. Casing	☐ foot mounted ☐ center-line mounted
66. Nozzle orientations (suction and discharge)	
67. Nozzle dimensions :	d_s, d_d L_s, L_d, nozzle positions (sketch etc.)
68. Inlet casing type :	☐ axial inlet nozzle ☐ inline pump ☐ radial
69. Outlet casing type :	☐ volute; ☐ annulus
70. Volute type:	☐ single volute; ☐ double volute
71. Volute shape of cross sections	☐ trapezoidal; ☐ rectangular ☐ circular
72. Volute design specifics:	
73. Volute surface roughness:	μm
74. Diffuser material	
75. Diffuser vane distance Volute cutwater distance	$d_3/d_2 =$ $d_z/d_2 =$
76. Number of diffuser vanes Number of return vanes	$z_{Le} =$ $z_{RV} =$
77. Gap A and overlap x_{ov}	
78. Seal design:	• type (annular, radial, diagonal, other): • clearance $\Delta D/D =$ • surface pattern (plain, serrated, etc.):
79. Radial forces; design criteria for volute casing pumps:	☐ shaft deflection at mechanical seal 50μm for $q^* > 0.5$ ☐ maximum shaft deflection at shut-off smaller than seal clearance ☐ other (specify):
80. Axial thrust balancing:	☐ none ☐ expeller vanes ☐ balance holes & rear shroud annular seal ☐ balance piston ☐ balance disk Axial bearing capacity: N
81. Bearing frame natural frequencies:	
82.	
83.	
84.	
85.	

References

1. Robinet, F., Gülich, J.F., Kaiser, T.: Vane pass vibrations—source, assessment and correction—a practical guide for centrifugal pumps. In: 16th International Pump Users Symposium, Houston, pp. 121–137 (1999)
2. Roark, R.J., Young, W.C.: Formulas for Stress and Strain, 5th edn. McGraw-Hill, New York (1986)

Bibliography

B.1. Pfleiderer, C.: Kreiselpumpen f. Flüssigkeiten u. Gase. 5. Aufl, Springer, Berlin (1961)
B.2. Stepanoff, A.J.: Radial- und Axialpumpen. Springer, Berlin (1959)
B.3. Traupel, W.: Thermische Turbomaschinen. 3. Aufl, Springer, Berlin (1977)
B.4. Anderson, H.H.: Centrifugal Pumps. Trade and Technical Press (1972)
B.5. KSB-Kreiselpumpen-Lexikon. 3. Aufl, KSB AG, Frankenthal (1989)
B.6. Raabe, J.: Hydro Power. VDI, Düsseldorf (1985)
B.7. Sigloch, H.: Strömungsmaschinen. 2. Aufl, Hanser, München (1993)
B.8. Spengler, H., et al.: Technisches Handbuch Pumpen. VEB Technik, Berlin (1976)
B.9. Lobanoff, V.S., Ross, R.R.: Centrifugal Pumps, 2nd edn. Gulf Publishing, Houston (1985)
B.10. Pumping Manual. 9th edn. Elsevier Advanced Technology, Oxford (1995)
B.11. Anderson, H.H.: Centrifugal Pumps and Allied Machinery, 4th edn. Elsevier Advanced Technology, Oxford (1994)
B.12. Faragallah, W.H.: (Hrsg) Pumpen als Turbinen. Faragallah, Sulzbach (1993)
B.13. Faragallah, W.H.: (Hrsg) Seitenkanal-Strömungsmaschinen. Faragallah, Sulzbach (1992)
B.14. Neumaier, R.: Hermetische Pumpen. Faragallah, Sulzbach (1994)
B.15. Karassik, I., Messina, J., Cooper, P., Heald, C.: Pump Handbook, 4th edn. McGraw-Hill, New York (2008)
B.16. Japikse, D., Marscher, W.D., Furst, R.B.: Centrifugal Pump Design and Performance. Concepts ETI, Norwich (1997)
B.17. Sulzer Centrifugal Pump Handbook. 3rd edn. Elsevier Advanced Technology, Oxford (2010)
B.18. Dicmas, J.L.: Vertical Turbine, Mixed Flow, and Propeller Pumps. McGraw-Hill, New York (1987)
B.19. Gülich, J.F., et al.: Feedpump operation and design guidelines. EPRI Report TR-102102 (June 1993)
B.20. Gülich, J.F., Egger, R.: Part load flow and hydraulic stability of centrifugal pumps. EPRI Report TR-100219 (March 1992)
B.21. Neumann, B.: The interaction between geometry and performance of a centrifugal pump. MEP, London (1991)
B.22. Troskolanski, A.T., Lazarkiewicz, S.: Kreiselpumpen, Berechnung und Konstruktion. Birkhäuser, Basel (1976)
B.23. de Kovats, A., Desmur, G.: Pumpen, Ventilatoren und Kompressoren. Braun, Karlsruhe (1968)
B.24. Wislicenus, G.F.: Preliminary design of turbo pumps and related machinery. NASA RP 1170 (1986)

J. F. Gülich, *Centrifugal Pumps*,
https://doi.org/10.1007/978-3-030-14788-4

B.25. Cumpsty, N.A.: Compressor Aerodynamics. Longman, Harlow, England (1989)
B.26. Grist, E.: Cavitation and the Centrifugal Pump. A Guide for Pump Users. Taylor & France, ISBN 1-56032-591-7 (1999)
B.27. Japikse, D., Baines, N.C.: Introduction to Turbomachinery. Oxford University Press (1994)
B.28. Sterling, S.I.H.I.: Pumpenhandbuch Grundlagen für Planung von Kreiselpumpenanlagen, 7. Aufl, 2000, Sterling SIHI GmbH, Itzehoe
B.29. Fogg, P.G.T., Gerrand, W.: Solubility of Gases in Liquids. Wiley, New York, ISBN 0 471 92925 5 (1991)
B.30. Wesche, W.: Radiale Kreiselpumpen. Springer, Berlin (2012)

Standards

N.1. DIN EN 12723 (formerly DIN 24260-1): Flüssigkeitspumpen; Kreiselpumpen und Kreiselpumpenanlagen. Begriffe, Formelzeichen, Einheiten. 2000-09
N.2. ISO 9906: 2012-05 Rotodynamic pumps – Hydraulic performance acceptance tests
N.3. DIN 1184-2: Schöpfwerke/Pumpwerke. Kreiselpumpen für den Betrieb mit Rechen. Richtlinien für die Planung. 1992
N.4. Standards for centrifugal, rotary & reciprocating pumps. Hydraulic Institute, 14th edn. Cleveland (1983)
N.5. IEC 60193: 1999 International code for model acceptance tests of storage pumps
N.6. ISO13709: 2010-05 (API Standard 610): Centrifugal pumps for petroleum, petrochemical and natural gas industries
N.7. Zustandsgrößen für Wasser und Wasserdampf in SI-Einheiten. Springer, Berlin (1989)
N.8. ISO 10816-3: 2009 Mechanical vibration - evaluation of machine vibration by measurements on non-rotating parts. Industrial machines with power greater than 15 kW (nicht spezifisch für Pumpen)
N.9. ISO 7919-3: 2008 Mechanical vibrations of non-reciprocating machines – Measurements on rotating shafts and evaluation criteria. Part 3: coupled industrial machines
N.10. ANSI/HI 9.8-2012 Standard: Pump Intake Design
N.11. ANSI/HI Standard 9.6.7-2004: Effect of liquid viscosity on pump performance. Also published as ISO/TR 17766:2005(E)
N.12. DIN EN 12502-1/-5: 2005-03 Korrosionsverhalten von metallischen Werkstoffen
N.13. DIN 50900, Teil 1 bis 5: Korrosion der Metalle (1993)
N.14. EN 10283: 2010-06 Korrosionsbeständiger Stahlguß
N.15. DIN-Taschenbuch 219: 2015-02 Korrosion u. Korrosionsschutz. Beurteilung, Prüfung, Schutzmaßnahmen. Normen, Technische Regeln, Beuth-Verlag
N.16. ISO 10816-7: 2009 Mechanical vibration – Evaluation of machine vibration by measurements on non-rotating parts. Part 7: Rotodynamic pumps for industrial application including measurements on rotating shafts
N.17. VDI 3839: Hinweise zur Messung und Interpretation der Schwingungen von Maschinen. Allgemeine Grundlagen. VDI-Richtlinie (2001)
N.18. DIN 45635: Geräuschmessung an Maschinen
N.19. Nichtrostender Stahlguß. Stahl-Eisen-Werkstoffblätter, SEW 410, 2017-12
N.20. ISO 5198: 1987 Centrifugal, mixed and axial pumps – Code for hydraulic performance tests – precision class
N.21. CCH 70-3: Pflichtenheft für die Abnahme von Stahlgußstücken für hydraulische Maschinen. Ausgabe 3, 1996, Groupe d' étude Cahier des charges hydrauliques, case postale 3, Ch 1297 Founex
N.22. ISO 5389: 2005: Turbocompressors – Performance test code
N.23. ANSI/HI Standard 1.1.-1.2-2000: Centrifugal Pumps

Index

J. F. Gülich, *Centrifugal Pumps*,
https://doi.org/10.1007/978-3-030-14788-4

<u>GPSR Compliance</u>

The European Union's (EU) General Product Safety Regulation (GPSR) is a set of rules that requires consumer products to be safe and our obligations to ensure this.

If you have any concerns about our products, you can contact us on ProductSafety@springernature.com

In case Publisher is established outside the EU, the EU authorized representative is:

Springer Nature Customer Service Center GmbH
Europaplatz 3
69115 Heidelberg, Germany

Batch number: 10770418

Printed by Printforce, the Netherlands